Seeley's
Principles of Anatomy & Physiology

ON TRACK... These interactive activities and quizzing keep you motivated and on track in mastering key concepts:

► **Animations** Access to hundreds of animations of key A & P processes will help you visualize and comprehend important concepts depicted in the text. The animations even include quiz questions to help ensure that you are retaining the information.

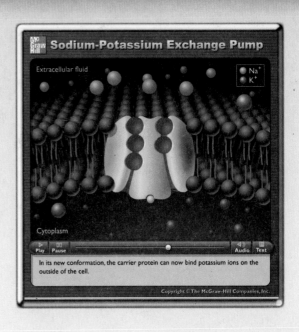

► **Test Yourself**
Take a chapter quiz at the ARIS website to gauge your mastery of chapter content. Each quiz is specially constructed to test your comprehension of key concepts. Immediate feedback explains incorrect responses. You can even e-mail your quiz results to your professor!

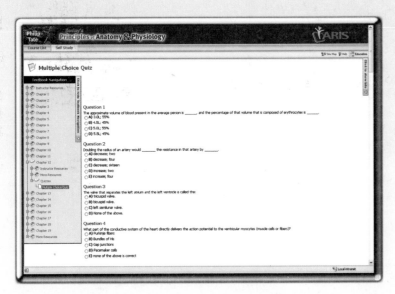

► **Learning Activities** Helpful and engaging learning experiences await you at the *Seeley's Principles of Anatomy & Physiology* ARIS site. In addition to interactive online quizzing and animations, each chapter offers relevant case study presentations, digital images for creating PowerPoints, vocabulary flash cards, and other activities designed to reinforce learning.

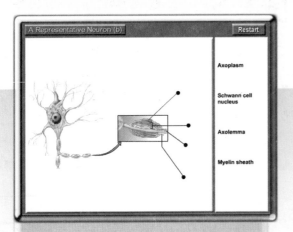

McGraw-Hill Higher Education

Seeley's Principles of Anatomy & Physiology

Philip Tate
Phoenix College

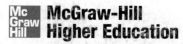
McGraw-Hill
Higher Education

Boston   Burr Ridge, IL   Dubuque, IA   New York   San Francisco   St. Louis
Bangkok   Bogotá   Caracas   Kuala Lumpur   Lisbon   London   Madrid   Mexico City
Milan   Montreal   New Delhi   Santiago   Seoul   Singapore   Sydney   Taipei   Toronto

McGraw-Hill
Higher Education

SEELEY'S PRINCIPLES OF ANATOMY & PHYSIOLOGY

Published by McGraw-Hill, a business unit of The McGraw-Hill Companies, Inc., 1221 Avenue of the
Americas, New York, NY 10020. Copyright © 2009 by The McGraw-Hill Companies, Inc. All rights reserved.
No part of this publication may be reproduced or distributed in any form or by any means, or stored in a
database or retrieval system, without the prior written consent of The McGraw-Hill Companies, Inc., including,
but not limited to, in any network or other electronic storage or transmission, or broadcast for distance
learning.

Some ancillaries, including electronic and print components, may not be available to customers outside the
United States.

 This book is printed on recycled, acid-free paper containing 10% postconsumer waste.

3 4 5 6 7 8 9 0 QPD/QPD 0 9

ISBN 978-0-07—337813—8
MHID 0-07—337813—5

Publisher: *Michelle Watnick*
Senior Sponsoring Editor: *James F. Connely*
Vice-President New Product Launches: *Michael Lange*
Senior Developmental Editor: *Kathleen R. Loewenberg*
Marketing Manager: *Lynn M. Breithaupt*
Lead Project Manager: *Mary E. Powers*
Senior Production Supervisor: *Laura Fuller*
Senior Freelance Design Coordinator: *Michelle D. Whitaker*
Cover Designer: *Christopher Reese*
(USE) Cover Image: © *Corey Rich/Getty Images*
Cover image: *An aerial dancer performs in a high angle environment while dangling from ropes.*
Senior Photo Research Coordinator: *John C. Leland*
Photo Research: *Jerry Marshall/pictureresearching.com*
Supplement Producer: *Mary Jane Lampe*
Compositor: *Aptara*
Typeface: *10.5/12 Minion*
Printer: *Quebecor World Dubuque, IA*

The credits section for this book begins on page C-1 and is considered an extension of the copyright page.

Library of Congress Cataloging-in-Publication Data

Tate, Philip.
  Seeley's principles of anatomy & physiology / Philip Tate. -- 1st ed.
    p. cm.
  "The hallmarks of Seeley, Stephens and Tate's Anatomy and Physiology are retained in this first edition, but
the material has been made more efficient and succinct."--Preface.
  Includes bibliographical references and index.
  ISBN 978-0-07-337813-8 — ISBN 0-07-337813-5 (hard copy : alk. paper)
1. Human physiology. 2. Human anatomy. I. Seeley, Rod R. Anatomy & physiology. II. Title. III. Title:
Seeley's principles of anatomy and physiology. IV. Title: Principles of anatomy & physiology.
QP34.5.T378 2009
612--dc22
                    2007049854

Brief Contents

Appendices

Contents

About the Author

Philip Tate earned a B.S. in zoology, a B.S. in mathematics, and an M.S. in ecology at San Diego State University; and a Doctor of Arts (D.A.) in biological education from Idaho State University. He is an award-winning instructor who has taught a wide spectrum of students at the four-year and community college levels. Phil has served as the annual conference coordinator, president-elect, president, and past president of the Human Anatomy and Physiology Society (HAPS). He presently teaches anatomy and physiology at Phoenix College in Phoenix, Arizona.

DEDICATION

To those who study, teach, and use the knowledge of human anatomy and physiology.

Preface

For Whom Is This Book Written?

Principles of Anatomy and Physiology is written for the two-semester anatomy and physiology course. The writing is comprehensive enough to provide the background necessary for those courses not requiring prerequisites, and yet is concise so as not to confuse and overwhelm the reader. Clear descriptions and exceptional illustrations combine with ample clinical information to help students develop a solid understanding of the concepts of anatomy and physiology and how that knowledge relates to the medical world.

Why a "Streamlined" Text?

Instructors: how often have you tried to cover all of the material in your current textbook, but in the end, had to skip a section or chapter? And students: are you struggling with knowing what parts of the book are necessary to retain and understand, and what is just "extra"? *Principles of Anatomy and Physiology* is designed to help students develop a basic understanding of the concepts of anatomy and physiology without overwhelming them with chatty anecdotes or in-depth coverage that may be too detailed for their needs. The hallmarks of Seeley, Stephens and Tate's *Anatomy and Physiology* are retained in this first edition, but the material has been made more efficient and succinct. For example, coverage of a particular topic that may span six to nine pages in another text could result in only four pages in this book because of combined art, limited boxed readings, and careful scrutiny of necessary detail.

> "The more chapters I read, the more I like the streamlined approach the author is taking in this text. A frequent comment from our students is that they have trouble wading through the details and figuring out which they are 'required' to know, as opposed to being supplementary."
>
> —Susan Spencer, *Mt. Hood Community College*

What's Different About This Anatomy and Physiology Book?

As mentioned earlier, the same popular features that the anatomy and physiology text by Seeley, Stephens, and Tate is known for can still be found in this "principles" textbook: realistic and beautifully rendered figures; problem-solving and critical thinking emphasis; and clear, straight-forward writing. So how is this textbook different? The explanations have been shortened, examples have been reduced to perhaps just one, multiple figures have been combined into one or two, and many boxed readings have been eliminated. The result is a shorter, simplified textbook that **covers all of the major points found in more lengthy texts,** but is easier to read and is more economical in price.

> "I really enjoyed the book. It was well written and I think students will like it. In this chapter, there was actually a lot more covered than I thought would be for a 'shorter' text."
>
> —Teresa Alvarez, *St. Louis Community College —Forest Park*

How Will This Book's Features Benefit Me?

Relevant clinical coverage means your students will "get" the connection to their future careers.

Examples of diseases, responses to exercise, clinical case studies, aging, and environmental conditions are all used to explain how our bodies function and to understand the consequences when systems do not operate normally. These conditions are also used to enhance comprehension of the relationship between structure and function. Brief, real-life **Case Studies** combined with **Clinical Asides,** the more in-depth **Clinical Relevance Readings,** and the **Systems Interactions** pages, provide a thorough clinical education that fully supports the textual material.

> "This is one place where you've outdone Tortora. Your clinical material is outstanding, and I like the fact that some of it is illustrated."
>
> —Beverly P. Kirk, *Northeast Mississippi Community College*

Integrated critical thinking will help students make progress in comprehension.

Because instructors realize that recall isn't enough and that learning needs to be developed and applied, Tate's *Anatomy and Physiology* employs a critical thinking and problem-solving approach throughout each chapter. No longer just a catch phrase, critical thinking is recognized as a necessity for those students entering health-related careers. At every opportunity, the author encourages students to apply information they have learned to practical "real life" scenarios. Abundant clinical content, step-by-step Process Figures, in-chapter Review and Predict Questions, macro-to-micro art, unique Homeostasis Summary Figures, cadaver images, and Bloom's Taxonomy based end-of-chapter questions all combine to make this a textbook that is focused on helping students test their comprehension and learn to solve problems.

> "Excellent critical thinking questions in this chapter! I normally do not assign homework to the students, but I may use some of the questions listed as applications during lecture."
>
> —Claudia Stanescu, *University of Arizona*

Exceptional art depicts important structure and function principles and offers another perspective to learning.

"Accuracy, Consistency, and Depth of Detail" are the guidelines for every piece of art in this textbook. Attractive and clearly presented, the visual program enhances comprehension in a number of ways: tables are often combined with illustrations, relevant photos are side-by-side with drawings, cadaver photos are included where appropriate, step-by-step Process Figures explain physiological processes, and the distinctive Homeostasis Summary Figures include explanations that are necessary to understand mechanisms and their roles in the maintenance of homeostasis. Images are rendered in a contemporary style and are coordinated so that colors and styles of structures in multiple figures are consistent with one another throughout the book.

> "The artwork is excellent. Several key processes (and a few anatomical features) are presented in more detail via figures, with brief discussions in the actual text. I prefer this since it allows the students to view the process as they follow along with the written steps. The number and placement of figures were great."
>
> —Paul Florence, *Jefferson Community and Technical College*

Clear and concise writing results in to-the-point language, helping students determine priority of importance.

Perhaps the central difference between this textbook and all other two-semester anatomy and physiology textbooks is its advantage of being briefer and less expensive while keeping all of the necessary content students need. How was this done? The text was rewritten to cover all essential learning objectives needed for clear and complete explanations, while reducing or eliminating additional in-depth coverage. This has been accomplished by the author being more efficient and simplifying where possible. All of the major concepts found in more lengthy texts have been made, but they are easier to absorb because they're shorter and more to the point.

> "The discussion of osmosis is one of the best I've read in a long time. Since this concept is important in the understanding of many other concepts (respiration, renal, capillary dynamics), I feel it should be well understood."
>
> —Lois Brewer Borek, *Georgia State University*

> "This chapter does a very good job tackling several very difficult topics including resting membrane potential and the acting potential. It does so directly and with minimal jargon. The text does an excellent job of meeting our student needs."
>
> —William Stewart, *Middle Tennessee State University*

Acknowledgments

First and foremost, it is my pleasure to acknowledge Rod Seeley and Trent Stephens. For over a quarter of a century we have collaborated to write eight editions of *Anatomy and Physiology* and six editions of *Essentials of Anatomy and Physiology*. These books are the foundation of *Principles of Anatomy and Physiology*. Rod and Trent are much more than colleagues—they are my mentors and my friends.

The encouragement and support of my family has been essential for the completion of this project. My mother, Billie, deserves special thanks and praise.

It is difficult to adequately acknowledge the contributions of all the people who have guided the book through its various stages of development. I wish to express my gratitude to the staff of McGraw-Hill for their assistance, especially Publisher Michelle Watnick and Senior Sponsoring Editor Jim Connely for encouraging me to undertake this project, and Senior Developmental Editor Kathy Loewenberg for her suggestions and help.

The efforts of many people are required to produce a modern textbook. The dedicated work of the entire project team is greatly appreciated. Thanks to Managing Editor Mary Powers, Senior Project Manager Brenda Trone, and Project Manager Carole Kuhn. Their efficiency and professionalism has lightened my load.

Thanks are gratefully offered to Copy Editor Debra DeBord for carefully polishing my words.

I also thank the illustrators who worked on the development and execution of the illustration program. I appreciate their contribution to the overall appearance and pedagogical value of the illustrations.

Finally, I sincerely thank the reviewers and the instructors who have provided me with excellent constructive criticism. The remuneration they received represents only a token payment for their efforts. To conscientiously review a textbook requires a true commitment and dedication to excellence in teaching. Their helpful criticisms and suggestions for improvement were significant contributions that I greatly appreciate. I acknowledge them by name in the next section. Special thanks to Angela Mick at Glendale Community College in Arizona, for her help in organizing the reviews.

—Philip Tate

Reviewers

Teresa Alvarez
St. Louis Community College Forest Park

Jerry D. Barton II
Tarrant County College–South Campus

J. Gordon Betts
Tyler Junior College

Lois Brewer Borek
Georgia State University

Danielle Desroches
William Paterson University of New Jersey

Martha R. Eshleman
Pulaski Technical College

Paul Florence
Jefferson Community & Technical College

Glenn M. Fox
Jackson Community College

Chaya Gopalan
St. Louis Community College

Virginia Irintcheva
Black Hawk College

Beverly P. Kirk
Northeast Mississippi Community College

Paul Luyster
Tarrant County College District

W. J. McCracken
Tallahassee Community College

Robert W. McMullen
Pikes Peak Community College

Scott Murdoch
Moraine Valley Community College

Gwen M. Niksic
University of Mary

Nathan O. Okia
Auburn University Montgomery

Reviewers for the eighth edition of *Anatomy and Physiology* by Seeley, Stephens, and Tate:

Terry A. Austin
Temple College

Gail Baker
LaGuardia Community College

David M. Bastedo
San Bernardino Valley College

Alease S. Bruce
University of Massachusetts—Lowell

Nishi S. Bryska
University of North Carolina—Charlotte

Patrick D. Burns
University of Northern Colorado

Brad Caldwell
Greenville Technical College

Ana Christensen
Lamar University

Nathan L. Collie
Texas Tech University

David T. Corey
Midlands Technical College

Ethel R. Cornforth
San Jacinto College—South Campus

Cara L. Davies
Ohio Northern University

Richard Doolin
Daytona Beach Community College

Kathryn A. Durham
Lorain County Community College

Adam Eiler
San Jacinto College—South

Lee F. Famiano
Cuyahoga Community College

Kathy E. Ferrell
Greenville Technical College

Edward R. Fliss
St. Louis Community College at Florissant Valley

Paul Florence
Jefferson Community College

Cliff Fontenot
Southeastern Louisiana University

Allan Forsman
East Tennessee State University

Ralph F. Fregosi
The University of Arizona

Paul Garcia
Houston Community College—Southwest

Chaya Gopalan
St. Louis Community College—Florissant Valley

Jean C. Jackson
Lexington Community College

Amy E. Jetton
Middle Tennessee State University

Jody E. Johnson
Arapahoe Community College

Sally Johnston
Community College of Southern Nevada

Beverly P. Kirk
Northeast Mississippi Community College

Michael Kopenits
Amarillo College

Karen K. McLellan
Indiana University—Purdue University Fort Wayne

Mara L. Manis
Hillsborough Community College

Glenn Merrick
Lake Superior College

Ralph R. Meyer
University of Cincinnati

Amy Griffin Ouchley
University of Louisiana—Monroe

Mark Paternostro
Pennsylvania College of Technology

Jennifer L. Regan
University of Southern Mississippi

Jackie Reynolds
Richland College

Tim Roye
San Jacinto College—South

Gerald Schafer
Mission College—Thailand

Marilyn Shannon
Indiana University—Purdue University Fort Wayne

Jeff S. Simpson
Metropolitan State College of Denver

Claudia Stanescu
University of Arizona

William Stewart
Middle Tennessee State University

Janis Thompson
Lorain County Community College

Katherine M. Van de Wal
Community College of Baltimore County, Essex

Jyoti R. Wagle
Houston Community College System—Central

Mark D. Womble
Youngstown State University

Linda S. Wooten
Bishop State Community College

Michelle Zurawski
Moraine Valley Community College

Focus Group Participants:

Pegge Alciatore
University of Louisiana—Lafayette

Sara Brenizer
Shelton State Community College

Kathy Bruce
Parkland College

Jorge Cortese
Durham Technical Community College

Juville Dario-Becker
Central Virginia Community College

Smruti Desai
Cy-Fair College

Don Steve Dutton
Amarillo College

Glen Early
Jefferson Community and Technical College

Clair Eckersell
Brigham Young University—Idaho

Jim Ezell
J. Sargeant Reynolds Community College

Deanna Ferguson
Gloucester County College

Pamela Fouche
Walters State Community College

Mary Fox
University of Cincinnati

Bagie George
Georgia Perimeter College—Lawrenceville Campus

Peter Germroth
Hillsborough Community College—Dale Mabry Campus

Richard Griner
Augusta State University

Mark Hubley
Prince George's Community College

Sister Carol Makravitz
Luzerne County Community College

Ronald Markle
Northern Arizona University

Carl McAllister
Georgia Perimeter College

Alfredo Munoz
University of Texas at Brownsville

Necia Nicholas
Calhoun Community College

Robyn O'Kane
LaGuardia Community College/CUNY

Karen Payne
Chattanooga State Technical Community College

Joseph Schiller
Austin Peay State University

Colleen Sinclair
Towson University

Mary Elizabeth Torrano
American River College

Cinnamon VanPutte
Southwestern Illinois College

Chuck Venglarik
Jefferson State Community College

Mark Wygoda
McNeese State University

Teaching and Learning Supplements

McGraw-Hill offers various tools and technology products to support *Seeley's Principles of Anatomy and Physiology*. Students can order supplemental study materials by contacting their campus bookstore. Instructors can obtain teaching aids by calling the McGraw-Hill Customer Service Department at 1-800-338-3987, visiting our A&P catalog at www.mhhe.com/ap, or contacting their local McGraw-Hill sales representative.

Anatomy & Physiology | REVEALED® 2.0

Anatomy & Physiology | REVEALED® 2.0 is a unique multimedia study aid designed to help students learn and review human anatomy using digital cadaver specimens. Dissections, animations, imaging, and self-tests all work together as an exceptional tool for the study of structure and function. Anatomy & Physiology| REVEALED® 2.0 includes:

- Integumentary System
- Skeletal and Muscular Systems
- Nervous System
- Cardiovascular, Respiratory, and Lymphatic Systems
- Digestive, Urinary, Reproductive, and Endocrine Systems
- Expanded physiology content
- Histology material

A new online version of Anatomy & Physiology | REVEALED® will be available soon! Visit www.aprevealed.com for more information.

Broyles Workbook to Accompany Anatomy & Physiology | Revealed©, Version 2.0

This workbook/study guide, by Robert Broyles, is designed to help students get the most they can out of the *Anatomy and Physiology Revealed* (APR), and out of their anatomy and physiology course. The Table of Contents closely follows the new CD set and is organized along the lines of a typical Anatomy and Physiology course. The individual exercises include dissection art from APR and also have review questions, tables, coloring exercises, terminology quiz questions, and reminders on key content.

ARIS Text Website

The ARIS website that accompanies this textbook includes tutorials, animations, practice quizzing, helpful Internet links and more—a whole semester's worth of study help for students. Instructors will find a complete electronic homework and course management system where they can create and share course materials and assignments with colleagues in just a few clicks of the mouse. Instructors can also edit questions, import their own content, and create announcements and/or due dates for assignments. ARIS offers automatic grading and reporting of easy-to-assign homework, quizzing, and testing. Check out www.aris.mhhe.com, select your subject and textbook, and start benefiting today!

New "Study on the Fly" Content!

Now students with portable media players can study anywhere, anytime.

- Audio tutorials, developed by topic and correlated to each chapter
- Audio quizzes
- Animations

Complete Set of Electronic Images and Assets for Instructors

Instructors, build instructional materials wherever, whenever, and however you want!

Part of the ARIS Website, the Presentation Center contains assets such as photos, artwork, animations, PowerPoints, and other media resources that can be used to create customized lectures, visually enhance tests and quizzes, and design compelling course websites or attractive printed support materials. All assets are copyrighted by McGraw-Hill Higher Education but can be used by instructors for classroom purposes. The visual resources in this collection include:

- **Art** Full-color digital files of all illustrations in the book can be readily incorporated into lecture presentations, exams, or custom-made classroom materials. In addition, all files are pre-inserted into blank PowerPoint slides for ease of lecture preparation.
- **Photos** The photos collection contains digital files of photographs from the text, which can be reproduced for multiple classroom uses.

- **Tables**   Every table that appears in the text has been saved in electronic form for use in classroom presentations and/or quizzes.
- **Animations**   Numerous full-color animations illustrating important anatomical structures or physiological processes are also provided. Harness the visual impact of concepts in motion by importing these files into classroom presentations or online course materials.
- **Lecture Outlines**   Specially prepared custom outlines for each chapter are offered in easy-to-use PowerPoint slides.

Online Computerized Test Bank

A comprehensive bank of test questions is provided within a computerized test bank powered by McGraw-Hill's flexible electronic testing program EZ Test Online. EZ Test Online allows instructors to create and access paper or online tests and quizzes in an easy-to-use program anywhere, at any time, without installing the testing software. Now, with EZ Test Online, instructors can select questions from multiple McGraw-Hill test banks or author their own, and then either print the test for paper distribution or give it online. Visit: www.eztestonline.com to learn more about creating and managing tests, online scoring and reporting, and support resources.

Laboratory Manual

The Anatomy and Physiology Laboratory Manual by Eric Wise of Santa Barbara City College is expressly written to coincide with the chapters of *Seeley's Principles of Anatomy and Physiology* by Phil Tate. This lab manual includes clear explanations of physiology experiments and computer simulations that serve as alternatives to frog experimentation. Each lab also includes an extensive set of review questions. Chapter Summary Data sheets at the end of appropriate exercises allow students to record answers from throughout the exercise on a single summary sheet that can be pulled out and handed in.

Electronic Books

If you, or your students, are ready for an alternative version of the traditional textbook, McGraw-Hill and VitalSource have partnered to introduce innovative and inexpensive electronic textbooks. By purchasing E-books from McGraw-Hill and VitalSource, students can save as much as 50% on selected titles delivered on the **most advanced** E-book platform available, VitalSource Bookshelf.

E-books from McGraw-Hill and VitalSource are *smart, interactive, searchable and portable*. VitalSource Bookshelf comes with a powerful suite of built-in tools that allow detailed searching, highlighting, note taking, and student-to-student or instructor-to-student note sharing. In addition, the media-rich E-book for *Principles of Anatomy and Physiology* integrates relevant animations and videos into the textbook content for a true multimedia learning experience.

E-books from McGraw-Hill and VitalSource will help students study smarter and quickly find the information they need while saving money. Instructors: contact your McGraw-Hill sales representative to discuss E-book packaging options.

Physiology Interactive Lab Simulations (Ph.I.L.S.) 3.0

This unique student study tool is the perfect way to reinforce key physiology concepts with powerful lab experiments. Created by Dr. Phil Stephens of Villanova University, the program offers 37 laboratory simulations that may be used to supplement or substitute for wet labs. Students can adjust variables, view outcomes, make predictions, draw conclusions, and print lab reports. The easy-to-use software offers the flexibility to change the parameters of the lab experiment— there is no limit to the amount of times an experiment can be repeated.

"MediaPhys" Tutorial

This physiology study aid offers detailed explanations, high-quality illustrations, and amazing animations to provide a thorough introduction to the world of physiology. MediaPhys is filled with interactive activities and quizzes to help reinforce physiology concepts that are often difficult to understand.

e-Instruction with CPS

The Classroom Performance System (CPS) is an interactive system that allows the instructor to administer in-class questions electronically. Students answer questions via hand-held remote control keypads (clickers), and their individual responses are logged into a grade book. Aggregated responses can be displayed in graphical form. Using this immediate feedback, the instructor can quickly determine if students understand the lecture topic, or if more clarification is needed. CPS promotes student participation, class productivity, and individual student confidence and accountability. Specially designed questions for e-Instruction to accompany *Principles of Anatomy and Physiology* are provided through the book's ARIS website.

Course Delivery Systems

In addition to McGraw-Hill's ARIS course management options, instructors can also design and control their course content with help from our partners WebCT, Blackboard, Top-Class, and eCollege. Course cartridges containing website content, online testing, and powerful student tracking features are readily available for use within these or any other HTML-based course management platforms.

Guided Tour

Streamlined and Complete—Introducing a Book Students Can Understand!

When polled, instructors ranked the following three goals highest in their assessment of what they wanted to achieve in their A & P course:

1. Improve Student Performance
2. Help My Students Apply Concepts and Think Critically
3. Help My Students Optimize Study Time

Seeley's Principles of Anatomy and Physiology by Phil Tate is written for these instructors. The text was carefully designed to give students everything they need in an Anatomy and Physiology textbook and nothing more. A foundation of clear, straight-forward writing; superior artwork; critical thinking based pedagogy; and relevant clinical coverage, have all built a textbook that supports the three goals above, and one that students really will understand.

Superior Artwork

A picture is worth a thousand words—especially when you're learning anatomy and physiology. Because words alone cannot convey the nuances of anatomy or the intricacies of physiology, Seeley's *Principles of Anatomy and Physiology* employs a dynamic program of full-color illustrations and photographs that support and further clarify the textual explanations. Brilliantly rendered and carefully reviewed for accuracy and consistency, the precisely labeled illustrations and photos provide concrete, visual reinforcement of important topics discussed throughout the text.

Realistic Anatomical Art

The anatomical figures in Seeley's *Anatomy and Physiology* have been carefully drawn to convey realistic, three-dimensional detail. Richly textured bones and artfully shaded muscles, organs, and vessels lend a sense of realism to the figures that helps students envision the appearance of actual structures within the body.

The colors used to represent different anatomical structures have been applied consistently throughout the book to help students easily identify structures in every figure.

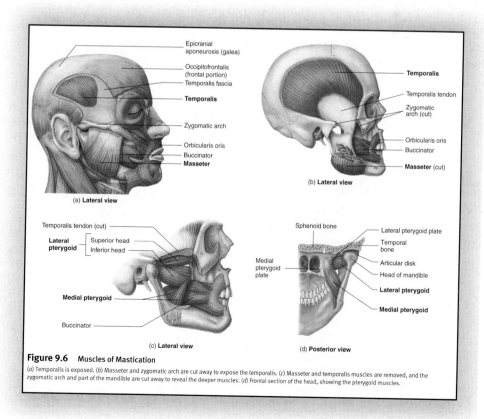

Figure 9.6   Muscles of Mastication

(*a*) Temporalis is exposed. (*b*) Masseter and zygomatic arch are cut away to expose the temporalis. (*c*) Masseter and temporalis muscles are removed, and the zygomatic arch and part of the mandible are cut away to reveal the deeper muscles. (*d*) Frontal section of the head, showing the pterygoid muscles.

Combination Art

Drawings are often paired with photographs to enhance visualization and comprehension of structures.

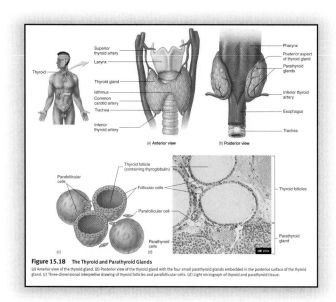

Figure 15.18 The Thyroid and Parathyroid Glands
(a) Anterior view of the thyroid gland. (b) Posterior view of the thyroid gland with the four small parathyroid glands embedded in the posterior surface of the thyroid gland. (c) Three-dimensional interpretive drawing of thyroid follicles and parafollicular cells. (d) Light micrograph of thyroid and parathyroid tissue.

Histology Micrographs

Light micrographs, as well as scanning and transmission electron micrographs, are used in conjunction with illustrations to present a true picture of anatomy and physiology from the cellular level.

Magnifications are indicated to help students estimate the size of structures shown in the photomicrographs.

Macro-to-Micro Art

Illustrations depicting complex structures or processes combine macroscopic and microscopic views to help students see the relationships between increasingly detailed drawings.

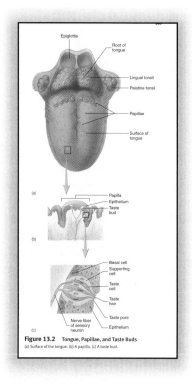

Figure 13.2 Tongue, Papillae, and Taste Buds
(a) Surface of the tongue. (b) A papilla. (c) A taste bud.

Atlas-Quality Cadaver Images

Clearly labeled photos of dissected human cadavers provide detailed views of anatomical structures, capturing the intangible characteristics of actual human anatomy that can be appreciated only when viewed in human specimens.

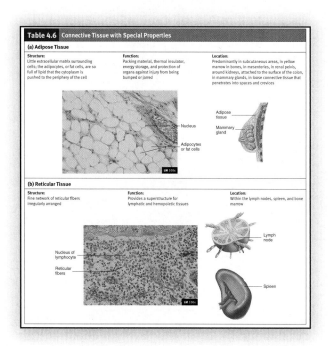

Table 4.6	Connective Tissue with Special Properties	
(a) Adipose Tissue		
Structure: Little extracellular matrix surrounding cells; the adipocytes, or fat cells, are so full of lipid that the cytoplasm is pushed to the periphery of the cell	**Function:** Packing material, thermal insulator, energy storage, and protection of organs against injury from being bumped or jarred	**Location:** Predominantly in subcutaneous areas, in yellow marrow in bones, in mesenteries, in renal pelvis, around kidneys, attached to the surface of the colon, in mammary glands, in loose connective tissue that penetrates into spaces and crevices
(b) Reticular Tissue		
Structure: Fine network of reticular fibers irregularly arranged	**Function:** Provides a superstructure for lymphatic and hemopoietic tissues	**Location:** Within the lymph nodes, spleen, and bone marrow

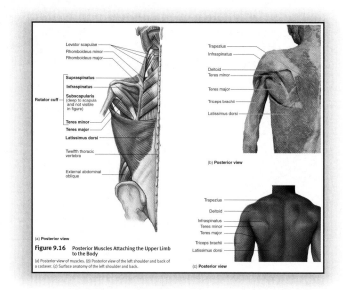

Figure 9.16 Posterior Muscles Attaching the Upper Limb to the Body
(a) Posterior view of muscles. (b) Posterior view of the left shoulder and back of a cadaver. (c) Surface anatomy of the left shoulder and back.

Specialized Figures Clarify Tough Concepts

Studying physiology does not have to be an intimidating task mired in memorization. *Seeley's Principles of Anatomy and Physiology* uses two special types of illustrations to help students not only learn the steps involved in specific processes, but also apply this knowledge as they predict outcomes in similar situations. **Process Figures** organize the key occurrences of physiological processes in an easy-to-follow format. **Homeostasis Summary Figures** detail the mechanisms of homeostasis by illustrating the means by which a system regulates a parameter within a narrow range of values.

Process Figures

Process Figures break down physiological processes into a series of smaller steps, allowing readers to build their understanding by learning each important phase.

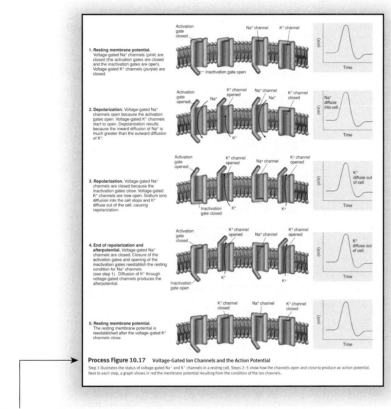

Process Figure 10.17   Voltage-Gated Ion Channels and the Action Potential

Step 1 illustrates the status of voltage-gated Na⁺ and K⁺ channels in a resting cell. Steps 2–5 show how the channels open and close to produce an action potential. Next to each step, a graph shows in *red* the membrane potential resulting from the condition of the ion channels.

Numbers are placed carefully in the art, permitting students to zero right in to where the action described in each step takes place.

Process Figures and Homeostasis Summary Figures are identified next to the figure number. The accompanying caption provides additional explanation.

Homeostasis Summary Figures

These specialized flowcharts illustrate the mechanisms that body systems employ to maintain homeostasis. Explanations within the figures help students better understand the process.

Changes caused by an increase of a variable outside its normal range are shown in the green boxes across the top.

The normal range for a given value is represented by the graphs in the center of each figure. Begin at the new yellow "Start" oval and follow the green arrows to learn about the chain of events triggered by an increase in the variable; or follow the red arrows for events resulting from a decrease in the variable.

Insulin levels increase:
- Increased blood glucose stimulates insulin secretion by the pancreas.
- Increased secretion of hormones associated with digestion, such as gastrin, secretin, and cholecystokinin, stimulate insulin secretion.
- Increased parasympathetic stimulation of the pancreas associated with digestion increases insulin secretion.

Follow the green arrows when blood glucose increases.

Blood glucose levels decrease:
- Most tissues increase glucose uptake (exceptions are the brain and liver, which do not depend on insulin for glucose uptake).
- The liver and skeletal muscle convert glucose to glycogen.
- Adipose tissue uses glucose to synthesize fats (triglycerides).

Blood glucose (normal range)

Start here

Blood glucose (normal range)

Blood glucose homeostasis is maintained.

Follow the red arrows when blood glucose decreases.

Changes caused by a decrease of a variable outside its normal range are shown in the red boxes across the bottom of the figure.

Insulin levels decrease:
- Decreased blood glucose inhibits insulin secretion by the pancreas.
- Low blood glucose levels increase epinephrine levels, which inhibits insulin secretion.
- Increased sympathetic stimulation of the pancreas during exercise decreases insulin secretion.

Blood glucose levels increase:
- Most tissues decrease glucose uptake, which makes glucose available for use by the brain.
- The liver breaks down glycogen to glucose and synthesizes glucose from amino acids. The glucose is released into the blood.
- Adipose tissue breaks down fats and releases fatty acids into the blood. The use of fatty acids by tissues as a source of energy spares glucose usage.
- The liver converts fatty acids into ketones. The use of ketones by tissues as a source of energy spares glucose usage.

Homeostasis Figure 15.26   Summary of Insulin Secretion Regulation

Relevant Clinical Content Puts Knowledge into Practice

Seeley's Principles of Anatomy and Physiology provides clinical examples to demonstrate the application of basic knowledge in a relevant clinical context. Exposure to clinical information is especially beneficial if students are planning on using their knowledge of anatomy and physiology in a health-related career.

Clinical Topics

Interesting clinical sidebars reinforce or expand upon the facts and concepts discussed within the narrative. Once students have learned a concept, applying that information in a clinical context shows them how their new knowledge can be put into practice.

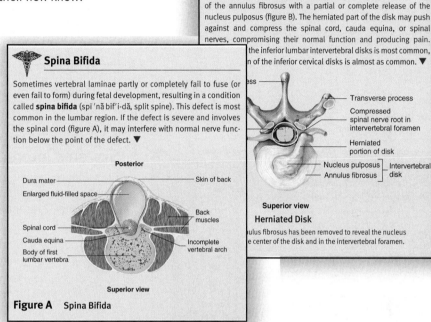

Herniated, or Ruptured, Intervertebral Disk

A **herniated**, or **ruptured**, **disk** results from the breakage or ballooning of the annulus fibrosus with a partial or complete release of the nucleus pulposus (figure B). The herniated part of the disk may push against and compress the spinal cord, cauda equina, or spinal nerves, compromising their normal function and producing pain. ...the inferior lumbar intervertebral disks is most common, ...n of the inferior cervical disks is almost as common. ▼

...ss

— Transverse process

Compressed spinal nerve root in intervertebral foramen

Herniated portion of disk

Nucleus pulposus ⎤ Intervertebral
Annulus fibrosus ⎦ disk

Superior view
Herniated Disk

...ulus fibrosus has been removed to reveal the nucleus ...e center of the disk and in the intervertebral foramen.

Spina Bifida

Sometimes vertebral laminae partly or completely fail to fuse (or even fail to form) during fetal development, resulting in a condition called **spina bifida** (spī′nă bif′i-dă, split spine). This defect is most common in the lumbar region. If the defect is severe and involves the spinal cord (figure A), it may interfere with normal nerve function below the point of the defect. ▼

Posterior

Dura mater —
Enlarged fluid-filled space —

— Skin of back

— Back muscles

Spinal cord —
Cauda equina —
Body of first lumbar vertebra —

— Incomplete vertebral arch

Superior view

Figure A   Spina Bifida

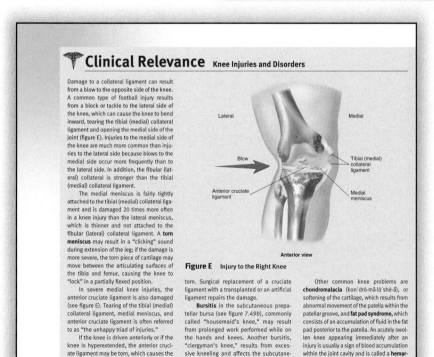

🜚 Clinical Relevance   Knee Injuries and Disorders

Damage to a collateral ligament can result from a blow to the opposite side of the knee. A common type of football injury results from a block or tackle to the lateral side of the knee, which can cause the knee to bend inward, tearing the tibial (medial) collateral ligament and opening the medial side of the joint (figure E). Injuries to the medial side of the knee are much more common than injuries to the lateral side because blows to the medial side occur more frequently than to the lateral side. In addition, the fibular (lateral) collateral is stronger than the tibial (medial) collateral ligament.

The medial meniscus is fairly tightly attached to the tibial (medial) collateral ligament and is damaged 20 times more often in a knee injury than the lateral meniscus, which is thinner and not attached to the fibular (lateral) collateral ligament. A **torn meniscus** may result in a "clicking" sound during extension of the leg; if the damage is more severe, the torn piece of cartilage may move between the articulating surfaces of the tibia and femur, causing the knee to "lock" in a partially flexed position.

In severe medial knee injuries, the anterior cruciate ligament is also damaged (see figure E). Tearing of the tibial (medial) collateral ligament, medial meniscus, and anterior cruciate ligament is often referred to as "the unhappy triad of injuries."

If the knee is driven anteriorly or if the knee is hyperextended, the anterior cruciate ligament may be torn, which causes the knee joint to be very unstable. If the knee is flexed and the tibia is driven posteriorly, the posterior cruciate ligament may be

Lateral

Medial

Blow

Anterior cruciate ligament

Tibial (medial) collateral ligament

Medial meniscus

Anterior view

Figure E   Injury to the Right Knee

torn. Surgical replacement of a cruciate ligament with a transplanted or an artificial ligament repairs the damage.

Bursitis in the subcutaneous prepatellar bursa (see figure 7.49b), commonly called "housemaid's knee," may result from prolonged work performed while on the hands and knees. Another bursitis, "clergyman's knee," results from excessive kneeling and affects the subcutaneous infrapatellar bursa. This type of bursitis is common among carpet layers and roofers.

Other common knee problems are **chondromalacia** (kon′drō-mă-lā′shē-ă), or softening of the cartilage, which results from abnormal movement of the patella within the patellar groove, and **fat pad syndrome**, which consists of an accumulation of fluid in the fat pad posterior to the patella. An acutely swollen knee appearing immediately after an injury is usually a sign of blood accumulation within the joint cavity and is called a **hemarthrosis** (hē′mar-thrō′sis, hem′ar-thrō′sis). A slower accumulation of fluid, "water on the knee," may be caused by bursitis.

Clinical Relevance

These in-depth boxed essays explore relevant topics of clinical interest. Subjects covered include pathologies, current research, sports medicine, exercise physiology, and pharmacology.

Case Studies

These specific, and yet brief, examples of how alternations of structure and/or function result in diseases helps students better understand the practical application of anatomy and physiology. The boxed summaries are placed strategically in the text so they can immediately start to see connections between learned concepts and realistic events.

CASE STUDY | Spinal Cord Injury

Harley was taking a curve on his motorcycle when he hit a patch of loose gravel on the side of the road. His motorcycle slid onto its side. Harley fell off and slid, as he was spun around, into a retaining wall. A passing motorist stopped at the wreck and saw that Harley was still conscious and breathing. The motorist (wisely) did not move Harley but dialed 911. The paramedics placed Harley on a back-support bed and rushed him to the hospital. Examination there revealed that Harley had suffered a fracture of his second lumbar vertebra. Methylprednisolone was administered to reduce inflammation around the spinal cord. His vertebrae were realigned and stabilized. A neurologist determined that the entire left half of Harley's spinal cord was destroyed at level L2. A lesion of the spinal cord that destroys half the cord (hemisection) at a specific level results in a very specific syndrome, called the Brown-Séquard syndrome.

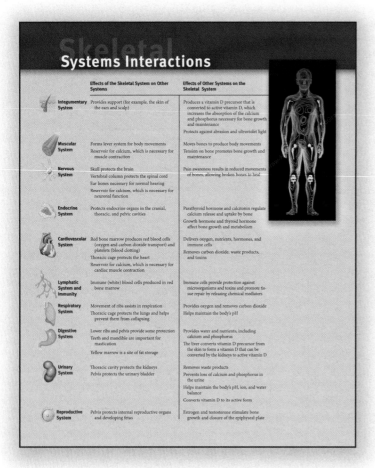

Systems Interactions

These pages appear at the end of every system chapter and summarize the important effects of all of the systems upon one another. Presented in table form, the page is clear and easy to read.

Critical Thinking Based Pedagogy

Learning anatomy and physiology is, in many ways, like learning a new language. Mastering the terminology is key to building a knowledge base. Once that base is established, students can learn to apply the knowledge and think critically. This is the basis of Bloom's taxonomy and the approach of the critical thinking learning system in *Seeley's Principles of Anatomy and Physiology.*

Building a Knowledge Base

Key terms are set in boldface where they are defined in the chapter, and most terms are included in the glossary at the end of the book. Pronunciation guides are provided for difficult words.

Because knowing the original meaning of a term can enhance understanding and retention, derivations of key words are given when they are relevant. Additionally, a handy list of prefixes, suffixes, and combining forms is printed on the inside back cover as a quick reference to help students identify commonly used word roots. A list of abbreviations used throughout the text is also provided.

as plants and animals. Molecules can combine to form **organelles** (or′gă-nelz, *organon,* a tool + *-elle,* small), which are the small structures that make up cells. For example, the nucleus contains the cell's hereditary information and mitochondria manufacture adenosine triphosphate (ATP), which is a molecule used by cells for a source of energy. Although cell types differ in their structure and function, they have many characteristics in common. Knowledge of these characteristics and their variations is essential to a basic understanding of anatomy and physiology. The cell is discussed in chapter 3.

5. *How are the olfactory bulbs related to the olfactory nerves and olfactory cortex?*
6. *Explain how smell can elicit various conscious and visceral responses.*

In-Chapter Section Reviews

Review questions at the end of sections *within* the chapter prompt students to test their understanding of key concepts. Readers can use them as a self-test to determine whether they have a sufficient grasp of the information before proceeding to the next section.

Chapter Summary

The summary outline briefly states the important facts and concepts covered in each chapter to provide a convenient "big picture" of the chapter content.

Review and Comprehension

These multiple-choice practice questions cover the main points of the chapter. Completing this self-test helps students gauge their mastery of the material. Answers are provided in Appendix E.

Critical Thinking

These innovative exercises encourage students to apply chapter concepts to solve problems. Answering these questions helps build their working knowledge of anatomy and physiology while developing reasoning and critical thinking skills. Answers are provided in Appendix F.

C R I T I C A L   T H I N K I N G

1. Given the observation that a tissue has more than one layer of cells lining a free surface, (a) list the possible tissue types that exhibit those characteristics, and (b) explain what additional observations need to be made to identify the tissue type.
2. A patient suffered from kidney failure a few days after he was exposed to a toxic chemical. A biopsy of his kidney indicated that many of the thousands of epithelium-lined tubules that make up the kidney had lost many of the simple cuboidal epithelial cells that normally line them, although most of the basement membranes appeared to be intact. Predict how likely this person is to recover fully.
3. What types of epithelium are likely to be found lining the trachea of a heavy smoker? Predict the changes that are likely to occur after he or she stops smoking for 1 or 2 years.

would be expected in the epithelium that is responsible for producing the digestive enzymes?
6. A tissue has the following characteristics: a free surface; a single layer of cells; narrow, tall cells; microvilli; many mitochondria; and goblet cells. Describe the tissue type and as many functions of the cell as possible.
7. Explain the consequences:
   a. if simple columnar epithelium replaced nonkeratinized stratified squamous epithelium that lines the mouth
   b. if tendons were dense elastic connective tissue instead of dense collagenous connective tissue
   c. if bones were made entirely of elastic cartilage
8. Antihistamines block the effect of a chemical mediator, histamine, that is released during the inflammatory response. Give an example

P R E D I C T   3

In tendons, collagen fibers are oriented parallel to the length of the tendon. In the skin, collagen fibers are oriented in many directions. What are the functional advantages of the fiber arrangements in tendons and in the skin?

Predict Questions

These innovative critical thinking questions encourage students to become active learners as they read. Predict Questions challenge the understanding of new concepts needed to solve a problem. Answers to the questions are provided at the end of the book, allowing students to evaluate their responses and to understand the logic used to arrive at the correct answer.

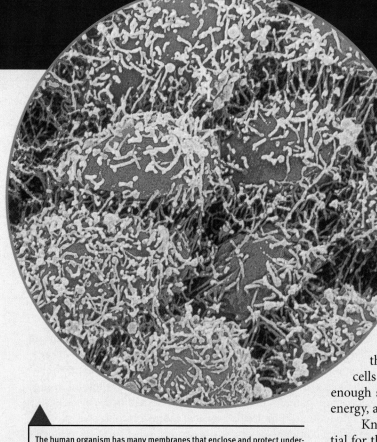

The human organism has many membranes that enclose and protect underlying structures. The colorized scanning electron micrograph shows the cells forming the peritoneum, a membrane covering abdominopelvic organs and the inside wall of the abdominopelvic cavity. These cells have many short, hairlike projections called microvilli. The microvilli increase the surface area of the cells, enabling them to secrete a slippery lubricating fluid that protects organs from friction as they rub against one another or the inside of the abdominopelvic wall.

Tools for Success

Anatomy & Physiology REVEALED®
aprevealed.com

A virtual cadaver dissection experience

Audio (MP3) and/or video (MP4) content, available at www.mhhe.aris.com

Introduction

Human anatomy and physiology is the study of the structure and function of the human body. The human body has many intricate parts with coordinated functions maintained by a complex system of checks and balances. For example, tiny collections of cells embedded in the pancreas affect the uptake and use of blood sugar in the body. Eating a candy bar results in an increase in blood sugar, which acts as a stimulus. Pancreatic cells respond to the stimulus by secreting insulin. Insulin moves into blood vessels and is transported to cells throughout the body, where it increases the movement of sugar from the blood into cells, thereby providing the cells with a source of energy and causing blood sugar levels to decrease.

Knowledge of the structure and function of the human body is the basis for understanding disease. In one type of diabetes mellitus, cells of the pancreas do not secrete adequate amounts of insulin. Not enough sugar moves into cells, which deprives them of a needed source of energy, and they malfunction.

Knowledge of the structure and function of the human body is essential for those planning a career in the health sciences. It is also beneficial to nonprofessionals because it helps with understanding overall health and disease, with evaluating recommended treatments, and with critically reviewing advertisements and articles.

▶This chapter defines **anatomy and physiology (p. 2)**. It also explains the body's **structural and functional organization (p. 2)** and provides an overview of the **characteristics of life (p. 4)** and **homeostasis (p. 7)**. Finally, the chapter presents **terminology and the body plan (p. 10)**.

Anatomy and Physiology

Anatomy (ă-nat′ŏ-mē) is the scientific discipline that investigates the structure of the body. The word *anatomy* means to dissect, or cut apart and separate, the parts of the body for study. Anatomy covers a wide range of studies, including the structure of body parts, their microscopic organization, and the processes by which they develop. In addition, anatomy examines the relationship between the structure of a body part and its function. Just as the structure of a hammer makes it well suited for pounding nails, the structure of body parts allows them to perform specific functions effectively. For example, bones can provide strength and support because bone cells surround themselves with a hard, mineralized substance. Understanding the relationship between structure and function makes it easier to understand and appreciate anatomy.

Systemic and regional anatomy are two basic approaches to the study of anatomy. **Systemic anatomy** is the study of the body by systems and is the approach taken in this and most other introductory textbooks. Examples of systems are the circulatory, nervous, skeletal, and muscular systems. **Regional anatomy** is the study of the organization of the body by areas. Within each region, such as the head, abdomen, or arm, all systems are studied simultaneously. It is the approach taken in most medical and dental schools.

Surface anatomy and anatomical imaging are used to examine the internal structures of a living person. **Surface anatomy** is the study of external features, such as bony projections, which serve as landmarks for locating deeper structures (for examples of external landmarks, see chapter 6). **Anatomical imaging** involves the use of x-rays, ultrasound, magnetic resonance imaging (MRI), and other technologies to create pictures of internal structures. Both surface anatomy and anatomical imaging provide important information useful in diagnosing disease.

Anatomical Anomalies

No two humans are structurally identical. For instance, one person may have longer fingers than another person. Despite this variability, most humans have the same basic pattern. Normally, we each have 10 fingers. **Anatomical anomalies** are structures that are unusual and different from the normal pattern. For example, some individuals have 12 fingers.

Anatomical anomalies can vary in severity from the relatively harmless to the life-threatening. For example, each kidney is normally supplied by one blood vessel, but in some individuals a kidney is supplied by two blood vessels. Either way, the kidney receives adequate blood. On the other hand, in the condition called "blue baby" syndrome certain blood vessels arising from the heart of an infant are not attached in their correct locations; blood is not effectively pumped to the lungs, resulting in tissues not receiving adequate oxygen.

Physiology (fiz-ē-ol′ō-jē, the study of nature) is the scientific discipline that deals with the processes or functions of living things. It is important in physiology to recognize structures as dynamic rather than static, or unchanging. The major goals of physiology are (1) to understand and predict the body's responses to stimuli and (2) to understand how the body maintains conditions within a narrow range of values in the presence of a continually changing environment.

Physiology is divided according to (1) the organisms involved or (2) the levels of organization within a given organism. **Human physiology** is the study of a specific organism, the human, whereas **cellular** and **systemic physiology** are examples of physiology that emphasize specific organizational levels. ▼

1. *Define anatomy and explain the importance of understanding the relationship between structure and function.*
2. *Define physiology and state two major goals of physiology.*
3. *Describe different ways or levels at which anatomy and physiology can be considered.*

Structural and Functional Organization

The body can be studied at six structural levels: the chemical, cell, tissue, organ, organ system, and organism (figure 1.1).

Chemical Level

The structural and functional characteristics of all organisms are determined by their chemical makeup. The **chemical** level of organization involves interactions between atoms, which are tiny building blocks of matter. Atoms can combine to form molecules, such as water, sugar, fats, proteins, and deoxyribose nucleic acid (DNA). The function of a molecule is related intimately to its structure. For example, collagen molecules are strong, ropelike fibers that give skin structural strength and flexibility. With old age, the structure of collagen changes, and the skin becomes fragile and is torn more easily. A brief overview of chemistry is presented in chapter 2.

Cell Level

Cells are the basic structural and functional units of organisms, such as plants and animals. Molecules can combine to form **organelles** (or′gă-nelz, *organon,* a tool + *-elle,* small), which are the small structures that make up cells. For example, the nucleus contains the cell's hereditary information and mitochondria manufacture adenosine triphosphate (ATP), which is a molecule used by cells for a source of energy. Although cell types differ in their structure and function, they have many characteristics in common. Knowledge of these characteristics and their variations is essential to a basic understanding of anatomy and physiology. The cell is discussed in chapter 3.

Tissue Level

A **tissue** (tish′ū, *texo,* to weave) is a group of similar cells and the materials surrounding them. The characteristics of the cells and surrounding materials determine the functions of the tissue. The

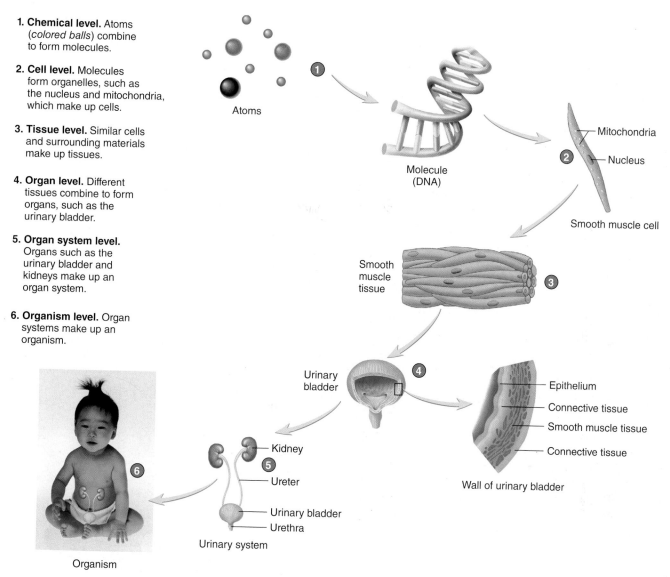

1. **Chemical level.** Atoms (*colored balls*) combine to form molecules.

2. **Cell level.** Molecules form organelles, such as the nucleus and mitochondria, which make up cells.

3. **Tissue level.** Similar cells and surrounding materials make up tissues.

4. **Organ level.** Different tissues combine to form organs, such as the urinary bladder.

5. **Organ system level.** Organs such as the urinary bladder and kidneys make up an organ system.

6. **Organism level.** Organ systems make up an organism.

Atoms

Molecule (DNA)

Mitochondria

Nucleus

Smooth muscle cell

Smooth muscle tissue

Urinary bladder

Epithelium

Connective tissue

Smooth muscle tissue

Connective tissue

Wall of urinary bladder

Kidney

Ureter

Urinary bladder

Urethra

Urinary system

Organism

Figure 1.1   Levels of Organization

Six levels of organization for the human body are the chemical, cell, tissue, organ, organ system, and organism levels.

many tissues that make up the body are classified into four primary tissue types: epithelial, connective, muscle, and nervous. Tissues are discussed in chapter 4.

Organ Level

An **organ** (or′găn, a tool) is composed of two or more tissue types that together perform one or more common functions. The urinary bladder, skin, stomach, eye, and heart are examples of organs (figure 1.2).

Organ System Level

An **organ system** is a group of organs classified as a unit because of a common function or set of functions. For example, the urinary system consists of the kidneys, ureter, urinary bladder, and urethra. The kidneys produce urine, which is transported by the ureters to the urinary bladder, where it is stored until eliminated from the body by passing through the urethra. In this book, the body is considered to have 11 major organ systems: the integumentary, skeletal, muscular, lymphatic, respiratory, digestive, nervous, endocrine, cardiovascular, urinary, and reproductive systems (figure 1.3).

The coordinated activity of the organ systems is necessary for normal function. For example, the digestive system takes in and processes food, which is carried by the blood of the cardiovascular system to the cells of the other systems. These cells use the food and produce waste products that are carried by the blood to the kidneys of the urinary system, which removes waste products from the blood. Because the organ systems are so interrelated, dysfunction of one organ system can have profound effects on other systems. For example, a heart attack can result in inadequate circulation of blood. Consequently, the organs of other systems, such as the brain and kidneys, can malfunction.

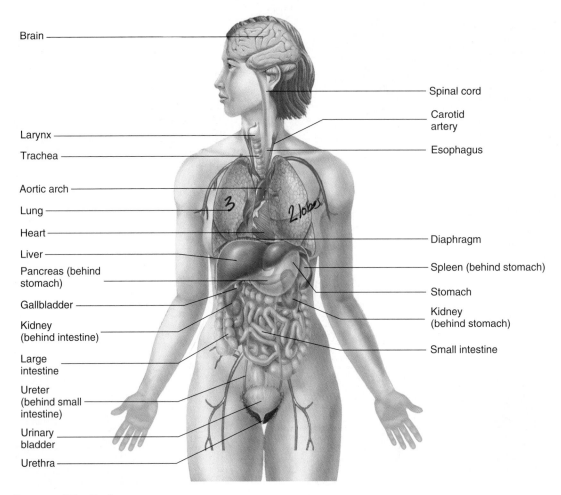

Figure 1.2   Organs of the Body

Organism Level

An **organism** is any living thing considered as a whole—whether composed of one cell, such as a bacterium, or trillions of cells, such as a human. The human organism is a complex of organ systems that are mutually dependent on one another. ▼

4. *From smallest to largest, list and define the body's six levels of organization.*
5. *What are the four primary tissue types?*
6. *Which two organ systems are responsible for regulating the other organ systems (see figure 1.3)? Which two are responsible for support and movement?*
7. *What are the functions of the integumentary, cardiovascular, lymphatic, respiratory, digestive, urinary, and reproductive systems (see figure 1.3)?*

P R E D I C T

In one type of diabetes, the pancreas (an organ) fails to produce insulin, which is a chemical normally made by pancreatic cells and released into the circulation. List as many levels of organization as you can in which this disorder could be corrected.

Characteristics of Life

Humans are organisms and have many characteristics in common with other organisms. The most important common feature of all organisms is life. Essential characteristics of life are organization, metabolism, responsiveness, growth, development, and reproduction.

1. **Organization** is the condition in which the parts of an organism have specific relationships to each other and the parts interact to perform specific functions. Living things are highly organized. All organisms are composed of one or more cells. Cells, in turn, are composed of highly specialized organelles, which depend on the precise functions of large molecules. Disruption of this organized state can result in loss of functions and death.

2. **Metabolism** (mĕ-tab′ō-lizm, *metabolē*, change) is the sum of the chemical and physical changes taking place in an organism. It includes an organism's ability to break down food molecules, which it uses as a source of energy and raw materials to synthesize its own molecules. Energy is also used when one part of a molecule moves relative to another

part, resulting in a change in shape of the molecule. Changes in molecular shape in turn can change the shape of cells, which can produce movements of the organism. Metabolism is necessary for vital functions, such as responsiveness, growth, development, and reproduction.

3. **Responsiveness** is the ability of an organism to sense changes in its external or internal environment and make the adjustments that help maintain its life. Responses include movement toward food or water and away from danger or poor environmental conditions.

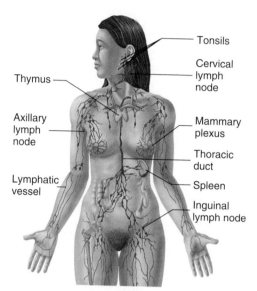

Lymphatic System

Removes foreign substances from the blood and lymph, combats disease, maintains tissue fluid balance, and transports fats from the digestive tract. Consists of the lymphatic vessels, lymph nodes, and other lymphatic organs.

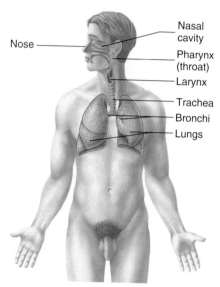

Respiratory System

Exchanges oxygen and carbon dioxide between the blood and air and regulates blood pH. Consists of the lungs and respiratory passages.

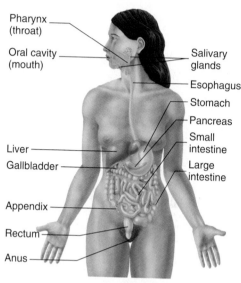

Digestive System

Performs the mechanical and chemical processes of digestion, absorption of nutrients, and elimination of wastes. Consists of the mouth, esophagus, stomach, intestines, and accessory organs.

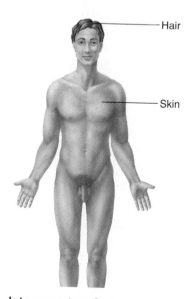

Integumentary System

Provides protection, regulates temperature, reduces water loss, and produces vitamin D precursors. Consists of skin, hair, nails, and sweat glands.

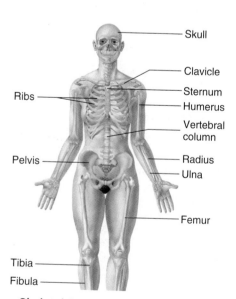

Skeletal System

Provides protection and support, allows body movements, produces blood cells, and stores minerals and fat. Consists of bones, associated cartilages, ligaments, and joints.

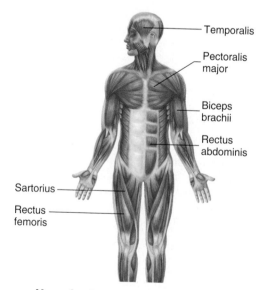

Muscular System

Produces body movements, maintains posture, and produces body heat. Consists of muscles attached to the skeleton by tendons.

Figure 1.3 Organ Systems of the Body

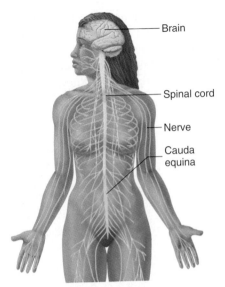

Nervous System

A major regulatory system that detects sensations and controls movements, physiologic processes, and intellectual functions. Consists of the brain, spinal cord, nerves, and sensory receptors.

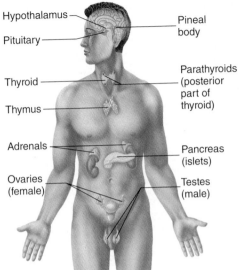

Endocrine System

A major regulatory system that influences metabolism, growth, reproduction, and many other functions. Consists of glands, such as the pituitary, that secrete hormones.

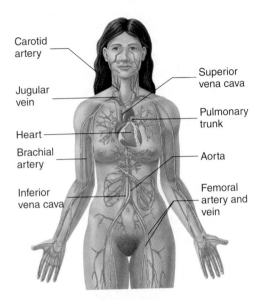

Cardiovascular System

Transports nutrients, waste products, gases, and hormones throughout the body; plays a role in the immune response and the regulation of body temperature. Consists of the heart, blood vessels, and blood.

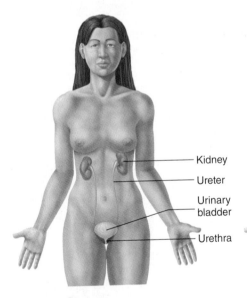

Urinary System

Removes waste products from the blood and regulates blood pH, ion balance, and water balance. Consists of the kidneys, urinary bladder, and ducts that carry urine.

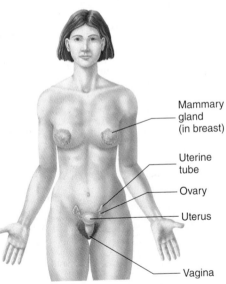

Female Reproductive System

Produces oocytes and is the site of fertilization and fetal development; produces milk for the newborn; produces hormones that influence sexual functions and behaviors. Consists of the ovaries, vagina, uterus, mammary glands, and associated structures.

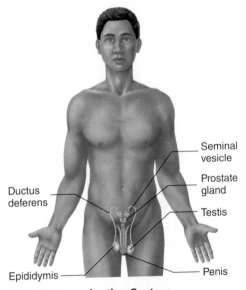

Male Reproductive System

Produces and transfers sperm cells to the female and produces hormones that influence sexual functions and behaviors. Consists of the testes, accessory structures, ducts, and penis.

Figure 1.3 *(continued)*

Organisms can also make adjustments that maintain their internal environment. For example, if body temperature increases in a hot environment, sweat glands produce sweat, which can lower body temperature back toward normal levels.

4. **Growth** results in an increase in size of all or part of the organism. It can result from an increase in cell number, cell size, or the amount of substance surrounding cells. For example, bones become larger as the number of bone cells increases and they surround themselves with bone matrix.

5. **Development** includes the changes an organism undergoes through time; it begins with fertilization and ends at death. The greatest developmental changes occur before birth, but many changes continue after birth, and some continue throughout life. Development usually involves growth, but it also involves differentiation. **Differentiation** is change in cell structure and function from generalized to specialized. For example, following fertilization, generalized cells specialize to become specific cell types, such as skin, bone, muscle, or nerve cells. These differentiated cells form the tissues and organs.

6. **Reproduction** is the formation of new cells or new organisms. Without reproduction of cells, growth and development are impossible. Without reproduction of the organism, the species becomes extinct.

Humans share many characteristics with other organisms, and much of the knowledge about humans has come from studying other organisms. For example, the study of bacteria (single-celled organisms) has provided much information about human cells, and great progress in open heart surgery was made possible by perfecting techniques on other mammals before attempting them on humans. Because other organisms are also different from humans, the ultimate answers to questions about humans can be obtained only from humans. ▼

8. *Describe six characteristics of life.*
9. *Why is it important to realize that humans share many, but not all, characteristics with other animals?*

Homeostasis

Homeostasis (hō′mē-ō-stā′sis, homeo-, the same + *stasis*, standing) is the existence and maintenance of a relatively constant environment within the body. A small amount of fluid surrounds most cells of the body. Normal cell functions depend on the maintenance of the cells' fluid environment within a narrow range of conditions, including temperature, volume, and chemical content. These conditions are called **variables** because their values can change. For example, body temperature is a variable that can increase in a hot environment or decrease in a cold environment.

Homeostatic mechanisms, such as sweating or shivering, normally maintain body temperature near an ideal normal value, or **set point** (figure 1.4). Note that these mechanisms are not able to maintain body temperature precisely at the set point. Instead, body temperature increases and decreases slightly around the

set point, producing a **normal range** of values. As long as body temperatures remain within this normal range, homeostasis is maintained.

The organ systems help control the body's internal environment so that it remains relatively constant. For example, the digestive, respiratory, circulatory, and urinary systems function together so that each cell in the body receives adequate oxygen and nutrients and so that waste products do not accumulate to a toxic level. If the fluid surrounding cells deviates from homeostasis, the cells do not function normally and can even die. Disease disrupts homeostasis and sometimes results in death.

Negative Feedback

Most systems of the body are regulated by **negative-feedback mechanisms,** which maintain homeostasis. *Negative* means that any deviation from the set point is made smaller or is resisted. Negative feedback does not prevent variation but maintains variation within a normal range.

Many negative-feedback mechanisms have three components: a **receptor,** which monitors the value of a variable; a **control center,** which receives information about the variable from the receptor, establishes the set point, and controls the effector; and an **effector,** which produces responses that change the value of the variable.

The maintenance of normal blood pressure is an example of a negative-feedback mechanism. Normal blood pressure is necessary for the movement of blood from the heart to tissues. The blood supplies the tissues with oxygen and nutrients and removes waste products, thus maintaining tissue homeostasis. Several negative-feedback mechanisms regulate blood pressure, and they are described more fully in chapters 17 and 18. One negative-feedback mechanism regulating blood pressure is described here. Receptors that monitor blood pressure are located within large blood vessels near the heart and head. A control center located in the brain receives signals sent through nerves from the receptors. The control center evaluates the information and sends signals through nerves to the heart. The heart is the effector, and the heart rate increases or decreases in

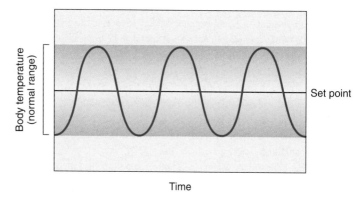

Figure 1.4   Homeostasis

Homeostasis is the maintenance of a variable, such as body temperature, around an ideal normal value, or set point. The value of the variable fluctuates around the set point to establish a normal range of values.

---

response to signals from the brain (figure 1.5). As heart rate increases, blood pressure increases, and as heart rate decreases, blood pressure decreases.

If blood pressure increases slightly, the receptors detect the increased blood pressure and send that information to the control center in the brain. The control center causes heart rate to decrease, resulting in a decrease in blood pressure. If blood pressure decreases slightly, the receptors inform the control center, which increases heart rate, thereby producing an increase in blood pressure (figure 1.6). As a result, blood pressure constantly rises and falls within a normal range of values.

PREDICT

Donating a pint of blood reduces blood volume, which results in a decrease in blood pressure (just as air pressure in a tire decreases as air is let out of the tire). What effect does donating blood have on heart rate? What would happen if a negative-feedback mechanism did not return the value of some parameter, such as blood pressure, to its normal range?

CASE STUDY | Orthostatic Hypotension

Molly is a 75-year-old widow who lives alone. For 2 days, she had a fever and chills and stayed mostly in bed. On rising to go to the bathroom, she felt dizzy, fainted, and fell to the floor. Molly quickly regained consciousness and managed to call her son, who took her to the emergency room, where a diagnosis of orthostatic hypotension was made.

Orthostasis literally means to stand and *hypotension* refers to low blood pressure. **Orthostatic hypotension** is a significant drop in blood pressure on standing. When a person changes position from lying down to standing, blood "pools" within the veins below the heart because of gravity, and less blood returns to the heart. Consequently, blood pressure decreases because the heart has less blood to pump.

PREDICT

Although orthostatic hypotension has many causes, in the elderly it can be due to age-related decreased neural and cardiovascular responses. Dehydration can result from decreased fluid intake while feeling ill and from sweating as a result of a fever. Dehydration can decrease blood volume and lower blood pressure, increasing the likelihood of orthostatic hypotension. Use figure 1.5 to answer the following:

a. Describe the normal response to a decrease in blood pressure on standing.
b. What happened to Molly's heart rate just before she fainted? Why did Molly faint?
c. How did Molly's fainting and falling to the floor assist in establishing homeostasis (assuming she was not injured)?

Although homeostasis is the maintenance of a normal range of values, this does not mean that all variables are maintained within the same narrow range of values at all times. Sometimes a deviation from the usual range of values can be beneficial. For example, during exercise the normal range for blood pressure differs from the range under resting conditions, and the blood pressure is significantly elevated (figure 1.7). Muscle cells require increased oxygen and nutrients and increased removal of waste products to support their increased level of activity during exercise. Increased blood pressure increases blood delivery to muscles, which maintains muscle cell homeostasis during exercise by increasing the delivery of oxygen and nutrients and the removal of waste products. ▼

10. *Define homeostasis, variable, set point, and normal range of values. If a deviation from homeostasis occurs, what mechanism restores it?*
11. *What are the three components of many negative-feedback mechanisms? How do they maintain homeostasis?*

PREDICT

Explain how negative-feedback mechanisms control respiratory rates when a person is at rest and when a person is exercising.

Positive Feedback

Positive-feedback mechanisms are *not* homeostatic and are rare in healthy individuals. *Positive* implies that, when a deviation from a normal value occurs, the response of the system is to make the deviation even greater (figure 1.8). Positive feedback therefore usually creates a cycle leading away from homeostasis and in some cases results in death.

Inadequate delivery of blood to cardiac (heart) muscle is an example of positive feedback. Contraction of cardiac muscle generates blood pressure and moves blood through blood vessels to tissues. A system of blood vessels on the outside of the heart provides cardiac muscle with a blood supply sufficient to allow normal contractions to occur. In effect, the heart pumps blood to itself. Just as with other tissues, blood pressure must be maintained to ensure adequate delivery of blood to cardiac muscle. Following extreme blood loss, blood pressure decreases to the point at which the delivery of blood to cardiac muscle is inadequate. As a result, cardiac muscle homeostasis is disrupted, and cardiac muscle does not function normally. The heart pumps less blood, which causes the blood pressure to drop even further. The additional decrease in blood pressure causes less blood delivery to cardiac muscle, and the heart pumps even less blood, which again decreases the blood pressure (figure 1.9). The process continues until the blood pressure is too low to sustain the cardiac muscle, the heart stops beating, and death results.

Following a moderate amount of blood loss (e.g., after donating a pint of blood), negative-feedback mechanisms result in an increase in heart rate and other responses that restore blood pressure. If blood loss is severe, however, negative-feedback mechanisms may not be able to maintain homeostasis, and the positive-feedback effect of an ever-decreasing blood pressure can develop.

1. Receptors monitor the value of a variable—in this case, receptors in the wall of a blood vessel monitor blood pressure.

2. Information about the value of the variable is sent to a control center. In this case, information is sent by nerves to the part of the brain responsible for regulating blood pressure.

3. The control center compares the value of the variable against the set point.

4. If a response is necessary to maintain homestasis, the control center causes an effector to respond. In this case, information is sent by nerves to the heart.

5. An effector produces a response that maintains homeostasis. In this case, changing heart rate changes blood pressure.

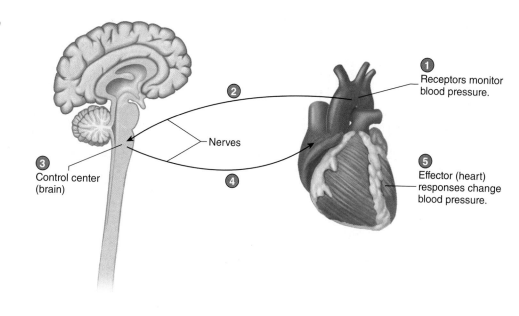

① Receptors monitor blood pressure.

② Nerves

③ Control center (brain)

④

⑤ Effector (heart) responses change blood pressure.

Process Figure 1.5 Negative-Feedback Mechanism: Blood Pressure

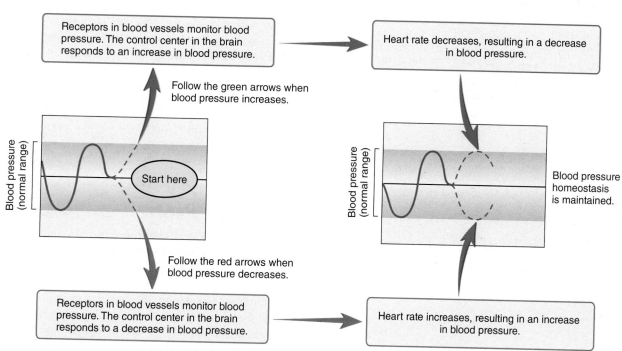

Receptors in blood vessels monitor blood pressure. The control center in the brain responds to an increase in blood pressure.

Heart rate decreases, resulting in a decrease in blood pressure.

Follow the green arrows when blood pressure increases.

Start here

Blood pressure (normal range)

Follow the red arrows when blood pressure decreases.

Receptors in blood vessels monitor blood pressure. The control center in the brain responds to a decrease in blood pressure.

Heart rate increases, resulting in an increase in blood pressure.

Blood pressure homeostasis is maintained.

Homeostasis Figure 1.6 Summary of Negative-Feedback Mechanism: Blood Pressure

Blood pressure is maintained within a normal range by negative-feedback mechanisms.

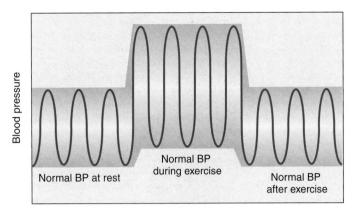

Figure 1.7 Changes in Blood Pressure During Exercise

During exercise, the demand for oxygen by muscle tissue increases. An increase in blood pressure (BP) results in an increase in blood flow to the tissues. The increased blood pressure is not an abnormal or a nonhomeostatic condition but is a resetting of the normal homeostatic range to meet the increased demand. The reset range is higher and broader than the resting range. After exercise ceases, the range returns to that of the resting condition.

Circumstances in which negative-feedback mechanisms are not adequate to maintain homeostasis illustrate a basic principle. Many disease states result from the failure of negative-feedback mechanisms to maintain homeostasis. The purpose of medical therapy is to overcome illness by assisting negative-feedback mechanisms. For example, a transfusion reverses a constantly decreasing blood pressure and restores homeostasis.

A few positive-feedback mechanisms do operate in the body under normal conditions, but in all cases they are eventually limited in some way. Birth is an example of a normally occurring positive-feedback mechanism. Near the end of pregnancy, the uterus is stretched by the baby's large size. This stretching, especially around the opening of the uterus, stimulates contractions of the uterine muscles. The uterine contractions push the baby

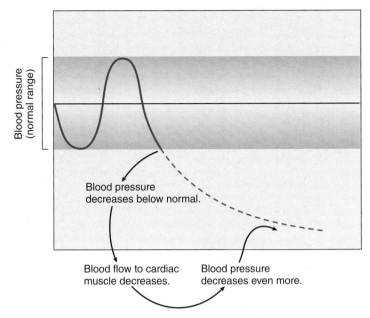

Figure 1.9 Example of Harmful Positive Feedback

A decrease in blood pressure below the normal range causes decreased blood flow to the heart. The heart is unable to pump enough blood to maintain blood pressure, and blood flow to the cardiac muscle decreases. Thus, the heart's ability to pump decreases further, and blood pressure decreases even more.

against the opening of the uterus, stretching it further. This stimulates additional contractions that result in additional stretching. This positive-feedback sequence ends only when the baby is delivered from the uterus and the stretching stimulus is eliminated. ▼

> 12. *Define positive feedback. Why are positive-feedback mechanisms often harmful? Give an example of a harmful and a beneficial positive-feedback mechanism.*

P R E D I C T   **5**

Is the sensation of thirst associated with a negative- or a positive-feedback mechanism? Explain. (*Hint:* What is being regulated when one becomes thirsty?)

Terminology and the Body Plan

When you begin to study anatomy and physiology, the number of new words may seem overwhelming. Learning is easier and more interesting if you pay attention to the origin, or **etymology** (et′ĕ-mol′o-jē), of new words. Most of the terms are derived from Latin or Greek and are descriptive in the original languages. For example, *foramen* is a Latin word for hole, and *magnum* means large. The foramen magnum is therefore a large hole in the skull through which the spinal cord attaches to the brain.

Words are often modified by adding a prefix or suffix. The suffix *-itis* means an inflammation; so appendicitis is an

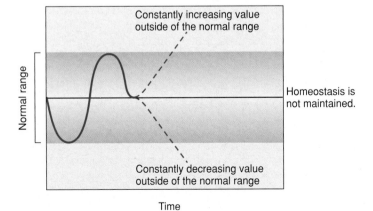

Figure 1.8 Positive Feedback

Deviations from the normal set point value cause an additional deviation away from that value in either a positive or negative direction.

inflammation of the appendix. As new terms are introduced in this text, their meanings are often explained. The glossary and the list of word roots, prefixes, and suffixes on the inside back cover of the textbook provide additional information about the new terms.

Body Positions

Anatomical position refers to a person standing erect with the face directed forward, the upper limbs hanging to the sides, and the palms of the hands facing forward (figure 1.10). A person is **supine** when lying face upward and **prone** when lying face downward.

The position of the body can affect the description of body parts relative to each other. In the anatomical position, the elbow is above the hand, but in the supine or prone position, the elbow and hand are at the same level. To avoid confusion, relational descriptions are always based on the anatomical position, no matter the actual position of the body. Thus, the elbow is always described as being above the wrist, whether the person is lying down or is even upside down. ▼

13. *What is the anatomical position in humans? Why is it important?*
14. *Define supine and prone.*

Directional Terms

Directional terms describe parts of the body relative to each other (see figure 1.10 and table 1.1). It is important to become familiar with these directional terms as soon as possible because you will see them repeatedly throughout the text. *Right* and *left* are retained as directional terms in anatomical terminology. *Up* is replaced by **superior,** *down* by **inferior.** In humans, *superior* is synonymous with **cephalic** (se-fal′ik), which means toward the head, because, when we are in the anatomic position, the head is the highest point. In humans, the term *inferior* is synonymous with **caudal** (kaw′dăl), which means toward the tail, which would be located at the end of the vertebral column if humans had tails. The terms *cephalic* and *caudal* can be used to describe directional movements on the trunk, but they are not used to describe directional movements on the limbs.

Proximal means nearest, whereas **distal** means distant. These terms are used to refer to linear structures, such as the limbs, in which one end is near some other structure and the other end is farther away. Each limb is attached at its proximal end to the body, and the distal end, such as the hand, is farther away.

In anatomical terminology, *front* is replaced by **anterior,** *back* by **posterior.** The word *anterior* means that which goes before, and **ventral** means belly. The anterior surface of the human body is therefore the ventral surface, or belly, because the

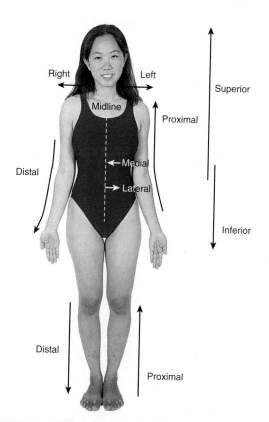
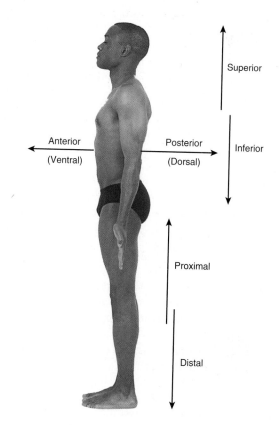

Figure 1.10   Directional Terms

All directional terms are in relation to a person in the anatomical position: a person standing erect with the face directed forward, the arms hanging to the sides, and the palms of the hands facing forward.

Table 1.1 Directional Terms for Humans

Terms	Etymology*	Definition	Example
Right		Toward the right side of the body	The right ear
Left		Toward the left side of the body	The left eye
Superior	L., higher	A structure above another	The chin is superior to the navel.
Inferior	L., lower	A structure below another	The navel is inferior to the chin.
Cephalic	G. *kephale*, head	Closer to the head than another structure (usually synonymous with *superior*)	The chin is cephalic to the navel.
Caudal	L. *cauda*, a tail	Closer to the tail than another structure (usually synonymous with *inferior*)	The navel is caudal to the chin.
Anterior	L., before	The front of the body	The navel is anterior to the spine.
Posterior	L. *posterus*, following	The back of the body	The spine is posterior to the breastbone.
Ventral	L. *ventr-*, belly	Toward the belly (synonymous with *anterior*)	The navel is ventral to the spine.
Dorsal	L. *dorsum*, back	Toward the back (synonymous with *posterior*)	The spine is dorsal to the breastbone.
Proximal	L. *proximus*, nearest	Closer to the point of attachment to the body than another structure	The elbow is proximal to the wrist.
Distal	L. *di-* plus *sto*, to stand apart or be distant	Farther from the point of attachment to the body than another structure	The wrist is distal to the elbow.
Lateral	L. *latus*, side	Away from the midline of the body	The nipple is lateral to the breastbone.
Medial	L. *medialis*, middle	Toward the midline of the body	The bridge of the nose is medial to the eye.
Superficial	L. *superficialis*, toward the surface	Toward or on the surface (not shown in figure 1.10)	The skin is superficial to muscle.
Deep	O.E. *deop*, deep	Away from the surface, internal (not shown in figure 1.10)	The lungs are deep to the ribs.

*Origin and meaning of the word: L., Latin; G., Greek; O.E., Old English.

belly "goes first" when we are walking. The word *posterior* means that which follows, and **dorsal** means back. The posterior surface of the body is the dorsal surface, or back, which follows as we are walking.

Medial means toward the midline, and **lateral** means away from the midline. The nose is located in a medial position in the face, and the eyes are lateral to the nose. The term **superficial** refers to a structure close to the surface of the body, and **deep** is toward the interior of the body. The skin is superficial to muscle and bone. ▼

15. *List two terms that in humans indicate* toward the head. *Name two terms that mean the opposite.*
16. *List two terms that indicate* the back *in humans. What two terms mean* the front?
17. *Define the following terms, and give the word that means the opposite:* proximal, lateral, *and* superficial.

P R E D I C T
Use as many directional terms as you can to describe the relationship between your kneecap and your heel.

Body Parts and Regions

A number of terms are used when referring to different regions or parts of the body (figure 1.11). The upper limb is divided into the arm, forearm, wrist, and hand. The **arm** extends from the shoulder to the elbow, and the **forearm** extends from the elbow to the wrist. The lower limb is divided into the thigh, leg, ankle, and foot. The **thigh** extends from the hip to the knee, and the **leg** extends from the knee to the ankle. Note that, contrary to popular usage, the terms *arm* and *leg* refer to only a part of the respective limb.

The central region of the body consists of the **head, neck, and trunk.** The trunk can be divided into the **thorax** (chest), **abdomen** (region between the thorax and pelvis), and **pelvis** (the inferior end of the trunk associated with the hips).

The abdomen is often subdivided superficially into four **quadrants** by two imaginary lines—one horizontal and one vertical—that intersect at the navel (figure 1.12a). The quadrants formed are the right-upper, left-upper, right-lower, and left-lower quadrants. In addition to these quadrants, the abdomen is sometimes subdivided into nine **regions** by four imaginary lines—two horizontal and two vertical. These four lines create

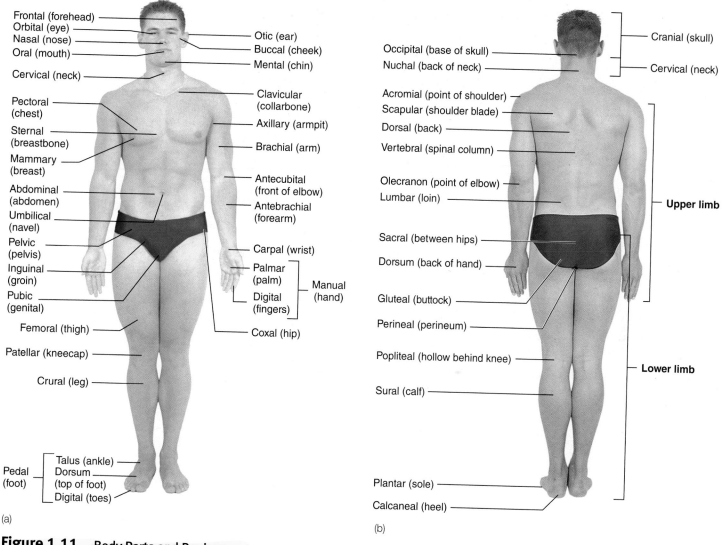

Figure 1.11   **Body Parts and Regions**

The anatomical and common (*in parentheses*) names are indicated for some parts and regions of the body. (*a*) Anterior view. (*b*) Posterior view.

an imaginary tic-tac-toe figure on the abdomen, resulting in nine regions: epigastric (ep-i-gas′trik), right and left hypo-chondriac (hī-pō-kon′drē-ak), umbilical (ŭm-bil′i-kăl), right and left lumbar (lŭm′bar), hypogastric (hī-pō-gas′trik), and right and left iliac (il′ē-ak) (figure 1.12*b*). Clinicians use the quadrants or regions as reference points for locating the underlying organs. For example, the appendix is located in the right-lower quadrant, and the pain of an acute appendicitis is usually felt there. ▼

18. *What is the difference between the arm and the upper limb and the difference between the leg and the lower limb?*

19. *Describe the quadrant and the nine-region methods of subdividing the abdominal region. What is the purpose of these subdivisions?*

Planes

At times it is conceptually useful to discuss the body in reference to a series of planes (imaginary flat surfaces) passing through it (figure 1.13). Sectioning the body is a way to "look inside" and observe the body's structures. A **sagittal** (saj′i-tăl) **plane** runs vertically through the body and separates it into right and left parts. The word *sagittal* literally means the flight of an arrow and refers to the way the body would be split by an arrow passing anteriorly to posteriorly. A **median plane** is a sagittal plane that passes through the midline of the body and divides it into equal right and left halves. A **transverse** (trans-vers′), or **horizontal, plane** runs parallel to the surface of the ground and divides the body into superior and inferior parts. A **frontal,** or **coronal** (kōr′ŏ-năl, kō-rō′nal), **plane** runs vertically from right to left and divides the body into anterior and posterior parts.

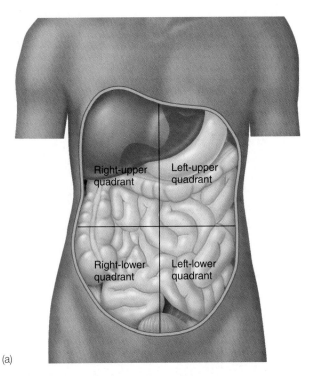

(a)

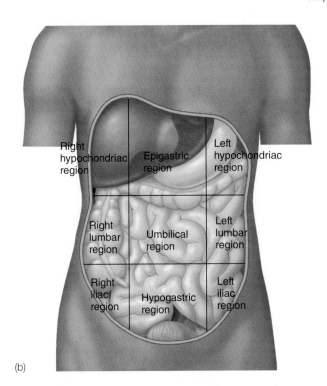

(b)

Figure 1.12 **Subdivisions of the Abdomen**

Lines are superimposed over internal organs to demonstrate the relationship of the organs to the subdivisions. (*a*) Abdominal quadrants consist of four subdivisions. (*b*) Abdominal regions consist of nine subdivisions.

Organs are often sectioned to reveal their internal structure (figure 1.14). A cut through the long axis of the organ is a **longitudinal section,** and a cut at a right angle to the long axis is a **cross,** or **transverse, section.** If a cut is made across the long axis at other than a right angle, it is called an **oblique section.** ▼

> 20. *Define the three planes of the body.*
> 21. *In what three ways can an organ be cut?*

Body Cavities

The body contains many cavities. Some of these cavities, such as the nasal cavity, open to the outside of the body, and some do not. The trunk contains three large cavities that do not open to the outside of the body: the thoracic cavity, the abdominal cavity, and the pelvic cavity (figure 1.15). The rib cage surrounds the **thoracic cavity,** and the muscular diaphragm separates it from the abdominal cavity. A median structure called the **mediastinum** (mē′dē-as-tī′nŭm, wall) divides the thoracic cavity into right and left parts. The mediastinum is a partition containing the heart, thymus, trachea, esophagus, and other structures. The two lungs are located on either side of the mediastinum.

Abdominal muscles primarily enclose the **abdominal cavity,** which contains the stomach, intestines, liver, spleen, pancreas, and kidneys. Pelvic bones encase the small space known as the **pelvic cavity,** where the urinary bladder, part of the large intestine, and the internal reproductive organs are housed. The abdominal and pelvic cavities are not physically separated and

sometimes are called the **abdominopelvic** (ab-dom′i-nō-pel′vik) **cavity.** ▼

> 22. *What structure separates the thoracic cavity from the abdominal cavity? The abdominal cavity from the pelvic cavity?*
> 23. *What structure divides the thoracic cavity into right and left parts?*

Serous Membranes

Serous (sēr′ŭs) **membranes** cover the organs of the trunk cavities and line the trunk cavities. Imagine an inflated balloon into which a fist has been pushed (figure 1.16). The fist represents an organ and the balloon represents a serous membrane. The inner balloon wall in contact with the fist (organ) represents the **visceral** (vis′er-ăl, organ) **serous membrane,** and the outer part of the balloon wall represents the **parietal** (pă-rī′ĕ-tăl, wall) **serous membrane.** The cavity, or space, between the visceral and parietal serous membranes is normally filled with a thin, lubricating film of **serous fluid** produced by the membranes. As an organ rubs against another organ or against the body wall, the serous fluid and smooth serous membranes function to reduce friction.

The thoracic cavity contains three serous membrane-lined cavities: a pericardial cavity and two pleural cavities. The **pericardial** (per-i-kar′dē-ăl, around the heart) **cavity** surrounds the heart (figure 1.17*a*). The visceral pericardium covers the heart, which is contained within a connective tissue sac lined with the parietal pericardium. The pericardial cavity is located between

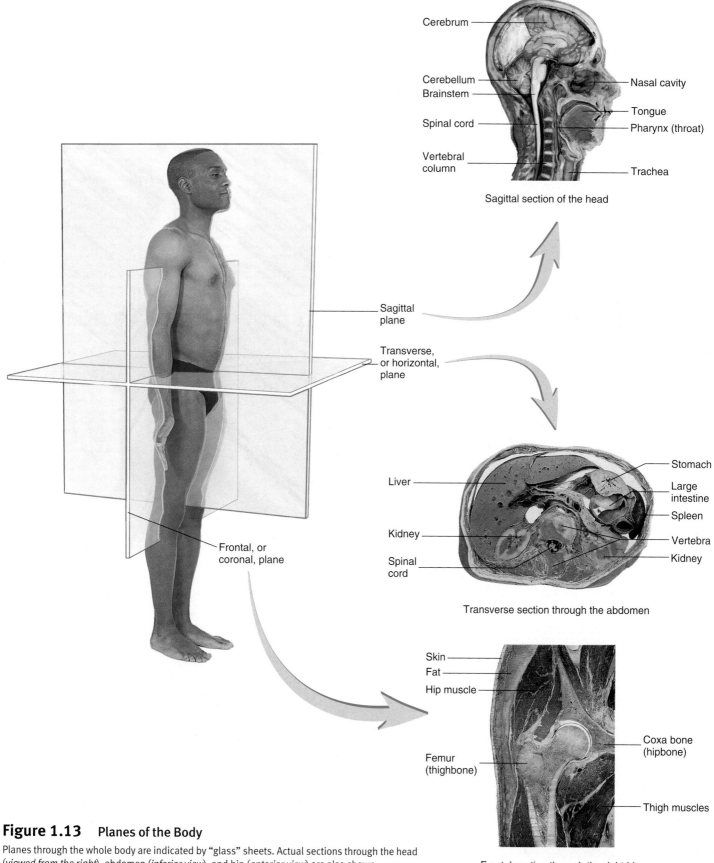

Sagittal section of the head

Transverse section through the abdomen

Frontal section through the right hip

Figure 1.13   Planes of the Body

Planes through the whole body are indicated by "glass" sheets. Actual sections through the head (*viewed from the right*), abdomen (*inferior view*), and hip (*anterior view*) are also shown.

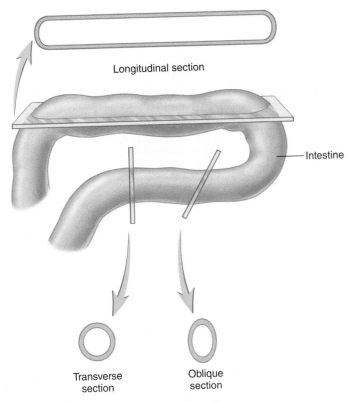

Longitudinal section

Intestine

Transverse
section

Oblique
section

Figure 1.14 **Planes Through an Organ**

Planes through the small intestine are indicated by "glass" sheets. The views
of the small intestine after sectioning are also shown. Although the small
intestine is basically a tube, the sections appear quite different in shape.

the visceral pericardium and the parietal pericardium and contains pericardial fluid.

A **pleural** (ploor´ăl, associated with the ribs) **cavity** surrounds
each lung, which is covered by visceral pleura (figure 1.17*b*).
Parietal pleura lines the inner surface of the thoracic wall, the lateral
surfaces of the mediastinum, and the superior surface of the
diaphragm. The pleural cavity is located between the visceral
pleura and parietal pleura and contains pleural fluid.

The abdominopelvic cavity contains a serous membrane-
lined cavity called the **peritoneal** (per´i-tō-nē´ăl, to stretch over)
cavity (figure 1.17*c*). Visceral peritoneum covers many of the
organs of the abdominopelvic cavity. Parietal peritoneum lines the
wall of the abdominopelvic cavity and the inferior surface of the
diaphragm. The peritoneal cavity is located between the visceral
peritoneum and parietal peritoneum and contains peritoneal fluid.

Inflammation of Serous Membranes

The serous membranes can become inflamed—usually as a result
of an infection. Pericarditis (per´i-kar-dī´tis) is inflammation of the
pericardium, pleurisy (ploor´i-sē) is inflammation of the pleura,
and peritonitis (per´i-tō-nī´tis) is inflammation of the peritoneum.
Visceral peritoneum covers the appendix, which is a small,
wormlike sac attached to the large intestine. An infection of
the appendix can rupture its wall, releasing bacteria into the
peritoneal cavity, resulting in peritonitis. Appendicitis is the
most common cause of emergency abdominal surgery in
children and it often leads to peritonitis.

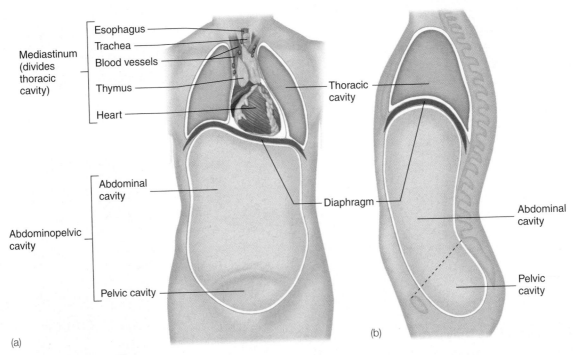

Esophagus
Trachea
Blood vessels

Mediastinum
(divides
thoracic
cavity)

Thymus

Heart

Thoracic
cavity

Abdominal
cavity

Abdominal
cavity

Abdominopelvic
cavity

Diaphragm

Pelvic cavity

Pelvic
cavity

(a)

(b)

Figure 1.15 **Trunk Cavities**

(*a*) Anterior view showing the major trunk cavities. The diaphragm separates the thoracic cavity from the abdominal cavity. The mediastinum, which includes the heart,
is a partition of organs dividing the thoracic cavity. (*b*) Sagittal section of the trunk cavities viewed from the left. The *dashed line* shows the division between the
abdominal and pelvic cavities. The mediastinum has been removed to show the thoracic cavity.

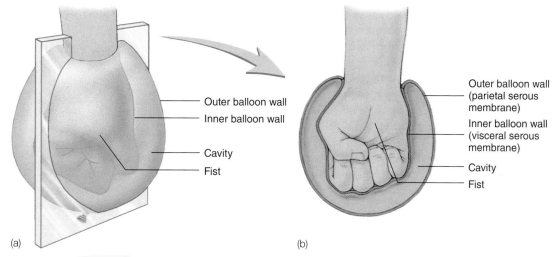

Figure 1.16 Serous Membranes

(a) Fist pushing into a balloon. A "glass" sheet indicates the location of a cross section through the balloon. (b) Interior view produced by the section in (a). The fist represents an organ, the walls of the balloon the serous membranes. The inner wall of the balloon represents a visceral serous membrane in contact with the fist (organ). The outer wall of the balloon represents a parietal serous membrane.

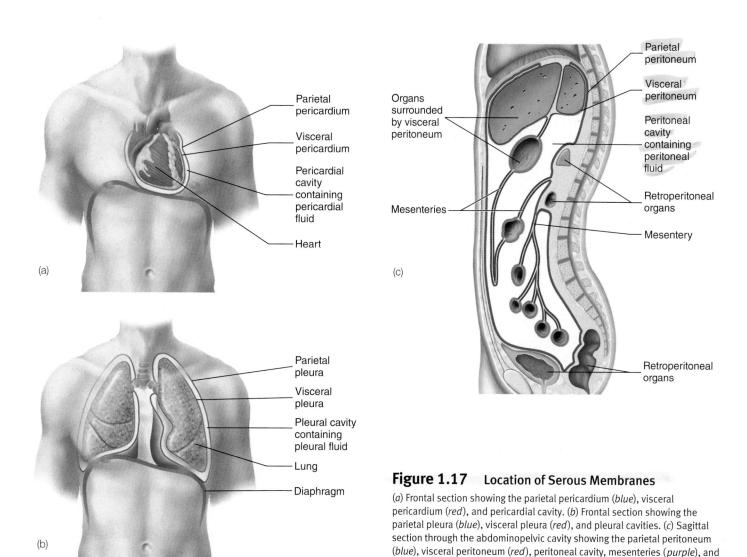

Figure 1.17 Location of Serous Membranes

(a) Frontal section showing the parietal pericardium (*blue*), visceral pericardium (*red*), and pericardial cavity. (b) Frontal section showing the parietal pleura (*blue*), visceral pleura (*red*), and pleural cavities. (c) Sagittal section through the abdominopelvic cavity showing the parietal peritoneum (*blue*), visceral peritoneum (*red*), peritoneal cavity, mesenteries (*purple*), and retroperitoneal organs.

Mesenteries (mes′en-ter-ēz), which consist of two layers of peritoneum fused together (see figure 1.17*c*), connect the visceral peritoneum of some abdominopelvic organs to the parietal peritoneum on the body wall or to the visceral peritoneum of other abdominopelvic organs. The mesenteries anchor the organs to the body wall and provide a pathway for nerves and blood vessels to reach the organs. Other abdominopelvic organs are more closely attached to the body wall and do not have mesenteries. Parietal peritoneum covers these other organs, which are said to be **retroperitoneal** (re′trō-per′i-tō-nē′ăl, *retro*, behind + peritoneum). The retroperitoneal organs include the kidneys, the adrenal glands, the pancreas, parts of the intestines, and the urinary bladder (see figure 1.14*c*) ▼

24. *Define serous membranes. Differentiate between parietal and visceral serous membranes. What is the function of the serous membranes?*
25. *Name the serous membrane-lined cavities of the trunk.*
26. *What are mesenteries? Explain their function.*
27. *What are retroperitoneal organs? List five examples.*

P R E D I C T

A bullet enters the left side of a male, passes through the left lung, and lodges in the heart. Name in order the serous membranes and their cavities through which the bullet passes.

S U M M A R Y

A knowledge of anatomy and physiology can be used to predict the body's responses to stimuli when healthy or diseased.

Anatomy and Physiology (p. 2)

1. Anatomy is the study of the structures of the body.
2. Systemic anatomy is the study of the body by organ systems. Regional anatomy is the study of the body by areas.
3. Surface anatomy uses superficial structures to locate deeper structures, and anatomical imaging is a noninvasive method for examining deep structures.
4. Physiology is the study of the processes and functions of the body. It can be approached according to the organism involved (e.g., human) or level of organization (e.g., cellular or systemic).

Structural and Functional Organization (p. 2)

1. The human body can be organized into six levels: chemical (atoms and molecules), cell, tissue (groups of similar cells and the materials surrounding them), organ (two or more tissues that perform one or more common functions), organ system (groups of organs with common functions), and organism.
2. The 11 organ systems are the integumentary, skeletal, muscular, nervous, endocrine, cardiovascular, lymphatic, respiratory, digestive, urinary, and reproductive systems (see figure 1.2).

Characteristics of Life (p. 4)

The characteristics of life include organization, metabolism, responsiveness, growth, development, and reproduction.

Homeostasis (p. 7)

1. Homeostasis is the existence and maintenance of a relatively constant internal environment.
2. Variables, such as body temperature, are maintained around a set point, resulting in a normal range of values.

Negative Feedback

1. Negative-feedback mechanisms maintain homeostasis.
2. Many negative-feedback mechanisms consist of a receptor, a control center, and an effector.

Positive Feedback

1. Positive-feedback mechanisms make deviations from normal even greater.
2. Although a few positive-feedback mechanisms normally exist in the body, most positive-feedback mechanisms are harmful.

Terminology and the Body Plan (p. 10)
Body Positions

1. A human standing erect with the face directed forward, the arms hanging to the sides, and the palms facing forward is in the anatomical position.
2. A person lying face upward is supine and face downward is prone.

Directional Terms

Directional terms always refer to the anatomical position, regardless of the body's actual position (see table 1.1).

Body Parts and Regions

1. The body can be divided into the upper limbs, lower limbs, head, neck, and trunk.
2. The abdomen can be divided superficially into four quadrants or nine regions that are useful for locating internal organs or describing the location of a pain.

Planes

1. A sagittal plane divides the body into left and right parts, a transverse plane divides the body into superior and inferior parts, and a frontal (coronal) plane divides the body into anterior and posterior parts.
2. A longitudinal section divides an organ along its long axis, a cross (transverse) section cuts an organ at a right angle to the long axis, and an oblique section cuts across the long axis at an angle other than a right angle.

Body Cavities

1. The thoracic cavity is bounded by the ribs and the diaphragm. The mediastinum divides the thoracic cavity into two parts.
2. The abdominal cavity is bounded by the diaphragm and the abdominal muscles.
3. The pelvic cavity is surrounded by the pelvic bones.

Serous Membranes

1. Serous membranes line the trunk cavities. The parietal part of a serous membrane lines the wall of the cavity, and the visceral part is in contact with the internal organs.

2. The serous membranes secrete serous fluid that fills the space between the parietal and visceral membranes. The serous membranes protect organs from friction.

3. The pericardial cavity surrounds the heart, the pleural cavities surround the lungs, and the peritoneal cavity surrounds certain abdominal and pelvic organs.

4. Mesenteries are parts of the peritoneum that hold the abdominal organs in place and provide a passageway for blood vessels and nerves to organs.

5. Retroperitoneal organs are located "behind" the parietal peritoneum. The kidneys, the adrenal glands, the pancreas, parts of the intestines, and the urinary bladder are examples of retroperitoneal organs.

R E V I E W   A N D   C O M P R E H E N S I O N

1. Physiology
   a. deals with the processes or functions of living things.
   b. is the scientific discipline that investigates the body's structures.
   c. is concerned with organisms and does not deal with different levels of organization, such as cells and systems.
   d. recognizes the static (as opposed to the dynamic) nature of living things.
   e. can be used to study the human body without considering anatomy.

2. The following are conceptual levels for considering the body.
   1. cell
   2. chemical
   3. organ
   4. organ system
   5. organism
   6. tissue

   Choose the correct order for these conceptual levels, from smallest to largest.
   a. 1,2,3,6,4,5
   b. 2,1,6,3,4,5
   c. 3,1,6,4,5,2
   d. 4,6,1,3,5,2
   e. 1,6,5,3,4,2

For questions 3–8, match each organ system with its correct function.
   a. regulates other organ systems
   b. removes waste products from the blood; maintains water balance
   c. regulates temperature; reduces water loss; provides protection
   d. removes foreign substances from the blood; combats disease; maintains tissue fluid balance
   e. produces movement; maintains posture; produces body heat

3. endocrine system

4. integumentary system

5. lymphatic system

6. muscular system

7. nervous system

8. urinary system

9. The characteristic of life that is defined as "the sum of the chemical and physical changes taking place in an organism" is
   a. development.
   b. growth.
   c. metabolism.
   d. organization.
   e. responsiveness.

10. Negative-feedback mechanisms
    a. make deviations from the set point smaller.
    b. maintain homeostasis.
    c. are associated with an increased sense of hunger the longer a person goes without eating.
    d. all of the above.

11. The following events are part of a negative-feedback mechanism.
    1. Blood pressure increases.
    2. The control center compares actual blood pressure to the blood pressure set point.
    3. The heart beats faster.
    4. Receptors detect a decrease in blood pressure.

    Choose the arrangement that lists the events in the order they occur.
    a. 1,2,3,4
    b. 1,3,2,4
    c. 3,1,4,2
    d. 4,2,3,1
    e. 4,3,2,1

12. Which of these statements concerning positive feedback is correct?
    a. Positive-feedback responses maintain homeostasis.
    b. Positive-feedback responses occur continuously in healthy individuals.
    c. Birth is an example of a normally occurring positive-feedback mechanism.
    d. When the cardiac muscle receives an inadequate supply of blood, positive-feedback mechanisms increase blood flow to the heart.
    e. Medical therapy seeks to overcome illness by aiding positive-feedback mechanisms.

13. The clavicle (collarbone) is _____ to the nipple of the breast.
    a. anterior
    b. distal
    c. superficial
    d. superior
    e. ventral

14. The term that means nearer the attached end of a limb is
    a. distal.
    b. lateral.
    c. medial.
    d. proximal.
    e. superficial.

15. Which of these directional terms are paired most appropriately as opposites?
    a. superficial and deep
    b. medial and proximal
    c. distal and lateral
    d. superior and posterior
    e. anterior and inferior

16. The part of the upper limb between the elbow and the wrist is called the
    a. arm.
    b. forearm.
    c. hand.
    d. inferior arm.
    e. lower arm.

17. A patient with appendicitis usually has pain in the _____ quadrant of the abdomen.
    a. left-lower
    b. right-lower
    c. left-upper
    d. right-upper

18. A plane that divides the body into anterior and posterior parts is a
    a. frontal (coronal) plane.
    b. sagittal plane.
    c. transverse plane.

19. The pelvic cavity contains the
    a. kidneys.
    b. liver.
    c. spleen.
    d. stomach.
    e. urinary bladder.

20. The lungs are
    a. part of the mediastinum.
    b. surrounded by the pericardial cavity.
    c. found within the thoracic cavity.
    d. separated from each other by the diaphragm.
    e. surrounded by mucous membranes.

21. Given these characteristics:
    1. reduce friction between organs
    2. line fluid-filled cavities
    3. line trunk cavities that open to the exterior of the body

Which of the characteristics describe serous membranes?
    a. 1,2   b. 1,3   c. 2,3   d. 1,2,3

22. Given these organ and cavity combinations:
    1. heart and pericardial cavity
    2. lungs and pleural cavity
    3. stomach and peritoneal cavity
    4. kidney and peritoneal cavity

Which of the organs is correctly paired with a space that surrounds that organ?
    a. 1,2   b. 1,2,3   c. 1,2,4   d. 2,3,4   e. 1,2,3,4

23. Which of these membrane combinations are found on the surface of the diaphragm?
    a. parietal pleura—parietal peritoneum
    b. parietal pleura—visceral peritoneum
    c. visceral pleura—parietal peritoneum
    d. visceral pleura—visceral peritoneum
24. Mesenteries
    a. are found in the pleural, pericardial, and abdominopelvic cavities.
    b. consist of two layers of peritoneum fused together.
    c. anchor organs, such as the kidneys and urinary bladder, to the body wall.
    d. are found primarily in body cavities that open to the outside.
    e. all of the above.
25. Which of the following organs is *not* retroperitoneal?
    a. adrenal glands   c. kidneys   e. stomach
    b. urinary bladder   d. pancreas

Answers in Appendix E

CRITICAL THINKING

1. A male has lost blood as a result of a gunshot wound. Even though bleeding has been stopped, his blood pressure is low and dropping and his heart rate is elevated. Following a blood transfusion, his blood pressure increases and his heart rate decreases. Which of the following statement(s) is (are) consistent with these observations?
   a. Negative-feedback mechanisms can be inadequate without medical intervention.
   b. The transfusion interrupted a positive-feedback mechanism.
   c. The increased heart rate after the gunshot wound and before the transfusion is a result of a positive-feedback mechanism.
   d. a and b
   e. a, b, and c
2. The anatomic position of a cat refers to the animal standing erect on all four limbs and facing forward. On the basis of the etymology of the directional terms, what two terms indicate movement toward the head? What two terms mean movement toward the back? Compare these terms with those referring to a human in the anatomic position.
3. Provide the correct directional term for the following statement: When a boy is standing on his head, his nose is _____ to his mouth.
4. Complete the following statements using the correct directional terms for a human being.
   a. The navel is _____ to the nose.
   b. The heart is _____ to the breastbone (sternum).
   c. The forearm is _____ to the arm.
   d. The ear is _____ to the brain.
5. Describe in as many directional terms as you can the relationship between your thumb and your elbow.
6. According to "Dear Abby," a wedding band should be worn closest to the heart, and an engagement ring should be worn as a "guard" on the outside. Should a woman's wedding band be worn proximal or distal to her engagement ring?
7. During pregnancy, which would increase more in size, the mother's abdominal or pelvic cavity? Explain.
8. Explain how an organ can be located within the abdominopelvic cavity but not be within the peritoneal cavity.
9. A woman falls while skiing and accidentally is impaled by her ski pole. The pole passes through the abdominal body wall and into and through the stomach, pierces the diaphragm, and finally stops in the left lung. List, in order, the serous membranes the pole pierces.
10. Can a kidney be removed without cutting through parietal peritoneum? Explain.

Answers in Appendix F

 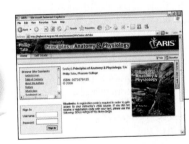

CHAPTER 2

The Chemical Basis of Life

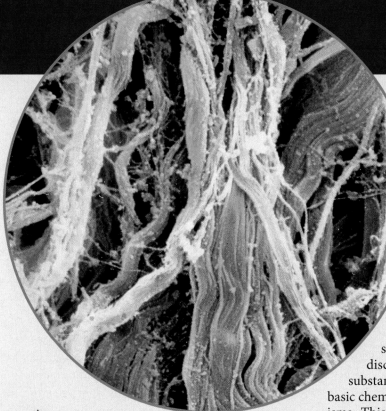

The chemical composition of the body's structures determines their function. This colorized scanning electron micrograph shows bundles of collagen fibers (brown) and elastic fibers (blue). The molecules forming collagen fibers are like tiny ropes, and collagen fibers make tissues strong yet flexible. The molecules forming elastic fibers resemble microscopic coiled springs. The elastic fibers allow tissues to be stretched and then recoil back to their original shape.

Introduction

Chemicals compose the structures of the body, and the interactions of chemicals with one another are responsible for the functions of the body. Nerve impulse generation, digestion, muscle contraction, and metabolism can be described in chemical terms. Many abnormal conditions and their treatments can also be explained in chemical terms. For example, Parkinson disease, which causes uncontrolled shaking movements, results from a shortage of a chemical called dopamine in certain nerve cells in the brain. It can be treated by giving patients another chemical that brain cells convert to dopamine.

▶A basic knowledge of chemistry is essential for understanding anatomy and physiology. **Chemistry** is the scientific discipline concerned with the atomic composition and structure of substances and the reactions they undergo. This chapter outlines some basic chemical principles and emphasizes their relationship to living organisms. This chapter outlines **basic chemistry (p. 22)**, **chemical reactions (p. 28)**, **acids and bases (p. 31)**, **inorganic chemistry (p. 32)**, and **organic chemistry (p. 33)**. It is not a comprehensive review of chemistry, but it does review some of the basic concepts. Refer back to this chapter when chemical phenomena are discussed later in the book.

Tools for Success

A virtual cadaver dissection experience

Audio (MP3) and/or video (MP4) content, available at www.mhhe.aris.com

Basic Chemistry

Matter, Mass, and Weight

All living and nonliving things are composed of **matter,** which is anything that occupies space and has mass. **Mass** is the amount of matter in an object, and **weight** is the gravitational force acting on an object of a given mass. For example, the weight of an apple results from the force of gravity "pulling" on the apple's mass.

P R E D I C T

The difference between mass and weight can be illustrated by considering an astronaut. How would an astronaut's mass and weight in outer space compare with his or her mass and weight on the earth's surface?

The international unit for mass is the **kilogram (kg),** which is the mass of a platinum–iridium cylinder kept at the International Bureau of Weights and Measurements in France. The mass of all other objects is compared with this cylinder. For example, a 2.2-lb lead weight or 1 liter (L) (1.06 qt) of water has a mass of approximately 1 kg. An object with 1/1000 the mass of the standard kilogram cylinder is defined to have a mass of 1 **gram (g).**

Chemists use a balance to determine the mass of objects. Although we commonly refer to weighing an object on a balance, we actually are "massing" the object because the balance compares objects of unknown mass with objects of known mass. When the unknown and known masses are exactly balanced, the gravitational pull of the earth on both of them is the same. Thus, the effect of gravity on the unknown mass is counteracted by the effect of gravity on the known mass. A balance produces the same results at sea level as on a mountaintop because it does not matter that the gravitational pull at sea level is stronger than on a mountaintop. It only matters that the effect of gravity on both the unknown and known masses is the same. ▼

> 1. *Define matter. How is the mass and the weight of an object different?*

Elements and Atoms

An **element** is the simplest type of matter with unique chemical and physical properties. A list of the elements commonly found in the human body is given in table 2.1. About 96% of the weight of the body results from the elements oxygen, carbon, hydrogen, and nitrogen. See appendix A for additional elements.

An **atom** (at′ŏm, indivisible) is the smallest particle of an element that has the characteristics of that element. An element is composed of atoms of only one kind. For example, the element carbon is composed of only carbon atoms, and the element oxygen is composed of only oxygen atoms.

An element, or an atom of that element, often is represented by a symbol. Usually the first letter or letters of the element's name are used—for example, C for carbon, H for hydrogen, Ca for calcium, and Cl for chlorine. Occasionally the symbol is taken from the Latin, Greek, or Arabic name for the element—for example, Na from the Latin word *natrium* is the symbol for sodium. ▼

> 2. *Define element and atom. What four elements are found in the greatest abundance in humans?*

Atomic Structure

The characteristics of matter result from the structure, organization, and behavior of atoms. Atoms are composed of subatomic particles, some of which have an electric charge. The three major types of subatomic particles are neutrons, protons, and electrons.

Table 2.1	Some Common Elements				
Element	**Symbol**	**Atomic Number**	**Mass Number**	**Percent in Human Body by Weight**	**Percent in Human Body by Number of Atoms**
Hydrogen	H	1	1	9.5	63.0
Carbon	C	6	12	18.5	9.5
Nitrogen	N	7	14	3.3	1.4
Oxygen	O	8	16	65.0	25.5
Sodium	Na	11	23	0.2	0.3
Phosphorus	P	15	31	1.0	0.22
Sulfur	S	16	32	0.3	0.05
Chlorine	Cl	17	35	0.2	0.03
Potassium	K	19	39	0.4	0.06
Calcium	Ca	20	40	1.5	0.31
Iron	Fe	26	56	Trace	Trace
Iodine	I	53	127	Trace	Trace

A **neutron** (noo′tron) has no electric charge, a **proton** (prō′ton) has one positive charge, and an **electron** (e-lek′tron) has one negative charge. The positive charge of a proton is equal in magnitude to the negative charge of an electron. The number of protons and electrons in each atom is equal, and the individual charges cancel each other. Therefore, each atom is electrically neutral.

Protons and neutrons form the **nucleus** at the center of an atom, and electrons are moving around the nucleus (figure 2.1). The nucleus accounts for 99.97% of an atom's mass, but only 1 ten-trillionth of its volume. Most of the volume of an atom is occupied by the electrons. Although it is impossible to know precisely where any given electron is located at any particular moment, the region where electrons are most likely to be found can be represented by an **electron cloud** (see figure 2.1). The darker the color in each small volume of the diagram, the greater the likelihood of finding an electron there at any given moment. ▼

> 3. *For each subatomic particle of an atom, state its charge and location. Which subatomic particles are most responsible for the mass and volume of an atom?*

Atomic Number and Mass Number

The **atomic number** of an element is equal to the number of protons in each atom. The atomic number is also the number of electrons in each atom because the number of electrons is equal to the number of protons in each atom. Each element is uniquely defined by the number of protons in the atoms of that element. For example,

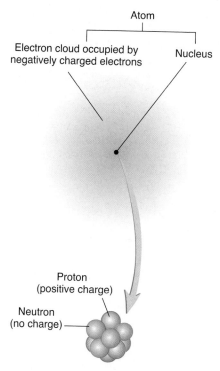

Figure 2.1 Model of an Atom

The tiny, dense nucleus consists of positively charged protons and uncharged neutrons. Most of the volume of an atom is occupied by rapidly moving, negatively charged electrons, which can be represented as an electron cloud. The probable location of an electron is indicated by the color of the electron cloud. The darker the color in each small part of the electron cloud, the more likely the electron is located there.

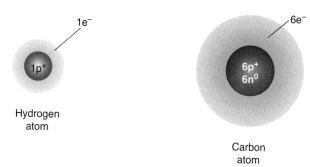

Figure 2.2 Hydrogen and Carbon Atoms

Within the nucleus, the number of positively charged protons (p^+) and uncharged neutrons (n^0) is indicated. The negatively charged electrons (e^-) are around the nucleus. Atoms are electrically neutral because the number of protons and electrons within an atom is equal.

only hydrogen atoms have one proton and only carbon atoms have six protons (figure 2.2; see table 2.1). There are 90 naturally occurring elements, but additional elements have been synthesized by altering atomic nuclei (see appendix A).

Protons and neutrons have about the same mass, and they are responsible for most of the mass of atoms. Electrons, on the other hand, have very little mass. The **mass number** of an element is the number of protons plus the number of neutrons in each atom. For example, the mass number for carbon is 12 because it has six protons and six neutrons.

P R E D I C T ②

The atomic number of fluorine is 9, and the mass number is 19. What is the number of protons, neutrons, and electrons in an atom of fluorine?

Isotopes (ī′sō-tōpz, *isos*, equal, + *topos*, part) are two or more forms of the same element that have the same number of protons but a different number of neutrons. Thus, isotopes have the *same* atomic number but *different* mass numbers. There are three isotopes of hydrogen: hydrogen, deuterium, and tritium. All three isotopes have an atomic number of 1 because they each have one proton. Hydrogen, with one proton and no neutrons, has a mass number of 1; deuterium, with one proton and one neutron, has a mass number of 2; and tritium, with one proton and two neutrons, has a mass number of 3. Isotopes can be denoted using the symbol of the element preceded by the mass number of the isotope. Thus, hydrogen is ^1H, deuterium is ^2H, and tritium is ^3H. ▼

> 4. *Which subatomic particles determine atomic number and mass number?*
> 5. *What are isotopes?*

Electrons and Chemical Bonding

The outermost electrons of an atom determine its chemical behavior. **Chemical bonding** occurs when the outermost electrons are transferred or shared between atoms. Two major types of chemical bonding are ionic and covalent bonding.

Ionic Bonding

An atom is electrically neutral because it has an equal number of protons and electrons. If an atom loses or gains electrons, the number

Protons, neutrons, and electrons are responsible for the chemical properties of atoms. They also have other properties that can be useful in a clinical setting. Radioactive isotopes have unstable nuclei that lose neutrons or protons. Several kinds of radiation can be produced when neutrons and protons, or the products formed by their breakdown, are released from the nucleus of the isotope. The radiation given off by some radioactive isotopes can penetrate and destroy tissues. Rapidly dividing cells are more sensitive to radiation than are slowly dividing cells. Radiation is used to treat cancerous (malignant) tumors because cancer cells divide rapidly. If the treatment is effective, few healthy cells are destroyed, but the cancerous cells are killed.

Radioactive isotopes also are used in medical diagnosis. The radiation can be detected, and the movement of the radioactive isotopes throughout the body can be traced. For example, the thyroid gland normally takes up iodine and uses it in the formation of thyroid hormones. Radioactive iodine can be used to determine if iodine uptake is normal in the thyroid gland.

Radiation can be produced in ways other than changing the nucleus of atoms. X-rays are a type of radiation formed when electrons lose energy by moving from a higher energy state to a lower one. X-rays are used in the examination of bones to determine if they are broken and of teeth to see if they have caries

(cavities). Mammograms, which are low-energy radiographs (x-ray films) of the breast, can be used to detect tumors because the tumors are slightly denser than normal tissue.

Computers can be used to analyze a series of radiographs, each made at a slightly different body location. The picture of each radiographic "slice" through the body is assembled by the computer to form a three-dimensional image. A **computed tomography** (tō-mog′ră-fē) **(CT) scan** is an example of this technique (figure A). CT scans are used to detect tumors and other abnormalities in the body.

Magnetic resonance imaging (MRI) is another method for looking into the body (figure B). The patient is placed into a very powerful magnetic field, which aligns the hydrogen nuclei. Radio waves given off by the hydrogen nuclei are monitored, and the data are used by a computer to make an image of the body. Because MRI detects hydrogen, it is very effective for visualizing soft tissues that contain a lot of water. MRI technology is used to detect tumors and other abnormalities in the body.

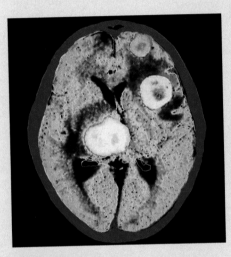

Figure A CT Scan

CT scan of the brain with iodine injection, showing three brain tumors (ovals) that have metastasized (spread) to the brain from cancer in the large intestine.

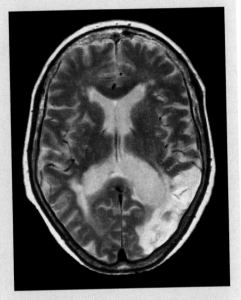

Figure B MRI

Colorized MRI brain scan showing a stroke. The whitish area in the lower right part of the MRI is blood that has leaked into the surrounding tissue.

---

of protons and electrons are no longer equal, and a charged particle called an **ion** (ī′on) is formed. After an atom loses an electron, it has one more proton than it has electrons and is positively charged. A sodium atom (Na) can lose an electron to become a positively charged sodium ion (Na^+) (figure 2.3*a*). After an atom gains an electron, it has one more electron than it has protons and is negatively charged. A chlorine atom (Cl) can accept an electron to become a negatively charged chloride ion (Cl^-).

Positively charged ions are called **cations** (kat′ī-onz), and negatively charged ions are called **anions** (an′īonz). Because oppositely charged ions are attracted to each other, cations tend to remain close to anions, which is called **ionic** (īon′ik) **bonding.** Thus, Na^+ and Cl^- are held together by ionic bonding to form an array of ions called sodium chloride (NaCl), or table salt (figure 2.3*b* and *c*).

Ions are denoted by using the symbol of the atom from which the ion was formed. The charge of the ion is indicated by a superscripted plus (+) or minus (−) sign. For example, a sodium ion is Na^+, and a chloride ion is Cl^-. If more than one electron has been lost or gained, a number is used with the plus or minus sign. Thus, Ca^{2+} is a calcium ion formed by the loss of two electrons. Some ions commonly found in the body are listed in table 2.2.

PREDICT ③

If an iron (Fe) atom loses three electrons, what is the charge of the resulting ion? Write the symbol for this ion.

Covalent Bonding

Covalent bonding results when atoms share one or more pairs of electrons. The resulting combination of atoms is called a **molecule.** An example is the covalent bond between two hydrogen atoms to form a hydrogen molecule (figure 2.4). Each hydrogen atom has

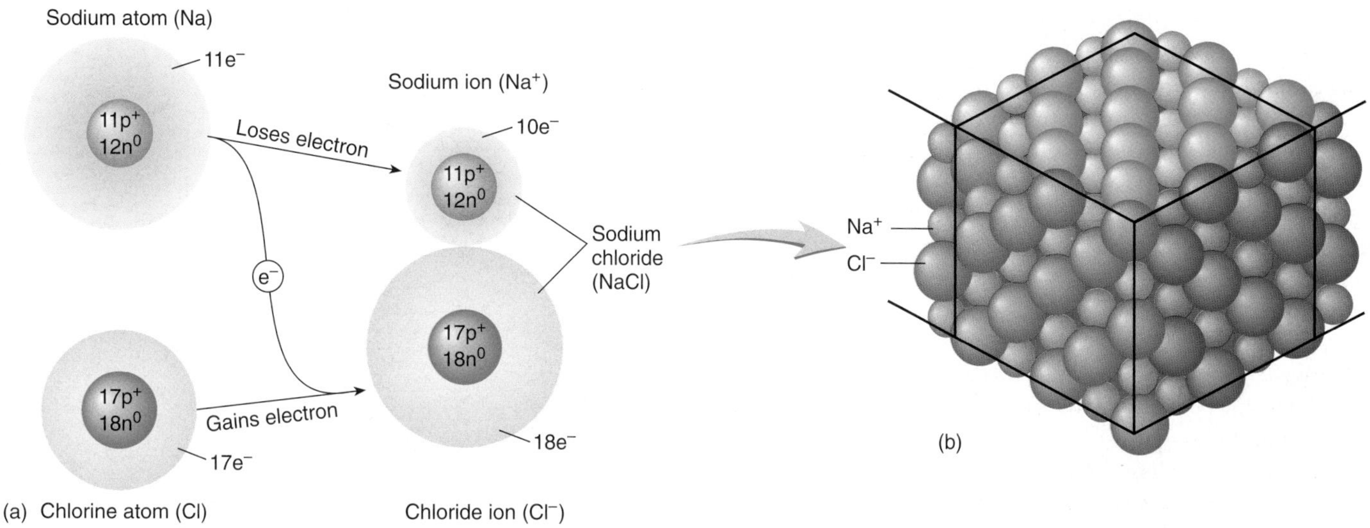

Sodium atom (Na)

Sodium ion (Na$^+$)

Loses electron

Gains electron

Chlorine atom (Cl)

Chloride ion (Cl$^-$)

Sodium chloride (NaCl)

Na$^+$

Cl$^-$

(b)

Figure 2.3 Ionic Bonding

(*a*) A sodium atom (Na) loses an electron to become a smaller positively charged ion, and a chlorine atom (Cl) gains an electron to become a larger negatively charged ion. The attraction between the oppositely charged ions results in ionic bonding and the formation of sodium chloride (NaCl). (*b*) The sodium ion (Na$^+$) and chloride ion (Cl$^-$) are organized to form a cube-shaped array. (*c*) A microphotograph of salt crystals reflects the cubic arrangement of the ions.

(c)

Table 2.2	Important Ions	
Common Ions	**Symbols**	**Significance***
Calcium	Ca^{2+}	Part of bones and teeth, blood clotting, muscle contraction, release of neurotransmitters
Sodium	Na$^+$	Membrane potentials, water balance
Potassium	K$^+$	Membrane potentials
Hydrogen	H$^+$	Acid–base balance
Hydroxide	OH$^-$	Acid–base balance
Chloride	Cl$^-$	Water balance
Bicarbonate	HCO$_3^-$	Acid–base balance
Ammonium	NH$_4^+$	Acid–base balance
Phosphate	PO$_4^{3-}$	Part of bones and teeth, energy exchange, acid–base balance
Iron	Fe^{2+}	Red blood cell formation
Magnesium	Mg^{2+}	Necessary for enzymes
Iodide	I$^-$	Present in thyroid hormones

*The ions are part of the structures or play important roles in the processes listed.

There is no interaction between the two hydrogen atoms because they are too far apart.

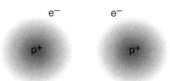

The positively charged nucleus of each hydrogen atom begins to attract the electron of the other.

A covalent bond is formed when the electrons are shared between the nuclei because the electrons are equally attracted to each nucleus.

Figure 2.4 Covalent Bonding

one electron. As the atoms get closer together, the positively charged nucleus of each atom begins to attract the electron of the other atom. At an optimal distance, the two nuclei mutually attract the two electrons, and each electron is shared by both nuclei. The two hydrogen atoms are now held together by a covalent bond.

A **single covalent bond** results when two atoms share a pair of electrons. A single covalent bond can be represented by a single line between the symbols of the atoms involved (for example, H—H). A **double covalent bond** results when two atoms share two pairs of electrons. When a carbon atom combines with two oxygen atoms to form carbon dioxide, two double covalent bonds are formed. Double covalent bonds are indicated by a double line between the atoms (O=C=O).

A **nonpolar covalent bond** occurs when electrons are shared equally between atoms, as in a hydrogen molecule (see figure 2.4). A **polar covalent bond** results when electrons are shared unequally between atoms, as in a water molecule (figure 2.5). The nucleus of the oxygen atom attracts a pair of electrons more strongly than does the nucleus of each hydrogen atom. Therefore, the electrons are shared unequally and tend to spend more time around the oxygen atom than around the hydrogen atoms. Thus,

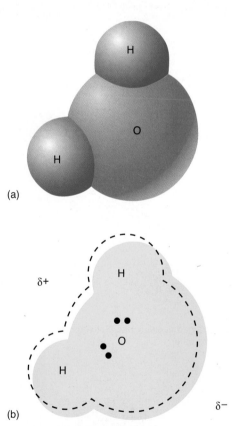

(a)

(b)

Figure 2.5 Polar Covalent Bonds

(*a*) A water molecule forms when two hydrogen atoms form covalent bonds with an oxygen atom. (*b*) Electron pairs (*indicated by the black dots*) are shared between the hydrogen atoms and oxygen. The dashed outline shows the expected location of the electron cloud if the electrons are shared equally. The electrons are shared unequally, as shown by the electron cloud (*yellow*) not coinciding with the dashed outline. Consequently, the oxygen side of the molecule has a slight negative charge (*indicated by* δ⁻), and the hydrogen side of the molecule has a slight positive charge (*indicated by* δ⁺).

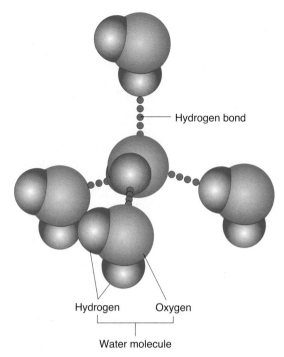

Hydrogen bond

Hydrogen   Oxygen

Water molecule

Figure 2.6 Hydrogen Bonds

The positive hydrogen part of one water molecule forms a hydrogen bond (*red dotted line*) with the negative oxygen part of another water molecule. As a result, hydrogen bonds hold the water molecules together.

the oxygen side of the molecule has a slight negative charge, compared with a slight positive charge on the hydrogen side. Molecules with this asymmetrical electric charge are called **polar molecules**, whereas molecules with a symmetrical electric charge are called **nonpolar molecules.**

Hydrogen Bonds

A polar molecule has a positive and a negative "end." A **hydrogen bond** is formed when the positive "end" of one polar molecule is weakly attracted to the negative "end" of another polar molecule. For example, water molecules are held together by hydrogen bonds when the positively charged hydrogen of one water molecule is weakly attracted to a negatively charged oxygen of another water molecule (figure 2.6). Although this attraction is called a hydrogen bond, it is not a chemical bond. The attraction between molecules resulting from hydrogen bonds is much weaker than ionic or covalent bonds.

Hydrogen bonds play an important role in determining the shape of complex molecules because the hydrogen bonds between different polar parts of a single large molecule hold the molecule in its normal three-dimensional shape (see "Proteins and Nucleic Acids," pp. 37–40). ▼

6. *Describe how ionic bonding occurs. Define an ion, a cation, and an anion.*
7. *Describe how covalent bonding occurs. What is a single and a double covalent bond?*
8. *What is the difference between nonpolar and polar covalent bonds? Between nonpolar and polar molecules?*

*9. Define hydrogen bond, and explain how hydrogen bonds
hold polar molecules, such as water, together. How do
hydrogen bonds affect the shape of a molecule?*

Molecules and Compounds

A molecule is formed when two or more atoms chemically combine to form a structure that behaves as an independent unit. The atoms that combine to form a molecule can be of the same type, such as two hydrogen atoms combining to form a hydrogen molecule. More typically, a molecule consists of two or more different types of atoms, such as two hydrogen atoms and an oxygen atom forming water. Thus, a glass of water consists of a collection of individual water molecules positioned next to one another.

A **compound** is a substance composed of two or more *different* types of atoms that are chemically combined. Not all molecules are compounds. For example, a hydrogen molecule is not a compound because it does not consist of different types of atoms. Some compounds are molecules and some are not. Covalent compounds, in which different types of atoms are held together by covalent bonds, are molecules because the sharing of electrons results in the formation of distinct, independent units. Water is an example of a substance that is a compound and a molecule.

On the other hand, ionic compounds, in which ions are held together by the force of attraction between opposite charges, are not molecules because they do not consist of distinct units. A piece of sodium chloride does not consist of individual sodium chloride molecules positioned next to one another. Instead, it is an organized array of individual Na^+ and individual Cl^- in which each charged ion is surrounded by several ions of the opposite charge (see figure 2.3b). Sodium chloride is an example of a substance that is a compound but not a molecule.

Molecules and compounds can be represented by the symbols of the atoms (or ions) forming the molecule or compound plus subscripts denoting the number of each type of atom (or ion). For example, glucose (a sugar) can be represented as $C_6H_{12}O_6$, indicating that glucose has 6 carbon, 12 hydrogen, and 6 oxygen atoms. ▼

*10. Distinguish between a molecule and a compound. Are all
molecules compounds? Are all compounds molecules?*

Dissociation

When ionic compounds dissolve in water, their ions **dissociate** (di-sō′sē-āt′), or separate, from each other because the positively charged ions are attracted to the negative ends of the water molecules, and the negatively charged ions are attracted to the positive ends of the water molecules. For example, when sodium chloride dissociates in water, the Na^+ and Cl^- separate, and water molecules surround and isolate the ions, keeping them in solution (figure 2.7).

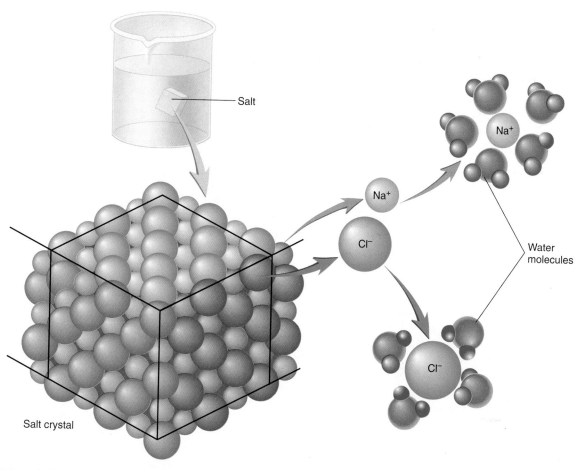

Figure 2.7   **Dissociation**

Sodium chloride (table salt) dissociating in water. The positively charged Na^+ are attracted to the negatively charged oxygen (*red*) end of the water molecule, and the negatively charged Cl^- are attracted to the positively charged hydrogen (*blue*) end of the water molecule.

When molecules dissolve in water, the molecules usually remain intact even though they are surrounded by water molecules. Thus, in a glucose solution, glucose molecules are surrounded by water molecules.

Ions that dissociate in water are sometimes called **electrolytes** (ē-lek′trō-lītz) because they have the capacity to conduct an electric current, which is the flow of charged particles. An electrocardiogram (ECG) is a recording of electric currents produced by the heart. These currents can be detected by electrodes on the surface of the body because the ions in the body fluids conduct electric currents. Molecules that do not dissociate form solutions that do not conduct electricity and are called **nonelectrolytes.** Pure water is a nonelectrolyte. ▼

11. *How do ionic compounds and molecules (covalent compounds) typically dissolve in water?*
12. *Distinguish between electrolytes and nonelectrolytes.*

Chemical Reactions

In a **chemical reaction,** the relationship between atoms, ions, molecules, or compounds is changed by forming or breaking chemical bonds. The substances that enter into a chemical reaction are called the **reactants,** and the substances that result from the chemical reaction are called the **products.**

A **synthesis reaction** occurs when two or more reactants combine to form a larger, more complex product. Synthesis reactions result in the production of the complex molecules of the human body from the basic "building blocks" obtained in food. For example, amino acids are basic building blocks that can combine to form a dipeptide (figure 2.8*a*). As the atoms rearrange, old chemical bonds are broken and new chemical bonds are formed.

As a result, the amino acids are joined by a covalent bond and water is formed. Synthesis reactions in which water is a product are called **dehydration** (water out) **reactions.**

A **decomposition reaction** occurs when reactants are broken down into smaller, less complex products. Decomposition reactions include the digestion of food molecules in the intestine and within cells. For example, a disaccharide is a type of carbohydrate that is broken down to form glucose (figure 2.8*b*). Note that this reaction requires that water be split into two parts and that each part be contributed to one of the new glucose molecules. Reactions that use water in this manner are called **hydrolysis** (hī-drol′i-sis, water dissolution) **reactions.**

All of the synthesis and decomposition reactions in the body are collectively defined as **metabolism.** ▼

13. *Define a chemical reaction, reactants, and products.*
14. *Describe synthesis and decomposition reactions, giving an example of each. What are dehydration and hydrolysis reactions?*
15. *Define metabolism.*

Reversible Reactions

A **reversible reaction** is a chemical reaction in which the reaction can proceed from reactants to products and from products to reactants. When the rate of product formation is equal to the rate of reactant formation, the reaction is said to be at **equilibrium.** At equilibrium the amount of the reactants relative to the amount of products remains constant.

The following analogy may help clarify the concept of reversible reactions and equilibrium. Imagine a trough containing water. The trough is divided into two compartments by a partition, but the partition contains holes that allow water to move freely

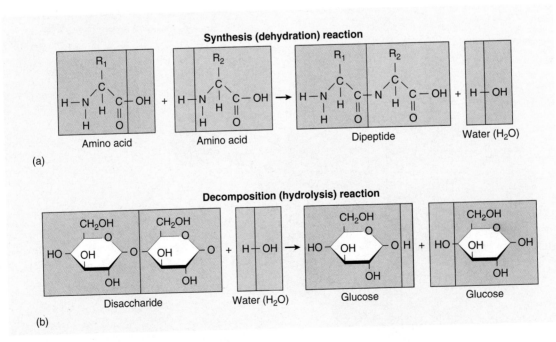

Figure 2.8 Synthesis and Decomposition Reactions

(*a*) Synthesis reaction in which two amino acids combine to form a dipeptide. This reaction is also a dehydration reaction because it results in the removal of a water molecule from the amino acids. (*b*) Decomposition reaction in which a disaccharide breaks apart to form glucose molecules. This reaction is also a hydrolysis reaction because it involves the splitting of a water molecule.

between the compartments. Because water can move in either direction, this is like a reversible reaction. Let the amount of water in the left compartment represent the amount of reactant and the amount of water in the right compartment represent the amount of product. At equilibrium the amount of reactant relative to the amount of product in each compartment is always the same because the partition allows water to pass between the two compartments until the level of water is the same in both compartments. If the amount of reactant is increased by adding water to the left compartment, water flows from the left compartment through the partition to the right compartment until the level of water is the same in both. Thus, the relative amounts of reactant and product are once again equal. Unlike this analogy, however, the amount of reactants relative to the products in most reversible reactions is not one to one. Depending on the specific reversible reaction, there can be one part reactant to two parts product, two parts reactants to one part product, or many other possibilities.

An important reversible reaction in the human body is the reaction between carbon dioxide (CO_2) and water (H_2O) to form hydrogen ions (H^+) and bicarbonate ions (HCO_3^-) (the reversibility of the reaction is indicated by two arrows pointing in opposite directions):

$$CO_2 + H_2O \rightleftarrows H^+ + HCO_3^-$$

If CO_2 is added to H_2O, the amount of CO_2 relative to the amount of H^+ increases. The reaction of CO_2 with H_2O produces more H^+, however, and the amount of CO_2 relative to the amount of H^+ returns to equilibrium. Conversely, adding H^+ results in the formation of more CO_2, and the equilibrium is restored.

Maintaining a constant level of H^+ in body fluids is necessary for the nervous system to function properly. This level can be maintained, in part, by controlling blood CO_2 levels. For example, slowing the respiration rate causes blood CO_2 levels to increase, which in turn causes an increase in H^+ concentration in the blood. ▼

16. Describe reversible reactions. What is meant by the equilibrium condition in reversible reactions?

P R E D I C T 4

If the respiration rate increases, carbon dioxide is removed from the blood. What effect does this have on blood hydrogen ion levels?

Energy and Chemical Reactions

Energy, unlike matter, does not occupy space and it has no mass. **Energy** is defined as the capacity to do **work**—that is, to move matter. Energy can be subdivided into potential energy and kinetic energy. **Potential energy** is stored energy that could do work but is not doing so. For example, a coiled spring has potential energy. It could push against an object and move the object, but as long as the spring does not uncoil, no work is accomplished. **Kinetic** (ki-net′ik, of motion) **energy** is energy caused by the movement of an object and is the form of energy that actually does work. An uncoiling spring pushing an object, causing it to move, is an example. When potential energy is released, it becomes kinetic energy, thus doing work.

Potential and kinetic energy can be found in many different forms. **Mechanical energy** is energy resulting from the position or movement of objects. Many of the activities of the human body, such as moving a limb, breathing, or circulating blood, involve mechanical energy. Other forms of energy are chemical energy, heat energy, electric energy, and electromagnetic (radiant) energy.

According to the law of conservation of energy, the total energy of the universe is constant. Therefore, energy is neither created nor destroyed. One type of energy can be changed into another, however. For example, as a moving object slows down and comes to rest, its kinetic energy is converted into heat energy by friction.

The **chemical energy** of a substance is a form of stored (potential) energy that results from the relative positions and interactions among its charged subatomic particles. Consider two balls attached by a relaxed spring. In order to push the balls together and compress the spring, energy must be put into this system. As the spring is compressed, potential energy increases. When the compressed spring expands, potential energy decreases. Similarly charged particles, such as two negatively charged electrons or two positively charged nuclei, repel each other. As similarly charged particles move closer together, their potential energy increases, much like the compression of a spring, and, as they move farther apart, their potential energy decreases. Chemical bonding is a form of potential energy because of the charges and positions of the subatomic particles bound together.

Chemical reactions are important because of the products they form and the energy changes that result as the relative position of subatomic particles changes. If the reactants of a chemical reaction contain more potential energy than the products, energy is released. For example, food molecules contain more potential energy than waste products. The difference in potential energy between food and waste products is used by living systems for many activities, such as growth, repair, movement, and heat production, as the potential energy in the food molecules changes into other forms of energy.

Adenosine triphosphate (ATP) (ă-den′ō-sēn trī-fos′fāt) is an important molecule involved with the transfer of energy in cells. In ATP, A stands for adenosine (adenine combined with ribose), T stands for tri- (or three), and P stands for a phosphate group (PO_4^{3-}). Thus, ATP consists of adenosine and three phosphate groups (figure 2.9). The phosphate group that reacts with ADP is often denoted as P_i, where the "i" indicates that the phosphate group is associated with an inorganic substance (see "Inorganic Chemistry," p. 32).

Potential energy is stored in the covalent bond between the second and third phosphate in ATP when ATP is synthesized from **adenosine diphosphate (ADP)** and a phosphate group. Energy must be added from another source when the products of a chemical reaction, such as ATP, contain more energy than the reactants (figure 2.10a). The energy released during the breakdown of food molecules is the source of energy for this kind of reaction in the body.

$$ADP + P_i + \text{Energy (from food molecules)} \rightarrow ATP$$

Energy is released during a chemical reaction when the potential energy in the chemical bonds of the reactants is greater than that of the products (figure 2.10b). The breakdown of ATP to adenosine diphosphate (ADP) and a phosphate group results in the release of energy and heat. The energy is used by cells for activities such as synthesizing new molecules or for muscle contraction.

$$ATP \rightarrow ADP + P_i + \text{Energy (used by cells)} + \text{Heat}$$

1. Adenosine is adenine, which is one of the nitrogenous bases in DNA, combined with the sugar ribose.

2. Adenosine diphosphate (ADP) is adenosine with two phosphate groups.

3. Adenosine triphosphate (ATP) is adenosine with three phosphates.

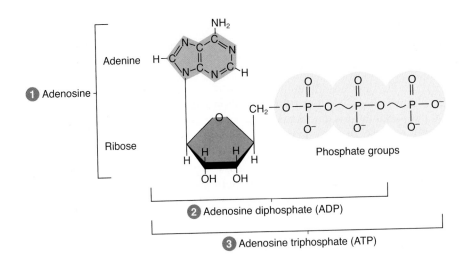

Figure 2.9   Adenosine Triphosphate (ATP) Molecule

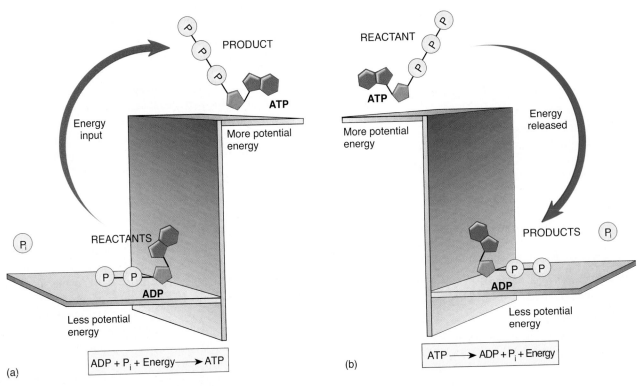

$$ADP + P_i + Energy \longrightarrow ATP$$

(a)

$$ATP \longrightarrow ADP + P_i + Energy$$

(b)

Figure 2.10   Energy and Chemical Reactions

In each figure the upper shelf represents a higher energy level, and the lower shelf represents a lower energy level. (*a*) Reaction in which energy is released as a result of the breakdown of ATP. (*b*) Reaction in which the input of energy is required for the synthesis of ATP.

P R E D I C T   5

Why does body temperature increase during exercise?

ATP is often called the energy currency of cells because it is capable of both storing and providing energy. The concentration of ATP is maintained within a narrow range of values, and essentially all energy-requiring chemical reactions stop when there is an inadequate quantity of ATP. ▼

17. *How is energy different from matter? How are potential and kinetic energy different from each other?*

18. *Define mechanical energy and chemical energy. How is chemical energy converted to mechanical energy and heat energy in the body?*

19. *Use ATP and ADP to illustrate the input or release of energy in chemical reactions. Why is ATP called the energy currency of cells?*

Rate of Chemical Reactions

The rate at which a chemical reaction proceeds is influenced by several factors, including how easily the substances react with one another, their concentrations, the temperature, and the presence of a catalyst.

Reactants

Reactants differ from one another in their ability to undergo chemical reactions. For example, iron corrodes much more rapidly than does stainless steel. For this reason, during its refurbishment the iron bars forming the skeleton of the Statue of Liberty were replaced with stainless steel bars.

Concentration

Within limits, the greater the concentration of reactants, the greater the rate at which a chemical reaction will occur because, as the concentration increases, the reacting molecules are more likely to come into contact with one another. For example, the normal concentration of oxygen inside cells enables it to come into contact with other molecules, producing the chemical reactions necessary for life. If the oxygen concentration decreases, the rate of chemical reactions decreases. A decrease in oxygen in cells can impair cell function and even result in cell death.

Temperature

The rate of chemical reactions also increases when the temperature is increased. When a person has a fever of only a few degrees, reactions occur throughout the body at a faster rate. The result is increased activity in most organ systems, such as increased heart and respiratory rates. When body temperature drops, the rate of reactions decreases. The clumsy movement of very cold fingers results largely from the reduced rate of chemical reactions in cold muscle tissue.

Catalysts

At normal body temperatures, most chemical reactions would take place too slowly to sustain life if it were not for the body's catalysts. A **catalyst** (kat′ă-list) is a substance that increases the rate of a chemical reaction, without itself being permanently changed or depleted. An **enzyme** (en′zīm) is a protein molecule that acts as a catalyst. Many of the chemical reactions that occur in the body require enzymes. Enzymes are considered in greater detail on p. 39. ▼

> 20. *Name three ways that the rate of a chemical reaction can be increased.*

Acids and Bases

An **acid** is a proton donor. Because a hydrogen atom without its electron is a proton, any substance that releases hydrogen ions in water is an acid. For example, hydrochloric acid (HCl) in the stomach forms hydrogen ions (H^+) and chloride ions (Cl^-):

$$HCl \rightarrow H^+ + Cl^-$$

A **base** is a proton acceptor. For example, sodium hydroxide (NaOH) forms sodium ions (Na^+) and hydroxide ions (OH^-). It is a base because the OH^- is a proton acceptor that binds with a H^+ to form water.

$$NaOH \rightarrow Na^+ + OH^-$$
$$H^+ \longrightarrow H_2O$$

Acids and bases are classified as strong or weak. Strong acids or bases dissociate almost completely when dissolved in water. Consequently, they release almost all of their H^+ or OH^-. The more completely the acid or base dissociates, the stronger it is. For example, HCl is a strong acid because it completely dissociates in water.

$$HCl \rightarrow H^+ + Cl^-$$
Not freely reversible

Weak acids or bases only partially dissociate in water. Consequently, they release only some of their H^+ or OH^-. For example, when acetic acid (CH_3COOH) is dissolved in water, some of it dissociates, but some of it remains in the undissociated form. An equilibrium is established between the ions and the undissociated weak acid.

$$CH_3COOH \rightleftarrows CH_3COO^- + H^+$$
Freely reversible

For a given weak acid or base, the amount of the dissociated ions relative to the weak acid or base is a constant.

The pH Scale

The **pH scale** (figure 2.11), which ranges from 0 to 14, indicates the H^+ concentration of a solution. Pure water is defined as a neutral

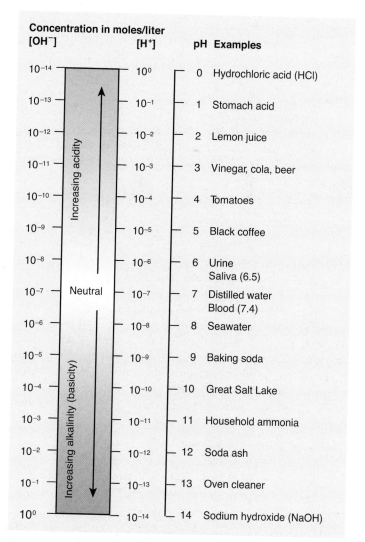

Figure 2.11 **The pH Scale**

A pH of 7 is considered neutral. Values less than 7 are acidic (the lower the number, the more acidic). Values greater than 7 are basic (the higher the number, the more basic). Representative fluids and their approximate pH values are listed.

solution. A **neutral solution** has an equal number of H^+ and OH^- and has a pH of 7.0. An **acidic solution** has a pH less than 7.0 and has a greater concentration of H^+ than OH^-. An **alkaline** (al′kă-līn), or **basic, solution** has a pH greater than 7.0 and has fewer H^+ than OH^-.

As the pH value becomes smaller, the solution is more acidic; as the pH value becomes larger, the solution is more basic. A change of one unit on the pH scale represents a 10-fold change in the H^+ concentration. For example, a solution of pH 6.0 has 10 times more H^+ than a solution with a pH of 7.0. Thus, small changes in pH represent large changes in H^+ concentration.

Acidosis and Alkalosis

The normal pH range for human blood is 7.35 to 7.45. The condition of **acidosis** (as-i-dō′sis) results if blood pH drops below 7.35. The nervous system becomes depressed, and the individual becomes disoriented and possibly comatose. **Alkalosis** (al-kă-lō′sis) results if blood pH rises above 7.45. The nervous system becomes overexcitable and the individual can be extremely nervous or have convulsions. Both acidosis and alkalosis can result in death.

Salts

A **salt** is a compound consisting of a positive ion other than a H^+ and a negative ion other than a OH^-. Salts are formed by the reaction of an acid and a base. For example, hydrochloric acid (HCl) combines with sodium hydroxide (NaOH) to form the salt sodium chloride (NaCl).

$$\underset{\text{(acid)}}{HCl} + \underset{\text{(base)}}{NaOH} \rightarrow \underset{\text{(salt)}}{NaCl} + \underset{\text{(water)}}{H_2O}$$

Buffers

The chemical behavior of many molecules changes as the pH of the solution in which they are dissolved changes. The survival of an organism depends on its ability to regulate body fluid pH within a narrow range. One way normal body fluid pH is maintained is through the use of buffers. A **buffer** (bŭf′er) is a chemical that resists changes in pH when either an acid or a base is added to a solution containing the buffer. When an acid is added to a buffered solution, the buffer binds to the H^+, preventing them from causing a decrease in the pH of the solution (figure 2.12).

PREDICT 6

If a base is added to a solution, would the pH of the solution increase or decrease? If the solution is buffered, what response from the buffer prevents the change in pH?

21. *Define acid and base. What is the difference between a strong acid or base and a weak acid or base?*
22. *Describe the pH scale. Define acidosis and alkalosis, and describe the symptoms of each.*
23. *What is a salt? What is a buffer, and why are buffers important to organisms?*

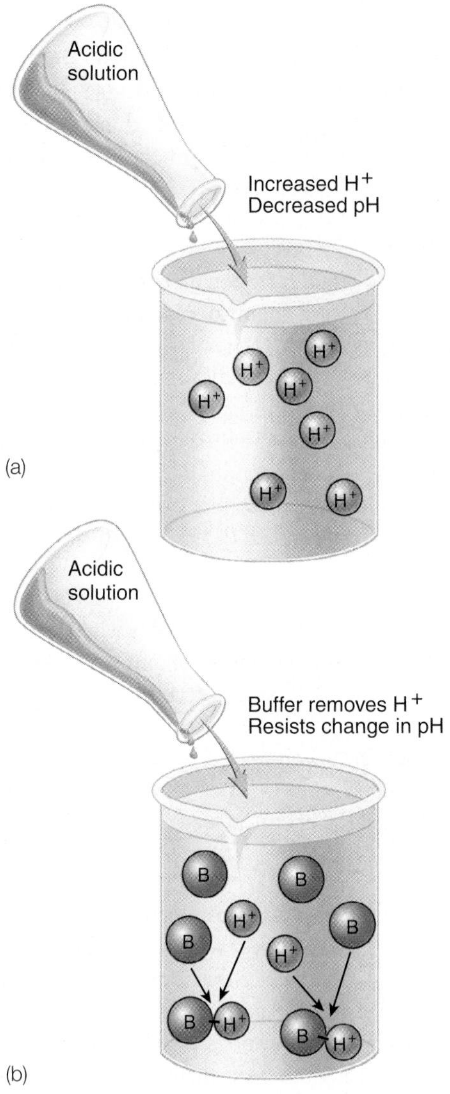

Figure 2.12 Buffers

(*a*) The addition of an acid to a nonbuffered solution results in an increase of H^+ and a decrease in pH. (*b*) The addition of an acid to a buffered solution results in a much smaller change in pH because the added H^+ bind to the buffer (symbolized by the letter "B").

Inorganic Chemistry

Originally it was believed that inorganic substances were those that came from nonliving sources and organic substances were those extracted from living organisms. As the science of chemistry developed, however, it became apparent that organic substances could be manufactured in the laboratory. As defined currently, **inorganic chemistry** deals with those substances that do not contain carbon, whereas **organic chemistry** is the study of carbon-containing substances. These definitions have a few exceptions. For example, carbon monoxide (CO), carbon dioxide (CO_2), and bicarbonate ion (HCO_3^-) are classified as inorganic molecules.

Oxygen and Carbon Dioxide

Oxygen (O_2) is an inorganic molecule consisting of two oxygen atoms bound together by a double covalent bond. About 21% of the gas in the atmosphere is oxygen, and it is essential for most living organisms. Oxygen is required by humans in the final step of a series of chemical reactions in which energy is extracted from food molecules to make ATP (see chapter 22).

Carbon dioxide (CO_2) consists of one carbon atom bound to two oxygen atoms. Each oxygen atom is bound to the carbon atom by a double covalent bond. Carbon dioxide is produced when food molecules, such as glucose, are metabolized within the cells of the body (see chapter 22). Once carbon dioxide is produced, it is eliminated from the cell as a metabolic by-product, transferred to the lungs by the blood, and exhaled during respiration. If carbon dioxide is allowed to accumulate within cells, it becomes toxic.

Water

Water (H_2O) is an inorganic molecule that consists of one atom of oxygen joined by polar covalent bonds to two atoms of hydrogen. It has many important properties for living organisms.

1. *Stabilizing body temperature.* Water can absorb large amounts of heat and remain at a stable temperature. Blood, which is mostly water, can transfer heat effectively from deep within the body to the body's surface. Blood is warmed deep in the body and then flows to the surface, where the heat is released. In addition, water evaporation in the form of sweat results in significant heat loss from the body.
2. *Protection.* Water is an effective lubricant. For example, tears protect the surface of the eye from the rubbing of the eyelids. Water also forms a fluid cushion around organs that helps protect them from damage. The cerebrospinal fluid that surrounds the brain is an example.
3. *Chemical reactions.* Most of the chemical reactions necessary for life do not take place unless the reacting molecules are dissolved in water. For example, sodium chloride must dissociate in water into Na^+ and Cl^- before they can react with other ions. Water also directly participates in many chemical reactions. For example, during the digestion of food, large molecules and water react to form smaller molecules.
4. *Transport.* Many substances dissolve in water and can be moved from place to place as the water moves. For example, blood transports nutrients, gases, and waste products within the body. ▼

24. *Define inorganic and organic chemistry.*
25. *What is the function of oxygen in living systems? Where does the carbon dioxide we breathe out come from?*
26. *List four functions that water performs in living organisms and give an example of each.*

Organic Chemistry

The ability of carbon to form covalent bonds with other atoms makes possible the formation of the large, diverse, complicated molecules necessary for life. A series of carbon atoms bound together by covalent bonds constitutes the "backbone" of many large molecules. Variation in the length of the carbon chains and the combination of atoms bound to the carbon backbone allow the formation of a wide

variety of molecules. For example, some protein molecules have thousands of carbon atoms bound by covalent bonds to one another or to other atoms, such as nitrogen, sulfur, hydrogen, and oxygen.

The four major groups of organic molecules essential to living organisms are carbohydrates, lipids, proteins, and nucleic acids. Each of these groups has specific structural and functional characteristics (table 2.3).

Carbohydrates

Carbohydrates are composed of carbon, hydrogen, and oxygen atoms. In most carbohydrates, for each carbon atom there are two hydrogen atoms and one oxygen atom. Note that this ratio of hydrogen atoms to oxygen atoms is two to one, the same as in water (H_2O). They are called carbohydrates because each carbon (carbo) is combined with the same atoms that form water (hydrated). For example, the chemical formula for glucose is $C_6H_{12}O_6$.

The smallest carbohydrates are **monosaccharides** (mon-ō-sak′ă-rīdz, one sugar), or simple sugars. Glucose (blood sugar) and fructose (fruit sugar) are important monosaccharide energy sources for many of the body's cells. Larger carbohydrates are formed by chemically binding monosaccharides together. For this reason, monosaccharides are considered the building blocks of carbohydrates. **Disaccharides** (dī-sak′ă-rīdz, two sugars) are formed when two monosaccharides join. For example, glucose and fructose combine to form the disaccharide sucrose (table sugar) (figure 2.13a). **Polysaccharides** (pol-ē-sak′ă-rīdz, many sugars) consist of many monosaccharides bound in long chains. Glycogen, or animal starch, is a polysaccharide of glucose (figure 2.13b). It is an energy-storage molecule. When cells containing glycogen need energy, the glycogen is broken down into individual glucose molecules, which can be used as energy sources. Plant starch, also a polysaccharide of glucose, can be ingested and broken down into glucose. Cellulose, another polysaccharide of glucose, is an important structural component of plant cell walls. Humans cannot digest cellulose, however, and it is eliminated in the feces, where the cellulose fibers provide bulk. ▼

27. *Name the basic building blocks of carbohydrates.*
28. *What are disaccharides and polysaccharides, and how are they formed?*
29. *Which carbohydrates are used for energy? What is the function of glycogen and cellulose in animals?*

Lipids

Lipids are substances that dissolve in nonpolar solvents, such as alcohol or acetone, but not in polar solvents, such as water. Lipids are composed mainly of carbon, hydrogen, and oxygen, but other elements, such as phosphorus and nitrogen, are minor components of some lipids. Lipids contain a lower proportion of oxygen to carbon than do carbohydrates.

Fats, phospholipids, eicosanoids, and steroids are examples of lipids. **Fats** are important energy-storage molecules. Energy from the chemical bonds of ingested foods can be stored in the chemical bonds of fat for later use as energy is needed. Fats also provide protection by surrounding and padding organs, and under-the-skin fats act as an insulator to prevent heat loss.

Table 2.3 Important Organic Molecules and Their Functions

Molecule	Building Blocks	Function	Examples
Carbohydrate	Monosaccharides	Energy	Monosaccharides (glucose, fructose) can be used as energy sources. Disaccharides (sucrose, lactose) and polysaccharides (starch, glycogen) must be broken down to monosaccharides before they can be used for energy. Glycogen (polysaccharide) is an energy-storage molecule in muscles and in the liver.
		Structure	Ribose forms part of RNA and ATP molecules, and deoxyribose forms part of DNA.
		Bulk	Cellulose forms bulk in the feces.
Lipid	Glycerol and fatty acids (for fats)	Energy	Fats can be stored and broken down later for energy; per unit of weight, fats yield twice as much energy as carbohydrates.
		Structure	Phospholipids and cholesterol are important components of plasma membranes.
		Regulation	Steroid hormones regulate many physiological processes. For example, estrogen and testosterone are responsible for many of the differences between males and females. Prostaglandins help regulate tissue inflammation and repair.
		Insulation	Fat under the skin prevents heat loss. Myelin surrounds nerve cells and electrically insulates the cells from one another.
Protein	Amino acids	Regulation	Enzymes control the rate of chemical reactions. Hormones regulate many physiological processes. For example, insulin affects glucose transport into cells.
		Structure	Collagen fibers form a structural framework in many parts of the body. Keratin adds strength to skin, hair, and nails.
		Energy	Proteins can be broken down for energy; per unit of weight, they yield the same energy as carbohydrates.
		Contraction	Actin and myosin in muscle are responsible for muscle contraction.
		Transport	Hemoglobin transports oxygen in the blood. Plasma proteins transport many substances in the blood.
		Protection	Antibodies and complement protect against microorganisms and other foreign substances.
Nucleic acid	Nucleotides	Regulation	DNA directs the activities of the cell.
		Heredity	Genes are pieces of DNA that can be passed from one generation to the next generation.
		Protein synthesis	RNA is involved in protein synthesis.

The building blocks of fats are **glycerol** (glis′er-ol) and **fatty acids** (figure 2.14). Glycerol is a three-carbon molecule with a **hydroxyl** (hī-drok′sil) **group** (—OH) attached to each carbon atom, and fatty acids consist of a carbon chain with a **carboxyl** (kar-bok′sil) **group** attached at one end. A carboxyl group consists of both an oxygen atom and a hydroxyl group attached to a carbon atom (—COOH).

$$\begin{array}{ccc} & \overset{\displaystyle O}{\underset{|}{\|}} & & \overset{\displaystyle O}{\|} \\ -\!\!\!&C\!-\!OH & \text{or} & HO\!-\!C\!-\!\! \end{array}$$

The carboxyl group is responsible for the acidic nature of the molecule because it releases hydrogen ions into solution.

Monoglycerides (mon-ō-glis′er-īdz) have one fatty acid, **diglycerides** (dī-glis′er-īdz) have two fatty acids, and **triglycerides** (trī-glis′er-īdz) have three fatty acids bound to glycerol (see figure 2.14). Triglycerides constitute 95% of the fats in the human body.

Fatty acids differ from one another according to the length and degree of saturation of their carbon chains. Most naturally occurring fatty acids contain 14–18 carbon atoms. A fatty acid is **saturated** if it contains only single covalent bonds between the carbon atoms (figure 2.15a). Sources of saturated fats include beef, pork, whole milk, cheese, butter, eggs, coconut oil, and palm oil. The carbon chain is **unsaturated** if it has one or more double covalent bonds (figure 2.15b). Because the double covalent bonds can occur anywhere along the carbon chain, many types of unsaturated fatty acids with an equal degree of unsaturation are possible. **Monounsaturated** fats, such as olive and peanut oils, have one double covalent bond between carbon atoms. **Polyunsaturated fats,** such as safflower, sunflower, corn, or fish oils, have two or more double covalent bonds between carbon atoms. Unsaturated fats are the best type of fats in the diet because, unlike saturated fats, they do not contribute to the development of cardiovascular disease.

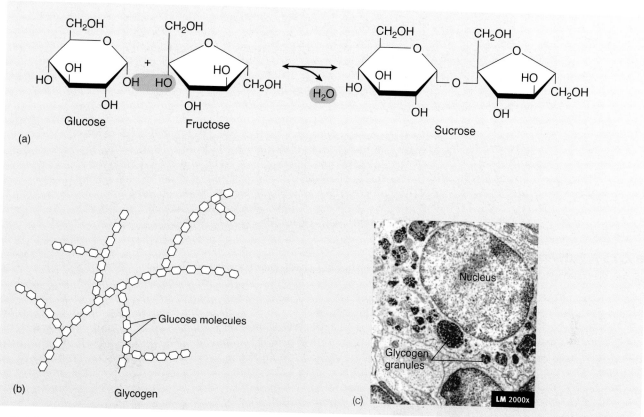

Figure 2.13 Carbohydrates

(a) Glucose and fructose are monosaccharides, which almost always form a ring-shaped molecule. Although not labeled with a "C," carbon atoms are located at the corners of the ring-shaped molecule. Glucose and fructose combine to form the disaccharide sucrose (table sugar). Glucose and fructose are held together by a covalent bond and a water molecule (H_2O) is given off. (b) Glycogen is a polysaccharide formed by combining many glucose molecules. (c) The photomicrograph shows glycogen granules in a liver cell.

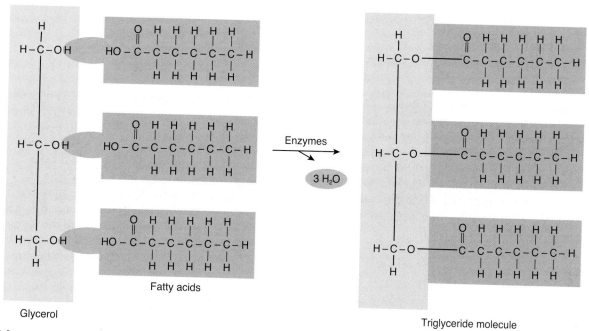

Figure 2.14 Triglyceride

Production of a triglyceride from one glycerol molecule and three fatty acids. One water molecule is given off for each covalent bond formed between a fatty acid molecule and glycerol.

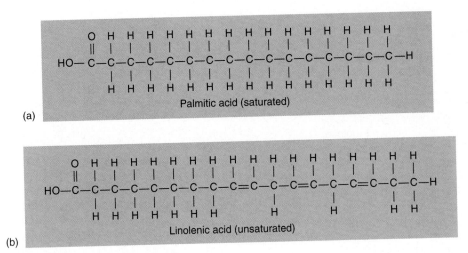

Figure 2.15   Fatty Acids

(*a*) Palmitic acid (saturated with no double bonds between the carbons). (*b*) Linolenic acid (unsaturated with three double bonds between the carbons).

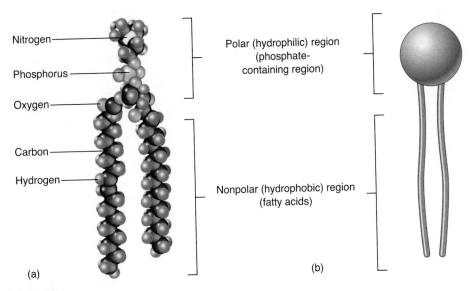

Figure 2.16   Phospholipids

(*a*) Molecular model of a phospholipid. (*b*) Simplified way in which phospholipids are often depicted.

Phospholipids are similar to triglycerides, except that one of the fatty acids bound to the glycerol is replaced by a molecule containing phosphorus (figure 2.16). They are polar at the end of the molecule to which the phosphate is bound and nonpolar at the other end. The polar end of the molecule is attracted to water and is said to be **hydrophilic** (water loving). The nonpolar end is repelled by water and is said to be **hydrophobic** (water fearing). Phospholipids are important structural components of cell membranes (see chapter 3).

The **eicosanoids** (ī′kō-să-noydz) are a group of important chemicals derived from fatty acids. They include **prostaglandins** (pros′tă-glan′dinz), **thromboxanes** (throm′bok-zānz), and **leukotrienes** (loo-kō-trī′ēnz). Eicosanoids are made in most cells and are important regulatory molecules. Among their numerous effects is their role in the response of tissues to injuries. Prostaglandins have been implicated in regulating the secretion of some hormones, blood clotting, some reproductive functions, and many other processes. Many of the therapeutic effects of aspirin and other anti-inflammatory drugs result from their ability to inhibit prostaglandin synthesis.

Steroids are composed of carbon atoms bound together into four ringlike structures. Important steroid molecules include cholesterol, bile salts, estrogen, progesterone, and testosterone (figure 2.17). Cholesterol is an important steroid because other molecules are synthesized from it. For example, bile salts, which increase fat absorption in the intestines, are derived from cholesterol, as are the reproductive hormones estrogen, progesterone, and testosterone. In addition, cholesterol is an important component of cell membranes. Although high levels of cholesterol in the blood increase the risk for cardiovascular disease, a certain amount of cholesterol is vital for normal function. ▼

30. *Give three functions of fats. What are the basic building blocks of fat? Name the most common type of fat in the human body.*
31. *What is the difference between a saturated and an unsaturated fat? Between monounsaturated and polyunsaturated fats?*

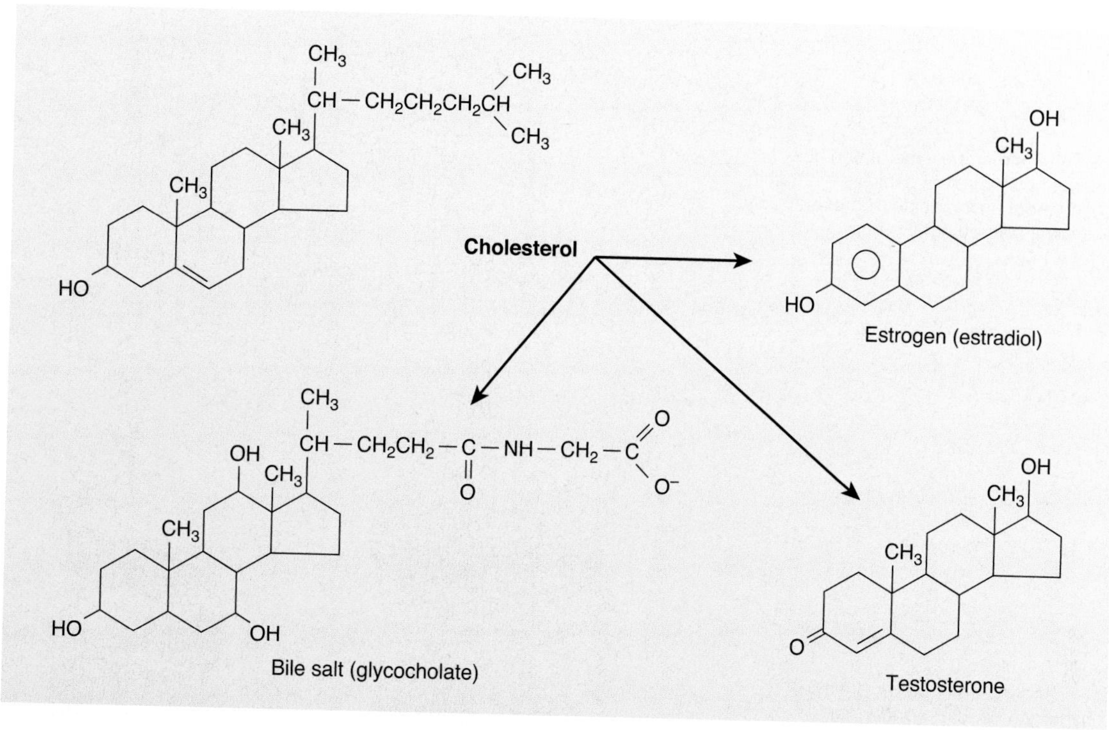

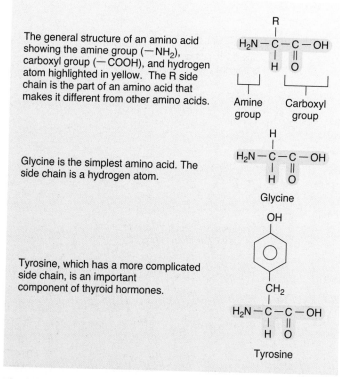

Figure 2.17 Steroids

Steroids are four-ringed molecules that differ from one another according to the groups attached to the rings. Cholesterol, the most common steroid, can be modified to produce other steroids.

32. *Describe the structure of phospholipids. Define hydrophilic and hydrophobic. Which end of a phospholipid is hydrophilic?*
33. *What are the functions of eicosanoids? Give three examples of eicosanoids.*
34. *Describe the structure of steroids. Why is cholesterol an important steroid?*

Proteins

All **proteins** contain carbon, hydrogen, oxygen, and nitrogen, and most have some sulfur. The building blocks of proteins are **amino** (ă-mē′nō) **acids,** which are organic acids containing an **amine** (ă-mēn′) **group** ($-NH_2$) and a carboxyl group, a hydrogen atom, and a side chain designated by the symbol R attached to the same carbon atom. The side chain can be a variety of chemical structures, and the differences in the side chains make the amino acids different from one another (figure 2.18). There are 20 basic types of amino acids.

A **peptide bond** is a covalent bond that binds two amino acids together (see figure 2.8). A **dipeptide** is two amino acids bound together by a peptide bond, a **tripeptide** is three amino acids bound together by peptide bonds, and a **polypeptide** is many amino acids bound together by peptide bonds. Proteins are polypeptides composed of hundreds of amino acids joined together to form a chain of amino acids.

The **primary structure** of a protein is determined by the sequence of the amino acids bound by peptide bonds (figure 2.19a). The potential number of different protein molecules is enormous because 20 different amino acids exist, and each amino acid can be located at any position along a polypeptide chain. The characteristics of the amino acids in a protein ultimately determine

the three-dimensional shape of the protein, and the shape of the protein determines its function. A change in one, or a few, amino acids in the primary structure can alter protein function, usually making the protein less or even nonfunctional.

The general structure of an amino acid showing the amine group ($-NH_2$), carboxyl group ($-COOH$), and hydrogen atom highlighted in yellow. The R side chain is the part of an amino acid that makes it different from other amino acids.

Amine group

Carboxyl group

Glycine is the simplest amino acid. The side chain is a hydrogen atom.

Glycine

Tyrosine, which has a more complicated side chain, is an important component of thyroid hormones.

Tyrosine

Figure 2.18 Amino Acids

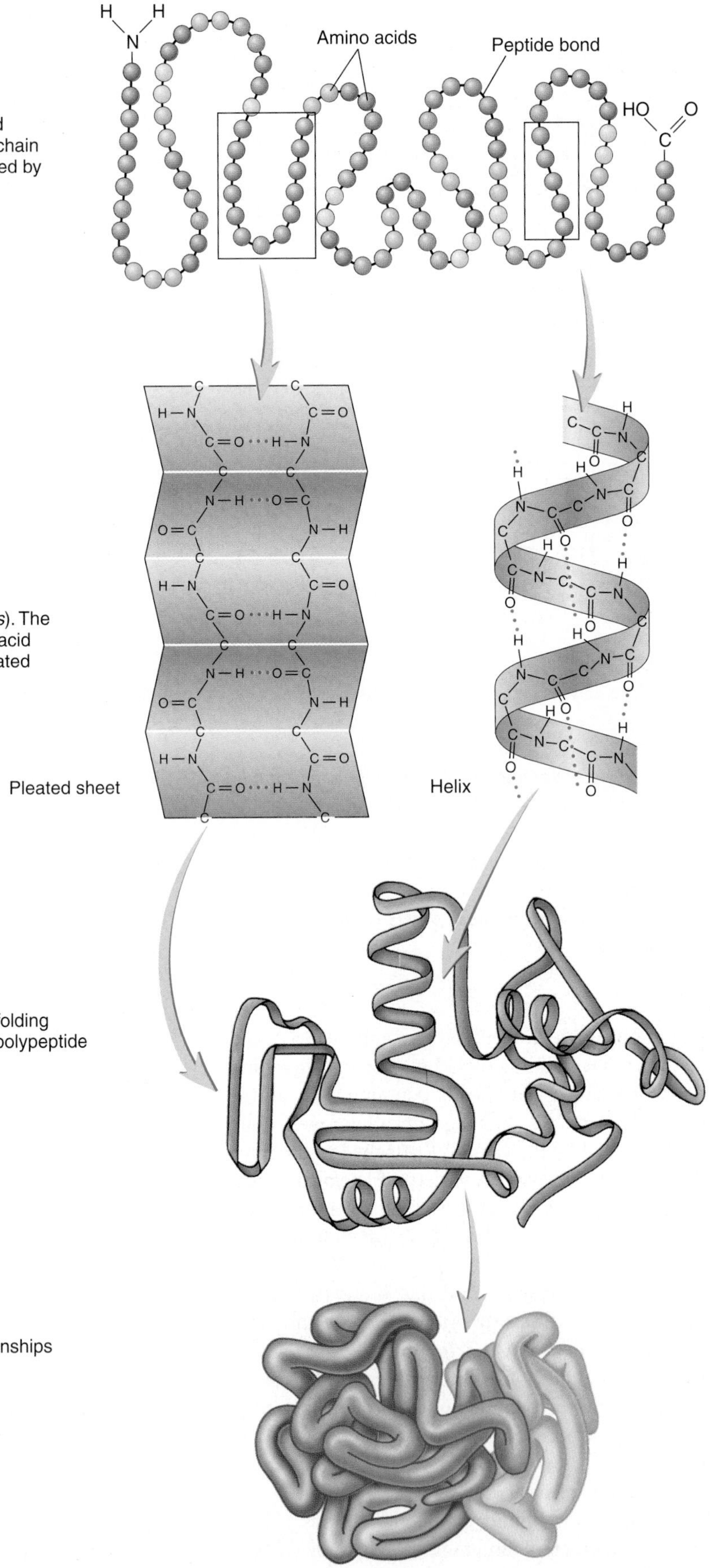

(a) Primary structure—the amino acid
sequence. A protein consists of a chain
of different amino acids (represented by
different colored spheres).

Amino acids

Peptide bond

(b) Secondary structure results from
hydrogen bonding (*dotted red lines*). The
hydrogen bonds cause the amino acid
chain to form helices (coils) or pleated
(folded) sheets.

Pleated sheet

Helix

(c) Tertiary structure with secondary folding
caused by interactions within the polypeptide
and its immediate environment

(d) Quaternary structure—the relationships
between individual subunits

Figure 2.19   Proteins

The **secondary structure** results from the folding or bending of the polypeptide chain caused by the hydrogen bonds between amino acids (figure 2.19*b*). Two common shapes that result are helices (coils) and pleated (folded) sheets. If the hydrogen bonds that maintain the shape of the protein are broken, the protein becomes nonfunctional. This change in shape is called **denaturation,** and it can be caused by abnormally high temperatures or changes in the pH of body fluids. An everyday example of denaturation is the change in the proteins of egg whites when they are cooked.

The **tertiary structure** results from the folding of the helices or pleated sheets (figure 2.19*c*). Some amino acids are quite polar and therefore form hydrogen bonds with water. The polar portions of proteins tend to remain unfolded, maximizing their contact with water, whereas the less polar regions tend to fold into a globular shape, minimizing their contact with water. The formation of covalent bonds between sulfur atoms of one amino acid and sulfur atoms of another amino acid located at a different place in the sequence of amino acids can also contribute to the tertiary structure of proteins. The tertiary structure determines the shape of a **domain,** which is a folded sequence of 100–200 amino acids within a protein. The functions of proteins occur at one or more domains. Therefore, changes in the primary or secondary structure that affect the shape of the domain can change protein function.

If two or more proteins associate to form a functional unit, the individual proteins are called subunits. The **quaternary** structure is the spatial relationships between the individual subunits (figure 2.19*d*).

Proteins perform many important functions. For example, enzymes are proteins that regulate the rate of chemical reactions, structural proteins provide the framework for many of the body's tissues, and muscles contain proteins that are responsible for muscle contraction. ▼

35. *Name and describe the structure of the building blocks of proteins. Define a peptide bond.*
36. *What determines the primary, secondary, tertiary, and quaternary structures of proteins?*
37. *Define denaturation and name two things that can cause it to occur.*

Enzymes

An **enzyme** (en′zīm) is a protein catalyst that increases the rate at which a chemical reaction proceeds without the enzyme being permanently changed. Enzymes increase the rate of chemical reactions by lowering the **activation energy,** which is the minimum energy necessary to start a chemical reaction. For example, heat in the form of a spark is required to start the reaction between oxygen and gasoline. Most of the chemical reactions that occur in the body have high activation energies, which are decreased by enzymes (figure 2.20). The lowered activation energies enable reactions to proceed at rates that sustain life.

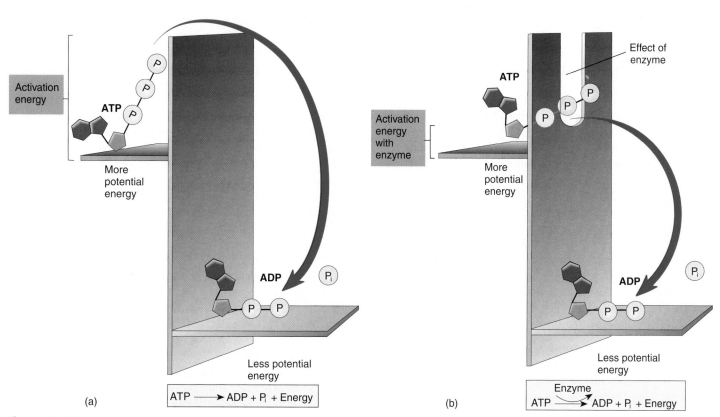

Figure 2.20   Activation Energy and Enzymes

(*a*) Activation energy is needed to change ATP to ADP. The upper shelf represents a higher energy level, and the lower shelf represents a lower energy level. The "wall" extending above the upper shelf represents the activation energy. Even though energy is given up moving from the upper to the lower shelf, the activation energy "wall" must be overcome before the reaction can proceed. (*b*) The enzyme lowers the activation energy, making it easier for the reaction to proceed.

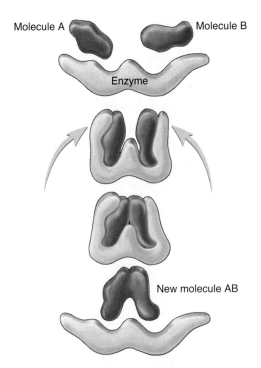

Molecule A   Molecule B

Enzyme

New molecule AB

Figure 2.21 **Enzyme Action**

The enzyme brings two reacting molecules together. After the reaction, the unaltered enzyme can be used again.

Consider this analogy, in which paper clips represent amino acids and your hands represent enzymes. Paper clips in a box only occasionally join together. Using your hands, however, a chain of paper clips can be rapidly formed. In a similar fashion, enzymes can quickly join amino acids into a chain, forming a protein. With an enzyme, the rate of a chemical reaction can take place more than a million times faster than without the enzyme.

The three-dimensional shape of enzymes is critical for their normal function. According to the **lock-and-key model** of enzyme action, the shape of an enzyme and that of the reactants allow the enzyme to bind easily to the reactants. Bringing the reactants very close to one another reduces the activation energy for the reaction. Enzymes are very specific for the reactions they control because the enzyme and the reactants must fit together. Thus, each enzyme controls only one type of chemical reaction. After the reaction takes place, the enzyme is released and can be used again (figure 2.21).

The chemical events of the body are regulated primarily by mechanisms that control either the concentration or the activity of enzymes. The rate at which enzymes are produced in cells or whether the enzymes are in an active or inactive form determines the rate of each chemical reaction. ▼

38. *What is an enzyme? How do enzymes affect activation energy and the speed of chemical reactions?*
39. *Describe the lock-and-key model of enzyme activity. Why are enzymes specific for the reactions they control?*

Nucleic Acids: DNA and RNA

Deoxyribonucleic (dē-oks′ē-rī′bō-noo-klē′ik) **acid (DNA)** is the genetic material of cells, and copies of DNA are transferred from one generation of cells to the next. DNA contains the information

that determines the structure of proteins. **Ribonucleic** (rī′bō-noo-klē′ik) **acid (RNA)** is structurally related to DNA, and three types of RNA also play important roles in protein synthesis. In chapter 3, the means by which DNA and RNA direct the functions of the cell are described.

The **nucleic** (noo-klē′ik, noo-klā′ik) **acids** are large molecules composed of carbon, hydrogen, oxygen, nitrogen, and phosphorus. Both DNA and RNA consist of basic building blocks called **nucleotides** (noo′klē-ō-tīdz). Each nucleotide is composed of a sugar (monosaccharide) to which a nitrogenous organic base and a phosphate group are attached (figure 2.22). The sugar is deoxyribose for DNA, ribose for RNA. The organic bases are thymine (thī′mēn, thī′min), cytosine (sī′tō-sēn), and uracil (ūr′ă-sil), which are single-ringed molecules, and adenine (ad′ĕ-nēn) and guanine (gwahn′ēn), which are double-ringed molecules.

DNA has two strands of nucleotides joined together to form a twisted, ladderlike structure called a double helix. The uprights of the ladder are formed by covalent bonds between the sugar molecules and phosphate groups of adjacent nucleotides. The rungs of the ladder are formed by the bases of the nucleotides of one upright connected to the bases of the other upright by hydrogen bonds. Each nucleotide of DNA contains one of the organic bases: adenine, thymine, cytosine, or guanine. **Complementary base pairs** are organic bases held together by hydrogen bonds. Adenine and thymine are complementary base pairs because the structure of these organic bases allows two hydrogen bonds to form between them. Cytosine and guanine are complementary base pairs because the structure of these organic bases allows three hydrogen bonds to form between them. The two strands of a DNA molecule are said to be complementary. If the sequence of bases in one DNA strand is known, the sequence of bases in the other strand can be predicted because of complementary base pairing.

The sequence of organic bases in DNA is a "code" that stores information used to determine the structures and functions of cells. A sequence of DNA bases that directs the synthesis of proteins or RNA molecules is called a **gene** (see chapter 3 for more information on genes). Genes determine the type and sequence of amino acids found in protein molecules. Because enzymes are proteins, DNA structure determines the rate and type of chemical reactions that occur in cells by controlling enzyme structure. The information contained in DNA, therefore, ultimately defines all cellular activities. Other proteins, such as collagen, that are coded by DNA determine many of the structural features of humans.

RNA has a structure similar to a single strand of DNA. Like DNA, four different nucleotides make up the RNA molecule, and the organic bases are the same, except that thymine is replaced with uracil. Uracil can bind only to adenine. ▼

40. *Name two types of nucleic acids. What are their functions?*
41. *What are the basic building blocks of nucleic acids? What kinds of sugar and bases are found in DNA? In RNA?*
42. *DNA is like a twisted ladder. What forms the uprights and rungs of the ladder?*
43. *Name the complementary base pairs in DNA and RNA.*
44. *Define gene and explain how genes determine the structures and functions of cells.*

1. The building blocks of nucleic acids are nucleotides, which consist of a phosphate group, a sugar, and a nitrogen base.

2. The phosphate groups connect the sugars to form two strands of nucleotides (*purple columns*).

3. Hydrogen bonds (*dotted red lines*) between the nucleotides join the two nucleotide strands together. Adenine binds to thymine and cytosine binds to guanine.

4. The two nucleotide strands coil to form a double-stranded helix.

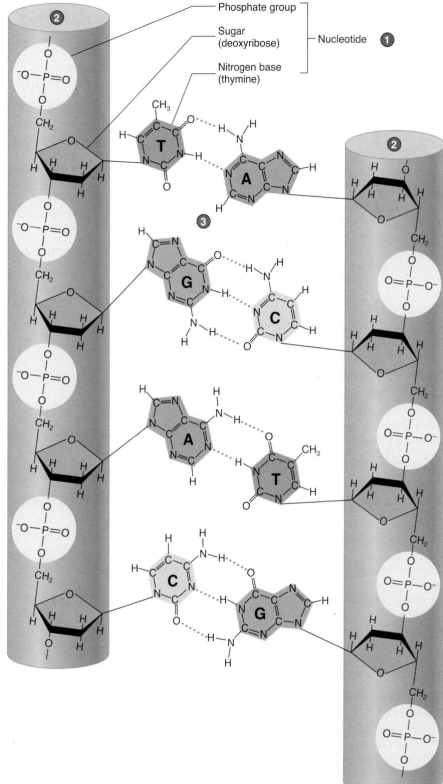

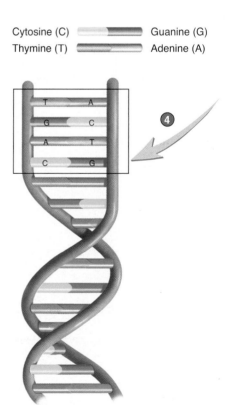

Figure 2.22   Structure of DNA

S U M M A R Y

Chemistry is the study of the composition and structure of substances and the reactions they undergo.

Basic Chemistry (p. 22)

Matter, Mass, and Weight

1. Matter is anything that occupies space.
2. Mass is the amount of matter in an object, and weight results from the gravitational attraction between earth and matter.

Elements and Atoms

1. An element is the simplest type of matter with unique chemical properties.
2. An atom is the smallest particle of an element that has the characteristics of that element. An element is composed of only one kind of atom.

Atomic Structure

1. Atoms consist of neutrons, positively charged protons, and negatively charged electrons.
2. Atoms are electrically neutral because the number of protons in atoms equals the number of electrons.
3. Protons and neutrons are found in the nucleus, and electrons, which are located around the nucleus, can be represented by an electron cloud.

Atomic Number and Mass Number

1. The atomic number is the unique number of protons in each atom of an element. The mass number is the number of protons and neutrons.
2. Isotopes are two or more forms of the same element that have the same number or protons but a different number of neutrons.

Electrons and Chemical Bonding

1. Ionic bonding results when an electron is transferred from one atom to another.
2. Cations are positively charged ions and anions are negatively charged ions.
3. Covalent bonding results when a pair of electrons are shared between atoms. Electrons are shared equally in nonpolar covalent bonds and can result in nonpolar molecules. Electrons are shared unequally in polar covalent bonds, producing polar molecules with asymmetric electrical charges.

Hydrogen Bonds

A hydrogen bond is the weak attraction that occurs between the oppositely charged regions of polar molecules. Hydrogen bonds are important in determining the three-dimensional structure of large molecules.

Molecules and Compounds

1. A molecule is two or more atoms chemically combined to form a structure that behaves as an independent unit.
2. A compound is two or more different types of atoms chemically combined. A compound can be a molecule (covalent compound) or an organized array of ions (ionic compound).
3. The kinds and numbers of atoms (or ions) in a molecule or compound can be represented by a formula consisting of the symbols of the atoms (or ions) plus subscripts denoting the number of each type of atom (or ion).

Dissociation

1. Dissociation is the separation of ions in an ionic compound by polar water molecules. Dissociated ions are called electrolytes because they can conduct electricity

2. Molecules do not dissociate in water and are called nonelectrolytes.

Chemical Reactions (p. 28)

1. A synthesis reaction is the combination of reactants to form a new, larger product. A dehydration reaction is a synthesis reaction in which water is a product.
2. A decomposition reaction is the breakdown of larger reactants into smaller products. A hydrolysis reaction is a decomposition reaction that uses water.
3. Metabolism is all of the synthesis and decomposition reactions in the body.

Reversible Reactions

1. In a reversible reaction, the reactants can form products, or the products can form reactants.
2. The amount of reactants relative to products is constant at equilibrium.

Energy and Chemical Reactions

1. Energy is the capacity to do work. Potential energy is stored energy that could do work, and kinetic energy does work by causing the movement of an object.
2. Energy can be neither created nor destroyed, but one type of energy can be changed into another.
3. Energy exists in chemical bonds as potential energy.
4. Energy is released in chemical reactions when the products contain less potential energy than the reactants. The energy can be lost as heat, can be used to synthesize molecules, or can do work.
5. Energy is absorbed in reactions when the products contain more potential energy than the reactants.
6. ATP stores and provides energy.

Rate of Chemical Reactions

1. The rate of chemical reactions increases when the concentration of the reactants increases, temperature increases, or a catalyst is present.
2. A catalyst (enzyme) increases the rate of chemical reactions without being altered permanently.

Acids and Bases (p. 31)

Acids are proton (hydrogen ion, H^+) donors, and bases are proton acceptors.

The pH Scale

1. A neutral solution has an equal number of H^+ and OH^- and a pH of 7.0.
2. An acidic solution has more H^+ than OH^- and a pH of less than 7.0.
3. A basic solution has fewer H^+ than OH^- and a pH greater than 7.0.

Salts

A salt is formed when an acid reacts with a base.

Buffers

Buffers are chemicals that resist changes in pH when acids or bases are added.

Inorganic Chemistry (p. 32)

Inorganic chemistry is mostly concerned with non-carbon-containing substances but does include such carbon-containing substances as carbon monoxide, carbon dioxide, and bicarbonate ion.

Oxygen and Carbon Dioxide

1. Oxygen is involved with the extraction of energy from food molecules.
2. Carbon dioxide is a by-product of the breakdown of food molecules.

Water

1. Water stabilizes body temperature.
2. Water provides protection by acting as a lubricant or cushion.
3. Water is necessary for many chemical reactions.
4. Water transports many substances.

Organic Chemistry   (p. 33)

Organic molecules contain carbon atoms bound together by covalent bonds.

Carbohydrates

1. Carbohydrates provide the body with energy.
2. Monosaccharides are the building blocks that form more complex carbohydrates, such as disaccharides and polysaccharides.

Lipids

1. Lipids are substances that dissolve in nonpolar solvents, such as alcohol or acetone, but not in polar solvents, such as water. Fats, phospholipids, eicosanoids, and steroids are examples of lipids.
2. Lipids provide energy (fats), are structural components (phospholipids), and regulate physiological processes (eicosanoids and steroids).
3. The building blocks of triglycerides (fats) are glycerol and fatty acids.

4. Fatty acids can be saturated (have only single covalent bonds between carbon atoms) or unsaturated (have one or more double covalent bonds between carbon atoms).

Proteins

1. Proteins regulate chemical reactions (enzymes), are structural components, and cause muscle contraction.
2. The building blocks of proteins are amino acids, which are joined by peptide bonds.
3. The number, kind, and arrangement of amino acids determine the primary structure of a protein. Hydrogen bonds between amino acids determine secondary structure, and hydrogen bonds between amino acids and water determine tertiary structure. Interactions between different protein subunits determine quaternary structure.
3. Denaturation of proteins disrupts hydrogen bonds, which changes the shape of proteins and makes them nonfunctional.
4. Enzymes speed up chemical reactions by lowering their activation energy. They bind to reactants according to the lock-and-key model.

Nucleic Acids: DNA and RNA

1. The basic unit of nucleic acids is the nucleotide, which is a monosaccharide with an attached phosphate and organic base.
2. DNA nucleotides contain the monosaccharide deoxyribose and the organic base adenine, thymine, guanine, or cytosine. DNA occurs as a double strand of joined nucleotides and is the genetic material of cells.
3. RNA nucleotides are composed of the monosaccharide ribose. The organic bases are the same as for DNA, except that thymine is replaced with uracil.

R E V I E W   A N D   C O M P R E H E N S I O N

1. The smallest particle of an element that still has the chemical characteristics of that element is a (an)
   a. electron.
   b. molecule.
   c. neutron
   d. proton.
   e. atom.

2. The number of electrons in an atom is equal to the
   a. atomic number.
   b. mass number.
   c. number of neutrons.
   d. isotope number.

3. A cation is a (an)
   a. uncharged atom.
   b. positively charged atom.
   c. negatively charged atom.
   d. atom that has gained an electron.
   e. both c and d.

4. A polar covalent bond between two atoms occurs when
   a. one atom attracts shared electrons more strongly than another atom.
   b. atoms attract electrons equally.
   c. an electron from one atom is completely transferred to another atom.
   d. a hydrogen atom is shared between two different atoms.

5. Table salt (NaCl) is
   a. an atom.
   b. organic.
   c. a molecule.
   d. a compound.

6. The weak attractive force between two water molecules forms a (an)
   a. covalent bond.
   b. hydrogen bond.
   c. ionic bond.
   d. compound.
   e. isotope.

7. Concerning chemical reactions,
   a. the formation of ADP is an example of a synthesis reaction.
   b. synthesis reactions in which water is a product are called hydrolysis reactions.
   c. basic building blocks from food are combined in decomposition reactions.
   d. the amount of reactants relative to products remains constant in reversible reactions.
   e. all of the above.

8. Potential energy
   a. is energy caused by movement of an object.
   b. is the form of energy that is actually doing work.
   c. includes energy within chemical bonds.
   d. can never be converted to kinetic energy.
   e. all of the above.

9. ATP
   a. is formed by the addition of a phosphate group to ADP.
   b. formation requires energy obtained from the breakdown of food molecules.
   c. stores potential energy.
   d. contains three phosphate groups.
   e. all of the above.

10. An *increase* in the speed of a chemical reaction occurs if
    a. the activation energy requirement is decreased.
    b. catalysts are increased.
    c. temperature increases.
    d. the concentration of the reactants increases.
    e. all of the above.

11. Water
    a. is composed of two oxygen atoms and one hydrogen atom.
    b. temperature increases dramatically when small amounts of heat are absorbed.
    c. is composed of polar molecules into which ionic substances dissociate.
    d. is a very small organic molecule.

12. A solution with a pH of 5 is _____ and contains _____ H^+ than a neutral solution.
    a. a base, more
    b. a base, less
    c. an acid, more
    d. an acid, less
    e. neutral, the same number of

13. A buffer
    a. slows down chemical reactions.
    b. speeds up chemical reactions.
    c. increases the pH of a solution.
    d. maintains a relatively constant pH.

14. Which of these is a carbohydrate?
    a. glycogen   c. steroid   e. polypeptide
    b. prostaglandin   d. DNA

15. The polysaccharide used for energy storage in the human body is
    a. cellulose.   c. lactose.   e. starch.
    b. glycogen.   d. sucrose.

16. The basic units or building blocks of triglycerides are
    a. simple sugars (monosaccharides).
    b. double sugars (disaccharides).
    c. amino acids.
    d. glycerol and fatty acids.
    e. nucleotides.

17. A _____ fatty acid has one double covalent bond between carbon atoms.
    a. cholesterol   c. phospholipid   e. saturated
    b. monounsaturated   d. polyunsaturated

18. The _____ structure of a protein results from the folding of the helices or pleated sheets.
    a. primary   c. tertiary
    b. secondary   d. quaternary

19. Which of these statements about enzymes is true?
    a. Enzymes are proteins.
    b. Enzymes lower the activation energy of chemical reactions.
    c. Enzymes increase the rate of chemical reactions without being permanently altered in the process.
    d. The shape of enzymes is critical for their normal function (lock-and-key model).
    e. all of the above.

20. DNA molecules
    a. contain genes.
    b. consist of a single strand of nucleotides.
    c. contain the nucleotide uracil.
    d. are held together when adenine binds to guanine and thymine binds to cytosine.

Answers in Appendix E

C R I T I C A L   T H I N K I N G

1. If an atom of iodine (I) gains an electron, what is the charge of the resulting ion? Write the symbol for this ion.

2. For each of the following chemical equations, determine if a synthesis reaction, a decomposition reaction, or dissociation has taken place:
   a. $HCl \rightarrow H^+ + Cl^-$
   b. Glucose + Fructose $\rightarrow$ Sucrose (table sugar)
   c. $2\,H_2O \rightarrow 2\,H_2 + O_2$

3. In terms of the energy in chemical bonds, explain why eating food is necessary for increasing muscle mass.

4. Given that the hydrogen ion concentration in a solution is based on the following reversible reaction:

   $$CO_2 + H_2O \leftrightarrows H^+ + HCO_3^-$$

   What happens to the pH of the solution when $NaHCO_3$ (sodium bicarbonate) is added to the solution? (*Hint:* The sodium bicarbonate dissociates to form Na^+ and HCO_3^-.)

5. A mixture of chemicals is warmed slightly. As a consequence, although little heat is added, the solution becomes very hot. Explain what happens to make the solution hot.

6. Two solutions, when mixed together at room temperature, produce a chemical reaction. When the solutions are boiled, however, and allowed to cool to room temperature before mixing, no chemical reaction takes place. Explain.

Answers in Appendix F

CHAPTER 3 Cell Structures and Their Functions

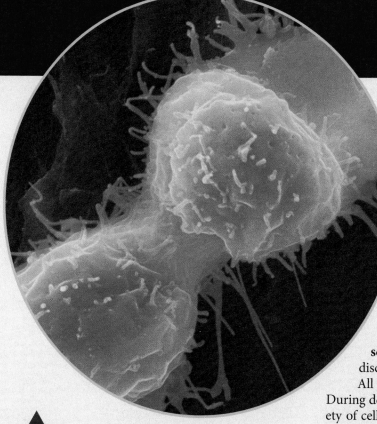

Colorized scanning electron micrograph (SEM) of a dividing cell. Cell division is one of the major features of organisms and is the basis of growth and repair of tissues and organs.

Tools for Success

apревealed.com

A virtual cadaver dissection experience

Audio (MP3) and/or video (MP4) content, available at www.mhhe.aris.com

Introduction

The cell is the basic living unit of all organisms. The simplest organisms consist of single cells, whereas humans are composed of trillions of cells. An average-sized cell is one-fifth the size of the smallest dot you can make on a sheet of paper with a sharp pencil. If each cell of the body were the size of a standard brick, the colossal human statue made from those bricks would be 6 miles high! **Light microscopes** allow us to visualize general features of cells. **Electron microscopes,** which achieve higher magnifications than light microscopes, are used to study the fine structure of cells. A **scanning electron microscope (SEM)** allows us to see features of the cell surface and the surfaces of internal structures. A **transmission electron microscope (TEM)** allows us to see "through" parts of the cell and thus to discover other aspects of cell structure.

All the cells of an individual originate from a single fertilized cell. During development, cell division and specialization give rise to a wide variety of cell types, such as nerve, muscle, bone, fat, and blood cells. Each cell type has important characteristics that are critical to the normal function of the body as a whole. Maintaining homeostasis is necessary to keep the trillions of cells that form the body functioning normally.

The study of cells is an important link between the study of chemistry in chapter 2 and tissues in chapter 4. Knowledge of chemistry makes it possible to understand cells because cells are composed of chemicals responsible for many of the characteristics of cells. Cells, in turn, determine the form and functions of the tissues of the body. In addition, a great many diseases and other human disorders have a cellular basis.

▶This chapter considers **cell organization and functions (p. 46),** the composition of the **plasma membrane (p. 48),** and **movement through the plasma membrane (p. 49).** The chapter then addresses the **cytoplasm (p. 58),** the **nucleus and cytoplasmic organelles (p. 59),** protein synthesis **(p. 65),** and **cell division (p. 69).** The chapter concludes with **differentiation (p. 72).**

Cell Organization and Functions

Each cell is a highly organized unit. The **plasma** (plaz′mă), or **cell, membrane** forms the outer boundary of the cell, through which the cell interacts with its external environment. Within cells, specialized structures called **organelles** (or′gă-nelz, little organs) perform specific functions (figure 3.1 and table 3.1). The nucleus is an organelle, usually located centrally in the cell. It contains the cell's genetic material and directs cell activities. **Cytoplasm** (sī′tō-plazm, *cyto-,* cell + *plasma,* a thing formed), located between the nucleus and plasma membrane, contains many organelles. The number and type of organelles within each cell determine the cell's specific structure and functions.

The main functions of the cell are

1. *Cell metabolism and energy use.* The chemical reactions that occur within cells are referred to collectively as **cell metabolism.** Energy released during metabolism is used for cell activities, such as the synthesis of new molecules and muscle contraction. Heat produced during metabolism helps maintain body temperature.
2. *Synthesis of molecules.* The structural and functional characteristics of cells are determined by the types of molecules they produce. Cells differ from each other because they synthesize different kinds of molecules, including lipids, proteins, and nucleic acids. For example, bone cells produce and secrete a mineralized material, making bone hard, whereas muscle cells contain proteins that enable the cells to shorten, resulting in contraction of muscles.
3. *Communication.* Cells produce and receive chemical and electric signals that allow them to communicate with one another. For example, nerve cells communicate with one another and with muscle cells, causing muscle cells to contract.
4. *Reproduction and inheritance.* Most cells contain a copy of all the genetic information of the individual. This genetic information ultimately determines the structural and functional characteristics of the cell. During the growth of an individual, cells divide to produce new cells, each containing the same genetic information. Specialized cells of the body, called gametes, are responsible for transmitting genetic information to the next generation. ▼

1. *Define plasma membrane, organelle, cytoplasm, and cell metabolism.*
2. *What are the main functions of the cell?*

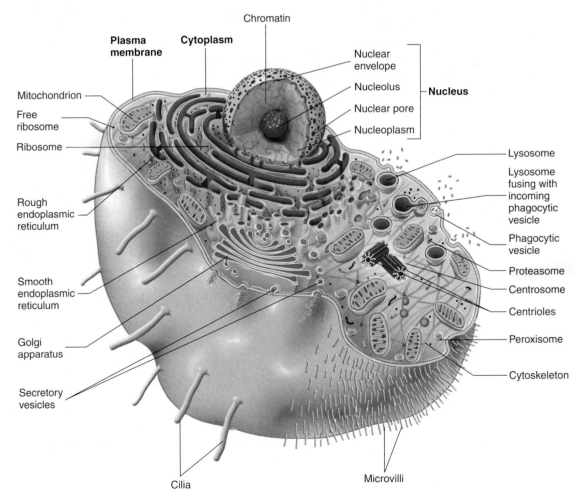

Figure 3.1   Cell

A generalized human cell showing the plasma membrane, nucleus, and cytoplasm with its organelles. Although no single cell contains all these organelles, many cells contain a large number of them.

Table 3.1 Summary of Cell Parts and Functions

Cell Parts	Structure	Function
Plasma Membrane	Lipid bilayer composed of phospholipids and cholesterol with proteins that extend across or are embedded in either surface of the lipid bilayer	Outer boundary of cells that controls the entry and exit of substances; attaches to other cells or intercellular molecules; part of intercellular communication and identification; catalyzes chemical reactions
Cytoplasm: Cytosol		
Fluid part	Water with dissolved ions and molecules; colloid with suspended proteins	Contains enzymes that catalyze the synthesis and breakdown of molecules
Cytoskeleton Microtubules	Hollow cylinders composed of protein; 25 nm in diameter	Support the cytoplasm and form centrioles, spindle fibers, cilia, and flagella
Actin filaments	Small fibrils of the protein actin; 8 nm in diameter	Provide structural support to cells, support microvilli, are responsible for cell movements
Intermediate filaments	Protein fibers; 10 nm in diameter	Provide structural support to cells
Cytoplasmic inclusions	Aggregates of molecules manufactured or ingested by the cell; may be membrane-bound	Function depends on the molecules: energy storage (lipids, glycogen), oxygen transport (hemoglobin), skin color (melanin), and others
Nucleus		
Nuclear envelope	Double membrane enclosing the nucleus; the outer membrane is continuous with the endoplasmic reticulum; nuclear pores extend through the nuclear envelope	Separates nucleus from cytoplasm; allows movement of materials into and out of nucleus
Chromatin	Dispersed, thin strands of DNA, histones, and other proteins; condenses to form chromosomes during cell division	DNA regulates protein (e.g., enzyme) synthesis and therefore the chemical reactions of the cell; DNA is the genetic, or hereditary, material.
Nucleolus	One or more dense bodies consisting of ribosomal RNA and proteins	Assembly site of large and small ribosomal subunits
Cytoplasmic Organelles		
Ribosome	Ribosomal RNA and proteins form large and small ribosomal subunits; attached to endoplasmic reticulum or free ribosomes are distributed throughout the cytoplasm	Site of protein synthesis
Rough endoplasmic reticulum	Membranous tubules and flattened sacs with attached ribosomes	Protein synthesis and transport to Golgi apparatus
Smooth endoplasmic reticulum	Membranous tubules and flattened sacs with no attached ribosomes	Manufactures lipids and carbohydrates; detoxifies harmful chemicals; stores calcium
Golgi apparatus	Flattened membrane sacs stacked on each other	Modifies proteins and lipids and packages them into vesicles for distribution (e.g., for internal use, secretion, or to become part of the plasma membrane)
Secretory vesicle	Membrane-bound sac pinched off Golgi apparatus	Carries proteins to cell surface for secretion
Lysosome	Membrane-bound vesicle pinched off Golgi apparatus	Contains digestive enzymes
Peroxisome	Membrane-bound vesicle	One site of lipid and amino acid degradation; breaks down hydrogen peroxide
Proteasomes	Tubelike protein complexes in the cytoplasm	Break down proteins in the cytoplasm
Mitochondria	Spherical, rod-shaped, or threadlike structures; enclosed by double membrane; inner membrane forms projections called cristae	Major site of ATP synthesis
Centrioles	Pair of cylindrical organelles in the centrosome, consisting of triplets of parallel microtubules	Centers for microtubule formation; determine cell polarity during cell division
Spindle fibers	Microtubules extending from the centrosome to chromosomes and other parts of the cell (i.e., aster fibers)	Assist in the separation of chromosomes during cell division
Cilia	Extensions of the plasma membrane containing parallel microtubules; 10 μm in length	Move materials over the surface of cells
Flagellum	Extension of the plasma membrane containing parallel microtubules; 55 μm in length	In humans, responsible for movement of sperm cells
Microvilli	Extension of the plasma membrane containing actin filaments	Increase surface area of the plasma membrane for absorption and secretion

Plasma Membrane

The plasma (plaz′mă) membrane is the outermost component of a cell. The plasma membrane encloses the cell, supports the cell contents, is a selective barrier that determines what moves into and out of the cell, and plays a role in communication between cells. Substances inside the cell are called **intracellular substances,** and those outside the cell are called **extracellular substances.** Sometimes extracellular substances are referred to as **intercellular,** meaning between cells.

The predominant lipids of the plasma membrane are phospholipids and cholesterol. **Phospholipids** readily assemble to form a **lipid bilayer,** which is a double layer of phospholipid molecules. The polar, phosphate-containing ends of the phospholipids are hydrophilic (water loving) and therefore face the water inside and outside the cell. The nonpolar, fatty acid ends of the phospholipids are hydrophobic (water fearing) and therefore face away from the water on either side of the membrane, toward the center of the double layer of phospholipids (figure 3.2). The double layer of phospholipids forms a lipid barrier between the inside and outside of the cell.

The modern concept of the plasma membrane, the **fluid-mosaic model,** suggests that the plasma membrane is highly flexible and can change its shape and composition through time. The lipid bilayer functions as a liquid in which other molecules, such as cholesterol and proteins, are suspended. Cholesterol within the phospholipid membrane gives it added strength and flexibility.

Protein molecules "float" among the phospholipid molecules and in some cases extend from the inner to the outer surface of the plasma membrane. Membrane proteins can function as marker molecules, attachment proteins, transport proteins, receptor proteins, or enzymes. The ability of membrane proteins to function depends on their three-dimensional shapes and their chemical characteristics.

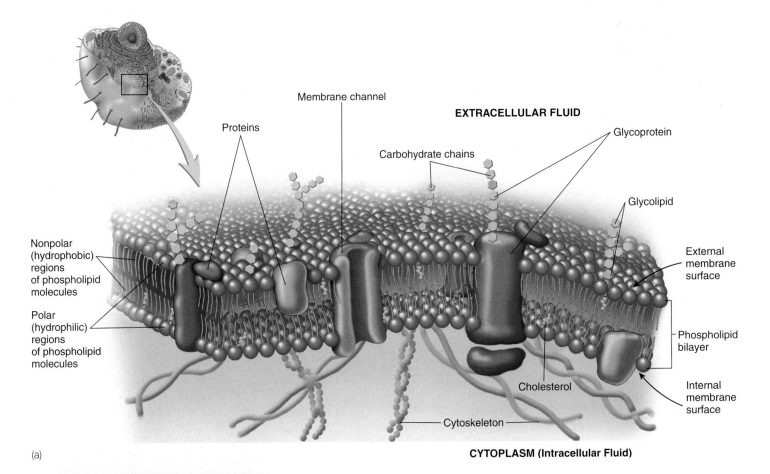

(a)

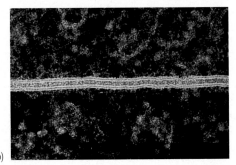

(b)

Figure 3.2   Plasma Membrane

(*a*) Fluid-mosaic model of the plasma membrane. The membrane is composed of a bilayer of phospholipids and cholesterol with proteins "floating" in the membrane. The nonpolar hydrophobic region of each phospholipid molecule is directed toward the center of the membrane and the polar hydrophilic region is directed toward the water environment either outside or inside the cell. (*b*) Colorized transmission electron micrograph of plasma membranes of two adjacent cells. Proteins at either surface of the lipid bilayer stain more readily than the lipid bilayer does and give each membrane the appearance of consisting of three parts: the two yellow, outer parts are proteins and the phospholipid heads, and the red, central part is the phospholipid tails and cholesterol.

Marker molecules are cell surface molecules that allow cells to identify one another or other molecules. They are mostly **glycoproteins,** which are proteins with attached carbohydrates, or **glycolipids,** which are lipids with attached carbohydrates. Marker molecules allow immune cells to distinguish between self-cells and foreign cells, such as bacteria or donor cells in an organ transplant. Intercellular recognition and communication are important because cells are not isolated entities.

Attachment proteins allow cells to attach to other cells or to extracellular molecules. Many attachment proteins also attach to intracellular molecules. **Cadherins** are proteins that attach cells to other cells. **Integrins** are proteins that attach cells to extracellular molecules. Because of their interaction with intracellular molecules, integrins also function in cellular communication.

Transport proteins extend from one surface of the plasma membrane to the other and move ions or molecules across the plasma membrane. Transport proteins include channel proteins, carrier proteins, and ATP-powered pumps (see "Movement Through the Plasma Membrane," next section). Channel proteins form **membrane channels,** which are like small pores extending from one surface of the plasma membranes to the other (see figure 3.2).

Receptor proteins are proteins or glycoproteins in the plasma membrane with an exposed **receptor site** on the outer cell surface, which can attach to specific chemical signals. Many receptors and the chemical signals they bind are part of intercellular communication systems that coordinate cell activities. One cell can release a chemical signal that diffuses to another cell and binds to its receptor. The binding acts as a signal that triggers a response. The same chemical signal has no effect on other cells lacking the specific receptor molecule.

Enzymes are protein catalysts which increase the rate of chemical reactions on either the inner or the outer surface of the plasma membrane. For example, some enzymes on the surface of cells in the small intestine promote the breakdown of dipeptides to form two single amino acids. ▼

3. *Define intracellular, extracellular, and intercellular.*
4. *How do the hydrophilic heads and hydrophobic tails of phospholipid molecules result in a lipid bilayer?*
5. *Describe the fluid-mosaic model of the plasma membrane. What is the function of cholesterol in plasma membranes?*
6. *List five kinds of plasma membrane proteins and state their functions.*
7. *Define glycolipid, glycoprotein, cadherin, and integrin.*

Movement Through the Plasma Membrane

Plasma membranes are **selectively permeable,** allowing some substances, but not others, to pass into or out of the cells. Intracellular material has a different composition from extracellular material, and the survival of cells depends on maintaining the difference. Substances such as enzymes, glycogen, and potassium ions are found at higher concentrations intracellularly; and Na^+, Ca^{2+}, and Cl^- are found in greater concentrations extracellularly. In addition,

nutrients must enter cells continually, and waste products must exit. Cells are able to maintain proper intracellular concentrations of ions and molecules because of the permeability characteristics of plasma membranes and their ability to transport certain ions and molecules. Rupture of the membrane, alteration of its permeability characteristics, or inhibition of transport processes disrupts the normal intracellular concentration of molecules and can lead to cell death.

Ions and molecules move across plasma membranes by diffusion, osmosis, mediated transport, and vesicular transport. ▼

8. *Define selectively permeable.*
9. *List four ways that substances move across the plasma membrane.*

Diffusion

Diffusion is the tendency for ions and molecules to move from an area of higher concentration to an area of lower concentration in a solution. Lipid-soluble molecules, such as oxygen, carbon dioxide, and steroid hormones, readily diffuse through plasma membranes by dissolving in the phospholipid bilayer. Most non-lipid-soluble molecules and ions do not pass through the phospholipid bilayer. Instead, they pass through transport proteins, such as membrane channels (figure 3.3). The normal intracellular concentrations of many substances depend on diffusion and some nutrients enter and some waste products leave cells by diffusion. For example, if the extracellular concentration of oxygen is reduced, not enough oxygen diffuses into the cell, and normal cell function cannot occur. Diffusion is also an important means of movement of substances through the extracellular and intracellular fluids in the body.

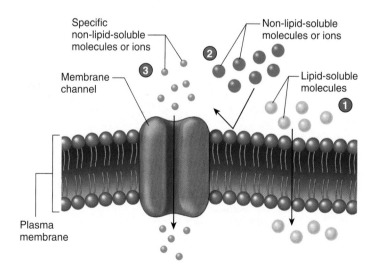

1. Lipid-soluble molecules diffuse directly through the plasma membrane.

2. Most non-lipid-soluble molecules and ions do not diffuse through the plasma membrane.

3. Some specific non-lipid-soluble molecules and ions pass through membrane channels or other transport proteins.

Process Figure 3.3   Movement Through the Plasma Membrane

A few terms need to be defined in order to better understand diffusion. A **solution** is any mixture of liquids, gases, or solids in which the substances are uniformly distributed with no clear boundary between the substances. For example, a salt solution consists of salt dissolved in water, air is a solution containing a variety of gases, and wax is a solid solution of several fatty substances. Solutions are often described in terms of one substance dissolving in another: The **solute** (sol′ūt) dissolves in the **solvent.** In a salt solution, water is the solvent and the dissolved salt is the solute. Sweat is a salt solution in which sodium chloride (NaCl) and other solutes are dissolved in water.

A concentration difference occurs when the concentration of a solute is greater at one point than at another point in a solvent. For example, when a sugar cube is placed in a beaker of still water, sugar dissolves into the water. There is a greater concentration of sugar near the cube than away from it (figure 3.4, step 1). A **concentration gradient** is the concentration difference between two points divided by the distance between the two points. Diffusion results from the constant, random motion of molecules and ions in a solution. Because the sugar molecules move randomly, like Ping-Pong balls in a lottery drawing, the chances are greater that they will move from a higher to a lower concentration than from a lower to a higher concentration. The sugar molecules are said to diffuse down, or with, their concentration gradient, from the area of higher sugar concentration to the area of lower sugar concentration (figure 3.4, step 2). At equilibrium, diffusion stops when the sugar molecules are uniformly distributed throughout the solution such that the random movement of the sugar molecules in any one direction is balanced by an equal movement in the opposite direction (figure 3.4, step 3).

The rate of diffusion is influenced by the magnitude of the concentration gradient. The greater the concentration gradient, the greater the number of solute particles moving from a higher to a lower solute concentration. The concentration gradient increases, or is said to be steeper, when the concentration difference between two points increases and/or the distance between them decreases. ▼

10. *Define diffusion, solution, solute, and solvent.*
11. *What causes diffusion and how does it stop?*
12. *What is the concentration gradient and how is it related to diffusion?*
13. *How is the rate of diffusion affected by an increased concentration gradient? How can the concentration gradient be increased?*

P R E D I C T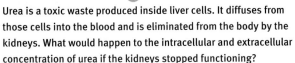

Urea is a toxic waste produced inside liver cells. It diffuses from those cells into the blood and is eliminated from the body by the kidneys. What would happen to the intracellular and extracellular concentration of urea if the kidneys stopped functioning?

Osmosis

Osmosis (os-mō′sis, a thrusting) is the diffusion of water (a solvent) across a selectively permeable membrane, such as the plasma membrane. Water can diffuse through the lipid bilayer of the plasma membrane. Rapid movement of water through the plasma membrane occurs through **water channels,** or **aquaporins,** in some cells, such as kidney cells. Osmosis is important to cells because large volume changes caused by water movement disrupt normal cell function.

A selectively permeable membrane allows water, but not all the solutes dissolved in the water, to diffuse through the membrane. Water diffuses across a selectively permeable membrane from a solution with a higher water concentration into a solution with a lower water concentration. Solution concentrations, however, are defined in terms of solute concentrations, not in terms of water concentration (see appendix C). For example, adding salt (solute) to distilled water produces a salt (NaCl) solution. The concentration of the salt solution increases as more and more salt is added to the water. Proportionately, as the concentration of salt increases, the concentration of water decreases. Therefore, water

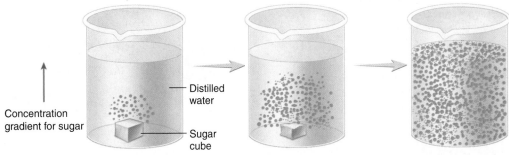

Concentration gradient for sugar

— Distilled water

— Sugar cube

1. When a sugar cube is placed into a beaker of water, there is a concentration gradient for sugar molecules from the sugar cube to the water that surrounds it. There is also a concentration gradient for water molecules from the water toward the sugar cube.

2. Sugar molecules (*green areas*) move down their concentration gradient (from an area of higher concentration toward an area of lower concentration) into the water.

3. Sugar molecules and water molecules are distributed evenly throughout the solution. Even though the sugar and water molecules continue to move randomly, an equilibrium exists, and no net movement occurs because no concentration gradient exists.

Process Figure 3.4   Diffusion

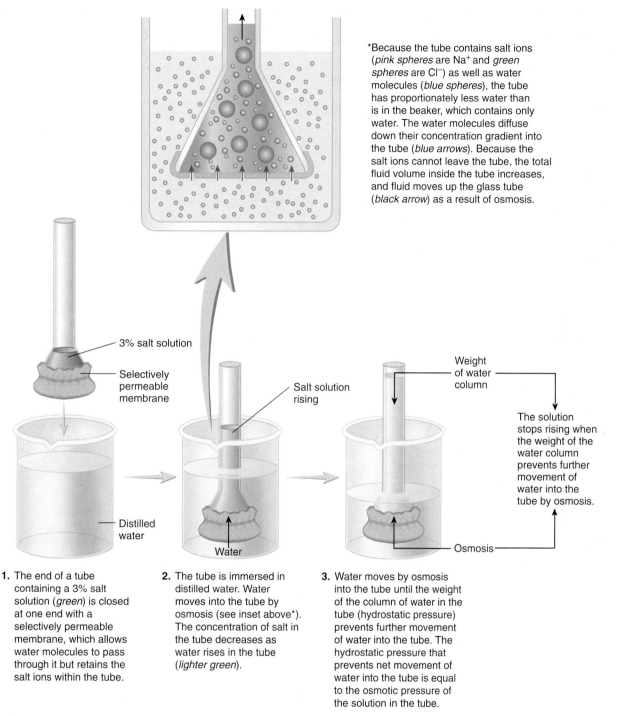

*Because the tube contains salt ions (*pink spheres* are Na+ and *green spheres* are Cl−) as well as water molecules (*blue spheres*), the tube has proportionately less water than is in the beaker, which contains only water. The water molecules diffuse down their concentration gradient into the tube (*blue arrows*). Because the salt ions cannot leave the tube, the total fluid volume inside the tube increases, and fluid moves up the glass tube (*black arrow*) as a result of osmosis.

Weight of water column

The solution stops rising when the weight of the water column prevents further movement of water into the tube by osmosis.

3% salt solution

Selectively permeable membrane

Salt solution rising

Distilled water

Water

Osmosis

1. The end of a tube containing a 3% salt solution (*green*) is closed at one end with a selectively permeable membrane, which allows water molecules to pass through it but retains the salt ions within the tube.

2. The tube is immersed in distilled water. Water moves into the tube by osmosis (see inset above*). The concentration of salt in the tube decreases as water rises in the tube (*lighter green*).

3. Water moves by osmosis into the tube until the weight of the column of water in the tube (hydrostatic pressure) prevents further movement of water into the tube. The hydrostatic pressure that prevents net movement of water into the tube is equal to the osmotic pressure of the solution in the tube.

Process Figure 3.5   Osmosis

diffuses from a less concentrated solution, which has fewer solute molecules but more water molecules, into a more concentrated solution, which has more solute molecules but fewer water molecules (figure 3.5 and table 3.2).

Osmotic pressure is the force required to prevent the movement of water by osmosis across a selectively permeable membrane. Osmotic pressure can be measured by placing a solution into a tube that is closed at one end by a selectively permeable membrane and immersing the tube in distilled water (see figure 3.5, step 1). Water molecules move by osmosis through the membrane into the tube, forcing the solution to move up the tube (see figure 3.5, step 2). As the solution rises, its weight produces **hydrostatic pressure,** which moves water out of the tube back into the distilled water surrounding the tube (see figure 3.5, step 3). Net movement of water into the tube stops when the hydrostatic pressure in the tube causes water to move out of the tube at the same rate that it diffuses into the tube by osmosis. The osmotic pressure of the solution in the tube is equal to the hydrostatic pressure that prevents net movement of water into the tube.

Table 3.2 Comparison of Membrane Transport Mechanisms

Transport Mechanism	Description	Substances Transported	Example
Diffusion	Random movement of molecules results in net movement from areas of higher to lower concentration.	Lipid-soluble molecules dissolve in the lipid bilayer and diffuse through it; ions and small molecules diffuse through membrane channels.	Oxygen, carbon dioxide, and lipids, such as steroid hormones, dissolve in the lipid bilayer.
Osmosis	Water diffuses across a selectively permeable membrane.	Water diffuses through the lipid bilayer or water channels.	Water moves from the intestines into the blood.
Facilitated diffusion	Carrier proteins combine with substances and move them across the plasma membrane; no ATP is used; substances are always moved from areas of higher to lower concentration	Some substances too large to pass through membrane channels and too polar to dissolve in the lipid bilayer are transported.	Glucose moves by facilitated diffusion into muscle cells and fat cells.
Active transport	ATP-powered pumps combine with substances and move them across the plasma membrane; ATP is required; substances can be moved from areas of lower to higher concentration.	Substances too large to pass through channels and too polar to dissolve in the lipid bilayer are transported; substances that are accumulated in concentrations higher on one side of the membrane than on the other are transported.	Ions, such as Na^+, K^+, and Ca^{2+}, are actively transported.
Secondary active transport	Ions are moved across the plasma membrane by active transport (ATP-powered pumps), which establishes an ion concentration gradient; ATP is required; ions then diffuse back down their concentration gradient assisted by carrier proteins, and another ion or molecule moves with the diffusion ion (symport) or in the opposite direction (antiport).	Some sugars, amino acids, and ions are transported.	There is a concentration gradient for Na^+ into intestinal epithelial cells. This gradient provides the energy for the symport of glucose. In many cells, H^+ are moved in the opposite direction of Na^+ (antiport).
Endocytosis	The plasma membrane forms a vesicle around the substances to be transported, and the vesicle is taken into the cell; ATP is required; in receptor-mediated endocytosis, specific substances are ingested.	Phagocytosis takes in cells and solid particles; pinocytosis takes in molecules dissolved in liquid.	Immune system cells called phagocytes ingest bacteria and cellular debris; most cells take in substances through pinocytosis.
Exocytosis	Materials manufactured by the cell are packaged in secretory vesicles that fuse with the plasma membrane and release their contents to the outside of the cell; ATP is required.	Proteins and other water-soluble molecules are transported out of cells.	Digestive enzymes, hormones, neurotransmitters, and glandular secretions are transported, and cell waste products are eliminated.

The greater the concentration of a solution, the greater is its osmotic pressure. This occurs because water moves from less concentrated solutions (less solute, more water) into more concentrated solutions (more solute, less water). The greater the concentration of a solution, the greater the tendency for water to move into the solution, and the greater the osmotic pressure must be to prevent that movement.

Three terms describe the relative osmotic concentration of solutions based on the number of solute particles, which can be ions, molecules, or a combination of ions and molecules. The number, not the type, of solute particles determines osmotic pressure. **Isosmotic** (ī′sos-mot′ik) solutions have the same concentration of solute particles and the same osmotic pressure. If one solution has a greater concentration of solute particles, and therefore a greater osmotic pressure, than another solution, the first solution is said to be **hyperosmotic** (hī′per-oz-mot′ik), compared with the more dilute solution. The more dilute solution, with the lower osmotic concentration and pressure, is

hyposmotic (hī-pos-mot′ik), compared with the more concentrated solution.

P R E D I C T
Solution A is hyperosmotic to solution B and, therefore, solution B is hyposmotic to solution A. If solutions A and B are separated by a selectively permeable membrane, will water move from the hyperosmotic solution into the hyposmotic solution, or vice versa? Explain.

Cells will swell, remain unchanged, or shrink when placed into a solution. When a cell is placed into a **hypotonic** (hī′pō-ton′ik, *hypo,* under + *tonos,* tone) solution, the solution usually has a lower concentration of solutes and a higher concentration of water than the cytoplasm of the cell. Water moves by osmosis into the cell, causing it to swell. If the cell swells enough, it can rupture, a process called **lysis** (lī′sis, loosening) (figure 3.6*a*). When a cell is immersed in an **isotonic** (ī′sō-ton′ik, *iso,* equal) solution, the concentrations of various solutes and water are the same on both sides of the plasma membrane. The cell therefore neither shrinks nor swells (figure 3.6*b*). In general, solutions injected into the circulatory system or into tissues must be isotonic because swelling or shrinking disrupts normal cell function and can lead to cell death. When a cell is immersed in a **hypertonic** (hī′per-ton′ik, *hyper,* above) solution, the solution usually has a higher concentration of solutes and a lower concentration of water than the cytoplasm of the cell. Water moves by osmosis from the cell into the hypertonic solution, resulting in cell shrinkage, or **crenation** (krē-nā′shŭn, *crena,* a notch) (figure 3.6*c*). ▼

14. *Define osmosis and osmotic pressure. As the concentration of a solution increases, what happens to its osmotic pressure and to the tendency for water to move into it?*

15. *Compare the osmotic pressure of isosmotic, hyperosmotic, and hyposmotic solutions.*

16. *Define isotonic, hypertonic, and hypotonic solutions. Which type of solution causes crenation of a cell and which type causes lysis?*

Mediated Transport

Most non-lipid-soluble molecules and ions do not readily pass through the phospholipid bilayer (see figure 3.3). Transport proteins move these substances across the plasma membrane. **Mediated transport** is the process by which transport proteins mediate, or assist in, the movement of ions and molecules across the plasma membrane. Mediated transport has three characteristics: specificity, competition, and saturation. **Specificity** means that each transport protein moves particular molecules or ions, but not others. For example, the transport protein that moves glucose does not move amino acids or ions. **Competition** occurs when similar molecules or ions can be moved by the transport protein. Although transport proteins exhibit specificity, a transport protein may transport very similar substances. The substance in the greater concentration or the substance for which the transport protein is the most specific is moved across the plasma membrane at the greater rate. **Saturation** means that the rate of movement of molecules or ions across the membrane is limited by the number of available transport proteins. As the concentration of a transported substance increases, more transport proteins become involved with transporting the substance and the rate at which the substance is moved across the plasma membrane increases. Once the concentration of the substance is increased so that all the transport proteins are in use, the rate of movement remains constant, even though the concentration of the substance increases further.

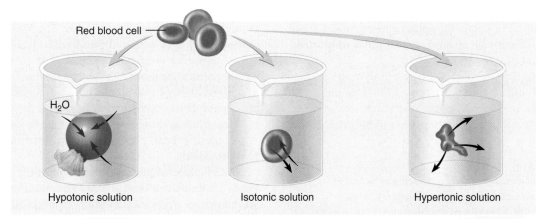

(a) A hypotonic solution with a low solute concentration results in swelling of the red blood cell placed into the solution. Water enters the cell by osmosis (*black arrows*), and the red blood cell lyses (bursts; *puff of red in the lower part of the cell*).

(b) An isotonic solution with a concentration of solutes equal to that inside the cell results in a normally shaped red blood cell. Water moves into and out of the cell at the same rate (*black arrows*), but there is no net water movement.

(c) A hypertonic solution, with a high solute concentration, causes shrinkage (crenation) of the red blood cell as water moves by osmosis out of the cell and into the hypertonic solution (*black arrows*).

Figure 3.6   Effects of Hypotonic, Isotonic, and Hypertonic Solutions on Red Blood Cells

Three types of transport proteins—channel proteins, carrier proteins (transporters), and ATP-powered pumps—are involved in mediated transport.

Channel Proteins

Channel proteins form membrane channels (see figure 3.2). **Ion channels** are membrane channels that transport ions. Ion channels are often thought of as simple tubes through which ions pass. Many ion channels, however, are more complex than once thought. It now appears that ions briefly bind to specific sites inside channels and that there is a change in the shape of those channels as ions are transported through them. The size and charge within a channel determine the channel's specificity. For example, Na^+ channels do not transport K^+ and vice versa. In addition, similar ions moving into and binding within an ion channel are in competition with each other. Furthermore, the number of ions moving into an ion channel can exceed the capacity of the channel, thus saturating the channel. Therefore, ion channels exhibit specificity, competition, and saturation.

Carrier Proteins

Carrier proteins, or **transporters,** are membrane proteins that move ions or molecules from one side of the plasma membrane to the other. The carrier proteins have specific binding sites to which ions or molecules attach on one side of the plasma membrane. The carrier proteins change shape to move the bound ions or molecules to the other side of the plasma membrane, where they are released (figure 3.7). The carrier protein then resumes its original shape and is available to transport more molecules.

Movement of ions or molecules *by carrier proteins* can be classified as uniport, symport, or antiport. **Uniport** is the movement of one specific ion or molecule across the membrane. **Symport** is the movement of two or more different ions or molecules in the same direction across the plasma membrane, whereas **antiport** is the movement of two or more different ions or molecules in opposite directions across the plasma membrane. Carrier proteins involved in these types of movement are called **uniporters, symporters,** and **antiporters,** respectively.

Movement of ions or molecules by a uniporter is often called **facilitated diffusion** (see figure 3.7). The uniporter facilitates, or helps in, the movement of the ion or molecule, and, as in diffusion, movement is from areas of higher to lower concentration.

P R E D I C T ③

The transport of glucose into most cells occurs by facilitated diffusion. Because diffusion occurs from a higher to a lower concentration, glucose cannot accumulate within these cells at a higher concentration than is found outside the cell. Once glucose enters cells, it is rapidly converted to other molecules, such as glucose phosphate or glycogen. What effect does this conversion have on the ability of cells to transport glucose?

ATP-Powered Pumps

ATP-powered pumps are transport proteins that use energy derived from the breakdown of adenosine triphosphate (ATP)

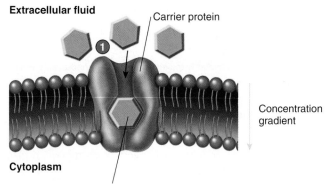

Extracellular fluid

Carrier protein

Concentration gradient

Cytoplasm

Transported molecule (glucose)

1. The carrier protein (uniporter) binds with a molecule, such as glucose, on the outside of the plasma membrane.

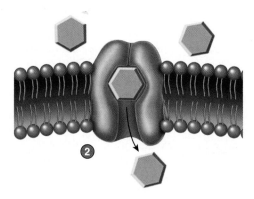

2. The carrier protein changes shape and releases the molecule on the inside of the plasma membrane.

Process Figure 3.7 Facilitated Diffusion

(see chapter 2) to move specific ions and molecules from one side of the plasma membrane to the other. The movement of ions and molecules by ATP-powered pumps is called **active transport.** Active transport is important because it can move substances against their concentration gradients—that is, from lower concentrations to higher concentrations. Consequently, active transport can accumulate substances on one side of the plasma membrane at concentrations many times greater than those on the other side. Active transport can also move substances from higher to lower concentrations.

ATP-powered pumps have binding sites, to which a specific ion or molecule can bind, as well as a binding site for ATP. The breakdown of ATP to adenosine diphosphate (ADP) provides energy that changes the shape of the protein, which moves the ion or molecule across the membrane. In some cases, the active transport mechanism can exchange one substance for another. For example, the **sodium–potassium (Na^+–K^+) pump** moves Na^+ out of cells and K^+ into cells (figure 3.8). Active transport requires energy in the form of ATP and, if ATP is not available, active transport stops.

In **secondary active transport,** the concentration gradient established by the active transport of one substance provides

1. Three sodium ions (Na⁺) and adenosine triphosphate (ATP) bind to the Na⁺–K⁺ pump, which is an ATP-powered pump.

2. The ATP breaks down to adenosine diphosphate (ADP) and a phosphate (P) and releases energy. That energy is used to power a shape change in the Na⁺–K⁺ pump. Phosphate remains bound to the Na⁺–K⁺–ATP binding site.

3. The Na⁺–K⁺ pump changes shape, and the Na⁺ are transported across the membrane.

4. The Na⁺ diffuse away from the Na⁺–K⁺ pump.

5. Two potassium ions (K⁺) bind to the Na⁺–K⁺ pump.

6. The phosphate is released from the Na⁺–K⁺ pump binding site.

7. The Na⁺–K⁺ pump resumes its original shape, transporting K⁺ across the membrane, and the K⁺ diffuse away from the pump. The Na⁺–K⁺ pump can again bind to Na⁺ and ATP.

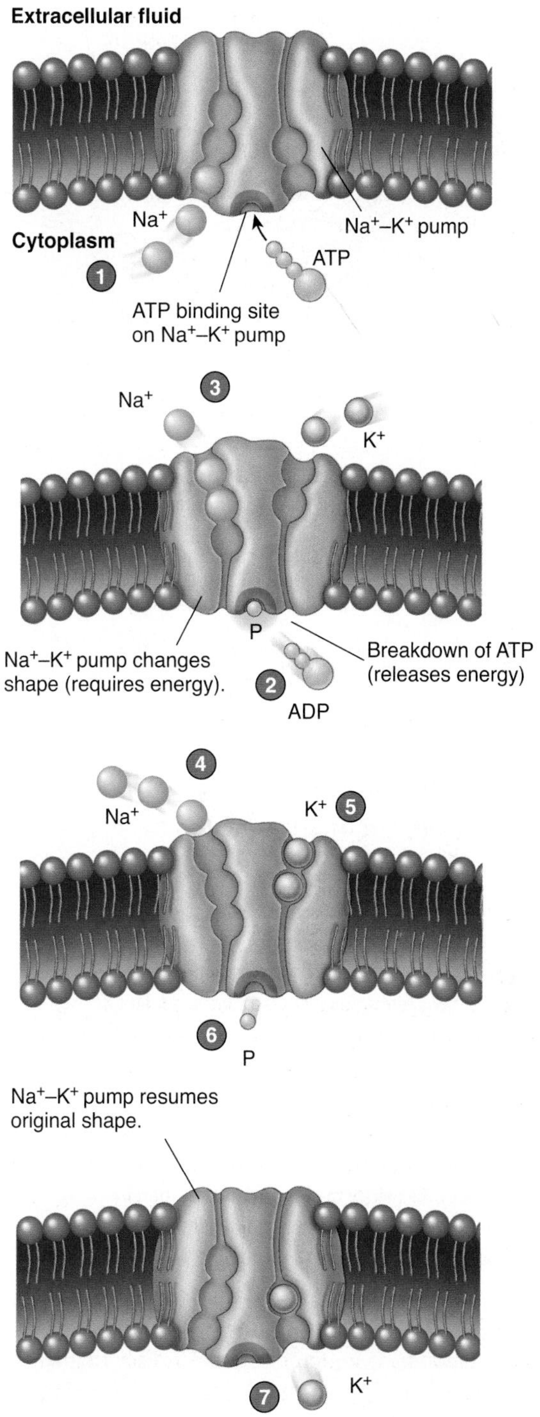

Extracellular fluid

Na⁺

Cytoplasm

ATP binding site on Na⁺–K⁺ pump

Na⁺–K⁺ pump

ATP

Na⁺

K⁺

Na⁺–K⁺ pump changes shape (requires energy).

P

Breakdown of ATP (releases energy)

ADP

Na⁺

K⁺

P

Na⁺–K⁺ pump resumes original shape.

K⁺

Process Figure 3.8   Sodium–Potassium Pump

the energy to move a second substance (figure 3.9). For example, Na⁺ are actively pumped out of a cell, establishing a higher concentration of Na⁺ outside the cell than inside. The movement of Na⁺ down its concentration gradient back into the cell provides the energy necessary to move glucose into the cell against its concentration gradient. In this example, a symporter moves Na⁺ and glucose into the cell together. Secondary active transport can also involve antiporters—for example, the movement

of Na⁺ into a cell coupled with the movement of H⁺ out of the cell. ▼

17. *What is mediated transport? Describe specificity, competition, and saturation as characteristics of mediated transport.*

18. *Name the types of transport proteins involved in mediated transport.*

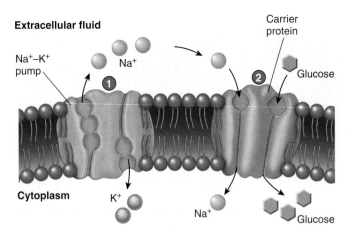

Extracellular fluid

Na⁺–K⁺ pump

Na⁺

Carrier protein

Glucose

Cytoplasm

K⁺

Na⁺

Glucose

1. A Na⁺–K⁺ pump (ATP-powered pump) maintains a concentration of Na⁺ that is higher outside the cell than inside.

2. Sodium ions move back into the cell through a carrier protein (symporter) that also moves glucose. The concentration gradient for Na⁺ provides energy required to move glucose against its concentration gradient.

Process Figure 3.9   Secondary Active Transport (Symport) of Na⁺ and Glucose

19. *Define uniport (facilitated diffusion), symport, and antiport.*
20. *What is active transport? Describe the operation of the (Na⁺–K⁺) pump.*
21. *What is secondary active transport?*

P R E D I C T   **4**

In cardiac (heart) muscle cells, the force of contraction increases as the intracellular Ca²⁺ concentration increases. Intracellular Ca²⁺ concentration is regulated in part by secondary active transport involving a Na⁺–Ca²⁺ antiporter. The movement of Na⁺ down their concentration gradient into the cell provides the energy to transport Ca²⁺ out of the cell against their concentration gradient. Digitalis, a drug often used to treat congestive heart failure, slows the active transport of Na⁺ out of the cell by the Na⁺–K⁺ pump, thereby increasing intracellular Na⁺ concentration. Should the heart beat more or less forcefully when exposed to this drug? Explain.

Vesicular Transport

A **vesicle** (ves′i-kl, a bladder) is a membrane-bound sac that surrounds substances within the cytoplasm of cells. **Vesicular transport** is the movement of materials by vesicles into, out of, or within cells. Vesicular transport into cells is called **endocytosis** (en′dō-sī-tō′sis, *endon*, within + *kytos*, cell + *osis*, condition) and includes phagocytosis and pinocytosis. Both phagocytosis and pinocytosis require energy in the form of ATP and, therefore, are active processes.

Phagocytosis (făg-ō-sī-tō′sis) literally means cell-eating and applies to endocytosis when solid particles, such as bacteria, cell debris, or foreign substances, are ingested. Phagocytosis is important in the elimination of harmful substances from the body by specialized cells called macrophages. In phagocytosis, a part of the plasma membrane extends around a particle and fuses so that the particle is surrounded by the membrane. That part of the

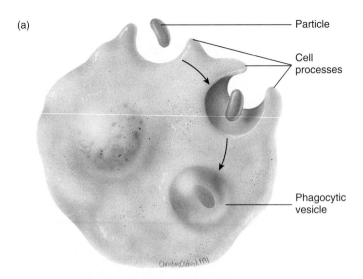

(a)

Particle

Cell processes

Phagocytic vesicle

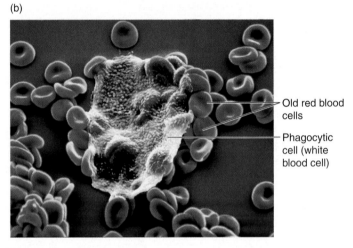

(b)

Old red blood cells

Phagocytic cell (white blood cell)

Figure 3.10   Phagocytosis

(*a*) Phagocytosis. (*b*) Scanning electron micrograph of phagocytosis of red blood cells.

membrane then pinches off to form a phagocytic vesicle containing the particle, which is inside the cell (figure 3.10).

Pinocytosis (pin′ō-sī-tō′sis) means cell-drinking and is the uptake of small droplets of extracellular fluid, and the materials contained in the fluid, by the formation of small endocytic vesicles. Pinocytosis often forms vesicles near the tips of deep invaginations of the plasma membrane. It is a common transport phenomenon in a variety of cell types and occurs in certain cells of the kidneys, epithelial cells of the intestines, cells of the liver, and cells that line capillaries.

Phagocytosis and pinocytosis can exhibit specificity. Cells that phagocytize bacteria do not phagocytize healthy cells and specific molecules are taken in by pinocytosis. In **receptor-mediated endocytosis,** the plasma membrane contains receptors (receptor proteins) that bind to specific molecules. When a specific molecule binds to its receptor, endocytosis is triggered, and the molecule and receptor are transported into the cell (figure 3.11). This mechanism increases the rate at which specific substances, such as cholesterol and insulin, are taken up by the cells.

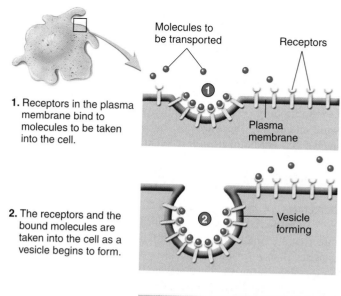

Molecules to be transported

Receptors

1. Receptors in the plasma membrane bind to molecules to be taken into the cell.

Plasma membrane

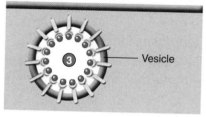

2. The receptors and the bound molecules are taken into the cell as a vesicle begins to form.

Vesicle forming

3. The vesicle fuses and separates from the plasma membrane.

Vesicle

Process Figure 3.11 **Receptor-Mediated Endocytosis**

Hypercholesterolemia

Hypercholesterolemia is a common genetic disorder affecting 1 in every 500 adults in the United States. It consists of a reduction in or absence of low-density lipoprotein (LDL) receptors on cell surfaces. This interferes with receptor-mediated endocytosis of LDL cholesterol. As a result of inadequate cholesterol uptake, cholesterol synthesis within these cells is not regulated, and too much cholesterol is produced. The excess cholesterol accumulates in blood vessels, resulting in atherosclerosis. Atherosclerosis can result in heart attacks or strokes.

Exocytosis (ek′sō-sī-tō′sis, *exo,* outside) is the movement of materials out of cells by vesicles (figure 3.12 and see table 3.2). **Secretory vesicles** accumulate materials for release from cells. The secretory vesicles move to the plasma membrane, where the vesicle membrane fuses with the plasma membrane, and the material in the vesicle is eliminated from the cell. Examples of exocytosis are the secretion of digestive enzymes by the pancreas, of mucus by the salivary glands, and of milk from the mammary glands. ▼

22. *Define vesicle, vesicular transport, and endocytosis. How do phagocytosis and pinocytosis differ from each other?*
23. *What is receptor-mediated endocytosis?*
24. *Describe and give examples of exocytosis.*

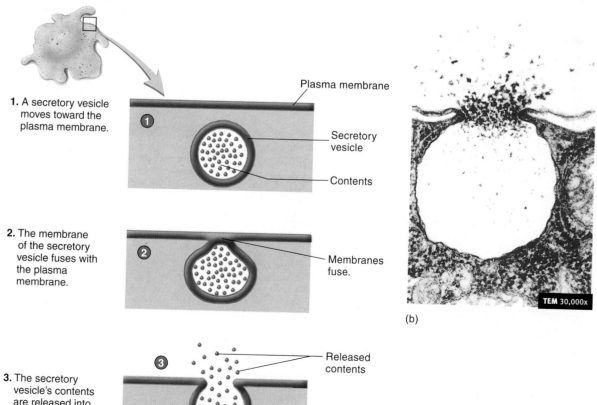

Plasma membrane

1. A secretory vesicle moves toward the plasma membrane.

Secretory vesicle

Contents

2. The membrane of the secretory vesicle fuses with the plasma membrane.

Membranes fuse.

3. The secretory vesicle's contents are released into the extracellular fluid.

Released contents

(a)

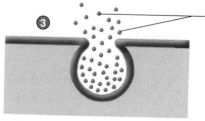

TEM 30,000x

(b)

Process Figure 3.12 **Exocytosis**

(*a*) Diagram of exocytosis. (*b*) Transmission electron micrograph of exocytosis.

Cytoplasm

Cytoplasm is the cellular material outside the nucleus but inside the plasma membrane. It is about half cytosol and half organelles.

Cytosol

Cytosol (sī′tō-sol) consists of a fluid portion, a cytoskeleton, and cytoplasmic inclusions. The fluid portion of cytosol is a solution and a colloid. The solution part contains dissolved ions and molecules. A **colloid** (kol′oyd) is a mixture in which large molecules or aggregates of atoms, ions, or molecules do not dissolve in a liquid but remain suspended in the liquid. Proteins, which are large molecules, and water form colloids. Many of the proteins in cytosol are enzymes that catalyze the breakdown of molecules for energy or the synthesis of sugars, fatty acids, nucleotides, amino acids, and other molecules.

Cytoskeleton

The **cytoskeleton** (sī-tō-skel′ĕ-ton) consists of proteins that support the cell, hold organelles in place, and enable the cell to change shape. The cytoskeleton consists of microtubules, actin filaments, and intermediate filaments (figure 3.13). **Microtubules** are hollow structures formed from protein subunits. They perform a variety of roles, such as helping provide support to the cytoplasm of cells, assisting in the process of cell division, and forming essential components of certain organelles, such as centrioles, spindle fibers, cilia, and flagella.

Actin filaments, or **microfilaments,** are small fibrils formed from protein subunits that form bundles, sheets, or networks in the cytoplasm of cells. Actin filaments support the plasma membrane and define the shape of the cell. Changes in cell shape involve the breakdown and reconstruction of actin filaments. These changes in shape allow some cells to move about. Muscle cells contain a large number of highly organized actin filaments, which are responsible for the muscle's ability to contract (see chapter 8).

Intermediate filaments are fibrils formed from protein subunits that are smaller in diameter than microtubules but larger in diameter than microfilaments. They provide mechanical support to the cell. For example, intermediate filaments support the extensions of nerve cells, which have a very small diameter but can be a meter in length.

Cytoplasmic Inclusions

The cytosol also contains **cytoplasmic inclusions,** which are aggregates of chemicals either produced by the cell or taken in by

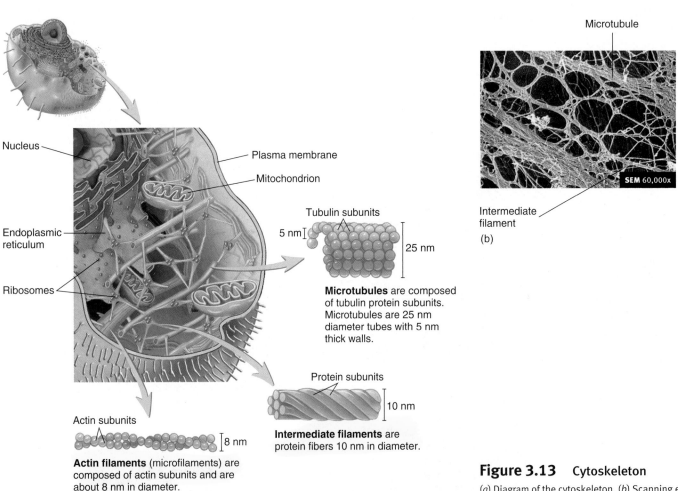

Nucleus

Endoplasmic reticulum

Ribosomes

Plasma membrane

Mitochondrion

Tubulin subunits

5 nm

25 nm

Microtubules are composed of tubulin protein subunits. Microtubules are 25 nm diameter tubes with 5 nm thick walls.

Protein subunits

10 nm

Intermediate filaments are protein fibers 10 nm in diameter.

Actin subunits

8 nm

Actin filaments (microfilaments) are composed of actin subunits and are about 8 nm in diameter.

(a)

Microtubule

SEM 60,000x

Intermediate filament

(b)

Figure 3.13   Cytoskeleton

(*a*) Diagram of the cytoskeleton. (*b*) Scanning electron micrograph of the cytoskeleton.

the cell. For example, lipid droplets or glycogen granules store energy-rich molecules and melanin, and carotene are pigments that color the skin. ▼

25. *Define cytoplasm and cytosol.*
26. *What are the functions of the cytoskeleton?*
27. *Describe and list the functions of microtubules, actin filaments, and intermediate filaments.*
28. *Define and give examples of cytoplasmic inclusions.*

The Nucleus and Cytoplasmic Organelles

Organelles are structures within cells that are specialized for particular functions, such as manufacturing proteins or producing ATP. Organelles can be thought of as individual workstations within the cell, each responsible for performing specific tasks. The nucleus is the largest organelle of the cell. The remaining organelles are referred to as cytoplasmic organelles (see table 3.1).

Nucleus

The **nucleus** (noo′klē-ŭs, a little nut, stone of fruit) is a large organelle, usually located near the center of the cell (see figure 3.1). All cells of the body have a nucleus at some point in their life cycle,

although some cells, such as red blood cells, lose their nuclei as they mature. Other cells, such as osteoclasts (a type of bone cell) and skeletal muscle cells, contain more than one nucleus. ▼

The nucleus consists of **nucleoplasm** surrounded by a **nuclear envelope** composed of two membranes separated by a space (figure 3.14). At many points on the surface of the nucleus, the inner and outer membranes come together to form **nuclear pores,** through which materials can pass into or out of the nucleus.

Deoxyribonucleic acid (DNA) is mostly found within the nucleus, although small amounts of DNA are also found within mitochondria. The genes that influence the structural and functional features of every individual are portions of DNA molecules. These sections of DNA molecules determine the structure of proteins. By determining the structure of proteins, genes direct cell structure and function.

Nuclear DNA and associated proteins are organized into discrete structures called **chromosomes** (krō′mō-sōmz, colored bodies). The proteins include **histones** (his′tōnz), which are important for the structural organization of DNA, and other proteins that regulate DNA function. Except when cells are dividing, the chromosomes are dispersed throughout the nucleus as delicate filaments referred to as **chromatin** (krō′ma-tin, colored material) (see figures 3.14*b* and 3.15). During cell division, the dispersed chromatin becomes densely coiled, forming compact chromosomes. Each chromosome consists of two **chromatids**

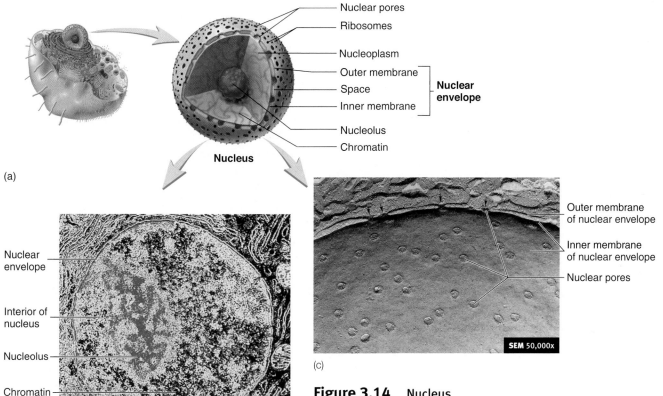

Figure 3.14   Nucleus

(*a*) The nuclear envelope consists of inner and outer membranes that become fused at the nuclear pores. The nucleolus is a condensed region of the nucleus not bound by a membrane and consisting mostly of RNA and protein. (*b*) Transmission electron micrograph of the nucleus. (*c*) Scanning electron micrograph showing the inner surface of the nuclear envelope and the nuclear pores.

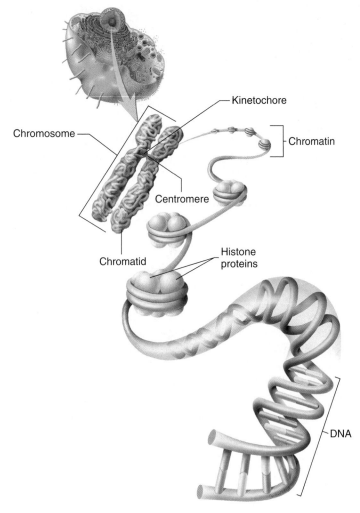

Figure 3.15   **Chromosome Structure**

DNA is associated with globular histone proteins and other DNA-binding proteins. DNA molecules and bound proteins are called chromatin. During cell division, the chromatin condenses so that individual structures, called chromosomes, become visible.

(krō′ma-tids), which are attached at the **centromere** (sen′trō-mĕr). The **kinetochore** (ki-nē′tō-kōr, ki-net′ō-kōr), a protein structure within the centromere, provides a point of attachment for microtubules during cell division (see "Centrioles and Spindle Fibers," p. 64). ▼

29. *Define organelles.*
30. *Describe the structure of the nucleus and nuclear envelope. What is the function of the nuclear pores?*
31. *What molecules are found in chromatin? When does chromatin become a chromosome?*

Nucleoli and Ribosomes

Nucleoli (noo-klē′ō-lī, sing. nucleolus, little nucleus) number from one to four per nucleus. They are rounded, dense, well-defined nuclear bodies with no surrounding membrane (see figure 3.14). The nucleolus produces **ribosomal ribonucleic** (rī′bō-noo-klē′ik) **acid (rRNA),** which combines with proteins produced in the cytoplasm to form large and small ribosomal subunits (figure 3.16). The proteins from the cytoplasm enter the nucleus through nuclear pores, and the ribosomal subunits move from the nucleus through the nuclear pores into the cytoplasm, where one large and one small subunit join to form a ribosome.

Ribosomes (rī′bō-sōmz) are the organelles where proteins are produced (see "Protein Synthesis," p. 65). **Free ribosomes** are not attached to any other organelles in the cytoplasm, whereas other ribosomes are attached to a network of membranes called the endoplasmic reticulum. Free ribosomes primarily synthesize proteins used inside the cell, whereas ribosomes attached to the endoplasmic reticulum usually produce proteins that are secreted from the cell. ▼

32. *What is a nucleolus?*
33. *What kinds of molecules combine to form ribosomes? Where are ribosomal subunits formed and assembled?*
34. *Compare the functions of free ribosomes and ribosomes attached to the endoplasmic reticulum.*

Rough and Smooth Endoplasmic Reticulum

The **endoplasmic reticulum** (en′dō-plas′mik re-tik′ū-lŭm, *rete,* a network) **(ER)** is a series of membranes forming sacs and tubules that extends from the outer nuclear membrane into the cytoplasm (figure 3.17). The interior spaces of those sacs and tubules are called **cisternae** (sis-ter′nē) and are isolated from the rest of the cytoplasm.

Rough ER is ER with ribosomes attached to it. A large amount of rough ER in a cell indicates that it is synthesizing large amounts of protein for export from the cell. On the other hand, ER without ribosomes is called **smooth ER.** Smooth ER is a site for lipid and carbohydrate synthesis, and it participates in the detoxification of chemicals within cells. In skeletal muscle cells, the smooth ER stores Ca^{2+}. ▼

35. *How is the endoplasmic reticulum related to the nuclear envelope? How are the cisternae of the endoplasmic reticulum related to the rest of the cytoplasm?*
36. *What are the functions of rough and smooth endoplasmic reticulum?*

Golgi Apparatus

The **Golgi** (gol′jē) **apparatus** (named for Camillo Golgi [1843–1926], an Italian histologist) consists of closely packed stacks of curved, membrane-bound sacs (figure 3.18). It collects, modifies, packages, and distributes proteins and lipids. Proteins produced at the ribosomes of the rough endoplasmic reticulum enter a cisterna of the endoplasmic reticulum. The ER then pinches off to form a small sac called a **transport vesicle.** The transport vesicle moves to the Golgi apparatus, fuses with its membrane, and releases the protein into its cisterna. The Golgi apparatus concentrates and, in some cases, chemically modifies the proteins by synthesizing and attaching carbohydrate molecules to the proteins to form glycoproteins or by attaching lipids to proteins to form lipoproteins. The proteins are then packaged into vesicles that pinch off from the margins of the Golgi apparatus and are distributed to various locations. Some vesicles contain enzymes that are used within the cell; some vesicles

1. Ribosomal proteins, produced in the cytoplasm, are transported through nuclear pores into the nucleolus.

2. rRNA, most of which is produced in the nucleolus, is assembled with ribosomal proteins to form small and large ribosomal subunits.

3. The small and large ribosomal subunits leave the nucleolus and the nucleus through nuclear pores.

4. The small and large subunits, now in the cytoplasm, combine with each other and with mRNA during protein synthesis.

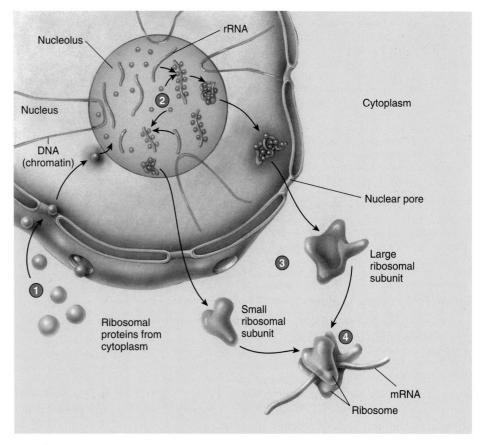

Process Figure 3.16 **Production of Ribosomes**

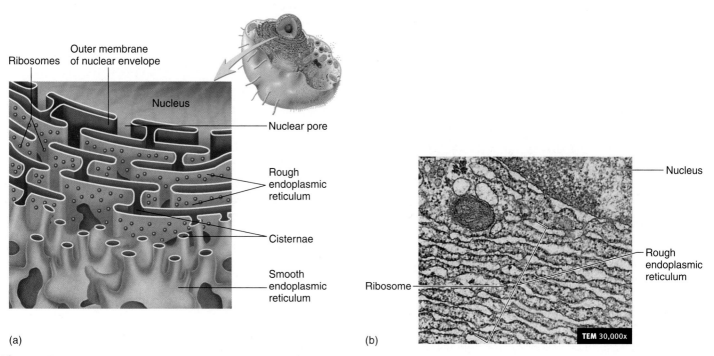

(a)

(b)

Figure 3.17 **Endoplasmic Reticulum**

(a) The endoplasmic reticulum is continuous with the nuclear envelope and occurs either as rough endoplasmic reticulum (with ribosomes) or as smooth endoplasmic reticulum (without ribosomes). (b) Transmission electron micrograph of the rough endoplasmic reticulum.

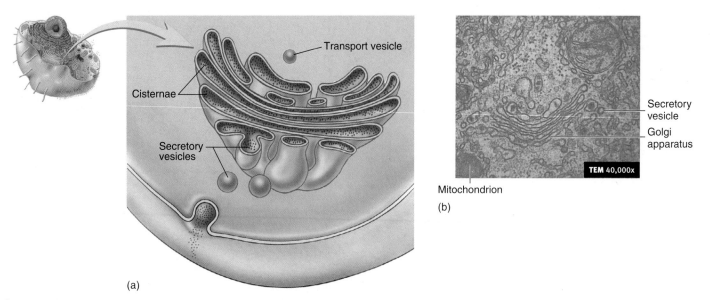

Figure 3.18   Golgi Apparatus

(*a*) The Golgi apparatus is composed of flattened, membranous sacs, containing cisternae, and resembles a stack of dinner plates or pancakes. (*b*) Transmission electron micrograph of the Golgi apparatus.

carry proteins to the plasma membrane, where the proteins are secreted from the cell by exocytosis; other vesicles contain proteins and lipids that become part of the plasma membrane.

The Golgi apparatus is present in larger numbers and is most highly developed in cells that secrete protein, such as the cells of the salivary glands or the pancreas. ▼

> 37. Describe the structure and function of the Golgi apparatus.
> 38. Describe the production of a protein at the endoplasmic reticulum and its distribution to the Golgi apparatus. Name three ways in which proteins are distributed from the Golgi apparatus.

Secretory Vesicles

Secretory vesicles are small, membrane-bound sacs that transport material produced in cells to the exterior of cells. Secretory vesicles pinch off from the Golgi apparatus and move to the surface of the cell (figure 3.19). Their membranes then fuse with the plasma membrane, and the contents of the vesicles are released to the exterior of the cell. In many cells, secretory vesicles accumulate in the cytoplasm and are released to the exterior when the cell receives a signal. For example, nerve cells release substances called neurotransmitters from secretory vesicles to communicate with other cells. ▼

> 39. Describe the formation of secretory vesicles and how their contents are released from cells.

Lysosomes and Peroxisomes

Lysosomes (lī′sō-sōmz, *lysis,* a loosening + *soma,* body) (figure 3.20) are membrane-bound vesicles formed from the Golgi apparatus. They contain a variety of enzymes that function as intracellular digestive systems. Vesicles formed by endocytosis may fuse with lysosomes. The enzymes within the lysosomes break down

the materials in the endocytic vesicle. For example, white blood cells phagocytize bacteria. Enzymes within lysosomes destroy the bacteria. Also, when tissues are damaged, ruptured lysosomes within the damaged cells release their enzymes and digest both healthy and damaged cells. The released enzymes are responsible for part of the resulting inflammation (see chapter 4).

Peroxisomes (per-ok′si-sōmz) are small, membrane-bound vesicles containing enzymes that break down fatty acids, amino acids, and hydrogen peroxide (H_2O_2). Hydrogen peroxide is a by-product of fatty acid and amino acid breakdown and can be toxic to the cell. The enzymes in peroxisomes break down hydrogen peroxide to water and oxygen. Cells active in detoxification, such as liver and kidney cells, have many peroxisomes. ▼

> 40. Describe the process by which lysosomal enzymes digest phagocytized materials.
> 41. What is the function of peroxisomes?

Proteasomes

Proteasomes (prō′tē-ă-sōmz) are tunnel-like structures, similar to channel protein, and are not bounded by membranes. They consist of large protein complexes that break down and recycle proteins within the cell. The inner surfaces of the tunnel have enzymatic regions that break down proteins. Smaller protein subunits close the ends of the tunnel and regulate which proteins are taken into it for digestion. ▼

> 42. What is the function of proteasomes?

Mitochondria

Mitochondria (mī′tō-kon′drē-ă, sing. mitochondrion, *mitos,* thread + *chondros,* granule) are bean-shaped, rod-shaped, or threadlike organelles with inner and outer membranes separated

1. Some proteins are produced at ribosomes on the surface of the rough endoplasmic reticulum and are transferred into the cisterna as they are produced.

2. The proteins are surrounded by a vesicle that forms from the membrane of the endoplasmic reticulum.

3. This transport vesicle moves from the endoplasmic reticulum to the Golgi apparatus, fuses with its membrane, and releases the proteins into its cisterna.

4. The Golgi apparatus concentrates and, in some cases, modifies the proteins into glycoproteins or lipoproteins.

5. The proteins are packaged into vesicles that form from the membrane of the Golgi apparatus.

6. Some vesicles, such as lysosomes, contain enzymes that are used within the cell.

7. Secretory vesicles carry proteins to the plasma membrane, where the proteins are secreted from the cell by exocytosis.

8. Some vesicles contain proteins that become part of the plasma membrane.

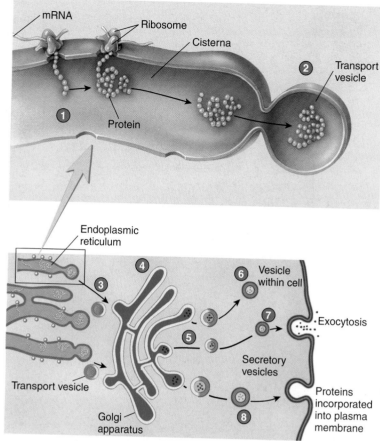

Process Figure 3.19 Function of the Golgi Apparatus

1. A vesicle forms around material outside the cell.

2. The vesicle is pinched off from the plasma membrane and becomes a separate vesicle inside the cell.

3. A lysosome is pinched off the Golgi apparatus.

4. The lysosome fuses with the vesicle.

5. The enzymes from the lysosome mix with the material in the vesicle, and the enzymes digest the material.

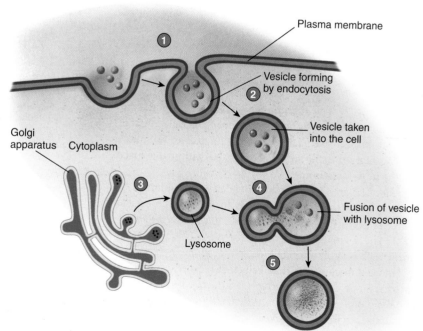

Process Figure 3.20 Action of Lysosomes

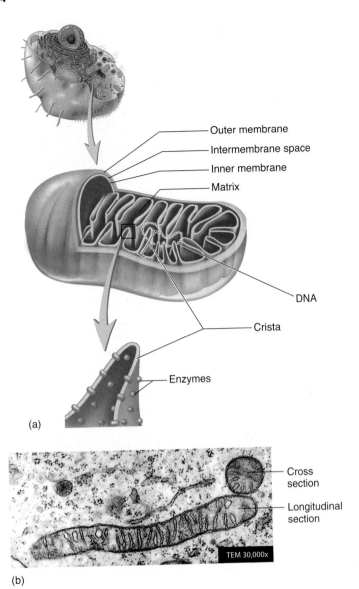

(a)

Outer membrane
Intermembrane space
Inner membrane
Matrix
DNA
Crista
Enzymes

Cross section
Longitudinal section

TEM 30,000x

(b)

Figure 3.21   Mitochondrion

(*a*) Typical mitochondrion structure. (*b*) Transmission electron micrograph of mitochondria in longitudinal and cross section.

by a space (figure 3.21). The outer membranes have a smooth contour, but the inner membranes have numerous folds called **cristae** (kris′tē), which project like shelves into the interior of the mitochondria. The **matrix** is the substance located within the space formed by the inner membrane.

Mitochondria are the major sites of adenosine triphosphate (ATP) production within cells. Enzymes located in the cristae and the matrix are responsible for the production of ATP (see chapter 22). The major energy source for most chemical reactions within cells is ATP, and cells with a large energy requirement have more mitochondria than cells that require less energy. Cells that carry out extensive active transport, which requires ATP, contain many mitochondria. When muscles enlarge as a result of exercise, the mitochondria increase in number within the muscle cells and provide the additional ATP required for muscle contraction. Increases in the number of mitochondria

result from the division of preexisting mitochondria. The information for making some mitochondrial proteins and for mitochondrial division is contained in a unique type of DNA within the mitochondria. This DNA is more like bacterial DNA than that of the cell's nucleus. ▼

43. Describe the structure of mitochondria.
44. What is the function of mitochondria? What enzymes are found on the cristae and in the matrix?

Centrioles and Spindle Fibers

The **centrosome** (sen′trō-sōm) is a specialized zone of cytoplasm close to the nucleus that is the center of microtubule formation. It contains two **centrioles** (sen′trē-ōlz), normally oriented perpendicular to each other (see figure 3.1). Each centriole is a small, cylindrical organelle composed of nine triplets, each consisting of three parallel microtubules joined together (figure 3.22).

Before cell division, the two centrioles double in number, the centrosome divides into two, and one centrosome, containing two centrioles, moves to each end of the cell. Microtubules called **spindle fibers** extend out in all directions from the centrosome. Eventually, spindle fibers from each centrosome bind to the kinetochore of all the chromosomes. During cell division, the spindle

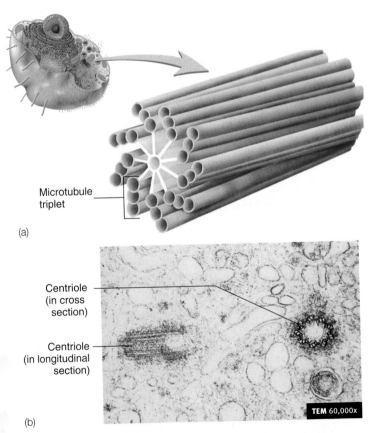

Microtubule triplet

(a)

Centriole (in cross section)

Centriole (in longitudinal section)

TEM 60,000x

(b)

Figure 3.22   Centriole

(*a*) Structure of a centriole, which comprises nine triplets of microtubules. Each triplet contains one complete microtubule fused to two incomplete microtubules. (*b*) Transmission electron micrograph of a pair of centrioles, which are normally located together near the nucleus. One is shown in cross section and one in longitudinal section.

Each cell is well adapted for the functions it performs, and the abundance of organelles in each cell reflects the function of the cell. For example, epithelial cells lining the larger-diameter respiratory passages secrete mucus and transport it toward the throat, where it is either swallowed or expelled from the body by coughing. Particles of dust and other debris suspended in the air become trapped in the mucus. The production and transport of mucus from the respiratory passages keep these passages clean. Cells of the respiratory system have abundant rough ER, Golgi apparatuses, secretory vesicles, and cilia. The ribosomes on the rough ER are the sites where proteins, a major component of mucus, are produced. The Golgi apparatuses package the proteins and other components of mucus into secretory vesicles, which move to the surface of the epithelial cells. The contents of the secretory vesicles are released onto the surface of the epithelial cells. Cilia on the cell surface then propel the mucus toward the throat.

In people who smoke, prolonged exposure of the respiratory epithelium to the irritation of tobacco smoke causes the respiratory epithelial cells to change in structure and function. The cells flatten and form several layers of epithelial cells. These flattened epithelial cells no longer contain abundant rough ER, Golgi apparatuses, secretory vesicles, or cilia. The altered respiratory epithelium is adapted to protect the underlying cells from irritation, but it can no longer secrete mucus, nor does it have cilia to move the mucus toward the throat to clean the respiratory passages. Extensive replacement of normal epithelial cells in respiratory passages is associated with chronic inflammation of the respiratory passages (bronchitis), which is common in people who smoke heavily. Coughing is a common manifestation of inflammation and mucus accumulation in the respiratory passages.

fibers facilitate the movement of chromosomes toward the two centrosomes (see "Cell Division," p. 69). ▼

> 45. What is the function of centrosomes?
> 46. What are spindle fibers? What is their function?

Cilia, Flagella, and Microvilli

Cilia (sĭl′ē-ă, sing. cilium, an eyelash) project from the surface of cells, are capable of moving (see figure 3.1), and vary in number from none to thousands per cell. Cilia have a cylindrical shape, contain specialized microtubules similar to the orientation in centrioles, and are enclosed by the plasma membrane. Cilia are numerous on surface cells that line the respiratory tract. Their coordinated movement moves mucus, in which dust particles are embedded, upward and away from the lungs. This action helps keep the lungs clear of debris.

Flagella (flă-jel′ă, sing. flagellum, a whip) have a structure similar to that of cilia but are much longer, and they usually occur only one per cell. Sperm cells each have one flagellum, which propels the sperm cell.

Microvilli (mī′krō-vil′ī, *mikros,* small + *villus,* shaggy hair) are specialized extensions of the plasma membrane that are supported by microfilaments (see figure 3.1), but they do not actively move like cilia and flagella. Microvilli are numerous on cells that have them, and they increase the surface area of those cells. They are abundant on the surface of cells that line the intestine, kidney, and other areas in which absorption or secretion is an important function. ▼

> 47. Contrast the structure and function of cilia and flagella.
> 48. Describe the structure and function of microvilli. How are microvilli different from cilia?

P R E D I C T

List the organelles that are common in cells that (a) synthesize and secrete proteins, (b) actively transport substances into cells, and (c) phagocytize foreign substances. Explain the function of each organelle you list.

Whole-Cell Activity

Interactions between organelles must be considered to understand how a cell functions. For example, the transport of many food molecules into the cell requires ATP and plasma membrane proteins. Most ATP is produced by mitochondria. ATP is required to transport amino acids across the plasma membrane. Amino acids are assembled to synthesize proteins, including the transport proteins of the plasma membrane and mitochondrial proteins. Information contained in DNA within the nucleus determines which amino acids are combined at ribosomes to form proteins. The mutual interdependence of cellular organelles is coordinated to maintain homeostasis within the cell and the entire body. The following sections, "Protein Synthesis" and "Cell Division," illustrate the interactions of organelles that result in a functioning cell.

Protein Synthesis

DNA contains the information that directs protein synthesis. The proteins form structural components of the cell; are enzymes, which regulate chemical reactions in the cell; and are secreted from the cell. DNA influences the structural and functional characteristics of the entire organism because it directs protein synthesis.

A DNA molecule consists of nucleotides joined together to form two nucleotide strands (see figure 2.22). The two strands are connected by complementary base pairs and resemble a ladder that is twisted around its long axis. The sequence of nucleotides in a DNA molecule is a method of storing information that is based on a triplet code. Three consecutive nucleotides, called **triplets,** form the words of the triplet code. Just as the sequence of letters "seedogrun" can be deciphered to mean "see dog run," a sequence of bases, such as "CATGAGTAG," has meaning, which is used to construct other DNA molecules, RNA molecules, or proteins. A molecular definition of a **gene** is all the triplets necessary to make a functional RNA molecule or protein. Each DNA molecule contains many different genes.

1. DNA contains the information necessary to produce proteins.

2. Transcription of one DNA strand results in mRNA, which is a complementary copy of the information in the DNA strand needed to make a protein.

3. The mRNA leaves the nucleus and goes to a ribosome.

4. Amino acids, the building blocks of proteins, are carried to the ribosome by tRNAs.

5. In the process of translation, the information contained in mRNA is used to determine the number, kinds, and arrangement of amino acids in the polypeptide chain.

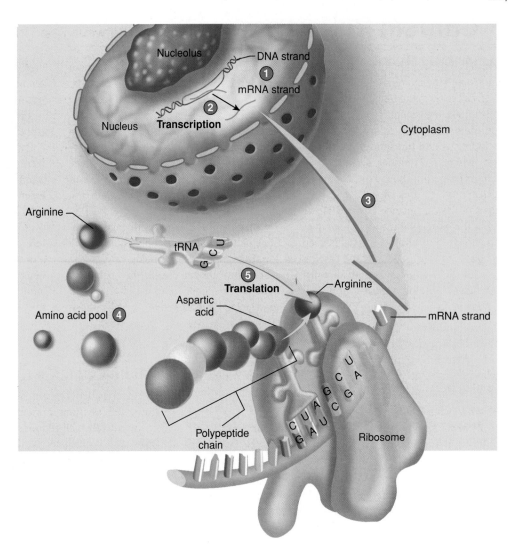

Process Figure 3.23   Overview of Protein Synthesis

Recall from chapter 2 that proteins consist of amino acids. The unique structural and functional characteristics of different proteins are determined by the kinds, numbers, and arrangement of their amino acids (see figure 2.19). The nucleotide sequence of a gene determines the amino acid sequence of a specific protein.

DNA directs the production of proteins in two steps—transcription and translation—which can be illustrated with an analogy. Suppose a cook wants a cake recipe that is found only in a reference book in the library. Because the book cannot be checked out, the cook makes a copy, or transcription, of the recipe. Later, in the kitchen, the information contained in the copied recipe is used to make the cake. The changing of something from one form to another (from recipe to cake) is called translation.

In this analogy, DNA is the reference book that contains many genes (recipes) for making different proteins. DNA, however, is too large a molecule to pass through the nuclear pores to go to the ribosomes where the proteins are prepared. Just as the reference book stays in the library, DNA remains in the nucleus. Through transcription the cell makes a copy of the gene (the recipe) necessary to make a particular protein (the cake). The copy, which is called **messenger RNA (mRNA),** travels from the nucleus to the ribosomes (the kitchen), where the information in the copy is used to construct a protein by means of translation. Of course, to turn a recipe into a cake, ingredients

are needed. The ingredients necessary to synthesize a protein are amino acids. Specialized transport molecules, called **transfer RNA (tRNA),** carry the amino acids to the ribosome (figure 3.23).

In summary, the synthesis of proteins involves transcription, making a copy of part of the information in DNA (a gene)—, and translation, converting that copied information into a protein. The details of transcription and translation are considered next. ▼

49. *Define a triplet in DNA. What is a gene?*
50. *What are the two steps in protein synthesis?*

Transcription

Transcription is the synthesis of mRNA, tRNA, and rRNA based on the nucleotide sequence in DNA (figure 3.24). Transcription occurs when a section of a DNA molecule unwinds and its complementary strands separate. One of the DNA strands serves as the template strand for the process of transcription. Nucleotides that form RNA align with the DNA nucleotides in the template strand by complementary base pairing. An adenine aligns with a thymine of DNA, a cytosine aligns with a guanine of DNA, and a guanine aligns with a cytosine of DNA. Instead of thymine, a uracil of RNA aligns with an adenine of DNA. Thus, the sequence of bases that aligns with the TCGA sequence of DNA is AGCU. This

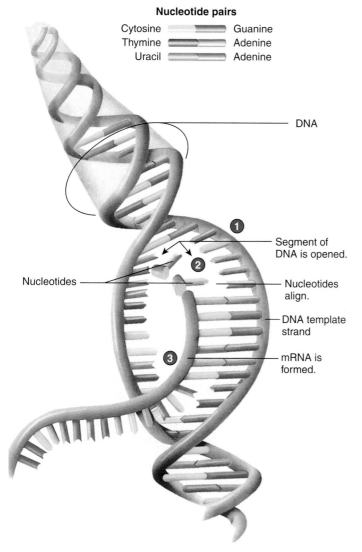

Nucleotide pairs

Cytosine ▬▬▬▬ Guanine
Thymine ▬▬▬▬ Adenine
Uracil ▬▬▬▬ Adenine

DNA

① Segment of
DNA is opened.

②

Nucleotides

Nucleotides
align.

DNA template
strand

③ mRNA is
formed.

1. The strands of the DNA molecule separate from each other. One DNA strand serves as a template for mRNA synthesis.

2. Nucleotides that will form mRNA pair with DNA nucleotides according to the base-pair combinations shown in the key at the top of the figure. Thus, the sequence of nucleotides in the template DNA strand (*purple*) determines the sequence of nucleotides in mRNA (*gray*). RNA polymerase (the enzyme is not shown) joins the nucleotides of mRNA together.

3. As nucleotides are added, an mRNA molecule is formed.

Process Figure 3.24 Formation of mRNA by Transcription of DNA

pairing relationship between nucleotides ensures that the information in DNA is transcribed correctly into RNA.

PREDICT
Given the following sequence of nucleotides of a DNA molecule, write the sequence of mRNA that is transcribed from it. What is the nucleotide sequence of the complementary strand of the DNA molecule? How does it differ from the nucleotide sequence of RNA?
DNA nucleotide sequence: CGTACG

After the DNA nucleotides pair up with the RNA nucleotides, an enzyme called **RNA polymerase** catalyzes reactions that form covalent bonds between the RNA nucleotides to form a long mRNA segment.

Posttranscriptional processing modifies mRNA before it leaves the nucleus to form the functional mRNA that is used in translation to produce a protein or part of a protein. Not all of the nucleotides in mRNA code for parts of a protein. The protein-coding region of mRNA contains sections, called **exons,** that code for parts of a protein as well as sections, called **introns,** that do not code for parts of a protein. The introns are removed from the mRNA and the exons are spliced together by enzymes called **spliceosomes.** The functional mRNA consists only of exons.

In a process called **alternative splicing,** various combinations of exons are incorporated into mRNA. Which exons, and how many exons, are used to make mRNA can vary between cells of different tissues, resulting in different mRNAs transcribed from the same gene. Alternative splicing allows a single gene to produce more than one specific protein. ▼

> 51. *Describe the formation of mRNA.*
> 52. *What is posttranscriptional processing and alternative splicing?*

Translation

The information contained in mRNA, called the **genetic code,** is carried in sets of three nucleotide units called **codons.** A codon specifies an amino acid during translation. For example, the codon GAU specifies the amino acid aspartic acid, and the codon CGA specifies arginine. Although there are only 20 different amino acids commonly found in proteins, 64 possible codons exist. Therefore, an amino acid can have more than one codon. The codons for arginine include CGA, CGG, CGU, and CGC. Furthermore, some codons act as signals during translation. AUG, which specifies methionine, also acts as a **start codon,** which signals the beginning of translation. UAA, UGA, and UAG act as **stop codons,** which signal the end of translation. Unlike the start codon, those codons do not specify amino acids. Therefore, the protein-coding region of an mRNA begins at the start codon and ends at a stop codon.

Translation is the synthesis of proteins in response to the codons of mRNA. In addition to mRNA, translation requires ribosomes and tRNA. Ribosomes consist of **ribosomal RNA (rRNA)** and proteins. Like mRNA, rRNA and tRNA are produced in the nucleus by transcription.

The function of tRNA is to match a specific amino acid to a specific codon of mRNA. To do this, one end of each kind of tRNA combines with a specific amino acid. Another part of the tRNA called the **anticodon** consists of three nucleotides and is complementary to a particular codon of mRNA. On the basis of the pairing relationships between nucleotides, the anticodon can combine only with its matched codon. For example, the tRNA that binds to aspartic acid has the anticodon CUA, which combines with the codon GAU of mRNA. Therefore, the codon GAU codes for aspartic acid.

The function of ribosomes is to align the codons of mRNA with the anticodons of tRNA molecules and then enzymatically join the amino acids of adjacent tRNA molecules. The mRNA moves through the ribosome one codon at a time. With each move, a new tRNA enters the ribosome and the amino acid is linked to the growing chain, forming a polypeptide. The step-by-step process of protein synthesis at the ribosome is described in detail in figure 3.25.

1. To start protein synthesis, a ribosome binds to mRNA. The ribosome also has two binding sites for tRNA, one of which is occupied by a tRNA with its amino acid. Note that the first codon to associate with a tRNA is AUG, the start codon, which codes for methionine. The codon of mRNA and the anticodon of tRNA are aligned and joined. The other tRNA binding site is open.

2. By occupying the open tRNA binding site, the next tRNA is properly aligned with mRNA and with the other tRNA.

3. An enzyme within the ribosome catalyzes a synthesis reaction to form a peptide bond between the amino acids. Note that the amino acids are now associated with only one of the tRNAs.

4. The ribosome shifts position by three nucleotides. The tRNA without the amino acid is released from the ribosome, and the tRNA with the amino acids takes its position. A tRNA binding site is left open by the shift. Additional amino acids can be added by repeating steps 2 through 4.

5. Eventually, a stop codon in the mRNA, such as UAA, ends the process of translation. At this point, the mRNA and polypeptide chain are released from the ribosome.

6. Multiple ribosomes attach to a single mRNA to form a polyribosome. As the ribosomes move down the mRNA, proteins attached to the ribosomes lengthen and eventually detach from the mRNA.

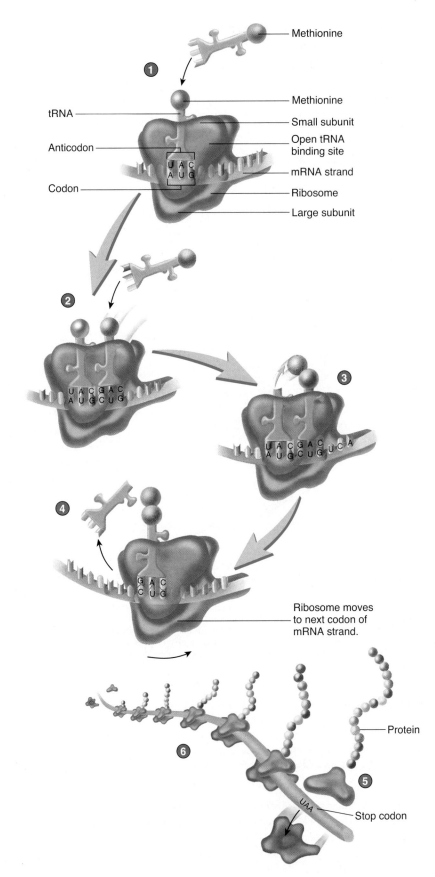

Process Figure 3.25   Translation of mRNA to Produce a Protein

After the initial part of mRNA is used by a ribosome, another ribosome can attach to the mRNA and begin to make a protein. The resulting cluster of ribosomes attached to the mRNA is called a **polyribosome** (see figure 3.25, step 6). Each ribosome in a polyribosome produces an identical protein, and polyribosomes are an efficient way to use a single mRNA molecule to produce many copies of the same protein.

Posttranslational processing is making modifications to proteins after they are produced. Many proteins are longer when first made than they are in their final, functional state. These proteins are called **proproteins,** and the extra piece of the molecule is cleaved off by enzymes to make the proprotein into a functional protein. Many proteins have side chains, such as polysaccharides, added to them following translation. Some proteins are composed of two or more amino acid chains that are joined after each chain is produced on separate ribosomes. ▼

53. Define genetic code, codon, anticodon, start codon, and stop codon.
54. What are the functions of tRNA and ribosomes in translation?
55. Describe posttranslational processing of proteins.

Cell Division

Cell division is the formation of two daughter cells from a single parent cell. The cytoplasm of the parent cell is divided in a process called cytokinesis. The nucleus of the parent cell can form the nuclei of the daughter cells by mitosis or meiosis. The new cells necessary for growth and tissue repair are formed through **mitosis** (mī-tō′sis, *mitos,* thread; refers to threadlike spindle fibers formed during mitosis). The sex cells, called **gametes** (gam′ētz), necessary for reproduction are formed through **meiosis** (mī-ō′sǐs, a lessening). In males, the gametes are **sperm cells;** in females, the gametes are **oocytes** (egg cells).

During mitosis and meiosis the DNA within the parent cell is distributed to the daughter cells. The DNA is found within chromosomes. The normal number of chromosomes in a somatic cell is called the **diploid** (dip′loyd, twofold) **number.** Somatic cells are cells of the body, except for gametes. The normal number of chromosomes in a gamete is the **haploid** (hap′loyd, single) **number.** In humans, the diploid number of chromosomes is 46 and the haploid number is 23. The 46 chromosomes are organized to form 23 pairs of chromosomes. Of the 23 pairs, one pair is the sex chromosomes, which consist of two **X chromosomes** if the person is a female or an X chromosome and a **Y chromosome** if the person is a male. The remaining 22 pairs of chromosomes are called **autosomes** (aw′tō-sōmz). The combination of sex chromosomes determines the individual's sex, and the autosomes determine most other characteristics.

Following mitosis, each daughter cell has the diploid number of chromosomes—they have the same type and amount of DNA as the parent cell. Because DNA determines cell structure and function, the daughter cells tend to have the same structure and perform the same functions as the parent cell. During development and cell differentiation, however, the functions of daughter cells

may differ from each other and from that of the parent cell (see "Differentiation," p. 72).

Following meiosis, the gametes have the haploid number of chromosomes   that is, half of the DNA as the parent cell. Furthermore, the DNA in each gamete differs from the DNA of the other gametes. Chapter 24 describes the details of meiosis.

Interphase is the period between cell divisions. The DNA in each chromosome replicates during interphase (figure 3.26). The two strands of DNA separate from each other, and each strand serves as a template for the production of a new strand of DNA. Nucleotides in the DNA of an old strand pair with nucleotides that are subsequently joined by enzymes to form a new strand of DNA.

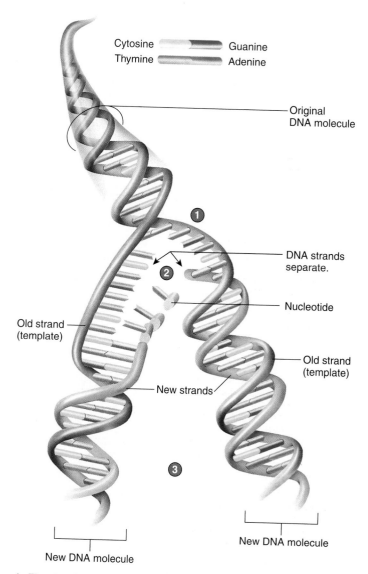

Cytosine Guanine
Thymine Adenine

Original DNA molecule

1

DNA strands separate.

2

Nucleotide

Old strand (template)

Old strand (template)

New strands

3

New DNA molecule

New DNA molecule

1. The strands of the DNA molecule separate from each other.
2. Each old strand (*dark purple*) functions as a template on which a new, complementary strand (*light purple*) is formed. The base-pairing relationship between nucleotides determines the sequence of nucleotides in the newly formed strands.
3. Two identical DNA molecules are produced.

Process Figure 3.26   Replication of DNA

Replication of DNA during interphase produces two identical molecules of DNA.

The sequence of nucleotides in the DNA template determines the sequence of nucleotides in the new strand of DNA because adenine pairs with thymine, and cytosine pairs with guanine. The two new strands of DNA combine with the two template strands to form two double strands of DNA.

PREDICT 7

Suppose a molecule of DNA separates, forming strands 1 and 2. Part of the nucleotide sequence in strand 1 is ATGCTA. From this template, what would be the sequence of nucleotides in the DNA replicated from strand 1 and strand 2?

Mitosis

The DNA replicated during interphase is dispersed as chromatin. During mitosis, the chromatin becomes very densely coiled to form compact chromosomes. These chromosomes are discrete bodies easily seen with a light microscope after staining the nuclei of dividing cells. Because the DNA has been replicated, each chromosome consists of two chromatids, which are attached at a single point called the centromere (figure 3.27). Each chromatid contains a DNA molecule. As two daughter cells are formed, the chromatids separate, and each is now called a chromosome. Each daughter cell receives one of the chromosomes. Thus, the daughter cells receive the same complement of chromosomes (DNA) and are genetically identical.

Mitosis follows interphase. For convenience of discussion, mitosis is divided into four phases: **prophase, metaphase, anaphase,** and **telophase** (tel′ō-fāz). Although each phase represents major events, mitosis is a continuous process, and no discrete jumps occur from one phase to another. Learning the characteristics associated with each phase is helpful, but a more important concept is that each daughter cell obtains the same number and type of chromosomes as the parent cell. The major events of mitosis are summarized in figure 3.28.

Cytokinesis

Cytokinesis (sī′tō-ki-nē′sis) is the division of the cytoplasm of the cell to produce two new cells. Cytokinesis begins in anaphase, continues through telophase, and ends in the following interphase (see figure 3.28). The first sign of cytokinesis is the formation of a **cleavage furrow,** which is an indentation of the plasma membrane. A contractile ring composed primarily of actin filaments pulls the plasma membrane inward, dividing the cell into halves. Cytokinesis is complete when the membranes of the halves separate at the cleavage furrow to form two separate cells. ▼

56. *How do the numbers of chromosomes in somatic cells and gametes differ from each other?*
57. *What are sex chromosomes and autosomes?*
58. *How does the DNA of daughter cells produced by mitosis and meiosis compare with that of the parent cell?*
59. *Define interphase. Describe the replication of DNA during interphase.*
60. *Define chromatin, chromatid, and chromosome.*
61. *Describe the events that occur during prophase, metaphase, anaphase, and telophase of mitosis.*
62. *Describe cytokinesis.*

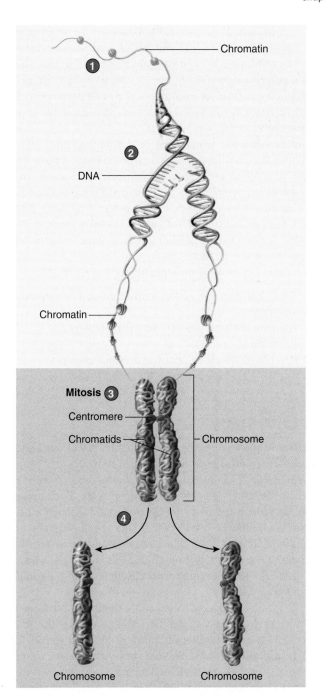

1. The DNA of a chromosome is dispersed as chromatin.
2. The DNA molecule unwinds and each strand of the molecule is replicated.
3. During mitosis the chromatin from each replicated DNA strand condenses to form a chromatid. The chromatids are joined at the centromere to form a single chromosome.
4. The chromatids separate to form two new, identical chromosomes. The chromosomes will unwind to form chromatin in the nuclei of the two daughter cells.

Process Figure 3.27 Replication of a Chromosome

1. **Interphase** is the time between cell divisions. DNA is found as thin threads of chromatin in the nucleus. DNA replication occurs during interphase. Organelles, other than the nucleus, duplicate during interphase.

2. In **prophase,** the chromatin condenses into chromosomes. Each chromosome consists of two chromatids joined at the centromere. The centrioles move to the opposite ends of the cell, and the nucleolus and the nuclear envelope disappear. Microtubules form near the centrioles and project in all directions. Some of the microtubules end blindly and are called astral fibers. Others, known as spindle fibers, project toward an invisible line called the equator and overlap with fibers from opposite centrioles.

3. In **metaphase,** the chromosomes align in the center of the cell in association with the spindle fibers. Some spindle fibers are attached to kinetochores in the centromere of each chromosome.

4. In **anaphase,** the chromatids separate, and each chromatid is then referred to as a chromosome. Thus, when the centromeres divide, the chromosome number is double, and there are two identical sets of chromosomes. The chromosomes, assisted by the spindle fibers, move toward the centrioles at each end of the cell. Separation of the chromatids signals the beginning of anaphase, and, by the time anaphase has ended, the chromosomes have reached the poles of the cell. Cytokinesis begins during anaphase as a cleavage furrow forms around the cell.

5. In **telophase,** migration of each set of chromosomes is complete. The chromosomes unravel to become less distinct chromatin threads. The nuclear envelope forms from the endoplasmic reticulum. The nucleoli form, and cytokinesis continues to form two cells.

6. Mitosis is complete, and a new interphase begins. The chromosomes have unraveled to become chromatin. Cell division has produced two daughter cells, each with DNA that is identical to the DNA of the parent cell.

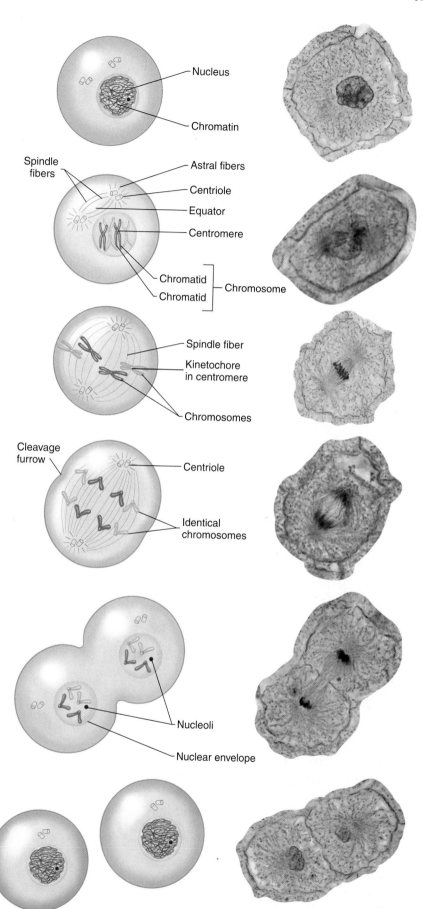

Process Figure 3.28 Mitosis and Cytokinesis

Differentiation

Differentiation is the process by which cells develop specialized structures and functions. The single cell formed by the union of a sperm cell and an oocyte during fertilization divides by mitosis to form two cells, which divide to form four cells, and so on (see chapter 25). The trillions of cells that ultimately make up the body of an adult stem from that single cell. Therefore, all the cells in an individual's body contain the same amount and type of DNA because they resulted from mitosis. Not all cells look and function alike, even though the genetic information contained in them is identical. Bone cells, for example, do not look like or function as muscle cells, nerve cells, or red blood cells.

During differentiation of a cell, some portions of DNA are active, but others are inactive. The active and inactive sections of DNA differ with each cell type. The portion of DNA responsible for the structure and function of a bone cell is different from that responsible for the structure and function of a muscle cell. Differentiation results from the selective activation and inactivation of segments of DNA. The mechanisms that determine which portions of DNA are active in any one cell type are not fully understood, but the resulting differentiation produces the many cell types that function together to make a person.

Eventually, as cells differentiate and mature, the rate at which they divide slows or even stops.

Cloning

Through the process of differentiation, cells become specialized to certain functions and are no longer capable of producing an entire organism if isolated. Over 30 years ago, however, it was demonstrated in frogs that if the nucleus is removed from a differentiated cell and is transferred to an oocyte with the nucleus removed, a complete normal frog can develop from that oocyte. This process, called **cloning,** demonstrated that during differentiation, genetic information is not irrevocably lost. Because mammalian oocytes are considerably smaller than frog oocytes, cloning of mammalian cells has been technically much more difficult. Dr. Ian Wilmut and his colleagues at the Roslin Institute in Edinburgh, Scotland, overcame those technical difficulties in 1996, when they successfully cloned the first mammal, a sheep. Since that time, many other mammalian species have been cloned.

63. How can cells that have the same DNA be structurally and functionally different from each other?

S U M M A R Y

Cell Organization and Functions (p. 46)

1. Cells are the basic unit of life, containing organelles, which perform specific functions.
2. The plasma membrane forms the outer boundary of the cell, the nucleus contains genetic material and directs cell activities, and cytoplasm is material between the nucleus and plasma membrane.
3. Cells metabolize and release energy, synthesize molecules, provide a means of communication, reproduce, and provide for inheritance.

Plasma Membrane (p. 48)

1. Intracellular substances are inside cells, whereas extracellular (intercellular) substances are between cells.
2. The plasma membrane is composed of a double layer of phospholipid molecules (lipid bilayer) in which proteins float (fluid-mosaic model). The proteins function as marker molecules, attachment proteins (cadherins and integrins), transport proteins, receptor proteins, and enzymes.

Movement Through the Plasma Membrane (p. 49)

Selectively permeable membranes allow some substances, but not others, to pass through.

Diffusion

1. Diffusion is the movement of a solute from an area of higher concentration to an area of lower concentration within a solvent. At equilibrium, there is a uniform distribution of molecules.

2. A solution is a mixture of liquids, gases or solids in which dissolved substances are uniformly distributed. A solute dissolves in a solvent to form a solution.
3. A concentration gradient is the concentration difference of a solute between two points divided by the distance between the points.

Osmosis

1. Osmosis is the diffusion of a solvent (water) across a selectively permeable membrane.
2. Osmotic pressure is the force required to prevent the movement of water across a selectively permeable membrane.
3. Isosmotic solutions have the same concentration, hyperosmotic solutions have a greater concentration, and hyposmotic solutions have a lesser concentration of solute particles than a reference solution.
4. In a hypotonic solution, cells swell (and can undergo lysis); in an isotonic solution, cells neither swell nor shrink; and in a hypertonic solution, cells shrink and undergo crenation.

Mediated Transport

1. Mediated transport is the movement of a substance across a membrane by means of a transport protein. The substances transported tend to be large, non-lipid-soluble molecules or ions.
2. Mediated transport exhibits specificity (selectiveness), competition (similar molecules or ions compete for a transport protein), and

saturation (rate of transport cannot increase because all transport proteins are in use).
3. Channel proteins form membrane channels (ion channels).
4. Carrier proteins bind to ions or molecules and transport them.
   - Uniport (facilitated diffusion) moves an ion or molecule down its concentration gradient.
   - Symport moves two or more ions or molecules in the same direction.
   - Antiport moves two or more ions or molecules in opposite directions.
5. ATP-powered pumps move ions or molecules against their concentration gradient using the energy from ATP.
6. Secondary active transport uses the energy of one substance moving down its concentration gradient to move another substance across the plasma membrane.

Vesicular Transport

1. Endocytosis is the movement of materials into cells by the formation of a vesicle.
   - Phagocytosis is the movement of solid material into cells.
   - Pinocytosis is the uptake of small droplets of liquids and the materials in them.
   - Receptor-mediated endocytosis involves plasma membrane receptors attaching to molecules that are then taken into the cell.
2. Exocytosis is the secretion of materials from cells by vesicle formation.

Cytoplasm   (p. 58)

The cytoplasm is the material outside the nucleus and inside the plasma membrane.

Cytosol

1. Cytosol consists of a fluid part (the site of chemical reactions), the cytoskeleton, and cytoplasmic inclusions.
2. The cytoskeleton supports the cell and enables cell movements. It consists of protein fibers (microtubules, actin filaments, and intermediate filaments).

The Nucleus and Cytoplasmic Organelles   (p. 59)

Organelles are subcellular structures specialized for specific functions.

Nucleus

1. The nuclear envelope consists of two separate membranes with nuclear pores.
2. DNA and associated proteins are found inside the nucleus. DNA is the hereditary material of the cell and controls the activities of the cell.
3. Between cell divisions DNA is organized as chromatin. During cell division chromatin condenses to form chromosomes consisting of two chromatids connected by a centromere.

Nucleoli and Ribosomes

1. Nucleoli consist of RNA and proteins and are the sites of ribosomal subunit assembly.
2. Ribosomes are the sites of protein synthesis.

Rough and Smooth Endoplasmic Reticulum

1. Rough ER is ER with ribosomes attached. It is a major site of protein synthesis.
2. Smooth ER does not have ribosomes attached. It is a major site of lipid synthesis.

Golgi Apparatus

The Golgi apparatus is a series of closely packed membrane sacs that collect, modify, package, and distribute proteins and lipids produced by the ER.

Secretory Vesicles

Secretory vesicles are membrane-bound sacs that carry substances from the Golgi apparatus to the plasma membrane, where the vesicle contents are released.

Lysosomes and Peroxisomes

Membrane-bound sacs containing enzymes include lysosomes and peroxisomes. Within the cell the lysosomes break down material within endocytic vesicles. Peroxisomes break down fatty acids, amino acids, and hydrogen peroxide.

Proteasomes

Proteasomes are large, multienzyme complexes, not bound by membranes, that digest selected proteins within the cell.

Mitochondria

1. Mitochondria are the major sites of the production of ATP, which is an energy source for cells.
2. The mitochondria have a smooth outer membrane and an inner membrane that is infolded to produce cristae.
3. Mitochondria contain their own DNA, can produce some of their own proteins, and can replicate independently of the cell.

Centrioles and Spindle Fibers

1. Centrioles are cylindrical organelles located in the centrosome, a specialized zone of the cytoplasm. The centrosome is the site of microtubule formation.
2. Spindle fibers are involved in the separation of chromosomes during cell division.

Cilia, Flagella, and Microvilli

1. Cilia move substances over the surface of cells.
2. Flagella are much longer than cilia and propel sperm cells.
3. Microvilli increase the surface area of cells and aid in absorption and secretion.

Whole-Cell Activity

The interactions between organelles must be considered for cell function to be fully understood. That function is reflected in the quantity and distribution of organelles.

Protein Synthesis   (p. 65)

1. DNA controls enzyme production and cell activity is regulated by enzymes (proteins).
2. A gene is all of the nucleotide triplets necessary to make a functional RNA molecule or protein.

Transcription

1. DNA unwinds and, through nucleotide pairing, produces mRNA.
2. During posttranscriptional processing, introns are removed and exons are spliced by spliceosomes.
3. Alternative splicing produces different combinations of exons, allowing one gene to produce more than one type of protein.

Translation

1. The mRNA moves through the nuclear pores to ribosomes.
2. Transfer RNA (tRNA), which carries amino acids, interacts at the ribosome with mRNA. The anticodons of tRNA bind to the codons of mRNA, and the amino acids are joined to form a protein.
3. During posttranslational processing, proproteins are modified by proenzymes.

Cell Division (p. 69)

1. Cell division that occurs by mitosis produces new cells for growth and tissue repair.
2. Cell division that occurs by meiosis produces gametes (sex cells). Sperm cells in males and oocytes (egg cells) in females are gametes.
3. Chromosomes
   - Somatic cells have a diploid number of chromosomes, whereas gametes have a haploid number. In humans, the diploid number is 46 (23 pairs) and the haploid number is 23.
   - Humans have 22 pairs of autosomal chromosomes and one pair of sex chromosomes. Females have the sex chromosomes XX and males have XY.
4. DNA replicates during interphase, the time between cell division.

Mitosis

1. Chromatin condenses to form chromosomes, consisting of two chromatids. The chromatids separate to form two chromosomes, which are distributed to each daughter cell.
2. Mitosis is divided into four stages:
   - *Prophase.* Chromatin condenses to become visible as chromosomes. Each chromosome consists of two chromatids joined at the centromere. Centrioles move to opposite poles of the cell, and astral fibers and spindle fibers form. Nucleoli disappear, and the nuclear envelope degenerates.
   - *Metaphase.* Chromosomes align at the equatorial plane.
   - *Anaphase.* The chromatids of each chromosome separate at the centromere. Each chromatid then is called a chromosome. The chromosomes migrate to opposite poles.
   - *Telophase.* Chromosomes unravel to become chromatin. The nuclear envelope and nucleoli reappear.

Cytokinesis

1. Cytokinesis is the division of the cytoplasm of the cell. It begins with the formation of the cleavage furrow during anaphase.
2. Cytokinesis is complete when the plasma membrane comes together at the equator, producing two new daughter cells.

Differentiation (p. 72)

Differentiation, the process by which cells develop specialized structures and functions, results from the selective activation and inactivation of DNA sections.

R E V I E W   A N D   C O M P R E H E N S I O N

1. In the plasma membrane, phospholipids
   a. form most of the bilayer.
   b. function as enzymes.
   c. bind cells together.
   d. allow cells to identify each other.
   e. all of the above.

2. Concerning diffusion,
   a. most non-lipid-soluble molecules and ions diffuse through the lipid bilayer.
   b. it stops when random movement of molecules and ions stops.
   c. it is the movement of molecules or ions from areas of lower concentration to areas of higher concentration.
   d. the greater the concentration gradient, the greater the rate of diffusion.
   e. it requires ATP.

3. Which of these statements about osmosis is true?
   a. Osmosis always involves a membrane that allows water and all solutes to diffuse through it.
   b. The greater the solute concentration, the smaller the osmotic pressure of a solution.
   c. Osmosis moves water from a solution with a greater solute concentration to a solution with a lesser solute concentration.
   d. The greater the osmotic pressure of a solution, the greater the tendency for water to move into the solution.
   e. Osmosis occurs because of hydrostatic pressure outside the cell.

4. If a cell is placed in a (an) _____ solution, lysis of the cell may occur.
   a. hypertonic   c. isotonic
   b. hypotonic   d. isosmotic

5. Container A contains a 10% salt solution, and container B contains a 20% salt solution. If the two solutions are connected, the net movement of water by diffusion is from _____ to _____ , and the net movement of salt by diffusion is from _____ to _____ .
   a. A, B; A, B   c. B, A; A, B
   b. A, B; B, A   d. B, A; B, A

6. Suppose that a woman ran a long-distance race in the summer. During the race, she lost a large amount of hyposmotic sweat. You would expect her cells to
   a. shrink.   b. swell.   c. stay the same.

7. Suppose that a man is doing heavy exercise in the hot summer sun. He sweats profusely. He then drinks a large amount of distilled water. After he drinks the water, you would expect his tissue cells to
   a. shrink.   c. remain the same.
   b. swell.

8. In mediated transport,
   a. the rate of transport is limited by the number of transport proteins.
   b. similar molecules may be moved by the same transport protein.
   c. each transport protein moves particular molecules or ions, but not others.
   d. all of the above.

9. Which of these statements about facilitated diffusion is true?
   a. In facilitated diffusion, net movement is down the concentration gradient.
   b. Facilitated diffusion requires the expenditure of energy.
   c. Facilitated diffusion does not require a carrier protein.
   d. Facilitated diffusion moves materials through membrane channels.
   e. Facilitated diffusion moves materials in vesicles.

10. The Na^+–K^+ pump
    a. requires ATP.
    b. moves Na^+ and K^+ down their concentration gradients.
    c. is an example of antiport.
    d. is a carrier protein.
    e. all of the above.

11. Which of these statements concerning the symport of glucose into cells is true?
    a. The Na^+–K^+ pump moves Na^+ into cells.
    b. The concentration of Na^+ outside cells is less than inside cells.
    c. A carrier protein moves Na^+ into cells and glucose out of cells.
    d. The concentration of glucose can be greater inside cells than outside cells.
    e. As Na^+ are actively transported into the cell, glucose is carried along.

12. A white blood cell ingests solid particles by forming vesicles. This describes the process of
    a. exocytosis.
    b. facilitated diffusion.
    c. secondary active transport.
    d. phagocytosis.
    e. pinocytosis.

13. Cytoplasm is found
    a. in the nucleus.
    b. outside the nucleus and inside the plasma membrane.
    c. outside the plasma membrane.
    d. everywhere in the cell.

14. Actin filaments
    a. are essential components of centrioles, spindle fibers, cilia, and flagella.
    b. are involved with changes in cell shape.
    c. are hollow tubes.
    d. are aggregates of chemicals taken in by cells.

15. Cylindrically shaped extensions of the plasma membrane that do not move, are supported with actin filaments, and may function in absorption or as sensory receptors are
    a. centrioles.   c. cilia.   e. microvilli.
    b. spindle fibers.   d. flagella.

16. Mature red blood cells cannot
    a. synthesize ATP.
    b. transport oxygen.
    c. synthesize new protein.
    d. use glucose as a nutrient.

17. A large structure, normally visible in the nucleus of a cell, where ribosomal subunits are produced is called a (an)
    a. endoplasmic reticulum.   c. nucleolus.
    b. mitochondrion.   d. lysosome.

18. A cell that synthesizes large amounts of protein for use outside the cell has a large
    a. number of cytoplasmic inclusions.
    b. amount of rough endoplasmic reticulum.
    c. amount of smooth endoplasmic reticulum.
    d. number of lysosomes.

19. After the ER produces a protein,
    a. a secretory vesicle caries it to the Golgi apparatus.
    b. a transport vesicle carries it to a lysosome.
    c. it can be modified by the Golgi apparatus.
    d. it can be modified by a proteasome.

20. Which of these organelles produces large amounts of ATP?
    a. nucleus   d. endoplasmic reticulum
    b. mitochondria   e. lysosomes
    c. ribosomes

21. A portion of an mRNA molecule that determines one amino acid in a polypeptide chain is called a (an)
    a. nucleotide.   c. codon.   e. intron.
    b. gene.   d. exon.

22. In which of these organelles is mRNA synthesized?
    a. nucleus   d. nuclear envelope
    b. ribosome   e. peroxisome
    c. endoplasmic reticulum

23. Which of the following structures bind to each other?
    a. DNA triplet—codon   d. adenine—thymine
    b. codon—anticodon   e. all of the above
    c. amino acid—tRNA

24. Concerning chromosomes,
    a. the diploid number of chromosomes is in gametes.
    b. humans have 23 chromosomes in somatic cells.
    c. males have two X chromosomes.
    d. they are replicated during interphase.

25. Given the following activities:
    1. repair
    2. growth
    3. gamete production
    4. differentiation
    Which of the activities is (are) the result of mitosis?
    a. 2   c. 1,2   e. 1,2,4
    b. 3   d. 3,4

Answers in Appendix E

CRITICAL THINKING

1. Given the setup in figure 3.5, what would happen to osmotic pressure if the membrane were not selectively permeable but instead allowed all solutes and water to pass through it?

2. The body of a man was found floating in the salt water of Grand-Pacific Bay, which has a concentration that is slightly greater than body fluids. When seen during an autopsy, the cells in his lung tissues were clearly swollen. Choose the most logical conclusion.
    a. He probably drowned in the bay.
    b. He may have been murdered elsewhere.
    c. He did not drown.

3. Why does a surgeon irrigate a surgical wound from which a tumor has been removed with sterile distilled water rather than with sterile isotonic saline?

4. Patients with kidney failure can be kept alive by dialysis, which removes toxic waste products from the blood. In a dialysis machine, blood flows past one side of a selectively permeable dialysis membrane, and dialysis fluid flows on the other side of the membrane. Small substances, such as ions, glucose, and urea, can pass through the dialysis membrane, but larger substances, such as proteins, cannot. If you wanted to use a dialysis machine to remove

only the toxic waste product urea from blood, what could you use for the dialysis fluid?

a. a solution that is isotonic and contains only protein
b. a solution that is isotonic and contains the same concentration of substances as blood, except for having no urea in it
c. distilled water
d. blood

5. Secretory vesicles fuse with the plasma membrane to release their contents to the outside of the cell. In this process the membrane of the secretory vesicle becomes part of the plasma membrane. Because small pieces of membrane are continually added to the plasma membrane, one would expect the plasma membrane to become larger and larger as secretion continues. The plasma membrane stays the same size, however. Explain how this happens.

6. Suppose that a cell has the following characteristics: many mitochondria, well-developed rough ER, well-developed Golgi apparatuses, and numerous vesicles. Predict the major function of the cell. Explain how each characteristic supports your prediction.

7. If you had the ability to inhibit mRNA synthesis with a drug, explain how you could distinguish between proteins released from secretory vesicles in which they had been stored and proteins released from cells in which they had been newly synthesized.

8. The proteins (hemoglobin) in red blood cells normally organize relative to one another, forming "stacks" of proteins, which are in part responsible for the normal shape of red blood cells. In sickle-cell anemia, proteins inside red blood cells do not stack normally. Consequently, the red blood cells become sickle-shaped and plug up small blood vessels. It is known that sickle-cell anemia is hereditary and results from changing one nucleotide for a different nucleotide within the gene that is responsible for producing the protein. Explain how this change results in an abnormally functioning protein.

Answers in Appendix F

CHAPTER 4 Tissues, Glands, and Membranes

This colorized scanning electron micrograph of simple columnar epithelium lining the uterine tube shows the columnar epithelial cells (*blue*) resting on a weblike basement membrane. Organelles of the epithelial cells synthesize a small volume of mucus and secrete it on their free surface. Cilia are the hairlike structures seen at the free surface of some of the epithelial cells. Cilia move the mucus produced by the epithelial cells over the surface of the cells.

Introduction

In some ways, the human body is like a complex machine, such as a car. Not all parts of a car can be made from a single type of material. Metal, capable of withstanding the heat of the engine, cannot be used for windows or tires. Similarly, the many parts of the human body are made of collections of specialized cells and the materials surrounding them. Muscle cells that contract to produce body movements have a structure and function different from that of epithelial cells that protect, secrete, or absorb.

Knowledge of tissue structure and function is important in understanding how individual cells are organized to form tissues and how tissues are organized to form organs, organ systems, and the complete organism. There is a relationship between the structure of each tissue type and its function and between the tissues in an organ and the organ's function. The structure and function of tissues are so closely related that you should be able to predict the function of a tissue when given its structure, and vice versa.

▶This chapter begins with brief discussions of **tissues and histology** (p. 78) and the development of **embryonic tissue** (p. 78) and then describes the structural and functional characteristics of the major tissue types: **epithelial tissue** (p. 78), **connective tissue** (p. 86), **muscle tissue** (p. 94), and **nervous tissue** (p. 96). In addition, the chapter provides an explanation of **membranes** (p. 97), **inflammation** (p. 97), **tissue repair** (p. 98), and **tissues and aging** (p. 101).

Tools for Success

A virtual cadaver dissection experience

Audio (MP3) and/or video (MP4) content, available at www.mhhe.aris.com

Tissues and Histology

Tissues (tish′ūz, to weave) are collections of similar cells and the substances surrounding them. The **extracellular matrix** is the substances surrounding the cells. The classification of tissue types is based on the structure of the cells, the composition of the **extracellular matrix,** and the functions of the cells. The four primary tissue types are epithelial, connective, muscle, and nervous tissue. This chapter emphasizes epithelial and connective tissues. Muscle and nervous tissues are considered in more detail in later chapters.

Development, growth, aging, trauma, and diseases result from changes in tissues. For example, enlargement of skeletal muscles occurs because skeletal muscle cells increase in size in response to exercise. Reduced elasticity of blood vessel walls in aging people results from slowly developing changes in connective tissue. Many tissue abnormalities, including cancer, result from changes in tissues.

Histology (his-tol′ō-jē, *histo-,* tissue + *-ology,* study) is the microscopic study of tissues. Much information about a person's health can be gained by examining tissues. For example, cancer is identified and classified based on characteristic changes that occur in tissues. A **biopsy** (bī′op-sē) is the process of removing tissue samples from patients surgically or with a needle for diagnostic purposes. In some cases, tissues are removed surgically and examined while the patient is still anesthetized. The results of the biopsy are used to determine the appropriate therapy. For example, the classification of cancer tissue determines the amount of tissue removed as part of breast or other types of cancer treatment. ▼

> 1. *Define tissue and name the four primary tissue types.*
> 2. *Define histology and biopsy.*

Embryonic Tissue

Approximately 13 or 14 days after fertilization, the cells that give rise to a new individual form a slightly elongated disk consisting of two layers called the **endoderm** (en′dō-derm) and the **ectoderm** (ek′tō-derm). Cells of the ectoderm then migrate between the two layers to form a third layer called the **mesoderm** (mez′ō-derm). The disk folds over to form a tube in which the ectoderm is the outer layer, the mesoderm is the middle layer, and the endoderm is the inner layer. The endoderm, mesoderm, and ectoderm are called **germ layers** because they give rise to all the tissues of the body (see chapter 25). The endoderm forms the lining of the digestive tract and its derivatives; the mesoderm forms tissues such as muscle, bone, and blood vessels; and the ectoderm forms the outermost layer of the skin. A portion of the ectoderm, called **neuroectoderm** (noor-ō-ek′tō-derm), becomes the nervous system (see chapter 25). Groups of cells that break away from the neuroectoderm during development, called **neural crest cells,** give rise to parts of the peripheral nerves (see chapter 11), skin pigment (see chapter 5), the medulla of the adrenal gland (see chapter 13), and many tissues of the face. ▼

> 3. *Name the three embryonic germ layers.*
> 4. *What adult structures are derived from the endoderm, mesoderm, ectoderm, neuroectoderm, and neural crest cells?*

Epithelial Tissue

Epithelium (ep-i-thē′lē-ŭm, pl. epithelia, ep-i-thē′lē-ă, *epi,* on + *thele,* covering or lining), or **epithelial tissue,** can be thought of as a protective covering of external and internal surfaces of the body. Characteristics common to most types of epithelium (figure 4.1) are

1. Epithelium covers surfaces of the body and forms glands that are derived developmentally from body surfaces. The body surfaces include the outside surface of the body and the lining of cavities such as the digestive tract, respiratory passages, and blood vessels.
2. Epithelium consists almost entirely of cells, with very little extracellular matrix between them.
3. Most epithelial tissues have one **free,** or **apical** (ap′i-kăl), **surface** not attached to other cells; a **lateral surface** attached to other epithelial cells; and a **basal surface.** The free surface often lines the lumen of ducts, vessels, or cavities. The basal surface of most epithelial tissues is attached to a **basement membrane,** a meshwork of proteins and other molecules. The basement membrane is secreted by the epithelial cells and by underlying connective tissue cells. It is like the adhesive on Scotch™ tape and helps attach the epithelial cells to the underlying tissues. The basement membrane also plays an important role in supporting and guiding cell migration during tissue repair. A few epithelial tissues, such as in lymphatic capillaries and liver sinusoids, do not have basement membranes.
4. Epithelium does not contain blood vessels. Epithelial cells exchange gases and nutrients by diffusion with blood vessels in underlying tissues. ▼

> 5. *List four characteristics common to most types of epithelium.*
> 6. *Define free (apical), lateral, and basal surfaces of epithelial cells. What is the basement membrane?*

Functions of Epithelia

The major functions of epithelia are

1. *Protecting underlying structures.* Examples are the outer layer of the skin and the epithelium of the oral cavity, which protect the underlying structures from abrasion.
2. *Acting as barriers.* Epithelium prevents the movement of many substances through the epithelial layer. For example, the epithelium of the skin is a barrier to water and reduces water loss from the body. The epithelium of the skin also prevents the entry of many toxic molecules and microorganisms into the body.
3. *Permitting the passage of substances.* Epithelium allows the movement of many substances through the epithelial layer. For example, oxygen and carbon dioxide are exchanged between the air and blood by diffusion through the epithelium in the lungs.
4. *Secreting substances.* Mucous glands, sweat glands, and the enzyme-secreting portions of the pancreas secrete their

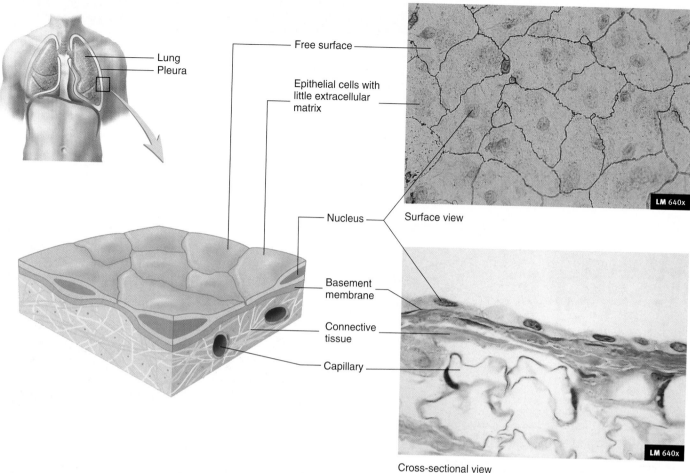

Figure 4.1 **Characteristics of Epithelium**

Surface and cross-sectional views of epithelium illustrate the following characteristics: little extracellular material between cells, a free surface, and a basement membrane attaching epithelial cells to underlying tissues. Capillaries in connective tissue do not penetrate the basement membrane. Nutrients, oxygen, and waste products must diffuse across the basement membrane between the capillaries and the epithelial cells.

products onto epithelial surfaces or into ducts that carry them to other areas of the body.

5. *Absorbing substances.* The plasma membranes of certain epithelial tissues contain carrier proteins (see chapter 3), which regulate the absorption of materials. For example, the epithelial cells of the intestine absorb digested food molecules, vitamins, and ions. ▼

7. **List five functions of epithelial tissue and give an example of each.**

Classification of Epithelia

Epithelia are classified according to the number of cell layers and the shape of the cells. **Simple epithelium** consists of a single layer of cells. **Stratified** (layers) **epithelium** consists of more than one layer of epithelial cells, with some cells sitting on top of other cells. The categories of epithelium based on cell shape are **squamous** (skwā′mŭs, flat and thin), **cuboidal** (cubelike), and **columnar** (tall and thin). In most cases, each epithelium is given two names. Examples include simple squamous, simple columnar, and stratified squamous epithelia. When epithelium is stratified, it is named according to the shape of the cells at the free surface. ▼

8. **What terms are used to describe the number of layers and the shapes of cells in epithelial tissue?**

Simple Epithelium

Simple squamous epithelium is a single layer of flat, thin cells (table 4.1*a*). It is often found where diffusion or filtration take place. Some substances easily pass through this thin layer of cells and other substances do not. For example, the respiratory passages end as small sacs called **alveoli** (al-vē′o-lī, sing. alveolus, hollow sac). The alveoli consist of simple squamous epithelium that allows oxygen from the air to diffuse into the body and carbon dioxide to diffuse out of the body into the air. Simple squamous epithelial tissue in the kidneys forms thin barriers through which small molecules, but not large ones, can pass. Small molecules, including water from blood, pass through these barriers as a major step in urine formation. Blood cells and large molecules, such as proteins, remain in the blood vessels of the kidneys.

Table 4.1 Simple Epithelium

(a) Simple Squamous Epithelium

Structure:
Single layer of flat, often hexagonal cells; the nuclei appear as bumps when viewed as a cross section because the cells are so flat

Function:
Diffusion, filtration, some secretion and some protection against friction

Location:
Alveoli of the lungs; lining (endothelium) of blood vessels, the heart, and lymphatic vessels; lining (mesothelium) of serous membranes of pericardial, pleural, and peritoneal cavities; portions of kidney tubules; small ducts; and inner surface of the eardrums

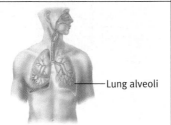
— Lung alveoli

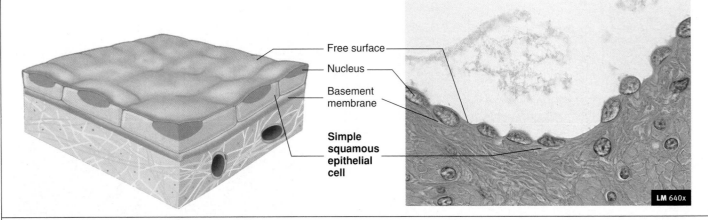

Free surface
Nucleus
Basement membrane
Simple squamous epithelial cell

LM 640x

(b) Simple Cuboidal Epithelium

Structure:
Single layer of cube-shaped cells; some cells have microvilli (kidney tubules) or cilia (terminal bronchioles of the lungs)

Function:
Active transport and facilitated diffusion result in secretion and absorption by cells of the kidney tubules; secretion by cells of glands and choroid plexuses; movement of particles embedded in mucus out of the terminal bronchioles by ciliated cells

Location:
Kidney tubules, glands and their ducts, choroid plexuses of the brain, lining of terminal bronchioles of the lungs, surfaces of the ovaries

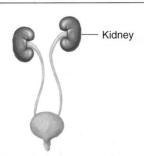
— Kidney

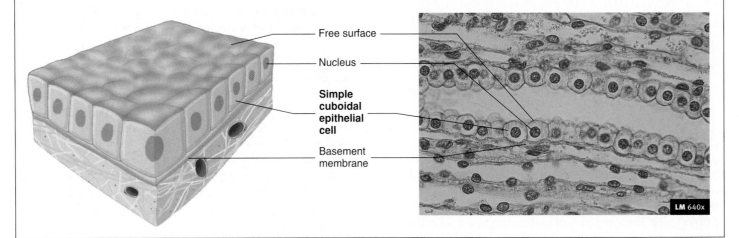

Free surface
Nucleus
Simple cuboidal epithelial cell
Basement membrane

LM 640x

Table 4.1 Simple Epithelium—Continued

(c) Simple Columnar Epithelium

Structure:
Single layer of tall, narrow cells; some cells have cilia (bronchioles of lungs, auditory tubes, uterine tubes, and uterus) or microvilli (intestines)

Function:
Movement of particles out of the bronchioles of the lungs by ciliated cells; partially responsible for the movement of oocytes through the uterine tubes by ciliated cells; secretion by cells of the glands, the stomach, and the intestine; absorption by cells of the intestine

Location:
Glands and some ducts, bronchioles of lungs, auditory tubes, uterus, uterine tubes, stomach, intestines, gallbladder, bile ducts, ventricles of the brain

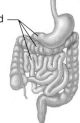

Lining of stomach and intestines

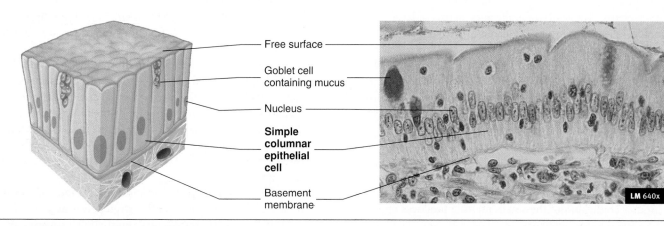

- Free surface
- Goblet cell containing mucus
- Nucleus
- **Simple columnar epithelial cell**
- Basement membrane

LM 640x

(d) Pseudostratified Columnar Epithelium

Structure:
Single layer of cells; some cells are tall and thin and reach the free surface, and others do not; the nuclei of these cells are at different levels and appear stratified; the cells are almost always ciliated and are associated with goblet cells that secrete mucus onto the free surface

Function:
Synthesize and secrete mucus onto the free surface and move mucus (or fluid) that contains foreign particles over the surface of the free surface and from passages

Location:
Lining of nasal cavity, nasal sinuses, auditory tubes, pharynx, trachea, bronchi of lungs

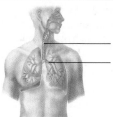

- Trachea
- Bronchus

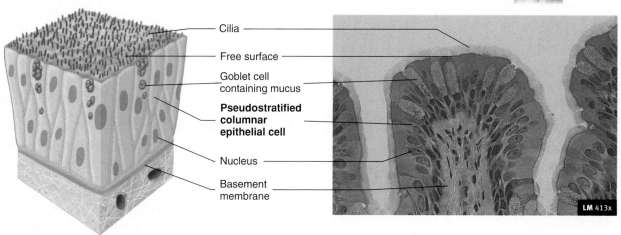

- Cilia
- Free surface
- Goblet cell containing mucus
- **Pseudostratified columnar epithelial cell**
- Nucleus
- Basement membrane

LM 413x

Simple squamous epithelium also prevents abrasion between organs in the pericardial, pleural, and peritoneal cavities (see chapter 1). The outer surfaces of organs are covered with simple squamous epithelium that secretes a slippery fluid. The fluid lubricates the surfaces between the organs, preventing damage from friction when the organs rub against one another.

Simple cuboidal epithelium is a single layer of cubelike cells (table 4.1*b*) that carry out active transport, facilitated diffusion, or secretion. Epithelial cells that secrete molecules such as proteins contain organelles that synthesize them. These cells have a greater volume than simple squamous epithelial cells and contain more cell organelles. The organelles of simple cuboidal cells that actively transport molecules into and out of the cells include mitochondria, which produce ATP, and organelles needed to synthesize the transport proteins. Transport of molecules across a layer of simple cuboidal epithelium can be regulated by the amount of ATP produced or by the type of transport proteins synthesized. The kidney tubules have large portions of their walls composed of simple cuboidal epithelium. These cuboidal epithelial cells secrete waste products into the tubules and reabsorb useful materials from the tubules as urine is formed. Some cuboidal epithelial cells have cilia that move mucus over the free surface or microvilli that increase the surface area for secretion and absorption.

Simple columnar epithelium is a single layer of tall, thin cells (table 4.1*c*). These large cells contain organelles that enable them to perform complex functions. For example, the simple columnar epithelium of the small intestine produces and secretes mucus and digestive enzymes. The mucus protects the lining of the intestine, and the digestive enzymes complete the process of digesting food. The columnar cells then absorb the digested foods by active transport, facilitated diffusion, or simple diffusion.

Pseudostratified (*pseudo,* false) **columnar epithelium** is a special type of simple epithelium (table 4.1*d*). The prefix *pseudo-* means false, so this type of epithelium appears to be stratified but is not. It consists of one layer of cells, with all the cells attached to the basement membrane. There is an appearance of two or more layers of cells because some of the cells are tall and reach the free surface, whereas others are short and do not reach the free surface. Pseudostratified columnar epithelium lines some glands and ducts; auditory tubes; and some of the respiratory passages, such as the nasal cavity, nasal sinuses, pharynx, trachea, and bronchi. Pseudostratified columnar epithelium secretes mucus, which covers its free surface. Cilia located on the free surface move the mucus and the debris that accumulates in it over the surfaces. For example, cilia of the respiratory passages move mucus toward the throat, where it is swallowed. ▼

> 9. *Name the four types of simple epithelium and give an example of each.*

Stratified Epithelium

Stratified squamous epithelium forms a thick epithelium because it consists of several layers of cells (table 4.2*a*). The deepest cells are cuboidal or columnar and are capable of dividing and producing new cells. As these newly formed cells are pushed to the surface, they become flat and thin. If cells at the surface are damaged or rubbed away, they are replaced by cells formed in the deeper layers.

Stratified squamous epithelium can be classified further as either nonkeratinized (moist) or keratinized, according to the condition of the outermost layer of cells. **Nonkeratinized (moist) stratified squamous epithelium** (see table 4.2*a*), found in areas such as the mouth, esophagus, rectum, and vagina, consists of living cells in the deepest and outermost layers. A layer of fluid covers the outermost layers of cells, which makes them moist. In contrast, **keratinized** (ker′ă-ti-nizd) **stratified squamous epithelium,** found in the skin (see chapter 5), consists of living cells in the deepest layers, and the outer layers are composed of dead cells containing the protein keratin. The dead, keratinized cells give the tissue a durable, moisture-resistant, dry character.

Stratified cuboidal epithelium consists of more than one layer of cuboidal epithelial cells (table 4.2*b*). This epithelial type is relatively rare and is found in sweat gland ducts, ovarian follicles, and the salivary glands. It functions in absorption, secretion, and protection.

Stratified columnar epithelium consists of more than one layer of epithelial cells, but only the surface cells are columnar in shape (table 4.2*c*). The deeper layers are irregular in shape or cuboidal. Like stratified cuboidal epithelium, stratified columnar epithelium is relatively rare. It is found in locations such as the mammary gland ducts, the larynx, and a portion of the male urethra. This epithelium carries out secretion, protection, and some absorption.

Transitional epithelium is a special type of stratified epithelium that can be greatly stretched (table 4.2*d*). In the unstretched state, transitional epithelium consists of five or more layers of cuboidal or columnar cells that often are dome-shaped at the free surface. As transitional epithelium is stretched, the cells change shape to a low cuboidal or squamous shape, and the number of cell layers decreases. Transitional epithelium lines cavities that can expand greatly, such as the urinary bladder. It also protects underlying structures from the caustic effects of urine. ▼

> 10. *Name the four types of stratified epithelium and give an example of each.*
> 11. *How do nonkeratinized (moist) stratified squamous epithelium and keratinized stratified squamous epithelium differ? Where is each type found?*

Structural and Functional Relationships

Cell Layers and Cell Shapes

The number of cell layers and the shape of the cells in a specific type of epithelium reflect the function the epithelium performs. Two important functions are controlling the passage of materials through the epithelium and protecting the underlying tissues. Simple epithelium, with its single layer of cells, is found in organs in which the principal function is the movement of materials. Examples include diffusion of gases across the wall of the alveoli of the lungs, secretion in glands, and nutrient absorption in the intestines. The movement of materials through a stratified epithelium is hindered by its many layers. Stratified epithelium is well adapted for its protective function. As the outer cell layers are damaged, they are replaced by cells from deeper layers. Stratified squamous

Table 4.2 Stratified Epithelium

(a) Stratified Squamous Epithelium

Structure:
Multiple layers of cells that are cuboidal in the basal layer and progressively flattened toward the surface; the epithelium can be nonkeratinized (moist) or keratinized; in nonkeratinized stratified squamous epithelium, the surface cells retain a nucleus and cytoplasm; in keratinized stratified epithelium, the cytoplasm of cells at the surface is replaced by a protein called keratin, and the cells are dead

Function:
Protection against abrasion, barrier against infection, reduction of water loss from the body

Location:
Keratinized—skin; nonkeratinized (moist)—mouth, throat, larynx, esophagus, anus, vagina, inferior urethra, cornea

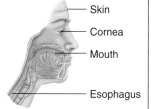

Skin
Cornea
Mouth
Esophagus

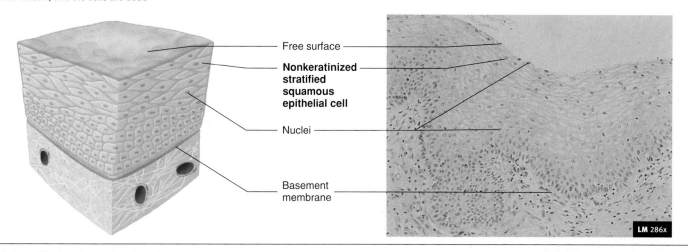

Free surface

Nonkeratinized stratified squamous epithelial cell

Nuclei

Basement membrane

LM 286x

(b) Stratified Cuboidal Epithelium

Structure:
Multiple layers of somewhat cube-shaped cells

Function:
Secretion, absorption, protection against infection

Location:
Sweat gland ducts, ovarian follicles, salivary gland ducts

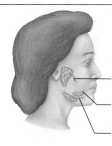

Parotid gland duct
Sublingual gland duct
Submandibular gland duct

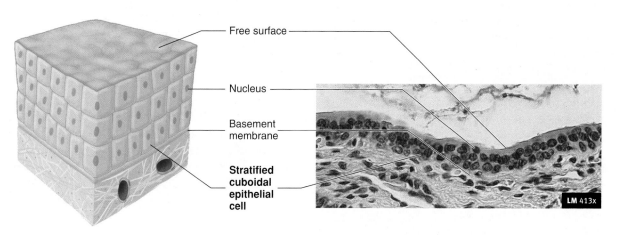

Free surface

Nucleus

Basement membrane

Stratified cuboidal epithelial cell

LM 413x

Table 4.2 Stratified Epithelium—Continued

(c) Stratified Columnar Epithelium

Structure:
Multiple layers of cells with tall, thin cells resting on layers of more cuboidal cells; the cells are ciliated in the larynx

Function:
Protection and secretion

Location:
Mammary gland ducts, larynx, a portion of the male urethra

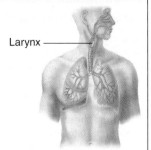

Larynx

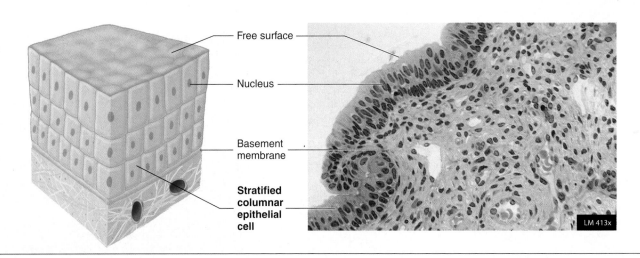

Free surface

Nucleus

Basement membrane

Stratified columnar epithelial cell

LM 413x

(d) Transitional Epithelium

Structure:
Stratified cells that appear cuboidal when the organ or tube is not stretched and squamous when the organ or tube is stretched by fluid

Function:
Accommodates fluctuations in the volume of fluid in organs or tubes; protects against the caustic effects of urine

Location:
Lining of urinary bladder, ureters, superior urethra

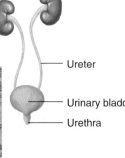

Ureter

Urinary bladder

Urethra

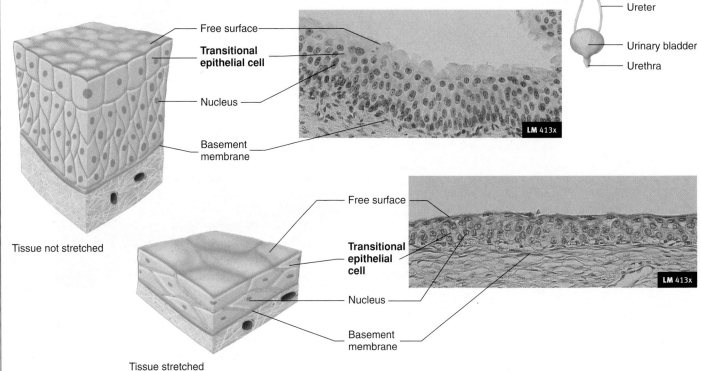

Free surface

Transitional epithelial cell

Nucleus

Basement membrane

Tissue not stretched

Free surface

Transitional epithelial cell

Nucleus

Basement membrane

LM 413x

Tissue stretched

epithelium is found in areas of the body where abrasion can occur, such as in the skin, mouth, throat, esophagus, anus, and vagina.

Differences in function are also reflected in cell shape. Cells are normally flat and thin when the function is diffusion, such as in the alveoli of the lungs. Cells with the major function of secretion or absorption are usually cuboidal or columnar. They are larger because they contain more organelles, which are responsible for the function of the cell. The stomach, for example, is lined with simple columnar epithelium. Some of the columnar cells are **goblet cells** (see table 4.1*c*), which secrete a clear, viscous material called **mucus** (mū′kŭs). The goblet cells contain abundant organelles responsible for the synthesis and secretion of mucus, such as ribosomes, endoplasmic reticulum, Golgi apparatuses, and secretory vesicles filled with mucus. The large amounts of mucus produced protects the stomach lining against the digestive enzymes and acid produced in the stomach. An ulcer, or irritation in the epithelium and underlying tissue, can develop if this protective mechanism fails. Simple cuboidal epithelial cells that secrete or absorb molecules, such as in the kidney tubules, contain many mitochondria, which produce the ATP required for active transport. ▼

> 12. **What functions would a single layer of epithelial cells be expected to perform? A stratified layer?**
> 13. **What cell shapes are typically found in locations where diffusion, secretion, or absorption are occurring?**

P R E D I C T ①
Explain the consequences of having (a) nonkeratinized stratified epithelium rather than simple columnar epithelium lining the digestive tract and (b) nonkeratinized stratified squamous epithelium rather than keratinized stratified squamous epithelium in the skin.

Free Cell Surfaces

Most epithelia have a free surface that is not in contact with other cells and faces away from underlying tissues. The characteristics of the free surface reflect the functions it performs. The free surface can be smooth, it can have microvilli or cilia, or it can be folded. **Smooth surfaces** reduce friction. For example, the lining of blood vessels is simple squamous epithelium with a smooth surface, which reduces friction as blood flows through the vessels.

Microvilli are cylindrical extensions of the cell membrane that increase the cell surface area (see chapter 3). Normally, many microvilli cover the free surface of each cell involved in absorption or secretion, such as the cells lining the small intestine.

Cilia (see chapter 3) propel materials along the surface of cells. The nasal cavity and trachea are lined with pseudostratified columnar ciliated epithelium. Intermixed with the ciliated cells are goblet cells (see table 4.1*d*). Dust and other materials are trapped in the mucus that covers the epithelium, and movement of the cilia propels the mucus with its entrapped particles to the back of the throat, where it is swallowed or coughed up. The constant movement of mucus helps keep the respiratory passages clean.

Transitional epithelium has a rather unusual plasma membrane specialization: More rigid sections of membrane are separated by very flexible regions in which the plasma membrane is folded. When transitional epithelium is stretched, the folded regions of the plasma membrane can unfold, like an accordion opening up. Transitional epithelium is specialized to expand. It is found in the urinary bladder, ureters, kidney pelvis, calyces of the kidney, and superior part of the urethra. ▼

> 14. **What is the function of an epithelial free surface that is smooth, has cilia, has microvilli, or is folded? Give an example of epithelium in which each surface type is found.**

Cell Connections

Epithelial cells are connected to one another in several ways (figure 4.2). **Tight junctions** bind adjacent cells together and form permeability barriers. Tight junctions prevent the passage of materials *between* epithelial cells because they completely surround each cell, similar to the way a belt surrounds the waist. Materials that pass through the epithelial layer must pass through the cells, which can regulate what materials cross the epithelial layer. Tight junctions are found in the lining of the intestines and most other simple epithelia. **Desmosomes** (dez′mō-sōmz, *desmos,* a band + *soma,* body) are mechanical links that bind cells together. Modified desmosomes, called **hemidesmosomes** (hem-ē-dez′mō-sōmz, *hemi,* one-half), also anchor cells to the basement membrane. Many desmosomes are found in epithelia subjected to stress, such as the stratified squamous epithelium of the skin. **Gap junctions** are small protein channels that allow small molecules and ions to pass from one epithelial cell to an adjacent one. Most epithelial cells are connected to one another by gap junctions, and it is believed that

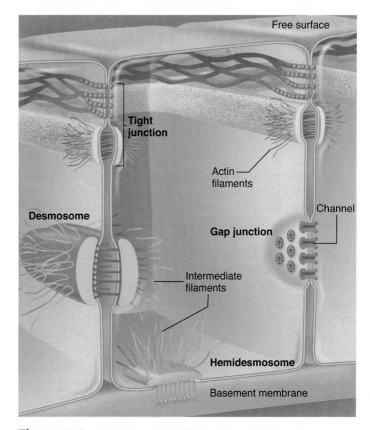

Figure 4.2   **Cell Connections**

Desmosomes anchor cells to one another and hemidesmosomes anchor cells to the basement membrane. Tight junctions bind adjacent cells together and prevent the passage of materials between the cells. Gap junctions allow adjacent cells to communicate with each other. Few cells have all of these different connections.

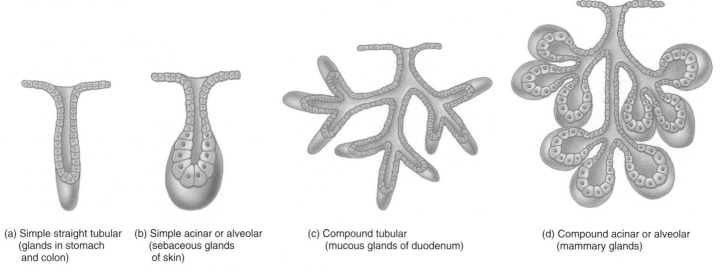

(a) Simple straight tubular (glands in stomach and colon)

(b) Simple acinar or alveolar (sebaceous glands of skin)

(c) Compound tubular (mucous glands of duodenum)

(d) Compound acinar or alveolar (mammary glands)

Figure 4.3   Structure of Exocrine Glands

The names of exocrine glands are based on the shapes of their secretory units and their ducts.

molecules or ions moving through the gap junctions act as communication signals to coordinate the activities of the cells. ▼

15. *Name the ways in which epithelial cells are bound to one another and to the basement membrane.*
16. *In addition to holding cells together, name an additional function of tight junctions. What is the general function of gap junctions?*

P R E D I C T   ②

If a simple epithelial type has well-developed tight junctions, explain how NaCl can be moved from one side of the epithelial layer to the other, what type of epithelium it is likely to be, and how the movement of NaCl causes water to move in the same direction.

Glands

A **gland** is a secretory structure. Most glands are multicellular, consisting of epithelium and a supporting network of connective tissue. Sometimes single goblet cells are classified as unicellular glands because they secrete mucus. Multicellular glands can be classified in several ways. **Endocrine** (en′dō-krin, *endo,* within) **glands** do not have ducts, and **exocrine** (ek′sō-krin, *exo,* outside +*krino,* to separate) **glands** do. Endocrine glands secrete hormones, which enter the blood and are carried to other parts of the body. Endocrine glands are so variable in their structure that they are not classified easily. They are described in chapter 15.

Exocrine glands release their secretions into ducts, which empty onto a surface or into a cavity. Examples of exocrine glands are sweat glands and mammary glands. The structure of the ducts is used to classify the glands (figure 4.3). Glands with one duct are called **simple,** and glands with ducts that branch repeatedly are called **compound.** Further classification is based on whether the ducts end in **tubules** (small tubes) or in saclike structures called **acini** (as′i-nī, grapes, suggesting a cluster of grapes or small sacs) or **alveoli** (al-vē′ō-lī, hollow sacs). Tubular glands can be classified as straight or coiled.

Exocrine glands can also be classified according to how products leave the cell. **Merocrine** (mer′ō-krin) **glands,** such as water-

producing sweat glands and the exocrine portion of the pancreas, secrete products with no loss of actual cellular material (figure 4.4a). Secretions are either actively transported or packaged in vesicles and then released by the process of exocytosis at the free surface of the cell. **Apocrine** (ap′ō-krin) **glands,** such as the milk-producing mammary glands, discharge fragments of the gland cells in the secretion (figure 4.4b). Products are retained within the cell, and portions of the cell are pinched off to become part of the secretion. Apocrine glands, however, also produce merocrine secretions. Most of the volume secreted by apocrine glands are merocrine in origin. **Holocrine** (hol′ō-krin) **glands,** such as sebaceous (oil) glands of the skin, shed entire cells (figure 4.4c). Products accumulate in the cytoplasm of each epithelial cell, the cell ruptures and dies, and the entire cell becomes part of the secretion. ▼

17. *Define gland. What is the difference between an endocrine and an exocrine gland?*
18. *Describe the classification scheme for multicellular exocrine glands on the basis of their duct systems.*
19. *Describe three ways in which exocrine glands release their secretions. Give an example of each method.*

Connective Tissue

Connective tissue is abundant, and it makes up part of every organ in the body. The major structural characteristic that distinguishes connective tissue from the other three tissue types is that it consists of cells separated from each other by abundant extracellular matrix.

Functions of Connective Tissue

Connective tissue structure is diverse, and it performs a variety of important functions:

1. *Enclosing and separating.* Sheets of connective tissue form capsules around organs, such as the liver and kidneys. Connective tissue also forms layers that separate tissues and organs. For example, connective tissues separate muscles, arteries, veins, and nerves from one another.

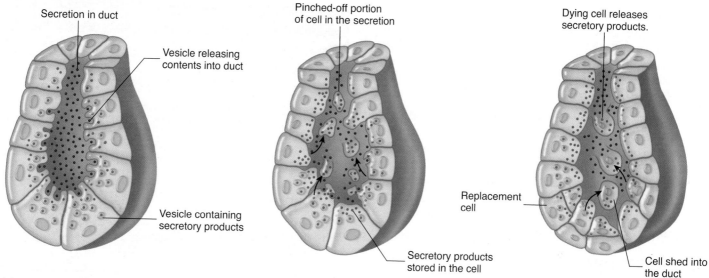

Figure 4.4 Exocrine Glands and Secretion Types

Exocrine glands are classified according to the type of secretion.

(a) Merocrine gland
Cells of the gland produce secretions by active transport or produce vesicles that contain secretory products, and the vesicles empty their contents into the duct through exocytosis.

(b) Apocrine gland
Secretory products are stored in the cell near the lumen of the duct. A portion of the cell near the lumen containing secretory products is pinched off the cell and joins secretions produced by a merocine process.

(c) Holocrine gland
Secretory products are stored in the cells of the gland. Entire cells are shed by the gland and become part of the secretion. The lost cells are replaced by other cells deeper in the gland.

2. *Connecting tissues to one another.* Tendons are strong cables, or bands, of connective tissue that attach muscles to bone, and ligaments are connective tissue bands that hold bones together.

3. *Supporting and moving.* Bones of the skeletal system provide rigid support for the body, and semirigid cartilage supports structures such as the nose, ears, and surfaces of joints. Joints between bones allow one part of the body to move relative to other parts.

4. *Storing.* Adipose tissue (fat) stores high-energy molecules, and bones store minerals, such as calcium and phosphate.

5. *Cushioning and insulating.* Adipose tissue (fat) cushions and protects the tissues it surrounds and provides an insulating layer beneath the skin that helps conserve heat.

6. *Transporting.* Blood transports substances throughout the body, such as gases, nutrients, enzymes, hormones, and cells of the immune system.

7. *Protecting.* Cells of the immune system and blood provide protection against toxins and tissue injury, as well as from microorganisms. Bones protect underlying structures from injury. ▼

20. *What is the major characteristic that distinguishes connective tissue from other tissues?*

21. *List the major functions of connective tissues, and give an example of a connective tissue that performs each function.*

Cells and Extracellular Matrix

The specialized cells of the various connective tissues produce the extracellular matrix. The names of the cells end with suffixes that identify the cell functions as blasts, cytes, or clasts. **Blast** (germ) cells produce the matrix, **cyte** (cell) cells maintain it, and **clast** (break) cells break it down for remodeling. For example, **fibroblasts** (fī′bro-blast, *fibra,* fiber) are cells that form fibers and ground substance in the extracellular matrix of fibrous connective tissue and fibrocytes are cells that maintain it. **Osteoblasts** (os′tē-ō-blasts, *osteo,* bone) form bone, **osteocytes** (os′tē-ō-sītz) maintain bone, and **osteoclasts** (os′tē-ō-klasts, broken) break down bone. Cells associated with the immune system are also found in connective tissue. **Macrophages** (mak′rō-fāj-ez, *makros,* large + *phago,* to eat) are large cells that are capable of moving about and ingesting foreign substances, including microorganisms found in the connective tissue. **Mast cells** are nonmotile cells that release chemicals, such as histamine, that promote inflammation.

The structure of the extracellular matrix gives connective tissue types most of their functional characteristics, such as the ability of bones and cartilage to bear weight, of tendons and ligaments to withstand tension, and of the skin's dermis to withstand punctures, abrasions, and other abuses. The extracellular matrix has three major components: (1) protein fibers, (2) ground substance consisting of nonfibrous protein and other molecules, and (3) fluid.

Three types of protein fibers help form most connective tissues. **Collagen** (kol′lă-jen, glue-producing) **fibers,** which resemble microscopic ropes, are flexible but resist stretching. **Reticular** (rē-tik′ū-lăr) **fibers** are very fine, short collagen fibers that branch to form a supporting network. **Elastic fibers** have a structure similar to that of coiled metal bed springs. After being stretched, elastic fibers can recoil to their original shape.

Ground substance is the shapeless background against which cells and collagen fibers are seen in the light microscope. Although ground substance appears shapeless, the molecules within the ground substance are highly structured. **Proteoglycans** (prō′tē-ō-glī′kanz, *proteo,* protein + *glycan,* polysaccharide) resemble the limbs of pine trees, with proteins forming the branches and polysaccharides forming the pine needles. This structure enables proteoglycans to hold large quantities of water between the polysaccharides. Like a water bed, tissues with large quantities of proteoglycans return to their original shape after being compressed or deformed. ▼

22. *Explain the differences among connective tissue cells that are termed blast, cyte, or clast cells.*
23. *What three components are found in the extracellular matrix of connective tissue?*
24. *Contrast the structure and characteristics of collagen, reticular, and elastin fibers.*
25. *What are proteoglycans? What characteristic is found in tissues with large quantities of proteoglycans?*

Classification of Connective Tissue

Embryonic connective tissue is called **mesenchyme** (mez′en-kīm). It is made up of irregularly shaped fibroblasts surrounded by abundant, semifluid extracellular matrix in which delicate collagen fibers are distributed. It forms in the embryo during the third and fourth weeks of development from mesoderm and neural crest cells (see chapter 25), and all adult connective tissue types develop from it. By 8 weeks of development, mesenchyme has become specialized to form the six major categories of connective tissue seen in adults (table 4.3). ▼

26. *What is mesenchyme? Where does it come from and what does it become?*

Loose Connective Tissue

Loose, or **areolar** (a-re′ō-lar, small areas), **connective tissue** has extracellular matrix consisting mostly of collagen and a few elastic fibers. These fibers are widely separated from one another

Table 4.3	Classification of Connective Tissues
Mesenchyme (embryonic connective tissue)	
Loose (areolar) connective tissue	

Dense connective tissue
  Dense, regular collagenous connective tissue
  Dense, regular elastic connective tissue
  Dense, irregular collagenous connective tissue
  Dense, irregular elastic connective tissue	

Special properties
  Adipose tissue
  Reticular tissue	

Cartilage
  Hyaline cartilage
  Fibrocartilage
  Elastic cartilage	

Bone
  Compact
  Cancellous	

Blood and hemopoietic tissue	

(table 4.4). The most common cells found in loose connective tissue are the fibroblasts, which are responsible for the production of the matrix. Loose connective tissue is widely distributed throughout the body and is the loose packing material of the body, filling the spaces between glands, muscles, and nerves. The basement membranes of epithelia often rest on loose connective tissue, and it attaches the skin to underlying tissues. ▼

27. *Describe the fiber arrangement in loose (areolar) connective tissue. What are the functions of this tissue type?*

Table 4.4 Loose Connective Tissue

Structure:
A fine network of fibers (mostly collagen fibers with a few elastic fibers) with spaces between the fibers; fibroblasts, macrophages, and lymphocytes are located in the spaces

Function:
Loose packing, support, and nourishment for the structures with which it is associated

Location:
Widely distributed throughout the body; substance on which epithelial basement membranes rest; packing between glands, muscles, and nerves; attaches the skin to underlying tissues

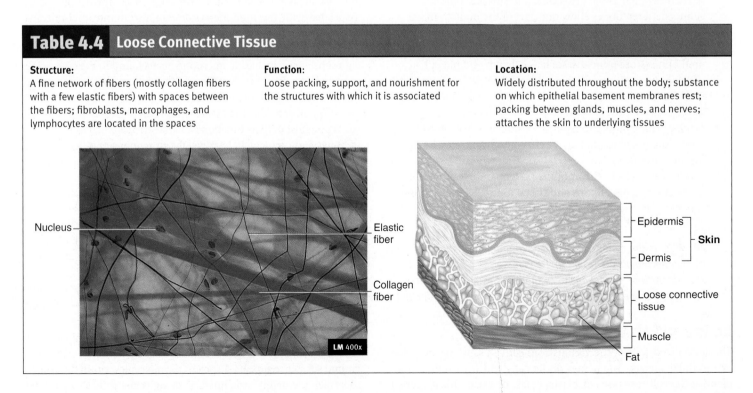

Dense Connective Tissue

Dense connective tissue has extracellular matrix that consists of densely packed fibers produced by fibroblasts. There are two major subcategories of dense connective tissue based on fiber orientation: regular and irregular. **Dense regular connective tissue** has protein fibers in the extracellular matrix that are oriented predominantly in one direction. **Dense regular collagenous connective tissue** (table 4.5a) has abundant collagen fibers. The collagen fibers resist stretching and give the tissue considerable strength in the direction of the fiber orientation. Dense regular collagenous

Table 4.5	Dense Connective Tissue

(a) Dense Regular Collagenous Connective Tissue

Structure:	Function:	Location:
Matrix composed of collagen fibers running in somewhat the same direction	Ability to withstand great pulling forces exerted in the direction of fiber orientation, great tensile strength and stretch resistance	Tendons (attach muscle to bone) and ligaments (attach bones to each other)

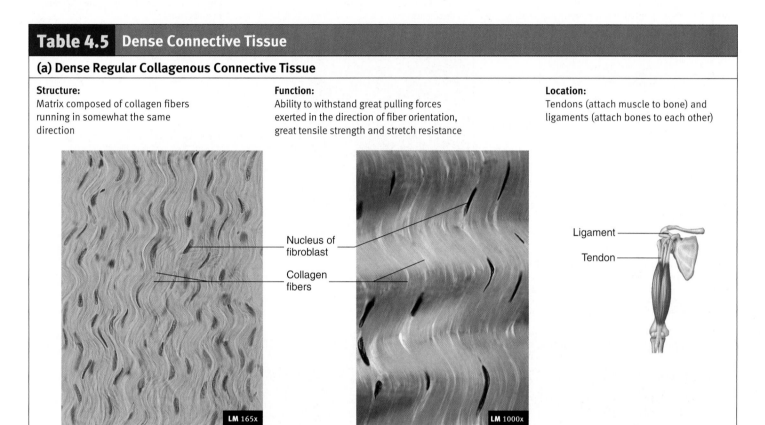

Nucleus of fibroblast

Collagen fibers

LM 165x

LM 1000x

Ligament

Tendon

(b) Dense Regular Elastic Connective Tissue

Structure:	Function:	Location:
Matrix composed of regularly arranged collagen fibers and elastin fibers	Capable of stretching and recoiling like a rubber band, with strength in the direction of fiber-orientation	Vocal folds and elastic ligaments between the vertebrae and along the dorsal aspect of the neck

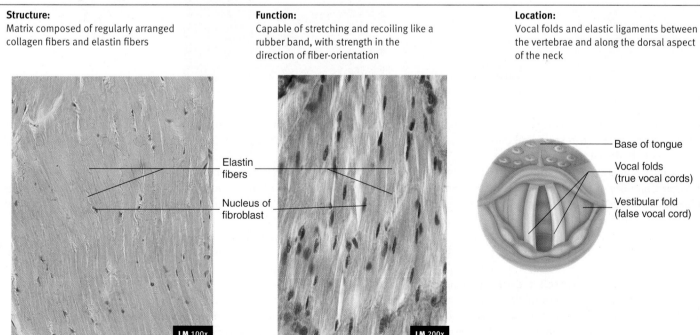

Elastin fibers

Nucleus of fibroblast

LM 100x

LM 200x

Base of tongue

Vocal folds (true vocal cords)

Vestibular fold (false vocal cord)

Table 4.5   Dense Connective Tissue—Continued

(c) Dense Irregular Collagenous Connective Tissue

Structure:
Matrix composed of collagen fibers that run in all -directions or in alternating planes of fibers oriented in a somewhat single direction

Function:
Tensile strength capable of withstanding stretching in all directions

Location:
Sheaths; most of the dermis of the skin; organ capsules and septa; outer covering of body tubes

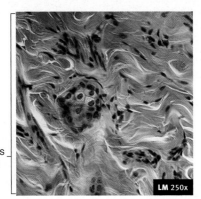

Epidermis

Dense irregular collagenous connective tissue of dermis

LM 100x

LM 250x

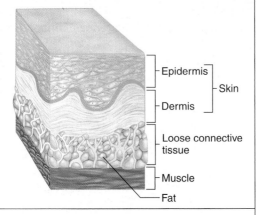

Epidermis ⎤
              ⎬ Skin
Dermis   ⎦

Loose connective tissue

Muscle

Fat

(d) Dense Irregular Elastic Connective Tissue

Structure:
Matrix composed of bundles and sheets of collagenous and elastin fibers oriented in multiple directions

Function:
Capable of strength with stretching and recoil in several directions

Location:
Elastic arteries

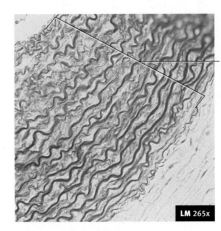

Dense irregular elastic connective tissue

LM 265x

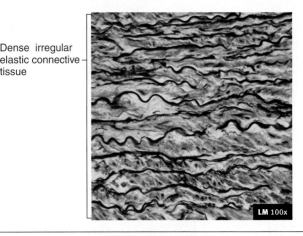

LM 100x

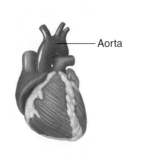

Aorta

---

connective tissue forms structures such as tendons, which connect muscles to bones (see chapter 9), and most ligaments, which connect bones to bones (see chapter 7).

Dense regular elastic connective tissue (table 4.5*b*) consists of parallel bundles of collagen fibers and abundant elastic fibers. When stretched, the tissue can shorten to its original length, much as an elastic band does. Dense regular elastic connective tissue forms some elastic ligaments, such as those in the vocal folds and the elastic ligaments holding the vertebrae (bones of the back) together.

Dense irregular connective tissue contains protein fibers arranged as a meshwork of randomly oriented fibers. Alternatively, the fibers within a given layer of dense irregular connective tissue can be oriented in one direction, whereas the fibers of adjacent layers are oriented at nearly right angles to that layer. Dense irregular

connective tissue forms sheets of connective tissue that have strength in many directions but less strength in any single direction than does regular connective tissue.

Dense irregular collagenous connective tissue (table 4.5*c*) forms most of the dermis of the skin, which is the tough, inner portion of the skin (see chapter 5) and of the connective tissue capsules that surround organs such as the kidney and spleen.

Dense irregular elastic connective tissue (table 4.5*d*) is found in the wall of elastic arteries. In addition to collagen fibers, oriented in many directions, there are abundant elastic fibers in the layers of this tissue. ▼

28. Structurally and functionally, what is the difference between dense regular connective tissue and dense irregular connective tissue?

29. *Name the two kinds of dense regular connective tissue, and give an example of each. Do the same for dense irregular connective tissue.*

P R E D I C T

In tendons, collagen fibers are oriented parallel to the length of the tendon. In the skin, collagen fibers are oriented in many directions. What are the functional advantages of the fiber arrangements in tendons and in the skin?

Connective Tissue with Special Properties

Adipose (ad′i-pōs, fat) **tissue** has an extracellular matrix with collagen and elastic fibers, but it is not a typical connective tissue because it has very little extracellular matrix. The individual fat cells, or **adipocytes** (ad′i-pō-sītz), are large and closely packed together (table 4.6a). Adipocytes are filled with lipids; they store energy. Adipose tissue under the skin also pads and protects parts of the body and acts as a thermal insulator. **Yellow bone marrow** is adipose tissue within cavities of bones.

| **Table 4.6** | **Connective Tissue with Special Properties** |

(a) Adipose Tissue

Structure:	Function:	Location:
Little extracellular matrix surrounding cells; the adipocytes, or fat cells, are so full of lipid that the cytoplasm is pushed to the periphery of the cell	Packing material, thermal insulator, energy storage, and protection of organs against injury from being bumped or jarred	Predominantly in subcutaneous areas, in yellow marrow in bones, in mesenteries, in renal pelvis, around kidneys, attached to the surface of the colon, in mammary glands, in loose connective tissue that penetrates into spaces and crevices

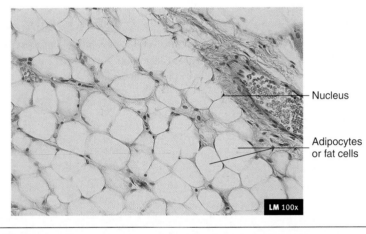
Nucleus

Adipocytes or fat cells

LM 100x

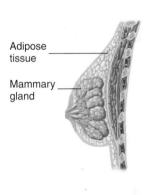
Adipose tissue

Mammary gland

(b) Reticular Tissue

Structure:	Function:	Location:
Fine network of reticular fibers irregularly arranged	Provides a superstructure for lymphatic and hemopoietic tissues	Within the lymph nodes, spleen, and bone marrow

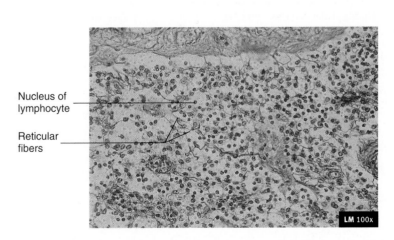
Nucleus of lymphocyte

Reticular fibers

LM 100x

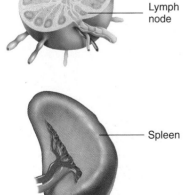
Lymph node

Spleen

Reticular tissue forms the framework of lymphatic tissue (table 4.6*b*), such as in the spleen and lymph nodes, as well as in bone marrow and the liver. It is characterized by a network of reticular fibers and reticular cells. **Reticular cells** produce the reticular fibers and remain closely attached to them. The spaces between the reticular fibers can contain a variety of other cells, such as dendritic cells, which are cells of the immune system, macrophages, and blood cells (see chapter 19). ▼

> 30. *What feature of the extracellular matrix distinguishes adipose tissue from other connective tissue types? What are the functions of adipose tissue?*
> 31. *What is the function of reticular tissue? Where is it found?*

Cartilage

Cartilage (kar′ti-lij, gristle) is composed of cartilage cells, or **chondrocytes** (kon′drō-sītz, cartilage cells), located in spaces called **lacunae** (lă-koo′nē, small spaces) within an extensive matrix (table 4.7*a*). Collagen in the matrix gives cartilage flexibility and strength. Cartilage is resilient because the proteoglycans of the matrix trap water, which makes the cartilage relatively rigid and enables it to spring back after being compressed. Cartilage provides support but, if bent or slightly compressed, it resumes its original shape. Cartilage heals slowly after an injury because blood vessels do not penetrate it. Thus, the cells and nutrients necessary for tissue repair do not easily reach the damaged area.

Hyaline (hī′ă-lin, clear or glassy) **cartilage** (see table 4.7*a*), the most abundant type of cartilage, has many functions. It covers the ends of bones where bones come together to form joints. In joints, hyaline cartilage forms smooth, resilient surfaces that can withstand repeated compression. Hyaline cartilage also forms the costal cartilages, which attach the ribs to the sternum (breastbone), the cartilage rings of the respiratory tract, and nasal cartilages.

Fibrocartilage (table 4.7*b*) has more collagen than does hyaline cartilage, and bundles of collagen fibers can be seen in the matrix. In addition to withstanding compression, it is able to resist pulling and tearing forces. It is found in the disks between vertebrae (bones of the back) and in some joints, such as the knee and temporomandibular (jaw) joints.

Elastic cartilage (table 4.7*c*) contains elastic fibers in addition to collagen and proteoglycans. The elastic fibers appear as coiled fibers among bundles of collagen fibers. Elastic cartilage is able to recoil to its original shape when bent. The external ear, epiglottis, and auditory tube contain elastic cartilage. ▼

> 32. *Describe the cells and matrix of cartilage. What are lacunae? Why does cartilage heal slowly?*
> 33. *How do hyaline cartilage, fibrocartilage, and elastic cartilage differ in structure and function? Give an example of each.*

P R E D I C T

One of several changes caused by rheumatoid arthritis in joints is the replacement of hyaline cartilage with dense irregular collagenous connective tissue. Predict the effect of replacing hyaline cartilage with fibrous connective tissue.

Bone

Bone is a hard connective tissue that consists of living cells and a mineralized matrix (table 4.8). Bone cells, or **osteocytes** (*osteo,*

Table 4.7 Connective Tissue: Cartilage

(a) Hyaline Cartilage

Structure:	Function:	Location:
Collagen fibers are small and evenly dispersed in the matrix, making the matrix appear transparent; the cartilage cells, or chondrocytes, are found in spaces, or lacunae, within the firm but flexible matrix	Allows growth of long bones; provides rigidity with some flexibility in the trachea, bronchi, ribs, and nose; forms rugged, smooth, yet somewhat flexible articulating surfaces; forms the embryonic skeleton	Growing long bones, cartiage rings of the respiratory system, costal cartilage of ribs, nasal cartilages, articulating surface of bones, embryonic skeleton

Chondrocyte in a lacuna

Nucleus

Matrix

LM 240x

Bone

Hyaline cartilage

Table 4.7 Connective Tissue: Cartilage—Continued

(b) Fibrocartilage

Structure:
Collagenous fibers similar to those in hyaline cartilage; the fibers are more numerous than in other cartilages and are arranged in thick bundles

Function:
Somewhat flexible and capable of withstanding considerable pressure; connects structures subjected to great pressure

Location:
Intervertebral disks, symphysis pubis, articular disks (e.g., knee and temporomandibular joints)

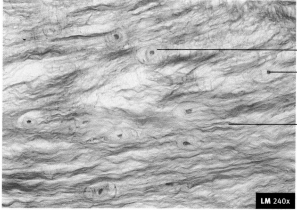

Chondrocyte in lacuna
Nucleus
Collagen fibers in matrix

LM 240x

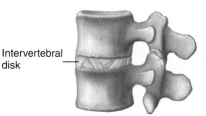

Intervertebral disk

(c) Elastic Cartilage

Structure:
Similar to hyaline cartilage, but matrix also contains elastic fibers

Function:
Provides rigidity with even more flexibility than hyaline cartilage because elastic fibers return to their original shape after being stretched

Location:
External ears, epiglottis, auditory tubes

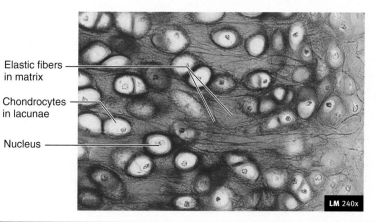

Elastic fibers in matrix
Chondrocytes in lacunae
Nucleus

LM 240x

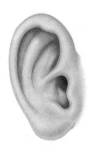

bone), are located within spaces in the matrix called lacunae. The strength and rigidity of the mineralized matrix enable bones to support and protect other tissues and organs of the body. **Compact bone** has more bone matrix than spaces. The bone matrix is organized into many thin layers of bone called **lamellae** (lă-mel′ē, sing. lă-mel′ă). **Cancellous** (kan′sē-lŭs), or **spongy, bone** has spaces between **trabeculae** (tră-bek′ū-lē, beams), or plates, of bone and therefore resembles a sponge. Compact and cancellous bone are considered in greater detail in chapter 6. ▼

34. *Describe the cells and matrix of bone. Differentiate between cancellous bone and compact bone.*

Blood and Hemopoietic Tissue

Blood is unique because the matrix is liquid, enabling blood cells to move through blood vessels (table 4.9). Some blood cells even leave the blood and wander into other tissues. The liquid matrix enables blood to flow rapidly through the body, carrying food, oxygen, waste products, and other materials. Blood is discussed more fully in chapter 16.

Hemopoietic (hē′mō-poy-et′ik) **tissue** produces red and white blood cells. Hemopoietic tissue is found as **red bone marrow,** a soft connective tissue in the cavities of bones. ▼

35. *What is the function of hemopoietic tissue? Where is it found?*

Table 4.8　Connective Tissue: Bone

Structure:
Hard, bony matrix predominates; spaces in the matrix contain osteocytes (not seen in this bone preparation); compact bone has more matrix than spaces and the matrix is organized into layers called lamellae; cancellous bone has more space than matrix

Function:
Provides great strength and support and protects internal organs; provides attachment site for muscles and ligaments; joints allow movements

Location:
Bones

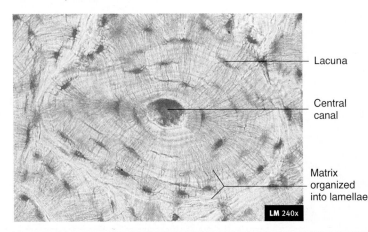

Lacuna

Central canal

Matrix organized into lamellae

LM 240x

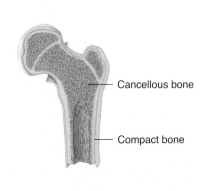

Cancellous bone

Compact bone

Table 4.9　Connective Tissue: Blood

Structure:
Blood cells and a fluid matrix

Function:
Transports oxygen, carbon dioxide, hormones, nutrients, waste products, and other substances; protects the body from infections and is involved in temperature regulation

Location:
Within blood vessels and heart; produced by the hemopoietic tissues (red bone marrow); white blood cells frequently leave the blood vessels and enter the interstitial spaces

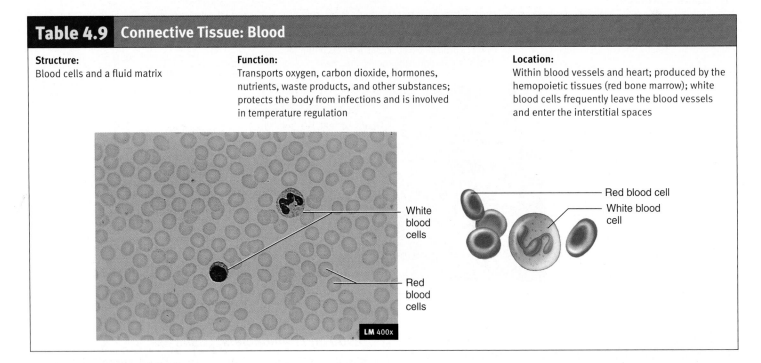

White blood cells

Red blood cells

Red blood cell

White blood cell

LM 400x

Muscle Tissue

The main characteristic of **muscle tissue** is its ability to contract, or shorten, making movement possible. Muscle contraction results from contractile proteins located within the muscle cells (see chapter 8). The length of muscle cells is greater than the diameter. Muscle cells are sometimes called **muscle fibers** because they often resemble tiny threads.

The three types of muscle tissue are skeletal, cardiac, and smooth muscle. **Skeletal muscle** is what normally is thought of as "muscle" (table 4.10*a*). It is the meat of animals and constitutes about 40% of a person's body weight. As the name implies, skeletal muscle attaches to the skeleton and enables body movement. Skeletal muscle is described as being under voluntary (conscious) control because one can purposefully cause skeletal muscle contraction to achieve specific body movements. However, the nervous system can cause skeletal muscles to contract without conscious involvement, such as during reflex movements and maintenance of muscle tone. Skeletal muscle cells tend to be long, cylindrical cells with several nuclei per cell.

Table 4.10　Muscle Tissue

(a) Skeletal Muscle

Structure:
Skeletal muscle cells or fibers appear striated (banded); cells are large, long, and cylindrical, with many nuclei located at the periphery

Function:
Movement of the body; under voluntary (conscious) control

Location:
Attaches to bone or other connective tissue

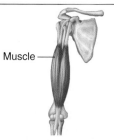

Muscle

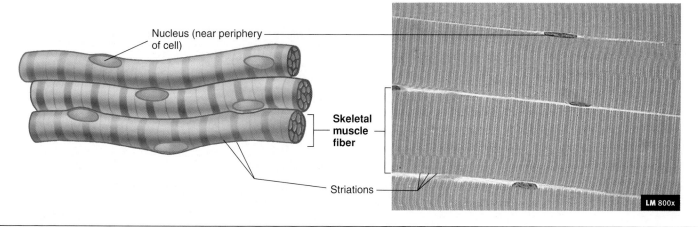

Nucleus (near periphery of cell)

Skeletal muscle fiber

Striations

LM 800x

(b) Cardiac Muscle

Structure:
Cardiac muscle cells are cylindrical and striated and have a single, centrally located nucleus; they are branched and connected to one another by intercalated disks, which contain gap junctions

Function:
Pumps the blood; under involuntary (unconscious) control

Location:
In the heart

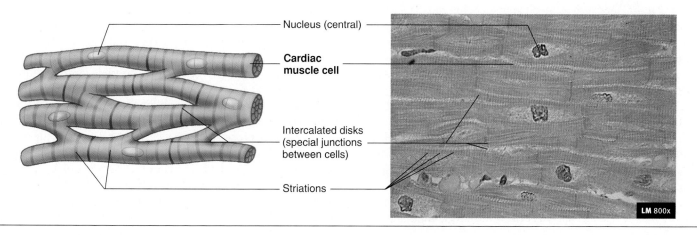

Nucleus (central)

Cardiac muscle cell

Intercalated disks (special junctions between cells)

Striations

LM 800x

Table 4.10 Muscle Tissue—Continued

(c) Smooth Muscle

Structure:
Smooth muscle cells are tapered at each end, are not striated, and have a single nucleus

Function:
Regulates the size of organs, forces fluid through tubes, controls the amount of light entering the eye, and produces "goose flesh" in the skin; under involuntary (unconscious) control

Location:
In hollow organs, such as the stomach and intestine

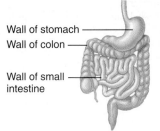

Wall of stomach
Wall of colon
Wall of small intestine

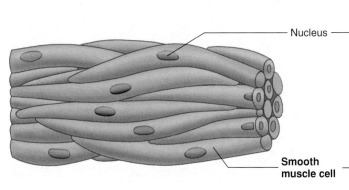

Nucleus

Smooth muscle cell

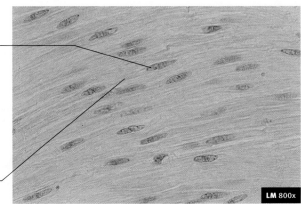

LM 800x

The nuclei of these cells are located near the periphery of the cell. Some skeletal muscle cells extend the length of an entire muscle. Skeletal muscle cells are **striated** (strī′āt-ed), or banded, because of the arrangement of contractile proteins within the cells (see chapter 8).

Cardiac muscle is the muscle of the heart; it is responsible for pumping blood (table 4.10*b*). It is under involuntary (unconscious) control, although one can learn to influence the heart rate by using techniques such as meditation and biofeedback. Cardiac muscle cells are cylindrical in shape but much shorter in length than skeletal muscle cells. Cardiac muscle cells are striated and usually have one nucleus per cell. They often are branched and connected to one another by **intercalated** (in-ter′kă-lā-ted, inserted between) **disks.** The intercalated disks, which contain specialized gap junctions, are important in coordinating the contractions of the cardiac muscle cells (see chapter 12).

Smooth muscle forms the walls of hollow organs (except the heart) and also is found in the skin and the eyes (table 4.10*c*). It is responsible for a number of functions, such as movement of food through the digestive tract and emptying of the urinary bladder. Like cardiac muscle, smooth muscle is controlled involuntarily. Smooth muscle cells are tapered at each end, have a single nucleus, and are not striated. ▼

36. *Functionally, what is unique about muscle tissue?*
37. *Compare the structure of skeletal, cardiac, and smooth muscle cells.*
38. *Which of the muscle types is under voluntary control?*
39. *What tasks does each muscle type perform?*

Nervous Tissue

Nervous tissue forms the brain, spinal cord, and nerves. It is responsible for coordinating and controlling many bodily activities. For example, the conscious control of skeletal muscles and the unconscious regulation of cardiac muscle are accomplished by nervous tissue. Awareness of ourselves and the external environment, emotions, reasoning skills, and memory are other functions performed by nervous tissue. Many of these functions depend on the ability of nervous tissue cells to communicate with one another and with the cells of other tissues by electric signals called **action potentials.**

Nervous tissue consists of neurons and support cells. The **neuron** (noor′on), or **nerve cell,** is responsible for the conduction of action potentials. It is composed of three parts (table 4.11). The **cell body** contains the nucleus and is the site of general cell functions. **Dendrites** (den′drītz, relating to a tree) and **axons** (ak′sonz) are nerve cell processes (extensions). Dendrites usually receive stimuli that lead to electric changes that either increase or decrease action potentials in the neuron's axon. There is usually one axon per neuron. Action potentials usually originate at the base of an axon where it joins the cell body and travel to the end of the axon. **Neuroglia** (noo-rog′lē-ă, *glia*, glue) are the support cells of the nervous system; they nourish, protect, and insulate the neurons. Nervous tissue is considered in greater detail in chapter 10. ▼

40. *What are action potentials?*
41. *Define and list the functions of the cell body, dendrites, and axons of a neuron.*
42. *What is the general function of neuroglia?*

Table 4.11   Neurons and Neuroglia

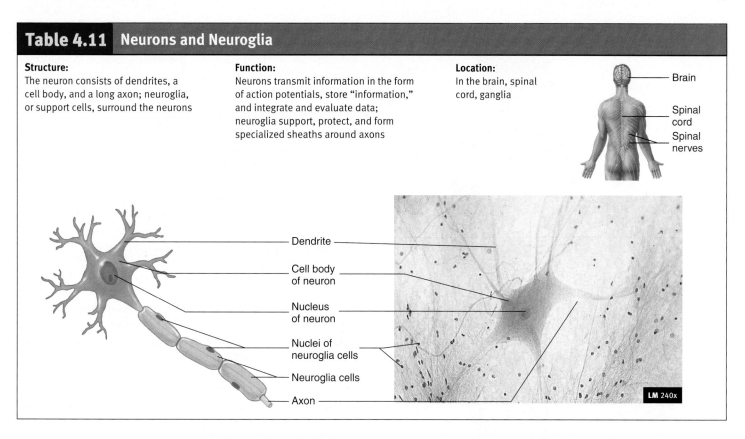

Structure:
The neuron consists of dendrites, a cell body, and a long axon; neuroglia, or support cells, surround the neurons

Function:
Neurons transmit information in the form of action potentials, store "information," and integrate and evaluate data; neuroglia support, protect, and form specialized sheaths around axons

Location:
In the brain, spinal cord, ganglia

Brain

Spinal cord

Spinal nerves

Dendrite

Cell body of neuron

Nucleus of neuron

Nuclei of neuroglia cells

Neuroglia cells

Axon

LM 240x

Membranes

A **membrane** is a thin sheet or layer of tissue that covers a structure or lines a cavity. Most membranes are formed from epithelium and the connective tissue on which it rests. The skin, or cutaneous, membrane (see chapter 5) is the external membrane. The three major categories of internal membranes are mucous membranes, serous membranes, and synovial membranes.

Mucous (mū′kŭs) **membranes** consist of various kinds of epithelium resting on a thick layer of loose connective tissue. They line cavities that open to the outside of the body, such as the digestive, respiratory, urinary, and reproductive tracts (figure 4.5). Many mucous membranes have goblet cells or mucous glands, which secrete mucus. The functions of mucous membranes vary, depending on their location, and include protection, absorption, and secretion.

A **serous** (sēr′ŭs, produces watery secretion) **membrane** consists of simple squamous epithelium resting on a delicate layer of loose connective tissue. Serous membranes line the pleural, peritoneal, and pericardial cavities within the trunk (see chapter 1 and figure 4.5). Although serous membranes do not contain glands, the epithelium of the serous membranes produce **serous fluid,** which lubricates the surfaces of serous membranes, making them slippery. Serous membranes protect the internal organs from friction and help hold them in place.

A **synovial** (si-nō′vē-ăl) **membrane** consists of modified connective tissue cells, either intermixed with part of the dense connective tissue of the joint capsule or separated from the capsule by areolar or adipose tissue. Synovial membranes line freely movable joints (see chapter 7). They produce **synovial fluid,** which lubricates the joint for smooth joint movement. ▼

43. *Compare mucous, serous, and synovial membranes according to the type of cavities they line and their secretions.*

Inflammation

The **inflammatory response,** or **inflammation** (*flamma*, flame) is a complex process involving cells and chemicals that occurs in response to tissue damage caused by physical, chemical, or biological agents. For example, a splinter in the skin or an infection can stimulate an inflammatory response (figure 4.6). **Mediators of inflammation** are chemicals, released or activated in injured or infected tissues, that promote inflammation. The mediators include **histamine** (his′tă-mēn), **kinins** (kī′ninz, to move), **prostaglandins** (pros-tă-glan′dinz), **leukotrienes** (lū-kō-trī′ēnz), and others.

Inflammation mobilizes the body's defenses and isolates and destroys microorganisms, foreign materials, and damaged cells so that tissue repair can proceed. Inflammation produces five major symptoms: redness, heat, swelling, pain, and disturbance of function. Although unpleasant, the processes producing the symptoms are usually beneficial. Redness and heat, similar to what occurs when a person blushes, result from the dilation of blood vessels induced by some mediators of inflammation. Dilation of blood vessels is beneficial because it increases the speed with which blood cells and other substances important for fighting infections and repairing the injury are brought to the injury site.

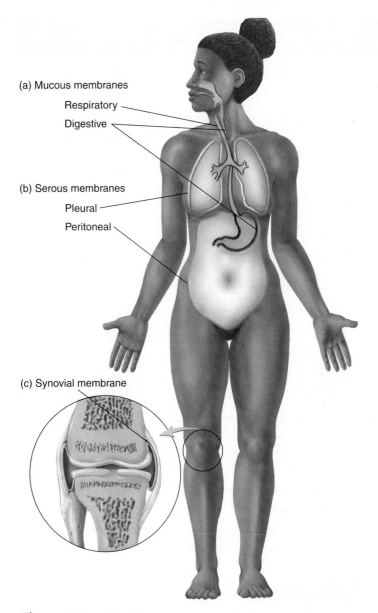

(a) Mucous membranes
   Respiratory
   Digestive

(b) Serous membranes
   Pleural
   Peritoneal

(c) Synovial membrane

Figure 4.5 **Membranes**

(*a*) Mucous membranes line cavities that open to the outside and often contain mucous glands, which secrete mucus. (*b*) Serous membranes line cavities that do not open to the exterior and do not contain glands but do secrete serous fluid. (*c*) Synovial membranes line cavities that surround synovial joints.

Swelling, or **edema** (e-dē′mǎ), is induced by mediators of inflammation that increase the permeability of blood vessels, allowing materials and blood cells to move out of the vessels and into the tissue, where they can deal directly with the injury. As proteins and other substances enter the tissues, water follows by osmosis, causing edema. **Neutrophils** (noo′trō-filz), a type of white blood cell, and macrophages are phagocytic cells that move to sites of infection, where they ingest bacteria and tissue debris.

Pain associated with inflammation is produced in several ways. Nerve cell endings are stimulated by direct damage and by some mediators of inflammation to produce pain sensations. In addition, the increased pressure in the tissue caused by edema and the accumulation of pus can cause pain.

Disturbance of function is caused by pain, limitation of movement resulting from edema, and tissue destruction. Disturbance of function can be adaptive because it warns the person to protect the injured area from further damage.

 Reducing Inflammation

Sometimes the inflammatory response lasts longer or is more intense than is desirable, and drugs are used to suppress the symptoms by inhibiting the synthesis, release, or actions of the mediators of inflammation. For example, the effects of histamine released in people with hay fever are suppressed by antihistamines. Aspirin and related drugs, such as ibuprofen and naproxen, are effective anti-inflammatory agents that relieve pain by preventing the synthesis of prostaglandins and related substances.

Chronic, or prolonged, inflammation results when the agent responsible for an injury is not removed or something else interferes with the process of healing. Infections of the lungs or kidneys usually result in a brief period of inflammation followed by repair. However, prolonged infections, or prolonged exposure to irritants, can result in chronic inflammation. Chronic inflammation caused by irritants, such as silica in the lungs, or abnormal immune responses can result in the replacement of normal tissue with fibrous connective tissue. Chronic inflammation of the stomach or small intestine may result in ulcers. The loss of normal tissue leads to the loss of normal organ functions. Consequently, chronic inflammation of organs such as the lungs, liver, or kidneys can lead to death. ▼

44. *What is the function of the inflammatory response?*
45. *Name five manifestations of the inflammatory response, and explain how each is produced.*
46. *What is chronic inflammation? Why is it bad?*

P R E D I C T   5

In some injuries, tissues are so severely damaged that areas exist where cells are killed and blood vessels are destroyed. For injuries such as these, where do the signs of inflammation, such as redness, heat, edema, and pain, occur?

Tissue Repair

Tissue repair is the substitution of viable cells for dead cells. It can occur by regeneration or replacement. In **regeneration,** the new cells are the same type as those that were destroyed, and normal function is usually restored. In **replacement,** a new type of tissue develops that eventually causes scar production and the loss of some tissue function.

Cells can be classified into three groups on the basis of their ability to divide and produce new cells. **Labile cells** continue to divide throughout life. Cells of the skin, mucous membranes, hemopoietic tissues, and lymphatic tissues are labile cells. Damage to labile cells can be repaired completely by regeneration. **Stable cells** do not actively divide after growth ceases, but they do retain the ability to divide after an injury. For example, connective tissue and glands, including the liver and pancreas, are capable of

1. A splinter in the skin causes damage and introduces bacteria. Mediators of inflammation are released or activated in injured tissues and adjacent blood vessels. Some blood vessels are ruptured, causing bleeding.

2. Mediators of inflammation cause blood vessels to dilate, causing the skin to become red. Mediators of inflammation also increase blood vessel permeability, and fluid leaves the blood vessels, producing swelling (*arrows*).

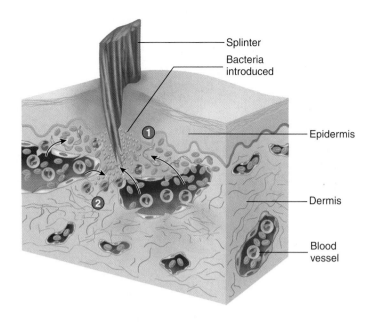

3. White blood cells (e.g., neutrophils and macrophages) leave the dilated blood vessels and move to the site of bacterial infection, where they begin to phagocytize bacteria and other debris.

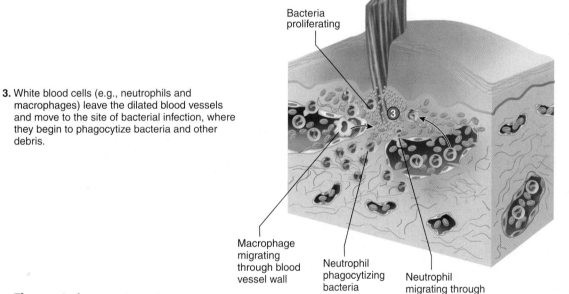

Process Figure 4.6   Inflammation

regeneration. Smooth muscle cells also readily regenerate following injury. **Permanent cells** have little or no ability to divide. Neurons and skeletal muscle cells are examples of permanent cells. If they are killed, they are usually replaced by connective tissue. Permanent cells, however, can usually recover from a limited amount of damage. For example, if the axon of a neuron is damaged, the neuron can grow a new axon. If the cell body is sufficiently damaged, however, the neuron dies and is replaced by connective tissue.

Some undifferentiated cells of the central nervous system can undergo mitosis and form functional neurons, although the degree to which mitosis occurs and its functional significance is not clear. Undifferentiated cells of skeletal and cardiac muscle also have very limited ability to regenerate in response to injury.

In addition to the type of cells involved, the severity of an injury can influence whether repair is by regeneration or replacement. Generally, the more severe the injury, the greater the likelihood that repair involves replacement.

Repair of the skin is an illustration of tissue repair (figure 4.7). When the edges of a wound are close together, the wound fills with blood, and a clot forms (see chapter 16). The **clot** contains a threadlike protein, fibrin, which binds the edges of the wound together and stops the bleeding. The surface of the clot dries to form a **scab,** which seals the wound and helps prevent infection.

An inflammatory response is activated to fight infectious agents in the wound and to help the repair process. Dilation of blood vessels brings blood cells and other substances to the injury area, and increased blood vessel permeability allows them to enter

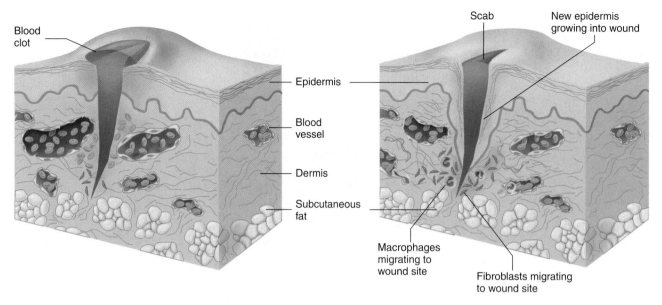

1. A fresh wound cuts through the epithelium (epidermis) and underlying connective tissue (dermis), and a clot forms.

2. Approximately 1 week after the injury, a scab is present, and epithelium (new epidermis) is growing into the wound.

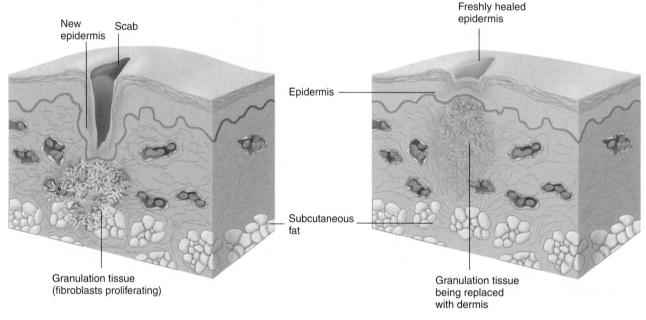

3. Approximately 2 weeks after the injury, the epithelium has grown completely into the wound, and granulation tissue has formed.

4. Approximately 1 month after the injury, the wound has completely closed, the scab has been sloughed, and the granulation tissue is being replaced with dermis.

Process Figure 4.7   Tissue Repair

the tissue. Neutrophils help fight infections by phagocytizing bacteria (see figure 4.6). **Pus** (pŭs) is an accumulation of dead neutrophils, other cells, and fluid.

The epithelium at the edge of the wound undergoes regeneration and migrates under the scab while the inflammatory response proceeds. Eventually, the epithelial cells from the edges meet, and the epithelium is restored. After the epithelium is repaired, the scab is sloughed off (shed).

Macrophages migrate to the injured tissue and remove dead neutrophils, cellular debris, and the decomposing clot.

Fibroblasts from the surrounding connective tissue migrate into the area, producing collagen and other extracellular matrix components. Capillaries grow from blood vessels at the edge of the wound and revascularize the area. The result is the replacement of the clot by a delicate connective tissue called **granulation (a**ppears granular**) tissue,** which consists of fibroblasts, collagen, and capillaries. Eventually, normal connective tissue replaces the granulation tissue. Sometimes a large amount of granulation tissue persists as a scar, which at first is bright red because of the vascularization of the tissue. The scar turns from

Cancer (kan′ser) is a malignant, spreading tumor, as well as the illness that results from it. A **tumor** (too′mŏr) is any swelling, although modern usage has limited the term to swellings that involve neoplastic tissue. **Neoplasm** (nē′ō-plazm) means new growth and refers to abnormal tissue growth, resulting in rapid cellular proliferation, which continues after the growth of normal tissue has stopped or slowed considerably. **Oncology** (ong-kol′ō-jē) is the study of tumors and the problems they cause. A neoplasm can be either **benign** (bē-nīn′, L. kind), not inclined to spread and not likely to become worse, or **malignant** (ma-lig′nănt, with malice or intent to cause harm), able to spread and become worse. Although benign tumors are usually less dangerous than malignant tumors, they can cause problems. As a benign tumor enlarges, it can compress surrounding tissues and impair their functions. In some cases, such as in some benign brain tumors, the result can be death.

Cells of malignant neoplasms, or cancer cells, differ from cells of normal tissues in several ways; the greater the degree to which they differ, the more dangerous they are. Cancer cells are more spherical because they do not adhere tightly to surrounding normal cells. They appear to be more embryonic, or less mature, than the normal tissue from which they arise. For example, a skin cancer cell is more spherical and softer than the stratified squamous epithelial cells of the skin. Cancer cells are also invasive. That is, they have the ability to squeeze into spaces and enter surrounding tissues. They secrete enzymes that cut paths through healthy tissue, so they are able to grow irregularly, sending processes in all directions. Cancer cells can dislodge; enter blood vessels, lymphatic vessels, or body cavities; and travel to distant sites, where they attach and invade tissues. The process by which cancer spreads to distant sites is called **metastasis** (me-tas′ta-sis). Cancer cells secrete substances that cause blood vessels to grow into the tumor and supply oxygen and nutrients. Cancer cells also produce a number of substances that can be found on their plasma membranes or are secreted. For example, prostate-specific antigen (PSA) is an enzyme, produced by prostate gland cells, that is involved with the liquefaction of semen. Normal blood levels of PSA are low. Prostate cancer cells secrete PSA in increasing amounts, and these proteins are released into the blood. Therefore, blood levels of PSA can be monitored to determine if a person is likely to have prostate cancer.

There are many types of cancer and special terms to name them. For example, a **carcinoma** (kar-si-nō′ma) is a cancer derived from epithelial tissue. **Basal cell** and **squamous cell carcinomas** are types of skin cancer derived from epithelial tissue. **Adenocarcinomas** (ad′ĕ-nō-kar-si-nō′maz) are derived from glandular epithelium. Most breast cancers are adenocarcinomas. A **sarcoma** (sar-kō′mă) is cancer derived from connective tissue. For example, an **osteosarcoma** (os′tē-ō-sar-kō′mă) is cancer of bone and a **chondrosarcoma** (kon′drō-sar-kō′mă) is cancer of cartilage.

Cancer therapy concentrates primarily on trying to confine and then kill the malignant cells. This goal is accomplished currently by killing the tissue with x-rays or lasers, by removing the tumor surgically, or by treating the patient with drugs that kill rapidly dividing cells or reduce the blood supply to the tumor. The major problem with current therapy is that some cancers cannot be removed completely by surgery or killed completely by x-rays or laser therapy. These treatments can also kill normal tissue adjacent to the tumor. Many drugs used in cancer therapy kill not only cancer tissue but also other rapidly growing tissues, such as bone marrow, where new blood cells are produced, and the lining of the intestinal tract. Loss of these tissues can result in anemia, caused by the lack of red blood cells, and nausea, caused by the loss of the intestinal lining.

A newer class of drugs eliminates these unwanted side effects. These drugs prevent blood vessel development, thus depriving the cancer tissue of a blood supply, rather than attacking dividing cells. Other normal tissues, in which cells divide rapidly, have well-established blood vessels and are therefore not affected by these drugs.

Promising anticancer therapies are also being developed in which the cells responsible for immune responses can be stimulated to recognize tumor cells and destroy them. A major advantage in such anticancer treatments is that the cells of the immune system can specifically attack the tumor cells and not other, healthy tissues.

---

red to white as collagen accumulates and the blood vessels decrease in number.

When the wound edges are far apart, the clot may not completely close the gap, and it takes much longer for the epithelial cells to regenerate and cover the wound. With increased tissue damage, the degree of the inflammatory response is greater, there is more cell debris for the phagocytes to remove, and the risk of infection is greater. Much more granulation tissue forms, and **wound contracture,** a result of the contraction of fibroblasts in the granulation tissue, pulls the edges of the wound closer together. Although wound contracture reduces the size of the wound and speeds healing, it can lead to disfiguring and debilitating scars. ▼

47. *Define tissue repair. Differentiate between tissue repair that occurs by regeneration and by replacement.*

48. *Compare labile cells, stable cells, and permanent cells. Give examples of each type. What is the significance of these cell types to tissue repair?*

49. *Describe the process of wound repair.*

50. *What are pus and granulation tissue? How does granulation tissue contribute to scars and wound contraction?*

P R E D I C T

Explain why it is advisable to suture large wounds.

Tissues and Aging

At the tissue level, age-related changes affect cells and the extracellular matrix produced by them. In general, cells divide more slowly in older than in younger people. The rate of red blood cell synthesis declines in the elderly. Injuries in the very young heal more rapidly and more completely than in older people, in part, because of the more rapid cell division. For example, a fracture in the femur of an infant is likely to heal quickly and eventually leave no evidence of the fracture in the bone. A similar fracture in an adult heals more slowly, and a scar, seen in radiographs of the bone, is likely to persist throughout life.

The consequences of changes in the extracellular matrix are important. Collagen fibers become more irregular in structure, even though they may increase in number. As a consequence, connective tissues with abundant collagen, such as tendons and ligaments, become less flexible and more fragile. Elastic fibers fragment, bind to Ca^{2+}, and become less elastic. Consequently, elastic connective tissues, such as elastic ligaments, become less elastic.

Reduced flexibility and elasticity of connective tissue is responsible for increased wrinkling of the skin, as well as an increased tendency for bones to break in older people.

The walls of arteries become less elastic because of changes in collagen and elastic fibers. Atherosclerosis results as plaques form in the walls of blood vessels, which contain collagen fibers, lipids, and calcium deposits. These changes result in reduced blood supply to tissues and increased susceptibility to blockage and rupture of arteries.

51. *What happens to the rate of cell division with age, and how does this affect tissue repair?*
52. *Describe the age-related changes in tissues with abundant collagen and elastic fibers.*

S U M M A R Y

Tissues and Histology (p. 78)

1. A tissue is a group of similar cells and the extracellular matrix surrounding them.
2. Histology is the study of tissues.
3. The four primary tissue types are epithelial, connective, muscle, and nervous tissue.

Embryonic Tissue (p. 78)

The primary tissue types are derived from the embryonic germ layers (endoderm, mesoderm, and ectoderm).

Epithelial Tissue (p. 78)

1. Epithelial tissue covers surfaces, has little extracellular material, usually has a basement membrane, and has no blood vessels.
2. Epithelial cells have a free, or apical, surface (not attached to other cells), a lateral surface (attached to other cells), and a basal surface (attached to the basement membrane).

Functions of Epithelia

General functions of epithelia include protecting underlying structures, acting as a barrier, permitting the passage of substances, secreting substances, and absorbing substances.

Classification of Epithelia

1. Epithelia are classified according to the number of cell layers and the shape of the cells.
2. Simple epithelium has one layer of cells, whereas stratified epithelium has more than one.
3. Pseudostratified columnar epithelium is simple epithelium that appears to have two or more cell layers.
4. Transitional epithelium is stratified epithelium that can be greatly stretched.

Structural and Functional Relationships

1. Simple epithelium is involved with diffusion, secretion, or absorption. Stratified epithelium serves a protective role. Squamous cells function in diffusion or filtration. Cuboidal or columnar cells, which contain more organelles, secrete or absorb.
2. A smooth, free surface reduces friction. Microvilli increase surface area, and cilia move materials over the cell surface.
3. Tight junctions bind adjacent cells together and form a permeability barrier.

4. Desmosomes mechanically bind cells together and hemidesmosomes mechanically bind cells to the basement membrane.
5. Gap junctions allow intercellular communication.

Glands

1. A gland is a single cell or a multicellular structure that secretes.
2. Exocrine glands have ducts, and endocrine glands do not.
3. Exocrine glands are classified by their ducts as simple versus compound and tubular versus acinar or alveolar.
4. Exocrine glands are classified by their mode of secretion as merocrine (no loss of cellular material), apocrine (part of the cell pinches off), and holocrine (entire cell is shed).

Connective Tissue (p. 86)
Functions of Connective Tissue

Connective tissues enclose and separate; connect tissues to one another; play a role in support and movement; and store, cushion, insulate, transport, and protect.

Cells and Extracellular Matrix

1. Blast cells form the matrix, cyte cells maintain it, and clast cells break it down.
2. Connective tissue has an extracellular matrix consisting of protein fibers, ground substance, and fluid.
3. Collagen fibers are flexible but resist stretching, reticular fibers form a fiber network, and elastic fibers recoil.
4. Proteoglycans in ground substance holds water, enabling connective tissues to return to their original shape after being compressed.

Classification of Connective Tissue

1. Mesenchyme is embryonic connective tissue that gives rise to six major categories of connective tissue.
2. Loose, or areolar, connective tissue is the "loose packing" material of the body, which fills the spaces between organs and holds them in place.
3. Adipose, or fat, tissue stores energy. Adipose tissue also pads and protects parts of the body and acts as a thermal insulator.
4. Dense connective tissue consists of matrix containing densely packed collagen fibers (tendons, ligaments, and dermis of the skin) or of matrix containing densely packed elastic fibers (elastic ligaments and in the walls of arteries).

5. Cartilage provides support and is found as hyaline cartilage (covers ends of bones and forms costal cartilages), fibrocartilage (disks between vertebrae), and elastic cartilage (external ear).
6. Bone has a mineralized matrix and forms most of the skeleton of the body. Compact bone has more matrix than spaces, whereas cancellous bone has more spaces than matrix.
7. Blood has a liquid matrix and is found in blood vessels. It is produced in hemopoietic tissue (red bone marrow).

Muscle Tissue (p. 94)

1. Muscle tissue is specialized to shorten, or contract.
2. The three types of muscle tissue are skeletal, cardiac, and smooth muscle.

Nervous Tissue (p. 96)

1. Nervous tissue is specialized to conduct action potentials (electric signals).
2. Neurons conduct action potentials, and neuroglia support the neurons.

Membranes (p. 97)

1. Mucous membranes line cavities that open to the outside of the body (digestive, respiratory, urinary, and reproductive tracts). They contain glands and secrete mucus.
2. Serous membranes line trunk cavities that do not open to the outside of the body (pleural, pericardial, and peritoneal cavities). They do not contain glands but do secrete serous fluid.
3. Synovial membranes line freely movable joints.

Inflammation (p. 97)

1. The function of the inflammatory response is to isolate and destroy harmful agents.
2. The inflammatory response produces five symptoms: redness, heat, swelling, pain, and disturbance of function.
3. Chronic inflammation results when the agent causing injury is not removed or something else interferes with the healing process.

Tissue Repair (p. 98)

1. Tissue repair is the substitution of viable cells for dead cells. Labile cells divide throughout life and can undergo regeneration. Stable cells do not ordinarily divide but can regenerate if necessary. Permanent cells have little or no ability to divide. If permanent cells are killed, repair is by replacement.
2. Tissue repair involves clot formation, inflammation, the formation of granulation tissue, and the regeneration or replacement of tissues. In severe wounds, wound contracture can occur.

Tissues and Aging (p. 101)

1. Cells divide more slowly as people age. Injuries heal more slowly.
2. Extracellular matrix containing collagen and elastic fibers become less flexible and less elastic. Consequently, skin wrinkles, elasticity in arteries is reduced, and bones break more easily.

R E V I E W   A N D   C O M P R E H E N S I O N

1. Epithelial tissue
   a. covers free body surfaces.
   b. lacks blood vessels.
   c. composes various glands.
   d. is anchored to connective tissue by a basement membrane.
   e. all of the above.

2. Which of these embryonic germ layers gives rise to muscle, bone, and blood vessels?
   a. ectoderm   b. endoderm   c. mesoderm

3. A tissue that covers a surface, is one cell layer thick, and is composed of flat cells is
   a. simple squamous epithelium.
   b. simple cuboidal epithelium.
   c. simple columnar epithelium.
   d. stratified squamous epithelium.
   e. transitional epithelium.

4. Epithelium composed of two or more layers of cells with only the deepest layer in contact with the basement membrane is
   a. stratified epithelium.   d. columnar epithelium.
   b. simple epithelium.   e. cuboidal epithelium.
   c. pseudostratified epithelium.

5. Simple squamous epithelium most likely
   a. performs phagocytosis.
   b. is involved with active transport.
   c. secretes many complex lipids and proteins.
   d. allows certain substances to diffuse across it.

6. Pseudostratified ciliated columnar epithelium can be found lining the
   a. digestive tract.
   b. respiratory passages (e.g., trachea).
   c. thyroid gland.
   d. kidney tubules.
   e. urinary bladder.

7. Which of these characteristics do *not* describe nonkeratinized stratified squamous epithelium?
   a. many layers of cells   d. found in the skin
   b. flat surface cells   e. outer layers covered by fluid
   c. surface cells

8. In parts of the body, such as the urinary bladder, where considerable expansion occurs, one can expect to find which type of epithelium?
   a. cuboidal   c. transitional   e. columnar
   b. pseudostratified   d. squamous

9. Epithelial cells with microvilli are most likely found
   a. lining blood vessels.   d. lining the small intestine.
   b. lining the lungs.   e. in the skin.
   c. lining the uterine tube.

10. A type of cell connection whose *only* function is to prevent the cells from coming apart is the
    a. desmosome.   b. gap junction.   c. tight junction.

11. The glands that do not have ducts and secrete their cellular products into the bloodstream are called _____ glands.
    a. apocrine   c. exocrine   e. merocrine
    b. endocrine   d. holocrine

12. A _____ gland has a duct that branches repeatedly, and the ducts end in saclike structures.
    a. simple tubular   d. simple acinar
    b. compound tubular   e. compound acinar
    c. simple coiled tubular

13. Glands that accumulate secretions and release them only when the individual secretory cells rupture and die are called _____ glands.
    a. apocrine   b. holocrine   c. merocrine

14. The fibers in dense connective tissue are produced by
    a. fibroblasts.   c. osteoblasts.   e. macrophages.
    b. adipocytes.   d. osteoclasts.

15. A tissue with a large number of collagen fibers organized parallel to each other would most likely be found in
    a. a muscle.
    b. a tendon.
    c. adipose tissue.
    d. a bone.
    e. cartilage.

16. Extremely delicate fibers that make up the framework for organs such as the liver, spleen, and lymph nodes are
    a. elastic fibers.
    b. reticular fibers.
    c. microvilli.
    d. cilia.
    e. collagen fibers.

17. Which of these types of connective tissue has the smallest amount of extracellular matrix?
    a. adipose
    b. bone
    c. cartilage
    d. loose connective tissue
    e. blood

18. Fibrocartilage is found
    a. in the cartilage of the trachea.
    b. in the rib cage.
    c. in the external ear.
    d. on the surface of bones in movable joints.
    e. between vertebrae.

19. A tissue in which cells are located in lacunae surrounded by a hard matrix organized as lamellae is
    a. hyaline cartilage.
    b. bone.
    c. nervous tissue.
    d. dense regular collagenous connective tissue.
    e. fibrocartilage.

20. Which of these characteristics apply to smooth muscle?
    a. striated, involuntary
    b. striated, voluntary
    c. unstriated, involuntary
    d. unstriated, voluntary

21. Which of these statements about nervous tissue is true?
    a. Neurons have cytoplasmic extensions called axons.
    b. Electric signals (action potentials) are conducted along axons.
    c. Neurons are nourished and protected by neuroglia.
    d. Dendrites receive electric signals and conduct them toward the cell body.
    e. All of the above are true.

22. The linings of the digestive, respiratory, urinary, and reproductive passages are composed of
    a. serous membranes.
    b. mucous membranes.
    c. synovial membranes.

23. Chemical mediators of inflammation
    a. cause blood vessels to constrict.
    b. decrease the permeability of blood vessels.
    c. initiate processes that lead to edema.
    d. help prevent clotting.
    e. decrease pain.

24. Which of these types of cells are labile?
    a. neurons
    b. skin
    c. liver
    d. pancreas

25. Permanent cells
    a. divide and replace damaged cells in replacement tissue repair.
    b. form granulation tissue.
    c. are responsible for removing scar tissue.
    d. are usually replaced by a different cell type if they are destroyed.
    e. are replaced during regeneration tissue repair.

Answers in Appendix E

C R I T I C A L   T H I N K I N G

1. Given the observation that a tissue has more than one layer of cells lining a free surface, (a) list the possible tissue types that exhibit those characteristics, and (b) explain what additional observations need to be made to identify the tissue type.

2. A patient suffered from kidney failure a few days after he was exposed to a toxic chemical. A biopsy of his kidney indicated that many of the thousands of epithelium-lined tubules that make up the kidney had lost many of the simple cuboidal epithelial cells that normally line them, although most of the basement membranes appeared to be intact. Predict how likely this person is to recover fully.

3. What types of epithelium are likely to be found lining the trachea of a heavy smoker? Predict the changes that are likely to occur after he or she stops smoking for 1 or 2 years.

4. The blood–brain barrier is a specialized epithelium in capillaries that prevents many materials from passing from the blood into the brain. What kind of cell connections would be expected in the blood–brain barrier?

5. One of the functions of the pancreas is to secrete digestive enzymes that are carried by ducts to the small intestine. How many cell layers and what cell shape, cell surface, and type of cell-to-cell connections would be expected in the epithelium that is responsible for producing the digestive enzymes?

6. A tissue has the following characteristics: a free surface; a single layer of cells; narrow, tall cells; microvilli; many mitochondria; and goblet cells. Describe the tissue type and as many functions of the cell as possible.

7. Explain the consequences:
   a. if simple columnar epithelium replaced nonkeratinized stratified squamous epithelium that lines the mouth
   b. if tendons were dense elastic connective tissue instead of dense collagenous connective tissue
   c. if bones were made entirely of elastic cartilage

8. Antihistamines block the effect of a chemical mediator, histamine, that is released during the inflammatory response. Give an example of when it could be harmful to use an antihistamine and an example of when it could be beneficial.

9. Granulation tissue and scars consist of dense irregular collagenous connective tissue. Vitamin C is required for collagen synthesis. Predict the effect of scurvy, which is a nutritional disease caused by vitamin C deficiency, on wound healing.

Answers in Appendix F

For additional study tools, go to: www.aris.mhhe.com.
Click on your course topic and then this textbook's author name/title. Register once for a semester's worth of interactive learning activities to help improve your grade!

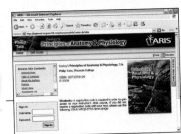

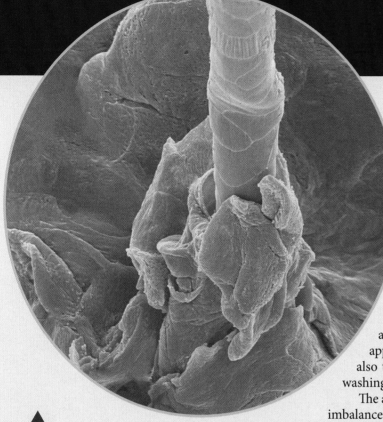

Skin, hair, glands, and nails form the integumentary system. This colorized scanning electron micrograph shows the shaft of a hair (*yellow*) protruding through the surface of the skin. Tightly bound epithelial cells form the hair shaft, and flat, scalelike epithelial cells form the skin's surface. Like the shingles on the roof of a house, the skin's tough cells protect underlying structures. Dandruff is excessive flaking of these epithelial cells from the scalp.

Tools for Success

 *A virtual cadaver dissection experience*

 Audio (MP3) and/or video (MP4) content, available at www.mhhe.aris.com

Introduction

The **integumentary** (in-teg-ū-men′tă-rē) **system** consists of the skin and accessory structures, such as hair, glands, and nails. *Integument* means covering, and the integumentary system is familiar to most people because it covers the outside of the body and is easily observed. In addition, humans are concerned with the appearance of the integumentary system. Skin without blemishes is considered attractive, whereas acne is a source of embarrassment for many people. The development of wrinkles and the graying or loss of hair are signs of aging that some people find unattractive. Because of these feelings, much time, effort, and money are spent on changing the appearance of the integumentary system. For example, people apply lotion to their skin, color their hair, and trim their nails. They also try to prevent sweating with antiperspirants and body odor with washing, deodorants, and perfumes.

The appearance of the integumentary system can indicate physiological imbalances in the body. Some disorders, such as acne or warts, affect just the integumentary system. Disorders of other parts of the body can be reflected there, and thus the integumentary system is useful for diagnosis. For example, reduced blood flow through the skin during a heart attack can cause a pale appearance, whereas increased blood flow as a result of fever can cause a flushed appearance. Also, the rashes of some diseases are very characteristic, such as the rashes of measles, chicken pox, and allergic reactions.

▶This chapter provides an overview of the **functions of the integumentary system (p. 106)** and an explanation of the **skin (p. 106)**, the **subcutaneous tissue (p. 111)**, and the **accessory skin structures (p. 112)**. A **summary of integumentary system functions (p. 116)**, **the integumentary system as a diagnostic aid (p. 118)**, **skin cancer (p. 118)**, and the **effects of aging on the integumentary system (p. 120)** are also presented.

Functions of the Integumentary System

The integumentary system consists of the skin, hair, glands, and nails. Although we are often concerned with how the integumentary system looks, it has many important functions that go beyond appearance. The major functions of the integumentary system are

1. *Protection.* The skin provides protection against abrasion and ultraviolet light. It also prevents the entry of microorganisms and prevents dehydration by reducing water loss from the body.
2. *Sensation.* The integumentary system has sensory receptors that can detect heat, cold, touch, pressure, and pain.
3. *Temperature regulation.* Body temperature is regulated by controlling blood flow through the skin and the activity of sweat glands.
4. *Vitamin D production.* When exposed to ultraviolet light, the skin produces a molecule that can be transformed into vitamin D.
5. *Excretion.* Small amounts of waste products are lost through the skin and in gland secretions. ▼

1. **What are the components of the integumentary system?**
2. **State five functions of the integumentary system.**

Skin

The skin is made up of two major tissue layers. The **epidermis** (ep-i-derm′is, on the dermis) is the most superficial layer of the skin; it consists of epithelial tissue (figure 5.1). The epidermis resists abrasion on the skin's surface and reduces water loss through the skin. The epidermis rests on the **dermis** (derm′is, skin), which is a layer of connective tissue. Depending on location, the dermis is 10 to 20 times thicker than the epidermis and is responsible for most of the structural strength of the skin. The strength of the dermis is seen in leather, which is produced from the hide (skin) of an animal. The epidermis is removed, and the dermis is preserved by tanning.

The skin rests on the **subcutaneous** (under the skin) **tissue,** which is a layer of loose connective tissue (see figure 5.1). The subcutaneous tissue is not part of the skin or the integumentary system, but it does connect the skin to underlying muscle or bone. Table 5.1 summarizes the structures and functions of the skin and subcutaneous tissue. ▼

3. *For the two layers of the skin, state their name, tissue type, and function.*
4. *What type of tissue is the subcutaneous layer and what is its function? How is subcutaneous tissue related to the integumentary system?*

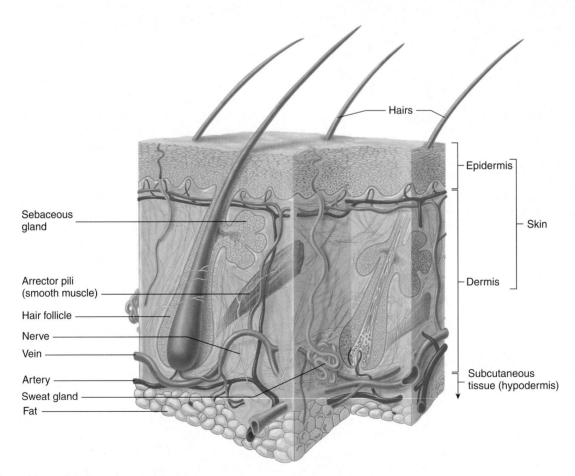

Figure 5.1   **Skin and Subcutaneous Tissue**

The skin, consisting of the epidermis and the dermis, is connected by the subcutaneous tissue to underlying structures. Note the accessory structures (hairs, glands, and arrector pili), some of which project into the subcutaneous tissue, and the large amount of fat in the subcutaneous tissue.

Table 5.1	Comparison of the Skin (Epidermis and Dermis) and Subcutaneous Tissue	
Parts	**Structure**	**Function**
Epidermis	Superficial part of skin; stratified squamous epithelium; composed of four or five strata	Barrier that prevents water loss and the entry of chemicals and microorganisms; protects against abrasion and ultraviolet light; produces vitamin D; gives rise to hair, nails, and glands
Stratum corneum	Most superficial strata of the epidermis; 25 or more layers of dead squamous cells	Provision of structural strength by keratin and protein envelope within cells; prevention of water loss by lipids surrounding cells; sloughing off of most superficial cells resists abrasion
Stratum lucidum	Three to five layers of dead cells; appears transparent; present in thick skin, absent in most thin skin	Dispersion of keratohyalin around keratin fibers
Stratum granulosum	Two to five layers of flattened, diamond-shaped cells	Production of keratohyalin granules; lamellar bodies release lipids from cells; cells die
Stratum spinosum	A total of 8 to 10 layers of many-sided cells	Production of keratin fibers; formation of lamellar bodies
Stratum basale	Deepest strata of the epidermis; single layer of cuboidal or columnar cells; basement membrane of the epidermis attaches to the dermis	Production of cells of the most superficial strata; melanocytes produce and contribute melanin, which protects against ultraviolet light
Dermis	Deep part of skin; connective tissue composed of two layers	Responsible for the structural strength and flexibility of the skin; the epidermis exchanges gases, nutrients, and waste products with blood vessels in the dermis
Papillary layer	Papillae project toward the epidermis; loose connective tissue	Brings blood vessels close to the epidermis; dermal papillae form fingerprints and footprints
Reticular layer	Mat of collagen and elastin fibers; dense irregular connective tissue	Main fibrous layer of the dermis; strong in many directions; forms cleavage lines
Subcutaneous tissue	Not part of the skin; loose connective tissue with abundant fat deposits	Attaches the dermis to underlying structures; fat tissue provides energy storage, insulation, and padding; blood vessels and nerves from the subcutaneous tissue supply the dermis

Epidermis

The epidermis is stratified squamous epithelium, and it is separated from the dermis by a basement membrane. The epidermis has no blood vessels and is nourished by diffusion from capillaries of the dermis (figure 5.2). Most cells of the epidermis are called **keratinocytes** (ke-rat′i-nō-sītz) because they produce a protein mixture called **keratin** (ker′ă-tin), which makes the cells hard. Keratinocytes are responsible for the ability of the epidermis to resist abrasion and reduce water loss. Other cells of the epidermis include **melanocytes** (mel′ă-nō-sītz), which contribute to skin color, **Langerhans cells,** which are part of the immune system (see chapter 22), and **Merkel cells,** which are specialized epidermal cells associated with the nerve endings responsible for detecting light touch and superficial pressure (see chapter 12).

Cells are produced by mitosis in the deepest layer of the epidermis. As new cells are formed, they push older cells to the surface, where they slough off, or **desquamate** (des′kwă-māt). The outermost cells in this stratified arrangement protect the cells underneath, and the deeper replicating cells replace cells lost from the surface. As they move from the deeper epidermal layers to the surface, the cells change shape and chemical composition. This process is called **keratinization** (ker′ă-tin-i-zā′shŭn) because the cells become filled with keratin. As keratinization proceeds, the epithelial cells die, producing an outer layer of dead, hard cells that resists abrasion and forms a permeability barrier.

Although keratinization is a continual process, distinct transitional stages called **strata** (sing. *stratum*) can be recognized as the cells change. From the deepest to the most superficial, the five strata are the stratum basale, stratum spinosum, stratum granulosum, stratum lucidum, and stratum corneum (see figure 5.2).

Stratum Basale

The **stratum basale** (bā′să-lē) is the deepest layer of the epidermis, consisting of a single layer of cuboidal or columnar cells (figure 5.3, step 1). Structural strength is provided by hemidesmosomes, which anchor the epidermis to the basement membrane, and by desmosomes, which hold the keratinocytes together (see chapter 4). Keratinocytes undergo mitotic divisions approximately every 19 days. One daughter cell becomes a new stratum basale cell and can divide again. The other cell is pushed toward the surface, a journey that takes about 40–56 days.

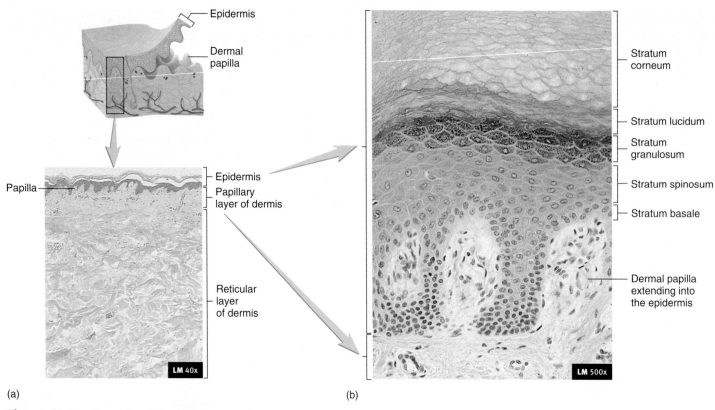

Figure 5.2   Dermis and Epidermis

(*a*) Photomicrograph of dermis covered by the epidermis. The dermis consists of the papillary and reticular layers. The papillary layer has projections, called papillae, that extend into the epidermis. (*b*) Higher-magnification photomicrograph of the epidermis resting on the papillary layer of the dermis. Note the strata of the epidermis.

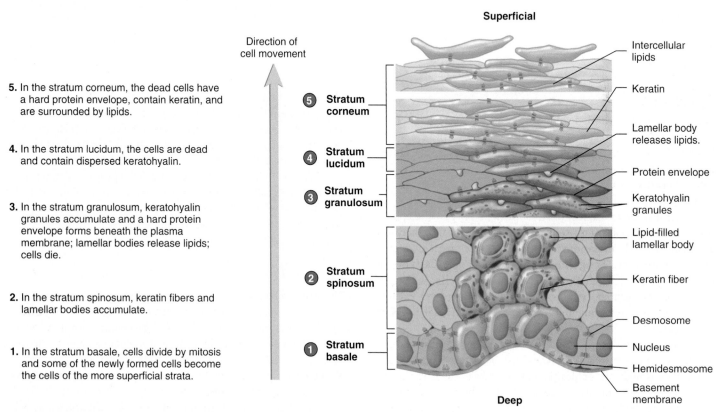

5. In the stratum corneum, the dead cells have a hard protein envelope, contain keratin, and are surrounded by lipids.

4. In the stratum lucidum, the cells are dead and contain dispersed keratohyalin.

3. In the stratum granulosum, keratohyalin granules accumulate and a hard protein envelope forms beneath the plasma membrane; lamellar bodies release lipids; cells die.

2. In the stratum spinosum, keratin fibers and lamellar bodies accumulate.

1. In the stratum basale, cells divide by mitosis and some of the newly formed cells become the cells of the more superficial strata.

Process Figure 5.3   Epidermal Layers and Keratinization

Stratum Spinosum

The **stratum spinosum** (spī-nō′sŭm) is superficial to the stratum basale and consists of 8–10 layers of many-sided cells (figure 5.3, step 2). As the cells in this stratum are pushed to the surface, they flatten. Keratin fibers and lipid-filled, membrane-bound organelles called **lamellar** (lam′ĕ-lăr, lă-mel′ar) **bodies** are formed inside the keratinocytes.

Stratum Granulosum

The **stratum granulosum** (gran-ū-lō′sŭm) consists of two to five layers of diamond-shaped cells (figure 5.3, step 3). This stratum derives its name from the nonmembrane-bound protein granules of **keratohyalin** (ker′ă-tō-hī′ă-lin), which accumulate in the cytoplasm of the cell. The lamellar bodies of these cells move to the plasma membrane and release their lipid contents into the intercellular space. Inside the cell, a protein envelope forms beneath the plasma membrane. In the most superficial layers of the stratum granulosum, the nucleus and other organelles degenerate, and the cell dies.

Stratum Lucidum

The **stratum lucidum** (loo′si-dŭm) is a thin, clear zone above the stratum granulosum, consisting of several layers of dead cells with indistinct boundaries (figure 5.3, step 4). Keratin fibers are present, but the keratohyalin, which was evident as granules in the stratum granulosum, has dispersed around the keratin fibers, and the cells appear somewhat transparent. The stratum lucidum is present in only a few areas of the body (see "Thick and Thin Skin," p. 109).

Stratum Corneum

The **stratum corneum** (kōr′nē-ŭm) is the most superficial layer of the epidermis, consisting of 25 or more layers of dead squamous cells joined by desmosomes (figure 5.3, step 5). Eventually, the desmosomes break apart, and the cells are desquamated from the surface of the skin. Excessive desquamation of the stratum corneum of the scalp is called dandruff.

The stratum corneum consists of **cornified cells,** which are dead cells, with a hard protein envelope, filled with the protein keratin. Keratin is a mixture of keratin fibers and keratohyalin. The envelope and the keratin are responsible for the structural strength of the stratum corneum. The type of keratin found in the skin is soft keratin. Another type of keratin, hard keratin, is found in nails and the external parts of hair. Cells containing hard keratin are more durable than cells with soft keratin, and they do not desquamate.

Surrounding the cells are the lipids released from lamellar bodies. The lipids are responsible for many of the permeability characteristics of the skin.

In skin subjected to friction or pressure, the number of layers in the stratum corneum greatly increases to produce a thickened area called a **callus** (kal′ŭs). The skin over bony prominences may develop a cone-shaped structure called a **corn.** The base of the cone is at the surface, but the apex extends deep into the epidermis, and pressure on the corn may be quite painful.

 Keratinization and Psoriasis

Psoriasis (sō-rī′ă-sis, the itch) is characterized by increased cell division in the stratum basale and abnormal keratin production. The result is a thicker-than-normal stratum corneum that sloughs to produce large, silvery scales. If the scales are scraped away, bleeding occurs from the blood vessels located in the dermis. Evidence suggests that the disease has a genetic component and that the immune system stimulates the increased cell divisions. Psoriasis is a chronic disease that can be controlled with drugs and phototherapy (UV light), but as yet it has no cure. ▼

5. *How does the epidermis receive nourishment?*
6. *From deepest to most superficial, name and describe the five strata of the epidermis. Name the stratum in which new cells form, the stratum from which cells desquamate, and the strata that have living versus dead cells.*
7. *Describe the structural features resulting from keratinization that make the epidermis structurally strong and resistant to water loss.*
8. *What are a callus and a corn?*

P R E D I C T
Some drugs are administered by applying them to the skin (e.g., a nicotine skin patch to help a person stop smoking). The drug diffuses through the epidermis to blood vessels in the dermis. What kinds of substances can pass easily through the skin by diffusion? What kinds have difficulty?

Thick and Thin Skin

When we say a person has thick or thin skin, we are usually referring metaphorically to the person's ability to take criticism. However, all of us in a literal sense have both thick and thin skin. Skin is classified as thick or thin on the basis of the structure of the epidermis. **Thick skin** has all five epithelial strata, and the stratum corneum has many layers of cells. Thick skin is found in areas subject to pressure or friction, such as the palms of the hands, the soles of the feet, and the fingertips.

Thin skin covers the rest of the body and is more flexible than thick skin. Each stratum contains fewer layers of cells than are found in thick skin; the stratum granulosum frequently consists of only one or two layers of cells, and the stratum lucidum generally is absent. Hair is found only in thin skin. ▼

9. *Compare the structure and location of thick skin and thin skin. Is hair found in thick or thin skin?*

Skin Color

Pigments in the skin, blood circulating through the skin, and the thickness of the stratum corneum together determine skin color. **Melanin** (mel′ă-nin) is the group of pigments responsible for skin, hair, and eye color. Most melanin molecules are brown to black pigments, but some are yellowish or reddish. Melanin provides protection against ultraviolet light from the sun.

Large amounts of melanin are found in certain regions of the skin, such as freckles, moles, nipples, areolae of the breasts, axillae,

1. Melanosomes are produced by the Golgi apparatus of the melanocyte.

2. Melanosomes move into melanocyte cell processes.

3. Epithelial cells phagocytize the tips of the melanocyte cell processes.

4. The melanosomes, which were produced inside the melanocytes, have been transferred to epithelial cells and are now inside them.

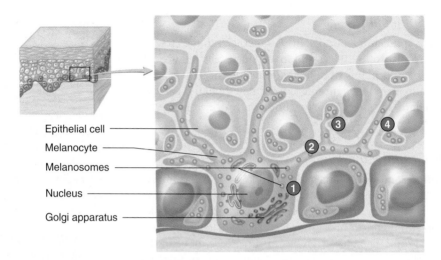

Epithelial cell

Melanocyte

Melanosomes

Nucleus

Golgi apparatus

Process Figure 5.4 Melanin Transfer from Melanocyte to Keratinocytes

Melanocytes make melanin, which is packaged into melanosomes and transferred to many keratinocytes.

and genitalia. Other areas of the body, such as the lips, palms of the hands, and soles of the feet, contain less melanin.

Melanin is produced by **melanocytes** (mel′ă-nō-sītz), irregularly shaped cells with many long processes that extend between the keratinocytes of the stratum basale and the stratum spinosum (figure 5.4). The Golgi apparatuses of the melanocytes package melanin into vesicles called **melanosomes** (mel′ă-nō-sōmz), which move into the cell processes of the melanocytes. Keratinocytes phagocytize (see chapter 3) the tips of the melanocyte cell processes, thereby acquiring melanosomes. Although all keratinocytes can contain melanin, only the melanocytes produce it.

Freckles, Moles, and Warts

Freckles are small, discrete areas of the epidermis with a uniform brown color. Freckles result from an increased amount of melanin in the stratum basale keratinocytes. There is no increase in the number of melanocytes. **Moles** are elevations of the skin that are variable in size and are often pigmented and hairy. A mole is an aggregation, or "nest," of melanocytes in the epidermis or dermis. They are a normal occurrence, and most people have 10–20 moles, which appear in childhood and enlarge until puberty. **Warts** are uncontrolled growths of the epidermis caused by the human *Papillomavirus*. Usually, the growths are benign and disappear spontaneously, or they can be removed by a variety of techniques. The viruses are transmitted to the skin by direct contact with contaminated objects or an infected person. They can also be spread by scratching.

Melanin production is determined by genetic factors, exposure to light, and hormones. Genetic factors are primarily responsible for the variations in skin color among different races and among people of the same race. The amount and types of melanin produced by the melanocytes, as well as the size, number, and distribution of the melanosomes, are genetically determined. Skin colors are not determined by the number of melanocytes because all races have essentially the same number. Although many genes are responsible for skin color, a single mutation (see chapter 25) can prevent the manufacture of melanin. **Albinism** (al′bi-nizm) is a recessive genetic trait that causes a deficiency or an absence of melanin. Albinos have fair skin, white hair, and unpigmented irises in the eyes.

Exposure to ultraviolet light darkens melanin already present and stimulates melanin production, resulting in tanning of the skin. The increased melanin provides additional protection against ultraviolet light.

During pregnancy, certain hormones, such as estrogen and melanocyte-stimulating hormone, cause an increase in melanin production in the mother, which in turn causes darkening of the nipples, areolae, and genitalia. The cheekbones, forehead, and chest also may darken, resulting in the "mask of pregnancy," and a dark line of pigmentation may appear on the midline of the abdomen.

Blood flowing through the skin imparts a reddish hue. **Erythema** (er-ĭ-thē′mă) is increased redness of the skin resulting from increased blood flow through the skin. An inflammatory response (see chapter 4) stimulated by infections, sunburn, allergic reactions, insect bites, or other causes can produce erythema, as can exposure to the cold, blushing, or flushing when angry or hot. A decrease in blood flow, such as occurs in shock, can make the skin appear pale. **Cyanosis** (sī-ă-nō′sis) is a bluish skin color caused by a decrease in the blood oxygen content.

Birthmarks are congenital (present at birth) disorders of the capillaries in the dermis of the skin. Usually, they are of concern only for cosmetic reasons. A strawberry birthmark is a mass of soft, elevated tissue that appears bright red to deep purple. In 70% of patients, strawberry birthmarks disappear spontaneously by age 7. Portwine stains appear as flat, dull red, or bluish patches that persist throughout life.

Carotene (kar′ō-tēn) is a yellow pigment found in plants, such as carrots and corn. Humans normally ingest carotene and use it as a source of vitamin A. Carotene is lipid-soluble, and, when large amounts of carotene are consumed, the excess accumulates in the stratum corneum and in the adipose cells of the dermis and subcutaneous tissue, causing the skin to develop a yellowish tint, which slowly disappears once carotene intake is reduced.

The location of pigments and other substances in the skin affects the color produced. If a dark pigment is located in the dermis or subcutaneous tissue, light reflected off the dark pigment can be scattered by collagen fibers of the dermis to produce a blue color. The same effect produces the blue color of the sky as light is reflected from dust particles in the air. The deeper within the dermis or subcutaneous tissue any dark pigment is located, the bluer the pigment appears because of the light-scattering effect of the overlying tissue. This effect causes the blue color of bruises and some superficial blood vessels. ▼

10. *What is the function of melanin? Which cells of the epidermis produce melanin? What happens to the melanin once it is produced?*
11. *How do genetic factors, exposure to light, and hormones determine the amount of melanin in the skin?*
12. *How do melanin, carotene, and blood affect skin color?*

P R E D I C T ②

Explain the differences in skin color between (a) the palms of the hands and the lips, (b) the palms of the hands of a person who does heavy manual labor and of one who does not, (c) the anterior and posterior surfaces of the forearm, and (d) the genitals and the soles of the feet.

Dermis

The dermis is connective tissue with fibroblasts, a few adipose cells, and macrophages. Collagen is the main connective tissue fiber, but elastin and reticular fibers are also present. Nerves, hair follicles, smooth muscles, glands, and lymphatic vessels extend into the dermis (see figure 5.1).

The dermis is divided into two layers (see figures 5.1 and 5.2): the superficial **papillary** (pap′i-lār-ē) **layer** and the deeper **reticular** (re-tik′ū-lăr) **layer.** The papillary layer derives its name from projections called **dermal papillae** (pă-pil′ē), which extend toward the epidermis (see figure 5.2). The papillary layer is loose connective tissue with thin fibers that are somewhat loosely arranged. The papillary layer also contains blood vessels that supply the overlying epidermis with nutrients, remove waste products, and aid in regulating body temperature.

The dermal papillae underlying thick skin are in parallel, curving ridges that shape the overlying epidermis into fingerprints and footprints. The ridges increase friction and improve the grip of the hands and feet. Everyone has unique fingerprints and footprints, even identical twins.

The reticular layer, which is composed of dense irregular connective tissue, is the main layer of the dermis. It is continuous with the subcutaneous tissue and forms a mat of irregularly arranged fibers that are resistant to stretching in many directions. The elastin and collagen fibers are oriented more in some directions than in others and produce **cleavage,** or **tension, lines** in the

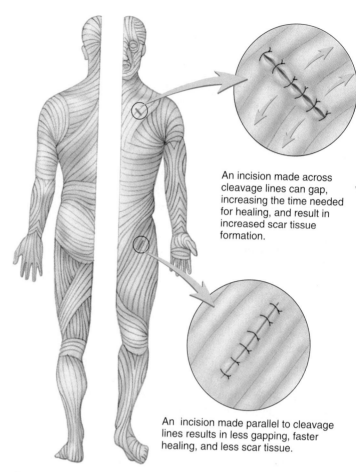

An incision made across cleavage lines can gap, increasing the time needed for healing, and result in increased scar tissue formation.

An incision made parallel to cleavage lines results in less gapping, faster healing, and less scar tissue.

Figure 5.5   Cleavage Lines
The orientation of collagen fibers produces cleavage, or tension, lines in the skin.

skin (figure 5.5). Knowledge of cleavage line directions is important because an incision made parallel to the cleavage lines is less likely to gap than is an incision made across them. The closer together the edges of a wound, the less likely is the development of infections and the formation of considerable scar tissue.

If the skin is overstretched, the dermis may rupture and leave lines that are visible through the epidermis. These lines of scar tissue, called **striae** (strī′ē), or **stretch marks,** can develop on the abdomen and breasts of a woman during pregnancy. ▼

13. *Name and compare the two layers of the dermis. Which layer is responsible for most of the structural strength of the skin?*
14. *What are cleavage lines and striae?*

Subcutaneous Tissue

Subcutaneous (under the skin) **tissue** attaches the skin to underlying bone and muscle and supplies it with blood vessels and nerves (see figure 5.1). Subcutaneous tissue is loose connective tissue with collagen and elastin fibers. The main types of cells within the subcutaneous tissue are fibroblasts, adipose cells, and macrophages. Subcutaneous tissue is not part of the integumentary system. An alternate name for subcutaneous tissue is the **hypodermis**

(hi-pō-der′mis, below the skin). *Subcutaneous tissue* is the term used clinically and by advanced texts.

Approximately half the body's stored fat is in the subcutaneous tissue, where it functions as a source of energy, insulation, and padding. The amount of fat in the subcutaneous tissue varies with age, sex, and diet. The subcutaneous tissue can be used to estimate total body fat. The skin is pinched at selected locations, and the thickness of the fold of skin and underlying subcutaneous tissue is measured. The thicker the fold, the greater the amount of total body fat. ▼

> 15. *What is the function of subcutaneous tissue? Name the types of tissue forming the subcutaneous tissue.*
> 16. *List the functions of the fat within the subcutaneous tissue.*

 Injections

There are three types of injections. An **intradermal injection,** such as for the tuberculin skin test, is an injection into the dermis. It is administered by drawing the skin taut and inserting a small needle at a shallow angle into the skin. A **subcutaneous injection** is an injection into the fatty tissue of the subcutaneous tissue, such as for an insulin injection. This injection is achieved by pinching the skin to form a "tent" into which a short needle is inserted. An **intramuscular injection** is an injection into a muscle deep to the subcutaneous tissue. It is accomplished by inserting a long needle at a 90-degree angle to the skin. Intramuscular injections are used for injecting most vaccines and certain antibiotics.

Accessory Skin Structures

Accessory skin structures are hair, smooth muscles called the arrector pili, glands, and nails.

Hair

The presence of **hair** is one of the characteristics common to all mammals; if the hair is dense and covers most of the body surface, it is called fur. In humans, hair is found everywhere on the skin except the palms, the soles, the lips, the nipples, parts of the external genitalia, and the distal segments of the fingers and toes.

By the fifth or sixth month of fetal development, delicate, unpigmented hair called **lanugo** (lă-noo′gō) has developed and covered the fetus. Near the time of birth, **terminal hairs,** which are long, coarse, and pigmented, replace the lanugo of the scalp, eyelids, and eyebrows. **Vellus** (vel′ŭs) **hairs,** which are short, fine, and usually unpigmented, replace the lanugo on the rest of the body. At puberty, terminal hair, especially in the pubic and axillary regions, replaces much of the vellus hair. The hair of the chest, legs, and arms is approximately 90% terminal hair in males, compared with approximately 35% in females. In males, terminal hairs replace the vellus hairs of the face to form the beard.

Hair Structure

The **shaft** of the hair protrudes above the surface of the skin, whereas the **root** is below the surface (figure 5.6). The base of the

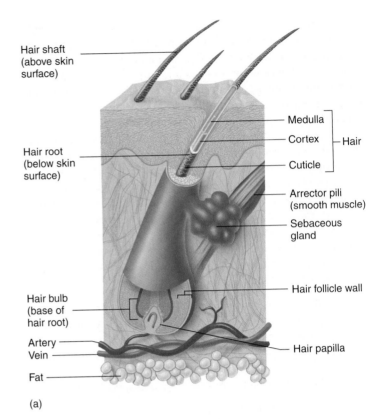

(a)

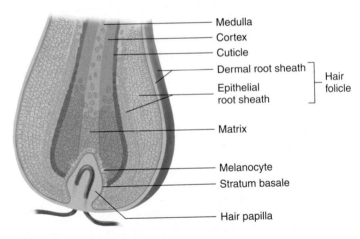

(b)

Figure 5.6   Hair Follicle

(*a*) Hair within a hair follicle. (*b*) Enlargement of the hair follicle wall and hair bulb.

root is expanded to form the **hair bulb.** Most of the root and the shaft of the hair are composed of columns of dead keratinized epithelial cells arranged in three concentric layers: the medulla, the cortex, and the cuticle. The **medulla** (me-dool′ă) is the central axis of the hair, consisting of two or three layers of cells containing soft keratin. The **cortex** forms the bulk of the hair and consists of cells containing hard keratin. The **cuticle** (kū′ti-kl) is a single layer of cells, with hard keratin, that forms the hair surface. The edges of the cuticle cells overlap like shingles on a roof. The raised edges help hold the hair in place. During textile manufacture, the sheep hair cuticles catch each other and hold the hairs together to form threads.

The **hair follicle** is a tubelike invagination of the epidermis into the dermis from which the hair develops. It consists of a **dermal root sheath** and an **epithelial root sheath** (see figure 5.6*b*). The dermal root sheath is the portion of the dermis that surrounds the epithelial root sheath. The epithelial root sheath is divided into an external and an internal part. At the opening of the follicle, the epithelial root sheath has all the strata found in thin skin. Deeper in the hair follicle, the number of cells decreases until at the hair bulb only the stratum basale is present. This has important consequences for the repair of the skin. If the epidermis and the superficial part of the dermis are damaged, the stratum basale in the undamaged part of the hair follicle can be a source of new epithelium.

The hair bulb is an expanded knob at the base of the hair root. Inside the hair bulb is a mass of undifferentiated epithelial cells, the **matrix,** which produces the hair (see figure 5.6*b*). The dermis of the skin projects into the hair bulb as a **hair papilla,** which contains blood vessels that provide nourishment to the cells of the matrix.

Hair Growth

Hair is produced in cycles that involve a **growth stage** and a **resting stage.** During the growth stage, hair is formed by matrix cells that differentiate, become keratinized, and die. The hair grows longer as cells are added at the base of the hair root. During the resting stage, there is no hair growth and the hair is held in the hair follicle. The duration of each stage depends on the hair—eyelashes grow for approximately 30 days and rest for 105 days, whereas scalp hairs grow for 3 years and rest for 1–2 years. At any given time, an estimated 90% of the scalp hairs are in the growing stage.

When the next growth stage begins, a new hair is formed, and the old hair falls out. A loss of approximately 100 scalp hairs per day is normal. Although loss of hair normally means that the hair is being replaced, loss of hair can be permanent. The most common kind of permanent hair loss is "pattern baldness." Hair follicles are lost, and the remaining hair follicles revert to producing vellus hair, which is very short, transparent, and for practical purposes invisible. Although more common and more pronounced in certain men, baldness can also occur in women. Genetic factors and the hormone testosterone are involved in causing pattern baldness.

Hair Color

Hair color is determined by varying amounts and types of melanin. The production and distribution of melanin by melanocytes occurs in the hair bulb by the same method as in the skin. With age, the amount of melanin in hair can decrease, causing the hair to become faded in color, or the hair can have no melanin and be white. Gray hair is usually a mixture of unfaded, faded, and white hairs.

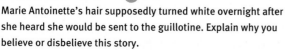

P R E D I C T 3
Marie Antoinette's hair supposedly turned white overnight after she heard she would be sent to the guillotine. Explain why you believe or disbelieve this story.

Muscles

Associated with each hair follicle are smooth muscle cells, the **arrector** (ă-rek′tōr, that which raises) **pili** (pī′lī, hair) (see figure 5.6*a*). Contraction of the arrector pili causes the hair to become more perpendicular to the skin's surface, or to "stand on end," and it produces a raised area of skin called "goose flesh." In animals with fur, contraction of the arrector pili is beneficial because it increases the thickness of the fur by raising the hairs. In the cold, the thicker layer of fur traps air and becomes a better insulator. The thickened fur can also make the animal appear larger and more ferocious, which might deter an attacker. It is unlikely that humans, with their sparse amount of hair, derive any important benefit from contraction of their arrector pili. ▼

17. When and where are lanugo, vellus, and terminal hairs found in the skin?
18. Define the root, shaft, and hair bulb of a hair. Describe the three layers of a hair.
19. Describe the parts of a hair follicle. How is the epithelial root sheath important in the repair of the skin?
20. In what part of a hair does growth take place? What are the stages of hair growth?
21. What determines the different shades of hair color?
22. What happens when the arrector pili muscles contract?

Glands

The major glands of the skin are the **sebaceous** (sē-bāshŭs) **glands** and the **sweat glands** (figure 5.7). Sebaceous glands are simple or compound alveolar glands located in the dermis. Most are connected by a duct to the superficial part of a hair follicle. Sebaceous

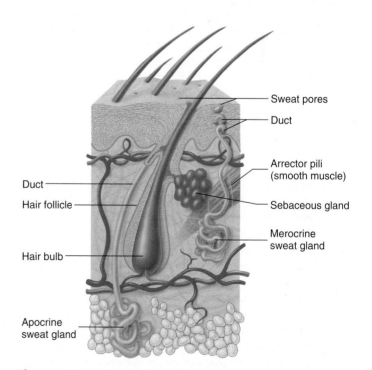

Figure 5.7 Glands of the Skin
Sebaceous and apocrine sweat glands empty into the hair follicle. Merocrine sweat glands empty onto the surface of the skin.

Clinical Relevance Burns

A **burn** is injury to a tissue caused by heat, cold, friction, chemicals, electricity, or radiation. Burns are classified according to the extent of surface area involved and the depth of the burn. For an adult, the surface area that is burned can be conveniently estimated by "the rule of nines," in which the body is divided into areas that are approximately 9%, or multiples of 9%, of the body surface area (BSA) (figure A). For younger patients, surface area relationships are different. For example, in an infant, the head and neck are 21% of BSA, whereas in an adult they are 9%. For burn victims younger than age 15, a table specifically developed for them should be consulted.

On the basis of depth, burns are classified as either partial-thickness or full-thickness burns (figure B). **Partial-thickness burns** are subdivided into first- and second-degree burns. **First-degree burns** involve only the epidermis and are red and painful, and slight edema (swelling) may occur. They can be caused by sunburn or brief exposure to hot or cold objects, and they heal in a week or so without scarring.

Second-degree burns damage the epidermis and the dermis. Minimal dermal damage causes redness, pain, edema, and blisters. Healing takes approximately 2 weeks, and no scarring results. If the burn goes deep into the dermis, however, the wound appears red, tan, or white; may take several months to heal; and might scar. In all second-degree burns, the epidermis

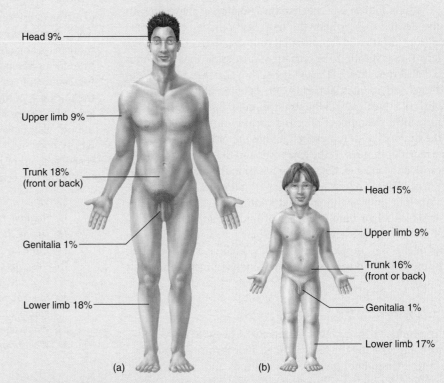

Head 9%

Upper limb 9%

Trunk 18% (front or back)

Genitalia 1%

Lower limb 18%

(a)

Head 15%

Upper limb 9%

Trunk 16% (front or back)

Genitalia 1%

Lower limb 17%

(b)

Figure A **The Rule of Nines**

(a) In an adult, surface areas can be estimated using the rule of nines: Each major area of the body is 9%, or a multiple of 9%, of the total body surface area. (b) In infants and children, the head represents a larger proportion of surface area. The rule of nines is not as accurate for children, as can be seen in this 5-year-old child.

regenerates from epithelial tissue in hair follicles and sweat glands, as well as from the edges of the wound.

Full-thickness burns are also called **third-degree burns.** The epidermis and dermis are completely destroyed, and deeper tissue may be involved. Third-degree burns

are often surrounded by first- and second-degree burns. Although the areas that have first- and second-degree burns are painful, the region of third-degree burn is usually painless because of destruction of sensory receptors. Third-degree burns appear white, tan, brown, black, or deep cherry red

---

glands produce **sebum,** an oily, white substance rich in lipids. The sebum lubricates the hair and the surface of the skin, which prevents drying and protects against some bacteria.

Merocrine (merō-krin) **sweat glands** are simple, coiled tubular glands located in almost every part of the skin and are most numerous in the palms and soles. They produce sweat, which is a secretion of mostly water with a few salts. Merocrine sweat glands have ducts that open onto the surface of the skin through sweat pores. When the body temperature starts to rise above normal levels, the sweat glands produce sweat, which evaporates and cools the body. Sweat can also be released in the palms, soles, axillae (armpits), and other places because of emotional stress. Emotional sweating is used in lie detector (polygraph) tests because sweat gland activity usually increases when a person tells a lie. Even small amounts of sweat can be detected

because the salt solution conducts electricity and lowers the electric resistance of the skin.

Apocrine (apō-krin) **sweat glands** are simple, coiled, tubular glands that produce a thick secretion rich in organic substances. They open into hair follicles, but only in the axillae and genitalia. Apocrine sweat glands become active at puberty because of the influence of sex hormones. Their organic secretion, which is essentially odorless when released, is quickly broken down by bacteria into substances responsible for what is commonly known as body odor. Many mammals use scent as a means of communication, and it has been suggested that the activity of apocrine sweat glands may be a sign of sexual maturity.

Other skin glands include the ceruminous glands and the mammary glands. The **ceruminous** (sĕ-roo′mi-nŭs) **glands** are modified merocrine sweat glands that produce **cerumen,** or earwax (see chapter 14). The **mammary glands** are modified apocrine

in color. Skin can regenerate in a third-degree burn only from the edges, and skin grafts are often necessary.

The depth of burns and the percent of BSA affected by the burns can be combined with other criteria to classify the seriousness of burns. A **major burn** is a third-degree burn over 10% or more of the BSA; a second-degree burn over 25% or more of the BSA; or a second- or third-degree burn of the hands, feet, face, genitals, or anal region. Facial burns are often associated with damage to the respiratory tract, and burns of joints often heal with scar tissue formation that limits movement. A **moderate burn** is a third-degree burn of 2%–10% of the BSA or a second-degree burn of 15%–25% of the BSA. A **minor burn** is a third-degree burn of less than 2% or a second-degree burn of less than 15% of the BSA.

Deep partial-thickness and full-thickness burns take a long time to heal and form scar tissue with disfiguring and debilitating wound contracture. Skin grafts are performed to prevent these complications and to speed healing. In a split skin graft, the epidermis and part of the dermis are removed from another part of the body and are placed over the burn. Interstitial fluid from the burned area nourishes the graft until its dermis becomes vascularized. At the graft donation site, part of the dermis is still present. The deep parts of hair follicles and sweat gland ducts are in this remaining dermis. These hair follicles and sweat gland ducts are a source of epithelial cells,

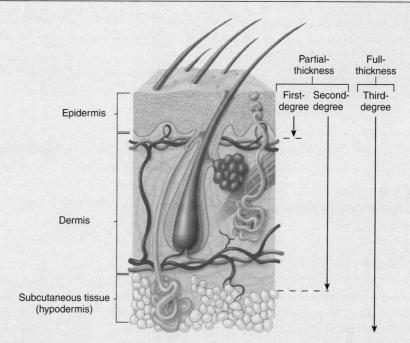

Figure B Burns

Parts of the skin damaged by burns of different degrees. Partial-thickness burns are subdivided into first-degree burns (damage to only the epidermis) and second-degree burns (damage to the epidermis and part of the dermis). Full-thickness, or third-degree, burns destroy the epidermis, dermis, and sometimes deeper tissues.

which form a new epidermis that covers the dermis. This is the same process of epidermis formation that occurs in superficial second-degree burns.

When it is not possible or practical to move skin from one part of the body to a burn site, artificial skin or grafts from human cadavers or pigs are used. These techniques are often unsatisfactory because the

body's immune system recognizes the graft as a foreign substance and rejects it. A solution to this problem is laboratory-grown skin. A piece of healthy skin from the burn victim is removed and placed into a flask with nutrients and hormones that stimulate rapid growth. The skin that is produced consists only of epidermis and does not contain glands or hair.

---

sweat glands located in the breasts. They produce milk. The structure and regulation of mammary glands are discussed in chapters 24 and 25. ▼

> 23. *What secretion is produced by the sebaceous glands? What is the function of the secretion?*
> 24. *Which glands of the skin are responsible for cooling the body? Which glands are involved with the production of body odor?*
> 25. *Where are ceruminous and mammary glands located, and what secretions do they produce?*

Nails

The distal ends of the digits of humans and other primates have nails, whereas reptiles, birds, and most mammals have claws or hooves. The

nail is a thin plate, consisting of layers of dead stratum corneum cells with hard keratin. The visible part of the nail is the **nail body,** and the part of the nail covered by skin is the **nail root** (figure 5.8). The lateral and proximal edges of the nail are held in place by a fold of skin called the **nail fold.** The **eponychium** (ep-ō-nik′ē-ŭm, *epi*, upon + *onyx*, nail), or **cuticle,** is stratum corneum from the nail fold that extends onto the nail body. The nail root extends distally from the **nail matrix.** The nail also attaches to the underlying **nail bed,** which is located distal to the nail matrix. The nail matrix and bed are epithelial tissue with a stratum basale that gives rise to the cells that form the nail. The nail matrix is thicker than the nail bed and produces most of the nail. A small part of the nail matrix, the **lunula** (loo′noo-lă, moon), can be seen through the nail body as a whitish, crescent-shaped area at the base of the nail (see figure 5.8*a*). The production of cells within the nail matrix results in growth of the nail. Unlike hair,

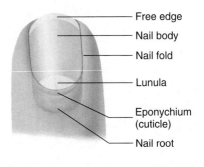

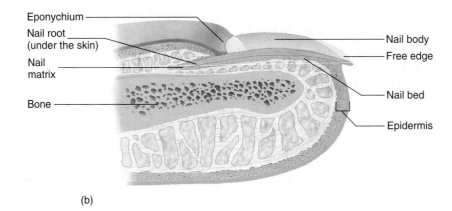

Figure 5.8   NAIL

(*a*) Dorsal view. (*b*) Lateral view of a sagittal section through the nail and finger.

Acne

Acne (aknē) is an inflammation of the hair follicles and sebaceous glands. Four factors are believed to be involved: hormones, sebum, abnormal keratinization, and the bacterium *Propionibacterium acnes*. The lesions of acne begin with the overproduction of epidermal cells in the hair follicle. These cells are shed from the wall of the hair follicle, and they stick to one another to form a mass of cells mixed with sebum that blocks the hair follicle. During puberty, hormones, especially testosterone, stimulate the sebaceous glands, and sebum production increases. Because both the testes and the ovaries produce testosterone, the effect is seen in males and females. An accumulation of sebum behind the blockage produces a whitehead. A blackhead develops when the accumulating mass of cells and sebum pushes through the opening of the hair follicle. Although there is general agreement that dirt is not responsible for the black color, the exact cause of the black color in blackheads is disputed. A pimple results if the wall of the hair follicle ruptures, forming an entry into the surrounding tissue. *P. acnes* and other bacteria stimulate an inflammatory response that results in the formation of a red pimple filled with pus. If tissue damage is extensive, scarring occurs.

nails grow continuously and do not have a resting stage. Nails grow at an average rate of 0.5–1.2 mm per day, and fingernails grow more rapidly than toenails. ▼

> 26. **Name the parts of a nail. Which part produces most of the nail? What is the lunula?**
> 27. **Do nails have growth stages?**

Summary of Integumentary System Functions

Protection

The integumentary system performs many protective functions:

1. The intact skin plays an important role in reducing water loss because its lipids act as a barrier to the diffusion of water.
2. The skin prevents the entry of microorganisms and other foreign substances into the body. Secretions from skin

glands also produce an environment unsuitable for some microorganisms.
3. The stratified squamous epithelium of the skin protects underlying structures against abrasion.
4. Melanin absorbs ultraviolet light and protects underlying structures from its damaging effects.
5. Hair provides protection in several ways: The hair on the head acts as a heat insulator, eyebrows keep sweat out of the eyes, eyelashes protect the eyes from foreign objects, and hair in the nose and ears prevents the entry of dust and other materials.
6. The nails protect the ends of the digits from damage and can be used in defense.

Sensation

The skin has receptors in the epidermis and dermis that can detect heat, cold, touch, pressure, and pain (see chapter 12). Although hair does not have a nerve supply, movement of the hair can be detected by sensory receptors around the hair follicle.

Temperature Regulation

Body temperature is affected by blood flow through the skin. When blood vessels (arterioles) in the dermis dilate, there is increased flow of warm blood from deeper structures to the skin and heat loss increases (figure 5.9, steps 1 and 2). Body temperature tends to increase as a result of exercise, fever, or an increase in environmental temperature. Homeostasis is maintained by the loss of excess heat. To counteract environmental heat gain or to get rid of excess heat produced by the body, sweat is produced. The sweat spreads over the surface of the skin; as it evaporates, heat is lost from the body.

When blood vessels in the dermis constrict, there is decreased flow of warm blood from deeper structures to the skin and heat loss decreases (figure 5.9, steps 3 and 4). If body temperature begins to drop below normal, heat can be conserved by a decrease in the diameter of dermal blood vessels.

Contraction of the arrector pili muscles causes hair to stand on end, but, with the sparse amount of hair covering the body, this does not significantly reduce heat loss in humans. Hair on the head, however, is an effective insulator. General temperature regulation is considered in chapter 22.

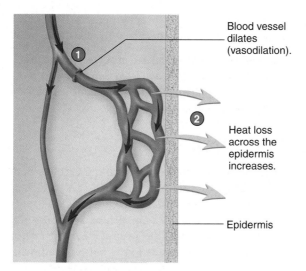

Blood vessel dilates (vasodilation).

Heat loss across the epidermis increases.

Epidermis

1. Blood vessel dilation results in increased blood flow toward the surface of the skin.
2. Increased blood flow beneath the epidermis results in increased heat loss (*gold arrows*).

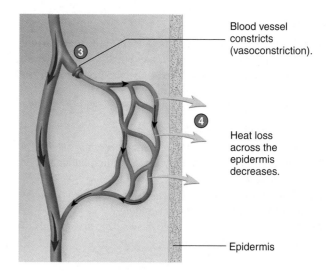

Blood vessel constricts (vasoconstriction).

Heat loss across the epidermis decreases.

Epidermis

3. Blood vessel constriction results in decreased blood flow toward the surface of the skin.
4. Decreased blood flow beneath the epidermis results in decreased heat loss.

Process Figure 5.9   Heat Exchange in the Skin

CASE STUDY | Frostbite

Billy was hiking in the mountains during the fall. Unexpectedly, a cold front moved in and the temperature dropped to well below freezing. Billy was unprepared for the temperature change, and he did not have a hat or ear muffs. As the temperature dropped, his ears and nose became pale in color. After a continued decrease in temperature and continued exposure to the cold, every 15–20 minutes his ears and nose turned red for 5–10 minutes, then became pale again. After several hours, Billy managed to hike back to the trail head. By then, he was very chilled and had no sensation in his ears or nose. As he looked in the rearview mirror of his car, he could see that the skin of his ears and nose had turned white. It took Billy 2 hours to drive to the nearest emergency room, where he was informed that the white-colored skin meant he had frostbite of his ears and nose. About 2 weeks later, the frostbitten skin peeled and, despite treatment with an antibiotic, Billy developed an infection of his right ear. Eventually, he recovered, but he lost part of his right ear.

P R E D I C T

Frostbite is the most common type of freezing injury. When skin temperature drops below 0°C (32°F), the skin freezes and ice crystal formation damages tissues.

a. Using figure 5.9, describe the mechanism responsible for Billy's ears and nose becoming pale. How is this beneficial when the ambient temperature is decreasing?
b. Explain what happened when Billy's ears and nose periodically turned red. How is this beneficial when the ambient temperature is very cold?
c. What is the significance of Billy's ears and nose turning and staying white?
d. Why is a person with frostbite likely to develop an infection of the affected part of the body?

Vitamin D Production

Vitamin D production begins when a molecule (7-dehydrocholesterol) in the skin is exposed to ultraviolet light and is converted into a vitamin D precursor molecule (cholecalciferol). The precursor is carried by the blood to the liver, where it is modified, and then to the kidneys, where the precursor is modified further to form active vitamin D (calcitriol). If exposed to enough ultraviolet light, humans can produce all the vitamin D they need. Many people need to ingest vitamin D, however, because clothing and indoor living reduce their exposure to ultraviolet light. Fatty fish (and fish oils) and vitamin D–fortified milk are the best sources of vitamin D. Eggs, butter, and liver contain small amounts of vitamin D but are not considered significant sources because too large a serving size is necessary to meet daily vitamin D requirements. Adequate levels of vitamin D are necessary because vitamin D stimulates calcium and phosphate uptake in the intestines. These substances are necessary for normal bone metabolism (see chapter 6) and normal muscle function (see chapter 8).

Excretion

Excretion is the removal of waste products from the body. In addition to water and salts, sweat contains a small amount of waste products, such as urea, uric acid, and ammonia. Even though large amounts of sweat can be lost from the body, the sweat glands do not play a significant role in the excretion of waste products. ▼

> 28. *How does the skin provide protection?*
> 29. *List the types of sensations detected by receptors in the skin.*
> 30. *How does the integumentary system help regulate body temperature?*
> 31. *What role does the skin play in the production of vitamin D? What are the functions of vitamin D?*
> 32. *What substances are excreted in sweat? Is the skin an important site of excretion?*

The Integumentary System as a Diagnostic Aid

The integumentary system is useful in diagnosis because it is observed easily and often reflects events occurring in other parts of the body. For example, cyanosis, a bluish color caused by decreased blood oxygen content, is an indication of impaired circulatory or respiratory function. A yellowish skin color, **jaundice** (jawn′dis), can occur when the liver is damaged by a disease such as viral hepatitis. Normally, the liver secretes bile pigments, which are products of the breakdown of worn-out red blood cells, into the intestine. Bile pigments are yellow, and their buildup in the blood and tissues can indicate an impairment of liver function.

Rashes and lesions in the skin can be symptoms of problems elsewhere in the body. For example, scarlet fever results from a bacterial infection in the throat. The bacteria release a toxin into the blood that causes a pink–red rash in the skin. The development of a rash can also indicate an allergic reaction to foods or drugs, such as penicillin.

The condition of the skin, hair, and nails is affected by nutritional status. In vitamin A deficiency the skin produces excess keratin and assumes a characteristic sandpaper texture, whereas in iron-deficiency anemia the nails lose their normal contour and become flat or concave (spoon-shaped).

Hair concentrates many substances that can be detected by laboratory analysis, and comparison of a patient's hair to a "normal" hair can be useful in certain diagnoses. For examples, lead poisoning results in high levels of lead in the hair. The use of hair

analysis as a screening test to determine the general health or nutritional status of an individual is unreliable, however. ▼

> 33. *What kinds of problems are indicated by cyanosis, jaundice, and rashes and lesions in the skin?*
> 34. *How can the integumentary system be used to determine nutritional status and exposure to toxins?*

Skin Cancer

Skin cancer is the most common type of cancer. Most skin cancers result from damage caused by the ultraviolet (UV) radiation in sunlight. Some skin cancers are induced by chemicals, x-rays, depression of the immune system, and inflammation, and some are inherited.

UV radiation damages the genes (DNA) in epidermal cells, producing mutations. If a mutation is not repaired, when a cell divides by mitosis the mutation is passed to one of the two daughter cells. An accumulation of mutations affecting oncogenes and tumor-suppressing genes in epidermal cells can lead to uncontrolled cell division and skin cancer (see "Cancer," chapter 4).

The amount of protective melanin in the skin affects the likelihood of developing skin cancer. Fair-skinned individuals, with less melanin, are genetically predisposed to develop skin cancer, compared with dark-skinned individuals, who have more melanin. Long-term or intense exposure to UV radiation also increases the risk of developing skin cancer. Thus, individuals who are older than 50, who have engaged in repeated recreational or occupational exposure to the sun, or who have experienced sunburn are at increased risk. The sites of development of most skin cancers are the parts of the body most exposed to sunlight, such as the face, neck, ears, and dorsum of the forearm and hand.

There are three types of skin cancer: basal cell carcinoma, squamous cell carcinoma, and melanoma (figure 5.10). **Basal cell carcinoma,** the most common type of skin cancer, arises from cells in the stratum basale. Basal cell carcinomas have a varied appearance. Some are open sores that bleed, ooze, or crust for several weeks. Others are reddish patches; shiny, pearly, or translucent bumps; or scarlike areas of shiny, taut skin. A physician should be consulted for the diagnosis and treatment of all skin cancers. Removal or destruction of the tumor cures most cases of basal cell carcinoma.

An **actinic keratosis** (ak-tin′ik ker-ă-tō′sis) is a small, scaly, crusty bump that arises on the surface of the skin. If untreated, about 2%–5% of actinic keratoses can progress to **squamous cell carcinoma,** the second most common type of skin cancer.

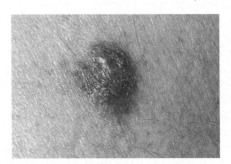
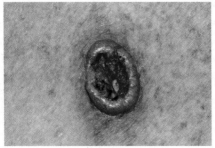
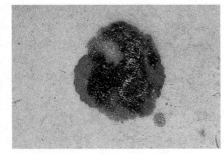

(a) Basal cell carcinoma   (b) Squamous cell carcinoma   (c) Melanoma

Figure 5.10   Cancer of the Skin

Systems Interactions

	Effects of the Integumentary System on Other Systems	Effects of Other Systems on the Integumentary System
Skeletal System	Produces a vitamin D precursor that is converted to active vitamin D, which increases the absorption of the calcium and phosphorus necessary for bone growth and maintenance Protects against abrasion and ultraviolet light	Provides support (for example, the skin of the ears and scalp)
Muscular System	Produces a vitamin D precursor that is converted to active vitamin D, which increases the absorption of the calcium necessary for muscle contraction Protects against abrasion and ultraviolet light	Moves skin for facial expressions Provides support Produces heat
Nervous System	Contains receptors for heat, cold, touch, pressure, and pain	Regulates body temperature by controlling the activity of sweat glands and blood vessels Stimulates arrector pili to contract
Endocrine System	Produces a vitamin D precursor that is converted in the liver and kidneys to active vitamin D, which functions as a hormone	Melanocyte-stimulating hormone stimulates increased melanin production
Cardiovascular System	Loss of heat from dermal vessels helps regulate body temperature	Delivers oxygen, nutrients, hormones, and immune cells Removes carbon dioxide, waste products, and toxins
Lymphatic System and Immunity	Forms a barrier to microorganisms Langerhans cells detect foreign substances	Immune cells provide protection against microorganisms and toxins and promote tissue repair by releasing chemical mediators Removes excess interstitial fluid
Respiratory System	Nasal hair prevents the entry of dust and other foreign materials	Provides oxygen and removes carbon dioxide Helps maintain the body's pH
Digestive System	Produces a vitamin D precursor that is converted in the liver and kidneys to active vitamin D, which increases the absorption of calcium and phosphorus in the small intestine	Provides nutrients and water
Urinary System	Reduces water loss Removes a very small amount of waste products in sweat	Removes waste products Helps maintain the body's pH, ion, and water balance
Reproductive System	Mammary glands produce milk Areolar glands help prevent chafing during nursing Covers external genitalia	Sex hormones increase sebum production (contributing to acne), increase apocrine gland secretion (contributing to body odor), and stimulate axillary and pubic hair growth Increases melanin production during pregnancy

Squamous cell carcinoma arises from cells in the stratum spinosum and can appear as a wartlike growth; a persistent, scaly red patch; an open sore; or an elevated growth with a central depression. Bleeding from these lesions can occur. Removal or destruction of the tumor cures most cases of squamous cell carcinoma.

Melanoma (mel′ă-nō′mă) is the least common, but most deadly, type of skin cancer, accounting for over 77% of the skin cancer deaths in the United States. Because they arise from melanocytes, most melanomas are black or brown, but occasionally a melanoma stops producing melanin and appears skin-colored, pink, red, or purple. About 40% of melanomas develop in preexisting moles. Treatment of melanomas when they are confined to the epidermis is almost always successful. If a melanoma invades the dermis and metastasizes to other parts of the body, it is difficult to treat and can be deadly.

Early detection and treatment of melanoma before it metastasizes can prevent death. Detection of melanoma can be accomplished by routine examination of the skin and the application of the **ABCDE rule,** in which *A* stands for asymmetry (one side of the lesion does not match the other side), *B* is for border irregularity (the edges are ragged, notched, or blurred), *C* is for color (pigmentation is not uniform), *D* is for diameter (greater than 6 mm), and *E* is for evolving (lesion changes over time). Evolving lesions change size, shape, elevation, or color; they may bleed, crust, itch, or become tender.

Limiting exposure to the sun and using sunscreens can reduce the likelihood of developing skin cancer in everyone. Two types of UV radiation play a role. Ultraviolet-B (UVB) (290–320 nm) radiation is the most potent for causing sunburn, is the main cause of basal and squamous cell carcinomas, and is a significant cause of melanoma. Ultraviolet-A (UVA) (320–400 nm) also contributes to skin cancer development, especially melanoma. It also penetrates into the dermis, causing wrinkling and leathering of the skin. A broad-spectrum sunscreen, which protects against both UVB and UVA, with a sun protection factor (SPF) of at least 15, is recommended by the Skin Cancer Foundation.

Someday increased protection against UV radiation may be achieved by stimulating tanning. Melanotan I is being tested. It is a synthetic version of melanocyte-stimulating hormone, which stimulates increased melanin production by melanocytes.

Sunless tanning stains the dead surface cells of the epidermis with dihydroxyacetone (DHA). At best, DHA provides a SPF of only 2 to 4. Although it does not provide adequate protection against UV radiation, it is beneficial in the sense that a tanned appearance is achieved without exposing the skin to the damaging effects of UV radiation. ▼

35. *What is the most common cause of skin cancer? Describe three types of skin cancer and the risks of each type.*
36. *What is the ABCDE rule?*
37. *What is the role of UVA and UVB in skin cancer and aging of the skin?*

Effects of Aging on the Integumentary System

As the body ages, the skin is more easily damaged because the epidermis thins and the amount of collagen in the dermis decreases. Skin infections are more likely and repair of the skin occurs more slowly. A decrease in the number of elastic fibers in the dermis and loss of fat from the subcutaneous tissue cause the skin to sag and wrinkle. A decrease in the activity of sweat glands and a decrease in the blood supply to the dermis result in a poor ability to regulate body temperature. The skin becomes drier as sebaceous gland activity decreases. The number of melanocytes generally decreases, but in some areas the number of melanocytes increases to produce **age spots.** Note that age spots are different from freckles, which are caused by increased melanin production. Gray or white hair also results because of a decrease in or a lack of melanin production. Skin that is exposed to sunlight shows signs of aging more rapidly than nonexposed skin, so avoiding overexposure to sunlight and using sun blockers is advisable. ▼

38. *Compared with young skin, why is aged skin more likely to be damaged, wrinkled, and dry?*
39. *Why is heat potentially dangerous to the elderly?*
40. *What are age spots?*

S U M M A R Y

Functions of the Integumentary System (p. 106)

1. The integumentary system consists of the skin, hair, glands, and nails.
2. The integumentary system protects us from the external environment. Other functions include sensation, temperature regulation, vitamin D production, and excretion of small amounts of waste products.

Skin (p. 106)
Epidermis

1. The epidermis is stratified squamous epithelium divided into five strata.
   - Cells are produced in the stratum basale.
   - The stratum corneum is many layers of dead, squamous cells containing keratin. The most superficial layers are sloughed.
2. Keratinization is the transformation of stratum basale cells into stratum corneum cells.
   - Structural strength results from keratin inside the cells and from desmosomes, which hold the cells together.

   - Permeability characteristics result from lipids surrounding the cells.

Thick and Thin Skin

1. Thick skin has all five epithelial strata.
2. Thin skin contains fewer cell layers per stratum, and the stratum lucidum is usually absent. Hair is found only in thin skin.

Skin Color

1. Melanocytes produce melanin inside melanosomes and then transfer the melanin to keratinocytes. The size and distribution of melanosomes determine skin color. Melanin production is determined genetically but can be influenced by ultraviolet light (tanning) and hormones.
2. Carotene, an ingested plant pigment, can cause the skin to appear yellowish.
3. Increased blood flow produces a red skin color, whereas a decreased blood flow causes a pale skin. Decreased oxygen content in the blood results in a bluish color called cyanosis.

Dermis

1. The dermis is connective tissue divided into two layers.
2. The papillary layer has projections called dermal papillae and is loose connective tissue that is well supplied with capillaries.
3. The reticular layer is the main layer. It is dense irregular connective tissue consisting mostly of collagen.

Subcutaneous Tissue (p. 111)

1. Located beneath the dermis, the subcutaneous tissue is loose connective tissue that contains collagen and elastin fibers.
2. Subcutaneous tissue attaches the skin to underlying structures and is a site of fat storage.

Accessory Skin Structures (p. 112)
Hair

1. Lanugo (fetal hair) is replaced near the time of birth by terminal hairs (scalp, eyelids, and eyebrows) and vellus hairs. At puberty, vellus hairs can be replaced with terminal hairs.
2. Hairs are columns of dead, keratinized epithelial cells. Each hair consists of a shaft (above the skin), root (below the skin), and hair bulb (site of hair cell formation).
3. Hairs have a growth stage and a resting stage.
4. Contraction of the arrector pili, which are smooth muscles, causes hair to "stand on end" and produces "goose flesh."

Glands

1. Sebaceous glands produce sebum, which oils the hair and the surface of the skin.
2. Merocrine sweat glands produce sweat, which cools the body.
3. Apocrine sweat glands produce an organic secretion that can be broken down by bacteria to cause body odor.
4. Other skin glands include ceruminous glands, which make cerumen (earwax), and mammary glands, which produce milk.

Nails

1. The nail is stratum corneum containing hard keratin.
2. The nail root is covered by skin, and the nail body is the visible part of the nail.
3. Nearly all of the nail is formed by the nail matrix, but the nail bed contributes.
4. The lunula is the part of the nail matrix visible through the nail body.

Summary of Integumentary System Functions (p. 116)
Protection

1. The skin protects against abrasion and ultraviolet light, prevents the entry of microorganisms, helps regulate body temperature, and prevents water loss.
2. Hair protects against abrasion and ultraviolet light and is a heat insulator.
3. Nails protect the ends of the digits.

Sensation

The skin contains sensory receptors for heat, cold, touch, pressure, and pain.

Temperature Regulation

1. Through dilation and constriction of blood vessels, the skin controls heat loss from the body.
2. Sweat glands produce sweat, which evaporates and lowers body temperature.

Vitamin D Production

1. Ultraviolet light stimulates the production of a precursor molecule in the skin that is modified by the liver and kidneys into vitamin D.
2. Vitamin D increases calcium uptake in the intestines.

Excretion

Skin glands remove small amounts of waste products but are not important in excretion.

The Integumentary System as a Diagnostic Aid (p. 118)

The integumentary system is easily observed and often reflects events occurring in other parts of the body (e.g., cyanosis, jaundice, rashes).

Skin Cancer (p. 118)

1. Basal cell carcinoma involves the cells of the stratum basale and is readily treatable.
2. Squamous cell carcinoma involves the cells of the stratum spinosum and can metastasize.
3. Malignant melanoma involves melanocytes, can metastasize, and is often fatal.

Effects of Aging on the Integumentary System (p. 120)

1. Blood flow to the skin is reduced, the skin becomes thinner, and elasticity is lost.
2. Sweat and sebaceous glands are less active, and the number of melanocytes decreases.

REVIEW AND COMPREHENSION

1. A layer of skin (where mitosis occurs) that replaces cells lost from the outer layer of the epidermis is the
   a. stratum corneum.
   b. stratum basale.
   c. stratum lucidum.
   d. reticular layer.
   e. subcutaneous tissue.

For questions 2–6, match the layer of the epidermis with the correct description or function:
   a. stratum basale
   b. stratum corneum
   c. stratum granulosum
   d. stratum lucidum
   e. stratum spinosum

2. Production of keratin fibers; formation of lamellar bodies

3. Desquamation occurs; 25 or more layers of dead squamous cells

4. Production of cells; melanocytes produce and contribute melanin; hemidesmosomes present

5. Production of keratohyalin granules; lamellar bodies release lipids; cells die

6. Dispersion of keratohyalin around keratin fibers; layer appears transparent; cells dead

7. In which of these areas of the body is thick skin found?
   a. back of the hand
   b. abdomen
   c. over the shin
   d. bridge of the nose
   e. heel of the foot

8. The function of melanin in the skin is to
   a. lubricate the skin.
   b. prevent skin infections.
   c. protect the skin from ultraviolet light.
   d. reduce water loss.
   e. help regulate body temperature.

9. Concerning skin color, which of these statements is *not* correctly matched?
   a. skin appears yellow—carotene present
   b. no skin pigmentation (albinism)—genetic disorder
   c. skin tans—increased melanin production

d. skin appears blue (cyanosis)—oxygenated blood
e. African-Americans darker than whites—more melanin in African-American skin

For questions 10–13, match the layer of the dermis with the correct description or function:

a. papillary layer   b. reticular layer

10. Layer of dermis closest to the epidermis

11. Layer of dermis responsible for most of the structural strength of the skin

12. Layer of dermis responsible for fingerprints and footprints

13. Layer of dermis responsible for cleavage lines and striae

14. After birth, the type of hair on the scalp, eyelids, and eyebrows is
a. lanugo.   b. terminal hair.   c. vellus hair.

15. Hair
a. is produced by the dermal root sheath.
b. consists of living keratinized epithelial cells.
c. is colored by melanin.
d. contains mostly soft keratin.
e. grows from the tip.

16. Given these parts of a hair and hair follicle:
1. cortex   4. epithelial root sheath
2. cuticle   5. medulla
3. dermal root sheath

Arrange the structures in the correct order from the outside of the hair follicle to the center of the hair.
a. 1,4,3,5,2   c. 3,4,2,1,5   e. 5,4,3,2,1
b. 2,1,5,3,4   d. 4,3,1,2,5

17. Concerning hair growth,
a. hair falls out of the hair follicle at the end of the growth stage.
b. most of the hair on the body grows continuously.
c. genetic factors and the hormone testosterone are involved in "pattern baldness."

d. eyebrows have a longer growth stage and resting stage than scalp hair.

18. Smooth muscles that produce "goose bumps" when they contract and are attached to hair follicles are called
a. arrector pili.   c. hair bulbs.
b. dermal papillae.   d. root sheaths.

For questions 19–21, match the type of gland with the correct description or function:
a. apocrine sweat gland   c. sebaceous gland
b. merocrine sweat gland

19. Alveolar glands that produce a white, oily substance; usually open into hair follicles

20. Coiled, tubular glands that secrete a fluid that cools the body; most numerous in the palms of the hands and soles of the feet

21. Secretions from these coiled, tubular glands are broken down by bacteria to produce body odor; found in the axillae and genitalia

22. The stratum corneum of the skin grows onto the nail body as the
a. eponychium.   c. lunula.   e. nail matrix.
b. hyponychium.   d. nail bed.

23. Most of the nail is produced by the
a. eponychium.   c. nail bed.   e. dermis.
b. hyponychium.   d. nail matrix.

24. The skin aids in maintaining the calcium and phosphate levels of the body at optimum levels by participating in the production of
a. vitamin A.   c. vitamin C.   e. keratin.
b. vitamin B.   d. melanin.

25. Which of these processes increase(s) heat loss from the body?
a. dilation of dermal blood vessels (arterioles)
b. constriction of dermal blood vessels (arterioles)   e. both b and c
c. increased sweating
d. both a and c

Answers in Appendix E

C R I T I C A L   T H I N K I N G

1. The skin of infants is more easily penetrated and injured by abrasion than that of adults. Based on this fact, which stratum of the epidermis is probably much thinner in infants than that in adults?

2. The rate of water loss from the skin of a hand was measured. Following the measurement, the hand was soaked in alcohol for 15 minutes. After all the alcohol had been removed from the hand, the rate of water loss was again measured. Compared with the rate of water loss before soaking the hand in alcohol, what difference, if any, would you expect in the rate of water loss after soaking the hand in alcohol?

3. Melanocytes are found primarily in the stratum basale of the epidermis. In reference to their function, why does this location make sense?

4. It has been several weeks since Goodboy Player has competed in a tennis match. After the match, he discovers that a blister has formed beneath an old callus on his foot, and the callus has fallen off. When he examines the callus, he discovers that it appears yellow. Can you explain why?

5. Harry Fastfeet, a white man, jogs on a cold day. What color would you expect his skin to be (a) just before starting to run, (b) during the run, and (c) 5 minutes after the run?

6. A woman has stretch marks on her abdomen, yet she states that she has never been pregnant. Is this possible?

7. The lips are muscular folds forming the anterior boundary of the oral cavity. A mucous membrane covers the lips internally and the skin of the face covers them externally. The vermillion border, which is the red part of the lips, is covered by keratinized epithelium that is the transition between the epithelium of the mucous membrane and the facial skin. The vermillion border can become chapped (dry and cracked), whereas the mucous membrane and the facial skin do not. Propose as many reasons as you can to explain why the vermillion border is more prone to drying than the mucous membrane or facial skin.

8. Why are your eyelashes not a foot long? Your fingernails?

9. Pulling on hair can be quite painful, yet cutting hair is not painful. Explain.

10. Given what you know about the cause of acne, propose some ways to prevent or treat the disorder.

11. A patient has an ingrown toenail, a condition in which the nail grows into the nail fold. Would cutting the nail away from the nail fold permanently correct this condition? Why or why not?

12. Consider the following statement: Dark-skinned children are more susceptible to rickets (insufficient calcium in the bones) than fair-skinned children. Defend or refute this statement.

Answers in Appendix F

**For additional study tools, go to: www.aris.mhhe.com.
Click on your course topic and then this textbook's author name/title. Register once for a semester's worth of interactive learning activities to help improve your grade!**

CHAPTER 6

Histology and Physiology of Bones

Introduction

Sitting, standing, walking, picking up a pencil, and taking a breath all involve the skeletal system. It is the structural framework that gives the body its shape and provides protection for internal organs and soft tissues. The term *skeleton* is derived from a Greek word meaning dried, indicating that the skeleton is the dried hard parts left after the softer parts are removed. Even with the flesh and organs removed, the skeleton is easily recognized as human. Despite its association with death, however, the skeletal system actually consists of dynamic, living tissues that are capable of growth, adapt to stress, and undergo repair after injury.

▶This chapter describes the **functions of the skeletal system (p. 124)**, provides an explanation of **cartilage (p. 124)**, and examines **bone histology (p. 125)**, **bone anatomy (p. 128)**, **bone development (p. 130)**, **bone growth (p. 132)**, **bone remodeling (p. 136)**, **bone repair (p. 139)**, **calcium homeostasis (p. 140)**, and the **effects of aging on the skeletal system (p. 141)**.

Colorized scanning electron micrograph of a tiny part of a bone called an osteon. The large opening is the space through which blood vessels bring blood to the bone. The surrounding bone matrix is organized into circular layers.

Tools for Success

A virtual cadaver dissection experience

Audio (MP3) and/or video (MP4) content, available at www.mhhe.aris.com

Functions of the Skeletal System

The skeletal system has four components: bones, cartilage, tendons, and ligaments. The functions of the skeletal system include

1. *Support.* Rigid, strong bone is well suited for bearing weight and is the major supporting tissue of the body. Cartilage provides a firm yet flexible support within certain structures, such as the nose, external ear, thoracic cage, and trachea. Ligaments are strong bands of fibrous connective tissue that attach to bones and hold them together.
2. *Protection.* Bone is hard and protects the organs it surrounds. For example, the skull encloses and protects the brain, and the vertebrae surround the spinal cord. The rib cage protects the heart, lungs, and other organs of the thorax.
3. *Movement.* Skeletal muscles attach to bones by tendons, which are strong bands of connective tissue. Contraction of the skeletal muscles moves the bones, producing body movements. Joints, which are formed where two or more bones come together, allow movement between bones. Smooth cartilage covers the ends of bones within some joints, allowing the bones to move freely. Ligaments allow some movement between bones but prevent excessive movements.
4. *Storage.* Some minerals in the blood are taken into bone and stored. Should blood levels of these minerals decrease, the minerals are released from bone into the blood. The principal minerals stored are calcium and phosphorus. Fat (adipose tissue) is also stored within bone cavities. If needed, the fats are released into the blood and used by other tissues as a source of energy.
5. *Blood cell production.* Many bones contain cavities filled with red bone marrow that gives rise to blood cells and platelets (see chapter 16). ▼

> 1. **Name the four components of the skeletal system. List the five functions of the skeletal system.**

Cartilage

The three types of cartilage are hyaline cartilage, fibrocartilage, and elastic cartilage (see chapter 4). Although each type of cartilage can provide support, hyaline cartilage is most intimately associated with bone. An understanding of hyaline cartilage is important because most of the bones in the body develop from it. In addition, the growth in length of bones and bone repair often involve the production of hyaline cartilage, followed by its replacement with bone.

Hyaline cartilage consists of specialized cells that produce a matrix surrounding the cells (figure 6.1). **Chondroblasts** (kon′drō-blastz; *chondro* is from the Greek word *chondrion,* cartilage) are cells that produce cartilage matrix. When the matrix surrounds a chondroblast, it becomes a **chondrocyte** (kon′drō-sīt), which occupies a space called a **lacuna** (lă-koo′nă) within the matrix. The matrix contains collagen, which provides strength, and proteoglycans, which make cartilage resilient by trapping water (see chapter 4).

The **perichondrium** (per-i-kon′drē-ŭm) is a double-layered connective tissue sheath covering most cartilage (see figure 6.1). The outer layer of the perichondrium is dense irregular connective tissue containing fibroblasts. The inner, more delicate layer has fewer fibers and contains chondroblasts. Blood vessels and nerves penetrate the outer layer of the perichondrium but do not enter the cartilage matrix so that nutrients must diffuse through the cartilage matrix to reach the chondrocytes. **Articular** (ar-tik′ū-lăr) **cartilage,** which is the cartilage covering the ends of bones where they come together to form joints, has no perichondrium, blood vessels, or nerves.

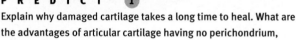

P R E D I C T   1

Explain why damaged cartilage takes a long time to heal. What are the advantages of articular cartilage having no perichondrium, blood vessels, or nerves?

Cartilage grows through appositional or interstitial growth (see figure 6.1). **Appositional growth** is the addition of new cartilage matrix on the surface of cartilage. Chondroblasts in the inner layer of the perichondrium lay down new matrix on the

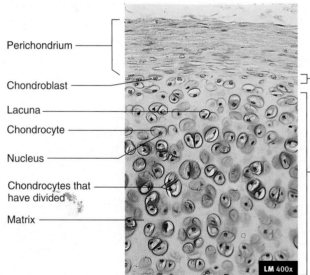

Perichondrium

Chondroblast

Lacuna

Chondrocyte

Nucleus

Chondrocytes that have divided

Matrix

LM 400x

Appositional growth
(new cartilage is added to the surface of the cartilage by chondroblasts from the inner layer of the perichondrium)

Interstitial growth
(new cartilage is formed within the cartilage by chondrocytes that divide and produce additional matrix)

Figure 6.1   Hyaline Cartilage

Photomicrograph of hyaline cartilage covered by perichondrium. Chondrocytes within lacunae are surrounded by cartilage matrix.

surface of the cartilage under the periosteum. When the chondroblasts are surrounded by matrix, they become chondrocytes in the new layer of cartilage. **Interstitial growth** is the addition of new cartilage matrix within cartilage. Chondrocytes within the tissue divide and add more matrix between the cells. The addition of new cells and matrix increases the thickness of the cartilage. ▼

> 2. Describe the structure of hyaline cartilage. Name two types of cartilage cells. What is a lacuna?
> 3. Describe the connective tissue and cells found in both layers of the perichondrium. How do nutrients from blood vessels in the perichondrium reach the chondrocytes?
> 4. Describe appositional and interstitial growth of cartilage.

Bone Histology

Bone consists of extracellular bone matrix and bone cells. The composition of the bone matrix is responsible for the characteristics of bone. The bone cells produce the bone matrix, become entrapped within it, and break it down so that new matrix can replace the old matrix.

Bone Matrix

By weight, mature bone matrix normally is approximately 35% organic and 65% inorganic material. The organic material consists primarily of collagen and proteoglycans. The inorganic material consists primarily of a calcium phosphate crystal called **hydroxyapatite** (hī-drok′sē-ap-ă-tīt), which has the molecular formula $Ca_{10}(PO_4)_6(OH)_2$.

The collagen and mineral components are responsible for the major functional characteristics of bone. Bone matrix might be said to resemble reinforced concrete. Collagen, like reinforcing steel bars, lends flexible strength to the matrix, whereas the mineral components, like concrete, give the matrix compression (weight-bearing) strength.

If all the mineral is removed from a long bone, collagen becomes the primary constituent, and the bone becomes overly flexible. On the other hand, if the collagen is removed from the bone, the mineral component becomes the primary constituent, and the bone is very brittle (figure 6.2).

Osteogenesis Imperfecta

Osteogenesis imperfecta (OI) (os′tē-ō-jen′ĕ-sis im-per-fek′tă, imperfect bone formation) is a rare disorder caused by any one of a number of faulty genes that results in either too little collagen formation or a poor quality of collagen. As a result, bone matrix has decreased flexibility and is more easily broken than normal bone. Osteogenesis imperfecta is also known as the "brittle bone" disorder. In mild forms of the disorder, children may appear normal except for a history of broken bones. It is important for children with OI to be properly diagnosed because broken bones can be associated with child abuse. Over a lifetime, the number of fractures can vary from a few to more than 100. In more severe forms of the disorder, fractures heal in poor alignment, resulting in bent limbs, short stature, curved spine, and small thorax.

P R E D I C T

In general, the bones of elderly people break more easily than the bones of younger people. Give as many possible explanations as you can for this observation.

Bone Cells

Bone cells are categorized as osteoblasts, osteocytes, and osteoclasts, which have different functions and origins.

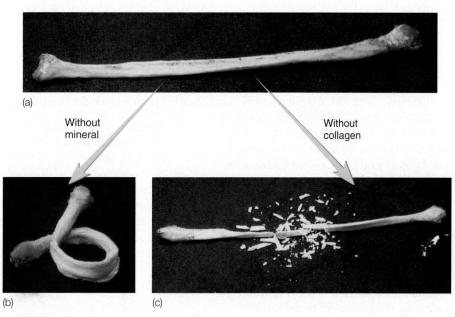

Figure 6.2   Effects of Changing the Bone Matrix

(*a*) Normal bone. (*b*) Demineralized bone, in which collagen is the primary remaining component, can be bent without breaking. (*c*) When collagen is removed, mineral is the primary remaining component, thus making the bone so brittle that it is easily shattered.

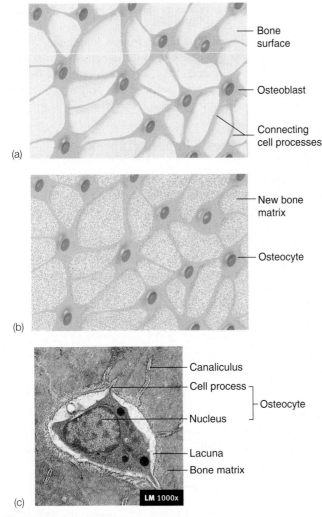

(a)

New bone matrix

Osteocyte

(b)

Canaliculus
Cell process
Nucleus
Osteocyte
Lacuna
Bone matrix

(c) LM 1000x

Figure 6.3   Ossification

(*a*) Osteoblasts on a preexisting surface, such as cartilage or bone. The cell processes of different osteoblasts join together. (*b*) Osteoblasts have produced bone matrix. The osteoblasts are now osteocytes. (*c*) Photomicrograph of an osteocyte in a lacuna with cell processes in the canaliculi.

Osteoblasts

Osteoblasts (os′tē-ō-blastz) produce new bone matrix. They have an extensive endoplasmic reticulum, numerous ribosomes, and Golgi apparatuses. Osteoblasts produce collagen and proteoglycans, which are packaged into vesicles by the Golgi apparatus and released from the cell by exocytosis. Osteoblasts also release **matrix vesicles,** which are membrane-bound sacs formed when the plasma membrane buds, or protrudes outward, and pinches off. The matrix vesicles concentrate Ca^{2+} and PO_4^{3-} and form needlelike hydroxyapatite crystals. When the hydroxyapatite crystals are released from the matrix vesicles, they act as templates, or "seeds," which stimulate further hydroxyapatite formation and mineralization of the matrix.

   Ossification (os′i-fi-kā′shŭn), or **osteogenesis** (os′tē-ō-jen′ĕ-sis), is the formation of bone by osteoblasts. Ossification occurs by appositional growth on the surface of previously existing bone or cartilage. For example, osteoblasts beneath the periosteum cover the surface of preexisting bone (figure 6.3). Elongated cell processes from osteoblasts connect to cell processes of other osteoblasts through gap junctions (see chapter 4). Bone matrix pro-

duced by the osteoblasts covers the older bone surface and surrounds the osteoblast cell bodies and processes. In a sense, the cell bodies and processes form a "mold" around which the matrix is formed during ossification. The result is a new layer of bone.

Osteocytes

Osteocytes (os′tē-ō-sītz) are mature bone cells that maintain the bone matrix. Once an osteoblast becomes surrounded by bone matrix, it becomes an osteocyte. The spaces occupied by the osteocyte cell bodies are called **lacunae** (lă-koo′nē), and the spaces occupied by the osteocyte cell processes are called **canaliculi** (kan-ă-lik′ū-lī, little canals) (see figure 6.3*c*). Bone differs from cartilage in that the processes of bone cells are in contact with one another through the canaliculi. Instead of diffusing through the mineralized matrix, nutrients and gases can pass through the small amount of fluid surrounding the cells in the canaliculi and lacunae or pass from cell to cell through the gap junctions connecting the cell processes.

Osteoclasts

Osteoclasts (os′tē-ō-klastz) are responsible for the **resorption,** or breakdown, of bone. They are large cells with several nuclei. Osteoclasts release H^+, which produce an acid environment necessary for the decalcification of the bone matrix. The osteoclasts also release enzymes that digest the protein components of the matrix. Through the process of endocytosis, some of the breakdown products of bone resorption are taken into the osteoclast.

Origin of Bone Cells

Connective tissue develops embryologically from mesenchymal cells (see chapter 4). Some of the mesenchymal cells become **stem cells,** which can replicate and give rise to more specialized cell types. **Osteochondral progenitor cells** are stem cells that can become osteoblasts or chondroblasts. Osteochondral progenitor cells are located in the inner layer of the perichondrium, the inner layer of the periosteum, and the endosteum. From these locations, they are a potential source of new osteoblasts or chondroblasts.

   Osteoblasts are derived from osteochondral progenitor cells, and osteocytes are derived from osteoblasts. Osteoclasts are derived from stem cells in red bone marrow (see chapter 16). The red bone marrow stem cells that give rise to a type of white blood cell, called a monocyte, also are the source of osteoclasts. ▼

5. *Name the components of bone matrix, and explain their contribution to bone flexibility and the ability of bones to bear weight.*
6. *What are the functions of osteoblasts, osteocytes, and osteoclasts?*
7. *Describe the formation of new bone by appositional growth. Name the spaces that are occupied by osteocyte cell bodies and cell processes.*
8. *What cells give rise to osteochondral progenitor cells? What kinds of cells are derived from osteochondral progenitor cells? What types of cells give rise to osteoclasts?*

Woven and Lamellar Bone

Bone tissue is classified as either woven or lamellar bone, according to the organization of collagen fibers within the bone matrix. In **woven**

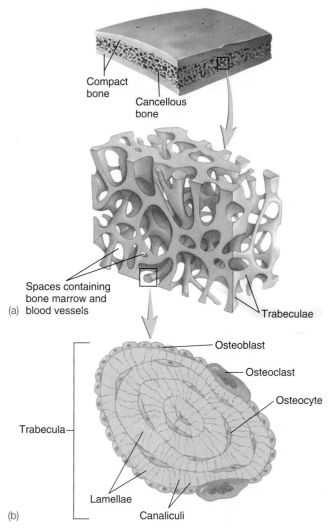

Compact bone

Cancellous bone

Spaces containing bone marrow and
(a) blood vessels

Trabeculae

Osteoblast

Osteoclast

Osteocyte

Trabecula

Lamellae

Canaliculi

(b)

Figure 6.4   **Cancellous Bone**

(*a*) Beams of bone, the trabeculae, surround spaces in the bone. In life, the spaces are filled with red or yellow bone marrow and with blood vessels. (*b*) Transverse section of a trabecula.

bone, the collagen fibers are randomly oriented in many directions. Woven bone is first formed during fetal development or during the repair of a fracture. After its formation, osteoclasts break down the woven bone and osteoblasts build new matrix. The process of removing old bone and adding new bone is called **remodeling.** It is an important process discussed later in this chapter (see "Bone Remodeling," p. 136). Woven bone is remodeled to form lamellar bone.

Lamellar bone is mature bone that is organized into thin sheets or layers approximately 3–7 micrometers (μm) thick called **lamellae** (lă-mel′ē). In general, the collagen fibers of one lamella lie parallel to one another, but at an angle to the collagen fibers in the adjacent lamellae. Osteocytes, within their lacunae, are arranged in layers sandwiched between lamellae.

Cancellous and Compact Bone

Bone, whether woven or lamellar, can be classified according to the amount of bone matrix relative to the amount of space present within the bone. Cancellous bone has less bone matrix and more space than compact bone, which has more bone matrix and less space than cancellous bone.

Cancellous bone (figure 6.4*a*) consists of thin rods or plates of interconnecting bone called **trabeculae** (tră-bek′ū-lē, beam). The trabeculae bear weight and help bones resist bending and stretching. If the stress on a bone is changed slightly (e.g., because of a fracture that heals improperly), the trabeculae realign with the new lines of stress. Between the trabeculae are spaces that, in life, are filled with bone marrow and blood vessels. Cancellous bone is sometimes called spongy bone because of its porous appearance.

Most trabeculae are thin (50–400 μm) and consist of several lamellae with osteocytes located in lacunae between the lamellae (figure 6.4*b*). Each osteocyte is associated with other osteocytes through canaliculi. Usually, no blood vessels penetrate the trabeculae, so osteocytes must obtain nutrients through their canaliculi. The surfaces of trabeculae are covered with a single layer of cells consisting mostly of osteoblasts with a few osteoclasts.

Compact bone (figure 6.5) is denser and has fewer spaces than cancellous bone. Blood vessels enter the substance of the bone itself, and the lamellae of compact bone are primarily oriented around those blood vessels. Vessels that run parallel to the long axis of the bone are contained within **central,** or **haversian** (ha-ver′shan), **canals. Concentric lamellae** are circular layers of bone matrix that surround the central canal. An **osteon** (os′tē-on), or **haversian system,** consists of a single central canal, its contents, and associated concentric lamellae and osteocytes. In cross section, an osteon resembles a circular target; the "bull's-eye" of the target is the central canal, and 4–20 concentric lamellae form the rings. Osteocytes are located in lacunae between the lamellar rings, and canaliculi connect lacunae, producing the appearance of minute cracks across the rings of the target.

The outer surfaces of compact bone are formed by **circumferential lamellae,** which are thin plates that extend around the bone (see figure 6.5). In between the osteons are **interstitial lamellae,** which are remnants of concentric or circumferential lamellae that were partially removed during bone remodeling.

Osteocytes receive nutrients and eliminate waste products through the canal system within compact bone. Blood vessels from the periosteum or medullary cavity enter the bone through **perforating,** or **Volkmann's, canals,** which run perpendicular to the long axis of the bone (see figure 6.5). Blood vessels from the perforating canals join blood vessels in the central canals. Nutrients in the blood vessels enter the central canals, pass into the canaliculi, and move through the cytoplasm of the osteocytes that occupy the canaliculi and lacunae to the most peripheral cells within each osteon. Waste products are removed in the reverse direction. ▼

9. *Distinguish between woven bone and lamellar bone. Where is woven bone found?*
10. *Describe the structure of cancellous bone. What are trabeculae, and what is their function? How do osteocytes within trabeculae obtain nutrients?*
11. *Describe the structure of compact bone. What is an osteon? Name three types of lamellae found in compact bone.*
12. *Trace the pathway nutrients must follow to go from blood vessels in the periosteum to osteocytes within osteons.*

P R E D I C T
Compact bone has perforating and central canals. Why isn't such a canal system necessary in cancellous bone?

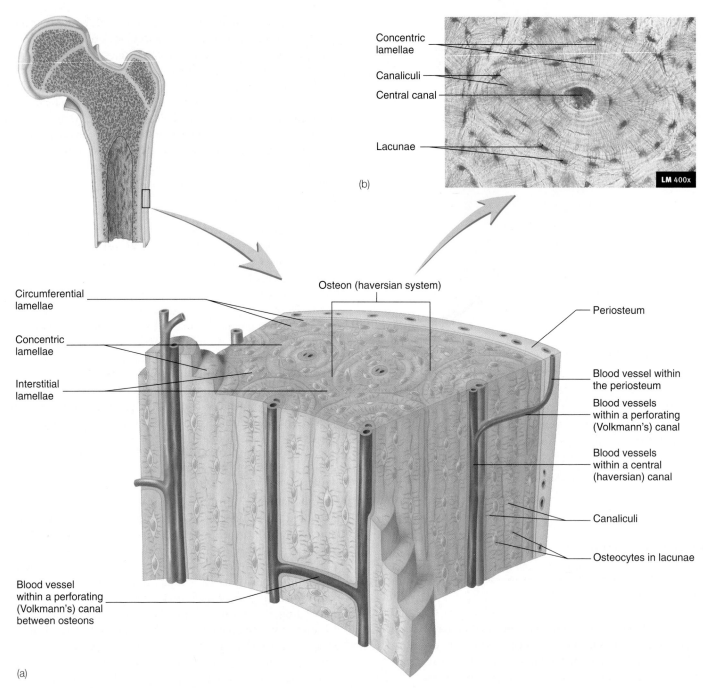

(b)

(a)

Figure 6.5   Compact Bone

(*a*) Compact bone consists mainly of osteons, which are concentric lamellae surrounding blood vessels within central canals. The outer surface of the bone is formed by circumferential lamellae, and bone between the osteons consists of interstitial lamellae. (*b*) Photomicrograph of an osteon.

Bone Anatomy

Bone Shapes

Individual bones are classified according to their shape as long, short, flat, or irregular. **Long bones** are longer than they are wide. Most of the bones of the upper and lower limbs are long bones. **Short bones** are about as wide as they are long. They are nearly cube-shaped or round and are exemplified by the bones of the wrist (carpals) and ankle (tarsals). **Flat bones** have a relatively thin, flattened shape and

are usually curved. Examples of flat bones are certain skull bones, the ribs, the breastbone (sternum), and the shoulder blades (scapulae). **Irregular bones,** such as the vertebrae and facial bones, have shapes that do not fit readily into the other three categories.

Structure of a Long Bone

The **diaphysis** (dī-af′i-sis), or shaft, of a long bone is composed primarily of compact bone (figure 6.6*a* and *b*). The end of a long bone is mostly cancellous bone, with an outer layer of compact

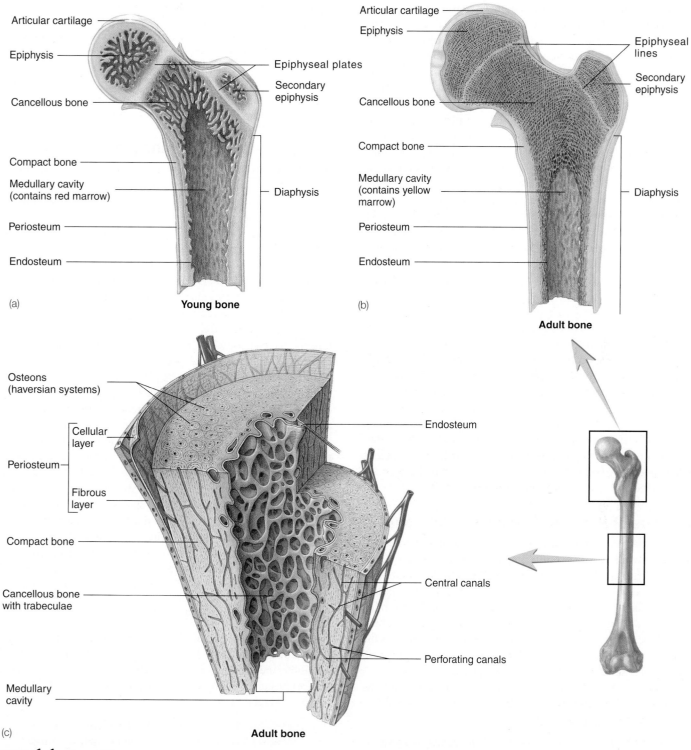

Figure 6.6   Long Bone

(a) Young long bone (the femur) showing epiphyses, epiphyseal plates, and diaphysis. The femur is unusual in that there are two epiphyses at the proximal end of the bone. (b) Adult long bone with epiphyseal lines. (c) Internal features of a portion of the diaphysis in (a).

bone. Within joints, the end of a long bone is covered with hyaline cartilage called **articular cartilage.** A long bone that is still growing has an **epiphyseal plate** (see figure 6.6a), or **growth plate,** composed of cartilage, between the diaphysis and the end of the bone, which is called an **epiphysis** (e-pif'i-sis, growing upon). The epiphyseal plate is the site of growth in bone length. When

bone growth stops, the cartilage of the epiphyseal plate is replaced by bone and is called an **epiphyseal line** (see figure 6.6b). Each long bone of the arm, forearm, thigh, and leg has one or more epiphyses (e-pif'i-sēz) on each end of the bone. Each long bone of the hand and foot has one epiphysis, which is located on the proximal or distal end of the bone.

The diaphysis of a long bone can have a large internal space called the **medullary cavity.** The cavities of cancellous bone and the medullary cavity are filled with marrow. **Red bone marrow** is the site of blood cell formation, and **yellow bone marrow** is mostly adipose tissue. In the fetus, the spaces within bones are filled with red bone marrow. The conversion of red bone marrow to yellow bone marrow begins just before birth and continues well into adulthood. Yellow bone marrow completely replaces the red bone marrow in the long bones of the limbs, except for some red bone marrow in the proximal part of the arm and thigh bones. Elsewhere, varying proportions of yellow and red bone marrow are found. In some locations, red bone marrow is completely replaced by yellow bone marrow; in others, there is a mixture of red and yellow bone marrow. For example, part of the hip bone (ilium) may contain 50% red bone marrow and 50% yellow bone marrow. Marrow from the hip bone is used as a source for donating red bone marrow because it is a large bone with more marrow than smaller bones and it is accessed relatively easily.

The **periosteum** (per-ē-os′tē-ŭm, *peri*, around + *osteon*, bone) is a connective tissue membrane that covers the outer surface of a bone (figure 6.6c). The outer fibrous layer is dense irregular collagenous connective tissue containing blood vessels and nerves. The inner cellular layer is a single layer of bone cells, which includes osteoblasts, osteoclasts, and osteochondral progenitor cells.

The **endosteum** (en-dos′tē-ŭm) is a single layer of cells lining the internal surfaces of all cavities within bones, such as the medullary cavity of the diaphysis and the smaller cavities in cancellous and compact bone (see figure 6.6c). The endosteum includes osteoblasts, osteoclasts, and osteochondral progenitor cells.

Structure of Flat, Short, and Irregular Bones

Flat bones contain an interior framework of cancellous bone sandwiched between two layers of compact bone (see figure 6.4a). Short and irregular bones have a composition similar to the ends of long bones. They have compact bone surfaces that surround a cancellous bone center with small spaces that usually are filled with marrow. Short and irregular bones are not elongated and have no diaphyses. Certain regions of these bones, however, such as the processes (projections) of irregular bones, possess epiphyseal growth plates and therefore have small epiphyses.

Some of the flat and irregular bones of the skull have air-filled spaces called **sinuses** (sī′nŭs-ĕz; see chapter 7), which are lined by mucous membranes. ▼

13. *List the four basic shapes of individual bones, and give an example of each.*
14. *Define the diaphysis, epiphysis, epiphyseal plate, and epiphyseal line of a long bone.*
15. *What are red bone marrow and yellow bone marrow? Where are they located in a child and in an adult?*
16. *Where are the periosteum and endosteum located? What types of cells are found in the periosteum and endosteum?*
17. *Compare the structure of long bones with the structure of flat, short, and irregular bones. How are compact bone and cancellous bone arranged in each?*

Bone Development

Bones develop in the fetus by intramembranous or endochondral ossification. Both processes involve the formation of bone matrix on preexisting connective tissue. Both methods initially produce woven bone, which is then remodeled. After remodeling, bone formed by intramembranous ossification cannot be distinguished from bone formed by endochondral ossification.

Intramembranous Ossification

Intramembranous (in′tră-mem′brā-nŭs, within membranes) **ossification** is the formation of bone within a connective tissue membrane. At about the eighth week of development, a connective tissue membrane has formed around the developing brain. Intramembranous ossification begins when osteochondral progenitor cells specialize to become osteoblasts. The osteoblasts lay down bone matrix on the collagen fibers of the connective tissue membrane, forming many, tiny trabeculae of woven bone (figure 6.7, step 1). The trabeculae enlarge as additional osteoblasts lay down bone matrix on their surfaces (figure 6.7, step 2). Cancellous bone forms as the trabeculae join together, resulting in an interconnected network of trabeculae separated by spaces. Red bone marrow develops within the spaces, and cells surrounding the developing bone specialize to form the periosteum. Osteoblasts from the periosteum lay down bone matrix to form an outer surface of compact bone (figure 6.7, step 3). Remodeling converts woven bone to lamellar bone and contributes to the final shape of the bone.

Centers of ossification are the locations in the membrane where ossification begins. The centers of ossification expand to form a bone by gradually ossifying the membrane. Thus, the centers of ossification have the oldest bone and the expanding edges the youngest bone. The larger membrane-covered spaces between the developing skull bones that have not yet been ossified are called **fontanels,** or soft spots (figure 6.8) (see chapter 7). The bones eventually grow together, and all the fontanels have usually closed by 2 years of age. Many skull bones, part of the mandible (lower jaw), and the diaphyses of the clavicles (collarbones) develop by intramembranous ossification.

Endochondral Ossification

Endochondral (en-dō-kon′drăl, within cartilage) is the formation of bone within cartilage. Bones of the base of the skull, part of the mandible, the epiphyses of the clavicles, and most of the remaining skeletal system develop through endochondral ossification (see figures 6.7 and 6.8).

1. Osteochondral progenitor cells specialize to become chondroblasts that produce a hyaline **cartilage model** having the approximate shape of the bone that will later be formed (figure 6.9, step 1). The cartilage model is surrounded by perichondrium, except where a joint will form connecting one bone to another bone.
2. Osteochondral progenitor cells within the perichondrium become osteoblasts that produce a **bone collar** around part of the outer surface of the diaphysis. The perichondrium becomes the periosteum once bone is produced (figure 6.9, step 2). Chondrocytes within the cartilage model **hypertrophy** (hī-per′trō-fē), or enlarge. **Calcified cartilage** is formed as the chondrocytes release matrix vesicles, which

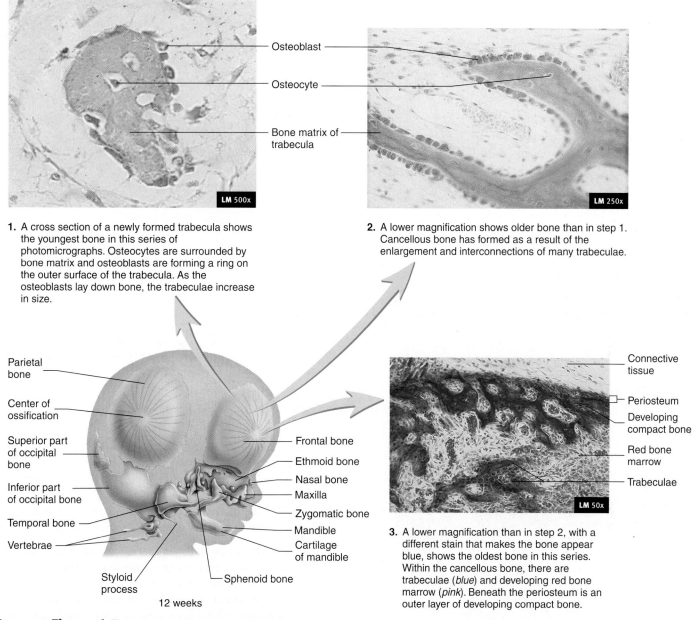

1. A cross section of a newly formed trabecula shows the youngest bone in this series of photomicrographs. Osteocytes are surrounded by bone matrix and osteoblasts are forming a ring on the outer surface of the trabecula. As the osteoblasts lay down bone, the trabeculae increase in size.

2. A lower magnification shows older bone than in step 1. Cancellous bone has formed as a result of the enlargement and interconnections of many trabeculae.

3. A lower magnification than in step 2, with a different stain that makes the bone appear blue, shows the oldest bone in this series. Within the cancellous bone, there are trabeculae (*blue*) and developing red bone marrow (*pink*). Beneath the periosteum is an outer layer of developing compact bone.

12 weeks

Process Figure 6.7   Intramembranous Ossification

The inset shows a 12-week-old fetus. Bones formed by intramembranous ossification are *yellow* and bones formed by endochondral ossification are *blue*. Intramembranous ossification starts at a center of ossification and expands outward. Therefore, the youngest bone is at the edge of the expanding bone, and the oldest bone is at the center of bones.

initiate the formation of hydroxyapatite crystals. The chondrocytes in the calcified cartilage eventually die, leaving enlarged lacunae with thin walls of calcified matrix.

3. Blood vessels grow into the calcified cartilage, bringing osteoblasts and osteoclasts from the periosteum. A **primary ossification center** forms as the osteoblasts lay down bone matrix on the surface of the calcified cartilage (figure 6.9, step 3). Cancellous bone is formed and remodeled. Red bone marrow forms within the spaces of the cancellous bone.

4. The cartilage model continues to grow and the bone collar and area of calcified cartilage enlarge (figure 6.9, step 4). Osteoclasts remove bone from the center of the diaphysis to form the medullary cavity. Within the epiphyses, calcified cartilage is formed.

5. Blood vessels grow into the calcified cartilage of each epiphysis, and a **secondary ossification center** forms as osteoblasts lay down bone matrix on the surface of the calcified cartilage (figure 6.9, step 5). The events occurring at the secondary ossification centers are the same as those occurring at the primary ossification centers, except that the spaces in the epiphyses do not enlarge to form a medullary cavity as in the diaphysis. Primary ossification centers appear during early fetal development, whereas secondary ossification centers appear in the proximal epiphysis of the femur, humerus, and tibia about 1 month before birth. A baby is considered full-term if one of these three ossification centers can be seen on radiographs at the time of birth. At about 18–20 years of age, the last

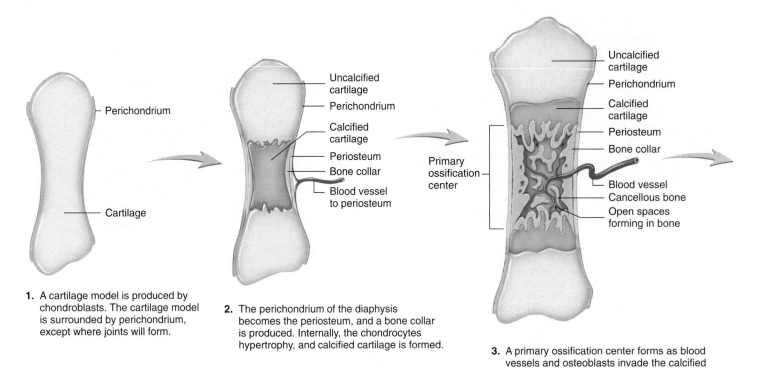

1. A cartilage model is produced by chondroblasts. The cartilage model is surrounded by perichondrium, except where joints will form.

2. The perichondrium of the diaphysis becomes the periosteum, and a bone collar is produced. Internally, the chondrocytes hypertrophy, and calcified cartilage is formed.

3. A primary ossification center forms as blood vessels and osteoblasts invade the calcified cartilage. The osteoblasts lay down bone matrix, forming cancellous bone.

Process Figure 6.9   Endochondral Ossification

Endochondral ossification begins with the formation of a cartilage model. See successive steps as indicated by the *blue arrows*.

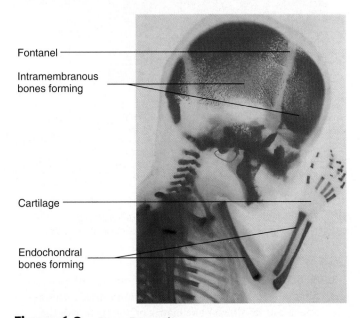

Figure 6.8   Bone Formation

An 18-week-old fetus showing intramembranous and endochondral ossification. Intramembranous ossification occurs at centers of ossification in the flat bones of the skull. Endochondral ossification has formed bones in the diaphyses of long bones. The ends of the long bones are still cartilage at this stage of development.

secondary ossification center appears in the medial epiphysis of the clavicle.

6. Replacement of cartilage by bone continues in the cartilage model until all the cartilage, except that in the epiphyseal plate and on articular surfaces, has been replaced by bone (figure 6.9, step 6).

7. In mature bone, cancellous and compact bone are fully developed and the epiphyseal plate has become the epiphyseal line. The only cartilage present is the articular cartilage at the ends of the bone (figure 6.9, step 7). ▼

18. *Describe the formation of cancellous and compact bone during intramembranous ossification. What are centers of ossification? What are fontanels?*

19. *For the process of endochondral ossification, describe the formation of these structures: cartilage model, bone collar, calcified cartilage, primary ossification center, medullary cavity, secondary ossification center, epiphyseal plate, epiphyseal line, and articular cartilage.*

20. *When do primary and secondary ossification centers appear during endochondral ossification?*

**P R E D I C T **

During endochondral ossification, calcification of cartilage results in the death of chondrocytes. However, ossification of the bone matrix does not result in the death of osteocytes. Explain.

Bone Growth

Unlike cartilage, bones cannot grow by interstitial growth. Bones increase in size only by appositional growth, the formation of new bone on the surface of older bone or cartilage. For example, trabeculae grow in size by the deposition of new bone matrix by osteoblasts onto the surface of the trabeculae (see figure 6.7, step 1).

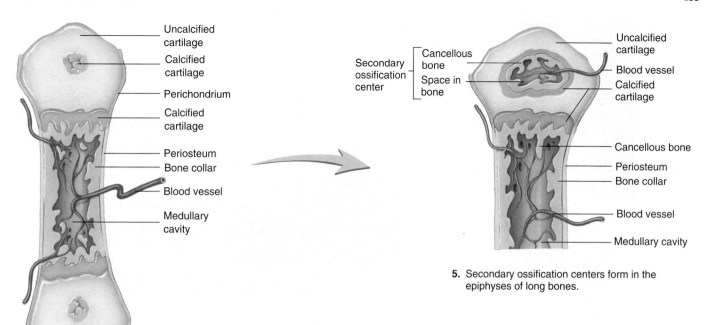

Uncalcified
cartilage

Calcified
cartilage

Perichondrium

Calcified
cartilage

Periosteum

Bone collar

Blood vessel

Medullary
cavity

Secondary
ossification
center

Cancellous
bone

Space in
bone

Uncalcified
cartilage

Blood vessel

Calcified
cartilage

Cancellous bone

Periosteum

Bone collar

Blood vessel

Medullary cavity

5. Secondary ossification centers form in the
epiphyses of long bones.

4. The process of bone collar formation, cartilage calcification,
and cancellous bone production continues. Calcified
cartilage begins to form in the epiphyses. A medullary
cavity begins to form in the center of the diaphysis.

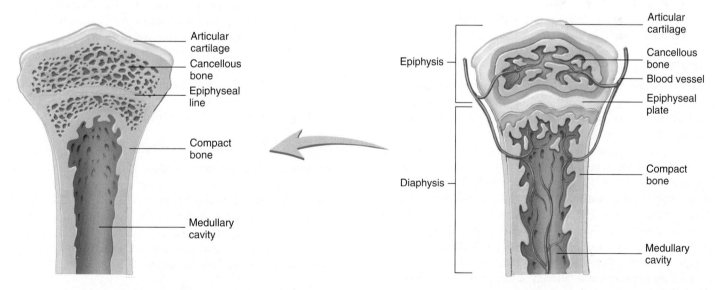

Articular
cartilage

Cancellous
bone

Epiphyseal
line

Compact
bone

Medullary
cavity

Epiphysis

Diaphysis

Articular
cartilage

Cancellous
bone

Blood vessel

Epiphyseal
plate

Compact
bone

Medullary
cavity

7. In a mature bone, the epiphyseal plate has
become the epiphyseal line and all the cartilage in the
epiphysis, except the articular cartilage, has become
bone.

6. The original cartilage model is almost completely
ossified. Unossified cartilage becomes the
epiphyseal plate and the articular cartilage.

Process Figure 6.9 (*continued*)

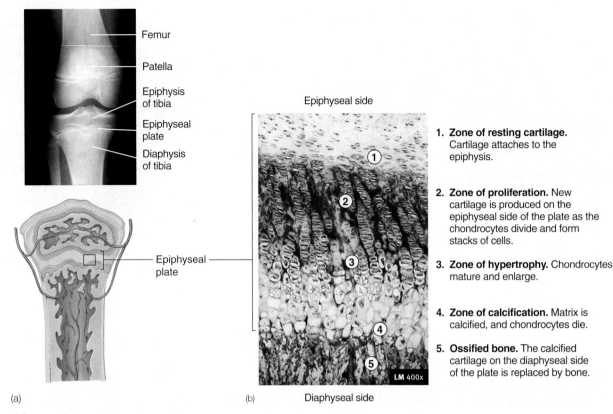

Femur

Patella

Epiphysis
of tibia

Epiphyseal
plate

Diaphysis
of tibia

Epiphyseal
plate

Epiphyseal side

1. **Zone of resting cartilage.**
   Cartilage attaches to the
   epiphysis.

2. **Zone of proliferation.** New
   cartilage is produced on the
   epiphyseal side of the plate as the
   chondrocytes divide and form
   stacks of cells.

3. **Zone of hypertrophy.** Chondrocytes
   mature and enlarge.

4. **Zone of calcification.** Matrix is
   calcified, and chondrocytes die.

5. **Ossified bone.** The calcified
   cartilage on the diaphyseal side
   of the plate is replaced by bone.

LM 400x

(a)   (b)   Diaphyseal side

Process Figure 6.10   Epiphyseal Plate

(a) Radiograph of the knee, showing the epiphyseal plate of the tibia (shinbone). Because cartilage does not appear readily on radiographs, the epiphyseal plate appears as a black area between the white diaphysis and the epiphyses. (b) Zones of the epiphyseal plate.

PREDICT

Explain why bones cannot undergo interstitial growth, as does cartilage.

Growth in Bone Length

Long bones and bony projections increase in length because of growth at the epiphyseal plate. Growth at the epiphyseal plate involves the formation of new cartilage by interstitial cartilage growth followed by appositional bone growth on the surface of the cartilage. The epiphyseal plate is organized into four zones (figure 6.10). The **zone of resting cartilage** is nearest the epiphysis and contains chondrocytes that do not divide rapidly. The chondrocytes in the **zone of proliferation** produce new cartilage through interstitial cartilage growth. The chondrocytes divide and form columns resembling stacks of plates or coins. In the **zone of hypertrophy,** the chondrocytes produced in the zone of proliferation mature and enlarge. The **zone of calcification** is very thin and contains hypertrophied chondrocytes and calcified cartilage matrix. The hypertrophied chondrocytes die and are replaced by osteoblasts from the endosteum. Through appositional bone growth, the osteoblasts deposit new bone matrix on the calcified cartilage matrix. The process of cartilage calcification and ossification in the epiphyseal plate occurs by the same basic process as the calcification and ossification of the cartilage model during endochondral bone formation.

As new cartilage cells form in the zone of proliferation, and, as these cells enlarge in the zone of hypertrophy, the overall length of the diaphysis increases (figure 6.11). The thickness of the epiphyseal

plate does not increase, however, because the rate of cartilage growth on the epiphyseal side of the plate is equal to the rate at which cartilage is replaced by bone on the diaphyseal side of the plate.

As the bones achieve normal adult size, growth in bone length ceases because the epiphyseal plate is ossified and becomes the epiphyseal line. This event, called **closure of the epiphyseal plate,** occurs between approximately 12 and 25 years of age, depending on the bone and the individual.

Growth at Articular Cartilage

Bone growth at the articular cartilage also increases the size of bones. The process of growth in articular cartilage is similar to that occurring in the epiphyseal plate, except that the chondrocyte columns are not as obvious. When bones reach their full size, the growth of articular cartilage and its replacement by bone cease. The articular cartilage, however, persists throughout life and does not become ossified as does the epiphyseal plate.

PREDICT

Growth at the epiphyseal plate stops when the epiphyseal cartilage becomes ossified. The articular cartilage, however, does not become ossified when growth of the epiphysis ceases. Explain why it is advantageous for the articular cartilage not to be ossified.

Growth in Bone Width

Long bones increase in width (diameter) and other bones increase in size or thickness through appositional bone growth beneath the

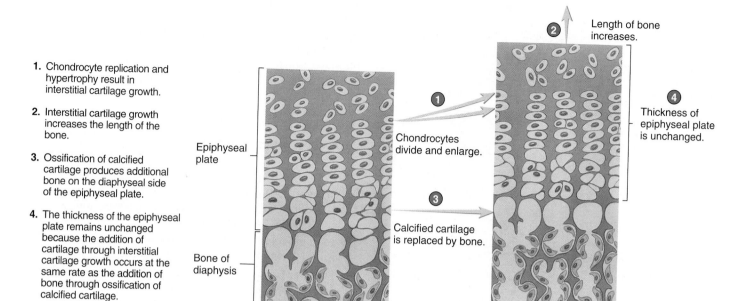

1. Chondrocyte replication and hypertrophy result in interstitial cartilage growth.

2. Interstitial cartilage growth increases the length of the bone.

3. Ossification of calcified cartilage produces additional bone on the diaphyseal side of the epiphyseal plate.

4. The thickness of the epiphyseal plate remains unchanged because the addition of cartilage through interstitial cartilage growth occurs at the same rate as the addition of bone through ossification of calcified cartilage.

Epiphyseal plate

Bone of diaphysis

Length of bone increases.

Chondrocytes divide and enlarge.

Calcified cartilage is replaced by bone.

Thickness of epiphyseal plate is unchanged.

Process Figure 6.11 Bone Growth in Length at the Epiphyseal Plate

periosteum. When bone growth in width is rapid, as in young bones or in bones during puberty, osteoblasts from the periosteum lay down bone to form a series of ridges with grooves between them (figure 6.12, step 1). The periosteum covers the bone ridges and extends down into the bottom of the grooves, and one or more blood vessels of the periosteum lie within each groove. As the osteoblasts continue to produce bone, the ridges increase in size, extend toward each other, and meet to change the groove into a tunnel (figure 6.12, step 2). The name of the periosteum in the tunnel changes to *endosteum* because the membrane now lines an internal bone surface. Osteoblasts from the endosteum lay down bone to form a concentric lamella (figure 6.12, step 3). The production of additional lamellae fills in the tunnel, encloses the blood vessel, and produces an osteon (figure 6.12, step 4).

When bone growth in width is slow, the surface of the bone becomes smooth as osteoblasts from the periosteum lay down even layers of bone to form circumferential lamellae. The circumferential lamellae are broken down during remodeling to form osteons (see "Bone Remodeling," p. 136). ▼

21. *Name and describe the events occurring in the four zones of the epiphyseal plate. Explain how the epiphyseal plate remains the same thickness while the bone increases in length.*
22. *Describe the process of growth at the articular cartilage. What happens to the epiphyseal plate and the articular - cartilage when bone growth ceases?*
23. *Describe how new osteons are produced as a bone increases in width.*

Factors Affecting Bone Growth

The potential shape and size of a bone and an individual's final adult height are determined genetically, but factors such as nutrition and hormones can greatly modify the expression of those genetic factors.

Nutrition

Inadequate intake of the materials necessary to support chondroblast and osteoblast activities, including the production of collagen and other matrix components, results in decreased cartilage and bone growth.

Vitamin D is necessary for the normal absorption of calcium from the intestines. The body can either synthesize or ingest vitamin D. Its rate of synthesis increases when the skin is exposed to sunlight (see chapter 5). Insufficient vitamin D in children causes **rickets,** a disease resulting from reduced mineralization of the bone matrix. Children with rickets can have bowed bones and inflamed joints. During the winter in northern climates, if children are not exposed to sufficient sunlight, they can take vitamin D as a dietary supplement to prevent rickets. The body's inability to absorb fats in which vitamin D is soluble can also result in vitamin D deficiency. This condition can occur in adults who suffer from digestive disorders and can be one cause of "adult rickets," or **osteomalacia** (os'tē-ō-mǎ-lā'shē-ǎ), which is a softening of the bones as a result of calcium depletion.

Hormones

Hormones are very important in bone growth. **Growth hormone** from the pituitary gland increases general tissue growth (see chapter 15), including overall bone growth, by stimulating interstitial cartilage growth and appositional bone growth. **Giantism** is a condition of abnormally increased height that usually results from excessive cartilage and bone formation at the epiphyseal plates of long bones. The most common type of giantism, **pituitary giantism,** results from excess secretion of pituitary growth hormone. The large stature of some individuals, however, can result from genetic factors rather than from abnormal levels of growth hormone. **Acromegaly** (ak-rō-meg'ǎ-lē) involves the growth of connective tissue, including bones, after the epiphyseal plates have

1. Osteoblasts beneath the periosteum lay down bone (*dark brown*) to form ridges separated by grooves. Blood vessels of the periosteum lie in the grooves.

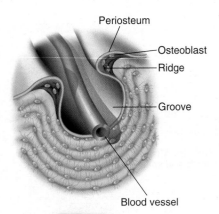

Periosteum
Osteoblast
Ridge
Groove
Blood vessel

2. The groove is transformed into a tunnel when the bone built on adjacent ridges meets. The periosteum of the groove becomes the endosteum of the tunnel.

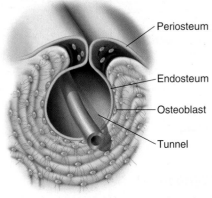

Periosteum
Endosteum
Osteoblast
Tunnel

3. Appositional growth by osteoblasts from the endosteum results in the formation of a new concentric lamella.

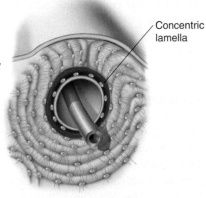

Concentric lamella

4. The production of additional concentric lamellae fills in the tunnel and completes the formation of the osteon.

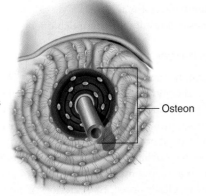

Osteon

Process Figure 6.12   Bone Growth in Width

Bones can increase in width by the formation of new osteons beneath the periosteum.

ossified. The effect mainly involves increased diameter or thickness of bones and is most strikingly apparent in the face and hands. Many pituitary giants also develop acromegaly later in life.

Dwarfism is a condition of abnormally decreased height. **Pituitary dwarfism** results when abnormally low levels of pituitary growth hormone affect the whole body, thus producing a small person who is normally proportioned.

 Achondroplasia

The most common type of dwarfism is not caused by a hormonal disorder. **Achondroplasia** (ā-kon-drō-plā′zē-ă), or **achondroplastic** (ā-kon-drō-plas′tik) **dwarfism** results in a person with a nearly normal-sized trunk and head but shorter than normal limbs. Achondroplasia is an autosomal-dominant trait (see chapter 25), caused by a mutation of a gene regulating bone growth. The normal effect of the gene is to slow bone growth by inhibiting chondrocyte division at the epiphyseal plate. Mutation of the gene results in a "gain of function," in which the normal inhibitory effect is increased, resulting in severely reduced bone growth in length. Approximately 80% of cases result from a spontaneous mutation of a gene during the formation of sperm cells or oocytes. Thus, the parents of most achondroplastic dwarfs are of normal height and proportions.

Sex hormones also influence bone growth. Estrogen (a class of female sex hormones) and testosterone (a male sex hormone) initially stimulate bone growth, which accounts for the burst of growth at puberty, when the production of these hormones increases. Both hormones also stimulate the ossification of epiphyseal plates, however, and thus the cessation of growth. Females usually stop growing earlier than males because estrogens cause a quicker closure of the epiphyseal plate than does testosterone. Because their entire growth period is somewhat shorter, females usually do not reach the same height as males. Decreased levels of testosterone or estrogen can prolong the growth phase of the epiphyseal plates, even though the bones grow more slowly. ▼

> 24. *How does vitamins D affect bone growth?*
> 25. *What causes pituitary giantism, acromegaly, pituitary dwarfism, and achondroplasia?*
> 26. *What effects do estrogen and testosterone have on bone growth? How do these effects account for the average height difference observed between men and women?*

**P R E D I C T   **

A 12-year-old female has an adrenal tumor that produces large amounts of estrogen. If untreated, what effect will this condition have on her growth for the next 6 months? On her height when she is 18?

Bone Remodeling

Just as we renovate or remodel our homes when they become outdated, when bone becomes old it is replaced with new bone in a process called **bone remodeling.** In this process, osteoclasts remove old bone and osteoblasts deposit new bone. Bone remodeling converts woven bone into lamellar bone, and it is involved in

Bone fractures are classified in several ways. The most commonly used classification involves the severity of the injury to the soft tissues surrounding the bone. An **open** (formerly called compound) fracture occurs when an open wound extends to the site of the fracture or when a fragment of bone protrudes through the skin. If the skin is not perforated, the fracture is called a **closed** (formerly called simple) fracture. If the soft tissues around a closed fracture are damaged, the fracture is called a **complicated fracture.**

Two other terms to designate fractures are **incomplete,** in which the fracture does not extend completely across the bone, and **complete,** in which the bone is broken into at least two fragments (figure A*a*). An incomplete fracture that occurs on the convex side of the curve of the bone is a **greenstick fracture. Hairline fractures** are incomplete fractures in which the two sections of bone do not separate; they are common in skull fractures.

Comminuted (kom'i-noo-ted) **fractures** are complete fractures in which the bone breaks into more than two pieces—

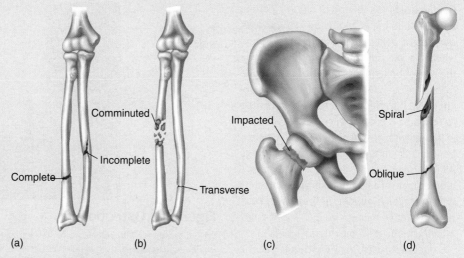

Figure A   Bone Fractures
(*a*) Complete and incomplete. (*b*) Transverse and comminuted. (*c*) Impacted. (*d*) Spiral and oblique.

usually two major fragments and a smaller fragment (figure A*b*). **Impacted fractures** are those in which one fragment is driven into the cancellous portion of the other fragment (figure A*c*).

Fractures are also classified according to the direction of the fracture within a bone. **Linear fractures** run parallel to the long axis

of the bone, and **transverse fractures** are at right angles to the long axis (see figure A*b*). **Spiral fractures** have a helical course around the bone, and **oblique fractures** run obliquely in relation to the long axis (figure A*d*). **Dentate fractures** have rough, toothed, broken ends, and **stellate fractures** have breakage lines radiating from a central point.

---

bone growth, changes in bone shape, the adjustment of the bone to stress, bone repair, and calcium ion (Ca^{2+}) regulation in the body. For example, as a long bone increases in length and diameter, the size of the medullary cavity also increases (figure 6.13). Otherwise, the bone would consist of nearly solid bone matrix and would be very heavy. The relative thickness of compact bone is maintained by the removal of bone on the inside by osteoclasts and the addition of bone to the outside by osteoblasts.

A **basic multicellular unit (BMU)** is a temporary assembly of osteoclasts and osteoblasts that travels through or across the surface of bone, removing old bone matrix and replacing it with new bone matrix. The average life span of a BMU is approximately 6 months, and BMU activity renews the entire skeleton every 10 years. The osteoclasts of a BMU break down bone matrix, forming a tunnel. Interstitial lamellae (see figure 6.5*a*) are remnants of osteons that were not completely removed when a BMU formed a tunnel. Blood vessels grow into the tunnel and osteoblasts of the BMU move in and lay down a layer of bone on the tunnel wall, forming a concentric lamella. Additional concentric lamellae are produced, filling in the tunnel from the outside to the inside, until an osteon is formed in which the center of the tunnel becomes a central canal containing blood vessels. In cancellous bone, the BMU removes bone matrix from the surface of a trabecula, forming a cavity, which the BMU then fills in with new bone matrix.

Mechanical Stress and Bone Strength

Remodeling, the formation of additional bone, alteration in trabecular alignment to reinforce the scaffolding, or other changes can modify the strength of the bone in response to the amount of stress applied to it. Mechanical stress applied to bone increases osteoblast activity in bone tissue, and removal of mechanical stress decreases osteoblast activity. Under conditions of reduced stress, such as when a person is bedridden or paralyzed, osteoclast activity continues at a nearly normal rate, but osteoblast activity is reduced, resulting in a decrease in bone density. In addition, pressure in bone causes an electrical change that increases the activity of osteoblasts. Applying weight (pressure) to a broken bone therefore speeds the healing process. Weak pulses of electric current applied to a broken bone sometimes are used clinically to speed the healing process. ▼

---

27. *What is bone remodeling? List the processes in which bone remodeling plays a role.*
28. *What is a BMU? Describe how BMUs produced osteons and interstitial lamellae.*

Osteoporosis (os'tē-ō-pō-rō'sis), or porous bone, results from a reduction in the overall quantity of bone tissue (figure B). It occurs when the rate of bone resorption exceeds the rate of bone formation. The loss of bone mass makes bones so porous and weakened that they become deformed and prone to fracture. The occurrence of osteoporosis increases with age. In both men and women, bone mass starts to decrease at about age 35 and continually decreases thereafter. Women can eventually lose approximately half, and men a quarter, of their cancellous bone. Osteoporosis is two and a half times more common in women than in men.

Osteoporosis has a strong genetic component. It is estimated that approximately 60% of a person's peak bone mass is genetically determined and that 40% is attributed to environmental factors, such as diet and physical activity. A woman whose mother has osteoporosis is more likely to develop osteoporosis than is a woman whose mother does not have the disorder. The genetic component of osteoporosis is complex and probably involves variations in a number of genes, such as those encoding vitamin D, collagen, calcitonin and estrogen receptors, and others.

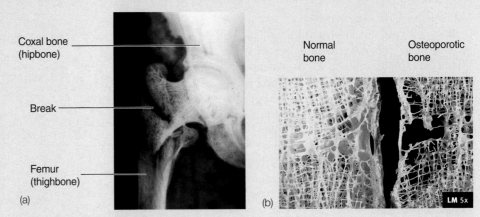

Figure B Osteoporosis

(*a*) Radiograph of a "broken hip," which is actually a break of the femur (thighbone) in the hip region. (*b*) Photomicrograph of normal bone and osteoporotic bone.

In postmenopausal women, the decreased production of the female sex hormone estrogen can cause osteoporosis. Estrogen is secreted by the ovaries, and it normally contributes to the maintenance of normal bone mass by inhibiting the stimulatory effects of PTH on osteoclast activity. Following menopause, estrogen production decreases, resulting in degeneration of cancellous bone, especially in the vertebrae of the spine and the bones of the forearm. Collapse of the vertebrae can cause a decrease in height or, in more severe cases, can produce kyphosis, or a "dowager's hump," in the upper back.

Conditions that result in decreased estrogen levels, other than menopause, can also cause osteoporosis. Examples include removal of the ovaries before menopause, extreme exercise to the point of amenorrhea (lack of menstrual flow), anorexia nervosa (self-starvation), and cigarette smoking.

In males, reduction in testosterone levels can cause loss of bone tissue.

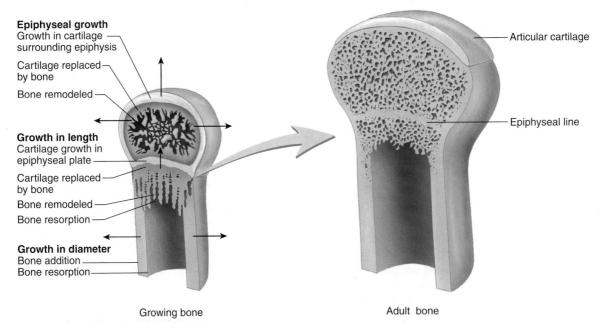

Growing bone

Adult bone

Figure 6.13   Remodeling of a Long Bone

The epiphysis enlarges and the diaphysis increases in length as new cartilage is formed and replaced by bone, which is remodeled. The diameter of the bone increases as a result of bone growth on the outside of the bone, and the size of the medullary cavity increases because of bone resorption.

Decreasing testosterone levels are usually less of a problem for men than decreasing estrogen levels are for women for two reasons. First, because males have denser bones than females, a loss of some bone tissue has less of an effect. Second, testosterone levels generally do not decrease significantly until after age 65, and even then the rate of decrease is often slow.

An overproduction of PTH, which results in overstimulation of osteoclast activity, can also cause osteoporosis.

Inadequate dietary intake or absorption of calcium can contribute to osteoporosis. Absorption of calcium from the small intestine decreases with age, and individuals with osteoporosis often have insufficient intake of calcium or vitamin D. Drugs that interfere with calcium uptake or use can also increase the risk for osteoporosis.

Finally, osteoporosis can result from inadequate exercise or disuse caused by fractures or paralysis. Significant amounts of bone are lost after 8 weeks of immobilization.

Treatments for osteoporosis are designed to reduce bone loss, increase bone formation, or both. Increased dietary calcium and vitamin D can increase calcium uptake and promote bone formation. Daily doses of 1000–1500 mg of calcium and 800 IU (20 µg) of vitamin D are recommended. Exercise, such as walking or using light weights, also appears to be effective not only in reducing bone loss but in increasing bone mass as well.

In postmenopausal women, **hormone replacement therapy (HRT)** with estrogen decreases osteoclast numbers. This reduces bone loss but does not result in an increase in bone mass because osteoclast activity still exceeds osteoblast activity. However, the use of HRT to prevent bone loss is now discouraged because of the results of a study sponsored by the Women's Health Initiative. The study examined HRT in over 16,000 women and found that HRT increased the risk for breast cancer, uterine cancer, heart attacks, strokes, and blood clots but decreased the risk for hip fractures and colorectal cancer. **Selective estrogen receptor modulators (SERMs)** are a class of drugs that bind to estrogen receptors. They may be able to protect against bone loss without increasing the risk for breast cancer. For example, raloxifene (ral-ox′ĭ-fēn) stimulates estrogen receptors in bone but inhibits them in the breast and uterus.

Calcitonin (Miacalcin), which inhibits osteoclast activity, is now available as a nasal spray. Calcitonin can be used to treat osteoporosis in men and women and has been shown to produce a slight increase in bone mass. **Statins** (stat′ins) are drugs that inhibit cholesterol synthesis; they also stimulate osteoblast activity, and there is some evidence that statins can reduce the risk for fractures. Alendronate (Fosamax) belongs to a class of drugs called **bisphosphonates** (bis-fos′fō-nāts). Bisphosphonates concentrate in bone; when osteoclasts break down bone, the bisphosphonates are taken up by the osteoclasts. The bisphosphonates interfere with certain enzymes, leading to inactivation and lysis of the osteoclasts. Alendronate increases bone mass and reduces fracture rates even more effectively than calcitonin. Slow-releasing sodium fluoride (Slow Fluoride) in combination with calcium citrate (Citracal) also appears to increase bone mass.

Early diagnosis of osteoporosis may lead to the use of more preventive treatments. Instruments that measure the absorption of photons (particles of light) by bone are currently used, of which dual-energy x-ray absorptiometry (DEXA) is considered the best.

Bone Repair

Bone is a living tissue that can undergo repair following damage to it. This process has four major steps.

1. *Hematoma formation* (figure 6.14, step 1). When bone is fractured, the blood vessels in the bone and surrounding periosteum are damaged, and a hematoma forms. A **hematoma** (hē-mă-tō′mă, hem-ă-tō′mă) is a localized mass of blood released from blood vessels but confined within an organ or a space. Usually, the blood in a hematoma forms a clot, which consists of fibrous proteins that stop the bleeding. Disruption of blood vessels in the central canals results in inadequate blood delivery to osteocytes, and bone tissue adjacent to the fracture site dies. Inflammation and swelling of tissues around the bone often occur following the injury.

2. *Callus formation* (figure 6.14, step 2). A **callus** (kal′ŭs) is a mass of tissue that forms at a fracture site, connecting the broken ends of the bone. The callus forms as the clot dissolves (see chapter 16) and is removed by macrophages. An **internal callus** forms *between* the ends of the broken bone, as well as in the marrow cavity if the fracture occurs in the diaphysis of a long bone. It consists of a dense fibrous network in which cartilage and woven bone forms. The **external callus** forms a collar of cartilage and woven bone *around* the opposing ends of the bone fragments. It helps stabilize the ends of the broken bone. In modern medical practice, stabilization of the bone is assisted with a cast or the surgical implantation of metal supports.

3. *Callus ossification* (figure 6.14, step 3). The fibers and cartilage of the internal and external calluses are ossified to produce woven, cancellous bone. The ossification process is similar to that which occurs during fetal bone development. Cancellous bone formation in the callus is usually complete 4–6 weeks after the injury. Immobilization of the bone is critical up to this time because movement can refracture the delicate new bone matrix.

4. *Bone remodeling* (figure 6.14, step 4). Filling the gap between bone fragments with an internal callus of woven bone is not the end of the repair process because woven bone is not as structurally strong as the original lamellar bone. Repair is not complete until the woven bone

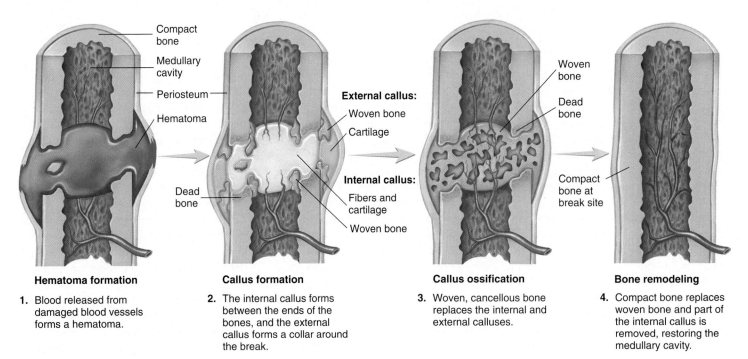

Compact bone
Medullary cavity
Periosteum
Hematoma

Dead bone

External callus:
Woven bone
Cartilage

Internal callus:
Fibers and cartilage
Woven bone

Woven bone
Dead bone

Compact bone at break site

Hematoma formation

1. Blood released from damaged blood vessels forms a hematoma.

Callus formation

2. The internal callus forms between the ends of the bones, and the external callus forms a collar around the break.

Callus ossification

3. Woven, cancellous bone replaces the internal and external calluses.

Bone remodeling

4. Compact bone replaces woven bone and part of the internal callus is removed, restoring the medullary cavity.

Process Figure 6.14   Bone Repair

of the internal callus and the dead bone adjacent to the fracture site are replaced by compact bone. In this compact bone, osteons from both sides of the break extend across the fracture line to "peg" the bone fragments together. This remodeling process takes time and may not be complete even after a year. As the internal callus is remodeled and becomes stronger, the external callus is reduced in size by osteoclast activity. The repaired zone usually remains slightly thicker than the adjacent bone, but the repair may be so complete that no evidence of the break remains. If the fracture occurred in the diaphysis of a long bone, remodeling also restores the medullary cavity. ▼

29. *How does breaking a bone result in hematoma formation?*
30. *Distinguish between the location and the composition of the internal and external callus.*
31. *Why is remodeling of the ossified callus necessary?*

 Uniting Broken Bones

Before the formation of compact bone between the broken ends of a bone can take place, the appropriate substrate must be present. Normally, this is the woven, cancellous bone of the internal callus. If formation of the internal callus is prevented by infections, bone movements, or the nature of the injury, nonunion of the bone occurs. This condition can be treated by surgically implanting an appropriate substrate, such as living bone from another site in the body or dead bone from cadavers. Other substrates have also been used. For example, a specific marine coral calcium phosphate is converted into a predominantly hydroxyapatite biomatrix that is very much like cancellous bone.

Calcium Homeostasis

Bone is the major storage site for calcium in the body, and the movement of Ca^{2+} into and out of bone helps determine blood Ca^{2+} levels, which is critical for normal muscle and nervous system function (see chapters 8 and 10). Calcium ions move into bone as osteoblasts build new bone and move out of bone as osteoclasts break down bone (figure 6.15). When osteoblast and osteoclast activity is balanced, the movement of Ca^{2+} into and out of a bone is equal.

When blood Ca^{2+} levels are too low, osteoclast activity increases. More Ca^{2+} are released by osteoclasts from bone into the blood than are removed by osteoblasts from the blood to make new bone. Consequently, a net movement of Ca^{2+} occurs from bone into blood, and blood Ca^{2+} levels increase. Conversely, if blood Ca^{2+} levels are too high, osteoclast activity decreases. Fewer Ca^{2+} are released by osteoclasts from bone into the blood than are taken from the blood by osteoblasts to produce new bone. As a result, a net movement of Ca^{2+} occurs from the blood to bone, and blood Ca^{2+} levels decrease.

Parathyroid hormone (PTH) from the parathyroid glands (see chapter 15) is the major regulator of blood Ca^{2+} levels. If the blood Ca^{2+} level decreases, the secretion of PTH increases, resulting in increased numbers of osteoclasts, which causes increased bone breakdown and increased blood Ca^{2+} levels (see figure 6.15). PTH also regulates blood Ca^{2+} levels by increasing Ca^{2+} uptake in the small intestine. Increased PTH promotes the formation of vitamin D in the kidneys, and vitamin D increases the absorption of Ca^{2+} from the small intestine. PTH also increases the reabsorption of Ca^{2+} from urine in the kidneys, which reduces Ca^{2+} lost in the urine.

When blood Ca^{2+} levels are too high, PTH levels decrease. As a result, the net movement of Ca^{2+} into bone increases because of

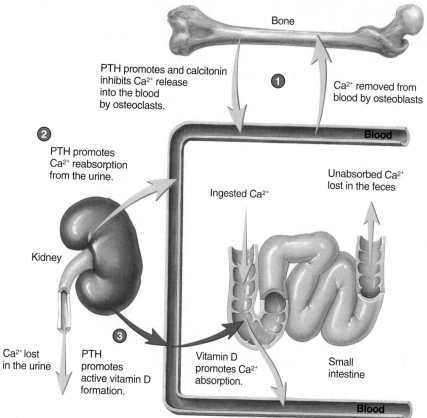

1. Osteoclasts break down bone and release Ca^{2+} into the blood, and osteoblasts remove Ca^{2+} from the blood to make bone. (*Blue arrows represent the movement of Ca^{2+}.*) Parathyroid hormone (PTH) regulates blood Ca^{2+} levels by indirectly stimulating osteoclast activity, resulting in increased Ca^{2+} release into the blood. Calcitonin plays a minor role in Ca^{2+} maintenance by inhibiting osteoclast activity.

2. In the kidneys, PTH increases Ca^{2+} reabsorption from the urine.

3. In the kidneys, PTH also promotes the formation of active vitamin D (*green arrows*), which increases Ca^{2+} absorption from the small intestine.

Labels in figure:
Bone
PTH promotes and calcitonin inhibits Ca^{2+} release into the blood by osteoclasts.
Ca^{2+} removed from blood by osteoblasts
Blood
PTH promotes Ca^{2+} reabsorption from the urine.
Unabsorbed Ca^{2+} lost in the feces
Ingested Ca^{2+}
Kidney
Ca^{2+} lost in the urine
PTH promotes active vitamin D formation.
Vitamin D promotes Ca^{2+} absorption.
Small intestine
Blood

Process Figure 6.15 Calcium Homeostasis

decreased osteoclast activity. Vitamin D levels decrease, resulting in less absorption of Ca^{2+} from the intestines, and reabsorption of Ca^{2+} from urine decreases, resulting in more Ca^{2+} lost in the urine.

Calcitonin (kal-si-tō'nin), secreted from the thyroid gland (see chapter 15), decreases osteoclast activity (see figure 6.15) and thus decreases blood calcium levels. Increased blood Ca^{2+} levels stimulate the thyroid gland to secrete calcitonin. PTH and calcitonin are described more fully in chapters 15 and 23. ▼

32. Name the hormone that is the major regulator of Ca^{2+} levels in the body. What stimulates the secretion of this hormone?

33. What are the effects of PTH on osteoclast numbers, the formation of vitamin D, and the reabsorption of Ca^{2+} from urine?

34. What stimulates calcitonin secretion? How does calcitonin affect osteoclast activity?

CASE STUDY | Bone Density

Henry is a 65-year-old man who was admitted to the emergency room after a fall. A radiograph confirmed that he had fractured the proximal part of his arm bone (surgical neck of the humerus). The radiograph also revealed that his bone matrix was not as dense as it should be for a man his age. A test for blood Ca^{2+} levels was normal. On

questioning, Henry confessed that he is a junk food addict with poor eating habits. He eats few vegetables and never consumes dairy products. In addition, Henry never exercises and seldom goes outdoors except at night.

P R E D I C T 8

Use your knowledge of bone physiology and figure 6.13 to answer the following questions.

a. Why is Henry more likely than most men his age to break a bone? *osteoporosis*

b. How has Henry's eating habits contributed to his low bone density? *no vitamins for bones*

c. Would Henry's PTH levels be lower than normal, normal, or higher than normal? *low levels of PTH*

d. What effect has Henry's nocturnal lifestyle had on his bone density? *no vitamin D from sun or dairy*

e. How has lack of exercise affected his bone density? *no stress on bones to strengthen*

Effects of Aging on the Skeletal System

The most significant age-related changes in the skeletal system affect the quality and quantity of bone matrix. Recall that a mineral (hydroxyapatite) in the bone matrix gives bone compression

Systems Interactions

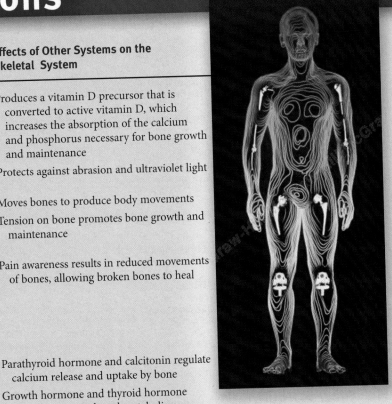

	Effects of the Skeletal System on Other Systems	Effects of Other Systems on the Skeletal System
Integumentary System	Provides support (for example, the skin of the ears and scalp)	Produces a vitamin D precursor that is converted to active vitamin D, which increases the absorption of the calcium and phosphorus necessary for bone growth and maintenance Protects against abrasion and ultraviolet light
Muscular System	Forms lever system for body movements Reservoir for calcium, which is necessary for muscle contraction	Moves bones to produce body movements Tension on bone promotes bone growth and maintenance
Nervous System	Skull protects the brain Vertebral column protects the spinal cord Ear bones necessary for normal hearing Reservoir for calcium, which is necessary for neuronal function	Pain awareness results in reduced movements of bones, allowing broken bones to heal
Endocrine System	Protects endocrine organs in the cranial, thoracic, and pelvic cavities	Parathyroid hormone and calcitonin regulate calcium release and uptake by bone Growth hormone and thyroid hormone affect bone growth and metabolism
Cardiovascular System	Red bone marrow produces red blood cells (oxygen and carbon dioxide transport) and platelets (blood clotting) Thoracic cage protects the heart Reservoir for calcium, which is necessary for cardiac muscle contraction	Delivers oxygen, nutrients, hormones, and immune cells Removes carbon dioxide, waste products, and toxins
Lymphatic System and Immunity	Immune (white) blood cells produced in red bone marrow	Immune cells provide protection against microorganisms and toxins and promote tissue repair by releasing chemical mediators
Respiratory System	Movement of ribs assists in respiration Thoracic cage protects the lungs and helps prevent them from collapsing	Provides oxygen and removes carbon dioxide Helps maintain the body's pH
Digestive System	Lower ribs and pelvis provide some protection Teeth and mandible are important for mastication Yellow marrow is a site of fat storage	Provides water and nutrients, including calcium and phosphorus The liver converts vitamin D precursor from the skin to form a vitamin D that can be converted by the kidneys to active vitamin D
Urinary System	Thoracic cavity protects the kidneys Pelvis protects the urinary bladder	Removes waste products Prevents loss of calcium and phosphorus in the urine Helps maintain the body's pH, ion, and water balance Converts vitamin D to its active form
Reproductive System	Pelvis protects internal reproductive organs and developing fetus	Estrogen and testosterone stimulate bone growth and closure of the epiphyseal plate

(weight-bearing) strength, but collagen fibers make the bone flexible. The bone matrix in an older bone is more brittle than in a younger bone because decreased collagen production results in a matrix that has relatively more mineral and fewer collagen fibers. With aging, the amount of matrix also decreases because the rate of matrix formation by osteoblasts becomes slower than the rate of matrix breakdown by osteoclasts.

Bone mass is at its highest around age 30, and men generally have denser bones than women because of the effects of testosterone and greater body weight. Race also affects bone mass. African-Americans and Hispanics have higher bone masses than whites and Asians. After age 35, both men and women have an age-related loss of bone of 0.3%–0.5% a year. This loss can increase by 10 times in women after menopause,

and women can have a bone loss of 3%–5% a year for approximately 5–7 years (see "Osteoporosis," pp. 138–139).

Significant loss of bone increases the likelihood of having bone fractures. For example, loss of trabeculae greatly increases the risk for compression fractures of the vertebrae (backbones) because the weight-bearing body of the vertebrae consists mostly of cancellous bone. In addition, loss of bone can cause deformity, loss of height, pain, and stiffness. For example, compression fractures of the vertebrae can cause an exaggerated curvature of the spine, resulting in a bent-forward, stooped posture. Loss of bone from the jaws can also lead to tooth loss. ▼

33. What effect does aging have on the quality and quantity of bone matrix?

S U M M A R Y

Functions of the Skeletal System (p. 124)

1. The skeletal system consists of bones, cartilage, tendons, and ligaments.
2. The skeletal system supports the body, protects the organs it surrounds, allows body movements, stores minerals and fats, and is the site of blood cell and platelet production.

Cartilage (p. 124)

1. Chondroblasts produce cartilage and become chondrocytes. Chondrocytes are located in lacunae surrounded by matrix.
2. The matrix of cartilage contains collagen fibers (for strength) and proteoglycans (trap water).
3. The perichondrium surrounds cartilage.
   - The outer layer contains fibroblasts.
   - The inner layer contains chondroblasts.
4. Cartilage grows by appositional and interstitial growth.

Bone Histology (p. 125)
Bone Matrix

1. Collagen provides flexible strength.
2. Hydroxyapatite provides compressional strength.

Bone Cells

1. Osteoblasts produce bone matrix and become osteocytes.
   - Osteoblasts connect to one another through cell processes and surround themselves with bone matrix to become osteocytes.
   - Osteocytes are located in lacunae and are connected to one another through canaliculi.
2. Osteoclasts break down bone.
3. Osteoblasts originate from osteochondral progenitor cells, whereas osteoclasts originate from stem cells in red bone marrow.
4. Ossification, the formation of bone, occurs through appositional growth.

Woven and Lamellar Bone

1. Woven bone has collagen fibers oriented in many different directions. It is remodeled to form lamellar bone.

2. Lamellar bone is arranged in thin layers, called lamellae, which have collagen fibers oriented parallel to one another.

Cancellous and Compact Bone

1. Cancellous bone has many spaces.
   - Lamellae combine to form trabeculae, beams of bone that interconnect to form a latticelike structure with spaces filled with bone marrow and blood vessels.
   - The trabeculae are oriented along lines of stress and provide structural strength.
2. Compact bone is dense with few spaces.
   - Compact bone consists of organized lamellae: Circumferential lamellae form the outer surface of compact bones; concentric lamellae surround central canals, forming osteons; interstitial lamellae are remnants of lamellae left after bone remodeling.
   - Canals within compact bone provide a means for the exchange of gases, nutrients, and waste products. From the periosteum or endosteum, perforating canals carry blood vessels to central canals, and canaliculi connect central canals to osteocytes.

Bone Anatomy (p. 128)
Bone Shapes

Individual bones can be classified as long, short, flat, or irregular.

Structure of a Long Bone

1. The diaphysis is the shaft of a long bone, and the epiphysis is separated from the diaphysis by the epiphyseal plate or line.
2. The epiphyseal plate is the site of bone growth in length. The epiphyseal plate becomes the epiphyseal line when all of its cartilage becomes bone.
3. The medullary cavity is a space within the diaphysis.
4. Red bone marrow is the site of blood cell production, and yellow bone marrow consists of fat.
5. The periosteum covers the outer surface of bone.
   - The outer layer contains blood vessels and nerves.
   - The inner layer contains osteoblasts, osteoclasts, and osteochondral progenitor cells.
6. The endosteum lines cavities inside bone and contains osteoblasts, osteoclasts, and osteochondral progenitor cells.

Structure of Flat, Short, and Irregular Bones

Flat, short, and irregular bones have an outer covering of compact bone surrounding cancellous bone.

Bone Development (p. 130)

Intramembranous Ossification

1. Some skull bones, part of the mandible, and the diaphyses of the clavicles develop from membranes.
2. Within the membrane at centers of ossification, osteoblasts produce bone along the membrane fibers to form cancellous bone.
3. Beneath the periosteum, osteoblasts lay down compact bone to form the outer surface of the bone.
4. Fontanels are areas of membrane that are not ossified at birth.

Endochondral Ossification

1. Most bones develop from a cartilage model.
2. The cartilage matrix is calcified, and chondrocytes die. Osteoblasts form bone on the calcified cartilage matrix, producing cancellous bone.
3. Osteoblasts build an outer surface of compact bone beneath the periosteum.
4. Primary ossification centers form in the diaphysis during fetal development. Secondary ossification centers form in the epiphyses.
5. Articular cartilage on the ends of bones and the epiphyseal plate is cartilage that does not ossify.

Bone Growth (p. 132)

1. Bones increase in size only by appositional growth, the adding of new bone on the surface of older bone or cartilage.
2. Trabeculae grow by appositional growth.

Growth in Bone Length

1. Epiphyseal plate growth involves the interstitial growth of cartilage followed by appositional bone growth on the cartilage.
2. Epiphyseal plate growth results in an increase in the length of the diaphysis and bony processes. Bone growth in length ceases when the epiphyseal plate becomes ossified and forms the epiphyseal line.

Growth at Articular Cartilage

1. Articular cartilage growth involves the interstitial growth of cartilage followed by appositional bone growth on the cartilage.
2. Articular cartilage growth results in larger epiphyses and an increase in the size of bones that do not have epiphyseal plates.

Growth in Bone Width

1. Appositional bone growth beneath the periosteum increases the diameter of long bones and the size of other bones.

2. Osteoblasts from the periosteum form ridges with grooves between them. The ridges grow together, converting the grooves into tunnels filled with concentric lamellae to form osteons.
3. Osteoblasts from the periosteum lay down circumferential lamellae, which can be remodeled.

Factors Affecting Bone Growth

1. Genetic factors determine bone shape and size. The expression of genetic factors can be modified.
2. Factors that alter the mineralization process or the production of organic matrix, such as deficiencies in vitamins D, can affect bone growth.
3. Growth hormone, estrogen, and testosterone stimulate bone growth.
4. Estrogen and testosterone cause closure of the epiphyseal plate.

Bone Remodeling (p. 136)

1. Remodeling converts woven bone to lamellar bone and allows bone to change shape, adjust to stress, repair itself, and regulate body calcium levels.
2. Basic multicellular units (BMUs) make tunnels in bone, which are filled with concentric lamellae to form osteons. Interstitial lamellae are remnants of bone not removed by BMUs

Bone Repair (p. 139)

1. Fracture repair begins with the formation of a hematoma.
2. The hematoma is replaced by an internal callus consisting of fibers and cartilage.
3. The external callus is a bone–cartilage collar that stabilizes the ends of the broken bone.
4. The internal and external calluses are ossified to become woven bone.
5. Woven bone is remodeled.

Calcium Homeostasis (p. 140)

PTH increases blood Ca^{2+} levels by increasing bone breakdown, Ca^{2+} absorption from the small intestine, and reabsorption of Ca^{2+} from the urine. Calcitonin decreases blood Ca^{2+} by decreasing bone breakdown.

Effects of Aging on the Skeletal System (p. 141)

1. With aging, bone matrix is lost and the matrix becomes more brittle.
2. Cancellous bone loss results from a thinning and a loss of trabeculae. Compact bone loss mainly occurs from the inner surface of bones and involves less osteon formation.
3. Loss of bone increases the risk for fractures and causes deformity, loss of height, pain, stiffness, and loss of teeth.

R E V I E W   A N D   C O M P R E H E N S I O N

1. The extracellular matrix of hyaline cartilage
   a. is produced by chondroblasts.
   b. contains collagen.
   c. contains proteoglycans.
   d. all of the above.

2. Chondrocytes are mature cartilage cells found within the _____, and they are derived from _____.
   a. perichondrium, fibroblasts
   b. perichondrium, chondroblasts
   c. lacunae, fibroblasts
   d. lacunae, chondroblasts

3. Which of these statements concerning cartilage is correct?
   a. Chondrocytes receive nutrients and oxygen from blood vessels in the matrix.
   b. Articular cartilage has a thick perichondrium layer.
   c. The perichondrium has both chondrocytes and osteocytes.
   d. Interstitial cartilage growth occurs when chondrocytes within the tissue add more matrix from the inside.

4. Which of these substances makes up the major portion of bone?
   a. collagen   d. osteocytes
   b. hydroxyapatite   e. osteoblasts
   c. proteoglycans

5. The flexible strength of bone occurs because of
   a. osteoclasts.
   b. ligaments.
   c. hydroxyapatite.
   d. collagen fibers.
   e. periosteum.

6. The prime function of osteoclasts is to
   a. prevent osteoblasts from forming.
   b. become osteocytes.
   c. break down bone.
   d. secrete calcium salts and collagen fibers.
   e. form the periosteum.

7. Osteochondral progenitor cells
   a. can become osteoblasts or chondroblasts.
   b. are derived from mesenchymal stem cells.
   c. are located in the perichondrium, periosteum, and endosteum.
   d. do not produce osteoclasts.
   e. all of the above.

8. Lamellar bone
   a. is mature bone.
   b. is remodeled to form woven bone.
   c. is the first type of bone formed during early fetal development.
   d. has collagen fibers randomly oriented in many directions.
   e. all of the above.

9. Central canals
   a. connect perforating canals to canaliculi.
   b. connect cancellous bone to compact bone.
   c. are where blood cells are produced.
   d. are found only in cancellous bone.
   e. are lined with periosteum.

10. The type of lamellae found in osteons is _____ lamellae.
    a. circumferential
    b. concentric
    c. interstitial

11. Cancellous bone consists of interconnecting rods or plates of bone called
    a. osteons.
    b. canaliculi.
    c. circumferential lamellae.
    d. a haversian system.
    e. trabeculae.

12. Yellow bone marrow is
    a. found mostly in children's bones.
    b. found in the epiphyseal plate.
    c. important for blood cell production.
    d. mostly adipose tissue.

13. The periosteum
    a. is a connective tissue membrane.
    b. covers the outer and internal surfaces of bone.
    c. contains only osteoblasts.
    d. has a single fibrous layer.

14. Given these events:
    1. Osteochondral progenitor cells become osteoblasts.
    2. Connective tissue membrane is formed.
    3. Osteoblasts produce woven bone.

    Which sequence best describes intramembranous bone formation?
    a. 1,2,3
    b. 1,3,2
    c. 2,1,3
    d. 2,3,1
    e. 3,2,1

15. Given these processes:
    1. Chondrocytes die.
    2. Cartilage matrix calcifies.
    3. Chondrocytes hypertrophy.
    4. Osteoblasts deposit bone.
    5. Chondroblasts produce hyaline cartilage.

    Which sequence best represents the order in which they occur during endochondral bone formation?
    a. 2,4,1,2,5
    b. 2,5,1,2,4
    c. 4,5,2,3,1
    d. 5,3,2,1,4
    e. 5,4,2,3,1

16. Intramembranous ossification
    a. occurs at the epiphyseal plate.
    b. is responsible for growth in diameter of a bone.
    c. gives rise to the flat bones of the skull.
    d. occurs within a hyaline cartilage model.
    e. produces articular cartilage in the long bones.

17. Growth in the length of a long bone occurs
    a. at the primary ossification center.
    b. beneath the periosteum.
    c. at the center of the diaphysis.
    d. at the epiphyseal plate.
    e. at the epiphyseal line.

18. During growth in length of a long bone, cartilage is formed and then ossified. The location of the ossification is the zone of
    a. calcification.
    b. hypertrophy.
    c. proliferation.
    d. resting cartilage.

19. Given these processes:
    1. An osteon is produced.
    2. Osteoblasts from the periosteum form a series of ridges.
    3. The periosteum becomes the endosteum.
    4. Osteoblasts lay down bone to produce a concentric lamella.
    5. Grooves are changed into tunnels.

    Which sequence best represents the order in which these processes occur during growth in width of a long bone?
    a. 1,4,2,3,5
    b. 2,5,3,4,1
    c. 3,4,2,1,5
    d. 4,2,1,5,3
    e. 5,4,2,1,3

20. Chronic vitamin D deficiency results in which of these consequences?
    a. Bones become brittle.
    b. The percentage of bone composed of hydroxyapatite increases.
    c. Bones become soft and pliable.
    d. Bone growth increases.

21. Estrogen
    a. stimulates a burst of growth at puberty.
    b. causes a later closure of the epiphyseal plate than does testosterone.
    c. causes a longer growth period in females than testosterone causes in males.
    d. tends to prolong the growth phase of the epiphyseal plates.
    e. all of the above.

22. Bone remodeling can occur
    a. when woven bone is converted into lamellar bone.
    b. as bones are subjected to varying patterns of stress.
    c. as a long bone increases in diameter.
    d. when new osteons are formed in compact bone.
    e. all of the above.

23. Given these processes:
    1. callus formation
    2. cartilage ossification
    3. hematoma formation
    4. remodeling of woven bone into compact bone

    Which sequence best represents the order in which the processes occur during repair of a fracture?
    a. 1,2,3,4
    b. 1,2,4,3
    c. 2,1,4,3
    d. 3,1,2,4
    e. 3,2,1,4

24. If the secretion of parathyroid hormone (PTH) increases, osteoclast activity _____ and blood Ca^{2+} levels _____.
    a. decreases, decrease
    b. decreases, increase
    c. increases, decrease
    d. increases, increase

25. Osteoclast activity is inhibited by
    a. calcitonin.
    b. growth hormone.
    c. parathyroid hormone.
    d. sex hormones.
    e. thyroid hormone.

Answers in Appendix E

C R I T I C A L   T H I N K I N G

1. When a person develops Paget disease, for unknown reasons the collagen fibers in the bone matrix run randomly in all directions. In addition, the amount of trabecular bone decreases. What symptoms would you expect to observe?

2. A 12-year-old boy fell while playing basketball. The physician explained that the head (epiphysis) of the femur was separated from the shaft (diaphysis). Although the bone was set properly, by the time the boy was 16 it was apparent that the injured lower limb was shorter than the normal one. Explain why this difference occurred.

3. When closure of the epiphyseal plate occurs, the cartilage of the plate is replaced by bone. Does this occur from the epiphyseal side of the plate or the diaphyseal side?

4. Assume that two patients have identical breaks in the femur (thighbone). If one is bedridden and the other has a walking cast, which patient's fracture heals faster? Explain.

5. Explain why running helps prevent osteoporosis in the elderly. Does the benefit include all bones or mainly those of the legs and spine?

6. Astronauts can experience a dramatic decrease in bone density while in a weightless environment. Explain how this happens, and suggest a way to slow the loss of bone tissue.

7. Would a patient suffering from kidney failure be more likely to develop osteomalacia or osteoporosis? Explain.

8. In some cultures, eunuchs were responsible for guarding harems, which are the collective wives of one male. Eunuchs are males who were castrated as boys. Castration removes the testes, the major site of testosterone production in males. Because testosterone is responsible for the sex drive in males, the reason for castration is obvious. As a side effect of this procedure, the eunuchs grew to above-normal heights. Can you explain why?

9. When a long bone is broken, blood vessels at the fracture line are severed. The formation of blood clots stops the bleeding. Within a few days, bone tissue on both sides of the fracture site dies. The bone dies back only a certain distance from the fracture line, however. Explain.

10. A patient has hyperparathyroidism because of a tumor in the parathyroid gland that produces excessive amounts of PTH. What effect does this hormone have on bone? Would the administration of large doses of vitamin D help the situation? Explain.

Answers in Appendix F

For additional study tools, go to: www.aris.mhhe.com. Click on your course topic and then this textbook's author name/title. Register once for a semester's worth of interactive learning activities to help improve your grade!

CHAPTER 7 Anatomy of Bones and Joints

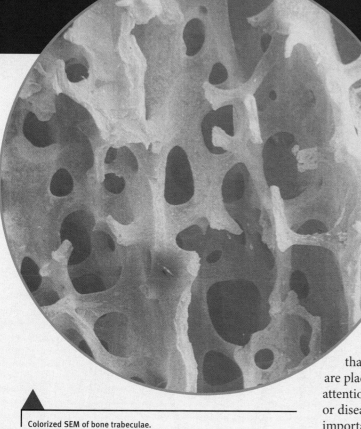

Colorized SEM of bone trabeculae.

Tools for Success

A virtual cadaver dissection experience

Audio (MP3) and/or video (MP4) content, available at www.mhhe.aris.com

Introduction

If the body had no skeleton, it would look somewhat like a poorly stuffed rag doll. Without a skeletal system, we would have no framework to help maintain shape and we would not be able to move normally. Bones of the skeletal system surround and protect organs, such as the brain and heart. Human bones are very strong and can resist tremendous bending and compression forces without breaking. Nonetheless, each year nearly 2 million Americans break a bone.

Muscles pull on bones to make them move, but movement would not be possible without joints between the bones. Humans would resemble statues, were it not for the joints between bones that allow bones to move once the muscles have provided the pull. Machine parts most likely to wear out are those that rub together and they require the most maintenance. Movable joints are places in the body where the bones rub together, yet we tend to pay little attention to them. Fortunately, our joints are self-maintaining, but damage to or disease of a joint can make movement very difficult. We realize then how important the movable joints are for normal function.

The skeletal system includes the bones, cartilage, ligaments, and tendons. To study skeletal gross anatomy, however, dried, prepared bones are used. This allows the major features of individual bones to be seen clearly without being obstructed by associated soft tissues, such as muscles, tendons, ligaments, cartilage, nerves, and blood vessels. The important relationships among bones and soft tissues should not be ignored, however.

▶This chapter includes a discussion of **general considerations of bones (p. 148)**. It then discusses the two categories of the named bones: the **axial skeleton (p. 149)** and the **appendicular skeleton (p. 165)**. **Articulations (p. 177)** are defined and the **classes of joints (p. 177)** are described. **Types of movements (p. 182)** are discussed, **a description of selected joints (p. 185)** is presented, and the chapter concludes with **the effects of aging on joints (p. 190)**.

General Considerations of Bones

The average adult skeleton has 206 bones (figure 7.1). Although this is the traditional number, the actual number of bones varies from person to person and decreases with age as some bones become fused. Bones can be categorized as paired or unpaired. A **paired bone** is two bones of the same type located on the right and left sides of the body, whereas an **unpaired bone** is a bone located on the midline of the body. For example, the bones of the upper and lower limbs are paired bones, whereas the bones of

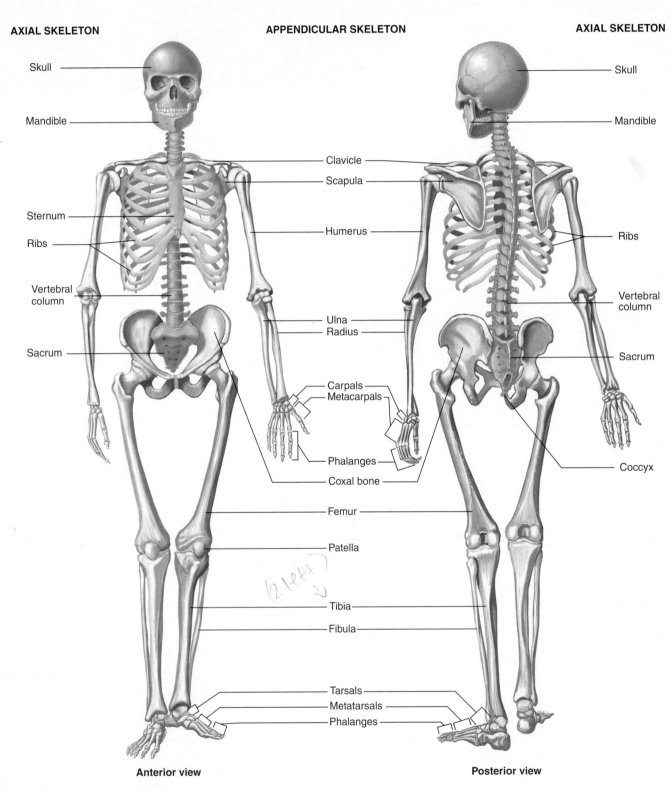

AXIAL SKELETON **APPENDICULAR SKELETON** **AXIAL SKELETON**

Skull — Mandible — Sternum — Ribs — Vertebral column — Sacrum

Clavicle — Scapula — Humerus — Ulna — Radius — Carpals — Metacarpals — Phalanges — Coxal bone — Femur — Patella — Tibia — Fibula — Tarsals — Metatarsals — Phalanges

Skull — Mandible — Ribs — Vertebral column — Sacrum — Coccyx

Anterior view **Posterior view**

Figure 7.1 Complete Skeleton

The skeleton is not shown in the anatomical position.

Table 7.1	General Anatomical Terms for Various Features of Bones

Term	Description
Body	Main part
Head	Enlarged (often rounded) end
Neck	Constriction between head and body
Margin or border	Edge
Angle	Bend
Ramus	Branch off the body (beyond the angle)
Condyle	Smooth, rounded articular surface *(bump)*
Facet	Small, flattened articular surface
Ridges	
Line or linea	Low ridge
Crest or crista	Prominent ridge *muscle attachments*
Spine	Very high ridge
Projections	
Process	Prominent projection *(for muscle attachment)*
Tubercle	Small, rounded bump
Tuberosity or tuber	Knob; larger than a tubercle
Trochanter	Tuberosities on the proximal femur
Epicondyle	Upon a condyle
Openings	
Foramen	Hole
Canal or meatus	Tunnel
Fissure	Cleft
Sinus or labyrinth	Cavity
Depressions	
Fossa	General term for a depression *shallow*
Notch	Depression in the margin of a bone
Groove or sulcus	Deeper, narrow depression

the vertebral column are unpaired bones. There are 86 paired and 34 unpaired bones.

Many of the anatomical terms used to describe the features of bones are listed in table 7.1. Most of these features are based on the relationship between the bones and associated soft tissues. If a bone has a **tubercle** (too′ber-kl, lump) or **process** (projection), such structures usually exist because a ligament or tendon was attached to that tubercle or process during life. If a bone has a **foramen** (fō-rā′men, pl. *foramina,* fō-ram′i-nă, a hole) in it, that foramen was occupied by something, such as a nerve or blood vessel. If a bone has a **condyle** (kon′dīl, knuckle), it has a smooth, rounded end, covered with articular cartilage (see chapter 6), that is part of a joint.

The skeleton is divided into the axial and appendicular skeletons. ▼

1. *How many bones are in an average adult skeleton? What are paired and unpaired bones?*
2. *How are lumps, projections, and openings in bones related to soft tissues?*

Axial Skeleton

The axial skeleton forms the upright axis of the body (see figure 7.1). It is divided into the skull, auditory ossicles, hyoid bone, vertebral column, and thoracic cage, or rib cage. The axial skeleton protects the brain, the spinal cord, and the vital organs housed within the thorax. ▼

3. *List the parts of the axial skeleton and its functions.*

Skull

The bones of the head form the **skull,** or **cranium** (krā′nē-ŭm). The 22 bones of the skull are divided into two groups: those of the braincase and those of the face. The **braincase** consists of 8 bones that immediately surround and protect the brain. The bones of the braincase are the paired parietal and temporal bones and the unpaired frontal, occipital, sphenoid, and ethmoid bones. The **facial bones** form the structure of the face. The 14 facial bones are the maxilla (2), zygomatic (2), palatine (2), lacrimal (2), nasal (2), inferior nasal concha (2), mandible (1), and vomer (1) bones. The frontal and ethmoid bones, which are part of the braincase, also contribute to the face. The **facial bones** support the organs of vision, smell, and taste. They also provide attachment points for the muscles involved in **mastication** (mas-ti-kā′shŭn, chewing), facial expression, and eye movement. The jaws (mandible and maxillae) hold the teeth (see chapter 21) and the temporal bones hold the **auditory ossicles,** or ear bones (see chapter 14).

The bones of the skull, except for the mandible, are not easily separated from each other. It is convenient to think of the skull, except for the mandible, as a single unit. The top of the skull is called the **calvaria** (kal-vā′rē-ă), or skullcap. It is usually cut off to reveal the skull's interior. Selected features of the intact skull are listed in table 7.2. ▼

4. *Name the bones of the braincase and the facial bones. What functions are accomplished by each group of bones?*

Superior View of the Skull

The skull appears quite simple when viewed from above (figure 7.2). The paired **parietal bones** are joined at the midline by the **sagittal suture,** and the parietal bones are connected to the **frontal bone** by the **coronal suture.**

P R E D I C T **1**
Explain the basis for the names *sagittal* and *coronal sutures.*

Posterior View of the Skull

The parietal bones are joined to the **occipital bone** by the **lambdoid** (lam′doyd, the shape resembles the Greek letter lambda) **suture** (figure 7.3). Occasionally, extra small bones called **sutural** (soo′choor-ăl), or wormian, **bones** form along the lambdoid suture. An **external occipital protuberance** is present on the posterior surface of the occipital bone. It can be felt through the scalp at the base of the head and varies considerably in size from person to person. The external occipital protuberance is the site of attachment of the **ligamentum nuchae** (noo′kē, nape of neck),

Table 7.2 Processes and Other Features of the Skull

Feature	Bone on Which Feature Is Found	Description
External Features		
Alveolar process	Mandible, maxilla	Ridges on the mandible and maxilla containing the teeth
Coronoid process	Mandible	Attachment point for the temporalis muscle
Horizontal plate	Palatine	The posterior third of the hard palate
Mandibular condyle	Mandible	Region where the mandible articulates with the temporal bone
Mandibular fossa	Temporal	Depression where the mandible articulates with the skull
Mastoid process	Temporal	Enlargement posterior to the ear; attachment site for several muscles that move the head
Nuchal lines	Occipital	Attachment points for several posterior neck muscles
Occipital condyle	Occipital	Point of articulation between the skull and the vertebral column
Palatine process	Maxilla	Anterior two-thirds of the hard palate
Pterygoid hamulus	Sphenoid	Hooked process on the inferior end of the medial pterygoid plate, around which the tendon of one palatine muscle passes; an important dental landmark
Pterygoid plates (medial and lateral)	Sphenoid	Bony plates on the inferior aspect of the sphenoid bone; the lateral pterygoid plate is the site of attachment for two muscles of mastication (chewing)
Styloid process	Temporal	Attachment site for three muscles (to the tongue, pharynx, and hyoid bone) and some ligaments
Temporal lines	Parietal	Where the temporalis muscle, which closes the jaw, attaches
Internal Features		
Crista galli	Ethmoid	Process in the anterior part of the cranial vault to which one of the connective tissue coverings of the brain (dura mater) connects
Petrous portion	Temporal	Thick, interior part of temporal bone containing the middle and inner ears and the auditory ossicles
Sella turcica	Sphenoid	Bony structure resembling a saddle in which the pituitary gland is located

an elastic ligament that extends down the neck and helps keep the head erect by pulling on the occipital region of the skull. **Nuchal lines** are a set of small ridges that extend laterally from the protuberance and are the points of attachment for several neck muscles.

Lateral View of the Skull

The parietal bone and the temporal bone form a large part of the side of the head (figure 7.4). The term *temporal* means related to time, and the temporal bone is so named because the hair of the temples is often the first to turn white, indicating the passage of

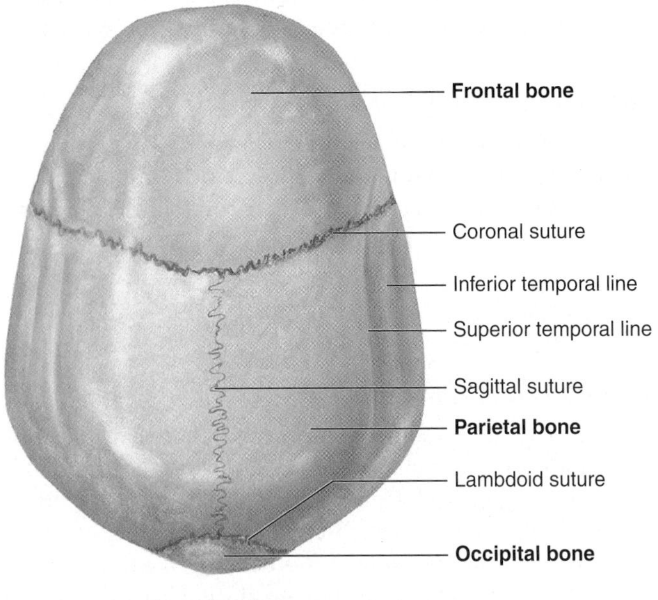

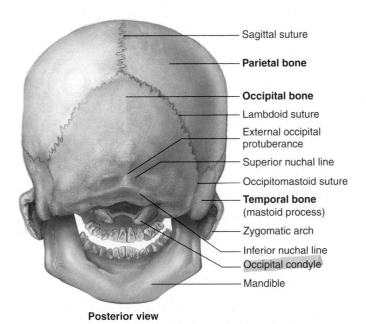

Superior view

Figure 7.2 Skull as Seen from the Superior View
The names of the bones are bolded.

Posterior view

Figure 7.3 Skull as Seen from the Posterior View
The names of the bones are bolded.

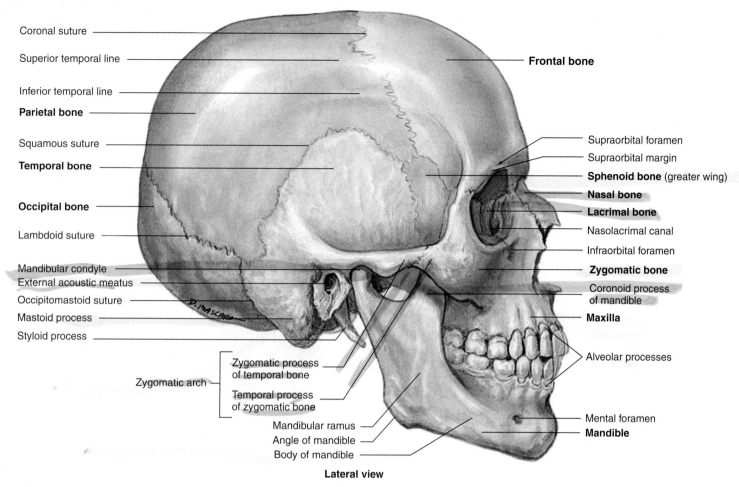

Coronal suture
Superior temporal line
Inferior temporal line
Parietal bone
Squamous suture
Temporal bone
Occipital bone
Lambdoid suture
Mandibular condyle
External acoustic meatus
Occipitomastoid suture
Mastoid process
Styloid process

Zygomatic arch

Zygomatic process of temporal bone
Temporal process of zygomatic bone

Mandibular ramus
Angle of mandible
Body of mandible

Frontal bone
Supraorbital foramen
Supraorbital margin
Sphenoid bone (greater wing)
Nasal bone
Lacrimal bone
Nasolacrimal canal
Infraorbital foramen
Zygomatic bone
Coronoid process of mandible
Maxilla
Alveolar processes
Mental foramen
Mandible

Lateral view

Figure 7.4   Lateral View of the Skull as Seen from the Right Side
The names of the bones are bolded.

time. The **squamous suture** joins the parietal and temporal bones. A prominent feature of the temporal bone is a large hole, the **external acoustic,** or **auditory, meatus** (mē-ā′tŭs, passageway or tunnel), which transmits sound waves toward the eardrum. Just posterior and inferior to the external auditory meatus is a large inferior projection, the **mastoid** (mas′toyd, resembling a breast) **process.** The process can be seen and felt as a prominent lump just posterior to the ear. The process is not solid bone but is filled with cavities called **mastoid air cells,** which are connected to the middle ear. Neck muscles involved in rotation of the head attach to the mastoid process. The superior and inferior **temporal lines** arch across the lateral surface of the parietal bone. They are attachment points of the temporalis muscle, one of the muscles of mastication.

The lateral surface of the **greater wing** of the **sphenoid** (sfē′noyd, wedge-shaped) **bone** is anterior to the temporal bone (see figure 7.4). Although appearing to be two bones, one on each side of the skull, the sphenoid bone is actually a single bone that extends completely across the skull. Anterior to the sphenoid bone is the **zygomatic** (zī′gō-mat′ik, a bar or yoke) **bone,** or cheekbone, which can be easily seen and felt on the face (figure 7.5).

The **zygomatic arch,** which consists of joined processes from the temporal and zygomatic bones, forms a bridge across the side of the skull (see figure 7.4). The zygomatic arch is easily felt on

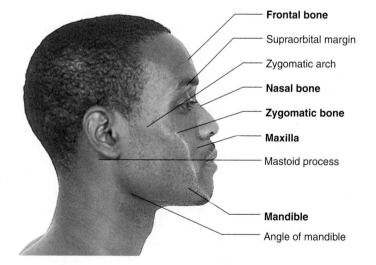

Frontal bone
Supraorbital margin
Zygomatic arch
Nasal bone
Zygomatic bone
Maxilla
Mastoid process
Mandible
Angle of mandible

Figure 7.5   Lateral View of Bony Landmarks on the Face
The names of the bones are bolded.

the side of the face, and the muscles on each side of the arch can be felt as the jaws are opened and closed.

The **maxilla** (mak-sil′ă), or upper jaw, is anterior to the zygomatic bone. The **mandible,** or lower jaw, is inferior to the maxilla

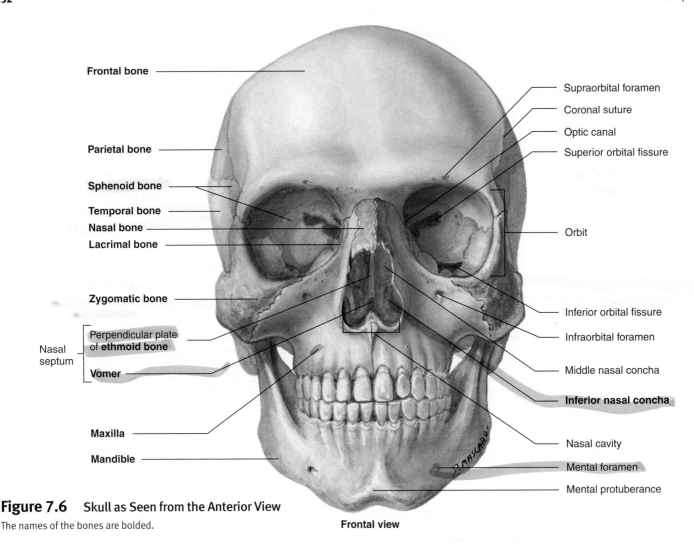

Figure 7.6 Skull as Seen from the Anterior View
The names of the bones are bolded.

(see figure 7.4). The mandible consists of two main parts: the **body** and the **ramus** (branch). The body and ramus join the **angle of the mandible.** The superior end of the ramus has a **mandibular condyle,** which articulates with the temporal bone, allowing movement of the mandible. The **coronoid** (kōr′ŏ-noyd, shaped like a crow's beak) **process** is the attachment site of the temporalis muscle to the mandible. The maxillae and mandible have **alveolar** (al-vē′ō-lǎr) **processes** with sockets for the attachment of the teeth.

Anterior View of the Skull

The major bones seen from the anterior view are the frontal bone (forehead), the zygomatic bones (cheekbones), the maxillae, and the mandible (figure 7.6). The teeth, which are very prominent in this view, are discussed in chapter 24. Many bones of the face can be easily felt through the skin of the face (figure 7.7).

   Two prominent cavities of the skull are the orbits and the nasal cavity (see figure 7.6). The **orbits** are so named because of the rotation of the eyes within them. The bones of the orbits (figure 7.8) provide protection for the eyes and attachment points for the muscles moving the eyes. The major portion of each eyeball is within the orbit, and the portion of the eye visible from the outside is relatively small. Each orbit contains blood vessels, nerves, and fat, as well as the eyeball and the muscles that move it.

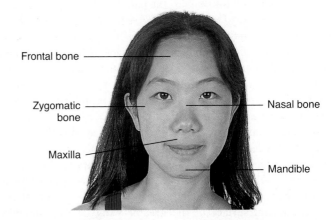

Figure 7.7 Anterior View of Bony Landmarks on the Face
The names of the bones are bolded.

   The orbit has several openings through which structures communicate between the orbit and other cavities (see figure 7.8). The largest of these are the **superior** and **inferior orbital fissures.** They provide openings through which nerves and blood vessels communicate with the orbit or pass to the face. The optic nerve, for the sense of vision, passes from the eye through the **optic canal** and enters the cranial cavity. The **nasolacrimal** (nā-zō-lak′ri-măl,

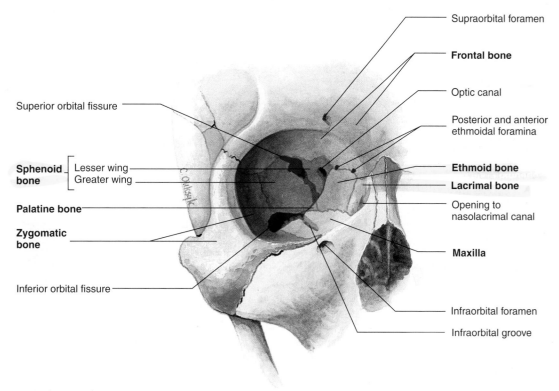

Superior orbital fissure

Sphenoid bone — Lesser wing
Greater wing

Palatine bone

Zygomatic bone

Inferior orbital fissure

Supraorbital foramen

Frontal bone

Optic canal

Posterior and anterior ethmoidal foramina

Ethmoid bone

Lacrimal bone

Opening to nasolacrimal canal

Maxilla

Infraorbital foramen

Infraorbital groove

Figure 7.8   Bones of the Right Orbit

The names of the bones are bolded.   **Anterior view**

nasus, nose + *lacrima,* tear) **canal** passes from the orbit into the nasal cavity. It contains a duct that carries tears from the eyes to the nasal cavity (see chapter 14).

The nasal cavity is divided into right and left halves by a **nasal septum** (sep′tŭm, wall) (see figure 7.6; figure 7.9). The bony part of the nasal septum consists primarily of the **vomer** (vō′mer, shaped like a plowshare) inferiorly and the **perpendicular plate** of the **ethmoid** (eth′moyd, sieve-shaped) **bone** superiorly. The anterior part of the nasal septum is formed by hyaline cartilage called **septal cartilage** (see figure 7.9*a*). The external part of the nose has some bone but is mostly hyaline cartilage (see figure 7.9*b*), which is absent in the dried skeleton.

 Deviated Nasal Septum

The nasal septum usually is located in the median plane until a person is 7 years old. Thereafter, it tends to deviate, or bulge slightly to one side. The septum can also deviate abnormally at birth or, more commonly, as a result of injury. Deviations can be severe enough to block one side of the nasal passage and interfere with normal breathing. The repair of severe deviations requires surgery.

P R E D I C T

A direct blow to the nose may result in a "broken nose." Using figures 7.6 and 7.9, list the bones most likely to be broken.

The lateral wall of the nasal cavity has three bony shelves, the **nasal conchae** (kon′kē, resembling a conch shell) (see figure 7.9*b*). The inferior nasal concha is a separate bone, and the middle and superior nasal conchae are projections from the ethmoid bone. The

conchae and the nasal septum increase the surface area in the nasal cavity, which promotes the moistening and warming of inhaled air and the removal of particles from the air by overlying mucous membranes.

Several of the bones associated with the nasal cavity have large, air-filled cavities within them called the **paranasal sinuses,** which open into the nasal cavity (figure 7.10). The sinuses decrease the weight of the skull and act as resonating chambers during voice production. Compare a normal voice with the voice of a person who has a cold and whose sinuses are "stopped up." The sinuses are named for the bones in which they are located and include the frontal, maxillary, and sphenoidal sinuses. The cavities within each ethmoid bone are collectively called the ethmoidal labyrinth because it consists of a maze of interconnected ethmoidal air cells. The ethmoidal labyrinth is often referred to as the ethmoidal sinuses, and is one of the paranasal sinuses.

Inferior View of the Skull

Seen from below with the mandible removed, the base of the skull is complex, with a number of foramina and specialized surfaces (figure 7.11 and table 7.3). The prominent **foramen magnum,** through which the spinal cord and brain are connected, is located in the occipital bone. The **occipital condyles,** located next to the foramen magnum, connect the skull to the vertebral column.

The major entry and exit points for blood vessels that supply the brain can be seen from this view. Blood is carried to the brain by the internal carotid arteries, which pass through the **carotid** (ka-rot′id, put to sleep) **canals,** and the vertebral arteries, which pass through the foramen magnum. Most blood leaves the brain through the internal jugular veins, which exit through the **jugular foramina** located lateral to the occipital condyles.

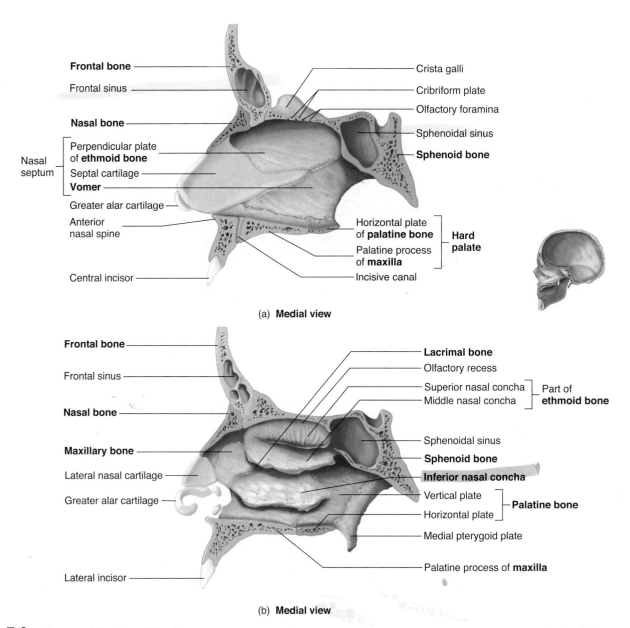

(a) **Medial view**

(b) **Medial view**

Figure 7.9   Bones of the Nasal Cavity

The names of the bones are bolded. (*a*) Nasal septum as seen from the left nasal cavity. (*b*) Right lateral nasal wall as seen from inside the nasal cavity with the nasal septum removed.

Two long, pointed **styloid** (stī′loyd, stylus- or pen-shaped) **processes** project from the inferior surface of the temporal bone (see figures 7.4 and 7.11). Muscles involved in movement of the tongue, hyoid bone, and pharynx attach to each process. The **mandibular fossa,** where the mandibular condyle articulates with the skull, is anterior to the mastoid process.

The posterior opening of the nasal cavity is bounded on each side by the vertical bony plates of the sphenoid bone: the **medial pterygoid** (ter′i-goyd, wing-shaped) **plates** and the **lateral pterygoid plates.** Muscles that help move the mandible attach to the lateral pterygoid plates (see chapter 9). The **vomer** forms most of the posterior portion of the nasal septum.

The **hard palate,** or **bony palate,** forms the floor of the nasal cavity. Sutures join four bones to form the hard palate; the palatine processes of the two maxillary bones form the anterior two-thirds of the palate, and the horizontal plates of the two palatine bones

form the posterior one-third of the palate. The tissues of the soft palate extend posteriorly from the hard palate. The hard and soft palates separate the nasal cavity from the mouth, enabling humans to chew and breathe at the same time.

 Cleft Lip or Palate

During development, the facial bones sometimes fail to fuse with one another. A **cleft lip** results if the maxillae do not form normally, and a **cleft palate** occurs when the palatine processes of the maxillae do not fuse with one another. A cleft palate produces an opening between the nasal and oral cavities, making it difficult to eat or drink or to speak distinctly. A cleft lip and cleft palate may also occur in the same person.

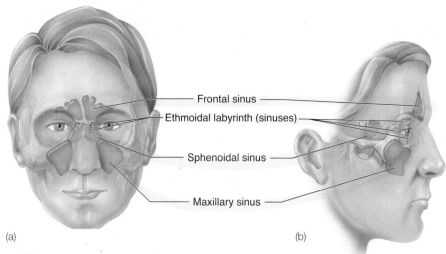

Figure 7.10 Paranasal Sinuses

(*a*) Anterior view. (*b*) Lateral view.

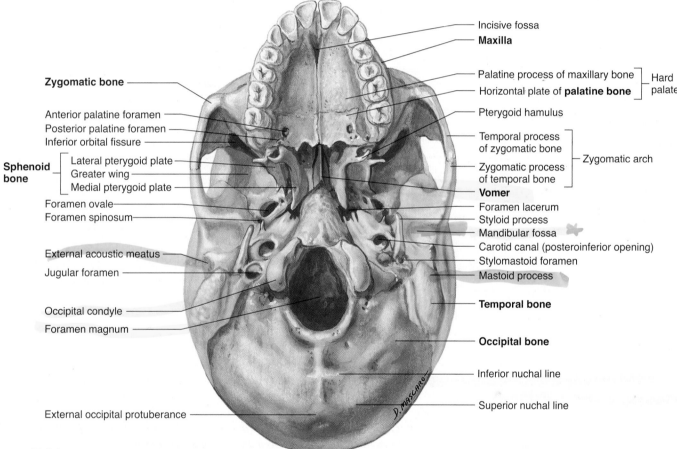

Figure 7.11 Inferior View of the Skull

The names of the bones are bolded.

Inferior view

Interior of the Cranial Cavity

The **cranial cavity** is the cavity in the skull occupied by the brain. When the floor of the cranial cavity is viewed from above with the calvaria cut away (figure 7.12), it can be divided into **anterior, middle, and posterior cranial fossae,** which are formed as the developing skull conforms to the shape of the brain.

The **crista galli** (kris′tă găl′ē, rooster's comb) of the ethmoid bone is a prominent ridge located in the center of the anterior fossa. It is a point of attachment for one of the **meninges** (mĕ-nin′jēz), a thick connective tissue membrane that supports and protects the brain (see chapter 11). On each side of the crista galli are the **cribriform** (krib′ri-fōrm, sievelike) **plates** of the ethmoid bone. The olfactory nerves extend from the cranial cavity into the roof of the nasal cavity through sievelike perforations in the cribriform plate called **olfactory foramina** (see chapter 14).

Table 7.3 Skull Foramina, Fissures, and Canals

Opening	Bone Containing the Opening	Structures Passing Through Openings
Carotid canal	Temporal	Carotid artery and carotid sympathetic nerve plexus
External acoustic meatus	Temporal	Sound waves en route to the eardrum
Foramen lacerum	Between temporal, occipital, and sphenoid	The foramen is filled with cartilage during life; the carotid canal and pterygoid canal cross its superior part but do not actually pass through it
Foramen magnum	Occipital	Spinal cord, accessory nerves, and vertebral arteries
Foramen ovale	Sphenoid	Mandibular division of trigeminal nerve
Foramen rotundum	Sphenoid	Maxillary division of trigeminal nerve
Foramen spinosum	Sphenoid	Middle meningeal artery
Hypoglossal canal	Occipital	Hypoglossal nerve
Incisive foramen (canal)	Between maxillae	Incisive nerve
Inferior orbital fissure	Between sphenoid and maxilla	Infraorbital nerve and blood vessels and zygomatic nerve
Infraorbital foramen	Maxilla	Infraorbital nerve
Internal acoustic meatus	Temporal	Facial nerve and vestibulocochlear nerve
Jugular foramen	Between temporal and occipital	Internal jugular vein, glossopharyngeal nerve, vagus nerve, and accessory nerve
Mandibular foramen	Mandible	Inferior alveolar nerve to the mandibular teeth
Mental foramen	Mandible	Mental nerve
Nasolacrimal canal	Between lacrimal and maxilla	Nasolacrimal (tear) duct
Olfactory foramina	Ethmoid	Olfactory nerves
Optic canal	Sphenoid	Optic nerve and ophthalmic artery
Stylomastoid foramen	Temporal	Facial nerve
Superior orbital fissures	Sphenoid	Oculomotor nerve, trochlear nerve, ophthalmic division of trigeminal nerve, abducens nerve, and ophthalmic veins
Supraorbital foramen or notch	Frontal	Supraorbital nerve and vessels

The sphenoid bone extends from one side of the skull to the other. The center of the sphenoid bone is modified into a structure resembling a saddle, the **sella turcica** (sel′ă tŭr′si-kă, Turkish saddle), which is occupied by the pituitary gland in life.

The **petrous** (rocky) **part** of the temporal bone is a thick, bony ridge lateral to the foramen magnum. It is hollow and contains the middle and inner ears. The auditory ossicles are located in the middle ear. An internal carotid artery enters the external opening of each carotid canal (see figure 7.11) and passes through the carotid canal, which runs anteromedially within the petrous part of the temporal bone. A thin plate of bone separates the carotid canal from the middle ear, making it possible for a person to hear his or her own heartbeat—for example, when frightened or after running.

Most of the foramina seen in the interior view of the skull, such as the foramen magnum and optic canals, can also be seen externally. A few foramina, such as the **internal acoustic meatus,**

do not open to the outside. The vestibulocochlear nerve for hearing and balance passes through the internal acoustic meatus and connects to the inner ear within the temporal bone. ▼

5. *Name the major sutures separating the frontal, parietal, occipital, and temporal bones.*
6. *Name the parts of the bones that connect the skull to the vertebral columnand that connect the mandible to the temporal bone.*
7. *Describe the bones and cartilage found in the nasal septum.*
8. *What is a sinus? What are the functions of sinuses? Give the location of the paranasal sinuses. Where else in the skull are there air-filled spaces?*
9. *Name the bones that form the hard palate. What is the function of the hard palate?*

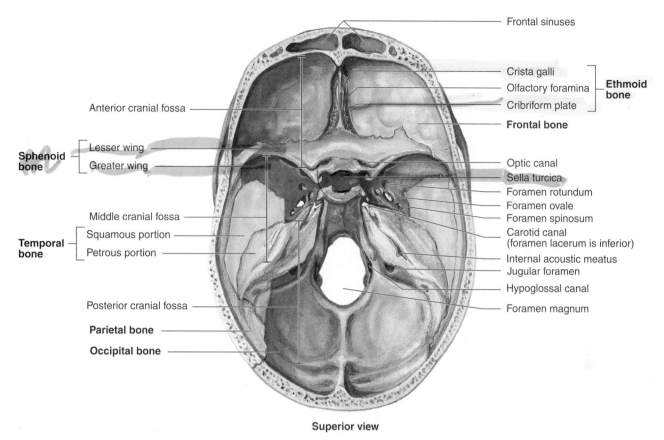

Superior view

Figure 7.12 Floor of the Cranial Cavity

The names of the bones are bolded. The roof of the skull has been removed, and the floor is seen from a superior view.

10. *Through what foramen does the brainstem connect to the spinal cord? Name the foramina that contain nerves for the senses of vision (optic nerve), smell (olfactory nerves), and hearing (vestibulocochlear nerve).*
11. *Name the foramina through which the major blood vessels enter and exit the skull.*
12. *List the places where the following muscles attach to the skull: neck muscles, throat muscles, muscles of mastication, muscles of facial expression, and muscles that move the eyeballs.*

Hyoid Bone

The **hyoid bone** (figure 7.13), which is unpaired, is often listed as part of the facial bones because it has a developmental origin in common with the bones of the face. It is not, however, part of the adult skull. The hyoid bone has no direct bony attachment to the skull. Instead, muscles and ligaments attach it to the skull, so the hyoid "floats" in the superior aspect of the neck just below the mandible. The hyoid bone provides an attachment point for some tongue muscles, and it is an attachment point for important neck muscles that elevate the larynx during speech or swallowing. ▼

13. *Where is the hyoid bone located and what does it do?*

Vertebral Column

The **vertebral** (*verto,* to turn) **column,** or backbone, is the central axis of the skeleton, extending from the base of the skull to slightly past the end of the pelvis (see figure 7.1). The vertebral column performs five major functions: (1) It supports the weight of the head and trunk, (2) it protects the spinal cord, (3) it allows spinal nerves to exit the spinal cord, (4) it provides a site for muscle attachment, and (5) it permits movement of the head and trunk.

The vertebral column usually consists of 26 individual bones, grouped into five regions (figure 7.14). Seven **cervical** (ser′vĭ-kal, neck) **vertebrae,** 12 **thoracic** (thō-ras′ik, chest) **vertebrae,** 5 **lumbar** (lŭm′bar, loin) **vertebrae,** 1 **sacral** (sā′krăl, sacred bone) **bone,** and 1 **coccygeal** (kok-sij′ē-ăl, shaped like a cuckoo's bill) **bone** make up the vertebral column. The cervical vertebrae are designated "C," thoracic "T," lumbar "L," sacral "S," and coccygeal "CO." A number after the letter indicates the number of the vertebra, from superior to inferior, within each vertebral region. The developing embryo has 33 or 34 vertebrae, but the 5 sacral vertebrae fuse to form 1 bone, and the 4 or 5 coccygeal bones usually fuse to form 1 bone.

The five regions of the adult vertebral column have four major curvatures (see figure 7.14). The primary thoracic and sacral curves appear during embryonic development and reflect the C-shaped curve of the embryo and fetus within the uterus. When the infant raises its head in the first few months after birth, a secondary curve, which is convex anteriorly, develops in the neck. Later, when the infant learns to sit and then walk, the lumbar portion of the column also becomes convex anteriorly.

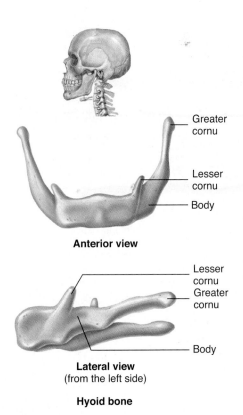

Greater
cornu

Lesser
cornu

Body

Anterior view

Lesser
cornu
Greater
cornu

Body

Lateral view
(from the left side)

Hyoid bone

Figure 7.13   Hyoid Bone

 Abnormal Spinal Curvatures

Lordosis (lōr-dō′sis, hollow back) is an exaggeration of the convex curve of the lumbar region, resulting in a swayback condition. **Kyphosis** (kī-fō′sis, hump back) is an exaggeration of the concave curve of the thoracic region, resulting in a hunchback condition. **Scoliosis** (skō′lē-ō′sis) is an abnormal lateral and rotational curvature of the vertebral column, which is often accompanied by secondary abnormal curvatures, such as kyphosis. ▼

14. *What are the functions of the vertebral column?*
15. *Name and give the number of the bones forming the vertebral column.*
16. *Describe the four major curvatures of the vertebral column and how they develop.*

General Plan of the Vertebrae

Each vertebra consists of a body, an arch, and various processes. The weight-bearing portion of the vertebra is the **body** (table 7.4a). The **vertebral arch** projects posteriorly from the body. Each vertebral arch consists of two **pedicles** (ped′ĭ-klz, feet), which are attached to the body, and two **laminae** (lam′i-nē, thin plates), which extend from the transverse processes to the spinous process. The vertebral arch and the posterior part of the body surround a large opening called the **vertebral foramen.** The vertebral foramina of adjacent vertebrae combine to form the **vertebral canal** (table 7.4b), which contains the spinal cord and cauda equina, which is a collection of spinal nerves (see

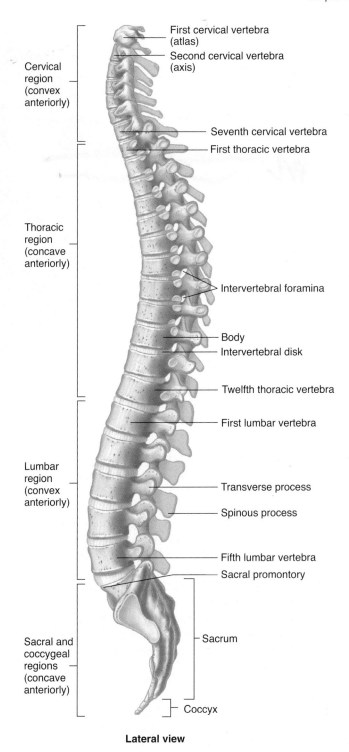

First cervical vertebra (atlas)
Second cervical vertebra (axis)

Cervical region (convex anteriorly)

Seventh cervical vertebra
First thoracic vertebra

Thoracic region (concave anteriorly)

Intervertebral foramina

Body
Intervertebral disk

Twelfth thoracic vertebra

First lumbar vertebra

Lumbar region (convex anteriorly)

Transverse process

Spinous process

Fifth lumbar vertebra
Sacral promontory

Sacral and coccygeal regions (concave anteriorly)

Sacrum

Coccyx

Lateral view

Figure 7.14   Complete Vertebral Column Viewed from the Left Side

chapter 11). The vertebral arches and bodies protect the spinal cord and cauda equina.

A **transverse process** extends laterally from each side of the arch between the lamina and pedicle, and a single **spinous process** is present at the junction between the two laminae (see table 7.4a). The spinous processes can be seen and felt as a series of lumps down the midline of the back (figure 7.15). The transverse and spinous processes are attachment sites for muscles moving the vertebral column.

Table 7.4 General Structure of a Vertebra

Feature	Description
Body	Disk-shaped; usually the largest part with flat surfaces directed superiorly and inferiorly; forms the anterior wall of the vertebral foramen; intervertebral disks are located between the bodies
Vertebral foramen	Hole in each vertebra through which the spinal cord passes; adjacent vertebral foramina form the vertebral canal
Vertebral arch	Forms the lateral and posterior walls of the vertebral foramen; possesses several processes and articular surfaces
Pedicle	Foot of the arch with one on each side; forms the lateral walls of the vertebral foramen
Lamina	Posterior part of the arch; forms the posterior wall of the vertebral foramen
Transverse process	Process projecting laterally from the junction of the lamina and pedicle; a site of muscle attachment
Spinous process	Process projecting posteriorly at the point where the two laminae join; a site of muscle attachment; strengthens the vertebral column and allows for movement
Articular processes	Superior and inferior projections containing articular facets where vertebrae articulate with each other; strengthen the vertebral column and allow for movement
Intervertebral notches	Form intervertebral foramina between two adjacent vertebrae through which spinal nerves exit the vertebral canal

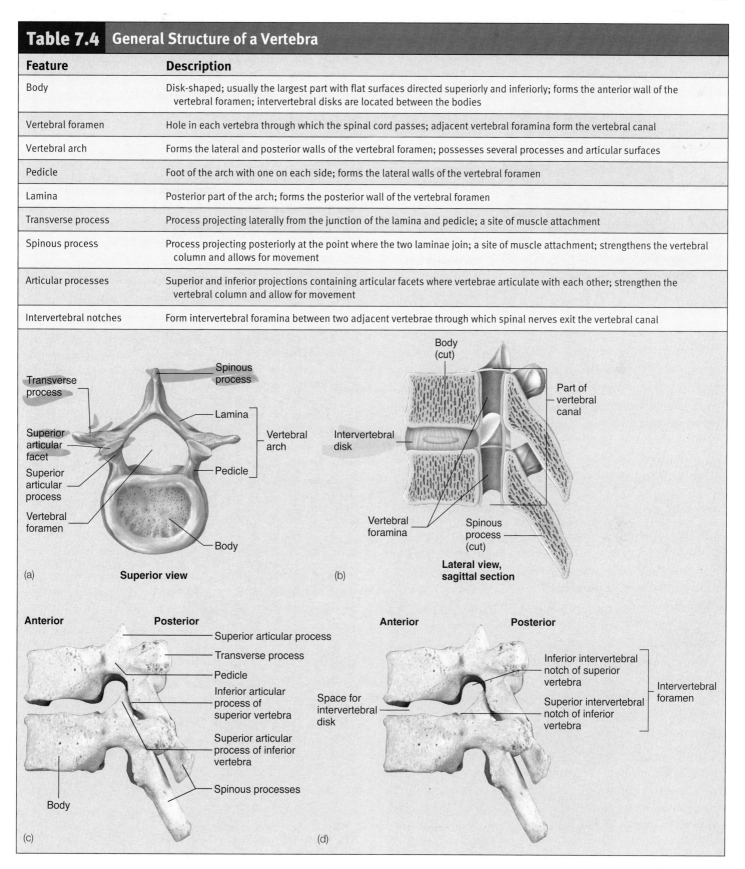

(a) **Superior view**

(b) **Lateral view, sagittal section**

(c)

(d)

Support and movement of the vertebral column are made possible by the articular processes. Each vertebra has two **superior** and two **inferior articular processes,** with the superior processes of one vertebra articulating with the inferior processes of the next superior vertebra (table 7.4c). Overlap of these processes helps hold the vertebrae together. Each articular process has a smooth **articular facet** (fas′et, little face), which allows movement between the processes (see table 7.4a).

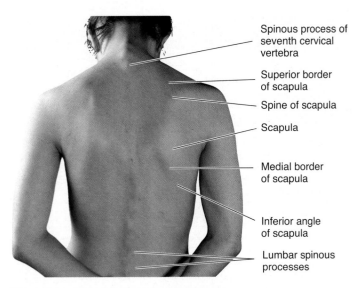

- Spinous process of seventh cervical vertebra
- Superior border of scapula
- Spine of scapula
- Scapula
- Medial border of scapula
- Inferior angle of scapula
- Lumbar spinous processes

Figure 7.15 Surface View of the Back Showing the Scapula and Vertebral Spinous Processes

Spinal nerves exit the vertebral canal through the **intervertebral foramina** (see table 7.4*d* and figure 7.14). Each intervertebral foramen is formed by **intervertebral notches** in the pedicles of adjacent vertebrae.

Spina Bifida

Sometimes vertebral laminae partly or completely fail to fuse (or even fail to form) during fetal development, resulting in a condition called **spina bifida** (spī′nă bif′i-dă, split spine). This defect is most common in the lumbar region. If the defect is severe and involves the spinal cord (figure A), it may interfere with normal nerve function below the point of the defect. ▼

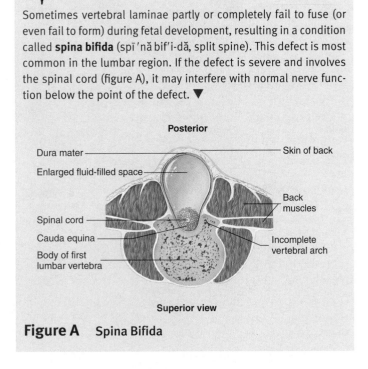

Posterior

- Dura mater
- Enlarged fluid-filled space
- Spinal cord
- Cauda equina
- Body of first lumbar vertebra
- Skin of back
- Back muscles
- Incomplete vertebral arch

Superior view

Figure A Spina Bifida

17. *What is the weight-bearing part of a vertebra?*
18. *Describe the structures forming the vertebral foramen and the vertebral canal. What structures are found within them?*

19. *What are the functions of the transverse and spinous processes?*
20. *Describe how superior and inferior articular processes help support and allow movement of the vertebral column.*
21. *Where do spinal nerves exit the vertebral column?*

Intervertebral Disks

Intervertebral disks are pads of fibrocartilage located between the bodies of adjacent vertebrae (figure 7.16). They act as shock absorbers between the vertebral bodies and allow the vertebral column to bend. The intervertebral disks consist of an external **annulus fibrosus** (an′ū-lŭs fī-brō′sŭs, fibrous ring) and an internal, gelatinous **nucleus pulposus** (pŭl-pō′sŭs, pulp). The disk becomes more compressed with increasing age so that the distance between vertebrae and therefore the overall height of the individual decreases. The annulus fibrosus also becomes weaker with age and more susceptible to herniation.

Herniated, or Ruptured, Intervertebral Disk

A **herniated**, or **ruptured, disk** results from the breakage or ballooning of the annulus fibrosus with a partial or complete release of the nucleus pulposus (figure B). The herniated part of the disk may push against and compress the spinal cord, cauda equina, or spinal nerves, compromising their normal function and producing pain. Herniation of the inferior lumbar intervertebral disks is most common, but herniation of the inferior cervical disks is almost as common. ▼

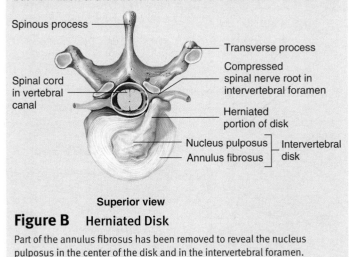

- Spinous process
- Transverse process
- Compressed spinal nerve root in intervertebral foramen
- Spinal cord in vertebral canal
- Herniated portion of disk
- Nucleus pulposus
- Annulus fibrosus
- Intervertebral disk

Superior view

Figure B Herniated Disk

Part of the annulus fibrosus has been removed to reveal the nucleus pulposus in the center of the disk and in the intervertebral foramen.

22. *What is the function of the intervertebral disks? Name the two parts of the disk.*

Regional Differences in Vertebrae

The vertebrae of each region of the vertebral column have specific characteristics that tend to blend at the boundaries between regions (figure 7.17 and table 7.5). The **cervical vertebrae** all have a **transverse foramen** in each transverse process through which the vertebral arteries extend toward the head.

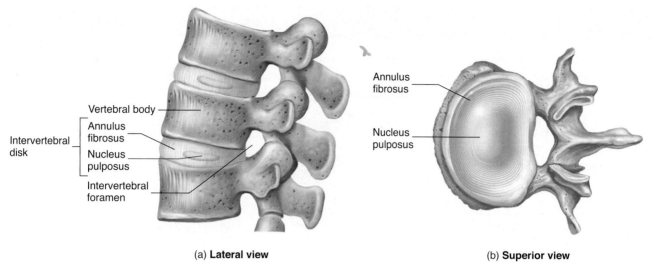

(a) **Lateral view**

Vertebral body

Intervertebral disk
- Annulus fibrosus
- Nucleus pulposus

Intervertebral foramen

(b) **Superior view**

Annulus fibrosus

Nucleus pulposus

Figure 7.16 Intervertebral Disk

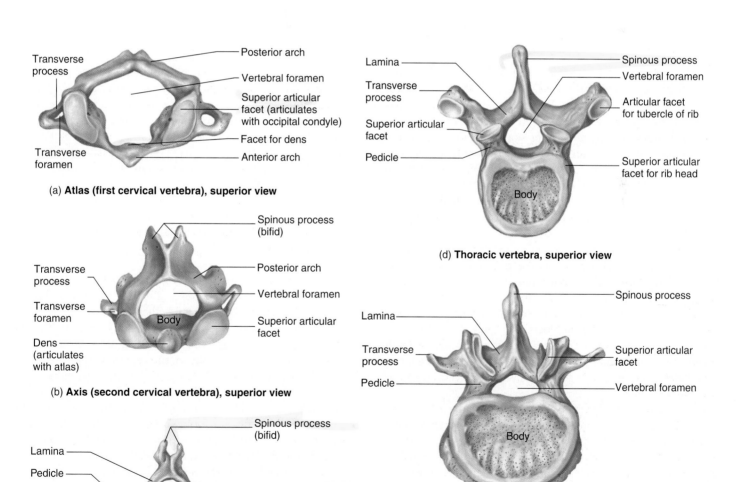

(a) **Atlas (first cervical vertebra), superior view**

Transverse process

Posterior arch

Vertebral foramen

Superior articular facet (articulates with occipital condyle)

Facet for dens

Anterior arch

Transverse foramen

(b) **Axis (second cervical vertebra), superior view**

Spinous process (bifid)

Transverse process

Transverse foramen

Body

Dens (articulates with atlas)

Posterior arch

Vertebral foramen

Superior articular facet

(c) **Fifth cervical vertebra, superior view**

Lamina

Pedicle

Transverse foramen

Transverse process

Body

Spinous process (bifid)

Vertebral foramen

Superior articular facet

(d) **Thoracic vertebra, superior view**

Lamina

Transverse process

Superior articular facet

Pedicle

Body

Spinous process

Vertebral foramen

Articular facet for tubercle of rib

Superior articular facet for rib head

(e) **Lumbar vertebra, superior view**

Lamina

Transverse process

Pedicle

Body

Spinous process

Superior articular facet

Vertebral foramen

Figure 7.17 Regional Differences in Vertebrae

Posterior is shown at the top of each illustration.

Table 7.5 Comparison of Vertebral Regions

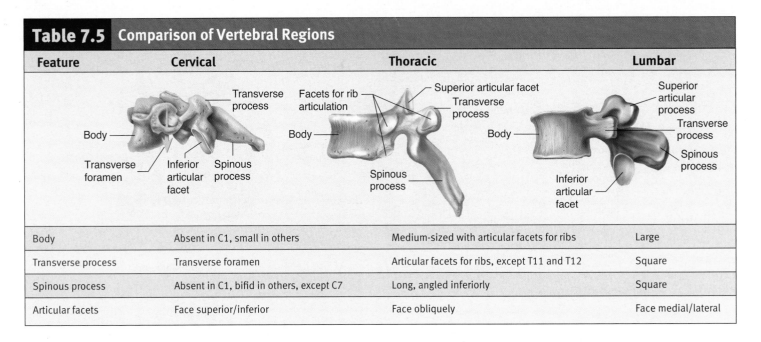

Feature	Cervical	Thoracic	Lumbar
Body	Absent in C1, small in others	Medium-sized with articular facets for ribs	Large
Transverse process	Transverse foramen	Articular facets for ribs, except T11 and T12	Square
Spinous process	Absent in C1, bifid in others, except C7	Long, angled inferiorly	Square
Articular facets	Face superior/inferior	Face obliquely	Face medial/lateral

The first cervical vertebra is called the **atlas** (see figure 7.17*a*) because it holds up the head, just as Atlas in classical mythology held up the world. The atlas has no body, but it has large superior articular facets where it articulates with the occipital condyles on the base of the skull. This joint allows the head to move in a "yes" motion or to tilt from side to side. The second cervical vertebra is called the **axis** (figure 7.17*b*) because it has a projection around which the atlas rotates to produce a "no" motion of the head. The projection is called the **dens** (denz, tooth-shaped) or **odontoid** (ō-don′toyd, tooth-shaped), **process**.

The atlas does not have a spinous process (see figure 7.17*a*). The spinous process of most cervical vertebrae end in two parts and are called **bifid** (bī′fid, split) **spinous processes** (see figure 7.17*b* and *c*). The spinous process of the seventh cervical vertebra is not bifid; it is often quite pronounced and often can be seen and felt as a lump between the shoulders (see figure 7.15). The most prominent spinous process in this area is called the **vertebra prominens.** This is usually the spinous process of the seventh cervical vertebra, but it may be that of the sixth cervical vertebra or even the first thoracic vertebra.

The **thoracic vertebrae** (see figure 7.14; figure 7.17*d*) have attachment sites for the ribs. The first 10 thoracic vertebrae have articular facets on their transverse processes, where they articulate with the tubercles of the ribs. Additional articular facets are on the superior and inferior margins of the body where the heads of the ribs articulate (see "Ribs and Costal Cartilages," p. 163). Thoracic vertebrae have long, thin spinous processes, which are directed inferiorly.

The **lumbar vertebrae** (see figure 7.14; figure 7.17*e*) have large, thick bodies and heavy, rectangular transverse and spinous processes. The superior articular facets face medially, and the

inferior articular facets face laterally. When the superior articular surface of one lumbar vertebra joins the inferior articulating surface of another lumbar vertebra, the arrangement tends to "lock" adjacent lumbar vertebrae together, giving the lumbar part of the vertebral column more stability and limiting rotation of the lumbar vertebrae. The articular facets in other regions of the vertebral column have a more "open" position, allowing for more movement but less stability.

P R E D I C T   ③
Cervical vertebrae have small bodies, whereas lumbar vertebrae have large bodies. Explain.

The five **sacral** (sā′krăl) **vertebrae** (see figure 7.14; figure 7.18) are fused into a single bone called the **sacrum** (sā′krŭm). Although the margins of the sacral bodies unite after the twentieth year, the interior of the sacrum is not ossified until midlife. The transverse processes fuse to form the lateral parts of the sacrum. The superior lateral part of the sacrum forms wing-shaped areas called the **alae** (ā′lē, wings). Much of the lateral surfaces of the sacrum are ear-shaped **auricular surfaces,** which join the sacrum to the pelvic bones. The spinous processes of the first four sacral vertebrae partially fuse to form projections, called the **median sacral crest.** The spinous process of the fifth sacral vertebra does not form, thereby leaving a **sacral hiatus** (hī-ā′tŭs), or gap, which exposes the sacral canal. The vertebral canal within the sacrum is called the **sacral canal.** The sacral hiatus is used to gain entry into the sacral canal to administer anesthetic injections—for example, just before childbirth. The anterior edge of the body of the first sacral vertebra bulges to form the **sacral promontory,** a landmark that separates the abdominal cavity from the pelvic cavity. The sacral promontory can be felt during a vaginal examination, and

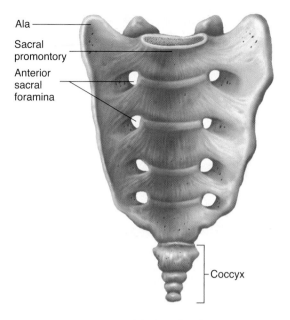

Ala

Sacral promontory

Anterior sacral foramina

Coccyx

(a) **Anterior view**

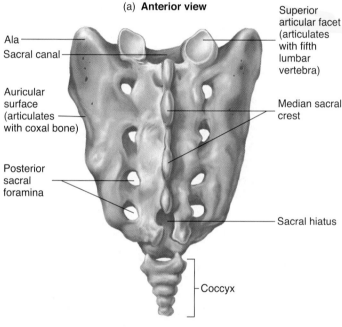

Ala

Sacral canal

Auricular surface (articulates with coxal bone)

Posterior sacral foramina

Superior articular facet (articulates with fifth lumbar vertebra)

Median sacral crest

Sacral hiatus

Coccyx

(b) **Posterior view**

Figure 7.18   Sacrum

it is used as a reference point during measurement to determine if the pelvic openings are large enough to allow for normal vaginal delivery of a baby.

The **coccyx** (kok′siks, shaped like a cuckoo's bill), or tailbone, usually consists of four more or less fused vertebrae (see figure 7.18). The vertebrae of the coccyx do not have the typical structure of most other vertebrae. They consist of extremely reduced vertebral bodies, without the foramina or processes, usually fused into a single bone. The coccyx is easily broken in a fall in

which a person sits down hard on a solid surface. Also, a mother's coccyx may be fractured during childbirth. ▼

> 23. *Describe the characteristics that distinguish the different types of vertebrae.*
> 24. *Describe the movements of the head produced by the atlas and axis.*

P R E D I C T
Which bone is the loneliest bone in the body?

Thoracic Cage

The **thoracic cage,** or **rib cage,** protects the vital organs within the thorax and forms a semirigid chamber that can increase and decrease in volume during breathing. It consists of the thoracic vertebrae, the ribs with their associated costal (rib) cartilages, and the sternum (figure 7.19*a*).

Ribs and Costal Cartilages

There are 12 pairs of **ribs,** which are numbered 1 through 12, starting with the most superior rib. All of the ribs articulate posteriorly with the thoracic vertebrae. **Costal cartilages** attach many of the ribs anteriorly to the sternum. Movement of the ribs relative to the vertebrae and the flexibility of the costal cartilages allow the thoracic cage to change shape during breathing.

The ribs are classified by their anterior attachments as true or false ribs. The **true ribs** attach directly through their costal cartilages to the sternum. The superior seven pairs of ribs are true ribs. The **false ribs** do not attach to the sternum. The inferior five pairs of ribs are false ribs. On each side, the three superior false ribs are joined by a common cartilage to the costal cartilage of the seventh true rib, which in turn is attached to the sternum. The two inferior pairs of false ribs are also called **floating ribs** because they do not attach to the sternum.

Most ribs have two points of articulation with the thoracic vertebrae (figure 7.19*b* and *c*). First, the **head** articulates with the bodies of two adjacent vertebrae and the intervertebral disk between them. The head of each rib articulates with the inferior articular facet of the superior vertebra and the superior articular facet of the inferior vertebra. Second, the **tubercle** articulates with the transverse process of the inferior vertebra. The **neck** is between the head and tubercle, and the **body,** or shaft, is the main part of the rib. The **angle** of the rib is located just lateral to the tubercle and is the point of greatest curvature.

⚕ **Rib Defects**

A **separated rib** is a dislocation between a rib and its costal cartilage. As a result of the dislocation, the rib can move, override adjacent ribs, and cause pain. Separation of the tenth rib is the most common.

The angle is the weakest part of the rib and may be fractured in a crushing accident, such as an automobile accident. Broken rib ends can damage internal organs, such as the lungs, spleen, liver, and diaphragm.

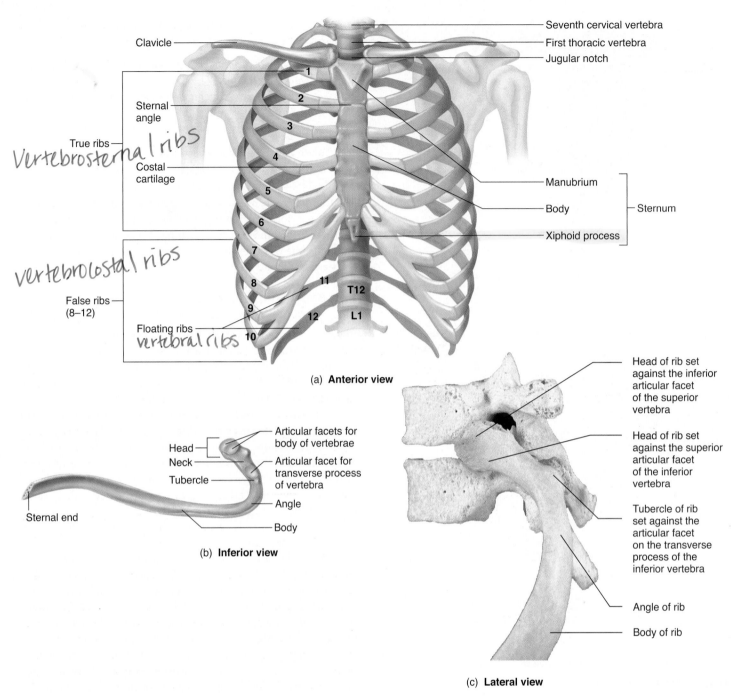

Clavicle

Seventh cervical vertebra
First thoracic vertebra
Jugular notch

Sternal angle

True ribs

Vertebrosternal ribs

Costal cartilage

Manubrium
Body — Sternum
Xiphoid process

vertebrocostal ribs

False ribs (8–12)

Floating ribs

vertebral ribs

T12
L1

(a) **Anterior view**

Head
Neck
Tubercle

Articular facets for body of vertebrae
Articular facet for transverse process of vertebra
Angle
Body

Sternal end

(b) **Inferior view**

Head of rib set against the inferior articular facet of the superior vertebra

Head of rib set against the superior articular facet of the inferior vertebra

Tubercle of rib set against the articular facet on the transverse process of the inferior vertebra

Angle of rib

Body of rib

(c) **Lateral view**

Figure 7.19 **Thoracic Cage**

(*a*) Entire thoracic cage as seen from an anterior view. (*b*) Typical rib, inferior view. (*c*) Photograph of two thoracic vertebrae and the proximal end of a rib, as seen from the left side, showing the relationship between the vertebra and the head and tubercle of the rib.

Sternum

The **sternum,** or breastbone, has three parts (see figure 7.19*a*): the **manubrium** (mă-noo′brē-ŭm, handle), the **body,** and the **xiphoid** (zi′foyd, sword) **process.** The sternum resembles a sword, with the manubrium forming the handle, the body forming the blade, and the xiphoid process forming the tip. At the superior end of the sternum, a depression, called the **jugular notch,** is located between the ends of the clavicles where they articulate with the manubrium of the sternum. The jugular notch can easily be found at the base of the neck (figure 7.20). A slight ridge, called the **sternal angle,** can be felt at the junction of the manubrium and the body of the sternum.

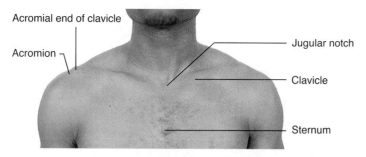

Figure 7.20   Surface Anatomy Showing Bones of the Upper Thorax

 Clinical Importance of the Sternum

The sternal angle is important clinically because the second rib is found lateral to it and can be used as a starting point for counting the other ribs. Counting ribs is important because they are landmarks used to locate structures in the thorax, such as areas of the heart. The sternum often is used as a site for taking red bone marrow samples because it is readily accessible. Because the xiphoid process of the sternum is attached only at its superior end, it may be broken during cardiopulmonary resuscitation (CPR) and then may lacerate the underlying liver. ▼

25. *What are the functions of the thoracic (rib) cage? Distinguish among true, false, and floating ribs, and give the number of each type.*
26. *Describe the articulation of the ribs with thoracic vertebrae.*
27. *Describe the parts of the sternum. What structures attach to the sternum?*

Appendicular Skeleton

The appendicular skeleton (see figure 7.1) consists of the bones of the **upper** and **lower limbs** and the **girdles** by which they are attached to the body. The term *girdle* means a belt or a zone and refers to the two zones, pectoral and pelvic, where the limbs are attached to the body. The pectoral girdle attaches the upper limbs to the body and allows considerable movement of the upper limbs. This freedom of movement allows the hands to be placed in a wide range of positions to accomplish their functions. The pelvic girdle attaches the lower limbs to the body, providing support while allowing movement. The pelvic girdle is stronger and attached much more firmly to the body than is the pectoral girdle, and the lower limb bones in general are thicker and longer than those of the upper limb.

Pectoral Girdle

The **pectoral** (pek′tŏ-răl), or **shoulder, girdle** consists of two **scapulae** (skap′ū-lăē), or shoulder blades, and two **clavicles** (klav′i-klz, key), or collarbones (see figure 7.1). Each humerus (arm bone) attaches to a scapula, which is connected by a clavicle to the sternum (figure 7.21). The scapula is a flat, triangular

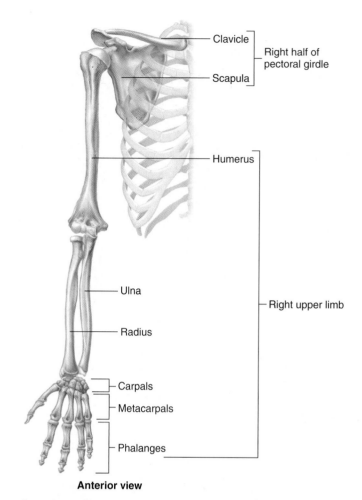

Figure 7.21   Bones of the Pectoral Girdle and Upper Limb

bone (figure 7.22) that can easily be seen and felt in a living person (see figure 7.15). The **glenoid** (glen′oyd) **cavity** is a depression where the humerus connects to the scapula. The scapula has three fossae where muscles extending to the arm are attached. The **scapular spine,** which runs across the posterior surface of the scapula, separates two of these fossae. The **supraspinous fossa** is superior to the spine and the **infraspinous fossa** is inferior to it. The **subscapular fossa** is on the anterior surface of the scapula. The **acromion** (ă-krō′mē-on, *akron,* tip + *omos,* shoulder) is an extension of the spine forming the point of the shoulder. The acromion forms a protective cover for the shoulder joint and is the attachment site for the clavicle and some of the shoulder muscles. The **coracoid** (kōr′ă-koyd, crow's beak) **process** curves below the clavicle and provides attachment for arm and chest muscles.

The clavicle is a long bone with a slight sigmoid (S-shaped) curve (figure 7.23) and is easily seen and felt in the living human (see figure 7.20). The **acromial (lateral) end** of the clavicle articulates with the acromion of the scapula, and the **sternal (medial) end** articulates with the manubrium of the sternum. The pectoral girdle's only attachment to the axial skeleton is at the sternum. Mobility of the upper limb is enhanced by movement of the scapula, which is possible because the clavicle can move relative to the

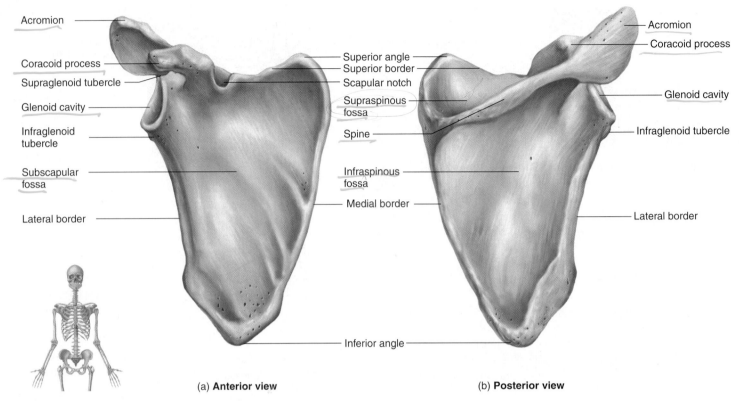

Acromion

Coracoid process

Supraglenoid tubercle

Glenoid cavity

Infraglenoid tubercle

Subscapular fossa

Lateral border

Superior angle
Superior border
Scapular notch
Supraspinous fossa
Spine
Infraspinous fossa
Medial border
Inferior angle

Acromion
Coracoid process

Glenoid cavity

Infraglenoid tubercle

Lateral border

(a) **Anterior view**

(b) **Posterior view**

Figure 7.22 Right Scapula

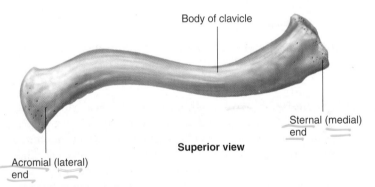

Body of clavicle

Sternal (medial) end

Superior view

Acromial (lateral) end

Figure 7.23 Right Clavicle

sternum. For example, feel the movement of the clavicle when shrugging the shoulders. ▼

28. *Name the bones that make up the pectoral girdle. Describe their functions.*
29. *What are the functions of the acromion and the coracoid process of the scapula?*

P R E D I C T ⑤
How does a broken clavicle change the position of the upper limb?

Upper Limb

The upper limb consists of the bones of the arm, forearm, wrist, and hand (see figure 7.21).

Arm

The arm is the part of the upper limb from the shoulder to the elbow. It contains only one bone, the **humerus** (figure 7.24). The humeral **head** articulates with the glenoid cavity of the scapula. The **anatomical neck,** around the head of the humerus, is where connective tissue holding the shoulder joint together attaches. The **surgical neck** is so named because it is a common fracture site that often requires surgical repair. If it becomes necessary to remove the humeral head because of disease or injury, it is removed down to the surgical neck. The **greater tubercle** and the **lesser tubercle** are sites of muscle attachment. The **intertubercular,** or **bicipital** (bī-sip′i-tăl), **groove** between the tubercles contains one tendon of the biceps brachii muscle. The **deltoid tuberosity** is located on the lateral surface of the humerus a little more than a third of the way along its length and is the attachment site for the deltoid muscle.

Condyles on the distal end of the humerus articulate with the two forearm bones. The **capitulum** (kă-pit′ū-lŭm, head-shaped) is very rounded and articulates with the radius. The **trochlea** (trok′lē-ă, spool) somewhat resembles a spool or pulley and articulates with the ulna. Proximal to the capitulum and the trochlea are the **medial** and **lateral epicondyles,** which are points of muscle attachment for the muscles of the forearm. They can be found as bony protuberances proximal to the elbow (figure 7.25).

Forearm

The forearm has two bones (figure 7.26). The **ulna** is on the medial (little finger) side of the forearm, whereas the **radius** is on the lateral (thumb) side of the forearm.

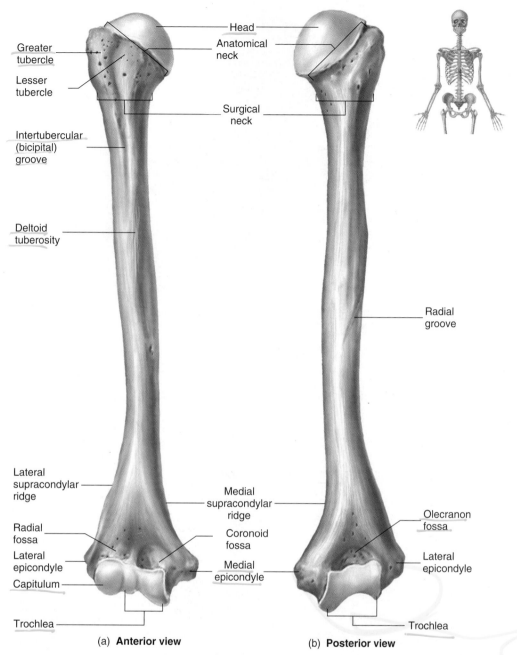

Figure 7.24   Right Humerus

The proximal end of the ulna has a C-shaped articular surface called the **trochlear** or **semilunar notch** that fits over the trochlea of the humerus, forming most of the elbow joint. The trochlear notch is bounded by two processes. The **olecranon** (ō-lek′ră-non, the point of the elbow) is the posterior process forming the tip of the elbow (see figure 7.25). It can easily be felt and is commonly referred to as "the elbow." Posterior arm muscles attach to the olecranon. The smaller, anterior process is the **coronoid** (kōr′ŏ-noyd, crow's beak) **process.**

The proximal end of the radius is the **head.** It is concave and articulates with the capitulum of the humerus. Movements of the radial head relative to the capitulum and of the trochlear notch relative to the trochlea allows the elbow to bend and straighten. The lateral surfaces of the radial head form a smooth cylinder where the radius rotates against the **radial notch** of the ulna. As the forearm supinates and pronates (see "Types of Movements," p. 182), the proximal end of the ulna stays in place and the radius rotates.

P R E D I C T

Explain the functions of the olecranon, coronoid, and radial fossae on the distal humerus (see figure 7.24).

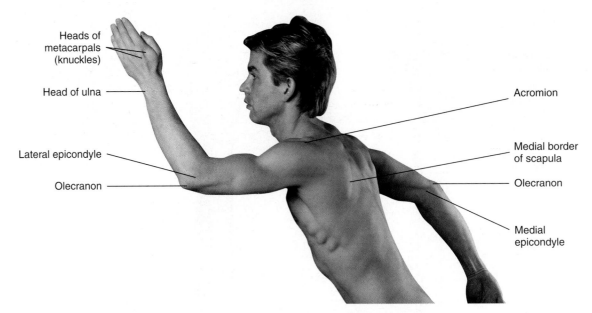

Heads of metacarpals (knuckles)

Head of ulna

Lateral epicondyle

Olecranon

Acromion

Medial border of scapula

Olecranon

Medial epicondyle

Figure 7.25   Surface Anatomy Showing Bones of the Pectoral Girdle and Upper Limb

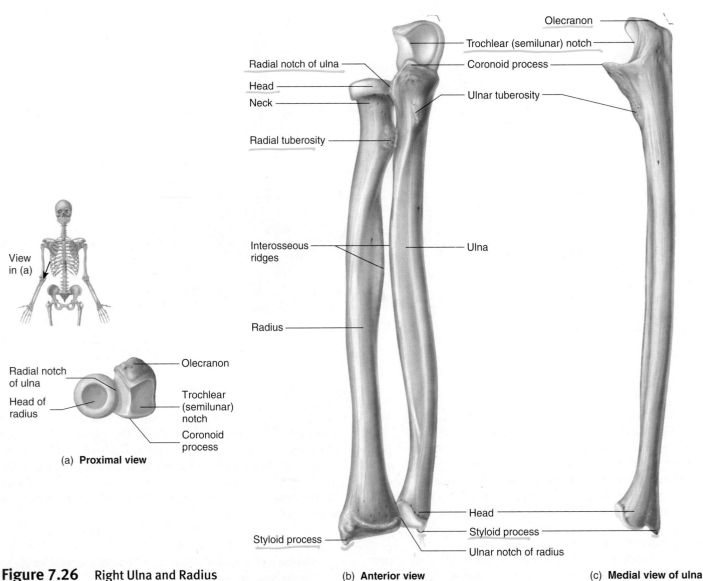

View in (a)

Radial notch of ulna

Head of radius

Olecranon

Trochlear (semilunar) notch

Coronoid process

(a) **Proximal view**

Olecranon

Trochlear (semilunar) notch

Radial notch of ulna

Head

Neck

Radial tuberosity

Coronoid process

Ulnar tuberosity

Interosseous ridges

Ulna

Radius

Head

Styloid process

Styloid process

Ulnar notch of radius

Figure 7.26   Right Ulna and Radius   (b) **Anterior view**   (c) **Medial view of ulna**

Just distal to the elbow joint, the **radial tuberosity** and the **ulnar tuberosity** are attachment sites for arm muscles.

The distal end of the ulna has a small **head,** which articulates with both the radius and the carpal (wrist) bones (see figure 7.26). The head can be seen as a prominent lump on the posterior, medial (ulnar) side of the distal forearm (see figure 7.25). The distal end of the radius, which articulates with the ulna and the carpals, is somewhat broadened. The ulna and radius have small **styloid** (stī′loyd, shaped like a stylus or writing instrument) **processes** to which ligaments of the wrist are attached.

Radius Fractures

The radius is the most commonly fractured bone in people over 50 years old. It is often fractured as the result of a fall on an outstretched hand, which results in posterior displacement of the hand. Typically, there is a complete transverse fracture of the radius 2.5 cm proximal to the wrist. The fracture is often comminuted, or impacted. Such a fracture is called a Colles fracture.

Wrist

The wrist is a relatively short region between the forearm and hand; it is composed of eight **carpal** (kar′păl) **bones** arranged into two rows of four each (figure 7.27). The proximal row of carpals, lateral to medial, includes the **scaphoid** (skaf′oyd, boat-shaped), **lunate** (loo′nāt, moon-shaped), **triquetrum** (trī-kwē′trŭm, trī-kwet′rŭm, three-cornered), and **pisiform** (pis′i-fōrm, pea-shaped). The distal row of carpals, from medial to lateral, includes the **hamate**

(ha′māt, hook), **capitate** (kap′i-tāt, head), **trapezoid** (trap′ĕ-zoyd, a four-sided geometric form with two parallel sides), and **trapezium** (tra-pē′zē-ŭm, a four-sided geometric form with no two sides parallel). A number of mnemonics have been developed to help students remember the carpal bones. The following mnemonic allows students to remember them in order from lateral to medial for the proximal row (top) and from medial to lateral (by the thumb) for the distal row: **S**o **L**ong **T**op **P**art, **H**ere **C**omes **T**he **T**humb—that is **S**caphoid, **L**unate, **T**riquetrum, **P**isiform, **H**amate, **C**apitate, **T**rapezoid, and **T**rapezium.

Carpal Tunnel Syndrome

The bones and ligaments on the anterior side of the wrist form a **carpal tunnel,** which does not have much "give." Tendons and nerves pass from the forearm through the carpal tunnel to the hand. Fluid and connective tissue can accumulate in the carpal tunnel as a result of inflammation associated with overuse or trauma. The inflammation can also cause the tendons in the carpal tunnel to enlarge. The accumulated fluid and enlarged tendons can apply pressure to a major nerve passing through the tunnel. The pressure on this nerve causes **carpal tunnel syndrome,** the symptoms of which are tingling, burning, and numbness in the hand

Hand

Five **metacarpals** (met′ă-kar′pălz, after the carpals) are attached to the carpal bones and constitute the bony framework of the hand (see figure 7.27). They are numbered 1 through 5, starting with the most lateral metacarpal, at the base of the thumb. The distal ends of the metacarpals help form the knuckles of the hand (see figure 7.25).

The five **digits** of each hand include one thumb and four fingers. The digits are also numbered 1 through 5, starting from the thumb. Each digit consists of small long bones called **phalanges** (fă-lan′jēz, sing. *phalanx,* a line or wedge of soldiers holding their spears, tips outward, in front of them). The thumb has two phalanges, called proximal and distal. Each finger has three phalanges designated proximal, middle, and distal. One or two **sesamoid** (ses′ă-moyd, resembling a sesame seed) **bones** (not shown in figure 7.27) often form near the junction between the proximal phalanx and the metacarpal of the thumb. Sesamoid bones are small bones located within some tendons, increasing their mechanical advantage where they cross joints.

P R E D I C T

Explain why a dried, articulated skeleton appears to have much longer "fingers" than are seen in a hand with the soft tissue intact. ▼

30. Distinguish between the anatomical and surgical necks of the humerus.
31. Name the important sites of muscle attachment on the humerus.
32. Give the points of articulation between the scapula, humerus, radius, ulna, and wrist bones.
33. What is the function of the radial and ulnar tuberosities? Of the styloid processes?

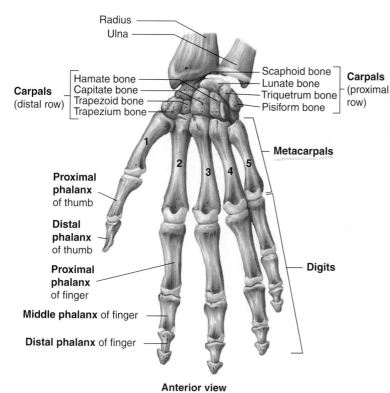

Figure 7.27 Bones of the Right Wrist and Hand

34. *Name the part of the ulna commonly referred to as "the elbow."*
35. *List the eight carpal bones.*
36. *What bones form the hand? The knuckles?*
37. *Name the phalanges in a thumb and in a finger.*

Pelvic Girdle

The pelvic girdle is the place of attachment for the lower limbs, it supports the weight of the body, and it protects internal organs (figure 7.28). The right and left **coxal** (kok′sul) **bones,** os coxae, or hipbones, join each other anteriorly and the **sacrum** posteriorly to form a ring of bone called the **pelvic girdle.** The **pelvis** (pel′vis, basin) includes the pelvic girdle and the coccyx (figure 7.29). Because the pelvic girdle is a complete bony ring, it provides more stable support but less mobility than the incomplete ring of the pectoral girdle. In addition, the pelvis in a woman protects a developing fetus and forms a passageway through which the fetus passes during delivery.

Each coxal bone is formed by three bones fused to one another to form a single bone (see figure 7.29). The **ilium** (il′ē-ŭm, groin) is the superior, the **ischium** (is′kē-ŭm, hip) is inferior and posterior, and the **pubis** (pū′bis, genital hair) is inferior and anterior. The coxal bones join anteriorly at the **symphysis** (sim′fi-sis, a coming together) **pubis,** or **pubic symphysis.** Posteriorly, each coxal bone joins the sacrum at the **sacroiliac joint.**

A fossa called the **acetabulum** (as-ĕ-tab′ū-lŭm, a shallow vinegar cup—a common household item in ancient times) is located on the lateral surface of each coxal bone (figure 7.30*a*). In a child, the joints between the ilium, ischium, and pubis can be

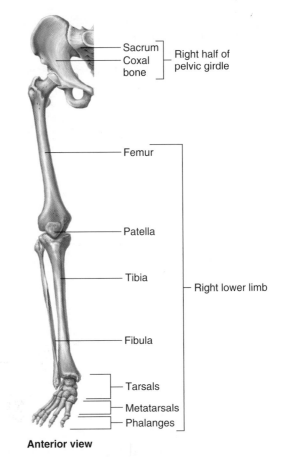

Figure 7.28   Bones of the Pelvic Girdle and Lower Limb

Labels (top to bottom): Sacrum, Coxal bone — Right half of pelvic girdle; Femur; Patella; Tibia; Fibula; Tarsals; Metatarsals; Phalanges — Right lower limb. **Anterior view**

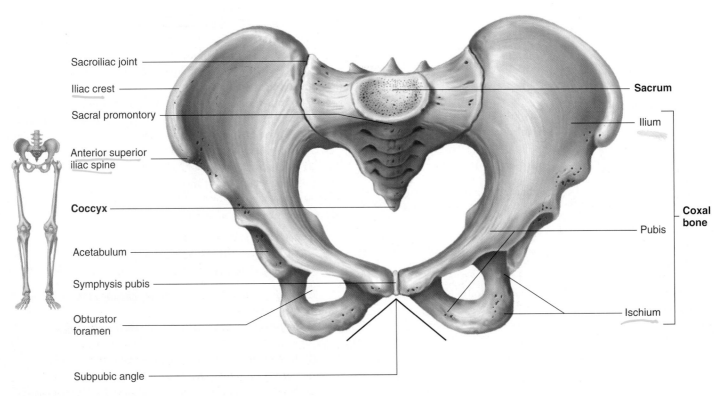

Figure 7.29   Pelvis

Labels: Sacroiliac joint, Iliac crest, Sacral promontory, Anterior superior iliac spine, Coccyx, Acetabulum, Symphysis pubis, Obturator foramen, Subpubic angle; Sacrum, Ilium, Pubis, Ischium, Coxal bone. **Anterosuperior view**

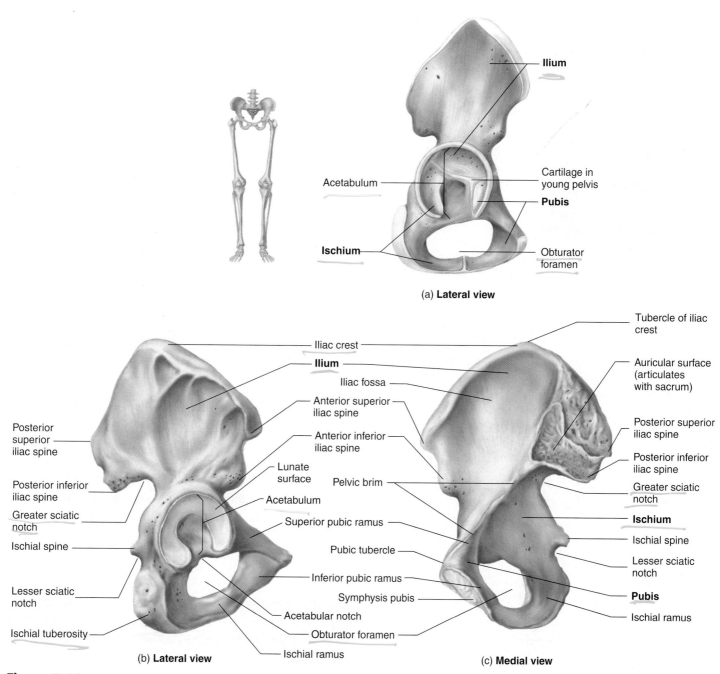

Figure 7.30 Coxal Bone

The three bones forming the coxal bone are bolded. (*a*) Right coxal bone of a growing child. Each coxal bone is formed by fusion of the ilium, ischium, and pubis. The three bones can be seen joining near the center of the acetabulum, separated by lines of cartilage. (*b*) Right coxal bone, lateral view. (*c*) Right coxal bone, medial view.

seen. The bones fuse together in some locations by the seventh or eighth year. Complete fusion within the acetabulum occurs between the sixteenth and eighteenth years. The acetabulum is the point of articulation of the lower limb with the pelvic girdle. The articular, **lunate surface** of the acetabulum is crescent-shaped and occupies only the superior and lateral aspects of the fossa (figure 7.30*b*). Inferior to the acetabulum is the large **obturator** (ob′too-rā-tŏr, to occlude or close up) **foramen.** In life, the obturator foramen is almost completely closed off by a connective tissue membrane, which separates the pelvic cavity from more superficial structures.

Despite its large size, only a few small blood vessels and nerves pass through the obturator foramen.

The superior portion of the ilium is called the **iliac crest** (see figure 7.30*b*; figure 7.30*c*). The crest ends anteriorly as the **anterior superior iliac spine** and posteriorly as the **posterior superior iliac spine.** The crest and anterior spine can be felt and even seen in thin individuals (figure 7.31). The anterior superior iliac spine is an important anatomical landmark used, for example, to find the correct location for giving gluteal injections into the hip. A dimple overlies the posterior superior iliac spine just superior to the buttocks.

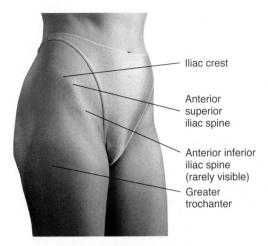

Iliac crest

Anterior
superior
iliac spine

Anterior inferior
iliac spine
(rarely visible)

Greater
trochanter

Figure 7.31   Surface Anatomy Showing an Anterolateral
View of the Coxal Bone and Femur

⚕ Gluteal Injections

The large gluteal (hip) muscles (see chapter 9) are a common site for intramuscular injections. Gluteal injections are made in the superolateral region of the hip (figure C) so as to avoid a large nerve (the sciatic nerve) (see chapter 11) located more posteriorly. The landmarks for such an injection are the anterior superior iliac spine and the tubercle of the iliac crest, which lies about one-third of the way along the iliac crest from anterior to posterior.

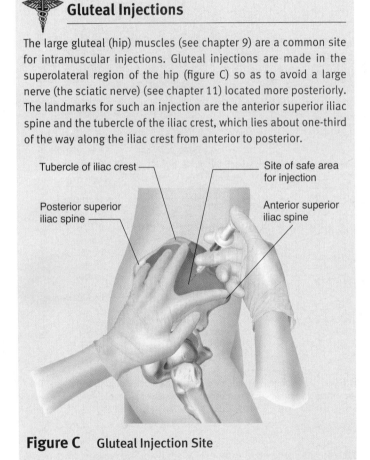

Tubercle of iliac crest

Site of safe area
for injection

Posterior superior
iliac spine

Anterior superior
iliac spine

Figure C   Gluteal Injection Site

Inferior to the anterior superior iliac spine is the **anterior inferior iliac spine** (see figure 7.30). The anterior iliac spines are attachment sites for anterior thigh muscles. Inferior to the superior posterior iliac spine are the **posterior inferior iliac spine, ischial spine,** and **ischial tuberosity.** The posterior iliac spines and ischial tuberosity are attachment sites for ligaments anchoring the coxal bone to the sacrum. The **auricular surface** of the ilium (see figure 7.30c) joins the auricular surface of the sacrum

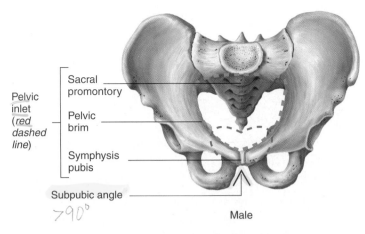

Pelvic
inlet
(*red
dashed
line*)

Sacral
promontory

Pelvic
brim

Symphysis
pubis

Subpubic angle

>90°

Male

(a) **Anterosuperior view**

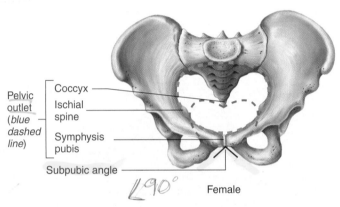

Pelvic
outlet
(*blue
dashed
line*)

Coccyx

Ischial
spine

Symphysis
pubis

Subpubic angle

∠90°

Female

(b) **Anterosuperior view**

Figure 7.32   Comparison of the Male and Female Pelvis

(*a*) Male. The pelvic inlet (*red dashed line*) and outlet (*blue dashed line*) are small, and the subpubic angle is less than 90 degrees. (*b*) Female. The pelvic inlet (*red dashed line*) and outlet (*blue dashed line*) are larger, and the subpubic angle is 90 degrees or greater.

Table 7.6	Differences Between the Male and Female Pelvis (See Figure 7.32)
Area	**Description**
General	In females, somewhat lighter in weight and wider laterally but shorter superiorly to inferiorly and less funnel-shaped; less obvious muscle attachment points in females than in males
Sacrum	Broader in females with the inferior part directed more posteriorly; the sacral promontory does not project as far anteriorly in females
Pelvic inlet	Heart-shaped in males; oval in females
Pelvic outlet	Broader and more shallow in females
Subpubic angle	Less than 90 degrees in males; 90 degrees or more in females
Ilium	More shallow and flared laterally in females
Ischial spines	Farther apart in females
Ischial tuberosities	Turned laterally in females and medially in males

(see figure 7.18) to form the sacroiliac joint. The ischial tuberosity is also an attachment site for posterior thigh muscles, and it is the part of the coxal bone on which a person sits.

The **greater sciatic notch** is superior to the ischial spine and the **lesser sciatic notch** is inferior to it (see figure 7.30). Nerves and blood vessels pass through the sciatic notches.

The pelvis can be thought of as having two parts divided by an imaginary plane passing from the sacral promontory to the pubic crest. The bony boundary of this plane is the **pelvic brim** (figure 7.32). The **false,** or **greater, pelvis** is superior to the pelvic brim and is partially surrounded by bone on the posterior and lateral sides. During life, the abdominal muscles form the anterior wall of the false pelvis. The **true pelvis** is inferior to the pelvic brim and is completely surrounded by bone. The superior opening of the true pelvis, at the level of the pelvic brim, is the **pelvic inlet.** The inferior opening of the true pelvis, bordered by the inferior margin of the pubis, the ischial spines and tuberosities, and the coccyx, is the **pelvic outlet.**

Comparison of the Male and Female Pelvis

The male pelvis usually is more massive than the female pelvis as a result of the greater weight and size of the male, but the female pelvis is broader and has a larger, more rounded pelvic inlet and outlet (see figure 7.32), consistent with the need to allow a fetus to

pass through these openings in the female pelvis during delivery. If the pelvic outlet is too small for normal delivery, delivery can be accomplished by **cesarean section,** which is the surgical removal of the fetus through the abdominal wall. Table 7.6 lists additional differences between the male and female pelvis. ▼

> 38. *Define the pelvic girdle. What bones fuse to form each coxal bone? Where and with what bones does each coxal bone articulate?*
> 39. *Name the important sites of muscle and ligament attachment on the pelvis.*
> 40. *Distinguish between the true pelvis and the false pelvis.*
> 41. *Describe the differences between a male and a female pelvis.*

Lower Limb

The lower limb consists of the bones of the thigh, leg, ankle, and foot (see figure 7.28).

Thigh

The thigh is the region between the hip and the knee. The thigh, like the arm, contains a single bone, called the **femur** (figure 7.33). The **head** of the femur articulates with the acetabulum of the coxal bone, and the **neck** of the femur connects the head to the **body** (shaft) of the femur. The **greater trochanter** (trō-kan′ter, runner)

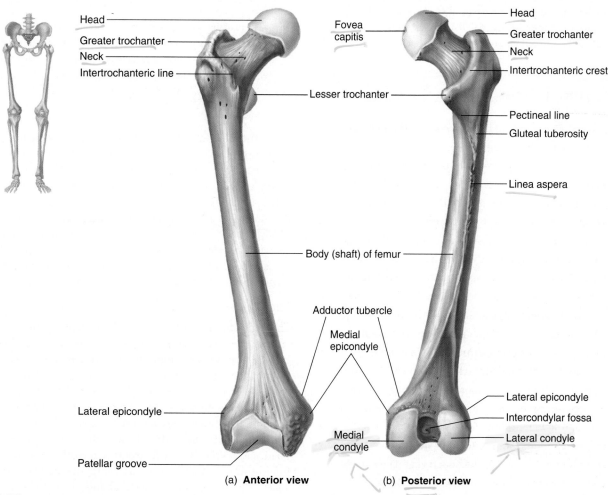

Figure 7.33   **Right Femur**

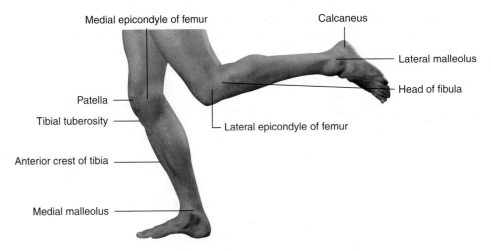

Figure 7.34   Surface Anatomy Showing Bones of the Lower Limb

and the **lesser trochanter** are attachment sites for muscles that fasten the hip to the thigh. The greater trochanter and its attached muscles form a bulge that can be seen as the widest part of the hips (see figure 7.31). The **pectineal line, gluteal tuberosity,** and the **linea aspera** are other muscle attachment sites. The distal end of the femur has **medial** and **lateral condyles** that articulate with the tibia (leg bone). Located proximally to the condyles are the **medial** and **lateral epicondyles,** important sites of ligament attachment. The epicondyles can be felt just proximal to the knee joint (figure 7.34). An **adductor tubercle,** to which muscles attach, is located just proximal to the medial epicondyle.

CASE STUDY | Fracture of the Femoral Neck

An 85-year-old woman who lived alone was found lying on her kitchen floor by her daughter, who had come to check on her mother. The woman could not rise, even with help, and when she tried she experienced extreme pain in her right hip. Her daughter immediately dialed 911 and paramedics took her mother to the hospital.

The elderly woman's hip was x-rayed in the emergency room, and it was determined that she had a fracture of the right femoral neck. A femoral neck fracture is commonly, but incorrectly, called a broken hip. Two days later, she received a partial hip replacement in which the head and neck of the femur, but not the acetabulum, were replaced. In falls involving femoral neck fracture, it is not always clear whether the fall caused the femoral neck to fracture or whether a fracture of the femoral neck caused the fall. Femoral neck fractures are among the most common injuries resulting in morbidity (disease) and mortality (death) in older adults. Four percent of women over age 85 experience femoral neck fractures each year. Only about 25% of victims fully recover from the injury. Despite treatment with anticoagulants and antibiotics, about 5% of patients with femoral neck fractures develop deep vein

thrombosis (blood clot) and about 5% develop wound infections, either of which can be life-threatening. Hospital mortality is 1%–7% among patients with femoral neck fractures, and nearly 20% of femoral neck fracture victims die within 3 months of the fracture.

PREDICT ⑧

Fracture of the femoral neck increases dramatically with age, and 81% of victims are women. The average age of such injury is 82. Why is the femoral neck so commonly injured? (*Hint:* See figure 7.1.) Why are elderly women most commonly affected?

The **patella** (figure 7.35), or kneecap, articulates with the **patellar groove** of the femur (see figure 7.33). It is a large sesamoid bone located within the tendon of the quadriceps femoris muscle group, which is the major muscle group of the anterior thigh. The patella holds the tendon away from the distal end of the femur and therefore changes the angle of the tendon between the quadriceps femoris muscle and the tibia, where the tendon attaches. This change in angle increases the force that can be applied from the muscle to the tibia. As a result of this increase in applied force, less muscle contraction force is required to move the tibia.

Leg

The leg is the part of the lower limb between the knee and the ankle. Like the forearm, it consists of two bones: the **tibia** (tibʹē-ă), or shinbone, and the **fibula** (fibʹū-lă, resembling a clasp or buckle) (figure 7.36). The tibia is by far the larger of the two and supports most of the weight of the leg. The rounded condyles of the femur rest on the flat **medial** and **lateral condyles** on the proximal end of the tibia. The **intercondylar eminence** is a ridge between the condyles. A **tibial tuberosity,** which is the attachment point for the quadriceps femoris muscle group, can easily be seen and felt just inferior to the patella (see figure 7.35). The **anterior crest** forms a sharp edge on the shin. The distal end of the tibia is enlarged to form the **medial malleolus** (ma-lēʹō-lŭs, mallet-shaped), which helps form the medial side of the ankle joint.

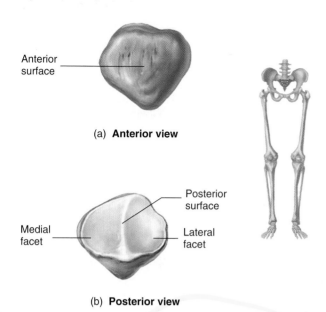

(a) **Anterior view**

Anterior surface

Posterior surface

Medial facet

Lateral facet

(b) **Posterior view**

Figure 7.35 Right Patella

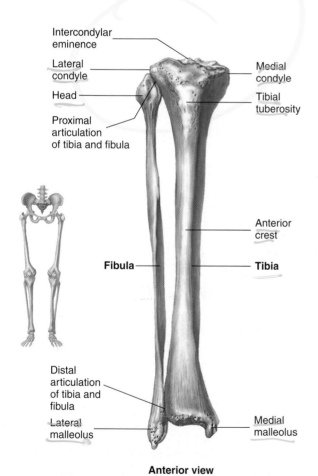

Intercondylar eminence

Lateral condyle

Head

Proximal articulation of tibia and fibula

Medial condyle

Tibial tuberosity

Anterior crest

Fibula

Tibia

Distal articulation of tibia and fibula

Lateral malleolus

Medial malleolus

Anterior view

Figure 7.36 Right Tibia And Fibula

The fibula does not articulate with the femur but has a small proximal **head** where it articulates with the tibia. The distal end of the fibula is also slightly enlarged as the **lateral malleolus** to create the lateral wall of the ankle joint. The lateral and medial malleoli can be felt and seen as prominent lumps on both sides of the ankle

(see figure 7.35). The thinnest, weakest portion of the fibula is just proximal to the lateral malleolus.

Foot

The proximal portion of the foot consists of seven **tarsal** (tar′săl, the sole of the foot) **bones** (figure 7.37). The tarsal bones are the **talus** (tā′lŭs, ankle bone), **calcaneus** (kal-kā′nē-ŭs, heel), **cuboid** (kū′boyd, cube-shaped), and **navicular** (nă-vik′yū-lăr, boat-shaped) bones and the medial, intermediate, and lateral **cuneiforms** (kū′nē-i-fōrmz, wedge-shaped). A mnemonic for the distal row of bones is **MILC**—that is, **M**edial, **I**ntermediate, and **L**ateral cuneiforms and the **C**uboid. The mnemonic for the proximal three bones is **No Thanks Cow**—that is, **N**avicular, **T**alus, and **C**alcaneus.

The talus articulates with the tibia and fibula to form the ankle joint. It is an unusual bone in that no muscles attach to it. The calcaneus forms the heel and is the attachment point for the large calf muscles.

 Fractures of the Malleoli

Turning the plantar surface of the foot outward so that it faces laterally is called eversion. Forceful eversion of the foot, such as when a person slips and twists the ankle or jumps and lands incorrectly on the foot, may cause the distal ends of the tibia and/or fibula to fracture (figure D*a*). When the foot is forcefully everted, the medial malleolus moves inferiorly toward the ground or floor and the talus slides laterally, forcing the medial and lateral malleoli to separate. The ligament holding the medial malleolus to the tarsal bones is stronger than the bones it connects, and often it does not tear as the malleoli separate. Instead, the medial malleolus breaks. Also, as the talus slides laterally, the force can shear off the lateral malleolus or, more commonly, can cause the fibula to break superior to the lateral malleolus. This type of injury to the tibia and fibula is often called a Pott fracture.

Turning the plantar surface of the foot inward so that it faces medially is called inversion. Forceful inversion of the foot can fracture the fibula just proximal to the lateral malleolus (figure D*b*). More often, because the ligament holding the medial malleolus to the tarsal bones is weaker than the bones it connects, the inversion of the foot causes a sprain in which ligaments are damaged.

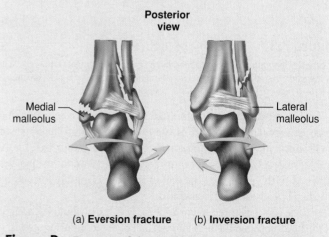

Posterior view

Medial malleolus

Lateral malleolus

(a) **Eversion fracture**

(b) **Inversion fracture**

Figure D Fracture of the Medial or Lateral Malleolus

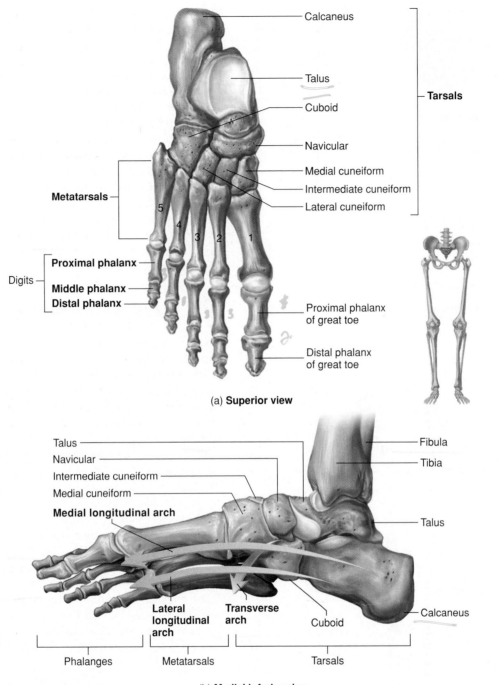

(a) **Superior view**

(b) **Medial inferior view**

Figure 7.37 Bones of the Right Foot

The medial longitudinal arch is formed by the calcaneus, the talus, the navicular, the cuneiforms, and three medial metatarsals. The lateral longitudinal arch is formed by the calcaneus, the cuboid, and two lateral metatarsals. The transverse arch is formed by the cuboid and cuneiforms.

The **metatarsals** and **phalanges** of the foot are arranged in a manner very similar to that of the metacarpals and phalanges of the hand, with the great toe analogous to the thumb (see figure 7.37). Small sesamoid bones often form in the tendons of muscles attached to the great toe. The ball of the foot is mainly formed by the distal heads of the metatarsals.

The foot has three major **arches,** which distribute the weight of the body between the heel and the ball of the foot during standing and walking (see figure 7.37b). As the foot is placed on the ground,

weight is transferred from the tibia and the fibula to the talus. From there, the weight is distributed first to the heel (calcaneus) and then through the arch system along the lateral side of the foot to the ball of the foot (head of the metatarsals). This effect can be observed when a person with wet, bare feet walks across a dry surface; the print of the heel, the lateral border of the foot, and the ball of the foot can be seen, but the middle of the plantar surface and the medial border leave no impression. The medial side leaves no mark because the arches on this side of the foot are higher than those on the lateral side. The shape of the arches is maintained by the configuration of the bones, the ligaments connecting them, and the muscles acting on the foot. ▼

> 42. *Name the bones of the thigh and leg.*
> 43. *Give the points of articulation among the pelvic girdle, femur, leg, and ankle.*
> 44. *What is the function of the greater trochanter and the lesser trochanter? The medial and lateral epicondyles?*
> 45. *Describe the function of the patella.*
> 46. *What is the function of the tibial tuberosity?*
> 47. *Name the seven tarsal bones. Which bones form the ankle joint? What bone forms the heel?*
> 48. *What bones form the ball of the foot? How many phalanges are in each toe?*
> 49. *List the three arches of the foot and describe their function.*

Articulations

An **articulation,** or **joint,** is a place where two or more bones come together. We usually think of joints as being movable, but that is not always the case. Many joints allow only limited movement, and others allow no apparent movement. The structure of a given joint is directly correlated with its degree of movement. Fibrous joints have much less movement than joints containing fluid and having smooth articulating surfaces.

Joints are commonly named according to the bones or portions of bones that are united at the joint, such as the temporomandibular joint between the temporal bone and the mandible. Some joints are simply given the Greek or Latin equivalent of the common name, such as the **cubital** (kū′bi-tăl, cubit, elbow or forearm) **joint** for the elbow joint. ▼

> 50. *What criteria are used to name joints?*

P R E D I C T
What is the name of the joint between the metacarpals and the phalanges?

Classes of Joints

The three major kinds of joints are classified structurally as fibrous, cartilaginous, and synovial. In this classification scheme, joints are categorized according to the major connective tissue type that binds the bones together and whether or not a fluid-filled joint capsule is present.

Fibrous Joints

Fibrous joints consist of two bones that are united by fibrous connective tissue, have no joint cavity, and exhibit little or no movement. Joints in this group are classified further as sutures, syndesmoses, or gomphoses (table 7.7), based on their structure.

Table 7.7	**Fibrous and Cartilaginous Joints**	
Class and Example of Joint	**Bones or Structures Joined**	**Movement**
Fibrous Joints		
Sutures		
Coronal	Frontal and parietal	None
Lambdoid	Occipital and parietal	None
Sagittal	The two parietal bones	None
Squamous	Parietal and temporal	Slight
Syndesmoses		
Radioulnar	Ulna and radius	Slight
Stylohyoid	Styloid process and hyoid bone	Slight
Stylomandibular	Styloid process and mandible	Slight
Tibiofibular	Tibia and fibula	Slight
Gomphoses		
Dentoalveolar	Tooth and alveolar process	Slight
Cartilaginous Joints		
Synchondroses		
Epiphyseal plate	Diaphysis and epiphysis of a long bone	None
Sternocostal	Anterior cartilaginous part of first rib; between rib and sternum	Slight
Sphenooccipital	Sphenoid and occipital	None
Symphyses		
Intervertebral	Bodies of adjacent vertebrae	Slight
Manubriosternal	Manubrium and body of sternum	None
Symphysis pubis	The two coxal bones	None except during childbirth
Xiphisternal	Xiphoid process and body of sternum	None

Sutures

Sutures (soo'choorz) are fibrous joints between the bones of the skull (see figure 7.2). The type of connective tissue in fibrous joints is dense regular collagenous connective tissue. The bones in a suture often have interlocking, fingerlike processes that add stability to the joint.

In a newborn, intramembranous ossification of the skull bones along their margins is incomplete. A large area of unossified membrane between some bones is called a **fontanel** (fon'tă-nel', little fountain, so named because the membrane can be seen to move with the pulse), or soft spot (figure 7.38). The unossified membrane makes the skull flexible during the birth process and allows for growth of the head after birth. The fontanels have usually ossified by 2 years of age.

The margins of bones within sutures are sites of continuous intramembranous bone growth, and many sutures eventually become ossified. For example, ossification of the suture between the two frontal bones occurs shortly after birth so that they usually form a single frontal bone in the adult skull. In most normal adults, the coronal, sagittal, and lambdoid sutures are not fused. In some very old adults, however, even these sutures become ossified. A **synostosis** (sin-os-tō'sis) results when two bones grow together across a joint to form a single bone.

P R E D I C T 10
Predict the result of a sutural synostosis that occurs prematurely in a child's skull before the brain has reached its full size.

Syndesmoses

A **syndesmosis** (sin'dez-mō'sis, to fasten or bind) is a fibrous joint in which the bones are farther apart than in a suture and are joined by ligaments. Some movement may occur at syndesmoses because of flexibility of the ligaments, such as in the radioulnar syndesmosis, which binds the radius and ulna together (figure 7.39).

Gomphoses

Gomphoses (gom-fō'sēz) consist of pegs held in place within sockets by fibrous tissue. The joints between the teeth and the sockets (alveoli) of the mandible and maxillae are gomphoses (figure 7.40). The connective tissue bundles between the teeth and their sockets are called **periodontal** (per'ē-ō-don'tăl) **ligaments.**

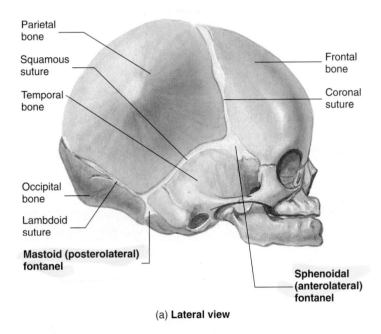

(a) **Lateral view**

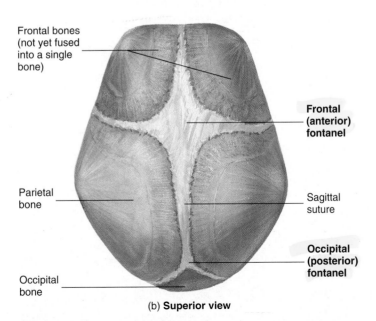

(b) **Superior view**

Figure 7.38 Fetal Skull Showing Fontanels and Sutures

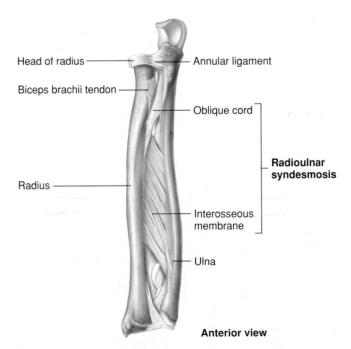

Anterior view

Figure 7.39 Right Radioulnar Syndesmosis

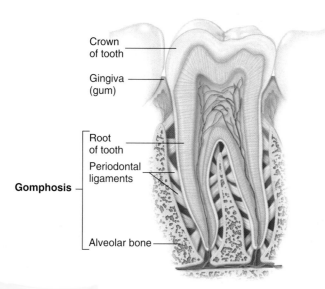

Crown
of tooth

Gingiva
(gum)

Root
of tooth

Periodontal
ligaments

Gomphosis

Alveolar bone

Figure 7.40 Gomphosis Between a Tooth and Alveolar
Bone of the Mandible

Gingivitis and Periodontal Disease

The **gingiva**, or gums, are the soft tissues covering the alveolar processes. Neglect of the teeth can result in **gingivitis**, an inflammation of the gingiva, often resulting from bacterial infection. Left untreated, gingivitis can spread to the tooth socket, resulting in **periodontal disease**, the leading cause of tooth loss in the United States. Periodontal disease involves an accumulation of plaque and bacteria, resulting in inflammation and the gradual destruction of the periodontal ligaments and the bone. Eventually, the teeth may become so loose that they come out of their sockets. Proper brushing, flossing, and professional cleaning to remove plaque can usually prevent gingivitis and periodontal disease. ▼

> 51. *Define fibrous joint, describe three types, and give an example of each.*
>
> 52. *What is a synostosis?*

Cartilaginous Joints

Cartilaginous joints unite two bones by means of either hyaline cartilage or fibrocartilage (see table 7.7).

Synchondroses

A **synchondrosis** (sin′kon-drō′sis, union through cartilage) consists of two bones joined by hyaline cartilage where little or no movement occurs. The joints between the ilium, ischium, and pubis before those bones fuse together are examples of synchondroses joints (see figure 7.30*a*). The epiphyseal plates of growing bones are synchondroses (see figure 6.5). Most synchondroses are temporary, with bone eventually replacing them to form synostoses. On the other hand, some synchondroses persist throughout life. An example is the sternocostal synchondrosis between the first rib and the sternum by way of the first costal cartilage (see figure 7.19). The remaining costal cartilages attach to the sternum by synovial joints (see "Synovial Joints," p. 179).

Symphyses

A **symphysis** (sim′fi-sis, a growing together) consists of fibrocartilage uniting two bones. Symphyses include the junction between the manubrium and the body of the sternum (see figure 7.19), the symphysis pubis (see figure 7.29), and the intervertebral disks (see figure 7.16). Some of these joints are slightly movable because of the somewhat flexible nature of fibrocartilage. ▼

> 53. *Define cartilaginous joints, describe two types, and give an example of each.*

Synovial Joints

Synovial (si-nō′vē-ăl, joint fluid, *syn,* coming together, *ovia,* resembling egg albumin) **joints** are freely movable joints that contain **synovial fluid** in a cavity surrounding the ends of articulating bones. Most joints that unite the bones of the appendicular skeleton are large synovial joints, whereas many of the joints that unite the bones of the axial skeleton are not. This pattern reflects the greater mobility of the appendicular skeleton, compared with the axial skeleton.

The articular surfaces of bones within synovial joints are covered with a thin layer of hyaline cartilage called **articular cartilage,** which provides a smooth surface where the bones meet (figure 7.41). In some synovial joints, a flat plate or pad of fibrocartilage, called an **articular disk,** is located between the articular cartilages of bones. Articular disks absorb and distribute the forces between the articular cartilages as the bones move. Examples of joints with articular disks are the temporomandibular, sternoclavicular, and acromioclavicular joints. A **meniscus** (mĕ-nis′kŭs, crescent-shaped) is an incomplete, crescent-shaped fibrocartilage pad found in joints such as the knee and wrist. A meniscus is much like an articular disk with a hole in the center.

A **joint capsule** (see figure 7.41) surrounds the ends of the bones forming synovial joints, forming a **joint cavity.** The capsule helps hold the bones together while allowing for movement. The joint capsule consists of two layers: an outer **fibrous capsule** and an inner **synovial membrane** (see figure 7.41). The fibrous capsule consists of dense irregular connective tissue and is continuous with the fibrous layer of the periosteum that covers the bones united at the joint. Portions of the fibrous capsule may thicken and the collagen fibers become regularly arranged to form ligaments. In addition, ligaments and tendons may be present outside the fibrous capsule, thereby contributing to the strength and stability of the joint while limiting movement in some directions.

The synovial membrane lines the joint cavity, except over the articular cartilage and articular disks. It is a thin, delicate membrane consisting of modified connective tissue cells. The membrane produces synovial fluid, which consists of a serum (blood fluid) filtrate and secretions from the synovial cells. Synovial fluid is a complex mixture of polysaccharides, proteins, fat, and cells. The major polysaccharide is hyaluronic acid, which provides much of the slippery consistency and lubricating qualities of synovial fluid. Synovial fluid forms an important thin, lubricating film that covers the surfaces of a joint.

P R E D I C T **11**
What would happen if a synovial membrane covered the articular cartilage?

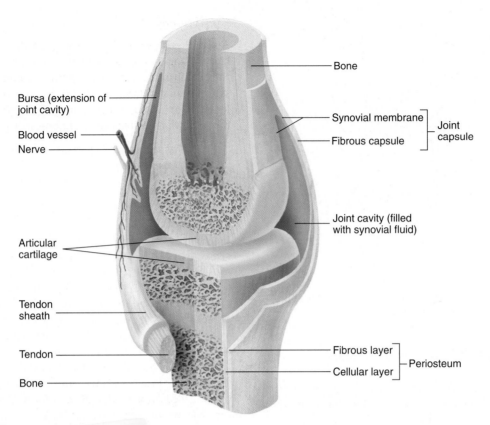

Figure 7.41   Structure of a Synovial Joint

In certain synovial joints, such as the shoulder and knee, the synovial membrane extends as a pocket, or sac, called a **bursa** (ber′să, pocket) for a distance away from the rest of the joint cavity (see figure 7.41). Bursae contain synovial fluid, providing a cushion between structures that otherwise would rub against each other, such as tendons rubbing on bones or other tendons. Some bursae, such as the subcutaneous olecranon bursae, are not associated with joints but provide a cushion between the skin and underlying bony prominences, where friction could damage the tissues. Other bursae extend along tendons for some distance, forming **tendon sheaths.** **Bursitis** (ber-sī′tis) is the inflammation of a bursa; it may cause considerable pain around the joint and restrict movement. ▼

54. *Describe the structure of a synovial joint. How do the different parts of the joint permit joint movement? What are articular disks and menisci?*

55. *Define bursa and tendon sheath. What is their function?*

Types of Synovial Joints

Synovial joints are classified according to the shape of the adjoining articular surfaces. The six types of synovial joints are plane, saddle, hinge, pivot, ball-and-socket, and ellipsoid (table 7.8). Movements at synovial joints are described as **uniaxial** (occurring around one axis), **biaxial** (occurring around two axes situated at right angles to each other), or **multiaxial** (occurring around several axes).

Plane, or **gliding, joints** consist of two opposed flat surfaces of about equal size in which a slight amount of gliding motion can occur between the bones. These joints are considered uniaxial because some rotation is also possible but is limited by ligaments and adjacent bone. Examples are the articular processes between vertebrae.

Pivot joints consists of a relatively cylindrical bony process that rotates within a ring composed partly of bone and partly of ligament. They are uniaxial joints that restrict movement to rotation around a single axis. The articulation between the head of the radius and the proximal end of the ulna is an example (see figures 7.26a and 7.39). The articulation between the dens and the atlas is another example (see figure 7.17a and b).

Hinge joints consist of a convex cylinder in one bone applied to a corresponding concavity in the other bone. They are uniaxial joints. Examples include the elbow and knee joints.

Ball-and-socket joints consist of a ball (head) at the end of one bone and a socket in an adjacent bone into which a portion of the ball fits. This type of joint is multiaxial, allowing a wide range of movement in almost any direction. Examples are the shoulder and hip joints.

Ellipsoid joints (or condyloid joints) are modified ball-and-socket joints. The articular surfaces are ellipsoid in shape (like a football). Ellipsoid joints are biaxial, because the shape of the joint limits its range of movement almost to a hinge motion in two axes and restricts rotation. An example is the atlantooccipital joint between the atlas and the occipital condyles, which allows a "yes" movement and a tilting, side-to-side movement of the head.

Saddle joints consist of two saddle-shaped articulating surfaces oriented at right angles to each other so that complementary surfaces articulate with each other. Saddle joints are biaxial joints. The carpometacarpal joint between the carpal (trapezium) and metacarpal of the thumb is a saddle joint. ▼

56. *On what basis are synovial joints classified? Describe the types of synovial joints, and give examples of each. What movements does each type of joint allow?*

Table 7.8 Types of Synovial Joints

Class and Example of Joint	Structures Joined	Movement
Plane		
Acromioclavicular	Acromion process of scapula and clavicle	Slight
Carpometacarpal	Carpals and metacarpals two through five	Multiple axes as a group
Costovertebral	Ribs and vertebrae	Slight
Intercarpal	Between carpals	Slight
Intermetatarsal	Between metatarsals	Slight
Intertarsal	Between tarsals	Slight
Intervertebral	Between articular processes of adjacent vertebrae	Slight
Sacroiliac	Between sacrum and coxa (complex joint with several planes and synchondroses)	Slight
Tarsometatarsal	Tarsals and metatarsals	Slight
Pivot		
Medial atlantoaxial	Atlas and axis	Rotation
Proximal radioulnar	Radius and ulna	Rotation
Distal radioulnar	Radius and ulna	Rotation
Hinge		
Cubital (elbow)	Humerus, ulna, and radius	One axis
Genu (knee)	Femur and tibia	One axis
Interphalangeal	Between phalanges	One axis
Talocrural (ankle)	Talus, tibia, and fibula	Multiple axes, one predominates
Ball-and-Socket		
Humeral (shoulder)	Scapula and humerus	Multiple axes
Coxal (hip)	Coxal bone and femur	Multiple axes
Ellipsoid		
Atlantooccipital	Atlas and occipital bone	Two axes
Metacarpophalangeal (knuckles)	Metacarpals and phalanges	Two axes
Metatarsophalangeal (ball of foot)	Metatarsals and phalanges	Two axes
Radiocarpal (wrist)	Radius and carpals	Multiple axes
Temporomandibular	Mandible and temporal bone	Multiple axes, one predominates
Saddle		
Carpometacarpal pollicis	Carpal and metacarpal of thumb	Two axes
Intercarpal	Between carpals	Slight
Sternoclavicular	Manubrium of sternum and clavicle	Slight

Plane

Pivot

Hinge

Ball-and-socket

Elipsoid

Saddle

Types of Movement

The types of movement occurring at a given joint are related to the structure of that joint. Some joints are limited to only one type of movement, whereas others permit movement in several directions. All the movements are described relative to the anatomical position. Because most movements are accompanied by movements in the opposite direction, they are listed in pairs.

Gliding Movements

Gliding movements are the simplest of all the types of movement. These movements occur in plane joints between two flat or nearly flat surfaces where the surfaces slide or glide over each other. These joints often give only slight movement, such as between carpal bones.

Angular Movements

Angular movements are those in which one part of a linear structure, such as the body as a whole or a limb, is bent relative to another part of the structure, thereby changing the angle between the two parts. Angular movements also involve the movement of a solid rod, such as a limb, that is attached at one end to the body so that the angle at which it meets the body changes. The most common angular movements are flexion and extension and abduction and adduction.

Flexion and Extension

Flexion and extension can be defined in a number of ways, but in each case exceptions to the definition exist. The literal definition for flexion is to bend; for extension, to straighten. This bending and straightening can easily be seen in the elbow (figure 7.42*a*).

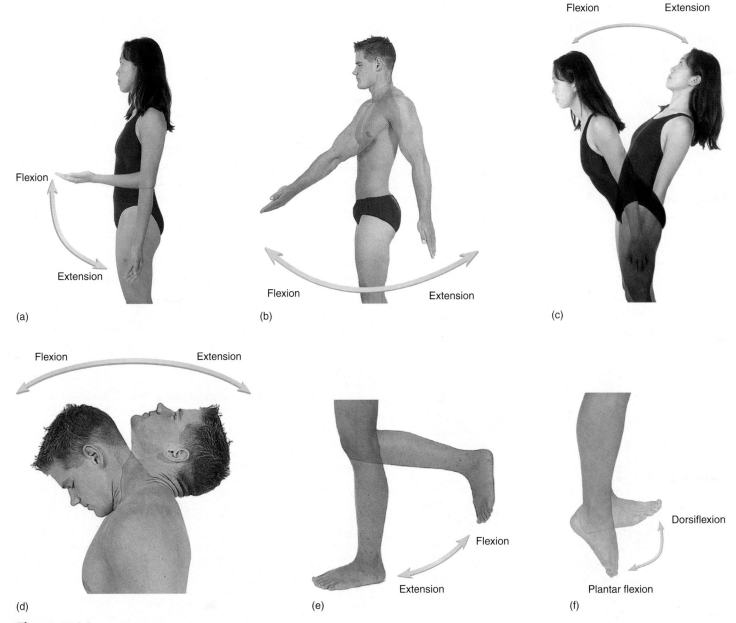

Figure 7.42   **Flexion and Extension**

Flexion and extension of (*a*) the elbow, (*b*) the shoulder, (*c*) the hip, (*d*) the neck, and (*e*) the knee. (*f*) Dorsiflexion and plantar flexion of the foot.

Flexion and extension can also be defined in relation to the coronal plane. Starting from the anatomical position, **flexion** is the movement of a body part anterior to the coronal plane, or in the anterior direction. **Extension** is the movement of a body part posterior to the coronal plane, or in the posterior direction (figure 7.42*b-d*).

The exceptions to defining flexion and extension according to the coronal plane are the knee and foot. At the knee, flexion moves the leg in a posterior direction and extension moves it in an anterior direction (figure 7.42*e*). Movement of the foot toward the plantar surface, such as when standing on the toes, is commonly called **plantar flexion;** movement of the foot toward the shin, such as when walking on the heels, is called **dorsiflexion** (figure 7.42*f*).

Hyperextension is usually defined as an abnormal, forced extension of a joint beyond its normal range of motion. For example, if a person falls and attempts to break the fall by putting out a hand, the force of the fall directed into the hand and wrist may cause hyperextension of the wrist, which may result in sprained joints or broken bones. Some health professionals, however, define hyperextension as the normal movement of structures, except the leg, into the space posterior to the anatomical position.

Abduction and Adduction

Abduction (to take away) is movement away from the median plane (midline of the body); **adduction** (to bring together) is movement toward the median plane. Moving the upper limbs away from the body, such as in the outward portion of doing jumping jacks, is abduction of the arm, and bringing the upper limbs back toward the body is adduction of the arm (figure 7.43*a*). Abduction of the fingers involves spreading the fingers apart, away from the midline of the hand, and adduction is bringing them back together (figure 7.43*b*). Abduction of the thumb moves it anteriorly, away from the palm. Abduction of the wrist is movement of the hand away from the midline of the body, and adduction of the wrist results in movement of the hand toward the midline of the body. Abduction of the neck tilts the head to one side and is commonly called lateral flexion of the neck. Bending at the waist to one side is usually called **lateral flexion** of the vertebral column, rather than abduction. ▼

> 57. *Define flexion and extension and demonstrate flexion and extension of the limbs and trunk. What is hyperextension?*
> 58. *Define plantar flexion and dorsiflexion.*
> 59. *Contrast abduction and adduction. Describe these movements for the arm, fingers, wrist, neck, and vertebral column. For what part of the body is the term lateral flexion used?*

Circular Movements

Circular movements involve the rotation of a structure around an axis or movement of the structure in an arc.

Rotation

Rotation is the turning of a structure around its long axis, such as the movement of the atlas around the axis when shaking the head "no." Medial rotation of the humerus with the forearm flexed brings the hand toward the body. Rotation of the humerus so that the hand moves away from the body is lateral rotation (figure 7.44).

Pronation and Supination

Pronation (prō-nā′shŭn) and **supination** (soo′pi-nā′shŭn) refer to the unique rotation of the forearm (figure 7.45). The word *prone* means lying facedown; the word *supine* means lying faceup. When the elbow is flexed to 90 degrees, pronation is rotation of the forearm so that the palm of the hand faces inferiorly. Supination is rotation of the forearm so that the palm faces superiorly. In pronation, the radius and ulna cross; in supination, they are in a parallel position. The head of the radius rotates against the radial notch of the ulna during supination and pronation (see figure 7.26*a*).

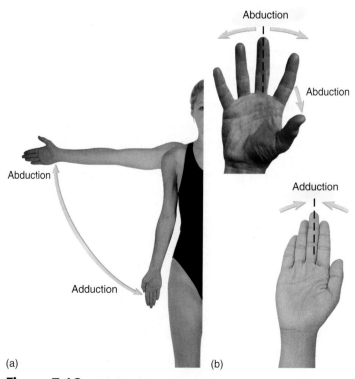

(a) (b)

Figure 7.43 **Abduction and Adduction**

Abduction and adduction of (*a*) the arm and (*b*) the fingers.

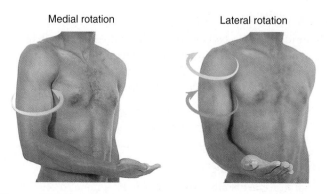

Medial rotation Lateral rotation

Figure 7.44 **Medial and Lateral Rotation of the Arm**

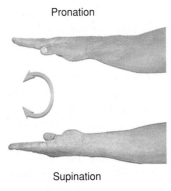

Figure 7.45 Pronation and Supination of the Forearm

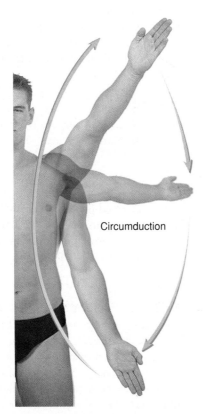

Figure 7.46 Circumduction

Circumduction

Circumduction is a combination of flexion, extension, abduction, and adduction (figure 7.46). It occurs at freely movable joints, such as the shoulder. In circumduction, the arm moves so that it describes a cone with the shoulder joint at the apex. ▼

> 60. *Distinguish among rotation, circumduction, pronation, and supination. Give an example of each.*

Special Movements

Special movements are those movements unique to only one or two joints; they do not fit neatly into one of the other categories.

Elevation and Depression

Elevation moves a structure superiorly; **depression** moves it inferiorly. The scapulae and mandible are primary examples. Shrugging the shoulders is an example of scapular elevation. Depression of the mandible opens the mouth, and elevation closes it.

Protraction and Retraction

Protraction consists of moving a structure in a gliding motion in an anterior direction. **Retraction** moves the structure back to the anatomical position or even more posteriorly. As with elevation and depression, the mandible and scapulae are primary examples. Pulling the scapulae back toward the vertebral column is retraction.

Excursion

Lateral excursion is moving the mandible to either the right or left of the midline, such as in grinding the teeth or chewing. **Medial excursion** returns the mandible to the neutral position.

Opposition and Reposition

Opposition is a unique movement of the thumb and little finger. It occurs when these two digits are brought toward each other across the palm of the hand. The thumb can also oppose the other digits, but those digits flex to touch the tip of the opposing thumb. **Reposition** is the movement returning the thumb and little finger to the neutral, anatomical position.

Inversion and Eversion

Inversion of the foot consists of turning the ankle so that the plantar surface of the foot faces medially, toward the opposite foot. **Eversion** of the foot is turning the ankle so that the plantar surface faces laterally (figure 7.47). Inversion of the foot is sometimes called supination, and eversion is called pronation.

Combination Movements

Most movements that occur in the course of normal activities are combinations of the movements named previously and are described by naming the individual movements involved in the combined movement. For example, when a person steps forward

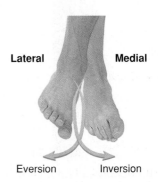

Figure 7.47 Inversion and Eversion of the Right Foot

at a 45-degree angle to the side, the movement at the hip is a combination of flexion and abduction. ▼

> 61. *Define the following jaw movements: protraction, retraction, lateral excursion, medial excursion, elevation, and depression.*
> 62. *Define opposition and reposition. Define inversion and eversion of the foot.*
> 63. *What are combination movements?*

P R E D I C T ⑫

What combination of movements is required at the shoulder and elbow joints for a person to move the right upper limb from the anatomical position to touch the right side of the head with the fingertips?

Description of Selected Joints

It is impossible in a limited space to describe all the joints of the body; therefore, only selected joints are described in this chapter, and they have been chosen because of their representative structure, important function, or clinical significance.

Temporomandibular Joint

The mandible articulates with the temporal bone to form the **temporomandibular joint (TMJ).** The mandibular condyle fits into the mandibular fossa of the temporal bone (figure 7.48*a*). A fibrocartilage articular disk is located between the mandible and the temporal bone, dividing the joint into superior and inferior joint cavities. The joint is surrounded by a fibrous capsule, to which the

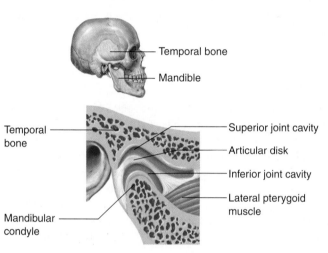

(a) **Sagittal section of temporomandibular joint**

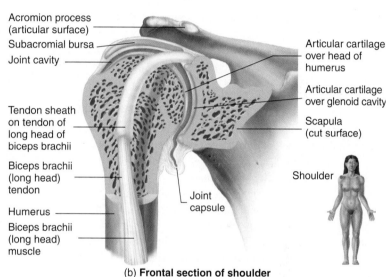

(b) **Frontal section of shoulder**

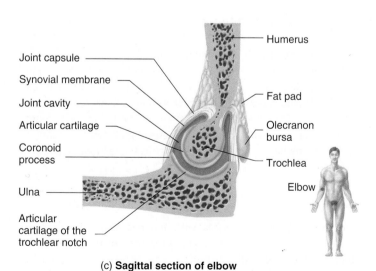

(c) **Sagittal section of elbow**

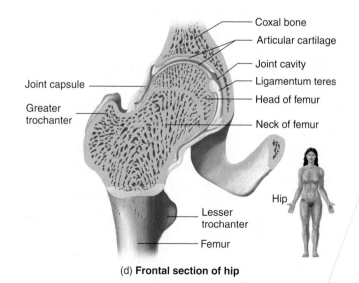

(d) **Frontal section of hip**

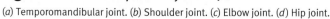

Figure 7.48 Examples of Synovial Joints

(*a*) Temporomandibular joint. (*b*) Shoulder joint. (*c*) Elbow joint. (*d*) Hip joint.

articular disk is attached at its margin, and it is strengthened by lateral and accessory ligaments.

The temporomandibular joint is a combination plane and ellipsoid joint, with the ellipsoid portion predominating. Depression of the mandible to open the mouth involves an anterior gliding motion of the mandibular condyle and articular disk relative to the temporal bone, which is about the same motion that occurs in protraction of the mandible; it is followed by a hinge motion between the articular disk and the mandibular condyle. The mandibular condyle is also capable of slight mediolateral movement, allowing excursion of the mandible.

 Temporomandibular Disorder

Temporomandibular disorder (TMD) is the second most common cause of orofacial pain, after toothache. TMD is broadly subdivided into muscle-related TMD and joint-related TMD. Muscle-related TMD is the more common form and is related to muscle hyperactivity and malalignment of the teeth. Muscle hyperactivity results in grinding of the teeth during sleep and jaw clenching during the day in a stressed person. Radiographs may reveal no obvious destructive changes of the joint.

Joint-related TMD can be caused by disk displacement, degeneration of the joint, arthritis, infections, and other causes. Abnormal movement of the disk can produce a characteristic popping or clicking sound. An abnormal position of the disk may make it impossible to open the mouth fully.

Treatment includes teaching the patient to reduce jaw movements that aggravate the problem and to reduce stress and anxiety. Physical therapy may help relax the muscles and restore function. Analgesic and anti-inflammatory drugs may be used, and oral splints may be helpful, especially at night.

Shoulder Joint

The **shoulder,** or **glenohumeral, joint** is a ball-and-socket joint (figure 7.48b) in which stability is reduced and mobility is increased, compared with the other ball-and-socket joint, the hip. Flexion, extension, abduction, adduction, rotation, and circumduction can all occur at the shoulder joint. The rounded head of the humerus articulates with the shallow glenoid cavity of the scapula. The rim of the glenoid cavity is built up slightly by a fibrocartilage ring, the **glenoid labrum,** to which the joint capsule is attached. A **subacromial bursa** is located near the joint cavity but is separated from the cavity by the joint capsule.

The stability of the joint is maintained primarily by ligaments and four muscles referred to collectively as the **rotator cuff,** which help hold the head of the humerus in the glenoid cavity. These muscles are discussed in more detail in chapter 9. The head of the humerus is also supported against the glenoid cavity by a tendon from the biceps brachii muscle. This tendon attaches to the supraglenoid tubercle (see figure 7.22a), crosses over the head of the humerus within the joint cavity (see figure 7.48b), and passes through the intertubercular groove (see figure 7.24a) to join the biceps brachii in the anterior arm.

 Shoulder Disorders

The most common traumatic shoulder disorders are **dislocation** and muscle or tendon **tears.** The shoulder is the most commonly dislocated joint in the body. The major ligaments cross the superior part of the shoulder joint, and no major ligaments or muscles are associated with the inferior side. As a result, dislocation of the humerus is most likely to occur inferiorly into the axilla. Because the axilla contains very important nerves and arteries, severe and permanent damage may occur when the humeral head dislocates inferiorly. The axillary nerve is the most commonly damaged (see chapter 11).

Chronic shoulder disorders include tendonitis (inflammation of tendons), bursitis (inflammation of bursae), and arthritis (inflammation of joints). Bursitis of the subacromial bursa can become very painful when the large shoulder muscle, called the deltoid muscle, compresses the bursa during shoulder movement.

Elbow Joint

The **elbow joint** (figure 7.48c) is a compound hinge joint consisting of the **humeroulnar joint,** between the humerus and ulna, and the **humeroradial joint,** between the humerus and radius. The **proximal radioulnar joint,** between the proximal radius and ulna, is also closely related. The shape of the trochlear notch and its association with the trochlea of the humerus limit movement at the elbow joint to flexion and extension. The rounded radial head, however, rotates in the radial notch of the ulna and against the capitulum of the humerus (see figure 7.26a), allowing pronation and supination of the hand.

The elbow joint is surrounded by a joint capsule and is reinforced by ligaments. A subcutaneous **olecranon bursa** covers the proximal and posterior surfaces of the olecranon process.

 Elbow Problems

Olecranon bursitis is an inflammation of the olecranon bursa. This inflammation can be caused by excessive rubbing of the elbow against a hard surface and is sometimes referred to as **student's elbow.** The radial head can become subluxated (partially dislocated) from the annular ligament of the radius (see figure 7.39). This condition is called **nursemaid's elbow.** If a child is lifted by one hand, the action may subluxate the radial head.

Hip Joint

The femoral head articulates with the relatively deep, concave acetabulum of the coxal bone to form the **coxal,** or **hip, joint** (figure 7.48d). The head of the femur is more nearly a complete ball than the articulating surface of any other bone of the body. The hip is capable of a wide range of movement, including flexion, extension, abduction, adduction, rotation, and circumduction.

Damage to a collateral ligament can result from a blow to the opposite side of the knee. A common type of football injury results from a block or tackle to the lateral side of the knee, which can cause the knee to bend inward, tearing the tibial (medial) collateral ligament and opening the medial side of the joint (figure E). Injuries to the medial side of the knee are much more common than injuries to the lateral side because blows to the medial side occur more frequently than to the lateral side. In addition, the fibular (lateral) collateral is stronger than the tibial (medial) collateral ligament.

The medial meniscus is fairly tightly attached to the tibial (medial) collateral ligament and is damaged 20 times more often in a knee injury than the lateral meniscus, which is thinner and not attached to the fibular (lateral) collateral ligament. A **torn meniscus** may result in a "clicking" sound during extension of the leg; if the damage is more severe, the torn piece of cartilage may move between the articulating surfaces of the tibia and femur, causing the knee to "lock" in a partially flexed position.

In severe medial knee injuries, the anterior cruciate ligament is also damaged (see figure E). Tearing of the tibial (medial) collateral ligament, medial meniscus, and anterior cruciate ligament is often referred to as "the unhappy triad of injuries."

If the knee is driven anteriorly or if the knee is hyperextended, the anterior cruciate ligament may be torn, which causes the knee joint to be very unstable. If the knee is flexed and the tibia is driven posteriorly, the posterior cruciate ligament may be

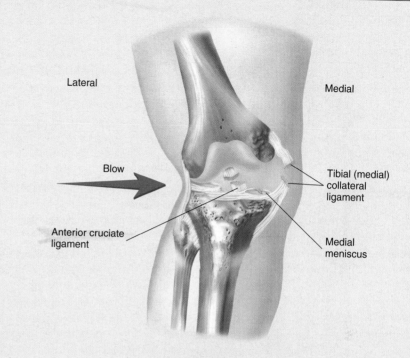

Figure E Injury to the Right Knee

torn. Surgical replacement of a cruciate ligament with a transplanted or an artificial ligament repairs the damage.

Bursitis in the subcutaneous prepatellar bursa (see figure 7.49b), commonly called "housemaid's knee," may result from prolonged work performed while on the hands and knees. Another bursitis, "clergyman's knee," results from excessive kneeling and affects the subcutaneous infrapatellar bursa. This type of bursitis is common among carpet layers and roofers.

Other common knee problems are **chondromalacia** (kon′drō-mă-lā′shē-ă), or softening of the cartilage, which results from abnormal movement of the patella within the patellar groove, and **fat pad syndrome,** which consists of an accumulation of fluid in the fat pad posterior to the patella. An acutely swollen knee appearing immediately after an injury is usually a sign of blood accumulation within the joint cavity and is called a **hemarthrosis** (hē′mar-thrō′sis, hem′ar-thrō′sis). A slower accumulation of fluid, "water on the knee," may be caused by bursitis.

The acetabulum is deepened and strengthened by a lip of fibrocartilage called the **acetabular labrum.** An extremely strong joint capsule, reinforced by several ligaments, extends from the rim of the acetabulum to the neck of the femur. The **ligament of the head of the femur** (round ligament of the femur) is located inside the hip joint between the femoral head and the acetabulum. This ligament does not contribute much toward strengthening the hip joint; however, it does carry a small nutrient artery to the head of the femur in about 80% of the population. The acetabular labrum, ligaments of the hip, and surrounding muscles make the hip joint much more stable but less mobile than the shoulder joint.

Hip Dislocation

Dislocation of the hip may occur when the hip is flexed and the femur is driven posteriorly, such as when a person sitting in an automobile is involved in an accident. The head of the femur usually dislocates posterior to the acetabulum, tearing the acetabular labrum, the fibrous capsule, and the ligaments. Fracture of the femur and the coxal bone often accompanies hip dislocation.

188

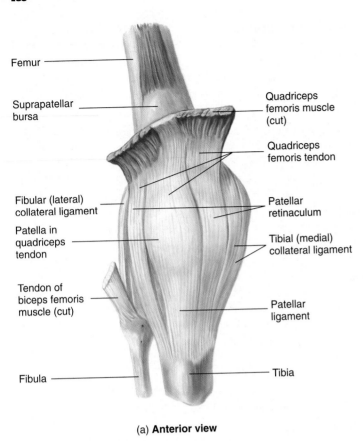

Femur

Suprapatellar bursa

Quadriceps femoris muscle (cut)

Quadriceps femoris tendon

Fibular (lateral) collateral ligament

Patellar retinaculum

Patella in quadriceps tendon

Tibial (medial) collateral ligament

Tendon of biceps femoris muscle (cut)

Patellar ligament

Fibula

Tibia

(a) **Anterior view**

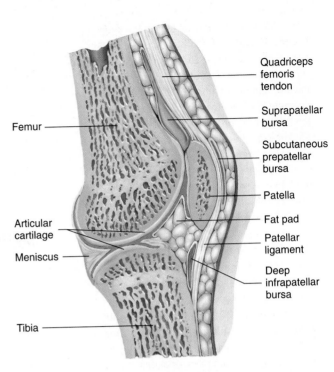

Femur

Quadriceps femoris tendon

Suprapatellar bursa

Subcutaneous prepatellar bursa

Patella

Articular cartilage

Meniscus

Fat pad

Patellar ligament

Deep infrapatellar bursa

Tibia

(b) **Sagittal section**

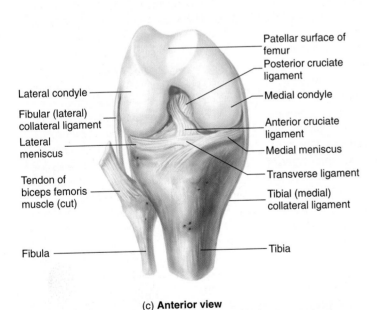

Patellar surface of femur

Posterior cruciate ligament

Lateral condyle

Medial condyle

Fibular (lateral) collateral ligament

Anterior cruciate ligament

Lateral meniscus

Medial meniscus

Tendon of biceps femoris muscle (cut)

Transverse ligament

Tibial (medial) collateral ligament

Fibula

Tibia

(c) **Anterior view**

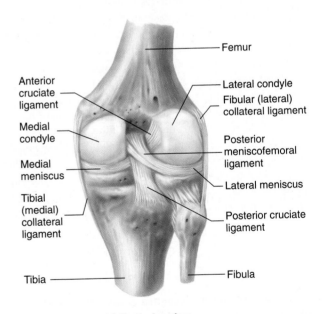

Femur

Anterior cruciate ligament

Lateral condyle

Fibular (lateral) collateral ligament

Medial condyle

Medial meniscus

Posterior meniscofemoral ligament

Tibial (medial) collateral ligament

Lateral meniscus

Posterior cruciate ligament

Tibia

Fibula

(d) **Posterior view**

Figure 7.49 Right Knee Joint

Knee Joint

At the knee, the femur joins the tibia and the patella and the fibula joins the tibia (figure 7.49). The patella is located within the tendon of the quadriceps femoris muscle. The **knee joint** traditionally is classified as a modified hinge joint located between the femur and the tibia. Actually, it is a complex ellipsoid joint that allows flexion, extension, and a small amount of rotation of the leg. The distal end of the femur has two large, rounded condyles with a deep intercondylar fossa between them (see figure 7.33b). The femur articulates with the proximal condyles of the tibia, which are flattened, with a crest called the intercondylar eminence between them (see figure 7.36). The **lateral and medial menisci** build up the margins of the tibial condyles and deepen their articular surfaces (see figure 7.49*b–d*).

Two ligaments extend between the tibia and the intercondylar fossa of the femur. The **anterior cruciate** (kroo′shē-āt, crossed) **ligament** attaches to the tibia anterior to the intercondylar eminence (see figure 7.49*c*) and prevents anterior displacement of the tibia relative to the femur. The **posterior cruciate ligament** attaches to the tibia posterior to the intercondylar eminence (see figure 7.49*d*) and prevents posterior displacement of the tibia. The **fibular (lateral) collateral ligament** and the **tibial (medial) collateral ligament** strengthen the sides of the knee joint and prevent the femur from tipping side to side on the tibia. Other ligaments and the tendons of the thigh muscles, which extend around the knee, also provide stability.

A number of bursae surround the knee (see figure 7.49*b*). The largest is the **suprapatellar bursa,** which is a superior extension of the joint capsule that allows for movement of the anterior thigh muscles over the distal end of the femur. Other knee bursae include the **subcutaneous prepatellar bursa** and the **deep infrapatellar bursa.**

Ankle Joint and Arches of the Foot

The distal tibia and fibula form a highly modified hinge joint with the talus called the **ankle,** or **talocrural** (tā′lō-kroo′răl), **joint.** The medial and lateral malleoli of the tibia and fibula, which form the medial and lateral margins of the ankle, are rather extensive, whereas the anterior and posterior margins are almost nonexistent (see figure 7.36). As a result, a hinge joint is created. Most of the weight from the leg is transferred from the tibia to the talus (see figure 7.37*b*). Ligaments extending from the lateral and medial malleoli attach to the tarsal bones, stabilizing the joint (figure 7.50). Dorsiflexion, plantar flexion, and limited inversion and eversion can occur at the ankle joint.

The arches of the foot (see figure 7.37*b*) are supported by ligaments (see figure 7.50). These ligaments hold the bones of the arches in their proper relationship and provide ties across the arch somewhat like a bowstring. As weight is transferred through the arch system, some of the ligaments are stretched, giving the foot

more flexibility and allowing it to adjust to uneven surfaces. When weight is removed from the foot, the ligaments recoil and restore the arches to their unstressed shape.

 Ankle Injury and Arch Problems

The ankle is the most frequently injured major joint in the body. A **sprained ankle** results when the ligaments of the ankle are torn partially or completely. The most common ankle injuries result from forceful inversion of the foot and damage to the lateral ligaments (figure F). Eversion of the foot and medial ligament damage is rare and usually involves fracture of the malleoli (see figure F).

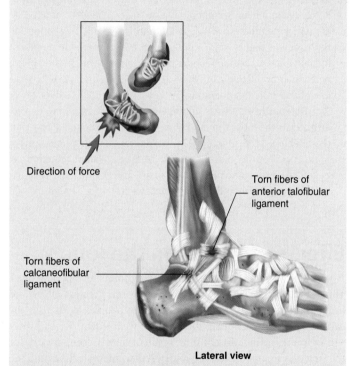

Direction of force

Torn fibers of calcaneofibular ligament

Torn fibers of anterior talofibular ligament

Lateral view

Figure F Injury of the Right Ankle

The arches of the foot normally form early in fetal life. Failure to form results in congenital **flat feet,** or fallen arches, a condition in which the arches, primarily the medial longitudinal arch (see figure 7.37), are depressed or collapsed. This condition is not always painful. Flat feet may also occur when the muscles and ligaments supporting the arch fatigue and allow the arch, usually the medial longitudinal arch, to collapse. During prolonged standing, the plantar calcaneonavicular ligament may stretch, flattening the medial longitudinal arch. The transverse arch may also become flattened. The strained ligaments can become painful. ▼

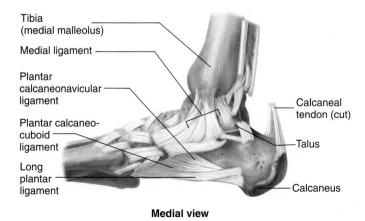

Tibia (medial malleolus)

Medial ligament

Plantar calcaneonavicular ligament

Plantar calcaneo-cuboid ligament

Long plantar ligament

Calcaneal tendon (cut)

Talus

Calcaneus

Medial view

Figure 7.50 Ligaments of the Right Ankle Joint

64. *For each of the following joints, name the bones of the joint, the specific part of the bones that form the joint, the type of joint, and the possible movement(s) at the joint: temporomandibular, shoulder, elbow, hip, knee, and ankle.*

Arthritis, an inflammation of any joint, is the most common and best known of the joint disorders, affecting 10% of the world's population and 14% of the U.S. population. There are over 37 million cases of arthritis in the United States alone. More than 100 types of arthritis exist. Classification is often based on the cause and progress of the arthritis. Its causes include infectious agents, metabolic disorders, trauma, and immune disorders. Mild exercise retards joint degeneration and enhances mobility. Swimming and walking are recommended for people with arthritis, but running, tennis, and aerobics are not recommended. Therapy depends on the type of arthritis but usually includes the use of anti-inflammatory drugs.

Osteoarthritis (OA), or **degenerative arthritis,** is the most common type of arthritis, affecting 10% of people in the United States (85% of those over age 70). OA may begin as a molecular abnormality in articular cartilage, with heredity and normal wear and tear of the joint important contributing factors. Slowed metabolic rates with increased age also seem to contribute to OA. Inflammation is usually secondary in this disorder. It tends to occur in the weight-bearing joints, such as the knees, and is more common in overweight individuals. OA is becoming more common in younger people as a result of increasing rates of childhood obesity.

Rheumatoid arthritis (RA) is the second most common type of arthritis. It affects about 3% of all women and about 1% of all men in the United States. It is a general connective tissue disorder that affects the skin, vessels, lungs, and other organs, but it is most pronounced in the joints. It is severely disabling and most commonly destroys small joints, such as those in the hands and feet. The initial cause is unknown but may involve a transient **infection** or an **autoimmune disease** (an immune reaction to one's own tissues) (see chapter 19) that develops against col-

lagen. A genetic predisposition may also exist. Whatever the cause, the ultimate course appears to be immunologic. People with classic RA have a protein, **rheumatoid factor,** in their blood. In RA, the synovial fluid and associated connective tissue cells proliferate, forming a **pannus** (clothlike layer), which causes the joint capsule to become thickened and which destroys the articular cartilage. In advanced stages, opposing joint surfaces can become fused.

As a result of recent advancements in biomedical technology, many joints of the body can be replaced by artificial joints. Joint replacement, called **arthroplasty,** was first developed in the late 1950s. One of the major reasons for its use is to eliminate unbearable pain in patients near ages 55 to 60 with joint disorders. Osteoarthritis is the leading disease requiring joint replacement and accounts for two-thirds of the patients. Rheumatoid arthritis accounts for more than half of the remaining cases.

Effects of Aging on the Joints

A number of changes occur within many joints as a person ages. Those that occur in synovial joints have the greatest effect and often present major problems for elderly people. With use, the cartilage covering articular surfaces can wear down. When a person is young, production of new, resilient matrix compensates for the wear. As a person ages, the rate of replacement declines and the matrix becomes more rigid, thus adding to its rate of wear. The production rate of lubricating synovial fluid also declines with age, further contributing to the wear of the articular cartilage. Many people also experience arthritis, an inflammatory degeneration of joints, with advancing age. In addition, the ligaments and tendons surrounding a joint shorten and become less flexible with age, resulting in a decrease in the range of motion of the joint. Furthermore, older people often experience a general decrease in activity, which causes the joints to become less flexible and their range of motion to decrease. ▼

> *65. What age-related factors contribute to cartilage wear in synovial joints? To loss of flexibility and loss of range of motion in synovial joints?*

S U M M A R Y

General Considerations of Bones (p. 148)

1. Bones are paired or unpaired.
2. Bones have processes, smooth surfaces, and holes that are associated with ligaments, muscles, joints, nerves, and blood vessels.

Axial Skeleton (p. 149)

The axial skeleton consists of the skull, hyoid bone, vertebral column, and thoracic cage.

Skull

1. The skull is composed of 22 bones.
   - The braincase protects the brain.
   - The facial bones protect the sensory organs of the head and are muscle attachment sites (mastication, facial expression, and eye muscles).

   - The mandible and maxillae hold the teeth, and the auditory ossicles, which function in hearing, are located inside the temporal bones.
2. The parietal bones are joined at the midline by the sagittal suture; they are joined to the frontal bone by the coronal suture, to the occipital bone by the lambdoid suture, and to the temporal bone by the squamous suture.
3. The external occipital protuberance is an attachment site for an elastic ligament. Nuchal lines are the points of attachment for neck muscles.
4. Several skull features are seen from a lateral view.
   - The external acoustic meatus transmits sound waves toward the eardrum.
   - Neck muscles attach to the mastoid process, which contains mastoid air cells.
   - The temporal lines are attachment points of the temporalis muscle.

- The zygomatic arch, from the temporal and zygomatic bones, forms a bridge across the side of the skull.
- The mandible articulates with the temporal bone.

5. Several skull features are seen from an anterior view.
   - The orbits contain the eyes.
   - The nasal cavity is divided by the nasal septum
   - Sinuses within bone are air-filled cavities. The paranasal sinuses, which connect to the nasal cavity, are the frontal, sphenoidal, and maxillary sinuses and the ethmoidal labyrinth.

6. Several features are on the inferior surface of the skull.
   - The spinal cord and brain are connected through the foramen magnum.
   - Occipital condyles are points of articulation between the skull and the vertebral column.
   - Blood reaches the brain through the internal carotid arteries, which pass through the carotid canals, and the vertebral arteries, which pass through the foramen magnum.
   - Most blood leaves the brain through the internal jugular veins, which exit through the jugular foramina.
   - Styloid processes provide attachment points for three muscles involved in movement of the tongue, hyoid bone, and pharynx.
   - The hard palate separates the oral cavity from the nasal cavity.

7. Several skull features are inside the cranial cavity.
   - The crista galli is a point of attachment for one of the meninges.
   - The olfactory nerves extend into the roof of the nasal cavity through the olfactory foramina of the cribriform plate.
   - The sella turcica is occupied by the pituitary gland.

Hyoid Bone

The hyoid bone, which "floats" in the neck, is the attachment site for throat and tongue muscles.

Vertebral Column

1. The vertebral column provides flexible support and protects the spinal cord.
2. The vertebral column has four major curvatures: cervical, thoracic, lumbar, and sacral/coccygeal. Abnormal curvatures are lordosis (lumbar), kyphosis (thoracic), and scoliosis (lateral).
3. A typical vertebra consists of a body, a vertebral arch, and various processes.
   - Part of the body and the vertebral arch (pedicle and lamina) form the vertebral foramen, which contains and protects the spinal cord.
   - The transverse and spinous processes are points of muscle and ligament attachment.
   - Vertebrae articulate with one another through the superior and inferior articular processes.
   - Spinal nerves exit through the intervertebral foramina.
4. Adjacent bodies are separated by intervertebral disks. The disk has a fibrous outer covering (annulus fibrosus) surrounding a gelatinous interior (nucleus pulposus).
5. Several types of vertebrae can be distinguished.
   - All seven cervical vertebrae have transverse foramina, and most have bifid spinous processes.
   - The 12 thoracic vertebrae have attachment sites for ribs and are characterized by long, downward-pointing spinous processes.
   - The five lumbar vertebrae have thick, heavy bodies and processes. Their superior articular facets face medially and their inferior articular facets face laterally.
   - The sacrum consists of five fused vertebrae and attaches to the coxal bones to form the pelvis.
   - The coccyx consists of four fused vertebrae attached to the sacrum.

Thoracic Cage

1. The thoracic cage (consisting of the ribs, their associated costal cartilages, and the sternum) protects the thoracic organs and changes volume during respiration.
2. Twelve pairs of ribs attach to the thoracic vertebrae. They are divided into seven pairs of true ribs and five pairs of false ribs. Two pairs of false ribs are floating ribs.
3. The sternum is composed of the manubrium, the body, and the xiphoid process.

Appendicular Skeleton   (p. 165)

The appendicular skeleton consists of the upper and lower limbs and the girdles that attach the limbs to the body.

Pectoral Girdle

1. The pectoral girdle consists of the scapulae and clavicles.
2. The scapula articulates with the humerus (at the glenoid cavity) and the clavicle (at the acromion). It is an attachment site for shoulder, back, and arm muscles.
3. The clavicle holds the shoulder away from the body and allows movement of the scapula, resulting in free movement of the arm.

Upper Limb

1. The arm bone is the humerus.
   - The humerus articulates with the scapula (head), the radius (capitulum), and the ulna (trochlea).
   - Sites of muscle attachment are the greater and lesser tubercles, the deltoid tuberosity, and the epicondyles.
2. The forearm contains the ulna and radius.
   - The ulna and radius articulate with each other and with the humerus and wrist bones.
   - The wrist ligaments attach to the styloid processes of the radius and ulna.
3. Eight carpal, or wrist, bones are arranged in two rows.
4. The hand consists of five metacarpal bones.
5. The phalanges are digital bones. Each finger has three phalanges, and the thumb has two phalanges.

Pelvic Girdle

1. The lower limb is attached solidly to the coxal bone and functions in support and movement.
2. The pelvic girdle consists of the right and left coxal bones and sacrum. Each coxal bone is formed by the fusion of the ilium, the ischium, and the pubis.
   - The coxal bones articulate with each other (symphysis pubis) and with the sacrum (sacroiliac joint) and the femur (acetabulum).
   - Muscles attach to the anterior iliac spines and the ischial tuberosity; ligaments attach to the posterior iliac spines, ischial spine, and ischial tuberosity.
   - The female pelvis has a larger pelvic inlet and outlet than the male pelvis.

Lower Limb

1. The thighbone is the femur.
   - The femur articulates with the coxal bone (head), the tibia (medial and lateral condyles), and the patella (patellar groove).
   - Sites of muscle attachment are the greater and lesser trochanters, as well as the adductor tubercle.
   - Sites of ligament attachment are the lateral and medial epicondyles.
2. The leg consists of the tibia and the fibula.
   - The tibia articulates with the femur, the fibula, and the talus. The fibula articulates with the tibia and the talus.
   - Tendons from the thigh muscles attach to the tibial tuberosity.

3. Seven tarsal bones form the proximal portion of the foot and five metatarsal bones form the distal portion.
4. The toes have three phalanges each, except for the big toe, which has two.
5. The bony arches transfer weight from the heels to the toes and allow the foot to conform to many different positions.

Articulations (p. 177)

1. An articulation, or joint, is a place where two bones come together.
2. Joints are named according to the bones or parts of bones involved.

Classes of Joints (p. 177)

Joints are classified according to function or type of connective tissue that binds them together and whether fluid is present between the bones.

Fibrous Joints

1. Fibrous joints are those in which bones are connected by fibrous tissue with no joint cavity. They are capable of little or no movement.
2. Sutures involve interdigitating bones held together by dense fibrous connective tissue. They occur between most skull bones.
3. Syndesmoses are joints consisting of fibrous ligaments.
4. Gomphoses are joints in which pegs fit into sockets and are held in place by periodontal ligaments (teeth in the jaws).
5. Some sutures and other joints can become ossified (synostosis).

Cartilaginous Joints

1. Synchondroses are joints in which bones are joined by hyaline cartilage. Epiphyseal plates are examples.
2. Symphyses are slightly movable joints made of fibrocartilage.

Types of Synovial Joints

1. Synovial joints are capable of considerable movement. They consist of the following:
   - Articular cartilage on the ends of bones provides a smooth surface for articulation. Articular disks and menisci can provide additional support.
   - A joint cavity is surrounded by a joint capsule of fibrous connective tissue, which holds the bones together while permitting flexibility, and a synovial membrane produces synovial fluid that lubricates the joint.
2. Bursae are extensions of synovial joints that protect skin, tendons, or bone from structures that could rub against them.
3. Synovial joints are classified according to the shape of the adjoining articular surfaces: plane (two flat surfaces), pivot (cylindrical projection inside a ring), hinge (concave and convex surfaces), ball-and-socket (rounded surface into a socket), ellipsoid (ellipsoid concave and convex surfaces), and saddle (two saddle-shaped surfaces).

Types of Movement (p. 182)

Gliding Movements

Gliding movements occur when two flat surfaces glide over one another.

Angular Movements

Angular movements include flexion and extension, plantar and dorsiflexion, and abduction and adduction.

Circular Movements

Circular movements include rotation, pronation and supination, and circumduction.

Special Movements

Special movements include elevation, depression, protraction, retraction, excursion, opposition, reposition, inversion and eversion.

Combination Movements

Combination movements involve two or more other movements.

Description of Selected Joints (p. 185)

Temporomandibular Joint

1. The temporomandibular joint is a complex gliding and hinge joint between the temporal and mandibular bones.
2. The temporomandibular joint is capable of elevation, depression, protraction, retraction, and lateral and medial excursion movements.

Shoulder Joint

1. The shoulder joint is a ball-and-socket joint between the head of the humerus and the glenoid cavity of the scapula. It is strengthened by ligaments and the muscles of the rotator cuff. The tendon of the biceps brachii passes through the joint capsule.
2. The shoulder joint is capable of flexion, extension, abduction, adduction, rotation, and circumduction.

Elbow Joint

1. The elbow joint is a compound hinge joint between the humerus, ulna, and radius.
2. Movement at this joint is limited to flexion and extension.

Hip Joint

1. The hip joint is a ball-and-socket joint between the head of the femur and the acetabulum of the coxal bone.
2. The hip joint is capable of flexion, extension, abduction, adduction, rotation, and circumduction.

Knee Joint

1. The knee joint is a complex ellipsoid joint between the femur and the tibia that is supported by many ligaments.
2. The joint allows flexion, extension, and slight rotation of the leg.

Ankle Joint and Arches of the Foot

1. The ankle joint is a special hinge joint of the tibia, fibula, and talus that allows dorsiflexion, plantar flexion, inversion, and eversion of the foot.
2. Ligaments of the foot arches hold the bones in an arch and transfer weight in the foot.

Effects of Aging on the Joints (p. 190)

With age, the connective tissue of the joints becomes less flexible and less elastic. The resulting joint rigidity increases the rate of wear in the articulating surfaces. The change in connective tissue also reduces the range of motion.

R E V I E W   A N D   C O M P R E H E N S I O N

1. The superior and middle nasal conchae are formed by projections of the
   a. sphenoid bone.
   d. palatine bone.
   b. vomer bone.
   e. ethmoid bone.
   c. palatine process of maxillae.

2. The perpendicular plate of the ethmoid and the form the nasal septum.
   a. palatine process of the maxilla
   d. nasal bone
   b. horizontal plate of the palatine
   e. lacrimal bone
   c. vomer

3. Which of these bones does not contain a paranasal sinus?
   a. ethmoid
   c. frontal
   e. maxilla
   b. sphenoid
   d. temporal

4. The mandible articulates with the skull at the
   a. styloid process.
   d. zygomatic arch.
   b. occipital condyle.
   e. medial pterygoid.
   c. mandibular fossa.

5. The nerves for the sense of smell pass through the
   a. cribriform plate.
   d. optic canal.
   b. nasolacrimal canal.
   e. orbital fissure.
   c. internal acoustic meatus.

6. The major blood supply to the brain enters through the
   a. foramen magnum.
   d. both a and b.
   b. carotid canals.
   e. all of the above.
   c. jugular foramina.

7. The site of the sella turcica is the
   a. sphenoid bone
   c. frontal bone.
   e. temporal bone.
   b. maxillae.
   d. ethmoid bone.

8. A herniated disk occurs when
   a. the annulus fibrosus ruptures.
   b. the intervertebral disk slips out of place.
   c. the spinal cord ruptures.
   d. too much fluid builds up in the nucleus pulposus.
   e. all of the above.

9. The weight-bearing portion of a vertebra is the
   a. vertebral arch.
   d. transverse process.
   b. articular process.
   e. spinous process.
   c. body.

10. Transverse foramina are found only in
    a. cervical vertebrae.
    c. lumbar vertebrae.
    e. the coccyx.
    b. thoracic vertebrae.
    d. the sacrum.

11. Articular facets on the bodies and transverse processes are found only on
    a. cervical vertebrae.
    c. lumbar vertebrae.
    e. the coccyx.
    b. thoracic vertebrae.
    d. the sacrum.

12. Which of these statements concerning ribs is true?
    a. The true ribs attach directly to the sternum with costal cartilage.
    b. There are five pairs of floating ribs.
    c. The head of the rib attaches to the transverse process of the vertebra.
    d. Floating ribs do not attach to vertebrae.

13. The point where the scapula and clavicle connect is the
    a. coracoid process.
    c. glenoid cavity.
    e. capitulum.
    b. styloid process.
    d. acromion.

14. The distal medial process of the humerus to which the ulna joins is the
    a. epicondyle.
    c. malleolus.
    e. trochlea.
    b. deltoid tuberosity.
    d. capitulum.

15. The bone/bones of the foot on which the tibia rests is (are) the
    a. talus.
    c. metatarsals.
    e. phalanges.
    b. calcaneus.
    d. navicular.

16. The projection on the coxal bone of the pelvic girdle that is used as a landmark for finding an injection site is the
    a. ischial tuberosity.
    d. posterior inferior iliac spine.
    b. iliac crest.
    e. ischial spine.
    c. anterior superior iliac spine.

17. When comparing the pectoral girdle with the pelvic girdle, which of these statements is true?
    a. The pectoral girdle has greater mass than the pelvic girdle.
    b. The pelvic girdle is more firmly attached to the body than the pectoral girdle.
    c. The pectoral girdle has the limbs more securely attached than the pelvic girdle.
    d. The pelvic girdle allows greater mobility than the pectoral girdle.

18. When comparing a male pelvis with a female pelvis, which of these statements is true?
    a. The pelvic inlet in males is larger and more circular.
    b. The subpubic angle in females is less than 90 degrees.
    c. The ischial spines in males are closer together.
    d. The sacrum in males is broader and less curved.

19. A site of muscle attachment on the proximal end of the femur is the
    a. greater trochanter.
    d. intercondylar eminence.
    b. epicondyle.
    e. condyle.
    c. greater tubercle.

20. The process that forms the large lateral bump in the ankle is the lateral
    a. malleolus.
    c. epicondyle.
    e. tubercle.
    b. condyle.
    d. tuberosity.

21. Given these types of joints:
    1. gomphosis
    3. symphysis
    5. syndesmosis
    2. suture
    4. synchondrosis
    Choose the types that are held together by fibrous connective tissue.
    a. 1,2,3
    c. 2,3,5
    e. 1,2,3,4,5
    b. 1,2,5
    d. 3,4,5

22. Which of these joints is not matched with the correct joint type?
    a. parietal bone to occipital bone—suture
    b. between the coxal bones—symphysis
    c. humerus and scapula—synovial
    d. shafts of the radius and ulna—synchondrosis
    e. teeth in alveolar process—gomphosis

23. In which of these joints are periodontal ligaments found?
    a. sutures
    c. symphyses
    e. gomphoses
    b. syndesmoses
    d. synovial

24. The intervertebral disks are an example of
    a. sutures.
    c. symphyses.
    e. gomphoses.
    b. syndesmoses.
    d. synovial joints.

25. Joints containing hyaline cartilage are called _____, and joints containing fibrocartilage are called _____.
    a. sutures, synchondroses
    d. synchondroses, symphyses
    b. syndesmoses, symphyses
    e. gomphoses, synchondroses
    c. symphyses, syndesmoses

26. The inability to produce the fluid that keeps most joints moist would likely be caused by a disorder of the
    a. cruciate ligaments.
    d. bursae.
    b. synovial membrane.
    e. tendon sheath.
    c. articular cartilage.

27. Which of these is not associated with synovial joints?
    a. perichondrium on the surface of articular cartilage
    b. fibrous capsule
    d. synovial fluid
    c. synovial membrane
    e. bursae

28. Assume that a sharp object penetrated a synovial joint. From this list of structures:
    1. tendon or muscle
    4. fibrous capsule (of joint capsule)
    2. ligament
    5. skin
    3. articular cartilage
    6. synovial membrane (of joint capsule)
    Choose the order in which they would most likely be penetrated.
    a. 5,1,2,6,4,3
    c. 5,1,2,6,3,4
    e. 5,1,2,4,6,3
    b. 5,2,1,4,3,6
    d. 5,1,2,4,3,6

29. Which of these do hinge joints and saddle joints have in common?
    a. Both are synovial joints.
    b. Both have concave surfaces that articulate with a convex surface.
    c. Both are uniaxial joints.
    d. Both a and b are correct.
    e. All of the above are correct.

30. Which of these joints is correctly matched with the type of joint?
    a. atlas to occipital condyle—pivot
    b. tarsals to metatarsals—saddle
    c. femur to coxal bone—ellipsoid
    d. tibia to talus—hinge
    e. scapula to humerus—plane

31. Once a doorknob is grasped, what movement of the forearm is necessary to unlatch the door—that is, turn the knob in a clockwise direction? (Assume using the right hand.)
    a. pronation
    b. rotation
    c. supination
    d. flexion
    e. extension

32. After the door is unlatched, what movement of the elbow is necessary to open it? (Assume the door opens in, and you are on the inside.)
    a. pronation
    b. rotation
    c. supination
    d. flexion
    e. extension

33. After the door is unlatched, what movement of the shoulder is necessary to open it? (Assume the door opens in, and you are on the inside.)
    a. pronation
    b. rotation
    c. supination
    d. flexion
    e. extension

34. When grasping a doorknob, the thumb and little finger undergo
    a. opposition.
    b. reposition.
    c. lateral excursion.
    d. medial excursion.
    e. dorsiflexion.

35. Spreading the fingers apart is
    a. rotation.
    b. depression.
    c. abduction.
    d. lateral excursion.
    e. flexion.

36. A runner notices that the lateral side of her right shoe is wearing much more than the lateral side of her left shoe. This could mean that her right foot undergoes more _____ than her left foot.
    a. eversion
    b. inversion
    c. plantar flexion
    d. dorsiflexion
    e. lateral excursion

37. For a ballet dancer to stand on her toes, her feet must
    a. evert.
    b. invert.
    c. lantar flex.
    d. dorsiflex.
    e. abduct.

38. A meniscus is found in the
    a. shoulder joint.
    b. elbow joint.
    c. hip joint.
    d. knee joint.
    e. ankle joint.

39. A lip (labrum) of fibrocartilage deepens the joint cavity of the
    a. temporomandibular joint.
    b. shoulder joint.
    c. elbow joint.
    d. knee joint.
    e. ankle joint.

40. Which of these structures helps stabilize the shoulder joint?
    a. rotator cuff muscles
    b. cruciate ligaments
    c. medial and lateral collateral ligaments
    d. articular disk
    e. all of the above

Answers in Appendix E

C R I T I C A L   T H I N K I N G

1. A patient has an infection in the nasal cavity. Name seven adjacent structures to which the infection could spread.

2. A patient is unconscious. Radiographic films reveal that the superior articular process of the atlas has been fractured. Which of the following could have produced this condition: falling on the top of the head or being hit in the jaw with an uppercut? Explain.

3. If the vertebral column is forcefully rotated, what part of the vertebra is most likely to be damaged? In what area of the vertebral column is such damage most likely?

4. An asymmetric weakness of the back muscles can produce which of the following: scoliosis, kyphosis, or lordosis? Which can result from pregnancy? Explain.

5. What might be the consequences of a broken forearm involving both the ulna and the radius when the ulna and radius fuse to each other during repair of the fracture?

6. Suppose you need to compare the length of one lower limb with the length of the other in an individual. Using bony landmarks, suggest an easy way to accomplish the measurements.

7. A decubitus ulcer is a chronic ulcer that appears in pressure areas of skin overlying a bony prominence in bedridden or otherwise immobilized patients. Where are likely sites for decubitus ulcers to occur?

8. Why are more women than men knock-kneed?

9. On the basis of bone structure of the lower limb, explain why it is easier to turn the foot medially (sole of the foot facing toward the midline of the body) than laterally. Why is it easier to cock the wrist medially than laterally?

10. Justin Time leaped from his hotel room to avoid burning to death in a fire. If he landed on his heels, what bone was likely fractured? Unfortunately for Justin, a 240-lb firefighter ran by and stepped heavily on the proximal part of Justin's foot (not the toes). What bones could have been broken?

11. For each of the following muscles, describe the motion(s) produced when the muscle contracts. It may be helpful to use an articulated skeleton.
    a. The biceps brachii muscle attaches to the coracoid process of the scapula (one head) and the radial tuberosity of the radius. Name two movements that the muscle accomplishes in the forearm.
    b. The rectus femoris muscle attaches to the anterior inferior iliac spine and the tibial tuberosity. How does contraction move the thigh? The leg?
    c. The supraspinatus muscle is located in and attached to the supraspinatus fossa of the scapula. Its tendon runs over the head of the humerus to the greater tubercle. When it contracts, what movement occurs at the glenohumeral (shoulder) joint?
    d. The gastrocnemius muscle attaches to the medial and lateral condyles of the femur and to the calcaneus. What movement of the leg results when this muscle contracts? Of the foot?

12. Crash McBang hurt his knee in an auto accident by ramming it into the dashboard. The doctor tested the knee for ligament damage by having Crash sit on the edge of a table with his knee flexed at a 90-degree angle. The doctor attempted to pull the tibia in an anterior direction (the anterior drawer test) and then tried to push the tibia in a posterior direction (the posterior drawer test). No unusual movement of the tibia occurred in the anterior drawer test but did occur during the posterior drawer test. Explain the purpose of each test, and tell Crash which ligament he has damaged. *Answers in Appendix F*

For additional study tools, go to: www.aris.mhhe.com. Click on your course topic and then this textbook's author name/title. Register once for a semester's worth of interactive learning activities to help improve your grade!

CHAPTER 8 Histology and Physiology of Muscles

Color-enhanced scanning electron micrograph of skeletal muscle fibers. Skeletal muscle fibers are very large, multinucleated cells that are long and uniform in diameter. The nuclei show up as egg-shaped bumps on the surface of the cell. Two nuclei have broken out of the middle cell. One is gone and the other is partially dislodged from the cell. The curly material in the upper right corner is connective tissue associated with the muscle fibers.

Tools for Success

Anatomy & Physiology **REVEALED**®
aprevealed.com — *A virtual cadaver dissection experience*

Audio (MP3) and/or video (MP4) content, available at www.mhhe.aris.com

Introduction

As a runner rounds the last corner of the track and sprints for the finish line, her arms and legs are pumping as she tries to reach her maximum speed. Her heart is beating rapidly and her breathing is rapid, deep, and regular. Blood is shunted away from digestive organs, and a greater volume is delivered to skeletal muscles to maximize the oxygen supply to them. These actions are accomplished by muscle tissue, the most abundant tissue of the body, and one of the most adaptable.

Movements of the limbs, the heart, and other parts of the body are made possible by muscle cells that function like tiny motors. Muscle cells use energy extracted from nutrient molecules much as motors use energy provided by electric current. The nervous system regulates and coordinates muscle cells so that smooth, coordinated movements are produced much as a computer regulates and coordinates several motors in robotic machines that perform assembly line functions.

▶ This chapter presents the **functions of the muscular system (p. 196)**, **general functional characteristics of muscle (p. 196)**, and **skeletal muscle structure (p. 197)**. The **sliding filament model (p. 200)** of muscle contraction is explained. The **physiology of skeletal muscle fibers (p. 202)**, the **physiology of skeletal muscle (p. 209)**, the **types of skeletal muscle fibers (p. 216)**, **muscular hypertrophy and atrophy (p. 218)**, and the **effects of aging on skeletal muscle (p. 219)** are presented. The structure and function of **smooth muscle (p. 219)** and **cardiac muscle (p. 222)** are introduced, but cardiac muscle is discussed in greater detail in chapter 17.

Functions of the Muscular System

Movements within the body are accomplished by cilia or flagella on the surface of some cells, by the force of gravity, or by the contraction of muscles. Most of the body's movements result from muscle contractions. As described in chapter 4, there are three types of muscle tissue: skeletal, smooth, and cardiac. The following are the major functions of muscles:

1. *Body movement.* Contraction of skeletal muscles is responsible for the overall movements of the body, such as walking, running, or manipulating objects with the hands.
2. *Maintenance of posture.* Skeletal muscles constantly maintain tone, which keeps us sitting or standing erect.
3. *Respiration.* Skeletal muscles of the thorax are responsible for the movements necessary for respiration.
4. *Production of body heat.* When skeletal muscles contract, heat is given off as a by-product. This released heat is critical to the maintenance of body temperature.
5. *Communication.* Skeletal muscles are involved in all aspects of communication, such as speaking, writing, typing, gesturing, and facial expression.
6. *Constriction of organs and vessels.* The contraction of smooth muscle within the walls of internal organs and vessels causes constriction of those structures. This constriction can help propel and mix food and water in the digestive tract, propel secretions from organs, and regulate blood flow through blood vessels.
7. *Heart beat.* The contraction of cardiac muscle causes the heart to beat, propelling blood to all parts of the body. ▼

1. **List the functions of skeletal, smooth, and cardiac muscles and explain how each is accomplished.**

General Functional Characteristics of Muscle

Muscle tissue is highly specialized to contract, or shorten, forcefully. The process of metabolism extracts energy from nutrient molecules. Part of that energy is used for muscle contraction, and the remainder is used for other cell processes or is released as heat.

Properties of Muscle

Muscle has four major functional properties: contractility, excitability, extensibility, and elasticity.

1. **Contractility** is the ability of muscle to shorten forcefully. When muscle contracts, it causes movement of the structures to which it is attached, or it may increase pressure inside hollow organs or vessels. Muscle contraction is an active process in which muscle cells generate the forces causing muscle to shorten. Muscle

Table 8.1 Comparison of Muscle Types

Features	Skeletal Muscle	Smooth Muscle	Cardiac Muscle
Location	Attached to bones	Walls of hollow organs, blood vessels, eyes, glands, and skin	Heart
Appearance (see table 4.10)			
Cell shape	Very long and cylindrical (1 mm–4 cm); extends the length of muscle fasciculi, which in some cases is the length of the muscle	Spindle-shaped (15–200 μm in length, 5–8 μm in diameter)	Cylindrical and branched (100–500 μm in length, 12–20 μm in diameter)
Nucleus	Multiple, peripherally located	Single, centrally located	Single, centrally located
Special cell–cell attachments	None	Gap junctions join some visceral smooth muscle cells together	Intercalated disks join cells to one another
Striations	Yes	No	Yes
Control	Voluntary and involuntary (reflexes)	Involuntary	Involuntary
Capable of spontaneous contraction	No	Yes (some smooth muscle)	Yes
Function	Body movement	Food movement through the digestive tract, emptying of the urinary bladder, regulation of blood vessel diameter, change in pupil size, contraction of many gland ducts, movement of hair, and many other functions	Pumps blood; contractions provide the major force for propelling blood through blood vessels

relaxation is a passive process and muscle elongation results from forces outside of the muscle, such as gravity, contraction of an opposing muscle, or the pressure of fluid in a hollow organ or vessel.

2. **Excitability** is the capacity of muscle to respond to a stimulus. Normally, skeletal muscle contracts as a result of stimulation by nerves. Smooth muscle and cardiac muscle can contract without outside stimuli, but they also respond to stimulation by nerves and hormones.

3. **Extensibility** means that muscle can be stretched beyond its normal resting length and is still able to contract.

4. **Elasticity** is the ability of muscle to recoil to its original resting length after it has been stretched.

Types of Muscle Tissue

Table 8.1 and "Muscle Tissue" in chapter 4 provide a comparison of the major characteristics of skeletal, smooth, and cardiac muscle. Skeletal muscle with its associated connective tissue constitutes about 40% of the body's weight and is responsible for locomotion, facial expressions, posture, respiratory movements, and many other body movements. The nervous system voluntarily, or consciously, controls the functions of the skeletal muscles.

Smooth muscle is the most widely distributed type of muscle in the body, and it has the greatest variety of functions. It is in the walls of hollow organs and tubes, the interior of the eye, the walls of blood vessels, and other areas. Smooth muscle performs a variety of functions, including propelling urine through the urinary tract, mixing food in the stomach and intestine, dilating and constricting the pupils, and regulating the flow of blood through blood vessels.

Cardiac muscle is found only in the heart, and its contractions provide the major force for moving blood through the circulatory system. Unlike skeletal muscle, cardiac muscle and many smooth muscles are **autorhythmic**; that is, they contract spontaneously at somewhat regular intervals, and nervous or hormonal stimulation is not always required for them to contract. Furthermore, unlike skeletal muscle, smooth muscle and cardiac muscle are not consciously controlled by the nervous system. Rather, they are controlled involuntarily, or unconsciously, by the autonomic nervous system and the endocrine system (see chapters 13 and 15). ▼

2. *Define contractility, excitability, extensibility, and elasticity of muscle tissue.*
3. *Describe the structure, function, location, and control of the three major muscle tissue types.*

Skeletal Muscle Structure

Skeletal muscles are composed of **skeletal muscle cells** associated with smaller amounts of connective tissue, blood vessels, and nerves. Each skeletal muscle cell is a single, long, cylindrical cell containing several nuclei located around the periphery of the fiber near the plasma membrane. A single fiber can extend from one end of a muscle to the other. In most muscles, the fibers range from approximately 1 mm to about 4 cm in length and from 10 μm to

100 μm in diameter. Skeletal muscle cells are often called **skeletal muscle fibers** because of their shape. ▼

4. ***Describe skeletal muscle fibers.***

Connective Tissue Coverings of Muscle

Fascia (fash′ē-ă) is a general term for connective tissue sheets within the body. **Muscular fascia** (formerly *deep fascia*) separates and compartmentalizes individual muscles or groups of muscles (figure 8.1). It consists of dense irregular collagenous connective tissue.

The **epimysium** (ep-i-mis′ē-ŭm, *epi,* upon + *mys,* muscle) is a connective tissue sheath of dense collagenous connective tissue

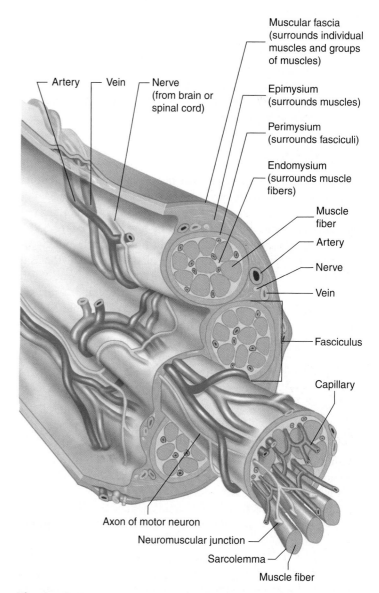

Figure 8.1 **Skeletal Muscle Structure: Connective Tissue, Innervation, and Blood Supply**

This figure shows the connective tissue surrounding a muscle, muscle fasciculi, and muscle fibers. Arteries, veins, and nerves course together through the connective tissue of muscles. They branch frequently as they approach individual muscle fibers. At the level of the perimysium, axons of neurons branch, and each branch extends to a muscle fiber.

surrounding individual muscles (see figure 8.1). A muscle is composed of **muscle fasciculi** (fă-sik′u-lī, *fascis,* bundle), which are bundles of muscle fibers surrounded by connective tissue called the **perimysium** (per′i-mis′ē-ŭm, *peri,* around + *mys,* muscle). The perimysium is an extension of the epimysium into the muscle and it connects with the **endomysium** (en′dō-mis′ē-ŭm, *endon,* within + *mys,* muscle), which is a layer of reticular fibers surrounding each muscle fiber.

Connective tissue associated with skeletal muscle is critical to its proper function. Muscle fibers are attached to the connective tissue. At the ends of muscles, the muscle connective tissue blends with other connective tissue, such as tendons (see "Dense Connective Tissue," chapter 4), that anchors the muscle to other structures (figure 8.2*a*). Thus, contraction of muscle fibers results in the movement of structures, such as bone and skin. Connective tissue also provides a pathway for nerves and blood vessels to reach muscle fibers (see figure 8.1). ▼

5. *Name the connective tissue that separates and compartmentalizes muscles.*
6. *Name the connective tissues that surrounds individual muscles, muscle fasciculi, and muscle fibers. What is a muscle fasciculus?*

Muscle Fibers

The plasma membrane of a muscle fiber is called the **sarcolemma** (sar′kō-lem′ă, *sarco,* flesh + *lemma,* husk). The many nuclei of each muscle fiber lie just inside the sarcolemma, whereas most of the interior of the fiber is filled with myofibrils (figure 8.2*b*). Other organelles, such as the numerous mitochondria and glycogen granules, are packed between the myofibrils. The cytoplasm without the myofibrils is called **sarcoplasm** (sar′kō-plazm). Each **myofibril** (mī-ō-fī′bril) is a threadlike structure approximately 1–3 μm in diameter that extends from one end of the muscle fiber to the other (figure 8.2*c*). Two kinds of protein filaments, called **actin** (ak′tin) and **myosin** (mī′ō-sin) **myofilaments** (mī-ō-fil′a-ments), are major components of myofibrils. The actin and myosin myofilaments form highly ordered units called **sarcomeres** (sar′kō-mērz), which are joined end to end to form the myofibrils (figure 8.2*d*).

Actin and Myosin Myofilaments

Actin myofilaments, or thin myofilaments, resemble two minute strands of pearls twisted together (figure 8.2*e*). Each strand of pearls is a **fibrous actin (F actin)** strand, and each pearl is a **globular actin (G actin).** Each G actin has an **active site,** to which myosin molecules can bind during muscle contraction. **Troponin** (trō′pō-nin, *trope,* a turning) molecules are attached at specific intervals along the actin myofilaments and have Ca^{2+} binding sites. Troponin is also attached to **tropomyosin** (trō-pō-mī′ō-sin) molecules located along the groove between the twisted strands of F actin. When Ca^{2+} is not bound to troponin, tropomyosin covers the active sites on G actin. When Ca^{2+} binds to troponin, tropomyosin moves, exposing the active sites.

Myosin myofilaments, or thick myofilaments, resemble bundles of minute golf clubs (figure 8.2*f*). Each golf club is a myosin molecule consisting of a head, a hinged region, and a rod. The head of the myosin molecule has a deep cleft where myosin can bind to the active site on G actin to form a **cross-bridge.** The head also has a **myosin ATPase,** which is an enzyme that promotes the breakdown of ATP. The energy from ATP enables movement of the head at the hinge region during contraction. The rods attach to each other and are arranged so that the heads of the myosin molecules are located at each end of the myosin myofilament.

Sarcomeres

Each sarcomere extends from one Z disk to an adjacent Z disk within a myofibril (see figure 8.2*d*; figure 8.3). A **Z disk** is a filamentous network of protein forming a disklike structure for the attachment of actin myofilaments. The arrangement of the actin myofilaments and myosin myofilaments gives the myofibril a banded, or striated, appearance. A light **I band** includes a Z disk and extends from each side of the Z disk to the ends of the myosin myofilaments. The I band on each side of the Z disk consists only of actin myofilaments. An **A band** extends the length of the myosin myofilaments within a sarcomere. The actin and myosin myofilaments overlap for some distance at both ends of the A band, producing a darker appearance. Each myosin myofilament is surrounded by six actin myofilaments where actin and myosin myofilaments overlap. In the center of each A band is a smaller band called the **H zone,** where the actin and myosin myofilaments do not overlap and only myosin myofilaments are present. A dark line, called the **M line,** is in the middle of the H zone and consists of delicate filaments that attach to the center of the myosin myofilaments. The M line helps hold the myosin myofilaments in place, similar to the way the Z disk holds actin myofilaments in place.

In addition to actin and myosin, there are other, less visible proteins within sarcomeres. These proteins help hold actin and myosin in place, and one of them accounts for muscle's ability to stretch (extensibility) and recoil (elasticity). **Titin** (tī′tin) (see figure 8.2*d*) is one of the largest known proteins, consisting of a single chain of nearly 27,000 amino acids. It attaches to a Z disk and extends along myosin myofilaments to the M line. The myosin myofilaments are attached to the titin molecules, which help hold them in position. Part of the titin molecule in the I band functions like a spring, allowing the sarcomere to stretch and recoil. ▼

7. *Describe the relationship among muscle fibers, myofibrils, myofilaments, and sarcomeres.*
8. *How do G actin, F actin, tropomyosin, and troponin combine to form an actin myofilament?*
9. *Name the parts of myosin molecules and how they combine to form a myosin myofilament.*
10. *What parts of actin and myosin myofilaments join to form a cross-bridge? What is the role of the myosin ATPase and hinge region during contraction?*
11. *How are Z disks, actin myofilaments, myosin myofilaments, and M lines arranged to form a sarcomere? Describe how this arrangement produces the I band, A band, and H zone.*

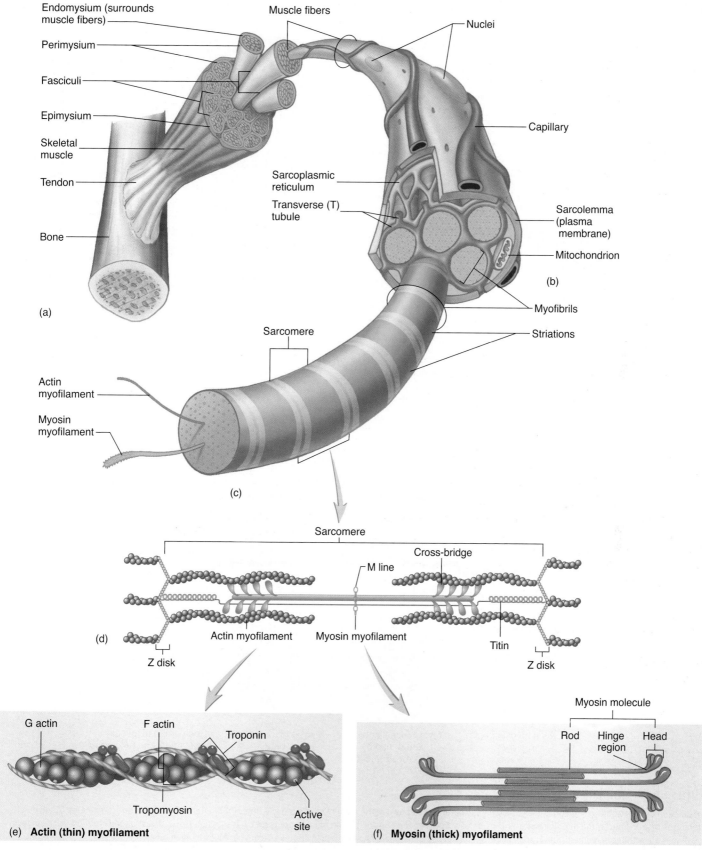

Figure 8.2 **Parts of a Muscle**

(a) Part of a muscle attached by a tendon to a bone. A muscle is composed of muscle fasciculi, each surrounded by perimysium. The fasciculi are composed of bundles of individual muscle fibers (muscle cells), each surrounded by endomysium. (b) Enlargement of one muscle fiber. The muscle fiber contains several myofibrils. (c) A myofibril extended out the end of the muscle fiber. The banding patterns of the sarcomeres are shown in the myofibril. (d) A single sarcomere of a myofibril is composed of actin myofilaments and myosin myofilaments. The Z disk anchors the actin myofilaments, and the myosin myofilaments are held in place by titin molecules and the M line. (e) Part of an actin myofilament is enlarged. (f) Part of a myosin myofilament is enlarged.

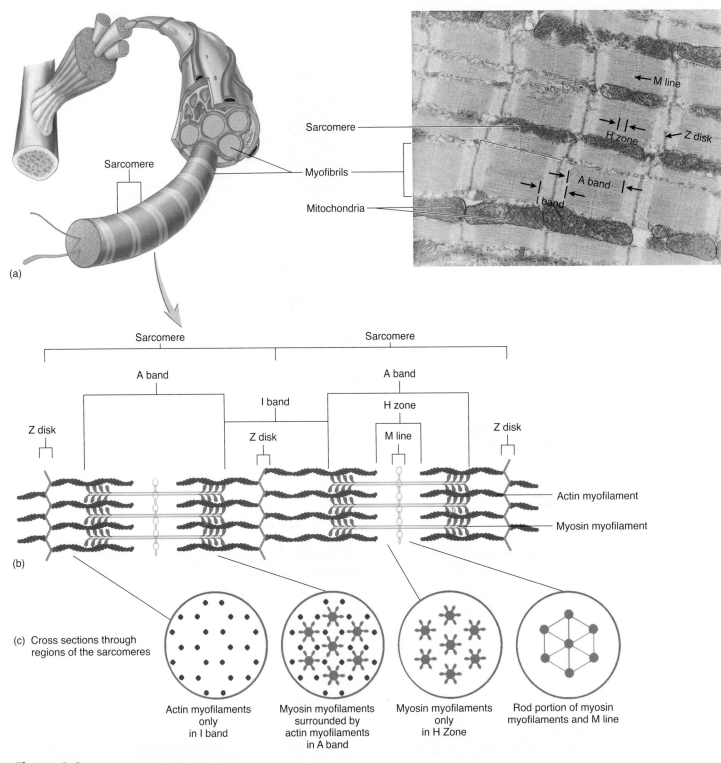

Figure 8.3 Organization of Sarcomeres

(a) Electron micrograph of a skeletal muscle. Several sarcomeres are shown in the myofibrils of a muscle fiber. (b) Diagram of two adjacent sarcomeres, depicting the structures responsible for the banding pattern. The I band is between the ends of myosin myofilaments on each side of a Z disk. The A band is formed by the myosin myofilaments within a sarcomere. The H zone is between the ends of the actin myofilaments within a sarcomere. Myosin myofilaments are attached to the M line. (c) Cross sections through regions of the sarcomeres show the arrangement of proteins in three dimensions.

Sliding Filament Model

The sarcomere is the basic structural and functional unit of skeletal muscle because it is the smallest portion of skeletal muscle capable of contracting. Sarcomeres shorten when the Z disks of sarcomeres are pulled closer together (figure 8.4). Within a sarcomere, actin myofilaments are attached to the Z disks. The Z disks are pulled together when the actin myofilaments of opposing Z disks move toward each other by sliding over myosin myofilaments. This is called the **sliding filament model** of muscle contraction. Actin and

1. Actin and myosin myofilaments in a relaxed muscle (*right*) and a contracted muscle (*#4 below*) are the same length. Myofilaments do not change length during muscle contraction.

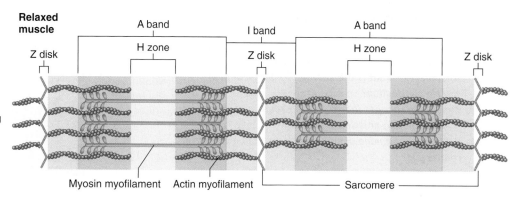

Relaxed muscle

Z disk | A band | I band | A band | Z disk

H zone | Z disk | H zone | Z disk

Myosin myofilament | Actin myofilament | Sarcomere

2. During contraction, actin myofilaments at each end of the sarcomere slide past the myosin myofilaments toward each other. As a result, the Z disks are brought closer together, and the sarcomere shortens.

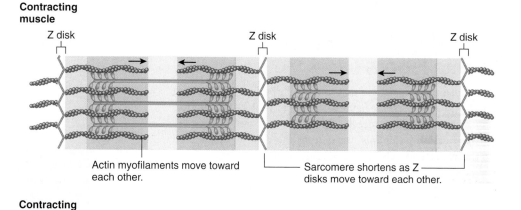

Contracting muscle

Z disk | Z disk | Z disk

Actin myofilaments move toward each other.

Sarcomere shortens as Z disks move toward each other.

3. As the actin myofilaments slide over the myosin myofilaments, the H zones (*yellow*) and the I bands (*blue*) narrow. The A bands, which are equal to the length of the myosin myofilaments, do not narrow, because the length of the myosin myofilaments does not change.

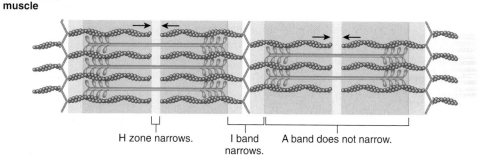

Contracting muscle

H zone narrows.

I band narrows.

A band does not narrow.

4. In a fully contracted muscle, the ends of the actin myofilaments overlap at the center of the sarcomere and the H zone disappears.

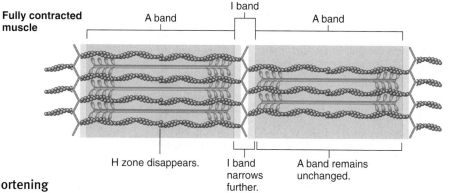

Fully contracted muscle

A band | I band | A band

H zone disappears.

I band narrows further.

A band remains unchanged.

Process Figure 8.4   Sarcomere Shortening

myosin myofilaments do not change length during contraction of skeletal muscle. Consequently, the A bands do not change in length, but the H zones and I bands shorten as actin myofilaments slide over myosin myofilaments.

When sarcomeres shorten, the myofibrils also shorten because the myofibrils consist of sarcomeres joined end to end.

The myofibrils extend the length of the muscle fibers and, when they shorten, the muscle fibers shorten. Muscle fibers are organized into muscle fasciculi, which are organized to form muscles (see figure 8.2). Therefore, when sarcomeres shorten, myofibrils, muscle fibers, muscle fasciculi, and muscles shorten to produce muscle contractions. During muscle relaxation, the sarcomeres lengthen as actin myofilaments slide back past myosin myofilaments and Z disks move farther apart. ▼

12. *How are Z disks involved in the shortening of sarcomeres? What is the sliding filament model of muscle contraction?*

13. *Explain the changes in width of the bands and zones of a myofibril during muscle contraction.*

14. *How does shortening of sarcomeres result in a muscle contracting?*

P R E D I C T ①
What happens to the width of the bands and zones in a myofibril when a muscle is stretched?

Physiology of Skeletal Muscle Fibers

Motor neurons are nerve cells that connect the brain and spinal cord to skeletal muscle fibers (see figure 8.1). The nervous system controls the contraction of skeletal muscles through these axons. Electric signals, called **action potentials,** travel from the brain or spinal cord along the axons to muscle fibers and cause them to contract.

Membrane Potentials

Plasma membranes are **polarized**—that is, there is an electric charge difference across each plasma membrane. Just as a battery has a negative and a positive end, or pole, the inside of the plasma membrane is negatively charged, compared with the outside. This charge difference across the plasma membrane of an unstimulated cell is called the **resting membrane potential.** Just as with a battery, the charge difference can be measured in units called volts. Any change in the charge difference results in a change in voltage across the plasma membrane. How the resting membrane potential is established is described in chapter 10.

Ion Channels

Once the resting membrane potential is established, action potentials can be produced. An action potential is a reversal of the resting membrane potential such that the inside of the plasma membrane becomes positively charged, compared with the outside. When a cell is stimulated, the permeability characteristics of the plasma membrane change as a result of the opening of certain ion channels. The diffusion of ions through these channels changes the charge across the plasma membrane and produces an action potential. Two types of gated ion channels play important roles in producing action potentials:

1. *Ligand-gated ion channels.* A **ligand** (lī′gand) is a molecule that binds to a receptor. A **receptor** is a protein or glycoprotein that has a receptor site to which a ligand can bind. **Ligand-gated ion channels** are channels with gates that open in response to a ligand binding to a receptor that is part of the ion channel. For example, the axons of nerve cells release ligands called **neurotransmitters** (noor′ō-trans-mit′erz). The motor neurons supplying skeletal muscle release the neurotransmitter **acetylcholine** (as-e-til-kō′1ēn) (ACH), which binds to ligand-gated Na^+ channels in the

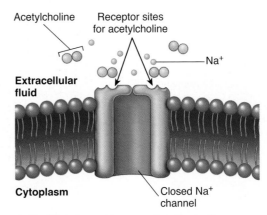

1. The Na^+ channel has receptor sites for the neurotransmitter acetylcholine. When the receptor sites are not occupied by acetylcholine, the Na^+ channel remains closed.

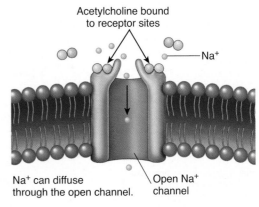

2. When two acetylcholine molecules bind to their receptor sites on the Na^+ channel, the channel opens to allow Na^+ to diffuse through the channel into the cell.

Process Figure 8.5   Ligand-Gated Ion Channel

membranes of the muscle fibers. As a result, the Na^+ channels open, allowing Na^+ to enter the cell (figure 8.5).

2. *Voltage-gated ion channels.* These channels are gated membrane channels that open and close in response to small voltage changes across the plasma membrane. When a nerve or muscle cell is stimulated, the charge difference across the plasma membrane changes, producing a voltage change that causes voltage-gated ion channels to open or close. Important voltage-gated ion channels are voltage-gated Na^+, K^+, and Ca^{2+} channels.

The concentration gradient for an ion determines whether that ion enters or leaves the cell after the ion channel, specific for that ion, opens. The Na^+–K^+ pump moves Na^+ to the outside of the cell and K^+ to the inside of the cell (see chapter 3). This results in a higher concentration of Na^+ in the extracellular fluid and a lower concentration of Na^+ in the intracellular fluid, and it results in a higher concentration of K^+ in the intracellular fluid and a lower concentration of K^+ in the extracellular fluid. Consequently, when voltage-gated Na^+ channels open, Na^+ move through them into the cell. In a similar fashion, when gated K^+ channels open, K^+ move out of the cell. ▼

15. *Define resting membrane potential and action potential.*

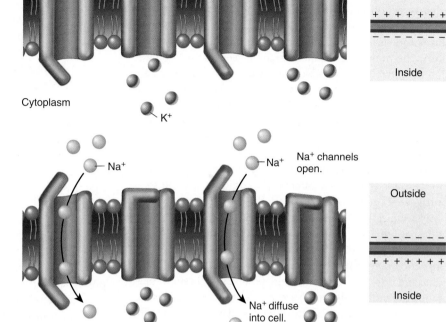

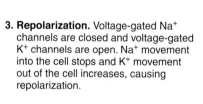

1. Resting membrane potential.
Voltage-gated Na⁺ channels (*pink*) and
voltage-gated K⁺ channels (*purple*) are
closed. The outside of the plasma
membrane is positively charged,
compared with the inside.

2. Depolarization. Voltage-gated Na⁺
channels are open. Depolarization
results because the inward
movement of Na⁺ makes the inside
of the membrane more positive.

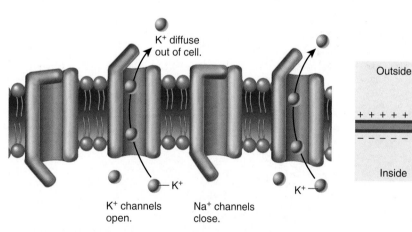

3. Repolarization. Voltage-gated Na⁺
channels are closed and voltage-gated
K⁺ channels are open. Na⁺ movement
into the cell stops and K⁺ movement
out of the cell increases, causing
repolarization.

Process Figure 8.6   Gated Ion Channels and the Action Potential

Step 1 illustrates the status of gated Na⁺ and K⁺ channels in a resting cell. Steps 2 and 3 show how the channels open and close to produce an action potential. Next to each step, the charge difference across the plasma membrane is illustrated.

16. *Name the two major categories of gated ion channels in the plasma membrane and give examples of each.*
17. *Describe the differences in Na⁺ and K⁺ concentrations across the plasma membrane. How does this affect the movement of Na⁺ and K⁺ into and out of cells?*

Action Potentials

Ligand-gated and voltage-gated ion channels allow cells to respond to stimuli and produce action potentials. When neurotransmitters, such as ACh, bind to ligand-gated Na⁺ channels, positively charged Na⁺ diffuse into cells (see figure 8.5), causing the inside of the plasma membrane to become more positive. This change in charge produces

a change in voltage called **depolarization.** Before a cell is stimulated, voltage-gated Na⁺ and K⁺ channels are closed (figure 8.6, step 1). If the change in voltage caused by the movement of Na⁺ into the cell is sufficient to reach a level called **threshold,** then voltage-gated Na⁺ channels open. When the voltage gated Na⁺ channels open, Na⁺ diffuse into the cell until the inside of the membrane becomes positive, compared with the outside of the membrane (figure 8.6, step 2). Thus, an action potential is a reversal of the charge across the plasma membrane. As the inside of the cell becomes positive, this voltage change causes voltage-gated Na⁺ channels to close and voltage-gated K⁺ channels to open (figure 8.6, step 3). Consequently, Na⁺ no longer move into the cell and K⁺ move out of the cell, causing the inside of the plasma membrane to become more negative and the outside more positive. The return of the membrane potential to its resting value is called **repolarization.** The action potential ends and the resting membrane potential is reestablished when the voltage-gated K⁺ channels close. The cell can now be stimulated again.

Action potentials occur according to the **all-or-none principle;** that is, action potentials either will not occur (the "none" part) or, if they do, are all the same (the "all" part). A **subthreshold stimulus** is too weak to initiate an action potential. A **threshold stimulus** is the minimum stimulus strength required to produce an action potential. Once an action potential begins, however, all of the ion channel changes responsible for producing the action potential proceed without stopping. A stronger-than-threshold stimulus results in the same action potential as a threshold stimulus because it, too, results in all of the ion channel changes necessary to produce an action potential. Consequently, all of the action potentials in a given cell are alike. An action potential can be compared to the flash system of a camera. If the shutter is depressed but not triggered (does not reach threshold), no flash results (the "none" part). Once the shutter is triggered (reaches threshold), the camera flashes (an action potential is produced), and each flash is the same brightness (the "all" part) as the previous flashes.

An action potential occurs in a very small area of the plasma membrane and does not affect the entire plasma membrane at one time. Action potentials can **propagate,** or travel, across the plasma membrane because an action potential produced at one location in the plasma membrane can stimulate the production of an action potential in an adjacent location. Thus, action potentials produced in the brain or spinal cord can propagate along nerve axons supplying skeletal muscles. ▼

18. **How can the opening of ligand-gated Na⁺ channels result in an action potential? What is threshold?**
19. **What happens to the charge across the plasma membrane during depolarization and repolarization? Describe the role of voltage-gated ion channels in these processes.**
20. **What is the all-or-none principle of action potentials? Define subthreshold and threshold stimuli.**
21. **What is propagation of an action potential?**

Neuromuscular Junction

Axons of motor neurons enter muscles and send out branches to several muscle fibers. Each axon branch forms a **neuromuscular junction,** or **synapse,** with a muscle fiber (see figure 8.1; figure 8.7). Within a synapse, an axon branch forms a cluster of enlarged axon endings called **presynaptic** (prē′si-nap′tik) **terminals.** Each presynaptic terminal rests in an invagination of the sarcolemma of the muscle fiber. The space between the presynaptic terminal and the muscle fiber is the **synaptic** (si-nap′tik) **cleft,** and the muscle plasma membrane in the area of the synapse is the **postsynaptic** (pōst-si-nap′tik) **membrane,** or **motor end-plate.**

Each presynaptic terminal contains numerous mitochondria and many small, spherical sacs called **synaptic vesicles,** which contain the neurotransmitter acetylcholine.

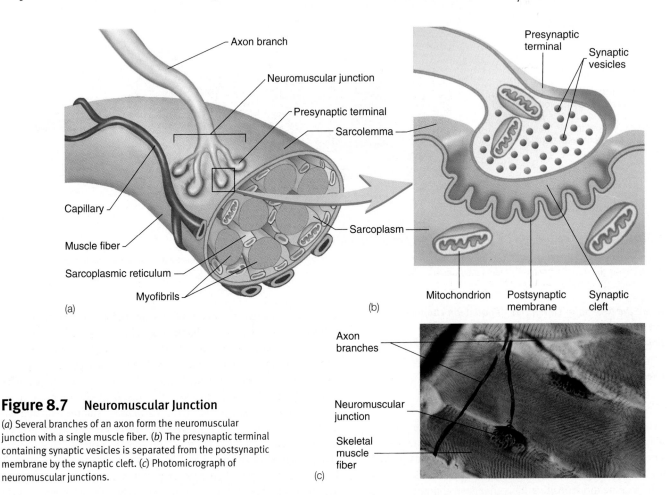

Figure 8.7   **Neuromuscular Junction**

(a) Several branches of an axon form the neuromuscular junction with a single muscle fiber. (b) The presynaptic terminal containing synaptic vesicles is separated from the postsynaptic membrane by the synaptic cleft. (c) Photomicrograph of neuromuscular junctions.

1. An action potential (*orange arrow*) arrives at the presynaptic terminal and causes voltage-gated Ca^{2+} channels in the presynaptic membrane to open.

2. Calcium ions enter the presynaptic terminal and initiate the release of the neurotransmitter acetylcholine (ACh) from synaptic vesicles.

3. ACh is released into the synaptic cleft by exocytosis.

4. ACh diffuses across the synaptic cleft and binds to ligand-gated Na^+ channels on the postsynaptic membrane.

5. Ligand-gated Na^+ channels open and Na^+ enter the postsynaptic cell, causing the postsynaptic membrane to depolarize. If depolarization passes threshold, an action potential is generated along the postsynaptic membrane.

6. ACh unbinds from the ligand-gated Na^+ channels, which then close.

7. The enzyme acetylcholinesterase, which is attached to the postsynaptic membrane, removes acetylcholine from the synaptic cleft by breaking it down into acetic acid and choline.

8. Choline is symported with Na^+ into the presynaptic terminal, where it can be recycled to make ACh. Acetic acid diffuses away from the synaptic cleft.

9. ACh is reformed within the presynaptic terminal using acetic acid generated from metabolism and from choline recycled from the synaptic cleft. ACh is then taken up by synaptic vesicles.

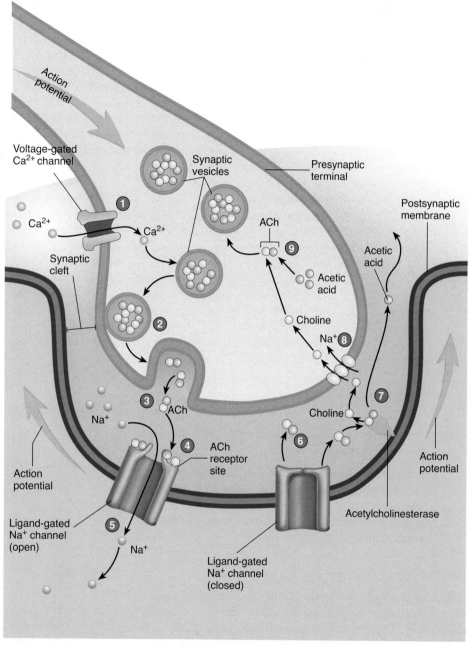

Process Figure 8.8   Function of the Neuromuscular Junction

When an action potential reaches the presynaptic terminal, it causes voltage-gated calcium ion (Ca^{2+}) channels in the plasma membrane of the axon to open (figure 8.8). Calcium ions diffuse into the presynaptic terminal because the concentration of Ca^{2+} is greater outside of the cell than inside. The Ca^{2+} cause a few synaptic vesicles to release acetylcholine by exocytosis from the presynaptic terminal into the synaptic cleft. The acetylcholine diffuses across the synaptic cleft and binds to ligand-gated Na^+ channels located within the postsynaptic membrane. The ligand-gated Na^+ channels open and Na^+ diffuse into the cell, causing depolarization and the production of an action potential, which leads to contraction of the muscle fiber (see "Excitation–Contraction Coupling," p. 206).

Acetylcholine released into the synaptic cleft is rapidly broken down to acetic acid and choline by the enzyme **acetylcholinesterase** (as′e-til-kō-lin-es′ter-ās) (see figure 8.8). Acetylcholinesterase keeps acetylcholine from accumulating within the synaptic cleft, where it would act as a constant stimulus at the postsynaptic terminal, producing many action potentials and continuous contraction of the muscle fiber. The release of acetylcholine and its rapid degradation in the synaptic cleft ensures that one presynaptic action potential yields only one postsynaptic action potential.

Choline molecules are absorbed by the presynaptic terminal by symport (see figures 3.9 and 8.8). Choline combines with the acetic acid produced within the cell to form acetylcholine. Recycling choline molecules requires less energy and is more rapid than

Neuromuscular-blocking drugs act at the neuromuscular junction to prevent or reduce action potential production in muscle fibers, thus preventing or reducing muscle contractions. **Presynaptic inhibition** prevents the synthesis or release of ACh from the presynaptic terminal. Botulinum toxin, produced by the bacterium *C. botulinum*, prevents ACh release. Improper canning of foods allows the bacteria to grow and produce the toxin. Ingestion of the toxin can lead to death because of paralysis of respiratory muscles. The most common cosmetic procedure in the United States is injection of botulinum toxin (BOTOX) to relax facial muscles causing wrinkles.

Postsynaptic inhibition prevents the production of action potentials in the postsynaptic membrane of muscle fibers. **Nondepolarizing blocking agents** prevent

ligand-gated Na$^+$ channel activity by binding to ACh receptors, which blocks ACh binding to the receptors, or by blocking the ion channel, which prevents Na$^+$ movement into the muscle fiber. The most famous drug of this type is curare, which is the name used for various dart/arrow tip poisons derived from plants in South America. Hunters use curare to paralyze wild animals used for food. Curare derivatives and other drugs are used to relax muscles, especially abdominal muscles, for surgery. During surgery anaesthetic drugs are used to render patients unconscious. Using non-depolarizing blocking agents in combination with anesthetic drugs can result in **anaesthesia awareness**, in which patients are conscious and can perceive pain during surgery but are unable to move and alert the medical staff of their condition.

Myasthenia gravis (mī-as-thē′nē-ă grăv′is) is an autoimmune disorder in which the immune system inappropriately attacks one's own cells as if they were foreign cells, such as bacteria. In myasthenia gravis, the immune response results in a decreased number of ligand-gated and voltage-gated Na$^+$ channels in neuromuscular junctions. Thus, there is a decreased ability of the nervous system to stimulate skeletal muscle fibers and there is a decreased ability of skeletal muscle fibers to generate action potentials. The result is muscular weakness. A class of drugs that includes neostigmine partially blocks the action of acetylcholinesterase and sometimes is used to treat myasthenia gravis. The drugs cause acetylcholine levels to increase in the synaptic cleft and combine more effectively with the remaining acetylcholine receptors.

---

completely synthesizing new acetylcholine molecules each time they are released from the presynaptic terminal. ▼

22. *Describe the neuromuscular junction. How does an action potential in the neuron produce an action potential in the muscle cell?*

23. *What is the importance of acetylcholinesterase in the synaptic cleft?*

Excitation–Contraction Coupling

Excitation–contraction coupling is the mechanism by which an action potential in the sarcolemma causes the contraction of a muscle fiber. The sarcolemma extends into the interior of the muscle fiber by tubelike invaginations called **transverse, or T, tubules.** The T tubules wrap around sarcomeres of myofibrils in the region where actin and myosin myofilaments overlap (figure 8.9). Suspended in the sarcoplasm between the T tubules is a highly specialized smooth endoplasmic reticulum called the **sarcoplasmic reticulum** (sar-kō-plaz′mik re-tik′ū-lŭm). Near the T tubules, the sarcoplasmic reticulum is enlarged to form **terminal cisternae** (sis-ter′nē). A T tubule and the two adjacent terminal cisternae together are called a **triad** (trī′ad). The sarcoplasmic reticulum actively transports Ca^{2+} into its lumen; thus, the concentration of Ca^{2+} is approximately 2000 times higher within the sarcoplasmic reticulum than in the sarcoplasm of a resting muscle.

Excitation–contraction coupling begins at the neuromuscular junction with the production of an action potential in the sarcolemma. The action potential is propagated along the sarcolemma and into the T tubules, where they cause gated Ca^{2+} channels in the terminal cisternae of the sarcoplasmic reticulum to open.

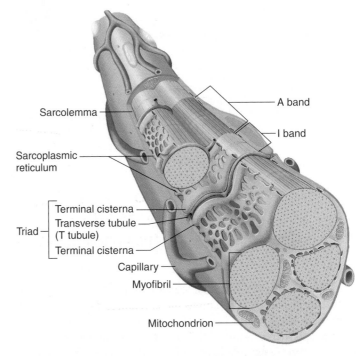

Figure 8.9   T Tubules and Sarcoplasmic Reticulum
A T tubule and the sarcoplasmic reticulum on each side of the T tubule form a triad.

When the Ca^{2+} channels open, Ca^{2+} rapidly diffuse into the sarcoplasm surrounding the myofibrils (figure 8.10).

Calcium ions bind to Ca^{2+} binding sites on the troponin molecules of the actin myofilaments. The combination of Ca^{2+}

1. An action potential produced at a presynaptic terminal in the neuromuscular junction is propagated along the sarcolemma of the skeletal muscle. The depolarization also spreads along the membrane of the T tubules.

2. The depolarization of the T tubule causes gated Ca^{2+} channels in the sarcoplasmic reticulum to open, resulting in an increase in the permeability of the sarcoplasmic reticulum to Ca^{2+}, especially in the terminal cisternae. Calcium ions then diffuse from the sarcoplasmic reticulum into the sarcoplasm.

3. Calcium ions released from the sarcoplasmic reticulum bind to troponin molecules. The troponin molecules bound to G actin molecules are released, causing tropomyosin to move, exposing the active sites on G actin.

4. Once active sites on G actin molecules are exposed, the heads of the myosin myofilaments bind to them to form cross-bridges.

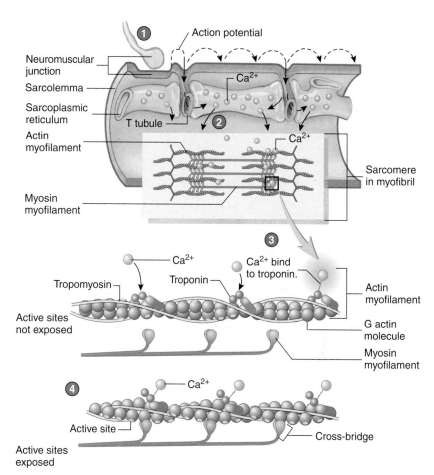

Process Figure 8.10 Action Potentials and Muscle Contraction

with troponin causes the troponin–tropomyosin complex to move deeper into the groove between the two F actin strands, which exposes active sites on the actin myofilaments. The heads of the myosin molecules then bind to the exposed active sites to form cross-bridges (see figure 8.10). Movement of the cross-bridges results in contraction. ▼

24. **What is a triad? Where are Ca^{2+} concentrated in a triad?**

25. **How does an action potential produced in the postsynaptic membrane of the neuromuscular junction eventually result in cross-bridge formation?**

Cross-Bridge Movement

The energy from one ATP molecule is required for cross-bridge movement. Before a myosin head binds to an active site on an actin myofilament, the head of the myosin molecule is in its cocked position. Attached to the head are ADP and phosphate, derived from the breakdown of ATP (figure 8.11, step 1). The energy from the breakdown of ATP is stored in the myosin head, which is cocked, just as the bar of a mousetrap is cocked when the trap is set. Just as a mouse nibbling on the cheese on a mousetrap sets off the trap, Ca^{2+} binding to troponin sets off a series of events resulting in the movement of myosin heads. When Ca^{2+} bind to troponin, tropomyosin moves, and the active sites on actin myofilaments are exposed. The head of a myosin molecule

binds to an active site, forming a cross-bridge and the phosphate is released from the head (figure 8.11, step 2). Energy stored in the myosin head is used to move the head at the hinge region of the myosin molecule, which is called the **power stroke.** Movement of the head causes the actin myofilament to slide past the myosin myofilament and ADP is released from the myosin head (figure 8.11, step 3). This movement of actin myofilaments results in muscle contraction (see "Sliding Filament Model," p. 200). ATP binding to the myosin head causes cross-bridge release and the myosin head separates from the actin (figure 8.11, step 4). ATP is broken down to ADP and phosphate, which remain attached to the myosin head (figure 8.11, step 5). Energy released from the breakdown of ATP is stored in the myosin head, which returns to its cocked position in the **recovery stroke** (figure 8.11, step 6). After the recovery stroke, the myosin head can form another cross-bridge at a different site on the actin myofilament, followed by movement of the head, release of the cross-bridge, and return of the head to its original position (see figure 8.11, steps 2–6). This process, called **cross-bridge cycling,** occurs many times during a single contraction.

Unlike the coordinated movements of a team of rowers, the myosin heads do not move in unison. Instead, their movements are more like those of people engaged in a game of "tug-of-war." While some people (myosin heads) are pulling on the rope (actin myofilament), other people let go of the rope and reach out to grab

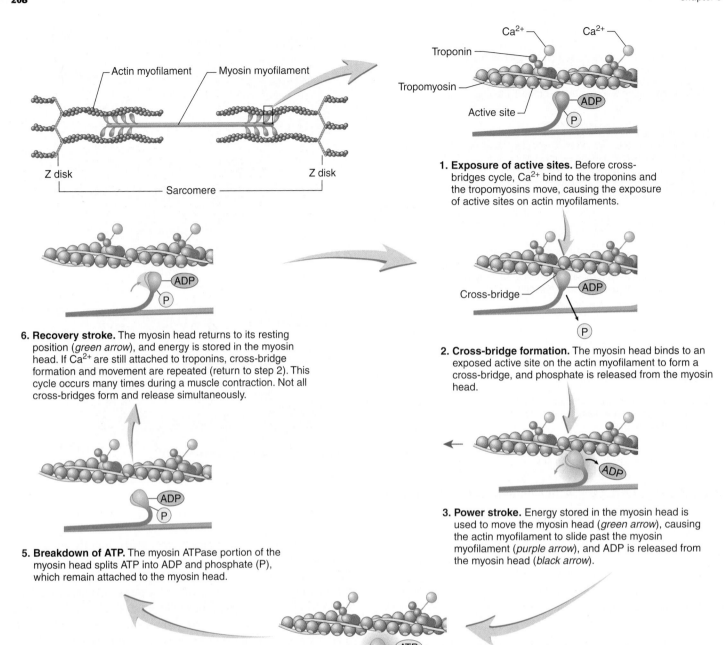

1. **Exposure of active sites.** Before cross-bridges cycle, Ca^{2+} bind to the troponins and the tropomyosins move, causing the exposure of active sites on actin myofilaments.

2. **Cross-bridge formation.** The myosin head binds to an exposed active site on the actin myofilament to form a cross-bridge, and phosphate is released from the myosin head.

3. **Power stroke.** Energy stored in the myosin head is used to move the myosin head (*green arrow*), causing the actin myofilament to slide past the myosin myofilament (*purple arrow*), and ADP is released from the myosin head (*black arrow*).

4. **Cross-bridge release.** ATP binds to the myosin head, causing it to detach from the actin.

5. **Breakdown of ATP.** The myosin ATPase portion of the myosin head splits ATP into ADP and phosphate (P), which remain attached to the myosin head.

6. **Recovery stroke.** The myosin head returns to its resting position (*green arrow*), and energy is stored in the myosin head. If Ca^{2+} are still attached to troponins, cross-bridge formation and movement are repeated (return to step 2). This cycle occurs many times during a muscle contraction. Not all cross-bridges form and release simultaneously.

Process Figure 8.11   Breakdown of ATP and Cross-Bridge Movement During Muscle Contraction

the rope. Thus, the actin myofilament moves while tension is maintained. ▼

> 26. **Describe the events occurring in one cross-bridge cycle. Define power stroke and recovery stroke.**

Muscle Relaxation

Excitation–contraction coupling results in the release of Ca^{2+} from the sarcoplasmic reticulum and the binding of Ca^{2+} to troponin. Relaxation occurs as a result of the active transport of Ca^{2+} back into the sarcoplasmic reticulum. As the Ca^{2+} concentration decreases in the sarcoplasm, Ca^{2+} diffuse away from the troponin molecules. The troponin–tropomyosin complex then reestablishes its position, which blocks the active sites on the actin molecules. Following cross-bridge release, new cross-bridges cannot form, and relaxation occurs.

ATP is required for contraction and relaxation. Energy from the breakdown of ATP is used to move myosin heads during contraction. Relaxation cannot occur without cross-bridge release, which requires ATP binding to myosin heads. Furthermore, relaxation requires the movement of Ca^{2+} into the sarcoplasmic reticulum, which is an active transport process requiring ATP. ▼

> 27. **What is the role of Ca^{2+} and ATP in muscle contraction and relaxation?**

P R E D I C T   2

Predict the consequences of having the following conditions develop in a muscle in response to a stimulus: (a) Na^+ cannot enter the skeletal muscle through voltage-gated Na^+ channels, (b) very little ATP is present in the muscle fiber before a stimulus is applied, and (c) adequate ATP is present within the muscle fiber, but action potentials occur at a frequency so great that Ca^{2+} is not transported back into the sarcoplasmic reticulum between individual action potentials.

Physiology of Skeletal Muscle

Muscle Twitch

A **muscle twitch** is the contraction of a muscle in response to a stimulus that causes an action potential in one or more muscle fibers. Even though the normal function of muscles is more complex, an understanding of the muscle twitch makes the function of muscles in living organisms easier to comprehend.

A hypothetical contraction of a single muscle fiber in response to a single action potential is illustrated in figure 8.12. The time between application of the stimulus to the motor neuron supplying the muscle fiber and the beginning of contraction is the **lag,** or **latent, phase;** the time during which contraction occurs is the **contraction phase;** and the time during which relaxation occurs is the **relaxation phase** (table 8.2). ▼

> 28. **Define the phases of a muscle twitch and describe the events responsible for each phase.**

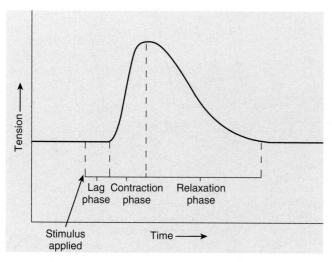

Figure 8.12   **Phases of a Muscle Twitch in a Single Muscle Fiber**

There is a short lag phase after stimulus application, followed by a contraction phase and a relaxation phase.

Table 8.2	Events That Occur during Each Phase of a Muscle Twitch*

Lag Phase

1. An action potential is propagated to the presynaptic terminal of the motor neuron, causing voltage-gated Ca^{2+} channels in the presynaptic terminal to open.

2. Calcium ions diffuse into the presynaptic terminal, causing several synaptic vesicles to release acetylcholine by exocytosis into the synaptic cleft.

3. Acetylcholine diffuses across the synaptic cleft and binds to acetylcholine receptor sites on ligand-gated Na^+ channels in the postsynaptic membrane of the sarcolemma, causing them to open.

4. Sodium ions diffuse into the muscle fiber, causing a local depolarization that exceeds threshold and produces an action potential in the sarcolemma.

5. Acetylcholine is rapidly degraded in the synaptic cleft to acetic acid and choline by acetylcholinesterase, thus limiting the length of time acetylcholine can stimulate the muscle fiber.

6. The action potential produced in the sarcolemma propagates into the T tubules, causing gated Ca^{2+} channels of the membrane of the sarcoplasmic reticulum to open.

7. Calcium ions diffuse from the sarcoplasmic reticulum into the sarcoplasm and bind to troponin; the troponin–tropomyosin complex changes its position and exposes active sites on the actin myofilaments.

Contraction Phase

8. Cross-bridges between actin molecules and myosin molecules form, move, release, and re-form many times, causing actin myofilaments to slide past myosin myofilaments; the sarcomeres shorten.

Relaxation Phase

9. Calcium ions are actively transported back into the sarcoplasmic reticulum.

10. The troponin–tropomyosin complexes move to inhibit cross-bridge formation.

11. The muscle fibers lengthen passively.

*Assuming that the process begins with a single action potential in the motor neuron.

Strength of Muscle Contraction

The strength of muscle contractions can vary from weak to strong. For example, the force generated by muscles to lift a feather is much less than the force required to lift a 25-pound weight. The force of contraction produced by a muscle is increased in two ways: (1) **multiple motor unit summation,** which involves increasing the number of muscle fibers contracting, and (2) **multiple-wave summation,** which involves increasing the force of contraction of the muscle fibers.

Multiple Motor Unit Summation

Within a muscle, the muscle fibers are functionally organized as motor units. A **motor unit** consists of a single motor neuron and all of the muscle fibers it innervates (figure 8.13). Muscle fibers, motor units, and muscles respond to stimulation by contracting. The response of muscles, however, is different from that of muscle fibers and motor units.

The all-or-none principle of action potentials (see "Action Potentials," p. 203) influences the response of a muscle fiber to

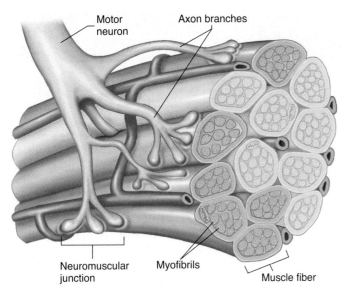

Figure 8.13   Motor Unit

A motor unit consists of a single motor neuron and all the muscle fibers its branches innervate.

stimulation. When brief electric stimuli of increasing strength are applied to the muscle fiber sarcolemma, one of the following events occur: (1) A subthreshold stimulus does not produce an action potential, and no muscle contraction occurs; (2) a threshold stimulus produces an action potential and results in contraction of the muscle fiber; or (3) a stronger-than-threshold stimulus produces an action potential of the same magnitude as the threshold stimulus and therefore produces an identical contraction. Thus, for a given condition, once an action potential is generated, the skeletal muscle fiber contracts to produce a constant force. If internal conditions change, the force of contraction can change as well. For example, muscle fatigue can result in a weaker force of contraction.

The response of motor units is also influenced by the all-or-none principle of action potentials. A subthreshold stimulus produces no action potential in the motor neuron and, therefore, no contraction of the muscle fibers of the motor unit. A threshold stimulus produces an action potential in the motor neuron, which then stimulates the production of action potentials in all of its muscle fibers, which contract. A stronger-than-threshold stimulus produces an action potential in the motor neuron and, therefore, the same contraction from all of the muscle fibers.

Whole muscles exhibit characteristics that are more complex than those of individual muscle fibers or motor units. Instead of responding in an all-or-none fashion, whole muscles respond to stimuli in a **graded fashion,** which means that the strength of the contractions can range from weak to strong.

A muscle is composed of many motor units, and the axons of the motor units combine to form a nerve. A whole muscle contracts with either a small force or a large force, depending on the number of motor units stimulated to contract. This relationship is called multiple motor unit summation because, as more and more (multiple) motor units are stimulated, the force of contraction increases (sums). Multiple motor unit summation resulting in graded responses can be demonstrated by applying brief electric stimuli of increasing strength to the nerve supplying a muscle (figure 8.14). A **subthreshold stimulus** is not strong enough to cause an action potential in any of the axons in a nerve and causes no contraction. As the stimulus strength increases, it eventually becomes a **threshold stimulus,** which is strong enough to produce an action potential in a single motor unit axon, causing all the muscle fibers of that motor unit to contract. Progressively stronger stimuli, called **submaximal stimuli,** produce action potentials in axons of additional motor

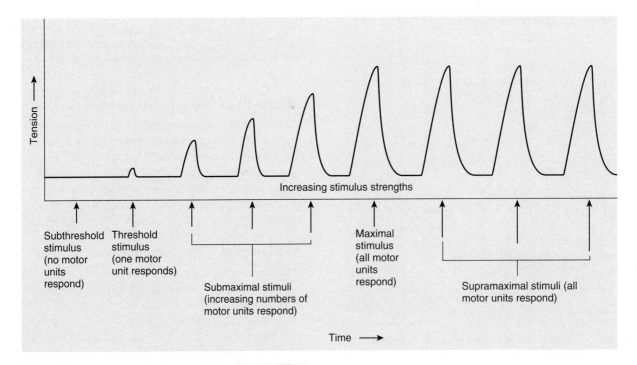

Figure 8.14   Multiple Motor Unit Summation in a Muscle

Multiple motor unit summation occurs as stimuli of increasing strength are applied to a nerve that innervates a muscle. The amount of tension (height of peaks) is influenced by the number of motor units responding.

units. A **maximal stimulus** produces action potentials in the axons of all motor units of that muscle. Consequently, even greater stimulus strengths, or **supramaximal stimuli,** have no additional effect. As the stimulus strength increases between threshold and maximum values, motor units are **recruited**—the number of motor units responding to the stimuli increases and the force of contraction produced by the muscle increases in a graded fashion.

Motor units in different muscles do not always contain the same number of muscle fibers. Muscles performing delicate and precise movements have motor units with a small number of muscle fibers, whereas muscles performing more powerful but less precise contractions have motor units with many muscle fibers. For example, in very delicate muscles, such as those that move the eye, the number of muscle fibers per motor unit can be less than 10, whereas, in the large muscles of the thigh, the number can be several hundred. ▼

> 29. *Why does a single muscle fiber either not contract or contract with the same force in response to stimuli of different strengths?*
> 30. *Why does a motor unit either not contract or contract with the same force in response to stimuli of different strengths?*
> 31. *How does increasing the strength of a stimulus cause a whole muscle to respond in a graded fashion?*
> 32. *Define multiple motor unit summation.*

P R E D I C T   3

In patients with poliomyelitis (pō′lē-ō-mī′ĕ-lī′tis), motor neurons are destroyed, causing loss of muscle function and even paralysis. Sometimes recovery occurs because of the formation of axon branches from the remaining motor neurons. These branches innervate the paralyzed muscle fibers to produce motor units with many more muscle fibers than usual, resulting in the recovery of muscle function. What effect would this reinnervation of muscle fibers have on the degree of muscle control in a person who has recovered from poliomyelitis?

Multiple-Wave Summation

Although an action potential triggers the contraction of a muscle fiber, the action potential is completed long before the contraction phase is completed. Action potentials occur in 1–2 milliseconds. A skeletal muscle twitch can take from 10 milliseconds up to 100 milliseconds (1 second).

Stimulus frequency is the number of times a motor neuron is stimulated per second. When stimulus frequency is low, there is time for the complete relaxation of muscle fibers between muscle twitches (figure 8.15, stimulus frequency 1). As stimulus frequency increases (figure 8.15, stimulus frequency 2), there is not enough time for muscle fibers to relax completely. Relaxation of a muscle fiber, however, is not required before a second action potential can stimulate a second contraction. One contraction can summate, or be added onto, a previous contraction. As a result, the overall force of contraction increases. In **incomplete tetanus** (tet′ă-nŭs), muscle fibers partially relax between the contractions (figure 8.15, stimuli frequencies 2–4) but, in **complete tetanus,** action potentials are produced so rapidly in muscle fibers that no relaxation occurs between them (figure 8.15, stimulus frequency 5). The increased tension resulting from increased frequency of stimulation is called **multiple-wave summation.**

Tetanus of a muscle caused by stimuli of increasing frequency can be explained by the effect of the action potentials on Ca^{2+}

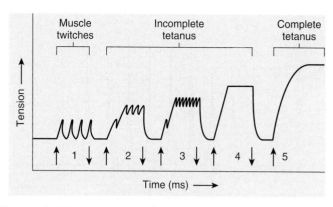

Figure 8.15   Multiple-Wave Summation

Stimuli 1–5 are stimuli of increasing frequency. For each stimulus, the up arrow indicates the start of stimulation, and the down arrow indicates the end of stimulation. Stimulus frequency 1 produces successive muscle twitches with complete relaxation between stimuli. Stimuli frequencies 2–4 do not allow complete relaxation between stimuli, resulting in incomplete tetanus. Stimulus frequency 5 allows no relaxation between stimuli, resulting in complete tetanus.

release from the sarcoplasmic reticulum. The first action potential causes Ca^{2+} release from the sarcoplasmic reticulum, the Ca^{2+} diffuse to the myofibrils, and contraction occurs. Relaxation begins as the Ca^{2+} are pumped back into the sarcoplasmic reticulum. If the next action potential occurs before relaxation is complete, two things happen. First, because not enough time has passed for all the Ca^{2+} to reenter the sarcoplasmic reticulum, Ca^{2+} levels around the myofibrils remain elevated. Second, the next action potential causes the release of additional Ca^{2+} from the sarcoplasmic reticulum. Thus, the elevated Ca^{2+} levels in the sarcoplasm produce continued contraction of the muscle fiber.

Another example of a graded response is **treppe** (trep′eh, staircase), which occurs in muscle that has rested for a prolonged period (figure 8.16). If the muscle is stimulated with a maximal stimulus at a low frequency, which allows complete relaxation between the stimuli, the contraction triggered by the second stimulus produces a slightly greater tension than the first. The contraction triggered by the third stimulus produces a contraction with a greater tension than the second. After only a few stimuli, the levels of tension produced by all the contractions are equal.

A possible explanation for treppe is an increase in Ca^{2+} levels around the myofibrils. The Ca^{2+} released in response to the first stimulus is not taken up completely by the sarcoplasmic reticulum before the second stimulus causes the release of additional Ca^{2+}, even though the muscle completely relaxes between the muscle twitches. As a consequence, during the first few contractions of the muscle, the Ca^{2+} concentration in the sarcoplasm increases slightly, making contraction more efficient because of the increased number of Ca^{2+} available to bind to troponin. Treppe achieved during warm-up exercises can contribute to improved muscle efficiency during athletic events. Factors such as increased blood flow to the muscle and increased muscle temperature are involved as well. Increased muscle temperature causes the enzymes responsible for muscle contraction to function at a more rapid rate. ▼

> 33. *Define multiple-wave summation, incomplete tetanus, and complete tetanus.*

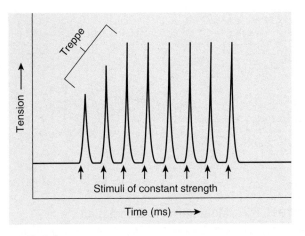

Figure 8.16 Treppe

When a rested muscle is stimulated repeatedly with maximal stimuli at a frequency that allows complete relaxation between stimuli, the second contraction produces a slightly greater tension than the first, and the third contraction produces greater tension than the second. After a few contractions, the levels of tension produced by all contractions are equal.

34. *Describe treppe.*
35. *In terms of Ca^{2+}, explain multiple-wave summation and treppe.*

CASE STUDY | Organophosphate Poisoning

John has a number of prize apple trees in his backyard. To prevent them from becoming infested with insects, he sprayed them with an organophosphate insecticide. He was in a rush to spray the trees before leaving town on vacation, and he failed to pay attention to the safety precautions on the packaging. He sprayed the trees without using any skin or respiratory protection. Soon he experienced severe stomach cramps, double vision, difficulty breathing, and spastic contractions of his skeletal muscles. His wife took John to the emergency room, where he was diagnosed with organophosphate poisoning. While in the emergency room, his physician administered a drug, and soon many of John's symptoms subsided.

Organophosphate insecticides exert their effects by binding to the enzyme acetylcholinesterase within synaptic clefts, rendering it ineffective. Thus, the organophosphate poison and acetylcholine "compete" for the acetylcholinesterase. As the organophosphate poison increases in concentration, less acetylcholinesterase is available to degrade acetylcholine. Organophosphate poisons affect synapses in which acetylcholine is the neurotransmitter, including skeletal muscle synapses and some smooth muscle synapses, such as in the walls of the stomach, intestines, and air passageways.

PREDICT 4

Organophosphate insecticides exert their effects by binding to the enzyme acetylcholinesterase within synaptic clefts, rendering it ineffective. Use figures 8.8 and 8.15 to help answer the following questions.

a. Explain the spastic contractions that occurred in John's skeletal muscles.
b. Propose as many mechanisms as you can by which a drug could counteract the effects of organophosphate poisoning.

Types of Muscle Contractions

In **isometric** (ī-sō-met′rik, equal distance) **contractions,** the length of the muscle does not change, but the amount of tension increases during the contraction process. Isometric contractions are responsible for the constant length of the postural muscles of the body, such as the muscles of the back. In **isotonic** (ī-sō-ton′ik, equal tension) **contractions,** the amount of tension produced by the muscle is constant during contraction, but the length of the muscle changes. Movements of the upper limbs or fingers are predominantly isotonic contractions. Most muscle contractions are a combination of isometric and isotonic contractions in which the muscles shorten some distance and the degree of tension increases.

Concentric (kon-sen′trik) **contractions** are isotonic contractions in which muscle tension increases and the muscle shortens. For example, flexing the elbow while holding a weight in the hand results from concentric contractions of arm muscles. **Eccentric** (ek-sen′trik) **contractions** are isotonic contractions in which tension is maintained as the muscle lengthens. For example, slowly lowering a weight held in the hand (extending the elbow) results from eccentric contractions of arm muscles. Thus, concentric contractions shorten muscles, whereas eccentric contractions resist lengthening of muscles.

Muscle Soreness Resulting from Exercise

Pain frequently develops after 1 or 2 days in muscles that are vigorously exercised, and the pain can last for several days. The pain is more common in untrained people who exercise vigorously. Eccentric contractions are of clinical interest because repetitive eccentric contractions, such as seen in the lower limbs of people who run downhill for long distances, cause more pain than concentric contractions. The pain produced appears to result from inflammation caused by damage to muscle fibers and connective tissue. Exercise schedules that alternate exercise with periods of rest, such as lifting weights every other day, provide time for the repair of muscle tissue.

Muscle tone is the constant tension produced by muscles of the body over long periods of time. Muscle tone is responsible for keeping the back and legs straight, the head in an upright position, and the abdomen from bulging. Muscle tone depends on a small percentage of the motor units in a muscle being stimulated at any point in time and out of phase with one another. The stimulation of the motor units results in incomplete or complete tetanus, and not single muscle twitches.

P R E D I C T   5

What types of muscle movements occur when a weight lifter lifts a weight above the head and then holds it there before lowering it?

Movements of the body are usually smooth and occur at widely differing rates—some very slowly and others quite rapidly. Most movements are produced by muscle contractions, but very few of the movements resemble the rapid contractions of individual muscle twitches. Smooth, slow contractions result from an increasing number of motor units contracting out of phase as the muscles shorten, as well as from a decreasing number of motor units contracting out of phase as muscles lengthen. Each motor unit exhibits either incomplete or complete tetanus, but, because the contractions are out of phase and because the number of motor units activated varies at each point in time, a smooth contraction results. Consequently, muscles are capable of contracting either slowly or rapidly, depending on the number of motor units stimulated and the rate at which that number increases or decreases. ▼

36. **Define isometric, isotonic, concentric, and eccentric contractions.**
37. **What is muscle tone, and how is it maintained?**
38. **How are smooth muscular contractions produced?**

Length Versus Tension

The number of cross-bridges formed during contraction determines the force of contraction. Just as many people pulling on a rope generates more force than a few people pulling, the force of contraction of a muscle increases with the number of cross-bridges formed. The number of cross-bridges that can form depends on the length of the muscle when it is stimulated to contract. Muscle length determines sarcomere length, which determines the amount of overlap between actin and myosin myofilaments. If a muscle is stretched to its optimum length, optimal overlap of the actin and myosin myofilaments takes place. When the muscle is stimulated, maximal cross-bridge formation results in maximal contraction (figure 8.17). If a muscle is stretched so that the actin and myosin myofilaments within the sarcomeres do not overlap or overlap to a very small extent, the muscle produces very little active tension when it is stimulated because few cross-bridges form. Also, if the muscle is not stretched at all, the myosin myofilaments touch each of the Z disks in each sarcomere, and very little contraction of the sarcomeres can occur.

Weight Lifters and Muscle Length

Weight lifters and others who lift heavy objects usually assume positions so that their muscles are stretched close to their optimum length before lifting. For example, the position a weight lifter assumes before power lifting stretches the upper limb and lower limb muscles to a near-optimum length for muscle contraction, and the stance a lineman assumes in a football game stretches most muscle groups in the lower limbs so they are near their optimum length for suddenly moving the body forward. ▼

39. **What effect does muscle length have on the force of contraction of a muscle? Explain how this occurs.**

Fatigue

Fatigue (fă-tēg′) is the decreased capacity to do work and the reduced efficiency of performance that normally follows a period of activity. **Psychologic fatigue** involves the central nervous system and is the most common type of fatigue. The muscles are capable of functioning, but the individual "perceives" that additional muscular work is not possible. A determined burst of activity in a tired athlete in response to pressure from a competitor is an example of how psychologic fatigue can be overcome.

Muscular fatigue results from ATP depletion. Without adequate ATP levels in muscle fibers, cross-bridges and ion transport do not function normally. As a consequence, the tension that a muscle is capable of producing declines. Fatigue in the lower limbs of marathon runners and in the upper and lower limbs of swimmers are examples.

As a result of extreme muscular fatigue, muscles occasionally become incapable of either contracting or relaxing—a condition called **physiologic contracture** (kon-trak′choor). When ATP levels are very low, active transport of Ca^{2+} into the sarcoplasmic reticulum slows and Ca^{2+} accumulate within the sarcoplasm. The Ca^{2+} bind to troponin, which promotes the formation of cross-bridges and contraction. With inadequate ATP, cross-bridge release does not occur, preventing relaxation.

Rigor Mortis

Rigor mortis (rig′er mōr′tĭs), the development of rigid muscles several hours after death, is similar to physiologic contracture. ATP production stops shortly after death, and ATP levels within muscle fibers decline. Because of low ATP levels, active transport of Ca^{2+} into the sarcoplasmic reticulum stops, and Ca^{2+} leak from the sarcoplasmic reticulum into the sarcoplasm. Calcium ions can also leak from the sarcoplasmic reticulum as a result of the breakdown of the sarcoplasmic reticulum membrane after cell death. As Ca^{2+} levels increase in the sarcoplasm, cross-bridges form. Without ATP, cross-bridges do not release and the muscles remain stiff until tissue degeneration occurs. ▼

40. **Describe two types of fatigue.**
41. **Define and explain the cause of physiologic contracture.**

Energy Sources

ATP is the immediate source of energy for muscle contractions. As long as adequate amounts of ATP are present, muscles can contract repeatedly for a long time. ATP must be synthesized continuously to sustain muscle contractions. The energy required to produce ATP comes from three sources: (1) creatine phosphate, (2) anaerobic respiration, and (3) aerobic respiration. Only the main points of anaerobic respiration and aerobic respiration are considered here (a more detailed discussion can be found in chapter 22).

Creatine Phosphate

During resting conditions, energy from aerobic respiration is used to synthesize **creatine** (krē′ă-tēn, krē′ă-tin) **phosphate.** Creatine

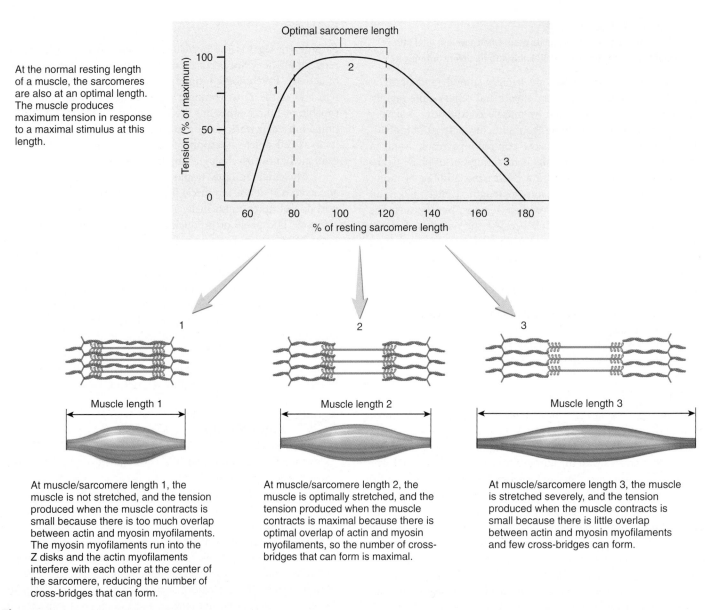

At the normal resting length of a muscle, the sarcomeres are also at an optimal length. The muscle produces maximum tension in response to a maximal stimulus at this length.

At muscle/sarcomere length 1, the muscle is not stretched, and the tension produced when the muscle contracts is small because there is too much overlap between actin and myosin myofilaments. The myosin myofilaments run into the Z disks and the actin myofilaments interfere with each other at the center of the sarcomere, reducing the number of cross-bridges that can form.

At muscle/sarcomere length 2, the muscle is optimally stretched, and the tension produced when the muscle contracts is maximal because there is optimal overlap of actin and myosin myofilaments, so the number of cross-bridges that can form is maximal.

At muscle/sarcomere length 3, the muscle is stretched severely, and the tension produced when the muscle contracts is small because there is little overlap between actin and myosin myofilaments and few cross-bridges can form.

Figure 8.17 **Muscle Length and Tension**

The length of a muscle before it is stimulated influences the force of contraction of the muscle. As the muscle changes length, the sarcomeres also change length.

phosphate accumulates in muscle cells and stores energy, which can be used to synthesize ATP. As ATP levels begin to fall, ADP reacts with creatine phosphate to produce ATP and creatine (figure 8.18a).

$$\text{ADP} + \text{Creatine phosphate} \rightarrow \text{Creatine} + \text{ATP}$$

This reaction is able to maintain ATP levels as long as creatine phosphate is available in the cell. During intense muscular contraction, however, creatine phosphate levels are quickly exhausted. ATP and creatine phosphate present in the cell provide enough energy to sustain maximum contractions for about 8–10 seconds.

Anaerobic Respiration

Anaerobic (an-ār-ō′bik) **respiration** does not require oxygen and results in the breakdown of glucose to yield ATP and lactic acid (figure 8.18b). For each molecule of glucose metabolized, a net production of two ATP molecules and two molecules of lactic acid

occurs. The first part of anaerobic respiration and aerobic respiration share an enzymatic pathway called **glycolysis** (glī-kol′-ī-sis). In glycolysis, a glucose molecule is broken down into two molecules of pyruvic acid. Two molecules of ATP are used in this process, but four molecules of ATP are produced, resulting in a net gain of two ATP molecules for each glucose molecule metabolized. In anaerobic respiration, the pyruvic acid is then converted to lactic acid. Unlike pyruvic acid, much of the lactic acid diffuses out of the muscle fibers into the bloodstream.

Anaerobic respiration is less efficient than aerobic respiration, but it is much faster, especially when oxygen availability limits aerobic respiration. By using many glucose molecules, anaerobic respiration can rapidly produce ATP for a short time. During short periods of intense exercise, such as sprinting, anaerobic respiration combined with the breakdown of creatine phosphate provides enough ATP to support intense muscle contraction for up to 3 minutes.

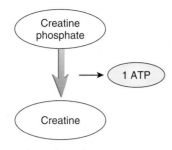

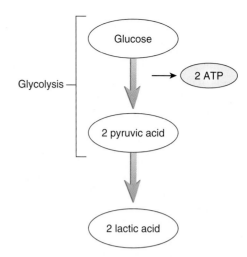

(a) **Creatine phosphate**
Energy source: creatine phosphate
Oxygen required: no
ATP yield: 1 per creatine phosphate
Duration: up to 10 seconds

(b) **Anaerobic respiration**
Energy source: glucose
Oxygen required: no
ATP yield: 2 per glucose molecule
Duration: up to 3 minutes

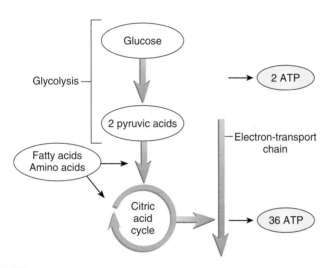

(c) **Aerobic respiration**
Energy source: glucose, fatty acids, amino acids
Oxygen required: yes
ATP yield: up to 38 per glucose molecule
Duration: hours

Figure 8.18   Energy Sources During Exercise

Glycogen stored in muscle fibers can be broken down to glucose (see chapter 2), which can then be used to produce more ATP. Anaerobic metabolism is ultimately limited by the depletion of glucose and a buildup of lactic acid within the muscle fiber.

Aerobic Respiration

Aerobic (ār-ō′bik) **respiration** requires oxygen and breaks down glucose to produce ATP, carbon dioxide, and water. Compared with anaerobic respiration, aerobic respiration is much more efficient. The metabolism of a glucose molecule by anaerobic respiration produces a net gain of 2 ATP molecules for each glucose molecule. In contrast, aerobic respiration can produce up to 38 ATP molecules for each glucose molecule (figure 8.18c). In addition, aerobic respiration uses a greater variety of molecules as energy sources, such as fatty acids and amino acids. Some glucose is used as an energy source in skeletal muscles, but fatty acids provide a more important source of energy during sustained exercise and during resting conditions.

In aerobic respiration, pyruvic acid is metabolized by chemical reactions within mitochondria. Two closely coupled sequences of reactions in mitochondria, called the **citric acid cycle** and the **electron-transport chain,** produce many ATP molecules (see chapter 22). Carbon dioxide molecules are produced and, in the

last step, oxygen atoms combine with hydrogen atoms to form water. Thus, carbon dioxide, water, and ATP are major end products of aerobic respiration. The following equation represents the aerobic respiration of one molecule of glucose:

$$\text{Glucose} + 6\,O_2 + 38\,\text{ADP} + 38\,P \rightarrow$$
$$6\,CO_2 + 6\,H_2O + \text{About 38 ATP}$$

Although aerobic respiration produces many more ATP molecules for each glucose molecule metabolized than does anaerobic respiration, the rate at which the ATP molecules are produced is slower. Resting muscles or muscles undergoing long-term exercise, such as long-distance running or other endurance exercises, depend primarily on aerobic respiration for ATP synthesis. At activity levels of 70% or greater of the maximum, ATP for muscle contraction is provided by aerobic and anaerobic respiration. ▼

42. *List the energy sources used to synthesize ATP for muscle contraction.*
43. *Contrast the efficiency of aerobic and anaerobic respiration. When is each type used by cells?*
44. *What is the function of creatine phosphate, and when is it used?*
45. *When do lactic acid levels increase in a muscle cell?*

Oxygen Deficit and Recovery Oxygen Consumption

There is a lag time between when exercise begins and when an individual begins to breathe more heavily because of the exercise. After exercise stops, there is a lag time before breathing returns to preexercise levels. These changes in breathing patterns reflect changes in oxygen consumption by muscle cells as they increase their demand for ATP supplied through aerobic respiration. The lack of increased oxygen consumption relative to increased activity at the onset of exercise creates an **oxygen deficit,** or **debt.** This deficit must be repaid during and after exercise once oxygen consumption catches up with the increased activity level. At the onset of exercise, the ATP required by muscles for the increased level of activity is primarily supplied by the creatine phosphate system and anaerobic respiration: two systems that can supply ATP relatively quickly and without a requirement for oxygen (see figure 8.18*a* and *b*). The ability of aerobic respiration to supply ATP at the onset of exercise lags behind that of the creatine phosphate system and anaerobic respiration. Consequently, there is a lag between the onset of exercise and the need for increased oxygen consumption.

The elevated oxygen consumption relative to activity level after exercise has ended is called **recovery oxygen consumption.** The oxygen is used in aerobic respiration to produce ATP. Some of the ATP is used to "repay" the oxygen deficit that was incurred at the onset of exercise, but most of the ATP is used to support processes that restore homeostasis resulting from the disturbances that occurred during exercise. Such disturbances include exercise-related increases in body temperature, changes in intra- and extracellular ion concentrations, and changes in metabolite and hormone levels. The duration of the recovery oxygen consumption generally lasts only minutes to hours, and it is dependent on the individual's physical conditioning and on the length and intensity of the exercise session. In extreme cases, such as following a marathon, recovery oxygen consumption can last as long as 15 hours. ▼

> 46. **Define oxygen deficit and recovery oxygen consumption. List the factors that contribute to them.**

P R E D I C T

Eric is a highly trained cross-country runner and his brother, John, is a computer programmer who almost never exercises. While the two brothers were working on a remodeling project in the basement of their house, the doorbell upstairs rang: A package they were both very excited about was being delivered. They raced each other up the stairs to the front door to see which one would get the package first. When they reached the front door, both were breathing heavily. However, John continued to breathe heavily for several minutes while Eric was opening the package. Explain why John continued to breathe heavily longer than Eric, even though they both had run the same distance.

Heat Production

The rate of metabolism in skeletal muscle differs before, during, and after exercise. As chemical reactions occur within cells, some energy is released in the form of heat. Normal body temperature

results in large part from this heat. Because the rate of chemical reactions increases in muscle fibers during contraction, the rate of heat production also increases, causing an increase in body temperature. After exercise, elevated metabolism resulting from recovery oxygen consumption helps keep the body temperature elevated. If the body temperature increases as a result of increased contraction of skeletal muscle, vasodilation of blood vessels in the skin and sweating speed heat loss and keep body temperature within its normal range (see chapter 22).

When body temperature declines below a certain level, the nervous system responds by inducing **shivering,** which involves rapid skeletal muscle contractions that produce shaking rather than coordinated movements. The muscle movement increases heat production up to 18 times that of resting levels, and the heat produced during shivering can exceed that produced during moderate exercise. The elevated heat production during shivering helps raise body temperature to its normal range. ▼

> 47. **How do muscles contribute to the heat responsible for body temperature before, during, and after exercise? What is accomplished by shivering?**

Types of Skeletal Muscle Fibers

There are three major types of skeletal muscle fibers: **slow-twitch oxidative (SO) fibers, fast-twitch glycolytic (FG) fibers, and fast-twitch oxidative glycolytic (FOG) fibers.** Many other names have been used for these fibers (table 8.3). Skeletal muscle fiber types can be described by their speed of contraction, fatigue resistance, and functions.

Speed of Contraction

The speed at which ATP is broken down on myosin heads affects the speed of cross-bridge cycling, which affects the speed of contraction. The speed at which ATP is broken down is determined by different types of ATPases on the myosin heads. Fast-twitch fibers have ATPases that break down ATP relatively rapidly, resulting in fast contractions, whereas slow-twitch fibers have an ATPase that breaks down ATP relatively slowly, resulting in slow contractions. In general, fast-twitch fibers contract approximately twice as fast as slow-twitch fibers. Within the fast-twitch group, FG fibers contract faster than FOG fibers.

P R E D I C T

Slow-twitch fibers relax more slowly than do fast-twitch fibers. Propose an explanation for the different relaxation speeds.

Fatigue Resistance

Fatigue-resistant fibers are able to contract for prolonged periods, whereas **fatigable fibers** cannot. Slow-twitch oxidative fibers are the most fatigue-resistant because they can produce many ATP to support muscle contraction. The term *oxidative* means that oxygen is required, and SO fibers rely upon aerobic (oxygen requiring) respiration to generate ATP. In support of aerobic respiration, SO fibers have many mitochondria and many capillaries, which take blood with oxygen to the muscle fibers.

Table 8.3	Characteristics of Skeletal Muscle Fiber Types		
	Slow-Twitch Oxidative (SO) Fibers	Fast-Twitch Oxidative Glycolytic (FOG) Fibers	Fast-Twitch Glycolytic (FG) Fibers
Myosin ATPase Activity	Slow	Fast	Fast
Metabolism	High aerobic capacity Low anaerobic capacity	Intermediate aerobic capacity High anaerobic capacity	Low aerobic capacity Highest anaerobic capacity
Fatigue Resistance	High	Intermediate	Low
Mitochondria	Many	Many	Few
Capillaries	Many	Many	Few
Myoglobin Content	High	High	Low
Glycogen Concentration	Low	High	High
Location Where Fibers Are Most Abundant	Generally in postural muscles and more in lower limbs than upper limbs	Generally predominate in lower limbs	Generally predominate in upper limbs
Functions	Maintenance of posture and performance in endurance activities	Endurance activities in endurance-trained muscles	Rapid, intense movements of short duration
Other Names	Type I fibers, slow fibers, red fibers	Type IIa fibers, fast fibers (fatigue resistant), red fibers	Type IIb fibers, fast fibers (fatigable), white fibers

They also contain large amounts of a dark pigment called **myoglobin** (mī-ō-glō′bin), which binds oxygen and acts as a muscle reservoir for oxygen when the blood does not supply an adequate amount. Myoglobin thus enhances the capacity of the muscle fibers to perform aerobic respiration. A muscle with predominately SO fibers appears red because of the myoglobin and rich blood supply.

Fast-twitch glycolytic fibers are the most fatigable fibers because they produce few ATP. The term *glycolytic* refers to glycolysis (see figure 8.18b), which produces ATP without oxygen. In support of anaerobic respiration, FG fibers have a high concentration of glycogen, from which glucose can be derived. FG fibers have few mitochondria, few capillaries, and little myoglobin. Consequently, a muscle with predominately FG fibers appears white.

Fast-twitch oxidative glycolytic fibers have fatigue resistance intermediate between SO and FG fibers, deriving ATP from oxidative and glycolytic processes. Compared to FG fibers, FOG fibers have more mitochondria, capillaries, and myoglobin. Compared with SO fibers, FOG fibers have more glycogen (see table 8.3).

P R E D I C T   8

Chickens can fly for only a few seconds, but they can stand and walk about all day. Explain.

Functions

Skeletal muscle fibers are specialized to perform a variety of tasks. SO fibers are well suited for maintaining posture and prolonged exercise because of their ability to produce ATP through aerobic respiration. FG fibers produce rapid, powerful contractions of

short duration. They readily fatigue because of their reliance on anaerobic respiration. FOG fibers can support moderate-intensity endurance exercises because of their oxidative and glycolytic capabilities. During low exercise intensities, primarily SO fibers are recruited. As exercise intensity increases, recruitment of SO fibers continues and there is progressively increased recruitment of FOG fibers and then FG fibers.

Humans exhibit no clear separation of slow-twitch and fast-twitch muscle fibers in individual muscles. Most muscles have both types of fibers, although the number of each type varies in a given muscle. The large postural muscles contain more slow-twitch fibers, whereas muscles of the upper limb contain more fast-twitch fibers. People who are good long-distance runners have a higher percentage of slow-twitch fibers, whereas good sprinters have a greater percentage of fast-twitch muscle fibers in their lower limbs. Athletes who are able to perform a variety of aerobic and anaerobic exercises tend to have a more balanced mixture of slow-twitch and fast-twitch muscle fibers.

The distribution of slow-twitch and fast-twitch muscle fibers in a given muscle is fairly constant for each individual and apparently is established developmentally. SO fibers are not converted to FG or FOG fibers, and vice versa. Aerobic exercise training, however, can convert FG fibers to FOG fibers, resulting in increased endurance. ▼

48. *Name the three major types of skeletal muscle fibers. Describe the factors responsible for the speed of contraction and fatigue resistance of each fiber type.*

49. *Explain the functions for which each fiber type is best adapted and how they are distributed in human muscles.*

Clinical Relevance   Duchenne Muscular Dystrophy

Duchenne muscular dystrophy (DMD) is usually identified in children around 3 years of age when their parents notice slow motor development with progressive weakness and muscle atrophy. Typically, muscular weakness begins in the hip muscles, which causes a waddling gait. Temporary enlargement of the calf muscles is apparent in 80% of cases. The enlargement is paradoxical because the muscle fibers are actually getting smaller, but there is an increase in fibrous connective tissue and fat between the muscle fibers. Wasting of the muscles with replacement by connective tissue results in shortened,

inflexible muscles called contractures. The contractures limit movements and can cause severe deformities of the skeleton. People with DMD are usually unable to walk by 10–12 years of age, and few live beyond age 20.

Mutations in the *dystrophin* (dis-trō′-fin) *(DMD)* gene are responsible for Duchenne muscular dystrophy. The dystrophin gene is responsible for producing a protein called **dystrophin.** Dystrophin plays a role in attaching actin myofilaments of myofibrils to, and regulating the activity of, other proteins in the sarcolemma. Dystrophin is thought to protect muscle

fibers against mechanical stress in a normal individual. In DMD, part of the dystrophin gene is missing, and the protein it produces is nonfunctional, resulting in progressive muscular weakness and muscle contractures. DMD is an X-linked recessive disorder (see chapter 25). Thus, although the gene is carried by females, DMD affects males almost exclusively. There are other types of muscular dystrophies that are less severe than DMD. Most of these are also inherited conditions, and all are characterized by degeneration of muscle fibers, leading to atrophy and replacement by connective tissues.

Muscular Hypertrophy and Atrophy

Muscular hypertrophy (hī-per′trō-fē) is an increase in the size of a muscle. During most of a person's life, hypertrophy of skeletal muscle results from changes in the size of muscle fibers because the number of muscle fibers is relatively constant. Muscles hypertrophy and increase in strength and endurance in response to exercise. As fibers increase in size, the number of myofibrils and sarcomeres increases within each muscle fiber. The number of nuclei in each muscle cell increases in response to exercise, but the nuclei of muscle cells cannot divide. New nuclei are added to muscle fibers because small satellite cells near skeletal muscle cells increase in number in response to exercise and then fuse with the skeletal muscle cells. The nuclei direct the synthesis of the myofibrils. Other elements, such as blood vessels, connective tissue, and mitochondria, also increase in response to exercise.

Aerobic exercise increases the vascularity of muscle and causes an enlargement of slow-twitch muscle fibers. Intense exercise resulting in anaerobic respiration, such as weight lifting, increases muscular strength and mass and results in an enlargement of fast-twitch muscle fibers more than slow-twitch muscle fibers.

The increased strength of trained muscle is greater than would be expected if that strength were based only on the change in muscle size. Part of the increase in strength results from the ability of the nervous system to recruit a large number of motor units simultaneously in a trained person to perform movements with better neuromuscular coordination. In addition, trained muscles usually are restricted less by excess fat. Metabolic enzymes increase in hypertrophied muscle fibers, resulting in a greater capacity for nutrient uptake and ATP production. Improved endurance in trained muscles is in part a result of improved metabolism, increased circulation to the exercising muscles, increased numbers of capillaries, more efficient respiration, and a greater capacity for the heart to pump blood.

Anabolic Steroids

Anabolic steroids (an-ă-bol′ik ster′oydz) are synthetic hormones related to testosterone, a reproductive hormone secreted by the testes. Anabolic steroids have been altered so that their reproductive effects are minimized but their effect on skeletal muscles is maintained. They increase muscle size and strength, but not muscle endurance. Some athletes have taken large doses of anabolic steroids to improve their athletic performance. Harmful side effects are associated with taking anabolic steroids, however, such as periods of irritability, acne, testicular atrophy and sterility, heart attack, stroke, and abnormal liver function. Most athletic organizations prohibit the use of anabolic steroids.

Muscular atrophy (at′rō-fē) is a decrease in muscle size. Individual muscle fibers decrease in size, and a progressive loss of myofibrils occurs. **Disuse atrophy** is muscular atrophy that results from a lack of muscle use. People with limbs in casts, bedridden people, and those who are inactive for other reasons experience disuse atrophy in the unused muscles. Disuse atrophy can be reversed by exercise. Extreme disuse of a muscle, however, results in permanent loss of muscle fibers. **Denervation** (dē-ner-vā′shŭn) **atrophy** results when nerves that supply skeletal muscles are severed. When motor neurons innervating skeletal muscle fibers are severed, the result is flaccid paralysis. If the muscle is reinnervated, muscle function is restored, and atrophy is stopped. If skeletal muscle is permanently denervated, however, it atrophies and exhibits permanent flaccid paralysis. Eventually, muscle fibers are replaced by connective tissue and the condition cannot be reversed. Muscle also atrophies with age (see "Effects of Aging on Skeletal Muscle," next section). ▼

50. **What factors contribute to an increase in muscle size, strength, and endurance?**
51. **How does aerobic versus anaerobic exercise affect muscles?**

P R E D I C T   9

Susan recently began racing her bicycle. Her training consists entirely of long rides on her bicycle at a steady pace. When she entered her first race, she was excited to find that she was able to keep pace with the rest of the riders. However, at the final sprint to the finish line, the other riders left her in their dust, and she finished in last place. Why was she unable to keep pace during the finishing sprint? As her coach, what advice would you give Susan about her training to prepare her better for her next race?

Effects of Aging on Skeletal Muscle

Several changes occur in aging skeletal muscle that reduce muscle mass, reduce stamina, and increase recovery time from exercise. There is a loss of muscle fibers as aging occurs, and the loss begins as early as 25 years of age. By 80 years of age, 50% of the muscle mass is gone, and this is due mainly to the loss of muscle fibers. In addition, fast-twitch muscle fibers decrease in number more rapidly than slow-twitch fibers. Most of the loss of strength and speed is due to the loss of fast-twitch muscle fibers. Weight-lifting exercises help slow the loss of muscle mass, but they do not prevent the loss of muscle fibers. Aging is associated with a decrease in the density of capillaries in skeletal muscles, and after exercise a longer period of time is required to recover.

Many of the age-related changes in skeletal muscle can be dramatically slowed if people remain physically active. As many people age, they assume a sedentary lifestyle. Age-related changes develop more rapidly in these people. It has been demonstrated that elderly people who are sedentary can become stronger and more mobile in response to exercise. ▼

> **52. Describe the changes in muscle that occur with aging.**

Smooth Muscle

Smooth muscle is distributed widely throughout the body and is more variable in function than skeletal and cardiac muscle (see table 8.1).

Types of Smooth Muscle

The two types of smooth muscle are visceral and multiunit smooth muscle. **Visceral** (vis′er-ăl) **smooth muscle** is located in the walls of most viscera (organs) of the body, including the digestive, reproductive, and urinary tracts. Visceral smooth muscle occurs in sheets and has numerous gap junctions (see chapter 4), which allow action potentials to pass directly from one cell to another. As a consequence, sheets of smooth muscle cells function as a unit, and a wave of contraction traverses the entire smooth muscle sheet. For this reason, visceral smooth muscle is also called **single-unit or unitary smooth muscle**.

Multiunit smooth muscle occurs as sheets, such as in the walls of blood vessels; as small bundles, such as in the arrector pili muscles and the iris of the eye; and as single cells, such as in the capsule of the spleen. Cells or groups of cells form many (multi)

units that function independently of each other because there are few gap junctions connecting the cells. ▼

> **53. Where are visceral and multiunit smooth muscle located? How do gap junctions explain the difference in contraction between these types of smooth muscle?**

Regulation of Smooth Muscle Contraction

Contraction of smooth muscle is involuntary. Multiunit smooth muscle usually contracts only when externally stimulated by nerves, hormones, or various substances. Visceral smooth muscle contraction can be autorhythmic or result from external stimulation. The resting membrane potential of autorhythmic smooth muscle cells fluctuates between slow depolarization and repolarization phases. As a result, smooth muscle cells can periodically and spontaneously generate action potentials that cause the smooth muscle cells to contract. For example, visceral smooth muscles of the digestive tract contract spontaneously and at relatively regular intervals, which mixes and moves the contents of the tract.

The autonomic nervous system (ANS) can stimulate or inhibit smooth muscle contraction (see chapter 13). For example, feces move out of the intestinal tract when the ANS causes smooth muscle of the rectum to contract and smooth muscle of the internal anal sphincter to relax.

Visceral smooth muscle has a different arrangement between neurons and smooth muscle fibers than does skeletal muscle. Instead of axon branches ending as presynaptic terminals supplying individual skeletal muscle fibers, the branches of axons supplying smooth muscle are a short distance from groups of smooth muscle cells. Each axon branch has many dilations called **varicosities,** each containing vesicles with neurotransmitters. Action potentials cause the vesicles to release neurotransmitters from the varicosities, and the neurotransmitters diffuse to several smooth muscle cells. Neurons release either acetylcholine or norepinephrine. When there are many layers of smooth muscle cells, axons may supply only the outer layer of cells. Action potentials produced in the outer layer are conducted through gap junctions to cells of the inner layer.

Multiunit smooth muscle has synapses more like those found in skeletal muscle tissue and there is greater control over individual smooth muscle cells.

Hormones are also important in regulating smooth muscle. For example, epinephrine, a hormone from the adrenal gland, helps regulate blood flow by changing the diameter of blood vessels. During exercise, epinephrine reduces blood flow to the stomach by stimulating the contraction of smooth muscle in stomach blood vessels and increases blood flow to skeletal muscles by decreasing smooth muscle contraction in skeletal muscle blood vessels.

The response of specific smooth muscle types to neurotransmitters, hormones, and other substances is presented in later chapters. ▼

> **54. List the ways in which multiunit and visceral smooth muscle are regulated.**
> **55. How are autorhythmic contractions produced in smooth muscle?**
> **56. What effect does the ANS have on smooth muscle contraction?**

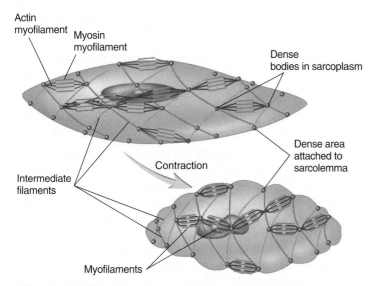

Actin myofilament
Myosin myofilament
Dense bodies in sarcoplasm
Dense area attached to sarcolemma
Contraction
Intermediate filaments
Myofilaments

Figure 8.19   Contractile Proteins in a Smooth Muscle Cell

Bundles of contractile myofilaments containing actin and myosin are anchored at one end to dense areas in the plasma membrane and at the other end, through dense bodies, to intermediate filament. The contractile myofilaments are oriented with the long axis of the cell; when actin and myosin slide over one another during contraction, the cell shortens.

57. *Describe the nerve supply to visceral and multiunit smooth muscle. How does this affect the control of groups of smooth muscle cells versus individual smooth muscle cells?*

Structure of Smooth Muscle

Smooth muscle cells are smaller than skeletal muscle cells and spindle-shaped, with a single nucleus located in the middle of the cell (see table 8.1). The actin and myosin myofilaments are not organized as sarcomeres. Consequently, there are no myofibrils and smooth muscle does not have a striated appearance. The actin and myosin myofilaments in smooth muscle are organized as loose bundles scattered throughout the cell (figure 8.19). The actin myofilaments are attached to **dense bodies** in the cytoplasm and to **dense areas** in the plasma membrane. Dense bodies and dense areas are considered to be equivalent to the Z disks in skeletal muscle. Noncontractile **intermediate filaments** also attach to the dense bodies and dense areas, forming an intracellular cytoskeleton. When actin myofilaments slide past the myosin myofilaments, the dense bodies and dense areas are pulled closer together, resulting in contraction of the smooth muscle cell (see figure 8.19). The dense areas of adjacent smooth muscle cells are connected so that the force of contraction is transmitted from one cell to the next.

Smooth muscle cells do not have T tubules and most have less sarcoplasmic reticulum for storing Ca^{2+} than does skeletal muscle. Nor do smooth muscle cells have troponin.

The plasma membrane of smooth muscle has Ca^{2+} channels and Ca^{2+} pumps. When the Ca^{2+} channels open, Ca^{2+} diffuse into the cell because there is a higher concentration of Ca^{2+} outside of cells than inside. The Ca^{2+} channels open in response to plasma membrane voltage changes; ligands, such as hormones, binding to their receptors; and mechanical changes in the plasma membrane,

such as stretch. The Ca^{2+} pumps actively transport Ca^{2+} out of cells, maintaining a low intracellular Ca^{2+} concentration. ▼

58. *Describe the structure of smooth muscle cells. How does movement of actin myofilaments past myosin myofilaments result in contraction?*
59. *Describe how Ca^{2+} move across the plasma membrane of smooth muscle.*

Smooth Muscle Contraction and Relaxation

Auto or external stimulation of smooth muscles can result in increased intracellular Ca^{2+} levels, which promotes contraction. The role of Ca^{2+} in smooth muscle differs from that in skeletal muscle cells, however, because there are no troponin molecules associated with the actin fibers of smooth muscle cells. Instead of binding to troponin, Ca^{2+} bind to a protein called **calmodulin** (kal-mod′ū-lin). Calmodulin with Ca^{2+} bound to it activates an enzyme called **myosin kinase** (kī′nās), which transfers a phosphate group from ATP to the head of a myosin molecule. Once the phosphate is attached to a myosin head, myosin ATPase can break down ATP, resulting in cross-bridge cycling, just as occurs in skeletal muscle. Cross-bridge cycling stops when another enzyme, called **myosin phosphatase** (fos′fă-tās), removes the phosphate from the myosin head. Active transport of Ca^{2+} out of the cell or into sarcoplasmic reticulum decreases Ca^{2+} levels, which decreases myosin kinase levels and the addition of phosphates to myosin heads. Relaxation occurs because fewer phosphates are added to myosin heads than are removed by myosin phosphatase. ▼

60. *Describe the role of Ca^{2+}, calmodulin, myosin kinase, myosin ATPase, and myosin phosphatase in smooth muscle contraction.*

Functional Properties of Smooth Muscle

Smooth muscle has several functional properties not seen in skeletal muscle:

1. Smooth muscle is capable of autorhythmic contractions. In some smooth muscle, **pacemaker cells** control the contraction of other smooth muscle cells by generating action potentials that spread to the other cells through gap junctions.
2. A typical smooth muscle cell contracts and relaxes approximately 30 times slower than a typical skeletal muscle cell. Although smooth muscle has sarcoplasmic reticulum, many of the Ca^{2+} that initiate contraction come from outside the cell. The latent phase is longer because it takes time for these Ca^{2+} to diffuse into the cell. The contraction phase is longer because of slower cross-bridge cycling. Cross-bridges form more slowly because the myosin ATPase in smooth muscle is slower than in skeletal muscle. Cross-bridges release very slowly when myosin phosphatase removes the phosphate from the myosin head while it is attached to actin. The relaxation phase is longer because the Ca^{2+} pumps that move Ca^{2+} out of the cell or into sarcoplasmic reticulum are slow.
3. **Smooth muscle tone** is the ability of smooth muscle to maintain tension for long periods while expending very little energy. Skeletal muscle uses 10 to 300 times more ATP than smooth muscle to maintain the same tension. Smooth

Systems Interactions

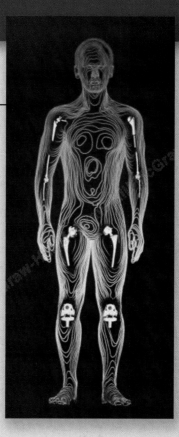

	Effects of the Muscular System on Other Systems	Effects of Other Systems on the Muscular System
Integumentary System	Moves skin for facial expressions Provides support Produces heat	Produces a vitamin D precursor that is converted to active vitamin D, which increases the absorption of the Ca²⁺ necessary for muscle contraction Protects against abrasion and ultraviolet light
Skeletal System	Moves bones, resulting in body movements Maintains body posture Tension on bone promotes bone growth and maintenance.	Forms lever system for body movements Reservoir for Ca²⁺, which is necessary for muscle contraction
Nervous System	Sensory receptors (for example, muscle spindles and Golgi tendon organs) provide information about body position.	Stimulates muscle contractions and maintains muscle tone Psychologic fatigue is the most common reason for stopping muscular exercise.
Endocrine System	Muscular exercise deacreases blood sugar, resulting in decreased insulin and increased glucagon secretion by the parcreas.	Hormones affect muscle growth, metabolism, and nutrient uptake
Cardiovascular System	Skeletal muscle "pump" helps move blood through blood vessels.	Delivers oxygen, nutrients, hormones, and immune cells Removes carbon dioxide, waste products, and toxins Transports lactic acid from muscle to the liver
Lymphatic System and Immunity	Skeletal muscle "pump" helps move lymph through lymphatic vessels.	Immune cells provide protection against microorganisms and toxins and promote tissue repair by releasing chemical mediators. Removes excess interstitial fluid
Respiratory System	Changes thoracic volume during breathing Controls tension on vocal cords during voice production	Provides oxygen and removes carbon dioxide Helps maintain the body's pH
Digestive System	Protects abdominal organs Is responsible for mastication and swallowing Controls voluntary defecation	Provides nutrients and water The liver converts lactic acid from muscle to glucose (oxygen deficit).
Urinary System	Controls voluntary urination Pelvic floor muscles support the urinary bladder.	Removes waste products Helps maintain the body's pH, ion, and water balance
Reproductive System	Pelvic floor muscles support internal reproductive organs, such as the uterus. Is responsible for ejaculation Cremaster muscles help regulate testes temperature. Abdominal muscles assist in delivery.	Sex hormones increase muscle growth.

muscle can maintain tension while using little ATP because cross-bridge cycling is slow and actin and myosin are bound together for much of the cycle as a result of the slow rate of cross-bridge release. The smooth muscle of blood vessels is in a tonic state of contraction, which helps maintain blood pressure. Loss of smooth muscle tone in blood vessels can cause a drop in blood pressure, resulting in dizziness, unconsciousness, and even death.

4. Contraction of smooth muscle in the walls of hollow organs maintains nearly the same pressure on the organ's contents despite changes in content volume. When the contents of an organ suddenly increase, the wall of the organ is stretched and pressure increases. Within a few seconds to minutes, however, the smooth muscle relaxes and pressure nearly returns to the original level. As cross-bridges slowly cycle, cross-bridges release, actin myofilaments slide apart because of the external force stretching the muscle, and the muscle elongates. As cross-bridges cycle, the number of cross-bridges in the stretched muscle remains approximately the same as before the muscle was stretched, resulting in approximately the same tension. Likewise, when organ contents suddenly decrease, pressure decreases. The smooth muscle responds by contracting and pressure returns to the original level. Maintaining a constant pressure despite volume changes allows additional materials to enter an organ and helps prevent the premature movement of materials out of an organ. For example, the urinary bladder can gradually fill, but pressure within the bladder does not increase to the point that urine is forced out.

5. The ability of smooth muscle to contract effectively remains relatively constant despite changes in muscle length. Smooth muscle can contract effectively even when stretched to more than two-thirds of its resting length, whereas skeletal muscle contracts effectively when stretched one-quarter to one-third of its resting length (see figure 8.17). The amount of stretch of smooth muscle in the walls of hollow organs, such as the stomach and urinary bladder, is determined by the organ's contents. When the organ is full, the smooth muscle can be greatly stretched, but, when the organ is nearly empty, the stretch is less. Despite changes in

organ contents, the smooth muscle can contract effectively to mix or move the contents.

The ability of smooth muscle to contract effectively despite changes in muscle length is related to the arrangement of actin and myosin myofilaments, which are scattered as bundles throughout the cells. The optimal overlap of some of the actin and myosin myofilament bundles occurs at one muscle length, whereas the optimal overlap of other actin and myosin bundles occurs at a different muscle length. In addition, there is no bare area in the middle of the myosin myofilament as occurs in skeletal muscle (see figure 8.2f), so there are more opportunities to form cross-bridges. ▼

61. *What are pacemaker cells?*
62. *Explain how smooth muscle contracts more slowly than skeletal muscle.*
63. *What is smooth muscle tone? Explain how it is produced.*
64. *How does smooth muscle maintain a constant pressure within hollow organs despite changes in the organ's volume?*
65. *Compare the ability of smooth muscle and skeletal muscle to contract when stretched.*

Cardiac Muscle

Cardiac muscle is found only in the heart; it is discussed in detail in chapter 17. Cardiac muscle tissue is striated, like skeletal muscle, but each cell usually contains one nucleus located near the center. Adjacent cells join together to form branching fibers by specialized cell-to-cell attachments called **intercalated** (in-ter′kă-lā-ted) **disks,** which have gap junctions that allow action potentials to pass from cell to cell. Some cardiac muscle cells are autorhythmic, and one part of the heart normally acts as the pacemaker. The action potentials of cardiac muscle are similar to those of nerve and skeletal muscle but have a much longer duration and refractory period. The depolarization of cardiac muscle results from the influx of both Na^+ and Ca^{2+} across the plasma membrane. The regulation of contraction in cardiac muscle by Ca^{2+} is similar to that of skeletal muscle. ▼

66. *Compare the structural and functional characteristics of cardiac muscle with those of skeletal muscle.*

S U M M A R Y

Functions of the Muscular System   (p. 196)

Muscle is responsible for movement of the arms, legs, heart, and other parts of the body; maintenance of posture; respiration; production of body heat; communication; constriction of organs and vessels; and heartbeat.

General Functional Characteristics of Muscle   (p. 196)
Properties of Muscle

1. Muscle exhibits contractility (shortens forcefully), excitability (responds to stimuli), extensibility (can be stretched and still contract), and elasticity (recoils to resting length).
2. Muscle tissue shortens forcefully but lengthens passively.

Types of Muscle Tissue

1. The three types of muscle are skeletal, smooth, and cardiac.
2. Skeletal muscle is responsible for most body movements, smooth muscle is found in the walls of hollow organs and tubes and moves substances through them, and cardiac muscle is found in the heart and pumps blood.

Skeletal Muscle Structure   (p. 197)

Skeletal muscle fibers are long, cylindrical cells, they contain several nuclei, and they appear striated.

Connective Tissue Coverings of Muscle

1. Muscular fascia separates and compartmentalizes individual or groups of muscles.

2. Muscles are composed of muscle fasciculi, which are composed of muscle fibers.
3. Epimysium surrounds muscles, perimysium surrounds muscle fasciculi (bundles), and endomysium surrounds muscle fibers.
4. The connective tissue of muscle blends with other connective tissue, such as tendons, which connect muscle to bone.
5. Muscle connective tissue provides a pathway for blood vessels and nerves to reach muscle fibers.

Muscle Fibers

1. A muscle fiber is a single cell consisting of a plasma membrane (sarcolemma), cytoplasm (sarcoplasm), several nuclei, and myofibrils.
2. Myofibrils are composed of two major protein fibers: actin and myosin.
   - Actin (thin) myofilaments consist of two strands of F actin (composed of G actin), tropomyosin, and troponin.
   - Myosin (thick) myofilaments consist of myosin molecules. Each myosin molecule has a head with an ATPase, which breaks down ATP; a hinge region, which enables the head to move; and a rod.
   - A cross-bridge is formed when a myosin head binds to the active site on G actin.
3. Actin and myosin are organized to form sarcomeres.
   - Sarcomeres are bound by Z disks that hold actin myofilaments.
   - Six actin myofilaments surround a myosin myofilament.
   - Myofibrils appear striated because of A bands and I bands.

Sliding Filament Model   (p. 200)

1. Actin and myosin myofilaments do not change in length during contraction.
2. Actin and myosin myofilaments slide past one another in a way that causes sarcomeres to shorten.
3. The I band and H zones become narrower during contraction, and the A band remains constant in length.

Physiology of Skeletal Muscle Fibers   (p. 202)
Membrane Potentials
The nervous system stimulates muscles to contract through electric signals called action potentials.

Membrane Potentials

1. Plasma membranes are polarized, which means there is a charge difference, called the resting membrane potential, across the plasma membrane.
2. The inside of plasma membranes is negative, compared with the outside in a resting cell.

Ion Channels

1. An action potential is a reversal of the resting membrane potential so that the inside of the plasma membrane becomes positive.
2. Ion channels are responsible for producing action potentials.
3. Ligand-gated and voltage-gated channels produce action potentials.

Action Potentials

1. Depolarization results from an increase in the permeability of the plasma membrane to Na^+. If depolarization reaches threshold, an action potential is produced.
2. The depolarization phase of the action potential results from many Na^+ channels opening.
3. The repolarization phase of the action potential occurs when the Na^+ channels close and K^+ channels open briefly.
4. Action potentials occur in an all-or-none fashion. A subthreshold stimulus produces no action potential, whereas threshold or stronger stimuli produce the same action potential.
5. Action potentials propagate, or travel, across plasma membranes.

Neuromuscular Junction

1. The presynaptic terminal of the axon is separated from the postsynaptic membrane of the muscle fiber by the synaptic cleft.
2. Action potentials cause Ca^{2+} channels in the presynaptic terminal to open. Calcium ions diffuse into the presynaptic terminal, stimulating the release of acetylcholine from the presynaptic terminal.
3. Acetylcholine binds to receptors of the postsynaptic membrane, thereby changing membrane permeability and producing an action potential, which stimulates muscle contraction.
4. After an action potential occurs, acetylcholinesterase splits acetylcholine into acetic acid and choline. Choline is transported into the presynaptic terminal to re-form acetylcholine.

Excitation–Contraction Coupling

1. Invaginations of the sarcolemma form T tubules, which wrap around the sarcomeres.
2. A triad is a T tubule and two terminal cisternae (an enlarged area of sarcoplasmic reticulum).
3. Action potentials move into the T tubule system, causing Ca^{2+} channels in the sarcoplasmic reticulum to open and release Ca^{2+}.
4. Calcium ions diffuse from the sarcoplasmic reticulum to the myofilaments and bind to troponin, causing tropomyosin to move and expose active sites on actin to myosin.
5. Contraction occurs when myosin heads bind to active sites on actin, myosin heads move at their hinge regions, and actin is pulled past the myosin.

Cross-Bridge Movement

1. ATP is required for the cycle of cross-bridge formation, movement, and release.
2. ATP is also required to transport Ca^{2+} into the sarcoplasmic reticulum and to maintain normal concentration gradients across the plasma membrane.

Muscle Relaxation

1. Calcium ions are transported into the sarcoplasmic reticulum.
2. Calcium ions diffuse away from troponin and tropomyosin moves, preventing further cross-bridge formation.

Physiology of Skeletal Muscle   (p. 209)
Muscle Twitch

1. A muscle twitch is the contraction of a muscle as a result of one or more muscle fibers contracting.
2. A muscle twitch has lag, contraction, and relaxation phases.

Strength of Muscle Contraction

1. For a given condition, a muscle fiber or motor unit contracts with a consistent force in response to each action potential.
2. For a whole muscle, stimuli of increasing strength result in graded contractions of increased force as more motor units are recruited (multiple motor unit summation).
3. Stimulus of increasing frequency increase the force of contraction (multiple-wave summation).
4. Incomplete tetanus is partial relaxation between contractions, and complete tetanus is no relaxation between contractions.
5. The force of contraction of a whole muscle increases with increased frequency of stimulation because of an increasing concentration of Ca^{2+} around the myofibrils.
6. Treppe is an increase in the force of contraction during the first few contractions of a rested muscle.

Types of Muscle Contraction

1. Isometric contractions cause a change in muscle tension but no change in muscle length.
2. Isotonic contractions cause a change in muscle length but no change in muscle tension.
3. Concentric contractions are isotonic contractions that cause muscles to shorten.
4. Eccentric contractions are isotonic contractions that enable muscles to resist an increase in length.
5. Muscle tone is the maintenance of a steady tension for long periods.
6. Asynchronous contractions of motor units produce smooth, steady muscle contractions.

Length Versus Tension

Muscle contracts with less than maximum force if its initial length is shorter or longer than optimum.

Fatigue

1. Fatigue, the decreased ability to do work, can be caused by the central nervous system (psychologic fatigue) and depletion of ATP in muscles (muscular fatigue).
2. Physiologic contracture (the inability of muscles to contract or relax) and rigor mortis (stiff muscles after death) result from inadequate amounts of ATP.

Energy Sources

1. Creatine phosphate
   - ATP is synthesized when ADP reacts with creatine phosphate to form creatine and ATP.
   - ATP from this source provides energy for a short time.
2. Anaerobic respiration
   - ATP is synthesized by anaerobic respiration and is used to provide energy for a short time at the beginning of exercise and during intense exercise.
   - Anaerobic respiration produces ATP less efficiently but more rapidly than aerobic respiration.
   - Lactic acid levels increase because of anaerobic respiration.
3. Aerobic respiration
   - Aerobic respiration requires oxygen.
   - Aerobic respiration produces energy for muscle contractions under resting conditions or during endurance exercises.

Oxygen Deficit and Recovery Oxygen Consumption

1. The oxygen deficit is the lack of oxygen consumed at the onset of exercise.
2. Recovery oxygen consumption is the oxygen used during aerobic respiration to restore homeostasis after exercise.

Heat Production

1. Heat is produced as a by-product of chemical reactions in muscles.
2. Shivering produces heat to maintain body temperature.

Types of Skeletal Muscle Fibers (p. 216)

The three main types of skeletal muscle fibers are slow-twitch oxidative (SO) fibers, fast-twitch glycolytic (FG) fibers, and fast-twitch oxidative glycolytic (FOG) fibers.

Speed of Contraction

SO fibers contract more slowly than FG and FOG fibers because they have slower myosin ATPases than FG and FOG fibers.

Fatigue Resistance

1. SO fibers are fatigue-resistant and rely on aerobic respiration. They have many mitochondria, a rich blood supply, and myoglobin.
2. FG fibers are fatigable. They rely on anaerobic respiration and have a high concentration of glycogen.
3. FOG fibers have fatigue resistance intermediate between SO and FG fibers. They rely on aerobic and anaerobic respiration.

Functions

1. SO fibers maintain posture and are involved with prolonged exercise.
2. FG fibers produce powerful contractions of short duration.
3. FOG fibers support moderate-intensity endurance exercises.
4. People who are good sprinters have a greater percentage of fast-twitch muscle fibers, and people who are good long-distance runners have a higher percentage of slow-twitch muscle fibers in their leg muscles.
5. Aerobic exercise can result in the conversion of FG fibers to FOG fibers.

Muscular Hypertrophy and Atrophy (p. 218)

1. Hypertrophy is an increase in the size of muscles due to an increase in the size of muscle fibers resulting from an increase in the number of myofibrils in the muscle fibers.
2. Aerobic exercise increases the vascularity of muscle and hypertrophy of slow-twitch fibers more than fast-twitch fibers. Intense, anaerobic exercise results in greater hypertrophy of fast-twitch than slow-twitch fibers.
3. Atrophy is a decrease in the size of muscle due to a decrease in the size of muscle fibers or a loss of muscle fibers.

Effects of Aging on Skeletal Muscle (p. 219)

Muscles atrophy because of an age-related loss of muscle fibers, especially fast-twitch fibers.

Smooth Muscle (p. 219)
Types of Smooth Muscle

1. Visceral smooth muscle fibers have many gap junctions and contract as a single unit.
2. Multiunit smooth muscle fibers have few gap junctions and function independently.

Regulation of Smooth Muscle

1. Smooth muscle contraction is involuntary.
   - Multiunit smooth muscle contracts when externally stimulated by nerves, hormones, or other substances.
   - Visceral smooth muscle contracts autorhythmically or when stimulated externally.
2. Hormones are important in regulating smooth muscle.

Structure of Smooth Muscle

1. Smooth muscle cells are spindle-shaped with a single nucleus.
2. Smooth muscle cells have actin and myosin myofilaments. The actin myofilaments are connected to dense bodies and dense areas. Smooth muscle is not striated.
3. Smooth muscle cells have no T tubule system and most have less sarcoplasmic reticulum than skeletal muscle cells. They do not have troponin.

Smooth Muscle Contraction and Relaxation

1. Calcium ions enter the cell to initiate contraction; Ca^{2+} bind to calmodulin and they activate myosin kinase, which transfers a phosphate group from ATP to myosin. When phosphate groups are attached to myosin, cross-bridges form.
2. Relaxation results when myosin phosphatase removes a phosphate group from the myosin molecule.

Functional Properties of Smooth Muscle

1. Pacemaker cells are autorhythmic smooth muscle cells that control the contraction of other smooth muscle cells.
2. Smooth muscle cells contract more slowly than skeletal muscle cells.
3. Smooth muscle tone is the ability of smooth muscle to maintain a steady tension for long periods with little expenditure of energy.
4. Smooth muscle in the walls of hollow organs maintain a relatively constant pressure on the organs' contents despite changes in content volume.

5. The force of smooth muscle contraction remains nearly constant despite changes in muscle length.

Cardiac Muscle (p. 222)

Cardiac muscle fibers are striated, have a single nucleus, are connected by intercalated disks (thus function as a single unit), and are capable of autorhythmicity.

REVIEW AND COMPREHENSION

1. Which of these is true of skeletal muscle?
   a. spindle-shaped cells
   b. under involuntary control
   c. many peripherally located nuclei per muscle cell
   d. forms the walls of hollow internal organs
   e. may be autorhythmic

2. The connective tissue sheath that surrounds a muscle fasciculus (bundle) is the
   a. perimysium.   c. epimysium.   e. external lamina.
   b. endomysium.   d. hypomysium.

3. Given these structures:
   1. whole muscle   3. myofilament   5. muscle fasciculus (bundle)
   2. muscle fiber (cell)   4. myofibril

   Choose the arrangement that lists the structures in the correct order from the largest to the smallest structure.
   a. 1,2,5,3,4   c. 1,5,2,3,4   e. 1,5,4,2,3
   b. 1,2,5,4,3   d. 1,5,2,4,3

4. Myofibrils
   a. are made up of many muscle fibers.
   b. are made up of many sarcomeres.
   c. contain sarcoplasmic reticulum.
   d. contain T tubules.

5. Myosin myofilaments are
   a. attached to the Z disk.   d. absent from the H zone.
   b. found primarily in the I band.   e. attached to the M line.
   c. thinner than actin myofilaments.

6. Which of these statements about the molecular structure of myofilaments is true?
   a. Tropomyosin has a binding site for Ca^{2+}.
   b. The head of the myosin molecule binds to an active site on G actin.
   c. ATPase is found on troponin.
   d. Troponin binds to the rod portion of myosin molecules.
   e. Actin molecules have a hingelike portion, which bends and straightens during contraction.

7. Which of these regions shortens during skeletal muscle contraction?
   a. A band   c. H zone   e. both b and c
   b. I band   d. both a and b

8. When the plasma membrane depolarizes during an action potential, the inside of the membrane
   a. becomes more negative than the outside of the membrane.
   b. becomes more positive than the outside of the membrane.
   c. is unchanged.

9. During repolarization of the plasma membrane,
   a. Na^+ move to the inside of the cell.
   b. Na^+ move to the outside of the cell.
   c. K^+ move to the inside of the cell.
   d. K^+ move to the outside of the cell.

10. Given these events:
    1. Acetylcholine is broken down into acetic acid and choline.
    2. Acetylcholine diffuses across the synaptic cleft.
    3. Action potential reaches the terminal branch of the motor neuron.
    4. Acetylcholine combines with a ligand-gated ion channel.
    5. Action potential is produced on the muscle fiber's sarcolemma.

    Choose the arrangement that lists the events in the order they occur at a neuromuscular junction.
    a. 2,3,4,1,5   c. 3,4,2,1,5   e. 5,1,2,4,3
    b. 3,2,4,5,1   d. 4,5,2,1,3

11. Acetylcholinesterase is an important molecule in the neuromuscular junction because it
    a. stimulates receptors on the presynaptic terminal.
    b. synthesizes acetylcholine from acetic acid and choline.
    c. stimulates receptors within the postsynaptic membrane.
    d. breaks down acetylcholine.
    e. causes the release of Ca^{2+} from the sarcoplasmic reticulum.

12. Given these events:
    1. Sarcoplasmic reticulum releases Ca^{2+}.
    2. Sarcoplasmic reticulum takes up Ca^{2+}.
    3. Calcium ions diffuse into myofibrils.
    4. Action potential moves down the T tubule.
    5. Sarcomere shortens.
    6. Muscle relaxes.

    Choose the arrangement that lists the events in the order they occur following a single stimulation of a skeletal muscle cell.
    a. 1,3,4,5,2,6   c. 4,1,3,5,2,6   e. 5,1,4,3,2,6
    b. 2,3,5,4,6,1   d. 4,2,3,5,1,6

13. Given these events:
    1. Calcium ions combine with tropomyosin.
    2. Calcium ions combine with troponin.
    3. Tropomyosin pulls away from actin.
    4. Troponin pulls away from actin.
    5. Tropomyosin pulls away from myosin.
    6. Troponin pulls away from myosin.
    7. Myosin binds to actin.

    Choose the arrangement that lists the events in the order they occur during muscle contraction.
    a. 1,4,7   b. 2,5,6   c. 1,3,7   d. 2,4,7   e. 2,3,7

14. Which of these events occurs during the lag (latent) phase of muscle contraction?
    a. cross-bridge movement
    b. active transport of Ca^{2+} into the sarcoplasmic reticulum
    c. Ca^{2+} binding to troponin
    d. sarcomere shortening
    e. ATP binding to myosin

15. With stimuli of increasing strength, which of these is capable of a graded response?
    a. nerve axon   c. motor unit
    b. muscle fiber   d. whole muscle

16. Considering the force of contraction of a skeletal muscle cell, multiple-wave summation occurs because of
    a. increased strength of action potentials on the plasma membrane.
    b. a decreased number of cross-bridges formed.

c. an increase in Ca^{2+} concentration around the myofibrils.
d. an increased number of motor units recruited.
e. increased permeability of the sarcolemma to Ca^{2+}.

17. A weight lifter attempts to lift a weight from the floor, but the weight is so heavy that he is unable to move it. The type of muscle contraction the weight lifter is using is mostly
   a. isometric.   c. isokinetic.   e. eccentric.
   b. isotonic.   d. concentric.

18. The length-tension curve illustrates
   a. how isometric contractions occur.
   b. that the greatest force of contraction occurs if a muscle is not stretched at all.
   c. that optimal overlap of actin and myosin produces the greatest force of contraction.
   d. that the greatest force of contraction occurs with little or no overlap of actin and myosin.

19. Given these conditions:
   1. low ATP levels
   2. little or no transport of Ca^{2+} into the sarcoplasmic reticulum
   3. release of cross-bridges
   4. Na^+ accumulation in the sarcoplasm
   5. formation of cross-bridges

   Choose the conditions that occur in both physiologic contracture and rigor mortis.
   a. 1,2,3   c. 1,2,3,4   e. 1,2,3,4,5
   b. 1,2,5   d. 1,2,4,5

20. Which of the following statements regarding energy sources for skeletal muscle is true?
   a. Creatine phosphate provides energy for endurance exercise.
   b. Anaerobic respiration provides more ATP per glucose molecule than does aerobic respiration.
   c. Aerobic respiration ends with the production of lactic acid.
   d. At the onset of exercise, aerobic respiration provides all of the ATP necessary for the increased activity.
   e. ATP is provided by aerobic and anaerobic respiration at exercise levels of 70% or greater of maximum.

21. Concerning different types of skeletal muscle fibers,
   a. FOG fibers are better at endurance-type activities than are SO fibers.
   b. FOG fibers are recruited before SO fibers during low-intensity exercise.
   c. FG fibers have more myoglobin than SO fibers.
   d. FG fibers have more glycogen than SO fibers.
   e. Aerobic training converts FG fibers into SO fibers.

22. Which of these increases the least as a result of muscle hypertrophy?
   a. number of sarcomeres
   b. number of myofibrils
   c. number of fibers
   d. blood vessels and mitochondria

23. Relaxation in smooth muscle occurs when
   a. myosin kinase attaches phosphate to the myosin head.
   b. Ca^{2+} bind to calmodulin.
   c. myosin phosphatase removes phosphate from myosin.
   d. Ca^{2+} channels open.
   e. Ca^{2+} are released from the sarcoplasmic reticulum.

24. Compared with skeletal muscle, visceral smooth muscle
   a. can have autorhythmic contractions.
   b. contracts more slowly.
   c. maintains tension for long periods with less expenditure of energy.
   d. can be stretched more and still retain the ability to contract forcefully.
   e. all of the above.

25. Which of these statements concerning aging and skeletal muscle is correct?
   a. There is a loss of muscle fibers with aging.
   b. Slow-twitch fibers decrease in number faster than fast-twitch fibers.
   c. There is an increase in the density of capillaries in skeletal muscle.
   d. The number of motor neurons remains constant.

Answers in Appendix E

C R I T I C A L   T H I N K I N G

1. Suppose that a poison causes death because of respiratory failure (the respiratory muscles relax but do not contract). Propose as many ways as possible that the toxin could cause respiratory failure.

2. A patient is thought to be suffering from either muscular dystrophy or myasthenia gravis. How would you distinguish between the two conditions?

3. Under certain circumstances, the actin and myosin myofilaments can be extracted from muscle cells and placed in a beaker. They subsequently bind together to form long filaments of actin and myosin. The addition of what cell organelle or molecule to the beaker would make the actin and myosin myofilaments unbind?

4. Design an experiment to test the following hypothesis: Muscle A has the same number of motor units as muscle B. Assume you could stimulate the nerves that innervate skeletal muscles with an electronic stimulator and monitor the tension produced by the muscles.

5. Explain what is happening at the level of individual sarcomeres when a person is using his or her biceps brachii muscle to hold a weight in a constant position. Contrast this with what is happening at the level of individual sarcomeres when an individual lowers the weight, as well as when the individual raises the weight.

6. Predict the shape of the length-tension curve for visceral smooth muscle. How does it differ from the length-tension curve for skeletal muscle?

7. Harvey Leche milked cows by hand each morning before school. One morning, he slept later than usual and had to hurry to get to school on time. As he was milking the cows as fast as he could, his hands became very tired, and for a short time he could neither release his grip nor squeeze harder. Explain what happened.

8. Predict and explain the response if the ATP concentration in a muscle that was exhibiting rigor mortis could be instantly increased.

9. Shorty McFleet noticed that his rate of respiration was elevated after running a 100 m race but was not as elevated after running slowly for a much longer distance. How would you explain this?

Answers in Appendix F

For additional study tools, go to: www.aris.mhhe.com.
Click on your course topic and then this textbook's author name/title. Register once for a semester's worth of interactive learning activities to help improve your grade!

CHAPTER 9 Gross Anatomy and Functions of Skeletal Muscles

Colorized SEM of skeletal muscle.

Tools for Success

A virtual cadaver dissection experience

Audio (MP3) and/or video (MP4) content, available at www.mhhe.aris.com

Introduction

Mannequins are rigid, expressionless, immobile recreations of the human form. They cannot walk or talk. One of the major characteristics of living human beings is our ability to move about. Without muscles, humans would be little more than mannequins. We would not be able to hold this book or turn its pages. We would not be able to blink, so our eyes would dry out. None of these inconveniences would bother us for long because we would not be able to breathe, either.

We use our skeletal muscles all the time. Postural muscles are constantly contracting to keep us sitting or standing upright. Respiratory muscles are constantly functioning to keep us breathing, even when we sleep. Communication of any kind requires skeletal muscles, whether we are writing, typing, or speaking. Even silent communication with hand signals or facial expression requires skeletal muscle function.

▶ This chapter explains the **general principles (p. 228)** of the muscular system and describes in detail the **head and neck muscles (p. 232)**, **trunk muscles (p. 239)**, **scapular and upper limb muscles (p. 244)**, and **hip and lower limb muscles (p. 253)**. The chapter concludes with a discussion of **bodybuilding (p. 261)**.

General Principles

This chapter is devoted to the description of the major named skeletal muscles. The structure and function of cardiac and smooth muscle are considered in other chapters. Most skeletal muscles extend from one bone to another and cross at least one joint. Muscle contraction causes most body movements by pulling one bone toward another across a movable joint. The **action** of a muscle is the movement accomplished when it contracts. Some muscles are not attached to bone at both ends. For example, some facial muscles attach to the skin, which moves as the muscles contract.

The two points of attachment of each muscle are its origin and insertion (figure 9.1). The **origin,** also called the **fixed end** or the **head,** is usually the most stationary end of a muscle. Some muscles have multiple heads; for example, the biceps brachii has two heads and the triceps brachii has three heads. The **insertion,** also called the **mobile end,** is usually the end of the muscle undergoing the greatest movement. A muscle can insert on more than one structure. Generally, the origin of a muscle is the proximal or medial end of a muscle, whereas the insertion is the distal or lateral end of a muscle. The part of the muscle between the origin and the insertion is the **belly.** Muscles are connected to bones and other structures by **tendons.** Tendons are long, cablelike structures; broad, sheetlike structures called **aponeuroses** (ap′ō-noo-rō′sēz, *apo,* from + *neuron,* sinew); or short, almost nonexistent structures.

Muscles are typically grouped so that the action of one muscle or group of muscles is opposed by that of another muscle or group of muscles. For example, the biceps brachii flexes the elbow and the triceps brachii extends the elbow. A muscle that accomplishes a certain movement, such as flexion, is called the **agonist** (ag′ō-nist, a contest). A muscle acting in opposition to an agonist is called an **antagonist** (an-tag′ō-nist). The biceps brachii is the agonist in elbow flexion, whereas the triceps brachii is the antagonist, which extends the elbow.

P R E D I C T
The triceps brachii is the agonist for extension of the elbow. What muscle is its antagonist?

Muscles also tend to function in groups to accomplish specific movements. For example, the deltoid, biceps brachii, and pectoralis major all help flex the shoulder. Furthermore, many muscles are members of more than one group, depending on the type of movement being considered. For example, the anterior part of the deltoid muscle functions with the flexors of the shoulder, whereas the posterior part functions with the extensors of the shoulder. Members of a group of muscles working together to produce a movement are called **synergists** (sin′er-jistz). For example, the biceps brachii and brachialis are synergists in elbow flexion. Among a group of synergists, if one muscle plays the major role in accomplishing the desired movement, it is the **prime mover.** The brachialis is the prime mover in flexing the elbow. **Fixators** are muscles that hold one bone in place relative to the body while a usually more distal bone is moved. Certain trunk muscles are fixators of the scapula. They hold the scapula in place while muscles attached to the scapula contract to move the humerus. ▼

> 1. **Define origin, insertion, and belly of a muscle; tendon and aponeurosis; agonist and antagonist; and synergist, prime mover, and fixator.**

Nomenclature

Muscles are named according to their location, size, shape, orientation of fasciculi, origin and insertion, number of heads, and function. Recognizing the descriptive nature of muscle names makes learning those names much easier.

1. *Location.* A pectoralis (chest) muscle is located in the chest, a gluteus (buttock) muscle is located in the buttock, and a brachial (arm) muscle is located in the arm.
2. *Size.* The gluteus maximus (large) is the largest muscle of the buttock, and the gluteus minimus (small) is the smallest. The pectoralis major (larger of two muscles) is larger than the pectoralis minor (smaller of two muscles). A longus (long) muscle is longer than a brevis (short) muscle.
3. *Shape.* The deltoid (triangular) muscle is triangular, a quadratus (quadrate) muscle is rectangular, and a teres (round) muscle is round.
4. *Orientation of fasciculi.* A rectus (straight) muscle has muscle fasciculi running straight with the axis of the structure to which the muscle is associated, whereas the fasciculi of an oblique muscle lie oblique to the longitudinal axis of the structure.
5. *Origin and insertion.* The sternocleidomastoid originates on the sternum and clavicle and inserts onto the mastoid process of the temporal bone. The brachioradialis originates in the arm (brachium) and inserts onto the radius.
6. *Number of heads.* A biceps (*bi,* two + *ceps,* head) muscle has two heads, and a triceps (*tri,* three + *ceps,* head) muscle has three heads.

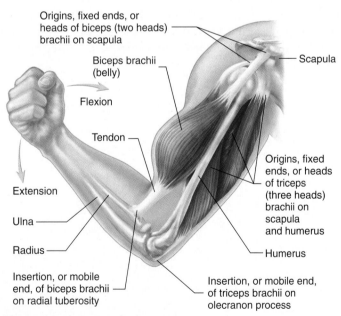

Origins, fixed ends, or heads of biceps (two heads) brachii on scapula

Biceps brachii (belly)

Scapula

Flexion

Tendon

Extension

Ulna

Radius

Origins, fixed ends, or heads of triceps (three heads) brachii on scapula and humerus

Humerus

Insertion, or mobile end, of biceps brachii on radial tuberosity

Insertion, or mobile end, of triceps brachii on olecranon process

Figure 9.1   Muscle Attachment

Muscles are attached to bones by tendons. The biceps brachii has two heads, which originate on the scapula. The triceps brachii has three heads, which originate on the scapula and humerus. The biceps brachii inserts onto the radial tuberosity and onto nearby connective tissue. The triceps brachii inserts onto the olecranon process of the ulna.

7. *Function.* An abductor moves a structure away from the midline, and an adductor moves a structure toward the midline. The masseter (a chewer) is a chewing muscle. ▼

> **2. List the criteria used to name muscles, and give an example of each.**

Movements Accomplished by Muscles

When muscles contract, the **pull** (P), or force, of muscle contraction is applied to levers, such as bones, resulting in movement of the levers (figure 9.2). A **lever** is a rigid shaft capable of turning about a hinge, or pivot point, called a **fulcrum** (F) and transferring a force applied at one point along the lever to a **weight** (W), or resistance, placed at another point along the lever. The joints function as fulcrums, the bones function as levers, and the muscles provide the pull to move the levers. Three classes of levers exist based on the relative positions of the levers, weights, fulcrums, and forces.

Class I Lever

In a **class I lever system,** the fulcrum is located between the pull and the weight (see figure 9.2*a*). A child's seesaw is an example of this type of lever. The children on the seesaw alternate between being the weight and the pull across a fulcrum in the center of the board. The head is an example of this type of lever in the body. The atlanto-occipital joint is the fulcrum, the posterior neck muscles provide the pull depressing the back of the head, and the face, which is elevated, is the weight. With the weight balanced over the fulcrum, only a small amount of pull is required to lift a weight. For example, only a very small shift in weight is needed for one child to lift the other on a seesaw. This system is quite limited, however, as

to how much weight can be lifted and how high it can be lifted. For example, consider what happens when the child on one end of the seesaw is much larger than the child on the other end.

Class II Lever

In a **class II lever system,** the weight is located between the fulcrum and the pull (see figure 9.2*b*). An example is a wheelbarrow; the wheel is the fulcrum and the person lifting on the handles provides the pull. The weight, or load, carried in the wheelbarrow is placed between the wheel and the operator. In the body, an example of a class II lever is depression of the mandible (as in opening the mouth; in this case, the whole system, the wheelbarrow and the person lifting it, is upside down).

Class III Lever

In a **class III lever system,** the most common type in the body, the pull is located between the fulcrum and the weight (see figure 9.2*c*). An example is a person using a shovel. The hand placed on the part of the handle closest to the blade provides the pull to lift the weight, such as a shovelful of dirt, and the hand placed near the end of the handle acts as the fulcrum. In the body, the action of the biceps brachii muscle (force) pulling on the radius (lever) to flex the elbow (fulcrum) and elevate the hand (weight) is an example of a class III lever. This type of lever system does not allow as great a weight to be lifted, but the weight can be lifted a greater distance. ▼

> **3. Using the terms fulcrum, lever, *and* force, *explain how contraction of a muscle results in movement. Define the three classes of levers, and give an example of each in the body.***

Muscle Anatomy

An overview of the superficial skeletal muscles is presented in figure 9.3. Muscles of the head, neck, trunk, and limbs are described in the following sections. Muscles often can cause more than one action. The discussion of muscle actions is sometimes simplified to emphasize the main actions. Consult the tables for more details.

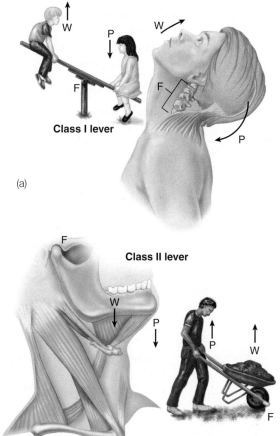

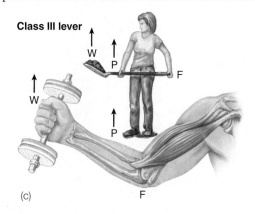

Figure 9.2 Lever Classes

(*a*) **Class I:** The fulcrum (*F*) is located between the weight (*W*) and the pull (*P*), or force. The pull is directed downward, and the weight, on the opposite side of the fulcrum, is lifted. In the example in the body, the fulcrum extends through several cervical vertebrae. (*b*) **Class II:** The weight (*W*) is located between the fulcrum (*F*) and the pull (*P*), or force. The upward pull lifts the weight. The movement of the mandible is easier to compare to a wheelbarrow if the head is considered upside down. (*c*) **Class III:** The pull (*P*), or force, is located between the fulcrum (*F*) and the weight (*W*). The upward pull lifts the weight.

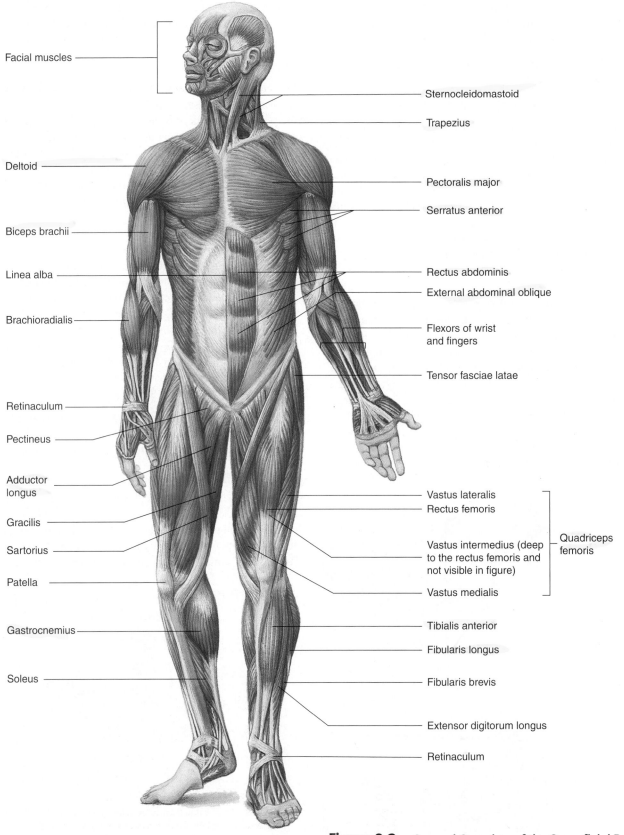

Facial muscles

Sternocleidomastoid

Trapezius

Deltoid

Pectoralis major

Serratus anterior

Biceps brachii

Linea alba

Rectus abdominis

External abdominal oblique

Brachioradialis

Flexors of wrist
and fingers

Tensor fasciae latae

Retinaculum

Pectineus

Adductor
longus

Vastus lateralis

Rectus femoris

Gracilis

Vastus intermedius (deep
to the rectus femoris and
not visible in figure)

Quadriceps
femoris

Sartorius

Patella

Vastus medialis

Gastrocnemius

Tibialis anterior

Fibularis longus

Soleus

Fibularis brevis

Extensor digitorum longus

Retinaculum

(a) **Anterior view**

Figure 9.3   General Overview of the Superficial Body
Musculature

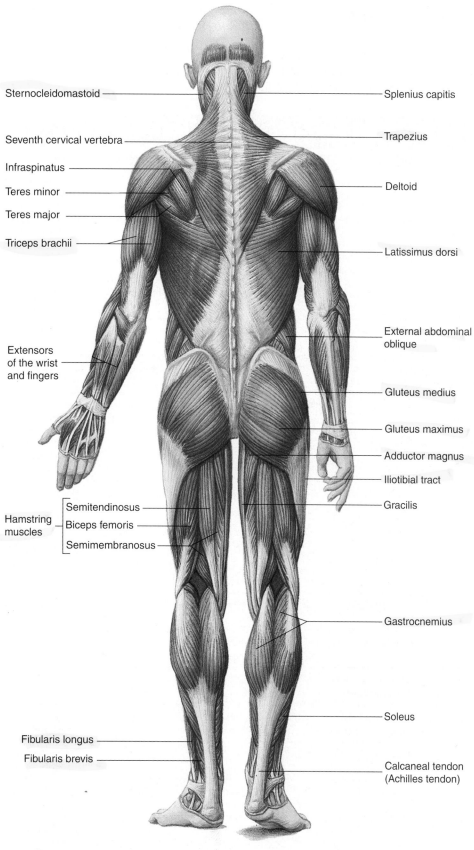

Sternocleidomastoid

Seventh cervical vertebra

Infraspinatus

Teres minor

Teres major

Triceps brachii

Extensors
of the wrist
and fingers

Hamstring
muscles
- Semitendinosus
- Biceps femoris
- Semimembranosus

Fibularis longus

Fibularis brevis

Splenius capitis

Trapezius

Deltoid

Latissimus dorsi

External abdominal
oblique

Gluteus medius

Gluteus maximus

Adductor magnus

Iliotibial tract

Gracilis

Gastrocnemius

Soleus

Calcaneal tendon
(Achilles tendon)

(b) **Posterior view**

Figure 9.3 (*continued*)

Head and Neck Muscles

The muscles of the head and neck involved in facial expression, mastication (chewing), movement of the tongue, and movement of the head and neck are emphasized in this chapter. Other head and neck muscles are considered in later chapters—movement of the eye and hearing in chapter 14, speech in chapter 20, and swallowing in chapter 21.

Facial Expression

The skeletal muscles of facial expression (table 9.1 and figure 9.4) are cutaneous muscles, which attach to and move the skin. Many

Table 9.1*	Muscles of Facial Expression (See Figures 9.4 and 9.5)			
Muscle	**Origin**	**Insertion**	**Nerve**	**Action**
Auricularis (aw-rik′ū-lăr′is) muscles	Aponeurosis over head and mastoid process	Ear	Facial	Move ear superiorly, posteriorly, and anteriorly
Buccinator (buk′sĭ-nā′tōr)	Mandible and maxilla	Orbicularis oris at angle of mouth	Facial	Retracts angle of mouth; flattens cheek
Corrugator supercilii (kōr′ŭ′gā′ter, soo′per-sil′ē-ī)	Nasal bridge and orbicularis oculi	Skin of eyebrow	Facial	Depresses medial portion of eyebrow; draws eyebrows together, as in frowning
Depressor anguli oris (dē-pres′ŏr ang′gū-lī ōr′is)	Lower border of mandible	Skin of lip near angle of mouth	Facial	Depresses angle of mouth, as in frowning
Depressor labii inferioris (dē-pres′ŏr lā′bē-ī in-fēr′ē-ōr-is)	Lower border of mandible	Skin of lower lip and orbicularis oris	Facial	Depresses lower lip, as in frowning
Levator anguli oris (lē-vā′tor, le-vē′ter ang′gū-lī ōr′is)	Maxilla	Skin at angle of mouth and orbicularis oris	Facial	Elevates angle of mouth, as in smiling
Levator labii superioris (lē-vā′tor, le-vā′ter lā′bē-ī sū-pēr′ē-ōr-is)	Maxilla	Skin and orbicularis oris of upper lip	Facial	Elevates upper lip, as in sneering
Levator labii superioris alaeque nasi (lē-vā′tor, le-vā′ter lā′bē-ī sū-pēr′ē-ōr-is ă-lak′ā nā′zī)	Maxilla	Ala at nose and upper lip	Facial	Elevates lateral side of nostril and upper lip
Levator palpebrae superioris (lē-vā′tor, le-vā′ter pal-pē′brē sū-pēr′ē-ōr-is) (not illustrated)	Lesser wing of sphenoid	Skin of eyelid	Oculomotor	Elevates upper eyelid
Mentalis (men-tā′lis)	Mandible	Skin of chin	Facial	Elevates and wrinkles skin over chin; protrudes lower lip, as in pouting
Nasalis (nā′ză-lis)	Maxilla	Bridge and ala of nose	Facial	Dilates nostril
Occipitofrontalis (ok-sip′i-tō-frŭn′tā′lis)	Occipital bone	Skin of eyebrow and nose	Facial	Moves scalp; elevates eyebrows
Orbicularis oculi (ōr-bik′ū-lā′ris ok′ū-lī)	Maxilla and frontal bones	Circles orbit and inserts near origin	Facial	Closes eyelids
Orbicularis oris (ōr-bik′ū-lā′ris ōr′is)	Nasal septum, maxilla, and mandible	Fascia and other muscles of lips	Facial	Closes lips
Platysma (plă-tiz′mă)	Fascia of deltoid and pectoralis major	Skin over inferior border of mandible	Facial	Depresses lower lip; wrinkles skin of neck and upper chest
Procerus (prō-sē′rŭs)	Bridge of nose	Frontalis	Facial	Creates horizontal wrinkles between eyes, as in frowning
Risorius (ri-sōr′ē-ŭs)	Platysma and masseter fascia	Orbicularis oris and skin at corner of mouth	Facial	Abducts angle of mouth, as in smiling
Zygomaticus major (zī′gō-mat′i-kŭs)	Zygomatic bone	Angle of mouth	Facial	Elevates and abducts upper lip, as in smiling
Zygomaticus minor (zī′gō-mat′i-kŭs)	Zygomatic bone	Orbicularis oris of upper lip	Facial	Elevates and abducts upper lip, as in smiling

*The tables in this chapter are to be used as references. As you study the muscular system, first locate the muscle on the figure, and then find its description in the corresponding table

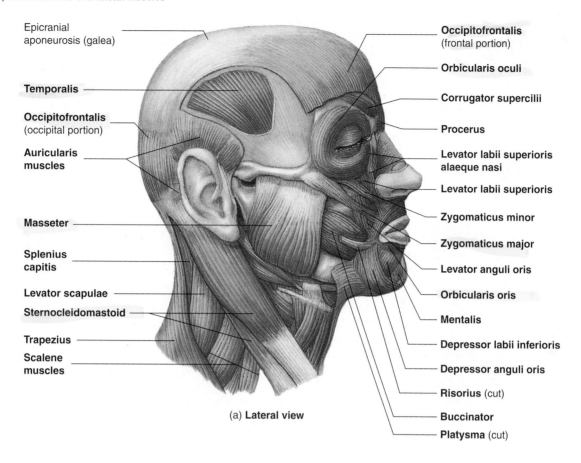

Epicranial
aponeurosis (galea)

Temporalis

Occipitofrontalis
(occipital portion)

Auricularis
muscles

Masseter

Splenius
capitis

Levator scapulae

Sternocleidomastoid

Trapezius

Scalene
muscles

Occipitofrontalis
(frontal portion)

Orbicularis oculi

Corrugator supercilii

Procerus

Levator labii superioris
alaeque nasi

Levator labii superioris

Zygomaticus minor

Zygomaticus major

Levator anguli oris

Orbicularis oris

Mentalis

Depressor labii inferioris

Depressor anguli oris

Risorius (cut)

Buccinator

Platysma (cut)

(a) Lateral view

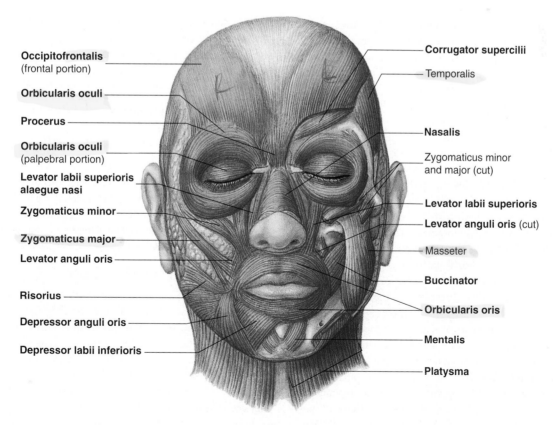

Occipitofrontalis
(frontal portion)

Orbicularis oculi

Procerus

Orbicularis oculi
(palpebral portion)

Levator labii superioris
alaegue nasi

Zygomaticus minor

Zygomaticus major

Levator anguli oris

Risorius

Depressor anguli oris

Depressor labii inferioris

Corrugator supercilii

Temporalis

Nasalis

Zygomaticus minor
and major (cut)

Levator labii superioris

Levator anguli oris (cut)

Masseter

Buccinator

Orbicularis oris

Mentalis

Platysma

(b) Anterior view

Figure 9.4   Muscle of the Head and Neck

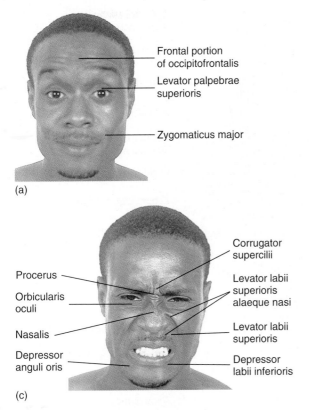

(a)

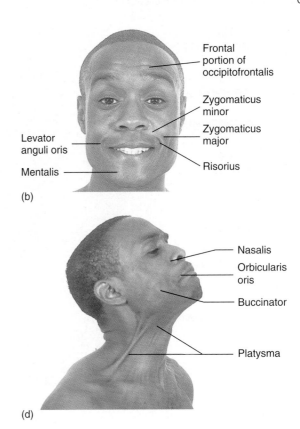

(b)

(c)

(d)

Figure 9.5   Surface Anatomy, Muscles of Facial Expression

animals have cutaneous muscles over the trunk that allow the skin to twitch to remove irritants, such as insects. In humans, cutaneous muscles are confined primarily to the face and neck. Facial expressions resulting from contraction of these muscles are used for nonverbal communication (figure 9.5).

Several muscles act on the skin around the eyes and eyebrows. The **occipitofrontalis** (ok-sip′i-tō-frŭn-tă′lis) raises the eyebrows and furrows the skin of the forehead. The **orbicularis oculi** (ōr-bik′ū-lā′ris ok′ū-lī, circular structure of the eye) closes the eyelids and causes "crow's-feet" wrinkles in the skin at the lateral corners of the eyes. The **levator palpebrae** (le-vā′ter pal-pē′brē, lifter of the eyelid) **superioris** raises the upper lids. A droopy eyelid on one side, called **ptosis** (tō′sis), usually indicates that the nerve to the levator palpebrae superioris, or the part of the brain controlling that nerve, has been damaged. The **corrugator supercilii** (kōr′ŭ-gā′ter, kōr′ŭ-gā′tōr soo′per-sil′ē-ī, wrinkle above the eyelid) draws the eyebrows inferiorly and medially, producing vertical corrugations (furrows) in the skin between the eyes.

Several muscles function in moving the lips and the skin surrounding the mouth (see figures 9.4 and 9.5). The **orbicularis oris** (ōr-bik′ū-lā′ris ōr′is, circular structure of the mouth) closes the mouth and the **buccinator** (buk′si-nā-tōr, trumpeter) flattens the cheek. They are called the kissing muscles because they pucker the lips. Smiling is accomplished by the **zygomaticus** (zī′gō-mat′i-kŭs) **major** and **minor,** the **levator anguli** (ang′gū-lī) **oris,** and the **risorius** (rī-sōr′ē-ŭs, laughter). Sneering is accomplished by the **levator labii** (lā′bē-ī, lip) **superioris** and frowning or pouting

by the **depressor anguli oris,** the **depressor labii inferioris,** and the **mentalis** (men-tā′lis, chin). If the mentalis muscles are well developed on each side of the chin, a chin dimple, where the skin is tightly attached to the underlying bone or other connective tissue, may appear between the two muscles. ▼

4. *What are cutaneous muscles? How are they used for communication?*
5. *What causes a dimple on the chin?*

P R E D I C T
Harry Wolf, a notorious flirt, on seeing Sally Gorgeous, raises his eyebrows, winks, whistles, and smiles. Name the facial muscles he uses to carry out this communication. Sally, thoroughly displeased with this exhibition, frowns and flares her nostrils in disgust. What muscles does she use (see table 9.1)?

Movement of the Mandible

Movement of the mandible is complex and is accomplished by the muscles of mastication and the hyoid muscles. Mastication is chewing. The **muscles of mastication** are the **temporalis** (tem′pŏ-rā′lis), **masseter** (mă-sē′ter, chewer), **lateral pterygoid** (ter′ĭ-goydz, wing-shaped) and **medial pterygoid** muscles (table 9.2 and figure 9.6). The temporalis and masseter muscles are powerful elevators of the mandible; they close the mouth to crush food between the teeth. Clench your teeth and feel the temporalis on the side of the head and the masseter on the side of the cheek. The medial and lateral pterygoids cause lateral and medial excursion of the mandible, moving it from side to side.

Table 9.2	Muscles of Mastication (See Figure 9.6)			
Muscle	**Origin**	**Insertion**	**Nerve**	**Action**
Temporalis (tem-pŏ-rā′lis)	Temporal fossa	Anterior portion of mandibular ramus and coronoid process	Mandibular division of trigeminal	Elevates and retracts mandible; involved in excursion
Masseter (ma′se-ter)	Zygomatic arch	Lateral side of mandibular ramus	Mandibular division of trigeminal	Elevates and protracts mandible; involved in excursion
Pterygoids (ter′i-goydz) Lateral	Lateral side of lateral pterygoid plate and greater wing of sphenoid	Condylar process of mandible and articular disk	Mandibular division of trigeminal	Protracts and depresses mandible; involved in excursion
Medial	Medial side of lateral pterygoid plate and tuberosity of maxilla	Medial surface of mandible	Mandibular division of trigeminal	Protracts and elevates mandible; involved in excursion

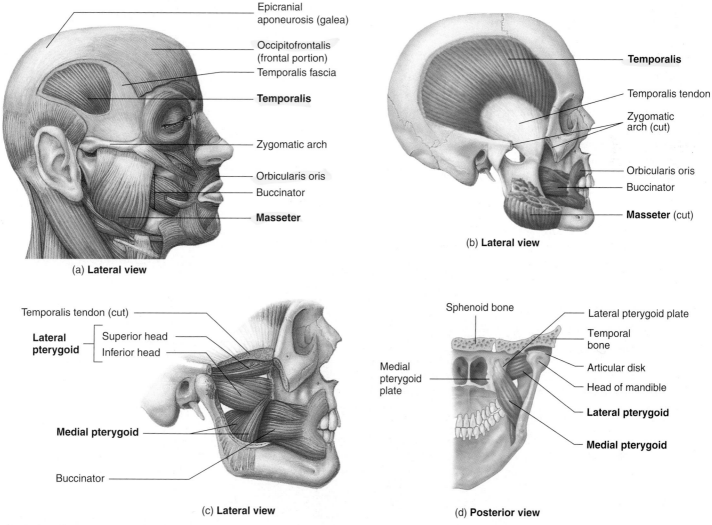

(a) **Lateral view**

(b) **Lateral view**

(c) **Lateral view**

(d) **Posterior view**

Figure 9.6 Muscles of Mastication

(a) Temporalis is exposed. (b) Masseter and zygomatic arch are cut away to expose the temporalis. (c) Masseter and temporalis muscles are removed, and the zygomatic arch and part of the mandible are cut away to reveal the deeper muscles. (d) Frontal section of the head, showing the pterygoid muscles.

Table 9.3 Hyoid Muscles (See Figure 9.7)

Muscle	Origin	Insertion	Nerve	Action
Suprahyoid Muscles				
Digastric (dī-gas'trik)	Mastoid process (posterior belly)	Mandible near midline (anterior belly)	Posterior belly—facial; anterior belly—mandibular division of trigeminal	Depresses and retracts mandible; elevates hyoid
Geniohyoid (jĕ-nī-ō-hī'oyd) (not illustrated)	Mental protuberance of mandible	Body of hyoid	Fibers of C1 and C2 with hypoglossal	Protracts hyoid; depresses mandible
Mylohyoid (mī'lō-hī'oyd)	Body of mandible	Hyoid	Mandibular division of trigeminal	Elevates floor of mouth and tongue; depresses mandible when hyoid is fixed
Stylohyoid (stī-lō-hī'oyd)	Styloid process	Hyoid	Facial	Elevates hyoid
Infrahyoid Muscles				
Omohyoid (ō-mō-hī'oyd)	Superior border of scapula	Hyoid	Upper cervical through ansa cervicalis	Depresses hyoid; fixes hyoid in mandibular depression
Sternohyoid (ster'nō-hī'oyd)	Manubrium and first costal cartilage	Hyoid	Upper cervical through ansa cervicalis	Depresses hyoid; fixes hyoid in mandibular depression
Sternothyroid (ster'nō-thī'royd)	Manubrium and first or second costal cartilage	Thyroid cartilage	Upper cervical through ansa cervicalis	Depresses larynx; fixes hyoid in mandibular depression
Thyrohyoid (thī-rō-hī'oyd)	Thyroid cartilage	Hyoid	Upper cervical, passing with hypoglossal	Depresses hyoid and elevates thyroid cartilage of larynx; fixes hyoid in mandibular depression

The **hyoid muscles** attach to the hyoid bone (table 9.3 and figure 9.7). The suprahyoid muscles are superior to the hyoid, whereas the infrahyoid muscles are inferior to it. Slight mandibular depression involves relaxation of the mandibular elevators and the pull of gravity. The **digastric** (two bellies) **muscles** are the most important muscles for opening the mouth. The hyoid muscles also move the hyoid or hold it in place. Along with other muscles, the hyoid muscle are involved with swallowing and movement of the larynx. To observe this movement, place your hand on your larynx (Adam's apple) and swallow. ▼

6. *Name the muscles primarily responsible for opening and closing the mandible and for lateral and medial excursion of the mandible.*

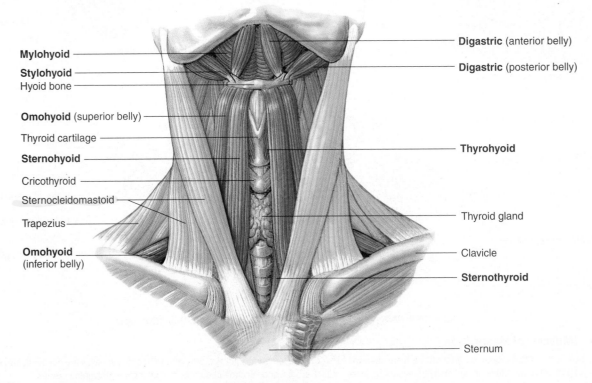

Figure 9.7 Hyoid Muscles   **Anterior superficial view**

Table 9.4	Tongue Muscles (See Figure 9.8)			
Muscle	**Origin**	**Insertion**	**Nerve**	**Action**
Intrinsic Muscles				
Longitudinal, transverse, and vertical (not illustrated)	Within tongue	Within tongue	Hypoglossal	Change tongue shape
Extrinsic Muscles				
Genioglossus (jĕ′nī-ō-glos′ŭs)	Mental protuberance of mandible	Tongue	Hypoglossal	Depresses and protrudes tongue
Hyoglossus (hī′ō-glos′ŭs)	Hyoid	Side of tongue	Hypoglossal	Retracts and depresses side of tongue
Styloglossus (stī′lō-glos′ŭs)	Styloid process of temporal bone	Tongue (lateral and inferior)	Hypoglossal	Retracts tongue
Palatoglossus (pal-ă-tō-glos′ŭs)	Soft palate	Tongue	Pharyngeal plexus	Elevates posterior tongue

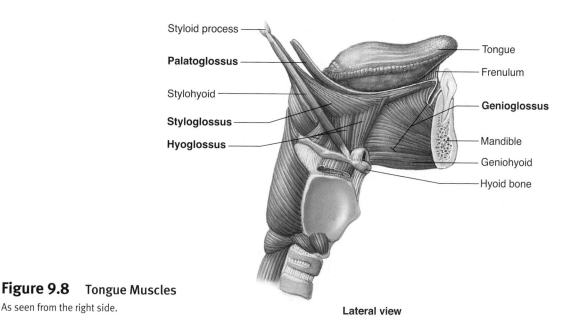

Styloid process

Palatoglossus

Stylohyoid

Styloglossus

Hyoglossus

Tongue

Frenulum

Genioglossus

Mandible

Geniohyoid

Hyoid bone

Lateral view

Figure 9.8 Tongue Muscles

As seen from the right side.

Tongue Movements

The tongue is very important in mastication and speech: (1) It moves food around in the mouth; (2) with the buccinator (see figure 9.6a), it holds food in place while the teeth grind it; (3) it pushes food up to the palate and back toward the pharynx to initiate swallowing; and (4) it changes shape to modify sound during speech. The tongue consists of intrinsic and extrinsic muscles (table 9.4 and figure 9.8). The **intrinsic muscles** are entirely within the tongue and are named for their fiber orientation. They are involved in changing the shape of the tongue. The **extrinsic muscles** are outside of the tongue but attached to it. They are named for their origin and insertion. The extrinsic muscles move the tongue about as a unit and also help change its shape.

 Tongue Rolling

Everyone can change the shape of the tongue, but not everyone can roll the tongue into the shape of a tube. This ability apparently is partially controlled genetically, but other factors are involved. In some cases, one of a pair of identical twins can roll the tongue but the other twin cannot. It is not known exactly what tongue muscles are involved in **tongue rolling,** and no anatomical differences are reported between tongue rollers and nonrollers. ▼

7. Define the intrinsic and extrinsic tongue muscles and describe the movements they produce.

Table 9.5 Muscles Moving the Head and Neck (See Figures 9.4a, 9.9, and 9.10)

Muscle	Origin	Insertion	Nerve	Action
Levator scapulae (lē-vă′tor, le-vă′ter skap′ŭ-lē)	Transverse processes of C1–C4	Superior angle of scapula	Dorsal scapular	laterally flexes neck; elevates, retracts, and rotates scapula
Scalene (skā′lēn) muscles	C2–C6	First and second ribs	Cervical and brachial plexuses	Flex, laterally flex, and rotate neck
Semispinalis capitis (sem′ē-spī-nă′lis ka′pĭ-tis)	C4–T6	Occipital bone	Dorsal rami of cervical nerves	Extends and rotates head
Splenius capitis (splē′nē-ŭs ka′pĭ-tis)	C4–T6	Superior nuchal line and mastoid process	Dorsal rami of cervical nerves	Extends, rotates, and laterally flexes head
Splenius cervicis (splē′nē-ŭs ser-vī′sis)	Spinous processes of C3–C5	Transverse processes of C1–C3	Dorsal rami of cervical nerves	Rotates and extends neck
Sternocleidomastoid (ster′nō-klī′dō-mas′toyd)	Manubrium and medial clavicle	Mastoid process and superior nuchal line	Accessory (cranial nerve XI)	One contracting alone: laterally flexes head and neck to same side and rotates the head and neck to opposite side Both contracting together: flexes neck
Trapezius (tra-pē′zē-ŭs)	Occipital protuberance, nuchal ligament, spinous processes of C7–T12	Clavicle, acromion, and scapular spine	Accessory (cranial nerve XI)	Extends and laterally flexes head and neck

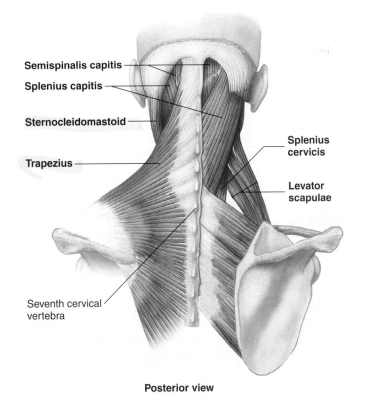

Semispinalis capitis

Splenius capitis

Sternocleidomastoid

Trapezius

Splenius cervicis

Levator scapulae

Seventh cervical vertebra

Posterior view

Figure 9.9 **Superficial Neck Muscles**

Head and Neck Movements

The head and neck are moved by posterior, lateral, and anterior neck muscles (table 9.5). The **trapezius** (tra-pē′zē-ŭs) and **splenius capitis** (splē′nē-ŭs ka′pĭ-tis) on the posterior neck are the major superficial muscles extending the head and neck (figure 9.9).

The lateral neck muscles are involved with rotation, flexion, and lateral flexion of the head and neck (see figure 9.4a). The **sternocleidomastoid** (ster′nō-klī′dō-mas′toyd) muscle is the prime mover of the lateral muscle group. Depending on what other neck muscles are doing, the sternocleidomastoid muscles are involved with different movements of the head and neck. While the head is held level, contraction of one sternocleidomastoid muscle helps rotate the head to the opposite side. Contraction of one sterno-cleidomastoid muscle can also laterally flex the head and neck to the same side and rotate the neck so that the chin turns to the opposite side, as occurs with an upward sideways glance. Contraction of both sternocleiodomastoid muscles results in flex-ion of the neck, as occurs when lifting the head off the ground while supine.

Deep neck muscle are located along the vertebral column. The deep posterior neck muscles (figure 9.10) extend, rotate, and laterally flex the head and neck, whereas the deep anterior neck muscles (not illustrated) flex the head.

Torticollis

Torticollis (tōr-ti-kol′is, twisted neck), or wry neck, is an abnormal twisting or bending of the neck. It can be caused by damage to the sternocleidomastoid muscle, resulting in fibrous tissue formation and contracture of the muscle. Damage to an infant's neck muscles because of interuterine positioning or a difficult birth may cause torticollis. In most cases, the sternocleidomastoid fibrosis resolves spontaneously. **Spasmodic torticollis** can occur in adults and may have many causes. It is characterized by intermittent contraction of neck muscles, especially the sternocleidomastoid and trapezius, resulting in rotation, flexion, and extension of the neck and elevation of the shoulder. ▼

8. *Name the major movements of the head caused by contraction of the trapezius, splenius capitis, and sternocleidomastoid muscles.*
9. *What movements of the head and neck are produced by the lateral and deep neck muscles?*

P R E D I C T
The parallel ridges or "grain" on the surface of muscles in muscle drawings represent muscle fasciculi (bundles). Shortening of the right sternocleidomastoid muscle rotates the head in which direction?

Trunk Muscles

Back Muscles

The muscles that extend, laterally flex, and rotate the vertebral column are divided into superficial and deep groups (table 9.6 and see figure 9.10). The superficial muscles are collectively called the **erector spinae** (spī′nē), which literally means the muscles that make the spine erect. The erector spine consists of three subgroups: the **iliocostalis** (il′ē-ō-kos-tā′lis), the **longissimus** (lon-gis′i-mŭs), and the **spinalis** (sp-ī-nā′lis). The longissimus group accounts for most of the muscle mass in the lower back. In general, the muscles of the erector spinae extend from the vertebrae to the ribs or from rib to rib, whereas the muscles of the deep group attach between the transverse and spinous processes of individual vertebra. ▼

10. *List the actions of the group of back muscles that attaches to the vertebrae or ribs (or both). What is the name of the superficial group?*

 Back Pain

Low back pain can result from poor posture, from being overweight, or from having a poor fitness level. A few changes may help: sitting and standing up straight; using a low-back support when sitting; losing weight; exercising, especially the back and abdominal muscles; and sleeping on your side on a firm mattress. Sleeping on your side all night, however, may be difficult because most people change position over 40 times during the night. ▼

Abdominal Wall Muscles

Muscles of the abdominal wall (table 9.7 and figure 9.11) flex, laterally flex, and rotate the vertebral column. Contraction of the abdominal muscles when the vertebral column is fixed decreases the volume of the abdominal cavity and the thoracic cavity and can aid in such functions as forced expiration, vomiting, defecation, urination, and childbirth. The crossing pattern of the abdominal muscles creates a strong anterior wall, which holds in and protects the abdominal viscera.

In a relatively muscular person with little fat, a vertical indentation extends from the xiphoid process of the sternum through the navel to the pubis. This tendinous area of the abdominal wall, called the **linea alba** (lin′ē-ă al′bă, white line), consists of white connective tissue rather than muscle. On each side of the linea alba is a **rectus abdominis** (rek′tŭs ab-dom′i-nis, *rectus,* straight) muscle, which flexes the vertebral column. **Tendinous intersections** cross the rectus abdominis at three or more locations, causing the abdominal wall of a well-muscled lean person to appear segmented. The **linea semilunaris** (sem-ē-loo-nar′is, a crescent- or half-moon-shaped line) is a slight groove in the abdominal wall running parallel to the lateral edge of each rectus abdominis muscle. Lateral to each rectus abdominis is three layers of muscle. From superficial to deep, these muscles are the **external abdominal oblique, internal abdominal oblique,** and **transversus abdominis** (trans-ver′sŭs ab-dom′i-nis) muscles. The fasciculi of these three muscle layers are oriented in different directions to one another. When these muscles contract, they flex, laterally flex, and rotate the vertebral column or compress the abdominal contents. The lateral abdominal wall muscles connect to the linea alba by broad, thin tendons called aponeuroses. These aponeuroses pass anterior and posterior to the rectus abdominis, enveloping it in connective tissue collectively called the **rectus sheath** (see figure 9.11*b*). ▼

11. *List the muscles of the anterior abdominal wall. What are their functions?*
12. *Define linea alba, tendinous intersection, linea semilunaris, and rectus sheath.*

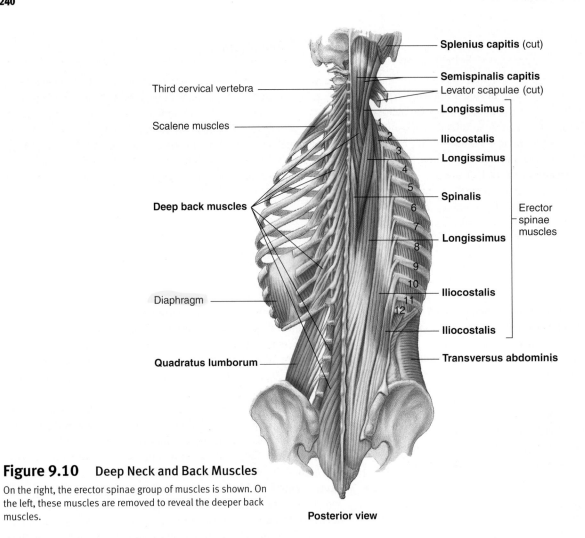

Third cervical vertebra

Scalene muscles

Deep back muscles

Diaphragm

Quadratus lumborum

— **Splenius capitis** (cut)

— **Semispinalis capitis**
— Levator scapulae (cut)

— **Longissimus**

— **Iliocostalis**
— **Longissimus**

— **Spinalis**

— **Longissimus**

— **Iliocostalis**

— **Iliocostalis**

— **Transversus abdominis**

1
2
3
4
5
6
7
8
9
10
11
12

Erector
spinae
muscles

Figure 9.10   **Deep Neck and Back Muscles**

On the right, the erector spinae group of muscles is shown. On
the left, these muscles are removed to reveal the deeper back
muscles.

Posterior view

Table 9.6	Muscles of the Back (See Figure 9.10)			
Muscle	**Origin**	**Insertion**	**Nerve**	**Action**
Erector spinae (ē-rek′tŏr, ē-rek′tŏr spī′nē; divides into three columns)				
Iliocostalis (il′ē-ō-kos-tā′lis)	Sacrum, ilium, and angles of ribs	Angles of ribs and transverse processes of vertebrae	Cervical, thoracic, and lumbar spinal nerves	Extends, laterally flexes, and rotates vertebral column; maintains posture
Longissimus (lon-gis′i-mŭs)	Sacrum, transverse processes of lumbar, thoracic, and lower cervical vertebrae	Transverse processes of vertebrae, angles of ribs, and mastoid process	Cervical, thoracic, and lumbar spinal nerves	Extends head, neck, and vertebral column; maintains posture
Spinalis (spī-nā′lis)	Spinous processes of T11–L2 and C7	Spinous processes of upper thoracic vertebrae and axis	Cervical and thoracic, spinal nerves	Extends neck and vertebral column; maintains posture
Deep back muscles	Spinous and transverse processes of each vertebra	Next superior spinous or transverse process, ribs, and occipital bone	Spinal nerves	Extends back and neck, laterally flexes and rotates vertebral column
Quadratus lumborum (kwah-drā′tŭs lŭm-bōr′ŭm)	Iliac crest and lower lumbar vertebrae	Twelfth rib and transverse processes of upper four lumbar vertebrae	Twelfth thoracic and upper lumbar spinal nerves	One acting alone: laterally flexes vertebral column and depresses twelfth rib Both contracting together: extend vertebral column

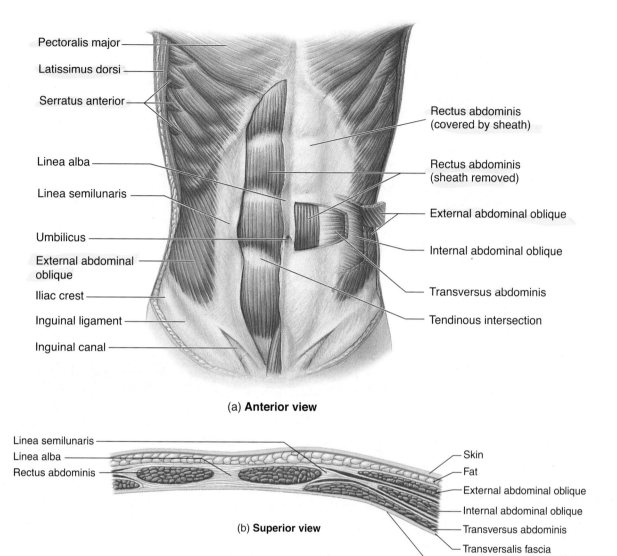

(a) **Anterior view**

(b) **Superior view**

Figure 9.11 Anterior Abdominal Wall Muscles

(a) Windows in the side reveal the various muscle layers. (b) Cross section superior to the umbilicus.

Table 9.7	Muscles of the Abdominal Wall (See Figure 9.11)			
Muscle	**Origin**	**Insertion**	**Nerve**	**Action**
Anterior				
Rectus abdominis (rek'tŭs ab-dom'i-nis)	Pubic crest and symphysis pubis	Xiphoid process and inferior ribs	Branches of lower thoracic	Flexes vertebral column; compresses abdomen
External abdominal oblique	Fifth to twelfth ribs	Iliac crest, inguinal ligament, and rectus sheath	Branches of lower thoracic	Flexes and rotates vertebral column; compresses abdomen; depresses thorax
Internal abdominal oblique	Iliac crest, inguinal ligament, and lumbar fascia	Tenth to twelfth ribs and rectus sheath	Lower thoracic	Flexes and rotates vertebral column; compresses abdomen; depresses thorax
Transversus abdominis (trans-ver'sŭs ab-dom'i-nis)	Seventh to twelfth costal cartilages, lumbar fascia, iliac crest, and inguinal ligament	Xiphoid process, linea alba, and pubic tubercle	Lower thoracic	Compresses abdomen

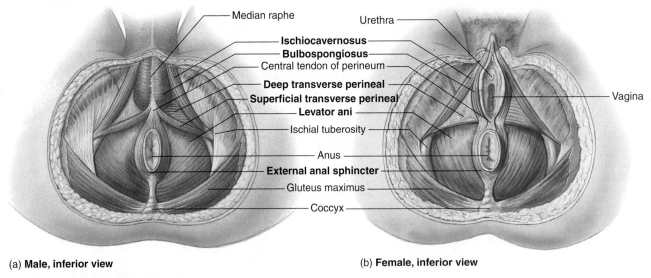

(a) **Male, inferior view**   (b) **Female, inferior view**

Figure 9.12   Muscles of the Pelvic Floor and Perineum

Table 9.8	Muscles of the Pelvic Floor and Perineum (See Figure 9.12)			
Muscle	**Origin**	**Insertion**	**Nerve**	**Action**
Bulbospongiosus (bul′bō-spŭn′jē-ō′sŭs)	Male—central tendon of perineum and median raphe of penis	Dorsal surface of penis and bulb of penis	Pudendal	Constricts urethra; erects penis
	Female—central tendon of perineum	Base of clitoris	Pudendal	Erects clitoris
Coccygeus (kok-si′jē-ŭs) (not illustrated)	Ischial spine	Coccyx	S3 and S4	Elevates and supports pelvic floor
Ischiocavernosus (ish′ē-ō-kav′er-nō′sŭs)	Ischial ramus	Corpus cavernosum	Perineal	Compresses base of penis or clitoris
Levator ani (lē-vā′tor, le-vā′ter ā′nī)	Posterior pubis and ischial spine	Sacrum and coccyx	Fourth sacral	Elevates anus; supports pelvic viscera
External anal sphincter (ā′năl sfingk′ter)	Coccyx	Central tendon of perineum	Fourth sacral and pudenda	Keeps orifice of anal canal closed
External urethral sphincter (ū-rē′thrăl sfingk′ter) (not illustrated)	Pubic ramus	Median raphe	Pudendal	Constricts urethra
Transverse perinei (pĕr′i-nē′ī)				
Deep	Ischial ramus	Median raphe	Pudendal	Supports pelvic floor
Superficial	Ischial ramus	Central perineal	Pudendal	Fixes central tendon

Pelvic Floor and Perineum

The pelvis is a ring of bone (see chapter 7) with an inferior opening that is closed by a muscular wall, the **pelvic floor,** through which the anus and the urogenital openings penetrate (table 9.8 and figure 9.12). Most of the pelvic floor is formed by the **coccygeus** (kok-si′jē-ŭs) muscle and the **levator ani** (a′nī) muscle, referred to jointly as the **pelvic diaphragm.** The area inferior to the pelvic floor is the **perineum** (per′i-nē′ŭm), which is somewhat diamond-shaped. The anterior half of the diamond is the urogenital triangle, and the posterior half is the anal triangle (see chapter 24).

The urogenital triangle contains the **external urethral sphincter,** which allows voluntary control of urination, and the **external anal sphincter,** which allows voluntary control of defecation. During pregnancy or delivery, the muscles of the pelvic diaphragm and perineum may be stretched or torn, changing the position of the urinary bladder and urethra. As a result, stress urinary incontinence may develop, in which coughing or physical exertion can cause dribbling of urine. ▼

13. What openings penetrate the pelvic floor muscles? Name the area inferior to the pelvic floor.

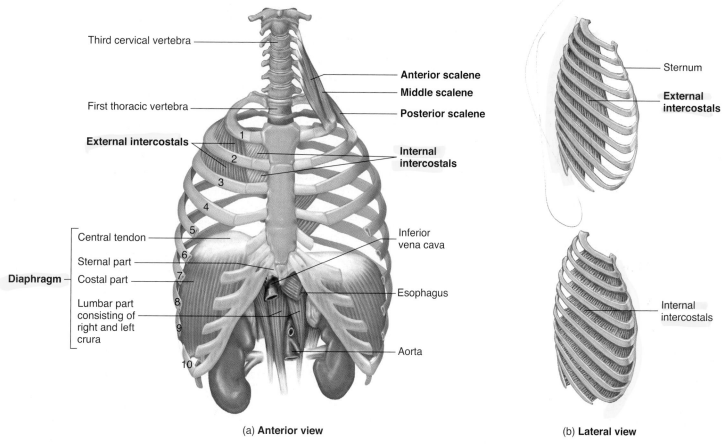

(a) **Anterior view**

(b) **Lateral view**

Figure 9.13   Muscles of the Thorax

Table 9.9	Muscles of the Thorax (See Figure 9.13)			
Muscle	**Origin**	**Insertion**	**Nerve**	**Action**
Diaphragm	Interior of ribs, sternum, and lumbar vertebrae	Central tendon of diaphragm	Phrenic	Inspiration; depresses floor of thorax
Intercostalis (in'ter-kos-ta'lis)				
External	Inferior margin of each rib	Superior border of next rib below	Intercostal	Inspiration; elevates ribs
Internal	Superior margin of each rib	Inferior border of next rib above	Intercostal	Expiration; depresses ribs
Scalene (skā-lēn) muscles	Transverse processes of C2–C6	First and second rib	Cervical and brachial plexuses	Elevates first and second rib

Thoracic Muscles

The muscles of the thorax are involved mainly in the process of breathing (table 9.9 and figure 9.13). Contraction and relaxation of these muscle results in changes in thoracic volume, which results in the movement of air into and out of the lungs (see chapter 20). The dome-shaped **diaphragm** (dī'ă-fram, a partition wall) causes the major change in thoracic volume during quiet breathing. When it contracts, the top of the diaphragm moves inferiorly, causing the volume of the thoracic cavity to increase. If the diaphragm or the phrenic nerve supplying it is severely damaged, the amount of air moving into and out of the lungs may be so small that the individual is likely to die without the aid of an artificial respirator.

Muscles associated with the ribs change thoracic volume by moving the ribs. The **external intercostals** (in′ter-kos′tŭlz, between ribs) increase thoracic volume by elevating the ribs during inspiration, whereas the **internal intercostals** decrease thoracic volume by depressing the ribs during forced expiration. When the vertebrae are fixed in position by neck muscles, the **scalene** (skā′lēn) muscles elevate the first two ribs. This movement is important when taking a deep breath. When the ribs are fixed in position by the intercostals, the scalenes can cause flexion, lateral flexion, and rotation of the neck (see table 9.5). Thus, the ends of a muscle can cause different movements. ▼

> 14. Describe how the diaphragm, intercostals, and scalenes change thoracic volume.

Scapular and Upper Limb Muscles

Scapular Movements

Trunk muscles form the major connections of the scapula to the body (table 9.10 and figure 9.14). Remember that the scapula is attached to the rest of the skeleton only by the clavicle (see chapter 7). The muscles attaching the scapula to thorax are the **trapezius** (tra-pē′zē-ŭs, a four-side figure with no sides parallel), **levator scapulae** (lē-vā′tor skap′ū-lē, lifter of the scapula), **rhomboideus** (rom-bō-id′ē-ŭs, rhomboid-shaped, an oblique parallelogram with unequal sides) **major** and **minor, serratus** (ser-ā′tŭs, serrated, referring to its appearance on the thorax) **anterior,** and **pectoralis** (pek′tō-ra′lis, chest) **minor.** These muscles move the scapula, permitting a wide range of movements of the upper limb, or act as fixators to hold the scapula firmly in position when the arm muscles contract. The trapezius forms the upper line from each shoulder to the neck, and the origin of the serratus anterior from the first eight or nine ribs can be seen along the lateral thorax. The serratus anterior inserts onto the medial border of the scapula (see figure 9.14c). ▼

> 15. Name six muscles that attach the scapula to the trunk. As a group, what is the function of these muscles?

Table 9.10 Muscles Acting on the Scapula (See Figure 9.14)

Muscle	Origin	Insertion	Nerve	Action
Levator scapulae (lē-vā′tor, le-vā′ter skap′ū-lē)	Transverse processes of C1–C4	Superior angle of scapula	Dorsal scapular	Elevates, retracts, and rotates scapula; laterally flexes neck
Pectoralis minor (pek′tō-ra′lis)	Third to fifth ribs	Coracoid process of scapula	Medial pectoral	Depresses scapula or elevates ribs
Rhomboideus (rom-bō-id′ē-ŭs)				
Major	Spinous processes of T1–T4	Medial border of scapula	Dorsal scapular	Retracts, rotates, and fixes scapula
Minor	Spinous processes of C6–C7	Medial border of scapula	Dorsal scapular	Retracts, slightly elevates, rotates, and fixes scapula
Serratus anterior (ser-ā′tūs)	First to eighth or ninth ribs	Medial border of scapula	Long thoracic	Rotates and protracts scapula; elevates ribs
Trapezius (tra-pē′zē-ŭs)	External occipital protuberance, ligamentum nuchae, and spinous processes of C7–T12	Clavicle, acromion, and scapular spine	Accessory and cervical plexus	Elevates, depresses, retracts, rotates, and fixes scapula; extends neck

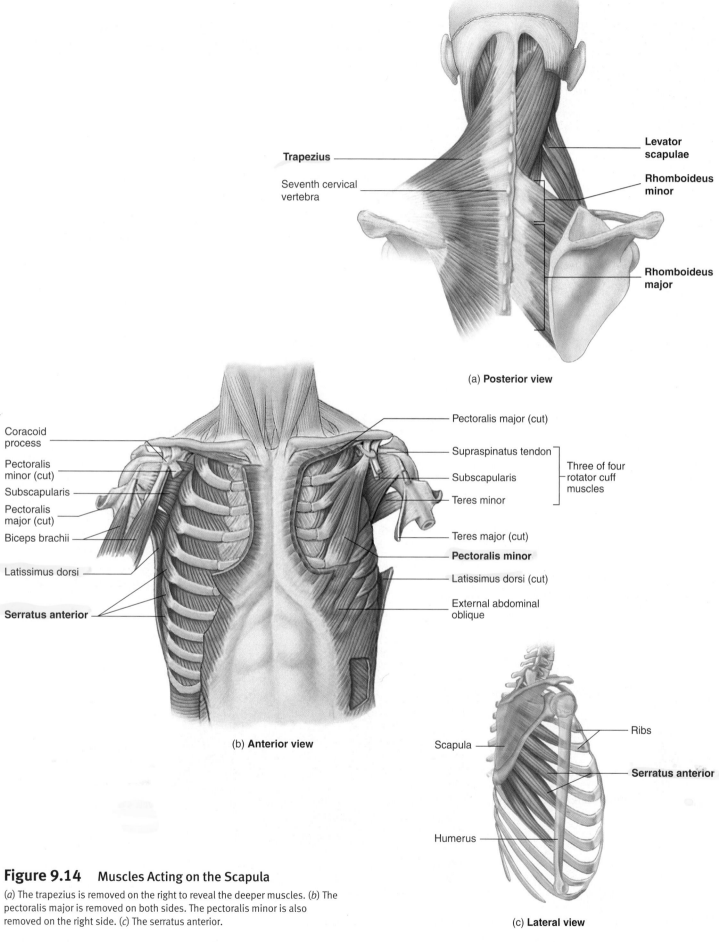

Trapezius

Seventh cervical vertebra

Levator scapulae

Rhomboideus minor

Rhomboideus major

(a) **Posterior view**

Coracoid process

Pectoralis minor (cut)

Subscapularis

Pectoralis major (cut)

Biceps brachii

Latissimus dorsi

Serratus anterior

Pectoralis major (cut)

Supraspinatus tendon ⎤
Subscapularis ⎥ Three of four rotator cuff muscles
Teres minor ⎦

Teres major (cut)

Pectoralis minor

Latissimus dorsi (cut)

External abdominal oblique

(b) **Anterior view**

Scapula

Ribs

Serratus anterior

Humerus

(c) **Lateral view**

Figure 9.14 Muscles Acting on the Scapula

(*a*) The trapezius is removed on the right to reveal the deeper muscles. (*b*) The pectoralis major is removed on both sides. The pectoralis minor is also removed on the right side. (*c*) The serratus anterior.

Table 9.11 Muscles Acting on the Arm (See Figures 9.15–9.17)

Muscle	Origin	Insertion	Nerve	Action
Coracobrachialis (kōr′ă-kō-brā-kē-ā′lis)	Coracoid process of scapula	Midshaft of humerus	Musculocutaneous	Adducts arm and flexes shoulder
Deltoid (del′toyd)	Clavicle, acromion, and scapular spine	Deltoid tuberosity	Axillary	Flexes and extends shoulder; abducts and medially and laterally rotates arm
Latissimus dorsi (lă-tis′i-mŭs dōr′sī)	Spinous processes of T7–L5; sacrum and iliac crest; inferior angle of scapula in some people	Medial crest of intertubercular groove	Thoracodorsal	Adducts and medially rotates arm; extends shoulder
Pectoralis major (pek′tō-rā′lis)	Clavicle, sternum, superior six costal cartilages, and external abdominal oblique aponeurosis	Lateral crest of intertubercular groove	Medial and lateral pectoral	Flexes shoulder; adducts and medially rotates arm; extends shoulder from flexed position
Teres major (ter′ēz, tēr-ēz)	Lateral border of scapula	Medial crest of intertubercular groove	Lower subscapular C5 and C6	Extends shoulder; adducts and medially rotates arm
Rotator Cuff				
Infraspinatus (in-fră-spī-nā′tŭs)	Infraspinous fossa of scapula	Greater tubercle of humerus	Suprascapular C5 and C6	Laterally rotates arm; holds head of humerus in place
Subscapularis (sŭb-skap-ū-lā′ris)	Subscapular fossa	Lesser tubercle of humerus	Upper and lower subscapular C5 and C6	Medially rotates arm; holds head of humerus in place
Supraspinatus (soo-pră-spī-nā′tŭs)	Supraspinous fossa	Greater tubercle of humerus	Suprascapular C5 and C6	Abducts arm; holds head of humerus in place
Teres minor (ter′ēz, tēr-ēz)	Lateral border of scapula	Greater tubercle of humerus	Axillary C5 and C6	Laterally rotates and adducts arm; holds head of humerus in place

Arm Movements

The muscles moving the arm are involved in flexion, extension, abduction, adduction, rotation, and circumduction (table 9.11). Muscles moving the arm have their origins on the trunk and scapula and insert on the arm. The trunk muscles moving the arm are the **pectoralis major** (figure 9.15) and the **latissimus dorsi** (lă-tis′i-mŭs dōr′sī, wide back) muscles (figure 9.16). Notice that the pectoralis major muscle is listed in table 9.11 as both a flexor and an extensor of the shoulder. It flexes the extended shoulder and extends the flexed shoulder. Try these movements and notice the position and action of the muscle.

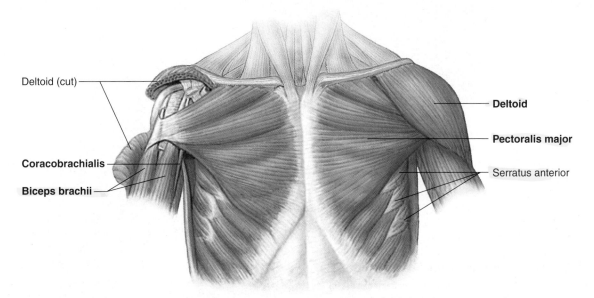

Figure 9.15 Anterior Muscles Attaching the Upper Limb to the Body

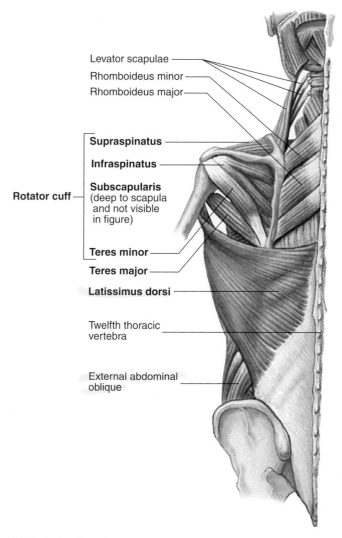

(a) **Posterior view**

Figure 9.16 **Posterior Muscles Attaching the Upper Limb to the Body**

(*a*) Posterior view of muscles. (*b*) Posterior view of the left shoulder and back of a cadaver. (*c*) Surface anatomy of the left shoulder and back.

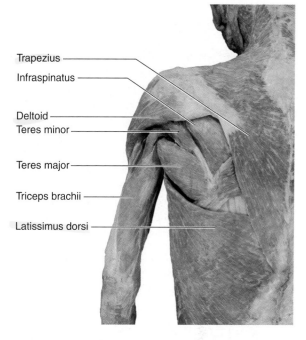

(b) **Posterior view**

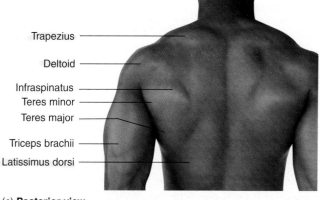

(c) **Posterior view**

Muscles located in the shoulder move the arm: The **deltoid** (del′toyd, resembling the Greek letter delta) muscle, **teres** (ter′ēz, tēr-ēz, round) **major,** and **rotator cuff muscles** (see figures 9.15 and 9.16). The deltoid muscle is like three muscles in one: The anterior fibers flex the shoulder, the lateral fibers abduct the arm, and the posterior fibers extend the shoulder.

The rotator cuff muscles are the **supraspinatus** (above the spine of the scapula), **infraspinatus** (below the spine of the scapula), **teres major,** and **subscapularis** (under the scapula). They can be remembered with the mnemonic **SITS: S**upraspinatus,

Infraspinatus, **T**eres minor, and **S**ubscapularis. The rotator cuff muscles form a cuff or cap over the proximal humerus (figure 9.17). Although they can cause movement of the arm, the primary function of the rotator cuff muscles is holding the head of the humerus in the glenoid cavity.

Muscles located in the arm move the arm: the **coracobrachialis** (kōr′ă-kō-brā-kē-ā′lis, attaches to the coracoid process and the arm), **biceps** (bī′seps, two heads) **brachii** (brā′kē-ī, arm), and the **triceps brachii** (trī′seps, three heads). The actions of the trunk, shoulder, and arm muscles on the shoulder and arm are summarized in table 9.12.

Several muscles acting on the arm can be seen very clearly in the living individual. The pectoralis major forms the upper chest, and the deltoids are prominent over the shoulders. The deltoid is a common site for administering injections.

Shoulder Pain

A **rotator cuff injury** involves damage to one or more of the rotator cuff muscles or their tendons. Athletes, such as tennis players and baseball pitchers, may tear their rotator cuffs. Such tears result in pain in the anterosuperior part of the shoulder. Older people may also develop such pain because of **degenerative tendonitis** of the rotator cuff. The supraspinatus tendon is the most commonly affected part of the rotator cuff in either trauma or degeneration, probably because it has a relatively poor blood supply. **Biceps tendinitis,** inflammation of the biceps brachii long head tendon, can also cause shoulder pain. This inflammation is commonly caused by throwing a baseball or football. Pain in the shoulder can also result from **subacromial bursitis,** which is inflammation of the subacromial bursa. ▼

16. *Name the muscles located in the trunk, shoulder, and arm that move the arm.*
17. *List the muscles forming the rotator cuff, and describe their function.*

P R E D I C T ④

A few muscles can act as their own antagonists because of the organization of their muscle fasciculi (bundles). For example, a muscle can cause both flexion and extension. Name three muscles acting on the arm and scapula that can be their own antagonist.

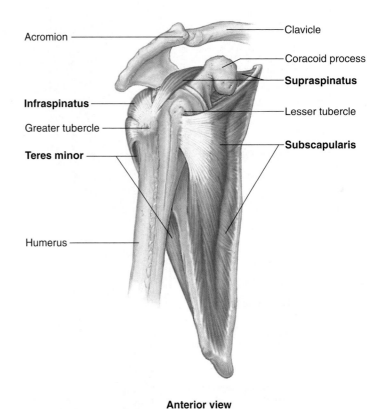

Anterior view

Figure 9.17   **Right Rotator Cuff Muscles**

Forearm Movements

Muscles of the arm and forearm produce flexion/extension of the elbow and pronation/supination of the forearm (table 9.13). The biceps brachii and the **brachialis** (brā′kē-al′is) are the large muscles of the anterior arm, whereas the triceps brachii forms the posterior arm (figure 9.18). When a person is called on to "make a muscle," it is likely the biceps brachii will be used. The brachialis lies deep to the biceps brachii and can be seen only as a mass on the lateral and medial sides of the arm. The biceps brachii and brachialis are often mistakenly perceived to be one muscle when they are not well defined.

The biceps brachii has two heads, the long and short heads, which attach to the scapula (see figure 9.18). The biceps brachii inserts on the radial tuberosity. The brachialis originates on the humerus and inserts on the ulnar tuberosity. Thus, the biceps brachii pulls on and moves the radius, whereas the brachialis pulls on and moves the ulna. The triceps brachii has three heads (see figure 9.18). The long head is attached to the scapula, whereas the medial and lateral heads are attached to the humerus

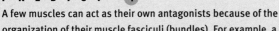

Table 9.12	**Summary of Muscle Actions on the Shoulder and Arm**				
Flexion	**Extension**	**Abduction**	**Adduction**	**Medial Rotation**	**Lateral Rotation**
Deltoid	Deltoid	Deltoid	Pectoralis major	Pectoralis major	Deltoid
Pectoralis major	Teres major	Supraspinatus	Latissimus dorsi	Teres major	Infraspinatus
Coracobrachialis	Latissimus dorsi		Teres major	Latissimus dorsi	Teres minor
Biceps brachii	Pectoralis major		Teres minor	Deltoid	
	Triceps brachii		Triceps brachii	Subscapularis	
			Coracobrachialis		

Table 9.13 Muscles Acting on the Forearm (See Figure 9.18)

Muscle	Origin	Insertion	Nerve	Action
Arm				
Biceps brachii (bī′seps brā′kē-ī)	Long head—supraglenoid tubercle Short head—coracoid process	Radial tuberosity and aponeurosis of biceps brachii	Musculocutaneous	Flexes shoulder and elbow; supinates forearm and hand
Brachialis (brā′kē-al′is)	Anterior surface of humerus	Ulnar tuberosity and coronoid process of ulna	Musculocutaneous and radial	Flexes elbow
Triceps brachii (trī′seps brā′kē-ī)	Long head—infraglenoid tubercle on the lateral border of scapula Lateral head—lateral and posterior surface of humerus Medial head—posterior humerus	Olecranon process of ulna	Radial	Extends elbow; extends shoulder and adducts arm
Forearm				
Anconeus (ang-kō′nē-ŭs)	Lateral epicondyle of humerus	Olecranon process and posterior ulna	Radial	Extends elbow
Brachioradialis (brā′kē-ō-rā′dē-al′is)	Lateral supracondylar ridge of humerus	Styloid process of radius	Radial	Flexes elbow
Pronator quadratus (prō-nā-ter, prō-nā-tōr kwah-drā′tŭs)	Distal ulna	Distal radius	Anterior interosseous	Pronates forearm (and hand)
Pronator teres (prō-nā-tōr ter′ēz, tēr-ēz)	Medial epicondyle of humerus and coronoid process of ulna	Radius	Median	Pronates forearm (and hand)
Supinator (soo′pi-nā-ter, soo′pi-nā-tōr)	Lateral epicondyle of humerus and ulna	Radius	Radial	Supinates forearm (and hand)

(see figure 9.1). All three heads insert on the olecranon process of the ulna.

Muscles that cross two joints can affect movements of both joints. The biceps brachii and the triceps brachii (long head) cross the shoulder and elbow joints and can cause both shoulder and elbow movements. Their role in elbow movements is usually emphasized.

Flexion of the elbow is accomplished by the brachialis, the biceps brachii, and the **brachioradialis** (brā′kē-ō-rā′dē-al′is, connected to the arm and radius bone) (see figure 9.18a). The brachialis is the prime mover for flexion of the elbow. The brachioradialis forms a bulge on the anterolateral side of the forearm just distal to the elbow (figure 9.19). If the elbow is forcefully flexed when the forearm is held midway between pronation and supination, the brachioradialis stands out clearly on the forearm. Extension of the elbow is accomplished primarily by the triceps brachii.

Supination of the forearm is accomplished by the **supinator** and biceps brachii. Pronation is a function of the **pronator quadratus** (kwah-drā′tŭs) and the **pronator teres.** The supinator and pronator muscles are located in the forearm (see figure 9.19). ▼

18. *List the muscles that cause flexion and extension of the elbow. Where are these muscles located?*

19. *Supination and pronation of the forearm are produced by what muscles? Where are these muscles located?*

P R E D I C T
It is easier to do a chin-up with the forearm supinated than when the forearm is pronated. Explain.

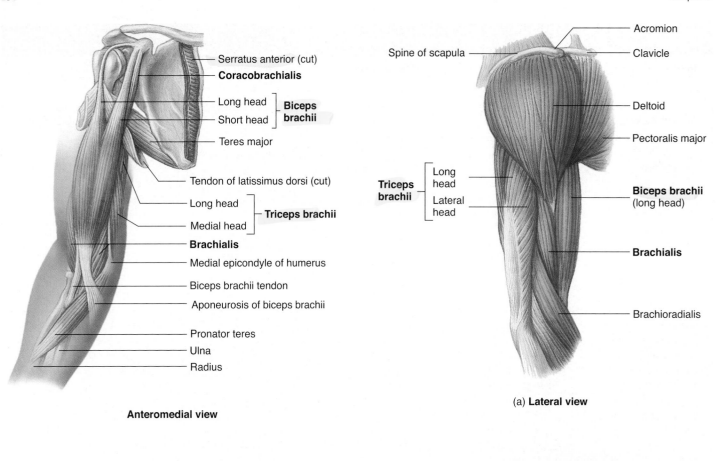

Serratus anterior (cut)

Coracobrachialis

Long head
Short head } **Biceps brachii**

Teres major

Tendon of latissimus dorsi (cut)

Long head
Medial head } **Triceps brachii**

Brachialis

Medial epicondyle of humerus

Biceps brachii tendon

Aponeurosis of biceps brachii

Pronator teres

Ulna

Radius

Anteromedial view

Acromion

Spine of scapula

Clavicle

Deltoid

Pectoralis major

Triceps brachii {
Long head
Lateral head

Biceps brachii (long head)

Brachialis

Brachioradialis

(a) **Lateral view**

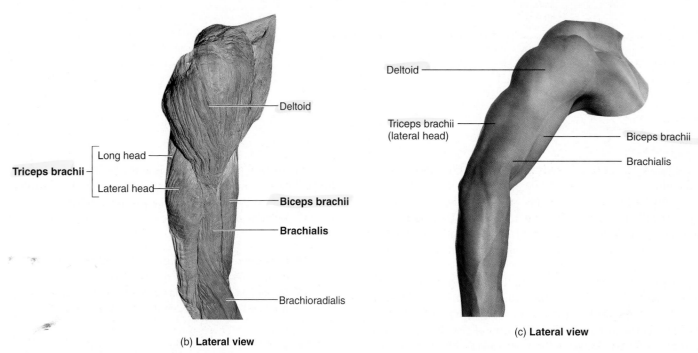

Deltoid

Triceps brachii {
Long head

Lateral head

Biceps brachii

Brachialis

Brachioradialis

(b) **Lateral view**

Deltoid

Triceps brachii (lateral head)

Biceps brachii

Brachialis

(c) **Lateral view**

Figure 9.18   Lateral Right Arm Muscles

(*a*) The right shoulder and arm. (*b*) The right shoulder and arm muscles of a cadaver. (*c*) Surface anatomy of the right shoulder and arm.

Table 9.14 Muscles of the Forearm Acting on the Wrist, Hand, and Fingers (See Figures 9.19 and 9.20)

Muscle	Origin	Insertion	Nerve	Action
Anterior Forearm				
Flexor carpi radialis (kar′pīrā-dē-ā′lis)	Medial epicondyle of humerus	Second and third metacarpals	Median	Flexes and abducts wrist
Flexor carpi ulnaris (kar′pīŭl-nā′ris)	Medial epicondyle of humerus and ulna	Pisiform, hamate, and fifth metacarpal	Ulnar	Flexes and adducts wrist
Flexor digitorum profundus (dij′i-tōr′ŭm prō-fŭn′dŭs)	Ulna	Distal phalanges of digits 2–5	Ulnar and median	Flexes fingers at metacarpophalangeal joints and interphalangeal joints and wrist
Flexor digitorum superficialis (dij′i-tōr′ŭm soo′per-fish-ē-ā′lis)	Medial epicondyle of humerus, coronoid process, and radius	Middle phalanges of digits 2–5	Median	Flexes fingers at interphalangeal joints and wrist
Flexor pollicis longus (pol′i-sis lon′gŭs)	Radius	Distal phalanx of thumb	Median	Flexes thumb
Palmaris longus (pawl-mār′is lon′gŭs)	Medial epicondyle of humerus	Palmar fascia	Median	Tenses palmar fascia; flexes wrist
Posterior Forearm				
Abductor pollicis longus (pol′i-sis lon′gŭs)	Posterior ulna and radius and interosseous membrane	Base of first metacarpal	Radial	Abducts and extends thumb; abducts wrist
Extensor carpi radialis brevis (kar′pī rā-dē-ā′lis brev′is)	Lateral epicondyle of humerus	Base of third metacarpal	Radial	Extends and abducts wrist
Extensor carpi radialis longus (kar′pī rā-dē-ā′lis lon′gus)	Lateral supracondylar ridge of humerus	Base of second metacarpal	Radial	Extends and abducts wrist
Extensor carpi ulnaris (kar′pī ŭl-nā′ris)	Lateral epicondyle of humerus and ulna	Base of fifth metacarpal	Radial	Extends and adducts wrist
Extensor digiti minimi (dij′i-tī min′i-mī)	Lateral epicondyle of humerus	Phalanges of fifth digit	Radial	Extends little finger and wrist
Extensor digitorum (dij′i-tōr′ŭm)	Lateral epicondyle of humerus	Extensor tendon expansion over phalanges of digits 2–5	Radial	Extends fingers and wrist
Extensor indicis (in′di-sis)	Ulna	Extensor tendon expansion over second digit	Radial	Extends forefinger and wrist
Extensor pollicis brevis (pol′i-sis brev′is)	Radius	Proximal phalanx of thumb	Radial	Extends and abducts thumb; abducts wrist
Extensor pollicis longus (pol′i-sis lon′gŭs)	Ulna	Distal phalanx of thumb	Radial	Extends thumb

Wrist, Hand, and Finger Movements

Forearm muscles moving the wrist, hand, and fingers (table 9.14) are divided into anterior (see figure 9.19) and posterior groups (figure 9.20). The superficial anterior forearm muscles have their origin on the medial epicondyle of the humerus, whereas the superficial posterior forearm muscles have their origin on the lateral epicondyle. The deep muscles originate on the radius and ulna.

The anterior forearm muscles are responsible for flexion of the wrist and fingers, whereas the posterior forearm muscles cause extension of the wrist and fingers. Both groups work together to cause abduction and adduction of the wrist.

 Tennis Elbow

Forceful, repetitive use of the forearm extensor muscles can damage them where they attach to the lateral epicondyle. This condition is often called **tennis elbow** because it can result from playing tennis. It is also called **lateral epicondylitis** because it can result from other sports and activities, such as shoveling snow.

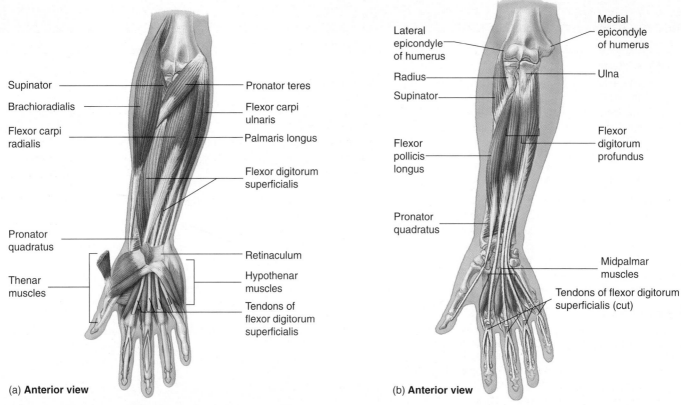

(a) **Anterior view**

(b) **Anterior view**

Figure 9.19   Anterior Right Forearm and Hand Muscles

(*a*) Anterior view of right forearm and intrinsic hand muscles. The retinaculum (see figure 9.2) has been removed. (*b*) Deep muscles of the right anterior forearm. The brachioradialis, pronator teres, flexor carpi radialis and ulnaris, palmaris longus, and flexor digitorum superficialis muscles are removed.

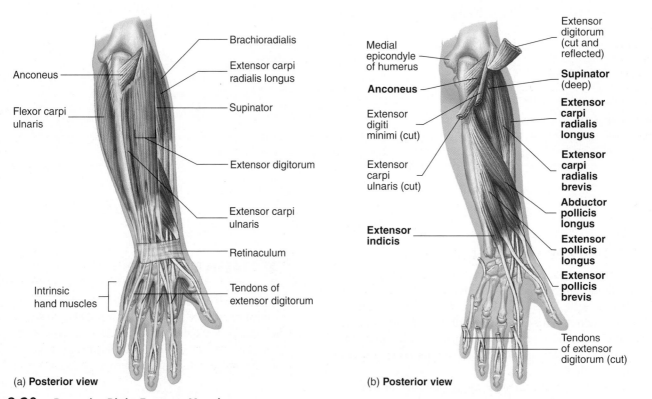

(a) **Posterior view**

(b) **Posterior view**

Figure 9.20   Posterior Right Forearm Muscles

(*a*) Posterior view. (*b*) Deep muscles of the right posterior forearm. The extensor digitorum, extensor digiti minimi, and extensor carpi ulnaris muscles are cut to reveal deeper muscles.

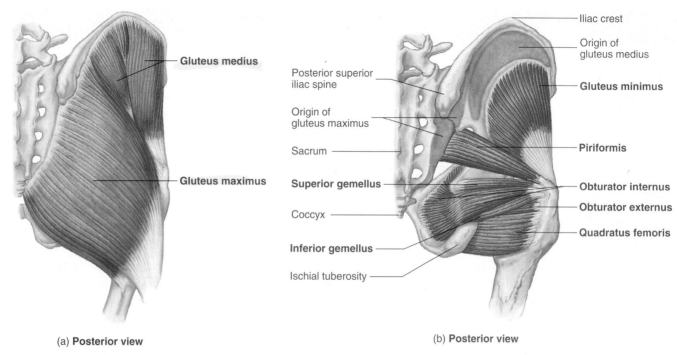

(a) **Posterior view**   (b) **Posterior view**

Figure 9.21   Right Gluteal and Deep Hip Muscles

(a) Right hip, superficial. (b) Gluteus maximus and medius are removed to reveal deeper muscles.

The tendons of the wrist flexors can be seen on the anterior surface of the forearm proximal to the wrist when the wrist is flexed against resistance. The tendon of the **flexor carpi radialis** (kar′pī-rā-dē-ā′lis) is an important landmark because the radial pulse can be felt just lateral to the tendon.

The **extrinsic hand muscles** are muscles located outside of the hand that control hand movements. These forearm muscles have tendons that extend into the hand. A strong band of fibrous connective tissue, the **retinaculum** (ret-i-nak′ū-lŭm, bracelet), covers the flexor and extensor tendons and holds them in place around the wrist so that they do not "bowstring" during muscle contraction (see figures 9.19a and 20a). The tendons of extrinsic hand muscles are readily seen on the posterior surface of the hand.

The **intrinsic hand muscles** are entirely within the hand and are divided into three groups. The **thenar** (thē′nar, palm of hand) **muscles** form a fleshy prominence at the base of the thumb on the palm called the thenar eminence, and the **hypothenar muscles** form a fleshy mass on the medial side of the palm called the hypothenar eminence (see figure 9.19a). The **midpalmar muscles** are in the middle of the palm (see figure 9.19b).

The extrinsic hand muscles are large and are responsible for powerful movement of the fingers, especially when grasping objects. They also assist the wrist flexors and extensors. The intrinsic

hand movements are responsible for abduction/adduction of the fingers and control of the thumb and little finger, including opposition/reposition movements. They also assist the extrinsic hand muscles in flexion/extension. ▼

20. *Contrast the location and actions of the anterior and posterior forearm muscles. Where is the origin of most of the muscles of these groups?*
21. *Contrast the location and actions of the extrinsic and intrinsic hand muscles. What is the retinaculum?*

Hip and Lower Limb Muscles

Thigh Movements

Several hip muscles originate on the coxal bone and insert onto the femur (table 9.15 and figure 9.21). They are the gluteal muscles, deep hip muscles, and anterior hip muscles. The **gluteus** (gloo-tē′ŭs, buttock) **maximus** forms most of the buttock. The gluteus maximus functions at its maximum force in extension of the hip when the hip is flexed at a 45-degree angle so that the muscle is optimally stretched, which accounts for both the sprinter's stance and the bicycle racing posture. The **gluteus medius,** a common site for injections, creates a smaller mass just superior and lateral

Table 9.15 Muscles Acting on the Thigh (See Figures 9.21 and 9.22)

Muscle	Origin	Insertion	Nerve	Action
Gluteal				
Gluteus maximus (gloo-tē'ŭs mak'si-mŭs)	Posterior surface of ilium, sacrum, and coccyx	Gluteal tuberosity of femur and iliotibial tract	Inferior gluteal	Extends hip; abducts and laterally rotates thigh
Gluteus medius (gloo-tē'ŭs mē'dē-ŭs)	Posterior surface of ilium	Greater trochanter of femur	Superior gluteal	Abducts and medially rotates thigh; tilts pelvis toward supported side
Gluteus minimus (gloo-tē'ŭs min-i-mŭs)	Posterior surface of ilium	Greater trochanter of femur	Superior gluteal	Abducts and medially rotates thigh; tilts pelvis toward supported side
Deep Hip				
Gemellus (jĕ-mel'ŭs)				
Inferior	Ischial tuberosity	Obturator internus tendon	L5 and S1	Laterally rotates and abducts thigh
Superior	Ischial spine	Obturator internus tendon	L5 and S1	Laterally rotates and abducts thigh
Obturator (ob'too-rā-tŏr)				
Externus (eks-ter'nŭs)	Inferior margin of obturator foramen	Greater trochanter of femur	Obturator	Laterally rotates thigh
Internus (in-ter'nŭs)	Interior margin of obturator foramen	Greater trochanter of femur	L5 and S1	Laterally rotates thigh
Piriformis (pir'i-fŏr'mis)	Sacrum and ilium	Greater trochanter of femur	S1 and S2	Laterally rotates and abducts thigh
Quadratus femoris (kwah'-drā'tŭs fem'ŏ-ris)	Ischial tuberosity	Intertrochanteric ridge of femur	L5 and S1	Laterally rotates thigh
Anterior Hip				
Iliopsoas (il'ē-ō-sō'as)				
Iliacus (il-ī'ă-kŭs)	Iliac fossa	Lesser trochanter of femur and capsule of hip joint	Lumbar plexus	Flexes hip
Psoas major (sō'as)	T12–L5	Lesser trochanter of femur	Lumbar plexus	Flexes hip

to the gluteus maximus. The **gluteus minimus** is deep to the gluteus medius. All of the gluteal muscles can abduct the thigh. The gluteus medius and minimus muscles help tilt the pelvis and maintain the trunk in an upright posture during walking and running. As a foot is raised off of the ground, that side of the pelvis is no longer supported from below. The gluteus minimus and medius on the opposite contract, which prevents the hip from sagging on the unsupported side.

The deep hip muscles and the gluteus maximus function as lateral thigh rotators. The gluteus medius and gluteus minimus are medial hip rotators (see table 9.15).

The anterior hip muscles (figure 9.22) are the **iliacus** (il-ī'ă-kŭs, related to the ilium) and the **psoas** (sō'as, loin muscle) **major.** They often are referred to as the **iliopsoas** (il'ē-ō-sō'as) because they share an insertion and produce the same movement.

Depending on which end of the muscle is fixed, flexion of the hip by the iliopsoas produces different movements. When the thigh is fixed, the trunk moves relative to the thigh. For example, the iliopsoas does most of the work when a person does sit-ups. When the trunk is fixed, the thigh moves, as when climbing stairs.

Some of the muscles located in the thigh originate on the coxal bone and can cause movement of the thigh when the trunk is fixed. The thigh muscles are grouped as the anterior, medial, and posterior compartments. The anterior thigh muscles flex the hip, the posterior thigh muscles extend the hip, and the medial thigh muscles adduct the thigh (table 9.16). ▼

22. Describe the movements produced by the gluteal muscles and the deep thigh rotators.
23. Name the muscle compartments of the thigh and state the hip movements produced by each compartment.

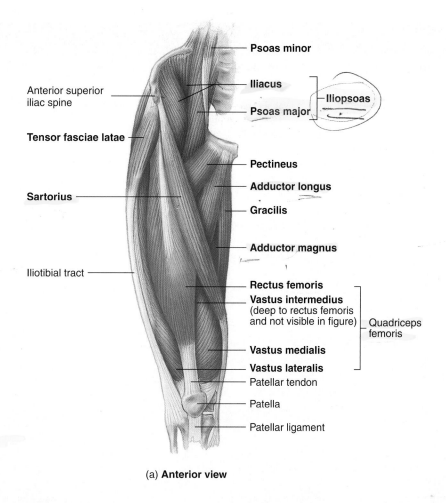

Psoas minor

Anterior superior iliac spine

Iliacus

Iliopsoas

Psoas major

Tensor fasciae latae

Pectineus

Adductor longus

Sartorius

Gracilis

Adductor magnus

Iliotibial tract

Rectus femoris

Vastus intermedius (deep to rectus femoris and not visible in figure)

Quadriceps femoris

Vastus medialis

Vastus lateralis

Patellar tendon

Patella

Patellar ligament

(a) **Anterior view**

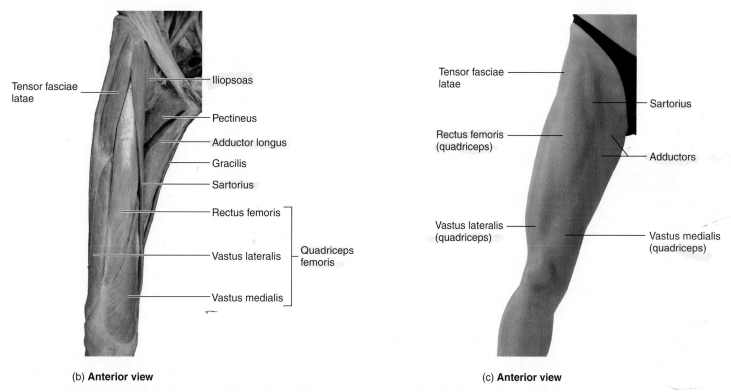

Tensor fasciae latae

Iliopsoas

Pectineus

Adductor longus

Gracilis

Sartorius

Rectus femoris

Vastus lateralis

Quadriceps femoris

Vastus medialis

(b) **Anterior view**

Tensor fasciae latae

Sartorius

Rectus femoris (quadriceps)

Adductors

Vastus lateralis (quadriceps)

Vastus medialis (quadriceps)

(c) **Anterior view**

Figure 9.22 **Right Anterior Hip and Thigh Muscles**

(a) Right anterior hip and thigh. (b) Photograph of the right anterior thigh muscles in a cadaver. (c) Surface anatomy of the right anterior thigh.

Table 9.16 Muscles of the Thigh (See Figures 9.22–9.24)

Muscle	Origin	Insertion	Nerve	Action
Anterior Compartment				
Quadriceps femoris (kwah′dri-seps fem′ŏ-ris)	Rectus femoris— anterior inferior iliac spine	Patella and onto tibial tuberosity through patellar ligament	Femoral	Extends knee; rectus femoris also flexes hip
	Vastus lateralis—greater trochanter and linea aspera of femur			
	Vastus intermedius— body of femur			
	Vastus medialis— linea aspera of femur			
Sartorius (sar-tōr′ē-ūs)	Anterior superior iliac spine	Medial side of tibial tuberosity	Femoral	Flexes hip and knee; rotates thigh laterally and leg medially
Tensor fasciae latae (ten′sōr fash′ē-ē lā′tē)	Anterior superior iliac spine	Through iliotibial tract to lateral condyle of tibia	Superior gluteal	Flexes hip; abducts and medially rotates thigh; stabilizes femur on tibia when standing
Medial Compartment				
Adductor brevis (a-dŭk′ter, a-dŭk′tōr brev′is)	Pubis	Pectineal line and linea aspera of femur	Obturator	Adducts, laterally rotates thigh; flexes hip
Adductor longus (a-dŭk′ter, a-dŭk′tōr lon′gŭs)	Pubis	Linea aspera of femur	Obturator	Adducts, laterally rotates thigh; flexes hip
Adductor magnus (a-dŭk′ter, a-dŭk′tōr mag′nŭs)	Adductor part: pubis and ischium	Adductor part: linea aspera of femur	Adductor part: obturator	Adductor part: adducts thigh and flexes hip
	Hamstring part: ischial tuberosity	Hamstring part: adductor tubercle of femur	Hamstring part: tibial	Hamstring part: extends hip
Gracilis (gras′i-lis)	Pubis near symphysis	Tibia	Obturator	Adducts thigh; flexes knee
Pectineus (pek′ti-nē′ŭs)	Pubic crest	Pectineal line of femur	Femoral and obturator	Adducts thigh; flexes hip
Posterior Compartment				
Biceps femoris (bī′seps fem′ŏ-ris)	Long head—ischial tuberosity	Head of fibula	Long head—tibial	Flexes knee; laterally rotates leg; extends hip
	Short head—femur		Short head— common fibular	
Semimembranosus (sem′ē-mem-bră-nō′sŭs)	Ischial tuberosity	Medial condyle of tibia and collateral ligament	Tibial	Flexes knee; medially rotates leg; tenses capsule of knee joint; extends hip
Semitendinosus (sem′ē-ten-di-nō′sŭs)	Ischial tuberosity	Tibia	Tibial	Flexes knee; medially rotates leg; extends hip

Leg Movements

The anterior thigh muscles are the **quadriceps** (four heads) **femoris** (fem′ŏ-ris), **sartorius** (sar-tōr′ē-ūs, tailor), and **tensor fasciae latae** (fash′ē-ē lā′tē) (see table 9.16 and figure 9.22). The quadriceps femoris is actually four muscles: the **rectus femoris,** the **vastus** (huge) **lateralis,** the **vastus medialis,** and the **vastus intermedius.** The quadriceps group extends the knee. The rectus femoris also flexes the hip because it crosses the hip joint.

The quadriceps femoris makes up the large mass on the anterior thigh (see figure 9.22c). The vastus lateralis sometimes is used as an injection site, especially in infants who may not have well-developed deltoid or gluteal muscles. The muscles of the

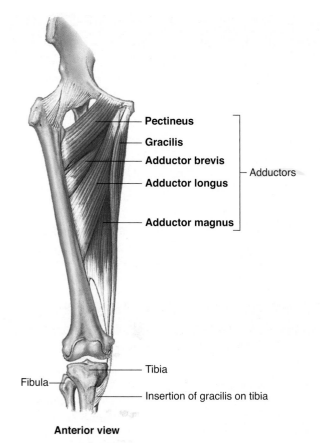

Figure 9.23   **Right Medial Thigh Muscles**

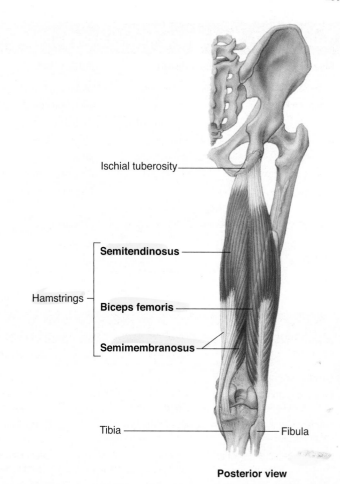

Figure 9.24   **Right Posterior Thigh Muscles**

Hip muscles are removed.

quadriceps femoris have a common insertion, the patellar tendon, on and around the patella. The patellar ligament is an extension of the patellar tendon onto the tibial tuberosity. The patellar ligament is the point that is tapped with a rubber hammer when testing the knee-jerk reflex in a physical examination.

The sartorius is the longest muscle of the body, crossing from the lateral side of the hip to the medial side of the knee (see figure 9.22). As the muscle contracts, it flexes the hip and knee and laterally rotates the thigh. This movement is the action required for crossing the legs. The term *sartorius* means tailor. The sartorius muscle is so-named because its action is to cross the legs, a common position traditionally preferred by tailors because they can hold their sewing in their lap as they sit and sew by hand.

The **tensor fasciae latae** (fash′ē-ē lā′tē) extends from the lateral side of the hip to the lateral condyle of the tibia through a connective tissue band, called the **iliotibial tract** (see figure 9.22*a*). The iliotibial tract is a thickening of the fascia lata, which is muscular fascia (see chapter 8) surrounding the thigh muscles like a supportive stocking. The tensor fasciae latae flexes the hip and helps stabilize the femur on the tibia when standing.

One of the medial thigh muscles, the **gracilis** (gras′i-lis, slender), flexes the knee. The medial thigh muscles (figure 9.23) are involved primarily in adduction of the thigh. Some of them cause flexion of the hip, extension of the hip, or lateral rotation of the thigh. An adductor strain, or "pulled groin," involves one or more of the adductor muscles, the adductor longus being the most commonly damaged. The damage usually occurs at the musculo-tendon junction, near its origin.

The posterior thigh muscles consist of the **biceps femoris, semimembranosus** (sem′ē-mem-bră-nō′sŭs), and **semitendinosus** (sem′ē-ten-di-nō′sŭs) (figure 9.24). They flex the knee. Their tendons are easily seen or felt on the medial and lateral posterior aspect of a slightly bent knee. The posterior thigh muscles are collectively called the **hamstrings** because in pigs these tendons can be used to suspend hams during curing. Some animals, such as wolves, often bring down their prey by biting through the hamstrings; therefore, "to hamstring" someone is to render the person helpless. A hamstring strain, or "pulled hamstring," results from tearing one or more of these muscles or their tendons. The biceps femoris is most commonly damaged.

Table 9.17 Summary of Muscle Actions on the Hip and Thigh

Flexion	Extension	Abduction	Adduction	Medial Rotation	Lateral Rotation
Iliopsoas	Gluteus maximus	Gluteus maximus	Adductor magnus	Gluteus medius	Gluteus maximus
Tensor fasciae latae	Semitendinosus	Gluteus medius	Adductor longus	Gluteus minimus	Obturator internus
	Semimembranosus	Gluteus minimus	Adductor brevis	Tensor fasciae latae	Obturator externus
Rectus femoris	Biceps femoris	Tensor fasciae	Pectineus		Superior gemellus
Sartorius	Adductor magnus	latae	Gracilis		Inferior gemellus
Adductor longus		Obturator internus			Quadratus femoris
Adductor brevis		Gemellus superior and inferior			Piriformis
Pectineus		Piriformis			Adductor magnus
					Adductor longus
					Adductor brevis

Table 9.18 Muscles of the Leg Acting on the Leg and Foot (See Figures 9.25 and 9.26)

Muscle	Origin	Insertion	Nerve	Action
Anterior Compartment				
Extensor digitorum longus (dij′i-tōr-ŭm lon′gŭs)	Lateral condyle of tibia and fibula	Four tendons to phalanges of four lateral toes	Deep fibular*	Extends four lateral toes; dorsiflexes and everts foot
Extensor hallucis longus (hal′i-sis lon′gŭs)	Middle fibula and interosseous membrane	Distal phalanx of great toe	Deep fibular*	Extends great toe; dorsiflexes and inverts foot
Tibialis anterior (tib-ē-a′lis)	Proximal, lateral tibia and interosseous membrane	Medial cuneiform and first metatarsal	Deep fibular*	Dorsiflexes and inverts foot
Fibularis tertius (peroneus tertius) (per′ō-nē′ŭs ter′shē-ŭs)	Fibula and interosseous membrane	Fifth metatarsal	Deep fibular*	Dorsiflexes and everts foot
Lateral Compartment				
Fibularis brevis (peroneus brevis) (fib-ū-lā′ris brev′is)	Inferior two-thirds of lateral fibula	Fifth metatarsal	Superficial fibular*	Everts and plantar flexes foot
Fibularis longus (peroneus longus) (fib-ū-lā′ris lon′gŭs)	Superior two-thirds of lateral fibula	First metatarsal and medial cuneiform	Superficial fibular*	Everts and plantar flexes foot
Posterior Compartment				
Superficial				
Gastrocnemius (gas-trok-nē′mē-ŭs)	Medial and lateral condyles of femur	Through calcaneal (Achilles) tendon to calcaneus	Tibial	Plantar flexes foot; flexes knee
Plantaris (plan-tār′is)	Femur	Through calcaneal tendon to calcaneus	Tibial	Plantar flexes foot; flexes knee
Soleus (sō-lē′ŭs)	Fibula and tibia	Through calcaneal tendon to calcaneus	Tibial	Plantar flexes foot
Deep				
Flexor digitorum longus (dij′i-tōr′ŭm lon′gŭs)	Tibia	Four tendons to distal phalanges of four lateral toes	Tibial	Flexes four lateral toes; flexes and inverts foot
Flexor hallucis longus (hal′i-sis lon′gŭs)	Fibula	Distal phalanx of great toe	Tibial	Flexes great toe; plantar flexes and inverts foot
Popliteus (pop-li-tē′ŭs)	Lateral femoral condyle	Posterior tibia	Tibial	Flexes knee; medially rotates leg
Tibialis posterior (tib-ē-a′lis)	Tibia, interosseous membrane, and fibula	Navicular, cuneiforms, cuboid, and second through fourth metatarsals	Tibial	Plantar flexes and inverts foot

*Also referred to as the peroneal nerve.

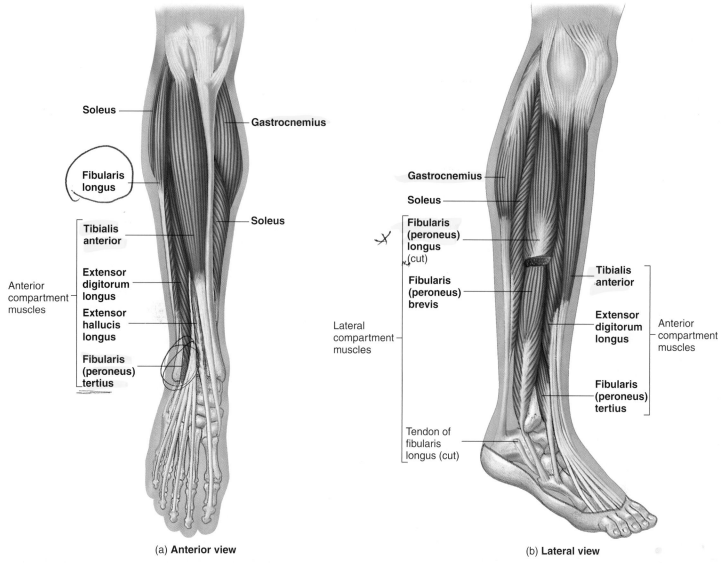

(a) **Anterior view**

(b) **Lateral view**

Figure 9.25 **Right Anterior and Lateral Leg Muscles**

(*a*) Anterior view of the right leg. (*b*) Lateral view of the right leg.

Table 9.17 summarizes muscle actions on the hip and thigh. ▼

24. *List the muscles of each thigh compartment and the individual action of each muscle.*
25. *How is it possible for thigh muscles to move both the thigh and the leg? Name six muscles that can do this.*

Foot and Toe Movements

The **extrinsic foot muscles** are muscles located outside of the foot that control foot movements (table 9.18). These leg muscles are divided into three compartments: anterior, posterior, and lateral. The anterior leg muscles (figure 9.25*a*) are extensor muscles involved in dorsiflexion and eversion or inversion of the foot and extension of the toes. The tendons of these muscles lie along the dorsum (top) of the foot.

 Shin Splints

Shin splints is a catchall term involving any one of the following four conditions associated with pain in the anterior portion of the leg:

1. Excessive stress on the tibialis anterior, resulting in pain along the origin of the muscle
2. Tibial periostitis, an inflammation of the tibial periosteum
3. Anterior compartment syndrome. During hard exercise, the anterior compartment muscles may swell with blood. The overlying fascia is very tough and does not expand; thus, the nerves and vessels are compressed, causing pain.
4. Stress fracture of the tibia 2–5 cm distal to the knee

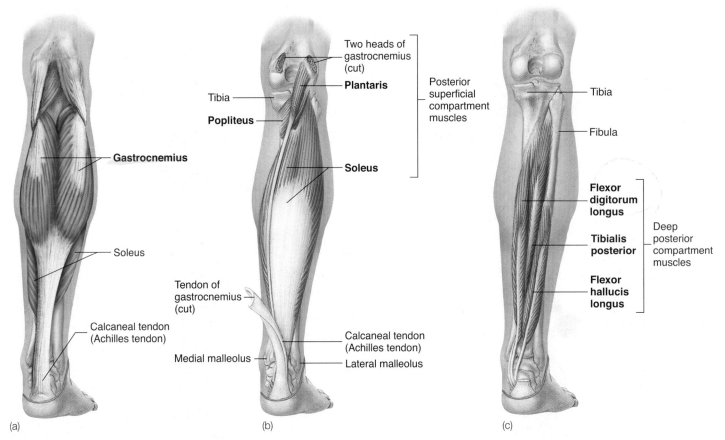

Figure 9.26   Right Posterior Leg Muscles

(*a*) Superficial muscles. (*b*) Posterior view of the right calf, superficial. Gastrocnemius is removed. (*c*) Posterior view of the right calf, deep. Gastrocnemius, plantaris, and soleus muscles are removed.

The lateral leg muscles (figure 9.25*b*) are primarily everters of the foot, but they also aid plantar flexion. The tendons of these two muscles passes posterior to the lateral malleolus to insert on the lateral side of the foot.

The posterior compartment is divided into superficial and deep groups (figure 9.26). The superficial muscles of the posterior compartment are powerful plantar flexors the foot. The **gastrocnemius** (gas-trok-nē′mē-ŭs, calf of the leg) and **soleus** (sō-lē′ŭs, sole of foot) form the bulge of the calf of the leg. They join with the small **plantaris** muscle to form the common **calcaneal** (kal-kā′nē-al), or Achilles, **tendon.** The **Achilles tendon** derives its name from a hero of Greek mythology. When Achilles was a baby, his mother dipped him into magic water, which made him invulnerable to harm everywhere the water touched his skin. His mother, however, held him by the heel and failed to submerge this part of his body under the water. Consequently, his heel was vulnerable and proved to be his undoing; he was shot in the heel with an arrow at the battle of Troy and died. Thus, saying that someone has an "Achilles heel" means that the person has a weak spot that can be attacked.

The tendons of the deep muscles of the posterior compartment pass posterior to the medial malleolus to insert on the plantar surface of foot bones. They plantar flex and invert the foot and flex the toes.

Intrinsic foot muscles, located within the foot itself, flex, extend, abduct, and adduct the toes. They are arranged in a manner similar to that of the intrinsic muscles of the hand.

 Plantar Fasciitis

The muscles in the plantar region of the foot are covered with thick fascia and the plantar aponeurosis. Running on a hard surface wearing poorly fitting or worn-out shoes can result in inflammation of the plantar aponeurosis. The pain resulting from **plantar fasciitis** occurs in the fascia over the heel and along the medial-inferior side of the foot. ▼

26. *What movements are produced by each of the three muscle compartments of the leg?*
27. *List the general actions performed by the intrinsic foot muscles.*

Clinical Relevance   Bodybuilding

Bodybuilding is a popular sport worldwide. Participants in this sport combine diet and specific weight training to develop maximum muscle mass and minimum body fat, with their major goal being a well-balanced, complete physique. An uninformed, untrained muscle builder can build some muscles and ignore others; the result is a disproportioned body. Skill, training, and concentration are required to build a well-proportioned, muscular body and to know which exercises build a large number of muscles and which are specialized to build certain parts of the body. Is the old adage "no pain, no gain" correct? Not really. Over exercising can cause small tears in muscles and soreness. Torn muscles are weaker, and it may take up to 3 weeks to repair the damage, even though the soreness may last only 5–10 days.

Bodybuilders concentrate on increasing skeletal muscle mass. Endurance tests conducted years ago demonstrated that the cardiovascular and respiratory abilities of bodybuilders were similar to the abilities of normal, healthy persons untrained in a sport. More recent studies, however, indicate that the cardiorespiratory fitness of bodybuilders is similar to that of other well-trained athletes. The difference between the results of the new studies and those of the older ones is attributed to modern bodybuilding techniques that

Anterior view

Lateral view

Figure 9.27   Bodybuilders

Name as many muscles as you can in these photographs.

include aerobic exercise and running, as well as "pumping iron."

Bodybuilding has its own language. Bodybuilders refer to the "lats," "traps," and "delts" rather than the latissimus dorsi, trapezius, and deltoids. The exercises also have special names, such as

"lat pulldowns," "preacher curls," and "triceps extensions."

Photographs of bodybuilders are very useful in the study of anatomy because they enable the easy identification of the surface anatomy of muscles that cannot usually be seen in untrained people (figure 9.27).

S U M M A R Y

General Principles   (p. 228)

1. The less movable end of a muscle attachment is the origin; the more movable end is the insertion.
2. An agonist causes a certain movement and an antagonist acts in opposition to the agonist.
3. Synergists are muscles that function together to produce movement.
4. Prime movers are mainly responsible for a movement. Fixators stabilize the action of prime movers.

Nomenclature

Muscles are named according to their location, size, shape, orientation of fasciculi, origin and insertion, number of heads, and function.

Muscle Anatomy

Muscles of the head, neck, trunk and limbs are presented.

Head and Neck Muscles   (p. 232)
Facial Expression

Origins of facial muscles are on skull bones or fascia; insertions are into the skin, causing movement of the facial skin, lips, and eyelids.

Mastication

The temporalis and masseter muscles elevate the mandible; gravity opens the jaw. The digastric depresses the mandible. The pterygoids move the mandible from side to side.

Tongue Movements

Intrinsic tongue muscles change the shape of the tongue; extrinsic tongue muscles move the tongue.

Head and Neck Muscles Movements

Origins of these muscles are mainly on the cervical vertebrae (except for the sternocleidomastoid). They cause flexion, extension, rotation, and lateral flexion of the head and neck.

Trunk Muscles   (p. 239)
Back Muscles

1. These muscles extend, laterally flex, and rotate the vertebral column. They also hold the vertebral column erect.
2. A superficial group of muscle, the erector spinae, runs from the pelvis to the skull, extending from the vertebrae to the ribs.
3. A deep group of muscles connects adjacent vertebrae.

Thoracic Muscles

1. Most respiratory movement is caused by the diaphragm.
2. Muscles attached to the ribs aid in respiration.

Pelvic Floor and Perineum

These muscles support the abdominal organs inferiorly.

Abdominal Wall Muscles

Abdominal wall muscles hold and protect abdominal organs and cause flexion, rotation, and lateral flexion of the vertebral column. They also compress the abdomen.

Scapular and Upper Limb Muscles   (p. 244)
Scapular Movements

Six muscles attach the scapula to the trunk and enable the scapula to function as an anchor point for the muscles and bones of the arm.

Arm Movements

Nine muscles attach the humerus to the scapula. Two additional muscles attach the humerus to the trunk. These muscles cause flexion and extension of the shoulder and abduction, adduction, rotation, and circumduction of the arm.

Forearm Movements

1. Flexion and extension of the elbow are accomplished by three muscles located in the arm and two in the forearm.

2. Supination and pronation are accomplished primarily by forearm muscles.

Wrist, Hand, and Finger Movements

1. Forearm muscles that originate on the medial epicondyle are responsible for flexion of the wrist and fingers. Muscles extending the wrist and fingers originate on the lateral epicondyle.
2. Extrinsic hand muscles are in the forearm. Intrinsic hand muscles are in the hand.

Hip and Lower Limb Muscles   (p. 253)
Thigh Movements

1. The gluteus maximus extends the hip. The gluteus medius and minimus help hold the hip level while walking or running.
2. Deep hip muscles laterally rotate the thigh.
3. Anterior hip muscles flex the hip.
4. The thigh can be divided into three compartments.
   - The anterior compartment muscles flex the hip.
   - The posterior compartment muscles extend the hip.
   - The medial compartment muscles adduct the thigh.

Leg Movements

1. Anterior thigh muscles extend the knee (quadriceps femoris), flex the knee (sartorius), and stabilize the knee (tensor fasciae latae).
2. Posterior thigh muscles flex the knee.
3. One medial thigh muscle flexes the knee (gracilis).

Foot and Toe Movements

1. The leg is divided into three compartments.
   - Muscles in the anterior compartment cause dorsiflexion, inversion, or eversion of the foot and extension of the toes.
   - Muscles of the lateral compartment plantar flex and evert the foot.
   - Muscles of the posterior compartment flex the leg, plantar flex and invert the foot, and flex the toes.
2. Intrinsic foot muscles flex or extend and abduct or adduct the toes.

Bodybuilding   (p. 261)

Bodybuilders combine diet and specific weight training to develop maximum muscle mass and minimum body fat.

R E V I E W   A N D   C O M P R E H E N S I O N

1. Muscles that oppose one another are
   a. synergists.   c. hateful.   e. fixators.
   b. levers.   d. antagonists.

2. The most movable attachment of a muscle is its
   a. origin.   c. fascia.
   b. insertion.   d. belly.

3. The muscle whose name means it is larger and round is the
   a. gluteus maximus.
   b. vastus lateralis.
   c. teres major.
   d. latissimus dorsi.
   e. adductor magnus.

4. Harry Wolf has just picked up his date for the evening. She is wearing a stunning new outfit. Harry shows his appreciation by moving his eyebrows up and down, winking, smiling, and finally kissing her. Given the muscles listed:
   1. zygomaticus
   2. levator labii superioris
   3. occipitofrontalis
   4. orbicularis oris
   5. orbicularis oculi

   In which order did Harry use these muscles?
   a. 2,3,4,1   c. 2,5,4,3   e. 3,5,2,4
   b. 2,5,3,1   d. 3,5,1,4

5. An aerial circus performer who supports herself only by her teeth while spinning around should have strong
   a. temporalis muscles.
   b. masseter muscles.
   c. buccinator muscles.
   d. both a and b.
   e. all of the above.

6. The digastric muscles
   a. depress the mandible.
   b. move the mandible from side to side.
   c. move the larynx during swallowing.
   d. are infrahyoid muscles.

7. The tongue curls and folds *primarily* because of the action of the
   a. extrinsic tongue muscles.
   b. intrinsic tongue muscles.

8. A prominent lateral muscle of the neck that can cause rotation and lateral flexion of the head is the
   a. digastric.
   b. mylohyoid.
   c. sternocleidomastoid.
   d. buccinator.
   e. platysma.

9. Which of these movements is *not* caused by contraction of the erector spinae muscles?
   a. flexion of the vertebral column
   b. lateral flexion of the vertebral column
   c. extension of the vertebral column
   d. rotation of the vertebral column

10. Which of these muscles is (are) responsible for flexion of the vertebral column (below the neck)?
    a. deep back muscles
    b. scalene muscles
    c. rectus abdominis
    d. both a and b
    e. all of the above

11. Given these muscles:
    1. external abdominal oblique
    2. internal abdominal oblique
    3. transversus abdominis

    Choose the arrangement that lists the muscles from most superficial to deepest.
    a. 1,2,3   c. 2,1,3   e. 3,1,2
    b. 1,3,2   d. 2,3,1

12. The diaphragm
    a. moves inferiorly, increasing thoracic volume.
    b. moves the ribs inferiorly during forced expiration.
    c. raises the superior two ribs when taking a deep breath.
    d. a and b.

13. Tendinous intersections
    a. attach the rectus abdominis muscles to the xiphoid process.
    b. divide the rectus abdominis muscles into segments.
    c. separate the rectus abdominis from the lateral abdominal wall muscles.
    d. are the central point of attachment for all the abdominal muscles.

14. Which of these muscles moves the scapula?
    a. deltoid
    b. latissimus dorsi
    c. pectoralis major
    d. pectoralis minor

15. Which of these muscles does *not* adduct the arm (humerus)?
    a. latissimus dorsi
    b. deltoid
    c. teres major
    d. pectoralis major
    e. coracobrachialis

16. Which of these muscles is most directly involved in abduction of the arm (humerus)?
    a. supraspinatus   c. teres minor   e. subscapularis
    b. infraspinatus   d. teres major

17. Which of these muscles would you expect to be especially well developed in a boxer known for his powerful jab (punching straight ahead)?
    a. biceps brachii
    b. brachialis
    c. trapezius
    d. triceps brachii
    e. supinator

18. Which of these muscles cause flexion of the elbow?
    a. biceps brachii
    b. brachialis
    c. brachioradialis
    d. a and b
    e. all of the above

19. The posterior group of forearm muscles is responsible for
    a. flexion of the wrist.
    b. flexion of the fingers.
    c. extension of the fingers.
    d. both a and b.
    e. all of the above.

20. The intrinsic hand muscles
    a. are divided into anterior and posterior groups.
    b. originate on the epicondyles of the humerus.
    c. cause abduction and adduction of the fingers.
    d. cause abduction and adduction of the wrist.
    e. all of the above

21. Which of these muscles can extend the hip?
    a. gluteus maximus
    b. gluteus medius
    c. gluteus minimus
    d. tensor fasciae latae
    e. sartorius

22. Given these muscles:
    1. iliopsoas
    2. rectus femoris
    3. sartorius

    Which of the muscles flex the hip?
    a. 1   c. 1,3   e. 1,2,3
    b. 1,2   d. 2,3

23. Which of these muscles is found in the medial compartment of the thigh?
    a. rectus femoris
    b. sartorius
    c. gracilis
    d. vastus medialis
    e. semitendinosus

24. Which of these is *not* a muscle that can flex the knee?
    a. biceps femoris
    b. gracilis
    c. sartorius
    d. vastus medialis

25. The _____ evert the foot, whereas the _____ plantar flex the foot.
    a. lateral leg muscles, gastrocnemius and soleus
    b. lateral leg muscles, anterior leg muscles
    c. gastrocnemius and soleus, lateral leg muscles
    d. gastrocnemius and soleus, anterior leg muscles

Answers in Appendix E

CRITICAL THINKING

1. For each of the following muscles, (1) describe the movement that the muscle produces and (2) name the muscles that act as synergists and antagonists for them: erector spinae and coracobrachialis.

2. Propose an exercise that would benefit each of the following muscles specifically: biceps brachii, triceps brachii, deltoid, rectus abdominis, quadriceps femoris, and gastrocnemius.

3. A patient was involved in an automobile accident in which the car was "rear-ended," resulting in whiplash injury of the head (hyperextension). What neck muscles might be injured in this type of accident? What is the easiest way to prevent such injury in an automobile accident?

4. During surgery, a branch of a patient's facial nerve was accidentally cut on one side of the face. As a result, after the operation, the lower eyelid and the corner of the patient's mouth drooped on that side of the face. What muscles were apparently affected?

5. When a person becomes unconscious, the tongue muscles relax and the tongue tends to retract or fall back and obstruct the airway.

Which tongue muscle is responsible? How can this be prevented or reversed?

6. The mechanical support of the head of the humerus in the glenoid cavity is weakest in the inferior direction. What muscles help prevent dislocation of the shoulder when a heavy weight, such as a suitcase, is carried?

7. How would paralysis of the quadriceps femoris of the left leg affect a person's ability to walk?

8. Speedy Sprinter started a 200 m dash and fell to the ground in pain. Examination of her right leg revealed the following symptoms: inability to plantar flex the foot against resistance, normal ability to evert the foot, dorsiflexion of the foot more than normal, and abnormal bulging of the calf muscles. Explain the nature of her injury.

Answers in Appendix F

For additional study tools, go to: www.aris.mhhe.com. Click on your course topic and then this textbook's author name/title. Register once for a semester's worth of interactive learning activities to help improve your grade!

CHAPTER 10

Functional Organization of Nervous Tissue

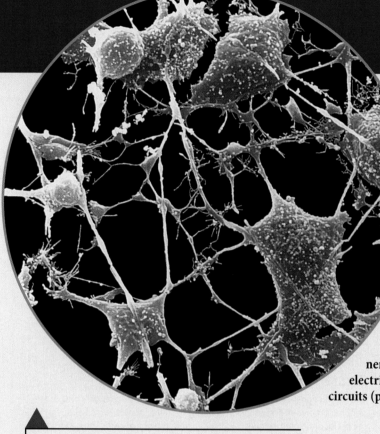

Introduction

A hungry man prepares to drink a cup of hot soup. He smells the aroma and anticipates the taste of the soup. Feeling the warmth of the cup in his hands, he quickly raises the cup to his lips and takes a sip. The soup is so hot that it burns his lips and tongue. He jerks the cup away from his lips and gasps in pain. None of these sensations, thoughts, emotions, and movements would be possible without the nervous system, which is responsible for sensations, mental activities, and control of the muscles and many glands. The nervous system is made up of the brain, spinal cord, nerves, and sensory receptors.

▶This chapter explains the **functions of the nervous system (p. 266)**, the **parts of the nervous system (p. 266)**, the **cells of the nervous system (p. 267)**, the **organization of nervous tissue (p. 271)**, **electric signals (p. 271)**, **the synapse (p. 283)**, and **neuronal pathways and circuits (p. 292)**.

This is a colorized scanning electron micrograph of a neuron network. The 10 or so cells that appear green in this photograph have yellow cell processes that reach out to almost touch one another. These contacts allow neurons of the nervous system to communicate. What is seen here is only a minute portion of the total connections among nerve cells. All the tiny processes branching off the larger processes contact processes of other cells.

Tools for Success

A virtual cadaver dissection experience

Audio (MP3) and/or video (MP4) content, available at www.mhhe.aris.com

Functions of the Nervous System

The nervous system is involved in most body functions. Some major functions of the nervous system are

1. *Sensory input.* Sensory receptors monitor numerous external and internal stimuli. Some stimuli result in sensations we are aware of, such as sight, vision, hearing, taste, smell, touch, pain, body position, and temperature. Other stimuli, such as blood pH, blood gases, or blood pressure, are processed at an unconscious level.

2. *Integration.* The brain and spinal cord are the major organs for processing sensory input and initiating responses. The input may produce an immediate response, may be stored as memory, or may be ignored.

3. *Control of muscles and glands.* Skeletal muscles normally contract only when stimulated by the nervous system. Thus, through the control of skeletal muscle, the nervous system controls the major movements of the body. The nervous system also participates in controlling cardiac muscle, smooth muscle, and many glands.

4. *Homeostasis.* The nervous system plays an important role in the maintenance of homeostasis. This function depends on the ability of the nervous system to detect, interpret, and respond to changes in internal and external conditions. In response, the nervous system can stimulate or inhibit the activities of other systems to help maintain a constant internal environment.

5. *Mental activity.* The brain is the center of mental activities, including consciousness, thinking, memory, and emotions. ▼

   1. Describe the major functions of the nervous system.

Parts of the Nervous System

The nervous system can be divided into the central and the peripheral nervous systems (figure 10.1). The **central nervous system (CNS)** consists of the brain and the spinal cord. The brain is located within the skull, and the spinal cord is located within the vertebral canal (see chapter 7).

The **peripheral nervous system (PNS)** is external to the CNS. It consists of sensory receptors and nerves. **Sensory receptors** are the endings of nerve cells or separate, specialized cells that detect temperature, pain, touch, pressure, light, sound, odors, and other stimuli. Sensory receptors are located in the skin, muscles, joints, internal organs, and specialized sensory organs, such as the eyes and ears. A **nerve** is a bundle of axons and their sheaths. Twelve pairs of **cranial nerves** originate from the brain, and 31 pairs of **spinal nerves** originate from the spinal cord (see figure 10.1).

Nerves transmit electric signals, called **action potentials.** The PNS is divided into two divisions (figure 10.2). The **sensory, or afferent, division** transmits action potentials to the CNS from sensory receptors, whereas the **motor, or efferent, division** transmits action potentials from the CNS to effector organs, such as muscles and glands.

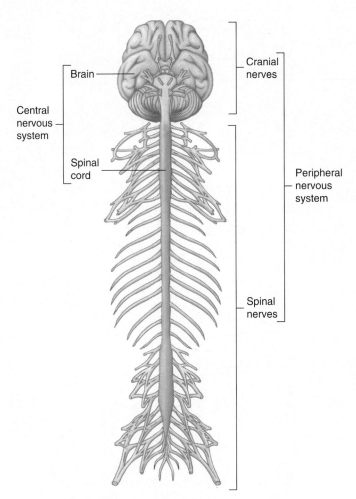

Figure 10.1 Nervous System

The central nervous system (CNS) consists of the brain and spinal cord. The peripheral nervous system (PNS) consists of cranial nerves, which arise from the brain, and spinal nerves, which arise from the spinal cord. The nerves, which are shown cut in the illustration, actually extend throughout the body.

The motor division can be further subdivided into the **somatic** (sō-mat'ik, bodily) **motor nervous system,** which transmits action potentials from the CNS to skeletal muscles, and the **autonomic** (aw-tō-nom'ik) **nervous system (ANS),** which transmits action potentials from the CNS to cardiac muscle, smooth muscle, and glands.

The ANS is subdivided into the sympathetic and the parasympathetic divisions and the enteric nervous system. In general, the **sympathetic division** is most active during physical activity, whereas the **parasympathetic division** regulates resting or vegetative functions, such as digesting food or emptying the urinary bladder. The **enteric nervous system (ENS)** is located in the digestive tract and can control the digestive tract independently of the CNS. See chapters 13 and 21 for details on the ANS.

The sensory division of the PNS detects stimuli and transmits information in the form of action potentials to the CNS (see figure 10.2). The CNS is the major site for processing information, initiating responses, and integrating mental processes. It is much like a computer, with the ability to receive input, process

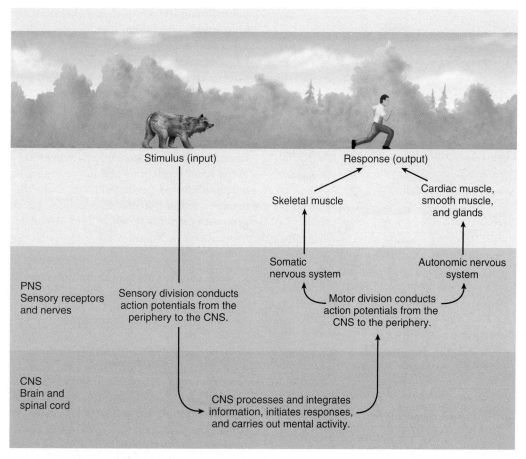

Figure 10.2   Organization of the Nervous System

The sensory division of the peripheral nervous system (PNS) detects stimuli and conducts action potentials to the central nervous system (CNS). The CNS interprets incoming action potentials and initiates action potentials that are conducted through the motor division to produce a response. The motor division is divided into the somatic nervous system and the autonomic nervous system.

and store information, and generate responses. The motor division of the PNS conducts action potentials from the CNS to muscles and glands. ▼

2. *Define CNS and PNS.*

3. *What is a sensory receptor and nerve?*

4. *Based on the direction they transmit action potentials, what are the two subcategories of the PNS?*

5. *Based on the structures they supply, what are the two subcategories of the motor division?*

6. *What are the subcategories of the ANS?*

7. *Compare the general functions of the CNS and the PNS.*

Cells of the Nervous System

The nervous system is made up of neurons and nonneural cells called glial cells.

Neurons

Neurons, or **nerve cells,** receive stimuli and transmit action potentials to other neurons or to effector organs. Each neuron consists of a cell body and two types of processes: dendrites and axons (figure 10.3).

Structure of Neurons

The cell body of neurons is called the **neuron cell body,** or **soma** (sō′mă, body). Each neuron cell body contains a single relatively large and centrally located nucleus with a prominent nucleolus (see figure 10.3). Extensive rough endoplasmic reticulum (ER) and Golgi apparatuses surround the nucleus, and a moderate number of mitochondria and other organelles are present. **Nissl** (nis′l) **substance,** which is located in the cell body and dendrites but not the axon, is aggregates of rough ER and free ribosomes. Nissl substance is the primary site of protein synthesis in neurons.

PREDICT   1
Predict the effect on the part of a severed axon that is no longer connected to its neuron cell body. Explain your prediction.

An **axon** (ak′son) is a long cell process extending from the neuron cell body. The **trigger zone,** which is the part of the neuron where the axon originates, is where nerve cells generate action potentials. Extending away from the trigger zone, each axon has a constant diameter and may vary in length from a few millimeters to more than a meter. Axons are often referred to as **nerve fibers** because of their shape. An axon can remain as a single structure or can branch to form collateral axons, or side

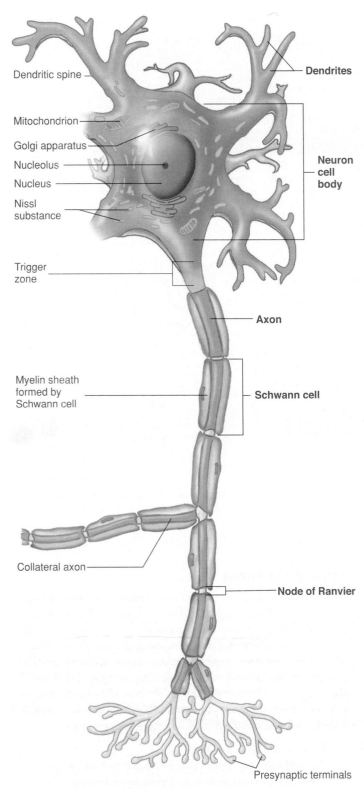

Figure 10.3 Neuron

The structural features of a neuron include a cell body and two types of cell processes: dendrites and an axon.

(Figure labels:)
Dendritic spine
Mitochondrion
Golgi apparatus
Nucleolus
Nucleus
Nissl substance
Trigger zone
Myelin sheath formed by Schwann cell
Collateral axon
Dendrites
Neuron cell body
Axon
Schwann cell
Node of Ranvier
Presynaptic terminals

neurotransmitters, which are chemicals that cross the synapse to stimulate or inhibit the postsynaptic cell (see "Chemical Synapses," p. 283).

Dendrites (den′drītz) are short, often highly branched cytoplasmic extensions that are tapered from their bases at the neuron cell body to their tips (see figure 10.3). Many dendrite surfaces have small extensions called **dendritic spines,** where axons of other neurons form synapses with the dendrites.

Dendrites are the input part of the neuron. When stimulated, they generate small electric currents, which are conducted to the neuron cell body and to the trigger zone, where an action potential can be generated. Axons are the output part of the neuron. Action potentials are conducted along the axon to the presynaptic terminal, where they stimulate the release of neurotransmitters. ▼

8. *Name the two types of cells forming the nervous system.*
9. *Compare the structure of axons and dendrites.*
10. *Define trigger zone, presynaptic terminal, and neurotransmitter.*
11. *Describe and give the function of a dendrite and an axon.*

Types of Neurons

Neurons are classified according to their function or structure. The functional classification is based on the direction in which action potentials are conducted. **Sensory,** or **afferent, neurons** conduct action potentials toward the CNS; **motor,** or **efferent, neurons** conduct action potentials away from the CNS toward muscles or glands. **Interneurons,** or **association neurons,** conduct action potentials from one neuron to another within the CNS.

The structural classification scheme is based on the number of processes that extend from the neuron cell body. The three major categories of neurons are multipolar, bipolar, and unipolar (figure 10.4).

Multipolar neurons have many dendrites and a single axon. Most of the neurons within the CNS and motor neurons are multipolar.

Bipolar neurons have two processes: one dendrite and one axon. Bipolar neurons are located in some sensory organs, such as in the retina of the eye and in the nasal cavity.

Unipolar neurons have a single process extending from the cell body. This process divides into two branches a short distance from the cell body. One branch extends to the CNS, and the other branch extends to the periphery and has dendritelike sensory receptors. The two branches function as a single axon. The sensory receptors respond to stimuli, resulting in the production of action potentials that are transmitted to the CNS. Most sensory neurons are unipolar neurons. ▼

12. *Describe the three types of neurons based on their function.*
13. *Describe the three types of neurons based on their structure, and give an example of where each type is found.*

Glial Cells of the CNS

Nonneural cells are called **glial** (glī′ăl, glē′ăl) **cells,** or **neuroglia** (noo-rog′lē-ă, nerve glue), and they account for over half of the

branches (see figure 10.3). Axons connect nerve cells to other nerve cells, muscle cells, and gland cells. The junction between a nerve cell and another cell is called a **synapse** and the ending of an axon in the synapse is called the **presynaptic terminal**. The presynaptic terminal has numerous vesicles containing

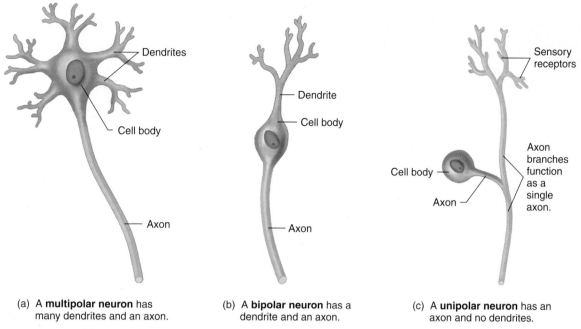

(a) A **multipolar neuron** has many dendrites and an axon.

(b) A **bipolar neuron** has a dendrite and an axon.

(c) A **unipolar neuron** has an axon and no dendrites.

Figure 10.4 Types of Neurons

Neurons are classified structurally by the number of cell processes extending from their cell bodies. Dendrites and sensory receptors are specialized to receive stimuli and axons are specialized to conduct action potentials.

brain's weight. Glial cells are the major supporting cells in the CNS; they participate in the formation of a permeability barrier between the blood and the neurons, phagocytize foreign substances, produce cerebrospinal fluid, and form myelin sheaths around axons. There are four types of CNS glial cells.

Astrocytes

Astrocytes (as′trō-sītz, *aster* is Greek, meaning star) are glial cells that are star-shaped because of cytoplasmic processes that extend from the cell body. These extensions widen and spread out to form foot processes, which cover the surfaces of blood vessels and neurons (figure 10.5). Astrocytes serve as the major supporting cells in the CNS.

Astrocytes play a role in regulating the extracellular composition of brain fluid. They release chemicals that promote the formation of tight junctions (see chapter 4) between the endothelial cells of capillaries. The endothelial cells with their tight junctions form the **blood–brain barrier,** which determines what substances can pass from the blood into the nervous tissue of the brain and spinal cord. The blood–brain barrier protects neurons from toxic substances in the blood, allows the exchange of nutrients and waste products between neurons and the blood, and prevents fluctuations in the composition of the blood from affecting the functions of the brain.

Astrocytes contribute to beneficial and detrimental responses to tissue damage in the CNS. Almost all injuries to CNS tissue induce **reactive astrocytosis,** in which astrocytes participate in walling off the injury site and limiting the spread of inflammation to the surrounding healthy tissue. Reactive scar-forming astrocytes also limit the regeneration of the axons of injured neurons.

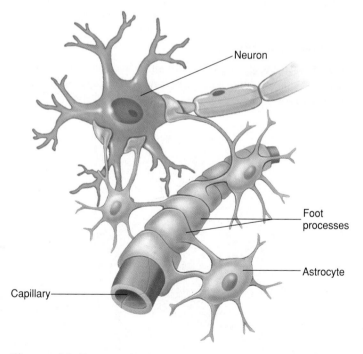

Figure 10.5 Astrocytes

Astrocyte processes form foot processes, which cover the surfaces of neurons, blood vessels, and the pia mater. The astrocytes provide structural support and play a role in regulating what substances from the blood reach the neurons.

Ependymal Cells

Ependymal (ep-en′di-măl) cells line the ventricles (cavities) of the brain and the central canal of the spinal cord (figure 10.6). Ependymal cells and blood vessels form the **choroid plexuses**

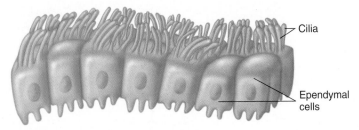

Figure 10.6   Ependymal Cells

Ciliated ependymal cells lining the ventricles of the brain and the central canal of the spinal cord help move cerebrospinal fluid.

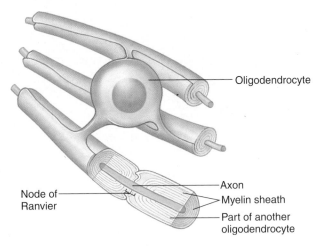

Figure 10.8   Oligodendrocytes

Extensions from oligodendrocytes form part of the myelin sheaths of several axons within the CNS.

(ko'royd plek'sŭs-ez), which are located within certain regions of the ventricles. The choroid plexuses secrete the cerebrospinal fluid that circulates through the ventricles of the brain (see chapter 11). The free surface of the ependymal cells frequently has patches of cilia that help move cerebrospinal fluid through the cavities of the brain.

Microglia

Microglia (mī-krog'lē-ă) are glial cells in the CNS that become mobile and phagocytic in response to inflammation. They phagocytize necrotic tissue, microorganisms, and other foreign substances that invade the CNS (figure 10.7). Numerous microglia migrate to areas damaged by infection, trauma, or stroke. A pathologist can identify these damaged areas in the CNS during an autopsy because large numbers of microglia are found in them.

Oligodendrocytes

Oligodendrocytes (ol'i-gō-den'drō-sītz) have cytoplasmic extensions that can surround axons. If the cytoplasmic extensions wrap many times around the axons, they form **myelin** (mī'ě-lin)

sheaths. A single oligodendrocyte can form myelin sheaths around portions of several axons (figure 10.8).

Glial Cells of the PNS

Schwann cells, or **neurolemmocytes** (noor-ō-lem'mō-sītz), are glial cells in the PNS that wrap around axons. If a Schwann cell wraps many times around an axon, it forms a myelin sheath. Unlike oligodendrocytes, however, each Schwann cell forms a myelin sheath around a portion of only one axon (figure 10.9).

Satellite cells surround neuron cell bodies in sensory ganglia (see figure 10.9). They provide support and nutrition to the neuron cell bodies, and they protect neurons from heavy metal poisons, such as lead and mercury, by absorbing them and reducing their access to the neuron cell bodies. ▼

14. *Which type of glial cell supports neurons and blood vessels and promotes the formation of the blood–brain barrier? What is the blood–brain barrier, and what is its function?*

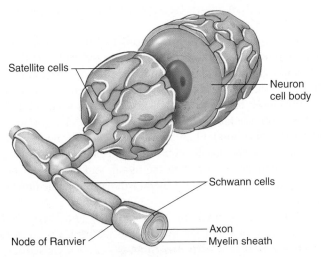

Figure 10.9   Glial Cells of the PNS

Neuron cell bodies are surrounded by satellite cells. Schwann cells form the myelin sheath of an axon within the PNS.

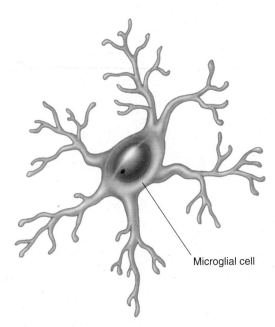

Figure 10.7   Microglia

Microglia are phagocytic cells within the CNS.

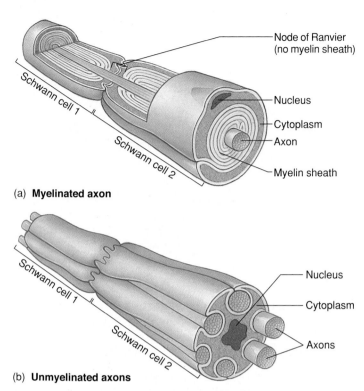

(a) Myelinated axon

- Node of Ranvier (no myelin sheath)
- Nucleus
- Cytoplasm
- Axon
- Myelin sheath
- Schwann cell 1
- Schwann cell 2

(b) Unmyelinated axons

- Nucleus
- Cytoplasm
- Axons
- Schwann cell 1
- Schwann cell 2

Figure 10.10 **Comparison of Myelinated and Unmyelinated Axons**

(*a*) Myelinated axon with two Schwann cells forming part of the myelin sheath around a single axon. Each Schwann cell surrounds part of one axon. (*b*) Unmyelinated axons with two Schwann cells surrounding several axons in parallel formation. Each Schwann cell surrounds part of several axons.

15. *Name the different kinds of glial cells responsible for the following functions: production of cerebrospinal fluid, phagocytosis, production of myelin sheaths in the CNS, production of myelin sheaths in the PNS, support of neuron cell bodies in the PNS.*

Myelinated and Unmyelinated Axons

Cytoplasmic extensions of the Schwann cells in the PNS and of the oligodendrocytes in the CNS surround axons to form either myelinated or unmyelinated axons. Myelin protects and electrically insulates axons from one another. In addition, action potentials travel along myelinated axons more rapidly than along unmyelinated axons (see "Propagation of Action Potentials," p. 280).

In **myelinated axons,** the plasma membrane of Schwann cells or oligodendrocytes repeatedly wraps around a segment of an axon to form the myelin sheath (figure 10.10*a*), which is white in color. Every 0.3–1.5 mm there is a bare area of the axon 2–3 μm in length called the **node of Ranvier** (ron′vē-ā). Although the axon at a node of Ranvier is not covered with myelin, Schwann cells or oligodendrocytes extend across the node and connect to each other.

Unmyelinated axons rest in invaginations of the Schwann cells or oligodendrocytes (figure 10.10*b*). The cell's plasma membrane surrounds each axon but does not wrap around it many times. Thus, each axon is surrounded by a series of cells, and each cell can simultaneously surround more than one unmyelinated axon. ▼

16. *Define myelin sheath and node of Ranvier.*
17. *How are myelinated and unmyelinated axons different from each other?*

Organization of Nervous Tissue

Nervous tissue is organized so that axons form bundles, and neuron cell bodies and their relatively short dendrites are grouped together. Bundles of parallel axons with their associated myelin sheaths, which are white in color, are called **white matter.** Collections of neuron cell bodies and unmyelinated axons are grayer in color and are called **gray matter.**

The axons that make up the white matter of the CNS form **nerve tracts,** which propagate action potentials from one area in the CNS to another. The gray matter of the CNS performs integrative functions or acts as relay areas in which axons synapse with the cell bodies of other neurons. The central area of the spinal cord is gray matter, and the outer surface of much of the brain consists of gray matter called **cortex.** Within the brain are other collections of gray matter called **nuclei.**

In the PNS, bundles of axons form nerves, which conduct action potentials to and from the CNS. Most nerves contain myelinated axons, but some consist of unmyelinated axons. Collections of neuron cell bodies in the PNS are called **ganglia**. ▼

18. *What are white matter and gray matter?*
19. *Define and state the location of nerve tracts, nerves, brain cortex, nuclei, and ganglia.*

Electric Signals

Like computers, humans depend on electric signals to communicate and process information. The electric signals produced by cells are called **action potentials.** They are an important means by which cells transfer information from one part of the body to another. For example, stimuli, such as light, sound, and pressure, act on specialized sensory cells in the eye, ear, and skin to produce action potentials, which are conducted from these cells to the spinal cord and brain. Action potentials originating within the brain and spinal cord are conducted to muscles and certain glands to regulate their activities.

The ability to perceive our environment, to perform complex mental activities, and to act depends on action potentials. For example, interpreting the action potentials received from sensory cells results in the sensations of sight, hearing, and touch. Complex mental activities, such as conscious thought, memory, and emotions, result from action potentials. The contraction of muscles and the secretion of certain glands occur in response to action potentials generated in them.

A basic knowledge of the electrical properties of cells is necessary for understanding many of the normal functions and pathologies of the body. These properties result from the ionic concentration differences across the plasma membrane and from the permeability characteristics of the plasma membrane.

Concentration Differences Across the Plasma Membrane

The concentration of a specific ion outside and inside a cell may be different. The concentration of sodium ions (Na^+), calcium ions (Ca^{2+}), and chloride ions (Cl^-) is much greater outside the cell than inside. The concentration of potassium ions (K^+) and negatively charged molecules, such as proteins, is much greater inside the cell than outside.

A steep concentration gradient (see chapter 3) exists for Na^+ from outside the cell to the inside because of the difference in Na^+ concentration across the plasma membrane. Also, a steep concentration gradient exists for K^+ from the inside to the outside of the cell. If membrane permeability permits, Na^+ diffuse down their concentration gradient into cells and K^+ diffuse out.

Differences in intracellular and extracellular concentrations of ions result primarily from (1) the Na^+–K^+ pump and (2) the permeability characteristics of the plasma membrane.

The Na⁺–K⁺ Pump

The differences in Na^+ and K^+ concentrations across the plasma membrane are maintained primarily by the action of the **Na⁺–K⁺ pump** (see figure 3.8). Through active transport, the Na^+–K^+ pump moves Na^+ and K^+ through the plasma membrane against their concentration gradients. Sodium ions are transported out of the cell, increasing the concentration of Na^+ outside the cell, and K^+ are transported into the cell, increasing the concentration of K^+ inside the cell. Approximately three Na^+ are transported out of the cell and two K^+ are transported into the cell for each ATP molecule used.

Permeability Characteristics of the Plasma Membrane

The plasma membrane is selectively permeable, thus allowing some, but not all, substances to pass through it (see chapter 3). Negatively charged proteins are synthesized inside the cell, and because of their large size and their solubility characteristics they cannot readily diffuse across the plasma membrane (figure 10.11).

Ions pass through the plasma membrane through ion channels. The two major types of ion channels are leak channels and gated ion channels. Leak channels are always open, whereas gated channels can open and close. Ions can pass through open channels, but not closed channels (see figure 10.11).

Leak Channels   Leak channels, or **nongated ion channels,** are always open and are responsible for the permeability of the plasma membrane to ions when the plasma membrane is unstimulated, or at rest. Each ion channel is specific for one type of ion, although the specificity is not absolute. The number of each type of leak channel in the plasma membrane determines the permeability characteristics of the resting plasma membrane to different types of ions. The plasma membrane is 50–100 times more permeable to K^+ and much less permeable to Na^+ because there are many more K^+ leak channels than Na^+ leak channels in the plasma membrane.

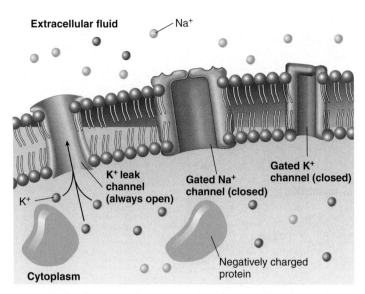

Figure 10.11   **Membrane Permeability and Ion Channels**

The plasma membrane is permeable to K^+ because K^+ leak channels are open. The permeability to Na^+ and K^+ can be regulated by gated Na^+ and K^+ channels. When the channels are closed, the passage of Na^+ and K^+ is prevented. Permeability to Na^+ and K^+ can increase, however, if the channels open. The membrane is not permeable to the negatively charged proteins inside the cell because they are too large to pass through membrane channels.

Gated Ion Channels   Gated ion channels open and close in response to stimuli. By opening and closing, these channels can change the permeability characteristics of the plasma membrane. The major types of gated ion channels are

1. *Ligand-gated ion channels.* A **ligand** is a molecule that binds to a receptor. A **receptor** is a protein or glycoprotein that has a **receptor site** to which a ligand can bind. Most receptors are located in the plasma membrane. **Ligand-gated ion channels** are receptors that have an extracellular receptor site and a membrane-spanning part that forms an ion channel. When a ligand binds to the receptor site, the ion channel opens or closes. For example, the neurotransmitter acetylcholine released from the presynaptic terminal of a neuron is a chemical signal that can bind to a ligand-gated Na^+ channel in the membrane of a muscle cell. As a result, the Na^+ channel opens, allowing Na^+ to enter the cell (see figure 8.5). Ligand-gated ion channels exist for Na^+, K^+, Ca^{2+}, and Cl^-, and these channels are common in tissues such as nervous and muscle tissues, as well as glands.

2. *Voltage-gated ion channels.* These channels open and close in response to small voltage changes across the plasma membrane. In an unstimulated cell, the inside of the plasma membrane is negatively charged relative to the outside. This charge difference can be measured in units called **millivolts** (mV) (1 mV = 1/1000 V). When a cell is stimulated, the permeability of the plasma membrane changes because gated ion channels open or close. The movement of ions into or out of the cell changes the charge difference across the plasma membrane, which causes voltage-gated ion channels to open or close. Voltage-gated channels specific

for Na⁺ and K⁺ are most numerous in electrically excitable tissues, but voltage-gated Ca^{2+} channels are also important, especially in smooth muscle and cardiac muscle cells (see chapters 8 and 17).

3. *Other gated ion channels.* Gated ion channels that respond to stimuli other than ligands or voltage changes are present in specialized electrically excitable tissues. Examples include touch receptors, which respond to mechanical stimulation of the skin, and temperature receptors, which respond to temperature changes in the skin. ▼

> **20.** *Describe the concentration differences for Na⁺ and K⁺ that exist across the plasma membrane.*
>
> **21.** *In what direction, into or out of cells, does the Na⁺–K⁺ pump move Na⁺ and K⁺?*
>
> **22.** *Define leak channels and gated ion channels. How are they responsible for the permeability characteristics of a resting versus a stimulated plasma membrane?*
>
> **23.** *Define ligand, receptor, and receptor site.*
>
> **24.** *What kinds of stimuli cause gated ion channels to open or close?*

Establishing the Resting Membrane Potential

The extracellular fluid is electrically neutral with no net overall charge because the number of positively charged ions in extracellular fluids is equal to the number of negatively charged ions. Similarly, the intracellular fluid is electrically neutral because the number of positively charged ions is equal to the number of negatively charged ions and proteins.

Even though extracellular and intracellular fluids are electrically neutral, there is a difference in charge across the plasma membrane. The immediate inside of the membrane is negative, compared with the immediate outside of the membrane (figure 10.12). The plasma membrane, therefore, is said to be **polarized** because there are opposite charges, or poles, across the membrane.

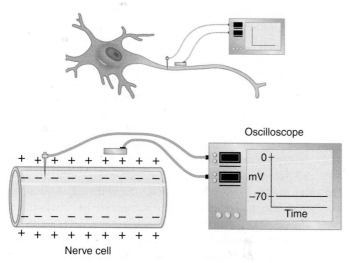

Figure 10.12   **Measuring Resting Membrane Potential**

The recording electrode is inside the cell; the reference electrode is outside. A potential difference of about −70 mV is recorded, with the inside of the plasma membrane negative with respect to the outside of the membrane.

The electric charge difference across the plasma membrane is called a **potential difference.** In an unstimulated, or resting, cell, the potential difference is called the **resting membrane potential.** It can be measured using a voltmeter or an oscilloscope connected to microelectrodes positioned just inside and outside the plasma membrane (see figure 10.12). The resting membrane potential of nerve cells is approximately −70 mV and of skeletal muscle cells is approximately −85 mV. The potential difference is reported as a negative number because the inside of the plasma membrane is negative, compared with the outside. The greater the charge difference across the plasma membrane, the greater the potential difference. A resting membrane potential of −85 mV has a greater charge difference than a resting membrane potential of −70 mV.

The resting membrane potential results from the permeability characteristics of the resting plasma membrane and the difference in concentration of ions between the intracellular and the extracellular fluids. The plasma membrane is somewhat permeable to K⁺ because of K⁺ leak channels. Positively charged K⁺ can therefore diffuse down their concentration gradient from inside to just outside the cell (figure 10.13). Negatively charged proteins and other molecules cannot diffuse through the plasma membrane with the K⁺. As K⁺ diffuse out of the cell, the loss of positive charges makes the inside of the plasma membrane more negative. Because opposite charges attract, the K⁺ are attracted back toward the cell. The accumulation of K⁺ just outside of the plasma membrane makes the outside of the plasma membrane positive relative to the inside. The resting membrane potential is an equilibrium that is established when the tendency for K⁺ to diffuse out of the cell, because of the K⁺ concentration gradient, is equal to the tendency for K⁺ to move into the cell because of the attraction of the positively charged K⁺ to negatively charged proteins and other molecules.

P R E D I C T

Given that tissue A has significantly more K⁺ leak channels than tissue B, which tissue has the larger resting membrane potential?

The major ionic influence on the resting membrane potential is due to the movement of K⁺ through leak channels. Other ions, such as Na⁺, have only a minor influence on the resting membrane potential because there are relatively few leak channels for them. Even after the resting membrane potential is established, however, cells gradually lose K⁺ and gain Na⁺. The large concentration gradients for Na⁺ and K⁺ would eventually disappear without the continuous activity of the Na⁺–K⁺ pump.

P R E D I C T

What would happen to the resting membrane potential if the Na⁺–K⁺ pump stopped? Explain.

The pump is also responsible for a small portion of the resting membrane potential, usually less than 15 mV, because it transports approximately three Na⁺ out of the cell and two K⁺ into the cell for each ATP molecule used. The outside of the plasma membrane becomes more positively charged than the inside because more positively charged ions are pumped out of the cell than are pumped into it.

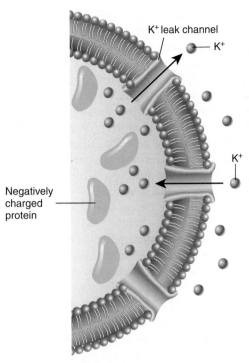

K⁺ leak channel

K⁺

Potassium ions diffuse out of the cell because there is a greater concentration of K⁺ inside than outside the cell.

The resting membrane potential is established when the movement of K⁺ out of the cell is equal to their movement into the cell.

K⁺

Potassium ions move into the cell because the positively charged ions are attracted to the negatively charged proteins and anions.

Negatively charged protein

Figure 10.13   Potassium Ions and the Resting Membrane Potential

The characteristics responsible for establishing and maintaining the resting membrane potential are summarized in table 10.1. ▼

25. *Define resting membrane potential. Is the outside of the plasma membrane positively or negatively charged relative to the inside?*

Table 10.1	Characteristics Responsible for the Resting Membrane Potential

1. The number of charged molecules and ions inside and outside the cell is nearly equal.
2. The concentration of K⁺ is higher inside than outside the cell, and the concentration of Na⁺ is higher outside than inside the cell.
3. The plasma membrane is 50–100 times more permeable to K⁺ than to other positively charged ions, such as Na⁺.
4. The plasma membrane is impermeable to large, intracellular, negatively charged molecules, such as proteins.
5. Potassium ions tend to diffuse across the plasma membrane from the inside to the outside of the cell.
6. Because negatively charged molecules cannot follow the positively charged K⁺, a small negative charge develops just inside the plasma membrane.
7. The negative charge inside the cell attracts positively charged K⁺. When the negative charge inside the cell is great enough to prevent additional K⁺ from diffusing out of the cell through the plasma membrane, an equilibrium is established.
8. The charge difference across the plasma membrane at equilibrium is reflected as a difference in potential, which is measured in millivolts (mV).
9. The Na⁺–K⁺ pump maintains the concentration gradients for Na⁺ and K⁺ and contributes to the resting membrane potential by pumping three Na⁺ out of the cell for each two K⁺ pumped into the cell.

26. *Explain the role of K⁺ and the Na⁺–K⁺ pump in establishing the resting membrane potential.*

Changing the Resting Membrane Potential

The resting membrane potential can decrease or increase (figure 10.14). **Depolarization** (dē-pō′lăr-i-zā′shŭn) is a decrease in the membrane potential in which the charge difference, or polarity, across the plasma membrane decreases. **Hyperpolarization** (hī′per-pō′lăr-i-zā′shŭn) is an increase in the membrane potential caused by an increase in the charge difference across the plasma membrane. The charge difference can be changed by decreasing or increasing the movement of ions across the plasma membrane. Ion movement can be changed by altering ion concentration gradients or ion permeability.

Potassium Ions

Changes in the K⁺ concentration gradient and K⁺ permeability can change the resting membrane potential. An increase in the extracellular concentration of K⁺ decreases the K⁺ concentration gradient because the concentration of K⁺ is lower outside than inside a cell. As a consequence, the tendency for K⁺ to diffuse out of the cell decreases, and a smaller negative charge inside the cell is required to oppose the diffusion of K⁺ out of the cell. At this new equilibrium, the smaller charge difference across the plasma membrane is a depolarization. On the other hand, a decrease in the extracellular concentration of K⁺ increases the K⁺ concentration gradient. As a result, the tendency for K⁺ to diffuse out of the cell increases, and a larger negative charge inside the cell is required to resist that diffusion. At this new equilibrium, the larger charge difference across the plasma membrane is a hyperpolarization.

Although K⁺ leak channels allow some K⁺ to diffuse across the plasma membrane, the resting membrane is not freely permeable

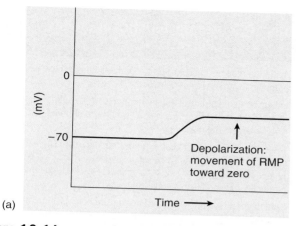

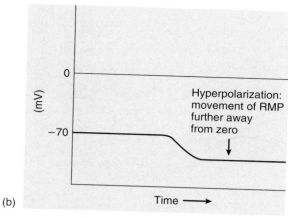

Figure 10.14 Depolarization and Hyperpolarization of the Resting Membrane Potential

(a) In depolarization, membrane potential decreases, becoming less negative. (b) In hyperpolarization, membrane potential increases, becoming more negative.

to K^+. There are gated K^+ channels in the plasma membrane, however (see figure 10.11). If gated K^+ channels open, membrane permeability to K^+ increases, and more K^+ diffuse out of the cell. The increased tendency for K^+ to diffuse out of the cell is opposed by the greater negative charge that develops inside the plasma membrane, resulting in hyperpolarization.

P R E D I C T   4

Does the resting membrane potential increase or decrease when the intracellular concentration of K^+ is increased by the injection of a potassium succinate solution into a cell? Explain.

Sodium Ions

In an unstimulated cell, the membrane is not very permeable to Na^+ because there are few Na^+ leak channels. Changes in the concentration of Na^+ on either side of the plasma membrane do not influence the resting membrane potential very much because of this low permeability. There are gated Na^+ channels in the plasma membrane, however (see figure 3.11). If gated Na^+ channels open, membrane permeability to Na^+ increases and Na^+ then diffuse into the cell because the concentration gradient for Na^+ is from the outside to the inside of the cell. As Na^+ diffuse into the cell, the inside of the membrane becomes more positive, resulting in depolarization.

Calcium Ions

Calcium ions alter membrane potentials in two ways: (1) by affecting voltage-gated Na^+ channels and (2) by entering cells through gated Ca^{2+} channels. Voltage-gated Na^+ channels are sensitive to changes in the extracellular concentration of Ca^{2+}. Positively charged Ca^{2+} in the extracellular fluid are attracted to the negatively charged groups of proteins in voltage-gated $Na+$ channels. If the extracellular concentration of Ca^{2+} decreases, these ions diffuse away from the voltage-gated Na^+ channels, causing the channels to open. If the extracellular concentration of Ca^{2+} increases, they bind to voltage-gated Na^+ channels, causing them to close. At the Ca^{2+} concentrations normally found in the extracellular fluid, only a small percentage of the voltage-gated Na^+ channels are open at any moment in an unstimulated cell.

P R E D I C T   5

Predict the effect of a decrease in the extracellular concentration of Ca^{2+} on the resting membrane potential.

Changes in the permeability of Ca^{2+} can change the resting membrane potential. If the plasma membrane becomes permeable to Ca^{2+} by the opening of gated Ca^{2+} channels, Ca^{2+} diffuse into the cell and the inside of the membrane becomes more positive, resulting in depolarization.

Chloride Ions

Changes in the permeability of Cl^- can change the resting membrane potential. If the permeability of the plasma membrane to Cl^- increases by the opening of gated Cl^- channels, then Cl^- diffuse into the cell and the inside of the membrane becomes more negative, resulting in hyperpolarization. ▼

27. **Define depolarization and hyperpolarization. How do alterations in the K^+ concentration gradient; changes in membrane permeability to K^+, Na^+, Ca^{2+}, or Cl^- and changes in extracellular Ca^{2+} concentration affect depolarization and hyperpolarization?**

Graded Potentials

A stimulus applied at one location on the plasma membrane of a cell normally causes a change in the resting membrane potential called a **graded potential.** Graded potentials are so called because the potential change can vary from small to large. Graded potentials are also called **local potentials** because they are confined to a small region of the plasma membrane. Graded potentials can result from (1) chemical signals binding to their receptors, (2) changes in the voltage across the plasma membrane, (3) mechanical stimulation, (4) temperature changes, or (5) spontaneous changes in membrane permeability.

A graded potential can be either a depolarization or a hyperpolarization (see figure 10.14). A change in membrane permeability to Na^+, K^+, or other ions can produce a graded potential. For example, if a stimulus causes gated Na^+ channels to open, the diffusion of a few Na^+ into cells results in depolarization. If a stimulus

276

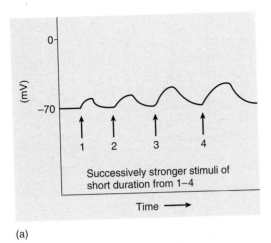

(a)

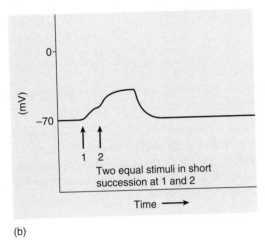

(b)

Figure 10.15 Graded Potentials

(a) Graded potentials are proportional to the stimulus strength. A weak stimulus applied briefly causes a small depolarization, which quickly returns to the resting membrane potential (1). Progressively stronger stimuli result in larger depolarizations (2 to 4). (b) A stimulus applied to a cell causes a small depolarization. When a second stimulus is applied before the depolarization disappears, the depolarization caused by the second stimulus is added to the depolarization caused by the first to result in a larger depolarization.

causes gated K^+ channels to open, the diffusion of a few K^+ out of the cell results in hyperpolarization.

The magnitude of graded potentials can vary from small to large, depending on the stimulus strength or on summation. For example, a weak stimulus can cause a few gated Na^+ channels to open. A few Na^+ diffuse into the cell and cause a small depolarization. A stronger stimulus can cause a greater number of gated Na^+ channels to open. A greater number of Na^+ diffusing into the cell causes a larger depolarization (figure 10.15a).

Local potentials can **summate** (sŭm-āt′), or add onto each other. **Summation** of graded potentials occurs when the effects produced by one graded potential are added onto the effects produced by another graded potential (figure 10.15b). For example, if a second stimulus is applied before the graded potential produced by the first stimulus has returned to the resting membrane potential, a larger depolarization results than would result from a single stimulus. The first stimulus causes gated Na^+ channels to open, and the second stimulus causes additional Na^+ channels to open.

Table 10.2	Characteristics of Graded Potentials

1. A stimulus causes ion channels to open, resulting in increased permeability of the membrane to Na^+, K^+, or Cl^-.

2. Increased permeability of the membrane to Na^+ results in depolarization. Increased permeability of the membrane to K^+ or Cl^- results in hyperpolarization.

3. The size of the graded potential is proportional to the strength of the stimulus. Graded potentials can also summate. Thus, a graded potential produced in response to several stimuli is larger than one produced in response to a single stimulus.

4. Graded potentials are conducted in a decremental fashion, meaning that their magnitude decreases as they spread over the plasma membrane. Graded potentials cannot be measured a few millimeters from the point of stimulation.

5. A depolarizing graded potential can cause an action potential.

As a result, more Na^+ diffuse into the cell, producing a larger graded potential. Summation is discussed in greater detail later in this chapter (see "Spatial and Temporal Summation," p. 290).

Graded potentials spread, or are conducted, over the plasma membrane in a decremental fashion. That is, they rapidly decrease in magnitude as they spread over the surface of the plasma membrane, much as a teacher talks to a large class. At the front of the class, the teacher's voice can be easily heard but, the farther away a student sits, the more difficult it is to hear. Normally, a graded potential cannot be detected more than a few millimeters from the site of stimulation. As a consequence, a graded potential cannot transfer information over long distances from one part of the body to another.

Graded potentials are important because of their effect on the generation of action potentials. The characteristics of graded potentials are summarized in table 10.2.

> **28. Define a graded potential. What does it mean to say a graded potential can summate and spreads in a decremental fashion?**

P R E D I C T 6

Given two cells that are identical in all ways except that the extracellular concentration of Na^+ is less for cell A than for cell B, how would the magnitude of the graded potential in cell A differ from that in cell B if stimuli of identical strength were applied to each?

Action Potentials

When a graded potential causes depolarization of the plasma membrane to a level called **threshold,** a series of permeability changes occurs that results in an action potential (figure 10.16). The action potential has a **depolarization phase,** in which the membrane potential moves away from the resting membrane potential and becomes more positive, and a **repolarization phase,** in which the membrane potential returns toward the resting membrane state and becomes more negative. After the repolarization phase, the plasma membrane may be slightly hyperpolarized for a short period called the **afterpotential.** An action potential is a large change in the membrane potential that propagates, without changing its magnitude, over long distances along the plasma

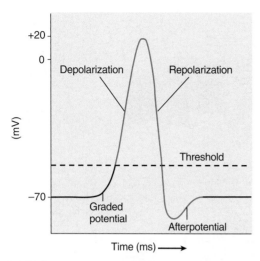

Figure 10.16 Action Potential

The action potential consists of a depolarization phase and a repolarization phase, often followed by a short period of hyperpolarization, called the afterpotential.

membrane. Thus, action potentials can transfer information from one part of the body to another. It generally takes 1–2 milliseconds (ms) (1 ms = 0.001 s) for an action potential to occur. The characteristics of action potentials are summarized in table 10.3.

All-or-None Principle

The generation of action potentials is dependent on graded potentials. Depolarizing graded potentials that reach threshold produce an action potential. Hyperpolarizing graded potentials can never reach threshold and do not produce action potentials. Thus, depolarizing graded potentials can potentially activate a cell by causing the production of an action potential, whereas hyperpolarizing graded potentials can prevent the production of an action potential.

Table 10.3	Characteristics of Action Potentials

1. Action potentials are produced when a graded potential reaches threshold.
2. Action potentials are all-or-none.
3. Depolarization is a result of increased membrane permeability to Na^+ and movement of Na^+ into the cell. Activation gates of the voltage-gated Na^+ channels open.
4. Repolarization is a result of decreased membrane permeability to Na^+ and increased membrane permeability to K^+, which stops Na^+ movement into the cell and increases K^+ movement out of the cell. The inactivation gates of the voltage-gated Na^+ channels close, and the voltage-gated K^+ channels open.
5. No action potential is produced by a stimulus, no matter how strong, during the absolute refractory period. During the relative refractory period, a stronger-than-threshold stimulus can produce an action potential.
6. Action potentials are propagated, and for a given axon or muscle fiber the magnitude of the action potential is constant.
7. Stimulus strength determines the frequency of action potentials.

The magnitude of a depolarizing graded potential affects the likelihood of generating an action potential. For example, a weak stimulus can produce a small depolarizing graded potential that does not reach threshold and therefore does not cause the production of an action potential. A stronger stimulus, however, can produce a larger depolarizing graded potential that reaches threshold, resulting in the production of an action potential.

Action potentials occur according to the **all-or-none principle.** If a stimulus produces a depolarizing graded potential that is large enough to reach threshold, all the permeability changes responsible for an action potential proceed without stopping and are constant in magnitude (the "all" part). If a stimulus is so weak that the depolarizing graded potential does not reach threshold, few of the permeability changes occur. The membrane potential returns to its resting level after a brief period without producing an action potential (the "none" part). An action potential can be compared to the flash system of a camera. When the shutter is triggered (reaches threshold), the camera flashes (an action potential is produced), and each flash is the same brightness (magnitude) (the "all" part) as previous flashes. If the shutter is depressed, but not triggered, no flash results (the "none" part).

Depolarization Phase

The change in charge across the plasma membrane caused by a graded potential causes increasing numbers of voltage-gated Na^+ channels to open for a brief time. As soon as a threshold depolarization is reached, many voltage-gated Na^+ channels open. Each voltage-gated Na^+ channel has two voltage-sensitive gates, called **activation** and **inactivation gates.** When the plasma membrane is at rest, the activation gates of the voltage-gated Na^+ channel are closed, and the inactivation gates are open (figure 10.17, step 1). Because the activation gates are closed, Na^+ cannot diffuse through the channels. When the graded potential reaches threshold, the change in the membrane potential causes many of the activation gates to open, and Na^+ can diffuse through the Na+ channels into the cell.

When the plasma membrane is at rest, voltage-gated K^+ channels, which have one gate, are closed (see figure 10.17, step 1). When the graded potential reaches threshold, the voltage-gated K^+ channels begin to open at the same time as the voltage-gated Na^+ channels, but they open more slowly (figure 10.17, step 2). Only a small number of voltage-gated K^+ channels are open, compared with the number of voltage-gated Na^+ channels, because the voltage-gated K^+ channels open slowly. Depolarization occurs because more Na^+ diffuse into the cell than K^+ diffuse out of it.

P R E D I C T   **7**

Predict the effect of a reduced extracellular concentration of Na^+ on the magnitude of the action potential in an electrically excitable cell.

Repolarization Phase

As the membrane potential approaches its maximum depolarization, the change in the potential difference across the plasma membrane causes the inactivation gates in the voltage-gated Na^+ channels to begin closing, and the permeability of the plasma

1. Resting membrane potential.
Voltage-gated Na⁺ channels (*pink*) are closed (the activation gates are closed and the inactivation gates are open). Voltage-gated K⁺ channels (*purple*) are closed.

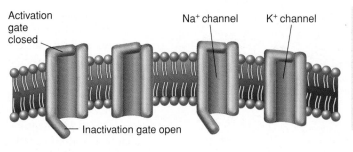
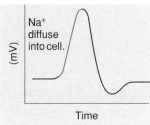

2. Depolarization. Voltage-gated Na⁺ channels open because the activation gates open. Voltage-gated K⁺ channels start to open. Depolarization results because the inward diffusion of Na⁺ is much greater than the outward diffusion of K⁺.

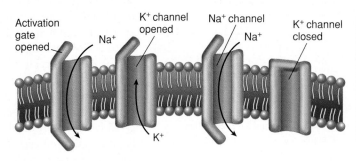
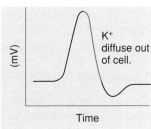

3. Repolarization. Voltage-gated Na⁺ channels are closed because the inactivation gates close. Voltage-gated K⁺ channels are now open. Sodium ions diffusion into the cell stops and K⁺ diffuse out of the cell, causing repolarization.

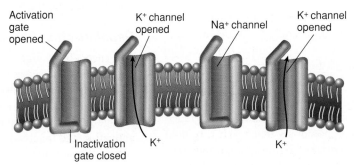

4. End of repolarization and afterpotential. Voltage-gated Na⁺ channels are closed. Closure of the activation gates and opening of the inactivation gates reestablish the resting condition for Na⁺ channels (see step 1). Diffusion of K⁺ through voltage-gated channels produces the afterpotential.

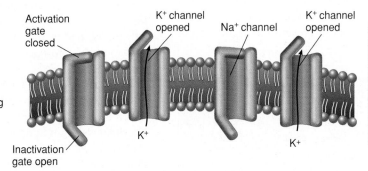
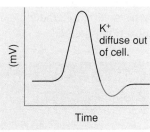

5. Resting membrane potential.
The resting membrane potential is reestablished after the voltage-gated K⁺ channels close.

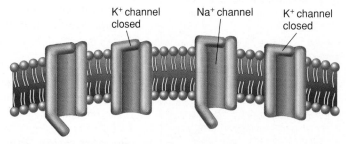
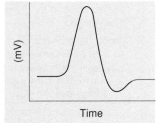

Process Figure 10.17   Voltage-Gated Ion Channels and the Action Potential

Step 1 illustrates the status of voltage-gated Na⁺ and K⁺ channels in a resting cell. Steps 2–5 show how the channels open and close to produce an action potential. Next to each step, a graph shows in *red* the membrane potential resulting from the condition of the ion channels.

membrane to Na^+ decreases. During the repolarization phase, the voltage-gated K^+ channels, which started to open along with the voltage-gated Na^+ channels, continue to open (figure 10.17, step 3). Consequently, the permeability of the plasma membrane to Na^+ decreases, and the permeability to K^+ increases. The decreased diffusion of Na^+ into the cell and the increased diffusion of K^+ out of the cell cause repolarization.

At the end of repolarization, the decrease in membrane potential causes the activation gates in the voltage-gated Na^+ channels to close and the inactivation gates to open. Although this change does not affect the diffusion of Na^+, it does return the voltage-gated Na^+ channels to their resting state (figure 10.17, step 4).

Afterpotential

In many cells, a period of hyperpolarization, or afterpotential, exists following each action potential. The afterpotential exists because the voltage-gated K^+ channels remain open for a short time (see figure 10.17, step 4). The increased K^+ permeability that develops during the repolarization phase of the action potential lasts slightly longer than the time required to bring the membrane potential back to its resting level. As the voltage-gated K^+ channels close, the original resting membrane potential is reestablished (figure 10.17, step 5).

During an action potential, a small number of Na^+ diffuse into the cell and a small number of K^+ diffuse out of the cell. The Na^+–K^+ pump restores normal resting ion concentrations by transporting these ions in the opposite direction of their movement during the action potential. That is, Na^+ are pumped out of the cell and K^+ are pumped into the cell. The Na^+–K^+ pump is too slow to have an effect on either the depolarization or the repolarization phase of individual action potentials. ▼

> **29. Define action potential. How do depolarizing and hyperpolarizing graded potentials affect the likelihood of generating an action potential?**
>
> **30. Explain the "all" and the "none" parts of the all-or-none principle of action potentials.**
>
> **31. What are the depolarization and repolarization phases of an action potential? Explain how changes in membrane permeability and the movement of Na^+ and K^+ cause each phase. What happens when the activation gates in the voltage-gated Na^+ channels open and the inactivation gates close?**
>
> **32. Describe the afterpotential and its cause.**

Refractory Period

Once an action potential is produced at a given point on the plasma membrane, the sensitivity of that area to further stimulation decreases for a time called the **refractory** (rē-frak′tōr-ē) **period.** The first part of the refractory period, during which complete insensitivity exists to another stimulus, is called the **absolute refractory period.** In many cells, it occurs from the beginning of the action potential until near the end of repolarization (figure 10.18). At the beginning of the action potential, depolarization occurs when the activation gates in the voltage-gated Na^+ channel open. At this time, the inactivation gates in the voltage-gated Na^+ channels are already open (see figure 10.17, step 2). Depolarization

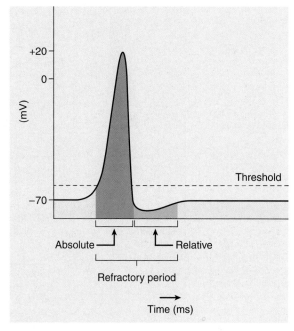

Figure 10.18 **Refractory Period**

The absolute and relative refractory periods of an action potential. In some cells, the absolute refractory period ends during the repolarization phase of the action potential.

ends as the inactivation gates close (see figure 10.17, step 3). As long as the inactivation gates are closed, further depolarization cannot occur. When the inactivation gates open and the activation gates close near the end of repolarization (see figure 10.17, step 4), once again it is possible to stimulate the production of another action potential.

The existence of the absolute refractory period guarantees that, once an action potential is begun, both the depolarization and the repolarization phases will be completed, or nearly completed, before another action potential can begin and that a strong stimulus cannot lead to prolonged depolarization of the plasma membrane. The absolute refractory period has important consequences for the rate at which action potentials can be generated and for the propagation of action potentials.

The second part of the refractory period, called the **relative refractory period,** follows the absolute refractory period. A stronger-than-threshold stimulus can initiate another action potential during the relative refractory period. Thus, after the absolute refractory period, but before the relative refractory period is completed, a sufficiently strong stimulus can produce another action potential. The relative refractory period ends when the voltage-gated K^+ channels close (see figure 10.17, step 5). ▼

> **33. What are the absolute and relative refractory periods? Relate them to the depolarization and repolarization phases of the action potential.**

P R E D I C T   8

Does a prolonged threshold stimulus or a prolonged stronger-than-threshold stimulus of the same duration produce the most action potentials? Explain.

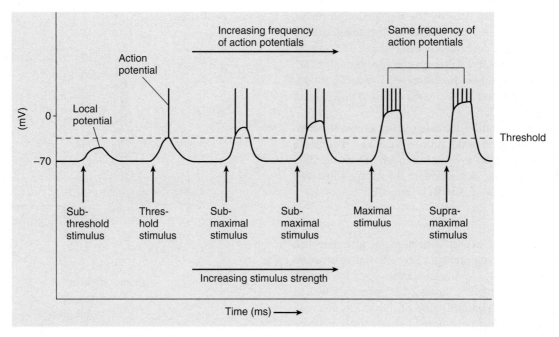

Figure 10.19 Stimulus Strength and Action Potential Frequency

From left to right, each stimulus in the figure is stronger than the previous one. As stimulus strength increases, the frequency of action potentials increases until a maximal rate is produced. Thereafter, increasing stimulus strength does not increase action potential frequency.

Action Potential Frequency

The **action potential frequency** is the number of action potentials produced per unit of time in response to a stimulus. Action potential frequency is directly proportional to stimulus strength and to the size of the graded potential. A **subthreshold stimulus** is any stimulus not strong enough to produce a graded potential that reaches threshold. Therefore, no action potential is produced (figure 10.19). A **threshold stimulus** produces a graded potential that is just strong enough to reach threshold and cause the production of a single action potential. A **maximal stimulus** is just strong enough to produce a maximum frequency of action potentials. A **submaximal stimulus** includes all stimuli between threshold and the maximal stimulus strength. For submaximal stimuli, the action potential frequency increases in proportion to the strength of the stimulus because the size of the graded potential increases with stimulus strength. A **supramaximal stimulus** is any stimulus stronger than a maximal stimulus. These stimuli cannot produce a greater frequency of action potentials than a maximal stimulus.

The duration of the absolute refractory period determines the maximum frequency of action potentials generated in an excitable cell. During the absolute refractory period, a second stimulus, no matter how strong, cannot stimulate an additional action potential. As soon as the absolute refractory period ends, however, it is possible for a second stimulus to cause the production of an action potential.

P R E D I C T 9

If the duration of the absolute refractory period of a nerve cell is 1 millisecond (ms), how many action potentials are generated by a maximal stimulus in 1 second?

The frequency of action potentials provides information about the strength of a stimulus. For example, a weak pain stimulus generates a low frequency of action potentials, whereas a stronger pain stimulus generates a higher frequency of action potentials. The ability to interpret a stimulus as mildly painful versus very painful depends, in part, on the frequency of action potentials generated by individual pain receptors. Communication regarding the strength of stimuli cannot depend on the magnitudes of action potentials because, according to the all-or-none principle, the magnitudes are always the same. The magnitudes of action potentials produced by weak and strong pain stimuli are the same.

The ability to stimulate muscle or gland cells also depends on action potential frequency. A low frequency of action potentials produces a weaker muscle contraction or less secretion than does a higher frequency. For example, a low frequency of action potentials in a muscle results in incomplete tetanus and a high frequency in complete tetanus (see chapter 8).

In addition to the frequency of action potentials, how long the action potentials are produced provides important information. For example, a pain stimulus of 1 second is interpreted differently than the same stimulus applied for 30 seconds. ▼

34. *Define action potential frequency. What two factors determine action potential frequency?*
35. *Define subthreshold, threshold, maximal, submaximal, and supramaximal stimuli. What determines the maximum frequency of action potential generation?*

Propagation of Action Potentials

An action potential occurs in a very small area of the plasma membrane and does not affect the entire membrane at one time. Action

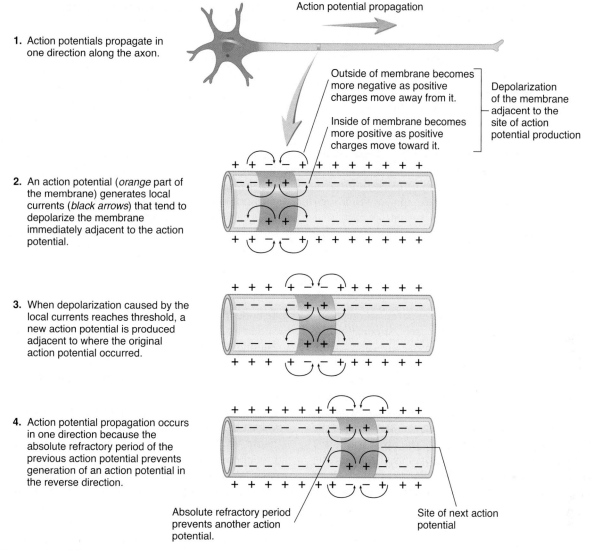

1. Action potentials propagate in one direction along the axon.

Action potential propagation

Outside of membrane becomes more negative as positive charges move away from it.

Inside of membrane becomes more positive as positive charges move toward it.

Depolarization of the membrane adjacent to the site of action potential production

2. An action potential (*orange* part of the membrane) generates local currents (*black arrows*) that tend to depolarize the membrane immediately adjacent to the action potential.

3. When depolarization caused by the local currents reaches threshold, a new action potential is produced adjacent to where the original action potential occurred.

4. Action potential propagation occurs in one direction because the absolute refractory period of the previous action potential prevents generation of an action potential in the reverse direction.

Absolute refractory period prevents another action potential.

Site of next action potential

Process Figure 10.20   Action Potential Propagation in an Unmyelinated Axon

potentials can, however, **propagate,** or spread, across the plasma membrane because an action potential produced at one location in the plasma membrane can stimulate the production of an action potential at an adjacent area of the plasma membrane. Note that an action potential does not actually move along an axon. Rather, an action potential at one location stimulates the production of another action potential at an adjacent location, which in turn stimulates the production of another, and so on, like a long row of toppling dominos in which each domino knocks down the next one. Each domino falls, but no one domino actually travels the length of the row.

In a neuron, action potentials are normally produced at the trigger zone and propagate in one direction along the axon (figure 10.20, step 1). The location at which the next action potential is generated is different for unmyelinated and myelinated axons (see figure 10.10). In an unmyelinated axon, the next action potential is generated immediately adjacent to the previous action potential. When an action potential is produced,

the inside of the membrane becomes more positive than the outside (figure 10.20, step 2). On the outside of the membrane, positively charged ions from the adjacent area are attracted to the negative charges at the site of the action potential. On the inside of the plasma membrane, positively charged ions at the site of the action potential are attracted to the adjacent negatively charged part of the membrane. The movement of positively charged ions is called an **ionic current,** or **local current.** As a result of the ionic current, the part of the membrane immediately adjacent to the action potential depolarizes. That is, the outside of the membrane immediately adjacent to the action potential becomes more negative because of the loss of positive charges and the inside becomes more positive because of the gain of positive charges. When the depolarization reaches threshold, an action potential is produced (figure 10.20, step 3).

If an action potential is initiated at one end of an axon, it is propagated in one direction down the axon. The absolute

Clinical Relevance   Examples of Abnormal Membrane Potentials

Several conditions are examples of the physiology of membrane potentials and the consequence of abnormal ones. **Hypokalemia** (hī-pō-ka-lē′mē-ă) is a lower than normal concentration of K+ in the blood or extracellular fluid. Reduced extracellular K+ concentrations cause hyperpolarization of the resting membrane potential (see figure 10.14b). Thus, a greater than normal stimulus is required to depolarize the membrane to its threshold level and to initiate action potentials in neurons, skeletal muscle, and cardiac muscle. Symptoms of hypokalemia include muscular weakness, an abnormal electrocardiogram, and sluggish reflexes. The symptoms result from the reduced sensitivity of the excitable tissues to stimulation. The causes of hypokalemia include potassium depletion during starvation, alkalosis, and certain kidney diseases.

Hypocalcemia (hī-pō-kal-sē′mē-ă) is a lower than normal concentration of Ca2+ in blood or extracellular fluid. Symptoms of hypocalcemia include nervousness and uncontrolled contraction of skeletal muscles, called **tetany** (tet′ă-nē). The symptoms are due to an increased membrane permeability to Na+ that results because low blood levels of Ca2+ cause voltage-gated Na+ channels in the membrane to open. Sodium ions diffuse into the cell, causing depolarization of the plasma membrane to threshold and initiating action potentials. The tendency for action potentials to occur spontaneously in nervous tissue and muscles accounts for the symptoms. A lack of dietary calcium, a lack of vitamin D, and a reduced secretion rate of a parathyroid gland hormone are conditions that cause hypocalcemia.

---

refractory period ensures one-way propagation of an action potential because it prevents the ionic current from stimulating the production of an action potential in the reverse direction (figure 10.20, step 4).

In a myelinated axon, an action potential is conducted from one node of Ranvier to another in a process called **saltatory conduction** (*saltare* is Latin, meaning to leap). An action potential at one node of Ranvier generates ionic currents that flow toward the next node of Ranvier (figure 10.21, step 1). The lipids within the membranes of the myelin sheath act as a layer of insulation, forcing the ionic currents to flow from one node of Ranvier to the next. In addition, voltage-gated Na+ channels are highly concentrated at the nodes of Ranvier. Therefore, the ionic current quickly flows to a node and stimulates the voltage-gated Na+ channels to open, resulting in the production of an action potential (figure 10.21, step 2).

The speed of action potential conduction along an axon depends on the myelination of the axon. Action potentials are conducted more rapidly in myelinated than unmyelinated axons because they are formed quickly at each successive node of Ranvier (figure 10.21, step 3), instead of being propagated more slowly through every part of the axon's membrane, as in unmyelinated axons (see figure 10.20). Action potential conduction in a myelinated fiber is like a grasshopper jumping; in an unmyelinated axon,

1. An action potential (*orange*) at a node of Ranvier generates local currents (*black arrows*). The local currents flow to the next node of Ranvier because the myelin sheath of the Schwann cell insulates the axon of the internode.

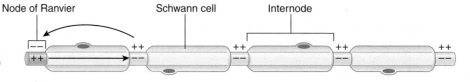

2. When the depolarization caused by the local currents reaches threshold at the next node of Ranvier, a new action potential is produced (*orange*).

3. Action potential propagation is rapid in myelinated axons because the action potentials are produced at successive nodes of Ranvier (*1–5*) instead of at every part of the membrane along the axon.

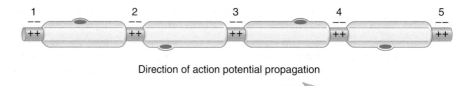

Direction of action potential propagation

Process Figure 10.21   Saltatory Conduction: Action Potential Propagation in a Myelinated Axon
The gaps between the Schwann cells are exaggerated for clarity.

it is like a grasshopper walking. The grasshopper (action potential) moves more rapidly by jumping. The generation of action potentials at nodes of Ranvier occurs so rapidly that as many as 30 successive nodes of Ranvier are simultaneously in some phase of an action potential.

The speed of action potential conduction is also affected by the thickness of the myelin sheath, which is determined by how many times oligodendrocytes or Schwann cells wrap around the axon. Heavily myelinated axons have a thicker myelin sheath and conduct action potentials more rapidly than lightly myelinated axons.

In addition to myelination, the diameter of axons affects the speed of action potential conduction. Large-diameter axons conduct action potentials more rapidly than small-diameter axons.

Combining axon size and myelination there are three major categories of axons. From fastest to slowest conducting speeds, they are heavily myelinated type A (largest diameter), lightly myelinated type B (medium diameter), and unmyelinated type C (small diameter). Most sensory and skeletal motor neurons have myelinated type A axons. Rapid response to the external environment is possible because of the rapid input of sensory information to the CNS and rapid output of action potentials to skeletal muscles. Type B and C fibers are primarily part of the ANS, which stimulates internal organs, such as the stomach, intestines, and heart. The responses necessary to maintain internal homeostasis, such as digestion, need not be as rapid as responses to the external environment. ▼

36. *What is an ionic current? How do ionic currents cause the propagation of action potentials in unmyelinated axons?*

37. *What prevents an action potential from reversing its direction of propagation?*

38. *Describe saltatory conduction of an action potential.*

39. *Compare the speed of action potential conduction based on amount of myelination and axon diameter.*

40. *Compare the functions of type A nerve fibers with those of type B and C nerve fibers.*

 ## Importance of Myelin Sheaths

Myelin sheaths begin to form late in fetal development. The process continues rapidly until the end of the first year after birth and continues more slowly thereafter. The development of myelin sheaths is associated with the infant's continuing development of more rapid and better-coordinated responses.

The importance of myelinated fibers is dramatically illustrated in diseases in which the myelin sheath is gradually destroyed. Action potential transmission is slowed, resulting in impaired control of skeletal and smooth muscles. In severe cases, complete blockage of action potential transmission can occur. Multiple sclerosis and some cases of diabetes mellitus result in myelin sheath destruction.

The Synapse

The **synapse** (sin′aps), which is the junction between two cells, is where two cells communicate with each other. The cell that transmits a signal toward a synapse is called the **presynaptic cell,** and the cell that receives the signal is called the **postsynaptic cell.** The average presynaptic neuron synapses with about 1000 other neurons, but the average postsynaptic neuron has up to 10,000 synapses. Some postsynaptic neurons in the part of the brain called the cerebellum have up to 100,000 synapses. There are two types of synapses: electrical and chemical.

Electrical Synapses

Electrical synapses are gap junctions (see chapter 4) that allow an ionic current to flow between adjacent cells. At these gap junctions, the membranes of adjacent cells are separated by a 2 nm gap spanned by tubular proteins called **connexons.** The movement of ions through the connexons generates an ionic current. Thus, an action potential in one cell produces an ionic current that generates an action potential in the adjacent cell almost as if the two cells had the same membrane. As a result, action potentials are conducted rapidly between cells, allowing the cells' activity to be synchronized. Electrical synapses are important in cardiac muscle and in many types of smooth muscle. Coordinated contractions of these muscle cells occur when action potentials in one cell propagate to adjacent cells because of electrical synapses (see chapters 8 and 17). ▼

41. *What is an electrical synapse? How are electrical synapses important in cardiac and smooth muscle?*

Chemical Synapses

The essential components of a **chemical synapse** are the presynaptic terminal, the synaptic cleft, and the postsynaptic membrane (figure 10.22). The **presynaptic terminal** is formed from the end of an axon, and the space separating the axon ending and the cell with which it synapses is the **synaptic cleft.** The membrane of the postsynaptic cell opposed to the presynaptic terminal is the **postsynaptic membrane.** Postsynaptic cells are typically other neurons, muscle cells, or gland cells.

Neurotransmitter Release

In chemical synapses, action potentials do not pass directly from the presynaptic terminal to the postsynaptic membrane. Instead, the action potentials in the presynaptic terminal cause the release of neurotransmitters from the terminal.

Presynaptic terminals are specialized to produce and release neurotransmitters. The major cytoplasmic organelles within presynaptic terminals are mitochondria and numerous membrane-bound **synaptic vesicles,** which contain neurotransmitters, such as acetylcholine (see figure 10.22). Each action potential arriving at the presynaptic terminal initiates a series of specific events, which result in the release of neurotransmitters. In response to an action potential, voltage-gated Ca^{2+} channels

When a nerve is cut, either healing or permanent interruption of the neural pathways occurs. The final outcome depends on the severity of the injury and on its treatment.

Several degenerative changes result when a nerve is cut (figure A). Within about 3–5 days, the axons in the part of the nerve distal to the cut break into irregular segments and degenerate. This occurs because the neuron cell body produces the substances essential to maintain the axon, and these substances have no way of reaching parts of the axon distal to the point of damage. Eventually, the distal part of the axon completely degenerates. As the axons degenerate, the myelin part of the Schwann cells around them also degenerates, and macrophages invade the area to phagocytize the myelin. The Schwann cells then enlarge, undergo mitosis, and finally form a column of cells along the regions once occupied by the axons. The columns of Schwann cells are essential for the growth of new axons. If the ends of the regenerating axons encounter a Schwann cell column, their rate of growth increases, and reinnervation of peripheral structures is likely. If the ends of the axons do not encounter the columns, they fail to reinnervate the peripheral structures.

The end of each regenerating axon forms several axonal sprouts. It normally takes about 2 weeks for the axonal sprouts to enter the Schwann cell columns. Only one of the sprouts from each severed neuron forms an axon, however. The other branches degenerate. After the axons grow through the Schwann cell columns, new myelin sheaths are formed, and the neurons reinnervate the structures they previously supplied.

Treatment strategies that increase the probability of reinnervation include bringing the ends of the severed nerve close together surgically. In some cases in which sections of nerves are destroyed as a result of trauma, nerve transplants are performed to replace damaged segments. The transplanted nerve eventually degenerates, but it does provide Schwann cell columns through which axons can grow.

The regeneration of damaged nerve tracts within the CNS is very limited and is poor in comparison with the regeneration of nerves in the PNS. In part, the difference may result from the oligodendrocytes, which exist only in the CNS. Each oligodendrocyte has several processes, each of which forms part of a myelin sheath. The cell bodies of the oligodendrocytes are a short distance from the axons they ensheathe, and fewer oligodendrocytes than Schwann cells are present. Consequently, when the myelin degenerates following damage, no column of cells remains in the CNS to act as a guide for the growing axons.

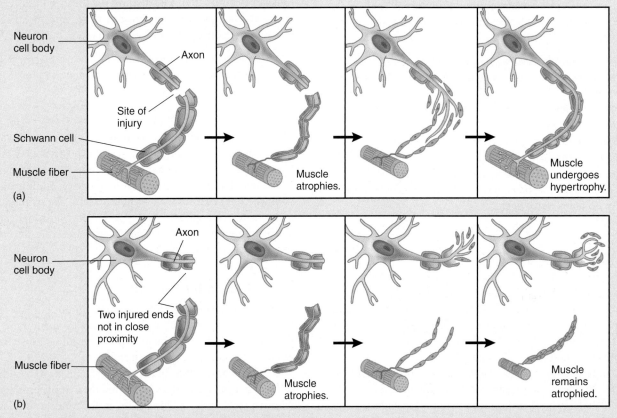

Figure A Changes That Occur in an Injured Nerve Fiber

(*a*) When the two ends of an injured nerve fiber are aligned in close proximity, healing and regeneration of the axon are likely to occur. Without stimulation from the nerve, the muscle is paralyzed and atrophies (shrinks in size). After reinnervation, the muscle can become functional and hypertrophy (increase in size).

(*b*) When the two ends of an injured nerve fiber are not aligned in close proximity, regeneration is unlikely to occur. Without innervation from the nerve, muscle function is completely lost, and the muscle remains atrophied.

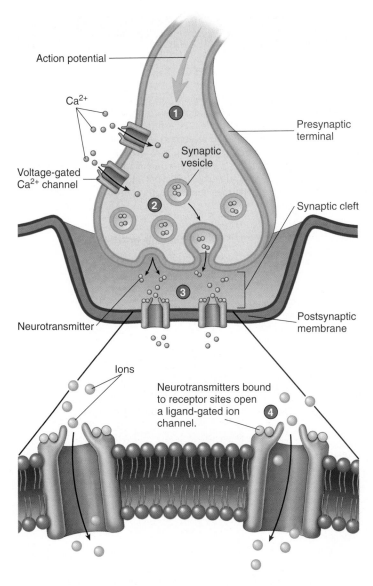

1. Action potentials arriving at the presynaptic terminal cause voltage-gated Ca^{2+} channels to open.

2. Calcium ions diffuse into the cell and cause synaptic vesicles to release neurotransmitters.

3. Neurotransmitters diffuse from the presynaptic terminal across the synaptic cleft.

4. Neurotransmitters combine with their receptor sites and cause ligand-gated ion channels to open. Ions diffuse into the cell (*shown in illustration*) or out of the cell (*not shown*) and cause a change in membrane potential.

Process Figure 10.22   Chemical Synapse

A chemical synapse consists of the end of a neuron (presynaptic terminal), a small space (synaptic cleft), and the postsynaptic membrane of another neuron or an effector cell, such as a muscle or gland cell.

open, and Ca^{2+} diffuse into the presynaptic terminal. These ions cause synaptic vesicles to fuse with the presynaptic membrane and release their neurotransmitters by exocytosis into the synaptic cleft.

Once neurotransmitters are released from the presynaptic terminal, they diffuse rapidly across the synaptic cleft, which is about 20 nm wide, and bind in a reversible fashion to specific receptors in the postsynaptic membrane (see figure 10.22). Depending on the receptor type, this binding produces a depolarizing or hyperpolarizing graded potential in the postsynaptic membrane. For example, the binding of acetylcholine to ligand-gated Na^+ channels causes them to open, allowing the diffusion of Na^+ into the postsynaptic cell. If the resulting depolarizing graded potential reaches threshold, an action potential is produced. On the other hand, the opening of K^+ or Cl^- channels results in a hyperpolarizing graded potential. ▼

42. *What are the three parts of a chemical synapse?*

43. *Describe the release of a neurotransmitter in a chemical synapse.*

P R E D I C T 10

Is an action potential transmitted fastest between cells connected by electrical synapses or chemical synapses? Explain.

Neurotransmitter Removal

Neurotransmitters have short-term effects on postsynaptic membranes because the neurotransmitter is rapidly destroyed or removed from the synaptic cleft. For example, in the neuromuscular junction, the neurotransmitter acetylcholine is broken down by the enzyme **acetylcholinesterase** (as′e-til-kō-lin-es′ter-ās) to acetic acid and choline (see figure 8.8). Choline is then transported back into the presynaptic terminal and reacts with acetyl-CoA to form acetylcholine. Thus, part of the neurotransmitter is recycled.

In most chemical synapses, the neurotransmitter is recycled intact. After the neurotransmitter norepinephrine is released into the synaptic cleft, most of it is transported back into the presynaptic terminal, where it is repackaged into synaptic vesicles for reuse. ▼

44. *Name two ways to stop the effect of a neurotransmitter on the postsynaptic membrane. Give an example of each way.*

Neurotransmitters and Neuromodulators

Several substances have been identified as neurotransmitters, and others are suspected neurotransmitters. **Neuromodulators** are substances released from neurons that can presynaptically or postsynaptically influence the likelihood that an action potential in the presynaptic terminal will result in the production of an action potential in the postsynaptic cell. For example, a neuromodulator that decreases the release of an excitatory neurotransmitter from a presynaptic terminal decreases the likelihood of action potential production in the postsynaptic cell. A list of neurotransmitters and neuromodulators is presented in table 10.4. ▼

45. *What is a neuromodulator?*

Table 10.4   Clinical Examples of Synaptic Function

Neurotransmitter/Neuromodulator	Clinical Examples

Acetylcholine

Structure:

Myasthenia Gravis

Myasthenia gravis is a disease in which the ability of skeletal muscle to respond to nervous system stimulation decreases, resulting in muscle weakness and even paralysis. Antibodies are proteins produced by the immune system that can attach to foreign substances, such as bacteria (see chapter 22). In myasthenia gravis, antibodies inappropriately attach to acetylcholine receptors. The antibodies link the receptors together, which cause them to be removed from the plasma membrane faster than normal, decreasing the number of receptors. The antibodies also stimulate immune responses that lead to destruction of the postsynaptic membrane, which decreases the number of Na^+ channels in the synapse. Thus, the ability of ACh to stimulate action potential production decreases because there are fewer ligand-gated ACh receptors, and the ability to generate an action potential is reduced because there are fewer voltage-gated Na^+ channels.

Site of release: CNS synapses, ANS synapses, and neuromuscular junctions

Effect: excitatory in the CNS and neuromuscular junctions; inhibitory or excitatory in ANS synapses

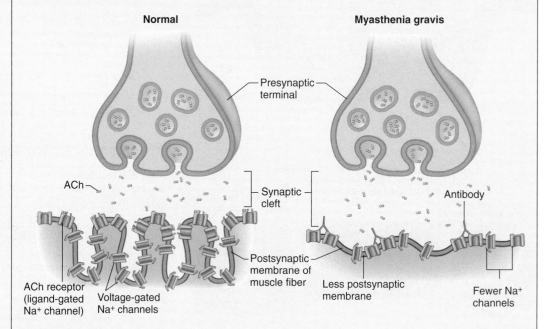

Biogenic Amines

Serotonin

Structure:

Site of release: CNS synapses

Effect: generally inhibitory

Antidepressant Therapy

Selective serotonin reuptake inhibitors (SSRIs), such as Prozac and Zoloft, are drugs commonly used to treat depression. They temporarily block serotonin transporters (symporters), which decreases serotonin transport back into presynaptic terminals, resulting in increased serotonin levels in synaptic clefts. In some people, the increased stimulation of the postsynaptic neuron by serotonin relieves depression.

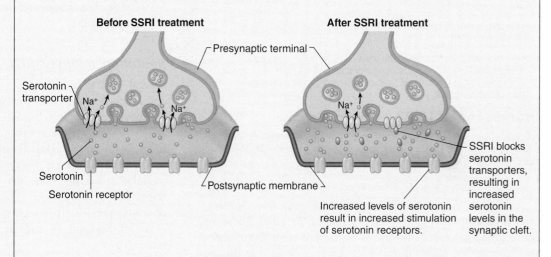

Table 10.4 **Clinical Examples of Synaptic Function—Continued**

Neurotransmitter/Neuromodulator	Clinical Examples
Serotonin (continued) *Dopamine* Structure: HO—, HO— (benzene ring) —CH₂—CH₂—NH₂ Site of release: selected CNS synapses; also found in some ANS synapses Effect: excitatory or inhibitory *Norepinephrine* Structure: OH HO—, HO— (benzene ring) —CH—CH₂—NH₂ Site of release: selected CNS synapses and some ANS synapses Effect: excitatory **Amino Acids** *Gamma-Amino Butyric Acid (GABA)* Structure: H₃N⁺—(chain)—C(=O)—O⁻ Site of release: CNS synapses Effect: inhibitory effect on postsynaptic neurons; some presynaptic inhibition occurs in the spinal cord	**Anxiety Disorders** SSRIs are also used to treat panic disorders, such as obsessive-compulsive disorder (OCD), a fact that makes researchers believe that OCD might be linked to abnormalities in serotonin function. **Hallucinogens** The hallucinogenic drug D-lysergic acid diethylamide (LSD) blocks serotonin transporters in specific areas of the brain and produces hallucinogenic effects. Other drugs, such as ecstasy, are also hallucinogens that block serotonin transporters. **Drug Addiction** Cocaine blocks dopamine transporters (symporters), which increases dopamine levels in synaptic clefts, resulting in overstimulation of postsynaptic neurons. Although moderate levels of dopamine can cause euphoria, high levels of dopamine can produce psychotic effects. **Parkinson Disease** Parkinson disease results from the destruction of dopamine-producing neurons, and it is characterized by tremors and decreased voluntary motor control. Parkinson disease is treated with the drug L-Dopa, which increases the production of dopamine in the presynaptic terminals of remaining neurons. Another treatment option is drugs that mimic the action of dopamine. **Attention-Deficit Hyperactivity Disorder (ADHD)** ADHD is often treated with drugs that increase the level of excitatory neurotransmitters, such as norepinephrine, in the synaptic clefts. This is achieved by using selective norepinephrine reuptake inhibitors (SNRIs) that block norepinephrine transporters (symporters) and increase the levels of norepinephrine in the synaptic clefts. **Amphetamines** Amphetamines are drugs with excitatory effects on the CNS. They increase the levels of norepinephrine and dopamine in synaptic clefts by either blocking the reuptake of these neurotransmitters or promoting their release from synaptic vesicles. The CNS effects of amphetamines include decreased appetite, increased alertness, and increased ability to concentrate and perform physical tasks. Amphetamines are used to treat ADHD, clinical depression, narcolepsy, and chronic fatigue syndrome. An overdose of amphetamines can cause insomnia, tremors, anxiety, and panic. **Barbiturates** Certain GABA receptors are ligand-gated channels that permit the inflow of Cl⁻ when stimulated. GABA produces an inhibitory (or a hyperpolarizing) effect by binding to these receptors and promoting Cl⁻ inflow. Barbiturates enhance the binding of GABA to their receptors, resulting in the prolonged inhibition of postsynaptic neurons. These drugs are used as sedatives and anesthetics and as a treatment for epilepsy, which is characterized by excessive neuronal discharge.

Without barbiturates

Cl⁻

GABA

GABA binding briefly opens Cl⁻ channel.

GABA receptor (GABA-gated Cl⁻ channel)

Empty barbiturate binding site

The inflow of Cl⁻ causes inhibition of the postsynaptic neuron.

With barbiturates

Barbiturate

Enhanced GABA binding opens Cl⁻ channel wider for an extended period.

Barbiturate binding enhances GABA binding.

Greater inflow of Cl⁻ causes prolonged inhibition of the postsynaptic neuron.

Table 10.4   Clinical Examples of Synaptic Function—Continued

Neurotransmitter/Neuromodulator	Clinical Examples
GABA (continued)	**Benzodiazepines** Benzodiazepines that are used in anti-anxiety drugs also have binding sites on certain GABA receptors. Their action is similar to that of barbiturates in that they enhance the binding of GABA to its receptor, producing an inhibitory effect. **Alcohol Dependence** Alcohol acts in ways similar to barbiturates to enhance the effect of GABA. As a result, the ligand-gated Cl$^-$ channel becomes more permeable to Cl$^-$, producing an inhibitory effect. Chronic consumption of alcohol renders the GABA receptor less sensitive to both alcohol and GABA, resulting in increased alcohol dependence and alcohol withdrawal symptoms, such as anxiety, tremors, and insomnia. Alcohol withdrawal symptoms are often treated with benzodiazepines.
Glycine Structure: Site of release: CNS synapses Effect: inhibitory	**Strychnine Poisoning** Glycine receptors are similar to GABA receptors in that they act as ligand-gated channels permitting the inflow of Cl$^-$. The poison strychnine blocks glycine receptors, which increases the excitability of certain neurons by preventing their inhibition. Strychnine poisoning results in powerful muscle contractions and convulsions. Tetanus of respiratory muscles can cause death.
Glutamate Structure: Site of release: CNS synapses; in areas of the brain that are involved in learning and memory Effect: excitatory	**Stroke and Excitotoxicity** Glutamate is the major excitatory neurotransmitter of the CNS. Glutamate receptors are ligand-gated Ca^{2+} channels. When stimulated, Ca^{2+} channels open, causing depolarization of postsynaptic membranes. Some glutamate is removed from the synapse by transporters in presynaptic terminals, whereas the bulk of it is removed by transporters (symporters) in neighboring astrocytes. When a person suffers a stroke, brain tissue is deprived of oxygen and ATP levels decrease. This causes the secondary active transport of glutamate by the glutamate transporters to fail temporarily. As a result, glutamate accumulates in the synaptic clefts and causes excessive stimulation of postsynaptic neurons. Excessive movement of Ca^{2+} into neurons activates a variety of destructive processes, which can cause cell death. 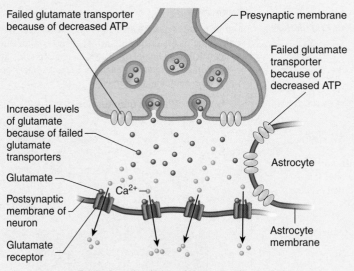 Excessive stimulation of postsynaptic neuron **Cognition** Glutamate is implicated in learning and memory. Drugs that target specific glutamate receptors are often used in the treatment of Alzheimer disease.

Table 10.4 Clinical Examples of Synaptic Function—Continued

Neurotransmitter/Neuromodulator	Clinical Examples
Purines *Adenosine* Structure: Site of release: CNS synapses; in the areas of the brain that are involved in learning and memory Effect: inhibitory	**Neuroprotective Agent** Adenosine acts both as a neurotransmitter and a neuromodulator. Adenosine receptors are linked to G-proteins (see figure 3.11). As a neurotransmitter, adenosine stimulates the opening of Cl^- and K^+ channels on postsynaptic membranes, thereby producing a hyperpolarizing effect. Acting as a neuromodulator, adenosine stimulates the closing of Ca^{2+} channels on presynaptic neurons, causing the inhibition of neurotransmitter release. Adenosine production greatly increase during a stroke. It prevents the release of glutamate from presynaptic vesicles, which reduces the level of glutamate in synaptic clefts. It also hyperpolarizes the postsynaptic membranes of glutamate synapses, thereby countering the excitatory effects of glutamate. As a result, the damaging effects of glutamate during a stroke are diminished. The possibility of using adenosine as an antistroke agent is being investigated. **Caffeine** Adenosine produces drowsiness, which is countered by caffeine, which blocks adenosine receptors and promotes alertness. Caffeine also promotes cognition by blocking adenosine's inhibitory effect on glutamate function.
Neuropeptides *Substance P* Structure: polypeptide (10 amino acids) Site of release: descending pain pathways Effect: excitatory *Endorphins* Structure: polypeptide (30 amino acids) Site of release: descending pain pathways Effect: inhibitory	**Pain Therapy** Substance P acts as a neurotransmitter and neuromodulator. The receptor for substance P is called a neurokinin receptor, which is linked to a G-protein complex (see figure 3.11). Drugs such as morphine reduce pain by blocking the release of substance P. **Opiates** Endorphins bind to endorphin receptors on presynaptic neurons and reduce pain by blocking the release of substance P. Endorphins also produce feelings of euphoria. Opiates such as morphine and heroin also bind to endorphin receptors, resulting in similar effects.
Gases *Nitric Oxide (NO)* Structure: N=O Site of release: CNS, nerves supplying the adrenal gland, penis Effect: excitatory	**Stroke Damage** During a stroke, rising glutamate levels act on postsynaptic neurons and cause the release of NO, which in high concentrations can be toxic to cells. Nitric oxide also diffuses out of postsynaptic neurons, enters neighboring cells, and damages them. **Treatment of Erectile Dysfunction** During sexual arousal, NO is released from nerves, causing the vasodilation of the blood vessels supplying the penis. Viagra, which is used to treat erectile dysfunction, acts by prolonging the effect of NO on these blood vessels (see chapter 24).

Excitatory and Inhibitory Postsynaptic Potentials

The combination of neurotransmitters with their specific receptors causes either depolarization or hyperpolarization of the postsynaptic membrane. When depolarization occurs, the response is stimulatory, and the graded potential is called an **excitatory postsynaptic potential (EPSP)** (figure 10.23*a*). EPSPs are important because the depolarization might reach threshold, thereby producing an action potential and a response from the cell. Neurons releasing neurotransmitters or neuromodulators that cause EPSPs are called **excitatory neurons.**

How Local Anesthetics Work

Awareness of pain can occur only if action potentials generated by sensory neurons stimulate the production of action potentials in CNS neurons. Local anesthetics, such as procaine (Novacain), act at their site of application to prevent pain sensations. They do so by blocking voltage-gated Na^+ channels, which prevents the propagation of action potentials along sensory neurons. Consequently, neurotransmitters are not released from the presynaptic terminals of the sensory neurons and EPSPs are not produced in CNS neurons.

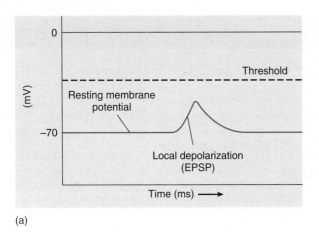

(a)

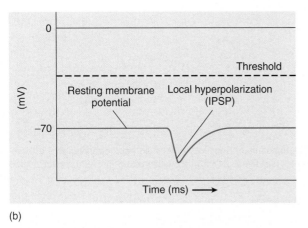

(b)

Figure 10.23 Postsynaptic Potentials

(a) Excitatory postsynaptic potential (EPSP) is closer to threshold. (b) Inhibitory postsynaptic potential (IPSP) is further from threshold.

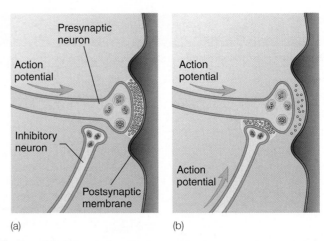

(a)   (b)

Figure 10.24 Presynaptic Inhibition at an Axoaxonic Synapse

(a) The inhibitory neuron of the axoaxonic synapse is inactive and has no effect on the release of neurotransmitter from the presynaptic terminal. (b) The release of a neuromodulator from the inhibitory neuron of the axoaxonic synapse reduces the amount of neurotransmitter released from the presynaptic terminal.

released from inhibitory neurons of axoaxonic synapses can reduce or eliminate pain sensations by inhibiting the release of neurotransmitter from the presynaptic terminals of sensory neurons (see figure 10.24). Enkephalins and endorphins can block voltage-gated Ca^{2+} channels. Consequently, when action potentials reach the presynaptic terminal, the influx of Ca^{2+} that normally stimulates neurotransmitter release is blocked.

In **presynaptic facilitation,** there is an increase in the amount of neurotransmitter released from the presynaptic terminal. For example, serotonin, released in certain axoaxonic synapses, functions as a neuromodulator that increases the release of neurotransmitter from the presynaptic terminal by causing voltage-gated Ca^{2+} channels to open. ▼

46. *Define and explain the production of EPSPs and IPSPs. Why are they important?*
47. *What is presynaptic inhibition and facilitation?*

Spatial and Temporal Summation

Within the CNS and in many PNS synapses, a single presynaptic action potential does not cause a graded potential (EPSP) in the postsynaptic membrane sufficient to reach threshold and produce an action potential. Instead, many presynaptic action potentials cause many graded potentials in the postsynaptic neuron. The graded potentials combine in summation at the trigger zone of the postsynaptic neuron, which is the normal site of action potential generation for most neurons. If summation results in a graded potential that exceeds threshold at the trigger zone, an action potential is produced. Action potentials are readily produced at the trigger zone because the concentration of voltage-gated Na^+ channels is approximately seven times greater there than at the rest of the cell body.

Two types of summation, called spatial summation and temporal summation, are possible. The simplest type of **spatial summation** occurs when two action potentials arrive simultaneously at two different presynaptic terminals that synapse with the same postsynaptic neuron. In the postsynaptic neuron, each action

When the combination of a neurotransmitter with its receptor results in hyperpolarization of the postsynaptic membrane, the response is inhibitory, and the local hyperpolarization is called an **inhibitory postsynaptic potential (IPSP)** (figure 10.23b). IPSPs are important because they decrease the likelihood of producing action potentials by moving the membrane potential further from threshold. Neurons releasing neurotransmitters or neuromodulators that cause IPSPs are called **inhibitory neurons.**

Presynaptic Inhibition and Facilitation

Many of the synapses of the CNS are **axoaxonic synapses,** meaning that the axon of one neuron synapses with the presynaptic terminal (axon) of another (figure 10.24). Through axoaxonic synapses, one neuron can release a neuromodulator that influences the release of a neurotransmitter from the presynaptic terminal of another neuron.

In **presynaptic inhibition,** there is a reduction in the amount of neurotransmitter released from the presynaptic terminal. For example, sensory neurons for pain can release neurotransmitters from their presynaptic terminals and stimulate the postsynaptic membranes of neurons in the brain or spinal cord. Awareness of pain occurs only if action potentials are produced in the postsynaptic membranes of the CNS neurons. Enkephalins and endorphins

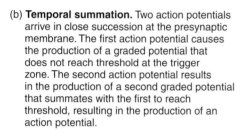

(a) **Spatial summation.** Action potentials 1 and 2 cause the production of graded potentials at two different dendrites. These graded potentials summate at the trigger zone to produce a graded potential that exceeds threshold, resulting in an action potential.

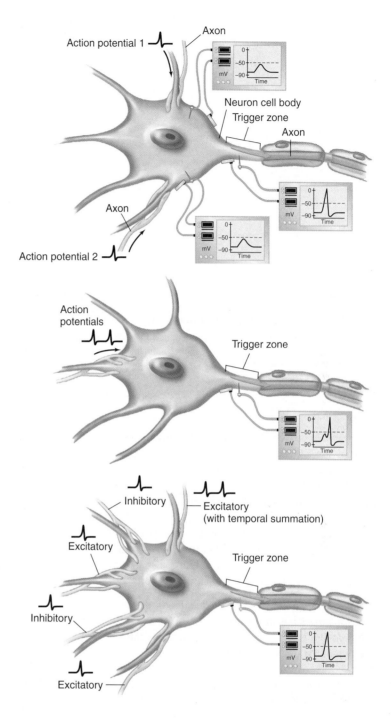

(b) **Temporal summation.** Two action potentials arrive in close succession at the presynaptic membrane. The first action potential causes the production of a graded potential that does not reach threshold at the trigger zone. The second action potential results in the production of a second graded potential that summates with the first to reach threshold, resulting in the production of an action potential.

(c) **Combined spatial and temporal summation with both excitatory postsynaptic potentials (EPSPs) and inhibitory postsynaptic potentials (IPSPs).** An action potential is produced at the trigger zone when the graded potentials produced as a result of the EPSPs and IPSPs summate to reach threshold.

Figure 10.25 Summation

potential causes a depolarizing graded potential that undergoes summation at the trigger zone. If the summated depolarization reaches threshold, an action potential is produced (figure 10.25a).

Temporal summation results when two or more action potentials arrive in very close succession at a single presynaptic terminal. The first action potential causes a depolarizing graded potential in the postsynaptic membrane that remains for a few milliseconds before it disappears, although its magnitude decreases through time. Temporal summation results when another action potential initiates another graded depolarization before the depolarization caused by the previous action potential returns to its resting value (see figure 10.15b). Subsequent action potentials cause depolarizations that

summate with previous depolarizations. If the summated depolarizing graded potentials reach threshold at the trigger zone, an action potential is produced in the postsynaptic neuron (figure 10.25b).

Excitatory and inhibitory neurons can synapse with the same postsynaptic neuron. Spatial summation of EPSPs and IPSPs occurs in the postsynaptic neuron, and whether a postsynaptic action potential is initiated or not depends on which type of graded potential has the greatest influence on the postsynaptic membrane potential (figure 10.25c). If the EPSPs (local depolarizations) cancel the IPSPs (local hyperpolarizations) and summate to threshold, an action potential is produced. If the IPSPs prevent the EPSPs from summating to threshold, no action potential is produced.

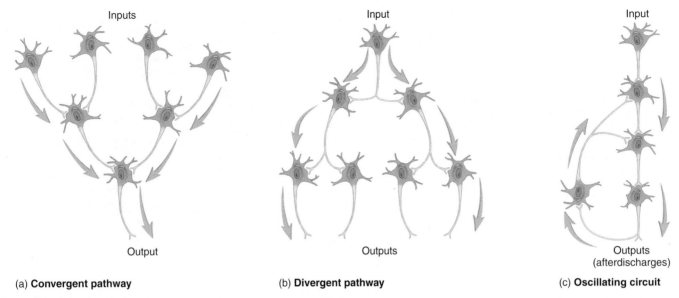

Inputs

Output

(a) **Convergent pathway**

Input

Outputs

(b) **Divergent pathway**

Input

Outputs
(afterdischarges)

(c) **Oscillating circuit**

Figure 10.26 Neuronal Pathways and Circuits

The direction of action potential propagation is represented by the *gold arrows*. (*a*) General model of a convergent pathway; many neurons converge and synapse with a smaller number of neurons. (*b*) General model of a divergent pathway; a few neurons synapse with a larger number of neurons. (*c*) Simple model of an oscillating circuit; input action potentials result in the production of a larger number of output action potentials because neurons within the circuit are repeatedly stimulated to produce action potentials.

The synapse is an essential structure for the process of integration carried out by the CNS. For example, action potentials propagated along axons from sensory organs to the CNS can produce a sensation, or they can be ignored. To produce a sensation, action potentials must be transmitted across synapses as they travel through the CNS to the cerebral cortex, where information is interpreted. Stimuli that do not result in action potential transmission across synapses are ignored because information never reaches the cerebral cortex. The brain can ignore large amounts of sensory information as a result of complex integration. ▼

48. *Define spatial and temporal summation. In what part of the neuron does summation take place?*
49. *How do EPSPs and IPSPs affect the likelihood that summation will result in an action potential?*

Neuronal Pathways and Circuits

The organization of neurons within the CNS varies from relatively simple to extremely complex patterns. The axon of a neuron can branch repeatedly to form synapses with many other neurons, and hundreds or even thousands of axons can synapse with the cell body and dendrites of a single neuron. Although their complexity varies, three basic patterns can be recognized: convergent pathways, divergent pathways, and oscillating circuits.

In **convergent pathways,** many neurons converge and synapse with a smaller number of neurons (figure 10.26*a*). Convergence makes it possible for different parts of the nervous system to activate or inhibit the activity of neurons. For example, one part of the nervous system can stimulate the neurons responsible for making a muscle contract, whereas another part can inhibit those neurons. Through summation, action potentials resulting in muscle contraction can be produced if the converging neurons stimulate the

production of more EPSPs than IPSPs. Conversely, muscle contraction is inhibited if the converging neurons stimulate the production of more IPSPs than EPSPs.

In **divergent pathways,** a smaller number of presynaptic neurons synapse with a larger number of postsynaptic neurons to allow information transmitted in one neuronal pathway to diverge into two or more pathways (figure 10.26*b*). Diverging pathways allow one part of the nervous system to affect more than one other part of the nervous system. For example, sensory input to the central nervous system can go to both the spinal cord and the brain.

Oscillating circuits have neurons arranged in a circular fashion, which allows action potentials entering the circuit to cause a neuron farther along in the circuit to produce an action potential more than once (figure 10.26*c*). This response is called **after-discharge,** and its effect is to prolong the response to a stimulus. Oscillating circuits are similar to positive-feedback systems. Once an oscillating circuit is stimulated, it continues to discharge until the synapses involved become fatigued or until they are inhibited by other neurons. Oscillating circuits play a role in neuronal circuits that are periodically active. Respiration may be controlled by an oscillating circuit that controls inspiration and another that controls expiration.

Neurons that spontaneously produce action potentials are common in the CNS and may activate oscillating circuits, which remain active awhile. The cycle of wakefulness and sleep may involve circuits of this type. Spontaneously active neurons can also influence the activity of other circuit types. The complex functions carried out by the CNS are affected by the numerous circuits operating together and influencing the activity of one another. ▼

50. *Diagram a convergent pathway, a divergent pathway, and an oscillating circuit, and describe what is accomplished in each.*

S U M M A R Y

Functions of the Nervous System (p. 266)

The nervous system detects external and internal stimuli (sensory input), processes and responds to sensory input (integration), controls body movements through skeletal muscles, maintains homeostasis by regulating other systems, and is the center for mental activities.

Parts of the Nervous System (p. 266)

1. The nervous system has two anatomical divisions.
   - The central nervous system (CNS) consists of the brain and spinal cord and is encased in bone.
   - The peripheral nervous system (PNS), the nervous tissue outside of the CNS, consists of sensory receptors and nerves.
2. The PNS has two divisions.
   - The sensory division transmits action potentials from sensory receptors to the CNS.
   - The motor division carries action potentials away from the CNS in cranial or spinal nerves.
3. The motor division has two subdivisions.
   - The somatic nervous system innervates skeletal muscle.
   - The autonomic nervous system (ANS) innervates cardiac muscle, smooth muscle, and glands.
4. The ANS is subdivided into the sympathetic division, which is most active during physical activity; the parasympathetic division, which regulates resting functions; and the enteric nervous system, which controls the digestive system.
5. The anatomical divisions perform different functions.
   - The PNS detects stimuli and transmits information to and receives information from the CNS.
   - The CNS processes, integrates, stores, and responds to information from the PNS.

Cells of the Nervous System (p. 267)

Neurons

1. Neurons receive stimuli and transmit action potentials.
2. Neurons have three components.
   - The cell body is the primary site of protein synthesis.
   - Dendrites are short, branched cytoplasmic extensions of the cell body that usually conduct electric signals toward the cell body.
   - An axon is a cytoplasmic extension of the cell body that transmits action potentials to other cells.

Types of Neurons

1. Multipolar neurons have several dendrites and a single axon. Interneurons and motor neurons are multipolar.
2. Bipolar neurons have a single axon and dendrite and are found as components of sensory organs.
3. Unipolar neurons have a single axon. Most sensory neurons are unipolar.

Glial Cells of the CNS

1. Glial cells are nonneural cells that support and aid the neurons of the CNS and PNS.
2. Astrocytes provide structural support for neurons and blood vessels. Astrocytes influence the functioning of the blood–brain barrier and process substances that pass through it. Astrocytes isolate damaged tissue and limit the spread of inflammation.
3. Ependymal cells line the ventricles and the central canal of the spinal cord. Some are specialized to produce cerebrospinal fluid.
4. Microglia phagocytize microorganisms, foreign substances, and necrotic tissue.

5. An oligodendrocyte forms myelin sheaths around the axons of several CNS neurons.

Glial Cells of the PNS

1. A Schwann cell forms a myelin sheath around part of the axon of a PNS neuron.
2. Satellite cells support and nourish neuron cell bodies within ganglia.

Myelinated and Unmyelinated Axons

1. Myelinated axons are wrapped by several layers of plasma membrane from Schwann cells (PNS) or oligodendrocytes (CNS). Spaces between the wrappings are the nodes of Ranvier. Myelinated axons conduct action potentials rapidly.
2. Unmyelinated axons rest in invaginations of Schwann cells (PNS) or oligodendrocytes (CNS). They conduct action potentials slowly.

Organization of Nervous Tissue (p. 271)

1. Nervous tissue can be grouped into white matter and gray matter.
   - White matter consists of myelinated axons; it propagates action potentials.
   - Gray matter consists of collections of neuron cell bodies or unmyelinated axons. Axons synapse with neuron cell bodies, which are functionally the site of integration in the nervous system.
2. White matter forms nerve tracts in the CNS and nerves in the PNS. Gray matter forms cortex and nuclei in the CNS and ganglia in the PNS.

Electric Signals (p. 271)

Electrical properties of cells result from the ionic concentration differences across the plasma membrane and from the permeability characteristics of the plasma membrane.

Concentration Differences Across the Plasma Membrane

1. The Na^+–K^+ pump moves ions by active transport. Potassium ions are moved into the cell, and Na^+ are moved out of it.
2. The concentration of K^+ and negatively charged proteins and other molecules is higher inside, and the concentrations of Na^+ and Cl^- are higher outside the cell.
3. Negatively charged proteins are synthesized inside the cell and cannot diffuse out of it.
4. The permeability of the plasma membrane to ions is determined by leak channels and gated ion channels.
   - Potassium ion leak channels are more numerous than Na^+ leak channels; thus, the plasma membrane is more permeable to K^+ than to Na^+ when at rest.
   - Gated ion channels in the plasma membrane include ligand-gated ion channels, voltage-gated ion channels, and other gated ion channels.

Establishing the Resting Membrane Potential

1. The resting membrane potential is a charge difference across the plasma membrane when the cell is in an unstimulated condition. The inside of the cell is negatively charged, compared with the outside of the cell.
2. The resting membrane potential is due mainly to the tendency of positively charged K^+ to diffuse out of the cell, which is opposed by the negative charge that develops inside the plasma membrane.

Changing the Resting Membrane Potential

1. Depolarization is a decrease in the resting membrane potential and can result from a decrease in the K^+ concentration gradient, a decrease in membrane permeability to K^+, an increase in membrane

permeability to Na^+, an increase in membrane permeability to Ca^{2+}, or a decrease in extracellular Ca^{2+} concentration.

2. Hyperpolarization is an increase in the resting membrane potential that can result from an increase in the K^+ concentration gradient, an increase in membrane permeability to K^+, an increase in membrane permeability to Cl^-, a decrease in membrane permeability to Na^+, or an increase in extracellular Ca^{2+} concentration.

Graded Potentials

1. A graded potential is a small change in the resting membrane potential that is confined to a small area of the plasma membrane.
2. An increase in membrane permeability to Na^+ can cause graded depolarization, and an increase in membrane permeability to K^+ or Cl^- can result in graded hyperpolarization.
3. The term graded potential is used because a stronger stimulus produces a *greater potential* change than a weaker stimulus.
4. Graded potentials can summate, or add together.
5. A graded potential decreases in magnitude as the distance from the stimulation increases.

Action Potentials

1. An action potential is a larger change in the resting membrane potential that spreads over the entire surface of the cell.
2. Threshold is the membrane potential at which a graded potential depolarizes the plasma membrane sufficiently to produce an action potential.
3. Action potentials occur in an all-or-none fashion. If action potentials occur, they are of the same magnitude, no matter how strong the stimulus.
4. Depolarization occurs as the inside of the membrane becomes more positive because Na^+ diffuse into the cell through voltage-gated ion channels.
5. Repolarization is a return of the membrane potential toward the resting membrane potential because voltage-gated Na^+ channels close and Na^+ diffusion into the cell slows to resting levels and because voltage-gated K^+ channels open and K^+ diffuse out of the cell.
6. The afterpotential is a brief period of hyperpolarization following repolarization.

Refractory Period

1. The absolute refractory period is the time during an action potential when a second stimulus, no matter how strong, cannot initiate another action potential.
2. The relative refractory period follows the absolute refractory period and is the time during which a stronger-than-threshold stimulus can evoke another action potential.

Action Potential Frequency

1. The strength of stimuli affects the frequency of action potentials.
   - A subthreshold stimulus produces only a graded potential.
   - A threshold stimulus causes a graded potential that reaches threshold and results in a single action potential.
   - A submaximal stimulus is greater than a threshold stimulus and weaker than a maximal stimulus. The action potential frequency increases as the strength of the submaximal stimulus increases.
   - A maximal or a supramaximal stimulus produces a maximum frequency of action potentials.
2. A low frequency of action potentials represents a weaker stimulus than a high frequency.

Propagation of Action Potentials

1. An action potential generates ionic currents, which stimulate voltage-gated Na^+ channels in adjacent regions of the plasma membrane to open, producing a new action potential.

2. In an unmyelinated axon, action potentials are generated immediately adjacent to previous action potentials.
3. In a myelinated axon, action potentials are generated at successive nodes of Ranvier.
4. Reversal of the direction of action potential propagation is prevented by the absolute refractory period.
5. Action potentials propagate most rapidly in myelinated, large-diameter axons.

The Synapse   (p. 283)

Electrical Synapses

1. Electrical synapses are gap junctions in which tubular proteins called connexons allow ionic currents to move between cells.
2. At an electrical synapse, an action potential in one cell generates an ionic current that causes an action potential in an adjacent cell.

Chemical Synapses

1. Anatomically, a chemical synapse has three components.
   - The enlarged ends of the axon are the presynaptic terminals containing synaptic vesicles.
   - The postsynaptic membranes contain receptors for the neurotransmitter.
   - The synaptic cleft, a space, separates the presynaptic and postsynaptic membranes.
2. An action potential arriving at the presynaptic terminal causes the release of a neurotransmitter, which diffuses across the synaptic cleft and binds to the receptors of the postsynaptic membrane.
3. The effect of the neurotransmitter on the postsynaptic membrane is stopped in two ways.
   - The neurotransmitter is broken down by an enzyme.
   - The neurotransmitter is taken up by the presynaptic terminal.
4. An excitatory postsynaptic potential (EPSP) is a depolarizing graded potential of the postsynaptic membrane. It can be caused by an increase in membrane permeability to Na^+.
5. An inhibitory postsynaptic potential (IPSP) is a hyperpolarizing graded potential of the postsynaptic membrane. It can be caused by an increase in membrane permeability to K^+ or Cl^-.
6. Presynaptic inhibition decreases neurotransmitter release. Presynaptic facilitation increases neurotransmitter release.

Spatial and Temporal Summation

1. Presynaptic action potentials through neurotransmitters produce graded potentials in postsynaptic neurons. The graded potential can summate to produce an action potential at the trigger zone.
2. Spatial summation occurs when two or more presynaptic terminals simultaneously stimulate a postsynaptic neuron.
3. Temporal summation occurs when two or more action potentials arrive in succession at a single presynaptic terminal.
4. Inhibitory and excitatory presynaptic neurons can converge on a postsynaptic neuron. The activity of the postsynaptic neuron is determined by the integration of the EPSPs and IPSPs produced in the postsynaptic neuron.

Neuronal Pathways and Circuits   (p. 292)

1. Convergent pathways have many neurons synapsing with a few neurons.
2. Divergent pathways have a few neurons synapsing with many neurons.
3. Oscillating circuits have collateral branches of postsynaptic neurons synapsing with presynaptic neurons.

R E V I E W   A N D   C O M P R E H E N S I O N

1. The peripheral nervous system includes
   a. the somatic nervous system.
   b. the brain.
   c. the spinal cord.
   d. nuclei.
   e. all of the above.

2. The part of the nervous system that controls smooth muscle, cardiac muscle, and glands is the
   a. somatic nervous system.
   b. autonomic nervous system.
   c. skeletal division.
   d. sensory division.

3. Neurons have cytoplasmic extensions that connect one neuron to another neuron. Given these structures:

   1. axon   3. dendritic spine
   2. dendrite   4. presynaptic terminal

   Choose the arrangement that lists the structures in the order they are found between two neurons.
   a. 1,4,2,3   c. 4,1,2,3   e. 4,3,2,1
   b. 1,4,3,2   d. 4,1,3,2

4. A neuron with many short dendrites and a single long axon is a _____ neuron.
   a. multipolar
   b. unipolar
   c. bipolar

5. Motor neurons and interneurons are _____ neurons.
   a. unipolar   c. multipolar
   b. bipolar   d. afferent

6. Cells found in the choroid plexuses that secrete cerebrospinal fluid are
   a. astrocytes.   c. ependymal cells.   e. Schwann cells.
   b. microglia.   d. oligodendrocytes.

7. Glial cells that are phagocytic within the central nervous system are
   a. oligodendrocytes.
   b. microglia.
   c. ependymal cells.
   d. astrocytes.
   e. Schwann cells.

8. Unmyelinated axons within nerves may have which of these associated with them?
   a. Schwann cells   c. oligodendrocytes
   b. nodes of Ranvier   d. all of the above

9. Action potentials are conducted more rapidly
   a. in small-diameter axons than in large-diameter axons.
   b. in unmyelinated axons than in myelinated axons.
   c. along axons that have nodes of Ranvier.
   d. all of the above.

10. Clusters of neuron cell bodies within the peripheral nervous system are
    a. ganglia.   c. nuclei.
    b. fascicles.   d. laminae.

11. Gray matter contains primarily
    a. myelinated fibers.
    b. neuron cell bodies.
    c. Schwann cells.
    d. oligodendrocytes.

12. Concerning concentration difference across the plasma membrane, there are
    a. more K^+ and Na^+ outside the cell than inside.
    b. more K^+ and Na^+ inside the cell than outside.
    c. more K^+ outside the cell than inside and more Na^+ inside the cell than outside.
    d. more K^+ inside the cell than outside and more Na^+ outside the cell than inside.

13. Compared with the inside of the resting plasma membrane, the outside surface of the membrane is
    a. positively charged.
    b. electrically neutral.
    c. negatively charged.
    d. continuously reversing so that it is positive 1 second and negative the next.
    e. negatively charged whenever the Na^+–K^+ pump is operating.

14. Leak channels
    a. open in response to small voltage changes.
    b. open when a chemical signal binds to its receptor.
    c. are responsible for the ion permeability of the resting plasma membrane.
    d. allow substances to move into the cell but not out.
    e. all of the above.

15. The resting membrane potential results when the tendency for _____ to diffuse out of the cell is balanced by their attraction to opposite charges inside the cell.
    a. Na^+
    b. K^+
    c. Cl^-
    d. negatively charged proteins

16. If the permeability of the plasma membrane to K^+ increases, resting membrane potential _____. This is called _____.
    a. increases, hyperpolarization
    b. increases, depolarization
    c. decreases, hyperpolarization
    d. decreases, depolarization

17. Decreasing the extracellular concentration of K^+ affects the resting membrane potential by causing
    a. hyperpolarization.   b. depolarization.   c. no change.

18. Which of these terms are correctly matched with their definition or description?
    a. depolarization: membrane potential becomes more negative
    b. hyperpolarization: membrane potential becomes more negative
    c. hypopolarization: membrane potential becomes more negative

19. Which of these statements about ion movement through the plasma membrane is true?
    a. Movement of Na^+ out of the cell requires energy (ATP).
    b. When Ca^{2+} bind to proteins in ion channels, the diffusion of Na^+ into the cell is inhibited.
    c. Specific ion channels regulate the diffusion of Na^+ through the plasma membrane.
    d. All of the above are true.

20. The *major* function of the Na^+–K^+ pump is to
    a. pump Na^+ into and K^+ out of the cell.
    b. generate the resting membrane potential.
    c. maintain the concentration gradients of Na^+ and K^+ across the plasma membrane.
    d. oppose any tendency of the cell to undergo hyperpolarization.

21. Graded potentials
    a. spread over the plasma membrane in decremental fashion.
    b. are not propagated for long distances.
    c. are confined to a small region of the plasma membrane.
    d. can summate.
    e. all of the above.

22. During the depolarization phase of an action potential, the permeability of the membrane
    a. to K^+ is greatly increased.
    b. to Na^+ is greatly increased.
    c. to Ca^{2+} is greatly increased.
    d. is unchanged.

23. During repolarization of the plasma membrane,
    a. Na^+ diffuse into the cell.
    b. Na^+ diffuse out of the cell.
    c. K^+ diffuse into the cell.
    d. K^+ diffuse out of the cell.

24. The absolute refractory period
    a. limits how many action potentials can be produced during a given period of time.
    b. prevents an action potential from starting another action potential at the same point on the plasma membrane.
    c. is the period of time when a strong stimulus can initiate a second action potential.
    d. both a and b.
    e. all of the above.

25. A subthreshold stimulus
    a. produces an afterpotential.
    b. produces a graded potential.
    c. causes an all-or-none response.
    d. produces more action potentials than a submaximal stimulus.

26. Neurotransmitter substances are stored in vesicles located in specialized portions of the
    a. neuron cell body.
    b. axon.
    c. dendrite.
    d. postsynaptic membrane.

27. In a chemical synapse,
    a. action potentials in the presynaptic terminal cause voltage-gated Ca^{2+} channels to open.
    b. neurotransmitters can cause ligand-gated Na^+ channels to open.
    c. neurotransmitters can be broken down by enzymes.
    d. neurotransmitters can be taken up by the presynaptic terminal.
    e. all of the above.

28. An inhibitory presynaptic neuron can affect a postsynaptic neuron by
    a. producing an IPSP in the postsynaptic neuron.
    b. hyperpolarizing the plasma membrane of the postsynaptic neuron.
    c. causing K^+ to diffuse out of the postsynaptic neuron.
    d. causing Cl^- to diffuse into the postsynaptic neuron.
    e. all of the above.

29. Summation
    a. is caused by combining two or more graded potentials.
    b. occurs at the trigger zone of the postsynaptic neuron.
    c. results in an action potential if it reaches the threshold potential.
    d. can occur when two action potentials arrive in close succession at a single presynaptic terminal.
    e. all of the above.

30. In convergent pathways,
    a. the response of the postsynaptic neuron depends on the summation of EPSPs and IPSPs.
    b. a smaller number of presynaptic neurons synapse with a larger number of postsynaptic neurons.
    c. information transmitted in one neuronal pathway can go into two or more pathways.
    d. all of the above.

Answers in Appendix E

C R I T I C A L   T H I N K I N G

1. Predict the consequence of a reduced intracellular K^+ concentration on the resting membrane potential.

2. A child eats a whole bottle of salt (NaCl) tablets. What effect does this have on action potentials?

3. Lithium ions reduce the permeability of plasma membranes to Na^+. Predict the effect lithium ions in the extracellular fluid would have on the response of a neuron to stimuli.

4. Some smooth muscle has the ability to contract spontaneously—that is, it contracts without any external stimulation. Propose an explanation for the ability of smooth muscle to contract spontaneously based on what you know about membrane potentials. Assume that an action potential in a smooth muscle cell causes it to contract.

5. Assume that you have two nerve fibers of the same diameter, but one is myelinated and the other is unmyelinated. The conduction of an action potential is most energy-efficient along which type of fiber? (*Hint:* ATP.)

6. The speed of action potential propagation and synaptic transmission decreases with aging. List possible explanations.

7. Students in a veterinary school are given the following hypothetical problem. A dog ingests organophosphate poison, and the students are responsible for saving the animal's life. Organophosphate poisons bind to and inhibit acetylcholinesterase. Several substances they could inject include the following: acetylcholine, curare (which blocks acetylcholine receptors), and potassium chloride. If you were a student in the class, what would you do to save the animal?

8. Strychnine blocks receptor sites for inhibitory neurotransmitter substances in the CNS. Explain how strychnine could produce tetanus in skeletal muscles.

9. The venom of many cobras contains a potent neurotoxin that binds to ligand-gated Na^+ channels, causing them to open. Unlike ACh, which binds to and then rapidly unbinds from ligand-gated Na^+ channels, the neurotoxin tends to remain bound to ligand-gated Na^+ channels. What effect would this neurotoxin have on the ability of the nervous system to stimulate skeletal muscle contraction? What effect would it have on the ability of skeletal muscle fibers to respond to stimulation?

10. Epilepsy is a chronic disorder characterized by seizures resulting from excessive production of action potentials by neurons in the brain. Channelopathies are genetic disorders caused by mutations in ion channel genes, which result in ion channels that do not function normally. What kind of Na^+ or Ca^{2+} channelopathies might contribute to epilepsy?

Answers in Appendix F

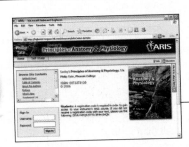

For additional study tools, go to: www.aris.mhhe.com.
Click on your course topic and then this textbook's author name/title. Register once for a semester's worth of interactive learning activities to help improve your grade!

Colorized SEM of nerve fascicles containing bundles of axons.

Tools for Success

A virtual cadaver dissection experience

Audio (MP3) and/or video (MP4) content, available at www.mhhe.aris.com

Introduction

The **central nervous system (CNS)** consists of the brain and spinal cord, with the division between these two parts of the CNS placed at the level of the foramen magnum. The **peripheral nervous system (PNS)** consists of sensory receptors and nerves outside the CNS. The PNS includes 12 pairs of cranial nerves and 31 pairs of spinal nerves.

The CNS receives sensory information, integrates and evaluates that information, stores some information, and initiates reactions. The PNS collects information from numerous sources both inside and outside the body and relays it through axons of sensory neurons to the CNS. Axons of motor neurons in the PNS relay information from the CNS to various parts of the body, primarily to muscles and glands, thereby regulating activity in those structures.

▶Although the CNS and PNS can be considered separately, the approach in this text is to consider parts of the CNS and their associated nerves together. Thus, the spinal cord and spinal nerves are described first, specifically the **spinal cord (p. 298)**, **reflexes (p. 302)**, and **nerves (p. 306)**. The chapter then describes the **brain (p. 314)**, specifically the **brainstem (p. 314)**, **cerebellum (p. 316)**, **diencephalon (p. 317)**, **cerebrum (p. 319)**, **meninges, ventricles, and cerebrospinal fluid (p. 322)**, **blood supply to the brain (p. 327)**, and **cranial nerves (p. 327)**.

Spinal Cord

The **spinal cord** is extremely important to the overall function of the nervous system. It is the major communication link between the brain and the PNS (spinal nerves) inferior to the head; it participates in the integration of incoming information and produces responses through reflex mechanisms.

General Structure

The spinal cord connects to the brain at the level of the foramen magnum and extends inferiorly in the vertebral canal to level L1–L2 of the vertebral column (figure 11.1). It is considerably shorter than the vertebral column because it does not grow as rapidly as the vertebral column during development. The spinal cord gives rise to 31 pairs of spinal nerves, which exit the vertebral column through intervertebral and sacral foramina (see figure 7.14).

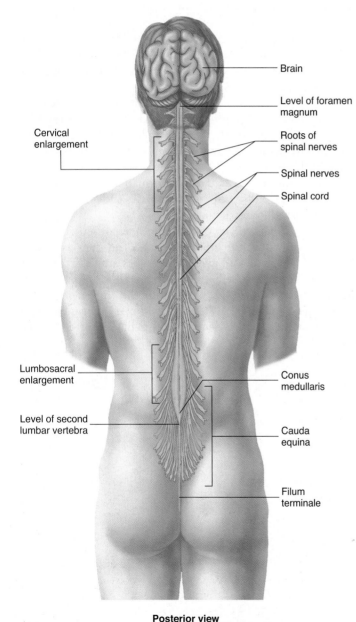

Cervical enlargement

Lumbosacral enlargement

Level of second lumbar vertebra

Brain

Level of foramen magnum

Roots of spinal nerves

Spinal nerves

Spinal cord

Conus medullaris

Cauda equina

Filum terminale

Posterior view

Figure 11.1   Spinal Cord and Spinal Nerve Roots

The spinal cord is not uniform in diameter throughout its length. The **cervical enlargement** in the inferior cervical region is where spinal nerves supplying the upper limbs arise. The **lumbosacral enlargement** in the inferior thoracic, lumbar, and superior sacral regions is the site where spinal nerves supplying the lower limbs arise.

Immediately inferior to the lumbosacral enlargement, the spinal cord tapers to form a conelike region called the **conus medullaris.** Nerves arising from the inferior lumbosacral enlargement and the conus medullaris extend inferiorly through the vertebral canal before exiting the vertebral column. The numerous roots (origins) of these nerves resemble the hairs in a horse's tail and are therefore called the **cauda** (kaw′dă, tail) **equina** (ē-kwī′nă, horse) (see figure 11.1). ▼

1. *At what level does the spinal cord begin and end?*
2. *How many pairs of spinal nerves arise from the spinal cord? How do they exit the vertebral column?*
3. *Describe the cervical and lumbar enlargements of the spinal cord, the conus medullaris, and the cauda equina.*

Meninges of the Spinal Cord

The spinal cord and brain are surrounded by connective tissue membranes called **meninges** (mĕ-nin′jēz) (figure 11.2). The most superficial and thickest membrane is the **dura mater** (doo′ră mā′ter, tough mother). The dura mater forms a sac, often called the **thecal** (the′kal, box) **sac,** which surrounds the spinal cord. The thecal sac attaches to the rim of the foramen magnum and ends at the level of the second sacral vertebra. The spinal dura mater is continuous with the dura mater surrounding the brain and the connective tissue surrounding the spinal nerves. The dura mater around the spinal cord is separated from the periosteum of the vertebral canal by the **epidural space,** which contains spinal nerve roots, blood vessels, areolar connective tissue, and fat. **Epidural anesthesia** of the spinal nerves is induced by injecting anesthetics into the epidural space. It is often given to women during childbirth.

The next deeper meningeal membrane is a very thin, wispy **arachnoid** (ă-rak′noyd, spiderlike—i.e., cobwebs) **mater.** The **subdural space** is between the dura mater and the arachnoid mater. It is a very small space containing a small amount of serous fluid.

The third, deepest, meningeal layer, the **pia** (pī′ă, affectionate) **mater** is bound very tightly to the surface of the spinal cord. Between the arachnoid mater and the pia mater is the **subarachnoid space,** which contains weblike strands of the arachnoid mater, blood vessels, and **cerebrospinal** (ser′ĕ-brō-spī-năl, sĕ-rē′brō-spī-nal) **fluid (CSF)** (see "Meninges, Ventricles, and Cerebrospinal Fluid," p. 322).

The subarachnoid space is relatively large compared with the spinal cord, which allows movement of the vertebral column without damaging the spinal cord. Lateral movement of the spinal cord is limited by 21 pairs of **denticulate** (den-tik′ū-lāt) **ligaments,** which are connective tissue septa extending from the lateral sides of the spinal cord to the dura mater (see figure 11.2b). Superior movement is limited by the **filum terminale** (fī′lŭm ter′mi-nal′ē), a connective tissue strand that anchors the conus medullaris and the thecal sac to the first coccygeal vertebra.

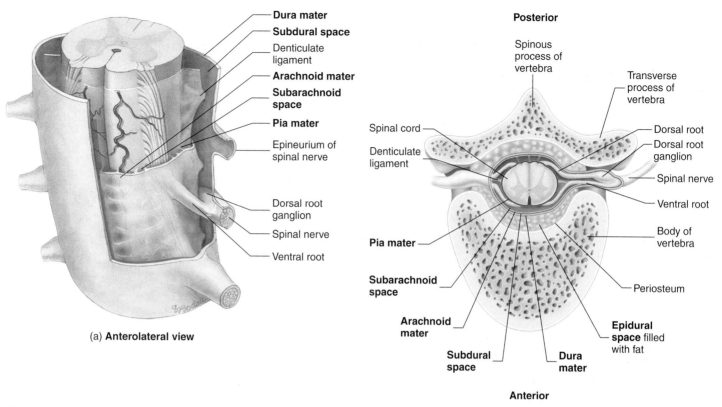

(a) **Anterolateral view**

Figure 11.2 Meningeal Membranes Surrounding the Spinal Cord

(b) **Superior view**

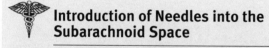

Introduction of Needles into the Subarachnoid Space

Several clinical procedures involve the insertion of a needle into the subarachnoid space inferior to the level of the second lumbar vertebra. The needle is introduced into either the L3/L4 or the L4/L5 intervertebral space. The needle does not contact the spinal cord because the spinal cord extends only approximately to the second lumbar vertebra of the vertebral column. The needle enters the subarachnoid space, which extends to level S2 of the vertebral column. The needle does not damage the nerve roots of the cauda equina located in the subarachnoid space because the needle quite easily pushes them aside.

In **spinal anesthesia,** or spinal block, drugs that block action potential transmission are introduced into the subarachnoid space to prevent pain sensations in the lower half of the body. A **spinal tap** is the removal of CSF from the subarachnoid space. A spinal tap may be performed to examine the CSF for infectious agents (meningitis), for the presence of blood (hemorrhage), or for the measurement of CSF pressure. A radiopaque substance may also be injected into this area, and a **myelogram** (radiograph of the spinal cord) may be taken to visualize spinal cord defects or damage. ▼

4. *From superficial to deep, name the meninges surrounding the spinal cord. What is found within the epidural, subdural, and subarachnoid spaces?*

5. *How is the spinal cord held within the vertebral canal?*

Cross Section of the Spinal Cord

The spinal cord is divided into right and left halves (figure 11.3). Internally, peripherally located white matter surrounds centrally located gray matter shaped like a butterfly. The white matter in each half of the spinal cord is organized into three **columns,** or **funiculi** (fū-nik′ū-lī, cord), called the **ventral** (anterior), **dorsal** (posterior), and **lateral columns.** Each column is subdivided into **tracts,** also called **fasciculi** (fă-sik′ū-lī, bundle) or **pathways.** The tracts consist of axons ascending to the brain or descending from the brain (see figure 11.3*b*). Axons within a given tract carry basically the same type of information, although they may overlap to some extent. For example, one ascending tract carries action potentials related to pain and temperature sensations, whereas another carries action potentials related to light touch (see chapter 12).

The gray matter consists of neuron cell bodies, dendrites, and axons. Each half of the central gray matter of the spinal cord consists of a relatively thin **posterior** (dorsal) **horn** and a larger **anterior** (ventral) **horn.** Small **lateral horns** exist in levels of the cord associated with the autonomic nervous system (see chapter 13).

The two halves of the spinal cord are connected by **gray** and **white commissures** (see figure 11.3*a*). The white and gray commissures contain axons that cross from one side of the spinal cord to the other. The **central canal** is in the center of the gray commissure.

Spinal nerves arise from numerous **rootlets** along the dorsal and ventral surfaces of the spinal cord (see figure 11.3*a*). The rootlets combine to form a **ventral root** and a **dorsal root** at each segment of the cord. The ventral and dorsal roots extend laterally from the spinal cord, passing through the subarachnoid space,

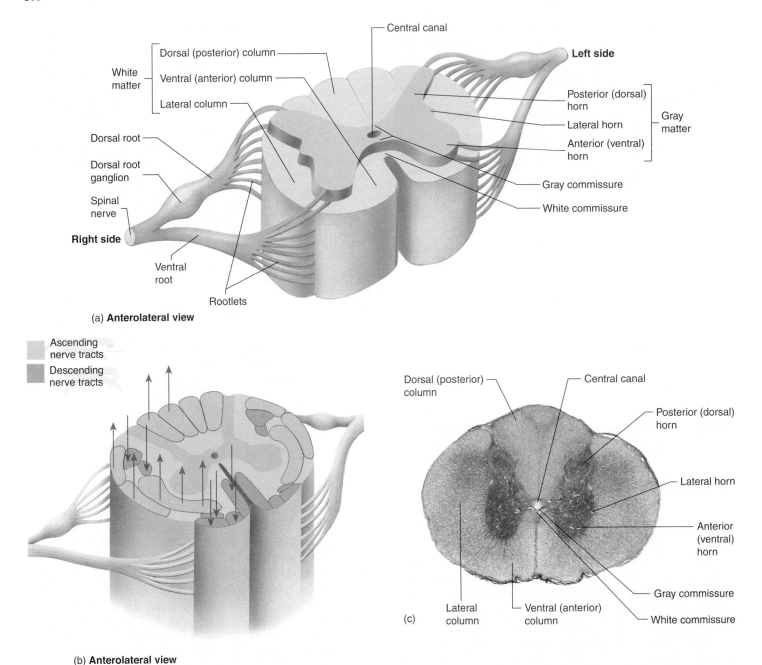

Figure 11.3   **Cross Section of the Spinal Cord**

(*a*) A segment of the spinal cord showing one dorsal and one ventral root on each side and the rootlets that form them. (*b*) Ascending and descending tracts in the spinal cord. Ascending nerve tracts are *blue*; descending nerve tracts are *pink*. The arrows indicate the direction of each pathway. (*c*) Photograph of a cross section through the midlumbar region. The *lighter tan areas* are white matter, where tracts are located. The *darker area* is gray matter, where neuron cell bodies are located.

piercing the arachnoid mater and dura mater, and joining one another to form a spinal nerve (see figure 11.2*b*).

Each dorsal root contains a ganglion, called the **dorsal root, or spinal, ganglion** (gang′glē-on, a swelling or knot). The dorsal root ganglia are collections of cell bodies of the sensory neurons forming the dorsal roots of the spinal nerves (figure 11.4). The axons of these unipolar neurons extend from various parts of the body and pass through spinal nerves to the dorsal root ganglia. The axons do not synapse in the dorsal root ganglion but pass through the dorsal root into the posterior horn of the spinal cord

gray matter. The axons either synapse with interneurons in the posterior horn or pass into the white matter and ascend or descend in nerve tracts of the spinal cord.

Motor neurons innervate muscles and glands. The cell bodies of the multipolar motor neurons are located in the anterior and lateral horns of the spinal cord gray matter (see figure 11.4). The cell bodies of somatic motor neurons are in the anterior horn and the cell bodies of autonomic motor neurons are in the lateral horn. Axons of the motor neurons form the ventral roots and pass into the spinal nerves. Thus, dorsal roots contain sensory axons, ventral

Damage to the spinal cord can disrupt ascending tracts from the spinal cord to the brain, resulting in the loss of sensation, and/or descending tracts from the brain to motor neurons in the spinal cord, resulting in the loss of motor functions. About 10,000 new cases of **spinal cord injury** occur each year in the United States. Automobile and motorcycle accidents are leading causes, followed by gunshot wounds, falls, and swimming accidents. Spinal cord injury is classified according to the vertebral level at which the injury occurred, whether the entire cord is damaged at that level or only a portion of the cord, and the mechanism of injury. Most spinal cord injuries occur in the cervical region or at the thoracolumbar junction and are incomplete. The primary mechanisms include concussion (an injury caused by a blow), contusion (an injury resulting in hemorrhage), and laceration (a tear or cut) and involve excessive flexion, extension, rotation, or compression of the vertebral column. The majority of spinal cord injuries are acute contusions of the cord due to bone or disk displacement into the cord and involve a combination of excessive directional movements, such as simultaneous flexion and compression.

At the time of spinal cord injury, two types of tissue damage occur: (1) primary, mechanical damage and (2) secondary, tissue damage. Secondary spinal cord damage, which begins within minutes of the primary damage, is caused by ischemia, edema, ion imbalances, the release of "excitotoxins" (such as glutamate), and inflammatory cell invasion. Secondary damage extends into a much larger region of the cord than the primary damage. It is the primary focus of current research in spinal cord injury. The only treatment for primary damage is prevention, such as wearing seat belts when riding in automobiles and not diving into shallow water. Once an accident occurs, however, little can be done about the primary damage. On the other hand, much of the secondary damage can be prevented or reversed.

Until the 1950s, many spinal cord injuries were ultimately fatal. Now, with quick treatment directed at the mechanisms of secondary tissue damage, much of the total damage to the spinal cord can be prevented. Treatment of the damaged spinal cord with large doses of anti-inflammatory steroids, such as methylprednisolone, within 8 hours of the injury can dramatically reduce the secondary damage to the cord. The objectives of this treatment are to reduce inflammation and edema. Current treatment includes ana-tomical realignment and stabilization of the vertebral column, decompression of the spinal cord, and administration of anti-inflammatory steroids. Rehabilitation is based on retraining the patient to use whatever residual connections exist across the site of damage.

It had long been thought that the spinal cord was incapable of regeneration following severe damage. However, it is now known that, following injury, most neurons of the adult spinal cord survive and begin to regenerate, growing about 1 mm into the site of damage, but then they regress to an inactive, atrophic state. In addition, fetuses and newborns exhibit considerable regenerative ability and functional improvement. A major block to adult spinal cord regeneration is the formation of a scar, consisting mainly of myelin and astrocytes, at the site of injury. Myelin in the scar is apparently the primary inhibitor of regeneration. Implantation of peripheral nerves, Schwann cells, or fetal CNS tissue can bridge the scar and stimulate some regeneration. Certain growth factors can also stimulate some regeneration. Current research continues to look for the right combination of chemicals and other factors to stimulate regeneration of the spinal cord following injury.

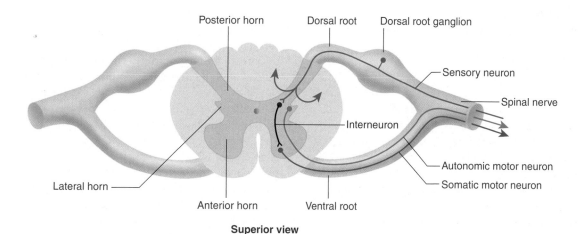

Superior view

Figure 11.4 Relationship of Sensory and Motor Neurons to the Spinal Cord

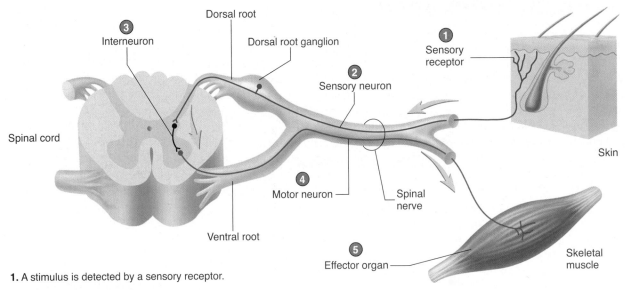

1. A stimulus is detected by a sensory receptor.

2. A sensory neuron conducts action potentials through the spinal nerve and dorsal root to the spinal cord.

3. In the spinal cord, the sensory neuron synapses with an interneuron.

4. The interneuron synapses with a motor neuron.

5. A motor neuron axon conducts action potentials through the ventral root and spinal nerve to an effector organ.

Process Figure 11.5   Reflex Arc
The parts of a reflex arc are labeled in the order in which action potentials pass through them.

roots contain motor axons, and spinal nerves have both sensory and motor axons. ▼

6. *Describe the arrangement of white and gray matter in the spinal cord.*

7. *What are tracts and commissures?*

8. *Where are the cell bodies of sensory, somatic motor, and autonomic motor neurons located in the gray matter?*

9. *Where do dorsal and ventral roots exit the spinal cord? What kinds of axons are in the dorsal and ventral roots and in the spinal nerves?*

P R E D I C T   1

Explain why the dorsal root ganglia are larger in diameter than the dorsal roots, and describe the direction of action potential propagation in the spinal nerves, dorsal roots, and ventral roots.

Reflexes

A **reflex** is a stereotypic, unconscious, involuntary response to a stimulus. *Stereotypic* means that the response to the stimulus is always more or less the same. For example, striking the patellar ligament with a rubber hammer stimulates a reflexive extension of the knee. The degree of extension may vary, but the knee never flexes. *Unconscious* means that no awareness is necessary for the reflex to occur, and *involuntary* means one cannot will the reflex to happen.

Reflexes maintain homeostasis. **Somatic reflexes** are mediated through the somatic motor nervous system and include responses that remove the body from painful stimuli that would cause tissue damage or keep the body from suddenly falling or moving because of external forces. **Autonomic reflexes** are mediated through the ANS and are responsible for maintaining variables within their normal ranges, such as maintaining relatively constant blood pressure, blood carbon dioxide levels, and water intake.

Reflexes are produced by reflex arcs. The **reflex arc** is the basic functional unit of the nervous system because it is the smallest, simplest portion capable of receiving a stimulus and producing a response. The reflex arc usually has five basic components: (1) a sensory receptor, (2) a sensory neuron, (3) an interneuron, (4) a motor neuron, and (5) an effector organ (figure 11.5). Most reflex arcs have at least one interneuron, but in a few reflex arcs the sensory neuron synapses directly with the motor neuron.

Reflexes occur unconsciously because reflex arcs are not part of the brain where consciousness occurs. Action potentials initiated by sensory receptors are transmitted along the axons of sensory neurons to the CNS, where the axons usually synapse with interneurons. Interneurons synapse with motor neurons, which send axons out of the spinal cord and through the PNS to muscles or glands, where the action potentials of the motor neurons cause these effector organs to respond.

Reflex arcs do not operate as isolated entities within the nervous system because of divergent and convergent pathways (see chapter 10). Diverging branches of the sensory neurons or interneurons in a reflex arc send action potentials along ascending nerve tracts to the brain (figure 11.6). For example, a pain stimulus can initiate a reflex that causes specific muscles to contract, resulting in the removal of the affected part of the body from the painful stimulus. Through divergence, the same pain stimulus results in action potentials going to parts of the brain where conscious awareness occurs, producing a sensation of pain. Although the

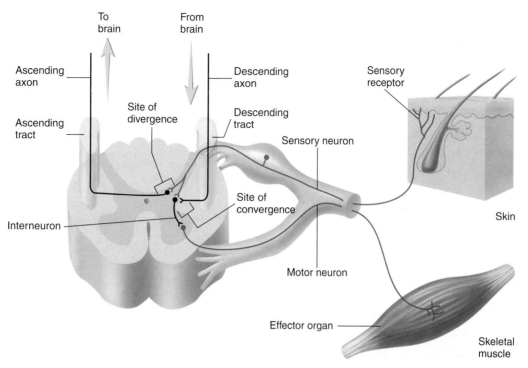

Figure 11.6   Spinal Reflex, with Ascending and Descending Axons

A diverging branch of a sensory neuron synapses with a neuron that extends to the brain in an ascending tract (*blue*). An axon from the brain extends to the spinal cord in a descending tract (*pink*) and synapses with an interneuron, influencing its action on the motor neuron.

reflex occurs unconsciously, there is eventually awareness of the painful stimulus.

Axons within descending tracts from the brain converge with neurons of reflex arcs (see figure 11.6). These descending neurons can change the sensitivity of the reflex arc. Descending excitatory neurons (see chapter 10) stimulate excitatory postsynaptic potentials (EPSPs), which increases the sensitivity of the reflex arc, making it more likely that a reflex occurs. Descending inhibitory neurons stimulate inhibitory postsynaptic potentials (IPSPs), which decrease the sensitivity of the reflex arc, making it less likely that a reflex occurs.

Through convergence, a motor neuron can be controlled in two ways—through a reflex arc or by the brain. For example, skeletal muscles can contract reflexively (unconsciously) or voluntarily (consciously).

P R E D I C T

Explain the advantage of removing your hand from a hot object reflexively versus consciously.

Many reflexes are integrated within the spinal cord; others are integrated within the brain. Some reflexes involve excitatory neurons and result in a response, such as when a muscle contracts. Other reflexes involve inhibitory neurons and result in the inhibition of a response, such as when a muscle is inhibited and relaxes. In addition, higher brain centers influence reflexes by either suppressing or exaggerating them. Major spinal cord reflexes include the stretch reflex, the Golgi tendon reflex, and the withdrawal reflex. ▼

> *10. Define a reflex and describe its characteristics. Give examples of somatic and autonomic reflexes.*

> *11. Name the parts of a reflex arc and explain how they produce a reflex.*
> *12. How do reflex arcs interact with ascending and descending tracts?*

Stretch Reflex

The simplest reflex is the **stretch reflex** (figure 11.7), a reflex in which muscles contract in response to a stretching force applied to them. The sensory receptor of this reflex is the **muscle spindle,** which consists of 3–10 small, specialized skeletal muscle fibers. The fibers are contractile only at their ends and are innervated by motor neurons. Sensory neurons innervate the noncontractile centers of the muscle spindle fibers. Axons of these sensory neurons extend to the spinal cord and synapse directly with motor neurons in the spinal cord, which in turn innervate the muscle in which the muscle spindle is embedded. Note that this reflex has no interneuron between the sensory neuron and the motor neuron.

Stretching a muscle also stretches muscle spindles located among the muscle fibers. The stretch stimulates the sensory neurons that innervate the centers of the muscle spindles. The increased frequency of action potentials carried to the spinal cord by sensory neurons stimulates the motor neurons in the spinal cord. The motor neurons transmit action potentials to skeletal muscle, causing a rapid contraction of the stretched muscle, which opposes the stretch of the muscle. The postural muscles demonstrate the adaptive nature of this reflex. If a person is standing upright and then begins to tip slightly to one side, the postural muscles associated with the vertebral column on the other

Sudden stretch of a muscle results in:

1. Muscle spindles detect stretch of the muscle.

2. Sensory neurons conduct action potentials to the spinal cord.

3. Sensory neurons synapse directly with motor neurons.

4. Stimulation of the motor neurons results in action potentials being conducted to the muscle, causing it to contract and resist being stretched. *Note:* The muscle that contracts is the muscle that is stretched.

Process Figure 11.7 Stretch Reflex

side are stretched. As a result, stretch reflexes are initiated in those muscles, which cause them to contract and reestablish normal posture.

Through divergence, collateral axons from the sensory neurons of the muscle spindles also synapse with neurons whose axons contribute to ascending tracts, which enable the brain to monitor the stretch of muscles (see figure 11.7). Through convergence, descending neurons within the spinal cord synapse with the neurons of the stretch reflex arc, modifying their activity. This activity is important in maintaining posture and in coordinating muscular activity.

The motor neurons innervating the ends of the muscle spindles are responsible for regulating the sensitivity of the muscle spindles to stretch. As a skeletal muscle contracts, the tension on the centers of muscle spindles within the muscle decreases because the muscle spindles passively shorten as the muscle shortens. This decrease in tension causes muscle spindles to be less sensitive to stretch. Sensitivity is maintained because motor neurons stimulate the ends of the muscle spindles to contract, which pulls on the center part of the muscle spindles and maintains the proper

tension. The activity of the muscle spindles help control posture, muscle tension, and muscle length.

Knee-Jerk Reflex

The **knee-jerk reflex,** or **patellar reflex,** is a classic example of the stretch reflex. When the patellar ligament is tapped, the tendons and muscles of the quadriceps femoris muscle group are stretched, activating a stretch reflex. Contraction of the muscles extends the leg, producing the characteristic knee-jerk response.

Clinicians use the knee-jerk to determine whether the parts of the brain controlling the sensitivity of muscle spindles are functioning normally. When the stretch reflex is greatly exaggerated or depressed, it can indicate that the neurons within the brain controlling the sensitivity of muscle spindles are overly active or suppressed. Absence of the stretch reflex can indicate that the descending tract from the brain or the reflex arc is not intact.

Intense stretch of a skeletal muscle results in:

1. Golgi tendon organs detect tension applied to a tendon.

2. Sensory neurons conduct action potentials to the spinal cord.

3. Sensory neurons synapse with inhibitory interneurons that synapse with motor neurons.

4. Inhibition of the motor neurons causes muscle relaxation, relieving the tension applied to the tendon. *Note:* The muscle that relaxes is attached to the tendon to which tension is applied.

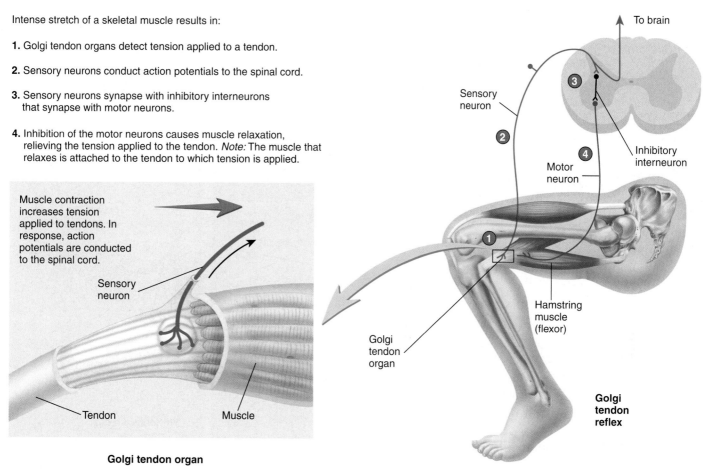

Muscle contraction increases tension applied to tendons. In response, action potentials are conducted to the spinal cord.

Sensory neuron

Tendon

Muscle

Golgi tendon organ

Process Figure 11.8   Golgi Tendon Reflex

To brain

Sensory neuron

Inhibitory interneuron

Motor neuron

Golgi tendon organ

Hamstring muscle (flexor)

Golgi tendon reflex

Golgi Tendon Reflex

The **Golgi tendon reflex** prevents contracting muscles from applying excessive tension to tendons. **Golgi tendon organs** are encapsulated nerve endings that have at their ends numerous terminal branches with small swellings associated with bundles of collagen fibers in tendons. The Golgi tendon organs are located within tendons near the muscle–tendon junction (figure 11.8). As a muscle contracts, the attached tendons are stretched, resulting in increased tension in the tendon. The increased tension stimulates action potentials in the sensory neurons from the Golgi tendon organs. Golgi tendon organs have a high threshold and are sensitive only to intense stretch.

The sensory neurons of the Golgi tendon organs synapse with inhibitory interneurons in the spinal cord. The interneurons synapse with motor neurons that innervate the muscle to which the Golgi tendon organ is attached. When a great amount of tension is applied to the tendon, the sensory neurons of the Golgi tendon organs are stimulated. The sensory neurons stimulate the interneurons to release inhibitory neurotransmitters, which inhibit the motor neurons of the associated muscle and cause it to relax. The sudden relaxation of the muscle reduces the tension

applied to the muscle and tendons and protects them from damage. A weight lifter who suddenly drops a heavy weight after straining to lift it does so, in part, because of the effect of the Golgi tendon reflex.

Tremendous amounts of tension can be applied to muscles and tendons in the legs. An athlete's Golgi tendon reflex can be inadequate to protect muscles and tendons from excessive tension. The large muscles and sudden movements of football players and sprinters can make them vulnerable to hamstring pulls and calcaneal (Achilles) tendon injuries.

Through divergence, collateral axons from the sensory neurons of the Golgi tendon organs also synapse with neurons whose axons contribute to ascending tracts, which enable the brain to monitor the tension generated by muscle contraction (see figure 11.8).

Withdrawal Reflex

The function of the **withdrawal,** or **flexor, reflex** is to remove a limb or another body part from a painful stimulus. The sensory receptors are pain receptors (see chapter 12). Action potentials from painful stimuli are conducted by sensory neurons to the

Stimulation of pain receptors results in:

1. Pain receptors detect a painful stimulus.

2. Sensory neurons conduct action potentials to the spinal cord.

3. Sensory neurons synapse with excitatory interneurons that synapse with motor neurons.

4. Excitation of the motor neurons results in contraction of the flexor muscles and withdrawal of the limb from the painful stimulus.

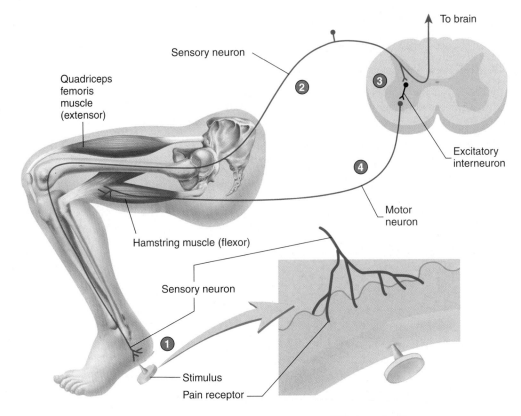

Process Figure 11.9   Withdrawal Reflex

spinal cord, where they synapse with excitatory interneurons, which in turn synapse with motor neurons (figure 11.9). The motor neurons stimulate muscles, usually flexor muscles, that remove the limb from the source of the painful stimulus. Collateral branches of the sensory neurons synapse with ascending fibers to the brain, providing conscious awareness of the painful stimuli.

Different reflex arcs often function together to produce a response. For example, when a reflex causes a muscle to contract, another reflex is initiated that causes the antagonists of the muscle to relax. At the same time that a withdrawal reflex causes the hamstrings to contract, another reflex causes the quadriceps to relax. At the same time that a stretch reflex causes the quadriceps to contract, another reflex causes the hamstrings to relax. ▼

> **13.** *Starting with their sensory receptors, name in the order stimulated the parts of stretch, Golgi tendon, and withdrawal reflex arcs.*
> **14.** *Describe the functional importance of responses produced by stretch, Golgi tendon, and withdrawal reflexes.*

Nerves

Structure of Nerves

Nerves consist of collections of axons, Schwann cells, and connective tissue in the PNS (figure 11.10). Each axon, or nerve fiber, and its Schwann cell sheath are surrounded by a delicate connective tissue layer, the **endoneurium** (en-dō-noo′rē-ŭm). A heavier connective tissue layer, the **perineurium** (per-i-noo′rē-ŭm), surrounds groups of axons to form nerve **fascicles** (fas′i-klz). A third layer of dense connective tissue, the **epineurium** (ep-i-noo′rē-ŭm), binds the nerve fascicles together to form a nerve. The connective tissue of the epineurium is continuous with the dura mater surrounding the CNS. An analogy of this relationship is a coat (the dura) and its sleeve (epineurium). The connective tissue layers of nerves make them tougher than the nerve tracts in the CNS. ▼

> **15.** *Describe the connective tissue layers within and surrounding spinal nerves.*

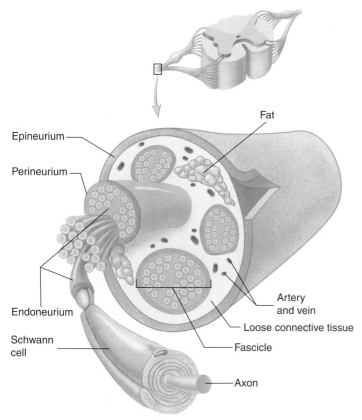

Figure 11.10 **Structure of a Nerve**

Nerve structure illustrating axons surrounded by various layers of connective tissue: epineurium surrounds the whole nerve, perineurium surrounds nerve-fascicles, and endoneurium surrounds Schwann cells and axons. Loose connective tissue also surrounds the nerve fascicles.

Spinal Nerves and Their Distribution

There are 31 pairs of spinal nerves, each identified by a letter and a number (figure 11.11). The letter indicates the region of the vertebral column from which the nerve emerges: C, cervical; T, thoracic; L, lumbar; and S, sacral. The single coccygeal nerve is often not designated, but when it is the symbol often used is Co. The number indicates the location in each region where the nerve emerges from the vertebral column, with the smallest number always representing the most superior origin. For example, the most superior nerve exiting from the thoracic region of the vertebral column is designated T1. The cervical nerves are designated C1–C8, the thoracic nerves T1–T12, the lumbar nerves L1–L5, and the sacral nerves S1–S5. Note that the number of each type of nerve matches the number of each type of vertebrae, except for the cervical nerves. Although there are only 7 cervical vertebrae, there are 8 pairs of cervical spinal nerves because 2 pairs of nerves are associated with the first cervical vertebra. The first pair of cervical nerves exit the spinal cord between the skull and the first cervical vertebra and the second pair exit through the intervertebral foramina between the first and second cervical vertebrae.

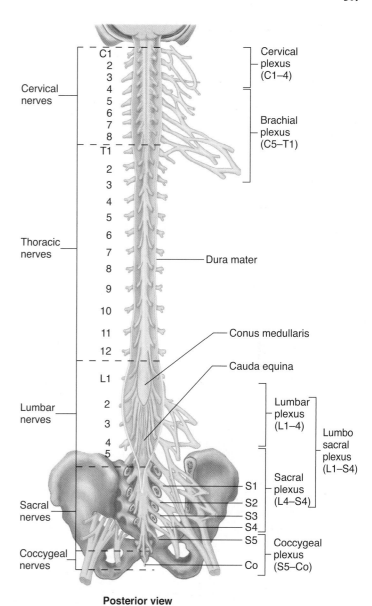

Posterior view

Figure 11.11 **Spinal Nerves and Plexuses**

The regional designations and the numbers of the spinal nerves are shown on the left. The plexuses formed by the spinal nerves are shown on the right.

The nerves arising from each region of the spinal cord and vertebral column supply specific regions of the body. Not surprisingly, the nerves supply skeletal muscles in a top to bottom pattern: the cervical nerves supply muscles of the head, neck, shoulders, and upper limbs; the thoracic nerves supply muscles of the upper limbs, thorax, vertebral column, and hips; the lumbar nerves supply muscles of the vertebral column, hips, and lower limbs; and the sacral nerves supply muscles of the lower limbs (figure 11.12*a*)

Each of the spinal nerves except C1 has a specific cutaneous sensory distribution. A **dermatome** (der-mă-tō′m) is the area of skin supplied with sensory innervation by a pair of spinal nerves.

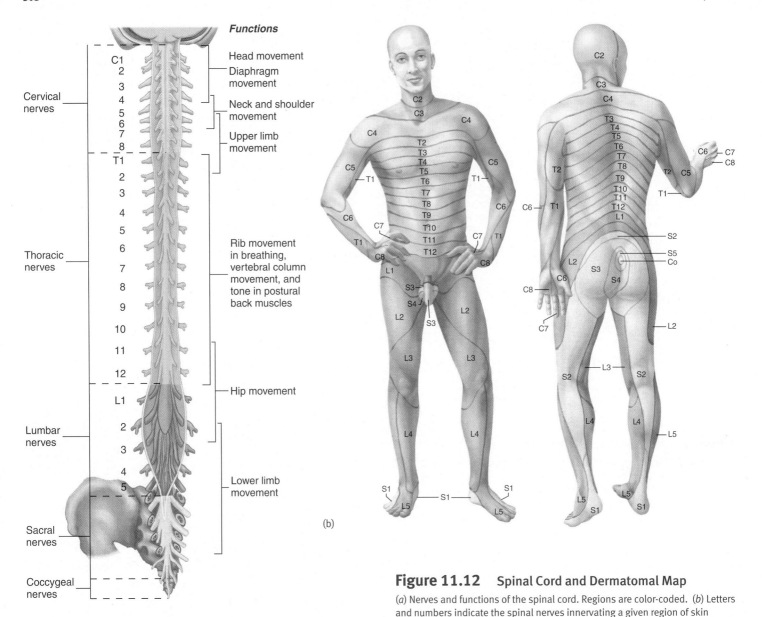

Functions

Head movement

Diaphragm movement

Neck and shoulder movement

Upper limb movement

Rib movement in breathing, vertebral column movement, and tone in postural back muscles

Hip movement

Lower limb movement

Cervical nerves

Thoracic nerves

Lumbar nerves

Sacral nerves

Coccygeal nerves

(a) **Posterior view**

(b)

Figure 11.12   Spinal Cord and Dermatomal Map

(*a*) Nerves and functions of the spinal cord. Regions are color-coded. (*b*) Letters and numbers indicate the spinal nerves innervating a given region of skin (dermatome).

A **dermatomal** (der-mă-tō′măl) **map** shows the distribution of all the dermatomes (figure 11.12*b*).

P R E D I C T
The dermatomal map is important in clinical considerations of nerve damage. Loss of sensation in a dermatomal pattern can provide valuable information about the location of nerve damage. Predict the possible site of nerve damage for a patient who suffered whiplash in an automobile accident and subsequently developed anesthesia (no sensations) in the left arm, forearm, and hand (see figure 11.12*b* for help). ▼

16. *How many spinal nerves of each type are there?*

17. *Describe the general pattern of distribution of spinal nerves to skeletal muscles.*

18. *What is a dermatome? Why are dermatomes clinically important?*

Spinal Nerves and Plexuses

Spinal nerves arise from the spinal cord as rootlets (figure 11.13). The rootlets combine to form the ventral and dorsal roots, which combine to form the spinal nerve. The spinal nerve divides to form branches called **rami** (rā′mī, branches). Each spinal nerve has a dorsal and a ventral ramus (rā′mŭs). Additional communicating

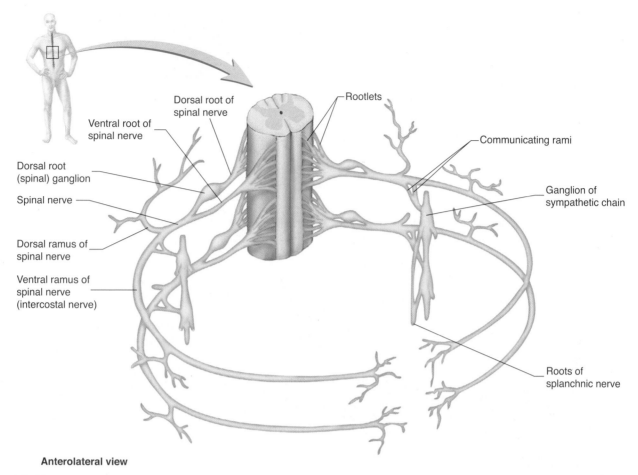

Dorsal root of
spinal nerve

Ventral root of
spinal nerve

Dorsal root
(spinal) ganglion

Spinal nerve

Dorsal ramus of
spinal nerve

Ventral ramus of
spinal nerve
(intercostal nerve)

Rootlets

Communicating rami

Ganglion of
sympathetic chain

Roots of
splanchnic nerve

Anterolateral view

Figure 11.13   Spinal Nerves

Typical thoracic spinal nerves. The figure shows dorsal and ventral roots, as well as dorsal, ventral, and communicating rami. Communicating rami connect to the sympathetic chain (see chapter 13).

rami are associated with the sympathetic division of the ANS (see chapter 13). The **dorsal rami** innervate most of the deep muscles of the dorsal trunk responsible for movement of the vertebral column. They also innervate the connective tissue and skin near the midline of the back.

The **ventral rami** are distributed in two ways. In the thoracic region, the ventral rami form **intercostal** (between ribs) **nerves** (see figure 11.13), which extend along the inferior margin of each rib and innervate the intercostal muscles and the skin over the thorax.

The ventral rami of the cervical, lumbar, sacral, and coccygeal spinal nerves form **plexuses** (plek′sŭs-ēz). The term *plexus* means braid and describes the organization produced by the intermingling of the nerves. The brachial plexus will be used to illustrate

how the spinal nerves intermingle within a plexus and then give rise to nerves that are distributed throughout the body. The brachial plexus is the most complicated plexus, and other plexuses have similar, but simpler, interconnections.

The **brachial plexus** originates from spinal nerves C5–T1 (figure 11.14). The ventral rami of the spinal nerves are called the **roots** of the plexus. These roots should not be confused with the dorsal and ventral roots from the spinal cord, which are more medial. The five roots join to form three **trunks,** which separate into six **divisions** and then join again to create three **cords** (posterior, lateral, and medial) from which five major **branches,** or nerves, emerge. The five major nerves are the **axillary, radial, musculocutaneous, ulnar,** and **median nerves.** They supply the upper limb.

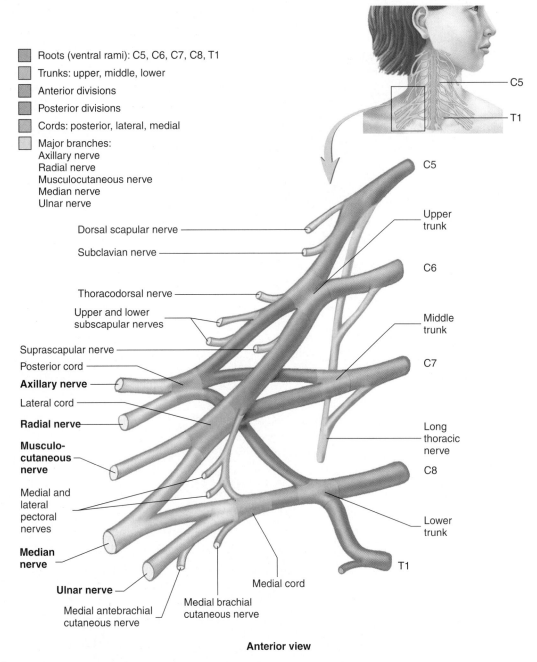

Roots (ventral rami): C5, C6, C7, C8, T1
Trunks: upper, middle, lower
Anterior divisions
Posterior divisions
Cords: posterior, lateral, medial
Major branches:
Axillary nerve
Radial nerve
Musculocutaneous nerve
Median nerve
Ulnar nerve

Dorsal scapular nerve
Subclavian nerve
Thoracodorsal nerve
Upper and lower subscapular nerves
Suprascapular nerve
Posterior cord
Axillary nerve
Lateral cord
Radial nerve
Musculo-cutaneous nerve
Medial and lateral pectoral nerves
Median nerve
Ulnar nerve
Medial antebrachial cutaneous nerve
Medial brachial cutaneous nerve
Medial cord

C5
Upper trunk
C6
Middle trunk
C7
Long thoracic nerve
C8
Lower trunk
T1

Anterior view

Figure 11.14 Brachial Plexus

The roots of the plexus are formed by the ventral rami of the spinal nerves C5–T1 and join to form an upper, middle, and lower trunk. Each trunk divides into anterior and posterior divisions. The divisions join together to form the posterior, lateral, and medial cords from which the major brachial plexus nerves arise. The major brachial plexus nerves include the axillary, radial, musculocutaneous, median, and ulnar nerves. These nerves innervate the muscles and skin of the upper limb.

The distribution of the radial nerve to skeletal muscles of the upper limb is shown in figure 11.15. In a similar fashion, nerves arising from plexuses are distributed to skeletal muscles throughout the body (table 11.1). Nerves arising from plexuses also supply the skin (figure 11.16). Referring to figure 11.14, note that axons of spinal nerves C5–T1 can reach the radial nerve through the interconnections of the brachial plexus. Thus, axons in the radial nerve arise from levels C5–T1 of the spinal cord. The cutaneous distribution of a nerve arising from a plexus is different from the cutaneous distribution (dermatome) arising from a spinal nerve because the plexus nerve consists of axons from more than one level of the spinal cord. ▼

19. Differentiate among rootlet, dorsal root, ventral root, and spinal nerve.

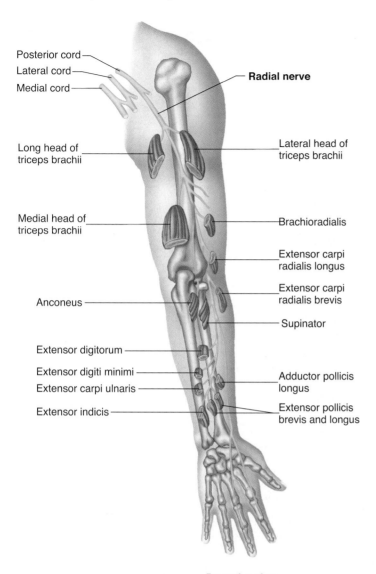

Posterior cord
Lateral cord
Medial cord

Radial nerve

Long head of
triceps brachii

Lateral head of
triceps brachii

Medial head of
triceps brachii

Brachioradialis

Extensor carpi
radialis longus

Anconeus

Extensor carpi
radialis brevis

Supinator

Extensor digitorum

Extensor digiti minimi

Extensor carpi ulnaris

Adductor pollicis
longus

Extensor indicis

Extensor pollicis
brevis and longus

Posterior view

Figure 11.15   Radial Nerve

The route of the radial nerve and the muscles it innervates are illustrated and listed.

20. *Contrast dorsal, ventral, and communicating rami of spinal nerves. What muscles do the dorsal rami innervate?*
21. *Describe the distribution of the ventral rami of the thoracic region.*
22. *What is a plexus? What happens to the axons of spinal nerves as they pass through a plexus?*

CASE STUDY | Cervical Rib Syndrome

Sarah, who is 26, noticed that over a period of time she experienced pain, tingling, and numbness in the ring finger and little finger of her right hand. She also had pain in her elbow, which radiated down the posteromedial portion of her forearm and hand. She made an appointment with her physician. After careful examination of Sarah's upper limb, her physician ordered a radiograph of her neck, which disclosed a right cervical rib attached to vertebra C7. Cervical ribs are not uncommon, occurring in about 1% of the population. Most people exhibit no symptoms, but symptoms may develop in some people. If the extra rib compresses the inferior roots of the brachial plexus (see figure 11.14), the condition is called **cervical rib syndrome.**

P R E D I C T

Use figures 11.12, 11.14, and 11.16 to answer these questions.
a. Name the brachial plexus nerves supplying the skin of the hand.
b. Damage to which of these nerves could produce the symptoms seen in Sarah's hand?
c. What made Sarah's physician suspect cervical rib syndrome rather than damage to an individual nerve?
d. Cervical rib syndrome can also affect muscles, producing muscle weakness and paralysis. Muscles supplied by what nerves could be affected by a cervical rib?

Major Spinal Nerve Plexuses

The ventral rami of the cervical, lumbar, sacral, and coccygeal spinal nerves form five major **plexuses** (plek′sŭs-ēz): the cervical, brachial, lumbar, sacral, and coccygeal plexuses (see figure 11.11).

Cervical Plexus

The **cervical plexus** is a relatively small plexus originating from spinal nerves C1–C4 (see figure 11.11). Branches derived from this plexus innervate superficial neck structures, including several of the muscles attached to the hyoid bone. The cervical plexus innervates the skin of the neck and posterior portion of the head. The **phrenic** (fren′ik) **nerve,** which originates from spinal nerves C3–C5, is derived from both the cervical and brachial plexus. The phrenic nerves descend along each side of the neck to enter the thorax. They descend along the sides of the mediastinum to reach the diaphragm, which they innervate. Contraction of the diaphragm is largely responsible for the ability to breathe.

P R E D I C T

Explain how damage to or compression of the right phrenic nerve affects the diaphragm. Describe the effect on breathing of a completely severed spinal cord at the level of C2 versus at the level of C6.

Brachial Plexus

The brachial plexus originates from spinal nerves C5–T1 (see figures 11.11 and 11.14). The five major nerves emerging from the

Table 11.1 — Major Nerves Arising from Spinal Nerve Plexuses

Nerve (Origin)	Muscles Innervated	Major Movements	Cutaneous Distribution
Cervical Plexus			
Phrenic (C1–C4)	Diaphragm	Inferior movement increases thoracic volume	None
Brachial Plexus			
Axillary (C5–C6)	Deltoid Teres minor	Abducts arm Laterally rotates arm	Inferior lateral shoulder
Radial (C5–T1)	Triceps brachii Brachialis (part) and brachioradialis Supinator Posterior extrinsic hand muscles	Extends elbow Flex elbow Supinates forearm Extend, abduct, and adduct wrist; extend fingers; extend and abduct thumb	Posterior surface of arm and forearm; lateral two-thirds of dorsum of hand
Musculocutaneous (C5–C7)	Coracobrachialis Biceps brachii Brachialis (most)	Flexes shoulder Flexes elbow and supinates forearm Flexes elbow	Lateral surface of forearm
Ulnar (C8–T1)	Anterior extrinsic hand muscles Intrinsic hand muscles	Flex and adduct wrist; flex distal phalanges of little and ring finger Adduct thumb; abduct and adduct fingers; oppose, flex, and extend little finger	Medial one-third of hand, little finger, and medial half of ring finger
Median (C5–T1)	Pronators Anterior extrinsic hand muscles Intrinsic hand muscles	Pronate forearm Flex and abduct wrist; flex distal phalanges of middle and little fingers; flex middle and proximal phalanges; flex thumb Oppose, flex, and abduct thumb	Lateral two-thirds of palm of hand; anterior surface of thumb, index, and middle fingers; dorsal tips of thumb, index, middle, and ring (lateral half) fingers
Lumbosacral Plexus			
Obturator (L2–L4)	Adductor muscles of medial thigh Gracilis	Adduct thigh Adducts thigh, flexes knee	Superior medial side of thigh
Femoral (L2–L4)	Pectineus Iliopsoas Sartorius Rectus femoris Vastus lateralis, vastus medialis, and vastus intermedius	Adducts thigh, flexes hip Flexes hip Flexes hip, flexes knee Flexes hip, extends knee Extend knee	Anterior and lateral thigh; medial leg and foot
Tibial (L4–S3)	Hamstrings (long head of biceps femoris, semitendinosus, and semimembranosus) Adductor magnus (hamstring part) Gastrocnemius and soleus Posterior extrinsic foot muscles Intrinsic foot muscles	Extend hip, flex knee Extends hip Plantar flex foot Plantar flex and invert foot; flex toes Flex, extend, abduct, and adduct toes	Heel and sole of foot; posterior surface of leg and lateral foot through the sural nerve
Common Fibular (L4–S2)	Biceps femoris (short head) Anterior extrinsic foot muscles Lateral extrinsic foot muscles Intrinsic foot muscle	Flexes knee Dorsiflex, invert, and evert foot; extend toes Planter flex and evert foot Extends toes	Anterior surface of leg and dorsum of foot; posterior surface of leg and lateral foot through the sural nerve

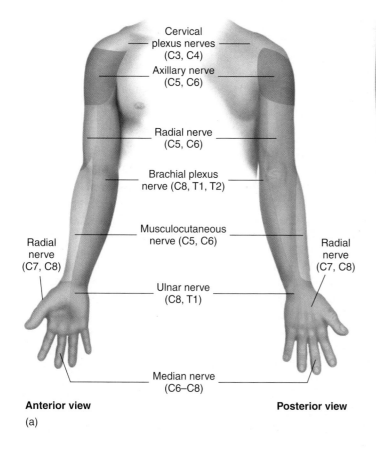

Cervical
plexus nerves
(C3, C4)

Axillary nerve
(C5, C6)

Radial nerve
(C5, C6)

Brachial plexus
nerve (C8, T1, T2)

Musculocutaneous
nerve (C5, C6)

Radial
nerve
(C7, C8)

Radial
nerve
(C7, C8)

Ulnar nerve
(C8, T1)

Median nerve
(C6–C8)

Anterior view

Posterior view

(a)

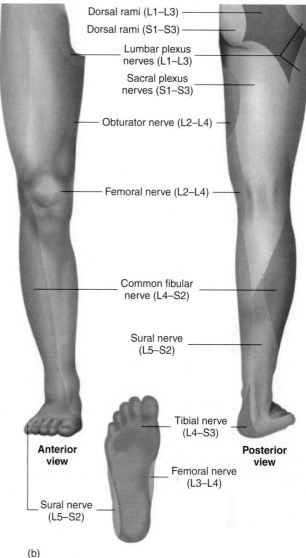

Dorsal rami (L1–L3)

Dorsal rami (S1–S3)

Lumbar plexus
nerves (L1–L3)

Sacral plexus
nerves (S1–S3)

Obturator nerve (L2–L4)

Femoral nerve (L2–L4)

Common fibular
nerve (L4–S2)

Sural nerve
(L5–S2)

Tibial nerve
(L4–S3)

**Anterior
view**

**Posterior
view**

Femoral nerve
(L3–L4)

Sural nerve
(L5–S2)

(b)

Figure 11.16 Cutaneous Distribution of Major Plexus Nerves

(a) Cutaneous distribution of cervical and brachial plexus nerves to the shoulder and upper limb. (b) Cutaneous distribution of lumbar and sacral plexus nerves to the hip and lower limb.

brachial plexus to supply the upper limb are the axillary, radial, musculocutaneous, ulnar, and median nerves (see table 11.1). Smaller nerves arising from the brachial plexus supply muscles and skin of the trunk and upper limb.

 Brachial Anesthesia

The entire upper limb can be anesthetized by injecting an anesthetic near the brachial plexus. This is called **brachial anesthesia.** The anesthetic can be injected between the neck and the shoulder posterior to the clavicle.

Lumbar and Sacral Plexuses

The **lumbar plexus** originates from the ventral rami of spinal nerves L1–L4 and the **sacral plexus** from L4 to S4. Because of their close, overlapping relationship and their similar distribution, however, the two plexuses often are considered together as a single **lumbosacral plexus** (L1–S4) (see figure 11.11). Four major nerves exit the lumbosacral plexus and enter the lower limb: the **obturator, femoral, tibial,** and **common fibular (peroneal)** (per-ō-nē′ăl) **nerves** (see table 11.1). Other lumbosacral nerves supply muscles of the lower back, hip, and lower abdomen and the skin of the hip and thigh (see figure 11.16).

The tibial and common fibular nerves originate from spinal segments L4–S3 and are bound together within a connective tissue sheath for the length of the thigh. The two nerves bound together are referred to jointly as the **sciatic** (sī-at′ik, hip) **nerve.** The sciatic nerve is by far the largest peripheral nerve in the body. It passes through the greater sciatic notch in the coxal bone and descends in the posterior thigh to the back of the knee, where the tibial and common fibular nerves separate from each other. The tibial nerve then supplies the posterior compartment of the leg and the common fibular supplies the anterior and lateral compartments. Branches of the two nerves combine to form the **sural nerve,** which supplies the skin on the posterior and lateral leg (see figure 11.16).

Radial Nerve Damage

The radial nerve lies near the humerus in the axilla. When crutches are used improperly, the crutch is pushed tightly into the axilla. This can damage the radial nerve by compressing it against the humerus, resulting in **crutch paralysis.** In this disorder, muscles innervated by the radial nerve lose their function. The major symptom of radial nerve damage is **wrist drop,** an inability to extend the wrist when the pronated forearm is held parallel to the ground. Consequently, the wrist drops into a flexed position.

The radial nerve can be permanently damaged by a fracture of the humerus in the proximal part of the arm. A sharp edge of the broken bone may cut the nerve, resulting in permanent paralysis unless the nerve is surgically repaired. Because of potential damage to the radial nerve, a broken humerus should be treated very carefully.

Ulnar Nerve Damage

The ulnar nerve is the most easily damaged of all the peripheral nerves, but such damage is almost always temporary. Slight damage to the ulnar nerve may occur where it passes posterior to the medial epicondyle of the humerus. The nerve can be felt just below the skin at this point, and, if this region of the elbow is banged against a hard object, temporary ulnar nerve damage may occur. This damage results in painful tingling sensations radiating down the ulnar side of the forearm and hand. Because of this sensation, this area of the elbow is often called the **funny bone** or **crazy bone.**

Median Nerve Damage

Damage to the median nerve occurs most commonly where it enters the wrist through the **carpal tunnel.** This tunnel is created by the concave organization of the carpal bones and the retinaculum on the anterior surface of the wrist. None of the connective tissue components of the carpal tunnel expand readily. The tendons passing through the carpal tunnel may become inflamed and enlarged as a result of repetitive movements. This inflammation can produce pressure within the carpal tunnel, thereby compressing the median nerve and resulting in numbness, tingling, and pain in the fingers. The thenar muscles, innervated by the median nerve, have reduced function, resulting in weakness in thumb flexion and opposition. This condition is referred to as **carpal tunnel syndrome.** Carpal tunnel syndrome is common among people who perform repetitive movements of the wrists and fingers, such as keyboard operators. Surgery is often required to relieve the pressure.

People attempting suicide by cutting the wrists commonly cut the median nerve proximal to the carpal tunnel.

Sciatic Nerve Damage

Sciatica produces severe spasms of throbbing or stabbing pain that radiates down the back of the thigh and leg. It is caused by inflammation or damage to the sciatic nerve. The most common cause is a herniated lumbar disk, resulting in pressure on the spinal nerves contributing to the lumbar plexus. Sciatica may also be produced by pressure from the uterus during pregnancy. Other causes of sciatica are nerve damage resulting from trauma or an improperly administered injection, mechanical stretching of the nerve during exertion, vitamin deficiency, and metabolic disorders, such as gout or diabetes.

If a person sits on a hard surface for a considerable time, the sciatic nerve may be compressed against the ischial portion of the coxal bone. When the person stands up, a tingling sensation, described as "pins and needles," can be felt throughout the lower limb, and the limb is said to have "gone to sleep."

Coccygeal Plexus

The **coccygeal** (kok-sij′ē-ăl) **plexus** is a very small plexus formed from the ventral rami of spinal nerve S5 and the coccygeal nerve (Co). This small plexus supplies the muscles of the pelvic floor and the skin over the coccyx. The dorsal rami of the coccygeal nerves also innervate some skin over the coccyx. ▼

23. Name the five major spinal nerve plexuses and the spinal nerves associated with each one.
24. Name the major nerves emerging from the cervical, brachial, and lumbosacral plexuses.
25. Describe the contribution of the tibial and common fibular nerves to the sciatic and sural nerves.
26. What structures are innervated by the coccygeal plexus?

Brain

The brain is that part of the CNS contained within the cranial cavity. It is the control center for many of the body's functions. The brain is much like a complex central computer but with additional functions that no computer can as yet match. Indeed, one goal in computer technology is to make computers that can function more like the human brain. The brain consists of the brainstem, the cerebellum, the diencephalon, and the cerebrum (figure 11.17). ▼

27. Name the parts of the brain.

Brainstem

The **brainstem** consists of the medulla oblongata, pons, and midbrain (see figure 11.17). It connects the spinal cord and cerebellum to the remainder of the brain, and 10 of the 12 pairs of cranial nerves arise from it. In general, the posterior part of the brainstem contains ascending tracts from the spinal cord, cerebellum, and cranial nerves, whereas the anterior part of the brainstem contains descending tracts involved with motor control. The brainstem contains several nuclei involved in vital body functions, such as the control of heart rate, blood pressure, and breathing. Damage to small areas of the brainstem can cause death, whereas damage to relatively large areas of the cerebrum or cerebellum often do not cause death. ▼

28. Name the parts of the brainstem.

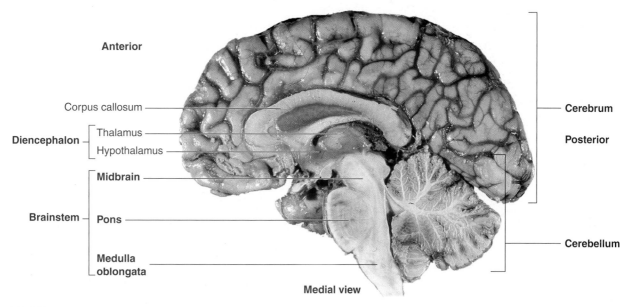

Anterior

Corpus callosum

Diencephalon — Thalamus
Hypothalamus

Midbrain

Brainstem — Pons

Medulla oblongata

Cerebrum

Posterior

Cerebellum

Medial view

Figure 11.17 Regions of the Right Half of the Brain

Right half of the brain as seen in a median section.

Medulla Oblongata

The **medulla oblongata** (ob′long-gă′tă, *oblongus,* rather long) is the most inferior portion of the brainstem (figure 11.18) and is continuous with the spinal cord. It extends from the level of the foramen magnum to the pons. In addition to ascending and descending nerve tracts, the medulla oblongata contains discrete nuclei with specific functions, such as the regulation of heart rate, blood vessel diameter, breathing, swallowing, vomiting, coughing, sneezing, hiccuping, balance, and coordination.

On the anterior surface, two prominent enlargements called **pyramids** extend the length of the medulla oblongata. They are called pyramids because they are broader near the pons and taper toward the spinal cord (see figure 11.18a). The pyramids are descending tracts involved in the conscious control of skeletal muscles. Near their inferior ends, most of the fibers of the descending tracts cross to the opposite side, or **decussate** (dē′kŭ-sāt, dē-kŭs′āt, to form an X, as in the Roman numeral X). This decussation accounts, in part, for the fact that each half of the brain controls the opposite half of the body.

Pons

The **pons** is the part of the brainstem just superior to the medulla oblongata (see figure 11.18). It contains ascending and descending nerve tracts, as well as several nuclei. Some of the nuclei in the pons relay information between the cerebrum and the cerebellum. The term *pons* means bridge, and it describes both the structure and the function of the pons. Not only is the pons a functional bridge between the cerebrum and cerebellum, but on the anterior surface it resembles an arched footbridge (see figure 11.17). Several nuclei of the medulla oblongata extend into the lower part of the pons, so that functions such as breathing, swallowing, and balance are controlled in the lower pons, as well as in the medulla oblongata. Other nuclei in the pons control functions such as chewing and salivation.

Midbrain

The **midbrain,** just superior to the pons, is the smallest region of the brainstem (see figure 11.18). The posterior part of the midbrain consists of four mounds collectively called the **corpora** (kōr′pōr-ă, bodies) **quadrigemina** (kwah′dri-jem′i-nă, four twins). Each mound is called a **colliculus** (ko-lik′ū-lŭs, hill). The two **inferior colliculi** are major relay centers for the auditory nerve pathways in the CNS. The two **superior colliculi** are involved in visual reflexes. Turning of the head toward a tap on the shoulder, a sudden loud noise, or a bright flash of light are reflexes controlled in the superior colliculi. The midbrain contains nuclei involved in the coordination of eye movements and in the control of pupil diameter and lens shape. The midbrain also has nuclei involved with the regulation of general body movements that are named according to their color. Two nuclei, each called a **substantia nigra** (sŭb-stan′shē-ă nī′gră, black substance), are dark gray or black in color because their neurons contain melanin granules. The paired **red nuclei** are pinkish in color as a result of their abundant blood supply. The rest of the midbrain consists largely of ascending tracts from the spinal cord to the cerebrum and descending tracts from the cerebrum to the spinal cord or cerebellum. The **cerebral peduncles** (pe-dŭng′klz, pē′dŭng-klz, the foot of a column) are descending motor tracts forming the anterior part of the midbrain.

Reticular Formation

Scattered throughout the brainstem is a group of nuclei collectively called the **reticular formation.** The reticular formation plays important regulatory functions in the brain. It is particularly involved in regulating cyclical motor functions, such as breathing, walking, and chewing. The reticular formation is a major component of the **reticular activating system,** which plays an important role in arousing and maintaining consciousness

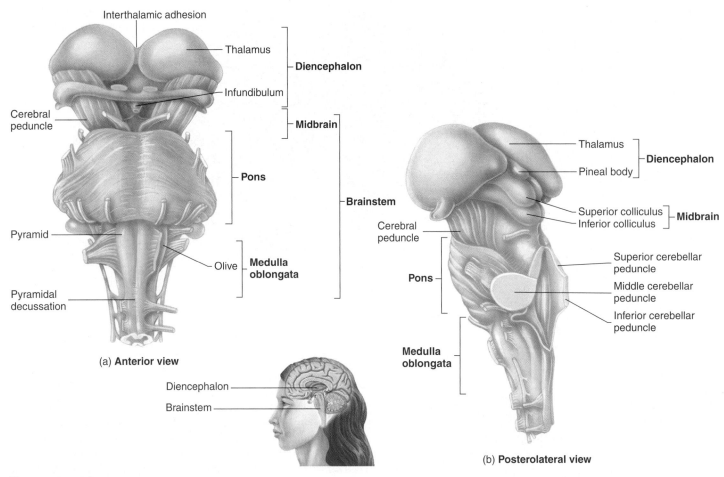

Figure 11.18   Brainstem and Diencephalon

and in regulating the sleep–wake cycle. Stimuli such as an alarm clock ringing, sudden bright lights, smelling salts, or cold water splashed on the face can arouse consciousness. Conversely, the removal of visual or auditory stimuli may lead to drowsiness or sleep. General anesthetics function by suppressing the reticular activating system. Damage to cells of the reticular formation can result in coma. ▼

> *28. Describe the major components of the medulla oblongata, pons, midbrain, and reticular formation. What are the general functions of each region?*

Cerebellum

The **cerebellum** (ser-e-bel′ŭm) is attached to the posterior brainstem (see figure 11.17). The term *cerebellum* means little brain. In many ways, the cerebellum is a smaller version of the main part of the human brain, the cerebrum. The cerebellum is attached to the brainstem by three large tracts called cerebellar peduncles (see figure 11.18b). The cerebellum connects with the midbrain by the **superior cerebellar peduncles,** the pons by the **middle cerebellar peduncles,** and the medulla oblongata by the **inferior cerebellar peduncles.**

The gray matter of the cerebellum consists of an outer cortex and nuclei deep within the cerebellum. The cerebellar cortex has ridges called **folia** (fō′lē-ă, leafs) (figure 11.19a). The white matter of the cerebellum consists of tracts collectively called the **arbor vitae** (ar′bōr vī′te, tree of life) because they resemble a tree when viewed in a median section. Many of the axons of the arbor vitae are the same axons forming the cerebellar peduncles. Thus, the arbor vitae connect the gray matter of the cerebellum to the rest of the CNS. For example, the terminal branches of the arbor vitae (tree of life) connect to the folia (leaves). The white matter of the cerebellum also connects the cerebellar cortex and cerebellar nuclei together.

The cerebellum consists of three parts: a small inferior part, the **flocculonodular** (flok′ū-lō-nod′ū-lăr, floccular, a tuft of wool) **lobe;** a narrow central **vermis** (worm-shaped); and two large **lateral hemispheres** (figure 11.19b and c).

The flocculonodular lobe, the simplest part of the cerebellum, helps control balance and eye movements. The vermis and medial portions of the lateral hemispheres are involved in the control of posture, locomotion, and fine motor coordination, thereby producing smooth, flowing movements. The major portions of the lateral hemispheres of the cerebellum function in concert with the

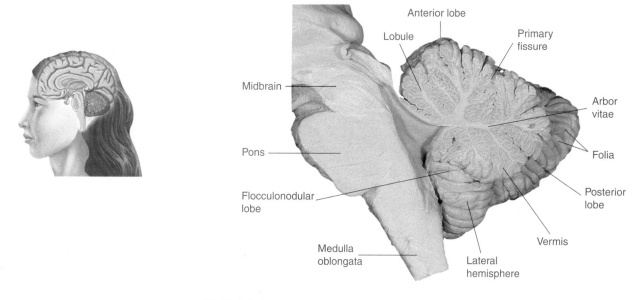

(a) **Medial view**

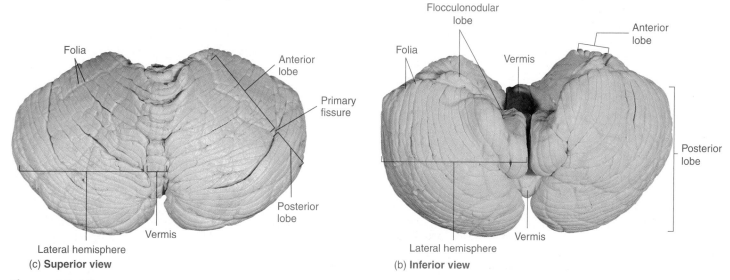

(c) **Superior view**

(b) **Inferior view**

Figure 11.19   Cerebellum

(*a*) Right half of the cerebellum and brainstem as seen in a median section. (*b*) Inferior view of the cerebellum. (*c*) Superior view of the cerebellum.

frontal lobes of the cerebral cortex in planning, practicing, and learning complex movements.

Each lateral hemisphere is divided by a **primary fissure** into an **anterior lobe** and a **posterior lobe.** The lobes are subdivided into **lobules,** which contain the folia. ▼

29. *Describe the connections of the cerebellum to the brainstem.*
30. *What parts of the cerebellum are gray and white matter?*
31. *What is the function of the arbor vitae?*
32. *Name the three major regions of the cerebellum and describe their major functions.*

Diencephalon

The **diencephalon** (dī-en-sef′ă-lon) is the part of the brain located between the brainstem and the cerebrum (see figure 11.17). Its main components are the thalamus, subthalamus, epithalamus, and hypothalamus.

Thalamus

The **thalamus** (thal′ă-mŭs,) (figure 11.20*a* and *b*) is by far the largest part of the diencephalon, constituting about four-fifths of its weight. It is shaped somewhat like a yo-yo, with two large,

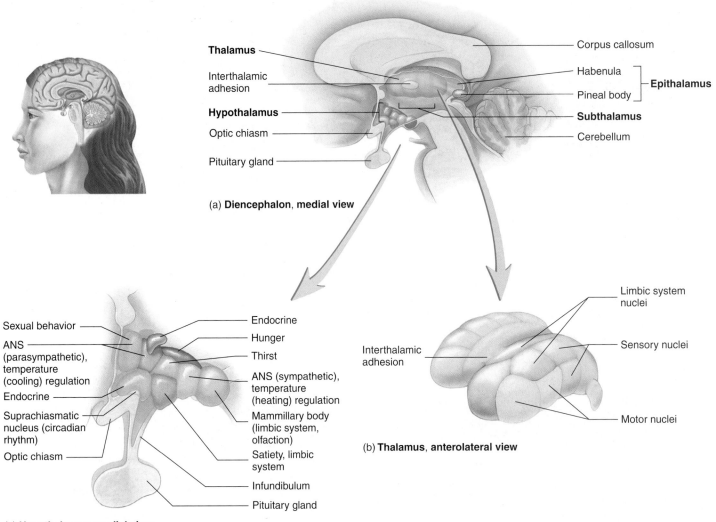

Figure 11.20   **Diencephalon**

(*a*) General overview of the right half of the diencephalon as seen in a median section. (*b*) Thalamus showing the nuclei. (*c*) Hypothalamus showing the nuclei and right half of the pituitary gland.

lateral portions connected in the center by a small stalk called the **interthalamic adhesion,** or **intermediate mass.** The thalamus is a collection of nuclei, each performing different functions. Some of the nuclei are involved with sensory input. All sensory input that reaches the cerebrum, except for the sense of smell, passes through the thalamus. Ascending axons carrying sensory information project to the thalamus, where they synapse with thalamic neurons in the nuclei. Thalamic neurons, in turn, send their axons to the cerebral cortex, where sensory input is localized and most awareness of sensory input occurs. For this reason, the thalamus is considered the **sensory relay center** of the brain. Auditory, visual, and other sensory input, such as pain, temperature, and touch, pass through the thalamus. The role of the thalamus in sensory input is considered in detail in chapter 12.

Some of the thalamic nuclei are involved with controlling skeletal muscles. They connect to, and interact with, other parts of

the brain that control skeletal muscle contraction, especially the motor areas of the cerebral cortex, the cerebellum, and the basal nuclei (see "Basal Nuclei," p. 321). The regulation of skeletal muscle contraction is considered in detail in chapter 12.

Some of the thalamic nuclei are involved with the limbic system and emotions. They connect different parts of the limbic system and influence mood and actions associated with strong emotions, such as fear and rage. They also register an unlocalized, uncomfortable perception of pain.

Subthalamus

The **subthalamus** is a small area immediately inferior to the thalamus (see figure 11.20*a*) that contains several ascending and descending tracts and the **subthalamic nuclei.** The subthalamic nuclei are associated with the basal nuclei and are involved in controlling motor functions.

Epithalamus

The **epithalamus** is a small area superior and posterior to the thalamus (see figure 11.20*a*). It consists of habenular nuclei and the pineal body. The **habenular** (hă-ben′ū-lăr) **nuclei** are influenced by the sense of smell and are involved in emotional and visceral responses to odors. The **pineal** (pin′ē-ăl) **body** is shaped somewhat like a pinecone, from which the name *pineal* is derived. The pineal body is an endocrine gland that appears to play a role in controlling the onset of puberty (see chapter 15). It also may influence the sleep–wake cycle and other biorhythms.

 Brain Sand in the Pineal

In about 75% of adults, the pineal body contains granules of calcium and magnesium salts called "brain sand." These granules can be seen on radiographs and are useful as landmarks in determining whether or not the pineal body has been displaced by a pathological enlargement of a part of the brain, such as a tumor or a hematoma.

Hypothalamus

The **hypothalamus** is the most inferior part of the diencephalon. Although a small portion of the brain, it connects to many other parts of the brain and spinal cord and is involved with autonomic, endocrine, emotional (limbic system), and basic body functions. The hypothalamus is a collection of nuclei. For simplicity, a few major functions of the nuclei are shown in figure 11.20*c*. The nuclei are actually more complicated, and a given nucleus can be involved with more than one function.

1. *Autonomic nervous system (ANS).* The hypothalamus is a major integrating center for controlling the autonomic nervous system, helping control heart rate, blood vessel diameter, urine release from the urinary bladder, and the movement of food through the digestive tract (see chapter 13).
2. *Endocrine system.* The hypothalamus is part of and helps regulate the endocrine system. A funnel-shaped stalk, the **infundibulum** (in-fŭn-dib′ū-lŭm, a funnel), extends from the floor of the hypothalamus to the pituitary gland, an important endocrine gland. Through the infundibulum, the hypothalamus regulates the pituitary gland's secretion of hormones, which influence functions as diverse as metabolism, reproduction, responses to stressful stimuli, and urine production (see chapter 15).
3. *Limbic system.* The hypothalamus is part of the limbic system and functions with other parts of the brain to affect mood, motivation, and emotions. Sexual behavior and pleasure, feeling relaxed, rage, and fear are related to hypothalamic functions. The hypothalamus is directly involved in stress-related and psychosomatic illnesses. The **mammillary** (mam′i-lār-ē, *mammilla,* nipple) **bodies** are collections of nuclei forming externally visible swellings on the posterior portion of the hypothalamus. They are involved in emotional responses to odors, olfactory reflexes, and memory.
4. *Basic body functions.* The hypothalamus is involved with many basic body functions. It plays a central role in the control of body temperature, activating sweat glands for cooling and shivering for heating the body. Hypothalamic nuclei are involved in the control of thirst, hunger, and sexual arousal. The **suprachiasmatic nucleus,** located above the optic chiasm (see chapter 14), functions as a "master clock" for the timing of circadian (ser-kā′dē-ăn, *circa,* about, *dies,* day) rhythms. Many events in our bodies wax and wane with a 24-hour cycle. For example, body temperature changes about a degree every day, increasing to a high in the late afternoon and low in the early morning. The sleep–wake cycle, feeding behavior, urinary output, and hormone secretion are other examples of circadian rhythms. The suprachiasmatic nucleus has connections to the retina of the eye and information about day length is used to set the clock. If a person is kept in constant light or darkness, the clock is not reset and a cycle is slightly more than 24 hours. Disruption of circadian rhythms has been related to jet lag, insomnia, fatigue, disorientation, and bipolar disorder. ▼

33. *Name the parts of the diencephalon.*
34. *Describe the structure of the thalamus. What are the general functions of the thalamic nuclei?*
35. *Where are the subthalamus and epithalamus located? What are their functions?*
36. *List four general functions of the hypothalamus and give examples of each.*
37. *What are the functions of the mammillary bodies and the suprachiasmatic nucleus?*

Cerebrum

The cerebrum (figure 11.21) is the part of the brain that most people think of when the term *brain* is mentioned. The cerebrum accounts for the largest portion of total brain weight, which is about 1200 g in females and 1400 g in males. Brain size is related to body size; larger brains are associated with larger bodies, not with greater intelligence.

The cerebrum is divided into left and right hemispheres by a **longitudinal fissure** (see figure 11.21*a*). The most conspicuous features on the surface of each hemisphere are numerous folds called **gyri** (jī′rī, sing. *gyrus,* circle), which greatly increase the surface area of the cortex. The grooves between the gyri are called **sulci** (sŭl′sī, sing. *sulcus,* a furrow or ditch). The **central sulcus,** which extends across the lateral surface of the cerebrum from superior to inferior, is located about midway along the length of the brain. The general pattern of the gyri is similar in all normal human brains, but some variation exists between individuals and even between the two hemispheres of the same cerebrum.

Each cerebral hemisphere is divided into lobes, which are named for the skull bones overlying each one (see figure 11.21). The **frontal lobe** is important in voluntary motor function, motivation, aggression, the sense of smell, and mood. The **parietal**

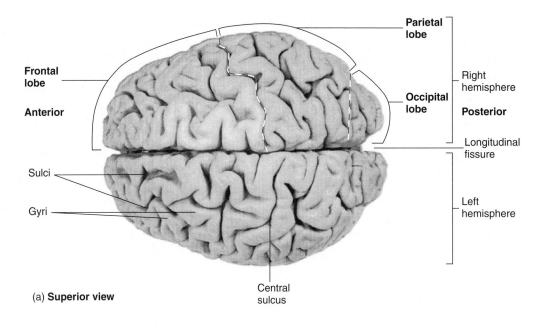

(a) **Superior view**

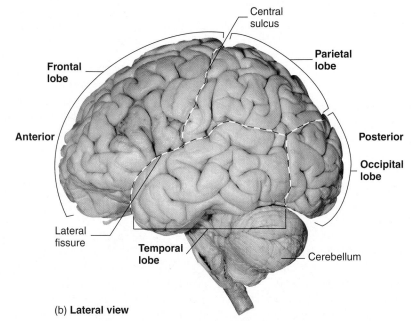

(b) **Lateral view**

Figure 11.21 Brain

(*a*) Superior view. (*b*) Lateral view of the left cerebral hemisphere.

lobe is the major center for the reception and evaluation of most sensory information, such as touch, pain, temperature, balance, and taste. The frontal and parietal lobes are separated by the central sulcus. The **occipital lobe** functions in the reception and integration of visual input and is not distinctly separate from the other lobes. The **temporal lobe** receives and evaluates input for smell and hearing and plays an important role in memory. Its anterior and inferior portions are referred to as the "psychic cortex," and they are associated with such brain functions as abstract thought and judgment. Most of the temporal lobe is

separated from the rest of the cerebrum by a **lateral fissure.** Deep within the lateral fissure is the **insula** (in′soo-lă, island), often referred to as a fifth lobe, which receives input for taste.

The gray matter on the outer surface of the cerebrum is the **cortex,** and clusters of gray matter deep inside the brain are nuclei. The white matter of the brain between the cortex and nuclei is the **cerebral medulla.** This term should not be confused with the medulla oblongata; *medulla* is a general term meaning the center of a structure. The cerebral medulla consists of tracts that connect areas of the cerebral cortex to each other or to other parts of the CNS

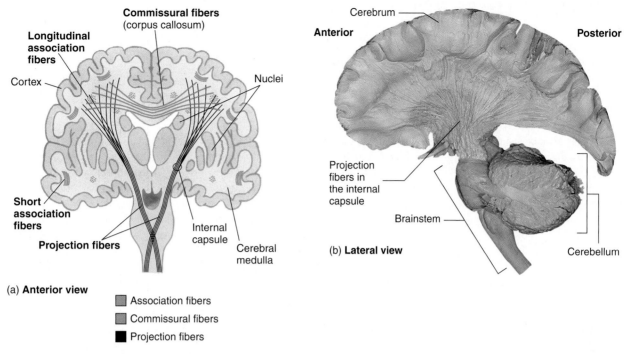

Figure 11.22　Cerebral Medullary Tracts

(*a*) Frontal section of the brain showing commissural, association, and projection fibers. (*b*) Photograph of the left cerebral hemisphere from a lateral view with the cortex and association fibers removed to reveal the projection fibers of the internal capsule deep within the brain.

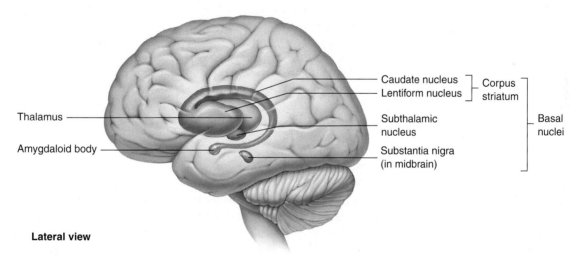

Figure 11.23　Basal Nuclei of the Left Hemisphere

"Transparent 3-D" drawing of the basal nuclei inside the left hemisphere.

(figure 11.22). The fibers in these tracts fall into three main categories. (1) **Association fibers** connect areas of the cerebral cortex within the same hemisphere. (2) **Commissural fibers** connect one cerebral hemisphere to the other. The largest bundle of commissural fibers connecting the two cerebral hemispheres is the **corpus callosum** (kōr′pŭs ka-lō′sŭm) (see figures 11.17 and 11.22). (3) **Projection fibers** are between the cerebrum and other parts of the brain and spinal cord. The projection fibers form the **internal capsule.** ▼

37. *Define gyri and sulci. What structures do the longitudinal fissure, central sulcus, and lateral fissure separate?*

38. *Define cerebral cortex and cerebral medulla.*
39. *Name the five lobes of the cerebrum, and describe their locations and functions.*
40. *List three categories of tracts in the cerebral medulla.*

Basal Nuclei

The **basal nuclei** are a group of functionally related nuclei located bilaterally in the inferior cerebrum, diencephalon, and midbrain (figure 11.23). These nuclei are involved in the control of motor functions (see chapter 12). The nuclei in the cerebrum are collectively

Head injuries are classified as open or closed. In an **open** injury the cranial cavity contents are exposed to the outside, whereas in a **closed** injury the cranial cavity remains intact. Closed injuries, which are more common than open injuries, involve the head striking a hard surface or an object striking the head. Such injuries may result in brain trauma. The main trauma may be **coup** (kŭ), occurring at the site of impact, or **contrecoup** (kon'tra-kŭ), occurring on the side of the brain opposite the impact. Contrecoup injuries result from movement of the brain within the skull.

The most common type of traumatic brain injury (75%–90%) is a **concussion,** which is an immediate, but transient, impairment of neural function, such as a loss of consciousness or blurred vision. In some cases, **postconcussion syndrome** occurs a short time after the injury. The syndrome includes increased muscle tension or migraine headaches, reduced alcohol tolerance, difficulty in learning new things, reduction in creativity, changes in motivation, fatigue, and personality changes. The symptoms may be gone in a month or may persist for as long as a year.

Traumatic brain injury can be diffuse or focal. **Diffuse brain injury** usually results from shaking, such as when a person shakes a child or when a person is thrown about in an automobile accident. As the name suggests, such injury is not localized to one place in the brain but involves damage to many small vessels and nerves, especially around the brainstem. **Focal brain injury** is a localized brain injury. It often results from direct impact injury to the brain or meninges. Focal trauma can produce contusions or hematomas. **Contusions,** or bruisings, are usually superficial and involve only the gyri.

A **hematoma** is a localized mass of blood released from blood vessels but confined within an organ or a space. An accumulated mass of blood can apply pressure to the brain, causing damage to brain tissue. Blood is toxic to brain tissue. Hematomas of the brain are classified according to where the bleeding occurs. **Epidural hematomas** result from an accumulation of blood in the epidural space and occur in about 1%–2% of major head injuries. **Subdural hematomas** result from an accumulation of blood in the subdural space and occur in

10%–20% of major head injuries. They most commonly involve tears of the cortical veins or dural venous sinuses, occur in the superior portion of the cranial cavity, and appear within hours of the head injury. **Chronic subdural hematomas,** which involve slow bleeding over weeks to months, are common in elderly people and in people who abuse alcohol. **Intracerebral hematomas** result from an accumulation of blood within the brain and occur in about 2%–3% of major head injuries. Intracerebral hematomas involve the damage of small vessels within the brain and are often associated with contusions.

A **stroke** is the sudden development of neurological defects due to an interruption of blood delivery to brain tissue or to bleeding in the brain. **Ischemic stroke** results from the blockage of blood vessels by a blood clot, and **hemorrhagic stroke** results from bleeding and hematoma formation. An **aneurysm** (an'ū-rizm) is a dilation, or ballooning, of an artery. The arteries around the brain are common sites for aneurysms, and hypertension can cause one of these "balloons" to burst or leak, casing a hemorrhagic stroke.

---

called the **corpus striatum** (kōr'pŭs strī-ā'tŭm, striped body) and include the **caudate** (kaw'dāt, having a tail) **nucleus** and **lentiform** (len'ti-fōrm, lens-shaped) **nucleus.** The **subthalamic nucleus** is located in the diencephalon and the **substantia nigra** is located in the midbrain. They function in conjunction with the caudate and lentiform nuclei in the control of movement. ▼

41. List the basal nuclei and state their general function.

Limbic System

Parts of the cerebrum and diencephalon are grouped together under the title **limbic** (lim'bik) **system** (figure 11.24). The limbic system plays a central role in basic survival functions, such as memory, reproduction, and nutrition. It is also involved in the emotional interpretation of sensory input and emotions in general. *Limbus* means border, and the term *limbic* refers to deep portions of the cerebrum that form a ring around the diencephalon. Structurally, the limbic system consists of (1) certain cerebral cortical areas, including the **cingulate** (sin'gū-lāt, to surround) **gyrus,** located along the inner surface of the longitudinal fissure just

above the corpus callosum, and the **parahippocampal gyrus,** located on the medial side of the temporal lobe; (2) various nuclei, such as anterior nuclei of the thalamus, the habenula in the epithalamus, and the **dentate nucleus** of the **hippocampus;** (3) the **amygdaloid body,** or **amygdala;** (4) the hypothalamus, especially the mammillary bodies; (5) the **olfactory cortex;** and (6) tracts connecting the various cortical areas and nuclei, such as the **fornix,** which connects the hippocampus to the thalamus and mammillary bodies. ▼

42. List the parts of the limbic system. What are the functions of the limbic system?

Meninges, Ventricles, and Cerebrospinal Fluid

Meninges

Three connective tissue membranes, the **meninges** (mĕ-nin'jēz), surround and protect the spinal cord (see figure 11.2) and the

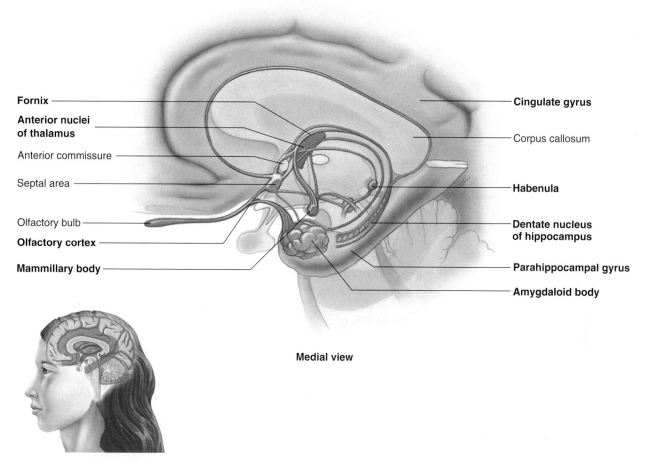

Fornix

Anterior nuclei
of thalamus

Anterior commissure

Septal area

Olfactory bulb

Olfactory cortex

Mammillary body

Cingulate gyrus

Corpus callosum

Habenula

Dentate nucleus
of hippocampus

Parahippocampal gyrus

Amygdaloid body

Medial view

Figure 11.24   Limbic System

Limbic system and associated structures as seen in a median section.

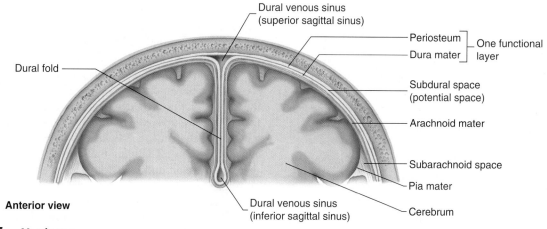

Dural venous sinus
(superior sagittal sinus)

Periosteum ⎤ One functional
Dura mater ⎦ layer

Dural fold

Subdural space
(potential space)

Arachnoid mater

Subarachnoid space

Pia mater

Dural venous sinus
(inferior sagittal sinus)

Cerebrum

Anterior view

Figure 11.25   Meninges

Frontal section of the head showing the meninges surrounding the brain.

brain (figure 11.25). The most superficial and thickest membrane is the **dura mater** (doo′ră mā′ter, tough mother). The dura mater is composed of dense irregular connective tissue. Within the vertebral canal, the dura mater is distinctly separate from the vertebrae, forming an **epidural space** (see figure 11.2). There is normally no epidural space within the cranial cavity because the dura mater is tightly attached to the cranial bones. An epidural space can develop, however, if an injury results in the accumulation of blood between the bone and dura mater. The dura mater within the cranial cavity consists of two layers. The outer layer is called the **periosteal dura,** and it is the inner periosteum of the cranial bones. The inner layer is the **meningeal dura.** The meningeal dura is continuous with the dura of the spinal cord. The meningeal dura is separated from the periosteal dura in several regions to form structures called dural folds and dural venous sinuses.

Dural folds are tough connective tissue partitions that extend into the major brain fissures. The dura mater and dural folds help hold the brain in place within the skull and keep it from moving around too freely. The largest dural fold lies in the longitudinal fissure and is anchored anteriorly to the crista galli of the ethmoid bone.

Dural venous sinuses are spaces that form where the two layers of the dura mater are separated from each other. The largest of these sinuses is the **superior sagittal sinus,** which runs along the median plane (see figure 11.25). All veins draining blood from the brain empty into dural venous sinuses. The dural venous sinuses subsequently drain into the internal jugular veins, which are the major veins that exit the cranial cavity to carry blood back to the heart (see chapter 19).

The next meningeal membrane is a very thin, wispy **arachnoid** (ă-rak′noyd, spiderlike—i.e., cobwebs) **mater.** The **subdural space** is between the dura mater and the arachnoid mater. It is a very small space containing a small amount of serous fluid.

The third meningeal layer, the **pia** (pī′ă, pē′ă, affectionate) **mater,** is bound very tightly to the surface of the brain. Between the arachnoid mater and the pia mater is the **subarachnoid space,** which contains weblike strands of arachnoid mater, the blood vessels supplying the brain, and cerebrospinal fluid. ▼

43. *Name the two layers of the dura mater. How are they related to the skull and the dura mater around the spinal cord?*
44. *What is a dural fold? How are dural venous sinuses formed?*
45. *Describe and list the contents of the dural sinuses, subdural space, and subarachnoid space.*

CASE STUDY | Traumatic Brain Injury

The body of an 80-year-old female was being examined by the hospital pathologist to determine the cause of death. Her medical records indicated no significant history of cardiovascular disease, stroke, Alzheimer disease, or cancer. She had been taken to the emergency room after having been found lying in her bathtub and not breathing. Bleeding over the occipital region of her scalp led the pathologist to hypothesize that she had slipped and fallen while getting into the bath and had hit the back of her head on the edge of the tub. Because the pathologist suspected a traumatic

brain injury as the cause of death, he focused most of his attention on the contents of the cranial cavity. The pathologist noted superficial bruising and bleeding of the scalp in the occipital region of the woman's head and then proceeded to open the cranial cavity. Inside the cranial cavity, the pathologist noted a large subdural hematoma above the right frontal lobe of the woman's brain. In addition, he noted that the brain had shifted as a result of the extensive bleeding such that the medulla oblongata had been pushed inferiorly through the foramen magnum into the vertebral canal. After completing a thorough inspection of the remainder of the woman's body, the pathologist determined the cause of death to be a subdural hematoma caused by a traumatic brain injury.

P R E D I C T 6

a. Explain why the subdural hematoma was found in the frontal region of the brain when the blow to the head occurred over the occipital region.
b. How did inferior movement of the medulla oblongata into the vertebral canal contribute to the woman's death?

Ventricles

Each cerebral hemisphere contains a relatively large cavity, the **lateral ventricle** (figure 11.26). The lateral ventricles meet at the midline just inferior to the corpus callosum. Where they come together, the wall of each ventricle is thin, forming a partition called the **septum pellucidum** (sep′tă pe-loo′sid-ŭm, pl. *septa pellucida,* translucent walls). Usually, the partitions are fused together and are referred to as the septa pellucida.

A smaller midline cavity, the **third ventricle,** is located in the center of the diencephalon between the two halves of the thalamus. If the thalamus is thought of as resembling a yo-yo, each lobe of the thalamus is half of the yo-yo and the interthalamic adhesion is the stick holding them together. The third ventricle is then the space around the stick where the yo-yo's string is wound up (see figure 11.26).

The two lateral ventricles connect to the third ventricle through two **interventricular foramina.** The lateral ventricles can be thought of as the first and second ventricles in the numbering scheme, but they are referred to as the lateral ventricles and are not numbered. The **fourth ventricle** is in the inferior part of the pons and the superior part of the medulla oblongata at the base of the cerebellum. The third ventricle connects with the fourth ventricle through a narrow canal, the **cerebral aqueduct,** which passes through the midbrain. The fourth ventricle is continuous with the central canal of the spinal cord, which extends nearly the full length of the cord. The fourth ventricle

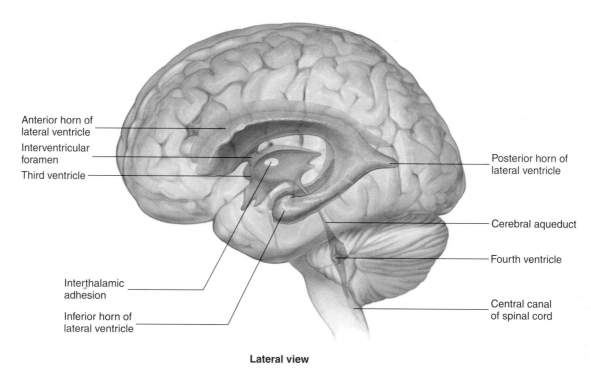

Figure 11.26 Ventricles of the Brain

"Transparent 3-D" drawing of the ventricles inside the brain viewed from the left.

connects with the subarachnoid space through three openings—a **median aperture** in the roof of the fourth ventricle and two **lateral apertures,** one in each lateral wall of the fourth ventricle. ▼

46. *Name the four ventricles of the brain, and describe their locations and the connections between them. What are the septa pellucida?*

47. *How is the fourth ventricle connected to the subarachnoid space?*

Cerebrospinal Fluid

Cerebrospinal (ser′ĕ-brō-spī-năl, sĕ-rē′brō-spī-năl) **fluid (CSF)** fills the ventricles, the subarachnoid space around the brain and spinal cord, and the central canal of the spinal cord. Approximately 23 mL of CSF fills the ventricles, and 117 mL fills the subarachnoid space. CSF provides a protective fluid cushion around the brain and spinal cord, protecting them from movements of the skull and vertebral column. It also provides some nutrients to CNS tissues.

CSF is produced by specialized neuroglia cells called ependymal cells (see chapter 10). Ependymal cells and associated blood vessels that produce CSF are collectively called the **choroid** (kō′royd, lacy) **plexuses** (plek′sŭs-ez). About 80%–90% of the CSF is produced by choroid plexuses within the lateral ventricles, with the remainder produced by choroid plexuses in the third and fourth ventricles.

CSF is a fluid similar to the fluid portion of blood with most of the proteins removed. CSF is formed through a variety of mechanisms. The majority of the fluid enters the ventricles by following a Na^+ concentration gradient. Ependymal cells of the choroid plexus actively transport Na^+ into the ventricles and water passively follows. Large molecules are transported by pinocytosis. The precise mechanisms for the transport of glucose and other substances into CSF remain unknown.

CSF produced from the blood in the choroid plexuses circulates through the ventricles and enters the subarachnoid space, from which it returns to the blood (figure 11.27). CSF passes from the lateral ventricles through the interventricular foramina into the third ventricle and then through the cerebral aqueduct into the fourth ventricle. Some CSF continues to flow inferiorly into the central canal of the spinal cord. The central canal, however, is a dead end, and CSF that flows into it eventually flows back into the fourth ventricle. CSF exits the fourth ventricle through the median and lateral apertures, enters the subarachnoid space, and flows throughout the subarachnoid space around the brain and spinal cord. Masses of arachnoid tissue, **arachnoid granulations,** penetrate into the dural venous sinuses, especially the superior sagittal sinus. CSF passes from the subarachnoid space into the blood of the dural venous sinuses through these granulations. From the dural venous sinuses, the blood drains into the internal jugular veins to enter the veins of the general circulation.

1. Cerebrospinal fluid (CSF) is produced by the choroid plexuses of each of the four ventricles (*inset*).

2. CSF from the lateral ventricles flows through the interventricular foramina to the third ventricle.

3. CSF flows from the third ventricle through the cerebral aqueduct to the fourth ventricle.

4. CSF exits the fourth ventricle through the lateral and median apertures and enters the subarachnoid space. Some CSF enters the central canal of the spinal cord.

5. CSF flows through the subarachnoid space to the arachnoid granulations in the superior sagittal sinus, where it enters the venous circulation (*inset*).

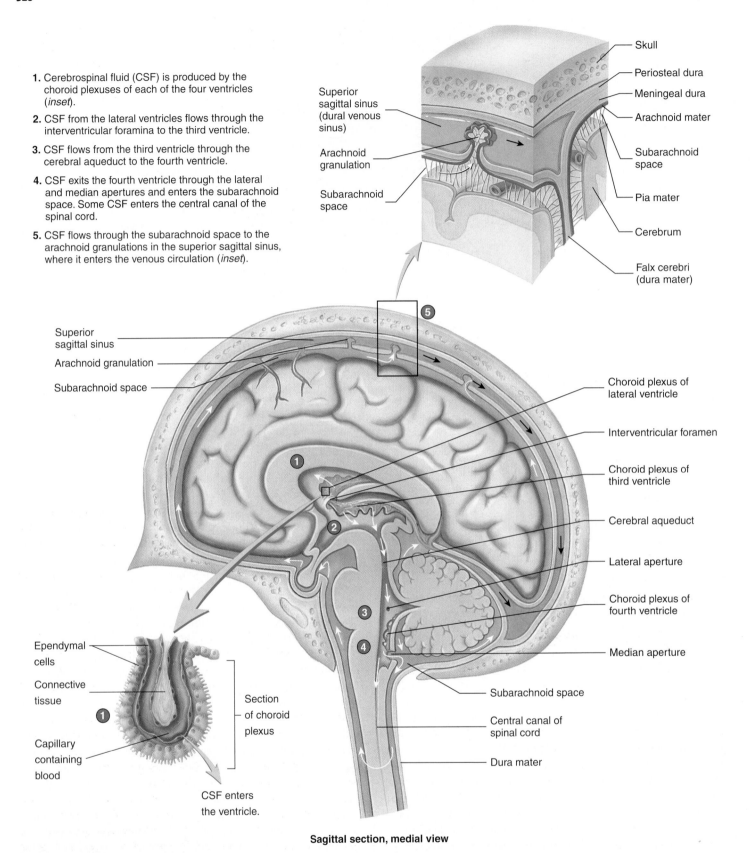

Sagittal section, medial view

Process Figure 11.27   Flow of CSF

CSF flowing through the ventricles and subarachnoid space is shown by *white arrows*. The *white arrows* going through the foramina in the wall and roof of the fourth ventricle depict the CSF entering the subarachnoid space. CSF passes back into the blood through the arachnoid granulations (*white* and *black arrows*), which penetrate the dural sinus. The *black arrows* show the direction of blood flow in the sinuses.

Hydrocephalus

CSF is produced at a rate of approximately 0.4 mL/min. If CSF returns to the blood at the same rate, the volume of CSF in the ventricles and subarachnoid space is constant. If the apertures of the fourth ventricle or the cerebral aqueduct are blocked, CSF can accumulate within the ventricles. This condition is called **internal,** or noncommunicating, **hydrocephalus,** and it results in increased CSF pressure. The production of CSF continues, even when the passages that normally allow it to exit the brain are blocked. Consequently, fluid builds inside the brain, causing pressure, which compresses the nervous tissue and dilates the ventricles. Compression of the nervous tissue usually results in irreversible brain damage. If the skull bones are not completely ossified when the hydrocephalus occurs, the pressure may also severely enlarge the head. The cerebral aqueduct may be blocked at the time of birth or may become blocked later in life because of a tumor growing in the brainstem.

Internal hydrocephalus can be successfully treated by placing a drainage tube (shunt) between the brain ventricles and abdominal cavity to eliminate the high internal pressures. There is some risk of infection being introduced into the brain through these shunts, however, and the shunts must be replaced as the person grows.

A subarachnoid hemorrhage may block the return of CSF to the circulation. If CSF accumulates in the subarachnoid space, the condition is called **external,** or communicating, **hydrocephalus.** In this condition, pressure is applied to the brain externally, compressing neural tissues and causing brain damage. ▼

48. Describe the production and circulation of CSF. Where does the CSF return to the blood?

Blood Supply to the Brain

The brain requires a tremendous amount of blood to maintain its normal functions. The brain has a very high metabolic rate and brain cells are not capable of storing high-energy molecules for any length of time. In addition, brain cells depend almost entirely on glucose as their energy source. Thus, the brain requires a constant blood supply to meet the demands of brain cells for both glucose and oxygen. Even though the brain accounts for only about 2% of the total weight of the body, it receives approximately 15%–20% of the blood pumped by the heart. Interruption of the brain's blood supply for only seconds can cause unconsciousness, and interruption of the blood supply for minutes can cause irreversible brain damage.

The major blood vessels supplying the brain are described in chapter 18. Branches of the arteries supplying the brain are located in the subarachnoid space. Smaller branches from the arteries in the subarachnoid space extend into the brain and quickly divide into capillaries. The epithelial cells of these capillaries are surrounded by the foot processes of astrocytes (see figure 10.5). The astrocytes promote the formation of tight junctions between the epithelial cells. The epithelial cells with their tight junctions form the **blood–brain barrier,** which regulates the movement of materials from the blood into the brain. Materials that would enter many tissues by passing between the epithelial cells of capillaries cannot pass through the blood–brain barrier because of the tight junctions. Most materials that enter the brain pass through the epithelial cells. Lipid-soluble substances, such as nicotine, ethanol, and heroin, can diffuse through the plasma membranes of the epithelial cells and enter the brain. Water-soluble molecules, such as amino acids and glucose, move across the plasma membranes of the epithelial cells by mediated transport (see chapter 3).

Drugs and the Blood–Brain Barrier

The permeability characteristics of the blood–brain barrier must be considered when developing drugs designed to affect the CNS. For example, Parkinson disease is caused by a lack of the neurotransmitter dopamine, which normally is produced by certain neurons of the brain. A lack of dopamine results in decreased muscle control and shaking movements. Administering dopamine is not helpful because dopamine cannot cross the blood–brain barrier. Levodopa (L-dopa), a precursor to dopamine, is administered instead because it can cross the blood–brain barrier. CNS neurons then convert levodopa to dopamine, which helps reduce the symptoms of Parkinson disease. ▼

49. What is the blood–brain barrier and what is its function?

Cranial Nerves

Names and Functions

The 12 pairs of cranial nerves arise from the brain. By convention, the cranial nerves are designated by Roman numerals (I–XII) from anterior to posterior (figure 11.28). The cranial nerves also have names. In numeric order, this mnemonic can help you remember their names: **O**n **O**ccasion **O**ur **T**rusty **T**ruck **A**cts **F**unny; **V**ery **G**ood **V**ehicle **A**ny**H**ow. The mnemonic corresponds to the **O**lfactory (I), **O**ptic (II), **O**culomotor (III), **T**rochlear (IV), **T**rigeminal (V), **A**bducent (VI), **F**acial (VII), **V**estibulocochlear (VIII), **G**lossopharyngeal (IX), **V**agus (X), **A**ccessory (XI), and **H**ypoglossal (XII) nerves.

Cranial nerve I (olfactory) is connected to the cerebrum and cranial nerve II (optic) is connected to the diencephalon. The remaining cranial nerves are connected to the brainstem. Cranial nerves III (oculomotor), IV (trochlear), and V (trigeminal) have nuclei in the midbrain; cranial nerves V (trigeminal), VI (abducent), VII (facial), VIII (vestibulocochlear), and IX (glossopharyngeal) have nuclei in the pons; and cranial nerves V (trigeminal), IX (glossopharyngeal), X (vagus), XI (accessory), and XII (hypoglossal) have nuclei in the medulla oblongata.

Table 11.2 describes the cranial nerves and their functions in detail. There are two general categories of cranial nerve function: sensory and motor. Sensory functions include the special senses, such as vision and hearing, and the more general senses, such as touch and pain in the face. The motor functions of the cranial nerves are subdivided into somatic motor and parasympathetic. Somatic motor cranial nerves innervate skeletal muscles in the head and neck. The parasympathetic division is part of the autonomic nervous system (see chapter 10). Parasympathetic cranial nerves innervate smooth muscle, cardiac muscle, and glands.

Anterior

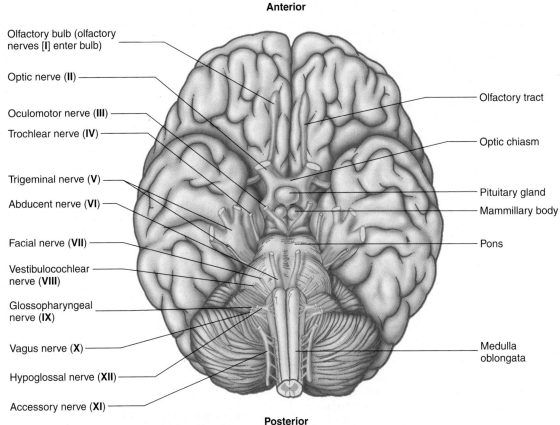

Olfactory bulb (olfactory nerves [I] enter bulb)

Optic nerve (**II**)

Oculomotor nerve (**III**)

Trochlear nerve (**IV**)

Trigeminal nerve (**V**)

Abducent nerve (**VI**)

Facial nerve (**VII**)

Vestibulocochlear nerve (**VIII**)

Glossopharyngeal nerve (**IX**)

Vagus nerve (**X**)

Hypoglossal nerve (**XII**)

Accessory nerve (**XI**)

Olfactory tract

Optic chiasm

Pituitary gland

Mammillary body

Pons

Medulla oblongata

Posterior

Figure 11.28 Inferior Surface of the Brain Showing the Origin of the Cranial Nerves

Inferior view

Cranial nerves can be grouped by function. Some cranial nerves have only one type of function, whereas others can perform two or more types of functions.

1. *Special senses.* Cranial nerve I (olfactory) for smell, II (optic) for vision, and VII (vestibulocochlear) for hearing and balance are sensory only nerves for the special senses. Cranial nerves VII (facial), IX (glossopharyngeal), and X (vagus) innervate the tongue for the sense of taste.
2. *General sensory.* Cranial nerve V (trigeminal) has the greatest general sensory function of all the cranial nerves and is the only cranial nerve involved in sensory cutaneous innervation of the head. All other cutaneous innervation comes from spinal nerves (see figure 11.12). Cranial nerve V (trigeminal) also provides sensory input from the oral and nasal cavities. Cranial nerves VII (facial), IX (glossopharyngeal), and X (vagus) provide general sensory input from the palate, pharynx, and larynx. Cranial nerve X (vagus) also provides sensory input from thoracic and abdominal organs.
3. *Somatic motor.* Cranial nerves III (oculomotor), IV (trochlear), and VI (abducent) innervate eye muscles that move the eye. Cranial nerve V (trigeminal) innervates muscles of mastication, VII (facial) innervates muscles of facial expression, XI (accessory nerve) innervates the trapezius and sternocleidomastoid muscles, and XII (hypoglossal) innervates the tongue muscles.
4. *Parasympathetic.* Cranial nerve III (oculomotor) innervates smooth muscle within the eye that regulates pupil diameter

and the shape of the lens. Cranial nerves VII (facial) and XI (glossopharyngeal) control secretions from salivary glands. Cranial nerve VII (facial nerve) also innervates lacrimal (tear) glands. Cranial nerve X (vagus) extends into the thoracic and abdominal cavities and is very important in regulating the functions of the heart, lungs, and digestive organs.

 Cranial Nerve Disorders

Trigeminal neuralgia, also called tic douloureux, involves the trigeminal nerve and consists of sharp bursts of pain in the face. This disorder often has a trigger point in or around the mouth, which, when touched, elicits the pain response in some other part of the face. The cause of trigeminal neuralgia is unknown.

Facial palsy (called Bell palsy) is a unilateral paralysis of the facial muscles. The affected side of the face droops because of the absence of muscle tone. Although the cause of facial palsy is often unknown, it can result from inflammation of the facial nerve or a stroke or tumor in the cerebral cortex or brainstem. The facial nerve passes from deep to superficial through the parotid gland, and temporary facial palsy can result from inflammation of the parotid gland. Temporary facial palsy can even result from extreme cold in the face, where the superficial branches of the facial nerve are located. ▼

50. Which cranial nerves are sensory only? With what sense is each of these nerves associated?

Table 11.2 Cranial Nerves and Their Functions

Cranial Nerve	Foramen or Fissure*	Function	Consequences of Lesions to Nerve
I: Olfactory	Cribriform plate	Sensory Special sense of smell	Inability to smell

Olfactory bulb

Olfactory tract (to cerebral cortex)

Cribriform plate of ethmoid bone

Fibers of olfactory nerves

Cranial Nerve	Foramen or Fissure*	Function	Consequences of Lesions to Nerve
II: Optic	Optic foramen	Sensory Special sense of vision	Blindness on the affected side

Eyeball

Pituitary gland

Mammillary body

Optic nerve

Optic chiasm

Optic tract

Cranial Nerve	Foramen or Fissure*	Function	Consequences of Lesions to Nerve
III: Oculomotor	Superior orbital fissure	Motor[†] and parasympathetic Motor to eye muscles (superior, medial, and inferior rectus; inferior oblique) and upper eyelid (levator palpebrae superioris) Proprioceptive from those muscles Parasympathetic to the sphincter of the pupil (causing constriction) and the ciliary muscle of the lens (causing accommodation)	Pupil dilation; eye deviates inferiorly and laterally from muscle paralysis, resulting in double vision; eyelid droops (ptosis); blurred vision from loss of accommodation

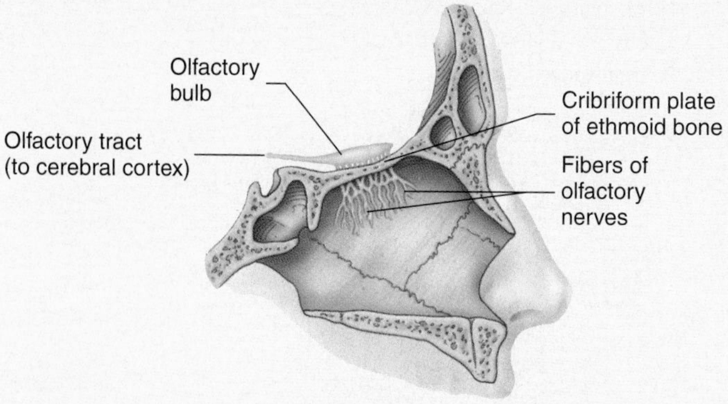

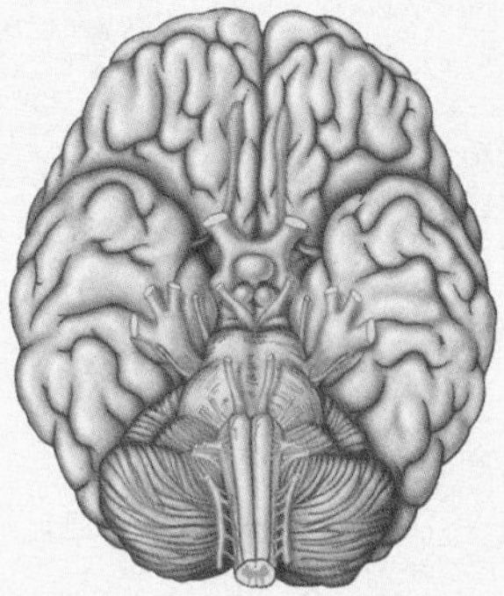

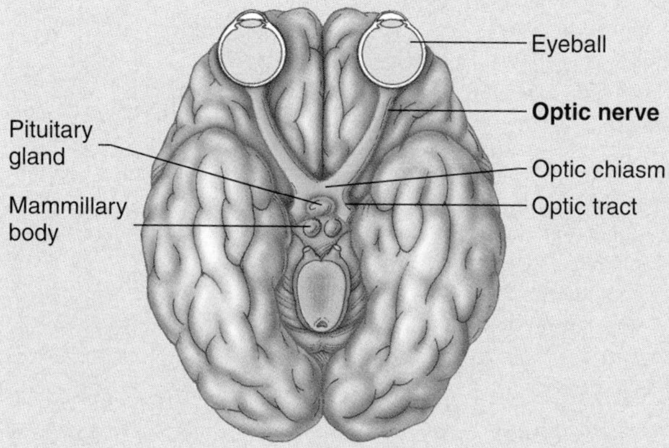

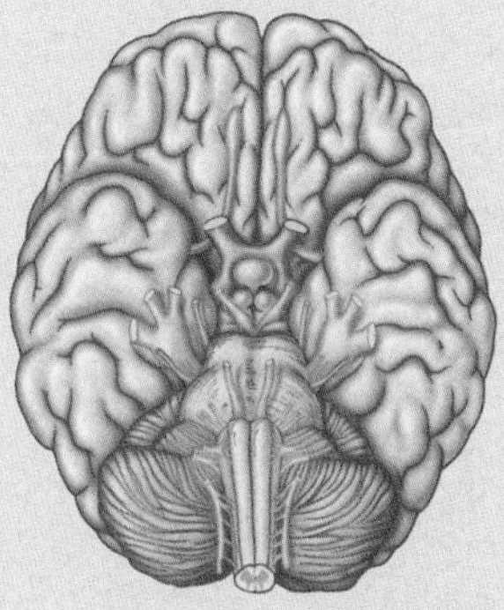

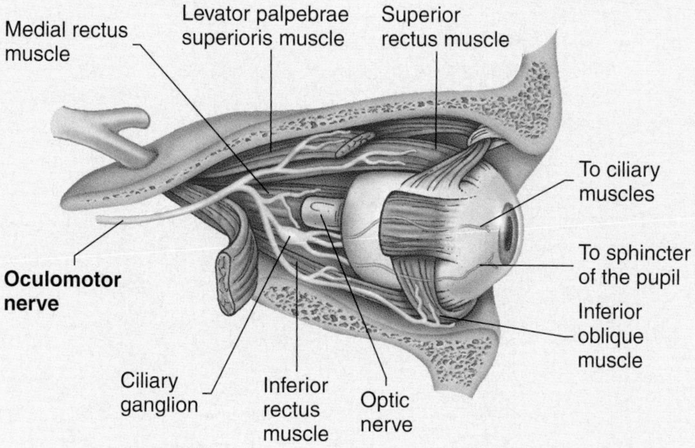

Medial rectus muscle

Levator palpebrae superioris muscle

Superior rectus muscle

To ciliary muscles

To sphincter of the pupil

Inferior oblique muscle

Oculomotor nerve

Ciliary ganglion

Inferior rectus muscle

Optic nerve

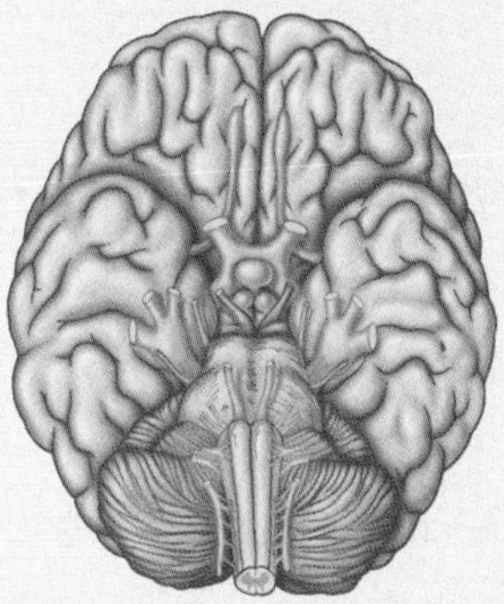

*Route of entry or exit from the skull.
[†]Proprioception is a sensory function, not a motor function; however, motor nerves to muscles also contain some proprioceptive afferent fibers from those muscles. Because proprioception is the only sensory information carried by some cranial nerves, these nerves still are considered "motor."

Table 11.2 Cranial Nerves and Their Functions—Continued

Cranial Nerve	Foramen or Fissure*	Function	Consequences of Lesions to Nerve
IV: Trochlear	Superior orbital fissure	Motor[†] Motor to one eye muscle (superior oblique) Proprioceptive from that muscle	Difficulty moving the eye inferior and lateral, which leads to double vision
V: Trigeminal Ophthalmic branch (V$_1$) Maxillary branch (V$_2$) Mandibular branch (V$_3$)	Superior orbital fissure Foramen rotundum Foramen ovale	Sensory Sensory from scalp, forehead, nose, upper eyelid, and cornea Sensory Sensory from palate, upper jaw, upper teeth and gums, nasopharynx, nasal cavity, skin and mucous membrane of cheek, lower eyelid, and upper lip Sensory and motor[†] Sensory from lower jaw, lower teeth and gums, anterior two-thirds of tongue, mucous membrane of cheek, lower lip, skin of cheek and chin, auricle, and temporal region Motor to muscles of mastication (masseter, temporalis, medial and lateral pterygoids), soft palate (tensor veli palatini), throat (anterior belly of digastric, mylohyoid), and middle ear (tensor tympani) Proprioceptive from those muscles All sympathetic and parasympathetic (III, VII, IX) axons in the head follow branches of the trigeminal nerve to their target tissues	Trigeminal neuralgia (see "Cranial Nerve Disorders," p. 328); intense pain along the course of a branch of the nerve; loss of tactile sensation in the face; weakness in biting or clenching jaw

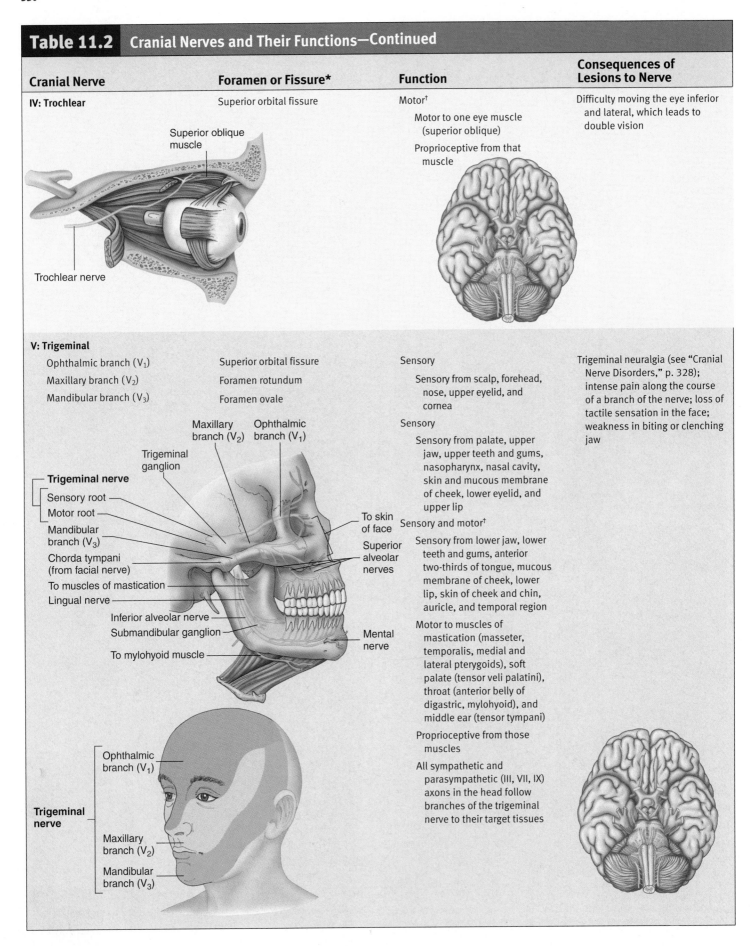

Superior oblique muscle

Trochlear nerve

Maxillary branch (V$_2$)   Ophthalmic branch (V$_1$)

Trigeminal ganglion

Trigeminal nerve

Sensory root

Motor root

Mandibular branch (V$_3$)

Chorda tympani (from facial nerve)

To muscles of mastication

Lingual nerve

Inferior alveolar nerve

Submandibular ganglion

To mylohyoid muscle

To skin of face

Superior alveolar nerves

Mental nerve

Ophthalmic branch (V$_1$)

Trigeminal nerve

Maxillary branch (V$_2$)

Mandibular branch (V$_3$)

Table 11.2   Cranial Nerves and Their Functions—Continued

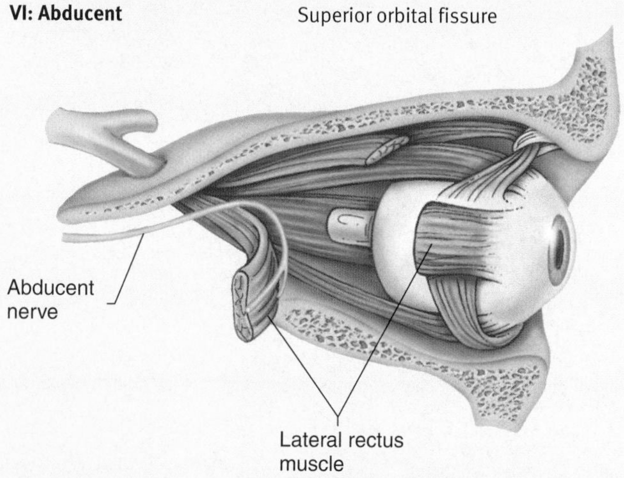

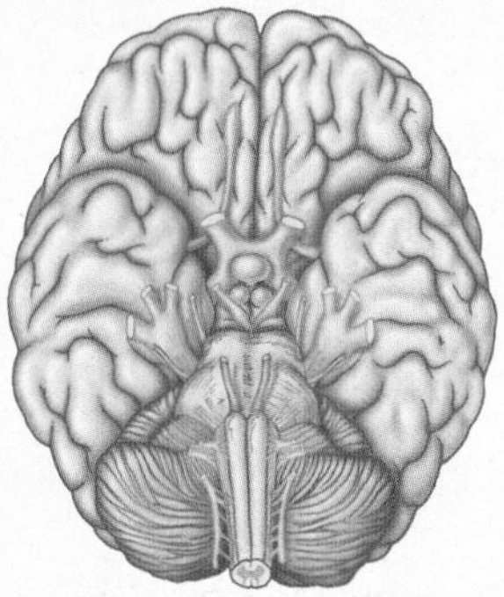

Cranial Nerve	Foramen or Fissure*	Function	Consequences of Lesions to Nerve

VI: Abducent	Superior orbital fissure	Motor[†]
		
		  Motor to one eye muscle (lateral rectus)
		
		  Proprioceptive from that muscle
			Eye deviates medially (adducts) causing double vision

VII: Facial	Internal auditory meatus
	
	Stylomastoid foramen
		Sensory, motor,[†] and parasympathetic
		
		  Sense of taste from anterior two-thirds of tongue, sensory from some of external ear and palate
		
		  Motor to muscles of facial expression, throat (posterior belly of digastric, stylohyoid), and middle ear (stapedius)
		
		  Proprioceptive from those muscles
		
		  Parasympathetic to submandibular and sublingual salivary glands, lacrimal gland, and glands of the nasal cavity and palate
			Facial palsy (see "Cranial Nerve Disorders," p. 328); loss of taste sensation on the anterior two-thirds of tongue; decreased salivation

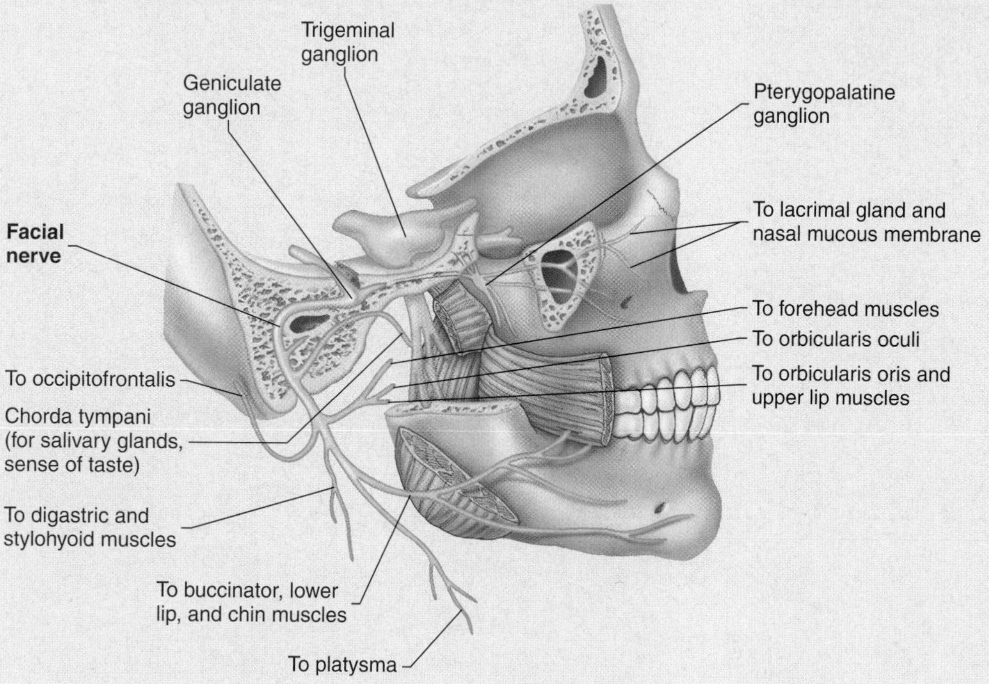

VI: Abducent

Abducent nerve

Lateral rectus muscle

VII: Facial

Trigeminal ganglion

Geniculate ganglion

Pterygopalatine ganglion

Facial nerve

To lacrimal gland and nasal mucous membrane

To forehead muscles

To orbicularis oculi

To occipitofrontalis

To orbicularis oris and upper lip muscles

Chorda tympani (for salivary glands, sense of taste)

To digastric and stylohyoid muscles

To buccinator, lower lip, and chin muscles

To platysma

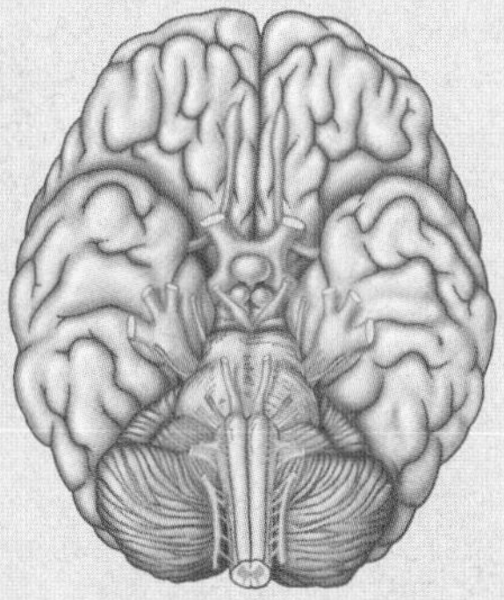

Table 11.2   Cranial Nerves and Their Functions—Continued

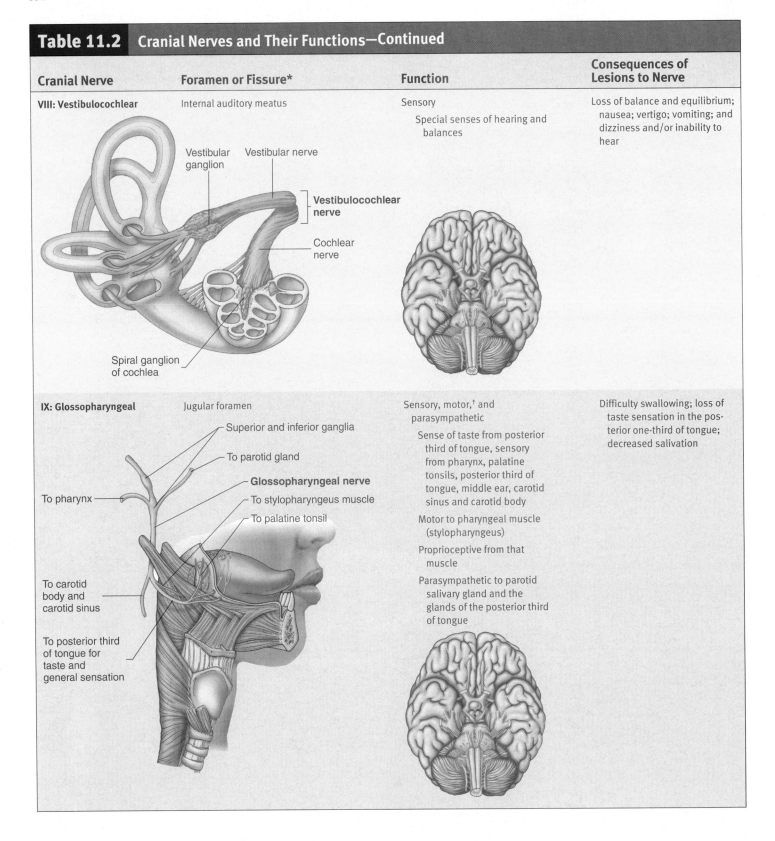

Cranial Nerve	Foramen or Fissure*	Function	Consequences of Lesions to Nerve
VIII: Vestibulocochlear	Internal auditory meatus	**Sensory** Special senses of hearing and balances	Loss of balance and equilibrium; nausea; vertigo; vomiting; and dizziness and/or inability to hear
IX: Glossopharyngeal	Jugular foramen	**Sensory, motor,[†] and parasympathetic** Sense of taste from posterior third of tongue, sensory from pharynx, palatine tonsils, posterior third of tongue, middle ear, carotid sinus and carotid body Motor to pharyngeal muscle (stylopharyngeus) Proprioceptive from that muscle Parasympathetic to parotid salivary gland and the glands of the posterior third of tongue	Difficulty swallowing; loss of taste sensation in the posterior one-third of tongue; decreased salivation

Labels in VIII image: Vestibular ganglion; Vestibular nerve; **Vestibulocochlear nerve**; Cochlear nerve; Spiral ganglion of cochlea

Labels in IX image: Superior and inferior ganglia; To parotid gland; **Glossopharyngeal nerve**; To stylopharyngeus muscle; To palatine tonsil; To pharynx; To carotid body and carotid sinus; To posterior third of tongue for taste and general sensation

Table 11.2 Cranial Nerves and Their Functions—Continued

Cranial Nerve	Foramen or Fissure*	Function	Consequences of Lesions to Nerve
X: Vagus	Jugular foramen	Sensory, motor,[†] and parasympathetic Sensory from inferior pharynx, larynx, thoracic and abdominal organs, sense of taste from posterior tongue Motor to soft palate, pharynx, intrinsic laryngeal muscles (voice production), and an extrinsic tongue muscle (palatoglossus) Proprioceptive from those muscles Parasympathetic to thoracic and abdominal viscera	Difficulty swallowing and/or hoarseness; uvula deviates away from side of the disfunction
XI: Accessory	Foramen magnum Jugular foramen	Motor[†] Motor to sternocleidomastoid and trapezius	Difficulty elevating the scapula or rotating the neck

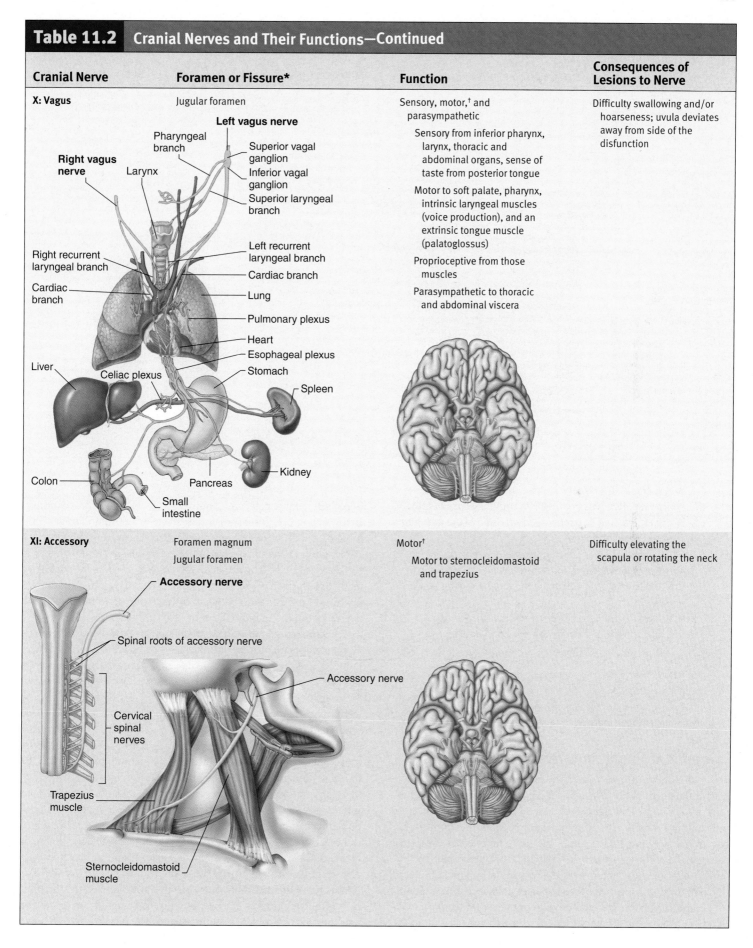

Left vagus nerve
Pharyngeal branch
Right vagus nerve
Larynx
Superior vagal ganglion
Inferior vagal ganglion
Superior laryngeal branch
Right recurrent laryngeal branch
Left recurrent laryngeal branch
Cardiac branch
Cardiac branch
Lung
Pulmonary plexus
Heart
Esophageal plexus
Liver
Celiac plexus
Stomach
Spleen
Colon
Pancreas
Kidney
Small intestine

Accessory nerve
Spinal roots of accessory nerve
Accessory nerve
Cervical spinal nerves
Trapezius muscle
Sternocleidomastoid muscle

Table 11.2 — Cranial Nerves and Their Functions—Continued

Cranial Nerve	Foramen or Fissure*	Function	Consequences of Lesions to Nerve
XII: Hypoglossal	Hypoglossal canal	Motor† Motor to intrinsic and extrinsic tongue muscles (styloglossus, hypoglossus, genioglossus) and throat muscles (thyrohyoid and geniohyoid) Proprioceptive from those muscles	When sticking out the tongue, it deviates toward the side of the damaged nerve

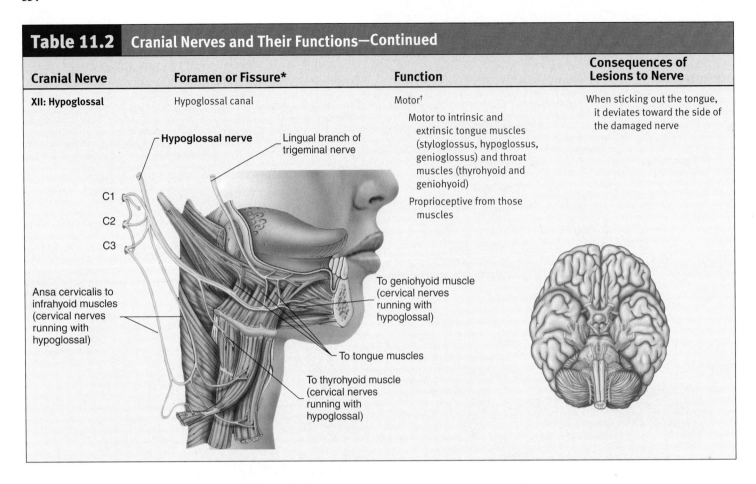

Hypoglossal nerve
Lingual branch of trigeminal nerve
C1
C2
C3
Ansa cervicalis to infrahyoid muscles (cervical nerves running with hypoglossal)
To geniohyoid muscle (cervical nerves running with hypoglossal)
To tongue muscles
To thyrohyoid muscle (cervical nerves running with hypoglossal)

51. *Which cranial nerves are involved with the sense of taste?*
52. *Which cranial nerve is sensory for the skin of the face, oral cavity, and nasal cavity?*
53. *Which cranial nerves control the muscles moving the eye, muscles of mastication, muscles of facial expression, trapezius and sternocleidomastoid muscles, and tongue muscles?*
54. *Which four cranial nerves have a parasympathetic function? Describe the function of each of these nerves.*

P R E D I C T
Unilateral damage to the hypoglossal nerve results in loss of tongue movement on one side, which is most obvious when the tongue is protruded. If the tongue is deviated to the right, is the left or right hypoglossal nerve damaged?

Trigeminal Nerve and Dental Anesthesia

Trigeminal means three twins, and the sensory distribution of the trigeminal nerve in the face is divided into three regions, each supplied by a branch of the nerve (see table 11.2). The three branches—**ophthalmic, maxillary,** and **mandibular**—arise directly from the trigeminal ganglion, which serves the same function as the dorsal root ganglia of the spinal nerves. Only the mandibular branch contains motor axons, which bypass the trigeminal ganglion, much as the ventral root of a spinal nerve bypasses a dorsal root ganglion.

The maxillary and mandibular branches are important in dentistry. The maxillary nerve supplies sensory innervation to the maxillary teeth, palate, and gingiva (jin′jĭ-vă, gum). The mandibular branch supplies sensory innervation to the mandibular teeth, tongue, and gingiva. The various nerves innervating the teeth are referred to as **alveolar nerves** (al-vē′ō-lăr, the sockets in which the teeth are located). The **superior alveolar nerves** to the maxillary teeth are derived from the maxillary branch of the trigeminal nerve, and the **inferior alveolar nerves** to the mandibular teeth are derived from the mandibular branch of the trigeminal nerve.

Dentists inject anesthetic to block sensory transmission by the alveolar nerves. The superior alveolar nerves are not usually anesthetized directly because they are difficult to approach with a needle. For this reason, the maxillary teeth are usually anesthetized locally by inserting the needle beneath the oral mucosa surrounding the teeth. The inferior alveolar nerve probably is anesthetized more often than any other nerve in the body. To anesthetize this nerve, the dentist inserts the needle somewhat posterior to the patient's last molar and extends the needle near where the mandibular branch of the trigeminal nerve enters the mandibular foramen.

Several nondental nerves are usually anesthetized during an inferior alveolar block. The mental nerve, which supplies cutaneous innervation to the anterior lip and chin, is a distal branch of the inferior alveolar nerve. When the inferior alveolar nerve is blocked, the mental nerve is blocked also, resulting in a numb lip and chin. Nerves lying near the point where the inferior alveolar nerve enters the mandible often are also anesthetized during inferior alveolar anesthesia. For example, the lingual nerve can be anesthetized to produce a numb tongue. The facial nerve lies some distance from the inferior alveolar nerve, but in rare cases anesthetic can diffuse far enough posteriorly to anesthetize that nerve. The result is a temporary facial palsy, with the injected side of the face drooping because of flaccid muscles, which disappears when the anesthesia wears off. If the facial nerve is cut by an improperly inserted needle, permanent facial palsy may occur. ▼

> **55.** Name the three branches of the trigeminal nerve.
> **56.** Describe the nerves involved in dental anesthesia. How can dental anesthesia result in a numb lip or tongue?

Brainstem Reflexes

Many of the brainstem reflexes are associated with cranial nerve function. The circuitry of most of these reflexes is too complex for our discussions. In general, these reflexes involve sensory input from the cranial nerves or spinal cord, as well as the motor output of the cranial nerves.

Turning of the eyes toward a flash of light, a sudden noise, and a touch on the skin are examples of brainstem reflexes. Moving the eyes to track a moving object is another complex brainstem reflex. Some of the sensory neurons from cranial nerve VIII (vestibulocochlear) form a reflex arc with neurons of cranial nerves V (trigeminal) and VII (facial), which send axons to muscles of the middle ear and dampen the effects of very loud, sustained noises on delicate inner ear structures (see chapter 14). Reflexes that occur during the process of chewing allow the jaws to react to foods of various hardness and protect the teeth from breakage. Both the sensory and motor components of the reflex arc are carried by cranial nerve V

(trigeminal). Reflexes involving input through cranial nerve V (trigeminal) and output through cranial nerve XII (hypoglossal) move the tongue about to position food between the teeth for chewing and then move the tongue out of the way so that it is not bitten.

Brainstem reflexes are used to determine if the brainstem is functioning. Brainstem reflexes are mediated by cranial nerve sensory nuclei, cranial nerve motor nuclei, and reticular formation nuclei. The presence of a reflex indicates that sensory input reaches the brainstem and the nuclei are functioning to produce a response. The absence of a reflex is taken to indicate damage to the nuclei involved in the reflex. It is also assumed that, if a nucleus is not functioning, it is likely that surrounding tissue has been damaged as well. Different reflexes are mediated through nuclei at different levels of the brainstem. For example, the pupillary reflex, in which the pupil constricts in response to light, is mediated through the midbrain; the corneal reflex, in which touching the cornea cause the eyelids to blink, is mediated through the pons; and the gag reflex, in which touching the back of the throat (pharynx) elicits a gag response, is mediated through the medulla oblongata. The brainstem is considered to be nonfunctional when reflexes at all levels of the brainstem are nonfunctional and there is no spontaneous breathing, which is mediated through the medulla oblongata. ▼

> **57.** Give examples of reflexes mediated through the brainstem.
> **58.** How are brainstem reflexes used to determine if the brainstem is functioning?

S U M M A R Y

Spinal Cord (p. 298)
General Structure

1. The spinal cord gives rise to 31 pairs of spinal nerves. The spinal cord has cervical and lumbosacral enlargements from which spinal nerves of the limbs originate.
2. The spinal cord is shorter than the vertebral column. Nerves from the end of the spinal cord form the cauda equina.

Meninges of the Spinal Cord

1. Three meningeal layers surround the spinal cord. From superficial to deep they are the dura mater, arachnoid mater, and pia mater.
2. The epidural space is between the periosteum of the vertebral canal and the dura mater, the subdural space is between the dura mater and the arachnoid mater, and the subarachnoid space is between the arachnoid mater and the pia mater.

Cross Section of the Spinal Cord

1. The spinal cord consists of peripheral white matter and central gray matter.
2. White matter is organized into columns (funiculi), which are subdivided into tracts (fasciculi or pathways), which consist of ascending and descending axons.
3. Gray matter is divided into horns.
   - The dorsal horns contain sensory axons that synapse with interneurons. The ventral horns contain the neuron cell bodies of somatic motor neurons, and the lateral horns contain the neuron cell bodies of autonomic motor neurons.

- The gray and white commissures connect each half of the spinal cord.
4. The dorsal root contains sensory axons, the ventral root has motor axons, and spinal nerves have sensory and motor axons.

Reflexes (p. 302)

1. Reflexes are stereotypic, unconscious, involuntary responses to stimuli.
2. Reflexes maintain homeostasis. Two general types of reflexes are somatic and autonomic reflexes.
3. A reflex arc is the functional unit of the nervous system.
   - Sensory receptors respond to stimuli and produce action potentials in sensory neurons.
   - Sensory neurons propagate action potentials to the CNS.
   - Interneurons in the CNS synapse with sensory neurons and with motor neurons.
   - Motor neurons carry action potentials from the CNS to effector organs.
   - Effector organs, such as muscles or glands, respond to the action potentials.
4. Convergent and divergent pathways interact with reflexes.
5. Reflexes are integrated within the brain and spinal cord. Higher brain centers can suppress or exaggerate reflexes.

Stretch Reflex

Muscle spindles detect the stretch of skeletal muscles and cause the muscle to shorten reflexively.

Golgi Tendon Reflex

Golgi tendon organs respond to increased tension within tendons and cause skeletal muscles to relax.

Withdrawal Reflex

Activation of pain receptors causes contraction of muscles and the removal of some part of the body from a painful stimulus.

Nerves (p. 306)
Structure of Nerves

Individual axons are surrounded by the endoneurium. Groups of axons, called fascicles, are bound together by the perineurium. The fascicles form the nerve and are held together by the epineurium.

Spinal Nerves and Their Distribution

1. Eight cervical, 12 thoracic, 5 lumbar, 5 sacral pairs, and 1 coccygeal pair make up the spinal nerves.
2. The distribution of spinal nerves to skeletal muscles has a top to bottom pattern in which superior nerves supply superior muscles and inferior nerves supply inferior muscles.
3. Spinal nerves have specific cutaneous distributions called dermatomes.

Spinal Nerves and Plexuses

1. Spinal nerves branch to form rami.
   - The dorsal rami supply the muscles and skin near the midline of the back.
   - The ventral rami in the thoracic region form intercostal nerves, which supply the thorax and upper abdomen. The remaining ventral rami join to form plexuses.
   - Communicating rami supply sympathetic nerves.
2. Axons from different levels of the spinal cord intermingle within plexuses and give rise to nerves that have axons from more than one level of the spinal cord.

Major Spinal Nerve Plexuses

1. The five major plexuses are the cervical (C1–C4), brachial (C5–T1), lumbar (L1–L4), sacral (L4–S4), and coccygeal (S5 and coccygeal nerve). The lumbar and sacral plexuses are often considered together as the lumbosacral plexus.
2. A major nerve of the cervical plexus is the phrenic nerve.
3. The major nerves of the brachial plexus are the axillary, radial, musculocutaneous, ulnar, and median nerves.
4. The major nerves of the lumbosacral plexus are the obturator, femoral, tibial, and common fibular nerves.
   - The tibial and common fibular nerves are bound together in the thigh to form the sciatic nerve.
   - Branches of the tibial and common fibular nerves form the sural nerve in the leg.

Brain (p. 314)

1. The brain is contained in the cranial cavity and is the control center for many of the body's functions.
2. The brain consists of the brainstem, cerebellum, diencephalon, and cerebrum.

Brainstem (p. 314)
Medulla Oblongata

1. The medulla oblongata is continuous with the spinal cord and contains ascending and descending tracts.
2. Medullary nuclei regulate the heart, blood vessels, breathing, swallowing, vomiting, coughing, sneezing, hiccuping, balance, and coordination.
3. The pyramids are tracts controlling voluntary muscle movement.

Pons

1. The pons is superior to the medulla.
2. Ascending and descending tracts pass through the pons. The pons connects the cerebrum and the cerebellum.
3. Pontine nuclei regulate breathing, swallowing, balance, chewing, and salivation.

Midbrain

1. The midbrain is superior to the pons.
2. The corpora quadrigemina consists of four colliculi. The two inferior colliculi are involved in hearing and the two superior colliculi in visual reflexes.
3. The substantia nigra and red nucleus help regulate body movements.
4. The cerebral peduncles are the major descending motor pathway.

Reticular Formation

1. The reticular formation consists of nuclei scattered throughout the brainstem.
2. The reticular system regulates cyclic motor functions, such as breathing, walking, and chewing.
3. The reticular activating system, which is part of the reticular formation, maintains consciousness and regulates the sleep–wake cycle.

Cerebellum (p. 316)

1. Gray matter forms the cortex and nuclei of the cerebellum. White matter forms the arbor vitae, which connects the cerebellum to the rest of the CNS and connects the cerebellar cortex and cerebellar nuclei.
2. The cerebellum has three parts.
   - The flocculonodular lobe controls balance and eye movements.
   - The vermis and medial part of the lateral hemispheres control posture, locomotion, and fine motor coordination.
   - Most of the lateral hemispheres are involved with the planning, practice, and learning of complex movements.

Diencephalon (p. 317)

The diencephalon is located between the brainstem and the cerebrum. It consists of the thalamus, subthalamus, epithalamus, and hypothalamus.

Thalamus

1. The thalamus consists of two lobes connected by the interthalamic adhesion. The thalamus functions as an integration center.
2. All sensory input that reaches the cerebrum, except for the sense of smell, synapses in the thalamus. Pain is registered in the thalamus.
3. The thalamus interacts with other parts of the brain to control motor activity.
4. The thalamus is involved with emotions and pain perception.

Subthalamus

The subthalamus is inferior to the thalamus and is involved in motor function.

Epithalamus

The epithalamus is superior and posterior to the thalamus and contains the habenular nuclei, which influence emotions through the sense of smell. The pineal body may play a role in the onset of puberty and the sleep–wake cycle.

Hypothalamus

1. The hypothalamus is inferior to the thalamus.

2. The hypothalamus is a major integrating center for the ANS, helping control heart rate, blood vessel diameter, urine release from the urinary bladder, and the movement of food through the digestive tract.
3. The hypothalamus regulates body temperature, hunger, satiety, thirst, and swallowing and interacts with the reticular activating system.
4. The hypothalamus regulates many endocrine functions (e.g., metabolism, reproduction, response to stress, and urine production) by controlling the pituitary gland.
5. The hypothalamus is involved with sensations (sexual pleasure, good feelings, rage, and fear).
6. The mammillary bodies are reflex centers for olfaction.

Cerebrum (p. 319)

1. The cortex of the cerebrum is folded into ridges called gyri and grooves called sulci, or fissures.
2. The longitudinal fissure divides the cerebrum into left and right hemispheres. Each hemisphere has five lobes.
   - The frontal lobes are involved in voluntary motor function, motivation, aggression, the sense of smell, and mood.
   - The parietal lobes contain the major sensory areas receiving sensory input, such as touch, pain, temperature, balance, and taste.
   - The occipital lobes contain the visual centers.
   - The temporal lobes evaluate smell and hearing input and are involved in memory, abstract thought, and judgment.
   - The insula is located deep within the lateral fissure and receives input for taste.
3. Gray matter forms the cortex and nuclei of the cerebrum. White matter forms the cerebral medulla, which consists of three types of tracts.
   - Association fibers connect areas of the cortex within the same hemisphere.
   - Commissural fibers connect the cerebral hemispheres.
   - Projection fibers connect the cerebrum to other parts of the brain and the spinal cord.

Basal Nuclei

1. Basal nuclei include the corpus striatum (caudate and lentiform nuclei), subthalamic nuclei, and substantia nigra.
2. The basal nuclei are important in controlling motor functions.

Limbic System

1. The limbic system includes parts of the cerebral cortex, basal nuclei, the thalamus, the hypothalamus, and the olfactory cortex.
2. The limbic system is involved in memory, reproduction, and nutrition. It is involved in the emotional interpretation of sensory input and emotions in general.

Meninges, Ventricles, and Cerebrospinal Fluid (p. 322)
Meninges

1. The brain and spinal cord are covered by the dura, arachnoid, and pia mater.

2. The dura mater attaches to the skull and has two layers that can separate to form dural folds and dural venous sinuses.
3. Beneath the arachnoid mater the subarachnoid space contains CSF that helps cushion the brain.
4. The pia mater attaches directly to the brain.

Ventricles

1. The lateral ventricles in the cerebrum are connected to the third ventricle in the diencephalon by the interventricular foramina.
2. The third ventricle is connected to the fourth ventricle in the pons by the cerebral aqueduct. The central canal of the spinal cord is connected to the fourth ventricle.
3. The fourth ventricle is connected to the subarachnoid space by the median and lateral apertures.

Cerebrospinal Fluid

1. CSF is produced from the blood in the choroid plexus of each ventricle by ependymal cells. CSF moves from the lateral to the third and then to the fourth ventricle.
2. From the fourth ventricle, CSF enters the subarachnoid space through three apertures.
3. CSF leaves the subarachnoid space through arachnoid granulations and returns to the blood in the dural venous sinuses.

Blood Supply to the Brain (p. 327)

1. The brain requires tremendous amounts of blood to function normally.
2. The blood–brain barrier is formed by the endothelial cells of the capillaries in the brain. It limits what substances enter brain tissue.

Cranial Nerves (p. 327)
Names and Functions

1. Cranial nerves are designated by Roman numerals (I–XII) or specific names.
2. The two types of general functions are sensory and motor. Sensory includes special senses and general senses. Motor includes somatic motor and parasympathetic.

Trigeminal Nerve and Dental Anesthesia

1. The trigeminal nerve (V) has three branches—ophthalmic, maxillary, and mandibular.
2. The superior alveolar nerve (branch of the maxillary) and the inferior alveolar nerve (branch of the mandibular) are used for dental anesthesia.

Brainstem Reflexes

1. Many reflexes are mediated through the brainstem.
2. The brainstem is considered to be nonfunctional when reflexes at all levels of the brainstem are nonfunctional and there is no spontaneous breathing, which is mediated through the medulla oblongata.

R E V I E W A N D C O M P R E H E N S I O N

1. The spinal cord extends from the
   a. medulla oblongata to the coccyx.
   b. level of the third cervical vertebra to the coccyx.
   c. level of the axis to the lowest lumbar vertebra.
   d. level of the foramen magnum to level L1–L2 of the vertebral column.
   e. axis to the sacral hiatus.

2. Axons of sensory neurons synapse with the cell bodies of interneurons in the _____ of spinal cord gray matter.
   a. anterior horn
   b. lateral horn
   c. posterior horn
   d. gray commissure
   e. lateral funiculi

3. Cell bodies for spinal sensory neurons are located in the
   a. anterior horn of spinal cord gray matter.
   b. lateral horn of spinal cord gray matter.
   c. posterior horn of spinal cord gray matter.
   d. dorsal root ganglia.
   e. posterior columns.

4. Given these components of a reflex arc:
   1. effector organ
   2. interneuron
   3. motor neuron
   4. sensory neuron
   5. sensory receptor

   Choose the correct order an action potential follows after a sensory receptor is stimulated.
   a. 5,4,3,2,1
   b. 5,4,2,3,1
   c. 5,3,4,1,2
   d. 5,2,4,3,1
   e. 5,3,2,1,4

5. A reflex response accompanied by the conscious sensation of pain is possible because of
   a. convergent pathways.
   b. divergent pathways.
   c. a reflex arc that contains only one neuron.
   d. sensory perception in the spinal cord.

6. Several of the events that occurred between the time that a physician struck a patient's patellar tendon with a rubber hammer and the time the quadriceps femoris contracted (knee-jerk reflex) are listed below:
   1. increased frequency of action potentials in sensory neurons
   2. stretch of the muscle spindles
   3. increased frequency of action potentials in the motor neurons
   4. stretch of the quadriceps femoris
   5. contraction of the quadriceps femoris

   Which of these lists most closely describes the sequence of events as they normally occur?
   a. 4,1,2,3,5
   b. 4,1,3,2,5
   c. 1,4,3,2,5
   d. 4,2,1,3,5
   e. 4,2,3,1,5

7. Which of these is a correct count of the spinal nerves?
   a. 9 cervical, 12 thoracic, 5 lumbar, 5 sacral, 1 coccygeal
   b. 8 cervical, 12 thoracic, 5 lumbar, 5 sacral, 1 coccygeal
   c. 7 cervical, 12 thoracic, 5 lumbar, 5 sacral, 1 coccygeal
   d. 8 cervical, 11 thoracic, 4 lumbar, 6 sacral, 1 coccygeal
   e. 7 cervical, 11 thoracic, 5 lumbar, 6 sacral, 1 coccygeal

8. Given these structures:
   1. dorsal ramus
   2. dorsal root
   3. plexus
   4. ventral ramus
   5. ventral root

   Choose the arrangement that lists the structures in the order that an action potential passes through them, given that the action potential originates in the spinal cord and propagates to a peripheral nerve.
   a. 2,1,3
   b. 2,3,1
   c. 3,4,5
   d. 5,3,4
   e. 5,4,3

9. Damage to the dorsal ramus of a spinal nerve results in
   a. loss of sensation.
   b. loss of motor function.
   c. both a and b.

10. A dermatome
    a. is the area of skin supplied by a pair of spinal nerves.
    b. exists for each spinal nerve except C1.
    c. can be used to locate the site of spinal cord or nerve root damage.
    d. all of the above.

11. A collection of spinal nerves that join together after leaving the spinal cord is called a
    a. ganglion.
    b. nucleus.
    c. projection nerve.
    d. plexus.

12. Which of these nerves arises from the cervical plexus?
    a. median   d. obturator
    b. musculocutaneous   e. ulnar
    c. phrenic

13. The sciatic nerve is actually two nerves combined within the same sheath. The two nerves are the
    a. femoral and obturator.
    b. femoral and gluteal.
    c. common fibular (peroneal) and tibial.
    d. common fibular (peroneal) and obturator.
    e. tibial and gluteal.

14. If a section is made that separates the brainstem from the rest of the brain, the cut is between the
    a. medulla oblongata and pons.
    b. pons and midbrain.
    c. midbrain and diencephalon.
    d. thalamus and cerebrum.
    e. medulla oblongata and spinal cord.

15. Important centers for heart rate, blood pressure, breathing, swallowing, coughing, and vomiting are located in the
    a. cerebrum.
    b. medulla oblongata.
    c. midbrain.
    d. pons.
    e. cerebellum.

16. In which of these parts of the brain does decussation of descending tracts involved in the conscious control of skeletal muscles occur?
    a. cerebrum
    b. diencephalon
    c. midbrain
    d. pons
    e. medulla oblongata

17. Important respiratory centers are located in the
    a. cerebrum.
    b. cerebellum.
    c. pons and medulla oblongata.
    d. midbrain.
    e. limbic system.

18. The cerebral peduncles are a major descending motor pathway found in the
    a. cerebrum.
    b. cerebellum.
    c. pons.
    d. midbrain.
    e. medulla oblongata.

19. The superior colliculi are involved in _____, whereas the inferior colliculi are involved in _____.
    a. hearing, visual reflexes
    b. visual reflexes, hearing
    c. balance, motor pathways
    d. motor pathways, balance
    e. respiration, sleep

20. The cerebellum communicates with other regions of the CNS through the
    a. flocculonodular lobe.
    b. cerebellar peduncles.
    c. vermis.
    d. lateral hemispheres.
    e. folia.

21. The major relay station for sensory input that projects to the cerebral cortex is the
    a. hypothalamus.
    b. thalamus.
    c. pons.
    d. cerebellum.
    e. midbrain.

22. Which part of the brain is involved with olfactory reflexes and emotional responses to odors?
    a. inferior colliculi
    b. superior colliculi
    c. mammillary bodies
    d. pineal body
    e. pituitary gland

23. Which of the following is a function of the hypothalamus?
    a. regulates autonomic nervous system functions
    b. regulates the release of hormones from the posterior pituitary
    c. regulates body temperature
    d. regulates food intake (hunger) and water intake (thirst)
    e. all of the above

24. Which of these cerebral lobes is important in voluntary motor function, motivation, aggression, sense of smell, and mood?
    a. frontal
    b. insula
    c. occipital
    d. parietal
    e. temporal

25. Fibers that connect areas of the cerebral cortex within the same hemisphere are
    a. projection fibers.
    b. commissural fibers.
    c. association fibers.
    d. all of the above.

26. The basal nuclei are located in the
    a. inferior cerebrum.
    b. diencephalon.
    c. midbrain.
    d. all of the above.

27. The most superficial of the meninges is a thick, tough membrane called the
    a. pia mater.
    b. dura mater.
    c. arachnoid mater.
    d. epidural mater.

28. Cerebrospinal fluid is produced by the _____, circulates through the ventricles, and enters the subarachnoid space. The cerebrospinal fluid leaves the subarachnoid space through the
    _____.
    a. choroid plexuses, arachnoid granulations
    b. arachnoid granulations, choroid plexuses
    c. dural venous sinuses, dura mater
    d. dura mater, dural venous sinuses

29. The cranial nerve i nvolved in chewing food is the
    a. trochlear (IV).
    b. trigeminal (V).
    c. abducent (VI).
    d. facial (VII).
    e. vestibulocochlear (VIII).

30. The cranial nerve involved in moving the tongue is the
    a. trigeminal (V).
    b. facial (VII).
    c. glossopharyngeal (IX).
    d. accessory (XI).
    e. hypoglossal (XII).

31. The cranial nerve involved in feeling a toothache is the
    a. trochlear (IV).
    b. trigeminal (V).
    c. abducent (VI).
    d. facial (VII).
    e. vestibulocochlear (VIII).

32. From this list of cranial nerves:
    1. olfactory (I)
    2. optic (II)
    3. oculomotor (III)
    4. abducent (VI)
    5. vestibulocochlear (VIII)
    Select the nerves that are sensory only.
    a. 1,2,3
    b. 2,3,4
    c. 1,2,5
    d. 2,3,5
    e. 3,4,5

33. From this list of cranial nerves:
    1. optic (II)
    2. oculomotor (III)
    3. trochlear (IV)
    4. trigeminal (V)
    5. abducent (VI)
    Select the nerves that are involved in moving the eyes.
    a. 1,2,3
    b. 1,2,4
    c. 2,3,4
    d. 2,4,5
    e. 2,3,5

34. From this list of cranial nerves:
    1. trigeminal (V)
    2. facial (VII)
    3. glossopharyngeal (IX)
    4. vagus (X)
    5. hypoglossal (XII)
    Select the nerves that innervate the salivary glands.
    a. 1,2
    b. 2,3
    c. 3,4
    d. 4,5
    e. 3,5

35. From this list of cranial nerves:
    1. oculomotor (III)
    2. trigeminal (V)
    3. facial (VII)
    4. vestibulocochlear (VIII)
    5. glossopharyngeal (IX)
    6. vagus (X)
    Select the nerves that are part of the parasympathetic division of the ANS.
    a. 1,2,4,5
    b. 1,3,5,6
    c. 1,4,5,6
    d. 2,3,4,5
    e. 2,3,5,6

Answers in Appendix E

CRITICAL THINKING

1. A patient loses all sense of feeling in the left side of the back, below the upper limb, and extending in a band around to the chest, as well as below the upper limb. All sensation on the right is normal. The line between normal and absent sensation is the anterior and posterior midline. Explain this condition.

2. A cancer patient has his left lung removed. To reduce the space remaining where the lung is removed, the diaphragm on the left side is paralyzed to allow the abdominal viscera to push the diaphragm upward. What nerve is cut? Where is a good place to cut it, and when would the surgery be done?

3. Based on sensory response to pain in the skin of the hand, how could you distinguish between damage to the ulnar, median, and radial nerves?

4. During a difficult delivery, the baby's arm delivered first. The attending physician grasped the arm and forcefully pulled it. Later a nurse observed that the baby could not abduct or adduct the medial four fingers and flexion of the wrist was impaired. What nerve was damaged?

5. Two patients are admitted to the hospital. According to their charts, both have herniated disks that are placing pressure on the roots of the sciatic nerve. One patient has pain in the buttocks and the posterior aspect of the thigh. The other patient experiences pain in the posterior and lateral aspects of the leg and the lateral part of the ankle and foot. Explain how the same condition, a herniated disk, could produce such different symptoms.

6. In an automobile accident, a woman suffers a crushing hip injury. For each of the following conditions, state what nerve is damaged.
   a. Unable to adduct the thigh
   b. Unable to extend the knee
   c. Unable to flex the knee
   d. Loss of sensation from the skin of the anterior thigh
   e. Loss of sensation from the skin of the medial thigh

7. A patient exhibits enlargement of the lateral and third ventricles, but no enlargement of the fourth ventricle. What do you conclude?

8. During a spinal tap of a patient, blood is discovered in the CSF. What does this finding suggest?

9. Describe a clinical test to evaluate the functioning of each of the 12 cranial nerves.

10. A baseball player was accidentally hit with a baseball, which struck him on the bridge of his nose. He suffered fractures of several facial bones as a result. Soon after sustaining the injury, he noticed he had lost his sense of smell. Explain how this probably happened.

Answers in Appendix F

CHAPTER 12 Integration of Nervous System Functions

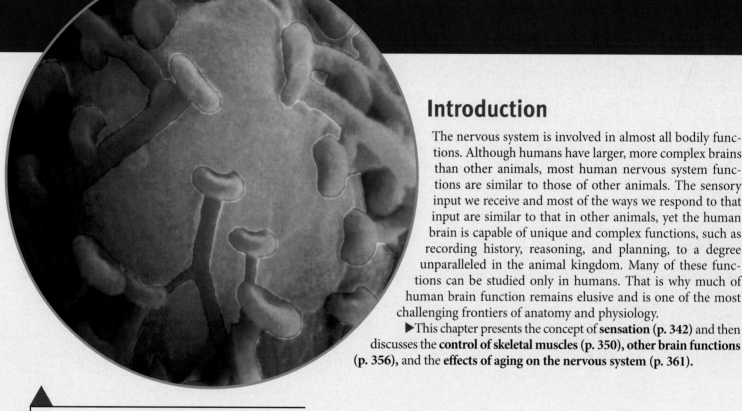

Introduction

The nervous system is involved in almost all bodily functions. Although humans have larger, more complex brains than other animals, most human nervous system functions are similar to those of other animals. The sensory input we receive and most of the ways we respond to that input are similar to that in other animals, yet the human brain is capable of unique and complex functions, such as recording history, reasoning, and planning, to a degree unparalleled in the animal kingdom. Many of these functions can be studied only in humans. That is why much of human brain function remains elusive and is one of the most challenging frontiers of anatomy and physiology.

▶This chapter presents the concept of **sensation (p. 342)** and then discusses the **control of skeletal muscles (p. 350), other brain functions (p. 356),** and the **effects of aging on the nervous system (p. 361).**

Colorized SEM of presynaptic terminals associated with a postsynaptic neuron.

Tools for Success

A virtual cadaver dissection experience

Audio (MP3) and/or video (MP4) content, available at www.mhhe.aris.com

Sensation

Sensation, or **perception,** is the conscious awareness of stimuli received by sensory receptors. The brain constantly receives action potentials from a wide variety of sensory receptors both inside and outside the body. Sensory receptors respond to stimuli by generating action potentials that are propagated along nerves to the spinal cord and brain. Most sensations result when action potentials reach the cerebral cortex. Some other parts of the brain are involved in sensation. For example, the thalamus and amygdaloid body are involved in the sensation of pain.

We are not consciously aware of much of the sensory information reaching the brain. For example, homeostasis is maintained by the unconscious monitoring and regulation of variables such as blood pressure, blood oxygen, and pH levels. Much of the information about body position and control of movements occurs at an unconscious level. In addition, we have selective awareness. We are more aware of sensations on which we focus our attention than on other sensations. If we were aware of all the sensory information that arrived at the cerebral cortex, we would probably not be able to function.

Even though sensory receptors are stimulated, we may not be aware of the stimulation because of **accommodation,** or **adaptation,** a decreased sensitivity to a continued stimulus. After exposure to a stimulus for a time, the response of some receptors or sensory pathways to a certain stimulus strength lessens from that which occurred when the stimulus was first applied. For example, when a person first gets dressed, tactile receptors and pathways relay information to the brain, resulting in awareness of clothes touching the skin. After a time, the action potentials from the skin decrease, and the clothes are ignored.

Sense is the ability to perceive stimuli. The senses are the means by which the brain receives information about the environment and the body. Historically, five senses were recognized: smell, taste, sight, hearing, and touch. Today, the senses are divided into two basic groups: general and special senses (figure 12.1). The **general senses** are those with receptors distributed over a large part of the body. They are divided into two groups: the somatic and the visceral senses. The **somatic senses,** which provide sensory information about the body and the environment, include touch, pressure, itch, vibration, proprioception, temperature, and pain. The **visceral senses,** which provide information about various internal organs, consist primarily of pain and pressure.

The **special senses** are more specialized in structure, have specialized nerve endings, and are localized to specific organs. The special senses are smell, taste, sight, hearing, and balance. Chapter 13 considers the special senses in detail.

Sensation requires the following steps:

1. Stimuli originating either inside or outside the body are detected by sensory receptors and action potentials are produced, which are propagated to the CNS by nerves (see "Sensory Receptors," next section).
2. Within the CNS, tracts conduct action potentials to the cerebral cortex and to other areas of the CNS (see "Sensory Tracts," p. 344).
3. Many action potentials reaching the cerebral cortex are ignored. Others are translated so the person becomes aware of the stimulus (see "Sensory Areas of the Cerebral Cortex," p. 346). ▼

1. *Define sensation and accommodation.*
2. *Define and give examples of general (somatic and visceral) and special senses.*

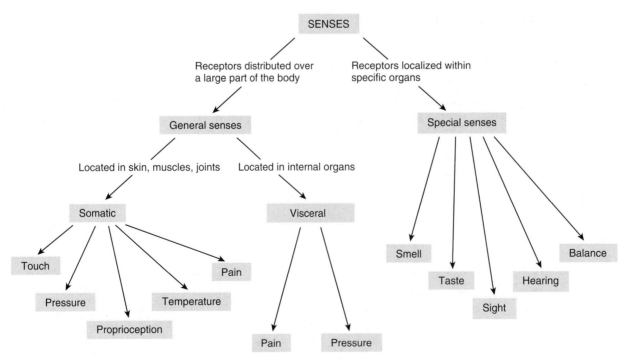

Figure 12.1   Classification of Senses

Sensory Receptors

Receptors are sensory nerve endings or specialized cells capable of responding to stimuli by developing action potentials. There are several types of receptors associated with both the special and general senses, each responding to a different type of stimulus. **Mechanoreceptors** (mek′ă-nō-rē-sep′tŏrz) respond to mechanical stimuli, such as the compression, bending, or stretching of receptors; **chemoreceptors** (kem′ō-rē-sep′tŏrz) respond to chemicals, such as odor molecules; **photoreceptors** (fō′tō-rē-sep′tŏrz) respond to light; **thermoreceptors** (ther′mō- rē-sep′tŏrz) respond to temperature changes; and **nociceptors** (nō′si-sep′tŏrs, *noceo,* to injure), or **pain receptors,** respond to painful mechanical, chemical, and thermal stimuli.

Many of the receptors for the general senses are associated with the skin (figure 12.2); others are associated with deeper structures, such as tendons, ligaments, and muscles. Structurally, the simplest and most common type of receptor nerve endings are **free nerve endings,** which are relatively unspecialized neuronal branches similar to dendrites. Free nerve endings are distributed throughout most parts of the body. Some free nerve endings respond to painful stimuli, some to temperature, some to itch, and some to movement.

Touch receptors are structurally more complex than free nerve endings, and many of them are enclosed by capsules (see figure 12.2). **Merkel** (mer′kĕl) **disks** are small, superficial nerve endings involved in detecting light touch and superficial pressure. **Hair follicle receptors,** associated with hairs, are involved in detecting light touch when the hair is bent. Light touch receptors are very sensitive but are not very discriminative, meaning that the point being touched cannot be precisely located. Receptors for

two-point discrimination are called **Meissner** (mīs′ner), or **tactile, corpuscles.** They are located just deep to the epidermis and are very specific in localizing tactile sensations. Deeper tactile receptors, called **Ruffini** (rū-fē′nē) **end organs,** play an important role in detecting continuous pressure in the skin. The deepest receptors in the dermis or subcutaneous tissue are called **Pacini** (pa-sin′ē), or **lamellated, corpuscles.** These receptors relay information concerning deep pressure and vibration.

Two-point discrimination, or fine touch, is the ability to detect simultaneous stimulation at two points on the skin. This sensation is important in evaluating the texture of objects. The distance between two points that a person can detect as separate points of stimulation differs for various regions of the body. Our fingertips, lips, and tongue have very good two-point discrimination because Meissner corpuscles are numerous and close together in these locations. Other parts of the body, such as the back, have fewer and more widely separated Meissner corpuscles and are less sensitive.

Proprioception (prō-prē-ō-sep′shun, perception of position) is the awareness of body position and movements. Proprioception provides information about the precise position and the rate of movement of various body parts, the weight of an object being held in the hand, and the range of movement of a joint. This information is involved in activities such as walking, climbing stairs, shooting a basketball, driving a car, eating, and writing. Although we can be consciously aware of body position and movements, much of this sensory information is processed at an unconscious level by the brain and spinal cord. Muscle spindles, Golgi tendon organs, Pacini corpuscles, and free nerve endings are the major receptors for proprioception. **Muscle spindles**

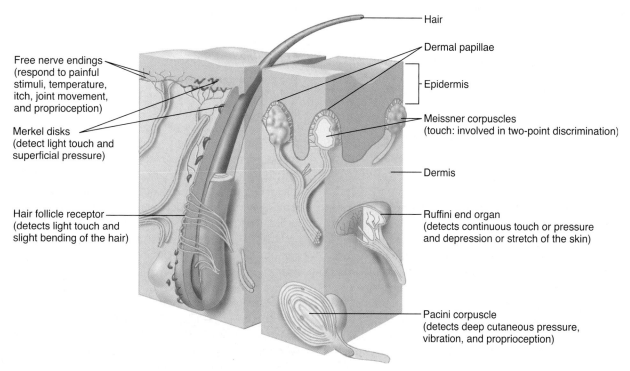

Figure 12.2 Sensory Nerve Endings in the Skin

Table 12.1 Ascending Spinal Pathways

Pathway	Information Transmitted
Anterolateral system	
Spinothalamic	Pain, temperature, light touch, pressure, tickle, and itch
Spinoreticular	Pain
Spinomesencephalic	Pain and touch
Dorsal column/ medial lemniscal system	Two-point discrimination, proprioception, pressure, and vibration
Trigeminothalamic	Pain, temperature, light touch, pressure, tickle, itch, two-point discrimination, proprioception, pressure, and vibration
Spinocerebellar	Proprioception

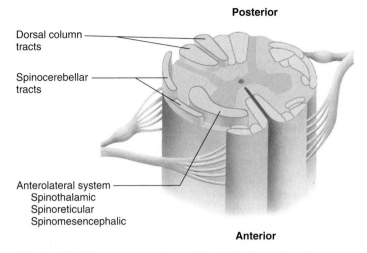

Posterior

Dorsal column tracts

Spinocerebellar tracts

Anterolateral system
Spinothalamic
Spinoreticular
Spinomesencephalic

Anterior

Figure 12.3 Cross Section of the Spinal Cord at the Cervical Level Depicting the Ascending Pathways

Ascending pathways are labeled on the left side of the figure only (*blue*), although they exist on both sides.

consist of 3–10 specialized skeletal muscle fibers. They are located in skeletal muscles and provide information about muscle length (see "Stretch Reflex," chapter 11). **Golgi tendon organs** are nerve endings associated with the fibers of a tendon near the junction between the muscle and tendon. They monitor the tension in tendons (see "Golgi Tendon Reflex," chapter 11). Pacini, or lamellated, corpuscles are complex receptors that resemble an onion (see figure 12.2). Pacini corpuscles in joints provide information about joint positions. ▼

3. *List five types of receptors based on the type of stimulus to which they respond.*
4. *List six types of receptors based on their structure and the sensations they detect.*
5. *Define two-point discrimination and proprioception. Which receptors are responsible for them?*

Sensory Tracts

The spinal cord and brainstem contain a number of sensory pathways that transmit action potentials from the periphery to various parts of the brain (table 12.1 and figure 12.3). Each pathway is involved with specific modalities, or types of sensory information. The neurons that make up each pathway are associated with specific types of sensory receptors. For example, thermoreceptors located in the skin generate action potentials that are propagated along the sensory pathway for pain and temperature, whereas Golgi tendon organs located in tendons generate action potentials that are propagated along sensory pathways involved with proprioception.

The names of many ascending pathways, or tracts, in the CNS indicate their origin and termination. Others indicate their location in the spinal cord. Each pathway usually is given a composite name in which the first half of the word indicates its origin and the second half indicates its termination. Ascending pathways therefore usually begin with the prefix *spino-,* indicating that they

originate in the spinal cord. For example, a spinocerebellar (spī′nō-ser-e-bel′ar) tract is one that originates in the spinal cord and terminates in the cerebellum. An example of location nomenclature is the dorsal column/medial lemniscal system, whose name is a combination of the pathway names in the spinal cord and brainstem. The specific function of each ascending tract, however, is not suggested by its name.

The two major ascending systems involved in the conscious perception of external stimuli are the anterolateral system and the dorsal column/medial lemniscal system. There are other pathways carrying sensory input of which we are not consciously aware.

Anterolateral System

The **anterolateral system** conveys cutaneous sensory information to the brain. It includes the spinothalamic, spinoreticular, and spinomesencephalic tracts. The **spinothalamic tracts** project to the thalamus, the sensory relay center of the brain (see chapter 11). Axons from the thalamus connect to the somatic sensory cortex, the part of the cerebral cortex where sensations are perceived (see "Sensory Areas of the Cerebral Cortex," p. 346). Thus, the spinothalamic tracts convey sensory information of which we are consciously aware, specifically pain and temperature information, as well as light touch and pressure and tickle and itch sensations. The **spinoreticular tracts** project to the reticular formation and the **spinomesencephalic tracts** project to the midbrain. They convey sensory information from pain and touch receptors of which we are not consciously aware.

The spinothalamic tracts are formed by three named neurons (figure 12.4). The **primary neuron** cell bodies of the spinothalamic tract are in the dorsal root ganglia. The primary neurons relay sensory input from the periphery to the posterior horn of the spinal cord, where they synapse with interneurons. The interneurons, which are not specifically named in the three-neuron sequence,

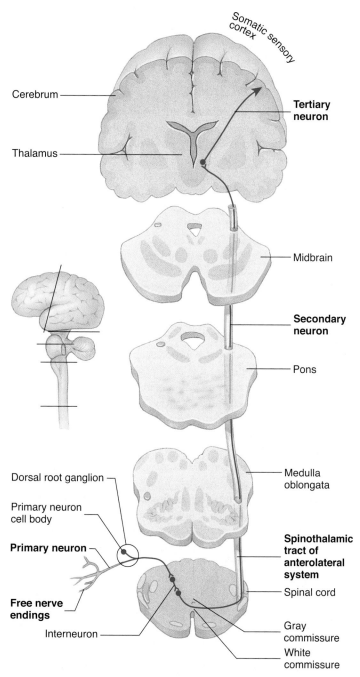

Figure 12.4 **Spinothalamic Tract of the Anterolateral System**

Primary neurons enter the spinal cord and synapse with interneurons that synapse with secondary neurons. The secondary neurons cross to the opposite side of the spinal cord, ascend to the thalamus, and synapse with tertiary neurons. The tertiary neurons connect to the somatic sensory cortex. Lines on the inset indicate levels of section.

synapse with secondary neurons. Axons from the **secondary neurons** cross to the opposite side of the spinal cord through the anterior portion of the gray and white commissures and enter the spinothalamic tract, where they ascend to the thalamus. The secondary neurons synapse with tertiary neurons in the thalamus. **Tertiary neurons** from the thalamus project to the somatic sensory cortex.

Sensory input transmitted by the anterolateral system that originates on one side of the body is perceived by the brain on the opposite side of the body because the secondary neurons cross over in the spinal cord. The term *contralateral* means relating to the opposite side. Each half of the cerebrum controls the contralateral body, and tracts on the opposite side of the body from their point of origin are said to be contralateral.

P R E D I C T **1**

As a result of an injury that interrupts the spinothalamic tract of the anterolateral system on the left side of the spinal cord, what sensations would be lost below the level of the injury and from which side of the body would they be lost?

Syringomyelia

Syringomyelia (sĭ-ring′gō-mī-ē′lē-ă) is a degenerative cavitation of the central canal of the spinal cord, often caused by a cord tumor. Symptoms include neuralgia (pain along the course of the nerve), paresthesia (abnormal sensations, burning, tingling, increased sensitivity to pain), specific loss of pain and temperature sensation, and paresis (partial paralysis). This defect is unusual in that it occurs in a distinct band that includes both sides of the body. Inferior to this band, sensations and motor functions are normal.

The defect causing syringomyelia occurs where the axons of secondary neurons of the anterolateral system cross over in the center of the spinal cord in commissural tracts. There is damage to crossing axons from both sides of the cord. As a result, there is loss of pain and temperature sensation on both sides of the body in a segment at the level of the cord damage. The loss of function is only at the level of damage to the cord. There is no loss of function below the level of the damage because the damage does not affect the ascending tracts in the anterolateral part of the cord.

Dorsal Column/Medial Lemniscal System

The **dorsal column/medial lemniscal** (lem-nis′kăl) **system** carries the sensations of two-point discrimination, proprioception, pressure, and vibration. This system is named for the dorsal column of the spinal cord and the medial lemniscus, which is the continuation of the dorsal column in the brainstem. The term *lemniscus* means ribbon and refers to the thin, ribbonlike appearance of the pathway as it passes through the brainstem.

Primary neurons of the dorsal column/medial lemniscal system are located in the dorsal root ganglia (figure 12.5). Axons of the primary neurons of the dorsal column/medial lemniscal system enter the spinal cord and ascend the spinal cord on the same side of the spinal cord on which they enter. The term *ipsilateral* means relating to the same side. Tracts on the same side of the body as their point of origin are said to be ipsilateral. The primary neurons synapse with secondary neurons located in the medulla oblongata. Axons from secondary neurons cross to the opposite side of the medulla oblongata through the decussations of the medial lemniscus and ascend through the medial lemniscus to synapse with tertiary neurons in the thalamus. Tertiary

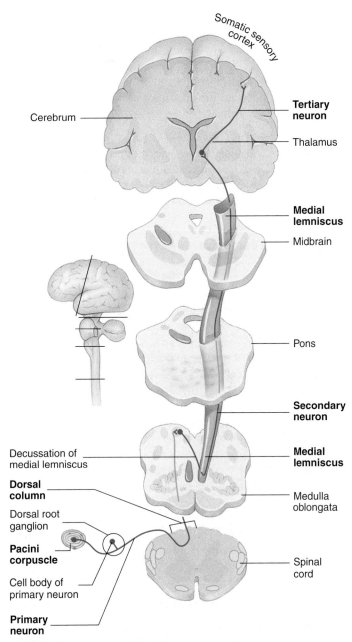

Figure 12.5   Dorsal Column/Medial Lemniscal System

Primary neurons enter the spinal cord, ascend ipsilaterally, and synapse with secondary neurons in the medulla oblongata. The secondary neurons cross to the opposite side of the spinal cord, ascend to the thalamus, and synapse with tertiary neurons. The tertiary neurons connect to the somatic sensory cortex. Lines on the inset indicate levels of section.

neurons from the thalamus project to the somatic sensory cortex.

Sensory input transmitted by the dorsal column/medial lemniscal system that originates on one side of the body is perceived by the brain on the opposite side of the body because the secondary neurons cross over in the medulla oblongata.

P R E D I C T

Two people, Bill and Mary, were involved in an accident and each experienced a loss of two-point discrimination, proprioception, and vibration on the right side of the body below the waist. It was determined that Bill had damage to his spinal cord as a result of the accident and that Mary had damage to her pons. Explain which side of the spinal cord was damaged in Bill and which side of the pons was damaged in Mary.

Trigeminothalamic Tract

As the fibers of the spinothalamic tracts pass through the brainstem, they are joined by fibers of the **trigeminothalamic tract.** The trigeminothalamic tract is made up primarily of afferent fibers from the trigeminal nerve. This tract carries the same sensory information as the spinothalamic tracts and dorsal column/medial lemniscal system but from the face, nasal cavity, and oral cavity, including the teeth. The trigeminothalamic tract is similar to the spinothalamic tracts and dorsal column/medial lemniscal system in that primary neurons from one side of the face synapse with secondary neurons, which cross to the opposite side of the brainstem. The secondary neurons synapse with tertiary neurons in the thalamus. Tertiary neurons from the thalamus project to the somatic sensory cortex.

Proprioception Tracts

The dorsal column/medial lemniscal system, the spinocerebellar tracts, and other tracts carry proprioceptive information to the brain. Proprioceptive input that reaches the cerebrum through the dorsal column/medial lemniscal system results in conscious proprioception, whereas input reaching the cerebellum results in unconscious proprioception. The cerebellum uses the proprioceptive information to compare actual movements with intended movements (see "Cerebellum," p. 355).

The **spinocerebellar tracts** project to the cerebellum by various routes, but in all cases proprioceptive input originating on one side of the body projects to the ipsilateral cerebellum. The spinocerebellar tracts provide unconscious proprioceptive input to the cerebellum from the lower limbs and trunk inferior to the midthorax. The dorsal column/medial lemniscal system receives proprioceptive input from the entire body, except for areas of the head supplied by the trigeminothalamic tract. Proprioceptive information carried by axons in the dorsal column and the trigeminothalamic tract can project to the contralateral cerebrum, resulting in conscious proprioception, or to the ipsilateral cerebrum, resulting in unconscious proprioception. ▼

6. *What sensory information is carried by the anterolateral and dorsal column/medial lemniscal systems? Describe where the neurons of these systems synapse and cross over from one side of the body to the opposite side.*
7. *What are the functions of the trigeminothalamic tracts?*
8. *Explain how the spinocerebellar tracts and the dorsal column/medial lemniscal system are involved with unconscious and conscious proprioception.*

Sensory Areas of the Cerebral Cortex

Figure 12.6 depicts a lateral view of the left cerebral cortex with some of its functional areas labeled. Sensory pathways project to specific regions of the cerebral cortex, called **primary sensory areas.** The primary sensory areas of the cerebral cortex must be intact for conscious perception, localization, and identification of a stimulus.

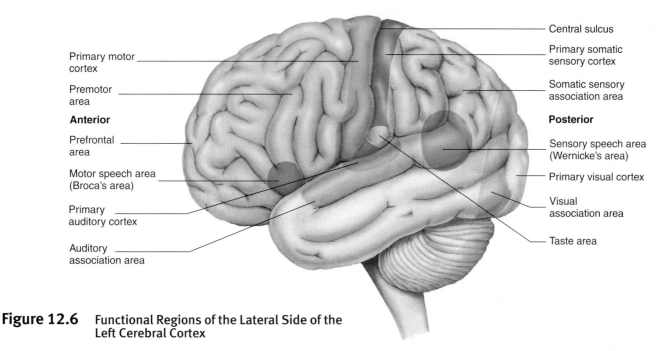

Central sulcus
Primary somatic sensory cortex
Somatic sensory association area

Posterior

Sensory speech area (Wernicke's area)
Primary visual cortex
Visual association area
Taste area

Primary motor cortex
Premotor area

Anterior

Prefrontal area
Motor speech area (Broca's area)
Primary auditory cortex
Auditory association area

Figure 12.6 Functional Regions of the Lateral Side of the Left Cerebral Cortex

The **primary visual cortex,** where portions of visual images are processed, is located in the occipital lobe. In the primary visual cortex, color, shape, and movement are processed separately rather than as a complete "color motion picture." The **primary auditory cortex,** where auditory stimuli are processed by the brain, is located in the temporal lobe. The terms *area* and *cortex* are often used interchangeably for the same functional region of the cerebral cortex. Taste is perceived in the **taste cortex** located in the insula (not shown in figure 12.6). The **olfactory cortex** (not shown in figure 12.6) is on the inferior surface of the temporal lobe and is where both conscious and unconscious responses to odor are initiated.

The central sulcus separates the frontal and parietal lobes of the brain (see figure 11.21). The gyrus immediately posterior to the central sulcus is the **postcentral gyrus**. Most of the postcentral gyrus is called the **somatic sensory cortex,** or **general sensory area.** Fibers carrying general sensory input in the anterolateral system and dorsal column/medial lemniscal system synapse in the thalamus, and thalamic neurons relay the information to the somatic sensory cortex.

The somatic sensory cortex is organized topographically relative to the general plan of the body (figure 12.7). The pattern of the somatic sensory cortex in each hemisphere is arranged in the form of an upside-down half homunculus (hō-mŭngk′ū-lŭs, little human) representing the opposite side of the body, with the feet located superiorly and the head located inferiorly. The size of various regions of the somatic sensory cortex is related to

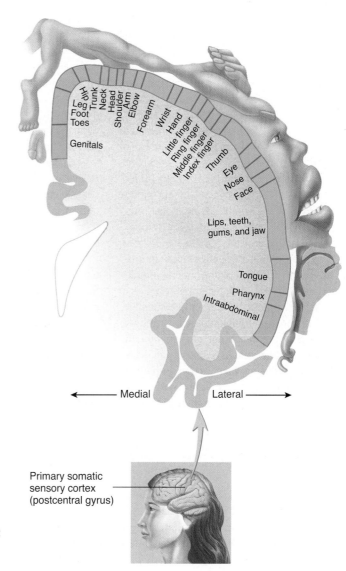

Medial

Lateral

Primary somatic sensory cortex (postcentral gyrus)

Figure 12.7 Topography of the Somatic Sensory Cortex

Cerebral cortex seen in frontal section on the left side of the brain. The figure of the body (homunculus) depicts the nerve distributions; the size of each body region shown indicates relative innervation. The cortex occurs on both sides of the brain but appears on only one side in this illustration. The inset shows the somatic sensory region of the left hemisphere (*green*).

Pain is a sensation characterized by a group of unpleasant and complex perceptual and emotional experiences that trigger autonomic, psychological, and somatic motor responses. There are two types of pain sensation: (1) sharp, well-localized, pricking, or cutting pain resulting from rapidly conducted action potentials carried by large-diameter, myelinated axons and (2) diffuse, burning, or aching pain resulting from action potentials that are propagated more slowly by smaller, less heavily myelinated axons (see chapter 10).

Superficial pain sensations in the skin are highly localized as a result of the simultaneous stimulation of the pain receptors of the anterolateral system and tactile receptors of the dorsal column/medial lemniscal system. Deep, or visceral, pain sensations are not highly localized because of the absence of tactile receptors in the deeper structures. Visceral pain stimuli are normally perceived as diffuse pain.

Dorsal column/medial lemniscal system neurons are involved in what is called the **gate-control theory** of pain control. Primary neurons of the dorsal column/medial lemniscal system send out collateral branches that synapse with interneurons in the posterior horn of the spinal cord. These interneurons have an inhibitory effect (see chapter 10) on secondary neurons of the spinothalamic tract. Thus, pain action potentials traveling through the spinothalamic tract can be suppressed by action potentials that originate in neurons of the dorsal column/medial lemniscal system. The arrangement may act as a "gate" for pain action potentials transmitted in the spinothalamic tract. Increased activity in the dorsal column/medial lemniscal system tends to close the gate, thereby reducing pain action potentials transmitted in the spinothalamic tract. Descending pathways from the cerebral cortex or other brain regions can also regulate this "gate."

The gate-control theory may explain the physiological basis for treatments that have been used to reduce the intensity of chronic pain: electric stimulation of dorsal column/medial lemniscal neurons, transcutaneous electric stimulation (applying a weak electric stimulus to the skin), acupuncture, massage, and exercise. The frequency of action potentials that are transmitted in the dorsal column/medial lemniscal system is increased when the skin is rubbed vigorously and when the limbs are moved and may explain why vigorously rubbing a large area around a source of pricking pain tends to reduce the intensity of the painful sensation. Exercise normally decreases the sensation of pain, and exercise programs are important components in the management of chronic pain not associated with illness. Action potentials initiated by acupuncture procedures may act through a gating mechanism in which the inhibition of action potentials in neurons that transmit pain action potentials upward in the spinal cord is influenced by activity in sensory cells that send collateral branches to the posterior horn.

Analgesics are drugs that reversibly relieve pain without loss of sensation or consciousness. Some analgesics block the transmission of pain in the spinal cord from primary neurons to neurons of the ascending pathways. Other analgesics function at the level of the cerebral cortex to modify pain perception. **Anesthetics** are drugs that reversibly depress nerve cell function, resulting in loss of sensation or consciousness. **Local anesthesia** is loss of sensation in a part of the body achieved by injecting anesthetics that block action potential transmission. **General anesthesia** is loss of consciousness achieved with anesthetics acting on the reticular formation.

Referred Pain

Referred pain is a painful sensation in a region of the body that is not the source of the pain stimulus. Most commonly, referred pain is sensed in the skin or other superficial structures when internal organs are damaged or inflamed (figure A). This occurs because sensory neurons from the superficial area to which the pain is referred and the neurons from the deeper, visceral area where the pain stimulation originates converge onto the same ascending neurons in the spinal cord. The brain cannot distinguish between the two sources of pain stimuli, and the painful sensation is referred to the most superficial structures innervated, such as the skin.

Referred pain is clinically useful in diagnosing the actual cause of the painful stimulus. For example, during a heart attack, pain receptors in the heart are stimulated when blood flow is blocked to some of the heart muscle. Heart attack victims, however, often do not feel the pain in the heart but instead feel what they perceive as cutaneous pain radiating from the left shoulder down the arm (see figure A).

Phantom Pain

Phantom pain occurs in people who have had appendages amputated or a structure, such as a tooth, removed. Many of these people perceive pain in the amputated structure as if it were still in place. If a neuron pathway that transmits action potentials is stimulated at any point along that pathway, action potentials are initiated and propagated toward the CNS. Integration results in the perception of pain that is projected to the site of the sensory receptors, even if those sensory receptors are no longer present. A similar phenomenon can be easily demonstrated by bumping the ulnar nerve as it crosses the elbow (the funny bone). A sensation of pain is often felt in

---

the number of sensory receptors in that area of the body. The density of sensory receptors is much greater in the face than in the legs; therefore, a greater area of the somatic sensory cortex contains sensory neurons associated with the face, and the homunculus has a disproportionately large face.

Cutaneous sensations, although integrated within the cerebrum, are perceived as though they were on the surface of the body.

This is called **projection** and indicates that the brain refers a cutaneous sensation to the superficial site at which the stimulus interacts with the sensory receptors. For example, when an injection is given in the gluteus medius, action potentials are generated and transmitted to the portion of the somatic sensory cortex corresponding to the hip. When neurons in the somatic sensory cortex are activated, there is a sensation of pain. Projection is the process by which the location

the fourth and fifth digits, even though the neurons were stimulated at the elbow.

A factor that may be important in phantom pain is the lack of touch, pressure, and proprioceptive impulses from the amputated limb. Those action potentials suppress the transmission of pain action potentials in the pain pathways, as explained by the gate-control theory of pain. When a limb is amputated, the inhibitory effect of sensory information is removed. As a consequence, the intensity of phantom pain may be increased. Another factor in phantom pain may be that the cerebral cortex retains an image of the amputated body part.

Chronic Pain

Chronic pain is long-lasting pain. Some chronic pain has a known cause, such as tissue damage, as in the case of arthritis. Other chronic pain cannot be associated with tissue damage and has no known cause.

Pain is important in warning us of potentially injurious conditions because pain receptors are stimulated when tissues are injured. Pain itself, however, can become a problem. Chronic pain, such as migraine headaches, localized facial pain, or back pain, can be very debilitating, and pain loses its value of providing information about the condition of the body. People suffering from chronic pain often feel helpless and hopeless, and they may become dependent on drugs. The pain can interfere with vocational pursuits, and many victims are unemployed or even housebound and socially isolated. They are easily frustrated or angered, and they suffer symptoms of major depression. These qualities are associated with what is called **chronic pain syndrome.** Over 2 million people in the United States at any given time suffer chronic pain sufficient to impair activity.

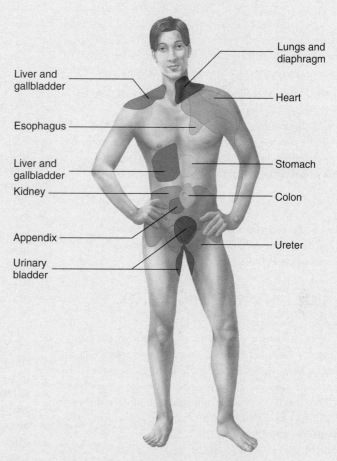

Figure A   Areas of Referred Pain on the Body Surface

Pain from the indicated internal organs is referred to the surface areas shown.

Chronic pain may originate with acute pain associated with an injury or may develop for no apparent reason. How sensory signals are processed in the thalamus and cerebrum may determine if the input is evaluated as only a discomfort, a minor pain, or a severe pain and how much distress is associated with the sensation. The brain actively regulates the amount of pain information that gets through to the level of perception, thereby suppressing much of the input. If this dampening system becomes less functional, pain perception may increase. Other nervous system factors, such as a loss of some sensory modalities from an area, or habituation of pain transmission, which may remain even after the stimulus is removed, may actually intensify otherwise normal pain sensations. Treatment often requires a multidisciplinary approach, including such interventions as surgery or psychotherapy. Some sufferers respond well to drug therapy, but some drugs, such as opiates, have a diminishing effect and may become addictive.

of the painful stimulus is determined. During brain surgery, direct stimulation of the somatic sensory cortex results in the perception of stimulation of the skin without the skin actually being stimulated.

Sensory Processing

Association areas are cortical areas immediately adjacent to the primary sensory areas that are involved in the process of recognition. Important association areas are the **visual association area, auditory association area**, and **somatic sensory association area** (see figure 12.7). Sensory information first reaches the primary sensory areas, where its basic components are interpreted. That information is passed to the association areas for further analysis. For example, action potentials originating in the retina of the eye reach the primary visual cortex, where the image is "perceived." Action potentials then pass from the primary visual cortex to the visual association area, where the present visual information is compared with past visual experience ("Have I seen this before?"). On the basis of this comparison, the visual association area "decides" whether or not the visual input is recognized and passes judgment concerning the significance of the input. For example, we generally pay less attention to people in a crowd we have never seen before than to someone we know. ▼

9. *Describe in the cerebral cortex the locations and functions of the primary sensory areas.*
10. *Describe the topographic organization of the general body plan in the somatic sensory cortex. Why are some areas of the body represented as larger than other areas?*
11. *Where are association areas located and what is their function?*

P R E D I C T ③

Using the visual association areas as an example, explain the general functions of the auditory association areas around the primary auditory cortex (see figure 12.7).

Control of Skeletal Muscles

The motor system of the brain and spinal cord is responsible for maintaining the body's posture and balance; for moving the trunk, head, limbs, and eyes; and for communicating through facial expressions and speech. Reflexes mediated through the spinal cord and brainstem (see chapter 11) are responsible for some body movements. They occur without conscious thought. **Voluntary movements,** on the other hand, are movements consciously activated to achieve a specific goal, such as walking or typing. Although consciously activated, the details of most voluntary movements occur automatically. After walking begins, it is not necessary to think about the moment-to-moment control of every muscle because neural circuits automatically control the limbs. After learning how to do complex tasks, such as typing, people can perform them relatively automatically.

Voluntary movements depend on lower and upper motor neurons. **Lower motor neurons** have axons that leave the central nervous system and extend through peripheral nerves to supply skeletal muscles. The cell bodies of lower motor neurons are located in the anterior horns of the spinal cord gray matter and in cranial nerve nuclei of the brainstem. Lower motor neurons are the neurons forming the motor units described in chapter 8.

The axons of **upper motor neurons** form tracts that directly or indirectly control the activities of lower motor neurons. The cell bodies of upper motor neurons are located in the cerebrum, brainstem, and cerebellum. Usually, upper motor neurons do not synapse directly with lower motor neurons. Instead, the upper motor neurons synapse with unnamed, short interneurons, which synapse with lower motor neurons.

Voluntary movements depend on the following:

1. The initiation of most voluntary movement begins with upper motor neurons in the cerebral cortex (see "Motor Areas of the Cerebral Cortex").
2. The axons of upper motor neurons form descending tracts that stimulate lower motor neurons, which stimulate skeletal muscles to contract (see "Motor Tracts," p. 351).

3. The cerebral cortex interacts with the basal nuclei and cerebellum in the planning, coordination, and execution of movements (see "Modifying and Refining Motor Activities," p. 355). ▼

12. *Distinguish between lower and upper motor neurons.*

Motor Areas of the Cerebral Cortex

The **precentral gyrus,** located immediately anterior to the central sulcus, is also called the **primary motor cortex** or **primary motor area** (see figure 12.6). Action potentials initiated in this region control many voluntary movements, especially the fine motor movements of the hands. Upper motor neurons are not confined to the precentral gyrus—only about 30% of them are located there. Another 30% are in the premotor area, and the rest are in the somatic sensory cortex.

The cortical functions of the precentral gyrus are arranged topographically according to the general plan of the body—similar to the topographic arrangement of the postcentral gyrus (figure 12.8). The neuron cell bodies controlling motor functions of the feet are in the most superior and medial portions of the precentral gyrus, whereas those for the face are in the inferior region. Muscle groups with many motor units are represented by relatively large areas of the precentral gyrus. For example, muscles performing precise movements, such as those controlling the hands and face, have many motor units, each of which has a small number of muscle fibers. Multiple motor unit summation (see chapter 8) can precisely control the force of contraction of these muscles because only a few muscle fibers at a time are recruited. Muscle groups with few motor units are represented by relatively small areas of the precentral gyrus, even if the muscles innervated are quite large. Muscles such as those controlling movements of the thigh and leg have proportionately fewer motor units than hand muscles but have many more and much larger muscle fibers per motor unit. They are less precisely controlled because the activation of a motor unit stimulates the contraction of many large muscle fibers.

The **premotor area,** located anterior to the primary motor cortex (see figure 12.6), is the staging area in which motor functions are organized before they are initiated in the motor cortex. For example, if a person decides to take a step, the neurons of the premotor area are stimulated first. The determination is made in the premotor area as to which muscles must contract, in what order, and to what degree. Action potentials are then passed to the

Apraxia

The premotor area must be intact for a person to carry out complex, skilled, or learned movements, especially ones related to manual dexterity—for example, a surgeon's use of a scalpel or a student's use of a pencil. Impairment in the performance of learned movements, called **apraxia** (ă-prak′sē-ă), can result from a lesion in the premotor area. Apraxia is characterized by hesitancy and reduced dexterity in performing these movements.

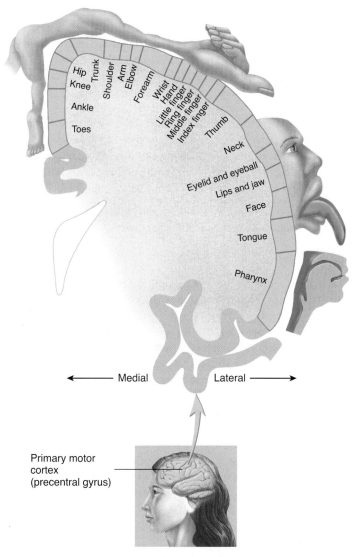

Figure 12.8 Topography of the Primary Motor Cortex

Cerebral cortex seen in frontal section on the left side of the brain. The figure of the body (homunculus) depicts the nerve distributions; the size of each body region shown indicates relative innervation. The cortex occurs on both sides of the brain but appears on only one side in this illustration. The inset shows the motor region of the left hemisphere (*pink*).

upper motor neurons in the motor cortex, which actually initiate the planned movements.

The motivation and the foresight to plan and initiate movements occur in the anterior portion of the frontal lobes, the **prefrontal area.** This is a region of association cortex that is well developed only in primates, especially in humans. It is involved in motivation and the regulation of emotional behavior and mood. The large size of this area in humans may account for our relatively well-developed forethought and motivation and for our emotional complexity. ▼

> 13. Where is the primary motor cortex located and what is its function? Why are some areas of the body represented as larger than other areas on the topographic map of the body on the primary motor cortex?
>
> 14. Where are the premotor and prefrontal areas of the cerebral cortex located and what is their function?

Motor Tracts

Motor tracts are pathways in the CNS that directly or indirectly affect the activities of lower motor neurons. The names of tracts are based on their origin and termination. Much like the names of ascending tracts, the prefix indicates its origin and the suffix indicates its destination. For example, the corticospinal tract is a motor tract that originates in the cerebral cortex and terminates in the spinal cord (table 12.2 and figure 12.9).

The internal capsule contains projection fibers, which form tracts connecting the cerebrum to other parts of the brain (see figure 11.22). Tracts of the internal capsule connect the cerebrum to the thalamus, basal nuclei, cerebellum, brainstem,

Table 12.2	Descending Spinal Pathways
Pathway	**Functions Controlled**
Direct	Conscious skilled movements
Corticospinal tract	Movements below the head, especially of the hands
Lateral	Movements of the neck, trunk, and limbs
Anterior	Movements of the neck and upper limbs
Corticobulbar tract	Movements of the head and face
Indirect	Conscious and unconscious movements
Rubrospinal	Coordination of fine hand movements
Reticulospinal	Posture adjustment, walking
Vestibulospinal	Maintenance of upright posture, balance

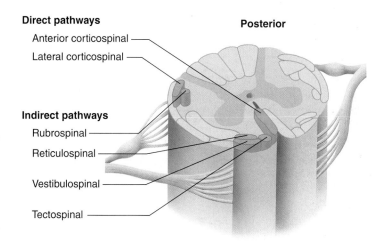

Direct pathways
Anterior corticospinal
Lateral corticospinal

Indirect pathways
Rubrospinal
Reticulospinal
Vestibulospinal
Tectospinal

Posterior

Anterior

Figure 12.9 Cross Section of the Spinal Cord at the Cervical Level Depicting the Descending Pathways

Descending pathways are labeled on the left side of the figure only (*pink*), though they exist on both sides.

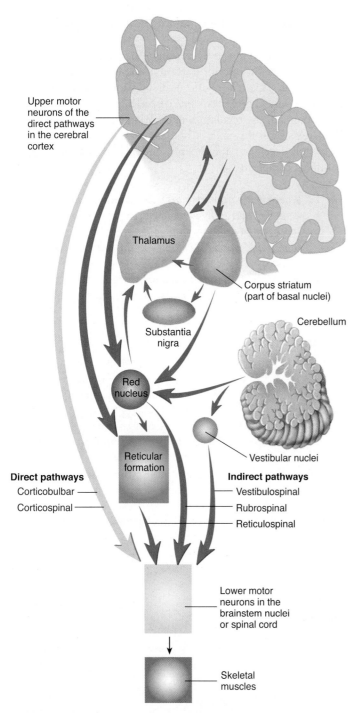

Upper motor neurons of the direct pathways in the cerebral cortex

Thalamus

Corpus striatum
(part of basal nuclei)

Cerebellum

Substantia
nigra

Red
nucleus

Reticular
formation

Vestibular nuclei

Direct pathways

Corticobulbar

Corticospinal

Indirect pathways

Vestibulospinal

Rubrospinal

Reticulospinal

Lower motor
neurons in the
brainstem nuclei
or spinal cord

Skeletal
muscles

Figure 12.10   Descending Pathways

The direct pathways (corticobulbar and corticospinal) are indicated by the *blue arrow*. The indirect pathways and their interconnections are indicated by the *red arrows*.

indirect pathways and spinal cord. The descending motor tracts are divided into two groups: direct pathways and indirect pathways (figure 12.10). The **direct pathways** arise from the cerebral motor cortex. Axons of upper motor neurons form tracts, which extend to lower motor neurons in the spinal cord or brainstem. The

indirect pathways arise from the cerebral motor cortex and the cerebellum and project to nuclei in the brainstem. Axons from these nuclei form tracts that project to lower motor neurons. The indirect pathways are so called because of the presence of these intermediate nuclei between the cerebral motor cortex and the cerebellum and lower motor neurons.

The direct pathways are involved in voluntary control of skeletal muscles and the maintenance of muscle tone. They are especially important for the control of distal limb and facial muscles. The indirect pathways are involved in voluntary control of skeletal muscles, especially of the trunk and proximal limbs. They are associated with overall body coordination and cerebellar function, such as posture and balance. ▼

15. What are direct and indirect pathways?

Direct Pathways

The direct pathways include groups of nerve fibers arrayed into two tracts: the corticospinal and the corticobulbar tracts. The **corticospinal tract** is involved in direct cortical control of movements below the head. It consists of axons of upper motor neurons located in the primary motor and premotor areas of the frontal lobes and the somatic sensory parts of the parietal lobes (figure 12.11). They descend through the internal capsules and the cerebral peduncles of the midbrain to the pyramids of the medulla oblongata. At the inferior end of the medulla 75%–85% of the corticospinal fibers cross to the opposite side of the CNS through the **pyramidal decussation,** which is visible on the anterior surface of the inferior medulla (see figure 11.18). The crossed fibers descend in the **lateral corticospinal tracts** of the spinal cord (see figure 12.11). The remaining 15%–25% descend uncrossed in the **anterior corticospinal tracts** and decussate near the level where they synapse with lower motor neurons. The anterior corticospinal tracts supply the neck and upper limbs, and the lateral corticospinal tracts supply all levels of the body.

Most of the corticospinal fibers synapse with interneurons in the lateral portions of the spinal cord central gray matter. The interneurons, in turn, synapse with the lower motor neurons of the anterior horn that innervate primarily distal limb muscles.

Damage to the corticospinal tracts results in clumsiness, weakness, and reduced muscle tone, but not in complete paralysis, even if the damage is bilateral. Experiments with monkeys have demonstrated that bilateral sectioning of the medullary pyramids results in difficulty using the distal parts of the limbs, especially the hands. The monkeys, however, still had use of axial and proximal limb muscles through the indirect pathways and could stand, walk, and climb. These and other data support the conclusion that the corticospinal system is superimposed over the indirect pathways and that it has many parallel functions. The main function of the direct pathways is to add speed and agility to voluntary movements, especially of the hands, and to provide a high degree of fine motor control, such as in the movements of individual fingers. Spinal cord lesions that affect both the direct and indirect pathways result in complete paralysis.

The **corticobulbar tracts** are involved in the direct cortical control of movements in the head and neck. Upper motor neurons that contribute to the corticobulbar tracts are in regions of the cortex similar to those of the corticospinal tracts and follow the same basic route as the corticospinal system down to the level of the brainstem. At that point, corticobulbar fibers terminate in the motor nuclei of some cranial nerves. Usually, the upper motor neurons synapse with interneurons, which synapse with lower motor neurons. The cranial nerves control the muscles of mastication, muscles of facial expression, trapezius, sternocleidomastoid, tongue muscles, and palatine, pharyngeal, and laryngeal muscles. Note that the corticobulbar tracts are not involved with the muscles controlling eye movements. The circuitry for controlling eye movements is complex, involving many parts of the brain, such as cerebral cortex motor areas, superior colliculi, reticular formation nuclei, vestibular nuclei, and the cerebellum. ▼

> 16. What two tracts form the direct pathways? Which parts of the body are supplied by each tract? Describe the location of the neurons in each tract, where they cross over, and where they synapse.
> 17. Describe the functions of the direct pathways. How are they related to the indirect pathways?

Indirect Pathways

The indirect pathways (see figure 12.10) originate with upper motor neurons of the cerebrum and cerebellum whose axons synapse in some intermediate nucleus rather than directly with lower motor neurons. Axons from the neurons in those nuclei form tracts that connect to lower motor neurons. In general, the indirect pathways decussate in the brainstem so that one side of the brain is controlling the opposite side of the body. Some indirect pathway fibers are distributed bilaterally. The major tracts of the indirect pathways are the reticulospinal, rubrospinal, and vestibulospinal tracts. The indirect pathways are involved with voluntary control of skeletal muscles and are very important in maintaining posture, holding of the limbs in position, and balance.

Neuron cell bodies of the **reticulospinal tract** (figure 12.12) are in the reticular formation of the pons and medulla oblongata. Their axons descend in the anterior portion of the lateral column and synapse with interneurons and lower motor neurons in the ventromedial portion of the spinal cord central gray matter. This tract is important for the voluntary control of skeletal muscles. It also maintains posture through the action of trunk and proximal upper and lower limb muscles during certain movements. For example, when a person who is walking lifts one foot off the ground, the weight of the body is shifted to the other limb.

Neurons of the **rubrospinal tract** begin in the red nucleus, which is located at the boundary between the diencephalon and midbrain (see figure 12.12). The tract decussates in the midbrain and descends in the lateral column of the spinal cord. The red nucleus receives input from both the motor cortex and the cerebellum. Lesions in the red nucleus result in intention, or action, tremors

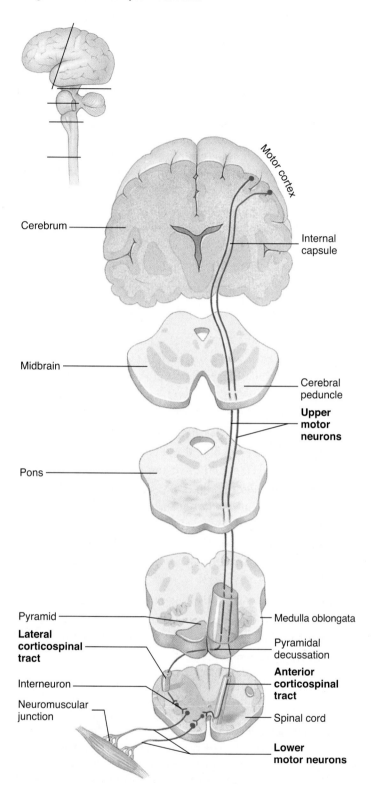

Figure 12.11 Direct Pathways

Upper motor neuron axons descend to the medulla oblongata. Most axons decussate in the medulla oblongata and descend in the lateral corticospinal tracts in the spinal cord. Some axons continue as the anterior corticospinal tracts and decussate in the spinal cord. Upper motor neurons synapse with interneurons that synapse with lower motor neurons. Lines on the inset indicate levels of section.

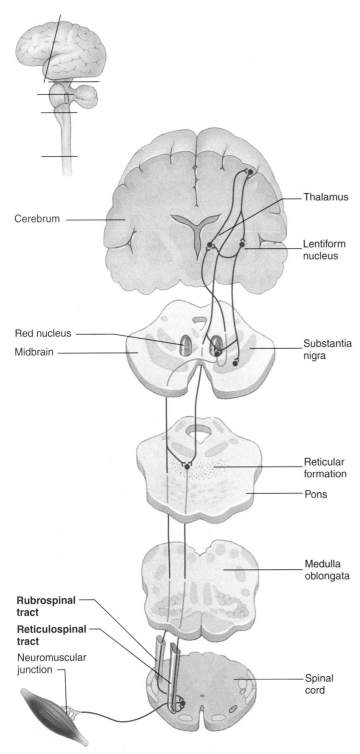

Figure 12.12   Indirect Pathways

Examples of indirect pathways are the reticulospinal and rubrospinal tracts. Neurons from the cerebrum project to neurons in brainstem nuclei. The axons of these neurons form the reticulospinal and rubrospinal tracts. Lines on the inset indicate levels of section.

similar to those seen in cerebellar lesions (see "Basal Nuclei and Cerebellar Disorders," p. 357). The function of the red nucleus, therefore, is related closely to cerebellar function. The rubrospinal tract is the one indirect tract that is very closely related to the

direct, corticospinal tract. It plays a role in coordinating fine motor movements in the distal part of the upper limbs. Damage to the rubrospinal tract impairs forearm and hand movements but does not greatly affect general body movements.

The **vestibulospinal tracts** originate in the vestibular nuclei, located in the medulla and pons. Their fibers preferentially influence neurons innervating extensor muscles in the trunk and proximal portion of the lower limbs and are involved primarily in the maintenance of upright posture and balance. The vestibular nuclei receive major input from the organs of balance in the inner ear (see chapter 13) and the cerebellum. ▼

> 18. *Name the structures and the tracts that form the indirect pathways. What functions do they control?*

CASE STUDY | Spinal Cord Injury

Harley was taking a curve on his motorcycle when he hit a patch of loose gravel on the side of the road. His motorcycle slid onto its side. Harley fell off and slid, as he was spun around, into a retaining wall. A passing motorist stopped at the wreck and saw that Harley was still conscious and breathing. The motorist (wisely) did not move Harley but dialed 911. The paramedics placed Harley on a back-support bed and rushed him to the hospital. Examination there revealed that Harley had suffered a fracture of his second lumbar vertebra. Methylprednisolone was administered to reduce inflammation around the spinal cord. His vertebrae were realigned and stabilized. A neurologist determined that the entire left half of Harley's spinal cord was destroyed at level L2. A lesion of the spinal cord that destroys half the cord (hemisection) at a specific level results in a very specific syndrome, called the Brown-Séquard syndrome.

P R E D I C T

Use figures 11.12, 12.3, 12.4, 12.5, 12.9, 12.11, and 12.12 to answer these questions.

a. What sensory tracts were damaged on the left side of the spinal cord? As a result, what sensations did Harley lose? On which side of his body did Harley lose sensations?

b. On the side of the body Harley lost sensation, which parts of the body lost sensation?

c. What motor tracts were damaged on the left side of the spinal cord? As a result, which side of his body was Harley no longer able to control voluntarily?

d. On the side of the body on which Harley lost motor control, which parts of the body were affected? Name the major nerves affected by Harley's injury.

Spinal Shock

Following injury to the spinal cord, reflexes below the level of injury are lost. This response is called **spinal shock.** Recall from chapter 11 that interneurons and motor neurons of reflex arcs are connected to the brain through descending tracts (see figure 11.6). Continual stimulation of reflex centers in the spinal cord through the reticulospinal, vestibulospinal, and corticospinal tracts helps neurons maintain a state of responsiveness to stimulation. That is, EPSPs move membrane potentials closer to threshold, increasing the excitability of neurons (see chapter 10). Loss of the facilitory input from the brain results in the depression of spinal cord reflex centers and loss of reflex activity. The presence or absence of specific reflexes following spinal cord injury is an indication of the severity and level of the injury. For example, the patellar (knee-jerk) reflex is an extension of the knee when the patellar ligament is struck with a rubber hammer (see figure 11.7). It is mediated through levels L2–L4 of the spinal cord. The plantar reflex is flexion of the toes when the lateral surface of the sole of the foot is stroked with a sharp object. It is mediated through levels S1–S2 of the spinal cord. The presence of the patellar reflex and the absence of the plantar reflex indicate injury to the spinal cord below level L4.

Within 2 weeks to several months, reflexes lost because of spinal shock are usually recovered. Sometimes recovery is not complete. The reasons for recovery from spinal shock are not well understood. Neurons deprived of normal facilitory input gradually return to their previous state of excitability. The formation of new synapses may play a role. The degeneration of descending neurons leaves vacant synaptic sites on neurons in spinal cord reflex centers. New synaptic connections by neurons of the reflex arc could fill these vacant sites and increase reflex sensitivity.

Damage to the spinal cord can also affect tracts and reflexes of the ANS. Spinal shock can result in the loss of reflexes necessary for urination and defecation to occur. Recovery from spinal shock results in reflexive urination and defecation.

Modifying and Refining Motor Activities

Basal Nuclei

The **basal nuclei** (see figure 11.23) are important in planning, organizing, and coordinating motor movements and posture. Complex neural circuits link the basal nuclei with each other, with the thalamus, and with the cerebral cortex. These connections form several feedback loops, some of which have excitatory effects on the cerebral motor cortex and some of which have inhibitory effects. The excitatory circuits facilitate muscle activity, especially at the beginning of a voluntary movement, such as rising from a sitting position or beginning to walk. The inhibitory circuits facilitate the actions of the stimulatory circuits by inhibiting muscle activity in antagonist muscles. Inhibitory circuits also decrease muscle tone when the body, limbs, and head are at rest, which eliminates "unwanted" movements. Basal nuclei disorders fall into two major categories: (1) decreased excitatory and inhibitory effects and (2) increased

excitatory and inhibitory effects (see "Basal Nuclei and Cerebellar Disorders," p. 357). ▼

19. What are the functions of the basal nuclei?

Cerebellum

The **cerebellum** is involved in balance, the coordination of fine motor movement, and the maintenance of muscle tone. If the cerebellum is damaged, fine motor movements become very clumsy and muscle tone decreases. The cerebellum (see figure 11.19) consists of three functional parts: the vestibulocerebellum, the spinocerebellum, and the cerebrocerebellum. The **vestibulocerebellum,** or **flocculonodular lobe,** receives direct input from the organs of balance in the inner ear (see chapter 13) and sends axons to the vestibular nuclei of the brainstem. It helps control balance, especially during movements, and it helps coordinate eye movement. The vestibulocerebellum also helps maintain muscle tone in postural muscles.

The **vermis** and medial portion of the **lateral hemisphere,** referred to jointly as the **spinocerebellum,** helps accomplish fine motor coordination of simple movements by means of its **comparator** function. Action potentials from the motor cortex descend into the spinal cord to initiate voluntary movements. At the same time, action potentials are carried from the motor cortex to the cerebellum to give the cerebellar neurons information representing the intended movement (figure 12.13). Simultaneously, action potentials from proprioceptive neurons ascend through the spinocerebellar tracts to the cerebellum. Proprioceptive neurons innervate the joints and tendons of the structure being moved, such as the elbow or knee, and provide information about the position of the body or body parts. These action potentials give the cerebellar neurons information from the periphery about the actual movements. The cerebellum compares the action potentials from the motor cortex with those from the moving structures. That is, it compares the intended movement with the actual movement. If a difference is detected, the cerebellum sends action potentials through the thalamus to the cerebral motor cortex and to the spinal cord to correct the discrepancy. The result is smooth and coordinated movement.

The comparator function coordinates simple movements, such as touching your nose. Rapid, complex movements, however, require much greater coordination and training. The **cerebrocerebellum** consists of the lateral two-thirds of the lateral hemispheres. It communicates with the motor, premotor, and prefrontal portions of the cerebral cortex to help in planning and practicing rapid, complex motor actions. Because of the cerebrocerebellum, with training a person can learn highly skilled and rapid movements that are accomplished more rapidly than can be accounted for by the comparator function of the cerebellum. In these cases, the cerebellum participates with the cerebrum in learning highly specialized movements, such as playing the piano or swinging a baseball bat. The cerebrocerebellum is also involved in cognitive functions, such as rhythm, conceptualization of time intervals, some word associations, and solutions to pegboard puzzles—tasks once thought to occur only in the cerebrum.

1. The motor cortex sends action potentials to lower motor neurons in the spinal cord.

2. Action potentials from the motor cortex inform the cerebellum of the intended movement.

3. Lower motor neurons in the spinal cord send action potentials to skeletal muscles, causing them to contract.

4. Proprioceptive signals from the skeletal muscles and joints to the cerebellum convey information concerning the status of the muscles and the structure being moved during contraction.

5. The cerebellum compares the information from the motor cortex to the proprioceptive information from the skeletal muscles and joints.

6. Action potentials from the cerebellum to the spinal cord modify the stimulation from the motor cortex to the lower motor neurons.

7. Action potentials from the cerebellum are sent to the motor cortex, which modify its motor activity.

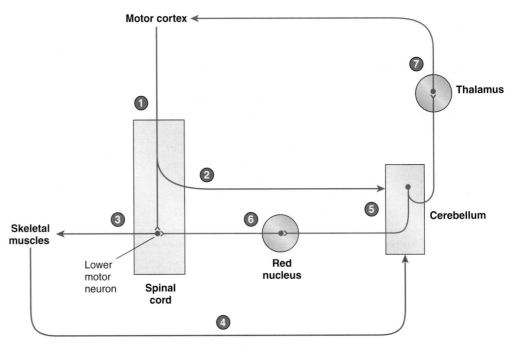

Process Figure 12.13   Cerebellar Comparator Function

There are two major cerebellar inputs: from the cerebral motor cortex and proprioception from the body. Tracts from the motor cortex descend through the internal capsule to synapse in pontine nuclei of the cerebral peduncles. Axons from the pontine nuclei cross the midline in the middle cerebellar peduncle and supply the contralateral cerebellum. Thus, the cerebral cortex on one side of the brain projects to the cerebellum on the opposite side. The cerebellum receives proprioception input ipsilaterally (see "Proprioception Tracts," p. 346).

There are two major cerebellar outputs: to the cerebral motor cortex and the spinal cord. Axons from the cerebellum pass through the superior cerebellar peduncles, which cross in the midbrain, and project to the contralateral thalamus. The thalamus connects to the cerebral motor cortex through the internal capsule. Thus, the cerebellum on one side of the brain projects to the cerebral motor cortex on the opposite side. Axons from the cerebellum synapse with brainstem nuclei of the indirect pathways, thus gaining access to lower motor neurons in the spinal cord. ▼

20. **What parts of the cerebellum contribute to the vestibulocerebellum, spinocerebellum, and cerebrocerebellum?**
21. **What is the function of the vestibulocerebellum?**
22. **Explain the comparator activities of the spinocerebellum.**
23. **Describe the role of the cerebrocerebellum in rapid and skilled motor movements, such as playing the piano.**
24. **Describe the major cerebellar inputs and outputs.**

PREDICT   5

If the right half of the cerebellum is injured, which side of the body exhibits defects associated with cerebellar function?

Other Brain Functions

The human brain is capable of many functions besides the awareness of sensory input and the control of skeletal muscles. Speech, mathematical and artistic abilities, sleep, and memory are other functions of the brain.

Speech

In most people, the speech area is in the left cerebral cortex. Two major cortical areas are involved in speech: **Wernicke's area** (sensory speech area), a portion of the parietal lobe, and **Broca's area** (motor speech area) in the inferior part of the frontal lobe (see figure 12.11). Wernicke's area is necessary for understanding and formulating coherent speech. Broca's area initiates the complex series of movements necessary for speech. The two areas are connected by an association tract known as the **arcuate fasciculus** (figure 12.14a).

For someone to speak a word that he or she sees, such as when reading out loud, the following sequence of events must take place. Action potentials from the eyes reach the primary visual cortex, where the word is seen. The word is then recognized in the visual association area and understood in parts of

Clinical Relevance Basal Nuclei and Cerebellar Disorders

Parkinson disease is caused by a dysfunction in the substantia nigra, resulting in decreased excitatory and inhibitory effects on the cerebral motor cortex by the basal nuclei circuits. Decreased excitatory effects result in a general lack of movement and difficulty in initiating movements, such as rising from a sitting position or initiating walking. Movements are not normal—for example, walking is accomplished by a slow, shuffling gait, and it is difficult to change directions. Basal nuclei circuits decrease muscle tone. Decreased inhibition of muscle tone results in increased muscle tone, causing muscular rigidity and loss of facial expression. It can also cause a **resting tremor,** called "pill-rolling," which consists of circular movement of the opposed thumb and index fingertips.

Parkinson disease results when neurons of the substantia nigra produce inadequate amounts of the inhibitory neurotransmitter dopamine. It cannot be treated by taking dopamine because dopamine cannot cross the blood–brain barrier. Parkinson disease is treated by taking levodopa (lē-vō-dō′pă) (L-dopa), a precursor to dopamine, which crosses the blood–brain barrier and is used to make dopamine. Sinemet, a combination of L-dopa and carbidopa (kar-bi-dō′pă) is even more effective than L-dopa alone because carbidopa is a decarboxylase inhibitor, which prevents the breakdown of L-dopa

before it reaches the brain. There are long-term, negative side effects with L-dopa, however. As a result, other drugs, such as ropinirole and pramipexole, which have fewer side effects, are now being used to treat the symptoms. Removal of a portion of the corpus striatum, or implantation of an electrical pulse generator to stimulate specific basal nuclei, is now being used to treat Parkinson disease effectively. Treatment by transplanting fetal tissues, or stem cells from adult tissues, capable of producing dopamine is under investigation.

Huntington disease and hemiballismus result from increased excitatory and inhibitory effects on the cerebral motor cortex by basal nuclei circuits. The increased excitatory effects result in involuntary movements and the increased inhibitory effects in decreased muscle tone. **Huntington disease** is a dominant hereditary disorder that begins in middle life, causing mental deterioration and progressive degeneration of the corpus striatum in affected individuals. It is characterized by a series of rapid, nearly continuous, involuntary movements of the limbs and facial muscles. Damage to a subthalamic nucleus can result in **hemiballismus** (hem-ē-bal-iz′mŭs), an uncontrolled, purposeless, and forceful throwing or flailing of one arm and one leg.

Cerebral palsy (pawl′zē) is a general term referring to defects in motor functions or coordination resulting from several

types of brain damage, which may be caused by abnormal brain development or birth-related injury. Some symptoms of cerebral palsy, such as increased muscle tension, are related to basal nuclei dysfunction. **Athetosis** (ath-ĕ-tō′sis), often one of the features of cerebral palsy, is characterized by slow, sinuous, aimless movements. When the face, neck, and tongue muscles are involved, grimacing, protrusion, and writhing of the tongue and difficulty in speaking and swallowing are characteristics.

Cerebellar lesions result in a spectrum of characteristic functional disorders. Movements tend to be jerky with overshooting. An example of overshooting is pointing past or deviating from a mark that one tries to touch with the finger. **Nystagmus** (nis-tag′mus), which is a constant motion of the eyes, may also occur. An **intention tremor** is a shaking or oscillatory movement that occurs when one tries to control a given movement. The more carefully one tries to control the movement, the greater the tremor becomes. For example, when a person with a cerebellar tremor attempts to drink a glass of water, the closer the glass comes to the mouth the shakier the movement becomes. An intention tremor is in direct contrast to basal nuclei tremors, in which the resting tremor largely or completely disappears during purposeful movement.

Wernicke's area. Then action potentials representing the word are conducted through association fibers that connect Wernicke's and Broca's areas. In Broca's area, the word is formulated as it will be spoken. Action potentials are then propagated to the premotor area, where the movements are programmed, and finally to the primary motor cortex, where the proper movements are triggered (figure 12.14b).

To repeat a word that has been heard is a similar process. The information passes from the ears to the primary auditory cortex and then passes to the auditory association area, where the sound is recognized as a word, and continues to Wernicke's area, where the word is understood. From Wernicke's area, it follows the same route as followed for speaking words that are seen. ▼

25. List the sequence of events that must occur for a person to say a word that is seen or heard.

P R E D I C T
Propose the sequence of events needed for a blindfolded person to name an object placed in the right hand.

Aphasia

Aphasia (ă-fā′zē-ă), absent or defective speech or language comprehension, results from a lesion in the language areas of the cortex. **Receptive aphasia** (Wernicke's aphasia), which includes defective auditory and visual comprehension of language, defective naming of objects, and repetition of spoken sentences, is caused by a lesion in Wernicke's area. **Expressive aphasia** (Broca's aphasia), caused by a lesion in Broca's area, is characterized by hesitant and distorted speech.

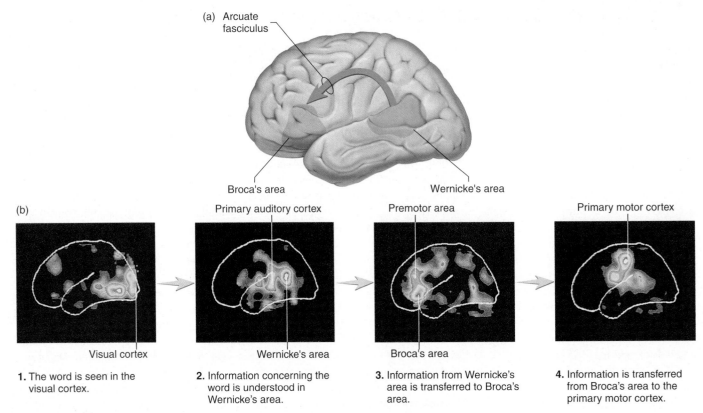

1. The word is seen in the visual cortex.

2. Information concerning the word is understood in Wernicke's area.

3. Information from Wernicke's area is transferred to Broca's area.

4. Information is transferred from Broca's area to the primary motor cortex.

Process Figure 12.14   **Demonstration of Cortical Activities During Speech**

(a) The arcuate fasciculus connects the two key areas involved in speech. (b) The figures show the pathway for reading and naming something that is seen, such as reading aloud. Positron emission tomography (PET) scans show the areas of the brain that are most active during various phases of speech. The highest level of brain activity is indicated in *red*, with successively lower levels represented by *yellow, green,* and *blue.*

Mathematical and Artistic Abilities

Some complex functions of the brain are not shared equally between the left and right cerebral hemispheres. The left cerebral hemisphere is more involved in such skills as mathematics and speech. The right hemisphere is involved in activities such as spatial perception, the recognition of faces, and musical ability. Dominance of one cerebral hemisphere over the other, for most functions, is probably not very important in most people because the two hemispheres are in constant communication through connections between the two hemispheres called **commissures** (kom′i-shūrz; a joining together). The largest of these commissures is the **corpus callosum** (kōr′pŭs kă-lō′sum; callous body), a broad band of tracts at the base of the longitudinal fissure (see figure 13.1).

Surgical cutting of the corpus callosum has been successful in treating a limited number of epilepsy cases. Severing the corpus callosum prevents electrical activity causing seizures from spreading from one cerebral hemisphere to the other. People with a cut corpus callosum function surprisingly normally. Under certain conditions, however, interesting functional defects are apparent in people who have had their corpus callosum severed. For example, if a patient with a severed corpus callosum is asked to reach behind a screen to touch one of several items with one hand without being able to see it and then is

asked to point out the same object with the other hand, the person cannot do it. Tactile information from the left hand enters the right somatic sensory cortex, but that information is not transferred to the left hemisphere, which controls the right hand. As a result, the left hemisphere cannot direct the right hand to the correct object. ▼

26. *What are the functions localized in the left cerebral hemisphere? In the right cerebral hemisphere?*
27. *Name the largest pathway that connects the right and left cerebral hemispheres.*

Brain Waves and Sleep

Electrodes placed on a person's scalp and attached to a recording device can record the electrical activity of the brain, producing an **electroencephalogram** (ē-lek′trō-en-sef′ă-lō-gram) **(EEG).** These electrodes are not sensitive enough to detect individual action potentials, but they can detect the simultaneous action potentials in large numbers of neurons. As a result, the EEG displays wavelike patterns known as **brain waves** (figure 12.15a). Most of the time, EEG patterns are irregular, with no particular pattern, because most of the brain's electrical activity is not synchronous. At other times, however, specific patterns can be detected. These regular patterns are classified as alpha, beta,

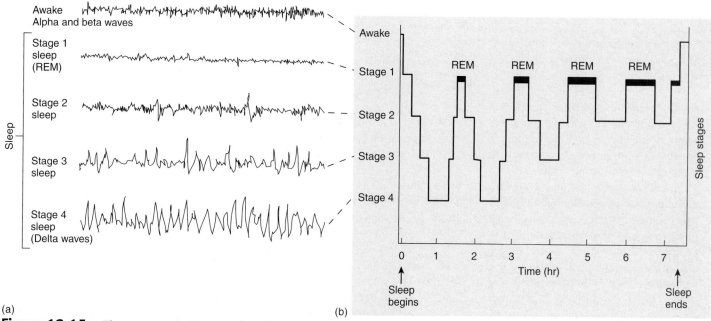

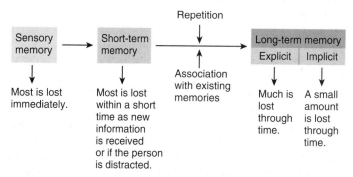

Figure 12.15 Electroencephalograms (EEGs) Showing Brain Waves

(*a*) EEG tracings when a person is awake and during four stages of sleep. (*b*) A typical night's sleep pattern in a young adult. The time spent in REM sleep is labeled and shown by *dark bars.*

theta, or delta waves. **Alpha waves** are observed in a normal person who is awake but in a quiet, resting state with the eyes closed. **Beta waves** have a higher frequency than alpha waves and occur during intense mental activity. **Theta waves** usually occur in children, but they can also occur in adults who are experiencing frustration or who have certain brain disorders. **Delta waves** occur in deep sleep, in infancy, and in patients with severe brain disorders.

Brain wave patterns vary during the four stages of sleep (figure 12.15*b*). A sleeping person arouses several times during a period of sleep. Dreaming occurs during periods when eye movement can be observed in a sleeping person, called **rapid eye movement (REM) sleep.**

Distinct types of EEG patterns can be detected in patients with specific brain disorders, such as epileptic seizures. Neurologists use these patterns to diagnose the disorders and determine the appropriate treatment. ▼

27. What is an EEG? What four conditions produce alpha, beta, theta, and delta waves, respectively?

Memory

Memory is divided into three major types: sensory, short-term, and long-term (figure 12.16). **Sensory memory** is the very brief retention of sensory input received by the brain while something is scanned, evaluated, and acted on. This type of memory lasts less than a second.

If a given piece of data held in sensory memory is considered valuable enough, it is moved into **short-term memory,** where it is retained for a few seconds to a few minutes. This memory is limited primarily by the number of bits of information, usually about seven or eight, that can be stored at any one time. More bits can be

stored when the numbers are grouped into specific segments separated by spaces. When new information is presented, or when the person is distracted, information previously stored in short-term memory is eliminated.

Long-term memory lasts for hours or longer. Some long-term memory lasts a lifetime. Two types of long-term memory exist: explicit and implicit. **Explicit memory** involves the retention of facts, such as names, dates, and places. Explicit memory is accessed by a gyrus of the temporal lobe called the **hippocampus** (hip-ō-kam′pus, shaped like a seahorse) and the **amygdaloid** (ă-mig′dă-loyd, almond-shaped) **body,** which is part of the limbic system. The hippocampus is involved in retrieving the actual memory, such as recalling a person's name, and the amygdaloid body is involved in the emotional overtones of that memory, such as feelings of like or dislike, and the recollection of good or bad memories associated with that person. A lesion in the temporal lobe affecting the hippocampus can prevent the brain from moving information from short-term to long-term memory. Emotion

Figure 12.16 Memory Processing

Nearly all **brain tumors** develop from neuroglia and not from neurons. Symptoms vary widely, depending on the location of the tumor, but include headaches, neuralgia (pain along the distribution of a peripheral nerve), paralysis, seizures, coma, and death. **Meningiomas** (mĕ-nin′jē-ō′mă̄z), tumors of the meninges, account for 25% of all primary intracranial tumors.

Stroke is a term meaning a sudden blow, suggesting the speed with which this type of defect can occur. It is also referred to clinically as a **cerebrovascular** (ser′ĕ-brō-vas′kū-lăr, sĕ-rē′brō-vas′kū-lăr) **accident (CVA).** A CVA may be caused by a **hemorrhage** (hem′ŏ-rij), bleeding into the tissue; by a clot, called a **thrombus** (throm′bus), in a blood vessel; by a piece of a clot, called an **embolus** (em′bō-lŭs) that has broken loose and floats through the circulation until it reaches and blocks a small vessel; or by **vasospasm** (vā′sō-spazm), constriction of the cerebral blood vessels. A hemorrhage, a thrombus, an embolism, or a vasospasm can result in a local area of cell death, called an **infarct** (in′farkt), caused by a lack of blood supply, surrounded by an area of cells that are secondarily affected. The symptoms depend on the location of the stroke and the size of the infarct, but they may include anesthesia (a lack of feeling) or paralysis on the side of the body opposite the cerebral infarct.

Alzheimer (ălz′hī-mer) **disease** is a severe type of mental deterioration, or dementia, usually affecting older people but occasionally affecting people younger than 60. It accounts for half of all dementias; the other half result from drug and alcohol abuse, infections, and strokes. Alzheimer disease is estimated to affect 10% of all people older than 65 and nearly half of those older than 85.

Alzheimer disease involves a general decrease in brain size that results from loss of neurons in the cerebral cortex. The gyri become more narrow, and the sulci widen. The frontal lobes and specific regions of the temporal lobes are affected most severely. The symptoms include general intellectual deficiency, memory loss, short attention span, moodiness, disorientation, and irritability.

Localized axonal enlargements, called **amyloid** (am′i-loyd) **plaques,** containing large amounts of β(beta)-amyloid protein, form in the cortex of patients with Alzheimer disease. There is some evidence that Alzheimer disease may have characteristics of a chronic inflammatory disease, and anti-inflammatory drugs have some effect in treating the disease. Estrogen, which affects some brain functions, such as emotion, memory, and cognition, may be involved in the disease.

The gene for β-amyloid protein has been mapped to chromosome 21 (see chapter 25), but this gene accounts for only a small portion of the cases. It is noteworthy that people with Down syndrome, which results from having three copies of chromosome 21 (trisomy 21), exhibit the cortical and other changes associated with Alzheimer disease.

The more common, late-onset form of the disease maps to chromosome 19. A protein, **apolipoprotein E** (ap′ō-lip-ō-prō′tēn) **(apo E),** which binds β-amyloid protein, has also been associated with Alzheimer disease. This protein maps to the same part of chromosome 19 as the late-onset form of Alzheimer disease. Apo E may also be involved in the regulation of yet another protein, called τ (tau), which is involved in microtubule formation inside neurons. If τ does not function properly, microtubules do

---

and mood apparently serve as gates in the brain and determine what is or is not stored in long-term explicit memory. The amygdaloid body is also a key to the development of fear, which also involves the prefrontal cortex and the hypothalamus.

Parts of explicit memory appear to be stored separately in various parts of the cerebrum, especially in the parietal lobe, much like storing items in separate "pigeonholes." Memories of people appear to be stored separately from memories of places. People's faces may be stored in yet other pigeonholes. Family members appear to be stored together. Items that are recognized by sight, such as an animal, are stored separately from items that are recognized by feel, such as tools. Damage, such as a stroke, to one part of the brain can remove certain memories without affecting others.

Retrieval of a complete memory requires accessing parts of the memory from different pigeonholes. A complex memory requires accessing and reassembling segments of memory each time the memory is recalled. The memory of an experience, for example, may be stored in at least four different pigeonholes. Where you were is stored in one place, who you were with in another, what happened in another, and how you felt in yet another place. The complexity of this process may be responsible for the changes in what is recalled over time. On occasion, parts of unrelated memories may be pulled out and put together incorrectly to create a "false memory." Much of what is stored as explicit memory is gradually lost through time.

Implicit memory involves the development of skills such as riding a bicycle or playing a piano. Implicit memory is stored primarily in the cerebellum and the premotor area of the cerebrum. Conditioned, or Pavlovian, reflexes are also implicit and can be eliminated in experimental animals by producing cerebellar lesions in the animals. The most famous example of a conditioned reflex is that of Ivan Pavlov's experiments with dogs. Each time he fed the dogs, a bell was rung; soon the dogs salivated when the bell rang, even if no food was presented. Only a small amount of implicit memory is lost through time.

A whole series of neurons and their pattern of activity, called a **memory engram,** or memory trace, probably is involved in the long-term retention of information, a thought, or an idea. Long-term memory may involve a physical change in neuron shape. Repetition of the information and association of the new information with existing memories assist in the transfer of information from short-term to long-term memory. ▼

29. Describe the three types of memory.
30. Distinguish between implicit and explicit memory.

not form normally, and the τ proteins become tangled within the neurons, decreasing their function. Nitric oxide production, which stimulates cerebral blood flow and memory in the brain, may help protect cerebral blood vessels and brain tissue from the toxic effects of β-amyloid protein.

Epilepsy is a group of brain disorders that have seizure episodes in common. The seizure, a sudden massive neuronal discharge, can be either partial or complete, depending on the amount of brain involved and whether or not consciousness is impaired. A seizure involves a change in sensation, consciousness, or behavior due to brief electrical discharge in the brain. Normally, a balance exists between excitation and inhibition in the brain. When this balance is disrupted by increased excitation or decreased inhibition, a seizure may result. The neuronal discharges may stimulate muscles innervated by the neurons involved, resulting in involuntary muscle contractions, or convulsions.

Headaches have a variety of causes, which can be grouped into two basic classes: extracranial and intracranial. Extracranial headaches can be caused by inflammation of the sinuses, dental irritations, temporomandibular joint disorders, ophthalmologic disorders, and tension in the muscles moving the head and neck. Intracranial headaches may result from inflammation of the brain or meninges, vascular problems, mechanical damage, or tumors. **Tension headaches** are extracranial muscle tension, stress headaches, consisting of a dull, steady pain in the forehead, temples, and neck or throughout the head. Tension headaches are associated with stress, fatigue, and posture. **Migraine headaches** (*migraine* means half a skull) occur in only one side of the head and appear to involve the abnormal dilation and constriction of blood vessels. They often start with distorted vision, shooting spots, and blind spots. Migraines consist of severe throbbing, pulsating pain. About 80% of migraine sufferers have a family history of the disorder, and women are affected four times more often than men. Those suffering migraines are usually women younger than 35. The severity and frequency usually decrease with age.

Dyslexia (dĭs-lek′sē-ă) is a defect in which the reading level is below that expected on the basis of an individual's overall intelligence. Most people with dyslexia have normal or above-normal intelligence quotients. It is three times more common in males than females. As many as 10% of males in the United States suffer from the disorder. The symptoms vary considerably from person to person and include the transposition of letters in a word, confusion between the letters *b* and *d*, and a lack of orientation in three-dimensional space. The brains of some dyslexics have abnormal cellular arrangements, including cortical disorganization and the appearance of bits of gray matter in medullary areas. Dyslexia apparently results from abnormal brain development.

Children with **attention-deficit disorder (ADD)** are easily distractible, have short attention spans, and may shift from one uncompleted task to another. Children with **attention-deficit/hyperactivity disorder (ADHD)** exhibit the characteristics of ADD, but they are also fidgety, have difficulty remaining seated and waiting their turn, engage in excessive talking, and commonly interrupt others. About 3% of all children exhibit ADHD, more boys than girls. Symptoms usually occur before age 7. The neurological basis of both ADD and ADHD is as yet unknown.

Effects of Aging on the Nervous System

As a person ages, sensory function gradually declines because the number of sensory neurons declines, the function of remaining neurons decreases, and CNS processing decreases. In the skin, free nerve endings and hair follicle receptors remain largely unchanged with age. Meissner corpuscles and Pacini corpuscles, however, decrease in number. The capsules of those that remain become thicker and structurally distorted and therefore exhibit reduced function. As a result of these changes in Meissner corpuscles and Pacini corpuscles, elderly people are less conscious of something touching or pressing on the skin, have a decreased sense of two-point discrimination, and have a more difficult time identifying objects by touch. These functional changes leave elderly people more prone to skin injuries and with a greater sense of isolation.

A loss of Pacini corpuscles also results in a decreased sense of position of the limbs and in the joints, which can affect balance and coordination. The functions of Golgi tendon organs and muscle spindles also decline with increasing age. As a result, information on the position, tension, and length of tendons and muscles decreases, resulting in additional reduction in the senses of movement, posture, and position, as well as reduced control and coordination of movement.

Other sensory neurons with reduced function include those that monitor blood pressure, thirst, objects in the throat, the amount of urine in the urinary bladder, and the amount of feces in the rectum. As a result, elderly people are more prone to high blood pressure, dehydration, swallowing and choking problems, urinary incontinence, and constipation or bowel incontinence.

There is also a general decline in the number of motor neurons. As many as 50% of the lower motor neurons in the lumbar region of the spinal cord may be lost by age 60. Muscle fibers innervated by the lost motor neurons are also lost, resulting in a general decline in muscle mass. The remaining motor units can compensate for some of the lost function. This, however, often results in a feeling that one must work harder to perform activities that were previously not so difficult. Loss of motor units also leads to more rapid fatigue, as the remaining units must perform compensatory work.

Systems Interactions

	Effects of the Nervous System on Other Systems	Effects of Other Systems on the Nervous System
Integumentary System	Regulates body temperature by controlling the activity of sweat glands and blood vessels Stimulates arrector pili to contract	Contains receptors for heat, cold, temperature, pain, pressure, and vibration
Skeletal System	Pain awareness results in reduced movement of bones, allowing broken bones to heal	Skull protects brain Vertebral column protects spinal cord Ear bones necessary for normal hearing Reservoir for calcium, which is necessary for neuronal function
Muscular System	Stimulates muscle contractions and maintains muscle tone	Sensory receptors (for example, muscle spindles and Golgi tendon organs) provide information about body position
Endocrine System	Controls the release of hormones from the hypothalamus, pituitary gland, and glands	Hormones affect neuron growth and metabolism
Cardiovascular System	Regulates heart rate and force of contraction Regulates blood vessel diameter	Delivers oxygen, nutrients, hormones, and immune cells Removes carbon dioxide, waste products, and toxins Cerebrospinal fluid and aqueous humor produced from, and returned to, the blood Contains receptors that monitor blood pressure, blood oxygen, and blood carbon dioxide levels
Lymphatic System and Immunity	The nervous system stimulates and inhibits immunity in ways that are not well understood	Immune cells provide protection against microorganisms and toxins and promote tissue repair by releasing chemical mediators
Respiratory System and	Regulates rate and depth of breathing	Provides oxygen and removes carbon dioxide Helps maintain the body's pH The lungs contain receptors that provide information on lung inflation
Digestive System	Regulates secretion from digestive glands and organs Controls mixing and movement of digestive tract contents	Provides nutrients and water
Urinary System	Controls emptying of the urinary bladder Regulates renal blood flow	Removes waste products Helps maintain the body's pH, ion, and water balance
Reproductive System	Stimulates sexual responses, such as erection and ejaculation in males and erection of the clitoris in females	Sex hormones influence CNS development and sexual behaviors

Reflexes slow as people age because both the generation and the conduction of action potentials and synaptic functions slow. The number of neurotransmitters and receptors declines. Age-related changes in the CNS also slow reflexes. The more complicated the reflex, the more it is affected by age. As reflexes slow, older people are less able to react automatically, quickly, and accurately to changes in internal and external conditions.

The size and weight of the brain decrease as a person ages. At least some of these changes result from the loss of neurons within the cerebrum. The remaining neurons can apparently compensate for much of this loss. In addition to loss of neurons, structural changes occur in the remaining neurons. Neuron plasma membranes become more rigid, the endoplasmic reticulum becomes more irregular in structure, neurofibrillar tangles develop in the cells, and amyloid plaques form in synapses. All these changes decrease the ability of neurons to function. Age-related changes in brain function include decreased voluntary movement, conscious sensations, reflexes, memory, and sleep. Short-term memory is decreased in most older people. This change varies greatly among individuals but, in general, such changes are slow until about age 60 and then become more rapid, especially after age 70. However, the total amount of memory loss is normally not great for most people. The most

difficult information for older people to assimilate is that which is unfamiliar and presented verbally and rapidly. Some of these problems may occur as older people are required to deal with new information in the face of existing, contradictory memories. Long-term memory appears to be unaffected or even improved in older people.

As with short-term memory, thinking, which includes problem solving, planning, and intelligence, in general declines slowly to age 60 but more rapidly thereafter. These changes, however, are slight and quite variable. Many older people show no change, and about 10% show an increase in thinking ability. Many of these changes are impacted by a person's background, education, health, motivation, and experience.

Among older people, more time is required to fall asleep, there are more periods of waking during the night, and the wakeful periods are of greater duration. Factors that can affect sleep include pain, indigestion, rhythmic leg movements, sleep apnea, decreased urinary bladder capacity, and circulatory problems. ▼

> *31. How does aging affect sensory function? How does loss of motor neurons affect muscle mass?*
> *32. Does aging always produce memory loss?*

S U M M A R Y

Sensation (p. 342)

1. The senses include general senses and special senses.
2. The general senses are divided into the somatic senses (touch, pressure, proprioception, temperature, and pain) and the visceral senses (pain and pressure).
3. The special senses are smell, taste, sight, hearing, and balance.
4. Sensation, or perception, is the conscious awareness of stimuli received by sensory receptors.
5. Sensation requires a stimulus, a receptor, the conduction of an action potential to the CNS, translation of the action potential, and processing of the action potential in the CNS so that the person is aware of the sensation.

Sensory Receptors

1. Receptors include mechanoreceptors, chemoreceptors, photoreceptors, thermoreceptors, and nociceptors.
2. Free nerve endings detect pain, temperature, itch, and movement (proprioception).
3. Merkel disks respond to light touch and superficial pressure.
4. Hair follicle receptors wrap around the hair follicle and are involved in the sensation of light touch when the hair is bent.
5. Meissner corpuscles, located deep to the epidermis, are responsible for two-point discriminative touch.
6. Ruffini end organs, located in the dermis, are involved in continuous touch or pressure.
7. Pacini corpuscles, located in the dermis and subcutaneous tissue, detect deep pressure and vibration. In joints, they serve a proprioceptive function.
8. Muscle spindles, located in skeletal muscle, are proprioceptors.
9. Golgi tendon organs, embedded in tendons, respond to changes in tension.

Sensory Tracts

1. Ascending pathways carry conscious and unconscious sensations. Each pathway carries specific sensory information because each pathway is associated with specific types of receptors.
2. In the anterolateral system,
   - The spinothalamic tract carries pain, temperature, light touch, pressure, tickle, and itch sensations.
   - The spinothalamic tracts are formed by primary neurons that enter the spinal cord and synapse with secondary neurons. The secondary neurons cross the spinal cord and ascend to the thalamus, where they synapse with tertiary neurons that project to the somatic sensory cortex.
   - The spinoreticular tracts ascend to the reticular formation and the spinomesencephalic tracts to the midbrain. They convey unconscious pain and touch information.
3. The dorsal column/medial lemniscal system carries the sensations of two-point discrimination, proprioception, pressure, and vibration. Primary neurons enter the spinal cord and ascend to the medulla, where they synapse with secondary neurons. Secondary neurons cross over and project to the thalamus. Tertiary neurons extend from there to the somatic sensory cortex.
4. The trigeminothalamic tract carries sensory information from the face, nose, and mouth.
5. Proprioception information carried to the cerebral cortex by the dorsal column/medial lemniscal system and trigeminothalamic tracts results in conscious proprioception. Proprioception information carried to the cerebellum by spinocerebellar tracts, the dorsal column/medial lemniscal system, and trigeminothalamic tracts results in unconscious proprioception.

Sensory Areas of the Cerebral Cortex

1. Sensory pathways project to primary sensory areas (visual, auditory, taste, olfactory, and somatic sensory) in the cerebral cortex.
2. Sensory areas are organized topographically in the somatic sensory cortex.

Sensory Processing

Association areas of the cerebral cortex process sensory input from the primary sensory areas.

Control of Skeletal Muscles   (p. 350)

1. Lower motor neurons are found in the cranial nuclei or the anterior horn of the spinal cord gray matter. Upper motor neurons are located in the cerebral cortex, brainstem, and cerebellum.
2. Upper motor neurons form tracts that directly or indirectly control the activities of lower motor neurons.

Motor Areas of the Cerebral Cortex

1. The primary motor cortex is the precentral gyrus and is organized topographically.
2. The premotor and prefrontal areas are staging areas for motor function.

Motor Tracts

1. The direct pathways arise from the cerebral cortex. Upper motor neurons extend to lower motor neurons in the brainstem and spinal cord. The indirect pathways arise from the cerebral cortex and cerebellum. Upper motor neurons extend to brainstem nuclei. Axons from the nuclei extend to lower motor neurons.
2. The corticospinal tracts control muscle movements below the head, especially of the distal limbs, where they make fine motor control of the fingers possible.
   - About 75%–85% of the upper motor neurons of the corticospinal tracts cross over in the medulla to form the lateral corticospinal tracts in the spinal cord.
   - The remaining upper motor neurons pass through the medulla to form the anterior corticospinal tracts, which cross over in the spinal cord.
   - The upper motor neurons of both tracts synapse with interneurons that then synapse with lower motor neurons in the spinal cord.
3. The corticobulbar tracts innervate the head muscles (except for muscles moving the eyes). Upper motor neurons synapse with interneurons that, in turn, synapse with lower motor neurons in the cranial nerve nuclei.
4. The indirect pathways include the reticulospinal, rubrospinal, and vestibulospinal tracts.
5. The indirect pathways are involved in conscious and unconscious trunk and proximal limb muscle movements, posture, and balance.

Modifying and Refining Motor Activities

1. Basal nuclei are important in planning, organizing, and coordinating motor movements and posture.
2. The cerebellum has three parts.
   - The vestibulocerebellum controls balance and eye movement.
   - The spinocerebellum corrects discrepancies between intended movements and actual movements.
   - The cerebrocerebellum can "learn" highly specific complex motor activities.
3. The cerebellum receives input from, and projects to, the contralateral cerebral motor cortex. The cerebellum receives proprioception input from the ipsilateral body.

Other Brain Functions   (p. 356)
Speech

1. Speech is located only in the left cortex in most people.
2. Wernicke's area comprehends and formulates speech.
3. Broca's area receives input from Wernicke's area and sends impulses to the premotor and motor areas, which cause the muscle movements required for speech.

Mathematical and Artistic Abilities

1. In most people, the left hemisphere is dominant for speech and analytical skills. The right hemisphere is dominant for spatial and musical abilities.
2. The right and left hemispheres are connected by commissures. The largest commissure is the corpus callosum, which allows the sharing of information between hemispheres.

Brain Waves and Sleep

1. Electroencephalograms (EEGs) record the electrical activity of the brain as brain waves.
2. Some brain disorders can be detected with EEGs.

Memory

At least three kinds of memory exist: sensory, short-term, and long-term.

Effects of Aging on the Nervous System   (p. 361)

1. There is a general decline in sensory and motor functions as a person ages.
2. Short-term memory is decreased in most older people.
3. Thinking ability does not decrease in most older people.

R E V I E W   A N D   C O M P R E H E N S I O N

1. Nociceptors respond to
   a. changes in temperature at the site of the receptor.
   b. compression, bending, or stretching of cells.
   c. painful mechanical, chemical, or thermal stimuli.
   d. light striking a receptor cell.
2. Which of these types of sensory receptors respond to pain, itch, tickle, and temperature?
   a. Merkel disks
   b. Meissner corpuscles
   c. Ruffini end organs
   d. free nerve endings
   e. Pacini corpuscles
3. Which of these types of sensory receptors are involved with proprioception?
   a. free nerve endings
   b. Golgi tendon organs
   c. muscle spindles
   d. Pacini corpuscles
   e. all of the above
4. Decreased sensitivity to a continued stimulus is called
   a. adaptation.
   b. projection.
   c. translation.
   d. conduction.
   e. phantom pain.

5. Secondary neurons in the spinothalamic tracts synapse with tertiary neurons in the
   a. medulla oblongata.
   b. gray matter of the spinal cord.
   c. cerebellum.
   d. thalamus.
   e. midbrain.

6. If the spinothalamic tract on the right side of the spinal cord is severed,
   a. pain sensations below the damaged area on the right side are eliminated.
   b. pain sensations below the damaged area on the left side are eliminated.
   c. temperature sensations are unaffected.
   d. neither pain sensations nor temperature sensations are affected.

7. Fibers of the dorsal column/medial lemniscal system
   a. carry the sensations of two-point discrimination, proprioception, pressure, and vibration.
   b. cross to the opposite side in the medulla oblongata.
   c. include secondary neurons that exit the medulla and synapse in the thalamus.
   d. all of the above.

8. Tertiary neurons in both the anterolateral system and the dorsal column/medial lemniscal system
   a. project to the somatic sensory cortex.
   b. cross to the opposite side in the medulla oblongata.
   c. are found in the spinal cord.
   d. connect to quaternary neurons in the thalamus.
   e. are part of a descending pathway.

9. General sensory inputs (pain, pressure, temperature) to the cerebrum end in the
   a. precentral gyrus.   d. corpus callosum.
   b. postcentral gyrus.   e. arachnoid mater.
   c. central sulcus.

10. Neurons from which area of the body occupy the greatest area of the somatic sensory cortex?
    a. foot   d. arm
    b. leg   e. face
    c. torso

11. A cutaneous nerve to the hand is severed at the elbow. The distal end of the nerve at the elbow is then stimulated. The person reports
    a. no sensation because the receptors are gone.
    b. a sensation only in the region of the elbow.
    c. a sensation "projected" to the hand.
    d. a vague sensation on the side of the body containing the cut nerve.

12. Which of these areas of the cerebral cortex is involved in the motivation and foresight to plan and initiate movements?
    a. primary motor cortex
    b. somatic sensory cortex
    c. prefrontal area
    d. premotor area
    e. basal nuclei

13. Which of these pathways is not an ascending (sensory) pathway?
    a. spinothalamic tract
    b. corticospinal tract
    c. dorsal column/medial lemniscal system
    d. trigeminothalamic tract
    e. spinocerebellar tract

14. The _____ tracts innervate the head muscles.
    a. corticospinal
    b. rubrospinal
    c. vestibulospinal
    d. corticobulbar
    e. dorsal column/medial lemniscal

15. Most fibers of the direct pathways
    a. decussate in the medulla oblongata.
    b. synapse in the pons.
    c. descend in the rubrospinal tract.
    d. begin in the cerebellum.

16. A person with a spinal cord injury is suffering from paresis (partial paralysis) in the right lower limb. Which of these pathways is probably involved?
    a. left lateral corticospinal tract
    b. right lateral corticospinal tract
    c. left dorsal column/medial lemniscal system
    d. right dorsal column/medial lemniscal system

17. Which of these tracts is a direct pathway?
    a. reticulospinal tract
    b. corticobulbar tract
    c. rubrospinal tract
    d. vestibulospinal tract

18. The indirect pathways are concerned with
    a. posture.
    b. trunk movements.
    c. proximal limb movements.
    d. all of the above.

19. A major effect of the basal nuclei is
    a. to act as a comparator for motor coordination.
    b. to decrease muscle tone and inhibit unwanted muscular activity.
    c. to affect emotions and emotional responses to odors.
    d. to modulate pain sensations.

20. Which of the parts of the cerebellum is correctly matched with its function?
    a. vestibulocerebellum—planning and learning rapid, complex movements
    b. spinocerebellum—comparator function
    c. cerebrocerebellum—balance
    d. none of the above

21. Given the following events:
    1. Action potentials from the cerebellum go to the motor cortex and spinal cord.
    2. Action potentials from the motor cortex go to lower motor neurons and the cerebellum.
    3. Action potentials from proprioceptors go to the cerebellum.

    Arrange the events in the order they occur in the cerebellar comparator function.
    a. 1,2,3   d. 2,3,1
    b. 1,3,2   e. 3,2,1
    c. 2,1,3

22. Given these areas of the cerebral cortex:
    1. Broca's area
    2. premotor area
    3. primary motor cortex
    4. Wernicke's area

    If a person hears and understands a word and then says the word out loud, in what order are the areas used?
    a. 1,4,2,3   d. 4,1,2,3
    b. 1,4,3,2   e. 4,1,3,2
    c. 3,1,4,2

23. The main connection between the right and left hemispheres of the cerebrum is the
    a. intermediate mass.
    b. corpus callosum.
    c. vermis.
    d. unmyelinated nuclei.
    e. thalamus.

24. Which of these activities is mostly associated with the left cerebral hemisphere in most people?
    a. sensory input from the left side of the body
    b. mathematics and speech
    c. spatial perception
    d. musical ability

25. Concerning long-term memory,
    a. explicit memory involves the development of skills, such as riding a bicycle.
    b. implicit memory involves the retention of facts, such as names, dates, or places.
    c. much of explicit memory is lost through time.
    d. explicit memory is stored primarily in the cerebellum and premotor area of the cerebrum.
    e. all of the above.

Answers in Appendix E

C R I T I C A L   T H I N K I N G

1. Describe all the sensations involved when a woman picks up an apple and bites into it. Explain which of those sensations are special and which are general. What types of receptors are involved? Which aspects of the taste of the apple are actually taste and which are olfaction?

2. Some student nurses are at a party. Because they love anatomy and physiology so much, they are discussing adaptation of the special senses. They make the following observations:
    a. When entering a room, an odor like brewing coffee is easily noticed. A few minutes later, the odor might be barely, if at all, detectable, no matter how hard one tries to smell it.
    b. When entering a room, the sound of a ticking clock can be detected. Later the sound is not noticed until a conscious effort is made to hear it. Then it is easily heard.

   Explain the basis for each of these observations.

3. A man has constipation, which causes distention and painful cramping in his colon. What kind of pain does he experience (local or diffuse) and where is it perceived? Explain.

4. A patient suffered a loss of two-point discrimination and proprioception on the right side of the body. Voluntary movement of muscles was not affected, and pain and temperature sensations were normal. Is it possible to conclude that the right side of the spinal cord was damaged?

5. A patient is suffering from the loss of two-point discrimination and proprioceptive sensations on the left side of the body resulting from a lesion in the thalamus. What tract is affected, and which side of the thalamus is involved?

6. A person in a car accident exhibits the following symptoms: extreme paresis on the right side, including the arm and leg; reduction of pain sensation on the left side; and normal tactile sensation on both sides. Which tracts are damaged? Where did the patient suffer tract damage?

7. A patient with a cerebral lesion exhibits a loss of fine motor control of the left hand, arm, forearm, and shoulder. All other motor and sensory functions appear to be intact. Describe the location of the lesion as precisely as possible.

8. Pamela is a 40-year-old woman experiencing limb weakness on her left side. A brain MRI shows lesions in her cerebral cortex. Which of the descending pathways are affected by these lesions? Would you expect the brain lesions to appear ipsilateral or contralateral to her limb weakness? Explain why.

9. A patient suffers brain damage in an automobile accident. It is suspected that the cerebellum is the part of the brain that is affected. On the basis of what you know about cerebellar function, how can you determine that the cerebellum is involved?

Answers in Appendix F

For additional study tools, go to: www.aris.mhhe.com. Click on your course topic and then this textbook's author name/title. Register once for a semester's worth of interactive learning activities to help improve your grade!

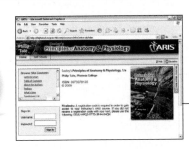

Photograph of an isolated cochlea from the inner ear.

Tools for Success

A virtual cadaver dissection experience

Audio (MP3) and/or video (MP4) content, available at www.mhhe.aris.com

Introduction

Historically, it was thought that we had just five senses: smell, taste, sight, hearing, and touch. Today we recognize many more. Some specialists suggest that there are at least 20, or perhaps as many as 40, different senses. Most of these senses are part of what was originally classified as "touch." These "general senses" were discussed in chapter 12. The sense of balance is now recognized as a "special sense," making a total of 5 special senses: smell, taste, sight, hearing, and balance.

Special senses are defined as senses with highly localized receptors that provide specific information about the environment. The sensations of smell and taste are closely related, both structurally and functionally, and they are both initiated by the interaction of chemicals with sensory receptors. The sense of vision is initiated by the interaction of light with sensory receptors. Both hearing and balance function in response to the interaction of mechanical stimuli with sensory receptors. Hearing occurs in response to sound waves, and balance occurs in response to gravity or motion.

▶This chapter describes **olfaction (p. 368)**, **taste (p. 369)**, the **visual system (p. 370)**, and **hearing and balance (p. 387)**. The chapter concludes with a look at the **effects of aging on the special senses (p. 399)**.

Olfaction

Olfaction (ol-fak′shŭn, to smell) is the sense of smell. It occurs in response to airborne molecules called **odorants** (ō′dŏr-ants, molecules with an odor) entering the nasal cavity.

Olfactory Epithelium and Bulb

A small superior part of the nasal cavity is lined with **olfactory epithelium** (figure 13.1). Ten million olfactory neurons are present within the olfactory epithelium. **Olfactory neurons** are bipolar neurons with dendrites extending to the epithelial surface of the nasal cavity. The ends of the dendrites are modified into bulbous

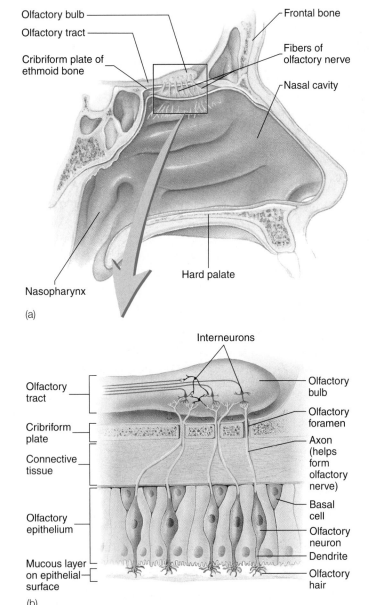

Interneurons

(b)

Figure 13.1   Olfactory Region, Epithelium, and Bulb

(*a*) The lateral wall of the nasal cavity (cut in sagittal section), showing the olfactory nerves, olfactory bulb, and olfactory tract. (*b*) The olfactory cells within the olfactory epithelium, the axons of olfactory neurons passing through the cribriform plate, and the fine structure of the olfactory bulb.

enlargements with long, specialized cilia, called **olfactory hairs,** which lie in a thin mucous film on the epithelial surface. The mucus keeps the nasal epithelium moist, traps and dissolves molecules, and facilitates the removal of molecules and particles from the olfactory epithelium.

Airborne odorants become dissolved in the mucus on the surface of the epithelium and bind to molecules on the membranes of the olfactory hairs called **olfactory receptors.** The odorants must first be dissolved in fluid in order to reach the olfactory receptors. When an odorant binds to its receptor, a G protein associated with the receptor is activated (G proteins are discussed in chapter 15). As a result of activation of the G protein, Na^+ and Ca^{2+} channels in the membrane open. The influx of ions into the olfactory hairs results in depolarization and the production of action potentials in the olfactory neurons. Once an odorant binds to its receptor, however, the receptor accommodates (see chapter 12) and does not respond to another odorant for some time.

P R E D I C T   1
Why is it helpful to sniff when trying to identify an odor?

The threshold for the detection of odors is very low, so very few odorants bound to an olfactory neuron can initiate an action potential. Methylmercaptan, which has a nauseating odor similar to that of rotten cabbage, is added to natural gas at a concentration of about one part per million. A person can detect the odor of about 1/25 billionth of a milligram of the substance and therefore is aware of the presence of the more dangerous but odorless natural gas.

There are approximately 1000 different odorant receptors, which can react to odorants of different sizes, shapes, and functional groups. The average person can distinguish among approximately 4000 different smells. It has been proposed that the wide variety of detectable odors are actually combinations of a smaller number of **primary odors.** The seven primary odors are camphoraceous (e.g., moth balls), musky, floral, pepperminty, ethereal (e.g., fresh pears), pungent, and putrid. It is very unlikely, however, that this list is an accurate representation of all primary odors, and some studies point to the possibility of as many as 50 primary odors.

 Odor Survey Results

The National Geographic Society conducted a smell survey in 1986, which is the largest sampling of its kind ever conducted. One and a half million people participated. Of six odors studied, 98%–99% of those responding could smell isoamyl acetate (banana), eugenol (cloves), mercaptans, and rose, but 29% could not smell galaxolide (musk), and 35% could not smell androsterone (contained in sweat). Of those responding to the survey, 1.2% could not smell at all, a disorder called **anosmia** (an-oz′mē-ă).

The primary olfactory neurons have the most exposed nerve endings of any neurons, and they are constantly being replaced. The entire olfactory epithelium, including the neurons, is lost about every 2 months as the olfactory epithelium degenerates and is lost from the surface. Lost olfactory cells are replaced by a proliferation of **basal cells** in the olfactory epithelium. This replacement of olfactory neurons is very unusual among neurons, most of which are permanent cells that have a very limited ability to replicate (see chapter 4). ▼

1. *Where are olfactory neurons and olfactory receptors located? What are odorants?*
2. *Describe the initiation of an action potential in an olfactory neuron.*
3. *How do the primary odors relate to the ability to smell many different odors?*
4. *What is the function of basal cells?*

Neuronal Pathways for Olfaction

Axons from olfactory neurons form the **olfactory nerves (I),** which pass through the olfactory foramina of the cribriform plate and enter the **olfactory bulb** (see figure 13.1*b*). There they synapse with interneurons that relay action potentials to the brain through the **olfactory tracts.** Each olfactory tract terminates in an area of the brain called the **olfactory cortex,** located within the temporal and frontal lobes. It is here that the sensation of smell is perceived. The olfactory cortex is part of the limbic system, projecting to the hypothalamus, amygdaloid bodies, and hippocampus of the cerebrum. Odors can produce strong emotional reactions, memories, and other responses. For example, the perfume or cologne of a loved one can evoke good feelings, and memories and food odors can cause salivation and hunger.

Within the olfactory bulb and olfactory cortex, feedback loops occur that tend to inhibit transmission of action potentials resulting from prolonged exposure to a given odorant. This feedback, plus the temporary decreased sensitivity at the level of the receptors, results in adaptation to a given odor. For example, if you enter a room that has an odor, you are aware of the odor, but you adapt to the odor and cannot smell it as well after the first few minutes. If you leave the room for some time and then reenter the room, the odor again seems more intense. ▼

5. *How are the olfactory bulbs related to the olfactory nerves and olfactory cortex?*
6. *Explain how smell can elicit various conscious and visceral responses.*

Taste

The sensory structures that detect **taste** stimuli are the **taste buds.** They are oval structures embedded in the epithelium of the tongue and mouth. Most taste buds are located in the epithelium of **papillae** (pă-pil′ē, nipples) which are small, raised areas on the surface of the tongue (figure 13.2). Taste buds are also located on other areas of the tongue, the palate, and even the lips and throat, especially in children. There are four types of papillae, classified according to their shape, and three of them have taste buds. The most numerous papillae on the surface of the tongue have no taste buds. They provide a rough surface, enabling the tongue to manipulate food more easily.

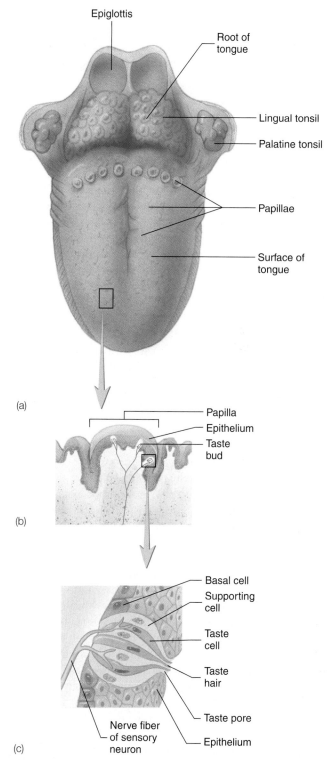

(a)

(b)

(c)

Figure 13.2 Tongue, Papillae, and Taste Buds
(*a*) Surface of the tongue. (*b*) A papilla. (*c*) A taste bud.

Histology of Taste Buds

Each of the 10,000 taste buds on a person's tongue consists of three major types of specialized epithelial cells. The sensory cells of each taste bud consist of about 50 **taste cells** (see figure 13.2). Each taste cell has several microvilli, called **taste hairs,** extending from its apex into a tiny opening in the epithelium called the **taste pore.** The remaining two nonsensory cell types are **basal cells** and **supporting cells.** Like olfactory cells, the cells of the taste buds are replaced continuously, each having a normal life span of about 10 days.

Function of Taste

Substances called **tastants** (tās′tants) dissolve in saliva, enter the taste pores, and by various mechanisms cause the taste cells to depolarize. For example, the diffusion of the positively charged ions Na^+ and H^+ into the taste cells causes depolarization. The binding of tastants to receptors on the taste hairs activates G proteins (see chapter 15), which results in depolarization by causing Ca^{2+} channels to open or K^+ channels to close. When the taste cells depolarize, they release neurotransmitters that diffuse to sensory neurons associated with the taste cells. The neurotransmitters stimulate action potentials in the sensory neurons, which are conducted to the brain, where the sense of taste is perceived.

Five **primary tastes** have been identified in humans: salty, sour, sweet, bitter, and umami (ū-ma′mē, a Japanese term loosely translated as savory). The salty taste is stimulated by Na^+; sour by acids; sweet by sugars, some other carbohydrates, and some proteins; bitter by alkaloids (bases); and umami by the amino acid glutamate and related compounds. The umami taste sensation is most intense when coupled with the salty taste, hence the popularity of adding salt to tomatoes, ketchup, soy sauce, and the food additive monosodium glutamate (MSG).

Even though only five primary tastes have been identified, humans can perceive a fairly large number of different tastes, presumably by combining the five basic taste sensations. As with olfaction, the specificity of the receptor molecules is not perfect. For example, artificial sweeteners have different chemical structures than the sugars they are designed to replace and some are many times more powerful than natural sugars in stimulating taste sensations.

Although all taste buds are able to respond to all five of the basic taste stimuli, each taste bud is usually most sensitive to one class of taste stimuli. Thresholds for the five primary tastes stimuli vary. Sensitivity for bitter substances is the highest. Many alkaloids are poisonous and the high sensitivity for bitter tastes may be protective because of the avoidance of bitter foods. On the other hand, humans tend to crave sweet, salty, and umami tastes, perhaps in response to the body's need for sugars, carbohydrates, proteins, and minerals.

Accommodation to taste stimuli occurs rapidly. This adaptation occurs at the level of the taste bud and within the CNS. Adaptation may begin within 1 or 2 seconds after a taste sensation is perceived, and complete adaptation may occur within 5 minutes.

Many of the sensations thought of as being taste are strongly influenced by olfactory sensations. This phenomenon can be demonstrated by pinching one's nose to close the nasal passages while trying to taste something. With olfaction blocked, it is difficult to distinguish between the taste of a piece of apple and the taste of potato. Much of the "taste" is lost by this action. This is one reason that a person with a cold has a reduced sensation of taste.

The tongue can detect other stimuli besides taste, such as temperature and texture, and they can influence the sensation of taste. Food served at the wrong temperature is perceived as distasteful. Stimulation of hot receptors by capsaicin or black pepper is interpreted as hot or spicy, and stimulation of cold receptors is perceived as fresh or minty.

Neuronal Pathways for Taste

Axons of the sensory neurons associated with taste cells combine and contribute to cranial nerves. Taste from the anterior two-thirds of the tongue is carried by a branch of the facial nerve (VII) called the **chorda tympani** (kōr′dă tim′pă-nē, so named because it crosses over the surface of the tympanic membrane of the middle ear). Taste from the posterior one-third of the tongue and the superior pharynx is carried by means of the glossopharyngeal nerve (IX). In addition to these two major nerves, the vagus nerve (X) carries a few fibers for taste sensation from the epiglottis.

These nerves extend from the taste buds to nuclei in the medulla oblongata. Fibers from these nuclei decussate and extend to the thalamus. Neurons from the thalamus project to the taste area of the cortex, which is at the extreme inferior end of the postcentral gyrus. ▼

> 7. *Where are taste buds located? Describe the structure of a taste bud.*
> 8. *Describe how tastants can cause the production of action potentials in sensory neurons.*
> 9. *What are the five primary tastes?*
> 10. *How is the sense of taste related to the sense of smell and other sensations, such as temperature?*
> 11. *Describe the neuronal pathway for taste.*

Visual System

The visual system includes the eyes, the accessory structures, and the sensory neurons that project to the cerebral cortex where action potentials conveying visual information are interpreted. The **eye** consists of the **eyeball,** or globe of the eye, and the optic nerve. Much of the information we obtain about the world around us is detected by the visual system. Our education is largely based on visual input and depends on our ability to read words and numbers. Visual input includes information about light and dark, movement, color, and hue.

Accessory Structures

Accessory structures protect, lubricate, and move the eye. They include the eyebrows, eyelids, conjunctiva, lacrimal apparatus, and extrinsic eye muscles.

Eyebrows

The **eyebrows** are curved lines of hair over the orbit (figure 13.3). They protect the eyes by preventing perspiration, which can irritate the eyes, from running down the forehead and into them, and they help shade the eyes from direct sunlight.

Eyelids

The **eyelids** are moveable folds covering the anterior surface of the eye when closed (see figure 13.3). The upper and lower eyelids meet at the **medial** and **lateral angles of the eye.** The medial angle contains a small, reddish-pink mound called the **caruncle**

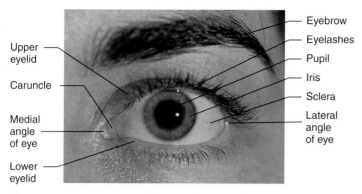

Eyebrow
Eyelashes
Pupil
Iris
Sclera
Lateral angle of eye

Upper eyelid
Caruncle
Medial angle of eye
Lower eyelid

Figure 13.3 The Left Eye and Its Accessory Structures

(kar′ŭng-kl, mound of tissue). The caruncle contains some modified sebaceous and sweat glands.

The **eyelids** consist of five layers of tissue (figure 13.4). From the outer to the inner surface, they are (1) a thin layer of skin on the external surface; (2) a thin layer of loose connective tissue; (3) a layer of skeletal muscle consisting of the orbicularis oculi and levator palpebrae superioris muscles; (4) a crescent-shaped layer of dense connective tissue called the **tarsal** (tar′săl) **plate,** which helps maintain the shape of the eyelid; and (5) the conjunctiva (described in the next section).

The eyelids, with their associated lashes, protect the eyes from foreign objects. If an object suddenly approaches the eye, the eyelids protect the eye by rapidly closing and then opening, a response called the blink reflex. Blinking, which normally occurs about 25 times per minute, also helps keep the eye lubricated by

spreading tears over the surface. Movements of the eyelids are a function of skeletal muscles. The orbicularis oculi muscle closes the lids, and the levator palpebrae superioris elevates the upper lid. The eyelids also help regulate the amount of light entering the eye.

Eyelashes are attached as a double or triple row of hairs to the free edges of the eyelids (see figure 13.3). **Ciliary glands** are modified sweat glands that open into the hair follicles of the eyelashes to keep them lubricated. When one of these glands becomes inflamed, it is called a **sty. Tarsal,** or **meibomian** (mī-bō′mē-an), **glands** are sebaceous glands near the inner margins of the eyelids. They produce **sebum** (sē′bŭm), an oily, semifluid substance that lubricates the lids and restrains tears from flowing over the margin of the eyelids. An infection or a blockage of a tarsal gland is called a **chalazion** (ka-lā′zē-on), or **meibomian cyst.**

Conjunctiva

The **conjunctiva** (kon-jŭnk-tī′vă, to bind together) (see figure 13.4) is a thin, transparent mucous membrane. It covers the inner surface of the eyelids and the anterior white surface of the eye. The conjunctiva reduces friction as the eyelids move over the surface of the eye. It also is a barrier to the entry of microorganisms.

⚕ Conjunctivitis

Conjunctivitis is an inflammation of the conjunctiva caused by an infection or another irritation. An example of conjunctivitis caused by a bacterium is **acute contagious conjunctivitis,** also called **pinkeye.**

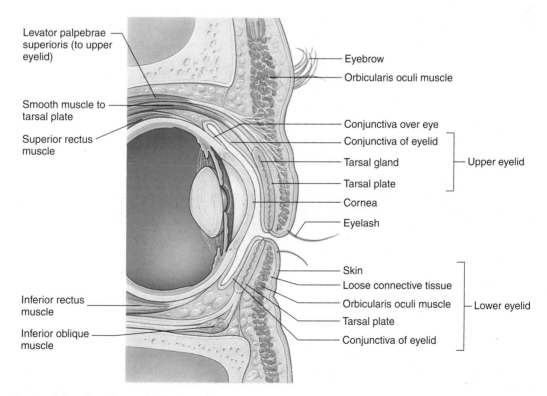

Levator palpebrae superioris (to upper eyelid)

Smooth muscle to tarsal plate

Superior rectus muscle

Inferior rectus muscle

Inferior oblique muscle

Eyebrow
Orbicularis oculi muscle

Conjunctiva over eye
Conjunctiva of eyelid
Tarsal gland
Tarsal plate
Cornea
Eyelash

Upper eyelid

Skin
Loose connective tissue
Orbicularis oculi muscle
Tarsal plate
Conjunctiva of eyelid

Lower eyelid

Figure 13.4 Sagittal Section Through the Eye, Showing Its Accessory Structures

1. Tears are produced in the lacrimal gland and exit the gland through several lacrimal ducts.

2. The tears pass over the surface of the eye.

3. Tears enter the lacrimal canaliculi.

4. Tears are carried through the nasolacrimal duct to the nasal cavity.

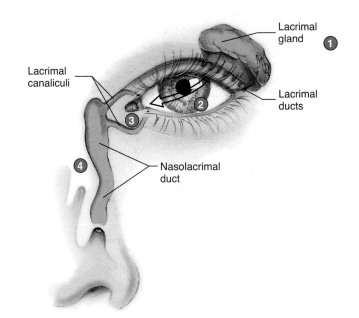

Process Figure 13.5   Lacrimal Apparatus

Lacrimal Apparatus

The **lacrimal** (lak′ri-măl, tear) **apparatus** (figure 13.5) consists of a lacrimal gland, lacrimal canaliculi, and a nasolacrimal duct. The lacrimal apparatus produces tears, releases the tears onto the surface of the eye, and removes excess tears from the surface.

The **lacrimal gland** is in the superolateral corner of the orbit and is innervated by parasympathetic fibers from the facial nerve (VII). The lacrimal gland produces tears, which leave the gland through several **lacrimal ducts** and pass over the anterior surface of the eyeball. The lacrimal gland produces tears constantly at the rate of about 1 mL/day to moisten the surface of the eye, lubricate the eyelids, and wash away foreign objects. Tears are mostly water, with some salts, mucus, and lysozyme, an enzyme that kills certain bacteria. Most of the fluid produced by the lacrimal glands evaporates from the surface of the eye, but excess tears are collected in the medial corner of the eye by small tubes called the **lacrimal canaliculi** (kan-ă-lik′ū-lī, little canal). One lacrimal canaliculus opens on the inner, medial surface of the upper eyelid, and the other lacrimal canaliculus opens on the inner, medial surface of

the lower eyelid. The lacrimal canaliculi connect to the **nasolacrimal duct,** which opens into the inferior meatus of the nasal cavity beneath the inferior nasal concha.

> **P R E D I C T**   **2**
> Explain why it is often possible to "taste" medications, such as eye drops, that have been placed into the eyes. Why does a person's nose "run" when he or she cries?

Extrinsic Eye Muscles

The **extrinsic eye muscles** attach to the outside of the eyeball and cause it to move (table 13.1 and figure 13.6). There are four rectus muscles, the **superior, inferior, medial, and lateral rectus muscles.** *Rectus* means straight, and the fibers of these muscles are nearly straight with the anterior-posterior axis of the eyeball. There are two oblique muscles, the **superior and inferior oblique muscles,** so named because they are at an angle to the axis of the eyeball.

The movements of the eye can be described graphically by a figure resembling the letter *H*. The clinical test for normal eye movement is therefore called the **H test.** A person's inability to

Table 13.1	Muscles Moving the Eye			
Muscle	**Origin**	**Insertion**	**Nerve**	**Action**
Rectus				
Inferior	Common tendinous ring	Sclera of eye	Oculomotor	Depresses and medially deviates gaze
Lateral	Common tendinous ring	Sclera of eye	Abducent	Laterally deviates gaze
Medial	Common tendinous ring	Sclera of eye	Oculomotor	Medially deviates gaze
Superior	Common tendinous ring	Sclera of eye	Oculomotor	Elevates and medially deviates gaze
Oblique				
Inferior	Orbital plate of maxilla	Sclera of eye	Oculomotor	Elevates and laterally deviates gaze
Superior	Common tendinous ring	Sclera of eye	Trochlear	Depresses and laterally deviates gaze

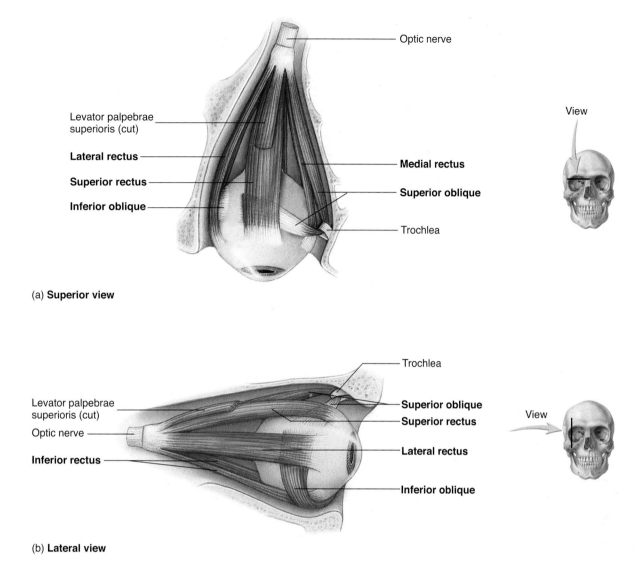

Figure 13.6 Extrinsic Eye Muscles of the Right Eye

move the eye toward one part of the *H* may indicate dysfunction of an extrinsic eye muscle or the cranial nerve to the muscle.

The superior oblique muscle is innervated by the trochlear nerve (IV). The nerve is so named because the superior oblique muscle goes around a little pulley, or trochlea, in the superomedial corner of the orbit. The lateral rectus muscle is innervated by the abducent nerve (VI), so named because the lateral rectus muscle abducts the eye. The other four extrinsic eye muscles are innervated by the oculomotor nerve (III). ▼

> **12. Describe and state the functions of the eyebrows, eyelids, conjunctiva, lacrimal apparatus, and extrinsic eye muscles.**

Anatomy of the Eye

The eyeball is composed of three layers. From superficial to deep they are the **fibrous layer, vascular layer,** and **retina** (figure 13.7). The fibrous layer consists of the sclera and cornea and the vascular layer consists of the choroid, ciliary body, and iris.

Fibrous Layer

The **sclera** (sklēr′ă, hard) is the firm, opaque, white, outer layer of the posterior five-sixths of the eyeball (see figure 13.7). It consists of dense collagenous connective tissue with elastic fibers. The sclera helps maintain the shape of the eyeball, protects its internal structures, and provides an attachment point for the extrinsic eye muscles. A small portion of the sclera can be seen as the "white of the eye" (see figure 13.3).

The sclera is continuous with the **cornea** (kōr′nē-ă, horn-like), the anterior sixth of the eyeball (see figure 13.7). The cornea is avascular and transparent, permitting light to enter the eye. The focusing system of the eye refracts, or bends, light and focuses it on the nervous layer (retina) (see "Focusing System of the Eye," p. 377). The cornea is responsible for most of the refraction of light entering the eye.

The cornea consists of a connective tissue matrix containing collagen, elastic fibers, and proteoglycans, with a layer of stratified squamous epithelium covering the outer surface and a layer of simple squamous epithelium on the inner surface. The outer

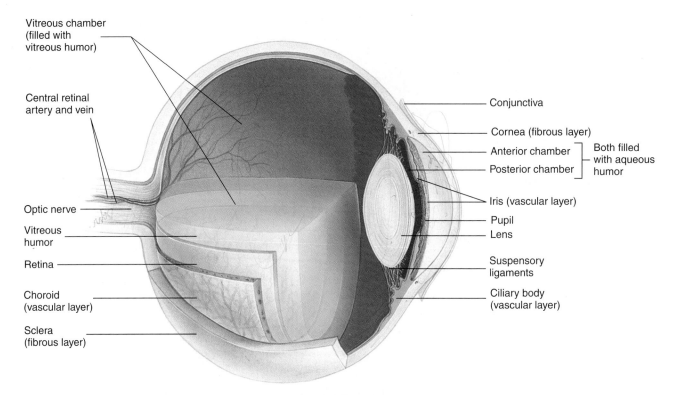

Vitreous chamber
(filled with
vitreous humor)

Central retinal
artery and vein

Optic nerve

Vitreous
humor

Retina

Choroid
(vascular layer)

Sclera
(fibrous layer)

Conjunctiva

Cornea (fibrous layer)

Anterior chamber ⎤ Both filled
⎦ with aqueous
Posterior chamber   humor

Iris (vascular layer)

Pupil

Lens

Suspensory
ligaments

Ciliary body
(vascular layer)

Figure 13.7   Sagittal Section of the Eye, Demonstrating the Layers of the Eyeball

epithelium is continuous with the conjunctiva over the sclera. The transparency of the cornea results from its low water content. In the presence of water, proteoglycans trap water and expand, which scatters light. In the absence of water, the proteoglycans decrease in size and do not interfere with the passage of light through the matrix.

 Cornea

The central part of the cornea receives oxygen from the outside air. Soft plastic contact lenses worn for long periods must therefore be permeable so that air can reach the cornea.

The most common eye injuries are cuts or tears of the cornea caused by foreign objects, stones or sticks, hitting the cornea. Extensive injury to the cornea may cause connective tissue deposition, thereby making the cornea opaque.

The cornea was one of the first organs to be transplanted. Several characteristics make it relatively easy to transplant: It is easily accessible and relatively easily removed; it is avascular and therefore does not require extensive circulation, as do other tissues; and it is less immunologically active and therefore less likely to be rejected than are other tissues. ▼

13. *What part of the eyeball is the sclera and cornea? What are their functions?*

P R E D I C T   3
Predict the effect of corneal inflammation on vision.

Vascular Layer

The vascular layer of the eye is so named because it contains most of the blood vessels of the eye. The posterior portion of the vascular layer, associated with the sclera, is the **choroid** (kō′royd, like a membrane) (see figure 13.7). This is a very thin structure consisting of a vascular network and many melanin-containing pigment cells so that it appears black in color. The black color absorbs light so that it is not reflected inside the eye. If light were reflected inside the eye, the reflection would interfere with vision. The interiors of cameras are black for the same reason.

The choroid is continuous anteriorly with the **ciliary** (sil′ē-ar-ē, like an eyelash) **body,** which consists of an outer **ciliary ring** and an inner group of **ciliary processes** (figure 13.8*a* and *b*). The ciliary ring and the base of the ciliary processes contain smooth muscle called **ciliary muscles. Suspensory ligaments** attach the ciliary ring and processes to the lens of the eye and contraction of the ciliary muscles can change the shape of the lens (see "Accommodation," p. 378). The ciliary process also produces aqueous humor (see "Chambers of the Eye," p. 376).

The ciliary body is continuous anteriorly with the **iris of the eye,** which is the "colored part" of the eye (see figures 13.3 and 13.8*a*). The iris is a contractile structure, consisting mainly of smooth muscle, surrounding an opening called the **pupil** (see figure 13.7). Light enters the eye through the pupil, and the iris regulates the amount of light by controlling the size of the pupil. The iris contains two groups of smooth muscles: a circular group called the **sphincter pupillae** (pū-pil′ē) and a radial group called the **dilator pupillae** (figure 13.8*c* and *d*). The sphincter pupillae are innervated by parasympathetic fibers from the oculomotor nerve (III). When

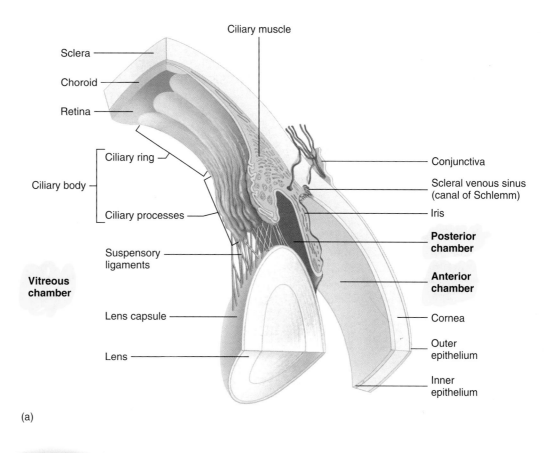

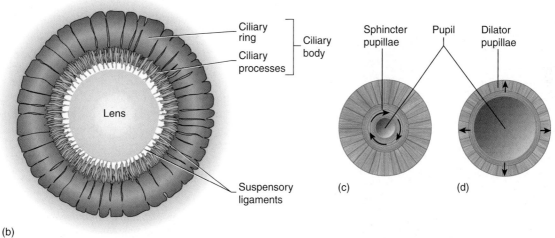

Figure 13.8   Ciliary Body, Lens, Iris, and Cornea

(*a*) Sagittal section of the eye. (*b*) The lens and ciliary body. (*c*) The sphincter pupillae muscles of the iris constrict the pupil. (*d*) The dilator pupillae muscles of the iris dilate the pupil.

they contract, the pupil constricts and less light enters the eye. The dilator pupillae are innervated by sympathetic fibers. When they contract, the pupil dilates and more light enters the eye. The ciliary muscles, sphincter pupillae, and dilator pupillae are sometimes referred to as the **intrinsic eye muscles.**

The color of the eye differs from person to person. A large amount of melanin in the iris causes it to appear brown or even black. Less melanin results in light brown, green, or gray irises. Even less melanin causes the eyes to appear blue. If there is no pigment in the iris, as in albinism, the iris is pink because blood vessels in the eye reflect light back to the iris. The genetics of eye color is quite complex. A gene on chromosome 15 and another on chromosome 19 affect eye color, but other genes, not yet identified, are postulated to explain complex eye colors and patterns.

Despite its appearance as a solid black disk (see figure 13.3), the pupil is not a solid structure but, rather, is an opening. It appears black because light entering the eye through the pupil is absorbed by the choroid and retina and is not reflected back out of the pupil. The absence of light is the color black. It is like looking down a hallway into a dark room. If a bright light is directed into the pupil, however, the reflected light is red because of the blood vessels on the surface of the retina. This is why the pupils of a person looking directly at a flash camera often appear red in a photograph. ▼

14. *Where is the choroid located and what is its function?*
15. *Describe the location and structure of the ciliary body. Give two functions of the ciliary body.*
16. *What is the iris and the pupil? How does the iris control the amount of light entering the eye through the pupil?*
17. *What determines the color of the iris?*

Retina

The **retina** is the inner layer of the eye, covering the inner surface of the eye posterior to the ciliary body (see figure 13.7). The retina has over 126 million photoreceptor cells, which respond to light (see "Structure and Function of the Retina," p. 379). As a result, action potentials are generated and conducted by the optic nerve (II) out of the eye to the cerebral cortex, where the sense of vision takes place.

When the posterior region of the retina is examined with an **ophthalmoscope** (of-thal′mō-skōp) (figure 13.9), several important features can be observed. Near the center of the posterior retina is a small, yellow spot approximately 4 mm in diameter, the **macula** (mak′ū-lă, a spot). In the center of the macula is a small pit, the **fovea** (fō′vē-ă) **centralis.** The fovea, followed by the macula, has the highest concentration of photoreceptor cells in the retina. Thus, they are the part of the retina most sensitive to light. Just medial to the macula

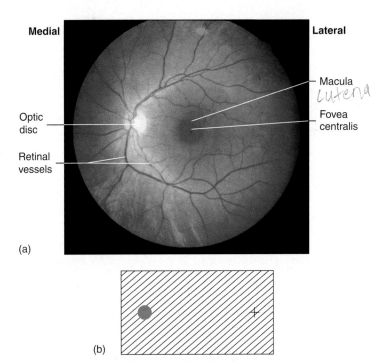

Medial   **Lateral**

Macula
Lutena

Optic disc

Fovea centralis

Retinal vessels

(a)

(b)

Figure 13.9   Ophthalmoscopic View of the Left Retina

(*a*) The posterior wall of the retina as seen when looking through the pupil. Notice the vessels entering the eye through the optic disc (the optic nerve) and the macula with the fovea centralis. (*b*) Demonstration of the blind spot. Close your right eye. Hold the figure in front of your left eye and stare at the +. Move the figure toward your eye. At a certain point, when the image of the spot is over the optic disc, the red spot seems to disappear.

is a white spot, the **optic disc,** through which the central retinal artery enters and the central retinal vein exits the eyeball. Branches from these vessels spread over the surface of the retina. The optic disc is also the place where axons from the neurons of the retina converge to form the optic nerve, which exits the posterior eye. The optic disc contains only axons and no photoreceptor cells. Therefore, it does not respond to light and is called the **blind spot** of the eye. See figure 13.9*b* for a simple test demonstrating the blind spot. ▼

18. *What is the macula, fovea centralis, and optic disc? How do they respond to light?*

Chambers of the Eye

The interior of the eye is divided into three chambers: **anterior chamber, posterior chamber,** and **vitreous** (vit′rē-ŭs, glassy) **chamber** (see figures 13.7 and 13.8*a*). The anterior and posterior chambers are located between the cornea and the lens. The iris separates the anterior chamber from the posterior chamber, which are continuous with each other through the pupil. The much larger vitreous chamber is posterior to the lens.

A fluid called **aqueous humor** fills the anterior and posterior chambers. Aqueous humor helps maintain intraocular pressure, which keeps the eyeball inflated and is largely responsible for maintaining the shape of the eyeball. The aqueous humor also refracts light and provides nutrition for the structures of the anterior chamber, such as the cornea, which has no blood vessels. Aqueous humor is produced by the ciliary processes as a blood filtrate and is returned to the circulation through a venous ring at the base of the cornea called the **scleral venous sinus** (canal of Schlemm) (shlem) (see figure 13.8*a*). The production and removal of aqueous humor result in the "circulation" of aqueous humor and maintenance of a constant intraocular pressure. **Glaucoma** (glaw-kō′mă) is an abnormal increase in intraocular pressure that results when the rate of production of aqueous humor exceeds its rate of removal.

A transparent, jellylike substance called **vitreous humor** fills the vitreous chamber. The vitreous humor helps maintain intraocular pressure and therefore the shape of the eyeball, and it holds the lens and retina in place. It also functions in the refraction of light in the eye. Vitreous humor is not produced as rapidly as is the aqueous humor, and its turnover is extremely slow. ▼

19. *Name the three chambers of the eye, describe their location, and state the substances in them.*

Lens

The **lens** is an avascular, transparent, biconvex disk located behind the pupil (see figure 13.7). *Biconvex* means that each side of the lens bulges outward. The lens is part of the focusing system of the eye and light passing through the lens is focused on the retina. The lens is a flexible structure, and changing the shape of the lens is involved with adjusting the focus of light.

The lens consists of a layer of cuboidal epithelial cells on its anterior surface and a posterior region of very long, columnar epithelial cells called **lens fibers.** Cells from the anterior epithelium proliferate and give rise to the lens fibers. The lens fibers lose their nuclei and other cellular organelles and accumulate a set of proteins called **crystallines.** A highly elastic, transparent **lens capsule** surrounds the lens (see figure 13.8*a*). The suspensory ligaments connect the lens capsule to the ciliary body. Through the lens capsule and the suspensory ligaments, the ciliary body can change the shape of the lens. ▼

> **20. Describe the location and structure of the lens of the eye. What is the function of the lens?**
>
> **21. What is the lens capsule and how is it connected to the ciliary body?**

Functions of the Complete Eye

The eye receives light and produces action potentials. When the brain interprets the action potentials, vision results.

Properties of Light

The **electromagnetic spectrum** is the entire range of wavelengths, or frequencies, of electromagnetic radiation. Gamma waves have the shortest wavelength and radio waves the longest wavelength (figure 13.10). **Visible light,** the portion of the electromagnetic spectrum that can be detected by the human eye, is a small part of the electromagnetic spectrum. Within the visible spectrum, each color has a different wavelength.

An important characteristic of light is that it can be refracted, or bent. As light passes from air to a denser substance, such as glass or water, its speed is reduced. If the surface of that substance is at an angle other than 90 degrees to the direction the light rays are traveling, the rays are bent as a result of variation in the speed of light as it encounters the new medium. This bending of light is called **refraction.** The greater the curvature of the surface, the greater is the refraction of light.

If the surface of a lens is concave, with the lens thinnest in the center, the light rays diverge as a result of refraction. If the surface is convex, with the lens thickest in the center, the light rays converge. As light rays converge, they finally reach a point at which

they cross. This point is called the **focal point (FP),** and causing light to converge is called **focusing.**

If light rays strike an object that is not transparent, they bounce off the surface. This phenomenon is called **reflection.** The images we see result from light reflected off of objects. ▼

> **22. Define visible light, refraction, focal point, focusing, and reflection.**

Focusing System of the Eye

Light entering the eye passes through the focusing system of the eye to strike the retina. The **focusing system of the eye,** which refracts light, is the cornea, aqueous humor, lens, and vitreous humor. Light passing through the focusing system is refracted, producing a focal point (figure 13.11*a*). No image is produced at the focal point, however. Past the focal point is a place where the image passing through the focusing system can be clearly seen. In a normal eye, the focused image falls on the retina as an inverted image. The image is inverted and reversed right to left because the light rays cross at the focal point.

 Visual Image Inversion

The image of the world focused on the retina is upside down because the visual image is inverted when it reaches the retina. The brain processes information from the retina so that the world is perceived the way "it really is." If a person wears glasses that invert the image entering the eye, he or she will see the world upside down for a few days, after which time the brain adjusts to the new input to set the world right side up again. If the glasses are then removed, another adjustment period is required before the world is made right by the brain.

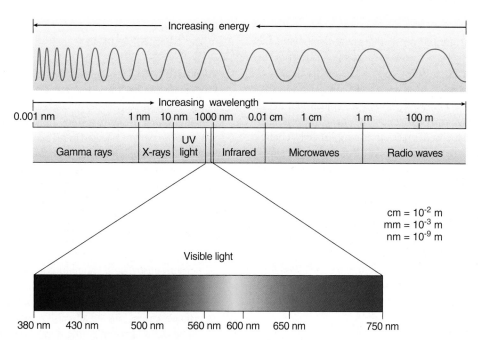

Figure 13.10   Electromagnetic Spectrum
Visible light is pulled out of the electromagnetic spectrum and expanded. The wavelengths of the various colors are also depicted.

Distant vision

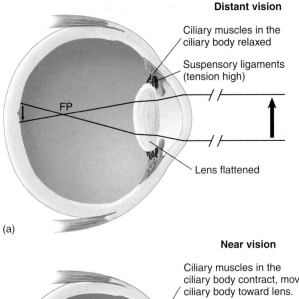

Ciliary muscles in the ciliary body relaxed

Suspensory ligaments (tension high)

FP

Lens flattened

(a)

Near vision

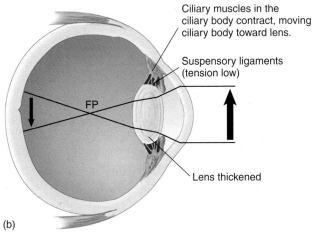

Ciliary muscles in the ciliary body contract, moving ciliary body toward lens.

Suspensory ligaments (tension low)

FP

Lens thickened

(b)

Figure 13.11   Focus and Accommodation by the Eye

The focal point (FP) is where light rays cross. (*a*) Distant vision. The lens is flattened, and the image is focused on the retina. (*b*) Accommodation for near vision. The lens is more rounded, and the image is focused on the retina.

The cornea and lens are the most important elements of the focusing system of the eye. The cornea is responsible for most of the refraction of light because the greatest contrast in media density is between the air and the cornea. The shape of the cornea and its distance from the retina are fixed, however, so that no adjustment in the location of the focused image can be made by the cornea. Fine adjustments to the location of the focused image are accomplished by changing the shape of the lens. Increasing the curvature of the lens increases the refraction of light, moving the focused image closer to the lens. Decreasing the curvature of the lens decreases the refraction of light, moving the focused image farther from the lens. In cameras, microscopes, and telescopes, focusing is not accomplished by changing lens shape. Instead, focusing is accomplished by moving the lens closer to or farther from the point at which the image will be focused. ▼

> **23. Name the parts of the focusing system of the eye. Which part does the most refraction of light? Which part can adjust the refraction?**

Distant and Near Vision

Distant vision occurs when looking at objects 20 feet or more from the eye, whereas **near vision** occurs when looking at objects that are less than 20 feet from the eye. In distant vision, the ciliary muscles

in the ciliary body are relaxed (see figure 13.11*a*). The suspensory ligaments, however, maintain elastic pressure on the lens, thereby keeping it relatively flat. The condition in which the lens is flattened so that nearly parallel rays from a distant object are focused on the retina is referred to as **emmetropia** (em-ĕ-trō′pē-ă, measure) and is the normal resting condition of the lens. The point at which the lens does not have to thicken for focusing to occur is called the **far point of vision** and normally is 20 feet or more from the eye.

When an object is brought closer than 20 feet to the eye, the image falling on the retina is no longer in focus. Three events occur to bring the image into focus on the retina: accommodation by the lens, constriction of the pupil, and convergence of the eyes.

1. *Accommodation.* When the eye focuses on a nearby object, the ciliary muscles contract as a result of parasympathetic stimulation from the oculomotor nerve (III). This sphincterlike contraction pulls the choroid toward the lens to reduce the tension on the suspensory ligaments. This allows the lens to assume a more spherical form because of its own elastic nature (figure 13.11*b*). The more spherical lens has a more convex surface, causing greater refraction of light, which brings the image back into focus on the retina. This process is called **accommodation.**

   As an object is brought closer and closer to the eye, accommodation becomes more and more difficult because the lens cannot become any more convex. At some point, the eye no longer can focus the object, and it is seen as a blur. The point at which this blurring occurs is called the **near point of vision,** which is usually about 2–3 inches from the eye for children, 4–6 inches for a young adult, 20 inches for a 45-year-old adult, and 60 inches for an 80-year-old adult. This increase in the near point of vision is called **presbyopia** (prez-bē-ō′pē-ă, elder eye). It occurs because the lens becomes more rigid with increasing age, which is why some older people say they could read with no problem if they only had longer arms.

 Vision Charts

Visual acuity is a measure of the sharpness or distinctiveness of vision. For example, the letters *e* and *o* have a similar shape and size. Visual acuity enables a person to tell the two letters apart. When a person's visual acuity is tested, a chart is placed 20 feet from the eye, and the person is asked to read a line of letters that is standardized for normal vision at 20 feet. If the person can correctly identify the letters in the line, the vision is considered to be 20/20, which means that the person can see at 20 feet what people with normal vision can see at 20 feet. If, on the other hand, the person can see letters only at 20 feet that people with normal vision can see at 40 feet, the vision is considered 20/40. They have less visual acuity.

2. *Pupil constriction.* When looking at a close-up object, the pupil diameter decreases, which increases the depth of focus. The **depth of focus** is the greatest distance through which an object can be moved and still remain in focus on the retina. The main factor affecting depth of focus is the size of the pupil. If the pupillary diameter is small, the depth of focus is greater than if the pupillary diameter is large. With a smaller pupillary opening, an object may therefore be moved slightly

nearer or farther from the eye without disturbing its focus. This is particularly important when viewing an object at close range because the interest in detail is much greater, and therefore the acceptable margin for error is smaller. When the pupil is constricted, the light entering the eye tends to pass more nearly through the center of the lens and is more accurately focused than light passing through the edges of the lens. Pupillary diameter also regulates the amount of light entering the eye. The smaller the pupil diameter, the less light entering the eye. As the pupil constricts during close vision, therefore, more light is required on the object being observed.

3. *Convergence.* Because the light rays entering the eyes from a distant object are nearly parallel, both pupils can pick up the light rays when the eyes are directed more or less straight ahead. As an object moves closer, however, the eyes must be rotated medially so that the object is kept focused on corresponding areas of each retina. Otherwise, the object appears blurry. This medial rotation of the eyes is accomplished by a reflex that stimulates the medial rectus muscle of each eye. This movement of the eyes is called convergence. Convergence can easily be observed. Have someone stand facing you. Have the person reach out one hand and extend an index finger as far in front of his or her face as possible. While the person keeps the gaze fixed on the finger, have the person slowly bring the finger in toward his or her nose until finally touching it. Notice the movement of the person's pupils. What happens? ▼

24. *Define distant and near vision. What is emmetropia and the far point of vision?*

25. *What is visual acuity and how is it tested?*
26. *Describe the process of accommodation.*
27. *What is the near point of vision and presbyopia?*
28. *What is depth of focus and how is it related to pupil diameter?*
29. *What is the benefit of the eyes converging as an object moves closer to the eyes?*

P R E D I C T ④
Explain how several hours of reading can cause eyestrain, or eye fatigue. Describe what structures are involved.

Structure and Function of the Retina

Leonardo da Vinci, in speaking of the eye, said, "Who would believe that so small a space could contain the images of all the universe?" The retina of each eyeball, which gives us the potential to see the whole world, is about the size and thickness of a postage stamp.

The retina consists of an outer, pigmented layer and an inner, neural layer (figure 13.12). The **pigmented layer** is a single layer of epithelial cells filled with melanin. Along with the choroid, the pigmented layer provides a black matrix that enhances visual acuity by isolating individual photoreceptors and reducing light scattering. This pigmentation is not strictly necessary for vision, however. People with albinism (lack of pigment) can see, although their visual acuity is reduced because of some light scattering.

The **neural layer** contains three layers of neurons: photoreceptor, bipolar, and ganglionic (see figure 13.12). The photoreceptor layer contains **rods** and **cones,** which are the photoreceptor cells that respond to light. The rods and cones synapse with **bipolar cells,** which in turn synapse with **ganglion cells.** Axons from

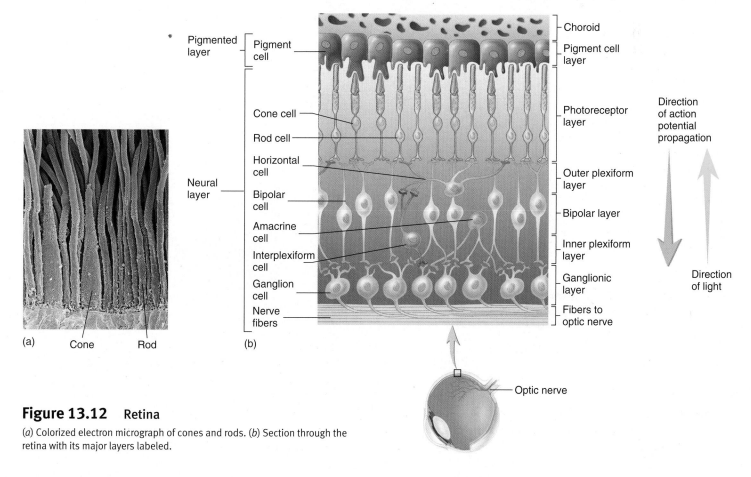

Figure 13.12   Retina

(*a*) Colorized electron micrograph of cones and rods. (*b*) Section through the retina with its major layers labeled.

the ganglion cells pass over the inner surface of the retina, converge at the optic disc (blind spot), and exit the eye as the optic nerve (II). The neural layers are separated by the plexiform (like a braid) layers. The outer plexiform layer is where the photoreceptor cells synapse with the bipolar cells and the inner plexiform layer is where the bipolar cells synapse with the ganglion cells. ▼

> 30. **What is the function of the pigmented retina and of the choroid?**
> 31. **Starting with a rod or cone, name the cells or structures that an action potential encounters while traveling to the optic nerve.**

Rods

Rods are responsible for noncolor vision and vision under conditions of reduced light. Even though rods are very sensitive to light, they cannot detect color, and sensory input reaching the brain from rods is interpreted by the brain as shades of gray. Rods are bipolar neurons with modified, light-sensitive dendrites, which are cylindrical in shape (figure 13.13). This rod-shaped photoreceptive part of the rods contains about 700 double-layered membranous discs. The discs contain **rhodopsin** (rō-dop′sin), a purple pigment consisting of the protein **opsin** covalently bound to a yellow photosensitive pigment called **retinal** (derived from vitamin A).

Exposure to light activates rhodopsin (figure 13.14). When light strikes retinal, it changes shape, which causes opsin to change shape, resulting in the activation of rhodopsin. Activated rhodopsin

stimulates a response in rods that results in vision. Rhodopsin is deactivated when retinal completely detaches from opsin. Retinal returns to its original shape, a process that requires energy from ATP. Retinal then reattaches to opsin, which changes shape, resulting in rhodopsin that can again be activated by light.

In the dark, rods continuously release the neurotransmitter glutamate, which inhibits the bipolar cells synapsing with the rods (see figure 13.12). The bipolar cells do not stimulate the ganglion cells and no action potentials are transmitted out the optic nerve. Thus, in the dark, the system is turned off. In the light, activated rhodopsin inhibits the production of glutamate by rods. Therefore, the bipolar cells become uninhibited, their membrane potentials reach threshold, and action potentials are produced. The bipolar cells stimulate the ganglion cells and action potentials are transmitted out the optic nerve. Thus, in the light, the system is turned on.

Light and **dark adaptation** is the adjustment of the eyes to changes in light intensity. Light adaptation occurs when the eyes are exposed to increased light intensity, such as when a person goes out of a darkened building into the sunlight. Dark adaptation occurs when the eyes are exposed to decreased light intensity, such as when a person goes from the sunlight into a building. Light and dark adaptation is accomplished by changes in the amount of available rhodopsin. In bright light, excess rhodopsin is broken down so that not as much is available to initiate action potentials, and the eyes become adapted to bright light. Conversely, in a dark room more rhodopsin is produced, making the retina more light-sensitive.

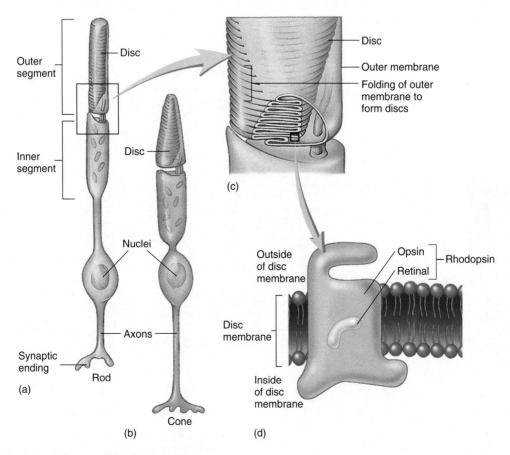

Figure 13.13   **Rods and Cones of the Retina**

(*a*) Rod cell. (*b*) Cone cell. (*c*) An enlargement of the discs in the outer segment of a rod. (*d*) An enlargement of one of the discs, showing rhodopsin in the membrane.

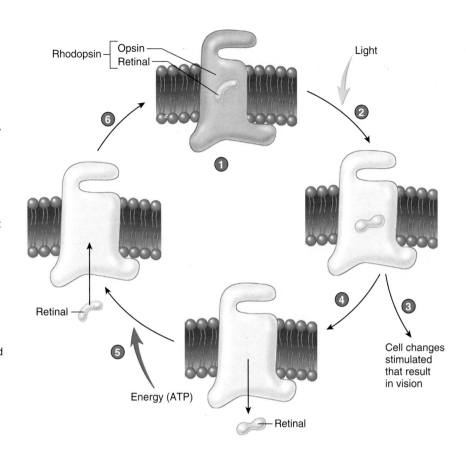

1. Retinal is attached inside opsin to make rhodopsin.

2. Light activates rhodopsin by causing retinal to change shape, which causes opsin to change shape.

3. Activated rhodopsin stimulates cell changes that result in vision.

4. Following rhodopsin activation, retinal detaches from opsin.

5. Energy from ATP is required to bring retinal back to its original form.

6. Retinal attaches to opsin to form rhodopsin (return to step 1).

Process Figure 13.14   Rhodopsin Cycle

Light and dark adaptation also involves pupil reflexes. The pupil enlarges in dim light to allow more light into the eye and contracts in bright light to allow less light into the eye. In addition, rod function decreases and cone function increases in light conditions, and vice versa during dark conditions. ▼

32. *What is the function of rods? Describe the structure of rods.*
33. *Describe the activation, deactivation, and reformation of rhodopsin after light strikes it.*
34. *How does the activation of rhodopsin result in action potential production in ganglion cells?*
35. *How does dark and light adaptation occur?*

P R E D I C T   5
If the breakdown of rhodopsin occurs rapidly and production is slow, do eyes adapt more rapidly to light or dark conditions?

Cones

Cones are responsible for color vision and visual acuity. Color is a function of the wavelength of light, and each color results from a certain wavelength of visible light. Cones require relatively bright light to function. As the light decreases, so does the color of objects that can be seen until, under conditions of very low illumination, the objects appear gray. This occurs because, as the light decreases, the number of cones responding to the light decreases but the number of rods increases.

Cones are bipolar photoreceptor cells with a conical, light-sensitive part that tapers slightly from base to apex (see figure 13.13b).

The outer segments of the cone cells, like those of the rods, consist of double-layered discs. The membranes of the discs contain a visual pigment, **iodopsin** (ī-ō-dop′sin), which consists of retinal combined with a photopigment opsin protein. There are three types of cones, based on the type of opsin they have, which determines the wavelength of light to which they are most responsive. **Blue cones** respond best to blue light, **green cones** respond best to green light, and **red cones** respond best to red light (figure 13.15). Note that the cones are named according to their light sensitivity, not their actual color.

As light of a given wavelength (color) strikes the retina, all cones containing photopigments capable of responding to that wavelength generate action potentials. Different proportions of cones respond to a given wavelength of light because of their overlap in pigment sensitivity (see figure 13.15). Color is interpreted in the visual cortex as combinations of sensory input originating from cones. For example, when orange light strikes the retina, 99% of the red cones respond, 42% of the green cones respond, and no blue cones respond. When yellow light strikes the retina, the response is shifted so that a greater number of green cones respond. The variety of combinations created allows humans to distinguish among several million gradations of light and shades of color. ▼

36. *What are the functions of cones? Describe the structure of cones.*
37. *What is iodopsin?*
38. *What are the three types of cone cells? How do they produce the colors we see?*

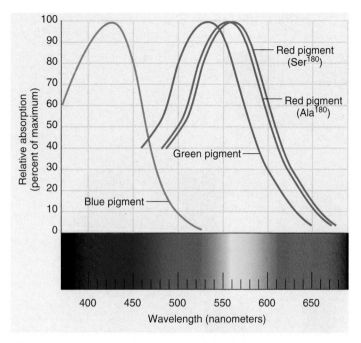

Figure 13.15 **Wavelengths to Which Each of the Three Visual Pigments Are Sensitive: Blue, Green, Red**

There are actually two forms of the red pigment. One form has a serine at position 180; the other has an alanine at position 180. Each red pigment has a slightly different wavelength sensitivity.

Seeing Red

Not everyone sees the same red. Two forms of the red photopigment are common in humans. Approximately 60% of people have the amino acid serine in position 180 of the red opsin protein, whereas 40% have alanine in that position. That subtle difference in the protein results in slightly different absorption characteristics (see figure 13.15). Even though we were each taught to recognize red when we see a certain color, we apparently do not see that color in quite the same way. This difference may contribute to people having different favorite colors.

Distribution of Rods and Cones in the Retina

Each eye has approximately 120 million rods and 6 or 7 million cones. The cones are most concentrated in the fovea centralis and the macula. The fovea centralis has approximately 35,000 cones and no rods. The rest of the macula has more cones than rods. Cones are involved in visual acuity, in addition to their role in color vision. When one is looking at an object directly in front of the eye, the focusing system of the eye places the image on the macula and fovea centralis. The high concentration of cones makes it possible to see fine details.

The rods are 10–20 times more plentiful than cones over most of the retina away from the macula. The high number of rods enables them to "collect" light and they are more important in low-light conditions. ▼

39. Describe the arrangement of cones and rods in the fovea centralis, the macula, and other parts of the retina.

Explain why at night a person may notice a movement "out of the corner of the eye" but, when the person tries to focus on the area of movement, it appears as though nothing is there.

Inner Layers of the Retina

Rod and cone cells differ in the way they interact with bipolar and ganglion cells. Several rods synapse with each bipolar cell, and several bipolar cells synapse with each ganglion cell (see figure 13.12). As a result of this convergence, stimulation of any of a large number of rods can result in stimulation of a ganglion cell. Propagation of action potentials from the ganglion cell to the visual cortex results in vision. Given the large number of rods in the retina, this arrangement allows an awareness of stimuli from very dim light sources. Visual acuity with this system is poor, however. From the brain's perspective, there is no way to determine which rods associated with a given ganglion cell are stimulated. Cones, on the other hand, exhibit little or no convergence on bipolar cells so that one cone cell may synapse with only one bipolar cell. This system reduces light sensitivity but enhances visual acuity because it increases the likelihood that the brain can determine which cone was stimulated.

Within the inner layers of the retina, interneurons modify the signals from the photoreceptor cells before the signal leaves the retina (see figure 13.12). **Horizontal cells** in the outer plexiform layer synapse with photoreceptor cells and bipolar cells. **Amacrine** (am′ă-krin) **cells** in the inner plexiform layer synapse with bipolar and ganglion cells. **Interplexiform cells** connect cells in the outer and inner plexiform layers, forming feedback loops. The interneurons are either excitatory or inhibitory on the cells with which they synapse. By increasing the signal from some photoreceptors and decreasing the signal from others, these interneurons increase the differences between boundaries, such as the edge of a dark object against a light background. ▼

40. How does the arrangement of cells in the retina explain the sensitivity of the rod system to dim light and the visual acuity of the cone system?
41. What is the function of interneurons in the retina?

Motion Pictures

Action potentials pass from the retina through the optic nerve at the rate of 20 to 25 per second. We "see" a given image for a fraction of a second longer than it actually appears. Motion pictures take advantage of these two facts. When still photographs are flashed on a screen at the rate of 24 frames per second, they appear to flow into each other, and a motion picture results.

Neuronal Pathways for Vision

The optic nerve (II) (figure 13.16) leaves the eye and exits the orbit through the optic foramen to enter the cranial cavity. Just anterior to the pituitary gland, the optic nerves are connected to each other at the **optic chiasm** (kī′azm, crossing). Ganglion cell axons from the nasal (medial) retina cross through the optic chiasm and project to the opposite side of the brain. Ganglion cell axons from

1. Each visual field is divided into temporal and nasal parts.

2. After passing through the lens, light from each half of a visual field projects to the opposite side of the retina.

3. An optic nerve consists of axons extending from the retina to the optic chiasm.

4. In the optic chiasm, axons from the nasal part of the retina cross and project to the opposite side of the brain. Axons from the temporal part of the retina do not cross.

5. An optic tract consists of axons that have passed through the optic chiasm (with or without crossing) to the thalamus.

6. The axons synapse in thalamic nuclei. Collateral branches of the axons in the optic tracts synapse in the superior colliculi.

7. An optic radiation consists of axons from thalamic neurons that project to the visual cortex in the occipital lobe.

8. The right part of each visual field (*dark green* and *light blue*) projects to the left side of the brain, and the left part of each visual field (*light green* and *dark blue*) projects to the right side of the brain.

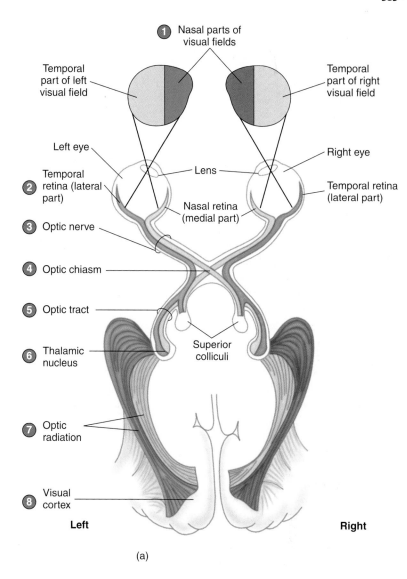

(a)

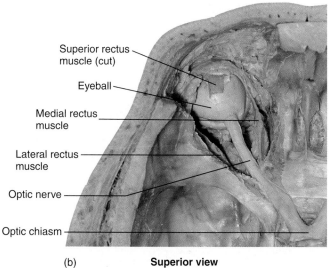

(b)   **Superior view**

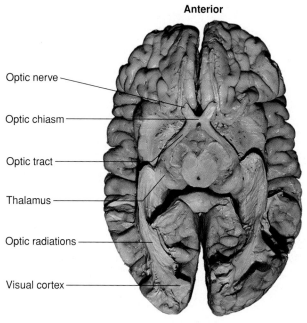

(c)   **Posterior**

Process Figure 13.16   Visual Pathways

(*a*) Visual pathways and their relationship to the visual fields. The left and right visual fields are shown separated from each other for clarity. They actually overlap. (*b*) Photograph of the eyeball, optic nerve, and optic chiasm (superior view). (*c*) Photograph of the optic nerves, tracts, and radiations (inferior view).

Myopia

Myopia (mī-ō′pē-ă), or **nearsightedness,** is the ability to see close objects clearly but distant objects appear blurry. Myopia is a defect of the eye in which the focusing system, the cornea and lens, is optically too powerful, or the eyeball is too long. As a result, images are focused in front of the retina (figure A*a*).

Myopia is corrected by a concave lens that spreads out the light rays coming to the eye so that, when the light is focused by the eye, it is focused on the retina (figure A*b*). Such lenses are called "minus" lenses.

Another technique for correcting myopia is **radial keratotomy** (ker′ă-tot′ō-mē), which consists of making a series of radiating cuts in the cornea. The cuts are intended to weaken the dome of the cornea slightly so that it becomes more flattened and eliminates the myopia. One problem with the technique is that it is difficult to predict exactly how much flattening will occur and

the surgery may overcorrect or undercorrect the problem. Another problem is that some patients are bothered by glare following radial keratotomy because the slits apparently do not heal evenly.

An alternative procedure is **lasix,** or **laser corneal sculpturing,** in which a thin portion of the cornea is etched away to make the cornea less convex. The advantage of this procedure is that the results can be more accurately predicted than those from radial keratotomy.

Hyperopia

Hyperopia (hī-per-ō′pē-ă), or **farsightedness,** is the ability to see distant objects clearly but close objects appear blurry. Hyperopia is a disorder in which the cornea and lens system is optically too weak or the eyeball is too short. As a result, the focused image is "behind" the retina when looking at a close object (figure A*c*). In hyperopia, the lens must accommodate to bring somewhat distant objects into focus,

which would not be necessary for a normal eye. Closer objects cannot be brought into focus because the lens cannot change shape enough to focus the image on the retina. Hyperopia is corrected by a convex lens that causes light rays to converge as they approach the eye and to focus on the retina (figure A*d*). Such lenses are called "plus" lenses.

Presbyopia

Presbyopia (prez-bē-ō′pē-ă) is the normal, presently unavoidable degeneration of the accommodation power of the eye that occurs as a consequence of aging. It occurs because the lens becomes hard and less flexible. The average age for onset of presbyopia is the midforties. Avid readers and people engaged in fine, close work may develop the symptoms earlier.

Presbyopia can be corrected by the use of "reading glasses," which are worn only for close work and are removed when the person wants to see at a distance. It is

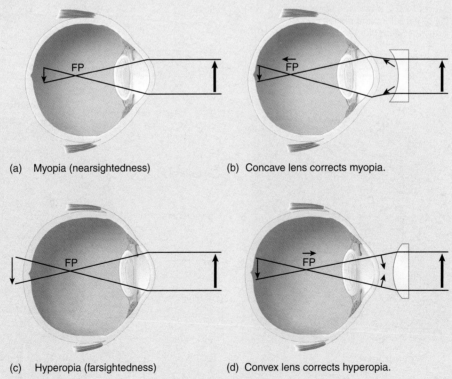

(a) Myopia (nearsightedness)

(b) Concave lens corrects myopia.

(c) Hyperopia (farsightedness)

(d) Convex lens corrects hyperopia.

Figure A Visual Disorders and Their Correction by Various Lenses

sometimes annoying to keep removing and replacing glasses because reading glasses hamper vision of only a few feet away. This problem may be corrected by the use of **bifocals,** which have a different lens in the top and the bottom, or by **progressive lenses,** in which the lens is graded.

Astigmatism

Astigmatism (ă-stig′mă-tizm) is a defect in which the cornea or lens is not uniformly curved and the image is not sharply focused. Glasses may be made to adjust for the abnormal curvature as long as the curvature is not too irregular. If the curvature of the cornea or lens is too irregular, the condition is difficult to correct.

Strabismus

Strabismus (stra-biz′mŭs) is a lack of parallelism of light paths through the eyes. Strabismus can involve one or both eyes, and the eyes may turn in (convergent) or out (divergent). The condition can result from abnormally weak eye muscles. In some cases, the image that appears on the retina of one eye may be considerably different from that appearing on the other eye. This problem is called **diplopia** (di-plō′pē-ă, double vision).

Retinal Detachment

Retinal detachment is a relatively common problem that can result in complete blindness. If a hole or tear occurs in the retina, fluid may accumulate between the neural and pigmented layers, thereby separating them. This separation, or detachment, may continue until the neural layer has become totally detached from the pigmented layer and has folded into a funnel-like form around the optic nerve. When the neural layer becomes separated from its nutrient supply in the choroid, it degenerates, and blindness follows. Causes of retinal detachment include a severe blow to the eye or head; a shrinking of the vitreous humor, which may occur with aging; and diabetes.

Color Blindness

Color blindness results from the dysfunction of one or more of the three photopigments involved in color vision. There may be a complete loss of color perception or only a decrease in perception. For example, in red-green color blindness, a person may be unable to distinguish between the colors red and green (figure B). Red-green color blindness occurs when mutations cause the red photopigment to respond more to green light, or the green photopigment to respond more to red light. Most forms of color blindness occur more frequently in males and are X-linked genetic traits (see chapter 25).

Night Blindness

Everyone sees less clearly in the dark than in the light. A person with **night blindness,** or **nyctalopia** (nik-tă-lō′pē-ă, obscure eye), however, may not see well enough in a dimly lit environment to function adequately. Night blindness results from loss of rod function. It can result from general retinal degeneration, such as occurs in retinitis pigmentosis or detached retinas. Temporary night blindness can result from a vitamin A deficiency because vitamin A is necessary to produce retinal. Patients with night blindness can be helped with electronic optical devices, including monocular pocket scopes and binocular goggles that electronically amplify light.

Glaucoma

Glaucoma (glaw-kō′mă, *glaukos,* blue-green) is a condition involving excessive pressure buildup in the aqueous humor. Glaucoma results from an interference with normal reentry of aqueous humor into the blood or from an overproduction of aqueous humor. The increased pressure within the eye can close off the blood vessels entering the eye and may destroy the retina or optic nerve, resulting in blindness. Everyone older than 40 should be checked every 2–3 years for glaucoma; those older than 40 who have relatives with glaucoma should have an annual checkup. Glaucoma is usually treated with eye drops, which do not cure the problem but keep it from advancing. In some cases, laser or conventional surgery may be used.

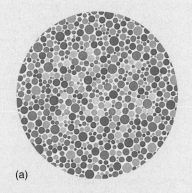 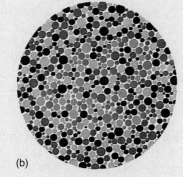

(a) (b)

Figure B   Color Blindness Charts

(*a*) A person with normal color vision can see the number 74, whereas a person with red-green color blindness sees the number 21. (*b*) A person with normal color vision can see the number 42. A person with red color blindness sees the number 2, and a person with green color blindness sees the number 4.

Reproduced from *Ishihara's Tests for Colour Blindness* published by Kanehara & Co., Ltd., Tokyo, Japan, but tests for color blindness cannot be conducted with this material. For accurate testing, the original plates should be used.

Cataract

A cataract (figure C*a*) is a clouding of the lens resulting from a buildup of proteins. The lens relies on the aqueous humor for its nutrition. Any loss of this nutrient source leads to degeneration of the lens and, ultimately, opacity of the lens (i.e., a cataract). A cataract may occur with advancing age, infection, or trauma.

A certain amount of lens clouding occurs in 65% of patients older than 50 and 95% of patients older than 65. The decision of whether to remove the cataract depends on the extent to which light passage is blocked. Over 400,000 cataracts are removed in the United States each year. Surgery to remove a cataract is actually the removal of the lens. The posterior portion of the lens capsule is left intact. Although the cornea can still accomplish light convergence, with the lens gone, the rays cannot be focused as well, and an artificial lens must be supplied to help accomplish focusing. In most cases, an artificial lens is implanted into the remaining portion of the lens capsule at the time that the natural lens is removed. The implanted lens helps restore normal vision, but glasses may be required for near vision.

Macular Degeneration

Macular degeneration (figure C*b*) is very common in older people. It does not cause total blindness but results in the loss of acute vision. This degeneration has a variety of causes, including hereditary disorders, infections, trauma, tumor, and most often poorly understood degeneration associated with aging. Because no satisfactory medical treatment has been developed, optical aids, such as magnifying glasses, are used to improve visual function.

Figure C Defects in Vision

Visual images as seen with various defects in vision. (*a*) Cataract. (*b*) Macular degeneration. (*c*) Retinitis pigmentosa. (*d*) Diabetic retinopathy.

Retinitis Pigmentosa

Retinitis pigmentosa (RP) is an inherited disorder characterized by the progressive degeneration of the retina, accompanied by the accumulation of pigment islands in the retina. Diagnosis of the disease can usually be made on the basis of family history and the presence of dark spots in the retina. RP results in the degeneration of rods and/or cones. The degeneration and death of photoreceptor cells result in progressive vision loss. In most forms of RP, called **rod-cone dystrophies,** the rods degenerate first. Progression of the disease includes night blindness and loss of peripheral vision as more rods are lost. Night blindness is often the first symptom of RP. Some patients retain a small amount of central vision throughout life (figure C*c*). Other forms of RP, called **cone-rod dystrophies,** first affect cones, resulting in loss of visual acuity and color vision.

Diabetes

Loss of visual function is one of the most common consequences of diabetes because a major complication of the disease is dysfunction of the peripheral circulation. Defective circulation to the eye may result in retinal degeneration or detachment. Diabetic retinopathy (figure C*d*) is one of the leading causes of blindness in the United States.

---

the temporal (lateral) retina pass through the optic chiasm and project to the brain on the same side of the body without crossing.

Beyond the optic chiasm, the axons form the **optic tracts** (see figure 13.16). Most of the optic tract axons terminate in the thalamus. Some axons do not terminate in the thalamus but separate from the optic tracts to terminate in the **superior colliculi,** the center for visual reflexes (see chapter 11). Neurons from the thalamus form the fibers of the **optic radiations,** which project to the **visual cortex** in the occipital lobe. Neurons of the visual cortex integrate the messages coming from the retina into a single message, translate that message into a mental image, and then transfer the image to other parts of the brain, where it is evaluated and either ignored or acted on.

The image seen by each eye is the **visual field** of that eye. The visual field of each eye can be divided into temporal (lateral) and nasal (medial) parts. The temporal part of a visual field projects onto the nasal retina, which projects to the visual cortex on the opposite side of the brain. The nasal part of a visual field projects onto the temporal retina, which projects to the same side of the brain. The nerve pathways are arranged in such a way that images entering the eye from the right part of each visual field (right temporal and left nasal) project to the left side of the brain. Conversely, the left part of each visual field (left temporal and right nasal) projects to the right side of the brain.

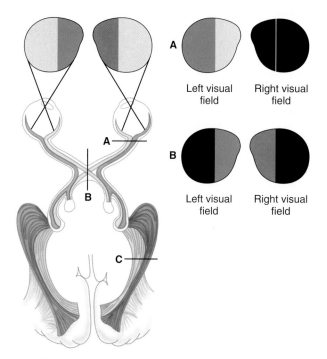

Figure 13.17 Lesions of the Visual Pathways

The lines at A, B, and C on the left indicate lesions in the visual pathways. The effects of lesions A and B on the visual fields are shown to the right. The *black areas* indicate what parts of the visual fields are defective.

PREDICT ⑦

While looking straight ahead, if a light flashes off to the right side, on which side of the brain is the visual cortex stimulated?

Figure 13.17 shows the visual pathways and potential sites of damage to them. If the right optic nerve is damaged, as in A, then vision is lost from the right visual field because action potentials from the eye cannot reach the brain. The loss of vision is indicated by the black color in the affected parts of the visual field.

The optic chiasm is just anterior to the pituitary gland. A pituitary tumor can expand and damage the axons where they cross in the optic chiasm, as in B in figure 13.17. The results are visual defects and loss of vision in the temporal halves of each visual field. This condition is called tunnel vision, and it can be an early sign of a pituitary tumor.

PREDICT ⑧

If the optic radiations are damaged at location C in figure 13.17, which parts of the visual fields would be affected?

Binocular vision is what is seen with two eyes at the same time. The visual field of one eye overlaps with the visual field of the other eye to produce binocular vision (figure 13.18). Where the visual fields do not overlap, there is **monocular vision,** that which is seen with one eye. Binocular vision is responsible for **depth perception,** the ability to distinguish between near and far objects and to judge their distance. When an object is seen with both eyes, the image of the object reaches the retina of one eye at a slightly different angle from that of the other. With experience, the brain can interpret these differences in angle so that distance can be judged quite accurately. ▼

42. Describe the neuronal pathway for vision.

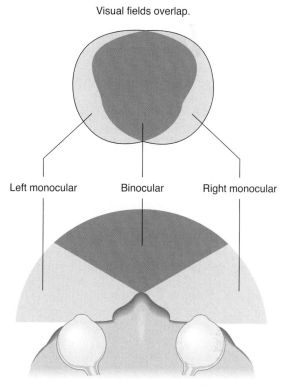

Figure 13.18 Binocular Vision

The left and right visual fields overlap to produce binocular vision. Monocular vision is where the visual fields do not overlap.

43. What is a visual field? How do the visual fields project to the brain?

44. Explain how binocular vision allows for depth perception.

Hearing and Balance

The organs of hearing and balance are divided into three parts: the external, middle, and inner ear (figure 13.19). The **external ear** is the part extending from the outside of the head to the **tympanic** (tim-pan′ik, drumlike) **membrane,** or **eardrum.** The **middle ear** is an air-filled chamber medial to the tympanic membrane. The **inner ear** is a set of fluid-filled chambers medial to the middle ear. The external and middle ears are involved in hearing only, whereas the inner ear functions in both hearing and balance. ▼

45. Name the three parts of the ear, and state their functions.

Auditory Structures and Their Functions

External Ear

The **auricle** (aw′ri-kl, ear) is the fleshy part of the external ear on the outside of the head. The auricle opens into the **external acoustic meatus** (mē-ā′tŭs, passage), a passageway that leads to the tympanic membrane. The auricle collects sound waves and directs them toward the external acoustic meatus, which transmits them to the tympanic membrane. The external acoustic meatus is lined with hairs and **ceruminous** (sĕ-roo′mi-nŭs, *cera,* wax) **glands,** which produce **cerumen** (sĕ-roo′men), a modified sebum

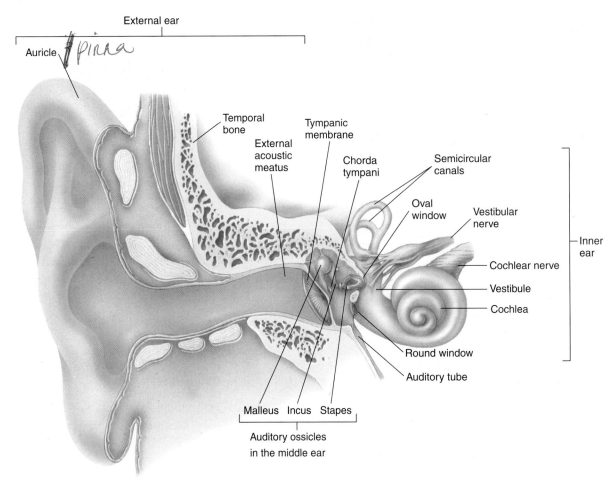

Figure 13.19   External, Middle, and Inner Ears

commonly called earwax. The hairs and cerumen help prevent foreign objects from reaching the delicate tympanic membrane.

The tympanic membrane is a thin membrane separating the external ear from the middle ear. It consists of a thin layer of connective tissue sandwiched between two epithelial layers. Sound waves reaching the tympanic membrane cause it to vibrate.

 Tympanic Membrane Rupture

Rupture of the tympanic membrane may result in hearing impairment. Foreign objects thrust into the ear, pressure, and infections of the middle ear can rupture the tympanic membrane. Sufficient differential pressure between the middle ear and the outside air can also rupture the tympanic membrane. This can occur in flyers, divers, or individuals who are hit on the side of the head by an open hand. ▼

46. From lateral to medial, name the parts of the external ear and describe their function.

Middle Ear

Medial to the tympanic membrane is the air-filled cavity of the middle ear (see figure 13.19). Two covered openings, the **oval window** and the **round window** on the medial side of the middle ear, connect the middle ear with the inner ear. The middle ear contains three **auditory ossicles** (os′i-klz, ear bones): the **malleus** (mal′ē-ŭs, hammer), **incus** (ing′kŭs, anvil), and **stapes** (stā′pēz, stirrup) (figure 13.20). These bones transmit vibrations from the tympanic membrane to the oval window. The malleus is attached to the medial surface of the tympanic membrane. The incus connects the malleus to the stapes. The base of the stapes is seated in the oval window and is surrounded by a flexible ligament.

Two small muscles in the middle ear help dampen vibrations of the auditory ossicles caused by loud noises. The **tensor tympani** (ten′sōr tim′păn-ē) muscle is attached to the malleus and is innervated by the trigeminal nerve (V). The **stapedius** (stā-pē′dē-ŭs) muscle is attached to the stapes and is innervated by the facial nerve (VII) (see figure 13.20).

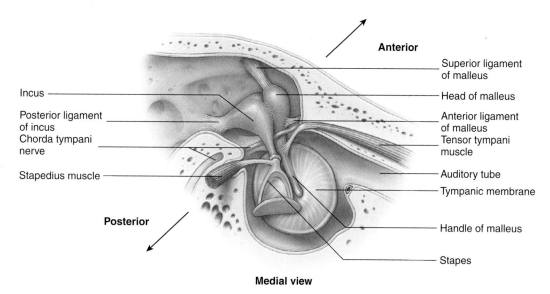

Medial view

Figure 13.20 Auditory Ossicles and Muscles of the Middle Ear

Medial view of the middle ear (as though viewed from the inner ear), showing the three auditory ossicles with their ligaments and the two muscles of the middle ear: the tensor tympani and the stapedius.

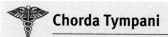

Chorda Tympani

The **chorda tympani** is a branch of the facial nerve carrying taste impulses from the anterior two-thirds of the tongue. It crosses through the middle ear between the malleus and incus (see figure 13.20). The chorda tympani has nothing to do with hearing but is just passing through. This nerve can be damaged, however, during ear surgery or by a middle ear infection, resulting in loss of taste sensation from the anterior two-thirds of the tongue on the side innervated by that nerve.

Two openings provide air passages from the middle ear. One passage opens into the **mastoid air cells** in the mastoid process of the temporal bone. The other passageway is the **auditory tube,** also called the **pharyngotympanic** (fă-ring′gō-tim-pan′ik) **tube** or the **eustachian** (ū-stā′shŭn) **tube.** The auditory tube opens into the pharynx and enables air pressure to be equalized between the outside air and the middle ear cavity.

When a person changes altitude, air pressure outside the tympanic membrane changes relative to the air pressure in the middle ear. With an increase in altitude, the pressure outside the tympanic membrane becomes less than the air pressure inside the middle ear and the tympanic membrane is pushed outward. With a decrease in altitude, the air pressure outside the ear becomes greater than in the middle ear and the tympanic membrane is pushed inward. Distortion of the tympanic membrane can make sounds seem muffled and stimulate pain. These symptoms can be relieved by opening the auditory tube to allow air to pass through the auditory tube to equalize air pressure. Swallowing, yawning, chewing, and holding the nose and mouth shut while gently trying to force air out of the lungs are methods used to open the auditory tube. ▼

47. Name the two membrane-covered openings that connect the middle ear to the inner ear.

48. Starting with the tympanic membrane, name in order the auditory ossicles connecting to the oval window.
49. What is the function of the muscles connected to the auditory ossicles?
50. What is the function of the auditory tube?

Inner Ear

The inner ear contains the sensory organs for hearing and balance. It consists of interconnecting, fluid-filled tunnels and chambers within the temporal bone called the **bony labyrinth** (lab′i-rinth, maze). The bony labyrinth is lined with **endosteum.** When the inner ear is drawn, it is the endosteum that is depicted (figure 13.21*a*). The bony labyrinth is divided into three regions: cochlea, vestibule, and semicircular canals. The **vestibule** (ves′ti-bool) and **semicircular canals** are involved primarily in balance, and the **cochlea** (kok′lē-ă, snail shell) is involved in hearing.

Inside the bony labyrinth is a similarly shaped but smaller set of membranous tunnels and chambers called the **membranous labyrinth** (figure 13.21*b*). The membranous labyrinth is filled with a clear fluid called **endolymph** (en′dō-limf, *lymph,* clear spring water), and the space between the membranous and bony labyrinths is filled with a fluid called **perilymph** (per′i-limf) (figure 13.21*c*). Perilymph is very similar to cerebrospinal fluid, but endolymph has a high concentration of potassium and a low concentration of sodium, which is opposite from perilymph and cerebrospinal fluid. The membranous labyrinth of the cochlea separates the bony labyrinth into two parts (figure 13.21*d*).

The cochlea is shaped like a snail shell—that is, a coiled tube. The **base** of the cochlea connects to the vestibule and the **apex** of the cochlea is the end of the coiled tube (figure 13.22*a*). The bony core of the cochlea, around which the tube coils, is shaped like a screw with threads called the **spiral lamina** (figure 13.22*b*). A Y-shaped, membranous complex divides the cochlea into three portions. The base of the Y is the spiral lamina. One branch of the Y is the **vestibular** (ves-tib′ū-lăr, entrance hall) **membrane,** and

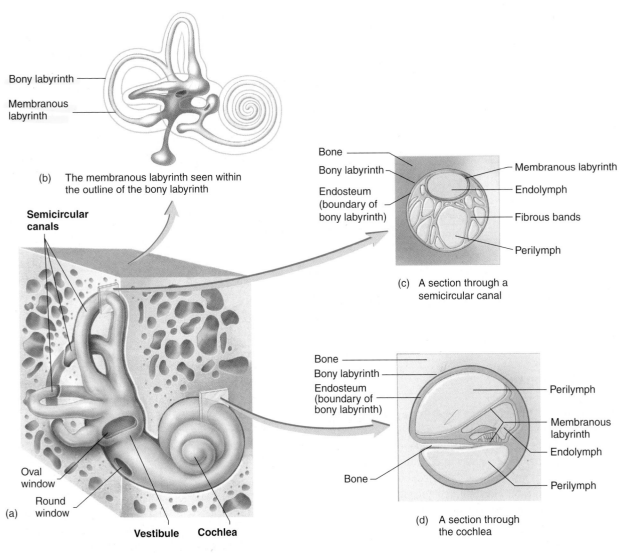

(b)   The membranous labyrinth seen within the outline of the bony labyrinth

Semicircular canals

Oval window

Round window

(a)

Vestibule   Cochlea

Bone
Bony labyrinth
Endosteum (boundary of bony labyrinth)

Membranous labyrinth
Endolymph
Fibrous bands
Perilymph

(c)   A section through a semicircular canal

Bone
Bony labyrinth
Endosteum (boundary of bony labyrinth)

Perilymph
Membranous labyrinth
Endolymph

Bone

Perilymph

(d)   A section through the cochlea

Figure 13.21   Inner Ear: Bony and Membranous Labyrinths

(*a*) The bony labyrinth in the temporal bone of the skull. (*b*) The membranous labyrinth within the bony labyrinth. (*c*) Cross section through a semicircular canal to show the relationship between the bony and membranous labyrinths. (*d*) Cross section through the cochlea to show the relationship between the bony and membranous labyrinths.

the other branch is the **basilar** (bas′i-lăr, base) **membrane.** The space between these membranes is called the **cochlear duct.** This complex is the membranous labyrinth, and it is filled with endolymph. If the Y is viewed lying on its right side, as in figure 13.22*b*, the space above the Y is called the **scala vestibuli** (skā′lă ves-tib′ū-lī, *scala,* stairway), and the space below the Y is called the **scala tympani** (tim-pa′nē). These two spaces are filled with perilymph.

The oval window connects the middle ear with the vestibule of the inner ear, which in turn connects to the scala vestibuli. The scala vestibuli extends the length of the cochlea and connects to the scala tympani at the apex of the cochlea. The opening connecting the two chambers is called the **helicotrema** (hel′i-kō-trē′mă, a hole at the end of a helix or spiral) (see figure 13.22*a*). The scala tympani extends from the apex of the cochlea to the round window, which connects to the middle ear. The round window is closed with a membrane.

The cells inside the cochlear duct are highly modified to form a structure called the **spiral organ,** or **organ of Corti,** which rests

on the basilar membrane (see figure 13.22*b* and *c*). The spiral organ contains supporting epithelial cells and specialized sensory cells called **hair cells,** which have hairlike projections at their apical ends (figure 13.22*d*). The projections are very long microvilli called **stereocilia.** Hair cells have no axons, but the basilar regions of each hair cell are covered by synaptic terminals of sensory neurons. Stimulation of these neurons by the hair cells results in the production of action potentials, which are transmitted to the brain.

The hair cells are arranged in four long rows extending the length of the cochlear duct. Each row contains 3500–4000 hair cells. The inner row consists of hair cells, called **inner hair cells,** which are the hair cells primarily responsible for hearing. The outer three rows contain **outer hair cells.** The tips of the longest stereocilia of the outer hair cells are embedded within an acellular gelatinous shelf called the **tectorial** (tek-tōr′ē-ăl) **membrane,** which is attached to the spiral lamina (see figure 13.22*d*). The outer hair cells are involved in regulating the tension of the basilar membrane and are separated from the inner hair cells by a gap in the basilar membrane.

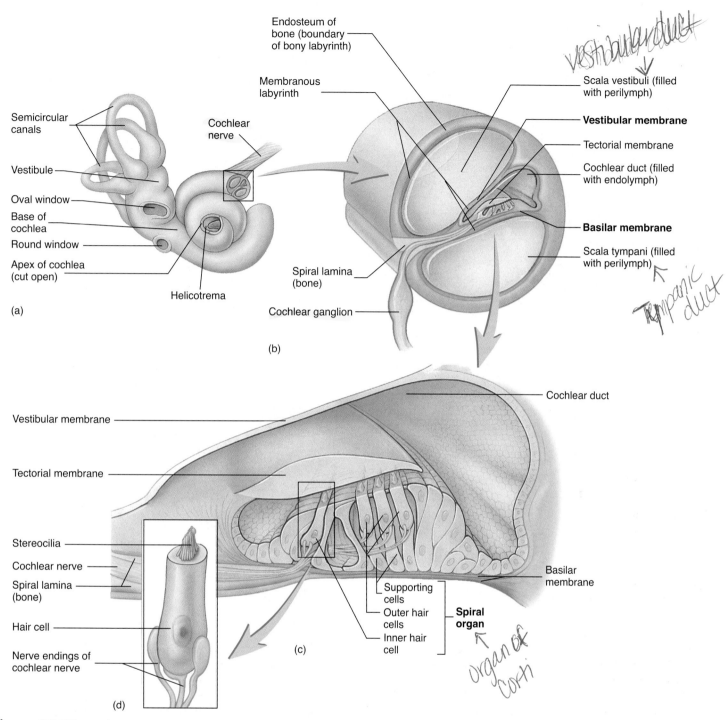

Figure 13.22 Structure of the Cochlea

(a) *The inner ear.* The outer surface (*gray*) is the endosteum lining the inner surface of the bony labyrinth. (b) *A cross section of the cochlea.* The outer layer is the endosteum lining the inner surface of the bony labyrinth. The membranous labyrinth is very small in the cochlea and consists of the vestibular and basilar membranes. The space between the membranous and bony labyrinths consists of two parallel tunnels: the scala vestibuli and scala tympani. (c) An enlarged section of the cochlear duct showing the spiral organ. (d) A greatly enlarged individual sensory hair cell.

The stereocilia of one inner hair cell form a conical group called a **hair bundle** (figure 13.23a). The length of each stereocilium within a hair bundle increases gradually from one side of the hair cell to the other. The stereocilia of an outer hair cell are arranged in a curved line (figure 13.23b).

A **tip link** connects the tip of each stereocilium in a hair bundle to the side of the next longer stereocilium (figure 13.24).

Each tip link is a **gating spring,** a pair of microtubule strands that attaches to the gate of a gated K^+ channel. The gated K^+ channels of hair cells open mechanically. As the stereocilia bend, the gating spring pulls the K^+ gate open. The response time for such a mechanism is very brief. The stereocilia are surrounded by perilymph, which has a high concentration of K^+. When gated K^+ channels open, positively charged K^+ move into the stereocilia,

(a)

(b)

Figure 13.23   Scanning Electron Micrograph of Cochlear Hair Cell Stereocilia

(*a*) The hair bundle of one inner hair cell. (*b*) The stereocilia of three outer hair cells.

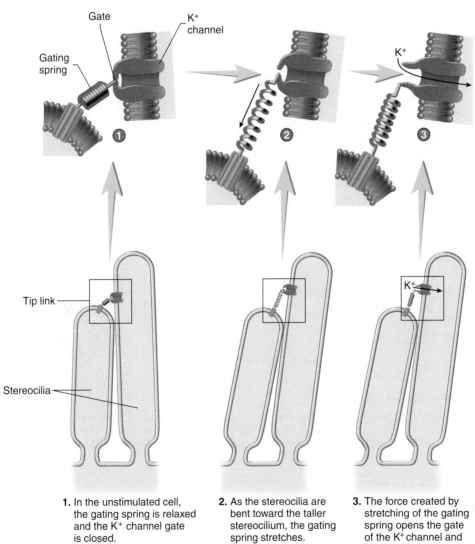

1. In the unstimulated cell, the gating spring is relaxed and the K$^+$ channel gate is closed.

2. As the stereocilia are bent toward the taller stereocilium, the gating spring stretches.

3. The force created by stretching of the gating spring opens the gate of the K$^+$ channel and K$^+$ enter the cell.

Process Figure 13.24   Action of the Gating Spring to Open a K$^+$ Channel When Two Stereocilia Bend

Diagram of two stereocilia connected by one tip link.

resulting in depolarization of the hair cells. (This is a rare instance in which an increase in K⁺ permeability of the plasma membrane of a cell results in depolarization.) The hair cells release the neurotransmitter glutamate, which results in the production of action potentials in the sensory neurons innervating the hair cells. ▼

51. *Describe the bony and membranous labyrinths. Name the three regions of the bony labyrinth.*
52. *Explain how the cochlea is divided into three compartments. Name the compartments and the membranes separating them. What fluid is found in each compartment?*
53. *Describe the structure of the spiral organ. Distinguish between inner and outer hair cells and their functions.*
54. *Explain how the bending of stereocilia results in the production of action potentials.*

Auditory Function

Properties of Sound

Vibration of matter, such as air, water, or a solid material, creates sound. No sound occurs in a vacuum. For example, when a tuning fork vibrates, it causes the surrounding air to vibrate. The air vibrations consist of bands of compressed air followed by bands of less compressed air (figure 13.25a). When these pressure changes are graphed through time, they have a wave form; hence, they are called sound waves. These sound waves are propagated through the air, somewhat as ripples are propagated over the surface of water. The **volume,** or loudness, of sound is a function of wave amplitude, or height, measured in decibels (figure 13.25b). The greater the amplitude, the louder the sound. The **pitch** of sound is a function of the wave frequency (figure 13.25c). It is measured in hertz (Hz), which is the number of waves or cycles per second. The higher the frequency, the higher the pitch. The normal range of human hearing is 20–20,000 Hz and 0 or more decibels (db). Sounds louder than 125 db are painful to the ear.

Human Speech and Hearing Impairment

The range of normal human speech is 250–8000 Hz. This is the range that is tested for the possibility of hearing impairment because it is the most important for communication.

Timbre (tim′br, tam′br) is the resonance quality or overtones of a sound. A smooth sigmoid curve is the image of a "pure" sound wave, but such a wave almost never exists in nature. The sounds made by musical instruments and the human voice are not

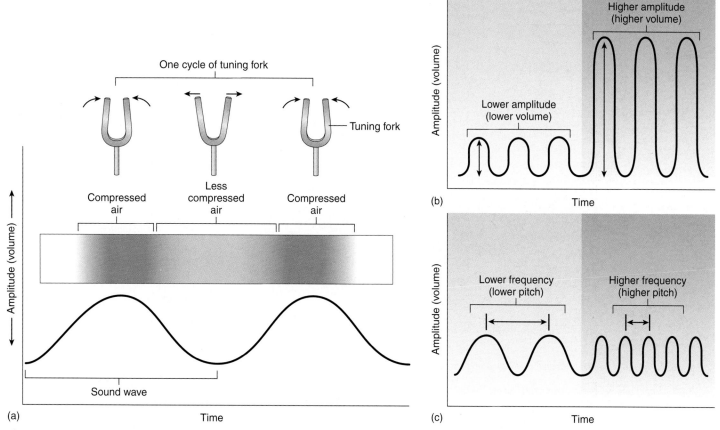

Figure 13.25   Sound Waves

(a) Vibration of objects, such as a tuning fork or vocal cords, produces sound. The movement of an object "into" the air compresses it (*blue bars*), whereas the movement of the object "away" from the air expands it, making it less compressed (*green bar*). The sigmoid waves correspond to the regions of more compressed air (peaks) and less compressed air (troughs). One cycle is the distance between peaks. (b) Low- and high-volume sound waves. Compare the relative lengths of the *arrows* indicating the wave height (amplitude). (c) Lower- and higher-pitch sound. Compare the relative number of peaks (frequency) within a given time interval (*between arrows*).

smooth sigmoid curves but, rather, are rough, jagged curves formed by numerous, superimposed curves of various amplitudes and frequencies. The roughness of the curve accounts for the timbre. Timbre allows one to distinguish between, for example, an oboe and a French horn playing a note at the same pitch and volume.

P R E D I C T   9
One dictionary definition of sound is "that which is heard." If a tree falls in the forest, does it make a sound?

External Ear

The auricle collects sound waves and directs them into the external acoustic meatus. Sound waves travel relatively slowly in air, 332 m/s, and a small time interval may elapse between the time a sound wave reaches one ear and the time it reaches the other. The brain can interpret this interval to determine the direction from which a sound is coming. Sound waves are conducted through the external acoustic meatus and strike the tympanic membrane, causing it to vibrate (figure 13.26).

Middle Ear

Vibration of the tympanic membrane causes vibration of the three auditory ossicles of the middle ear, and by this mechanical linkage vibration is transferred to the oval window (see figure 13.26). The oval window is approximately 20 times smaller than the tympanic membrane. The mechanical force of vibration is amplified about 20-fold as it passes from the tympanic membrane through the auditory ossicles to the oval window because of this size difference. This amplification is necessary because more force is required to cause vibration in a liquid, such as the perilymph of the inner ear, than is required in air.

The tensor tympani and stapedius muscles are attached to auditory ossicles (see figure 13.20). Excessively loud sounds cause these muscles to reflexively contract and dampen the movement of the auditory ossicles. This **sound attenuation reflex** protects the delicate inner ear structures from damage by loud noises. The sound attenuation reflex responds most effectively to low-frequency sounds and can reduce by a factor of 100 the energy reaching the oval window. The reflex is too slow to prevent damage from a sudden noise, such as a gunshot, and it cannot function effectively for longer than about 10 minutes, in response to prolonged noise.

Inner Ear

As the stapes vibrates, it produces sound waves in the perilymph of the scala vestibuli (see figure 13.26). Vibrations of the perilymph are transmitted through the vestibular membrane and cause simultaneous vibrations of the endolymph. The mechanical effect is

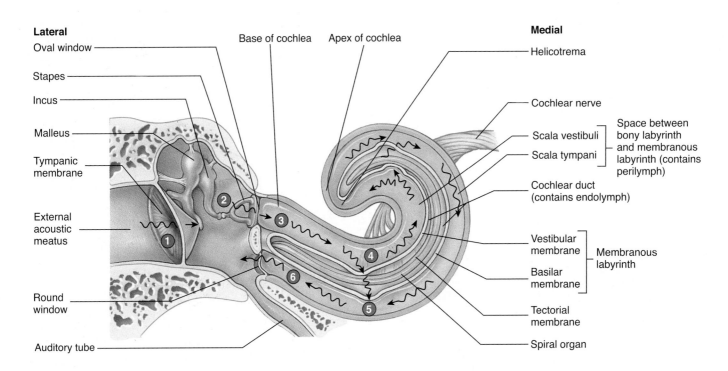

1. Sound waves strike the tympanic membrane and cause it to vibrate.

2. Vibration of the tympanic membrane causes the three bones of the middle ear—the malleus, incus, and stapes—to vibrate.

3. Vibration of the stapes in the oval window causes the perilymph in the scala vestibuli to vibrate.

4. Vibration of the perilymph causes the vestibular membrane to vibrate, which causes vibrations in the endolymph.

5. Vibration of the endolymph causes displacement of the basilar membrane. Movement of the basilar membrane is detected in the hair cells of the spiral organ, which are attached to the basilar membrane. Vibrations of the basilar membrane are transferred to the perilymph of the scala tympani.

6. Vibrations in the perilymph of the scala tympani are transferred to the round window, where they are dampened.

Process Figure 13.26   Effect of Sound Waves on Cochlear Structures

The cochlea is drawn uncoiled for clarity.

as though the perilymph and endolymph were a single fluid because the vestibular membrane is very thin. Vibration of the endolymph causes distortion of the basilar membrane, which is most important to hearing. As the basilar membrane moves, the hair cells resting on it move relative to the tectorial membrane, which remains stationary. The inner hair cell stereocilia are bent as they move against the tectorial membrane, resulting in the release of neurotransmitter and the production of action potentials in sensory neurons.

The basilar membrane is not uniform throughout its length (figure 13.27). The membrane is narrower and denser near the oval window and wider and less dense near the apex of the cochlea. The

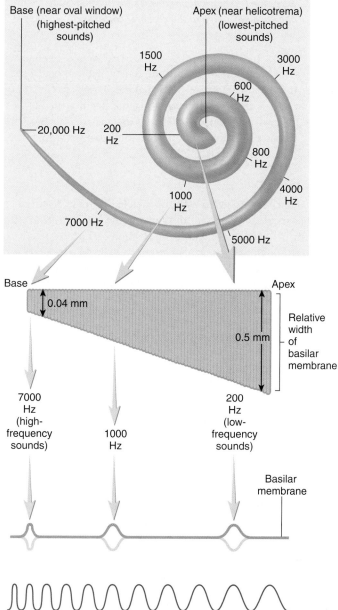

Figure 13.27 Effect of Sound Waves on Points Along the Basilar Membrane

The basilar membrane is thinnest at the base of the cochlea and thickest at the apex. The frequency of sound (Hz) affects which part of the basilar membrane vibrates. Higher-pitched sounds cause the basilar membrane near the base of the cochlea to vibrate, whereas lower-pitched sounds cause the basilar membrane to vibrate near the apex.

various regions of the membrane can be compared to the strings in a piano, which has strings of varying length and thickness. As a result of this organization, sounds with higher pitches cause the basilar membrane nearer the oval window to distort maximally, whereas sounds with lower pitches cause the basilar membrane nearer the apex of the cochlea to distort maximally. Depending on which hair cells are stimulated along the length of the basilar membrane, the brain interprets the pitch and timbre of sounds.

Sound volume, or loudness, is a function of sound wave amplitude. As the volume of sound increases, the vibration of the basilar membrane increases, the stimulation of hair cells increases, and the production of action potentials increases. The brain interprets the higher frequency of action potentials as a louder sound.

The outer hair cells are involved in regulating the tension of the basilar membrane, thereby increasing the sensitivity of the inner ear to sounds. Stimulation of the inner hair cells by the nervous system stimulates the contraction of actin filaments within the hair cells, causing them to shorten. This adjustment in the height of the outer hair cells, attached to both the basilar membrane and the tectorial membrane, fine tunes the tension of the basilar membrane and the distance between the basilar membrane and tectorial membrane.

Sound waves in the perilymph of the scala vestibuli are also transmitted the length of the scala vestibuli and through the helicotrema into the perilymph of the scala tympani (see figure 13.26). This transmission of sound waves is probably of little consequence because the helicotrema is very small. Vibration of the basilar membrane produces most of the sound waves in the perilymph of the scala tympani. Sound waves in the scala tympani perilymph cause vibration of the membrane of the round window. Vibration of the round window membrane is important to hearing because it acts as a mechanical release for sounds waves within the scala tympani. If this window were solid, it would reflect the sound waves, which would interfere with and dampen later sound waves. The round window also allows the relief of pressure in the perilymph because fluid is not compressible, thereby preventing compression damage to the spiral organ.

 Loud Noises and Hearing Loss

Prolonged or frequent exposure to excessively loud noises can cause degeneration of the spiral organ at the base of the cochlea, resulting in high-frequency deafness. The actual amount of damage can vary greatly from person to person. High-frequency loss can cause a person to miss hearing consonants in a noisy setting. Loud music, amplified to 120 db, can impair hearing. The defects may not be detectable on routine diagnosis, but they include decreased sensitivity to sound in specific narrow frequency ranges and a decreased ability to discriminate between two pitches. Loud music, however, is not as harmful as is the sound of a nearby gunshot, which is a sudden sound occurring at 140 db. The sound is too sudden for the attenuation reflex to protect the inner ear structures, and the intensity is great enough to cause auditory damage. In fact, gunshot noise is the most common recreational cause of serious hearing loss. ▼

55. *Starting with the auricle, trace a sound wave into the inner ear to the point at which action potentials are generated in sensory neurons.*

56. *What is the sound attenuation reflex?*

57. *How are we able to distinguish between sounds of different pitch and volume?*

58. *What is the function of the round window?*

P R E D I C T   10

Explain why it is much easier to perceive subtle musical tones when music is played somewhat softly, as opposed to very loudly.

Neuronal Pathways for Hearing

The axons of the sensory neurons supplying hair cells form the **cochlear nerve** (see figure 13.22d). These sensory neurons are bipolar neurons, and their cell bodies are in the **cochlear,** or **spiral, ganglion,** located in the bony core of the cochlea. The cochlear nerve joins the vestibular nerve to become the **vestibulocochlear nerve (VIII),** which traverses the internal acoustic meatus and enters the cranial cavity. The special senses of hearing and balance are both transmitted by the vestibulocochlear (VIII) nerve. The term *vestibular* refers to the vestibule of the inner ear, which is involved in balance. The term *cochlear* refers to the cochlea of the inner ear, which is involved in hearing. The vestibulocochlear nerve functions as two separate nerves carrying information from two separate but closely related structures.

The auditory pathways within the CNS are very complex, with both crossed and uncrossed tracts. Unilateral CNS damage therefore usually has little effect on hearing. The neurons from the cochlear ganglion synapse with CNS neurons in the **cochlear nucleus** in the medulla oblongata. These neurons in turn either synapse in or pass through the **superior olivary nucleus** in the medulla oblongata. Neurons terminating in the superior olivary

nucleus may synapse with efferent neurons returning to the cochlea to modulate pitch perception. Nerve fibers from the superior olivary nucleus also project to the trigeminal (V) nucleus, which controls the tensor tympani, and the facial (VII) nucleus, which controls the stapedius muscle. This is part of the sound attenuation reflex pathway. Ascending neurons from the superior olivary nucleus synapse in the **inferior colliculi,** and neurons from there project to the **thalamus,** where they synapse with neurons that project to the cortex. These neurons terminate in the **auditory cortex** (see chapter 12). Neurons from the inferior colliculi also project to the **superior colliculi,** where reflexes that turn the head and eyes in response to loud sounds are initiated.

59. *Describe the neuronal pathways for hearing from the cochlear nerve to the cerebral cortex.*

Balance

The sense of balance, or equilibrium, has two components: static balance and dynamic balance. **Static balance** is associated with the vestibule and is involved in evaluating the position of the head relative to gravity. **Dynamic balance** is associated with the semicircular canals and is involved in evaluating changes in the direction and rate of head movements.

Static Balance

Static balance is associated with the **utricle** (oo′tri-kl, leather bag) and the **saccule** (sak′ūl, sac) of the vestibule (figure 13.28a). It is primarily involved in evaluating the position of the head relative to gravity, although the system also responds to linear acceleration or deceleration, such as when a person is in a car that is increasing or decreasing speed.

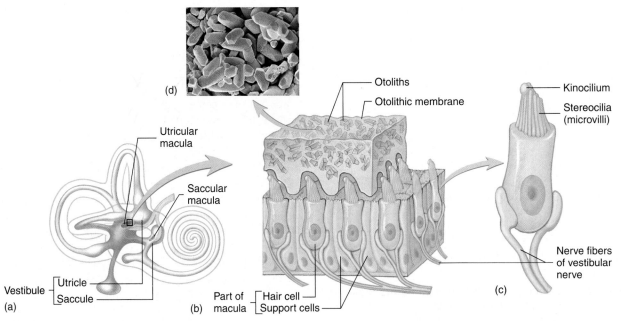

Figure 13.28   Structure of the Macula

(*a*) Vestibule showing the location of the utricular and saccular maculae. (*b*) Enlargement of the utricular macula, showing hair cells and otoliths in the macula. (*c*) An enlarged hair cell, showing the kinocilium and stereocilia. (*d*) Colorized scanning electron micrograph of otoliths.

Most of the utricular and saccular walls consist of simple cuboidal epithelium. The utricle and saccule, however, each contain a specialized patch of epithelium about 2–3 mm in diameter called the **macula** (mak′ū-lă) (figure 13.28*b*). The macula of the utricle is oriented parallel to the base of the skull, and the macula of the saccule is perpendicular to the base of the skull.

The maculae resemble the spiral organ and consist of hair cells, sensory neurons, and supporting cells. The "hairs" of the hair cells consist of numerous microvilli, called **stereocilia,** and one cilium, called a **kinocilium** (kī-nō-sil′ē-ŭm) (figure 13.28*c*). The stereocilia and kinocilium are embedded in the **otolithic** (ō′tō-lith-ik, ear stones) **membrane,** which is a gelatinous mass weighted with crystals of calcium carbonate and protein called **otoliths** (ō′tō-liths) (see figure 13.28*b* and *d*). The otolithic membrane moves in response to gravity, bending the hair cells and initiating action potentials in their associated sensory neurons. The stereocilia function much as do the stereocilia of cochlear hair cells, with tip links connected to gated K^+ channels. Deflection

of the hairs toward the kinocilium results in depolarization of the hair cell.

The hair cells are constantly being stimulated at a low level by the presence of the otolith-weighted covering of the macula. When the otolithic membrane moves in response to gravity, the pattern and intensity of hair cell stimulation change (figure 13.29). The change in action potentials produced is translated by the brain into specific information about head position or linear acceleration/deceleration. Much of this information is not perceived consciously but is dealt with subconsciously. The body responds by making subtle tone adjustments in the muscles of the back and neck, which are intended to restore the head to its proper neutral, balanced position.

Dynamic Balance

Dynamic balance is associated with the semicircular canals and is involved in evaluating movements of the head. There are three **semicircular canals** placed at nearly right angles to one another

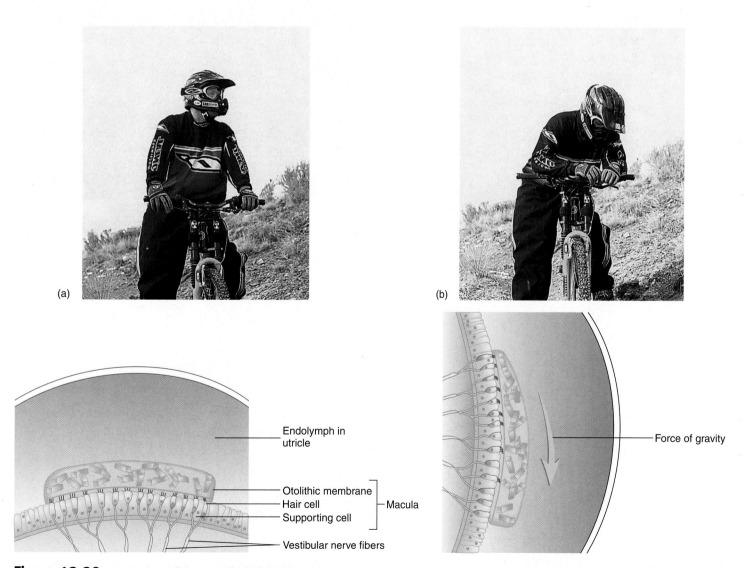

Figure 13.29   **Function of the Vestibule in Maintaining Balance**

(*a*) In an upright position, the otolithic membrane does not move. (*b*) As the position of the head changes, such as when a person bends over, gravity causes the otolithic membrane to move.

(figure 13.30a). One semicircular canal lies nearly in the transverse plane, one in the frontal plane, and one in the sagittal plane. The arrangement of the semicircular canals enables a person to detect movement in all directions.

The base of each semicircular canal is expanded into an **ampulla** (see figure 13.30a). Within each ampulla, the epithelium is specialized to form a **crista ampullaris** (kris′tă am-pū-lar′ŭs) (figure 13.30b). This specialized sensory epithelium is structurally and functionally very similar to the sensory epithelium of the maculae. Each crista consists of a ridge or crest of epithelium with a curved, gelatinous mass, the **cupula** (koo′poo-lă), suspended over the crest. The hairlike processes of the

crista hair cells, which are stereocilia similar to those in the maculae and cochlear hair cells, are embedded in the cupula (figure 13.30c). The cupula contains no otoliths and therefore does not respond to gravitational pull. Instead, the cupula is a float that is displaced by endolymph movements within the semicircular canals.

As the head begins to move, the endolymph does not move at the same rate as the semicircular canals, which are part of the temporal bone (figure 13.31). This difference causes displacement of the cupula in a direction opposite to that of the movement of the head. To appreciate this effect, imagine holding a feather (the cupula) in your hand (the crista ampullaris).

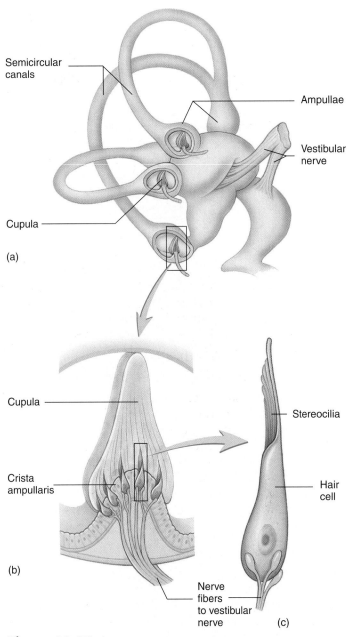

(a)

(b)

(c)

Cupula — Stereocilia — Hair cell — Crista ampullaris — Nerve fibers to vestibular nerve — Semicircular canals — Ampullae — Vestibular nerve

Figure 13.30   Semicircular Canals

(a) Semicircular canals showing the location of the crista ampullaris in the ampullae of the semicircular canals. (b) Enlargement of the crista ampullaris, showing the cupula and hair cells. (c) Enlargement of a hair cell.

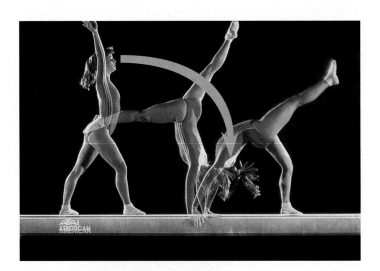

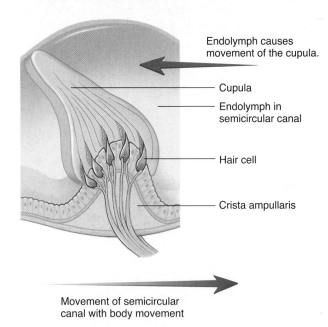

Endolymph causes movement of the cupula.

Cupula — Endolymph in semicircular canal — Hair cell — Crista ampullaris

Movement of semicircular canal with body movement

Figure 13.31   Function of the Semicircular Canals

As a person begins to move, the semicircular canals (crista ampullaris) begin to move with the body (*blue arrow*), but the endolymph tends to remain stationary relative to the movement. The cupula is displaced by the endolymph in a direction opposite to the direction of movement (*red arrow*).

If you rapidly move your hand, the feather bends over in the direction opposite the movement as it drags through the air (endolymph). Bending of the stereocilia results in the stimulation of sensory neurons. The brain interprets the direction of head movement based on which hair cells in which crista ampullaris are stimulated.

The semicircular canals detect changes in the rate of movement rather than movement alone because displacement of the cupula is most intense when the rate of head movement changes rapidly. As with the static balance, the information the brain obtains regarding dynamic balance is largely subconscious.

If a person continually spins in one direction, the endolymph of the semicircular canals begins to move and catches up with the cupula, and stimulation is stopped. If spinning suddenly stops, the endolymph continues to move because of its momentum, causing displacement of the cupula in the same direction as the head had been moving. The brain interprets this movement of the cupula to mean the head is moving in the opposite direction of the spin, even though the head is no longer moving. This is why a person has a feeling of moving even after he or she has stopped spinning.

Motion Sickness and Space Sickness

Motion sickness consists of nausea, weakness, vomiting, and incapacitation brought on by movement. It is caused by stimulation of the semicircular canals during motion, such as in a boat, an automobile, or an airplane; on a swing; or on an amusement park ride. The brain compares sensory input from the semicircular canals, eyes, and proprioceptors in the lower limbs. Perceived differences in the input may result in motion sickness.

Space sickness is a balance disorder occurring in zero gravity and resulting from unfamiliar sensory input to the brain. The brain must adjust to these unusual signals, or severe symptoms, such as headaches and dizziness, may result. Space sickness is unlike motion sickness in that motion sickness results from an excessive stimulation of the brain, whereas space sickness results from too little stimulation as a result of weightlessness. ▼

60. *Define static and dynamic balance. Which parts of the inner ear are responsible for them?*
61. *Describe the structure of maculae in the saccule and utricle. Explain how movement of the otolithic membrane is translated into perception of head position.*
62. *What is the function of the semicircular canals? Describe the crista ampullaris and its mode of operation.*

Neuronal Pathways for Balance

The axons of the sensory neurons supplying hair cells of the maculae and crista ampullaris form the **vestibular nerve** (see figure 13.28c and 13.30c). These sensory neurons are bipolar neurons, and their cell bodies are in the **vestibular ganglion,** a swelling of the vestibulocochlear nerve located in the internal acoustic meatus. The sensory neurons terminate in the **vestibular nucleus**

within the medulla oblongata. Axons run from this nucleus to numerous areas of the CNS, such as the spinal cord, the cerebellum, the cerebral cortex, and the nuclei controlling the extrinsic eye muscles.

Balance is a complex process not simply confined to one type of input. In addition to vestibular sensory input, the vestibular nucleus receives input from proprioceptive neurons throughout the body, as well as from the visual system. People are asked to close their eyes while balance is evaluated in a sobriety test because alcohol affects the proprioceptive and vestibular components of balance (cerebellar function) to a greater extent than it does the visual portion.

Reflex pathways exist between the dynamic balance part of the vestibular system and the nuclei controlling the extrinsic eye muscles (oculomotor, trochlear, and abducent). A reflex pathway allows the maintenance of visual fixation on an object while the head is in motion. This function can be demonstrated by spinning a person around about 10 times in 20 seconds, stopping him or her, and observing eye movements. A slight oscillatory movement of the eyes, called **nystagmus** (nis-tag′mŭs), occurs. The eyes track in the direction of motion and return with a rapid recovery movement before repeating the tracking motion. ▼

63. Describe the neuronal pathways for balance.

Effects of Aging on the Special Senses

Elderly people experience only a slight loss in the ability to detect odors. However, the ability to identify specific odors correctly is decreased, especially in men over age 70.

In general, the sense of taste decreases as people age. The number of sensory receptors decreases and the brain's ability to interpret taste sensations declines.

The lenses of the eyes lose flexibility as a person ages because the connective tissue of the lenses becomes more rigid. Consequently, there is first a reduction in and then an eventual loss of the lenses' ability to change shape. This condition, called presbyopia, is the most common age-related change in the eyes.

The most common visual problem in older people requiring medical treatment, such as surgery, is the development of cataracts. Macular degeneration is the second most common defect, glaucoma is third, and diabetic retinopathy is fourth.

The number of cones decreases, especially in the fovea centralis. These changes cause a gradual decline in visual acuity and color perception.

As people age, the number of hair cells in the cochlea decreases, leading to age-related hearing loss, called **presbyacusis** (prez′bē-ă-koo′sis). This decline does not occur equally in both ears, however. As a result, because direction is determined by comparing sounds coming into each ear, elderly people may experience a decreased ability to localize the origin of certain sounds. In some people, this leads to a general sense of disorientation. In addition, CNS defects in the auditory pathways can result in difficulty understanding sounds with echoes or background noise. Such a

Hearing Impairment

The term *hearing-impaired* refers to any type or degree of hearing loss; the hearing loss can be conductive, sensorineural, or a combination of both. **Conduction deafness** involves a mechanical deficiency in the transmission of sound waves from the external ear to the spiral organ. The spiral organ and neuronal pathways for hearing function normally. Conductive hearing loss often can be treated—for example, by removing earwax blocking the external acoustic meatus or by replacing or repairing the auditory ossicles. If the degree of conductive hearing loss does not justify surgical treatment, or if treatment does not resolve the hearing loss, a hearing aid may be beneficial because the amplified (louder) sound waves it produces are transmitted through the conductive blockage and may provide normal stimulation to the spiral organ.

Sensorineural hearing loss involves the spiral organ or neuronal pathways. Sound waves are transmitted normally to the spiral organ, but the nervous system's ability to respond to the sound waves is impaired. Hearing aids are commonly used by people with sensorineural hearing loss because the amplified sound waves they produce result in greater than normal stimulation of the spiral organ, helping overcome the perception of reduced sound volume. Sound clarity also improves with sound amplification but may never be perceived as normal.

The term *deaf* refers to sensorineural hearing loss so profound that the sense of hearing is nonfunctional, with or without amplification, for ordinary purposes of life. Stimulation of the spiral organ or hearing nerve pathways can help deaf people hear. One approach involves the direct stimulation of the cochlear nerve by action potentials. The mechanism consists of a microphone for picking up sound waves; a microelectronic processor for converting the sound into electrical signals; a transmission system for relaying the signals to the inner ear; and a long, slender electrode that is threaded into the cochlea. This electrode delivers electrical signals directly to the cochlear nerve.

Otosclerosis

Otosclerosis (ō′tō-sklē-rō′sis) is an ear disorder in which spongy bone grows over the oval window and immobilizes the stapes, leading to progressive loss of hearing. This disorder can be surgically corrected by breaking away the bony growth and the immobilized stapes. During surgery, the stapes is replaced by a small rod connected by a fat pad or a synthetic membrane to the oval window at one end and to the incus at the other end.

Tinnitus

Tinnitus (ti-ni′tŭs) consists of phantom noises, such as ringing, clicking, whistling, buzzing, or booming, in the ears. These noises may occur as a result of disorders in the middle or inner ear or along the central neuronal pathways. Tinnitus is a common problem, affecting 17% of the world's population. It is treated primarily by training people to ignore the sounds.

Meniere Disease

Meniere disease is the most common disease involving dizziness from the inner ear. Its cause is unknown but it appears to involve a fluid abnormality in one (usually) or both ears. Symptoms include vertigo, hearing loss, tinnitus, and a feeling of "fullness" in the affected ear. Treatment includes a low-salt diet and diuretics (water pills). Symptoms may also be treated with medications for motion sickness.

---

deficit makes it difficult for elderly people to understand rapid or broken speech.

With age, the number of hair cells in the saccule, utricle, and ampullae decreases. The number of otoliths also declines. As a result, elderly people experience a decreased sensitivity to gravity, acceleration, and rotation. Because of these decreases, elderly people experience dizziness (instability) and vertigo (a feeling of spinning). They often feel that they cannot maintain posture and are prone to falling. ▼

64. *Explain the changes in taste, vision, hearing, and balance that occur with aging.*

S U M M A R Y

Olfaction (p. 368)

Olfaction is the sense of smell.

Olfactory Epithelium and Bulb

1. Olfactory neurons in the olfactory epithelium are bipolar neurons. Their distal ends have olfactory hairs.
2. The olfactory hairs have receptors that respond to dissolved substances. There are approximately 1000 different odorant receptors.
3. The receptors activate G proteins, which results in ion channels opening and depolarization.
4. At least 7 (perhaps 50) primary odors exist. The olfactory neurons have a very low threshold and accommodate rapidly.

Neuronal Pathways for Olfaction

1. Axons from the olfactory neurons extend as olfactory nerves to the olfactory bulb, where they synapse with interneurons. Axons from these cells form the olfactory tracts, which connect to the olfactory cortex.
2. The olfactory bulbs and cortex accommodate to odors.

Taste (p. 369)

Taste buds usually are associated with papillae.

Histology of Taste Buds

1. Taste buds consist of taste cells, basilar cells, and supporting cells.
2. The taste cells have taste hairs that extend into taste pores.

Function of Taste

1. Receptors on the hairs detect dissolved substances.
2. Five basic types of taste exist: sour, salty, bitter, sweet, and umami.

Neuronal Pathways for Taste

1. The facial nerve carries taste sensations from the anterior two-thirds of the tongue, the glossopharyngeal nerve from the posterior one-third of the tongue, and the vagus nerve from the epiglottis.
2. The neural pathways for taste extend from the medulla oblongata to the thalamus and to the cerebral cortex.

Visual System (p. 370)
Accessory Structures

1. The eyebrows prevent perspiration from entering the eyes and help shade the eyes.
2. The eyelids consist of five tissue layers. They protect the eyes from foreign objects and help lubricate the eyes by spreading tears over their surface.
3. The conjunctiva covers the inner eyelid and the anterior part of the eye.
4. Lacrimal glands produce tears, which flow across the surface of the eye. Excess tears enter the lacrimal canaliculi and reach the nasal cavity through the nasolacrimal canal. Tears lubricate and protect the eye.
5. The extrinsic eye muscles move the eyeball.

Anatomy of the Eye

1. The fibrous layer is the outer layer of the eyeball. It consists of the sclera and cornea.
   - The sclera is the posterior four-fifths of the eyeball. It is white connective tissue that maintains the shape of the eyeball and provides a site for muscle attachment.
   - The cornea is the anterior one-fifth of the eye. It is transparent and refracts light that enters the eye.
2. The vascular layer is the middle layer of the eyeball.
   - The black choroid prevents the reflection of light inside the eye.
   - The iris is smooth muscle regulated by the autonomic nervous system. It controls the amount of light entering the pupil.
   - The ciliary muscles control the shape of the lens. They are smooth muscles regulated by the autonomic nervous system. The ciliary process produces aqueous humor.
3. The retina is the inner layer of the eyeball and contains neurons sensitive to light.
   - The macula (fovea centralis) is the area of greatest sensitivity to light.
   - The optic disc is the location through which nerves exit and blood vessels enter the eye. It has no photosensory cells and is therefore the blind spot of the eye.
4. The eyeball has three chambers: anterior, posterior, and vitreous.
   - The anterior and posterior chambers are filled with aqueous humor, which circulates and leaves by way of the scleral venous sinus.
   - The vitreous chamber is filled with vitreous humor.
5. The lens is held in place by the suspensory ligaments, which are attached to the lens capsule and ciliary body.

Functions of the Complete Eye

1. Visible light is the portion of the electromagnetic spectrum that humans can see.
2. When light travels from one medium to another, it can bend, or refract. Light striking a concave surface refracts outward (divergence). Light striking a convex surface refracts inward (convergence).
3. Converging light rays meet at the focal point and are said to be focused.
4. The cornea, aqueous humor, lens, and vitreous humor all refract light. The cornea is responsible for most of the convergence, whereas the lens can adjust the convergence by changing shape.
   - Relaxation of the ciliary muscles causes the lens to flatten, producing the emmetropic eye.
   - Contraction of the ciliary muscles causes the lens to become more spherical. This change in lens shape enables the eye to focus on objects that are less than 20 feet away, a process called accommodation.
5. The far point of vision is the distance at which the eye no longer has to change shape to focus on an object. The near point of vision is the closest an object can come to the eye and still be focused.
6. The pupil becomes smaller during accommodation, increasing the depth of focus.
7. The eyes converge (rotate medially) when looking at close-up objects.

Structure and Function of the Retina

1. The pigmented layer of the retina provides a black backdrop for increasing visual acuity.
2. The rods and the cones synapse with bipolar cells that in turn synapse with ganglion cells, which form the optic nerves.
3. Rods are responsible for noncolor vision and vision in low illumination (night vision).
   - A pigment, rhodopsin, is split by light into retinal and opsin, eventually resulting in action potentials going to the brain.
   - Light adaptation is caused by a reduction of rhodopsin; dark adaptation is caused by rhodopsin production.
4. Cones are responsible for color vision and visual acuity.
   - Cones are of three types, each with a different type of iodopsin photopigment. The pigments are most sensitive to blue, red, and green light.
   - Perception of many colors results from mixing the ratio of the different types of cones that are active at a given moment.
5. Most visual images are focused on the fovea centralis and macula. The fovea centralis has a very high concentration of cones. In the rest of the macula there are more cones than rods. Mostly rods are in the periphery of the retina.
6. Bipolar and ganglion cells in the retina can modify information sent to the brain.
7. Interneurons in the inner layers of the retina enhance contrast between the edges of objects.

Neuronal Pathways for Vision

1. Ganglion cell axons form the optic nerve, optic chiasm, and optic tracts. They extend to the thalamus, where they synapse. From there, neurons form the optic radiations that project to the visual cortex.
2. Neurons from the nasal visual field (temporal retina) of one eye and the temporal visual field (nasal retina) of the opposite eye project to the same cerebral hemisphere. Axons from the nasal retina cross in the optic chiasm, and axons from the temporal retina remain uncrossed.
3. Depth perception is the ability to judge relative distances of an object from the eyes and is a property of binocular vision. Binocular vision results because a slightly different image is seen by each eye.

Hearing and Balance (p. 387)

The osseous labyrinth is a canal system within the temporal bone that contains perilymph and the membranous labyrinth. Endolymph is inside the membranous labyrinth.

Auditory Structures and Their Functions

1. The external ear consists of the auricle and external acoustic meatus.
2. The middle ear connects the external and inner ears.
   - The tympanic membrane is stretched across the external acoustic meatus.
   - The malleus, incus, and stapes connect the tympanic membrane to the oval window of the inner ear.
   - The auditory tube connects the middle ear to the pharynx and equalizes pressure.
   - The middle ear is connected to the mastoid air cells.
3. The inner ear has three parts: the semicircular canals; the vestibule, which contains the utricle and the saccule; and the cochlea.
4. The cochlea is a spiral-shaped canal within the temporal bone.
   - The cochlea is divided into three compartments by the vestibular and basilar membranes. The scala vestibuli and scala tympani contain perilymph. The cochlear duct contains endolymph and the spiral organ.
   - The spiral organ consists of inner hair cells and outer hair cells, which attach to the tectorial membrane.

Auditory Function

1. Pitch is determine by the frequency of sound waves and volume by the amplitude of sound waves. Timbre is the resonance quality (overtones) of sound.
2. Hearing involves the following:
   - Sound waves are funneled by the auricle down the external acoustic meatus, causing the tympanic membrane to vibrate.
   - The tympanic membrane vibrations are passed along the auditory ossicles to the oval window of the inner ear.
   - Movement of the stapes in the oval window causes the perilymph, vestibular membrane, and endolymph to vibrate, producing movement of the basilar membrane.
   - Movement of the basilar membrane causes bending of the stereocilia of inner hair cells in the spiral organ.
   - Bending of the stereocilia pulls on gating springs, which open K^+ channels.
   - Potassium ions entering the hair cell result in depolarization of the cell.
   - Depolarization causes the release of glutamate, generating action potentials in the sensory neurons associated with hair cells.
   - The round window dissipates sound waves and protects the inner ear from pressure buildup.

Neuronal Pathways for Hearing

1. Axons from the vestibulocochlear nerve synapse in the medulla. Neurons from the medulla project axons to the inferior colliculi, where they synapse. Neurons from this point project to the thalamus and synapse. Thalamic neurons extend to the auditory cortex.
2. Efferent neurons project to cranial nerve nuclei responsible for controlling muscles that dampen sound in the middle ear.

Balance

1. Static balance evaluates the position of the head relative to gravity and detects linear acceleration and deceleration.
   - The utricle and saccule in the inner ear contain maculae. The maculae consist of hair cells with the hairs embedded in an otolithic membrane consisting of a gelatinous mass and crystals called otoliths.
   - The otolithic membrane moves in response to gravity.
2. Dynamic balance evaluates movements of the head.
   - Three semicircular canals at right angles to one another are present in the inner ear. The ampulla of each semicircular canal contains the crista ampullaris, which has hair cells with hairs embedded in a gelatinous mass, the cupula.
   - When the head moves, endolymph within the semicircular canal moves the cupula.

Neuronal Pathways for Balance

1. Axons from the maculae and the cristae ampullares extend to the vestibular nucleus of the medulla. Fibers from the medulla run to the spinal cord, cerebellum, cortex, and nuclei that control the extrinsic eye muscles.
2. Balance also depends on proprioception and visual input.

Effects of Aging on the Special Senses   (p. 399)

Elderly people experience a decline in function of all the special functions: olfaction, taste, vision, hearing, and balance. These declines can result in loss of appetite, visual impairment, disorientation, and risk of falling.

R E V I E W   A N D   C O M P R E H E N S I O N

1. Olfactory neurons
   a. have projections called cilia.
   b. have axons that combine to form the olfactory nerves.
   c. connect to the olfactory bulb.
   d. have receptors that react with odorants dissolved in fluid.
   e. all of the above.

2. Taste cells
   a. are found only on the tongue.
   b. extend through tiny openings called taste buds.
   c. have no axons but release neurotransmitters when stimulated.
   d. have axons that extend directly to the taste area of the cerebral cortex.

3. Which of these is *not* one of the basic tastes?
   a. spicy   d. umami
   b. salty   e. sour
   c. bitter

4. Tears
   a. are released onto the surface of the eye near the medial corner of the eye.
   b. in excess are removed by the scleral venous sinus.
   c. in excess can cause a sty.
   d. can pass through the nasolacrimal duct into the oral cavity.
   e. contain water, salts, mucus, and lysozyme.

5. The fibrous layer of the eye includes the
   a. conjunctiva.   d. iris.
   b. sclera.   e. retina.
   c. choroid.

6. The ciliary body
   a. attaches to the lens by suspensory ligaments.
   b. produces the vitreous humor.
   c. is part of the iris of the eye.
   d. is part of the sclera.
   e. all of the above.

7. Given these structures:
   1. lens
   2. aqueous humor
   3. vitreous humor
   4. cornea

   Choose the arrangement that lists the structures in the order that light entering the eye encounters them.
   a. 1,2,3,4
   b. 1,4,2,3
   c. 4,1,2,3
   d. 4,2,1,3
   e. 4,3,2,1

8. Aqueous humor
   a. is the pigment responsible for the black color of the choroid.
   b. exits the eye through the scleral venous sinus.
   c. is produced by the iris.
   d. can cause cataracts if overproduced.
   e. is composed of proteins called crystallines.

9. Contraction of the smooth muscle in the ciliary body causes the
   a. lens to flatten.
   b. lens to become more spherical.
   c. pupil to constrict.
   d. pupil to dilate.

10. Given these events:
    1. medial rectus contracts
    2. lateral rectus contracts
    3. pupils dilate
    4. pupils constrict
    5. lens of the eye flattens
    6. lens of the eye becomes more spherical

    Assume you are looking at an object 30 feet away. If you suddenly look at an object that is 1 foot away, which events occur?
    a. 1,3,6
    b. 1,4,5
    c. 1,4,6
    d. 2,3,6
    e. 2,4,5

11. Given these neurons in the retina:
    1. bipolar cells
    2. ganglion cells
    3. photoreceptor cells

    Choose the arrangement that lists the correct order of the cells encountered by light as it enters the eye and travels toward the pigmented retina.
    a. 1,2,3
    b. 1,3,2
    c. 2,1,3
    d. 2,3,1
    e. 3,1,2

12. Which of these photoreceptor cells is correctly matched with its function?
    a. rods—vision in bright light
    b. rods—visual acuity
    c. cones—color vision

13. Concerning dark adaptation,
    a. the amount of rhodopsin increases.
    b. the pupils constrict.
    c. it occurs more rapidly than light adaptation.
    d. all of the above.

14. In the retina, there are cones that are most sensitive to a particular color. Given this list of colors:
    1. red
    2. yellow
    3. green
    4. blue

    Indicate which colors correspond to specific types of cones.
    a. 2,3
    b. 3,4
    c. 1,2,3
    d. 1,3,4
    e. 1,2,3,4

15. Given these areas of the retina:
    1. macula
    2. fovea centralis
    3. optic disc
    4. periphery of the retina

    Choose the arrangement that lists the areas according to the density of cones, starting with the area that has the highest density of cones.
    a. 1,2,3,4
    b. 1,3,2,4
    c. 2,1,4,3
    d. 2,4,1,3
    e. 3,4,1,2

16. Concerning axons in the optic nerve from the right eye,
    a. they all go to the right occipital lobe.
    b. they all go to the left occipital lobe.
    c. they all go to the thalamus.
    d. most go to the thalamus but some go to the superior colliculus.
    e. some go to the right occipital lobe, but some go to the left occipital lobe.

17. A person with an abnormally powerful focusing system is _____ and uses a _____ to correct his or her vision.
    a. nearsighted, concave lens
    b. nearsighted, convex lens
    c. farsighted, concave lens
    d. farsighted, convex lens

18. Which of these structures is found within or is a part of the external ear?
    a. oval window
    b. auditory tube
    c. ossicles
    d. external acoustic meatus
    e. cochlear duct

19. Given these auditory ossicles:
    1. incus
    2. malleus
    3. stapes

    Choose the arrangement that lists the auditory ossicles in order from the tympanic membrane to the inner ear.
    a. 1,2,3
    b. 1,3,2
    c. 2,1,3
    d. 2,3,1
    e. 3,2,1

20. The spiral organ is found within the
    a. cochlear duct.
    b. scala vestibuli.
    c. scala tympani.
    d. vestibule.
    e. semicircular canals.

21. An increase in the loudness of sound occurs as a result of an increase in the _____ of the sound wave.
    a. frequency
    b. amplitude
    c. resonance
    d. both a and b

22. Given these structures:
    1. perilymph
    2. endolymph
    3. vestibular membrane
    4. basilar membrane

    Choose the arrangement that lists the structures in the order sound waves coming from the outside encounter them in producing sound.
    a. 1,3,2,4
    b. 1,4,2,3
    c. 2,3,1,4
    d. 2,4,1,3
    e. 3,4,2,1

23. Interpretation of different sounds is possible because of the ability of the _____ to vibrate at different frequencies and stimulate the _____.
    a. vestibular membrane, vestibular nerve
    b. vestibular membrane, spiral organ
    c. basilar membrane, vestibular nerve
    d. basilar membrane, spiral organ

24. Which structure is a specialized receptor found within the utricle?
    a. macula
    b. crista ampullaris
    c. spiral organ
    d. cupula

25. Damage to the semicircular canals affects the ability to detect
    a. sound.
    b. the position of the head relative to the ground.
    c. the movement of the head in all directions.
    d. all of the above.

Answers in Appendix E

CRITICAL THINKING

1. An elderly man with normal vision develops cataracts. He is surgically treated by removing the lenses of his eyes. What kind of glasses would you recommend he wear to compensate for the removal of his lenses?

2. Some animals have a reflective area in the choroid called the tapetum lucidum. Light entering the eye is reflected back instead of being absorbed by the choroid. What would be the advantage of this arrangement? The disadvantage?

3. Perhaps you have heard someone say that eating carrots is good for the eyes. What is the basis for this claim?

4. On a camping trip, Jean Tights rips her pants. That evening, she is going to repair the rip. As the sun goes down, the light becomes dimmer and dimmer. When she tries to thread the needle, it is obvious that she is not looking directly at the needle but is looking a few inches to the side. Why does she do this?

5. A man stares at a black clock on a white wall for several minutes. Then he shifts his view and looks at only the blank white wall.

Although he is no longer looking at the clock, he sees a light clock against a dark background. Explain what is happening.

6. Describe the results of a lesion of the left optic tract.

7. Professional divers are subject to increased pressure as they descend to the bottom of the ocean. Sometimes this pressure can lead to damage to the ear and loss of hearing. Describe the normal mechanisms that adjust for changes in pressure, suggest some conditions that might interfere with pressure adjustment, and explain how the increased pressure might cause loss of hearing.

8. What effect does facial nerve damage have on hearing?

9. If a vibrating tuning fork is placed against the mastoid process of the temporal bone, the vibrations are perceived as sound, even if the external acoustic meatus is plugged. Explain how this happens.

Answers in Appendix F

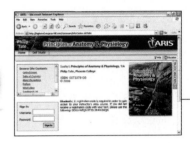

CHAPTER 14 Autonomic Nervous System

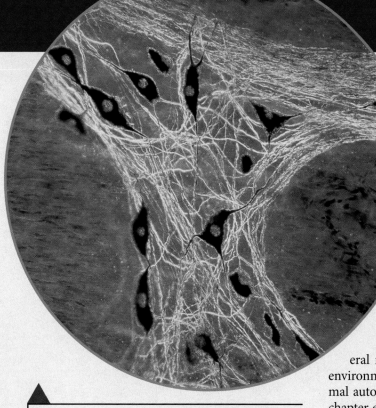

Light photomicrograph from a section of the small intestine, showing the nerve cells of the enteric plexus. These nerve cells regulate the contraction of smooth muscle and the secretion of glands within the intestinal wall.

Tools for Success

A virtual cadaver dissection experience

Audio (MP3) and/or video (MP4) content, available at www.mhhe.aris.com

Introduction

During a picnic on a sunny spring day, it is easy to concentrate on the delicious food and the pleasant surroundings. The maintenance of homeostasis requires no conscious thought. The autonomic nervous system (ANS) helps keep body temperature at a constant level by controlling the activity of sweat glands and the amount of blood flowing through the skin. The ANS helps regulate the complex activities necessary for the digestion of food. The movement of absorbed nutrients to tissues is possible because the ANS controls heart rate, which helps maintain the blood pressure necessary to deliver blood to tissues. Without the ANS, all of the activities necessary to maintain homeostasis would be overwhelming.

▶A functional knowledge of the ANS enables you to predict general responses to a variety of stimuli, explain responses to changes in environmental conditions, comprehend symptoms that result from abnormal autonomic functions, and understand how drugs affect the ANS. This chapter examines the autonomic nervous system by **contrasting the somatic and autonomic nervous systems (p. 406);** describing the **anatomy of the autonomic nervous system (p. 406),** the **physiology of the autonomic nervous system (p. 411),** and the **regulation of the autonomic nervous system (p. 416);** and examining **functional generalizations about the autonomic nervous system (p. 418).**

Contrasting the Somatic and Autonomic Nervous Systems

The peripheral nervous system (PNS) is composed of sensory and motor neurons. Sensory neurons carry action potentials from the periphery to the central nervous system (CNS), and motor neurons carry action potentials from the CNS to the periphery. Motor neurons are either somatic motor neurons, which innervate skeletal muscle, or autonomic motor neurons, which innervate smooth muscle, cardiac muscle, and glands.

Although axons of autonomic, somatic, and sensory neurons are in the same nerves, the proportion varies from nerve to nerve. For example, nerves innervating smooth muscle, cardiac muscle, and glands, such as the vagus nerves, consist primarily of axons of autonomic motor neurons and sensory neurons. Nerves innervating skeletal muscles, such as the sciatic nerves, consist primarily of axons of somatic motor neurons and sensory neurons. Some cranial nerves, such as the olfactory, optic, and vestibulocochlear nerves, are composed entirely of axons of sensory neurons.

The cell bodies of somatic motor neurons are in the CNS, and their axons extend from the CNS to skeletal muscle (figure 14.1a). The ANS, on the other hand, has two neurons in a series extending between the CNS and the organs innervated (figure 14.1b). The first neurons of the series are called **preganglionic neurons.** Their cell bodies are located in the CNS within either the brainstem or the lateral part of the spinal cord gray matter, and their axons extend to autonomic ganglia located outside the CNS. The **autonomic ganglia** contain the cell bodies of the second neurons of the series, which are called **postganglionic neurons.** The preganglionic neurons synapse with the postganglionic neurons in the autonomic ganglia. The axons of the postganglionic neurons extend from autonomic ganglia to effector organs, where they synapse with their target tissues.

Many movements controlled by the somatic nervous system are conscious, whereas ANS functions are unconsciously controlled. The effect of somatic motor neurons on skeletal muscle is always excitatory, but the effect of the ANS on target tissues can be excitatory or inhibitory. For example, after a meal, the ANS can stimulate stomach activities but, during exercise, the ANS can inhibit those activities. Table 14.1 summarizes the differences between the somatic nervous system and the ANS.

Sensory neurons are not classified as somatic or autonomic. These neurons propagate action potentials from sensory receptors to the CNS and can provide information for reflexes mediated through the somatic nervous system or the ANS. For example, stimulation of pain receptors can initiate somatic reflexes, such as the withdrawal reflex, and autonomic reflexes, such as an increase in heart rate. Although some sensory neurons primarily affect somatic functions and others primarily influence autonomic functions, functional overlap makes attempts to classify sensory neurons as either somatic or autonomic meaningless. ▼

1. *Contrast the somatic nervous system with the ANS for each of the following:*
   a. *the number of neurons between the CNS and effector organ*
   b. *the location of neuron cell bodies*
   c. *the structures each innervates*
   d. *inhibitory or excitatory effects*
   e. *conscious or unconscious control*

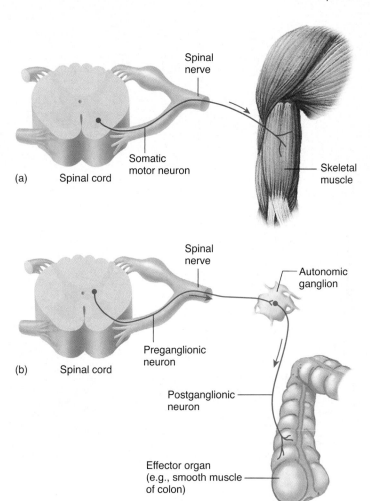

Figure 14.1 **Organization of Somatic and Autonomic Nervous System Neurons**

(*a*) The cell body of the somatic neuron is in the CNS, and its axon extends to the skeletal muscle. (*b*) The cell body of the preganglionic neuron is in the CNS, and its axon extends to the autonomic ganglion and synapses with the postganglionic neuron. The postganglionic neuron extends to and synapses with its effector organ.

2. *Why are sensory neurons not classified as somatic or autonomic?*
3. *Define preganglionic neuron, postganglionic neuron, and autonomic ganglia.*

Anatomy of the Autonomic Nervous System

The ANS is subdivided into the **sympathetic** and the **parasympathetic divisions** and the **enteric** (en-ter′ik, bowels) **nervous system (ENS).** The sympathetic and parasympathetic divisions differ structurally in (1) the location of their preganglionic neuron cell bodies within the CNS and (2) the location of their autonomic ganglia.

The enteric nervous system is a complex network of neuron cell bodies and axons within the wall of the digestive tract. An important part of this network is sympathetic and parasympathetic neurons. For this reason, the enteric nervous system is considered to be part of the ANS.

Table 14.1	Comparison of the Somatic and Autonomic Nervous Systems	
Features	**Somatic Nervous System**	**Autonomic Nervous System**
Target tissues	Skeletal muscle	Smooth muscle, cardiac muscle, and glands
Regulation	Controls all conscious and unconscious movements of skeletal muscle	Unconscious regulation, although influenced by conscious mental functions
Response to stimulation	Skeletal muscle contracts.	Target tissues are stimulated or inhibited.
Neuron arrangement	One neuron extends from the central nervous system (CNS) to skeletal muscle.	There are two neurons in series; the preganglionic neuron extends from the CNS to an autonomic ganglion, and the postganglionic neuron extends from the autonomic ganglion to the target tissue.
Neuron cell body location	Neuron cell bodies are in motor nuclei of the cranial nerves and in the ventral horn of the spinal cord.	Preganglionic neuron cell bodies are in autonomic nuclei of the cranial nerves and in the lateral part of the spinal cord; postganglionic neuron cell bodies are in autonomic ganglia.
Number of synapses	One synapse between the somatic motor neuron and the skeletal muscle	Two synapses: first is in the autonomic ganglia; second is at the target tissue
Axon sheaths	Myelinated	Preganglionic axons are myelinated; postganglionic axons are unmyelinated.
Neurotransmitter substance	Acetylcholine	Acetylcholine is released by preganglionic neurons; either acetylcholine or norepinephrine is released by postganglionic neurons.
Receptor molecules	Receptor molecules for acetylcholine are nicotinic.	In autonomic ganglia, receptor molecules for acetylcholine are nicotinic; in target tissues, receptor molecules for acetylcholine are muscarinic, whereas receptor molecules for norepinephrine are either α- or β-adrenergic.

Sympathetic Division

Cell bodies of sympathetic preganglionic neurons are in the lateral horns of the spinal cord gray matter between the first thoracic (T1) and the second lumbar (L2) segments (figure 14.2). The sympathetic division is sometimes called the **thoracolumbar division** because of the location of the preganglionic cell bodies. The axons of the preganglionic neurons pass through the ventral roots of spinal nerves T1–L2, course through the spinal nerves for a short distance, leave these nerves, and project to sympathetic ganglia. There are two types of sympathetic ganglia: sympathetic chain ganglia and collateral ganglia. **Sympathetic chain ganglia** are ganglia connected to each other to form a chain. A set of sympathetic chain ganglia is located along the left and right sides of the vertebral column. They are also called **paravertebral** (alongside the vertebral column) **ganglia** because of their location. Although the sympathetic division originates in the thoracic and lumbar vertebral regions, the sympathetic chain ganglia extend into the cervical and sacral regions. As a result of ganglia fusion during fetal development, there are typically 3 pairs of cervical ganglia, 11 pairs of thoracic ganglia, 4 pairs of lumbar ganglia, and 4 pairs of sacral ganglia. The **collateral** (meaning accessory) **ganglia** are unpaired ganglia located in the abdominopelvic cavity. They are also called **prevertebral** (in front of the vertebral column) **ganglia** because they are anterior to the vertebral column.

The axons of preganglionic neurons are small in diameter and myelinated. The short connection between a spinal nerve and a sympathetic chain ganglion through which the preganglionic axons pass is called a **white ramus communicans** (rā′mĭs kŏ-mū′ni-kans, pl. *rami communicantes,* rā′mī kŏ-mū-ni-kan′tēz) because of the whitish color of the myelinated axons (figure 14.3).

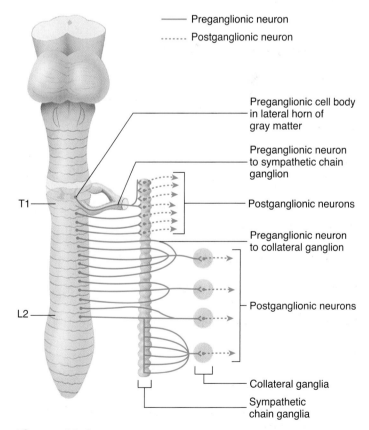

— Preganglionic neuron
...... Postganglionic neuron

Preganglionic cell body in lateral horn of gray matter

Preganglionic neuron to sympathetic chain ganglion

Postganglionic neurons

Preganglionic neuron to collateral ganglion

Postganglionic neurons

Collateral ganglia

Sympathetic chain ganglia

T1

L2

Figure 14.2   Sympathetic Division

The location of sympathetic preganglionic (*solid blue*) and postganglionic (*dotted blue*) neurons. The preganglionic cell bodies are in the lateral gray matter of the thoracic and lumbar parts of the spinal cord. The cell bodies of the postganglionic neurons are within the sympathetic chain ganglia or within collateral ganglia.

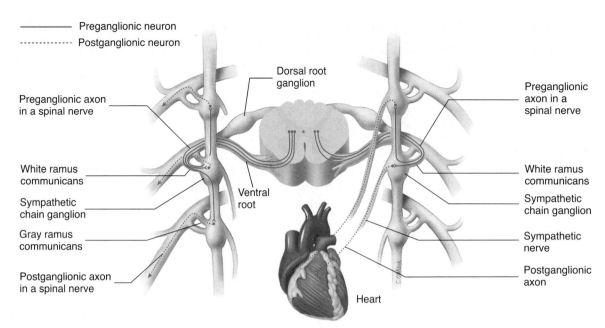

—— Preganglionic neuron
········· Postganglionic neuron

Dorsal root ganglion

Preganglionic axon in a spinal nerve

White ramus communicans

Sympathetic chain ganglion

Gray ramus communicans

Postganglionic axon in a spinal nerve

Ventral root

Preganglionic axon in a spinal nerve

White ramus communicans

Sympathetic chain ganglion

Sympathetic nerve

Postganglionic axon

Heart

(a) Preganglionic axons from a spinal nerve pass through a white ramus communicans into a sympathetic chain ganglion. Some axons synapse with a postganglionic neuron at the level of entry; others ascend or descend to other levels before synapsing. Each postganglionic axon exits the sympathetic chain through a gray ramus communicans and enters a spinal nerve.

(b) Part (*b*) is like part (*a*), except that each postganglionic neuron exits a sympathetic chain ganglion through a sympathetic nerve.

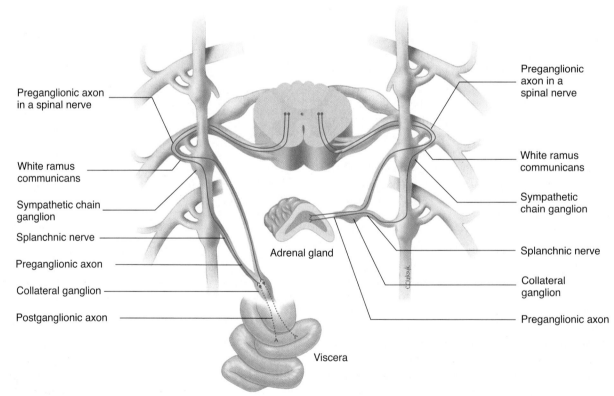

Preganglionic axon in a spinal nerve

White ramus communicans

Sympathetic chain ganglion

Splanchnic nerve

Preganglionic axon

Collateral ganglion

Postganglionic axon

Adrenal gland

Viscera

Preganglionic axon in a spinal nerve

White ramus communicans

Sympathetic chain ganglion

Splanchnic nerve

Collateral ganglion

Preganglionic axon

(c) Preganglionic neurons do not synapse in the sympathetic chain ganglia, but exit in splanchnic nerves and extend to a collateral ganglion, where they synapse with postganglionic neurons.

(d) Part (*d*) is like part (*c*), except that the preganglionic axons extend to the adrenal medulla, where they synapse with specialized adrenal medullary cells.

Figure 14.3 Routes Taken by Sympathetic Axons

Preganglionic axons are illustrated as *solid lines* and postganglionic axons as *dashed lines*.

Sympathetic axons exit the sympathetic chain ganglia by the following four routes:

1. *Spinal nerves* (see figure 14.3*a*). Preganglionic axons synapse with postganglionic neurons in sympathetic chain ganglia. They can synapse at the same level that the preganglionic axons enter the sympathetic chain, or they can pass superiorly or inferiorly through one or more ganglia and synapse with postganglionic neurons in a sympathetic chain ganglion at a different level. Axons of the postganglionic neurons pass through a **gray ramus communicans** and reenter a spinal nerve. Postganglionic axons are not myelinated, thereby giving the gray ramus communicans its grayish color. All spinal nerves receive postganglionic axons from a gray ramus communicans. The postganglionic axons then project through the spinal nerve to smooth muscle and glands located in the skin and skeletal muscles.

2. *Sympathetic nerves* (see figure 14.3*b*). Preganglionic axons enter the sympathetic chain and synapse in a sympathetic chain ganglion at the same or a different level with postganglionic neurons. The postganglionic axons leaving the sympathetic chain ganglion form **sympathetic nerves,** which supply organs in the thoracic cavity.

3. *Splanchnic* (splangk′nik) *nerves* (see figure 14.3*c*). Some preganglionic axons enter sympathetic chain ganglia and, without synapsing, exit at the same or a different level to form **splanchnic nerves.** Those preganglionic axons extend to collateral ganglia, where they synapse with postganglionic neurons. Axons of the postganglionic neurons leave the collateral ganglia through small nerves that extend to effector organs in the abdominopelvic cavity.

4. *Innervation to the adrenal gland* (see figure 14.3*d*). The splanchnic nerve innervation to the adrenal glands is different from other ANS nerves because it consists of only preganglionic neurons. Axons of the preganglionic neurons do not synapse in sympathetic chain ganglia or in collateral ganglia. Instead, the axons pass through those ganglia and synapse with cells in the adrenal medulla. The **adrenal medulla** is the inner portion of the adrenal gland; it consists of specialized cells derived during embryonic development from neural crest cells (see chapter 25), which are the same population of cells that gives rise to the postganglionic cells of the ANS. Adrenal medullary cells are round, have no axons or dendrites, and are divided into two groups. About 80% of the cells secrete **epinephrine** (ep′i-nef′rin), also called **adrenaline** (ă-dren′ă-lin), and about 20% secrete **norepinephrine** (nōr′ep-i-nef′rin), also called **noradrenaline** (nōr-ă-dren′ă-lin). Stimulation of these cells by preganglionic axons causes the release of epinephrine and norepinephrine. These substances circulate in the blood and affect all tissues having receptors to which they can bind. The general response to epinephrine and norepinephrine released from the adrenal medulla is to prepare the individual for physical activity. Secretions of the adrenal medulla are considered hormones because they are released into the general circulation and travel some

distance to the tissues in which they have their effect (see chapter 15). ▼

4. *Where are the cell bodies of sympathetic preganglionic neurons located?*
5. *What types of axon (preganglionic or postganglionic, myelinated or unmyelinated) are found in white and gray rami communicantes?*
6. *Where do preganglionic neurons synapse with postganglionic neurons that are found in spinal and sympathetic nerves?*
7. *Where do preganglionic axons that form splanchnic nerves (except those to the adrenal gland) synapse with postganglionic neurons?*
8. *What is unusual about the splanchnic nerve innervation to the adrenal gland? What do the specialized cells of the adrenal medulla secrete, and what is the effect of these substances?*

Parasympathetic Division

The cell bodies of parasympathetic preganglionic neurons are either within cranial nerve nuclei in the brainstem or within the lateral parts of the gray matter in the sacral region of the spinal cord from S2 to S4 (figure 14.4). For that reason, the parasympathetic division is sometimes called the **craniosacral** (krā′nē-ō-sā′krăl) **division.**

Axons of the parasympathetic preganglionic neurons from the brain are in **cranial nerves III, VII, IX,** and **X** and from the spinal cord in **pelvic splanchnic nerves.** The preganglionic axons course through these nerves to **terminal ganglia,** where they synapse with postganglionic neurons. The axons of the postganglionic neurons extend relatively short distances from the terminal ganglia to the effector organs. The terminal ganglia are either near or embedded within the walls of the organs innervated by the parasympathetic neurons. Many of the parasympathetic ganglia are small but some, such as those in the wall of the digestive tract, are large.

Table 14.2 summarizes the structural differences between the sympathetic and parasympathetic divisions. ▼

9. *Where are the cell bodies of parasympathetic preganglionic neurons located? In what structure do parasympathetic preganglionic neurons synapse with postganglionic neurons?*
10. *What nerves are formed by the axons of parasympathetic preganglionic neurons?*

Enteric Nervous System

The enteric nervous system consists of nerve plexuses within the wall of the digestive tract (see chapter 21). The plexuses have contributions from three sources: (1) sensory neurons that connect the digestive tract to the CNS, (2) ANS motor neurons that connect the CNS to the digestive tract, and (3) enteric neurons, which are confined to the enteric plexuses. The CNS is capable of monitoring the digestive tract and controlling its smooth muscle and glands through autonomic reflexes (see "Regulation of the

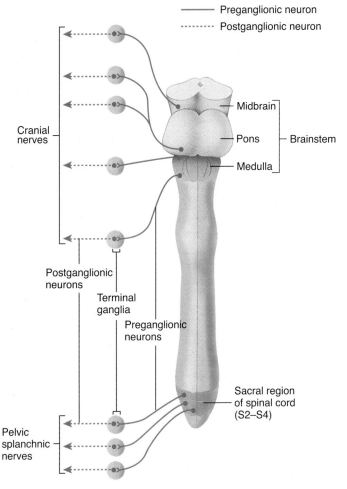

Figure 14.4   Parasympathetic Division

The location of parasympathetic preganglionic (*solid red*) and postganglionic (*dotted red*) neurons. The preganglionic neuron cell bodies are in the brainstem and the lateral gray matter of the sacral part of the spinal cord, and the postganglionic neuron cell bodies are within terminal ganglia.

Autonomic Nervous System," p. 416). For example, stretch of the digestive tract is detected by sensory neurons and action potentials are transmitted to the CNS. In response, the CNS sends action potentials to glands in the digestive tract, causing them to secrete.

There are several major types of enteric neurons: (1) Enteric sensory neurons can detect changes in the chemical composition of the contents of the digestive tract or detect stretch of the digestive tract wall, (2) enteric motor neurons can stimulate or inhibit smooth muscle contraction and gland secretion, and (3) enteric interneurons connect enteric sensory and motor neurons to each other. Enteric neurons are capable of monitoring and controlling the digestive tract independently of the CNS through local reflexes (see "Regulation of the Autonomic Nervous System," p. 416). For example, stretch of the digestive tract is detected by enteric sensory neurons, which stimulate enteric interneurons. The enteric interneurons stimulate enteric motor neurons, which stimulate glands to secrete. Although the enteric nervous system is capable of controlling the activities of the digestive tract completely independently of the CNS, normally the two systems work together. ▼

11. *What is the enteric nervous system and where is it located?*
12. *How does the CNS monitor and control the digestive tract?*
13. *Name three major types of enteric neurons. How do the enteric neurons monitor and control the digestive tract?*

P R E D I C T   1
What type of ganglia (chain ganglia, collateral ganglia, or terminal ganglia) are found in the enteric plexus? What type (preganglionic or postganglionic) of sympathetic and parasympathetic axons contribute to the enteric plexus?

Distribution of Autonomic Nerve Fibers

The ANS supplies smooth muscle, cardiac muscle, and glands. When naming the effector organs supplied by the ANS, it is the smooth muscle, cardiac muscle, or glands of the organ that is supplied. For example, smooth muscle is part of the walls of blood

Table 14.2	Comparison of the Sympathetic and Parasympathetic Divisions	
Features	**Sympathetic Division**	**Parasympathetic Division**
Location of preganglionic cell body	Lateral horns of spinal cord gray matter (T1–L2)	Brainstem and lateral parts of spinal gray matter (S2–S4)
Outflow from the CNS	Spinal nerves Sympathetic nerves Splanchnic nerves	Cranial nerves Pelvic splanchnic nerves
Ganglia	Sympathetic chain ganglia along spinal cord for spinal and sympathetic nerves; collateral ganglia for splanchnic nerves	Terminal ganglia near or on effector organ
Number of postganglionic neurons for each preganglionic neuron	Many (much divergence)	Few (less divergence)
Relative length of neurons	Short preganglionic Long postganglionic	Long preganglionic Short postganglionic

vessels, controlling blood vessel diameter and the flow of blood through vessels. All organs supplied by blood vessels are innervated by the ANS. To say that the ANS supplies skeletal muscle does not mean that it innervates skeletal muscle fibers. Instead, it means that the ANS innervates smooth muscle in the blood vessels of skeletal muscle. To say that the ANS supplies the stomach means that it innervates smooth muscle and glands in the wall of the stomach.

Sympathetic and parasympathetic axons intermingle as they extend to effector organs. **Autonomic nerve plexuses** are complex, interconnected neural networks formed by neurons of the sympathetic and parasympathetic divisions. Plexuses typically are on the outer surfaces of blood vessels and following the route of blood vessels is a major means by which autonomic axons are distributed throughout the body to effector organs. Sensory axons from the effector organs pass to the CNS through the autonomic nerve plexuses.

Sympathetic Division

Sympathetic axons pass from the sympathetic chain ganglia to their target tissues through spinal, sympathetic, and splanchnic nerves. Spinal nerves supply smooth muscle and glands in the skin and skeletal muscles of most of the body (see figure 11.12).

Sympathetic nerves supply the parts of the head and neck not supplied by spinal nerves. Most of the sympathetic nerve supply to the head and neck is derived from the superior cervical sympathetic chain ganglion (figure 14.5). Sympathetic axons join cranial nerves and are distributed to effector organs. Sympathetic nerves also supply thoracic organs, such as the lungs and heart.

Splanchnic nerves mainly supply organs in the abdominopelvic organs. The preganglionic axons from sympathetic chain ganglia T5 and below synapse with collateral ganglia in the abdominopelvic cavity. The largest collateral ganglia are the **celiac** (sē′lē-ak), **superior mesenteric** (mez-en-ter′ik), **inferior mesenteric,** and **hypogastric ganglia.**

Parasympathetic Division

Parasympathetic outflow is through cranial and pelvic splanchnic nerves (see figure 14.5). Cranial nerve III supplies smooth muscle in the eye, and cranial nerves VII and IX supply the salivary and lacrimal glands. Cranial nerve X supplies thoracic organs, abdominal organs, and the proximal large intestine. The pelvic splanchnic nerves supply organs of the pelvic cavity and the distal large intestine. ▼

14. *What is an autonomic nerve plexus?*
15. *Describe the distribution of the sympathetic division through spinal, sympathetic, and splanchnic nerves.*
16. *Describe the distribution of the parasympathetic division through cranial and pelvic splanchnic nerves.*

P R E D I C T
Starting in the small intestine and ending with the ganglia where their cell bodies are located, trace the route for sensory axons passing alongside sympathetic axons. Name all of the nerves, ganglia, and so on that the sensory axon passes through. Also trace the route for sensory neurons passing alongside parasympathetic axons.

Physiology of the Autonomic Nervous System

Neurotransmitters

Sympathetic and parasympathetic nerve endings secrete one of two neurotransmitters. If the neuron secretes acetylcholine, it is a **cholinergic** (kol-in-er′jik) **neuron;** if it secretes norepinephrine (or epinephrine), it is an **adrenergic** (ad-rĕ-ner′jik) **neuron.** Adrenergic neurons are so named because at one time it was believed that they secreted adrenaline, which is another name for epinephrine. All preganglionic neurons of the sympathetic and parasympathetic divisions and all postganglionic neurons of the parasympathetic division are cholinergic. Almost all postganglionic neurons of the sympathetic division are adrenergic, but a few postganglionic neurons that innervate thermoregulatory sweat glands are cholinergic (figure 14.6).

In recent years, substances in addition to the regular neurotransmitters have been extracted from ANS neurons. These substances include nitric oxide; fatty acids, such as eicosanoids; peptides, such as gastrin, somatostatin, cholecystokinin, vasoactive intestinal peptide, enkephalins, and substance P; and monoamines, such as dopamine, serotonin, and histamine. The specific role that many of these compounds play in the regulation of the ANS is unclear, but they appear to function as either neurotransmitters or neuromodulator substances (see chapter 11).

Receptors

Receptors for acetylcholine and norepinephrine are located in the plasma membrane of certain cells (table 14.3). The combination of neurotransmitter and receptor functions as a signal to cells, causing them to respond. Depending on the type of cell, the response is excitatory or inhibitory.

Cholinergic Receptors

Cholinergic receptors are receptors to which acetylcholine binds. They have two major, structurally different forms. **Nicotinic** (nik-ō-tin′ik) **receptors** bind to nicotine, an alkaloid substance found in tobacco, and **muscarinic** (mŭs-kă-rin′ik) **receptors** bind to muscarine, an alkaloid extracted from some poisonous mushrooms. Although nicotine and muscarine are not naturally in the human body, they demonstrate the differences between the two classes of cholinergic receptors. Nicotine binds to nicotinic receptors but not to muscarinic receptors, whereas muscarine binds to muscarinic receptors but not to nicotinic receptors. On the other hand, nicotinic and muscarinic receptors are very similar because acetylcholine binds to and activates both types of receptors.

The membranes of all postganglionic neurons in autonomic ganglia and the membranes of skeletal muscle cells have nicotinic receptors. The membranes of effector cells that respond to acetylcholine released from postganglionic neurons have muscarinic receptors (see figure 14.6).

P R E D I C T
Would structures innervated by the sympathetic division or the parasympathetic division be affected after the consumption of nicotine? After the consumption of muscarine? Explain.

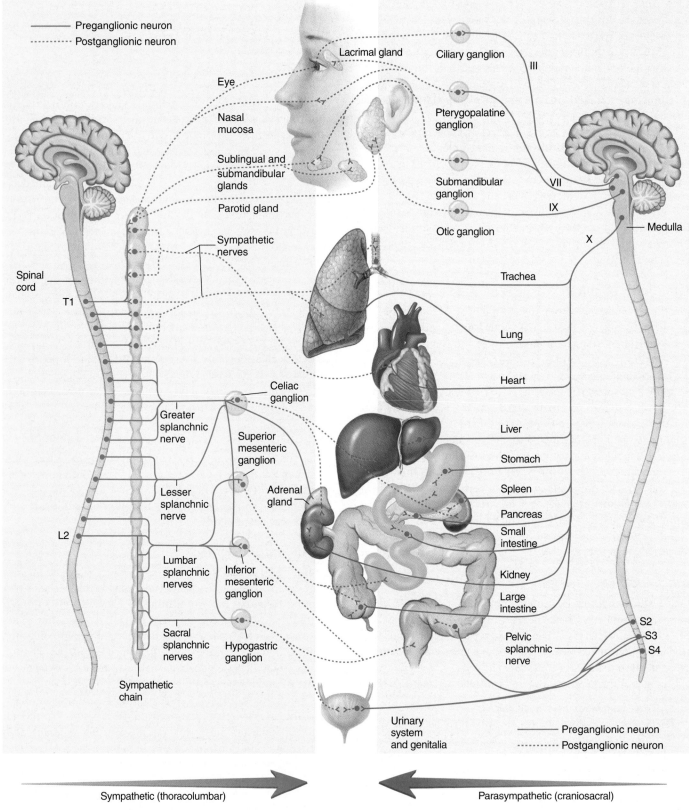

Figure 14.5 **Innervation of Organs by the ANS**

Preganglionic fibers are indicated by *solid lines*, and postganglionic fibers are indicated by *dashed lines*.

Sympathetic division

Most target tissues innervated by the sympathetic division have adrenergic receptors. When norepinephrine (NE) binds to adrenergic receptors, some target tissues are stimulated, and others are inhibited. For example, smooth muscle cells in blood vessels are stimulated to constrict, and stomach glands are inhibited.

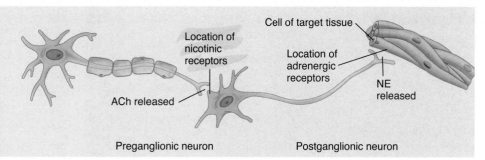

Sympathetic division

Some sympathetic target tissues, such as sweat glands, have muscarinic receptors, which respond to acetylcholine (ACh). Stimulation of sweat glands results in increased sweat production.

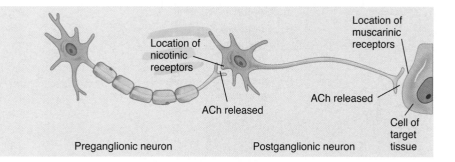

Parasympathetic division

All parasympathetic target tissues have muscarinic receptors. The general response to ACh is excitatory, but some target tissues, such as the heart, are inhibited.

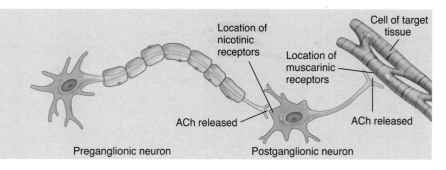

Figure 14.6 Location of ANS Receptors

Nicotinic receptors are on the cell bodies of both sympathetic and parasympathetic postganglionic cells in the autonomic ganglia. *Abbreviations: NE, norepinephrine; ACh, acetylcholine.*

Acetylcholine binding to nicotinic receptors has an excitatory effect because it results in the direct opening of Na⁺ channels and the production of action potentials. When acetylcholine binds to muscarinic receptors, the cell's response is mediated through G proteins (see chapter 15). The response is either excitatory or inhibitory, depending on the target tissue in which the receptors are found. For example, acetylcholine binds to muscarinic receptors in cardiac muscle, thereby reducing heart rate, and acetylcholine binds to muscarinic receptors in smooth muscle cells of the stomach, thus increasing its rate of contraction. ▼

17. *Define cholinergic and adrenergic neurons. Which neurons of the ANS are cholinergic and adrenergic?*

18. *Name the two major subtypes of cholinergic receptors. Where are they located? When acetylcholine binds to each subtype, does it result in an excitatory or inhibitory cell response?*

Adrenergic Receptors

Adrenergic receptors are receptors to which norepinephrine or epinephrine bind. They are located in the plasma membranes of target tissues innervated by the sympathetic division (see figure 14.6). The response of cells to norepinephrine or epinephrine binding to adrenergic receptors is mediated through G proteins (see chapter 15). Depending on the target tissue, the activation of G proteins can result in excitatory or inhibitory responses.

Adrenergic receptors are subdivided into two major categories: **alpha (α) receptors** and **beta (β) receptors.** Epinephrine has a greater effect than norepinephrine on most α and β receptors. The main subtypes for alpha receptors are α₁- and α₂-adrenergic receptors and for beta receptors are β₁- and β₂-adrenergic receptors.

Table 14.3	Effects of the Sympathetic and Parasympathetic Divisions on Various Tissues	
Organ	**Sympathetic Effects and Receptor Type***	**Parasympathetic Effects and Receptor Type***
Adipose tissue	Fat breakdown and release of fatty acids (α_2, β_1)	None
Arrector pili muscle	Contraction (α_1)	None
Blood (platelets)	Increased coagulation (α_2)	None
Blood vessels		
Arterioles (carry blood to tissues)		
Digestive organs	Constriction (α_1)	None
Heart	Constriction (α_1), dilation (β_2)[†]	None
Kidneys	Constriction (α_1, α_2), dilation (β_1, β_2)	None
Lungs	Constriction (α_1), dilation (β_2)	None
Skeletal muscle	Constriction (α_1), dilation (β_2)	None
Skin	Constriction (α_1, α_2)	None
Veins (carry blood away from tissues)	Constriction (α_1, α_2), dilation (β_2)	
Eye		
Ciliary muscle	Relaxation for far vision (β_2)	Contraction for near vision (m)
Pupil	Dilated (α_1)[‡]	Constriction (m)[‡]
Gallbladder	Relaxation (β_2)	Contraction (m)
Glands		
Adrenal	Release of epinephrine and norepinephrine (n)	None
Gastric	Decreased gastric secretion (α_2)	Increased gastric secretion (m)
Lacrimal	Slight tear production (α)	Increased tear secretion (m)
Pancreas	Decreased insulin secretion (α_2)	Increased insulin secretion (m)
	Decreased exocrine secretion (α)	Increased exocrine secretion (m)
Salivary	Constriction of blood vessels and slight production of a thick, viscous saliva (α_1)	Dilation of blood vessels and thin, copious saliva (m)
Sweat		
Apocrine	Thick, organic secretion (m)	None
Merocrine	Watery sweat from most of the skin (m), sweat from the palms and soles (α_1)	None
Heart	Increased rate and force of contraction (β_1, β_2)	Decreased rate of contraction (m)
Liver	Glucose released into blood (α_1, β_2)	None
Lungs	Dilated air passageways (β_2)	Constricted air passageways (m)
Metabolism	Increased up to 100% (α, β)	None
Sex organs	Ejaculation (α_1), erection[§]	Erection (m)
Skeletal muscles	Breakdown of glycogen to glucose (β_2)	None
Stomach and intestines		
Wall	Decreased tone (α_1, α_2, β_2)	Increased motility (m)
Sphincter	Increased tone (α_1)	Decreased tone (m)
Urinary bladder		
Wall (detrusor)	None	Contraction (m)
Neck of bladder	Contraction (α_1)	Relaxation (m)
Internal urinary sphincter	Contraction (α_1)	Relaxation (m)

*Receptor subtypes are indicated. The receptors are α_1- and α_2-adrenergic, β_1- and β_2-adrenergic, nicotinic cholinergic (n), and muscarinic cholinergic (m).

[†]Normally, blood flow increases through coronary arteries because of increased demand by cardiac tissue for oxygen (local control of blood flow is discussed in chapter 21). In experiments that isolate the coronary arteries, sympathetic nerve stimulation, acting through α-adrenergic receptors, causes vasoconstriction. The β-adrenergic receptors are relatively insensitive to sympathetic nerve stimulation but can be activated by epinephrine released from the adrenal gland and by drugs. As a result, coronary arteries vasodilate.

[‡]Contraction of the radial muscles of the iris causes the pupil to dilate. Contraction of the circular muscles causes the pupil to constrict (see chapter 15).

[§]Decreased stimulation of alpha receptors by the sympathetic division can cause vasodilation of penile blood vessels, resulting in an erection.

Some drugs that affect the ANS have important therapeutic value in treating certain diseases because they can increase or decrease activities normally controlled by the ANS. Chemicals that affect the ANS are also found in medically hazardous substances, such as tobacco and insecticides.

Direct-acting and indirect-acting drugs influence the ANS. Direct-acting drugs bind to ANS receptors to produce their effects. For example, **agonists,** or **stimulating agents,** bind to specific receptors and activate them, and **antagonists,** or **blocking agents,** bind to specific receptors and prevent them from being activated. The main topic of this discussion is direct-acting drugs. It should be noted, however, that some indirect-acting drugs produce a stimulatory effect by causing the release of neurotransmitters or by preventing the metabolic breakdown of neurotransmitters. Other indirect-acting drugs produce an inhibitory effect by preventing the biosynthesis or release of neurotransmitters.

Drugs That Bind to Nicotinic Receptors

Drugs that bind to nicotinic receptors and activate them are **nicotinic agents.** Although these agents have little therapeutic value and are mainly of interest to researchers, nicotine is medically important because of its presence in tobacco. Nicotinic agents bind to the nicotinic receptors on all postganglionic neurons within autonomic ganglia and produce stimulation. Responses to nicotine are variable and depend on the amount taken into the body. Because nicotine stimulates the postganglionic neurons of both the sympathetic and parasympathetic divisions, much of the variability of its effects results from the opposing actions of these divisions. For example, in response to the nicotine contained in a cigarette, the heart rate may either increase or decrease. Heart rate rhythm tends to become less regular as a result of the simultaneous actions on the sympathetic division, which increase the heart rate, and on the parasympathetic division, which decrease the heart rate. Blood pressure tends to increase because of the constriction of blood vessels, which are almost exclusively innervated by sympathetic neurons. In addition to its influence on the ANS, nicotine also affects the CNS; therefore, not all of its effects can be explained on the basis of action on the ANS. Nicotine is extremely toxic, and small amounts can be lethal.

Drugs that bind to and block nicotinic receptors are called **ganglionic blocking agents** because they block the effect of acetylcholine on both parasympathetic and sympathetic postganglionic neurons. The effect of these substances on the sympathetic division, however, overshadows the effect on the parasympathetic division. For example, trimethaphan camsylate (trī-meth′ă-fan kam′sil-āt), used to treat high blood pressure, blocks the sympathetic stimulation of blood vessels, causing the blood vessels to dilate, which decreases blood pressure. Ganglionic blocking agents have limited uses because they affect both sympathetic and parasympathetic ganglia. Whenever possible, more selective drugs, which affect receptors of target tissues, are now used.

Drugs That Bind to Muscarinic Receptors

Drugs that bind to and activate muscarinic receptors are **muscarinic,** or **parasympathomimetic** (par-ă-sim′pă-thō-mi-met′ik), **agents.** These drugs activate the muscarinic receptors of target tissues of the parasympathetic division and the muscarinic receptors of sweat glands, which are innervated by the sympathetic division. Muscarine causes increased sweating, increased secretion of glands in the digestive system, decreased heart rate, constriction of the pupils, and contraction of respiratory, digestive, and urinary system smooth muscles. Bethanechol (be-than′ē-kol) chloride is a parasympathomimetic agent used to stimulate the urinary bladder following surgery because the general anesthetics used for surgery can temporarily inhibit a person's ability to urinate.

Drugs that bind to and block the action of muscarinic receptors are **muscarinic,** or **parasympathetic, blocking agents.** For example, the activation of muscarinic receptors causes the constriction of air passageways. Ipratropium (i-pră-trō′pē-ŭm) is used to treat chronic obstructive pulmonary disease because it blocks muscarinic receptors, which promotes the relaxation of air passageways. Atropine (at′rō-pēn) is used to block parasympathetic reflexes associated with the surgical manipulation of organs.

Drugs That Bind to Alpha and Beta Receptors

Drugs that activate adrenergic receptors are **adrenergic,** or **sympathomimetic** (sim′pă-thō-mi-met′ik), **agents.** Drugs such as phenylephrine (fen-il-ef′rin) stimulate alpha receptors, which are numerous in the smooth muscle cells of certain blood vessels, especially in the digestive tract and the skin. These drugs increase blood pressure by causing vasoconstriction. On the other hand, albuterol (al-bū′ter-ol) is a drug that selectively activates beta receptors in bronchiolar smooth muscle. β-adrenergic-stimulating agents are sometimes used to dilate bronchioles in respiratory disorders such as asthma.

Drugs that bind to and block the action of alpha receptors are **α-adrenergic-blocking agents.** For example, prazosin (pră′zō-sin) hydrochloride is used to treat hypertension. By binding to alpha receptors in the smooth muscle of blood vessel walls, prazosin hydrochloride blocks the normal effects of norepinephrine released from sympathetic postganglionic neurons. Thus, the blood vessels relax, and blood pressure decreases.

Propranolol (prō-pran′ō-lōl) is an example of a **β-adrenergic-blocking agent.** These drugs are sometimes used to treat high blood pressure, some types of cardiac arrhythmias, and patients recovering from heart attacks. Blockage of the beta receptors within the heart prevents sudden increases in heart rate and thus decreases the probability of arrhythmic contractions.

Future Research

Present knowledge of the ANS is more complicated than the broad outline presented here. In fact, each of the major receptor types has subtype receptors. For example, α-adrenergic receptors are divided into the following subgroups: α_{1A}-, α_{1B}-, α_{2A}-, and α_{2B}-adrenergic receptors. The exact number of subtypes in humans is not yet known; however, their existence suggests the possibility of designing drugs that affect only one subtype. For example, a drug that affects the blood vessels of the heart but not other blood vessels might be developed. Such drugs could produce specific effects yet would not produce undesirable side effects because they would act only on specific target tissues.

CASE STUDY | Eye Drops

Sally is a 50-year-old woman with diabetes. Every year, she has her eyes examined by her ophthalmologist because damage to the retina can be associated with diabetes. In addition to checking her vision using an eye chart, her doctor examines the insides of her eyes using an ophthalmoscope (see figure 13.9a). To see inside Sally's eyes better, the doctor applies pupil-dilating eye drops that cause the pupils to enlarge. The diameter of a typical pupil in normal room light is approximately 3–4 mm, whereas a dilated pupil is 7–8 mm. Because Sally has complained in the past about light sensitivity as a result of the pupil-dilating eye drops, after the eye exam the doctor applies pupil-constricting eye drops to cause the pupil diameter to decrease.

P R E D I C T

a. Review the anatomy of the iris of the eye in chapter 13 (see p. 374). Do the radial muscles or circular muscles of the iris cause the pupil to dilate?

b. Which division of the ANS controls the radial muscles? The circular muscles?

c. Four types of drugs act on the receptors of target tissues of the ANS (see Clinical Relevance "Influence of Drugs on the Autonomic Nervous System," p. 415). Which of these drugs could explain dilation of Sally's pupils? (*Hint:* See table 14.3.)

d. A side effect of the pupil-dilating eye drops is blurred vision—that is, an inability to see close-up objects clearly. Based on this observation, which type of drug is in the pupil-dilating eye drops?

e. Which type of drug could be in the pupil-constricting eye drops that reversed the dilation of Sally's pupils?

f. A side effect of the pupil-constricting eye drops is blood-shot eyes. Based on this observation, which type of drug is in the pupil-constricting eye drops?

Adrenergic receptors can be stimulated in two ways: by the nervous system and by epinephrine and norepinephrine released from the adrenal gland. Sympathetic postganglionic neurons release norepinephrine, which stimulates adrenergic receptors within synapses (see figure 14.6). For example, blood vessels are stimulated to contract through the release of norepinephrine at synapses. Epinephrine and norepinephrine released from the adrenal glands and carried to effector organs by the blood can bind to adrenergic receptors located in the plasma membrane away from synapses. For example, during exercise epinephrine and norepinephrine bind to β_2 receptors and cause blood vessel dilation in skeletal muscles.

Dopamine and the Treatment of Shock

Norepinephrine is produced from a precursor molecule called dopamine. Certain sympathetic neurons release dopamine, which binds to dopamine receptors. Dopamine is structurally similar to norepinephrine and it binds to beta receptors. Dopamine hydrochloride has been used successfully to treat circulatory shock because it can bind to dopamine receptors in kidney blood vessels. The resulting vasodilation increases blood flow to the kidneys and prevents kidney damage. At the same time, dopamine can bind to beta receptors in the heart, causing stronger contractions. ▼

19. Where are adrenergic receptors located? Name the two major types of adrenergic receptors.

20. In what two ways are adrenergic receptors stimulated?

Regulation of the Autonomic Nervous System

Much of the regulation of structures by the ANS occurs through autonomic reflexes, but input from the cerebrum, the hypothalamus, and other areas of the brain allows conscious thoughts and actions, emotions, and other CNS activities to influence autonomic functions. Without the regulatory activity of the ANS, an individual has limited ability to maintain homeostasis.

Autonomic reflexes, like other reflexes, involve sensory receptors, sensory neurons, interneurons, motor neurons, and effector cells (figure 14.7) (see chapter 11). For example, **baroreceptors** (stretch receptors) in the walls of large arteries near the heart detect changes in blood pressure, and sensory neurons transmit information from the baroreceptors through the glossopharyngeal and vagus nerves to the medulla oblongata. Interneurons in the medulla oblongata integrate the information, and action potentials are produced in autonomic neurons that extend to the heart. If baroreceptors detect a change in blood pressure, autonomic reflexes change heart rate, which returns blood pressure to normal. A sudden increase in blood pressure initiates a parasympathetic reflex, which inhibits cardiac muscle cells and reduces heart rate, thus bringing blood pressure down toward its normal value. Conversely, a sudden decrease in blood pressure initiates a sympathetic reflex, which stimulates the heart to increase its rate and force of contraction, thus increasing blood pressure.

P R E D I C T

Sympathetic neurons stimulate sweat glands in the skin. Predict how they control body temperature during exercise and during exposure to cold temperatures.

Other autonomic reflexes participate in the regulation of blood pressure (see chapter 18). For example, numerous sympathetic neurons transmit a low but relatively constant frequency of action potentials that stimulate blood vessels throughout the body, keeping them partially constricted. If the vessels constrict further,

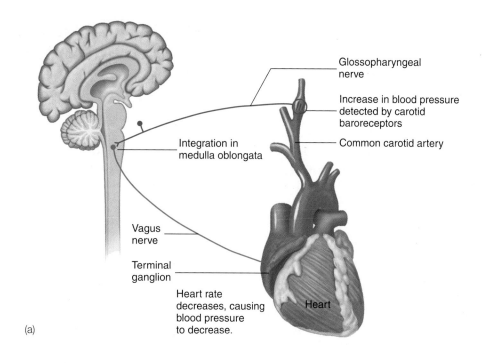

(a)

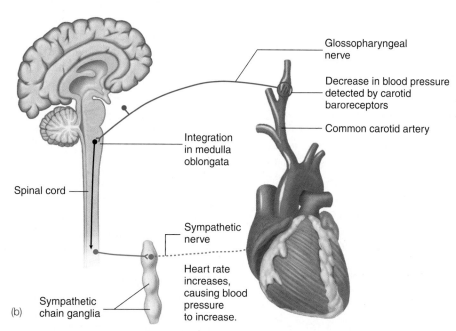

(b)

Figure 14.7 Autonomic Reflexes

Sensory input from the carotid baroreceptors is sent along the glossopharyngeal nerves to the medulla oblongata. The input is integrated in the medulla, and motor output is sent to the heart. (*a*) *Parasympathetic reflex*. Increased blood pressure results in increased stimulation of the heart by the vagus nerves, which increases inhibition of the heart and lowers heart rate. (*b*) *Sympathetic reflex*. Decreased blood pressure results in increased stimulation of the heart by sympathetic nerves, which in turn increases stimulation of the heart and increases heart rate and the force of contraction.

blood pressure increases; if they dilate, blood pressure decreases. Thus, altering the frequency of action potentials delivered to blood vessels along sympathetic neurons can either raise or lower blood pressure.

P R E D I C T

How would sympathetic reflexes that control blood vessels respond to a sudden decrease and a sudden increase in blood pressure?

The brainstem and the spinal cord contain important autonomic reflex centers responsible for maintaining homeostasis. The hypothalamus, however, is in overall control of the ANS. Almost any type of autonomic response can be evoked by stimulating a part of the hypothalamus, which in turn stimulates ANS centers in the brainstem or spinal cord. Although there is overlap, stimulation of the posterior hypothalamus produces sympathetic responses, whereas stimulation of the anterior hypothalamus produces parasympathetic responses. In addition, the hypothalamus monitors and controls body temperature (see figure 11.20*c*).

The hypothalamus has connections with the cerebrum and is an important part of the limbic system, which plays an important role in emotions. The hypothalamus integrates thoughts and emotions to produce ANS responses. Pleasant thoughts of a delicious banquet initiate increased secretion by salivary glands and by glands within the stomach and increased smooth muscle contractions within the digestive system. These responses are controlled by parasympathetic neurons. Emotions such as anger increase blood pressure by increasing heart rate and constricting blood vessels through sympathetic stimulation.

The enteric nervous system is involved with autonomic and local reflexes that regulate the activity of the digestive tract. For example, in an ANS reflex, sensory neurons detecting stretch of the digestive tract wall send action potentials to the CNS. In response, the CNS sends action potentials out the ANS, causing smooth muscle in the digestive tract wall to contract.

The neurons of the enteric nervous system also operate independently of the CNS to produce local reflexes. A **local reflex** does not involve the CNS, but it does produce an involuntary, unconscious, stereotypic response to a stimulus. For example, sensory neurons not connected to the CNS detect stretch of the digestive tract wall. These sensory neurons send action potentials through the enteric plexuses to motor neurons, causing smooth muscle contraction or relaxation. See chapter 21 for more information on local reflexes.

 ### Effects of Spinal Cord Injury on ANS Functions

Spinal cord injury can damage nerve tracts, resulting in the loss of sensation and motor control below the level of the injury. Spinal cord injury also interrupts the control of autonomic neurons by ANS centers in the brain. For the parasympathetic division, effector organs innervated through the sacral region of the spinal cord are affected, but most effector organs still have normal parasympathetic function because they are innervated by the vagus nerve. For the sympathetic division, brain control of sympathetic neurons is lost below the site of the injury. The higher the level of injury, the greater the number of body parts affected.

Immediately after spinal cord injury, spinal cord reflexes below the level of the injury are lost, including ANS reflexes (see "Spinal Shock," chapter 12). With time, the reflex centers in the spinal cord become functional again. This recovery is particularly important for reflexes involving urination and defecation. Autonomic reflexes mediated through the vagus nerves or the enteric nervous system are not affected by spinal cord injury. ▼

---

21. *Name the components of an autonomic reflex. Describe the autonomic reflex that maintains blood pressure by altering heart rate or the diameter of blood vessels.*
22. *What part of the CNS stimulates ANS reflexes and integrates thoughts and emotions to produce ANS responses?*
23. *Define local reflex. Explain how the enteric nervous system operates to produce local reflexes.*

Functional Generalizations about the Autonomic Nervous System

Generalizations can be made about the function of the ANS on effector organs, but most of the generalizations have exceptions.

Stimulatory versus Inhibitory Effects

Both divisions of the ANS produce stimulatory and inhibitory effects. For example, the parasympathetic division stimulates contraction of the urinary bladder and inhibits the heart, causing a decrease in heart rate. The sympathetic division causes vasoconstriction by stimulating smooth muscle contraction in blood vessel walls and produces dilation of lung air passageways by inhibiting smooth muscle contraction in the walls of the passageways. Thus, it is *not* true that one division of the ANS is always stimulatory and the other is always inhibitory.

Dual Innervation

Most organs that receive autonomic neurons are innervated by both the parasympathetic and the sympathetic divisions (see figure 14.5). The gastrointestinal tract, heart, urinary bladder, and reproductive tract are examples (see table 14.3). Dual innervation of organs by both divisions of the ANS is not universal, however. Sweat glands and blood vessels, for example, are innervated by sympathetic neurons almost exclusively. In addition, most structures receiving dual innervation are not regulated equally by both divisions. For example, parasympathetic innervation of the gastrointestinal tract is more extensive and exhibits a greater influence than does sympathetic innervation.

Opposite Effects

When a *single* structure is innervated by both autonomic divisions, the two divisions usually produce opposite effects on the structure. As a consequence, the ANS is capable of both increasing and decreasing the activity of the structure. In the gastrointestinal tract, for example, parasympathetic stimulation increases secretion from glands, whereas sympathetic stimulation decreases secretion. In a few instances, however, the effect of the two divisions is not clearly opposite. For example, both divisions of the ANS increase salivary secretion: The parasympathetic division initiates the production of a large volume of thin, watery saliva, and the sympathetic division causes the secretion of a small volume of viscous saliva.

Cooperative Effects

One autonomic division can coordinate the activities of *different* structures. For example, the parasympathetic division stimulates the pancreas to release digestive enzymes into the small intestine and stimulates contractions to mix the digestive enzymes with food within the small intestine. These responses enhance the digestion and absorption of the food.

Both divisions of the ANS can act together to coordinate the activity of *different* structures. In the male, the parasympathetic division initiates erection of the penis, and the sympathetic division

Biofeedback takes advantage of electronic instruments or other techniques to monitor and change subconscious activities, many of which are regulated by the ANS. Skin temperature, heart rate, and brain waves are monitored electronically. By watching the monitor and using biofeedback techniques, a person can learn how to reduce heart rate and blood pressure consciously and regulate blood flow in the limbs. For example, people claim that they can prevent the onset of migraine headaches or reduce their intensity by learning to dilate blood vessels in the skin of their forearms and hands. Increased blood vessel dilation increases skin temperature, which is correlated with a decrease in the severity of the migraine. Some people use biofeedback methods to relax by learning to reduce their heart rate or change the pattern of their brain waves. The severity of some stomach ulcers, high blood pressure, anxiety, and depression may be reduced by using biofeedback techniques.

Meditation is another technique that influences autonomic functions. Although numerous claims about the value of meditation include improving one's spiritual well-being, consciousness, and holistic view of the universe, it has been established that meditation does influence autonomic functions. Meditation techniques are useful in some people in reducing heart rate, blood pressure, severity of ulcers, and other symptoms that are frequently associated with stress.

The fight-or-flight response occurs when an individual is subjected to stress, such as a threatening, frightening, embarrassing, or exciting situation. Whether a person confronts or avoids a stressful situation, the nervous system and the endocrine system are involved either consciously or unconsciously. The autonomic part of the fight-or-flight response results in a general increase in sympathetic activity, including heart rate, blood pressure, sweating, and other responses, that prepare the individual for physical activity. The fight-or-flight response is adaptive because it also enables the individual to resist or move away from a threatening situation.

---

stimulates the release of secretions from male reproductive glands and helps initiate ejaculation in the male reproductive tract.

General versus Localized Effects

The sympathetic division has a more general effect than the parasympathetic division because activation of the sympathetic division often causes secretion of both epinephrine and norepinephrine from the adrenal medulla. These hormones circulate in the blood and stimulate effector organs throughout the body. Because circulating epinephrine and norepinephrine can persist for a few minutes before being broken down, they can also produce an effect for a longer time than the direct stimulation of effector organs by postganglionic sympathetic axons.

The sympathetic division diverges more than the parasympathetic division. Each sympathetic preganglionic neuron synapses with many postganglionic neurons, whereas each parasympathetic preganglionic neuron synapses with about two postganglionic neurons. Consequently, stimulation of sympathetic preganglionic neurons can result in greater stimulation of an effector organ.

Sympathetic stimulation often activates many different kinds of effector organs at the same time as a result of CNS stimulation or epinephrine and norepinephrine release from the adrenal medulla. It is possible, however, for the CNS to selectively activate effector organs. For example, vasoconstriction of cutaneous blood vessels in a cold hand is not always associated with an increased heart rate or other responses controlled by the sympathetic division.

Functions during Activity versus at Rest

In cases in which both sympathetic and parasympathetic neurons innervate a single organ, the sympathetic division has a major influence under conditions of physical activity or stress, whereas the parasympathetic division tends to have a greater influence under resting conditions. The sympathetic division does play a major role during resting conditions, however, by maintaining blood pressure and body temperature.

In general, the sympathetic division decreases the activity of organs not essential for the maintenance of physical activity and shunts blood and nutrients to structures that are active during physical exercise. This is sometimes referred to as the **fight-or-flight response** (see Clinical Relevance "Biofeedback, Meditation, and the Fight-or-Flight Response," on this page). Typical responses produced by the sympathetic division during exercise include

1. Increased heart rate and force of contraction increase blood pressure and the movement of blood.
2. As skeletal or cardiac muscle contracts, oxygen and nutrients are used and waste products are produced. During exercise, a decrease in oxygen and nutrients and an accumulation of waste products are stimuli that cause the vasodilation of muscle blood vessels (see "Local Control," chapter 18). Vasodilation is beneficial because it increases blood flow, bringing needed oxygen and nutrients and removing waste products. Too much vasodilation, however, can cause a decrease in blood pressure that decreases blood flow. Increased stimulation of skeletal muscle blood vessels by sympathetic nerves during exercise causes vasoconstriction, which prevents a drop in blood pressure (see chapter 18).
3. Increased heart rate and force of contraction potentially increase blood flow through tissues. Vasoconstriction of blood vessels in tissues not involved in exercise, such as abdominopelvic organs, reduces blood flow through them, thus making more blood available for the exercising tissues.
4. Dilation of air passageways increases air flow into and out of the lungs.

5. The availability of energy sources increases. Skeletal muscle cells and liver cells (hepatocytes) are stimulated to break down glycogen to glucose. Skeletal muscle cells use the glucose and liver cells release it into the blood for use by other tissues. Fat cells (adipocytes) break down triglycerides and release fatty acids into the blood, which are used as an energy source by skeletal and cardiac muscle.

6. As exercising muscles generate heat, body temperature increases. Vasodilation of blood vessels in the skin brings warm blood close to the surface, where heat is lost to the environment. Sweat gland activity increases, resulting in increased sweat production, and evaporation of the sweat removes additional heat.

7. The activities of organs not essential for exercise decrease. For example, the process of digesting food slows as digestive glands decrease their secretions and the contractions of smooth muscle that mix and move food through the gastrointestinal tract decrease.

Increased activity of the parasympathetic division is generally consistent with resting conditions. The parasympathetic division regulates digestion by increasing the secretions of glands, promoting the mixing of food with digestive enzymes and bile, and moving materials through the digestive tract. Defecation and urination are also controlled by the parasympathetic division.

Increased parasympathetic stimulation lowers heart rate, which lowers blood pressure, and constricts air passageways, which decreases air movement through them. ▼

24. *What kinds of effects, excitatory or inhibitory, are produced by the sympathetic and parasympathetic divisions?*
25. *Give two exceptions to the generalization that organs are innervated by both divisions of the ANS.*
26. *When a single organ is innervated by both ANS divisions, do they usually produce opposite effects?*
27. *Explain how the ANS coordinates the activities of different organs.*
28. *Which ANS division produces the most general effects? How does this happen?*
29. *Use the fight-or-flight response to describe the responses produced by the sympathetic division.*

P R E D I C T **7**
Bethanechol (be-than′ĕ-kol) chloride is a drug that binds to muscarinic receptors. Explain why this drug can be used to promote emptying of the urinary bladder. Which of the following side effects would you predict: abdominal cramps, asthmatic attack, decreased tear production, decreased salivation, dilation of the pupils, or sweating?

S U M M A R Y

Contrasting the Somatic and Autonomic Nervous Systems (p. 406)

1. The cell bodies of somatic neurons are located in the CNS, and their axons extend to skeletal muscles, where they have an excitatory effect that usually is controlled consciously.
2. The cell bodies of the preganglionic neurons of the ANS are located in the CNS and extend to ganglia, where they synapse with postganglionic neurons. The postganglionic axons extend to smooth muscle, cardiac muscle, or glands and have an excitatory or inhibitory effect, which usually is controlled unconsciously.

Anatomy of the Autonomic Nervous System (p. 406)
Sympathetic Division

1. Preganglionic cell bodies are in the lateral horns of the spinal cord gray matter from T1 to L2.
2. Preganglionic axons pass through the ventral roots to the white rami communicantes to the sympathetic chain ganglia. From there, four courses are possible:
   - Preganglionic axons synapse (at the same or a different level) with postganglionic neurons, which exit the ganglia through the gray rami communicantes and enter spinal nerves.
   - Preganglionic axons synapse (at the same or a different level) with postganglionic neurons, which exit the ganglia through sympathetic nerves.
   - Preganglionic axons pass through the chain ganglia without synapsing to form splanchnic nerves. Preganglionic axons then synapse with postganglionic neurons in collateral ganglia.

   - Preganglionic axons synapse with the cells of the adrenal medulla.

Parasympathetic Division

1. Preganglionic cell bodies are in nuclei in the brainstem or the lateral parts of the spinal cord gray matter from S2 to S4.
   - Preganglionic axons from the brain pass to ganglia through cranial nerves.
   - Preganglionic axons from the sacral region pass through the pelvic splanchnic nerves to the ganglia.
2. Preganglionic axons pass to terminal ganglia within the wall of or near the organ that is innervated.

Enteric Nervous System

1. The enteric nerve plexus is within the wall of the digestive tract.
2. The enteric plexus consists of sensory neurons, ANS motor neurons, and enteric neurons.

Distribution of Autonomic Nerve Fibers

1. Sympathetic, parasympathetic, and sensory neurons intermingle in autonomic nerve plexuses.
2. Sympathetic axons reach organs through spinal, sympathetic, and splanchnic nerves.
3. Parasympathetic axons reach organs through cranial and pelvic splanchnic nerves.
4. Sensory neurons run alongside sympathetic and parasympathetic neurons within nerves and nerve plexuses.

Physiology of the Autonomic Nervous System (p. 411)
Neurotransmitters

1. Acetylcholine is released by cholinergic neurons (all preganglionic neurons, all parasympathetic postganglionic neurons, and some sympathetic postganglionic neurons).
2. Norepinephrine is released by adrenergic neurons (most sympathetic postganglionic neurons).

Receptors

1. Acetylcholine binds to nicotinic receptors (found in all postganglionic neurons) and muscarinic receptors (found in all parasympathetic and some sympathetic effector organs).
2. Norepinephrine and epinephrine bind to alpha and beta receptors (found in most sympathetic effector organs).
3. Activation of nicotinic receptors is excitatory, whereas activation of muscarinic, alpha, and beta receptors is either excitatory or inhibitory.

Regulation of the Autonomic Nervous System (p. 416)

1. Autonomic reflexes control most of the activity of visceral organs, glands, and blood vessels.

2. Autonomic reflex activity can be influenced by the hypothalamus and higher brain centers.
3. The sympathetic and parasympathetic divisions can influence the activities of the enteric nervous system through autonomic reflexes. The enteric nervous system can function independently of the CNS through local reflexes.

Functional Generalizations About the Autonomic Nervous System (p. 418)

1. Both divisions of the ANS produce stimulatory and inhibitory effects.
2. Most organs are innervated by both divisions. Usually, each division produces an opposite effect on a given organ.
3. Either division alone or both working together can coordinate the activities of different structures.
4. The sympathetic division produces more generalized effects than the parasympathetic division.
5. Sympathetic activity generally prepares the body for physical activity, whereas parasympathetic activity is more important for resting conditions.

R E V I E W   A N D   C O M P R E H E N S I O N

1. Given these phrases:
   1. neuron cell bodies in the nuclei of cranial nerves
   2. neuron cell bodies in the lateral gray matter of the spinal cord (S2–S4)
   3. two synapses between the CNS and effector organs
   4. regulates smooth muscle

   Which of the phrases are true for the autonomic nervous system?
   a. 1,3   c. 1,2,3   e. 1,2,3,4
   b. 2,4   d. 2,3,4

2. Given these structures:
   1. gray ramus communicans
   2. white ramus communicans
   3. sympathetic chain ganglion

   Choose the arrangement that lists the structures in the order an action potential passes through them from a spinal nerve to an effector organ.
   a. 1,2,3   c. 2,1,3   e. 3,2,1
   b. 1,3,2   d. 2,3,1

3. Given these structures:
   1. collateral ganglion
   2. sympathetic chain ganglion
   3. white ramus communicans
   4. splanchnic nerve

   Choose the arrangement that lists the structures in the order an action potential travels through them on the way from a spinal nerve to an effector organ.
   a. 1,3,2,4   c. 3,1,4,2   e. 4,3,1,2
   b. 1,4,2,3   d. 3,2,4,1

4. The white ramus communicans contains
   a. preganglionic sympathetic fibers.
   b. postganglionic sympathetic fibers.
   c. preganglionic parasympathetic fibers.
   d. postganglionic parasympathetic fibers.

5. The cell bodies of the postganglionic neurons of the sympathetic division are located in the
   a. sympathetic chain ganglia.   d. dorsal root ganglia.
   b. collateral ganglia.   e. both a and b.
   c. terminal ganglia.

6. Splanchnic nerves
   a. are part of the parasympathetic division.
   b. have preganglionic neurons that synapse in the collateral ganglia.
   c. exit from the cervical region of the spinal cord.
   d. travel from the spinal cord to the sympathetic chain ganglia.
   e. all of the above.

7. Which of the following statements regarding the adrenal gland is true?
   a. The parasympathetic division stimulates the adrenal gland to release acetylcholine.
   b. The parasympathetic division stimulates the adrenal gland to release epinephrine.
   c. The sympathetic division stimulates the adrenal gland to release acetylcholine.
   d. The sympathetic division stimulates the adrenal gland to release epinephrine.

8. The parasympathetic division
   a. is also called the craniosacral division.
   b. has preganglionic axons in cranial nerves.
   c. has preganglionic axons in pelvic splanchnic nerves.
   d. has ganglia near or in the wall of effector organs.
   e. all of the above.

9. Which of these is *not* a part of the enteric nervous system?
   a. ANS motor neurons
   b. neurons located only in the digestive tract
   c. sensory neurons
   d. somatic neurons

10. Sympathetic axons reach organs through all of the following *except*
   a. abdominopelvic nerve plexuses.
   b. head and neck nerve plexuses.
   c. thoracic nerve plexuses.
   d. pelvic splanchnic nerves.
   e. spinal nerves.

11. Which of these cranial nerves does *not* contain parasympathetic fibers?
   a. oculomotor (III)   d. trigeminal (V)
   b. facial (VII)   e. vagus (X)
   c. glossopharyngeal (IX)

12. Which of the following statements concerning the preganglionic neurons of the ANS is true?
    a. All parasympathetic preganglionic neurons secrete acetylcholine.
    b. Only parasympathetic preganglionic neurons secrete acetylcholine.
    c. All sympathetic preganglionic neurons secrete norepinephrine.
    d. Only sympathetic preganglionic neurons secrete norepinephrine.

13. A cholinergic neuron
    a. secretes acetylcholine.
    b. has receptors for acetylcholine.
    c. secretes norepinephrine.
    d. has receptors for norepinephrine.
    e. secretes both acetylcholine and norepinephrine.

14. When acetylcholine binds to nicotinic receptors,
    a. the cell's response is mediated by G proteins.
    b. the response can be excitatory or inhibitory.
    c. Na⁺ channels open.
    d. the binding occurs at the effector organ.
    e. all of the above.

15. Nicotinic receptors are located in
    a. postganglionic neurons of the parasympathetic division.
    b. postganglionic neurons of the sympathetic division.
    c. membranes of skeletal muscle cells.
    d. both a and b.
    e. all of the above.

16. The activation of α and β receptors
    a. can produce an excitatory or inhibitory response.
    b. can be caused by the sympathetic division.
    c. can be caused by epinephrine released from the adrenal gland.
    d. can be caused by norepinephrine.
    e. all of the above.

17. The sympathetic division
    a. is always stimulatory.
    b. is always inhibitory.
    c. is usually under conscious control.
    d. generally opposes the actions of the parasympathetic division.
    e. both a and c.

18. A sudden increase in blood pressure
    a. initiates a sympathetic reflex that decreases heart rate.
    b. initiates a local reflex that decreases heart rate.
    c. initiates a parasympathetic reflex that decreases heart rate.
    d. both a and b.
    e. both b and c.

19. Which of these structures is innervated almost exclusively by the sympathetic division?
    a. gastrointestinal tract
    b. heart
    c. urinary bladder
    d. reproductive tract
    e. blood vessels

20. Which of these is expected if the sympathetic division is activated?
    a. Secretion of watery saliva increases.
    b. Tear production increases.
    c. Air passageways dilate.
    d. Glucose release from the liver decreases.
    e. All of the above are true.

Answers in Appendix E

CRITICAL THINKING

1. When a person is startled or sees a "pleasurable" object, the pupils of the eyes may dilate. What division of the ANS is involved in this reaction? Describe the nerve pathway involved.

2. Reduced secretion from salivary and lacrimal glands could indicate damage to what nerve?

3. In a patient with Raynaud disease, blood vessels in the skin of the hand may become chronically constricted, thereby reducing blood flow and producing gangrene. These vessels are supplied by nerves that originate at levels T2 and T3 of the spinal cord and eventually exit through the first thoracic and inferior cervical sympathetic ganglia. Surgical treatment for Raynaud disease severs this nerve supply. At which of the following locations would you recommend that the cut be made: white rami of T2–T3, gray rami of T2–T3, or spinal nerves T2–T3? Explain.

4. Patients with diabetes mellitus can develop autonomic neuropathy, which is damage to parts of the autonomic nerves. Given the following parts of the ANS—vagus nerve, oculomotor nerve, splanchnic nerve, pelvic splanchnic nerve, outflow of gray ramus—match the part with the symptom it would produce if the part were damaged:
    a. impotence
    b. subnormal sweat production
    c. stomach muscles relaxed and delayed emptying of the stomach
    d. diminished pupil reaction (constriction) to light
    e. bladder paralysis with urinary retention

5. Explain why methacholine, a drug that acts like acetylcholine, is effective for treating tachycardia (heart rate faster than normal). Which of the following side effects would you predict: increased salivation, dilation of the pupils, sweating, or difficulty in taking a deep breath?

6. A patient has been exposed to the organophosphate pesticide malathion, which inactivates acetylcholinesterase. Which of the following symptoms would you predict: blurring of vision, excess tear formation, frequent or involuntary urination, pallor (pale skin), muscle twitching, or cramps? Would atropine, a muscarinic blocking agent, be an effective drug to treat the symptoms? Explain.

7. Epinephrine is routinely mixed with local anesthetic solutions. Why?

8. A drug blocks the effect of the sympathetic division on the heart. Careful investigation reveals that, after administration of the drug, normal action potentials are produced in the sympathetic preganglionic and postganglionic neurons. Also, injection of norepinephrine produces a normal response in the heart. Explain, in as many ways as you can, the mode of action of the unknown drug.

9. Make a list of the responses controlled by the ANS in (a) a person who is extremely angry and (b) a person who has just finished eating and is relaxing.

Answers in Appendix F

For additional study tools, go to: www.aris.mhhe.com.
Click on your course topic and then this textbook's author name/title. Register once for a semester's worth of interactive learning activities to help improve your grade!

CHAPTER 15 Endocrine System

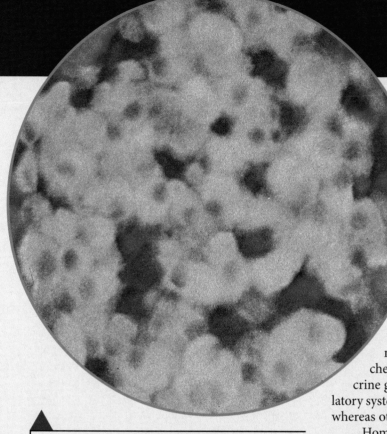

Light micrograph of a pancreatic islet showing insulin-secreting beta cells (*green*) and the glucagon-secreting cells (*red*).

Tools for Success

A virtual cadaver dissection experience

Audio (MP3) and/or video (MP4) content, available at www.mhhe.aris.com

Introduction

The nervous and endocrine systems are the two major regulatory systems of the body. Together, they regulate and coordinate the activities of essentially all other body structures. The nervous system functions something like telephone messages sent along many telephone wires to their specific destinations. It transmits information in the form of action potentials along the axons of nerve cells. Chemical signals in the form of neurotransmitters are released at synapses between neurons and the cells they control. The endocrine system is more like satellite radio or television signals broadcast widely so that every radio or television set, with its receiver adjusted properly, can receive the signals. It sends information to the cells it controls in the form of chemical signals, called hormones, which are released from endocrine glands. Hormones are carried to all parts of the body by the circulatory system. Cells that are able to recognize the hormones respond to them, whereas other cells do not.

Homeostasis depends on the precise regulation of the organs and organ systems of the body. Together, the nervous and endocrine systems regulate and coordinate the activity of nearly all other body structures. When either the nervous or the endocrine system fails to function properly, conditions can rapidly deviate from homeostasis. Disorders of the endocrine system can result in diseases such as insulin-dependent diabetes and Addison disease. Early in the 1900s, people who developed those diseases died. No effective treatments were available for those and other diseases of the endocrine system, such as diabetes insipidus, Cushing syndrome, and many reproductive abnormalities. Advances have been made in understanding the endocrine system, so the outlook for people with these and other endocrine diseases has improved.

▶This chapter provides an **overview of the endocrine system (p. 424)**, listing its functions, comparing the nervous and endocrine systems, emphasizing the role of the endocrine system in the maintenance of homeostasis, and illustrating the means by which the endocrine system regulates the functions of cells. Next, the chapter profiles the **pituitary gland and hypothalamus (p. 433), thyroid gland (p. 439), parathyroid glands (p. 444), adrenal glands (p. 445), pancreas (p. 450), hormonal regulation of nutrients (p. 453), testes and ovaries (p. 454), pineal body (p. 454), other endocrine organs (p. 455),** and **hormonelike substances (p. 457).** The chapter concludes with a look at the **effects of aging on the endocrine system (p. 457).**

Overview of the Endocrine System

Functions of the Endocrine System

The main regulatory functions of the endocrine system are the following:

1. *Metabolism and tissue maturation.* The endocrine system regulates the rate of metabolism and influences the maturation of tissues, such as those of the nervous system.
2. *Ion regulation.* The endocrine system helps regulate blood pH, as well as Na^+, K^+, and Ca^{2+} concentrations in the blood.
3. *Water balance.* The endocrine system regulates water balance by controlling the solute concentration of the blood.
4. *Immune system regulation.* The endocrine system helps control the production of immune cells.
5. *Heart rate and blood pressure regulation.* The endocrine system helps regulate the heart rate and blood pressure and helps prepare the body for physical activity.
6. *Control of blood glucose and other nutrients.* The endocrine system regulates blood glucose levels and other nutrient levels in the blood.
7. *Control of reproductive functions.* The endocrine system controls the development and functions of the reproductive systems in males and females.
8. *Uterine contractions and milk release.* The endocrine system regulates uterine contractions during delivery and stimulates milk release from the breasts in lactating females. ▼

> **1. List eight regulatory functions of the endocrine system.**

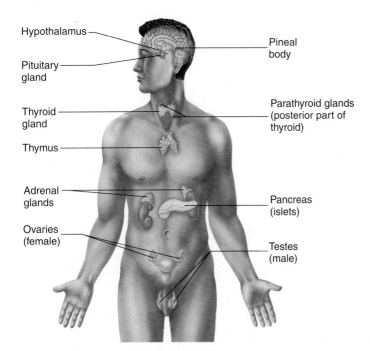

Figure 15.1   **Major Endocrine Glands**

Hypothalamus

Pituitary gland

Thyroid gland

Thymus

Adrenal glands

Ovaries (female)

Pineal body

Parathyroid glands (posterior part of thyroid)

Pancreas (islets)

Testes (male)

General Characteristics of the Endocrine System

The **endocrine system** is composed of **endocrine glands,** which are ductless glands secreting chemical signals into the circulatory system (figure 15.1). In contrast, exocrine glands have ducts that carry their secretions to surfaces (see chapter 4). The term *endocrine* (en′dō-krin) is derived from the Greek words *endo,* meaning within, and *krinō,* to separate. The term implies that cells of endocrine glands produce chemical signals within the glands that influence tissues separated from the glands by some distance.

The chemical signals secreted by endocrine glands are called **hormones** (hōr′mōnz), a term derived from the Greek word *hormon,* meaning to set into motion. Thus, hormones stimulate responses from cells. Hormones can be chemically categorized as belonging to the "protein group" or the "lipid group." The protein group includes proteins, glycoproteins, short sequences of amino acids called polypeptides, and derivatives of amino acids. The lipid group includes steroids and fatty acid derivatives.

 Lipid- and Water-Soluble Hormones in Medicine

Specific hormones are given as treatments for certain illnesses. Hormones that are soluble in lipids, such as steroids, can be taken orally because they can diffuse across the wall of the stomach and intestine into the circulatory system. Examples include the synthetic estrogen and progesterone-like hormones in birth control pills and steroids that reduce the severity of inflammation, such as prednisone (pred′ni-sōn). In contrast to lipid-soluble hormones, protein hormones cannot diffuse across the wall of the intestine because they are not lipid-soluble. Furthermore, protein hormones are not transported across the wall of the intestine because they are broken down to individual amino acids by the digestive system. The normal structure of a protein hormone is therefore destroyed, and its physiological activity is lost. Consequently, protein hormones must be injected rather than taken orally. The most commonly administered protein hormone is insulin, which is prescribed for the treatment of diabetes mellitus.

Traditionally, a hormone is defined as a chemical signal, or **ligand,** that (1) is produced in minute amounts by a collection of cells; (2) is secreted into the interstitial fluid; (3) enters the circulatory system, where it is transported some distance; and (4) acts on specific tissues, called **target tissues,** at another site in the body to influence the activity of those tissues in a specific fashion. All hormones exhibit most components of this traditional definition, but some components do not apply to every hormone.

Both the endocrine system and the nervous system regulate the activities of structures in the body, but they do so in different ways. The hormones secreted by most endocrine glands can be described as **amplitude-modulated signals,** which consist mainly of increases or decreases in the concentration of hormones in the body fluids (figure 15.2a). The responses to hormones either

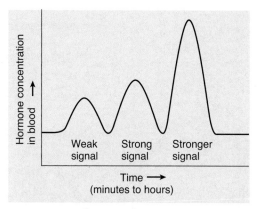

(a) **Amplitude-modulated system.** The concentration of the hormone determines the strength of the signal and the magnitude of the response. For most hormones, a small concentration of a hormone is a weak signal and produces a small response, whereas a larger concentration is a stronger signal and results in a greater response.

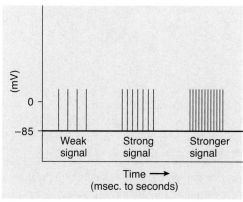

(b) **Frequency-modulated system.** The strength of the signal depends on the frequency, not the size, of the action potentials. All action potentials are the same size in a given tissue. A low frequency of action potentials is a weak stimulus, and a higher frequency is a stronger stimulus.

Figure 15.2 Regulatory Systems

increase or decrease as a function of the hormone concentration. A small amount of a hormone is a weak stimulus, producing a small response, whereas a large amount of the hormone is a stronger stimulus, producing a stronger response. For example, antidiuretic hormone released from the posterior pituitary gland acts on the kidney to decrease the volume of urine produced. The amount of urine produced is a function of the amount of hormone released.

The all-or-none action potentials carried along axons can be described as **frequency-modulated signals** (figure 15.2b), which vary in frequency but not in amplitude. A low frequency of action potentials is a weak stimulus, whereas a high frequency of action potentials is a stronger stimulus. For example, action potentials carried by axons to a muscle cause the muscle to contract. A low frequency of action potentials produces a weaker contraction than does a high frequency (see "Multiple-Wave Summation," chapter 10).

The responses of the endocrine system are usually slower and of longer duration than those of the nervous system. Hormones can persist for seconds, minutes, and even many hours. Action potentials occur in milliseconds and are rapidly initiated and stopped. Some endocrine responses, however, are more rapid than some neural responses, and some endocrine responses have a shorter duration than some neural responses. In addition, a few hormones act as both amplitude- and frequency-modulated signals, in which the concentrations of the hormones and the frequencies at which the increases in hormone concentrations occur are important.

At one time, the endocrine system was believed to be relatively independent and different from the nervous system; however, an intimate relationship between these systems is now recognized, and the two systems cannot be completely separated either anatomically or functionally. Some neurons secrete hormones and some neurons directly innervate endocrine glands and influence their secretory activity. Conversely, some hormones secreted by endocrine glands affect the nervous system and markedly influence its activity. ▼

2. *Define endocrine system, endocrine gland, and hormone.*
3. *Contrast the endocrine system and the nervous system for the following: amplitude versus frequency modulation, speed and duration of target cell response.*
4. *Explain why, despite their differences, the nervous and endocrine systems cannot be completely separated.*

Control of Secretion Rate

The secretion of hormones is usually controlled by negative-feedback mechanisms (see chapter 1). Negative-feedback mechanisms keep the body functioning within a narrow range of values consistent with life. For example, insulin is a hormone that regulates the concentration of blood glucose, or blood sugar. When blood glucose levels increase after a meal, insulin is secreted. Insulin acts on several target tissues and causes them to take up glucose, causing blood glucose levels to decline. As blood glucose levels begin to fall, however, the rate at which insulin is secreted falls also. As insulin levels fall, the rate at which glucose is taken up by the tissues decreases, keeping the blood glucose levels from declining too much. This negative-feedback mechanism counteracts increases and decreases in blood glucose levels to maintain homeostasis.

Hormone secretion is regulated in three ways. The secretion of some hormones is regulated by one of these methods, whereas the secretion of other hormones can be regulated by two or even all of these methods:

1. *Blood levels of chemicals.* The secretion of some hormones is directly controlled by the blood levels of certain chemicals. For example, insulin secretion is controlled by blood glucose levels. When blood glucose levels increase, insulin secretion increases, and vice versa.
2. *Nervous system.* The secretion of some hormones is controlled by the nervous system. An example is epinephrine, which is released from the adrenal medulla as a result of nervous system stimulation.

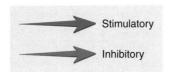

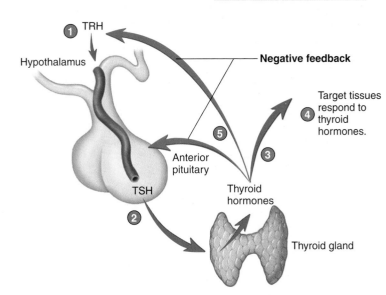

1. Thyrotropin-releasing hormone (TRH) is released from neurons in the hypothalamus and travels in the blood to the anterior pituitary gland.

2. TRH stimulates the release of thyroid-stimulating hormone (TSH) from the anterior pituitary gland. TSH travels in the blood to the thyroid gland.

3. TSH stimulates the secretion of thyroid hormones from the thyroid gland into the blood.

4. Thyroid hormones act on tissues to produce responses.

5. Thyroid hormones also have a negative-feedback effect on the hypothalamus and the anterior pituitary to inhibit both TRH secretion and TSH secretion. The negative feedback helps keep blood thyroid hormone levels within a narrow range.

Process Figure 15.3   Hormonal Regulation of Hormone Secretion

Hormones can stimulate or inhibit the secretion of other hormones.

3. *Hormones.* The secretion of some hormones is controlled by other hormones. Figure 15.3 illustrates how **thyrotropin-releasing hormone (TRH)** from the hypothalamus of the brain stimulates the secretion of **thyroid-stimulating hormone (TSH)** from the anterior pituitary gland, which in turn stimulates the secretion of **thyroid hormones** from the thyroid gland. It is important to maintain normal hormone levels because the response of tissues to hormones is amplitude-modulated. Too much or too little hormone results in too much or too little response to maintain homeostasis. The concentration of many hormones, such as thyroid hormones, is maintained within a normal range by a negative-feedback mechanism (see figure 15.3). As TRH levels increase, TSH levels increase, which causes thyroid hormone levels to increase. Therefore, the response of tissues to thyroid hormones increases. As the concentration of thyroid hormones increases, however, they inhibit the secretion of TRH and TSH. A decrease in TRH also causes a decrease in TSH. As TSH levels decrease, thyroid hormone levels decrease, and the response of tissues to thyroid hormones decreases.

P R E D I C T

Suppose that a person has a pituitary gland tumor that causes an overproduction of TSH. Would the blood levels of TRH and thyroid hormones be normal, higher than normal, or lower than normal? Explain.

Some hormones are in the circulatory system at relatively constant levels, some change suddenly in response to certain stimuli, and others change in relatively constant cycles. For example, thyroid hormones in the blood vary within a small range of concentrations that remain relatively constant over long periods of

time. Epinephrine is released in large amounts in response to stress or physical exercise; thus, its concentration can change suddenly. Reproductive hormones increase and decrease in a cyclic fashion in women during their reproductive years. ▼

5. *Describe and give examples of the three major patterns by which hormone secretion is regulated.*

6. *Describe three patterns of hormone secretion and give an example of each.*

Transport and Excretion

Hormones in blood plasma are transported either as unbound hormones or hormones bound to plasma proteins called **binding proteins.** The hormone molecules bound to binding proteins act as a reservoir for the hormone. If the amount of unbound hormone begins to decrease, some of the bound hormones are released from the binding proteins so that the concentration of unbound hormones decreases less than it otherwise would have. Consequently, hormones that bind to binding proteins can remain in the blood at a relatively constant level for longer periods of time than unbound hormones.

The concentration of hormones in the blood decreases for two major reasons: (1) They leave the blood to reach target tissues or (2) they are excreted. Hormones are removed from the blood primarily by excretion by the kidneys into urine and by the liver into bile.

Different hormones remain in the blood for different lengths of time. Water-soluble hormones, such as proteins, glycoproteins, epinephrine, and norepinephrine, remain for seconds to minutes because they are degraded rapidly by enzymes. Hormones that are rapidly removed from the blood generally regulate activities that have a rapid onset and a short duration. Hormones that are lipid-soluble, such as the steroid hormones and thyroid hormones secreted by the thyroid gland, commonly circulate in the blood in combination with binding proteins. The rate at which hormones are degraded, or are eliminated from the circulation, is greatly reduced when the hormones bind to binding proteins. Consequently, hormone levels remain relatively constant through time, having a prolonged, consistent regulatory effect. ▼

7. *What are binding proteins? How do they help maintain the level of hormones in the blood?*
8. *What are the two major reasons hormone blood concentrations decrease?*
9. *What kinds of hormones are removed from the blood rapidly and which kinds persist?*

P R E D I C T

What effect would a decrease in the concentration of a hormone's binding protein have on the concentration of the hormone in the blood?

Interaction of Hormones with Their Target Tissues

Hormones bind to proteins or glycoproteins called **receptors.** The portion of each protein or glycoprotein molecule where a hormone binds is called a **receptor site,** or **binding site.** The shape and chemical characteristics of each receptor site allow only a specific type of chemical signal to bind to it (figure 15.4). The tendency for each type of chemical signal to bind to a specific type of receptor, and not to others, is called **specificity.** Insulin therefore binds to insulin receptors but not to receptors for growth hormone. Some hormones, however, can bind to a number of different receptors that are closely related. For example, epinephrine can bind to more than one type of epinephrine receptor. Hormone receptors have a high affinity for the hormones that bind to them, so only a small concentration of a given hormone results in a significant number of receptors with hormones bound to them.

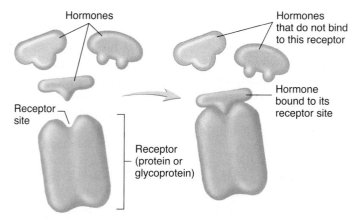

Figure 15.4   **Receptors and Specificity of Receptor Sites**

Hormones bind to receptor molecules. The shape and chemical characteristics of each receptor site allow certain hormones that have a compatible shape and compatible chemical characteristics to bind to it, not others. This relationship is called specificity.

Hormones are secreted and distributed throughout the body by the circulatory system, but the presence or absence of specific receptor molecules in cells determines which cells will or will not respond to each hormone. For example, there are receptors for thyroid-stimulating hormone (TSH) in cells of the thyroid gland, but there are no such receptors in most other cells of the body. Consequently, cells of the thyroid gland produce a response when exposed to TSH, but cells without receptor molecules do not respond to it.

Drugs with structures similar to specific hormones may compete with those hormones for their receptors (see chapter 3). A drug that binds to a hormone receptor and activates it is called an **agonist** for that hormone. A drug that binds to a hormone receptor and inhibits its action is called an **antagonist** for that hormone. For example, drugs exist that compete with epinephrine for its receptor. Some of these drugs, called epinephrine agonists, activate epinephrine receptors and others, called epinephrine antagonists, inhibit them. ▼

10. *What characteristics of a hormone receptor make it specific for one type of hormone?*
11. *Why do only certain cells respond to a given hormone?*
12. *What effects do drug agonists and antagonists have on hormone receptors?*

Classes of Receptors

The two major categories of hormone receptors are membrane-bound receptors and intracellular receptors. **Membrane-bound receptors** are receptors that extend across the plasma membrane and have their receptor sites exposed to the outer surface of the plasma membrane (figure 15.5*a*). When a hormone binds to a receptor on the outside of the plasma membrane, the receptor initiates a response inside the cell. Hormones that bind to membrane-bound receptors are large molecules and water-soluble molecules that cannot pass through the plasma membrane. Hormones that bind to membrane-bound receptors are proteins, glycoproteins, polypeptides, and some smaller molecules, such as epinephrine and norepinephrine.

Intracellular receptors are receptors in the cytoplasm or in the nucleus of the cell (figure 15.5*b*). Hormones that diffuse through the plasma membrane bind to intracellular receptors and the hormone-receptor complex then initiates the cell's response. These hormones are lipid-soluble and relatively small. Thyroid hormones and steroid hormones, such as testosterone, estrogen, progesterone, aldosterone, and cortisol, are examples.

The combination of a hormone with its receptor can stimulate a response in several ways (figure 15.6). Hormones binding to membrane-bound receptors can activate G proteins or intracelluar enzymes. The activated G protein or enzyme then causes changes leading to the cell's response. Hormones combining with intracellular receptors activate genes, promoting the synthesis of new proteins or enzymes responsible for the cell's response. ▼

13. *Define membrane-bound receptor and intracellular receptor. What types of hormones bind to them?*
14. *What are the major ways in which a hormone binding to a receptor produces a response in a cell?*

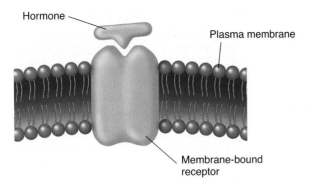

(a) Hormones that are large and water-soluble interact with membrane-bound receptors that extend across the plasma membrane and are exposed on the outer surface of the plasma membrane. The portion of the receptor inside the cell produces the response.

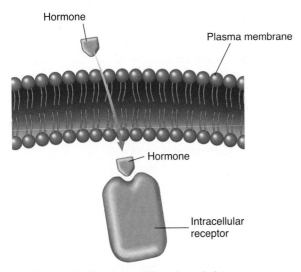

(b) Small lipid-soluble hormones diffuse through the plasma membrane and combine with intracellular receptors. The combination of hormones and intracellular receptors produces a response.

Figure 15.5   Membrane-Bound and Intracellular Receptors

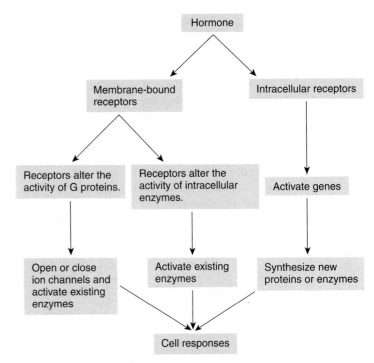

Figure 15.6   Overview of Responses to Hormones Binding to Their Receptors

Membrane-Bound Receptors That Activate G Proteins

Many membrane-bound receptors produce responses through the action of G proteins. **G proteins** consist of three subunits; from the largest to smallest, they are called **alpha (α), beta (β), and gamma (γ).** The G proteins are so named because one of the subunits binds to a guanine nucleotide. In the inactive state, a **guanine diphosphate (GDP)** molecule is bound to the α subunit of each G protein (figure 15.7, step 1).

G proteins can bind with receptors at the inner surface of the plasma membrane. After a hormone binds to the receptor on the outside of a cell, the receptor changes shape. As a result, a G protein combines with the receptor, and GDP is released from the α subunit. **Guanine triphosphate (GTP),** which is more abundant than GDP in the cytoplasm, binds to the α subunit, thereby activating it (figure 15.7, step 2). The G protein separates from the

receptor and the activated α subunit separates from the β and γ subunits (figure 15.7, step 3). The activated α subunit can alter the activity of molecules within the plasma membrane or inside the cell, thus producing cellular responses. After a short time, the activated α subunit is turned off because a phosphate group is removed from GTP, converting it to GDP (figure 15.7, step 4). The α subunit then recombines with the β and γ subunits (see figure 15.7, step 1).

Examples of hormones that activate G proteins are luteinizing hormone, follicle-stimulating hormone, thyroid-stimulating hormone, adrenocorticotropic hormone, oxytocin, antidiuretic hormone, calcitonin, parathyroid hormone, glucagon, and epinephrine.

Some activated α subunits of G proteins can combine with ion channels, causing them to open or close (figure 15.8). For example, activated α subunits can open Ca^{2+} channels in smooth muscle cells, resulting in the movement of Ca^{2+} into those cells. The Ca^{2+} function as intracellular mediators. **Intracellular mediators** are ions or molecules that either enter the cell or are synthesized in the cell and regulate enzyme activities inside the cell. The Ca^{2+} combine with calmodulin (kal-mod′ū-lin) molecules, and the calcium–calmodulin complexes activate enzymes that cause contraction of the smooth muscle cells.

Activated α subunits of G proteins can also alter the activity of enzymes inside the cell. For example, activated α subunits can alter the activity of **adenylate cyclase** (a-den′i-lāt sī′klās), which converts ATP to **cyclic adenosine monophosphate (cAMP)** and two inorganic phosphate groups (PP$_i$) (figure 15.9). The cAMP is an intracellular mediator that binds to protein kinases and activates them. **Protein kinases** are enzymes that regulate the activity of other enzymes by attaching phosphates to them, a process called **phosphorylation.** Depending on the enzyme, phosphorylation increases or decreases the activity of the enzyme. The amount of time cAMP is present to produce a response in a cell is limited.

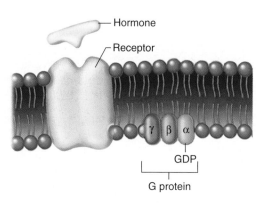

1. The hormone is not bound to its receptor. The G protein consists of three subunits with GDP attached to the α subunit.

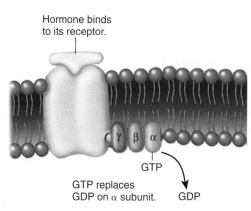

2. After the hormone binds to its membrane-bound receptor, the receptor is altered and the G protein binds to it. GTP replaces GDP on the α subunit of the G protein.

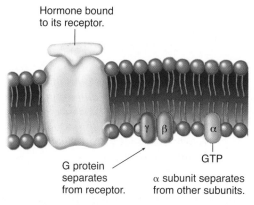

3. The G protein separates from the receptor and the α subunit with GTP bound to it separates from the γ and β subunits. The α subunit is the active component of the G protein.

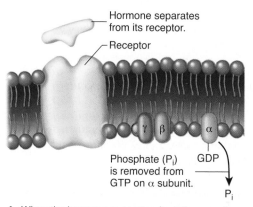

4. When the hormone separates from the receptor, additional G proteins are no longer activated. Inactivation of the α subunit occurs when phosphate (Pᵢ) is removed from the GTP, leaving GDP bound to the α subunit. The α subunit will combine with the β and γ subunits (go back to step 1).

Process Figure 15.7   Membrane-Bound Receptors and G Proteins

An enzyme in the cytoplasm, called **phosphodiesterase** (fos′fō-dī-es′ter-ās), breaks down cAMP to **adenosine monophosphate (AMP).** The response of the cell is terminated after cAMP levels are reduced below a certain level.

Cyclic AMP acts as an intracellular mediator in many cell types. The response in each cell type is different because the enzymes activated by cAMP in each cell type are different. For example, glucagon combines with receptors on the surface of liver cells, activating G proteins and causing an increase in cAMP synthesis, which increases the release of glucose from liver cells. In contrast, luteinizing hormone combines with receptors on the surface of cells of the ovary, activating G proteins and increasing cAMP synthesis. The major response to the increased cAMP is ovulation.

The combination of hormones with their receptors does not always result in increased cAMP synthesis. There are common intracellular mediators other than Ca^{2+} and cAMP, such as **cyclic guanine** (gwahn′ēn) **monophosphate (cGMP), diacylglycerol** (dī′as-il-glis′er-ol) **(DAG),** and **inositol** (in-ō′si-tōl, in-ō′si-tol)

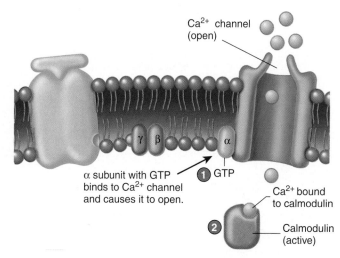

1. The α subunit, with GTP bound to it, combines with the Ca^{2+} channel, and the combination causes the Ca^{2+} channel to open.

2. The Ca^{2+} diffuse into the cell and combine with calmodulin. The combination of Ca^{2+} with calmodulin produces the cell's response to the hormone.

Process Figure 15.8   G Proteins and Ion Channels

1. After the hormone binds to its receptor, the G protein is activated (see figure 15.7).

2. The activated α subunit, with GTP bound to it, binds to and activates an adenylate cyclase enzyme so that it converts ATP to cAMP and two inorganic phosphates (PPi).

3. The cAMP activates protein kinase enzymes, which phosphorylate specific enzymes, activating them. The chemical reactions catalyzed by the activated enzymes produce the cell's response.

4. Phosphodiesterase enzymes inactivate cAMP by converting cAMP to AMP.

Glucagon bound to its receptor.

①

α subunit of G protein with GTP bound to it.

② GTP

ATP

Adenylate cyclase catalyzes the formation of cAMP.

cAMP + PPi

③

Protein kinase

cAMP is an intracellular mediator that activates protein kinases.

Phosphodiesterase inactivates cAMP.

④

AMP (inactive)

Response
Protein kinases phosphorylate specific enzymes and activate them.

Process Figure 15.9   G Proteins and Enzymes

triphosphate (IP₃). In some cell types, the combination of hormones with their receptors causes the G proteins to inhibit the synthesis of intracellular mediators. ▼

15. *Explain how the combination of a hormone and its receptor can alter the G proteins on the inner surface of the plasma membrane. Which activated subunit of the G protein alters the activity of molecules inside the plasma membrane or inside the cell?*
16. *Describe how G proteins can alter the permeability of the plasma membrane and how they can alter the synthesis*

of an intracellular mediator molecule, such as cAMP. Give examples.
17. *What limits the activity of intracellular mediator molecules, such as cAMP?*

Membrane-Bound Receptors That Directly Alter the Activity of Intracellular Enzymes

Some hormones bind to membrane-bound receptors and activate already existing intracellular enzymes. The intracellular enzymes can be separate molecules or they can be part of the membrane-

1. The hormone binds with its receptor.

2. At the inner surface of the plasma membrane, guanylate cyclase is activated to produce cGMP from GTP.

3. Cyclic GMP is an intracellular mediator that alters the activity of other intracellular enzymes to produce the response of the cell.

4. Phosphodiesterase is an enzyme that converts cGMP to inactive GMP.

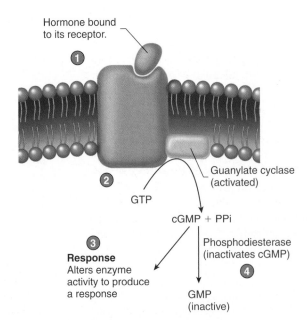

Hormone bound to its receptor.

①

②

GTP

Guanylate cyclase (activated)

cGMP + PPi

③

Response
Alters enzyme activity to produce a response

Phosphodiesterase (inactivates cGMP)

④

GMP (inactive)

Process Figure 15.10   Membrane-Bound Receptor That Directly Activates Enzymes

bound receptor. The activated enzyme either increases or decreases the synthesis of intracellular mediators, or it results in the phosphorylation of intracellular proteins. The intracellular mediators or phosphorylated proteins activate processes that produce the response of cells to the chemical signals. Examples of hormones that activate already existing enzymes are insulin, growth hormone, prolactin, and atrial natriuretic hormone.

Cyclic guanine monophosphate (cGMP) is an intracellular mediator that is synthesized in response to a hormone binding with a membrane-bound receptor (figure 15.10). The hormone binds to its receptor, and the combination activates an enzyme called **guanylate cyclase** (gwahn'i-lāt sī'klās), located at the inner surface of the plasma membrane. The guanylate cyclase enzyme converts guanine triphosphate (GTP) to cGMP and two inorganic phosphate groups (PP_i). The cGMP molecules then combine with specific enzymes in the cytoplasm of the cell and activate them. The activated enzymes, in turn, produce the cell's response to the hormone. For example, atrial natriuretic hormone combines with its receptor in the plasma membrane of kidney cells. The result is an increase in the rate of cGMP synthesis at the inner surface of the plasma membranes. Cyclic GMP influences the action of enzymes in the kidney cells, which increases the rate of Na^+ and water excretion by the kidney (see chapter 23). The amount of time the cGMP is present to produce a response in the cell is limited. Phosphodiesterase breaks down cGMP to GMP. Consequently, the length of time a hormone increases cGMP synthesis and has an effect on a cell is brief after the hormone is no longer present.

Some hormones bind to membrane-bound receptors, and the portion of the receptor on the inner surface of the plasma membrane acts as a phosphorylase enzyme that phosphorylates several specific proteins. Some of the phosphorylated proteins are part of the membrane-bound receptor, and others are in the cytoplasm of the cell (figure 15.11). The phosphorylated proteins influence the activity of other enzymes in the cytoplasm of the cell. For example, insulin binds to its membrane-bound receptor, resulting in the phosphorylation of parts of the receptor on the inner surface of the plasma membrane and the phosphorylation of certain other intracellular proteins. The phosphorylated proteins produce the cells' responses to insulin.

Hormones that stimulate the synthesis of an intracellular mediator molecule often produce rapid responses. This is possible because the mediator influences already existing enzymes and causes a **cascade effect,** which results when a few mediator molecules activate several enzymes, and each of the activated enzymes in turn activates several other enzymes that produce the final response. Thus, an amplification system exists in which a few molecules, such as cAMP, cGMP, or phosphorylated proteins, can control the activity of many enzymes within a cell (figure 15.12). ▼

> 18. *Describe how a hormone can combine with a membrane-bound receptor, change enzyme activity inside the cell, and increase phosphorylation of intracellular proteins. Give examples.*
>
> 19. *Explain the cascade effect for the intracellular mediator model of hormone action. Does the intracellular mediator mechanism produce a slow or rapid response?*

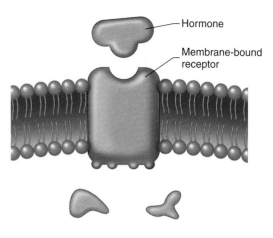

Unphosphorylated receptor and proteins

1. The membrane-bound receptor and other proteins have sites that can be phosphorylated at the inner surface of the cell membrane. When the receptor is not bound to a hormone, these sites remain unphosphorylated.

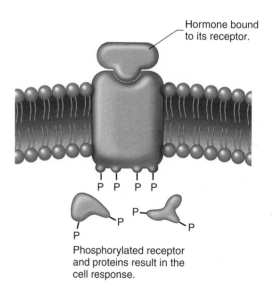

Phosphorylated receptor and proteins result in the cell response.

2. The hormone binds to its receptor site on the outside of the cell membrane. The receptor acts as an enzyme that phosphorylates the sites on the receptor and associated proteins. These phosphorylated proteins produce a response inside the cell.

Figure 15.11   Membrane-Bound Receptors That Phosphorylate Intracellular Proteins

P R E D I C T

When smooth muscle cells in the airways of the lungs contract, breathing can be very difficult, whereas breathing is easy if the smooth muscle cells are relaxed. During asthma attacks, the smooth muscle cells in the airways of the lungs strongly contract. Some of the drugs used to treat asthma increase cAMP in smooth muscle cells. Explain as many ways as possible how these drugs work.

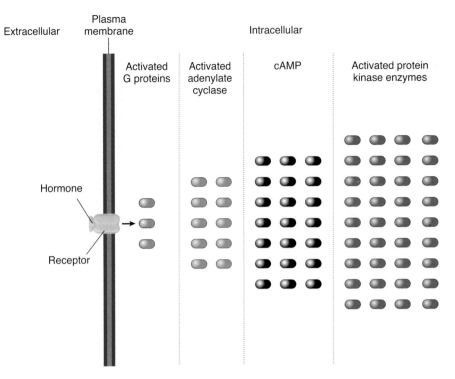

Figure 15.12 Cascade Effect

The combination of a hormone with a membrane-bound receptor activates several G proteins. The G proteins, in turn, activate many inactive adenylate cyclase enzymes, which cause the synthesis of a large number of cAMP molecules. The large number of cAMP molecules, in turn, activate many inactive protein kinase enzymes, which produce a rapid and amplified response.

1. The lipid-soluble hormone diffuses through the plasma membrane and enters the cytoplasm of the cell.

2. The hormone combines with a receptor in the cytoplasm (or in the nucleus).

3. The hormone–receptor complex interacts with DNA and increases the synthesis of specific messenger RNA (mRNA) molecules.

4. The mRNA passes from the nucleus to the cytoplasm.

5. In the cytoplasm of the cell, the mRNA combines with ribosomes, and new proteins are synthesized.

6. The new proteins produce the response of the cell to the hormone.

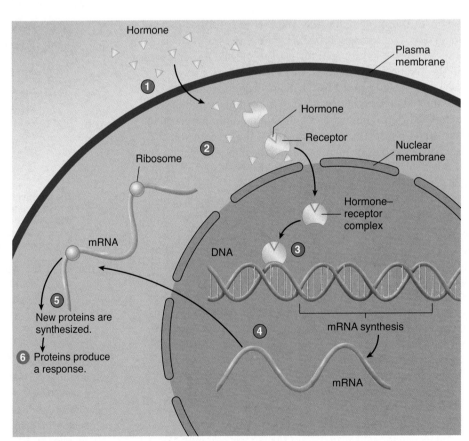

Process Figure 15.13 Intracellular Receptors

Intracellular Receptors

Lipid-soluble hormones diffuse cross the plasma membrane into the cytoplasm or into the nucleus (figure 15.13). Some intracellular receptors are located in the cytoplasm. Once a hormone binds to its receptor, the hormone–receptor complex diffuses into the nucleus and binds to DNA. Other intracellular receptors are located in the nucleus. A hormone diffuses into the nucleus and binds to its receptor, and the hormone–receptor complex then binds to DNA.

Receptors that interact with DNA have specific "fingerlike" projections that interact with specific parts of a DNA molecule. The combination of the hormone–receptor complex with DNA increases the synthesis of specific **messenger ribonucleic acid (mRNA)** molecules. The mRNA then moves to the cytoplasm and increases the synthesis of specific proteins at the ribosomes. The newly synthesized proteins produce the cell response to the hormone. For example, testosterone from the testes and estrogen from the ovaries stimulate the synthesis of proteins that are responsible for the secondary sex characteristics of males and females. Other hormones that activate intracellular receptors are aldosterone, cortisol, thyroid hormones, and vitamin D.

Cells that synthesize new proteins in response to hormonal stimuli normally have a latent period of several hours between the time the hormones bind to their receptors and the time responses are observed. During this latent period, mRNA and new proteins are synthesized. Hormone–receptor complexes normally are degraded within the cell, limiting the length of time hormones influence the cells' activities, and the cells slowly return to their previous functional states. ▼

20. **Describe how a hormone that crosses the plasma membrane interacts with its receptor and how it alters the rate of protein synthesis. Why is there normally a latent period between the time hormones bind to their receptors and the time responses are observed?**
21. **What finally limits the processes activated by the intracellular receptor mechanism?**

P R E D I C T

Of membrane-bound receptors and intracellular receptors, which is better adapted for mediating a response with a rapid onset and a short duration and which is better for mediating a response that lasts a considerable length of time? Explain.

Pituitary Gland and Hypothalamus

The **pituitary** (pi-too′i-tār-rē) **gland,** or **hypophysis** (hī-pof′i-sis, an undergrowth), secretes nine major hormones that regulate numerous body functions and the secretory activity of several other endocrine glands.

The **hypothalamus** (hī′pō-thal′ă-mŭs, under the thalamus) of the brain and the pituitary gland are major sites where the nervous and endocrine systems interact (figure 15.14a). The hypothalamus regulates the secretory activity of the pituitary

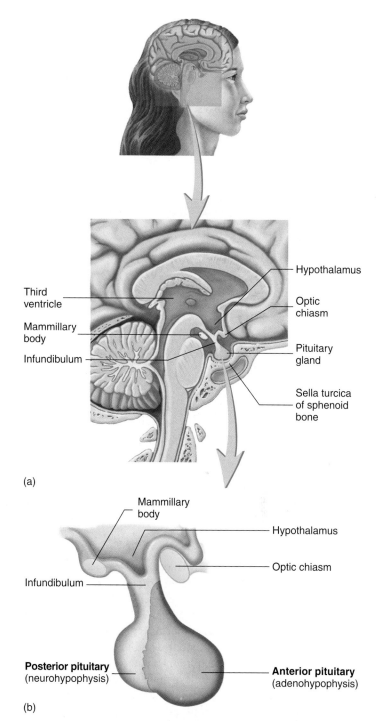

(a)

(b)

Figure 15.14 Hypothalamus and Pituitary Gland

(a) A median section of the head through the pituitary gland showing the location of the hypothalamus of the brain and the pituitary gland. (b) The pituitary gland is divided into the anterior pituitary and the posterior pituitary. The infundibulum connects the hypothalamus to the posterior pituitary.

gland. Hormones, sensory information that enters the central nervous system, and emotions, in turn, influence the activity of the hypothalamus.

Structure of the Pituitary Gland

The pituitary gland is roughly 1 cm in diameter, weighs 0.5–1.0 g, and rests in the sella turcica of the sphenoid bone (see figure 15.14a).

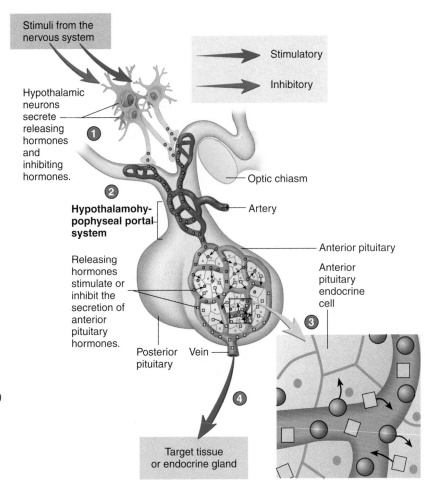

1. Stimuli within the nervous system increase or decrease the secretion of releasing hormones and inhibiting hormones (*blue balls*) from neurons of the hypothalamus.

2. Releasing hormones and inhibiting hormones pass through the hypothalamohypophyseal portal system to the anterior pituitary.

3. Releasing hormones and inhibiting hormones leave capillaries, bind to membrane-bound receptors, and stimulate or inhibit the release of hormones (*yellow squares*) from anterior pituitary cells.

4. Anterior pituitary hormones (*yellow squares*) are carried in the blood to their target tissues (*green arrow*), which, in some cases, are other endocrine glands.

Process Figure 15.15 Hypothalamic Regulation of the Anterior Pituitary

It is located inferior to the hypothalamus and is connected to it by a stalk of tissue called the **infundibulum** (in-fŭn-dib′ū-lŭm).

The pituitary gland is divided functionally into two parts: the posterior and anterior pituitary (figure 15.14*b*). The **posterior pituitary,** or **neurohypophysis** (noor′ō-hī-pof′i-sis, *neuro-* refers to the nervous system) is continuous with the brain. It is formed during embryonic development from an outgrowth of the inferior part of the brain in the area of the hypothalamus (see chapter 25). The outgrowth forms the infundibulum, and the distal end of the infundibulum enlarges to form the posterior pituitary.

The **anterior pituitary,** or **adenohypophysis** (ad′ĕ-nō-hī-pof′i-sis, *adeno-* means gland) arises as an outgrowth of the roof of the embryonic oral cavity. The outgrowth extends toward the posterior pituitary, loses its connection with the oral cavity, and becomes the anterior pituitary.

Relationship of the Pituitary to the Brain

Portal vessels are blood vessels that begin in a primary capillary network, extend some distance, and end in a secondary capillary network. The **hypothalamohypophyseal** (hī′pō-thal′ă-mō-hī′pō-fīz′ē-ăl) **portal system** is one of two major portal systems. The other is the hepatic portal system (see chapter 18). The hypothal-amohypophyseal portal system extends from the hypothalamus to the anterior pituitary (figure 15.15). The primary capillary network in the hypothalamus is supplied with blood from arteries that deliver blood to the hypothalamus. From the primary capillary network, the hypothalamohypophyseal portal vessels carry blood to a secondary capillary network in the anterior pituitary. Veins from the secondary capillary network eventually merge with the general circulation.

Hormones, produced and secreted by neurons of the hypothalamus, enter the primary capillary network and are carried to the secondary capillary network. There the hormones leave the blood and act on cells of the anterior pituitary. They act either as **releasing hormones,** increasing the secretion of anterior pituitary hormones, or as **inhibiting hormones,** decreasing the secretion of anterior pituitary hormones. Each releasing hormone stimulates and each inhibiting hormone inhibits the production and secretion of a specific hormone by the anterior pituitary. In response to the releasing hormones, anterior pituitary cells secrete hormones that enter the secondary capillary network and are carried by the general circulation to their target tissues. Thus, the hypothalamohypophyseal portal system provides a means by which the hypothalamus, using hormones as chemical signals, regulates the secretory activity of the anterior

Table 15.1 Hormones of the Hypothalamus

Hormones	Structure	Target Tissue	Response
Growth hormone–releasing hormone (GHRH)	Peptide	Anterior pituitary cells that secrete growth hormone	Increased growth hormone secretion
Growth hormone–inhibiting hormone (GHIH), or somatostatin	Small peptide	Anterior pituitary cells that secrete growth hormone	Decreased growth hormone secretion
Thyrotropin-releasing hormone (TRH)	Small peptide	Anterior-pituitary cells that secrete thyroid-stimulating hormon	Increased thyroid-stimulating hormone secretion
Corticotropin-releasing hormone (CRH)	Peptide	Anterior pituitary cells that secrete adrenocorticotropic hormone	Increased adrenocorticotropic hormone secretion
Gonadotropin-releasing hormone (GnRH)	Small peptide	Anterior pituitary cells that secrete luteinizing hormone and follicle-stimulating hormone	Increased secretion of luteinizing hormone and follicle-stimulating hormone
Prolactin-inhibiting hormone (PIH)	Unknown (possibly dopamine)	Anterior pituitary cells that secrete prolactin	Decreased prolactin secretion
Prolactin-releasing hormone (PRH)	Unknown	Anterior pituitary cells that secrete prolactin	Increased prolactin secretion

pituitary (see figures 11.20c and 15.15). The major hormones of the hypothalamus are listed in table 15.1 and discussed later with the hormones they regulate.

There is no portal system to carry hypothalamic hormones to the posterior pituitary. Hormones released from the posterior pituitary are produced by neurosecretory cells with their cell bodies located in the hypothalamus. The axons of these cells extend from the hypothalamus through the infundibulum into the posterior pituitary and form a nerve tract called the **hypothalamohypophyseal tract** (figure 15.16). Hormones produced in the hypothalamus pass down these axons in tiny vesicles and are stored in secretory vesicles in the enlarged ends of the axons. Action potentials originating in the neuron cell bodies in the hypothalamus are propagated along the axons to the axon terminals in the posterior pituitary. The action potentials cause the release of hormones from the axon terminals, and they enter the circulatory system. ▼

22. *Where is the pituitary gland located? Contrast the embryonic origin of the anterior pituitary and the posterior pituitary.*
23. *Define portal system. Describe the hypothalamo-hypophyseal portal system. How does the hypothalamus regulate the secretion of the anterior pituitary hormones?*
24. *Describe the hypothalamohypophyseal tract, including the production of hormones in the hypothalamus and their release from the posterior pituitary.*

P R E D I C T
Surgical removal of the posterior pituitary in experimental animals results in marked symptoms, but these symptoms associated with hormone shortage are temporary. Explain these results.

Hormones of the Pituitary Gland

Posterior Pituitary Hormones

The posterior pituitary stores and secretes two polypeptide hormones called antidiuretic hormone and oxytocin. A separate population of cells secretes each hormone.

Antidiuretic (an′tē-dī-ū-ret′ik) **hormone (ADH)** is so named because it prevents (*anti-*) the output of large amounts of urine (*diuresis*). ADH binds to membrane-bound receptors and increases water reabsorption by kidney tubules. This results in less water loss from the blood into the urine, and urine volume decreases. ADH can also cause blood vessels to constrict when released in large amounts. Consequently, it is sometimes called **vasopressin** (vă-sō-pres′in, *vaso,* blood vessel + *pressum,* to press down). The role of ADH in regulating blood osmolality, blood volume, and blood pressure is discussed in chapters 18 and 23.

Oxytocin (ok′sĭ-tō′sin, swift birth) binds to membrane-bound receptors and causes contraction of the smooth muscle cells of the uterus and milk ejection, or milk "let-down," from the breasts in lactating women. Oxytocin plays an important role in the expulsion of the fetus from the uterus during delivery by stimulating uterine smooth muscle contraction. Commercial preparations of oxytocin are given under certain conditions to assist in childbirth and to constrict uterine blood vessels following childbirth. Oxytocin also causes the contraction of uterine smooth muscle in nonpregnant women, primarily during menses and sexual intercourse. The uterine contractions help expel the uterine epithelium and small amounts of blood during menses and can participate in the movement of sperm cells through the uterus after sexual intercourse. The role of oxytocin in the reproductive system is described in greater

1. Stimuli within the nervous system stimulate hypothalamic neurons to either increase or decrease their action potential frequency.

2. Action potentials are carried by axons of the hypothalamic neurons through the hypothalamohypophyseal tract to the posterior pituitary. The axons of neurons store hormones in the posterior pituitary.

3. In the posterior pituitary gland, action potentials cause the release of hormones (*red balls*) from axon terminals into the circulatory system.

4. The hormones pass through the circulatory system and influence the activity of their target tissues (*green arrow*).

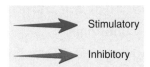

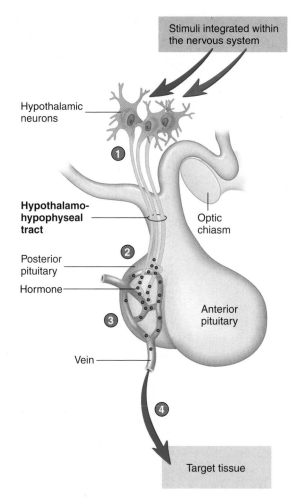

Process Figure 15.16 Hypothalamic Regulation of the Posterior Pituitary

Table 15.2	Hormones of the Pituitary Gland		
Hormones	**Structure**	**Target Tissue**	**Response**
Posterior Pituitary			
Antidiuretic hormone (ADH)	Small peptide	Kidney	Increased water reabsorption (less water is lost in the form of urine)
Oxytocin	Small peptide	Uterus; mammary glands	Increased uterine contractions; increased milk expulsion from mammary glands; unclear function in males
Anterior Pituitary			
Growth hormone (GH), or somatotropin	Protein	Most tissues	Increased growth in tissues; increased amino acid uptake and protein synthesis; increased breakdown of lipids and release of fatty acids from cells; increased glycogen synthesis and increased blood glucose levels; increased somatomedin production
Thyroid-stimulating hormone (TSH)	Glycoprotein	Thyroid gland	Increased thyroid hormone secretion
Adrenocorticotropic hormone (ACTH)	Peptide	Adrenal cortex	Increased glucocorticoid hormone secretion
Melanocyte-stimulating hormone (MSH)	Peptide	Melanocytes in the skin	Increased melanin production in melanocytes to make the skin darker in color
Luteinizing hormone (LH)	Glycoprotein	Ovaries in females; testes in males	Ovulation and progesterone production in ovaries; testosterone synthesis and support for sperm cell production in testes
Follicle-stimulating hormone (FSH)	Glycoprotein	Follicles in ovaries in females; seminiferous tubes in males	Follicle maturation and estrogen secretion in ovaries; sperm cell production in testes
Prolactin	Protein	Ovaries and mammary glands in females	Milk production in lactating women; increased response of follicle to LH and FSH; unclear function in males

Clinical Relevance Growth Hormone Disorders

Several pathological conditions are associated with GH. Tumors in the hypothalamus or anterior pituitary can cause hypersecretion or hyposecretion of GH. The synthesis of structurally abnormal GH, the liver's inability to produce somatomedins, and the lack of functional GH receptors in target tissues can also cause growth problems.

Giantism is a condition of abnormally increased height that usually results from excessive cartilage and bone formation at the epiphyseal plates of long bones (figure A). The most common type of giantism, **pituitary giantism,** results from excess secretion of GH. The large stature of some individuals, however, can result from genetic factors rather than from abnormal levels of GH. **Acromegaly** (ak-rō-meg′ă-lē, large extremity) is caused by excess GH secretion in adults, and many pituitary giants develop acromegaly later in life. The GH stimulates the growth of connective tissue, including bones. Bones in adults can increase in diameter and thickness, but not in length because the epiphyseal plates have ossified. The effects of acromegaly are most apparent in the face and hands. Hypersecretion of GH can also cause elevated blood glucose levels and may eventually lead to diabetes mellitus.

Dwarfism, the condition in which a person is abnormally short, is the opposite of giantism. **Pituitary dwarfism** results when abnormally low levels of GH affect the whole body, thus producing a small person who is normally proportioned. **Achondroplasia** (ā-kon-drō-plā′zē-ă), or **achondroplastic** (ā-kon-drō-plas′tik) **dwarfism,** is the most common type of dwarfism; it produces a person with a nearly normal-sized trunk and head but shorter than normal limbs. Achondroplasia is a genetic disorder, not a hormonal disorder (see "Achondroplasia," chapter 6).

Modern genetic engineering has provided a source of human GH for people who produce inadequate quantities. Human genes for GH have been successfully introduced into bacteria using genetic engineering techniques. The gene in the bacteria causes GH synthesis, and the GH can be extracted from the medium in which the bacteria are grown.

Figure A Effect of Overproduction and Underproduction of Growth Hormone on Height

detail in chapter 24. Little is known about the effect of oxytocin in males. ▼

> 25. What effect does ADH have on urine production?
> 26. What is the role of oxytocin in uterine contractions and milk ejection?

Anterior Pituitary Hormones

Releasing and inhibiting hormones that pass from the hypothalamus through the hypothalamohypophyseal portal system to the anterior pituitary influence anterior pituitary secretions. The hormones secreted are proteins, glycoproteins, or polypeptides. They are transported in the circulatory system and bind to membrane-bound receptor molecules on their target cells. For the most part, each hormone is secreted by a separate cell type.

The major hormones of the anterior pituitary, their target tissues, and their effects on target tissues are listed in table 15.2. Anterior pituitary hormones include growth hormone, thyroid-stimulating hormone, adrenocorticotropic hormone and related substances, luteinizing hormone, follicle-stimulating hormone, and prolactin. In this chapter, growth hormone is discussed in the

next section, and adrenocorticotropic hormone is discussed with the adrenal glands. Luteinizing hormone, follicle-stimulating hormone, and prolactin are involved with the reproductive system and are discussed in chapter 24. ▼

> 27. What kinds of chemicals are anterior pituitary hormones? How do they activate their target tissues?
> 28. List the hormones secreted by the anterior pituitary.

Growth Hormone Growth hormone (GH) stimulates the growth of most tissues and, through its effect on the epiphyseal plates of bones, GH plays a role in determining how tall a person becomes (see chapter 6). GH promotes the protein synthesis necessary for growth by increasing the movement of amino acids into cells and promoting their incorporation into proteins. It also decreases the breakdown of proteins.

GH plays an important role in regulating blood nutrient levels between meals and during periods of fasting. GH increases lipolysis, the breakdown of lipids. Fatty acids released from fat cells into the blood circulate to other tissues and are used as an energy source. The use of fatty acids as an energy source "spares" the use of blood glucose, helping maintain blood sugar levels. In

1. Stress and decreased blood glucose levels increase the release of growth hormone–releasing hormone (GHRH) and decrease the release of growth hormone–inhibiting hormone (GHIH).

2. GHRH and GHIH travel through the hypothalamohypophyseal portal system to the anterior pituitary.

3. Increased GHRH and reduced GHIH act on the anterior pituitary and result in increased GH secretion.

4. GH acts on target tissues.

5. Increasing GH has a negative-feedback effect on the hypothalamus, resulting in decreased GHRH and decreased GHIH release.

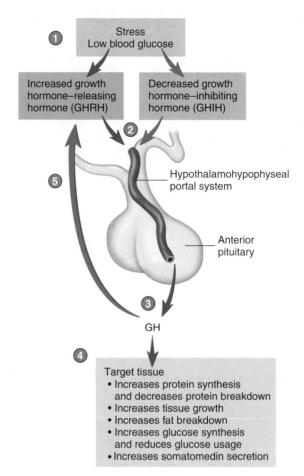

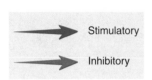

Process Figure 15.17   Control of Growth Hormone (GH) Secretion

addition, GH increases glucose synthesis by the liver, which releases glucose into the blood. Thus, through its effects on adipose tissue and the liver, GH maintains or increases blood sugar levels.

GH has indirect effects on some tissues by stimulating the production of polypeptides called **somatomedins** (sō′mă-tō-mē′dinz), primarily by the liver but also by skeletal muscle and other tissues. Somatomedins circulate in the blood, stimulating growth in cartilage and bone and increasing the synthesis of protein in skeletal muscles. The best known somatomedins are two polypeptide hormones produced by the liver called **insulin-like growth factor I** and **II** because of the similarity of their structure to insulin.

Two hormones released from the hypothalamus regulate the secretion of GH (figure 15.17). Growth hormone–releasing hormone (GHRH) stimulates the secretion of GH, whereas growth hormone–inhibiting hormone (GHIH) inhibits the secretion of GH. Stimuli that influence GH secretion act on the hypothalamus to increase or decrease the secretion of the releasing and inhibiting hormones. Low blood glucose levels and stress stimulate the secretion of GH, and high blood glucose levels inhibit the secretion of GH. An increase in certain amino acids stimulates increased GH secretion.

Most people have a rhythm of growth hormone secretion, with daily peak levels occurring during deep sleep. Growth hormone secretion also increases during periods of fasting and prolonged exercise. Blood growth hormone levels do not become greatly elevated during periods of rapid growth, although children tend to have somewhat higher blood levels of growth hormone than do adults. In addition to growth hormone, genetics, nutrition, and sex hormones influence growth. ▼

29. *What effects does GH have on protein synthesis, lipid breakdown, and blood glucose levels?*
30. *What stimulates somatomedin production, where is it produced, and what are its effects?*
31. *What effects do stress and blood glucose levels have on GH secretion?*

P R E D I C T

Mr. Hoops has a son who wants to be a basketball player almost as much as Mr. Hoops wants him to be one. Mr. Hoops knows a little bit about growth hormone and asks his son's doctor if she would prescribe some for his son, so he can grow taller. What do you think the doctor tells Mr. Hoops?

Thyroid Gland

The **thyroid** (thī′royd, shield) **gland** consists of two lobes connected by a narrow band called the **isthmus** (is′mŭs, a constriction). The lobes are located on each side of the trachea, just inferior to the larynx (figure 15.18a). The thyroid gland is one of the largest endocrine glands, with a weight of approximately 20 g. It is highly vascular and appears more red than its surrounding tissues.

The thyroid gland contains numerous **follicles,** which are small spheres whose walls are composed of a single layer of cuboidal epithelial cells (figure 15.18c and d). Each thyroid follicle is filled with proteins, called **thyroglobulin** (thī-rō-glob′ū-lin), which are synthesized and secreted by the cells of the thyroid follicles. Large amounts of the thyroid hormones are stored in the thyroid follicles as part of the thyroglobulin molecules.

Between the follicles, a delicate network of loose connective tissue contains scattered **parafollicular** (par-ă-fo-lik′ū-lăr) **cells.**

Calcitonin (kal-si-tō′nin) is secreted from the parafollicular cells and plays a role in reducing the concentration of Ca^{2+} in the body fluids when Ca^{2+} levels become elevated. ▼

> *32. Where is the thyroid gland located? Describe the follicles and the parafollicular cells within the thyroid gland. What hormones do they produce?*

Thyroid Hormones

The thyroid hormones are **triiodothyronine** (trī-ī′ō-dō-thī′rō-nēn) **(T3)** and **tetraiodothyronine** (tet′ră-ī′ō-dō-thī′rō-nēn) **(T4).** Another name for T4 is **thyroxine** (thī-rok′sēn, thī-rok′sin) (table 15.3). T_3 constitutes 90% of thyroid gland secretions and T_4 10%. Although calcitonin is secreted by the parafollicular cells of the thyroid gland, T_3 and T_4 are considered to be the thyroid hormones because they are more clinically important and because they are secreted from the thyroid follicles.

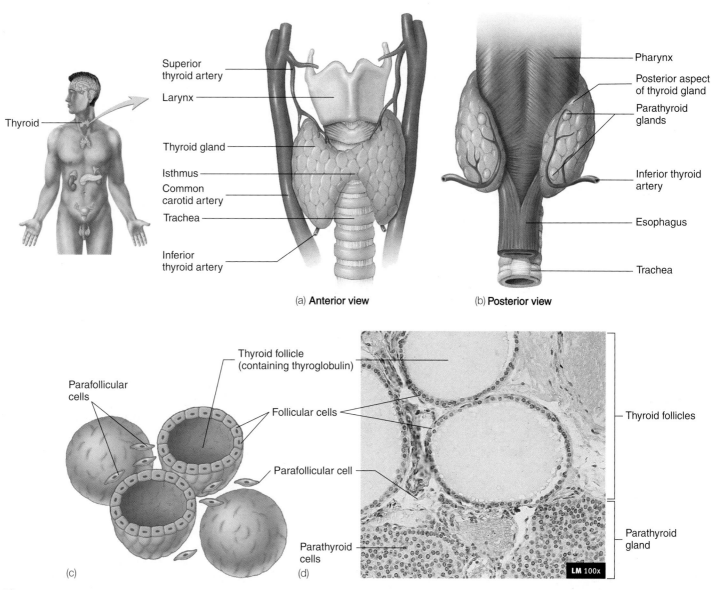

(a) **Anterior view**

(b) **Posterior view**

(c)

(d)

Figure 15.18   The Thyroid and Parathyroid Glands

(a) Anterior view of the thyroid gland. (b) Posterior view of the thyroid gland with the four small parathyroid glands embedded in the posterior surface of the thyroid gland. (c) Three-dimensional interpretive drawing of thyroid follicles and parafollicular cells. (d) Light micrograph of thyroid and parathyroid tissue.

Table 15.3 Major Endocrine Glands and Their Hormones

Hormones	Structure	Target Tissue	Response
Thyroid Gland			
Thyroid Follicles Thyroid hormones (T3 and T4)	Amino acid derivative	Most cells of the body	Increased metabolic rate; essential for normal process of growth and maturation
Parafollicular Cells Calcitonin	Polypeptide	Bone	Decreased rate of breakdown of bone by osteoclasts; prevention of a large increase in blood Ca^{2+} levels
Parathyroid Gland			
Parathyroid hormone	Polypeptide	Bone; kidney; small intestine	Increased rate of breakdown of bone by osteoclasts; increased reabsorption of Ca^{2+} in kidneys; increased absorption of Ca^{2+} from small intestine; increased vitamin D synthesis; increased blood Ca^{2+} levels
Adrenal Medulla			
Epinephrine primarily; norepinephrine	Amino acid derivatives	Heart; blood vessels; liver; fat cells	Increased cardiac output; increased blood flow to skeletal muscles and increased blood flow to heart; vasoconstriction of blood vessels, especially visceral and skin blood vessels; increased release of glucose and fatty acids into blood; in general, preparation for physical activity
Adrenal Cortex			
Glucocorticoids (cortisol)	Steroids	Most tissues	Increased protein and fat breakdown; increased glucose production; inhibition of immune response and decreased inflammation
Mineralocorticoids (aldosterone)	Steroids	Kidney	Increased Na^+ reabsorption and K^+ and H^+ excretion; enhanced water reabsorption
Androgens	Steroids	Many tissues	Minor importance in males; in females, development of some secondary sexual characteristics, such as axillary and pubic hair
Pancreas			
Insulin	Protein	Especially liver; skeletal muscle; fat tissue	Increased uptake and use of glucose and amino acids
Glucagon	Polypeptide	Primarily liver	Increased breakdown of glycogen; release of glucose into the circulatory system
Testis			
Testosterone	Steroid	Most cells	Aids in spermatogenesis; development of genitalia; maintenance of functional reproductive organs; secondary sex characteristics; sexual behavior
Ovary			
Estrogens	Steroids	Most cells	Uterine and mammary gland development and function; maturation of genitalia; secondary sex characteristics; sexual behavior and menstrual cycle
Progesterone	Steroid	Most cells	Uterine and mammary gland development and function; maturation of genitalia; secondary sex characteristics; menstrual cycle
Pineal Body			
Melatonin	Amino acid derivative	At least the hypothalamus	Inhibition of gonadotropin-releasing hormone secretion, thereby inhibiting reproduction; significance is not clear in humans; may help regulate sleep–wake cycles
Thymus			
Thymosin	Peptide	Immune tissues	Development and function of the immune system

T₃ and T₄ Synthesis

Thyroid-stimulating hormone (TSH) from the anterior pituitary stimulates thyroid hormone synthesis and secretion. TSH causes an increase in the synthesis of T_3 and T_4, which are then stored inside the thyroid follicles as part of thyroglobulin. TSH also causes T_3 and T_4 to be released from thyroglobulin and enter the circulatory system. An adequate amount of iodine in the diet is required for thyroid hormone synthesis because iodine is a

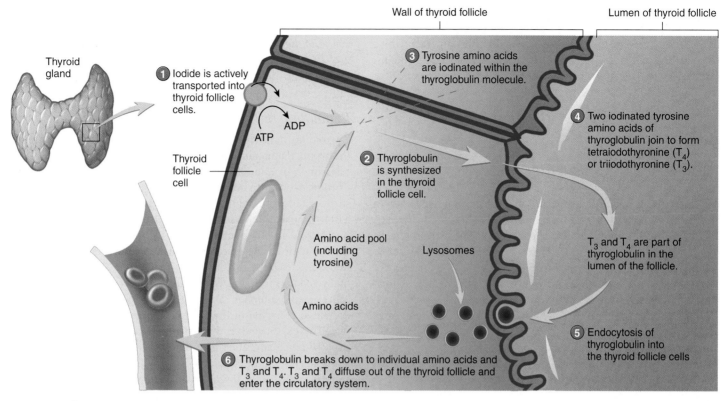

Wall of thyroid follicle

Lumen of thyroid follicle

Thyroid gland

1 Iodide is actively transported into thyroid follicle cells.

Thyroid follicle cell

ATP

ADP

2 Thyroglobulin is synthesized in the thyroid follicle cell.

Amino acid pool (including tyrosine)

Amino acids

3 Tyrosine amino acids are iodinated within the thyroglobulin molecule.

4 Two iodinated tyrosine amino acids of thyroglobulin join to form tetraiodothyronine (T_4) or triiodothyronine (T_3).

Lysosomes

T_3 and T_4 are part of thyroglobulin in the lumen of the follicle.

5 Endocytosis of thyroglobulin into the thyroid follicle cells

6 Thyroglobulin breaks down to individual amino acids and T_3 and T_4. T_3 and T_4 diffuse out of the thyroid follicle and enter the circulatory system.

Process Figure 15.19   Synthesis of Thyroid Hormones

The numbered steps describe the synthesis and secretion of T_3 and T_4 from the thyroid gland.

component of T_3 and T_4. The following events in the thyroid follicles result in T_3 and T_4 synthesis and secretion (figure 15.19):

1. Iodide ions (I^-) are taken up by thyroid follicle cells by active transport. The active transport of the I^- is against a concentration gradient of approximately 30-fold in healthy individuals. TSH promotes the active transport of I^-.
2. Thyroglobulins, which contain numerous tyrosine amino acids, are synthesized within the cells of the follicles.
3. The I^- are oxidized to form iodine (I), and either one or two iodine atoms are bound to each of the tyrosine amino acids of thyroglobulin. This occurs close to the time the thyroglobulin molecules are secreted by the process of exocytosis into the lumen of the follicle. As a result, the secreted thyroglobulin contains many iodinated tyrosine amino acids with either one iodine atom (monoiodotyrosine) or two iodine atoms (diiodotyrosine).
4. In the lumen of the follicle, two diiodotyrosine molecules of thyroglobulin combine to form tetraiodothyronine (T_4), which has four iodine atoms. Also, one monoiodotyrosine and one diiodotyrosine molecule combine to form triiodothyronine (T_3), which has three iodine atoms. Large amounts of T_3 and T_4 are stored within the thyroid follicles as part of thyroglobulin. A reserve sufficient to supply thyroid hormones for approximately 2–4 months is stored in this form.

5. Thyroglobulin is taken into the thyroid follicle cells by endocytosis, where lysosomes fuse with the endocytotic vesicles.
6. Proteolytic enzymes break down thyroglobulin to release T_3 and T_4. They are lipid-soluble, so T_3 and T_4 diffuse through the plasma membranes of the follicular cells into the interstitial fluids and finally into the capillaries of the thyroid gland. The remaining amino acids of thyroglobulin are used again to synthesize more thyroglobulin.

Transport in the Blood

Thyroid hormones are transported in combination with plasma proteins in the circulatory system. Approximately 70%–75% of the circulating thyroid hormones are bound to **thyroxine-binding globulin (TBG)**, which is synthesized by the liver, and 20%–30% are bound to other plasma proteins, including albumen. Thyroid hormones, bound to these plasma proteins, form a large reservoir of circulating thyroid hormones. Thyroid hormones are converted to other compounds and excreted in the urine.

Effects of Thyroid Hormones

Thyroid hormones interact with their target tissues in a fashion similar to that of the steroid hormones. They readily diffuse through plasma membranes into the cytoplasm of cells. Within cells, they bind to receptor molecules in the nuclei. Thyroid hormones combined with their receptor molecules interact with DNA in the nuclei to influence genes and initiate new protein synthesis. The newly synthesized proteins within the target cells mediate the cells' response to thyroid hormones. It takes up to a week after the administration of thyroid hormones for a maximal response to develop, and new protein synthesis occupies much of that time.

Thyroid hormones affect nearly every tissue in the body, but not all tissues respond identically. Metabolism is primarily affected in some tissues, and growth and maturation are influenced in others. The normal rate of metabolism for an individual depends on an adequate supply of thyroid hormone, which increases the rate at which glucose, fat, and protein are metabolized. The increased rate of metabolism produces heat. Thyroid hormones increase the activity of N^+–K^+ pumps, which helps increase the body temperature as ATP molecules are broken down. Thyroid hormones also alter the number and activity of mitochondria, resulting in greater ATP synthesis and heat production. The metabolic rate can increase 60%–100% when blood thyroid hormones are elevated. Low levels of thyroid hormones lead to the opposite effect. Maintaining normal body temperature depends on an adequate amount of thyroid hormones.

Normal growth and maturation of organs also depend on thyroid hormones. For example, bone, hair, teeth, connective tissue, and nervous tissue require thyroid hormones for normal growth and development. Both normal growth and normal maturation of the brain require thyroid hormones.

Regulation of Thyroid Hormone Secretion

Thyroid hormone secretion is regulated by hormones produced in the hypothalamus and anterior pituitary (figure 15.20). Thyrotropin-releasing hormone (TRH) is produced in the hypothalamus. Chronic exposure to cold increases TRH secretion, whereas stress, such as starvation, injury, and infections, decreases TRH secretion. TRH stimulates TSH secretion from the anterior pituitary. Small fluctuations in blood levels of TSH occur on a daily basis, with a small nocturnal increase. TSH stimulates the secretion of thyroid hormones from the thyroid gland. TSH also increases the synthesis of thyroid hormones, as well as causing an increase in thyroid gland cell size and number. Decreased blood levels of TSH lead to decreased secretion of thyroid hormones and thyroid gland atrophy. Thyroid hormones have a negative- feedback effect on the hypothalamus and anterior pituitary gland. As thyroid hormone levels increase in the circulatory system, they inhibit TRH and TSH secretion. Also, if the thyroid gland is removed or if the secretion of

1. TRH is released from neurons within the hypothalamus. It passes through the hypothalamohypophyseal portal system to the anterior pituitary.

2. TRH causes cells of the anterior pituitary to secrete TSH, which passes through the general circulation to the thyroid gland.

3. TSH causes increased release of T_3 and T_4 into the general circulation.

4. T_3 and T_4 act on target tissues to produce a response.

5. T_3 and T_4 also have an inhibitory effect on the secretion of TRH from the hypothalamus and TSH from the anterior pituitary.

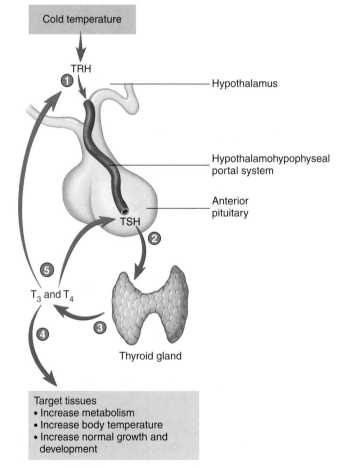

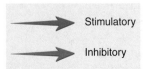

Process Figure 15.20   Regulation of Thyroid Hormone (T_3 and T_4) Secretion

Clinical Relevance   Hypothyroidism and hyperthyroidism

Hypothyroidism is reduced or no secretion of thyroid hormones. It can be caused by inadequate TSH stimulation of the thyroid gland, an inability of the thyroid gland to produce thyroid hormones, or the surgical removal or destruction of the thyroid gland for various reasons. Damage to the pituitary gland can result in decreased TSH secretion. Tumors and inadequate blood delivery to the pituitary because of blood loss during childbirth are causes of pituitary insufficiency. Lack of iodine in the diet can result in decreased thyroid hormone levels because iodine is necessary for the synthesis of thyroid hormones. Damage to the thyroid gland by drugs, chemicals, or an autoimmune disease (Hashimoto disease) can also reduce thyroid hormone production.

Hyposecretion of thyroid hormones decreases the rate of metabolism. Low body temperature, weight gain, reduced appetite, reduced heart rate, reduced blood pressure, decreased muscle tone, constipation, drowsiness, and apathy are major symptoms. **Myxedema** (mik-se-dē′mă), a swelling of the face and body as a result of mucoprotein deposits, can occur. If hyposecretion of thyroid hormones occurs during development, there is a decreased metabolic rate, abnormal nervous system development, abnormal growth, and abnormal maturation of tissues. The consequence is a mentally retarded person of short stature and distinctive form called a **cretin** (krē′tin).

Hyperthyroidism is an abnormally increased secretion of thyroid hormones. After diabetes mellitus, the most common endocrine disorder is a type of hyperthyroidism called **Graves disease.** It is an autoimmune disorder that produces a spe-cific immunoglobulin, called **thyroid stimulating immunoglobulin (TSI).** The structure of TSI is similar to the structure of TSH and it can bind to TSH receptors, resulting in continual stimulation of the thyroid gland and overproduction of thyroid hormones. TSH-secreting tumors and viral infections are other causes of hyperthyroidism. **Thyroid storm** is the sudden release of large amounts of thyroid hormones. It can be caused by surgery, stress, infections, or unknown causes.

Many of the symptoms of hypersecretion of thyroid hormones are the opposite of what occurs with hyposecretion of thyroid hormones. Hypersecretion of thyroid hormones increases the rate of metabolism. High body temperature, weight loss, increased appetite, rapid heart rate, elevated blood pressure, muscle tremors, diarrhea, insomnia, and hyperactivity are major symptoms. **Exophthalmos** (ek-sof-thal′mos) is a bulging of the eyes that often accompanies hyperthyroidism and is caused by the deposition of excess connective tissue proteins behind the eyes (figure B*a*). The excess tissue makes the eyes move anteriorly; consequently, they appear to be larger than normal.

Goiter (goy′ter, throat) is a chronic enlargement of the thyroid gland not due to a tumor (figure B*b*). Goiter eventually develops with chronic hypersecretion of thyroid hormones. TSI in Graves disease or elevated TSH produce by pituitary tumors results in continual overstimulation of the thyroid gland. Thyroid hormone synthesis increases and thyroid gland cells increase in size and number, producing goiter.

Hypothyroidism caused by an iodine deficiency in the diet can also cause goiter.

(a) Exophthalmos

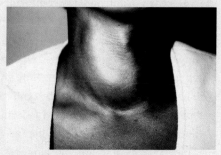

(b) Goiter

Figure B   Person with Hyperthyroidism

Without adequate iodine to synthesize thyroid hormones, blood levels of thyroid hormones decrease. The reduced negative feedback of thyroid hormones on the anterior pituitary and hypothalamus results in elevated TSH secretion. TSH causes increased thyroid gland cell size and number and increased thyroglobulin synthesis, even though there is not enough iodine to synthesize thyroid hormones. Historically, goiters were common in people from areas where the soil was depleted of iodine. Consequently, plants grown in these areas, called goiter belts, had little iodine in them and caused iodine-deficient diets. Iodized salt has nearly eliminated iodine-deficiency goiters. However, it remains a problem in some developing countries.

thyroid hormones declines, TSH levels in the blood increase dramatically. ▼

33. *Starting with the uptake of iodide by the follicles, describe the production and secretion of thyroid hormones.*
34. *How are the thyroid hormones transported in the blood?*
35. *By what mechanism do thyroid hormones alter the activities of their target tissues?*
36. *What are the target tissues of thyroid hormones? What effects are produced by thyroid hormones?*
37. *Starting in the hypothalamus, explain how chronic exposure to cold and stress can affect thyroid hormone production.*
38. *Describe the causes and effects of hyposecretion and hypersecretion of thyroid hormones.*

Josie owns a business, has several employees, and works hard to manage her business and make time for her family. Over several months, she slowly recognized that she felt warm when others did not; she sweated excessively and her skin was often flushed. She often felt as if her heart were pounding, she was much more nervous than usual, and it was difficult for her to concentrate. She began to feel weak and lose weight, even though her appetite was greater than normal. Her family recognized some of these changes and that her eyes seemed larger than usual. They encouraged her to see her physician. Based on the symptoms, her physician suspected that Josie was suffering from hyperthyroidism. A blood sample was taken and the results indicated that her blood levels of thyroid hormones were elevated and her blood levels of TRH and TSH were very low. In addition, a specific immunoglobulin, called thyroid-stimulating immunoglobulin (TSI), was present in significant concentrations in her blood. The structure of TSI is very similar to the structure of TSH. The physician concluded that Josie was suffering from Graves disease.

Josie was treated with radioactive iodine (^{131}I) atoms, which were actively transported into Josie's thyroid cells, where they destroyed a substantial portion of the thyroid gland. Subsequently, Josie had to take thyroid hormones in the form of a pill to keep her blood levels of thyroid hormones within their normal range of values.

P R E D I C T

a. Prior to treatment, explain why Josie's blood levels of thyroid hormones were elevated.

b. Prior to treatment, why were her TRH and TSH levels lower than normal?

c. After the ^{131}I treatment, why are her thyroid hormone levels lower than normal?

d. After the ^{131}I treatment, predict what will happen to Josie's TRH and TSH levels.

e. Why will Josie have to take thyroid hormone pills for the rest of her life? What effect will that have on her TRH and TSH levels?

Calcitonin

In addition to secreting thyroid hormones, the thyroid gland secretes a hormone called **calcitonin** (kal-si-tō′nin) produced by the parafollicular cells (see figure 15.18c). Calcitonin secretion is directly regulated by blood Ca^{2+} levels. As blood Ca^{2+} concentration increases, calcitonin secretion increases, and, as blood Ca^{2+} concentration decreases, calcitonin secretion decreases. Calcitonin binds to membrane-bound receptors of osteoclasts and inhibits them, which reduces the rate of bone matrix breakdown and the release of Ca^{2+} from bone into the blood. Calcitonin may prevent blood Ca^{2+} levels from becoming overly elevated following a meal that contains a high concentration of Ca^{2+}.

Calcitonin helps prevent elevated blood Ca^{2+} levels, but a lack of calcitonin secretion does not result in a prolonged increase in blood Ca^{2+} levels. Other mechanisms controlling blood Ca^{2+} levels, such as parathyroid hormone and vitamin D, are able to compensate for the lack of calcitonin secretion. ▼

39. *How is calcitonin secretion regulated?*

40. *What effect does calcitonin have on osteoclasts and blood calcium levels?*

Parathyroid Glands

The **parathyroid** (par-ă-thī′royd) **glands** are usually embedded in the posterior part of each lobe of the thyroid gland (figure 15.18b). Usually, four parathyroid glands are present, with their cells organized in densely packed masses or cords rather than in follicles.

The parathyroid glands secrete a polypeptide hormone called **parathyroid hormone (PTH),** which is essential for the regulation of blood Ca^{2+} levels (see table 15.3). PTH is much more important than calcitonin in regulating blood Ca^{2+} levels. PTH regulates blood Ca^{2+} levels by affecting Ca^{2+} release from bones, Ca^{2+} excretion by the kidneys, and vitamin D formation by the kidneys, which promotes Ca^{2+} absorption by the small intestine (see figure 6.13, p. 139).

PTH increases the release of Ca^{2+} from bones into blood by increasing the number of osteoclasts in bone, which results in increased bone breakdown. PTH promotes an increase in osteoclast numbers by stimulating stem cells in red bone marrow to differentiate and become osteoclasts. The effect of PTH on osteoclast formation, however, is indirect. PTH binds to its receptors on osteoblasts, stimulating them. The osteoblasts, through surface molecules and released chemicals, stimulate osteoclast stem cells to become osteoclasts.

In the kidneys, PTH increases Ca^{2+} reabsorption from the urine into the blood so that less calcium leaves the body in urine. PTH also increases the formation of active vitamin D in the kidneys. The vitamin D is carried by the blood to epithelial cells of the small intestine, where it promotes the synthesis of Ca^{2+} transport proteins. PTH increases blood Ca^{2+} levels by increasing the rate of active vitamin D formation, which in turn increases the rate of Ca^{2+} absorption in the intestine.

PTH secretion is directly regulated by blood Ca^{2+} levels. As blood Ca^{2+} concentration increases, PTH secretion decreases; as blood Ca^{2+} concentration decreases, PTH secretion increases (figure 15.21). This regulation keeps blood Ca^{2+} levels fluctuating within a normal range of values. ▼

41. *Where are the parathyroid glands located, and what hormone do they produce?*

42. *Name three ways by which PTH regulates blood Ca^{2+} levels. What stimulus can cause an increase in PTH secretion?*

P R E D I C T

Predict the effect of an inadequate dietary intake of calcium on PTH secretion and on target tissues for PTH.

Hypoparathyroidism is an abnormally low rate of PTH secretion resulting from injury to, or surgical removal of, the parathyroid glands. The low blood levels of PTH result in a reduced rate of bone breakdown and reduced vitamin D formation. As a result, blood Ca^{2+} levels decrease. In response to low blood Ca^{2+} levels, nerves and muscles become excitable and produce spontaneous action potentials (see chapter 10). Tetanus of skeletal muscles (see chapter 8) produces muscle cramps. Severe tetanus of the diaphragm and other respiratory muscles can stop breathing, resulting in death.

Hyperparathyroidism is an abnormally high rate of PTH secretion. **Primary hyperparathyroidism** usually results from a tumor of a parathyroid gland. The elevated blood levels of PTH increase bone breakdown and elevate blood Ca^{2+} levels. As a result, bones can become soft, deformed, and easily fractured. In addition, the elevated blood Ca^{2+} levels make nerve and muscle less excitable, resulting in fatigue and muscle weakness. The excess Ca^{2+} can be deposited in soft tissues of the body, and kidney stones can result. The Ca^{2+} deposits in soft tissues cause inflammation.

Secondary hyperparathyroidism is caused by conditions that cause long-term hyperstimulation of the parathyroid glands, such as low Ca^{2+} levels or tissue resistance to PTH. Chronic renal failure is the most common cause. Damage to the kidneys results in inadequate vitamin D production, which causes decreased Ca^{2+} absorption from the small intestine. The elevated PTH promotes increased breakdown of bone.

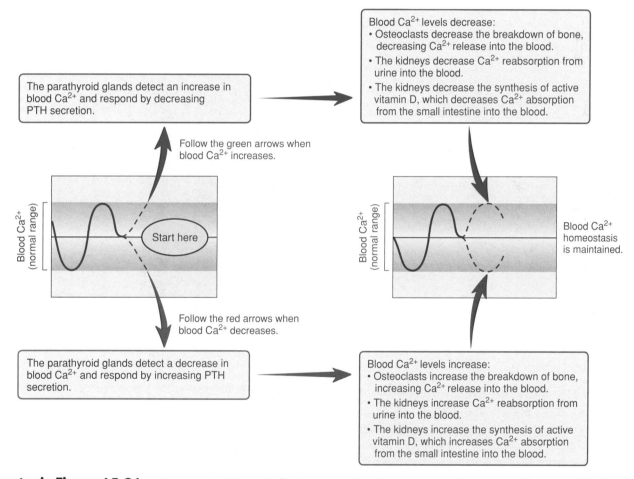

Homeostasis Figure 15.21   Summary of Blood Ca^{2+} Concentration Regulation by Parathyroid Hormone (PTH)

Adrenal Glands

The **adrenal** (ă-drē′năl, *ad*, near + *ren,* kidney) **glands,** or **suprarenal** (above the kidney) **glands,** are two small glands that are located superior to each kidney (figure 15.22). Each adrenal gland has an inner part, called the **adrenal medulla** (marrow or middle),

and an outer part, called the **adrenal cortex** (bark or outer). The cortex has three indistinct layers: the **zona glomerulosa** (glō-mār′ū-lōs-ă), the **zona fasciculata** (fa-sik′ū-lă-tă), and the **zona reticularis** (re-tik′ū-lăr′is). The medulla and the three layers of the cortex are structurally and functionally specialized. In many ways, an adrenal gland is four glands in one.

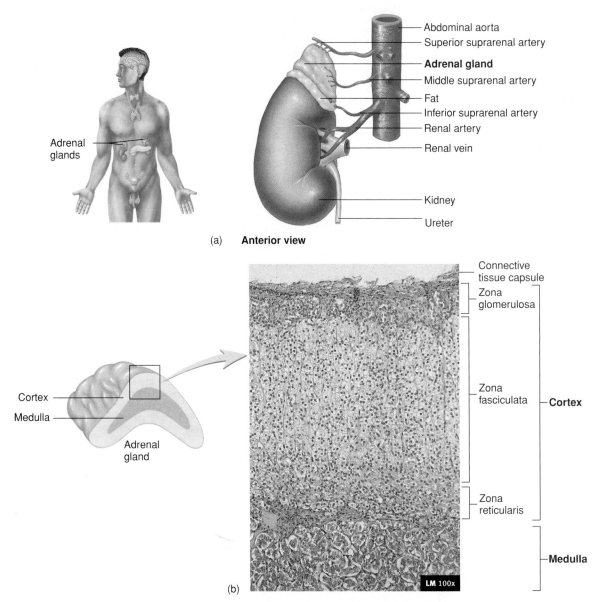

Figure 15.22 **Adrenal Glands**

(a) An adrenal gland is at the superior pole of each kidney. (b) The adrenal glands have an outer cortex and an inner medulla. The cortex is surrounded by a connective tissue capsule and consists of three layers: the zona glomerulosa, the zona fasciculata, and the zona reticularis.

Hormones of the Adrenal Medulla

Approximately 80% of the hormones released from the adrenal medulla is **epinephrine** (ep′i-nef′rin, *epi,* upon + *nephros,* kidney), or **adrenaline** (ă-dren′ă-lin, from the adrenal gland). The remaining 20% is **norepinephrine** (nōr′ep-i-nef′rin) (see table 15.3). The adrenal medulla consists of cells derived from the same cells that give rise to postganglionic sympathetic neurons, which secrete norepinephrine. Epinephrine is derived from norepinephrine.

The adrenal medulla and the sympathetic division function together to prepare the body for physical activity, producing the "fight-or-flight" response (see chapter 14). Some of the major effects of the hormones released from the adrenal medulla are the following:

1. Increased breakdown of glycogen to glucose in the liver, the release of the glucose into the blood, and the release of fatty acids from fat cells. The glucose and fatty acids are used as energy sources to maintain the body's increased rate of metabolism.
2. Increased heart rate, which increases blood pressure and blood delivery to tissues
3. Increased vasodilation of blood vessels of the heart and skeletal muscle, which increases blood flow to the organs responsible for increased physical activity. The hormones increase vasoconstriction of blood vessels to the internal organs and skin, which decreases blood flow to organs not directly involved in physical activity.
4. Increased metabolic rate of several tissues, especially skeletal muscle, cardiac muscle, and nervous tissue

The release of adrenal medullary hormones primarily occurs in response to stimulation by sympathetic neurons because the

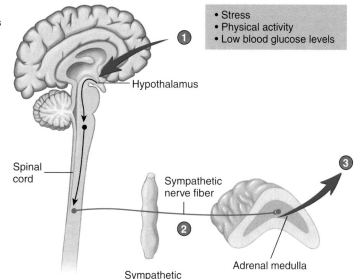

1. Stress, physical activity, and low blood glucose levels act as stimuli to the hypothalamus, resulting in increased sympathetic activity.

2. An increased frequency of action potentials conducted through the sympathetic division of the autonomic nervous system stimulates the adrenal medulla to secrete epinephrine and some norepinephrine into the circulatory system.

3. Epinephrine and norepinephrine act on their target tissues to produce responses.

- Stress
- Physical activity
- Low blood glucose levels

- Increase the release of glucose from the liver into the blood
- Increase the release of fatty acids from adipose tissue into the blood
- Increase heart rate
- Decrease blood flow through blood vessels of internal organs
- Increase blood flow through blood vessels of skeletal muscle and the heart
- Increase blood pressure
- Decrease the function of visceral organs
- Increase the metabolic rate of skeletal muscles

Hypothalamus

Spinal cord

Sympathetic nerve fiber

Adrenal medulla

Sympathetic chain

Process Figure 15.23 Regulation of Adrenal Medullary Secretions

Stress, physical exercise, and low blood glucose levels cause increased sympathetic activity, which increases epinephrine and norepinephrine secretion from the adrenal medulla.

adrenal medulla is a specialized part of the autonomic nervous system. Several conditions, including exercise, emotional excitement, injury, stress, and low blood glucose levels, lead to the release of adrenal medullary hormones (figure 15.23).

 Adrenal Tumors

The two major disorders of the adrenal medulla are both tumors. **Pheochromocytoma** (fē′ō-krō′mō-sī-tō′mă) is a benign tumor; **neuroblastoma** (noor′ō-blas-tō′mă) is a malignant tumor. Symptoms, resulting from the release of large amounts of epinephrine and norepinephrine, include hypertension (high blood pressure), sweating, nervousness, pallor, and tachycardia (rapid heart rate). The high blood pressure results from the effect of these hormones on the heart and blood vessels and is correlated with an increased chance of heart disease and stroke. ▼

43. *Where are the adrenal glands located?*
44. *Name two hormones secreted by the adrenal medulla, and list the effects of these hormones.*
45. *List conditions that can stimulate the secretion of adrenal medullary hormones.*
46. *Name the two major disorders of the adrenal medulla and the symptoms they produce.*

Hormones of the Adrenal Cortex

The adrenal cortex secretes three hormone types: **mineralocorticoids** (min′er-al-ō-kōr′ti-koydz), **glucocorticoids** (gloo-kō-kōr′ti-koydz), and **androgens** (an′drō-jenz) (see table 15.3). All are similar in structure in that they are steroids, highly specialized

lipids that are derived from cholesterol. Because they are lipid-soluble, they are not stored in the adrenal gland cells but diffuse from the cells as they are synthesized. Adrenal cortical hormones are transported in the blood in combination with specific plasma proteins; they are metabolized in the liver and excreted in the bile and urine. The hormones of the adrenal cortex bind to intracellular receptors and stimulate the synthesis of specific proteins that are responsible for producing the cell's responses.

Mineralocorticoids

The major secretory products of the zona glomerulosa are the mineralocorticoids. The mineralocorticoids are so named because they are steroids, produced by the adrenal cortex, that affect the "minerals" Na⁺, K⁺, and H⁺. **Aldosterone** (al-dos′ter-ōn) is produced in the greatest amounts, although other, closely related mineralocorticoids are also secreted. Aldosterone increases the rate of Na⁺ reabsorption by the kidneys, thereby increasing blood Na⁺ levels. Sodium reabsorption can result in increased water reabsorption by the kidneys and an increase in blood volume, providing ADH is also secreted. Increased blood volume can increase blood pressure.

Aldosterone increases K⁺ and H⁺ excretion into the urine by the kidneys, thereby decreasing blood levels of K⁺ and H⁺. When aldosterone is secreted in high concentrations, it can result in abnormally low blood levels of K⁺ and H⁺. The reduction in H⁺ can cause alkalosis, an abnormally elevated pH of body fluids. The details of the effects of aldosterone and the mechanisms controlling aldosterone secretion are discussed along with kidney functions in chapter 23 and with the cardiovascular system in chapter 18.

PREDICT

What effect do high and low blood levels of aldosterone have on nerve and muscle function (*hint:* blood K⁺ levels)?

Glucocorticoids

The zona fasciculata of the adrenal cortex primarily secretes glucocorticoids. The glucocorticoids are so named because they are steroids produced by the adrenal cortex that affect glucose metabolism. The major glucocorticoid is **cortisol** (kōr′ti-sol). The target tissues and responses to the glucocorticoids are numerous. The two major types of responses to glucocorticoids are classified as metabolic and anti-inflammatory. Cortisol increases the breakdown of protein and fat and increases their conversion to forms that can be used as energy sources by the body. For example, cortisol causes proteins in skeletal muscles to be broken down to amino acids, which are then released into the circulatory system. Cortisol acts on the liver, causing it to convert amino acids to glucose, which is released into the blood or stored as glycogen. Thus, cortisol increases blood sugar levels. Cortisol also acts on adipose tissue, causing fat stored in fat cells to be broken down to fatty acids, which are released into the circulation. The glucose and fatty acids released into the circulatory system are taken up by tissues and used as sources of energy.

Glucocorticoids decrease the intensity of the inflammatory and immune responses by decreasing both the number of white blood cells and the secretion of inflammatory chemicals from tissues. **Cortisone** (kōr′ti-sōn), a steroid closely related to cortisol, is often given as a medication to reduce inflammation that occurs in response to injuries. It is also given to reduce the immune and inflammatory responses that occur as a result of allergic reactions or diseases resulting from abnormal immune responses, such as rheumatoid arthritis or asthma.

In response to stressful conditions, cortisol is secreted in larger than normal amounts. Cortisol aids the body in responding to stressful conditions by providing energy sources for tissues. If stressful conditions are prolonged, however, immunity can be suppressed enough to make the body susceptible to infections.

Cortisol secretion is regulated through the hypothalamus and anterior pituitary gland (figure 15.24). Stress and low blood glucose levels stimulate increased **corticotropin-releasing hormone (CRH)** from the hypothalamus. CRH stimulates increased **adrenocorticotropic** (ă-drē′nō-kōr′ti-kō-trō′pik) **hormone (ACTH)** secretion from the anterior pituitary gland. ACTH stimulates increased cortisol secretion from the adrenal gland. ACTH is also necessary to maintain the secretory activity of the *entire* adrenal cortex, which rapidly atrophies without this hormone. Cortisol inhibits CRH and ACTH secretion. This negative-feedback loop is important in maintaining blood cortisol levels within a narrow range of concentrations.

PREDICT 10

Cortisone, a drug similar to cortisol, is sometimes given to people to reduce inflammation. Taking cortisone for long periods of time can damage the adrenal cortex. Explain how this damage can occur.

1. Cortiocotropin-releasing hormone (CRH) is released from hypothalamic neurons in response to stress or low blood glucose and passes, by way of the hypothalamohypophyseal portal system, to the anterior pituitary.

2. CRH stimulates the secretion of adrenocorticotropic hormone (ACTH) from the anterior pituitary.

3. ACTH stimulates the secretion of cortisol from the adrenal cortex.

4. Cortisol acts on target tissues, resulting in increased fat and protein breakdown, increased glucose levels, and anti-inflammatory effects.

5. Cortisol has a negative-feedback effect because it inhibits CRH release from the hypothalamus and ACTH secretion from the anterior pituitary.

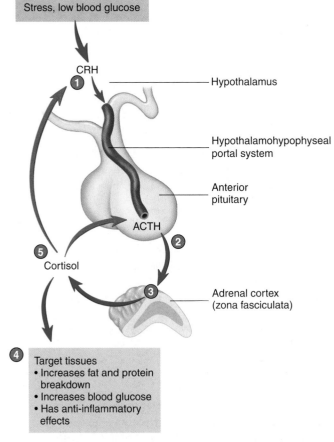

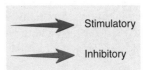

Process Figure 15.24   Regulation of Cortisol Secretion

Several pathologies are associated with hyposecretion and hypersecretion of adrenal cortex hormones.

Hyposecretion

Chronic adrenocortical insufficiency, often called **Addison disease,** results from abnormally low levels of aldosterone and cortisol in the blood. The cause of many cases of chronic adrenocortical insufficiency is unknown, but it frequently results from an autoimmune disease in which the body's defense mechanisms inappropriately destroy the adrenal cortex. Other causes are infections and tumors that damage the adrenal cortex. Prolonged treatment with glucocorticoids, which suppresses pituitary gland function, can also cause chronic adrenocortical insufficiency. Symptoms resulting from low levels of aldosterone include hyponatremia (decreased blood Na$^+$), hyperkalemia (elevated K$^+$), acidosis (elevated H$^+$), tremors and tetanus of skeletal muscle (due to elevated K$^+$), and increased urine production (due to a loss of Na$^+$ and water in the urine). As water is lost in the urine, blood volume decreases, causing blood pressure to decrease. Reduced blood pressure is the most critical manifestation of Addison disease and requires immediate treatment.

Symptoms resulting from low levels of cortisol include hypoglycemia (low blood sugar), weight loss (due to inadequate protein and fat metabolism), and in many cases increased pigmentation of the skin. An understanding of how ACTH is produced is necessary to understand how low cortisol levels can cause increased skin pigmentation. ACTH is one of several peptide hormones derived from a precursor protein called **proopiomelanocortin** (prō-ō′pē-ō-mel′ă-nō-kōr′tin). This precursor protein is broken apart to form ACTH and melanocyte-stimulating hormone (MSH), which binds to melanocytes in the skin and increases skin pigmentation (see chapter 5). Thus, when ACTH production increases, so does MSH production. When cortisol levels are low, there is a decrease in inhibition of the anterior pituitary by cortisol and ACTH and MSH production increases.

Hypersecretion

Aldosteronism (al-dos′ter-on-izm) is caused by an excess production of aldosterone. Primary aldosteronism results from an adrenal cortex tumor, and secondary aldosteronism occurs when an extraneous factor, such as an overproduction of renin, a substance produced by the kidneys, increases aldosterone secretion. Major symptoms of aldosteronism are hypernatremia (elevated blood Na$^+$), hypokalemia (decreased K$^+$), alkalosis (decreased H$^+$), and high blood pressure due to the retention of water and Na$^+$ by the kidneys.

Cushing syndrome is a disorder characterized by the hypersecretion of cortisol and androgens and possibly by excess aldosterone production. Most cases are caused by excess ACTH production by nonpituitary tumors, which usually result from a type of lung cancer. Some cases of increased ACTH secretion do result from pituitary tumors. Sometimes adrenal tumors or unidentified causes can be responsible for hypersecretion of the adrenal cortex without increases in ACTH secretion. Elevated secretion of glucocorticoids results in hyperglycemia (high blood glucose levels) and depression of the immune system. Destruction of tissue protein causes muscle atrophy and weakness, osteoporosis, bruising (weak capillary walls), thin skin, and impaired wound healing. Fats accumulate in the face, neck, and trunk (figure C).

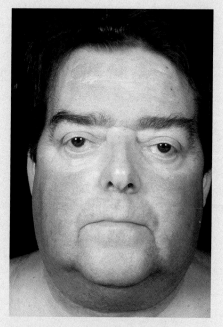

Figure C Male Patient with Cushing Syndrome

Adrenogenital (ă-drē′nō-jen′i-tăl) **syndrome** results from the hypersecretion of androgens from the adrenal cortex. If the condition occurs in childhood, secondary sex characteristics develop early in boys, and girls are masculinized. If the condition develops before birth in females, the external genitalia can be masculinized to the extent that the infant's reproductive structures are neither clearly female nor male. Hypersecretion of adrenal androgens in male children before puberty results in rapid and early development of the reproductive system. If not treated, early sexual development and short stature result. The short stature results from the effect of androgens on skeletal growth (see chapter 6). In women, the partial development of male secondary sex characteristics, such as facial hair and a masculine voice, occurs.

Adrenal Androgens

The zona reticularis secretes **androgens** (an′drō-jenz, *andros,* male), which are named for their ability to stimulate the development of secondary male sexual characteristics (see chapter 24). Adrenal androgens are converted by peripheral tissues to the more potent androgen testosterone. Small amounts of androgens are secreted from the adrenal cortex in both males and females. In adult males, most androgens are secreted by the testes. The effects of adrenal androgens in males are negligible, in comparison with testosterone secreted by the testes. In adult females, androgens stimulate pubic and axillary hair growth and sexual drive. If the secretion of sex hormones from the adrenal cortex is abnormally high, exaggerated male characteristics develop in both males and females. This condition is most apparent in females and in males before puberty, when the effects are not masked by the secretion of androgens by the testes. ▼

47. Describe the three layers of the adrenal cortex, and name the hormones produced by each layer.

48. *What effect does increased aldosterone have on blood Na⁺, K⁺, and H⁺ levels?*

49. *Describe the metabolic and anti-inflammatory effects produced by an increase in cortisol secretion.*

50. *Starting in the hypothalamus, describe how stress or low blood sugar levels can stimulate cortisol release.*

51. *What effects do adrenal androgens have on males and females?*

Pancreas

The **pancreas** (pan′krē-us) is an elongated organ approximately 15 cm long and weighing approximately 85–100 g. It lies posterior to the stomach and is partially surrounded by a loop of the first part of the small intestine (figure 15.25). The pancreas is both an exocrine gland and an endocrine gland. The exocrine portion consists of **acini** (as′i-nī), which produce pancreatic juice, and a duct system, which carries the pancreatic juice to the small intestine (see chapter 24). The endocrine part consists of **pancreatic islets** (islets of Langerhans), which secrete hormones that enter the circulatory system.

Between 500,000 and 1 million pancreatic islets are dispersed among the ducts and acini of the pancreas. Within the islets, **alpha (α) cells** secrete **glucagon** and **beta (β) cells** secrete **insulin.** Nerves from both divisions of the autonomic nervous system innervate the pancreatic islets, and a well-developed capillary network surrounds each islet.

Effect of Insulin and Glucagon on Their Target Tissues

Insulin

Insulin increases the uptake of glucose and amino acids by target cells. Once insulin binds to its receptors, the receptors cause specific proteins in the membrane to become phosphorylated. Part of the cell's response to insulin is to increase the number of transport proteins in the membrane of cells for glucose and amino acids.

The major target tissues of insulin are the liver, adipose tissue, the skeletal muscles, and the satiety center within the hypothalamus of the brain. The **satiety** (sa′-tī-ĕ-tē) **center** is a collection of neurons in the hypothalamus that controls appetite (see chapter 11). Unlike the satiety center, most of the nervous system does not depend on insulin for the uptake of glucose. Insulin is very important for the normal functioning of the nervous system, however, because insulin regulates blood glucose levels. If blood glucose levels are not maintained within a normal range, the brain malfunctions because glucose is its primary energy source.

When insulin levels increase, the movement of glucose and amino acids into cells increases. Glucose molecules that are not immediately used as an energy source are stored as glycogen in the liver, skeletal muscle, and other tissues, or they are converted to fat in adipose tissue. Amino acids are used to synthesize proteins.

When insulin levels decrease, the opposite effects are observed. The movement of glucose and amino acids into tissues

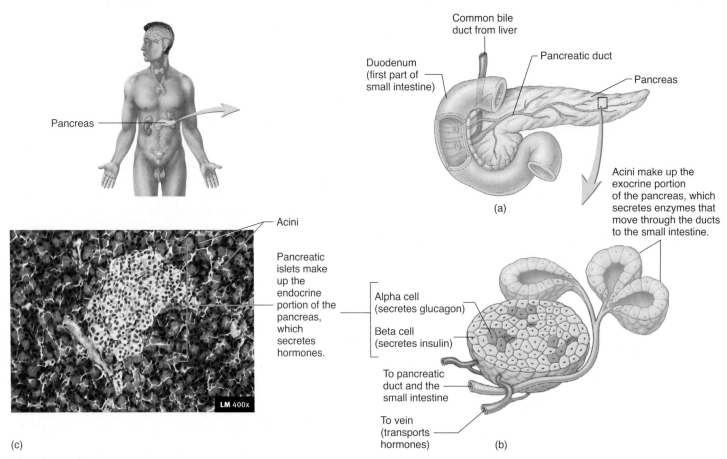

Figure 15.25 **Pancreas**

(*a*) Anterior view of the pancreas. (*b*) A pancreatic islet consists of clusters of specialized cells among the acini of the exocrine portion of the pancreas. (*c*) Light micrograph of pancreatic tissue. The stain used for this slide does not distinguish between alpha and beta cells.

slows, glycogen is broken down to glucose and released from the liver, adipose tissue releases fatty acids, and proteins are broken down into amino acids, especially in skeletal muscle. The amino acids are released into the blood, taken up by the liver, and used to synthesize glucose, which is released into the blood.

When insulin levels decrease, the liver uses fatty acids to make **acetoacetic** (as′e-tō-a-sē′tik) **acid,** which is converted to **acetone** (as′e-tōn) and **β-hydroxybutyric** (bā′tă hī-drok′sē-bū-tir′ik) **acid.** These three substances collectively are referred to as **ketone** (kē′tōn) **bodies.** The liver releases the ketone bodies into the blood, from which other tissues take them up and use them as a source of energy. The ketone bodies are smaller, more readily used "packets" of energy than are fatty acids. Ketone bodies, however, are acids that can adversely affect blood pH if too many of them are produced (see "Diabetes mellitus," p. 452). In addition, when insulin levels are low, the liver releases cholesterol and triglycerides into the blood.

Glucagon

Glucagon increases blood sugar and ketone levels. Glucagon primarily influences the liver, although it has some effect on skeletal muscle and adipose tissue. The pancreas secretes glucagon into the hepatic portal system, which carries blood to the liver from the pancreas and intestines (see chapter 18). Glucagon binds to membrane-bound receptors, activates G proteins, and increases cAMP synthesis. In general, glucagon causes the breakdown of glycogen to glucose and increases glucose synthesis from amino acids. The release of glucose from the liver increases blood glucose levels.

Regulation of Pancreatic Hormone Secretion

Insulin

Blood levels of nutrients, neural stimulation, and hormones control the secretion of insulin. Elevated blood levels of glucose directly affect the β cells and stimulate insulin secretion. Low blood levels of glucose directly inhibit insulin secretion. Thus, blood glucose levels play a major role in the regulation of insulin secretion. Certain amino acids also stimulate insulin secretion by acting directly on the β cells. After a meal, when glucose and amino acid levels increase in the circulatory system, insulin secretion increases. During periods of fasting, when blood glucose levels are low, the rate of insulin secretion declines (figure 15.26).

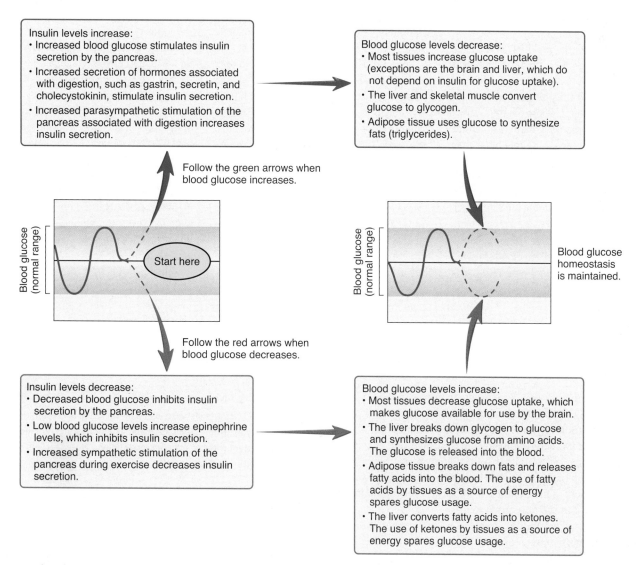

Homeostasis Figure 15.26   Summary of Insulin Secretion Regulation

Clinical Relevance   Diabetes mellitus

Diabetes mellitus results from the inadequate secretion of insulin or the inability of tissues to respond to insulin. As a result, blood sugar levels increase. Approximately 7% of Americans have diabetes and one-third of them are unaware that they have the disease. The two major types of diabetes are type 1 and type 2 diabetes. **Type 1 diabetes mellitus,** also called **insulin-dependent diabetes mellitus (IDDM),** results from diminished or absent insulin secretion. It affects approximately 5%–10% of people with diabetes mellitus and most commonly occurs in young people. Type 1 diabetes mellitus develops as a result of autoimmune destruction of the pancreatic islets, and symptoms appear after approximately 90% of the islets have been destroyed. Heredity plays a role in the condition, although the initiation of pancreatic islet destruction may involve a viral infection of the pancreas.

Type 2 diabetes mellitus, also called **noninsulin-dependent diabetes mellitus (NIDDM),** results from insulin resistance, the inability of tissues to respond normally to insulin. It affects approximately 90%–95% of people who have diabetes mellitus and usually develops in people older than 40–45 years of age, although the age of onset varies considerably. People with type 2 diabetes mellitus have a reduced number of functional receptors for insulin, or one or more of the enzymes activated by the insulin receptor are defective. Heredity influences the likelihood of developing type 2 diabetes, but it is not as important a risk factor as for type 1 diabetes.

Gestational diabetes mellitus results from insulin resistance. Hormones produced by the placenta, such as estrogens, progesterone, and human placental lactogen, increase insulin resistance. Gestational diabetes occurs in approximately 4% of pregnant women. Usually, following delivery the gestational diabetes goes away. Women who have had gestational diabetes, however, may develop type 2 diabetes years later.

Hyperglycemia is abnormally high blood glucose levels. It can develop in patients with untreated diabetes mellitus for two reasons: (1) Glucose is not entering cells normally and (2) glucose is being synthesized because of increased glucagon, cortisol, and epinephrine levels.

Glucose tolerance tests are used to diagnose diabetes mellitus. In general, the test involves feeding the patient a large amount of glucose after a period of fasting. Blood samples are collected for a few hours, and a sustained increase in blood glucose levels strongly indicates that the person is suffering from diabetes mellitus.

Polyuria and polydipsia are major symptoms of untreated diabetes mellitus related to hyperglycemia. **Polyuria** (pol-ē-ū′rē-ă) is increased urine volume. When blood glucose levels are very high, glucose molecules enter the kidney tubules and increase the solute concentration of the urine. Normally, urine does not contain glucose, and the presence of glucose in the urine means it is very likely that a person has diabetes. Water is attracted to the glucose because of osmosis and urine volume

increases. **Polydipsia** (pol-ē-dip′sē-ă) is increased thirst. The loss of water and the high blood glucose levels increase the osmotic concentration of blood, which increases the sensation of thirst.

Type 1 and type 2 diabetes can have different effects on body weight. Before the treatment of diabetes with insulin, untreated type 1 diabetics appeared to starve to death in spite of eating a large amount of food. In the absence of insulin, the satiety center cannot detect the presence of glucose in the extracellular fluid, even when high levels are present. Intense hunger results in excessive eating, called **polyphagia** (pol-ē-fā′jē-ă, much eating). Despite high blood glucose levels, tissues break down amino acids and fats for energy, resulting in tissue wasting.

In patients with type 2 diabetes mellitus, obesity is common, although not universal. Elevated blood glucose levels cause fat cells to convert glucose to fat, even though the rate at which adipose cells take up glucose is impaired.

Three potentially life-threatening conditions are associated with untreated diabetes mellitus: diabetic ketoacidosis, hyperglycemic hyperosmolar state, and insulin shock. **Diabetic ketoacidosis (DKA)** is the triad of hyperglycemia, ketosis, and acidosis. **Ketosis** (kē-tō′sis) is the presence of excess ketone bodies in the blood. When insulin levels are low and glucagon and epinephrine levels are high, there is increased utilization of fatty acids for energy, resulting in the excess production

---

The autonomic nervous system also controls insulin secretion. The parasympathetic stimulation of digestive system organs is associated with food intake. Parasympathetic stimulation of the pancreas increases its secretion of insulin and digestive enzymes. Sympathetic stimulation inhibits insulin secretion and helps prevent a rapid fall in blood glucose levels during periods of physical activity or excitement. This response is important for maintaining normal functioning of the nervous system.

Gastrointestinal hormones involved with the regulation of digestion, such as gastrin, secretin, and cholecystokinin (see chapter 21), increase insulin secretion.

PREDICT

Explain why the increase in insulin secretion in response to parasympathetic stimulation and gastrointestinal hormones is consistent with the maintenance of blood glucose levels in the circulatory system.

Glucagon

Low blood glucose levels stimulate glucagon secretion, and high blood glucose levels inhibit it. Certain amino acids and sympathetic stimulation also increase glucagon secretion. After a high-protein

of ketone bodies by the liver. The presence of excreted ketone bodies in the urine and in expired air ("acetone breath") suggests that a person has diabetes mellitus. Ketone bodies are acids, and an increase in ketone bodies can produce **acidosis,** an abnormally acidic blood pH (see chapter 23). Type 1 diabetics are most likely to develop DKA because they do not produce enough insulin to prevent ketosis.

Hyperglycemic hyperosmolar state (HHS) consists of very elevated blood sugar levels. It is most likely to develop in type 2 diabetics who have enough insulin to prevent ketosis, but not enough insulin to prevent hyperglycemia. Over time, blood sugar levels can gradually increase to very high levels, especially if fluid intake does not match fluid loss caused by polyuria. In DKA and HHS, elevated blood sugar levels cause the dehydration of tissues as water moves from the tissues into the sugar-concentrated blood. The dehydration of brain cells causes lethargy, fatigue, periods of irritability, disorientation, and coma.

Insulin shock occurs when there is too much insulin relative to the amount of blood glucose. Too much insulin, too little food intake after an injection of insulin, or increased metabolism of glucose due to excess exercise by a diabetic patient can cause insulin shock. The high levels of insulin cause target tissues to take up glucose at a very high rate. As a result, blood glucose levels rapidly fall to a low level. The nervous system depends on glucose as its major source of energy and neurons malfunction when blood glucose levels become too low. The result can be disorientation, confusion, and coma. The rapid decrease in blood glucose levels stimulates epinephrine secretion. Epinephrine causes pale skin (vasoconstriction), rapid heart rate, and increased nervous system excitability.

Diabetes has many serious, long-term consequences. Damage to the retina of the eye causes impaired vision and blindness. Damage to the kidneys results in kidney failure. Depending on the part of the nervous system affected, there can be sensations of pain, coldness, tingling, or burning; muscle weakness and paralysis; loss of bladder and bowel control; and impotence. There is an increased risk for cardiovascular disease associated with atherosclerosis because low insulin levels promote the release of cholesterol and triglycerides from the liver. Decreased blood delivery to tissues increases the risk for infections and results in poor wound healing, which can lead to amputation.

Keeping blood glucose levels within a normal range appears to reduce the risk of developing the severe, long-term consequences of diabetes. Taking insulin injections (figure D), monitoring blood glucose levels, and following a strict diet to keep blood glucose levels within a normal range of values are the major treatments for type 1 diabetes mellitus. In some people with type 2 diabetes mellitus, insulin production eventually decreases because pancreatic islet cells atrophy and type 1 diabetes mellitus develops. Approximately 25%–

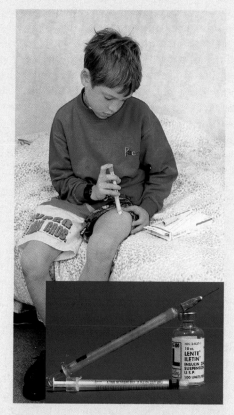

Figure D   Ten-Year-Old Boy Giving Himself Insulin

30% of patients with type 2 diabetes mellitus take insulin, 50% take oral medication to increase insulin secretion and increase the efficiency of glucose utilization, and the remainder control blood glucose levels with exercise and diet.

meal, amino acids increase both insulin and glucagon secretion. Insulin causes target tissues to accept the amino acids for protein synthesis, and glucagon increases the process of glucose synthesis from amino acids in the liver. ▼

52. *Where is the pancreas located? Describe the exocrine and endocrine parts of this gland and the secretions produced by each portion.*
53. *What hormones are produced by alpha (α) and beta (β) cells of the pancreas?*
54. *Name the major target tissues for insulin and glucagon.*
55. *What effects are produced by an increase and decrease of insulin? Describe three ways in which insulin secretion is regulated.*
56. *What effects are produced by an increase and a decrease of glucagon? How is glucagon secretion regulated?*

Hormonal Regulation of Nutrients

Two situations—after a meal and during exercise—can illustrate how several hormones function together to regulate blood nutrient levels. After a meal and under resting conditions, increasing blood glucose levels and parasympathetic stimulation elevate insulin secretion, but the secretion of glucagon, cortisol, GH, and epinephrine is reduced. Increased insulin levels promote the uptake of glucose, amino acids, and fats by target tissues. Substances not immediately used for cell metabolism are stored. Glucose is converted to glycogen in skeletal muscle and the liver, and it is used for fat synthesis in adipose tissue and the liver. The rapid uptake and storage of glucose prevent too large an increase in blood glucose levels. Amino acids are incorporated into proteins, and fats ingested as part of the meal are stored in adipose tissue and the liver. If the meal is high in protein, a small amount of glucagon is secreted, thereby increasing the rate at which the liver uses amino acids to form glucose.

Within 1–2 hours after the meal, absorption of digested materials from the gastrointestinal tract declines and blood glucose levels decline. As a result, the secretion of insulin decreases, but the secretion of glucagon, GH, cortisol, and epinephrine increases. As a result, the rate of glucose entry into the target tissues for insulin decreases. Glycogen, stored in cells, is converted back to glucose and is used as an energy source. The liver converts glycogen to glucose and synthesizes glucose from amino acids derived from the breakdown of proteins. The liver releases this glucose into the blood. The decreased uptake of glucose by most tissues, combined with its release from the liver, helps maintain blood glucose at levels necessary for normal brain function. Cells that use less glucose start using more fats. Adipose tissue releases fatty acids, and the liver uses fatty acids to produce ketones, which are released into the blood. Tissues take up fatty acids and ketones from the blood and use them for energy. Fatty acids and ketones are major sources of energy for most tissues when blood glucose levels are low.

The interactions of insulin, GH, glucagon, epinephrine, and cortisol are excellent examples of negative-feedback mechanisms. When blood glucose levels are high, these hormones cause the rapid uptake and storage of glucose, amino acids, and fats. When blood glucose levels are low, they cause the release of glucose and a switch to fat and protein metabolism as a source of energy for most tissues.

During exercise, skeletal muscles require energy to support the contraction process (see chapter 8). Although metabolism of intracellular nutrients can sustain muscle contraction for a short time, additional energy sources are required during prolonged activity. Sympathetic nervous system activity, which increases during exercise, stimulates the release of epinephrine from the adrenal medulla and of glucagon from the pancreas. These hormones induce the conversion of glycogen to glucose in the liver and the release of glucose into the blood, thus providing skeletal muscles with a source of energy.

During sustained activity, glucose released from the liver and other tissues is not adequate to support muscle activity, and a danger exists that blood glucose levels will become too low to support brain function. A decrease in insulin prevents the uptake of glucose by most tissues, thus conserving glucose for the brain. Epinephrine, glucagon, cortisol, and GH cause an increase of fatty acids in the blood. Because GH increases protein synthesis and it slows the breakdown of proteins, muscle proteins are not used as an energy source. Consequently, glucose metabolism decreases and fat metabolism in skeletal muscles increases. At the end of a long race, for example, muscles rely to a large extent on fat metabolism for energy. ▼

57. *Describe the hormonal effects after a meal that result in the movement of nutrients into cells and their storage. Describe the hormonal effects that later cause the release of stored materials for use as energy.*
58. *During exercise, how does sympathetic nervous system activity regulate blood glucose levels? Name five hormones that interact to ensure that both the brain and muscles have adequate energy sources.*

P R E D I C T   **12**

Explain why long-distance runners may not have much of a "kick" left when they try to sprint to the finish line.

Testes and Ovaries

The testes of the male and the ovaries of the female secrete sex hormones, in addition to producing sperm cells or oocytes (see figure 15.1 and table 15.3). The main sex hormone produced in the male is **testosterone** (tes′tos′tĕ-rōn), which is secreted by the testes. It is responsible for the growth and development of the male reproductive structures, muscle enlargement, the growth of body hair, voice changes, and the male sexual drive.

In the female, two classes of sex hormones are secreted by the ovaries: **estrogen** (es′trō-jen) and **progesterone** (prō-jes′ter-ōn). Together these hormones contribute to the development and function of female reproductive structures and other female sex characteristics. These characteristics include enlargement of the breasts and distribution of fat, which influences the shape of the hips, breasts, and thighs. The female menstrual cycle is controlled by the cyclical release of estrogen and progesterone from the ovaries.

The control of the hormones that regulate reproductive functions is discussed in greater detail in chapter 24. ▼

59. *Name the major hormones secreted by the testes and ovaries. Describe some of the effects of these hormones.*

Pineal Body

The **pineal** (pin′ē-ăl, pinecone) **body** is a small, pinecone-shaped structure located superior and posterior to the thalamus of the brain (see chapter 11). The pineal body produces a hormone called **melatonin** (mel-ă-tōn′in), which inhibits the functions of the reproductive system in some animals. Melatonin decreases the secretion of gonadotropin–releasing hormone (GnRH) from the hypothalamus. GnRH decreases the secretion of luteinizing hormone and follicle-stimulating hormone from the anterior pituitary. Without these anterior pituitary hormones, the testes do not produce testosterone and the ovaries do not produce estrogen and progesterone. Consequently, the development of reproductive structures and behavior does not occur. The function of the pineal body in humans is not clear, but melatonin may play an important role in the onset of puberty. Tumors that destroy the pineal body correlate with early sexual development, and tumors that result in pineal hormone secretion correlate with retarded development of the reproductive system. It is not clear, however, if the pineal body controls the onset of puberty.

The amount of light detected by the eyes regulates the rate of melatonin secretion. The axons of some neurons in the retina pass from the optic chiasm to the suprachiasmatic nucleus in the hypothalamus (see figure 11.20c), which influences the pineal body through sympathetic neurons (figure 15.27). Increased light exposure inhibits melatonin secretion, whereas darkness allows melatonin secretion. Melatonin is sometimes called the "hormone of darkness" because its production increases in the dark. In many animals, longer day length (shorter nights) causes a decrease in melatonin secretion, whereas shorter day length (longer nights) causes an increase in melatonin secretion. For example, in animals that breed in the spring, increased day length results in decreased melatonin secretion. With decreased inhibition of the hypothalamus by melatonin, sex hormone production increases, which promotes the development of reproductive structures and behavior. In the fall, decreased day length results in increased melatonin secretion, decreased sex hormone production, atrophy of reproductive organs, and a cessation of reproductive behavior. Thus, reproductive development and behavior is tied to seasonal changes in day length. Humans are not seasonal breeders, but human do secrete larger amounts of melatonin at night than in the daylight. Melatonin may effect circadian rhythms (see chapter 11). For example, melatonin effects the sleep-wake cycle by increasing the tendency to sleep. ▼

60. Where is the pineal body located? Name the hormone it produces and its possible effects.
61. What effect does light exposure have on melatonin secretion?

Other Endocrine Organs

The **thymus** (thī'mŭs) is in the neck and superior to the heart in the thorax; it secretes a hormone called **thymosin** (thī'mō-sin) (see table 15.3). Both the thymus and thymosin play an important role in the development and maturation of the immune system, as discussed in chapter 19.

Several hormones, such as **gastrin** (gas'trin), **secretin** (se-krē'tin), and **cholecystokinin** (kō'lē-sis-tō-kī'nin), are released from the gastrointestinal tract. They regulate digestive functions by influencing the activity of the stomach, intestines, liver, and pancreas. They are discussed in chapter 21.

The kidneys secrete a hormone in response to reduced oxygen levels in the kidney. The hormone is called **erythropoietin** (ĕ-rith'rō-poy'ĕ-tin, *erythro* refers to red blood cells). It acts on red bone marrow to increase the production of red blood cells (see chapter 16).

In pregnant women, the placenta is an important source of hormones that maintain pregnancy and stimulate breast

1. Light entering the eye stimulates neurons in the retina of the eye to produce action potentials.

2. Action potentials are transmitted to the hypothalamus in the brain.

3. Action potentials from the hypothalamus are transmitted through the sympathetic division to the pineal body.

4. A decrease in light (dark) results in increased sympathetic stimulation of the pineal gland and increased melatonin secretion. An increase in light results in decreased sympathetic stimulation of the pineal gland and decreased melatonin secretion.

5. Melatonin inhibits GnRH secretion from the hypothalamus and may help regulate sleep cycles.

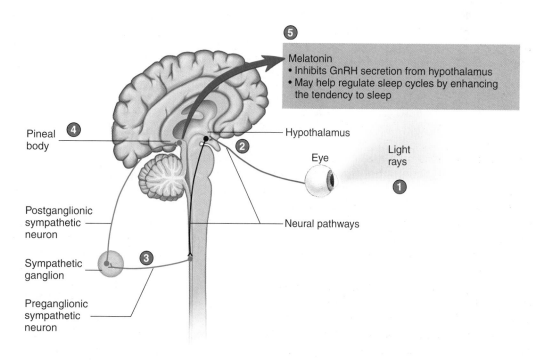

Process Figure 15.27 Regulation of Melatonin Secretion from the Pineal Body

Light entering the eye inhibits and dark stimulates the release of melatonin from the pineal body.

Endocrine

Systems Interactions

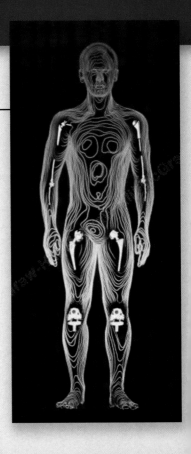

	Effects of the Endocrine System on Other Systems	Effects of Other Systems on the Endocrine System
Integumentary System	Sex hormones increase sebum production (contributing to acne), increase apocrine gland secretion (contributing to body odor), and stimulate axillary and pubic hair growth Melanocyte-stimulating hormone increases melanin production	Produces a vitamin D precursor that is converted in the liver and kidneys to active vitamin D, which functions as a hormone
Skeletal System	Parathyroid hormone and calcitonin regulate calcium release and uptake by bone Estrogen and testosterone stimulate bone growth and closure of the epiphyseal plate Growth hormone and thyroid hormone affect bone growth and metabolism	Protects endocrine organs in the cranial, thoracic, and pelvic cavities
Muscular System	Hormones affect muscle development, growth, and metabolism	Muscular exercise stimulates endorphin release
Nervous System	Hormones affect neuron development, growth, and metabolism	Controls the release of hormones from the hypothalamus, posterior pituitary, and adrenal glands
Cardiovascular System	Epinephrine and norepinephrine increase heart rate and force of contraction and change blood vessel diameter Hormones regulate blood pressure Erythropoietin stimulates red blood cell formation	Transports hormones released from endocrine tissues Delivers oxygen, nutrients, and immune cells Removes carbon dioxide, waste products, and toxins
Lymphatic System and Immunity	Thymosin is necessary for immune cell (T cell) maturation in the thymus gland Hormones affect immune cell functions	Lymph transports hormones Immune cells provide protection against microorganisms and toxins and promote tissue repair by releasing chemical mediators Removes excess interstitial fluid
Respiratory System	Epinephrine causes dilation of air passage ways (bronchioles)	Provides oxygen and removes carbon dioxide Helps maintain the body's pH
Digestive System	Regulates secretion from digestive glands and organs Controls mixing and movement of digestive tract contents	Provides nutrients and water
Urinary System	ADH and aldosterone regulate fluid and electrolyte balance	Removes waste products Helps maintain the body's pH, ion, and water balance
Reproductive System	Stimulates the onset of puberty and sexual characteristics Stimulates gamete formation Promotes uterine contractions for delivery Makes possible and regulates milk production	Testosterone, estrogen, and inhibin regulate the release of hormones from the hypothalamus and pituitary gland

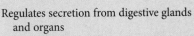

development. These hormones include estrogen, progesterone, and **human chorionic** (kō-rē-on′ik) **gonadotropin** (gō′nad-ō-trō′pin), which is similar in structure and function to LH. These hormones are essential to the maintenance of pregnancy (see chapter 24). ▼

> 62. **What hormones are secreted by the thymus, gastrointestinal tract, kidneys, and placenta?**

Hormonelike Substances

Autocrine (aw′tō-krin, *autos*, self + *krinō*, to separate) **agents** are chemical signals released by cells that have a local effect on the same cell type from which the autocrine agent is released. Examples of autocrine chemical signals include a group of related chemical mediators called **eicosanoids** (ī′kō-să-noydz), which are derived from fatty acids. The eicosanoids include **prostaglandins** (pros′-stă-glandinz), **thromboxanes** (throm′bok-zānz), **prostacyclins** (pros-tă-sī′klinz), and **leukotrienes** (loo′kō-trī′ēnz).

Paracrine (par′ă-krin, *para*, alongside + crine) **agents** are chemical signals released by cells that have a local effect on different cell types from which the paracrine agent is released. Examples of paracrine chemical signals include growth factors, clotting factors, and histamine.

Autocrine and paracrine chemical signals differ from hormones in that they are not secreted from discrete endocrine glands, they have local effects rather than systemic effects, or they have functions that are not understood adequately to explain their role in the body.

The schemes used to classify chemicals on the basis of their functions are useful, but they do not indicate that a specific molecule always performs as the same type of chemical signal in every place it is found. The distinction between autocrine and paracrine is not absolute because some chemical signals, such as prostaglandins, have both autocrine and paracrine functions. Furthermore, some of these chemicals can also act as hormones. Testosterone produced in the testes has a paracrine effect on the development of sperm cells, but it is released into the blood and has an endocrine effect on skeletal muscle development. ▼

> 63. **Define autocrine and paracrine chemical signals.**

Effects of Aging on the Endocrine System

Age-related changes of the endocrine system include a gradual decrease in the secretion of some, but not all, endocrine glands. Some of the decreases in secretion may be due to a decrease in physical activity as people age.

GH secretion decreases as people age, and the decrease is greatest in people who do not exercise. It may not occur in older people who exercise regularly. Decreasing GH levels may explain some of the gradual decrease in bone and muscle mass and some of the increase in adipose tissue in many elderly people. Administering GH to slow or prevent the consequences of aging has not been established to be effective, however, and unwanted side effects are possible.

A decrease in melatonin secretion may influence age-related changes in sleep patterns.

The secretion of thyroid hormones decreases slightly with age. Age-related damage to the thyroid gland by the immune system can occur, and this happens in women more than in men. Approximately 10% of elderly women have some reduction in thyroid hormone secretion.

Parathyroid hormone secretion does not appear to decrease with age. Blood levels of Ca^{2+} may decrease slightly because of reduced dietary calcium intake and vitamin D levels. The greatest risk is a loss of bone matrix as parathyroid hormone increases to maintain blood levels of Ca^{2+} within their normal range.

Reproductive hormone secretion gradually declines in elderly men, and women experience menopause. These age-related changes are described in chapter 24.

There are no age-related decreases in the ability to regulate blood glucose levels. There is an age-related tendency to develop type 2 diabetes mellitus in those who have a familial tendency to do so, and it is correlated with age-related increases in body weight.

Thymosin from the thymus decreases with age. Fewer immature lymphocytes are able to mature and become functional, and the immune system becomes less effective in protecting the body. There is an increased susceptibility to infection and to cancer. ▼

> 64. **Describe age-related changes in the secretion of the following and the consequences of those changes: GH, melatonin, thyroid hormones, reproductive hormones, and thymosin. Name one hormone that does not appear to decrease with age.**

S U M M A R Y

Overview of the Endocrine System   (p. 424)
Functions of the Endocrine System

The main regulatory functions include water balance, uterine contractions and milk release, metabolism and tissue maturation, ion regulation, heart rate and blood pressure regulation, control of blood glucose and other nutrients, immune system regulation, and control of reproductive functions.

General Characteristics of the Endocrine System

1. Endocrine glands produce hormones that are released into the interstitial fluid and diffuse into the blood. Hormones act on target tissues, producing specific responses.
2. The protein group of hormones includes hormones that are proteins, glycoproteins, polypeptides, and amino acid derivatives. The lipid group of hormones includes hormones that are steroids and fatty acid derivatives.

3. Generalizations about the differences between the endocrine and nervous systems include the following: (a) The endocrine system is amplitude-modulated, whereas the nervous system is frequency-modulated, and (b) the response of target tissues to hormones is usually slower and of longer duration than that to neurons.

Control of Secretion Rate

1. Negative-feedback mechanisms, which maintain homeostasis, control the secretion of most hormones.
2. Hormone secretion from an endocrine tissue is regulated by one or more of three mechanisms: changes in the extracellular concentration of a nonhormone substance, stimulation by the nervous system, or stimulation by a hormone from another endocrine tissue.
3. Hormones are secreted at a constant rate, suddenly in response to stimuli, or at a cyclic rate.

Transport and Excretion

1. Hormones are dissolved in plasma or bind to plasma proteins.
2. Water-soluble hormones, such as proteins, epinephrine, and norepinephrine, do not bind to plasma proteins or readily diffuse out of the blood. Instead, they are quickly broken down by enzymes or are taken up by tissues. These hormones regulate activities that have a rapid onset and a short duration.
3. Lipid-soluble hormones and thyroid hormones are not quickly removed from the blood. They produce a prolonged effect.
4. Hormones leave the blood to reach target tissues or are excreted by the kidneys or liver.

Interaction of Hormones with Their Target Tissues

1. Target tissues have receptor molecules that are specific for a particular hormone.
2. Only cells with a receptor for a hormone respond to the hormone.

Classes of Receptors

1. Membrane-bound receptors span plasma membranes. They bind to water-soluble or large-molecular-weight hormones.
2. Intracellular receptors are in the cell cytoplasm or nucleus. They bind to lipid-soluble hormones.
3. A hormone binding to a membrane-bound receptor can activate G proteins.
   - The α subunit of the G protein can bind to ion channels and cause them to open or close.
   - The α subunit of the G protein can change the rate of synthesis of intracellular mediator molecules, such as cAMP.
4. A hormone binding to a membrane-bound receptor can directly activate intracellular enzymes, which in turn synthesizes intracellular mediators, such as cGMP, or adds a phosphate group to intracellular enzymes, which alters their activity.
5. Intracellular mediator mechanisms are rapid-acting because they act on already existing enzymes and produce a cascade effect.
   - Hormones bind with the intracellular receptor, and the receptor–hormone complex activates genes. Consequently, DNA is activated to produce mRNA. The mRNA initiates the production of certain proteins (enzymes) that produce the response of the target cell to the hormone.
   - Intracellular receptor mechanisms are slow-acting because time is required to produce the mRNA and the protein.

Pituitary Gland and Hypothalamus   (p. 433)

1. The pituitary gland secretes at least nine hormones that regulate numerous body functions and other endocrine glands.
2. The hypothalamus regulates pituitary gland activity through hormones and action potentials.

Structure of the Pituitary Gland

1. The posterior pituitary develops from the floor of the brain and connects to the hypothalamus by the infundibulum.
2. The anterior pituitary develops from the roof of the mouth.

Relationship of the Pituitary to the Brain

1. The hypothalamohypophyseal portal system connects the hypothalamus and the anterior pituitary.
   - Hormones are produced in hypothalamic neurons.
   - Through the portal system, the hormones inhibit or stimulate hormone production in the anterior pituitary.
2. The hypothalamohypophyseal tract connects the hypothalamus and the posterior pituitary.
   - Hormones are produced in hypothalamic neurons.
   - The hormones move down the axons of the tract and are secreted from the posterior pituitary.

Hormones of the Pituitary Gland

1. Antidiuretic hormone (ADH) promotes water retention by the kidneys.
2. Oxytocin promotes uterine contractions during delivery and causes milk ejection in lactating women.
3. Growth hormone (GH) stimulates growth in most tissues and is a regulator of metabolism.
   - GH stimulates the uptake of amino acids and their conversion into proteins and stimulates the breakdown of fats and the synthesis of glucose.
   - GH stimulates the production of somatomedins; together, they promote bone and cartilage growth.
   - GH secretion increases in response to low blood glucose, stress, and an increase in certain amino acids.
   - GH is regulated by growth hormone–releasing hormone (GHRH) and growth hormone–inhibiting hormone (GHIH) from the hypothalamus.

Thyroid Gland   (p. 439)

1. The thyroid gland is just inferior to the larynx.
2. The thyroid gland is composed of small, hollow balls of cells called follicles, which contain thyroglobulin.
3. Parafollicular cells are scattered throughout the thyroid gland.

Thyroid Hormones

1. Thyroid hormone (T_3 and T_4) synthesis occurs in thyroid follicles.
   - Iodide ions are taken into the follicles by active transport, are oxidized, and are bound to tyrosine molecules in thyroglobulin.
   - Thyroglobulin is secreted into the follicle lumen. Tyrosine molecules with iodine combine to form T_3 and T_4 thyroid hormones.
   - Thyroglobulin is taken into the follicular cells and is broken down; T_3 and T_4 diffuse from the follicles to the blood.
2. Thyroid hormones are transported in the blood.
   - Thyroid hormones bind to thyroxine-binding globulin and other plasma proteins. The plasma proteins prolong the time that thyroid hormones remain in the blood.

3. T_3 and T_4 bind with intracellular receptor molecules and initiate new protein synthesis.
4. T_3 and T_4 affect nearly every tissue in the body.
   - T_3 and T_4 increase the rate of glucose, fat, and protein metabolism in many tissues, thus increasing body temperature.
   - Normal growth of many tissues is dependent on T_3 and T_4.
5. Thyrotropin-releasing hormone (TRH) and thyroid-stimulating hormone (TSH) regulate T_3 and T_4 secretion.
   - TRH from the hypothalamus increases TSH secretion. TRH increases as a result of chronic exposure to cold and decreases as a result of food deprivation, injury, and infections.
   - Increased TSH from the anterior pituitary increases T_3 and T_4 secretion.
   - T_3 and T_4 inhibit TSH and TRH secretion.

Calcitonin

1. The parafollicular cells secrete calcitonin.
2. An increase in blood calcium levels stimulates calcitonin secretion.
3. Calcitonin decreases blood Ca^{2+} levels by inhibiting osteoclasts.

Parathyroid Glands (p. 444)

1. The parathyroid glands are embedded in the thyroid gland.
2. Parathyroid hormone (PTH) increases blood Ca^{2+} levels.
   - PTH stimulates an increase in osteoclast numbers, resulting in increased breakdown of bone.
   - PTH promotes Ca^{2+} reabsorption by the kidneys and the formation of active vitamin D by the kidneys.
   - Active vitamin D increases calcium absorption by the intestine.
3. A decrease in blood Ca^{2+} levels stimulates PTH secretion.

Adrenal Glands (p. 445)

1. The adrenal glands are near the superior poles of the kidneys.
2. The adrenal medulla arises from the same cells that give rise to postganglionic sympathetic neurons.
3. The adrenal cortex is divided into three layers: the zona glomerulosa, the zona fasciculata, and the zona reticularis.

Hormones of the Adrenal Medulla

1. Epinephrine accounts for 80% and norepinephrine for 20% of the adrenal medulla hormones.
2. The adrenal medulla hormones prepare the body for physical activity. They increase blood glucose levels, increase the use of glycogen and glucose by skeletal muscle, increase heart rate and force of contraction, and cause vasoconstriction in the skin and viscera and vasodilation in skeletal and cardiac muscle.
3. Release of adrenal medulla hormones is mediated by the sympathetic division of the ANS in response to emotions, injury, stress, exercise, and low blood glucose levels.

Hormones of the Adrenal Cortex

1. The zona glomerulosa secretes the mineralocorticoids, especially aldosterone. Aldosterone acts on the kidneys to increase Na^+ and to decrease K^+ and H^+ levels in the blood.
2. The zona fasciculata secretes glucocorticoids, especially cortisol.
   - Cortisol increases fat and protein breakdown, increases glucose synthesis from amino acids, decreases the inflammatory response, and is necessary for the development of some tissues.
   - Low blood glucose levels and stress stimulate corticotropin-releasing hormone (CRH) secretion from the hypothalamus. CRH stimulates adrenocorticotropic hormone (ACTH) secretion from the anterior pituitary, which stimulates cortisol secretion.

3. The zona reticularis secretes androgens. In females, androgens stimulate axillary and pubic hair growth and sexual drive.

Pancreas (p. 450)

1. The pancreas is located posterior to the stomach along the small intestine.
2. The exocrine portion of the pancreas consists of a complex duct system, which ends in small sacs, called acini, that produce pancreatic digestive juices.
3. The endocrine portion consists of the pancreatic islets. Each islet is composed of alpha cells, which secrete glucagon, and beta cells, which secrete insulin.

Effect of Insulin and Glucagon on Their Target Tissues

1. Insulin's target tissues are the liver, adipose tissue, muscle, and the satiety center in the hypothalamus. The nervous system is not a target tissue, but it does rely on blood glucose levels maintained by insulin.
2. Insulin increases the uptake of glucose and amino acids by cells. Glucose is used for energy, stored as glycogen, or converted into fats. Amino acids are used to synthesize proteins.
3. Low levels of insulin promote the formation of ketone bodies by the liver.
4. Glucagon's target tissue is mainly the liver.
5. Glucagon causes the breakdown of glycogen to glucose and the synthesis of glucose from amino acids. The liver releases glucose into the blood.

Regulation of Pancreatic Hormone Secretion

1. Insulin secretion increases because of elevated blood glucose levels, an increase in some amino acids, parasympathetic stimulation, and gastrointestinal hormones. Sympathetic stimulation decreases insulin secretion.
2. Glucagon secretion is stimulated by low blood glucose levels, certain amino acids, and sympathetic stimulation.
3. Somatostatin inhibits insulin and glucagon secretion.

Hormonal Regulation of Nutrients (p. 453)

1. After a meal, the following events take place.
   - High glucose levels stimulate insulin secretion but inhibit glucagon, cortisol, GH, and epinephrine secretion.
   - Insulin increases the uptake of glucose, amino acids, and fats, which are used for energy or are stored.
   - Some time after the meal, blood glucose levels drop. Insulin levels decrease and glucagon, GH, cortisol, and epinephrine levels increase. Insulin levels decrease, and glucose is released from tissues.
   - The liver releases glucose into the blood, and the use of glucose by most tissues, other than nervous tissue, decreases.
   - Adipose tissue releases fatty acids and ketones, which most tissues use for energy.
2. During exercise, the following events occur.
   - Sympathetic activity increases epinephrine and glucagon secretion, causing a release of glucose from the liver into the blood.
   - Low blood sugar levels, caused by the uptake of glucose by skeletal muscles, stimulate epinephrine, glucagon, GH, and cortisol secretion, causing an increase in fatty acids and ketones in the blood, all of which are used for energy.

Testes and Ovaries (p. 454)

The testes secrete testosterone and the ovaries secrete estrogens and progesterone. These are sex hormones, responsible for the development and functioning of sexual organs.

Pineal Body (p. 454)

The pineal body produces melatonin, which can inhibit reproductive
maturation and may regulate sleep–wake cycles.

Other Endocrine Organs (p. 455)

1. The thymus produces thymosin, which is involved in the
   development of the immune system.
2. The gastrointestinal tract produces gastrin, secretin, and
   cholecystokinin, which regulate digestive functions.
3. The kidneys produce erythropoietin, which stimulates red blood cell
   production.
4. The placenta secretes human chorionic gonadrotropin, which is
   essential for the maintenance of pregnancy.

Hormonelike Substances (p. 457)

1. Autocrine agents are chemical signals that locally affect cells of the
   same type as the cell producing the autocrine agent.
2. Paracrine agents are chemical signals that locally affect cells of a
   different type than the cell producing the paracrine agent.

Effects of Aging on the Endocrine System (p. 457)

There is a gradual decrease in the secretion rate of most, but not all,
hormones. Some decreases are secondary to gradual decreases in
physical activity.

REVIEW AND COMPREHENSION

1. When comparing the endocrine system and the nervous system,
   generally speaking, the endocrine system
   a. is faster-acting than the nervous system.
   b. produces effects that are of shorter duration.
   c. uses amplitude-modulated signals.
   d. produces more localized effects.
   e. relies less on chemical signals.

2. Given this list of molecule types:
   1. nucleic acid derivatives
   2. fatty acid derivatives
   3. polypeptides
   4. proteins
   5. phospholipids
   Which could be hormone molecules?
   a. 1,2,3   c. 1,2,3,4   e. 1,2,3,4,5
   b. 2,3,4   d. 2,3,4,5

3. Which of these regulates the secretion of a hormone from an
   endocrine tissue?
   a. other hormones
   b. negative-feedback mechanisms
   c. nonhormone substance in the blood
   d. the nervous system
   e. all of the above

4. Hormones are released into the blood
   a. at relatively constant levels.
   b. in large amounts in response to a stimulus.
   c. in a cyclic fashion.
   d. all of the above.

5. Given these observations:
   1. A hormone affects only a specific tissue (not all tissues).
   2. A tissue can respond to more than one hormone.
   3. Some tissues respond rapidly to a hormone, whereas others take
      many hours to respond.
   Which of these observations can be explained by the characteristics
   of hormone receptors?
   a. 1   c. 2,3   e. 1,2,3
   b. 1,2   d. 1,3

6. A hormone
   a. can function as an enzyme.
   b. is also a G protein.
   c. can bind to a receptor.
   d. is an intracellular mediator.
   e. all of the above.

7. Given these events:
   1. GTP is converted to GDP.
   2. The α subunit separates from the β and γ units.
   3. GDP is released from the α subunit.
   List the order in which the events occur after a hormone binds to a
   membrane-bound receptor.
   a. 1,2,3   c. 2,3,1   e. 3,1,2
   b. 1,3,2   d. 3,2,1

8. Given these events:
   1. The α subunit of a G protein interacts with Ca^{2+} channels.
   2. Calcium ions diffuse into the cell.
   3. The α subunit of a G protein is activated.
   Choose the arrangement that lists the events in the order they
   occur after a hormone combines with a receptor on a smooth
   muscle cell.
   a. 1,2,3   c. 2,1,3   e. 3,2,1
   b. 1,3,2   d. 3,1,2

9. Given these events:
   1. cAMP is synthesized.
   2. The α subunit of G protein is activated.
   3. Phosphodiesterase breaks down cAMP.
   Choose the arrangement that lists the events in the order they occur
   after a hormone binds to a receptor.
   a. 1,2,3   c. 2,1,3   e. 3,2,1
   b. 1,3,2   d. 2,3,1

10. When a hormone binds to an intracellular receptor,
    a. DNA produces mRNA.
    b. G proteins are activated.
    c. the receptor–hormone complex causes ion channels to open or
       close.
    d. the cell's response is faster than when a hormone binds to a
       membrane-bound receptor.
    e. the hormone is usually a large, water-soluble molecule.

11. The pituitary gland
    a. develops from the floor of the brain.
    b. develops from the roof of the mouth.
    c. is stimulated by hormones produced in the midbrain.
    d. secretes only three major hormones.
    e. both a and b.

12. The hypothalamohypophyseal portal system
    a. contains one capillary bed.
    b. carries hormones from the anterior pituitary to the body.
    c. carries hormones from the posterior pituitary to the body.

d. carries hormones from the hypothalamus to the anterior pituitary.

e. carries hormones from the hypothalamus to the posterior pituitary.

13. Hormones secreted from the posterior pituitary
    a. are produced in the anterior pituitary.
    b. are transported to the posterior pituitary within axons.
    c. include GH and TSH.
    d. are steroids.
    e. all of the above.

14. Oxytocin is responsible for
    a. preventing milk release from the mammary glands.
    b. preventing goiter.
    c. causing contraction of the uterus.
    d. maintaining normal calcium levels.
    e. increasing metabolic rate.

15. Growth hormone
    a. increases the usage of glucose.
    b. increases the breakdown of lipids.
    c. decreases the synthesis of proteins.
    d. decreases the synthesis of glycogen.
    e. all of the above.

16. T_3 and T_4
    a. require iodine for their production.
    b. are made from the amino acid tyrosine.
    c. are transported in the blood bound to thyroxine-binding globulin.
    d. all of the above.

17. Which of these conditions most likely occurs if a healthy person receives an injection of T_3 and T_4?
    a. The secretion rate of TSH declines.
    b. The person develops symptoms of hypothyroidism.
    c. The person develops hypercalcemia.
    d. The person secretes more TRH.

18. Which of these occurs as a response to a thyroidectomy (removal of the thyroid gland)?
    a. increased calcitonin secretion   c. decreased TRH secretion
    b. increased T_3 and T_4 secretion   d. increased TSH secretion

19. Calcitonin
    a. is secreted by the parathyroid glands.
    b. levels increase when blood calcium levels decrease.
    c. causes blood calcium levels to decrease.
    d. insufficiency results in weak bones and tetany.

20. If parathyroid hormone levels increase, which of these conditions is expected?
    a. Osteoclast numbers are increased.
    b. Calcium absorption from the small intestine is inhibited.
    c. Calcium reabsorption from the urine is inhibited.
    d. Less active vitamin D is formed in the kidneys.
    e. All of the above are true.

21. The adrenal medulla
    a. produces steroids.
    b. has cortisol as its major secretory product.
    c. decreases its secretions during exercise.
    d. is formed from a modified portion of the sympathetic division of the ANS.
    e. all of the above.

22. If aldosterone secretions increase,
    a. blood potassium levels increase.
    b. blood hydrogen levels increase.
    c. acidosis results.
    d. blood sodium levels decrease.
    e. blood volume increases.

23. Glucocorticoids (cortisol)
    a. increase the breakdown of fats.
    b. increase the breakdown of proteins.
    c. increase blood glucose levels.
    d. decrease inflammation.
    e. all of the above.

24. The release of cortisol from the adrenal cortex is regulated by other hormones. Which of these hormones is correctly matched with its origin and function?
    a. CRH—secreted by the hypothalamus; stimulates the adrenal cortex to secrete cortisol
    b. CRH—secreted by the anterior pituitary; stimulates the adrenal cortex to secrete cortisol
    c. ACTH—secreted by the hypothalamus; stimulates the adrenal cortex to secrete cortisol
    d. ACTH—secreted by the anterior pituitary; stimulates the adrenal cortex to produce cortisol

25. Within the pancreas, the pancreatic islets produce
    a. insulin.   d. both a and b.
    b. glucagon.   e. all of the above.
    c. digestive enzymes.

26. Insulin increases
    a. the uptake of glucose by its target tissues.
    b. the breakdown of protein.
    c. the breakdown of fats.
    d. glycogen breakdown in the liver.
    e. all of the above.

27. Which of these tissues is least affected by insulin?
    a. adipose tissue   c. skeletal muscle   e. liver
    b. heart   d. brain

28. Glucagon
    a. primarily affects the liver.
    b. causes glycogen to be stored.
    c. causes blood glucose levels to decrease.
    d. decreases fat metabolism.
    e. all of the above.

29. When blood glucose levels increase, the secretion of which of these hormones increases?
    a. glucagon   c. GH   e. epinephrine
    b. insulin   d. cortisol

30. Melatonin
    a. is produced by the posterior pituitary.
    b. production increases as day length increases.
    c. inhibits the development of the reproductive system.
    d. increases GnRH secretion from the hypothalamus.
    e. decreases the tendency to sleep.

Answers in Appendix E

C R I T I C A L   T H I N K I N G

1. If the effect of a hormone on a target tissue is through a membrane-bound receptor that has a G protein associated with it, predict the consequences if a genetic disease causes the α subunit of the G protein to have a structure that prevents it from binding to GTP.

2. For a hormone that binds to a membrane-bound receptor and has cAMP as the intracellular mediator, predict and explain the consequences if a drug that strongly inhibits phosphodiesterase is taken.

3. An increase in thyroid hormones causes an increase in metabolic rate. If liver disease results in reduced production of the plasma proteins to which thyroid hormones normally bind, what is the effect on metabolic rate? Explain.

4. How can you determine whether a hormone-mediated response resulted from the membrane-bound receptor mechanism or the intracellular receptor mechanism?

5. The hypothalamohypophyseal portal system connects the hypothalamus with the anterior pituitary. Why is such a special circulatory system advantageous?

6. A patient complains of headaches and visual disturbances. A casual glance reveals that the patient's finger bones are enlarged in diameter, a heavy deposition of bone exists over the eyes, and the patient has a prominent jaw. The doctor tells you that the headaches and visual disturbances result from increased pressure within the skull and that the patient is suffering from a pituitary tumor that is affecting hormone secretion. Name the hormone that is causing the problem, and explain the increase in pressure and the visual disturbances.

7. Most laboratories have the ability to determine blood levels of TSH, T_3, and T_4. Given that ability, design a method of determining whether hyperthyroidism in a patient results from a pituitary abnormality or from the production of a nonpituitary thyroid-stimulatory substance.

8. An anatomy and physiology instructor asks two students to predict a patient's response to chronic vitamin D deficiency. One student claims that the person would suffer from hypocalcemia and the symptoms associated with that condition. The other student claims that calcium levels would remain within their normal range, although at the low end of the range, and that bone resorption would occur to the point that advanced osteomalacia might be seen. With whom do you agree, and why?

9. Given the ability to measure blood glucose levels, design an experiment that distinguishes among a person with diabetes, a healthy person, and a person who has a pancreatic tumor that secretes large amounts of insulin.

10. Diabetes mellitus can result from a lack of insulin, which results in hyperglycemia. Adrenal diabetes and pituitary diabetes also produce hyperglycemia. What hormones produce the last two conditions?

11. A patient arrives in an unconscious condition. A medical emergency bracelet reveals that he has diabetes. The patient can be in diabetic coma or insulin shock. How can you tell which, and what treatment do you recommend for each condition?

12. A patient exhibits polydipsia (thirst), polyuria (excess urine production), and urine with a low specific gravity (contains few ions and no glucose). If you wanted to reverse the symptoms, would you administer insulin or ADH? Explain.

Answers in Appendix F

For additional study tools, go to: www.aris.mhhe.com. Click on your course topic and then this textbook's author name/title. Register once for a semester's worth of interactive learning activities to help improve your grade!

CHAPTER 16 Blood

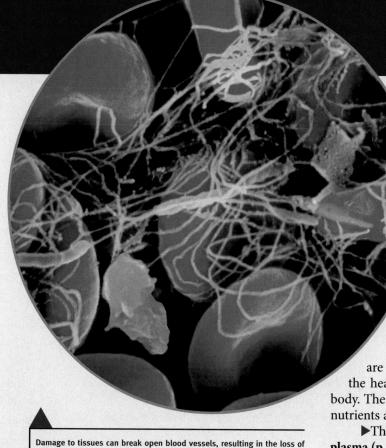

Damage to tissues can break open blood vessels, resulting in the loss of blood. This colorized scanning electron micrograph shows a blood clot, which prevents blood loss. The blue particles are platelets, which are activated when tissues are damaged. Activated platelets promote the formation of protein fibers called fibrin, which are seen here as yellow strands. The fibrin forms a network that captures red blood cells (red discs) and prevents their loss from the body.

Introduction

Historically, many cultures around the world, both ancient and modern, shared beliefs in the magical qualities of blood. Blood is considered the "essence of life" because the uncontrolled loss of it can result in death. Blood was also thought to define character and emotions. People of a noble bloodline were described as "blue bloods," whereas criminals were considered to have "bad" blood. It was said that anger caused the blood to "boil," and fear resulted in blood "curdling." The scientific study of blood reveals characteristics as fascinating as any of these fantasies. Blood performs many functions essential to life and often can reveal much about our health.

Cells require constant nutrition and waste removal because they are metabolically active. The cardiovascular system, which consists of the heart, blood vessels, and blood, connects the various tissues of the body. The heart pumps blood through blood vessels, and the blood delivers nutrients and picks up waste products.

▶This chapter explains the **functions and composition of blood (p. 464)**, **plasma (p. 465)**, and the **formed elements (p. 465)** of blood. **Preventing blood loss (p. 470)**, **blood grouping (p. 475)**, and **diagnostic blood tests (p. 479)** are also described.

Tools for Success

A virtual cadaver dissection experience

Audio (MP3) and/or video (MP4) content, available at www.mhhe.aris.com

Functions and Composition of Blood

The heart pumps blood through blood vessels, which extend throughout the body. Blood helps maintain homeostasis in several ways:

1. *Transport of gases, nutrients, and waste products.* Oxygen enters blood in the lungs and is carried to cells. Carbon dioxide, produced by cells, is carried in the blood to the lungs, from which it is expelled. The blood transports ingested nutrients, ions, and water from the digestive tract to cells, and the blood transports the waste products of cells to the kidneys for elimination.

2. *Transport of processed molecules.* Many substances are produced in one part of the body and transported in the blood to another part, where they are modified. For example, the precursor to vitamin D is produced in the skin (see chapter 5) and transported by the blood to the liver and then to the kidneys for processing into active vitamin D. Active vitamin D is transported in the blood to the small intestines, where it promotes the uptake of calcium. Another example is lactic acid produced by skeletal muscles during anaerobic respiration (see chapter 8). The blood carries lactic acid to the liver, where it is converted into glucose.

3. *Transport of regulatory molecules.* The blood carries many of the hormones and enzymes that regulate body processes from one part of the body to another.

4. *Regulation of pH and osmosis.* Buffers (see chapter 2), which help keep the blood's pH within its normal limits of 7.35–7.45, are in the blood. The osmotic composition of

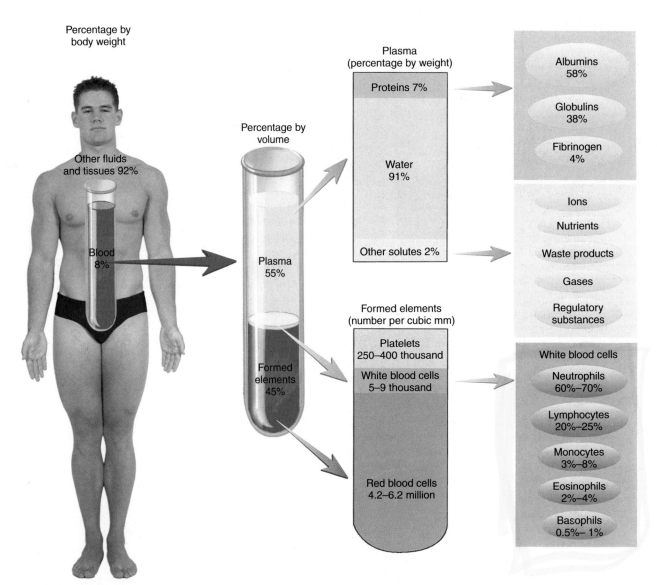

Figure 16.1   Composition of Blood

Approximate values for the components of blood in a normal adult.

Table 16.1 Composition of Plasma

Plasma Components	Functions and Examples
Water	Acts as a solvent and suspending medium for blood components
Proteins	Maintain osmotic pressure (albumin), destroy foreign substances (antibodies and complement), transport molecules (albumin, globulins), and form clots (fibrinogen)
Ions	Involved in osmotic pressure (sodium and chloride ions), membrane potentials (sodium and potassium ions), and acid–base balance (hydrogen, hydroxide, and bicarbonate ions)
Nutrients	Source of energy and "building blocks" of more complex molecules (glucose, amino acids, triglycerides)
Gases	Involved in aerobic respiration (oxygen and carbon dioxide)
Waste products	Break down products of protein metabolism (urea and ammonia salts), red blood cells (bilirubin), and anaerobic respiration (lactic acid)
Regulatory substances	Catalyze chemical reactions (enzymes) and stimulate or inhibit many body functions (hormones)

blood is also critical for maintaining normal fluid and ion balance.

5. *Maintenance of body temperature.* Blood is involved with body temperature regulation because warm blood is transported from the interior to the surface of the body, where heat is released from the blood.

6. *Protection against foreign substances.* The cells and chemicals of the blood make up an important part of the immune system, protecting against foreign substances, such as microorganisms and toxins.

7. *Clot formation.* Blood clotting protects against excessive blood loss when blood vessels are damaged. When tissues are damaged, the blood clot that forms is also the first step in tissue repair and the restoration of normal function (see chapter 4).

Blood is a type of connective tissue, consisting of cells and cell fragments surrounded by a liquid matrix. The cells and cell fragments are the **formed elements,** and the liquid is the **plasma** (plaz′mă, something formed). The formed elements make up about 45%, and plasma makes up about 55% of the total blood volume (figure 16.1). The total blood volume in the average adult is about 4–5 L in females and 5–6 L in males. Blood makes up about 8% of the total weight of the body. ▼

> 1. **List the ways that blood helps maintain homeostasis in the body.**
> 2. **Define formed elements and plasma of blood.**

Plasma

Plasma is a pale yellow fluid that consists of about 91% water; 7% proteins; and 2% other substances, such as ions, nutrients, gases, and waste products (see figure 16.1 and table 16.1). Plasma proteins include albumin, globulins, and fibrinogen. **Albumin** (al-bū′min, the white of egg) makes up 58% of the plasma proteins.

Although the osmotic pressure (see chapter 3) of blood results primarily from sodium chloride, albumin makes an important contribution. The water balance between blood and tissues is determined by the movement of water into and out of the blood by osmosis. **Globulins** (glob′ū-linz, globule) account for 38% of the plasma proteins. Some globulins, such as antibodies and complement, are part of the immune system (see chapter 19). Other globulins and albumin function as transport molecules because they bind to molecules, such as hormones (see chapter 15), and carry them in the blood throughout the body. Some globulins are clotting factors, which are necessary for the formation of blood clots. **Fibrinogen** (fī-brin′ō-jen, *fibra,* fiber + *gen,* produce) is a clotting factor that constitutes 4% of plasma proteins. Activation of clotting factors results in the conversion of fibrinogen into **fibrin** (fī′brin), a threadlike protein that forms blood clots (see "Blood Clotting," p. 473). **Serum** (ser′um, whey) is plasma without the clotting factors. ▼

> 3. **What are the functions of albumin, globulins, and fibrinogen in plasma? What other substances are found in plasma?**

Formed Elements

About 95% of the volume of the formed elements consists of **red blood cells (RBCs),** or **erythrocytes** (ĕ-rith′rō-sītz, erythro-, red + *kytos,* cell). Red blood cells transport oxygen and carbon dioxide. Most of the remaining 5% of the volume of the formed elements consists of **white blood cells (WBCs),** or **leukocytes** (loo′kō-sītz, leuko-, white + *kytos,* cell). There are five types of white blood cells, which are involved in immunity. **Platelets** (plāt′letz), or **thrombocytes** (throm′bō-sītz, thrombo-, clot + *kytos,* cell) are tiny cell fragments making a negligible contribution to plasma volume. Platelets are involved in blood clotting. Red blood cells are 700 times more numerous than white blood cells and 17 times

Table 16.2 — Formed Elements of the Blood

Cell Type	Illustration	Description	Function
Red Blood Cell		Biconcave disc; no nucleus; contains hemoglobin, which colors the cell red; 7.5 µm in diameter	Transports oxygen and carbon dioxide
White Blood Cells		Spherical cells with a nucleus	Five types of white blood cells, each with specific functions
Granulocytes			
Neutrophil		Nucleus with two to four lobes connected by thin filaments; cytoplasmic granules stain a light pink or reddish purple; 10–12 µm in diameter	Phagocytizes microorganisms and other substances
Basophil		Nucleus with two indistinct lobes; cytoplasmic granules stain blue-purple; 10–12 µm in diameter	Releases histamine, which promotes inflammation, and heparin, which prevents clot formation
Eosinophil		Nucleus often bilobed; cytoplasmic granules stain orange-red or bright red; 11–14 µm in diameter	Releases chemicals that reduce inflammation; attacks certain worm parasites
Agranulocytes			
Lymphocyte		Round nucleus; cytoplasm forms a thin ring around the nucleus; 6–14 µm in diameter	Produces antibodies and other chemicals responsible for destroying microorganisms; contributes to allergic reactions, graft rejection, tumor control, and regulation of the immune system
Monocyte		Nucleus round, kidney-shaped, or horseshoe-shaped; contains more cytoplasm than does lymphocyte; 12–20 µm in diameter	Phagocytic cell in the blood; leaves the blood and becomes a macrophage, which phagocytizes bacteria, dead cells, cell fragments, and other debris within tissues
Platelet		Cell fragment surrounded by a plasma membrane and containing granules; 2–4 µm in diameter	Forms platelet plugs; releases chemicals necessary for blood clotting

more numerous than platelets. The formed elements of the blood are outlined and illustrated in table 16.2. ▼

> 4. *Name the three general types of formed elements in the blood and state their major functions.*

Production of Formed Elements

The process of blood cell production is called **hematopoiesis** (hē′mă-tō-poy-ē′sis, hemato-, blood + *poiēsis,* a making). In the fetus, hematopoiesis occurs in several tissues, such as the liver, thymus, spleen, lymph nodes, and red bone marrow. After birth, hematopoiesis is confined primarily to red bone marrow, but some white blood cells are produced in lymphatic tissues (see chapter 19).

All the formed elements of blood are derived from a single population of cells called **stem cells,** or **hemocytoblasts.** These stem cells differentiate to give rise to different cell lines, each of which ends with the formation of a particular type of formed element (figure 16.2). The development of each cell line is regulated by specific proteins called **growth factors.** That is, the types of formed element derived from the stem cells and how many formed elements are produced are determined by the growth factors.

Stem Cells and Cancer Therapy

Many cancer therapies attack dividing cells, such as those found in tumors. An undesirable side effect, however, can be the destruction of nontumor cells that divide rapidly, such as the stem cells and their derivatives in red bone marrow. After treatment for cancer, growth factors are used to stimulate the rapid regeneration of the red bone marrow. Although not a treatment for the cancer itself, the use of growth factors can speed recovery from the side effects of cancer therapy.

Some types of leukemia and genetic immune deficiency diseases can be treated with a bone marrow transplant containing blood stem cells. To avoid problems of tissue rejection, families with a history of these disorders can freeze the umbilical cord blood of their newborn children. The cord blood contains many stem cells and can be used instead of a bone marrow transplant. ▼

> 5. *Define hematopoiesis. Where does it occur?*
> 6. *What is a stem cell? What is the function of growth factors?*

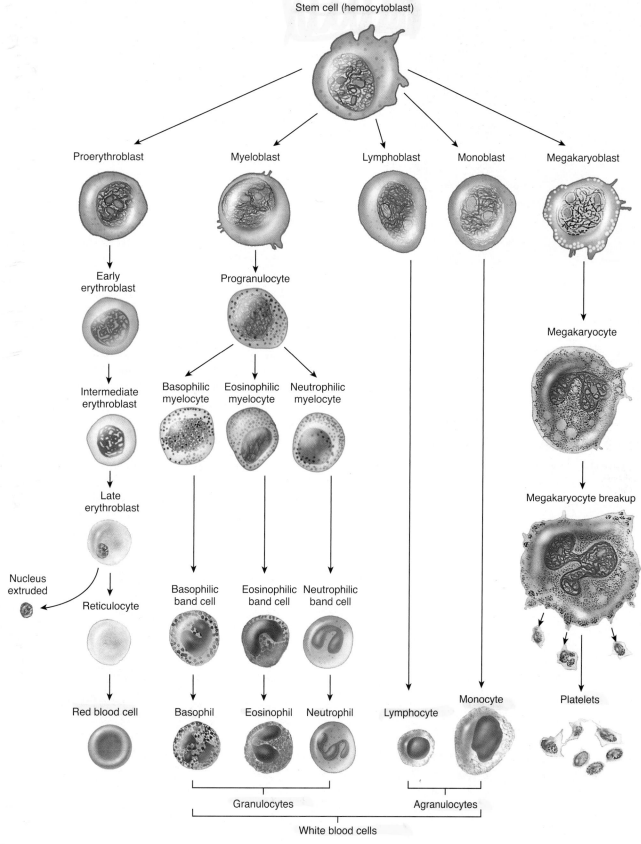

Figure 16.2 Hematopoiesis

Stem cells give rise to the cell lines that produce the formed elements.

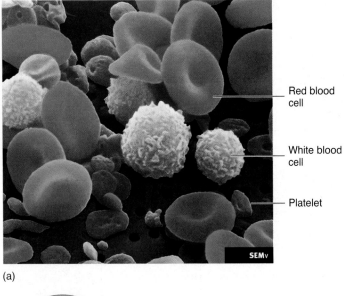

(a)

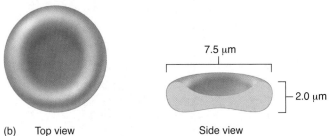

7.5 μm

2.0 μm

(b)   Top view   Side view

Figure 16.3   Red Blood Cells and White Blood Cells

(a) Scanning electron micrograph of formed elements: red blood cells (*red doughnut shapes*) and white blood cells (*yellow*). (b) Shape and dimensions of a red blood cell.

Red Blood Cells

Normal red blood cells are disc-shaped cells with edges that are thicker than the center of the cell (figure 16.3). The biconcave shape increases the surface area of the red blood cell, compared

with a flat disc of the same size. The greater surface area increases the movement of gases into and out of the red blood cells. In addition, a red blood cell can bend or fold around its thin center, decreasing its size and enabling it to pass more easily through small blood vessels.

During their development, red blood cells lose their nuclei and most of their organelles. Consequently, they are unable to divide. Red blood cells live for about 120 days in males and 110 days in females. ▼

> 7. *How does the shape of red blood cells contribute to their ability to exchange gases and move through blood vessels?*

Hemoglobin

The main component of a red blood cell is the pigmented protein **hemoglobin** (hē-mō-glō′bin, hemo-, blood + *globus*, a ball), which accounts for about a third of the cell's volume and is responsible for its red color. Hemoglobin consists of four polypeptide chains and four heme groups. Each polypeptide chain, called a **globin** (glō′bin), is bound to one **heme** (hēm). Each heme is a red-pigment molecule containing one **iron atom** (figure 16.4). When hemoglobin is exposed to oxygen, one oxygen molecule can become associated with each heme group. When most of the hemes are carrying oxygen, blood has a bright red appearance, as in arterial blood or blood exposed to the air in a wound. When the oxygen content of blood decreases, as in venous blood, the hemes without oxygen cause the blood to appear a darker red color.

Because iron is necessary for oxygen transport, it is not surprising that two-thirds of the body's iron is found in hemoglobin. Small amounts of iron are required in the diet to replace the small amounts lost in the urine and feces. Women need more dietary iron than men do because women lose iron as a result of menstruation. Dietary iron is absorbed into the circulation from the upper part of the intestinal tract. Stomach acid and vitamin C in food increase the absorption of iron by converting ferric iron (Fe^{3+}) to ferrous iron (Fe^{2+}), which is more readily absorbed.

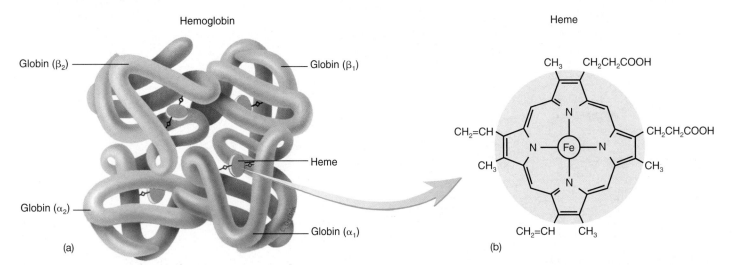

Figure 16.4   Hemoglobin

(a) Hemoglobin consists of four globins and four hemes. There are two alpha (α) globins and two beta (β) globins. A heme is associated with each globin.
(b) Each heme contains one iron atom.

Hemoglobin transports carbon dioxide, which does not combine with the iron atoms but is attached to amino groups of the globin molecules. ▼

8. Describe the two basic parts of a hemoglobin molecule. Which part is associated with iron? What gases are transported by each part?

Transport of Oxygen and Carbon Dioxide

The primary functions of red blood cells are to transport oxygen from the lungs to the various tissues of the body and to assist in the transport of carbon dioxide from the tissues to the lungs. Approximately 98.5% of the oxygen in the blood is transported in combination with the hemoglobin in the red blood cells, and the remaining 1.5% is dissolved in the water part of the plasma.

 Carbon Monoxide and Oxygen Transport

Carbon monoxide is a gas produced by the incomplete combustion of hydrocarbons, such as gasoline. It binds to the iron in hemoglobin about 210 times as readily as does oxygen and does not tend to unbind. As a result, the hemoglobin bound to carbon monoxide no longer transports oxygen. Nausea, headache, unconsciousness, and death are possible consequences of prolonged exposure to carbon monoxide.

Carbon dioxide is transported in the blood in three major ways: Approximately 7% is transported as carbon dioxide dissolved in the plasma, approximately 23% is transported in combination with hemoglobin, and 70% is transported in the form of bicarbonate ions. The bicarbonate ions (HCO_3^-) are produced when carbon dioxide (CO_2) and water (H_2O) combine to form carbonic acid (H_2CO_3), which dissociates to form hydrogen (H^+) and bicarbonate ions. The combination of carbon dioxide and water is catalyzed by an enzyme, **carbonic anhydrase,** which is located primarily within red blood cells. ▼

$$CO_2 + H_2O \underset{}{\overset{\text{Carbonic anhydrase}}{\rightleftarrows}} H_2CO_3 \rightleftarrows H^+ + HCO_3^-$$

Carbon   Water   Carbonic   Hydrogen   Bicarbonate
dioxide   acid   ion   ion

9. Give the percentage for each of the ways that oxygen and carbon dioxide are transported in the blood. What is the function of carbonic anhydrase?

Life History of Red Blood Cells

Under normal conditions, about 2.5 million red blood cells are destroyed every second. Fortunately, new red blood cells are produced as rapidly as old red blood cells are destroyed. The process by which new red blood cells are produced is called **erythropoiesis** (ĕ-rith′rō-poy-ē′sis), and the time required for the production of a single red blood cell is about 4 days.

Stem cells form **proerythroblasts** (prō-ĕ-rith′rō-blastz, pro-, before + erythro-, red + *blastos,* germ), which give rise to the red blood cell line (see figure 16.2). Red blood cells are the final

cells produced from a series of cell divisions. After each cell division, the newly formed cells change and become more like mature red blood cells. For example, following one of these cell divisions, the newly formed cells manufacture large amounts of hemoglobin. After the final cell division, the nucleus is lost from the cells, to produce immature red blood cells called **reticulocytes** (re-tik′ū-lō-sītz). Reticulocytes are released from the red bone marrow into the circulating blood, which normally consists of mature red blood cells and 1%–3% reticulocytes. Within 1 to 2 days, reticulocytes become mature red blood cells when the ribosomes responsible for the manufacture of hemoglobin degenerate.

P R E D I C T
What does an elevated reticulocyte count indicate? How would the reticulocyte count change during the week after a person had donated a unit (about 500 mL) of blood?

The process of cell division requires the B vitamins folate and B_{12}, which are necessary for the synthesis of DNA (see chapter 3). Iron is required for the production of hemoglobin. Consequently, a lack of folate, vitamin B_{12}, or iron can interfere with normal red blood cell production.

Red blood cell production is stimulated by low blood oxygen levels. Typical causes of low blood oxygen are decreased numbers of red blood cells, decreased or defective hemoglobin, diseases of the lungs, high altitude, inability of the cardiovascular system to deliver blood to tissues, and increased tissue demands for oxygen, such as during endurance exercises.

Low blood oxygen levels increase red blood cell production by increasing the formation of the glycoprotein **erythropoietin** (ĕ-rith-rō-poy′ĕ-tin, erythrocyte + *poiēsis,* a making) by the kidneys (figure 16.5). Erythropoietin stimulates red bone marrow to

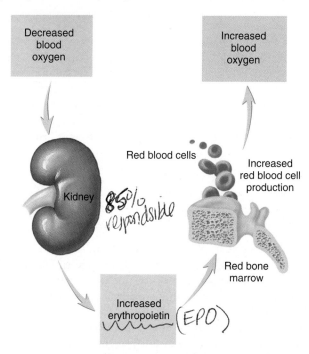

Figure 16.5   Red Blood Cell Production

In response to decreased blood oxygen, the kidneys release erythropoietin into the general circulation. The increased erythropoietin stimulates red blood cell production in the red bone marrow. This process increases blood oxygen levels.

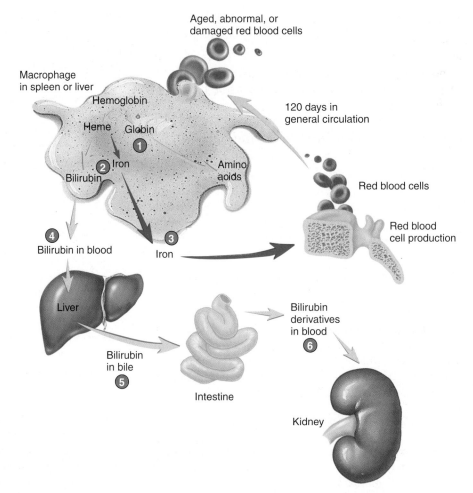

1. The globin chains of hemoglobin are broken down to individual amino acids (*pink arrow*) and are metabolized or used to build new proteins.

2. Iron is released from the heme of hemoglobin. The heme is converted into bilirubin.

3. Iron is transported in the blood to the red bone marrow and used in the production of new hemoglobin (*green arrows*).

4. Bilirubin (*blue arrow*) is transported in the blood to the liver.

5. Bilirubin is excreted as part of the bile into the small intestine.

6. Bilirubin derivatives contribute to the color of feces or are reabsorbed from the intestine into the blood and excreted from the kidneys in the urine.

Process Figure 16.6 Hemoglobin Breakdown

Hemoglobin is broken down in macrophages, and the breakdown products are used or excreted.

produce more red blood cells. Thus, when oxygen levels in the blood decrease, the production of erythropoietin increases, which increases red blood cell production. The increased number of red blood cells increases the ability of the blood to transport oxygen. This mechanism returns blood oxygen levels to normal and maintains homeostasis by increasing the delivery of oxygen to tissues. Conversely, if blood oxygen levels increase, less erythropoietin is released, and red blood cell production decreases.

P R E D I C T 2

Cigarette smoke produces carbon monoxide. If a nonsmoker smoked a pack of cigarettes a day for a few weeks, what would happen to the number of red blood cells in the person's blood? Explain.

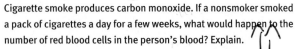

Old, abnormal, or damaged red blood cells are removed from the blood by macrophages located in the spleen and liver (figure 16.6). Within the macrophage the globin part of the molecule is broken down into amino acids that are metabolized or reused to produce other proteins. The iron released from heme is transported in the blood to the red bone marrow and is used to produce new hemoglobin. Only small amounts of iron are required in the daily diet because the iron is recycled. The heme molecules are converted to **bile pigments,** which are normally taken up by the liver and released into the small intestine as part of the bile (see chapter 21). Bile pigments are so named because they give bile its

color. The main bile pigment is **bilirubin** (bil-i-roo′bin, bili-, bile + *ruber,* red), a yellow pigment molecule. After it enters the intestine, bilirubin is converted by bacteria into other pigments. Some of these pigments give feces their brown color, whereas others are absorbed from the intestine into the blood, modified by the kidneys, and excreted in the urine, contributing to the characteristic yellow color of urine.

Jaundice (jawn′dis, yellow) is a yellowish staining of the skin and sclerae caused by a buildup of bile pigments in the circulation and interstitial fluids. Any process that causes increased destruction of red blood cells can cause jaundice, such as damage by toxins, genetic defects in red blood cell plasma membranes, infections, and immune reactions. Other causes of jaundice include dysfunction or destruction of liver tissue and blockage of the duct system that drains bile from the liver (see chapter 21). ▼

10. *Define erythropoiesis. What are proerythroblasts and reticulocytes?*
11. *What is erythropoietin, where is it produced, what causes it to be produced, and what effect does it have on red blood cell production?*
12. *Where are red blood cells removed from the blood? List the three breakdown products of hemoglobin and explain what happens to them.*

White Blood Cells

White blood cells, or **leukocytes,** are spherical cells that lack hemoglobin. White blood cells form a thin, white layer of cells between plasma and red blood cells when the components of blood are separated from each other. They are larger than red blood cells, and each has a nucleus (see table 16.2). Although white blood cells are components of the blood, the blood serves primarily as a means to transport these cells to other tissues of the body. White blood cells can leave the blood and move by **ameboid** (ă-mē′boyd, like an ameba) **movement** through the tissues. In this process, the cell projects a cytoplasmic extension that attaches to an object. Then the rest of the cell's cytoplasm flows into the extension. Two functions of white blood cells are (1) to protect the body against invading microorganisms and (2) to remove dead cells and debris from the tissues by phagocytosis.

Each white blood cell type is named according to its appearance in stained preparations. Those containing large cytoplasmic granules are **granulocytes** (gran′ū-lō-sītz, granulo-, granular + *kytos,* cell), and those with very small granules that cannot be easily seen with the light microscope are **agranulocytes** (ă-gran′ū-lō-sītz, a-, without).

There are three kinds of granulocytes: neutrophils, basophils, and eosinophils. **Neutrophils** (noo′trō-filz, neutro-, neutral + *philos,* loving), the most common type of white blood cells, have small cytoplasmic granules that stain with both acidic and basic dyes (figure 16.7). Their nuclei are commonly lobed, with the number of lobes varying from two to four. Neutrophils usually remain in the blood for a short time (10–12 hours), move into other tissues, and phagocytize microorganisms and other foreign substances. Neutrophils also secrete a class of enzymes called **lysozymes** (lī′sō-zīmz), which are capable of destroying certain bacteria. **Pus** is an accumulation of dead neutrophils, cell debris, and fluid at sites of infections.

Basophils (bā′sō-filz, baso-, base + *philos,* loving), the least common of all white blood cells, contain large cytoplasmic granules that stain blue or purple with basic dyes (see table 16.2). Basophils release **histamine** and other chemicals that promote inflammation (see chapters 4 and 19). They also release **heparin,** which prevents the formation of clots.

Eosinophils (ē-ō-sin′o-filz, eosin, an acidic dye + *philos,* loving) contain cytoplasmic granules that stain bright red with eosin, an acidic stain. Many have a two-lobed nucleus (see table 16.2). Eosinophils release chemicals that reduce inflammation. Their numbers are elevated in people with allergies. In addition, chemicals from eosinophils are involved with the destruction of certain worm parasites.

There are two kinds of agranulocytes: lymphocytes and monocytes. **Lymphocytes** (lim′fō-sītz, lympho-, lymph + *kytos,* cell) are the smallest of the white blood cells (see figure 16.7). The cytoplasm of lymphocytes consists of only a thin, sometimes imperceptible ring around the nucleus. Lymphocytes originate in red bone marrow and migrate through the blood to lymphatic tissues, where they can proliferate and produce more lymphocytes. There are several types of lymphocytes, and they play an important role in immunity (see chapter 19). For example, B cells can be stimulated by bacteria or toxins to divide and form cells that produce proteins called **antibodies.** Antibodies can attach to bacteria and activate mechanisms that result in the destruction of the bacteria. T cells protect against viruses and other intracellular microorganisms by attacking and destroying the cells in which they are found. In addition, T cells are involved in the destruction of tumor cells and tissue graft rejections.

Monocytes (mon′ō-sītz, mono-, one + *kytos,* cell) are the largest of the white blood cells (see table 16.2). After they leave the blood and enter tissues, monocytes enlarge and become **macrophages** (mak′rō-fă-jez, macro-, large + *phagō,* to eat), which phagocytize bacteria, dead cells, cell fragments, and any other debris within the tissues. In addition, macrophages can break down phagocytized foreign substances and present the processed substances to lymphocytes, which results in activation of the lymphocytes (see chapter 19). ▼

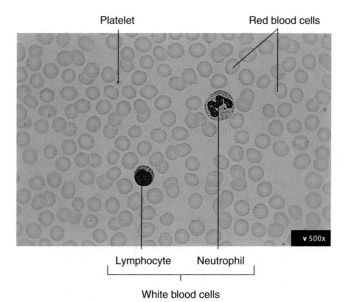

Platelet   Red blood cells

Lymphocyte   Neutrophil

White blood cells

Figure 16.7   **Standard Blood Smear**

A thin film of blood is spread on a microscope slide and stained. The white blood cells have pink-colored cytoplasm and purple-colored nuclei. The red blood cells do not have nuclei. The center of a red blood cell appears whitish because light more readily shines through the thin center of the disc than through the thicker edges. The platelets are purple cell fragments.

13. **What are the two major functions of white blood cells? Describe ameboid movement.**
14. **Describe the morphology of the five types of white blood cells.**
15. **Name the two white blood cells that function primarily as phagocytic cells. Define lysozymes.**
16. **Which white blood cell reduces the inflammatory response? Which white blood cell releases histamine and promotes inflammation?**
17. **B and T cells are examples of what type of white blood cell? How do these cells protect against bacteria and viruses?**

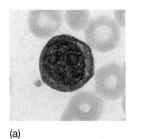
(a)

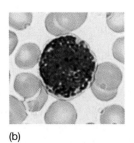
(b)

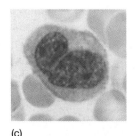
(c)

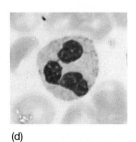
(d)

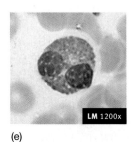
(e) LM 1200x

Figure 16.8 Identification of White Blood Cells
See Predict question 3.

Platelets

Platelets (plăt′letz), or **thrombocytes** (throm′bō-sītz), are minute fragments of cells, each consisting of a small amount of cytoplasm surrounded by a cell membrane (see figure 16.7). They are produced in the red bone marrow from **megakaryocytes** (meg-ă-kar′ē-ō-sītz, mega-, large + *karyon,* nucleus + *kytos,* cell), which are large cells (see figure 16.2). Small fragments of these cells break off and enter the blood as platelets, which play an important role in preventing blood loss. This prevention is accomplished in two ways: (1) the formation of platelet plugs, which seal holes in small vessels, and (2) the formation of clots, which help seal off larger wounds in the vessels. ▼

18. *What is a platelet? How are platelets formed?*
19. *What are the two major roles of platelets in preventing blood loss?*

Preventing Blood Loss

When a blood vessel is damaged, blood can leak into other tissues and interfere with normal tissue function, or blood can be lost from the body. A small amount of blood loss from the body can be tolerated, and new blood is produced to replace it. If a large amount of blood is lost, death can occur. Fortunately, when a blood vessel is damaged, vascular spasm, platelet plug formation, and blood clotting minimize the loss of blood.

Vascular Spasm

Vascular spasm is an immediate but temporary constriction of a blood vessel resulting from a contraction of smooth muscle within the wall of the vessel. This constriction can close small vessels completely and stop the flow of blood through them. Damage to blood vessels can activate nervous system reflexes that cause vascular spasms. Chemicals also produce vascular spasms. For example, platelets release **thromboxanes** (throm′bok-zānz), which are derived from certain prostaglandins, and endothelial (epithelial) cells lining blood vessels release the peptide **endothelin** (en-dō′thē-lin). ▼

20. *What is a vascular spasm? Name two factors that produce it. What is the source of thromboxanes and endothelin?*

Platelet Plugs

A **platelet plug** is an accumulation of platelets that can seal up a small break in a blood vessel. Platelet plug formation is very important in

1. Platelet adhesion occurs when von Willebrand factor connects collagen and platelets.

2. The platelet release reaction results in the release of ADP, thromboxanes, and other chemicals that activate other platelets.

3. Platelet aggregation occurs when fibrinogen receptors on activated platelets bind to fibrinogen, connecting the platelets to one another. A platelet plug is formed by the accumulating mass of platelets.

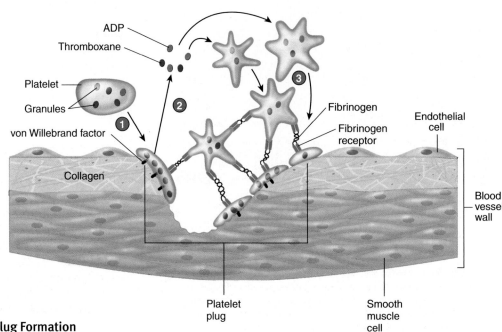

Process Figure 16.9 Platelet Plug Formation

maintaining the integrity of the circulatory system because small tears occur in the smaller vessels and capillaries many times each day, and platelet plug formation quickly closes them. People who lack the normal number of platelets tend to develop numerous small hemorrhages in their skin and internal organs.

The formation of a platelet plug can be described as a series of steps, but in actuality many of these steps occur at the same time. **Platelet adhesion** results in platelets sticking to collagen exposed by blood vessel damage (figure 16.9, step 1). Most platelet adhesion is mediated through **von Willebrand factor,** which is a protein produced and secreted by blood vessel endothelial cells. Von Willebrand factor forms a bridge between collagen and platelets by binding to platelet surface receptors and collagen. After platelets adhere to collagen, they become activated, change shape, and release chemicals. In the **platelet release reaction,** platelets release chemicals, such as adenosine diphosphate (ADP) and thromboxane (figure 16.9, step 2). ADP and thromboxane bind to their respective receptors on the surfaces of platelets, resulting in the activation of the platelets. These activated platelets also release ADP and thromboxanes, which activates more platelets. Thus, a cascade of chemical release activates many platelets. As platelets become activated, they express surface receptors called **fibrinogen receptors,** which can bind to fibrinogen, a plasma protein. In **platelet aggregation,** fibrinogen forms bridges between the fibrinogen receptors of numerous platelets, resulting in the formation of a platelet plug (figure 16.9, step 3). ▼

21. *What is the function of a platelet plug? Describe the process of platelet plug formation.*

Blood Clotting

Vascular spasms and platelet plugs alone are not sufficient to close large tears or cuts. When a blood vessel is severely damaged, **blood clotting,** or **coagulation** (kō-ag-ū-lā'shŭn), results in the formation of a blood clot. A **blood clot** is a network of threadlike protein fibers, called **fibrin,** that traps blood cells, platelets, and fluid (see chapter opener, p. 463).

The formation of a blood clot depends on a number of proteins, called **clotting factors.** Most clotting factors are manufactured in the liver, and many of them require vitamin K for their synthesis. In addition, many of the chemical reactions of clot formation require Ca^{2+} and the chemicals released from platelets. Low levels of vitamin K, low levels of Ca^{2+}, low numbers of platelets, or reduced synthesis of clotting factors because of liver dysfunction can seriously impair the blood-clotting process.

1. The extrinsic pathway of clotting starts with thromboplastin, which is released outside of the plasma in damaged tissue.

2. The intrinsic pathway of clotting starts when inactive factor XII, which is in the plasma, is activated by coming into contact with a damaged blood vessel.

3. Activation of the extrinsic or intrinsic clotting pathway results in the production of activated factor X.

4. Activated factor X, factor V, phospholipids, and Ca^{2+} form prothrombinase.

5. Prothrombin is converted to thrombin by prothrombinase.

6. Fibrinogen is converted to fibrin (the clot) by thrombin.

7. Thrombin activates clotting factors, promoting clot formation and stabilizing the fabrin clot.

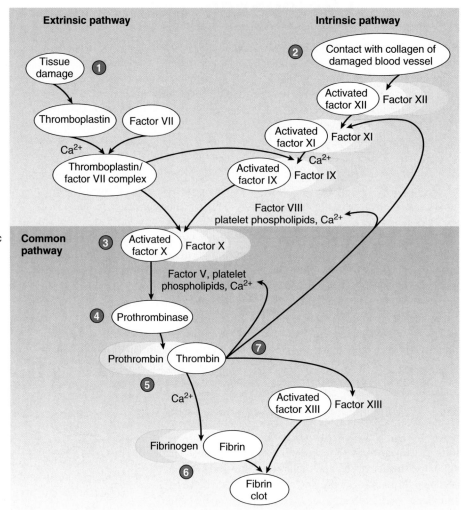

Process Figure 16.10 Clot Formation

A sequence of chemical reactions occurs as activated coagulation factors (*white ovals*) activate inactive coagulation factors (*yellow ovals*). Clot formation begins through the extrinsic or intrinsic pathway. The common pathway starts with factor X and results in a fibrin clot.

Normally, the clotting factors are in an inactive state and do not cause clotting. After injury, the clotting factors are activated to produce a clot. The activation of clotting proteins begins with the extrinsic and intrinsic pathways (figure 16.10). These pathways converge to form the common pathway, which results in the formation of a fibrin clot.

Extrinsic Pathway

The extrinsic pathway is so named because it begins with chemicals that are outside of, or extrinsic to, the blood (see figure 16.10). Damaged tissues release a mixture of lipoproteins and phospholipids called **thromboplastin** (throm-bō-plas′tin), also known as **tissue factor (TF),** or factor III. Thromboplastin, in the presence of Ca^{2+}, forms a complex with factor VII, which activates factor X, which is the beginning of the common pathway.

Intrinsic Pathway

The intrinsic pathway is so named because it begins with chemicals that are inside, or intrinsic to, the blood (see figure 16.10). Damage to blood vessels can expose collagen in the connective tissue

beneath the epithelium lining the blood vessel. When plasma factor XII comes into contact with collagen, factor XII is activated and it stimulates factor XI, which in turn activates factor IX. Activated factor IX joins with factor VIII, platelet phospholipids, and Ca^{2+} to activate factor X, which is the beginning of the common pathway.

Common Pathway

On the surface of platelets, activated factor X, factor V, platelet phospholipids, and Ca^{2+} combine to form **prothrombinase.** Prothrombinase converts the soluble plasma protein **prothrombin** into the enzyme **thrombin.** Thrombin converts the soluble plasma protein fibrinogen into the insoluble protein **fibrin.** Fibrin forms the fibrous network of the clot.

At each step of the clotting process, each clotting factor activates many additional clotting factors. Consequently, a large quantity of clotting factors are activated, resulting in the formation of a clot. ▼

22. *What is a blood clot and what is its function?*
23. *What are blood clotting factors? Where are most of them produced? How is vitamin K involved?*

24. *How do the extrinsic and intrinsic pathways begin? With what clotting factor do they end?*
25. *Starting with the formation of prothrombinase, describe the events that result in clot formation.*

Control of Clot Formation

Without control, clotting would spread from the point of its initiation throughout the entire circulatory system. The blood contains several **anticoagulants** (an′tē-kō-ag′ū-lantz), which prevent clotting factors from forming clots. **Antithrombin** (an-tē-throm′bin) and **heparin** (hep′ă-rin), for example, inactivate thrombin. Without thrombin, fibrinogen is not converted to fibrin, and no clot forms. Normally, there are enough anticoagulants in the blood to prevent clot formation. At an injury site, however, the activation of clotting factors is very rapid. Enough clotting factors are activated that the anticoagulants can no longer prevent a clot from forming. Away from the injury site there are enough anticoagulants to prevent clot formation from spreading.

The Danger of Unwanted Clots

When platelets encounter damaged or diseased areas of blood vessels or heart walls, an attached clot, called a **thrombus** (throm′bus), can form. A thrombus that breaks loose and begins to float through the circulation is called an **embolus** (em′bō-lŭs). Both thrombi and emboli can result in death if they block vessels that supply blood to essential organs, such as the heart, brain, or lungs. Abnormal blood clotting can be prevented or hindered by the injection of anticoagulants, such as heparin, which acts rapidly. Coumadin (koo′mă-din), or warfarin (war′fă-rin), acts more slowly than heparin. Warfarin prevents clot formation by suppressing the production of vitamin K–dependent clotting factors by the liver. ▼

26. *What is the function of anticoagulants in the blood? How do antithrombin and heparin prevent clot formation?*
27. *Define thrombus and embolus, and explain why they are dangerous.*

Clot Retraction and Fibrinolysis

After a clot has formed, it begins to condense into a more compact structure by a process known as **clot retraction.** Platelets contain the contractile proteins actin and myosin, which operate in a fashion similar to that of the actin and myosin in muscle (see chapter 8). Platelets form small extensions that attach to fibrin through surface receptors. Contraction of the extensions pulls on the fibrin and is responsible for clot retraction. **Serum** (sēr′ŭm), which is plasma without the clotting factors, is squeezed out of the clot during clot retraction.

Retraction of the clot pulls the edges of the damaged blood vessel together, helping stop the flow of blood, reducing the probability of infection, and enhancing healing. The damaged vessel is repaired by the movement of fibroblasts into the damaged area and

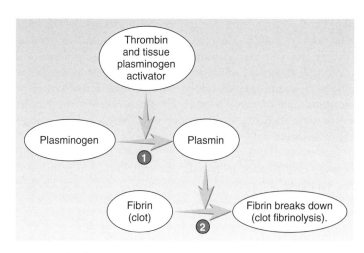

1. Thrombin and tissue plasminogen activator convert inactive plasminogen into plasmin.

2. Plasmin breaks down the fibrin in a blood clot, resulting in clot fibrinolysis.

Process Figure 16.11   Fibrinolysis

the formation of new connective tissue. In addition, epithelial cells around the wound divide and fill in the torn area (see chapter 4).

Clots are dissolved by a process called **fibrinolysis** (fi-bri-nol′-i-sis, fibrino-, fiber- + *lysis,* dissolution) (figure 16.11). **Plasminogen** (plaz-min′ō-jen, plasmin- + *gen,* produce) is an inactive plasma protein produced by the liver. Thrombin, other clotting factors activated during clot formation, and **tissue plasminogen activator (t-PA)** released from surrounding tissues can stimulate the conversion of plasminogen to its active form, **plasmin** (plaz′min). Over a period of a few days, plasmin slowly breaks down the fibrin.

Dissolving Clots

A heart attack can result from blockage by a clot of blood vessels that supply blood to the heart. One treatment for a heart attack is to inject into the blood chemicals that activate plasmin. Unlike aspirin and anticoagulant therapies, which are used to prevent heart attacks, the strategy in using plasmin activators is to dissolve the clot quickly and restore blood flow to cardiac muscle, thus reducing damage to tissues. **Streptokinase** (strep-tō-kīn′ās), a bacterial enzyme, and t-PA, produced through genetic engineering, have been used successfully to dissolve clots. ▼

28. *Describe clot retraction. What is serum?*
29. *What is fibrinolysis? How does it occur?*

Blood Grouping

If large quantities of blood are lost during surgery or in an accident, a person can go into shock and die unless a transfusion or an infusion is performed. A **transfusion** is the transfer of blood or blood components from one individual to another. When large quantities of blood are lost, red blood cells must be replaced so that the blood's oxygen-carrying capacity is restored. An **infusion** is the introduction of a fluid other than blood, such as a saline or glucose solution, into the blood. In many cases, the return of blood volume to normal levels is all that is necessary to prevent shock. Eventually, the body produces red blood cells to replace those that were lost.

Early attempts to transfuse blood were often unsuccessful because they resulted in **transfusion reactions,** which included the clumping of blood cells, rupture of blood cells, and clotting within blood vessels. It is now known that transfusion reactions are caused by interactions between antigens and antibodies (see chapter 19). In brief, the surfaces of red blood cells have molecules called **antigens** (an′ti-jenz, anti-, body + -*gen,* producing), and the plasma has proteins called **antibodies** (an′te-bod-ēz, anti-, against + body, a thing). Antibodies are very specific, meaning that each antibody can combine only with a certain antigen. When the antibodies in the plasma bind to the antigens on the surface of the red blood cells, they form molecular bridges that connect the red blood cells together. As a result, **agglutination** (ă-gloo-ti-nā′shŭn, *ad,* to + *gluten,* glue), or clumping of the cells, occurs. The combination of the antibodies with the antigens also can initiate reactions that cause **hemolysis** (hē-mol′i-sis, hemo-, blood + *lysis,* destruction), or rupture of the red blood cells. The debris formed from the ruptured red blood cells can trigger clotting within small blood vessels. As a result of these changes, tissue damage and death can occur.

The antigens on the surface of red blood cells have been categorized into **blood groups.** Although many blood groups are recognized, the ABO and Rh blood groups are the most important for transfusion reactions. ▼

30. *Define transfusion, infusion, and transfusion reaction.*
31. *What are antigens and antibodies? How do they cause agglutination, hemolysis, and blood clotting?*
32. *What are blood groups?*

ABO Blood Group

The **ABO blood group** system is used to categorize human blood. ABO antigens appear on the surface of the red blood cells. Type A blood has type A antigens, type B blood has type B antigens, type AB blood has both types of antigens, and type O blood has neither A nor B antigens (figure 16.12). In addition, plasma from type A blood contains anti-B antibodies, which act against type B antigens, whereas plasma from type B blood contains anti-A antibodies, which act against type A antigens. Type AB blood has neither type of antibody, and type O blood has both anti-A and anti-B antibodies.

The ABO blood types are not found in equal numbers. In the United States, the distribution is type O, 47%; type A, 41%; type B, 9%; and type AB, 3%. Among African-Americans the distribution is type O, 46%; type A, 27%; type B, 20%; and type AB, 7%.

Antibodies do not normally develop against an antigen unless the body is exposed to that antigen. One possible explanation for the production of anti-A and/or anti-B antibodies is that type A or B antigens on bacteria or food in the digestive tract stimulate the formation of antibodies against antigens that are different from one's own antigens. In support of this explanation is the observation that anti-A and anti-B antibodies are not found in the blood until about 2 months after birth. For example, an infant with

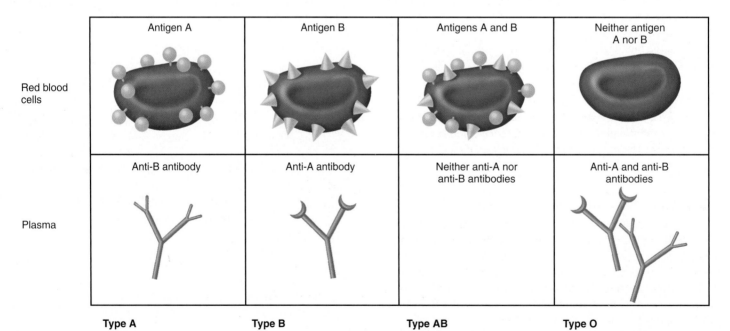

Type A
Red blood cells with type A surface antigens and plasma with anti-B antibodies

Type B
Red blood cells with type B surface antigens and plasma with anti-A antibodies

Type AB
Red blood cells with both type A and type B surface antigens, and neither anti-A nor anti-B plasma antibodies

Type O
Red blood cells with neither type A nor type B surface antigens, but both anti-A and anti-B plasma antibodies

Figure 16.12 **ABO Blood Groups**
For simplicity, only parts of the anti-A and anti-B antibodies are illustrated. Each antibody has five identical, Y-shaped arms (see IgM in chapter 19).

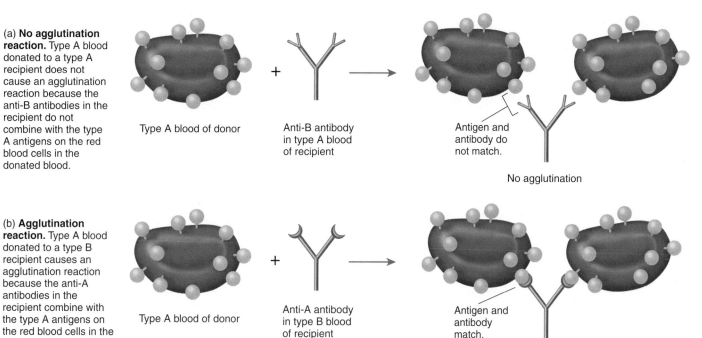

(a) **No agglutination reaction.** Type A blood donated to a type A recipient does not cause an agglutination reaction because the anti-B antibodies in the recipient do not combine with the type A antigens on the red blood cells in the donated blood.

Type A blood of donor

Anti-B antibody in type A blood of recipient

Antigen and antibody do not match.

No agglutination

(b) **Agglutination reaction.** Type A blood donated to a type B recipient causes an agglutination reaction because the anti-A antibodies in the recipient combine with the type A antigens on the red blood cells in the donated blood.

Type A blood of donor

Anti-A antibody in type B blood of recipient

Antigen and antibody match.

Agglutination

Figure 16.13   Agglutination Reaction

For simplicity, only parts of the anti-A and anti-B antibodies are illustrated. Each antibody has five identical, Y-shaped arms (see IgM in chapter 19).

type A blood produces anti-B antibodies against the B antigens on bacteria or food. An infant with A antigens does not produce antibodies against the A antigen on bacteria or food because mechanisms exist in the body to prevent the production of antibodies that would react with the body's own antigens (see chapter 19).

A **donor** is a person who gives blood, and a **recipient** is a person who receives blood. Usually, a recipient can receive blood from a donor if they both have the same blood type. For example, a person with type A blood can receive blood from a person with type A blood. There is no ABO transfusion reaction because the recipient has no anti-A antibodies against the type A antigen. On the other hand, if type A blood were donated to a person with type B blood, a transfusion reaction would occur because the person with type B blood has anti-A antibodies against the type A antigen, and agglutination would result (figure 16.13).

Historically, people with type O blood have been called universal donors because they usually can give blood to the other ABO blood types without causing an ABO transfusion reaction. Their red blood cells have no ABO surface antigens and, therefore, do not react with the recipient's anti-A or anti-B antibodies. For example, if type O blood is given to a person with type A blood, the type O red blood cells do not react with the anti-B antibodies in the recipient's blood. In a similar fashion, if type O blood is given to a person with type B blood, there is no reaction with the recipient's anti-A antibodies.

The term *universal donor* is misleading, however. There are two ways in which the transfusion of type O blood can produce a transfusion reaction. First, mismatching blood groups other than the ABO blood group can cause a transfusion reaction. Second, antibodies in the donor's blood can react with antigens in the recipient's blood. For example, type O blood has anti-A and anti-B antibodies. If type O

blood is transfused into a person with type A blood, the anti-A antibodies (in the type O blood) react against the A antigens (on the red blood cells in the type A blood). Usually such reactions are not serious because the antibodies in the donor's blood are diluted in the large volume of the recipient's blood, and few reactions take place.

Blood banks separate donated blood into several products, such as packed red blood cells; plasma; platelets; and cryoprecipitate, which contains von Willebrand factor, clotting factors, and fibrinogen. This process allows the donated blood to be used by multiple recipients, each of whom may need only one of the blood components. Type O packed red blood cells are unlikely to cause an ABO transfusion reaction when given to a person with a different blood type because it has very little plasma with anti-A and anti-B antibodies. ▼

33. *What kinds of antigens and antibodies are found in each of the four ABO blood types?*

34. *Why is a person with type O blood considered to be a universal donor?*

P R E D I C T   4

Historically, people with type AB blood were called universal recipients. What is the rationale for this term? Explain why the term is misleading.

Rh Blood Group

Another important blood group is the **Rh blood group,** so named because it was first studied in the rhesus monkey. People are Rh-positive if they have a certain Rh antigen (the D antigen) on the surface of their red blood cells, and people are Rh-negative if they do not have this Rh antigen. About 85% of white people and 95%

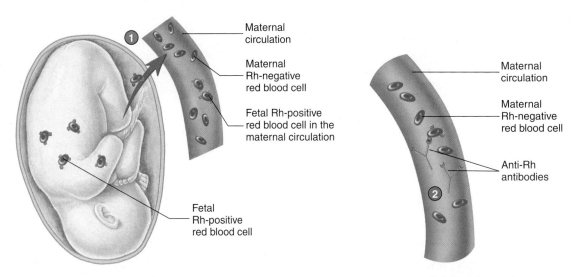

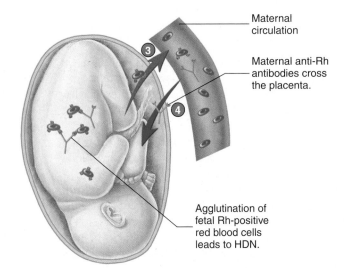

1. Before or during delivery, Rh-positive red blood cells from the fetus enter the blood of an Rh-negative woman through a tear in the placenta.

2. The mother is sensitized to the Rh antigen and produces anti-Rh antibodies. Because this usually happens after delivery, there is no effect on the fetus in the first pregnancy.

3. During a subsequent pregnancy with an Rh-positive fetus, Rh-positive red blood cells cross the placenta, enter the maternal circulation, and stimulate the mother to produce antibodies against the Rh antigen. Antibody production is rapid because the mother has been sensitized to the Rh antigen.

4. The anti-Rh antibodies from the mother cross the placenta, causing agglutination and hemolysis of fetal red blood cells, and hemolytic disease of the newborn (HDN) develops.

Figure 16.14 **Hemolytic Disease of The Newborn (HDN)**

of African-Americans are Rh-positive. The ABO blood type and the Rh blood type usually are designated together. For example, a person designated as A positive is type A in the ABO blood group and Rh-positive. The rarest combination in the United States is AB negative, which occurs in less than 1% of all Americans.

Antibodies against the Rh antigens do not develop unless an Rh-negative person is exposed to Rh-positive red blood cells. This can occur through a transfusion or by the transfer of blood across the placenta to a mother from her fetus. When an Rh-negative person receives a transfusion of Rh-positive blood, the recipient becomes sensitized to the Rh antigens and produces anti-Rh antibodies. If the Rh-negative person is unfortunate enough to receive a second transfusion of Rh-positive blood after becoming sensitized, a transfusion reaction results.

Rh incompatibility can pose a major problem in some pregnancies when the mother is Rh-negative and the fetus is Rh-positive. If fetal blood leaks through the placenta and mixes with the mother's blood, the mother becomes sensitized to the Rh antigen (figure 16.14). The mother produces anti-Rh antibodies that cross the placenta and cause agglutination and hemolysis of fetal red blood cells. This disorder is called **hemolytic** (hē-mō-lit′ik) **disease of the newborn (HDN),** or **erythroblastosis fetalis** (ĕ-rith′rō-blas-tō′sis fē-ta′lis), and it can be fatal to the fetus. In the mother's first pregnancy, there is often no problem. The leakage of fetal blood is usually the result of a tear in the placenta that takes place either late in the pregnancy or during delivery. Thus, there is not enough time for the mother to produce enough anti-Rh antibodies to harm the fetus. If sensitization occurs, however, it can cause problems in a subsequent pregnancy in two ways. First, once a woman is sensitized and produces anti-Rh antibodies, she may continue to produce the antibodies throughout her life. Thus, in a subsequent pregnancy, anti-Rh antibodies may already be present. Second, and especially dangerous in a subsequent pregnancy with an Rh-positive fetus, if any leakage of fetal blood into the mother's blood occurs, she rapidly produces large amounts of anti-Rh antibodies, and HDN develops. Therefore, the levels of anti-Rh antibodies in the mother should be tested. If they are too high, the fetus should be tested to determine the severity of the HDN. In severe cases, a transfusion to replace lost red blood cells can be performed through the umbilical cord, or the baby can be delivered if mature enough.

Prevention of HDN is often possible if the Rh-negative mother is given an injection of a specific type of antibody preparation, called $Rh_o(D)$ immune globulin (RhoGAM). The injection can be given during the pregnancy, before delivery, or immediately after each delivery, miscarriage, or abortion. The injection contains antibodies against Rh antigens. The injected antibodies bind to the Rh antigens of any fetal red blood cells that may have entered the mother's blood. This treatment inactivates the fetal Rh antigens and prevents sensitization of the mother. However, if sensitization of the mother has already occurred, the treatment is ineffective. ▼

35. *Define Rh-positive.*
36. *What Rh blood types must the mother and fetus have before HDN can occur? How is HDN harmful to the fetus?*
37. *Describe the events that lead to the development of HDN. Why doesn't HDN usually develop in the first pregnancy?*
38. *How can HDN be prevented?*

CASE STUDY | Treatment of Hemolytic Disease of the Newborn

Billy was born with HDN. He was treated with exchange transfusion, phototherapy, and erythropoietin. An exchange transfusion replaced Billy's blood with donor blood. In this procedure, as the donor's blood was transfused into Billy, his blood was withdrawn. Phototherapy, in which blood passing through the skin is exposed to blue or white lights, results in the breakdown of bilirubin to less toxic compounds that are removed by the newborn's liver. During fetal development, the increased rate of red blood cell destruction caused by the mother's anti-Rh antibodies results in lower than normal numbers of red blood cells, a condition called anemia (ă-nē′mē-ă). It also results in increased levels of bilirubin. Although high levels of bilirubin can damage the brain by killing nerve cells, this is not usually a problem in the fetus because the bilirubin is removed by the placenta. Following birth, bilirubin levels can increase because of the continued lysis of red blood cells and the inability of the newborn's liver to handle the large bilirubin load.

PREDICT 5

Answer the following questions about Billy's treatment for HDN.
a. What was the purpose of giving Billy a transfusion?
b. What was the benefit of an exchange transfusion?
c. Explain the reason for giving Billy erythropoietin.
d. Just before birth, would Billy's erythropoietin levels have been higher or lower than those of a fetus without HDN?
e. After birth, but before treatment, would Billy's erythropoietin levels have increased or decreased?
f. When treating HDN with an exchange transfusion, should the donor's blood be Rh-positive or Rh-negative? Explain.
g. Does giving an Rh-positive newborn a transfusion of Rh-negative blood change the newborn's blood type? Explain.

Diagnostic Blood Tests

Type and Crossmatch

To prevent transfusion reactions, blood is typed and crossmatched. **Blood typing** determines the ABO and Rh blood groups of the blood sample. Typically, the cells are separated from the serum. The cells are tested with known antibodies to determine the type of antigen on the cell surface. For example, if a patient's blood cells agglutinate when mixed with type A antibodies but do not agglutinate when mixed with type B antibodies, it is concluded that the cells have type A antigen. In a similar fashion, the serum is mixed with known cell types (antigens) to determine the type of antibodies in the serum.

The International Society of Blood Transfusion recognizes 29 blood groups, including the ABO and Rh groups. Because any of these blood groups can cause a transfusion reaction, a crossmatch is performed. In a **crossmatch,** the donor's blood cells are mixed with the recipient's serum, and the donor's serum is mixed with the recipient's cells. The donor's blood is considered safe for transfusion only if no agglutination occurs in either crossmatch.

Complete Blood Count

The **complete blood count (CBC)** is an analysis of the blood that provides much information. It consists of a red blood count, hemoglobin and hematocrit measurements, a white blood count, and a differential white blood count.

Red Blood Count

Blood cell counts are usually done electronically with a machine, but they can be done manually with a microscope. A normal **red blood count (RBC)** for a male is 4.6–6.2 million red blood cells per microliter (μL) of blood, and for a female it is 4.2–5.4 million μL of blood. A microliter is equivalent to 1 cubic millimeter (mm^3) or 10^{-6} L, and one drop of blood is approximately 50 μL. **Erythrocytosis** (ĕ-rith′rō-sī-tō′sis) is an overabundance of red blood cells (see "Disorders of the Blood," p. 780).

Erythrocytosis (ĕ-rith′rō-sī-tō′sis) is an overabundance of red blood cells, resulting in increased blood viscosity, reduced flow rates, and, if severe, plugging of the capillaries. **Relative erythrocytosis** results from decreased plasma volume, such as that caused by dehydration, diuretics, and burns. **Primary erythrocytosis,** often called **polycythemia vera** (pol′ē-sī-thē′mē-ă ve′ra), is a stem cell defect of unknown cause that results in the overproduction of red blood cells, granulocytes, and platelets. Erythropoietin levels are low and the spleen can be enlarged. **Secondary erythrocytosis (polycythemia)** results from a decreased oxygen supply, such as that which occurs at high altitudes, in chronic obstructive pulmonary disease, and in congestive heart failure. The resulting decrease in oxygen delivery to the kidneys stimulates erythropoietin secretion and causes an increase in red blood cell production. In primary and secondary erythrocytosis, the increased number of red blood cells increases blood viscosity and blood volume. There can be clogging of capillaries and the development of hypertension.

Anemia

Anemia is a deficiency of normal hemoglobin in the blood, resulting from a decreased number of red blood cells, a decreased amount of hemoglobin in each red blood cell, or both. Anemia can also be the result of abnormal hemoglobin production.

Anemia reduces the blood's ability to transport oxygen. People with anemia suffer from a lack of energy and feel excessively tired and listless. They may appear pale and quickly become short of breath with only slight exertion.

Red blood cell production also can be lower than normal as a result of nutritional deficiencies. **Iron-deficiency anemia** results from a deficient intake or absorption of iron or from excessive iron loss. Consequently, not enough hemoglobin is produced, the number of red blood cells decreases, and the red blood cells that are manufactured are smaller than normal.

Folate deficiency can also cause anemia. Inadequate amounts of folate in the diet are the usual cause of folate deficiency, with the disorder developing most often in the poor, in pregnant women, and in chronic alcoholics. Because folate helps in the synthesis of DNA, a folate deficiency results in fewer cell divisions and, therefore, decreased red blood cell production. A deficiency in folate during pregnancy is also associated with birth disorders called neural tube defects, such as spina bifida.

Another type of nutritional anemia is **pernicious** (per-nish′ŭs) **anemia,** which is caused by inadequate vitamin B_{12}. A 2- to 3-year supply of vitamin B_{12} can be stored in the liver. Inadequate amounts of vitamin B_{12} can result in decreased red blood cell production because vitamin B_{12} is important for folate synthesis. Although inadequate levels of vitamin B_{12} in the diet can cause pernicious anemia, the usual cause is insufficient absorption of the vitamin. Normally, the stomach produces **intrinsic factor,** a protein that binds to vitamin B_{12}. The combined molecules pass into the small intestine, where intrinsic factor facilitates the absorption of the vitamin. Without adequate levels of intrinsic factor, insufficient vitamin B_{12} is absorbed, and pernicious anemia develops. Most cases of pernicious anemia probably result from an autoimmune disease in which the body's immune system damages the stomach cells that produce intrinsic factor.

One general cause of anemia is insufficient production of red blood cells. **Aplastic** (ā-plas′tik) **anemia** is caused by an inability of the red bone marrow to produce red blood cells and, often, white blood cells and platelets. It is usually acquired as a result of damage to the stem cells in red marrow by chemicals such as benzene, drugs such as certain antibiotics and sedatives, or radiation.

Another general cause of anemia is the loss or destruction of red blood cells. **Hemorrhagic** (hem-ŏ-raj′ik) **anemia** results from a loss of blood, such as can result from trauma, ulcers, or excessive menstrual bleeding. Chronic blood loss, in which small amounts of blood are lost over a period of time, can result in iron-deficiency anemia. **Hemolytic** (hē-mō-lit′ik) **anemia** is a disorder in which red blood cells rupture or are destroyed at an excessive rate. It can be caused by inherited defects in the red blood cells. For example, one kind of inherited hemolytic anemia results from a defect in the cell membrane that causes red blood cells to rupture easily. Many kinds of hemolytic anemia result from damage to the red blood cells by drugs, snake venom, artificial heart valves, autoimmune disease, or hemolytic disease of the newborn.

Anemia can result from a reduced rate of synthesis of the globin chains in hemoglobin. **Thalassemia** (thal-ă-sē′mē-ă) is a hereditary disease found in people of Mediterranean, Asian, and African ancestry. If hemoglobin production is severely depressed, death usually occurs before age 20. In less severe cases, thalassemia produces a mild anemia.

Some anemias are caused by defective hemoglobin production. **Sickle-cell disease** is a hereditary disease that results in the formation of an abnormal

---

Hemoglobin Measurement

A hemoglobin measurement determines the amount of hemoglobin in a given volume of blood, usually expressed as grams of hemoglobin per 100 mL of blood. The normal hemoglobin measurement for a male is 14–18 grams (g)/100 mL of blood, and for a female it is 12 to 16 g/100 mL of blood. An abnormally low hemoglobin measurement is an indication of **anemia**, which is either a reduced number of red blood cells or a reduced amount of hemoglobin in each red blood cell (see "Disorders of the Blood," above).

Hematocrit Measurement

The percentage of total blood volume composed of red blood cells is the **hematocrit** (hē′mă-tō-krit, hem′a-tō-krit). One way to determine hematocrit is to place blood in a tube and spin the

hemoglobin. When blood oxygen levels decrease, as when oxygen diffuses away from the hemoglobin in tissue capillaries, the abnormal hemoglobin molecules join together, causing a change in red blood cell shape. When blood oxygen levels increase, as in the lungs, the abnormal hemoglobin molecules separate and red blood cells can resume their normal shape.

The major consequence of sickle-cell disease is tissue damage resulting from reduced blood flow through tissues because sickle-shaped red blood cells become lodged in capillaries. The most common symptom is pain, often severe pain, associated with the tissues deprived of blood. Spleen and liver enlargement, kidney and lung damage, and stroke can occur. Priapism (prī-ă-pizm), a prolonged, painful erection due to venous blockage, can develop in men. In severe cases, there is so much abnormal hemoglobin production that the disease is usually fatal before age 30.

Sickle-shaped red blood cells are also likely to rupture, and they have a life span of 10–20 days, compared with 120 days for normal red blood cells. Rupture of red blood cells can result in hemolytic anemia.

Sickle-cell disease is an autosomal recessive disorder (see chapter 25), which means only individuals who have two mutated genes express the disease. Individuals who have one normal and one abnormal hemoglobin gene produce sufficient amounts of normal hemoglobin that their red blood cells do not usually become sickle-shaped. Individuals with one normal and one abnormal hemoglobin gene are **carriers** (see chapter 25) and are said to have **sickle-cell trait.** They usually do not have disease symptoms, but can pass the abnormal gene to their children. Under stressful situations, such as low oxygen levels at high altitude, severe infections, or exhaustion, complications caused by sickling can develop.

Sickle-cell disease is an example of **balanced polymorphism,** in which the carrier has a better ability to survive under certain circumstances than a person who has normal hemoglobin or a person with sickle-cell disease. Carriers with sickle-cell trait are healthier than those with sickle-cell disease. They also have increased resistance to malaria, compared with individuals who have only normal hemoglobin. In the United States, sickle-cell disease and sickle-cell trait are found mostly in people of African descent.

Leukemia

Leukemia (loo-kē′mē-ă) is a cancer in which abnormal production of one or more of the white blood cell types occurs. Because these cells are usually immature or abnormal and lack normal immunological functions, people with leukemia are very susceptible to infections. The excess production of white blood cells in the red marrow can also interfere with red blood cell and platelet formation and thus leads to anemia and bleeding.

Disseminated Intravascular Coagulation

Disseminated intravascular coagulation (DIC) is a complex disorder involving clotting throughout the vascular system followed by bleeding. Normally, excessive clotting is prevented by anticoagulants. DIC can develop when these control mechanisms are overwhelmed. Many conditions can cause DIC by overstimulating blood clotting. Examples include massive tissue damage, such as burns, and alteration of the lining of blood vessels caused by infections or snake bites. If DIC occurs slowly, the predominant effect is thrombosis and the blockage of blood vessels. If DIC occurs rapidly, massive clot formation occurs, quickly using up available blood clotting factors and platelets. The result is continual bleeding around wounds, intravenous lines, and catheters, as well as internal bleeding. The best therapy for DIC is to treat and stop whatever condition is stimulating blood clotting.

Von Willebrand Disease

Von Willebrand disease is the most common inherited bleeding disorder, occurring as frequently as 1 in 1000 individuals. Von Willebrand factor helps platelets adhere to collagen and become activated. In von Willebrand disease, platelet plug formation and the contribution of activated platelets to blood clotting are impaired. Treatments for von Willebrand disease include injections of von Willebrand factor and the administration of drugs that increase von Willebrand factor levels in the blood.

Hemophilia

Hemophilia (hē-mō-fil′ē-ă) is a genetic disorder in which clotting is abnormal or absent. It is most often found in people from northern Europe and their descendants. Hemophilia is most often a sex-linked trait, and it occurs almost exclusively in males (see chapter 25). There are several types of hemophilia, each the result of a deficiency or dysfunction of a clotting factor. Treatment of hemophilia involves injection of the missing clotting factor taken from donated blood or produced by genetic engineering.

tube in a centrifuge. The formed elements are heavier than the plasma and are forced to one end of the tube. White blood cells and platelets form a thin, whitish layer, called the **buffy coat,** between the plasma and the red blood cells (figure 16.15). The red blood cells account for 40%–52% of the total blood volume in males and 38%–48% in females. The hematocrit measurement is affected by the number and size of red blood cells because it is based on volume. For example, a decreased hematocrit can result from a decreased number of normal-sized red blood cells or a normal number of small red blood cells. The average size of a red blood cell is calculated by dividing the hematocrit by the red blood cell count. A number of disorders cause red blood cells to be smaller or larger than normal. For example, inadequate iron in the diet can impair hemoglobin

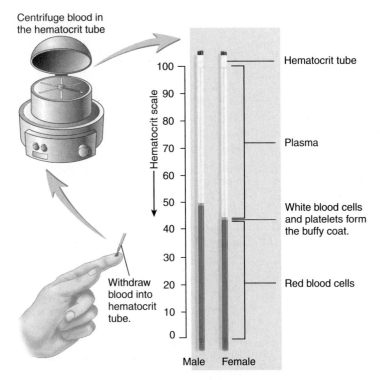

Centrifuge blood in the hematocrit tube

Hematocrit tube

Plasma

White blood cells and platelets form the buffy coat.

Red blood cells

Withdraw blood into hematocrit tube.

Male   Female

Figure 16.15   Hematocrit

Blood is withdrawn into a capillary tube and placed in a centrifuge. The blood is separated into plasma, red blood cells, and a small amount of white blood cells and platelets, which rest on the red blood cells. The hematocrit measurement is the percentage of the total blood volume that is red blood cells. It does not measure the white blood cells and platelets. Normal hematocrits for a male and a female are shown.

production. Consequently, red blood cells do not fill up with hemoglobin during their formation, and they remain smaller than normal.

White Blood Count

A **white blood count (WBC)** measures the total number of white blood cells in the blood. There are normally 5000–9000 white blood cells per microliter of blood. **Leukopenia** (loo-kō-pē′nē-ă) is a lower than normal WBC; it often indicates decreased production or destruction of the red marrow. Radiation, drugs, tumors, viral infections, or a deficiency of the vitamins folate or B_{12} can cause leukopenia. **Leukocytosis** (loo′kō-sī-tō′sis) is an abnormally high WBC. Bacterial infections often cause leukocytosis by stimulating neutrophils to increase in number. **Leukemia** (loo-kē′mē-ă), a cancerous tumor of the red marrow, can cause leukocytosis, but the white blood cells do not function normally.

Differential White Blood Count

A **differential white blood count** determines the percentage of each of the five kinds of white blood cells in the white blood cell count. Normally, neutrophils account for 60%–70%, lymphocytes 20%–25%, monocytes 3%–8%, eosinophils 2%–4%, and basophils 0.5%–1% of all white blood cells. For example, in bacterial infections, the neutrophil count is often greatly increased, whereas in allergic reactions the eosinophil and basophil counts are elevated.

Clotting

Two measurements that test the ability of the blood to clot are the platelet count and the prothrombin time.

Platelet Count

A normal **platelet count** is 250,000–400,000 platelets per microliter of blood. **Thrombocytopenia** (throm′bō-sī-tō-pē′nē-ă) is a condition in which the platelet count is greatly reduced, resulting in chronic bleeding through small vessels and capillaries. It can be caused by decreased platelet production as a result of hereditary disorders, a lack of vitamin B_{12}, drug therapy, or radiation therapy.

Prothrombin Time

Prothrombin time is a measure of how long it takes for the blood to start clotting, which is normally 9–12 seconds. Prothrombin time is determined by adding thromboplastin to whole plasma. Thromboplastin is a chemical released from injured tissues that starts the process of clotting (see figure 16.10). Prothrombin time is officially reported as the International Normalized Ratio (INR), which standardizes the time it takes to clot on the basis of the slightly different thromboplastins used by different labs. Because many clotting factors have to be activated to form fibrin, a deficiency of any one of them can cause an abnormal prothrombin time. Vitamin K deficiency, certain liver diseases, and drug therapy can cause an increased prothrombin time.

Blood Chemistry

The composition of materials dissolved or suspended in the plasma can be used to assess the functioning of many of the body's systems. For example, high blood glucose levels can indicate that the pancreas is not producing enough insulin, high blood urea nitrogen (BUN) is a sign of reduced kidney function, increased bilirubin can indicate liver dysfunction, and high cholesterol levels can indicate an increased risk of developing cardiovascular disease. A number of blood chemistry tests are routinely done when a blood sample is taken, and additional tests are available. ▼

> 39. For each of the following tests, define the test and give an example of a disorder that would cause an abnormal test result:
>   a. red blood count
>   b. hemoglobin measurement
>   c. hematocrit measurement
>   d. white blood count
>   e. differential white blood count
>   f. platelet count
>   g. prothrombin time
>   h. blood chemistry tests

P R E D I C T   **6**
When a patient complains of acute pain in the abdomen, his physician suspects appendicitis, which is a bacterial infection of the appendix. What blood test could provide supporting evidence for the diagnosis?

S U M M A R Y

Functions and Composition of Blood (p. 464)

1. Blood transports gases, nutrients, waste products, processed molecules, and regulatory molecules.
2. Blood regulates pH, fluid, and ion balance.
3. Blood is involved with temperature regulation and protects against foreign substances such as microorganisms and toxins.
4. Blood clotting prevents fluid and cell loss and is part of tissue repair.
5. Blood is a connective tissue consisting of plasma and formed elements. Total blood volume is approximately 5 L.

Plasma (p. 465)

1. Plasma is mostly water (91%) and contains proteins, such as albumin (maintains osmotic pressure), globulins (function in transport and immunity), fibrinogen (is involved in clot formation), and hormones and enzymes (are involved in regulation).
2. Plasma contains ions, nutrients, waste products, and gases.

Formed Elements (p. 465)

The formed elements include red blood cells (erythrocytes), white blood cells (leukocytes), and platelets (cell fragments).

Production of Formed Elements

Formed elements arise in red bone marrow from stem cells (hemocytoblasts).

Red Blood Cells

1. Red blood cells are biconcave discs containing hemoglobin and carbonic anhydrase.
   - A hemoglobin molecule consists of four heme and four globin molecules. The heme molecules transport oxygen. Iron is required for oxygen transport.
   - The globin molecules transport carbon dioxide.
2. Carbon dioxide is transported dissolved in plasma (7%), bound to hemoglobin (23%), and as bicarbonate ions (70%).
3. Carbonic anhydrase in red blood cells promotes the formation of bicarbonate ions.
4. Erythropoiesis is the production of red blood cells.
   - Stem cells in red bone marrow give rise to proerythroblasts. Through a series of cell division and differentiation, cells accumulate hemoglobin, lose their nuclei, and are released into the blood as reticulocytes, which become red blood cells.
   - In response to low blood oxygen, the kidneys produce erythropoietin, which stimulates erythropoiesis.
5. Worn-out red blood cells are phagocytized by macrophages in the spleen or liver. Hemoglobin is broken down, iron and amino acids are reused, and heme becomes bilirubin that is secreted in bile.

White Blood Cells

1. White blood cells protect the body against microorganisms and remove dead cells and debris.
2. Five types of white blood cells exist.
   - Neutrophils are small phagocytic cells.
   - Eosinophils reduce inflammation.
   - Basophils release histamine and are involved with increasing the inflammatory response.
   - Lymphocytes are important in immunity, including the production of antibodies.

- Monocytes leave the blood, enter tissues, and become large phagocytic cells called macrophages.

Platelets

Platelets are cell fragments involved with preventing blood loss.

Preventing Blood Loss (p. 472)
Vascular Spasm

Blood vessels constrict in response to injury, resulting in decreased blood flow.

Platelet Plugs

1. Platelets repair minor damage to blood vessels by forming platelet plugs.
   - In platelet adhesion, platelets bind to collagen in damaged tissues.
   - In the platelet release reaction, platelets release chemicals that activate additional platelets.
   - In platelet aggregation, platelets bind to one another to form a platelet plug.
2. Platelets also release chemicals (ADP and thromboxanes) involved with blood clotting.

Blood Clotting

1. Blood clotting, or coagulation, is the formation of a clot (a network of protein fibers called fibrin).
2. Blood clotting begins with the extrinsic or intrinsic pathway. Both pathways end with the production of activated factor X.
   - The extrinsic pathway begins with the release of thromboplastin from damaged tissues.
   - The intrinsic pathway begins with the activation of factor XII.
3. Activated factor X, factor V, phospholipids, and Ca^{2+} form prothrombinase.
4. Prothrombin is converted to thrombin by prothrombinase.
5. Fibrinogen is converted to fibrin by thrombin. The insoluble fibrin forms the clot.

Control of Clot Formation

Anticoagulants in the blood, such as antithrombin and heparin, prevent clot formation.

Clot Retraction and Fibrinolysis

1. Clot retraction results from the contraction of platelets, which pull the edges of damaged tissue closer together.
2. Serum, which is plasma minus fibrinogen and some clotting factors, is squeezed out of the clot.
3. Thrombin and tissue plasminogen activator activate plasmin, which dissolves fibrin (fibrinolysis).

Blood Grouping (p. 475)

1. Blood groups are determined by antigens on the surface of red blood cells.
2. In transfusion reactions, antibodies can bind to red blood cell antigens, resulting in agglutination or hemolysis of red blood cells.

ABO Blood Group

1. Type A blood has A antigens, type B blood has B antigens, type AB blood has A and B antigens, and type O blood has neither A nor B antigens.

2. Type A blood has anti-B antibodies, type B blood has anti-A antibodies, type AB blood has neither A nor B antibodies, and type O blood has both anti-A and anti-B antibodies.
3. Mismatching the ABO blood group can result in transfusion reactions.

Rh Blood Group

1. Rh-positive blood has certain Rh antigens (the D antigen), whereas Rh-negative blood does not.
2. Antibodies against the Rh antigen are produced when an Rh-negative person is exposed to Rh-positive blood.
3. The Rh blood group is responsible for hemolytic disease of the newborn, which can occur when the fetus is Rh-positive and the mother is Rh-negative.

Diagnostic Blood Tests (p. 479)
Type and Crossmatch

1. Blood typing determines the ABO and Rh blood groups of a blood sample.

2. A crossmatch tests for agglutination reactions between donor and recipient blood.

Complete Blood Count

The complete blood count consists of the following: red blood count (million/μL), hemoglobin measurement (grams of hemoglobin per 100 mL of blood), hematocrit measurement (percent volume of red blood cells), white blood count (million/μL), and differential white blood count (the percentage of each type of white blood cell).

Clotting

Platelet count and prothrombin time measure the ability of the blood to clot.

Blood Chemistry

The composition of materials dissolved or suspended in plasma (e.g., glucose, urea nitrogen, bilirubin, and cholesterol) can be used to assess the functioning and status of the body's systems.

REVIEW AND COMPREHENSION

1. Which of these is a function of blood?
   a. clot formation
   b. protection against foreign substances
   c. maintenance of body temperature
   d. regulation of pH and osmosis
   e. all of the above

2. Which of these is *not* a component of plasma?
   a. oxygen
   b. sodium ions
   c. platelets
   d. water
   e. urea

3. Which of these plasma proteins plays an important role in maintaining the osmotic concentration of the blood?
   a. albumin
   b. fibrinogen
   c. platelets
   d. hemoglobin
   e. globulins

4. Red blood cells
   a. are the least numerous formed element in the blood.
   b. are phagocytic cells.
   c. are produced in the yellow marrow.
   d. do not have a nucleus.
   e. all of the above.

5. Given these ways of transporting carbon dioxide in the blood:
   1. as bicarbonate ions
   2. combined with blood proteins
   3. dissolved in plasma

   Choose the arrangement that lists them in the correct order from largest to smallest percentage of carbon dioxide transported.
   a. 1,2,3
   b. 1,3,2
   c. 2,3,1
   d. 2,1,3
   e. 3,1,2

6. Which of these components of a red blood cell is correctly matched with its function?
   a. heme group of hemoglobin—oxygen transport
   b. globin portion of hemoglobin—carbon dioxide transport
   c. carbonic anhydrase—carbon dioxide transport
   d. all of the above

7. Each hemoglobin molecule can become associated with _____ oxygen molecules.
   a. one
   b. two
   c. three
   d. four
   e. an unlimited number of

8. Which of these substances is *not* required for normal red blood cell production?
   a. folate   c. iron
   b. vitamin K   d. vitamin B_{12}

9. Erythropoietin
   a. is produced mainly by the heart.
   b. inhibits the production of red blood cells.
   c. production increases when blood oxygen decreases.
   d. production is inhibited by testosterone.
   e. all of the above.

10. Which of these changes occurs in the blood in response to the initiation of a vigorous exercise program?
    a. increased erythropoietin production
    b. increased number of reticulocytes
    c. decreased bilirubin formation
    d. both a and b
    e. all of the above

11. Which of the components of hemoglobin is correctly matched with its fate following the destruction of a red blood cell?
    a. heme—reused to form a new hemoglobin molecule
    b. globin—broken down into amino acids
    c. iron—mostly secreted in bile
    d. all of the above

12. If you lived near sea level and were training for a track meet in Denver (5280 ft elevation), you would want to spend a few weeks before the meet training at
    a. sea level.
    b. an altitude similar to Denver's.
    c. a facility with a hyperbaric chamber.
    d. any location—it does not matter.

13. The blood cells that inhibit inflammation are
    a. eosinophils.
    b. basophils.
    c. neutrophils.
    d. monocytes.
    e. lymphocytes.

14. The most numerous type of white blood cell, whose primary function is phagocytosis, is
    a. eosinophils.
    b. basophils.
    c. neutrophils.
    d. monocytes.
    e. lymphocytes.

15. Monocytes
    a. are the smallest white blood cells.
    b. increase in number during chronic infections.
    c. give rise to neutrophils.
    d. produce antibodies.

16. The white blood cells that release large amounts of histamine and heparin are
    a. eosinophils.
    b. basophils.
    c. neutrophils.
    d. monocytes.
    e. lymphocytes.

17. The smallest white blood cells, which include B cells and T cells, are
    a. eosinophils.
    b. basophils.
    c. neutrophils.
    d. monocytes.
    e. lymphocytes.

18. Platelets
    a. are derived from megakaryocytes.
    b. are cell fragments.
    c. have surface molecules that attach to collagen.
    d. play an important role in clot formation.
    e. all of the above.

19. Given these processes in platelet plug formation:
    1. platelet adhesion
    2. platelet aggregation
    3. platelet release reaction

    Choose the arrangement that lists the processes in the correct order after a blood vessel is damaged.
    a. 1,2,3
    b. 1,3,2

    c. 3,1,2
    d. 3,2,1
    e. 2,3,1

20. A constituent of blood plasma that forms the network of fibers in a clot is
    a. fibrinogen.
    b. tissue factor.
    c. platelets.
    d. thrombin.
    e. prothrombinase.

21. Given these chemicals:
    1. activated factor XII
    2. fibrinogen
    3. prothrombinase
    4. thrombin

    Choose the arrangement that lists the chemicals in the order they are used during clot formation.
    a. 1,3,4,2
    b. 2,3,4,1
    c. 3,2,1,4
    d. 3,1,2,4
    e. 3,4,2,1

22. The extrinsic pathway
    a. begins with the release of thromboplastin (tissue factor).
    b. leads to the production of activated factor X.
    c. requires Ca^{2+}.
    d. all of the above.

23. The chemical involved in the breakdown of a clot (fibrinolysis) is
    a. antithrombin.
    b. fibrinogen.
    c. heparin.
    d. plasmin.
    e. sodium citrate.

24. A person with type A blood
    a. has anti-A antibodies.
    b. has type B antigens.
    c. will have a transfusion reaction if given type B blood.
    d. all of the above.

25. Rh-negative mothers who receive a RhoGAM injection are given that injection to
    a. initiate the synthesis of anti-Rh antibodies in the mother.
    b. initiate anti-Rh antibody production in the baby.
    c. prevent the mother from producing anti-Rh antibodies.
    d. prevent the baby from producing anti-Rh antibodies.

Answers in Appendix E

CRITICAL THINKING

1. Why is it advantageous for clot formation to involve molecules on the surface of activated platelets?

2. In hereditary hemolytic anemia, massive destruction of red blood cells occurs. Would you expect the reticulocyte count to be above or below normal? Explain why one of the symptoms of the disease is jaundice. In 1910, it was discovered that hereditary hemolytic

anemia can be successfully treated by removing the spleen. Explain why this treatment is effective.

3. Red Packer, a physical education major, wanted to improve his performance in an upcoming marathon race. About 6 weeks before the race, 500 mL of blood was removed from his body, and the formed elements were separated from the plasma. The formed

elements were frozen, and the plasma was reinfused into his body. Just before the competition, the formed elements were thawed and injected into his body. Explain why this procedure, called blood doping or blood boosting, would help Red's performance. Can you suggest any possible bad effects?

4. Chemicals such as benzene and chloramphenicol can destroy red bone marrow and cause aplastic anemia. What symptoms develop as a result of the lack of (a) red blood cells, (b) platelets, and (c) white blood cells?

5. Some people habitually use barbiturates to depress feelings of anxiety. Barbiturates cause hypoventilation, which is a slower than normal rate of breathing, because they suppress the respiratory centers in the brain. What happens to the red blood count of a habitual user of barbiturates? Explain.

6. What blood problems would you expect to observe in a patient after total gastrectomy (removal of the stomach)? Explain.

7. According to an old saying, "good food makes good blood." Name three substances in the diet that are essential for "good blood." What blood disorders develop if these substances are absent from the diet?

8. Why do many anemic patients have gray feces? (*Hint:* The feces is lacking its normal coloration.)

9. Reddie Popper has a cell membrane defect in her red blood cells that makes them more susceptible to rupturing. Her red blood cells are destroyed faster than they can be replaced. Are her RBC, hemoglobin, hematocrit, and bilirubin levels below normal, normal, or above normal? Explain.

Answers in Appendix F

For additional study tools, go to: www.aris.mhhe.com. Click on your course topic and then this textbook's author name/title. Register once for a semester's worth of interactive learning activities to help improve your grade!

CHAPTER 17 The Heart

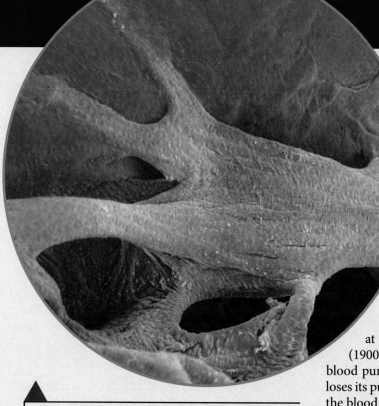

Introduction

People often refer to the heart as if it were the seat of certain strong emotions. A very determined person may be described as having "a lot of heart," and a person who has been disappointed romantically can be described as having a "broken heart." A popular holiday in February not only dramatically distorts the heart's anatomy but also attaches romantic emotions to it. The heart is a muscular organ that is essential for life because it pumps blood through the body. Emotions are a product of brain function, not heart function.

Like a pump that forces water to flow through a pipe, the heart contracts forcefully to pump blood through the blood vessels of the body. At rest, the heart of a healthy adult pumps blood at the rate of approximately 5 L/min (1.3 gal/min) or 7200 L/day (1900 gal/day). During short periods of vigorous exercise, the amount of blood pumped by the heart can increase to 20 L/min or more. If the heart loses its pumping ability for even a few minutes, however, blood flow through the blood vessels stops, and the life of the individual is in danger.

▶This chapter describes the **functions of the heart (p. 488)**, the **location, shape, and size of the heart (p. 488)**, the **anatomy of the heart (p. 489)**, the **histology of the heart (p. 497)**, and the **electrical activity of the heart (p. 498)**. The **cardiac cycle (p. 503)**, the **mean arterial blood pressure (p. 508)**, the **regulation of the heart (p. 509)**, and **the heart and homeostasis (p. 510)** are described. The chapter ends with the **effects of aging on the heart (p. 513)**.

This is a colorized scanning electron micrograph of Purkinje fibers of the heart, which are specialized cardiac muscle cells that conduct action potentials more rapidly than do other cardiac muscle fibers and have a reduced ability to contract. Purkinje fibers make up much of the conducting system of the heart.

Tools for Success

A virtual cadaver dissection experience

Audio (MP3) and/or video (MP4) content, available at www.mhhe.aris.com

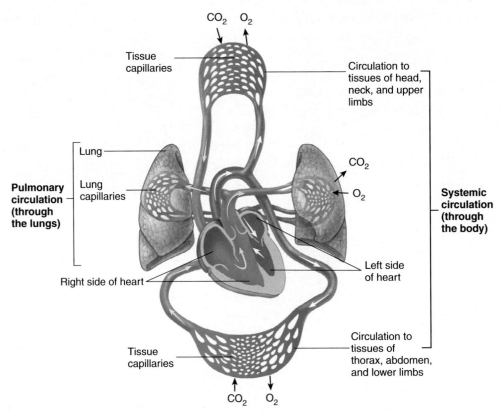

Figure 17.1 Systemic and Pulmonary Circulation
The circulatory system consists of the pulmonary and systemic circulations. The right side of the heart pumps blood through vessels to the lungs and back to the left side of the heart through the pulmonary circulation. The left side of the heart pumps blood through vessels to the tissues of the body and back to the right side of the heart through the systemic circulation.

Functions of the Heart

The following are the functions of the heart:

1. *Generating blood pressure.* Contractions of the heart generate blood pressure, which is required for blood flow through the blood vessels.
2. *Routing blood.* The heart is two pumps, moving blood through the pulmonary and systemic circulations. The **pulmonary** (pŭl′mō-nār-ē) **circulation** is the flow of blood from the heart through the lungs back to the heart. Blood in the pulmonary circulation picks up oxygen and releases carbon dioxide in the lungs. The **systemic circulation** is the flow of blood from the heart through the body back to the heart. Blood in the systemic circulation delivers oxygen and picks up carbon dioxide in the body's tissues (figure 17.1).
3. *Regulating blood supply.* Changes in the rate and force of heart contraction match blood flow to the changing metabolic needs of the tissues during rest, exercise, and changes in body position. ▼

1. **List the three functions of the heart.**
2. **Define the pulmonary and systemic circulations and describe the gas exchange taking place in each of them.**

Location, Shape, and Size of the Heart

It is important for clinical reasons to know the location and shape of the heart in the thoracic cavity. This knowledge allows a person to accurately place a stethoscope to hear the heart sounds, place chest leads to record an **electrocardiogram** (ē-lek-trō-kar′dē-ō-gram) (**ECG** or **EKG**), or administer effective **cardiopulmonary resuscitation** (kar′dē-ō-pŭl′mo-nār-ē rē-sŭs′i-tā-shun) (**CPR**).

The heart is located in the thoracic cavity between the lungs. The heart, trachea, esophagus, and associated structures form a midline partition, the **mediastinum** (me′dē-as-tī′nŭm) (see figure 1.15). The adult heart is shaped like a blunt cone and is approximately the size of a closed fist. It is larger in physically active adults than in less active but otherwise healthy adults. The blunt, rounded point of the cone is the **apex;** the larger, flat part at the opposite end of the cone is the **base** (figure 17.2).

The heart lies obliquely in the mediastinum, with its base directed posteriorly and slightly superiorly and the apex directed anteriorly and slightly inferiorly (see figure 17.2). The apex is also directed to the left so that approximately two-thirds of the heart's mass lies to the left of the midline of the sternum. The base of the heart is located deep to the sternum and extends to

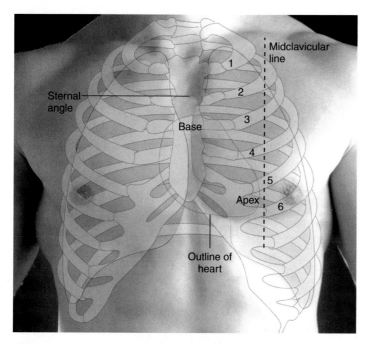

Figure 17.2   Location of the Heart in the Thorax

The heart is located in the thoracic cavity, deep and slightly to the left of the sternum. The sternal angle, the joint between the manubrium and body of the sternum, can be used as a landmark for counting the ribs, which are numbered in the figure. The base of the heart extends superiorly to the level of the second intercostal space (between ribs 2 and 3), and the apex of the heart is located deep to the fifth intercostal space (between ribs 5 and 6), approximately 7–9 cm to the left of the sternum where the midclavicular line intersects with the fifth intercostal space.

the level of the second intercostal space. The apex is located deep to the fifth intercostal space, approximately 7–9 centimeters (cm) to the left of the sternum near the midclavicular line, which is a perpendicular line that extends down from the middle of the clavicle. ▼

> 3. *Describe the location, shape, and approximate size of the heart.*

Anatomy of the Heart

Pericardium

The heart is surrounded by the **pericardial** (*peri,* around + *cardio,* heart) **cavity,** which is formed by the **pericardium** (per-i-kar′dē-ŭm), or **pericardial sac** (figure 17.3). The pericardium consists of two layers, an outer fibrous pericardium and an inner serous pericardium. The **fibrous pericardium** is tough, fibrous connective tissue that anchors the heart within the mediastinum. The **serous pericardium** is simple squamous epithelium overlying a layer of loose connective tissue and fat. The part of the serous pericardium lining the fibrous pericardium is the **parietal pericardium,** whereas the part covering the heart surface is the **visceral pericardium.** The **pericardial cavity,** located between the visceral and parietal pericardia, is filled with a thin layer of **pericardial fluid** produced by the serous pericardium. The pericardial fluid helps reduce friction as the heart moves within the pericardial sac.

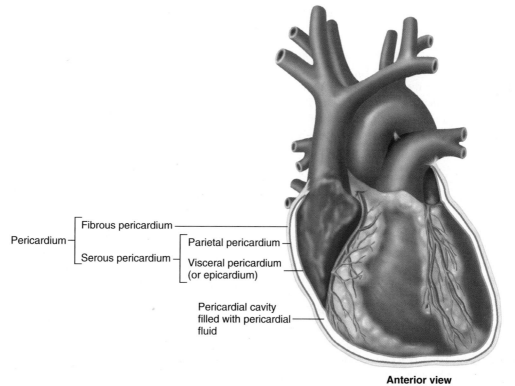

Anterior view

Figure 17.3   Heart in the Pericardium

The heart is located in the pericardium, which consists of an outer fibrous pericardium and an inner serous pericardium. The serous pericardium has two parts: The parietal pericardium lines the fibrous pericardium, and the visceral pericardium (epicardium) covers the surface of the heart. The pericardial cavity, between the parietal and visceral pericardia, is filled with a small amount of pericardial fluid.

Pericarditis and Cardiac Tamponade

Pericarditis (per'i-kar-dī'tis) is an inflammation of the serous pericardium. The cause is frequently unknown, but it can result from infection, diseases of connective tissue, or damage due to radiation treatment for cancer. It can be extremely painful, with sensations of pain referred to the back and to the chest, which can be confused with the pain of a heart attack. Pericarditis can result in a small amount of fluid accumulation within the pericardial sac.

Cardiac tamponade (tam-pŏ-nād', a pack or plug) is a potentially fatal condition in which fluid or blood accumulates in the pericardial sac. The fluid compresses the heart from the outside. The heart is a powerful muscle, but it relaxes passively. When it is compressed by fluid within the pericardial sac, it cannot expand when the cardiac muscle relaxes. Consequently, the heart cannot fill with blood during relaxation, which makes it impossible for it to pump blood. Cardiac tamponade can cause a person to die quickly unless the fluid is removed from the pericardial cavity. Causes of cardiac tamponade include rupture of the heart wall following a heart attack, rupture of blood vessels in the pericardium after a malignant tumor invades the area, damage to the pericardium resulting from radiation therapy, and trauma such as occurs in a traffic accident. ▼

4. *What is the pericardium? Describe its parts and their functions.*

Heart Wall

The heart wall is composed of three layers of tissue: the epicardium, the myocardium, and the endocardium (figure 17.4). The **epicardium** (ep-i-kar'dē-ŭm, upon the heart) is a thin serous membrane forming the smooth outer surface of the heart. It consists of simple squamous epithelium overlying a layer of loose connective tissue and fat. The epicardium and the visceral pericardium are two terms for the same structure. When considering the pericardium, it is called visceral pericardium; when considering the heart wall, it is called epicardium.

The thick middle layer of the heart, the **myocardium** (mī-ō-kar'dē-ŭm, heart muscle), is composed of cardiac muscle cells and is responsible for contractions of the heart chambers. The inner surface of the heart chambers is the **endocardium** (en-dō-kar'dē-ŭm, within the heart), which consists of simple squamous epithelium over a layer of connective tissue. The smooth endocardium allows blood to move easily through the heart.

The surfaces of the interior walls of the ventricles are modified by ridges and columns of cardiac muscle called **trabeculae** (trǎ-bek'ū-lē, beams) **carneae** (kar'nē-ē, flesh). Smaller, muscular ridges called **pectinate** (pek'ti-nāt, hair comb-shaped) **muscles** are also found in portions of the atria. ▼

5. *Describe the three layers of the heart, and state their functions.*
6. *Name the muscular ridges found on the interior walls of the ventricles and atria.*

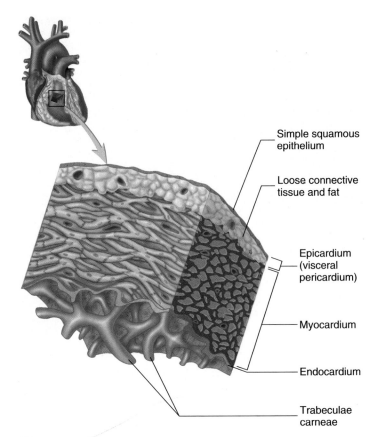

Figure 17.4   Heart Wall

Part of the wall of the heart has been removed, enlarged, and rotated so that its inner surface is visible. The enlarged section illustrates the epicardium (visceral pericardium), myocardium, and endocardium.

External Anatomy

The heart consists of four chambers: two **atria** (ā'trē-ǎ, entrance chamber) and two **ventricles** (ven'tri-klz, under side). The right and left atria are located at the base of the heart, and the right and left ventricles extend from the base of the heart toward the apex (figure 17.5a). Flaplike **auricles** (aw'ri-klz, ears) are extensions of the atria that can be seen anteriorly between each atrium and ventricle (figure 17.5b). A **sulcus** (sul'kus, ditch) is a groove on the surface of the heart containing blood vessels and fat. Several large sulci mark the boundaries of the heart chambers. A **coronary** (kōr'o-nār-ē, circling like a crown) **sulcus** extends around the heart, separating the atria from the ventricles. Two sulci indicate the division between the right and left ventricles. The **anterior interventricular sulcus** extends inferiorly from the coronary sulcus on the anterior surface of the heart, and the **posterior interventricular sulcus** extends inferiorly from the coronary sulcus on the posterior surface of the heart (figure 17.5c).

Six large veins carry blood to the heart (see figure 17.5a and c): The **superior vena cava** and **inferior vena cava** carry blood from the body to the right atrium, and four **pulmonary** (pǔl'mō-nār-ē, lung) **veins** carry blood from the lungs to the left atrium. Two arteries, the **pulmonary trunk** and the **aorta** (ā-ōr'tǎ), exit the heart. The pulmonary trunk, arising from the right ventricle, splits into the right and left **pulmonary arteries,**

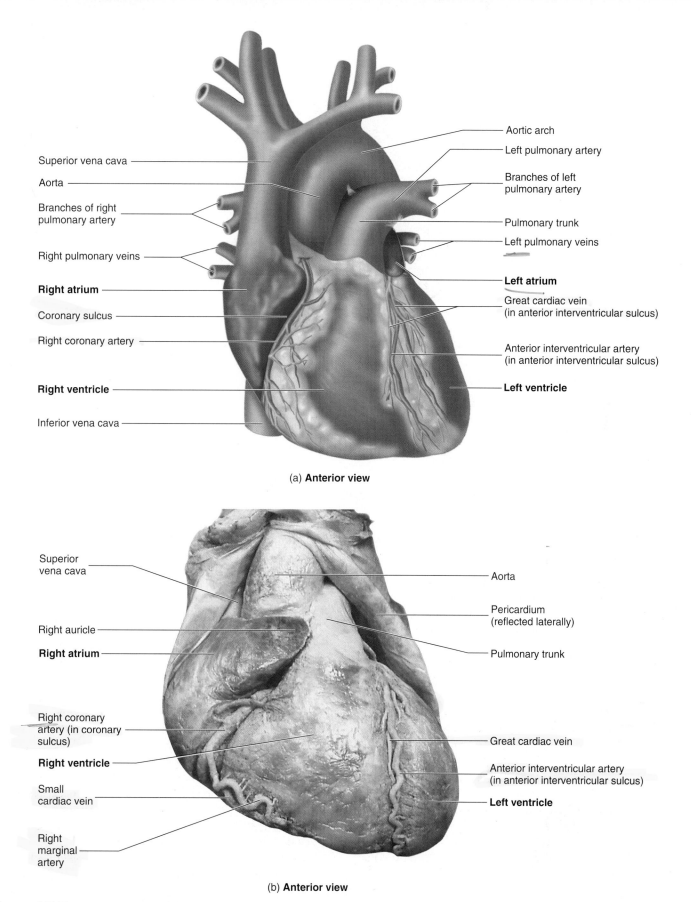

Superior vena cava

Aorta

Branches of right
pulmonary artery

Right pulmonary veins

Right atrium

Coronary sulcus

Right coronary artery

Right ventricle

Inferior vena cava

Aortic arch

Left pulmonary artery

Branches of left
pulmonary artery

Pulmonary trunk

Left pulmonary veins

Left atrium

Great cardiac vein
(in anterior interventricular sulcus)

Anterior interventricular artery
(in anterior interventricular sulcus)

Left ventricle

(a) **Anterior view**

Superior
vena cava

Right auricle

Right atrium

Right coronary
artery (in coronary
sulcus)

Right ventricle

Small
cardiac vein

Right
marginal
artery

Aorta

Pericardium
(reflected laterally)

Pulmonary trunk

Great cardiac vein

Anterior interventricular artery
(in anterior interventricular sulcus)

Left ventricle

(b) **Anterior view**

Figure 17.5 Surface View of the Heart

(a) The two atria (right and left) are located superiorly, and the two ventricles (right and left) are located inferiorly. The superior vena cava and inferior vena cava enter the right atrium. The pulmonary veins enter the left atrium. The pulmonary trunk exits the right ventricle, and the aorta exits the left ventricle. (b) Photograph of the anterior surface of the heart.

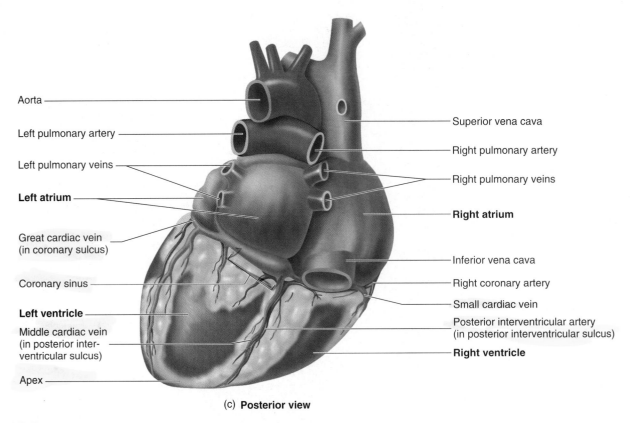

Aorta

Left pulmonary artery

Left pulmonary veins

Left atrium

Great cardiac vein
(in coronary sulcus)

Coronary sinus

Left ventricle

Middle cardiac vein
(in posterior inter-
ventricular sulcus)

Apex

Superior vena cava

Right pulmonary artery

Right pulmonary veins

Right atrium

Inferior vena cava

Right coronary artery

Small cardiac vein

Posterior interventricular artery
(in posterior interventricular sulcus)

Right ventricle

(c) **Posterior view**

Figure 17.5 (*continued*)

(*c*) The two atria (right and left) are located superiorly, and the two ventricles (right and left) are located inferiorly. The superior vena cava and inferior vena cava enter the right atrium, and the four pulmonary veins enter the left atrium. The pulmonary trunk divides, forming the left and right pulmonary arteries.

which carry blood to the lungs. The aorta, arising from the left ventricle, carries blood to the rest of the body. ▼

7. *Name the chambers of the heart and describe their location as seen on the outside of the heart.*
8. *Name the major blood vessels that enter and leave the heart. Which chambers of the heart do they enter or exit?*

Heart Chambers

With the anterior wall of the heart removed, the four chambers of the heart can be seen in figure 17.6. The atria receive blood from veins. The right atrium has three major openings through which blood enters: The openings from the superior vena cava and the inferior vena cava receive blood from the body, and the opening of the coronary sinus receives blood from the heart itself. The left atrium has four openings that receive blood from the four pulmonary veins from the lungs. The two atria are separated from each other by the **interatrial septum.** A slight, oval depression, the **fossa ovalis** (fos′ă ō-va′lis), on the right side of the septum marks the former location of the **foramen ovale** (ō-va′lē), an opening between the right and left atria in the embryo and the fetus. This opening allows blood to flow from the right to the left atrium in the fetus to bypass the pulmonary circulation (see chapter 25).

The atria open into the ventricles, and each ventricle has one large, superiorly placed outflow route near the midline of the heart. The **right ventricle** opens into the pulmonary trunk, and the **left ventricle** opens into the aorta (see figure 17.6). The two

ventricles are separated from each other by the **interventricular septum.** ▼

9. *Describe the openings through which blood enters the atria. What structure separates the atria from each other?*
10. *What is the fossa ovalis and the foramen ovale?*
11. *Describe the openings through which blood leaves the ventricles. What structure separates the ventricles from each other?*

Heart Valves

The heart valves are formed by folds of endocardium that consist of a double layer of endocardium with connective tissue in between. The valves allow blood to flow into and out of the ventricles but prevent the backflow of blood. This ensures one-way flow of blood through the heart.

An **atrioventricular (AV) valve** is located between each atrium and its ventricle. The AV valve between the right atrium and the right ventricle has three cusps and is called the **tricuspid valve** (see figures 17.6 and 17.7*a*). The AV valve between the left atrium and left ventricle has two cusps and is called the **bicuspid,** or **mitral** (resembling a bishop's miter, a two-pointed hat) **valve** (see figures 17.6 and 17.7*b*).

Each ventricle contains cone-shaped, muscular pillars called **papillary** (pap′ĭ-lār-ē, nipple- or pimple-shaped) **muscles.** These muscles are attached by thin, strong connective tissue strings called **chordae tendineae** (kōr′dē ten′di-nē-ē, heart strings) to

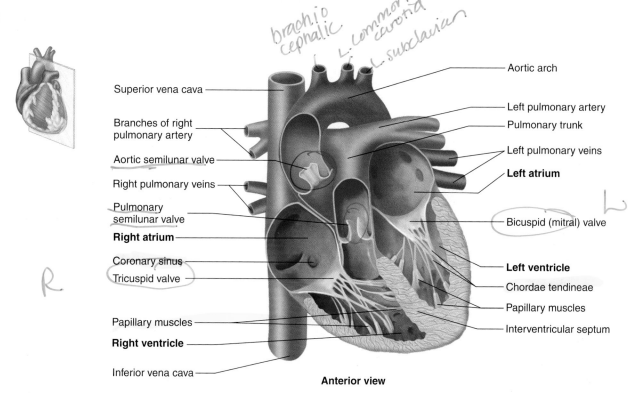

brachio cephalic *L. common carotid* *L. subclavian*

Superior vena cava

Branches of right pulmonary artery

Aortic semilunar valve

Right pulmonary veins

Pulmonary semilunar valve

Right atrium

Coronary sinus

Tricuspid valve

R

Papillary muscles

Right ventricle

Inferior vena cava

Aortic arch

Left pulmonary artery

Pulmonary trunk

Left pulmonary veins

Left atrium

Bicuspid (mitral) valve

Left ventricle

Chordae tendineae

Papillary muscles

Interventricular septum

Anterior view

Figure 17.6 **Internal Anatomy of the Heart**
The heart is cut in a frontal plane to show the internal anatomy.

the free margins of the cusps of the atrioventricular valves (see figures 17.6 and 17.7*a*).

A **semilunar** (half-moon–shaped) valve is located at the base of the large blood vessels carrying blood away from the ventricles. The **aortic semilunar valve** is in the aorta and the **pulmonary semilunar valve** is in the pulmonary trunk. Each valve consists of three pocketlike semilunar cusps (see figures 17.6 and 17.7*b*).

The left ventricle will be used to illustrate the operation of the heart valves. The right ventricular valves work in the same way. When the left ventricle relaxes, blood pushes the cusps of the bicuspid valve toward the ventricle, the valve opens, and blood flows into the ventricle. The chordae tendineae are slack when the

valve is open (figure 17.8, step 1). When the ventricle is relaxed, blood in the aorta flows back toward the ventricle and enters the pockets of the semilunar cusps, causing them to fill and expand. The semilunar cusps meet at the midline of the aorta and the semilunar valve closes, preventing the backflow of blood into the ventricle (figure 17.8, step 2).

When the left ventricle contracts, blood pushes the cusps of the bicuspid valve back toward the atrium. When the cusps meet, the valve is closed, preventing the backflow of blood into the atrium. The papillary muscles contract and the chordae tendineae are drawn taut, preventing the cusps of the valve from moving backwards into the atria (figure 17.8, step 3). When the ventricle contracts, blood pushes on the cusps of the semilunar valves, pushing

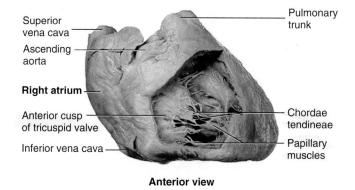

Superior vena cava

Ascending aorta

Right atrium

Anterior cusp of tricuspid valve

Inferior vena cava

Pulmonary trunk

Chordae tendineae

Papillary muscles

Anterior view

(a) View of the tricuspid valve, the chordae tendineae, and the papillary muscles.

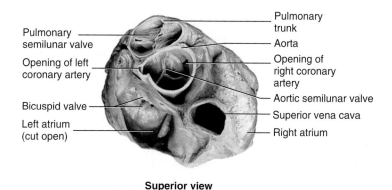

Pulmonary semilunar valve

Opening of left coronary artery

Bicuspid valve

Left atrium (cut open)

Pulmonary trunk

Aorta

Opening of right coronary artery

Aortic semilunar valve

Superior vena cava

Right atrium

Superior view

(b) A superior view of the heart valves. Note the three cusps of each semilunar valve meeting to prevent the backflow of blood.

Figure 17.7 **Heart Valves**

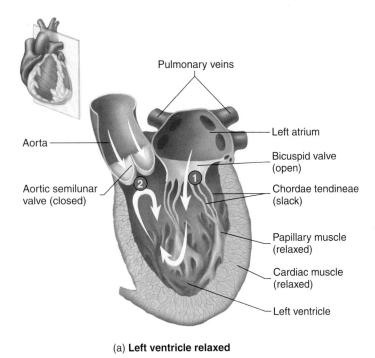

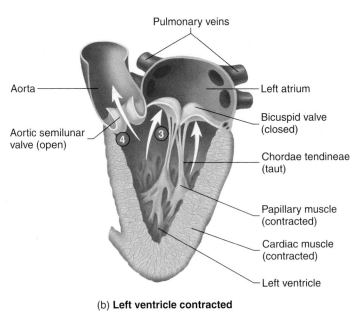

(a) **Left ventricle relaxed**

(b) **Left ventricle contracted**

1. The bicuspid valve is open. The cusps of the valve are pushed by the blood into the ventricle, and blood flows into the ventricle. The chordae tendineae are relaxed.

2. The aortic semilunar valve is closed. The cusps of the valve overlap as they are pushed by the blood in the aorta toward the ventricle. Backflow of blood into the ventricle is prevented.

3. The bicuspid valve is closed. The cusps of the valves overlap as they are pushed by the blood toward the left atrium. The chordae tendineae are tensed. Backflow of blood into the atrium is prevented.

4. The aortic semilunar valve is open. The cusps of the valve are pushed by the blood toward the aorta, and blood flows into the aorta.

Process Figure 17.8   Function of the Heart Valves

blood out of them. As the cusps empty, they flatten against the wall of the aorta and the semilunar valve opens (figure 17.8, step 4). ▼

> 12. *Name the heart valves, give their location, and describe their structure.*
> 13. *Describe how the heart valves operate, allowing blood to flow into and out of a ventricle, but preventing the backflow of blood.*

Route of Blood Flow Through the Heart

The route of blood flow through the heart is depicted in figure 17.9. Even though blood flow through the heart is described for the right and then the left side of the heart, it is important to understand that both atria and both ventricles contract at the same time. Thus, the events described for the right side of the heart take place at the same time as the events on the left side of the heart.

Blood enters the right atrium from the systemic circulation through the superior and inferior venae cavae and from heart muscle through the coronary sinus. Blood flows from the right atrium through the tricuspid valve into the right ventricle. Contraction of the right ventricle pushes blood through the pulmonary semilunar valves into the pulmonary trunk, which divides to form the left and right **pulmonary arteries. The pulmonary arteries** carry blood to the lungs and four pulmonary veins return blood to the left atrium. Blood flows from the left atrium through the bicuspid valve into the left ventricle. Contraction of the left ventricle pushes blood through the aortic semilunar valves into the

aorta. Blood flowing through the aorta is distributed to the body and to the coronary arteries supplying the heart. ▼

> 14. *Starting at the venae cavae and ending at the aorta, describe the flow of blood through the heart.*

Congenital Conditions Affecting the Heart

Congenital (occurring at birth) **heart disease** is the result of abnormal development of the heart. A **septal defect** is a hole in a septum between the left and right sides of the heart. The hole may be in the interatrial or interventricular septum. These defects allow blood to flow from one side of the heart to the other and, as a consequence, greatly reduce the pumping effectiveness of the heart.

Patent (to lie open) **ductus arteriosus** (dŭk'tŭs artēr'ē-ō-sŭs) results when a blood vessel called the **ductus arteriosus,** which is present in the fetus, fails to close after birth. The ductus arteriosus extends between the pulmonary trunk and the aorta. It allows blood to pass from the pulmonary trunk to the aorta, thus bypassing the lungs. This is normal before birth because the lungs are not functioning. If the ductus arteriosus fails to close after birth, however, blood flows in the opposite direction, from the aorta to the pulmonary trunk. As a consequence, blood flows through the lungs under a higher pressure and damages them. In addition, the amount of work required of the left ventricle to maintain an adequate systemic blood pressure increases.

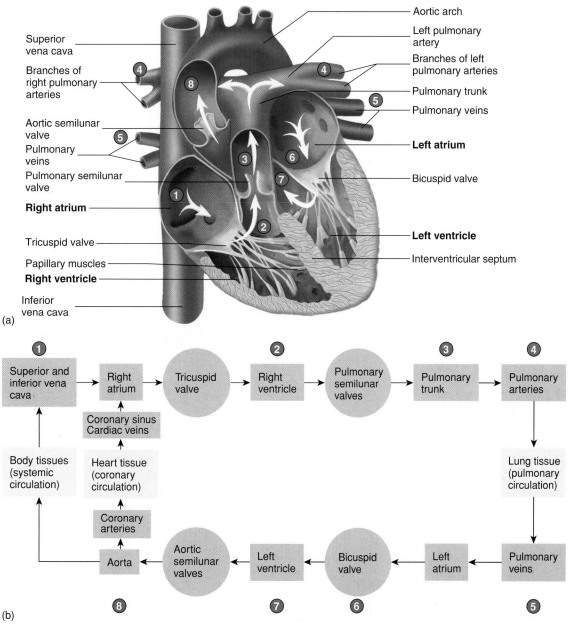

Figure 17.9 Blood Flow Through the Heart

(*a*) Frontal section of the heart, revealing the four chambers and the direction of blood flow through the heart. The numbers identify the direction of blood flow.
(*b*) Diagram listing in order the structures through which blood flows in the systemic, pulmonary, and coronary circulations. The heart valves are indicated by *circles*: deoxygenated blood (*blue*), oxygenated blood (*red*).

Blood Supply to the Heart

Cardiac muscle in the wall of the heart is thick and metabolically very active. The **coronary circulation** supplies cardiac muscle with blood (see figure 17.9*b*). Two coronary arteries originate from the base of the aorta, just above the aortic semilunar valves (figure 17.10*a*). The **left coronary artery** originates on the left side of the aorta. It has two major branches. The **anterior interventricular artery** lies in the anterior interventricular sulcus, and the **circumflex artery** extends around the coronary sulcus on the left to the posterior surface of the heart. The branches of the left coronary artery supply the left atrium, most of the left ventricle, and part of the right anterior ventricle. The **right coronary artery** originates

on the right side of the aorta. It extends around the coronary sulcus on the right to the posterior surface of the heart and gives rise to the **posterior interventricular artery,** which lies in the posterior interventricular sulcus. The right coronary artery and its branches supply the right atrium, most of the right ventricle, and part of the left posterior ventricle.

P R E D I C T

Predict the effect on the heart of a blood clot that completely blocks the flow of blood through the anterior interventricular artery if the blood clot forms in (1) the anterior interventricular artery just after it branches from the coronary artery and (2) the anterior interventricular artery at the apex of the heart.

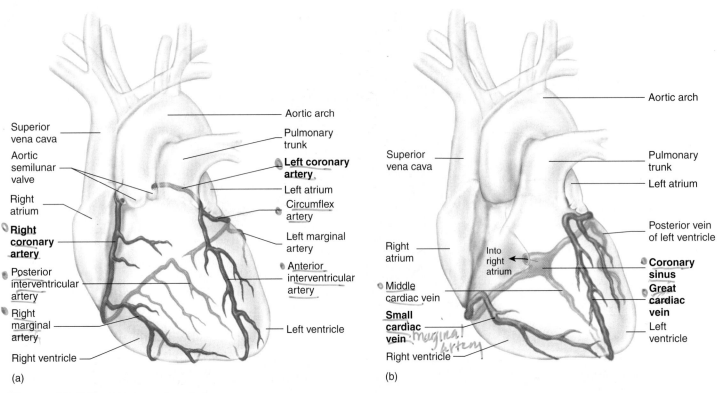

Figure 17.10   Coronary Circulation

(*a*) Arteries supplying blood to the heart. The arteries of the anterior surface are seen directly and are darker in color; the arteries of the posterior surface are seen through the heart and are lighter in color. (*b*) Veins draining blood from the heart. The veins of the anterior surface are seen directly and are darker in color; the veins of the posterior surface are seen through the heart and are lighter in color.

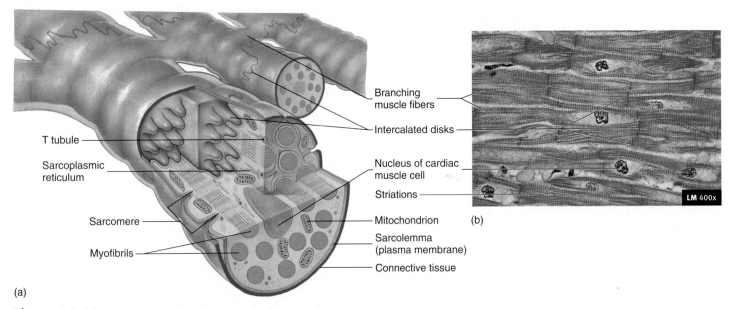

Figure 17.11   Histology of the Heart

(*a*) Cardiac muscle cells are branching cells with centrally located nuclei. As in skeletal muscle, sarcomeres join end-to-end to form myofibrils, and mitochondria provide ATP for contraction. The cells are joined to one another by intercalated disks. Gap junctions in the intercalated disks allow action potentials to pass from one cardiac muscle cell to the next. Sarcoplasmic reticulum and T tubules are visible but are not as numerous as they are in skeletal muscle. (*b*) A light micrograph of cardiac muscle tissue. The cardiac muscle fibers appear to be striated because of the arrangement of the individual myofilaments.

Clinical Relevance   Disorders of Coronary Arteries and Their Treatments

When a blood clot, or **thrombus** (throm′bŭs, a clot), suddenly blocks a coronary blood vessel, a **heart attack,** or **coronary thrombosis** (throm′bō-sis), occurs. The area that has been cut off from its blood supply suffers from a lack of oxygen and nutrients and dies if the blood supply is not quickly reestablished. The region of dead heart tissue is called an **infarct** (in′farkt), or **myocardial infarction.** People who are at risk for coronary thromboses can reduce the likelihood of heart attack by taking small amounts of aspirin or other drugs daily, which inhibit thrombus formation (see chapter 16).

Aspirin is also administered to many people who are exhibiting clear symptoms of a heart attack. In some cases, it is possible to treat heart attacks with enzymes, such as **streptokinase** (strep-tō-kī′nās) or **tissue plasminogen** (plaz-min′o-jen) **activator (t-PA),** which break down blood clots. One of the enzymes is injected into the circulatory system of a heart attack patient, where it reduces or removes the blockage in the coronary artery. If the clot is broken down quickly, the blood supply to cardiac muscle is reestablished, and the heart may suffer little permanent damage.

Coronary arteries can become blocked more gradually by **atherosclerotic** (ath′er-ō-skler-ot′ik, athero, pasty material + sclerosis, hardness) **lesions.** These thickenings in the walls of arteries can contain deposits that are high in cholesterol and other lipids. The lesions narrow the lumen (opening) of the arteries, thus restricting blood flow (figure A). The ability of cardiac muscle to

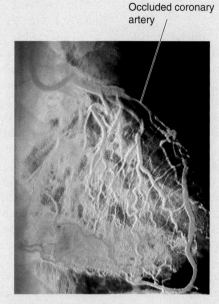

Figure A   Angiogram

An angiogram is a picture of a blood vessel. It is usually obtained by placing a catheter into a blood vessel and injecting a dye that can be detected with x-rays. Note the partially occluded (blocked) coronary blood vessel in this angiogram, which has been computer-enhanced to show colors.

function is reduced when it is deprived of an adequate blood supply. The person suffers from fatigue and often pain in the area of the chest and usually in the left arm with the slightest exertion. The pain is called **angina pectoris** (an-jī′nă, pain, pek′tō-ris, in the chest). Rest and drugs, such as nitroglycerin, frequently relieve angina pectoris. Nitroglycerin dilates the blood vessels, including the coronary arteries. Consequently, the drug increases the oxy-

gen supply to cardiac muscle. Nitroglycerin also reduces the heart's workload because the heart has to pump blood against a smaller pressure when peripheral arteries are dilated.

Angioplasty (an′jē-ō-plas-tē) is a surgical procedure in which a small balloon is threaded through the aorta and into a coronary artery. After the balloon has entered a partially blocked coronary artery, it is inflated, flattening the atherosclerotic deposits against the vessel wall and opening the blocked blood vessel. This technique improves the function of cardiac muscle in patients suffering from inadequate blood flow to the cardiac muscle through the coronary arteries. Some controversy exists about its effectiveness, at least in some patients, because dilation of the coronary arteries can be reversed within a few weeks or months and because blood clots can form in coronary arteries following angioplasty. Small, rotating blades and lasers are also used to remove lesions from coronary vessels, or a small coil device, called a **stent,** is placed in the vessels to hold them open following angioplasty.

A **coronary bypass** is a surgical procedure that relieves the effects of obstructions in the coronary arteries. The technique involves taking healthy segments of blood vessels from other parts of the patient's body and using them to bypass, or create an alternative path around, obstructions in the coronary arteries. The technique is common for those who suffer from severe blockage of parts of the coronary arteries.

The **cardiac veins** drain blood from the cardiac muscle. Their pathways are nearly parallel to the coronary arteries and most drain blood into the **coronary sinus,** a large vein located within the coronary sulcus on the posterior aspect of the heart. Blood flows from the coronary sinus into the right atrium (figure 17.10*b*). Some small cardiac veins drain directly into the right atrium. ▼

> 15. *Starting at the left ventricle and ending at the right atrium, describe the flow of blood through the coronary circulation.*

Histology of the Heart
Fibrous Skeleton of the Heart

The **fibrous skeleton of the heart** consists of a plate of fibrous connective tissue between the atria and ventricles. This connective

tissue plate forms **fibrous rings** around the atrioventricular and semilunar valves and provides a solid support for them. The fibrous connective tissue plate also provides a rigid site for attachment of the cardiac muscles and serves as electrical insulation between the atria and the ventricles. ▼

> 16. *Describe and list the functions of the heart skeleton.*

Cardiac Muscle

Cardiac muscle cells are elongated, branching cells containing one, or occasionally two, centrally located nuclei (figure 17.11). The cardiac muscle cells contain actin and myosin myofilaments organized to form **sarcomeres,** which are joined end-to-end to form **myofibrils** (see chapter 8). The actin and myosin myofilaments are responsible for muscle contraction, and their organization gives cardiac muscle a striated (banded) appearance, much like that of skeletal muscle. However, the striations are less regularly arranged and less numerous than is the case in skeletal muscle.

Cardiac muscle cells have **transverse (T) tubules,** which extend into the interior of the cells. Associated with the T tubules is **sarcoplasmic reticulum** but, compared with skeletal muscle, it is not as regularly arranged and there are no dilated terminal cisternae. As in skeletal muscle, action potentials propagate down T tubules and stimulate the release of Ca^{2+} from the sarcoplasmic reticulum. The Ca^{2+} initiate contraction by binding to troponin (see chapter 8). In addition to the Ca^{2+} from the sarcoplasmic reticulum, cardiac muscle cells also require an influx of extracellular Ca^{2+} for normal contractions.

Adenosine triphosphate (ATP) provides the energy for cardiac muscle contraction, and cardiac muscle cells have more mitochondria than skeletal muscle cells. Cardiac muscle cells rely on aerobic respiration to produce the ATP necessary for contractions. Accordingly, they have an extensive blood supply for the delivery of oxygen and nutrients. Cardiac muscle cells have a limited ability to produce ATP through anaerobic respiration. If blood delivery is inadequate to supply the oxygen needs of cardiac muscle cells, they quickly fatigue.

P R E D I C T   **2**

Under resting conditions, most of the ATP produced in cardiac muscle is derived from the metabolism of fatty acids. During heavy exercise, however, cardiac muscle cells use lactic acid as an energy source. Explain why this arrangement is an advantage.

Cardiac muscle cells are bound end-to-end and laterally to adjacent cells by specialized cell-to-cell contacts called **intercalated** (in-ter′kă-lā-ted, insertion between two others) **disks** (see figure 17.11). The membranes of the intercalated disks are highly folded, and the adjacent cells fit together, greatly increasing contact between them and preventing cells from pulling apart. Gap junctions within the intercalated disks form **electrical synapses** (see chapter 10), which allow action potentials to pass easily from one cell to another. Action potentials stimulate cardiac muscle cells to contract. Cardiac muscle cells connected by intercalated disks contract as a single unit because action potentials are rapidly propagated between the cells. The highly coordinated contractions of the heart depend on this functional characteristic. ▼

17. *Compare cardiac muscle to skeletal muscle for sarcomeres, myofibrils, sarcoplasmic reticulum, T tubules, mitochondria, and ATP production.*
18. *Explain how cardiac muscle cells contract as a single unit.*

Electrical Activity of the Heart
Action Potentials

Like action potentials in skeletal muscle and neurons, those in cardiac muscle exhibit depolarization followed by repolarization. In cardiac muscle, however, a period of slow repolarization greatly prolongs the action potential (figure 17.12). In contrast to action potentials in skeletal muscle, which take less than 2 milliseconds (ms) to complete, action potentials in cardiac muscle take approximately 200 to 500 ms to complete.

In cardiac muscle, each action potential consists of a **depolarization phase** followed by a rapid but partial **early repolarization phase.** This is followed by a longer period of slow repolarization, called the **plateau phase.** At the end of the plateau phase, a more rapid **final repolarization phase** takes place. During the final repolarization phase, the membrane potential achieves its maximum degree of repolarization (see figure 17.12).

Opening and closing of voltage-gated ion channels are responsible for the changes in the permeability of the plasma membrane that produce action potentials. The depolarization phase of the action potential results from three permeability changes. **Sodium ion channels** open, increasing the permeability of the plasma membrane to Na^+. Sodium ions then diffuse into the cell, causing depolarization. This depolarization causes K^+ channels to close quickly, decreasing the permeability of the plasma membrane to K^+. The decreased diffusion of K^+ out of the cell causes further depolarization. **Calcium ion channels** slowly open, increasing the permeability of the plasma membrane to Ca^{2+}. Calcium ions then diffuse into the cell and cause depolarization. It is not until the plateau phase that most of the Ca^{2+} channels open.

Early repolarization occurs when the Na^+ channels close and a small number of K^+ **channels** open. Diffusion of Na^+ into the cell stops, and there is some movement of K^+ out of the cell, which produces a small repolarization.

The plateau phase occurs as Ca^{2+} channels continue to open, and the diffusion of Ca^{2+} into the cell counteracts the potential change produced by the diffusion of K^+ out of the cell. The plateau phase ends and final repolarization begins as the Ca^{2+} channels close, and many K^+ channels open. Diffusion of Ca^{2+} into the cell decreases and diffusion of K^+ out of the cell increases. These changes cause the membrane potential to repolarize during the final repolarization phase. ▼

19. *For cardiac muscle action potentials, describe ion movement during the depolarization, early repolarization, plateau, and final repolarization phases.*

Refractory Periods

Cardiac muscle, like skeletal muscle, has absolute and relative refractory periods associated with its action potentials. The **absolute refractory** (rē-frak′tōr-ē) **period** begins with depolarization and extends into the final repolarization phase. During the absolute refractory period, the cardiac muscle cell is completely insensitive to further stimulation. The relative refractory period occurs during the last part of the final repolarization phase. During the **relative refractory period,** the cell is sensitive to stimulation, but a greater stimulation than normal is required to cause an action potential. The absolute refractory period is much longer in cardiac muscle than in skeletal muscle because of the long duration of the plateau phase in cardiac muscle. The long absolute refractory period ensures that contraction and most of relaxation are complete before another action potential can be initiated. This

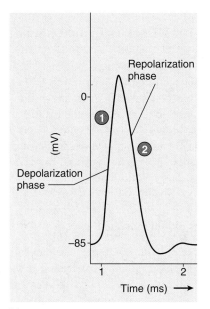

(a)

Permeability changes during an action potential in skeletal muscle:
1. **Depolarization phase**
   - Voltage-gated Na$^+$ channels open.
   - Voltage-gated K$^+$ channels begin to open.

2. **Repolarization phase**
   - Voltage-gated Na$^+$ channels close.
   - Voltage-gated K$^+$ channels continue to open.
   - Voltage-gated K$^+$ channels close at the end of repolarization and return the membrane potential to its resting value.

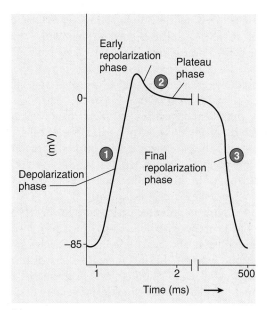

(b)

Permeability changes during an action potential in cardiac muscle:
1. **Depolarization phase**
   - Voltage-gated Na$^+$ channels open.
   - Voltage-gated K$^+$ channels close.
   - Voltage-gated Ca^{2+} channels begin to open.

2. **Early repolarization and plateau phases**
   - Voltage-gated Na$^+$ channels close.
   - Some voltage-gated K$^+$ channels open, causing early repolarization.
   - Voltage-gated Ca^{2+} channels are open, producing the plateau by slowing further repolarization.

3. **Final repolarization phase**
   - Voltage-gated Ca^{2+} channels close.
   - Many voltage-gated K$^+$ channels open.

Process Figure 17.12   Comparison of Action Potentials in Skeletal and Cardiac Muscle

(*a*) An action potential in skeletal muscle consists of depolarization and repolarization phases. (*b*) An action potential in cardiac muscle consists of depolarization, early repolarization, plateau, and final repolarization phases. Cardiac muscle does not repolarize as rapidly as skeletal muscle (*indicated by the break in the curve*) because of the plateau phase.

prevents tetanus in cardiac muscle (see figure 8.15) and is responsible for rhythmic contractions. ▼

> 20. **What causes a long absolute refractory period in cardiac muscle? What does the length of the refractory period prevent?**

P R E D I C T

Why is it important to prevent tetanic contractions in cardiac muscle but not in skeletal muscle?

Autorhythmicity of Cardiac Muscle

The heart is said to be **autorhythmic** (aw'tō-rith'mik) because it stimulates itself (*auto*) to contract at regular intervals (*rhythmic*). If the heart is removed from the body and maintained under physiological conditions with the proper nutrients and temperature, it will continue to beat autorhythmically for a long time.

Some cardiac muscle cells have the ability to generate action potentials spontaneously. When these action potentials propagate to other cardiac muscle cells, the action potentials stimulate them to contract. The **sinoatrial (SA) node** is a collection of cardiac muscle cells capable of spontaneously generating action potentials (see "Conducting System of the Heart," next section). The SA node is called the pacemaker of the heart because it can spontaneously generate action potentials faster than any other part of the heart. Action potentials generated by the SA node cause the heart to contract and thus determine **heart rate,** the number of beats (contractions) per minute (bpm).

As soon as an action potential ends, the production of the next action potential begins with a slowly developing local potential, called the **prepotential.** Sodium ions enter the cell through Na$^+$ leak channels, causing a small depolarizing voltage change across the plasma membrane. As a result of the depolarization, voltage-gated Ca^{2+} channels open, and the movement of Ca^{2+} into the pacemaker cells causes further depolarization. This depolarization causes additional voltage-gated Ca^{2+} channels to open. When the prepotential reaches threshold, many voltage-gated Ca^{2+} channels open. Unlike other cardiac muscle cells, the movement of Ca^{2+} into the pacemaker cells is primarily responsible for the depolarization phase of the action potential. Repolarization occurs, as in other cardiac muscle cells, when the Ca^{2+} channels close and the K$^+$ channels open.

The duration of the prepotential determines the heart rate. As the duration of the prepotential decreases, the time between action potentials decreases and heart rate increases. For example, epinephrine and norepinephrine increase the heart rate by opening voltage-gated Ca^{2+} channels, which decreases the duration of the prepotential by increasing the movement of Ca^{2+} into cells. ▼

21. *Define autorhythmic and heart rate.*
22. *Why is the SA node called the pacemaker of the heart?*
23. *Describe the production of the prepotential. How does the duration of the prepotential in the SA node determine heart rate?*

Drugs That Block Calcium Channels

Calcium channel blockers are drugs that block the movement of Ca^{2+} through voltage-gated Ca^{2+} channels. Consequently, the duration of the prepotential increases and heart rate slows. Calcium channel blockers also reduce the amount of work performed by the heart because fewer Ca^{2+} enter cardiac muscle cells to activate the contractile mechanism. Various cardiac disorders, including tachycardia (rapid heart rate) and certain arrhythmias, are treated with calcium channel blockers.

Conducting System of the Heart

Effective pumping of blood through the heart depends on coordinated contraction of the atria and ventricles. The atria contract, moving blood into the ventricles. Then the ventricles contract, moving blood to the lungs and body. The **conducting system of the heart,** which consists of specialized cardiac muscle cells, stimulates the atria and ventricles to contract by relaying action potentials through the heart (figure 17.13).

The sinoatrial (SA) node is located medial to the opening of the superior vena cava. Action potentials originating in the SA node spread over the right and left atria, causing them to contract.

1. Action potentials originate in the sinoatrial (SA) node (the pacemaker) and travel across the wall of the atrium (*arrows*) from the SA node to the atrioventricular (AV) node.

2. Action potentials pass through the AV node and along the atrioventricular (AV) bundle, which extends from the AV node, through the fibrous skeleton, into the interventricular septum.

3. The AV bundle divides into right and left bundle branches, and action potentials descend to the apex of each ventricle along the bundle branches.

4. Action potentials are carried by the Purkinje fibers from the bundle branches to the ventricular walls.

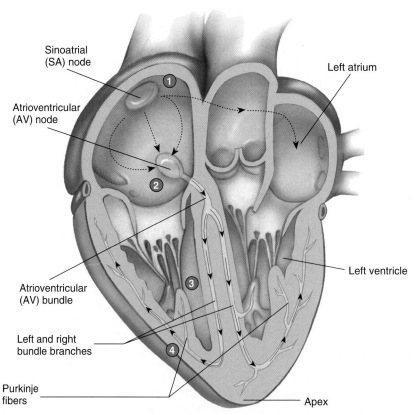

Process Figure 17.13   Conducting System of the Heart

Clinical Relevance Cardiac Arrhythmias

Dysfunction of the conducting system of the heart produces **cardiac arrhythmia,** which is an abnormal or loss of normal rhythm (table 17.1). For example, **tachycardia** is a resting heart rate greater than 100 bpm, **bradycardia** is a resting heart rate lower than 60 bpm, and an **arrhythmia** is an irregular heart rate that speeds up and slows down.

The normal relationship between atrial contraction and ventricular contraction can be disrupted by heart blocks. The SA node is the pacemaker of the heart, but other cells of the conduction system also are capable of producing action potentials spontaneously. An **ectopic focus** (eek-top′ik fō′kŭs, pl. *foci,* fō′sī) is any part of the heart other than the SA node that generates a heartbeat. For example, in an **SA node block,** the SA node no longer functions as the pacemaker of the heart. Instead, another part of

Table 17.1	Major Cardiac Arrhythmias	
Condition	**Symptoms**	**Possible Causes**
Abnormal Heart Rhythms		
Tachycardia	Heart rate in excess of 100 bpm	Elevated body temperature, excessive sympathetic stimulation, toxic conditions
Bradycardia	Heart rate less than 60 bpm	Increased stroke volume in athletes, excessive vagus nerve stimulation, nonfunctional SA node, carotid sinus syndrome
Sinus arrhythmia	Heart rate varies as much as 5% during respiratory cycle and up to 30% during deep respiration	Cause not always known; occasionally caused by ischemia, inflammation, or cardiac failure
Paroxysmal atrial tachycardia	Sudden increase in heart rate to 95–150 bpm for a few seconds or even for several hours; P waves precede every QRS complex; P wave inverted and superimposed on T wave	Excessive sympathetic stimulation, abnormally elevated permeability of cardiac muscle to Ca^{2+}
Atrial flutter	As many as 300 P waves/min and 125 QRS complexes/min; resulting in two or three P waves (atrial contractions) for every QRS complex (ventricular contraction)	Ectopic beats in the atria
Atrial fibrillation	No P waves, normal QRS and T waves, irregular timing, ventricles are constantly stimulated by atria, reduced ventricle filling; increased chance of fibrillation	Ectopic beats in the atria
Ventricular tachycardia	Frequently causes fibrillation	Often associated with damage to AV node or ventricular muscle
Heart Blocks		
SA node block	No P waves, low heart rate resulting from AV node acting as the pacemaker, normal QRS complexes and T waves	Ischemia, tissue damage resulting from infarction; cause sometimes is unknown
AV node blocks		
First-degree	PQ interval greater than 0.2 s	Inflammation of AV bundle
Second-degree	PQ interval 0.25–0.45 s; some P waves trigger QRS complexes and others do not; examples of 2:1, 3:1, and 3:2 P wave/QRS complex ratios	Excessive vagus nerve stimulation, AV node damage
Complete heart block	P wave dissociated from QRS complex, atrial rhythm about 100 bpm, ventricular rhythm less than 40 bpm	Ischemia of AV node or compression of AV bundle
Premature Contractions		
Premature atrial contractions	Occasional shortened intervals between one contraction and the succeeding contraction; frequently occurs in healthy people	Excessive smoking, lack of sleep, or too much caffeine
Premature ventricular contractions (PVCs)	Prolonged QRS complex, exaggerated voltage because only one ventricle may depolarize, possible inverted T wave, increased probability of fibrillation	Ectopic beat in ventricles, lack of sleep, too much coffee, irritability; occasionally occurs with coronary thrombosis

(continued)

the heart, such as the AV node, becomes the pacemaker. The resulting heart rate is much slower than normal. In an **AV block,** the transmission of action potentials through the AV node is slowed or blocked completely. Depending on the severity of the block, the time between atrial and ventricular contractions increases, the atria contract two or three times for each ventricular contraction, or the atria and ventricles contract independently of each other.

Cardiac muscle can also act as if there were thousands of pacemakers, each making a very small portion of the heart contract rapidly and independently of all other areas. This condition is called **fibrillation** (fĭ-bri-lā′shŭn), and it reduces the output of the heart to only a few milliliters of blood per minute when it occurs in the ventricles. Death of the individual results in a few minutes unless fibrillation of the ventricles is stopped. To stop the process of fibrillation, defibrillation is used, in which a strong electrical shock is applied to the chest region. The electrical shock causes the simultaneous depolarization of most cardiac muscle fibers. Following depolarization, the SA node can recover and produce action potentials before any other area of the heart. Consequently, the normal pattern of action potential generation and the normal rhythm of contraction can be reestablished.

An **artificial pacemaker** is an instrument placed beneath the skin; it is equipped with an electrode that extends to the heart. An artificial pacemaker provides an electrical stimulus to the heart at a set frequency. Artificial pacemakers are used in patients in whom the natural pacemaker of the heart does not produce a heart rate high enough to sustain normal physical activity. Modern electronics has made it possible to design artificial pacemakers that can increase the heart rate as physical activity increases. In addition, special artificial pacemakers can defibrillate the heart if it becomes arrhythmic. It is likely that rapid development of electronics for artificial pacemakers will further increase the degree to which the pacemakers can regulate the heart.

Some of these action potentials reach the **atrioventricular (AV)** (ā-trē-ō-ven′trik′-ū′lăr) **node,** located in the lower portion of the right atrium. Action potentials propagate slowly through the AV node before moving on to stimulate the ventricles to contract. The slow rate of action potential conduction in the AV node delays the transmission of action potentials to the ventricles for approximately 0.11 sec. This delay allows the atria to complete their contraction before the ventricles contract.

The AV node gives rise to the **atrioventricular (AV) bundle,** which passes through a small opening in the fibrous skeleton of the heart to reach the interventricular septum. The fibrous skeleton electrically separates the atria and ventricles so that action potentials spreading through the atria do not cause the ventricles to contract at the same time as the atria. Thus, only action potentials transmitted through the conducting system of the heart normally cause the ventricles to contract, allowing atria and ventricular contractions to be coordinated.

The AV bundle divides at the interventricular septum to form the **right** and **left bundle branches,** which extend to the apex of the heart. Many small bundles of **Purkinje** (pŭr-kĭn′jē) **fibers** pass from the tips of the right and left bundle branches to the apex of the heart and then extend superiorly to the cardiac muscle of the ventricular walls. The AV bundle, the bundle branches, and the Purkinje fibers are composed of specialized cardiac muscle fibers that conduct action potentials very rapidly. Consequently, ventricular contraction begins at the apex and progresses superiorly throughout the ventricles. This action pushes blood superiorly out the pulmonary trunk and aorta.

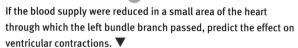

P R E D I C T 4

If the blood supply were reduced in a small area of the heart through which the left bundle branch passed, predict the effect on ventricular contractions. ▼

24. *What is the function of the conducting system of the heart?*
25. *What is accomplished by the delay of action potential transmission through the AV node?*

26. *What is the benefit of the electrical isolation of the atria from the ventricles by the fibrous skeleton of the heart?*
27. *What is the effect of the rapid transmission of action potentials through the AV bundle, the bundle branches, and the Purkinje fibers on ventricular contraction?*

Electrocardiogram

Action potentials conducted through the heart as it contracts and relaxes produce electric currents that can be measured at the surface of the body. Electrodes placed on the surface of the body and attached to a recording device can detect the small electrical changes resulting from the action potentials in all of the cardiac muscle cells. The record of these electrical events is an **electrocardiogram** (**ECG** or **EKG**) (figure 17.14). Note that ECGs record the summed effect of all the action potentials in the heart. They do not record individual action potentials.

A normal ECG consists of a P wave, a QRS complex, and a T wave (see figure 17.14). The **P wave** results from depolarization of the atrial myocardium, and the beginning of the P wave precedes the onset of atrial contraction. The **QRS complex** consists of three individual waves: the Q, R, and S waves. The QRS complex results from depolarization of the ventricles, and the beginning of the QRS complex precedes ventricular contraction. The **T wave** represents repolarization of the ventricles, and the beginning of the T wave precedes ventricular relaxation. A wave representing repolarization of the atria cannot be seen because it occurs during the QRS complex.

The time between the beginning of the P wave and the beginning of the QRS complex is the **PQ interval,** commonly called the **PR interval** because the Q wave is very small. During the PQ interval the atria contract and begin to relax. At the end of the PQ interval the ventricles begin to depolarize.

The **QT interval** extends from the beginning of the QRS complex to the end of the T wave and represents the length of time required for ventricular depolarization and repolarization.

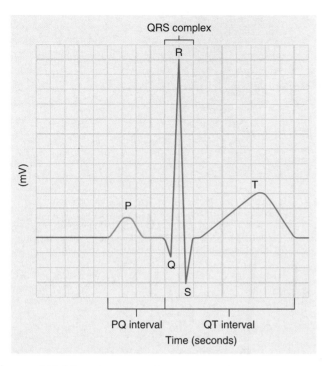

Figure 17.14 Electrocardiogram

The major waves and intervals of an electrocardiogram. Each thin horizontal line on the ECG recording represents 1 mV, and each thin vertical line represents 0.04 second.

 The ECG as a Diagnostic Tool

The ECG is not a direct measurement of mechanical events in the heart, and neither the force of contraction nor the blood pressure can be determined from it. Each deflection in the ECG record, however, indicates an electrical event within the heart and correlates with a subsequent mechanical event. Consequently, it is an extremely valuable diagnostic tool in identifying a number of cardiac abnormalities, particularly because it is painless, is easy to record, and does not require surgical procedures. Abnormal heart rates or rhythms; abnormal conduction pathways, such as blockages in the conduction pathways; hypertrophy or atrophy of portions of the heart; and the approximate location of damaged cardiac muscle can be determined from the analysis of an ECG. ▼

28. *What does an ECG measure? Name the waves produced by an ECG, and state what events occur during each wave.*

Cardiac Cycle

The heart can be viewed as two separate pumps represented by the right and left halves of the heart. Each pump consists of a primer pump—the atrium—and a power pump—the ventricle. The atria act as primer pumps because they complete the filling of the ventricles with blood, and the ventricles act as power pumps because they produce the major force that causes blood to flow through the pulmonary and systemic circulations. The term **cardiac cycle**

refers to the repetitive pumping process that begins with the onset of cardiac muscle contraction and ends with the beginning of the next contraction (figure 17.15). Pressure changes produced within the heart chambers as a result of cardiac muscle contraction are responsible for blood movement because blood moves from areas of higher pressure to areas of lower pressure.

Overview of Systole and Diastole

Atrial systole (sis′tō-lē, a contracting) refers to contraction of the two atria. **Ventricular systole** refers to contraction of the two ventricles. **Atrial diastole** (dī-as′tō-lē, dilation) refers to relaxation of the two atria, and **ventricular diastole** refers to relaxation of the two ventricles. When the terms **systole** and **diastole** are used without reference to the atria or ventricles, they refer to ventricular contraction or relaxation. The ventricles contain more cardiac muscle than the atria and produce far greater pressures, which force blood to circulate throughout the vessels of the body.

Just before systole begins, the atria and ventricles are relaxed, the ventricles are filled with blood, the semilunar valves are closed, and the AV valves are open. As systole begins, contraction of the ventricles increases ventricular pressures, causing blood to flow toward the atria and close the AV valves. As contraction proceeds, ventricular pressures continue to rise, but no blood flows from the ventricles because all the valves are closed. Thus, ventricular volume does not change, even though the ventricles are contracting (see figure 17.15, step 1).

P R E D I C T
As the ventricles contract while the AV and semilunar valves are closed, is the cardiac muscle contracting isotonically or isometrically?

As the ventricles continue to contract, ventricular pressures become greater than the pressures in the pulmonary trunk and aorta. As a result, the semilunar valves are pushed open and blood flows from the ventricles into those arteries (see figure 17.15, step 2).

As diastole begins, the ventricles relax and ventricular pressures decrease below the pressures in the pulmonary trunk and aorta. Blood begins to flow back toward the ventricles, causing the semilunar valves to close (see figure 17.15, step 3). With closure of the semilunar valves, all the heart valves are closed and no blood flows into the relaxing ventricles.

Throughout systole and the beginning of diastole, the atria relax and blood flows into them from the veins. When ventricular pressures become lower than atrial pressures, the AV valves open and blood flows from the atria into the relaxed ventricles (see figure 17.15, step 4). Most ventricular filling occurs as a result of this passive flow of blood during the first two-thirds of diastole. The remainder of ventricular filling occurs when the atria contract and push blood into the ventricles (see figure 17.15, step 5). During exercise, atrial contraction is more important for ventricular filling because, as heart rate increases, less time is available for passive ventricular filling. ▼

29. *Define systole and diastole.*
30. *Describe the opening and closing of the AV and semilunar valves during the cardiac cycle.*
31. *When does most ventricular filling occur?*

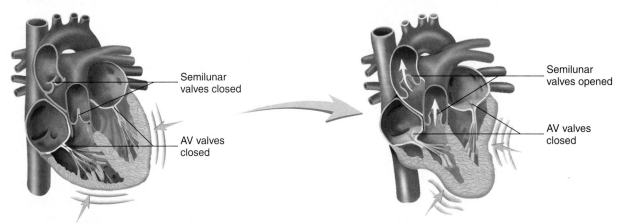

1. *Systole: AV valves close.* Ventricular contraction causes ventricular pressure to increase and causes the AV valves to close, which is the beginning of ventricular systole. The semilunar valves were closed in the previous diastole.

2. *Systole: semilunar valves open.* Continued ventricular contraction causes a greater increase in ventricular pressure, which pushes blood out of the ventricles, causing the semilunar valves to open. Blood is ejected from the ventricles.

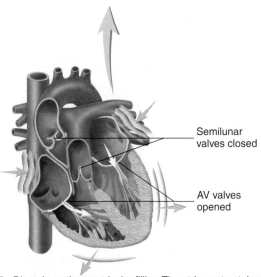

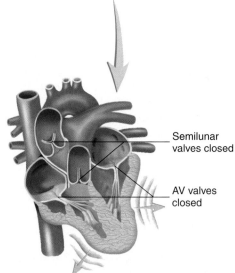

5. *Diastole: active ventricular filling.* The atria contract, increasing atrial pressure and completing ventricular filling while the ventricles are relaxed.

3. *Diastole: semilunar valves close.* As the ventricles begin to relax at the beginning of ventricular diastole, blood flowing back from the aorta and pulmonary trunk toward the relaxing ventricles causes the semilunar valves to close. Note that the AV valves are closed also.

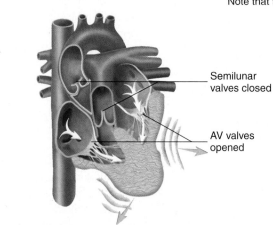

4. *Diastole: AV valves open.* As ventricular relaxation continues, the AV valves open and blood flows from the atria into the relaxing ventricles, accounting for most of the ventricular filling.

Process Figure 17.15   Cardiac Cycle

The cardiac cycle is a repeating series of contraction and relaxation that moves blood through the heart (*AV* = atrioventricular).

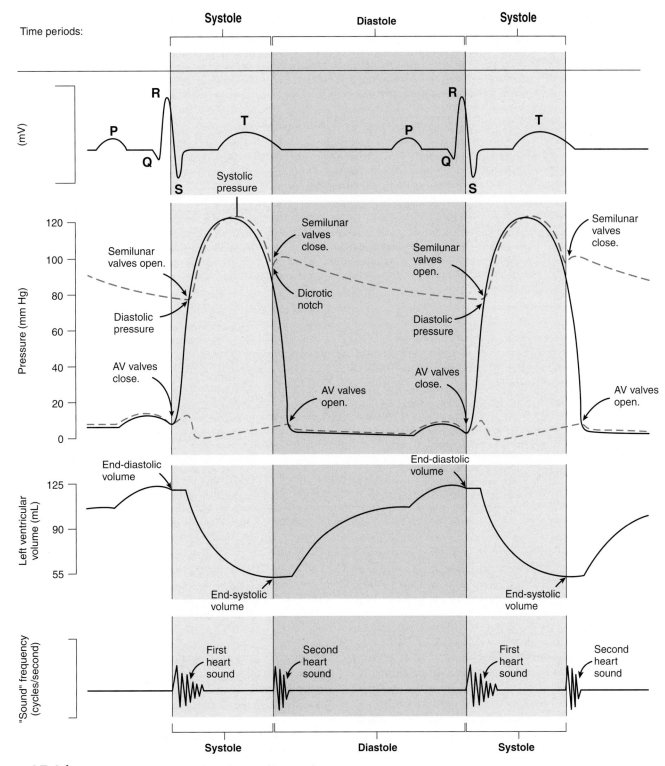

Figure 17.16 Events Occurring During the Cardiac Cycle

The cardiac cycle is divided into systole and diastole (see *top of figure*). Within systole and diastole, four graphs are presented. From *top* to *bottom*, the electrocardiogram; pressure changes for the left atrium (*blue line*), left ventricle (*black line*), and aorta (*red line*); left ventricular volume curve; and heart sounds are illustrated.

Events Occurring During Ventricular Systole

Figure 17.16 displays the main events of the cardiac cycle in graphic form and should be examined from top to bottom for systole and diastole. An ECG indicates the electrical events that cause contraction and relaxation of the atria and ventricles. The pressure graph shows the pressure changes within the left atrium, left ventricle, and aorta resulting from atrial and ventricular contraction and relaxation. Although pressure changes in the right side of the heart are not shown, they are similar to those in the left side, only lower. The volume graph presents the changes in left ventricular

Table 17.2	Summary of Events of the Cardiac Cycle for the Left Atrium and Ventricle (See Figure 17.16)	
	Ventricular Systole	**Ventricular Diastole**
ECG	The QRS complex is completed and the ventricles are stimulated to contract. The T wave begins.	The T wave is completed and the ventricles relax. Then the P wave stimulates the atria to contract, after which they relax.
Ventricular pressure curve (*black*)	Pressure increases rapidly as a result of left ventricular contraction. When left ventricular pressure exceeds aortic pressure, blood pushes the aortic semilunar valve open. Continued contraction increases ventricular pressure to a peak value of 120 mm Hg. Ventricular pressure then decreases as blood flows out of the left ventricle into the aorta.	Ventricular pressure decreases rapidly to nearly zero as the left ventricle relaxes.
Aortic pressure curve (*red*)	As ventricular contraction forces blood into the aorta, pressure in the aorta increases to its highest value (120 mm Hg), called the systolic pressure.	Ventricular pressure decreases below aortic pressure. Blood flows back toward the left ventricle and the aortic semilunar valve closes. As blood flows out of the aorta toward the body, elastic recoil of the aorta prevents a sudden decrease in pressure. Just before the aortic semilunar valve opens, pressure in the aorta decreases to its lowest value (80 mm Hg), called the diastolic pressure.
Atrial pressure curve (*blue*)	Atrial pressure increases slightly as contraction of the left ventricle pushes blood through the aorta and toward the left atrium. After closure of the bicuspid valve, pressure drops in the left atrium as it relaxes, then increases as blood flows into the left atrium from the four pulmonary veins.	After the bicuspid valve opens, pressure decreases slightly as blood flows into the left ventricle. At the end of ventricular diastole, contraction of the left atrium increases the pressure slightly.
Volume graph	Blood pushes the aortic semilunar valve open, blood is ejected from the left ventricle, and ventricular volume decreases. The amount of blood left in the ventricle is the end-systolic volume.	Blood flows from the left atrium into the left ventricle, accounting for 70% of ventricular filling. Near the end of ventricular diastole, contraction of the left atrium pushes blood into the left ventricle, completing ventricular filling. The amount of blood in the filled ventricle is the end-diastolic volume.
Sound graph	As contraction of the ventricles pushes blood toward the atria, the AV valves close, preventing the flow of blood into the atria and producing the first heart sound.	As blood flows back toward the heart, the semilunar valves close, preventing the flow of blood into the ventricles and producing the second heart sound.

volume as blood flows into and out of the left ventricle as a result of the pressure changes. The sound graph records the closing of valves caused by blood flow. Table 17.2 summarizes the events of systole and diastole.

Ventricular depolarization produces the QRS complex and initiates contraction of the ventricles. Ventricular pressure rapidly increases, resulting in closure of the AV valves. During the previous ventricular diastole, the ventricles were filled with blood, which is called the **end-diastolic volume.** At first, ventricular volume does not change because all the heart valves are closed. As soon as ventricular pressures exceed the pressures in the aorta and pulmonary trunk, the semilunar valves open. The aortic semilunar valve opens at approximately 80 mm Hg ventricular pressure, whereas the pulmonary semilunar valve opens at approximately 8 mm Hg. Although the pressures are different, both valves open at nearly the same time. As blood flows from the ventricles, the left ventricular pressure continues to climb to approximately 120 mm Hg, and the right ventricular

pressure increases to approximately 22 mm Hg. The higher pressure generated by the left ventricle is necessary to move blood through the larger systemic circulation. A lower pressure is adequate to move blood through the smaller pulmonary circulation (see figure 17.1)

PREDICT
Which ventricle has the thickest wall? Why is it important for each ventricle to pump approximately the same volume of blood?

At first, blood flows rapidly out of the ventricles. Toward the end of systole, very little blood flow occurs, which causes the ventricular pressure to decrease despite continued ventricular contraction. As systole ends, the volume of blood remaining in the ventricle is called the **end-systolic volume.** ▼

32. *Describe the relationship among the QRS complex, increased ventricular pressure, and decreased ventricular volume.*
33. *Define end-diastolic volume and end-systolic volume.*

Events Occurring During Ventricular Diastole

Ventricular repolarization produces the T wave and the ventricles relax. The already decreasing ventricular pressure falls very rapidly as the ventricles suddenly relax. When the ventricular pressures fall below the pressures in the aorta and pulmonary trunk, the recoil of the elastic arterial walls, which were stretched during the period of ejection, forces the blood to flow back toward the ventricles, thereby closing the semilunar valves. Ventricular volume does not change at this time because all the heart valves are closed.

When ventricular pressure drops below atrial pressure, the atrioventricular valves open and blood flows from the area of higher pressure in the veins and atria toward the area of lower pressure in the relaxed ventricles, which decreases to nearly 0 mm Hg. Approximately 70% of ventricular filling occurs during the first two-thirds of diastole.

P R E D I C T   7

Fibrillation is abnormal, rapid contractions of different parts of the heart that prevent the heart muscle from contracting as a single unit. Explain why atrial fibrillation does not immediately cause death but ventricular fibrillation does.

Depolarization of the SA node generates action potentials that spread over the atria, producing the P wave and stimulating both atria to contract (atrial systole). The atria contract during the last one-third of diastole and complete ventricular filling. Under most conditions, the atria function primarily as reservoirs, and the ventricles can pump sufficient blood to maintain homeostasis even if the atria do not contract at all. During exercise, however, the heart pumps 300%–400% more blood than during resting condition. It is under these conditions that the pumping action of the atria becomes important in maintaining the pumping efficiency of the heart. ▼

34. Describe the relationship among the T wave, decreased ventricular pressure, and increased ventricular volume.
35. Describe the relationship among the P wave, increased ventricular pressure, and increased ventricular volume.

Aortic Pressure Curve

As the ventricles contract and push blood into the aorta, aortic pressure increases to a maximum value called the **systolic pressure,** which is approximately 120 mm Hg (see figure 17.16). Even though ventricular pressure drops below aortic pressure as the ventricle relaxes, the forward momentum of the ejected blood maintains aortic pressure and prevents the backflow of blood into the ventricle. The higher pressure in the aorta eventually slows and then reverses the flow of blood, which moves back toward the ventricle, closing the aortic semilunar valve. Elastic recoil of the aorta prevents a large increase in aortic pressure and moves blood through the systemic circulation. Elastic recoil also moves blood back toward the left ventricle and pressure within the aorta increases slightly, producing a **dicrotic** (dī-krot′ik) **notch** in the aortic pressure curve (see figure 17.16). The term *dicrotic* means double-beating; when increased pressure caused by recoil is large, a double pulse can be felt.

Aortic pressure gradually falls throughout the rest of ventricular diastole as blood flows through the peripheral vessels. Aortic pressure decreases to a minimum value called **diastolic pressure,** which is approximately 80 mm Hg. The blood pressure

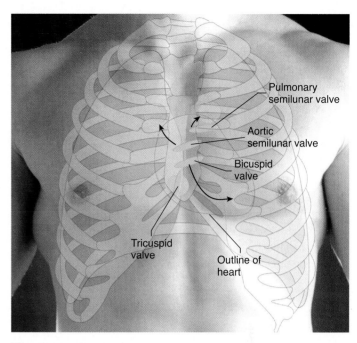

Figure 17.17   Location of the Heart Valves in the Thorax
Surface markings of the heart in the male. The positions of the four heart valves are indicated by *blue ellipses,* and the sites where the sounds of the valves are best heard with the stethoscope are indicated by *pink circles.*

in the aorta fluctuates between systolic and diastolic pressure. Blood pressure measurements performed for clinical purposes reflect the pressure changes that occur in the aorta rather than in the left ventricle (see chapter 18). Arteriosclerosis is a hardening of the arteries that occurs with age. The decreased elasticity of the aorta results in less expansion of the aorta when blood is injected into it, producing high blood pressure. ▼

36. Explain the production in the aorta of systolic pressure, diastolic pressure, and the dicrotic notch.

Heart Sounds

Distinct sounds are heard when a stethoscope (steth′ō-skōp, view the chest) is used to listen to the heart (figure 17.17). The **first heart sound** is a low-pitched sound, often described as a "lubb" sound. It is caused by vibration of the AV valves and surrounding fluid as the valves close at the beginning of ventricular systole (see figure 17.16).

The **second heart sound** is a higher-pitched sound often described as a "dupp" sound. It results from closure of the aortic and pulmonary semilunar valves, at the beginning of ventricular diastole (see figure 17.16). Systole is, therefore, approximately the time between the first and second heart sounds. Diastole, which lasts somewhat longer, is approximately the time between the second heart sound and the next first heart sound.

Abnormal heart sounds called **murmurs** are usually a result of faulty valves. For example, an **incompetent valve** fails to close tightly and blood leaks through the valve when it is closed. A murmur caused by an incompetent valve makes a swishing or gurgling sound immediately after closure of the valve. When the opening of a valve is narrowed, or **stenosed** (sten′ozd, a narrowing), a rushing sound precedes closure of the stenosed valve.

P R E D I C T

If normal heart sounds are represented by lubb-dupp, lubb-dupp, what do heart sounds represented by lubb-duppshhh, lubb-duppshhh represent (assume that shhh represents an abnormal sound)? Heart sounds represented by shhhlubb-dupp, shhhlubb-dubb? Be specific as to type of murmur and type of valve affected.

 ## Heart Murmurs

Inflammation of the heart valves, resulting from conditions such as rheumatic fever, can cause valves to become incompetent or stenosed. In addition, myocardial infarctions that make papillary muscles nonfunctional can cause bicuspid or tricuspid valves to be incompetent. Age-related changes in the connective tissue of the heart valves can produce defective valves. Heart murmurs also result from congenital abnormalities in the hearts of infants. For example, septal defects in the heart and patent ductus arteriosus result in distinct heart murmurs.

Incompetent valves increase the workload of the heart because the backflow of blood into a heart chamber increases the volume of blood in the chamber. Thus, the chamber must pump more blood with each contraction. Stenosed valves increase the workload of the heart because a stronger contraction is needed to push blood through the narrowed valve. Increased workload of the heart causes hypertrophy of the affected chamber and can lead to heart failure.

An incompetent bicuspid valve allows blood to flow back into the left atrium from the left ventricle during ventricular systole. This increases the pressure in the left atrium and pulmonary veins, which results in pulmonary edema. Also, the stroke volume of the left ventricle is reduced, which causes a decrease in systolic blood pressure. Similarly, an incompetent tricuspid valve allows blood to flow back into the right atrium and systemic veins, causing edema in the periphery.

An incompetent aortic semilunar valve allows blood to flow from the aorta into the left ventricle during diastole. Thus, the left ventricle fills with blood to a greater degree than normal. The ejection of the greater volume of blood into the aorta causes a greater than normal systolic pressure. Diastolic pressure, however, decreases below normal as blood rapidly leaks back into the ventricle.

Stenosis of the bicuspid valve prevents the flow of blood into the left ventricle, causing blood to back up in the left atrium and the lungs, resulting in edema in the lungs. Stenosis of the tricuspid valve causes blood to back up in the right atrium and systemic veins, causing edema in the periphery.

Surgical repair or replacement of defective valves is possible. Substitute valves made of synthetic materials such as plastic or Dacron are effective; valves transplanted from pigs are also used. ▼

37. *What produces the first and second heart sounds?*
38. *Name two types of valve problems that cause heart murmurs.*

Mean Arterial Blood Pressure

Blood pressure is necessary for blood movement and, therefore, is critical to the maintenance of homeostasis. Blood flows from areas of higher pressure to areas of lower pressure. For example, during one cardiac cycle, blood flows from the higher pressure in the aorta, resulting from contraction of the left ventricle, toward the lower pressure in the relaxed right atrium.

Mean arterial pressure (MAP) is the average pressure in the aorta. It is proportional to **cardiac output (CO)** times **peripheral resistance (PR).**

$$MAP = CO \times PR$$

Cardiac output is the amount of blood pumped by the heart per minute, and peripheral resistance is the total resistance against which blood must be pumped. Changes in cardiac output and peripheral resistance can alter mean arterial pressure. Cardiac output is discussed in this chapter, and peripheral resistance is explained in chapter 21.

Cardiac output is equal to heart rate (HR) times stroke volume (SV).

$$CO = HR \times SV$$

Heart rate is the number of times the heart beats (contracts) per minute, and **stroke volume** is the volume of blood pumped during each heartbeat (cardiac cycle). Stroke volume is equal to end-diastolic volume minus end-systolic volume. At rest, end-diastolic volume, the volume of blood in the relaxed ventricle just before it contracts, is 125 mL. End-systolic volume, the volume of blood left in the ventricle after it contracts, is 55 mL. The stroke volume is therefore 70 mL (125 − 55).

To better understand stroke volume, imagine that you are rinsing out a sponge under a running water faucet. As you relax your hand, the sponge fills with water; as your fingers contract, water is squeezed out of the sponge; and, after you have squeezed it, some water is left in the sponge. In this analogy, the amount of water you squeeze out of the sponge (stroke volume) is the difference between the amount of water in the sponge when your hand is relaxed (end-diastolic volume) and the amount that is left in the sponge after you squeeze it (end-systolic volume).

Stroke volume can be increased by increasing end-diastolic volume or by decreasing end-systolic volume. During exercise, end-diastolic volume increases because of an increase in **venous return,** which is the amount of blood returning to the heart from the peripheral circulation. End-systolic volume decreases because the heart contracts more forcefully and ejects more blood. For example, end-diastolic volume can increase to 145 mL and end-diastolic volume can decrease to 30 mL. Therefore, stroke volume increases to 115 mL (145 − 30) during exercise.

Under resting conditions, the heart rate is approximately 72 **beats per minute (bpm),** and the stroke volume is approximately 70 mL/beat, although these values can vary considerably from person to person. The cardiac output is therefore

$$\begin{aligned} CO &= HR \times SV \\ &= 72 \text{ bpm} \times 70 \text{ mL/beat} \\ &= 5040 \text{ mL/min (approximately 5 L/min)} \end{aligned}$$

During exercise, heart rate can increase to 190 bpm, and the stroke volume can increase to 115 mL. Consequently, cardiac output is

$$CO = 190 \text{ bpm} \times 115 \text{ mL/beat}$$
$$= 21,850 \text{ mL/min (approximately 22 L/min)}$$

Cardiac reserve is the difference between maximum cardiac output and cardiac output when a person is at rest. The greater a person's cardiac reserve, the greater his or her capacity for doing exercise. Lack of exercise and cardiovascular diseases can reduce cardiac reserve and affect a person's quality of life. Exercise training can greatly increase cardiac reserve by increasing maximum cardiac output. In well-trained athletes, stroke volume during exercise can increase to over 200 mL/beat, resulting in cardiac outputs of 40 L/min or more. ▼

39. *Define mean arterial pressure, cardiac output, and peripheral resistance. What is the role of mean arterial pressure in causing blood flow?*
40. *Define stroke volume and venous return. State two ways to increase stroke volume.*
41. *What is cardiac reserve? How does exercise training influence cardiac reserve?*

Regulation of the Heart

To maintain homeostasis, the amount of blood pumped by the heart must vary dramatically. For example, during exercise cardiac output can increase several times over resting values. Intrinsic and extrinsic regulatory mechanisms control cardiac output.

Intrinsic Regulation of the Heart

Intrinsic regulation of the heart modifies stroke volume through the normal functional characteristics of cardiac muscle cells. It does not depend on neural or hormonal regulation. According to **Starling's law of the heart,** as the resting length of cardiac muscle fibers increases, the force of contraction they produce increases. Past a certain length, however, the force of contraction decreases. This is similar to the length-tension relationship seen in skeletal muscle (see figure 8.17). Normally, cardiac muscle fibers are not stretched past the point at which they can contract with a maximal force.

The amount of blood in the ventricles at the end of ventricular diastole (end-diastolic volume) determines the degree to which cardiac muscle fibers are stretched. Venous return is the amount of blood that returns to the heart, and the degree to which the ventricular walls are stretched at the end of diastole is called **preload.** If venous return increases, the heart fills to a greater volume and stretches the cardiac muscle fibers, producing an increased preload. In response to the increased preload, cardiac muscle fibers contract with a greater force (Starling's law). The greater force of contraction causes an increased volume of blood to be ejected from the heart, resulting in an increased stroke volume, which increases cardiac output.

As a result of Starling's law of the heart, the amount of blood entering the heart (venous return) is equal to the amount of blood leaving the heart (cardiac output). When venous return increases, preload, force of contraction, stroke volume, and cardiac output increase. Conversely, when venous return decreases, preload, force of contraction, stroke volume, and cardiac output decrease.

Starling's law of the heart has a major influence on cardiac output because venous return is influenced by many conditions. For example, during exercise, blood vessels in skeletal muscles dilate, which increases blood delivery to the muscles. As the blood flows through the muscles and returns to the heart, venous return increases. Increased venous return results in an increased preload, stroke volume, and cardiac output. This is beneficial because an increased cardiac output is needed during exercise to deliver blood to exercising skeletal muscles.

Consequences of Heart Failure

Heart failure usually results from a progressive weakening of the heart muscle in elderly people, but it can occur in young people. A failing heart gradually enlarges and eventually does not adequately pump blood because further stretching of the cardiac muscle fibers does not increase the stroke volume of the heart. Consequently, blood backs up in the veins. **Right heart failure** causes blood to back up in the veins of the systemic circulation. Filling of the veins with blood causes fluid to accumulate in tissues, producing swelling, or edema, especially in the legs and feet. **Left heart failure** causes blood to back up in the veins of the pulmonary circulation. Filling of these veins causes edema in the lungs, which makes breathing difficult.

Afterload refers to the pressure against which the ventricles must pump blood. People suffering from hypertension have an increased afterload because they have an elevated aortic pressure during contraction of the ventricles. The heart must do more work to pump blood from the left ventricle into the aorta, which increases the workload on the heart and can eventually lead to heart failure. A reduced afterload decreases the work the heart must do. People who have a lower blood pressure have a reduced afterload and develop heart failure less often than people who have hypertension. The afterload, however, influences cardiac output less than preload influences it. Aortic blood pressure must increase to more than 170 mm Hg before it hampers the ability of a healthy ventricle to pump blood. ▼

42. *Define intrinsic regulation of the heart.*
43. *State Starling's law of the heart. How does venous return affect preload? How does preload affect cardiac output?*
44. *Define afterload, and describe its effect on the pumping effectiveness of the heart.*

Extrinsic Regulation of the Heart

Extrinsic regulation of the heart modifies heart rate and stroke volume through neural and hormonal mechanisms. Neural control of the heart results from sympathetic and parasympathetic reflexes, and the major hormonal control comes from epinephrine and norepinephrine secreted from the adrenal medulla.

The heart is autorhythmic with its own inherent heart rate (see "Autorhythmicity of Cardiac Muscle," p. 499). The intrinsic heart rate can be modified by hyperpolarization or depolarization of cardiac muscle cell plasma membranes. Hyperpolarization moves the membrane potential further from threshold, increasing the duration of the prepotential and slowing heart rate. Depolarization moves the membrane potential closer to threshold, decreasing the duration of the prepotential and increasing heart rate.

The parasympathetic division supplies the heart through the **vagus nerves,** which primarily innervate the SA and AV nodes. The postganglionic neurons of the vagus nerves release acetylcholine, which causes ligand-gated K^+ channels to open. Increased movement of K^+ out of cardiac muscle cells causes hyperpolarization, and heart rate slows. At rest, the parasympathetic division tonically stimulates the heart and depresses heart rate. Without parasympathetic stimulation, the resting heart rate would be approximately 100 beats/minute. During exercise, withdrawal of parasympathetic stimulation (removal of the inhibitory effect) contributes to an increase in heart rate.

The sympathetic division supplies the heart through sympathetic nerves, called **cardiac nerves,** which arise from the inferior cervical and upper thoracic sympathetic chain ganglia. The cardiac nerves innervate the SA node, the AV node, and the myocardium of the atria and ventricles. The postganglionic neurons of cardiac nerves release norepinephrine that binds to membrane-bound receptors, activating G proteins and causing Ca^{2+} channels to open. The movement of Ca^{2+} into cardiac muscle cells causes depolarization, and heart rate speeds up. The movement of Ca^{2+} into cardiac muscle cells also increases their force of contraction. In response to strong sympathetic stimulation, the heart rate can increase to 250 or, occasionally, 300 bpm. Stronger contractions cause the heart to empty to a greater extent by decreasing end-systolic volume, which increases stroke volume.

P R E D I C T

What effect does sympathetic stimulation have on stroke volume if the venous return remains constant?

 Beta-Adrenergic Blocking Agents

Adrenergic receptors respond to norepinephrine and epinephrine. The two subtypes of adrenergic receptors are alpha and beta receptors (see chapter 14). Cardiac muscle has beta receptors. **Beta-adrenergic** (bā-tă ad-rĕ-ner′jik) **blocking agents** are drugs that block the actions of beta receptors. They are used to reduce the rate and strength of cardiac muscle contractions, thus reducing the oxygen demand of the heart. Beta-adrenergic blocking agents are often used to treat people who suffer from rapid heart rates, certain types of arrhythmias, and hypertension.

Epinephrine and norepinephrine released from the adrenal medulla can markedly influence the pumping effectiveness of the heart. Epinephrine has essentially the same effect on cardiac muscle as norepinephrine and, therefore, increases the rate and force of heart contractions. The secretion of epinephrine and norepinephrine from the adrenal medulla is controlled by sympathetic stimulation of the adrenal medulla and occurs in response to increased physical activity, emotional excitement, and stressful conditions. ▼

45. Define extrinsic regulation of the heart.

46. *For each division of the ANS, name the nerves supplying the heart and the neurotransmitters they release. Describe the effects of the neurotransmitters on cardiac muscle plasma membrane potentials.*
47. *What effect do parasympathetic and sympathetic stimulation have on heart rate, force of contraction, and stroke volume?*
48. *Name the two main hormones that affect the heart. Where are they produced, what effects do they have on the heart, and what causes their release?*

The Heart and Homeostasis

The pumping of blood by the heart plays an important role in the maintenance of homeostasis. Blood pressure generated by the heart moves blood throughout the circulatory system. Adequate blood delivery ensures that the metabolic needs of tissues are met. The heart's activity must be regulated because the metabolic activities of the tissues change under such conditions as rest and exercise.

Effect of Blood Pressure

Baroreceptor (bar′ō-rē-sep′ter, bar′ō-rē-sep′tōr) **reflexes** detect changes in blood pressure and produce changes in heart rate and in the force of contraction. The sensory receptors of the baroreceptor reflexes are stretch receptors called **baroreceptors.** They are in the walls of certain large arteries, such as the internal carotid arteries and the aorta, and they measure blood pressure (figure 17.18).

Within the medulla oblongata of the brain is a **cardioregulatory center,** which receives and integrates action potentials from the baroreceptors. The cardioregulatory center controls the action potential frequency in sympathetic and parasympathetic nerve fibers that extend from the brain and spinal cord to the heart. The cardioregulatory center also controls the release of epinephrine and norepinephrine from the adrenal gland.

When the blood pressure increases, the walls of blood vessels are stretched, and the baroreceptors are stimulated. An increased frequency of action potentials is sent along the nerve fibers to the medulla oblongata of the brain. This prompts the cardioregulatory center to increase parasympathetic stimulation and to decrease sympathetic stimulation of the heart. As a result, the heart rate and stroke volume decrease, causing blood pressure to decline (figure 17.19).

When the blood pressure decreases, there is less stimulation of the baroreceptors. A lower frequency of action potentials is sent to the medulla oblongata of the brain. The cardioregulatory center responds by increasing sympathetic stimulation of the heart and decreasing parasympathetic stimulation. Consequently, the heart rate and stroke volume increase (see figure 17.19). Withdrawal of parasympathetic stimulation is primarily responsible for increases in heart rate up to approximately 100 bpm. Larger increases in heart rate, especially during exercise, result from sympathetic stimulation.

The baroreceptor reflexes provide fast-acting but short-term regulation of blood pressure. The baroreceptors help maintain a constant blood pressure despite changes in body position, such as from lying down to standing up. During exercise, blood pressure increases, and the baroreceptor reflexes help main the elevated blood pressure. The baroreceptors, however, do not maintain blood pressure over the long term and are not involved with chronic hypertension.

1. Sensory (*green*) neurons carry action potentials from baroreceptors and carotid body chemoreceptors to the cardioregulatory center. Chemoreceptors in the medulla oblongata also influence the cardioregulatory center.

2. The cardioregulatory center controls the frequency of action potentials in the parasympathetic (*red*) neurons extending to the heart through the vagus nerves. The parasympathetic neurons decrease the heart rate.

3. The cardioregulatory center controls the frequency of action potential in the sympathetic (*blue*) neurons. The sympathetic neurons extend through the cardiac nerves and increase the heart rate and the stroke volume.

4. The cardioregulatory center influences the frequency of action potentials in the sympathetic (*blue*) neurons extending to the adrenal medulla. The sympathetic neurons increase the secretion of epinephrine and some norepinephrine into the general circulation. Epinephrine and norepinephrine increase the heart rate and stroke volume.

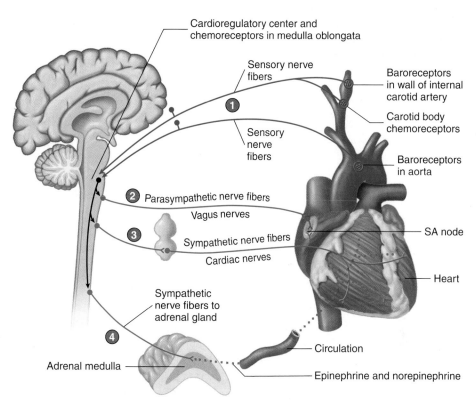

Process Figure 17.18 Baroreceptor and Chemoreceptor Reflexes

Sensory (*green*) nerves carry action potentials from sensory receptors to the medulla oblongata. Sympathetic (*blue*) and parasympathetic (*red*) nerves exit the spinal cord or medulla oblongata and extend to the heart to regulate its function. Epinephrine and norepinephrine from the adrenal gland also help regulate the heart's action (*SA* = sinoatrial).

The **adrenal medullary mechanism** is the release of epinephrine and norepinephrine from the adrenal gland. If the decrease in blood pressure is large, sympathetic stimulation of the adrenal medulla also increases. The epinephrine and norepinephrine secreted by the adrenal medulla increase the heart rate and stroke volume, also causing the blood pressure to increase toward its normal value (see figure 17.19).

P R E D I C T

In response to a severe hemorrhage, blood pressure lowers, the heart rate increases dramatically, and the stroke volume lowers. If low blood pressure activates reflexes that increase sympathetic stimulation of the heart, why is the stroke volume low?

Emotions integrated in the cerebrum of the brain can influence the heart. Excitement, anxiety, and anger can affect the cardioregulatory center, resulting in increased sympathetic stimulation of the heart and an increased cardiac output. Depression, on the other hand, can increase parasympathetic stimulation of the heart, causing a slight reduction in cardiac output.

▼

49. *Define baroreceptor reflex, baroreceptor, and cardioregulatory center.*
50. *How does the nervous system detect and respond to an increase and a decrease in blood pressure?*
51. *What effects do emotions have on cardiac output?*

CASE STUDY | Aortic Valve Stenosis

Norma is a 62-year-old woman who had rheumatic fever when she was 12 years old. She has had a heart murmur since then. Norma went to her doctor, complaining of fatigue; dizziness, especially on rising from a sitting or lying position; and pain in her chest when she exercises. Her doctor listened to Norma's heart sounds and determined she has a systolic murmur (see "Abnormal Heart Sounds," p. 507). Norma's blood pressure (90/65 mm Hg) and heart rate (55 beats/min) were lower than normal. Norma's doctor referred her to a cardiologist, who did additional tests. An electrocardiogram indicated she has left ventricular hypertrophy, and imaging techniques confirmed the left ventricular hypertrophy and a stenosed aortic semilunar valve. The cardiologist explained to Norma that the rheumatic fever she had as a child damaged her aortic semilunar valve and that the valve's condition had gradually become worse. The cardiologist recommended surgical replacement of

(continued)

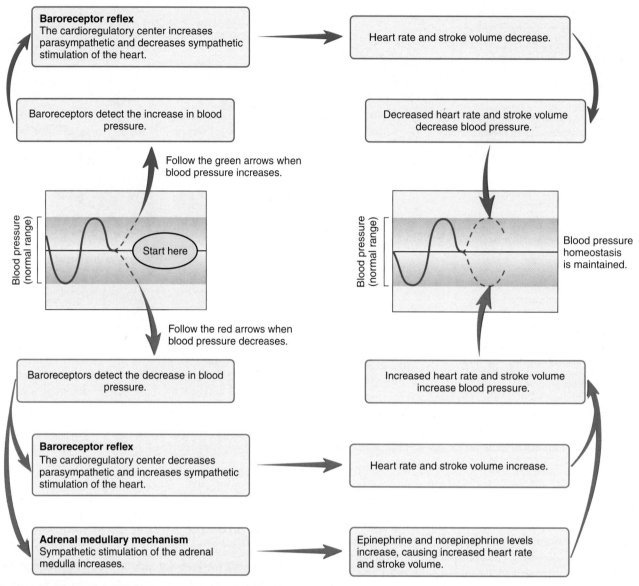

Homeostasis Figure 17.19 Summary of the Baroreceptor Reflex

The baroreceptor reflex maintains homeostasis in response to changes in blood pressure.

Norma's aortic semilunar valve, or she would likely develop heart failure.

P R E D I C T

a. What effect does Norma's stenosed valve have on stroke volume?

b. Norma has left ventricular hypertrophy, which means the left ventricle is enlarged and has thicker walls than normal. Explain how that condition developed.

c. Explain Norma's low blood pressure.

d. Explain why Norma becomes dizzy on rising from a sitting or lying position (*Hint:* venous return).

e. Predict how Norma's heart rate changes on standing (see figure 17.19).

f. Norma experiences chest pain when she exercises, a condition called angina pectoris (see p. 497). Why doesn't she feel this pain at rest?

Effect of pH, Carbon Dioxide, and Oxygen

Chemoreceptors (kē′mō-rē-sep′tors) are sensory receptors responding to chemicals, such as oxygen, carbon dioxide, and H$^+$. Chemoreceptors primarily sensitive to blood oxygen levels are found in the carotid and aortic bodies (figure 17.18). These small structures are located near large arteries close to the brain and heart, and they monitor blood flowing to the brain and to the rest of the body. There are also chemoreceptors sensitive to changes in carbon dioxide and pH levels in the medulla oblongata. Only when oxygen levels, carbon dioxide levels, or pH significantly deviate from normal are reflexes activated that affect the cardiovascular system. The chemoreceptor reflexes are not important for the normal regulation of the heart. They are more important in the regulation of blood vessel constriction (see chapter 18) and respiration (see chapter 20).

Chemoreceptors in the heart respond to chemicals, such as bradykinin and prostaglandins, released when cardiac muscle is deprived of oxygen. Axons from these sensory neurons extend to the spinal cord through sympathetic nerves and ascend the spinal cord in the spinothalamic tracts. There is considerable convergence with neurons of the spinothalamic tract, which could explain the referred pain of angina pectoris or a heart attack to the chest wall and arms (see chapter 10). ▼

52. Define chemoreceptor. How important are they for the regulation of the heart, blood vessels, and respiratory system?
53. Describe the role of chemoreceptors in the referred pain of angina pectoris or a heart attack.

Effect of Ions and Body Temperature

Changes in the extracellular concentration of K^+ and Ca^{2+}, which influence other electrically excitable tissues, also affect cardiac muscle function. The extracellular levels of Na^+ rarely deviate enough from the normal value to affect the function of cardiac muscle significantly.

Increased extracellular K^+ levels cause the heart rate and stroke volume to decrease. The excess K^+ in the extracellular fluid cause partial depolarization of the resting membrane potential, resulting in a decreased amplitude of action potentials. The decreased amplitude decreases the rate at which action potentials are conducted through the AV node, producing an AV node block. Ectopic foci and fibrillation can occur. The reduced action potential amplitude also results in fewer Ca^{2+} entering the sarcoplasm of the cell, which decreases the strength of cardiac muscle contraction.

Decreased extracellular K^+ levels cause the heart rate to decrease, but they do not affect the stroke volume. A decrease in extracellular K^+ results in a decrease in the heart rate because the resting membrane potential is hyperpolarized, which increases the duration of the prepotential.

Increased extracellular Ca^{2+} levels cause the stroke volume to increase and the heart rate to decrease. An increase in the extracellular concentration of Ca^{2+} produces an increase in the force of cardiac contraction because of a greater influx of Ca^{2+} into the sarcoplasm during action potential generation. Elevated plasma Ca^{2+} levels have an indirect effect on heart rate because they reduce the frequency of action potentials in nerve fibers, thus reducing sympathetic and parasympathetic stimulation of the heart. Generally, elevated blood Ca^{2+} levels reduce the heart rate.

Decreased extracellular Ca^{2+} levels decrease the stroke volume and increase the heart rate, although the effect is imperceptible until blood Ca^{2+} levels are reduced to approximately one-tenth of their normal value. A decrease in the extracellular concentration of Ca^{2+} produces a decrease in the force of cardiac contraction because of a decreased influx of Ca^{2+} into the sarcoplasm during action potential generation. Reduced extracellular Ca^{2+} levels also cause Na^+ channels to open, which allows Na^+ to diffuse more readily into the cell, resulting in depolarization and action potential generation. Reduced Ca^{2+} levels, however, usually cause death as a result of tetany of skeletal muscles before they decrease enough to markedly influence the heart's function.

Body temperature affects metabolism in the heart the way it affects other tissues. Elevated body temperature increases the heart rate, and reduced body temperature slows the heart rate. For example, during fever the heart rate is usually elevated. During heart surgery the body temperature is sometimes intentionally lowered to slow the heart rate and metabolism. ▼

54. What effects do an increased and a decreased extracellular concentration of K^+ and Ca^{2+} have on heart rate and stroke volume?
55. What effect does temperature have on heart rate?

Effects of Aging on the Heart

Gradual changes in the function of the heart are associated with aging. These changes are minor under resting conditions, but they become more obvious during exercise and in response to age-related diseases.

By age 70, cardiac output often decreases by approximately one-third. Because of the decrease in the cardiac reserve, many elderly people are limited in their ability to respond to emergencies, infections, blood loss, and stress.

Aging cardiac muscle requires a greater amount of time to contract and relax. Thus, there is a decrease in the maximum heart rate. This can be roughly predicted by the following formula: Maximum heart rate = 220 − age of the individual.

There is an age-related increase in cardiac arrhythmias as a consequence of a decrease in the number of cardiac cells in the SA node and because of the replacement of the cells of the AV bundle.

Hypertrophy (enlargement) of the left ventricle is a common age-related change. This appears to result from a gradual increase in the pressure in the aorta (afterload) against which the left ventricle must pump blood. The increased pressure in the aorta results from a gradual decrease in arterial elasticity, resulting in an increased stiffness of the aorta and other large arteries. The enlarged left ventricle has a reduced ability to pump blood, which can cause pulmonary edema. Consequently, there is an increased tendency for people to feel out of breath when they exercise strenuously.

Age-related changes occur in the connective tissue of the heart valves. The connective tissue becomes less flexible, and Ca^{2+} deposits develop in the valves. As a result, there is an increased tendency for the aortic semilunar valve to become stenosed or incompetent.

The development of coronary artery disease and heart failure also is age-related. Approximately 10% of elderly people over age 80 have heart failure, and a major contributing factor is coronary heart disease. Advanced age, malnutrition, chronic infections, toxins, severe anemias, hyperthyroidism, and hereditary factors can lead to heart failure.

Exercise has many beneficial effects on the heart. Regular aerobic exercise improves the functional capacity of the heart at all ages, provided there are no conditions that cause the heart's increased workload to be harmful. ▼

56. What happens to cardiac reserve with age?
57. How does maximum heart rate change with age? How does aging affect the conducting system of the heart?
58. Explain how age-related changes affect the function of the left ventricle.
59. Describe how increasing age affects the heart valves.
60. How does coronary artery disease and heart failure affect the aged?

Functions of the Heart (p. 488)

The heart produces the force that creates blood pressure, which causes blood to circulate. The heart routes blood to the pulmonary and systemic circulations and regulates blood delivery.

Location, Shape, and Size of the Heart (p. 488)

1. The heart is located in the mediastinum deep to the sternum and deep to the left second to fifth intercostal spaces.
2. The heart is shaped like a blunt cone, with an apex and a base. It is approximately the size of a fist.

Anatomy of the Heart (p. 489)

The heart consists of two atria and two ventricles.

Pericardium

1. The pericardium is a sac surrounding the heart and consisting of the fibrous pericardium and the serous pericardium.
2. The fibrous pericardium helps hold the heart in place.
3. The serous pericardium reduces friction as the heart beats. It consists of the following parts:
   - The parietal pericardium lines the fibrous pericardium.
   - The visceral pericardium covers the exterior surface of the heart.
   - The pericardial cavity lies between the parietal pericardium and visceral pericardium and is filled with pericardial fluid.

Heart Wall

1. The heart wall has three layers:
   - The outer epicardium (visceral pericardium) provides protection against the friction of rubbing organs.
   - The middle myocardium is responsible for contraction.
   - The inner endocardium reduces the friction resulting from the passage of blood through the heart.
2. The ventricles have ridges called trabeculae carneae.
3. The inner surfaces of the atria are mainly smooth. The auricles have raised areas called musculi pectinati.

External Anatomy

1. Each atrium has a flap called an auricle.
2. The coronary sulcus separates the atria from the ventricles. The interventricular grooves separate the right and left ventricles.
3. The inferior vena cava, superior vena cava, and coronary sinus enter the right atrium. The four pulmonary veins enter the left atrium.
4. The pulmonary trunk exits the right ventricle, and the aorta exits the left ventricle.

Heart Chambers

1. The interatrial septum separates the atria from each other. The fossa ovalis is the former location of the foramen ovalis through which blood bypassed the lungs in the fetus.
2. The interventricular septum separates the ventricles.

Heart Valves

1. The tricuspid valve separates the right atrium and ventricle. The bicuspid valve separates the left atrium and ventricle. The chordae tendineae attach the papillary muscles to the atrioventricular valves.
2. The semilunar valves separate the aorta and pulmonary trunk from the ventricles.

Route of Blood Flowing Through the Heart

1. Blood from the body flows through the right atrium into the right ventricle and then to the lungs.
2. Blood returns from the lungs to the left atrium, enters the left ventricle, and is pumped back to the body.

Blood Supply to the Heart

1. Coronary arteries branch off the aorta to supply the heart.
2. Blood returns from the heart tissues to the right atrium through the coronary sinus and cardiac veins.

Histology of the Heart (p. 497)

Fibrous Skeleton of the Heart

The fibrous heart skeleton supports the openings of the heart, provides a point of attachment for heart muscle, and electrically insulates the atria from the ventricles.

Cardiac Muscle

1. Cardiac muscle cells are branched and have a centrally located nucleus. Actin and myosin are organized to form sarcomeres.
2. The T tubules and sarcoplasmic reticulum are not as organized as in skeletal muscle. Normal contractions depend on extracellular Ca^{2+}.
3. Cardiac muscle cells rely on aerobic respiration for ATP production. They have many mitochondria and are well supplied with blood vessels.
4. Cardiac muscle cells are joined by intercalated disks, which allow action potentials to move from one cell to the next. Thus, cardiac muscle cells function as a unit.

Electrical Activity of the Heart (p. 498)

Action Potentials

1. After depolarization and partial repolarization, a plateau phase is reached, during which the membrane potential only slowly repolarizes.
2. The opening and closing of voltage-gated ion channels produce the action potential.
   - The movement of Na^+ through Na^+ channels causes depolarization.
   - During depolarization, K^+ channels close and Ca^{2+} channels begin to open.
   - Early repolarization results from closure of the Na^+ channels and the opening of some K^+ channels.
   - The plateau exists because Ca^{2+} channels remain open.
   - The rapid phase of repolarization results from the closure of the Ca^+ channels and the opening of many K^+ channels.

Refractory Periods

1. During the absolute refractory period, cardiac muscle cells are insensitive to further stimulation. During the relative refractory period, stronger than normal stimulation can produce an action potential.
2. Cardiac muscle has a prolonged depolarization and thus a prolonged absolute refractory period, which allows time for the cardiac muscle to relax before the next action potential causes a contraction.

Autorhythmicity of Cardiac Muscle

1. Cardiac muscle cells are autorhythmic because of the spontaneous development of a prepotential. The SA node is the pacemaker of the heart.

2. The prepotential results from the movement of Na^+ and Ca^{2+} into the SA node cells.
3. The duration of the prepotential determines heart rate.

Conducting System of the Heart

1. The SA node and the AV node are in the right atrium.
2. The AV node is connected to the bundle branches in the interventricular septum by the AV bundle.
3. The bundle branches give rise to Purkinje fibers, which supply the ventricles.
4. The SA node initiates action potentials, which spread across the atria and cause them to contract.
5. Action potentials are slowed in the AV node, allowing the atria to contract and blood to move into the ventricles. Then the action potentials travel through the AV bundles and bundle branches to the Purkinje fibers, causing the ventricles to contract, starting at the apex.

Electrocardiogram

1. The ECG records only the electrical activities of the heart.
   - Depolarization of the atria produces the P wave.
   - Depolarization of the ventricles produces the QRS complex. Repolarization of the atria occurs during the QRS complex.
   - Repolarization of the ventricles produces the T wave.
2. Based on the magnitude of the ECG waves and the time between waves, ECGs can be used to diagnose heart abnormalities.

Cardiac Cycle (p. 503)
Overview of Systole and Diastole

1. The cardiac cycle is repetitive contraction and relaxation of the heart chambers.
2. Atrial systole is contraction of the atria, and systole is contraction of the ventricles. Atrial diastole is relaxation of the atria, and diastole is relaxation of the ventricles.
3. During systole, the AV valves close, pressure increases in the ventricles, the semilunar valves are forced to open, and blood flows into the aorta and pulmonary trunk.
4. At the beginning of diastole, pressure in the ventricles decreases. The semilunar valves close to prevent backflow of blood from the aorta and pulmonary trunk into the ventricles.
5. When the pressure in the ventricles is lower than in the atria, the AV valves open and blood flows from the atria into the ventricles.
6. During atrial systole, the atria contract and complete the filling of the ventricles.

Events Occurring During Ventricular Systole

1. Ventricular depolarization produces the QRS complex and initiates contraction of the ventricles, which increases ventricular pressure. The AV valves close, the semilunar valves open, and blood is ejected from the heart.
2. The volume of blood in a ventricle just before it contracts is the end-diastolic volume. The volume of blood after contraction is the end-systolic volume.

Events Occurring During Ventricular Diastole

1. Ventricular repolarization produces the T wave and the ventricles relax. Blood flowing back toward the relaxed ventricles closes the semilunar valves. The AV valves open and blood flows into the ventricles.
2. Approximately 70% of ventricular filling occurs when blood flows from the higher pressure in the veins and atria to the lower pressure in the relaxed ventricles.

3. Atrial depolarization produces the P wave; the atria contract and complete ventricular filling.

Aortic Pressure Curve

1. Contraction of the ventricles forces blood into the aorta. The maximum pressure in the aorta is the systolic pressure.
2. Elastic recoil of the aorta maintains pressure in the aorta and produces the dicrotic notch.
3. Blood pressure in the aorta falls as blood flows out of the aorta. The minimum pressure in the aorta is the diastolic pressure.

Heart Sounds

1. Closure of the atrioventricular valves produces the first heart sound.
2. Closure of the semilunar valves produces the second heart sound.

Mean Arterial Blood Pressure (p. 508)

1. Mean arterial pressure is the average blood pressure in the aorta. Adequate blood pressure is necessary to ensure delivery of blood to the tissues.
2. Mean arterial pressure is proportional to cardiac output (amount of blood pumped by the heart per minute) times peripheral resistance (total resistance to blood flow through blood vessels).
3. Cardiac output is equal to heart rate times stroke volume.
4. Stroke volume, the amount of blood pumped by the heart per beat, is equal to end-diastolic volume minus end-systolic volume.
   - Venous return is the amount of blood returning to the heart. Increased venous return increases stroke volume by increasing end-diastolic volume.
   - Increased force of contraction increases stroke volume by decreasing end-systolic volume.
5. Cardiac reserve is the difference between resting and exercising cardiac output.

Regulation of the Heart (p. 509)
Intrinsic Regulation

1. Intrinsic regulation modifies stroke volume through the functional characteristics of cardiac muscle cells.
2. Starling's law of the heart describes the relationship between preload and the stroke volume of the heart. An increased preload causes the cardiac muscle fibers to contract with a greater force and produce a greater stroke volume.
3. Afterload is the pressure against which the ventricles must pump blood.

Extrinsic Regulation

1. Extrinsic regulation modifies heart rate and stroke volume through nervous and hormonal mechanisms.
2. The cardioregulatory center in the medulla oblongata regulates the parasympathetic and sympathetic nervous control of the heart.
3. Parasympathetic stimulation is supplied by the vagus nerve.
   - Parasympathetic stimulation decreases heart rate.
   - Postganglionic neurons secrete acetylcholine, which increases membrane permeability to K^+. Hyperpolarization of the plasma membrane increases the duration of the prepotential.
4. Sympathetic stimulation is supplied by the cardiac nerves.
   - Sympathetic stimulation increases heart rate and the force of contraction (stroke volume).
   - Postganglionic neurons secrete norepinephrine, which increases membrane permeability to Ca^{2+}. Depolarization of the plasma membrane decreases the duration of the prepotential.

5. Epinephrine and norepinephrine are released into the blood from the adrenal medulla as a result of sympathetic stimulation. They increase the rate and force of heart contraction.

The Heart and Homeostasis (p. 508)
Effect of Blood Pressure

1. Baroreceptors monitor blood pressure and the cardioregulatory center modifies heart rate and stroke volume.
2. In response to a decrease in blood pressure, the baroreceptor reflexes increase heart rate and stroke volume. When blood pressure increases, the baroreceptor reflexes decrease heart rate and stroke volume.

Effect of pH, Carbon Dioxide, and Oxygen

1. Carotid body and aortic chemoreceptor receptors monitor blood oxygen levels.
2. Medullary chemoreceptors monitor blood pH and carbon dioxide levels.
3. Chemoreceptors are not important for the normal regulation of the heart, but are important in the regulation of respiration and blood vessel constriction.

Effect of Ions and Body Temperature

1. Increased extracellular K^+ decrease heart rate and stroke volume. Decreased extracellular K^+ decrease heart rate.
2. Increased extracellular Ca^{2+} increase stroke volume and decrease heart rate. Decreased extracellular Ca^{2+} levels produce the opposite effect.
3. Heart rate increases when body temperature increases, and it decreases when body temperature decreases.

Effects of Aging on the Heart (p. 513)

1. Aging results in gradual changes in the function of the heart, which are minor under resting conditions but are more significant during exercise.
2. Some age-related changes to the heart are the following.
   - Decreased cardiac output and heart rate
   - Increased cardiac arrhythmias
   - Hypertrophy of the left ventricle
   - Development of stenosed or incompetent valves
   - Development of coronary artery disease and heart failure
3. Exercise improves the functional capacity of the heart at all ages.

REVIEW AND COMPREHENSION

1. The fibrous pericardium
   a. is in contact with the heart.
   b. is a serous membrane.
   c. is also known as the epicardium.
   d. forms the outer layer of the pericardial sac.
   e. all of the above.

2. The bulk of the heart wall is
   a. epicardium.
   b. pericardium.
   c. myocardium.
   d. endocardium.
   e. exocardium.

3. Which of these structures returns blood to the right atrium?
   a. coronary sinus
   b. inferior vena cava
   c. superior vena cava
   d. both b and c
   e. all of the above

4. The valve located between the right atrium and the right ventricle is the
   a. aortic semilunar valve.
   b. pulmonary semilunar valve.
   c. tricuspid valve.
   d. bicuspid (mitral) valve.

5. The papillary muscles
   a. are attached to chordae tendineae.
   b. are found in the atria.
   c. contract to close the foramen ovale.
   d. are attached to the semilunar valves.
   e. surround the openings of the coronary arteries.

6. Given these blood vessels:
   1. aorta
   2. inferior vena cava
   3. pulmonary trunk
   4. pulmonary vein

Choose the arrangement that lists the vessels in the order a red blood cell would encounter them in going from the systemic veins to the systemic arteries.
   a. 1,3,4,2
   b. 2,3,4,1
   c. 2,4,3,1
   d. 3,2,1,4
   e. 3,4,2,1

7. Which of these does *not* correctly describe the skeleton of the heart?
   a. electrically insulates the atria from the ventricles
   b. provides a rigid source of attachment for the cardiac muscle
   c. reinforces or supports the valve openings
   d. is composed mainly of bone

8. Cardiac muscle has
   a. sarcomeres.
   b. a sarcoplasmic reticulum.
   c. transverse tubules.
   d. many mitochondria.
   e. all of the above.

9. Action potentials pass from one cardiac muscle cell to another
   a. through gap junctions.
   b. by a special cardiac nervous system.
   c. because of the large voltage of the action potentials.
   d. because of the plateau phase of the action potentials.
   e. by neurotransmitters.

10. During the transmission of action potentials through the conducting system of the heart, there is a temporary delay at the
   a. bundle branches..
   b. Purkinje fibers.
   c. AV node.
   d. SA node
   e. AV bundle.

11. Given these structures of the conduction system of the heart:
    1. atrioventricular bundle
    2. AV node
    3. bundle branches
    4. Purkinje fibers
    5. SA node

    Choose the arrangement that lists the structures in the order an action potential passes through them.
    a. 2,5,1,3,4
    b. 2,5,3,1,4
    c. 2,5,4,1,3
    d. 5,2,1,3,4
    e. 5,2,4,3,1

12. Purkinje fibers
    a. are specialized neurons.
    b. conduct action potentials rapidly.
    c. conduct action potentials through the atria.
    d. connect between the SA node and the AV node.
    e. ensure that ventricular contraction starts at the base of the heart.

13. T waves on an ECG represent
    a. depolarization of the ventricles.
    b. repolarization of the ventricles.
    c. depolarization of the atria.
    d. repolarization of the atria.

14. The greatest amount of ventricular filling occurs during
    a. the first two-thirds of diastole.
    b. the middle one-third of diastole.
    c. the last two-thirds of diastole.
    d. ventricular systole.

15. While the semilunar valves are open during a normal cardiac cycle, the pressure in the left ventricle is
    a. greater than the pressure in the aorta.
    b. less than the pressure in the aorta.
    c. the same as the pressure in the left atrium.
    d. less than the pressure in the left atrium.

16. The pressure within the left ventricle fluctuates between
    a. 120 and 80 mm Hg.
    b. 120 and 0 mm Hg.
    c. 80 and 0 mm Hg.
    d. 20 and 0 mm Hg.

17. Pressure in the aorta is at its lowest
    a. at the time of the first heart sound.
    b. at the time of the second heart sound.
    c. just before the AV valves open.
    d. just before the semilunar valves open.

18. Stroke volume is the
    a. amount of blood pumped by the heart per minute.
    b. difference between end-diastolic and end-systolic volume.
    c. difference between the amount of blood pumped at rest and that pumped at maximum output.
    d. amount of blood pumped from the atria into the ventricles.

19. Cardiac output is
    a. blood pressure times peripheral resistance.
    b. peripheral resistance times heart rate.
    c. heart rate times stroke volume.
    d. stroke volume times blood pressure.
    e. blood pressure minus peripheral resistance.

20. The first heart sound, the "lubb" sound, is caused by the
    a. closing of the AV valves.
    b. closing of the semilunar valves.
    c. blood rushing out of the ventricles.
    d. filling of the ventricles.
    e. ventricular contraction.

21. Increased venous return results in
    a. increased stroke volume.
    b. increased cardiac output.
    c. decreased heart rate.
    d. both a and b.

22. Parasympathetic nerve fibers are found in the _____ nerves and release _____ at the heart.
    a. cardiac, acetylcholine
    b. cardiac, norepinephrine
    c. vagus, acetylcholine
    d. vagus, norepinephrine

23. Increased parasympathetic stimulation of the heart
    a. increases the force of ventricular contraction.
    b. increases the rate of depolarization in the SA node.
    c. decreases the heart rate.
    d. increases cardiac output.

24. Because of the baroreceptor reflex, when normal arterial blood pressure decreases, the
    a. heart rate increases.
    b. stroke volume decreases.
    c. frequency of afferent action potentials from baroreceptors increases.
    d. cardioregulatory center stimulates parasympathetic neurons.
    e. all of the above.

25. A decrease in blood pH and an increase in blood carbon dioxide levels result in
    a. increased heart rate.
    b. increased stroke volume.
    c. increased sympathetic stimulation of the heart.
    d. increased cardiac output.
    e. all of the above.

Answers in Appendix E

C R I T I C A L   T H I N K I N G

1. Is blood flow through the coronary circulation greatest during systole or diastole? Explain.

2. A patient has tachycardia. Would you recommend a drug that prolongs or shortens the plateau of cardiac muscle cell action potentials?

3. A friend tells you that her son had an ECG, and it revealed that he has a slight heart murmur. Are you convinced that he has a heart murmur? Explain.

4. Compare the rate of blood flow out of the ventricles between the first and second heart sounds of the same beat with the rate of blood flow out of the ventricles between the second heart sound of one beat and the first heart sound of the next beat.

5. Explain why it is sufficient to replace the ventricles, but not the atria, in artificial heart transplantation.

6. What happens to cardiac output following the ingestion of a large amount of fluid?

7. At rest, the cardiac output of athletes can be equal to that of nonathletes, but the heart rate of athletes is lower than that of nonathletes. At maximum exertion, the maximum heart rates of athletes and nonathletes can be equal, but the cardiac output of athletes is greater than that of nonathletes. Explain.

8. Predict the effect on Starling's law of the heart if the parasympathetic (vagus) nerves to the heart are cut.

9. Predict the effect on heart rate if the sensory nerve fibers from the baroreceptors are cut.

10. A doctor lets you listen to a patient's heart with a stethoscope at the same time that you feel the patient's pulse. Once in a while, you hear two heartbeats very close together, but you feel only one pulse beat. Later, the doctor tells you that the patient has an ectopic focus in the right atrium. Explain why you hear two heartbeats very close together. The doctor also tells you that the patient exhibits a pulse deficit (i.e., the number of pulse beats felt is fewer than the number of heartbeats heard). Explain why a pulse deficit occurs.

11. During shock caused by loss of blood, the blood pressure may fall dramatically, although the heart rate is elevated. Explain why blood pressure falls despite the increase in heart rate.

Answers in Appendix F

CHAPTER 18 Blood Vessels and Circulation

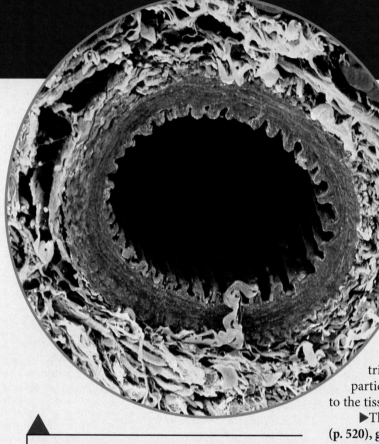

Color-enhanced scanning electron micrograph of an artery. The inner surface of the artery appears to be scalloped, or folded, because the circular smooth muscle cells in the wall of the artery partially contract when tissues are prepared for microscopic examination.

Tools for Success

A virtual cadaver dissection experience

Audio (MP3) and/or video (MP4) content, available at www.mhhe.aris.com

Introduction

The peripheral circulatory system comprises two sets of blood vessels: systemic and pulmonary vessels. **Systemic vessels** transport blood through all parts of the body from the left ventricle and back to the right atrium. **Pulmonary vessels** transport blood from the right ventricle through the lungs and back to the left atrium. The systemic vessels and the pulmonary vessels together constitute the **peripheral circulation.**

Complex urban water systems seem rather simple when compared with the intricacy and coordinated functions of blood vessels. The heart is the pump providing the major force causing blood to circulate. The blood vessels are the pipes carrying blood to within two or three cell diameters of nearly all of the trillions of cells that make up the body. In addition, the blood vessels participate in the regulation of blood pressure and help direct blood flow to the tissues that are most active.

▶This chapter explains the **functions of the peripheral circulation (p. 520), general features of blood vessel structure (p. 520), pulmonary circulation (p. 525), systemic circulation: arteries (p. 525), systemic circulation: veins (p. 535), physiology of circulation (p. 544), control of blood flow (p. 549),** and **regulation of mean arterial pressure (p. 551).** The chapter concludes with **examples of cardiovascular regulation (p. 557).**

Functions of the Peripheral Circulation

The heart provides the major force that causes blood to circulate, and the peripheral circulation has five functions:

1. *Carries blood.* Blood vessels carry blood from the heart to all the tissues of the body and back to the heart.
2. *Exchanges nutrients, waste products, and gases with tissues.* Nutrients and oxygen diffuse from blood vessels to cells in all areas of the body. Waste products and carbon dioxide diffuse from the cells to blood vessels.
3. *Transports substances.* Hormones, components of the immune system, molecules required for blood clotting, enzymes, nutrients, gases, waste products, and other substances are transported in the blood to all areas of the body.
4. *Helps regulate blood pressure.* The blood vessels and the heart work together to regulate blood pressure within a normal range of values.
5. *Directs blood flow to tissues.* The blood vessels direct blood to tissues when increased blood flow is required to maintain homeostasis. ▼

> 1. *List the functions of the peripheral circulation.*

General Features of Blood Vessels

Arteries (ar′ter-ēz, resembling a windpipe) are blood vessels that carry blood away from the heart. The heart pumps blood into large, elastic arteries that branch repeatedly to form many, progressively smaller arteries. As they become smaller, the arteries undergo a gradual transition from having walls that contain a large amount of elastic tissue and a smaller amount of smooth muscle to having walls with a smaller amount of elastic tissue and a relatively large amount of smooth muscle. Although the arteries form a continuum from the largest to the smallest branches, they normally are classified as (1) elastic arteries, (2) muscular arteries, or (3) arterioles.

Blood flows from arterioles into **capillaries** (kap′i-lār-ēz, resembling fine hair). Most of the exchange that occurs between the blood and interstitial fluids occurs across the walls of capillaries. Their walls are the thinnest of all the blood vessels, blood flows through them slowly, and a greater number of them exist than any other blood vessel type.

From the capillaries, blood flows into the venous system. **Veins** (vānz) are blood vessels that carry blood from the capillaries toward the heart. Compared with arteries, the walls of the veins are thinner and contain less elastic tissue and fewer smooth muscle cells. The veins increase in diameter and decrease in number, and their walls increase in thickness as they project toward the heart. They are classified as (1) venules, (2) small veins, or (3) medium or large veins.

Blood vessel walls consist of three layers, except in capillaries and some venules. The relative thickness and composition of each

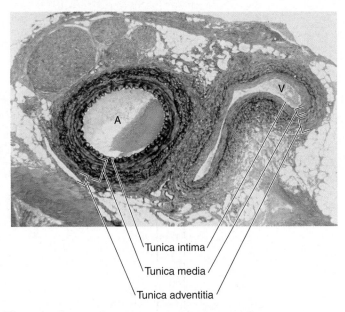

Figure 18.1   **Photomicrograph of an Artery and a Vein**

The typical structure of a medium-sized artery (A) and vein (V). Note that the artery has a thicker wall than the vein. The tunica intima, tunica media, and tunica adventitia make up the walls of the blood vessels. The predominant layer in the wall of the artery is the tunica media, with its circular layers of smooth muscle. The predominant layer in the wall of the vein is the tunica adventitia, and the tunica media is thinner.

layer vary with the type and diameter of the blood vessel. From the inner to the outer wall of the blood vessels, the layers, or **tunics** (too′niks), are (1) the tunica intima, (2) the tunica media, and (3) the tunica adventitia (figures 18.1 and 18.2).

The **tunica intima** (too′ni-kă in′ti-mă, a coat + *intima*, innermost) consists of an endothelium composed of simple squamous epithelial cells, a basement membrane, and a small amount of connective tissue. In muscular arteries, the tunica intima also contains a layer of thin, elastic connective tissue. The **tunica media,** or middle layer, consists of smooth muscle cells arranged circularly around the blood vessel. It also contains variable amounts of elastic and collagen fibers, depending on the size and type of the vessel. In muscular arteries, there is a layer of elastic connective tissue at the outer margin of the tunica media. The **tunica adventitia** (ad-ven-tish′ă, a coat + *adventicius*, to come from abroad) is composed of connective tissue. It is a denser connective tissue adjacent to the tunica media that becomes loose connective tissue toward the outer portion of the blood vessel wall.

Not surprisingly, the structure of all blood vessel walls is not the same. Modifications of the general pattern, such as differences in connective tissue and smooth muscle, result in different functional capabilities. ▼

> 2. *Define artery and vein.*
> 3. *Name, in order, all the types of blood vessels, starting at the heart, going into the tissues, and returning to the heart.*
> 4. *Name the three layers of a blood vessel. What kinds of tissue are in each layer?*

(a) **Elastic arteries.** The tunica media is mostly elastic connective tissue. Elastic arteries recoil when stretched, which prevents blood pressure from falling rapidly.

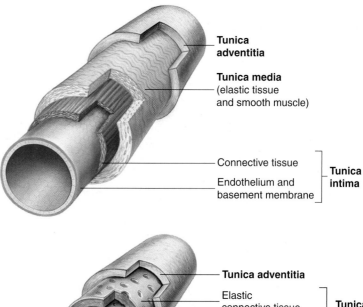

Tunica adventitia

Tunica media (elastic tissue and smooth muscle)

Connective tissue

Endothelium and basement membrane

Tunica intima

(b) **Muscular Arteries.** The tunica media is a thick layer of smooth muscle. Muscular arteries regulate blood flow to different regions of the body.

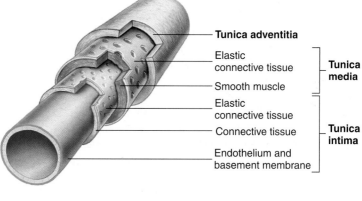

Tunica adventitia

Elastic connective tissue

Tunica media

Smooth muscle

Elastic connective tissue

Connective tissue

Endothelium and basement membrane

Tunica intima

(c) **Medium and Large Veins.** All three tunics are present. The tunica media is thin but can regulate vessel diameter because blood pressure in the venous system is low. The predominant layer is the tunica adventitia.

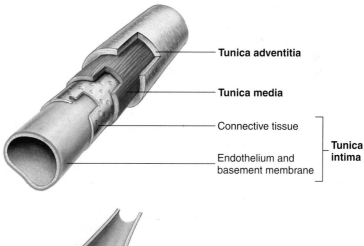

Tunica adventitia

Tunica media

Connective tissue

Endothelium and basement membrane

Tunica intima

(d) Folds in the endothelium form the valves of veins, which allow blood to flow toward the heart but not in the opposite direction.

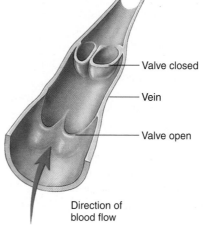

Valve closed

Vein

Valve open

Direction of blood flow

Figure 18.2 Structural Comparison of Blood Vessel Types

Arteries

Arteries conduct blood under high pressure and have thick, strong walls. **Elastic arteries** are the largest-diameter arteries and have the thickest walls (see figure 18.2a). A greater proportion of their walls is elastic tissue, and a smaller proportion is smooth muscle, compared with other arteries. Elastic arteries are stretched when the ventricles of the heart pump blood into them, which reduces blood pressure as a large volume of blood is suddenly ejected into them. The elastic recoil of the elastic arteries prevents blood pressure from falling rapidly and maintains blood flow while the ventricles are relaxed.

The **muscular arteries** include medium-sized and small-diameter arteries. The walls of medium-sized arteries are relatively thick, compared with their diameter. Most of the thickness of the wall results from smooth muscle cells of the tunica media (see figure 18.2b). Medium-sized arteries are frequently called **distributing arteries** because the smooth muscle tissue enables these vessels to control blood flow to different regions of the body. Contraction of the smooth muscle in blood vessels, which is called **vasoconstriction** (vā′sō-kon-strik′shŭn), decreases blood vessel diameter and blood flow. Relaxation of the smooth muscle in blood vessels, which is called **vasodilation** (vā′sō-dī-lā′shŭn), increases blood vessel diameter and blood flow.

Small arteries have about the same structure as medium-sized arteries, except that small arteries have a smaller diameter and their walls are thinner.

Arterioles (ar′ter-ē′ōlz) transport blood from small arteries to capillaries and are the smallest arteries in which the three tunics can be identified. The tunica media consists of only one or two layers of circular smooth muscle cells. Just before arterioles connect to capillaries, their continuous layer of smooth muscle becomes isolated smooth muscle cells encircling the arteriole at scattered locations along their walls. Arterioles control the amount of blood reaching capillaries. ▼

5. *Describe the function of elastic arteries, muscular (distributing) arteries, and arterioles.*
6. *Define vasoconstriction and vasodilation.*

Capillaries

Capillary walls consist of **endothelium** (*endon,* within + *thēlē,* nipple), which is a layer of simple squamous epithelium surrounded by a delicate loose connective tissue (figure 18.3). The thin walls of capillaries facilitate diffusion between the capillaries and surrounding cells. Each capillary is 0.5–1 millimeter (mm) long. Capillaries branch without changing their diameter, which is approximately the same as the diameter of a red blood cell (7.5 μm).

Blood flows from arterioles into a network of capillaries called a **capillary bed** (figure 18.4). The arterioles connect to thoroughfare channels or capillaries. **Thoroughfare channels** extend in a relatively direct fashion from arterioles to venules. Blood flow through thoroughfare channels is relatively continuous. Several capillaries branch from the thoroughfare channels. Smooth muscle cells called **precapillary sphincters,** which are located at the origin of capillaries from thoroughfare channels and arterioles, regulate blood flow into capillaries. Blood flow through capillaries is intermittent and slower than in thoroughfare channels.

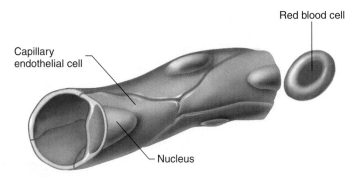

Figure 18.3 Capillary

The capillary wall is a thin layer of simple squamous epithelium, which facilitates the exchange of gases, nutrients, and waste products between the blood and tissues.

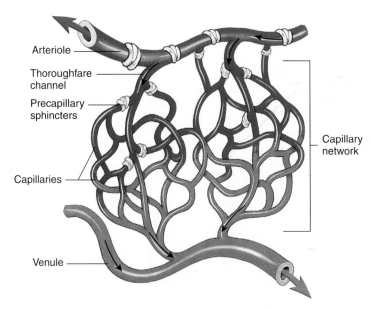

Figure 18.4 Capillary Bed

An arteriole gives rise to a capillary bed, which has numerous branches. Isolated smooth muscle encircles the arteriole and smooth muscle cells, called precapillary sphincters, encircle capillaries as they branch from the arteriole. The smooth muscle regulates the flow of blood, which decreases when the smooth muscle contracts and increases when it relaxes. Blood flows from capillaries into venules. *Red* represents oxygenated blood, *blue* represents deoxygenated blood.

Capillary beds in the skin have many more thoroughfare channels than capillary beds in cardiac or skeletal muscle. Capillaries in the skin function in thermoregulation, and heat loss results from the flow of a large volume of blood through the thoroughfare channels. In muscle, however, nutrient and waste product exchange is the major function of the capillaries. ▼

7. *What is the function of capillaries and thoroughfare channels? What structure regulates the flow of blood through capillaries?*

Veins

Blood flows from capillaries into **venules** (ven′oolz), which transport blood from capillaries to small veins. The venules connected to capillaries are tubes with a diameter slightly larger than that of capillaries and are composed of endothelium resting on a delicate connective tissue layer. As the venules converge, they become larger in diameter, with scattered smooth muscle cells in their walls. Eventually, venules have a tunic media, with one or two layers of smooth muscle, and a tunic adventitia.

Small veins collect blood from venules and deliver it to **medium-sized veins,** which deliver the blood to **large veins.** Veins return blood to the heart under low pressure. They have relatively thin walls with large diameters. The three thin but distinctive tunics make up the wall of the veins. The predominant layer is the outer tunica adventitia, which consists primarily of dense collagen fibers (see figures 18.2c). Veins expand more easily than arteries, and the connective tissue of the tunica adventitia determines the degree to which they can distend. Although their walls have less smooth muscle than arteries, they can effectively vasoconstrict and vasodilate because of the low blood pressure.

Veins having diameters greater than 2 mm contain **valves,** which allow blood to flow toward the heart but not in the opposite direction. Preventing the backflow of blood is necessary because the pressure moving blood in the veins is low. Each valve consists of folds in the tunica intima that form two flaps, which are shaped like and function like the semilunar valves of the heart. There are many valves in medium-sized veins (see figure 18.2d). There are more valves in the veins of the lower limbs than in the veins of the upper limbs. This prevents the flow of blood toward the feet in response to the pull of gravity.

Varicose Veins

Varicose (văr′ĭ-kōs) **veins** are permanently dilated veins in which the valves in the veins do not prevent the backflow of blood. They are most common in the veins of the lower limbs. As a consequence of varicose veins, venous pressure is greater than normal and can result in edema. Some people have a genetic tendency for the development of varicose veins, which is encouraged by conditions that increase the pressure in veins, causing them to stretch. Examples include standing in place for prolonged periods and pregnancy. Standing in place allows the pressure of the blood to stretch the veins, and pregnancy allows compression of the veins in the pelvis by the enlarged uterus, resulting in increased venous pressure in the veins that drain the lower limbs. Blood in the veins can become so stagnant that the blood clots. The clots, called **thromboses** (throm-bō′sēz), can result in inflammation of the veins, a condition called **phlebitis** (fle-bī′tis, *phlebo*, vein). If the condition becomes severe enough, the blocked veins can prevent blood flow through capillaries that are drained by the veins. The lack of blood flow can lead to tissue death, or **necrosis** (ně-krō′sis, death), and infection of the tissue with anaerobic bacteria, a condition called **gangrene** (gang′grēn, an eating sore). In addition, fragments of the clots can dislodge and travel through the veins to the lungs, where they can cause severe damage. Fragments of thromboses that dislodge and float in the blood are called **emboli** (em′bō-lī, plugs). ▼

8. *How do the veins function as a reservoir for blood?*
9. *What is the function of valves in veins?*

Aging of the Arteries

The walls of all arteries undergo changes as they age, although some arteries change more rapidly than others and some individuals are more susceptible to change. The most significant change occurs in the large elastic arteries, such as the aorta; the large arteries that carry blood to the brain; and the coronary arteries. The age-related changes described here refer to these blood vessel types. Changes in muscular arteries do occur, but they are less dramatic and they often do not result in the disruption of normal blood vessel function.

Arteriosclerosis (ar-tēr′ē-ō-skler-o′sis, hardening of the arteries) consists of degenerative changes in arteries that make them less elastic. These changes occur in many individuals, and they become more severe with advancing age. Arteriosclerosis greatly increases resistance to blood flow. Therefore, advanced arteriosclerosis reduces the normal circulation of blood and greatly increases the work performed by the heart.

Arteriosclerosis involves general thickening of the tunica intima and the tunica media. For example, when arteriosclerosis is associated with hypertension, there is an increase in smooth muscle and elastic and collagen fibers in the arterial walls. Arteriosclerosis in some older people can involve arteries, primarily of the lower limbs, in which calcium deposits form in the tunica media of the arteries with little or no encroachment on their lumens.

Atherosclerosis (ath′er-ō-skler-ō′sis) is the deposition of material in the walls of arteries to form distinct plaques. It affects primarily medium-sized and larger arteries, including the coronary arteries. The plaques form as macrophages containing cholesterol accumulate in the tunica intima, and smooth muscle cells of the tunica media proliferate (figure 18.5). After the plaques enlarge, they consist of smooth muscle cells; leukocytes; lipids, including cholesterol; and, in the largest plaques, fibrous connective tissue and calcium deposits. The plaques narrow the lumens of blood vessels and make their walls less elastic. Atherosclerotic plaques can become so large that they severely restrict, or block, blood flow through arteries, and the plaques are sites where thromboses and emboli form.

Some investigators propose that arteriosclerosis may not be a pathological process. Instead, it may be an aging or a wearing-out

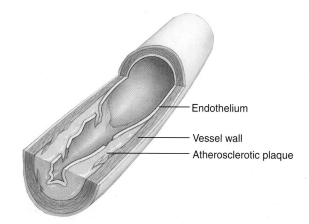

Figure 18.5 **Atherosclerotic Plaque in an Artery**
Atherosclerotic plaques develop within the tissue of the artery wall.

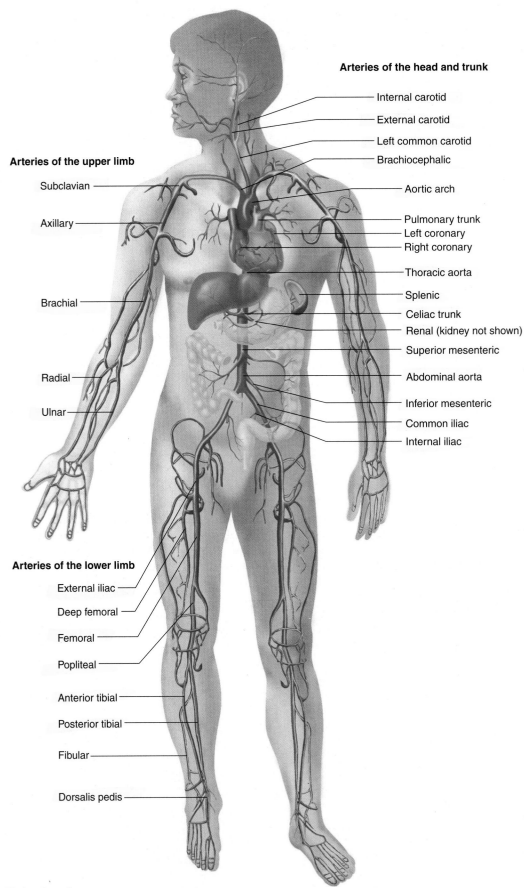

Arteries of the head and trunk

- Internal carotid
- External carotid
- Left common carotid
- Brachiocephalic
- Aortic arch
- Pulmonary trunk
- Left coronary
- Right coronary
- Thoracic aorta
- Splenic
- Celiac trunk
- Renal (kidney not shown)
- Superior mesenteric
- Abdominal aorta
- Inferior mesenteric
- Common iliac
- Internal iliac

Arteries of the upper limb

- Subclavian
- Axillary
- Brachial
- Radial
- Ulnar

Arteries of the lower limb

- External iliac
- Deep femoral
- Femoral
- Popliteal
- Anterior tibial
- Posterior tibial
- Fibular
- Dorsalis pedis

Figure 18.6 Major Arteries

The arteries carry blood from the heart to the tissues of the body.

process. Evidence also suggests that arteriosclerosis may result from inflammation, which, in some cases, may be the result of an autoimmune disease. Atherosclerosis has been studied extensively, and there are many risk factors associated with the development of atherosclerotic plaques. These factors include being at an advanced age, being a male, being a postmenopausal woman, having a family history of atherosclerosis, smoking cigarettes, having hypertension, having diabetes mellitus, having increased blood LDL and cholesterol levels, being overweight, leading a sedentary lifestyle, and having high blood triglyceride levels. Avoiding the environmental factors that influence atherosclerosis slows the development of atherosclerotic plaques. In some cases, the severity of the plaques can be reduced. For example, treatments, such as regulating blood glucose levels in people with diabetes mellitus and taking drugs that lower blood cholesterol levels, especially in people with high blood cholesterol levels that cannot be controlled by dietary changes, can provide some protection. ▼

> 10. *Describe the changes that occur in arteries due to arteriosclerosis and atherosclerosis. In which vessels do the most significant changes occur?*

Pulmonary Circulation

Blood from the right ventricle is pumped into the **pulmonary** (pŭl′mō-nār-ē, relating to the lungs) **trunk** (see chapter 17). This short vessel branches into the **right** and **left pulmonary arteries,** which extend to the right and left lungs, respectively. Poorly oxygenated blood is carried by these arteries to the pulmonary capillaries in the lungs, where oxygen is taken up by the blood and carbon dioxide is released. Blood rich in oxygen flows from the lungs to the left atrium. Four **pulmonary veins,** two from each lung, carry the oxygenated blood to the left atrium. ▼

> 11. *For the vessels of the pulmonary circulation, give their starting point, ending point, and function.*

Systemic Circulation: Arteries

Oxygenated blood entering the heart from the pulmonary veins passes through the left atrium into the left ventricle and from the left ventricle into the aorta. Blood flows from the aorta to all parts of the body (figure 18.6).

Aorta

All **arteries** of the systemic circulation are derived either directly or indirectly from the **aorta** (ā-ōr′tă, to lift up), which is usually considered in three parts: the ascending aorta, the aortic arch, and the descending aorta. The descending aorta is divided further into a thoracic aorta and an abdominal aorta (figure 18.7).

The part of the aorta that passes superiorly from the left ventricle is called the **ascending aorta.** The right and left **coronary arteries** arise from the base of the ascending aorta and supply blood to the heart (see chapter 17).

The aorta then arches posteriorly and to the left as the **aortic arch.** Three major branches, which carry blood to the head and upper limbs, originate from the aortic arch: the brachiocephalic artery, the left common carotid artery, and the left subclavian artery.

 Trauma and the Aorta

Trauma that ruptures the aorta is almost immediately fatal. Trauma can also lead to an **aneurysm** (an′ū-rizm), however, which is a bulge caused by a weakened spot in the aortic wall. If the weakened aortic wall leaks blood slowly into the thorax, the aneurysm must be corrected surgically. Also, once the aneurysm is formed it is likely to enlarge and becomes more likely to rupture. The majority of traumatic aortic arch ruptures occur during automobile accidents and result from the great force with which the body is thrown into the steering wheel, the dashboard, or other objects. Waist-type safety belts alone do not prevent this type of injury as effectively as shoulder-type safety belts and air bags.

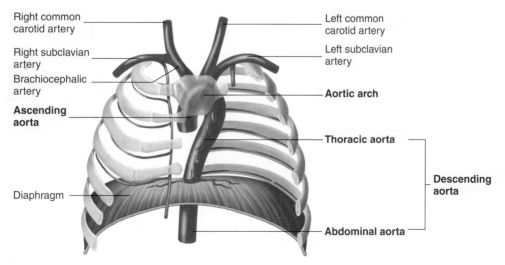

Figure 18.7   Aorta

The aorta has three parts: the ascending aorta, the aortic arch, and the descending aorta. The descending aorta consists of the thoracic and abdominal aortae.

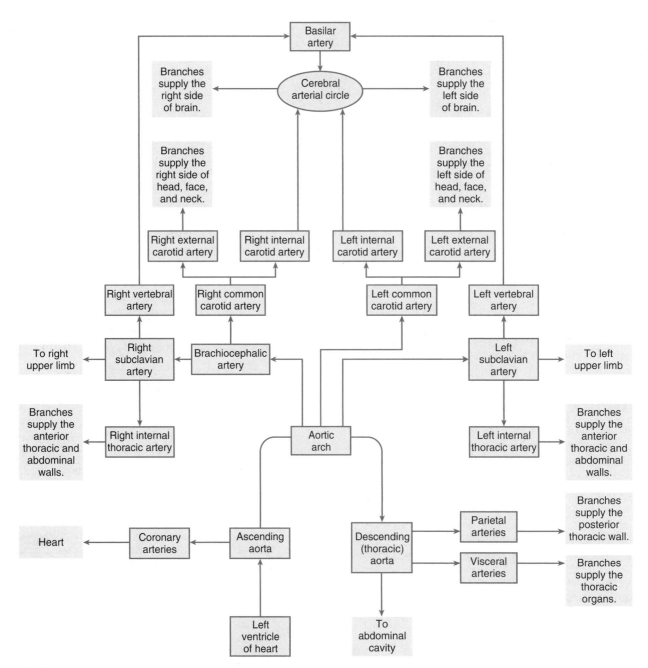

Figure 18.8 Major Arteries of the Head and Thorax

The **descending aorta** is the longest part of the aorta. It extends through the thorax and abdomen to the upper margin of the pelvis. The **thoracic** (thō-ras′ik) **aorta** is the portion of the descending aorta located in the thorax. The **abdominal aorta** is the part of the descending aorta inferior to the diaphragm. ▼

12. *Name the parts of the aorta. How does the aorta end?*
13. *Name the arteries that branch from the aorta to supply the heart.*

Arteries to the Head and the Neck

Figure 18.8 is a flow chart illustrating the relationship between the aorta and arteries supplying the head and thorax. The first vessel to branch from the aortic arch is the **brachiocephalic** (brā′kē-ō-se-

fal′ik, arm and head) **artery** (see figure 18.7). It is a very short artery, and it branches at the level of the clavicle to form the **right common carotid** (ka-rot′id, to put to sleep) **artery,** which transports blood to the right side of the head and neck, and the **right subclavian** (sŭb-klā′vē-an, under the clavicle) **artery,** which transports blood to the right upper limb (figure 18.9).

There is no brachiocephalic artery on the left side of the body. The second and third branches of the aortic arch are the **left common carotid artery,** which transports blood to the left side of the head and neck, and the **left subclavian artery,** which transports blood to the left upper limb (see figure 18.7).

The common carotid arteries extend superiorly along each side of the neck to the angle of the mandible, where they branch into **external** and **internal carotid arteries** (see figure 18.9). The

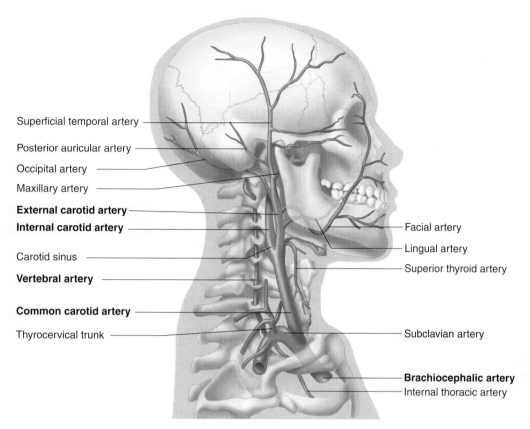

Superficial temporal artery
Posterior auricular artery
Occipital artery
Maxillary artery
External carotid artery
Internal carotid artery
Carotid sinus
Vertebral artery
Common carotid artery
Thyrocervical trunk

Facial artery
Lingual artery
Superior thyroid artery
Subclavian artery
Brachiocephalic artery
Internal thoracic artery

Figure 18.9 **Arteries of the Head and Neck**

The brachiocephalic artery, the right common carotid artery, and the right vertebral artery are the major arteries that supply the head and neck. The right common carotid artery branches from the brachiocephalic artery, and the vertebral artery branches from the subclavian artery.

external carotid arteries have several branches that supply the structures of the neck, face, nose, and mouth. The base of each internal carotid artery is slightly dilated to form a **carotid sinus,** which contains baroreceptors important in monitoring blood pressure (see chapter 17). The internal carotid arteries pass through the carotid canals and contribute to the **cerebral arterial circle** (circle of Willis) at the base of the brain (figure 18.10). The blood vessels that supply blood to most of the cerebrum branch from the cerebral arterial circle.

P R E D I C T

The term *carotid* means to put to sleep, implying that, if the carotid arteries are blocked for several seconds the patient can lose consciousness. Interruption of the blood supply for a few minutes can result in permanent brain damage. What is the physiological significance of atherosclerosis in the carotid arteries?

Some of the blood to the brain is supplied by the **vertebral** (ver′tĕ-brăl) **arteries,** which branch from the subclavian arteries (see figure 18.9) and pass to the head through the transverse foramina of the cervical vertebrae. The vertebral arteries then pass into the cranial cavity through the foramen magnum. Branches of the vertebral arteries supply blood to the spinal cord, as well as to the vertebrae, muscles, and ligaments in the neck.

Within the cranial cavity, the vertebral arteries unite to form a single, **basilar** (bas′i-lăr, relating to the base of the brain) **artery**

located along the anterior, inferior surface of the brainstem (see figure 18.10). The basilar artery gives off-branches that supply blood to the pons, cerebellum, and midbrain. It also forms right and left branches that contribute to the cerebral arterial circle. Most of the blood supply to the brain is through the internal carotid arteries; however, not enough blood is supplied to the brain to maintain life if either the carotid arteries or the vertebral arteries are blocked.

Stroke

A **stroke** is a sudden neurological disorder, often caused by a decreased blood supply to a part of the brain. It can occur as a result of a **thrombosis** (throm-bō′sis, a stationary clot), an **embolism** (em′bō-lizm, a floating clot that becomes lodged in smaller vessels), or a **hemorrhage** (hem′ŏ-rij, a rupture or leaking of blood from vessels). Any one of these conditions can result in a loss of blood supply or in trauma to a part of the brain. As a result, the tissue normally supplied by the arteries becomes **necrotic** (nĕ-krot′ik, dead). The affected area is called an **infarct** (in′farkt, to stuff into, an area of cell death). ▼

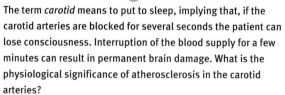

14. *Name the arteries that branch from the aorta to supply the head and neck.*

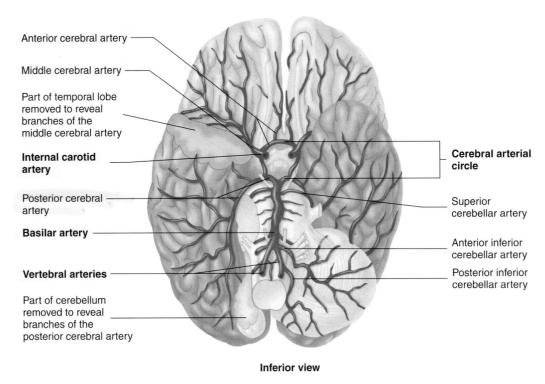

Anterior cerebral artery

Middle cerebral artery

Part of temporal lobe
removed to reveal
branches of the
middle cerebral artery

**Internal carotid
artery**

Posterior cerebral
artery

Basilar artery

Vertebral arteries

Part of cerebellum
removed to reveal
branches of the
posterior cerebral artery

**Cerebral arterial
circle**

Superior
cerebellar artery

Anterior inferior
cerebellar artery

Posterior inferior
cerebellar artery

Inferior view

Figure 18.10   Cerebral Arterial Circle

The blood supply to the brain is carried by the internal carotid and vertebral arteries. The vertebral arteries join to form the basilar artery. Branches of the internal carotid arteries and basilar artery supply blood to the brain and complete a circle of arteries around the pituitary gland and the base of the brain called the cerebral arterial circle (circle of Willis).

15. *Name the arteries that supply the cerebral arterial circle. Why is blood delivery to the cerebral arterial circle important?*

Arteries of the Upper Limb

The arteries of the upper limb are named differently as they pass into different body regions, even though no major branching occurs (figures 18.11 and 18.12). The subclavian artery becomes the **axillary** (ak′sil-ār-ē, refers to the axillary area) **artery** at the outer border of the first rib. The axillary artery passes to the axilla (armpit) and becomes the **brachial** (brā′kē-ăl, relating to the arm) **artery** at the superior border of the teres major muscle. The brachial artery lies medial to the proximal humerus, but it gradually curves laterally and crosses the anterior elbow. Blood pressure measurements are normally taken from the brachial artery at the elbow.

The brachial artery divides at the elbow into **radial** and **ulnar arteries,** which supply blood to the forearm and hand. The radial artery is the artery most commonly used for taking a pulse. The pulse can be detected conveniently on the thumb (radial) side of the anterior surface of the wrist.

The ulnar and radial arteries give rise to two arches within the palm of the hand. The **superficial palmar arch** is formed by the ulnar artery and is completed by connecting to the radial artery. The **deep palmar arch** is formed by the radial artery and is completed by connecting to the ulnar artery.

Digital (dij′i-tăl, relating to the digits—the fingers and the thumb) **arteries** branch from each of the two palmar arches and

unite to form single arteries on the medial and lateral sides of each digit. ▼

16. *Name the arteries that branch from the aorta to supply the upper limbs.*
17. *List, in order, the arteries that travel through the axilla and upper limb to the digits.*

Thoracic Aorta and Its Branches

The branches of the thoracic aorta can be divided into two groups: the **visceral** (vis′er-ăl, refers to internal organs) **arteries** supply the thoracic organs, and the **parietal arteries** (pă-rī′ě-tăl, wall) supply the thoracic wall (see figure 18.8 and figure 18.13*b*). The visceral arteries supply the pericardium, the esophagus, the trachea, and the bronchi and bronchioles of the lungs. Even though a large quantity of blood flows to the lungs through the pulmonary arteries, the bronchi and bronchioles require a separate oxygenated blood supply through small bronchial arteries from the thoracic aorta.

The major parietal arteries are the **posterior intercostal** (inter-kos′tăl, *inter-* + *costa*, between the ribs) **arteries,** which arise from the thoracic aorta and extend between the ribs. They supply intercostal muscles, the vertebrae, the spinal cord, and the deep muscles of the back. The **superior phrenic** (fren′ik, diaphragm) **arteries** supply the diaphragm.

The **internal thoracic arteries** are branches of the subclavian arteries (see figure 18.13*b*). They descend along the internal surface of the anterior thoracic wall and give rise to branches called the

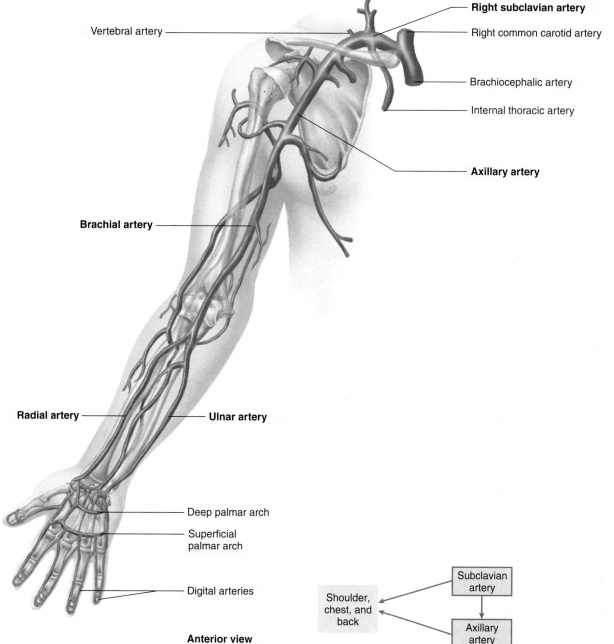

Vertebral artery

Right subclavian artery

Right common carotid artery

Brachiocephalic artery

Internal thoracic artery

Axillary artery

Brachial artery

Radial artery

Ulnar artery

Deep palmar arch

Superficial palmar arch

Digital arteries

Anterior view

Figure 18.11 Arteries of the Shoulder and Upper Limb

The arteries of the right upper limb and their branches: the right brachiocephalic, subclavian, axillary, radial, and ulnar arteries and their branches.

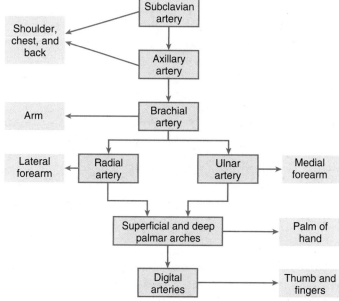

Subclavian artery

Shoulder, chest, and back

Axillary artery

Arm

Brachial artery

Lateral forearm

Radial artery

Ulnar artery

Medial forearm

Superficial and deep palmar arches

Palm of hand

Digital arteries

Thumb and fingers

Figure 18.12 Major Arteries of the Shoulder and Upper Limb

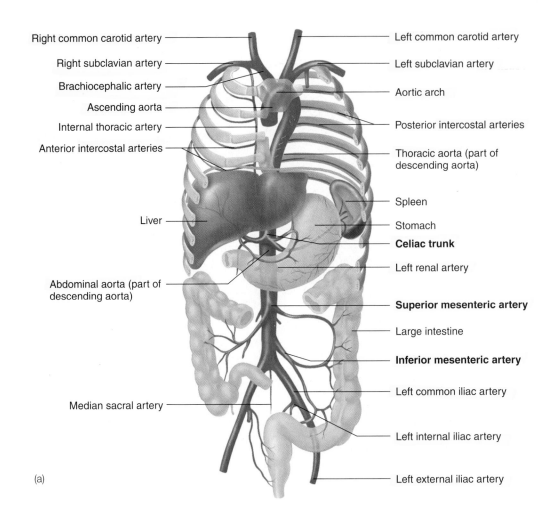

Right common carotid artery — Left common carotid artery
Right subclavian artery — Left subclavian artery
Brachiocephalic artery — Aortic arch
Ascending aorta —
Internal thoracic artery — Posterior intercostal arteries
Anterior intercostal arteries — Thoracic aorta (part of descending aorta)
Spleen
Liver — Stomach
Celiac trunk
Left renal artery
Abdominal aorta (part of descending aorta) — **Superior mesenteric artery**
Large intestine
Inferior mesenteric artery
Left common iliac artery
Median sacral artery — Left internal iliac artery
Left external iliac artery
(a)

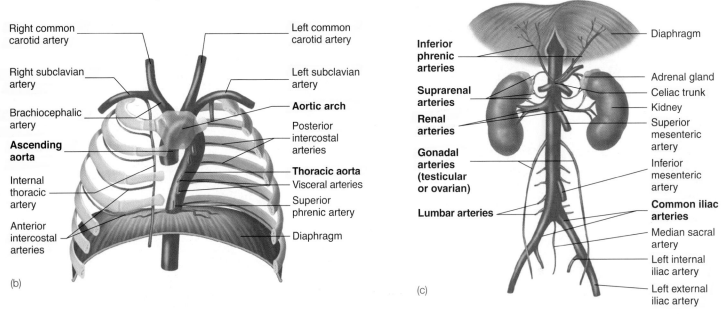

Right common carotid artery — Left common carotid artery
Right subclavian artery — Left subclavian artery
Brachiocephalic artery — **Aortic arch**
Ascending aorta — Posterior intercostal arteries
Internal thoracic artery — **Thoracic aorta**
Visceral arteries
Anterior intercostal arteries — Superior phrenic artery
Diaphragm
(b)

Inferior phrenic arteries — Diaphragm
Suprarenal arteries — Adrenal gland
Celiac trunk
Kidney
Renal arteries — Superior mesenteric artery
Gonadal arteries (testicular or ovarian) — Inferior mesenteric artery
Common iliac arteries
Lumbar arteries — Median sacral artery
Left internal iliac artery
Left external iliac artery
(c)

Figure 18.13   Branches of the Aorta

(a) Branches of the aortic arch supply the head and upper limbs. The thoracic aorta carries blood to the thorax and the abdominal aorta carries blood to the abdomen. Unpaired arteries supplying abdominal organs are bolded. (b) Branches of the thoracic aorta supply the thoracic organs and wall. (c) Branches of the abdominal aorta supply the abdominopelvic organs and wall. Pair arteries are bolded.

anterior intercostal arteries, which extend between the ribs to supply the anterior chest wall. The anterior and posterior intercostal arteries connect with each other approximately midway between the ends of the ribs. ▼

18. *Name the two types of branches arising from the thoracic aorta. What structures are supplied by each group?*

19. *Name two routes by which blood is delivered to the lungs.*
20. *Name two routes by which blood is delivered to the thoracic wall.*

Abdominal Aorta and Its Branches

Figure 18.14 is a flow chart illustrating the relationship between the aorta and the arteries supplying the abdomen and pelvis. The

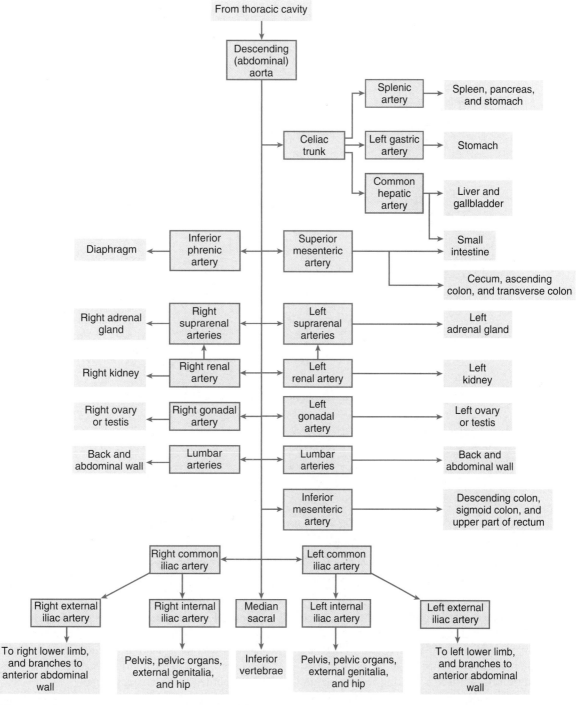

Figure 18.14 Major Arteries of the Abdomen and Pelvis

Visceral arteries include those that are unpaired (celiac trunk, superior mesenteric, inferior mesenteric) and those that are paired (renal, suprarenal, testicular, ovarian). Parietal arteries include inferior phrenic, lumbar, and median sacral.

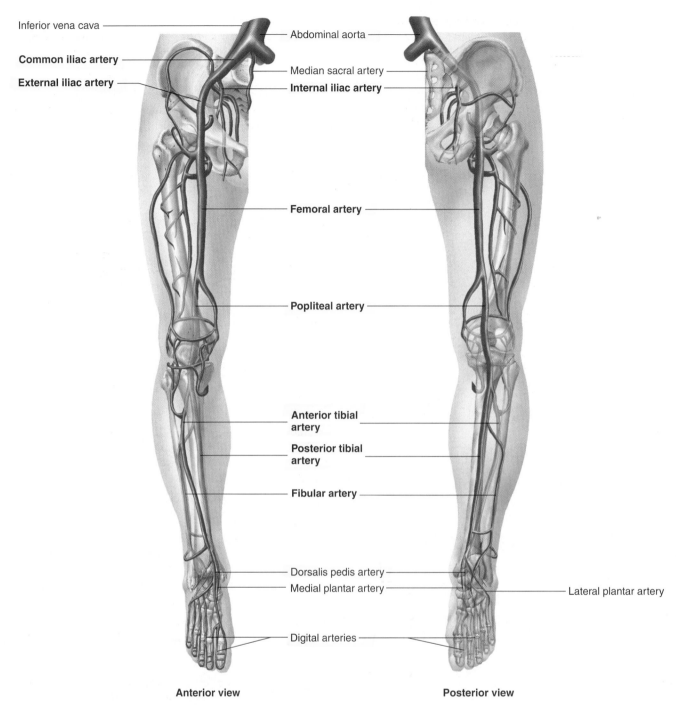

Figure 18.15 Arteries of the Pelvis and Lower Limb

The internal and external iliac arteries and their branches. The internal iliac artery supplies the pelvis and hip, and the external iliac artery supplies the lower limb through the femoral artery.

branches of the abdominal aorta, like those of the thoracic aorta, can be divided into visceral and parietal arteries. The visceral arteries are divided into paired and unpaired arteries. There are three major unpaired branches: the **celiac** (sē′lē-ak, belly) **trunk,** the **superior mesenteric** (mez-en-ter′ik, relating to membranes attached to the intestine) **artery,** and the **inferior mesenteric artery** (figure 18.13*a*). The celiac trunk gives rise to three branches: The **splenic** (splen′ik) **artery** supplies the spleen, pancreas, and stomach; the **left gastric artery** supplies the stomach; and the

common hepatic artery supplies the liver, gallbladder, and small intestine (see figure 18.14). The superior mesenteric artery supplies blood to the small intestine and the proximal portion of the large intestine, and the inferior mesenteric artery supplies blood to the remainder of the large intestine.

There are three paired visceral branches of the abdominal aorta. The **renal** (rē′nal, kidney) **arteries** supply the kidneys, and the **suprarenal** (sū′pră-rē′năl, superior to the kidney) **arteries** supply the adrenal glands. The **testicular arteries** supply the testes in males, and the **ovarian arteries** supply the ovaries in females (figure 18.13*c*).

The parietal branches of the abdominal aorta supply the diaphragm and abdominal wall. The **inferior phrenic arteries** supply the diaphragm, the **lumbar** (between the ribs and pelvis) **arteries** supply the lumbar vertebrae and back muscles, and the **median sacral** (area of sacrum) **artery** supplies the inferior vertebrae (see figure 18.13*c*). ▼

> 21. *Name the three unpaired visceral branches of the abdominal aorta.*
> 22. *Name the three branches of the celiac trunk and list the organs they supply.*
> 23. *What organs are supplied by the superior and inferior mesenteric arteries?*
> 24. *Name the three paired visceral branches of the abdominal aorta and list the parts of the body they supply.*
> 25. *Name the parietal branches of the abdominal aorta and list the parts of the body they supply.*

Arteries of the Pelvis

The abdominal aorta divides at the level of the fifth lumbar vertebra into two **common iliac** (il′ē-ak, relating to the flank area) **arteries** (see figures 18.13 and 18.14). Each common iliac artery divides to form an **external iliac artery,** which enters a lower limb, and an **internal iliac artery,** which supplies the pelvic area. Visceral branches of the internal iliac artery supply organs such as the

urinary bladder, rectum, uterus, and vagina. Parietal branches supply blood to the walls and floor of the pelvis; the lumbar, gluteal, and proximal thigh muscles; and the external genitalia. ▼

> 26. *Name the arteries that branch from the common iliac arteries to supply the lower limbs and pelvic area.*
> 27. *List the organs of the pelvis that are supplied by visceral and parietal branches.*

Arteries of the Lower Limb

Like the arteries of the upper limb, the arteries of the lower limb are named differently as they pass into different body regions, even though there are no major branches (figures 18.15 and 18.16). The external iliac artery passes under the inguinal ligament and becomes the **femoral** (fem′ŏ-răl, relating to the thigh) **artery** in the thigh. The femoral artery extends inferiorly along the anterior medial thigh and passes posteriorly through an opening in the adductor magnus muscle to become the **popliteal** (pop-lit′ē-ăl, pop-li-tē′ăl, ham, the hamstring area posterior to the knee) **artery.** The popliteal artery traverses the back of the knee and divides just inferior to the knee, forming the **anterior tibial artery** and the **posterior tibial artery.** The anterior tibial artery passes anteriorly through the leg and descends to the ankle, becoming the **dorsalis pedis artery** on the dorsum of the foot. The posterior tibial artery gives off the **fibular,** or **peroneal, artery,** which supplies the lateral leg and foot. The posterior tibial artery descends to the ankle and gives rise to **medial** and **lateral plantar** (plan′tăr, the sole of

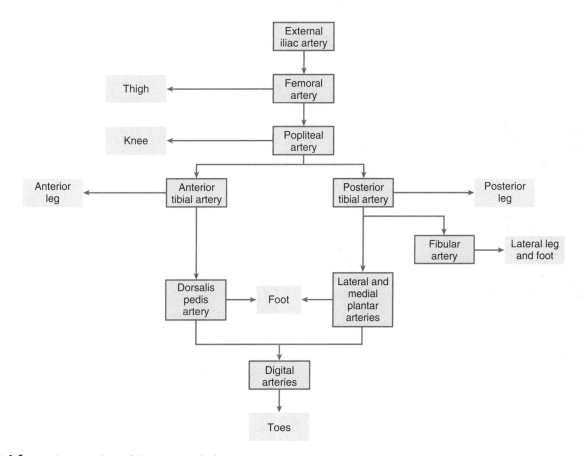

Figure 18.16 Major Arteries of the Lower Limb

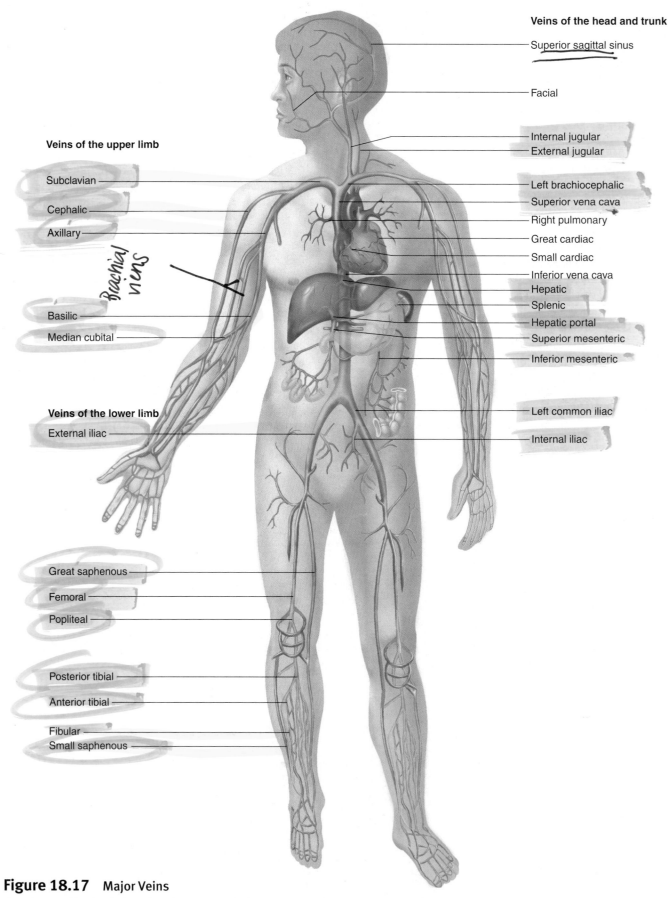

Veins of the head and trunk

Superior sagittal sinus

Facial

Internal jugular

External jugular

Veins of the upper limb

Subclavian

Cephalic

Axillary

Left brachiocephalic

Superior vena cava

Right pulmonary

Great cardiac

Small cardiac

Inferior vena cava

Hepatic

Splenic

Brachial veins

Basilic

Median cubital

Hepatic portal

Superior mesenteric

Inferior mesenteric

Veins of the lower limb

External iliac

Left common iliac

Internal iliac

Great saphenous

Femoral

Popliteal

Posterior tibial

Anterior tibial

Fibular

Small saphenous

Figure 18.17 Major Veins

The veins carry blood to the heart from the tissues of the body.   **Anterior view**

the foot) **arteries,** which supply the sole of the foot. The dorsalis pedis artery and the medial and lateral plantar arteries supply the **digital arteries** to the toes.

The Femoral Triangle

The femoral triangle is found in the superior and medial area of the thigh. Its margins are formed by the inguinal ligament, the medial margin of the sartorius muscle, and the lateral margin of the adductor longus muscle. The femoral artery, vein, and nerve pass though the femoral triangle. A pulse in the femoral artery can be detected in the area of the femoral triangle, and it is an area that is susceptible to serious traumatic injuries that can result in hemorrhage and nerve damage. In addition, pressure can be applied to this area to help prevent bleeding from wounds in more inferior areas of the lower limb. The femoral triangle is also an important access point for certain medical procedures. ▼

> 28. List, in order, the arteries that travel from the aorta to the digits of the lower limbs.

Systemic Circulation: Veins

The **superior vena cava** (vē′nă kā′vă, venous cave) returns blood from the head, neck, thorax, and upper limbs to the right atrium of the heart, and the **inferior vena cava** returns blood from the abdomen, pelvis, and lower limbs to the right atrium (figure 18.17). The **coronary sinus** and **cardiac veins** return blood from the walls of the heart to the right atrium (see chapter 17).

In a very general way, the smaller veins follow the same course as the arteries and many are given the same names. The veins, however, are more numerous and more variable. The larger veins often follow a very different course and have names different from the arteries.

The three major types of veins are the superficial veins, deep veins, and sinuses. The superficial veins of the limbs are, in general, larger than the deep veins, whereas in the head and trunk the opposite is the case. Venous sinuses occur primarily in the cranial cavity and the heart. ▼

> 29. Name the veins that return blood to the heart from each of the major body areas.

Veins of the Head and Neck

Dural venous sinuses are spaces within the dura mater surrounding the brain (see chapter 11). Blood from the brain and cerebrospinal fluid from the subarachnoid space drain into the dural venous sinuses. The **internal jugular** (jŭg′ū-lar, neck) **veins** are formed primarily as the continuation of the dural venous sinuses (figure 18.18). The internal jugular veins also drain the anterior head, face, and neck. The internal jugular veins join the **subclavian veins** on each side of the body to form the **brachiocephalic veins.** The

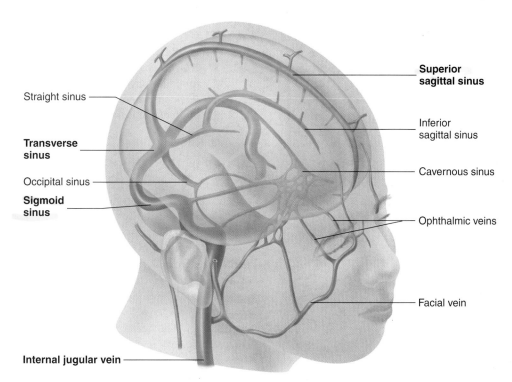

Figure 18.18 **Venous Sinuses Associated with the Brain**

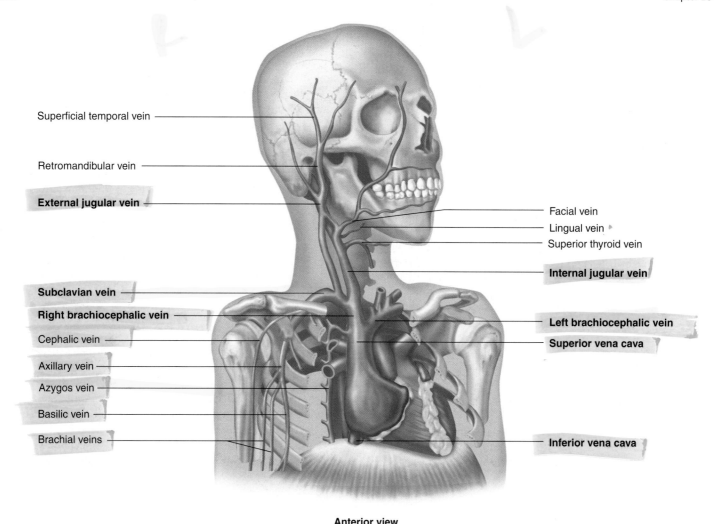

Superficial temporal vein

Retromandibular vein

External jugular vein

Facial vein

Lingual vein

Superior thyroid vein

Internal jugular vein

Subclavian vein

Right brachiocephalic vein

Cephalic vein

Axillary vein

Azygos vein

Basilic vein

Brachial veins

Left brachiocephalic vein

Superior vena cava

Inferior vena cava

Anterior view

Figure 18.19 Veins of the Head and Neck

The right brachiocephalic vein and its tributaries. The major veins draining the head and neck are internal and external jugular veins.

brachiocephalic veins join to form the superior vena cava (figures 18.19 and 18.20).

The **external jugular veins** are smaller and more superficial than the internal jugular veins, and they drain blood primarily from the posterior head and neck. The external jugular veins drain into the subclavian veins.

Facial Pimples and Meningitis

Because venous communication exists between the facial veins and venous sinuses through the ophthalmic veins, infections can be introduced into the cranial cavity through this route. A superficial infection of the face on either side of the nose can enter the facial vein. The infection can then pass through the ophthalmic veins to the venous sinuses and result in meningitis. For this reason, people are warned not to aggravate pimples or boils on the face on either side of the nose. ▼

30. *What are dural venous sinuses? To what large veins do the dural venous sinuses connect?*

31. *Which veins join to form the brachiocephalic veins? To form the superior vena cava?*

32. *What parts of the head and neck are drained by the internal and external jugular veins?*

Veins of the Upper Limb

The veins of the upper limbs (figures 18.21 and 18.22) can be divided into superficial and deep groups. The two major superficial veins are the **basilic** (ba-sil′ik, toward the base of the arm) and **cephalic** (se-fal′ik, toward the head) **veins,** which are responsible for draining most of the blood from the upper limbs. The basilic vein arises from superficial veins of the medial, posterior hand and curves around the forearm to the anterior elbow. From there, it extends medially along the arm and passes deep to the axilla. The basilic vein becomes the **axillary vein** at the inferior border of the teres minor. The axillary vein becomes the subclavian vein at the outer border of the first rib.

The cephalic vein arises from superficial veins of the lateral posterior hand and curves around the forearm to the anterior elbow. From there, it extends laterally along the arm to the

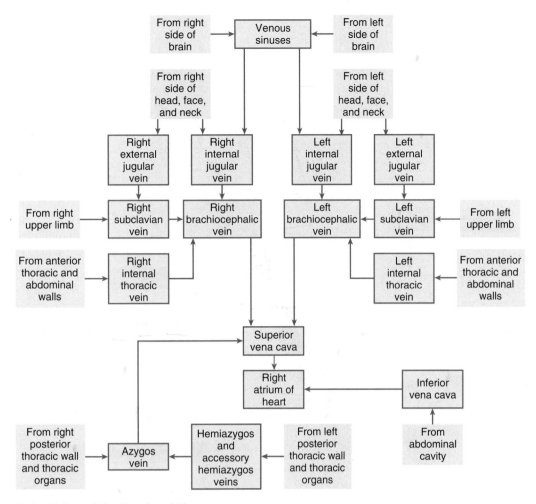

Figure 18.20 Major Veins of the Head and Thorax

shoulder and passes deep to join the axillary vein (see figures 18.21 and 18.22).

The **median cubital** (kū′bi-tăl, pertaining to the elbow) **vein** is a variable vein that usually connects the cephalic vein with the basilic vein at the anterior elbow (see figures 18.21 and 18.22). In many people, this vein is quite prominent and is used as a site for drawing blood. There is, however, considerable variation in the veins at the elbow. The median cubital can be replaced by other tributaries of the cephalic and basilic veins.

The deep veins, which drain the deep structures of the upper limbs, follow the same course as the arteries and are named for the arteries they accompany. **Digital veins** drain into **deep** and **superficial palmar arches,** which give rise to **radial** and **ulnar veins.** They usually are paired, with one small vein lying on each side of their corresponding artery. Each member of a pair has numerous connections with the other member and with the superficial veins. The radial and ulnar veins empty into the **brachial veins,** which accompany the brachial artery and empty into the axillary vein (see figures 18.21 and 18.22). ▼

33. *Describe two routes by which blood returns through superficial veins from the hand to the subclavian vein.*

34. *Where is the median cubital vein located, and how is it used clinically?*

35. *Describe two routes by which blood returns through deep veins from the hand to the subclavian vein.*

Veins of the Thorax

Three major veins return blood from the thorax to the superior vena cava: the right and left brachiocephalic veins and the **azygos** (az-ī′gos, az′i-gos, unpaired) **vein** (figure 18.23). Blood from the posterior thoracic wall is collected by **posterior intercostal veins,** which drain into the azygos vein on the right and the **hemiazygos vein** or **accessory hemiazygos vein** on the left. The hemiazygos and accessory hemiazygos veins empty into the azygos vein, which drains into the superior vena cava (see figures 18.20 and 18.23). Blood drains from the anterior thoracic wall by way of the **anterior intercostal veins.** These veins empty into the **internal thoracic veins,** which empty into the brachiocephalic veins (see figure 18.20). ▼

36. *List the three major veins that return blood from the thorax to the superior vena cava.*

37. *By what routes does blood return to the superior vena cava from the posterior and anterior thoracic walls?*

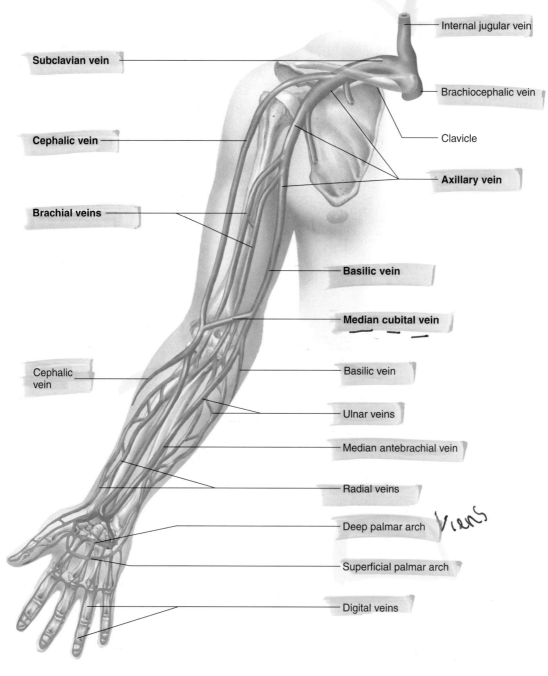

Internal jugular vein

Subclavian vein

Brachiocephalic vein

Clavicle

Cephalic vein

Axillary vein

Brachial veins

Basilic vein

Median cubital vein

Cephalic vein

Basilic vein

Ulnar veins

Median antebrachial vein

Radial veins

Deep palmar arch

Superficial palmar arch

Digital veins

Anterior view

Figure 18.21 Veins of the Shoulder and Upper Limb

The subclavian vein and its tributaries. The major veins draining the superficial structures of the limb are the cephalic and basilic veins. The brachial veins drain the deep structures.

Veins of the Abdomen and Pelvis

Blood from the posterior abdominal wall drains into the **ascending lumbar veins.** These veins are continuous superiorly with the hemiazygos on the left and the azygos on the right (see figure 18.23). Blood from the rest of the abdomen, pelvis, and lower limbs returns to the heart through the inferior vena cava (figure 18.24). The gonads (testes and ovaries), kidneys, and adrenal glands are the only abdominal organs outside the pelvis that drain

directly into the inferior vena cava. The **internal iliac veins** drain the pelvis and join the **external iliac veins** from the lower limbs to form the **common iliac veins.** The common iliac veins combine to form the inferior vena cava (figure 18.25).

Hepatic Portal System

A **portal** (pōr′tăl, *porta,* gate) **system** is a vascular system that begins and ends with capillary beds and has no pumping mechanism,

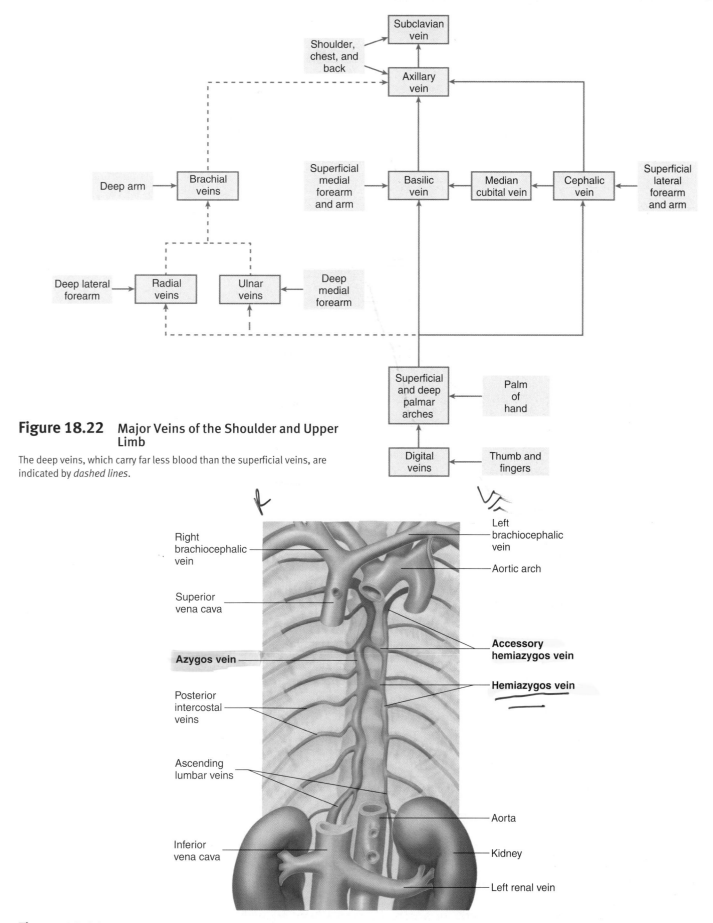

Figure 18.22 Major Veins of the Shoulder and Upper Limb

The deep veins, which carry far less blood than the superficial veins, are indicated by *dashed lines*.

Figure 18.23 Veins of the Thorax

Anterior view of the azygos, hemiazygos, and accessory hemiazygos veins and their tributaries.

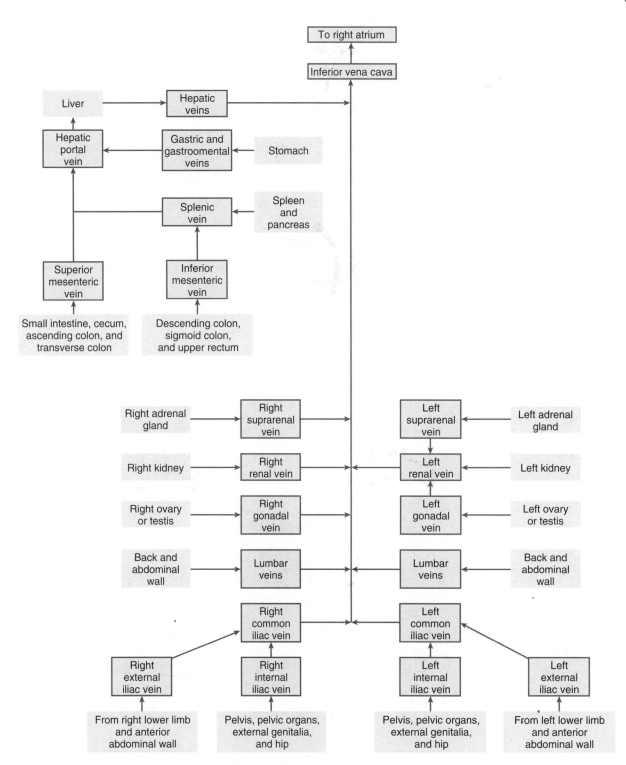

Figure 18.24   Major Veins of the Abdomen and Pelvis

such as the heart, between them. The **hepatic** (he-pat′ik, relating to the liver) **portal system** (figure 18.26) carries blood through veins from capillaries within most of the abdominal viscera, such as the stomach, intestines, and spleen, to capillaries in the liver. Nutrients absorbed from the stomach or intestines are delivered to the liver by

the hepatic portal system. The liver stores or modifies the nutrients so they can be used by other cells of the body. The liver also converts toxic substances absorbed from the stomach or intestines into nontoxic substances. The **hepatic portal vein,** the largest vein of the system, is formed by the union of the **superior mesenteric vein,**

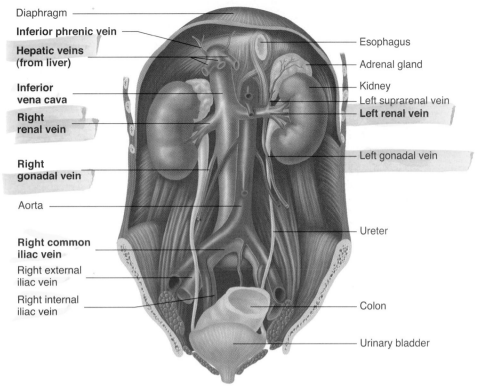

Figure 18.25   Inferior Vena Cava and Its Tributaries

The hepatic veins transport blood to the inferior vena cava from the hepatic portal system, which ends as a series of blood sinusoids in the liver (see figure 18.26).

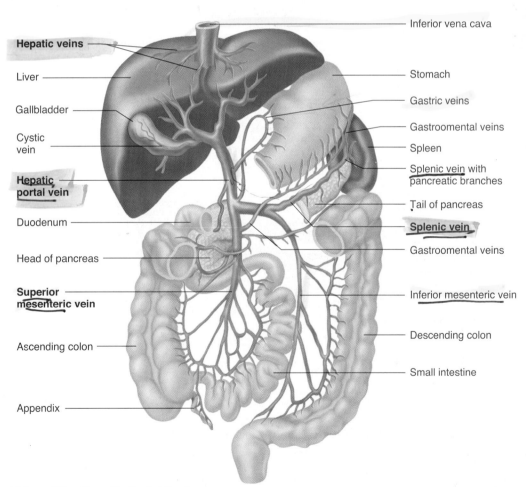

Figure 18.26   Veins of the Hepatic Portal System

The hepatic portal system begins as capillary beds in the stomach, pancreas, spleen, small intestine, and large intestine. The veins of the hepatic portal system converge on the hepatic portal vein, which carries blood to a series of capillaries (sinusoids) in the liver. Hepatic veins carry blood from capillaries in the liver to the inferior vena cava (see figure 18.25).

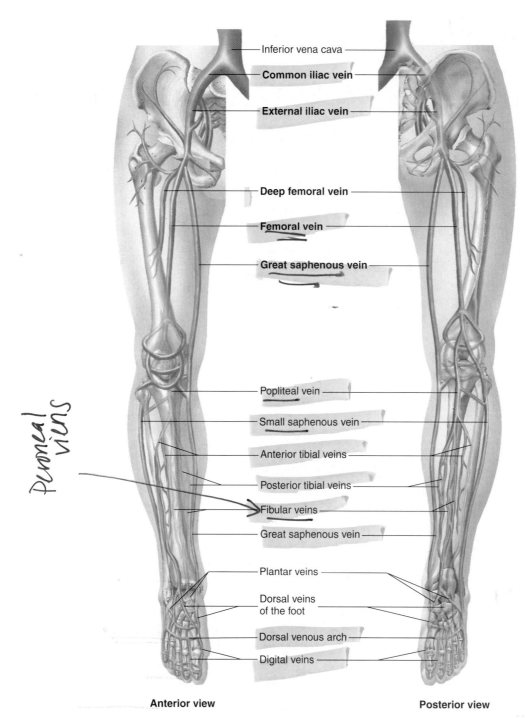

Figure 18.27 Veins of the Pelvis and Lower Limb

The right common iliac vein and its tributaries.

which drains the small intestine, and the **splenic vein,** which drains the spleen. The splenic vein receives the **inferior mesenteric vein,** which drains part of the large intestine, and the **pancreatic veins,** which drain the pancreas. The hepatic portal vein also receives veins from the stomach (see figures 18.24 and 18.26).

The hepatic portal vein enters the liver and divides many times to form capillaries supplying liver cells. Blood from the liver

flows into **hepatic veins,** which join the inferior vena cava (see figure 18.26). ▼

38. *Explain the three ways that blood from the abdomen returns to the heart.*
39. *List the vessels that carry blood from the abdominal organs to the hepatic portal vein.*

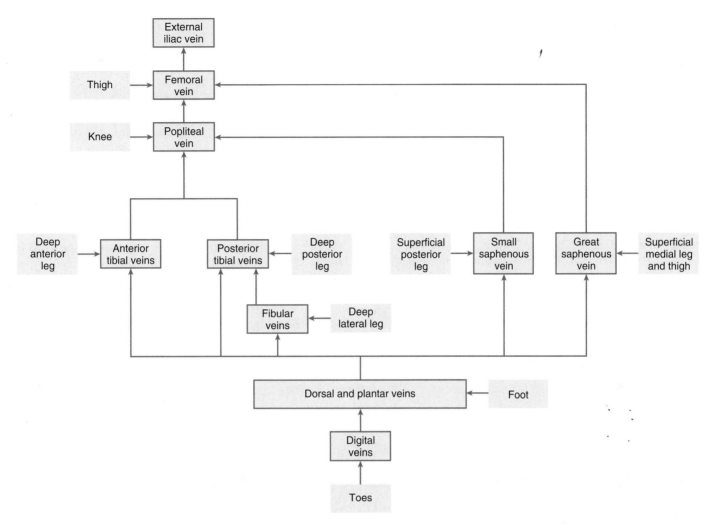

Figure 18.28 Major Veins of the Lower Limb

Veins of the Lower Limb

The veins of the lower limb, like those of the upper limb, consist of deep and superficial groups (figures 18.27 and 18.28). The digital veins drain the toes, and the dorsal and plantar veins drain the foot. The **anterior** and **posterior tibial veins** are paired and accompany the anterior and posterior tibial arteries. They unite just inferior to the knee to form the single **popliteal vein,** which ascends through the thigh and becomes the **femoral vein.** The femoral vein becomes the external iliac vein. **Fibular,** or **peroneal** (per-ō-nē′ăl), **veins** also are paired in each leg and accompany the fibular arteries. They empty into the posterior tibial veins just before those veins contribute to the popliteal vein.

The superficial veins consist of the great and small saphenous veins. The **great saphenous** (să-fē′nŭs, visible) **vein,** the longest vein of the body, originates over the dorsal and medial side of the foot and ascends along the medial side of the leg and thigh to empty into the femoral vein. The **small saphenous vein** begins over the lateral side of the foot and ascends along the posterior leg to the back of the knee, where it empties into the popliteal vein.

Blood Vessels Used for Coronary Bypass Surgery

The great saphenous vein often is removed surgically and used in coronary bypass surgery. Portions of the saphenous vein are grafted to create a route of blood flow that bypasses blocked portions of the coronary arteries. The circulation interrupted by the removal of the saphenous vein flows through other veins of the lower limb. The internal thoracic artery is also used for coronary bypasses. The distal end of the artery is freed and attached to a coronary artery at a point that bypasses the blocked portion of the coronary artery. This technique appears to be better because the internal thoracic artery does not become blocked as fast as the saphenous vein. ▼

40. *Describe three routes by which blood returns through deep veins from the foot to the external iliac vein.*
41. *Describe two routes by which blood returns through superficial veins from the foot to the external iliac vein.*

Physiology of Circulation

The function of the circulatory system is to maintain adequate blood flow to all tissues. An adequate blood flow maintains homeostasis by providing nutrients and oxygen to tissues and removing the waste products of metabolism from the tissues. Blood flows through blood vessels primarily as a result of the pressure produced by contractions of the heart's ventricles.

Blood Pressure

Blood pressure is a measure of the force blood exerts against the blood vessel walls. In arteries, blood pressure values exhibit a cycle dependent on the rhythmic contractions of the heart. When the ventricles contract, blood is forced into the arteries, and the pressure reaches a maximum, called the **systolic pressure.** When the ventricles relax, blood pressure in the arteries falls to a minimum value, called the **diastolic pressure.** The standard unit for measuring blood pressure is millimeters of mercury (mm Hg). If the blood pressure is 100 mm Hg, the pressure is great enough to lift a column of mercury 100 mm.

The **auscultatory** (aws-kŭl′tă-tō-rē, to listen) **method** of determining blood pressure is used under most clinical conditions (figure 18.29). A blood pressure cuff connected to a **sphygmomanometer** (sfig′mō-mă-nom′ĕ-ter, *sphygmic*, relating to the pulse + *manometer*, instrument for measuring pressure) is placed around the patient's arm, and a **stethoscope** (steth′ō-skōp, *stēthos*, chest + *skopeō*, to view) is placed over the brachial artery. The blood pressure cuff is then inflated until the brachial artery is completely blocked. Because no blood flows through the constricted area, no sounds can be heard through the stethoscope at this point. The pressure in the cuff is then gradually lowered. As soon as the pressure in the cuff declines below the systolic pressure, blood flows through the constricted area each time the left ventricle contracts. The blood flow is turbulent immediately downstream from the constricted area. This turbulence produces vibrations in the blood and surrounding tissues that can be heard through the stethoscope. These sounds are called **Korotkoff** (Kō-rot′kof, Nikolai Korotkoff, Russian physician [1874–1920]) **sounds,** and the pressure at which the first Korotkoff sound is heard is the systolic pressure.

As the pressure in the blood pressure cuff is lowered still more, the Korotkoff sounds change tone and loudness. When the pressure has dropped until the brachial artery is no longer constricted and blood flow is no longer turbulent, the sound

1. No sound is heard because there is no blood flow when the cuff pressure is high enough to keep the brachial artery closed.

2. **Systolic pressure** is the pressure at which a Korotkoff sound is first heard. When cuff pressure decreases and is no longer able to keep the brachial artery closed during systole, blood is pushed through the partially opened brachial artery to produce turbulent blood flow and a sound. The brachial artery remains closed during diastole.

3. As cuff pressure continues to decrease, the brachial artery opens even more during systole. At first, the artery is closed during diastole, but, as cuff pressure continues to decrease, the brachial artery partially opens during diastole. Turbulent blood flow during systole produces Korotkoff sounds, although the pitch of the sounds changes as the artery becomes more open.

4. **Diastolic pressure** is the pressure at which the sound disappears. Eventually, cuff pressure decreases below the pressure in the brachial artery and it remains open during systole and diastole. Nonturbulent flow is reestablished and no sounds are heard.

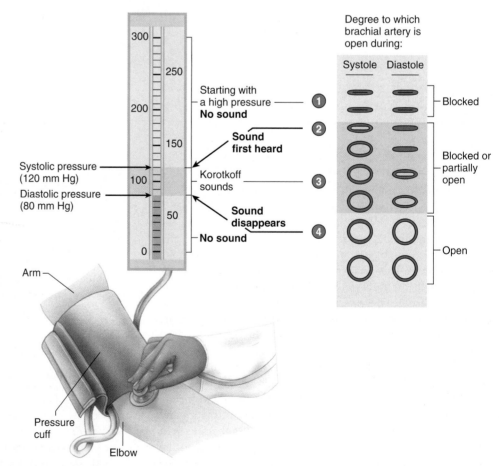

Process Figure 18.29   Blood Pressure Measurement

disappears completely. The pressure at which the Korotkoff sounds disappear is the diastolic pressure. The brachial artery remains open during systole and diastole, and continuous, nonturbulent blood flow is reestablished.

The systolic pressure is the maximum pressure produced in the large arteries. It is also a good measure of the maximum pressure within the left ventricle. The diastolic pressure is close to the lowest pressure within the large arteries. During relaxation of the left ventricle, the aortic semilunar valve closes, trapping the blood that was ejected during ventricular contraction in the aorta. The pressure in the ventricles falls to 0 mm Hg during ventricular relaxation. The blood trapped in the elastic arteries is compressed by the recoil of the elastic arteries, however, and the pressure falls more slowly, reaching the diastolic pressure (see figure 17.16). ▼

> **42. Define blood pressure. Describe how systolic and diastolic blood pressure can be measured.**

Blood Flow Through a Blood Vessel

Blood flow through a blood vessel is the volume of blood that passes through the vessel per unit of time. The blood flow in a vessel can be described by the following equation:

$$\text{Flow} = \frac{P_1 - P_2}{R}$$

where P_1 and P_2 are the pressures in the vessel at points one and two, respectively, and R is the resistance to flow. Blood always flows from an area of higher pressure to an area of lower pressure and, the greater the pressure difference, the greater the rate of flow. For example, the average blood pressure in the aorta (P_1) is greater than the blood pressure in the relaxed right atrium (P_2). Therefore, blood flows from the aorta to tissues and from tissues to the right atrium. If the heart should stop contracting, the pressure in the aorta would become equal to that in the right atrium and blood would no longer flow.

The flow of blood, resulting from a pressure difference between the two ends of a blood vessel, is opposed by a resistance to flow. As the resistance increases, blood flow decreases; as the resistance decreases, blood flow increases. Factors that affect resistance can be represented as follows:

$$\text{Resistance} = \frac{128vl}{\pi d^4}$$

where v is the viscosity of blood, l is the length of the vessel, and d is the diameter of the vessel. The **diameter** of a round vessel is the distance from one side of the vessel through the center of the vessel to the opposite side. Both 128 and π are constants and, for practical purposes, the length of the blood vessel is constant. Thus, the diameter of the blood vessel and the viscosity of the blood determine resistance.

When the equation for resistance is combined with the equation for flow, the following relationship, called **Poiseuille's** (pwah-zuh'yes) **law** results:

$$\text{Flow} = \frac{P_1 - P_2}{R} = \frac{\pi(P_1 - P_2)d^4}{128vl}$$

According to Poiseuille's law, a small change in the diameter of a vessel dramatically changes the resistance to flow, and therefore the amount of blood that flows through the vessel, because the diameter is raised to the fourth power. Vasoconstriction decreases the diameter of a vessel, increases the resistance to flow, and decreases the blood flow through the vessel. For example, decreasing the diameter of a vessel by half increases the resistance to flow 16-fold and decreases flow 16-fold. Vasodilation increases the diameter of a vessel, decreases resistance to flow, and increases blood flow through the vessel.

Viscosity (vis-kos'i-tē) is a measure of the resistance of a liquid to flow. A common means for reporting the viscosity of liquids is to consider the viscosity of distilled water as 1 and to compare the viscosity of other liquids with it. Using this procedure, whole blood normally has a viscosity of 3.0–4.5. As the viscosity of a liquid increases, the pressure required to force it to flow increases. It takes 3.0–4.5 times as much pressure to move whole blood through a tube at the same rate as water.

The viscosity of blood is influenced largely by **hematocrit** (hē'mă-tō-krit, hem'ă-tō-krit), which is the percentage of the total blood volume composed of red blood cells (see chapter 16). Increasing the number of red blood cells or decreasing plasma volume increases hematocrit; decreasing the number of red blood cells or increasing plasma volume decreases viscosity. As the hematocrit changes, the viscosity of blood changes logarithmically. Blood with a hematocrit of 45% has a viscosity about three times that of water, whereas blood with a very high hematocrit of 65% has a viscosity about seven to eight times that of water. Viscosity above its normal range of values increases the workload on the heart; if this workload is great enough, heart failure can result. ▼

> **43. Define blood flow and resistance.**
> **44. Describe the relationship among blood flow, blood pressure, and resistance.**
> **45. According to Poiseuille's law, what effect do viscosity, blood vessel diameter, and blood vessel length have on resistance? On blood flow?**
> **46. Define viscosity, and state the effect of hematocrit on viscosity.**

P R E D I C T ②
Use Poiseuille's law to explain the effect of the following on blood flow: (a) vasoconstriction of blood vessels in the skin in response to cold exposure, (b) vasodilation of the blood vessels in the skin in response to an elevated body temperature, (c) erythrocytosis, which results in a greatly increased hematocrit, and (d) dehydration.

Blood Flow Through the Body

The equation for blood flow can be used to describe blood flow through the body. The volume of blood flowing through the body per minute is **cardiac output (CO),** which is the volume of blood pumped per minute by the left ventricle (see chapter 17). Thus, flow (F) in the equation for blood flow is CO. Contrac-tions of the left ventricle maintain a **mean arterial pressure (MAP)** of 100 mm Hg in the aorta. Blood flows from the aorta to the relaxed right atrium, which has a pressure near 0 mm Hg. Thus, $P_1 - P_2$ is MAP − 0, or MAP. **Peripheral resistance (PR)** is the sum of all the resistances to blood flow in all of the blood vessels in the body. Thus, R in the flow equation becomes PR.

$$\text{Flow} = \frac{P_1 - P_2}{R} = CO = \frac{\text{MAP}}{\text{PR}}$$

Rearranging the terms of the equation,

$$\text{MAP} = CO \times PR$$

Thus, maintaining adequate blood pressure, which is necessary for blood delivery to tissues, depends on CO and PR. Peripheral resistance is maintained and regulated through the sympathetic division of the autonomic nervous system (ANS). Continual sympathetic stimulation of the smooth muscle in the walls of blood vessels keeps them in a state of partial constriction called **vasomotor tone.** Changing the amount of sympathetic stimulation changes vasomotor tone and peripheral resistance. Increased sympathetic stimulation of blood vessel smooth muscle causes vasoconstriction by decreasing blood vessel diameter. Increased vasoconstriction increases vasomotor tone and the resistance to blood flow, which increases peripheral resistance. Conversely, decreased sympathetic stimulation results in vasodilation, decreased vasomotor tone, and decreased peripheral resistance.

P R E D I C T   ③
What effect does vasoconstriction and vasodilation have on MAP? Explain.

The aortic pressure fluctuates between a systolic pressure of 120 mm Hg and a diastolic pressure of 80 mm Hg (figure 18.30).

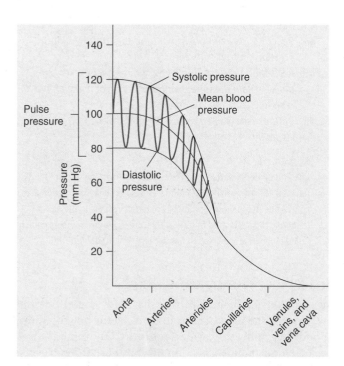

Figure 18.30   **Blood Pressure in the Major Blood Vessel Types**

Blood pressure fluctuations between systole and diastole are damped in small arteries and arterioles. There are no large fluctuations in blood pressure in capillaries and veins. The largest drop in blood pressure occurs in the arterioles and capillaries.

MAP is slightly less than the average of systolic and diastolic pressures because diastole lasts longer than systole. MAP is approximately 70 mm Hg at birth, is slightly less than 100 mm Hg from adolescence to middle age, and reaches 110 mm Hg in healthy older persons, but it can be as high as 130 mm Hg.

Blood pressure falls progressively as blood flows from arteries through the capillaries and veins to about 0 mm Hg by the time blood is returned to the right atrium (see figure 18.30). The greater the resistance to blood flow in a blood vessel, the more rapidly the pressure decreases as blood flows through it. The most rapid decline in blood pressure occurs in the arterioles and then in capillaries because their small diameters increase the resistance to blood flow. In addition, the pressure is damped, in that the difference between the systolic and diastolic pressures is decreased in the small-diameter vessels. By the time blood reaches the capillaries, there is no variation in blood pressure, and only a steady pressure of about 30 mm Hg remains. ▼

47. *Describe the relationship among blood pressure, cardiac output, and peripheral resistance.*
48. *Define vasomotor tone. How does vasoconstriction and vasodilation affect peripheral resistance?*
49. *Describe the changes in resistance and blood pressure as blood flows through the aorta to the superior and inferior venae cavae.*

Pulse Pressure and Vascular Compliance

Pulse pressure is the difference between the systolic and diastolic pressures. If a person has a systolic pressure of 120 mm Hg and a diastolic pressure of 80 mm Hg, the pulse pressure is 40 mm Hg. Two major factors influence pulse pressure: stroke volume and vascular compliance. When the stroke volume increases, the systolic pressure increases more than the diastolic pressure, causing the pulse pressure to increase.

Vascular compliance (kom-plī′ans) is the tendency for blood vessel volume to increase as the blood pressure increases. The more easily the vessel wall stretches the greater is its compliance, whereas the less easily the vessel wall stretches the smaller is its compliance. By analogy, it is easier to blow up a thin-walled (more compliant) balloon than a thick-walled (less compliant) balloon. As vascular compliance decreases, pulse pressure increases. Arteriosclerosis in older people results in less elastic arteries, which decreases compliance. The decreased compliance causes the pressure in the aorta to rise more rapidly and to a greater degree during systole. Thus, for a given stroke volume, systolic pressure and pulse pressure are higher. Arteriosclerosis increases the amount of work performed by the heart because the left ventricle must produce a greater pressure to eject the same amount of blood into a less elastic artery. In severe cases, the increased workload on the heart leads to heart failure.

Ejection of blood from the left ventricle into the aorta produces a **pulse,** or pressure wave, which travels rapidly along the arteries. The rate of transmission of the pulse wave is much more rapid than the flow of blood. A pulse wave takes approximately 0.1 sec to travel from the aorta to the radial artery, whereas a drop of blood takes approximately 8 sec to make the same journey.

A pulse can be felt at locations where large arteries are close to the surface of the body (figure 18.31). It is helpful to know the

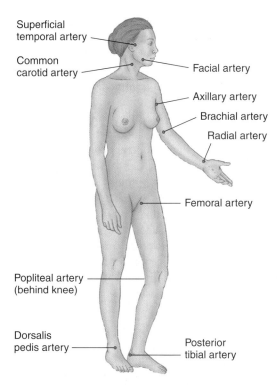

Superficial
temporal artery

Common
carotid artery

Facial artery

Axillary artery

Brachial artery

Radial artery

Femoral artery

Popliteal artery
(behind knee)

Dorsalis
pedis artery

Posterior
tibial artery

Figure 18.31 Major Points at Which the Pulse Can Be Monitored

Each pulse point is named after the artery on which it occurs.

major locations where the pulse can be detected because monitoring the pulse is important clinically. Heart rate, heart rhythm, and other characteristics can be determined by feeling the pulse. For example, a weak pulse usually indicates a decreased stroke volume or increased constriction of the arteries. ▼

> 50. *Define pulse pressure and vascular compliance. How do stroke volume and vascular compliance affect pulse pressure?*
> 51. *What is a pulse?*

P R E D I C T

A weak pulse occurs in response to premature beats of the heart and during cardiovascular shock due to hemorrhage. A stronger than normal pulse occurs in a person who has received too much saline intravenously and in a healthy exercising person. Explain the causes for the changes in the pulse under these conditions.

Blood Pressure and the Effect of Gravity

Hydrostatic pressure is the pressure, or weight, produced by a column of fluid due to the effects of gravity. When a person is standing, the blood pressure in a blood vessel in the foot results from the pressure generated by contraction of the heart plus the hydrostatic pressure produced by the "column" of blood above the foot. The hydrostatic pressure at the foot adds approximately 90 mm Hg to the pressures recorded in figure 18.30. Blood pressure is approximately 0 mm Hg in the relaxed right atrium, and it averages approximately 100 mm Hg in the aorta. In the foot, pressure in a vein is approximately 90 mm Hg and in an artery is approximately 190 mm Hg.

When a person changes position from lying down to standing, the blood pressure in the veins of the lower limbs increases. The compliance of veins is approximately 24 times greater than the compliance of arteries because of the structure of their walls. The increased blood pressure causes the distensible (compliant) veins to expand, but has little effect on the arteries. Venous return decreases because less blood is returning to the heart as the veins are filling with blood. As venous return decreases, cardiac output and blood pressure decrease. If negative feedback mechanisms do not compensate and cause blood pressure to increase, there is inadequate delivery of blood to the brain. Homeostasis of the brain is disrupted and dizziness or fainting can occur. ▼

> 52. *Define hydrostatic pressure. How does it affect blood pressure in the feet of a standing person?*
> 53. *Do veins or arteries have greater compliance?*
> 54. *What effect does standing have on venous return and cardiac output?*

P R E D I C T

If a person faints upon standing, explain why falling to a horizontal position restores homeostasis.

Capillary Exchange and Regulation of Interstitial Fluid Volume

There are approximately 10 billion capillaries in the body. The heart and blood vessels all maintain blood flow through those capillaries and support **capillary exchange,** which is the movement of substances between capillaries and the interstitial fluids of tissues. Capillary exchange is the process by which cells receive everything they need to survive and to eliminate metabolic waste products. If blood flow through capillaries is not adequate to maintain capillary exchange, cells cannot survive.

Substances leave the blood and enter the interstitial fluid of tissues by passing through or between the endothelial cells of capillaries.

1. *Passage through endothelial cells.* Substances can pass through endothelial cells by crossing their plasma membranes. Chapter 3 describes four ways that substances pass through plasma membranes: diffusion, osmosis, mediated transport, and vesicular transport. For example, in vesicular transport, endothelial cells take in substances from the blood by pinocytosis. The vesicles formed cross the cell and release their contents into the interstitial fluid by exocytosis. Some endothelial cells are penetrated by large pores, called **fenestrae** (fe-nes′trē, windows), through which substances can pass through the cells.
2. *Passage between endothelial cells.* In a typical capillary, there are intercellular spaces approximately 6–7 nm wide between cells through which substances can pass.

Lipid-soluble molecules cross capillary walls by diffusing through the plasma membranes of the endothelial cells of the capillaries. Examples include oxygen, carbon dioxide, steroid hormones, and fatty acids.

The permeability of capillaries to water-soluble substances, such as glucose and amino acids, varies tremendously. In a typical capillary the space between endothelial cells allows the passage of most substances, except for proteins. In the kidneys, there are many fenestrae, which also allow the passage of most substances, except for proteins. The numerous fenestrae allow large amounts of substances to move rapidly from the blood into kidney tubules, where urine is produced. In the liver, the spaces between the endothelial cells are large enough to allow proteins to pass through them. The capillaries in the brain forming the blood–brain barrier (see chapter 11) have tight junctions between cells (see chapter 3) and few molecules pass between them. In these capillaries, mediated transport processes move water-soluble substances across the capillary walls.

Diffusion and filtration are the primary means by which most substances cross capillary walls. Nutrients, oxygen, and hormones diffuse from a higher concentration in capillaries to a lower concentration in the interstitial fluid. Waste products, including carbon dioxide, diffuse from a higher concentration in the interstitial fluid to a lower concentration in the capillaries.

Osmosis is the diffusion of water across a selectively permeable membrane. The capillary wall acts as a selectively permeable membrane that prevents proteins from moving from the capillary into the interstitial fluid but allows water to move across the wall of the capillary. There is a higher concentration of proteins in the blood than in the interstitial fluid. Consequently, water diffuses from the less concentrated interstitial fluid, which has fewer proteins but more water molecules, into the more concentrated blood, which has more proteins but fewer water molecules (see chapter 3). Materials dissolved or suspended in the water that can pass through the capillary wall accompany the water into the blood.

Filtration is the movement of fluid through a partition containing small holes. The fluid movement results from the pressure or weight of the fluid pushing against the partition, and the fluid moves from the side of the partition with the greater pressure to the side with the lower pressure. The fluid and substances small enough to pass through the holes move through the partition, but substances larger than the holes do not pass through it. For example, in a car, oil but not dirt particles passes through an oil filter. In capillaries, blood pressure moves fluid through the spaces between endothelial cells or through fenestrae.

Blood pressure forces fluid out of capillaries and osmosis moves fluid into them. The balance between these two forces determines whether or not fluid leaves or enters capillaries. At the arterial end of the capillary, the movement of fluid out of the capillary caused by blood pressure is greater than the movement of fluid into the capillary as a result of osmosis. Consequently, there is a net movement of fluid out of the capillary (figure 18.32).

At the venous end of the capillary, blood pressure is lower than at the arterial end because of the resistance to blood flow through the capillary. Consequently, the movement of fluid into the capillary caused by osmosis is greater than the movement of fluid out of the capillary resulting from blood pressure, and there is a net movement of fluid into the capillary (see figure 18.32).

Approximately nine-tenths of the fluid that leaves the capillary at the arterial end reenters the capillary at its venous end. The remaining one-tenth of the fluid enters the lymphatic capillaries and is eventually returned to the blood (see chapter 19).

1. At the arterial end of the capillary, the movement of fluid out of the capillary due to blood pressure (*green arrow*) is greater than the movement of fluid into the capillary due to osmosis (*orange arrow*).

2. At the venous end of the capillary, the movement of fluid into the capillary due to osmosis (*orange arrow*) is greater than the movement of fluid out of the capillary due to blood pressure (*green arrow*).

3. Approximately nine-tenths of the fluid (*blue arrow*) that leaves the capillary at its arterial end reenters the capillary at its venous end. About one-tenth of the fluid passes into the lymphatic capillaries.

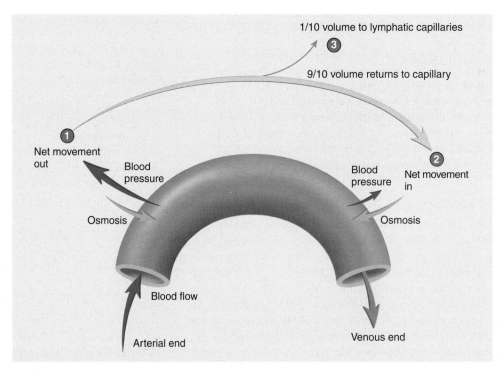

Process Figure 18.32   Fluid Exchange Across the Walls of Capillaries

Edema

Edema, or swelling, of tissues results from increased interstitial fluid caused by a disruption of the normal inwardly and outwardly directed pressures across the capillary walls. For example, inflammation results in an increase in the permeability of capillaries. Proteins, mainly albumen, leak out of the capillaries into the interstitial fluid. As the concentration difference for proteins decreases between the blood and the interstitial fluid, less water moves by osmosis into capillaries. Consequently, more fluid passes out the arterial end of capillaries and less fluid passes into the venous ends of capillaries. The lymphatic capillaries cannot carry all of the fluid away, and edema results.

Standing still for long periods can cause edema. Standing increases the hydrostatic pressure in the lower limbs, especially the feet. Consequently, blood pressure at the arterial end of capillaries increases and more fluid leaves the blood and enters the tissues. Blood pressure also increases at the venous end of the capillaries, which decreases the net movement of fluid from the tissues into the blood (see figure 18.32). Edema results when the fluid buildup exceeds the ability of the lymphatic capillaries to remove fluid. Up to 15%–20% of the total blood volume can pass through the walls of the capillaries into the interstitial fluids of the lower limbs during 15 minutes of standing still. Movement of the limbs helps prevent edema because the compression of veins and lymphatic vessels by skeletal muscles helps move fluid out of tissues. ▼

55. *Define capillary exchange. What are the two basic ways in which capillary exchange takes place? What are fenestrae?*
56. *How do lipid-soluble and water-soluble substances pass through capillaries? Explain the basis for the wide variability of capillary permeability to water-soluble substances.*
57. *Define osmosis and filtration. Explain how they account for the movement of fluid out of and into capillaries.*
58. *How much of the fluid leaving the arterial ends of capillaries reenters the blood at the venous ends of capillaries? What happens to the excess fluid entering tissues?*

P R E D I C T   6

Explain how a loss of protein molecules in urine through the kidneys and protein starvation result in edema. Explain why people who are suffering from edema in the legs are told to keep them elevated.

Control of Blood Flow

Blood flow is highly controlled and matched closely to the metabolic needs of tissues. Mechanisms that control blood flow to and through tissues are classified as (1) local control and (2) nervous and hormonal control.

Local Control

Blood flow at the tissue level is regulated by arterioles and precapillary sphincters. The arterioles control the amount of blood reaching the capillary beds, and the precapillary sphincters control the flow of blood through capillaries (see figure 18.4). The arterioles and the precapillary sphincters are regulated by local control. The arterioles are also regulated by nervous and hormonal control (see next section).

Local control is the response of vascular smooth muscle to changes in tissue gases, nutrients, and waste products levels; it does not involve the nervous system or hormones. Arterioles vasodilate and precapillary sphincters relax when oxygen levels decrease or, to a lesser degree, when glucose, amino acids, fatty acids, and other nutrients decrease. An increase in carbon dioxide and lactic acid or a decrease in pH also causes arterioles to vasodilate and precapillary sphincters to relax. For example, during exercise, the metabolic needs of skeletal muscle increase dramatically, and the by-products of metabolism are produced at a more rapid rate. Local control increases blood flow to match the metabolic needs of the tissue. Conversely, when metabolic needs are low, the vasoconstriction of arterioles and contraction of precapillary sphincters reduce blood flow.

P R E D I C T   7

A student has been sitting for a short time with her legs crossed. After getting up to walk out of class, she notices a red blotch on the back of one of her legs. On the basis of what you know about the local control of blood flow, explain why this happens.

The long-term regulation of blood flow through tissues is matched closely to the metabolic requirements of the tissue. If the metabolic activity of a tissue increases and remains elevated for an extended period, the diameter and the number of capillaries in the tissue increase, and local blood flow increases. The increased density of capillaries in the well-trained skeletal muscles of athletes, compared with that in poorly trained skeletal muscles, is an example. ▼

59. *What is the role of arterioles and precapillary sphincters in matching blood flow through capillaries to the metabolic needs of tissues?*
60. *How is long-term regulation of blood flow through tissues accomplished?*

Nervous and Hormonal Control

Blood flow through arterioles, arteries, and veins is regulated by nervous and hormonal mechanisms. An area of the lower pons and upper medulla oblongata, called the **vasomotor center,** continually transmits a low frequency of action potentials through sympathetic fibers to the smooth muscle of blood vessels, except for the precapillary sphincters (figure 18.33). As a consequence, the peripheral blood vessels, especially the arterioles, are continually in a partially constricted state, a condition called vasomotor tone. Increased sympathetic stimulation of a blood vessel causes vasoconstriction, and decreased stimulation causes vasodilation.

Areas throughout the pons, midbrain, and diencephalon can stimulate or inhibit the vasomotor center. For example, the hypothalamus can exert either strong excitatory or inhibitory effects on the vasomotor center. Increased body temperature detected by temperature receptors in the hypothalamus causes vasodilation of blood vessels in the skin (see chapter 5). The cerebral cortex also can either excite or inhibit the vasomotor center. For example, action potentials that originate in the cerebral cortex during periods of emotional excitement activate hypothalamic centers, which in turn increase vasomotor tone.

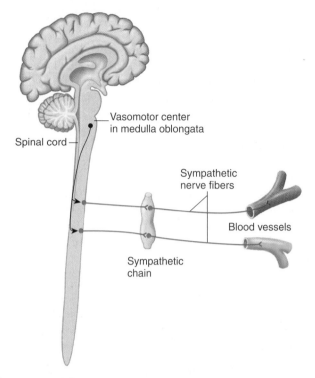

Figure 18.33 Nervous Regulation of Blood Vessels

Most blood vessels are innervated by sympathetic nerve fibers. The vasomotor center within the medulla oblongata plays a major role in regulating the frequency of action potentials in nerve fibers that innervate blood vessels.

The neurotransmitter for the sympathetic fibers is norepinephrine, which binds to α-adrenergic receptors on blood vessel smooth muscle cells to cause vasoconstriction. Sympathetic action potentials also cause the release of epinephrine and norepinephrine into the blood from the adrenal medulla. These hormones are transported in the blood to all parts of the body. In most blood vessels, they bind to α-adrenergic receptors and cause vasoconstriction. In skeletal and cardiac muscle, epinephrine causes vasodilation. There are large numbers of β-adrenergic receptors in addition to α-adrenergic receptors in skeletal and cardiac muscle. Epinephrine binding to β-adrenergic receptors promotes vasodilation. The overall effect is vasodilation because the effect of activating the large number of β-adrenergic receptors outweighs the effect of activating the α-adrenergic receptors. ▼

> 61. *In what two ways is blood flow through arteries and veins regulated?*
> 62. *Where is the vasomotor center located, and how does it control blood flow?*
> 63. *What effects do norepinephrine and epinephrine have on vasoconstriction?*

Arteries

Although precapillary sphincters control the flow of blood through capillaries, they can control only the blood that is delivered to the tissues. The muscular arteries and arterioles control blood delivery by vasoconstricting and vasodilating. The muscular arteries control the flow of blood to large areas of the body, such as the gastrointes-

tinal tract or the limbs. For example, following a meal, muscular arteries supplying the gastrointestinal tract vasodilate, increasing blood flow to the gastrointestinal tract. Arterioles control the flow of blood to tissues by supplying the capillaries. ▼

> 64. *How do the muscular arteries and arterioles control the flow of blood? What parts of the body do they affect?*

P R E D I C T

Raynaud syndrome is a condition in which arterioles, primarily in the fingers, undergo exaggerated vasoconstriction in response to exposure to cold or emotions. Would the fingers appear red or white? Would the precapillary sphincters be relaxed or contracted?

Veins

The compliance of veins is approximately 24 times greater than the compliance of arteries because of the structure of their walls. Even though the pressure in veins is less than in arteries, the diameter of veins is usually larger than that of corresponding arteries because they are so compliant. Consequently, veins act as storage areas, or reservoirs, for blood. The small to large veins hold five times as much blood as the comparable arteries.

Cardiac output depends on venous return, the amount of blood entering the heart. According to Starling's law, as venous return increases, the heart contracts more forcefully, causing stroke volume and cardiac output to increase (see chapter 17). The factors that affect flow in the veins are, therefore, of great importance to the overall function of the cardiovascular system. Venous return is affected by the following.

1. *Vasoconstriction and vasodilation of veins.* The veins are a reservoir for blood. The sympathetic division can control the amount of blood flowing from the reservoir to the heart. Vasoconstriction of veins pushes more blood toward the heart, increasing venous return and cardiac output. Vasodilation of veins decreases venous return and cardiac output because less blood flows to the heart as the dilated veins are filling with blood.
2. *Blood volume.* As blood volume increases, venous return and cardiac output increases. Conversely, decreased blood volume results in decreased venous return and cardiac output.
3. *Valves and the skeletal muscle pump.* Blood pressure in the venous system is low and blood tends to flow backwards in the veins when a person is standing. Valves in the veins prevent the backflow of blood. The **skeletal muscle pump** compresses the veins when skeletal muscles contract. Compression of the veins pushes blood toward the heart because the valves prevent the backflow of blood. The action of the skeletal muscle pump is important during exercise. ▼

> 65. *Do veins or arteries have greater compliance?*
> 66. *How does venous return affect cardiac output?*
> 67. *State three ways in which venous return is increased.*

P R E D I C T

What effect does a rapid loss of blood have on MAP? Explain. Propose a mechanism by which the veins could oppose this change.

Hypertension, or high blood pressure, affects approximately 20% of the human population at sometime in their lives. Generally, a person is considered hypertensive if the systolic blood pressure is greater than 140 mm Hg and the diastolic pressure is greater than 90 mm Hg. Current methods of evaluation, however, take into consideration diastolic and systolic blood pressures in determining whether a person is suffering from hypertension. In addition, normal blood pressure is age-dependent, so classification of an individual as hypertensive depends on the person's age.

Chronic hypertension has an adverse effect on the function of both the heart and the blood vessels. Hypertension requires the heart to work harder than normal. This extra work leads to hypertrophy of the cardiac muscle, especially in the left ventricle, and can lead to heart failure. Hypertension also increases the rate at which arteriosclerosis develops. Arteriosclerosis, in turn, increases the probability that blood clots, or thromboses (throm'bō-sēz), will form and that blood vessels will rupture. Common medical problems associated with hypertension are cerebral hemorrhage, coronary infarction, hemorrhage of renal blood vessels, and poor vision caused by burst blood vessels in the retina.

Some conditions leading to hypertension are a decrease in functional kidney mass, excess aldosterone or angiotensin production, and increased resistance to blood flow in the renal arteries. All of these conditions cause an increase in total blood volume, which causes cardiac output to increase. Increased cardiac output forces blood to flow through tissue capillaries at a higher than needed rate. In response, the precapillary sphincters constrict, which increases peripheral resistance. The increased cardiac output and peripheral resistance increase blood pressure.

Although these conditions result in hypertension, roughly 90% of the diagnosed cases of hypertension are **idiopathic,** or **essential, hypertension,** which means the cause of the condition is unknown. Drugs that dilate blood vessels (called vasodilators), drugs that increase the rate of urine production (called diuretics), and drugs that decrease cardiac output normally are used to treat essential hypertension. The vasodilator drugs increase the rate of blood flow through the kidneys and thus increase urine production, and the diuretics also increase urine production. Increased urine production reduces blood volume, which reduces blood pressure. Substances that decrease cardiac output, such as β-adrenergic–blocking agents, decrease the heart rate and force of contraction. In addition to these treatments, low-salt diets normally are recommended to reduce the amount of sodium chloride and water absorbed from the intestine into the bloodstream.

Regulation of Mean Arterial Pressure

Blood flow to all areas of the body depends on the maintenance of an adequate pressure in the arteries. As long as arterial blood pressure is adequate, local control of blood flow through tissues is appropriately matched to their metabolic needs. If blood pressure is too low, the metabolic needs of tissues are not met. If blood pressure is too high, blood vessels and the heart can be damaged.

Blood flow through the circulatory system is determined by the cardiac output (*CO*), which is equal to the heart rate (*HR*) times the stroke volume (*SV*), and peripheral resistance (*PR*), which is the resistance to blood flow in all the blood vessels (see p. 545):

$$\text{MAP} = CO \times PR \quad \text{or} \quad \text{MAP} = HR \times SV \times PR$$

This equation expresses the effect of heart rate, stroke volume, and peripheral resistance on blood pressure. An increase in any one of them results in an increase in blood pressure. Conversely, a decrease in any one of them produces a decrease in blood pressure. The mechanisms that control blood pressure do so by changing peripheral resistance, heart rate, or stroke volume. Regulatory mechanisms that control blood volume also affect blood pressure because stroke volume depends on the amount of blood entering the heart. For example, an increase in blood volume increases venous return, which increases stroke volume.

Two major types of control systems operate to achieve these responses: (1) those that respond in the short term and (2) those that respond in the long term. The regulatory mechanisms that control pressure on a short-term basis respond quickly but begin to lose their capacity to regulate blood pressure a few hours to a few days after blood pressure is maintained at higher or lower values. This occurs because sensory receptors adapt to the altered pressures. The long-term regulation of blood pressure is controlled primarily by mechanisms that influence kidney function, and those mechanisms do not adapt rapidly to altered blood pressures. ▼

> 68. *How does heart rate, stroke volume, and peripheral resistance affect mean arterial pressure?*
>
> 69. *Contrast the short-term and long-term regulation of blood pressure.*

Short-Term Regulation of Blood Pressure

The short-term, rapidly acting mechanisms controlling blood pressure are the baroreceptor reflexes, the adrenal medullary mechanism, and chemoreceptor reflexes. Some of these reflex mechanisms operate on a minute-to-minute basis and help regulate blood pressure within a narrow range of values. Some of them respond primarily to emergency situations.

Baroreceptor Reflexes

Baroreceptor reflexes help regulate blood pressure by modifying heart rate and stroke volume (see chapter 17). In addition, they regulate blood pressure by changing peripheral resistance. **Baroreceptors** are sensory receptors that respond to stretch in arteries caused by increased blood pressure. They are scattered along the walls of most of the large arteries of the neck and thorax, and there are many in the carotid sinus at the base of the internal carotid artery and in the walls of the aortic arch (figure 18.34). Action potentials are transmitted from the baroreceptors to the medulla oblongata along sensory nerve fibers in cranial nerves.

A sudden increase in blood pressure stretches the artery walls and increases action potential frequency in the baroreceptors. The increased action potential frequency delivered to the vasomotor and cardioregulatory centers in the medulla oblongata causes responses that lower the blood pressure. One major response is vasodilation, resulting in decreased vasomotor tone and peripheral resistance. Other responses, controlled by the cardioregulatory center, are an increase in the parasympathetic stimulation of the heart, which decreases the heart rate, and a decrease in sympathetic stimulation of the heart, which reduces the stroke volume. The decreased heart rate, stroke volume, and peripheral resistance lower the blood pressure toward its normal value.

A sudden decrease in blood pressure results in a decreased action potential frequency in the baroreceptors. The decreased action potential frequency delivered to the vasomotor and cardioregulatory centers in the medulla oblongata produces responses that raise blood pressure. Vasoconstriction increases vasomotor tone and peripheral resistance. Increased sympathetic stimulation of the heart increases the heart rate and stroke volume. The increased peripheral resistance, heart rate, and stroke volume raise the blood pressure toward its normal value.

The baroreceptor reflexes regulate blood pressure on a moment-to-moment basis. When a person rises rapidly from a sitting or lying position to a standing position, hydrostatic pressure in the lower limbs increases, resulting in decreased venous return, cardiac output, and blood pressure (see "Blood Pressure and the Effect of Gravity," p. 547). This reduction in blood pressure can be so great that blood flow to the brain is reduced enough to cause dizziness or even loss of consciousness. The falling blood pressure activates the baroreceptor reflexes, which reestablish normal blood pressure within a few seconds. In a healthy person, there may be no awareness that blood pressure has dropped momentarily. ▼

70. **Define baroreceptor. Describe the response of the baroreceptor reflex when blood pressure increases and decreases.**

P R E D I C T 10

When a person does a headstand, what happens to blood pressure in the head? Explain. What happens to heart rate when doing a headstand? Explain.

1. Baroreceptors in the carotid sinus and aortic arch monitor blood pressure.

2. Action potentials are conducted by the glossopharyngeal and vagus nerves to the cardioregulatory and vasomotor centers in the medulla oblongata.

3. Increased parasympathetic stimulation of the heart decreases the heart rate.

4. Increased sympathetic stimulation of the heart increases the heart rate and stroke volume.

5. Increased sympathetic stimulation of blood vessels increases vasoconstriction.

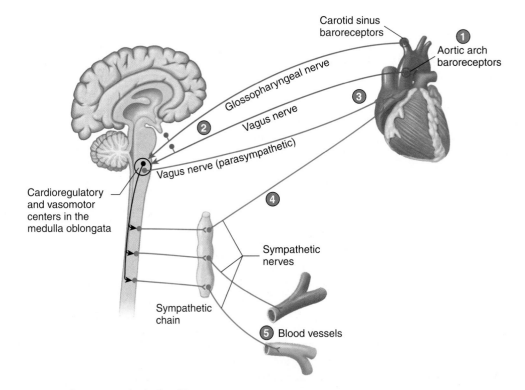

Process Figure 18.34 **Baroreceptor Reflex Control of Blood Pressure**

An increase in blood pressure increases parasympathetic stimulation of the heart and decreases sympathetic stimulation of the heart and blood vessels, resulting in a decrease in blood pressure. A decrease in blood pressure decreases parasympathetic stimulation of the heart and increases sympathetic stimulation of the heart and blood vessels, resulting in an increase in blood pressure.

Carotid Sinus Syndrome

Occasionally, the application of pressure to the carotid arteries in the upper neck results in a dramatic decrease in blood pressure. This condition, called the **carotid sinus syndrome,** is most common in patients in whom arteriosclerosis of the carotid artery is advanced. In such patients, a tight collar can apply enough pressure to the region of the carotid sinuses to stimulate the baroreceptors. The increased action potentials from the baroreceptors initiate reflexes that result in a decrease in vasomotor tone and an increase in parasympathetic action potentials to the heart. As a result of the decreased peripheral resistance and heart rate, blood pressure decreases dramatically. As a consequence, blood flow to the brain decreases to such a low level that the person becomes dizzy or may even faint. People suffering from this condition must avoid applying external pressure to the neck region. If the carotid sinus becomes too sensitive, a treatment for this condition is surgical destruction of the innervation to the carotid sinuses.

The baroreceptor reflexes do not change the average blood pressure in the long run. The baroreceptors adapt within 1–3 days to any new sustained blood pressure to which they are exposed. If the blood pressure is elevated for more than a few days, the baroreceptors adapt to the elevated pressure and the baroreceptor reflexes do not reduce the blood pressure to its original value. This adaptation is common in people who have hypertension.

CASE STUDY | A Venous Thrombosis

Harry is a 55-year-old college professor who teaches a night class in a small town about 50 miles from his home. One night, as he walked to his car after class, Harry noticed that his right leg was uncomfortable. By the time he reached home, about 90 minutes later, the calf of his right leg had become very swollen. When he extended his knee and plantar flexed his foot, the pain in his right leg increased. Harry thought this might be a serious condition, so he drove to the emergency room.

In the emergency room, a Doppler test, which monitors the flow of blood through blood vessels, was performed on Harry's right leg. The test confirmed that a thrombus had formed in one of the deep veins of his right leg. His pain and edema were consistent with the presence of a venous thrombosis.

Harry was admitted to the hospital and his physician prescribed intravenous (IV) heparin. About 4 A.M., Harry experienced an increase in his respiratory rate and his breathing became labored. He experienced pain in his chest and back, and there was a decrease in his arterial

oxygen levels. In response to these changes, Harry's physician increased the amount of heparin. The chest pain and changes in Harry's respiratory movements improved over the next 24 hours. The next day, a CT scan was performed and pulmonary emboli were identified, but infarctions of the lung were not apparent. The edema in Harry's leg also slowly improved.

Harry was told that he would have to remain in the hospital for several days. Heparin was continued for several days and then oral coumadin was prescribed. Frequent blood samples were taken to determine Harry's prothrombin time. After about a week, Harry was released from the hospital, but his physician prescribed oral coumadin for at least several months. Harry was required to have his prothrombin time (see chapter 16) checked periodically.

P R E D I C T

a. Explain why edema and pain developed in response to a thrombus in one of the deep veins of Harry's right leg.
b. If a thrombus in the posterior tibial vein gave rise to an embolus, name in order the parts of the circulatory system the embolus would pass through before lodging in a blood vessel in the lungs. Explain why the lungs are the most likely places the embolus would lodge.
c. Predict the effect of pulmonary emboli on the right ventricle's ability to pump blood.
d. Predict the effect of pulmonary emboli on blood oxygen levels, on the left ventricle's ability to pump blood, and on systemic blood pressure. What responses would be activated by this change in blood pressure?
e. Explain why Harry's physician prescribed heparin and coumadin, and explain why coumadin was prescribed long after the venous thrombosis and lung emboli were dissolved.

Adrenal Medullary Mechanism

Stimuli that result in increased sympathetic stimulation of the heart and blood vessels also cause increased stimulation of the adrenal medulla. The adrenal medulla responds by releasing epinephrine and norepinephrine into the blood (figure 18.35). The main effect of epinephrine and norepinephrine is to increase heart rate and stroke volume. Epinephrine causes vasodilation in skeletal muscle and cardiac muscle and vasoconstriction in the skin and kidneys. Norepinephrine increases vasoconstriction everywhere. The overall effect of epinephrine is to decrease peripheral resistance slightly, whereas norepinephrine increases peripheral resistance.

The adrenal medullary mechanism is short-term and rapid-acting. It responds within seconds to minutes and is usually active for minutes to hours. Other hormonal mechanisms are long-term

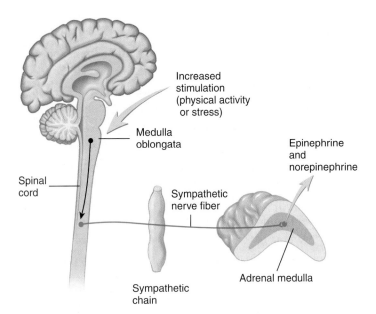

Figure 18.35 **Adrenal Medullary Mechanism**

Stimuli that increase sympathetic stimulation of the heart and blood vessels also result in increased sympathetic stimulation of the adrenal medulla and result in epinephrine and some norepinephrine secretion.

and slow-acting. They respond within minutes to hours and continue to function for many hours to days. ▼

> 71. *Describe the effects of epinephrine and norepinephrine on heart rate, stroke volume, and peripheral resistance.*

Chemoreceptor Reflexes

The **chemoreceptor** (kē′mō-rē-sep′tor) **reflexes** help maintain homeostasis when oxygen tension in the blood decreases or when carbon dioxide and H^+ concentrations increase (figure 18.36). **Chemoreceptors** are sensory receptors responding to chemicals, such as oxygen, carbon dioxide, and pH. **Peripheral chemoreceptors** are found in the **carotid bodies,** structures located near the carotid sinuses, and in the **aortic bodies,** located near the aortic arch. **Central chemoreceptors** are in the medulla oblongata.

The peripheral chemoreceptors are most sensitive to oxygen. They act under emergency conditions and do not regulate the cardiovascular system under resting conditions. For example, when blood pressure drops significantly, blood oxygen levels decrease because of inadequate circulation of blood through the lungs. The peripheral chemoreceptors stimulate the vasomotor center, resulting in vasoconstriction that maintains or increases blood pressure.

The central chemoreceptors are most sensitive to changes in carbon dioxide and pH. They also act under emergency conditions or unusual conditions. For example, when blood pressure drops significantly, blood carbon dioxide levels increase because of inadequate circulation of blood through the lungs. The central chemoreceptors stimulate the vasomotor center, resulting in vasoconstriction that maintains or increases blood pressure. Heart rate and force of contraction also increase. This response is called the **CNS ischemic response.** It is activated only when blood pressure is very low and is a last-ditch effort to maintain blood pressure. If the vasomotor center fails because of inadequate

1. Chemoreceptors in the carotid and aortic bodies monitor blood O_2, CO_2, and pH.

2. Chemoreceptors in the medulla oblongata monitor blood CO_2 and pH.

3. Decreased blood O_2, increased CO_2, and decreased pH decrease parasympathetic stimulation of the heart, which increases the heart rate.

4. Decreased blood O_2, increased CO_2, and decreased pH increase sympathetic stimulation of the heart, which increases the heart rate and force of contraction.

5. Decreased blood O_2, increased CO_2, and decreased pH increase sympathetic stimulation of blood vessels, which increases vasoconstriction.

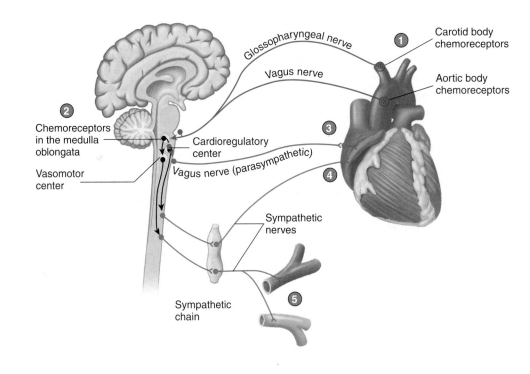

Process Figure 18.36 **Chemoreceptor Reflex Control of Blood Pressure**

An increase in blood CO_2 and a decrease in pH and O_2 result in vasoconstriction and increased heart rate and force of contraction. A decrease in blood CO_2 and an increase in blood pH result in vasodilation and decreased heart rate and force of contraction.

delivery of blood, vasomotor tone decreases, blood pressure drops, and death results.

The peripheral and central chemoreceptor reflexes are more important in the regulation of respiration than in the regulation of the cardiovascular system (see chapter 20). Low oxygen, increased carbon dioxide, and decreased pH can stimulate increased rate and depth of respiration. The increased respiratory activity stimulates reflexes that cause heart rate and stroke volume to increase. Thus, there is a match between respiratory and circulatory activities. ▼

72. Define peripheral and central chemoreceptors.
73. Describe the response of the peripheral chemoreceptor reflex to decreased oxygen.
74. Describe the CNS ischemic response to increased carbon dioxide.
75. How are chemoreceptor reflexes involved in matching respiratory and circulatory activities?

Long-Term Regulation of Blood Pressure

The regulation of blood volume and concentration by the kidneys and the movement of fluid across the wall of blood vessels play a central role in the long-term regulation of blood pressure. Some of the long-term regulatory mechanisms begin to respond in minutes, but they continue to function for hours, days, or longer. They adjust the blood pressure precisely and keep it within a narrow range of values for years. The major regulatory mechanisms are the renin-angiotensin-aldosterone mechanism, vasopressin (ADH) mechanism, atrial natriuretic mechanism, and fluid shift mechanism.

Renin-Angiotensin-Aldosterone Mechanism

The **renin-angiotensin-aldosterone mechanism** helps regulate blood pressure by changing peripheral resistance and blood volume. In response to reduced blood flow, the kidneys release an enzyme called **renin** (rē′nin, *rena,* kidney) into the circulatory system (figure 18.37). Renin acts on the blood protein **angiotensinogen** (an′jē-ō-ten-sin′ō-jen) to produce **angiotensin I** (an-jē-ō-ten′sin, *angio,* blood vessel + *tensus,* to stretch). Another enzyme, called **angiotensin-converting enzyme,** found in large amounts in organs such as the lungs, acts on angiotensin I to convert it to its most active form, called **angiotensin II.** Angiotensin II is a potent vasoconstrictor substance. Thus, in response to a reduced blood pressure, the release of renin by the kidney increases peripheral resistance, which causes blood pressure to increase toward its normal value.

Angiotensin II also acts on the adrenal cortex to increase the secretion of **aldosterone** (al-dos′ter-ōn). Aldosterone acts on the kidneys, causing them to conserve Na^+ and water. As a result, the volume of water lost from the blood into the urine is reduced. The decrease in urine volume results in less fluid loss from the body, which maintains blood volume (see figure 18.37). An adequate blood volume is essential for the maintenance of normal venous return to the heart and therefore for the maintenance of blood pressure.

ACE Inhibitors and Hypertension

Angiotensin-converting enzyme (ACE) inhibitors are a class of drugs that inhibit angiotensin-converting enzyme, which converts angiotensin I to angiotensin II. These drugs were first identified as components of the venom of pit vipers. Subsequently, several ACE inhibitors were synthesized. ACE inhibitors are effective in lowering blood pressure in many people who suffer from hypertension and are commonly administered to people to combat hypertension.

Vasopressin (ADH) Mechanism

The **vasopressin (ADH) mechanism** works in harmony with the renin-angiotensin-aldosterone mechanism in response to changes in blood pressure (figure 18.38). Baroreceptors are sensitive to changes in blood pressure, and decreases in blood pressure detected by the baroreceptors result in the release of vasopressin (vā-sō-pres′in, vas-ō-pres′in), or antidiuretic hormone (ADH), from the posterior pituitary. Blood pressure must decrease substantially before this mechanism is activated. For example, extensive burns decrease blood volume and blood pressure because of plasma loss at the burn site.

Neurons of the hypothalamus are sensitive to changes in the solute concentration of the plasma. Even small increases in the plasma concentration of solutes directly stimulate hypothalamic neurons that increase ADH secretion (see figure 18.38 and chapter 23). Increases in the concentration of the plasma, such as during dehydration, stimulate ADH secretion.

ADH acts directly on blood vessels to cause vasoconstriction, although it is not as potent as other vasoconstrictor agents. Within minutes after a rapid and substantial decline in blood pressure, ADH is released in sufficient quantities to help reestablish normal blood pressure. ADH also decreases the rate of urine production by the kidneys, thereby helping maintain blood volume and blood pressure.

Atrial Natriuretic Mechanism

A polypeptide called **atrial natriuretic** (ā′trē-ăl nā′trē-ū-ret′ik) **hormone** is released from cells in the atria of the heart. A major stimulus for its release is increased venous return, which stretches atrial cardiac muscle cells. Atrial natriuretic hormone acts on the kidneys to increase the rate of urine production and Na^+ loss in the urine. It also dilates arteries and veins. Loss of water and Na^+ in the urine causes the blood volume to decrease, which decreases venous return, and vasodilation results in a decrease in peripheral resistance. These effects cause a decrease in blood pressure.

The renin-angiotensin-aldosterone, vasopressin (ADH), and atrial natriuretic mechanisms work simultaneously to help regulate blood pressure by controlling urine production by the kidneys. If blood pressure drops below 50 mm Hg, the volume of urine produced by the kidneys is reduced to nearly zero. If blood pressure is increased to 200 mm Hg, the urine volume produced is approximately six to eight times greater than normal. The hormonal mechanisms that regulate blood pressure in the long term are summarized in figure 18.39.

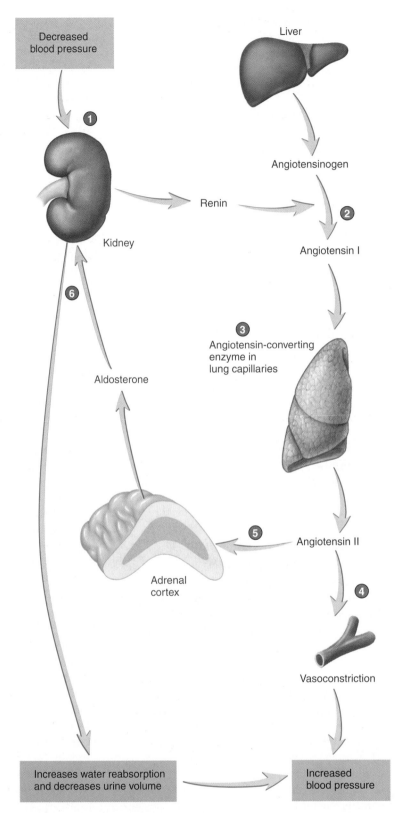

1. Decreased blood pressure is detected by the kidneys, resulting in increased renin secretion.

2. Renin converts angiotensinogen, a protein secreted from the liver, to angiotensin I.

3. Angiotensin-converting enzyme in the lungs converts angiotensin I to angiotensin II.

4. Angiotensin II is a potent vasoconstrictor, resulting in increased blood pressure.

5. Angiotensin II stimulates the adrenal cortex to secrete aldosterone.

6. Aldosterone acts on the kidneys to increase Na^+ reabsorption. As a result, urine volume decreases and blood volume increases, resulting in increased blood pressure.

Process Figure 18.37 Renin-Angiotensin-Aldosterone Mechanism

Decreased blood pressure is detected by the kidney, resulting in increased renin secretion. The result is vasoconstriction, increased water reabsorption, and decreased urine volume. These changes maintain blood pressure.

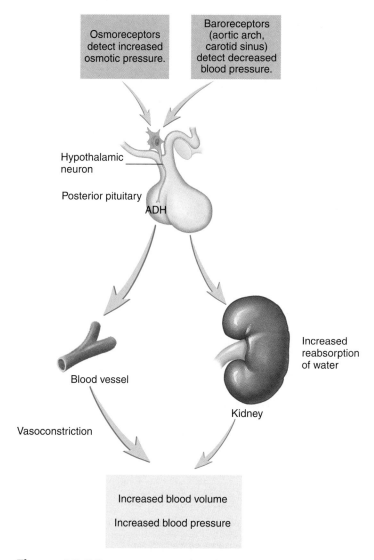

Figure 18.38 Vasopressin (ADH) Mechanism

Increases in osmolality of blood or decreases in blood pressure result in ADH secretion. ADH increases water reabsorption by the kidney, and large amounts of ADH result in vasoconstriction. These changes maintain blood pressure.

Fluid Shift Mechanism

The **fluid shift mechanism** begins to act within a few minutes but requires hours to achieve its full functional capacity. It plays a very important role when dehydration develops over several hours, or when a large volume of saline is administered over several hours. The fluid shift mechanism occurs in response to small changes in pressures across capillary walls. As blood pressure increases, some fluid is forced from the capillaries into the interstitial spaces. The movement of fluid into the interstitial spaces helps prevent the development of high blood pressure. As blood pressure falls, interstitial fluid moves into capillaries, which resists a further decline in blood pressure. The fluid shift mechanism is a powerful method through which blood pressure is maintained because the interstitial volume acts as a reservoir, and it is in equilibrium with the large volume of intercellular fluid. ▼

> *76. For each of these hormones—renin, angiotensin, aldosterone, antidiuretic hormone, and atrial natriuretic*

hormone—state where the hormone is produced and what effects it has on the circulatory system.

77. What is fluid shift, and what does it accomplish?

P R E D I C T **12**

Explain the differences in the mechanisms that regulate blood pressure in response to hemorrhage that results in the rapid loss of a large volume of blood, compared with hemorrhage that results in the loss of the same volume of blood but over a period of several hours.

Examples of Cardiovascular Regulation

Exercise

During exercise, blood flow through tissues is changed dramatically. Blood flow through exercising skeletal muscles can be 15–20 times greater than through resting muscles. Local, nervous, and hormonal regulatory mechanisms are responsible for the increased blood flow.

When skeletal muscle is resting, only 20%–25% of the capillaries in the skeletal muscle are open, whereas during exercise most or all of the capillaries are open. As muscular activity increases, oxygen levels in muscle tissue decreases and carbon dioxide and lactic acid levels increase. Local control of blood flow in the tissues results in arteriole vasodilatation and precapillary sphincter relaxation. Blood flow through capillaries increases, providing muscle cells with needed oxygen and nutrients and removing waste products. Peripheral resistance decreases because of the arteriolar vasodilation and the relaxation of precapillary sphincters. The decrease in peripheral resistance can produce a decrease in mean arterial pressure, which can jeopardize blood delivery. Increased cardiac output and increased vasoconstriction, however, not only prevent a decrease but also cause an increase in mean arterial pressure.

Cardiac output increases because of increased venous return, stroke volume, and heart rate. Venous return increases because of (1) an increased flow of blood through muscle capillaries, (2) the vasoconstriction of veins, and (3) the skeletal muscle pump. The increased venous return increases stroke volume and cardiac output (Starling's law). In addition, decreased parasympathetic stimulation, increased sympathetic stimulation, epinephrine, and norepinephrine cause heart rate and stroke volume to increase. As a consequence, blood pressure usually increases by 20–60 mm Hg, which helps sustain the increased blood flow through skeletal muscle blood vessels.

Despite the increased cardiac output during exercise, resistance to blood flow in skeletal muscles can decrease so much that hypotension could develop. Vasoconstriction counters this effect by increasing peripheral resistance. Sympathetic stimulation and norepinephrine cause vasoconstriction in the skin, kidneys, gastrointestinal tract, and nonexercising muscles. An added benefit of vasoconstriction in these organs is rerouting blood away from them to the exercising muscle. Vasoconstriction of arterioles in exercising muscle also helps maintain peripheral resistance. Although this counteracts local control and reduces blood flow to muscle tissue, it ensures adequate blood pressure for the delivery of blood.

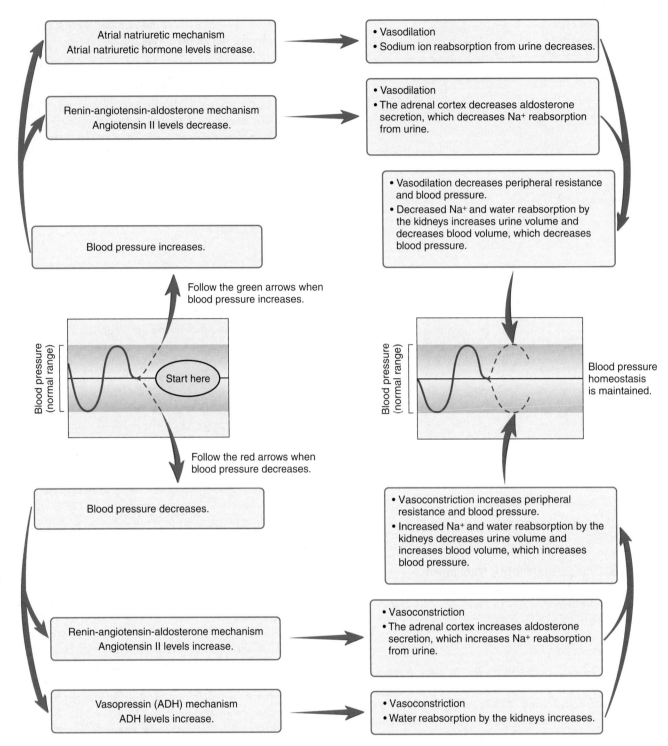

Homeostasis Figure 18.39   Summary of the Control of Blood Pressure by Long-Term (Slow-Acting) Hormonal Mechanisms

For more information on the renin-angiotensin-aldosterone mechanism see figure 18.37 (p. 556); on the vasopressin (ADH) mechanism, see figure 18.38 (p. 557).

In response to sympathetic stimulation, some decrease in blood flow through the skin can occur at the beginning of exercise. As body temperature increases in response to the increased muscular activity, however, temperature receptors in the hypothalamus are stimulated. As a result, action potentials in sympathetic nerve fibers causing vasoconstriction decrease, resulting in vasodilation of blood vessels in the skin. As a consequence, the skin turns a red or pinkish color, and a great deal of excess heat is lost as blood flows through the dilated blood vessels. As vasodilation occurs, sweat glands in the skin are stimulated to increase sweat production. Heat is removed from the skin by the evaporation of sweat.

The overall effect of exercise on circulation is to greatly increase blood flow by increasing blood pressure and decreasing peripheral resistance. Resistance to blood flow decreases in exercising

Systems Interactions

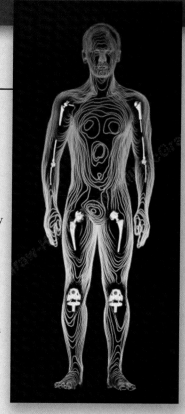

	Effects of the Cardiovascular System on Other Systems	Effects of Other Systems on the Cardiovascular System
Integumentary System	Delivers oxygen, nutrients, hormones, and immune cells Removes carbon dioxide, waste products, and toxins	Loss of heat from dermal vessels helps regulate body temperature
Skeletal System	Delivers oxygen, nutrients, hormones, and immune cells Removes carbon dioxide, waste products, and toxins	Red bone marrow produces red blood cells (oxygen and carbon dioxide transport) and platelets (blood clotting) Thoracic cage protects the heart Is a reservoir for calcium, which is necessary for cardiac muscle contraction
Muscular System	Delivers oxygen, nutrients, hormones, and immune cells Removes carbon dioxide, waste products, and toxins Removes lactic acid and heat from muscle	Skeletal muscle "pump" helps move blood through blood vessels
Nervous System	Delivers oxygen, nutrients, hormones, and immune cells Removes carbon dioxide, waste products, and toxins Blood–brain barrier formed by capillary endothelial cells Cerebrospinal fluid and aqueous humor produced from, and returned to, the blood	Regulates heart rate and force of contraction Regulates blood vessel diameter
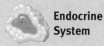 **Endocrine System**	Delivers oxygen, nutrients, hormones, and immune cells Removes carbon dioxide, waste products, and toxins	Epinephrine and norepinephrine increase heart rate and force of contraction and change blood vessel diameter Renin-angiotensin-aldosterone mechanism, antidiuretic hormone, and atrial natriuretic hormone regulate blood pressure
Lymphatic System and Immunity	Receives lymph (fluid) from tissues Delivers oxygen, nutrients, hormones, and immune cells Removes carbon dioxide, waste products, and toxins	Spleen removes damaged red blood cells and helps recycle iron Immune cells provide protection against microorganisms and toxins and promote tissue repair by releasing chemical mediators
Respiratory System	Delivers oxygen, nutrients, hormones, and immune cells Removes carbon dioxide, waste products, and toxins	Regulates blood pressure by converting angiotensin I to angiotensin II Provides oxygen and removes carbon dioxide Helps maintain body's pH
Digestive System	Delivers oxygen, nutrients, hormones, and immune cells Removes carbon dioxide, waste products, and toxins Receives nutrients and water	Provides nutrients, including iron (for hemoglobin production) Vitamin B12 and folic acid (for cell production) and vitamin K (for clotting proteins) Absorbs water and ions necessary to maintain blood volume and pressure
Urinary System	Maintenance of blood pressure necessary for urine formation Delivers oxygen, nutrients, hormones, and immune cells Removes carbon dioxide, waste products, and toxins	Removes waste products Helps maintain the body's pH, ion, and water balance Is a long-term regulator of blood volume and blood pressure
Reproductive System	Delivers oxygen, nutrients, hormones, and immune cells Removes carbon dioxide, waste products, and toxins	Estrogen may slow the development of atherosclerosis

muscle but increases elsewhere for a net decrease in peripheral resistance. In addition, the increased blood flow is routed to skeletal and cardiac muscle, while blood flow through other organs is maintained at a value adequate to supply their metabolic needs. ▼

78. *How does local control affect blood flow in exercising muscle? What effect does this have on peripheral resistance?*
79. *What causes blood pressure to increase during exercise?*
80. *How is heat loss through the skin regulated?*

Circulatory Shock

Circulatory shock is inadequate blood flow throughout the body. As a consequence, tissues suffer damage resulting from a lack of oxygen. Severe shock may damage vital body tissues and lead to death.

There are several causes of circulatory shock, but hemorrhagic shock resulting from excessive blood loss exemplifies the general characteristics of shock. If hemorrhagic shock is not severe, blood pressure decreases only a moderate amount. Under these conditions, the mechanisms that normally regulate blood pressure reestablish normal pressure and blood flow. The baroreceptor reflexes produce strong sympathetic responses, resulting in intense vasoconstriction and increased heart rate and force of contraction. The adrenal medullary mechanism increases heart rate and force of contraction.

As a result of the reduced blood flow through the kidneys, increased amounts of renin are released. The elevated renin level results in a greater rate of angiotensin II formation, causing vasoconstriction and increased aldosterone release from the adrenal cortex. Aldosterone, in turn, promotes water and salt retention by the kidneys. In response to reduced blood pressure, antidiuretic hormone (ADH) is released from the posterior pituitary gland, and ADH causes vasoconstriction and enhances the retention of water by the kidneys. An intense sensation of thirst leads to increased water intake, which helps restore the normal blood volume. The fluid shift mechanism also helps maintain blood volume and blood pressure.

In mild cases of shock, the baroreceptor reflexes can be adequate to compensate for blood loss until the blood volume is restored, but in more severe cases all the mechanisms are required

to sustain life. Peripheral and central chemoreceptor reflexes increase vasoconstriction, heart rate, and force of contraction.

In even more severe cases of shock, the regulatory mechanisms are not adequate to compensate for the effects of shock. As a consequence, a positive-feedback cycle begins to develop in which the blood pressure regulatory mechanisms lose their ability to control the blood pressure, shock worsens, and the effectiveness of the regulatory mechanisms deteriorates even further. The positive-feedback cycle proceeds until death occurs or until treatment, such as a transfusion, terminates the cycle. Several types of shock are classified by the cause of the condition:

1. **Hypovolemic shock** is the result of reduced blood volume. **Hemorrhagic shock** is caused by internal or external bleeding. **Plasma loss shock** results from a loss of plasma, such as from severely burned areas. **Interstitial fluid loss shock** is reduced blood volume resulting from a loss of interstitial fluid, as in diarrhea, vomiting, and dehydration.
2. **Neurogenic shock** is caused by vasodilation in response to emotional upset or anesthesia.
3. **Anaphylactic shock** is caused by an allergic response, resulting in the release of inflammatory substances that cause vasodilation and an increase in capillary permeability. Large amounts of fluid then move from capillaries into the interstitial spaces.
4. **Septic shock,** or **"blood poisoning,"** is caused by infections that result in the release of toxic substances into the circulatory system, which depress the activity of the heart and lead to vasodilation and increased capillary permeability.
5. **Cardiogenic shock** results from a decrease in cardiac output caused by events that decrease the heart's ability to function. Heart attack (myocardial infarction) is a common cause of cardiogenic shock. Fibrillation of the heart, which can be initiated by stimuli such as cardiac arrhythmias or exposure to electrical shocks, also results in cardiogenic shock. ▼

81. *Describe how various mechanisms help regulate blood pressure and blood volume during shock.*
82. *Describe the types of shock as classified by the causes of shock.*

S U M M A R Y

Functions of the Peripheral Circulation (p. 520)

The peripheral circulation carries blood; exchanges nutrients, waste products, and gases with tissues; helps regulate blood pressure; and directs blood flow to tissues.

General Features of Blood Vessels (p. 520)

1. Arteries carry blood away from the heart toward capillaries, where exchange between the blood and interstitial fluids occur. Veins carry blood from the capillaries toward the heart.
2. Blood flows from the heart through elastic arteries, muscular arteries, and arterioles to the capillaries.
3. Blood returns to the heart from the capillaries through venules, small veins, and large veins.
4. Blood vessels, except for capillaries, have three layers.
   - The inner tunica intima consists of endothelium (simple squamous epithelium), basement membrane, and internal elastic lamina.

   - The middle tunica media contains circular smooth muscle and elastic and collagen fibers.
   - The outer tunica adventitia is connective tissue.
5. The thickness and the composition of the layers vary with blood vessel type and diameter.

Arteries

1. Large elastic arteries are thick-walled with large diameters. The tunica media has many elastic fibers and little smooth muscle.
2. Muscular (distributing) arteries are thick-walled with small diameters. The tunica media has abundant smooth muscle and some elastic fibers.
3. Arterioles are the smallest arteries. The tunica media consists of one or two layers of smooth muscle cells and a few elastic fibers.

Capillaries

1. Capillaries consist only of endothelium.

2. A capillary bed is a network of capillaries.
3. Thoroughfare channels carry blood from arterioles to venules. Blood can pass rapidly through thoroughfare channels.
4. Precapillary sphincters regulate the flow of blood into capillaries.

Veins

1. Venules connect to capillaries and are like capillaries, except they are larger in diameter.
2. Large venules and all veins have all three layers.
3. Valves prevent the backflow of blood in the veins.

Aging of the Arteries

1. Arteriosclerosis results from a loss of elasticity in the aorta, large arteries, and coronary arteries.
2. Atherosclerosis is the deposition of materials in arterial walls to form plaques.

Pulmonary Circulation (p. 525)

The pulmonary circulation moves blood to and from the lungs. The pulmonary trunk arises from the right ventricle and divides to form the pulmonary arteries, which project to the lungs. From the lungs, four pulmonary veins return blood to the left atrium.

Systemic Circulation: Arteries (p. 525)

Arteries carry blood from the left ventricle of the heart to all parts of the body.

Aorta

1. The aorta leaves the left ventricle to form the ascending aorta, aortic arch, and descending aorta (consisting of the thoracic and abdominal aortae).
2. Coronary arteries branch from the aorta and supply the heart.

Arteries to the Head and the Neck

1. The brachiocephalic, left common carotid, and left subclavian arteries branch from the aortic arch to supply the head and the upper limbs. The brachiocephalic artery divides to form the right common carotid and the right subclavian arteries. The vertebral arteries branch from the subclavian arteries.
2. The common carotid arteries and the vertebral arteries supply the head.
   - The common carotid arteries divide to form the external carotids, which supply the face and mouth, and the internal carotids, which supply the brain.
   - The vertebral arteries join within the cranial cavity to form the basilar artery, which supplies the brain.
   - The internal carotids and basilar arteries contribute to the cerebral arterial circle.

Arteries of the Upper Limb

1. The subclavian artery continues (without branching) as the axillary artery and then as the brachial artery. The brachial artery divides into the radial and ulnar arteries.
2. The radial artery supplies the deep palmar arch, and the ulnar artery supplies the superficial palmar arch. Both arches give rise to the digital arteries.

Thoracic Aorta and Its Branches

The thoracic aorta has visceral branches that supply the thoracic organs and parietal branches that supply the thoracic wall.

Abdominal Aorta and Its Branches

1. The abdominal aorta has visceral branches that supply the abdominal organs and parietal branches that supply the abdominal wall.

2. The visceral branches are paired and unpaired. The unpaired arteries supply the stomach, spleen, and liver (celiac trunk); the small intestine and upper part of the large intestine (superior mesenteric); and the lower part of the large intestine (inferior mesenteric). The paired arteries supply the kidneys, adrenal glands, and gonads.

Arteries of the Pelvis

1. The common iliac arteries arise from the abdominal aorta, and the internal iliac arteries branch from the common iliac arteries.
2. The visceral branches of the internal iliac arteries supply the pelvic organs, and the parietal branches supply the pelvic wall and floor and the external genitalia.

Arteries of the Lower Limb

1. The external iliac arteries branch from the common iliac arteries.
2. The external iliac artery continues (without branching) as the femoral artery and then as the popliteal artery. The popliteal artery divides to form the anterior and posterior tibial arteries.
3. The posterior tibial artery gives rise to the fibular (peroneal) and plantar arteries. The plantar arteries form the plantar arch, from which the digital arteries arise.

Systemic Circulation: Veins (p. 535)

1. The three major veins returning blood to the heart are the superior vena cava (head, neck, thorax, and upper limbs), the inferior vena cava (abdomen, pelvis, and lower limbs), and the coronary sinus (heart).
2. Veins are of three types: superficial veins, deep veins, and sinuses.

Veins of the Head and Neck

1. The internal jugular veins drain the dural venous sinuses and the veins of the anterior head, face, and neck.
2. The external jugular veins and the vertebral veins drain the posterior head and neck.

Veins of the Upper Limb

1. The deep veins are the small ulnar and radial veins of the forearm, which join the brachial veins of the arm. The brachial veins drain into the axillary vein.
2. The superficial veins are the basilic, cephalic, and median cubital.
   - The basilic vein becomes the axillary vein, which then becomes the subclavian vein. The cephalic vein drains into the axillary vein.
   - The median cubital connects the basilic and cephalic veins at the elbow.

Veins of the Thorax

The left and right brachiocephalic veins and the azygos veins return blood to the superior vena cava.

Veins of the Abdomen and Pelvis

1. Ascending lumbar veins from the abdomen join the azygos and hemiazygos veins.
2. Veins from the kidneys, adrenal glands, and gonads directly enter the inferior vena cava.
3. Veins from the stomach, intestines, spleen, and pancreas connect with the hepatic portal vein. The hepatic portal vein transports blood to the liver for processing. Hepatic veins from the liver join the inferior vena cava.

Veins of the Lower Limb

1. The deep veins are the fibular (peroneal), anterior tibial, posterior tibial, popliteal, femoral, and external iliac veins.
2. The superficial veins are the small and great saphenous veins.

Physiology of Circulation   (p. 544)
Blood Pressure

1. Blood pressure is a measure of the force exerted by blood against the blood vessel wall. Blood moves through vessels because of blood pressure.
2. Blood pressure can be measured by listening for Korotkoff sounds produced by turbulent flow in arteries as pressure is released from a blood pressure cuff.

Blood Flow Through a Blood Vessel

1. Blood flow is the amount of blood that moves through a vessel in a given period. Blood flow is directly proportional to pressure differences and is inversely proportional to resistance.
2. Resistance is the sum of all the factors that inhibit blood flow. Resistance increases when blood vessels become smaller and viscosity increases.
3. Viscosity is the resistance of a liquid to flow. Most of the viscosity of blood results from red blood cells. The viscosity of blood increases when the hematocrit increases or plasma volume decreases.

Blood Flow Through the Body

1. Mean arterial pressure equals cardiac output times peripheral resistance.
2. Vasomotor tone is a state of partial contraction of blood vessels. Vasoconstriction increases vasomotor tone and peripheral resistance, whereas vasodilation decreases vasomotor tone and peripheral resistance.
3. Blood pressure averages 100 mm Hg in the aorta and drops to 0 mm Hg in the right atrium. The greatest drop occurs in the arterioles and capillaries.

Pulse Pressure and Vascular Compliance

1. Pulse pressure is the difference between systolic and diastolic pressures. Pulse pressure increases when stroke volume increases or vascular compliance decreases.
2. Vascular compliance is a measure of the change in volume of blood vessels produced by a change in pressure.
3. Pulse pressure waves travel through the vascular system faster than the blood flows. Pulse pressure can be used to take the pulse.

Blood Pressure and the Effect of Gravity

In a standing person, hydrostatic pressure caused by gravity increases blood pressure below the heart and decreases pressure above the heart.

Capillary Exchange and Regulation of Interstitial Fluid Volume

1. Capillary exchange occurs through or between endothelial cells.
2. Diffusion, which includes osmosis, and filtration are the primary means of capillary exchange.
3. Filtration moves materials out of capillaries and osmosis moves them into capillaries.
4. A net movement of fluid occurs from the blood into the tissues. The fluid gained by the tissues is removed by the lymphatic system.

Control of Blood Flow   (p. 549)

Blood flow through tissues is highly controlled and matched closely to the metabolic needs of tissues.

Local Control

1. Local control is the response of vascular smooth muscle to changes in tissue gases, nutrients, and waste products.
2. If the metabolic activity of a tissue increases, the diameter and number of capillaries in the tissue increase over time.

Nervous and Hormonal Control

1. The sympathetic nervous system (vasomotor center in the medulla) controls blood vessel diameter. Other brain areas can excite or inhibit the vasomotor center.
2. Epinephrine and norepinephrine cause vasoconstriction in most tissues. Epinephrine causes vasodilation in skeletal and cardiac muscle.
3. The muscular arteries and arterioles control the delivery of blood to tissues.
4. The veins are a reservoir for blood.
5. Venous return to the heart increases because of the vasoconstriction of veins, an increased blood volume, and the skeletal muscle pump (with valves).

Regulation of Mean Arterial Pressure   (p. 551)

Mean arterial pressure (MAP) is proportional to cardiac output times peripheral resistance.

Short-Term Regulation of Blood Pressure

1. Baroreceptors are sensory receptors sensitive to stretch.
   - Baroreceptors are located in the carotid sinuses and the aortic arch.
   - The baroreceptor reflex changes peripheral resistance, heart rate, and stroke volume in response to changes in blood pressure.
2. Epinephrine and norepinephrine are released from the adrenal medulla as a result of sympathetic stimulation. They increase heart rate, stroke volume, and vasoconstriction.
3. Peripheral chemoreceptor reflexes respond to decreased oxygen, leading to increased vasoconstriction.
4. Central chemoreceptors respond to high carbon dioxide or low pH levels in the medulla, leading to increased vasoconstriction, heart rate, and force of contraction (CNS ischemic response).

Long-Term Regulation of Blood Pressure

1. Through the renin-angiotensin-aldosterone mechanism, renin is released by the kidneys in response to low blood pressure. Renin promotes the production of angiotensin II, which causes vasoconstriction and an increase in aldosterone secretion. Aldosterone helps maintain blood volume by decreasing urine production.
2. The vasopressin (ADH) mechanism causes ADH release from the posterior pituitary in response to a substantial decrease in blood pressure. ADH causes vasoconstriction and helps maintain blood volume by decreasing urine production.
3. The atrial natriuretic mechanism causes atrial natriuretic hormone release from the cardiac muscle cells when atrial blood pressure increases. It stimulates an increase in urinary production, causing a decrease in blood volume and blood pressure.
4. The fluid shift mechanism causes fluid shift, which is a movement of fluid from the interstitial spaces into capillaries in response to a decrease in blood pressure to maintain blood volume.

Examples of Cardiovascular Regulation   (p. 557)
Exercise

1. Local control mechanisms increase blood flow through exercising muscles, which lowers peripheral resistance.
2. Cardiac output increases because of increased venous return, stroke volume, and heart rate.
3. Vasoconstriction in the skin, the kidneys, the gastrointestinal tract, and skeletal muscle (nonexercising and exercising) increases peripheral resistance, which helps prevent a drop in blood pressure.
4. Blood pressure increases despite an overall decrease in peripheral resistance because of increased cardiac output.

Circulatory Shock

1. Baroreceptor reflexes and the adrenal medullary response increase blood pressure.
2. The renin-angiotensin-aldosterone mechanism and the vasopressin mechanism increase vasoconstriction and blood volume. The fluid shift mechanism increases blood volume.
3. In severe shock, the chemoreceptor reflexes increase vasoconstriction, heart rate, and force of contraction.
4. In severe shock, despite negative-feedback mechanisms, a positive-feedback cycle of decreasing blood pressure can cause death.

R E V I E W   A N D   C O M P R E H E N S I O N

1. Given these blood vessels:
   1. arteriole
   2. capillary
   3. elastic artery
   4. muscular artery
   5. vein
   6. venule

   Choose the arrangement that lists the blood vessels in the order a red blood cell passes through them as it leaves the heart, travels to a tissue, and returns to the heart.
   a. 3,4,2,1,5,6
   b. 3,4,1,2,6,5
   c. 4,3,1,2,5,6
   d. 4,3,2,1,6,5
   e. 4,2,3,5,1,6

2. Given these structures:
   1. arteriole
   2. precapillary sphincter
   3. thoroughfare channel

   Choose the arrangement that lists the structures in the order a red blood cell encounters them as it passes through a tissue.
   a. 1,3,2
   b. 2,1,3
   c. 2,3,1
   d. 3,1,2
   e. 3,2,1

3. In which of these blood vessels are elastic fibers present in the largest amounts?
   a. large arteries
   b. medium-sized arteries
   c. arterioles
   d. venules
   e. large veins

4. Comparing and contrasting arteries and veins, the veins have
   a. thicker walls.
   b. a greater amount of smooth muscle than arteries.
   c. a tunica media, but arteries do not.
   d. valves, but arteries do not.
   e. all of the above.

5. Given these blood vessels:
   1. aorta
   2. inferior vena cava
   3. pulmonary arteries
   4. pulmonary veins

   Which vessels carry oxygen-rich blood?
   a. 1,3
   b. 1,4
   c. 2,3
   d. 2,4
   e. 3,4

6. Given these arteries:
   1. basilar
   2. common carotid
   3. internal carotid
   4. vertebral

   Which arteries have *direct* connections with the cerebral arterial circle (circle of Willis)?
   a. 1,2
   b. 2,4
   c. 1,3
   d. 3,4
   e. 2,3

7. Given these blood vessels:
   1. axillary artery
   2. brachial artery
   3. brachiocephalic artery
   4. radial artery
   5. subclavian artery

   Choose the arrangement that lists the vessels in order, from the aorta to the right hand.
   a. 2,5,4,1
   b. 5,2,1,4
   c. 5,3,1,4,2
   d. 3,5,1,2,4
   e. 4,5,1,2,3

8. A branch of the aorta that supplies the liver, stomach, and spleen is the
   a. celiac trunk.
   b. common iliac.
   c. inferior mesenteric.
   d. superior mesenteric.
   e. renal.

9. Given these arteries:
   1. common iliac
   2. external iliac
   3. femoral
   4. popliteal

   Choose the arrangement that lists the arteries in order, from the aorta to the knee.
   a. 1,2,3,4
   b. 1,2,4,3
   c. 2,1,3,4
   d. 2,1,4,3
   e. 3,1,2,4

10. Given these veins:
    1. brachiocephalic
    2. internal jugular
    3. superior vena cava
    4. dural venous sinus

    Choose the arrangement that lists the veins in order, from the brain to the heart.
    a. 1,2,4,3
    b. 2,4,1,3
    c. 2,4,3,1
    d. 4,2,1,3
    e. 4,2,3,1

11. Blood returning from the arm to the subclavian vein could pass through which of these veins?
    a. cephalic
    b. basilic
    c. brachial
    d. both a and b
    e. all of the above

12. Given these blood vessels:
    1. inferior mesenteric vein
    2. superior mesenteric vein
    3. hepatic portal vein
    4. hepatic vein

    Choose the arrangement that lists the vessels in order, from the small intestine to the inferior vena cava.
    a. 1,3,4
    b. 1,4,3
    c. 2,3,4
    d. 2,4,3
    e. 3,1,4

13. Given these veins:
    1. small saphenous
    2. great saphenous
    3. fibular (peroneal)
    4. posterior tibial

    Which are superficial veins?
    a. 1,2
    b. 1,3
    c. 2,3
    d. 2,4
    e. 3,4

14. If you could increase any of these factors that affect blood flow by twofold, which one would cause the greatest increase in blood flow?
    a. blood viscosity
    b. pressure gradient
    c. vessel diameter
    d. vessel length

15. Pulse pressure
    a. is the difference between the systolic and diastolic pressures.
    b. increases when stroke volume increases.
    c. increases as vascular compliance decreases.
    d. all of the above.

16. Local direct control of blood flow through a tissue
    a. results from vasoconstriction and vasodilation of arterioles.
    b. results from relaxation and contraction of precapillary sphincters.
    c. occurs in response to a buildup in carbon dioxide in the tissues.
    d. occurs in response to a decrease in oxygen in the tissues.
    e. all of the above.

17. An increase in mean arterial pressure can result from
    a. an increase in peripheral resistance.
    b. an increase in heart rate.
    c. an increase in stroke volume.
    d. all of the above.

18. In response to an increase in mean arterial pressure, the baroreceptor reflex causes
   a. an increase in sympathetic nervous system activity.
   b. a decrease in peripheral resistance.
   c. stimulation of the vasomotor center.
   d. vasoconstriction.
   e. an increase in cardiac output.

19. When blood oxygen levels markedly decrease, the chemoreceptor reflex causes
   a. peripheral resistance to decrease.
   b. mean arterial blood pressure to increase.
   c. vasomotor tone to decrease.
   d. vasodilation.
   e. all of the above.

20. When blood pressure is suddenly decreased a small amount (10 mm Hg), which of these mechanisms is activated to restore blood pressure to normal levels?
   a. chemoreceptor reflexes
   b. baroreceptor reflexes
   c. CNS ischemic response
   d. all of the above

21. A sudden release of epinephrine from the adrenal medulla
   a. increases heart rate.
   b. increases stroke volume.
   c. causes vasodilation in skeletal and cardiac muscle blood vessels.
   d. all of the above.

22. When blood pressure decreases,
   a. renin secretion increases.
   b. angiotensin II formation decreases.
   c. aldosterone secretion decreases.
   d. all of the above.

23. In response to a decrease in blood pressure,
   a. ADH secretion increases.
   b. the kidneys decrease urine production.
   c. blood volume increases.
   d. all of the above.

24. A patient is found to have severe arteriosclerosis of his renal arteries, which has reduced renal blood pressure. Which of these is consistent with that condition?
   a. hypotension
   b. hypertension
   c. decreased vasomotor tone
   d. exaggerated sympathetic stimulation of the heart
   e. both a and c

25. During exercise, the blood flow through skeletal muscle may increase up to 20-fold. However, the cardiac output does not increase that much. This occurs because of
   a. vasoconstriction in the viscera.
   b. vasoconstriction in the skin (at least temporarily).
   c. vasodilation of skeletal muscle blood vessels.
   d. both a and b.
   e. all of the above.

Answers in Appendix E

CRITICAL THINKING

1. For each of the following destinations, name all the arteries that a red blood cell would encounter if it started its journey in the left ventricle.
   a. posterior interventricular groove of the heart
   b. anterior neck to the brain (give two ways)
   c. posterior neck to the brain (give two ways)
   d. external skull
   e. tip of the fingers of the left hand (what other blood vessel would be encountered if the trip were through the right upper limb?)
   f. anterior compartment of the leg
   g. liver
   h. small intestine
   i. urinary bladder

2. For each of the following starting places, name all the veins that a red blood cell would encounter on its way back to the right atrium.
   a. anterior interventricular groove of the heart (give two ways)
   b. venous sinus near the brain
   c. external posterior of skull
   d. hand (return deep and superficial)
   e. foot (return deep and superficial)
   f. stomach
   g. kidney
   h. left inferior wall of the thorax

3. In a study of heart valve functions, it is necessary to inject a dye into the right atrium of the heart by inserting a catheter into a blood vessel and moving the catheter into the right atrium. What route would you suggest? If you wanted to do this procedure into the left atrium, what would you do differently?

4. In endurance-trained athletes, the hematocrit can be lower than normal because plasma volume increases more than red blood cell numbers increase. Explain why this condition would be beneficial.

5. Observe a large vein on the back of the hand (seen best on an older person with less connective tissue around the veins). What happens to the shape of the vein while raising and lowering the hand? Explain.

6. A very short nursing student is asked to measure the blood pressure of a very tall person. She decides to measure the blood pressure at the level of the tall person's foot while he is standing. What artery does she use? After taking the blood pressure, she decides that the tall person is suffering from hypertension because the systolic pressure is 210 mm Hg. Is her diagnosis correct? Why or why not?

7. Mr. D. was suffering from severe cirrhosis of the liver and hepatitis. He developed severe edema over a period of time. Explain how decreased liver function can result in edema.

8. A patient is suffering from edema in the lower right limb. Explain why massage helps remove the excess fluid.

9. One cool evening, Skinny Dip jumps into a hot Jacuzzi. Predict what will happen to Skinny's heart rate.

10. It rarely occurs that a person congenitally has no sympathetic control over his or her cardiovascular system. If such a person exercises, what happens to heart rate, stroke volume, peripheral resistance, and blood pressure?

Answers in Appendix F

For additional study tools, go to: www.aris.mhhe.com. Click on your course topic and then this textbook's author name/title. Register once for a semester's worth of interactive learning activities to help improve your grade!

CHAPTER 19 Lymphatic System and Immunity

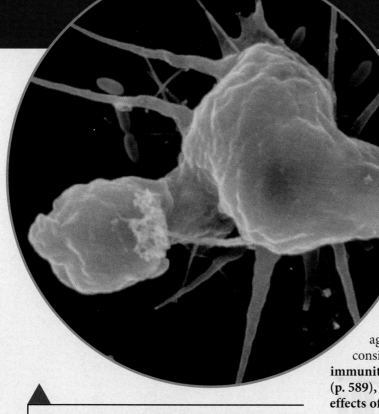

The blue rods in this colorized scanning electron micrograph are bacteria (*E. coli*) that commonly cause urinary tract infections. The yellow cell with extended cell processes is a macrophage. This large, motile cell can capture, phagocytize, and destroy the bacteria. The immune system provides protection against microorganisms.

Introduction

One of the basic themes of life is that many organisms consume, or use, other organisms to survive. For example, deer graze on grasses and wolves feed on deer. A parasite lives on or in another organism, called the host. The host provides the parasite with the conditions and food necessary for survival. For example, hookworms can live in the sheltered environment of the human intestine, where they feed on blood. Humans are host to many kinds of organisms, including microorganisms, such as bacteria, viruses, fungi, and protozoans; insects; and worms. Often, parasites harm humans, causing disease and sometimes death. However, our bodies have ways to resist or destroy harmful parasites. The lymphatic system and immunity are the body's defense systems against threats arising from inside and outside the body. This chapter considers the **lymphatic system (p. 566), immunity (p. 572), innate immunity (p. 574), adaptive immunity (p. 578), immune interactions (p. 589), immunotherapy (p. 589), acquired immunity (p. 591),** and the **effects of aging on the lymphatic system and immunity (p. 596).**

Tools for Success

A virtual cadaver dissection experience

Audio (MP3) and/or video (MP4) content, available at www.mhhe.aris.com

Lymphatic System

The **lymphatic** (lim-fat′ik) **system** includes lymph, lymphatic vessels, lymphatic tissue, lymphatic nodules, lymph nodes, tonsils, the spleen, and the thymus (figure 19.1).

Functions of the Lymphatic System

The lymphatic system is part of the body's defense system against microorganisms and other harmful substances. In addition, it helps maintain fluid balance in tissues and absorb fats from the digestive tract.

1. *Fluid balance.* Approximately 30 L of fluid pass from the blood capillaries into the interstitial spaces each day, whereas only 27 L pass from the interstitial spaces back into the blood capillaries. If the extra 3 L of fluid were to remain in the interstitial spaces, edema would result, causing tissue damage and eventual death. Instead, the 3 L of fluid enters the lymphatic capillaries, where the fluid is called **lymph** (limf, clear spring water), and it passes through the lymphatic vessels back to the blood (see chapter 18). In addition to water, lymph contains solutes derived from two sources: (1) Substances in plasma, such as ions, nutrients, gases, and some proteins, pass from blood capillaries into the interstitial fluid and become part of the lymph and (2) substances derived from cells, such as hormones, enzymes, and waste products, are also found in the lymph.

2. *Fat absorption.* The lymphatic system absorbs fats and other substances from the digestive tract (see chapter 21). Lymphatic vessels called **lacteals** (lak′tē-ălz) are located in the lining of the small intestine. Fats enter the lacteals and pass through the lymphatic vessels to the venous circulation. The lymph passing through these lymphatic vessels, called **chyle** (kīl), has a milky appearance because of its fat content.

3. *Defense.* Microorganisms and other foreign substances are filtered from lymph by lymph nodes and from blood by the spleen. In addition, lymphocytes and other cells are capable of destroying microorganisms and other foreign substances. ▼

1. **List the parts of the lymphatic system, and list the three main functions of the lymphatic system.**
2. **Define lymph. How is it formed?**
3. **What is the function of lacteals?**

Lymphatic Vessels

The lymphatic system, unlike the circulatory system, does not circulate fluid to and from tissues. Instead, the lymphatic system carries fluid in one direction, from tissues to the circulatory system. Fluid moves from blood capillaries into the interstitial spaces. Most of the fluid returns to the blood, but some of the fluid moves from the interstitial spaces into lymphatic capillaries to become lymph (figure 19.2*a*). The **lymphatic capillaries** are tiny, closed-ended vessels consisting of simple squamous epithelium.

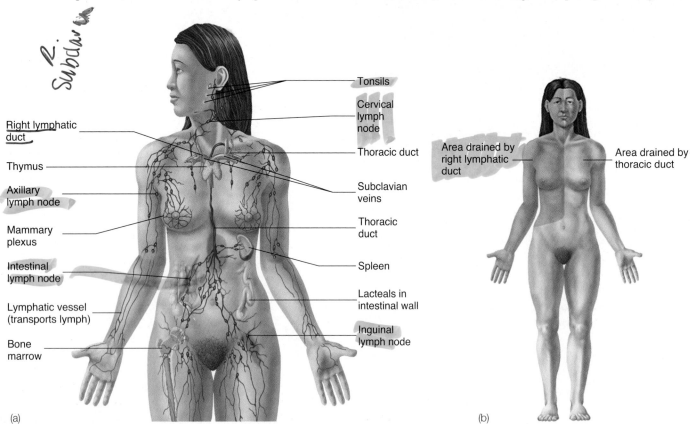

Right lymphatic duct

Thymus

Axillary lymph node

Mammary plexus

Intestinal lymph node

Lymphatic vessel (transports lymph)

Bone marrow

Tonsils

Cervical lymph node

Thoracic duct

Subclavian veins

Thoracic duct

Spleen

Lacteals in intestinal wall

Inguinal lymph node

Area drained by right lymphatic duct

Area drained by thoracic duct

(a)

(b)

Figure 19.1 **Lymphatic System and Lymph Drainage**

(*a*) The major lymphatic organs and vessels are shown. (*b*) Lymph from the uncolored areas drains through the thoracic duct. Lymph from the brown areas drains through the right lymphatic duct.

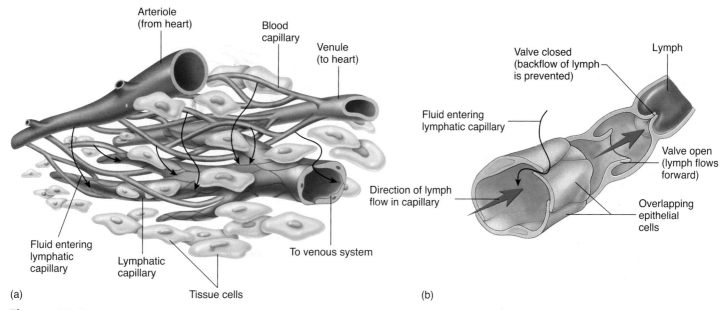

Figure 19.2 Lymph Formation and Movement

(a) Movement of fluid from blood capillaries into tissues and from tissues into lymphatic capillaries to form lymph. (b) The overlap of epithelial cells of the lymphatic capillary allows easy entry of fluid but prevents movement back into the tissue. Valves, located farther along in lymphatic vessels, also ensure the one-way flow of lymph.

Lymphatic capillaries differ from blood capillaries in that they lack a basement membrane and the cells of the simple squamous epithelium slightly overlap and are attached loosely to one another (figure 19.2b). Two things occur as a result of this structure. First, the lymphatic capillaries are far more permeable than blood capillaries, and nothing in the interstitial fluid is excluded from the lymphatic capillaries. Second, the lymphatic capillary epithelium functions as a series of one-way valves that allow fluid to enter the capillary but prevent it from passing back into the interstitial spaces.

Lymphatic capillaries are in most tissues of the body. Exceptions are the central nervous system, bone marrow, and tissues without blood vessels, such as the epidermis and cartilage. A superficial group of lymphatic capillaries drains the dermis and subcutaneous tissue, and a deep group drains muscle, viscera, and other deep structures.

The lymphatic capillaries join to form larger **lymphatic vessels,** which resemble small veins (see figure 19.2b). Small lymphatic vessels have a beaded appearance because of one-way valves that are similar to the valves of veins. When a lymphatic vessel is compressed, the valves prevent the backward movement of lymph. Consequently, compression of lymphatic vessels causes lymph to move forward through them. Three factors cause compression of the lymphatic vessels: (1) the periodic contraction of smooth muscle in the lymphatic vessel wall, (2) the contraction of surrounding skeletal muscle during activity, and (3) pressure changes in the thorax during respiration.

The lymphatic vessels converge and eventually empty into the blood at two locations in the body. Lymphatic vessels from the upper right limb and the right half of the head, neck, and chest empty into the right subclavian vein. These lymphatic vessels often converge to form a short duct, 1 cm in length, called the **right lymphatic duct,** which connects to the subclavian vein (see figure

19.1). Lymphatic vessels from the rest of the body enter the **thoracic duct,** which empties into the left subclavian vein. The thoracic duct is the largest lymphatic vessel. It is approximately 38–45 cm in length, extending from the twelfth thoracic vertebra to the base of the neck. ▼

4. *Describe the structure of a lymphatic capillary. Why is it easy for fluid and other substances to enter a lymphatic capillary?*
5. *What is the function of the valves in lymphatic vessels? Name three mechanisms responsible for the movement of lymph through the lymphatic vessels.*
6. *Where does lymph return to the blood? What areas of the body are drained by the right lymphatic and thoracic ducts?*

Lymphatic Tissue and Organs

The **lymphatic organs** include the tonsils, lymph nodes, spleen, and thymus. Lymphatic organs contain **lymphatic tissue,** which consists primarily of lymphocytes, but it also includes macrophages, dendritic cells, reticular cells, and other cell types. **Lymphocytes** are a type of white blood cell (see chapter 16). The two types of lymphocytes are **B cells** and **T cells.** They originate from red bone marrow and are carried by the blood to lymphatic organs and other tissues. When the body is exposed to microorganisms or other foreign substances, the lymphocytes divide, increase in number, and are part of the immune response that destroys microorganisms and foreign substances. Lymphatic tissue also has very fine collagen fibers, called **reticular fibers,** which are produced by **reticular cells.** Lymphocytes and other cells attach to these fibers. When lymph or blood filters through lymphatic organs, the fiber network traps microorganisms and other particles in the fluid.

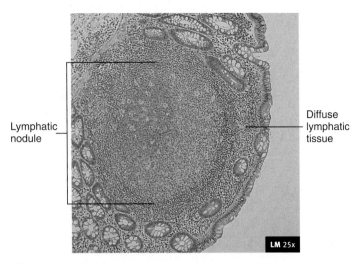

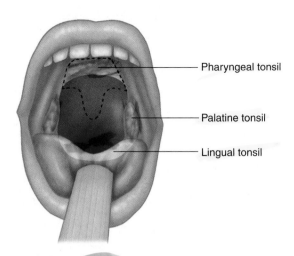

Pharyngeal tonsil

Palatine tonsil

Lingual tonsil

Lymphatic nodule

Diffuse lymphatic tissue

LM 25x

Figure 19.3 Diffuse Lymphatic Tissue and Lymphatic Nodule

Diffuse lymphatic tissue surrounding a lymphatic nodule in the small intestine.

Figure 19.4 Tonsils

Anterior view of the oral cavity, showing the tonsils. Part of the palate is removed (*dotted line*) to show the pharyngeal tonsil.

Lymphatic tissue surrounded by a connective tissue capsule is said to be encapsulated, whereas lymphatic tissue without a capsule is called nonencapsulated. Lymphatic organs with a capsule include lymph nodes, the spleen, and the thymus. Nonencapsulated lymphatic tissue includes diffuse lymphatic tissue, lymphatic nodules, and the tonsils.

Diffuse Lymphatic Tissue and Lymphatic Nodules

Diffuse lymphatic tissue has no clear boundary, blends with surrounding tissues, and contains dispersed lymphocytes, macrophages, and other cells (figure 19.3). It is located deep to mucous membranes, around lymphatic nodules, and within the lymph nodes and spleen.

Lymphatic nodules are denser arrangements of lymphatic tissue organized into compact, somewhat spherical structures, ranging in size from a few hundred microns to a few millimeters or more in diameter (see figure 19.3). Lymphatic nodules are numerous in the loose connective tissue of the digestive, respiratory, urinary, and reproductive systems. **Peyer's patches** are aggregations of lymphatic nodules found in the distal half of the small intestine and the appendix. Lymphatic nodules are also found within lymph nodes and the spleen.

Tonsils

Tonsils are large groups of lymphatic nodules and diffuse lymphatic tissue located deep to the mucous membranes within the pharynx (throat). The tonsils provide protection against bacteria and other potentially harmful material entering the pharynx from the nasal or oral cavities. In adults, the tonsils decrease in size and eventually may disappear.

There are three groups of tonsils (figure 19.4). The **palatine** (pal′ă-tīn, palate) **tonsils** are located on each side of the posterior opening of the oral cavity. They usually are referred to as "the tonsils." The **pharyngeal** (fă-rin′jē-ăl) **tonsil** is located near the internal opening of the nasal cavity. When the pharyngeal tonsil is enlarged, it is commonly referred to as the **adenoid** (ad′ē-noid,

glandlike) or **adenoids.** An enlarged pharyngeal tonsil can interfere with normal breathing. The **lingual** (ling′gwăl, tongue) **tonsil** is on the posterior surface of the tongue.

  Tonsillectomy and Adenoidectomy

Enlarged tonsils make swallowing difficult, and infected tonsils stimulate inflammation that causes a sore throat. An enlarged adenoid restricts air flow, causing snoring and sleeping with the mouth open. In addition, an enlarged and/or infected adenoid is associated with chronic middle ear infections because the openings to the auditory tubes are located next to the adenoid. Chronic middle ear infections are associated with loss of hearing, which in turn affects speech development. A **tonsillectomy** (ton′si-lek′tō-mē) is the removal of the palatine tonsils, and an **adenoidectomy** (ad′ē-noy-dek′tō-mē) is the removal of the adenoid. Both procedures performed at the same time is called a T&A. ▼

7. *What are the functions of lymphocytes and reticular fibers in lymphatic tissue?*
8. *Define diffuse lymphatic tissue, lymphatic nodule, and Peyer's patches.*
9. *Describe the structure, function, and location of the three groups of tonsils.*

Lymph Nodes

Lymph nodes are small, round, or bean-shaped structures, ranging in size from 1–25 mm long. They are distributed along the course of the lymphatic vessels (see figure 19.1). They filter the lymph, removing bacteria and other materials. In addition, lymphocytes congregate, function, and proliferate within lymph nodes.

Approximately 450 lymph nodes are found throughout the body. **Superficial lymph nodes** are in the subcutaneous tissue, and

Clinical Relevance Disorders of the Lymphatic System

It is not surprising that many infectious diseases produce symptoms associated with the lymphatic system, because the lymphatic system is involved with the production of lymphocytes that fight infectious diseases, as well as filtering blood and lymph to remove microorganisms. **Lymphadenitis** (lim-fad′ĕ-nī′tis) is an inflammation of the lymph nodes, causing them to become enlarged and tender. It is an indication that microorganisms are being trapped and destroyed within the lymph nodes. **Lymphangitis** (lim-fan-jī′tis) is an inflammation of the lymphatic vessels. This often results in visible red streaks in the skin that extend away from the site of infection. If microorganisms pass through the lymphatic vessels and lymph nodes to reach the blood, **septicemia** (sep-ti-sē′mē-ă), or blood poisoning, can result.

A **lymphoma** (lim-fō′mă) is a neoplasm (tumor) of lymphatic tissue that is almost always malignant. Lymphomas are usually divided into two groups: **Hodgkin disease** and all other lymphomas, which are called **non-Hodgkin lymphomas.** The different types of lymphomas are diagnosed based on their histological appearance and cell of origin.

Typically, a lymphoma begins as an enlarged, painless mass of lymph nodes. Enlargement of the lymph nodes can compress surrounding structures and produce complications. Immunity is depressed, and the patient has an increased susceptibility to infections. Fortunately, treatment with drugs and radiation is effective for many people who suffer from lymphomas.

Bubonic plague and elephantiasis are diseases of the lymphatic system. **Bubonic** (bū-bon′ik) **plague** is caused by bacteria that are transferred to humans from rats by the bite of the rat flea. The bacteria localize in the lymph nodes, causing them to enlarge. The term *bubonic* is derived from a Greek word referring to the groin, because the disease often causes the inguinal lymph nodes of the groin to swell. Without treatment, septicemia followed rapidly by death occurs in 70%–90% of those infected. In the sixth, fourteenth, and nineteenth centuries, the bubonic plague killed large numbers of people in Europe. Fortunately, there are relatively few cases today.

Elephantiasis (el-ĕ-fan-tī′ă-sis) is caused by long, slender roundworms. The adult worms lodge in the lymphatic vessels and can block lymph flow. Consequently, edema develops and a limb can become permanently swollen and enlarged. The resemblance of an affected limb to that of an elephant's leg is the basis for the name of the disease. The offspring of the adult worms pass through the lymphatic system into the blood. They can be transferred from an infected person to other humans by mosquitoes.

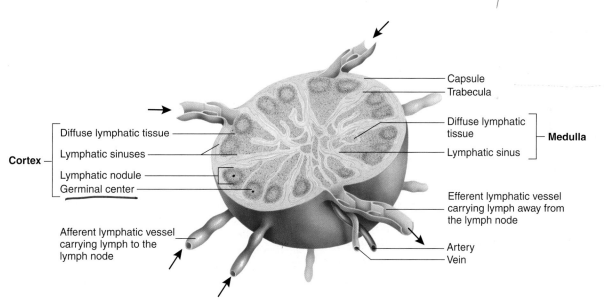

Figure 19.5   Lymph Node

Arrows indicate the direction of lymph flow. As lymph moves through the sinuses, phagocytic cells remove foreign substances. The germinal centers are sites of lymphocyte production.

deep lymph nodes are everywhere else. There are three superficial groups of lymph nodes on each side of the body: inguinal nodes in the groin, axillary nodes in the axilla (armpit), and cervical nodes in the neck.

A dense connective tissue **capsule** surrounds each lymph node (figure 19.5). Extensions of the capsule, called **trabeculae** (tră-bek′ū-lē), subdivide lymph nodes into compartments containing diffuse lymphatic tissue, lymphatic nodules, and lymphatic sinuses. **Lymphatic sinuses** are spaces between lymphatic tissue containing macrophages on a network of reticular fibers. The outer **cortex** consists of lymphatic nodules separated by diffuse lymphatic tissue and lymphatic sinuses. The inner **medulla** is organized into branching, irregular strands of diffuse lymphatic tissue separated by lymphatic sinuses.

Lymph nodes are the only structures to filter lymph. They have **afferent lymphatic vessels,** which carry lymph to the lymph nodes, where it is filtered, and **efferent lymphatic vessels,** which carry lymph away from the nodes (see figure 19.5). As lymph moves through the lymph nodes, two functions are performed. One function is activation of the immune system. Microorganisms or other foreign substances in the lymph can stimulate lymphocytes in the lymphatic tissue to divide. The lymph nodules containing the rapidly dividing lymphocytes are called **germinal centers.** The newly produced lymphocytes are released into the lymph and eventually reach the blood, where they circulate and enter other lymphatic tissues. The lymphocytes are part of the adaptive immune response (see p. 578) that destroys microorganisms and foreign substances. Another function of the lymph nodes is the removal of microorganisms and foreign substances from the lymph by macrophages. ▼

> 10. *Where are lymph nodes found? Describe the parts of a lymph node, and explain how lymph flows through a lymph node.*
>
> 11. *What are the functions of lymph nodes? What is a germinal center?*

CASE STUDY | Lymphedema

Cindy, a 40-year-old woman, had been diagnosed with breast cancer. Before removing the cancerous tumor from her left breast, her surgeon injected a dye and a radioactive tracer, called technetium-99, at the tumor site. The dye and tracer enabled the surgeon to find sentinel lymph nodes, which are the lymph nodes closest to the tumor. The sentinel lymph nodes were sampled for cancer. When cancer cells were found in all of them, the surgeon removed the axillary lymph nodes from under Cindy's left arm.

A few days after the surgery, Cindy noticed that the skin on her left arm felt tight and the arm felt heavy. In addition, her wedding ring was tighter than usual. Cindy had an abnormal accumulation of fluid in her upper limb, called **lymphedema** (limf′e-dē′mă), caused by the removal of her axillary lymph nodes. In the United States, the most common cause of lymphedema is the removal or damage of lymph nodes and vessels by cancer surgery or radiation treatment. Approximately 10%–20% of women who have their axillary lymph nodes removed develop lymphedema.

P R E D I C T

a. What is the rationale for testing sentinel lymph nodes for cancer?

b. What was the rationale for removing Cindy's axillary lymph nodes?

c. Why does removing the axillary lymph nodes result in lymphedema?

d. Exercise can help reduce lymphedema. Explain.

e. A compression bandage or garment can help reduce lymphedema. Explain.

f. In intermittent pneumatic pump compression therapy, the pressure of a garment enclosing a limb increases and decreases periodically. In addition, the pressure increases sequentially from the distal part to the proximal part of a limb. How does this therapy help reduce lymphedema?

Spleen

The **spleen** (splēn) is roughly the size of a clenched fist, and it is located in the left, superior corner of the abdominal cavity (figure 19.6*a*). The spleen has an outer **capsule** of dense connective tissue and a small amount of smooth muscle. **Trabeculae** from the capsule divide the spleen into small, interconnected compartments containing two specialized types of lymphatic tissue called **white pulp** and **red pulp.** Approximately one-fourth of the volume of the spleen is white pulp and three-fourths is red pulp.

Branches of the **splenic** (splen′ik) **artery** enter the spleen and their branches follow the various trabeculae into the spleen to supply white pulp within the compartments. Blood flows from white pulp into red pulp. Veins from the red pulp converge, forming the splenic vein, which exits the spleen.

White pulp is diffuse lymphatic tissue and lymphatic nodules surrounding the arteries within the spleen. Red pulp is associated with the veins (figure 19.6*b*). It consists of the splenic cords and the venous sinuses. The **splenic cords** are a network of reticular cells that produce reticular fibers. The spaces between the reticular cells are occupied by macrophages and blood cells that have come from the capillaries. The **venous sinuses** are enlarged capillaries between the splenic cords. They are unusual in that they have large, intercellular slits in their walls (figure 19.6*c*).

Most white pulp capillaries and red pulp venous sinuses are separated by a small gap, although a few are directly connected to each other. Most blood entering the spleen flows quickly through it by passing across the gap between white and red pulp capillaries. Some blood, however, enters the gap and then flows into the fibrous network of the splenic cords. This blood slowly percolates through the fibrous network, eventually passing through intercellular slits in the walls of the venous sinuses to enter the blood (see figure 19.6*c*).

The spleen destroys defective red blood cells and detects and responds to foreign substances in the blood. As red blood cells age, they lose their ability to bend and fold. Consequently, the cells can rupture as they pass slowly through the fibrous network of the splenic cords or the intercellular slits of the venous sinuses. Splenic macrophages then phagocytize the cellular debris. Foreign substances in the blood passing through the spleen can stimulate an immune response because of the presence in the white pulp of lymphocytes (see "Adaptive Immunity," p. 578).

The spleen also functions as a blood reservoir, holding a small volume of blood. For example, during exercise splenic volume can be reduced by 40%–50%. The resulting small increase in

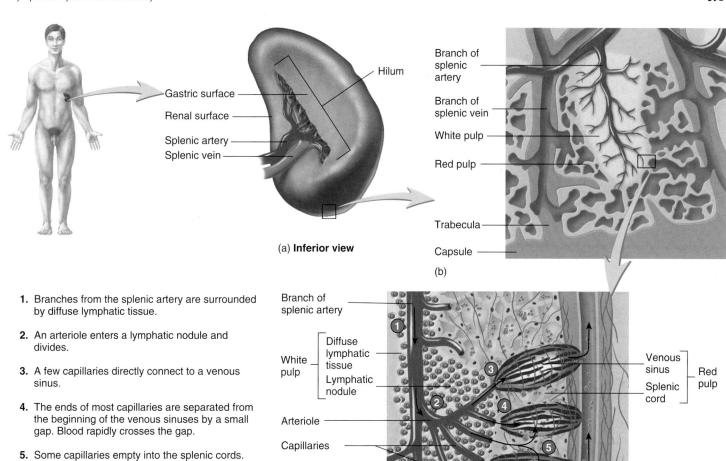

Hilum

Gastric surface
Renal surface
Splenic artery
Splenic vein

(a) **Inferior view**

Branch of splenic artery
Branch of splenic vein
White pulp
Red pulp
Trabecula
Capsule

(b)

1. Branches from the splenic artery are surrounded by diffuse lymphatic tissue.

2. An arteriole enters a lymphatic nodule and divides.

3. A few capillaries directly connect to a venous sinus.

4. The ends of most capillaries are separated from the beginning of the venous sinuses by a small gap. Blood rapidly crosses the gap.

5. Some capillaries empty into the splenic cords. Blood percolates through the splenic cords and passes through the walls of the venous sinuses.

6. The venous sinuses connect to a branch of the splenic vein.

Branch of splenic artery
White pulp — Diffuse lymphatic tissue / Lymphatic nodule
Arteriole
Capillaries
Reticular cell
Space

Venous sinus / Splenic cord — Red pulp
Trabecula
Branch of splenic vein

(c)

Process Figure 19.6   Spleen

(a) Inferior view of the spleen. (b) Section showing the arrangement of arteries, veins, white pulp, and red pulp. White pulp is associated with arteries, and red pulp is associated with veins. (c) Blood flow through white and red pulp.

circulating red blood cells can promote better oxygen delivery to muscles during exercise or emergency situations.

 Ruptured Spleen

Although the spleen is protected by the ribs, it is often ruptured in traumatic abdominal injuries. A ruptured spleen can cause severe bleeding, shock, and death. Surgical intervention may stop the bleeding. Cracks in the spleen are repaired using sutures and blood-clotting agents. Mesh wrapped around the spleen can hold it together. A **splenectomy** (splē-nek′tō-mē), removal of the spleen, may be necessary if these techniques do not stop the bleeding. Other lymphatic organs and the liver compensate for the loss of the spleen's functions. ▼

12. Where is the spleen located? Describe white and red pulp.

13. Where does blood enter and exit the red pulp?
14. What are three functions of the spleen?

Thymus

The **thymus** (thī′mŭs, sweetbread) is a bilobed gland (figure 19.7) located in the superior mediastinum, the partition dividing the thoracic cavity into left and right parts. The thymus increases in size until the first year of life, after which it remains approximately the same size until 60 years of age, after which it decreases in size. Although the size of the thymus is fairly constant throughout much of life, by 40 years of age much of the thymic lymphatic tissue has been replaced with adipose tissue.

Each lobe of the thymus is surrounded by a thin connective tissue **capsule. Trabeculae** extend from the capsule into the substance of the gland, dividing it into **lobules.** Unlike other lymphatic tissue, which has a fibrous network of reticular fibers, the framework of thymic tissue consists of epithelial cells. The

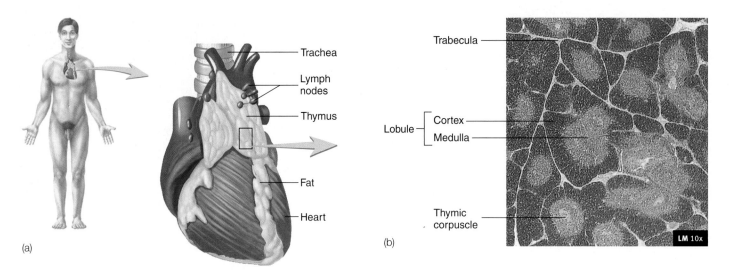

Figure 19.7 Thymus
(a) Location and shape of the thymus. (b) Histology of thymic lobules, showing outer cortex and inner medulla.

processes of the epithelial cells are joined by desmosomes, and the cells form small, irregularly shaped compartments filled with lymphocytes. Near the capsule and trabeculae, the lymphocytes are numerous and form dark-staining areas of the lobules called the **cortex.** A lighter-staining, central portion of the lobules, called the **medulla,** has fewer lymphocytes. The medulla also contains rounded epithelial structures, called **thymic corpuscles** (Hassall corpuscles), whose function is unknown.

The thymus is the site of the maturation of T cells. Large numbers of lymphocytes are produced in the thymus, but most degenerate. The surviving thymic lymphocytes enter the blood and travel to other lymphatic tissues. These lymphocytes are capable of reacting to foreign substances, but they normally do not react to and destroy healthy body cells (see "Origin and Development of Lymphocytes," p. 579). ▼

15. *Where is the thymus located? Describe its structure and function.*

Overview of the Lymphatic System

Figure 19.8 summarizes the parts of the lymphatic system and their functions. Lymphatic capillaries and vessels remove fluid from tissues and absorb fats from the small intestines. Lymph nodes filter lymph, and the spleen filters blood.

Figure 19.8 also illustrates two types of lymphocytes, the B and T cells. Pre-B cells originate and mature in red bone marrow to become B cells. Pre-T cells are produced in red bone marrow and migrate to the thymus, where they mature to become T cells. B cells from red bone marrow and T cells from the thymus circulate to, and populate, other lymphatic tissues.

B and T cells are responsible for much of immunity. In response to infections, B and T cells increase in number and circulate to lymphatic and infected tissues. How B and T cells protect the body is discussed in the section "Adaptive Immunity," p. 578.

Immunity

Immunity is the ability to resist damage from foreign substances, such as microorganisms; harmful chemicals; and internal threats, such as cancer cells. Immunity is categorized as **innate immunity** and **adaptive immunity,** although the two systems are fully integrated in the body. In innate immunity, the body recognizes and destroys certain foreign substances, but the response to them is the same each time the body is exposed to them. In adaptive immunity, the body recognizes and destroys foreign substances, but the response to them improves each time the foreign substance is encountered.

Specificity and memory are characteristics of adaptive immunity but not innate immunity. **Specificity** is the ability of adaptive immunity to recognize a particular substance. For example, innate immunity can act against bacteria in general, whereas adaptive immunity can distinguish among various kinds of bacteria. **Memory** is the ability of adaptive immunity to "remember" previous encounters with a particular substance. As a result, the response is faster and stronger and lasts longer.

In innate immunity, each time the body is exposed to a substance, the response is the same because specificity and memory of previous encounters are not present. For example, each time a bacterial cell is introduced into the body, it is phagocytized with the same speed and efficiency. Innate immunity is present to some degree in all multicellular organisms.

In adaptive immunity, the response during the second exposure is faster and stronger than the response to the first exposure because the immune system remembers the bacteria from the first exposure. For example, following the first exposure to the bacteria, the body can take many days to destroy them. During this time, the bacteria damage tissues and produce the symptoms of disease. Following the second exposure to the same bacteria, the response is rapid and effective. Bacteria are destroyed before any symptoms

(a) Lymphatic capillaries remove fluid from tissues. The fluid becomes lymph (see figure 19.2*a*).

(b) Lymph flows through lymphatic vessels, which have valves that prevent the backflow of lymph (see figure 19.2*b*).

(c) Lymph nodes filter lymph (see figure 19.5) and are sites where lymphocytes respond to infections, etc.

(d) Lymph enters the thoracic duct or the right lymphatic duct.

(e) Lymph enters the blood.

(f) Lacteals in the small intestine absorb fats, which enter the thoracic duct.

(g) Chyle, which is lymph containing fats, enters the blood.

(h) The spleen (see figure 19.6) filters blood and is a site where lymphocytes respond to infections, etc.

(i) Lymphocytes (pre-B and pre-T cells) originate from stem cells in the red bone marrow (see figure 19.11). The pre-B cells become mature B cells in the red bone marrow and are released into the blood. The pre-T cells enter the blood and migrate to the thymus.

(j) The thymus (see figure 19.7) is where pre-T cells derived from red bone marrow increase in number and become mature T cells that are released into the blood (see figure 19.11).

(k) B and T cells from the blood enter and populate all lymphatic tissues. These lymphocytes can remain in tissues or pass through them and return to the blood. B and T cells can also respond to infections, etc., by dividing and increasing in number. Some of the newly formed cells enter the blood and circulate to other tissues.

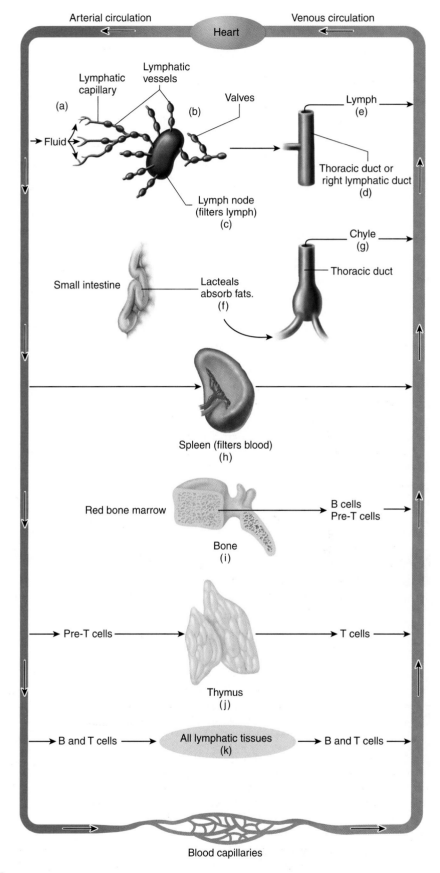

Figure 19.8 Overview of the Lymphatic System

develop, and the person is said to be **immune.** Adaptive immunity is unique to vertebrates.

Innate and adaptive immunity are intimately linked. Innate immunity is required for the initiation and regulation of the adaptive immune response. ▼

16. *Define immunity, specificity, and memory.*
17. *What are the differences between innate and adaptive immunity?*

Innate Immunity

Innate immunity is accomplished by mechanical mechanisms, chemical mediators, cells, and the inflammatory response.

Mechanical Mechanisms

Mechanical mechanisms prevent the entry of microorganisms and chemicals into the body in two ways: (1) The skin and mucous membranes form barriers that prevent their entry, and (2) tears, saliva, and urine wash them from the surfaces of the body. Microorganisms cannot cause a disease if they cannot get into the body.

Chemical Mediators

Chemical mediators are molecules responsible for many aspects of innate immunity (table 19.1). Some chemical mediators on the surface of cells, such as lysozyme, sebum, and mucus, kill microorganisms or prevent their entry into the cells. Other chemical mediators, such as histamine, complement, and eicosanoids (e.g., prostaglandins and leukotrienes), promote inflammation by causing vasodilation, increasing vascular permeability, attracting white blood cells, and stimulating phagocytosis. **Cytokines** (si'tō-kīnz) are proteins or peptides secreted by cells that bind to receptors on cell surfaces, stimulating a response. They usually bind to receptors on neighboring cells, but sometimes they bind to receptors on the secreting cell. Cytokines regulate the intensity and duration of immune responses and stimulate the proliferation and differentiation of cells. Examples of cytokines are interferons, interleukins, and lymphokines.

Complement

Complement is a group of about 20 proteins found in plasma. Normally, complement proteins circulate in the blood in an inactive, nonfunctional form. They become activated in the **complement cascade,** a series of reactions in which each component of the series activates the next component (figure 19.9). The complement cascade begins through either the classical pathway or the alternative pathway. The **classical pathway** is part of adaptive immunity, discussed on p. 585.

The **alternative pathway,** part of innate immunity, is initiated when the complement protein C3 becomes spontaneously active. Activated C3 normally is quickly inactivated by proteins on the surface of the body's cells. Activated C3 can combine with some foreign substances, such as part of a bacterial cell or virus. It can become stabilized and cause activation of the complement cascade.

Activated complement proteins provide protection in several ways (see figure 19.9). They can form a **membrane attack complex (MAC),** which produces a channel through the plasma membrane. The movement of water through the channels results in the lysis (rupture) of cells. Complement proteins can also attach to the surface of bacterial cells and stimulate macrophages to

Table 19.1	Chemical Mediators of Innate Immunity and Their Functions			
Chemical	**Description**	**Chemical**	**Description**
Surface chemicals	Lysozymes (in tears, saliva, nasal secretions, and sweat) lyse cells; acid secretions (sebum in the skin and hydrochloric acid in the stomach) prevent microbial growth or kill microorganisms; mucus on the mucous membranes traps microorganisms until they can be destroyed.	Complement	Complement is a group of plasma proteins that increase vascular permeability, stimulate the release of histamine, activate kinins, lyse cells, promote phagocytosis, and attract neutrophils, monocytes, macrophages, and eosinophils.
Histamine	Histamine is an amine released from mast cells, basophils, and platelets; histamine causes vasodilation, increases vascular permeability, stimulates gland secretions (especially mucus and tear production), causes smooth muscle contraction of airway passages (bronchioles) in the lungs, and attracts eosinophils.	Prostaglandins	Prostaglandins are a group of lipids (PGEs, PGFs, thromboxanes, and prostacyclins), some of which cause smooth muscle relaxation and vasodilation, increase vascular permeability, and stimulate pain receptors.
Kinins	Kinins are polypeptides derived from plasma proteins; kinins cause vasodilation, increase vascular permeability, stimulate pain receptors, and attract neutrophils.	Leukotrienes	Leukotrienes are a group of lipids, produced primarily by mast cells and basophils, that cause prolonged smooth muscle contraction (especially in the lung bronchioles), increase vascular permeability, and attract neutrophils and eosinophils.
Interferons	Interferons are proteins, produced by most cells, that interfere with virus production and infection.	Pyrogens	Pyrogens are chemicals, released by neutrophils, monocytes, and other cells, that stimulate fever production.

Abbreviations: PGE = prostaglandin E; PGF = prostaglandin F.

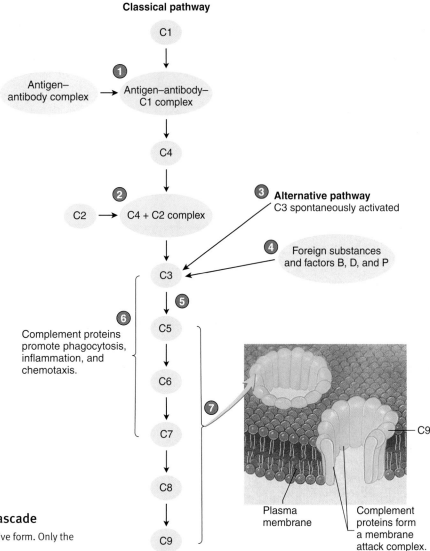

1. The classical pathway begins when an antigen–antibody complex binds to C1. The C1–antigen–antibody complex activates C4.

2. Activated C4 forms a complex with C2 that activates C3.

3. The alternate pathway begins when C3 is spontaneously activated.

4. Foreign substances and factors B, D, and P stabilize activated C3.

5. Once C3 is activated, the classical and alternate pathways are the same. C3 activates C5, C5 activates C6, C6 activates C7, C7 activates C8, and C8 activates C9.

6. Activated C3–C7 promote phagocytosis, inflammation, and chemotaxis (attracts cells).

7. Activated C5–C9 combine to form a membrane attack complex (MAC), which forms a channel through the plasma membrane (only C9 of MAC is shown).

Process Figure 19.9 Complement Cascade

Complement proteins circulate in the blood in an inactive form. Only the activated proteins are shown.

phagocytize the bacteria. In addition, complement proteins attract immune system cells to sites of infection and promote inflammation.

Interferons

Interferons (in-ter-fēr′onz) are proteins that protect the body against viral infection. When a virus infects a cell, the infected cell produces viral nucleic acids and proteins, which are assembled into new viruses. The new viruses are released from the infected cell to infect other cells. Viral infections are harmful to the body because infected cells usually stop their normal functions or die during viral replication. Fortunately, viruses often stimulate infected cells to produce interferons. Interferons do not protect the cell that produces them. Instead, interferons bind to the surface of neighboring cells, where they stimulate those cells to produce antiviral proteins. These antiviral proteins inhibit viral reproduction by preventing the production of new viral nucleic acids and proteins. Some interferons also play a role in the activation of immune cells, such as macrophages and natural killer cells.

Treating Viral Infections and Cancer with Interferons

Interferons may play a role in controlling cancers because some cancers are induced by viruses. Interferons activate macrophages and natural killer cells (a type of lymphocyte), which attack tumor cells. Through genetic engineering, interferons are produced in sufficient quantities for clinical use; along with other therapies, they have been effective in treating certain viral infections and cancers. For example, interferons are used to treat hepatitis C, a viral disorder that can cause cirrhosis and cancer of the liver, and to treat genital warts, caused by the herpes virus. Interferons are also approved for the treatment of Kaposi sarcoma, a cancer that can develop in AIDS patients. ▼

18. *List the four main components of innate immunity.*
19. *Name two mechanical mechanisms that prevent the entry of microorganisms into the body.*

Table 19.2 Immune System Cells and Their Primary Functions

Cell	Primary Function	Cell	Primary Function
Innate Immunity Neutrophil	Phagocytosis and inflammation; usually the first cell to leave the blood and enter infected tissues	**Adaptive Immunity** B cell	After activation, differentiates to become plasma cell or memory B cell
Monocyte	Leaves the blood and enters tissues to become a macrophage	Plasma cell	Produces antibodies that are directly or indirectly responsible for the destruction of the antigen
Macrophage	Most effective phagocyte; important in later stages of infection and in tissue repair; located throughout the body to "intercept" foreign substances; processes antigens; involved in the activation of B and T cells	Memory B cell	Quick and effective response to an antigen against which the immune system has previously reacted; responsible for immunity
Basophil	Motile cell that leaves the blood, enters tissues, and releases chemicals that promote inflammation	Cytotoxic T cell	Responsible for the destruction of cells by lysis or by the production of cytokines
Mast cell	Nonmotile cell in connective tissues that promotes inflammation through the release of chemicals	Delayed hypersensitivity T cell	Produces cytokines that promote inflammation
		Helper T cell	Activates B and effector T cells
Eosinophil	Enters tissues from the blood and releases chemicals that inhibit inflammation	Suppressor T cell	Inhibits B and effector T cells
		Memory T cell	Quick and effective response to an antigen against which the immune system has previously reacted; responsible for adaptive immunity
Natural killer cell	Lyses tumor and virus-infected cells	Dendritic cell	Processes antigen and is involved in the activation of B and T cells

20. *What is complement? In what two ways is it activated? How does complement provide protection?*
21. *What are interferons? How do they provide protection against viruses?*

Cells

White blood cells and the cells derived from them are the most important cellular components of the immune system (table 19.2). White blood cells are produced in red bone marrow and lymphatic tissue and are released into the blood, where they are transported throughout the body. Chemicals released from microorganisms or damaged tissues attract the white blood cells, and they leave the blood and enter affected tissues. Important chemicals known to attract white blood cells include complement, leukotrienes, kinins (kī′ninz), and histamine. The movement of white blood cells toward these chemicals is called **chemotaxis** (kem-ō-tak′sis, kē-mō-tak′sis).

Phagocytosis (fag-ō-sī-tō′sis) is the endocytosis and destruction of particles by cells called **phagocytes** (see figure 3.10). The particles can be microorganisms or their parts, foreign substances, or dead cells from the body. The most important phagocytic cells are neutrophils and macrophages.

Neutrophils

Neutrophils are small, phagocytic cells. Approximately 126 billion neutrophils per day leave the blood and pass through the wall of the gastrointestinal tract, where they provide phagocytic protection. The neutrophils are then eliminated as part of the feces. Neutrophils are usually the first cells to enter infected tissues in large numbers.

They release chemical signals, such as cytokines and chemotactic factors, that increase the inflammatory response by recruiting and activating other immune cells. Neutrophils often die after a single phagocytic event.

Neutrophils also release lysosomal enzymes that kill microorganisms and cause tissue damage and inflammation. **Pus** is an accumulation of dead neutrophils, dead microorganisms, debris from dead tissue, and fluid.

Macrophages

Macrophages are monocytes that leave the blood, enter tissues, enlarge about fivefold, and increase their number of lysosomes and mitochondria. They are large, phagocytic cells that outlive neutrophils, and they can ingest more and larger phagocytic particles than neutrophils can. Macrophages usually accumulate in tissues after neutrophils and are responsible for most of the phagocytic activity in the late stages of an infection, including the cleanup of dead neutrophils and other cellular debris. In addition to their phagocytic role, macrophages produce a variety of chemicals, such as interferons, prostaglandins, and complement, that enhance the immune response.

Macrophages are beneath the free surfaces of the body, such as the skin (dermis), subcutaneous tissue, mucous membranes, and serous membranes, and around blood and lymphatic vessels. In these locations, macrophages provide protection by trapping and destroying microorganisms entering the tissues. If microbes do gain entry to the blood, macrophages are waiting within the blood vessels of the spleen, bone marrow, and liver to phagocytize them. Microbes that enter lymph are phagocytized as the lymph filters through lymph nodes. Sometimes macrophages are given specific names—for instance, dust cells in the lungs, Kupffer cells in the liver, and microglia in the central nervous system.

Basophils, Mast Cells, and Eosinophils

Basophils are motile white blood cells that can leave the blood and enter infected tissues. **Mast cells,** which are derived from red bone marrow, are nonmotile cells in connective tissue, especially near capillaries. Like macrophages, mast cells are located at potential points of entry of microorganisms into the body, such as the skin, lungs, gastrointestinal tract, and urogenital tract.

Basophils and mast cells can be activated through innate immunity (e.g., by complement) or through adaptive immunity (see "Effects of Antibodies," p. 585). When activated, they release chemicals, such as histamine and leukotrienes. The chemicals produce an inflammatory response or activate other mechanisms, such as smooth muscle contraction in the lungs.

Eosinophils release enzymes that break down chemicals released by basophils and mast cells. Thus, as inflammation is initiated, mechanisms are activated that contain and reduce the inflammatory response. This process is similar to the blood clotting system, in which clot prevention and removal mechanisms are activated while the clot is being formed (see chapter 16). In patients with parasitic infections or allergic reactions with much inflammation, eosinophil numbers greatly increase. Eosinophils also secrete enzymes that effectively kill some parasites.

Natural Killer (NK) Cells

Natural killer (NK) cells account for up to 15% of lymphocytes. NK cells recognize classes of cells, such as tumor cells or virus-infected cells in general, rather than specific tumor cells or cells infected by a specific virus. For this reason and because NK cells do not exhibit a memory response, they are classified as part of innate immunity. NK cells use a variety of methods to kill their target cells, including the release of chemicals that damage plasma membranes, causing the cells to lyse. ▼

22. *Define chemotaxis and phagocytosis.*
23. *What are the functions of neutrophils and macrophages? What is pus?*
24. *What effects are produced by the chemicals released from basophils, mast cells, and eosinophils?*
25. *Describe the function of NK cells.*

Inflammatory Response

The **inflammatory response** to injury involves many of the chemicals and cells previously discussed. Most inflammatory responses are very similar, although some details can vary, depending on the intensity of the response and the type of injury. A bacterial infection is used here to illustrate an inflammatory response (figure 19.10). The bacteria or damage to tissues causes the release or activation of chemical mediators, such as histamine, prostaglandins, leukotrienes, complement, and kinins. The chemicals produce several effects: (1) vasodilation

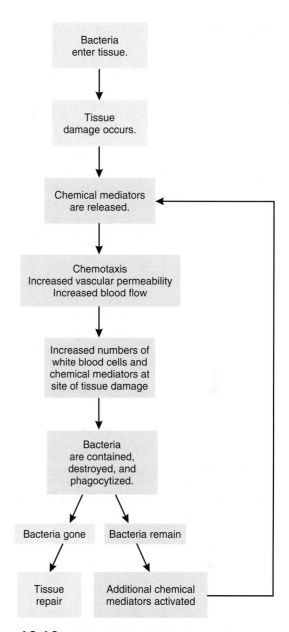

Figure 19.10 **Inflammatory Response**

Bacteria cause tissue damage and the release of chemical mediators, which initiate inflammation and phagocytosis, resulting in the destruction of the bacteria. If any bacteria remain, additional chemical mediators are activated. After all the bacteria have been destroyed, the tissue is repaired.

increases blood flow, which brings phagocytes and other white blood cells to the area; (2) chemotaxis of phagocytes, which leave the blood and enter the tissue; and (3) increased vascular permeability, allowing fibrinogen and complement to enter the tissue from the blood. Fibrinogen is converted to fibrin (see chapter 16), which isolates the infection by walling off the infected area. Complement further enhances the inflammatory response and attracts additional phagocytes. This process of releasing chemical mediators and attracting phagocytes and other white blood cells continues until the bacteria are destroyed. Phagocytes remove microorganisms and dead tissue, and the damaged tissues are repaired.

Inflammation can be localized or systemic. **Local inflammation** is an inflammatory response confined to a specific area of the body. Symptoms of local inflammation include redness, heat, swelling, pain, and loss of function. Redness, heat, and swelling result from increased blood flow and increased vascular permeability. Pain is caused by swelling and by chemical mediators acting on pain receptors. Loss of function results from tissue destruction, swelling, and pain.

Systemic inflammation is an inflammatory response that occurs in many parts of the body. In addition to the local symptoms at the sites of inflammation, three additional features can be present. First, red bone marrow produces and releases large numbers of neutrophils, which promote phagocytosis. Second, **pyrogens** (pī′rō-jenz, fire-producing), chemicals released by lymphocytes, macrophages, neutrophils, and other cells, stimulate fever production. Pyrogens affect the body's temperature-regulating mechanism in the hypothalamus, heat is conserved, and body temperature increases. Fever promotes the activities of the immune system, such as phagocytosis, and inhibits the growth of some microorganisms. Third, in severe cases of systemic inflammation, increased vascular permeability is so widespread that large amounts of fluid are lost from the blood into the tissues. The decreased blood volume can cause shock and death. ▼

26. **Describe the events that take place during an inflammatory response.**
27. **What are the symptoms of local and systemic inflammation?**

Adaptive Immunity

Adaptive immunity can recognize, respond to, and remember a particular substance. Substances that stimulate adaptive immunity are called **antigens** (an′ti-jenz). Antigens are divided into two groups: foreign antigens and self-antigens. **Foreign antigens** are not produced by the body but are introduced from outside it. Components of bacteria, viruses, and other microorganisms are examples of foreign antigens that cause disease. Pollen, animal dander (scaly, dried skin), feces of house dust mites, foods, and drugs are also foreign antigens and can trigger an overreaction of the immune system in some people, called an **allergic reaction.** Transplanted tissues and organs that contain foreign antigens result in the rejection of the transplant.

Self-antigens are molecules produced by the body that stimulate an adaptive immune system response. The response to self-antigens can be beneficial or harmful. For example, the recognition of tumor antigens can result in tumor destruction, whereas **autoimmune disease** can result when self-antigens stimulate unwanted tissue destruction.

Adaptive immunity can be divided into antibody-mediated immunity and cell-mediated immunity. **Antibody-mediated immunity** involves proteins called **antibodies,** which are found in fluids outside of cells, such as blood, interstitial fluid, and lymph. **B cells** give rise to cells that produce antibodies. **Cell-mediated immunity** involves **T cells.** Several subpopulations of T cells exist,

Table 19.3 Comparison of Innate and Adaptive Immunity

Characteristics	Innate Immunity	Adaptive Immunity Antibody-Mediated Immunity	Cell-Mediated Immunity
Primary cells	Neutrophils, eosinophils, basophils, mast cells, monocytes, and macrophages	B cells	T cells
Origin of cells	Red bone marrow	Red bone marrow	Red bone marrow
Site of maturation	Red bone marrow (neutrophils, eosinophils, basophils, and monocytes) and tissues (mast cells and macrophages)	Red bone marrow	Thymus
Location of mature cells	Blood, connective tissue, and lymphatic tissue	Blood and lymphatic tissue	Blood and lymphatic tissue
Primary secretory products	Histamine, kinins, complement, prostaglandins, leukotrienes, and interferons	Antibodies	Cytokines
Primary actions	Inflammatory response and phagocytosis	Protection against extracellular antigens (bacteria, toxins, parasites, and viruses outside cells)	Protection against intracellular antigens (viruses, intracellular bacteria, and intracellular fungi) and tumors: regulates antibody-mediated immunity and cell-mediated immunity responses (helper T and suppressor T cells)
Hypersensitivity reactions	None	Immediate hypersensitivity (e.g., hay fever, asthma, hives, and anaphylaxis)	Delayed hypersensitivity (e.g., contact hypersensitivity)

each of which is responsible for a particular aspect of cell-mediated immunity. For example, **effector T cells,** such as **cytotoxic T cells** and **delayed hypersensitivity T cells,** are responsible for producing the effects of cell-mediated immunity. **Regulatory T cells,** such as **helper T cells** and **suppressor T cells,** can promote or inhibit the activities of both antibody-mediated immunity and cell-mediated immunity.

Table 19.3 summarizes and contrasts the main features of innate and adaptive immunity. ▼

> 28. *Define antigen. Distinguish between a foreign antigen and a self-antigen.*
> 29. *What are allergic reactions and autoimmune diseases?*
> 30. *What types of lymphocytes are responsible for antibody-mediated and cell-mediated immunity?*

Origin and Development of Lymphocytes

All blood cells, including lymphocytes, are derived from stem cells in the red bone marrow (see chapter 16). Some stem cells give rise to pre-T cells, which migrate through the blood to the thymus, where they divide and are processed into T cells (see figures 19.8 and 19.11). The thymus produces hormones, such as thymosin, which stimulates T-cell maturation. Other stem cells produce pre-B cells, which are processed in the red bone marrow into B cells. A **positive selection** process results in the survival of pre-B and pre-T cells that are capable of an immune response. Cells that are incapable of an immune response die. A **negative selection** process eliminates or suppresses lymphocytes acting against self-antigens, thereby preventing the destruction of one's own cells. Although the negative selection process occurs mostly during

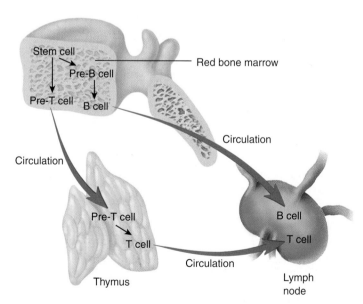

Figure 19.11 **Origin and Processing of B and T Cells**

Pre-B and pre-T cells originate from stem cells in red bone marrow. The pre-B cells remain in the red bone marrow and become B cells. The pre-T cells circulate to the thymus, where they become T cells. Both B and T cells circulate to other lymphatic tissues, such as lymph nodes, where they can divide and increase in number in response to antigens.

prenatal development, it continues throughout life (see "Inhibition of Lymphocytes," p. 583).

The B and T cells that can respond to antigens are composed of small groups of identical lymphocytes called **clones.** Although each clone can respond only to a particular antigen, such a large number of clones exist that the immune system can react to most molecules.

B cells are released from red bone marrow, T cells are released from the thymus, and both types of cells move through the blood to lymphatic tissue. There are approximately five T cells for every B cell in the blood. These lymphocytes live for a few months to many years and continually circulate between the blood and the lymphatic tissues. Antigens can come into contact with and activate lymphocytes, resulting in cell divisions that increase the number of lymphocytes that can recognize the antigen. ▼

> 31. *Describe the origin and development of B and T cells.*
> 32. *Distinguish between positive and negative lymphocyte selection. What are lymphocyte clones?*
> 33. *What happens to lymphocytes after they are produced in red bone marrow or the thymus?*

Activation of Lymphocytes

The specialized B- or T-cell clones can respond to antigens and produce an adaptive immune response. For the adaptive immune response to occur, lymphocytes first have to recognize the antigen. After recognition, the lymphocytes in a clone divide, greatly increasing the number of lymphocytes capable of responding to the antigen.

Antigenic Determinants and Antigen Receptors

Lymphocytes do not recognize an entire antigen. **Antigenic determinants,** or **epitopes** (ep′i-tōps), are specific regions of a given antigen recognized by a lymphocyte, and each antigen has many different antigenic determinants. All the lymphocytes of a given clone have, on their surfaces, identical proteins called **antigen receptors,** which combine with a specific antigenic determinant. The immune system response to an antigen with a particular antigenic determinant is similar to the lock-and-key model for enzymes (see chapter 2), and any given antigenic determinant can combine only with a specific antigen receptor. The antigen receptors on B cells are called **B-cell receptors** and those on T cells are called **T-cell receptors.**

Major Histocompatibility Complex Molecules

Although some antigens bind to their receptors and directly activate B cells and some T cells, most lymphocyte activation involves glycoproteins on the surfaces of cells called **major histocompatibility complex (MHC) molecules.** MHC molecules are attached to plasma membranes, and they can bind to foreign and self-antigens.

MHC class I molecules are found on nucleated cells; they display antigens produced inside the cells on their surfaces (figure 19.12*a*). This is necessary because the immune system cannot directly respond to an antigen inside a cell. For example, viruses

1. Foreign proteins or self-proteins within the cytosol are broken down into fragments that are antigens.
2. Antigens are transported into the rough endoplasmic reticulum.
3. Antigens combine with MHC class I molecules.
4. The MHC class I/antigen complex is transported to the Golgi apparatus, packaged into a vesicle, and transported to the plasma membrane.
5. Foreign antigens combined with MHC class I molecules stimulate cell destruction.
6. Self-antigens combined with MHC class I molecules do not stimulate cell destruction.

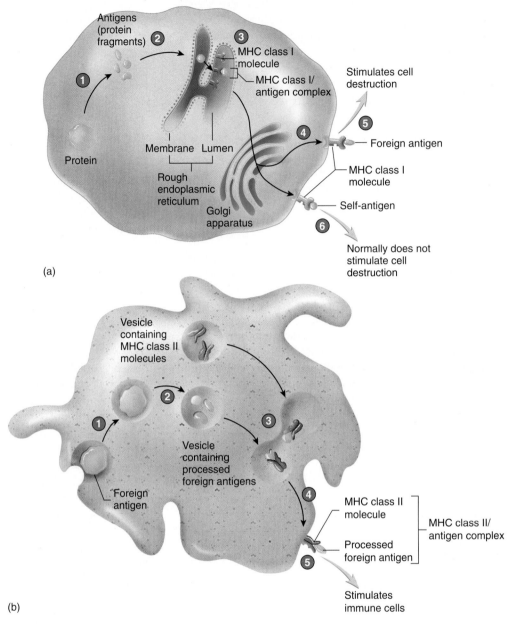

(a)

1. A foreign antigen is ingested by endocytosis and is within a vesicle.
2. The antigen is broken down into fragments to form processed foreign antigens.
3. The vesicle containing the processed foreign antigens fuses with vesicles produced by the Golgi apparatus that contain MHC class II molecules. Processed foreign antigens and MHC class II molecules combine.
4. The MHC class II/antigen complex is transported to the plasma membrane.
5. The displayed MHC class II/antigen complex can stimulate immune cells.

(b)

Process Figure 19.12   Antigen Processing

(a) Foreign proteins, such as viral proteins, or self-proteins in the cytosol are processed and presented at the cell surface by MHC class I molecules. (b) Foreign antigens are taken into an antigen-presenting cell, processed, and presented at the cell surface by MHC class II molecules.

reproduce inside cells, forming viral proteins that are foreign antigens. Some of these viral proteins are broken down in the cytoplasm. The protein fragments enter the rough endoplasmic reticulum and combine with MHC class I molecules to form complexes that move through the Golgi apparatus to be distributed on the surface of the cell (see chapter 3).

MHC class I/antigen complexes on the surface of cells can bind to T-cell receptors on the surface of T cells. This combination is a signal that activates T cells. Activated T cells can destroy infected cells, which effectively stops viral replication (see "Cell-Mediated Immunity," p. 588). Thus, the MHC class I/antigen complex functions as a signal, or "red flag," that prompts the immune

system to destroy the displaying cell. In essence, the cell is displaying a sign that says, "Kill me!" This process is said to be **MHC-restricted** because both the antigen and the individual organism's own MHC molecule are required.

P R E D I C T   2

In mouse A, T cells can respond to virus X. If these T cells are transferred to mouse B, which is infected with virus X, will the T cells respond to the virus? Explain.

The same process that moves foreign protein fragments to the surface of cells can also inadvertently transport self-protein fragments (see figure 19.12a). As part of normal protein metabolism, cells continually break down old proteins and synthesize new ones. Some self-protein fragments that result from protein breakdown can combine with MHC class I molecules and be displayed on the surface of the cell, thus becoming self-antigens. Normally, the immune system does not respond to self-antigens in combination with MHC molecules because the lymphocytes that could respond have been eliminated or inactivated.

MHC class II molecules are found on **antigen-presenting cells,** which include B cells, macrophages, monocytes, and dendritic cells. **Dendritic** (den-drit′ik) **cells** are large, motile cells with long, cytoplasmic extensions, and they are scattered throughout most tissues (except the brain), with their highest concentrations in lymphatic tissues and the skin. Dendritic cells in the skin are often called **Langerhans cells.**

Antigen-presenting cells can take in foreign antigens by endocytosis (figure 19.12*b*). Within the endocytotic vesicle, the antigen is broken down into fragments to form processed antigens. Vesicles from the Golgi apparatus containing MHC class II molecules combine with the endocytotic vesicles. The MHC class II molecules and processed antigens combine, and the MHC class II/antigen complexes are transported to the surface of the cell, where they are displayed to other immune cells.

MHC class II/antigen complexes on the surface of cells can bind to T-cell receptors on the surface of T cells. The presentation of antigens using MHC class II molecules is MHC-restricted because both the antigen and the individual's own MHC class II molecule are required. Unlike MHC class I molecules, however, this display does not result in the destruction of the antigen-presenting cell. Instead, the MHC class II/antigen complex is a "rally around the flag" signal that stimulates other immune system cells to respond to the antigen. The displaying cell is like Paul Revere, who spread the alarm for the militia to arm and organize. The militia then went out and killed the enemy. For example, when the lymphocytes of the B-cell clone that can recognize the antigen come into contact with the MHC class II/antigen complex, they are stimulated to divide. The activities of these lymphocytes, such as the production of antibodies, then result in the destruction of the antigen. ▼

34. *Define antigenic determinant and antigen receptor. How are they related to each other?*
35. *What type of antigens are displayed by MHC class I and II molecules?*
36. *What type of cells display MHC class I and II antigen complexes, and what happens as a result?*
37. *Define MHC-restricted.*

P R E D I C T 3

How does elimination of an antigen stop the production of antibodies?

Lymphocyte Proliferation

Before exposure to an antigen, the number of lymphocytes in a clone is too small to produce an effective response against the antigen. Exposure to an antigen results in an increase in lymphocyte number. The first step in this process is an increase in the number of helper T cells. This is necessary because the increased number of helper T cells responding to the antigen can find and stimulate B or effector T cells. The second step is an increase in the number of B or effector T cells. This is necessary because these cells are responsible for the immune response that destroys the antigen.

Antigen-presenting cells use MHC class II molecules to present processed antigens to helper T cells (figure 9.13, steps 1 and 2). Only the helper T cells with the T-cell receptors that can bind to the antigen respond. The combination of an MHC class II/antigen complex with an antigen receptor is usually only the first signal necessary to produce a response from a B or T cell. In many cases, **costimulation** by additional signals is also required. Costimulation is accomplished by cytokines, such as **interleukin-1,** and by molecules attached to the surface of cells (figure 19.13, step 3). Table 19.4 lists important cytokines and their functions.

Lymphocytes have other surface molecules besides MHC molecules that help bind cells together and stimulate a response. For example, a molecule called B7 on macrophages must bind with a molecule called CD28 on helper T cells before the helper T cells can respond to the antigen presented by the macrophage (figure 19.13, step 4). Helper T cells have a glycoprotein called CD4, which helps to connect helper T cells to the macrophage by binding to MHC molecules. For this reason, helper T cells are sometimes referred to as **CD4, or T4, cells.** In a similar fashion, cytotoxic T cells are sometimes called **CD8, or T8, cells** because they have a glycoprotein called CD8, which helps connect cytotoxic T cells to cells displaying MHC molecules.

The helper T cell responds to antigen presented by the MHC molecule and costimulation by producing the cytokine **interleukin-2** and **interleukin-2 receptors** (figure 19.13, step 5). Interleukin-2 binds to the receptors and stimulates the helper T cell to divide (figure 19.13, step 6). The helper T cells produced by this division can again be presented with antigen by macrophages and again be stimulated to divide. Thus, the number of helper T cells is greatly increased (figure 19.13, step 7).

Inhibiting and Stimulating Immunity

Decreasing the production or activity of cytokines can suppress the immune system. For example, cyclosporine, a drug used to prevent the rejection of transplanted organs, inhibits the production of interleukin-2. Conversely, genetically engineered interleukins can be used to stimulate the immune system. Administering interleukin-2 has promoted the destruction of cancer cells in some cases by increasing the activities of T cells.

It is important for the number of helper T cells to increase because helper T cells are necessary for the activation of most B or T cells (figure 19.13, step 8). For example, B cells have receptors that can recognize antigens. Most B cells, however, do not respond to antigens without stimulation from helper T cells. This process begins when a B cell takes in by receptor-mediated endocytosis the same kind of antigen that stimulated the helper

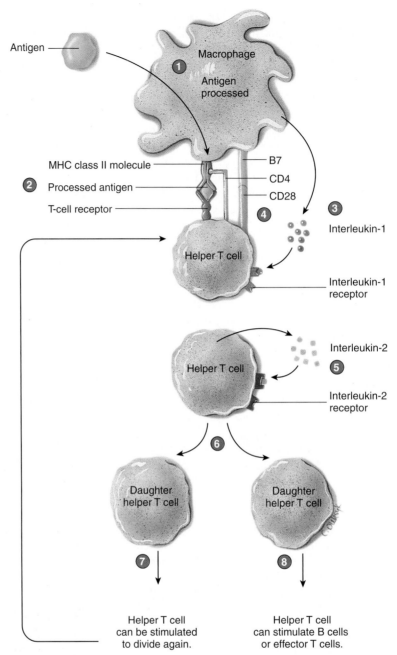

1. Antigen-presenting cells, such as macrophages, take in, process, and display antigens on the cell's surface (see figure 19.12).

2. The antigen is bound to an MHC class II molecule, which functions to present the processed antigen to a T-cell receptor of a helper T cell for recognition.

3. There is costimulation by the cytokine interleukin-1 released from the macrophage.

4. There is costimulation of the helper T cell by CD4 and other surface molecules (e.g., B7/CD28).

5. Interleukin-1 attaches to interleukin receptors and stimulates the helper T cell to secrete the cytokine interleukin-2 and to produce interleukin-2 receptors.

6. The helper T cell stimulates itself to divide when interleukin-2 binds to interleukin-2 receptors.

7. The "daughter" helper T cells resulting from this division can be stimulated to divide again if they are exposed to the same antigen that stimulated the "parent" helper T cell. This greatly increases the number of helper T cells.

8. The increased number of helper T cells can facilitate the activation of B cells or effector T cells.

Process Figure 19.13   Proliferation of Helper T Cells

An antigen-presenting cell (macrophage) stimulates helper T cells to divide and produce cytokines.

T cell (figure 19.14, step 1). The antigen is processed by the B cell and presented on the B-cell surface by an MHC class II molecule (figure 19.14, step 2). A helper T cell is stimulated when it binds to the MHC class II/antigen complex (figure 19.14, step 3). There is also costimulation involving CD4 and interleukins (figure 19.14, steps 4 and 5). The result is the division of the B cell (figure 19.14, step 6). Many of the resulting daughter cells differentiate to become specialized cells, called **plasma cells,** which produce antibodies (figure 19.14, step 7). The increased number of cells, each producing antibodies, can

produce an immune response that destroys the antigens (see "Effects of Antibodies," p. 585). ▼

38. *What is costimulation? State two ways in which it can happen.*

39. *Why are helper T cells sometimes called CD4 or T4 cells? Why are cytotoxic T cells sometimes called CD8 or T8 cells?*

40. *Describe how antigen-presenting cells stimulate an increase in the number of helper T cells. Why is this necessary?*

41. *Describe how helper T cells stimulate an increase in the number of B or T cells. Why is this necessary?*

Table 19.4	Cytokines and Their Functions
Cytokine*	**Description**
Interferon alpha (IFNα)	Prevents viral replication and inhibits cell growth; secreted by virus-infected cells
Interferon beta (IFNβ)	Prevents viral replication, inhibits cell growth, and decreases the expression of major histocompatibility complex (MHC) class I and II molecules; secreted by virus-infected fibroblasts
Interferon gamma (IFNγ)	About 20 different proteins that activate macrophages and natural killer (NK) cells, stimulate adaptive immunity by increasing the expression of MHC class I and II molecules, and prevent viral replication; secreted by helper T, cytotoxic T, and NK cells
Interleukin-1 (IL-1)	Costimulation of B and T cells, promotes inflammation through prostaglandin production, and induces fever acting through the hypothalamus (pyrogen); secreted by macrophages, B cells, and fibroblasts
Interleukin-2 (IL-2)	Costimulation of B and T cells, activation of macrophages and NK cells; secreted by helper T cells
Interleukin-4 (IL-4)	Plays a role in allergic reactions by activation of B cells, resulting in the production of immunoglobulin E (IgE); secreted by helper T cells
Interleukin-5 (IL-5)	Part of the response against parasites by stimulating eosinophil production; secreted by helper T cells
Interleukin-8 (IL-8)	Chemotactic factor that promotes inflammation by attracting neutrophils and basophils; secreted by macrophages
Interleukin-10 (IL-10)	Inhibits the secretion of interferon gamma and interleukins; secreted by suppressor T cells
Interleukin-15 (IL-15)	Promotes inflammation, activates memory T cells and natural killer cells
Lymphotoxin	Kills target cells; secreted by cytotoxic T cells
Perforin	Makes a hole in the membrane of target cells, resulting in lysis of the cell; secreted by cytotoxic T cells
Tumor necrosis factor α (TNFα)	Activates macrophages and promotes fever (pyrogen); secreted by macrophages

*Some cytokines were named according to the laboratory test first used to identify them; however, these names rarely are a good description of the actual function of the cytokine.

Inhibition of Lymphocytes

Tolerance is a state of unresponsiveness of lymphocytes to a specific antigen. Although foreign antigens can induce tolerance, the most important function of tolerance is to prevent the immune system from responding to self-antigens. The need to maintain tolerance and to avoid the development of autoimmune disease is obvious. Tolerance can be induced in several ways.

1. *Deletion of self-reactive lymphocytes.* During prenatal development and after birth, stem cells in red bone marrow and the thymus give rise to immature lymphocytes that develop into mature lymphocytes capable of an immune response. When immature lymphocytes are exposed to antigens, instead of responding in ways that result in the elimination of the antigen, they respond by dying. Because immature lymphocytes are exposed to self-antigens, this process eliminates self-reactive lymphocytes. In addition, immature lymphocytes that escape deletion during their development and become mature, self-reacting lymphocytes can still be deleted in ways that are not clearly understood.

2. *Prevention of the activation of lymphocytes.* For the activation of lymphocytes to take place, two signals are usually required: (1) the MHC/antigen complex binding

with an antigen receptor and (2) costimulation. Preventing either of these events stops lymphocyte activation. For example, blocking, altering, or deleting an antigen receptor prevents activation. **Anergy** (an′er-jē), which means without working, is a condition of inactivity in which a B or T cell does not respond to an antigen. Anergy develops when an MHC/antigen complex binds to an antigen receptor and no costimulation occurs. For example, if a helper T cell encounters a self-antigen on a cell that cannot provide costimulation, the T cell is turned off. It is likely that only antigen-presenting cells can provide costimulation.

3. *Activation of suppressor T cells.* **Suppressor T cells** are a poorly understood group of T cells that are defined by their ability to suppress immune responses. It is likely that suppressor T cells are subpopulations of helper T cells and cytotoxic T cells. The suppressor (helper) T cells release suppressive cytokines, or the suppressor (cytotoxic) T cells kill antigen-presenting cells. ▼

42. What is tolerance? List three ways it is accomplished.

Antibody-Mediated Immunity

Exposure of the body to an antigen can result in the activation of B cells and the production of antibodies. The antibodies bind to

1. Before a B cell can be activated by a helper T cell, the B cell must phagocytize and process the same antigen that activated the helper T cell. The antigen binds to a B-cell receptor, and both the receptor and antigen are taken into the cell by endocytosis.

2. The B cell uses an MHC class II molecule to present the processed antigen to the helper T cell.

3. The T-cell receptor binds to the MHC class II/antigen complex.

4. There is costimulation of the B cell by CD4 and other surface molecules.

5. There is costimulation by interleukins (cytokines) released from the helper T cell.

6. The B cell divides, and the resulting daughter cells divide, and so on, eventually producing many cells (only two are shown here) that recognize the same antigen.

7. Many of the daughter cells differentiate to become plasma cells, which produce antibodies. Antibodies are part of the immune response that eliminates the antigen.

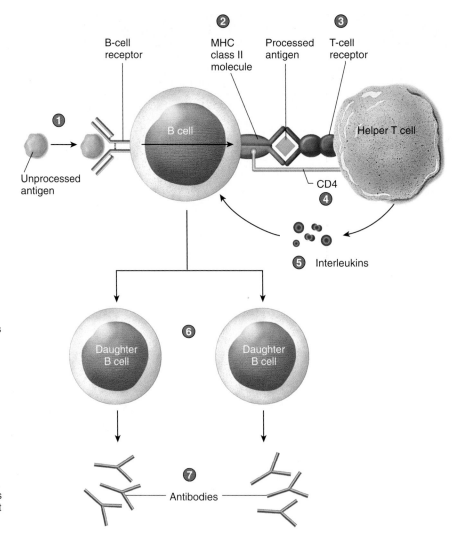

Process Figure 19.14 **Proliferation of B Cells**

A helper T cell stimulates a B cell to divide and produce antibodies.

the antigens, and through several different mechanisms the antigens can be destroyed. Because antibodies are in body fluids, antibody-mediated immunity is effective against extracellular antigens, such as bacteria, viruses (when they are outside cells), and toxins. Antibody-mediated immunity can also cause immediate hypersensitivity reactions (see "Immune System Problems of Clinical Significance," p. 592).

Structure of Antibodies

Antibodies are proteins produced in response to an antigen. They are Y-shaped molecules consisting of four polypeptide chains: two identical heavy chains and two identical light chains (figure 19.15). The end of each "arm" of the antibody is the **variable region,** which is the part of the antibody that combines with the antigen. The variable region of a particular antibody can only join with a particular antigen. This is similar to the lock-and-key model of enzymes. The rest of the antibody is the **constant region,** which

has several functions. For example, the constant region can activate complement, or it can attach the antibody to cells such as macrophages, basophils, and mast cells.

Antibodies make up a large portion of the proteins in plasma. Most plasma proteins can be separated into albumin and alpha, beta, and gamma globulin portions. Antibodies are called **gamma globulins** (glob′ū-linz, globule) because they are found mostly in the gamma globulin part of plasma. Antibodies are also called **immunoglobulins (Ig)** because they are globulin proteins involved in immunity. The five general classes of immunoglobulins are denoted IgG, IgM, IgA, IgE, and IgD (table 19.5). ▼

43. *Antibody-mediated immunity is effective against what kinds of antigens?*

44. *What are the functions of the variable and constant regions of an antibody? List the five classes of antibodies, and state their functions.*

Uses of Monoclonal Antibodies

Each type of **monoclonal antibody** is a pure antibody preparation that is specific for only one antigen. When the antigen is injected into a laboratory animal, it activates a B-cell clone against the antigen. The B cells are removed from the animal and fused with tumor cells. The resulting hybridoma cells have two ideal characteristics: They divide to form large numbers of cells, and the cells of a given clone produce only one kind of antibody.

Monoclonal antibodies are used for determining pregnancy and for diagnosing diseases, such as gonorrhea, syphilis, hepatitis, rabies, and cancer. These tests are specific and rapid because the monoclonal antibodies bind only to the antigen being tested. Monoclonal antibodies have been called "magic bullets" because they may someday be used to treat cancer by delivering drugs to cancer cells (see "Immunotherapy," p. 589).

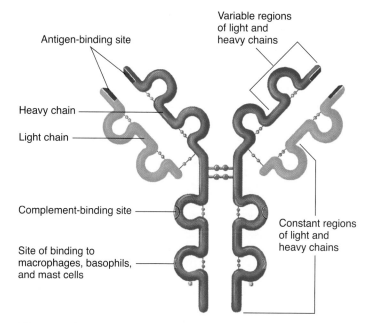

Figure 19.15 **Structure of an Antibody**

Antibodies consist of two heavy and two light polypeptide chains. The variable region of the antibody binds to the antigen. The constant region of the antibody can activate the classical pathway of the complement cascade. The constant region can also attach the antibody to the plasma membrane of cells such as macrophages, basophils, or mast cells.

Effects of Antibodies

Antibodies can affect antigens either directly or indirectly. Direct effects occur when a single antibody binds to an antigen and inactivates the antigen, or when many antigens are bound together and are inactivated by many antibodies (figure 19.16*a* and *b*). The ability of antibodies to join antigens together is the basis for many clinical tests, such as blood typing because, when enough antigens are bound together, they form visible clumps.

Most of the effectiveness of antibodies results from indirect effects (figure 19.16*c* to *e*). After an antibody has attached by its variable region to an antigen, the constant region of the antibody can activate other mechanisms that destroy the antigen. For example, the constant region of antibodies can activate complement proteins through the classical pathway (see figure 19.9). Activated

complement stimulates inflammation, attracts white blood cells through chemotaxis, and lyses bacteria. When an antigen combines with the antibody, the constant region triggers a release of inflammatory chemicals from mast cells and basophils. Finally, macrophages can attach to the constant region of the antibody and phagocytize both the antibody and the antigen. ▼

45. Describe the direct and indirect ways in which antibodies affect antigens.

Table 19.5	Classes of Antibodies and Their Functions		
Antibody	**Total Serum Antibody (%)**	**Description**	**Structure**
IgG	80–85	Activates complement and promotes phagocytosis; can cross the placenta and provide immune protection to the fetus and newborn; responsible for Rh reactions, such as hemolytic disease of the newborn	
IgM	5–10	Activates complement and acts as an antigen-binding receptor on the surface of B cells; responsible for transfusion reactions in the ABO blood system; often the first antibody produced in response to an antigen	
IgA	15	Secreted into saliva, into tears, and onto mucous membranes to provide protection on body surfaces; found in colostrum and milk to provide immune protection to newborns	
IgE	0.002	Binds to mast cells and basophils and stimulates the inflammatory response	
IgD	0.2	Functions as antigen-binding receptors on B cells	

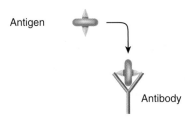

(a) **Inactivates the antigen.** An antibody binds to an antigen and inactivates it.

(b) **Binds antigens together.** Antibodies bind several antigens together.

(c) **Activates the complement cascade.** An antigen binds to an antibody. As a result, the antibody can activate complement proteins, which can produce inflammation, chemotaxis, and lysis.

(d) **Initiates the release of inflammatory chemicals.** An antibody binds to a mast cell or basophil. When an antigen binds to the antibody, it triggers a release of chemicals that cause inflammation.

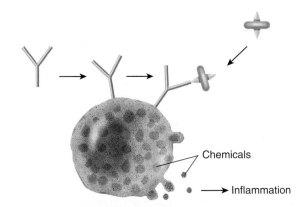

(e) **Facilitates phagocytosis.** An antibody binds to an antigen and then to a macrophage, which phagocytizes the antibody and antigen.

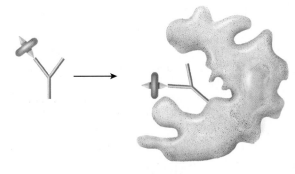

Figure 19.16 Effects of Antibodies

Antibodies directly affect antigens by inactivating the antigens or binding the antigens together. Antibodies indirectly affect antigens by activating other mechanisms through the constant region of the antibody. Indirect mechanisms include activation of complement, increased inflammation resulting from the release of inflammatory chemicals from mast cells or basophils, and increased phagocytosis resulting from antibody attachment to macrophages.

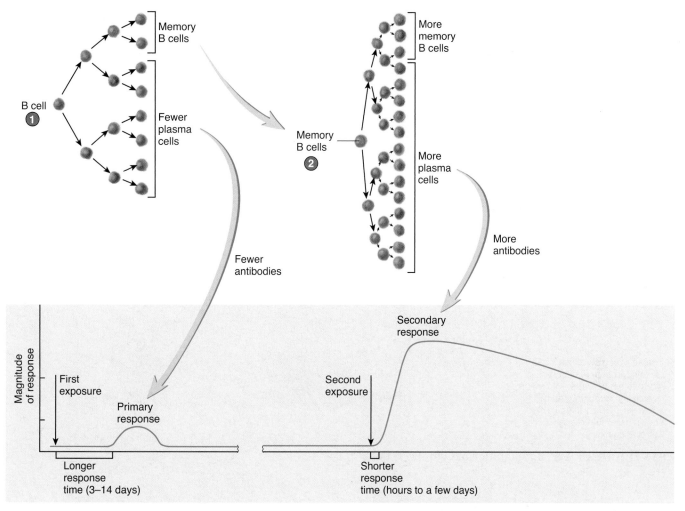

1. Primary response. The primary response occurs when a B cell is first activated by an antigen. The B cell proliferates to form plasma cells and memory cells. The plasma cells produce antibodies.

2. Secondary response. The secondary response occurs when another exposure to the same antigen causes the memory cells to rapidly form plasma cells and additional memory cells. The secondary response is faster and produces more antibodies than the primary response.

Process Figure 19.17   Antibody Production

Antibody Production

The production of antibodies after the first exposure to an antigen is different from that following a subsequent exposure. The **primary response** results from the first exposure of a B cell to an antigen (figure 19.17, step 1). When the antigen binds to the antigen-binding receptor on the B cell, the B cell undergoes several divisions to form plasma cells and memory B cells. **Plasma cells** produce antibodies. The primary response normally takes 3–14 days to produce enough antibodies to be effective against the antigen. In the meantime, the individual usually develops disease symptoms because the antigen has had time to cause tissue damage.

Memory B cells are responsible for the **secondary,** or **memory, response,** which occurs when the immune system is exposed to an antigen against which it has already produced a primary response (figure 19.17, step 2). When exposed to the antigen, the

B memory cells quickly divide to form plasma cells, which rapidly produce antibodies. The secondary response provides better protection than the primary response for two reasons: (1) The time required to start producing antibodies is less (hours to a few days) and (2) more plasma cells and antibodies are produced. As a consequence, the antigen is quickly destroyed, no disease symptoms develop, and the person is immune.

The memory response also includes the formation of new memory cells, which provide protection against additional exposures to a specific antigen. Memory cells are the basis of adaptive immunity. After destruction of the antigen, plasma cells die, the antibodies they released are degraded, and antibody levels decline to the point where they can no longer provide adequate protection. Memory cells persist for many years, however, probably for life in some cases. If memory cell production is not stimulated, or if the memory cells produced are short-lived, it is possible to have repeated infections of

1. An MHC class I molecule displays an antigen, such as a viral protein, on the surface of a target cell.

2. Activation of a cytotoxic T cell begins when a T-cell receptor binds to the MHC class I/antigen complex.

3. There is costimulation of the cytotoxic T cell by CD8 and other surface molecules.

4. There is costimulation by cytokines, such as interleukin-2, released from helper T cells.

5. The activated cytotoxic T cell divides, and the resulting daughter cells divide, and so on, eventually producing many cytotoxic T cells (only two are shown here) that recognize the same antigen.

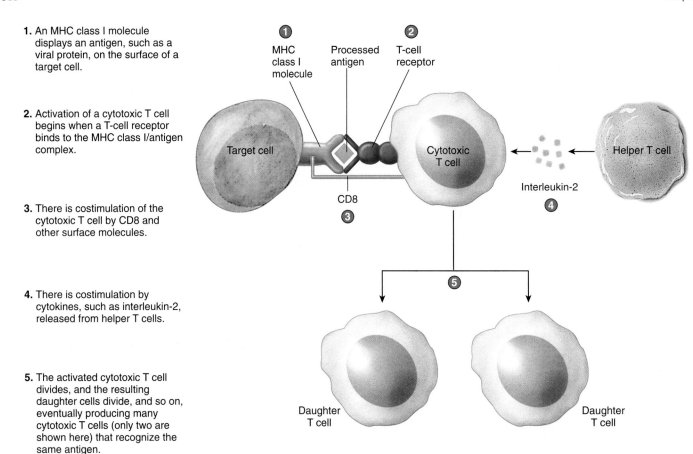

Process Figure 19.18   Proliferation of Cytotoxic T Cells

the same disease. For example, the same cold virus can cause the common cold more than once in the same person. ▼

46. *What are plasma cells and memory cells, and what are their functions?*
47. *What are the primary and secondary antibody responses? Why doesn't the primary response prevent illness but the secondary response does?*

P R E D I C T
One theory for long-lasting immunity assumes that humans are continually exposed to the disease-causing agent. Explain how this exposure can produce lifelong immunity.

Cell-Mediated Immunity

Cell-mediated immunity is a function of cytotoxic T cells and is most effective against microorganisms that live inside the cells of the body. All viruses and some bacteria, fungi, and parasites are examples of intracellular microorganisms. Cell-mediated immunity is also involved with some allergic reactions, the control of tumors, and graft rejections.

Cell-mediated immunity is essential for fighting viral infections. When viruses infect cells, they direct the cells to make new viruses, which are then released and infect other cells. Thus, cells are turned into virus manufacturing plants. While inside the cell, viruses have a safe haven from antibody-mediated immunity because antibodies cannot cross the plasma membrane. Cell-

mediated immunity fights viral infections by destroying virally infected cells. When viruses infect cells, some viral proteins are broken down and become processed antigens that are combined with MHC class I molecules and displayed on the surface of the infected cell (figure 19.18, step 1). Cytotoxic T cells can distinguish between virally infected cells and noninfected cells because the T cell receptor can bind to the MHC class I/viral antigen complex, which is not present on uninfected cells.

The T-cell receptor binding with the MHC molecule/antigen complex is a signal for activating cytotoxic T cells (figure 19.18, step 2). Costimulation by other surface molecules, such as CD8, also occurs (figure 19.18, step 3). Helper T cells provide costimulation by releasing cytokines, such as interleukin-2, which stimulates activation and cell division of cytotoxic T cells (figure 19.18, step 4). Unlike their interactions with macrophages and B cells, however, helper T cells do not connect to cytotoxic T cells through MHC molecule/antigen complexes or other surface molecules.

Increasing the number of helper T cells (figure 19.18, step 5) results in greater stimulation of cytotoxic T cells. In cell-mediated responses, helper T cells are activated and stimulated to divide in the same fashion as in antibody-mediated responses (see figure 19.13).

After cytotoxic T cells are activated by an antigen on the surface of a target cell, they undergo a series of divisions to produce additional cytotoxic T cells and memory T cells (figure 19.19). The cytotoxic T cells are responsible for the cell-mediated immune response, and the **memory T cells** provide a secondary response and long-lasting immunity in the same fashion as memory B cells.

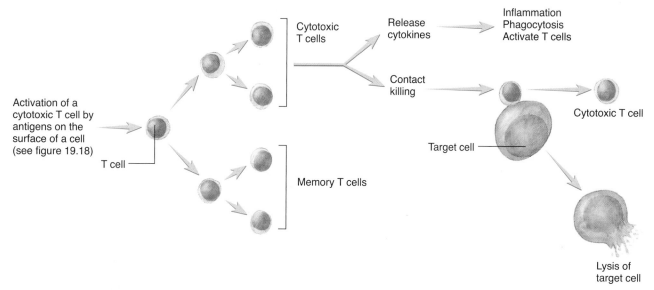

Figure 19.19 Stimulation and Effects of T Cells

When activated, cytotoxic T cells form many cytotoxic T cells and memory T cells. The cytotoxic T cells release cytokines that promote the destruction of the antigen or cause the lysis of target cells, such as virus-infected cells, tumor cells, or transplanted cells. The memory T cells are responsible for the secondary response.

Cytotoxic T Cells

Cytotoxic T cells have two main effects: They lyse cells and they produce cytokines (see figure 19.19). Cytotoxic T cells can come into contact with other cells and cause them to lyse. Virus-infected cells have viral antigens, tumor cells have tumor antigens, and tissue transplants have foreign antigens on their surfaces that can stimulate cytotoxic T-cell activity. A cytotoxic T cell binds to a target cell and releases chemicals that cause the target cell to lyse. The major method of lysis involves a protein called **perforin,** which is similar to the complement protein C9 (see figure 19.9). Perforin forms a channel in the plasma membrane of the target cell through which water enters the cell, causing lysis. The cytotoxic T cell then moves on to destroy additional target cells.

In addition to lysing cells, cytotoxic T cells release cytokines that activate additional components of the immune system. For example, one important function of cytokines is the recruitment of cells, such as macrophages. These cells are then responsible for phagocytosis and inflammation.

P R E D I C T 5

In patients with acquired immunodeficiency syndrome (AIDS), helper T cells are destroyed by a viral infection. The patients can die of pneumonia caused by an intracellular fungus (*Pneumocystis carinii*) or from Kaposi sarcoma, which consists of tumorous growths in the skin and lymph nodes. Explain what is happening.

Delayed Hypersensitivity T Cells

Delayed hypersensitivity T cells respond to antigens by releasing cytokines. Consequently, they promote phagocytosis and inflammation, especially in allergic reactions (see "Immune System Problems of Clinical Significance," p. 592). For example, poison ivy antigens can be processed by Langerhans cells in the skin, which present the antigen to delayed hypersensitivity

T cells, resulting in an intense inflammatory response with redness and itching. ▼

48. *Cell-mediated immunity is effective against what kinds of antigens?*
49. *How do intracellular microorganisms and helper T cells stimulate cytotoxic T cells?*
50. *How is long-lasting immunity achieved in cell-mediated immunity?*
51. *State the two main responses of cytotoxic T cells.*
52. *What kind of immune response is produced by delayed hypersensitivity T cells?*

Immune Interactions

Although the immune response can be described in terms of innate, antibody-mediated, and cell-mediated immunity, only one immune system really exists. These categories are artificial divisions used to emphasize particular aspects of immunity. Actually, immune responses often involve components of more than one type of immunity (figure 19.20). For example, although adaptive immunity can recognize and remember specific antigens, once recognition has occurred, many of the events that lead to the destruction of the antigen are innate immunity activities, such as inflammation and phagocytosis. ▼

53. *Describe how interactions among innate, antibody-mediated, and cell-mediated immunity can eliminate an antigen.*

Immunotherapy

Knowledge of the basic ways that the immune system operates has produced two fundamental benefits: (1) an understanding of the cause and progression of many diseases and (2) the development

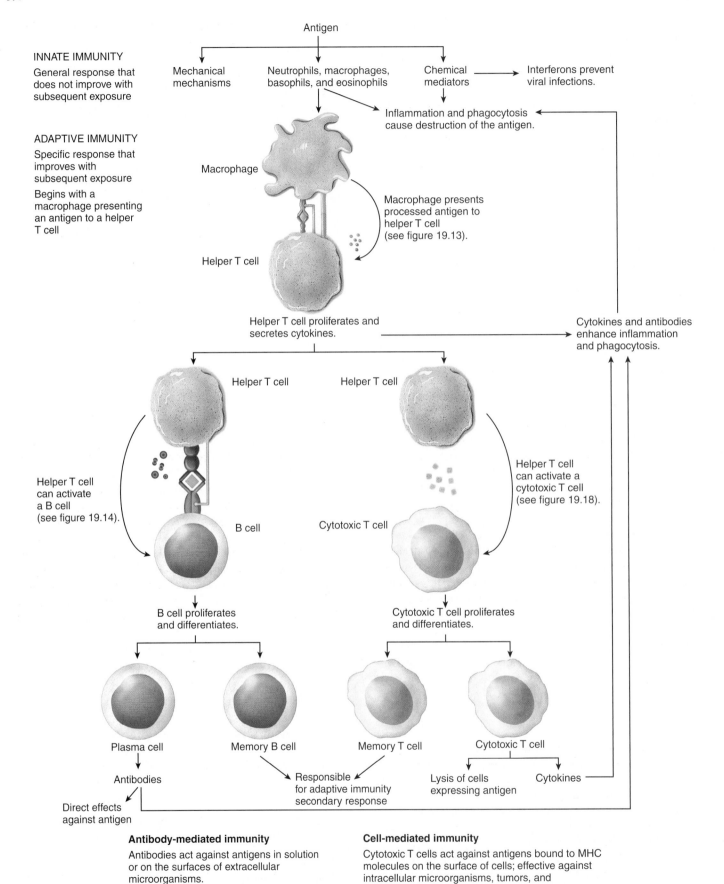

INNATE IMMUNITY
General response that does not improve with subsequent exposure

ADAPTIVE IMMUNITY
Specific response that improves with subsequent exposure

Begins with a macrophage presenting an antigen to a helper T cell

Antigen

Mechanical mechanisms

Neutrophils, macrophages, basophils, and eosinophils

Chemical mediators

Interferons prevent viral infections.

Inflammation and phagocytosis cause destruction of the antigen.

Macrophage

Helper T cell

Macrophage presents processed antigen to helper T cell (see figure 19.13).

Helper T cell proliferates and secretes cytokines.

Cytokines and antibodies enhance inflammation and phagocytosis.

Helper T cell

Helper T cell

Helper T cell can activate a B cell (see figure 19.14).

B cell

Cytotoxic T cell

Helper T cell can activate a cytotoxic T cell (see figure 19.18).

B cell proliferates and differentiates.

Cytotoxic T cell proliferates and differentiates.

Plasma cell

Memory B cell

Memory T cell

Cytotoxic T cell

Antibodies

Responsible for adaptive immunity secondary response

Lysis of cells expressing antigen

Cytokines

Direct effects against antigen

Antibody-mediated immunity
Antibodies act against antigens in solution or on the surfaces of extracellular microorganisms.

Cell-mediated immunity
Cytotoxic T cells act against antigens bound to MHC molecules on the surface of cells; effective against intracellular microorganisms, tumors, and transplanted cells.

Figure 19.20   Immune Interactions

The major interactions and responses of innate and adaptive immunity to an antigen.

or proposed development of effective methods to prevent, stop, or even reverse diseases.

Immunotherapy treats disease by altering immune system function or by directly attacking harmful cells. Some approaches attempt to boost immune system function in general. For example, administering cytokines or other agents can promote inflammation and the activation of immune cells, which can help in the destruction of tumor cells. On the other hand, sometimes inhibiting the immune system is helpful. For example, multiple sclerosis is an autoimmune disease in which the immune system treats self-antigens as foreign antigens, thereby destroying the myelin that covers axons. Interferon beta (IFNβ) blocks the expression of MHC molecules that display self-antigens and is used to treat multiple sclerosis.

Some immunotherapy takes a more specific approach. For example, vaccination can prevent many diseases (see "Acquired Immunity," this page). The ability to produce monoclonal antibodies may result in therapies that are effective for treating tumors. If an antigen unique to tumor cells can be found, then monoclonal antibodies could be used to deliver radioactive isotopes, drugs, toxins, enzymes, or cytokines that can kill the tumor cell or can activate the immune system to kill the cell. Unfortunately, no antigen on tumor cells has been found that is not also found on normal cells. Nonetheless, this approach may be useful if damage to normal cells is minimal.

One problem with monoclonal antibody delivery systems is that the immune system recognizes the monoclonal antibody as a foreign antigen. After the first exposure, a memory response quickly destroys the monoclonal antibodies, rendering the treatment ineffective. In a process called **humanization,** the monoclonal antibodies are modified to resemble human antibodies. This approach has allowed monoclonal antibodies to sneak past the immune system.

The use of monoclonal antibodies to treat tumors is mostly in the research stage of development, but a few clinical trials are now yielding promising results. For example, monoclonal antibodies with radioactive iodine (^{131}I) have caused the regression of B-cell lymphomas and have produced few side effects. Herceptin is a monoclonal antibody that binds to a growth factor that is overexpressed in 25%–30% of primary breast cancers. The antibodies "tag" cancer cells, which are then lysed by natural killer cells.

Herceptin slows disease progression and increases survival time, but it is not a cure for breast cancer.

Many other approaches for immunotherapy are being studied, and the development of treatments that use the immune system are certain to increase in the future. ▼

54. *What is immunotherapy? Give examples.*

Neuroendocrine Regulation of Immunity

An intriguing possibility for reducing the severity of diseases or even curing them is to use neuroendocrine regulation of immunity. The nervous system regulates the secretion of hormones, such as cortisol, epinephrine, endorphins, and enkephalins, for which lymphocytes have receptors. For example, cortisol released during times of stress inhibits the immune response. In addition, most lymphatic tissues, including some individual lymphocytes, receive sympathetic innervation. That a neuroendocrine connection exists with the immune system is clear. The question we need to answer is, can we use this connection to control our own immunotherapy?

Acquired Immunity

It is possible to acquire adaptive immunity in four ways: active natural, active artificial, passive natural, and passive artificial immunity (figure 19.21). The terms *natural* and *artificial* refer to the method of exposure. Natural exposure implies that contact with an antigen or antibody occurs as part of everyday living and is not deliberate. Artificial exposure, also called **immunization,** is a deliberate introduction of an antigen or antibody into the body.

The terms *active* and *passive* indicate whether an individual's immune system is directly responding to the antigen. When an individual is naturally or artificially exposed to an antigen, an adaptive immune system response can occur that produces antibodies. This is called **active immunity** because the individual's own immune system is the cause of the immunity. **Passive immunity** occurs when another

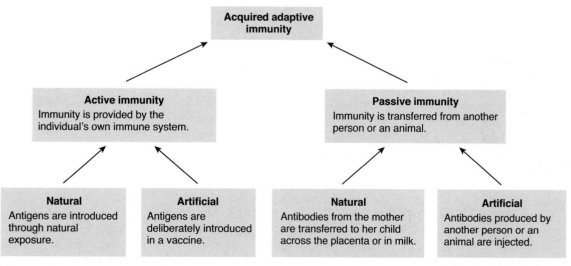

Figure 19.21 **Ways to Acquire Adaptive Immunity**

Allergy

An **allergy,** or **hypersensitivity reaction,** is a harmful response to an antigen that does not stimulate an adaptive immune response in most people. Immune and allergic reactions involve the same mechanisms, and the differences between them are unclear. Both require exposure to an antigen and stimulation of antibody-mediated or cell-mediated immunity. If immunity to the antigen is established, later exposure to the antigen results in an immune response that eliminates the antigen, and no symptoms appear. In allergic reactions, the antigen is called the **allergen** (al'er-jen, allergy + -gen, producing), and later exposure to the allergen stimulates much the same processes that occur during a normal immune response. The processes that eliminate the allergen, however, also produce undesirable side effects such as a strong inflammatory reaction, which can be more harmful than beneficial.

Immediate hypersensitivities produce symptoms within a few minutes of exposure to the allergen and are caused by antibodies. The reaction takes place rapidly because the antibodies are already present because of prior exposure to the allergen. For example, in people with **hay fever,** the allergens, usually plant pollens, are inhaled and absorbed through the respiratory mucous membrane. The combination of the allergen with antibodies stimulates mast cells to release inflammatory chemicals, such as histamine (see figure 19.16d). The resulting localized inflammatory response produces swelling and excess mucus production. In **asthma** (az'mă), resulting from an allergic reaction, the allergen combines with antibodies on mast cells or basophils in the lungs. As a result, these cells release inflammatory chemicals, such as leukotrienes and histamine. The chemicals cause constriction of smooth muscle in the walls of the tubes that transport air throughout the lungs. Consequently, less air flows into and out of the lungs, and the patient has difficulty breathing. **Urticaria** (er'ti-kar'i-ă, to burn), or **hives,** is a skin rash or localized swelling that can be caused by an ingested allergen. **Anaphylaxis** (an'ă-fī-lak'sis, an'ă-fī-lak'sis, ana, away from + phylaxis, protection) is a systemic allergic reaction, often resulting from insect stings or drugs, such as penicillin. The chemicals released from mast cells and basophils cause systemic vasodilation, increased vascular permeability, a drop in blood pressure, and possibly death. Transfusion reactions and hemolytic disease of the newborn (see chapter 16) are also examples of immediate hypersensitivity reactions.

Delayed hypersensitivities take hours to days to develop and are caused by T cells. It takes some time for this reaction to develop because it takes time for the T cells to move by chemotaxis to the allergen. It also takes time for the T cells to release cytokines that attract other immune system cells involved with producing inflammation. The most common type of delayed hypersensitivity reactions result from contact of an allergen with the skin or mucous membranes. For example, poison ivy, poison oak, soaps, cosmetics, and drugs can cause a delayed hypersensitivity reaction. The allergen is absorbed by epithelial cells, which are then destroyed by T cells, causing inflammation and tissue destruction. Although itching can be intense, scratching is harmful because it damages tissues and causes additional inflammation.

Autoimmune Disease

In **autoimmune disease,** the immune system incorrectly treats self-antigens as foreign antigens. Autoimmune disease operates through the same mechanisms as hypersensitivity reactions except that the reaction is stimulated by self-antigens. Examples of autoimmune diseases are thrombocytopenia, lupus erythematosus, rheumatoid arthritis, rheumatic fever, diabetes mellitus (type I), and myasthenia gravis.

Immunodeficiency

Immunodeficiency is a failure of some part of the immune system to function properly. It can be congenital (present at birth) or acquired. Congenital immunodeficiencies usually involve failure of the fetus to form adequate numbers of B cells, T cells, or both. **Severe combined immunodeficiency (SCID),** in which both B cells and T cells fail to form, is probably the best known. Unless the person suffering from SCID is kept in a sterile environment or is provided with a compatible bone marrow transplant, death from infection results.

Acquired immunodeficiency can result from many different causes. For example, inadequate protein in the diet inhibits protein synthesis and, therefore, antibody levels decrease. Immunity can be depressed as a result of stress, illness, or drugs, such as those used to prevent graft rejection. Diseases such as leukemia cause an overproduction of lymphocytes that do not function properly.

Acquired immunodeficiency syndrome (AIDS) is a life-threatening disease caused by the **human immunodeficiency virus (HIV).** HIV is transmitted from an infected to a noninfected person in body fluids, such as blood, semen, or vaginal secretions. The major methods of transmission are through unprotected intimate sexual contact, with contaminated needles used by intravenous drug users, through tainted blood products, and from a pregnant woman to her fetus. Present evidence indicates that household, school, and work contacts do not result in transmission.

HIV infection begins when a protein on the surface of the virus, called gp120, binds to a CD4 molecule on the surface of a cell. The CD4 molecule is found primarily on helper T cells, and it normally enables helper T cells to adhere to other lymphocytes—for example, during the process of antigen presentation. Certain monocytes, macrophages, neurons, and neuroglial cells also have CD4 molecules. Once attached to the CD4 molecules, the virus injects its genetic material (RNA) and enzymes into the cell and begins to replicate. Copies of the virus are manufactured using the organelles and materials within the cell. Replicated viruses escape from the cell and infect other cells.

Following infection by HIV, within 3 weeks to 3 months, many patients develop mononuculeosis-like symptoms, such as fever, sweats, fatigue, muscle and joint aches, headache, sore throat, diarrhea, rash, and swollen lymph nodes. Within 1–3 weeks, these symptoms disappear as the immune system responds to the virus by producing antibodies and activating cytotoxic T cells that kill HIV-infected cells. The immune system is not able to eliminate HIV completely, however, and by about 6 months a kind of "set point" is achieved in which the virus continues to replicate at a low but steady rate. This chronic stage of infection lasts, on the average, 8–10 years, and the infected person feels good and exhibits few, if any, symptoms.

Although helper T cells are infected and destroyed during the chronic stage of HIV infection, the body responds by producing large numbers of helper T cells. Nonetheless, over a period of years the HIV numbers gradually increase and helper T cell numbers decrease. Normally, approximately 1200 helper T cells are present per cubic millimeter of blood. An HIV-infected person is considered to have AIDS when one or more of the following conditions appear: The helper T cell count falls below 200 cells/mm³, an opportunistic infection occurs, or Kaposi sarcoma develops.

Opportunistic infections involve organisms that normally do not cause disease but can do so when the immune system is depressed. Without helper T cells, cytotoxic T- and B-cell activation is impaired, and adaptive resistance is suppressed. Examples of opportunistic infections include pneumocystis (noo-mō-sis′tis) pneumonia (caused by an intracellular fungus, *Pneumocystis carinii*), tuberculosis (caused by an intracellular bacterium, *Mycobacterium tuberculosis*), syphilis (caused by a sexually transmitted bacterium, *Treponema pallidum*), candidiasis (kan-di-dī′ă-sis, a yeast infection of the mouth or vagina caused by *Candida albicans*), and protozoans that cause severe, persistent diarrhea. Kaposi sarcoma is a type of cancer that produces lesions in the skin, lymph nodes, and visceral organs. Also associated with AIDS are symptoms resulting from the effects of HIV on the nervous system, including motor retardation, behavioral changes, progressive dementia, and possibly psychosis.

No cure for AIDS has yet been discovered. Management of AIDS can be divided into two categories: (1) the management of secondary infections or malignancies associated with AIDS and (2) the treatment of HIV. In order for HIV to replicate, the viral RNA is used to make viral DNA, which is inserted into the host cell's DNA. The inserted viral DNA directs the production of new viral RNA and proteins, which are assembled to form new HIV. Key steps in the replication of HIV require viral enzymes. **Reverse transcriptase** promotes the formation of viral DNA from viral RNA, and **integrase** (in′te-grās) inserts the viral DNA into the host cell's DNA. A viral **protease** (prō′tē-ās) breaks large viral proteins into smaller proteins, which are incorporated into the new HIV.

Blocking the activity of HIV enzymes can inhibit the replication of HIV. The first effective treatment of AIDS was the drug azidothymidine (AZT) (az′i-dō-thī′mi-dēn), also called zidovudine (zī-dō′voo-dēn). AZT is a **reverse transcriptase inhibitor,** which prevents HIV RNA from producing viral DNA. AZT can delay the onset of AIDS but does not appear to increase the survival time of AIDS patients. However, the number of babies who contract AIDS from their HIV-infected mothers can be dramatically reduced by giving AZT to the mothers during pregnancy and to the babies following birth.

Protease inhibitors are drugs that interfere with viral proteases. The current treatment for suppressing HIV replication is **highly active antiretroviral therapy (HAART).** This therapy involves the use of drugs from at least two classes of antivirals. Treatment may involve the combination of three drugs, such as two reverse transcriptase inhibitors and one protease inhibitor. It is less likely that HIV will develop resistance to all three drugs. This strategy has proven very effective in reducing the death rate from AIDS and partially restoring health in some individuals.

Still in the research stage are **integrase inhibitors,** which prevent the insertion of viral DNA into the host cell's DNA. Perhaps someday integrase inhibitors will be part of a combination drug therapy for AIDS.

Another advance in AIDS treatment is a test for measuring **viral load,** which measures the number of viral RNA molecules in a milliliter of blood. The actual level of HIV is one-half the RNA count because each HIV has two RNA strands. Viral load is a good predictor of how soon a person will develop AIDS. If the viral load is high, the onset of AIDS is much sooner than if it is low. It is also possible to detect developing viral resistance by an increase in viral load. In response, a change in drug dose or type may slow viral replication. Current treatment guidelines are to keep viral load below 500 RNA molecules per milliliter of blood.

An effective treatment for AIDS is not a cure. Even if viral load decreases to the point that the virus is undetected in the blood, the virus still remains in cells throughout the body. The virus may eventually mutate and escape drug suppression. The long-term goal for dealing with AIDS is to develop a vaccine that prevents HIV infection.

People with HIV/AIDS can now live for many years because of improved treatment. HIV/AIDS is therefore being viewed increasingly as a chronic disease, not as a death sentence. A multidisciplinary team of occupational therapists, physical therapists, nutritionists/dieticians, psychologists, infectious disease physicians, and others can work together to manage patients with HIV/AIDS to help them have a better quality of life.

Tumor Control

According to the concept of **immune surveillance,** the immune system detects tumor cells and destroys them before a tumor can form. T cells, NK cells, and macrophages are involved in the destruction of tumor cells. Immune surveillance may exist for some forms of cancer caused by viruses. The immune response appears to be directed more against the viruses, however, than against tumors in general. Only a few cancers are known to be caused by viruses in humans. For most tumors the response of the immune system may be ineffective and too late.

Transplantation

The genes that code for the production of the MHC molecules are generally called the **major histocompatibility complex genes.** *Histocompatibility* refers to the ability of tissues (*histo*) to get along (compatibility) when tissues are transplanted from one individual to another. In humans, the MHC genes are often referred to as **human leukocyte antigen (HLA) genes** because they were first identified in leukocytes. There are millions of possible combinations of the

continued

HLA genes, and it is very rare for two individuals (except identical twins) to have the same set of HLA genes. The closer the relationship between two individuals, the greater the likelihood of sharing the same HLA genes.

The immune system can distinguish between self- and foreign cells because self-cells have self-HLAs, whereas foreign cells have foreign HLAs. Rejection of a graft is caused by a normal immune response to foreign HLAs.

Graft rejection can occur in two different directions. In **host-versus-graft rejection,** the recipient's immune system recognizes the donor tissue as foreign and rejects the transplant. In a **graft-versus-host rejection,** the donor tissue (e.g., bone marrow) recognizes the recipient's tissue as foreign and the transplant rejects the recipient, causing destruction of the recipient's tissue and possibly death.

To reduce graft rejection, a tissue match is performed. Only tissue with HLAs similar to the recipient's have a chance of being accepted. An exact match is possible only for a graft from one part to another part of the same person, or between identical twins. For all other graft situations, drugs, such as cyclosporine (sī-klō-spōr′in), that suppress the immune system must be administered throughout the patient's life to prevent graft rejection. Unfortunately, the person then has a drug-produced immunodeficiency and is more susceptible to infections.

---

person or an animal develops antibodies and the antibodies are transferred to a nonimmune individual. For example, infants can acquire antibodies from mother's milk. This is called passive immunity because the nonimmune individual did not produce the antibodies.

How long the immunity lasts differs for active and passive immunity. Active immunity can persist for a few weeks (e.g., common cold) to a lifetime (e.g., whooping cough and polio). Immunity can be long-lasting if enough B or T memory cells are produced and persist to respond to later antigen exposure. Passive immunity is not long-lasting because the individual does not produce his or her own memory cells. Because active immunity can last longer than passive immunity, it is the preferred method. Passive immunity is preferred, however, when immediate protection is needed.

Active Natural Immunity

Active natural immunity results from natural exposure to an antigen, such as a disease-causing microorganism that stimulates an individual's immune system to respond against the antigen. Symptoms of the disease usually develop because the person is not immune during the first exposure. Interestingly, exposure to an antigen does not always produce symptoms. Many people exposed to the poliomyelitis virus at an early age have an immune system response and produce poliomyelitis antibodies, yet they do not exhibit any disease symptoms.

Active Artificial Immunity

In **active artificial immunity,** an antigen is deliberately introduced into an individual to stimulate the immune system. This process is **vaccination** (vak′si-nā-shŭn), and the introduced antigen is a **vaccine** (vak′sēn, vak-sēn′, *vaccinus,* relating to a cow). Injection of the vaccine is the usual mode of administration. Examples of injected vaccinations are the DTP injection against diphtheria, tetanus, and pertussis (whooping cough) and the MMR injection against mumps, measles, and rubella (German measles). Sometimes the vaccine is ingested, as in the oral poliomyelitis vaccine (OPV).

The vaccine usually consists of a part of a microorganism, a dead microorganism, or a live, altered microorganism. The antigen has been changed so that it will stimulate an immune response but will not cause the symptoms of disease. Active artificial immunity is the preferred method of acquiring adaptive immunity because it produces long-lasting immunity without disease symptoms.

P R E D I C T   6

In some cases, a booster shot is used as part of a vaccination procedure. A booster shot is another dose of the original vaccine given some time after the original dose was administered. Why are booster shots given?

Passive Natural Immunity

Passive natural immunity results from the transfer of antibodies from a mother to her child across the placenta before birth. During her life, the mother has been exposed to many antigens, either naturally or artificially, and she has antibodies against many of these antigens. These antibodies protect the mother and the developing fetus against disease. Some of the antibodies (IgG) can cross the placenta and enter the fetal blood. Following birth, the antibodies provide protection for the first few months of the baby's life. Eventually, the antibodies are broken down, and the baby must rely on his or her own immune system. If the mother nurses her baby, antibodies (IgA) in the mother's milk may also provide some protection for the baby.

Passive Artificial Immunity

Achieving **passive artificial immunity** usually begins with vaccinating an animal, such as a horse. After the animal's immune system responds to the antigen, antibodies (sometimes T cells) are removed from the animal and injected into the individual requiring immunity. In some cases, a human who has developed immunity through natural exposure or vaccination is used as a source of antibodies. Passive artificial immunity provides immediate protection for the individual receiving the antibodies and is therefore preferred when time might not be available for the individual to develop his or her own active immunity. This technique provides only temporary immunity, however, because the antibodies are used or eliminated by the recipient.

Antiserum is the general term used for serum, which is plasma minus the clotting factors, that contains antibodies responsible for passive artificial immunity. Antisera are available against microorganisms that cause diseases, such as rabies, hepatitis, and measles; bacterial toxins, such as tetanus, diphtheria,

Systems Interactions

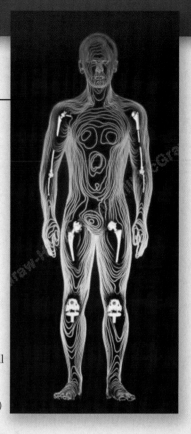

	Effects of the Lymphatic System and Immunity on Other Systems	Effects of Other Systems on the Lymphatic System and Immunity
Integumentary System	Immune cells and chemicals provide protection against microorganisms and toxins and promote tissue repair by releasing chemical mediators Removes excess interstitial fluid	Forms barrier to microorganisms Dendritic cells detect foreign substances
Skeletal System	Immune cells and chemicals provide protection against microorganisms and toxins and promote tissue repair by releasing chemical mediators	Lymphocytes are produced in red bone marrow Lower ribs protect the spleen
Muscular System	Immune cells and chemicals provide protection against microorganisms and toxins and promote tissue repair by releasing chemical mediators Removes excess interstitial fluid	Skeletal muscle "pump" helps move lymph through lymphatic vessels
Nervous System	Immune cells and chemicals provide protection against microorganisms and toxins and promote tissue repair by releasing chemical mediators	The nervous system stimulates and inhibits immunity in ways that are not well understood
Endocrine System	Lymph distributes hormones Immune cells and chemicals provide protection against microorganisms and toxins and promote tissue repair by releasing chemical mediators Removes excess interstitial fluid	Thymosin necessary for immune cell (T cell) maturation in the thymus Hormones affect immune cell functions
Cardiovascular System	The spleen removes damaged red blood cells and helps recycle iron for hemoglobin synthesis Immune cells and chemicals provide protection against microorganisms and toxins and promote tissue repair by releasing chemical mediators	Receives lymph from tissues Moves immune cells and chemicals to sites of infection Delivers oxygen, nutrients, and hormones Removes carbon dioxide, waste products, and toxins
Respiratory System	Immune cells and chemicals provide protection against microorganisms and toxins and promote tissue repair by releasing chemical mediators	Provides oxygen and removes carbon dioxide Helps maintain the body's pH
Digestive System	Carries absorbed fats to blood Immune cells and chemicals provide protection against microorganisms and toxins and promote tissue repair by releasing chemical mediators Removes excess interstitial fluid	Provides nutrients and water
Urinary System	Immune cells and chemicals provide protection against microorganisms and toxins and promote tissue repair by releasing chemical mediators	Removes waste products Helps maintain the body's pH, ion, and water balance
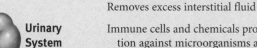 **Reproductive System**	Immune cells and chemicals provide protection against microorganisms and toxins and promote tissue repair by releasing chemical mediators Removes excess interstitial fluid	Blood–testes barrier isolates and protects sperm cells from the immune system

and botulism; and venoms from poisonous snakes and black widow spiders. ▼

> **55.** *Distinguish between natural and artificial immunity and between active and passive immunity.*
>
> **56.** *State four general ways of acquiring adaptive immunity. Which two provide the longest-lasting immunity?*

Effects of Aging on the Lymphatic System and Immunity

Aging appears to have little effect on the ability of the lymphatic system to remove fluid from tissues, absorb fats from the digestive tract, or remove defective red blood cells from the blood.

Aging also seems to have little direct effect on the ability of B cells to respond to antigens, and the number of circulating B cells remains stable in most individuals. With age, thymic tissue is replaced with adipose tissue, and the ability to produce new, mature T cells in the thymus is eventually lost. Nonetheless, the number of T cells remains stable in most individuals due to the replication (not maturation) of T cells in lymphatic tissues. In many individuals, however, there is a decreased ability of helper T cells to proliferate in response to antigens. Thus, antigen exposure produces fewer helper T cells, which results in less stimulation of B cells and effector T cells. Consequently, both antibody-mediated immunity and cell-mediated immunity responses to antigens decrease.

Primary and secondary antibody responses decrease with age. More antigen is required to produce a response, the response is slower, less antibody is produced, and fewer memory cells result. Thus, the ability to resist infections and develop immunity decreases. It is recommended that vaccinations be given well before age 60 because these declines are most evident after age 60. Vaccinations, however, can be beneficial at any age, especially if the individual has reduced resistance to infection.

The ability of cell-mediated immunity to resist intracellular pathogens decreases with age. For example, the elderly are more susceptible to influenza (flu) and should be vaccinated every year. Some pathogens cause disease but are not eliminated from the body. With age, a decrease in immunity can result in reactivation of the pathogen. For example, the virus that causes chicken pox in children can remain latent within nerve cells, even though the disease seems to have disappeared. Later in life, the virus can leave the nerve cells and infect skin cells, causing painful lesions known as herpes zoster, or shingles.

Autoimmune disease occurs when immune responses destroy otherwise healthy tissue. There is very little increase in the number of new-onset autoimmune diseases in the elderly. However, the chronic inflammation and immune responses that begin earlier in life have a cumulative, damaging effect. ▼

> **57.** *What effect does aging have on the major functions of the lymphatic system?*
>
> **58.** *Describe the effects of aging on B cells and T cells. Give examples of how they affect antibody-mediated immunity and cell-mediated immunity responses.*

S U M M A R Y

Lymphatic System   (p. 566)

The lymphatic system consists of lymph, lymphatic vessels, lymphatic tissue, lymphatic nodules, lymph nodes, tonsils, the spleen, and the thymus.

Functions of the Lymphatic System

The lymphatic system maintains fluid balance in tissues, absorbs fats from the small intestine, and defends against microorganisms and foreign substances.

Lymphatic Vessels

1. Lymphatic vessels carry lymph away from tissues.
2. Lymphatic capillaries lack a basement membrane and have loosely overlapping epithelial cells. Fluids and other substances easily enter the lymphatic capillary.
3. Lymphatic capillaries join to form lymphatic vessels.
   - Lymphatic vessels have valves that ensure the one-way flow of lymph.
   - Contraction of lymphatic vessel smooth muscle, skeletal muscle action, and thoracic pressure changes move the lymph.
4. Lymph from the right thorax, the upper-right limb, and the right side of the head and the neck enters the right subclavian vein (often through the right lymphatic duct).
5. Lymph from the lower limbs, pelvis, and abdomen; the left thorax; the upper-left limb; and the left side of the head and the neck enters left subclavian vein through the thoracic duct.

Lymphatic Tissue and Organs

1. Lymphatic tissue is reticular connective tissue that contains lymphocytes and other cells.
2. Lymphatic tissue can be surrounded by a capsule (lymph nodes, spleen, thymus).
3. Lymphatic tissue can be nonencapsulated (diffuse lymphatic tissue, lymphatic nodules, tonsils).
4. Diffuse lymphatic tissue consists of dispersed lymphocytes and has no clear boundaries.
5. Lymphatic nodules are small aggregates of lymphatic tissue (e.g., Peyer's patches in the small intestines).
6. The tonsils
   - The tonsils are large groups of lymphatic nodules in the oral cavity and nasopharynx.
   - The three groups of tonsils are the palatine, pharyngeal, and lingual tonsils.
7. Lymph nodes
   - Lymphatic tissue in the node is organized into the cortex and the medulla. Lymphatic sinuses extend through the lymphatic tissue.
   - Substances in lymph are removed by phagocytosis, or they stimulate lymphocytes (or both).
   - Lymphocytes leave the lymph node and circulate to other tissues.

8. The spleen
   - The spleen is in the left superior side of the abdomen.
   - Foreign substances stimulate lymphocytes in the white pulp.
   - Foreign substances and defective red blood cells are removed from the blood by phagocytes in the red pulp.
   - The spleen is a limited reservoir for blood.
9. The thymus
   - The thymus is a gland in the superior mediastinum and is divided into a cortex and a medulla.
   - Lymphocytes in the cortex are separated from the blood by epithelial cells.
   - Lymphocytes produced in the cortex migrate through the medulla, enter the blood, and travel to other lymphatic tissues, where they can proliferate.

Overview of the Lymphatic System

See figure 19.8.

Immunity (p. 572)

1. Immunity is the ability to resist the harmful effects of microorganisms and other foreign substances.
2. Adaptive immunity exhibits specificity and memory, whereas innate immunity does not.

Innate Immunity (p. 574)

Mechanical Mechanisms

Mechanical mechanisms prevent the entry of microbes (skin and mucous membranes) or remove them (tears, saliva, and mucus).

Chemical Mediators

1. Chemical mediators promote phagocytosis and inflammation.
2. Complement can be activated by either the classical or the alternative pathway. Complement lyses cells, increases phagocytosis, attracts immune system cells, and promotes inflammation.
3. Interferons prevent viral replication. Interferons are produced by virally infected cells and move to other cells, which are then protected.

Cells

1. Chemotaxis is the ability of white blood cells to move to tissues that release certain chemicals.
2. Phagocytosis is the ingestion and destruction of materials.
3. Neutrophils are small phagocytic cells.
4. Macrophages are large phagocytic cells.
   - Macrophages can engulf more than neutrophils can.
   - Macrophages in connective tissue protect the body at locations where microbes are likely to enter, and macrophages clean blood and lymph.
5. Basophils and mast cells release chemicals that promote inflammation.
6. Eosinophils release enzymes that reduce inflammation.
7. Natural killer cells lyse tumor cells and virus-infected cells.

Inflammatory Response

1. The inflammatory response can be initiated in many ways.
   - Chemical mediators cause vasodilation and increase vascular permeability, which allows the entry of other chemical mediators.
   - Chemical mediators attract phagocytes.
   - The amount of chemical mediators and phagocytes increases until the cause of the inflammation is destroyed. Then the tissue undergoes repair.
2. Local inflammation produces the symptoms of redness, heat, swelling, pain, and loss of function. Symptoms of systemic inflammation include an increase in neutrophil numbers, fever, and shock.

Adaptive Immunity (p. 578)

1. Antigens are large molecules that stimulate an adaptive immune system response.
2. Foreign antigens are not produced by the body, but self-antigens are.
3. B cells are responsible for antibody-mediated immunity. T cells are involved with cell-mediated immunity.

Origin and Development of Lymphocytes

1. B cells and T cells originate in red bone marrow. B cells are processed in bone marrow and T cells are processed in the thymus.
2. Positive selection ensures the survival of lymphocytes that can react against antigens, and negative selection eliminates lymphocytes that react against self-antigens.
3. A clone is a group of identical lymphocytes that can respond to a specific antigen.
4. B cells and T cells move to lymphatic tissue from their processing sites. They continually circulate from one lymphatic tissue to another.

Activation of Lymphocytes

1. The antigenic determinant is the specific part of the antigen to which the lymphocyte responds. The antigen receptor (T-cell receptor or B-cell receptor) on the surface of lymphocytes combines with the antigenic determinant.
2. MHC class I molecules display antigens on the surface of nucleated cells, resulting in the destruction of the cells.
3. MHC class II molecules display antigens on the surface of antigen-presenting cells, resulting in the activation of immune cells.
4. MHC antigen complex and costimulation are usually necessary to activate lymphocytes. Costimulation involves cytokines and certain surface molecules.
5. Antigen-presenting cells stimulate the proliferation of helper T cells, which stimulate the proliferation of B or T effector cells.

Inhibition of Lymphocytes

1. Tolerance is suppression of the immune system's response to an antigen.
2. Tolerance is produced by the deletion of self-reactive cells, by the prevention of lymphocyte activation, and by suppressor T cells.

Antibody-Mediated Immunity

1. Antibodies are proteins.
   - The variable region of an antibody combines with the antigen. The constant region activates complement or binds to cells.
   - Five classes of antibodies exist: IgG, IgM, IgA, IgE, and IgD.
2. Antibodies affect the antigen in many ways.
   - Antibodies bind to the antigen and interfere with antigen activity or bind the antigens together.
   - Antibodies act increase phagocytosis by binding to the antigen and to macrophages.
   - Antibodies can activate complement through the classical pathway.
   - Antibodies attach to mast cells or basophils and cause the release of inflammatory chemicals when the antibody combines with the antigen.
3. The primary response results from the first exposure to an antigen. B cells form plasma cells, which produce antibodies and memory B cells.
4. The secondary response results from exposure to an antigen after a primary response, and memory B cells quickly form plasma cells and additional memory B cells.

Cell-Mediated Immunity

1. Cells infected with intracellular microorganisms process antigens that combine with MHC class I molecules.

2. Cytotoxic T cells are stimulated to divide, producing more cytotoxic T cells and memory T cells, when MHC class I/antigen complexes are presented to T-cell receptors. Cytokines released from helper T cells also stimulate cytotoxic T cells.
3. Cytotoxic T cells lyse virus-infected cells, tumor cells, and tissue transplants.
4. Cytotoxic T cells produce cytokines, which promote phagocytosis and inflammation.

Immune Interactions   (p. 589)

Innate immunity, antibody-mediated immunity, and cell-mediated immunity can function together to eliminate an antigen.

Immunotherapy   (p. 589)

Immunotherapy stimulates or inhibits the immune system to treat diseases.

Acquired Immunity   (p. 591)

Active Natural Immunity

Active natural immunity results from natural exposure to an antigen.

Active Artificial Immunity

Active artificial immunity results from deliberate exposure to an antigen.

Passive Natural Immunity

Passive natural immunity results from the transfer of antibodies from a mother to her fetus or baby.

Passive Artificial Immunity

Passive artificial immunity results from the transfer of antibodies (or cells) from an immune animal to a nonimmune animal.

Effects of Aging on the Lymphatic System and Immunity   (p. 596)

1. Aging has little effect on the ability of the lymphatic system to remove fluid from tissues, absorb fats from the digestive tract, or remove defective red blood cells from the blood.
2. Decreased helper T cell proliferation results in decreased antibody-mediated immunity and cell-mediated immunity responses to antigens.
3. The primary and secondary antibody responses decrease with age.
4. The ability to resist intracellular pathogens decreases with age.

R E V I E W   A N D   C O M P R E H E N S I O N

1. The lymphatic system
   a. removes excess fluid from tissues.
   b. absorbs fats from the digestive tract.
   c. defends the body against microorganisms and other foreign substances.
   d. all of the above.
2. Which of the following statements is true?
   a. Lymphatic vessels do not have valves.
   b. Lymphatic vessels empty into lymph nodes.
   c. Lymph from the right-lower limb passes into the right subclavian vein.
   d. Lymph from the thoracic duct empties into the right subclavian vein.
   e. All of the above are true.
3. The tonsils
   a. consist of large groups of lymphatic nodules.
   b. protect against bacteria.
   c. can become chronically infected.
   d. decrease in size in adults.
   e. all of the above.
4. Lymph nodes
   a. filter lymph.
   b. are where lymphocytes divide and increase in number.
   c. contain a network of reticular fibers.
   d. contain lymphatic sinuses.
   e. all of the above.
5. Which of these statements about the spleen is *not* correct?
   a. The spleen has white pulp associated with the arteries.
   b. The spleen has red pulp associated with the veins.
   c. The spleen destroys defective red blood cells.
   d. The spleen is surrounded by trabeculae located outside the capsule.
   e. The spleen is a limited reservoir for blood.
6. The thymus
   a. increases in size in adults.
   b. produces lymphocytes that move to other lymphatic tissue.
   c. is located in the abdominal cavity.
   d. all of the above.
7. Which of these is an example of innate immunity?
   a. Tears and saliva wash away microorganisms.
   b. Basophils release histamine and leukotrienes.
   c. Neutrophils phagocytize a microorganism.
   d. The complement cascade is activated.
   e. All of the above are correct.
8. Neutrophils
   a. enlarge to become macrophages.
   b. account for most of the dead cells in pus.
   c. are usually the last type of cell to enter infected tissues.
   d. are usually located in lymph nodes.
9. Macrophages
   a. are large, phagocytic cells that outlive neutrophils.
   b. develop from mast cells.
   c. often die after a single phagocytic event.
   d. have the same function as eosinophils.
   e. all of the above.
10. Which of these cells is the most important in the release of histamine, which promotes inflammation?
   a. monocyte
   b. macrophage
   c. eosinophil
   d. mast cell
   e. natural killer cell
11. Which of these conditions does *not* occur during the inflammatory response?
   a. the release of histamine and other chemical mediators
   b. the chemotaxis of phagocytes
   c. the entry of fibrinogen into tissues from the blood
   d. the vasoconstriction of blood vessels
   e. the increased permeability of blood vessels
12. Antigens
   a. are foreign substances introduced into the body.
   b. are molecules produced by the body.
   c. stimulate an adaptive immune system response.
   d. all of the above.

13. B cells
    a. are processed in the thymus.
    b. originate in red bone marrow.
    c. once released into the blood remain in the blood.
    d. are responsible for cell-mediated immunity.
    e. all of the above.

14. MHC molecules
    a. are glycoproteins.
    b. attach to the plasma membrane.
    c. can bind to foreign and self-antigens.
    d. may form an MHC/antigen complex that activates T cells.
    e. all of the above.

15. Antigen-presenting cells can
    a. take in foreign antigens.
    b. process antigens.
    c. use MHC class II molecules to display the antigens.
    d. stimulate other immune system cells.
    e. all of the above.

16. Which of these participates in costimulation?
    a. cytokines
    b. complement
    c. antibodies
    d. histamine
    e. natural killer cells

17. Helper T cells
    a. respond to antigens from macrophages.
    b. respond to cytokines from macrophages.
    c. stimulate B cells with cytokines.
    d. all of the above.

18. The most important function of tolerance is to
    a. increase lymphocyte activity.
    b. increase complement activation.
    c. prevent the immune system from responding to self-antigens.
    d. prevent excessive immune system response to foreign antigens.
    e. process antigens.

19. The variable region of the antibody molecule
    a. makes the antibody specific for a given antigen.
    b. enables the antibody to activate complement.
    c. enables the antibody to attach to basophils and mast cells.
    d. forms the body of the Y-shaped molecule.
    e. all of the above.

20. The largest percentage of antibodies in the blood consists of
    a. IgA.
    b. IgD.
    c. IgE.
    d. IgG.
    e. IgM.

21. Antibodies
    a. prevent antigens from binding together.
    b. promote phagocytosis.
    c. inhibit inflammation.
    d. block complement activation.

22. The secondary antibody response
    a. is slower than the primary response.
    b. produces fewer antibodies than the primary response.
    c. prevents disease symptoms from occurring.
    d. occurs because of cytotoxic T cells.

23. The type of lymphocyte that is responsible for the secondary antibody response is the
    a. memory B cell.
    b. B cell.
    c. T cell.
    d. helper T cell.

24. Antibody-mediated immunity
    a. works best against intracellular antigens.
    b. regulates the activity of T cells.
    c. cannot be transferred from one person to another person.
    d. is responsible for immediate hypersensitivity reactions.

25. The activation of cytotoxic T cells can result in the
    a. lysis of virus-infected cells.
    b. production of cytokines.
    c. production of memory T cells.
    d. all of the above.

Answers in Appendix E

C R I T I C A L   T H I N K I N G

1. If the thymus of an experimental animal is removed immediately after its birth, the animal exhibits the following characteristics: (a) It is more susceptible to infections, (b) it has decreased numbers of lymphocytes in lymphatic tissue, and (c) its ability to reject grafts is greatly decreased. Explain these observations.

2. If the thymus of an adult experimental animal is removed, the following observations can be made: (a) No immediate effect occurs and (b) after 1 year the number of lymphocytes in the blood decreases, the ability to reject grafts decreases, and the ability to produce antibodies decreases. Explain these observations.

3. Adjuvants are substances that slow but do not stop the release of an antigen from an injection site into the blood. Suppose injection A of a certain amount of antigen is given without an adjuvant and injection B of the same amount of antigen is given with an adjuvant that causes the release of antigen over a period of 2–3 weeks. Does injection A or B result in the greater amount of antibody production? Explain.

4. An infant appears to be healthy until about 9 months of age. Then he develops severe bacterial infections, one after another. Fortunately, the infections are successfully treated with antibiotics.

When infected with the measles and other viral diseases, the infant recovers without unusual difficulty. Explain the different immune responses to these infections. Why did it take so long for this disorder to become apparent? (*Hint:* Consider IgG.)

5. A baby is born with severe combined immunodeficiency disease (SCID). In an attempt to save her life, a bone marrow transplant is performed. Explain how this procedure might help. Unfortunately, there is a graft rejection, and the baby dies. Explain what happened.

6. A patient has many allergic reactions. As part of the treatment scheme, doctors decide to try to identify the allergen that stimulates the immune system's response. A series of solutions, each containing an allergen that commonly causes a reaction, is composed. Each solution is injected into the skin at different locations on the patient's back. The following results are obtained: (a) At one location, the injection site becomes red and swollen within a few minutes; (b) at another injection site, swelling and redness appear 2 days later; and (c) no redness or swelling develops at the other sites. Explain what happened for each observation by describing what part of the immune system was involved and what caused the redness and swelling.

7. Ivy Hurtt developed a poison ivy rash after a camping trip. Her doctor prescribed a cortisone ointment to relieve the inflammation. A few weeks later, Ivy scraped her elbow, which became inflamed. Because she had some of the cortisone ointment left over, she applied it to the scrape. Explain why the ointment was or was not a good idea for the poison ivy and for the scrape.

8. Suzy Withitt has just had her ears pierced. To her dismay, she finds that, when she wears inexpensive (but tasteful) jewelry, by the end of the day there is an inflammatory (allergic) reaction to the metal in the jewelry. Is this because of antibodies or cytokines?

9. Tetanus is caused by bacteria that enter the body through wounds in the skin. The bacteria produce a toxin, which causes spastic muscle contractions. Death often results from the failure of the respiration muscles. A patient goes to the emergency room after stepping on a nail. If the patient has been vaccinated against tetanus, the patient is given a tetanus booster shot, which consists of the toxin altered so that it is harmless. If the patient has never been vaccinated against tetanus, the patient is given an antiserum shot against tetanus. Explain the rationale for this treatment strategy. Sometimes both a booster and an antiserum shot are given, but at different locations of the body. Explain why this is done and why the shots are given in different locations.

Answers in Appendix F

For additional study tools, go to: www.aris.mhhe.com.
Click on your course topic and then this textbook's author name/title. Register once for a semester's worth of interactive learning activities to help improve your grade!

CHAPTER 20 Respiratory System

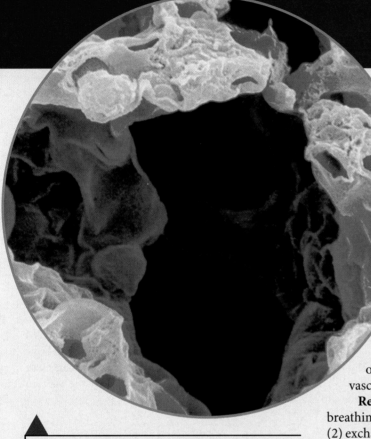

The central space in this colorized scanning electron micrograph of the lungs is a tiny air passageway called an alveolar duct. Air from outside the body passes through a series of ever smaller tubes to reach the alveolar duct. The even smaller spaces around the alveolar duct are alveoli, which are small chambers where gas exchange takes place between the air and the blood.

Tools for Success

 *A virtual cadaver dissection experience*

 *Audio (MP3) and/or video (MP4) content, available at www.mhhe.aris.com*

Introduction

From our first breath at birth, the rate and depth of our breathing are unconsciously matched to our activities, whether studying, sleeping, talking, eating, or exercising. We can voluntarily stop breathing, but within a few seconds we must breathe again. Breathing is so characteristic of life that, along with the pulse, it is one of the first things we check for to determine if an unconscious person is alive.

Breathing is necessary because all living cells of the body require oxygen and produce carbon dioxide. The respiratory system allows the exchange of these gases between the air and the blood, and the cardiovascular system transports them between the lungs and the cells of the body. The capacity to carry out normal activity is reduced without healthy respiratory and cardiovascular systems.

Respiration includes the following processes: (1) ventilation, or breathing, which is the movement of air into and out of the lungs; (2) exchange of oxygen and carbon dioxide between the air in the lungs and the blood; (3) transport of oxygen and carbon dioxide in the blood; and (4) exchange of oxygen and carbon dioxide between the blood and the tissues. The term *respiration* is also used in reference to cell metabolism, which is discussed in chapter 22.

▶This chapter explains the **functions of the respiratory system (p. 602)**, the **anatomy and histology of the respiratory system (p. 602)**, **ventilation (p. 613)**, **measurement of lung function (p. 617)**, **gas exchange in the lungs (p. 619)**, **oxygen and carbon dioxide transport in the blood (p. 620)**, **regulation of ventilation (p. 626)**, and **respiratory adaptations to exercise (p. 631)**. The chapter concludes by looking at the **effects of aging on the respiratory system (p. 631)**.

Functions of the Respiratory System

Respiration is necessary because all living cells of the body require oxygen and produce carbon dioxide. The respiratory system assists in gas exchange and performs other functions as well:

1. *Gas exchange.* The respiratory system allows oxygen from the air to enter the blood and carbon dioxide to leave the blood and enter the air. The cardiovascular system transports oxygen from the lungs to the cells of the body and carbon dioxide from the cells of the body to the lungs. Thus, the respiratory and cardiovascular systems work together to supply oxygen to all cells and to remove carbon dioxide.
2. *Regulation of blood pH.* The respiratory system can alter blood pH by changing blood carbon dioxide levels.
3. *Voice production.* Air movement past the vocal folds makes sound and speech possible.
4. *Olfaction.* The sensation of smell occurs when airborne molecules are drawn into the nasal cavity.
5. *Protection.* The respiratory system provides protection against some microorganisms by preventing their entry into the body and by removing them from respiratory surfaces. ▼

   1. Describe the functions of the respiratory system.

Anatomy and Histology of the Respiratory System

The respiratory system consists of the external nose, the nasal cavity, the pharynx, the larynx, the trachea, the bronchi, and the lungs (figure 20.1). Although air frequently passes through the oral cavity, it is considered to be part of the digestive system instead of the respiratory system. The **upper respiratory tract** consists of the external nose, nasal cavity, pharynx, and associated structures, and the **lower respiratory tract** consists of the larynx, trachea, bronchi, and lungs. These terms are not official anatomical terms, however, and there are several alternate definitions. For example, one alternate definition places the larynx in the upper respiratory tract. The diaphragm and the muscles of the thoracic and abdominal walls are responsible for respiratory movements. ▼

   2. Define upper and lower respiratory tract.

Nose

The **nasus** (nā′sŭs), or **nose,** consists of the external nose and the nasal cavity. The **external nose** is the visible structure that forms a prominent feature of the face. The largest part of the external nose is composed of hyaline cartilage plates, and the bridge of the nose consists of bones (see figure 7.9*b*).

   The **nasal cavity** extends from the nares to the choanae (figure 20.2). The **nares** (nā′res, sing. naris), or **nostrils,** are the external openings of the nasal cavity and the **choanae** (kō′an-ē) are the openings into the pharynx. The anterior part of the nasal cavity, just inside each naris, is the **vestibule** (ves′ti-bool, entry room). The vestibule is lined with stratified squamous epithelium, which is continuous with the stratified squamous epithelium of the skin. The **hard palate** (pal′ăt) is a bony plate covered by a mucous membrane that forms the floor of the nasal cavity. It separates the nasal cavity from the oral cavity. The **nasal septum** is a partition of bone and cartilage dividing the nasal cavity into right and left parts (see figure 7.9*a*). A deviated nasal septum occurs when the septum bulges to one side.

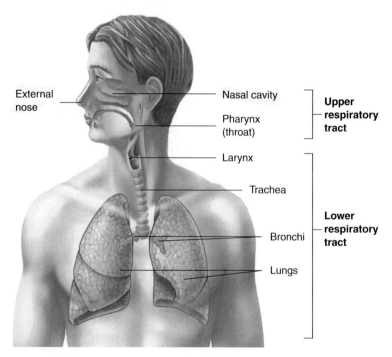

Figure 20.1   Respiratory System
The upper respiratory tract consists of the external nose, nasal cavity, and pharynx (throat). The lower respiratory tract consists of the larynx, trachea, bronchi, and lungs.

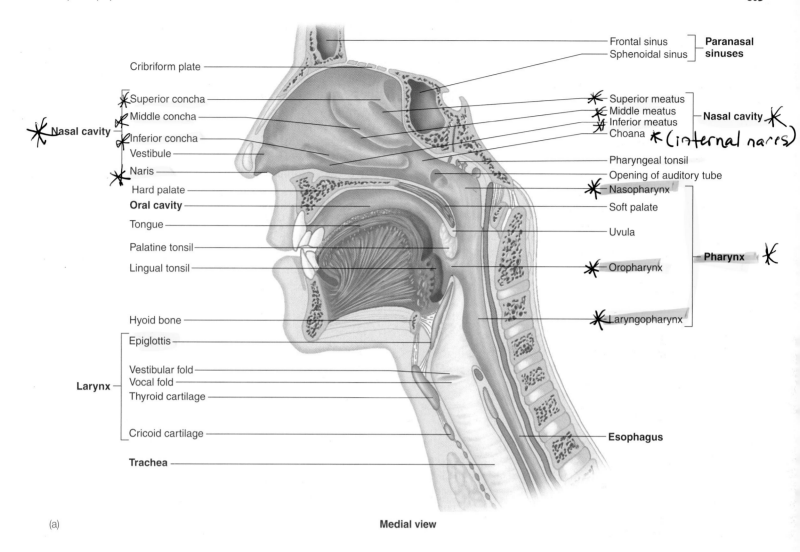

Cribriform plate

Superior concha
Middle concha
Nasal cavity
Inferior concha
Vestibule
Naris
Hard palate
Oral cavity
Tongue
Palatine tonsil
Lingual tonsil
Hyoid bone

Larynx
Epiglottis
Vestibular fold
Vocal fold
Thyroid cartilage
Cricoid cartilage

Trachea

Frontal sinus — **Paranasal sinuses**
Sphenoidal sinus

Superior meatus
Middle meatus — **Nasal cavity**
Inferior meatus
Choana ✶ (internal nares)

Pharyngeal tonsil
Opening of auditory tube
Nasopharynx
Soft palate
Uvula — **Pharynx**
Oropharynx
Laryngopharynx

Esophagus

(a)   **Medial view**

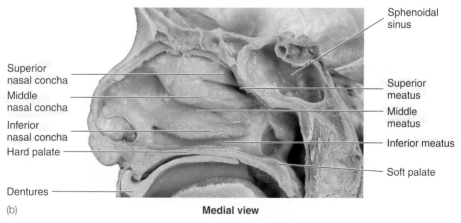

Superior nasal concha
Middle nasal concha
Inferior nasal concha
Hard palate
Dentures

Sphenoidal sinus
Superior meatus
Middle meatus
Inferior meatus
Soft palate

(b)   **Medial view**

Figure 20.2   Nasal Cavity and Pharynx

(a) Sagittal section through the nasal cavity and pharynx. (b) Photograph of sagittal section of the head.

Three bony ridges, called **conchae** (kon′kē, resembling a conch shell), are present on the lateral walls on each side of the nasal cavity. Beneath each concha is a passageway called a **meatus** (mē-ā′tŭs, a tunnel or passageway). Within the superior and middle meatus are openings from the various **paranasal sinuses** (see figure 7.10), and the opening of a **nasolacrimal** (nā-zō-lak′ri-măl) **duct** is within each inferior meatus (see figure 13.5). Sensory receptors for the sense of smell are found in the superior part of the nasal cavity (see figure 13.1).

Sinusitis

Sinusitis (sī-nŭ-sī′tis) is an inflammation of the mucous membrane of any sinus, especially of one or more paranasal sinuses. Viral infections, such as the common cold, can cause mucous membranes to become inflamed, to swell, and to produce excess mucus. As a result, the sinus opening into the nasal cavity can be partially or completely blocked. Mucus accumulates within the sinus, which can promote the development of a bacterial infection. Treatments consist of taking antibiotics and promoting sinus drainage with decongestants, hydration, and steam inhalation. Sinusitis can also result from swelling caused by allergies or by polyps that obstruct a sinus opening into the nasal cavity.

Air enters the nasal cavity through the nares. Just inside the nares the epithelial lining is composed of stratified squamous epithelium containing coarse hairs. The hairs trap some of the large particles of dust suspended in the air. The rest of the nasal cavity is lined with pseudostratified columnar epithelial cells containing cilia and many mucus-producing goblet cells (see chapter 4). Mucus produced by the goblet cells also traps debris in the air. The cilia sweep the mucus posteriorly to the pharynx, where it is swallowed. As air flows through the nasal cavities, it is humidified by moisture from the mucous epithelium and is warmed by blood flowing through the superficial capillary networks underlying the mucous epithelium. ▼

> 3. Describe the structures of the nasal cavity. How is air cleaned, humidified, and warmed as it passes through the nasal cavity?

P R E D I C T

Explain what happens to your throat when you sleep with your mouth open, especially when your nasal passages are plugged as a result of having a cold. Explain what may happen to your lungs when you run a long way in very cold weather while breathing rapidly through your mouth.

Pharynx

The **pharynx** (far′ingks, throat) is the common passageway of both the digestive and the respiratory systems. It receives air from the nasal cavity and receives air, food, and drink from the oral cavity. Inferiorly, the pharynx is connected to the respiratory system at the larynx and to the digestive system at the esophagus. The pharynx is divided into three regions: the nasopharynx, the oropharynx, and the laryngopharynx (see figure 20.2).

The **nasopharynx** (nā′zō-far′ingks) is located posterior to the choanae and superior to the **soft palate,** which is an incomplete muscle and connective tissue partition separating the nasopharynx from the oropharynx. The **uvula** (ū′vū-lă, a grape) is the posterior extension of the soft palate. The soft palate prevents swallowed materials from entering the nasopharynx and nasal cavity. The nasopharynx is lined with a mucous membrane containing pseudostratified ciliated columnar epithelium with goblet cells. Debris-laden mucus from the nasal cavity is moved through the nasopharynx and swallowed. Two auditory tubes from the middle ears open into the nasopharynx (see figures 13.19 and 20.2a). Air passes through them to equalize air pressure between the atmosphere and the middle ears. The posterior surface of the nasopharynx contains the **pharyngeal tonsil,** or **adenoid** (ad′ĕ-noyd), which helps defend the body against infection (see figure 19.4). An enlarged pharyngeal tonsil can interfere with normal breathing and the passage of air through the auditory tubes.

The Sneeze Reflex

The **sneeze reflex** dislodges foreign substances from the nasal cavity. Sensory receptors detect the foreign substances, and action potentials are conducted along the trigeminal nerves to the medulla oblongata in which the reflex is triggered. During the sneeze reflex, the uvula and the soft palate are depressed so that rapidly flowing air from the lungs is directed primarily through the nasal passages, although a considerable amount passes through the oral cavity.

Some people have a photic sneeze reflex, in which exposure to bright light, such as the sun, can stimulate a sneeze reflex. The pupillary reflex causes the pupils to constrict in response to bright light. It is speculated that the complicated "wiring" of the pupillary and sneeze reflexes is intermixed in some people so that, when bright light activates a pupillary reflex, it also activates a sneeze reflex.

The **oropharynx** (ōr′ō-far′ingks) extends from the soft palate to the epiglottis, and the oral cavity opens into the oropharynx. Thus, air, food, and drink all pass through the oropharynx. Moist stratified squamous epithelium lines the oropharynx and protects it against abrasion. The **palatine** (pal′ă-tīn) **tonsils** are located in the lateral walls near the border of the oral cavity and the oropharynx. The **lingual tonsil** is located on the surface of the posterior part of the tongue (see figure 19.4).

The **laryngopharynx** (lă-ring′gō-far-ingks) extends from the tip of the epiglottis to the esophagus and passes posterior to the larynx. Food and drink pass through the laryngopharynx to the esophagus. A small amount of air is usually swallowed with the food and drink. Swallowing too much air can cause excess gas in the stomach and belching. The laryngopharynx is lined with moist stratified squamous epithelium. ▼

> 4. Name the three parts of the pharynx. With what structures does each part communicate?

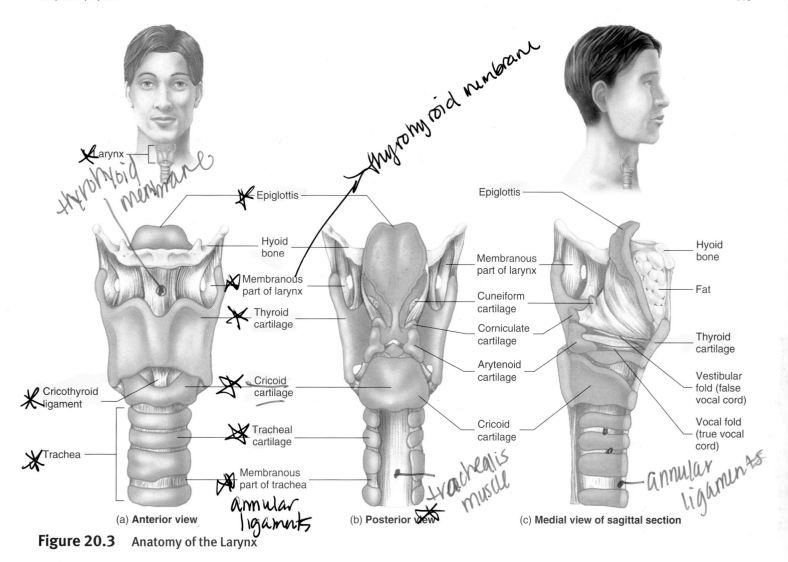

(a) **Anterior view**   (b) **Posterior view**   (c) **Medial view of sagittal section**

Figure 20.3   Anatomy of the Larynx

Larynx

The **larynx** (lar′ingks) is located in the anterior part of the throat and extends from the base of the tongue to the trachea (see figure 20.2*a*). It is a passageway for air between the pharynx and the trachea. The larynx is connected by membranes and/or muscles superiorly to the hyoid bone and consists of an outer casing of nine cartilages connected to one another by muscles and ligaments (figure 20.3). Three of the nine cartilages are unpaired, and six of them form three pairs.

The largest of the cartilages is the unpaired **thyroid** (shield, refers to the shape of the cartilage) **cartilage,** or Adam's apple. The most inferior cartilage of the larynx is the unpaired **cricoid** (krī′koyd, ring-shaped) **cartilage,** which forms the base of the larynx on which the other cartilages rest. The third unpaired cartilage is the **epiglottis** (ep-i-glot′is, on the glottis). It is attached to the thyroid cartilage and projects superiorly as a free flap toward the tongue. The epiglottis differs from the other cartilages in that it consists of elastic rather than hyaline cartilage.

The six paired cartilages consist of three cartilages on each side of the posterior part of the larynx (see figure 20.3*b*). The superior cartilage on each side is the **cuneiform** (kū′nē-i-fōrm,

wedge-shaped) **cartilage,** the middle cartilage is the **corniculate** (kōr-nik′ū-lāt, horn-shaped) **cartilage,** and the inferior cartilage is the **arytenoid** (ar-i-tē′noyd, ladle-shaped) **cartilage.**

Two pairs of ligaments extend from the anterior surface of the arytenoid cartilages to the posterior surface of the thyroid cartilage. The superior ligaments are covered by a mucous membrane called the **vestibular folds,** or **false vocal cords** (see figure 20.3*c* and figure 20.4).

The inferior ligaments are covered by a mucous membrane called the **vocal folds,** or **true vocal cords** (see figure 20.4). The vocal folds and the opening between them are called the **glottis** (glot′is). The vestibular folds and the vocal folds are lined with stratified squamous epithelium. The remainder of the larynx is lined with pseudostratified ciliated columnar epithelium. An inflammation of the mucosal epithelium of the vocal folds is called **laryngitis** (lar-in-jī′tis).

The larynx prevents the entry of swallowed materials into the lower respiratory tract and regulates the passage of air into and out of the lower respiratory tract. During swallowing, the epiglottis tips posteriorly until it lies below the horizontal plane and covers the opening into the larynx (see "Swallowing," chapter 21). Thus,

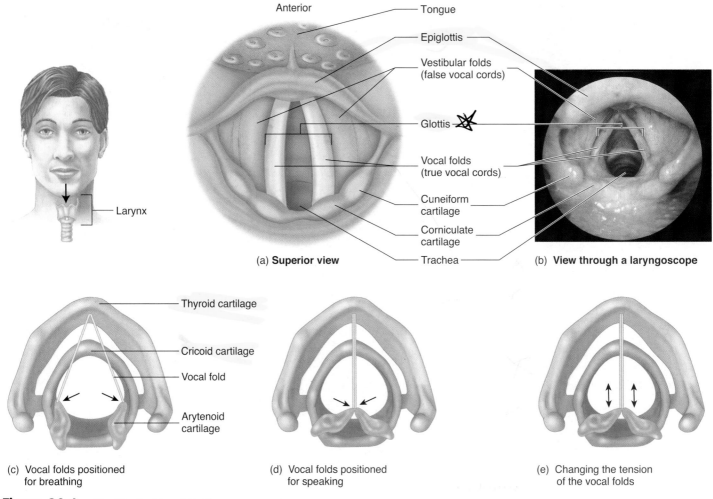

(a) **Superior view**

(b) **View through a laryngoscope**

(c) Vocal folds positioned for breathing

(d) Vocal folds positioned for speaking

(e) Changing the tension of the vocal folds

Figure 20.4 Vestibular Vocal Folds

Arrows show the direction of viewing the vestibular and vocal folds. (*a*) Relationship of the vestibular folds to the vocal folds and the laryngeal cartilages. (*b*) Laryngoscopic view of the vestibular and vocal folds. (*c*) Lateral rotation of the arytenoid cartilages moves the vocal folds laterally for breathing. (*d*) Medial rotation of the arytenoid cartilages moves the vocal folds medially for speaking. (*e*) Anterior/posterior movement of the arytenoid cartilages changes the length and tension of the vocal folds, changing the pitch of sounds (see text).

food and liquid slide over the epiglottis toward the esophagus. The most important event for preventing the entry of materials into the larynx, however, is the closure of the vestibular and vocal folds. That is, the vestibular folds move medially and come together, as do the vocal folds. The closure of the vestibular and vocal folds can also prevent the passage of air, as when a person holds his or her breath or increases air pressure within the lungs prior to coughing or sneezing.

The vocal folds are the primary source of sound production. Air moving past the vocal folds causes them to vibrate and produce sound. The greater the amplitude of the vibration, the louder the sound. The force of air moving past the vocal folds determines the amplitude of vibration and the loudness of the sound. The frequency of vibrations determines pitch, with higher-frequency vibrations producing higher-pitched sounds and lower-frequency vibrations producing lower-pitched sounds. Variations in the length of the vibrating segments of the vocal folds affect the frequency of the vibrations. Higher-pitched tones

are produced when only the anterior parts of the folds vibrate, and progressively lower tones result when longer sections of the folds vibrate. Most males have lower-pitched voices than females because males usually have longer vocal folds. The sound produced by the vibrating vocal folds is modified by the tongue, lips, teeth, and other structures to form words. A person whose larynx has been removed because of carcinoma of the larynx can produce sound by swallowing air and causing the esophagus to vibrate.

Movement of the arytenoid and other cartilages is controlled by skeletal muscles, thereby changing the position and length of the vocal folds. When a person is only breathing, lateral rotation of the arytenoid cartilages abducts the vocal folds, which allows greater movement of air (see figure 20.4c). Medial rotation of the arytenoid cartilages adducts the vocal folds, places them in position for producing sounds, and changes the tension on them (see figure 20.4d). Anterior movement of the arytenoid cartilages decreases the length and tension of the vocal folds, lowering pitch. Posterior movement of

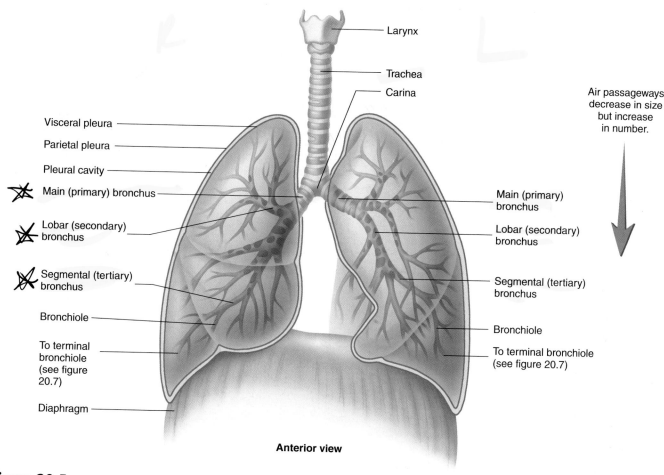

Larynx

Trachea

Carina

Air passageways decrease in size but increase in number.

Visceral pleura

Parietal pleura

Pleural cavity

Main (primary) bronchus

Lobar (secondary) bronchus

Segmental (tertiary) bronchus

Bronchiole

To terminal bronchiole (see figure 20.7)

Diaphragm

Main (primary) bronchus

Lobar (secondary) bronchus

Segmental (tertiary) bronchus

Bronchiole

To terminal bronchiole (see figure 20.7)

Anterior view

Figure 20.5 Anatomy of the Tracheobronchial Tree and Lungs

Drawing of the trachea and lungs, showing the branching of the bronchi to form the tracheobronchial tree. Each lung is surrounded by a pleural cavity, formed by the visceral and parietal pleurae.

the arytenoid cartilages increases the length and tension of the vocal folds, increasing pitch (figure 20.4*e*). ▼

5. *Name the unpaired and paired cartilages of the larynx.*
6. *Distinguish between the vestibular and vocal folds.*
7. *How is the entry of swallowed materials into the larynx prevented?*
8. *How are sounds of different loudness and pitch produced by the vocal folds?*

Trachea

The **trachea** (trā′kē-ă), or windpipe, is a membranous tube attached to the larynx (see figure 20.3). The trachea has an inside diameter of 12 mm and a length of 10–12 cm, descending from the larynx to the level of the fifth thoracic vertebra (figure 20.5). It consists of dense regular connective tissue and smooth muscle reinforced with 15–20 C-shaped pieces of hyaline cartilage. The cartilages support the anterior and lateral sides of the trachea. They protect the trachea and maintain an open passageway for air. The posterior wall of the trachea is devoid of cartilage; it contains an elastic ligamentous membrane and bundles of smooth muscle called the **trachealis**

(trā′kē-ā-lis) **muscle.** The esophagus lies immediately posterior to the cartilage-free posterior wall of the trachea.

 The Cough Reflex

The function of the **cough reflex** is to dislodge foreign substances from the trachea. Sensory receptors detect the foreign substances, and action potentials are conducted along the vagus nerves to the medulla oblongata in which the cough reflex is triggered. During coughing, contraction of trachealis muscle decreases the diameter of the trachea. As a result, air moves rapidly through the trachea, which helps expel mucus and foreign substances. Also, the uvula and soft palate are elevated so that air passes primarily through the oral cavity.

P R E D I C T **2**

Explain what happens to the shape of the trachea when a person swallows a large mouthful of food. Why is this change of shape advantageous?

The mucous membrane lining the trachea consists of pseudostratified ciliated columnar epithelium with numerous goblet cells. The goblet cells produce mucus, which traps inhaled foreign particles. The cilia move the mucus and foreign particles into the larynx, from which they enter the pharynx and are swallowed. Constant, long-term irritation to the trachea, such as occurs in smokers, can cause the tracheal epithelium to become moist stratified squamous epithelium that lacks cilia and goblet cells. Consequently, the normal function of the tracheal epithelium is lost.

 Establishing Airflow

In cases of extreme emergency when the upper air passageway is blocked by a foreign object to the extent that the victim cannot breathe, quick reaction is required to save the person's life. The **Heimlich maneuver** is designed to force such an object out of the air passage by the sudden application of pressure to the abdomen. The person who performs the maneuver stands behind the victim, with his or her arms under the victim's arms and hands over the victim's abdomen between the navel and the rib cage. With one hand formed into a fist and the other hand over it, both hands are suddenly pulled toward the abdomen with an accompanying upward motion. This maneuver, if done properly, forces air up the trachea and dislodges most foreign objects.

There are other ways to establish airflow, but they should be performed only by trained medical personnel. **Intubation** is the insertion of a tube into an opening, a canal, or a hollow organ. A tube can be passed through the mouth or nose into the pharynx and then through the larynx to the trachea.

Sometimes it is necessary to make an opening through which to pass the tube. The preferred point of entry in emergency cases is through the membrane between the cricoid and thyroid cartilages, a procedure referred to as a **cricothyrotomy** (krī'kō-thī-rot'ō-mē).

A **tracheostomy** (trā'-kē-os'tō-mē, *tracheo-* + *stoma*, mouth) is an operation to make an opening into the trachea, usually between the second and third cartilage rings. Usually, the opening is intended to be permanent, and a tube is inserted into the trachea to allow airflow and provide a way to remove secretions. The term **tracheotomy** (trā-kē-ot'ō-mē, *tracheo-* + *tome*, incision) refers to the actual cutting into the trachea. Sometimes the terms *tracheostomy* and *tracheotomy* are used interchangeably. It is not advisable to enter the air passageway through the trachea in emergency cases because arteries, nerves, and the thyroid gland overlie the anterior surface of the trachea.

Main Bronchi

The trachea divides to form two smaller tubes called **main,** or **primary, bronchi** (brong'kī, sing. *bronchus*, brong'kŭs, windpipe), each of which extends to a lung (see figure 20.5). The most inferior tracheal cartilage forms a ridge called the **carina** (kă-rī'nă), which separates the openings into the main bronchi. The carina is an important radiological landmark. In addition, the mucous membrane of the carina is very sensitive to mechanical stimulation, and materials reaching the carina stimulate a powerful cough reflex.

Materials in the air passageways inferior to the carina do not usually stimulate a cough reflex.

The left main bronchus is more horizontal than the right main bronchus because it is displaced by the heart (see figure 20.5). Foreign objects that enter the trachea usually lodge in the right main bronchus, because it is more vertical than the left main bronchus and therefore more in direct line with the trachea.

Lungs and the Tracheobronchial Tree

The **lungs** are the principal organs of respiration. Each lung is cone-shaped, with its base resting on the diaphragm and its apex extending superiorly to a point about 2.5 cm above the clavicle (figure 20.6). The right lung has three **lobes** called the superior, middle, and inferior lobes. The left lung has two lobes called the superior and inferior lobes. The lobes of the lungs are separated by deep, prominent fissures on the surface of the lung. Each lobe is divided into **bronchopulmonary segments** separated from one another by connective tissue septa, but these separations are not visible as surface fissures. Individual diseased bronchopulmonary segments can be surgically removed, leaving the rest of the lung relatively intact, because major blood vessels and bronchi do not cross the septa. There are 9 bronchopulmonary segments in the left lung and 10 in the right lung.

The main bronchi branch many times to form the **tracheobronchial** (trā'kē-ō-brong'kē-ăl) **tree** (see figure 20.5). Each main bronchus divides into lobar bronchi as they enter their respective lungs (see figure 20.6). The **lobar (secondary) bronchi,** two in the left lung and three in the right lung, conduct air to each lobe. The lobar bronchi in turn give rise to **segmental (tertiary) bronchi,** which extend to the bronchopulmonary segments of the lungs. The bronchi continue to branch many times, finally giving rise to **bronchioles** (brong'kē-ōlz), which subdivide numerous times to give rise to **terminal bronchioles.** There are approximately 16 generations of branching from the trachea to the terminal bronchioles.

The terminal bronchioles divide to form **respiratory bronchioles** (figure 20.7), which have a few attached alveoli. **Alveoli** (al-vē'ō-lī, hollow cavity) are small, air-filled chambers where gas exchange between the air and blood takes place. There are approximately 300 million alveoli in the two lungs. As the respiratory bronchioles divide to form smaller respiratory bronchioles, the number of attached alveoli increases. The respiratory bronchioles give rise to **alveolar** (al-vē'ō-lăr) **ducts,** which are like long, branching hallways with many open doorways. The "doorways" open into alveoli, which become so numerous that the alveolar duct wall is little more than a succession of alveoli. Approximately seven generations of branching occur from the terminal bronchioles to the alveolar ducts.

Two types of cells form the alveolar wall (figure 20.8*a*). **Type I pneumocytes** are thin squamous epithelial cells that form 90% of the alveolar surface. Most gas exchange between alveolar air and the blood takes place through these cells. **Type II pneumocytes** are round or cube-shaped secretory cells that produce surfactant, which makes it easier for the alveoli to expand during inspiration (see "Lung Recoil," p. 614). The tissue surrounding the alveoli contains elastic fibers, which allow the alveoli to expand during inspiration and recoil during expiration (see figure 20.7).

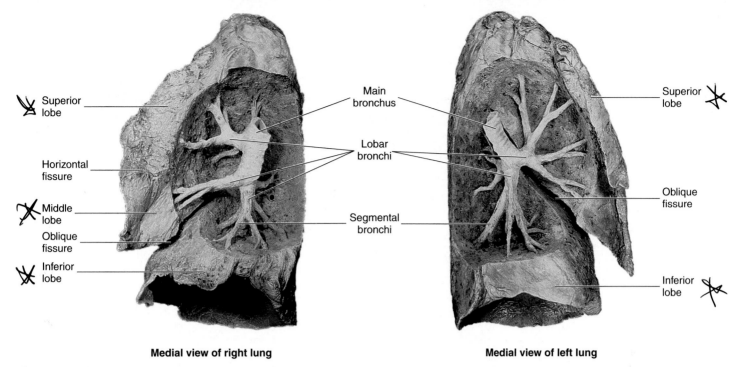

Superior lobe

Horizontal fissure

Middle lobe

Oblique fissure

Inferior lobe

Main bronchus

Lobar bronchi

Segmental bronchi

Superior lobe

Oblique fissure

Inferior lobe

Medial view of right lung

Medial view of left lung

Figure 20.6   Lungs, Lung Lobes, and Bronchi

The right lung is divided into three lobes by the horizontal and oblique fissures. The left lung is divided into two lobes by the oblique fissure. A main bronchus supplies each lung, a lobar bronchus supplies each lung lobe, and segmental bronchi supply the bronchopulmonary segments (not visible).

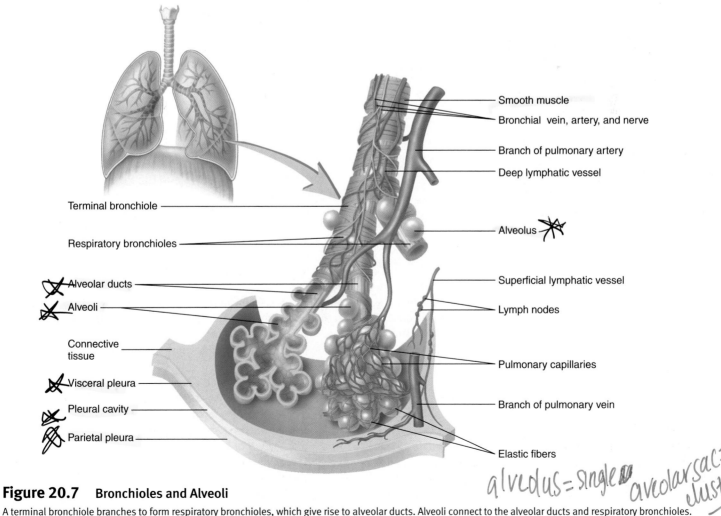

Smooth muscle

Bronchial vein, artery, and nerve

Branch of pulmonary artery

Deep lymphatic vessel

Terminal bronchiole

Respiratory bronchioles

Alveolus

Alveolar ducts

Alveoli

Superficial lymphatic vessel

Lymph nodes

Connective tissue

Pulmonary capillaries

Visceral pleura

Pleural cavity

Branch of pulmonary vein

Parietal pleura

Elastic fibers

alveolus = single
alveolar sac = cluster

Figure 20.7   Bronchioles and Alveoli

A terminal bronchiole branches to form respiratory bronchioles, which give rise to alveolar ducts. Alveoli connect to the alveolar ducts and respiratory bronchioles.

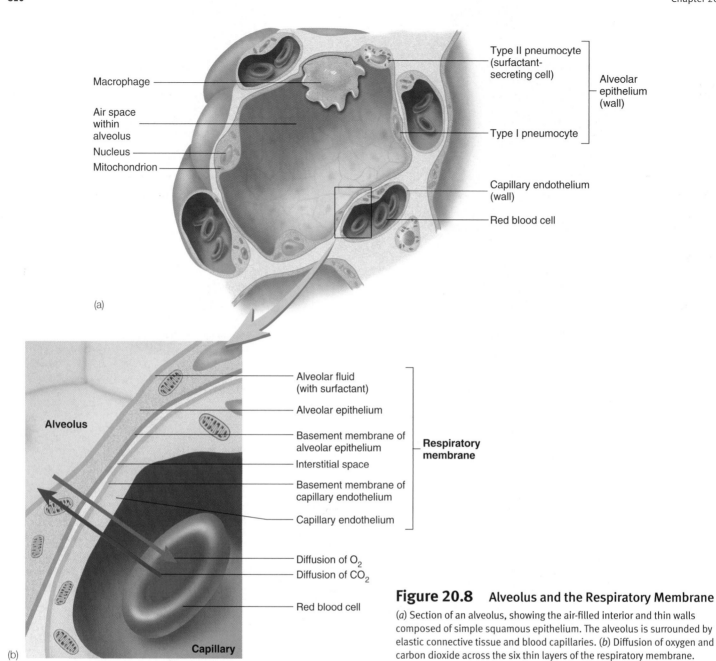

Figure 20.8   **Alveolus and the Respiratory Membrane**

(*a*) Section of an alveolus, showing the air-filled interior and thin walls composed of simple squamous epithelium. The alveolus is surrounded by elastic connective tissue and blood capillaries. (*b*) Diffusion of oxygen and carbon dioxide across the six thin layers of the respiratory membrane.

As the air passageways of the lungs become smaller, the structure of their walls changes. The amount of cartilage decreases and the amount of smooth muscle increases, until at the terminal bronchioles the walls have a prominent smooth muscle layer, but no cartilage. Relaxation and contraction of the smooth muscle within the bronchi and bronchioles can change the diameter of the air passageways. For example, during exercise the diameter can increase, thus increasing the volume of air moved.

As the air passageways of the lungs become smaller, the lining of their walls also changes. The lining changes from pseudostratified ciliated columnar epithelium in the trachea and bronchi to ciliated simple cuboidal epithelium in the terminal bronchioles. This ciliated epithelium is a mucus–cilia escalator that traps debris in the air and removes it from the respiratory system. The respiratory bronchioles have a simple cuboidal epithelium, and the alveolar ducts and alveoli consist of simple squamous epithelium, which facilitates the diffusion of gases. Although this epithelium is not ciliated, debris from the air is removed by macrophages that move over the surfaces of the cells. ▼

9. **Distinguish among a lung, a lung lobe, and a bronchopulmonary segment. How are they related to the branches of the tracheobronchial tree?**
10. **Name the two types of cells in the alveolar wall, and state their functions.**
11. **Describe the arrangement of cartilage and smooth muscle in the tracheobronchial tree.**
12. **Describe the epithelium of the tracheobronchial tree. How is debris removed from the tracheobronchial tree?**

Clinical Relevance Asthma and Cystic Fibrosis

Asthma (az′mă) is a disease characterized by abnormally increased constriction of the bronchi and bronchioles in response to various stimuli, resulting in a narrowing of the air passageways and decreased ventilation efficiency. Symptoms include rapid, shallow breathing; wheezing; coughing; and shortness of breath. In contrast to many other respiratory disorders, however, the symptoms of asthma typically reverse either spontaneously or with therapy.

The exact cause or causes of asthma are unknown, but asthma and allergies are more common in some families. Multiple genes contribute to a person's susceptibility to asthma; genes on chromosomes 5, 6, 11, 12, and 14 have all been linked to asthma. Although no definitive pathological feature or diagnostic test for asthma has yet been discovered, three important features of the disease are chronic airway inflammation, airway hyperreactivity, and airflow obstruction. The inflammatory response results in tissue damage, edema, and mucus buildup, which can block airflow through the bronchi and bronchioles. Airway hyperreactivity results in greatly increased contraction of the smooth muscle in the bronchi and bronchioles in response to a stimulus. As a result of airway hyperreactivity, the diameter of the airway decreases, and resistance to airflow increases. The effects of inflammation and airway hyperreactivity combine to cause airflow obstruction.

Many cases of asthma appear to be associated with a chronic inflammatory response. The number of immune cells in the bronchi and bronchioles, including mast cells, eosinophils, neutrophils, macrophages, and lymphocytes, increases. Inflammation appears to be linked to airway hyperreactivity by some chemical mediators released by immune cells (e.g., leukotrienes, prostaglandins, and interleukins), which increase the airway's sensitivity to stimulation and cause smooth muscle contraction.

The stimuli that prompt airflow obstruction in asthma vary from one individual to another. Stimuli that cause an asthma attack include allergens, smoke, aspirin, ibuprofen, and strenuous exercise, especially in cold weather. Treatment of asthma involves avoiding the causative stimulus and administering drug therapy. Steroids and mast-stabilizing agents are used to reduce airway inflammation. Theophylline (thē-of′i-lēn, thē-of′i-lin) and β-adrenergic agents (see chapter 14) are commonly used to stimulate bronchiolar dilation.

Cystic fibrosis (CF) is a disease characterized by frequent, serious respiratory infections and thick, sticky mucus in the lungs and digestive tract. It is the most common lethal genetic disorder in Caucasians. Cystic fibrosis is inherited as an autosomal-recessive gene on chromosome 7 (see chapter 25). In CF, a mutated

gene results in the production of a defective chloride ion (Cl^-) channel. In addition to transporting Cl^-, the chloride channel inhibits the activity of Na^+ channels. Exactly how this inhibition is accomplished is not completely understood. The chloride channel usually acts as a "brake" on the Na^+ channels, thereby reducing the amount of Na^+ normally reabsorbed. In cystic fibrosis, the chloride channel is either nonfunctional or marginally functional. Consequently, the normal inhibition of the Na^+ channels is lost or greatly diminished and Na^+ absorption increases dramatically. Water follows the Na^+ by osmosis, resulting in increased water loss from the mucus lining the respiratory passageways. The mucus becomes thicker than normal, the ciliated epithelium of the air passageways cannot remove it, and the air passageways become blocked with mucus.

People with CF must undergo **chest physical therapy,** also called **chest clapping** or **chest percussion.** This involves manually pounding the back and chest for 30 to 40 minutes three or four times daily to dislodge mucus trapped in the chest. Automated chest clappers are preferred by some CF patients. Antibiotics may be prescribed to help control lung infections. Mucus-thinning drugs, such as Pulmozyme, and bronchodilators can be inhaled to improve mucus clearance and open airways.

P R E D I C T

Based on function, the respiratory passageways can be subdivided into the conducting and respiratory zones. The conducting zone functions as a passageway for the exchange of air between the outside of the body, and the respiratory zone is where gas exchange between the air and blood takes place. Name in order the parts of the conducting and respiratory zones, starting with the nose and ending with the alveoli.

Respiratory Membrane

The **respiratory membrane** of the lungs is where gas exchange between the air and blood takes place. It is formed mainly by the alveolar walls and surrounding pulmonary capillaries (figure 20.8*b*), but there is some contribution by the respiratory bronchioles and alveolar ducts. The respiratory membrane is very thin to facilitate

the diffusion of gases. It consists of

1. a thin layer of fluid lining the alveolus
2. the alveolar epithelium composed of simple squamous epithelium
3. the basement membrane of the alveolar epithelium
4. a thin interstitial space
5. the basement membrane of the capillary endothelium
6. the capillary endothelium composed of simple squamous epithelium ▼

13. List the parts of the respiratory membrane.

Pleura

The lungs are contained within the thoracic cavity, but each lung is surrounded by a separate **pleural** (ploor′ăl, relating to the ribs)

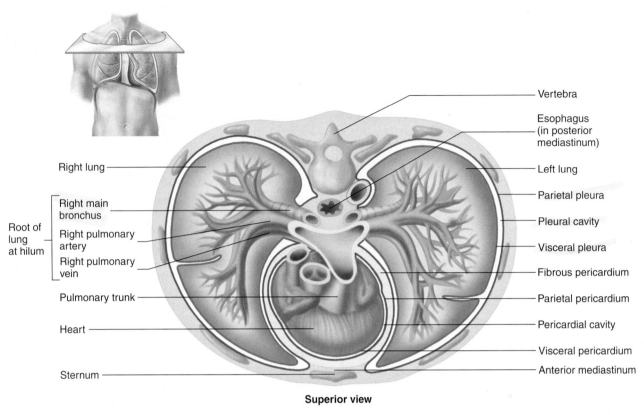

Superior view

Figure 20.9   Pleural Cavities and Membranes

Transverse section of the thorax, showing the relationship of the pleural cavities to the thoracic organs. Each lung is surrounded by a pleural cavity. The parietal pleura lines the wall of each pleural cavity, and the visceral pleura covers the surface of the lungs. The space between the parietal and visceral pleurae is small and filled with pleural fluid.

cavity formed by the pleural serous membranes (figure 20.9). The **parietal pleura** covers the inner thoracic wall, diaphragm, and mediastinum. The parietal pleura is continuous with the **visceral pleura,** which covers the surface of the lung.

The pleural cavity is filled with pleural fluid, which is produced by the pleural membranes. The pleural fluid does two things: (1) It acts as a lubricant, allowing the parietal and visceral pleural membranes to slide past each other as the lungs and the thorax change shape during respiration, and (2) it helps hold the parietal and visceral pleural membranes together. The pleural fluid acts like a thin film of water between two sheets of glass (the visceral and parietal pleurae); the glass sheets can slide over each other easily, but it is difficult to separate them. ▼

> 14. *Name the pleurae of the lungs. What are the functions of the pleural fluid?*

P R E D I C T
Pleurisy is an inflammation of the pleural membranes. Explain why this condition is so painful, especially when a person takes deep breaths.

Blood Supply

Blood that has passed through the lungs and picked up oxygen is called **oxygenated blood,** and blood that has passed through the tissues and released some of its oxygen is called **deoxygenated**

blood. Two blood flow routes to the lungs exist. The major route is the **pulmonary circulation,** which takes deoxygenated blood to the lungs, where it is oxygenated (see chapter 18). The deoxygenated blood flows through pulmonary arteries to pulmonary capillaries, becomes oxygenated, and returns to the heart through pulmonary veins. The **bronchial circulation** takes oxygenated blood to the tissues of the bronchi down to the respiratory bronchioles (see figure 20.7). The oxygenated blood flows from the thoracic aorta through bronchial arteries to capillaries, where oxygen is released. Deoxygenated blood from the proximal part of the bronchi returns to the heart through the bronchial veins and the azygos venous system (see chapter 18). More distally, the venous drainage from the bronchi enters the pulmonary veins. Thus, the oxygenated blood returning from the alveoli in the pulmonary veins is mixed with a small amount of deoxygenated blood returning from the bronchi. ▼

> 15. *What are the two major routes of blood flow to and from the lungs? What is the function of each route?*

Lymphatic Supply

The lungs have two lymphatic supplies (see figure 20.7). The **superficial lymphatic vessels** are deep to the visceral pleura; they drain lymph from the superficial lung tissue and the visceral pleura. The **deep lymphatic vessels** follow the bronchi; they drain lymph from the bronchi and associated connective tissues. No lymphatic vessels are located in the walls of the alveoli. Both the superficial and deep lymphatic vessels exit the lung at the main bronchi.

Phagocytic cells within the lungs phagocytize carbon particles and other debris from inspired air and move them to the lymphatic vessels. In older people, the surface of the lungs can appear gray to black because of the accumulation of these particles, especially if the person smoked or lived most of his or her life in a city with air pollution. Cancer cells from the lungs can spread to other parts of the body through the lymphatic vessels. ▼

16. Describe the lymphatic supply of the lungs.

Ventilation

Ventilation, or **breathing,** is the process of moving air into and out of the lungs. There are two phases of ventilation: (1) **inspiration,** or **inhalation,** is the movement of air into the lungs; (2) **expiration,** or **exhalation,** is the movement of air out of the lungs. Changes in thoracic volume, which produce changes in air pressure within the lungs, are responsible for ventilation. ▼

17. Define ventilation, inspiration, and expiration.

Changing Thoracic Volume

Muscles associated with the ribs are responsible for ventilation (figure 20.10). The **muscles of inspiration** are the diaphragm and muscles that elevate the ribs and sternum, such as the external intercostals, pectoralis minor, and scalenes. The **diaphragm** (dī′a-fram, partition) is a large dome of skeletal muscle that separates the thoracic cavity from the abdominal cavity. The **muscles of expiration** are muscles that depress the ribs and sternum, such as the abdominal muscles and the internal intercostals.

At the end of a normal, quiet expiration, the respiratory muscles are relaxed (see figure 20.10*a*). During quiet inspiration, contraction of the diaphragm causes the top of the dome to move inferiorly, which increases the volume of the thoracic cavity. The largest change in thoracic volume results from movement of the diaphragm. Contraction of the external intercostals also elevates the ribs and sternum (see figure 20.10*b*), which increases thoracic volume by increasing the diameter of the thoracic cage.

P R E D I C T   5
During inspiration, the abdominal muscles relax. How is this advantageous?

Expiration during quiet breathing occurs when the diaphragm and external intercostals relax and the elastic properties of the thorax and lungs cause a passive decrease in thoracic volume.

There are several differences between normal, quiet breathing and labored breathing. During labored breathing, all of the inspiratory muscles are active and they contract more forcefully than during quiet breathing, causing a greater increase in thoracic volume (see figure 20.10*b*). During labored breathing, forceful contraction of the internal intercostals and the abdominal muscles produces a more rapid and greater decrease in

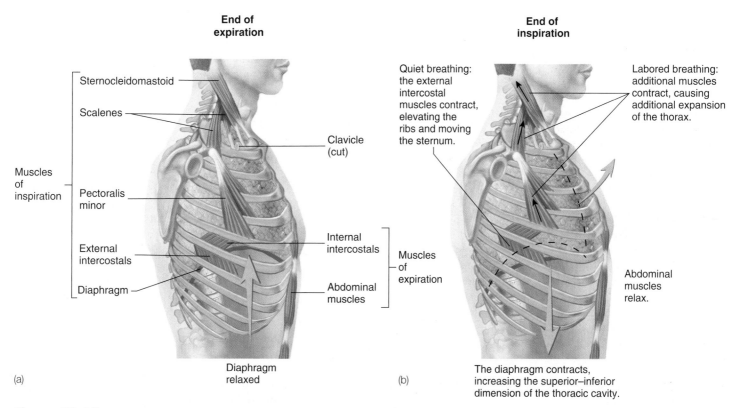

End of
expiration

End of
inspiration

Sternocleidomastoid

Scalenes

Clavicle
(cut)

Muscles
of
inspiration

Pectoralis
minor

External
intercostals

Diaphragm

Internal
intercostals

Muscles
of
expiration

Abdominal
muscles

Quiet breathing:
the external
intercostal
muscles contract,
elevating the
ribs and moving
the sternum.

Labored breathing:
additional muscles
contract, causing
additional expansion
of the thorax.

Abdominal
muscles
relax.

(a)

Diaphragm
relaxed

(b)

The diaphragm contracts,
increasing the superior–inferior
dimension of the thoracic cavity.

Figure 20.10   **Effect of the Muscles of Respiration on Thoracic Volume**

(*a*) Muscles of respiration at the end of expiration. (*b*) Muscles of respiration at the end of inspiration.

thoracic volume than would be produced by the passive recoil of the thorax and lungs. ▼

18. *List the muscles of respiration and describe their role in quiet inspiration and expiration. How does this change during labored breathing?*

Pressure Changes and Airflow

The physics of airflow in tubes, such as the ones that make up the respiratory passages, is the same as that of the flow of blood in blood vessels (see chapter 18). Thus, the following relationships hold:

$$F = \frac{P_1 - P_2}{R}$$

where *F* is airflow (milliliters per minute) in a tube, P_1 is pressure at point one, P_2 is pressure at point two, and *R* is resistance to airflow.

The flow of air into and out of the lungs is governed by three physical principles:

1. *Air flows from areas of higher to lower pressure.* If the pressure is higher at one end of a tube (P_1) than at the other (P_2), air flows from the area of higher pressure toward the area of lower pressure. The greater the pressure difference, the greater the rate of airflow. Air flows through the respiratory passages because of pressure differences between the outside of the body and the alveoli inside the body. These pressure differences are produced by changes in thoracic volume.

2. *Changes in volume result in changes in pressure.* As the volume of a container increases, the pressure within the container decreases. As the volume of a container decreases, the pressure within the container increases. This inverse relationship between volume and pressure is called **Boyle's law.** The muscles of respiration change thoracic volume and therefore pressure within the thoracic cavity.

P R E D I C T ⑥
What happens to pressure in the thoracic cavity when the muscles of inspiration contract?

3. *Changes in tube diameter result in changes in resistance.* According to Poiseuille's law (see chapter 18), the resistance to airflow is proportional to the diameter (*d*) of a tube raised to the fourth power (d^4). Thus, a small change in diameter results in a large change in resistance, which greatly decreases airflow. For example, asthma results in the release of inflammatory chemicals, such as leukotrienes, that cause severe constriction of the bronchioles. Emphysema increases airway resistance because the bronchioles are obstructed as a result of inflammation and because damaged bronchioles collapse during expiration. Tumors increase resistance by growing into and blocking air passageways. ▼

19. *How do pressure differences and resistance affect airflow through a tube?*

20. *What happens to the pressure within a container when the volume of the container increases?*

P R E D I C T
If resistance to airflow increases because of disease, how can adequate airflow be maintained?

Airflow into and Out of Alveoli

The volume and pressure changes responsible for one cycle of inspiration and expiration can be described as follows:

1. At the end of expiration, **alveolar pressure,** which is the air pressure within the alveoli, is equal to **atmospheric pressure,** which is the air pressure outside the body. There is no movement of air into or out of the lungs because alveolar pressure and atmospheric pressure are equal (figure 20.11, step 1).

2. During inspiration, contraction of the muscles of inspiration increases the volume of the thoracic cavity. The increased thoracic volume causes the lungs to expand, resulting in an increase in alveolar volume (see "Changing Alveolar Volume," p. 616). As the alveolar volume increases, alveolar pressure becomes less than atmospheric pressure, and air flows from outside the body through the respiratory passages to the alveoli (figure 20.11, step 2).

3. At the end of inspiration, the thorax and alveoli stop expanding. When the alveolar pressure and atmospheric pressure become equal, airflow stops (figure 20.11, step 3).

4. During expiration, the thoracic volume decreases, producing a decrease in alveolar volume. Consequently, alveolar pressure increases above the air pressure outside the body, and air flows from the alveoli through the respiratory passages to the outside (figure 20.11, step 4).

As expiration ends, the decrease in thoracic volume stops and the process repeats, beginning at step 1. ▼

21. *Explain how changes in alveolar volume cause air to move into and out of the lungs.*

Lung Recoil

During quiet expiration, thoracic volume and lung volume decrease because of passive recoil of the thoracic wall and lungs. The recoil of the thoracic wall results from the elastic properties of the thoracic wall tissues. **Lung recoil** is the tendency for an expanded lung to decrease in size, due to (1) the elastic fibers in the connective tissue of the lungs and (2) surface tension of the film of fluid that lines the alveoli. **Surface tension** exists because the oppositely charged ends of water molecules attract each other (see chapter 2). As the water molecules pull together, they also pull on the alveolar walls, causing the alveoli to recoil and become smaller.

Two factors keep the lungs from collapsing: (1) surfactant and (2) pressure in the pleural cavity. ▼

22. *Name two things that cause the lungs to recoil. Name two things that keep the lungs from collapsing.*

Surfactant

Surfactant (ser-fak′tănt, <u>surf</u>ace <u>act</u>ing <u>ag</u>ent) is a mixture of lipoprotein molecules produced by secretory cells of the alveolar epithelium. The surfactant molecules form a single layer on the surface of the thin fluid layer lining the alveoli, reducing surface tension.

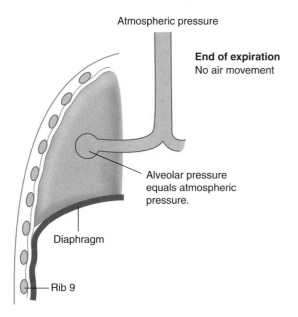

Atmospheric pressure

End of expiration
No air movement

Alveolar pressure equals atmospheric pressure.

Diaphragm

Rib 9

1. At the end of expiration, alveolar pressure is equal to atmospheric pressure and there is no air movement.

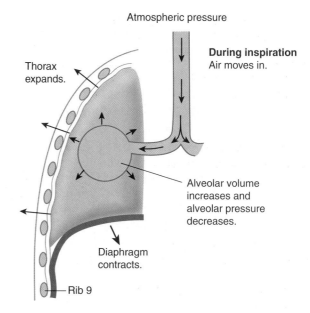

Atmospheric pressure

During inspiration
Air moves in.

Thorax expands.

Alveolar volume increases and alveolar pressure decreases.

Diaphragm contracts.

Rib 9

2. During inspiration, increased thoracic volume results in increased alveolar volume and decreased alveolar pressure. Air moves into the lungs because atmospheric pressure is greater than alveolar pressure.

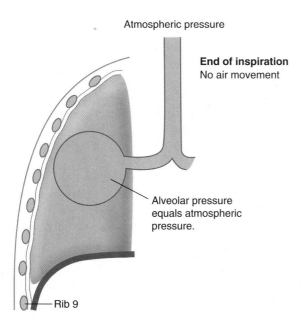

Atmospheric pressure

End of inspiration
No air movement

Alveolar pressure equals atmospheric pressure.

Rib 9

3. At the end of inspiration, alveolar pressure is equal to atmospheric pressure and there is no air movement.

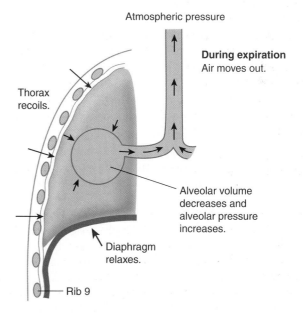

Atmospheric pressure

During expiration
Air moves out.

Thorax recoils.

Alveolar volume decreases and alveolar pressure increases.

Diaphragm relaxes.

Rib 9

4. During expiration, decreased thoracic volume results in decreased alveolar volume and increased alveolar pressure. Air moves out of the lungs because alveolar pressure is greater than atmospheric pressure.

Process Figure 20.11   Alveolar Pressure Changes During Inspiration and Expiration

The combined space of all the alveoli is represented by a large "bubble." The alveoli are actually microscopic and cannot be seen in the illustration.

Without surfactant, the surface tension causing the alveoli to recoil can be 10 times greater than when surfactant is present. Thus, surfactant greatly reduces the tendency of the lungs to collapse.

Infant Respiratory Distress Syndrome

Surfactant is not produced in adequate quantities until about the seventh month of gestation. Thereafter, the amount produced increases as the fetus matures. In premature infants, **infant respiratory distress syndrome (IRDS)**, or **hyaline** (hī′ă-lin, glass) **membrane disease,** is caused by too little surfactant. It is common, especially for infants delivered before the seventh month of pregnancy. Cortisol can be given to pregnant women who are likely to deliver prematurely, because it crosses the placenta into the fetus and stimulates surfactant synthesis.

If too little surfactant has been produced by the time of birth, the lungs tend to collapse, and a great deal of energy must be exerted by the muscles of respiration to keep the lungs inflated; even then, inadequate ventilation occurs. Without specialized treatment, most babies with this condition die soon after birth as a result of inadequate ventilation of the lungs and fatigue of the respiratory muscles. Treatment strategies include forcing enough oxygen-rich air into the lungs to inflate them and administering surfactant. ▼

> **23. What is surfactant? What effect does it have on water surface tension and the tendency for the lungs to collapse?**

Pleural Pressure

Pleural pressure is the pressure in the pleural cavity. When pleural pressure is less than alveolar pressure, the alveoli tend to expand. This principle can be understood by considering a balloon. The balloon expands when the pressure outside it is lower than the pressure inside. This pressure difference is normally achieved by increasing the pressure inside the balloon when a person forcefully blows into it. This pressure difference, however, can also be achieved by decreasing the pressure outside the balloon. For example, if the balloon is placed in a chamber from which air is removed, the pressure around the balloon becomes lower than atmospheric pressure, and the balloon expands. The lower the pressure outside the balloon, the greater the tendency for the higher pressure inside the balloon to cause it to expand. In a similar fashion, decreasing pleural pressure can result in expansion of the alveoli.

Normally, the alveoli are expanded because of a negative pleural pressure that is lower than alveolar pressure. At the end of a normal expiration, alveolar pressure is equal to atmospheric pressure and pleural pressure is slightly less than atmospheric pressure—that is, it is subatmospheric. The subatmospheric pleural pressure results from a "suction effect" caused by fluid removal by the lymphatic system and by lung recoil. As the lungs recoil, the visceral and parietal pleurae tend to be pulled apart. Normally, the lungs do not pull away from the thoracic wall because pleural fluid holds the

visceral and parietal pleurae together. Nonetheless, this pull decreases pressure in the pleural cavity, an effect that can be appreciated by putting water on the palms of the hands and putting them together. A sensation of negative pressure is felt as the hands are gently pulled apart.

When pleural pressure is lower than alveolar pressure, the alveoli tend to expand. This expansion is opposed by the tendency of the lungs to recoil. Therefore, the alveoli expand when the pleural pressure is low enough that lung recoil is overcome. If the pleural pressure is not low enough to overcome lung recoil, then the alveoli collapse.

Pneumothorax

A **pneumothorax** (noo-mō-thōr′aks) is the introduction of air into the pleural cavity. Air can enter by an external route when a sharp object, such as a bullet or broken rib, penetrates the thoracic wall, or air can enter the pleural cavity by an internal route if alveoli at the lung surface rupture, such as can occur in a patient with emphysema. When the pleural cavity is connected to the outside by such openings, air moves into the pleural cavity because air moves from the higher atmospheric pressure to the lower subatmospheric pressure in the pleural cavity. When air moves into the pleural cavity, the pressure in the pleural cavity increases and becomes equal to the atmospheric pressure outside the body. Thus, pleural pressure is also equal to alveolar pressure because pressure in the alveoli at the end of expiration is equal to atmospheric pressure outside the body. When pleural pressure and alveolar pressure are equal, there is no tendency for the alveoli to expand, lung recoil is unopposed, and the lungs collapse. A pneumothorax can occur in one lung while the lung on the opposite side remains inflated because the two pleural cavities are separated by the mediastinum. ▼

> **24. Define pleural pressure. What causes pleural pressure to be lower than alveolar pressure?**
> **25. What happens to alveolar volume when pleural pressure decreases?**
> **26. How does an opening in the chest wall cause a lung to collapse?**

P R E D I C T
Treatment of a pneumothorax involves closing the opening into the pleural cavity that caused the pneumothorax. Then a tube is placed into the pleural cavity. In order to inflate the lung, should this tube pump in air under pressure (as in blowing up a balloon) or should the tube apply suction? Explain.

Changing Alveolar Volume

Changes in alveolar volume result in the changes in alveolar pressure that are responsible for the movement of air into and out of the lungs (see figure 20.11). Alveolar volume changes result from changes in pleural pressure. For example, during inspiration, pleural pressure decreases and the alveoli expand. The decrease in pleural pressure occurs for two reasons:

1. Increasing the volume of the thoracic cavity results in a decrease in pleural pressure because of the effect of changing volume on pressure.
2. As the lungs expand, lung recoil increases, resulting in an increased suction effect and a lowering of pleural pressure. The increased lung recoil of the stretched lung is similar to the increased force generated in a stretched rubber band.

The events of inspiration and expiration can be summarized as follows:

1. During inspiration, pleural pressure decreases because of increased thoracic volume and increased lung recoil. As pleural pressure decreases, alveolar volume increases, alveolar pressure decreases, and air flows into the lungs.
2. During expiration, pleural pressure increases because of decreased thoracic volume and decreased lung recoil. As pleural pressure increases, alveolar volume decreases, alveolar pressure increases, and air flows out of the lungs. ▼

> **27. During inspiration, what causes pleural pressure to decrease? What effect does this have on alveolar pressure and air movement?**
>
> **28. During expiration, what causes pleural pressure to increase? What effect does this have on alveolar pressure and air movement?**

Measurement of Lung Function

A variety of measurements can be used to assess lung function. Each of these tests compares a subject's measurements with a normal range. These measurements can be used to diagnose diseases, track the progress of diseases, and track recovery from diseases.

Compliance of the Lungs and Thorax

Compliance is a measure of the ease with which the lungs and thorax expand. The compliance of the lungs and thorax is the volume by which they increase for each unit of pressure change in alveolar pressure. It is usually expressed in liters (volume of air) per centimeter of water (pressure), and for a normal person the compliance of the lungs and thorax is 0.13 L/cm H_2O. That is, for every 1 cm H_2O change in alveolar pressure, the volume changes by 0.13 L.

The greater the compliance, the easier it is for a change in pressure to cause expansion of the lungs and thorax. For example, one possible result of emphysema is the destruction of elastic lung tissue. This reduces the elastic recoil force of the lungs, thereby making expansion of the lungs easier and resulting in a higher

than normal compliance. A lower than normal compliance means that it is harder to expand the lungs and thorax. Conditions that decrease compliance include the deposition of inelastic fibers in lung tissue (pulmonary fibrosis), the collapse of the alveoli (infant respiratory distress syndrome and pulmonary edema), an increased resistance to airflow caused by airway obstruction (asthma, bronchitis, and lung cancer), and deformities of the thoracic wall that reduce the ability of the thoracic volume to increase (kyphosis and scoliosis).

Effects of Decreased Compliance

Pulmonary diseases can markedly affect the total amount of energy required for ventilation, as well as the percentage of the total amount of energy expended by the body. Diseases that decrease compliance can increase the energy required for breathing up to 30% of the total energy expended by the body. ▼

> **29. Define compliance. What is the effect on lung expansion when compliance increases or decreases?**

Pulmonary Volumes and Capacities

Spirometry (spī-rom′ĕ-trē, *spiro,* to breathe + *metron,* to measure) is the process of measuring volumes of air that move into and out of the respiratory system, and a **spirometer** (spī-rom′ĕ-ter) is a device used to measure these pulmonary volumes. The four pulmonary volumes and representative values (figure 20.12) for a young adult male follow:

1. **Tidal volume** is the volume of air inspired or expired with each breath. At rest, quiet breathing results in a tidal volume of approximately 500 mL.
2. **Inspiratory reserve volume** is the amount of air that can be inspired forcefully after inspiration of the tidal volume (approximately 3000 mL at rest).
3. **Expiratory reserve volume** is the amount of air that can be forcefully expired after expiration of the tidal volume (approximately 1100 mL at rest).
4. **Residual volume** is the volume of air still remaining in the respiratory passages and lungs after the most forceful expiration (approximately 1200 mL).

The tidal volume increases when a person is more active. An increase in tidal volume causes a decrease in the inspiratory and expiratory reserve volumes because the maximum volume of the respiratory system does not change from moment to moment.

Pulmonary capacities are the sum of two or more pulmonary volumes (see figure 20.12). Some pulmonary capacities follow:

1. **Inspiratory capacity** is the tidal volume plus the inspiratory reserve volume, which is the amount of air that a person can inspire maximally after a normal expiration (approximately 3500 mL at rest).

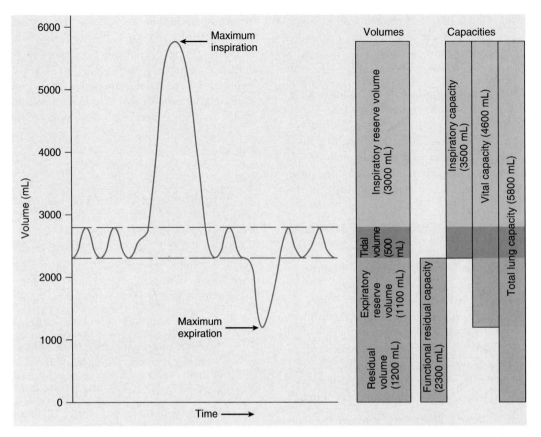

Figure 20.12 Lung Volumes and Capacities

The tidal volume in the figure is the tidal volume during resting conditions.

2. **Functional residual capacity** is the expiratory reserve volume plus the residual volume, which is the amount of air remaining in the lungs at the end of a normal expiration (approximately 2300 mL at rest).
3. **Vital capacity** is the sum of the inspiratory reserve volume, the tidal volume, and the expiratory reserve volume, which is the maximum volume of air that a person can expel from the respiratory tract after a maximum inspiration (approximately 4600 mL).
4. **Total lung capacity** is the sum of the inspiratory and expiratory reserve volumes plus the tidal volume and the residual volume (approximately 5800 mL).

Factors such as sex, age, body size, and physical conditioning cause variations in respiratory volumes and capacities from one individual to another. For example, the vital capacity of adult females is usually 20%–25% less than that of adult males. The vital capacity reaches its maximum amount in young adults and gradually decreases in the elderly. Tall people usually have a greater vital capacity than short people, and thin people have a greater vital capacity than obese people. Well-trained athletes can have a vital capacity 30%–40% above that of untrained people. In patients whose respiratory muscles are paralyzed by spinal cord injury or diseases such as poliomyelitis or muscular

dystrophy, vital capacity can be reduced to values not consistent with survival (less than 500–1000 mL). Factors that reduce compliance also reduce vital capacity.

The **forced expiratory vital capacity** is a simple and clinically important pulmonary test. The individual inspires maximally and then exhales maximally into a spirometer as rapidly as possible. The volume of air expired at the end of the test is the person's vital capacity. The spirometer also records the volume of air that enters it per second. The **forced expiratory volume in 1 second (FEV$_1$)** is the amount of air expired during the first second of the test. In some conditions, the vital capacity may not be dramatically affected, but how rapidly air is expired can be greatly decreased. Airway obstruction, caused by asthma, the collapse of bronchi in emphysema, or a tumor, and disorders that reduce the ability of the lungs or chest wall to deflate, such as pulmonary fibrosis, silicosis, kyphosis, and scoliosis, can cause a decreased FEV$_1$. ▼

30. *Define tidal volume, inspiratory reserve volume, expiratory reserve volume, and residual volume.*
31. *Define inspiratory capacity, functional residual capacity, vital capacity, and total lung capacity.*
32. *What is forced expiratory volume in 1 second, and why is it clinically important?*

Minute Ventilation and Alveolar Ventilation

Minute ventilation is the total amount of air moved into and out of the respiratory system each minute; it is equal to tidal volume times respiratory rate. **Respiratory rate,** or **respiratory frequency,** is the number of breaths taken per minute. Minute ventilation averages approximately 6 L/min because resting tidal volume is approximately 500 mL and respiratory rate is approximately 12 breaths per minute.

Although minute ventilation measures the amount of air moving into and out of the lungs per minute, it is not a measure of the amount of air available for gas exchange because gas exchange takes place mainly in the alveoli and to a lesser extent in the alveolar ducts and respiratory bronchioles. The part of the respiratory system where gas exchange does not take place is called the **dead space.** The nasal cavity, pharynx, larynx, trachea, bronchi, bronchioles, and terminal bronchioles form 150 mL of dead space. Nonfunctional alveoli also contribute to the dead space. In a healthy person, there are few nonfunctional alveoli.

Emphysema and Dead Space

Disease can decrease gas exchange in the alveoli and increase dead space. In patients with emphysema, alveolar walls degenerate and small alveoli combine to form larger alveoli. The result is fewer alveoli, but alveoli with an increased volume and decreased surface area. Although the enlarged alveoli are still ventilated, surface area is inadequate for complete gas exchange, and the dead space increases.

During inspiration, much of the inspired air fills the dead space first before reaching the alveoli and, thus, is unavailable for gas exchange. The volume of air available for gas exchange per minute is called **alveolar ventilation ($\dot{V}_A$),** and it is calculated as follows:

$$\dot{V}_A = f(V_T - V_D)$$

where $\dot{V}_A$ is alveolar ventilation (milliliters per minute), f is respiratory rate (frequency, breaths per minute), V_T is tidal volume (milliliters per respiration), and V_D is dead space (milliliters per respiration). ▼

33. Define minute ventilation, respiratory rate, dead space, and alveolar ventilation.
34. How does dead space affect alveolar ventilation?

P R E D I C T 9

What is the alveolar ventilation of a resting person with a tidal volume of 500 mL, a dead space of 150 mL, and a respiratory rate of 12 breaths per minute? If the person exercises and tidal volume increases to 4000 mL, dead space increases to 300 mL as a result of dilation of the respiratory passageways, and respiratory rate increases to 24 breaths per minute, what is the alveolar ventilation? How is the change in alveolar ventilation beneficial for doing exercise?

Gas Exchange in the Lungs

Ventilation supplies atmospheric air to the alveoli. The next step in the process of respiration is the diffusion of gases between alveoli and blood in the pulmonary capillaries.

Partial Pressure

Gases diffuse from higher concentrations toward lower concentrations. One measurement of the concentration of gases is **partial pressure,** which is the pressure exerted by a gas in a mixture of gases. According to **Dalton's law,** in a mixture of gases, the part of the total pressure resulting from each type of gas is determined by the percentage of the total volume represented by each gas type. For example, if the total pressure of all gases in a mixture of gases is 760 millimeters of mercury (mm Hg), which is the atmospheric pressure at sea level, and 21% of the mixture is made up of oxygen, then the partial pressure for oxygen is 160 mm Hg (0.21 × 760 mm Hg = 160 mm Hg). If the composition of air is 0.04% carbon dioxide at sea level, the partial pressure for carbon dioxide is 0.3 mm Hg (0.0004 × 760 = 0.3 mm Hg) (table 20.1). It is traditional to designate the partial pressure of individual gases in a mixture with a capital P followed by the symbol for the gas. Thus, the partial pressure of oxygen is P_{O_2}, and carbon dioxide is P_{CO_2}. ▼

35. According to Dalton's law, what is the partial pressure of a gas?

Diffusion of Gases into and Out of Liquids

Gas molecules diffuse from the air into a liquid, or from a liquid into the air, because of partial pressure gradients. For example, oxygen moves from the air into the thin layer of fluid lining the alveolus when P_{O_2} in the air is greater than in the fluid. According to **Henry's law,** the amount of gas that can dissolve in a liquid is equal to the partial pressure of the gas times its solubility coefficient. The solubility coefficient is a measure of how easily the gas dissolves in the liquid. In water, the solubility coefficient for oxygen is 0.024; for carbon dioxide, it is 0.57. Thus, carbon dioxide is approximately 24 times more soluble in water than is oxygen. ▼

36. According to Henry's law, how does the partial pressure and solubility of a gas affect its concentration in a liquid?

P R E D I C T 10

As a SCUBA diver descends, the pressure of the water on the body prevents normal expansion of the lungs. To compensate, the diver breathes pressurized air, which has a greater pressure than air at sea level. What effect does the increased pressure have on the amount of gas dissolved in the diver's body fluids? A SCUBA diver who suddenly ascends to the surface from a great depth can develop decompression sickness (the bends), in which bubbles of nitrogen gas form. The expanding bubbles damage tissues or block blood flow through small blood vessels. Explain the development of the bubbles.

Diffusion of Gases Through the Respiratory Membrane

The factors that influence the rate of gas diffusion through the respiratory membrane are (1) the partial pressure gradient of the gas across the membrane, (2) the diffusion coefficient of the gas in the substance of the membrane, (3) the thickness of the membrane, (4) the surface area of the membrane.

Partial Pressure Gradient

The partial pressure gradient of a gas across the respiratory membrane is the difference between the partial pressure of the gas in the alveoli and the partial pressure of the gas in the blood of the pulmonary capillaries. Oxygen diffuses from the alveoli into the pulmonary capillaries because the P_{O_2} in the alveoli is greater than that in the pulmonary capillaries. In contrast, carbon dioxide diffuses from the pulmonary capillaries into the alveoli because the P_{CO_2} is greater in the pulmonary capillaries than in the alveoli (see figure 20.8b).

The partial pressure gradient for oxygen and carbon dioxide can be increased by increasing alveolar ventilation. The greater volume of atmospheric air exchanged with the alveoli raises alveolar P_{O_2}, lowers alveolar P_{CO_2}, and thus promotes gas exchange. Conversely, inadequate ventilation causes a lower than normal partial pressure gradient for oxygen and carbon dioxide, resulting in inadequate gas exchange.

Diffusion Coefficient

A **diffusion coefficient** is a measure of how easily a gas diffuses into and out of a liquid or tissue, taking into account the solubility coefficient of the gas in the liquid and the size of the gas molecule (molecular weight). If the diffusion coefficient of oxygen is assigned a value of 1, the relative diffusion coefficient of carbon dioxide is 20, which means carbon dioxide diffuses through the respiratory membrane about 20 times more readily than oxygen does.

When the respiratory membrane becomes progressively damaged as a result of disease, its capacity for allowing the movement of oxygen into the blood is often impaired enough to cause death from oxygen deprivation before the diffusion of carbon dioxide is dramatically reduced. If life is being maintained by extensive oxygen therapy, which increases the concentration of oxygen in the lung alveoli, the reduced capacity for the diffusion of carbon dioxide across the respiratory membrane can result in substantial increases in carbon dioxide in the blood.

Respiratory Membrane Thickness

Fluid accumulation within alveoli increases the thickness of the fluid lining the alveoli, which increases the thickness of the respiratory membrane. If the thickness of the respiratory membrane increases two or three times, the rate of gas exchange markedly decreases. Pulmonary edema caused by failure of the left side of the heart is the most common cause of an increase in the thickness of the respiratory membrane. Conditions that result in inflammation of the lung tissues, such as tuberculosis,

pneumonia, or advanced silicosis, can also cause fluid accumulation within the alveoli.

Surface Area

In a healthy adult, the total surface area of the respiratory membrane is approximately 70 m^2 (approximately the floor area of a 25- by 30-foot room). Under resting conditions, a decrease in the surface area of the respiratory membrane to one-third or one-fourth of normal can significantly restrict gas exchange. During strenuous exercise, even small decreases in the surface area of the respiratory membrane can adversely affect gas exchange. A decreased surface area for gas exchange results from the surgical removal of lung tissue, the destruction of lung tissue by cancer, or the degeneration of the alveolar walls by emphysema. ▼

37. Describe four factors that affect the diffusion of gases through the respiratory membrane. Give examples of diseases that decrease diffusion by altering these factors.
38. Does oxygen or carbon dioxide diffuse more easily through the respiratory membrane?

Oxygen and Carbon Dioxide Transport in the Blood

Once oxygen diffuses through the respiratory membrane into the blood, most of it combines reversibly with hemoglobin, and a smaller amount dissolves in the plasma. Hemoglobin transports oxygen from the pulmonary capillaries through the blood vessels to the tissue capillaries, where some of the oxygen is released. The oxygen diffuses from the blood to tissue cells, where it is used in aerobic respiration.

Cells produce carbon dioxide during aerobic metabolism. The carbon dioxide diffuses from the cells into the tissue capillaries. Once carbon dioxide enters the blood, it is transported in three ways: dissolved in the plasma, in combination with hemoglobin, and in the form of bicarbonate ions (HCO_3^-).

Oxygen Partial Pressure Gradients

The P_{O_2} within the alveoli averages approximately 104 mm Hg, whereas the P_{O_2} in blood flowing into the pulmonary capillaries is approximately 40 mm Hg (figure 20.13). Consequently, oxygen diffuses down its partial pressure gradient from the alveoli into the pulmonary capillary blood. By the time blood flows through the first third of the pulmonary capillary beds, an equilibrium has been achieved, and the P_{O_2} in the blood is 104 mm Hg, which is equivalent to the P_{O_2} in the alveoli.

Blood leaving the pulmonary capillaries has a P_{O_2} of 104 mm Hg, but blood leaving the lungs in the pulmonary veins has a P_{O_2} of approximately 95 mm Hg. This decrease in the P_{O_2} occurs because the blood from the pulmonary capillaries mixes with deoxygenated blood called **shunted blood.** One source of shunted blood is the deoxygenated blood from the bronchial circulation (see "Blood Supply," p. 612). The other source is deoxygenated blood from the pulmonary circulation. Normally, 1%–2% of cardiac output is shunted blood.

Table 20.1   Partial Pressures of Gases at Sea Level

Gases	Dry Air		Humidified Air		Alveolar Air		Expired Air	
	mm Hg	%	mm Hg	%	mm Hg	%	mm Hg	%
Nitrogen	597.5	78.62	563.4	74.09	569.0	74.9	566.0	74.5
Oxygen	158.4	20.84	149.3	19.67	104.0	13.6	120.0	15.7
Carbon dioxide	0.3	0.04	0.3	0.04	40.0	5.3	27.0	3.6
Water vapor	0.0	0.00	47.0	6.20	47.0	6.2	47.0	6.2

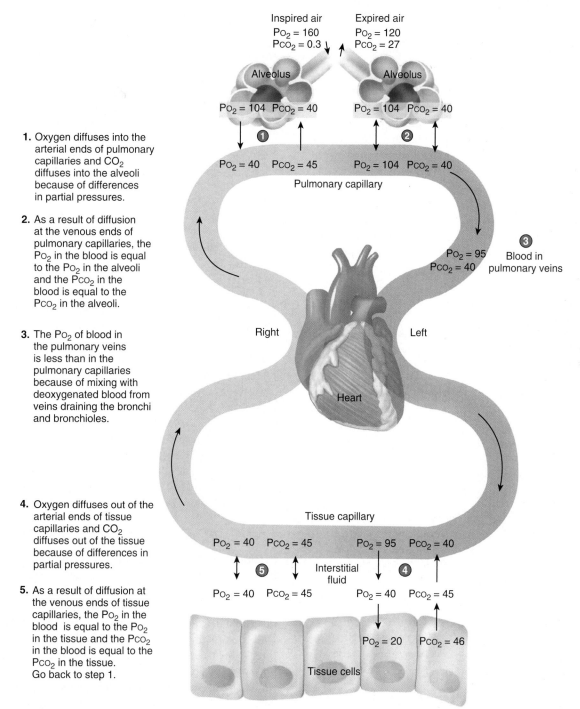

1. Oxygen diffuses into the arterial ends of pulmonary capillaries and CO_2 diffuses into the alveoli because of differences in partial pressures.

2. As a result of diffusion at the venous ends of pulmonary capillaries, the P_{O_2} in the blood is equal to the P_{O_2} in the alveoli and the P_{CO_2} in the blood is equal to the P_{CO_2} in the alveoli.

3. The P_{O_2} of blood in the pulmonary veins is less than in the pulmonary capillaries because of mixing with deoxygenated blood from veins draining the bronchi and bronchioles.

4. Oxygen diffuses out of the arterial ends of tissue capillaries and CO_2 diffuses out of the tissue because of differences in partial pressures.

5. As a result of diffusion at the venous ends of tissue capillaries, the P_{O_2} in the blood is equal to the P_{O_2} in the tissue and the P_{CO_2} in the blood is equal to the P_{CO_2} in the tissue. Go back to step 1.

Process Figure 20.13   Gas Exchange

Oxygen and carbon dioxide partial pressure gradients between the alveoli and the pulmonary capillaries and between the tissues and the tissue capillaries are responsible for gas exchange.

Disorders That Increase Shunted Blood

Any condition that decreases gas exchange between the alveoli and the blood can increase the amount of shunted blood. For example, obstruction of the bronchioles in conditions such as asthma can decrease ventilation beyond the obstructed areas. The result is a large increase in shunted blood because the blood flowing through the pulmonary capillaries in the obstructed area remains unoxygenated. In pneumonia or pulmonary edema, a buildup of fluid in the alveoli results in poor gas diffusion and less oxygenated blood.

The blood that enters the arterial end of the tissue capillaries has a P_{O_2} of approximately 95 mm Hg. The P_{O_2} of the interstitial fluid, in contrast, is close to 40 mm Hg and is probably near 20 mm Hg in the individual cells. Oxygen diffuses from the tissue capillaries to the interstitial fluid and from the interstitial fluid into the cells of the body, where it is used in aerobic metabolism. Because the cells use oxygen continuously, a constant partial pressure gradient exists for oxygen from the tissue capillaries to the cells.

Carbon Dioxide Partial Pressure Gradients

Carbon dioxide is continually produced as a by-product of cellular respiration, and a partial pressure gradient is established from tissue cells to the blood within the tissue capillaries. The intracellular P_{CO_2} is approximately 46 mm Hg, and the interstitial fluid P_{CO_2} is approximately 45 mm Hg. At the arterial end of the tissue capillaries, the P_{CO_2} is close to 40 mm Hg. As blood flows through the tissue capillaries, carbon dioxide diffuses from a higher P_{CO_2} to a lower P_{CO_2} until an equilibrium in P_{CO_2} is established. At the venous end of the capillaries, blood has a P_{CO_2} of 45 mm Hg (see figure 20.13).

After blood leaves the venous end of the capillaries, it is transported through the cardiovascular system to the lungs. At the arterial end of the pulmonary capillaries, the P_{CO_2} is 45 mm Hg. Because the P_{CO_2} is approximately 40 mm Hg in the alveoli, carbon dioxide diffuses from the pulmonary capillaries into the alveoli. At the venous end of the pulmonary capillaries, the P_{CO_2} has again decreased to 40 mm Hg. ▼

39. Describe the partial pressures of oxygen and carbon dioxide in the alveoli, lung capillaries, tissue capillaries, and tissues. How do these partial pressures account for the movement of oxygen and carbon dioxide between air and blood and between blood and tissues?

40. What is shunted blood?

P R E D I C T **11**

During exercise, the movement of oxygen into skeletal muscle cells and the movement of carbon dioxide out of skeletal muscle cells increases. Explain how this happens.

Hemoglobin and Oxygen Transport

Approximately 98.5% of the oxygen transported in the blood from the lungs to the tissues is transported in combination with hemoglobin in red blood cells, and the remaining 1.5% is dissolved in the water part of the plasma. The combination of oxygen with hemoglobin is reversible. In the pulmonary capillaries, oxygen binds to hemoglobin; in the tissue spaces, oxygen diffuses away from hemoglobin and enters the tissues.

Effect of P_{O_2}

The **oxygen–hemoglobin dissociation curve** describes the percent saturation of hemoglobin in the blood at different blood P_{O_2} values. Hemoglobin is 100% saturated with oxygen when four oxygen molecules are bound to each hemoglobin molecule in the blood. There are four heme groups in a hemoglobin molecule (see chapter 16), and an oxygen molecule is bound to each heme group. Hemoglobin is 50% saturated with oxygen when there is an average of two oxygen molecules bound to each hemoglobin molecule.

The P_{O_2} in the blood leaving the pulmonary capillaries is normally 104 mm Hg. At that partial pressure, hemoglobin is 98% saturated (figure 20.14a). Decreases in the P_{O_2} in the pulmonary capillaries have a relatively small effect on hemoglobin saturation, as shown by the fairly flat shape of the upper part of the oxygen–hemoglobin dissociation curve. Even if the blood P_{O_2} decreases from 104 mm Hg to 60 mm Hg, hemoglobin is still 90% saturated. Hemoglobin is very effective at picking up oxygen in the lungs, even if the P_{O_2} in the pulmonary capillaries decreases significantly.

In a resting person, the normal blood P_{O_2} leaving the tissue capillaries is 40 mm Hg and hemoglobin is 75% saturated. Thus, 23% (98% − 75%) of the oxygen picked up in the lungs is released from hemoglobin and diffuses into the tissues (figure 20.14b). The 75% of oxygen still bound to the hemoglobin is an oxygen reserve, which can be released if blood P_{O_2} decreases further. In the tissues, a relatively small change in blood P_{O_2} results in a relatively large change in hemoglobin saturation, as shown by the steep slope of the oxygen–hemoglobin dissociation curve. For example, during vigorous exercise, the P_{O_2} in skeletal muscle capillaries can decline to levels as low as 15 mm Hg because of the increased use of oxygen during aerobic respiration in skeletal muscle cells (see chapter 8). At a P_{O_2} of 15 mm Hg, hemoglobin is only 25% saturated, resulting in the release of 73% (98% − 25%) of the oxygen picked up in the lungs (figure 20.14c). Thus, as tissues use more oxygen, hemoglobin releases more oxygen to those tissues.

Effect of pH, P_{CO_2}, and Temperature

In addition to P_{O_2}, blood pH, P_{CO_2}, and temperature influence the degree to which oxygen binds to hemoglobin (figure 20.15). As the pH of the blood declines, the amount of oxygen bound to hemoglobin at any given P_{O_2} also declines. This occurs because decreased pH results from an increase in H^+, and the H^+ combine with the protein part of the hemoglobin molecule and change its three-dimensional structure, causing a decrease in the hemoglobin's ability to bind oxygen. Conversely, an increase in blood pH results in an increase in hemoglobin's ability to bind oxygen. The effect of pH on the oxygen–hemoglobin dissociation curve is called the **Bohr effect,** after its discoverer, Christian Bohr.

An increase in P_{CO_2} also decreases hemoglobin's ability to bind oxygen because of the effect of carbon dioxide on pH. Within

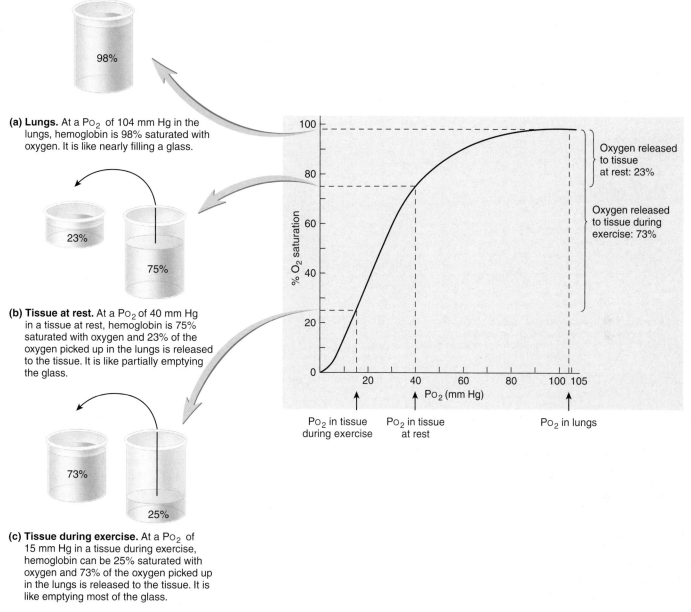

(a) Lungs. At a P_{O_2} of 104 mm Hg in the lungs, hemoglobin is 98% saturated with oxygen. It is like nearly filling a glass.

(b) Tissue at rest. At a P_{O_2} of 40 mm Hg in a tissue at rest, hemoglobin is 75% saturated with oxygen and 23% of the oxygen picked up in the lungs is released to the tissue. It is like partially emptying the glass.

(c) Tissue during exercise. At a P_{O_2} of 15 mm Hg in a tissue during exercise, hemoglobin can be 25% saturated with oxygen and 73% of the oxygen picked up in the lungs is released to the tissue. It is like emptying most of the glass.

Figure 20.14 Oxygen–Hemoglobin Dissociation Curve

The oxygen–hemoglobin dissociation curve shows the percent saturation of hemoglobin as a function of P_{O_2}. The ability of hemoglobin to pick up oxygen in the lungs and release it in the tissues is like a glass filling and emptying.

red blood cells, an enzyme called **carbonic anhydrase** catalyzes this reversible reaction.

$$\underset{\text{Carbon dioxide}}{CO_2} + \underset{\text{Water}}{H_2O} \underset{\text{Carbonic anhydrase}}{\rightleftarrows} \underset{\text{Carbonic acid}}{H_2CO_3} \rightleftarrows \underset{\text{Hydrogen ion}}{H^+} + \underset{\text{Bicarbonate ion}}{HCO_3^-}$$

As carbon dioxide levels increase, more H^+ are produced, and the pH declines. As carbon dioxide levels decline, the reaction proceeds in the opposite direction, resulting in a decrease in H^+ concentration and an increase in pH. Thus, changes in carbon dioxide levels indirectly produce a Bohr effect by altering pH. In addition, carbon dioxide can directly affect hemoglobin's ability to

bind oxygen, to a small extent. When carbon dioxide binds to the α- and β-globin chains of hemoglobin (see chapter 16), hemoglobin's ability to bind with oxygen decreases.

P R E D I C T   12

Predict how the pH of the blood in a sprinter changes during a 400-meter race. How does the pH change affect O_2 delivery to skeletal muscles?

As blood passes through tissue capillaries, carbon dioxide enters the blood from the tissues. As a consequence, blood carbon dioxide levels increase, pH decreases, and hemoglobin has less affinity for oxygen in the tissue capillaries. Therefore, a greater amount of oxygen is released in the tissue capillaries than would be

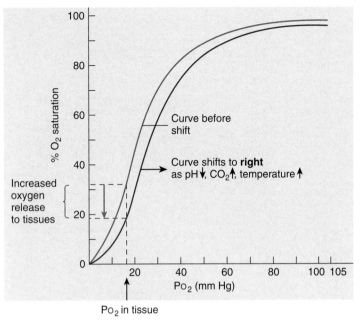

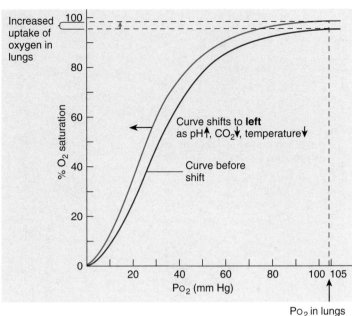

(a) **In the tissues, the oxygen–hemoglobin dissociation curve shifts to the right.** As pH decreases, P_{CO_2} increases, or temperature increases, the curve (*red*) shifts to the right (*blue*), resulting in an increased release of oxygen.

(b) **In the lungs, the oxygen–hemoglobin dissociation curve shifts to the left.** As pH increases, P_{CO_2} decreases, or temperature decreases, the curve (*blue*) shifts to the left (*red*), resulting in an increased ability of hemoglobin to pick up oxygen.

Figure 20.15   **Effects of Shifting the Oxygen–Hemoglobin Dissociation Curve**

released if carbon dioxide were not present. When blood is returned to the lungs and passes through the pulmonary capillaries, carbon dioxide leaves the capillaries and enters the alveoli. As a result, carbon dioxide levels in the pulmonary capillaries are reduced, pH increases, and hemoglobin's affinity for oxygen increases.

An increase in temperature also decreases oxygen's tendency to remain bound to hemoglobin. Elevated temperatures resulting from increased metabolism, therefore, increase the amount of oxygen released into the tissues by hemoglobin. In less metabolically active tissues in which the temperature is lower, less oxygen is released from hemoglobin.

When hemoglobin's affinity for oxygen decreases, the oxygen–hemoglobin dissociation curve is shifted to the right, and hemoglobin releases more oxygen (see figure 20.15*a*). During exercise, when carbon dioxide and acidic substances, such as lactic acid, accumulate and the temperature increases in the tissue spaces, the oxygen–hemoglobin curve shifts to the right. Under these conditions, as much as 75%–85% of the oxygen is released from the hemoglobin. In the lungs, however, the curve shifts to the left because of the lower carbon dioxide levels, lower temperature, and lower lactic acid levels. Hemoglobin's affinity for oxygen, therefore, increases, and it becomes easily saturated (see figure 20.15*b*).

During resting conditions, approximately 5 mL of oxygen are transported to the tissues in each 100 mL of blood, and cardiac output is approximately 5000 mL/min. Consequently, 250 mL of oxygen are delivered to the tissues each minute. During exercise, this value can increase up to 15 times. Oxygen transport can be increased threefold because of a greater degree of oxygen release from hemoglobin in the tissue capillaries, and the rate of oxygen transport is increased another five times because of the increase in cardiac output. Consequently, the volume of oxygen delivered to the tissues can

be as high as 3750 mL/min (15×250 mL/min). Highly trained athletes can increase this volume to as high as 5000 mL/min. ▼

41. *Name two ways that oxygen is transported in the blood, and state the percentage of total oxygen transport for which each is responsible.*

42. *How does the oxygen–hemoglobin dissociation curve explain the uptake of oxygen in the lungs and the release of oxygen in tissues?*

43. *What is the Bohr effect? How is it related to blood carbon dioxide?*

44. *Why is it advantageous for the oxygen–hemoglobin dissociation curve to shift to the left in the lungs and to the right in tissues?*

P R E D I C T   13

In carbon monoxide (CO) poisoning, CO binds to hemoglobin, thereby decreasing the uptake of oxygen by hemoglobin. In addition, when CO binds to hemoglobin, the oxygen–hemoglobin dissociation curve shifts to the left. What are the consequences of this shift on the ability of tissues to get oxygen? Explain.

Transport of Carbon Dioxide

Carbon dioxide is transported in the blood in three ways: approximately 7% as carbon dioxide dissolved in the plasma, approximately 70% in the form of HCO_3^- dissolved in plasma and in red blood cells, and approximately 23% bound to hemoglobin.

Carbon Dioxide Exchange in Tissues

Carbon dioxide diffuses from tissues into the plasma of blood (figure 20.16*a*). Most of the carbon dioxide diffuses into red blood

1. In the tissues, carbon dioxide (CO_2) diffuses into the plasma and into red blood cells. Some of the carbon dioxide remains in the plasma.

2. In red blood cells, carbon dioxide reacts with water (H_2O) to form carbonic acid (H_2CO_3) in a reaction catalyzed by the enzyme carbonic anhydrase (CA).

3. Carbonic acid dissociates to form bicarbonate ions (HCO_3^-) and hydrogen ions (H^+).

4. In the chloride shift, as HCO_3^- diffuse out of the red blood cells, electrical neutrality is maintained by the diffusion of chloride ions (Cl^-) into them.

5. Oxygen (O_2) is released from hemoglobin (Hb). Oxygen diffuses out of red blood cells and plasma into the tissue.

6. Hydrogen ions combine with hemoglobin, which promotes the release of oxygen from hemoglobin (Bohr effect).

7. Carbon dioxide combines with hemoglobin. Hemoglobin that has released oxygen readily combines with carbon dioxide (Haldane effect).

(a) **Gas exchange in the tissues**

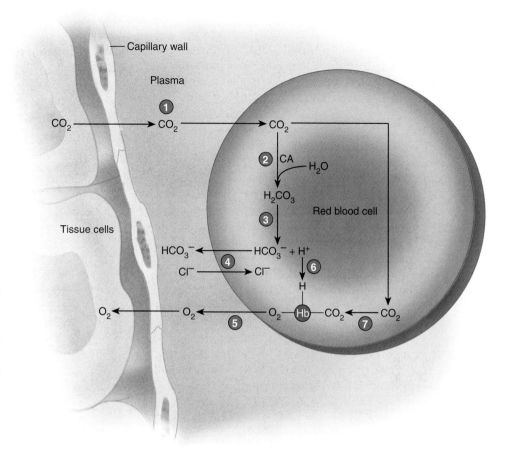

1. In the lungs, carbon dioxide (CO_2) diffuses from red blood cells and plasma into the alveoli.

2. Carbonic anhydrase catalyzes the formation of CO_2 and H_2O from H_2CO_3.

3. Bicarbonate ions and H^+ combine to replace H_2CO_3.

4. In the chloride shift, as HCO_3^- diffuse into red blood cells, electrical neutrality is maintained by the diffusion of chloride ions (Cl^-) out of them.

5. Oxygen diffuses into the plasma and into red blood cells. Some of the oxygen remains in the plasma. Oxygen binds to hemoglobin.

6. Hydrogen ions are released from hemoglobin, which promotes the uptake of oxygen by hemoglobin (Bohr effect).

7. Carbon dioxide is released from hemoglobin. Hemoglobin that is bound to oxygen readily releases carbon dioxide (Haldane effect).

(b) **Gas exchange in the lungs**

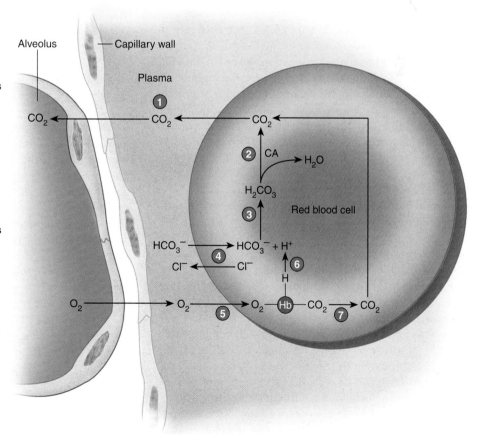

Process Figure 20.16 Gas Exchange

cells, but approximately 7% of it is transported dissolved in the plasma. Carbon dioxide inside red blood cells reacts with water to form carbonic acid, a reaction catalyzed by carbonic anhydrase. Carbonic acid then dissociates to form HCO_3^- and H^+. Approximately 70% of blood carbon dioxide is transported in the form of HCO_3^-.

Removing HCO_3^- from inside the red blood cells promotes carbon dioxide transport because, as HCO_3^- concentration decreases, more carbon dioxide combines with water to form additional HCO_3^- and H^+ (see "Reversible Reactions," chapter 2). In a process called **chloride shift** (see figure 20.16a, step 4), antiporters exchange Cl^- for HCO_3^-. This exchange maintains electrical balance in the red blood cells and plasma as HCO_3^- diffuse out of, and Cl^- diffuse into, red blood cells.

Hydrogen ions bind to hemoglobin (see figure 20.16a, step 6). As a result, three effects are produced: (1) The transport of carbon dioxide increases because, as H^+ concentration decreases, more carbon dioxide combines with water to form additional HCO_3^- and H^+; (2) the pH inside the red blood cells does not decrease because hemoglobin is a buffer, preventing an increase in H^+ concentration; and (3) the affinity of hemoglobin for oxygen decreases. Hemoglobin releases oxygen in tissue capillaries because of decreased P_{O_2} (see figure 20.14). Hemoglobin's decreased affinity for oxygen results in a shift of the oxygen–hemoglobin curve to the right (the Bohr effect) (see figure 20.15a) and an increase in the release of oxygen from hemoglobin.

Approximately 23% of blood carbon dioxide is transported bound to hemoglobin. Many carbon dioxide molecules bind in a reversible fashion to the α- and β-globin chains of hemoglobin molecules (see figure 20.16a, step 7). Carbon dioxide's ability to bind to hemoglobin is affected by the amount of oxygen bound to hemoglobin. The smaller the amount of oxygen bound to hemoglobin, the greater the amount of carbon dioxide that can bind to it, and vice versa. This relationship is called the **Haldane effect.** In tissues, as hemoglobin releases oxygen, the hemoglobin has an increased ability to pick up carbon dioxide.

Carbon Dioxide Exchange in the Lungs

Carbon dioxide diffuses from red blood cells and plasma into the alveoli (figure 20.16b). As carbon dioxide levels in the red blood cells decrease, carbonic acid is converted to carbon dioxide and water. In response, HCO_3^- join with H^+ to form carbonic acid. As HCO_3^- and H^+ concentrations decrease because of this reaction, HCO_3^- enter red blood cells in exchange for Cl^- and H^+ are released from hemoglobin. Hemoglobin picks up oxygen in pulmonary capillaries because of increased P_{O_2} (see figure 20.14). The release of H^+ from hemoglobin increases hemoglobin's affinity for oxygen, resulting in a shift of the oxygen–hemoglobin curve to the left (Bohr effect) (see figure 20.15b). Oxygen from the alveoli diffuses into the pulmonary capillaries and into the red blood cells, and it binds with hemoglobin. Carbon dioxide is released from hemoglobin and diffuses out of the red blood cells into the alveoli. As hemoglobin binds to oxygen, it more readily releases carbon dioxide (Haldane effect).

Carbon Dioxide and Blood pH

Blood pH refers to the pH in plasma, not inside red blood cells. In plasma, carbon dioxide can combine with water to form carbonic acid, a reaction that is catalyzed by carbonic anhydrase on the surface of capillary endothelial cells. The carbonic acid then dissociates to form HCO_3^- and H^+. Thus, as plasma carbon dioxide levels increase, H^+ levels increase and blood pH decreases. An important function of the respiratory system is to regulate blood pH by changing plasma carbon dioxide levels (see chapter 23). Hyperventilation decreases plasma carbon dioxide, and hypoventilation increases it. ▼

45. What is the effect of lowering HCO_3^- concentrations inside red blood cells on carbon dioxide transport? What is the chloride shift, and what does it accomplish?
46. Name three effects produced by H^+ binding to hemoglobin.
47. What is the Haldane effect?
48. What effect does blood carbon dioxide level have on blood pH?

P R E D I C T 14
What effect does hyperventilation and holding one's breath have on blood pH? Explain.

Regulation of Ventilation

Respiratory Areas in the Brainstem

The classic view of respiratory areas held that distinct inspiratory and expiratory centers were located in the brainstem. This view is now known to be too simplistic. Although neurons involved with respiration are aggregated in certain parts of the brainstem, neurons that are active during inspiration are intermingled with neurons that are active during expiration.

The **medullary respiratory center** consists of two **dorsal respiratory groups,** each forming a longitudinal column of cells located bilaterally in the dorsal part of the medulla oblongata, and two **ventral respiratory groups,** each forming a longitudinal column of cells located bilaterally in the ventral part of the medulla oblongata (figure 20.17). The dorsal respiratory groups are primarily responsible for stimulating contraction of the diaphragm. The ventral respiratory groups are primarily responsible for stimulating the external intercostal, internal intercostal, and abdominal muscles. A part of the ventral respiratory group, the **pre-Bötzinger complex,** is believed to establish the basic rhythm of respiration.

The **pontine respiratory group** is a collection of neurons in the pons (see figure 20.17). It has connections with the medullary respiratory center and appears to play a role in the switching between inspiration and expiration. ▼

49. Name the three respiratory groups and describe their main functions.

Generation of Rhythmic Ventilation

One explanation for the generation of rhythmic ventilation involves the integration of stimuli that start and stop inspiration:

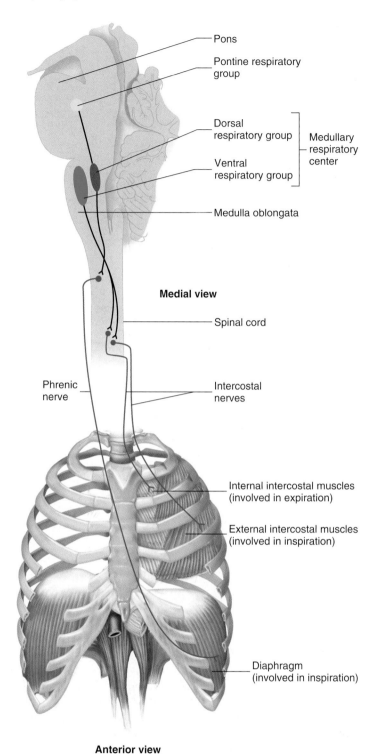

Medial view

Pons

Pontine respiratory group

Dorsal respiratory group

Ventral respiratory group

Medullary respiratory center

Medulla oblongata

Spinal cord

Phrenic nerve

Intercostal nerves

Internal intercostal muscles (involved in expiration)

External intercostal muscles (involved in inspiration)

Diaphragm (involved in inspiration)

Anterior view

Figure 20.17   Respiratory Structures in the Brainstem

The relationship of respiratory structures to each other and to the nerves innervating the muscles of respiration.

1. *Starting inspiration.* Neurons in the medullary respiratory center spontaneously establish the basic rhythm of respiration. The medullary respiratory center constantly receives stimulation from receptors that monitor blood gas levels, blood temperature, and the movements of muscles and joints. In addition, stimulation from the parts of the

brain concerned with voluntary respiratory movements and emotions can occur. Inspiration starts when the combined input from all these sources causes the production of action potentials in the neurons that stimulate respiratory muscles.

2. *Increasing inspiration.* Once inspiration begins, more and more neurons are gradually activated. The result is progressively stronger stimulation of the respiratory muscles, which lasts for approximately 2 seconds.

3. *Stopping inspiration.* The neurons stimulating the muscles of respiration also stimulate the neurons in the medullary respiratory center that are responsible for stopping inspiration. The neurons responsible for stopping inspiration also receive input from the pontine respiratory group, stretch receptors in the lungs, and probably other sources. When these inhibitory neurons are activated, they cause the neurons stimulating respiratory muscles to be inhibited. Relaxation of respiratory muscles results in expiration, which lasts approximately 3 seconds. The next inspiration begins again at step 1. ▼

50. How is rhythmic ventilation generated?

Cerebral and Limbic System Control

Through the cerebral cortex, it is possible to consciously or unconsciously increase or decrease the rate and depth of the respiratory movements (figure 20.18). For example, during talking or singing, air movement is controlled to produce sounds, as well as to facilitate gas exchange.

Apnea (ap′nē-ă) is the absence of breathing. A person may stop breathing voluntarily. As the period of voluntary apnea increases, greater and greater urge to breathe develops. That urge is primarily associated with increasing P_{CO_2} levels in the arterial blood. Finally, the P_{CO_2} reaches levels that cause the respiratory center to override the conscious influence from the cerebrum.

Emotions acting through the limbic system of the brain can also affect the respiratory center (see figure 20.18). For example, strong emotions can cause hyperventilation or produce the sobs and gasps of crying. ▼

51. Describe cerebral and limbic system control of ventilation.

Chemical Control of Ventilation

The respiratory system maintains blood oxygen and carbon dioxide concentrations and blood pH within a normal range of values. A deviation in any of these parameters from their normal range has a marked influence on respiratory movements. The effect of changes in oxygen and carbon dioxide concentrations and in pH is superimposed on the neural mechanisms that establish rhythmic ventilation.

Chemoreceptors

Chemoreceptors are specialized neurons that respond to changes in chemicals in solution. The chemoreceptors involved in the regulation of respiration respond to changes in hydrogen ion

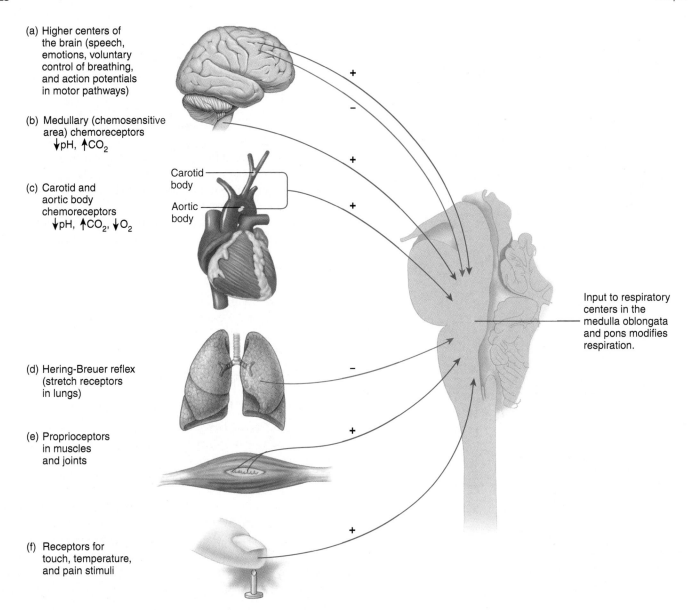

(a) Higher centers of the brain (speech, emotions, voluntary control of breathing, and action potentials in motor pathways)

(b) Medullary (chemosensitive area) chemoreceptors
↓pH, ↑CO_2

Carotid body

Aortic body

(c) Carotid and aortic body chemoreceptors
↓pH, ↑CO_2, ↓O_2

Input to respiratory centers in the medulla oblongata and pons modifies respiration.

(d) Hering-Breuer reflex (stretch receptors in lungs)

(e) Proprioceptors in muscles and joints

(f) Receptors for touch, temperature, and pain stimuli

Figure 20.18   Major Regulatory Mechanisms of Ventilation

A plus sign indicates an increase in ventilation, and a minus sign indicates a decrease in ventilation.

concentrations, changes in P_{O_2}, or both (see figure 20.18 and figure 20.19). **Central chemoreceptors** are located bilaterally and ventrally in the **chemosensitive area** of the medulla oblongata, and they are connected to the respiratory center. **Peripheral chemoreceptors** are found in the carotid and aortic bodies. These structures are small vascular sensory organs, which are encapsulated in connective tissue and located near the carotid sinuses and the aortic arch (see chapter 18). The respiratory center is connected to the carotid body chemoreceptors through the glossopharyngeal nerve (IX) and to the aortic body chemoreceptors by the vagus nerve (X).

Effect of pH

The chemosensitive area of the medulla oblongata and the carotid and aortic bodies respond to changes in blood pH. The chemo-

sensitive area responds indirectly to changes in blood pH, whereas the carotid and aortic bodies respond directly. Hydrogen ions do not easily cross the blood–brain barrier (see chapter 11) to affect the chemosensitive area, but they do easily cross from the blood to the carotid and aortic bodies.

The chemosensitive area detects changes in blood pH through changes in blood carbon dioxide, which easily diffuses across the blood–brain barrier. For example, if blood carbon dioxide levels increase, carbon dioxide diffuses across the blood–brain barrier into the cerebrospinal fluid. The carbon dioxide combines with water to form carbonic acid, which dissociates into H^+ and HCO_3^-. The increased concentration of H^+ lowers the pH and stimulates the chemosensitive area, which then stimulates the respiratory center, resulting in a greater rate and depth of breathing. Consequently, carbon dioxide levels decrease

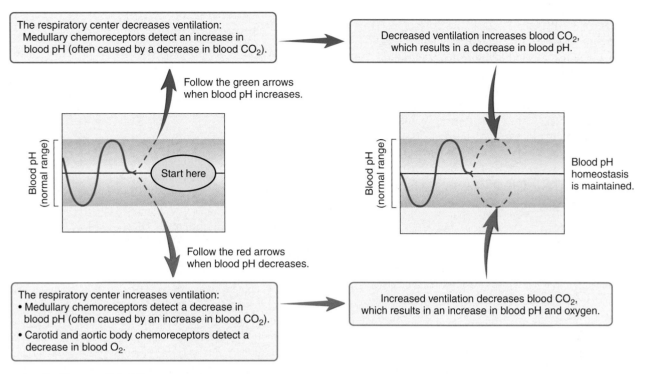

Homeostasis Figure 20.19 Summary of the Regulation of Blood pH and Gases

as carbon dioxide is eliminated from the body and blood pH increases to normal levels.

Maintaining body pH levels within normal parameters is necessary for the proper functioning of cells. Because changes in carbon dioxide levels can change pH, the respiratory system plays an important role in acid–base balance. For example, if blood pH decreases, the respiratory center is stimulated, resulting in the elimination of carbon dioxide and an increase in blood pH back to normal levels. Conversely, if blood pH increases, the respiratory rate decreases and carbon dioxide levels increase, causing blood pH to decrease back to normal levels. The role of the respiratory system in maintaining pH is considered in greater detail in chapter 23.

CASE STUDY | Asthma

Will is an 18-year-old track athlete in seemingly good health. Despite suffering from a slight cold, Will went jogging one morning with his running buddy, Al. After a few minutes of exercise, Will felt that he could hardly get enough air. Even though he stopped jogging, he continued to breathe rapidly and wheeze forcefully. Because his condition was not improving, Al took him to the emergency room of a nearby hospital.

The emergency room doctor used a stethoscope to listen to air movement in Will's lungs and noted that movement was poor. In addition, she ordered an arterial blood gas measurement for Will. He had a P_{O_2} of 60 mm Hg

and a P_{CO_2} of 30 mm Hg. Although Will had no previous history of asthma, the emergency room doctor was convinced that he was having an asthma attack.

Asthma is characterized by airway inflammation, which episodically results in shortness of breath, coughing, and wheezing due to bronchoconstriction. Bronchoconstriction decreases compliance, which makes expansion of the lungs more difficult. An asthma attack can be provoked by viral infections, exercise, or exposure to environmental irritants, such as pollen or cigarette smoke (see "Asthma and Cystic Fibrosis," p. 611).

P R E D I C T

a. Are Will's arterial blood gas values above or below normal (see figure 20.13)?

b. Why did the asthma attack cause Will to breathe more rapidly (see figure 20.19)?

c. Why did the asthma attack cause Will to wheeze forcefully?

d. Did Will's rapid, forceful wheezing restore homeostasis? Explain.

e. Explain Will's arterial blood P_{O_2} and P_{CO_2} values.

f. Is Will's blood pH lower or higher than normal? What effect does this blood pH normally have on respiration rate? Why didn't that happen?

g. Explain how β-adrenergic agents (see "The influence of Drugs on the Autonomic Nervous System," chapter 14) or inhaled glucocorticoids (see chapter 15) can help Will.

Effect of Carbon Dioxide

Blood carbon dioxide levels are a major regulator of respiration during resting conditions and conditions when the carbon dioxide levels are elevated—for example, during intense exercise. Even a small increase in carbon dioxide in the circulatory system triggers a large increase in the rate and depth of respiration. An increase in P_{CO_2} of 5 mm Hg, for example, causes an increase in ventilation of 100%. A greater than normal amount of carbon dioxide in the blood is called **hypercapnia** (hī-per-kap′nē-ă). Conversely, lower than normal carbon dioxide levels, a condition called **hypocapnia** (hī-pō-kap′nē-ă), result in periods in which respiratory movements are reduced or do not occur.

P R E D I C T 16

Explain why a person who breathes rapidly and deeply (hyperventilates) for several seconds experiences a short period during which respiration does not occur (apnea) before normal breathing resumes.

The chemoreceptors in the chemosensitive area of the medulla oblongata and in the carotid and aortic bodies respond to changes in carbon dioxide because of the effects of carbon dioxide on blood pH (see figure 20.19). The chemosensitive area in the medulla oblongata is far more important for the regulation of P_{CO_2} and pH than are the carotid and aortic bodies. The carotid and aortic bodies are responsible for, at most, 15%–20% of the total response to changes in P_{CO_2} or pH. During intense exercise, however, the carotid bodies respond more rapidly to changes in blood pH than does the chemosensitive area of the medulla.

Effect of Oxygen

Changes in P_{O_2} can affect respiration (see figure 20.19), although P_{CO_2} levels detected by the chemosensitive area are responsible for most of the changes in respiration. A decrease in oxygen levels below normal values is called **hypoxia** (hī-pok′sē-ă). Within a normal range of P_{O_2} levels, the effect of oxygen on the regulation of respiration is small. Only after arterial P_{O_2} decreases to approximately 50% of its normal value does it begin to have a large stimulatory effect on respiratory movements.

At first, it is somewhat surprising that small changes in P_{O_2} do not cause changes in respiratory rate. Consideration of the oxygen–hemoglobin dissociation curve, however, provides an explanation. Because of the S shape of the curve, at any P_{O_2} above 80 mm Hg nearly all of the hemoglobin is saturated with oxygen. Consequently, until P_{O_2} levels change significantly, the oxygen-carrying capacity of the blood is unaffected.

The carotid and aortic body chemoreceptors respond to decreased P_{O_2} with increased stimulation of the respiratory center, which can keep it active despite decreasing oxygen levels. If P_{O_2} decreases sufficiently, however, the respiratory center can fail to function, resulting in death.

 Importance of Reduced P_{O_2}

Carbon dioxide is much more important than oxygen as a regulator of normal alveolar ventilation, but under certain circumstances a reduced P_{O_2} in the arterial blood plays an important stimulatory role. During conditions of shock in which blood pressure is very low, the P_{O_2} in arterial blood can drop to levels sufficiently low to strongly stimulate carotid and aortic body sensory receptors. At high altitudes where atmospheric pressure is low, the P_{O_2} in arterial blood can also drop to levels sufficiently low to stimulate carotid and aortic bodies. Although P_{O_2} levels in the blood are reduced, the respiratory system's ability to eliminate carbon dioxide is not greatly affected by low atmospheric pressure. Thus, blood carbon dioxide levels become lower than normal because of the increased alveolar ventilation initiated in response to low P_{O_2}.

In people with emphysema, the destruction of the respiratory membrane results in decreased oxygen movement into the blood. The resulting low arterial P_{O_2} levels stimulate an increased rate and depth of respiration. At first, arterial P_{CO_2} levels may be unaffected by the reduced surface area of the respiratory membrane because carbon dioxide diffuses across the respiratory membrane 20 times more readily than does oxygen. However, if alveolar ventilation increases to the point that carbon dioxide exchange increases above normal, arterial carbon dioxide becomes lower than normal. More severe emphysema, in which the surface area of the respiratory membrane is reduced to a minimum, can decrease carbon dioxide exchange to the point that elevated arterial carbon dioxide occurs. ▼

52. *Define central and peripheral chemoreceptors. Which are most important for the regulation of blood pH and carbon dioxide?*
53. *Define hypercapnia and hypocapnia.*
54. *What effect does a decrease in blood pH or carbon dioxide have on respiratory rate?*
55. *Define hypoxia. Why must arterial P_{O_2} change significantly before it affects respiratory rate?*

Hering-Breuer Reflex

The **Hering-Breuer** (her′ing-broy′er) **reflex** limits the degree to which inspiration proceeds and prevents overinflation of the lungs (see figure 20.18). As the lungs fill with air, stretch receptors in the lungs are stimulated. Action potentials from the lung stretch receptors propagate to the medulla oblongata, where they inhibit the respiratory center neurons and cause expiration. In infants, the Hering-Breuer reflex plays an important role in regulating the basic rhythm of breathing and in preventing overinflation of the lungs. In adults, however, the reflex is important only when the tidal volume is large, such as during heavy exercise. ▼

56. *Describe the Hering-Breuer reflex and its function.*

Effect of Exercise on Ventilation

The mechanisms by which ventilation is regulated during exercise are controversial, and no one factor can account for all of the observed responses. Ventilation during exercise is divided into two phases:

1. *Ventilation increases abruptly.* At the onset of exercise, ventilation immediately increases. This initial increase can be as much as 50% of the total increase that occurs during exercise. The immediate increase in ventilation occurs too quickly to be explained by changes in metabolism or blood gases. At the same time that motor cortex stimulates skeletal muscle contractions, collateral branches from motor pathways stimulate the respiratory center. In addition, input from proprioceptors in muscles and joints stimulate the respiratory center (see figure 20.18).

2. *Ventilation increases gradually.* After the immediate increase in ventilation, a gradual increase occurs and levels off within 4–6 minutes after the onset of exercise. Factors responsible for the immediate increase in ventilation may play a role in the gradual increase as well.

Despite large changes in oxygen consumption and carbon dioxide production during exercise, the *average* arterial P_{O_2}, P_{CO_2}, and pH remain constant and close to resting levels as long as the exercise is aerobic (see chapter 8). This suggests that changes in blood gases and pH do not play an important role in regulating ventilation during aerobic exercise. During exercise, however, the values of arterial P_{O_2}, P_{CO_2}, and pH rise and fall more than at rest. Thus, even though their average values do not change, their oscillations may be a signal for helping control ventilation.

The highest level of exercise that can be performed without causing a significant change in blood pH is called the **anaerobic threshold.** If the exercise intensity is high enough to exceed the anaerobic threshold, skeletal muscles produce and release lactic acid into the blood. The resulting change in blood pH stimulates the carotid bodies, resulting in increased ventilation. In fact, ventilation can increase so much that arterial P_{CO_2} decreases below resting levels and arterial P_{O_2} increases above resting levels. ▼

57. *What mechanisms regulate ventilation at the onset of exercise and during exercise? What is the anaerobic threshold?*

Other Modifications of Ventilation

The activation of touch, thermal, and pain receptors can also affect the respiratory center (see figure 20.18). For example, irritants in the nasal cavity can initiate a sneeze reflex, and irritants in the lungs can stimulate a cough reflex. An increase in body temperature can stimulate increased ventilation. ▼

58. *Give examples of sensory receptor stimulation that alter respiration.*

PREDICT 17
Describe the respiratory response when cold water is splashed onto a person. In the past, newborn babies were sometimes swatted on the buttocks. Explain the rationale for this procedure.

Respiratory Adaptations to Exercise

In response to training, athletic performance increases because the cardiovascular and respiratory systems become more efficient at delivering oxygen and picking up carbon dioxide. Ventilation in most individuals does not limit performance because ventilation can increase to a greater extent than does cardiovascular function.

After training, vital capacity increases slightly and residual volume decreases slightly. Tidal volume at rest and during submaximal exercise does not change. At maximal exercise, however, tidal volume increases. After training, the respiratory rate at rest or during submaximal exercise is slightly lower than in an untrained person, but at maximal exercise respiratory rate is generally increased.

Minute ventilation is affected by the changes in tidal volume and respiratory rate. After training, minute ventilation is essentially unchanged or slightly reduced at rest and is slightly reduced during submaximal exercise. Minute ventilation is greatly increased at maximal exercise. For example, an untrained person's minute ventilation of 120 L/min can increase to 150 L/min after training. Increases to 180 L/min are typical of highly trained athletes. ▼

59. *What effect does training have on resting, submaximal, and maximal tidal volumes and on minute ventilation?*

Effects of Aging on the Respiratory System

Almost all aspects of the respiratory system are affected by aging. Even though vital capacity, maximum ventilation rates, and gas exchange decrease with age, the elderly can engage in light to moderate exercise because the respiratory system has a large reserve capacity.

With age, mucus accumulates within the respiratory passageways. The mucus–cilia escalator is less able to move the mucus because it becomes more viscous and because the number of cilia and their rate of movement decrease. As a consequence, the elderly are more susceptible to respiratory infections and bronchitis.

Vital capacity decreases with age because of a decreased ability to fill the lungs (decreased inspiratory reserve volume) and a decreased ability to empty the lungs (decreased expiratory reserve volume). As a result, maximum minute ventilation rates decrease, which in turn decreases the ability to perform intense exercise. These changes are related to weakening of respiratory muscles and to stiffening of cartilage and ribs.

Bronchitis (brong-kī'tis) is an inflammation of the bronchi caused by irritants, such as cigarette smoke, air pollution, or infections. The inflammation results in swelling of the mucous membrane lining the bronchi, increased mucus production, and decreased movement of mucus by cilia. Consequently, the diameter of the bronchi is decreased, and ventilation is impaired. Bronchitis can progress to emphysema.

Emphysema (em-fi-zē'mă) results in the destruction of the alveolar walls. Many smokers have both bronchitis and emphysema, which are often referred to as **chronic obstructive pulmonary disease (COPD).** Chronic inflammation of the bronchioles, usually caused by cigarette smoke or air pollution, probably initiates emphysema. Narrowing of the bronchioles restricts air movement, and air tends to be retained in the lungs. Coughing to remove accumulated mucus increases pressure in the alveoli, resulting in the rupture and destruction of alveolar walls. Loss of alveolar walls has two important consequences. The respiratory membrane has a decreased surface area, which decreases gas exchange, and loss of elastic fibers decreases the lung's ability to recoil and expel air. Symptoms of emphysema include shortness of breath and enlargement of the thoracic cavity. Treatment involves removing the sources of irritants (e.g., stopping smoking), promoting the removal of bronchial secretions, using bronchodilators, retraining people to breathe so that expiration of air is maximized, and using antibiotics to prevent infections. The progress of emphysema can be slowed, but no cure exists.

Alpha-1 antitrypsin (AAT) deficiency is a type of emphysema resulting from defects of the AAT gene located on chromosome 14. Although cigarette smoking is the major risk factor for emphysema, approximately 1%–2% of emphysema cases are due to a deficiency of AAT. As part of the inflammatory response, neutrophils and macrophages release **proteases,** which are enzymes that break down proteins. Proteases in the lungs provide protection against some bacteria and foreign substances. Too much protease activity, however, can be harmful because it results in the breakdown of lung tissue proteins, especially elastin in elastic fibers. AAT, which is synthesized in the liver, is a **protease inhibitor.** Normally, AAT inhibits protease activity, preventing the destruction of lung tissue. Excess protease production stimulated by cigarette smoke, however, can cause lung damage, leading to emphysema. In addition to the usual treatment for emphysema, AAT deficiency can be treated with drugs, such as danazol and tamoxifen, that stimulate increased AAT production in the liver. In addition, individuals may receive intravenous infusions of AAT, a process called **alpha-1 antitrypsin augmentation.**

Adult respiratory distress syndrome (ARDS) is caused by damage to the respiratory membrane. The damage stimulates an inflammatory response, which further damages the respiratory membrane. Water, ions, and proteins leave the blood and enter alveoli. Surfactant in the alveoli is reduced as surfactant-producing cells are damaged and surfactant present in the alveoli is diluted. The fluid-filled alveoli reduce gas exchange and make it more difficult for the lungs to expand. ARDS usually develops rapidly following an injurious event, such as an infection, inhalation of smoke from a fire, inhalation of toxic fumes, trauma, aspiration of gastric content associated with gastric reflux, or circulatory shock. Even with oxygen inhalation therapy, the mortality rate is high.

Pulmonary fibrosis is the replacement of lung tissue with fibrous connective tissue, thereby making the lungs less elastic and breathing more difficult. Exposure to asbestos, silica (silicosis), or coal dust is the most common cause.

Lung, or **bronchiogenic, cancer** arises from the epithelium of the respiratory tract. Cancers arising from tissues other than respiratory epithelium are not called lung cancer, even though they occur in the lungs. Lung cancer is the most common cause of cancer death in males and females in the United States, and most cases occur in smokers or those exposed to secondhand smoke. Because of the rich lymph and blood supply in the lungs, cancer in the lung can readily spread to other parts of the lung or body. In addition, the disease is often advanced before symptoms become severe enough for the victim to seek medical aid. Typical symptoms include coughing, sputum production, and blockage of the airways. Treatments include removal of part or all of the lung, chemotherapy, and radiation. Promising new, early detection tests are being explored. These include blood tests for a lung cancer–specific protein and sputum DNA tests that detect genetic abnormalities.

Residual volume increases with age as the alveolar ducts and many of the larger bronchioles increase in diameter. This increases the dead space, which decreases the amount of air available for gas exchange. In addition, gas exchange across the respiratory membrane is reduced because parts of the alveolar walls are lost, which decreases the surface area available for gas exchange, and the remaining walls thicken, which decreases diffusion of gases. A gradual increase in resting tidal volume with age compensates for these changes. ▼

60. *Why are the elderly more likely to develop respiratory infections and bronchitis?*

61. *Why do vital capacity, alveolar ventilation, and the diffusion of gases across the respiratory membrane decrease with age?*

Respiratory

Systems Interactions

	Effects of the Respiratory System on Other Systems	Effects of Other Systems on the Respiratory System
Integumentary System	Provides oxygen and removes carbon dioxide Helps maintain the body's pH balance	Nasal hair prevents the entry of dust and other foreign materials
Skeletal System	Provides oxygen and removes carbon dioxide Helps maintain the body's pH balance	Movement of ribs assists in respiration Thoracic cage protects the lungs and helps prevent them from collapsing
Muscular System	Provides oxygen and removes carbon dioxide Helps maintain the body's pH balance	Changes thoracic volume during breathing Controls tension on vocal cords during voice production
Nervous System	Provides oxygen and removes carbon dioxide Helps maintain the body's pH balance	Centers in the pons and medulla regulate the rate and depth of breathing Chemoreceptors in the medulla oblongata, carotid bodies, and aortic bodies monitor blood carbon dioxide, pH, and oxygen levels Mechanoreceptors in the lungs help regulate lung volumes by detecting lung stretch
Endocrine System	Provides oxygen and removes carbon dioxide Helps maintain the body's pH balance	Epinephrine causes dilation of bronchioles
Cardiovascular System	Angiotensin-converting enzyme converts angiotensin I to angiotensin II Provides oxygen and removes carbon dioxide Helps maintain the body's pH balance	Delivers oxygen, nutrients, hormones, and immune cells Removes carbon dioxide, waste products, and toxins
Lymphatic System and Immunity	Provides oxygen and removes carbon dioxide Helps maintain the body's pH balance	Immune cells provide protection against microorganisms and toxins and promote tissue repair by releasing chemical mediators
Digestive System	Provides oxygen and removes carbon dioxide Helps maintain the body's pH balance	Provides nutrients and water
Urinary System	Provides oxygen and removes carbon dioxide Helps maintain the body's pH balance	Removes waste products
Reproductive System	Provides oxygen and removes carbon dioxide Helps maintain the body's pH balance	Sexual arousal increases respiration

Respiration includes the movement of air into and out of the lungs, the exchange of gases between the air and the blood, the transport of gases in the blood, and the exchange of gases between the blood and tissues.

Functions of the Respiratory System (p. 602)

The major functions of the respiratory system are gas exchange, regulation of blood pH, voice production, olfaction, and protection against some microorganisms.

Anatomy and Histology of the Respiratory System (p. 602)

Nose

1. The nose consists of the external nose and the nasal cavity.
2. The bridge of the nose is bone, and most of the external nose is cartilage.
3. Openings of the nasal cavity
   - The nares open to the outside, and the choanae lead to the pharynx.
   - The paranasal sinuses and the nasolacrimal duct open into the nasal cavity.
4. Parts of the nasal cavity
   - The nasal cavity is divided by the nasal septum.
   - The anterior vestibule contains hairs that trap debris.
   - The nasal cavity is lined with pseudostratified ciliated columnar epithelium that traps debris and moves it to the pharynx.
   - The superior part of the nasal cavity contains the olfactory epithelium.
5. The nasal cavity is a passageway for air; it cleans, warms, and humidifies air.

Pharynx

1. The nasopharynx joins the nasal cavity through the choanae and contains the openings to the auditory tube and the pharyngeal tonsils.
2. The oropharynx joins the oral cavity and contains the palatine and lingual tonsils.
3. The laryngopharynx opens into the larynx and the esophagus.

Larynx

1. Cartilage
   - Three unpaired cartilages exist. The thyroid cartilage and cricoid cartilage form most of the larynx. The epiglottis covers the opening of the larynx during swallowing.
   - Six paired cartilages exist. The vocal folds attach to the arytenoid cartilages.
2. The epiglottis, vestibular folds, and vocal cords prevent swallowed materials from entering the larynx.
3. The vocal folds produce sounds.

Trachea

1. The trachea connects the larynx to the main bronchi.
2. The trachealis muscle regulates the diameter of the trachea.

Main Bronchi

The trachea divides to form two main bronchi, which go to the lungs.

Lungs and Tracheobronchial Tree

1. The body contains two lungs, which are divided into lobes and bronchopulmonary segments.

2. The main bronchi supply the lungs.
   - The main bronchi divide to form lobar bronchi, which supply lung lobes.
   - The lobar bronchi divide to form segmental bronchi, which supply bronchopulmonary segments.
   - The segmental bronchi divide to form bronchioles, which divide to form terminal bronchioles.
3. The trachea to the terminal bronchioles is a passageway for air movement.
   - The area from the trachea to the terminal bronchioles is ciliated to facilitate the removal of inhaled debris.
   - Cartilage helps hold the tube system open (from the trachea to the bronchioles).
   - Smooth muscle controls the diameter of the tubes (terminal bronchioles).
4. Terminal bronchioles divide to form respiratory bronchioles, which give rise to alveolar ducts. Air-filled chambers called alveoli open into the respiratory bronchioles and alveolar ducts.
5. Gas exchange occurs from the respiratory bronchioles to the alveoli.

Respiratory Membrane

The components of the respiratory membrane are a film of water, the walls of the alveolus and the capillary, and an interstitial space.

Pleura

The pleural membranes surround the lungs and provide protection against friction.

Blood Supply

1. Deoxygenated blood is transported to the lungs through the pulmonary arteries, and oxygenated blood leaves through the pulmonary veins.
2. Oxygenated blood is mixed with a small amount of deoxygenated blood from the bronchi.

Lymphatic Supply

The superficial and deep lymphatic vessels drain lymph from the lungs.

Ventilation (p. 613)

Changing Thoracic Volume

1. Contraction of the diaphragm increases thoracic volume.
2. Muscles can elevate the ribs and increase thoracic volume or can depress the ribs and decrease thoracic volume.

Pressure Differences and Airflow

1. Air moves from an area of higher pressure to an area of lower pressure.
2. Pressure is inversely related to volume (Boyle's law).
3. Changes in tube diameter result in changes in resistance to airflow.

Airflow into and Out of Alveoli

1. Inspiration results when atmospheric pressure is greater than alveolar pressure.
2. Expiration results when atmospheric pressure is less than alveolar pressure.

Lung Recoil

1. Lung recoil causes alveoli to collapse.
   - Lung recoil results from elastic fibers and water surface tension.
   - Surfactant reduces water surface tension.
2. Pleural pressure is the pressure in the pleural cavity.
   - A pleural pressure lower than alveolar pressure can cause the alveoli to expand.
   - Pneumothorax is an opening between the pleural cavity and the air that causes a loss of pleural pressure.

Changing Alveolar Volume

Changes in thoracic volume cause changes in pleural pressure, resulting in changes in alveolar volume, alveolar pressure, and airflow.

Measurement of Lung Function (p. 617)
Compliance of the Lungs and Thorax

1. Compliance is a measure of lung expansion caused by alveolar pressure.
2. Reduced compliance means that it is more difficult than normal to expand the lungs.

Pulmonary Volumes and Capacities

1. Four pulmonary volumes exist: tidal volume, inspiratory reserve volume, expiratory reserve volume, and residual volume.
2. Pulmonary capacities are the sum of two or more pulmonary volumes and include inspiratory capacity, functional residual capacity, vital capacity, and total lung capacity.
3. The forced expiratory vital capacity measures vital capacity as the individual exhales as rapidly as possible.

Minute Ventilation and Alveolar Ventilation

1. The minute ventilation is the total amount of air moved in and out of the respiratory system per minute.
2. Dead space is the part of the respiratory system in which gas exchange does not take place.
3. Alveolar ventilation is how much air per minute enters the parts of the respiratory system in which gas exchange takes place.

Gas Exchange in the Lungs (p. 619)
Partial Pressure

Partial pressure is the contribution of a gas to the total pressure of a mixture of gases (Dalton's law).

Diffusion of Gases into and Out of Liquids

The concentration of a dissolved gas in a liquid is determined by its pressure and by its solubility coefficient (Henry's law).

Diffusion of Gases Through the Respiratory Membrane

The rate of diffusion of gases through the respiratory membrane depends on its thickness, the diffusion coefficient of the gas, the surface area of the membrane, and the partial pressure of the gases in the alveoli and the blood.

Oxygen and Carbon Dioxide Transport in the Blood (p. 620)
Oxygen Partial Pressure Gradients

1. Oxygen moves from the alveoli ($P_{O_2} = 104$ mm Hg) into the blood ($P_{O_2} = 40$ mm Hg). Blood is almost completely saturated with oxygen when it leaves the capillary.
2. The P_{O_2} in the blood decreases ($P_{O_2} = 95$ mm Hg) because of mixing with deoxygenated blood called shunted blood.
3. Oxygen moves from the tissue capillaries ($P_{O_2} = 95$ mm Hg) into the tissues ($P_{O_2} = 40$ mm Hg).

Carbon Dioxide Partial Pressure Gradients

1. Carbon dioxide moves from the tissues ($P_{CO_2} = 45$ mm Hg) into tissue capillaries ($P_{CO_2} = 40$ mm Hg).
2. Carbon dioxide moves from the pulmonary capillaries ($P_{CO_2} = 45$ mm Hg) into the alveoli ($P_{CO_2} = 40$ mm Hg).

Hemoglobin and Oxygen Transport

1. Oxygen is transported by hemoglobin (98.5%) and is dissolved in plasma (1.5%).
2. The oxygen–hemoglobin dissociation curve shows that hemoglobin is almost completely saturated when P_{O_2} is 80 mm Hg or above. At lower partial pressures, the hemoglobin releases oxygen.
3. A shift of the oxygen–hemoglobin dissociation curve to the right because of a decrease in pH (Bohr effect), an increase in carbon dioxide, or an increase in temperature results in a decrease in hemoglobin's ability to hold oxygen.
4. A shift of the oxygen–hemoglobin dissociation curve to the left because of an increase in pH (Bohr effect), a decrease in carbon dioxide, or a decrease in temperature results in an increase in hemoglobin's ability to hold oxygen.

Transport of Carbon Dioxide

1. Carbon dioxide is transported dissolved in plasma (7%), as HCO_3^- dissolved in plasma and in red blood cells (70%), and bound to hemoglobin (23%).
2. In tissue capillaries, the following occur.
   - Carbon dioxide combines with water inside red blood cells to form carbonic acid, which dissociates to form HCO_3^-. Decreasing HCO_3^- concentrations promote carbon dioxide transport.
   - The chloride shift is the exchange of Cl^- for HCO_3^- between plasma and red blood cells.
   - Hydrogen ions binding to hemoglobin promote carbon dioxide transport, prevent a change in pH in red blood cells, and produce a Bohr effect.
   - In the Haldane effect, the smaller the amount of oxygen bound to hemoglobin, the greater the amount of carbon dioxide bound to it, and vice versa.
3. In pulmonary capillaries, the events occurring in the tissue capillaries are reversed.

Regulation of Ventilation (p. 626)
Respiratory Areas in the Brainstem

1. The medullary respiratory center consists of the dorsal and ventral respiratory groups.
   - The dorsal respiratory groups stimulate the diaphragm.
   - The ventral respiratory groups stimulate the intercostal and abdominal muscles.
2. The pontine respiratory group is involved with switching between inspiration and expiration.

Generation of Rhythmic Ventilation

1. Neurons in the medullary respiratory center establish the basic rhythm of respiration.
2. When stimuli from receptors or other parts of the brain exceed a threshold level, inspiration begins.
3. As respiratory muscles are stimulated, neurons that stop inspiration are stimulated. When the stimulation of these neurons exceeds a threshold level, inspiration is inhibited.

Cerebral and Limbic System Control

Respiration can be voluntarily controlled and can be modified by emotions.

Chemical Control of Ventilation

1. Carbon dioxide is the major regulator of respiration. An increase in carbon dioxide or a decrease in pH can stimulate the chemosensitive area, causing a greater rate and depth of respiration.
2. Oxygen levels in the blood affect respiration when a 50% or greater decrease from normal levels exists. Decreased oxygen is detected by receptors in the carotid and aortic bodies, which then stimulate the respiratory center.

Hering-Breuer Reflex

Stretch of the lungs during inspiration can inhibit the respiratory center and contribute to a cessation of inspiration.

Effect of Exercise on Ventilation

1. Collateral fibers from motor neurons and from proprioceptors stimulate the respiratory centers.
2. Chemosensitive mechanisms fine-tune the effects produced through the motor neurons and proprioceptors.

Other Modifications of Ventilation

Touch, thermal, and pain sensations can modify ventilation.

Respiratory Adaptations to Exercise (p. 631)

Tidal volume, respiratory rate, minute ventilation, and gas exchange between the alveoli and blood remain unchanged or slightly lower at rest or during submaximal exercise but increase at maximal exercise.

Effects of Aging on the Respiratory System (p. 631)

1. The ability to remove mucus from the respiratory passageways decreases with age.
2. Vital capacity and maximum minute ventilation decrease with age because of a weakening of respiratory muscles and decreased thoracic cage compliance.
3. Residual volume and dead space increase because of the increased diameter of respiratory passageways. As a result, alveolar ventilation decreases.
4. An increase in resting tidal volume compensates for decreased alveolar ventilation, loss of alveolar walls (surface area), and thickening of alveolar walls.

R E V I E W   A N D   C O M P R E H E N S I O N

1. The nasal cavity
   a. has openings for the paranasal sinuses.
   b. has a vestibule, which contains the olfactory epithelium.
   c. is connected to the pharynx by the nares.
   d. has passageways called conchae.
   e. is lined with squamous epithelium, except for the vestibule.

2. The nasopharynx
   a. is lined with moist stratified squamous epithelium.
   b. contains the pharyngeal tonsil.
   c. opens into the oral cavity.
   d. extends to the tip of the epiglottis.
   e. is an area through which food, drink, and air normally pass.

3. The larynx
   a. connects the oropharynx to the trachea.
   b. has three unpaired and six paired cartilages.
   c. contains the vocal folds.
   d. contains the vestibular folds.
   e. all of the above.

4. Terminal bronchioles branch to form
   a. the alveolar duct.
   b. alveoli.
   c. bronchioles.
   d. respiratory bronchioles.

5. During quiet expiration, the
   a. abdominal muscles relax.
   b. diaphragm moves inferiorly.
   c. external intercostal muscles contract.
   d. thorax and lungs passively recoil.
   e. all of the above.

6. The parietal pleura
   a. covers the surface of the lung.
   b. covers the inner surface of the thoracic cavity.
   c. is the connective tissue partition that divides the thoracic cavity into right and left pleural cavities.
   d. covers the inner surface of the alveoli.
   e. is the membrane across which gas exchange occurs.

7. Contraction of the bronchiolar smooth muscle has which of these effects?
   a. A smaller pressure gradient is required to get the same rate of airflow, compared with normal bronchioles.
   b. It increases airflow through the bronchioles.
   c. It increases resistance to airflow.
   d. It increases alveolar ventilation.

8. During expiration, the alveolar pressure is
   a. less than the pleural pressure.
   b. greater than the atmospheric pressure.
   c. less than the atmospheric pressure.
   d. unchanged.

9. The lungs do not normally collapse because of
   a. surfactant.
   b. pleural pressure.
   c. elastic recoil.
   d. both a and b.

10. Immediately after the creation of an opening through the thorax into the pleural cavity,
    a. air flows through the hole and into the pleural cavity.
    b. air flows through the hole and out of the pleural cavity.
    c. air flows neither out nor in.
    d. the lung protrudes through the hole.

11. Compliance of the lungs and thorax
    a. is the volume by which the lungs and thorax change for each unit change of alveolar pressure.
    b. increases in emphysema.
    c. decreases because of lack of surfactant.
    d. all of the above.

12. Given these lung volumes:
    1. tidal volume = 500 mL
    2. residual volume = 1000 mL
    3. inspiratory reserve volume = 2500 mL
    4. expiratory reserve volume = 1000 mL
    5. dead space = 1000 mL

The vital capacity is
a. 3000 mL. c. 4000 mL. e. 6000 mL.
b. 3500 mL. d. 5000 mL.

13. The alveolar ventilation is the
 a. tidal volume times the respiratory rate.
 b. minute ventilation plus the dead space.
 c. amount of air available for gas exchange in the lungs.
 d. vital capacity divided by the respiratory rate.
 e. inspiratory reserve volume times minute ventilation.

14. If the total pressure of a gas is 760 mm Hg and its composition is 20% oxygen, 0.04% carbon dioxide, 75% nitrogen, and 5% water vapor, the partial pressure of oxygen is
 a. 15.2 mm Hg.
 b. 20 mm Hg.
 c. 118 mm Hg.
 d. 152 mm Hg.
 e. 740 mm Hg.

15. The rate of diffusion of a gas across the respiratory membrane increases as the
 a. respiratory membrane becomes thicker.
 b. surface area of the respiratory membrane decreases.
 c. partial pressure gradient of the gas across the respiratory membrane increases.
 d. diffusion coefficient of the gas decreases.
 e. all of the above.

16. The partial pressure of carbon dioxide in the venous blood is
 a. greater than in the tissue spaces.
 b. less than in the tissue spaces.
 c. less than in the alveoli.
 d. less than in arterial blood.

17. Oxygen is mostly transported in the blood
 a. dissolved in plasma.
 b. bound to blood proteins.
 c. within HCO_3^-.
 d. bound to the heme portion of hemoglobin.

18. The oxygen–hemoglobin dissociation curve is adaptive because it
 a. shifts to the right in the pulmonary capillaries and to the left in the tissue capillaries.
 b. shifts to the left in the pulmonary capillaries and to the right in the tissue capillaries.
 c. does not shift.

19. Carbon dioxide is mostly transported in the blood
 a. dissolved in plasma.
 b. bound to blood proteins.
 c. within HCO_3^-.

d. bound to the heme portion of hemoglobin.
e. bound to the globin portion of hemoglobin.

20. When blood passes through the tissues, the hemoglobin in blood is better able to combine with carbon dioxide because of the
 a. Bohr effect.
 b. Haldane effect.
 c. chloride shift.
 d. Boyle effect.
 e. Dalton effect.

21. The chloride shift
 a. promotes the transport of carbon dioxide in the blood.
 b. occurs when Cl^- replace HCO_3^- within red blood cells.
 c. maintains electrical neutrality in red blood cells and the plasma.
 d. all of the above.

22. Which of these parts of the brainstem is correctly matched with its main function?
 a. ventral respiratory groups—stimulate the diaphragm
 b. dorsal respiratory groups—limit inflation of the lungs
 c. pontine respiratory group—switches between inspiration and expiration
 d. all of the above

23. The chemosensitive area
 a. stimulates the respiratory center when blood carbon dioxide levels increase.
 b. stimulates the respiratory center when blood pH increases.
 c. is located in the pons.
 d. stimulates the respiratory center when blood oxygen levels increase.
 e. all of the above.

24. Blood oxygen levels
 a. are more important than carbon dioxide levels in the regulation of respiration.
 b. need to change only slightly to cause a change in respiration.
 c. are detected by sensory receptors in the carotid and aortic bodies.
 d. all of the above.

25. At the onset of exercise, respiration rate and depth increase primarily because of
 a. increased blood carbon dioxide levels.
 b. decreased blood oxygen levels.
 c. decreased blood pH levels.
 d. input to the respiratory center from the cerebral motor cortex and proprioceptors.

Answers in Appendix E

CRITICAL THINKING

1. What effect does rapid (respiratory rate equals 24 breaths per minute), shallow (tidal volume equals 250 mL per breath) breathing have on minute ventilation, alveolar ventilation, and alveolar P_{O_2} and P_{CO_2}?

2. A person's vital capacity is measured while standing and while lying down. What difference, if any, in the measurement do you predict and why?

3. Ima Diver wanted to do some underwater exploration. She did not want to buy expensive SCUBA equipment, however. Instead, she bought a long hose and an inner tube. She attached one end of the hose to the inner tube so that the end was always out of the water, and she inserted the other end of the hose in her mouth and went diving. What happened to her alveolar ventilation and why? How can she compensate for this change? How does diving affect thoracic compliance and the work of ventilation?

4. The bacteria that cause gangrene (*Clostridium perfringens*) are anaerobic microorganisms that do not thrive in the presence of oxygen. Hyperbaric oxygenation (HBO) treatment places a person in a chamber that contains oxygen at three to four times normal atmospheric pressure. Explain how HBO helps in the treatment of gangrene.

5. Cardiopulmonary resuscitation (CPR) has replaced older, less efficient methods of sustaining respiration. The back-pressure/arm-lift method is one such technique that is no longer used. This procedure is performed with the victim lying face down. The rescuer presses firmly on the base of the scapulae for several seconds and then grasps the arms and lifts them. The sequence is then repeated. Explain why this procedure results in ventilation of the lungs.

6. A technique for artificial respiration is mouth-to-mouth resuscitation. The rescuer takes a deep breath, blows air into the victim's mouth, and

then lets air flow out of the victim. The process is repeated. Explain the following: (1) Why do the victim's lungs expand? (2) Why does air move out of the victim's lungs? (3) What effect do the P_{O_2} and the P_{CO_2} of the rescuer's air have on the victim?

7. The left phrenic nerve supplies the left side of the diaphragm and the right phrenic nerve supplies the right side. Damage to the left phrenic nerve results in paralysis of the left side of the diaphragm. During inspiration, does the left side of the diaphragm move superiorly, move inferiorly, or stay in place?

8. Suppose that the thoracic wall is punctured at the end of a normal expiration, producing a pneumothorax. Does the thoracic wall move inward, move outward, or not move?

9. During normal, quiet respiration, when does the maximum rate of diffusion of oxygen in the pulmonary capillaries occur? When does the maximum rate of diffusion of carbon dioxide occur?

10. Predict what would happen to tidal volume if the vagus nerves were cut, the phrenic nerves were cut, or the intercostal nerves were cut.

11. You and your physiology instructor are trapped in an overturned ship. To escape, you must swim under water a long distance. You tell your instructor it would be a good idea to hyperventilate before making the escape attempt. Your instructor calmly replies, "What good would that do, since your pulmonary capillaries are already 100% saturated with oxygen?" What would you do and why?

Answers in Appendix F

For additional study tools, go to: www.aris.mhhe.com. Click on your course topic and then this textbook's author name/title. Register once for a semester's worth of interactive learning activities to help improve your grade!

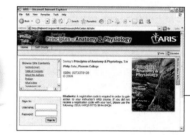

CHAPTER 21 Digestive System

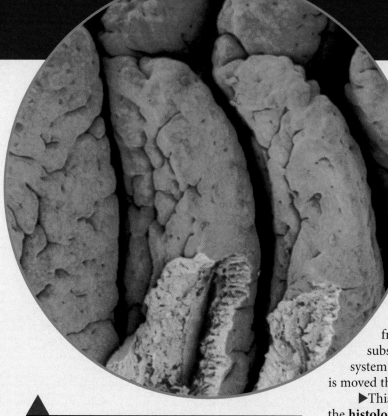

This colorized scanning electron micrograph of the interior surface of the small intestine shows villi. The surface of the villi is *blue*. The *orange* color shows two places where the surface has been broken open to reveal the interior of the villi. The small intestine is the portion of the digestive system where nutrients are absorbed from the food we eat. This absorption is facilitated by the fingerlike villi, which greatly enlarge the internal surface area of the small intestine.

Introduction

Every cell of the body needs nourishment, yet most cells cannot leave their position in the body and travel to a food source, so the food must be converted to a usable form and delivered. The digestive system, with the help of the circulatory system, acts as a gigantic "meals on wheels," providing nourishment to over a hundred trillion "customer" cells in the body. It also has its own quality control and waste disposal system.

The digestive system provides the body with water, electrolytes, and other nutrients. To do this, the digestive system is specialized to ingest food, propel it through the digestive tract, digest it, and absorb water, electrolytes, and other nutrients from the lumen of the gastrointestinal tract. Once these useful substances are absorbed, they are transported through the circulatory system to cells, where they are used. The undigested portion of the food is moved through the digestive tract and eliminated through the anus.

▶This chapter presents the **functions of the digestive system (p. 640),** the **histology of the digestive tract (p. 640),** and the **peritoneum (p. 641).** The anatomy and physiology of each section of the digestive tract and its accessory structures are then presented—the **oral cavity (p. 643), pharynx (p. 647),** and **esophagus (p. 648)**—along with a section on **swallowing (p. 648),** the **stomach (p. 650), small intestine (p. 656), liver and gallbladder (p. 659), pancreas (p. 664),** and **large intestine (p. 667). Digestion, absorption, and transport (p. 671)** of nutrients are then discussed, along with the **effects of aging on the digestive system (p. 676).**

Tools for Success

A virtual cadaver dissection experience

Audio (MP3) and/or video (MP4) content, available at www.mhhe.aris.com

Functions of the Digestive System

The functions of the digestive system are to

1. *Take in food.* Food and water are taken into the body through the mouth.
2. *Break down the food.* The food that is taken into the body is broken down during the process of digestion from complex molecules to smaller molecules that can be absorbed. Digestion consists of mechanical digestion, which involves the chewing of food and the mixing of food with digestive tract secretions, and chemical digestion, which is accomplished by digestive enzymes that are secreted along the digestive tract.
3. *Absorb nutrients.* The small molecules that result from digestion are absorbed through the walls of the intestine for use in the body. Water, electrolytes, and other nutrients, such as vitamins and minerals, are also absorbed.
4. *Eliminate wastes.* Undigested material, such as fiber from food, plus waste products excreted into the digestive tract are eliminated in the feces. ▼

   1. Describe the functions of the digestive system.

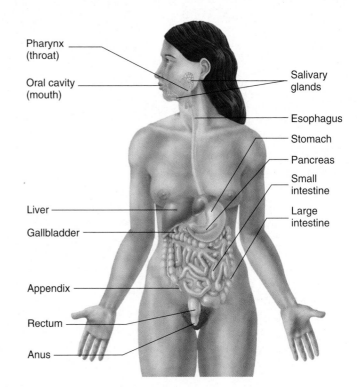

Figure 21.1 Digestive System

Histology of the Digestive Tract

The digestive system consists of the **digestive tract,** a tube extending from the mouth to the anus, plus associated organs, which secrete fluids into the digestive tract. The term **gastrointestinal (GI)** (gas′trō-in-tes′tin-ăl) tract technically refers only to the stomach and intestines but is often used as a synonym for *digestive tract.* The inside of the digestive tract is continuous with the outside environment, where it opens at the mouth and anus. Nutrients cross the wall of the digestive tract to enter the circulation.

The digestive tract consists of the oral cavity, pharynx, esophagus, stomach, small intestine, large intestine, and anus. Accessory glands are associated with the digestive tract (figure 21.1). The salivary glands empty into the oral cavity, and the liver and pancreas are connected to the small intestine.

Tunics

Various parts of the digestive tract are specialized for different functions, but nearly all parts consist of four **tunics,** or layers: the mucosa, submucosa, muscularis, and serosa or adventitia (figure 21.2). These will be described in order from the inside of the tube.

1. The innermost tunic is mucous membrane called the **mucosa** (mū-kō′să, a tissue producing mucus). The mucosa consists of **mucous epithelium,** a loose connective tissue called the **lamina propria,** and a thin smooth muscle layer, the **muscularis mucosa.** The epithelium in the mouth, esophagus, and anus resists abrasion, and epithelium in the stomach and intestine absorbs and secretes.
2. The **submucosa** lies just outside the mucosa. It is a thick layer of loose connective tissue containing nerves, blood vessels, and small glands. The submucosa has glands with

ducts that extend through the mucosa to empty into the digestive tract. An extensive network of nerve cell processes forms a **submucosal plexus** (network), which regulates secretion from the glands.

3. The next tunic is the **muscularis,** which in most parts of the digestive tube consists of an inner layer of **circular smooth muscle** and an outer layer of **longitudinal smooth muscle.** Another nerve plexus, the **myenteric plexus,** lies between the two muscle layers. The myenteric plexus is much more extensive than the submucosal plexus and controls the **motility,** or movements, of the intestinal tract. Together the nerve plexuses of the submucosa and muscularis compose the **enteric** (en-tĕr′ik, relating to the intestine) **plexus.**
4. The fourth, or outermost, layer of the digestive tract is either a serosa or an adventitia. Some regions of the digestive tract are covered by peritoneum, and other regions are not. The peritoneum, which is a smooth epithelial layer, and its underlying connective tissue are referred to histologically as the **serosa.** In regions of the digestive tract not covered by peritoneum, the digestive tract is covered by a connective tissue layer called the **adventitia** (ad′ven-tish′ă, foreign, coming from outside), which is continuous with the surrounding connective tissue. These areas include the esophagus and the retroperitoneal organs (see "Peritoneum"). ▼

   2. What are the major layers of the digestive tract? How do the serosa and adventitia differ?
   3. Describe the enteric plexus. In what layers of the digestive tract are the submucosal and myenteric plexuses found?

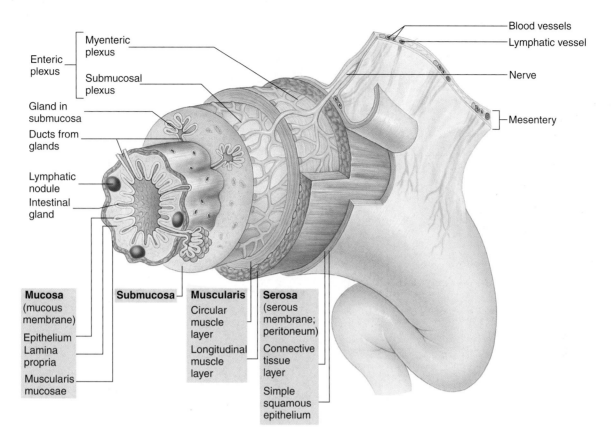

Figure 21.2 Digestive Tract Histology

The four tunics are the mucosa, submucosa, muscularis, and serosa or adventitia. Glands may exist along the digestive tract as part of the epithelium, as glands within the submucosa, or as large glands outside the digestive tract.

Enteric Nervous System

The **enteric nervous system (ENS)** consists of nerve plexuses within the wall of the digestive tract. The plexuses are composed of sensory neurons connecting the digestive tract to the central nervous system (CNS), autonomic nervous system (ANS) motor neurons connecting the CNS to the digestive tract, and enteric neurons, which are only in the enteric nerve plexus. There are more neurons in the enteric plexus than in the spinal cord. Most of the nervous control of the GI tract is local, occurring as a result of activities within the enteric plexus. Some of the nervous control is more general, mediated largely by the parasympathetic division of the ANS through the vagus nerves. The sympathetic division of the ANS also exerts control through sympathetic nerves (see chapter 14).

There are three major types of enteric neurons: (1) Enteric sensory neurons detect changes in the chemical composition of the digestive tract contents or detect mechanical changes, such as stretch of the digestive tract wall; (2) enteric motor neurons stimulate or inhibit smooth muscle contraction and glandular secretion in the digestive system; (3) enteric interneurons connect enteric sensory and motor neurons.

The enteric plexus controls activities within specific, short regions of the digestive tract through **local reflexes.** For example, enteric sensory neurons detect stretch of the digestive tract and stimulate enteric interneurons. The enteric interneurons stimulate enteric motor neurons, which stimulate smooth muscle in the muscularis, resulting in contractions that resist the stretch. The ENS is capable of controlling the complex movements, secretions, and blood flow of the GI tract, without any outside influences. Although the ENS can control the activities of the digestive tract independent of the CNS, normally the two systems work together. For example, autonomic innervation from the CNS influences ENS activity. ▼

4. *What three types of neurons are in the enteric nervous system?*
5. *What are the kinds of enteric neurons? What is a local reflex?*

Peritoneum

The body walls and organs of the abdominopelvic cavity are covered with **serous membranes.** These membranes are very smooth and secrete a serous fluid, which provides a lubricating film between the layers of membranes. These membranes and fluid reduce friction as organs move within the abdominopelvic cavity. The serous membrane that covers the organs is the **visceral peritoneum** (per′i-tō-nē′ŭm, to stretch over), and the one that covers the interior surface of the body wall is the **parietal peritoneum** (figure 21.3). The **peritoneal cavity** is between the visceral peritoneum and parietal peritoneum.

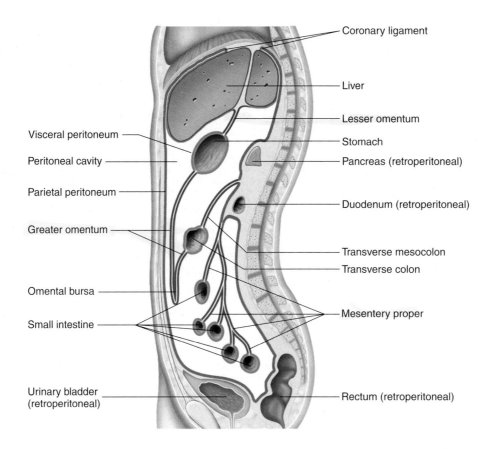

Coronary ligament

Liver

Visceral peritoneum

Lesser omentum

Stomach

Peritoneal cavity

Pancreas (retroperitoneal)

Parietal peritoneum

Duodenum (retroperitoneal)

Greater omentum

Transverse mesocolon

Transverse colon

Omental bursa

Mesentery proper

Small intestine

Urinary bladder
(retroperitoneal)

Rectum (retroperitoneal)

(a) **Medial view**

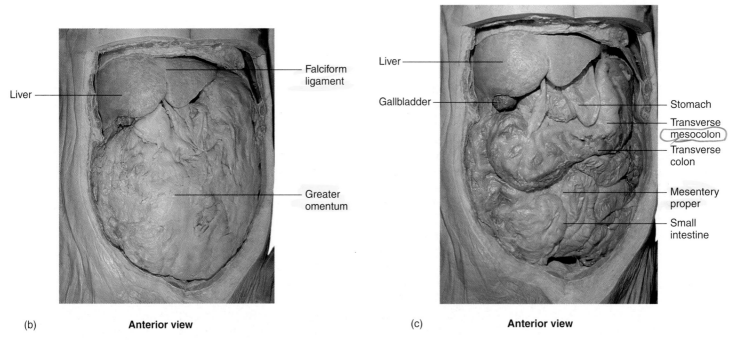

Liver

Falciform
ligament

Greater
omentum

Liver

Gallbladder

Stomach

Transverse
mesocolon

Transverse
colon

Mesentery
proper

Small
intestine

(b)   **Anterior view**

(c)   **Anterior view**

Figure 21.3   Peritoneum and Mesenteries

(*a*) Sagittal section through the trunk, showing the peritoneum and mesenteries associated with some abdominopelvic organs. The parietal peritoneum lines the abdominopelvic cavity (*blue*), and the visceral peritoneum covers abdominal organs (*red*). Retroperitoneal organs are behind the parietal peritoneum. The mesenteries are membranes (*purple*) that connect organs to each other and to the body wall. (*b*) Photograph of the abdomen of a cadaver, with the greater omentum in place. (*c*) Photograph of the abdomen of a cadaver, with the greater omentum removed to reveal the underlying viscera.

Peritonitis

Peritonitis is a potentially life-threatening inflammation of the peritoneal membranes. The inflammation can result from chemical irritation by substances, such as bile, that have escaped from a damaged digestive tract, or it can result from infection originating in the digestive tract, such as when the appendix ruptures. An accumulation of excess serous fluid in the peritoneal cavity, called **ascites** (ă-sī′tēz), can occur in peritonitis. Ascites may also accompany starvation, alcoholism, or liver cancer.

Mesenteries (mes′en-ter′ēz, middle intestine) are connective tissue sheets holding many of the organs in place within the abdominopelvic cavity (see figure 21.3a). The mesenteries consist of two layers of serous membranes with a thin layer of loose connective tissue between them. They provide a route by which vessels and nerves can pass from the body wall to the organs. Other abdominal organs lie against the abdominal wall, have no mesenteries, and are referred to as **retroperitoneal** (re′trō-per′i-tō-nē′ăl, behind the peritoneum). The retroperitoneal organs include the duodenum, pancreas, ascending colon, descending colon, rectum, kidneys, adrenal glands, and urinary bladder.

The mesentery connecting the lesser curvature of the stomach to the liver and diaphragm is called the **lesser omentum** (ō-men′tŭm, membrane of the bowels), and the mesentery extending as a fold from the greater curvature and then to the transverse colon is called the **greater omentum**. The greater omentum is unusual in that it is a long, double fold of mesentery that extends inferiorly from the stomach before looping back to the transverse colon to create a cavity, or pocket, called the **omental bursa** (ber′să, pocket) (see figure 21.3a). A large amount of fat accumulates in the greater omentum, and it is sometimes referred to as the "fatty apron" (see figure 21.3b).

The **coronary ligament** attaches the liver to the diaphragm. Unlike other mesenteries, the coronary ligament has a wide space in the center, the bare area of the liver, where no peritoneum exists. The **falciform ligament** attaches the liver to the anterior abdominal wall (see figure 21.3b).

Although *mesentery* is a general term referring to the serous membranes attached to the abdominal organs, it is also used specifically to refer to the mesentery associated with the small intestine, sometimes called the **mesentery proper**. The **transverse mesocolon** extends from the transverse colon to the posterior body wall (see figure 21.3c). ▼

6. *Where are visceral peritoneum and parietal peritoneum found? What is a retroperitoneal organ?*
7. *Define mesentery. Name and describe the location of the mesenteries in the abdominal cavity.*

Oral Cavity

The **oral cavity** (figure 21.4), or mouth, is the part of the digestive tract bounded by the lips anteriorly, the cheeks laterally, the palate superiorly, and a muscular floor inferiorly. The oral cavity connects posteriorly to the pharynx through a space called the **fauces**

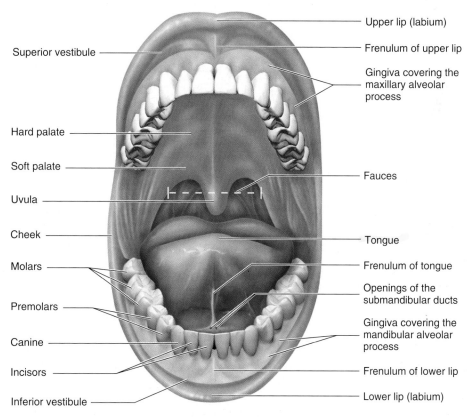

Superior vestibule

Hard palate

Soft palate

Uvula

Cheek

Molars

Premolars

Canine

Incisors

Inferior vestibule

Upper lip (labium)

Frenulum of upper lip

Gingiva covering the maxillary alveolar process

Fauces

Tongue

Frenulum of tongue

Openings of the submandibular ducts

Gingiva covering the mandibular alveolar process

Frenulum of lower lip

Lower lip (labium)

Anterior view

Figure 21.4 Oral Cavity

(faw′sēz, throat). The oral cavity is divided into two regions: (1) the **vestibule** (ves′ti-bool, entry) is the space between the lips or cheeks and the alveolar processes, which contain the teeth; and (2) the **oral cavity proper,** which lies medial to the alveolar processes. The oral cavity is lined with moist stratified squamous epithelium, which protects against abrasion. ▼

   8. *Define fauces, vestibule, and oral cavity proper.*

Lips and Cheeks

The **lips,** or **labia** (lā′bē-ă) (see figure 21.4), are muscular structures formed mostly by the **orbicularis oris** (ōr-bik′ū-lā′ris ōr′is) **muscle** (see chapter 9) and connective tissue. The keratinized stratified epithelium of the skin covering the lips is thinner than the epithelium of the surrounding skin (see chapter 5). Consequently, it is more transparent and the color from the underlying blood vessels give the lips a reddish pink to dark red appearance. At the internal margin of the lips, the epithelium is continuous with the moist stratified squamous epithelium of the mucosa in the oral cavity.

One or more **labial frenula** (fren′ū-lă, bridle), which are mucosal folds, extend from the alveolar processes of the maxilla to the upper lip and from the alveolar process of the mandible to the lower lip.

The **cheeks** form the lateral walls of the oral cavity. They consist of an interior lining of moist stratified squamous epithelium and an exterior covering of skin. The substance of the cheek includes the **buccinator muscle** (see chapter 9), which flattens the cheek against the teeth, and the **buccal fat pad,** which rounds out the profile on the side of the face.

The lips and cheeks are important in the processes of mastication (chewing) and speech. They help manipulate food within the mouth and hold it in place while the teeth crush or tear it. They also help form words during the speech process. A large number of the muscles of facial expression are involved in lip movement. They are listed in chapter 9. ▼

   9. *What muscles form the substance of the lips and cheeks? What are the functions of the lips and cheeks?*

Palate and Palatine Tonsils

The **palate** (see figure 21.4) consists of two parts: an anterior bony part, the **hard palate** (see chapter 7), and a posterior part, the **soft palate,** which consists of skeletal muscle and connective tissue. The **uvula** (ū′vū-lă, a grape) is the projection from the posterior edge of the soft palate. The palate is important in the swallowing process; it prevents food from passing into the nasal cavity.

The **palatine tonsils** are located in the lateral wall of the fauces (see chapter 19). ▼

   10. *What are the hard and soft palates? Where is the uvula found? What is the function of the palate?*

Tongue

The **tongue** is a large, muscular organ that occupies most of the oral cavity. The major attachment of the tongue is in the posterior

part of the oral cavity. The anterior part of the tongue is relatively free. There is an anterior attachment to the floor of the mouth by a thin fold of tissue called the **frenulum** (fren′ū-lŭm, *frenum,* bridle) (see figure 21.4). The muscles associated with the tongue are described in chapter 9.

The tongue moves food in the mouth and, in cooperation with the lips and cheeks, holds the food in place during mastication. It also plays a major role in the process of swallowing. The tongue is a major sensory organ for taste (see chapter 13), as well as being one of the major organs of speech. ▼

   11. *List the functions of the tongue.*

Lipid-Soluble Drugs

Drugs that are lipid-soluble and can diffuse through the plasma membranes of the oral cavity can be quickly absorbed into the circulation. An example is nitroglycerin, which is a vasodilator used to treat angina pectoris. The drug is placed under the tongue, where, in less than 1 minute, it dissolves and passes through the very thin oral mucosa into the lingual veins.

Teeth

Adults have 32 **teeth,** which are distributed in two **dental arches:** the maxillary arch and the mandibular arch. The teeth in the right and left halves of each dental arch are roughly mirror images of each other. As a result, the teeth are divided into four quadrants: right upper, left upper, right lower, and left lower. The teeth in each quadrant include one central and one lateral **incisor,** one **canine,** first and second **premolars,** and first, second, and third **molars** (figure 21.5*a*). The third molars are called **wisdom teeth** because they usually appear in a person's late teens or early twenties, when the person is thought to be old enough to have acquired some wisdom.

Impacted Wisdom Teeth

In some people with small dental arches, the third molars may not have room to erupt into the oral cavity and remain embedded within the jaw. Embedded wisdom teeth are referred to as impacted. They may cause pain or irritation and their surgical removal is often necessary.

The teeth of the adult mouth are **permanent,** or **secondary, teeth.** Most of them are replacements for **deciduous** (dē-sid′ū-ŭs, those that fall out), or **primary, teeth,** which are lost during childhood (figure 21.5*b*). The deciduous teeth erupt (the crowns appear within the oral cavity) between about 6 months and 24 months of age. The permanent teeth begin replacing the deciduous teeth at about 5 years and the process is complete by about 11 years.

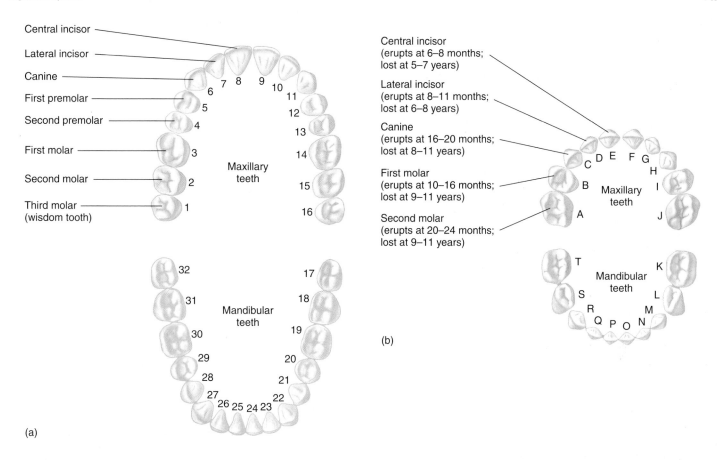

Central incisor
Lateral incisor
Canine
First premolar
Second premolar
First molar
Second molar
Third molar (wisdom tooth)

Maxillary teeth

Mandibular teeth

(a)

Central incisor (erupts at 6–8 months; lost at 5–7 years)
Lateral incisor (erupts at 8–11 months; lost at 6–8 years)
Canine (erupts at 16–20 months; lost at 8–11 years)
First molar (erupts at 10–16 months; lost at 9–11 years)
Second molar (erupts at 20–24 months; lost at 9–11 years)

Maxillary teeth

Mandibular teeth

(b)

Figure 21.5   Teeth

(*a*) Permanent teeth. (*b*) Deciduous teeth. Dental professionals have developed a "universal" numbering and lettering system for convenience in identifying individual teeth.

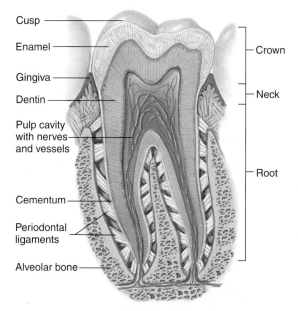

Cusp
Enamel
Gingiva
Dentin
Pulp cavity with nerves and vessels
Cementum
Periodontal ligaments
Alveolar bone

Crown
Neck
Root

Figure 21.6   Molar Tooth in Place in the Alveolar Bone

The tooth consists of a crown and root. The root is covered with cementum, and the tooth is held in the socket by periodontal ligaments. Nerves and vessels enter and exit the tooth through the tip of the root.

Each tooth consists of a **crown** with one or more **cusps** (points), a **neck,** and a **root** (figure 21.6). The center of the tooth is a **pulp cavity,** which is filled with blood vessels, nerves, and connective tissue called **pulp.** The pulp cavity within the root is called the **root canal.** The nerves and blood vessels of the tooth enter and exit the pulp through holes at the tips of the roots. The pulp cavity is surrounded by a living, cellular, calcified tissue called **dentin.** The dentin of the tooth crown is covered by an extremely hard, nonliving, acellular substance called **enamel,** which protects the tooth against abrasion and acids produced by bacteria in the mouth. The surface of the dentin in the root is covered with a bonelike substance called **cementum,** which helps anchor the tooth in the jaw.

The teeth are set in **alveoli** (al-vē′ō-lī, sockets) along the alveolar processes of the mandible and maxilla. The **gingiva** (jin′ji-vă, gums) is dense fibrous connective tissue and stratified squamous epithelium covering the alveolar processes (see figure 21.4). **Periodontal** (per′ē-ō-don′tăl, around a tooth) **ligaments** secure the teeth in the alveoli.

Dental Diseases

The formation of **dental caries** (kār′ēz), or tooth decay, is the result of the breakdown of enamel by acids produced by bacteria on the tooth surface. Enamel is nonliving and cannot repair itself. Consequently, a dental filling is necessary to prevent further damage. If the decay reaches the pulp cavity, with its rich supply of nerves, toothache pain may result. In some cases in which decay has reached the pulp cavity, it is necessary to perform a dental procedure called a root canal, which consists of removing the pulp from the tooth. **Periodontal disease** is inflammation and degeneration of the periodontal ligaments, gingiva, and alveolar bone. This disease is the most common cause of tooth loss in adults.

The teeth play an important role in mastication and a role in speech. ▼

12. *Name the kinds of teeth. What are permanent and deciduous teeth?*
13. *Name the three parts of a tooth. What are dentin, enamel, cementum, and pulp?*

Mastication

Food taken into the mouth is **masticated,** or **chewed,** by the teeth. The incisors and the canines primarily cut and tear food, whereas the premolars and molars primarily crush and grind it. Mastication breaks large food particles into smaller ones, which have a much larger total surface area. Mastication increases the efficiency of digestion because digestive enzymes digest food molecules only at the surface of the food particles. The muscles of mastication (see chapter 9) move the mandible. ▼

14. *Define mastication. What does it accomplish?*

Salivary Glands

Salivary (sal′i-vār-ē) **glands** produce **saliva** (să-lī′vă), which is a mixture of **serous** (watery) and **mucous** fluids. There are many small, coiled tubular salivary glands located deep to the epithelium of the tongue, palate, cheeks, and lips. There are three pairs of large, compound alveolar glands: the parotid, submandibular, and sublingual glands (figure 21.7). The largest of the salivary glands, the **parotid** (pă-rot′id, beside the ear) **glands,** are serous glands located just anterior to each ear. Parotid ducts enter the oral cavity adjacent to the second upper molars. Minerals secreted in the saliva of the parotid salivary glands tend to accumulate on the surface of the second upper molar.

Mumps

Inflammation of the parotid gland is called **parotiditis** (pă-rot-i-dī′tis). **Mumps,** which is caused by a virus, is the most common type of parotiditis.

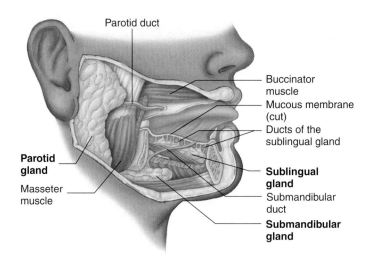

Figure 21.7 Salivary Glands

The large salivary glands are the parotid glands, the submandibular glands, and the sublingual glands.

The **submandibular** (sŭb-man-dib′ū-lăr, below the mandible) **glands** produce more serous than mucous secretions. Each gland can be felt as a soft lump along the inferior border of the mandible. The submandibular ducts open into the oral cavity on each side of the frenulum of the tongue (see figure 21.4). In certain people, if the mouth is opened and the tip of the tongue is elevated, saliva can squirt out of the mouth from the ducts of these glands.

The **sublingual** (sŭb-ling′gwăl, below the tongue) **glands,** the smallest of the three paired salivary glands, produce primarily mucous secretions. They lie immediately below the mucous membrane in the floor of the oral cavity. Each sublingual gland has 10–12 small ducts opening onto the floor of the oral cavity.

Saliva is secreted at the rate of approximately 1 liter (L) per day. The serous part of saliva, produced mainly by the parotid and submandibular glands, contains a digestive enzyme called **salivary amylase** (am′il-ās, starch-splitting enzyme) (table 21.1), which breaks the covalent bonds between glucose molecules in starch and other polysaccharides to produce the disaccharides maltose and isomaltose. Maltose and isomaltose have a sweet taste; thus, the digestion of polysaccharides by salivary amylase enhances the sweet taste of food. Food spends very little time in the mouth. Consequently, only about 5% of the total carbohydrates humans absorb are digested in the mouth.

The mucous secretions of the submandibular and sublingual glands contain a large amount of **mucin** (mū′sin), a proteoglycan that gives a lubricating quality to the secretions of the salivary glands.

Salivary gland secretion is regulated primarily by the autonomic nervous system, with parasympathetic stimulation being the most important. Salivary secretions increase in response to a variety of stimuli, such as tactile stimulation in the oral cavity and certain tastes, especially sour. Higher brain centers can stimulate parasympathetic activity and thus increase the activity of the salivary glands in response to the thought of food, to odors, or to the sensation of hunger. Sympathetic stimulation increases the mucus content of

Table 21.1	Functions of Digestive Tract and Accessory Organs Secretions	
Fluid, Enzyme, or Hormone	**Source**	**Function**
Mouth		
Saliva	Salivary glands	Moistens and lubricates food
Salivary amylase	Salivary glands	Digests starch (conversion to maltose and isomaltose)
Lingual lipase	Salivary glands	Begins lipid digestion ($< 10\%$)
Lysozyme	Salivary glands	Weak antibacterial action
Esophagus		
Mucus	Mucous glands	Lubricates esophagus; protects esophageal lining from abrasion and allows food to move easily through the esophagus
Stomach		
Hydrochloric acid	Gastric glands (parietal cells)	Converts pepsinogen into pepsin, kills bacteria
Pepsin	Gastric glands (chief cells)	Activated pepsinogen, digests protein into smaller peptide chains
Gastric lipase	Gastric glands (chief cells)	Digests a minor amount of lipid
Mucus	Mucous cells	Protects stomach lining from stomach acids and digestive enzymes
Intrinsic factor	Gastric glands (parietal cells)	Binds to vitamin B_{12}, aiding in its absorption
Gastrin	Gastric glands (endocrine cells)	Increases stomach secretions and motility
Small Intestine and Associated Glands		
Bile salts	Liver	Emulsify fats, form micelles, contain waste products (bilirubin, cholesterol)
Bicarbonate ions	Pancreas. liver	Neutralize stomach acid
Pancreatic amylase	Pancreas	Digests starch
Pancreatic lipase	Pancreas	Digests lipid
Trypsin	Pancreas	Activated trypsinogen, digests proteins
Chymotrypsin	Pancreas	Activated chymotrypsinogen, digests proteins
Carboxypeptidase	Pancreas	Activated procarboxypeptidase, digests proteins
Nucleases	Pancreas	Digest nucleic acid
Sucrase	Small intestine	Digests sucrose
Lactase	Small intestine	Digests lactose
Maltase	Small intestine	Digests maltose
Isomaltase	Small intestine	Digests isomaltose
Lipase	Small intestine	Digests lipid
Peptidases	Small intestine	Digest polypeptide
Mucus	Duodenal glands and goblet cells	Protects duodenum from stomach acid and digestive enzymes
Secretin	Small intestine	Inhibits gastric secretions, stimulates secretion of aqueous component (bicarbonate ions) of pancreatic juice, increases bile secretion (bicarbonate ions), decreases gastric motility
Cholecystokinin	Small intestine	Inhibits gastric secretion, stimulates secretion of enzymatic component of pancreatic juice, decreases gastric motility, stimulates gallbladder contraction and sphincter relaxation

saliva. When a person becomes frightened and the sympathetic division of the autonomic nervous system is stimulated, the person may have a dry mouth with thick mucus. ▼

15. *Name and give the location of the three largest salivary glands.*
16. *What is the function of salivary amylase and mucin in saliva?*

Pharynx

The **pharynx** was described in detail in chapter 20. It consists of three parts: the nasopharynx, oropharynx, and laryngopharynx. Normally, only the oropharynx and laryngopharynx transmit food. The **oropharynx** communicates with the nasopharynx superiorly, the larynx and **laryngopharynx** inferiorly, and the mouth anteriorly. The laryngopharynx extends from the oropharynx to the esophagus and is posterior to the larynx. The posterior walls of the oropharynx and laryngopharynx consist of three muscles: the superior, middle, and inferior **pharyngeal constrictors,** which are arranged like three stacked flowerpots, one inside the other. The oropharynx and the laryngopharynx are lined with moist stratified squamous epithelium, and the nasopharynx is lined with ciliated pseudostratified columnar epithelium. ▼

> 17. *Name the three parts of the pharynx. What are the pharyngeal constrictors?*

P R E D I C T

Explain the functional significance of the differences in epithelial types among the three pharyngeal regions.

Esophagus

The **esophagus** (ē-sof′ă-gŭs, gullet) is a muscular tube, lined with moist stratified squamous epithelium, extending from the pharynx to the stomach. It is about 25 centimeters (cm) long and lies anterior to the vertebrae and posterior to the trachea within the mediastinum. It passes through the diaphragm and ends at the stomach. The esophagus transports food from the pharynx to the stomach.

The esophagus has thick walls consisting of the four tunics: the mucosa, submucosa, muscularis, and adventitia. The muscularis is different from other parts of the digestive tract because the superior part of the esophagus consists of skeletal muscle. An **upper esophageal sphincter** at the superior end of the esophagus and a **lower esophageal sphincter** at the lower end of the esophagus regulate the movement of materials into and out of the esophagus. Numerous mucous glands produce a thick, lubricating mucus that coats the inner surface of the esophagus.

⚕ Hiatal Hernia

A **hiatus** (hī-ā′tŭs, to yawn) is an opening. The esophageal hiatus is the opening in the diaphragm through which the esophagus passes to join the stomach. A **hernia** (her′nē-ă) is a protrusion of a part or structure through the tissues normally containing it. A **hiatal hernia** occurs when the esophageal hiatus widens and part of the stomach extends through the opening into the thorax. The hernia can decrease some of the pressure in the lower esophageal sphincter, allowing gastroesophageal reflux (movement of stomach contents back into the esophagus) and subsequent esophagitis (inflammation of the esophagus) to occur. Hiatal herniation can also compress blood vessels in the stomach mucosa when the stomach extends through the hernia, which can lead to gastritis (inflammation of the stomach) or ulcer formation. Esophagitis, gastritis, and ulcers are very painful. ▼

> 18. *Where is the esophagus located and what is its function? Describe the muscles of the esophageal wall and the esophageal sphincters.*

Swallowing

Swallowing, or **deglutition** (dē-gloo-tish′ŭn), can be divided into three separate phases: the voluntary phase, the pharyngeal phase, and the esophageal phase. During the **voluntary phase,** a **bolus,** or mass of food, is formed in the mouth. The bolus is pushed by the tongue against the hard palate, forcing the bolus toward the posterior part of the mouth and into the oropharynx (figure 21.8, step 1).

The **pharyngeal phase** of swallowing is a reflex that is initiated when a bolus of food stimulates tactile receptors in the oropharynx (figure 21.8, steps 2–4). This phase of swallowing begins with the elevation of the soft palate, which closes the passage between the nasopharynx and oropharynx. The pharynx elevates to receive the bolus of food from the mouth. The three **pharyngeal constrictor muscles** then contract in succession, forcing the food through the pharynx. At the same time, the upper esophageal sphincter relaxes, and food is pushed into the esophagus.

During the pharyngeal phase, the vestibular folds and vocal cords close, the **epiglottis** (ep-i-glot′is, on the glottis) is tipped posteriorly so that it covers the opening into the larynx, and the larynx is elevated. These movements of the larynx prevent food from passing through the opening into the larynx.

The **esophageal phase** (figure 21.8, step 5) of swallowing, which takes about 5–8 seconds, is responsible for moving food from the pharynx to the stomach. Muscular contractions in the wall of the esophagus occur in **peristaltic** (per-i-stal′tik, *peri,* around + *stalsis,* constriction) **waves** (figure 21.9). A wave of relaxation of the circular esophageal muscles precedes the bolus of food down the esophagus, and a wave of strong contraction of the circular muscles follows and propels the bolus through the esophagus. The peristaltic contractions associated with swallowing cause relaxation of the lower esophageal sphincter in the esophagus as the peristaltic waves approach the stomach.

The presence of food in the esophagus stimulates the enteric plexus, which controls the peristaltic waves through local reflexes. The presence of food in the esophagus also stimulates tactile receptors, which send afferent impulses to the medulla oblongata through the vagus nerves. Motor impulses, in turn, pass along the vagal efferent fibers to the striated and smooth muscles within the esophagus, thereby stimulating their contractions and reinforcing the peristaltic contractions.

⚕ Swallowing and Gravity

Gravity assists the movement of material through the esophagus, especially when liquids are swallowed. The peristaltic contractions that move material through the esophagus are sufficiently forceful, however, to allow a person to swallow even while doing a headstand or floating in the zero-gravity environment of space. ▼

> 19. *What are the three phases of deglutition? List sequentially the processes involved in the last two phases, and describe how they are regulated.*

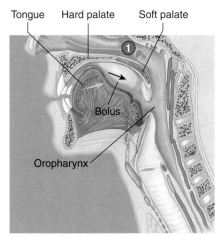

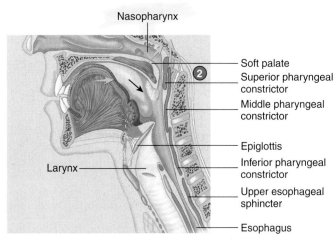

1. During the voluntary phase, a bolus of food (*yellow*) is pushed by the tongue against the hard and soft palates and posteriorly toward the oropharynx (*blue arrow* indicates tongue movement; *black arrow* indicates movement of the bolus). *Tan:* bone, *purple:* cartilage, *red:* muscle.

2. During the pharyngeal phase, the soft palate is elevated, closing off the nasopharynx. The pharynx and larynx are elevated (*blue arrows* indicate muscle movement).

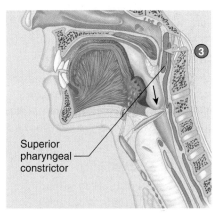

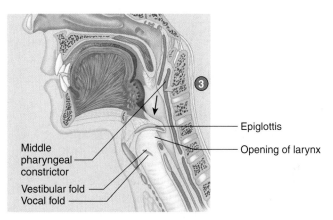

3. Successive constriction of the pharyngeal constrictors from superior to inferior (*blue arrows*) forces the bolus through the pharynx and into the esophagus. As this occurs, the vestibular and vocal folds meet medially to close the passage of the larynx. The epiglottis is bent down over the opening of the larynx largely by the force of the bolus pressing against it.

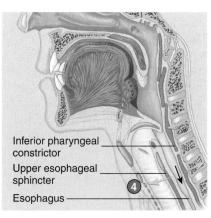

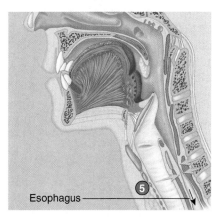

4. As the inferior pharyngeal constrictor contracts, the upper esophageal sphincter relaxes (outwardly directed *blue arrows*), allowing the bolus to enter the esophagus.

5. During the esophageal phase, the bolus is moved by peristaltic contractions of the esophagus toward the stomach (inwardly directed *blue arrows*).

Process Figure 21.8   Three Phases of Swallowing (Deglutition)

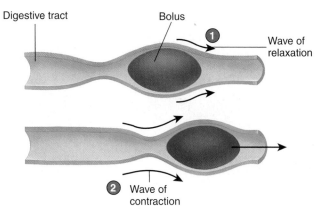

Digestive tract   Bolus

1. A wave of circular smooth muscle relaxation moves ahead of the bolus of food, allowing the digestive tract to expand.

2. A wave of contraction of the circular smooth muscles behind the bolus of food propels it through the digestive tract.

Wave of relaxation

Wave of contraction

Process Figure 21.9 Peristalsis

20. *Explain how peristaltic waves move a bolus through the digestive tract.*

P R E D I C T **2**
Sometimes food becomes "stuck" in the esophagus. How does sipping a liquid help the food become "unstuck?"

Stomach

The **stomach** is an enlarged segment of the digestive tract in the left superior part of the abdomen (see figure 21.1).

Anatomy and Histology of the Stomach

The opening from the esophagus into the stomach is the **gastroesophageal,** or **cardiac** (located near the heart), **opening.** The region of the stomach around the cardiac opening is the **cardiac part** (figure 21.10*a*). The lower esophageal sphincter, also called the **cardiac sphincter,** surrounds the cardiac opening. The most superior part of the stomach is the **fundus** (fŭn′dŭs, the bottom of a round-bottomed leather bottle). The largest part of the stomach is the **body,** which turns to the right, forming a **greater curvature** on the left, and a **lesser curvature** on the right. The body narrows to form the funnel-shaped **pyloric** (pī-lōr′ik, gatekeeper) **part** of the stomach, which joins the small intestine. The opening from the stomach into the small intestine is the **pyloric orifice,** which is surrounded by a relatively thick ring of smooth muscle called the **pyloric sphincter.** The pyloric sphincter helps regulate the movement of gastric contents into the small intestine.

The muscular layer of the stomach is different from other regions of the digestive tract in that it consists of three layers: an outer longitudinal layer, a middle circular layer, and an inner oblique layer. These muscular layers produce a churning action in the stomach, important in the digestive process. The submucosa and mucosa of the stomach are thrown into large folds called **rugae** (roo′gē, wrinkles) (see figure 21.10*a*) when the stomach is empty. These folds allow the mucosa and submucosa to stretch, and the folds disappear as the stomach is filled.

The stomach is lined with simple columnar epithelium. The epithelium forms numerous, tubelike **gastric pits,** which are the openings for the **gastric glands** (figure 21.10*b* and *c*). The epithelial cells of the stomach are of five types. **Surface mucous cells** are on the inner surface of the stomach and line the gastric pits. Those cells produce mucus, which coats and protects the stomach lining. The remaining cell types are in the gastric glands. They are **mucous neck cells,** which produce mucus; **parietal cells,** which produce hydrochloric acid and intrinsic factor; **chief cells,** which produce **pepsinogen** (pep-sin′ō-jen); and **endocrine cells,** which produce regulatory hormones. There are several types of endocrine cells. For example, **G cells** secrete gastrin and **enterochromaffin-like cells** produce histamine. ▼

21. *Describe the parts of the stomach.*
22. *How is the muscular layer of the stomach different from the rest of the digestive tract? What are rugae?*
23. *What are gastric pits and gastric glands? Name the types of cells in the stomach and the secretions they produce.*

Secretions of the Stomach

The secretions of the stomach are called **gastric juice.** As food enters the stomach, it is mixed with gastric juice to become a semifluid mixture called **chyme** (kīm, juice). Although some digestion of chyme and a small amount of absorption occur in the stomach, the stomach is primarily a storage and mixing chamber for ingested food.

Secretions from the gastric glands include mucus, hydrochloric acid, pepsinogen, intrinsic factor, and gastrin (see table 21.1). A thick layer of **mucus** lubricates and protects the epithelial cells of the stomach wall from the damaging effect of the acidic chyme and pepsin. Irritation of the stomach mucosa stimulates the secretion of a greater volume of mucus. **Hydrochloric acid** produces a pH of about 2.0 in the stomach. **Pepsinogen** is a precursor of the enzyme pepsin. Hydrochloric acid converts pepsinogen to the active enzyme **pepsin** (pep′sin, *pepsis,* digestion), which digests proteins by breaking them down into smaller peptides. A **peptide** is two or more amino acids united by a peptide bond (see chapter 2). Pepsin exhibits optimum enzymatic activity at a pH of about 2.0. The low pH also kills microorganisms. **Intrinsic** (in-trin′sik) **factor** binds with vitamin B_{12} and makes it more readily absorbed in the small intestine. Vitamin B_{12} is important in deoxyribonucleic acid (DNA) synthesis and is important to red blood cell production. **Gastrin** (gas′trin, *gaster,* belly or stomach) and **histamine** help regulate stomach secretions.

P R E D I C T **3**
Peptides in the stomach stimulate parietal cell and chief cell secretion. Explain how this is a negative-feedback mechanism.

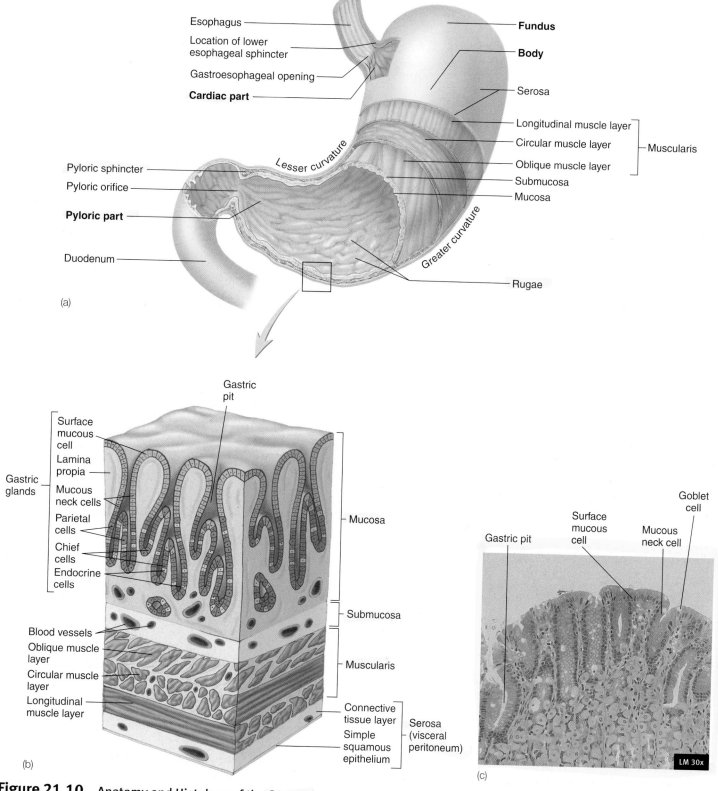

Esophagus
Location of lower esophageal sphincter
Gastroesophageal opening
Cardiac part
Fundus
Body
Serosa
Longitudinal muscle layer
Circular muscle layer
Oblique muscle layer
Muscularis
Submucosa
Mucosa
Pyloric sphincter
Pyloric orifice
Pyloric part
Lesser curvature
Greater curvature
Duodenum
Rugae

(a)

Gastric pit

Surface mucous cell
Lamina propia
Gastric glands
Mucous neck cells
Parietal cells
Chief cells
Endocrine cells
Mucosa
Submucosa
Muscularis
Blood vessels
Oblique muscle layer
Circular muscle layer
Longitudinal muscle layer
Connective tissue layer
Simple squamous epithelium
Serosa (visceral peritoneum)

(b)

Gastric pit
Surface mucous cell
Mucous neck cell
Goblet cell
LM 30x

(c)

Figure 21.10 Anatomy and Histology of the Stomach

(*a*) Cutaway section reveals muscular layers and internal anatomy. (*b*) Section of the stomach wall that illustrates its histology, including several gastric pits and glands. (*c*) Photomicrograph of gastric glands.

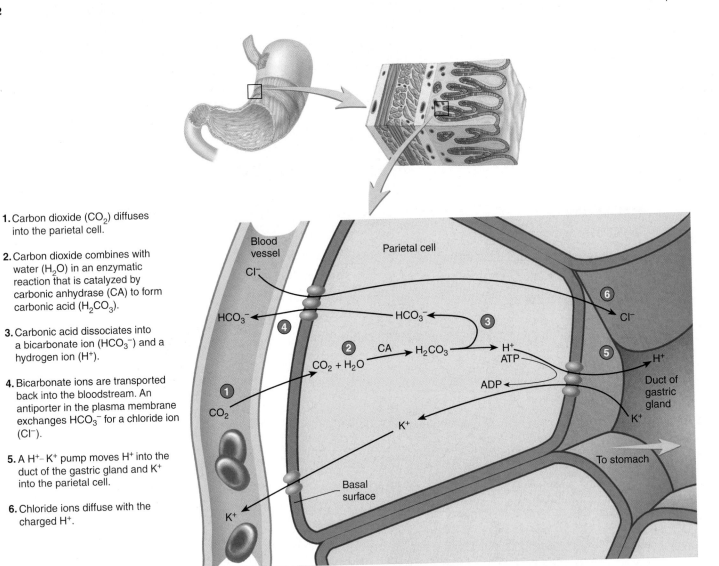

1. Carbon dioxide (CO_2) diffuses into the parietal cell.

2. Carbon dioxide combines with water (H_2O) in an enzymatic reaction that is catalyzed by carbonic anhydrase (CA) to form carbonic acid (H_2CO_3).

3. Carbonic acid dissociates into a bicarbonate ion (HCO_3^-) and a hydrogen ion (H^+).

4. Bicarbonate ions are transported back into the bloodstream. An antiporter in the plasma membrane exchanges HCO_3^- for a chloride ion (Cl^-).

5. A H^+–K^+ pump moves H^+ into the duct of the gastric gland and K^+ into the parietal cell.

6. Chloride ions diffuse with the charged H^+.

Process Figure 21.11   Hydrochloric Acid Production by Parietal Cells in the Gastric Glands of the Stomach

 Gastroesophageal Reflux Disease

Gastroesophageal reflux disease (GERD), or heartburn, is a painful or burning sensation in the chest associated with the reflux of acidic chyme into the esophagus. The pain is usually short-lived but may be confused with the pain of an ulcer or a heart attack. Overeating, eating fatty foods, lying down immediately after a meal, consuming too much alcohol or caffeine, smoking, and wearing extremely tight clothing can all cause heartburn. A hiatal hernia can also cause heartburn, especially in older people. Drugs that neutralize gastric acid or reduce gastric acid production can effectively treat GERD (see "Treatment of Excess Acid Secretion," p. 655).

Parietal cells produce hydrochloric acid (figure 21.11). Hydrogen ions are derived from carbon dioxide and water, which enter the parietal cell from its basal surface, which is the side opposite the lumen of the gastric pit. Once inside the cell, carbonic anhydrase catalyzes the reaction between carbon dioxide and water to form carbonic acid. Some of the carbonic acid molecules then dissociate to form H^+ and HCO_3^-. The H^+ are actively transported across the mucosal surface of the parietal cell into the lumen of the stomach by a H^+-K^+ exchange pump, often called a **proton pump.** Although H^+ are actively transported against a steep concentration gradient, Cl^- diffuse with the H^+ from the cell through the plasma membrane. Diffusion of Cl^- with the positively charged H^+ reduces the amount of energy needed to transport the H^+ against both a concentration gradient and an electrical gradient. Bicarbonate ions diffuse down their concentration gradient from the parietal cell into the extracellular fluid. As they diffuse, HCO_3^- are exchanged for Cl^- through an antiporter, and the Cl^- move into the cell. ▼

24. *What is gastric juice and chyme?*
25. *What is the function of mucus, hydrochloric acid, pepsinogen, intrinsic factor, gastrin, and histamine?*
26. *Describe the production of hydrochloric acid. What is a proton pump?*

P R E D I C T **4**

Explain why a slight increase in blood pH may occur following a heavy meal. The elevated pH of blood, especially in the veins that carry blood away from the stomach, is called the postenteric alkaline tide.

Regulation of Stomach Secretion

Approximately 2–3 L of gastric secretions (gastric juice) are produced each day. Nervous and hormonal mechanisms regulate gastric secretions. The neural mechanisms involve central nervous system (CNS) reflexes integrated within the medulla oblongata. Higher brain centers can influence these reflexes. Local reflexes are integrated within the enteric plexus in the wall of the digestive tract and do not involve the CNS. Hormones produced by the stomach and intestine help regulate stomach secretions.

The regulation of stomach secretion is divided into three phases: cephalic, gastric, and gastrointestinal. The cephalic phase can be viewed as the "get started" phase, when stomach secretions are increased in anticipation of incoming food. This is followed by the gastric "go for it" phase, when most of the stimulation of secretion occurs. Finally, the gastrointestinal phase is the "slow down" phase, during which stomach secretion decreases.

1. **Cephalic phase.** The cephalic phase prepares the stomach to receive food. The sensations of the taste and smell of food, the stimulation of tactile receptors during the process of chewing and swallowing, and pleasant thoughts of food stimulate the centers within the medulla oblongata that influence gastric secretions (figure 21.12a). Action potentials are sent from the medulla along parasympathetic neurons within the vagus (X) nerves to the stomach. Within the stomach wall, the preganglionic neurons stimulate the postganglionic neurons in the enteric plexus. The postganglionic neurons stimulate secretory activity in the cells of the stomach mucosa, causing the release of mucus, hydrochloric acid, pepsinogen, intrinsic factor, histamine, and gastrin. The gastrin enters the circulation and is carried back to the stomach, where it stimulates additional secretory activity.

   Acetylcholine, released by postganglionic cells, stimulates the secretion of hydrochloric acid from parietal cells, gastrin from G cells, and histamine from enterochromaffin-like cells. The gastrin stimulates parietal cells to release additional hydrochloric acid. In addition, gastrin stimulates enterochromaffin-like cells to release histamine, which stimulates parietal cells to secrete hydrochloric acid. Acetylcholine, histamine, and gastrin working together cause a greater secretion of hydrochloric acid than any of them does separately. Of the three, histamine has the greatest stimulatory effect.

2. **Gastric phase.** The greatest volume of gastric secretions is produced during the gastric phase of gastric regulation. The presence of food in the stomach initiates the gastric phase (figure 21.12b). The primary stimuli are distention of the stomach and the presence of amino acids and peptides in the stomach.

   Distention of the stomach wall results in the stimulation of mechanoreceptors. Action potentials generated by these receptors initiate reflexes that involve both the CNS and the ENS. These reflexes result in acetylcholine release and the cascade of events in the cephalic phase. Peptides, produced by the action of pepsin on proteins, stimulate the secretion of gastrin, which in turn stimulates additional hydrochloric acid secretion. Moderate amounts of alcohol or caffeine in the stomach also stimulate gastrin secretion.

   When the pH of the stomach contents falls below 2.0, increased gastric secretion produced by distention of the stomach is blocked. This negative-feedback mechanism limits the secretion of gastric juice.

3. **Gastrointestinal phase.** The entrance of chyme into the duodenum of the small intestine stimulates neural and hormonal mechanisms that inhibit stomach secretions (figure 21.12c). This reduces the acidity of chyme, making it easier for pancreatic and liver secretions to neutralize the chyme, which is required for the digestion of food by pancreatic enzymes and for the prevention of peptic ulcer formation.

   When the pH of the chyme entering the duodenum drops to 2.0 or below, the inhibitory influence of the intestinal phase is greatest. Acidic chyme in the duodenum inhibits CNS stimulation and initiates local reflexes that inhibit gastric secretion. Acidic chyme also stimulates the duodenum to release the hormone **secretin** (se-krē′tin), which enters the blood and is carried to the stomach, where it inhibits gastric secretion. Fatty acids and certain other lipids in the duodenum initiate the release of the hormone **cholecystokinin** (kō′lē-sis-tō-kī′nin), **(CCK)** which also inhibits gastric secretion. ▼

27. *Name the three phases of stomach secretion. In general terms, what does each phase accomplish?*
28. *What stimuli initiate each phase of stomach secretion?*
29. *Describe three ways in which parietal cells are stimulated to secrete hydrochloric acid.*
30. *Describe the effects of ANS and local reflexes on gastric secretion during each phase of stomach secretion.*
31. *What effect do gastrin, secretin, and cholecystokinin have on gastric secretions?*

Cephalic Phase

1. The taste, smell, or thought of food or tactile sensations of food in the mouth stimulate the medulla oblongata (*green arrows*).

2. Parasympathetic action potentials are carried by the vagus nerves to the stomach (*pink arrow*), where enteric plexus neurons are activated.

3. Postganglionic neurons stimulate secretion by parietal and chief cells and stimulate gastrin and histamine secretion by endocrine cells.

4. Gastrin is carried through the circulation back to the stomach (*purple arrow*), where, along with histamine, it stimulates secretion.

Gastric Phase

1. Distention of the stomach stimulates mechanoreceptors (stretch receptors) and activates a parasympathetic reflex. Action potentials generated by the mechanoreceptors are carried by the vagus nerves to the medulla oblongata (*green arrow*).

2. The medulla oblongata increases action potentials in the vagus nerves that stimulate secretions by parietal and chief cells and stimulate gastrin and histamine secretion by endocrine cells (*pink arrow*).

3. Distention of the stomach also activates local reflexes that increase stomach secretions (*orange arrow*).

4. Gastrin is carried through the circulation back to the stomach (*purple arrow*), where, along with histamine, it stimulates secretion.

Gastrointestinal Phase

1. Chyme in the duodenum with a pH less than 2.0 or containing fat digestion products (lipids) inhibits gastric secretions by three mechanisms (2–4).

2. Chemoreceptors in the duodenum are stimulated by H^+ (low pH) or lipids. Action potentials generated by the chemoreceptors are carried by the vagus nerves to the medulla oblongata (*green arrow*), where they inhibit parasympathetic action potentials (*pink arrow*), thereby decreasing gastric secretions.

3. Local reflexes activated by H^+ or lipids also inhibit gastric secretion (*orange arrows*).

4. Secretin and cholecystokinin produced by the duodenum (*brown arrows*) decrease gastric secretions in the stomach.

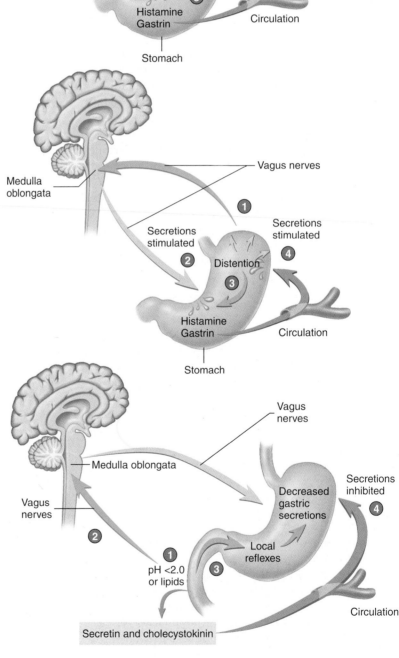

Process Figure 21.12   Phases of Stomach Secretion

Treatment of Excess Acid Secretion

Antacids are bases, such as $CaCO_3$, $Al(OH)_3$, and $Mg(OH)_2$, that neutralize gastric acid when ingested. They are fast acting but have relatively short duration of effect. Antacids are effective for up to 30 minutes on an empty stomach and neutralize acid for 2 to 3 hours when taken with food. Long-term self-treatment with antacids should be avoided because the cause of excess acid production should be determined. Also, there are more effective treatments.

H_2-receptor antagonists are drugs that block the histamine receptor on parietal cells. Recall that histamine, gastrin, and acetylcholine stimulate parietal cells to secrete acid and that histamine has the greatest stimulatory effect. H_2 receptors are the type of histamine receptor on parietal cells. Cimetidine (Tagamet), ranitidine (Zantac), and famotidine (Pepcid) are H_2-receptor antagonists that bind reversibly to H_2 receptors. They suppress 24-hour gastric acid secretion approximately 70%. The H_2 receptors are different from the H_1 receptors involved in allergic reactions. Antihistamines that block allergic reactions do not affect histamine-mediated gastric acid secretion and vice versa.

Proton pump inhibitors bind irreversibly with the H^+-K^+ exchange pump in parietal cells. The pump is inactivated and acid secretion does not resume until a new pump molecule is manufactured and inserted into the plasma membrane. Proton pump inhibitors effectively reduce acid secretion for 24 to 48 hours.

Movements of the Stomach

Two types of stomach movement occur: mixing waves and peristaltic waves (figure 21.13). Both types of movements result from smooth muscle contractions in the stomach wall. The contractions occur about every 20 seconds and proceed from the body of the stomach toward the pyloric sphincter. Relatively weak contractions result in **mixing waves,** which thoroughly mix ingested food with stomach secretions to form chyme. The more fluid part of the chyme is pushed toward the pyloric sphincter, whereas the more solid center moves back toward the body of the stomach. Stronger contractions result in **peristaltic waves,** which force the chyme toward and through the pyloric sphincter. The pyloric sphincter usually remains closed because of mild tonic contraction. Each peristaltic contraction is sufficiently strong to pump a few milliliters of chyme through the pyloric opening and into the duodenum. Roughly 20% of the stomach contractions are peristaltic waves, and 80% are mixing waves.

If the stomach empties too fast, the efficiency of digestion and absorption is reduced, and acidic gastric contents dumped into the duodenum may damage its lining. If the rate of emptying is too slow, the highly acidic contents of the stomach may damage the stomach wall and reduce the rate at which nutrients are digested and absorbed. Stomach emptying is regulated to prevent these two extremes. The neural mechanisms that stimulate stomach secretions also are involved with increasing stomach motility. The major stimulus is distention of the stomach wall. Increased stomach motility increases stomach emptying. Conversely, the hormonal and neural mechanisms associated with the duodenum that decrease gastric secretions also inhibit gastric motility and increase constriction of the pyloric sphincter.

Pregame Meal

A meal of polysaccharide carbohydrates (starch and glycogen) is considered the best meal before engaging in a sporting activity. Polysaccharides help provide a steady source of glucose and have the fastest clearance time, typically 1 hour, from the stomach. For comparison, a meal containing both carbohydrates and proteins takes 3 hours to clear from the stomach, and a meal heavy with fats and proteins takes up to 6 hours. A major reason for the fast clearance of carbohydrates is that they do not increase cholecystokinin release, which is a major inhibitor of stomach emptying.

Hunger Contractions

Hunger contractions are peristaltic contractions that approach tetanic contractions for periods of about 2–3 minutes. Low blood glucose levels cause the contractions to increase and become sufficiently strong to create uncomfortable sensations called **hunger pangs.** Hunger pangs usually begin 12–24 hours after a meal or in less time for some people. If nothing is ingested, they reach their maximum intensity within 3–4 days and then become progressively weaker.

Vomiting

Vomiting is the ejection of the stomach's contents through the esophagus and out the mouth. Vomiting can result from excessive irritation or overdistention of the GI tract. **Reversed peristalsis,** which is peristaltic contractions moving GI tract contents backwards, precedes vomiting. Intestinal contents accumulating in the duodenum and stomach stimulate the vomiting center in the medulla oblongata. As a result, (1) contractions of the stomach begin to push the gastric contents into the esophagus as the lower esophageal sphincter relaxes; (2) a deep breath is taken and the vestibular and vocal folds close the opening of the larynx; (3) the hyoid bone and larynx are elevated, opening the upper esophageal sphincter; (4) the soft palate elevates, closing the connection between the oropharynx and nasopharynx; (5) the diaphragm and abdominal muscles are forcefully contracted, strongly compressing the stomach and increasing the intragastric pressure; (6) the lower esophageal sphincter relaxes completely; and (7) the gastric contents are forced out of the stomach, through the esophagus and oral cavity, to the outside. ▼

32. *What are two kinds of stomach movements? How are stomach movements regulated by hormones and nervous control?*
33. *What are hunger contractions?*
34. *Describe the events of vomiting.*

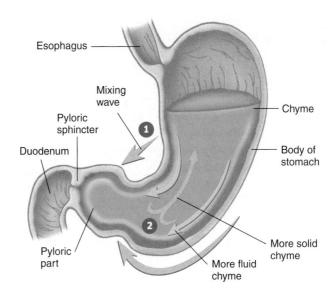

1. A mixing wave initiated in the body of the stomach progresses toward the pyloric sphincter (*pink arrows directed inward*).

2. The more fluid part of the chyme is pushed toward the pyloric sphincter (*blue arrows*), whereas the more solid center of the chyme squeezes past the peristaltic constriction back toward the body of the stomach (*orange arrow*).

3. Peristaltic waves (*purple arrows*) move in the same direction and in the same way as the mixing waves but are stronger.

4. Again, the more fluid part of the chyme is pushed toward the pyloric region (*blue arrows*), whereas the more solid center of the chyme squeezes past the peristaltic constriction back toward the body of the stomach (*orange arrow*).

5. Peristaltic contractions force a few milliliters of the mostly fluid chyme through the pyloric opening into the duodenum (*small red arrows*). Most of the chyme, including the more solid portion, is forced back toward the body of the stomach for further mixing (*yellow arrow*).

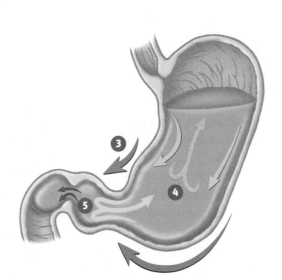

Process Figure 21.13 Movements in the Stomach

Small Intestine

The **small intestine** consists of three parts: the duodenum, the jejunum, and the ileum (figure 21.14). The entire small intestine is about 6 m long (range: 4.6–9 m). The duodenum is about 25 cm long (*duodenum* means 12, suggesting that it is 12 inches long), the jejunum is about 2.5 m long, and the ileum is about 3.5 m long. Two major accessory glands, the liver and the pancreas, are associated with the duodenum. The small intestine is where the greatest amount of digestion and absorption occur.

Anatomy and Histology of the Small Intestine

Duodenum

The **duodenum** (doo-ō-dē′nŭm, doo-od′ĕ-nŭm) nearly completes a 180-degree arc as it curves within the abdominal cavity (figure 21.15a), and the head of the pancreas lies within this arc.

The duodenum begins with a short, superior part, which is where it exits the pylorus of the stomach, and ends in a sharp bend, which is where it joins the jejunum. Ducts from the liver and pancreas open into the duodenum.

The surface of the duodenum has several modifications that increase its surface area about 600-fold to allow for more efficient digestion and absorption of food. The mucosa and submucosa form a series of folds called the **circular folds** (see figure 21.15b), which run perpendicular to the long axis of the digestive tract. Tiny, fingerlike projections of the mucosa form numerous **villi** (vil′ī, shaggy hair), which are 0.5–1.5 mm in length (see figure 21.15c). Each villus is covered by simple columnar epithelium and contains a blood capillary network and a lymphatic capillary called a **lacteal** (lak′tē-ăl) (see figure 21.15d). The blood capillary network and the lacteal are very important in transporting absorbed nutrients. Most of the cells that make up the surface of the villi have numerous cytoplasmic extensions (about 1 μm long)

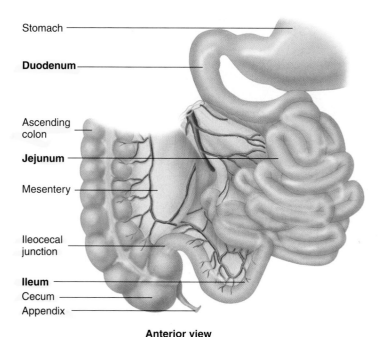

Stomach

Duodenum

Ascending colon

Jejunum

Mesentery

Ileocecal junction

Ileum

Cecum

Appendix

Anterior view

Figure 21.14   Small Intestine

called **microvilli,** which further increase the surface area (see figure 21.15*e*).

The mucosa of the duodenum is simple columnar epithelium with four major cell types: (1) **Absorptive cells,** which have microvilli, produce digestive enzymes and absorb digested food; (2) **goblet cells** produce a protective mucus; (3) **granular cells** (Paneth cells) may help protect the intestinal epithelium from bacteria; and (4) **endocrine cells** produce regulatory hormones. The epithelial cells are produced within tubular invaginations of the mucosa, called **intestinal glands,** at the base of the villi. The absorptive and goblet cells migrate from the intestinal glands to cover the surface of the villi and eventually are shed from its tip. The granular and endocrine cells remain in the bottom of the glands. The submucosa of the duodenum contains coiled, tubular mucous glands, called **duodenal glands,** which open into the base of the intestinal glands.

Jejunum and Ileum

The **jejunum** (jĕ-joo′nŭm) and **ileum** (il′ē-ŭm) are similar in structure to the duodenum, except that a gradual decrease occurs in the diameter of the small intestine, the thickness of the intestinal wall, the number of circular folds, and the number of villi as one progresses through the small intestine. The duodenum and jejunum are the major sites of nutrient absorption, although some absorption occurs in the ileum. Diffuse lymphatic tissue and lymphatic nodules, called **Peyer's patches,** are numerous in the mucosa and submucosa of the ileum. Lymphatic tissue in the digestive tract initiates immune responses against microorganisms that enter the mucosa from ingested food (see chapter 19).

The junction between the ileum and the large intestine is the **ileocecal junction** (see figure 21.14). It has a ring of smooth muscle, the **ileocecal sphincter,** and a one-way **ileocecal valve.** ▼

> 35. *Name and describe the three parts of the small intestine.*

> 36. *What are the circular folds, villi, and microvilli in the small intestine? What are their functions?*
> 37. *Name the four types of cells found in the duodenal mucosa, and state their functions.*
> 38. *What are intestinal glands and duodenal glands?*
> 39. *What is the function of lymphatic tissue in the digestive tract?*

Secretions of the Small Intestine

The mucosa of the small intestine produces secretions that primarily contain mucus, electrolytes, and water. Intestinal secretions lubricate and protect the intestinal wall from the acidic chyme and the action of digestive enzymes. They also keep the chyme in the small intestine in a liquid form to facilitate the digestive process. The intestinal mucosa produces most of the secretions that enter the small intestine, but the secretions of the liver and the pancreas also enter the small intestine and play essential roles in the process of digestion (see table 21.1).

The duodenal glands and goblet cells secrete large amounts of mucus. This mucus protects the wall of the intestine from the irritating effects of acidic chyme and from the digestive enzymes that enter the duodenum from the pancreas. The vagus nerve, secretin, and chemical or tactile irritation of the duodenal mucosa stimulate secretion from the duodenal glands. Goblet cells produce mucus in response to the tactile and chemical stimulation of the mucosa.

The microvilli of absorptive cells have enzymes bound to their free surfaces that play a significant role in the final steps of digestion. **Disaccharidases** (dī-sak′ă-rid-ās-ez) break down disaccharides, such as maltose and isomaltose, into monosaccharides. **Peptidases** (pep′ti-dās-ez) break the peptide bonds in proteins to form amino acids. Although these enzymes are not secreted into the intestine, the large surface area of the microvilli brings these enzymes into contact with the intestinal contents. Small molecules, which are breakdown products of digestion, are absorbed through the microvilli and enter the circulatory or lymphatic system. ▼

> 40. *What is the main secretion of duodenal glands and from goblet cells?*
> 41. *List the enzymes of the small intestine wall and give their functions.*

Movement of the Small Intestine

Mixing and propulsion of chyme are the primary mechanical events that occur in the small intestine. **Segmental contractions** are propagated for only short distances and mix intestinal contents (figure 21.16). **Peristaltic contractions** proceed along the length of the intestine for variable distances and cause the chyme to move along the small intestine (see figure 21.9).

Local reflexes are the most important regulator of contractions in the small intestine. Distention of the intestinal wall, amino acids, and low pH stimulate contractions. Parasympathetic stimulation increases contractions but is not as important for the intestines as for the stomach.

The ileocecal sphincter at the juncture of the ileum and the large intestine remains mildly contracted most of the time. CNS and local reflexes control the ileocecal sphincter. When the cecum is full, for example, increased constriction of the sphincter prevents

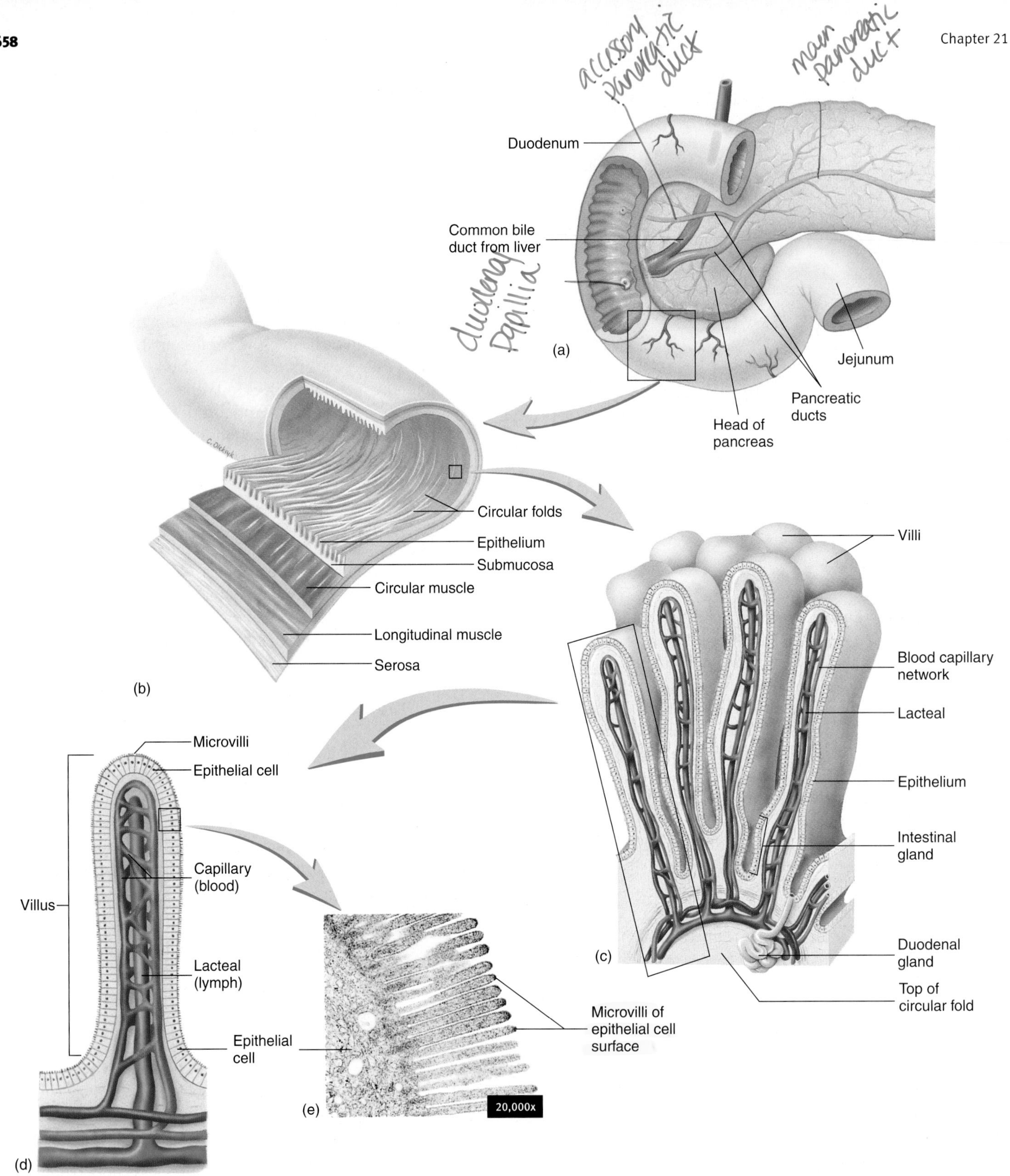

Figure 21.15 **Anatomy and Histology of the Duodenum**

(*a*) Ducts from the liver and pancreas empty into the duodenum. (*b*) Wall of the duodenum, showing the circular folds. (*c*) The villi on a circular fold. (*d*) A single villus, showing the lacteal and capillary network. (*e*) Transmission electron micrograph of microvilli on the surface of a villus.

1. A secretion introduced into the digestive tract or food within the tract begins in one location.

2. Segments of the digestive tract alternate between contraction and relaxation.

3. Material (*brown*) in the intestine is spread out in both directions from the site of introduction.

4. The secretion or food is spread out in the digestive tract and becomes more diffuse (*lighter color*) through time.

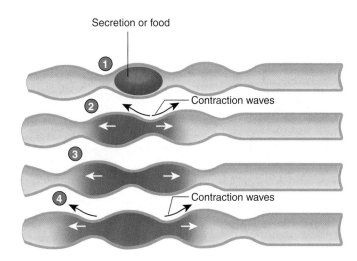

Process Figure 21.16   Segmental Contractions

chyme from entering. The ileocecal valve allows chyme to move from the ileum into the large intestine but tends to prevent movement from the large intestine back into the ileum. ▼

> 42. *What are two kinds of movement of the small intestine? How are they regulated?*
> 43. *What is the function of the ileocecal sphincter and valve?*

Liver and Gallbladder

Anatomy and Histology of the Liver

The **liver** is the largest internal organ of the body, weighing about 1.36 kg (3 pounds); it is in the right-upper quadrant of the abdomen, tucked against the inferior surface of the diaphragm (figure 21.17). Historically, the liver was divided into four lobes based on superficial structures. Anteriorly, the boundary between the **left** and **right lobes** is marked by the falciform ligament. Inferiorly, the left lobe is separated from the **quadrate** and **caudate lobes** by the lesser omentum. The **porta hepatis** (gate of the liver) on the inferior surface of the liver is where the various vessels, ducts, and nerves enter and exit the liver. The porta hepatis separates the quadrate and caudate lobes. The **gallbladder** is a small sac on the inferior surface of the liver that stores bile. The gallbladder marks the division between the right and quadrate lobes. The inferior vena cava marks the division between the right and caudate lobes.

A connective tissue capsule and visceral peritoneum cover the liver, except for the **bare area,** which is a small area on the diaphragmatic surface that lacks a visceral peritoneum and is surrounded by the coronary ligament (see figure 21.17*b* and *c*). At the porta hepatis, the connective tissue capsule sends a branching network of septa (walls) into the substance of the liver to provide its main support. Vessels, nerves, and ducts follow the connective tissue branches throughout the liver.

It is now known that the external division of the liver into lobes has nothing to do with its internal organization. Internally, the liver is divided into eight **segments** based on the distribution of blood vessels and ducts transporting bile.

 Liver Segments

Knowledge of the liver segments allows surgeons to safely remove parts of the liver. For example, the part of the liver containing a tumor. The remaining segments can regenerate and the liver will regrow to its original size in 6 to 12 months. A liver divided by segments can be transplanted into more than one person, helping more people who need a liver transplant. Also, using part of an adult liver for a liver transplant into a child achieves a better size match.

The porta hepatis contains the **hepatic artery** (he-pa′tik, associated with the liver) and **hepatic portal vein,** which carry blood to the liver, and the **right** and **left hepatic ducts,** which conduct bile toward the duodenum (see figure 21.17*b*). Connective tissue septa divide the liver segments into many hexagon-shaped **lobules** with a **portal triad** at each corner (figure 21.18). The triads are so named because they contain three structures derived from the porta hepatis: branches of the hepatic artery, hepatic portal vein, and hepatic ducts. The hepatic artery branches and the hepatic portal branches join enlarged capillaries called **hepatic sinusoids.** The wall of the hepatic sinusoids consists of simple squamous epithelium (endothelium) and **hepatic phagocytic cells** (Kupffer cells). The hepatic sinusoids join a **central vein** located in the center of the lobule. **Hepatic cords** radiate out from the central vein of each lobule like the spokes of a wheel, surrounding the hepatic sinusoids. The hepatic cords are composed of **hepatocytes,** the functional cells of the liver. A cleftlike lumen, the **bile canaliculus** (kan-ă-lik′ū-lŭs, little canal), lies between the hepatocytes within each cord. The bile canaliculi join the hepatic duct branches in the portal triad. ▼

> 44. *Describe the four lobes of the liver. What are the porta hepatis and the bare area?*
> 45. *What are liver segments and liver lobules?*
> 46. *What are a portal triad, hepatic sinusoid, central vein, hepatic cord, and bile canaliculus?*

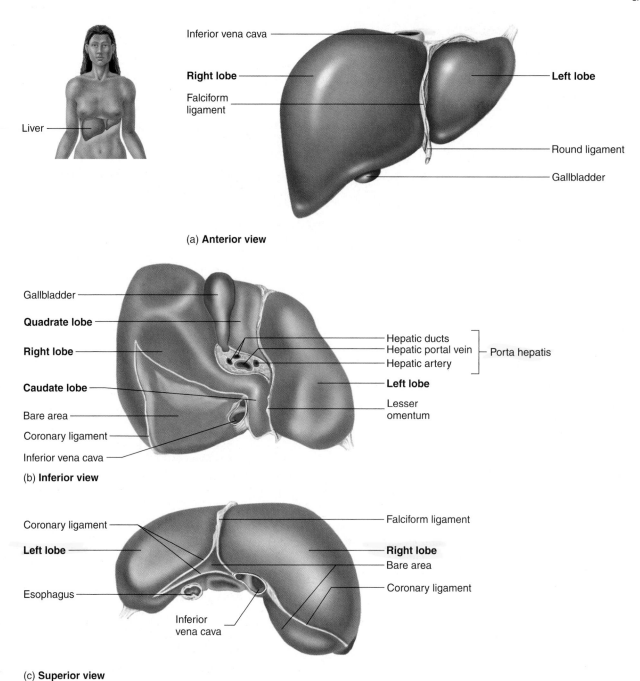

(a) **Anterior view**

(b) **Inferior view**

(c) **Superior view**

Figure 21.17 Liver

Functions of the Liver

The liver performs important digestive and excretory functions, stores and processes nutrients, synthesizes new molecules, and detoxifies harmful chemicals.

Bile Production

The liver produces and secretes about 600–1000 mL of bile each day. Bile contains no digestive enzymes, but it plays a role in digestion because it neutralizes and dilutes gastric acid and emulsifies fats. The pH of chyme as it leaves the stomach is too low for the normal function of pancreatic enzymes. Bicarbonate ions in bile help neutralize the acidic chyme and bring the pH up to a level at which pancreatic enzymes can function. **Bile salts** emulsify fats, changing large lipid droplets into much smaller droplets (see p. 672). Bile contains excretory products, such as the bile pigment bilirubin (see chapter 16). Bile also contains cholesterol, fats, fat-soluble hormones, and lecithin.

Storage

Hepatocytes can remove sugar from the blood and store it in the form of **glycogen.** They can also store fat, vitamins (A, B_{12}, D, E, and

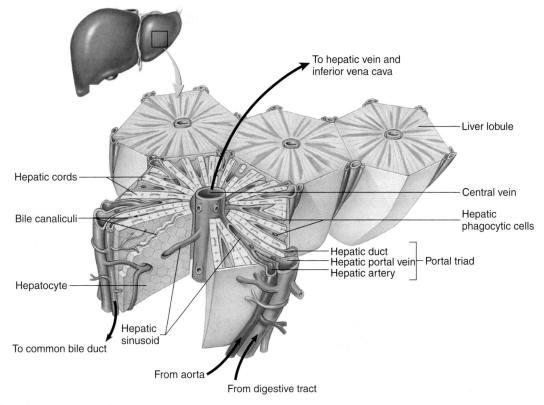

To hepatic vein and inferior vena cava

Liver lobule

Hepatic cords

Bile canaliculi

Central vein

Hepatic phagocytic cells

Hepatic duct
Hepatic portal vein — Portal triad
Hepatic artery

Hepatocyte

Hepatic sinusoid

To common bile duct

From aorta

From digestive tract

Figure 21.18   Histology of the Liver

The liver consists of hexagon-shaped lobules with a portal triad at each corner. A central vein is located in the center of each lobule.

K), copper, and iron. This storage function is usually short-term, and the amount of stored material in the hepatocytes and the cell size fluctuate during a given day.

Hepatocytes help control blood sugar levels within very narrow limits. If a large amount of sugar were to enter the general circulation after a meal, it would increase the osmolality of the blood and produce hyperglycemia. This is prevented because the blood from the intestine passes to the liver, where glucose and other substances are removed from the blood by hepatocytes, stored, and secreted back into the circulation when needed.

Nutrient Interconversion

The interconversion of nutrients is another important function of the liver. Ingested nutrients are not always in the proportion needed by the tissues. If this is the case, the liver can convert some nutrients into others. If, for example, a person is on a diet that is excessively high in protein, an oversupply of amino acids and an undersupply of lipids and carbohydrates may be delivered to the liver. The hepatocytes break down the amino acids and cycle many of them through metabolic pathways so they can be used to produce adenosine triphosphate, lipids, and glucose (see chapter 22).

Hepatocytes also transform substances that cannot be used by most cells into more readily usable substances. For example, ingested fats are combined with choline and phosphorus in the liver to produce phospholipids, which are essential components of plasma membranes. The liver converts lactic acid, produced by exercising muscle, to glucose.

Detoxification

Many ingested substances are harmful to the cells of the body. In addition, the body itself produces many by-products of metabolism that, if accumulated, are toxic. The liver forms a major line of defense against many of these harmful substances. It detoxifies many substances by altering their structure to make them less toxic or to make their elimination easier. Ammonia, for example, a by-product of amino acid metabolism, is toxic and is not readily removed from the circulation by the kidneys. Hepatocytes remove ammonia from the circulation and convert it to urea, which is less toxic than ammonia. The urea is secreted into the circulation and is then eliminated by the kidneys in the urine. Other substances are removed from the circulation and excreted by the hepatocytes into the bile.

Phagocytosis

Hepatic phagocytic cells (Kupffer cells) in the wall of hepatic sinusoids phagocytize "worn-out" and dying red and white blood cells, some bacteria, and other debris that enters the liver through the circulation.

Synthesis

The liver can produce its own new compounds. It produces many blood proteins, such as albumins, fibrinogen, globulins, heparin, and clotting factors, which are released into the circulation. ▼

> 47. *Explain and give examples of the major functions of the liver.*

Hepatitis and Cirrhosis

Strictly defined, **hepatitis** is an inflammation of the liver. Infectious hepatitis can be caused by viruses, bacteria, fungi, or parasites, whereas noninfectious hepatitis can be caused by medications, toxins, and autoimmune disorders. In the United States, hepatitis viruses A, B, C, and D cause most cases of infectious hepatitis. Many individuals infected by hepatitis virus are asymptomatic. Others develop fever, abdominal discomfort, nausea, loss of appetite, fatigue, and other symptoms. Liver damage can occur, resulting in jaundice and loss of liver function as liver cells die and are replaced with scar tissue. Severe damage and liver failure cause death. Chronic hepatitis B infections also lead to the development of liver cancer.

Cirrhosis (sir-rō′sis) of the liver involves the death of hepatocytes and their replacement by fibrous connective tissue. The liver becomes pale in color (the term *cirrhosis* means a tawny or orange condition) because of the presence of excess white connective tissue. It also becomes firmer, and the surface becomes nodular. The loss of hepatocytes eliminates the function of the liver, often resulting in jaundice. The buildup of connective tissue can impede blood flow through the liver. Cirrhosis frequently develops in alcoholics and can develop as a result of biliary obstruction, hepatitis, or nutritional deficiencies.

Blood Flow Through the Liver

The hepatic artery takes oxygen-rich blood to the liver, which supplies hepatocytes with oxygen (figure 21.19, step 1). The hepatic portal vein carries blood from the digestive tract to the liver (figure 21.19, step 2). Portal blood vessels are blood vessels that begin in a primary capillary bed, extend some distance, and end in a second capillary bed. The **hepatic portal system** is one of two major portal systems. The other is the hypothalamohypophyseal portal system in the brain (see chapter 15).

The hepatic portal system begins with capillaries in the digestive tract and ends with the hepatic sinusoids (capillaries) in the liver. The hepatic portal vein has blood that is oxygen-poor because oxygen left the blood in the capillaries of the digestive tract. The blood, however, contains materials, such as nutrients, absorbed from the digestive tract. These absorbed materials leave the blood in the capillaries of the liver and are processed (see "Functions of the Liver," p. 660).

The hepatic artery and hepatic portal vein enter the porta hepatis and branch many times to supply the portal triads. The hepatic artery and hepatic portal vein branches in a triad empty into a hepatic sinusoid. Thus, the hepatocytes surrounding the hepatic sinusoid receive oxygen and substances absorbed from the digestive tract. The mixed blood flows toward the center of each lobule into a central vein. The central veins from all the lobules unite to form the **hepatic veins,** which carry blood out of the liver to the inferior vena cava (figure 21.19, step 3).

In the fetus, the liver does not process absorbed nutrients from the digestive tract because the fetus derives the nutrients from the mother's blood at the placenta. In the adult, the round ligament (ligamentum teres) is the remnant of the fetal blood vessel carrying blood from the placenta to the liver (see figure 21.17a). The ligamentum venosum is the remnant of the fetal blood vessel through which blood bypasses the liver (see chapter 25). ▼

> 48. *What is the hepatic portal system?*
> 49. *Describe the oxygen and nutrient levels in the blood of the hepatic artery and hepatic portal vein.*
> 50. *Starting with the hepatic artery and hepatic portal vein, describe in order all the structures through which blood flows to reach the inferior vena cava.*
> 51. *What is the round ligament (ligamentum teres) and the ligamentum venosum?*

Bile Transport

Bile, produced by the hepatocytes, flows through the bile canaliculi to the hepatic duct branches in the portal triads. The hepatic ducts converge and empty into the right and left hepatic ducts, which transport bile out of the liver (figure 21.19, steps 4 and 5). The right and left hepatic ducts unite to form a single **common hepatic duct** (figure 21.20). The common hepatic duct is joined by the **cystic** (sis′tik, *kystis,* bladder) **duct** from the gallbladder to form the **common bile duct.** The **gallbladder** is a small sac on the inferior surface of the liver that stores and concentrates bile. The common bile duct joins the pancreatic duct at the **hepatopancreatic ampulla** (hĕ-pat′ō-pan-crē-at′ik am-pul′lă), an enlargement where the hepatic and pancreatic ducts come together. The hepatopancreatic ampulla empties into the duodenum at the **major duodenal papilla** (see figure 21.18). Smooth muscle sphincters surround the common bile duct, hepatopancreatic ampulla, and pancreatic duct. ▼

> 52. *Starting with a hepatocyte, list in order all the structures through which bile flows to reach the duodenum.*

Gallbladder and Bile Storage

The **gallbladder** is a saclike structure on the inferior surface of the liver that is about 8 cm long and 4 cm wide (see figure 21.20). Three tunics form the gallbladder wall: (1) an inner mucosa folded into rugae that allow the gallbladder to expand; (2) a muscularis, which is a layer of smooth muscle that allows the gallbladder to contract; and (3) an outer covering of serosa. The cystic duct connects the gallbladder to the common bile duct.

Bile is continually secreted by the liver and flows through the cystic duct to the gallbladder, where 40–70 mL of bile can be stored. While the bile is in the gallbladder, water and electrolytes are absorbed, and bile salts and pigments become as much as 5–10 times more concentrated than they were when secreted by the liver. Contraction of the gallbladder moves the stored bile into the duodenum.

1. The hepatic artery carries oxygen-rich blood from the aorta through the porta of the liver. Hepatic artery branches become part of the portal triads. Blood from the hepatic artery branches enters the hepatic sinusoids and supplies hepatocytes in the hepatic cords with oxygen.

2. The hepatic portal vein carries nutrient-rich, oxygen-poor blood from the intestines through the porta of the liver. Hepatic portal vein branches become part of the portal triads. Blood from the hepatic portal vein branches enters the hepatic sinusoids and supplies hepatocytes in the hepatic cords with nutrients.

3. Blood in the hepatic sinusoids that comes from the hepatic artery and hepatic portal vein picks up plasma proteins, processed molecules, and waste products produced by the hepatocytes of the hepatic cords. The hepatic sinusoids empty into central veins. The central veins connect to hepatic veins, which connect to the inferior vena cava.

4. Bile produced by hepatocytes in the hepatic cords enters bile canaliculi, which connect to hepatic duct branches that are part of the portal triads.

5. The hepatic duct branches converge to form the left and right hepatic ducts, which carry bile out the porta of the liver.

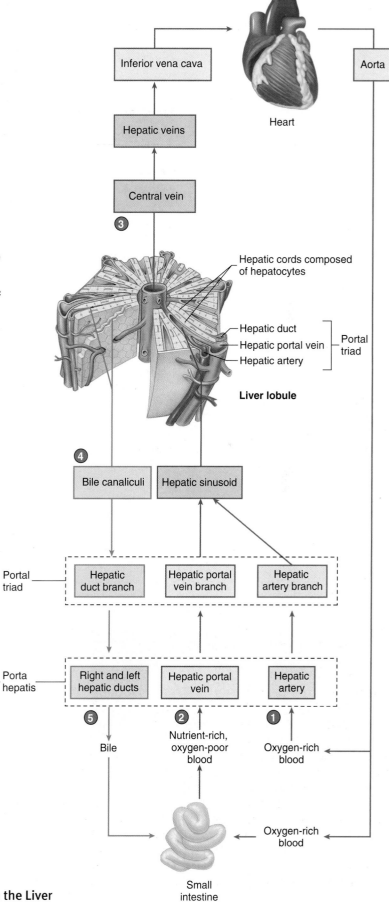

Process Figure 21.19   **Blood and Bile Flow Through the Liver**

1. The hepatic ducts, which carry bile from the liver lobes, combine to form the common hepatic duct.

2. The common hepatic duct combines with the cystic duct from the gallbladder to form the common bile duct.

3. The common bile duct and the pancreatic duct combine to form the hepatopancreatic ampulla.

4. The hepatopancreatic ampulla empties bile and pancreatic secretions into the duodenum at the major duodenal papilla.

5. The accessory pancreatic duct empties pancreatic secretions into the duodenum at the minor duodenal papilla.

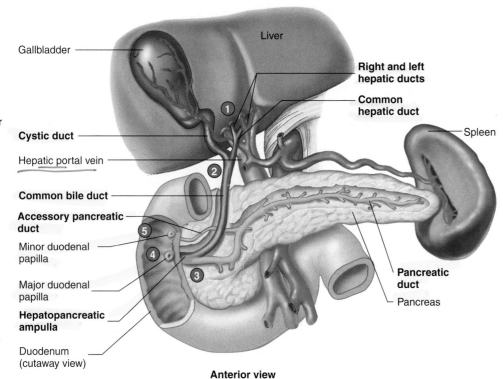

Process Figure 21.20 Liver, Gallbladder, Pancreas, and Duct System

Gallstones

Cholesterol, secreted by the liver, may precipitate in the gallbladder to produce **gallstones** (figure A). Cholesterol is not soluble in water and is ordinarily kept in solution by bile salts. Gallstones can form when there is excess cholesterol in the bile due to a high-cholesterol diet or when cholesterol is overly concentrated in the gallbladder. Occasionally, a gallstone passes out of the gallbladder and enters the cystic duct, blocking the release of bile. Such a condition interferes with normal digestion, and the gallstone often must be removed surgically. If the gallstone moves far enough down the duct, it can also block the pancreatic duct, resulting in pancreatitis. ▼

Figure A Gallstones

53. Describe the three tunics of the gallbladder wall.
54. How does bile reach the gallbladder? What is the function of the gallbladder?

Regulation of Bile Secretion and Release

Secretin released from the duodenum stimulates bile secretion, primarily by increasing the water and bicarbonate ion content of bile (figure 21.21). Cholecystokinin released from the duodenum stimulates the gallbladder to contract and sphincters of the bile duct and hepatopancreatic ampulla to relax. To a lesser degree, parasympathetic stimulation through the vagus nerves cause the gallbladder to contract. Thus, large amounts of concentrated bile move rapidly into the duodenum.

Bile salts also increase bile secretion through a positive-feedback system. Over 90% of bile salts are reabsorbed in the ileum and carried in the blood back to the liver, where they contribute to further bile secretion. The loss of bile salts in the feces is reduced by this recycling process. ▼

55. How do secretin and bile salts affect bile secretion?
56. How do cholecystokinin and parasympathetic stimulation affect the movement of bile from the gallbladder to the duodenum?

P R E D I C T 5
Regulation of the secretion of bile salts by bile salts is a positive-feedback mechanism. How does this positive-feedback mechanism get turned off?

Pancreas

Anatomy and Histology of the Pancreas

The **pancreas** is located retroperitoneal, posterior to the stomach in the inferior part of the left-upper quadrant (see figure 21.1). It consists of a **head,** located within the curvature of the duodenum (figure

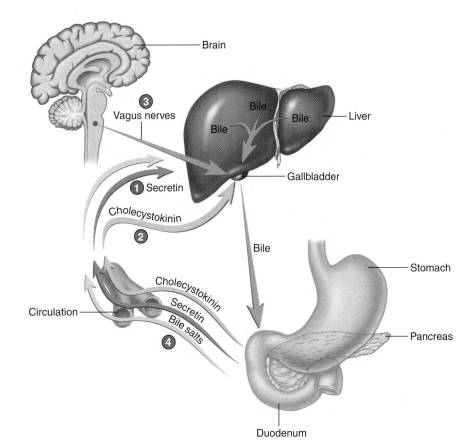

1. Secretin, produced by the duodenum (*purple arrows*) and carried through the circulation to the liver, stimulates bicarbonate secretion into bile (*green arrows inside the liver*).

2. Cholecystokinin, produced by the duodenum (*pink arrows*) and carried through the circulation to the gallbladder, stimulates the gallbladder to contract and sphincters to relax, thereby releasing bile into the duodenum (*green arrow outside the liver*).

3. Vagus nerve stimulation (*red arrow*) causes the gallbladder to contract, thereby releasing bile into the duodenum.

4. Bile salts stimulate bile secretion. Over 90% of bile salts are reabsorbed in the ileum and returned to the liver (*blue arrows*), where they stimulate additional secretion of bile salts.

Process Figure 21.21   Control of Bile Secretion and Release

21.22*a*), a **body,** and a **tail,** which extends to the spleen (see figure 21.20). The pancreas is a complex organ composed of endocrine and exocrine tissues performing several functions. The endocrine part of the pancreas consists of **pancreatic islets** (islets of Langerhans). The islet cells produce the hormones insulin and glucagon, which enter the blood. These hormones are very important in controlling blood levels of nutrients, such as glucose and amino acids (see chapter 15).

The exocrine part of the pancreas is a compound acinar gland (see chapter 4). The **acini** (as'i-nī, grapes) produce digestive enzymes. Clusters of acini are connected by small ducts, which join to form larger ducts, and the larger ducts join to form the **pancreatic duct.** The pancreatic duct joins the common bile duct at the hepatopancreatic ampulla and empties into the duodenum at the major duodenal papilla. An **accessory pancreatic duct,** present in most people, opens at the **minor duodenal papilla.** A smooth muscle sphincter surrounds the pancreatic duct where it enters the hepatopancreatic ampulla and there is a smooth muscle sphincter at the minor duodenal papilla. ▼

57. *Describe the location of the pancreas and its parts.*
58. *Describe the endocrine and exocrine parts of the pancreas and their secretions.*
59. *How do pancreatic secretions reach the duodenum?*

Pancreatic Secretions

Pancreatic juice, the exocrine secretion of the pancreas, moves through the pancreatic ducts to the small intestine, where it functions in digestion. Pancreatic juice has an aqueous and an enzymatic component. The **aqueous component** is produced principally by columnar epithelial cells that line the smaller ducts of the pancreas. Bicarbonate ions are a major part of the aqueous component, and they neutralize the acidic chyme that enters the small intestine from the stomach. The increased pH caused by pancreatic secretions in the duodenum stops pepsin digestion but provides the proper environment for the function of pancreatic enzymes.

The **enzymatic component** consists of enzymes produced by the acinar cells of the pancreas. These enzymes are important for the digestion of all major classes of food. Without the enzymes produced by the pancreas, carbohydrates, lipids, and proteins are not adequately digested (see table 21.1).

The proteolytic pancreatic enzymes digest proteins. The major proteolytic enzymes are **trypsin, chymotrypsin,** and **carboxypeptidase.** They are secreted in inactive forms, whereas many of the other enzymes are secreted in active form. They are secreted in their inactive forms as trypsinogen, chymotrypsinogen, and procarboxypeptidase and are activated by the removal of certain peptides from the larger precursor proteins. If these enzymes were produced in their active forms, they would digest the tissues producing them.

Pancreatic juice also contains **pancreatic amylase,** which continues the polysaccharide digestion initiated in the oral cavity; **pancreatic lipases,** which break down lipids; and **nucleases** (noo'klē-ās-ez), which reduce DNA and ribonucleic acid to their component nucleotides. ▼

60. *What is in the two components of pancreatic juice? Where are these components produced?*
61. *What are the enzymes present in pancreatic juice? Explain the function of each.*

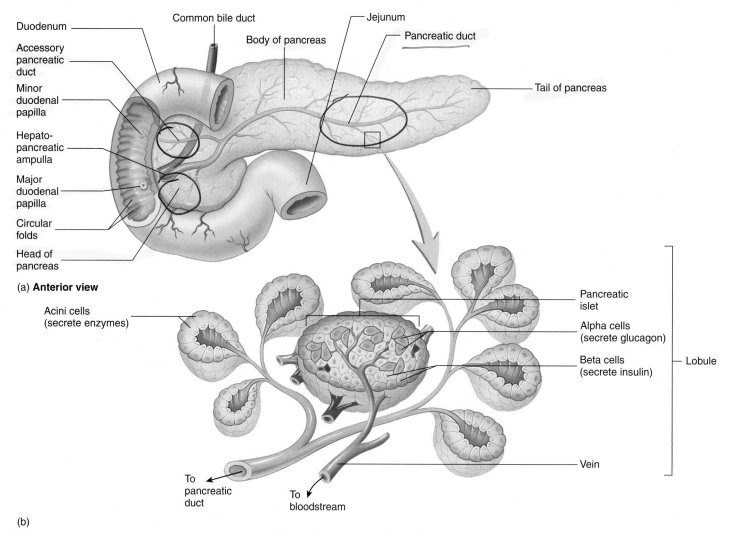

Figure 21.22 Anatomy and Histology of the Duodenum and Pancreas

(*a*) The head of the pancreas lies within the duodenal curvature, with the pancreatic duct emptying into the duodenum. (*b*) Histology of the pancreas, showing both the acini and the pancreatic duct system.

 Pancreatitis and Pancreatic Cancer

Pancreatitis is an inflammation of the pancreas that occurs quite commonly. Pancreatitis involves the release of pancreatic enzymes within the pancreas itself, which digest pancreatic tissue. It can result from alcoholism, the use of certain drugs, pancreatic duct blockage, cystic fibrosis, viral infection, or pancreatic cancer. Symptoms range from mild abdominal pain to systemic shock and coma.

Cancer of the pancreas can obstruct the pancreatic duct and the common bile duct, resulting in jaundice. Pancreatic cancer may not be detected until the tumor has become fairly large, and it can become so large as to block off the pyloric part of the stomach.

Regulation of Pancreatic Secretion

Both hormonal and neural mechanisms control the exocrine secretions of the pancreas (figure 21.23). An acidic chyme in the duodenum stimulates the release of secretin. Secretin stimulates the secretion of the aqueous component of pancreatic juice, which contains a large amount of bicarbonate ions. The bicarbonate ions increase the pH of chyme in the duodenum so that the duodenum is not damaged by the low pH. In addition, pancreatic and microvilli enzymes do not function at a low pH.

P R E D I C T
Explain why secretin production in response to acidic chyme and its stimulation of bicarbonate ion secretion is a negative-feedback mechanism.

Cholecystokinin stimulates the secretion of the enzymatic component of pancreatic juice. The major stimulus for the release of cholecystokinin is the presence of fatty acids and other lipids in the duodenum. Cholecystokinin also causes the sphincters of the pancreatic duct and hepatopancreatic ampulla to relax.

Parasympathetic stimulation through the vagus (X) nerves also stimulates the secretion of pancreatic juices rich in pancreatic

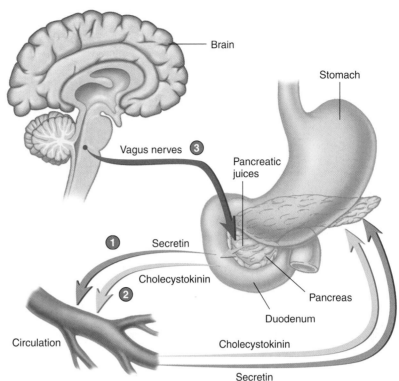

1. Secretin (*purple arrows*) released from the duodenum stimulates the pancreas to release a watery secretion rich in HCO_3^-.

2. Cholecystokinin (*pink arrows*) released from the duodenum causes the pancreas to release a secretion rich in digestive enzymes.

3. Parasympathetic stimulation from the vagus nerve (*red arrow*) causes the pancreas to release a secretion rich in digestive enzymes.

Process Figure 21.23 **Control of Pancreatic Secretion**

enzymes, and sympathetic impulses inhibit secretion. The effect of vagal stimulation on pancreatic juice secretion is greatest during the cephalic and gastric phases of stomach secretion. ▼

> **62. What stimulates the release of the aqueous and enzymatic components of pancreatic juice?**

Large Intestine

The **large intestine** is the portion of the digestive tract extending from the ileocecal junction to the anus. It consists of the cecum, colon, rectum, and anal canal. While in the colon, chyme is converted to feces. The absorption of water and salts, secretion of mucus, and extensive action of microorganisms are involved in the formation of feces, which the colon stores until the feces are eliminated by the process of defecation.

Anatomy and Histology of the Large Intestine

Cecum

The **cecum** (sē'kŭm, blind) is the proximal end of the large intestine. It is where the large and small intestines meet at the ileocecal junction. The cecum is a sac extending inferiorly about 6 cm past the ileocecal junction (figure 21.24). Attached to the cecum is a small, blind tube about 9 cm long called the **vermiform** (ver'mifōrm, worm-shaped) **appendix.** The walls of the appendix contain many lymphatic nodules.

Appendicitis

Appendicitis is an inflammation of the appendix; it usually occurs because of obstruction. Secretions from the appendix cannot pass the obstruction and accumulate, causing enlargement and pain. Bacteria in the area can cause infection. Symptoms include sudden abdominal pain, particularly in the right-lower quadrant of the abdomen, along with a slight fever, loss of appetite, constipation or diarrhea, nausea, and vomiting. If the appendix bursts, the infection can spread throughout the peritoneal cavity, causing **peritonitis,** with life-threatening results. Each year, 500,000 people in the United States suffer from appendicitis. An **appendectomy** is removal of the appendix.

McBurney's point is located on the abdomen in the right-lower quadrant approximately one-third of the distance along a line from the anterior superior iliac spine to the umbilicus. McBurney's point is over the most common location of the attachment of the appendix to the cecum. Pain produced by applying pressure to McBurney's point is suggestive of appendicitis. However, many cases of appendicitis do not exhibit pain at McBurney's point and other conditions can cause pain at McBurney's point.

Colon

The **colon** (kō'lon) is about 1.5–1.8 m long and consists of four parts: the ascending colon, the transverse colon, the descending colon, and the sigmoid colon (see figure 21.24). The **ascending**

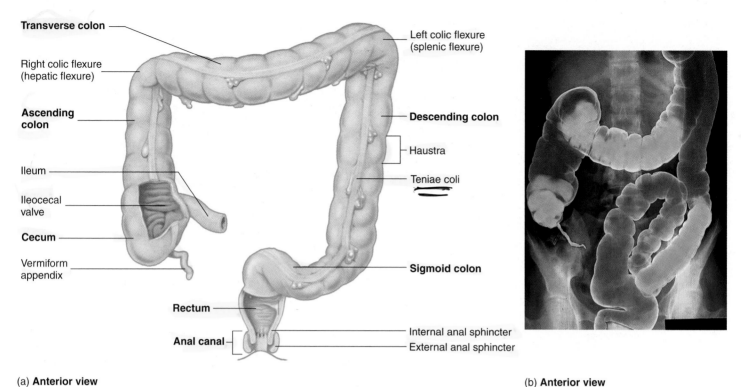

(a) Anterior view

(b) Anterior view

Figure 21.24 Large Intestine

(a) The large intestine consists of the cecum, colon, rectum, and anal canal. The teniae coli are along the length of the colon. (b) Radiograph of the large intestine following a barium enema.

colon extends superiorly from the cecum to the liver, where it turns to the left, forming the right colic flexure (hepatic flexure). The **transverse colon** extends from the right colic flexure to the spleen, where it turns inferiorly, forming the left colic flexure (splenic flexure). The **descending colon** extends from the left colic flexure to the pelvis, where it becomes the sigmoid colon. The **sigmoid colon** forms an S-shaped tube that extends medially and then inferiorly into the pelvic cavity and ends at the rectum.

The circular muscle layer of the colon is complete, but the longitudinal muscle layer is incomplete. The longitudinal layer does not completely envelop the intestinal wall but forms three bands, called the **teniae coli** (tē′nē-ē kō′lī, a band or tape along the colon), that run the length of the colon (see figure 21.24). **Haustra** (haw′stră, to draw up) are pouches formed in the colon wall along its length, giving the colon a puckered appearance.

The mucosal lining of the large intestine consists of simple columnar epithelium. This epithelium is not formed into folds or villi like that of the small intestine but has numerous, straight, tubular glands called **crypts.** The crypts are somewhat similar to the intestinal glands of the small intestine, with three cell types— absorptive, goblet, and granular cells. The major difference is that, in the large intestine, goblet cells predominate and the other two cell types are greatly reduced in number.

Rectum

The **rectum** is a straight, muscular tube that begins at the termination of the sigmoid colon and ends at the anal canal (see figure 21.24). The mucosal lining of the rectum is simple columnar epithelium, and the muscular tunic is relatively thick, compared with the rest of the digestive tract.

Anal Canal

The last 2–3 cm of the digestive tract is the **anal canal** (see figure 21.24). It begins at the inferior end of the rectum and ends at the **anus,** the inferior opening of the digestive tract. The smooth muscle layer of the anal canal is even thicker than that of the rectum and forms the **internal anal sphincter** at the superior end of the anal canal. Skeletal muscle forms the **external anal sphincter** at the inferior end of the canal. The epithelium of the superior part of the anal canal is simple columnar and that of the inferior part is stratified squamous.

 Hemorrhoids

Hemorrhoids are enlarged rectal veins. Enlargement is associated with increased pressure in the veins, as occurs when straining to have a bowel movement (constipation), pregnancy, and liver disease. Internal hemorrhoids are enlarged rectal veins in the anal canal. They are usually painless but often bleed following a bowel movement. The most common sign of internal hemorrhoids is blood on toilet paper. Prolapse, or extrusion, of an internal hemorrhoid through the anus can be painful. External hemorrhoids are enlarged rectal veins around the anus. The development of a blood clot in an external hemorrhoid can be painful. Treatments for hemorrhoids include increasing the bulk (indigestible fiber) in the diet, taking sitz baths, and using hydrocortisone suppositories. Surgery may be necessary if the condition is extreme and does not respond to other treatments. ▼

63. Describe the parts of the large intestine and the colon. What are teniae coli, haustra, and crypts?

64. Explain the difference in structure between the internal anal sphincter and the external anal sphincter.

Secretions of the Large Intestine

The mucosa of the colon has numerous goblet cells scattered along its length and numerous crypts lined almost entirely with goblet cells. Little enzymatic activity is associated with secretions of the colon because mucus is the major secretory product. Mucus lubricates the wall of the colon and helps the fecal matter stick together. Tactile stimuli and irritation of the wall of the colon trigger local enteric reflexes that increase mucous secretion. Parasympathetic stimulation also increases the secretory rate of the goblet cells.

The feces that leave the digestive tract consist of water, solid substances (e.g., undigested food), microorganisms, and sloughed-off epithelial cells.

Numerous microorganisms inhabit the colon. They reproduce rapidly and ultimately constitute about 30% of the dry weight of the feces. Some bacteria in the intestine synthesize vitamin K, which is passively absorbed in the colon. Bacterial actions in the colon produce gases called **flatus** (flā'tŭs, blowing). The amount of flatus depends partly on the bacterial population in the colon and partly on the type of food consumed. For example, beans, which contain certain complex carbohydrates, are well known for their flatus-producing effect. ▼

65. What is the major secretory product of the large intestine? How are secretions regulated in the large intestine?

66. What is the role of microorganisms in the colon?

Movement in the Large Intestine

Normally, 18–24 hours are required for material to pass through the large intestine, in contrast to the 3–5 hours required for the movement of chyme through the small intestine. Simultaneous contractions of the circular muscles and teniae coli of the colon wall cause constriction and shortening of the colon, resulting in haustra formation. As the colon constricts and bulges, its contents are mixed. In addition, the formation of haustra proceeds toward the anus, resulting in the slow, analward movement of colon contents. Local reflexes regulate haustra formation.

Three or four times each day, the circular muscles in large parts of the transverse and descending colon undergo several strong peristaltic contractions, called **mass movements.** Each mass movement contraction extends over 20 cm of the colon and propels the colon contents a considerable distance toward the anus. Mass movements are stimulated by irritation or distention of the colon, local reflexes in the enteric plexus, and intense parasympathetic stimulation. The gastrocolic and duodenocolic reflexes are local reflexes that can stimulate mass movements (figure 21.25, steps 1 and 2). The **gastrocolic reflex** is initiated in the stomach and the **duodenocolic reflex** is initiated in the duodenum. The thought or smell of food, distention of the stomach, and the movement of chyme into the duodenum can stimulate these reflexes. Mass movements are most common about 15 minutes after breakfast. They usually persist for 10–30 minutes and then stop for perhaps half a day.

Distention of the rectal wall by feces stimulates the **defecation reflex,** which consists of local and parasympathetic reflexes. Local reflexes cause weak contractions, whereas parasympathetic reflexes cause strong contractions and are normally responsible for most of the defecation reflex. Local reflexes cause weak contractions of the distal colon and rectum and relaxation of the internal anal sphincter (figure 21.25, steps 3 and 4). Defecation requires that the contractions that move feces toward the anus be coordinated with the relaxation of the internal and external anal sphincters.

Action potentials produced in response to the distention travel along sensory nerve fibers to the defecation center in the conus medullaris of the spinal cord (S2–S4), where motor action potentials are initiated that reinforce peristaltic contractions in the lower colon and rectum. Action potentials from the spinal cord also cause the internal anal sphincter to relax (figure 21.25, steps 5 and 6).

Action potentials from the sacral spinal cord ascend to the brain, where parts of the brainstem and hypothalamus inhibit or facilitate reflex activity in the spinal cord (figure 21.25, steps 7 and 8). In addition, action potentials ascend to the cerebrum, where awareness of the need to defecate is realized. The external anal sphincter is composed of skeletal muscle and is under conscious cerebral control. If this sphincter is relaxed voluntarily, feces are expelled. On the other hand, increased contraction of the external anal sphincter prevents defecation (figure 21.25, step 9). The defecation reflex persists for only a few minutes and quickly declines. Generally, the reflex is reinitiated after a period that may be as long as several hours. Mass movements in the colon are usually the reason for the reinitiation of the defecation reflex.

Contraction of the internal and external anal sphincters prevents defecation. Resting sphincter pressure results from tonic muscle contractions, mostly of the internal anal sphincter. In response to increased abdominal pressure, reflexes mediated through the spinal cord cause contractions of the external anal sphincter. Thus, the untimely expulsion of feces during coughing or exertion is avoided.

Defecation can be initiated by voluntary actions that stimulate a defecation reflex. This "straining" includes a large inspiration of air, followed by closure of the larynx and forceful contraction of the abdominal muscles. As a consequence, the pressure in the abdominal cavity increases and forces feces into the rectum. Stretch of the rectum initiates a defecation reflex and input from the brain overrides the reflexive contraction of the external anal sphincter stimulated by increased abdominal pressure. The increased abdominal pressure also helps push feces through the rectum.

On Being "Regular"

The importance of the regularity of defecation has been greatly overestimated. Many people have the misleading notion that a daily bowel movement is critical for good health. As with many other body functions, what is "normal" differs from person to person. Whereas many people defecate one or more times per day, some normal, healthy adults defecate on the average only every other day. A defecation rate of only twice per week, however, is usually described as constipation. Habitually postponing defecation when the defecation reflex occurs can lead to constipation and may eventually result in desensitization of the rectum so that the defecation reflex is greatly diminished. ▼

67. How are the contents of the colon mixed? What are mass movements?

1. The thought or smell of food, distention of the stomach, and the movement of chyme into the duodenum can stimulate the gastrocolic and duodenocolic reflexes (*green arrows*).

2. The gastrocolic and duodenocolic reflexes stimulate mass movements in the colon, which propel the contents of the colon toward the rectum (*orange arrow*).

3. Distention of the rectum by feces stimulates local defecation reflexes. These reflexes cause contractions of the colon and rectum (*brown arrow*), which move feces toward the anus.

4. Local reflexes cause relaxation of the internal anal sphincter (*brown arrow*).

5. Distention of the rectum by feces stimulates parasympathetic reflexes. Action potentials are propagated to the defecation reflex center located in the spinal cord (*yellow arrow*).

6. Action potentials stimulate contraction of the colon and rectum and relaxation of the internal anal sphincter (*purple arrows*).

7. Action potentials are propagated through ascending nerve tracts to the brain (*blue arrow*).

8. Descending nerve tracts from the brain regulate the defecation reflex center (*pink arrow*).

9. Action potentials from the brain control the external anal sphincter (*purple arrow*).

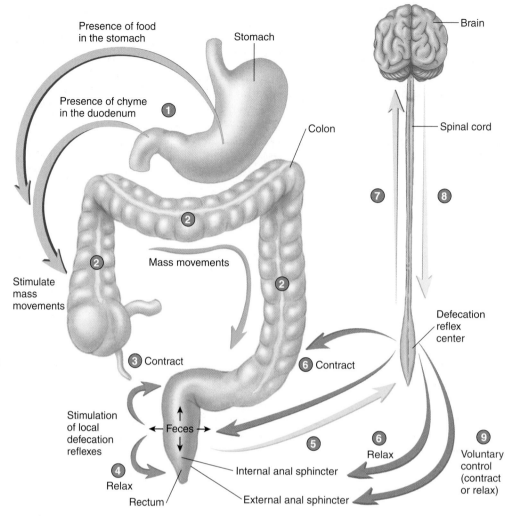

Process Figure 21.25   Control of Defecation

---

68. *Describe gastrocolic and duodenocolic reflexes.*
69. *What stimulus initiates the defecation reflex? Describe the role of local and parasympathetic reflexes in defecation.*
70. *How are the internal anal sphincter and the external anal sphincter controlled during defecation?*

CASE STUDY | Spinal Cord Injury and Defecation

Dan, a 17-year-old male, was driving home late at night from a ski trip when he missed a sharp curve and crashed. He suffered traumatic injury at the T11 level of the spinal cord, with complete paralysis of both lower limbs. Dan also has incontinence and is no longer able to control his bowel movements. A loss of the ability to control defecation commonly affects the quality of life of most spinal cord injury patients. About 10,000 new spinal cord injuries occur per year in the United States.

About 80% of those injuries involve men, usually in their late teens or twenties. The most common cause is motor vehicle accidents, followed by violence, falls, and sports.

The spinal cord is required for a normal defecation reflex and voluntary control of the external anal sphincter (see figure 21.25). In regard to defecation, spinal cord injuries can be divided into two groups based on the level of injury: those that occur above the conus medullaris and those that damage the conus medullaris where the defecation reflex center is located. Immediately following a spinal cord injury, there is a loss of reflexes below the level of the injury, called **spinal shock** (see chapter 12). The reflexes usually become functional again, however, and the defecation reflex is usually depressed for a few weeks but eventually returns.

Digestion, Absorption, and Transport

Digestion is the breakdown of food to molecules that are small enough to be absorbed into the circulation. **Mechanical digestion** breaks large food particles down into smaller ones. **Chemical digestion** is the breaking of covalent chemical bonds in organic molecules by digestive enzymes. Carbohydrates are broken down into monosaccharides, fats are broken down into fatty acids and monoglycerides, and proteins are broken down into amino acids.

Absorption is the uptake of digestive tract contents by the lining of the digestive tract. A few chemicals, such as nitroglycerin, can be absorbed through the thin mucosa of the oral cavity below the tongue. Some small molecules, such as alcohol and aspirin, can diffuse through the stomach epithelium into the circulation. Most absorption occurs in the duodenum and jejunum, although some absorption occurs in the ileum. **Transport** is the distribution of nutrients throughout the body. ▼

71. *Define digestion, mechanical digestion, chemical digestion, absorption, and transport.*

Carbohydrates

Ingested **carbohydrates** consist primarily of polysaccharides, such as starches; disaccharides, such as sucrose (table sugar) and lactose (milk sugar); and monosaccharides, such as glucose and fructose (the sugar found in many fruits). A minor amount of carbohydrate digestion begins in the oral cavity with the partial digestion of starches by **salivary amylase** (am′il-ās), which splits off the disaccharides maltose and isomaltose from starch. Starch digestion continues in the stomach until the food is well mixed with acid, which inactivates salivary amylase. Carbohydrate digestion is resumed in the intestine by **pancreatic amylase.** A series of **disaccharidases** that are bound to the microvilli of the intestinal epithelium digest disaccharides into monosaccharides (table 21.2).

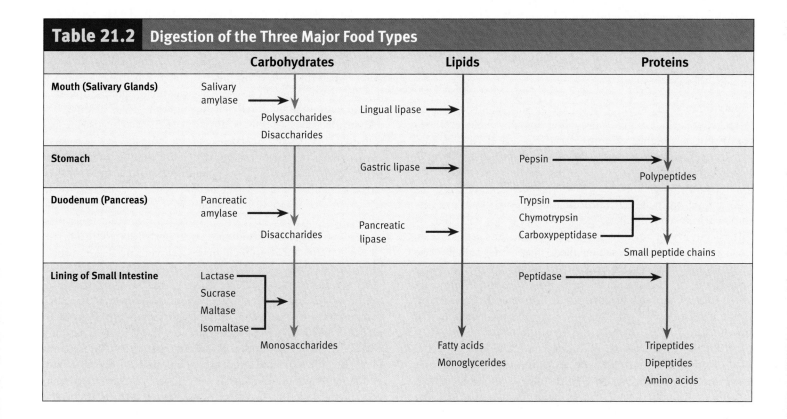

Table 21.2	Digestion of the Three Major Food Types		
	Carbohydrates	**Lipids**	**Proteins**
Mouth (Salivary Glands)	Salivary amylase → Polysaccharides Disaccharides	Lingual lipase →	
Stomach		Gastric lipase →	Pepsin → Polypeptides
Duodenum (Pancreas)	Pancreatic amylase → Disaccharides	Pancreatic lipase →	Trypsin Chymotrypsin Carboxypeptidase → Small peptide chains
Lining of Small Intestine	Lactase Sucrase Maltase Isomaltase → Monosaccharides	Fatty acids Monoglycerides	Peptidase → Tripeptides Dipeptides Amino acids

Lactose Intolerance

Lactose intolerance is the inability to digest the lactose sugar found in milk and other dairy products. Adults in most of the world are lactose intolerant, although infants are not. Why can infants digest milk when their parents cannot? The reason is that many adults lack the enzyme lactase. Lactase is on the surface of absorptive cells in the intestinal mucosa; it digests the disaccharide lactose down to two monosaccharides, glucose and galactose. The consequence of lactose intolerance is diarrhea due to fluid loss as water follows lactose through the GI tract. In addition, a considerable amount of gas is generated from lactose metabolism by bacteria in the large intestine.

Some monosaccharides, such as glucose and galactose, are taken up into intestinal epithelial cells by symport, powered by a Na^+ gradient (figure 21.26). The Na^+ gradient is generated by a Na^+–K^+ pump. Other monosaccharides, such as fructose, are taken up by facilitated diffusion. Once inside the intestinal epithelial cell, monosaccharides are transferred by facilitated diffusion to the capillaries of the intestinal villi (see figure 21.15) and are carried by the hepatic portal system to the liver (see figure 21.19), where the nonglucose sugars are converted to glucose. Glucose enters the body's cells through facilitated diffusion. The rate of glucose transport into most types of cells is greatly influenced by **insulin** and may increase 10-fold in its presence (see chapter 15). ▼

72. *Describe the enzymatic digestion of carbohydrates to monosaccharides. State the part of the digestive tract where each step in the process occurs.*
73. *Describe the transport mechanisms by which monosaccharides are absorbed.*

Lipids

Lipids are molecules that are insoluble or only slightly soluble in water. They include triglycerides, phospholipids, cholesterol, steroids, and fat-soluble vitamins. **Triglycerides** (trī-glis′er-īdz) consist of three fatty acids and one glycerol molecule covalently bonded together.

The first step in lipid digestion is **emulsification** (ē-mŭl′si-fi-kā′shŭn), which is the transformation of large lipid droplets into much smaller droplets. Emulsification is accomplished by bile salts in bile secreted by the liver. The bile salts act like soap, breaking up lipids and making them more water-soluble. The enzymes that digest lipids, such as pancreatic lipase, are water-soluble and can digest the lipids only by acting at the surface of the droplets. The emulsification process increases the surface area of the lipid exposed to the digestive enzymes by decreasing the droplet size.

Lipase (lip′ās) digests lipid molecules. The primary products of lipase digestion are fatty acids and monoglycerides. A minor amount of lingual lipase is secreted in the oral cavity, is swallowed with food, and digests lipids in the stomach. The stomach also produces very small amounts of gastric lipase. Less than 10% of lipid digestion takes place in the stomach. The vast majority of lipase is secreted by the pancreas (see table 21.2).

Cystic Fibrosis

Cystic Fibrosis is a hereditary disorder that affects both the respiratory and digestive systems. It results from defective chloride channels, which cause cells to produce thick, viscous mucous secretions. Blockage of the pancreatic ducts often occurs so that the pancreatic digestive enzymes are prevented from reaching the duodenum. Consequently, fat digestion, which depends on pancreatic enzymes, is slowed or even stopped. Fats and fat-soluble vitamins are not absorbed, resulting in vitamin A, D, E, and K deficiencies. These deficiencies can cause night blindness, skin disorders, rickets, and excessive bleeding. Therapy consists of administering the vitamins and reducing dietary fat intake.

Once lipids are digested in the intestine, bile salts aggregate around the small droplets to form **micelles** (mi-selz′, mī-selz′, a small morsel) (figure 21.27). The hydrophobic ends of the bile salts are directed toward the free fatty acids and monoglycerides at the center of the micelle; the hydrophilic ends are directed outward toward the water environment. When a micelle comes into contact with the epithelial cells of the small intestine, the lipid contents of the micelle pass by means of simple diffusion through the plasma membrane of the epithelial cells. The bile salts are not absorbed until they reach the epithelium of the distal ileum.

Once inside the intestinal epithelial cells, the fatty acids and monoglycerides are recombined to form triglycerides. These, and other lipids, are packaged inside a protein coat within the epithelial cells of the intestinal villi. The packaged lipids, called **chylomicrons** (kī-lō-mi′kron, *chylo,* juice + *micros,* small), leave the epithelial cells and enter the lacteals. Lacteals are lymphatic capillaries located within the intestinal villi (see figure 21.15). Chylomicrons enter the lacteals rather than the blood capillaries because the lacteals lack a basement membrane and are more permeable to large particles, such as chylomicrons, which are about 0.3 mm in diameter. Lymph containing large amounts of absorbed lipid is called **chyle** (kīl, milky lymph). The lymphatic system carries the chyle to the blood (see chapter 19). Chylomicrons are transported to the liver, where the lipids are stored, converted into other molecules, or used as energy. Lipids are also transported to adipose tissue, where they are stored until an energy source is needed elsewhere in the body.

Lipids are not soluble in water, so they are transported in the blood in combination with proteins. **Lipoproteins** are any compound or complex containing lipid and protein. Fatty acids released from adipose tissue bind to the plasma protein albumin and are transported to other tissues as lipoproteins. More complex assemblies of lipids and proteins consist of phospholipids and proteins surrounding lipids. The phospholipids and proteins have charged groups facing outward, which attracts water and

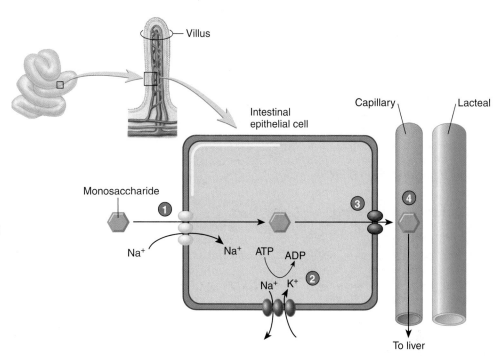

Monosaccharide transport

1. Monosaccharides are absorbed by symport with Na^+ into intestinal epithelial cells.

2. Symport is driven by a sodium gradient established by a Na^+–K^+ pump.

3. Monosaccharides move out of the intestinal epithelial cells by facilitated diffusion.

4. Monosaccharides enter the capillaries of the intestinal villi and are carried through the hepatic portal vein to the liver.

Process Figure 21.26   Transport of Monosaccharides Across the Intestinal Epithelium

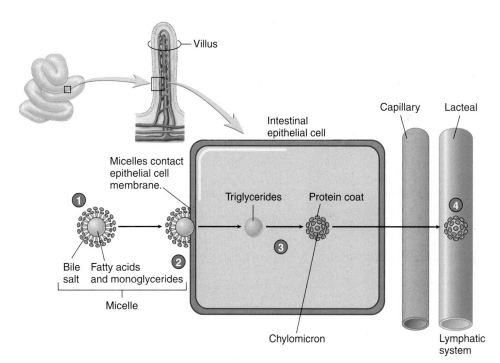

Lipid transport

1. Bile salts surround fatty acids and monoglycerides to form micelles.

2. Micelles attach to the cell membranes of intestinal epithelial cells, and the fatty acids and monoglycerides pass by simple diffusion into the intestinal epithelial cells.

3. Within the intestinal epithelial cell, the fatty acids and monoglycerides are converted to triglycerides; proteins coat the triglycerides to form chylomicrons, which move out of the intestinal epithelial cells by exocytosis.

4. The chylomicrons enter the lacteals of the intestinal villi and are carried through the lymphatic system to the general circulation.

Process Figure 21.27   Transport of Lipids Across the Intestinal Epithelium

makes the lipoprotein soluble in water. The internally located lipids are separated from the water by the phospholipids and proteins.

Complex lipoprotein assemblies are classified according to their densities. The greater the amount of protein relative to the amount of lipid, the denser is the lipoprotein. Chylomicrons, which are made up of 99% lipid and only 1% protein, have an extremely low density. The other major transport lipoproteins are **very-low-density lipoproteins (VLDL),** which are 92% lipid and 8% protein, **low-density lipoproteins (LDL),** which are 75% lipid and 25% protein, and **high-density lipoproteins (HDL),** which are 55% lipid and 45% protein (figure 21.28).

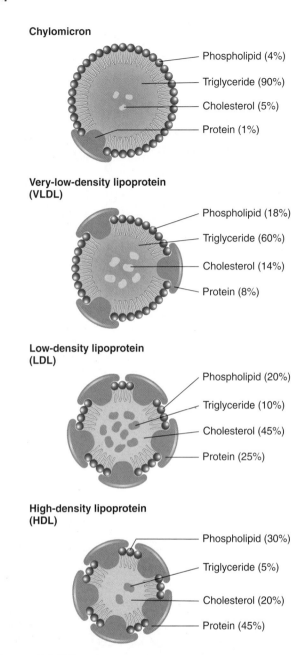

Figure 21.28　Lipoproteins

packaged into HDL, which are manufactured by the liver. The HDL transport the cholesterol back to the liver for recycling or excretion in bile.

 Cholesterol and Cardiovascular Disease

Cholesterol levels in the blood are of great concern to many adults. Cholesterol levels of less than 200 mg/dL are considered desirable, whereas cholesterol levels of 200 mg/dL or higher are considered to be too high. People with high blood cholesterol levels run a much greater risk for heart disease and stroke than people with low cholesterol levels. People with high levels should seek advice from their physician, reduce their intake of foods rich in cholesterol and other fats, and increase their level of exercise. Some people with high cholesterol levels may have to take medication to reduce their cholesterol levels.

LDL is considered to be "bad" because, when in excess, it deposits cholesterol in arterial walls. On the other hand, HDL is considered to be "good" because it transports cholesterol from the tissues via blood to the liver for removal from the body in the bile. A high HDL/LDL ratio is related to a lower risk for heart disease. Aerobic exercise is one way to elevate blood levels of HDL. ▼

74. *What is emulsification of lipids? How is it accomplished and how does it promote lipid digestion?*
75. *Describe the role of micelles, chylomicrons, VLDL, LDL, and HDL in the absorption and transport of lipids in the body.*

Proteins

Proteins are long chains of amino acids found in most of the plant and animal products we eat. In the stomach, the enzyme **pepsin** breaks down proteins, producing shorter amino acid chains called **polypeptides.** Only about 10%–20% of the total ingested protein is digested by pepsin in the stomach. The remaining proteins and polypeptide chains leave the stomach and enter the small intestine. The pancreatic enzymes **trypsin, chymotrypsin,** and **carboxypeptidase** continue the digestive process and produce small peptide chains. **Peptidases,** which are enzymes bound to the microvilli of the small intestine, break down the small peptide chains into tripeptides (three amino acids), dipeptides (two amino acids), and single amino acids (see table 21.2).

Dipeptides and tripeptides enter intestinal epithelial cells by a symport mechanism powered by a Na^+ concentration gradient (figure 21.29). Acidic and most neutral amino acids enter by symport with a Na^+ gradient, whereas basic amino acids enter the epithelial cells by facilitated diffusion. The total amount of each amino acid that enters the intestinal epithelial cells as dipeptides or tripeptides is considerably more than the amount that enters as single amino acids. Once inside the cells, **dipeptidases** and **tripeptidases** split the dipeptides and tripeptides into their component amino acids. Individual amino acids then leave the epithelial cells and enter the hepatic portal system, which transports them to the liver (see figure 21.29). The amino acids may

Most of the lipid taken into or manufactured in the liver leaves the liver in the form of VLDL. Most of the triglycerides are removed from the VLDL to be stored in adipose tissue; as a result, VLDL becomes LDL. Cells have surface **LDL receptors,** which bind to LDL proteins. The cells then take in the LDL by receptor-mediated endocytosis. Thus, phospholipids and cholesterol are delivered to cells.

Cells not only take in cholesterol and other lipids from LDL but also make their own cholesterol. When the combined intake and manufacture of cholesterol exceeds a cell's needs, a negative-feedback system reduces the amount of LDL receptors and cholesterol manufactured by the cell. Excess lipids in cells are also

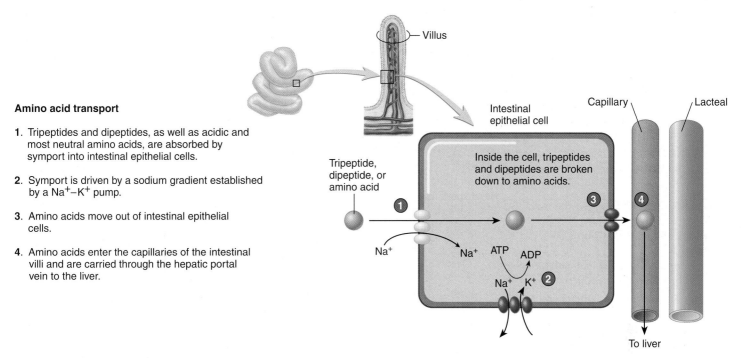

Amino acid transport

1. Tripeptides and dipeptides, as well as acidic and most neutral amino acids, are absorbed by symport into intestinal epithelial cells.

2. Symport is driven by a sodium gradient established by a Na^+–K^+ pump.

3. Amino acids move out of intestinal epithelial cells.

4. Amino acids enter the capillaries of the intestinal villi and are carried through the hepatic portal vein to the liver.

Process Figure 21.29 Amino Acid Transport Across the Intestinal Epithelium

be modified in the liver or released into the blood and distributed throughout the body.

Amino acids are actively transported into the various cells of the body. This transport is stimulated by growth hormone and insulin. Most amino acids are used as building blocks to form new proteins (see chapter 2), but some amino acids may be used for energy. The body cannot store excess amino acids. Instead, they are partially broken down and used to synthesize glycogen or fat, which can be stored. The body can store only small amounts of glycogen, so most of the excess amino acids are converted to fat. ▼

76. *Describe the enzymatic digestion of proteins to tripeptides, dipeptides, and amino acids. State the part of the digestive tract where each step in the process occurs.*
77. *Describe the transport mechanisms by which tripeptides, dipeptides, and amino acids are absorbed.*

Water and Ions

Approximately 9 L of water enters the digestive tract each day (figure 21.30). We ingest about 2 L in food and drink, and the remaining 7 L is from digestive secretions. Approximately 92% of that water is absorbed in the small intestine, about 7% is absorbed in the large intestine, and about 1% leaves the body in the feces. Water can move in either direction by osmosis across the wall of the gastrointestinal tract. The direction of its movement is determined by osmotic gradients across the epithelium. When the chyme is diluted, water moves out of the intestine into the blood; when the chyme is concentrated, water moves into the lumen of the small intestine.

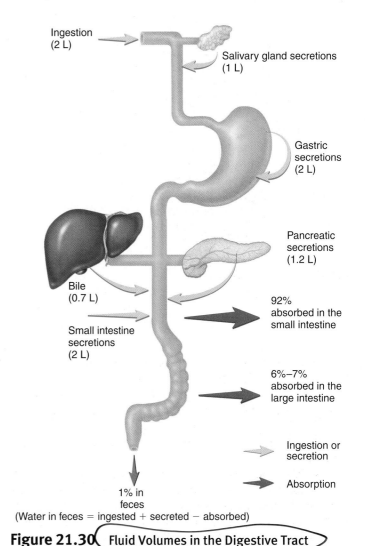

(Water in feces = ingested + secreted − absorbed)

Figure 21.30 Fluid Volumes in the Digestive Tract

Peptic Ulcer

Approximately 5%–12% of the population is affected by **peptic ulcers,** which are lesions in the lining of the digestive tract. For many years it was thought that stress, diet, smoking, or alcohol caused excess acid secretion in the stomach, resulting in ulcers. Although these factors can contribute to ulcers, it is now clear that the root cause of most ulcers is an infection of a specific bacterium—*Helicobacter pylori.* The bacterium may also be involved in many cases of gastritis and gastric cancer.

Most bacteria cannot survive in the stomach. Hence, *H. pylori* is one of the most pervasive of human pathogens because it inhabits a niche without competition. It is estimated that well over half of the world's population is infected with *H. pylori.* The infection rate in the United States is about 1% per year of age—for example, 30% of all 30-year-olds are infected. In developing countries, nearly all people over age 25 are infected. Only about 15%–20% of infected people show gastric problems attributed to *H. pylori.* What triggers the bacterial infection to become symptomatic is a major unanswered question. It seems likely that *H. pylori* infection and conditions that elevate acid secretion or damage the digestive tract wall contribute to the development of an ulcer. The best therapy for ulcers involving *H. pylori* is a combination of antibiotics and drugs that reduce the acid content of the stomach.

Peptic ulcers can occur in the duodenum and stomach. **Duodenal ulcers** are in duodenum, usually near the pyloric sphincter. Approximately 80% of all peptic ulcers are duodenal ulcers. The most common contributing factor to developing duodenal ulcers is the oversecretion of gastric juices relative to the degree of mucous and alkaline protection of the small intestine. People who experience severe anxiety over a long period are the most prone to developing duodenal ulcers. They often have a rate of gastric secretion between meals that is as much as 15 times the normal amount. **Gastric ulcers** are in the stomach. In some patients with gastric ulcers, normal or even low levels of gastric acid secretion often occur. The stomachs of these patients, however, have reduced resistance to their own acid. Such inhibited resistance can result from excessive ingestion of alcohol or aspirin, which directly damages the mucosa. Reflux of duodenal contents into the pylorus can also cause gastric ulcers. In this case, bile, which is present in the reflux, has a detergent effect that reduces gastric mucosal resistance to acid, as well as to bacteria.

Prostaglandins normally stimulate increased mucus production and inhibit acid production in the stomach. Some drugs used to treat arthritis inhibit prostaglandin synthesis, which reduces inflammation. A side effect of taking prostaglandin inhibitors is irritation and ulcers of the mucosa caused by decreased mucus production and increased acid production.

Inflammation of the Intestines

Enteritis is inflammation of the intestine, and **colitis** is inflammation of the colon. Both can result from an infection, chemical irritation, or an unknown cause. **Inflammatory bowel disease (IBD)** is the general name given to either Crohn disease (regional enteritis) or ulcerative colitis. IBD occurs at a rate in Europe and North America of approximately 4 to 8 new cases per 100,000 people per year, which is much higher than in Asia and Africa. Males and females are affected about equally. IBD is of unknown cause, but infectious, autoimmune, and hereditary factors have been implicated.

Crohn disease involves localized inflammatory degeneration, which can occur anywhere along the digestive tract but most commonly involves the distal ileum and proximal colon. The degeneration involves the entire thickness of the digestive tract wall. The intestinal wall often becomes thickened, constricting the lumen, with ulcerations and fissures in the damaged areas. The disease causes diarrhea, abdominal pain, fever, and weight loss. Treatment centers around anti-inflammatory drugs, but other treatments, including the avoidance of foods that increase symptoms and even surgery, are used.

Ulcerative colitis is limited to the mucosa of the large intestine and rectum. The involved mucosa exhibits inflammation, including edema, vascular congestion, hemorrhage, and the accumulation of plasma cells, lymphocytes, neutrophils, and eosinophils. Patients may experience abdominal

---

Epithelial cells actively transport Na^+, K^+, Ca^{2+}, and Mg^{2+} from the intestine. Vitamin D is required for the transport of Ca^{2+}. Negatively charged Cl^- move passively through the wall of the duodenum and jejunum with the positively charged Na^+, but Cl^- are actively transported from the ileum. ▼

78. *Describe the movement of water through the intestinal wall.*
79. *How are ions absorbed?*

Effects of Aging on the Digestive System

As a person ages, the connective tissue layers of the digestive tract, the submucosa and serosa, tend to thin. The blood supply to the digestive tract decreases. There is also a decrease in the number of smooth muscle cells in the muscularis, resulting in decreased motility in the digestive tract. In addition, goblet cells within the

pain, fever, malaise, fatigue, and weight loss, as well as diarrhea and hemorrhage. In rare cases, severe diarrhea and hemorrhage require transfusions. Treatment includes the use of anti-inflammatory drugs and, in some cases, the avoidance of foods that increase symptoms.

Irritable Bowel Syndrome

Irritable bowel syndrome (IBS), also called spastic colon, is a disorder of unknown cause in which intestinal mobility is abnormal. The disorder accounts for over half of all referrals to gastroenterologists. Male and female children are affected equally, but women are affected twice as often as men.

IBS patients experience abdominal pain mainly in the left-lower quadrant, especially after eating. They also have alternating bouts of constipation and diarrhea. There is no specific histopathology in the digestive tracts of IBS patients. There are no anatomical abnormalities, no indication of infection, and no sign of metabolic causes. Patients with IBS appear to exhibit greater than normal levels of psychological stress or depression and show increased contractions of the esophagus and small intestine during times of stress. There is a high familial incidence. Some patients present with a history of traumatic events, such as physical or sexual abuse. Treatments include psychiatric counseling and stress management, diets with increased fiber and limited gas-producing foods, and loose clothing. In some patients, drugs that reduce parasympathetic stimulation of the digestive system are useful.

Malabsorption Syndrome

Malabsorption syndrome (sprue) is a spectrum of disorders of the small intestine that results in abnormal nutrient absorption. One type of malabsorption, called **celiac disease,** or **gluten-sensitive enteropathy,** results from an immune response to gluten, which is present in certain types of grains and involves the destruction of newly formed epithelial cells in the intestinal glands. These cells fail to migrate to the villi surface, the villi become blunted, and the surface area decreases. As a result, the intestinal epithelium is less capable of absorbing nutrients. Another type of malabsorption, tropical malabsorption, is apparently caused by bacteria, although no specific bacterium has been identified.

Colon Cancer

Colon cancer is the second leading cause of cancer-related deaths in the United States; it accounts for 55,000 deaths a year. Susceptibility to colon cancer can be familial; however, a correlation exists between colon cancer and lifestyle factors, including diets low in fiber and high in fat and red or processed meats. Screening for colon cancer includes testing the stool for blood content and performing a colonoscopy, which allows physicians to see into the colon.

Constipation

Constipation is the slow movement of feces through the large intestine. The feces often become dry and hard because of increased fluid absorption during the extended time they are retained in the large intestine. Constipation often results after a prolonged time of inhibiting normal defecation reflexes. A change in habits (such as travel), dehydration, depression, disease, metabolic disturbances, certain medications, pregnancy, or dependency on laxatives can cause constipation. Irritable bowel syndrome can also cause constipation. Constipation can also occur with diabetes, kidney failure, colon nerve damage, or spinal cord injuries or as a result of an obstructed bowel. Of greatest concern, the obstruction could be caused by colon cancer.

Diarrhea

Diarrhea is an abnormally frequent discharge of fluid feces. Increased secretion into the digestive tract can cause diarrhea. Infections and irritation of the digestive tract can cause the intestinal mucosa to secrete large amounts of mucus and electrolytes into the colon. Water follows by osmosis. Although such discharge increases fluid and electrolyte loss, it also moves the infected feces out of the intestine more rapidly and speeds recovery from the disease. Diarrhea can also occur when there is inadequate time to reabsorb the normal amount of water from the feces. Increased motility decrease the time feces spend in the colon. For example, emotional stress can increase parasympathetic stimulation of the digestive tract and cause diarrhea.

mucosa secrete less mucus. Glands along the digestive tract, such as the gastric glands, the liver, and the pancreas, also tend to secrete less with age. These changes by themselves do not appreciably decrease the function of the digestive system.

Through the years the digestive tract, like the skin and lungs, is directly exposed to materials from the outside environment. Some of those substances can cause mechanical damage to the digestive tract and others may be toxic to the tissues. The digestive tract of elderly people becomes less and less protected from these outside influences because the connective tissue of the digestive tract becomes thin with age and because the protective mucous covering is reduced. In addition, the mucosa of elderly people tends to heal more slowly following injury. The liver's ability to detoxify certain chemicals tends to decline, the ability of the hepatic phagocytic cells to remove particulate contaminants decreases, and the liver's ability to store glycogen decreases. ▼

80. Describe the general effects of aging on the digestive tract.

Digestive
Systems Interactions

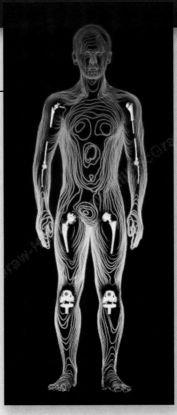

	Effects of the Digestive System on Other Systems	Effects of Other Systems on the Digestive System
Integumentary System	Provides nutrients and water	Produces a vitamin D precursor that is converted in the liver and kidneys to active vitamin D, which increases the absorption of calcium and phosphorus in the small intestine
Skeletal System	Provides water and nutrients, including calcium and phosphorus The liver modifies vitamin D precursor produced in the skin	Lower ribs and pelvis provide some protection Teeth and mandible are important for mastication
Muscular System	Provides nutrients and water The liver converts lactic acid from muscle to glucose	Protects abdominal organs Is responsible for mastication and swallowing Controls voluntary defecation
Nervous System	Provides nutrients and water	Regulates secretion from digestive glands and organs Controls mixing and movement of digestive tract contents Hypothalamus is responsible for hunger, thirst, and satiety sensations
Endocrine System	Provides nutrients and water	Regulates secretion from digestive glands and organs Controls mixing and movement of digestive tract contents
Cardiovascular System	Provides nutrients, including iron (for hemoglobin production), vitamin B_{12} and folic acid (for cell production), and vitamin K (for clotting proteins) Absorbs water and ions necessary to maintain blood volume and pressure The liver produces clotting proteins Heme is converted to bilirubin and excreted in bile	Delivers oxygen, nutrients, hormones, and immune cells Removes carbon dioxide, waste products, and toxins
Lymphatic System and Immunity	Provides nutrients and water	Carries absorbed fats to the blood Immune cells provide protection against microorganisms and toxins and promote tissue repair by releasing chemical mediators Removes excess interstitial fluid
Respiratory System	Provides nutrients and water	Provides oxygen and removes carbon dioxide Helps maintain the body's pH
Urinary System	Provides nutrients and water The liver converts vitamin D precursor from the skin to a form of vitamin D that can be converted by the kidneys to active vitamin D	Removes waste products Helps maintain the body's pH, ion, and water balance
Reproductive System	Provides nutrients and water Supports fetal development	Developing fetus can crowd digestive organs

S U M M A R Y

Functions of the Digestive System (p. 640)

The digestive system takes in food, breaks down food, absorbs nutrients, and eliminates wastes.

Histology of the Digestive Tract (p. 640)

1. The digestive system consists of a digestive tract and its associated accessory organs.
   - The digestive tract consists of the oral cavity, pharynx, esophagus, stomach, small intestine, large intestine, and anus.
   - Accessory organs, such as the salivary glands, liver, gallbladder, and pancreas, are located along the digestive tract.
2. The digestive tract is composed of four tunics: mucosa, submucosa, muscularis, and serosa or adventitia.
   - The mucosa consists of a mucous epithelium, a lamina propria, and a muscularis mucosae.
   - The submucosa is a connective tissue layer containing the submucosal plexus (part of the enteric plexus), blood vessels, and small glands.
   - The muscularis consists of an inner layer of circular smooth muscle and an outer layer of longitudinal smooth muscle. The myenteric plexus is between the two muscle layers.
   - The serosa or adventitia forms the outermost layer of the digestive tract.

Enteric Nervous System

1. The enteric nervous system (ENS) consists of nerve plexuses within the wall of the digestive tract.
2. Nervous regulation involves local reflexes in the ENS and CNS reflexes.

Peritoneum (p. 641)

1. The peritoneum is a serous membrane that lines the abdominopelvic cavity and organs.
2. Mesenteries are peritoneum that extends from the body wall to many of the abdominopelvic organs.
3. Retroperitoneal organs are located behind the peritoneum.

Oral Cavity (p. 643)

The oral cavity includes the vestibule and oral cavity proper.

Lips and Cheeks

The lips and cheeks are involved in facial expression, mastication, and speech.

Palate and Palatine Tonsils

1. The roof of the oral cavity is divided into the hard and soft palates.
2. The palatine tonsils are located in the lateral wall of the fauces.

Tongue

The tongue is involved in speech, taste, mastication, and swallowing.

Teeth

1. Twenty deciduous teeth are replaced by 32 permanent teeth.
2. The types of teeth are incisors, canines, premolars, and molars.
3. A tooth consists of a crown, a neck, and a root.
4. The root is composed of dentin. Within the dentin of the root is the pulp cavity, which is filled with pulp, blood vessels, and nerves. The crown is dentin covered by enamel.
5. Periodontal ligaments hold the teeth in the alveoli.

Mastication

Mastication breaks large food particles into small ones.

Salivary Glands

1. Salivary glands produce serous and mucous secretions.
2. The three pairs of large salivary glands are the parotid, submandibular, and sublingual.

Pharynx (p. 647)

The pharynx consists of the nasopharynx, oropharynx, and laryngopharynx.

Esophagus (p. 648)

1. The esophagus connects the pharynx to the stomach. The upper and lower esophageal sphincters regulate movement.
2. Mucous glands produce a lubricating mucus.

Swallowing (p. 648)

1. During the voluntary phase of swallowing, a bolus of food is moved by the tongue from the oral cavity to the pharynx.
2. The pharyngeal phase is a reflex caused by the stimulation of stretch receptors in the pharynx.
   - The soft palate closes the nasopharynx, and the epiglottis, vestibular folds, and vocal folds close the opening into the larynx.
   - Pharyngeal muscles move the bolus to the esophagus.
3. The esophageal phase is a reflex initiated by the stimulation of stretch receptors in the esophagus. A wave of contraction (peristalsis) moves the food to the stomach.

Stomach (p. 650)

Anatomy and Histology of the Stomach

1. The openings of the stomach are the gastroesophageal opening (to the esophagus) and the pyloric orifice (to the duodenum).
2. The wall of the stomach consists of an external serosa, a muscle layer (longitudinal, circular, and oblique), a submucosa, and simple columnar epithelium (surface mucous cells).
3. Rugae are the folds in the stomach when it is empty.
4. Gastric pits are the openings to the gastric glands, which contain mucous neck cells, parietal cells, chief cells, and endocrine cells.

Secretions of the Stomach

1. Gastric juice is stomach secretions. Chyme is ingested food mixed with gastric juice.
   - Mucus protects the stomach lining.
   - Pepsinogen is converted to pepsin, which digests proteins.
   - Hydrochloric acid promotes pepsin activity and kills microorganisms.
   - Intrinsic factor is necessary for vitamin B_{12} absorption.
   - Gastrin and histamine regulate stomach secretions.
2. A proton pump (H^+–K^+ exchange pump) moves H^+ out of parietal cells.
3. There are three phases of stomach secretion.
   - The sight, smell, taste, or thought of food initiates the cephalic phase. Nerve impulses from the medulla stimulate hydrochloric acid, pepsinogen, gastrin, and histamine secretion.
   - Distention of the stomach, which stimulates gastrin secretion and activates CNS and local reflexes that promote secretion, initiates the gastric phase.

- Acidic chyme, which enters the duodenum and stimulates neuronal reflexes and the secretion of hormones (secretin, cholecystokinin) that inhibit gastric secretions, initiates the gastrointestinal phase.

Movements of the Stomach

1. Mixing waves mix the stomach contents with stomach secretions to form chyme.
2. Peristaltic waves move the chyme into the duodenum.

Small Intestine (p. 656)

The small intestine is divided into the duodenum, jejunum, and ileum.

Anatomy and Histology of the Small Intestine

1. Circular folds, villi, and microvilli greatly increase the surface area of the intestinal lining.
2. Absorptive, goblet, granular, and endocrine cells are in intestinal glands. Duodenal glands produce mucus.

Secretions of the Small Intestine

1. Mucus protects against digestive enzymes and gastric acids.
2. Digestive enzymes (disaccharidases and peptidases) are bound to the intestinal wall.
3. Chemical or tactile irritation, vagal stimulation, and secretin stimulate intestinal secretion.

Movement of the Small Intestine

1. Segmental contractions mix intestinal contents. Peristaltic contractions move materials distally.
2. Distension of the intestinal wall, local reflexes, and the parasympathetic nervous system stimulate contractions. Distension of the cecum initiates a reflex that stimulates contraction of the ileocecal sphincter.

Liver and Gallbladder (p. 659)

Anatomy and Histology of the Liver

1. The liver has four external lobes: right, left, caudate, and quadrate. Internally, the liver is divided into eight segments.
2. Liver segments are divided into lobules.
   - The hepatic cords are composed of columns of hepatocytes separated by the bile canaliculi.
   - The sinusoids are enlarged spaces filled with blood and lined with endothelium and hepatic phagocytic cells.

Functions of the Liver

1. The liver produces bile, which contains bile salts that emulsify fats.
2. The liver stores and processes nutrients, produces new molecules, and detoxifies molecules.
3. Hepatic phagocytic cells phagocytize red blood cells, bacteria, and other debris.
4. The liver produces blood components.

Blood Flow Through the Liver

1. Branches of the hepatic artery and the hepatic portal vein in the portal triads empty into hepatic sinusoids.
2. Hepatic sinusoids empty into central veins, which join to form the hepatic veins, which leave the liver.

Bile Transport

1. Bile canaliculi collect bile from hepatocytes and join the small hepatic ducts in the portal triads.

2. Small hepatic ducts converge to form the right and left hepatic ducts, which exit the liver.
3. The right and left hepatic ducts join to form the common hepatic duct.
4. The cystic duct from the gallbladder joins the common hepatic duct to form the common bile duct.
5. The common bile duct and pancreatic duct join the hepatopancreatic ampulla, which opens into the duodenum at the major duodenal papilla.

Gallbladder and Bile Storage

1. The gallbladder is a small sac on the inferior surface of the liver.
2. The gallbladder stores and concentrates bile.

Regulation of Bile Secretion

1. Secretin increases bile secretion (water and bicarbonate ions).
2. Cholecystokinin stimulates gallbladder contraction and relaxation of the sphincters of the bile duct and hepatopancreatic ampulla.

Pancreas (p. 664)
Anatomy and Histology of the Pancreas

1. The pancreas is an endocrine and an exocrine gland. Its exocrine function is the production of digestive enzymes.
2. The pancreas is divided into lobules that contain acini. The acini connect to a duct system that eventually forms the pancreatic duct.
3. The pancreatic duct joins the hepatopancreatic ampulla. The accessory pancreatic duct empties into the duodenum at the minor duodenal papilla.

Pancreatic Secretions

1. The aqueous component of pancreatic juice is produced by the small pancreatic ducts and contains bicarbonate ions.
2. The enzymatic component of pancreatic juice is produced by the acini and contains enzymes that digest carbohydrates, lipids, and proteins.

Regulation of Bile Secretion and Release

1. Secretin stimulates the release of the aqueous component, which neutralizes acidic chyme.
2. Cholecystokinin stimulates the secretion of the enzymatic component and relaxation of the sphincters of the pancreatic duct and hepatopancreatic ampulla.
3. Parasympathetic stimulation increases, and sympathetic stimulation decreases, secretion of the enzymatic component.

Large Intestine (p. 667)
Anatomy and Histology of the Large Intestine

1. The cecum forms a blind sac at the junction of the small and large intestines. The vermiform appendix is a blind tube off the cecum.
2. The ascending colon extends from the cecum superiorly to the right colic flexure. The transverse colon extends from the right to the left colic flexure. The descending colon extends inferiorly to join the sigmoid colon.
3. The sigmoid colon is an S-shaped tube that ends at the rectum.
4. Longitudinal smooth muscles of the large intestine wall are arranged into bands, called teniae coli. Haustra are pouches.
5. The mucosal lining of the large intestine is simple columnar epithelium with mucus-producing crypts.
6. The rectum is a straight tube that ends at the anus.
7. An internal anal sphincter (smooth muscle) and an external anal sphincter (skeletal muscle) are in the wall of the anal canal.

Secretions of the Large Intestine

1. Mucus protects the intestinal lining.
2. Microorganisms are responsible for vitamin K production, gas production, and much of the bulk of feces.

Movement in the Large Intestine

1. Haustra formation mixes the colon's contents and moves them slowly toward the anus.
2. Mass movements are strong peristaltic contractions that occur three or four times a day.
3. Defecation is the elimination of feces. Reflex activity moves feces through the internal anal sphincter. Voluntary activity regulates movement through the external anal sphincter.

Digestion, Absorption, and Transport (p. 671)

1. Digestion (mechanical and chemical) is the breakdown of organic molecules into their component parts.
2. Absorption is the uptake of digestive tract contents and transport is the distribution of nutrients throughout the body.

Carbohydrates

1. Carbohydrates include starches, glycogen, sucrose, lactose, glucose, and fructose.
2. Polysaccharides are broken down into monosaccharides by a number of different enzymes.
3. Monosaccharides are taken up by intestinal epithelial cells by symport that is powered by a Na^+ gradient or by facilitated diffusion.
4. The monosaccharides are carried to the liver, where the nonglucose sugars are converted to glucose.
5. Glucose is transported to the cells that require energy.
6. Glucose enters the cells through facilitated diffusion.
7. Insulin influences the rate of glucose transport.

Lipids

1. Lipids include triglycerides, phospholipids, steroids, and fat-soluble vitamins.
2. Emulsification is the transformation of large lipid droplets into smaller droplets and is accomplished by bile salts.
3. Lipase digests lipid molecules to form fatty acids and monoglyceride.
4. Micelles form around lipid digestion products and move to epithelial cells of the small intestine, where the products pass into the cells by simple diffusion.

5. Within the epithelial cells, free fatty acids are combined with monoglyceride to form triglyceride.
6. Proteins coat triglycerides, phospholipids, and cholesterol to form chylomicrons.
7. Chylomicrons enter lacteals within intestinal villi and are carried through the lymphatic system to the bloodstream.
8. Triglyceride is stored in adipose tissue, converted into other molecules, or used as energy.
9. Lipoproteins include chylomicrons, VLDL, LDL, and HDL.
10. LDL transports cholesterol to cells, and HDL transports it from cells to the liver.
11. LDL are taken into cells by receptor-mediated endocytosis, which is controlled by a negative-feedback mechanism.

Proteins

1. Pepsin in the stomach breaks proteins into smaller polypeptide chains.
2. Proteolytic enzymes from the pancreas produce small peptide chains.
3. Peptidases, bound to the microvilli of the small intestine, break down peptides.
4. Tripeptides, dipeptides, and amino acids are absorbed by symport that is powered by a Na^+ gradient.
5. Amino acids are transported to the liver, where the amino acids can be modified or released into the bloodstream.
6. Amino acids are actively transported into cells under the stimulation of growth hormone and insulin.
7. Amino acids are used as building blocks or for energy.

Water and Ions

1. Water can move in either direction across the wall of the small intestine, depending on the osmotic gradients across the epithelium.
2. Epithelial cells actively transport Na^+, K^+, Ca^{2+}, and Mg^{2+} from the intestine.
3. Chloride ions move passively through the wall of the duodenum and jejunum but are actively transported from the ileum.

Effects of Aging on the Digestive System (p. 676)

The mucous layer, the connective tissue, the muscles, and the secretions all tend to decrease as a person ages. These changes make an older person more open to infections and toxic agents.

REVIEW AND COMPREHENSION

1. Which layer of the digestive tract is in direct contact with the food that is consumed?
   a. mucosa   c. serosa
   b. muscularis   d. submucosa

2. The enteric plexus is found in the
   a. submucosa layer.   c. serosa layer.   e. all of the above.
   b. muscularis layer.   d. both a and b.

3. The tongue
   a. holds food in place during mastication.
   b. plays a major role in swallowing.
   c. helps form words during speech.
   d. is a major sense organ for taste.
   e. all of the above.

4. Dentin
   a. forms the surface of the crown of the teeth.
   b. holds the teeth to the periodontal ligaments.
   c. is found in the pulp cavity.
   d. makes up most of the structure of the teeth.
   e. is harder than enamel.

5. The number of premolar deciduous teeth is
   a. 0.   c. 4.   e. 12.
   b. 2.   d. 8.

6. Which of these glands does *not* secrete saliva into the oral cavity?
   a. submandibular glands   c. sublingual glands
   b. pancreas   d. parotid glands

7. The portion of the digestive tract in which digestion begins is the
   a. oral cavity.
   b. esophagus.
   c. stomach.
   d. duodenum.
   e. jejunum.

8. During swallowing (deglutition),
   a. the movement of food results primarily from gravity.
   b. the swallowing center in the medulla oblongata is activated.
   c. food is pushed into the oropharynx during the pharyngeal phase.
   d. the soft palate closes off the opening into the larynx.

9. The stomach
   a. has large folds in the submucosa and mucosa called rugae.
   b. has two layers of smooth muscle in the muscularis layer.
   c. opening from the esophagus is the pyloric orifice.
   d. has an area closest to the duodenum called the fundus.
   e. all of the above.

10. Which of these stomach cell types is *not* correctly matched with its function?
    a. surface mucous cells—produce mucus
    b. parietal cells—produce hydrochloric acid
    c. chief cells—produce intrinsic factor
    d. endocrine cells—produce regulatory hormones

11. HCl
    a. is an enzyme.
    b. creates the acid condition necessary for pepsin to work.
    c. is secreted by the small intestine.
    d. activates salivary amylase.
    e. all of the above.

12. Why doesn't the stomach digest itself?
    a. The stomach wall is not composed of protein, so it is not affected by proteolytic enzymes.
    b. The digestive enzymes of the stomach are not strong enough to digest the stomach wall.
    c. The lining of the stomach wall has a protective layer of epithelial cells.
    d. The stomach wall is protected by large amounts of mucus.

13. Which of these hormones stimulates stomach secretions?
    a. cholecystokinin
    b. insulin
    c. gastrin
    d. secretin

14. Which of these phases of stomach secretion is correctly matched?
    a. cephalic phase—the largest volume of secretion is produced
    b. gastric phase—gastrin secretion is inhibited by distension of the stomach
    c. gastric phase—initiated by chewing, swallowing, or thinking of food
    d. gastrointestinal phase—stomach secretions are inhibited

15. Which of these structures increase the mucosal surface of the small intestine?
    a. circular folds
    b. villi
    c. microvilli
    d. length of the small intestine
    e. all of the above

16. Given these parts of the small intestine:
    1. duodenum   2. ileum   3. jejunum

    Choose the arrangement that lists the parts in the order food encounters them as it passes from the stomach through the small intestine.
    a. 1,2,3
    b. 1,3,2
    c. 2,1,3
    d. 2,3,1
    e. 3,1,2

17. Which cells in the small intestine have digestive enzymes attached to their surfaces?
    a. mucous cells
    b. goblet cells
    c. endocrine cells
    d. absorptive cells

18. The hepatic sinusoids
    a. receive blood from the hepatic artery.
    b. receive blood from the hepatic portal vein.
    c. empty into the central veins.
    d. all of the above.

19. Given these ducts:
    1. common bile duct
    2. common hepatic duct
    3. cystic duct
    4. hepatic ducts

    Choose the arrangement that lists the ducts in the order bile passes through them when moving from the bile canaliculi of the liver to the small intestine.
    a. 3,4,2
    b. 3,2,1
    c. 3,4,1
    d. 4,1,2
    e. 4,2,1

20. Which of these might occur if a person suffers from a severe case of hepatitis that impairs liver function?
    a. Fat digestion is difficult.
    b. By-products of hemoglobin breakdown accumulate in the blood.
    c. Plasma proteins decrease in concentration.
    d. Toxins in the blood increase.
    e. All of the above occur.

21. The gallbladder
    a. produces bile.
    b. stores bile.
    c. contracts and releases bile in response to secretin.
    d. contracts and releases bile in response to sympathetic stimulation.
    e. both b and c.

22. The aqueous component of pancreatic secretions
    a. is secreted by the pancreatic islets.
    b. contains HCO_3^-.
    c. is released primarily in response to cholecystokinin.
    d. passes directly into the blood.
    e. all of the above.

23. Given these structures:
    1. ascending colon
    2. descending colon
    3. sigmoid colon
    4. transverse colon

    Choose the arrangement that lists the structures in the order that food encounters them as it passes between the small intestine and the rectum.
    a. 1,2,3,4
    b. 1,4,2,3
    c. 2,3,1,4
    d. 2,4,1,3
    e. 3,4,1,2

24. Which of these is *not* a function of the large intestine?
    a. absorption of fats
    b. absorption of water and salts
    c. chyme converted to feces
    d. production of mucus
    e. all of the above

25. Defecation
    a. can be initiated by stretch of the rectum.
    b. can occur as a result of mass movements.
    c. involves local reflexes.
    d. involves parasympathetic reflexes mediated by the spinal cord.
    e. all of the above.

26. Which of these structures produces enzymes that digest carbohydrates?
    a. salivary glands
    b. pancreas
    c. lining of the small intestine
    d. both a and b
    e. all of the above

27. Bile
    a. is an important enzyme for the digestion of fats.
    b. is made by the gallbladder.
    c. contains breakdown products from hemoglobin.
    d. emulsifies fats.
    e. both c and d.

28. Micelles are
    a. lipids surrounded by bile salts.
    b. produced by the pancreas.
    c. released into lacteals.
    d. stored in the gallbladder.
    e. reabsorbed in the colon.

29. If the thoracic duct were tied off, which of these classes of nutrients would *not* enter the circulatory system at their normal rate?
    a. amino acids
    b. glucose
    c. lipids
    d. fructose
    e. nucleotides

30. Which of these lipoprotein molecules transports excess lipids from cells back to the liver?
    a. high-density lipoprotein (HDL)
    b. low-density lipoprotein (LDL)
    c. very-low-density lipoprotein (VLDL)

Answers in Appendix E

C R I T I C A L T H I N K I N G

1. While anesthetized, patients sometimes vomit. Given that the anesthetic eliminates the swallowing reflex, explain why it is dangerous for an anesthetized patient to vomit.

2. Achlorhydria is a condition in which the stomach stops producing hydrochloric acid and other secretions. What effect would achlorhydria have on the digestive process? On red blood cell count?

3. Victor experienced the pain of a duodenal ulcer during final examination week. Describe the possible reasons. Explain what habits could have caused the ulcer, and recommend a reasonable remedy.

4. Gallstones sometimes obstruct the common bile duct. What are the consequences of such a blockage?

5. The bowel (colon) occasionally can become impacted. Given what you know about the functions of the colon and the factors that determine the movement of substances across the colon wall, predict the effect of the impaction on the contents of the colon above the point of impaction.

6. The bacterium *Vibrio cholerae* produces cholera toxin, which activates a chloride channel in the intestinal epithelium. In contrast, mutations that inactivate the same channel cause cystic fibrosis. Explain how increased chloride channel activity causes severe diarrhea, whereas decreased activity causes cystic fibrosis.

7. Discuss why the most effective oral rehydration therapy is water with sodium and glucose instead of water alone or water with fructose.

Answers in Appendix F

For additional study tools, go to: www.aris.mhhe.com.
Click on your course topic and then this textbook's author name/title. Register once for a semester's worth of interactive learning activities to help improve your grade!

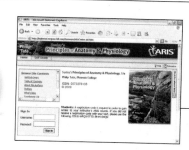

CHAPTER 22

Nutrition, Metabolism, and Temperature Regulation

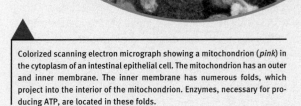

Colorized scanning electron micrograph showing a mitochondrion (*pink*) in the cytoplasm of an intestinal epithelial cell. The mitochondrion has an outer and inner membrane. The inner membrane has numerous folds, which project into the interior of the mitochondrion. Enzymes, necessary for producing ATP, are located in these folds.

Introduction

We are often more concerned with the taste of food than with its nutritional value when choosing from a menu or when selecting food to prepare. Knowing about nutrition is important, however, because the food we eat provides us with the energy and the building blocks necessary to synthesize new molecules. What happens if we do not obtain enough vitamins, or if we eat too much sugar and fats? Health claims about foods and food supplements bombard us every day. Which ones are ridiculous, and which ones have merit? A basic understanding of nutrition can answer these and other questions so that we can develop a healthy diet. It also allows us to know which questions currently do not have good answers.

▶This chapter explains **nutrition (p. 686)**, **metabolism (p. 695)**, **carbohydrate metabolism (p. 696)**, **lipid metabolism (p. 703)**, **protein metabolism (p. 704)**, the **interconversion of nutrient molecules (p. 705)**, **metabolic states (p. 707)**, **metabolic rate (p. 707)**, and **body temperature regulation (p. 710)**.

Tools for Success

A virtual cadaver dissection experience

Audio (MP3) and/or video (MP4) content, available at www.mhhe.aris.com

Nutrition

Nutrition (noo-trish′ŭn, to nourish) is the process by which the body obtains and uses certain components of food. The process includes digestion, absorption, transportation, and cell metabolism. Nutrition is also the evaluation of food and drink requirements for normal body function.

Nutrients

Nutrients are the chemicals taken into the body that are used to produce energy, to provide building blocks for new molecules, and to function in other chemical reactions. Some important substances in food, such as nondigestible plant fibers, are not nutrients. Nutrients are divided into six major classes: carbohydrates, proteins, lipids, vitamins, minerals, and water. Carbohydrates, lipids, and proteins are the major organic nutrients, and they are broken down by enzymes into their individual components during digestion. Many of these subunits are broken down further to supply energy, whereas others are used as building blocks for making new carbohydrates, lipids, and proteins. Carbohydrates, lipids, proteins, and water are required in fairly substantial quantities, whereas vitamins and minerals are required in only small amounts. Vitamins, minerals, and water are taken into the body without being digested.

Essential nutrients are nutrients that must be ingested because the body cannot manufacture them or adequate amounts of them. The essential nutrients include certain amino acids, certain fatty acids, most vitamins, minerals, water, and a minimum amount of carbohydrates. The term *essential* does not mean, however, that the body requires only the essential nutrients. Other nutrients are necessary but, if they are not part of the diet, they can be synthesized from the essential nutrients. Most of this synthesis takes place in the liver, which has a remarkable ability to transform and manufacture molecules. ▼

> 1. *Define nutrient and essential nutrient, and list the six major classes of nutrients.*

Kilocalories

The body uses the energy stored within the chemical bonds of certain nutrients. A **calorie** (kal′ō-rē, **cal**) is the amount of energy (heat) necessary to raise the temperature of 1 g of water 1°C. A **kilocalorie** (kil′ō-kal-ō-rē, **kcal**) is 1000 calories and is used to express the larger amounts of energy supplied by foods and released through metabolism.

A kilocalorie is often called a Calorie (with a capital C). Unfortunately, this usage has resulted in confusion between the terms *calorie* (with a lowercase c) and *Calorie* (with a capital C). It is common practice on food labels and in nutrition books to use *calorie* when *Calorie* (*kilocalorie*) is the proper term.

Most of the kilocalories supplied by food come from carbohydrates, proteins, or fats. For each gram of carbohydrate or protein that the body metabolizes, about 4 kcal of energy are released. Fats contain more energy per unit of weight than carbohydrates and proteins and yield about 9 kcal/g. Table 22.1 lists the kilocalories supplied by some typical foods. A typical diet in the United States consists of 50%–60% carbohydrates, 35%–45% fats, and 10%–15% protein. Table 22.1 also lists the carbohydrate, fat, and protein composition of some foods. ▼

> 2. *Define kilocalorie, and state the number of kilocalories supplied by a gram of carbohydrate, lipid, and protein.*

MyPyramid

Every 5 years, the U.S. Department of Health and Human Services (HHS) and the Department of Agriculture (USDA) jointly make their recommendations on what Americans should eat to be healthy. The latest recommendations, "The Dietary Guidelines for Americans 2005," were published in January 2005. Unlike the previous, single food guide pyramid, there are now 12 pyramids, which take into account a person's age, sex, and activity level. Thus, you can pick the pyramid, called MyPyramid, that best describes you (www.mypyramid.gov). All of the new pyramids have the same form (figure 22.1). Six colored bands represent the approximate, recommended proportions of grains (orange), vegetables (green), fruits (red), fats and oils (yellow), milk and milk products (blue), and meat and beans (purple). A balanced diet includes a variety of foods from each of the major food groups. Variety is necessary because no one food contains all of the nutrients necessary for health. Moderation is indicated by the narrowing of each food group from bottom to top. The wider base represents foods with little or no solid fats, added sugars, or caloric sweeteners. The climbing stick figure stresses the importance of daily exercise.

Recommendations and Criticisms of MyPyramid

The 2005 MyPyramid guidelines make several important recommendations: Weight control and exercise are emphasized. Moderation, how much we eat, is at least as important as what we eat. Reducing intake by 50–100 kilocalories per day may prevent weight gain in many people. Thirty minutes a day of exercise is good for the heart, but even more is needed for the waist. There are good and bad fats. We should limit the intake of saturated fats and *trans* fats, while using unsaturated fats. Carbohydrates are good, especially whole grains, but sugars and highly refined grains should be avoided. We should consume foods rich in vitamins, minerals, and fiber. Nine servings a day of fruits and vegetables and three cups a day of low-fat or fat-free milk or yogurt, or three servings of other dairy products, are recommended.

Although the MyPyramid guidelines are considered a positive step forward, there are some criticisms. The guidelines permit half of the carbohydrates in the diet to be refined starch, such as white bread, white rice, and chips. The body's response to these carbohydrates is similar to its response to sugar. The guidelines recommend protein sources that are "lean, low-fat, or fat-free." Thus, the guidelines do not distinguish between lean red meat and other sources of protein, such as poultry, fish, and beans, which have less saturated fats and more unsaturated fats. The guidelines' recommendation for dairy products is intended to increase calcium intake to help prevent osteoporosis. Supplements of calcium and vitamin D, along with exercise, are effective preventive measures that do not increase daily intake of kilocalories. Also, many people are lactose intolerant and cannot consume dairy products (see chapter 21). ▼

> 3. *List the six food groups shown in MyPyramid. How is the moderation of eating and exercise indicated in MyPyramid?*

Table 22.1 Food Consumption

Food	Quantity	Food Energy (kcal)	Carbohydrate (g)	Fat (g)	Protein (g)
Dairy Products					
Whole milk (3.3% fat)	1 cup	150	11	8	8
Low-fat milk (2% fat)	1 cup	120	12	5	8
Butter	1 tablespoon	100	—	12	—
Grain					
Bread, white enriched	1 slice	75	24	1	2
Bread, whole-wheat	1 slice	65	14	1	3
Fruit					
Apple	1	80	20	1	—
Banana	1	100	26	—	1
Orange	1	65	16	—	1
Vegetables					
Corn, canned	1 cup	140	33	1	4
Peas, canned	1 cup	150	29	1	8
Lettuce	1 cup	5	2	—	—
Celery	1 cup	20	5	—	1
Potato, baked	1 large	145	33	—	4
Meat, Fish, and Poultry					
Lean ground beef (10% fat)	3 ounces	185	—	10	23
Shrimp, french fried	3 ounces	190	9	9	17
Tuna, canned	3 ounces	170	—	7	24
Chicken breast, fried	3 ounces	160	1	5	26
Bacon	2 slices	85	—	8	4
Hot dog	1	170	1	15	7
Fast Foods					
McDonald's Egg McMuffin	1	327	31	15	19
McDonald's Big Mac	1	563	41	33	26
Taco Bell's beef burrito	1	466	37	21	30
Arby's roast beef	1	350	32	15	22
Pizza Hut Super Supreme	1 slice	260	23	13	15
Long John Silver's fish	2 pieces	366	21	22	22
Dairy Queen malt, large	1	840	125	28	22
Desserts					
Chocolate chip cookie	4	200	29	9	2
Apple pie	1 piece	345	51	15	3
Dairy Queen cone, large	1	340	52	10	10
Beverage					
Cola soft drink	12 ounces	145	37	—	—
Beer	12 ounces	144	13	—	1
Wine	3-1/2 ounces	73	2	—	—
Hard liquor (86 proof)	1-1/2 ounces	105	—	—	—
Miscellaneous					
Egg	1	80	1	6	6
Mayonnaise	1 tablespoon	100	—	11	—
Sugar	1 tablespoon	45	12	—	—

Figure 22.1 MyPyramid

The pyramid suggests the approaches to a healthy diet: Eat various foods (different-colored bands), eat different amounts of each food type (band width), eat in moderation (bands narrow from bottom to top), use fats and sugars sparingly (the wide base stands for foods with little or no solid fats or added sugars), and exercise (stick figure climbing stairs).

Source: U.S. Department of Agriculture.

Carbohydrates

Sources in the Diet

Carbohydrates include monosaccharides, disaccharides, and polysaccharides (see chapter 2). Most of the carbohydrates humans ingest come from plants. An exception is lactose (milk sugar), which is found in animal and human milk.

The most common monosaccharides in the diet are glucose and fructose. Plants capture the energy in sunlight and use that energy to produce glucose, which can be found in vegetables. Fructose (fruit sugar) and galactose are isomers of glucose. **Isomers** are molecules that have the same number and kinds of atoms, but the atoms are in different positions within the molecules. Glucose is found in vegetables and fructose is found in fruits, berries, honey, and high-fructose corn syrup, which is used to sweeten soft drinks and desserts. Galactose is usually found in milk.

The disaccharide sucrose (table sugar) is what most people think of when they use the term *sugar*. Sucrose is a glucose and a fructose molecule joined together (see figure 2.13), and its principal sources are sugarcane, sugar beets, maple sugar, and honey. Maltose (malt sugar), derived from germinating cereals, is a combination of two glucose molecules, and lactose (in milk) consists of a glucose and a galactose molecule.

The **complex carbohydrates** are the polysaccharides: starch, glycogen, and cellulose. These polysaccharides consist of many glucose molecules bound together to form long chains. Starch is an energy-storage molecule in plants and is found primarily in vegetables, fruits, and grains. Glycogen is an energy-storage molecule in animals and is located primarily in muscle and in the liver. By the time meats have been processed, they contain little, if any, glycogen because it is used up by the dying muscle cells (see "Anaerobic Respiration," p. 699). Cellulose forms cell walls, which surround plant cells.

Uses in the Body

During digestion, polysaccharides and disaccharides are split into monosaccharides, which are absorbed into the blood (see chapter 21). Humans can digest starch and glycogen because they can break the bonds between the glucose molecules of starch and glycogen. Humans are unable to digest cellulose because they cannot break the bonds between its glucose molecules. Instead, cellulose provides fiber, or "roughage," thereby increasing the bulk of feces, making it easier to defecate.

The liver converts fructose, galactose, and other monosaccharides absorbed by the blood into glucose. Glucose, whether absorbed from the digestive tract or synthesized in the liver, is an energy source used to produce **adenosine triphosphate (ATP)** molecules (see "Anaerobic Respiration," p. 699 and "Aerobic Respiration," p. 700). Blood glucose levels are carefully regulated because the brain relies almost entirely on glucose for its energy (see chapter 18).

If excess amounts of glucose are present, the glucose is converted into glycogen, which is stored in muscle and in the liver. Because cells can store only a limited amount of glycogen, any additional glucose is converted into fat, which is stored in adipose tissue. Glycogen can be rapidly converted back to glucose when energy is needed. For example, during exercise muscles convert glycogen to glucose and in-between meals the liver helps maintain blood sugar levels by converting glycogen to glucose, which is released into the blood.

In addition to being used as a source of energy, sugars have other functions. They form part of deoxyribonucleic acid (DNA), ribonucleic acid (RNA), and ATP molecules (see chapter 2); they also combine with proteins to form glycoproteins, such as the glycoprotein receptor molecules on the outer surface of the plasma membrane (see chapter 3).

Recommended Amounts

According to the Dietary Guidelines Advisory Committee, the **Acceptable Macronutrient Distribution Range (AMDR)** for carbohydrates is 45%–65% of total kilocalories. Although a minimum level of carbohydrates is not known, it is assumed that amounts of 100 g or less per day result in overuse of the body's proteins and fats for energy sources. The use of proteins for energy can result in the breakdown of muscle tissue because muscles are primarily protein. The extensive use of fats as an energy source can result in acidosis (see chapter 15).

Complex carbohydrates are recommended in the diet because starchy foods often contain other valuable nutrients, such as vitamins and minerals, and because the slower rate of digestion and absorption of complex carbohydrates does not result in large increases and decreases in blood glucose levels, as the consumption of large amounts of simple sugars does. Foods containing large amounts of simple sugars, such as soft drinks and candy, are rich in carbohydrates, but they have few other nutrients. For example, a typical soft drink is mostly sucrose, containing 9 teaspoons of sugar per 12-ounce container. In excess, the consumption of these kinds of foods usually results in obesity and tooth decay. ▼

4. *What are the most common monosaccharides in the diet? What are sucrose, maltose, and lactose?*

5. *Give three examples of complex carbohydrates. How does the body use them?*

6. How does the body use glucose and other monosaccharides?

7. What quantities of carbohydrate should be ingested daily?

Lipids

Sources in the Diet

About 95% of the lipids in the human diet are **triglycerides** (trī-glis′er-īdz). Triglycerides, which are sometimes called **triacylglycerols** (trī-as′il-glis′er-olz), consist of three fatty acids attached to a glycerol molecule (see figure 2.14). Triglycerides are often referred to as fats or oils. Fats are solid at room temperature, whereas oils are liquid. Fats and oils can be categorized as saturated or unsaturated. **Saturated fats and oils** have only single covalent bonds between the carbon atoms of their fatty acids (see figure 2.15a). They are found in the fats of meats (e.g., beef, pork), dairy products (e.g., whole milk, cheese, butter), eggs, coconut oil, and palm oil. **Unsaturated fats and oils** have one or more double covalent bonds between the carbon atoms of their fatty acids (see figure 2.15b). **Monounsaturated** fats have one double bond and **polyunsaturated** fats have two or more double bonds. Monounsaturated fats include olive and peanut oils; polyunsaturated fats are in fish, safflower, sunflower, and corn oil.

Unsaturated fatty acids can be classified according to the location of their first double bond from the omega (methyl) end of the fatty acid. The first double bond of an omega-3 fatty acid starts three carbon atoms after the omega end, an omega-6 fatty acid after six carbons, and an omega-9 fatty acid after nine carbons (see figure 2.15b).

Saturating Fats

Solid fats, such as shortening and margarine, work better than liquid oils do for preparing some foods, such as pastries. Polyunsaturated vegetable oils can be changed from a liquid to a solid by making them more saturated—that is, by decreasing the number of double covalent bonds in their polyunsaturated fatty acids. To saturate an unsaturated oil, hydrogen gas is bubbled through it. As hydrogen binds to the fatty acids, double covalent bonds are converted to single covalent bonds to produce a change in molecular shape that solidifies the oil. The more saturated the product, the harder it becomes at room temperature.

Unprocessed polyunsaturated fats are found mostly in the **cis form,** which means the hydrogen atoms are on the same side of the carbon–carbon double bond in their fatty acids (see figure 2.15b). During hydrogenation, some of the hydrogen atoms are transferred to the opposite side of the double bond to make the **trans form,** in which one hydrogen atom is on one side of the double bond and another is on the opposite side. Processed foods and oils account for most of the *trans* fats in the American diet, although some *trans* fats occur naturally in food from animal sources. *Trans* fatty acids raise the concentration of low-density lipoproteins and lower the concentration of high-density lipoproteins in the blood (see chapter 21). These changes are associated with a greater risk for cardiovascular disease.

The remaining 5% of lipids include cholesterol and phospholipids, such as **lecithin** (les′i-thin, *lekithos,* egg yolk). **Cholesterol** is a steroid (see figure 2.17) found in high concentrations in liver and egg yolks, but it is also present in whole milk, cheese, butter, and meats. Cholesterol is not found in plants. Phospholipids are major components of plasma membranes, and they are found in a variety of foods, such as egg yolks.

Uses in the Body

Triglycerides are important sources of energy that are used to produce ATP molecules. A gram of triglyceride delivers more than twice as many kilocalories as a gram of carbohydrate. Some cells, such as skeletal muscle cells, derive most of their energy from triglycerides.

After a meal, excess triglycerides that are not immediately used are stored in adipose tissue or the liver. Later, when energy is required, the triglycerides are broken down, and their fatty acids are released into the blood, where they are taken up and used by various tissues. In addition to storing energy, adipose tissue surrounds and pads organs, and under the skin adipose tissue is an insulator, which prevents heat loss.

Cholesterol is an important molecule with many functions in the body. It can be either obtained in food or manufactured by the liver and most other tissues. Cholesterol is a component of the plasma membrane, and it can be modified to form other useful molecules, such as bile salts and steroid hormones. Bile salts are necessary for fat digestion and absorption. Steroid hormones include the sex hormones estrogen, progesterone, and testosterone, which regulate the reproductive system.

The **eicosanoids** (ī′kō-să-noydz), which include prostaglandins and leukotrienes, are derived from fatty acids. The molecules are involved in activities such as inflammation, blood clotting, tissue repair, and smooth muscle contraction. Phospholipids, such as lecithin, are part of the plasma membrane and are used to construct the myelin sheath around the axons of nerve cells. Lecithin is also found in bile and helps emulsify fats.

Recommended Amounts

The AMDR for fats is 20%–35% for adults, 25%–35% for children and adolescents 4 to 18 years of age, and 30%–35% for children 2 to 3 years of age. Saturated fats should be 10% of total kilocalories, or as low as possible. Most dietary fat should come from sources of polyunsaturated and monounsaturated fats. Cholesterol should be limited to 300 mg (the amount in one egg yolk) or less per day and *trans* fat consumption should be as low as possible. These guidelines reflect the belief that excess amounts of fats, especially saturated fats, *trans* fats, and cholesterol, contribute to cardiovascular disease. The typical American diet derives 35%–45% of its kilocalories from fats, indicating that most Americans need to reduce fat consumption.

Most of the lecithin consumed in the diet is broken down in the digestive tract. The liver can manufacture all of the lecithin necessary to meet the body's needs, so it is not necessary to consume lecithin supplements.

Alpha-linolenic (lin-ō-len′ik) **acid** is an omega-3 fatty acid, and **linoleic** (lin-ō-lē′ik) **acid** is an omega-6 fatty acid. They are essential fatty acids, which must be ingested because humans lack the enzymes necessary to synthesize them. Other fatty acids, such as omega-9 fatty acids, can be synthesized from the essential fatty acids. Seeds, nuts, and legumes are good sources of alpha-linolenic and linoleic acids. Alpha-linolenic acid is in the green leaves of plants, and linoleic acid is found in grains.

 Fatty Acids and Blood Clotting

The essential fatty acids are used to synthesize prostaglandins that affect blood clotting. Linoleic acid can be converted to **arachidonic** (ă-rak-i-don′ik) **acid,** which is an omega-6 fatty acid. The arachidonic acid is used to produce thromboxanes, which increase blood clotting. Alpha-linolenic acid can be converted to **eicosapentaenoic** (ī′kō-să-pen-tă-nō′ik) **acid (EPA)** and **docasahexaenoic** (dō′kō-să-heks-ă-nō′ik) **acid (DHA),** which are omega-3 fatty acids. They can be used to synthesize prostaglandins that decrease blood clotting. Individuals who consume foods rich in EPA and DHA, such as herring, salmon, tuna, and sardines, increase the synthesis of prostaglandins from EPA and DHA. Individuals who eat these fish two or more times per week have a lower risk for heart attack than those who do not, possibly because of reduced blood clotting. EPA and DHA are also known to reduce blood triglyceride levels. Those who do not like to eat fish can take fish oil supplements as a source of EPA and DHA. Flaxseed is a source of alpha-linolenic acid, from which EPA and DHA can be synthesized. Whether or not flaxseed can provide the same benefits as EPA and DHA is under investigation. Individuals who have bleeding disorders, take anticoagulants, or anticipate surgery should follow their physician's advice regarding the use of these supplements because they can increase the risk for bleeding and hemorrhagic stroke. ▼

8. *What is the major source of lipids in the diet? What are other sources?*
9. *Define saturated and unsaturated fats.*
10. *How does the body use triglycerides, cholesterol, prostaglandins, and lecithin?*
11. *Describe the recommended dietary intake of lipids. List the essential fatty acids.*

Proteins

Sources in the Diet

Proteins are chains of amino acids (see chapter 2). Proteins in the body are constructed of 20 kinds of amino acids, which are divided into two groups: essential and nonessential. The body cannot synthesize **essential amino acids,** so they must be obtained in the diet. The nine essential amino acids are histidine, isoleucine, leucine, lysine, methionine, phenylalanine, threonine, tryptophan, and valine. Although the **nonessential amino acids** are necessary to construct our proteins, they are nonessential in the sense that it is not necessary to ingest them because they can be synthesized from the essential amino acids.

A **complete protein** food contains adequate amounts of all nine essential amino acids, whereas an **incomplete protein** food does not. Examples of complete proteins are meat, fish, poultry, milk, cheese, and eggs; incomplete proteins include leafy green vegetables, grains, and legumes (peas and beans).

Uses in the Body

The body uses essential and nonessential amino acids to synthesize proteins. Proteins perform numerous functions, as the following examples illustrate. Collagen provides structural strength in connective tissue, as does keratin in the skin, and the combination of actin and myosin makes muscle contraction possible. Enzymes regulate the rate of chemical reactions, and protein hormones regulate many physiological processes (see chapter 15). Proteins in the blood prevent changes in pH (buffers), promote blood clotting (coagulation factors), and transport oxygen and carbon dioxide in the blood (hemoglobin). Transport proteins (see chapter 3) move materials across plasma membranes, and other proteins in the plasma membrane function as receptor molecules. Antibodies, lymphokines, and complement are part of the immune system response that protects against microorganisms and other foreign substances.

The body also uses proteins as a source of energy, yielding the same amount of energy as carbohydrates. If excess proteins are ingested, the energy in the proteins can be stored by converting their amino acids into glycogen or fats.

Recommended Amounts

The AMDR for protein is 10%–35% of total kilocalories. If two incomplete proteins, such as rice and beans, are ingested, each can provide the amino acids lacking in the other. Thus, a correctly balanced vegetarian diet can provide all of the essential amino acids.

When protein intake is adequate, the synthesis and breakdown of proteins in a healthy adult occur at the same rate. The amino acids of proteins contain nitrogen, so saying that a person is in **nitrogen balance** means that the nitrogen content of ingested protein is equal to the nitrogen excreted in urine and feces. A starving person is in negative nitrogen balance because the nitrogen gained in the diet is less than that lost by excretion. In other words, when proteins are broken down for energy, more nitrogen is lost than is replaced in the diet. A growing child or a healthy pregnant woman, on the other hand, is in positive nitrogen balance because more nitrogen is going into the body to produce new tissues than is lost by excretion. ▼

12. *Distinguish between essential and nonessential amino acids and between complete and incomplete protein foods.*
13. *Describe some of the functions performed by proteins in the body.*
14. *What is the AMDR of proteins? Define nitrogen balance.*

Vitamins

Vitamins (vīt′ă-minz, life-giving chemicals) are organic molecules that exist in minute quantities in food and are essential to normal metabolism (table 22.2). **Essential vitamins** cannot be produced by the body and must be obtained through the diet. Because no

Table 22.2 Principal Vitamins

Vitamin	Fat (F)- or Water (W)- Soluble	Source	Function	Symptoms of Deficiency	Reference Daily Intake (RDI)*
A (retinol)	F	From provitamin beta carotene found in yellow and green vegetables: preformed in liver, egg yolk, butter, and milk	Necessary for rhodopsin synthesis, normal health of epithelial cells, and bone and tooth growth	Rhodopsin deficiency, night blindness, retarded growth, skin disorders, and increase in infection risk	900 RE†
B$_1$ (thiamine)	W	Yeast, grains, and milk	Involved in carbohydrate and amino acid metabolism, necessary for growth	Beriberi—muscle weakness (including cardiac muscle), neuritis, and paralysis	1.2 mg
B$_2$ (riboflavin)	W	Green vegetables, liver, wheat germ, milk, and eggs	Component of flavin adenine dinucleotide; involved in citric acid cycle	Eye disorders and skin cracking, especially at corners of the mouth	1.3 mg
B$_3$ (niacin)	W	Fish, liver, red meat, yeast, grains, peas, beans, and nuts	Component of nicotinamide adenine dinucleotide; involved in glycolysis and citric acid cycle	Pellagra—diarrhea, dermatitis, and nervous system disorder	16 mg
Pantothenic acid	W	Liver, yeast, green vegetables, grains, and intestinal bacteria	Constituent of coenzyme-A, glucose production from lipids and amino acids, and steroid hormone synthesis	Neuromuscular dysfunction and fatigue	5 mg
Biotin	W	Liver, yeast, eggs, and intestinal bacteria	Fatty acid and nucleic acid synthesis, movement of pyruvic acid into citric acid cycle	Mental and muscle dysfunction, fatigue, and nausea	30 µg
B$_6$ (pyridoxine)	W	Fish, liver, yeast, tomatoes, and intestinal bacteria	Involved in amino acid metabolism	Dermatitis, retarded growth, and nausea	1.7 mg
Folate	W	Liver, leafy green vegetables, and intestinal bacteria	Nucleic acid synthesis, hematopoiesis, prevent birth defects	Macrocytic anemia (enlarged red blood cells) and neural tube defects	0.4 mg
B$_{12}$ (cobalamins)	W	Liver, red meat, milk, and eggs	Necessary for red blood cell production, some nucleic acid and amino acid metabolism	Pernicious anemia and nervous system disorders	2.4 µg
C (ascorbic acid)	W	Citrus fruit, tomatoes, and green vegetables	Collagen synthesis, general protein metabolism	Scurvy—defective bone formation and poor wound healing	90 mg
D (cholecalciferol, ergosterol)	F	Fish liver oil, enriched milk, and eggs; provitamin D converted by sunlight to cholecalciferol in the skin	Promotes calcium and phosphorus use, normal growth and bone and teeth formation	Rickets—poorly developed, weak bones, osteomalacia; and bone reabsorption	10 µg‡
E (tocopherol, tocotrienols)	F	Wheat germ, cottonseed, palm, and rice oils; grain; liver; and lettuce	Prevents the oxidation of cell membranes and DNA	Hemolysis of red blood cells	15 mg
K (phylloquinone)	F	Alfalfa, liver, spinach, vegetable oils, cabbage, and intestinal bacteria	Required for synthesis of a number of clotting factors	Excessive bleeding due to retarded blood clotting	120 µg

*RDIs for people over 4 years of age; IU = international units.
†Retinol equivalents (RE). 1 retinol equivalent = 1 µg retinol or 6 µg beta carotene.
‡As cholecalciferol, 1 µg cholecalciferol = 40 IU vitamin D.

single food item or nutrient class provides all the essential vitamins, it is necessary to maintain a balanced diet by eating a variety of foods. The absence of an essential vitamin in the diet can result in a specific deficiency disease. A few vitamins, such as vitamin K, are produced by intestinal bacteria, and a few can be formed by the body from substances called provitamins. A **provitamin** is a part of a vitamin that can be assembled or modified by the body into a functional vitamin. Beta carotene is an example of a provitamin that can be modified by the body to form vitamin A. The other provitamins are **7-dehydrocholesterol** (dē-hī′dro-kō-les′ter-ol), which can be converted to vitamin D, and **tryptophan** (trip′tō-fan), which can be converted to niacin.

Vitamins are not broken down by catabolism but are used by the body in their original or slightly modified forms. After the chemical structure of a vitamin is destroyed, its function is usually lost. The chemical structure of many vitamins is destroyed by heat, such as when food is overcooked.

Many vitamins function as **coenzymes,** which combine with enzymes to make the enzymes functional. Without coenzymes and their enzymes, many chemical reactions would occur too slowly to support good health and even life. For example, vitamins B_2 and B_3, biotin (bī′ō-tin), and pantothenic (pan-tō-then′ik) acid are critical for some of the chemical reactions involved in the production of ATP. Folate (fō′lāt) and vitamin B_{12} are involved in nucleic acid synthesis. Vitamins A, B_1, B_6, B_{12}, C, and D are necessary for growth. Vitamin K is necessary for the synthesis of proteins involved in blood clotting (see table 22.2).

Vitamins are either fat-soluble or water-soluble. **Fat-soluble vitamins,** such as vitamins A, D, E, and K, dissolve in lipids. They are absorbed from the intestine along with lipids. Some of them can be stored in the body for a long time. Fat-soluble vitamins can accumulate in the body to the point of toxicity because they can be stored. **Water-soluble vitamins,** such as the B vitamins and vitamin C, dissolve in water. They are absorbed from the water in the intestinal tract and typically remain in the body only a short time before being excreted in the urine.

Vitamins were discovered at the beginning of the twentieth century. They were found to be associated with certain foods known to protect people from diseases such as rickets and beriberi. In 1941, the first Food and Nutrition Board established the **Recommended Dietary Allowances (RDAs),** which are the nutrient intakes sufficient to meet the needs of nearly all people in certain age and gender groups. RDAs were established for different-aged males and females, starting with infants and continuing on to adults. RDAs are also set for pregnant and lactating women. The RDAs have been reevaluated every 4–5 years and updated when necessary on the basis of new information.

The RDAs establish a minimum intake of vitamins and minerals that should protect almost everyone (97%) in a given group from diseases caused by vitamin or mineral deficiencies. Although personal requirements can vary, the RDAs are a good benchmark. The further dietary intake is below the RDAs, the more likely a nutritional deficiency will occur. On the other hand, the consumption of too large a quantity of some vitamins and minerals can have harmful effects. For example, the long-term ingestion of 3–10 times the RDA for vitamin A can cause bone and muscle pain, skin disorders, hair loss, and increased liver size. The long-term

consumption of 5–10 times the RDA of vitamin D can result in the deposition of calcium in the kidneys, heart, and blood vessels, and the regular consumption of more than 2 g of vitamin C daily can cause stomach inflammation and diarrhea.

 Free Radicals and Antioxidants

Free radicals are molecules, produced as part of normal metabolism, that are missing an electron. Free radicals can replace the missing electron by taking an electron from cell molecules, such as fats, proteins, or DNA, resulting in damage to the cell. Damage from free radicals may contribute to aging and certain diseases, such as atherosclerosis and cancer. The loss of an electron from a molecule is called oxidation. **Antioxidants** are substances that prevent the oxidation of cell components by donating an electron to free radicals. Examples of antioxidants include beta carotene (provitamin A), vitamin C, and vitamin E.

Many studies have been done to determine whether or not taking large doses of antioxidants is beneficial. Although future research may suggest otherwise, the consensus among scientists establishing the RDAs is that the best evidence presently available does not support claims that taking large doses of antioxidants prevents chronic disease or otherwise improves health. On the other hand, the amount of antioxidants normally found in a balanced diet that includes fruits and vegetables rich in antioxidants, combined with the complex mix of other chemicals found in food, can be beneficial. ▼

15. *What are vitamins, essential vitamins, and provitamins? Name the water-soluble vitamins and the fat-soluble vitamins.*
16. *List some of the functions of vitamins.*
17. *What are Recommended Dietary Allowances (RDAs)? Why are they useful?*

P R E D I C T ①
What would happen if vitamins were broken down during the process of digestion rather than being absorbed intact into the circulation?

Minerals

Minerals (min′er-ălz) are inorganic nutrients that are necessary for normal metabolic functions. Based on the amount of the mineral required in the diet for good health, the minerals are divided into two groups. The daily requirement for **major minerals** is 100 mg or more daily, whereas for **trace minerals** it is less than 100 mg daily. The requirement for some trace minerals is unknown. Minerals constitute about 4%–5% of total body weight and are components of coenzymes, a few vitamins, hemoglobin, and other organic molecules. Minerals are involved in a number of important functions, such as establishing resting membrane potentials and generating action potentials, adding mechanical strength to bones and teeth, combining with organic molecules, and acting as coenzymes, buffers, and regulators of osmotic pressure. Table 22.3 lists important minerals and their functions.

Table 22.3 — Important Minerals

Mineral	Function	Symptoms of Deficiency	Reference Daily Intake (RDI)*
Calcium	Bone and teeth formation, blood clotting, muscle activity, and nerve function	Spontaneous action potential generation in neurons and tetany	1300 mg
Chlorine	Blood acid–base balance, hydrochloric acid production in stomach	Acid–base imbalance	2.3 g†
Chromium	Associated with enzymes in glucose metabolism	Unknown	35 µg
Cobalt	Component of vitamin B_{12}, red blood cell production	Anemia	Unknown
Copper	Hemoglobin and melanin production, electron-transport system	Anemia and loss of energy	0.9 mg
Fluorine	Provides extra strength in teeth, prevents dental caries	No real pathology	4 mg
Iodine	Thyroid hormone production, maintenance of normal metabolic rate	Goiter and decrease in normal metabolism	150 µg
Iron	Component of hemoglobin, ATP production in electron-transport system	Anemia, decreased oxygen transport, and energy loss	18 mg
Magnesium	Coenzyme constituent, bone formation, and muscle and nerve function	Increased nervous system irritability, vasodilation, and arrhythmias	420 mg
Manganese	Hemoglobin synthesis, growth, and activation of several enzymes	Tremors and convulsions	2.3 mg
Molybdenum	Enzyme component	Unknown	45 µg
Phosphorus	Bone and teeth formation, energy transfer (ATP), and component of nucleic acids	Loss of energy and cellular function	1250 mg
Potassium	Muscle and nerve function	Muscle weakness, abnormal electrocardiogram, and alkaline urine	4.7 g
Selenium	Component of many enzymes	Unknown	55 µg
Sodium	Osmotic pressure regulation and nerve and muscle function	Nausea, vomiting, exhaustion, and dizziness	1.5 g†
Sulfur	Component of hormones, several vitamins, and proteins	Unknown	Unknown
Zinc	Component of several enzymes, carbon dioxide transport and metabolism, and protein metabolism	Deficient carbon dioxide transport and deficient protein metabolism	11 mg

*RDIs for people over 4 years of age, except for sodium.
†3.8 g sodium chloride (table salt).

Minerals are ingested by themselves or in combination with organic molecules, and they are obtained from animal and plant sources. Mineral absorption from plants, however, can be limited because the minerals tend to bind to plant fibers. Refined breads and cereals have hardly any minerals or vitamins because they are lost in the processing of the seeds used to make them. The seeds are crushed and the outer parts of the seeds, which contain most of their minerals and vitamins, are removed. The inner part of the seeds, which has few minerals and vitamins, is used to make the refined breads and cereals. Minerals and vitamins are often added to refined breads and cereals to compensate for their loss during the refinement process.

A balanced diet can provide all the vitamins and minerals required for good health for most people. Some nutritionists, however, recommend taking a once-a-day multiple vitamin and mineral supplement as insurance because many people do not have a balanced diet. ▼

18. *What are minerals? List some of the important functions of minerals.*

Vegetarian Diet

Plants alone can provide all of the protein required for good health. In order to get adequate amounts of the essential amino acids, a variety of protein sources, such as grains and legumes, should be consumed.

The Dietary Guidelines for Americans recommend that vegan diets be supplemented with vitamin B_{12}, vitamin D, calcium, iron, and zinc. This is especially important for children and pregnant and lactating women. Plant sources do not supply vitamin B_{12} or sufficient amounts of vitamin D, although the body can produce vitamin D with adequate exposure to sunlight (see chapter 5). Calcium is found in leafy green vegetables and nuts. Iron and zinc are in whole grains, nuts, and legumes. However, these minerals are either in low amounts or not easily absorbed.

Daily Values

Daily Values are dietary reference values that appear on food labels to help consumers plan a healthy diet. However, not all possible Daily Values are required to be listed on food labels. Daily Values are based on two other sets of reference values: Reference Daily Intakes and Daily Reference Values. The **Reference Daily Intakes (RDIs)** are based on the 1968 RDAs for certain vitamins and minerals. RDIs have been set for four categories of people: infants, toddlers, people over 4 years of age, and pregnant or lactating women. Generally, the RDIs are set to the highest 1968 RDA value of an age category. For example, the highest RDA for iron in males over 4 years of age is 10 mg/day and for females over 4 years of age is 18 mg/day. Thus, the RDI for iron is set at 18 mg/day. The **Daily Reference Values (DRVs)** are set for total fat, saturated fat, cholesterol, total carbohydrate, dietary fiber, sodium, potassium, and protein.

Having two standards on food labels, RDIs for vitamins and minerals and DRVs for other nutrients, was thought to be more confusing for consumers than having one standard. Therefore, the RDIs and DRVs were combined to form the Daily Values.

The Daily Values appearing on food labels are based on a 2000 kcal reference diet, which approximates the weight maintenance requirements of postmenopausal women, women who exercise moderately, teenage girls, and sedentary men (figure 22.2). On large food labels, additional information is listed based on a daily intake of 2500 kcal, which is adequate for young men.

The Daily Values for energy-producing nutrients are determined as a percentage of daily kilocaloric intake: 60% for carbohydrates, 30% for total fats, 10% for saturated fats, and 10% for proteins. The Daily Value for fiber is 11.5 g for each 1000 kcal of intake. The Daily Values for a nutrient in a 2000 kcal/day diet can be calculated on the basis of the recommended daily percentage of the nutrient and the kilocalories in a gram of the nutrient. For example, carbohydrates should be 60% of a 2000 kcal/day diet, or 1200 kcal/day (0.60 × 2000). Since there are 4 kcal in a gram of carbohydrate, the Daily Value for carbohydrate is 300 g/day (1200/4).

The Daily Values for some nutrients are the uppermost limits considered desirable because of the link between these nutrients and certain diseases. Thus, the Daily Values for total fats are less than 65 g; saturated fats, less than 20 g; and cholesterol, less than 300 mg because of their association with increased risk for heart

Nutrition Facts

Serving Size 1 oz. (28g/About 32 chips)
Servings Per Container 2.5

Amount Per Serving

Calories 160	Calories from Fat 90

	% Daily Value*
Total Fat 10g	**16%**
Saturated Fat 1.5g	**7%**
Cholesterol 0mg	**0%**
Sodium 170mg	**7%**
Total Carbohydrate 15g	**5%**
Dietary Fiber 1g	**4%**
Sugars less than 1g	
Protein 2g	

Vitamin A 0%	•	Vitamin C 0%
Calcium 2%	•	Iron 0%

* Percent Daily Values are based on a 2,000 calorie diet. Your daily values may be higher or lower depending on your calorie needs:

	Calories:	2,000	2,500
Total Fat	Less than	65g	80g
Sat Fat	Less than	20g	25g
Cholesterol	Less than	300mg	300mg
Sodium	Less than	2,400mg	2,400mg
Total Carbohydrate		300g	375g
Dietary Fiber		25g	30g

Calories per gram:
Fat 9 • Carbohydrate 4 • Protein 4

Figure 22.2 Food Label

disease. The Daily Value for sodium is less than 2400 mg because of its association with high blood pressure in some people.

For a particular food, the Daily Value is used to calculate the **Percent Daily Value (% Daily Value)** for some of the nutrients in one serving of the food (see figure 22.2). For example, if a serving of food has 3 g of fat and the Daily Value for total fat is 65 g, the % Daily Value is 5% (3/65 = 0.05, or 5%). The Food and Drug Administration (FDA) requires % Daily Values to be on food labels so that the public has useful and accurate dietary information.

P R E D I C T   2
One serving of a food has 30 g of carbohydrate. What % Daily Value for carbohydrate is on the food label for this food?

The % Daily Values for nutrients related to energy consumption are based on a 2000 kcal/day diet. For people who maintain their weight on a 2000 kcal/day diet, the total of the % Daily Values for each of these nutrients should add up to no more than 100%. For individuals consuming more or fewer kilocalories per day than 2000 kcal, however, the total of the % Daily Values can be more or fewer than 100%. For example, for a person consuming 2200 kcal/day, the total of the % Daily Values for each of these nutrients should be no more than 110% because 2200/2000 = 1.10, or 110%.

P R E D I C T   3
Suppose a person consumes 1800 kcal/day. What total % Daily Values for energy-producing nutrients is recommended?

When using the % Daily Values of a food to determine how the amounts of certain nutrients in the food fit into the overall diet, the number of servings in a container or package needs to be considered. For example, suppose a small (2.25-ounce) bag of corn chips has a % Daily Value of 16% for total fat. One might suppose that eating the bag of chips accounts for 16% of total fat for the day. The bag, however, contains 2.5 servings. Therefore, if all the chips in the bag are consumed, they account for 40% (16% × 2.5) of the maximum recommended total fat. ▼

19. What are the Reference Daily Intakes and the Daily Reference Values? When combined, what reference set of values is established?
20. Define % Daily Values. The % Daily Values appearing on food labels is based on how many kilocalories per day?

Metabolism

Metabolism (mě-tab′ō-lizm, change) is the total of all the chemical reactions that occur in the body. It consists of **catabolism** (kă-tab′ō-lizm), the energy-releasing process by which large molecules are broken down into smaller molecules, and **anabolism** (ă-nab′ō-lizm), the energy-requiring process by which small molecules are joined to form larger molecules.

Catabolism begins during the process of digestion and is concluded within individual cells. The energy derived from catabolism is used to drive anabolic reactions and processes such as active transport and muscle contraction. Anabolism occurs in all the cells of the body as they divide to form new cells, maintain their own intracellular structure, and produce molecules, such as hormones, neurotransmitters, and extracellular matrix molecules for export.

Large nutrient molecules, such as carbohydrates, lipids, and proteins, are broken down by digestion into smaller molecules, such as glucose, amino acids, and fatty acids, which are absorbed from the digestive tract into the blood (see chapter 21). These smaller molecules are taken into cells, they are catabolized, and the energy from them is used to combine adenosine diphosphate (ADP) and an inorganic phosphate group (P_i) to form ATP (figure 22.3):

$$ADP + P_i + Energy \rightarrow ATP$$

The energy in small nutrient molecules is used to produce many ATP molecules, each of which stores a small amount of energy. The smaller amount of energy in each of the many ATP molecules is more readily available for use in cells than is the larger amount of energy stored in nutrient molecules. ATP is often called the energy currency of the cell because, when it is spent, or broken down to ADP, energy becomes available for use by the cell. If a quarter represents an ATP molecule, then a $20 bill is analogous to a small nutrient molecule. The quarter (ATP) can be used in various vending machines (chemical reactions), but the $20 bill (nutrient molecule) cannot.

The chemical reactions responsible for the transfer of energy from the chemical bonds of nutrient molecules to ATP molecules involve oxidation–reduction reactions.

An **oxidation–reduction reaction** is a chemical reaction in which there is an exchange of electrons between the reactants. For example, when sodium and chlorine react to form sodium chloride, the sodium atom loses an electron, and the chlorine atom gains an electron (see figure 2.3). The loss of an electron by an atom is called **oxidation,** and the gain of an electron is called **reduction.** The transfer of the electron can be complete, resulting in an ionic bond, or it can be a partial transfer, resulting in a covalent bond. These reactions are called oxidation–reduction reactions because the complete or partial loss of an electron by one atom is accompanied by the gain of that electron by another atom. Synthesis and decomposition

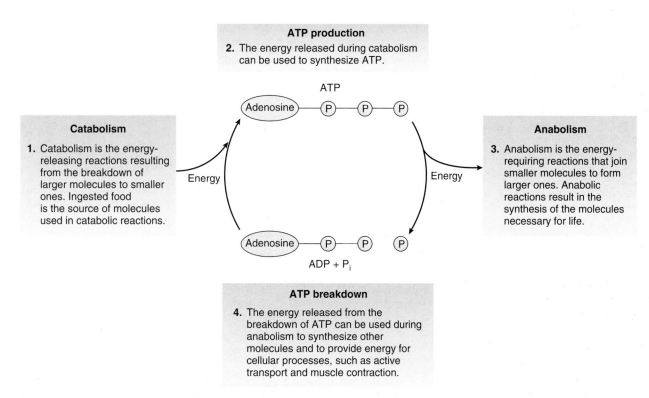

Process Figure 22.3 ATP Coupling of Catabolism and Anabolism
Energy from catabolism is required to form ATP from ADP and phosphate (P). Energy and a phosphate are given off when ATP is converted back to ADP.

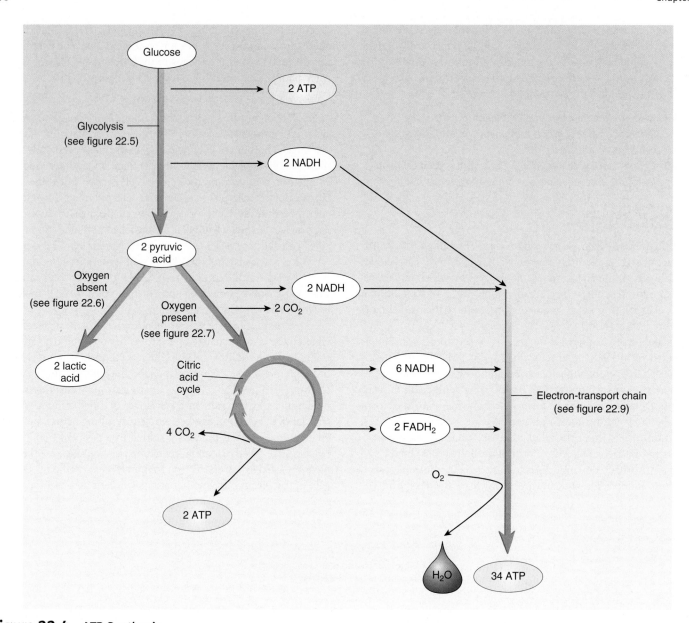

Figure 22.4 ATP Synthesis

Glycolysis, the citric acid cycle, and the electron-transport chain.

reactions can be oxidation–reduction reactions. Thus, a chemical reaction can be described in more than one way.

A nutrient molecule has many hydrogen atoms covalently bonded to the carbon atoms that form the "backbone" of the molecule. Nutrient molecules have many electrons because a hydrogen atom is a H^+ (proton) and an electron. Therefore, nutrient molecules are highly reduced. When a H^+ and an associated electron are lost from the nutrient molecule, the molecule loses energy and becomes oxidized. The energy in the electron is used to synthesize ATP. The major events of ATP synthesis are summarized in figure 22.4. ▼

> 21. *Define metabolism, catabolism, and anabolism. How is the energy derived from catabolism used to drive anabolic reactions?*
> 22. *Define oxidation–reduction reactions. How does the removal of hydrogen atoms from nutrient molecules result in a loss of energy from the nutrient molecule?*

Carbohydrate Metabolism

Monosaccharides are the breakdown products of carbohydrate digestion. Of these, glucose is the most important as far as cellular metabolism is concerned. Glucose is transported in the circulation to all the tissues of the body, where it is used as a source of energy. Any excess glucose in the blood following a meal can be used to form **glycogen** (glī′kō-jen, *glyks,* sweet), or it can be partially broken down and the components used to form fat. Glycogen is a short-term energy-storage molecule, which can be stored by the body only in limited amounts, whereas fat is a long-term energy-storage molecule that can be stored in the body in large amounts. Most of the body's glycogen is in skeletal muscle and in the liver.

Glycolysis

Carbohydrate metabolism begins with **glycolysis** (glī-kol′i-sis), which is a series of chemical reactions in the cytosol that results

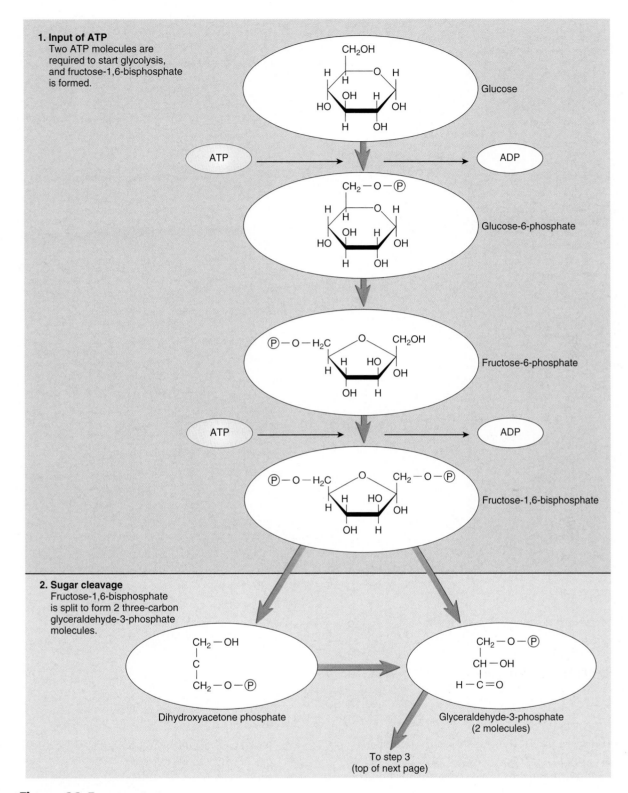

1. Input of ATP
Two ATP molecules are
required to start glycolysis,
and fructose-1,6-bisphosphate
is formed.

Glucose

ATP → ADP

Glucose-6-phosphate

Fructose-6-phosphate

ATP → ADP

Fructose-1,6-bisphosphate

2. Sugar cleavage
Fructose-1,6-bisphosphate
is split to form 2 three-carbon
glyceraldehyde-3-phosphate
molecules.

Dihydroxyacetone phosphate

Glyceraldehyde-3-phosphate
(2 molecules)

To step 3
(top of next page)

Process Figure 22.5   Glycolysis

The chemical reactions of glycolysis take place in the cytosol.

in the breakdown of glucose into two **pyruvic** (pī-roo′vik) **acid** molecules (figure 22.5).

Glycolysis is divided into four phases:

1. *Input of ATP.* The first steps in glycolysis require the input of energy in the form of two ATP molecules. A phosphate

group is transferred from ATP to the glucose molecule, a process called **phosphorylation** (fos′fōr-i-lā′shŭn), to form glucose-6-phosphate. The glucose-6-phosphate atoms are rearranged to form fructose-6-phosphate, which is then converted to fructose-1,6-bisphosphate by the addition of another phosphate group from another ATP.

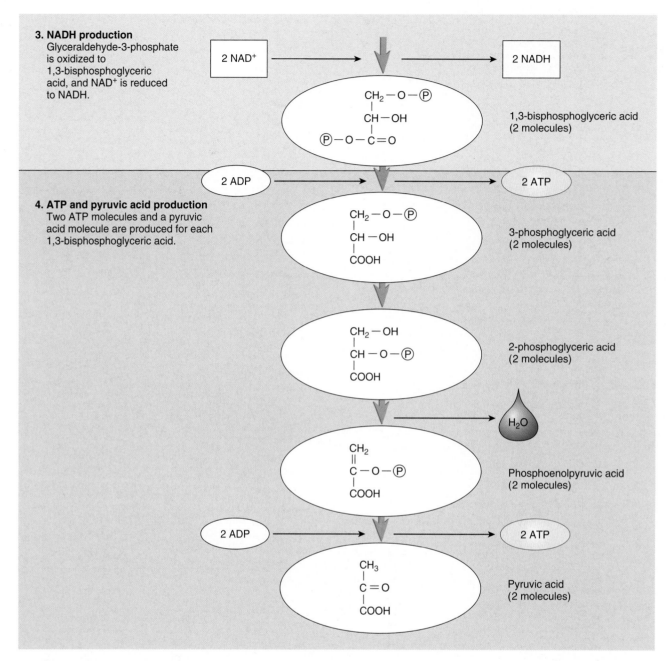

3. NADH production
Glyceraldehyde-3-phosphate is oxidized to 1,3-bisphosphoglyceric acid, and NAD⁺ is reduced to NADH.

2 NAD⁺ → → 2 NADH

1,3-bisphosphoglyceric acid (2 molecules)

2 ADP → → 2 ATP

4. ATP and pyruvic acid production
Two ATP molecules and a pyruvic acid molecule are produced for each 1,3-bisphosphoglyceric acid.

3-phosphoglyceric acid (2 molecules)

2-phosphoglyceric acid (2 molecules)

H_2O

Phosphoenolpyruvic acid (2 molecules)

2 ADP → → 2 ATP

Pyruvic acid (2 molecules)

Process Figure 22.5 *(continued)*

2. *Sugar cleavage.* Fructose-1,6-bisphosphate is cleaved into two three-carbon molecules, glyceraldehyde (glis-er-al′dĕ-hīd)-3-phosphate and dihydroxyacetone (dī′hī-drok-sē-as′e-tōn) phosphate. Dihydroxyacetone phosphate is rearranged to form glyceraldehyde-3-phospha te; consequently, two molecules of glyceraldehyde-3-phosphate result.

3. *NADH production.* Each glyceraldehyde-3-phosphate molecule is oxidized (loses two electrons) to form 1,3-bisphosphoglyceric (biz′phos-fo-gli′sēr′ik) acid, and **nicotinamide adenine** (nik-ō-tin′ă-mīd ad′ĕ-nēn) **dinucleotide (NAD⁺)** is reduced (gains two electrons) to

NADH. Glyceraldehyde-3-phosphate also loses two H⁺, one of which binds to NAD⁺.

$$NAD^+ + 2\,e^- + 2\,H^+ \rightarrow NADH + H^+$$

NAD⁺ is the oxidized form of nicotinamide adenine dinucleotide, and NADH is the reduced form. NADH is a carrier molecule with two high-energy electrons (e⁻) that can be used to produce ATP molecules through the electron-transport chain (see "Electron-Transport Chain," p. 702).

4. *ATP and pyruvic acid production.* The last four steps of glycolysis produce two ATP molecules and one pyruvic acid molecule from each 1,3-bisphosphoglyceric acid molecule.

Table 22.4	ATP Production from One Glucose Molecule	
Process	**Product**	**Total ATP Produced***
Glycolysis	4 ATP	2 ATP (4 ATP produced minus 2 ATP to start)
	2 NADH	6 ATP (or 4 ATP; see text)
Acetyl-CoA production	2 NADH	6 ATP
Citric acid cycle	2 ATP	2 ATP
	6 NADH	18 ATP
	2 FADH₂	4 ATP
Total		38 ATP (or 36 ATP; see text)

*NADH and FADH₂ are used in the production of ATP in the electron-transport chain.
Abbreviations: ATP = adenosine triphosphate, NADH = reduced nicotinamide adenine dinucleotide, FADH₂ = reduced flavin adenine diphosphate, acetyl-CoA = acetyl coenzyme A.

The events of glycolysis are summarized in table 22.4. Each glucose molecule that enters glycolysis forms two glyceraldehyde-3-phosphate molecules at the sugar cleavage phase. Each glyceraldehyde-3-phosphate molecule produces two ATP molecules, one NADH molecule, and one pyruvic acid molecule. Each glucose molecule, therefore, forms four ATP, two NADH, and two pyruvic acid molecules. Because the start of glycolysis requires the input of two ATP molecules, however, the final yield of each glucose molecule is two ATP, two NADH, and two pyruvic acid molecules (see figure 22.4).

If the cell has adequate amounts of oxygen, the NADH and pyruvic acid molecules are used in aerobic respiration to produce ATP. In the absence of sufficient oxygen, they are used in anaerobic respiration. ▼

> 23. Describe the four phases of glycolysis. What are the products of glycolysis?
> 24. What determines whether the pyruvic acid produced in glycolysis is used in aerobic or anaerobic respiration?

Anaerobic Respiration

Anaerobic (an-ār-ō′bik) **respiration** is the breakdown of glucose in the absence of oxygen to produce two molecules of **lactic** (lak′tik) **acid** and two molecules of ATP (figure 22.6). The ATP thus produced is a source of energy during activities, such as intense exercise, when insufficient oxygen is delivered to tissues. Anaerobic respiration can be divided into two phases:

1. *Glycolysis.* Glucose undergoes several reactions to produce two pyruvic acid molecules and two NADH. There is also a net gain of two ATP molecules.

1. Glycolysis converts glucose to two pyruvic acid molecules. The many reactions within the pathway are not shown. There is a net gain of two ATP and two NADH from glycolysis.

2. Anaerobic respiration, which does not require oxygen, includes glycolysis and converts the two pyruvic acid molecules produced by glycolysis to two lactic acid molecules. This conversion requires energy, which is derived from the NADH generated in glycolysis.

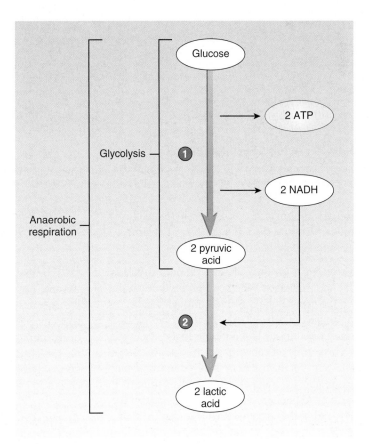

Process Figure 22.6 Glycolysis and Anaerobic Respiration

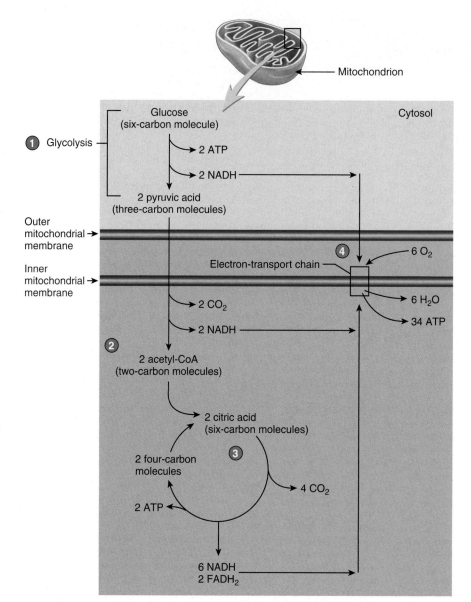

1. Glycolysis in the cytosol converts glucose to two pyruvic acid molecules and produces two ATP and two NADH. The NADH can go to the electron-transport chain in the inner mitochondrial membrane.

2. The two pyruvic acid molecules produced in glycolysis are converted to two acetyl-CoA molecules, producing two CO_2 and two NADH. The NADH can go to the electron-transport chain.

3. The two acetyl-CoA molecules enter the citric acid cycle, which produces four CO_2, six NADH, two $FADH_2$, and two ATP. The NADH and $FADH_2$ can go to the electron-transport chain.

4. The electron-transport chain uses NADH and $FADH_2$ to produce 34 ATP. This process requires O_2, which combines with H^+ to form H_2O.

Process Figure 22.7   Aerobic Respiration

Aerobic respiration involves four phases: (1) glycolysis, (2) acetyl-CoA formation, (3) the citric acid cycle, and (4) the electron-transport chain. The number of carbon atoms in a molecule is indicated after the molecule's name. As glucose is broken down, the carbon atoms from glucose are incorporated into carbon dioxide.

2. *Lactic acid formation.* Pyruvic acid is converted to lactic acid, a reaction that requires the input of energy from the NADH produced in glycolysis.

Lactic acid is released from the cells that produce it and is transported by the blood to the liver. When oxygen becomes available, the lactic acid in the liver can be converted through a series of chemical reactions into glucose. The glucose then can be released from the liver and transported in the blood to cells that use glucose as an energy source. This process of converting lactic acid to glucose is called the **Cori cycle.** Some of the reactions involved in converting lactic acid into glucose require the input of energy derived from ATP that is produced by aerobic respiration. The oxygen necessary for the synthesis of the ATP is part of the **oxygen deficit** (see chapter 8). ▼

25. Describe the two phases of anaerobic respiration. How many ATP molecules are produced by anaerobic respiration?

26. What happens to the lactic acid produced in anaerobic respiration when oxygen becomes available?

Aerobic Respiration

Aerobic (ār-ō′bik) **respiration** is the breakdown of glucose in the presence of oxygen to produce carbon dioxide, water, and 38 ATP molecules. Most of the ATP molecules required to sustain life are produced through aerobic respiration, which can be considered in four phases. The first phase of aerobic respiration, as in anaerobic respiration, is glycolysis. The remaining phases are acetyl-CoA formation, the citric acid cycle, and the electron-transport chain (figure 22.7).

Acetyl-CoA Formation

In the second phase of aerobic respiration, pyruvic acid moves from the cytosol into a mitochondrion, which is separated into

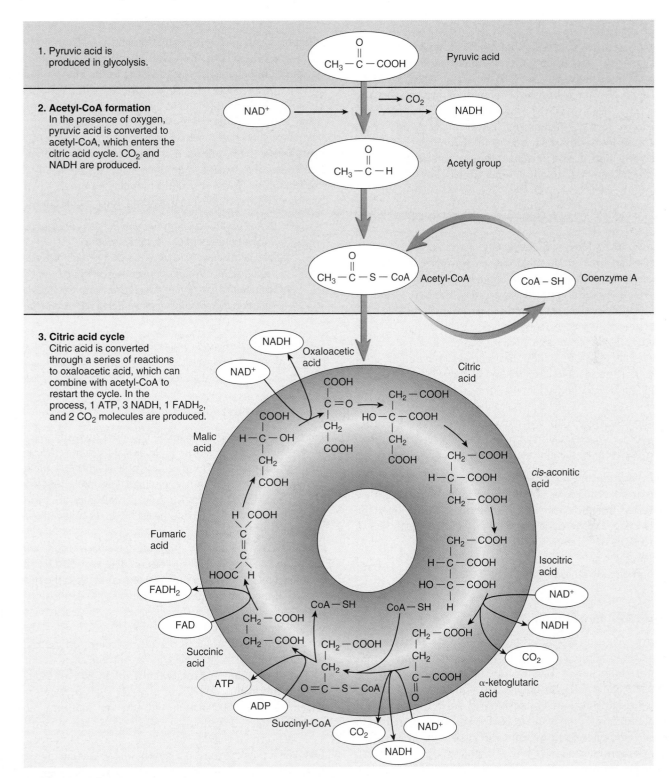

Process Figure 22.8 Acetyl-CoA and the Citric Acid Cycle

Pyruvic acid from the cytosol is converted to acetyl-CoA in mitochondria. The acetyl-CoA enters the citric acid cycle.

inner and outer compartments by the inner mitochondrial membrane. Within the inner compartment, enzymes remove a carbon and two oxygen atoms from the three-carbon pyruvic acid molecule to form carbon dioxide and a two-carbon acetyl (as'e-til) group (figure 22.8). Energy is released in the reaction and is used to reduce NAD^+ to NADH. The acetyl group combines with coenzyme A (CoA) to form acetyl-CoA. For each two pyruvic acid molecules from glycolysis, two acetyl-CoA molecules, two carbon dioxide molecules, and two NADH are formed (see figure 22.4).

Citric Acid Cycle

The third phase of aerobic respiration is the **citric acid cycle,** which is named after the six-carbon citric acid molecule formed in the first step of the cycle (see figure 22.8). It is also called the Krebs cycle after its discoverer, British biochemist Sir Hans Krebs. The citric acid cycle begins with the production of citric acid from the combination of acetyl-CoA and a four-carbon molecule called oxaloacetic (ok′să-lō-ă-sē′tik) acid. A series of reactions occurs, resulting in the formation of another oxaloacetic acid, which can start the cycle again by combining with another acetyl-CoA. During the reactions of the citric acid cycle, three important events occur:

1. *ATP production.* For each citric acid molecule, one ATP is formed.
2. *NADH and FADH₂ production.* For each citric acid molecule, three NAD^+ molecules are converted to NADH molecules, and one flavin (flā′vin) adenine dinucleotide (FAD) molecule is converted to $FADH_2$. The NADH and $FADH_2$ molecules are electron carriers that enter the electron-transport chain and are used to produce ATP.
3. *Carbon dioxide production.* Each six-carbon citric acid molecule at the start of the cycle becomes a four-carbon oxaloacetic acid molecule at the end of the cycle. Two carbon and four oxygen atoms from the citric acid molecule are used to form two carbon dioxide molecules. Thus, some of the carbon and oxygen atoms that make up food molecules, such as glucose, are eventually eliminated from the body as carbon dioxide. Humans literally breathe out part of the food they eat.

For each glucose molecule that begins aerobic respiration, two pyruvic acid molecules are produced in glycolysis, and they are converted into two acetyl-CoA molecules, which enter the citric acid cycle. To determine the number of molecules produced from glucose by the citric acid cycle, two "turns" of the cycle must be counted; the results are two ATP, six NADH, two $FADH_2$, and four carbon dioxide molecules (see figure 22.4).

Electron-Transport Chain

The fourth phase of aerobic respiration involves the **electron-transport chain** (figure 22.9), which is a series of electron carriers in the inner mitochondrial membrane. Electrons are transferred from NADH and $FADH_2$ to the electron-transport carriers, and H^+ are released from NADH and $FADH_2$. After the loss of the electrons and the H^+, the oxidized NAD^+ and FAD are reused to transport additional electrons from the citric acid cycle to the electron-transport chain.

The electrons released from NADH and $FADH_2$ pass from one electron carrier to the next through a series of oxidation–reduction reactions. Three of the electron carriers also function as proton pumps, which move the H^+ from the inner mitochondrial compartment into the outer mitochondrial compartment. Each proton pump accepts an electron, uses some of the electron's energy to export a H^+, and passes the electron to the next electron carrier. The last electron carrier in the series collects the electrons and combines them with oxygen and H^+ to form water.

$$1/2\ O_2 + 2\ H^+ + 2\ e^- \rightarrow H_2O$$

Without oxygen to accept the electrons, the reactions of the electron-transport chain cease, effectively stopping aerobic respiration.

The H^+ released from NADH and $FADH_2$ are moved from the inner mitochondrial compartment to the outer mitochondrial compartment by active transport. As a result, the concentration of H^+ in the outer compartment exceeds that of the inner compartment, and H^+ diffuse back into the inner compartment. The H^+ pass through certain channels formed by an enzyme called **ATP synthase.** As the H^+ diffuse down their concentration gradient, they lose energy that is used to produce ATP. This process is called the **chemiosmotic** (kem-ē-os-mot′ik) **model** because the chemical formation of ATP is coupled to a diffusion force similar to osmosis. ▼

27. *Define aerobic respiration, and list its products. Describe the four phases of aerobic respiration.*
28. *Why is the citric acid cycle a cycle? What molecules are produced as a result of the citric acid cycle?*
29. *What is the function of the electron-transport chain? Describe the chemiosmotic model of ATP production.*

P R E D I C T 4

Many poisons function by blocking certain steps in the metabolic pathways. For example, cyanide blocks the last step in the electron-transport chain. Explain why this blockage causes death.

Summary of ATP Production

For each glucose molecule, aerobic respiration produces a net gain of 38 ATP molecules: 2 from glycolysis, 2 from the citric acid cycle, and 34 from the NADH molecules and $FADH_2$ molecules that pass through the electron-transport chain (see table 22.4). For each NADH molecule formed, three ATP molecules are produced by the electron-transport chain; for each $FADH_2$ molecule, two ATP molecules are produced.

The number of ATP molecules produced from each glucose molecule can be 36 ATP molecules. The two NADH molecules produced by glycolysis in the cytosol cannot cross the inner mitochondrial membrane; thus, their electrons are donated to a shuttle molecule, which carries the electrons to the electron-transport chain. Depending on the shuttle molecule, each glycolytic NADH molecule can produce 2 or 3 ATP molecules. In skeletal muscle and the brain, 2 ATP molecules are produced for each NADH molecule, resulting in a total number of 36 ATP molecules; however, in the liver, kidneys, and heart, 3 ATP molecules are produced for each NADH molecule, and the total number of ATP molecules formed is 38.

Six carbon dioxide molecules are produced in aerobic respiration. Water molecules are reactants in some of the chemical reactions of aerobic respiration and products in others. Six water molecules are used, but 12 are formed, for a net gain of 6 water molecules. Thus, aerobic respiration can be summarized as follows: ▼

$$C_6H_{12}O_6 + 6\ O_2 + 6\ H_2O + 38\ ADP + 38\ P_i \rightarrow$$
$$6\ CO_2 + 12\ H_2O + 38\ ATP$$

30. *In aerobic respiration, how many ATP molecules are produced from one molecule of glucose through glycolysis, the citric acid cycle, and the electron-transport chain?*

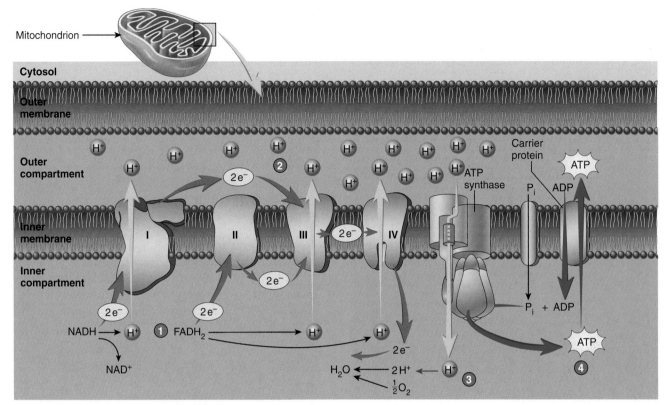

1. NADH or FADH$_2$ transfer their electrons to the electron-transport chain.

2. As the electrons move through the electron-transport chain, some of their energy is used to pump H$^+$ into the outer compartment, resulting in a higher concentration of H$^+$ in the outer than in the inner compartment.

3. The H$^+$ diffuse back into the inner compartment through special channels (ATP synthase) that couple the H$^+$ movement with the production of ATP. The electrons, H$^+$, and O$_2$ combine to form H$_2$O.

4. ATP is transported out of the inner compartment by a carrier protein that exchanges ATP for ADP. A different carrier protein moves phosphate into the inner compartment.

Process Figure 22.9 **Electron-Transport Chain**

The electron-transport chain in the mitochondrial inner membrane consists of four protein complexes (*purple*; numbered I to IV) with carrier proteins. As electrons are transferred from one carrier protein to another, they lose energy that is used to move H$^+$ out of the inner compartment. Hydrogen ions move back into the inner compartment through special channels (ATP synthase; *green*), which produce ATP. Carrier proteins (*brown*) move ATP out of and ADP and Pi into the inner compartment.

31. *Why is the total number of ATP produced in aerobic respiration listed as 38 or 36?*

 Quantity of ATP Produced from Glucose

The number of ATP molecules produced per glucose molecule is a theoretical number that assumes that two H$^+$ are necessary for the formation of each ATP. If more than two are required, the efficiency of aerobic respiration decreases. In addition, it costs energy to get ADP and phosphates into the mitochondria and to get ATP out. Considering all these factors, each glucose molecule yields about 25 ATP molecules instead of 38 ATP molecules.

Lipid Metabolism

Lipids are the body's main energy-storage molecules. In a healthy person, lipids are responsible for about 99% of the body's energy storage, and glycogen accounts for about 1%. Although proteins are used as an energy source, they are not considered storage molecules because the breakdown of proteins normally involves the loss of molecules that perform other functions.

Lipids are stored primarily as triglycerides in adipose tissue. There is constant synthesis and breakdown of triglycerides; thus, the fat present in adipose tissue today is not the same fat that was there a few weeks ago. Between meals, when triglycerides are broken down in adipose tissue, some of the fatty acids produced are released into the blood, where they are called **free fatty acids.** Other tissues, especially skeletal muscle and the liver, use the free fatty acids as a source of energy.

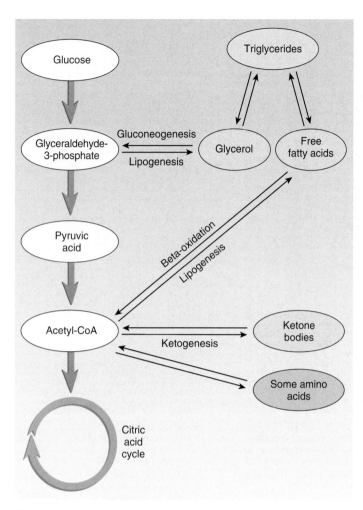

Figure 22.10 Lipid Metabolism

Triglyceride is broken down into glycerol and fatty acids. Glycerol enters glycolysis to produce ATP. The fatty acids are broken down by beta-oxidation into acetyl-CoA, which enters the citric acid cycle to produce ATP. Acetyl-CoA can also be used to produce ketone bodies (ketogenesis). Lipogenesis is the production of lipids. Glucose is converted to glycerol, and amino acids are converted to acetyl-CoA molecules. Acetyl-CoA molecules can combine to form fatty acids. Glycerol and fatty acids join to form triglycerides.

The metabolism of fatty acids occurs by **beta-oxidation,** a series of reactions in which two carbon atoms at a time are removed from the end of a fatty acid chain to form acetyl-CoA. The process of beta-oxidation continues to remove two carbon atoms at a time until the entire fatty acid chain is converted into acetyl-CoA molecules. Acetyl-CoA can enter the citric acid cycle and be used to generate ATP (figure 22.10).

Acetyl-CoA is also used in **ketogenesis** (kē-tō-jen′ě-sis), the formation of ketone bodies. In the liver, when large amounts of acetyl-CoA are produced, not all of the acetyl-CoA enters the citric acid cycle. Instead, two acetyl-CoA molecules combine to form a molecule of acetoacetic (as′e-tō-a-sē′tik) acid, which is converted mainly into β-hydroxybutyric (hī-drōk′sē-bū-tir′ik) acid and a smaller amount of acetone (as′e-tōn). Acetoacetic acid, β-hydroxybutyric acid, and acetone are called **ketone** (kē′tōn) **bodies;** they are released into the blood, where they travel to other

tissues, especially skeletal muscle. In these tissues, the ketone bodies are converted back into acetyl-CoA, which enters the citric acid cycle to produce ATP.

 The Danger of Excessive Amounts of Ketones

Small amounts of ketone bodies in the blood are normal and beneficial. An excessive production of ketone bodies is called **ketosis** (kē-tō′sis). If the increased number of acidic ketone bodies exceeds the capacity of the body's buffering systems, acidosis, a decrease in blood pH, can occur (see chapter 23). Conditions that increase fat metabolism can increase the rate of ketone body formation. Examples are starvation, diets consisting of proteins and fats with few carbohydrates, and untreated diabetes mellitus (see chapter 15). Ketone bodies are excreted by the kidneys and diffuse into the alveoli of the lungs. Ketone bodies in the urine and "acetone breath" are characteristic of untreated diabetes mellitus. ▼

32. *Define beta-oxidation, and explain how it results in ATP production.*
33. *What are ketone bodies, how are they produced, and for what are they used?*

Protein Metabolism

Once absorbed into the body, amino acids are quickly taken up by cells, especially in the liver. Amino acids are used to synthesize needed proteins (see chapter 3) or as a source of energy (figure 22.11). Unlike glycogen and triglycerides, amino acids are not stored in the body.

The synthesis of nonessential amino acids usually begins with keto acids (figure 22.12). A keto acid can be converted into an amino acid by replacing its oxygen with an amine group. Usually, this conversion is accomplished by transferring an amine group from an amino acid to the keto acid, a reaction called **transamination** (trans-am′i-nā′shŭn). For example, α-ketoglutaric acid (a keto acid) reacts with an amino acid to form glutamic acid (an amino acid, figure 22.13*a*). Most amino acids can undergo transamination to produce glutamic acid. The glutamic acid is used as a source of an amine group to construct most of the nonessential amino acids. A few nonessential amino acids are formed by other chemical reactions from the essential amino acids.

Amino acids can be used as a source of energy. In **oxidative deamination** (dē-am-i-nā′shŭn) or **deaminization** (dē-am′i-nizā′shŭn), an amine group is removed from an amino acid (usually glutamic acid), leaving ammonia and a keto acid (figure 22.13*b*). In the process, NAD$^+$ is reduced to NADH, which can enter the electron-transport chain to produce ATP. Ammonia is toxic to cells. An accumulation of ammonia to toxic levels is prevented because the liver converts it into urea, which is carried by the blood to the kidneys, where the urea is eliminated (figure 22.13*c*).

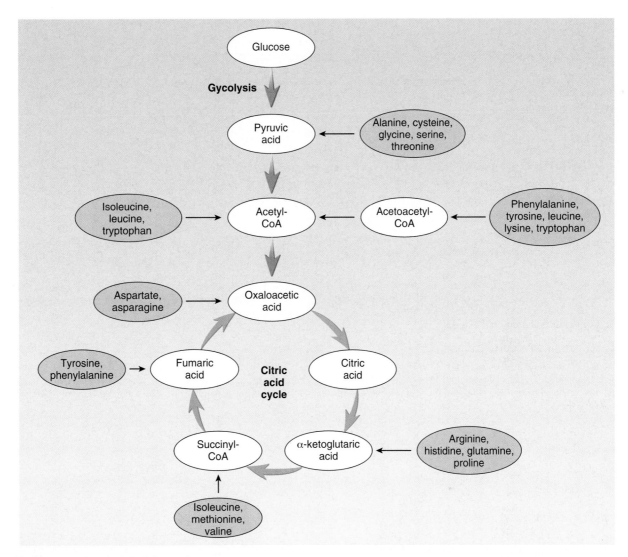

Figure 22.11   Amino Acid Metabolism

Various entry points for amino acids (*tan ovals*) into carbohydrate metabolism.

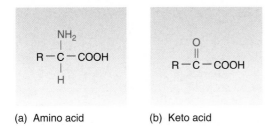

(a) Amino acid   (b) Keto acid

Figure 22.12   General Formulas of an Amino Acid and a Keto Acid

(*a*) Amino acid with a carboxyl group (—COOH), an amine group (NH₂), a hydrogen atom (H), and a group called "R," which represents the rest of the molecule. (*b*) Keto acid with a double-bonded oxygen replacing the amine group and the hydrogen atom of the amino acid.

Amino acids are also used as a source of energy by converting them into the intermediate molecules of carbohydrate metabolism (see figure 22.11). These molecules are then metabolized to yield ATP. The conversion of an amino acid often begins with a trans-amination or oxidative deamination reaction, in which the amino acid is converted into a keto acid (see figure 22.13). The keto acid enters the citric acid cycle or is converted into pyruvic acid or acetyl-CoA. ▼

34. What is accomplished by transamination and oxidative deamination?

35. How are proteins (amino acids) used to produce energy?

Interconversion of Nutrient Molecules

Blood glucose enters most cells by facilitated diffusion and is immediately converted to glucose-6-phosphate, which cannot recross the plasma membrane (figure 22.14*a*). Glucose-6-phosphate then continues through glycolysis to produce ATP. If, however, excess glucose is present (e.g., after a meal), it is used to form glycogen through a process called **glycogenesis** (glī-kō-jen′ĕ-sis).

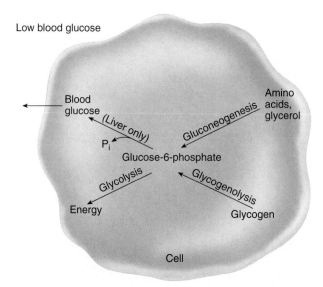

Figure 22.13 Amino Acid Reactions

(*a*) Transamination reaction in which an amine group is transferred from an amino acid to a keto acid to form a different amino acid. (*b*) Oxidative deamination reaction in which an amino acid loses an amine group to become a keto acid and to form ammonia. In the process, NADH, which can be used to generate ATP, is formed. (*c*) Ammonia is converted to urea in the liver. The actual conversion of ammonia to urea is more complex, involving a number of intermediate reactions that constitute the urea cycle.

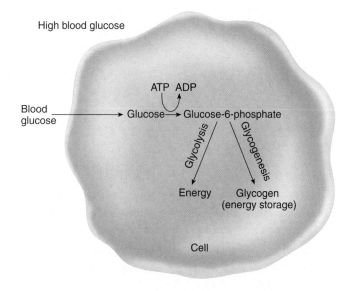

(a) When blood glucose levels are high, glucose enters the cell and is phosphorylated to form glucose-6-phosphate, which can enter glycolysis or glycogenesis.

(b) When blood glucose levels drop, glucose-6-phosphate can be produced through glycogenolysis or gluconeogenesis. Glucose-6-phosphate can enter glycolysis, or the phosphate group can be removed in liver tissue and glucose released into the blood.

Figure 22.14 Interconversion of Nutrient Molecules

Most of the body's glycogen is contained in skeletal muscle and the liver.

Once glycogen stores, which are quite limited, are filled, glucose and amino acids are used to synthesize lipids, a process called **lipogenesis** (lip-ō-jen′ĕ-sis) (see figure 22.10). Glucose molecules can be used to form glyceraldehyde-3-phosphate and acetyl-CoA. Amino acids can also be converted to acetyl-CoA. Glyceraldehyde-3-phosphate is converted to glycerol, and the two-carbon acetyl-CoA molecules are joined together to form fatty acid chains. Glycerol and three fatty acids then combine to form triglycerides.

Alcoholism and Cirrhosis of the Liver

Enzymes in the liver convert ethanol (beverage alcohol) into acetyl-CoA; in the process, two NADH molecules are produced. The NADH molecules enter the electron-transport chain and are used to produce ATP molecules. Each gram of ethanol provides 7 kcal of energy. The high level of NADH in the cell resulting from the metabolism of ethanol inhibits the production of NADH by glycolysis and by the citric acid cycle. Consequently, sugars and amino acids are not broken down but are converted into fats, which accumulate in the liver. Chronic alcohol abuse can, therefore, result in **cirrhosis** (sir-rō'sis) **of the liver,** which involves fat deposition, cell death, inflammation, and scar tissue formation. Death can occur because the liver is unable to carry out its normal functions.

When glucose is needed, glycogen can be broken down into glucose-6-phosphate through a set of reactions called **glycogenolysis** (glī'kō-jĕ-nol'i-sis) (figure 22.14*b*). In skeletal muscle, glucose-6-phosphate continues through glycolysis to produce ATP. The liver can use glucose-6-phosphate for energy or can convert it to glucose, which diffuses into the blood. The liver can release glucose, but skeletal muscle cannot because it lacks the necessary enzymes to convert glucose-6-phosphate into glucose.

The release of glucose from the liver is necessary to maintain blood glucose levels between meals. Maintaining these levels is especially important to the brain, which normally uses only glucose for an energy source and consumes about two-thirds of the total glucose used each day. When liver glycogen levels are inadequate to supply glucose, it is synthesized from molecules other than carbohydrates, such as amino acids from proteins and glycerol from triglycerides, in a process called **gluconeogenesis** (gloo'kō-nē-ō-jen'ĕ-sis). Most amino acids can be converted into citric acid cycle molecules, acetyl-CoA, or pyruvic acid (see figure 22.11). Through a series of chemical reactions, these molecules are converted into glucose. Glycerol enters glycolysis by becoming glyceraldehyde-3-phosphate. ▼

36. *Define glycogenesis, lipogenesis, glycogenolysis, and gluconeogenesis.*

Metabolic States

The absorptive and postabsorptive states are the two major metabolic states of the body. The regulation of these states is discussed in chapter 15 (see "Hormonal Regulation of Nutrients" p. 453). The **absorptive state** is the period immediately after a meal when nutrients are being absorbed through the intestinal wall into the circulatory and lymphatic systems (figure 22.15).

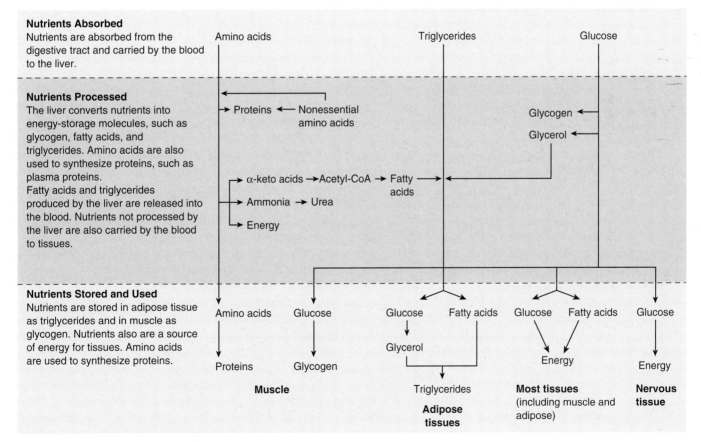

Figure 22.15 Events of the Absorptive State
Absorbed molecules, especially glucose, are used as sources of energy. Molecules not immediately needed for energy are stored: Glucose is converted to glycogen or triglycerides, triglycerides are deposited in adipose tissue, and amino acids are converted to triglycerides or carbohydrates.

The absorptive state usually lasts about 4 hours after each meal, and the cells use most of the glucose that enters the circulation for the energy they require. The remainder of the glucose is converted into glycogen or fats. Most of the absorbed fats are deposited in adipose tissue. Many of the absorbed amino acids are used by cells in protein synthesis, some are used for energy, and others enter the liver and are converted into fats or carbohydrates.

The **postabsorptive state** occurs late in the morning, late in the afternoon, or during the night after each absorptive state is concluded. It is vital to the body's homeostasis that normal blood glucose levels be maintained, especially for normal functioning of the brain. During the postabsorptive state, blood glucose levels are maintained by the conversion of other molecules to glucose (figure 22.16). The first source of blood glucose during the postabsorptive state is the glycogen stored in the liver. This glycogen supply, however, can provide glucose for only about 4 hours. The glycogen stored in skeletal muscles can also be used during times of vigorous exercise. As glycogen stores are depleted, fats are used as an energy source. The glycerol from triglycerides can be converted to glucose. The fatty acids from fat can be converted to acetyl-CoA, moved into the citric acid cycle, and used as a source of energy to produce ATP. In the liver, acetyl-CoA is used to produce ketone bodies, which other tissues use for energy. The use of fatty acids as an energy source partly eliminates the need to use glucose for energy, resulting in reduced glucose removal from the blood and the maintenance of blood glucose levels at homeostatic levels. Proteins can also be used as a source of glucose or can be used for energy production, again sparing the use of blood glucose. ▼

37. *What happens to glucose, fats, and amino acids during the absorptive state?*
38. *Why is it important to maintain blood glucose levels during the postabsorptive state? Name three sources of this glucose.*

Metabolic Rate

Metabolic rate is the total amount of energy produced and used by the body per unit of time. A molecule of ATP exists for less than 1 minute before it is degraded back to ADP and inorganic phosphate. For this reason, ATP is produced in cells at about the same rate as it is used. Thus, in examining metabolic rate, ATP production and use can be roughly equated. Metabolic rate is usually estimated by measuring the amount of oxygen used per minute because most ATP production involves the use of oxygen. One liter of oxygen consumed by the body is assumed to produce 4.825 kcal of energy.

The daily input of energy should equal the metabolic expenditure of energy; otherwise, a person will gain or lose weight. For a typical 23-year-old, 70 kg (154-pound) male to maintain his weight, the daily input should be 2700 kcal/day; for a typical 58 kg (128-pound) female of the same age, 2000 kcal/day is necessary. A pound of body fat provides about 3500 kcal. Reducing kilocaloric intake by 500 kcal/day can result in the loss of 1 pound of fat per week. Clearly, adjusting kilocaloric input is an important way to control body weight.

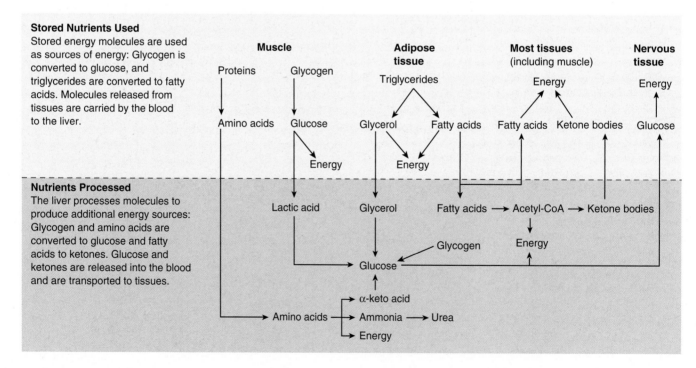

Figure 22.16   Events of the Postabsorptive State

Stored energy molecules are used as sources of energy: Glycogen is converted to glucose; triglycerides are broken down to fatty acids, some of which are converted to ketones; and proteins are converted to glucose.

Proportion of Fat in the Diet and Body Weight

Not only the number of kilocalories ingested but also the proportion of fat in the diet has an effect on body weight. To convert dietary fat into body fat, 3% of the energy in the dietary fat is used, leaving 97% for storage as fat deposits. On the other hand, the conversion of dietary carbohydrate to fat requires 23% of the energy in the carbohydrate, leaving just 77% as body fat. If two people have the same kilocaloric intake, the one with the higher proportion of fat in his or her diet is more likely to gain weight because fewer kilocalories are used to convert the dietary fat into body fat.

Metabolic energy is used in three ways: for basal metabolism, for the thermic effect of food, and for muscular activity.

Basal Metabolic Rate

The **basal metabolic rate (BMR)** is the energy needed to keep the resting body functional. It is the metabolic rate calculated in expended kilocalories per square meter of body surface area per hour. BMR is determined by measuring the oxygen consumption of a person who is awake but restful and has not eaten for 12 hours. The liters of oxygen consumed are then multiplied by 4.825 because each liter of oxygen used results in the production of 4.825 kcal of energy. A typical BMR for a 70 kg (154-pound) male is 38 kcal/m^2/h.

In the average person, basal metabolism accounts for about 60% of energy expenditure. Basal metabolism supports active-transport mechanisms, muscle tone, maintenance of body temperature, beating of the heart, and other activities. A number of factors can affect the BMR. Muscle tissue is metabolically more active than adipose tissue, even at rest. Younger people have a higher BMR than older people because of increased cell activity, especially during growth. Fever can increase BMR 7% for each degree Fahrenheit increase in body temperature. During dieting or fasting, greatly reduced kilocaloric input can depress BMR, which apparently is a protective mechanism to prevent weight loss. Thyroid hormones can increase BMR on a long-term basis, and epinephrine can increase BMR on a short-term basis. Males have a greater BMR than females because men have proportionately more muscle tissue and less adipose tissue than women do. During pregnancy, a woman's BMR can increase 20% because of the metabolic activity of the fetus.

Thermic Effect of Food

The second component of metabolic energy is the assimilation of food. When food is ingested, the accessory digestive organs and the intestinal lining produce secretions, the motility of the digestive tract increases, active transport increases, and the liver is involved in the synthesis of new molecules. The energy cost of these events is called the **thermic effect of food,** and it accounts for about 10% of the body's energy expenditure.

Muscular Activity

Muscular activity consumes about 30% of the body's energy. Physical activity resulting from skeletal muscle movement requires the expenditure of energy. In addition, energy must be provided for the increased contraction of the heart and muscles of respiration. The number of kilocalories used in an activity depends almost entirely on the amount of muscular work performed and on the duration of the activity. Despite the fact that studying can make a person feel tired, intense mental concentration produces little change in BMR.

Energy loss through muscular activity is the only component of energy expenditure that a person can reasonably control. A comparison of the number of kilocalories gained from food versus the number of kilocalories lost in exercise reveals why losing weight can be difficult. For example, if brisk walking uses 225 kcal/h, it takes 20 minutes of brisk walking to burn off the 75 kcal in one slice of bread (75/225 = 0.33 h). Research suggests that a combination of appropriate physical activity and appropriate kilocaloric intake is the best approach to maintaining a healthy body composition and weight.

PREDICT
If watching TV uses 95 kcal/h, how long does it take to burn off the kilocalories in one cola or beer (see table 22.1)? If jogging at a pace of 6 mph uses 580 kcal/h, how long does it take to use the kilocalories in one cola or beer? ▼

39. *Define metabolic rate.*
40. *What is BMR? What factors can alter BMR?*
41. *What is the thermic effect of food?*
42. *BMR, the thermic effect of food, and muscular activity each accounts for what percent of total energy expenditure?*
43. *How are kilocaloric input and output adjusted to maintain body weight?*

Body Temperature Regulation

Humans can maintain a relatively constant internal body temperature despite changes in the temperature of the surrounding environment. The maintenance of a constant body temperature is very important to homeostasis. Most enzymes are very temperature-sensitive, functioning only in narrow temperature ranges. Environmental temperatures are too low for normal enzyme function, and the heat produced by metabolism helps maintain body temperature at a steady, elevated level that is high enough for normal enzyme function.

Free energy is the total amount of energy liberated by the complete catabolism of food. It is usually expressed in terms of kilocalories (kcal) per mole of food consumed. For example, the complete catabolism of 1 mole of glucose (168 g, see appendix C) releases 686 kcal of free energy. About 43% of the total energy released by catabolism is used to produce ATP and to accomplish biological work, such as anabolism, muscular contraction, and other cellular activities. The remaining energy is lost as **heat.**

PREDICT
Explain why we become warm during exercise and why we shiver when it is cold.

Starvation

Starvation results from the inadequate intake of nutrients or the inability to metabolize or absorb nutrients. It has a number of causes, such as prolonged fasting, anorexia, deprivation, and disease. No matter what the cause, starvation follows the same course and consists of three phases. The events of the first two phases occur even during relatively short periods of fasting or dieting, but the third phase occurs only in prolonged starvation and can end in death.

During the first phase of starvation, blood glucose levels are maintained through the production of glucose from glycogen, proteins, and fats. At first, glycogen is broken down into glucose; however, only enough glycogen is stored in the liver to last a few hours. Thereafter, blood glucose levels are maintained by the breakdown of proteins and fats. Fats are decomposed into fatty acids and glycerol. Fatty acids can be used as a source of energy, especially by skeletal muscle, thus decreasing the use of glucose by tissues other than the brain. Glycerol can be used to make a small amount of glucose, but most of the glucose is formed from the amino acids of proteins. In addition, some amino acids can be used directly for energy.

In the second phase, which can last for several weeks, fats are the primary energy source. The liver metabolizes fatty acids into ketone bodies, which can be used as a source of energy. After about a week of fasting, the brain begins to use ketone bodies, as well as glucose, for energy. This usage decreases the demand for glucose, and the rate of protein breakdown diminishes but does not stop. In addition, the proteins not essential for survival are used first.

The third phase of starvation begins when the fat reserves are depleted and a switch to proteins as the major energy source takes place. Muscles, the largest source of protein in the body, are rapidly depleted. At the end of this phase, proteins essential for cellular functions are broken down, and cell function degenerates.

In addition to weight loss, the symptoms of starvation include apathy, listlessness, withdrawal, and increased susceptibility to infectious disease. Few people die directly from starvation because they usually die of an infectious disease first. Other signs of starvation include changes in hair color, flaky skin, and massive edema in the abdomen and lower limbs, causing the abdomen to appear bloated.

During starvation, the body's ability to consume normal volumes of food also decreases. Foods high in bulk but low in protein content often cannot reverse the process of starvation. Intervention involves feeding the starving person low-bulk food that provides ample proteins and kilocalories and is fortified with vitamins and minerals. Starvation also results in dehydration, and rehydration is an important part of intervention. Even with intervention, a victim may be so affected by disease or weakness that he or she cannot recover.

Obesity

Obesity is the presence of excess body fat, resulting from the ingestion of more food than is necessary for the body's energy needs. Obesity can be defined on the basis of body weight, body mass index, or percent body fat. "Desirable body weight" is listed in the Metropolitan Life Insurance Table (1983) and indicates, for any height, the weight that is associated with a maximum life span. Being overweight is defined as weighing 10% to 20% more than the "desirable weight," and being obese is defined as weighing 20% or more than the "desirable weight." Body mass index (BMI) can be calculated by dividing a person's weight (Wt) in kilograms by the square of his or her height (Ht) in meters: $BMI = Wt/Ht^2$. A BMI greater than 25–27 is overweight, and a value greater than 30 is defined as obese. About 10% of people in the United States have a BMI of 30 or greater. In terms of the percent of the total body weight contributed by fat, 15% body fat or less in men and 25% body fat or less in women is associated with reduced health risks. Obesity is defined to be more than 25% body fat in men and 30%–35% in women.

Obesity is classified according to the number and size of fat cells. The greater the amount of lipids stored in the fat cells, the larger their size. In **hyperplastic obesity,** a greater than normal number of fat cells occurs, and they are larger than normal. This type of obesity, which is associated with massive obesity, begins at an early age. In nonobese children, the number of fat cells triples or quadruples between birth and 2 years of age and then remains relatively stable until puberty, when a further increase in the number occurs. In obese children, however, between 2 years of age and

---

The average normal body temperature usually is considered 37°C (98.6°F) when measured orally and 37.6°C (99.7°F) when measured rectally. Rectal temperature comes closer to the true core body temperature, but an oral temperature is more easily obtained in older children and adults and therefore is the preferred measure.

Heat can be exchanged with the environment in a number of ways (figure 22.17). **Radiation** is the loss of heat as infrared radiation, a type of electromagnetic radiation. For example, the coals in a fire give off radiant heat, which can be felt some distance away from the fire. **Conduction** is the exchange of heat between objects in direct contact with each other, such as the bottoms of the feet and the floor. **Convection** is a transfer of heat between the body and the air or water. A cool breeze results in the movement of air over the body and loss of heat from the body. **Evaporation** is the conversion of water from a liquid to a gas, a process that requires heat. The evaporation of 1 g of water from the body's surface results in the loss of 580 cal of heat.

Body temperature is maintained by balancing heat gain with heat loss. When heat gain equals heat loss, body temperature is maintained. If heat gain exceeds heat loss, body temperature increases; if heat loss exceeds heat gain, body temperature decreases. Heat gain occurs through metabolism and the muscular contractions of shivering, whereas heat loss occurs through evaporation. Heat gain or loss can occur by radiation, conduction, or convection, depending on skin temperature and the environmen-

puberty, an increase also occurs in the number of fat cells.

Hypertrophic obesity results from a normal number of fat cells that have increased in size. This type of obesity is more common, is associated with moderate obesity or being "overweight," and typically develops in adults. People who were thin or of average weight and quite active when they were young become less active as they become older. They begin to gain weight between ages 20 and 40, and, although they no longer use as many kilocalories, they still take in the same amount of food as when they were younger. The unused kilocalories are turned into fat, causing fat cells to increase in size. At one time, it was believed that the number of fat cells did not increase after adulthood. It is now known that the number of fat cells can increase in adults. Apparently, if all the existing fat cells are filled to capacity with lipids, new fat cells are formed to store the excess lipids. Once fat cells are formed, however, dieting and weight loss do not result in a decrease in the number of fat cells—instead, they become smaller in size as their lipid content decreases.

The distribution of fat in obese individuals varies. Fat can be found mainly in the upper body, such as in the abdominal region, or it can be associated with the hips and buttocks. These distribution differences are clinically significant because upper body obesity is associated with an increased likelihood of developing diabetes mellitus, cardiovascular disease, stroke, and death.

In some cases, a specific cause of obesity can be identified. For example, a tumor in the hypothalamus can stimulate overeating. In most cases, however, no specific cause is apparent. In fact, obesity occurs for many reasons, and obesity in an individual can have more than one cause. Obesity seems to have a genetic component and, if one or both parents are obese, their children are more likely to be obese also. Environmental factors, such as eating habits, however, can also play an important role. For example, adopted children can exhibit the obesity of their adoptive parents. In addition, psychological factors, such as overeating as a means for dealing with stress, can contribute to obesity.

The regulation of body weight is actually a matter of regulating body fat because most changes in body weight reflect changes in the amount of fat in the body. According to the "set point" theory of weight control, the body maintains a certain amount of body fat. If the amount decreases below or increases above this level, mechanisms are activated to return the amount of body fat to its normal value.

It is a common belief that the main cause of obesity is overeating. Certainly, for obesity to occur, at sometime energy intake must have exceeded energy expenditure. A comparison of the kilocaloric intake of obese and of lean individuals at their usual weights, however, reveals that, on a per kilogram basis, obese people consume fewer kilocalories than lean people.

When people lose a large amount of weight, their feeding behavior changes. They become hyperresponsive to external food cues, think of food often, and cannot get enough to eat without gaining weight. This behavior is typical of both lean and obese individuals who are below their relative set point for weight. Other changes, such as a decrease in basal metabolic rate, take place in a person who has lost a large amount of weight. Most of this decrease in BMR probably results from a decrease in muscle mass associated with weight loss. In addition, some evidence exists that energy lost through exercise and the thermic effect of food are also reduced.

Thus, a person who has lost a large amount of weight has an increased appetite and a decreased ability to expend energy. It is no surprise that only a small percentage of obese people maintain weight loss over the long term. Instead, the typical pattern is one of repeated cycles of weight loss followed by a rapid regain of the lost weight.

The message emerging from current research is that body weight results from many complicated genetic and metabolic factors that go awry in many ways. Obesity is being regarded as a chronic condition that may respond to medication in much the same way that diabetes does. Nonetheless, medication is only part of the story. Drugs can help, but eating less and exercising more is still necessary for optimal health.

tal temperature. If the skin temperature is lower than the environmental temperature, heat is gained; however, if skin temperature is higher than the environmental temperature, heat is lost.

The difference in temperature between the body and the environment determines the amount of heat exchanged between the environment and the body. The greater the temperature difference, the greater the rate of heat exchange. Control of the temperature difference is used to regulate body temperature. For example, if environmental temperature is very cold, as on a cold winter day, a large temperature difference exists between the body and the environment, and a large loss of heat occurs. The heat loss can be decreased by behaviorally selecting a warmer environment— for example, by going inside a heated house. Heat loss can also be decreased by insulating the exchange surface, such as by putting on extra clothes. Physiologically, temperature difference can be controlled through the dilation and constriction of blood vessels in the skin. When these blood vessels dilate, they bring warm blood to the surface of the body, raising skin temperature; conversely, vasoconstriction decreases blood flow and lowers skin temperature.

P R E D I C T
Explain why vasoconstriction of the skin's blood vessels on a cool day is beneficial.

When the environmental temperature is greater than body temperature, vasodilation brings warm blood to the skin, causing an increase in skin temperature, which decreases heat gain from

Figure 22.17   Heat Exchange

Heat exchange between a person and the environment occurs by convection, radiation, evaporation, and conduction. *Arrows* show the direction of net heat gain or loss in this environment.

the environment. At the same time, evaporation carries away excess heat to prevent heat gain and overheating.

Body temperature regulation is an example of a negative-feedback system that is controlled by a set point. A small area in the anterior part of the hypothalamus detects slight increases in body temperature through changes in blood temperature (figure 22.18). As a result, mechanisms are activated that cause heat loss, such as vasodilation and sweating, and body temperature decreases. A small area in the posterior hypothalamus can detect slight decreases in body temperature and can initiate heat gain by increasing muscular activity (shivering) and vasoconstriction.

Under some conditions, the set point of the hypothalamus is actually changed. For example, during a fever, the set point is raised, heat-conserving and heat-producing mechanisms are stimulated, and body temperature increases. In recovery from a fever, the set point is lowered to normal, heat loss mechanisms are initiated, and body temperature decreases. ▼

44. *Define free energy. How much of the free energy is lost as heat from the body?*
45. *What are four ways that heat is exchanged between the body and the environment?*
46. *How is body temperature behaviorally and physiologically maintained in a cold and in a hot environment?*
47. *How does the hypothalamus regulate body temperature?*

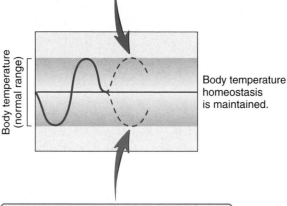

Homeostasis Figure 22.18   Summary of Temperature Regulation

Hyperthermia

Hyperthermia, an elevated body temperature, develops when heat gain exceeds the body's ability to lose heat. Hyperthermia can result from exercise, exposure to hot environments, fever, or anesthesia.

Exercise increases body temperature because of the heat produced as a by-product of muscle activity (see chapter 8). Normally, vasodilation and increased sweating prevent harmful body temperature increases. In a hot, humid environment, the evaporation of sweat is decreased, and exercise levels have to be reduced to prevent overheating.

Exposure to a hot environment normally results in the activation of heat loss mechanisms, and body temperature is maintained at normal levels, a negative-feedback mechanism. Prolonged exposure to a hot environment, however, can result in **heat exhaustion.** The normal negative-feedback mechanisms for controlling body temperature are operating, but they are unable to prevent an increase in body temperature above normal levels. Heavy sweating results in dehydration, decreased blood volume, decreased blood pressure, and increased heart rate. Individuals suffering from heat exhaustion have a wet, cool skin because of the heavy sweating. They usually feel weak, dizzy, and nauseated. Treatment includes reducing heat gain by moving to a cooler environment, ceasing activity to reduce the heat produced by muscle metabolism, and restoring blood volume by drinking fluids.

Heat stroke is more severe than heat exhaustion because it results from a breakdown in the normal negative-feedback mechanisms of temperature regulation. If the temperature of the hypothalamus becomes too high, it no longer functions appropriately. Sweating stops, and the skin becomes dry and flushed. The person becomes confused, irritable, or even comatose. In addition to the treatment for heat exhaustion, heat loss from the skin should be increased. This can be accomplished by increasing evaporation from the skin by applying wet cloths or by increasing conductive heat loss by immersing the person in a cool bath.

Fever is the development of a higher than normal body temperature following the invasion of the body by microorganisms or other foreign substances. Lymphocytes, neutrophils, and macrophages release chemicals called **pyrogens** (pī′rō-jenz). Examples of pyrogens include certain interleukins, interferons, and tissue necrosis factor. Pyrogens increase the synthesis of prostaglandins, which stimulate a rise in the temperature set points of the hypothalamus. Consequently, body temperature and metabolic rate increase. Fever is believed to be beneficial because it speeds up the chemical reactions of the immune system (see chapter 19) and inhibits the growth of some microorganisms. Although beneficial, body temperatures greater than 41°C (106°F) can be harmful. Aspirin, nonsteroidal anti-inflammatory drugs (NSAIDs), and acetaminophen lower body temperature by inhibiting the synthesis of prostaglandins.

Malignant hyperthermia is an inherited muscle disorder. Certain drugs used to induce general anesthesia for surgery cause sustained, uncoordinated muscle contractions in individuals with this disorder. Consequently, body temperature increases.

Therapeutic hyperthermia is an induced local or general body increase in temperature. It is a treatment sometimes used on tumors and infections.

Hypothermia

If heat loss exceeds the body's ability to produce heat, body temperature decreases below normal levels. **Hypothermia** is a decrease in body temperature to 35°C (95°F) or below. Hypothermia usually results from prolonged exposure to cold environments. At first, normal negative-feedback mechanisms maintain body temperature. Heat loss is decreased by constricting blood vessels in the skin, and heat production is increased by shivering. If body temperature decreases despite these mechanisms, hypothermia develops. The individual's thinking becomes sluggish, and movements are uncoordinated. Heart, respiratory, and metabolic rates decline, and death results unless body temperature is restored to normal. Rewarming should occur at a rate of a few degrees per hour.

Frostbite is damage to the skin and deeper tissues, resulting from prolonged exposure to the cold. Damage results from direct cold injury to cells, injury from ice crystal formation, and reduced blood flow to affected tissues. The fingers, toes, ears, nose, and cheeks are most commonly affected. Damage from frostbite can range from redness and discomfort to loss of the affected part. The best treatment is immersion in a warm-water bath. Rubbing the affected area and local, dry heat should be avoided.

Therapeutic hypothermia is sometimes used to slow metabolic rate during surgical procedures, such as heart surgery. Because the metabolic rate is decreased, the tissues do not require as much oxygen as normal and are less likely to be damaged.

$$\boxed{\text{S ~ U ~ M ~ M ~ A ~ R ~ Y}}$$

Nutrition (p. 686)

Nutrition is the taking in and use of food.

Nutrients

1. Nutrients are the chemicals used by the body. They are carbohydrates, lipids, proteins, vitamins, minerals, and water.
2. Essential nutrients are nutrients that must be ingested because the body cannot manufacture them or is unable to manufacture adequate amounts of them.

Kilocalories

1. A calorie (cal) is the heat (energy) necessary to raise the temperature of 1 g of water 1°C. A kilocalorie (kcal), or Calorie (Cal), is 1000 calories.
2. A gram of carbohydrate or protein yields 4 kcal, and a gram of fat yields 9 kcal.

MyPyramid

The MyPyramid guide recommends the amounts of different food types and fiber necessary for good health, based on a person's age, sex, and physical activity.

Carbohydrates

1. Carbohydrates are ingested as monosaccharides (glucose, fructose), disaccharides (sucrose, maltose, lactose), and polysaccharides (starch, glycogen, cellulose).
2. Monounsaturated fats and oils have one double bond, and polyunsaturated fats and oils have two or more double bonds.
3. Most unprocessed polyunsaturated oils occur in the *cis* form, whereas hydrogenated polyunsaturated oils are in the *trans* form.
4. Polysaccharides and disaccharides are converted to glucose. Glucose can be used for energy or stored as glycogen or fats.
5. The Acceptable Macronutrient Distribution Range (AMDR) for carbohydrates is 45%–65% of total kilocalories.

Lipids

1. Lipids are ingested as triglycerides (95%) or cholesterol and phospholipids (5%).
2. Triglycerides are used for energy or are stored in adipose tissue. Cholesterol forms other molecules, such as steroid hormones. Cholesterol and phospholipids are part of the plasma membrane.
3. The AMDR for lipids is 20%–35%.

Proteins

1. Proteins are ingested and broken down into amino acids.
2. Proteins perform many functions: protection (antibodies), regulation (enzymes, hormones), structure (collagen), muscle contraction (actin and myosin), transportation (hemoglobin, transport proteins), and receptors.
3. The AMDR for protein is 10%–35% of total kilocalories.

Vitamins

1. Many vitamins function as coenzymes or as parts of coenzymes.
2. Most vitamins are not produced by the body and must be obtained in the diet. Some vitamins can be formed from provitamins.
3. Vitamins are classified as either fat-soluble or water-soluble.
4. Recommended Dietary Allowances (RDAs) are a guide for estimating the nutritional needs of groups of people based on their age, sex, and other factors.

Minerals

1. Minerals are necessary for normal metabolism, add mechanical strength to bones and teeth, function as buffers, and are involved in osmotic balance.
2. The daily requirement for major minerals is 100 mg or more daily, whereas for trace minerals it is less than 100 mg daily.

Daily Values

1. Daily Values are dietary references that can be used to help plan a healthy diet.
2. Daily Values for vitamins and minerals are based on Reference Daily Intakes, which are generally the highest 1968 RDA values of age categories.
3. Daily Values are based on Daily Reference Values.
   - The Daily Reference Values for energy-producing nutrients (carbohydrates, total fat, saturated fat, and proteins) and dietary fiber are recommended percentages of the total kilocalories ingested daily for each nutrient.
   - The Daily Reference Values for total fats, saturated fats, cholesterol, and sodium are the uppermost limit considered desirable because of their link to diseases.
4. The % Daily Value is the percent of the recommended Daily Value of a nutrient found in one serving of a particular food.

Metabolism (p. 695)

1. Metabolism consists of catabolism and anabolism. Catabolism is the breaking down of molecules and gives off energy. Anabolism is the building up of molecules and requires energy.
2. The energy in carbohydrates, lipids, and proteins is used to produce ATP through oxidation–reduction reactions.

Carbohydrate Metabolism (p. 696)

Glycolysis

Glycolysis is the breakdown of glucose into two pyruvic acid molecules. Also produced are two NADH molecules and two ATP molecules.

Anaerobic Respiration

1. Anaerobic respiration is the breakdown of glucose in the absence of oxygen into two lactic acid and two ATP molecules.
2. Lactic acid can be converted to glucose (Cori cycle) using aerobically produced ATP (oxygen debt).

Aerobic Respiration

1. Aerobic respiration is the breakdown of glucose in the presence of oxygen to produce carbon dioxide, water, and 38 (or 36) ATP molecules.
2. The first phase is glycolysis, which produces two ATP, two NADH, and two pyruvic acid molecules.
3. The second phase is the conversion of the two pyruvic acid molecules into two molecules of acetyl-CoA. These reactions also produce two NADH and two carbon dioxide molecules.
4. The third phase is the citric acid cycle, which produces two ATP, six NADH, two $FADH_2$, and four carbon dioxide molecules.
5. The fourth phase is the electron-transport chain. The high-energy electrons in NADH and $FADH_2$ enter the electron-transport chain and are used in the synthesis of ATP and water.

Lipid Metabolism (p. 702)

1. Adipose triglycerides are broken down and released as free fatty acids.

2. Free fatty acids are taken up by cells and broken down by beta-oxidation into acetyl-CoA.
   - Acetyl-CoA can enter the citric acid cycle.
   - Acetyl-CoA can be converted into ketone bodies.

Protein Metabolism (p. 704)

1. New amino acids are formed by transamination, the transfer of an amine group to a keto acid.
2. Amino acids are used to synthesize proteins. If used for energy, ammonia is produced as a by-product of oxidative deamination. Ammonia is converted to urea and is excreted.

Interconversion of Nutrient Molecules (p. 705)

1. Glycogenesis is the formation of glycogen from glucose.
2. Lipogenesis is the formation of lipids from glucose and amino acids.
3. Glycogenolysis is the breakdown of glycogen to glucose.
4. Gluconeogenesis is the formation of glucose from amino acids and glycerol.

Metabolic States (p. 707)

1. In the absorptive state, nutrients are used as energy or stored.
2. In the postabsorptive state, stored nutrients are used for energy.

Metabolic Rate (p. 707)

Metabolic rate is the total energy expenditure per unit of time, and it has three components.

Basal Metabolic Rate

Basal metabolic rate is the energy used at rest. It is 60% of the metabolic rate.

Thermic Effect of Food

The thermic effect of food is the energy used to digest and absorb food. It is 10% of the metabolic rate.

Muscular Activity

Muscular energy is used for muscle contraction. It is 30% of the metabolic rate.

Body Temperature Regulation (p. 710)

1. Body temperature is a balance between heat gain and heat loss.
   - Heat is produced through metabolism.
   - Heat is exchanged through radiation, conduction, convection, and evaporation.
2. The greater the temperature difference between the body and the environment, the greater the rate of heat exchange.
3. Body temperature is regulated by a set point in the hypothalamus.

R E V I E W   A N D   C O M P R E H E N S I O N

1. Which of these statements concerning kilocalories is true?
   a. A kilocalorie is the amount of energy required to raise the temperature of 1 g of water 1°C.
   b. There are 9 kcal in a gram of protein.
   c. There are 4 kcal in a gram of fat.
   d. A pound of body fat contains 3500 kcal.

2. Complex carbohydrates include
   a. sucrose.
   b. milk sugar (lactose).
   c. starch, an energy storage molecule in plants.
   d. all of the above.

3. What type of nutrient is recommended as the primary energy source in the diet?
   a. carbohydrates
   b. fats
   c. proteins
   d. cellulose

4. A source of monounsaturated fats is
   a. fat associated with meat.
   b. egg yolks.
   c. whole milk.
   d. fish oil.
   e. olive oil.

5. A complete protein food
   a. provides the daily amount (grams) of protein recommended for a healthy diet.
   b. can be used to synthesize the nonessential amino acids.
   c. contains all 20 amino acids.
   d. includes beans, peas, and leafy green vegetables.

6. Concerning vitamins,
   a. most can be synthesized by the body.
   b. they are normally broken down before they can be used by the body.
   c. A, D, E, and K are water-soluble vitamins.
   d. many function as coenzymes.

7. Minerals
   a. are inorganic nutrients.
   b. compose about 4%–5% of total body weight.
   c. act as buffers and osmotic regulators.
   d. are components of enzymes.
   e. all of the above.

8. Glycolysis
   a. is the breakdown of glucose to two pyruvic acid molecules.
   b. requires the input of two ATP molecules.
   c. produces two NADH molecules.
   d. does not require oxygen.
   e. all of the above.

9. Anaerobic respiration _____ oxygen and produces _____ energy (ATP) for the cell than aerobic respiration.
   a. does not require, more
   b. does not require, less
   c. requires, more
   d. requires, less

10. Which of these reactions takes place in both anaerobic and aerobic respiration?
    a. glycolysis
    b. citric acid cycle
    c. electron-transport chain
    d. acetyl-CoA formation

11. The molecule that moves electrons from the citric acid cycle to the electron-transport chain is
    a. tRNA.   d. NADH.
    b. mRNA.   e. pyruvic acid.
    c. ADP.

12. The production of ATP molecules by the electron-transport chain is accompanied by the synthesis of
    a. alcohol.   d. lactic acid.
    b. water.   e. glucose.
    c. oxygen.

13. The carbon dioxide you breathe out comes from
    a. glycolysis.
    b. the electron-transport chain.
    c. anaerobic respiration.
    d. the food you eat.

14. Lipids are
    a. stored primarily as triglycerides.
    b. synthesized by beta-oxidation.
    c. broken down by oxidative deamination.
    d. all of the above.

15. Amino acids
    a. are classified as essential or nonessential.
    b. can be synthesized in a transamination reaction.
    c. can be used as a source of energy.
    d. can be converted to keto acids.
    e. all of the above.

16. Ammonia is
    a. a by-product of lipid metabolism.
    b. formed during ketogenesis.
    c. converted into urea in the liver.
    d. produced during lipogenesis.
    e. converted to keto acids.

17. The conversion of amino acids and glycerol into glucose is called
    a. gluconeogenesis.
    b. glycogenolysis.
    c. glycogenesis.
    d. ketogenesis.

18. Which of these events takes place during the absorptive state?
    a. Glycogen is converted into glucose.
    b. Glucose is converted into fats.
    c. Ketones are produced.
    d. Proteins are converted into glucose.

19. The major use of energy by the body is in
    a. basal metabolism.
    b. physical activity.
    c. the thermic effect of food.

20. The loss of heat resulting from the loss of water from the body's surface is
    a. radiation.
    b. evaporation.
    c. conduction.
    d. convection.

Answers in Appendix E

C R I T I C A L   T H I N K I N G

1. One serving of a food has 2 g of saturated fat. What % Daily Value for saturated fat would appear on a food label for this food? (See the bottom of figure 22.2 for information needed to answer this question.)

2. An active teenage boy has a daily intake of 3000 kcal/day. What is the maximum amount (weight) of total fats he should consume, according to the Daily Values?

3. If the teenager in question 2 eats a serving of food that has a total fat content of 10 g/serving, what is his % Daily Value for total fat?

4. Suppose the food in question 3 is in a package that lists a serving size of 1/2 cup, with 4 servings in the package. If the teenager eats half of the contents of the package (1 cup), how much of his % Daily Value does he consume?

5. Why does a vegetarian usually have to be more careful about his or her diet than a person who includes meat in the diet?

6. Explain why a person suffering from copper deficiency feels tired all the time.

7. Some people claim that occasionally fasting for short periods can be beneficial. How can fasts be damaging?

8. Why can some people lose weight on a 1200 kcal/day diet but others cannot?

9. Lotta Bulk, a muscle builder, wanted to increase her muscle mass. Knowing that proteins are the main components of muscle, she began a high-protein diet in which most of her daily kilocalories were supplied by proteins. She also exercised regularly with heavy weights. After 3 months of this diet and exercise program, Lotta increased her muscle mass, but not any more than her friend, who did the same exercises but did not have a high-protein diet. Explain what happened. Was Lotta in positive or negative nitrogen balance?

10. On learning that sweat evaporation results in the loss of calories, an anatomy and physiology student enters a sauna in an attempt to lose weight. He reasons that a liter (about a quart) of water weighs 1000 g, which is equivalent to 580,000 cal, or 580 kcal, of heat when lost as sweat. Instead of reducing his diet by 580 kcal/day, if he loses a liter of sweat every day in the sauna, he believes he will lose about a pound of fat a week. Will this approach work? Explain.

11. Thyroid hormone is known to increase the activity of the Na^+–K^+ pump. If a person produced excess amounts of thyroid hormone, what effect would this have on basal metabolic rate, body weight, and body temperature? How would the body attempt to compensate for the changes in body weight and temperature?

12. In some diseases, an infection causes a high fever, resulting in a crisis state. The person is on the way to recovery when body temperature begins to return to normal. If you were looking for symptoms in a person who had just passed through the crisis state, would you look for a dry, pale skin or a wet, flushed skin? Explain.

Answers in Appendix F

CHAPTER  Urinary System and Body Fluids

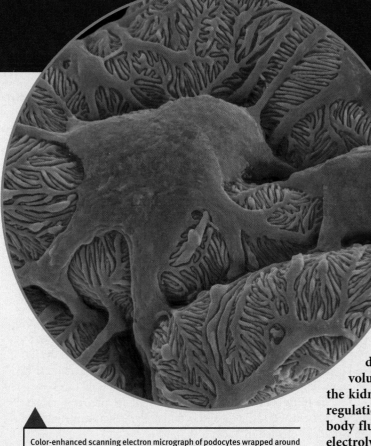

▲ Color-enhanced scanning electron micrograph of podocytes wrapped around the glomerular capillaries. These structures play a major role in filtration, one of the important steps in urine formation.

Introduction

The kidneys make up the body's main purification system. They control the composition of blood by removing waste products, many of which are toxic, and conserving useful substances. The kidneys help control blood volume and consequently play a role in regulating blood pressure. The kidneys also play an essential role in regulating blood pH. Approximately one-third of one kidney is all that is needed to maintain homeostasis. Even after extensive damage, the kidneys can still perform their life-sustaining functions. If the kidneys are damaged further, however, death results unless specialized medical treatment is administered.

▶This chapter explains the **functions of the urinary system (p. 718)**, **kidney anatomy and histology (p. 718)**, **urine production (p. 723)**, **hormonal regulation of urine concentration and volume (p. 734)**, **urine movement (p. 739)**, and the **effects of aging on the kidneys (p. 742)**. The chapter also considers **body fluids (p. 743)**, the **regulation of intracellular fluid composition (p. 744)**, the **regulation of body fluid concentration and volume (p. 745)**, the **regulation of specific electrolytes in the extracellular fluid (p. 747)**, and the **regulation of acid–base balance (p. 752)**.

Tools for Success

 *A virtual cadaver dissection experience*

 Audio (MP3) and/or video (MP4) content, available at www.mhhe.aris.com

Functions of the Urinary System

The **urinary system** consists of two kidneys, which produce urine; two ureters, which carry urine from the kidneys to the urinary bladder; a single, midline urinary bladder, which stores urine; and a single urethra, which carries urine from the urinary bladder to the outside of the body (figure 23.1).

The kidneys are the major excretory organs of the body. The skin, liver, lungs, and intestines eliminate some waste products, but, if the kidneys fail to function, these other excretory organs cannot adequately compensate. The following functions are performed by the kidneys:

1. *Excretion.* The kidneys filter blood, and a large volume of filtrate is produced. Large molecules, such as proteins and blood cells, are retained in the blood, whereas smaller molecules and ions enter the filtrate. Most of the filtrate volume is reabsorbed back into the blood, along with useful molecules and ions. Metabolic wastes, toxic molecules, and excess ions remain in a small volume of filtrate. Additional waste products are secreted into the filtrate and the result is urine formation.
2. *Regulation of blood volume and pressure.* The kidneys play a major role in controlling the extracellular fluid volume in the body by producing either a large volume of dilute urine or a small volume of concentrated urine. Consequently, the kidneys regulate blood volume and blood pressure.

3. *Regulation of the concentration of solutes in the blood.* The kidneys help regulate the concentration of the major ions, such as Na^+, Cl^-, K^+, Ca^{2+}, HCO_3^-, and HPO_4^{2-}.
4. *Regulation of extracellular fluid pH.* The kidneys secrete variable amounts of H^+ to help regulate the extracellular fluid pH.
5. *Regulation of red blood cell synthesis.* The kidneys secrete a hormone, erythropoietin, which regulates the synthesis of red blood cells in bone marrow (see chapter 16).
6. *Vitamin D synthesis.* The kidneys play an important role in controlling blood levels of Ca^{2+} by regulating the synthesis of vitamin D (see chapter 6). ▼

> 1. List the functions performed by the kidneys, and briefly describe each.

Kidney Anatomy and Histology

Location and External Anatomy of the Kidneys

The **kidneys** are bean-shaped and each is about the size of a tightly clenched fist. They lie behind the peritoneum on the posterior abdominal wall on each side of the vertebral column (figure 23.2). Structures that are behind the peritoneum are said to be **retroperitoneal** (re′trō-per′i-tō-nē′ăl). The **renal** (kidney) **capsule** is a layer of fibrous connective tissue surrounding each kidney. A thick layer of adipose tissue, in turn, surrounds the renal capsule. This fat acts as a shock absorber, cushioning the kidneys against mechanical shock. The **renal fascia** is a thin layer of connective tissue surrounding the fat. Numerous connective tissue strands cross the fat and connect the renal fascia to the renal capsule. The renal fascia helps anchor the kidneys and fat to the abdominal wall.

The **hilum** (hī′lŭm) is a small area where the renal artery and nerves enter, and the renal vein and ureter exit, the medial side of the kidney (figure 23.3). ▼

> 2. Describe the shape, size, and location of the kidneys.
> 3. Describe the renal capsule and the structures surrounding the kidney.
> 4. What is the hilum?

Internal Anatomy and Histology of the Kidneys

The kidney is divided into an outer **cortex** and an inner **medulla,** which surrounds the renal sinus (see figure 23.3). The **renal sinus** is a cavity connected to the hilum. It contains blood vessels, urine-collecting chambers that empty into the ureter, and fat.

The **renal columns** are extensions of the renal cortex into the renal medulla. Between the renal columns are the cone-shaped **renal pyramids.** The bases of the pyramids form the boundary between the cortex and the medulla. The tips of the pyramids, the **renal papillae,** point toward the renal sinus.

Minor calyces (kal′i-sēz, cup of flower) are funnel-shaped chambers into which the renal papillae extend. The minor calyces of several pyramids merge to form larger funnels, the **major calyces.** Each kidney contains 8–20 minor calyces and 2 or 3 major calyces. The major calyces converge to form an enlarged chamber called the

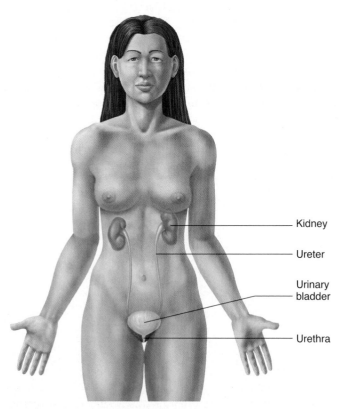

Figure 23.1   Urinary System

The urinary system consists of two kidneys, two ureters, a urinary bladder, and a urethra.

Kidney

Ureter

Urinary bladder

Urethra

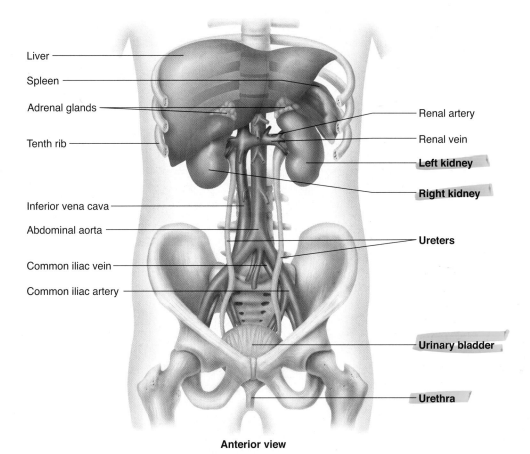

Liver
Spleen
Adrenal glands
Tenth rib
Inferior vena cava
Abdominal aorta
Common iliac vein
Common iliac artery

Renal artery
Renal vein
Left kidney
Right kidney
Ureters
Urinary bladder
Urethra

Anterior view

(a) The kidneys are located in the abdominal cavity, with the right kidney just below the liver and the left kidney below the spleen. A ureter extends from each kidney to the urinary bladder within the pelvic cavity. An adrenal gland is located at the superior pole of each kidney.

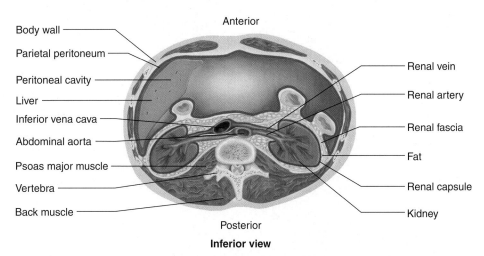

Anterior

Body wall
Parietal peritoneum
Peritoneal cavity
Liver
Inferior vena cava
Abdominal aorta
Psoas major muscle
Vertebra
Back muscle

Renal vein
Renal artery
Renal fascia
Fat
Renal capsule
Kidney

Posterior
Inferior view

(b) The kidneys are located behind the parietal peritoneum. Fat surrounds the kidneys. A connective tissue layer, the renal fascia, anchors the kidney to the abdominal wall. The renal arteries extend from the abdominal aorta to each kidney, and the renal veins extend from the kidneys to the inferior vena cava.

Figure 23.2 Anatomy of the Urinary System

renal pelvis, which is surrounded by the renal sinus. The renal pelvis narrows into a small-diameter tube, the **ureter,** which exits the kidney at the hilum and connects to the urinary bladder. Urine formed within the kidney flows from the renal papillae into the minor calyces. From the minor calyces, urine flows into the major calyces, collects in the renal pelvis, and then leaves the kidney through the ureter.

The functional unit of the kidney is the **nephron** (nef′ron, kidney), and there are approximately 1.3 million of them in each kidney. Each nephron consists of a **renal corpuscle;** a **proximal convoluted tubule;** a **loop of Henle,** or nephronic loop; and a **distal convoluted tubule** (figure 23.4). The renal corpuscle consists of Bowman's capsule and the glomerulus. **Bowman's capsule**

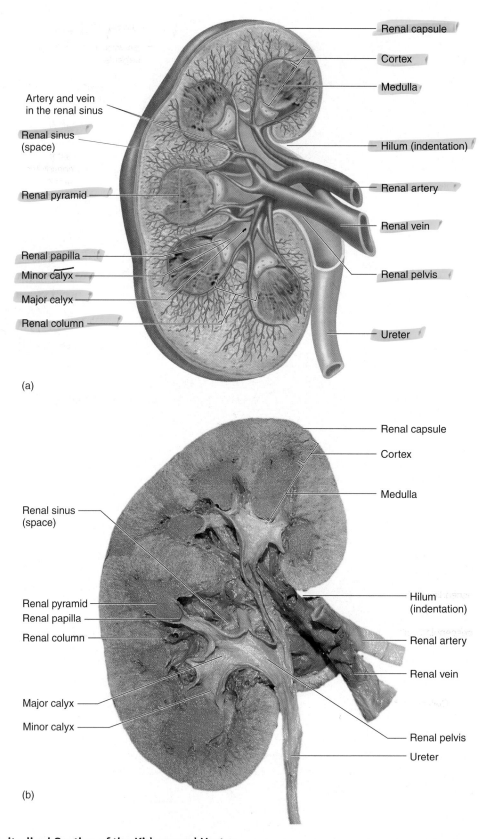

Figure 23.3 Longitudinal Section of the Kidney and Ureter

(*a*) The cortex forms the outer part of the kidney, and the medulla forms the inner part. A central cavity called the renal sinus contains the renal pelvis. The renal columns of the kidney project from the cortex into the medulla and separate the pyramids. (*b*) Photograph of a longitudinal section of a human kidney and ureter.

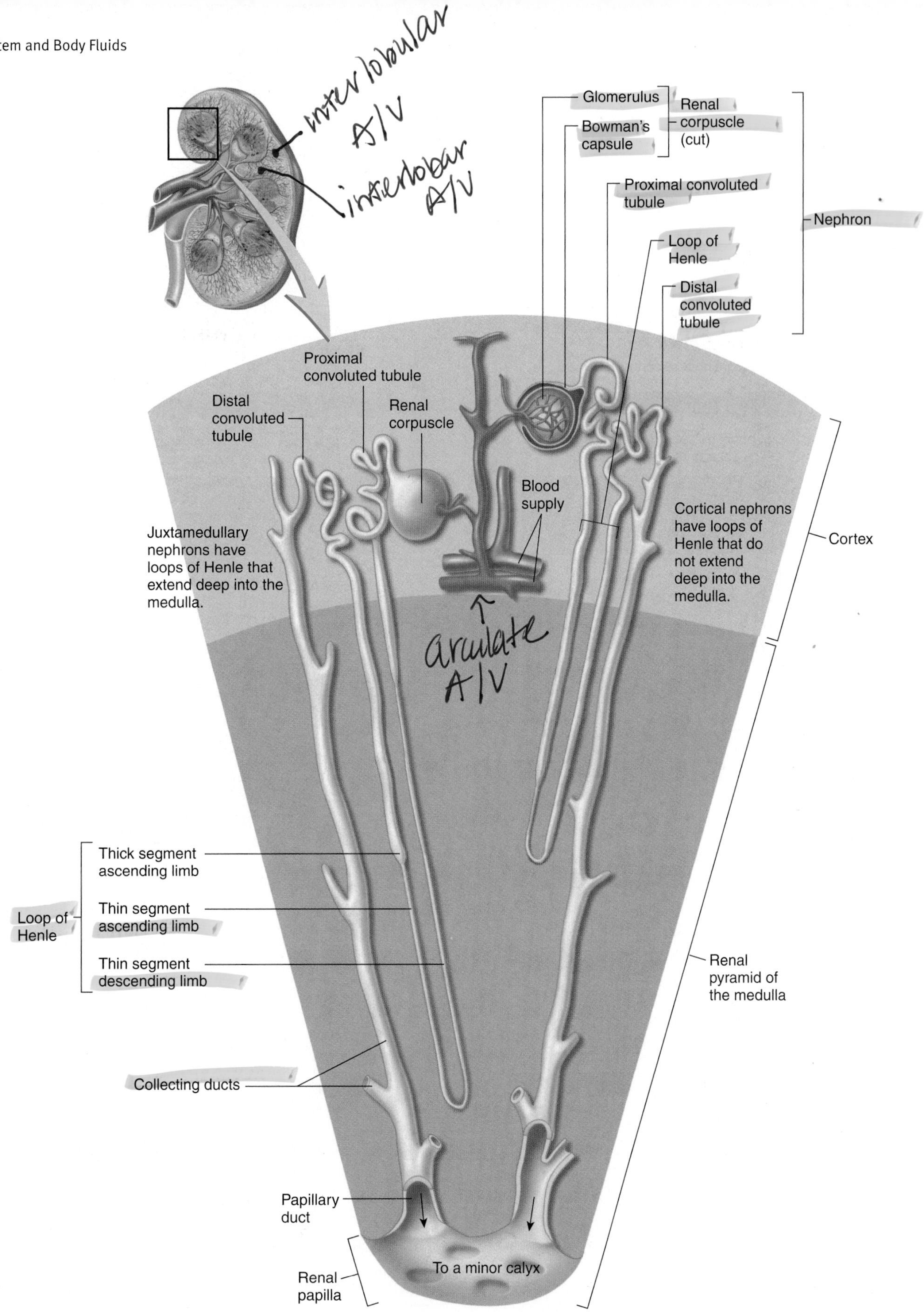

Figure 23.4   Functional Unit of the Kidney—the Nephron

A nephron consists of a renal corpuscle, proximal convoluted tubule, loop of Henle, and distal convoluted tubule. The distal convoluted tubule empties into a collecting duct. Juxtamedullary nephrons (those near the medulla of the kidney) have loops of Henle that extend deep into the medulla of the kidney, whereas other nephrons do not. Collecting ducts undergo a transition to larger-diameter papillary ducts near the tip of the renal papilla. The papillary ducts empty into a minor calyx.

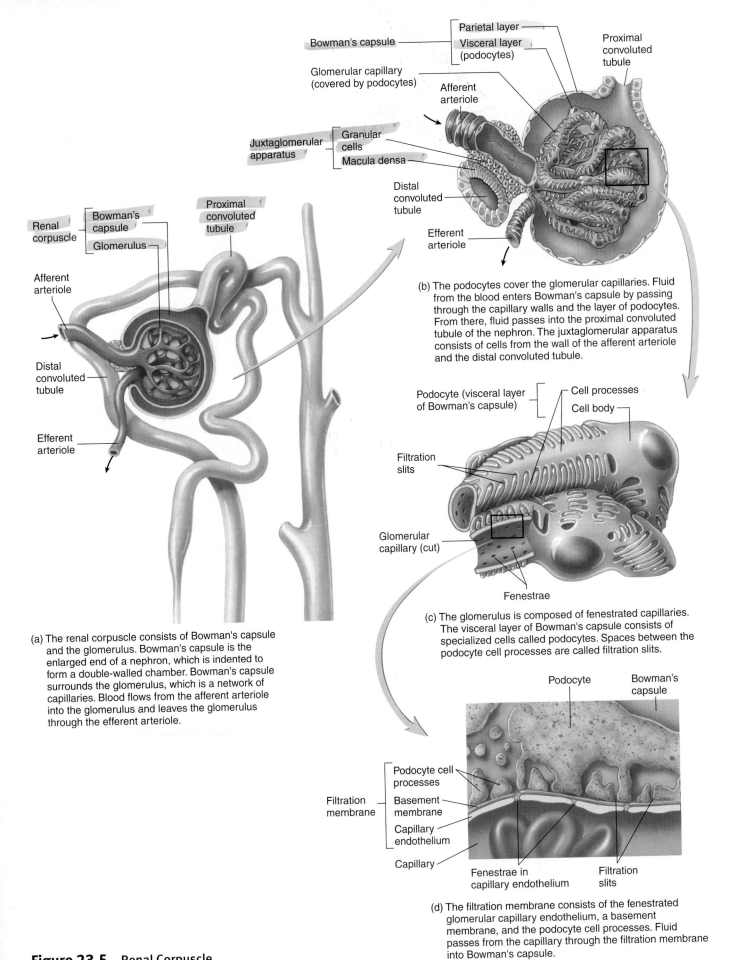

Parietal layer
Bowman's capsule
Visceral layer (podocytes)
Proximal convoluted tubule
Glomerular capillary (covered by podocytes)
Afferent arteriole
Juxtaglomerular apparatus
Granular cells
Macula densa
Distal convoluted tubule
Efferent arteriole

Renal corpuscle
Bowman's capsule
Glomerulus
Proximal convoluted tubule
Afferent arteriole
Distal convoluted tubule
Efferent arteriole

(a) The renal corpuscle consists of Bowman's capsule and the glomerulus. Bowman's capsule is the enlarged end of a nephron, which is indented to form a double-walled chamber. Bowman's capsule surrounds the glomerulus, which is a network of capillaries. Blood flows from the afferent arteriole into the glomerulus and leaves the glomerulus through the efferent arteriole.

(b) The podocytes cover the glomerular capillaries. Fluid from the blood enters Bowman's capsule by passing through the capillary walls and the layer of podocytes. From there, fluid passes into the proximal convoluted tubule of the nephron. The juxtaglomerular apparatus consists of cells from the wall of the afferent arteriole and the distal convoluted tubule.

Podocyte (visceral layer of Bowman's capsule)
Cell processes
Cell body
Filtration slits
Glomerular capillary (cut)
Fenestrae

(c) The glomerulus is composed of fenestrated capillaries. The visceral layer of Bowman's capsule consists of specialized cells called podocytes. Spaces between the podocyte cell processes are called filtration slits.

Podocyte
Bowman's capsule
Filtration membrane
Podocyte cell processes
Basement membrane
Capillary endothelium
Capillary
Fenestrae in capillary endothelium
Filtration slits

(d) The filtration membrane consists of the fenestrated glomerular capillary endothelium, a basement membrane, and the podocyte cell processes. Fluid passes from the capillary through the filtration membrane into Bowman's capsule.

Figure 23.5   Renal Corpuscle

is the enlarged end of the nephron surrounding the **glomerulus** (glō-mār′ū-lŭs, *glomus,* a ball of yarn), which is a network of capillaries. Fluid from the blood in the glomerulus enters Bowman's capsule and then flows into the proximal convoluted tubule. From there, it flows into the loop of Henle. Each loop of Henle has a descending limb, which extends toward the renal sinus, and an ascending limb, which extends back toward the cortex. The fluid flows through the ascending limb of the loop of Henle to the distal convoluted tubule. Many distal convoluted tubules empty into a **collecting duct,** which carries the fluid from the cortex, through the medulla. Many collecting ducts empty into a **papillary duct,** and the papillary ducts empty their contents into a minor calyx.

The renal corpuscle and convoluted tubules are in the renal cortex (see figure 23.4). The collecting duct and loop of Henle enter the medulla. Approximately 15% of the nephrons, called **juxtamedullary** (next to the medulla) **nephrons,** have loops of Henle that extend deep into the medulla of the kidney. The other 85% of nephrons, called **cortical nephrons,** have loops of Henle that do not extend deep into the medulla.

Bowman's capsule is indented to form a double-walled chamber (figure 23.5). The **parietal layer** of Bowman's capsule is simple squamous epithelium forming the outer wall. The **visceral layer** consists of cells called **podocytes** (pod′ō-sīts, *pod,* foot + *kytos,* a hollow cell), which form the inner wall of the capsule. The podocytes wrap around the capillaries of the glomerulus. The cavity of Bowman's capsule opens into the proximal convoluted tubule, which carries fluid away from the capsule.

The endothelium of the glomerular capillaries has pores, called **fenestrae** (fe-nes′trē), and the podocytes have numerous cell processes with gaps between them, called **filtration slits.** The endothelium of the glomerular capillaries, the podocytes, and the basement membrane between them form a **filtration membrane** (see figure 23.5d). In the first step of urine formation, fluid, consisting of water and solutes smaller than proteins, passes from the blood in the glomerular capillaries through the filtration membrane into Bowman's capsule. The fluid that passes across the filtration membrane is called the **filtrate.**

The proximal convoluted tubules, the thick segment of the loops of Henle, the distal convoluted tubules, and the collecting ducts consist of simple cuboidal epithelium. The cuboidal epithelial cells have microvilli and many mitochondria. These portions of the nephron actively transport molecules and ions across the wall of the nephron. The thin segments of the descending and ascending limbs of the loops of Henle have very thin walls made up of simple squamous epithelium. Water and solutes pass through the walls of these portions of the nephron by diffusion. The thin segment of the descending limb of the loop of Henle is permeable to water and solutes, and the thin segment of the ascending limb is permeable to solutes, but not to water. ▼

5. *What are the renal cortex, renal medulla, and renal sinus?*
6. *Describe the relationships among the renal papillae, calyces, renal pelvis, and ureter.*
7. *What is the functional unit of the kidney? Name its parts.*
8. *What is the relationship among nephrons, collecting ducts, papillary ducts, and minor calyces?*

9. *Where in the cortex and the medulla are the parts of a nephron located? Distinguish between cortical and juxtamedullary nephrons.*
10. *Describe the structure of Bowman's capsule, the glomerulus, and the filtration membrane.*

Arteries and Veins of the Kidneys

A **renal artery** branches off the abdominal aorta and enters the renal sinus of each kidney (figure 23.6a). Within the kidney, there is repeated branching of the artery, with the branches becoming smaller and smaller. These branches pass along the sides and bases of the renal pyramids, project into the cortex, and give rise to **afferent arterioles** supplying the glomerular capillaries inside Bowman's capsule (figure 23.6b). **Efferent arterioles** arise from the glomerular capillaries and carry blood away from the glomeruli. After each efferent arteriole exits the glomerulus, it gives rise to a plexus of capillaries, called the **peritubular capillaries,** around the proximal and distal convoluted tubules. The **vasa recta** (vā′să rek′tă, straight vessels) are specialized portions of the peritubular capillaries that extend deep into the medulla of the kidney and surround the loops of Henle and collecting ducts. The peritubular capillaries, which include the vasa recta, join small veins in the cortex. These veins join larger veins that run parallel to the arteries. In the renal sinus, the veins converge to form the renal vein, which exits the kidney.

A structure called the **juxtaglomerular** (jŭks′-tă-glō-mer′ū-lăr, close to the medulla) **apparatus** is formed where the distal convoluted tubule comes into contact with the afferent arteriole next to Bowman's capsule (see figure 23.5b). The juxtaglomerular apparatus is composed of the granular cells and the macula densa. The **granular cells** are specialized smooth muscle cells in the wall of the afferent arteriole that secrete renin. The **macula densa** is specialized epithelial cells of the distal convoluted tubule that monitor the flow of filtrate. The juxtaglomerular apparatus plays an important role in blood pressure regulation (see Regulation of Renin Secretion p. 738). ▼

11. *Describe the blood supply for the kidney.*
12. *What is the juxtaglomerular apparatus?*

Urine Production

The nephron is called the **functional unit of the kidney** because it is the smallest structural component of the kidney capable of producing urine. Three major processes are essential for urine formation: filtration, tubular reabsorption, and tubular secretion (figure 23.7). All three are essential for the regulation of body fluid composition. **Filtration** is the movement of water and small solutes across the filtration membrane as a result of a pressure difference.

Tubular reabsorption is the removal of water and solutes from the filtrate back into the blood. In general, most of the water and useful solutes are reabsorbed, whereas waste products, excess solutes, and a small amount of water are not.

Tubular secretion is the addition of solutes across the walls of the nephron into the filtrate. Consequently, urine consists of substances that are filtered across the filtration membrane and

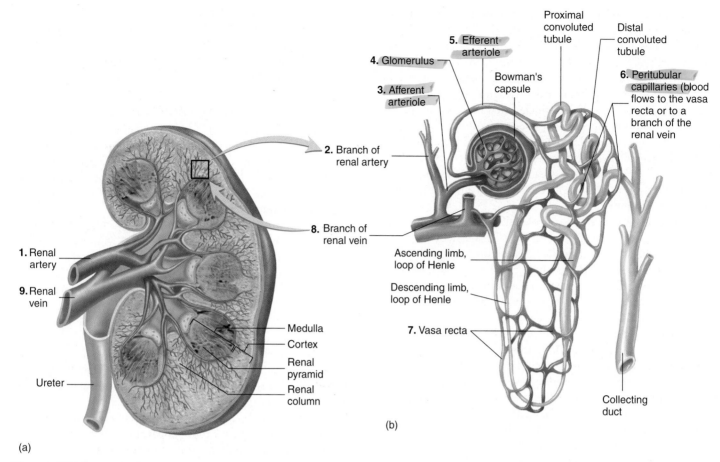

Figure 23.6   Blood Flow Through the Kidney

(a) Blood flow through the larger arteries and veins of the kidney. (b) Blood flow through the arteries, capillaries, and veins that provide circulation to the nephrons.

Urine formation results from the following three processes:

1. Filtration

Filtration (*blue arrow*) is the movement of materials across the filtration membrane into Bowman's capsule to form filtrate.

2. Tubular reabsorption

Solutes are reabsorbed (*purple arrow*) across the wall of the nephron into the interstitial fluid by transport processes, such as active transport and cotransport.
    Water is reabsorbed (*orange arrow*) across the wall of the nephron by osmosis. Water and solutes pass from the interstitial fluid into the peritubular capillaries.

3. Tubular secretion

Solutes are secreted (*green arrow*) across the wall of the nephron into the filtrate.

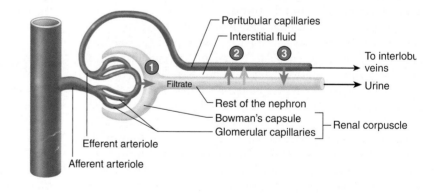

Process Figure 23.7   Urine Formation

Clinical Relevance
Hematuria and Glomerulonephritis

Blood cells do not normally appear in the urine because the filtration membrane does not permit their passage. **Hematuria** (hē-mă-too′rē-ă, hem-ă-too′rē-ă) is the presence of blood in the urine. The source of blood cells in the urine can be inside or outside the kidney. Damage to the filtration membrane or other areas of the kidney can result in hematuria. **Glomerular nephritis** (glō-măr′ū-lăr ne-frī′tis) results from inflammation of the filtration membrane within the renal corpuscle. It is characterized by an increased permeability of the filtration membrane and the accumulation of numerous white blood cells in the area. As a consequence, a high concentration of plasma proteins enters the

filtrate and a greater than normal urine volume is produced. As damage to the filtration membrane increases, red and white blood cells pass through the filtration membrane.

Acute glomerular nephritis often occurs 1–3 weeks after a severe bacterial infection, such as streptococcal sore throat or scarlet fever. Antigen–antibody complexes associated with the disease become deposited in the filtration membrane and cause inflammation. This acute inflammation normally subsides after several days.

Chronic glomerular nephritis is long-term and usually progressive. The filtration membrane thickens and eventually is replaced by connective tissue. Although in

the early stages chronic glomerular nephritis resembles the acute form, in the advanced stages many of the renal corpuscles have been replaced by fibrous connective tissue, and the kidney eventually ceases to function.

Blood can be added to the urine just before or after it leaves the kidneys. Kidney stones or tumors in the renal pelvis, ureter, urinary bladder, prostate, or urethra can cause hematuria. Infections of the urinary tract, resulting in inflammation of the urinary bladder (cystitis), of the prostate gland (prostatitis, pros-tă-tī′tis), and of the urethra (urethritis, ū-rē-thrī′tis) can also cause hematuria.

---

those that are secreted from the peritubular capillaries into the nephron, minus the substances that are reabsorbed (table 23.1). ▼

> **13. Describe the three general processes involved in the production of urine.**

Filtration

An average of 21% of the blood pumped by the heart each minute flows through the kidneys. Of the total volume of blood plasma that flows through the glomerular capillaries, about 19% passes through the filtration membrane into Bowman's capsule to become filtrate. The amount of filtrate produced each minute is called the

Table 23.1	Concentrations of Major Substances in Urine (Average Values)	
Substance	**Plasma**	**Urine**
Water (L/day)		1.4
Organic molecules (mg/dL)		
Protein	3900–5000	0*
Glucose	100	0
Urea	26	1820
Uric acid	3	42
Creatinine	1	196
Ions (mEq/L)		
Na⁺	142	128
K⁺	5	60
Cl⁻	103	134
HCO₃⁻	28	14
Specific gravity (g/ml)†		1.005–1.030
pH		4.5–8.0

*Trace amounts of protein can be found in the urine.
†The specific gravity increases as the concentration of solutes in urine increases.

glomerular filtration rate (GFR). The GFR is approximately 125 mL/min, or about 180 L/day. Approximately 99% of the filtrate volume is reabsorbed as it passes through the nephron, and less than 1% becomes urine. Thus, a healthy person produces approximately 1–2 L of urine each day. ▼

> **14. Define glomerular filtration rate.**

Filtration Membrane

The filtration membrane allows some substances, but not others, to pass from the blood into Bowman's capsule. Water and solutes of small size readily pass through the openings of the filtration membrane but blood cells, platelets, and proteins, which are too large to pass through the filtration membrane, do not enter Bowman's capsule. Albumin, a small blood protein with a diameter slightly less than the openings in the filtration membrane, enters the filtrate in very small amounts. Consequently, the filtrate contains no blood cells or platelets and only a small amount of protein. ▼

> **15. Name the things that do and do not pass through the filtration membrane.**

P R E D I C T
Hemoglobin has a smaller diameter than albumin, but very little hemoglobin normally passes from the blood into the filtrate. Explain why. Under what circumstances would large amounts of hemoglobin enter the filtrate?

Filtration Pressure

The formation of filtrate depends on the **filtration pressure (FP)**, which is a pressure gradient that forces fluid from the glomerular capillary across the filtration membrane into Bowman's capsule (figure 23.8). The filtration pressure results from a force moving fluid into Bowman's capsule from the glomerular capillary minus a force moving fluid out of Bowman's capsule into the glomerular capillary. Filtration moves fluid into Bowman's capsule and osmosis moves fluid out of it.

1. Fluid moves by filtration into Bowman's capsule because the glomerular capillary pressure (50 mm Hg) is greater than the capsular pressure (10 mm Hg).

2. Fluid moves by osmosis out of Bowman's capsule. The blood colloid osmotic pressure (30 mm Hg) results from the concentration of proteins in the blood.

3. Filtration pressure (blue arrow) is the difference between the filtration force moving fluid into Bowman's capsule and the osmotic force moving fluid out.

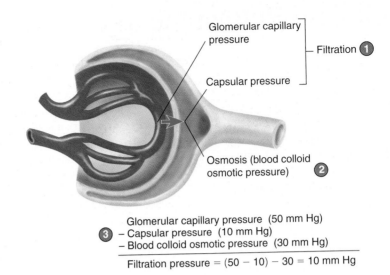

Glomerular capillary pressure } Filtration **1**

Capsular pressure

Osmosis (blood colloid osmotic pressure) **2**

3 Glomerular capillary pressure (50 mm Hg)
 – Capsular pressure (10 mm Hg)
 – Blood colloid osmotic pressure (30 mm Hg)

Filtration pressure = (50 − 10) − 30 = 10 mm Hg

Process Figure 23.8 Filtration Pressure

$$\text{Filtration pressure} = \text{Force in} - \text{Force out}$$
$$= \text{Filtration} - \text{Osmosis}$$

Filtration is the movement of fluid through a partition, such as the filtration membrane, that allows fluid and smaller substances to pass through the partition but prevents the passage of larger substances. The fluid movement results from a pressure difference across the partition, with fluid moving from the side of the partition with the greater pressure to the side with the lower pressure. The **glomerular capillary pressure (GCP)** is the blood pressure in the glomerular capillary, and the **capsular pressure (CP)** is the pressure produced by filtrate already inside Bowman's capsule. The GCP is 50 mm Hg and the CP is 10 mm Hg. Therefore, the force moving fluid into Bowman's capsule by filtration is 40 mm Hg (50 − 10).

Osmosis is the diffusion of water across a selectively permeable membrane. The filtration membrane acts as a selectively permeable membrane that prevents proteins from moving into Bowman's capsule but allows the passage of water and solutes. The concentration of most substances in the plasma and filtrate is the same, except for proteins. The plasma has proteins and the filtrate has so few proteins that the protein concentration is considered to be zero. Consequently, water moves by osmosis from the less concentrated filtrate (no proteins, more water) into the more concentrated blood (proteins, less water). The osmotic force moving fluid out of Bowman's capsule is called the **blood colloid osmotic pressure (BCOP)** because proteins in water form a colloid. A colloid consists of large molecules, such as proteins, suspended in a fluid. The BCOP is approximately 30 mm Hg. Therefore, filtration pressure is 10 mm Hg (see figure 23.8).

$$FP = \text{Force in} - \text{Force out}$$
$$= \text{Filtration} - \text{Osmosis}$$
$$= (GCP - CP) - BCOP$$
$$= (50 - 10) - 30$$
$$= 40 - 30$$
$$= 10 \text{ mm Hg}$$

Changes in filtration pressure have a direct effect on the glomerular filtration rate and urine production. For example, as filtration pressure increases, the glomerular filtration rate increases. As more filtrate is produced, urine volume increases, unless there is a compensating increase in tubular reabsorption. Changes in filtration pressure result from changes in glomerular capillary pressure and changes in protein concentrations in the blood and filtrate.

The glomerular capillary pressure is affected by the diameter of the efferent and afferent arterioles. The glomerular capillary pressure is much higher than that in most capillaries because of the resistance to blood flow through the efferent arterioles, which have a small diameter. As the diameter of a vessel decreases, resistance to flow through the vessel increases (see Poiseuille's law, chapter 18). Also, the pressure upstream from the point of decreased vessel diameter is higher than the pressure downstream from the point of decreased diameter. For example, in the extreme case of a completely closed vessel, pressure is high upstream from the constriction and falls to zero downstream from the constriction.

Glomerular capillary pressure can be changed by vasoconstriction and vasodilation of the afferent arterioles. Vasoconstriction of the afferent arteriole decreases (downstream) glomerular capillary pressure, whereas vasodilation increases it.

Changes in blood or filtrate protein concentrations change filtration pressure because the movement of fluid out of Bowman's capsule by osmosis is primarily due to the difference in protein concentration between the blood and the filtrate. It is usually assumed that the filtrate has no proteins because very few proteins cross the filtration membrane. If the permeability of the filtration membrane to proteins increases, however, the proteins entering Bowman's capsule can affect the movement of fluid by osmosis. The presence of proteins in Bowman's capsule produces a **capsular colloid osmotic pressure (CCOP)** that counteracts the blood colloid osmotic pressure. The result is less movement of fluid by osmosis out of Bowman's capsule and an increase in filtration pressure. ▼

16. *What is filtration pressure? How is it related to filtration and osmosis across the filtration membrane?*
17. *Define glomerular capillary pressure, capsular pressure, and blood colloid osmotic pressure.*
18. *How do changes in filtration pressure affect the glomerular filtration rate and urine production?*
19. *Why is glomerular capillary pressure higher than in other capillaries? What effect does vasoconstriction and vasodilation of the afferent arterioles have on glomerular capillary pressure?*
20. *How do changes in blood and plasma proteins affect filtration pressure?*

Regulation of Glomerular Filtration Rate

Autoregulation and sympathetic stimulation regulate the glomerular filtration rate and renal blood flow.

Autoregulation

Autoregulation is the maintenance of a relatively stable glomerular filtration rate over a wide range of blood pressure. The glomerular filtration rate increases only slightly as blood pressure increases from 90 to 180 mm Hg. Autoregulation is achieved through constriction and dilation of the afferent arterioles. As blood pressure increases, the afferent arterioles constrict, preventing a large increase in glomerular capillary pressure, filtration pressure, and glomerular filtration rate. Conversely, a decrease in blood pressure results in dilation of the afferent arterioles, preventing a large decrease in glomerular capillary pressure, filtration pressure, and glomerular filtration rate.

Autoregulation prevents large changes in the glomerular filtration rate, which prevents large changes in urine production. Despite autoregulation, changes in blood pressure do cause slight changes in the glomerular filtration rate. These changes are stimuli affecting the secretion of renin from the juxtaglomerular apparatus (see "Regulation of Renin Secretion," p. 738). The slight changes in glomerular filtration rate also cause changes in urine volume, which affect blood volume. Increased glomerular filtration rate increases urine volume and decreases blood volume, whereas decreased glomerular filtration rate decreases urine volume and increases blood volume.

Stretch of the afferent arteriole and the flow of filtrate past the juxtaglomerular apparatus stimulate autoregulation. As blood pressure increases, the smooth muscle in the wall of the afferent arteriole is stretched, which stimulates the smooth muscle to contract. The macula densa of the juxtaglomerular apparatus monitors the rate of filtrate flow through the nephron. When the glomerular filtration rate increases, more filtrate flows past the macula densa, which sends a signal to the afferent arteriole, causing it to contract. As a result, the glomerular filtration rate decreases. ▼

21. *What is autoregulation and how is it achieved? Describe two mechanisms responsible for autoregulation.*

Sympathetic Stimulation

Sympathetic stimulation constricts the small arteries and afferent arterioles, thereby decreasing renal blood flow and filtrate formation. In response to severe stress or circulatory shock, intense sympathetic stimulation reduces renal blood flow. This helps maintain homeostasis by maintaining blood pressure at levels adequate to sustain blood flow to organs, such as the heart and brain. A prolonged decrease in renal blood, however, can damage the kidneys. Consequently, shock should be treated quickly. ▼

22. *What effect does sympathetic stimulation have on renal blood flow?*

P R E D I C T
Karl was pouring gasoline from a small can into his lawn mower when it ignited. He experienced third-degree burns on his chest, neck, arms, and hands. In the hospital emergency room, his blood pressure was found to be in the low normal range and his heart rate was very high. His skin appeared very pale. There was a delay of a couple of hours from the time of the accident until Karl was treated and an IV was started. During that time he produced almost no urine. After IV treatment, urine production resumed. Explain these responses.

Tubular Reabsorption

Approximately 180 L of filtrate are produced each day. Given that plasma volume is approximately 3 L, that means the entire plasma volume is filtered 60 times a day, yet only 1–2 L of urine are produced each day. Tubular reabsorption, the movement of water and solutes from the filtrate back into the blood, makes this possible. Almost all of the water and useful solutes are reabsorbed, whereas waste products, excess solutes, and a small amount of water are not. The reabsorbed substances enter the peritubular capillaries and vasa recta and flow through the renal veins to enter the general circulation (see figure 23.6b). ▼

23. *Where do solutes and water reabsorbed by the nephrons go?*

Medullary Concentration Gradient

Solutions in the body are complex, containing many different substances in different amounts. For simplicity, the ions and molecules in a solution are referred to as particles. When considering osmosis, the number of particles, whether of one type or a mixture of different types, determines the concentration of a solution. An **osmole (Osm)** is 6.022×10^{23} particles in a kilogram (kg) of water (see Appendix C for an explanation of this number). The concentration of particles in body fluids is so low that the measurement **milliosmole (mOsm)**, 1/1000 of an osmole, is used. Most body fluids have an osmotic concentration of approximately 300 mOsm/kg.

Water moves by osmosis from a solution with a lower osmolality (fewer particles, more water) to a solution with a higher osmolality (more particles, less water). For example, water moves by osmosis from a solution of 300 mOsm/kg toward a solution of 1200 mOsm/kg.

There is a **medullary concentration gradient** extending from the bases of the renal pyramids to their tips. The interstitial fluid concentration is 300 mOsm/kg in the cortical region of the kidney. Moving from the cortex into the medulla, the interstitial fluid becomes progressively more concentrated until it achieves a maximum concentration of 1200 mOsm/kg at the tip of the renal pyramid. This is more concentrated than seawater. ▼

24. *Define milliosmole. How does the osmolality of solutions affect the movement of water by osmosis?*
25. *Describe the medullary concentration gradient.*

Reabsorption in the Proximal Convoluted Tubule

Most reabsorbed substances pass through the cells that make up the wall of the nephrons. The cells have an **apical membrane,** which is in contact with the filtrate, and a **basolateral membrane,** which forms the base and sides of the cell and is in contact with the interstitial fluid between the nephron and peritubular capillaries (figure 23.9). Substances reabsorbed through cells must pass through the apical and basolateral membranes. Some reabsorbed substances pass between the nephron cells. which are connected by **tight junctions** (see chapter 4). In some parts of the nephron, such as the proximal convoluted tubule, the tight

junctions are "leaky," allowing the passage of significant amounts of the filtrate.

The proximal convoluted tubule is the primary site for the reabsorption of solutes and water. A Na^+–K^+ pump moves Na^+ out of tubule cells and K^+ into tubule cells by active transport. As a result, the concentration of Na^+ inside the tubule cell is lower than the concentration of Na^+ in the filtrate. This concentration gradient for Na^+ is responsible for the symport of many ions and molecules from the filtrate into the tubule cells.

Carrier proteins that transport ions, amino acids, glucose, and other solutes are located within the apical membrane. Each of these carrier proteins binds specifically to one of the substances to be transported and to Na^+. The concentration gradient for Na^+ provides the source of energy that moves the Na^+ and the other ions or molecules bound to the carrier protein from the filtrate into the tubule cell. Once the symported molecules are inside the tubule cell, they cross the basolateral membrane of the tubule cell by facilitated diffusion or symport.

Some solutes also diffuse between the cells from the filtrate into the interstitial fluid. The concentration of these solutes increases as other solutes and water leave the filtrate and enter tubule cells. As the concentration of these solutes increases above the concentration in the interstitial fluid, they diffuse between the

1. In the basolateral membrane, Na^+-K^+ pumps establish a concentration gradient for Na^+.

2. Sodium ions cross the apical membrane by antiport in exchange for H^+ (see figure 23.27, step 2).

3. Glucose, amino acids, and other organic molecules cross the apical membrane by symport and the basolateral membrane by facilitated diffusion.

4. Chloride ions and other anions follow the positively charged Na^+.

5. Water follows Na^+ and Cl^- by osmosis.

6. Many ions (K^+, Ca^{2+}, Mg^{2+}, Cl^-), urea, and water pass between cells.

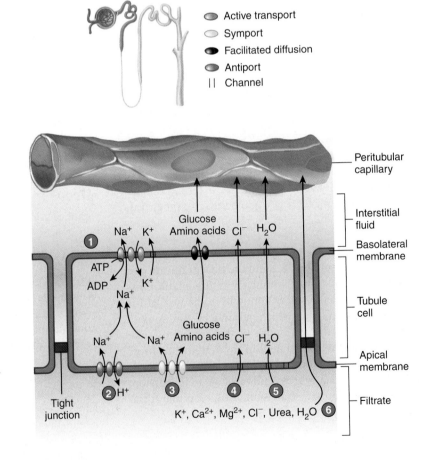

Process Figure 23.9 Reabsorption in the Proximal Convoluted Tubule

tubule cells. Some K^+, Ca^{2+}, Mg^{2+}, Cl^-, and urea diffuse between the cells of the proximal convoluted tubule wall from the filtrate to the interstitial fluid. Reabsorption of these solutes by diffusion occurs, even though the same ions are also reabsorbed by symport processes.

Reabsorption of solutes in the proximal convoluted tubule is extensive, and the tubule is permeable to water. As solute molecules are transported from the nephron to the interstitial fluid, water moves by osmosis in the same direction (see figure 23.9). By the time the filtrate reaches the end of the proximal convoluted tubule, 65% of its water and NaCl (and other solutes) has been reabsorbed. The concentration of the filtrate in the proximal convoluted tubule remains about the same as that of the interstitial fluid (300 mOsm/kg) because proportional amounts of water and solutes are reabsorbed. ▼

26. *Define apical and basolateral membranes of nephron cells.*
27. *Explain how solutes and water are reabsorbed through and between tubule cells.*

Reabsorption in the Loop of Henle

The descending limb of the loop of Henle concentrates the filtrate by adding solutes and removing water. The thin segment of the descending limb of the loop of Henle is moderately permeable to urea, sodium, and most other ions and is highly permeable to water. The loop of Henle descends into the medulla of the kidney, where the interstitial fluid concentration is very high. As the filtrate passes through the thin segment of the descending limb, some solutes diffuse into it, and water moves out of it by osmosis. By the time the filtrate has reached the tip of the loop of Henle, the concentration of the filtrate is equal to the high concentration of the interstitial fluid (1200 mOsm/L) and the volume of the filtrate has been reduced by another 15%.

The thin segment of the ascending limb of the loop of Henle dilutes the filtrate by *passively* removing solutes. The thin segment of the ascending limb removes solutes from the filtrate by diffusion but is impermeable to water. The ascending limb is surrounded by interstitial fluid, which becomes less concentrated toward the cortex. As the filtrate flows through the thin segment of the ascending limb, solutes diffuse out of the filtrate into the interstitial fluid, which dilutes the filtrate. The water content of the filtrate does not change because the ascending limb is impermeable to water.

The thick segment of the ascending limb of the loop of Henle dilutes the filtrate by *actively* removing solutes. The thick segment of the ascending limb removes solutes from the filtrate by secondary active transport but is impermeable to water. A Na^+–K^+ pump actively transports Na^+ across the basolateral membrane into the interstitial fluid, establishing a concentration gradient for Na^+ (figure 23.10). A symporter moves Na^+, K^+, and Cl^- across the apical membrane. Chloride ions and K^+ enter the interstitial

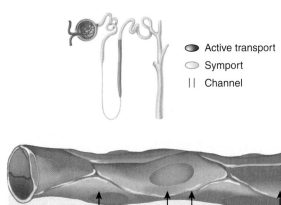

1. In the basolateral membrane, Na^+–K^+ pumps establish a concentration gradient for Na^+.

2. Symporters move Na^+, K^+, and $2Cl^-$ across the apical membrane.

3. Water does not follow the symported ions because the apical membrane is impermeable to water.

4. Ions, such as K^+, Ca^{2+}, Mg^{2+}, and Cl^-, pass between cells.

5. Potassium ions and Cl^- pass through the basolateral membrane by symport and diffusion.

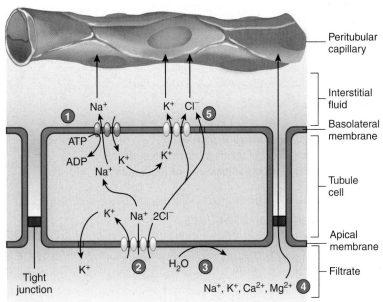

Process Figure 23.10 **Reabsorption in the Thick Segment of the Ascending Limb of the Loop of Henle**

fluid by symport and diffusion. Sodium and other ions also pass between cells. Water does not follow the solutes because the ascending limb is impermeable to water. The removal of solutes, but not water, dilutes the filtrate.

By the time the fluid reaches the distal convoluted tubule, the filtrate concentration is reduced to 100 mOsm/kg, and 25% of the NaCl has been reabsorbed. The filtrate entering the distal convoluted tubule is more dilute than the surrounding interstitial fluid, which has an osmolality of 300 mOsm/kg. ▼

> 28. Describe the movement of solutes and water in the descending and ascending limbs of the loop of Henle.

Reabsorption in the Distal Convoluted Tubule and Collecting Duct

Most of the water (80%) and solutes (90%) in the filtrate have been reabsorbed by the time the filtrate reaches the distal convoluted tubules and collecting ducts. The volume and concentration of the urine are determined by the amount of water and solutes reabsorbed in the distal convoluted tubules and collecting ducts. This reabsorption is under hormonal control.

Reabsorption of solutes occurs in the distal convoluted tubules and collecting ducts. A Na^+–K^+ pump actively transports Na^+ across the basolateral membrane into the interstitial fluid, establishing a concentration gradient for Na^+. Sodium ions and Cl^- enter the apical membrane by symport, and the Cl^- diffuse across the basolateral membrane. The distal convoluted tubules and collecting ducts reabsorb approximately 9%–10% of the Na^+ and Cl^- in the filtrate. The reabsorption of Na^+ is regulated by the hormone **aldosterone** (see "Renin–Angiotensin–Aldosterone," p. 736).

Water moves by osmosis from the distal convoluted tubules and collecting ducts. The permeability of the distal convoluted tubules and collecting ducts to water is under the control of **antidiuretic hormone (ADH)** control (see "Antidiuretic Hormone," p. 734). When ADH increases, the distal convoluted tubule and collecting duct are permeable to water. The water moves out of the filtrate and a small volume of very concentrated urine is produced. Another 19% of the filtrate can be reabsorbed, leaving less than 1% of the filtrate as urine. When ADH decreases, the distal convoluted tubule and collecting duct are impermeable to water. The water remains in the filtrate and a large volume of dilute urine is produced. ▼

> 29. How are Na^+ and Cl^- reabsorbed from the filtrate in the distal convoluted tubules and collecting ducts?
> 30. How does the permeability of the distal convoluted tubules and collecting ducts affect the volume and concentration of urine?

Changes in the Concentration of Solutes in the Nephron

Waste products eliminated in the urine are in a higher concentration in the urine than in the blood. Urea enters the glomerular filtrate and is in the same concentration there as in the blood. As the volume of filtrate decreases in the nephron, the concentration of urea increases because renal tubules are not as permeable to urea as they are to water. Approximately 50% of the urea is passively reabsorbed in the nephron, although about 99% of the water is reabsorbed. Urate ions, creatinine, sulfates, phosphates, and nitrates are reabsorbed, but not to the same extent as water. They become more concentrated in the filtrate as the volume of the filtrate becomes smaller. These substances are toxic if they accumulate in the body, so their accumulation in the filtrate and elimination in urine help maintain homeostasis. ▼

> 31. How do waste products, such as urea, become concentrated in the urine?

Conjugation and Excretion

Some drugs, environmental pollutants, and other foreign substances that gain access to the circulatory system are reabsorbed from the nephron. These substances are usually lipid-soluble, nonpolar compounds. They enter the glomerular filtrate and are reabsorbed passively by a process similar to that by which urea is reabsorbed. Because these substances are passively reabsorbed within the nephron, they are not rapidly excreted. The liver cells attach other molecules to them by a process called **conjugation** (kon-jŭ-gā′shŭn), which converts them to more water-soluble molecules. These more water-soluble substances enter the filtrate but do not pass as readily through the wall of the nephron, are not reabsorbed from the renal tubules, and consequently are more rapidly excreted in the urine. One of the important functions of the liver is to convert nonpolar toxic substances to more water-soluble forms, thus increasing the rate at which they are excreted in the urine.

Tubular Secretion

Tubular secretion is the addition of substances to the filtrate. By-products of metabolism that become toxic in high concentrations, drugs or molecules not normally produced by the body, and other substances are secreted into the nephron from the peritubular capillaries. As with tubular reabsorption, tubular secretion can be active or passive. For example, H^+, K^+, creatinine, histamine, and penicillin are actively transported into the nephron, whereas ammonia diffuses into the filtrate. The secretion of H^+ (see "Renal Regulation of Acid–Base Balance," p. 756) and K^+ (see "Regulation of Potassium Ions," p. 749) is discussed later in the chapter. ▼

> 32. List examples of substances secreted into the filtrate. Describe the way in which they are transported.
> 33. Describe how and where H^+ and K^+ are secreted.

Summary of Changes in Filtrate Volume and Concentration

Approximately 180 L of filtrate enter the proximal convoluted tubules daily (figure 23.11, step 1). Most of the useful solutes that pass through the filtration membrane into Bowman's capsule are reabsorbed by secondary active transport in the proximal convoluted tubule. Filtrate volume and NaCl (solutes) are reduced by

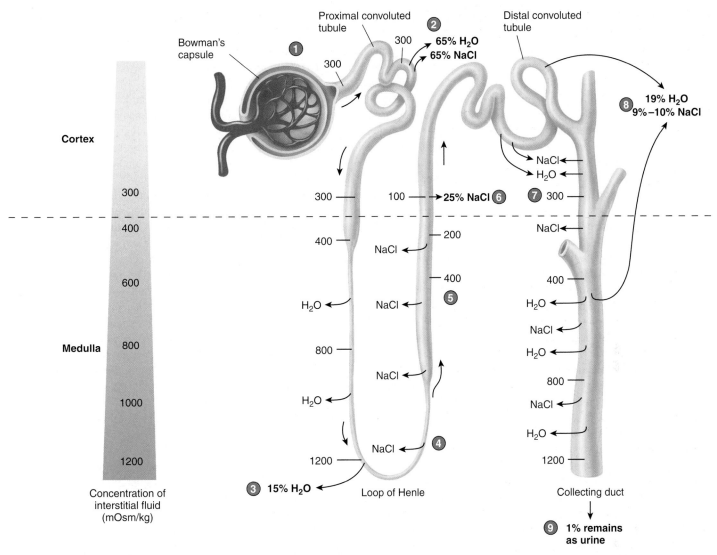

Process Figure 23.11 Summary of Urine Concentrating Mechanism

1. Approximately 180 L of filtrate enters the nephrons each day (see figure 23.8). The filtrate concentration is 300 mOsm/kg.

2. Approximately 65% of the water and NaCl in the original filtrate is reabsorbed in the proximal convoluted tubule (see figure 23.9). The filtrate concentration is 300 mOsm/kg.

3. Approximately 15% of the water is reabsorbed in the thin segment of the descending limb of the loop of Henle. At the tip of the renal pyramid, filtrate concentration is 1200 mOsm/kg, which is equal to the interstitial fluid concentration.

4. The thin segment of the ascending limb of the loop of Henle is not permeable to water. Sodium chloride diffuses out of the thin segment.

5. The thick segment of the ascending limb of the loop of Henle is not permeable to water. Sodium ions are actively transported into the interstitial fluid and Cl⁻ follow by diffusion (see figure 23.10).

6. The volume of the filtrate does not change as it passes through the ascending limb, but the concentration is greatly reduced. By the time the filtrate reaches the cortex, the concentration is 100 mOsm/kg and an additional 25% of NaCl has been reabsorbed.

7. The distal convoluted tubules and collecting ducts reabsorb water and NaCl.

8. If ADH is present, water moves by osmosis from the less concentrated filtrate into the more concentrated interstitial fluid. By the time the filtrate reaches the tip of the renal pyramid, an additional 19% of water and 9%–10% of NaCl is reabsorbed.

9. One percent or less of the filtrate remains as urine when ADH is present.

65%. The osmolality of the interstitial fluid and the filtrate is maintained at about 300 mOsm/kg (figure 23.11, step 2).

The filtrate passes into the descending limbs of the loops of Henle, which are permeable to water and solutes. As the descending limbs penetrate deep into the medulla of the kidney, the surrounding interstitial fluid has a progressively greater osmolality. Water diffuses out of the nephrons as solutes slowly diffuse into

them. By the time the filtrate reaches the deepest part of the loops of Henle, its volume has been reduced by an additional 15% of the original volume and its osmolality has increased to that of the surrounding interstitial fluid. At the tips of the renal pyramids, filtrate concentration can be 1200 mOsm/kg (figure 23.11, step 3).

The thin and thick segments of the ascending limbs of the loops of Henle are impermeable to water but permeable to solutes.

Diuretics (dī-ū-ret′iks) are agents that increase the rate of urine formation. Although the definition is simple, the effect can be achieved by different physiological mechanisms. Diuretics are used to treat disorders such as hypertension and several types of edema caused by congestive heart failure, cirrhosis of the liver, and other anomalies. The use of diuretics can lead to complications, however, including dehydration and electrolyte imbalances.

Many diuretics reduce the reabsorption of Na^+ from the filtrate, which promotes the loss of Na^+, Cl^-, and water in the urine. **Carbonic anhydrase** (kar-bon′ik an-hī′drās) **inhibitors** reduce the uptake of Na^+ and HCO_3^- from the filtrate by inhibiting H^+ secretion into the filtrate. Carbonic anhydrase is an enzyme that reversibly promotes the formation of carbonic acid (H_2CO_3) from carbon dioxide and water. The carbonic acid dissociates to form H^+ and HCO_3^-. In the filtrate, carbonic anhydrase associated with the microvilli of tubule cells converts H^+ and HCO_3^- to carbon dioxide and water (see figure 23.27, step 4). This effectively removes the HCO_3^- from the filtrate. The carbon dioxide diffuses into tubule cells, where carbonic anhydrase promotes the formation of H^+ and HCO_3^-. The H^+ are secreted into the nephron by an antiporter in exchange for

Na^+. Carbonic anhydrase inhibitors reduce the production of H^+, which results in reduced uptake of Na^+ by antiport from the filtrate. Consequently, less water leaves the filtrate by osmosis and urine volume increases. The diuretic effect is useful in treating conditions such as glaucoma and altitude sickness.

Loop diuretics inhibit the $Na^+-K^+-2Cl^-$ symporters in the apical membrane of cells in the ascending thick segment of the loop of Henle (see figure 23.10). Loop diuretics are used to treat pulmonary edema and edema associated with congestive heart failure, cirrhosis of the liver, and renal failure. **Thiazide diuretics** inhibit the Na^+-Cl^- symporters in the distal convoluted tubules and collecting ducts. These diuretics are given to some people who have hypertension.

A possible side effect of diuretic drugs that inhibit Na^+ reabsorption is an increase in the excretion of K^+ in the urine. Increased delivery of Na^+ to the distal convoluted tubules and collecting ducts results in increased movement of Na^+ out of the filtrate and increased movement of K^+ into the filtrate. The increased loss of K^+ in the urine can cause hypokalemia. **Potassium-sparing diuretics** act on the distal convoluted tubules and the collecting ducts to prevent the reabsorption of Na^+, which reduces K^+ loss and therefore preserves, or

"spares," these ions. Some potassium-sparing diuretic drugs act by blocking aldosterone receptors, which results in less antiport of Na^+ and K^+ (see figure 23.15). Other potassium-sparing diuretics block Na^+ channels through which Na^+ diffuse into cells. Potassium-sparing diuretics are often used in combination with other diuretics.

Osmotic diuretics freely pass by filtration into the filtrate, but they undergo limited reabsorption by the nephron. These diuretics increase urine volume by elevating the osmotic concentration of the filtrate, thus reducing the amount of water moving by osmosis out of the nephron. Urea, mannitol, and glycerine have been used as osmotic diuretics. Although they are not commonly used, they are effective in treating people who are suffering from cerebral edema and edema in acute renal failure (see "Renal Failure," p. 742).

Xanthines (zan′thēnz), including caffeine and related substances, act as diuretics partly because they increase renal blood flow and the rate of glomerular filtrate formation. They also influence the nephron by decreasing Na^+ and Cl^- reabsorption.

Alcohol acts as a diuretic, although it is not used clinically for that purpose. It inhibits ADH secretion from the posterior pituitary and results in increased urine volume.

The movement of solutes, but not water, across the wall of the ascending limbs causes the osmolality of the filtrate to decrease from 1200 to about 100 mOsm/kg by the time the filtrate again reaches the cortex of the kidney. The volume of the filtrate does not change as it passes through the ascending limbs, but 25% of the NaCl is reabsorbed (figure 23.11, steps 4–6).

The distal convoluted tubules and the collecting ducts reabsorb water and NaCl (figure 23.11, step 7). When ADH is present, an additional 19% of the filtrate is reabsorbed, leaving 1% of the original filtrate as urine. The concentration of the urine can be as high as 1200 mOsm/kg, which is the concentration of the interstitial fluid at the tip of the renal pyramid (figure 23.11, steps 8 and 9). In the absence of ADH, a greater volume of urine is produced, and it can be as dilute as 65 mOsm/kg because of the reabsorption of NaCl. ▼

> 34. **Describe how the filtrate volume and concentration change as it flows through the nephrons and collecting ducts.**

Maintaining the Medullary Concentration Gradient

The kidneys' ability to concentrate urine depends on maintaining the medullary concentration gradient because the urine will become only as concentrated as the interstitial fluid at the tip of the renal pyramid. Maintenance of the high solute concentration in the kidney medulla depends on the addition of solutes to the interstitial fluid and the removal of reabsorbed solutes and water from the nephron by the vasa recta.

Addition of Solutes

The addition of solutes, without water, by the thin and thick segments of the ascending limbs of the loops of Henle is the main mechanism increasing the medullary interstitial fluid concentration (see figures 23.10 and 23.11). The thick segments of the ascending limbs actively pump Na^+ into the interstitial fluid and other ions follow by symport or by passing between tubule cells.

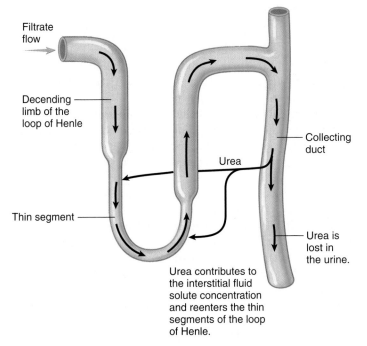

Figure 23.12 **Urea Cycling and the Medullary Concentration Gradient**

Urea cycles from the medullary collecting duct to the descending and ascending thin segments of the loop of Henle. The addition of urea to the interstitial fluid contributes to the medullary concentration gradient.

The thin segments of the ascending limbs add solutes to the interstitial fluid by diffusion.

Urea (ū-rē′ă) molecules are responsible for a substantial part of the high osmolality in the medulla of the kidney (figure 23.12). The medullary collecting ducts and the descending and ascending thin segments of the loops of Henle are permeable to urea. As water is reabsorbed from the collecting ducts, the concentration of urea in the filtrate increases until it is higher than in the surrounding interstitial fluid. Approximately 50% of the urea reaching the collecting ducts moves by facilitated diffusion out of them into the interstitial fluid, which increases the interstitial fluid concentration. Around the thin segments, the concentration of urea in the interstitial fluid is higher than in the filtrate and urea moves into the filtrate by facilitated diffusion. As urea cycles from the medullary collecting ducts into the thin segments, it contributes to the medullary concentration gradient. ▼

> **35.** *In what locations are solutes added to the medullary interstitial fluid?*
> **36.** *Describe the cycling of urea.*

Removal of Reabsorbed Solutes and Water

The vasa recta has two functions: (1) It carries oxygen and nutrients to nephron cells in the medulla and removes waste products, and (2) it removes solutes and water reabsorbed by the nephrons and collecting ducts without destroying the medullary concentration gradient.

Blood flows through the vasa recta without destroying the medullary concentration gradient because the vasa recta is a countercurrent mechanism. A **countercurrent mechanism** consists of two parallel tubes through which fluid flows in opposite directions. The vasa recta has a descending limb, through which blood flows toward the tip of the renal pyramid, and an ascending limb, which returns blood in the opposite direction to the cortex. As blood flows through the descending limb, it becomes more concentrated because it loses water to, and gains solutes from, the more concentrated interstitial fluid. At the tip of the renal pyramid, the concentration of the blood is 1200 mOsm and is equal to that of the interstitial fluid (figure 23.13, steps 1–3).

1. The concentration of the blood entering the descending limb of the vasa recta is 300 mOsm/kg.

2. The concentration of blood in the descending limb increases as water is lost and solutes are gained.

3. The concentration of the blood is equal to the surrounding interstitial fluid (1200 mOsm/kg).

4. The concentration of blood in the ascending limb decreases as water is gained and solutes are lost. The medullary concentration gradient is maintained by the exchange of water and solutes between the descending and ascending limbs.

5. Solutes reabsorbed by nephrons and collecting ducts enters the descending limb.

6. Water reabsorbed by nephrons and collecting ducts enter the ascending limb.

7. The concentration and volume of the blood leaving the ascending limb are slightly greater than those entering the descending limb because of solutes and water gained from the nephrons and collecting ducts.

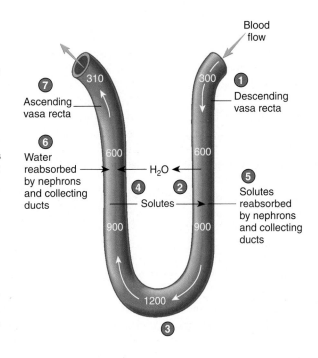

Figure 23.13 **Blood Flow Through the Vasa Recta**

If the descending limb exited the tip of the renal pyramid, the medullary concentration gradient would be diluted because of the addition of water from the blood and the removal of solutes by the blood. Instead of exiting the tip of the renal pyramid, however, the vasa recta turns and the ascending limb carries blood back to the cortex. As blood flows through the ascending limb, it becomes less concentrated because it gains water from, and loses solutes to, the less concentrated interstitial fluid. The medullary concentration gradient is preserved because of the exchange of water and solutes between the descending and ascending limbs of the vasa recta. The water entering the interstitial fluid from the descending limb is removed by the ascending limb, and the solutes removed from the interstitial fluid by the descending limb are replaced from the ascending limb (figure 23.13, step 4).

Blood passing through the vasa recta picks up solutes and water reabsorbed by the nephrons and collecting ducts. Solutes reabsorbed by nephrons and collecting ducts enter the descending limb, and water reabsorbed by nephrons and collecting ducts enters the ascending limb. Slightly more water and slightly more solutes are carried from the medulla by the vasa recta than enter it. Thus, the composition of the blood at both ends of the vasa recta is nearly the same, with the volume and osmolality being slightly greater as the blood once again reaches the cortex (figure 23.13, steps 5–7). ▼

37. *What are the two functions of the vasa recta?*
38. *What is a countercurrent mechanism?*
39. *How does the vasa recta remove reabsorbed solutes and water while maintaining the medullary concentration gradient?*

Hormonal Regulation of Urine Concentration and Volume

When a large volume of water is consumed, it is necessary to eliminate the excess without losing large amounts of electrolytes or other substances essential for the maintenance of homeostasis. The kidneys' response is to produce a large volume of dilute urine. On the other hand, when drinking water is not available, producing a large volume of dilute urine would lead to rapid dehydration. When water intake is restricted, the kidneys produce a small volume of concentrated urine that conserves water and contains sufficient waste products to prevent their accumulation in the circulatory system.

The hormonal mechanisms involved in regulating urine concentration and volume are the antidiuretic hormone (ADH) mechanism, the renin–angiotensin–aldosterone mechanism, and the atrial natriuretic hormone (ANH) mechanism. The ADH mechanism is more sensitive to changes in blood osmolality, and the renin–angiotensin–aldosterone and ANH mechanisms are more sensitive to changes in blood pressure. In addition to hormonal mechanisms, autoregulation and sympathetic stimulation affect urine production (see "Regulation of Glomerular Filtration Rate," p. 727).

Antidiuretic Hormone

Antidiuretic hormone (ADH) is secreted from the posterior pituitary gland. Neurons with cell bodies primarily in the hypothalamus have axons that extend to the posterior pituitary gland in the hypothalamohypophyseal tract (see figure 15.16). ADH is produced in the cell bodies of these neurons and transported to their axon endings in the posterior pituitary gland, from which ADH is released into the circulatory system. **Osmoreceptor cells** in the hypothalamus are very sensitive to even slight changes in the osmolality of the interstitial fluid. If the osmolality of the blood and interstitial fluid increases, these cells stimulate the ADH-secreting neurons. Action potentials are then propagated along the axons of the ADH-secreting neurons to the posterior pituitary gland, and ADH is released. Reduced osmolality of the interstitial fluid within the hypothalamus inhibits ADH secretion from the posterior pituitary gland (see figure 18.38).

ADH increases the permeability of the distal convoluted tubules and collecting ducts to water by stimulating an increase in the number of aquaporins in the apical plasma membranes of their cells. **Aquaporins** are water channels through which water passes by osmosis. There are different forms of aquaporins. Aquaporin-2 molecules are found in the apical membranes. Their numbers in the apical membranes increase in response to ADH stimulation and they regulate water movement out of the filtrate into cells. Aquaporin-3 and aquaporin-4 molecules are in the basolateral membranes. Water passes out of cells to the blood through these aquaporins, which are insensitive to ADH stimulation.

In cells that have not been exposed to ADH, the aquaporin-2 molecules are found in the membranes of vesicles in the cytoplasm. When ADH binds to its receptors, it activates a G protein mechanism, causing in increased cAMP synthesis. As a result, the vesicles containing aquaporin-2 molecules insert them into the apical plasma membrane. Water moves from the filtrate into the tubule cells by osmosis and exits the cells into the interstitial fluid by passing through aquaporin-3 and aquaporin-4 molecules (figure 23.14).

ADH regulates the osmolality of the body's fluids by controlling the amount of water entering or leaving the blood in the kidneys. Consider the following situation to understand how this works. Given a 10% salt solution in a container, such as a pan on a stove, it is possible to decrease the concentration of the solution by adding pure water to it. It is also possible to increase the concentration of the solution by boiling the water in the pan, thus removing water from the solution by evaporation. Similarly, the kidneys decrease blood osmolality by increasing water reabsorption, and they increase blood osmolality by decreasing water reabsorption.

The amount of water reabsorbed by the kidneys affects the concentration and volume of the urine.

Formation of Concentrated and Dilute Urine

ADH regulates the permeability of the ends of the distal convoluted tubules and the collecting ducts to water. Depending on the levels of ADH, water permeability can be high, low, or everything

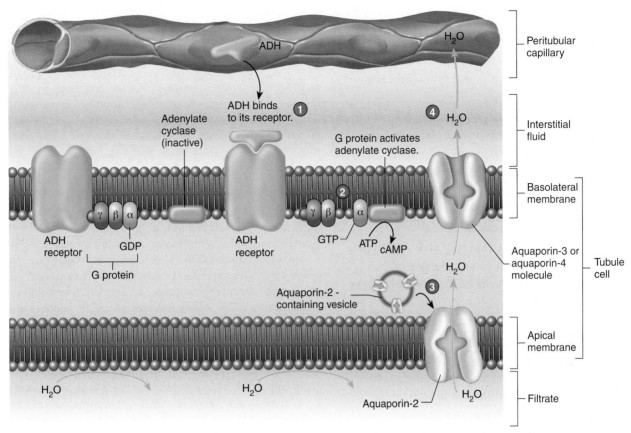

cAMP increases aquaporin-2 to increase permeability of the membrane to H_2O.

1. ADH moves from the peritubular capillaries and binds to ADH receptors in the plasma membranes of the distal convoluted tubule cells and the collecting duct cells.

2. When ADH binds to its receptor, a G protein mechanism is activated, which in turn activates adenylate cyclase.

3. Adenylate cyclase increases the rate of cAMP synthesis. Cyclic AMP promotes the insertion of aquaporin-2 containing cytoplasmic vesicles into the apical membranes of the distal convoluted tubules and collecting ducts, thereby increasing their permeability to water. Water then moves by osmosis out of the distal convoluted tubules and collecting ducts into the tubule cells through the aquaporin-2 water channels.

4. Water exits the tubule cells and enters the interstitial fluid through aquaporin-3 and aquaporin-4 water channels in the basolateral membranes.

Process Figure 23.14 Effect of Antidiuretic Hormone (ADH) on Water Movement

in between. When ADH levels are high, water permeability is high, water reabsorption is high, and a small volume of concentrated urine is produced. By the time the filtrate has reached the end of the collecting ducts, the osmolality of the filtrate is approximately 1200 mOsm/kg and most of the water in the filtrate has been reabsorbed (see figure 23.11).

When ADH levels are low, water permeability is low, water reabsorption is low, and a large volume of dilute urine is produced. Filtrate entering the distal convoluted tubules has a concentration of 100 mOsm/kg (see figure 23.11). The removal of Na^+ and other

solutes from the filtrate can reduce the urine concentration to 65 mOsm/kg. In a healthy person, even when the kidneys produce dilute urine, beneficial substances are retained, and toxic substances and excess water are eliminated.

P R E D I C T

Drugs that increase urine volume are called diuretics. Some diuretics inhibit the active transport of Na^+ in the nephron. Explain how these diuretic drugs can cause increased urine volume.

CASE STUDY | Diabetes Insipidus

Billy is a newborn. Not long after Billy arrived home from the hospital, his parents noticed that his diapers were excessively wet hour after hour throughout the day and night. After the first week, Billy's parents took him to the doctor because he had a slight fever and had vomited, even though he had not eaten for several feedings. Billy was irritable, but his parents thought this was due to a virus causing the vomiting. The doctor took a blood sample, which indicated that Billy had high blood Na$^+$ levels. Subsequently, the doctor ordered a water deprivation test, during which plasma levels of ADH were monitored. Based on the test results, the doctor diagnosed Billy's condition to be **nephrogenic diabetes insipidus (NDI)**.

The term *diabetes* refers to a disease state characterized by an excess production of urine. The word *insipidus* implies the production of a clear, tasteless, dilute urine. People who have diabetes insipidus often produce 10–20 L of urine per day and develop major problems, such as dehydration and ion imbalances. Diabetes insipidus is a relatively rare disease that occurs in two varieties: central and nephrogenic diabetes insipidus. **Central diabetes insipidus (CDI)** is caused by a failure of ADH secretion. It can be of unknown cause or be caused by pituitary tumors, head trauma, cranial surgery, or genetic abnormalities. **Nephrogenic diabetes insipidus (NDI)** results when ADH secretion is normal but the ADH receptor, or the response to ADH, in the kidney is abnormal. In most cases, NDI results from an inherited condition that affects the function of the ADH receptor or aquaporin-2 molecules. Billy's NDI results from an abnormal ADH receptor.

The treatment of NDI includes ensuring a plentiful supply of water, following a low-sodium and sometimes a low-protein diet, and using thiazide diuretics (Na$^+$ ion reabsorption inhibitors) in combination with a potassium-sparing diuretic.

P R E D I C T **4**

Use your knowledge of kidney physiology and figure 23.14 to answer the following questions.

a. Why does Billy have high blood Na$^+$ levels and dilute urine?

b. Predict how Billy's ADH plasma levels will change during the water deprivation test, given a diagnosis of NDI. If Billy had CDI, how would his ADH plasma levels change?

c. How would a genetic abnormality that produces abnormal aquaporin-2 molecules result in excessive urine production?

d. Even though mutations of aquaporin-3 and aquaporin-4 in the collecting duct have not been described in the literature, if mutations occurred that resulted in a reduced number of these aquaporins in the cells of the collecting ducts, what effect would they have on urine volume and concentration? Would ADH be an effective treatment?

Regulation of ADH Secretion

The primary stimulus for ADH secretion is a change in blood osmolality. Small changes in blood osmolality cause changes in ADH secretion. When blood osmolality increases, ADH secretion increases and the kidneys reabsorb more water. The reabsorption of water decreases blood osmolality, decreases urine volume, and increases urine concentration. Conversely, when blood osmolality decreases, ADH secretion decreases and the kidneys reabsorb less water, resulting in increased blood osmolality and a large volume of dilute urine.

ADH secretion also changes in response to significant changes in blood pressure of 5%–10% or more through baroreceptor reflexes (see figure 18.38). A significant decrease in blood pressure stimulates increased ADH secretion, which promotes water reabsorption and increased blood volume. The increased blood volume helps increase blood pressure. Conversely, when blood pressure increases significantly, ADH secretion decreases, water reabsorption decreases, blood volume decreases, and blood pressure decreases. ▼

40. *Where is ADH produced and secreted? What are osmoreceptors?*

41. *Describe the role of ADH and aquaporins in water reabsorption in the kidney.*

42. *What effect does ADH have on blood osmolality and volume? On urine concentration and volume?*

43. *How do changes in blood osmolality and blood pressure affect ADH secretion?*

Renin–Angiotensin–Aldosterone

Renin (rē′nin, ren′in) is an enzyme secreted by the granular cells of the juxtaglomerular apparatus. Renin enters the general circulation and acts on a protein produced by the liver called **angiotensinogen** (an′jē-ō-ten-sin′ō-jen). Renin removes amino acids from angiotensinogen to produce **angiotensin** (an′jē-ō-ten′sin) **I**. **Angiotensin-converting enzyme (ACE)** is found in capillary beds in organs such as the lungs. ACE removes amino acids from angiotensin I to produce **angiotensin II.** Angiotensin II stimulates the secretion of **aldosterone,** which is a steroid hormone secreted by the cortical cells of the adrenal glands (figure 23.15).

Aldosterone passes through the circulatory system from the adrenal glands to the cells in the distal convoluted tubules and the collecting ducts, where it diffuses through the plasma membranes and binds to intracellular receptors. The combination of aldosterone molecules with their receptor molecules increases synthesis of the transport proteins that increase the movement of Na$^+$ across the basolateral and apical membranes of nephron cells. As a result, the

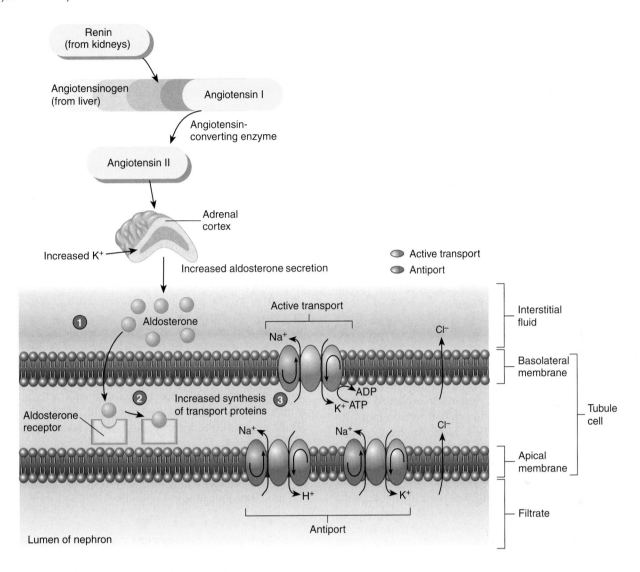

1. Aldosterone secreted from the adrenal cortex enters cells of the distal convoluted tubule.

2. Aldosterone binds to intracellular receptors and increases the synthesis of transport proteins of the apical and basolateral membranes.

3. Newly synthesized transport proteins increase the rate at which Na^+ are absorbed and K^+ and H^+ are secreted. Chloride ions move with the Na^+ because they are attracted to the positive charge of Na^+.

Process Figure 23.15   Effect of Aldosterone on Ion Movement

rate of Na^+ and Cl^- transport out of the filtrate back into the blood increases (see figure 23.15).

The water content of the body is controlled by regulating the Na^+ content of the body. Consider the following situation to understand how this works. Given a 10% salt solution in a container, if salt is added to the solution, its osmolality increases but there is little increase in volume. If, however, a cup of 10% salt solution is added to the container, its volume increases but its osmolality stays the same. Thus, adding salt with just the right amount of water increases volume but does not change osmolality. Similarly, removing salt with just the right amount of water decreases volume and maintains osmolality. For example, scooping out a cup of solution from the container decreases the volume of the solution in the container but does not change its osmolality.

The renin–angiotensin–aldosterone mechanism regulates blood volume by controlling the total Na^+ content of the body. Increased aldosterone promotes increased Na^+ reabsorption, and Cl^- follow the Na^+. As Na^+ and Cl^- are reabsorbed, water follows by osmosis. Given that ADH maintains a constant blood osmolality, increased Na^+ reabsorption increases water reabsorption and blood volume, without changing blood osmolality. Conversely, decreased aldosterone decreases water reabsorption and blood volume. ▼

44. *Starting with renin, describe the events leading to increased secretion of aldosterone from the adrenal gland.*

45. *What are the effects of aldosterone on Na^+ and Cl^- transport?*

46. *How does regulating the body's Na^+ content regulate the body's water content? How does aldosterone affect blood volume?*

Regulation of Renin Secretion

Although the renin–angiotensin–aldosterone mechanism regulates the body's Na$^+$ content, the primary variable stimulating renin secretion is not the body's Na$^+$ content or plasma Na$^+$ levels but, rather, is systemic blood pressure. This makes sense when the relationship between the body's Na$^+$ content and blood pressure is understood. Approximately 90%–95% of the osmolality of the body's fluids is due to Na$^+$ and its accompanying anion, Cl$^-$. Given that fluid osmolality is maintained at a constant level over the long run, if the body's Na$^+$ content increases, then the body's fluid volume increases, which increases blood volume and blood pressure. If the body's Na$^+$ content decreases, the body's fluid volume, blood volume, and blood pressure decrease. Changes in blood pressure normally indicate a change in the body's Na$^+$ content because the body's Na$^+$ content, blood volume, and blood pressure normally change together.

A decrease in blood pressure causes increased renin secretion, whereas an increase in blood pressure results in decreased renin secretion. For example, a decrease in blood pressure results in increased renin secretion, aldosterone secretion, Na$^+$ reabsorption, blood volume, and blood pressure.

Changes in blood pressure alter renin secretion through intrinsic and extrinsic mechanisms. The intrinsic mechanisms occur within the kidney and operate even in a transplanted kidney. The same stimuli involved with autoregulation are involved with the intrinsic regulation of renin by the kidneys. As blood pressure decreases, the granular cells in the wall of the afferent arteriole are less stretched, which stimulates increased renin secretion. The macula densa of the juxtaglomerular apparatus monitors the rate of filtrate flow through the nephron. When blood pressure decreases, the glomerular filtration rate decreases slightly despite autoregulation. Consequently, slightly less filtrate flows past the macula densa, which sends a signal to the granular cells, stimulating them to increase renin secretion. Conversely, when blood pressure increases, renin secretion decreases.

Extrinsic regulation of renin secretion occurs through the sympathetic division, which innervates the juxtaglomerular apparatus and renal blood vessels. A decrease in blood pressure is detected by baroreceptors, resulting in increased sympathetic stimulation of granular cells and increased renin secretion, whereas an increase in blood pressure causes decreased sympathetic stimulation and decreased renin secretion. Sympathetic stimulation also indirectly affects renin secretion by altering renal blood flow and afferent arteriole diameter. For example, increased sympathetic stimulation results in decreased renal blood flow and vasoconstriction of the afferent arterioles, which decreases glomerular filtration rate. A decreased glomerular filtration rate stimulates the macula densa, causing increased renin secretion.

P R E D I C T

Angiotensin II is a potent vasoconstricting substance that increases peripheral resistance, causing blood pressure to increase (see chapter 18). Angiotensin II also increases aldosterone secretion, salt appetite, the sensation of thirst, and ADH secretion. How is this adaptive when blood pressure decreases?

Aldosterone secretion is regulated by renin secretion through the production of angiotensin II. Aldosterone secretion is also controlled through plasma K$^+$ levels. Increases in blood K$^+$ levels act directly on the adrenal cortex to stimulate aldosterone secretion, whereas decreases in blood K$^+$ levels decrease aldosterone secretion (see "Regulation of Potassium Ions," p. 749). ▼

> **47. What is the primary stimulus for renin secretion? How does the renin–angiotensin–aldosterone mechanism affect Na$^+$ reabsorption, blood volume, and blood pressure?**
> **48. Describe intrinsic and extrinsic regulation of renin secretion.**
> **49. How is aldosterone secretion regulated?**

Atrial Natriuretic Hormone

Atrial natriuretic (nā′trē-ū-ret′ik) **hormone (ANH)** is a polypeptide hormone secreted from cardiac muscle cells in the right atrium of the heart when blood volume in the right atrium increases and stretches the cardiac muscle cells (figure 23.16). Atrial natriuretic hormone inhibits Na$^+$ reabsorption in the kidney tubules. ANH also inhibits ADH secretion from the posterior pituitary gland. Consequently, increased ANH secretion increases the volume of urine produced and lowers blood volume and blood pressure. Atrial natriuretic hormone also dilates arteries and veins, which reduces peripheral resistance and lowers blood pressure. Thus, a decrease occurs in venous return and blood volume in the right atrium. ▼

> **50. Where is atrial natriuretic hormone produced, and what effect does it have on urine production, blood volume, and blood pressure?**

P R E D I C T

A man eats a bag of salty potato chips. Assuming the salty potato chips make him thirsty and he drinks some water, what happens to his urine concentration and volume? Describe the hormonal mechanisms involved.

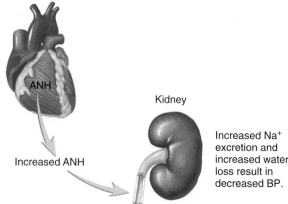

Figure 23.16   ANH and the Regulation of Na$^+$ and Water
Increased blood pressure in the right atrium of the heart causes increased secretion of atrial natriuretic hormone (ANH), which increases Na$^+$ excretion and water loss in the urine.

Urine Movement

Anatomy and Histology of the Ureters, Urinary Bladder, and Urethra

The **ureters** are small tubes that carry urine from the renal pelvis of each kidney to the posterior inferior portion of the urinary bladder (see figure 23.2).

The **urinary bladder** is a hollow, muscular container that lies in the pelvic cavity just posterior to the symphysis pubis. In males, the urinary bladder is just anterior to the rectum; in females, it is just anterior to the vagina and inferior and anterior to the uterus. The urinary bladder stores urine, and its size depends on the quantity of urine present. The urinary bladder can hold from a few milliliters (mL) to a maximum of about 1000 mL of urine.

The **urethra** is a tube that exits the urinary bladder inferiorly and anteriorly. It carries urine to the outside of the body. In males, the urethra extends to the end of the penis. In females, it opens into the vestibule anterior to the vaginal opening (see chapter 24). The female urethra is approximately 4 cm in length, whereas the male urethra is approximately 20 cm.

PREDICT **7**

Cystitis (sis-tī′tis) is inflammation of the urinary bladder. It typically results from infections that often occur when bacteria from outside the body enter the bladder. Are males or females more prone to cystitis? Explain.

The linings of the ureters, urinary bladder, and urethra are mucous membranes consisting of epithelium and an underlying layer of connective tissue, the lamina propria (figure 23.17). The epithelium of the ureters and urinary bladder is transitional epithelium, which is specialized to stretch (see chapter 4) as the ureters transport urine and the urinary bladder expands to store urine. Transitional epithelium lines the superior urethra and nonkeratinized stratified squamous epithelium lines the inferior urethra.

The walls of the ureter and urinary bladder are composed of layers of smooth muscle and connective tissue (see figure 23.17). Regular waves of smooth muscle contractions in the ureters

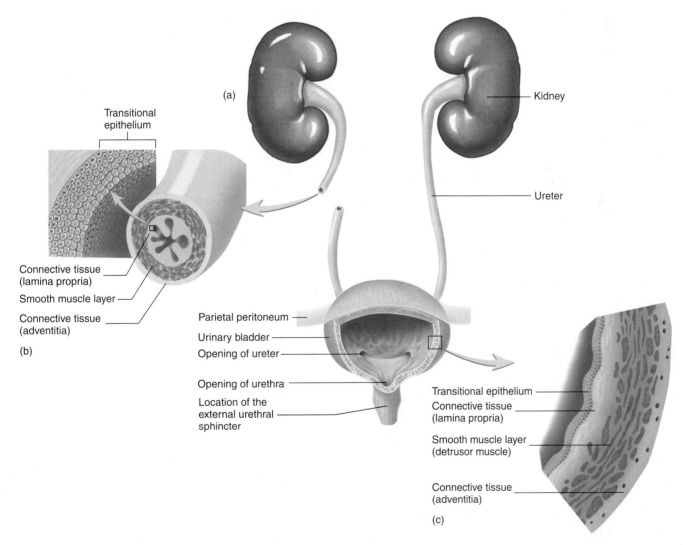

Figure 23.17 Ureters and Urinary Bladder

(a) Ureters extend from the pelvis of the kidney to the urinary bladder. (b) The walls of the ureters and the urinary bladder are lined with transitional epithelium, which is surrounded by a connective tissue layer (lamina propria), smooth muscle layers, and a fibrous adventitia. (c) Section through the wall of the urinary bladder.

produce the force that causes urine to flow from the kidneys to the urinary bladder. The wall of the urinary bladder is much thicker than the wall of a ureter. This thickness is caused by layers composed primarily of smooth muscle, sometimes called the **detrusor** (dē-troo′ser). Contractions of the detrusor force urine to flow from the bladder through the urethra. The urethral wall contains smooth muscle superiorly, but it disappears inferiorly.

Elastic fibers at the junction of the urethra and urinary bladder keep urine from passing through the urethra until the urinary bladder pressure increases. At the junction of the urinary bladder and urethra, the smooth muscle of the bladder wall and urethra form the **internal urethral sphincter** in males. No well-defined internal urethral sphincter is found in females. The internal urethral sphincter of males is under involuntary control. Contraction of the internal urethral sphincter during ejaculation prevents semen from entering the urinary bladder and keeps urine from flowing through the urethra. The **external urethral sphincter** is skeletal muscle in the urethral wall surrounding the urethra as it extends through the pelvic floor (see figure 23.17). The external urethral sphincter is under voluntary control and regulates the flow of urine through the urethra.

Urinary Bladder Cancer

In the United States, urinary bladder cancer affects more than 60,000 new patients each year and is among the 10 most common cancers in men and women. Half the diagnosed cases of urinary bladder cancer can be attributed to cigarette smoking, even 10 years or more after cessation of smoking.

When bladder cancer is detected early (the cancer is confined to the bladder), the survival rate is 94%, whereas, if it is detected late (after it has spread to other areas), the survival rate is 6%. Unfortunately, early detection of urinary bladder cancer is especially challenging due to its rapid growth rate. Frequently, blood in the urine is a symptom but, because this symptom is also associated with other, less serious problems, it tends to be ignored.

Scientists are investigating ways to detect urinary bladder cancer early and noninvasively. Currently, urinary bladder cancer tests screen for abnormal cells; the bladder is visually examined with a catheter in a process called **cystoscopy** (sis-tos′kŏ-pē). New tests that hold promise include a urine test that measures the levels of an enzyme called telomerase, present in nearly all human cancer cells. Measurements of telomerase levels are especially promising for the detection of urinary bladder cancer because telomerase levels are measurable earlier in urinary bladder cancer than in many other cancers. ▼

51. *What are the functions of the ureters, urinary bladder, and urethra? Describe the epithelium and smooth muscle forming their walls.*
52. *What is the function of the internal and external urethral sphincters? Are they under voluntary or involuntary control?*

Urine Flow Through the Nephron and Ureters

Capsular pressure averages 10 mm Hg in Bowman's capsule and nearly 0 mm Hg in the renal pelvis. This pressure gradient forces the filtrate to flow from Bowman's capsule through the nephron into the renal pelvis. Urine moves through the ureters as a result of peristaltic contractions. The peristaltic waves progress from the region of the renal pelvis to the urinary bladder and occur from every few seconds to every 2–3 minutes. Parasympathetic stimulation increases their frequency, and sympathetic stimulation decreases it.

The peristaltic ureter contractions can generate pressures in excess of 50 mm Hg, which moves urine into the urinary bladder. Where the ureters penetrate the urinary bladder, they course obliquely through the bladder wall. Pressure inside the urinary bladder compresses that part of the ureter to prevent the backflow of urine.

When no urine is present in the urinary bladder, internal pressure is about 0 mm Hg. When the volume is 100 mL of urine, pressure is elevated to only 10 mm Hg. Pressure increases slowly as bladder volume increases to approximately 300 mL, but above 400 mL the pressure rises rapidly.

Kidney Stones

Kidney stones are hard objects usually found in the renal pelvis of the kidney. They are normally 2–3 mm in diameter, with a smooth or a jagged surface. About 1% of all autopsies reveal kidney stones, and many of the stones occur without causing symptoms. The symptoms associated with kidney stones occur when a stone passes into the ureter, resulting in intense referred pain down the back, side, and groin area. The ureter contracts around the stone, causing the stone to irritate the epithelium and produce bleeding. Kidney stones can also block the ureter, cause ulceration in the ureter, and increase the probability of bacterial infections.

About 65% of all kidney stones are composed of calcium oxalate mixed with calcium phosphate, 15% are magnesium ammonium phosphate, and 10% are uric acid or cystine. The cause of kidney stones is usually obscure. Predisposing conditions include a concentrated urine and an abnormally high calcium concentration in the urine, although the cause of the high calcium concentration is usually unknown. Magnesium ammonium phosphate stones are often found in people with recurrent kidney infections, and uric acid stones often occur in people suffering from gout. Severe kidney stones must be removed surgically. **Lithotripsy** (lith′ō-trip-sē), the use of instruments that pulverize kidney stones with ultrasound or lasers, however, has replaced most traditional surgical procedures. ▼

53. *What causes urine to flow from Bowman's capsule to the ureter? What causes urine to move through the ureter to the urinary bladder?*

Micturition Reflex

The **micturition** (mik-choo-rish′ŭn) **reflex** is emptying of the urinary bladder activated by stretch of the urinary bladder wall. As the bladder fills with urine, pressure increases and stretch receptors in the wall of the bladder are stimulated. Action potentials are

1. Urine in the urinary bladder stretches the bladder wall.

2. Action potentials produced by stretch receptors are carried along pelvic nerves (*green line*) to the sacral region of the spinal cord.

3. Action potentials are carried by parasympathetic nerves (*red line*) to contract the smooth muscles of the urinary bladder.

4. Ascending pathways carry an increased frequency of action potentials up the spinal cord to the pons and cerebrum when the urinary bladder becomes stretched. This increases the conscious urge to urinate.

5. Descending pathways carry action potentials to the sacral region of the spinal cord to tonically inhibit the micturition reflex, preventing automatic urination when the bladder is full. Descending pathways facilitate the reflex when stretch of the urinary bladder produces the conscious urge to urinate. This reinforces the micturition reflex.

6. The brain voluntarily controls the external uretheral sphincter through somatic motor nerves (*purple*), causing the sphincter to relax or constrict.

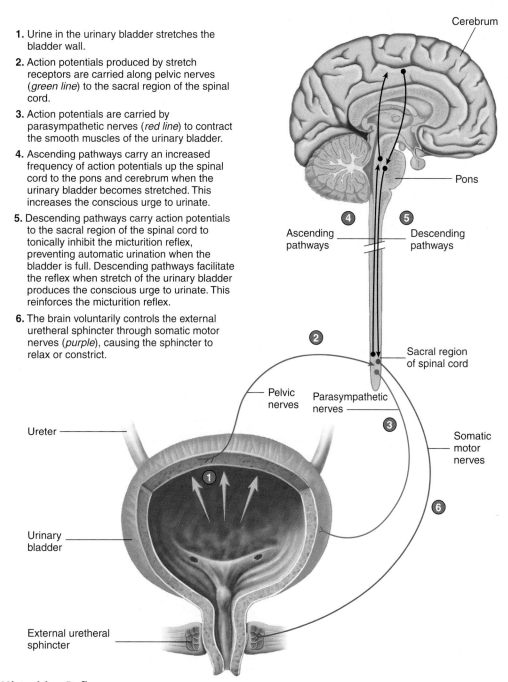

Figure 23.18   **Micturition Reflex**

conducted from the bladder to the sacral region of the spinal cord through the pelvic nerves. Integration of the reflex occurs in the spinal cord, and action potentials are conducted along parasympathetic nerve fibers to the urinary bladder. Parasympathetic action potentials cause the urinary bladder to contract (figure 23.18). The external urethral sphincter is tonically contracted as a result of stimulation from the somatic motor nervous system. A strong micturition reflex can inhibit that tonic contraction, causing the external urethral sphincter to relax. This is the normal way in which the urinary bladder empties in infants before they learn bladder control.

The micturition reflex occurs automatically, but it can be inhibited or facilitated by higher centers in the brain. The higher brain centers prevent micturition by sending action potentials

through the spinal cord to decrease the intensity of the autonomic reflex that stimulates urinary bladder contractions and to stimulate nerve fibers that keep the external urethral sphincter contracted. The ability to inhibit micturition develops at the age of 2–3 years.

When the desire to urinate exists, the higher brain centers alter action potentials sent to the spinal cord to facilitate the micturition reflex and relax the external urethral sphincter (see figure 23.18). Awareness of the need to urinate occurs because stretch of the urinary bladder stimulates sensory nerve fibers that increase action potentials carried to the brain by ascending tracts in the spinal cord. Irritation of the urinary bladder or the urethra by bacterial infections or by other conditions can also initiate the urge to urinate, even though the bladder is nearly empty.

Clinical Relevance Renal Failure

Renal failure can result from any condition that interferes with kidney function. **Acute renal failure** occurs when kidney damage is extensive and leads to the accumulation of urea in the blood and to acidosis. In complete renal failure, death can occur in 1–2 weeks. Acute renal failure can result from acute glomerular nephritis, or it can be caused by damage to or blockage of the renal tubules. Some poisons, such as mercuric ions or carbon tetrachloride, which are common to certain industrial processes, cause necrosis of the nephron epithelium. If the damage does not interrupt the basement membrane surrounding the nephrons, extensive regeneration can occur within 2–3 weeks. Severe ischemia associated with circulatory shock resulting from sympathetic vasoconstriction of the renal blood vessels can cause necrosis of the epithelial cells of the nephron.

Chronic renal failure results when so many nephrons are permanently damaged that the nephrons that remain functional cannot adequately compensate. **Diabetic nephropathy** (ne-frop′ă-thē) is a disease of the kidney associated with diabetes mellitus, and it is the principal cause of chronic renal failure. It damages renal glomeruli and

ultimately results in the destruction of functional nephrons through progressive scar tissue formation, which is mediated in part by an inflammatory response. Chronic renal failure can also result from chronic glomerular nephritis, trauma to the kidneys, the absence of kidney tissue caused by congenital abnormalities, tumors, urinary tract obstruction by kidney stones, damage resulting from pyelonephritis (inflammation of the renal pelvis), and severe arteriosclerosis of the renal arteries.

In chronic renal failure, the glomerular filtration rate is dramatically reduced, and the kidney is unable to excrete excess excretory products, including electrolytes and metabolic waste products. The accumulation of solutes in the body fluids causes water retention and edema. Potassium levels in the extracellular fluid are elevated, and acidosis occurs because the distal convoluted tubules and collecting ducts cannot excrete sufficient quantities of K^+ and H^+. Acidosis, elevated potassium levels in the body fluids, and the toxic effects of metabolic waste products cause mental confusion, coma, and finally death when chronic renal failure is severe.

Hemodialysis (hē′mō-dī-al′i-sis) is used when a person is suffering from severe acute

or chronic kidney failure. The procedure substitutes for the excretory functions of the kidney. Hemodialysis is based on blood flow through tubes composed of a selectively permeable membrane. Blood is usually taken from an artery, passed through the tubes of the dialysis machine, and then returned to a vein (figure A). On the outside of the dialysis tubes is a fluid, called dialysis fluid, that contains the same concentration of solutes as normal plasma except for the metabolic waste products. As a consequence, the metabolic wastes diffuse from the blood to the dialysis fluid. The dialysis membrane has pores that are too small to allow plasma proteins to pass through them and, because the dialysis fluid contains the same beneficial solutes as the plasma, the net movement of these substances is zero.

Peritoneal (per′i-tō-nē′ăl) **dialysis** is sometimes used to treat people suffering from kidney failure. The principles by which peritoneal dialysis works are the same as those for hemodialysis. The dialysis fluid flows through a tube inserted into the peritoneal cavity, where the visceral peritoneum and parietal peritoneum act as the dialysis membrane. Waste products diffuse from the

Automatic, Noncontracting, and Hyperexcitable Urinary Bladder

If the spinal cord is damaged above the sacral region, no micturition reflex exists for a time; however, if the urinary bladder is emptied frequently, the micturition reflex eventually becomes adequate to cause it to empty. Some time is generally required for the micturition reflex integrated within the spinal cord to begin to operate. A typical micturition reflex can exist, but there is no conscious control over its onset or duration. This condition is called the **automatic bladder.**

Damage to the sacral region of the spinal cord or to the nerves that carry action potentials between the spinal cord and the urinary bladder can result in failure of the urinary bladder to contract, although the external urethral sphincter is relaxed. As a result, the micturition reflex cannot occur. The bladder fills to capacity, and urine is forced in a slow dribble through the external urethral sphincter.

In elderly people and in patients with damage to the brainstem or spinal cord, a loss of inhibitory action potentials to the sacral region of the spinal cord can occur. Without inhibition, the sacral centers are hyperexcitable, and even a small amount of urine in the bladder can elicit an uncontrollable micturition reflex. ▼

54. Describe the micturition reflex. How is voluntary control of micturition accomplished?

Effects of Aging on the Kidneys

Aging causes a gradual decrease in the size of the kidneys, beginning as early as age 20, becoming obvious by age 50, and continuing until death. The loss of kidney size appears to be related to changes in the blood vessels of the kidney. The amount of blood flowing through the kidneys gradually decreases. Small arteries, including the afferent and efferent arterioles, become irregular and twisted. Functional glomeruli are lost. By age 80, 40% of the glomeruli are not functioning. Some nephrons and collecting ducts become thicker, shorter, and more irregular in structure. The capacity to secrete and absorb declines, and whole nephrons stop functioning. The kidney's ability to concentrate urine gradually declines. Eventually, changes in the kidney increase the risk for dehydration because of the kidney's reduced ability to produce a concentrated urine. There is also a decreased ability to eliminate uric acid, urea, creatine, and toxins from the blood.

An age-related loss of responsiveness to ADH and to aldosterone occurs. The kidneys decrease renin secretion. A reduced ability to participate in vitamin D synthesis occurs, which contributes to Ca^{2+} deficiency, osteoporosis, and bone fractures.

Recall that one-third of one kidney is required to maintain homeostasis. The additional kidney tissue beyond this constitutes a reserve capacity. The age-related changes in the kidney cause a

blood vessels beneath the peritoneum, across the peritoneum, and into the dialysis fluid.

Kidney transplants are sometimes performed on people who suffer from severe renal failure. Usually, the donor has suffered an accidental death and had granted permission to have his or her kidneys used for trans-plantation. An attempt is made to match the immune characteristics of the donor and recipient to reduce the tendency for the recipient's immune system to reject the transplanted kidney. Even with careful matching, however, recipients have to take medication for the rest of their lives to suppress their immune systems so that rejection is less likely. The major cause of kidney transplant failure is rejection by the recipient's immune system. In most cases, the transplanted kidney functions well, and the tendency for the recipient's immune system to reject the transplanted kidney can be controlled.

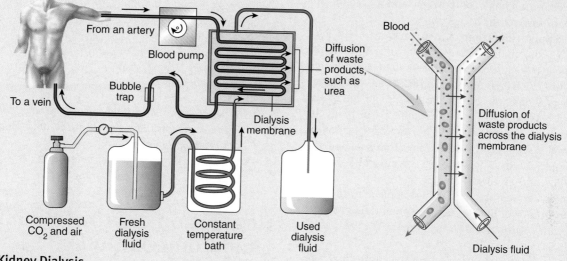

Figure A   **Kidney Dialysis**

During kidney dialysis, blood flows through a system of tubes composed of a selectively permeable membrane. Dialysis fluid flows in the opposite direction on the outside of the dialysis tubes. The composition of the dialysis fluid is similar to that of normal blood, except the concentration of waste products is very low. Waste products, such as urea, diffuse from the blood into the dialysis fluid. There is no net exchange of substances, such as Na^+ and K^+ that have the same concentrations in the blood and dialysis fluid.

reduction in the kidney's reserve capacity. As the functional kidney mass is reduced substantially in older people, high blood pressure, atherosclerosis, and diabetes have greater adverse effects. ▼

> *55. Describe the effect of aging on the kidneys. Why do the kidneys gradually decrease in size?*

Body Fluids

Life depends on many complex and highly regulated chemical reactions, all of which occur in water. Many of these reactions are catalyzed by enzymes that can function only within a narrow range of conditions. Changes in the total amount of water, the pH, or the concentration of specific electrolytes can alter the chemical reactions on which life depends. Homeostasis requires the maintenance of these parameters within a narrow range of values, and the failure to maintain homeostasis can result in illness or death.

The kidneys, along with the respiratory, integumentary, and gastrointestinal systems, regulate water volume, electrolyte concentrations, and pH. The nervous and endocrine systems coordinate the activities of these systems.

For an adult male, approximately 60% of the total body weight consists of water. For an adult female, approximately 50% of the total body weight is water (table 23.2). The fraction of the

Table 23.2	Approximate Volumes of Body Fluid Compartments*				
				Extracellular Fluid	
	Total Body Water	**Intracellular Fluid**	**Plasma**	**Interstitial Fluid**	**Total**
Infants	75	45	4	26	30
Adult Males	60	40	5	15	20
Adult Females	50	35	5	10	15

*Expressed as percentage of body weight.

body's weight composed of water decreases as the amount of adipose tissue increases because the water content of adipose tissue is relatively low. A smaller percentage of the body weight of the adult female consists of water because females generally have a greater percentage of body fat than males.

For people of all ages and body compositions, the two major fluid compartments are the intracellular and extracellular fluid compartments. The **intracellular fluid compartment** includes the fluid inside the several trillion cells of the body. The intracellular fluid from all cells has a similar composition, and it accounts for approximately two-thirds of the body's water.

The **extracellular fluid compartment** includes all the fluid outside the cells, constituting approximately one-third of the body's water. The extracellular fluid compartment is divided into subcompartments. The largest subcompartments are the plasma within blood vessels and the interstitial fluid. The interstitial fluid is the fluid between cells and, for simplicity, includes the lymph in the lymphatic vessels. A small portion of the extracellular fluid volume is separated by membranes into subcompartments. These special subcompartments contain fluid with a composition different from that of the remainder of the extracellular fluid. Included among the subcompartments are the aqueous and vitreous humor of the eye, cerebrospinal fluid, synovial fluid in joint cavities, serous fluid in the body cavities, and fluid secreted by glands.

Although the fluid contained in each extracellular fluid subcompartment differs somewhat in composition from that in the others, continuous and extensive exchange occurs between the subcompartments. Water diffuses from one subcompartment to another, and small molecules and ions are either transported or diffuse freely between them. Large molecules, such as proteins, are much more restricted in their movement because of the permeability characteristics of the membranes that separate the fluid subcompartments.

 Edema

Edema is an example of a fluid shift from the plasma to the interstitial fluid. Edema commonly results from an increase in the permeability of the capillary walls due to inflammation, which allows proteins to diffuse from the plasma into the interstitial fluid. Water moves in the same direction by osmosis. Edema can also result from a change in the hydrostatic pressure across capillary walls. An increased hydrostatic pressure in capillaries due to the blockage of veins or heart failure forces fluid from plasma into the interstitial spaces (see chapter 18). ▼

56. *Define intracellular fluid and extracellular fluid compartments. Describe the extracellular fluid subcompartments.*
57. *Describe the movement of water, ions, and molecules between body fluid compartments.*

Regulation of Intracellular Fluid Composition

The composition of intracellular fluid is substantially different from that of extracellular fluid. The intracellular fluid contains a relatively high concentration of ions, such as K^+, magnesium (Mg^{2+}), phosphate (PO_4^{3-}), and sulfate ions (SO_4^{2-}), compared with the extracellular fluid. It has a lower concentration of Na^+, Ca^{2+}, Cl^-, and HCO_3^- than that of the extracellular fluid. The

1. Large organic molecules, such as proteins, which cannot cross the plasma membrane, are synthesized inside cells and influence the concentration of solutes inside the cells.

2. The transport of ions across the plasma membrane, such as Na^+, K^+, and Ca^{2+}, influences the concentration of ions inside and outside the cell.

3. An electric charge difference across the plasma membrane influences the distribution of ions inside and outside the cell.

4. The distribution of water inside and outside the cell is determined by osmosis.

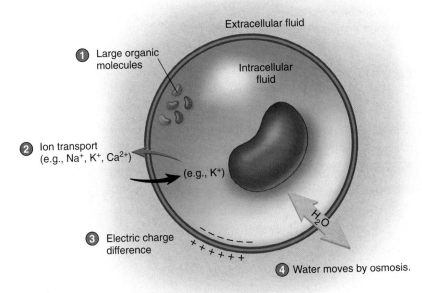

Process Figure 23.19 Regulation of Intracellular Fluid

concentration of protein in the intracellular fluid is also greater than that in the extracellular fluid.

Plasma membranes, which separate the two compartments, are selectively permeable—they are relatively impermeable to proteins and other large molecules and have limited permeability to smaller molecules and ions. Consequently, most large molecules synthesized within cells, such as proteins, remain within the intracellular fluid. Some substances, such as electrolytes, are actively transported across the plasma membrane, and their concentrations in the intracellular fluid are determined by the transport processes and by the electric charge difference across the plasma membrane (figure 23.19).

Water movement across the plasma membrane occurs by osmosis. Thus, the net movement of water is affected by changes in the concentration of solutes in the extracellular and intracellular fluids. For example, as dehydration develops, the concentration of solutes in the extracellular fluid increases, resulting in the movement of water by osmosis from the intracellular fluid into the extracellular fluid. If dehydration is severe, enough water moves from the intracellular fluid to cause the cells to function abnormally. If water intake increases after a period of dehydration, the concentration of solutes in the extracellular fluids decreases, which results in the movement of water back into the cells. ▼

58. Compare the composition of intracellular and extracellular fluids.
59. What factors determine the composition of intracellular fluid? How does water move between the intracellular and extracellular compartments?

Regulation of Body Fluid Concentration and Volume

Water Input

The body's water content is regulated so that the total volume of water in the body remains constant. Thus, the volume of water taken into the body is equal to the volume lost each day. The total volume of water entering the body each day is 1500–3000 mL. Approximately 90% of that volume comes from ingested fluids and the water within food, and approximately 10% is derived from the water produced during cellular metabolism (table 23.3).

Table 23.3	**Summary of Water Intake and Loss**
Sources of Water	**Routes by Which Water Is Lost**
Ingestion (90%)	Urine (61%)
Cellular metabolism (10%)	Evaporation (35%)
	Perspiration
	Insensible
	Sensible
	Respiratory passages
	Feces (4%)

The sensation of thirst is influenced by habit and by social settings. The sensation of thirst, however, is primarily stimulated by an increase in the osmolality of the extracellular fluids and from a reduction in plasma volume. These stimuli are part of regulatory mechanisms involved with maintaining blood osmolality and pressure. **Osmoreceptors** within the **thirst center** of the hypothalamus can detect an increased extracellular fluid osmolality and initiate activity in neural circuits, resulting in a conscious sensation of thirst (figure 23.20).

Baroreceptors can also influence the sensation of thirst. When they detect a substantial decrease in blood pressure, action potentials are conducted to the brain along sensory neurons to influence the sensation of thirst. Decreased blood pressure also stimulates increased renin secretion, which leads to increased angiotensin II production. Angiotensin II stimulates increased thirst. Low blood pressure associated with hemorrhagic shock, for example, is correlated with an intense sensation of thirst.

When people who are dehydrated drink water, they eventually consume a quantity sufficient to reduce the osmolality of the extracellular fluid to its normal value. They do not normally consume the water all at once. Instead, they drink intermittently until the proper osmolality of the extracellular fluid is established. The thirst sensation is temporarily reduced after the ingestion of small amounts of liquid. At least two factors are responsible for this temporary interruption of the thirst sensation. First, when the oral mucosa becomes wet after it has been dry, sensory neurons conduct action potentials to the thirst center of the hypothalamus and temporarily decrease the sensation of thirst. Second, consumed fluid increases the gastrointestinal tract volume, and stretch of the gastrointestinal wall initiates sensory action potentials in stretch receptors. The sensory neurons conduct action potentials to the thirst center of the hypothalamus, where they temporarily suppress the sensation of thirst. Because the absorption of water from the gastrointestinal tract requires time, mechanisms that temporarily suppress the sensation of thirst prevent the consumption of extreme volumes of fluid that would exceed the amount required to reduce blood osmolality. A longer-term suppression of the thirst sensation results when the extracellular fluid osmolality and blood pressure are within their normal ranges.

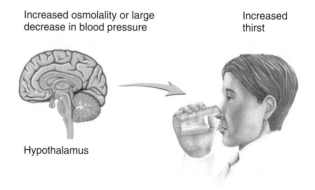

Figure 23.20 Effect of Blood Osmolality and Pressure on Water Intake

Increased blood osmolality affects hypothalamic neurons, and large decreases in blood pressure affect baroreceptors in the aortic arch, carotid sinuses, and atrium. As a result of these stimuli, an increase in thirst results, which increases water intake. Increased water intake reduces blood osmolality.

Learned behavior can be very important in avoiding periodic dehydration through the consumption of fluids either with or without food, even though blood osmolality is not reduced. The volume of fluid ingested by a healthy person usually exceeds the minimum volume required to maintain homeostasis, and the kidneys eliminate the excess water in urine. ▼

> **60.** *What are the sources of the body's water?*
>
> **61.** *List three factors that increase thirst. Name two factors that inhibit the sense of thirst.*

Water Output

Water loss from the body occurs through three routes (see table 23.3). The greatest amount of water, approximately 61%, is lost through the urine. Approximately 35% of water loss occurs through evaporation of water from respiratory passages and the skin. Approximately 4% is lost in the feces.

The volume of water lost through the respiratory system depends on the temperature and humidity of the air, body temperature, and the volume of air expired. Water lost through the diffusion and evaporation of water from the skin is called **insensible perspiration** and it plays a role in heat loss. For each degree that the body temperature rises above normal, an increased volume of 100–150 mL of water is lost each day in the form of insensible perspiration.

Sweat, or **sensible perspiration,** is secreted by the sweat glands; in contrast to insensible perspiration, it contains solutes. Sweat resembles extracellular fluid in its composition, with sodium chloride as the major component, but it also contains some potassium, ammonia, and urea. The volume of fluid lost as sweat is negligible for a person at rest in a cool environment. The volume of sweat produced is determined primarily by neural mechanisms regulating body temperature, although some sweat is produced as a result of sympathetic stimulation in response to stress. During exercise, elevated environmental temperature, or fever, the volume increases substantially and plays an important role in heat loss. Sweat losses of 8–10 L/day have been measured in outdoor workers in the summer.

Relatively little water is lost by way of the digestive tract. Although the total volume of fluid secreted into the gastrointestinal tract is large, nearly all the fluid is reabsorbed under normal conditions (see figure 21.30). Severe vomiting and diarrhea, however, are exceptions and can result in a large volume of fluid loss.

The kidneys are the primary organs that regulate the composition and volume of body fluids by controlling the volume and concentration of water excreted in the form of urine. Urine production varies greatly, ranging from a small volume of concentrated urine to a large volume of dilute urine in response to the mechanisms that regulate the body's water content. The mechanisms that respond to changes in extracellular fluid osmolality and extracellular fluid volume keep the total body water levels within a narrow range of values. ▼

> **62.** *Describe three routes for the loss of water from the body. Contrast insensible and sensible perspiration.*
>
> **63.** *What are the primary organs that regulate the composition and volume of body fluids?*

P R E D I C T
Mary Thon runs several miles each day. List the mechanisms through which water loss changes during her run.

Regulation of Extracellular Fluid Osmolality

Adding water to, or removing water from, body fluids maintains their osmolality between 285 and 300 mOsm/kg.

An increase in the osmolality of the extracellular fluid triggers thirst and antidiuretic hormone (ADH) secretion. Consumed water is absorbed from the intestine and enters the extracellular fluid. ADH acts on the distal convoluted tubules and collecting ducts of the kidneys to increase the reabsorption of water from the filtrate. The increase in the amount of water entering the extracellular fluid causes a decrease in osmolality (figure 23.21). The ADH and thirst mechanisms are sensitive to even small changes in extracellular fluid osmolality and the response is fast (from minutes to a few hours). Larger increases in extracellular fluid osmolality, such as during dehydration, result in an even greater increase in thirst and ADH secretion.

A decrease in extracellular fluid osmolality inhibits thirst and ADH secretion. Less water is consumed and reabsorbed from the filtrate in the kidneys. Consequently, more water is lost as a large volume of dilute urine. The result is an increase in the osmolality of the extracellular fluid (see figure 23.21). For example, consumption of a large volume of water in a beverage results in reduced extracellular fluid osmolality. This results in reduced ADH secretion, less reabsorption of water from the filtrate in the kidneys, and the production of a large volume of dilute urine. This response occurs quickly enough that the osmolality of the extracellular fluid is maintained within a normal range of values. ▼

> **64.** *What two mechanisms are triggered by an increase in the osmolality of the extracellular fluid?*
>
> **65.** *Describe how thirst and ADH regulate extracellular fluid osmolality when extracellular fluid osmolality increases and decreases.*

Regulation of Extracellular Fluid Volume

The volume of extracellular fluid can increase or decrease even if the osmolality of the extracellular fluid is maintained within a narrow range of values. Neural and hormonal mechanisms regulate extracellular fluid volume (figure 23.22).

When blood volume (pressure) decreases, regulatory mechanisms increase blood volume (pressure). Blood volume is increased by increasing the sensation of thirst, decreasing glomerular filtration rate, and increasing the reabsorption of Na$^+$ and water in the distal convoluted tubules and collecting ducts. Increasing the sensation of thirst results in drinking fluids and the addition of water to the extracellular fluid compartment. Decreasing glomerular filtration rate decreases urine production and conserves water. Increasing the reabsorption of Na$^+$ and water results in the movement of Na$^+$ and water into the blood (see figure 23.22).

When blood volume (pressure) increases, regulatory mechanisms decrease blood volume (pressure). Blood volume is decreased by decreasing the sensation of thirst, increasing glomerular filtration rate, and decreasing the reabsorption of Na$^+$ and water in the distal convoluted tubules and collecting ducts. Decreasing the sensation of thirst results in the decreased input of water to the

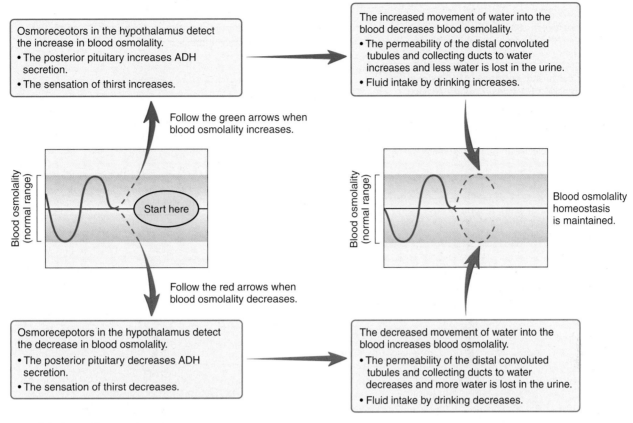

Homeostasis Figure 23.21 Summary of Blood Osmolality Regulation

extracellular fluid compartment. Increasing glomerular filtration rate increases urine production, which increases water loss. Decreasing the reabsorption of Na$^+$ and water results in the increased loss of Na$^+$ and water in the urine (see figure 23.22). ▼

66. *What neural and hormonal mechanisms regulate blood volume?*

67. *Describe how thirst, glomerular filtration rate, and the reabsorption of Na$^+$ and water regulate blood volume (pressure) when blood volume (pressure) decreases and increases.*

P R E D I C T ❾

In response to hemorrhagic shock, the kidneys produce a small volume of concentrated urine. Describe the mechanisms responsible. How do they help maintain homeostasis?

Regulation of Specific Electrolytes in the Extracellular Fluid

Electrolytes (ē-lek′trō-līts) are ions or molecules with an electric charge. The major extracellular ions are Na$^+$, Cl$^-$, K$^+$, Ca^{2+}, Mg^{2+}, and phosphate ions. The ingestion of electrolytes in food and water adds them to the body, whereas organs—such as the kidneys and, to a lesser degree, the liver, skin, and lungs—remove them from the body. The concentrations of electrolytes in the extracellular fluid are regulated so that they do not change unless the individual is growing, gaining weight, or losing weight. The

regulation of each electrolyte involves the coordinated participation of several organ systems.

Regulation of Sodium Ions

Sodium ions (Na$^+$) are the major cations (positively charged ions) in the extracellular fluid. In the United States, the quantity of Na$^+$ ingested each day is 20–30 times the amount needed. Less than 0.5 g is required to maintain homeostasis, but the average individual ingests approximately 10–15 g of sodium chloride daily. Regulation of the Na$^+$ content in the body usually involves the excretion of excess quantities of Na$^+$. When the Na$^+$ intake is very low, however, the mechanisms for conserving Na$^+$ in the body are effective.

The kidneys are the major route by which Na$^+$ are excreted. Aldosterone and ANH regulate Na$^+$ excretion (see figures 23.17 and 23.18). Aldosterone stimulates the reabsorption of Na$^+$ from filtrate, whereas ANH stimulates increased Na$^+$ excretion. Aldosterone is by far more important than ANH in Na$^+$ regulation. Reabsorption of Na$^+$ from the distal convoluted tubules and collecting ducts is very efficient, and few Na$^+$ are lost in the urine when aldosterone is present. As little as 0.1 g of sodium is excreted in the urine each day in the presence of high blood levels of aldosterone. When aldosterone is absent, Na$^+$ reabsorption in the nephron is greatly reduced, and as much as 30–40 g of sodium can be lost in the urine daily.

Sodium ions are also excreted from the body in **sweat.** Normally, only a small quantity of Na$^+$ is lost each day in the form of sweat, but the amount increases during heavy exercise in a warm environment. The loss of Na$^+$ in sweat is rarely physiologically significant.

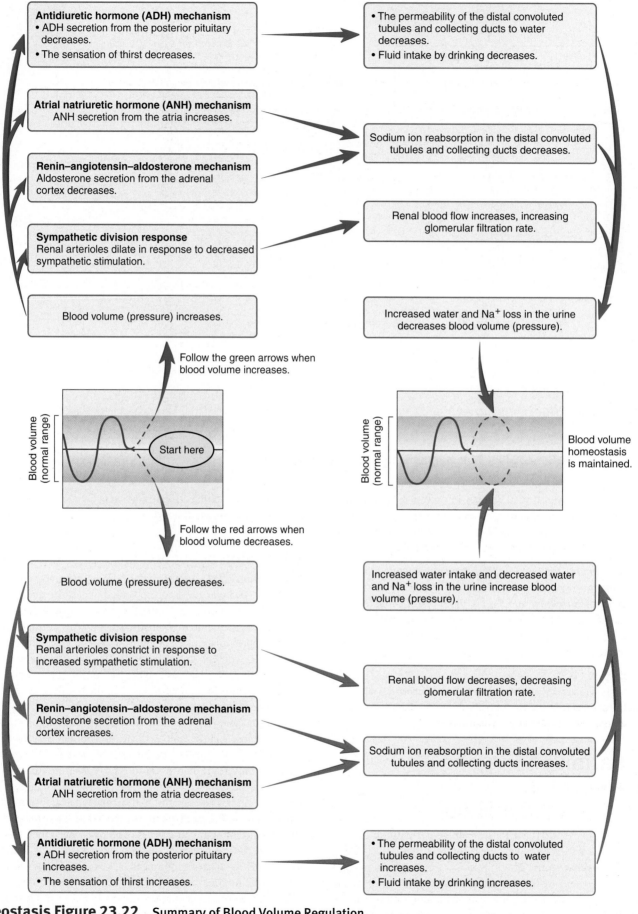

Antidiuretic hormone (ADH) mechanism
• ADH secretion from the posterior pituitary decreases.
• The sensation of thirst decreases.

• The permeability of the distal convoluted tubules and collecting ducts to water decreases.
• Fluid intake by drinking decreases.

Atrial natriuretic hormone (ANH) mechanism
ANH secretion from the atria increases.

Sodium ion reabsorption in the distal convoluted tubules and collecting ducts decreases.

Renin–angiotensin–aldosterone mechanism
Aldosterone secretion from the adrenal cortex decreases.

Sympathetic division response
Renal arterioles dilate in response to decreased sympathetic stimulation.

Renal blood flow increases, increasing glomerular filtration rate.

Blood volume (pressure) increases.

Increased water and Na$^+$ loss in the urine decreases blood volume (pressure).

Follow the green arrows when blood volume increases.

Blood volume (normal range)

Start here

Blood volume (normal range)

Blood volume homeostasis is maintained.

Follow the red arrows when blood volume decreases.

Blood volume (pressure) decreases.

Increased water intake and decreased water and Na$^+$ loss in the urine increase blood volume (pressure).

Sympathetic division response
Renal arterioles constrict in response to increased sympathetic stimulation.

Renal blood flow decreases, decreasing glomerular filtration rate.

Renin–angiotensin–aldosterone mechanism
Aldosterone secretion from the adrenal cortex increases.

Sodium ion reabsorption in the distal convoluted tubules and collecting ducts increases.

Atrial natriuretic hormone (ANH) mechanism
ANH secretion from the atria decreases.

Antidiuretic hormone (ADH) mechanism
• ADH secretion from the posterior pituitary increases.
• The sensation of thirst increases.

• The permeability of the distal convoluted tubules and collecting ducts to water increases.
• Fluid intake by drinking increases.

Homeostasis Figure 23.22 Summary of Blood Volume Regulation

For more information on the renin–angiotensin–aldosterone mechanism, see figure 23.15; on the atrial natriuretic hormone mechanism, see figure 23.16; on the antidiuretic hormone mechanism, see figure 23.14; on the thirst mechanism, see figure 23.20.

Table 23.4	Consequences of Abnormal Plasma Levels of Sodium Ions
Major Causes	**Symptoms**
Hyponatremia	
Inadequate dietary intake of sodium rarely causes symptoms—can occur in those on low-sodium diets and those taking diuretics.	Lethargy, confusion, apprehension, seizures, and coma
Extrarenal losses—vomiting, prolonged diarrhea, and burns	When accompanied by reduced blood volume—reduced blood pressure, tachycardia, and decreased urine output
Dilution—intake of large water volume after excessive sweating	When accompanied by increased blood volume—weight gain, edema, and distention of veins
Hyperglycemia, which attracts water into the circulatory system but reduces the concentration of Na^+	
Hypernatremia	
High dietary sodium rarely causes symptoms.	Thirst, fever, dry mucous membranes, restlessness; most serious symptoms—convulsions and pulmonary edema
Administration of hypertonic saline solutions (e.g., sodium bicarbonate treatment for acidosis)	
Oversecretion of aldosterone (i.e., aldosteronism)	When occurring with an increased water volume—weight gain, edema, elevated blood pressure, and bounding pulse
Water loss (e.g., because of fever, respiratory infections, diabetes insipidus, diabetes mellitus, and diarrhea)	

Approximately 90%–95% of the osmotic pressure of the extracellular fluid results from Na^+ and the anions (negatively charged ions) associated with them. Mechanisms that influence Na^+ concentrations in the extracellular fluid also influence the extracellular fluid volume and blood pressure because Na^+ have such a large effect on the osmotic pressure of the extracellular fluid (see figure 23.22).

Hyponatremia (hī′pō-nă-trē′mē-ă) is a below-normal plasma Na^+ concentration, and **hypernatremia** (hī′per-nă-trē′mē-ă) is an above-normal plasma Na^+ concentration. The major causes of hyponatremia and hypernatremia and their symptoms are listed in table 23.4. ▼

68. *How do aldosterone and ANH affect the concentration of Na^+ in the urine?*
69. *What role does sweating play in Na^+ balance?*
70. *How does extracellular Na^+ affect blood pressure?*
71. *Define hypernatremia and hyponatremia and list their effects.*

Regulation of Chloride Ions

The predominant anions in the extracellular fluid are **chloride ions (Cl^-)**. The electrical attraction of anions and cations makes it difficult to separate these charged particles. Consequently, the regulatory mechanisms that influence the concentration of cations in the extracellular fluid also influence the concentration of anions. The mechanisms that regulate Na^+, K^+, and Ca^{2+} levels in the body are important in influencing Cl^- levels. The mechanisms that regulate extracellular Na^+ concentration are the most important because Na^+ are the dominant cation in the extracellular fluid. ▼

72. *What mechanisms regulate Cl^- concentrations?*

Regulation of Potassium Ions

Potassium ions pass freely through the filtration membrane of the renal corpuscle. They are reabsorbed in the proximal convoluted tubules (see figure 23.9) and secreted in the distal convoluted tubules and collecting ducts (see figure 23.15). Potassium ion secretion into the distal convoluted tubule and collecting duct is highly regulated and primarily responsible for controlling the extracellular concentration of K^+.

Aldosterone plays a major role in regulating the concentration of K^+ in the extracellular fluid (see figures 23.15 and 23.23). Dehydration, circulatory system shock resulting from plasma loss, and tissue damage due to injuries, such as severe burns, all cause extracellular K^+ to become more concentrated than normal. In response, aldosterone secretion from the adrenal cortex increases and causes K^+ secretion to increase.

If the K^+ concentration in the extracellular fluid becomes reduced, aldosterone secretion from the adrenal cortex decreases. In response, the rate of K^+ secretion by the kidney is reduced (see figure 23.23).

Hypokalemia (hī′pō-ka-lē′mē-ă) is an abnormally low level of K^+ in the extracellular fluid, and **hyperkalemia** (hī′per-kă-lē′mē-ă) is an abnormally high level of K^+ in the extracellular fluid. The major causes of hypokalemia and hyperkalemia and their symptoms are listed in table 23.5. ▼

73. *Where are K^+ secreted in the nephron?*
74. *How is the secretion of K^+ regulated?*
75. *What effect does an increase or a decrease in extracellular K^+ concentration have on resting membrane potential?*
76. *Define hyperkalemia and hypokalemia and list their effects.*

Regulation of Calcium Ions

The kidneys, intestinal tract, and bones are important in maintaining extracellular Ca^{2+} levels (figure 23.24). Almost 99% of total body calcium is contained in bone. Part of the extracellular Ca^{2+} regulation involves the regulation of Ca^{2+} deposition into and resorption from bone (see chapter 6). Long-term regulation of Ca^{2+} levels, however, depends on maintaining a balance between Ca^{2+} absorption across the wall of the intestinal tract and Ca^{2+} excretion by the kidneys.

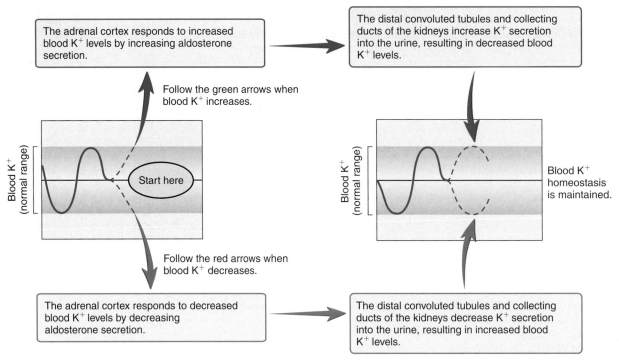

Homeostasis Figure 23.23   Summary of Blood K⁺ Regulation

Table 23.5	Consequences of Abnormal Concentrations of Potassium Ions
Major Causes	**Symptoms**
Hypokalemia	
Alkalosis (K⁺ shift into cell in exchange for H⁺)	Symptoms are mainly due to hyperpolarization of membranes.
Insulin administration (promotes cellular uptake of K⁺)	Decreased neuromuscular excitability—skeletal muscle weakness
Reduced K⁺ intake (especially with anorexia nervosa and alcoholism)	Decreased tone in smooth muscle
Increased renal loss (excessive aldosterone secretion, improper use of diuretics, and kidney diseases that result in reduced ability to reabsorb Na⁺)	Cardiac muscle—delayed ventricular repolarization, bradycardia, and atrioventricular block
Hyperkalemia	
Movement of K⁺ from intracellular to extracellular fluid resulting from cell trauma (e.g., burns or crushing injuries) and alterations in plasma membrane permeability (e.g., acidosis, insulin deficiency, and cell hypoxia)	Mild hyperkalemia (caused mainly by partial depolarization of plasma membranes):
	Increased neuromuscular irritability and restlessness
Decreased renal excretion of K⁺ (e.g., from decreased secretion of aldosterone in persons with Addison disease)	Intestinal cramping and diarrhea
	Electrocardiogram—alterations, including rapid repolarization with narrower and taller T waves and shortened QT intervals
	Severe hyperkalemia (caused mainly by partial depolarization of plasma membranes severe enough to hamper action potential conduction): muscle weakness, loss of muscle tone, and paralysis
	Electrocardiogram—alterations, including changes caused by reduced rate of action potential conduction (e.g., depressed ST segment, prolonged PR interval, wide QRS complex, arrhythmias, and cardiac arrest)

Parathyroid hormone (PTH) is secreted by the parathyroid glands (see figure 15.18*b*). PTH increases extracellular Ca^{2+} concentration. The rate of PTH secretion is regulated by the extracellular Ca^{2+} concentration (see figure 23.24). An elevated Ca^{2+} concentration inhibits, and a reduced Ca^{2+} concentration stimulates, the secretion of PTH. PTH causes osteoclasts to degrade bone and release Ca^{2+} into the body fluids. PTH also increases the rate of Ca^{2+} reabsorption from the urine by the kidneys and promotes the synthesis of active vitamin D by the kidneys (see chapter 5). Active vitamin D increases Ca^{2+}

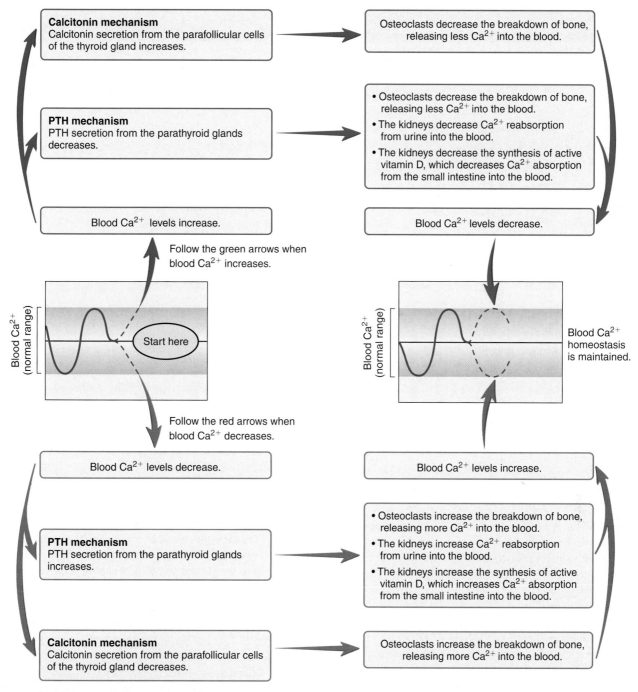

Homeostasis Figure 23.24 Summary of Blood Ca²⁺ Regulation

For more information on the PTH and calcitonin mechanisms, see figure 6.20.

uptake from the small intestine. Vitamin D can also be obtained in the food.

Calcitonin (kal-si-tō′nin) is secreted by the thyroid gland (see figure 15.18). Calcitonin reduces the blood Ca²⁺ concentration when it is too high. An elevated blood Ca²⁺ concentration causes the thyroid gland to secrete calcitonin, and a low blood Ca²⁺ concentration inhibits calcitonin secretion (see figure 23.24). Calcitonin inhibits osteoclasts, which reduces the rate at which bone is broken down and decreases the release of Ca²⁺ from bone.

Hypocalcemia (hī′pō-kal-sē′mē-ă) is a below-normal level of Ca²⁺ in the extracellular fluid, and **hypercalcemia** (hī′per-kal-

sē′mē-ă) is an above-normal level of Ca²⁺ in the extracellular fluid. Table 23.6 lists the major causes and symptoms of hypocalcemia and hypercalcemia. ▼

77. *Describe three ways by which parathyroid hormone regulates the extracellular Ca²⁺ concentration.*
78. *Describe the role of calcitonin in the regulation of extracellular Ca²⁺ concentration.*
79. *Define hypercalcemia and hypocalcemia and list their effects.*

Table 23.6	Consequences of Abnormal Concentrations of Calcium
Major Causes	**Symptoms**
Hypocalcemia	
Nutritional deficiencies	Symptoms due mainly to increased permeability of plasma membranes to Na$^+$
Vitamin D deficiency	Increase in neuromuscular excitability—confusion, muscle spasms, hyperreflexia, and intestinal cramping
Decreased parathyroid hormone secretion	
Malabsorption of fats (reduced vitamin D absorption)	Severe neuromuscular excitability—convulsions, tetany, and inadequate respiratory movements
	Electrocardiogram—prolonged QT interval (prolonged ventricular depolarization)
Bone tumors that increase Ca^{2+} deposition	Reduced absorption of phosphate from the intestine
Hypercalcemia	
Excessive parathyroid hormone secretion	Symptoms due mainly to decreased permeability of plasma membranes to Na$^+$
Excess vitamin D	Loss of membrane excitability—fatigue, weakness, lethargy, anorexia, nausea, and constipation
	Electrocardiogram—shortened QT segment and depressed T waves
	Kidney stones

Regulation of Other Ions

Other ions, such as phosphate ions (PO_4^{3-}), sulfate ions (SO_4^{2-}), and magnesium ions (Mg^{2+}) are reabsorbed by active transport in the kidneys. The rate of reabsorption is slow so that, if the concentration of these ions in the filtrate exceeds the nephron's ability to reabsorb them, the excess is excreted into the urine. As long as the concentration of these ions is low, nearly all of them are reabsorbed by active transport.

Regulation of Acid–Base Balance

Hydrogen ions affect the activity of enzymes and interact with many electrically charged molecules. Consequently, most chemical reactions within the body are highly sensitive to the H$^+$ concentration of the fluid in which they occur. The maintenance of the H$^+$ concentration within a narrow range of values is essential for normal metabolic reactions. The H$^+$ concentration is determined by acids and bases in the body.

Acids are substances that release H$^+$ into a solution; **bases** bind to H$^+$ and remove them from solution. Many bases release hydroxide ions (OH$^-$), which react with H$^+$ to form water (H$_2$O). Acids and bases are categorized as either strong or weak. Strong acids and strong bases completely dissociate to form ions in solution. For example, hydrochloric acid is a strong acid that completely dissociates to form H$^+$ and Cl$^-$.

$$HCl \rightarrow H^+ + Cl^-$$

Therefore, for every HCl added to a solution, one H$^+$ is added.

Weak acids and bases do not dissociate completely in solution. Most of the weak acid or base molecules remain intact and only a few of them dissociate. For example, carbonic acid (H$_2$CO$_3$) is a weak acid. When it is dissolved in water, the following equilibrium is established in which *some* of the carbonic acid molecules dissociate to form bicarbonate ions (HCO$_3^-$) and H$^+$.

$$H_2CO_3 \leftrightarrows HCO_3^- + H^-$$

Even though many carbonic acid molecules are added to a solution, only a few of them dissociate. Therefore, there is little change

in solution H$^+$ concentration and pH. Weak acids are common in living systems, and they play important roles in preventing large changes in body fluid pH.

Buffer systems, the respiratory system, and the kidneys work together and play essential roles in the regulation of acid–base balance (figure 23.25). When pH changes, buffer systems respond almost instantaneously, the respiratory system responds within a few minutes, and the kidneys respond within hours to days. The kidneys have the greatest capacity to return pH to its precise range of normal values. ▼

80. *Define acid and base.*
81. *Describe weak acids and bases. Why are weak acids important in living systems?*
82. *Compare the rate at which the buffer system, respiratory system, and kidneys control body fluid pH. Which system has the greatest capacity to control body fluid pH?*

Buffer Systems

Buffers (bŭf′erz) resist changes in the pH of a solution (see chapter 2). Buffers within body fluids stabilize the pH by chemically binding to excess H$^+$ when H$^+$ are added to a solution or by releasing H$^+$ when the H$^+$ concentration in a solution begins to fall.

Several important buffer systems function together to resist changes in the pH of body fluids. The carbonic acid/bicarbonate buffer system, proteins, phosphate compounds, and ammonia function as buffers. ▼

83. *Define buffer.*

Carbonic Acid/Bicarbonate Buffer System

The carbonic acid/bicarbonate buffer system depends on the equilibrium between H$_2$CO$_3$ and the H$^+$ and bicarbonate (HCO$_3^-$). When an acid is added to a solution containing the carbonic acid/bicarbonate buffer system, a large proportion of the H$^+$ released from the acid binds to HCO$_3^-$ to form H$_2$CO$_3$, and only a small percentage remains as free H$^+$. Thus, a large decrease in pH is resisted by the carbonic acid/bicarbonate buffer system.

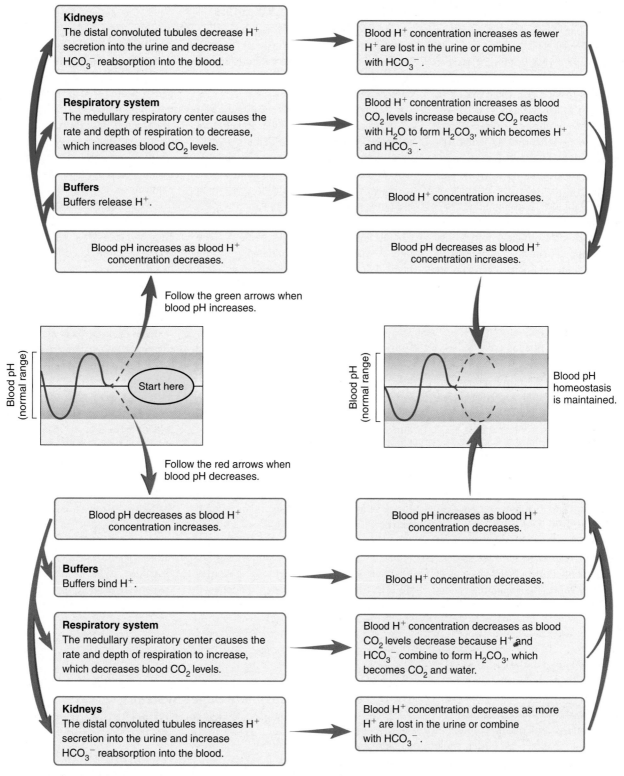

Kidneys
The distal convoluted tubules decrease H$^+$ secretion into the urine and decrease HCO$_3^-$ reabsorption into the blood.

Blood H$^+$ concentration increases as fewer H$^+$ are lost in the urine or combine with HCO$_3^-$.

Respiratory system
The medullary respiratory center causes the rate and depth of respiration to decrease, which increases blood CO$_2$ levels.

Blood H$^+$ concentration increases as blood CO$_2$ levels increase because CO$_2$ reacts with H$_2$O to form H$_2$CO$_3$, which becomes H$^+$ and HCO$_3^-$.

Buffers
Buffers release H$^+$.

Blood H$^+$ concentration increases.

Blood pH increases as blood H$^+$ concentration decreases.

Blood pH decreases as blood H$^+$ concentration increases.

Follow the green arrows when blood pH increases.

Start here

Follow the red arrows when blood pH decreases.

Blood pH homeostasis is maintained.

Blood pH decreases as blood H$^+$ concentration increases.

Blood pH increases as blood H$^+$ concentration decreases.

Buffers
Buffers bind H$^+$.

Blood H$^+$ concentration decreases.

Respiratory system
The medullary respiratory center causes the rate and depth of respiration to increase, which decreases blood CO$_2$ levels.

Blood H$^+$ concentration decreases as blood CO$_2$ levels decrease because H$^+$ and HCO$_3^-$ combine to form H$_2$CO$_3$, which becomes CO$_2$ and water.

Kidneys
The distal convoluted tubules increases H$^+$ secretion into the urine and increase HCO$_3^-$ reabsorption into the blood.

Blood H$^+$ concentration decreases as more H$^+$ are lost in the urine or combine with HCO$_3^-$.

Homeostasis Figure 23.25   Summary of Acid–Base Balance Regulation

For more information on the respiratory system, see figure 23.26; on the kidneys, see figure 23.28.

When a base is added to a solution containing the carbonic acid/bicarbonate buffer system, the base binds to H$^+$. As the H$^+$ are removed by the base, additional H$^+$ are formed from H$_2$CO$_3$, which prevents a large increase in pH.

The carbonic acid/bicarbonate buffer system plays an important role in regulating the extracellular pH. It quickly responds to the addition of substances such as carbon dioxide or lactic acid produced by increased metabolism during exercise (see chapter 20) and increased fatty acid and ketone body production during periods of elevated fat metabolism (see chapter 22). It also responds to the addition of basic substances, such as large amounts of NaHCO$_3$ consumed as an antacid.

The normal pH value of the body fluids is between 7.35 and 7.45. When the pH value of body fluids is below 7.35, the condition is called **acidosis** (as-i-dō′sis); when the pH is above 7.45, it is called **alkalosis** (al′kă-lō′sis).

Metabolism produces acidic products that lower the pH of the body fluids. For example, CO_2 is a by-product of metabolism, and CO_2 combines with water to form H_2CO_3. Also, lactic acid is a product of anaerobic metabolism, protein metabolism produces phosphoric and sulfuric acids, and lipid metabolism produces fatty acids. These acidic substances must continuously be eliminated from the body to maintain pH homeostasis. The failure to eliminate the acidic products of metabolism results in acidosis. The excess elimination of acidic products of metabolism results in alkalosis.

The major effect of acidosis is depression of the central nervous system. When blood pH falls below 7.35, the central nervous system malfunctions and the individual becomes disoriented and possibly comatose as the condition worsens.

A major effect of alkalosis is hyperexcitability of the nervous system. Peripheral nerves are affected first, resulting in spontaneous nervous stimulation of muscles. Spasms, tetanic contractions, and possibly extreme nervousness or convulsions result. Severe alkalosis can cause death as a result of tetany of the respiratory muscles.

Although buffers help resist changes in the pH of body fluids, the respiratory system and the kidneys regulate the pH of the body fluids. Malfunctions in either the respiratory system or the kidneys can result in acidosis or alkalosis.

Acidosis and alkalosis are categorized by the cause of the condition. **Respiratory acidosis** or **respiratory alkalosis** results from abnormalities in the respiratory system. **Metabolic acidosis** or **metabolic alkalosis** results from all causes other than abnormal respiratory functions.

Inadequate ventilation of the lungs causes respiratory acidosis (table A). The rate at which CO_2 is eliminated from the body fluids through the lungs falls. This increases the concentration of CO_2 in the body fluids. As CO_2 levels increase, CO_2 reacts with water to form H_2CO_3. Carbonic acid forms H^+ and HCO_3^-. The increase in H^+ concentration causes the pH of the body fluids to decrease. If the pH of the body fluids falls below 7.35, the symptoms of respiratory acidosis become apparent.

Buffers help resist a decrease in pH, and the kidneys help compensate for the failure of the lungs to prevent respiratory acidosis by increasing the rate at which they secrete H^+ into the filtrate and reabsorb HCO_3^-. The capacity of buffers to resist changes in pH can be exceeded, however, and a period of 1–2 days is required for the kidneys to become maximally functional. Thus, the kidneys are not effective if respiratory acidosis develops quickly. The kidneys are very effective if respiratory acidosis develops slowly or if it lasts long enough for the kidneys to respond. For example, the kidneys cannot compensate for respiratory acidosis occurring in response to a severe asthma attack that begins quickly and subsides within hours. If, however, respiratory acidosis results from emphysema, which develops over a long time, the kidneys play a significant role in helping compensate.

Respiratory alkalosis results from hyperventilation of the lungs (see table A). This increases the rate at which CO_2 is eliminated from the body fluids and results in a decrease in the concentration of CO_2 in the body fluids. As CO_2 levels decrease, H^+ react with HCO_3^- to form H_2CO_3. The H_2CO_3 form H_2O and CO_2. The resulting decrease in the concentration of H^+ causes the pH of the body fluids to increase. If the pH of body fluids increases above 7.45, the symptoms of respiratory alkalosis become apparent.

The kidneys help compensate for respiratory alkalosis by decreasing the rate of H^+ secretion into the filtrate and the rate of HCO_3^- reabsorption. If an increase in pH occurs, the kidneys need 1–2 days to compensate. Thus, the kidneys are not effective if respiratory alkalosis develops quickly. They are very effective, however, if respiratory alkalosis develops slowly. For example, the kidneys are not effective in compensating for respiratory alkalosis that occurs in response to hyperventilation triggered by emotions,

Protein Buffer System

Intracellular proteins and plasma proteins form a large pool of protein molecules that act as buffer molecules. They provide approximately three-fourths of the buffer capacity of the body. Hemoglobin in red blood cells is one of the most important intracellular proteins. Other intracellular molecules, such as histone proteins associated with nucleic acids, also act as buffers. The capacity of proteins to function as buffers is due to the functional groups of amino acids. The carboxyl (—COOH) groups act as weak acids and the amine (—NH₂) groups act as weak bases. When the H^+ concentration increases, H^+ bind to the amine groups and, when the H^+ concentration decreases, H^+ are released from the carboxyl groups.

Phosphate Buffer System

Dihydrogen phosphate ($H_2PO_4^-$) is a weak acid that dissociates to form H^+ and **hydrogen phosphate (HPO_4^{2-}).**

$$H_2PO_4^- \rightleftarrows H^+ + HPO_4^{2-}$$

Hydrogen phosphate is an important non-bicarbonate buffer in the filtrate (see "Adding or Deleting HCO_3^-," p. 757). Phosphate-containing molecules, such as DNA, RNA, and ATP, are important intracellular buffers.

Ammonia Buffer System

Ammonium (NH_4^+) is a weak acid that dissociates to form H^+ and **ammonia (NH_3).**

$$NH_4^+ \rightleftarrows H^+ + NH_3$$

which usually begins quickly and subsides within minutes or hours. If alkalosis results, however, from staying at a high altitude over a 2- or 3-day period, the kidneys play a significant role in helping compensate.

Metabolic acidosis results from all conditions that decrease the pH of the body fluids below 7.35, with the exception of conditions resulting from altered function of the respiratory system (see table A). As H^+ accumulate in the body fluids, buffers first resist a decline in pH. If the buffers cannot compensate for the increase in H^+, the respiratory center helps regulate body fluid pH. The reduced pH stimulates the respiratory center, which causes hyperventilation. During hyperventilation, CO_2 is eliminated at a greater rate. The elimination of CO_2 also eliminates excess H^+ and helps maintain the pH of the body fluids within a normal range.

If metabolic acidosis persists for many hours and if the kidneys are functional, the kidneys can also help compensate for metabolic acidosis by secreting H^+ at a greater rate and increasing the rate of HCO_3^- reabsorption. The symptoms of metabolic acidosis appear if the respiratory and renal systems are not able to maintain the pH of the body fluids within its normal range.

Metabolic alkalosis results from all conditions that increase the pH of the body fluids above 7.45, with the exception of conditions resulting from altered function of the respiratory system. As H^+ decrease in

the body fluids, buffers first resist an increase in pH. If the buffers cannot compensate for the decrease in H^+, the respiratory center helps regulate body fluid pH. The increased pH inhibits respiration. Reduced respiration allows CO_2 to accumulate in the

body fluids. Carbon dioxide reacts with water to produce H_2CO_3. If metabolic alkalosis persists for several hours and if the kidneys are functional, the kidneys reduce the rate of H^+ secretion to help reverse alkalosis (see table A).

Table A	Acidosis and Alkalosis

Acidosis

Respiratory Acidosis

Reduced elimination of CO_2 from the body fluids

Asphyxia

Hypoventilation (e.g., impaired respiratory center function due to trauma, tumor, shock, or renal failure)

Advanced asthma

Severe emphysema

Metabolic Acidosis

Elimination of large amounts of HCO_3^- resulting from mucous secretion (e.g., severe diarrhea and vomiting of lower intestinal contents)

Direct reduction of body fluid pH as acid is absorbed (e.g., ingestion of acidic drugs, such as aspirin)

Production of large amounts of fatty acids and other acidic metabolites, such as ketone bodies (e.g., untreated diabetes mellitus)

Inadequate oxygen delivery to tissue, resulting in anaerobic respiration and lactic acid buildup (e.g., exercise, heart failure, or shock)

Alkalosis

Respiratory Alkalosis

Reduced CO_2 levels in the extracellular fluid (e.g., hyperventilation due to emotions)

Decreased atmospheric pressure reduces oxygen levels, which stimulates the chemoreceptor reflex (e.g., causes hyperventilation at high altitudes)

Metabolic Alkalosis

Elimination of H^+ and reabsorption of HCO_3^- in the stomach or kidney (e.g., severe vomiting or formation of acidic urine in response to excess aldosterone)

Ingestion of alkaline substances (e.g., large amounts of sodium bicarbonate)

Ammonia is an important non-bicarbonate buffer in the filtrate (see "Adding or Deleting HCO_3^-," p. 757). ▼

84. Describe the buffer systems of the body.

Respiratory Regulation of Acid–Base Balance

The respiratory system regulates acid–base balance by influencing the carbonic acid/bicarbonate buffer system. Carbon dioxide (CO_2) reacts with water (H_2O) to form carbonic acid (H_2CO_3), which dissociates to form H^+ and HCO_3^-.

$$H_2O + CO_2 \rightleftarrows H_2CO_3 \rightleftarrows H^- + HCO_3^-$$

This reaction is in equilibrium. As carbon dioxide concentration increases, carbon dioxide combines with water. The higher the concentration of carbon dioxide, the greater the amount of H_2CO_3 formed. The H_2CO_3 molecules then dissociate to form H^+ and

HCO_3^-, which decreases blood pH. If carbon dioxide levels decline, however, the reaction shifts in the opposite direction so that many H^+ and HCO_3^- combine to form H_2CO_3, which then forms carbon dioxide and water. Thus, H^+ and HCO_3^- decrease in the solution, which increases blood pH.

The reaction between carbon dioxide and water is catalyzed by an enzyme, **carbonic anhydrase,** which is found on the surface of capillary epithelial cells (figure 23.26). This enzyme does not influence equilibrium but accelerates the rate at which the reaction proceeds in either direction so that equilibrium is achieved quickly.

Decreases in body fluid pH, regardless of the cause, stimulate neurons in the respiratory center in the brainstem and cause the rate and depth of ventilation to increase. The increased rate and depth of ventilation cause carbon dioxide to be eliminated from the body through the lungs at a greater rate, and the concentration of carbon dioxide in the body fluids decreases. As carbon

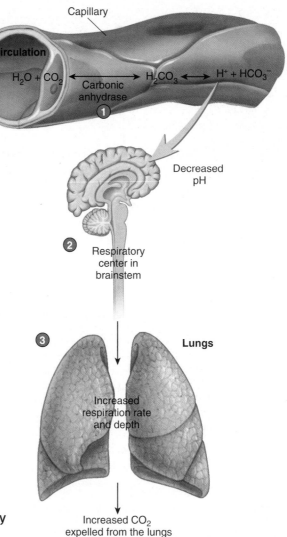

1. Carbon dioxide reacts with H_2O to form H_2CO_3. An enzyme, carbonic anhydrase, found in red blood cells and on the surface of blood vessel epithelium, catalyzes the reaction. Bicarbonate ion dissociates to form H^+ and HCO_3^-. An equilibrium is quickly established.

2. A decreased pH of the extracellular fluid stimulates the respiratory center and causes an increased rate and depth of respiration.

3. An increased rate and depth of respiration causes CO_2 to be expelled from the lungs, thus reducing the extracellular CO_2 levels. As CO_2 levels decrease, the extracellular concentration of H^+ decreases, and the extracellular fluid pH increases.

Process Figure 23.26 Respiratory Regulation of Body Fluid Acid–Base Balance

dioxide levels decline, the concentration of H^+ decreases, which increases body fluid pH (see figure 23.26).

Increases in body fluid pH, regardless of the cause, inhibit neurons in the respiratory center in the brainstem and cause the rate and depth of ventilation to decrease. The decreased rate and depth of ventilation cause less carbon dioxide to be eliminated from the body through the lungs. The concentration of carbon dioxide in the body fluids increases because carbon dioxide is continually produced as a by-product of metabolism in all tissues. As carbon dioxide levels increase, the concentration of H^+ increases, which decreases body fluid pH. ▼

85. *What effect does an increase or a decrease in blood CO_2 levels have on blood pH? Explain how this happens.*
86. *What effect does decreased or increased pH have on respiration? How does this change in respiration affect blood pH?*

PREDICT ⑩

Under stressful conditions, some people hyperventilate. What effect does the rapid rate of ventilation have on blood pH? Explain why a person who is hyperventilating may benefit from breathing into a paper bag.

Renal Regulation of Acid–Base Balance

The key to understanding acid–base regulation by the kidneys is total-body HCO_3^- content. If H^+ are added to the blood, each H^+ combines with a HCO_3^- to form H_2CO_3, which becomes carbon dioxide and water. The carbon dioxide is eliminated by the respiratory system, which maintains constant carbon dioxide levels. Thus, when pH decreases (H^+ concentration increases), for each H^+ entering the blood a HCO_3^- is lost. In order to maintain the total-body HCO_3^- content, the HCO_3^- must be replaced. That replacement takes place in the kidneys.

Conversely, if H^+ are removed from the blood, carbon dioxide and water combine to form H_2CO_3, which becomes H^+ and HCO_3^-. The carbon dioxide is replaced by the respiratory system. Thus, when pH increases (H^+ concentration decreases), for each H^+ lost from the blood a HCO_3^- is gained. In order to maintain the total-body HCO_3^- content, the HCO_3^- must be lost. That loss takes place in the kidneys.

Kidney regulation of total-body HCO_3^- content involves two processes: (1) recovering filtered HCO_3^- and (2) adding HCO_3^- to, or deleting HCO_3^- from, the blood.

Recovering Filtered HCO₃⁻

Bicarbonate ions freely pass through the filtration membrane, and the concentration of HCO_3^- in the filtrate is the same as in the blood. If this HCO_3^- were lost in the urine, total-body HCO_3^- content would quickly be depleted and the ability to regulate blood pH would be lost. In order to maintain the status quo, all of the filtered HCO_3^- must be recovered. This recovery process is not a conventional process in which HCO_3^- are *directly* transported across the apical membranes of tubules cells. Instead, it involves the production *inside* tubule cells of H^+ and HCO_3^- from carbon dioxide and water, followed by the secretion of H^+ into the filtrate and the transport of HCO_3^- into the blood.

Approximately 80% of HCO_3^- reabsorption occurs in the proximal convoluted tubules, 15% in the ascending limb of the loop of Henle and distal convoluted tubules, and 5% in the collecting ducts. Carbonic anhydrase within tubule cells promotes carbon dioxide and water to form H_2CO_3, which becomes H^+ and HCO_3^- (figure 23.27). The H^+ are then secreted from the cell. Depending on the cell, different methods of secreting H^+ are used. In the proximal convoluted tubules, an antiporter exchanges H^+ for Na^+ across the apical membrane of the cells. This is the major means by which Na^+ are reabsorbed from the filtrate.

The secreted H^+ combine with filtered HCO_3^- to form H_2CO_3. Carbonic anhydrase associated with the microvilli of the tubule cells promotes the conversion of H_2CO_3 to carbon dioxide and water. The carbon dioxide diffuses into the tubule cell and is used to produce additional H^+ and HCO_3^-. A filtered HCO_3^- is removed from the filtrate when it combines with a secreted H^+, and the secreted H^+ does not change the pH of the filtrate because it combines with filtered HCO_3^- to become water.

For every H^+ secreted into the filtrate, a HCO_3^- is added to the blood. The movement of carbon dioxide into the tubule cell results in the production of H^+ and HCO_3^-. The H^+ are secreted into the filtrate and the HCO_3^- and Na^+ are symported across the basolateral membrane and diffuse to the peritubular capillaries. Through H^+ secretion, all of the filtered HCO_3^- can be recovered from the filtrate.

Adding or Deleting HCO₃⁻

The process of H^+ secretion that results in the reabsorption of HCO_3^- and the maintenance of total-body HCO_3^- content can also be used to lose H^+ from the body and add *new* HCO_3^- to the blood. The key difference between reabsorbing filtered HCO_3^- and generating new HCO_3^- is the fate of the secreted H^+. By the time the filtrate reaches the collecting duct, almost all of the HCO_3^- has been recovered. Hydrogen ions secreted into the collecting duct can combine with non-bicarbonate bases and be eliminated in the urine. The HCO_3^- produced inside the collecting duct from carbon dioxide and water are added to the blood. They represent a gain of *new* HCO_3^-

1. Carbon dioxide and water combine to form H_2CO_3 that dissociates to form H^+ and HCO_3^-.

2. Hydrogen ions are secreted into the filtrate in exchange for Na^+ by antiport.

3. Hydrogen ions combine with HCO_3^- to form H_2CO_3. The H^+ and HCO_3^- have been removed from the filtrate.

4. Carbonic acid becomes carbon dioxide and water.

5. Carbon dioxide diffuses into the tubule cell. The carbon dioxide can generate additional H^+ and HCO_3^-.

6. Bicarbonate ions are transported with Na^+ into the interstitial fluid by symport. They diffuse to the peritubular capillary.

7. For each H^+ secreted into the filtrate, a HCO_3^- has been removed from the filtrate (step 3) and a HCO_3^- has been added to the blood. Thus, HCO_3^- is removed from the filtrate and added to the blood.

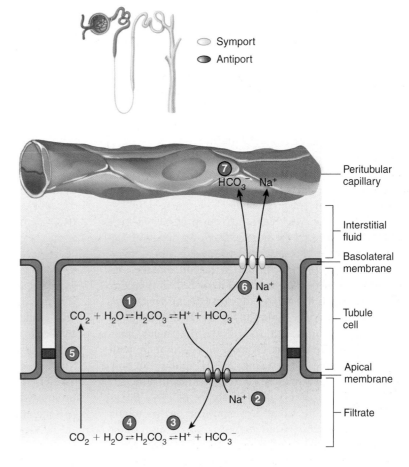

Process Figure 23.27 Reabsorption of HCO₃⁻

As the extracellular pH decreases, the rate of H^+ secretion and HCO_3^- reabsorption increase.

1. When blood pH decreases because of increased H^+, the H^+ combine with HCO_3^- to form H_2CO_3. The H^+ and HCO_3^- have been removed from the blood.

2. Carbonic acid becomes carbon dioxide and water.

3. Carbon dioxide diffuses into the tubule cell. Carbon dioxide and water combine to form H_2CO_3, which dissociates to form H^+ and HCO_3^-.

4. A H^+ pump actively transports H^+ into the filtrate.

5. Hydrogen ions can combine with filtered HPO_4^{2-} to form $H_2PO_4^-$. The H^+ are eliminated from the body in the urine.

6. Hydrogen ions can combine with NH_3 to form NH_4^+. The H^+ are eliminated from the body in the urine.

7. Bicarbonate ions are exchanged for Cl^- by antiport.

8. For each H^+ secreted into the filtrate (step 4), a new HCO_3^- has been added to the blood. The new HCO_3^- replaces, the HCO_3^- lost when H^+ were added to the blood (step 1).

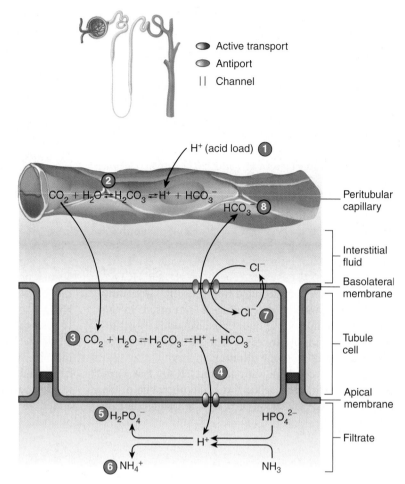

Process Figure 23.28 **Hydrogen Ion Secretion**

The secretion of H^+ into the filtrate decreases filtrate pH. As the concentration of H^+ increases in the filtrate, the ability of tubule cells to secrete additional H^+ becomes limited. Buffering the H^+ in the filtrate (steps 5 and 6) decreases their concentration and enables tubule cells to secrete additional H^+.

because they are generated in addition to the HCO_3^- produced when HCO_3^- are reabsorbed from the filtrate.

Distal convoluted tubule and collecting duct cells directly regulate acid balance by increasing the rate of H^+ secretion into the filtrate and the rate of new HCO_3^- formation (figure 23.28). A decrease in blood pH occurs when H^+ are added to the blood. The H^+ combine with HCO_3^- to form H_2CO_3, which increases blood pH back toward normal levels by removing the H^+. This process, however, eliminates a HCO_3^-, which has to be replaced. The H_2CO_3 is converted to carbon dioxide and water. The carbon dioxide diffuses into tubule cells and is converted to H_2CO_3, which dissociates to form H^+ and HCO_3^-. A H^+ pump actively secretes H^+ into the filtrate. Hydrogen phosphate (HPO_4^{2-}) and ammonia (NH_3) are the major non-bicarbonate bases in the filtrate. Hydrogen ions combine with them and are eliminated in the urine. The H^+ pump can only reduce filtrate pH to approximately 4.5. It is essential that the H^+ be removed by buffers; otherwise, the ability of the kidneys to eliminate H^+ would be reduced. Meanwhile, in the tubule cells the HCO_3^- produced from carbon dioxide and water are exchanged for Cl^- by an antiporter in the basolateral membrane. The HCO_3^- enter the blood and replace the HCO_3^- lost when H^+ were added to the blood.

Protein metabolism in the liver also adds new HCO_3^- to the blood. Proteins are broken down to amino acids, which are converted into urea and the amino acid glutamine. The urea and glutamine are released into the blood and become part of the filtrate. The urea is eliminated in the urine, but the glutamine is taken up by proximal convoluted tubule cells and converted into NH_4^+ and HCO_3^-. The HCO_3^- enter the blood and are a source of new HCO_3^- (as opposed to HCO_3^- recovered from the filtrate). Approximately 50% of the new HCO_3^- produced daily is derived from glutamine. The amount of HCO_3^- generated from glutamine can be regulated. When pH decreases, liver cells increase their production of glutamine and proximal convoluted tubules increase their breakdown of glutamine. As a result, the amount of HCO_3^- generated from glutamine can increase to over 10 times normal levels when the blood becomes very acidic and decrease to zero when the blood is very alkaline. The NH_4^+ produced from glutamine are released into the filtrate. Thus, the NH_4^+ in the filtrate are derived from two sources, both of which are involved with generating new HCO_3^-: the buffering of H^+ by NH_3 and the metabolism of glutamine.

Distal convoluted tubule and collecting duct cells directly regulate acid balance by decreasing the rate of H^+ secretion into the filtrate and the rate of new HCO_3^- formation. An increase in blood pH occurs when a base is added to the blood because H^+

Systems Interactions

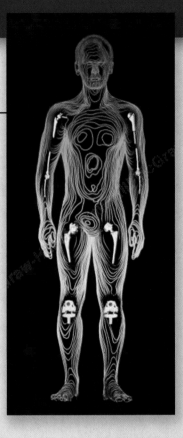

	Effects of the Urinary System on Other Systems	Effects of Other Systems on the Urinary System
Integumentary System	Removes waste products Helps maintain the body's pH, ion, and water balance	Prevents water loss Removes very small amount of waste products
Skeletal System	Removes waste products Helps maintain the body's pH, ion, and water balance Kidneys activate vitamin D, which increases the absorption of the calcium and phosphorus necessary for bone growth and maintenance	Thoracic cage protects the kidneys Pelvis protects the urinary bladder
Muscular System	Removes waste products Helps maintain the body's pH, ion, and water balance	Controls voluntary urination Pelvic floor muscles support the urinary bladder
Nervous System	Removes waste products Helps maintain the body's pH, ion, and water balance	Controls emptying of the bladder Regulates renal blood flow
Endocrine System	Removes waste products Helps maintain the body's pH, ion, and water balance	Antidiuretic hormone, renin–angiotensin–aldosterone mechanism, and atrial natriuretic hormone regulate blood volume and electrolyte balance
 **Cardiovascular System**	Removes waste products Helps maintain the body's pH, ion, and water balance Erythropoietin from the kidneys stimulates red blood cell production	Maintenance of blood pressure is necessary for filtration Delivers oxygen, nutrients, hormones, and immune cells Removes carbon dioxide, waste products, and toxins
Lymphatic System and Immunity	Removes waste products Helps maintain the body's pH, ion, and water balance	Immune cells provide protection against microorganisms and toxins and promote tissue repair by releasing chemical mediators
Respiratory System	Regulates pH, which is a major stimulus affecting respiration rate and depth Removes waste products Helps maintain the body's ion and water balance	Provides oxygen and removes carbon dioxide Helps maintain the body's pH by regulating carbon dioxide levels
Digestive System	Removes waste products Helps maintain the body's pH, ion, and water balance	Provides nutrients The liver converts vitamin D precursor from the skin to a form of vitamin D that can be converted by the kidneys to active vitamin D
Reproductive System	Removes waste products Helps maintain the body's pH, ion, and water balance Urethra common passageway for sperm cells and urine in males	Developing fetus compresses the urinary bladder

combine with the base and are removed from solution. When H^+ levels decrease, additional H^+ and HCO_3^- are produced from carbon dioxide and water. The result is the addition of HCO_3^-, which must be eliminated. The kidneys accomplish this task in two ways. First, they allow some of the filtered HCO_3^- to pass through to the urine. Second, some of the cells in the collecting duct can actively secrete HCO_3^- formed from carbon dioxide and water. These cells have Cl^-–HCO_3^- antiporters in their apical membranes and H^+ pumps in their basolateral membranes and operate in the opposite fashion of cells that secrete H^+ into the filtrate. ▼

87. *For each H^+ added to, or removed from, the blood, what happens to HCO_3^-?*

88. *Describe the process by which tubule cells recover filtered HCO_3^-. What is the relationship between a H^+ secreted into the filtrate and a HCO_3^- entering the blood?*

89. *Describe the role of non-bicarbonate bases and glutamine in adding new HCO_3^- to the blood.*

90. *Name two ways in which the kidneys eliminate HCO_3^-.*

P R E D I C T 11

Mr. Puffer suffers from severe emphysema. Gas exchange in his lungs is not adequate, and he must have a supply of oxygen. His blood carbon dioxide level is elevated. Nevertheless, his blood pH is close to normal. Explain. Also explain how his blood bicarbonate ion (HCO_3^-) level is likely to differ from normal values.

S U M M A R Y

Functions of the Urinary System (p. 718)

1. The kidneys produce urine, the ureters transport urine to the urinary bladder, the urinary bladder stores urine, and the urethra transports urine to the outside of the body.
2. The kidneys eliminates waste; regulate blood volume, blood pressure, ion concentration, and pH; and are involved with red blood cell and vitamin D production.

Kidney Anatomy and Histology (p. 718)

Location and External Anatomy of the Kidneys

1. The kidneys lie behind the peritoneum on the posterior abdominal wall on each side of the vertebral column.
2. The kidney is surrounded by a renal capsule and fat and is held in place by the renal fascia.
3. The hilum, on the medial side of each kidney, is where blood vessels and nerves enter and exit the kidney.

Internal Anatomy and Histology of the Kidneys

1. The two layers of the kidney are the cortex and the medulla.
   - The renal columns extend into the medulla between the renal pyramids.
   - The tips of the renal pyramids project to the minor calyces.
2. The minor calyces open into the major calyces, which open into the renal pelvis. The renal pelvis leads to the ureter.
3. The functional unit of the kidney is the nephron. The parts of a nephron are the renal corpuscle, the proximal convoluted tubule, the loop of Henle, and the distal convoluted tubule.
   - The renal corpuscle is Bowman's capsule and the glomerulus. Fluid leaves the blood in the glomerulus and enters Bowman's capsule.
   - The nephron empties through the distal convoluted tubule into a collecting duct.
   - The collecting ducts empty into papillary ducts, which empty into minor calyces.
4. Bowman's capsule has an outer parietal layer and an inner visceral layer consisting of podocytes.
5. The filtration membrane consists of the endothelium of glomerular capillaries (with fenestrae), a basement membrane, and podocytes (with filtration slits).

Arteries and Veins of the Kidneys

1. The renal artery enters the kidney and branches many times, forming afferent arterioles, which supply the glomeruli.
2. Efferent arteries from the glomeruli supply the peritubular capillaries and vasa recta.
3. The peritubular capillaries and vasa recta join small veins that converge to form the renal vein, which exits the kidney.
4. The juxtaglomerular apparatus consists of the granular cells of the afferent arteriole and the macula densa (part of the distal convoluted tubule).

Urine Production (p. 723)

Urine is produced by the processes of filtration, tubular reabsorption, and tubular secretion.

Filtration

1. The glomerular filtration rate is the amount of filtrate produce per minute.
2. The filtrate is plasma minus blood cells, platelets, and blood proteins. Most (99%) of the filtrate is reabsorbed.
3. Filtration pressure is responsible for filtrate formation.
   - Filtration pressure is glomerular capillary pressure minus capsule pressure minus blood colloid osmotic pressure.
   - Filtration pressure changes are primarily caused by changes in glomerular capillary pressure.

Regulation of Glomerular Filtration Rate

1. Autoregulation dampens systemic blood pressure changes by altering afferent arteriole diameter.
2. Sympathetic stimulation decreases renal blood flow and afferent arteriole diameter.

Tubular Reabsorption

1. There is a medullary concentration gradient from the cortex (300 mOsm/kg) to the tip of the renal pyramids (1200 mOsm/kg).
2. The apical membrane of tubule cells are in contact with filtrate, whereas the basolateral membranes are in contact with interstitial fluid.
3. Filtrate is reabsorbed by diffusion, facilitated diffusion, active transport, symport, and antiport from the nephron and collecting ducts into the peritubular capillaries and vasa recta.
   - The proximal convoluted tubule reabsorbs 65% of filtrate water and NaCl (solutes).
   - The descending limb of the loop of Henle reabsorbs 15% of filtrate water.
   - The ascending limb of the loop of Henle reabsorbs 25% of filtrate NaCl.

■ The distal convoluted tubules and collecting ducts reabsorb up to 19% of filtrate water and 9%–10% of filtrate water.

4. Waste products and toxic substances are concentrated in the urine.

Tubular Secretion

1. Substances are secreted in the proximal or distal convoluted tubules and the collecting ducts.
2. Hydrogen ions, K^+, and some substances not produced in the body are secreted by antiport mechanisms.

Summary of Changes in Filtrate Volume and Concentration (p. 731)

Maintaining the Medullary Concentration Gradient

1. Maintaining the medullary concentration gradient is necessary for the production of concentrated urine.
2. The addition of solutes increases the medullary interstitial fluid concentration.
   ■ The ascending limb of the loop of Henle adds NaCl, but not water.
   ■ Urea cycles between the collecting ducts and the thin segments of the loop of Henle.
3. The vasa recta is a countercurrent mechanism that removes reabsorbed water and solutes without disturbing the medullary concentration gradient.

Hormonal Regulation of Urine Concentration and Volume (p. 734)

Antidiuretic Hormone

1. Antidiuretic hormone (ADH) is secreted by the posterior pituitary and increases water permeability in the distal convoluted tubules and collecting ducts by stimulating the insertion of aquaporin-2 molecules into apical membranes.
2. ADH regulates blood osmolality by altering water reabsorption.
   ■ An increase in blood osmolality or a significant decrease in blood pressure stimulates increased ADH secretion, which increases water reabsorption. As a result, blood osmolality decreases, blood volume and blood pressure increase, urine concentration increases, and urine volume decreases.
   ■ A decrease in blood osmolality or a significant increase in blood pressure stimulates decreased ADH secretion, which decreases water reabsorption. As a result, blood osmolality increases, blood volume and blood pressure decrease, urine concentration decreases, and urine volume increases.

Renin–Angiotensin–Aldosterone

1. Renin, produced by the kidneys, causes the conversion of angiotensinogen to angiotensin I. Angiotensin-converting enzyme converts angiotensin I into angiotensin II, which stimulates aldosterone secretion from the adrenal cortex.
2. Aldosterone affects Na^+ and Cl^- transport in the nephron and collecting ducts by stimulating an increase in transport proteins.
3. Aldosterone regulates the body's water content by regulating the body's Na^+ content (assuming that ADH maintains blood osmolality).
   ■ A decrease in blood pressure results in increased renin secretion, aldosterone secretion, Na^+ reabsorption, blood volume, and blood pressure.
   ■ An increase in blood pressure results in decreased renin secretion, aldosterone secretion, Na^+ reabsorption, blood volume, and blood pressure.

Atrial Natriuretic Hormone

1. Atrial natriuretic hormone (ANH) is produced by the heart when blood pressure increases.

2. ANH inhibits Na^+ reabsorption in the kidneys, resulting in increased urine volume and decreased blood volume and blood pressure.
3. ANH inhibits ADH secretion and dilates arteries and veins.

Urine Movement (p. 739)

Anatomy and Histology of the Ureters, Urinary Bladder, and Urethra

1. The walls of the ureter and urinary bladder consist of the epithelium, the lamina propria, a muscular coat, and a fibrous adventitia.
   ■ The transitional epithelium permits changes in size.
   ■ Contraction of the smooth muscle moves urine.
2. The urethra is lined with transitional and stratified squamous epithelium.
   ■ Males have an internal urethral sphincter of smooth muscle that prevents retrograde ejaculation of semen.
   ■ An external urethral sphincter of skeletal muscle allows voluntary control of urination.

Urine Flow Through the Nephron and Ureters

1. A pressure gradient causes urine to flow from Bowman's capsule to the ureters.
2. Peristalsis moves urine through the ureters.

Micturition Reflex

1. Stretch of the urinary bladder stimulates a reflex that causes the urinary bladder to contract and inhibits the external urethral sphincter.
2. Higher brain centers can stimulate or inhibit the micturition reflex.
3. Voluntary relaxation of the external urethral sphincter permits urination and contraction prevents it.

Effects of Aging on the Kidneys (p. 742)

1. There is a gradual decrease in the size of the kidney.
2. The decrease in kidney size is associated with a decrease in renal blood flow.
3. The number of functional nephrons decreases.
4. Renin secretion and vitamin D synthesis decrease.
5. The nephron's ability to secrete and absorb declines.

Body Fluids (p. 743)

1. Intracellular fluid is inside cells.
2. Extracellular fluid is outside cells and includes interstitial fluid and plasma.

Regulation of Intracellular Fluid Composition (p. 744)

1. Substances used or produced inside the cell and substances exchanged with the extracellular fluid determine the composition of intracellular fluid.
2. Intracellular fluid is different from extracellular fluid because the plasma membrane regulates the movement of materials.
3. The difference between intracellular and extracellular fluid concentrations determines water movement.

Regulation of Body Fluid Concentration and Volume (p. 745)

Water Input

1. Water is ingested (90%) or produced in metabolism (10%).
2. Habit and social setting influence thirst. An increase in extracellular osmolality or a decrease in blood pressure stimulates the sense of thirst.
3. Wetting of the oral mucosa or stretch of the gastrointestinal tract inhibits thirst.

Water Output

1. Water is lost through evaporation from the respiratory system and the skin (insensible perspiration and sweat) (35%).
2. Water loss into the gastrointestinal tract normally is small (4%).
3. The kidneys are the primary regulator of water excretion (61%).

Regulation of Extracellular Fluid Osmolality

1. An increase in extracellular fluid osmolality stimulates thirst and ADH secretion, resulting in increased fluid intake and increased water reabsorption in the kidneys.
2. A decrease in extracellular osmolality inhibits thirst and decreases ADH secretion, resulting in decreased fluid intake and decreased water reabsorption in the kidneys.

Regulation of Extracellular Fluid Volume

1. Decreased extracellular fluid volume results in increased aldosterone secretion, decreased ANH secretion, increased ADH secretion, and increased sympathetic stimulation of the afferent arterioles. Consequently, thirst increases, glomerular filtration rate decreases, and the reabsorption of Na^+ and water increases.
2. Increased extracellular fluid volume results in decreased aldosterone secretion, increased ANH secretion, decreased ADH secretion, and decreased sympathetic stimulation of the afferent arterioles. Consequently, thirst decreases, glomerular filtration rate increases, and the reabsorption of Na^+ and water decreases.

Regulation of Specific Electrolytes in the Extracellular Fluid   (p. 747)

Electrolytes are ions or molecules with a charge.

Regulation of Sodium Ions

1. The kidneys are the major route by which Na^+ is excreted.
2. Sodium ion excretion is regulated by aldosterone and ANH.
3. Sodium ions are lost in sweat.
4. Sodium ions and association anions are responsible for 90%–95% of extracellular osmotic pressure.

Regulation of Chloride Ions

Chloride ions are the dominant negatively charged ions in extracellular fluid. They are regulated by the mechanisms regulating cations.

Regulation of Potassium Ions

1. Potassium ions are reabsorbed from the proximal convoluted tubule and secreted into the distal convoluted tubule.
2. Aldosterone increases the amount of K^+ secreted.

Regulation of Calcium Ions

1. Parathyroid hormone increases extracellular Ca^{2+} levels. It increases osteoclast activity, increases calcium reabsorption from the kidneys, and stimulates active vitamin D production.
3. Vitamin D stimulates Ca^{2+} uptake in the intestines.
4. Calcitonin decreases extracellular Ca^{2+} levels by inhibiting osteoclasts.

Regulation of Acid–Base Balance   (p. 752)

1. Acids release H^+ into solution, and bases remove them.
2. Buffers respond almost instantaneously to changes in pH, whereas the respiratory system takes minutes and the kidneys may take hours to days. The kidneys have the greatest ability to regulate pH precisely.

Buffer Systems

1. A buffer resists changes in pH.
   - When H^+ are added to a solution, the buffer removes them.
   - When H^+ are removed from a solution, the buffer replaces them.
2. Carbonic acid/bicarbonate, proteins, phosphate compounds, and ammonia are important buffers.

Respiratory Regulation of Acid–Base Balance

1. Respiratory regulation of pH is achieved through the carbonic acid/bicarbonate buffer system.
   - As carbon dioxide levels increase, pH decreases.
   - As carbon dioxide levels decrease, pH increases.
   - Carbon dioxide levels and pH affect the respiratory centers.
2. The pH affects the respiratory centers. Hypoventilation increases blood carbon dioxide levels, and hyperventilation decreases blood carbon dioxide levels.
   - Decreased pH increases the rate and depth of respiration, which lowers carbon dioxide levels and increases blood pH.
   - Increased pH decreases the rate and depth of respiration, which increases carbon dioxide levels and decreases blood pH.

Renal Regulation of Acid–Base Balance

1. For each H^+ added to the blood, a HCO_3^- is removed; for each H^+ removed from the blood, a HCO_3^- is added.
2. For each H^+ secreted into the filtrate, a HCO_3^- is removed from the filtrate and a HCO_3^- is added to the blood.
3. Hydrogen phosphate (HPO_4^{2-}) and ammonia (NH_3) are the major non-bicarbonate bases in the filtrate.
   - When H^+ combine with HPO_4^{2-} and NH_3, the filtrate is buffered, allowing additional H^+ to be secreted.
   - For each H^+ that combines with HPO_4^{2-} and NH_3, a new HCO_3^- is added to the blood.
4. The metabolism of glutamine produces new HCO_3^-.

R E V I E W   A N D   C O M P R E H E N S I O N

1. Which of these is *not* a general function of the kidneys?
   a. regulation of blood volume
   b. regulation of solute concentration in the blood
   c. regulation of the pH of the extracellular fluid
   d. regulation of vitamin A synthesis
   e. regulation of red blood cell synthesis

2. Given these structures:
   1. fat   2. renal capsule   3. renal fascia
   Choose the order in which they would be encountered from superficial to deep.
   a. 1,2,3   c. 2,3,1   e. 3,2,1
   b. 2,1,3   d. 3,1,2

3. The cortex of the kidney contains the
   a. hilum.   d. renal pyramids.
   b. glomeruli.   e. renal pelvis.
   c. perirenal fat.

4. Given these structures:
   1. major calyx   3. renal papilla
   2. minor calyx   4. renal pelvis

   Choose the arrangement that lists the structures in order as urine leaves the collecting duct and travels to the ureter.
   a. 1,4,2,3   d. 4,1,3,2
   b. 2,3,1,4   e. 4,3,2,1
   c. 3,2,1,4

5. The juxtaglomerular cells of the _____ and the macula densa cells of the _____ form the juxtaglomerular apparatus.
   a. afferent arteriole, proximal convoluted tubule
   b. afferent arteriole, distal convoluted tubule
   c. efferent arteriole, proximal convoluted tubule
   d. efferent arteriole, distal convoluted tubule

6. Given these structures:
   1. basement membrane
   2. fenestra
   3. filtration slit

   Choose the arrangement that lists the structures in the order a molecule of glucose encounters them as the glucose passes through the filtration membrane to enter Bowman's capsule.
   a. 1,2,3   c. 2,3,1   e. 3,2,1
   b. 2,1,3   d. 3,1,2

7. Given these blood vessels:
   1. afferent arteriole   3. glomerulus
   2. efferent arteriole   4. peritubular capillaries

   Choose the correct order as blood passes from an interlobular artery to an interlobular vein.
   a. 1,2,3,4   c. 2,1,4,3   e. 4,3,1,2
   b. 1,3,2,4   d. 3,2,4,1

8. Kidney function is accomplished by which of these means?
   a. filtration   d. both a and b
   b. secretion   e. all of the above
   c. reabsorption

9. If the glomerular capillary pressure is 45 mm Hg, the capsule pressure is 10 mm Hg, and the blood colloid osmotic pressure within the glomerulus is 25 mm Hg, the filtration pressure is
   a. −10 mm Hg.   d. 75 mm Hg.
   b. 10 mm Hg.   e. 80 mm Hg.
   c. 30 mm Hg.

10. Which of these conditions reduces filtration pressure in the glomerulus?
    a. elevated blood pressure
    b. constriction of the afferent arterioles
    c. decreased plasma protein in the glomerulus
    d. dilation of the afferent arterioles
    e. decreased capsule pressure

11. If blood pressure increases by 50 mm Hg,
    a. the afferent arterioles constrict.
    b. glomerular capillary pressure increases by 50 mm Hg.
    c. GFR increases dramatically.
    d. all of the above.

12. The greatest volume of water is reabsorbed from the nephron by the
    a. proximal convoluted tubule.   c. distal convoluted tubule.
    b. loop of Henle.   d. collecting duct.

13. Water leaves the nephron by
    a. active transport.
    b. filtration into the capillary network.
    c. osmosis.
    d. facilitated diffusion.
    e. symport.

14. Reabsorption of most solute molecules from the proximal convoluted tubule is linked to the active transport of Na^+
    a. across the apical membrane and out of the cell.
    b. across the apical membrane and into the cell.
    c. across the basolateral membrane and out of the cell.
    d. across the basolateral membrane and into the cell.

15. Which of these ions is used to symport amino acids, glucose, and other solutes through the apical membrane of nephron epithelial cells?
    a. K^+ c. Cl^- e. Mg^{2+}
    b. Na^+ d. Ca^{2+}

16. Which of the following contributes to the formation of a hyperosmotic environment in the medulla of the kidney?
    a. the symport of Na^+, K^+, and Cl^- out of the ascending limb of the loop of Henle
    b. the active transport of Na^+ out of the ascending limb of the loop of Henle
    c. the cycling of urea between the collecting ducts and the thin segments of the loops of Henle
    d. all of the above

17. At which of these sites is the osmolality lowest?
    a. glomerular capillary
    b. proximal convoluted tubule
    c. tip of the loop of Henle
    d. initial section of the distal convoluted tubule
    e. collecting duct

18. Increased aldosterone causes
    a. increased reabsorption of Na^+.
    b. decreased blood volume.
    c. decreased reabsorption of Cl^-.
    d. increased permeability of the distal convoluted tubule to water.
    e. increased volume of urine.

19. Granular cells secrete
    a. ADH.   c. aldosterone.
    b. angiotensin.   d. renin.

20. ADH governs the
    a. Na^+ pump of the proximal convoluted tubules.
    b. water permeability of the loop of Henle.
    c. Na^+ pump of the vasa recta.
    d. water permeability of the distal convoluted tubules and collecting ducts.
    e. Na^+ reabsorption in the proximal convoluted tubule.

21. A decrease in blood osmolality results in
    a. increased ADH secretion.
    b. increased permeability of the collecting ducts to water.
    c. decreased urine osmolality.
    d. decreased urine output.
    e. all of the above.

22. The urinary bladder
    a. is made up of skeletal muscle.
    b. is lined by simple columnar epithelium.
    c. is connected to the outside of the body by the ureter.
    d. is located in the pelvic cavity.
    e. has two urethras and one ureter attached to it.

23. Given these events:
    1. loss of voluntary control of urination
    2. loss of the sensation or desire to urinate
    3. loss of reflex emptying of the urinary bladder

    Which of these events occurs following transection of the spinal cord at level L5?
    a. 1   c. 1,2   e. 1,2,3
    b. 2   d. 2,3

24. The sensation of thirst increases when
    a. the levels of angiotensin II increase.
    b. the osmolality of the blood decreases.
    c. blood pressure increases.
    d. renin secretion decreases.

25. Insensible perspiration
    a. is lost through sweat glands.
    b. results in heat loss from the body.
    c. increases when ADH secretion increases.
    d. results in the loss of solutes, such as Na^+ and Cl^-.

26. The composition and volume of body fluid are regulated primarily by the
    a. skin.   c. kidneys.   e. spleen.
    b. lungs.   d. heart.

27. Which of these conditions *decreases* extracellular fluid volume?
    a. increased ADH secretion
    b. decreased ANH secretion
    c. decreased aldosterone secretion
    d. increased sympathetic stimulation of the kidneys

28. Which of these results in an increased blood Na$^+$ concentration?
    a. a decrease in ADH secretion
    b. a decrease in aldosterone secretion
    c. an increase in ANH
    d. a decrease in renin secretion

29. Which of these mechanisms is the most important for regulating blood osmolality?
    a. ADH
    b. renin–angiotensin–aldosterone
    c. ANH
    d. parathyroid hormone

30. Calcium ion concentration in the blood decreases when
    a. vitamin D levels are lower than normal.
    b. calcitonin secretion decreases.
    c. parathyroid hormone secretion increases.
    d. all of the above.

31. An acid
    a. solution has a pH greater than 7.
    b. is a substance that releases H$^+$ into a solution.
    c. is considered weak if it completely dissociates in water.
    d. all of the above.

32. Buffers
    a. release H$^+$ when pH increases.
    b. resist changes in the pH of a solution.
    c. include the proteins of the blood.
    d. all of the above.

33. Which of these is *not* a buffer system in the body?
    a. sodium chloride buffer system
    b. carbonic acid/bicarbonate buffer system
    c. phosphate buffer system
    d. protein buffer system

34. Which of these systems regulating blood pH is the fastest-acting?
    a. respiratory b. kidney

35. High levels of bicarbonate ions in the urine indicate
    a. a low level of H$^+$ secretion into the urine.
    b. that the kidneys are causing blood pH to increase.
    c. that urine pH is decreasing.
    d. all of the above.

Answers in Appendix E

C R I T I C A L   T H I N K I N G

1. Propose as many ways as you can to decrease the GFR.

2. If only a very small amount of urea were present in the interstitial fluid of the kidney instead of its normal concentration, how would it affect the kidney's ability to concentrate urine?

3. Research has shown that mammals with kidneys having relatively thicker medullas have the ability to produce more concentrated urine than humans. Explain why this is so.

4. Design a kidney that can produce hyposmotic urine, which is less concentrated than plasma, or hyperosmotic urine, which is more concentrated than plasma, by the active transport of water instead of Na$^+$. Assume that the anatomical structure of the kidney is the same as that in humans. Feel free to change anything else you choose.

5. Ethyl alcohol inhibits ADH secretion. Suppose a person drinks two shots of vodka. What effect does this have on urine production?

6. To relax after an anatomy and physiology examination, Mucho Gusto goes to a local bistro and drinks 2 quarts of low-sodium, non-alcoholic beer. What effect does this beer have on urine concentration and volume? Explain the mechanisms involved.

7. Some patients with hypertension are kept on a low-salt (low-sodium) diet. Propose an explanation for this therapy.

8. In patients with diabetes mellitus, not enough insulin is produced; as a consequence, blood glucose levels increase. If blood glucose levels rise high enough, the kidneys are unable to absorb the glucose from the glomerular filtrate and glucose "spills over" into the urine. What effect does this glucose have on urine concentration and volume? How does the body adjust to the excess glucose in the urine?

9. A patient suffering from a tumor in the hypothalamus produces excessive amounts of ADH, a condition called syndrome of inappropriate ADH (SIADH) production. For this patient, the excessive ADH production is chronic and has persisted for many months. A student nurse keeps a fluid intake–output record on the patient. She is surprised to find that fluid intake and urinary output are normal. What effect was she expecting? Can you explain why urinary output is normal?

10. A patient exhibits elevated urine ammonium ions (NH$_4$$^+$) and increased rate of respiration. Does she have metabolic acidosis or metabolic alkalosis?

11. Swifty Trotts has an enteropathogenic *Escherichia coli* infection that produces severe diarrhea. What does this diarrhea do to his blood pH, urine pH, and respiratory rate?

12. As part of a physiology experiment, an anatomy and physiology student is asked to breathe through a 3-foot-long glass tube. What effect does this action have on his blood pH, urine pH, and respiratory rate?

13. A young girl is suspected of having epilepsy and therefore is prone to having convulsions. Based on your knowledge of acid–base balance and respiration, propose a hypothetical experiment that might suggest that the girl is susceptible to convulsions.

14. Hardy Explorer has climbed to the top of a mountain. To celebrate, he drinks a glass of whiskey. Alcohol stimulates hydrochloric acid secretion in the stomach. What do you expect to happen to Hardy's respiratory rate and the pH of his urine?

Answers in Appendix F

CHAPTER 24 Reproductive System

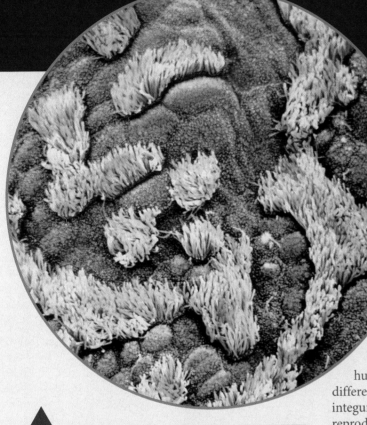

Color-enhanced scanning electron micrograph of the ciliated epithelial surface of a uterine tube. The cilia help move the ovulated oocyte from the ovary to the uterus.

Tools for Success

A virtual cadaver dissection experience

Audio (MP3) and/or video (MP4) content, available at www.mhhe.aris.com

Introduction

Reproduction is an essential characteristic of living organisms, and functional male and female reproductive systems are necessary for individuals to reproduce. The male reproductive system produces sperm cells and can transfer them to the female. The female reproductive system produces oocytes and can receive sperm cells, one of which may unite with an oocyte. The female reproductive system is then intimately involved with nurturing the development of a new individual until birth and usually for some considerable time after birth.

Even though some individuals do not reproduce, their reproductive systems play important roles. The reproductive system controls the development of the structural and functional differences between males and females, and it has profound effects on human behavior. Not only are male and female reproductive structures different, but also the differences that exist among other systems, such as the integumentary, muscular, and skeletal systems, result from the effects of the reproductive system on them.

Although the male and female reproductive systems have striking differences, they also have a number of similarities. Many reproductive organs of males and females are derived from the same embryological structures (see chapter 25). In addition, some hormones are the same in males and females, even though they act in very different ways.

▶This chapter discusses the **functions of the reproductive system (p. 766)**, **meiosis (p. 766)**, the **anatomy of the male reproductive system (p. 767)**, the **physiology of male reproduction (p. 776)**, the **anatomy of the female reproductive system (p. 781)**, the **physiology of female reproduction (p. 790)**, and the **effects of aging on the reproductive system (p. 799)**.

Functions of the Reproductive System

The male reproductive system consists of the testes, epididymis, ducts, accessory glands, and penis (figure 24.1*a*). It performs the following functions:

1. *Production of sperm cells.* The reproductive system produces male sex cells, or sperm cells, in the testes.
2. *Sustaining and transfer of the sperm cells to the female.* The duct and gland system provides an environment in which the sperm cells mature, provides nutrients for the sperm cells produced in the testes, and transports the sperm cells from the testes through the penis, which is a specialized organ that deposits the sperm cells in the female reproductive system.
3. *Production of male sex hormones.* Hormones produced by the male reproductive system control the development of the reproductive system itself and of the male body form. These hormones are also essential for the normal function of the reproductive system and reproductive behavior.

The female reproductive system consists of the ovaries, uterine tubes, uterus, vagina, and mammary glands (figure 24.1*b*). It performs the following functions:

1. *Production of female sex cells.* The reproductive system produces female sex cells, or oocytes, in the ovaries.
2. *Reception of sperm cells from the male.* The female reproductive system includes structures that receive sperm cells from the male and transport the sperm cells to the site of fertilization.
3. *Nurturing the development of and providing nourishment for the new individual.* The female reproductive system nurtures the development of a new individual in the uterus until birth and provides nourishment in the form of milk after birth.
4. *Production of female sex hormones.* Hormones produced by the female reproductive system control the development of the reproductive system itself and of the female body form. These hormones are also essential for the normal function of the reproductive system and reproductive behavior. ▼

1. **List the functions of the male and female reproductive systems.**

Meiosis

Gametes (gam′ētz), or sex cells, have half the number of chromosomes as other cells in the body. Sperm cells are the male gametes and oocytes are the female gametes. **Meiosis** (mī-ō′sis) is

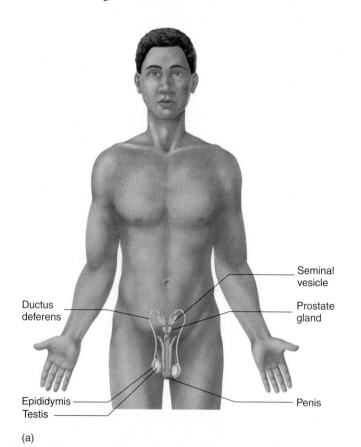

(a)

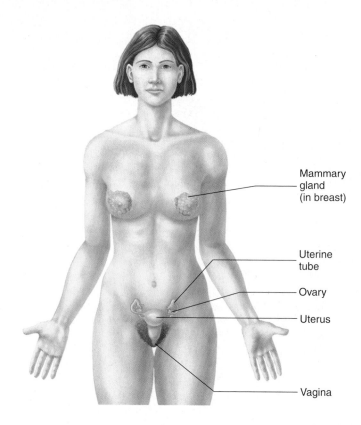

(b)

Figure 24.1   Major Organs of the Reproductive System

(*a*) The male reproductive system: testes, epididymis, ductus deferens, seminal vesicles, prostate gland, and penis. (*b*) The female reproductive system: ovaries, uterine tubes, uterus, vagina, and mammary glands.

a kind of cell division that produces gametes. It consists of two consecutive nuclear divisions without replication of the genetic material between the divisions. Four daughter cells are produced, and each has half as many chromosomes as the parent cell. ▼

> 2. **Define gamete and meiosis.**

Chromosomes

Chromosomes contain most of the cell's DNA, which has the genes determining the structural and functional features of every individual (see chapter 3). The normal chromosome number in human cells is 46. This number is called a **diploid** (dip′loyd), or a **2n, number** of chromosomes. The chromosomes consist of 23 pairs, each of which is called a **homologous** (hŏ-mol′ō-gŭs) **pair.** The homologous pairs consist of 22 autosomal pairs, which are all of the chromosomes except the sex chromosomes, and 1 pair of sex chromosomes. The sex chromosomes are an X and a Y chromosome in males and two X chromosomes in females. One chromosome of each homologous pair is from the male parent, and the other is from the female parent. The chromosomes of each homologous pair look alike, except for the sex chromosomes, and they contain genes for the same traits.

In sperm cells and oocytes, the number of chromosomes is 23. This number is called a **haploid** (hap′loyd), or **n, number** of chromosomes. Each gamete contains one chromosome from each of the homologous pairs. Reduction in the number of chromosomes in sperm cells or oocytes to an *n* number is important. When a sperm cell and an oocyte fuse to form a fertilized egg, each provides an *n* number of chromosomes, which reestablishes a *2n* number of chromosomes. If meiosis did not take place, each time fertilization occurred the number of chromosomes in the fertilized oocyte would double. The extra chromosomal material would be lethal to the developing offspring.

Sperm cells have 22 autosomal chromosomes and either an X or a Y chromosome. Oocytes contain 1 autosomal chromosome from each of the 22 homologous pairs and an X chromosome. During fertilization, when a sperm cell fuses with an oocyte, the normal number of 46 chromosomes, consisting of 23 pairs of chromosomes, is reestablished. The sex of the baby is determined by the sperm cell that fertilizes the oocyte. The sex is male if the sperm cell that fertilizes the oocyte carries a Y chromosome, female if the sperm cell carries an X chromosome. ▼

> 3. **Define diploid and haploid. Why is it necessary for gametes to have a haploid number of chromosomes?**
> 4. **What is a homologous pair of chromosomes?**
> 5. **Which sex chromosomes are found in males and females?**

Meiotic Divisions

The two divisions of meiosis are called **meiosis I** and **meiosis II.** Like mitosis, each division of meiosis has prophase, metaphase, anaphase, and telophase (see figure 3.28). Distinct differences exist, however, between mitosis and meiosis.

Before meiosis begins, all the chromosomes are duplicated. At the beginning of meiosis, each of the 46 chromosomes consists of two **chromatids** (krō′mă-tid) connected by a **centromere** (sen′trō-mēr) (figure 24.2, step 1). Each chromatid is a replicated chromosome. Homologous chromosomes align as pairs in a process called **synapsis** (figure 24.2, step 2). Occasionally, part of a chromatid of one homologous chromosome breaks off and is exchanged with part of another chromatid from the other homologous chromosome. This exchange of genetic material is called **crossing over.** Crossing over allows the exchange of genetic material between maternal and paternal chromosomes.

The chromosomes align along the center of the cell (figure 24.2, step 3). For each pair of homologous chromosomes, however, the side of the cell on which the maternal or paternal chromosome is located is random. The way the chromosomes align during synapsis results in the **random assortment** of maternal and paternal chromosomes in the daughter cells during meiosis. Crossing over and the random assortment of maternal and paternal chromosomes are responsible for the diversity in the genetic makeup of sperm cells and oocytes. We are unique individuals because of this genetic diversity.

The aligned chromosomes separate and move to opposite sides of the cell. New nuclei form and the cell completes division of the cytoplasm to form two cells (figure 24.2, steps 4 and 5). When meiosis I is complete, each daughter cell has 1 chromosome from each of the homologous pairs. Each of the 23 chromosomes in each daughter cell consists of two chromatids joined by a centromere.

It is during the first meiotic division that the chromosome number is reduced from a *2n* number (46 chromosomes, or 23 pairs) to an *n* number (23 chromosomes, or 1 from each homologous pair). The first meiotic division is therefore called a **reduction division.**

The second meiotic division is similar to mitosis (figure 24.2, steps 6–9). The chromosomes, each consisting of two chromatids, line up near the middle of the cell. Then the chromatids separate at the centromere, and each daughter cell receives one of the chromatids from each chromosome. When the centromere separates, each of the chromatids is called a chromosome. Consequently, each of the four daughter cells produced by meiosis contains 23 chromosomes. ▼

> 6. **Describe the process by which chromosome number is reduced by half in meiosis.**
> 7. **What is the significance of crossing over and random assortment of chromosomes?**

Anatomy of the Male Reproductive System

The male reproductive system consists of the testes (sing. *testis*), a series of ducts, accessory glands, and supporting structures. The ducts include the epididymides (sing. *epididymis*), ductus deferentia (sing. *deferens,* also *vas deferens*), and urethra. Accessory glands include the seminal vesicles, prostate gland, and bulbourethral glands. Supporting structures include the scrotum and penis

First Meiotic Division (Meiosis I)

1. Early prophase I
The duplicated chromosomes become visible chromatids (shown separated for emphasis; they actually are so close together that they appear as a single strand).

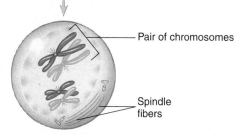

Centromere

Chromosomes

Nucleus

Chromatids

Centrioles

2. Middle prophase I
Homologous chromosomes synapse. Crossing over may occur at this stage.

Pair of chromosomes

Spindle fibers

3. Metaphase I
Homologous chromosomes align at the center of the cell. Random assortment of homologous chromosomes occurs.

Equatorial plane

4. Anaphase I
Homologous chromosomes move apart to opposite sides of the cell.

5. Telophase I
New nuclei form, and the cell divides.

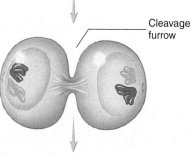

Cleavage furrow

Prophase II (top of next column)

Second Meiotic Division (Meiosis II)
(continued from the bottom of previous column)

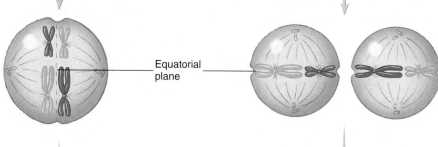

6. Prophase II
Each chromosome consists of two chromatids.

7. Metaphase II
Chromosomes align along the center of the cell.

8. Anaphase II
Chromatids separate and each is now called a chromosome.

9. Telophase II
New nuclei form around the chromosomes. The cells divide to form four daughter cells with a haploid (*n*) number of chromosomes.

Process Figure 24.2   Meiosis

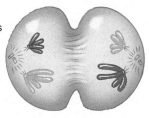

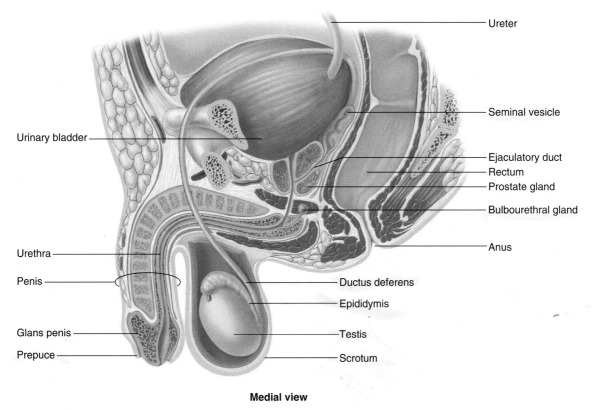

Figure 24.3 Male Reproductive Structures

Sagittal section of the male pelvis showing the male reproductive structures.

(figure 24.3). Collectively, all of these structures are called the **male genitalia**, or **male genitals**. The **external male genitalia** consist of the scrotum and penis, and the **internal male genitalia** are the testes, epididymis, ducts, and glands. ▼

> **8.** *List the structures that make up the male reproductive system. What are the male genitalia?*

Scrotum

The **scrotum** (skrō′tum) is a saclike structure containing the testes. It is divided into right and left internal compartments by an incomplete connective tissue septum. Externally the scrotum consists of skin. Beneath the skin is a layer of loose connective tissue and a layer of smooth muscle, called the **dartos** (dar′tōs, to skin) **muscle.**

When the scrotum is exposed to cool temperatures, the dartos muscle contracts, causing the skin of the scrotum to become firm and wrinkled and reducing the overall size of the scrotum. At the same time, extensions of abdominal muscles connected to the testes, called **cremaster** (krē-mas′ter, to hang) **muscles,** contract. Consequently, the testes are pulled nearer to the body, and their temperature is raised. During warm weather or exercise, the dartos and cremaster muscles relax, the skin of the scrotum becomes loose and thin, and the testes descend away from the body, which lowers their temperature. The response of the dartos and cremaster muscles is important in the regulation of temperature in the testes.

If the testes become too warm or too cold, normal sperm cell development does not occur. ▼

> **9.** *Describe the structure of the scrotum.*
> **10.** *Describe the role of the dartos and cremaster muscles in temperature regulation of the testes.*

Testes

Testicular Histology

The **testes** (tes′tēz) are small, ovoid organs, each about 4–5 cm long, within the scrotum (see figure 24.3). They are both exocrine and endocrine glands. Sperm cells form a major part of the exocrine secretions of the testes, and testosterone is the major endocrine.

The **tunica albuginea** (al-bū-jin′ē-ă, white) is connective tissue forming the outer capsule of each testis (figure 24.4). Extensions of the tunica albuginea form incomplete **septa** (sep′tă) dividing the testis into 300–400 cone-shaped **lobules. Seminiferous** (sem′i-nif′er-ŭs, seed carriers) **tubules** are coiled tubes located in the lobules. The uncoiled, combined length of the seminiferous tubules in both testes is nearly half a mile. The seminiferous tubules are the site of sperm cell production. **Interstitial** (in-ter-stish′ăl) **cells,** or **cells of Leydig,** are clusters of cells within delicate connective tissue surrounding the seminiferous tubules. They are endocrine cells that secrete testosterone.

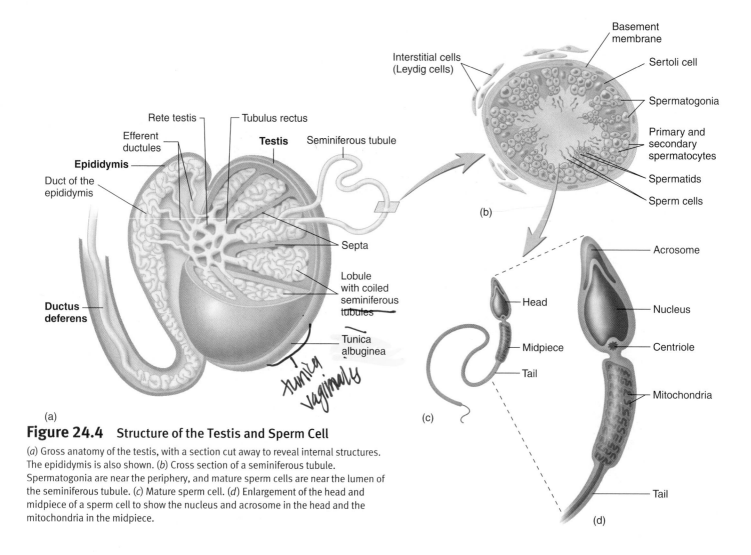

Figure 24.4 **Structure of the Testis and Sperm Cell**

(*a*) Gross anatomy of the testis, with a section cut away to reveal internal structures. The epididymis is also shown. (*b*) Cross section of a seminiferous tubule. Spermatogonia are near the periphery, and mature sperm cells are near the lumen of the seminiferous tubule. (*c*) Mature sperm cell. (*d*) Enlargement of the head and midpiece of a sperm cell to show the nucleus and acrosome in the head and the mitochondria in the midpiece.

The seminiferous tubules empty into a set of short, straight tubules, the **tubuli recti,** which in turn empty into a tubular network called the **rete** (rĕ′tē, net) **testis.** The rete testis empties into 15–20 tubules called **efferent ductules** (dŭk′tools). They have a ciliated pseudostratified columnar epithelium, which helps move sperm cells out of the testis. The efferent ductules pierce the tunica albuginea to exit the testis. ▼

11. *Describe the covering and connective tissue of the testis.*
12. *Where are the seminiferous tubules and interstitial cells located? What are their functions?*
13. *Describe the tubuli recti, rete testis, and efferent ductules.*

Descent of the Testes

By 8 weeks following fertilization, the testes have developed as retroperitoneal organs in the abdominopelvic cavity (figure 24.5, step 1). The **gubernaculum** (goo′ber-nak′ū-lŭm) is a column of mesenchymal tissue connecting each testis to the anterolateral abdominal wall at the site of the future inguinal canal. An inguinal canal is an oblique passageway through the anterior abdominal wall. The **deep inguinal ring** is the opening into the inguinal canal from the abdominal cavity, and the **superficial**

inguinal ring is the opening of the canal to the outside of the abdominal wall.

The testes descend (figure 24.5, step 2) and come to rest near the deep inguinal ring. The gubernaculum extends through the inguinal canal and swelling of the gubernaculum enlarges the inguinal canal. Between 7 and 9 months of development, the testes pass through the inguinal canals, a journey taking about 3 days, and end up in the scrotum (figure 24.5, step 3). In females, the ovaries also develop in the abdomen and descend, but they stop their descent in the pelvic cavity and do not enter the inguinal canals.

An outpocketing of the peritoneum, called the **process vaginalis** (vaj′i-nă-lis), precedes each testis into the scrotum (see figure 24.5, steps 2 and 3). The superior part of each process vaginalis usually becomes obliterated, and the inferior part remains as a small, closed sac, the **tunica** (too′ni-kă) **vaginalis** (figure 24.5, step 4). The tunica vaginalis is a serous membrane surrounding most of the testis in much the same way that the pericardium surrounds the heart. The visceral layer of the tunica vaginalis covers the anterior surface of the testis, and the parietal layer lines the scrotum. A small amount of fluid within the tunica vaginalis allows the testes to move with little friction.

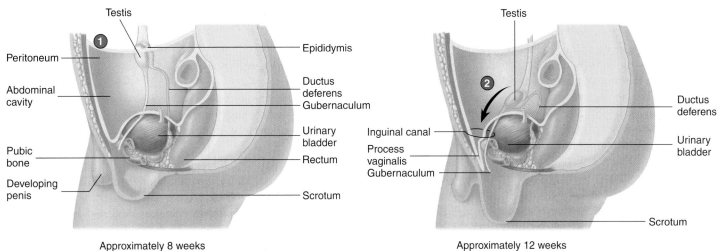

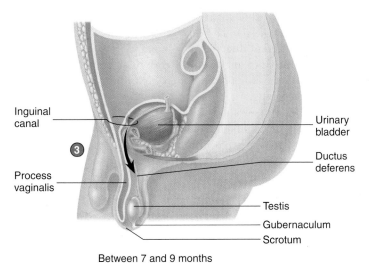

Approximately 8 weeks

(1) Each testis forms as a retroperitoneal structure near the level of the kidney.

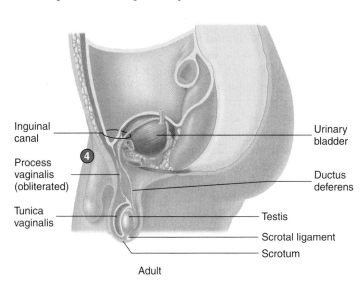

Approximately 12 weeks

(2) The testis beneath the parietal peritoneum is descending toward the inguinal canal. The process vaginalis extends into the inguinal canal alongside the gubernaculum.

Between 7 and 9 months

(3) The testis follows the process vaginalis and descends through the inguinal canal into the scrotum.

Adult

(4) Between birth and adulthood, the process vaginalis is obliterated and its inferior portion becomes the tunica vaginalis. The gubernaculum becomes the scrotal ligament.

Process Figure 24.5   Descent of the Testes

⚕ Cryptorchidism and Inguinal Hernia

Cryptorchidism (krip-tōr′ki-dizm) is the failure of one or both of the testes to descend into the scrotum. Cryptorchidism occurs in 3% of full-term newborns and 30% of premature births. By 6 months to a year, cryptorchidism has decreased to 1%. Later in life, an undescended testis does not produce sperm cells because of the higher temperature of the abdominal cavity.

A **hernia** (her′nē-ă) is the protrusion of part of an organ through the tissues normally containing it. An **inguinal hernia** is the protrusion of an abdominopelvic organ, usually the intestine, through an opening in the abdominal wall in the inguinal region. An inguinal hernia can pass partially or completely through the inguinal canal and can even extend to the testis. Inguinal hernias can be

quite painful and even very dangerous, especially if a portion of the small intestine is compressed so that its blood supply is cut off. Fortunately, inguinal hernias can be repaired surgically. Males are much more prone to inguinal hernias than are females because a male's inguinal canals are larger and weakened because the testes pass through them on their way into the scrotum.

An **indirect inguinal hernia** enters the inguinal canal through the deep inguinal ring. In most cases, the process vaginalis is not obliterated and the hernia protrudes into it. A **direct inguinal hernia** passes through a weakened or ruptured abdominal wall medial to the deep inguinal ring. Frequently, the hernia enters the inguinal canal. ▼

14. *When and how do the testes descend into the scrotum?*
15. *Describe the tunica vaginalis.*

Spermatogenesis

Before puberty, the testes remain relatively simple and unchanged from the time of their initial development. The interstitial cells are not prominent, and the seminiferous tubules are small and not yet functional. At the time of puberty, the interstitial cells increase in number and size, the seminiferous tubules enlarge, and spermatogenesis begins.

Spermatogenesis (sper′mă-tō-jen′ĕ-sis) is the formation of sperm cells. **Spermatogonia** (sper′mă-tō-gō′nē-ă, primitive sperm cell) are the cells that give rise to sperm cells. They are located in the periphery of the seminiferous tubules and divide by mitosis (figure 24.6a). Some daughter cells produced from these mitotic divisions remain as spermatogonia and continue to divide by mitosis. Other daughter cells form **primary spermatocytes** (sper′mă-tō-sītz, sperm cell), which divide by meiosis (see "Meiotic Divisions," p. 767).

A primary spermatocyte contains 46 chromosomes, each consisting of two chromatids. Each primary spermatocyte passes through the first meiotic division to produce two **secondary spermatocytes.** The first meiotic division is a reduction division and the secondary spermatocytes now have 23 chromosomes (two chromatids). Each secondary spermatocyte undergoes a second meiotic division to produce two smaller cells called **spermatids** (sper′mă-tidz). Each spermatid has 23 chromosomes, each derived from a chromatid in the secondary spermatocyte. After the second meiotic division, the spermatids undergo major structural changes to form sperm cells (see figure 24.6a). Much of the cytoplasm of the spermatids is eliminated, and each spermatid develops a head, midpiece, and flagellum (tail) to become a **sperm cell,** or **spermatozoon** (see figures 24.4c and d). The head contains the nucleus with its chromosomes. The leading end of the head has a cap, the **acrosome** (ak′rō-sōm), which contains enzymes that are released during the process of fertilization and are necessary for the sperm cell to penetrate the oocyte. The flagellum is similar to a cilium (see chapter 3), and movement of microtubules past one another within the tail causes the tail to move and propel the sperm cell forward. The midpiece has large numbers of mitochondria, which produce the ATP necessary for microtubule movement.

At the end of spermatogenesis, the developing sperm cells are located around the lumen of the seminiferous tubules with their tails directed toward the center of the lumen. Finally, sperm cells are released into the lumen of the seminiferous tubules (see figures 24.4b and 24.6).

In addition to the developing sperm cells, **sustentacular** (sŭs-ten-tak′ū-lăr), or **Sertoli** (sēr-tō′lē, named for an Italian histologist), **cells** are in the seminiferous tubules. The sustentacular cells are also sometimes referred to as **nurse cells.** Sustentacular cells are large cells that extend from the periphery to the lumen of the seminiferous tubule (figure 24.6b). They nourish the developing sperm cells and probably produce, together with the interstitial cells, a number of hormones, such as androgens, estrogen, and inhibins. In addition, tight junctions between the sustentacular cells form a **blood–testis barrier** between spermatogonia and sperm cells (see figure 24.6b). It isolates the sperm cells from the immune system. This barrier is necessary because developing sperm cells form surface antigens that could stimulate an immune response, resulting in their destruction. ▼

> 16. Starting with spermatogonia, describe the production of sperm cells.
> 17. Name the parts of a sperm cell and state their functions.
> 18. Where are sustentacular cells located and what is their function? What is the blood–testis barrier?

Ducts

After their release into the seminiferous tubules, the sperm cells pass through the tubuli recti to the rete testis. From the rete testis, they pass through the efferent ductules to the epididymis.

Epididymis

The efferent ductules from each testis extend to a comma-shaped structure on the posterior side of the testis called the **epididymis** (ep-i-did′i-mis, pl. *epididymides,* ep-i-di-dim′i-dēz, on the twin, which refers to the paired, or twin, testes) (see figure 24.4a). Each epididymis consists of a head, a body, and a long tail (figure 24.7). The head contains the very convoluted ends of the efferent ductules, which empty into a single convoluted tube, the **duct of the epididymis.** The body and tail consist of the duct of the epididymis, which if unraveled would be several meters long.

The final maturation of the sperm cells occurs within the epididymis. It takes 12–16 days for sperm cells to travel through the epididymis. As they pass through the epididymis the acrosome matures, the ability to fertilize an oocyte develops, and the flagella become capable of movement. ▼

> 19. Name the ducts located in the epididymis.
> 20. List the changes that occur in sperm cells while in the epididymis.

Ductus Deferens and Ejaculatory Duct

The duct of the epididymis becomes the **ductus deferens,** or **vas deferens,** which emerges from the tail of the epididymis, ascends along the posterior side of the testis, and joins the spermatic cord. Each **spermatic cord** consists of the ductus deferens, testicular artery and veins, lymphatic vessels, and testicular nerve. The **coverings of the spermatic cord** surround the spermatic cord. They consist of the **external spermatic fascia** (fash′ē-ă); the **cremaster muscle,** an extension of the muscle fibers of the internal abdominal oblique muscle of the abdomen; and the **internal spermatic fascia** (see figure 24.7a).

Each ductus deferens extends, in the spermatic cord, through the abdominal wall by way of the inguinal canal. Each ductus deferens then crosses the lateral wall of the pelvic cavity and loops behind the posterior surface of the urinary bladder to approach the prostate gland (see figure 24.3). The total length of the ductus deferens is about 45 cm. The end of the ductus deferens enlarges to form the **ampulla of the ductus deferens** (see figure 24.7a). The wall of the ductus deferens contains smooth muscle. Peristaltic contractions of the smooth muscle propel the sperm cells from the epididymis through the ductus deferens.

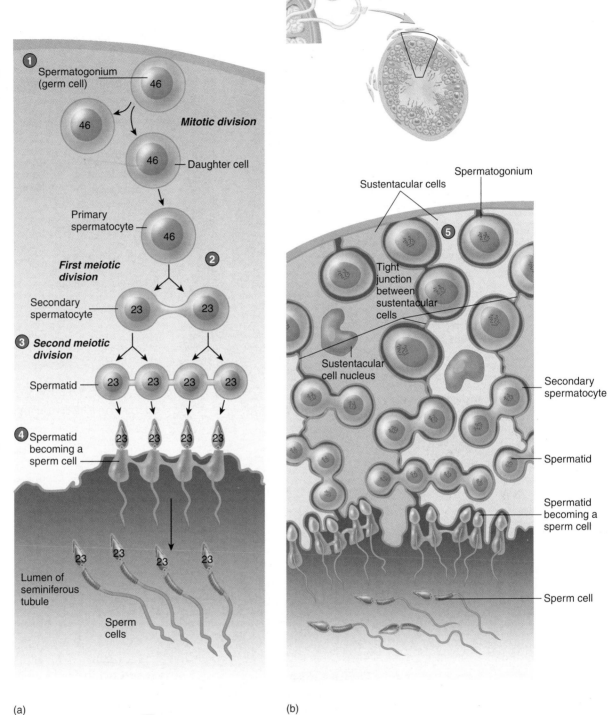

1. Spermatogonia are the cells from which sperm cells arise. The spermatogonia divide by mitosis. One daughter cell remains a spermatogonium that can divide again by mitosis. One daughter cell becomes a primary spermatocyte.

2. The primary spermatocyte divides by meiosis to form secondary spermatocytes.

3. The secondary spermatocytes divide by meiosis to form spermatids.

4. The spermatids differentiate to form sperm cells.

5. Sustentacular cells are between the spermatogonia and developing sperm cells.

Seminiferous tubule

① Spermatogonium (germ cell) 46

Mitotic division

46

46 — Daughter cell

Primary spermatocyte 46

② *First meiotic division*

Secondary spermatocyte 23 23

③ *Second meiotic division*

Spermatid 23 23 23 23

④ Spermatid becoming a sperm cell 23 23 23 23

23 23 23 23

Lumen of seminiferous tubule

Sperm cells

Sustentacular cells

Spermatogonium

⑤

Tight junction between sustentacular cells

Sustentacular cell nucleus

Secondary spermatocyte

Spermatid

Spermatid becoming a sperm cell

Sperm cell

(a)

(b)

Process Figure 24.6   Spermatogenesis

(*a*) Meiosis during spermatogenesis. A section of the seminiferous tubule illustrating the process of meiosis and sperm cell formation. (*b*) The sustentacular cells (shown in alternating dark and light shades) extend from the periphery to the lumen of the seminiferous tubules. The tight junctions that form between adjacent sustentacular cells form the blood–testis barrier. Spermatogonia are peripheral to the blood–testis barrier, and spermatocytes are central to it.

Adjacent to the ampulla of each ductus deferens is a sac-shaped gland called the seminal vesicle. A short duct from the seminal vesicle joins the ampulla of the ductus deferens to form the **ejaculatory** (ē-jak′ū-lă-tōr-ē) **duct.** The ejaculatory ducts are approximately 2.5 cm long. These ducts project into the prostate gland and end by opening into the urethra (see figures 24.3 and 24.7*a*). ▼

21. Describe the route by which the ductus deferens extends from the testis to the prostate gland. What is the ejaculatory duct?

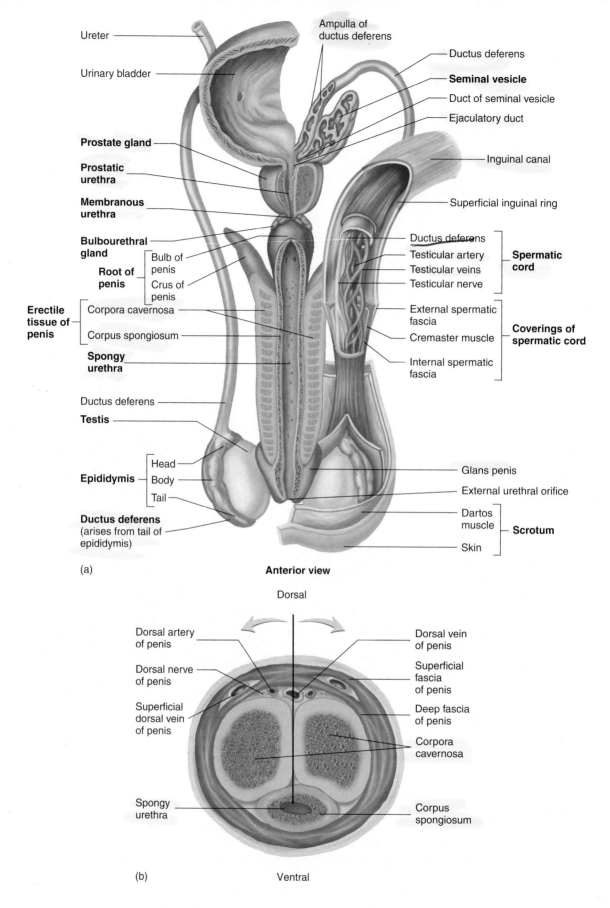

Ureter

Urinary bladder

Prostate gland

Prostatic urethra

Membranous urethra

Bulbourethral gland

Root of penis
- Bulb of penis
- Crus of penis

Erectile tissue of penis
- Corpora cavernosa
- Corpus spongiosum

Spongy urethra

Ductus deferens

Testis

Epididymis
- Head
- Body
- Tail

Ductus deferens
(arises from tail of epididymis)

Ampulla of ductus deferens

Ductus deferens

Seminal vesicle

Duct of seminal vesicle

Ejaculatory duct

Inguinal canal

Superficial inguinal ring

Ductus deferens
Testicular artery   **Spermatic cord**
Testicular veins
Testicular nerve

External spermatic fascia   **Coverings of spermatic cord**
Cremaster muscle
Internal spermatic fascia

Glans penis

External urethral orifice

Dartos muscle   **Scrotum**
Skin

(a)   **Anterior view**

Dorsal

Dorsal artery of penis

Dorsal nerve of penis

Superficial dorsal vein of penis

Spongy urethra

Dorsal vein of penis

Superficial fascia of penis

Deep fascia of penis

Corpora cavernosa

Corpus spongiosum

(b)   Ventral

Figure 24.7 Male Reproductive Structures

(a) Testes, epididymis, ductus deferens, and glands of the male reproductive system. The penis is cut open along its dorsal side. (b) Cross section of the penis, showing principal nerves, arteries, and veins along the dorsum of the penis. The *line* and *arrows* depict the manner in which (a) is cut and laid open.

22. List the components of the spermatic cord and the coverings of the spermatic cord.

Urethra

The **male urethra** (ū-rē′thră) is about 20 cm long and extends from the urinary bladder to the distal end of the penis (see figure 24.7a). The urethra can be divided into three parts: the **prostatic urethra,** which passes through the prostate gland; the **membranous urethra,** which passes through the floor of the pelvis and is surrounded by the external urethral sphincter; and the **spongy urethra,** which extends the length of the penis and opens at its end.

The urethra is a passageway for both urine and male reproductive fluids. Many minute, mucus-secreting glands are located in the epithelial lining of the urethra. ▼

23. Define the parts of the urethra. What is the function of the urethra?

Penis

The **penis** contains three columns of erectile tissue (see figure 24.7a). Engorgement of this erectile tissue with blood causes the penis to enlarge and become firm, a condition called an **erection.** The penis is the male organ of copulation, transferring sperm cells from the male to the female. The **corpora cavernosa** (kōr′pōr-ă kav-er-nos′ă) are two columns of erectile tissue forming the dorsal portion and the sides of the penis. The **corpus spongiosum** (kōr′pŭs spŭn′jē-ō′sŭm) surrounds the spongy urethra and forms the ventral portion of the penis. The distal corpus spongiosum expands to form a cap, the **glans penis,** over the distal end of the penis. At the base of the penis, the corpus spongiosum expands to form the **bulb of the penis.** Each corpus cavernosum expands to form a **crus** (kroos, pl. *crura,* kroo′ră) **of the penis,** which attaches to a coxal bone. The **root of the penis** is the combined bulb and crura of the penis.

Skin is loosely attached to the connective tissue that surrounds the erectile columns in the shaft of the penis. The skin is firmly attached at the base of the glans penis, and a thinner layer of skin tightly covers the glans penis. The skin of the penis, especially the glans penis, is well supplied with sensory receptors. A loose fold of skin called the **prepuce** (prē′poos), or **foreskin,** covers the glans penis (see figure 24.3).

 Circumcision

Circumcision (ser-kŭm-sizh′ŭn) is the surgical removal of the prepuce, usually near the time of birth. Compelling medical reasons for circumcision do not appear to exist. Uncircumcised males have a higher incidence of penile cancer, but the underlying cause appears to be related to chronic infections and poor hygiene. In the few cases in which the prepuce is "too tight" to be moved over the glans penis, circumcision can be necessary to avoid chronic infections and maintain normal circulation. ▼

24. Define erection. Describe the erectile tissues of the penis.
25. Define glans penis, root of penis, and prepuce (foreskin).

Accessory Glands

Seminal Vesicles

Each of the two **seminal vesicles** (sem′i-năl ves′i-klz) is a sac-shaped gland located next to the ampulla of the ductus deferens (see figures 24.3 and 24.7a). Each gland is about 5 cm long and tapers into a short excretory duct that joins the ampulla of the ductus deferens to form the ejaculatory duct. The seminal vesicles have a capsule containing fibrous connective tissue and smooth muscle cells.

Prostate Gland

The **prostate** (pros′tāt, one standing before) **gland** is about the size and shape of a walnut (about 4 cm long and 2 cm wide) and surrounds the urethra and the two ejaculatory ducts at the base of the urinary bladder (see figures 24.3 and 24.7a). The prostate gland has a capsule and partitions, which consist of connective tissue and smooth muscle. Epithelial cells lining the partitions secrete prostatic fluid. Fifteen to 30 small prostatic ducts carry these secretions into the prostatic urethra.

P R E D I C T **1**

The prostate gland can enlarge for several reasons, including infections, tumor, and old age. The detection of enlargement or changes in the prostate is an important way to detect prostatic cancer. Suggest a way other than surgery that the prostate gland can be examined by palpation for any abnormal changes.

 Prostate Disorders

Benign prostatic hyperplasia (BPH) is a noncancerous enlargement of the prostate gland. BPH affects approximately one-third of men over the age of 50 and up to 90% of men by age 80. BPH can be treated with drugs that inhibit enlargement of the prostate and reduce smooth muscle tension in the prostate, urethra, and urinary bladder. Various surgical procedures are available to remove the prostate. Most involve gaining access to the prostate gland through the urethra, but sometimes open abdominal surgery is performed. **Transurethral resection of the prostate (TURP)** is a surgical procedure that cuts away the prostate gland. Other methods of destroying the prostate gland include lasers, microwaves, radiofrequency energy, ultrasound waves, and electricity.

Cancer of the prostate is the second most common cause of cancer in men in the United States; it is less common than lung cancer but more common than colon cancer. Prostate-specific antigen (PSA) is a protein produced by the prostate gland, and small amounts of PSA leak into the blood and can be detected with a simple blood test. Blood PSA levels often increase in men with prostate cancer. An annual examination of the prostate and a PSA test are recommended for men over the age of 50. If a biopsy confirms prostate cancer, treatment options include radiation, chemotherapy, and surgery. Treatment can result in incontinence and an inability to have an erection because of nerve damage.

Bulbourethral Glands

The **bulbourethral** (bŭl′bō-ū-rē′thrăl) **glands** are a pair of small, mucus-secreting glands located near the membranous part of the urethra (see figures 24.3 and 24.7a). In young males, each is about the size of a pea, but they decrease in size with age and are almost impossible to see in old men. A single duct from each gland enters the urethra.

Semen

Semen (sē′men) is a mixture of sperm cells and secretions from the male reproductive glands. Sperm cells and testicular secretions account for about 4% of semen volume. The bulbourethral glands produce less than 1% of the semen, the seminal vesicles produce about 65% of the semen, and the prostate gland contributes about 30%. The mucous secretion of the bulbourethral glands lubricates the urethra, helps neutralize the contents of the normally acidic urethra, provides a small amount of lubrication during intercourse, and helps reduce the acidity in the vagina.

The thick, mucuslike secretion of the seminal vesicles contains the sugar fructose and other nutrients that provide nourishment to sperm cells. The seminal vesicle secretions also contain proteins that weakly clot the semen after ejaculation, making it a sticky, jellylike fluid. Prostaglandins, which stimulate smooth muscle contractions, are present in high concentrations in the secretions of the seminal vesicles and can cause contractions of the female reproductive tract, which help transport sperm cells through the female reproductive tract.

The thin, milky secretions of the prostate have an alkaline pH and help neutralize the acidic urethra, as well as the acidic secretions of the testes, the seminal vesicles, and the vagina. The increased pH is important for normal sperm cell function. The movement of sperm cells is not optimal until the pH is increased to between 6.0 and 6.5. In contrast, the secretions of the vagina have a pH between 3.5 and 4.0. Prostatic secretions also contain proteolytic enzymes, such as prostate-specific antigen (PSA), which break down the clotting proteins of the seminal vesicles and make the semen more liquid.

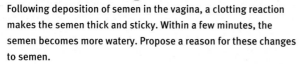

P R E D I C T ❷

Following deposition of semen in the vagina, a clotting reaction makes the semen thick and sticky. Within a few minutes, the semen becomes more watery. Propose a reason for these changes to semen.

Normal sperm cell counts in the semen range from 75 to 400 million sperm cells per milliliter of semen, and a normal ejaculation usually consists of about 2–5 mL of semen. Sperm cells become motile after ejaculation once they are mixed with secretions of the male accessory glands and with secretions of the female reproductive tract. Most of the sperm cells (millions) are expended in moving the general group of sperm cells through the female reproductive system. Enzymes carried in the acrosome of each sperm cell help digest a path through the mucoid fluids of the female reproductive tract and through materials surrounding the oocyte. Once the acrosomal fluid is depleted from a sperm cell, the sperm cell is no longer capable of fertilization. ▼

26. State where the seminal vesicles, prostate gland, and bulbourethral glands empty into the male reproductive duct system.
27. Define semen. By volume, what are the contributions of various sources to semen?
28. What is the function of the secretions of each of the accessory glands?
29. What is a normal sperm count?

Physiology of Male Reproduction

The male reproductive system depends on both hormonal and neural mechanisms to function normally. Hormones control the development of reproductive structures, the development of secondary sexual characteristics, spermatogenesis, and, in part, sexual behavior. The mature neural mechanisms are primarily involved in controlling the sexual act and in the expression of sexual behavior.

Regulation of Sex Hormone Secretion

The hypothalamus of the brain, the anterior pituitary gland, and the testes (figure 24.8) produce hormones that influence the male reproductive system. **Gonadotropin-releasing** (gō′nad-ō-trō′pin, *gonad* + *trophe*, nourishment) **hormone (GnRH)** is released from neurons in the hypothalamus and passes to the anterior pituitary gland (table 24.1). GnRH causes cells in the anterior pituitary gland to secrete two hormones, **luteinizing** (loo′tē-ĭ-nīz-ing) **hormone (LH)** and **follicle-stimulating hormone (FSH),** into the blood (see table 24.1). LH and FSH are named for their functions in females, but they are also essential reproductive hormones in males.

LH binds to the interstitial cells in the testes and causes them to secrete testosterone. LH was once referred to as interstitial cell–stimulating hormone (ICSH) because it stimulates interstitial cells of the testes to secrete testosterone, but it was later discovered to be identical to LH found in the female. Consequently, it is now simply called LH. FSH binds primarily to sustentacular cells in the seminiferous tubules and promotes sperm cell development. It also increases the secretion of a hormone called **inhibin** (in-hib′in, to inhibit).

Testosterone has a negative-feedback effect on the secretion of GnRH from the hypothalamus, as well as on LH and FSH from the anterior pituitary gland. Inhibin has a negative-feedback effect on the secretion of FSH from the anterior pituitary gland.

For GnRH to stimulate the secretion of large quantities of LH and FSH, the anterior pituitary must be exposed to a series of brief increases and decreases in GnRH. Chronically elevated GnRH levels in the blood cause the anterior pituitary cells to become insensitive to stimulation by GnRH molecules, and little LH or FSH is secreted.

1. GnRH from the hypothalamus stimulates the secretion of LH and FSH from the anterior pituitary.

2. LH stimulates testosterone secretion from the interstitial cells.

3. FSH stimulates sustentacular cells of the seminiferous tubules to increase spermatogenesis and to secrete inhibin.

4. Testosterone has a stimulatory effect on the sustentacular cells of the seminiferous tubules, and it has a stimulatory effect on the development of sex organs and secondary sex characteristics.

5. Testosterone has a negative-feedback effect on the hypothalamus and pituitary to reduce LH and FSH secretion.

6. Inhibin has a negative-feedback effect on the anterior pituitary to reduce FSH secretion.

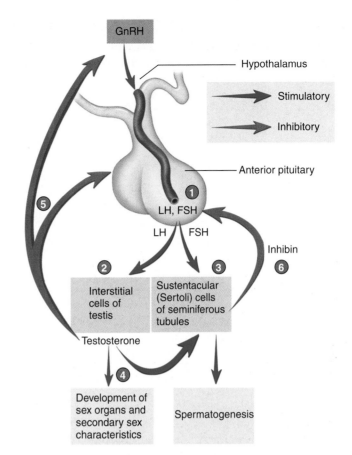

Process Figure 24.8   Regulation of Reproductive Hormone Secretion in Males

 GnRH Pulses and Infertility

GnRH can be produced synthetically, and, if administered in small amounts in frequent pulses or surges, it can be useful in treating males who are infertile. GnRH can also inhibit reproduction because the chronic administration of it can sufficiently reduce LH and FSH levels to prevent sperm cell production in males or ovulation in females. ▼

30. *Where are GnRH, FSH, LH, and inhibin produced? What effects do they produce?*
31. *Where is testosterone produced?*
32. *Describe the regulation of testosterone secretion.*

Puberty

Human chorionic (kō-rē-on'ik) **gonadotropin (hCG)** is a hormone secreted by the placenta. Human chorionic gonadotropin is structurally similar to LH and stimulates the synthesis and secretion of testosterone by the fetal testes before birth. After birth, however, no source of stimulation is present, and the testes of the newborn baby atrophy slightly and secrete only small amounts of testosterone until puberty, which normally begins when a boy is 12–14 years old and is mostly completed by age 18.

Puberty (pū'ber-tē) is the age at which individuals become capable of sexual reproduction. Before puberty, small amounts of testosterone inhibit GnRH, LH, and FSH secretion. At puberty, the hypothalamus becomes much less sensitive to the inhibitory effect of testosterone, and the rate of GnRH, LH, and FSH secretion increases. Elevated FSH levels promote sperm cell formation, and elevated LH levels cause the interstitial cells to secrete larger amounts of testosterone. Testosterone still has a negative-feedback effect on the hypothalamus and anterior pituitary gland, but GnRH, LH, and FSH secretion occurs at substantially higher levels. ▼

33. *What changes in hormone production occur at puberty?*

Effects of Testosterone

Testosterone is the major male hormone secreted by the interstitial cells of the testes. It is classified as an **androgen** (*andros* is Greek for male human being) because it stimulates the development of male reproductive structures (see chapter 25) and male secondary sexual characteristics. The testes secrete other androgens, but they are produced in smaller concentrations and are less potent than testosterone. In addition, the testes secrete small amounts of estrogen and progesterone. Small amounts of testosterone are also produced by the adrenal cortex (see chapter 15).

Before birth, testosterone stimulates the development of the male reproductive system and is necessary for the descent of the

Table 24.1 — Major Reproductive Hormones

Hormone	Source	Target Tissue	Response
Males			
Gonadotropin-releasing hormone (GnRH)	Hypothalamus	Anterior pituitary	Stimulates secretion of LH and FSH
Luteinizing hormone (LH) (also called interstitial cell-stimulating hormone [ICSH] in males)	Anterior pituitary	Interstitial cells in the testes	Stimulates synthesis and secretion of testosterone
Follicle-stimulating hormone (FSH)	Anterior pituitary	Seminiferous tubules (sustentacular cells)	Supports spermatogenesis
Testosterone	Leydig cells in the testes	Testes and body tissues	Supports spermatogenesis, stimulates development and maintenance of reproductive organs, causes development of secondary sexual characteristics
		Anterior pituitary and hypothalamus	Inhibits GnRH, LH, and FSH secretion through negative feedback
Females			
Gonadotropin-releasing hormone (GnRH)	Hypothalamus	Anterior pituitary	Stimulates production of LH and FSH
Luteinizing hormone (LH)	Anterior pituitary	Ovaries	Causes follicles to complete maturation and undergo ovulation; causes ovulated follicle to become the corpus luteum; stimulates the primary oocyte to divide
Follicle-stimulating hormone (FSH)	Anterior pituitary	Ovaries	Causes follicles to begin development
Prolactin	Anterior pituitary	Mammary glands	Stimulates milk secretion following parturition
Estrogen	Follicles of ovaries	Uterus	Causes proliferation of endometrial cells
		Mammary glands	Causes development of the mammary glands (especially duct systems)
		Anterior pituitary and hypothalamus	Has a positive-feedback effect before ovulation, resulting in increased LH and FSH secretion: has a negative-feedback effect with progesterone on the hypothalamus and anterior pituitary after ovulation, resulting in decreased LH and FSH secretion
		Other tissues	Causes development of secondary sexual characteristics
Progesterone	Corpus luteum of ovaries	Uterus	Causes hypertrophy of endometrial cells and secretion of fluid from uterine glands
		Mammary glands	Causes development of the mammary glands (especially alveoli)
		Anterior pituitary	Has a negative-feedback effect with estrogen on the hypothalamus and anterior pituitary after ovulation, resulting in decreased LH and FSH secretion
		Other tissues	Causes development of secondary sexual characteristics
Oxytocin*	Posterior pituitary	Uterus and mammary glands	Causes contraction of uterine smooth muscle during intercourse and childbirth; causes contraction of myoepithelial cells in the breast, resulting in milk letdown in lactating women
Human chorionic gonadotropin (HCG)	Placenta	Corpus luteum of ovaries	Maintains the corpus luteum and increases its rate of progesterone secretion during the first one-third (first trimester) of pregnancy; increases testosterone production in testes of male fetuses

*Covered in chapter 25.

testes. During puberty, testosterone promotes the development of the **male primary sexual characteristics,** which include the ability to produce sperm cells and the enlargement and differentiation of the male genitalia. Testosterone also promotes the development of **male secondary sexual characteristics,** which are the structural and behavioral changes, other than in the reproductive organs, that develop at puberty and distinguish adult males from females. Secondary sexual characteristics in males include hair distribution and growth, skin texture, changes in the larynx, increased metabolism, and skeletal muscle and bone development. After puberty, testosterone maintains the adult structure of the male genitalia, accessory sex glands, and secondary sexual characteristics.

Testosterone stimulates hair growth in the following regions: (1) the pubic area and extending up the linea alba, (2) the legs, (3) the chest, (4) the axillary region, (5) the face, and (6) the back. It causes vellus hair to be converted to terminal hair, which is more pigmented and coarser.

Male Pattern Baldness

Some men have a genetic tendency for **male pattern baldness,** which develops in response to testosterone and other androgens. When testosterone levels increase at puberty, the density of the hair on the top of the head begins to decrease. Baldness usually reaches its maximum rate of development when the individual is in the third or fourth decade of life.

Testosterone causes the texture of the skin to become rougher or coarser. The quantity of melanin in the skin also increases, making the skin darker. Testosterone increases the rate of secretion from the sebaceous glands, contributing to the development of acne (see p. 116).

Testosterone causes hypertrophy of the larynx and reduced tension on the vocal folds. The structural changes can first result in a voice that is difficult to control, but ultimately the voice reaches its normal masculine quality.

Testosterone has a general stimulatory effect on metabolism so that males have a slightly higher metabolic rate than females. The red blood cell count is increased by nearly 20% as a result of the effects of testosterone on erythropoietin production. Testosterone promotes protein synthesis in most tissues; as a result, skeletal muscle mass increases at puberty. The average percentage of the body weight composed of skeletal muscle is greater for males than for females because of the effect of androgens.

Synthetic Androgens and Muscle Mass

Some athletes, especially weight lifters, ingest synthetic androgens in an attempt to increase muscle mass. The side effects of large doses of androgens are often substantial and include testicular atrophy, kidney damage, liver damage, heart attack, and stroke. The use of synthetic androgens is highly discouraged by the medical profession and is a violation of the rules of most athletic organizations.

Testosterone causes rapid bone growth and increases the deposition of calcium in bone, resulting in an increase in height. The growth in height is limited, however, because testosterone also stimulates ossification of the epiphyseal plates of the long bones (see chapter 6). Males who mature sexually at an earlier age grow rapidly but reach their maximum height earlier. Males who mature sexually at a later age do not exhibit a rapid period of growth, but they grow for a longer period and can become taller than those who mature sexually at an earlier age. ▼

> *34. Describe the effects of testosterone on male primary and secondary sexual characteristics.*

Male Sexual Behavior and the Male Sexual Act

Testosterone is required to initiate and maintain male sexual behavior. Testosterone enters cells within the hypothalamus and the surrounding areas of the brain and influences their function, resulting in sexual behavior. Male sexual behavior may depend in part, however, on the conversion of testosterone to other steroids, such as estrogen, in the cells of the brain.

P R E D I C T
Predict the effect on male primary and secondary sexual characteristics and sexual behavior if the testes fail to produce normal amounts of testosterone at puberty.

The blood levels of testosterone remain relatively constant throughout the lifetime of a male from puberty until about 40 years of age. Thereafter, the levels slowly decline to about 20% of this value by 80 years of age, causing a slow decrease in sex drive and fertility.

The male sexual act is a complex series of reflexes that result in erection of the penis, secretion of mucus into the urethra, emission, and ejaculation. **Emission** (ē-mish'ŭn, *emitto,* to send out) is the movement of sperm cells, mucus, prostatic secretions, and seminal vesicle secretions into the urethra. **Ejaculation** (ē-jak'ū-lā'shŭn) is the forceful expulsion of the secretions that have accumulated in the urethra to the exterior. Sensations, normally interpreted as pleasurable, occur during the male sexual act and result in an intense sensation called an **orgasm** (ōr'gazm, *orgaō,* to swell or be excited), or **climax.** In males, orgasm is closely associated with ejaculation, although they are separate functions and do not always occur simultaneously. A phase called **resolution** occurs after ejaculation. During resolution the penis becomes flaccid, an overall feeling of satisfaction exists, and the male is unable to achieve erection and a second ejaculation. ▼

> *35. What is the role of testosterone in male sexual behavior?*
> *36. Define emission, ejaculation, orgasm, and resolution.*

Erection, Emission, and Ejaculation

The male sexual act is mediated through the brain and spinal cord. Psychic stimuli, such as sight, sound, odor, or thoughts, can affect the brain and cause a psychic erection. Ejaculation while sleeping (nocturnal emission) is a relatively common event in young males and is thought to be triggered by psychic stimuli associated with dreaming. Psychic stimuli can also inhibit the sexual act, and thoughts that are not sexual tend to decrease the effectiveness of the male sexual act.

Action potentials from the brain affect sexual functions through parasympathetic centers (S2–S4) and sympathetic centers (T10–T12) in the spinal cord. Tactile stimulation can generate action potentials affecting these centers. Rhythmic massage of the penis, especially the glans, and surrounding tissues, such as the scrotal, anal, and pubic regions, is an important source of sensory action potentials that can activate reflexes causing erection, emission, and ejaculation. As a result of this type of stimulation, action potentials also ascend the spinal cord to the cerebrum to produce conscious sexual sensations. Action potentials from the brain then reinforce the spinal cord reflexes. Action potentials from the brain, however, are not absolutely required for the culmination of the male sexual act, which can occur in men with spinal cord injury if the appropriate reflex centers are not damaged.

Erection is the first major component of the male sexual act. Parasympathetic nerves from the sacral region of the spinal cord stimulate the arteries that supply blood to the erectile tissues to dilate. At the same time, other arteries of the penis constrict to shunt blood to the erectile tissues. Blood then fills small venous sinuses called **sinusoids** in the erectile tissue, causing the erectile tissue to expand. Expansion of the erectile tissue compresses veins and reduces blood flow from the penis, which contributes to the development and maintenance of the erection.

The nerve fibers causing dilation of penile blood vessels release acetylcholine and **nitric oxide (NO)** as neurotransmitter substances. Acetylcholine binds to muscarinic receptors and activates a G protein mechanism that causes smooth muscle relaxation. Nitric oxide diffuses into the smooth muscle cells of blood vessels and combines with the enzyme guanylate cyclase, which converts guanosine triphosphate (GTP) to cyclic guanosine monophosphate (cGMP). The cGMP causes smooth muscle cells to relax and blood vessels to dilate (figure 24.9).

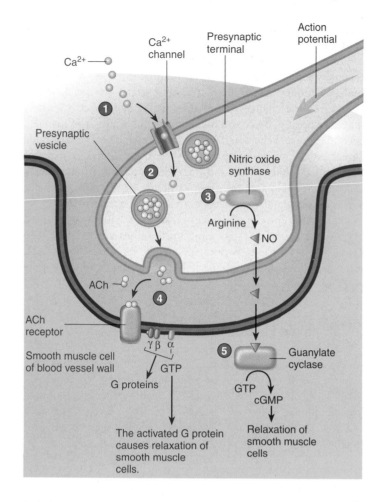

1. Action potentials in parasympathetic neurons cause voltage-gated Ca^{2+} channels to open and Ca^{2+} diffuse into the presynaptic terminals.
2. Calcium ions initiate the release of acetylcholine (ACh) from presynaptic vesicles.
3. Calcium ions also activate nitric oxide synthase, which promotes the synthesis of nitric oxide (NO) from arginine.
4. ACh binds to ACh receptors on the smooth muscle cells and activates a G protein mechanism. The activated G protein causes the relaxation of smooth muscle cells and erection of the penis.
5. NO binds to guanylate cyclase enzymes and activates them. The activated enzymes convert GTP to cGMP, which causes relaxation of the smooth muscle cells and erection of the penis.

Process Figure 24.9 Neural Control of Erection

 Treatment of Erectile Dysfunction

Erectile dysfunction (ED), or **impotence** (im′pŏ-tens, lack of power), is the failure to achieve or maintain an erection. ED can be caused by physical or psychological factors, many of which are treatable. There are many physical causes of ED, including neurological disorders, such as spinal cord and brain injuries; vascular disease or trauma that reduces blood flow to the penis; hormonal disorders resulting in decreased testosterone; aging; side effects of medications; and lifestyle factors, such as alcohol use.

Erection can be achieved in some people by oral medications, such as sildenafil (Viagra), tadalafil (Cialis), or verdenafil (Livitra), or by the injection of specific drugs into the base of the penis. These drugs increase blood flow into the erectile tissue of the penis, resulting in erection for many minutes. Sildenafil blocks the activity of an enzyme that converts cGMP to GMP. This allows cGMP to accumulate in smooth muscle cells in the arteries of erectile tissues and causes them to relax. This response is effective in enhancing erection of the penis in males. Sildenafil's action is not specific to erectile tissue of the penis. It also causes vasodilation in other tissues and can increase the workload of the heart.

Emission is the accumulation of sperm cells and secretions of the prostate gland and seminal vesicles in the urethra. Sympathetic centers (T10–T12) in the spinal cord, which are stimulated as the level of sexual tension increases, control emission. Sympathetic action potentials cause peristaltic contractions of the reproductive ducts and stimulate the seminal vesicles and the prostate gland to release their secretions. Consequently, semen accumulates in the prostatic urethra and produces sensory action potentials that pass through the pudendal nerves to the spinal cord. Integration of these action potentials results in both sympathetic and somatic motor output. Sympathetic action potentials cause constriction of the internal urethral sphincter of the urinary

Causes of Male Infertility

Infertility (in-fer-til′i-tē) is a reduced or absent ability to produce offspring. The most common cause of infertility in males is a low sperm cell count. If the sperm cell count drops to below 20 million sperm cells per milliliter, the male is usually infertile.

A decreased sperm cell count can occur because of damage to the testes as a result of trauma, radiation, cryptorchidism, or infections, such as mumps. A **varicocele** (var′i-kō-sēl) is an abnormal dilation of a vein in the spermatic cord. It can result from incompetent or absent valves in testicular veins, from thrombi, or from tumors. There is a decrease in testicular blood flow and decreased spermatogenesis.

Reduced sperm cell counts can also result from inadequate secretion of luteinizing hormone and follicle-stimulating hormone, which can be caused by hypothyroidism, trauma to the hypothalamus, infarctions of the hypothalamus or anterior pituitary gland, and tumors. Decreased testosterone secretion also reduces the sperm cell count.

Fertility is reduced if the sperm cell count is normal but sperm cell structure is abnormal. Abnormal sperm cell structure can be due to chromosomal abnormalities or abnormal genes. Reduced sperm cell motility also results in infertility. A major cause of reduced sperm cell motility is anti-sperm antibodies, produced by the immune system, which bind to sperm cells.

Fertility can sometimes be achieved by collecting several ejaculations and concentrating the sperm cells. The more concentrated sperm cells can then be introduced into the female's reproductive tract, a process called **artificial insemination** (in-sem-i-nā′shŭn).

Causes of Female Infertility

The causes of infertility in females include malfunctions of the uterine tubes, reduced hormone secretion from the pituitary or ovary, and interruption of implantation. Adhesions from pelvic inflammatory conditions caused by a variety of infections can cause the blockage of one or both uterine tubes and is a relatively common

cause of infertility in women. Reduced ovulation can result from inadequate secretion of LH and FSH, which can be caused by hypothyroidism, trauma to the hypothalamus, infarctions of the hypothalamus or anterior pituitary gland, and tumors. Interruption of implantation may result from uterine tumors or conditions causing abnormal ovarian hormone secretion.

Endometriosis (en′dō-mē-trē-ō′sis), a condition in which endometrial tissue is found in abnormal locations, reduces fertility. Generally, endometriosis is thought to result from some endometrial cells passing from the uterus through the uterine tubes into the pelvic cavity. The endometrial cells invade the peritoneum of the pelvic cavity. Periodic inflammation of the areas where the endometrial cells have invaded occurs because the endometrium is sensitive to estrogen and progesterone. Endometriosis is a cause of abdominal pain associated with menstruation and it can reduce fertility.

---

bladder so that semen and urine are not mixed. Somatic motor action potentials are sent to skeletal muscle surrounding the bulb of the penis, stimulating several rhythmic contractions that cause ejaculation. ▼

> 37. Describe the role of the brain and spinal cord centers in the male sexual act. What are psychic stimuli?
> 38. What changes in blood flow in the penis result in an erection?
> 39. By what mechanisms do acetylcholine and nitric oxide cause vasodilation?
> 40. Describe the process of emission and ejaculation.

Anatomy of the Female Reproductive System

The female reproductive organs consist of the ovaries, uterine tubes, uterus, vagina, external genitalia, and mammary glands. The internal reproductive organs of the female are within the pelvis between the urinary bladder and the rectum (figure 24.10). The uterus and the vagina are in the midline, with the ovaries to each side of the uterus. A group of ligaments holds the internal reproductive organs in place. The most conspicuous is the **broad**

ligament, an extension of the peritoneum that spreads out on both sides of the uterus and to which the ovaries and uterine tubes are attached (figure 24.11). ▼

> 41. List the organs of the female reproductive system. What is the broad ligament?

Ovaries

The two **ovaries** (ō′var-ēz) are small organs about 2–3.5 cm long and 1–1.5 cm wide (see figure 24.11). The **suspensory ligament** extends from each ovary to the lateral body wall, and the **ovarian ligament** attaches the ovary to the superior margin of the uterus. In addition, an extension of the broad ligament supports the ovaries. The ovarian arteries, veins, and nerves traverse the suspensory ligament and enter the ovary. ▼

> 42. Name and describe the ligaments that hold the ovaries in place.

Ovarian Histology

A layer of visceral peritoneum covers the surface of the ovary. Immediately under the visceral peritoneum is a capsule of dense fibrous connective tissue, the **tunica albuginea** (al-bū-jin′ē-ă). The denser, outer part of the ovary is called the **cortex** and the

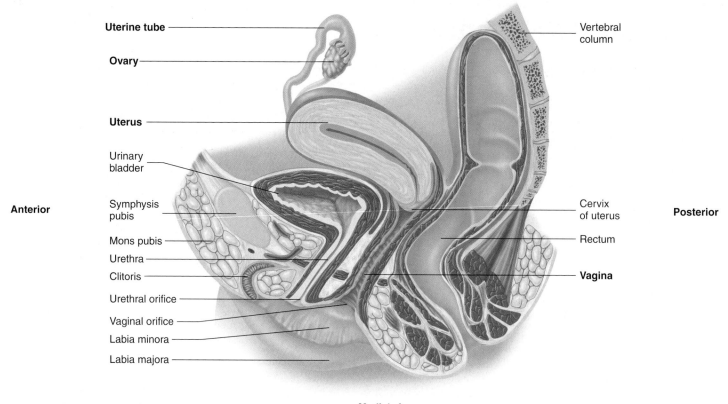

Medial view

Figure 24.10 Sagittal Section of the Female Pelvis

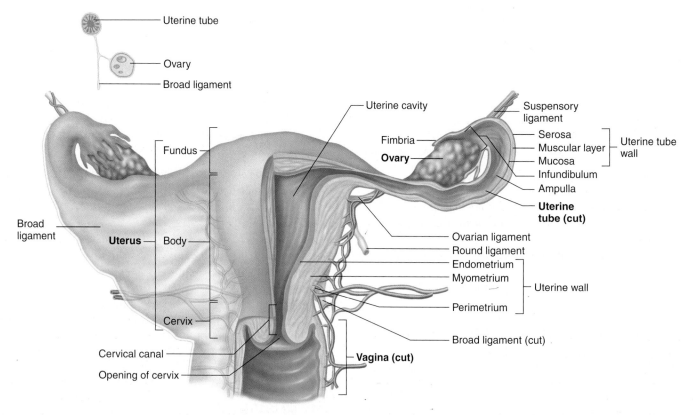

Anterior view

Figure 24.11 Uterus, Vagina, Uterine Tubes, Ovaries, and Supporting Ligaments

The uterus and uterine tubes are cut in section (on the left side), and the vagina is cut to show the internal anatomy. The inset shows the relationships among the ovary, the uterine tube, and the broad ligament.

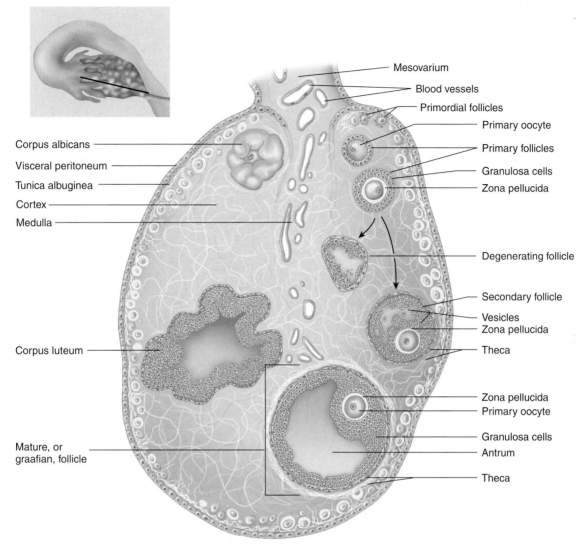

Figure 24.12 Histology of the Ovary

The ovary is sectioned to illustrate its internal structure (the inset shows plane of section). Ovarian follicles from each major stage of development are shown.

looser, inner part is called the **medulla** (figure 24.12). Blood vessels, lymphatic vessels, and nerves enter the medulla. Numerous small vesicles called ovarian follicles, each of which contains an oocyte, are distributed throughout the cortex. ▼

43. Describe the coverings and structure of the ovary.

Oocyte Development and Fertilization

The development of oocytes begins in the fetus. By the fourth month of development, the ovaries contain 5 million **oogonia** (ō-ō-gō′nē-ă, *oon,* egg + *gone,* generation), the cells from which oocytes develop (figure 24.13). Oogonia can form after birth from stem cells, but the extent to which this occurs, and how long it occurs, is not clear. By the time of birth, many of the oogonia have degenerated, and the remaining have begun meiosis and are now called **primary oocytes.** Meiosis stops, however, during the first meiotic division at a stage called prophase I. There are about 2 million primary oocytes at birth, but their numbers decrease to around 300,000 to 400,000 by puberty. Only about 400 primary oocytes will complete development and give rise to the secondary oocytes that are released from the ovaries during ovulation.

Ovulation (ov′ū-lā′shŭn, ō′vū-lā′shŭn) is the release of a **secondary oocyte** from an ovary (see figure 24.13). Just before ovulation, the primary oocyte completes the first meiotic division to produce a secondary oocyte and a **polar body.** The first meiotic division is a reduction division and the secondary oocyte and the polar body each have 23 chromosomes (two chromatids). Unlike meiosis in males, cytoplasm is not split evenly between the two cells. Most of the cytoplasm of the primary oocyte remains with the secondary oocyte. The cytoplasm contains organelles, such as mitochondria, and nutrients that increase the viability of the secondary oocyte. The polar body degenerates or divides to form two polar bodies, which degenerate. The secondary oocyte begins the second meiotic division, but it stops in metaphase II.

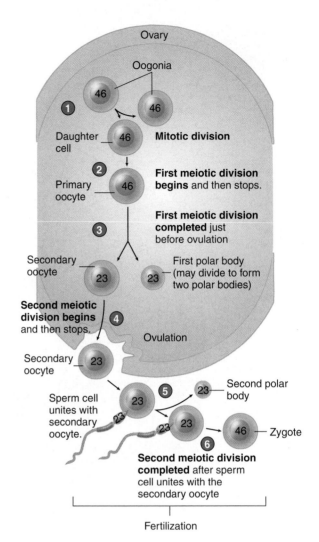

1. Oogonia are the cells from which oocytes arise. The oogonia divide by mitosis to produce other oogonia and primary oocytes.

2. Five million oocytes may be produced by the fourth month of prenatal life. Primary oocytes begin the first meiotic division but stop at prophase I. All of the primary oocytes remain in this state until puberty.

3. The first meiotic division is completed in a single mature follicle just before ovulation during each menstrual cycle. A secondary oocyte and the first polar body result from the unequal division of the cytoplasm.

4. The secondary oocyte begins the second meiotic division but stops at metaphase II.

5. The second meiotic division is completed after ovulation and after a sperm cell unites with the secondary oocyte. A secondary oocyte and a second polar body are formed.

6. Fertilization is completed after the nuclei of the secondary oocyte and the sperm cell unite. The resulting cell is called a zygote.

Process Figure 24.13   Maturation and Fertilization of the Oocyte

After ovulation, the secondary oocyte may be fertilized by a sperm cell (see figure 24.13). **Fertilization** (fer′til-i-zā-shŭn) begins when a sperm cell penetrates into the cytoplasm of a secondary oocyte and ends with the formation of a zygote. After the sperm cell penetrates the secondary oocyte, the secondary oocyte completes the second meiotic division to form two cells, each containing 23 chromosomes derived from the chromatids in the secondary oocyte. One of these cells has very little cytoplasm and is another polar body that degenerates. In the other, larger cell, the 23 chromosomes from the female join with the 23 chromosomes from the sperm cell to complete fertilization. The single cell produced as a result of fertilization is called a **zygote** (zī′gōt, *zygotos,* yolked). The zygote has 23 pairs of chromosomes for a total of 46 chromosomes. One member of each pair came from the father and one member came from the mother. The zygote divides by mitosis to form two cells, which divide to form four cells, and so on. An **embryo** (em′brē-ō) is the developing human between the time of fertilization and 8 weeks of development. Approximately 7 or 8 days after ovulation, the embryo attaches to the wall of the uterus and becomes embedded within it. This process is called **implantation.** The embryo continues development within the uterus and becomes a **fetus** (fē′tus), which is a developing human from 8 weeks to birth (see chapter 25). ▼

44. *Starting with oogonia, describe the formation of secondary oocytes by meiosis. What are polar bodies?*

45. *Define ovulation, fertilization, zygote, implantation, embryo, and fetus.*

46. *How many pairs of chromosomes are in a zygote and where do they come from?*

Follicle Development

A **primordial follicle** is a primary oocyte surrounded by a single layer of flat cells, called **granulosa cells** (figure 24.14). Beginning with puberty, hormonal changes periodically stimulate follicle development. Primordial follicles are converted to **primary follicles** when the oocyte enlarges and the single layer of granulosa cells becomes enlarged and cuboidal. Subsequently, several layers of granulosa cells form and a layer of clear material is deposited around the primary oocyte called the **zona pellucida** (zō′nă pellū′sid-dă, *zone,* girdle + *pellucidus,* passage of light).

1. The primordial follicle consists of a primary oocyte surrounded by a single layer of squamous granulosa cells.

2. A primordial follicle becomes a primary follicle as the granulosa cells become enlarged and cuboidal.

3. The primary follicle enlarges. Granulosa cells form more than one layer of cells. A zona pellucida forms around the primary oocyte.

4. A secondary follicle forms when fluid-filled vesicles (spaces) develop among the granulosa cells and a well-developed theca becomes apparent around the granulosa cells.

5. A mature follicle forms when the fluid-filled vesicles form a single antrum. When a follicle becomes fully mature, it is enlarged to its maximum size, a large antrum is present, and the primary oocyte is surrounded by a mass of granulosa cells. Just prior to ovulation, the first meiotic division is completed to produce a secondary oocyte and a polar body (see figure 24.13).

6. During ovulation the secondary oocyte is released from the follicle. Granulosa cells surrounding the secondary oocyte form the corona radiata.

7. Following ovulation, the granulosa cells divide rapidly and enlarge to form the corpus luteum.

8. When the corpus luteum degenerates, it forms the corpus albicans.

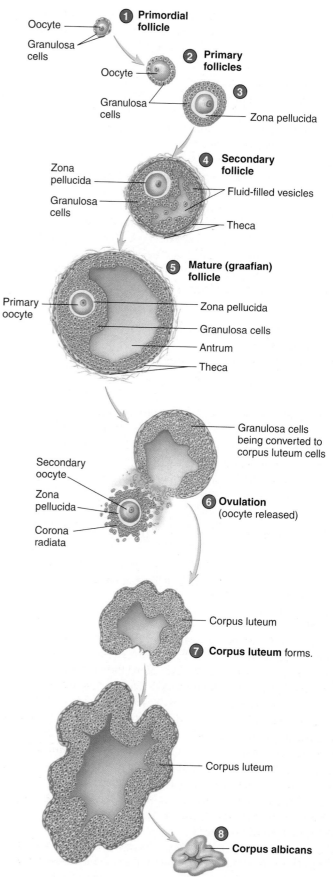

Process Figure 24.14   Maturation of the Follicle and Oocyte

The primary follicle enlarges and becomes a **secondary follicle** as fluid-filled spaces called **vesicles** form among the granulosa cells, and a capsule called the **theca** (thē′kă, a box) forms around the follicle (see figure 24.14).

The secondary follicle continues to enlarge and becomes a **mature,** or **graafian** (graf′ē-ăn, named for Dutch histologist Reijnier de Graaf [1641]), **follicle** when the vesicles fuse to form a single, fluid-filled chamber called the **antrum** (an′trŭm). The primary oocyte is pushed off to one side and lies in a mass of granulosa cells.

The mature follicle forms a lump on the surface of the ovary, and the primary oocyte divides to form a secondary oocyte. During ovulation, the mature follicle ruptures and the secondary oocyte surrounded by some granulosa cells is released into the peritoneal cavity. The granulosa cells resemble a crown radiating from the oocyte and are thus called the **corona radiata.**

In most cases, only one of the follicles that begin to develop forms a mature follicle and undergoes ovulation. The other follicles degenerate. ▼

47. *Distinguish among primordial, primary, secondary, and mature follicles.*
48. *What is released from the ovary by ovulation?*

Fate of the Follicle

After ovulation, the remaining cells of the ruptured follicle are transformed into a glandular structure called the **corpus luteum** (kōr′pŭs, body; loo′tē-ŭm, yellow) (see figure 24.14). If pregnancy occurs, the corpus luteum enlarges and remains throughout pregnancy as the **corpus luteum of pregnancy.** If pregnancy does not occur, the corpus luteum remains functional for about 10–12 days and then degenerates to become the **corpus albicans** (al′bī-kanz, white body). The corpus albicans shrinks and eventually disappears after several months or even years. ▼

49. *What is the corpus luteum? What happens to the corpus luteum if fertilization occurs? If fertilization does not occur?*

Uterine Tubes

There are two **uterine tubes,** also called **fallopian** (fa-lō′pē-an) **tubes** or **oviducts** (ō′vi-dŭkts). Each uterine tube is associated with an ovary and extends to the uterus along the superior margin of the broad ligament (see figure 24.11). The uterine tube opens directly into the peritoneal cavity to receive the oocyte from the ovary. The **infundibulum** (in-fŭn-dib′ū-lŭm, funnel) is the funnel-shaped end of the uterine tube. The opening of the infundibulum is surrounded by long, thin processes called **fimbriae** (fim′brē-ē, fringe). The inner surfaces of the fimbriae consist of a ciliated mucous membrane. The part of the uterine tube that is nearest the infundibulum is called the **ampulla.** It is the widest and longest part of the tube and accounts for about 7.5–8 cm of the total 10 cm length of the tube.

The wall of each uterine tube consists of three layers (see figure 24.11). The outer **serosa** is formed by the visceral peritoneum, the middle **muscular layer** consists of longitudinal and circular smooth muscle cells, and the inner **mucosa** is a mucous membrane of simple ciliated columnar epithelium. Muscular contractions and

movement of the cilia move the oocyte, or the developing embryo, through the uterine tubes toward the uterus. ▼

50. *Describe the parts of the uterine tube and the three layers of the uterine wall.*
51. *How are the uterine tubes involved in moving the oocyte or the zygote?*

Uterus

The **uterus** (ū′ter-ŭs) is the size and shape of a medium-sized pear—about 7.5 cm long and 5 cm wide (see figures 24.10 and 24.11). It is oriented in the pelvic cavity, with the larger, rounded part directed superiorly. The part of the uterus superior to the entrance of the uterine tubes is the **fundus** (fŭn′dŭs), and the main part of the uterus is the **body.** The **cervix** (ser′viks, neck) is the narrower, inferior part of the uterus. Internally, the **uterine cavity** in the fundus and uterine body continues through the cervix as the **cervical canal,** which opens into the vagina.

Disorders of the Uterus

Uterine **leiomyomas** (lī′-ō-mī′măs), also called uterine fibroids, are enlarged masses of smooth muscle in the myometrium, and they are one of the most common disorders of the uterus. They are the most frequent tumor in women, affecting one of every four. Three-fourths of the women with this condition, however, experience no symptoms. The enlarged mass compresses the uterine lining (endometrium), resulting in ischemia, inflammation, and frequent and severe menstruations. Abdominal cramping because of strong uterine contractions can be present. Constant menstruation is a frequent manifestation of these tumors, and it is one of the most common reasons that women elect to have the uterus removed, a procedure called a **hysterectomy** (his-ter-ek′tō-mē).

Cancer of the cervix is relatively common and fortunately can be detected and treated. Early in the development of cervical cancer, the cells of the cervix change in a characteristic way. This change can be observed by examining a cell sample microscopically. The most common technique is to obtain a **Papanicolaou (Pap) smear,** which is named after a U.S. physician who developed the technique. Pap smears have a reliability of 90% for detecting cervical cancer. A cervical cancer vaccine (Gardasil) prevents cervical cancer by providing protection against infection by the human papillomavirus, which can cause cancer and genital warts.

The major ligaments holding the uterus in place are the **broad ligament,** the **round ligaments** (see figure 24.11), and the **uterosacral ligaments.** The broad ligament is a peritoneal fold extending from the lateral margin of the uterus to the wall of the pelvis on either side. It also ensheaths the ovaries and the uterine tubes. The round ligaments extend from the uterus through the inguinal canals to the labia majora, and the uterosacral ligaments attach the lateral wall of the uterus to the sacrum. In addition to these ligaments that support the uterus, much support is provided inferiorly to the uterus by skeletal muscles of the pelvic floor. If ligaments that support the uterus or muscles of the pelvic floor are weakened, such as in childbirth, the uterus can extend inferiorly into the vagina, a condition called a **prolapsed uterus.** Severe cases require surgical correction.

The uterine wall is composed of three layers: the perimetrium, myometrium, and endometrium (see figure 24.11). The outer **perimetrium** (per-i-mē′trē-ŭm), or **serous layer,** is visceral peritoneum. The middle **myometrium** (mī′ō-mē′trē-ŭm), or **muscular layer,** consists of a thick layer of smooth muscle. The myometrium accounts for the bulk of the uterine wall and is the thickest layer of smooth muscle in the body. The inner **endometrium** (en′dō-mē′trē-ŭm), or **mucous membrane,** consists of a simple columnar epithelial lining and connective tissue. Simple tubular glands, called endometrial glands, are formed by folds of the endometrium. The superficial part of the endometrium is sloughed off during menstruation.

Columnar epithelial cells line the cervical canal, which contains **cervical mucous glands.** The mucus fills the cervical canal and acts as a barrier to substances that could pass from the vagina into the uterus. Near ovulation, the consistency of the mucus changes, making the passage of sperm cells from the vagina into the uterus easier. ▼

> 52. *Name the parts and cavities of the uterus.*
> 53. *Name the major ligaments holding the uterus in place.*
> 54. *Describe the layers of the uterine wall.*

Vagina

The **vagina** (vă-jī′nă) is the female organ of copulation and receives the penis during intercourse. It also allows menstrual flow and childbirth. The vagina is a tube about 10 cm long that extends from the uterus to the outside of the body (see figures 24.10 and 24.11). The superior, domed part of the vagina, the **fornix** (fōr′niks, domed), is attached to the sides of the cervix so that a part of the cervix extends into the vagina.

The wall of the vagina consists of an outer, muscular layer and an inner mucous membrane. The muscular layer is smooth muscle and contains many elastic fibers. Thus, the vagina can increase in size to accommodate the penis during intercourse, and it can stretch greatly during childbirth. The mucous membrane is moist stratified squamous epithelium that forms a protective surface layer. Lubricating fluid passes through the vaginal epithelium into the vagina during intercourse.

The **vaginal orifice** is the opening of the vagina to the outside. In young females, the vaginal orifice is covered by a thin mucous membrane called the **hymen** (hī′men, membrane). The hymen can completely close the vaginal orifice, in which case it must be removed to allow menstrual flow. More commonly, the hymen is perforated by one or several holes. The openings in the hymen are usually greatly enlarged during the first sexual intercourse. The hymen can also be perforated or torn at some earlier time in a young female's life during a variety of activities, including strenuous exercise. The condition of the hymen is therefore not a reliable indicator of virginity. ▼

> 55. *Where is the vagina located?*
> 56. *Describe the layers of the vaginal wall.*
> 57. *What is the hymen?*

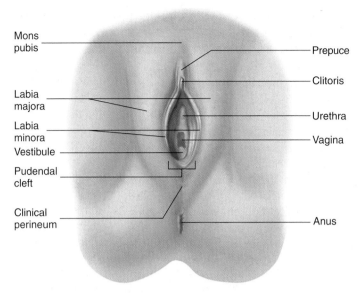

Figure 24.15 Female External Genitalia

External Genitalia

The **female genitalia** can be divided into the internal and external female genitalia. The **internal female genitalia** are the ovaries, uterine tubes, uterus, and vagina. The **external female genitalia** are also called the **external female genitals, vulva** (vŭl′vă, a wrapper or covering), or **pudendum** (pū-den′dŭm, to feel ashamed). The external female genitalia consist of the vestibule and its surrounding structures (figure 24.15). The **vestibule** (ves′ti-bool, entrance court) is the space into which the vagina and urethra open. The urethra opens just anterior to the vagina. The vestibule is bordered by a pair of thin, longitudinal skin folds called the **labia minora** (lā′bē-ă, lips; mī-nō′ră, small). A small, erectile structure called the **clitoris** (klit′ŏ-ris, klī′tŏ-ris) is located in the anterior margin of the vestibule. The two labia minora unite over the clitoris to form a fold of skin called the **prepuce.**

The clitoris is usually less than 2 cm in length and consists of a shaft and a distal glans. Well supplied with sensory receptors, it initiates and intensifies levels of sexual tension. The clitoris contains two erectile structures, the **corpora cavernosa.** The bases of the corpora cavernosa expand to form the **crura of the clitoris,** which attach the clitoris to the coxal bones. The corpora cavernosa of the clitoris are comparable to the corpora cavernosa of the penis, and they become engorged with blood as a result of sexual excitement. In most women, this engorgement results in an increase in the diameter, but not the length, of the clitoris. With increased diameter, the clitoris makes better contact with the prepuce and surrounding tissues and is more easily stimulated.

Erectile tissue that corresponds to the corpus spongiosum of the male lies deep to the vaginal orifice along the lateral sides of the vagina. Each erectile body is called a **bulb of the vestibule.** Like other erectile tissue, it becomes engorged with blood and is more sensitive during sexual arousal. Expansion of the bulbs causes narrowing of the vaginal orifice and produces better contact of the vagina with the penis during intercourse.

Fibrocystic Changes in the Breast

Fibrocystic changes in the breast are benign changes. They include the formation of fluid-filled cysts, hyperplasia of the duct system of the breast, and the deposition of fibrous connective tissue. These changes occur in approximately 10% of women who are less than 21 years of age, 25% of women in their reproductive years, and 50% of women who are postmenopausal. The cause of the condition is not known. Major manifestations are breast pain, especially during the secretory phase of the menstrual cycle and continuing until menstruation. Some evidence suggest that some women with certain types of duct hyperplasia, when associated with a family history of breast cancer, have an increased likelihood of developing breast cancer.

Breast Cancer

Breast cancer is a serious, often fatal, disease in women. It is the most common cancer in North American women. Greater than 75% of the breast cancer cases occur in women older than 50 years of age, and 85% involve cancer of the epithelium of the ducts of the mammary glands (ductal carcinoma).

The younger a woman's age when her first child is born, the lower her risk of developing breast cancer. However, the risk for breast cancer is not lower in young women who become pregnant but whose pregnancies do not go full-term. Also, women who have never given birth are at a greater risk than those who have given birth. It appears that differentiation of the breast epithelium caused by the first pregnancy decreases the risk for cancer because it reduces the amount of undifferentiated epithelium in the breast.

A reduced cancer risk because of early menopause, or removal of the ovaries, is probably due to the decreased exposure of the breast epithelium to estrogen and progesterone. An increased duration of the use of postmenopausal estrogen and progesterone is correlated with an increased incidence of cancer. There is a decrease in breast cancer incidence with the use of drugs that block the effect of estrogen receptors.

A number of environmental factors have been associated with the development of breast cancer. Exposure to ionizing radiation, especially during adolescence and during pregnancy when epithelial cells are dividing more rapidly than at other times, is correlated with breast cancer. High fat intake and obesity are also correlated with breast cancer. A number of environmental factors mimic estrogen, such as components of plastics, fuels, pharmaceuticals, and chlorine-based chemicals (such as DDT, PCBs, and chlorofluorocarbons), and these may increase the incidence of breast cancer.

On each side of the vestibule, between the vaginal orifice and the labia minora, is an opening of the duct of the **greater vestibular gland.** Additional small mucous glands—the **lesser vestibular glands,** or **paraurethral glands**—are located near the clitoris and urethral opening. They produce a lubricating fluid that helps maintain the moistness of the vestibule.

Lateral to the labia minora are two prominent, rounded folds of skin called the **labia majora** (see figure 24.15). Subcutaneous fat is primarily responsible for the prominence of the labia majora. The two labia majora unite anteriorly in an elevation over the symphysis pubis called the **mons pubis** (monz, mound, pū′bis). The lateral surfaces of the labia majora and the surface of the mons pubis are covered with coarse hair. The medial surfaces are covered with numerous sebaceous and sweat glands. The space between the labia majora is called the **pudendal** (pū-den′dăl) **cleft.** Most of the time, the labia majora are in contact with each other across the midline, closing the pudendal cleft and covering the deeper structures within the vestibule. ▼

58. *What are the female genitalia?*
59. *What erectile tissue is in the clitoris and bulb of the vestibule?*
60. *What is the function of the clitoris and bulb of the vestibule?*
61. *Describe the labia minora, prepuce, labia majora, pudendal cleft, and mons pubis.*
62. *Where are the greater and lesser vestibular glands located? What is their function?*

Perineum

The region between the vagina and the anus is the **clinical perineum** (per′i-nē′um, area between the thighs) (see figure 24.15). The skin and muscle of this region can tear during childbirth. To prevent such tearing, an incision called an **episiotomy** (e-piz-ē-ot′ō-mē, pubic region + tōme, incision) is sometimes made in the clinical perineum. Traditionally, this clean, straight incision was thought to result in less injury, less trouble in healing, and less pain. However, many studies indicate that there is less injury and pain when no episiotomy is performed. ▼

63. *What is the clinical perineum?*
64. *Define and give the purpose of an episiotomy.*

Mammary Glands

The **mammary glands** are modified sweat glands that produce milk. The mammary glands are located within the **mammae** (mam′ē), or **breasts** (figure 24.16). Externally, the breasts of both males and females have a raised **nipple** surrounded by a circular, pigmented **areola** (ă-rē′ō-lă). The areolae normally have a slightly bumpy surface caused by the presence of modified sweat glands called **areolar glands.** Secretions from these glands protect the nipple and the areola from chafing during nursing. The nipples contain smooth muscle; when the smooth muscle contracts, the nipple becomes erect. The smooth muscle contracts in response to stimuli, such as touch, cold, and sexual arousal.

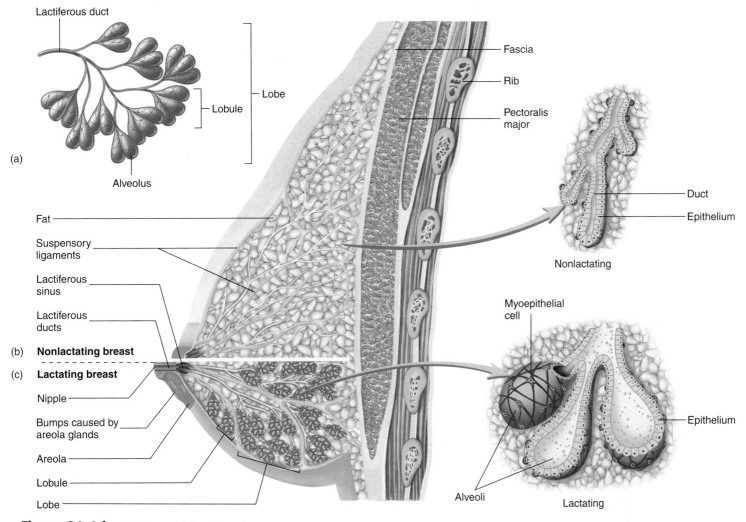

(a)

Lactiferous duct

Lobe

Lobule

Alveolus

Fat

Suspensory ligaments

Lactiferous sinus

Lactiferous ducts

(b) **Nonlactating breast**

(c) **Lactating breast**

Nipple

Bumps caused by areola glands

Areola

Lobule

Lobe

Fascia

Rib

Pectoralis major

Duct

Epithelium

Nonlactating

Myoepithelial cell

Epithelium

Alveoli

Lactating

Figure 24.16 Anatomy of the Breast

(a) Lactiferous ducts divide to supply lobules, which form lobes. (b) In the nonlactating breast, only the duct system is present. (c) In the milk-producing breast, the ends of the ducts of the mammary glands have secretory sacs, called alveoli, which produce milk. Surrounding the alveoli are myoepithelial cells, which can contract, causing the movement of milk out of the alveoli.

In prepubescent children, the general structure of the male and female breasts is similar, and both males and females possess a rudimentary duct system. The female breasts begin to enlarge during puberty, under the influence of estrogen and progesterone. Some males also experience a minor and temporary enlargement of the breasts at puberty. The breasts of a male can become permanently enlarged, however, a condition called **gynecomastia** (gī′nĕ-kō-mas′tē-ă). Causes of gynecomastia include hormonal imbalances and the abuse of anabolic steroids.

Each adult female mammary gland usually consists of 15–20 glandular **lobes** covered by a considerable amount of adipose tissue. It is primarily this superficial fat that gives the breast its form. The lobes are subdivided into **lobules.** In the milk-producing breast, milk-producing sacs called **alveoli** form at the end of ducts in the lobules (see figure 24.16a and c). **Myoepithelial cells** surrounding the alveoli can contract to expel milk from them. In nonmilk-producing breasts, the lobule

ducts end without alveoli (see figure 24.16b). The small lobule ducts converge to form a single duct that exits each lobule. Ducts from the lobules converge to form a single **lactiferous duct** exiting each lobe. Each lactiferous duct opens independently to the surface of the nipple. A **lactiferous sinus** is a spindle-shaped expansion of the lactiferous duct that can accumulate milk when it is produced.

The breasts are supported by the **suspensory ligaments,** which extend from the fascia over the pectoralis major muscles to the skin over the breasts and prevent them from excessive sagging (see figure 24.16b). In older adults, the suspensory ligaments can weaken and elongate, increasing the tendency for the breasts to sag. ▼

65. *Describe the anatomy of the mammary glands.*
66. *Describe the route taken by a drop of milk from its site of production to the outside of the body.*
67. *What are suspensory ligaments?*

Physiology of Female Reproduction

Female reproduction is under the control of hormonal and nervous regulation.

Puberty

The first signs of puberty typically appear between 11 and 13 years of age in girls, and the process is largely completed by age 16. Puberty in females is marked by the first episode of menstrual bleeding, which is called **menarche** (me-nar′kē, *mēn,* month + *archē,* beginning).

Before puberty, GnRH, LH, FSH, estrogen, and progesterone are secreted in very small amounts. Estrogen and progesterone from the ovaries have a strong negative-feedback effect on the hypothalamus and pituitary gland. At puberty, not only are GnRH, LH, and FSH secreted in greater quantities than before puberty but also the adult pattern is established, in which a cyclic pattern of hormone secretion occurs.

During puberty, increased estrogen and progesterone promote the development of the **female primary sexual characteristics,** which include the ability to produce oocytes and the enlargement and differentiation of the female genitalia. Hormones also promote the development of **female secondary sexual characteristics,** which are the structural and behavioral changes, other than in the reproductive organs, that develop at puberty and distinguish adult females from males. For example, fat is deposited in the breasts and around the hips, the pelvis widens, and the sex drive develops. Pubic and axillary hair forms and terminal hair develops in other parts of the body, but to a lesser extent than in males. ▼

> 68. *Define menarche. What changes in hormone secretion occur during puberty?*
> 69. *Describe the female primary and secondary sexual characteristics.*

Menstrual Cycle

The **menstrual cycle** consists of the periodic changes occurring in the ovaries and uterus of a sexually mature, nonpregnant female that result in the production of a secondary oocyte and prepare the uterus for implantation. Typically, the menstrual cycle is about 28 days long, although it can be as short as 18 days in some women and as long as 40 days in others (figure 24.17).

The menstrual cycle is subdivided into phases based on the timing of events in the uterus. Day 1 of the menstrual cycle is marked by **menstruation** (men-stroo-ā′shŭn), which is the discharge of blood and part of the endometrium from the uterus. Menstruation is also called **menses** and **menstrual bleeding.** The **menstrual phase** is the time between the beginning and the end of menstruation, which is typically days 1–5 of the cycle.

Ovulation occurs on about day 14 of a 28–day menstrual cycle, although the timing of ovulation varies from individual to individual and varies within a single individual from one menstrual cycle to the next. The **proliferation phase** is the time between the end of menstruation and ovulation. During the proliferation phase, the number of cells in the endometrium rapidly increases and the endometrium becomes thicker. The **secretory phase** is the time between ovulation and the beginning of the next menstruation. During the secretory phase, the endometrium produces secretions and becomes thicker.

Gonadotropin-releasing hormone produced by the hypothalamus stimulates the release of FSH and LH from the anterior pituitary (see figure 24.17). These hormones affect follicle development, ovulation, and corpus luteum development in the ovaries. The follicles and corpus luteum produce estrogen and progesterone, which regulate changes in the endometrium in the uterus (table 24.2). ▼

> 70. *Define the menstrual cycle.*
> 71. *What is the length of a typical menstrual cycle? On which day does ovulation occur?*
> 72. *Define the phases of the menstrual cycle.*

Menstrual Phase

FSH from the anterior pituitary is mainly responsible for initiating the development of primary follicles, and as many as 25 begin to mature during each cycle (see figure 24.17). The follicles that start to develop in response to FSH may not ovulate during the same cycle in which they begin to mature, but they may ovulate one or two cycles later.

Menstruation begins because of a decrease in estrogen and progesterone production in the previous cycle. Decreased blood delivery to the endometrium results in cell death, the breakdown of the superficial endometrium, and some bleeding from the damaged tissue. Uterine contractions expel the menstrual fluid from the uterus through the cervix and into the vagina. ▼

> 73. *How many follicles normally start to mature during a cycle? What keeps them all from maturing?*

Proliferative Phase and Ovulation

Early in the proliferative phase, the release of GnRH from the hypothalamus increases, which stimulates the production and release of a small amount of FSH and LH by the anterior pituitary. FSH and LH stimulate follicular growth and estrogen secretion. LH stimulates the thecal cells to produce androgens, which diffuse from these cells to the granulosa cells. FSH stimulates the granulosa cells to convert the androgens to estrogen.

Estrogen produced by the maturing follicles stimulates cell divisions in the endometrium. The epithelial cells remaining after menstruation rapidly divide and replace the cells that were sloughed during the previous cycle. A relatively uniform layer of low cuboidal endometrial cells is produced. It later becomes columnar and is folded to form tubular **spiral glands.** Blood vessels called **spiral arteries** project through the delicate connective tissue that separates the individual spiral glands to supply nutrients to the endometrial cells (see figure 24.17). Estrogen also makes the uterus more sensitive to progesterone by stimulating the synthesis of progesterone receptors within the uterine cells.

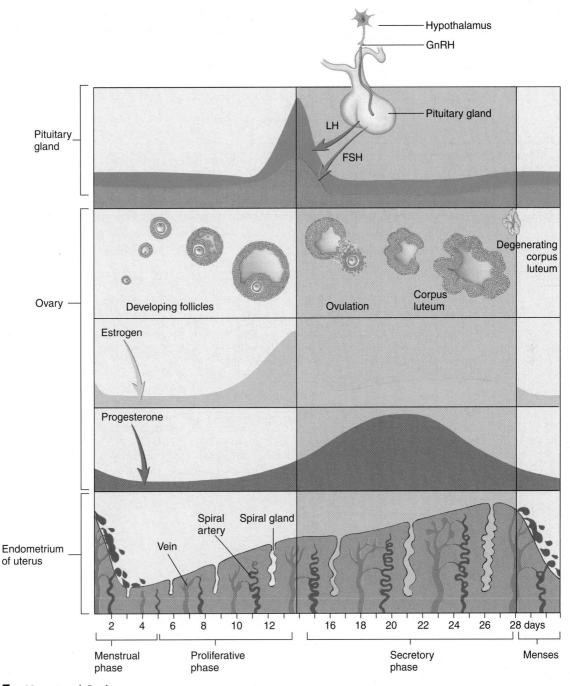

Figure 24.17   Menstrual Cycle

The various lines depict the changes in blood hormone levels, the development of the follicles, and the changes in the endometrium during the cycle.

As estrogen levels increase, they have a negative-feedback effect on the secretion of LH and FSH by the anterior pituitary. Inhibin produced by the follicles also has a negative-feedback effect on FSH secretion. When estrogen levels increase to a critical level, however, they begin to have a positive-feedback effect on LH and FSH release from the anterior pituitary. Consequently, LH and FSH secretion increases rapidly and in large amounts just before ovulation. The rapid increase in blood levels of LH is called the **LH surge,** and the increase in FSH is called the **FSH surge**

(see figure 24.17). The LH surge causes the primary oocyte to complete the first meiotic division and become a secondary oocyte, stimulates ovulation, and causes the ovulated follicle to become the corpus luteum. The FSH surge initiates the development of follicles that develop and may ovulate during later ovarian cycles. Although several follicles begin to mature during each cycle, normally only one is ovulated. The remaining follicles degenerate. Larger and more mature follicles appear to secrete estrogen and other substances that have an inhibitory effect on

Table 24.2   Events During the Menstrual Cycle

Menses (Day 1 to Day 5 of the Menstrual Cycle)

Pituitary gland	The rate of FSH and LH secretion is low.
Ovary	The rate of estrogen and progesterone secretion is low after degeneration of the corpus luteum produced during the previous menstrual cycle.
Uterus	In response to declining progesterone levels, the endometrial lining of the uterus sloughs off, resulting in menses followed by repair of the endometrium.

Proliferative Phase (Day 5 Until Ovulation on Day 14)

Pituitary gland	The rate of FSH and LH secretion is low during most of the proliferative phase; FSH and LH secretion increases near the end of the proliferative phase in response to increasing estrogen secretion from the ovaries.
Ovary	Developing follicles secrete increasing amounts of estrogen, especially near the end of the proliferative phase; increasing FSH and LH levels cause additional estrogen secretion from the ovaries near the end of the proliferative phase.
Uterus	Estrogen causes endometrial cells of the uterus to divide. The endometrium of the uterus thickens and spiral glands and spiral arteries form. Estrogen causes the cells of the uterus to be more sensitive to progesterone by increasing the number of progesterone receptors in uterine tissues.

Ovulation (Day 14)

Pituitary gland	The rate of FSH and LH secretion increases rapidly just before ovulation in response to increasing estrogen levels. Increasing FSH and LH levels stimulate estrogen secretion, resulting in a positive-feedback cycle.
Ovary	LH stimulates division of the primary oocyte, causes final maturation of a mature follicle, and initiates the process of ovulation. FSH acts on immature follicles and causes several of them to begin to enlarge.
Uterus	The endometrial cells continue to divide in response to estrogen.

Secretory Phase (Day 14 to Day 28)

Pituitary gland	Estrogen and progesterone reach levels high enough to inhibit FSH and LH secretion from the pituitary gland.
Ovary	After ovulation, the follicle is converted to the corpus luteum; the corpus luteum secretes large amounts of progesterone and smaller amounts of estrogen from shortly after ovulation until about day 24 or 25. If fertilization does not occur, the corpus luteum degenerates after about day 25, and the rate of progesterone secretion rapidly declines to low levels.
Uterus	In response to progesterone, the endometrial cells enlarge, the endometrial layer thickens, and the glands of the endometrium reach their greatest degree of development; the endometrial cells secrete a small amount of fluid. After progesterone levels decline, the endometrium begins to degenerate.

other, less mature follicles. Estrogen levels are at their highest value at ovulation. ▼

74. *Describe the role of FSH and LH in estrogen production. What effect does estrogen have on the endometrium?*
75. *How are the LH and FSH surges produced? List the effects of these hormones.*

Secretory Phase and the Next Cycle

Shortly after ovulation, the follicle's production of estrogen decreases (see figure 24.17). The remaining granulosa and thecal cells of the ovulated follicle are converted to corpus luteum cells and begin to secrete progesterone and some estrogen. After the corpus luteum forms, progesterone levels become much higher than before ovulation. The increased progesterone and estrogen have a negative-feedback effect on GnRH release from the hypothalamus. As a result, GnRh, FSH, and LH decrease to low levels after ovulation.

The increased progesterone and estrogen stimulate the endometrium, which thickens, becomes more vascular, and produces secretions. Estrogen causes the endometrial cells and, to a lesser degree, the myometrial cells to divide and increase in number. Progesterone binds to progesterone receptors, which have increased in number as a result of estrogen stimulation. Progesterone causes endometrial and myometrial cells to increase in size and causes the endometrial cells to become secretory. The secretions are rich in glycogen and lipids and are a source of nourishment for the embryo. Estrogen increases the tendency of the smooth muscle cells of the uterus to contract in response to stimuli, but progesterone inhibits smooth muscle contractions. When progesterone levels increase while estrogen levels are low, contractions of the uterine smooth muscle are reduced.

If fertilization and implantation do not occur, the cells of the corpus luteum begin to atrophy after day 25 or 26, and the blood levels of progesterone and estrogen decrease rapidly (see figure 24.17). As a consequence, the uterine lining begins to degenerate as the spiral arteries constrict, but occasionally they dilate. Constriction of the spiral arteries deprives cells of their blood supply. The cells die and all but the most basal part of the endometrium breaks up and is sloughed into the uterine lumen. Dilation of the spiral arteries releases small amounts of blood. The sloughed endometrium, mucous secretions, and a small amount of blood make up the menstrual fluid. Decreases in progesterone levels and increases in inflammatory substances that stimulate myometrial smooth muscle cells cause uterine contractions, which expel the menstrual fluid from the uterus. Thus, the next cycle begins. The decrease in progesterone and estrogen also decreases the inhibition of the hypothalamus and anterior pituitary, so the cycle of hormone secretion begins again.

If fertilization and implantation occur, the developing placenta begins to secrete the LH-like substance human chorionic gonadotropin (hCG), which keeps the corpus luteum from degenerating (see chapter 25). As a result, blood levels of estrogen and progesterone do not decrease, and menses does not occur. ▼

76. Describe the effects of increased and decreased progesterone and estrogen on the endometrium during the secretory phase.
77. Where is hCG produced, and what effect does it have on the ovary?

CASE STUDY | Oral Contraceptives

Mary and Sam are 20-year-olds attending college. They plan to marry in May and return to school to complete their senior year. Mary and Sam want very much to have children, but not before they graduate and become established in their careers. After discussing their options with their physician, they decide that oral contraception will work best for them. Their physician recommends a combined oral contraceptive pill consisting of synthetic progesterone and estrogen. There are 28 pills in a package, of which 21 pills contain progesterone and estrogen and 7 pills are placebo, or sugar, pills.

PREDICT 4

Refer to figure 24.17 to answer these questions.
a. How does taking progesterone and estrogen prevent pregnancy?
b. Why are progesterone and estrogen taken for only 21 days?
c. Since the sugar pill has no pharmacological effect, what is the reason for taking a pill every day?
d. Before the wedding, Mary has to decide when to start taking the pill. The two alternatives are the first day of her period or the first Sunday after her period starts. Why should she start at, or near, the beginning of her cycle? What difference does it make if she is a first-day starter or a Sunday starter?

Menstrual Cramps and Amenorrhea

Menstrual cramps are the result of strong contractions of the myometrium occurring before and during menstruation. In some women, menstrual cramps are extremely uncomfortable. Prostaglandins, produced as part of an inflammatory response when the endometrium breaks down, can stimulate smooth muscle contractions. Many women can alleviate painful menstruation by taking drugs, such as aspirinlike drugs, that inhibit prostaglandin synthesis just before the onset of menstruation.

The absence of a menstrual cycle is called **amenorrhea** (ă-men-ō-rē′ă, without menses). If the pituitary gland does not function properly because of abnormal development, the woman will not begin to menstruate at puberty. This condition is called **primary amenorrhea.** In contrast, if a woman has had normal menstrual cycles and later stops menstruating, the condition is called **secondary amenorrhea.** Anorexia can cause secondary amenorrhea because lack of food causes the hypothalamus of the brain to decrease GnRH secretion to levels so low that the menstrual cycle cannot occur. Many female athletes, such as ballet dancers and gymnasts, who have rigorous training schedules coupled with inadequate food intake have secondary amenorrhea. Head trauma and tumors that affect the hypothalamus can also result in lack of GnRH secretion.

Secondary amenorrhea can be also be caused by disorders of the pituitary gland and ovaries. For example, pituitary tumors decrease FSH and LH secretion; autoimmune diseases attack the ovaries, causing decreased estrogen and progesterone secretion; and **polycystic ovarian disease** produces large amounts of androgens that suppress the menstrual cycle.

Female Sexual Behavior and the Female Sexual Act

Sex drive in females, like sex drive in males, depends on hormones. The adrenal gland, liver, and other tissues convert steroids, such as progesterone, to androgens. Androgens and possibly estrogen affect cells in the brain, especially in the hypothalamus, to influence sexual behavior. Androgens and estrogen alone do not control sex drive, however. For example, sex drive cannot be predictably increased simply by injecting these hormones into healthy women or men. Psychological factors also affect sexual behavior. For example, after removal of the ovaries or after menopause, many women report having an increased sex drive because they no longer fear pregnancy.

The sensory and motor neural pathways involved in controlling female sexual responses are similar to those found in the male. During sexual excitement, erectile tissue within the clitoris and around the vaginal orifice becomes engorged with blood as a result of parasympathetic stimulation. The mucous glands within the vestibule, especially the vestibular glands, secrete small amounts of mucus. Large amounts of mucuslike fluid are also extruded into the vagina through its wall, although no well-developed mucous glands are within the vaginal wall. These secretions provide

Many methods are used to prevent or terminate pregnancy (figure A), including methods that prevent fertilization (contraception), prevent implantation of the developing embryo (IUDs), or remove the implanted embryo or fetus (abortion). Many of these techniques are quite effective when done properly and used consistently (table A).

Behavioral Methods

Abstinence, or refraining from sexual intercourse, is a sure way to prevent pregnancy when practiced consistently. It is not an effective method when used only occasionally.

Coitus interruptus (kō′i-tŭs int-ĕ-rŭp′tŭs), or **withdrawal,** is removal of the penis from the vagina just before ejaculation. This is a very unreliable method of preventing pregnancy because it requires perfect awareness and willingness to withdraw the penis at the correct time. It also ignores the fact that some sperm cells are found in preejaculatory emissions.

The **rhythm method** requires abstaining from sexual intercourse near the time of ovulation. A major factor in the success of this method is the ability to predict accurately the time of ovulation. Although the rhythm method provides some protection against becoming pregnant, it has a relatively high rate of failure because of both the inability to predict the time of ovulation and the failure to abstain from intercourse around the time of ovulation.

Barrier Methods

A **condom** (kon′dom) is a sheath of animal membrane, rubber, or latex. Placed over the erect penis, it is a barrier device because the semen is collected within the condom instead of within the vagina. Condoms also provide protection against sexually transmitted diseases. A **vaginal condom** also acts as a barrier device. The vaginal condom can be placed into the vagina by the woman before sexual intercourse.

Methods to prevent sperm cells from reaching the oocyte once they are in the vagina include the use of a diaphragm, cervical cap, and spermicidal agents. The **diaphragm** and the **cervical cap** are flexible latex domes that are placed over the cervix within

Table A — Effectiveness of Various Methods for Preventing Pregnancy

Technique	Optimal Effectiveness (%)
Abortion	100
Sterilization	100
Combination (estrogen and progesterone) pill	99.9
Intrauterine device	98
Minipill (low dose of estrogen and progesterone)	99
Condom plus spermicide	99
Condom alone	97
Diaphragm plus spermicide	97
Foam	97
Rhythm	97

the vagina, where they prevent the passage of sperm cells from the vagina through the cervical canal of the uterus. The diaphragm is a larger, shallow latex cup and the cervical cap is a smaller, thimble-shaped latex cup. The most commonly used **spermicidal agents** are foams or creams that kill sperm cells. They are inserted into the vagina before sexual intercourse. **Spermicidal douches** (dūsh′ez), which remove and kill sperm cells, are sometimes used. Spermicidal douches used alone are not very effective.

Lactation

Lactation (lak-tā′shŭn) often prevents the menstrual cycle for a few months after childbirth. Action potentials sent to the hypothalamus in response to suckling inhibit GnRH release from the hypothalamus. Reduced GnRH reduces LH, which prevents ovulation. Despite continual lactation, the menstrual cycle eventually resumes. Relying on lactation to prevent pregnancy is not consistently effective because ovulation normally precedes menstruation.

Chemical Methods

Synthetic estrogen and progesterone in **oral contraceptives** (birth control pills) effectively suppress fertility in females (see "Oral Contraceptives," p. 793). Over the years, the dose of estrogen and progesterone in birth control pills has been reduced. The current lower-dose birth control pills have fewer side effects than earlier dosages. There is an increased risk for heart attack or stroke in

females using oral contraceptives who smoke or who have a history of hypertension or coagulation disorders. For most females, the pill is effective and has a minimum frequency of complications, until at least age 35.

The **minipill** is an oral contraceptive that contains only synthetic progesterone. It reduces and thickens mucus of the cervix, which prevents sperm cells from reaching the egg. It also prevents embryos from implanting in the uterus.

Progesterone-like chemicals, such as medroxyprogesterone (med-rok′-sē-prō-jes′ter-ōn) (Depo-Provera), which are injected intramuscularly and slowly released into the circulatory system, can act as effective contraceptives. Injected progesterone-like chemicals can provide protection from pregnancy for approximately 1 month, depending on the amount injected. Thin silastic tubes containing these chemicals can be implanted beneath the skin, usually in the upper arm; the chemicals are slowly released into the circulatory system. The implants can be effective for up to 5 years. Menstruation does not normally occur in women using these techniques while the progesterone levels are elevated.

The **patch** is an adhesive skin patch containing synthetic estrogen and progesterone. It is worn on the lower abdomen, buttocks, or upper body. The **vaginal contraceptive ring** is inserted into the vagina and releases synthetic estrogen and progesterone.

A drug called **RU486,** or **mifepristone** (mif′pris-tōn), blocks the action of progesterone, causing the endometrium of the uterus to slough off as it does at the time of menstruation. It can, therefore, be used to induce menstruation and reduce the possibility of implantation when sexual intercourse has occurred near the time of ovulation. It can also be used to terminate pregnancies.

Morning-after pills, similar in composition to birth control pills, are available. Doubling the number of birth control pills after sexual intercourse within 3 days and again after 12 more hours is sometimes recommended. These techniques can be used after intercourse, but they are only about 75% effective. The elevated blood levels of estrogen and progesterone may inhibit the increase in LH that causes ovulation, may alter the rate at which the embryo is transported through the uterine tube to the uterus, or may inhibit implantation.

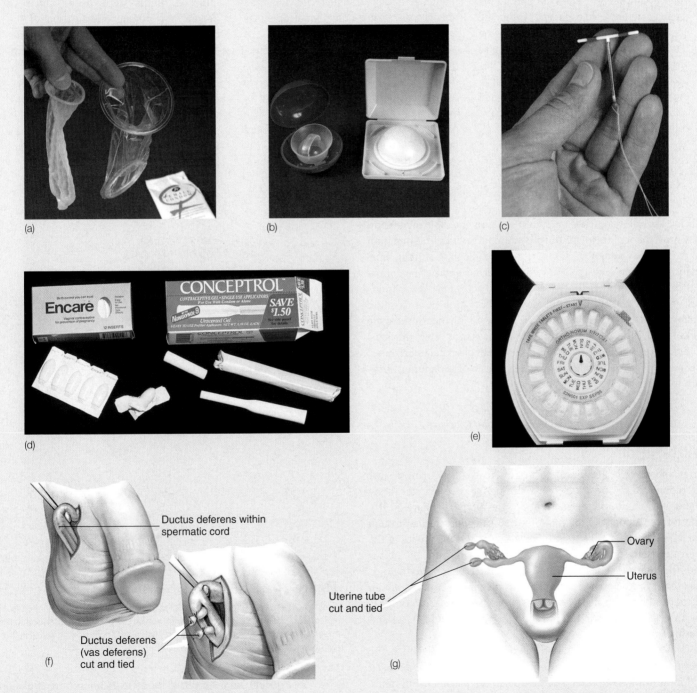

Figure A Contraceptive Devices and Techniques

(a) Condom. (b) Cervical cap and diaphragm used with spermicidal jelly. (c) Intrauterine device (IUD). (d) Spermicidal foam. (e) Oral contraceptives. (f) Vasectomy. (g) Tubal ligation.

Clinical Relevance Control of Pregnancy

continued

Surgical Methods

Vasectomy (va-sek'tō-mē) is a common method used to render males incapable of fertilization without affecting the performance of the sexual act. Vasectomy is a surgical procedure used to cut and tie the ductus deferens from each testis within the scrotal sac. In the majority of cases the procedure is permanent, but in some cases vasectomy can be surgically reversed. This procedure prevents sperm cells from passing through the ductus deferens and becoming part of the ejaculate. The sperm cells are reabsorbed in the epididymis.

A common method of permanent birth control in females is **tubal ligation** (lī-gā'shŭn), a procedure in which the uterine tubes are tied and cut or clamped through an incision made through the wall of the abdomen. This procedure closes off the pathway between the sperm cells and the oocyte. **Laparoscopy** (lap-ă-ros'kŏ-pē), in which an instrument is inserted into the abdomen through a small incision, is commonly used so that only small openings are required to perform the operation.

In some cases, pregnancies are terminated by surgical procedures called **abortions.** The most common method for performing abortions is the insertion of an instrument through the cervix into the uterus. The instrument scrapes the endometrial surface while a strong suction is applied. The endometrium and the embed-ded embryo are disrupted and sucked out of the uterus. This technique is normally used only in pregnancies that have progressed less than 3 months.

Prevention of Implantation

Intrauterine devices (IUDs) are inserted into the uterus through the cervix to prevent the normal implantation of the developing embryonic mass within the endometrium. Some early IUD designs produced serious side effects, such as perforation of the uterus; as a result, many IUDs were removed from the market. Data indicate, however, that IUDs are effective in preventing pregnancy. The effective IUDs include those containing copper and progesterone.

---

lubrication, allowing easy entry of the penis into the vagina and easy movement of the penis during intercourse. Tactile stimulation of the female's genitalia, along with psychological stimuli, normally triggers an orgasm. The vaginal, uterine, and perineal muscles contract rhythmically. After the sexual act, a period of resolution characterized by an overall sense of satisfaction and relaxation occurs. Females are sometimes receptive to further immediate stimulation, however, and can experience successive orgasms. Although orgasm is a pleasurable component of sexual intercourse, it is not necessary for females to experience an orgasm for fertilization to occur. ▼

> 78. *What determines sex drive in females?*
> 79. *Is an orgasm required for fertilization to occur?*

Fertilization

After sperm cells are ejaculated into the vagina during sexual intercourse, they are transported through the cervix, the body of the uterus, and the uterine tubes to the ampulla (figure 24.18). The forces responsible for the movement of sperm cells through the female reproductive tract involve the swimming ability of the sperm cells and possibly the muscular contractions of the uterus and the uterine tubes. During sexual intercourse, oxytocin is released from the posterior pituitary of the female, and the semen introduced into the vagina contains prostaglandins. Both of these hormones stimulate smooth muscle contractions in the uterus and uterine tubes.

While passing through the female reproductive tract, the sperm cells undergo **capacitation** (kă-pas'i-tā'shŭn), the removal and modification of proteins of the sperm cell plasma membranes. Following capacitation, acrosomal enzymes of some sperm cells are released as the sperm cells move through the uterus and oviducts. The acrosomal enzymes allow penetration of the cervical mucus, corona radiata, zona pellucida, and oocyte plasma membrane.

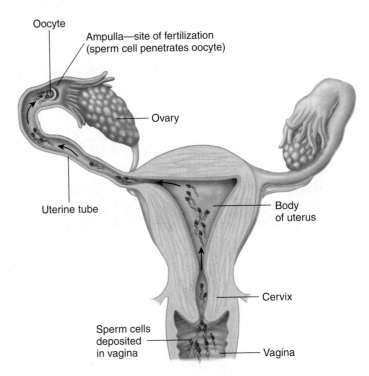

Figure 24.18 Sperm Cell Movement

Sperm cells are deposited into the vagina as part of the semen when the male ejaculates. Sperm cells pass through the cervix, the body of the uterus, and the uterine tube. Fertilization normally occurs when the secondary oocyte is in the ampulla (upper one-third of the uterine tube).

The oocyte can be fertilized for up to 24 hours after ovulation, and some sperm cells remain viable in the female reproductive tract for up to 6 days, although most of them have degenerated after 24 hours. For fertilization to occur successfully, sexual intercourse must therefore occur between 5 days before and 1 day after ovulation.

Fertilization usually takes place in the ampulla. By 7 or 8 days after ovulation, which is day 21 or 22 of the average menstrual cycle, the embryo has reached the uterus. The endometrium of the uterus has reached its maximum thickness and secretory activity in response to estrogen and progesterone.

Ectopic Pregnancy

An **ectopic pregnancy** results if the embryo implants anywhere other than in the uterine cavity. The most common site of ectopic pregnancy is the uterine tube. Implantation in the uterine tube eventually is fatal to the fetus and can cause the tube to rupture. In some cases, implantation occurs in the mesenteries of the abdominal cavity, and the fetus can develop normally but must be delivered by cesarean section. ▼

80. *Describe the transport of sperm cells through the female reproductive system.*
81. *What is capacitation of sperm cells?*
82. *When is fertilization possible? Where does fertilization usually take place?*

Menopause

When a woman is 40–50 years old, menstrual cycles become less regular, and ovulation often does not occur. Eventually, menstrual cycles stop completely. **Menopause** (men′ō-pawz) is the cessation of menstrual cycles, and **perimenopause** is the time from the onset of irregular cycles to their complete cessation. Perimenopause often lasts 3 to 5 years.

The major cause of menopause is age-related changes in the ovaries. The number of follicles remaining in the ovaries of menopausal women is small. In addition, the follicles that remain become less sensitive to stimulation by FSH and LH. As the ovaries become less responsive to stimulation by FSH and LH, fewer mature follicles and corpora lutea are produced. Gradual changes occur in women in response to the reduced amount of estrogen and progesterone produced by the ovaries (table 24.3).

During perimenopause, some women experience "hot flashes," irritability, fatigue, anxiety, temporary decrease in libido, and occasionally severe emotional disturbances. Many of these symptoms can be treated successfully with hormone replacement therapy (HRT), which usually consists of small amounts of estrogen, or estrogen in combination with progesterone. Although estrogen therapy has been successful, in many women it prolongs the symptoms associated with menopause. A potential side effect of HRT is a slightly increased possibility of the development of breast cancer, uterine cancer, heart attacks, strokes, and blood clots. HRT does slow the decrease in bone density that can become severe in some women after menopause and decreases the risk of developing colorectal cancer. ▼

83. *Define menopause and perimenopause.*
84. *What causes the changes that lead to menopause?*

P R E D I C T
After menopause, would a woman's levels of GnRH, FSH, and LH be higher or lower than before perimenopause?

Table 24.3	Possible Changes Caused by Decreased Ovarian Hormone Secretion in Postmenopausal Women
Affected Structures and Functions	**Changes**
Menstrual cycle	Five to 7 years before menopause, the cycle becomes more irregular; finally, the number of cycles in which ovulation does not occur increases, and corpora lutea do not develop.
Uterine tubes	Little change occurs.
Uterus	Irregular menstruation is gradually followed by no menstruation; the chance of cystic glandular hypertrophy of the endometrium increases; the endometrium finally atrophies, and the uterus becomes smaller.
Vagina and external genitalia	The dermis and epithelial lining become thinner; the vulva becomes thinner and less elastic; the labia majora become smaller; pubic hair decreases; the vaginal epithelium produces less glycogen; vaginal pH increases; reduced secretion leads to dryness; the vagina is more easily inflamed and infected.
Skin	The epidermis becomes thinner; melanin synthesis increases.
Cardiovascular system	Hypertension and atherosclerosis occur more frequently.
Vasomotor instability	Hot flashes and increased sweating are correlated with the vasodilation of cutaneous blood vessels; hot flashes are not caused by abnormal FSH and LH secretion but are related to decreased estrogen levels.
Sex drive	Temporary changes, such as either decreases or increases in sex drive, are often associated with the onset of menopause.
Fertility	Fertility begins to decline approximately 10 years before the onset of menopause; by age 50, almost all oocytes and follicles have been lost; the loss is gradual, and no increased follicular degeneration is associated with the onset of menopause.

Systems Interactions

	Effects of the Reproductive System on Other Systems	Effects of Other Systems on the Reproductive System
Integumentary System	Sex hormones increase sebum production (contributing to acne), increase apocrine gland secretion (contributing to body odor), and stimulate axillary and pubic hair growth Increases melanin production during pregnancy	Mammary glands produce milk Areolar glands help prevent chafing during nursing Covers external genitalia
Skeletal System	Estrogen and testosterone stimulate bone growth and closure of the epiphyseal plate	Pelvis protects internal reproductive organs and developing fetus
Muscular System	Sex hormones increase muscle growth	Pelvic floor muscles support internal reproductive organs, such as the uterus Responsible for ejaculation Cremaster muscles help regulate testes temperature Abdominal muscles assist in delivery
Nervous System	Sex hormones influence CNS development and sexual behaviors	Stimulates sexual responses, such as erection and ejaculation in males and erection of the clitoris in females
Endocrine System	Testosterone, estrogen, and inhibin regulate the release of hormones from the hypothalamus and pituitary gland	Stimulates the onset of puberty and sexual characteristics Stimulates gamete formation Promotes uterine contractions for delivery Makes possible and regulates milk production
Cardiovascular System	Estrogen may slow the development of atherosclerosis	Delivers oxygen, nutrients, hormones, and immune cells Removes carbon dioxide, waste products, and toxins Vasodilation necessary for erection
Lymphatic System and Immunity	Blood–testes barrier isolates and protects sperm cells from the immune system	Immune cells provide protection against microorganisms and toxins and promote tissue repair by releasing chemical mediators Removes excess interstitial fluid
Respiratory System	Sexual arousal increases respiration	Provides oxygen and removes carbon dioxide Helps maintain the body's pH
Digestive System	Developing fetus crowds digestive organs	Provides nutrients
Urinary System	Developing fetus compresses the urinary bladder	Removes waste products Helps maintain the body's pH, ion, and water balance Urethra common passageway for sperm cells and urine in males

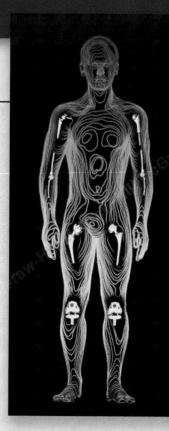

Effects of Aging on the Reproductive System

Benign prostatic enlargement is common in men over 50 years of age. A major consequence of prostatic enlargement is blockage of the prostatic urethra. The frequency of prostate cancer also increases as men age and is a significant cause of death in men. There is a decrease in the rate of sperm cell production and an increase in the number of abnormal sperm cells. However, sperm cell production does not stop, and it remains adequate for fertility for most men.

Although sexual activity is often maintained in men and women as they age, there is usually a gradual decrease in the frequency of sexual intercourse. There is an increased tendency for erectile dysfunction to occur as men age.

The most significant age-related change in females is menopause. By age 50, few viable follicles remain in the ovaries. As a result, there is a decrease in the estrogen and progesterone produced by the ovaries. The uterus decreases in size, and the endometrium decreases in thickness. The times between menstruations become irregular and longer. Finally, menstruation stops. The vaginal wall becomes thinner and less elastic. There is less lubrication of the vagina and the epithelial lining is more fragile. There is an increased tendency for vaginal infections.

Approximately 10% of all women will have breast cancer. The incidence of breast cancer is greatest between 45 and 65 years of age, and the incidence is greater for those who have a family history of breast cancer. The frequency of cancer of the uterus and cancer of the uterine cervix increases between 50 and 65 years of age. Ovarian cancer also increases in frequency in older women. Occasionally, a prolapsed uterus develops when the ligaments of the uterus allow it to descend and protrude into the vagina. ▼

85. List the major age-related changes that occur in males and females.

S U M M A R Y

Functions of the Reproductive System (p. 766)

1. The male reproductive system produces sperm cells, provides nutrients for the sperm cells and secretions, transfers the sperm cells to the female, and makes male sex hormones.
2. The female reproductive system produces oocytes, receives sex cells from the male, provides nourishment for the developing individual before and after birth, and produces female sex hormones.

Meiosis (p. 766)

Chromosomes

1. The diploid number of chromosomes in humans is 46, consisting of 23 pairs of homologous chromosomes.
2. A male has the sex chromosomes XY and a female XX.
3. The haploid number of chromosomes in humans is 23.

Meiotic Divisions

1. A reduction division decreases the number of chromosomes from the diploid to the haploid number.
2. Crossing over and random assortment are responsible for the genetic diversity of sperm cells and oocytes.

Anatomy of the Male Reproductive System (p. 767)

The male reproductive system includes the testes, ducts, accessory glands, and supporting structures. Collectively, all of these structures are called the male genitalia.

Scrotum

1. The scrotum is a sac containing the testes.
2. The dartos and cremaster muscles help regulate testicular temperature.

Testes

1. The tunica albuginea is the outer connective tissue capsule of the testes.

2. The testes are divided by septa into compartments that contain the seminiferous tubules and the interstitial cells.
3. The seminiferous tubules join the tubuli recti, which lead to the rete testis. The rete testis opens into the efferent ductules of the epididymis.
4. During development, the testes pass from the abdominal cavity through the inguinal canal to the scrotum.
5. A tunic vaginalis covers each testis, protecting against friction.

Spermatogenesis

1. Sperm cells (spermatozoa) are produced in the seminiferous tubules.
2. Spermatogonia divide (mitosis) to form primary spermatocytes.
3. Primary spermatocytes divide (first division of meiosis) to form secondary spermatocytes, which divide (second division of meiosis) to form spermatids.
4. Spermatids develop an acrosome and a flagellum to become sperm cells.
5. Sustentacular cells nourish the sperm cells, form a blood–testes barrier, and produce hormones.

Ducts

1. Efferent ductules extend from the testes into the head of the epididymis and join the duct of the epididymis.
2. The epididymis is a coiled tube system located on the testis that is the site of sperm cell maturation. It consists of a head, body, and tail.
3. The ductus deferens passes from the epididymis into the abdominal cavity.
4. The end of the ductus deferens, called the ampulla, and the seminal vesicle join to form the ejaculatory duct, which connects to the prostatic urethra.
5. The prostatic urethra extends from the urinary bladder through the prostate gland to the membranous urethra.
6. The membranous urethra extends through the pelvic floor and becomes the spongy urethra, which continues through the penis.

7. The spermatic cord consists of the ductus deferens, blood and lymphatic vessels, and nerves.
8. Coverings of the spermatic cord consist of the external spermatic fascia, cremaster muscle, and internal spermatic fascia.
9. The spermatic cord passes through the inguinal canal into the abdominal cavity.

Penis

1. The penis consists of erectile tissue.
   - The two corpora cavernosa form the dorsum and the sides of the penis.
   - The corpus spongiosum forms the ventral part and the glans penis.
2. The bulb of the penis and the crura form the root of the penis and the crura attach the penis to the coxal bones.
3. The prepuce covers the glans penis.

Accessory Glands

1. The seminal vesicles empty into the ejaculatory ducts.
2. The prostate gland consists of glandular and muscular tissue and empties into the prostatic urethra.
3. The bulbourethral glands are mucous glands that empty into the spongy urethra.
4. Semen is a mixture of gland secretions and sperm cells.
   - The testicular secretions contain sperm cells.
   - The bulbourethral glands produce mucus, which neutralizes the acidic pH of the urethra.
   - The seminal vesicle fluid contains fructose, clotting proteins, and prostaglandins.
   - The prostate secretions make the seminal fluid more pH-neutral. Proteolytic enzymes break down clotting proteins.

Physiology of Male Reproduction   (p. 776)

Normal function of the male reproductive system depends on hormonal and neural mechanisms.

Regulation of Sex Hormone Secretion

1. GnRH stimulates LH and FSH release from the anterior pituitary.
   - LH stimulates the interstitial cells to produce testosterone.
   - FSH stimulates sperm cell formation.
2. Inhibin, produced by sustentacular cells, inhibits FSH secretion.

Puberty

1. Before puberty, small amounts of testosterone inhibit GnRH release.
2. During puberty, testosterone does not completely suppress GnRH release, resulting in increased production of FSH, LH, and testosterone.

Effects of Testosterone

1. Interstitial cells produce testosterone.
2. Testosterone causes the development of male sex organs in the embryo and stimulates the descent of the testes.
3. Testosterone causes enlargement of the genitalia and is necessary for sperm cell formation.
4. Other effects of testosterone occur.
   - Hair growth stimulation (pubic area, axilla, and beard) and inhibition (male pattern baldness) occur.
   - Increased skin thickness and melanin and sebum production occur.
   - Enlargement of the larynx and deepening of the voice occur.
   - Increased protein synthesis (muscle), bone growth, blood cell synthesis, and blood volume occur.
   - Metabolic rate increases.

Male Sexual Behavior and the Male Sexual Act

1. Testosterone is required for normal sex drive.
2. The male sexual act includes erection, emission, ejaculation, orgasm, and resolution.
3. Stimulation of the sexual act can be psychic or tactile.

Anatomy of the Female Reproductive System   (p. 781)

The female reproductive system includes the ovaries, uterine tubes, uterus, vagina, external genitalia, and summary glands.

Ovaries

1. The suspensory ligament, ovarian ligament, and broad ligament hold the ovary in place.
2. The visceral peritoneum covers the surface of the ovaries.
3. The ovary has an outer capsule (tunica albuginea) and is divided internally into a cortex (contains follicles) and a medulla (receives blood and lymphatic vessels and nerves).
4. Oocyte development and fertilization
   - Oogonia proliferate and become primary oocytes that are in prophase I of meiosis.
   - A primary oocyte continues meiosis and produces a secondary oocyte, which is in metaphase II of meiosis, and a polar body, which degenerates or divides to form two polar bodies.
   - Ovulation is the release of a secondary oocyte from an ovary.
   - Fertilization is the joining of a sperm cell and a secondary oocyte to form a zygote.
   - An embryo is the developing human between the time of fertilization and 8 weeks of development. A fetus is the developing human from 8 weeks to birth.
5. Follicle development
   - Primordial follicles are surrounded by a single layer of flat granulosa cells.
   - Primary follicles are primary oocytes surrounded by a zona pellucida and cuboidal granulosa cells.
   - The primary follicles become secondary follicles as granulosa cells increase in number and fluid begins to accumulate in the vesicles. The granulosa cells increase in number, and theca cells form around the secondary follicles.
   - Mature follicles have an antrum.
6. Ovulation occurs when the follicle swells and ruptures and the secondary oocyte is released from the ovary. The corona radiata surround the oocyte.
7. Fate of the follicle
   - The mature follicle becomes the corpus luteum.
   - If pregnancy occurs, the corpus luteum persists. If no pregnancy occurs, it becomes the corpus albicans.

Uterine Tubes

1. The uterine tubes extend from the ovaries to the uterus.
   - The ovarian end of the uterine tube is expanded as the infundibulum. The opening of the infundibulum is surrounded by fimbriae.
   - The ampulla is the widest, longest part of the uterine tube.
2. The uterine tube consists of an outer serosa, a middle muscular layer, and an inner mucosa with simple ciliated columnar epithelium.
3. Muscular contractions and cilia move the oocyte through the uterine tube.

Uterus

1. The uterus consists of the fundus, body, and cervix. The uterine cavity and the cervical canal are the spaces formed by the uterus.
2. The uterus is held in place by the broad, round, and uterosacral ligaments.

3. The wall of the uterus consists of the perimetrium (visceral peritoneum), the myometrium (smooth muscle), and the endometrium (mucous membrane).

Vagina

1. The vagina is the female organ of copulation. It connects the uterus (cervix) to the vestibule.
2. The vagina consists of a layer of smooth muscle and an inner lining of moist stratified squamous epithelium.
3. The hymen covers the vaginal orifice.

External Genitalia

1. The external female genitalia consist of the vestibule and its surrounding structures.
2. The vestibule is the space into which the vagina and the urethra open.
3. Erectile tissue is associated with the vestibule.
   - The two corpora cavernosa form the clitoris.
   - The corpora spongiosa form the bulbs of the vestibule.
4. The labia minora are folds that cover the vestibule and form the prepuce.
5. The greater and lesser vestibular glands produce a mucous fluid.
6. When closed, the labia majora cover the labia minora.
   - The mons pubis is an elevated fat deposit superior to the labia majora.
   - The pudendal cleft is a space between the labia majora.

Perineum

The clinical perineum is the region between the vagina and the anus.

Mammary Glands

1. The mammary glands are modified sweat glands located in the breasts.
2. The areola surrounds the nipple.
   - The mammary glands consist of glandular lobes and adipose tissue.
   - The lobes consist of lobules that have milk-producing alveoli.
   - The lobes connect to the nipple through the lactiferous ducts.
3. Suspensory ligaments support the breasts.

Physiology of Female Reproduction   (p. 790)

Puberty

1. Puberty begins with the first menstrual bleeding (menarche).
2. Puberty begins when GnRH, FSH, LH, estrogen, and progesterone levels increase.
3. Increased estrogen and progesterone promote the development of the female primary and secondary sexual characteristics.

Menstrual Cycle

1. The menstrual cycle consists of the periodic changes occurring in the ovaries and uterus of a sexually mature, nonpregnant female that result in the production of a secondary oocyte and prepare the uterus for implantation.
2. The menstrual phase is the time between the beginning and the end of menstruation (days 1–5).
   - Menstruation is the discharge of blood and part of the endometrium from the uterus.
   - Menstruation begins because of a decrease in progesterone and estrogen from the previous cycle.
3. The proliferation phase is the time between the end of menstruation and ovulation (days 6–14).
   - FSH and LH stimulate follicular growth and estrogen production.
   - Estrogen stimulates epithelial cells in the endometrium to multiply. The endometrium becomes thicker and spiral glands and arteries develop.
   - The LH surge stimulates completion of the first meiotic division by the primary oocyte, ovulation, and formation of the corpus luteum.
   - The FSH surge stimulates follicle development. Mature follicles inhibit the development of less mature follicles.
4. The secretory phase is the time between ovulation and the beginning of menstruation (days 14–28).
   - Estrogen stimulates cell division in the endometrium.
   - Progesterone stimulates the spiral glands to produce a secretion rich in glycogen and lipids and inhibits uterine contractions.
   - If fertilization does not occur, menses begins and the corpus luteum becomes the corpus albicans.
   - If fertilization occurs, hCG stimulates the corpus luteum to persist.

Female Sexual Behavior and the Female Sexual Act

1. Female sex drive is partially influenced by androgens (produced by the adrenal gland) and steroids (produced by the ovaries).
2. Events of the female sexual act including the following.
   - The erectile tissue of the clitoris and the bulbs of the vestibule become filled with blood.
   - The vestibular glands secrete mucus, and the vagina extrudes a mucuslike substance.
   - Orgasm and resolution occur.

Fertilization

1. Intercourse must take place 5 days before to 1 day after ovulation if fertilization is to occur.
2. Sperm cell transport to the ampulla depends on the ability of the sperm cells to swim and possibly on contractions of the uterus and the uterine tubes.
3. Implantation of the developing embryo into the uterine wall occurs when the uterus is most receptive.

Menopause

1. Menopause is the cessation of menstrual cycles.
2. Perimenopause is the time between the beginning of irregular menstrual cycles and menopause.

Effects of Aging on the Reproductive System   (p. 799)

1. The prostate gland enlarges, and there is an age-related increase in prostatic cancer.
2. There is decreased sperm cell production and increased production of abnormal sperm cells.
3. Erectile dysfunction increases.
4. The most significant age-related change in females is menopause.
5. The uterus decreases in size and the vaginal wall thins.
6. There is an age-related increase in breast, uterine, and ovarian cancer.

1. If an adult male walked into a swimming pool of cold water, which of these muscles would be expected to contract?
   a. cremaster muscle
   b. dartos muscle
   c. gubernaculum
   d. prepuce muscle
   e. both a and b

2. Testosterone is produced in the
   a. interstitial cells.
   b. seminiferous tubules of the testes.
   c. anterior lobe of the pituitary.
   d. sperm cells.

3. Early in development (4 months after fertilization), the testes
   a. are found in the abdominal cavity.
   b. move through the inguinal canal.
   c. produce a membrane that becomes the scrotum.
   d. produce sperm cells.
   e. all of the above.

4. The site of spermatogenesis in the male is the
   a. ductus deferens.
   b. seminiferous tubules.
   c. epididymis.
   d. rete testis.
   e. efferent ductule.

5. The location of final maturation and storage of sperm cells before their ejaculation is the
   a. seminal vesicles.
   b. seminiferous tubules.
   c. glans penis.
   d. epididymis.
   e. sperm bank.

6. Given these structures:
   1. ductus deferens
   2. efferent ductule
   3. epididymis
   4. ejaculatory duct
   5. rete testis

   Choose the arrangement that lists the structures in the order a sperm cell passes through them from the seminiferous tubules to the urethra.
   a. 2,3,5,4,1
   b. 2,5,3,4,1
   c. 3,2,4,1,5
   d. 3,4,2,1,5
   e. 5,2,3,1,4

7. Concerning the penis,
   a. the membranous urethra passes through the corpora cavernosa.
   b. the glans penis is formed by the corpus spongiosum.
   c. the penis contains four columns of erectile tissue.
   d. the crus of the penis is part of the corpus spongiosum.
   e. the bulb of the penis is covered by the prepuce.

8. Given these glands:
   1. prostate gland
   2. bulbourethral gland
   3. seminal vesicle

   Choose the arrangement that is in the order that a sperm cell would encounter their secretions.
   a. 1,2,3
   b. 2,1,3
   c. 2,3,1
   d. 3,1,2
   e. 3,2,1

9. Which of these glands is correctly matched with the function of its secretions?
   a. bulbourethral gland—neutralizes acidic contents of the urethra
   b. seminal vesicles—contain large amounts of fructose, which nourishes the sperm cells
   c. prostate gland—contains proteins that cause clotting of the semen
   d. all of the above

10. LH in the male stimulates
    a. the development of the seminiferous tubules.
    b. spermatogenesis.
    c. testosterone production.
    d. both a and b.
    e. all of the above.

11. Which of these factors causes a decrease in GnRH release?
    a. decreased inhibin
    b. increased testosterone
    c. decreased FSH
    d. decreased LH

12. In the male, before puberty
    a. FSH levels are higher than after puberty.
    b. LH levels are higher than after puberty.
    c. GnRH release is inhibited by testosterone.
    d. all of the above.

13. Testosterone
    a. stimulates the development of terminal hairs.
    b. decreases red blood cell count.
    c. prevents closure of the epiphyseal plate.
    d. decreases blood volume.
    e. all of the above.

14. Which of these is consistent with erection of the penis?
    a. parasympathetic stimulation
    b. dilation of arterioles
    c. engorgement of sinusoids with blood
    d. compression of veins
    e. all of the above

15. The first polar body
    a. is normally formed before fertilization.
    b. is normally formed after fertilization.
    c. normally receives most of the cytoplasm.
    d. is larger than an oocyte.
    e. both a and b.

16. After ovulation, the mature follicle collapses, taking on a yellowish appearance to become the
    a. degenerating follicle.
    b. corpus luteum.
    c. corpus albicans.
    d. tunica albuginea.
    e. corona radiata.

17. The ampulla of the uterine tube
    a. is the opening of the uterine tube into the uterus.
    b. has long, thin projections called fimbriae.
    c. is lined with ciliated columnar epithelium.
    d. all of the above.

18. The layer of the uterus that undergoes the greatest change during the menstrual cycle is the
    a. perimetrium.
    b. hymen.
    c. endometrium.
    d. myometrium.
    e. broad ligament.

19. Concerning the vagina,
    a. its wall consists of skeletal muscle.
    b. the vaginal orifice is covered by the hymen.
    c. it is lined with simple squamous epithelium.
    d. all of the above.

20. During sexual excitement, which of these structures fills with blood and causes the vaginal orifice to narrow?
    a. bulb of the vestibule
    b. clitoris
    c. mons pubis
    d. labia majora
    e. prepuce

21. Given these vestibular–perineal structures:
    1. vaginal orifice
    2. clitoris
    3. urethral orifice
    4. anus

    Choose the arrangement that lists the structures in their proper order from the anterior to the posterior aspect.
    a. 2,3,1,4
    b. 2,4,3,1
    c. 3,1,2,4
    d. 3,1,4,2
    e. 4,2,3,1

22. Concerning the breasts,
    a. lactiferous ducts open on the areola.
    b. each lactiferous duct supplies an alveolus.
    c. they are attached to the pectoralis major muscles by suspensory ligaments.
    d. even before puberty, the female breast is quite different from the male breast.

23. The major secretory product of the mature follicle is
    a. estrogen.
    b. progesterone.
    c. LH.
    d. FSH.
    e. relaxin.

24. In the average adult female, ovulation occurs at day _____ of the menstrual cycle.
    a. 1
    b. 7
    c. 14
    d. 21
    e. 28

25. Which of these processes or phases in the menstrual cycle of the human female occur at the same time?
    a. maximal LH secretion and menstruation
    b. early follicular development and the secretory phase of the uterus

    c. degeneration of the corpus luteum and an increase in ovarian progesterone secretion
    d. ovulation and menstruation
    e. increased estrogen production and the proliferation phase

26. During the proliferative phase of the menstrual cycle, one would normally expect
    a. the highest levels of estrogen that occur during the menstrual cycle.
    b. the mature follicle to be present in the ovary.
    c. an increase in the thickness of the endometrium.
    d. both a and b.
    e. all of the above.

27. During the secretory phase of the menstrual cycle,
    a. the highest levels of progesterone occur.
    b. there is a follicle present in the ovary that is ready to undergo ovulation.
    c. the endometrium reaches its greatest degree of development.
    d. both a and b.
    e. both a and c.

28. The cause of menses in the menstrual cycle appears to be
    a. increased progesterone secretion from the ovary, which produces blood clotting.
    b. increased estrogen secretion from the ovary, which stimulates the muscles of the uterus to contract.
    c. decreased progesterone secretion by the ovary.
    d. decreased production of oxytocin, causing the muscles of the uterus to relax.

29. A woman with a 28-day menstrual cycle is most likely to become pregnant as a result of coitus on days
    a. 1–3.
    b. 5–8.
    c. 9–14.
    d. 15–20.
    e. 21–24.

30. Menopause
    a. develops when follicles become less responsive to FSH and LH.
    b. results from elevated estrogen levels in 40- to 50-year-old women.
    c. occurs because too many follicles develop during each cycle.
    d. results when follicles develop but contain no oocytes.
    e. occurs because FSH and LH levels decline.

Answers in Appendix E

C R I T I C A L   T H I N K I N G

1. If a 20-year-old male were castrated, what would happen to the levels of GnRH, FSH, LH, and testosterone in his blood? What effect would these hormonal changes have on his sexual characteristics and behavior?

2. If a 9-year-old boy were castrated, what would happen to the levels of GnRH, FSH, LH, and testosterone in his blood? What effect would these hormonal changes have on his sexual characteristics and behavior?

3. Suppose you want to produce a birth control pill for men. On the basis of what you know about the male hormonal system, what do you want the pill to do? Discuss any possible side effects of the pill.

4. If the ovaries are removed from a 20-year-old woman, what happens to the levels of GnRH, FSH, LH, estrogen, and progesterone in her blood? What side effects do these hormonal changes have on her sexual characteristics and behavior?

5. If the ovaries are removed from a postmenopausal woman, what happens to the levels of GnRH, FSH, LH, estrogen, and progesterone in her blood? What symptoms do you expect to observe?

6. A study divides healthy women into two groups (A and B). Both groups are composed of women who have been sexually active for at least 2 years and are not pregnant at the beginning of the experiment. The subjects weigh about the same amount, and none

smokes cigarettes, although some drink alcohol occasionally. Group A women receive a placebo in the form of a sugar pill each morning during their menstrual cycles. Group B women receive a pill containing estrogen and progesterone each morning of their menstrual cycles. Then plasma LH levels are measured before, during, and after ovulation. The results are as follows:

Group	4 Days Before Ovulation	The Day of Ovulation	4 Days After Ovulation
A	18 mg/100 mL	300 mg/100 mL	17 mg/100 mL
B	21 mg/100 mL	157 mg/100 mL	15 mg/100 mL

The number of pregnancies in group A is 37/100 women/year. The number of pregnancies in group B is 1.5/100 women/year. What conclusion can you reach on the basis of these data? Explain the mechanism involved.

7. A woman who is taking birth control pills that consist of only progesterone experiences the hot flash symptoms of menopause. Explain why.

8. Predict the effect on the endometrium of maintaining high progesterone levels in the circulatory system, including the time during which estrogen normally increases following menstruation.

Answers in Appendix F

CHAPTER 25 Development and Genetics

Colorized scanning electron micrograph of a zygote (120 μm in diameter) with two polar bodies attached.

Tools for Success

A virtual cadaver dissection experience

Audio (MP3) and/or video (MP4) content, available at www.mhhe.aris.com

Introduction

The life span is usually considered the period between birth and death; however, the 9 months before birth are a critical part of a person's existence. What happens in these 9 months profoundly affects the rest of a person's life. Although most people develop normally and are born without defects, approximately 3 out of every 100 people are born with a birth defect so severe that it requires medical attention during the first year of life. Later in life, many more people discover previously unknown problems, such as the tendency to develop asthma, certain brain disorders, or cancer.

Genetics is the study of heredity—that is, those characteristics inherited by children from their parents. Although the environment can influence gene expression, people's physical characteristics and abilities are largely determined by their genetic makeup. Many of a person's abilities, susceptibility to disease, and even life span are influenced by the genes they inherit from their parents. An understanding of genetics is an important tool for medical professionals. Patients are asked questions about their family medical history to help diagnose many diseases. The family medical history also allows doctors to determine the probability that patients will develop certain diseases, such as heart disease or cancer, and to suggest preventive measures. Recent advances in the field of genetics have shown how genes influence health and have provided new methods for treating certain diseases.

▶This chapter discusses **prenatal development (p. 806), labor (p. 826), the newborn (p. 828), lactation (p. 831),** and **genetics (p. 832).**

Prenatal Development

The **prenatal** (prē-nā′tăl, before birth) **period** is the period from conception until birth, and it is divided into three parts: (1) approximately the first 2 weeks of development, during which the primitive germ layers are formed; (2) from about the second to the end of the eighth week of development, during which the major organ systems come into existence; and (3) the last 30 weeks of the prenatal period, during which the organ systems grow and become more mature.

The developing human from the time of fertilization to the end of the eighth week of development is called an **embryo** (em′brē-ō, to swell). From 8 weeks to birth, the developing human is called a **fetus** (fē′tus, offspring).

Malformations

During the first 2 weeks of development, the embryo is quite resistant to outside influences that may cause malformations. Factors that adversely affect the embryo at this age are more likely to kill it. Between 2 weeks and the next 4–7 weeks (depending on the structure considered), the embryo is more sensitive to outside influences that cause malformations than at any other time.

The medical community in general uses the mother's **last menstrual period (LMP)** to calculate the **clinical age** of the unborn child. Most embryologists, on the other hand, use **developmental age,** which begins with fertilization, to describe the timing of developmental events. It is assumed that developmental age is 14 days less than clinical age because fertilization occurs approximately 14 days after LMP. The times presented in this chapter are based on developmental age. ▼

1. *Describe the three parts of the prenatal period. Give the length of time for each part.*
2. *Define clinical age and postovulatory age.*

Fertilization

Fertilization (fer′til-i-zā′shŭn, *fero,* to bear + process) (figure 25.1) is the union of a sperm cell and a secondary oocyte to produce a single cell called the **zygote** (zī′gōt, having a yoke) (see figure 24.13). The **corona radiata,** which consists of granulosa cells expelled from the follicle with the secondary oocyte (see figure 24.14), is a barrier to the sperm cells reaching the oocyte. The sperm cells are propelled through the loose matrix between the follicular cells of the corona radiata by the action of their flagella. The **zona pellucida** is an extracellular membrane, comprised mostly of glycoproteins, between the corona radiata and the oocyte. The **acrosome** is a cap on the head of the sperm cell containing enzymes (see figure 24.4*d*). When the acrosome comes into contact with the zona pellucida, it triggers the release and activation of digestive enzymes that break down the zona pellucida. The first sperm cell through the zona pellucida attaches to

the oocyte plasma membrane and enters the oocyte. In response, the oocyte plasma membrane depolarizes and the oocyte releases chemicals that denature the glycoproteins in the zona pellucida and cause it to expand, creating a space. These changes prevent more than one sperm cell from entering the secondary oocyte.

The entrance of a sperm cell into the secondary oocyte stimulates the second meiotic division, and the second polar body is formed (see figure 24.13). The haploid nucleus formed by the second meiotic division is called the **female pronucleus.** Meanwhile, the haploid nucleus of the sperm cell (see figure 24.6) separates from the sperm head and becomes the male pronucleus, which swells as its chromosomes "unpack." The male pronucleus and female pronucleus meet in the center of the cell and fuse, which completes the process of fertilization and restores the diploid number of chromosomes. ▼

3. *Describe the process by which only one sperm cell enters an oocyte.*
4. *How does fertilization result in a zygote with a diploid number of chromosomes?*

Cloning

Fusion of the male and female pronuclei forms a new diploid nucleus in the zygote. Alternatively, the nucleus can be removed from an oocyte and the diploid nucleus from another cell can be introduced to form a zygote. This process of introducing a nucleus from a cell into an oocyte, whose own nucleus has been removed, is called **cloning.** The first cloned mammal was a sheep named Dolly in Scotland in 1996.

Early Cell Division

About 18–36 hours after fertilization, the zygote divides to form two cells (figure 25.2). Those two cells divide to form four cells, which divide to form eight, and so on. Even though the number of cells increases, the size of each cell decreases so that the total mass of cells remains about the same size as the zygote. These cells have the ability to develop into a wide range of tissues. As a result, the total number of cells can be decreased, increased, or reorganized during this period without affecting normal development.

In the very early stages of development (days 1–4), the cells are said to be **totipotent** (tō-tip′ō-tent, whole-powered), meaning each cell has the potential to give rise to any tissue type necessary for development. At this point, if a cell separates from the embryo, the totipotent cell can give rise to another individual, as in identical twins. However, the cells of the developing embryo soon undergo **differentiation,** or specialization. Once differentiation occurs, the dividing cells of the embryo are referred to as **pluripotent** (ploo-rip′ō-tent, multiple-powered), which means that any cell has the ability to develop into a wide range of tissues, but not all the tissues necessary for development, as is the case with totipotent cells.

1. Many sperm cells attach to the corona radiata of a secondary oocyte.

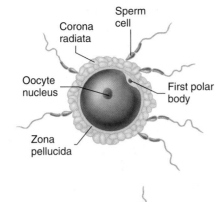

Sperm cell

Corona radiata

Oocyte nucleus

First polar body

Zona pellucida

2. One sperm cell attaches to the zona pellucida, and enzymes in the acrosome digest through the zona pellucida.

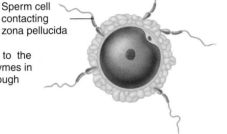

Sperm cell contacting zona pellucida

3. The head of one sperm cell penetrates the zona pellucida and oocyte plasma membrane to enter the oocyte cytoplasm. Changes in the zona pellucida (moving away from the oocyte) form a space and prevent additional sperm cells from entering the oocyte.

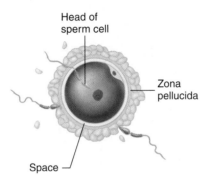

Head of sperm cell

Zona pellucida

Space

4. In response to the entry of the contents of the sperm head into the oocyte, the oocyte nucleus moves to one side of the oocyte, where it completes the second meiotic division and gives off a second polar body. When the second meiosis is complete, the oocyte nucleus, now called the female pronucleus, moves back toward the center of the oocyte.

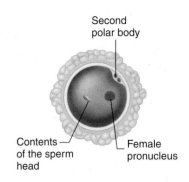

Second polar body

Contents of the sperm head

Female pronucleus

5. The contents of the sperm head enlarge and become the male pronucleus.

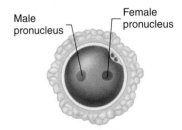

Male pronucleus

Female pronucleus

6. The two pronuclei fuse to form a single nucleus. Fertilization is complete and a zygote results.

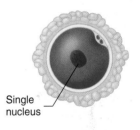

Single nucleus

Process Figure 25.1   Fertilization

Twins

In rare cases, following early cell divisions, the cells separate and develop to form two individuals, called identical, or **monozygotic, twins.** Identical twins have identical genetic information in their cells. Other mechanisms that occur a little later in development can also cause identical twins. Occasionally, a woman ovulates two or more secondary oocytes at the same time. Fertilization of two oocytes by different sperm cells results in fraternal, or **dizygotic, twins.** Multiple ovulations can occur naturally or be stimulated by the injection of drugs. These drugs are sometimes used to treat certain forms of infertility and can result in multiple births in women undergoing the treatment.

Once the dividing embryo is a solid ball of 12 or more cells, it is called a **morula** (mōr′oo-lă, mōr′ū-lă, mulberry) (see figure 25.2). Four or 5 days after ovulation, the morula consists of about

32 cells. Near this time, a fluid-filled cavity called the **blastocele** (blas′tō-sēl) begins to appear approximately in the center of the cellular mass. The hollow sphere that results is called a **blastocyst** (blas′tō-sist) (see figure 25.2). A single layer of cells, the **trophoblast** (trof′ō-blast, trō′fō-blast, feeding layer), surrounds most of the blastocele, but at one end of the blastocyst the cells are several layers thick. The thickened area is the **inner cell mass.** Some of the cells of the inner cell mass give rise to the **embryo proper,** from which all of the tissues of the individual are derived. For simplicity, the embryo proper is referred to as the embryo. The trophoblast and some of the inner cell mass cells give rise to **extraembryonic tissue,** which is tissue outside of the embryo. The extraembryonic tissue becomes structures, such as the placenta, that nourish and support the embryo. ▼

5. *Define totipotent, pluripotent, morula, blastocyst, trophoblast, inner cell mass, embryo proper, and extraembryonic tissue.*

Embryo Transfer

In a small number of women, normal pregnancy is not possible because of an anatomical or physiological condition. In 87% of these cases, the uterine tubes are incapable of transporting the zygote to the uterus or of allowing sperm cells to reach the oocyte. In vitro fertilization and embryo transfer have made pregnancy possible in hundreds of such women since 1978. **In vitro fertilization** involves removing secondary oocytes from a woman, placing the oocytes into a petri dish, and adding sperm cells to the dish, allowing fertilization and early development to occur in vitro, which means in glass. **Embryo transfer** involves the removal of the developing embryo from the petri dish and introduction of the embryo into the uterus of a recipient female.

For in vitro fertilization and embryo transfer to be accomplished, a woman is first injected with a drug similar in structure to follicle stimulating hormone (FSH). While follicles are developing, a drug that blocks gonadotropin-releasing hormone (GnRH) is administered to block the luteinizing hormone (LH) surge that would cause ovulation. After the follicles have had time to develop, an LH-like substance is given to promote ovulation. Just before the follicles rupture, the secondary oocytes are surgically removed from the ovary. The oocytes are then incubated in a dish and maintained at body temperature for 6 hours. Then sperm cells are added to the dish.

After 24–48 hours, several of the embryos are transferred to the uterus. Several embryos are transferred because only a few of them survive. Implantation and subsequent development then proceed in the uterus as they would for natural implantation. The woman is usually required to lie perfectly still for several hours after the embryos have been introduced into the uterus to prevent possible expulsion before implantation can occur. It is not fully understood why such expulsion does not occur in natural fertilization and implantation.

The success rate of embryo transfer is dependent on many factors, including the reason for infertility, the age of the mother, and the expertise of the physician. The success rate at the best U.S. clinics is as high as 72%. Many multiple births have occurred following embryo transfer because of the practice of introducing more than one embryo into the uterus in an attempt to increase the success rate as much as possible.

Stem Cells

During growth and development, many cells differentiate for a particular function and many lose the ability to divide. Stem cells, however, are cells that do not become fully specialized and retain the ability to undergo mitosis and differentiation. Stem cells have the potential to treat many diseases by replacing dysfunctional cells with normal cells. For example, blood stem cells, found in the red bone marrow, are harvested from a donor and introduced into a leukemia patient. The stem cells provide a new source of normal blood cells for that individual. The question for many other diseases, however, is whether a source of stem cells can be isolated and used to treat them.

Adult stem cells are stem cells obtained from adult tissues. Adult stem cells are usually limited as to the type of cell they can produce. For example, adult liver stem cells can give rise to many cell types found in the liver, but they cannot give rise to other tissue cell types, such as blood cells. The most versatile stem cell is the totipotent zygote because it can give rise to all cell types. However, the zygote also gives rise to placental tissue. The pluripotent cells of the inner cell mass that give rise to the embryo (proper) are called **embryonic stem cells.** These cells are also very versatile because they can give rise to all tissues of a new individual, but they cannot give rise to the extraembryonic tissues necessary for implantation and development of the placenta.

Embryonic stem cells were first isolated in 1998; since then, many people have suggested the great potential of these cells for treating many diseases. One problem, however, lies in the ability to stimulate embryonic stem cells artificially to differentiate into the right type of cell. Scientists are still in the early stages of determining the biochemical processes of how embryonic stem cells communicate to allow for the proper development of all tissues in multicellular organisms.

Adult stem cells have had limited success in treating certain diseases. Stem cells obtained from adult blood may be used to address the complications associated with coronary heart disease, particularly the growth of new vessels around areas of blockage. It appears that adult stem cells are not effective in treating other diseases, such as type I diabetes mellitus and Parkinson disease. For these cases, research suggests that embryonic stem cells have a greater potential to provide a longer-lasting cure.

Important ethical issues are involved in the use of embryonic stem cells. To harvest embryonic cells, an embryo must be destroyed. Opponents of embryonic stem cell use argue that a zygote is a living human and the destruction of the zygote or cells derived from the zygote is unacceptable. Their argument is that the sanctity of human life extends well into the prenatal period to include the earliest stages of human development. Proponents of embryonic stem cell use argue that the definition of a human does not extend to the zygote, and the sanctity and quality of human life for those suffering from debilitating, often fatal, and currently incurable diseases are worth the price. Furthermore, proponents argue that the zygotes providing the embryonic stem cells were produced for some other reason, such as infertility treatment, and will more than likely be destroyed.

The ethical debate over the use of embryos to obtain stem cells probably will not end. The debate, however, may become less important because of the discovery of a way of obtaining stem cells without using embryos. In 2007, scientists reprogrammed adult cells to become stem cells that have the same capabilities as embryonic stem cells. This feat was accomplished using viruses to insert into adult cells four genes known to be active in early development, followed by treatment with protein factors. The advantage of this technique is that it does not require embryos and tissues grown from such stem cells can probably be transplanted into the donor with little risk of graft rejection (see chapter 19). A possible disadvantage is that the viruses used are know to induce cancer. Much has yet to be learned about this process and how to best coax stem cells to become specific tissues.

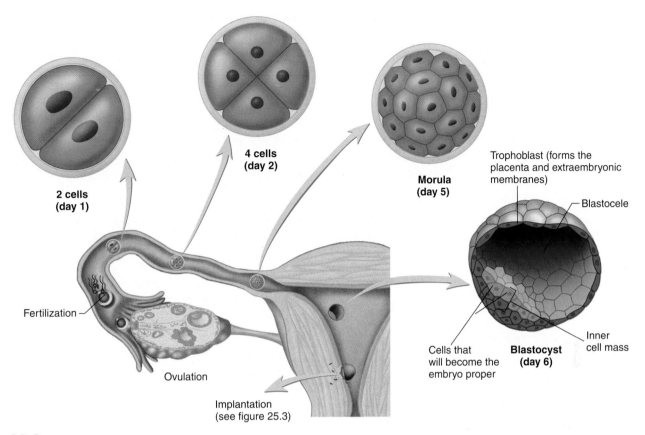

**2 cells
(day 1)**

**4 cells
(day 2)**

**Morula
(day 5)**

Trophoblast (forms the
placenta and extraembryonic
membranes)

Blastocele

Inner
cell mass

Fertilization

Ovulation

Cells that
will become the
embryo proper

**Blastocyst
(day 6)**

Implantation
(see figure 25.3)

Figure 25.2 Blastocyst

Green cells are trophoblastic, and *orange* cells will form the embryo proper.

Implantation of the Blastocyst and Development of the Placenta

All the early events in development, from the first cell division to formation of the blastocyst, occur as the embryo moves from the site of fertilization in the ampulla of the uterine tube to the uterus. About 7 days after fertilization, the blastocyst attaches to the uterine wall and begins the process of **implantation,** which is the burrowing of the blastocyst into the uterine wall.

As the blastocyst invades the uterine wall, the trophoblast gives rise to two layers of cells (figure 25.3). The outer layer is the **syncytiotrophoblast** (sin-sish′ē-ō-trō′fō-blast) and the inner layer is the **cytotrophoblast** (sī-tō-trof′ō-blast). The syncytiotrophoblast is a **syncytium,** which is a collection of cells that fuse together to form a larger, multinucleated structure. The syncytiotrophoblast invades the endometrium, breaking down glands and blood vessels. The nutrients in the glandular secretions provide nourishment for the developing embryo. As the blood vessel walls break down, cavities called **lacunae** filled with maternal blood form. As maternal blood flows through the lacunae, nutrients and gases are exchanged with the embryo.

The **placenta** is the organ of nutrient and waste product exchange between the mother and the embryo. The maternal part of the placenta is formed from the endometrium and contains the lacunae filled with maternal blood. The **chorion** (kō′rē-on), which is the embryonic portion of the placenta, is formed from the syncytiotrophoblast, cytotrophoblast, and extraembryonic mesenchyme (see figure 25.3). Branches of the chorion, called **chorionic** (kō-rē-on′ik) **villi,** protrude into the endometrium and are surrounded by the lacunae, which eventually merge together to form the **intervillous space.** As the chorionic villi form, the syncytiotrophoblast grows into the endometrium like so many roots. As a finger slides into a glove, the cytotrophoblast grows into the syncytiotrophoblast. Extraembryonic mesenchyme grows into the cytotrophoblast, and embryonic blood vessels form in the mesenchyme. In the mature placenta, the cytotrophoblast disappears so that the embryonic blood supply is separated from the maternal blood supply by a thin layer of syncytiotrophoblast, a basement membrane, extraembryonic mesenchyme, and the embryonic capillary wall (figure 25.4). Gases, nutrients, and waste products are exchanged between maternal and fetal blood, but there is no mixing of blood.

The chorion secretes **human chorionic gonadotropin** (gō′nad-ō-trō′pin) **(hCG),** which is transported in the blood to the maternal ovary and causes the corpus luteum to remain functional. The secretion of hCG begins shortly after implantation, increases rapidly, and reaches a peak about 8 or 9 weeks after fertilization. Subsequently, hCG levels decline to a lower level and are maintained at a low level throughout the remainder of the

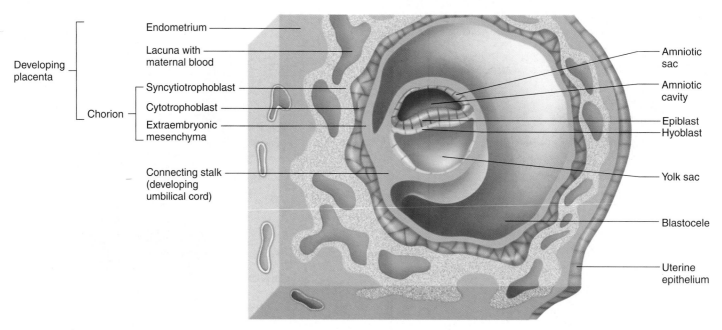

Process Figure 25.3   Formation of the Placenta

Implantation of the blastocyst and invasion of the trophoblast to form the placenta. The epiblast and hyoblast separate the amniotic cavity and yolk sac.

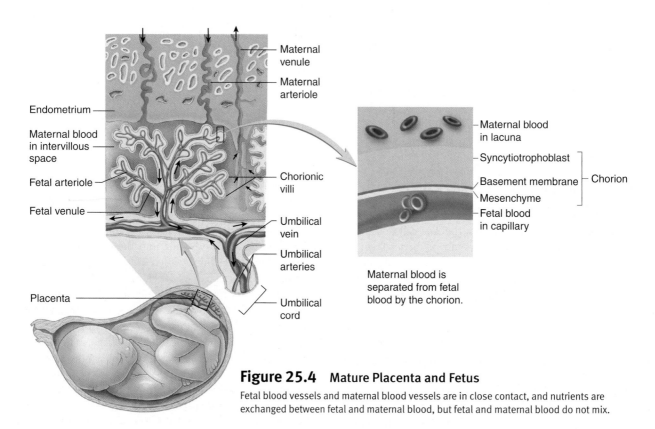

Figure 25.4   Mature Placenta and Fetus

Fetal blood vessels and maternal blood vessels are in close contact, and nutrients are exchanged between fetal and maternal blood, but fetal and maternal blood do not mix.

pregnancy (figure 25.5). Most pregnancy tests are designed to detect hCG in either urine or blood.

The estrogen and progesterone secreted by the corpus luteum (see chapter 24) are essential for the maintenance of the endometrium for the first 3 months of pregnancy. After the placenta forms, it also begins to secrete estrogens and progesterone.

By the third month of pregnancy, the placenta has become an endocrine gland that secretes sufficient quantities of estrogen and progesterone to maintain pregnancy, and the corpus luteum is no longer needed. Estrogen and progesterone levels increase in the mother's blood throughout pregnancy.

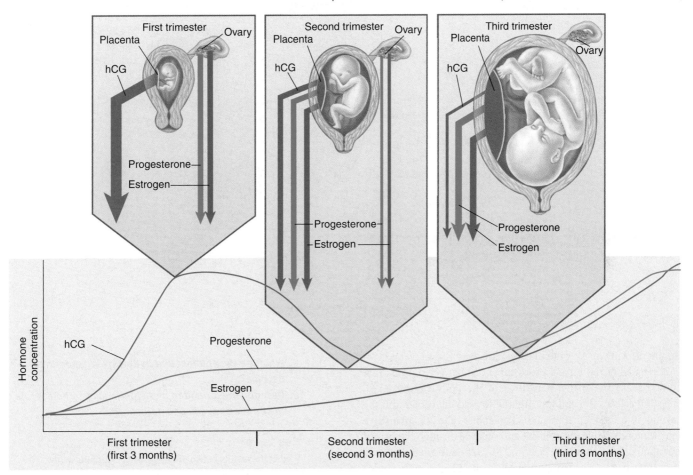

1. Human chorionic gonadotropin (hCG) increases until it reaches a maximum concentration near the end of the first 3 months of pregnancy and then decreases to a low level thereafter.

2. Progesterone continues to increase until it levels off near the end of pregnancy. Early in pregnancy, progesterone is produced by the corpus luteum in the ovary; later production shifts to the placenta.

3. Estrogen levels increase slowly throughout pregnancy, but they increase more rapidly as the end of pregnancy approaches. Early in pregnancy, estrogen is produced only in the ovary; later production shifts to the placenta.

Process Figure 25.5   **Changes in Hormone Secretion and Concentrations During Pregnancy**

The thickness of the arrows is proportional to the amount of hormone secreted from the placenta and ovaries.

 Placental Problems

Usually, the blastocyst implants in or near the fundus of the uterus. If the blastocyst implants near the cervix, a condition called **placenta previa** (prē′vē-ă) occurs. In this condition, as the placenta grows, it may extend partially or completely across the internal cervical opening. As the fetus and placenta continue to grow and the uterus stretches, the region of the placenta over the cervical opening may tear, and hemorrhaging may occur. **Abruptio** (ab-rŭp′shē-ō) **placentae** is a tearing away of a normally positioned placenta from the uterine wall accompanied by hemorrhaging. Both of these conditions can result in miscarriage and can be life-threatening to the mother. ▼

6. *Define implantation, syncytium, chorion, chorionic villi, and intervillous space.*
7. *Describe the role of hCG in maintaining pregnancy.*

Formation of the Germ Layers

After implantation, a new cavity called the **amniotic** (am-nē-ot′ik) **cavity** forms inside the inner cell mass. An extraembryonic membrane, a layer of cells called the **amniotic sac,** or **amnion** (am′nē-on), forms the boundary of the amniotic cavity (see figure 24.3). **Amniotic fluid** fills the amniotic sac, which eventually enlarges to surround the developing embryo, providing it with a protective fluid environment, the "bag of waters."

Formation of the amniotic cavity causes part of the inner cell mass nearest the blastocele to separate as a flat disk of tissue with two layers, called the **embryonic disk.** The **epiblast** is the layer adjacent to the amniotic cavity and the **hyoblast** is on the side of the disk opposite the amnion. The epiblast becomes the embryo and the hyoblast gives rise to extraembryonic tissue. An extraembryonic membrane forms a third cavity, the **yolk sac,** inside the blastocele.

About 13 or 14 days after fertilization, the embryonic disk becomes a slightly elongated, oval structure. This phase of

1. Cells in the epiblast move toward the primitive streak and migrate through the streak (*blue arrows*).

2. Cells of the epiblast that migrate through the primitive streak become mesodermal cells.

3. The mesoderm (*red*) lies between the ectoderm (*blue*) and endoderm (*yellow*). Cells of the epiblast that migrate through the primitive streak become endodermal cells.

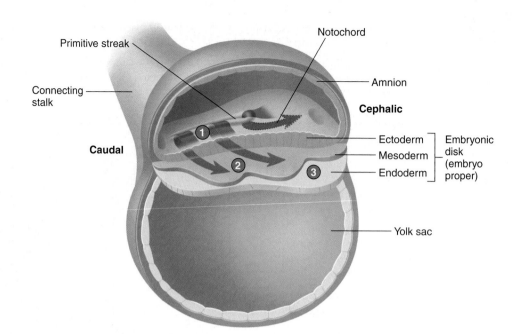

Process Figure 25.6 Embryonic Disk with a Primitive Streak

The head of the embryo will develop over the notochord.

development, known as **gastrulation,** involves the movement of cells and results in the formation of three distinct germ layers that will give rise to the many structures of the body (figure 25.6). Proliferating cells of the epiblast migrate toward the center and the caudal end of the disk, forming a thickened line called the **primitive streak.** Some cells leave the epiblast and migrate through the primitive streak to displace hyoblast cells and form the **endoderm** (en′dō-derm, inside layer). Other migrating cells collect between the endoderm and epiblast to form the **mesoderm** (mez′ō-derm, middle layer). The cells remaining in the epiblast become the **ectoderm** (ek′tō-derm, outside layer). These three germ layers, the ectoderm, mesoderm, and endoderm, are the beginning of the embryo proper. All tissues of the adult can be traced to them (table 25.1).

A specialized group of cells at the cephalic end of the primitive streak moves from one end of the primitive streak to the other and, in some as yet unknown way, organizes the embryo. A cord-like structure called the **notochord** (nō′tō-kōrd, *notos,* back + *chorde,* cord or string) is formed by these cells as they move down the primitive streak. The notochord marks the central axis of the developing embryo (see figure 25.6).

The development of the germ layers and the subsequent development of the organ systems is heavily dependent on cell communication. Some of the cell communication depends on direct cell–cell contact, whereas other communication depends on diffusible molecules, such as **growth factors.** Two important families of growth factors are epidermal growth factors (EGF) and fibroblast growth factors (FGF).▼

> 8. Describe the formation of the amniotic sac, embryonic disk, and yolk sac.

> 9. What are the epiblast and hyoblast? What do they become?
> 10. Describe the formation of the germ layers and the role of the primitive streak and notochord.

P R E D I C T   1

Predict the results of two primitive streaks forming in one embryonic disk. What happens if the two primitive streaks are touching each other?

Neural Tube and Neural Crest Formation

At about 18 days after fertilization, the ectoderm overlying the notochord thickens to form the **neural plate** (figure 25.7). The lateral edges of the plate begin to rise like two ocean waves coming together. These edges are called the **neural folds,** and a **neural groove** lies between them. The neural folds begin to meet in the midline and fuse into a neural tube, which has completely closed by day 26. The cells of the **neural tube** are called **neuroectoderm** (noor-ō-ek′tō-derm) (see table 25.1). Neuroectoderm becomes the brain, the spinal cord, and parts of the peripheral nervous system. If the neural tube fails to close, major defects of the central nervous system can result.

As the neural folds come together and fuse, a population of cells breaks away from the neuroectoderm all along the crests of the folds. Most of these **neural crest cells** become part of the peripheral nervous system or become melanocytes of the skin. In the head, neural crest cells also contribute to the skull, the dentin of teeth, blood vessels, and general connective tissue.

Table 25.1   Germ Layer Derivatives

Ectoderm	Mesoderm	Endoderm
Epidermis of skin	Dermis of skin	Lining of gastrointestinal tract
Tooth enamel	Circulatory system	Lining of lungs
Lens and cornea of eye	Parenchyma of glands	Lining of hepatic, pancreatic, and other exocrine ducts
Outer ear	Muscle	Urinary bladder
Nasal cavity	Bones (except facial)	Thymus
Anterior pituitary	Microglia	Thyroid
Neuroectoderm	Kidneys	Parathyroid
Brain and spinal cord		Tonsils
Somatic motor neurons		
Preganglionic autonomic neurons		
Neuroglia cells (except microglia)		
Posterior pituitary		
Neural crest cells		
Melanocytes		
Sensory neurons		
Postganglionic autonomic neurons		
Adrenal medulla		
Facial bones		
Teeth (dentin, pulp, and cementum) and gingiva		
A few skeletal muscles in head		

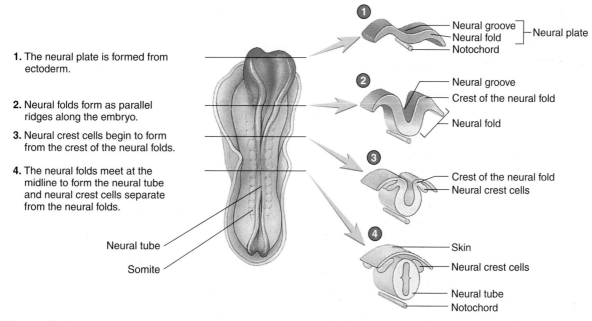

1. The neural plate is formed from ectoderm.

2. Neural folds form as parallel ridges along the embryo.

3. Neural crest cells begin to form from the crest of the neural folds.

4. The neural folds meet at the midline to form the neural tube and neural crest cells separate from the neural folds.

Neural tube

Somite

Neural groove
Neural fold — Neural plate
Notochord

Neural groove
Crest of the neural fold
Neural fold

Crest of the neural fold
Neural crest cells

Skin
Neural crest cells
Neural tube
Notochord

Process Figure 25.7   Formation of the Neural Tube

The neural folds come together in the midline and fuse to form a neural tube. This fusion begins in the center and moves both cranially and caudally. The embryo shown is about 21 days after fertilization. The insets to the *right* show progressive closure of the neural tube. The *lines* indicate locations of cross sections.

Neural Tube Defects

Anencephaly (an'en-sef'ă-lē, no brain) is a birth defect in which much of the brain fails to form. It results when the neural tube fails to close in the region of the head. A baby born with anencephaly cannot survive. **Spina bifida** (spī'nă bi'fi-dă, split spine) is a general term describing defects of the spinal cord, vertebral column, or both (figure A). Spina bifida can range from a simple defect with no clinical manifestations and with one or more vertebral spinous processes split or missing to a more severe defect that results in paralysis of the limbs or the bowels and bladder, depending on where the defect occurs.

The inclusion of **folic acid,** the B vitamin folate, in a woman's diet during the early stages of her pregnancy significantly reduces the risk for neural tube defects in her developing embryo.▼

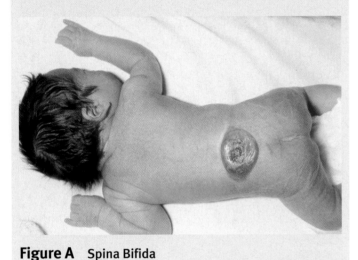

Figure A   Spina Bifida

> 11. Describe the formation of the neural tube and neural crest cells.

Somite Formation

As the neural tube develops, the mesoderm immediately adjacent to the tube forms distinct segments called **somites** (sō'mītz) (see figure 25.7). In the head, the first few somites never become clearly divided but develop into indistinct, segmented structures called **somitomeres.** The somites and somitomeres eventually give rise to part of the skull, the vertebral column, and skeletal muscle. Most of the head muscles are derived from the somitomeres.▼

> 12. What are a somite and a somitomere?

Formation of the Gut and Body Cavities

At the same time that its neural tube is forming, the embryo itself is becoming a tube along the upper part of the yolk sac. The **foregut** and **hindgut** develop as the cephalic and caudal ends of the yolk sac are separated from the main yolk sac. This is the beginning of the digestive tract (figure 25.8a). The developing digestive tract pinches off from the yolk sac as a tube but remains attached in the center to the yolk sac by a yolk stalk.

The ends of the foregut and hindgut (figure 25.8b) are in close relationship to the overlying ectoderm and form membranes called the oropharyngeal membrane and the cloacal membrane, respectively. The **oropharyngeal membrane** opens to form the mouth, and the **cloacal membrane** opens to form the urethra and anus. Thus, the digestive tract becomes a tube that opens to the outside at both ends.

A considerable number of **evaginations** (ē-vaj-i-nā'shŭnz, outpocketings) occur along the early digestive tract (figure 25.8c). The first to form is the allantois (see figure 25.8b), part of which will form the urinary bladder. Other evaginations develop into structures such as the anterior pituitary, the thyroid gland, the lungs, the liver, the pancreas, and the urinary bladder. At the same time, solid bars of tissue known as **branchial** (brang'kē-ăl, gill) **arches** (see figure 25.8c and figure 25.9) form along the lateral sides of the head, and the sides of the foregut expand as pockets between the branchial arches. The central expanded foregut is called the **pharynx,** and the pockets along both sides of the pharynx are called **pharyngeal pouches.** Adult derivatives of the pharyngeal pouches include the auditory tubes, tonsils, thymus, and parathyroids.

At about the same time the gut is developing, a series of isolated cavities forms in the mesoderm; they fuse together to form the **celom** (sē'lom) (see figure 25.8b and c), or body cavity. Eventually, the celom separates into three distinct adult cavities, the **pericardial, pleural, and peritoneal cavities** (see chapter 1).▼

> 13. Describe the formation of the gut and body cavities.

Limb Bud Development

Arms and legs first appear at about 28 days as **limb buds** (see figure 25.9). As the limb buds elongate, limb tissues are laid down in a proximal-to-distal sequence. For example, in the upper limb, the arm is formed before the forearm, which is formed before the hand.▼

> 14. What is a limb bud? What is meant by a proximal-to-distal growth sequence of the limb?

Development of the Face

The face develops by the fusion of five embryonic structures: the **frontonasal process,** which forms the forehead, nose, and midportion of the upper jaw and lip; two **maxillary processes,** which form the lateral parts of the upper jaw and lip; and two **mandibular processes,** which form the lower jaw and lip (figure 25.10, step 1). **Nasal placodes** (plak'ōdz), which develop at the lateral margins of the frontonasal process, develop into the nose and the center of the upper jaw and lip (figure 25.10, step 2).

As the brain enlarges and the face matures, the nasal placodes approach each other in the midline. The medial edges of the placodes fuse to form the midportion of the upper jaw and lip (figure 25.10, steps 3–5). This part of the frontonasal process is between the two maxillary processes, which are expanding toward the midline, and fuses with them to form the upper jaw and lip, known as the **primary palate** (see figure 25.10, step 4).

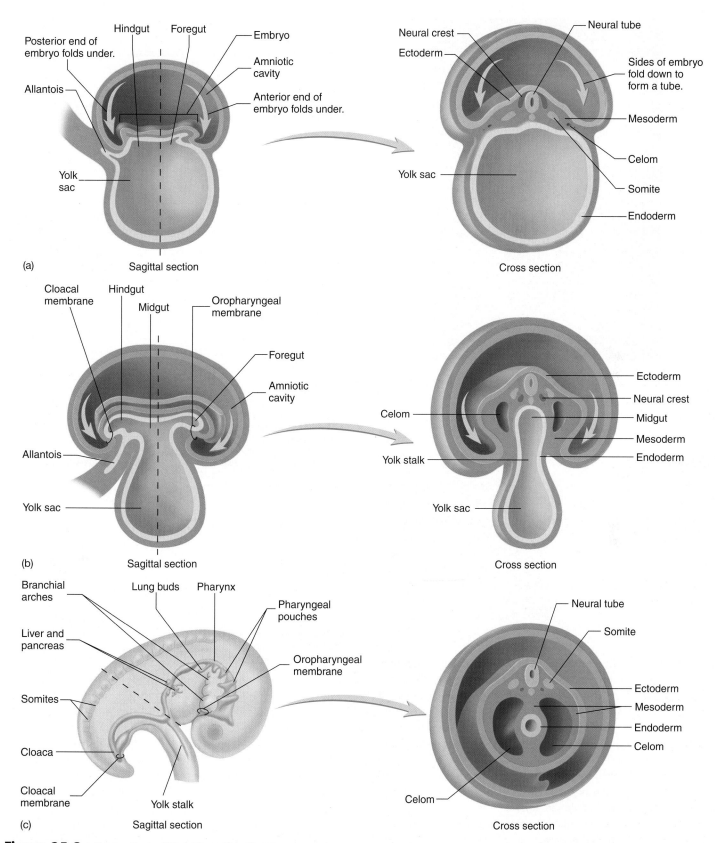

Figure 25.8 Formation of the Digestive Tract

Blue arrows show the folding of the digestive tract into a tube. *Dotted lines* show the plane of the section from which cross sections were taken. (*a*) Twenty days after fertilization. (*b*) Twenty-five days after fertilization. (*c*) Thirty days after fertilization. Evaginations are identified along the pharynx and digestive tract.

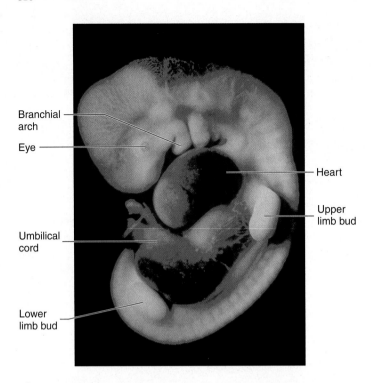

Branchial arch
Eye
Umbilical cord
Lower limb bud
Heart
Upper limb bud

Figure 25.9 Human Embryo 35 Days After Fertilization

Cleft Lip

A **cleft lip** results from failure of the frontonasal and one or both maxillary processes to fuse (see figure 25.10). Cleft lips usually occur to one side (or both sides) because the frontonasal process is a midline structure that normally fuses with the two lateral maxillary processes during formation of the primary palate. The cleft can vary in severity from a slight indentation in the lip to a fissure that extends from the mouth to the nares (nostril).

At about the same time that the primary palate is forming, the lateral edges of the nasal placodes fuse with the maxillary processes to close off the groove extending from the mouth to the eye (see figure 25.10, steps 4 and 5). On rare occasions, these structures fail to meet, resulting in a facial cleft extending from the mouth to the eye. The inferior margins of the maxillary processes fuse with the superior margins of the mandibular processes to decrease the size of the mouth. All of the previously described fusions and the growth of the brain give the face a recognizably "human" appearance by about 50 days.

The roof of the mouth, known as the **secondary palate,** forms as vertical shelves that swing to a horizontal position and begin to fuse with each other at about 56 days of development. Fusion of the entire palate is not completed until about 90 days. If the secondary palate does not fuse, a midline fissure in the roof of the mouth, called a **cleft palate,** results. A cleft palate can range in severity from a slight cleft of the uvula (see figure 21.4) to a fissure

1. 28 days after fertilization
The face develops from five processes: frontonasal (*blue*), two maxillary (*yellow*), and two mandibular (*orange;* already fused).

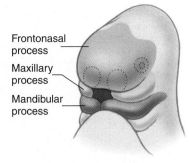

Frontonasal process
Maxillary process
Mandibular process

2. 33 days after fertilization
Nasal placodes appear in the frontonasal process.

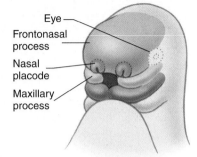

Eye
Frontonasal process
Nasal placode
Maxillary process

3. 40 days after fertilization
Maxillary processes extend toward the midline. The nasal placodes also move toward the midline and fuse with the maxillary processes to form the jaw and lip.

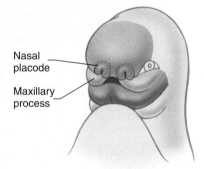

Nasal placode
Maxillary process

4. 48 days after fertilization
Continued growth brings structures more toward the midline.

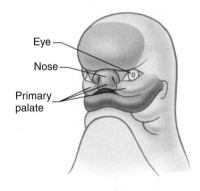

Eye
Nose
Primary palate

5. 14 weeks after fertilization
Colors show the contributions of each process to the adult face.

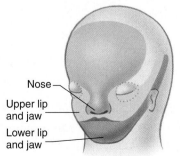

Nose
Upper lip and jaw
Lower lip and jaw

Process Figure 25.10 Development of the Face

extending the entire length of the palate. A cleft lip and cleft palate can occur together, forming a continuous fissure.▼

> **15. Describe the processes involved in the formation of the face. What clefts can result if these processes fail to fuse?**

Development of the Organ Systems

The major organ systems appear and begin to develop during the embryonic period. The period between 14 and 60 days is therefore called the period of **organogenesis** (ōr′gă-nō-jen′ĕ-sis) (table 25.2).

Integumentary System

The **epidermis** of the skin is derived from ectoderm, and the **dermis** is derived from mesoderm, or from neural crest cells, in the case of the face. Nails, hair, and glands develop from the epidermis (see chapter 5). Melanocytes and sensory receptors in the skin are derived from neural crest cells.▼

> **16. List the developmental sources of the epidermis, dermis, nails, hair, glands, melanocytes, and sensory receptors of the skin.**

Skeletal System

The skeleton is derived from mesoderm or neural crest cells and forms through intramembranous or endochondral bone formation (see chapter 6). The bones of the face develop from neural crest cells, whereas the rest of the skull, the vertebral column, and the ribs develop from somite- or somitomere-derived mesoderm. The appendicular skeleton develops from limb bud mesoderm.▼

> **17. From what developmental sources is the skeletal system derived?**

Skeletal Muscle

Myoblasts (mī′ō-blastz) are the early, embryonic cells that give rise to skeletal muscle fibers. They migrate from somites or somitomeres to sites of future muscle development. The myoblasts continue to divide and begin to fuse and form multinucleated cells called **myotubes,** which enlarge to become the muscle fibers of the skeletal muscles. Shortly after myotubes form, nerves grow into the area and innervate the developing muscle fibers. After the basic form of each muscle is established, an increase in the number of muscle fibers causes continued muscle growth. The total number of muscle fibers is established before birth and remains relatively constant thereafter. Muscle enlargement after birth results from an increase in the size of individual fibers.▼

> **18. Describe the formation of skeletal muscles from myoblasts.**

Nervous System

The nervous system is derived from the neural tube and neural crest cells (see figure 25.7). Neural tube closure begins at about 21 days of development in the upper cervical region and proceeds into the head and down the spinal cord. Soon after the neural tube has closed at about 25 days of development, the part of the neural tube that becomes the brain begins to expand and develops a series of pouches (figure 25.11). The pouch walls become the various portions of the adult brain, and the pouch cavities become fluid-filled **ventricles** (ven′tri-klz). The ventricles are continuous with each other and with the **central canal** of the spinal cord. The neural tube develops flexures that cause the brain to be oriented almost 90 degrees to the spinal cord.

Three brain regions can be identified in the early embryo: a forebrain, or **prosencephalon** (pros-en-sef′ă-lon); a midbrain, or **mesencephalon** (mez-en-sef′ă-lon); and a hindbrain, or **rhombencephalon** (rom-ben-sef′ă-lon). During development, the forebrain divides into the **telencephalon** (tel-en-sef′ă-lon), which becomes the cerebrum, and the **diencephalon** (dī-en-sef′ă-lon). The midbrain remains as a single structure, but the hindbrain divides into the **metencephalon** (met′en-sef′ă-lon), which becomes the pons and cerebellum, and the **myelencephalon** (mī′el-en-sef′ă-lon), which becomes the medulla oblongata (see figure 25.11).

The neuron cell bodies of somatic motor neurons and preganglionic neurons of the autonomic nervous system, which provide axons to the peripheral nervous system, are located within the neural tube. Sensory neurons and postganglionic neurons of the autonomic nervous system are derived from neural crest cells.

 Alcohol and Cigarette Smoke

A number of drugs and other chemicals are known to affect the embryo and fetus during development. The two most common are alcohol and cigarette smoke. Alcohol consumption can result in **fetal alcohol syndrome,** which includes decreased mental function. Though excessive alcohol consumption, such as alcoholism and binge drinking, are known to cause fetal alcohol syndrome, research results are inconsistent about the effects of lower levels of consumption. Most physicians recommend that alcohol not be consumed at all during pregnancy. Exposure of the fetus to **cigarette smoke** throughout pregnancy can stunt the physical growth and mental development of the fetus.▼

> **19. Name the five divisions of the neural tube and the part of the brain that each division becomes.**
>
> **20. What do the cavities of the neural tube become in the adult brain?**

Special Senses

The **olfactory bulbs** and **nerves** develop as an evagination from the telencephalon. The eyes develop as evaginations from the diencephalon. Each evagination elongates to form an **optic stalk,** and a bulb called the **optic vesicle** develops at its terminal end (see figure 25.11). The optic vesicle reaches the side of the head and stimulates the overlying ectoderm to thicken into a **lens.** The sensory part of the ear appears as an ectodermal thickening, or placode, that invaginates and pinches off from the overlying ectoderm.▼

> **21. Describe the development of the special senses.**

Table 25.2 Development of the Organ Systems

	Age (Days Since Fertilization)					
	1–5	**6–10**	**11–15**	**16–20**	**21–25**	**26–30**
General Features	Fertilization Morula Blastocyst	Blastocyst implants	Primitive streak Three germ layers	Neural plate	Neural tube closed	Limb buds and other "buds" appear.
Integumentary System			Ectoderm Mesoderm			Melanocytes from neural crest
Skeletal System			Mesoderm		Neural crest (will form facial bones)	Limb buds
Muscular System			Mesoderm	Somites begin to form.		Somites all present
Nervous System			Ectoderm	Neural plate	Neural tube complete Neural crest Eyes and ears begin.	Lens begins to form.
Endocrine System			Ectoderm Mesoderm Endoderm	Thyroid begins to develop.		Parathyroids appear.
Cardiovascular System			Mesoderm	Blood islands form. Two heart tubes	Single-tubed heart begins to beat.	Interatrial septum begins to form.
Lymphatic System			Mesoderm			Thymus appears.
Respiratory System			Mesoderm Endoderm		Diaphragm begins to form.	Trachea forms as single bud. Lung buds (primary bronchi)
Digestive System			Mesoderm Endoderm		Neural crest (will form tooth dentin) Foregut and hindgut form.	Liver and pancreas appear as buds. Tongue bud appears.
Urinary System			Mesoderm Endoderm		Pronephros develops. Allantois appears.	Mesonephros appears.
Reproductive System			Mesoderm Endoderm		Primordial germ cells on yolk sac	Mesonephros appears. Genital tubercle forms.

Age (Days Since Fertilization)					
31–35	**36–40**	**41–45**	**46–50**	**51–55**	**56–60**
Hand and foot plates on limbs	Fingers and toes appear. Lips formed Embryo 15 mm	External ear forming Embryo 20 mm	Embryo 25 mm	Limbs elongate to a more adult relationship. Embryo 35 mm	Face is distinctly human in appearance.
Sensory receptors appear in skin.		Collagen fibers clearly present in skin		Extensive sensory endings in skin	
Mesoderm condensation in areas of future bone	Cartilage in site of future humerus	Cartilage in site of future ulna and radius	Cartilage in site of hand and fingers		Ossification begins in clavicle and then in other bones.
Muscle precursor cells enter limb buds.			Functional muscle		Nearly all muscles appear in adult form.
Nerve processes enter limb buds.		External ear forming Olfactory nerves begin to form.		Semicircular canals in inner ear complete	Eyelids form. Cochlea in inner ear complete
Pituitary appears as evaginations from brain and mouth.	Gonadal ridges form. Adrenal glands forming		Pineal body appears.	Thyroid gland in adult position and attachment to tongue lost	Anterior pituitary loses its connection to the mouth.
Interventricular septum begins to form.		Interventricular septum complete	Interatrial septum complete but still has opening until birth		
Large lymphatic vessels form in neck.	Spleen appears.			Adult lymph pattern forms.	
Secondary bronchi to lobes form.	Tertiary bronchi to bronchopulmonary segments form.		Tracheal cartilage begins to form.		
Oropharyngeal membrane ruptures.		Secondary palate begins to form. Tooth buds begin to form.			Secondary palate begins to fuse (fusion complete by 90 days).
Metanephros begins to develop.				Mesonephros degenerates.	Anal portion of cloacal membrane ruptures.
	Gonadal ridges form.	Primordial germ cells enter gonadal ridges.	Paramesonephric ducts appear.		Uterus forming Beginning of differentiation of external genitalia in male and female

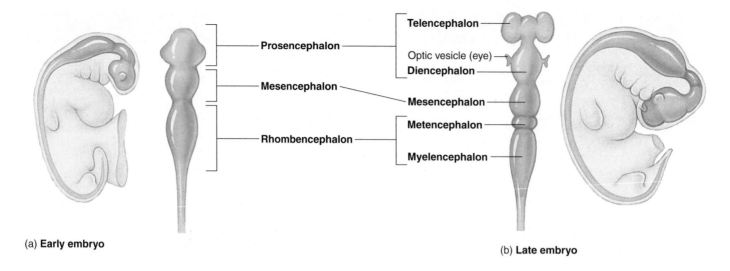

(a) **Early embryo**

(b) **Late embryo**

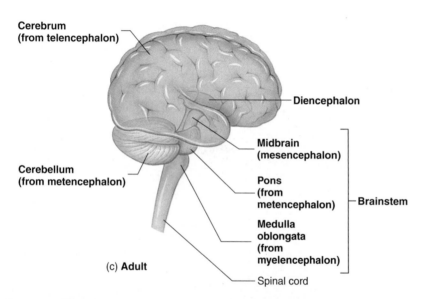

(c) **Adult**

Process Figure 25.11 Development of the Brain

(*a*) The neural tube forms three pouches. The walls of the pouches become parts of the brain (note color coding) and the lumen of the tube becomes the ventricles and central canal of the spinal cord (*b*) Two of the pouches divide, resulting in five pouches. (*c*) The adult brain.

Endocrine System

An evagination from the floor of the diencephalon forms the **posterior pituitary gland.** The **anterior pituitary gland** develops from an evagination of ectoderm in the roof of the embryonic oral cavity and grows toward the floor of the brain. It eventually loses its connection with the oral cavity and becomes attached to the posterior pituitary gland (see chapter 15).

The **thyroid gland** originates as an evagination from the floor of the pharynx in the region of the developing tongue and moves into the lower neck, eventually losing its connection with the pharynx. The **parathyroid glands,** which are derived from the third and fourth pharyngeal pouches, migrate inferiorly and become associated with the thyroid gland.

The **adrenal medulla** arises from neural crest cells and consists of specialized postganglionic neurons of the sympathetic

division of the autonomic nervous system (see chapter 14). The **adrenal cortex** is derived from mesoderm.

The **pancreas** originates as two evaginations from the duodenum, which come together to form a single gland.▼

> 22. *Explain the formation of the following endocrine glands: anterior pituitary, posterior pituitary, thyroid, parathyroid, adrenal medulla, adrenal cortex, and pancreas.*

Circulatory System

The heart develops from two endothelial tubes (figure 25.12, step 1), which fuse into a single, midline heart tube (figure 25.12, step 2). Blood vessels and blood cells form from blood islands on the surface of the yolk sac and inside the embryo. **Blood islands** are small masses of mesoderm that become blood vessels on the

1. **20 days after fertilization**
   At this age, the heart consists of two parallel tubes.

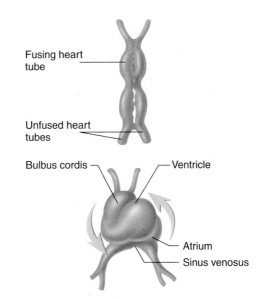

Fusing heart tube

Unfused heart tubes

2. **22 days after fertilization**
   The two parallel tubes have fused to form one tube. This tube bends as it elongates (*blue arrows* suggest the direction of bending) within the confined space of the pericardium.

Bulbus cordis — Ventricle

Atrium

Sinus venosus

3. **31 days after fertilization**
   The interatrial septum (septum primum, *green*) and the interventricular septum grow toward the center of the heart.

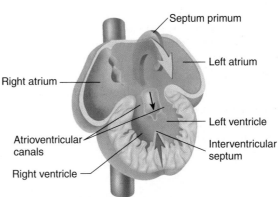

Septum primum

Left atrium

Right atrium

Left ventricle

Atrioventricular canals

Right ventricle

Interventricular septum

4. **35 days after fertilization**
   The interventricular septum is nearly complete. A foramen opens in the septum primum (*green*) as the septum secundum begins to form (*blue*).

Septum secundum

Septum primum

Foramen

Interventricular septum

5. **The final embryonic condition of the interatrial septum**
   Blood from the right atrium can flow through the foramen ovale into the left atrium. After birth, as blood begins to flow in the other direction, the flaplike septum primum is forced against the septum secundum, closing the foramen ovale.

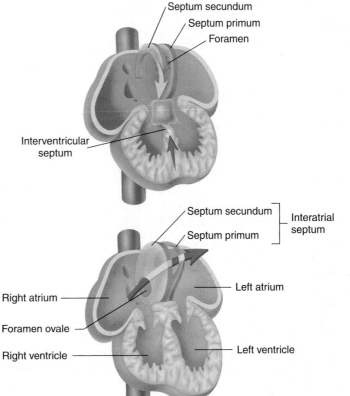

Septum secundum

Septum primum

Interatrial septum

Right atrium

Left atrium

Foramen ovale

Right ventricle

Left ventricle

Process Figure 25.12   Development of the Heart

outside and blood cells on the inside. These islands expand and fuse to form the circulatory system. A series of dilations appears along the length of the primitive heart tube, and four major regions can be identified: the **sinus venosus,** the site where blood enters the heart; a single **atrium;** a single **ventricle;** and the **bulbus cordis,** where blood exits the heart (see figure 25.12, step 2).

The elongating heart, confined within the pericardium, becomes bent into a loop, the apex of which is the ventricle (see figure 25.12, step 2). The major chambers of the heart, the atrium and the ventricle, expand rapidly. The right part of the sinus venosus becomes absorbed into the atrium, and the bulbus cordis is absorbed into the ventricle. The embryonic sinus venosus initiates contraction at one end of the tubular heart. Later in development, part of the sinus venosus becomes the sinoatrial node, which is the adult pacemaker.

P R E D I C T

What would happen if the sinus venosus did not contract before other areas of the primitive heart?

The **interatrial septum,** which separates the two atria in the adult heart, is formed from two parts: the **septum primum** (primary septum) and the **septum secundum** (secondary septum). An opening in the interatrial septum called the **foramen ovale** (ō-val'ē) connects the two atria and allows blood to flow from the right to the left atrium in the embryo and fetus. An **interventricular septum** (figure 25.12, steps 3–5) develops, which divides the single ventricle into two chambers.

⚕ Heart Defects

If the septum secundum fails to grow far enough or if the foramen in the septum primum becomes too large, an **atrial septal defect (ASD)** occurs, allowing blood to flow from the left atrium to the right atrium in the newborn. If the interventricular septum does not grow enough to completely separate the ventricles, a **ventricular septal defect (VSD)** results. VSDs are more common than ASDs. Both ASDs and VSDs result in abnormal heart sounds called **heart murmurs.** Blood passes through the ASD and VSD from the left to right side of the heart. The right side of the heart usually hypertrophies. In many cases, septal defects are not serious. In severe cases of VSD (Eisenmenger syndrome), the increased pressure in the pulmonary blood vessels and a decreased blood flow through the systemic blood vessels result in pulmonary edema, cyanosis (a bluish color due to deficient blood oxygenation), or heart failure.▼

---

23. *Explain the process whereby a one-chambered heart becomes a four-chambered heart.*

Respiratory System

The lungs begin to develop as a single midline evagination from the foregut in the region of the future esophagus. This evagination branches to form two **lung buds** (figure 25.13, step 1). The lung buds elongate and branch, forming the main bronchi that project to the lungs. The main bronchi branch, forming the lobar bronchi, which project to the lobes of the lungs (figure 25.13, step 2). The

---

1. **28 days after fertilization**
A single bud forms and divides into two buds, which will become the lungs and main bronchi.

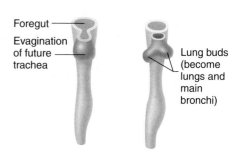

Foregut
Evagination of future trachea
Lung buds (become lungs and main bronchi)

2. **32 days after fertilization**
Main bronchi branch to form lobar bronchi, which supply the lung lobes.

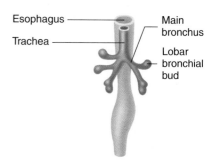

Esophagus
Trachea
Main bronchus
Lobar bronchial bud

3. **35 days after fertilization**
Lobar bronchi branch to form segmental bronchi, which supply the bronchopulmonary segments.

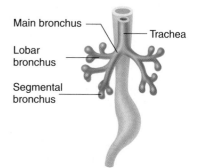

Main bronchus
Trachea
Lobar bronchus
Segmental bronchus

4. **50 days after fertilization**
The branching will continue to eventually form the extensive respiratory passages in the lungs.

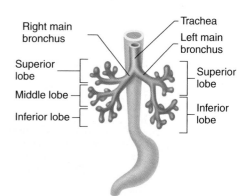

Right main bronchus
Trachea
Left main bronchus
Superior lobe
Superior lobe
Middle lobe
Inferior lobe
Inferior lobe

Process Figure 25.13   **Development of the Lung**

lobar bronchi branch, forming the segmental bronchi, which project to the bronchopulmonary segments of the lungs (figure 25.13, step 3). This branching continues (figure 25.13, step 4) until, by the end of the sixth month, about 17 generations of branching have occurred. Even after birth, some branching continues as the lungs grow larger, and in the adult about 23 generations of branches have been established.▼

24. Describe the formation of the bronchi and lungs.

Urinary System

The kidneys develop from mesoderm located between the somites and the lateral part of the embryo. About 21 days after fertilization, the mesoderm in the cervical region differentiates into a structure called the **pronephros** (the most forward or earliest kidney) (figure 25.14a), which consists of a duct and simple tubules connecting the duct to the open celomic cavity. This type of kidney is the functional adult kidney in some lower chordates, but it is probably not functional in the human embryo and soon disappears.

The **mesonephros** (middle kidney) (see figure 25.14a) is a functional organ in the embryo. It consists of a duct, which is a caudal extension of the pronephric duct, and a number of minute tubules, which are smaller and more complex than those of the pronephros. One end of each tubule opens into the mesonephric duct, and the other end forms a glomerulus (see chapter 23).

As the mesonephros is developing, the caudal end of the hindgut begins to enlarge to form the **cloaca** (klō-ā′kă, sewer), the common junction of the digestive, urinary, and genital systems (figure 25.14b). A **urorectal septum** divides the cloaca into two parts: a digestive part called the **rectum** and a urogenital part called the **urethra** (figure 24.14c). The cloaca has two tubes associated with it: the hindgut and the **allantois** (ă-lan′tō-is, sausage), which is a blind tube extending into the umbilical cord (see figures 25.9 and 25.14). The part of the allantois nearest the cloaca enlarges to form the urinary bladder, and the remainder, which is from the bladder to the umbilicus, forms the median umbilical ligament (figure 25.14d).

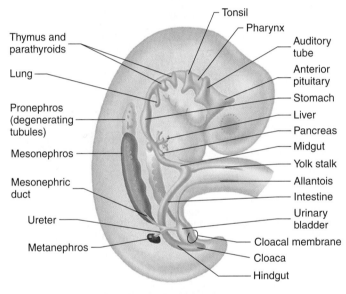

(a) The three parts of the developing kidney: pronephros, mesonephros, metanephros

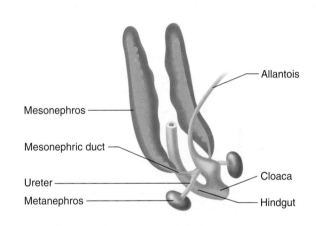

(b) The metanephros (adult kidney) enlarges as the mesonephros degenerates.

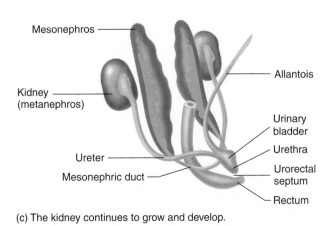

(c) The kidney continues to grow and develop.

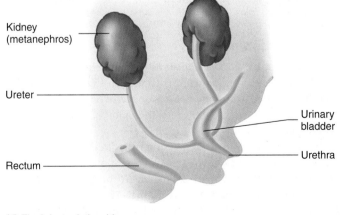

(d) Final duct relationships

Figure 25.14 **Development of the Kidney and Urinary Bladder**

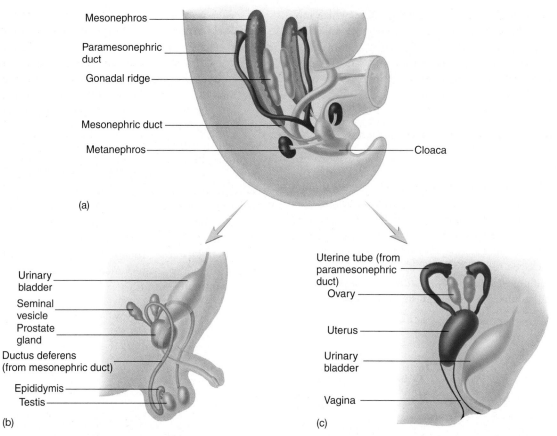

Figure 25.15 Development of the Reproductive System

(*a*) Indifferent stage. (*b*) The male, under the influence of male hormones, develops a ductus deferens from the mesonephric duct, and the paramesonephric duct degenerates. (*c*) The female, without male hormones, develops a uterus and uterine tubes from the paramesonephric duct, and the mesonephros disappears.

The mesonephric duct extends caudally as it develops and eventually joins the cloaca. At the point of junction, another tube, the **ureter,** begins to form. Its distal end enlarges and branches to form the duct system of the adult kidney, called the **metanephros** (last kidney), which takes over the function of the degenerating mesonephros.▼

> 25. *Describe the development of the pronephros, mesonephros, and metanephros in the development of the kidneys.*

Reproductive System

The male and female gonads appear as **gonadal ridges** along the ventral border of each mesonephros (figure 25.15*a*). **Primordial germ cells,** destined to become oocytes or sperm cells, form on the surface of the yolk sac, migrate into the embryo, and enter the gonadal ridge.

In the female, the ovaries descend from their original position in the abdominal cavity to a position in the pelvic cavity. In the male, the testes descend even farther. As the testes reach the anteroinferior abdominal wall, two tunnels called the **inguinal canals** form through the abdominal musculature. The testes pass through these canals, leaving the abdominal cavity and coming to lie within the **scrotum** (see figure 24.5). Descent of

the testes through the canals begins about 7 months after conception, and the testes enter the scrotum about 1 month before the infant is born.

Cryptorchidism

In approximately 3% of male children, one or both testes fail to enter the scrotum. This condition is called undescended testes, or **cryptorchidism** (krip-tōr'ki-dizm). Because testosterone is required for the testes to descend into the scrotum, cryptorchidism is often the result of inadequate testosterone secreted by the fetal testes. If neither testis descends and the defect is not corrected, the male will be infertile because the slightly higher temperature of the body cavity, compared with that of the scrotal sac, causes the spermatogonia to degenerate. Cryptorchidism is an important risk factor for testicular cancer. Cryptorchidism is treated with hormone therapy or surgery.

Paramesonephric ducts (also called müllerian ducts) begin to develop just lateral to the **mesonephric ducts** (also called wolffian ducts) and grow inferiorly to meet one another, where they enter the cloaca as a single, midline tube.

In male embryos, testosterone is secreted by the testes. It causes the mesonephric duct system to enlarge and differentiate to form the epididymis, ductus deferens, seminal vesicles, and prostate gland (figure 25.15*b*). Müllerian-inhibiting hormone, which the testes also secrete, causes the paramesonephric müllerian ducts to degenerate. In female embryos, neither testosterone nor müllerian-inhibiting hormone is secreted. As a result, the mesonephric duct system atrophies, and the paramesonephric duct system develops to form the uterine tubes, the uterus, and part of the vagina (figure 25.15*c*).

Like the other sexual organs, the external genitalia begin as the same structures in the male and female and then diverge (figure 25.16). An enlargement called the **genital tubercle** develops in the groin of the embryo. **Genital folds** develop on each side of a **urethral groove**, and **labioscrotal swellings** develop lateral to the folds.

In the male, under the influence of testosterone, the genital tubercle and the genital folds close over the urethral groove to form the penis. If this closure does not proceed all the way to the end of the penis, a defect known as **hypospadias** (hī′pō-spā′dē-ăs) results. The testes move into the labioscrotal swellings, which become the scrotum of the male.

In the female, in the absence of testosterone, the genital tubercle becomes the clitoris. The urethral groove disappears; genital folds do not fuse. As a result, the urethra opens somewhat posterior to the clitoris but anterior to the vaginal opening. The unfused genital folds become the labia minora, and the labioscrotal folds become the labia majora.▼

26. *Describe the effect of hormones on the development of the male and female reproductive systems.*
27. *Compare the male and female structures formed from each of the following: genital tubercle, genital folds, and labioscrotal swellings.*

P R E D I C T ③
How would the failure to produce müllerian-inhibiting hormone affect the development of the internal reproductive system and external genitalia in a male embryo?

Growth of the Fetus

The embryo becomes a **fetus** approximately 60 days after fertilization (figure 25.17). In the embryo, the organ systems are developing, whereas in the fetus the organs grow and mature. The fetus

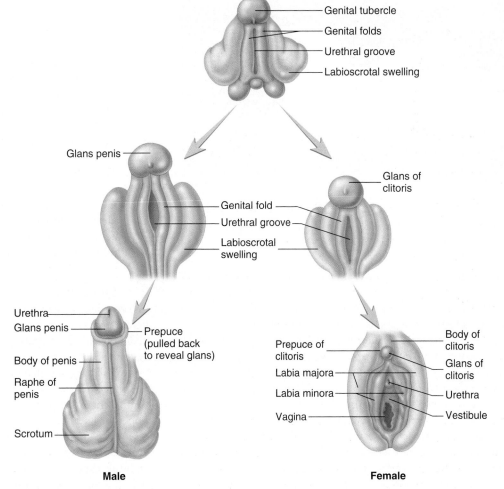

1. At approximately 5 weeks after fertilization, the genital tubercle, genital folds, and labioscrotal swellings are the same for the male and female.

2. By 10 weeks of development, the male penile structures are somewhat larger than those of the female clitoris. The urethral groove is still open in the male and female.

3. Near term, the general adult condition is achieved. The labioscrotal swellings become the scrotum in the male, the labia majora in the female. The genital folds fuse in the male to form the body of the penis, whereas they form the labia minora in the female. The urethral groove in the male is closed by the raphe of the penis. It remains open in the female as the vestibule.

Process Figure 25.16   **Development of the External Genitalia**

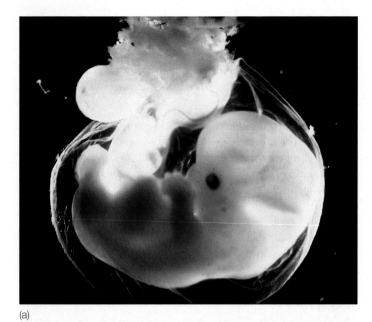

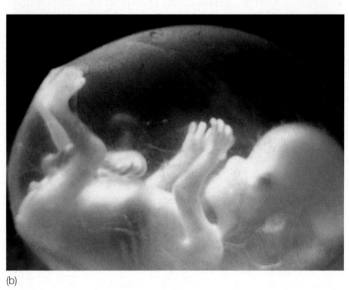

Figure 25.17 **Embryos and Fetuses at Different Ages**
(*a*) Embryo 50 days after fertilization. (*b*) Fetus 3 months after fertilization.

grows on average from about 3 cm and 2.5 g (0.09 oz) at 8 weeks to 50 cm and 3300 g (7 lb, 4 oz) at the end of pregnancy. The growth during the fetal period represents more than a 15-fold increase in length and a 1400-fold increase in weight.

Fine, soft hair called **lanugo** (la-noo′gō) covers the fetus, and a waxy coat of sloughed epithelial cells called **vernix caseosa** (ver′niks kā-se-ō′să) protects the fetus from the somewhat toxic nature of the amniotic fluid formed by the accumulation of waste products from the fetus.

Subcutaneous fat that accumulates in the older fetus and newborn provides a nutrient reserve, helps insulate the baby, and aids the baby in sucking by strengthening and supporting the cheeks so that negative pressure can be developed in the oral cavity.

Peak body growth occurs late in gestation, but, as placental size and blood supply limits are approached, the growth rate slows.

Growth of the placenta essentially stops at about 35 weeks, thus restricting further intrauterine growth.▼

28. What major events distinguish embryonic development from fetal development?

Labor

Labor, childbirth, or **parturition** (par-toor-ish′ŭn) is the process by which a baby is born. Physicians usually calculate the **gestation period,** or length of the pregnancy, as 280 days (40 weeks) from the last menstrual period (LMP) to the date of delivery of the infant.

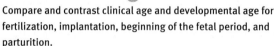

P R E D I C T **4**
Compare and contrast clinical age and developmental age for fertilization, implantation, beginning of the fetal period, and parturition.

Labor results in the expulsion of the fetus and placenta from the uterus. Although labor may differ greatly from woman to woman and from one pregnancy to another for the same woman, it can usually be divided into three stages (figure 25.18):

1. The **first stage** begins with the onset of regular uterine contractions and extends until the cervix dilates to a diameter about the size of the fetus's head (10 cm). This stage takes approximately 24 hours, but it may be as short as a few minutes in some women who have had more than one child. During this stage of labor, the amniotic sac ruptures, releasing the amniotic fluid. This event is commonly referred to as when the "water breaks."
2. The **second stage** lasts from the time of maximum cervical dilation until the baby exits the vagina. This stage may last from 1 minute to up to 1 hour. During this stage, contraction of the woman's abdominal muscles assists the uterine contractions.
3. The **third stage,** often called the placental stage, involves the expulsion of the placenta from the uterus. Contractions of the uterus cause the placenta to tear away from the wall of the uterus. Some bleeding occurs because of the intimate contact between the placenta and the uterus; however, bleeding normally is restricted because uterine smooth muscle contractions compress the blood vessels to the placenta.

During the 4 or 5 weeks following delivery, the uterus becomes much smaller, but it remains somewhat larger than it was before pregnancy. A vaginal discharge composed of small amounts of blood and degenerating endometrium can persist for several days after delivery.

The precise signal that triggers labor is unknown, but many of the factors that support it have been identified. Before labor begins, the progesterone concentration in the maternal circulation is at its highest level. Progesterone has an inhibitory effect on uterine smooth muscle cells. Near the end of pregnancy, however, estrogen levels rapidly increase in the maternal circulation, and the excitatory influence of estrogens on uterine smooth muscle cells overcomes the inhibitory influence of progesterone (see figure 25.5).

Amniocentesis (am'nē-ō-sen-tē'sis) is the removal of amniotic fluid from the amniotic cavity (figure B). As the fetus develops, it expels molecules of various types, as well as living cells, into the amniotic fluid. These molecules and cells can be collected and analyzed. A number of normal conditions can be evaluated, and a number of metabolic disorders can be detected by analyzing the types of molecules that the fetus expels. The cells collected by amniocentesis can be grown in culture, and additional metabolic disorders can be evaluated. Chromosome analysis, called a **karyotype** (see p. 833), can also be performed on the cultured cells. Amniocentesis is most commonly performed at 13–16 weeks after fertilization.

Fetal tissue samples can also be obtained by **chorionic villus sampling,** in which a probe is introduced into the uterine cavity through the cervix and a small piece of chorion removed. This technique has an advantage over amniocentesis in that it can be used earlier in development, as early as 8–10 weeks after fertilization. Furthermore, cells can be used directly for analyses such as karyotyping, rather than having to culture cells, as in amniocentesis.

One of the molecules normally produced by the fetus and released into the amniotic fluid is **α-fetoprotein.** If the fetus has tissues exposed to the amniotic fluid that are normally covered by skin, an

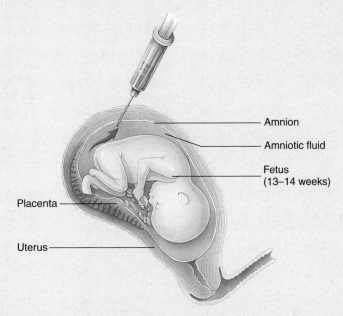

Figure B Removal of Amniotic Fluid for Amniocentesis

excessive amount of α-fetoprotein is lost into the amniotic fluid. For example, failure of the neural tube to close results in the exposure of neural tissue to amniotic fluid, and failure of the abdominal wall to fully form exposes abdominal organs to amniotic fluid.

Some of the metabolic by-products from the fetus, such as α-fetoprotein, can enter the maternal blood. In some cases, the by-products are processed and passed to the maternal urine. The levels of these

fetal products can then be measured in the mother's blood or urine.

The fetus can be seen within the uterus by ultrasound, which uses sound waves that are bounced off the fetus like sonar and then analyzed and enhanced by computer.

Fetal heart rate can be detected with an ultrasound stethoscope by the tenth week after fertilization and with a conventional stethoscope by 20 weeks. The normal fetal heart rate is 140 bpm (normal range is 110–160).

The stress of the confined space of the uterus and the limited oxygen supply resulting from a more rapid increase in the size of the fetus than in the size of the placenta increase the rate of adrenocorticotropic hormone (ACTH) secretion by the fetus's anterior pituitary gland (figure 25.19). ACTH causes the fetal adrenal cortex to produce glucocorticoids, which travel to the placenta, where they decrease the rate of progesterone secretion and increase the rate of estrogen synthesis. In addition, prostaglandin synthesis is initiated. Prostaglandins strongly stimulate uterine contractions.

During labor, stretch of the uterine cervix initiates nervous reflexes that cause oxytocin to be released from the woman's posterior pituitary gland. Oxytocin stimulates uterine contractions, which move the fetus farther into the cervix, causing further stretch. Thus, a positive-feedback mechanism is established in which stretch stimulates oxytocin release and oxytocin causes fur-

ther stretch. This positive-feedback system stops after delivery, when the cervix is no longer stretched.

Oxytocin

Occasionally, synthetic oxytocin (pitocin) is administered to women during labor to increase the force of uterine contractions. Caution must be exercised in the use of this drug, however, so that tetanic-like contractions, which would drastically reduce the blood flow through the placenta, do not occur.

Progesterone inhibits oxytocin release and decreases the number of oxytocin receptors. Decreased progesterone levels in the maternal circulation results in increased oxytocin secretion

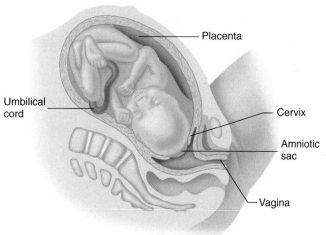

1. **First stage:** Begins with the onset of regular uterine contractions. The cervix begins to dilate. The first stage commonly lasts 8–24 hours.

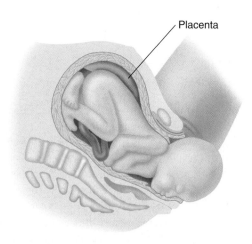

2. **Second stage:** Extends from maximal cervical dilation until the baby exits the vagina. It lasts from a few minutes to an hour.

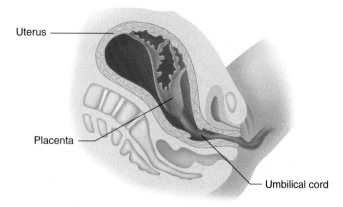

3. **Third stage:** The placenta is expelled.

Process Figure 25.18 Labor

and an increase in the number of oxytocin receptors in the uterus. In addition, estrogens make the uterus more sensitive to oxytocin stimulation by increasing the synthesis of receptor sites for oxytocin. Estrogen may also increase the formation of gap junctions between myometrial cells, thereby increasing the contractility of the uterus. Some evidence suggests that oxytocin also stimulates prostaglandin synthesis in the uterus. All these events support the development of strong uterine contractions.▼

29. List the stages of labor, and indicate when each stage begins and its approximate length of time.
30. Describe the hormonal changes that take place before and during delivery. How is stretch of the cervix involved in delivery?

P R E D I C T 5
A woman is having an extremely prolonged labor. From her anatomy and physiology course, she remembers the role of Ca^{2+} in muscle contraction and asks the doctor to give her a Ca^{2+} injection to speed the delivery. Explain why the doctor would or would not do as she requests.

The Newborn

The newborn, or **neonate** (nē′ō-nāt, newborn), experiences several dramatic changes at the time of birth. The major and earliest changes are the separation of the infant from the maternal circulation and the transfer from a fluid to a gaseous environment.

Respiratory and Circulatory Changes

Fetal blood circulation is different from the adult pattern for several reasons:

1. The fetus is connected to the mother through the placenta. Oxygen-rich, nutrient-rich blood is carried from the placenta to the fetus by the **umbilical vein,** and oxygen-poor, nutrient-poor blood is returned to the placenta by the **umbilical arteries** (figure 25.20, steps 4 and 5).
2. The umbilical vein joins the left branch of the hepatic portal vein opposite the origin of the **ductus venosus,** which is a short vessel connecting to the inferior vena cava (figure 25.20, step 3). Approximately 50% of the blood flowing through the umbilical vein bypasses the liver by flowing through the ductus venosus to the inferior vena cava. The other 50% flows through the liver with blood from the fetal hepatic portal system.
3. The fetus is not breathing and blood bypasses the lungs in two ways. Approximately 28% of the blood entering the fetal heart bypasses the pulmonary circuit by flowing from the right atrium into the left atrium through an opening in the atrial septum called the **foramen ovale** (figure 25.20, step 2). The remainder enters the right ventricle and exits the heart through the pulmonary trunk. Approximately 61% of the blood entering the fetal heart passes through the **ductus arteriosus,** which connects the left pulmonary artery with the aorta (figure 25.20, step 1). Thus, only 11% of the blood entering the fetal heart flows through the fetal lungs.

1. The fetal hypothalamus secretes a releasing hormone, corticotropin-releasing hormone (CRH), which stimulates adrenocorticotropic hormone (ACTH) secretion from the pituitary. The fetal pituitary secretes ACTH in greater amounts near parturition.

2. ACTH causes the fetal adrenal gland to secrete greater quantities of adrenal cortical steroids.

3. Adrenal cortical steroids travel in the umbilical blood to the placenta.

4. In the placenta, the adrenal cortical steroids cause progesterone synthesis to level off and estrogen and prostaglandin synthesis to increase, making the uterus more excitable.

5. The stretching of the uterus produces action potentials that are transmitted to the brain through ascending pathways.

6. Action potentials stimulate the secretion of oxytocin from the posterior pituitary.

7. Oxytocin causes the uterine smooth muscle to contract.

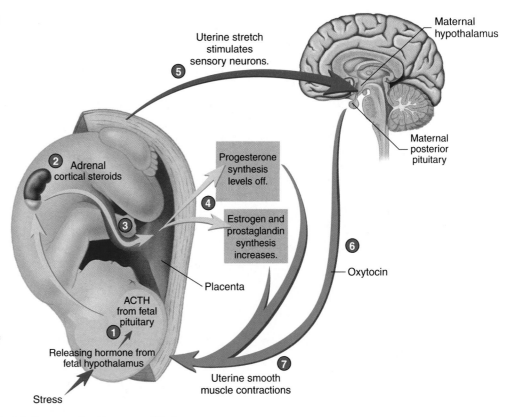

Process Figure 25.19 **Factors That Influence the Process of Labor**

Although the precise control of labor is unknown, these changes appear to play a role.

The large, forced gasps of air that occur when the infant cries at the time of delivery help inflate the lungs. The initial inflation of the lungs causes important changes in the circulatory system. Expansion of the lungs reduces the resistance to blood flow through the lungs, resulting in increased blood flow through the pulmonary arteries. Consequently, more blood flows from the right atrium to the right ventricle and into the pulmonary arteries, and less blood flows from the right atrium through the foramen ovale to the left atrium. In addition, an increased volume of blood returns from the lungs through the pulmonary veins to the left atrium, which increases the pressure in the left atrium. The increased left atrial pressure and decreased right atrial pressure, resulting from decreased pulmonary resistance, forces blood against the septum primum (see figure 25.12, step 5), causing the foramen ovale to close. This action functionally completes the separation of the heart into two pumps: the right side of the heart and the left side of the heart. The closed foramen ovale becomes the **fossa ovalis** (figure 25.21, step 2).

The ductus arteriosus closes off within 1 or 2 days after birth. This closure occurs because of the sphincterlike constriction of the artery and is probably stimulated by local changes in blood oxygen content and blood pressure. Once closed, the ductus arteriosus is replaced by connective tissue and is known as the **ligamentum arteriosum** (figure 25.21, step 1).

Patent Ductus Arteriosus

If the ductus arteriosus does not close completely, it is said to be **patent.** This is a serious birth defect, resulting in marked elevation in pulmonary blood pressure because blood flows from the left ventricle to the aorta, through the ductus arteriosus, to the pulmonary arteries. If not corrected, it can lead to irreversible degenerative changes in the heart and lungs.

When the umbilical cord is tied and cut, no more blood flows through the umbilical vein and arteries, and they degenerate. The remnant of the umbilical vein becomes the **ligamentum teres,** or **round ligament,** of the liver, and the ductus venosus becomes the **ligamentum venosum.** The remnants of the umbilical arteries become the **cords of the umbilical arteries** (figure 25.21, steps 3–5).▼

31. *Describe fetal circulation. How is it different from adult circulation?*

32. *What changes take place in the newborn's circulatory system shortly after birth? What does each of the following become: foramen ovale, ductus arteriosus, umbilical vein, umbilical arteries, and ductus venosus?*

1. Blood bypasses the lungs by flowing from the pulmonary trunk through the ductus arteriosus to the aorta.

2. Blood also bypasses the lungs by flowing from the right to the left atrium through the foramen ovale.

3. Blood bypasses the liver sinusoids by flowing through the ductus venosus.

4. Oxygen-rich blood is returned to the fetus from the placenta by the umbilical vein.

5. Oxygen-poor blood is carried from the fetus to the placenta through the umbilical arteries.

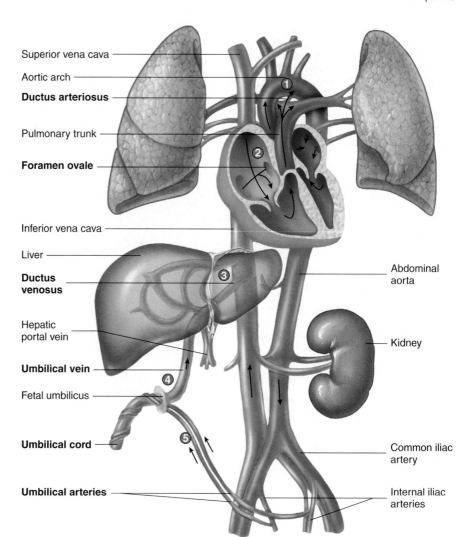

Process Figure 25.20   Fetal Circulation

Digestive Changes

When a baby is born, it is suddenly separated from its source of nutrients provided by the maternal circulation. Because of this separation and the stress of birth and new life, the neonate usually loses 5%–10% of its total body weight during the first few days of life. Although the digestive system of the fetus becomes somewhat functional late in development, it is still very immature, compared with that of an adult, and can digest only a limited number of food types.

Late in gestation, the fetus swallows amniotic fluid from time to time. Shortly after birth, this swallowed fluid plus cells sloughed from the mucosal lining, mucus produced by intestinal mucous glands, and bile from the liver pass from the digestive tract as a greenish anal discharge called **meconium** (mē-kō′nē-ŭm).

The pH of the stomach at birth is nearly neutral because of the presence of swallowed alkaline amniotic fluid. Within the first 8 hours of life, a striking increase in gastric acid secretion occurs, causing the stomach pH to decrease. Maximum acidity is reached at 4–10 days, and the pH gradually increases for the next 10–30 days.

The neonatal liver is functionally immature and lacks adequate amounts of the enzyme required in the production of bilirubin. This enzyme system usually develops within 2 weeks after birth in a healthy neonate, but, because this enzyme system is not fully developed at birth, some full-term babies temporarily develop jaundice, with elevated blood levels of bilirubin. Jaundice often occurs in premature babies.

The newborn digestive system is capable of digesting lactose (milk sugar) from the time of birth. The pancreatic secretions are sufficiently mature for a milk diet, but the digestive system only gradually develops the ability to digest more solid foods over the first year or two. New foods should be introduced gradually during the first 2 years. It is also advised that only one new food be introduced at a time into the infant's diet so that, if an allergic reaction occurs, the cause is more easily determined.

Amylase secretion by the salivary glands and the pancreas remains low until after the first year. Lactase activity in the small intestine is high at birth but declines during infancy, although the levels still exceed those in adults. Lactase activity is lost in many adults (see chapter 21).▼

33. *What changes take place in the newborn's digestive system shortly after birth?*
34. *Define meconium. Why does jaundice often develop after birth?*

1. When air enters the lungs, blood is forced through the pulmonary arteries to the lungs. The ductus arteriosus closes and becomes the ligamentum arteriosum (*gray*).

2. The foramen ovale closes and becomes the fossa ovalis. Blood can no longer flow from the right to the left atrium.

3. The ductus venosus degenerates and becomes the ligamentum venosum (*gray*).

4. The umbilical arteries and vein are cut. The umbilical vein becomes the round ligament of the liver (*gray*).

5. The umbilical arteries also degenerate and become the cords of the umbilical arteries (*gray*).

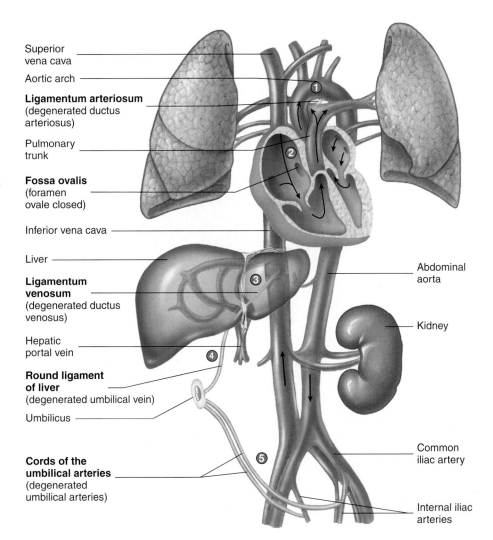

Superior vena cava
Aortic arch
Ligamentum arteriosum (degenerated ductus arteriosus)
Pulmonary trunk
Fossa ovalis (foramen ovale closed)
Inferior vena cava
Liver
Ligamentum venosum (degenerated ductus venosus)
Hepatic portal vein
Round ligament of liver (degenerated umbilical vein)
Umbilicus
Cords of the umbilical arteries (degenerated umbilical arteries)
Abdominal aorta
Kidney
Common iliac artery
Internal iliac arteries

Process Figure 25.21 Circulatory Changes Following Birth

Congenital Disorders

The term *congenital* means present at birth and **congenital disorders** are abnormalities commonly referred to as birth defects. Approximately 15% of all congenital disorders have a known genetic cause, and approximately 70% of all birth defects are of unknown cause. The remaining 15% are the result of environmental causes or a combination of environmental and genetic causes. In the case of environmental causes, the birth defect results from damage to the fetus during development. Agents that cause birth defects are called **teratogens** (ter′ă-tō-jenz). For example, fetal alcohol syndrome results when a pregnant woman drinks alcohol, which crosses the placenta and damages the fetus. The baby is born with a smaller than normal head and mental retardation and may exhibit other birth defects. Researchers are working to identify various teratogens. With this information, women can avoid known teratogens and reduce the risk for birth defects.▼

    35. What are congenital disorders? What causes these disorders?

    36. What is a teratogen? Give an example.

Lactation

Lactation (lak-tā′shŭn, suckle) is the production of milk by the mammary glands within the breasts. It normally occurs in females after delivery and may continue for 2 or 3 years, or even longer, provided suckling occurs often and regularly.

    During pregnancy, the high concentration and continuous presence of estrogens and progesterone cause expansion of the duct system and secretory units of the mammary glands. The ducts grow and branch repeatedly to form an extensive network. The size of the breasts increases substantially throughout pregnancy because of increased fat deposition. Estrogen is primarily responsible for breast growth during pregnancy, but normal development of the breast does not occur without the presence of several other hormones. Progesterone causes development of the breasts' secretory alveoli, which enlarge but do not usually secrete milk during pregnancy. The other hormones are growth hormone, prolactin, thyroid hormones, glucocorticoids, and insulin. The placenta secretes a growth hormonelike substance (human somatotropin) and a prolactin-like substance (human placental lactogen), and these substances help support the development of the breasts.

1. Stimulation of the nipple by the baby's suckling initiates action potentials in sensory neurons that connect with the hypothalamus.

2. In response, the hypothalamus stimulates the posterior pituitary to release oxytocin and the anterior pituitary to release prolactin.

3. Oxytocin stimulates milk release from the breast. Prolactin stimulates additional milk production.

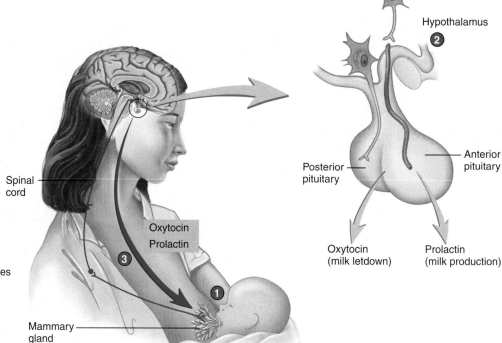

Spinal cord

Oxytocin
Prolactin

Mammary gland

Hypothalamus

Posterior pituitary

Anterior pituitary

Oxytocin (milk letdown)

Prolactin (milk production)

Process Figure 25.22   Hormonal Control of Lactation

Prolactin, which is produced by the anterior pituitary gland, is the hormone responsible for milk production. Before parturition, high levels of estrogen stimulate an increase in prolactin production. Milk production is inhibited during pregnancy, however, because high levels of estrogen and progesterone inhibit the effect of prolactin on the mammary glands. After parturition, estrogen, progesterone, and prolactin levels decrease, and, with lower estrogen and progesterone levels, prolactin stimulates milk production. Despite a decrease in the basal levels of prolactin, a reflex response produces surges of prolactin release. During suckling, mechanical stimulation of the breasts initiates nerve impulses that reach the hypothalamus, causing the secretion of **prolactin-releasing factor (PRF)** and inhibiting the release of **prolactin-inhibiting factor (PIF).** Consequently, prolactin levels temporarily increase and stimulate milk production (figure 25.22). Prolactin results in the production of milk to be used in the next nursing period because it takes time to produce milk.

For the first few days after birth, the mammary glands secrete **colostrum** (kō-los′trŭm), a very nutritious substance that contains little fat and less lactose than milk. The breasts may secrete some colostrum during pregnancy. Eventually, milk with a higher fat and lactose content is produced. Colostrum and milk not only provide nutrition but also contain antibodies (see chapter 19) that help protect the nursing baby from infections.

Repeated stimulation of prolactin release, by suckling, makes nursing (breast-feeding) possible for several years. If nursing stops, however, within a few days the ability to produce prolactin ceases, and milk production stops.

At the time of nursing, stored milk is released as a result of a reflex response. Mechanical stimulation of the breasts produces nerve impulses that cause the release of oxytocin from the posterior pituitary, which stimulates cells surrounding the alveoli to contract. Milk is then released from the breasts, a process called **milk letdown** (see figure 25.22). In addition, higher brain centers can stimulate oxytocin release, and such things as hearing an infant cry can result in milk letdown.▼

37. *What hormones are involved in preparing the breast for lactation? Describe the events involved in milk production and milk letdown. What is colostrum?*

P R E D I C T   6
While nursing her baby, a woman notices that she develops "uterine cramps." Explain what is happening.

Genetics

Genetics is the study of heredity; that is, the characteristics children inherit from their parents. Many of a person's abilities, susceptibility to disease, and even life span are influenced by heredity.

Chromosomes

Deoxyribonucleic (dē-oks′ē-rī′bō-noo-klē′ic) **acid (DNA)** molecules and their associated proteins become visible as densely stained bodies, called **chromosomes** (krō′mō-sōmz, colored bodies), during cell division (see chapter 3). **Somatic** (sō-mat′ik) **cells,**

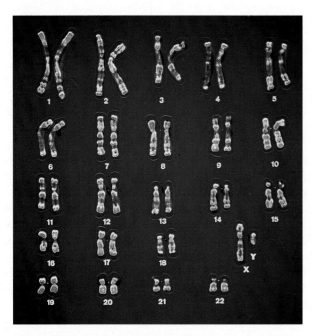

Figure 25.23 Human Karyotype

The 23 pairs of chromosomes in humans consist of 22 pairs of autosomal chromosomes (numbered 1–22) and 1 pair of sex chromosomes. The autosomal chromosome pairs are numbered in order from the largest to smallest. This karyotype is of a male and has an X and a Y sex chromosome. A female karyotype would have two X chromosomes.

all the cells of the body except the sex cells, contain 23 pairs of chromosomes, or 46 total chromosomes. The sex cells, or **gametes** (gam′ētz), contain 23 unpaired chromosomes.

A **karyotype** (kar′ē-ō-tīp, *karyon,* nucleus + *typos,* model) is a display of the chromosomes in a somatic cell (figure 25.23). There are 22 pairs of **autosomal** (aw-tō-sō′măl) **chromosomes,** which are all the chromosomes but the sex chromosomes, and there is one pair of **sex chromosomes.** A normal female has two **X chromosomes** (XX) in each somatic cell, whereas a normal male has one X and one **Y chromosome** (XY).

 Sex Chromosome Abnormalities

There is a wide range of sex chromosome abnormalities. The presence of a Y chromosome makes a person male, and the absence of a Y chromosome makes a person female, regardless of the number of X chromosomes. The following combinations, therefore, are female: XO (O means a chromosome is missing) (Turner syndrome), XX, XXX, and XXXX. Any combinations that include a Y are male: XY, XXY, XXXY, and XYY. A YO condition is lethal, because the genes on the X chromosome are necessary for survival. Secondary sexual characteristics are sometimes underdeveloped in both the XXX female and the XXY male (called Klinefelter syndrome), and additional X chromosomes (XXXX or XXXY) are often associated with some degree of mental retardation.

Gametes are produced by **meiosis** (mī-ō′sis, a lessening) (see chapter 24). Meiosis is called a reduction division because the number of chromosomes in the gametes is half the number in the somatic cells. When a sperm cell and an oocyte fuse during fertilization, each contributes one-half of the chromosomes necessary to produce new somatic cells. Half of an individual's genetic makeup therefore comes from the father, half from the mother.

During meiosis, the chromosomes are distributed in such a way that each gamete receives only one chromosome from each **homologous** (hŏ-mol′ō-gŭs) pair of chromosomes (see chapter 24). Homologous chromosomes contain the same compliment of genetic information. The inheritance of sex illustrates, in part, how chromosomes are distributed during gamete formation and fertilization. During meiosis and gamete formation, the pair of sex chromosomes separates so that each oocyte receives one of a homologous pair of X chromosomes, whereas each sperm cell receives either an X chromosome or a Y chromosome (figure 25.24). When a sperm cell fertilizes an oocyte to form a single cell, the sex of the individual is determined randomly. If the oocyte is fertilized by a sperm cell with a Y chromosome, a male results; if the oocyte is fertilized by a sperm cell with an X chromosome, a female results. Estimating the probability of any given zygote being male or female is much like flipping a coin. When all the possible combinations of sperm

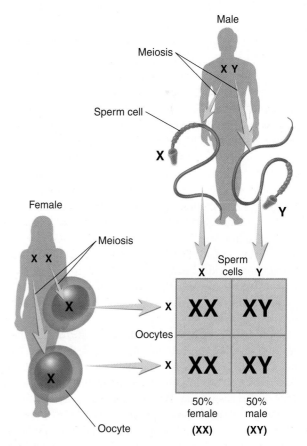

Figure 25.24 Inheritance of Sex

The female produces oocytes containing one X chromosome, whereas the male produces sperm cells with either an X or a Y chromosome. There are four possible combinations of an oocyte with a sperm cell, half of which produce females and half of which produce males.

cells with oocytes are considered, half the individuals should be female and the other half should be male. ▼

38. *What are somatic cells and gametes?*
39. *What is a karyotype? Define autosomal and sex chromosomes and state the number of each in humans.*
40. *What are homologous chromosomes? How are they distributed in meiosis?*

Genes

The functional unit of heredity is the gene. Each **gene** consists of a certain portion of a DNA molecule, but not necessarily a continuous stretch of DNA. Each chromosome contains thousands of genes. Both chromosomes of a given pair contain similar, but not necessarily identical, genes. Similar genes on homologous chromosomes are called **alleles** (ă-lēlz′, *allelon,* reciprocally). If the two allelic genes are identical, the person is **homozygous** (hō-mō-zī′gŭs) for the trait specified by that gene. If the two alleles are slightly different, the person is **heterozygous** (het′er-ō-zī′gŭs) for the trait. All the genes in one homologous set of 23 chromosomes in one individual are called the **genome.**

Phenylketonuria

The importance of genes is dramatically illustrated by situations in which the alteration of a single gene results in a genetic disorder. For example, in **phenylketonuria** (fen′il-kē′tō-nū′rē-ă) **(PKU)** the gene responsible for producing an enzyme that converts the amino acid phenylalanine to the amino acid tyrosine is defective. Phenylalanine therefore accumulates in the blood and is eventually converted to harmful substances that can cause mental retardation. Reducing phenylalanine in the diet can prevent the development of mental retardation.

Through the processes of meiosis, gamete formation, and fertilization the distribution of genes received from each parent is essentially random. This random distribution is influenced by several factors, however. For example, all of the genes on a given chromosome are **linked;** that is, they tend to be inherited as a set rather than as individual genes because chromosomes, not individual genes, segregate during meiosis. Also during meiosis, however, homologous chromosomes may exchange genetic information by **crossing over.**

Furthermore, segregation errors can occur during meiosis. As the chromosomes separate during meiosis, the two members of a homologous pair may not segregate. As a result, one of the daughter cells receives both chromosome pairs and the other daughter cell receives none. When the gametes are fertilized, the resulting zygote has either 47 chromosomes or 45 chromosomes, rather than the normal 46. When this condition results in an abnormal autosomal chromosome number, it is usually, but not always, lethal and is one reason for a high rate of early embryo loss. **Down syndrome,** or **trisomy 21,** in which there are three #21 chromosomes is one of the few autosomal trisomies that is not lethal. In contrast to autosomal trisomies, sex chromosome

trisomies (see "Sex Chromosome Abnormalities," p. 833) are not usually lethal. ▼

41. *Define gene, allele, homozygous, heterozygous, and genome.*
42. *What effect do linked genes, crossing over, and segregation errors have on the distribution of genes?*

Dominant and Recessive Genes

Most human genetic traits are recognized because defective alleles for those traits exist in the population. For example, on chromosome 11 is a gene that produces an enzyme necessary for the synthesis of melanin, the pigment responsible for skin, hair, and eye color (see chapter 5). An abnormal allele, however, produces a defective enzyme not capable of catalyzing one of the steps in melanin synthesis. If a given person inherits two defective alleles, a homozygous condition, the person is unable to produce melanin and therefore lacks normal pigment. This condition is referred to as **albinism** (al′bi-nizm, *albo,* white) (figure 25.25).

For many genetic traits, the effects of one allele for that trait can mask the effect of another allele for the same trait. For example, a person who is heterozygous for the melanin-producing enzyme gene has one normal gene for melanin production and one defective gene for melanin production. One copy of the gene and its resulting enzymes are enough to make normal melanin. As a result, the person who is heterozygous produces melanin and appears normal. In this case, the allele that produces the normal enzyme is said to be **dominant,** whereas the allele producing the abnormal enzyme is **recessive.** By convention, dominant traits are indicated by uppercase letters and recessive traits are indicated by lowercase letters. In this example, the letter *A* designates the dominant normal, pigmented condition, and the letter *a* the recessive albino condition.

The possible combinations of dominant and recessive alleles for normal melanin production versus albanism are *AA* (homozygous dominant, normal), *Aa* (heterozygous, normal), and *aa* (homozygous recessive, albino). The alleles a person has for a given trait are called the **genotype** (jen′ō-tīp). The person's appearance is called the **phenotype** (fē′nō-tīp). A person with the genotype *AA* or *Aa* would have the phenotype of normal pigmentation, whereas

Figure 25.25   Albinism

Photograph of an albino male with his normally pigmented father.

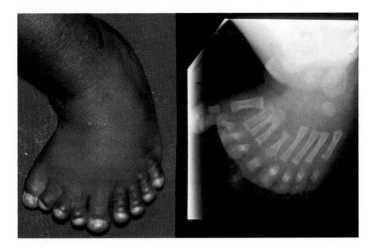

Figure 25.26   Polydactyly

Polydactyly, having extra fingers or toes, is determined by a dominant gene.

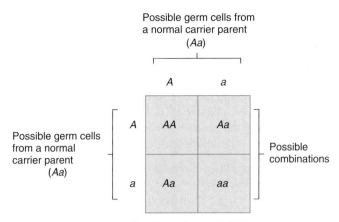

Possible outcome in children:
$\frac{1}{4}$ AA (normal) : $\frac{1}{2}$ Aa (normal carrier) : $\frac{1}{4}$ aa (albino)

Figure 25.27   Inheritance of a Recessive Trait: Albinism

A represents the normal pigmented condition, and *a* represents the recessive unpigmented condition. The figure shows a Punnett square of a mating between two normal carriers.

a person with the genotype *aa* would have the phenotype of albinism. Note that the recessive trait is expressed only when no allele for the dominant trait is present. The possible genotypes and phenotypes for albinism are

Alleles	Genotype	Phenotype
AA	Homozygous dominant	Normal pigmentation
Aa	Heterozygous	Normal pigmentation
aa	Homozygous recessive	Albino

Not all dominant traits are the normal condition, and not all recessive traits are abnormal. In many cases, the dominant trait is abnormal. For example, a person with **polydactyly** (pol-ē-dak′ti-lē) has extra fingers or toes (figure 25.26). One allele for polydactyly, which results in extra fingers or toes, is dominant over the recessive, normal allele, which results in the normal number of fingers or toes.

PREDICT   7
Given that polydactyly is a dominant trait, list all the possible genotypes and phenotypes for polydactyly. Use the letters *D* and *d* for the genotypes.

The inheritance of dominant and recessive traits can be determined if the genotypes of the parents are known. For example, if an albino person (*aa*) mates with a heterozygous normal person (*Aa*), the probability is that half of the children will be albino (*aa*), and half will be normal heterozygous carriers (*Aa*). If two carriers (*Aa*) mate, the probability is that 1/4 will be homozygous dominant (*AA*), 1/4 will be homozygous recessive (*aa*), and 1/2 will be heterozygous (*Aa*). Such a probability can be easily determined by the use of a table called a **Punnett square** (figure 25.27). A **carrier** is a heterozygous person with an abnormal recessive gene but with a normal phenotype because he or she also has a normal dominant allele for that gene. ▼

43. *Define dominant and recessive traits.*
44. *What is the difference between genotype and phenotype?*
45. *What is a Punnett square?*
46. *Define carrier.*

PREDICT   8
If a carrier for albinism mates with a homozygous normal person, what is the likelihood that any of their children will be albinos? Explain.

Multiple Alleles

Alleles can exist in many forms, called **multiple alleles.** An individual has only two alleles for a given gene, one on each homologous chromosome. At the population level, however, there can be many forms of an allele.

Differences in alleles arise by mutation, in which the DNA nucleotide sequence is altered. Many alleles are possible in a population because any change in the nucleotide sequence of DNA, even of one base pair, potentially produces a different allele. A different form of an allele is called an **allelic variant**, a **mutated allele (gene)**, or a **polymorphism** (many forms).

Allelic variants can result in either no effect on the phenotype or minor to major phenotypic changes. Allelic variants can encode for different sequences of amino acids in proteins. Recall from chapter 2 that the sequence of amino acids in a protein affects the shape of the protein, including the protein's domain (see p. 39), which is the functional part of the protein. A mutated allele that causes an amino acid change in a protein that does not significantly change protein shape may not affect the phenotype. The greater the effect of an amino acid change on a protein's shape, however, the greater the effect on the protein's function, and the greater the effect on the phenotype. Millions of allelic variants have been identified. The significance of most of them is unknown.

The allelic variant present at a locus can affect a person's phenotype. For example, PKU is an autosomal recessive trait with multiple alleles. On chromosome 12 is a gene that encodes for an enzyme that converts the amino acid phenylalanine to the amino acid tyrosine. If phenylalanine is not converted to tyrosine, high levels of phenylalanine can accumulate, leading to brain cell damage. The more defective the enzyme, the greater the accumulation of phenylalanine and the greater the damage to the brain, in most cases. Over 400 disease-causing allelic variants of the PKU gene are known. The severity of PKU depends in part on which two allelic variants are present. The less severe of the two alleles determines the severity of the disorder. Symptoms of untreated PKU can vary from profound mental retardation to nearly normal mental abilities. Thus, even though PKU is an autosomal recessive trait, the expression of the recessive condition exhibits considerable variability because of multiple alleles. ▼

    47. What are multiple alleles? How can they affect phenotype?

Other Types of Gene Expression

In some cases, the dominant allele does not completely mask the effects of the recessive allele. This is called **incomplete dominance.** An example of incomplete dominance is **sickle-cell disease,** in which the hemoglobin produced by the gene is abnormal (see chapter 16). The result is sickle-shaped red blood cells, which are likely to stick in capillaries and tend to rupture more easily than normal red blood cells. The normal hemoglobin allele (*S*) is dominant over the sickle-cell allele (*s*). A normal person (*SS*) has normal hemoglobin, and person with sickle-cell anemia (*ss*) has abnormal hemoglobin. A person who is heterozygous (*Ss*) has half normal hemoglobin and half abnormal hemoglobin and usually has only a few sickle-shaped red blood cells. This condition is called **sickle-cell trait.**

In **codominance,** two alleles at the same locus are expressed so that separate, distinguishable phenotypes occur at the same time. ABO blood types are an example of codominance. A gene with three alleles on chromosome 9 encodes for enzymes that add sugar molecules to certain carbohydrates found on the surface of red blood cells. The carbohydrates are part of glycoproteins, called A and B antigens (see chapter 16). These carbohydrates consist of different sugars joined together. Traditionally, in designating alleles for blood type, the A and B alleles are superscripted to a capital letter *I* and the O allele is designated with the lowercase letter *i*. The I^A allele encodes for an enzyme that adds a particular sugar to the ends of the carbohydrate, producing the A antigen. The I^B allele encodes for a different enzyme that adds a different sugar to the ends of the carbohydrate, producing the B antigen. The *i* allele encodes for no functional enzyme and therefore is recessive to A and B. The possible genotypes and phenotypes for the ABO blood group are

Genotype	Phenotype
$I^A I^A$ or $I^A i$	**Type A blood (A antigen only)**
$I^B I^B$ or $I^B i$	**Type B blood (B antigen only)**
$I^A I^B$	**Type AB blood (A and B antigens)**
ii	**Type O blood (neither A nor B antigen)**

Type A blood results from the expression of only the I^A allele, type B blood from the expression of only the I^B allele, and type O blood from the expression of neither the I^A allele nor the I^B allele. Type AB blood illustrates codominance and results from the expression of both the I^A allele and the I^B allele at the same time.

Many traits, called **polygenic** (pol-ē-jen′ik) **traits,** are determined by the expression of multiple genes on different chromosomes. Examples are a person's height, intelligence, eye color, and skin color. Polygenic traits typically are characterized by having a great amount of variability. For example, there are many different shades of eye color and skin color.

Multifactorial traits are traits affected by various combinations of genes and the environment. In addition to genes, height is affected by nutrition, and skin color by exposure to the sun. Knowledge of environmental effects can be used to improve our genetic potential and to prevent harmful effects. For example, a healthy diet can promote growth or help prevent diabetes, and not smoking can reduce the risk of developing cancer. PKU causes damage to the brain because of a buildup of phenylalanine. Reducing phenylalanine in the diet can prevent the development of mental retardation. ▼

    48. Define incomplete dominance, codominance, polygenic traits, and multifactorial traits. Give an example of each.

Sex-Linked Traits

Traits affected by genes on the sex chromosomes are called **sex-linked traits.** Most sex-linked traits are **X-linked;** that is, they are on the X chromosome. Only a few **Y-linked** traits exist, however, largely because the Y chromosome is very small. An example of an X-linked trait is **hemophilia A** (classic hemophilia), in which the person is unable to produce one of the clotting factors (see chapter 16). Consequently, clotting is impaired and persistent bleeding can occur either spontaneously or as a result of an injury. Hemophilia A is a recessive trait and the allele for the trait is located on the X chromosome. Traditionally, X-linked genes are designated by letters superscripted to the letter *X*. The possible genotypes and phenotypes for hemophilia A are

Genotype	Phenotype
$X^H X^H$ or $X^H X^h$	**Normal female**
$X^h X^h$	**Hemophiliac female**
$X^H Y$	**Normal male**
$X^h Y$	**Hemophiliac male**

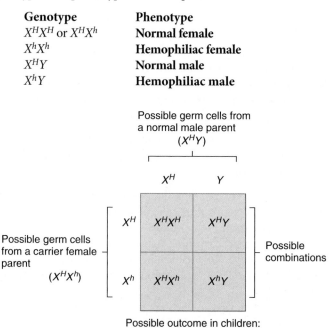

Figure 25.28   **Inheritance of an X-Linked Trait: Hemophilia**

X^H represents the normal X chromosome condition with all clotting factors, and X^h represents the X chromosome lacking an allele for one clotting factor. The figure shows a Punnett square of a mating between a normal male and a normal carrier female.

Table 25.3 Genetic Disorders

Disorder	Description
Dominant Traits	
Achondroplasia	Dwarfism characterized by shortening of the upper and lower limbs
Huntington chorea	Severe degeneration of the basal nuclei and frontal cerebral cortex; characterized by purposeless movements and mental deterioration; onset is usually between 40 and 50 years of age
Hypercholesterolemia	Elevated blood cholesterol levels that contribute to atherosclerosis and cardiovascular disease
Marfan syndrome	Abnormal connective tissue, resulting in increased height, elongated digits, and weakness in the aortic wall
Neurofibromatosis	Small, pigmented lesions (café-au-lait spots) in the skin and disfiguring tumors (noncancerous) caused by the proliferation of Schwann cells along nerves
Osteogenesis imperfecta	Abnormal collagen synthesis, resulting in brittle bones that break repeatedly
Recessive Traits	
Albinism	Lack of an enzyme necessary to produce the pigment melanin; characterized by lack of skin, hair, and eye coloration
Cystic fibrosis	Impaired transport of chloride ions across plasma membranes; resulting in excessive production of thick mucus, which blocks the respiratory and gastrointestinal tract; the most common fatal genetic disorder
Phenylketonuria	Lack of the enzyme necessary to convert the amino acid phenylalanine to the amino acid tyrosine; an accumulation of phenylalanine leads to mental retardation
Severe combined immune deficiency	Inability to form the white blood cells (B cells, T cells, and phagocytes) necessary for an immune system response
Sickle-cell disease	Inability to produce normal hemoglobin, resulting in abnormally shaped red blood cells that clog capillaries or rupture
Tay-Sachs disease	Lack of the enzyme necessary to break down certain fatty substances; an accumulation of fatty substances impairs action potential propagation, resulting in deterioration of mental and physical functions and death by 3–4 years of age
Thalassemia	Decreased rate of hemoglobin synthesis, resulting in anemia, enlargement of the spleen, increased cell numbers in red bone marrow, and congestive heart failure
Sex-Linked Traits	
Duchenne muscular dystrophy	Deletion or alteration of part of the X chromosome, resulting in progressive weakness and wasting of muscles
Hemophilia	Most commonly, failure to produce blood clotting factors, caused by a recessive gene, resulting in prolonged bleeding
Red-green color blindness	Most commonly, deficiency in functional green-sensitive cones, caused by a recessive gene; inability to distinguish between red and green colors
Chromosomal Disorders	
Down syndrome	Caused by having three chromosomes 21, resulting in mental retardation, short stature, and poor muscle tone
Klinefelter syndrome	Caused by two or more X chromosomes in a male (XXY), resulting in small testes, sterility, and development of femalelike breasts
Turner syndrome	Caused by having only one X chromosome, resulting in immature uterus, lack of ovaries, and short stature

Note that a female must have both recessive alleles to exhibit hemophilia, whereas a male, because he has only one X chromosome, has hemophilia if he has one recessive allele. An example of the inheritance of hemophilia is illustrated in figure 25.28. If a woman who is a carrier for hemophilia mates with a man who does not have hemophilia, none of their daughters but half of their sons will have hemophilia. Half of their daughters will be carriers. ▼

49. What is a sex-linked trait?

P R E D I C T ⑨

Predict the probability of a girl with Turner syndrome (see "Sex Chromosome Abnormalities," p. 833) having hemophilia if her mother is a carrier for hemophilia.

Genetic Disorders

Genetic disorders are caused by abnormalities in a person's genetic makeup—that is, in his or her DNA (table 25.3). They may involve a single gene or an entire chromosome. Some genetic disorders result from a **mutation** (mū-tā′shŭn, to change), a change in a gene that usually involves a change in the nucleotides composing the DNA (see chapter 2). Mutations occur by chance or can be caused by chemicals, radiation, or viruses. Once a mutation has occurred, the abnormal trait can be passed from one generation to the next.

Cancer is a tumor resulting from uncontrolled cell divisions. **Oncogenes** (ong′kō-jēnz) are genes associated with cancer. Many oncogenes are actually control genes involved in regulating cell

Clinical Relevance The Human Genome Project

The human genome consists of all the genes found in one homologous set of human chromosomes. It is estimated that humans have 25,000–30,000 genes. A **genomic map** is a description of the DNA nucleotide sequences of the genes and their locations on the chromosomes (figure C). Celera, a private corporation working on the **Human Genome Project,** announced on February 12, 2001, that it had completed sequencing 99% of the genome. Online Mendelian Inheritance in Man (OMIM) is a database that contains information about many human genes and genetic disorders (www.ncbi.nlm.nih.gov/omim).

Armed with a knowledge of the human genome and its effects on a person's physical, mental, and behavioral abilities, medicine and society will be transformed in many ways. Medicine, for example, will shift emphasis from the curative to the preventive. The disorders or diseases a person is likely to develop can be prevented or their

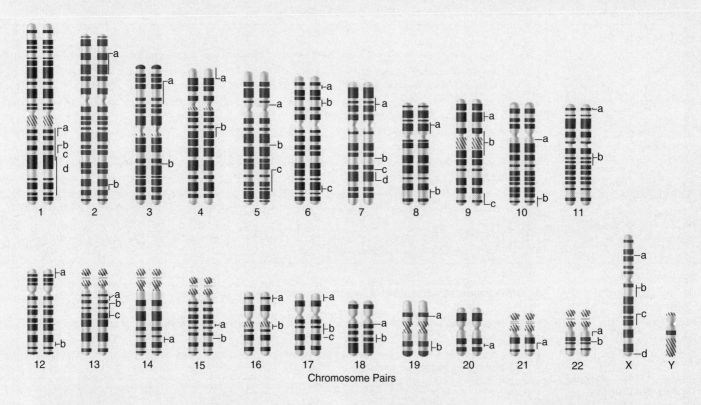

Chromosome Pairs

1. a. Gaucher disease
   b. Prostate cancer
   c. Glaucoma
   d. Alzheimer disease*
2. a. Familial colon cancer*
   b. Waardenburg syndrome
3. a. Lung cancer
   b. Retinitis pigmentosa*
4. a. Huntington chorea
   b. Parkinson disease
5. a. Cockayne syndrome
   b. Familial polyposis of the colon
   c. Asthma
6. a. Spinocerebellar ataxia
   b. Diabetes*
   c. Epilepsy*

7. a. Diabetes*
   b. Osteogenesis imperfecta
   c. Cystic fibrosis
   d. Obesity*
8. a. Werner syndrome
   b. Burkitt lymphoma
9. a. Malignant melanoma
   b. Friedreich ataxia
   c. Tuberous sclerosis
10. a. Multiple endocrine neoplasia, type 2
    b. Gyrate atrophy
11. a. Sickle-cell disease
    b. Multiple endocrine neoplasia
12. a. Zellweger syndrome
    b. Phenylketonuria (PKU)

13. a. Breast cancer*
    b. Retinoblastoma
    c. Wilson disease
14. a. Alzheimer disease*
15. a. Marfan syndrome
    b. Tay-Sachs disease
16. a. Polycystic kidney disease
    b. Crohn disease*
17. a. Tumor-suppressor protein
    b. Breast cancer*
    c. Osteogenesis imperfecta
18. a. Amyloidosis
    b. Pancreatic cancer*
19 a. Familial hypercholesterolemia
    b. Myotonic dystrophy

20. a. Severe combined immunodeficiency
21. a. Amyotrophic lateral sclerosis*
22. a. DiGeorge syndrome
    b. Neurofibromatosis, type 2
X a. Duchenne muscular dystrophy
    b. Menkes syndrome
    c. X-linked severe combined immunodeficiency
    d. Factor VIII deficiency (hemophilia A)

*Gene responsible for only some cases.

Figure C Human Genomic Map
Representative genetic defects mapped to date. The bars and lines indicate the location of the genes listed for each chromosome.

severity lessened. When prevention is not possible, knowledge of the enzymes or other molecules involved in a disorder may result in new drugs and techniques that can compensate for the genetic disorder. Knowledge of the genes involved in a disorder may result in **gene therapy,** or **genetic engineering,** that repairs or replaces defective genes, resulting in cures of or treatments for genetic disorders.

Despite the great promise of benefits from the Human Genome Project, the knowledge produced has raised a number of ethical and legal questions for society. Should a person's genomic information be public knowledge? Should persons with a genome that predisposes them to cancer or behavioral disorders be barred from certain types of employment or be refused medical insurance because they are a high risk? Can a person demand to know a prospective mate's genome? Should parents know the genome of their fetus and be allowed to make decisions regarding abortion based on this knowledge? Should the same genetic-engineering techniques that provide alteration of the genome to cure genetic disorders be used to create genomes that are deemed to be superior? Such questions raise the specter of genetic discrimination.

division and differentiation in the embryo and fetus. A change in an oncogene or in the regulation of an oncogene can result in uncontrolled cell division and the development of cancerous tumors. The normal control of oncogenes involves other genes, called **tumor-suppressor genes.** Cancer can occur when a mutation activates an oncogene or inactivates a tumor-suppressor gene. An accumulation of several mutations is necessary for cancer to occur. It is believed that exposure to ionizing radiation or to certain chemicals called **carcinogens** (kar-sin′ō-jenz) can induce such mutations and thereby initiate the development of cancer. For example, chemicals in cigarette smoke are known to cause lung cancer.

A change in cells that results in cancer is not usually inherited. Nonetheless, there may be a genetic basis that allows cancer development, especially under the right environmental conditions. In this sense, the inheritance of cancer and other abnormalities has been described as **genetic susceptibility,** or **genetic predisposition.** For example, if a woman's close relatives, such as her mother or sister, have breast cancer, she has a greater than average risk of developing it herself. Similar genetic susceptibilities have been found for diabetes mellitus, schizophrenia, and other disorders. ▼

50. *What is a mutation?*
51. *Describe the role of oncogenes and tumor-suppressor genes in cancer.*
52. *What is genetic susceptibility?*

Genetic Counseling

Genetic counseling includes predicting the possible results of matings involving carriers of harmful genes and talking to parents or prospective parents about the possible outcomes and treatments of a genetic disorder. With this knowledge, prospective parents can make informed decisions about having children.

A first step in genetic counseling is to attempt to determine the genotype of the individuals involved. A family tree, or **pedigree,** provides historical information about family members (figure 25.29)—for example, in the case of a dominant trait, such as Huntington chorea, a neurological disorder. Sometimes by knowing the phenotypes of relatives it is possible to determine a person's genotype. As part of the process of collecting information, a karyotype can be prepared. For some genetic disorders, the amount of a given substance, such as an enzyme, produced by a carrier can be tested. For example, carriers for cystic fibrosis produce more salt in their sweat than is normal.

Sometimes, it is suspected that a fetus has a genetic abnormality. Fetal cells can be tested by amniocentesis, which takes cells floating in the amniotic fluid, or chorionic villus sampling, which takes cells from the fetal side of the placenta (see chapter 24). ▼

53. *How are pedigrees, karyotypes, chemical tests, amniocentesis, and chorionic villus sampling used in genetic counseling?*

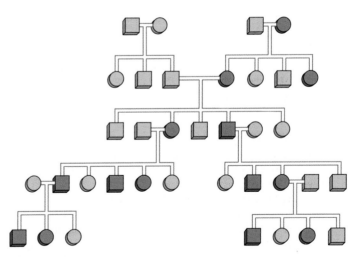

Figure 25.29   Pedigree of a Simple Dominant Trait

Males are indicated by squares, females by circles. Affected people are indicated in ***purple***. The horizontal line between symbols represents a mating. The symbols connected to the mating line by vertical and horizontal lines represent the children resulting from the mating in order of birth from left to right. Matings not related to the pedigree are not shown.

Prenatal Development (p. 806)

1. Prenatal development is divided into three parts: formation of the germ layers, formation of the organ systems, and growth and maturation.
2. The developing human from fertilization to the end of the eighth week of development is called an embryo. From 8 weeks to birth, the developing human is called a fetus.
3. Postovulatory age is 14 days less than clinical age.

Fertilization

1. Fertilization, the union of the secondary oocyte and sperm cell, results in a zygote.
2. A sperm cell must pass through the corona radiata, zona pellucida, and oocyte plasma membrane for fertilization to occur.
3. Only one sperm cell enters a secondary oocyte because of changes in the plasma membrane and zona pellucida.
4. Fusion of the male and female pronuclei produces a diploid nucleus.

Early Cell Division

1. In the very early stages of development the cells are totipotent, meaning each cell can give rise to any tissue necessary for development.
2. The cells of the early embryo are pluripotent (capable of making any cell of the body).
3. The product of fertilization undergoes divisions until it becomes a mass of cells, called a morula, and then a hollow ball of cells, called a blastocyst.
   - The trophoblast and some inner cell mass cells give rise to extra-embryonic tissues, such as the placenta.
   - Some inner cell mass cells give rise to the embryo proper.

Implantation of the Blastocyst and Development of the Placenta

1. Implantation is the burrowing of the blastocyst into the uterine wall about 7 days after fertilization.
   - The syncytiotrophoblast, cytotrophoblast, and extraembryonic mesenchyme form the chorion (with chorionic villi), which is the embryonic part of the placenta.
   - The endometrium forms the maternal part of the placenta.
   - Maternal and fetal blood do not mix, but exchange of gases, nutrients, and waste products occurs through the chorion.
2. The chorion produces hCG, which stimulates the corpus luteum to produce progesterone and estrogen.

Formation of the Germ Layers

1. The amniotic cavity and yolk sac form in the inner cell mass.
2. The embryonic disk, consisting of the epiblast and hyoblast, is between the amniotic cavity and yolk sac.
   - The epiblast gives rise to the embryo proper.
   - The hyoblast gives rise to extraembryonic tissue.
3. Gastrulation is the formation of a primitive streak, resulting in the production of ectoderm, mesoderm, and endoderm, the germ layers from which all the tissues of the body are derived.

Neural Tube and Neural Crest Formation

The nervous system develops from a neural tube that forms in the ectodermal surface of the embryo and from neural crest cells derived from the developing neural tube.

Somite Formation

Segments, called somites, that develop along the neural tube give rise to the musculature, vertebral column, and ribs.

Formation of the Gut and Body Cavities

1. The gastrointestinal tract develops as the developing embryo closes off part of the yolk sac.
2. The celom develops from small cavities that fuse within the embryo.

Limb Bud Development

The limbs develop from proximal to distal as outgrowths called limb buds.

Development of the Face

The face develops from the fusion of five major tissue processes.

Development of the Organ Systems

1. The skin develops from the ectoderm (epithelium), mesoderm and neural crest (dermis), and neural crest (melanocytes).
2. The skeletal system develops from mesoderm or neural crest cells.
3. Muscle develops from myoblasts, which migrate from somites or somitomeres.
4. The brain and spinal cord develop from the neural tube, and the peripheral nervous system develops from the neural tube and the neural crest cells.
5. The special senses develop mainly as neural tube or neural crest cell derivatives.
6. Many endocrine organs develop mainly as evaginations of the brain or digestive tract.
7. The heart develops as two tubes fuse into a single tube, which bends and develops septa to form four chambers.
8. The peripheral circulation develops from mesoderm as blood islands become hollow and fuse to form a network.
9. The lungs form as evaginations of the digestive tract. These evaginations undergo repeated branching.
10. The urinary system develops in three stages—pronephros, mesonephros, and metanephros—from the head to the tail of the embryo. The ducts join the allantois, part of which becomes the urinary bladder.
11. The reproductive system develops in conjunction with the urinary system. The presence or absence of certain hormones is very important to sexual development.

Growth of the Fetus

1. The embryo becomes a fetus at 60 days.
2. The fetal period is from day 60 to birth. It is a time of rapid growth.

Labor (p. 826)

1. The total length of gestation is 280 days (clinical age).
2. Uterine contractions force the baby out of the uterus during labor.
3. Increased estrogen levels and decreased progesterone levels help initiate parturition.
4. Fetal glucocorticoids act on the placenta to decrease progesterone synthesis and to increase estrogen and prostaglandin synthesis.
5. Stretch of the uterus and decreased progesterone levels stimulate oxytocin secretion, which stimulates uterine contraction.

The Newborn (p. 828)

The newborn baby experiences several dramatic changes when it is separated from the maternal circulation that provided nutrients, oxygen, and waste removal before parturition.

Respiratory and Circulatory Changes

1. The foramen ovale closes, separating the two atria.
2. The ductus arteriosus closes and becomes the ligamentum arteriosum. Blood no longer flows between the pulmonary trunk and the aorta.
3. The ductus venosus closes and becomes the ligamentum venosum. Blood no longer bypasses the liver.
4. The umbilical vein closes and becomes the round ligament of the liver.
5. The umbilical arteries become the cords of the umbilical arteries.

Digestive Changes

1. Meconium is a mixture of cells from the digestive tract, amniotic fluid, bile, and mucus excreted by the newborn.
2. The stomach begins to secrete acid.
3. The liver does not form adult bilirubin for the first 2 weeks.
4. Lactose can be digested, but other foods must be gradually introduced.

Congenital Disorders

1. Congenital disorders are abnormalities present at birth.
2. Teratogens are environmental agents that cause some congenital disorders.

Lactation (p. 831)

1. Estrogen, progesterone, and other hormones stimulate the growth of the breasts during pregnancy.
2. Suckling stimulates prolactin and oxytocin synthesis. Prolactin stimulates milk production, and oxytocin stimulates milk letdown.

Genetics (p. 832)

Chromosomes

1. Humans have 46 chromosomes in 23 pairs (homologous chromosomes).
2. Males have the sex chromosomes XY and females have XX.
3. During gamete formation, the chromosomes of each pair of homologous chromosomes separate; therefore, half of a person's genetic makeup comes from the father and half from the mother.

Genes

1. Genes are the functional units of heredity. Similar genes on homologous chromosomes are alleles.
   - A person is homozygous for a trait if he or she has two identical alleles for the trait.
   - A person is heterozygous for a trait if he or she has two different alleles for the trait.
2. The random distribution of genes is altered by linkage, cross over, and segregation errors.
3. Dominant genes mask the effects of recessive genes.
4. Genotype is a person's genetic makeup and phenotype is a person's appearance.
5. Punnett squares determine the probable genotypes of offspring.
6. A carrier is a person with a normal phenotype but is heterozygous for an abnormal recessive gene.
7. Multiple alleles are different forms of a gene.
8. In incomplete dominance, the heterozygote expresses a trait that is intermediate between the two homozygous traits.
9. In codominance, neither gene is dominant or recessive, but both are fully expressed.
10. Polygenic traits result from the expression of multiple genes.
11. Multifactorial traits are traits affected by various combinations of genes and the environment.
12. Sex-linked traits result from genes on the sex chromosomes.

Genetic Disorders

1. A mutation is a change in the DNA.
2. Some genetic disorders result from an abnormal distribution of chromosomes during gamete formation.
3. Oncogenes and tumor-suppressor genes are associated with cancer.
4. Genetic predisposition makes it more likely a person will develop a disorder.

Genetic Counseling

1. A pedigree (family history) can be used to determine the risk of having children with a genetic disorder.
2. A person's karyotype, chemical test, amniocentesis, and chorionic villi sampling can be used to determine a person's genotype.

R E V I E W   A N D   C O M P R E H E N S I O N

1. The major development of organ systems takes place in the
   a. first 2 weeks of development.
   b. third to eighth week of development.
   c. ninth to twentieth week of development.
   d. last 30 weeks of development.

2. Given these structures:
   1. blastocyst   2. morula   3. zygote
   Choose the arrangement that lists the structures in the order in which they are formed during development.
   a. 1,2,3   c. 2,3,1   e. 3,2,1
   b. 1,3,2   d. 3,1,2

3. The embryo proper develops from the
   a. inner cell mass.   c. blastocele.
   b. trophoblast.   d. yolk sac.

4. The placenta
   a. develops from the trophoblast.
   b. allows maternal blood to mix with embryonic blood.
   c. invades the lacunae of the embryo.
   d. all of the above.

5. The embryonic disk
   a. forms between the amniotic cavity and the yolk sac.
   b. contains the primitive streak.
   c. becomes a three-layered structure.
   d. all of the above.

6. The brain develops from
   a. ectoderm.   b. endoderm.   c. mesoderm.

7. Most of the skeletal system develops from
   a. ectoderm.   b. endoderm.   c. mesoderm.

8. Given these structures:
   1. neural crest   2. neural plate   3. neural tube
   Choose the arrangement that lists the structure in the order in which they form during development.
   a. 1,2,3   c. 2,1,3   e. 3,2,1
   b. 1,3,2   d. 2,3,1

9. The somites give rise to the
   a. circulatory system.   c. lungs.   e. brain.
   b. skeletal muscle.   d. kidneys.

10. The pericardial cavity forms from
    a. evagination of the early gastrointestinal tract.
    b. the neural tube.
    c. the celom.
    d. the branchial arches.
    e. pharyngeal pouches.

11. The parts of the limbs develop
    a. in a proximal-to-distal sequence.
    b. in a distal-to-proximal sequence.
    c. at approximately the same time.
    d. before the primitive streak is formed.

12. Concerning development of the face,
    a. the face develops by the fusion of five embryonic structures.
    b. the maxillary processes normally meet at the midline to form the lip.
    c. the primary palate starts as two vertical shelves that join to form the roof of the mouth.
    d. clefts of the secondary palate normally occur to one side of the midline.

13. Concerning the development of the heart,
    a. the heart develops from a single tube, which results from the fusion of two tubes.
    b. the SA node develops in the wall of the sinus venosus.
    c. the foramen ovale lets blood flow from the right atrium to the left atrium.
    d. the bulbus cordis is absorbed into the ventricle.
    e. all of the above.

14. Given these structures:
    1. mesonephros
    2. metanephros
    3. pronephros

    Choose the arrangement that lists the structures in the order in which they form during development.
    a. 1,2,3   c. 2,3,1   e. 3,2,1
    b. 1,3,2   d. 3,1,2

15. A study of the early embryo indicates that the glans penis of the male develops from the same embryonic structure as which of these female structures?
    a. labia majora   c. clitoris   e. urinary bladder
    b. uterus   d. vagina

16. Which hormone causes the differentiation of sex organs in the developing male fetus?
    a. FSH
    b. LH
    c. testosterone
    d. estrogen and progesterone
    e. GnRH

17. The onset of labor may be a result of
    a. increased estrogen secretion by the placenta.
    b. increased glucocorticoid secretion by the fetus.
    c. increased secretion of oxytocin.
    d. stretch of the uterus.
    e. all of the above.

18. Following birth,
    a. the ductus arteriosus closes.
    b. the pH of the stomach increases.
    c. the fossa ovalis becomes the foramen ovale.
    d. blood flow through the pulmonary arteries decreases.
    e. all of the above.

19. The hormone involved in milk production is
    a. oxytocin.   c. estrogen.   e. ACTH.
    b. prolactin.   d. progesterone.

20. Which of these most appropriately predicts the consequences of removing the sensory neurons from the areola of a lactating rat (or human)?
    a. Blood levels of oxytocin decrease.
    b. Blood levels of prolactin decrease.
    c. Milk production and letdown decrease.
    d. All of the above occur.

21. A gene is
    a. the functional unit of heredity.
    b. a certain portion of a DNA molecule.
    c. a part of a chromosome.
    d. all of the above.

22. Which of these terms is correctly matched with its definition?
    a. autosome—an X or a Y chromosome
    b. phenotype—the genetic makeup of an individual
    c. allele—similar genes on homologous chromosomes
    d. heterozygous—having two identical genes for a trait
    e. recessive—a trait expressed when the genes are heterozygous

23. Which of these genotypes is heterozygous?
    a. *DD*   c. *dd*
    b. *Dd*   d. both a and c

24. The ABO blood group is an example of
    a. dominant versus recessive alleles.
    b. incomplete dominance.
    c. codominance.
    d. a polygenic trait.
    e. sex-linked inheritance.

25. Assume that a trait is determined by an X-linked dominant gene. If the mother exhibits the trait, but the father does not, then their
    a. sons are more likely than their daughters to exhibit the trait.
    b. daughters are more likely than their sons to exhibit the trait.
    c. sons and daughters are equally likely to exhibit the trait.

Answers in Appendix E

CRITICAL THINKING

1. Triploidy is the presence of three sets of chromosomes in a cell. Though rare, triploidy does occur in humans. What failures during fertilization could lead to triploidy?

2. A woman is told by her physician that she is pregnant and that she is 44 days past her LMP. Approximately how many days has the embryo been developing, and what developmental events are occurring?

3. A high fever can prevent neural tube closure. If a woman has a high fever approximately 35–45 days after her LMP, what kinds of birth defects may be seen in the developing embryo?

4. If the limb bud is damaged during embryonic development when the limb bud is about one-half grown, what kinds of birth defects might be expected? Describe the anatomy of the affected structure.

5. What are the results of exposing a female embryo to high levels of testosterone while she is developing?

6. A woman goes into labor during the thirtieth week of her pregnancy. What would be the effect of administering progesterone at this stage?

7. When a woman nurses, milk letdown can occur in the breast that is not being suckled. Explain how this response happens.

8. Dimpled cheeks are inherited as a dominant trait. Is it possible for two parents, each of whom has dimpled cheeks, to have a child without dimpled cheeks? Explain.

9. The ability to roll the tongue to form a "tube" results from a dominant gene. Suppose that a woman and her son can both roll their tongues, but her husband cannot. Is it possible to determine if the husband is the father of her son?

10. The ABO antigens are a group of molecules found on the surface of red blood cells. Can an individual who has blood type AB be a parent of a child with blood type O? Why or why not?

11. A woman who does not have hemophilia marries a man who has the disorder. Determine the genotype of both parents if half of their children have hemophilia.

Answers in Appendix F

For additional study tools, go to: www.aris.mhhe.com.
Click on your course topic and then this textbook's author name/title. Register once for a semester's worth of interactive learning activities to help improve your grade!

Appendix A

Periodic Table of the Elements

The legend box indicates: Metals, Metalloids, Nonmetals. A sample element box shows Atomic number (9), symbol F, name Fluorine, and Atomic mass (19.00).

1 1A																	18 8A
1 **H** Hydrogen 1.008	2 2A											13 3A	14 4A	15 5A	16 6A	17 7A	2 **He** Helium 4.003
3 **Li** Lithium 6.941	4 **Be** Beryllium 9.012											5 **B** Boron 10.81	6 **C** Carbon 12.01	7 **N** Nitrogen 14.01	8 **O** Oxygen 16.00	9 **F** Fluorine 19.00	10 **Ne** Neon 20.18
11 **Na** Sodium 22.99	12 **Mg** Magnesium 24.31	3 3B	4 4B	5 5B	6 6B	7 7B	8 / 8B	9 / 8B	10 / 8B	11 1B	12 2B	13 **Al** Aluminum 26.98	14 **Si** Silicon 28.09	15 **P** Phosphorus 30.97	16 **S** Sulfur 32.07	17 **Cl** Chlorine 35.45	18 **Ar** Argon 39.95
19 **K** Potassium 39.10	20 **Ca** Calcium 40.08	21 **Sc** Scandium 44.96	22 **Ti** Titanium 47.88	23 **V** Vanadium 50.94	24 **Cr** Chromium 52.00	25 **Mn** Manganese 54.94	26 **Fe** Iron 55.85	27 **Co** Cobalt 58.93	28 **Ni** Nickel 58.69	29 **Cu** Copper 63.55	30 **Zn** Zinc 65.39	31 **Ga** Gallium 69.72	32 **Ge** Germanium 72.59	33 **As** Arsenic 74.92	34 **Se** Selenium 78.96	35 **Br** Bromine 79.90	36 **Kr** Krypton 83.80
37 **Rb** Rubidium 85.47	38 **Sr** Strontium 87.62	39 **Y** Yttrium 88.91	40 **Zr** Zirconium 91.22	41 **Nb** Niobium 92.91	42 **Mo** Molybdenum 95.94	43 **Tc** Technetium (98)	44 **Ru** Ruthenium 101.1	45 **Rh** Rhodium 102.9	46 **Pd** Palladium 106.4	47 **Ag** Silver 107.9	48 **Cd** Cadmium 112.4	49 **In** Indium 114.8	50 **Sn** Tin 118.7	51 **Sb** Antimony 121.8	52 **Te** Tellurium 127.6	53 **I** Iodine 126.9	54 **Xe** Xenon 131.3
55 **Cs** Cesium 132.9	56 **Ba** Barium 137.3	57 **La** Lanthanum 138.9	72 **Hf** Hafnium 178.5	73 **Ta** Tantalum 180.9	74 **W** Tungsten 183.9	75 **Re** Rhenium 186.2	76 **Os** Osmium 190.2	77 **Ir** Iridium 192.2	78 **Pt** Platinum 195.1	79 **Au** Gold 197.0	80 **Hg** Mercury 200.6	81 **Tl** Thallium 204.4	82 **Pb** Lead 207.2	83 **Bi** Bismuth 209.0	84 **Po** Polonium (210)	85 **At** Astatine (210)	86 **Rn** Radon (222)
87 **Fr** Francium (223)	88 **Ra** Radium (226)	89 **Ac** Actinium (227)	104 **Rf** Rutherfordium (257)	105 **Db** Dubnium (260)	106 **Sg** Seaborgium (263)	107 **Bh** Bohrium (262)	108 **Hs** Hassium (265)	109 **Mt** Meitnerium (266)	110 **Ds** Darmstadtium (269)	111 **Rg** Roentgenium (272)	112	(113)	114	(115)	116	(117)	(118)

Lanthanide series:

58 **Ce** Cerium 140.1	59 **Pr** Praseodymium 140.9	60 **Nd** Neodymium 144.2	61 **Pm** Promethium (147)	62 **Sm** Samarium 150.4	63 **Eu** Europium 152.0	64 **Gd** Gadolinium 157.3	65 **Tb** Terbium 158.9	66 **Dy** Dysprosium 162.5	67 **Ho** Holmium 164.9	68 **Er** Erbium 167.3	69 **Tm** Thulium 168.9	70 **Yb** Ytterbium 173.0	71 **Lu** Lutetium 175.0
90 **Th** Thorium 232.0	91 **Pa** Protactinium (231)	92 **U** Uranium 238.0	93 **Np** Neptunium (237)	94 **Pu** Plutonium (242)	95 **Am** Americium (243)	96 **Cm** Curium (247)	97 **Bk** Berkelium (247)	98 **Cf** Californium (249)	99 **Es** Einsteinium (254)	100 **Fm** Fermium (253)	101 **Md** Mendelevium (256)	102 **No** Nobelium (254)	103 **Lr** Lawrencium (257)

The 1–18 group designation has been recommended by the International Union of Pure and Applied Chemistry (IUPAC) but is not yet in wide use.

Each element of the periodic table has a box that contains the name of the element and its chemical symbol, atomic number, and atomic mass. The atomic number is the number of protons in an element. Each element has a unique number of protons and therefore a unique atomic number.

There are 90 naturally occurring elements. Scientists have been able to create new elements by changing the number of protons in the nuclei of existing elements. Protons, neutrons, or electrons from one atom are accelerated to very high speeds and then smashed into the nucleus of another atom. The resulting changes in the nucleus produce a new element with a new atomic number. These artificially produced elements are usually unstable, and they quickly convert back to more stable elements. The synthetic elements are technetium (Tc, atomic number 43), promethium (Pm, atomic number 61), and all the elements with an atomic number of 93 or higher. An element with an atomic number of 112 has the highest number officially recognized by the International Union of Pure and Applied Chemistry, but elements with higher atomic numbers have reportedly been made.

The modern periodic table of the elements lists the known elements in order of their atomic masses. The boxes are organized into a grid of horizontal rows, called periods, and vertical columns, called groups. Within a period, the elements are listed in order of increasing atomic number from left to right. Elements in a period have different chemical properties, whereas elements in a group have similar chemical properties.

Individual atoms have very little mass. A hydrogen atom has a mass of 1.67×10^{-24} g (see appendix B for an explanation of the scientific notation of numbers). To avoid using such small numbers, a system of relative atomic mass is used. In this system, a **unified atomic mass unit (u), or dalton (Da),** is 1/12 the mass of ^{12}C, a carbon atom with six protons and six neutrons. Thus, ^{12}C has an atomic mass of exactly 12 u. A naturally occurring sample of carbon, however, contains mostly ^{12}C but also a small quantity of other carbon isotopes, such as ^{13}C, which has six protons and seven neutrons. The **atomic mass** of an element is the *average* mass of its naturally occurring isotopes, taking into account the relative abundance of each isotope. For example, the atomic mass of the element carbon is 12.01 u (see table 2.1), which is slightly more than 12 u because of the additional mass of the small amount of other carbon isotopes. Because the atomic mass is an average, a sample of carbon can be treated as if all the carbon atoms had an atomic mass of 12.01 u.

Appendix B

Scientific Notation

Very large numbers with many zeros, such as 1,000,000,000,000,000, or very small numbers, such as 0.0000000000000001, are very cumbersome to work with. Consequently, the numbers are expressed in a kind of mathematical shorthand known as scientific notation. Scientific notation has the following form:

$$M \times 10^n$$

where n specifies how many times the number M is raised to the power of 10. The exponent n has two meanings, depending on its sign. If n is positive, M is multiplied by 10 n times. For example, if $n = 2$ and $M = 1.2$, then

$$1.2 \times 10^2 = 1.2 \times 10 \times 10 = 120$$

In other words, if n is positive, the decimal point of M is moved to the right n times. In this case the decimal point of 1.2 is moved two places to the right.

$$1.20.$$

If n is negative, M is divided by 10 n times.

$$1.2 \times 10^{-2} = \frac{1.2}{(10 \times 10)} = \frac{1.2}{100} = 0.012$$

In other words, if n is negative, the decimal point of M is moved to the left n times. In this case, the decimal point of 1.2 is moved two places to the left.

$$0.01.2$$

If M is the number 1.0, it often is not expressed in scientific notation. For example, 1.0×10^2 is the same thing as 10^2, and 1.0×10^{-2} is the same thing as 10^{-2}.

Two common examples of the use of scientific notation in chemistry are Avogadro's number and pH. Avogadro's number, 6.023×10^{23}, is the number of atoms in 1 molar mass of an element. Thus,

$$6.023 \times 10^{23} = 602,300,000,000,000,000,000,000$$

which is a very large number of atoms.

The pH scale is a measure of the concentration of hydrogen ions in a solution. A neutral solution has 10^{-7} moles of hydrogen ions per liter. In other words,

$$10^{-7} = 0.0000001$$

which is a very small amount (1 ten-millionth of a gram) of hydrogen ions.

Appendix C

Solution Concentrations

Physiologists often express solution concentration in terms of percent, molarity, molality, and equivalents.

Percent

The weight-volume method of expressing percent concentrations states the weight of a solute in a given volume of solvent. For example, to prepare a 10% solution of sodium chloride, 10 g of sodium chloride is dissolved in a small amount of water (solvent) to form a salt solution. Then additional water is added to the salt solution to form 100 mL of salt solution. Note that the sodium chloride is dissolved in water and then diluted to the required volume. The sodium chloride is not dissolved directly in 100 mL of water.

Molarity

Molarity determines the number of moles of solute dissolved in a given volume of solvent. A **mole (mol)** of a substance contains Avogadro's number of entities, such as atoms, ions, or molecules. **Avogadro's number** is the number of atoms in exactly 12 g of ^{12}C (carbon atoms with six protons and six neutrons). This enormous number is 6.022×10^{23}. The **molar mass** of a substance is the mass of one mole of the substance expressed in grams. Because 12 g of ^{12}C is used as the standard, the atomic mass of an entity expressed in unified atomic mass units (see appendix A) is the same as the molar mass expressed in grams. Thus, carbon atoms have an atomic mass of 12.01 u, and 12.01 g of carbon have Avogadro's number (1 mol) of carbon atoms.

Just as a grocer sells eggs in lots of 12 (a dozen), a chemist groups atoms in lots of 6.022×10^{23} (Avogadro's number, 1 mol). The molar mass is used as a convenient way to determine the number of atoms in a sample of an element. For example, 1.008 g of hydrogen (1 mol) has the same number of atoms as 12.01 g of carbon (1 mol).

A 1 molar (1 M) solution is made by dissolving 1 mole (mol) of a substance in enough water to make 1 L of solution. For example, 1 mol of sodium chloride solution is made by dissolving 58.44 g of sodium chloride in enough water to make 1 L of solution. One mol of glucose solution is made by dissolving 180.2 g of glucose in enough water to make 1 L of solution.

Molality

Although 1 M solutions have the same number of solute molecules, they do not have the same number of solvent (water) molecules. Because 58.5 g of sodium chloride occupies less volume than 180 g of glucose, the sodium chloride solution has more water molecules. **Molality** is a method of calculating concentrations that takes into account the number of solute and solvent molecules. A 1 molal solution (1 m) is 1 mol of a substance dissolved in 1 kg of water. Thus, all 1-molal solutions have the same number of solvent molecules.

Osmolality

Osmosis is the diffusion of water across a selectively permeable membrane. Osmosis is determined by the concentration of solute particles in a solution, where a particle can be an atom, an ion, a molecule, or any combination of atoms, ions, or molecules (see chapter 3). That is, the number of solute particles, not the kind, determines the movement of water by osmosis. When sodium chloride, which is an ionic compound, is dissolved in water, it dissociates to form two ions, a sodium cation (Na^+) and a chloride anion (Cl^-). Glucose does not dissociate when dissolved in water, however, because it is a molecule. Thus, the sodium chloride solution contains twice as many particles as the glucose solution (one Na^+ and one Cl^- for each glucose molecule). To report the concentration of these substances in a way that reflects the number of

particles in a given mass of solvent, the concept of **osmolality** is used. An **osmole** (Osm) is the molality of a solution corrected for the degree of dissociation of the solute. Thus, 1 mol of sodium chloride in 1 kg of water is a 2 Osm solution because sodium chloride dissociates to form two ions, whereas 1 mol of glucose in 1 kg of water is a 1 Osm solution.

The osmolality of a solution is a reflection of the number, not the type, of particles in a solution. Thus, a 1 Osm solution contains 1 Osm of particles per kilogram of solvent, but the particles may be all one type or a complex mixture of different types. The kind of solute, however, affects the number of solute particles because the dissociation of solutes can vary between 0 (e.g., glucose) and 1.0 (e.g., NaCl).

The concentration of particles in body fluids is so low that the measurement milliosmole (mOsm), 1/1000 of an osmole, is used. Most body fluids have an osmotic concentration of approximately 300 mOsm and consist of many different ions and molecules. The osmotic concentration of body fluids is important because it influences the movement of water into or out of cells (see chapter 3).

Equivalents

Equivalents are a measure of the concentrations of ionized substances. One equivalent (Eq) is 1 mol of an ionized substance multiplied by the absolute value of its charge. For example, 1 mol of NaCl dissociates into 1 mol of Na^+ and 1 mol of Cl^-. Thus, there is 1 Eq of Na^+ (1 mol × 1) and 1 Eq of Cl^- (1 mol × 1). One mole of $CaCl_2$ dissociates into 1 mol of Ca^{2+} and 2 mol of Cl^-. Thus, there are 2 Eq of Ca^{2+} (1 mol × 2) and 2 Eq of Cl^- (2 mol × 1). In an electrically neutral solution, the equivalent concentration of positively charged ions is equal to the equivalent concentration of the negatively charged ions. One milliequivalent (mEq) is 1/1000 of an equivalent.

Appendix D

pH

Pure water weakly dissociates to form small numbers of hydrogen and hydroxide ions:

$$H_2O \longleftrightarrow H^+ + OH^-$$

At 25°C, the concentration of both hydrogen ions and hydroxide ions is 10^{-7} mol/L. Any solution that has equal concentrations of hydrogen and hydroxide ions is considered **neutral.** A solution is an **acid** if it has a higher concentration of hydrogen ions than hydroxide ions, and a solution is a **base** if it has a lower concentration of hydrogen ions than hydroxide ions. In any aqueous solution (at 25°C), the hydrogen ion concentration $[H^+]$ times the hydroxide ion concentration $[OH^-]$ is a constant that is equal to 10^{-14}.

$$[H^+] \times [OH^-] = 10^{-14}$$

Consequently, as the hydrogen ion concentration decreases, the hydroxide ion concentration increases, and vice versa—for example,

	$[H^+]$	$[OH^-]$
Acidic solution	10^{-3}	10^{-11}
Neutral solution	10^{-7}	10^{-7}
Basic solution	10^{-12}	10^{-2}

Although the acidity or basicity of a solution could be expressed in terms of either hydrogen or hydroxide ion concentration, it is customary to use hydrogen ion concentration. The pH of a solution is defined as

$$pH = -\log_{10}(H^+)$$

Thus, a neutral solution with 10^{-7} mol of hydrogen ions per liter has a pH of 7.

$$\begin{aligned} pH &= -\log_{10}(H^+) \\ &= -\log_{10}(10^{-7}) \\ &= -(-7) \\ &= 7 \end{aligned}$$

In simple terms, to convert the hydrogen ion concentration to the pH scale, the exponent of the concentration (e.g., -7) is used, and it is changed from a negative to a positive number. Thus, an acidic solution with 10^{-3} mol of hydrogen ions/L has a pH of 3, whereas a basic solution with 10^{-12} hydrogen ions/L has a pH of 12.

Appendix E

Answers to Review and Comprehension Questions

Chapter One

1. a; 2. b; 3. a; 4. c; 5. d; 6. e; 7. a; 8. b; 9. c; 10. d; 11. d; 12. c; 13. d; 14. d; 15. a; 16. b; 17. b; 18. a; 19. e; 20. c; 21. a; 22. b; 23. a; 24. b; 25. e

Chapter Two

1. e; 2. a; 3. b; 4. a; 5. d; 6. b; 7. d; 8. c; 9. e; 10. e; 11. c; 12. c; 13. d; 14. a; 15. b; 16. d; 17. b; 18. b; 19. e; 20. a

Chapter Three

1. a; 2. d; 3. d; 4. a; 5. b; 6. a; 7. b; 8. d; 9. a; 10. a; 11. d; 12. d; 13. b; 14. b; 15. e; 16. c; 17. c; 18. b; 19. c; 20. b; 21. c; 22. a; 23. e; 24. d; 25. c

Chapter Four

1. e; 2. c; 3. a; 4. a; 5. d; 6. b; 7. d; 8. c; 9. d; 10. a; 11. b; 12. e; 13. b; 14. a; 15. b; 16. b; 17. a; 18. e; 19. b; 20. c; 21. e; 22. b; 23. c; 24. b; 25. d

Chapter Five

1. b; 2. e; 3. b; 4. a; 5. c; 6. d; 7. e; 8. c; 9. d; 10. a; 11. b; 12. a; 13. d; 14. b; 15. c; 16. c; 17. c; 18. a; 19. c; 20. b; 21. a; 22. a; 23. d; 24. c.; 25. d

Chapter Six

1. d; 2. d; 3. d; 4. b; 5. d; 6. c; 7. e; 8. a; 9. a; 10. b; 11. e; 12. d; 13. a; 14. c; 15. d; 16. c; 17. d; 18. a; 19. b; 20. c; 21. a; 22. e; 23. d; 24. d; 25. a

Chapter Seven

1. e; 2. c; 3. d; 4. c; 5. a; 6. d; 7. a; 8. a; 9. c; 10. a; 11. b; 12. a; 13. d; 14. e; 15. a; 16. c; 17. b; 18. c; 19. a; 20. a; 21. e; 22. d; 23. e; 24. c; 25. d; 26. b; 27. a; 28. e; 29. a; 30. e; 31. c; 32. d; 33. e; 34. a; 35. c; 36. b; 37. c; 38. d; 39. b; 40. a

Chapter Eight

1. c; 2. a; 3. d; 4. b; 5. e; 6. b; 7. e; 8. b; 9. d; 10. b; 11. d; 12. c; 13. e; 14. c; 15. d; 16. c; 17. a; 18. c; 19. b; 20. e; 21. d; 22. c; 23. c; 24. e; 25. a

Chapter Nine

1. d; 2. b; 3. c; 4. d; 5. d; 6. a; 7. b; 8. c; 9. a; 10. c; 11. a; 12. a; 13. b; 14. d; 15. b; 16. a; 17. d; 18. e; 19. c; 20. c; 21. a; 22. e; 23. c; 24. d; 25. a

Chapter Ten

1. a; 2. b; 3. b; 4. a; 5. c; 6. c; 7. b; 8. a; 9. c; 10. a; 11. b; 12. d; 13. a; 14. c; 15. b; 16. a; 17. a; 18. b; 19. d; 20. c; 21. e; 22. b; 23. d; 24. d; 25. b; 26. b; 27. e; 28. e; 29. e; 30. a

Chapter Eleven

1. d; 2. c; 3. d; 4. b; 5. b; 6. d; 7. b; 8. e; 9. a; 10. d; 11. d; 12. c; 13. c; 14. c; 15. b; 16. e; 17. c; 18. d; 19. b; 20. b; 21. b; 22. c; 23. e; 24. a; 25. c; 26. d; 27. b; 28. a; 29. b; 30. e; 31. b; 32. c; 33. e; 34. b; 35. b

Chapter Twelve

1. c; 2. d; 3. e; 4. a; 5. d; 6. b; 7. d; 8. a; 9. b; 10. e; 11. c; 12. c; 13. b; 14. d; 15. a; 16. b; 17. b; 18. d; 19. b; 20. b; 21. d; 22. d; 23. b; 24. b; 25. c

Chapter Thirteen

1. e; 2. c; 3. a; 4. e; 5. b; 6. a; 7. d; 8. b; 9. b; 10. c; 11. c; 12. c; 13. a; 14. d; 15. c; 16. d; 17. a; 18. d; 19. c; 20. a; 21. b; 22. a; 23. d; 24. a; 25. c

Chapter Fourteen

1. e; 2. d; 3. d; 4. a; 5. e; 6. b; 7. d; 8. e; 9. d; 10. d; 11. d; 12. a; 13. a; 14. c; 15. e; 16. e; 17. d; 18. c; 19. e; 20. c

Chapter Fifteen

1. c; 2. b; 3. e; 4. d; 5. e; 6. c; 7. d; 8. d; 9. c; 10. a; 11. e; 12. d; 13. b; 14. c; 15. b; 16. d; 17. a; 18. d; 19. c; 20. a; 21. d; 22. e; 23. e; 24. d; 25. d; 26. a; 27. d; 28. a; 29. b; 30. c

Chapter Sixteen

1. e; 2. c; 3. a; 4. d; 5. a; 6. d; 7. d; 8. b; 9. c; 10. d; 11. b; 12. b; 13. a; 14. e; 15. b; 16. b; 17. e; 18. e; 19. b; 20. a; 21. a; 22. d; 23. d; 24. c; 25. c

Chapter Seventeen

1. d; 2. c; 3. e; 4. c; 5. a; 6. b; 7. d; 8. e; 9. a; 10. c; 11. d; 12. b; 13. b; 14. a; 15. a; 16. b; 17. d; 18. b; 19. c; 20. a; 21. d; 22. c; 23. c; 24. a; 25. e

Chapter Eighteen

1. b; 2. a; 3. a; 4. d; 5. b; 6. c; 7. d; 8. a; 9. a; 10. d; 11. e; 12. c; 13. a; 14. c; 15. d; 16. e; 17. d; 18. b; 19. b; 20. b; 21. d; 22. a; 23. d; 24. b; 25. e

Chapter Nineteen

1. d; 2. b; 3. e; 4. e; 5. d; 6. b; 7. e; 8. b; 9. a; 10. d; 11. d; 12. d; 13. b; 14. e; 15. e; 16. a; 17. d; 18. c; 19. a; 20. d; 21. b; 22. c; 23. a; 24. d; 25. d

Chapter Twenty

1. a; 2. b; 3. e; 4. d; 5. d; 6. b; 7. c; 8. b; 9. d; 10. a; 11. d; 12. c; 13. c; 14. d; 15. c; 16. b; 17. d; 18. b; 19. c; 20. b; 21. d; 22. c; 23. a; 24. c; 25. d

Chapter Twenty-One

1. a; 2. d; 3. e; 4. d; 5. a; 6. b; 7. a; 8. b; 9. a; 10. c; 11. b; 12. d; 13. c; 14. d; 15. e; 16. b; 17. d; 18. d; 19. e; 20. e; 21. b; 22. e; 23. b; 24. a; 25. e; 26. e; 27. e; 28. a; 29. c; 30. a

Chapter Twenty-Two

1. d; 2. c; 3. a; 4. e; 5. b; 6. d; 7. e; 8. e; 9. b; 10. a; 11. d; 12. b; 13. d; 14. a; 15. e; 16. c; 17. a; 18. b; 19. a; 20. b

Chapter Twenty-Three

1. d; 2. d; 3. b; 4. c; 5. b; 6. b; 7. b; 8. e; 9. b; 10. b; 11. a; 12. a; 13. c; 14. c; 15. b; 16. d; 17. d; 18. a; 19. d; 20. d; 21. c; 22. d; 23. c; 24. a; 25. b; 26. c; 27. c; 28. a; 29. a; 30. a; 31. b; 32. d; 33. a; 34. a; 35. a

Chapter Twenty-Four

1. e; 2. a; 3. a; 4. b; 5. d; 6. e; 7. b; 8. d; 9. d; 10. c; 11. b; 12. c; 13. a; 14. e; 15. a; 16. b; 17. c; 18. c; 19. b; 20. a; 21. a; 22. c; 23. a; 24. c; 25. e; 26. e; 27. e; 28. c; 29. c; 30. a

Chapter Twenty-Five

1. b; 2. e; 3. a; 4. a; 5. d; 6. a; 7. c; 8. c; 9. b; 10. c; 11. a; 12. a; 13. e; 14. d; 15. c; 16. c; 17. c; 18. a; 19. b; 20. d; 21. d; 22. c; 23. b; 24. c; 25. c

Appendix F

Answers to Critical Thinking Questions

Chapter 1

1. Answer *e* is correct. Positive-feedback mechanisms result in movement away from homeostasis and are usually harmful. The continually decreasing blood pressure is an example. Negative-feedback mechanisms result in a return to homeostasis. The elevated heart rate is a negative-feedback mechanism that attempts to return blood pressure to a normal value. In this case, the negative-feedback mechanism was inadequate to restore homeostasis, and medical intervention (a transfusion) was necessary.
2. In the cat, cephalic and anterior are toward the head; dorsal and superior are toward the back. In humans, cephalic and superior are toward the head; dorsal and posterior are toward the back.
3. When a boy is standing on his head, his nose is superior to his mouth. Directional terms refer to a person in the anatomical position, not to the body's current position.
4. a. Inferior
   b. Posterior (dorsal) or deep
   c. Distal or inferior
   d. Lateral
5. The thumb is lateral and distal (inferior) to the elbow. The elbow is medial and proximal (superior) to the thumb.
6. The wedding band should be worn proximal to the engagement ring.
7. The uterus is located in the pelvic cavity. The pelvic cavity, however, is surrounded by the bones of the pelvis and does not increase in size during pregnancy. Instead, as the fetus grows the expanding uterus must move into the abdominal cavity, thereby crowding abdominal organs and dramatically increasing the size of the abdominal cavity.
8. There are two ways. First, the visceral peritoneum wraps around organs. Thus, the peritoneal cavity surrounds the organ, but the organ is not inside the peritoneal cavity. The peritoneal cavity contains only peritoneal fluid. Second, retroperitoneal organs are in the abdominopelvic cavity, but they are between the wall of the abdominopelvic cavity and the parietal peritoneal membrane.
9. After the pole passes through the abdominal wall, it pierces the parietal peritoneum. In passing through the stomach, it penetrates the visceral peritoneum, the stomach itself, and the visceral peritoneum on the other side of the stomach. Because the diaphragm is lined inferiorly by parietal peritoneum and superiorly by parietal pleura, these are the next two membranes pierced. The pole then passes through the pleural space and visceral pleura to enter the lung.
10. The kidneys are located in the abdominal cavity but are retroperitoneal. When a person is lying prone, it is possible to cut through the posterior abdominal wall and remove a kidney without cutting through parietal peritoneum.

Chapter 2

1. Because atoms are electrically neutral, the iodine atom has the same number of protons and electrons. The gain of an electron means the iodine ion has one more electron than protons and therefore a charge of minus one. The correct symbol is I^{-1}.
2. a. Dissociation
   b. Synthesis
   c. Decomposition
3. Muscle contains proteins. To increase muscle mass, proteins must be synthesized from amino acids. The synthesis of molecules in living organisms requires the input of energy. That energy comes from the potential energy stored in the chemical bonds of food molecules, which is released during the decomposition of food molecules.
4. The sodium bicarbonate dissociates to form sodium ions (Na^+) and bicarbonate ions (HCO_3^-). Because this is a reversible reaction, The HCO_3^- added to the solution bind with H^+ to form carbon dioxide and water. The decrease in H^+ causes the pH to increase (become more basic).
5. The slight amount of heat functioned as activation energy and started a chemical reaction. The reaction released energy, especially heat, which caused the solution to become very hot.
6. There was most likely at least one enzyme present that was required for one of the initial reactions. Boiling denatured any enzymes present in the solution. Without enzymes, the reactions will not occur.

Chapter 3

1. If the membrane is freely permeable, the solutes in the tube diffuse from the tube (higher concentration of solutes) into the beaker (lower concentration of solutes) until equal amounts of solutes exist inside the tube and beaker (i.e., equilibrium). In a similar fashion, water in the beaker diffuses from the beaker (higher concentration of water) into the tube (lower concentration of water) until equal amounts of water are inside the tube and beaker. Consequently, the solution concentrations inside the tube and beaker are the same because they both contain the same amount of solutes and water. Under these conditions, no net movement of water into the tube occurs. This simple experiment demonstrates that osmosis and osmotic pressure require a membrane that is selectively permeable.
2. *B* is the most logical conclusion. Swollen lung tissue suggests that the tissues had been submerged in a hypotonic solution, such as fresh water. Because the bay contains salt water, which is slightly hypertonic to blood, it is unlikely he drowned in the bay. It is more likely he drowned in fresh water and was later placed in the bay. Although this is the most logical conclusion of those given, there are possibilities other than murder. For example, he may have accidentally drowned in a fresh-water stream and then been washed into the bay.
3. The cells within the wound swell with water and lyse by the introduction of a hypotonic solution. This kills potentially metastatic cells that may still be present in the wound.
4. Answer *b* is correct. Because the solution is isotonic, there is no exchange of water. Because the solution contains the same concentration of all substances except that it has no urea, only a net movement of urea occurs across the membrane.
5. Because the plasma membrane stays the same size, even though small pieces of membrane from secretory vesicles are continually added to the plasma membrane, one must conclude that some process is removing small pieces of membrane at the same rate that they are added. The plasma membrane is constantly being recycled.
6. The well-developed rough endoplasmic reticulum is indicative of protein synthesis, and a well-developed Golgi apparatus is indicative of secretion. It is likely that this cell synthesizes and secretes proteins.
7. Because the drug inhibits mRNA synthesis, protein synthesis is stopped. If the cell releases proteins as they are synthesized, the rate of protein secretion dramatically decreases following the administration of the drug. On the other hand, if the cell releases stored proteins, the rate of secretion at first is normal and then gradually declines.
8. The sickle-cell gene for producing the protein has a different nucleotide than the normal gene for producing the protein. Because the nucleotides are the set of instructions for making the protein, the change means the instructions contain an error. When mRNA copies the instructions from the faulty DNA, the error is also copied. Consequently, when the amino acids are joined together to form the protein at the ribosome, the instructions are wrong and the protein is incorrectly assembled. A substitution of a single nucleotide results in a different amino acid in the chain of amino acids that make up the protein. This change alters the shape of the protein. Just as the function of enzymes depends on their shape, the functions of other proteins depends on their shape. The incorrectly manufactured proteins in the red blood cells have the wrong shape and do not stack correctly. As a result of the abnormal stacking, the red blood cells have an abnormal, sickle shape.

Chapter 4

1. The tissue is epithelial tissue, as it is lining a free surface, and the epithelium is stratified because it consists of more than one layer. The types of stratified epithelium are stratified squamous, stratified cuboidal, stratified columnar, and

transitional epithelium. The structure of the cells in the surface layers enables the determination of a specific tissue type. Flat cells in the surface layer indicate stratified squamous epithelium. Cuboidal cells in the surface layer indicate stratified cuboidal epithelium, and columnar cells in the surface layer point to stratified columnar epithelium. The surface cells of transitional epithelium are roughly cuboidal with cubelike or columnar cells beneath them. When transitional epithelium is stretched, the surface cells are still roughly cuboidal, but underlying layers can be somewhat flattened.

2. In general, epithelial cells undergo cell division (mitosis) in response to injury, and the newly produced cells replace the damaged cells. If the basement membrane is destroyed, however, nothing is present to provide scaffolding for the newly formed epithelial cells. Without the basement membrane, there is no effective way for the newly formed epithelial cells to repair a structure, such as a kidney tubule. Since the basement membranes appear to be mostly present, the person is likely to survive and the kidney will regain most of its ability to function.

3. Pseudostratified columnar ciliated epithelium is found in the trachea. It produces mucus that traps dust and debris in air. The cilia move the mucus with entrapped dust and debris to the throat, where it is swallowed. In heavy smokers the pseudostratified columnar epithelium is replaced by stratified squamous epithelium that serves a protective function against the irritating materials in the smoke. Unfortunately, this type of epithelium is not ciliated, so removal of foreign materials from the trachea is more difficult. If a person doesn't smoke for 2 years, then the original pseudostratified columnar ciliated epithelium replaces the stratified squamous epithelium and debris removal resumes.

4. Tight junctions prevent the passage of materials between the epithelial cells.

5. A secretory epithelium is generally a simple epithelium. To manufacture large amounts of enzymes, a cuboidal or columnar cell with the appropriate organelles such as rough endoplasmic reticulum and Golgi apparatuses would be expected. The pancreas is formed of simple columnar epithelium. The epithelium has microvilli on its free surface, which increase the surface area and facilitate secretion. The tight junctions that connect the epithelial cells of the pancreatic glands and duct system to each other prevent damage to the underlying tissues (by the action of pancreatic digestive enzymes).

6. Because this tissue has a free surface and consists of a single layer of cells, it is simple epithelium. Because the cells are narrow and tall, it is simple columnar epithelium. The microvilli increase the surface area of the free surface of the epithelial cells and the mitochondria synthesize ATP. Therefore, this cell type is probably involved in active transport of solutes. The goblet cells indicate that it also secretes mucus. These characteristics are consistent with the epithelium lining the small intestine.

7. a. The stratified squamous epithelium that lines the mouth provides protection. Replacement of it with simple columnar epithelium makes the lining of the mouth much more susceptible to damage because the single layer of epithelial cells is easier to damage.
   b. Tendons attach muscles to bone. When muscles contract, muscles pull on the tendons and, because the tendons are attached to bone, the bones move. If the tendons contained elastic tissue, the tendons would be more like elastic bands. The muscles would contract and stretch the tendons. Not all of the force of muscle contraction would be transferred to the bones to cause them to move.
   c. If bones were made of elastic cartilage, they would be much more flexible and could bend and then return to their original shape. They would not be rigid structures, like bone, that support our weight and result in efficient movement.

8. Chemical mediators of inflammation normally produce beneficial responses, such as dilation of small-diameter blood vessels and increased blood vessel permeability. Blocking these effects could reduce the ability of the body to deal with harmful agents, such as bacteria. On the other hand, antihistamines also reduce many of the unpleasant symptoms of inflammation, making the patient more comfortable and this can be considered beneficial. Antihistamines are commonly taken to prevent allergy symptoms that are often an overreaction of the inflammatory response to foreign substances, such as pollen.

9. Collagen synthesis is required for granulation tissue and scar formation. If collagen synthesis does not occur because of a lack of vitamin C, or if collagen synthesis is slowed, wound healing does not occur or is slower than normal. One might expect that the density of collagen fibers in a scar is reduced and the scar is not as durable as a normal scar.

Chapter 5

1. The stratum corneum, the outermost layer of the skin, consists of many rows of flat, dead epithelial cells. The many rows of cells, which are continuously shed and replaced, are responsible for the protective nature of the integument. In infants, there are fewer rows of cells, resulting in skin that is more easily damaged than that in adults.

2. Alcohol is a solvent that dissolves lipids. It removes the lipids from the skin, especially in the stratum corneum. The rate of water loss increases after soaking the hand in alcohol because of the removal of the lipids that normally prevent water loss.

3. Melanocytes produce melanin, which protects underlying tissue from ultraviolet radiation. Therefore, we expect melanocytes to be as superficial as possible. Also, melanin production varies, depending on exposure to the sun. Response to stimulation is a characteristic of living cells. Thus, melanocytes should be found in the most superficial living layer of the epidermis, the stratum basale.

4. Carotene, a yellow pigment from ingested plants, accumulates in lipids. The stratum corneum of a callus has more layers of cells than other noncallused parts of the skin, and the cells in each layer are surrounded by lipids. The carotene in the lipids makes the callus appear yellow.

5. When first exposed to the cold temperature just before starting the run, the blood vessels in the skin constrict to conserve heat. This produces a pale skin color. After a person has been running for awhile, as a result of the excess heat generated by the exercise, the blood vessels in the skin dilate. This results in heat loss and helps prevent overheating. Increased blood flow through the skin causes it to turn red. After the run, the body still has excess heat to eliminate, so the skin remains red for some time.

6. Yes, the skin (dermis) can be overstretched due to obesity or rapid growth.

7. The vermillion border is covered by keratinized epithelium that is transitional between the nonkeratinized stratified epithelium of the mucous membrane and the keratinized stratified epithelium of the facial skin. The mucous membrane has mucous glands, which secrete mucus (see chapter 4). The mucus helps keep the inner surface of the lips moist. In addition, the inner surface of the lips is "sealed off" from the outside air most of the time and is moistened by saliva. The keratinized stratified epithelium of the skin has a stratum corneum, with lipids that reduce water loss. The skin also has sebaceous glands, which produce sebum that reduces water loss. The skin of the vermillion border is not moistened by mucus or saliva and it does not have sebaceous glands. Without sebaceous glands, the surface of the vermillion border is not protected against drying by sebum. Also, the vermillion border is not as heavily keratinized as facial skin—that is, there are fewer cell layers with surrounding lipids. For these reasons, the vermillion border dries out more easily than the mucous membrane of the oral cavity or the skin of the face.

8. Eyelashes have a short growth stage (30 days) and are therefore short. Fingernails grow continuously but are short because they are cut, broken off, or worn down.

9. The hair follicle, but not the hair, is surrounded with nerve endings that can detect movement or pulling of the hair. The hair is dead, keratinized epithelium, so cutting the hair is not painful.

10. Several methods have some degree of success in treating acne. (a) Kill the bacteria. One effective agent is benzoyl peroxide, found in some acne medications. (b) Prevent blockage of the hair follicle. A vitamin A derivative (tretinoin; Retin-A) has proven effective in keeping the follicular epithelial cells and sebum from building up and closing off the hair follicle. (c) Unplug the follicle. Some sulfur compounds (e.g., Acnederm) speed up peeling of the skin and thus unplug the follicle.

11. Probably not, because, following removal of the nail from the nail fold, it may grow back into the nail fold and the ingrown toenail may reoccur. One solution is to remove the small part of the nail responsible for the ingrown toenail. Prior to this drastic approach, sterile gauze can be placed between the nail and the nail fold to force the nail away from the nail fold. After the nail fold is healed, the gauze can be removed.

12. Rickets is a disease of children resulting from inadequate vitamin D intake. With inadequate vitamin D, there is insufficient absorption of calcium from the intestine, resulting in soft bones. If adequate vitamin D is ingested, rickets is prevented, whether one is dark- or fair-skinned. However, if dietary vitamin D is inadequate, when the skin is exposed to ultraviolet light, 7-dehydrocholesterol is converted into cholecalciferol, which can be converted to vitamin D. Dark-skinned children are more susceptible to rickets because the additional melanin in their skin screens out the ultraviolet light and they produce less vitamin D.

Chapter 6

1. The injury separated the head (epiphysis) from the shaft (diaphysis) at the epiphyseal plate, which is cartilage. Because the epiphyseal plate is the site of bone elongation, damage to the epiphyseal plate can interfere with bone elongation, resulting in a shortened limb.

2. Normally, bone matrix and bone trabeculae are organized to be strongest along lines of stress. Random organization of the collagen fibers of bone matrix results in weaker bones. In addition, the reduced amount of trabecular bone makes the bone weaker. Fractures of the bone can occur when the weakened bone is subjected to stress.

3. Replacement of cartilage of the epiphyseal plate by bone normally occurs on the diaphyseal side of the plate. As growth ceases, the cartilage cells stop dividing and producing new cartilage. Replacement of cartilage with bone continues from the diaphyseal side, and eventually all of the cartilage of the plate is bone.

4. Mechanical stress applied to bone stimulates osteoblast activity, so the patient with a walking cast should heal faster.

5. Osteoporosis is depletion of bone matrix that results when more bone is destroyed than is formed. Because mechanical stress stimulates bone formation (osteoblast activity), running helps prevent osteoporosis in the bones being stressed. This includes the bones of the lower limbs and the spine.

6. The loss of bone density results because the bones are not bearing weight in the weightless environment. Therefore, osteoblasts are not sufficiently stimulated and bone resorption exceeds bone building. Bone loss can be slowed by stressing the bones using exercises against resistance, such as cycling.

7. The kidneys are the site of production of active vitamin D (see chapter 5), which is needed for Ca^{2+} absorption in the small intestine. Kidney failure can result in inadequate vitamin D production, too little uptake of Ca^{2+}, and therefore osteomalacia.

8. Testosterone normally causes a growth spurt at puberty, followed by slower growth and closure of the epiphyseal plate. Without testosterone, growth is slower but proceeds longer, resulting in a taller than normal person.

9. Blood vessels in central canals run parallel to the long axis of the bone, and perforating canals run at approximately a right angle to the central canals. Thus, perforating canals connect to central canals, which allows blood vessels in the perforating canals to connect with blood vessels in the central canals. After a fracture, blood flow through the central canals stops back to the point where the blood vessels in the central canals connect to the blood vessels in the perforating canals. The regions of bone on both sides of the fracture associated with this lack of blood delivery die.

10. Hyperparathyroidism stimulates increased bone breakdown and could cause osteitis fibrosa cystica, a condition in which the bone is eaten away as Ca^{2+} are released from the bone. The result can be a deformed bone that is likely to fracture. Vitamin D therapy might help because vitamin D promotes an increase in blood Ca^{2+} levels and therefore increased deposition of Ca^{2+} in bone.

Chapter 7

1. An infection in the nasal cavity could spread to adjacent cavities and fossae, including the paranasal sinuses: (1) frontal, (2) maxillary, (3) ethmoidal, and (4) sphenoidal; (5) the orbit (through the nasolacrimal duct); (6) the cranial cavity (through the cribriform plate); and (7) the throat (through the posterior opening of the nasal cavity).

2. Falling on the top of the head could drive the occipital condyles into the superior articulating processes of the atlas, causing a fracture. An uppercut to the jaw would slightly lift the occipital condyles away from the superior articulating processes of the atlas and usually does not result in a fractured atlas. Such a blow to the jaw can, however, fracture the temporal bone where it articulates with the mandible.

3. Forceful rotation of the vertebral column is most likely to damage the articular processes, especially in the lumbar region, where the articular processes tend to prevent excessive rotation (the superior articular processes face medially and the inferior articular processes face laterally).

4. Weaker back muscles on one side can cause the vertebral column to bend laterally (scoliosis) toward the opposite side. Lordosis can result from pregnancy. As the fetus causes the abdomen to move anteriorly, the thorax and head tend to pull posteriorly, to restore the center of gravity. This posture increases the lumbar curvature. The same effect can be seen in people who are "pot-bellied."

5. If the ulna and radius become fused, the radius can no longer rotate relative to the ulna. As a result, most of the rotation of the forearm and hand is lost.

6. Measure from the anterior superior iliac spine (a "stationary" point relative to the limb, which can be easily found as a surface landmark) to the lateral malleolus. The inferior side of the foot can also be used, if the person is standing on a flat surface. A defect of the foot or ankle may occur, however, in which the ankle on one side is elevated. If the length of the thigh is the only part to be measured, measure to the lateral epicondyle.

7. Decubitus ulcers form over bony prominences where the bone is close to the overlying skin and where the body contacts the bed when lying down. Such sites are the back and front of the skull and the cheeks (over zygomatic bones), acromion, scapula, olecranon, coccyx, greater trochanter, lateral epicondyle of the femur, patella, lateral malleolus, and calcaneus.

8. Women's hips are wider than men's. As the knees are positioned toward the midline, the slope of the femur from its proximal end toward its distal end is greater in women. As a result, more women than men tend to be knock-kneed.

9. The lateral malleolus extends farther distally than does the medial malleolus, thus making it more difficult to turn the foot laterally than to turn it medially. The styloid process of the radius extends farther distally than the styloid process of the ulna, thus making it more difficult to cock the wrist toward the thumb (laterally) than toward the little finger (medially).

10. Landing on the heels could fracture the calcaneus. Heavy objects, such as the firefighters, landing on the dorsal surface of the foot could fracture the metatarsals or even the tarsals.

11. a. flexion and supination
    b. flexion of the hip and extension of the knee
    c. abduction of the arm at the shoulder
    d. flexion of the knee and plantar flexion of the foot

12. The anterior drawer test determines the integrity of the anterior cruciate ligament, and the posterior drawer test determines the integrity of the posterior cruciate ligament. Unusual movement during the posterior drawer test indicates damage to the posterior cruciate ligament.

Chapter 8

1. The toxin could prevent nerves from stimulating contraction of respiratory muscles. Possibilities include blocking Ca^{2+} channels in the presynaptic terminals so that they do not respond to action potentials, preventing the synthesis or release of ACh from the presynaptic terminals, inactivating ACh, increasing the activity of acetylcholinesterase, blocking ACh from binding to its receptor, blocking ligand-gated Na^+ channels in the postsynaptic membranes, blocking voltage-gated Na^+ channels in the postsynaptic membranes, and blocking Ca^{2+} channels in the sarcoplasmic reticulum. The toxin could also prevent ATP synthesis or utilization.

2. Muscular dystrophy results from gradual atrophy of skeletal muscle fibers and their replacement with connective tissue. Myasthenia gravis results from the degeneration of the receptors for acetylcholine on the postsynaptic membranes of skeletal muscle cells. If an inhibitor of acetylcholinesterase is administered, the result should be an increase in the concentration of acetylcholine in the nerve muscle synapse. Thus, more acetylcholine is available to bind to acetylcholine receptors. In people suffering from myasthenia gravis, the increased concentration of acetylcholine in the synapse allows acetylcholine to bind a greater percentage of the acetylcholine receptors present and causes the muscle contractions to increase in strength. In people who have muscular dystrophy, the muscle contractions do not increase in strength because muscle atrophy is the cause of the weakness. The additional acetylcholine in the neuromuscular synapse has no effect on the weakened muscle fibers.

3. Placing sarcoplasmic reticulum from skeletal muscle cells into the beaker would remove Ca^{2+} from the solution because sarcoplasmic reticulum transports Ca^{2+} from the solution into the sarcoplasmic reticulum. ATP would have to be added for two reasons: (1) The sarcoplasmic reticulum actively transports Ca^{2+} and therefore requires ATP and (2) ATP must bind to the heads of the myosin molecules before the myosin heads can release from the active sites on the actin molecules.

4. Start with a subthreshold stimulus and increase the stimulus strength by very small increments. Apply the stimulus to the nerve of muscle A and muscle B. If the number of motor units is the same for both preparations, each time

the stimulus strength is increased the degree of tension produced by the muscles will also increase to the same degree in each muscle. If one muscle has more motor units than the other, the muscle with the greater number of motor units will exhibit a greater number of separate increases in tension, and the magnitude of the increases in tension will be smaller than those seen in the muscle with fewer motor units.

5. While the weight is being held steady, the cross-bridges are pulling the Z disks closer together, but the external load (the weight) is pulling the sarcomeres apart with equal force. Because the internal and external forces are equal, the cross-bridges are producing enough force to hold the weight steady, but not enough to shorten the muscle, so the sarcomeres remain the same length. When the individual lowers the weight, the cross-bridges are producing less force than the weight. Thus, each time a cross-bridge detaches from actin, the thin filaments "slip" and the sarcomeres lengthen. When the individual raises the weight, the cross-bridges are producing more force than the external load. Thus, the cross-bridges collectively are able to produce enough force to pull the Z disks closer together and the sarcomeres get shorter.

6. The shape of an length-tension curve for skeletal muscle can be seen in figure 8.17. In contrast, a length-tension curve is much flatter for smooth muscle. That is, for each increase in the length of a smooth muscle fiber, there is little change in the tension produced. Smooth muscle has the ability to increase in length without much increase in the tension produced by the smooth muscle cells.

7. During intense exercise, it is possible to experience physiological contracture. Being unable to either contract or relax the muscles for a short time while exercising suggests the existence of physiological contracture.

8. The muscles would contract. ATP would be available to bind to the myosin heads, thus allowing myosin molecules to be released from actin molecules. The cross-bridges would immediately re-form, and complete cross-bridge cycling would result in contraction of the muscle fibers. As long as Ca^{2+} were present at high concentrations in the sarcoplasm, contraction of the muscles would occur. If the sarcoplasmic reticulum were intact, ATP would be available to drive the active transport of Ca^{2+} into the sarcoplasmic reticulum. As the Ca^{2+} decreased in the sarcoplasm, relaxation would result. If the sarcoplasmic reticulum were not intact, however, and could not transport Ca^{2+} into the sarcoplasmic reticulum as fast as they leaked out, the muscle would remain contracted until it fatigued.

9. During the 100 m race, Shorty depended on ATP produced by anaerobic metabolism. That produced an oxygen deficit at the end of the run, which resulted in an elevated rate of respiration for a time. During the longer and slower run, most of the ATP for muscle contractions was produced by aerobic respiration, and very little oxygen deficit developed. Prolonged aerobic respiration is required to "pay back" the oxygen deficit. Shorty's rate of respiration was, therefore, prolonged after the 100 m race but not after the longer but slower run.

Chapter 9

1.
Muscle	Action	Synergist	Antagonist
Erector spinae	Extends vertebral column	Interspinales Multifidus Semispinalis thoracis	Most anterior abdominal muscles
Coraco-brachialis	Adducts arm	Latissimus dorsi Pectoralis major Teres major Teres minor	Deltoid Supra-spinatus
	Flexes arm	Deltoid (anterior) Pectoralis major Biceps brachii	Deltoid (posterior) Latissimus dorsi Teres major Teres minor Infra-spinatus Sub-scapularis Triceps brachii

2. Biceps brachii: pull-ups with hands supinated; Triceps brachii: push-ups
Deltoid: abduction of the arms to shoulder height, with weights in the hands (abduction past shoulder height involves mostly scapular rotation by the trapezius)
Rectus abdominis: sit-ups to 45 degrees (sit-ups past 45 degrees involve mostly the iliopsoas)
Quadriceps femoris: extending the legs against a force
Gastrocnemius: plantar flexion of the feet against a force, such as toe raises with a weight on the shoulders

3. The muscles that flex the head also oppose extension of the neck. In an accident causing hyperextension of the neck, these muscles could be stretched and torn. The muscles involved could include the sternocleidomastoid and deep anterior neck muscles. Automobile headrests are designed so that, if adjusted correctly, the back of the head hits the headrest during a rear-end accident, thereby preventing hyperextension of the neck.

4. The only muscle that elevates the lower eyelid is the orbicularis oculi, which "closes the eye." With this muscle not functioning, the lower eyelid would droop. The levator anguli oris, which elevates the angle of the mouth, was also apparently affected, allowing the corner of the mouth to droop. The zygomaticus major may also have been affected, as it inserts onto the corner of the mouth (see figure 9.3).

5. The genioglossus muscle protrudes the tongue. If it becomes relaxed, or paralyzed, the tongue may fall back and obstruct the airway. This can be prevented or reversed by pulling forward and down on the mandible, thus opening the mouth. The genioglossus originates on the mental protuberance of the mandible. As the mandible is pulled down and forward, the genioglossus is pulled forward with the mandible, thus pulling the tongue forward also.

6. The rotator cuff muscles are the primary muscles holding the head of the humerus in the glenoid cavity, especially the supraspinatus. In fact, a torn rotator cuff, which usually involves a tear of the supraspinatus muscle, often results in dislocation of the shoulder.

7. With the quadriceps femoris paralyzed, the leg could not be extended, and the lower limb could not bear weight unless the knee were passively extended, such as by pushing back on the distal end of the thigh with the hand. Walking would be almost impossible, except by taking very small steps and by pushing back on the knee with each step, or by bracing the knee in an extended position.

8. Speedy has ruptured the calcaneal tendon, and the gastrocnemius and soleus muscles have retracted, thereby causing the abnormal bulging of the calf muscles. Because the major plantar flexors are no longer connected to the calcaneus, the runner cannot plantar flex the foot, and the foot is abnormally dorsiflexed because the antagonists have been disconnected.

Chapter 10

1. A reduced intracellular concentration of K^+ causes depolarization of the resting membrane potential. Because the intracellular concentration of K^+ is reduced, the concentration gradient for K^+ from the inside to the outside of the plasma membrane is also reduced. Thus, the rate at which K^+ diffuse out of the cell is reduced, and a smaller charge difference across the plasma membrane is required to oppose the diffusion of the K^+ out of the cell. Therefore, the potential difference across the plasma membrane is reduced, and the cell is depolarized.

2. Because the plasma membrane is much less permeable to Na^+ than to K^+, changes in the extracellular concentration of Na^+ affect the resting membrane potential less than do changes in the extracellular concentration of K^+. Therefore, increases in extracellular Na^+ have a minimal effect on the resting membrane potential. Because the membrane is much more permeable to Na^+ during the action potential, the elevated concentration of Na^+ in the extracellular fluid results in Na^+ diffusing into the cell at a more rapid rate during the action potential, resulting in a greater degree of depolarization during the depolarization phase of the action potential.

3. Because lithium ions reduce the permeability of plasma membranes to Na^+, the Na^+ channels in the plasma membrane tend to remain closed. A normal stimulus causes Na^+ channels to open, allowing Na^+ to diffuse into the cell, thereby resulting in depolarization. The cell is less sensitive to stimuli because the membrane is less permeable to Na^+.

4. Smooth muscle cells contract spontaneously in response to spontaneous depolarizations that produce action potentials. One way action potentials can be produced spontaneously is if membrane permeability to Na^+ spontaneously increases. As a result, a few Na^+ enter the smooth muscle cells and cause a small depolarizing graded potential. The small depolarization can cause voltage-gated Na^+ channels to open, which results in further depolarization, thereby stimulating additional Na^+ voltage-gated ion channels to open. This positive-feedback cycle can continue until the plasma membrane is depolarized to its threshold level and an action potential is produced.

5. Action potential conduction along a myelinated nerve fiber is more energy-efficient

because the action potential is propagated by saltatory conduction, which produces action potentials at the nodes of Ranvier. Compared with an unmyelinated nerve fiber, only a small portion of the myelinated neuron's membrane has action potentials. Thus, there is less flow of Na^+ into the neuron (depolarization) and less flow of K^+ out of the neuron (repolarization). Consequently, the Na^+–K^+ pump has to move fewer ions in order to restore ion concentrations. Because the Na^+–K^+ pump requires ATP, myelinated axons use less ATP than unmyelinated axons.

6. With aging, there is a decrease in the amount of myelin surrounding axons, which decreases the speed of action potential propagation. At synapses, there is also an increase in the time it takes for action potentials in the presynaptic terminal to cause the production of action potentials in the postsynaptic membrane. It is believed this results from a reduced release of neurotransmitter by the presynaptic terminal and a reduced number of receptors in the postsynaptic membrane.

7. Organophosphates inhibit acetylcholinesterase, thereby causing an increase in acetylcholine in the synaptic cleft, leading to overproduction of action potentials, tetanus of muscles, and possible death resulting from respiratory failure (see chapter 8). Curare is the best antidote because it blocks the effect of acetylcholine and counteracts the organophosphate. Too much curare, however, can cause flaccid paralysis of the respiratory muscles. Injecting acetylcholine would make the effect of the organophosphate worse. Potassium chloride causes depolarization of muscle cell membranes, thereby making them more sensitive to acetylcholine.

8. If the motor neurons supplying skeletal muscle are innervated by both excitatory and inhibitory neurons, blocking the activity of the inhibitory neurons with strychnine results in overstimulation of the motor neurons by the excitatory neurons.

9. When the neurotoxin binds to ligand-gated Na^+ channels in the postsynaptic membrane of a skeletal muscle fiber, they open and Na^+ enter the cell, producing graded potentials. When the graded potentials reach threshold, an action potential is produced, stimulating the muscle fiber to contract. The neurotoxin tends to remain bound to the ligand-gated Na^+ channels, however, which prevents ACh from binding. Thus, the nervous system's ability to stimulate the muscle fiber decreases as more and more neurotoxin binds to ligand-gated Na^+ channels. Because the ligand-gated Na^+ channels with bound neurotoxin remain open, Na^+ continue to enter the muscle fiber, causing its resting membrane potential to depolarize. Eventually, the membrane becomes so depolarized that it is unresponsive to stimulation. Death from a cobra bite usually occurs because of paralysis of respiratory muscles.

10. A Na^+ channelopathy in which Na^+ channels open more readily than normal or stay open longer than normal could cause an increased production of action potentials. A Ca^{2+} channelopathy in which more Ca^{2+} than normal enter the presynaptic terminal could result in increased release of neurotransmitter from the presynaptic terminal and thus an increased production of action potentials.

Chapter 11

1. A condition in which a patient loses all sense of feeling in the left side of the back, below the upper limb, and extending in a band around to the chest, also below the upper limb, but all sensation on the right is normal, suggests that the patient's dorsal roots have been damaged on the left side adjacent to the part of the spinal cord supplying that part of the body.

2. The phrenic nerve is cut in the thorax, and the surgery is performed while the lung is being removed.

3. The ulnar nerve supplies the medial third of the hand, little finger, and medial half of the ring finger. The median nerve supplies the lateral two-thirds of the palm and thumb and the surface of the index, middle, and lateral half of the ring finger. The radial nerve supplies the lateral two-thirds of the dorsum of the hand.

4. Pulling on the upper limb when it is raised over the head can damage the lower brachial plexus—in this case, the origin of the ulnar nerve. The ulnar nerve innervates muscles that abduct/adduct the fingers and flex the wrist.

5. The sciatic nerve has rootlets from L4 to S3. Depending on the rootlet compressed, pain can be felt in different locations.

6. a. obturator nerve
   b. femoral nerve
   c. sciatic (tibial) nerve
   d. obturator nerve
   e. obturator nerve, some from femoral nerve

7. Enlargement of the lateral and third ventricles, without enlargement of the fourth ventricle, suggests a blockage between the third and fourth ventricles in the cerebral aqueduct. This defect, called aqueductal stenosis, is a common congenital problem.

8. Blood in the CSF taken through a spinal tap indicates the presence of blood in the subarachnoid space and suggests that the patient has a damaged blood vessel in the subarachnoid space.

9. I: Have the person describe the smell of something placed under his or her nose; II: test vision; III: test eye movements; IV: test the ability to move the eye down and out; V: test the sense of feeling in the face; VI: test the ability to move the eye to the side; VII: test for the ability to taste an item on the front of the tongue and check for facial expression; VIII: test the person's ability to hear; IX: test the ability to swallow; X: test the ability to swallow; XI: test the ability of the person to turn the head and shrug the shoulders; XII: test the ability to protrude the tongue.

10. The olfactory nerves (CN I) travel through the olfactory foramina within the cribriform plate of the ethmoid bone as they pass from the nasal epithelium to the olfactory bulbs (see table 11.2). In this case, the cribriform plate of the ethmoid bone was probably fractured, severing the connections of the olfactory nerves to the olfactory bulb and resulting in a loss of the sense of smell.

Chapter 12

1. The first sensations that occur when a woman picks up an apple and bites into it are visual (special), tactile (general), and proprioceptive (general). The woman holds the apple in her hand and looks at it. The tactile sensations from mechanoreceptors in the hand tell her that the apple is firm and smooth. The proprioceptive sensations originating in the joints of the hand tell the woman the size and shape of the apple. Visual input also tells her the size of the apple and that it has a smooth surface, as well as its color. As the woman bites into the apple and begins to chew, proprioceptive sensations from the teeth and jaw provide information on how widely the jaws must be opened to accommodate the bite and how hard to bite down. Tactile sensations originating in the tongue and cheeks tell her the location of the bite of apple and its texture as it is moved about in the mouth. Taste and olfactory sensations (special, chemoreceptor) provide information about the taste of the apple.

2. a. The most likely explanation is that the olfactory neurons accommodate and no longer respond to odor stimulus.
   b. The fact that one can hear the sound when one tries indicates that the sensory receptors for sound have not accommodated and are still able to detect the sound stimulus. Many action potentials arriving in the brain are prevented from causing conscious perception, until we consciously "pay attention" to the stimulus. For example, you may not be paying attention to general conversations in a crowded room or hall, until someone says your name. The sound of your name leaps out of the surrounding babble, and you are suddenly interested in what was being said by the person who spoke your name.

3. Constipation, with painful distention and cramping of the colon, results in the sensation of diffuse pain. Deep, visceral pain is not highly localized because few mechanoreceptors are present in deeper structures, such as the colon. The pain is perceived as occurring in the skin over the lower central portion of the abdomen (in the hypogastric region) because it is referred to that location because of converging CNS pathways (see figure A).

4. It is possible that the dorsal column/medial lemniscal system within the right side of the spinal cord is damaged, and that the corticospinal tracts for motor control and the spinothalamic tracts for pain and temperature were not damaged. However, it is also possible that the dorsal column/medial lemniscal system is damaged within the medulla oblongata, where neurons synapse and cross over to the left side of the brain, or within the tracts on the left side that ascend from the medulla oblongata to the thalamus. Another possibility is damage to the cerebral cortex on the left side. Additional information is needed to determine exactly where the injury is located.

5. The fibers of the dorsal column/medial lemniscal system carry two-point discrimination and proprioceptive information. Primary neurons from the left side of the body ascend the spinal cord in the dorsal column and synapse with secondary neurons in the medulla oblongata. The secondary neurons cross over in the upper medulla and ascend through the right side of the pons to the thalamus. A patient suffering from a loss of two-point discrimination and proprioception on the left side of the body as a result of a lesion in the medial lemniscal system in the thalamus has a lesion in the right side of the thalamus.

6. The damaged tracts are the lateral corticospinal tract, controlling motor functions on the right side of the body, and the lateral spinothalamic tract for pain and temperature sensations from the left side of the body. Damage to these tracts in the right side of the spinal cord produces the observed symptoms because, in the cord the lateral spinothalamic tract crosses over at the level of entry and is therefore located on the opposite side of the cord from its peripheral nerve endings, whereas the corticospinal tract lies on the same side of the cord as its target muscles.

7. The right cerebral cortex controls the left side of the body. The motor cortex has a topographic representation of the opposite side of the body, with the hand, forearm, arm, and shoulder located approximately in the center of the precentral gyrus. The lesion is therefore in the center of the right precentral gyrus of the cerebrum. Some grosser control of the left-upper limb may still exist because of the indirect pathways.

8. The descending pathway affected is most likely the corticospinal tract. The lesions in the cerebral cortex would have to be contralateral to the affected side. This is because descending pathways arising from the cerebral cortex cross over at the medulla oblongata (lateral corticospinal tract) and spinal cord (anterior corticospinal tract) before they synapse with lower motor neurons.

9. Damage to the cerebellum can result in decreased muscle tone, balance impairment, a tendency to overshoot when reaching for or touching something, and an intention tremor. These symptoms are opposite to those seen with basal ganglia dysfunction. Cerebellar dysfunction exhibits symptoms very similar to those seen in an inebriated person, and the same tests can be applied, such as having the person touch his or her nose or walk a straight line.

Chapter 13

1. The lens of the eye is biconvex and causes light rays to converge. If the lens is removed, the replacement lens should also cause light rays to converge. A biconvex lens or a lens with a single convex surface would work. Bifocals or trifocals can also be recommended because of the loss of accommodation.

2. The tapetum lucidum can increase the sensitivity of the eye to light, which can be an advantage when light levels are low. Photoreceptors can be stimulate twice: once as light passes through the photoreceptor layer of the retina on its way to the tapetum lucidum (part of the pigmented retina), again when the light is reflected back. The disadvantage of a tapetum lucidum is a decrease in visual acuity because the same light image can stimulate different photoreceptors, which can cause blurring of vision.

3. Carrots contain vitamin A (retinoic acid), which can be used to form retinal. Retinal and opsin combine to form rhodopsin, which is found in rods. Rhodopsin is necessary for rods to respond to low levels of light. Lack of vitamin A can result in lack of rhodopsin and night blindness.

4. When Jean looks a few inches to the side, the image of the needle and thread is projected to the periphery of the retina, where there is the highest concentration of rods. The rods function better than cones in low-light intensities. If Jean looks directly at the needle and thread, their image falls on the macula, which has few rods and mostly cones, which do not function well in dim light. By looking to the side, however, she is using a part of the retina where the photoreceptor cells are not as densely packed as in the macula, and the image is fuzzy rather than sharp.

5. This phenomenon is called a negative afterimage. While a man is staring at the clock, the darkest portion of the image (the black clock) causes dark adaptation in part of the retina. That is, part of the retina becomes more sensitive to light. At the same time, the lightest part of the image (the white wall) causes light adaptation in the rest of the retina, and that part of the retina becomes less sensitive to light. When the man looks at a black wall, the dark-adapted portion of the retina, which is more sensitive to light, produces more action potentials than does the light-adapted part of the retina. Consequently, he perceives a light clock against a darker background.

6. A lesion of the left optic tract results in visual loss in right temporal and right nasal visual fields.

7. Normally, as external pressure changes, the auditory tubes open to allow an equalization of pressure between the middle ear and the external environment. As a diver descends, external water pressure on the tympanic membrane increases. This is countered by increased air pressure from the SCUBA tank. The same effect occurs in reverse when a diver ascends. As long as the acoustic tube is open to allow equalization of pressure, all is well. If the auditory tube does not open, however, the pressure difference can rupture the tympanic membrane, resulting in impairment of hearing. Having a cold can increase the likelihood of damage to the tympanic membrane. A cold can cause inflammation of the auditory tube. Swelling and blockage of the tube by mucus can prevent equalization of pressure between the middle ear and the external environment.

8. The stapedius muscle, attached to the stapes, is innervated by the facial nerve (VII). Loss of facial nerve function eliminates part of the sound attenuation reflex, although not all of it because the tensor tympani muscle, innervated by the trigeminal nerve, is still functional. A reduction in the sound attenuation reflex results in loud sounds being excessively loud in the affected ear. A reduced reflex can also leave the ear more susceptible to damage by prolonged loud sounds.

9. Normally, airborne sounds cause the tympanic membrane to vibrate, resulting in the movement of the middle ear ossicles and the production of sound waves in the perilymph of the scala vestibuli. Vibration of the skull bones can also cause vibration of the perilymph in the scala vestibuli.

Chapter 14

1. The sympathetic division of the ANS is responsible for dilation of the pupil. Preganglionic fibers from the upper thoracic region of the spinal cord pass through spinal nerves (T1 and T2), into the white rami communicantes, and into the sympathetic chain ganglia. The preganglionic fibers ascend the sympathetic chain and synapse with postganglionic neurons in the superior cervical sympathetic chain ganglia. The axons of the postganglionic neurons leave the sympathetic chain ganglia as small nerves that project to the iris of the eye.

2. Reduced salivary and lacrimal gland secretions can indicate damage to the facial nerves, which innervate the submandibular, sublingual, and lacrimal glands. The glossopharyngeal nerves innervate the parotid glands.

3. Cutting the preganglionic fibers in the white rami of T2–T3 is the best way to eliminate innervation of the blood vessels in the skin. Cutting the gray rami at levels T2–T3 is inappropriate because the postganglionic fibers that innervate the hand blood vessels exit from the first thoracic and inferior cervical sympathetic chain ganglia. Cutting the spinal nerves is inappropriate because it eliminates all sensory and motor functions to the area supplied.

4. a. pelvic splanchnic nerves
   b. gray rami
   c. vagus nerves
   d. oculomotor nerve
   e. pelvic splanchnic nerves

5. The parasympathetic division innervates the heart through the vagus nerves. The postganglionic nerve fibers of the vagus nerves release acetylcholine, which reduces heart rate. Methacholine can bind to the same receptors as acetylcholine and reduce heart rate. Side effects result from stimulating other parasympathetic effector organs. For example, stimulating the salivary glands results in increased salivation. Dilation of the pupils and sweating are effects expected from sympathetic stimulation. The muscles of respiration are not regulated by the ANS, but they do respond to acetylcholine through somatic neurons. Methacholine would be expected to make contractions of respiratory muscles more likely.

6. Inactivation of acetylcholinesterase results in a buildup of acetylcholine in synapses and overstimulation of muscarinic receptors. One would expect mostly parasympathetic effects because the effects of acetylcholine are enhanced: blurring of vision as a result of contraction of ciliary muscles, excess tear formation because of overstimulation of the lacrimal glands, and frequent or involuntary urination because of overstimulation of the urinary bladder. Pallor resulting from vasoconstriction in the skin is a sympathetic effect that would not be expected because skin blood vessels respond to norepinephrine. Muscle twitching or cramps of skeletal muscles might occur because they normally respond to acetylcholine. Atropine, a muscarinic blocking agent, can be used to treat exposure to malathion.

7. Epinephrine causes vasoconstriction and confines the drug to the site of administration. This increases the drug's duration of action locally and decreases its systemic effects. Vasoconstriction also reduces bleeding if a dry field (an area clear of blood on its surface) is required.

8. Because normal action potentials are produced, the drug does not act at the synapse between the preganglionic and postganglionic neurons. Because injected norepinephrine works, sympathetic receptors in the heart are functioning and are not affected by the drug. Therefore, the drug must somehow affect the postganglionic neurons. Possibly it inhibits neurotransmitter production or release from the postganglionic neurons.

9. a. Responses in a person who is extremely angry are primarily controlled by the sympathetic division of the ANS. These responses include increased heart rate and blood pressure, decreased blood flow to the internal organs, increased blood flow to skeletal muscles, decreased contractions of the intestinal smooth muscle, flushed skin in the face and neck region, and dilation of the pupils of the eyes.

   b. For a person who has just finished eating and is now relaxing, the parasympathetic reflexes are more important than sympathetic reflexes. The blood pressure and heart rate are at normal resting levels, the blood flow to the internal organs is greater, contractions of smooth muscle in the intestines are greater, and secretions that achieve digestion are more active. If the urinary bladder or the colon becomes distended, autonomic reflexes that result in urination or defecation can result. Blood flow to the skeletal muscles is reduced.

Chapter 15

1. When the hormone binds to its receptor, the receptor changes shape and a G protein binds to the receptor and GDP detaches from the α subunit of the G protein. Normally, GTP would then attach to the α subunit. If this does not happen because of the genetic disease, then the α subunit is not activated and there is no response to the hormone.

2. Phosphodiesterase causes the conversion of cAMP to AMP, thus reducing the concentration of cAMP. A drug that inhibits phosphodiesterase, therefore, increases the amount of cAMP in cells where cAMP is produced. Therefore, an inhibitor of phosphodiesterase increases the response of a tissue to a hormone that has cAMP as an intracellular mediator.

3. If liver disease results in a decrease of plasma proteins to which thyroid hormones bind, higher than normal concentrations of free (unbound) thyroid hormones occur in the circulatory system. Because of the higher than normal concentration of thyroid hormones that are unbound, the responses to thyroid hormones increase. In addition, the half-life of the thyroid hormones is shortened. Thus, as thyroid hormone secretion increases, the concentration of thyroid hormone also increases. As the thyroid hormone secretion decreases, the concentration of thyroid hormone also decreases. Thyroid hormones fluctuate in concentration in the circulatory system more than normal.

4. Usually, membrane-bound receptor mechanisms respond quickly, and the hormone's effect is brief. Intracellular receptor mechanisms usually take a long time (several hours) to respond, and their effects last much longer. If the hormone is large and water-soluble, it is probably functioning through a membrane-bound receptor mechanism; if the hormone is lipid-soluble, it is probably an intracellular receptor mechanism. If you have the ability to monitor the concentration of a suspected intracellular mediator activated by the membrane-bound receptor and it increases in response to the hormone, or if you can inhibit the synthesis of an intracellular mediator and it prevents the target cells' response to the hormone, it is a membrane-bound receptor mechanism. If you can inhibit the synthesis

of mRNA and this inhibits the action of the hormone, or if you can measure an increase in mRNA synthesis in response to the hormone, then the mechanism is an intracellular receptor mechanism.

5. The hypothalamohypophysial portal system allows releasing and inhibiting hormones, which are secreted by neurons in the hypothalamus, to be carried directly from the hypothalamus to the anterior pituitary gland. Consequently, the releasing and inhibiting hormones are not diluted or destroyed by the enzymes, which are abundant in the kidneys, liver, lungs, and general circulation, before they reach the anterior pituitary. Also, the time it takes for releasing and inhibiting hormones to reach the anterior pituitary is less than if they were secreted into the general circulation.

6. The symptoms are consistent with acromegaly, which is a consequence of elevated GH secretion after the epiphyses have closed. Increased GH causes enlarged finger bones, the growth of bony ridges over the eyes, and increased growth of the jaw. The GH also causes bone deposition on the inner surface of skull bones, which increases the pressure inside the skull. The visual disturbances result from pressure on the optic chiasm. The pituitary gland rests in the sella turcica of the sphenoid bone next to the optic chiasm. As the pituitary gland enlarges because of the tumor, it presses on the optic chiasm.

7. If hyperthyroidism results from a pituitary abnormality, laboratory tests should show elevated TSH levels in the circulatory system in addition to elevated thyroid hormone levels. If hyperthyroidism results from the production of a nonpituitary thyroid-stimulating substance, laboratory tests should also show elevated thyroid hormone levels, but TSH levels will be low because of the negative-feedback effects of thyroid hormones on the hypothalamus and pituitary gland.

8. The second student is correct. Low levels of vitamin D reduce Ca^+ uptake in the gastrointestinal tract, which results in a decreased blood Ca^+ levels. As blood Ca^+ levels decrease, the rate of PTH secretion increases. Parathyroid hormone increases bone breakdown, which maintains blood calcium levels, even if vitamin D deficiency exists for a prolonged time. Osteomalacia results because of the increased bone breakdown necessary to maintain normal blood Ca^+ levels.

9. A glucose tolerance test can distinguish between these conditions. The person would consume glucose after a period of fasting. Over the next few hours, the blood glucose levels in a healthy person would increase and then return to fasting levels. The blood glucose levels would always remain within the normal range, however. In a person with diabetes, the blood glucose levels would increase to above normal levels and would remain elevated for several hours. In a person who secreted large amounts of insulin, blood glucose levels would increase, and then they would decrease to below normal levels within a relatively short time.

10. Adrenal diabetes results from the elevated and uncontrolled secretion of glucocorticoid hormones, such as cortisol, from the adrenal gland. Elevated levels of glucocorticoids cause elevated blood glucose levels and symptoms similar to those of diabetes mellitus. Pituitary diabetes results from elevated secretion of GH from the anterior pituitary. Elevated GH levels cause an increase in blood glucose levels

and, therefore, produce symptoms similar to diabetes mellitus. Prolonged elevation of both glucocorticoids and growth hormone secretion can lead to the development of diabetes mellitus if the insulin-secreting cells of the pancreatic islets degenerate because of the prolonged need to secrete insulin in response to the elevated blood glucose levels.

11. Because the person is a diabetic and probably is taking insulin, the condition is more likely to be insulin shock than a diabetic coma. To confirm the condition, however, a blood sample should be taken. If the condition is due to a diabetic coma, the blood glucose levels will be elevated. If the condition is due to insulin shock, the blood glucose levels will be below normal. In the case of insulin shock, glucose can be administered intravenously. In the case of diabetic coma, insulin should be administered. An isotonic solution containing insulin can be administered to reduce the osmolality of the extracellular fluid.

12. Low levels of insulin or insulin resistance, as in diabetes mellitus, can result in the production of a large volume of urine containing glucose. Low levels of ADH decrease water reabsorption by the kidneys, resulting in the production of a large volume of urine with few ions and no glucose. Polyuria and polydipsia are consistent with low levels of insulin and ADH. The administration of ADH is recommended.

Chapter 16

1. Platelets become activated at sites of tissue damage, where it is advantageous to form a clot to stop bleeding.

2. Because of the rapid destruction of the red blood cells, we would expect erythropoiesis to increase in an attempt to replace the lost red blood cells. The reticulocyte count would therefore be above normal. Jaundice is a symptom of hereditary hemolytic anemia because the destroyed red blood cells release hemoglobin, which is converted into bilirubin. Removal of the spleen cures the disease because the spleen is the major site of red blood cell destruction.

3. Blood doping increases the number of red blood cells in the blood, thereby increasing its oxygen-carrying capacity. The increased number of red blood cells also makes it more difficult for the blood to flow through the blood vessels, increasing the heart's workload.

4. Symptoms resulting from decreased red blood cells are associated with a decreased ability of the blood to carry oxygen: shortness of breath, weakness, fatigue, and pallor. Symptoms resulting from decreased platelets are associated with a decreased ability to form platelet plugs and clots: small areas of hemorrhage in the skin (petechiae), bruises, and decreased ability to stop bleeding. Symptoms resulting from decreased white blood cells include an increased susceptibility to infections.

5. Hypoventilation results in decreased blood oxygen levels, which stimulate erythropoiesis. Therefore, the number of red blood cells increases and produces secondary polycythemia.

6. Removal of the stomach removes intrinsic factor, which is necessary for vitamin B_{12} absorption. Therefore, the patient develops pernicious anemia. Lack of stomach acid can decrease iron absorption in the small intestine and results in iron-deficiency anemia.

7. Vitamin B_{12} and folic acid are necessary for blood cell division. Lack of these vitamins results in pernicious anemia. Iron is necessary for the production of hemoglobin. Lack of iron results in iron-deficiency anemia. Vitamin K is necessary for the production of many blood clotting factors. Lack of vitamin K can greatly increase blood clotting time, resulting in excessive bleeding.

8. The anemia results from too little hemoglobin. Because there is less hemoglobin, less hemoglobin is broken down into bilirubin. Consequently, less bilirubin is excreted as part of the bile into the small intestine. With decreased bilirubin in the intestine, bacteria produce fewer of the pigments that normally color the feces.

9. Reddie Popper has hemolytic anemia. The RBC is lower than normal because the red blood cells are being destroyed faster than they are being replaced. With fewer red blood cells, hemoglobin and hematocrit levels are lower than normal. Bilirubin levels are above normal because of the breakdown of the hemoglobin released from the ruptured red blood cells.

Chapter 17

1. During systole, the cardiac muscle in the right and left ventricles contracts, which compresses the coronary arteries. During diastole, the cardiac muscle of the ventricles relaxes and blood flow through the coronary arteries increases. The diastolic pressure is sufficient to cause blood to flow through coronary arteries during diastole.

2. A drug that prolongs the plateau phase of the action potential slows heart rate because it prolongs the time each action potential exists and increases the absolute refractory period. A drug that shortens the plateau phase of the action potential increases heart rate because it shortens the length of time each action potential exists and decreases the absolute refractory period.

3. An ECG measures the electrical activity of the heart and does not indicate a slight heart murmur. Heart murmurs are detected by listening to the heart sounds. The boy may have a heart murmur, but the mother does not understand the basis for making such a diagnosis.

4. Most of the ventricular contraction occurs between the first and second heart sounds of the same beat. Between the first and second heart sounds, blood therefore is ejected from the ventricles into the pulmonary trunk and the aorta. Between the second heart sound of one beat and the first heart sound of the next beat, the ventricles are relaxing and the semilunar valves are closed. No blood passes from the ventricles into the aorta or pulmonary trunk during that period.

5. Atrial contractions complete ventricular filling, but atrial contractions are not primarily responsible for ventricular filling. Therefore, even if the atria are fibrillating, blood can still flow into the ventricles and ventricular contractions can occur. As long as the ventricles contract rhythmically, the heart can pump an adequate amount of blood, even though the atria are not effective pumps. If the ventricles fail to contract forcefully and rhythmically, however, they cannot function as pumps. Thus, the stroke volume will become too low to maintain adequate blood flow to tissues.

Therefore, atrial transplants are not essential but ventricle transplants are.

6. Consuming a large amount of fluid increases the total volume of the blood, at least until the mechanisms that regulate blood volume decrease the blood volume to normal values. If the blood volume is increased, it causes an increase in venous return to the heart. Because of Starling's law of the heart, the increased venous return results in an increased stroke volume. Mechanisms that regulate blood pressure, such as the baroreceptor reflex, would prevent a large increase in blood pressure. Therefore, there may be an increased stroke volume but the blood pressure would not increase dramatically.

7. Cardiac output is influenced by the heart rate and stroke volume (CO = SV × HR). An athlete's cardiac output can be equal to a nonathlete's cardiac output while they are both at rest, even though the athlete's heart rate is lower than the nonathlete's heart rate because the stroke volume of the athlete's heart is greater than the stroke volume of the nonathlete's heart. Athletic training causes a gradual hypertrophy (enlargement) of the heart. Therefore, athletes can maintain a cardiac output that is equal to a nonathlete's because they have an increased stroke volume, but a decreased heart rate. During exercise the athlete's heart rate and stroke volume increase. The stroke volume of the athlete's heart is much greater than the stroke volume of the nonathlete's. Therefore, the cardiac output of the athlete's heart is greater than the cardiac output of the nonathlete's during exercise, even though they have same maximum heart rate.

8. Starling's law of the heart is an intrinsic regulatory mechanism, whereas parasympathetic innervation of the heart is a component of the extrinsic regulation of the heart. Cutting the vagus nerve does not significantly affect the ability of Starling's law of the heart to operate.

9. Cutting sensory nerve fibers from baroreceptors would reduce the frequency of action potentials delivered to the medulla oblongata from the baroreceptors. This results because the normal blood pressure stimulates the baroreceptors. An increase in blood pressure increases the action potential frequency and a decrease in blood pressure decreases the action potential frequency. Because cutting the sensory nerve fibers decreases the action potential frequency, this acts as a signal to the medulla oblongata that a decrease in blood pressure has occurred, even though the blood pressure has not decreased. The medulla oblongata responds by increasing sympathetic action potentials and reducing parasympathetic action potentials delivered to the heart. Consequently, the heart rate increases.

10. The two heartbeats occurring close together can be heard through the stethoscope, because the heart valves open and close normally during each of the heartbeats even if they are close together. The second heartbeat, however, produces a greatly reduced stroke volume because there is not enough time for the ventricles to fill with blood between the first and second contractions of the heart. Thus, the preload is reduced. Because the preload is reduced, the second heartbeat has a greatly reduced stroke volume. The reduced stroke volume fails to produce a normal pulse. The

pulse deficit, therefore, results from the reduced stroke volume of the second of the two beats that are very close together.

11. Venous return declines markedly in hemorrhagic shock because of the loss of blood volume. With decreased venous return, stroke volume decreases (Starling's law of the heart). The decreased stroke volume results in a decreased cardiac output, which produces a decreased blood pressure. In response to the decreased blood pressure, the baroreceptor reflex causes an increase in heart rate in an attempt to restore normal blood pressure. However, with inadequate venous return the increased heart rate is not able to restore normal blood pressure.

Chapter 18

1. a. aorta, left coronary artery, circumflex artery, posterior interventricular artery or aorta, right coronary artery, posterior interventricular artery
   b. aorta, brachiocephalic artery, right common carotid artery, right internal carotid artery or aorta, left common carotid artery, left internal carotid artery
   c. aorta, brachiocephalic artery, right subclavian artery, right vertebral artery, basilar artery or aorta, left subclavian artery, left vertebral artery, basilar artery
   d. aorta, left or right common carotid artery, left or right external carotid artery
   e. aorta, left subclavian artery, axillary artery, brachial artery, radial or ulnar artery, deep or superficial palmar arch, digital artery (on the right: the brachiocephalic artery would be included)
   f. aorta, common iliac artery, external iliac artery, femoral artery, popliteal artery, anterior tibial artery
   g. aorta, celiac artery, common hepatic artery
   h. aorta, superior mesenteric artery, intestinal branches
   i. aorta, left or right internal iliac artery

2. a. great cardiac vein, coronary sinus or anterior cardiac vein
   b. transverse sinus, sigmoid sinus, internal jugular vein, brachiocephalic vein, superior vena cava
   c. retromandibular vein, external jugular vein, subclavian vein, brachiocephalic vein, superior vena cava
   d. deep: vein of hand, radial or ulnar vein, brachial vein, axillary vein, subclavian vein, brachiocephalic vein, superior vena cava superficial: vein of hand, radial or ulnar vein, cephalic or basilic vein, axillary vein, subclavian vein, brachiocephalic vein, superior vena cava
   e. deep: vein of foot, dorsalis veins of foot, anterior tibial vein, popliteal vein, femoral vein, external iliac vein, common iliac vein, inferior vena cava superficial: vein of foot, great saphenous vein, external iliac vein, common iliac vein, inferior vena cava; or vein of foot, small saphenous vein, popliteal vein, femoral vein, external iliac vein, common iliac vein, inferior vena cava
   f. gastric vein or gastroepiploic vein, hepatic portal vein, hepatic sinusoids, hepatic vein, inferior vena cava
   g. renal vein, inferior vena cava
   h. hemiazygos vein or accessory hemiazygos vein, azygos vein, superior vena cava

3. A superficial vessel is easiest, such as the right cephalic or basilic vein. The catheter is passed through the cephalic (or brachial) vein and the superior vena cava to the right atrium. Because the pulmonary veins are not readily accessible, dye is not normally placed directly into them. Instead, the dye is placed in the right atrium using the procedure just described. The dye passes from the right atrium into the right ventricle, the pulmonary arteries, the lungs, the pulmonary veins, and the left atrium. If the catheter has to be placed in the left atrium, it can be inserted through an artery, such as the femoral artery, and passed via the aorta to the left ventricle and then into the left atrium.

4. The viscosity of the blood is affected primarily by the hematocrit. As hematocrit increases, the viscosity of the blood increases logarithmically so that even a small increase in hematocrit results in a large increase in viscosity. Greater force is therefore not needed to cause blood to flow through the blood vessels. With the increased blood volume, blood flow through vessels is adequate without an increase in viscosity.

5. The vein becomes larger when the hand is lowered and smaller when the hand is raised. Lowering the hand increases the hydrostatic pressure in the vein. The vein expands and fills with blood because veins are compliant and easily distended. Raising the hand decreases the hydrostatic pressure and the vein holds less blood and becomes smaller.

6. The nursing student's diagnosis was incorrect. Blood pressure measurements are normally made in either the right or the left arm, both of which are close to the level of the heart. Blood pressure taken in the leg is influenced by pressure created by the pumping action of the heart, but the effect of gravity on the blood, as it flows into the leg, also influences the blood pressure in a substantial way. In this case, gravity increases blood pressure from about 120 mm Hg for the systolic pressure to 210 mm Hg.

7. Decreased liver function includes a decrease in the synthesis of plasma proteins, such as blood clotting factors (see chapter 16). Consequently, the concentration of plasma proteins decreases. Less water moves by osmosis into the capillaries at the venous ends and the result is edema.

8. Veins and lymphatic vessels have one-way valves in them. Massage creates a cycle of increasing and decreasing pressure to the veins, which rhythmically compresses them. The compression of the veins forces fluid to move out of the limb through both veins and lymphatic vessels. The movement of fluid through the veins lowers the pressure within the venous end of the capillary. Thus, the forces that move fluid into the capillaries at their venous ends are greater and they move more interstitial fluid into the capillaries. Compression of the lymphatic capillaries also causes more lymphatic fluid to move into the lymphatic vessels. Because there is less fluid in the limb, the edema decreases.

9. The hot Jacuzzi increases Skinny's skin and body temperature. As a result, the blood vessels of the skin dilate. Because the blood vessels dilate, peripheral resistance decreases, causing the blood pressure to decrease. The baroreceptors of the carotid sinus and aortic arch detect the decrease in blood pressure and send action potentials to the cardioregulatory center in the medulla oblongata. As a result,

parasympathetic stimulation of the heart decreases and sympathetic stimulation to the heart increases. Heart rate and force of contraction increase, which elevates the blood pressure back to within its normal range of values.

10. Heart rate increases to approximately 100 beats/minute because of withdrawal of parasympathetic stimulation. Heart rate could increase another 20 beats/minute because of the adrenal medullary mechanism. Lack of sympathetic stimulation of the heart results in less force of contraction, so end-diastolic volume should remain unchanged. Stroke volume increases, however, because end-systolic volume increases as a result of increased venous return. Venous return increases because blood flow through exercising muscles increases as a result of local control mechanisms. The increased venous return increases stroke volume (Starling's law). The increased heart rate and stroke volume tend to increase blood pressure. The vasodilation effects in exercising skeletal muscle, however, lower peripheral resistance. Without sympathetic stimulation of blood vessels, the normal vasoconstriction seen in exercise does not occur. Consequently, peripheral resistance drops. The decrease in peripheral resistance is so large that it more than cancels the effects of increased heart rate and stroke volume. The result is a drop in blood pressure.

Chapter 19

1. Normally, T cells are processed in the thymus and then migrate to other lymphatic tissues. Without the thymus, this processing is prevented. Because there are normally five T cells for every one B cell, the number of lymphocytes is greatly reduced. The loss of T cells results in an increased susceptibility to infection and an inability to reject grafts because of the loss of cell-mediated immunity. In addition, since helper T cells are involved with the activation of B cells, antibody-mediated immunity is also depressed.

2. That there is no immediate effect indicates there is a reservoir of T cells in the lymphatic tissue. As the reservoir is depleted through time, the number of lymphocytes decreases and cell-mediated immunity is depressed, the animal is more susceptible to infections, and the ability to reject grafts decreases. The ability to produce antibodies decreases because of the loss of helper T cells that are normally involved with the activation of B cells.

3. Injection B results in the greatest amount of antibody production. At first, the antigen causes a primary response. A few weeks later, the slowly released antigen causes a secondary response, resulting in a greatly increased production of antibodies. Injection A does not cause a secondary response because all of the antigen is eliminated by the primary response.

4. The infant's antibody-mediated immunity is not functioning properly, whereas his cell-mediated immunity is working properly. This explains the susceptibility to extracellular bacterial infections and the resistance to intracellular viral infections. It took so long to become apparent because IgG from the mother crossed the placenta and provided the infant with protection. The infant began to get sick after these antibodies degraded.

5. Bone marrow is the source of the lymphocytes responsible for adaptive immunity. If successful, the transplanted bone marrow starts producing lymphocytes and the baby has a functioning immune response. In this case, there is a graft versus host rejection in which the lymphocytes in the transplanted red marrow mount an immune attack against the baby's tissues, resulting in death.

6. At the first location, an antibody-mediated response results in an immediate hypersensitivity reaction, which produces inflammation. Most likely, the response resulted from IgE antibodies. At the second location, a cell-mediated response results in a delayed hypersensitivity reaction, which produces inflammation. This probably involves the release of cytokines and the lysis of cells. At the other locations, there is neither an antibody-mediated nor a cell-mediated response.

7. The ointment is a good idea for the poison ivy, which causes a delayed hypersensitivity reaction—for example, too much inflammation. For the scrape, it is a bad idea because a normal amount of inflammation is beneficial and helps fight infection in the scrape.

8. Because both antibodies and cytokines produce inflammation, the fact that the metal in the jewelry results in inflammation is not enough information to answer the question. However, the fact that it took most of the day (many hours) to develop the reaction indicates a delayed hypersensitivity reaction and therefore cytokines.

9. If the patient has already been vaccinated, the booster shot stimulates a memory (secondary) response and rapid production of antibodies against the toxin. If the patient has never been vaccinated, vaccinating now is not effective because there is not enough time for the patient to develop his or her own primary response. Therefore, antiserum is given to provide immediate, but temporary, protection. Sometimes both are given: The antiserum provides short-term protection and the tetanus vaccine stimulates the patient's immune system to provide long-term protection. If the shots are given at the same location in the body, the antiserum (antibodies against the tetanus toxin) can cancel the effects of the tetanus vaccine (tetanus toxin altered to be nonharmful).

Chapter 20

1. Minute respiratory volume is equal to the respiration rate times the tidal volume. With a respiration rate of 12 breaths per minute and a tidal volume of 500 mL per breath, normal minute ventilation is 6000 mL/min (12×500). Rapid (24 breaths per minute), shallow (250 mL per breath) breathing results in the same minute ventilation—that is, 6000 mL/min (24×250). Alveolar ventilation rate (V_A) is the respiratory rate (frequency; f) times the difference between the tidal volume (V_T) and dead space (V_D).

$$V_A = f(V_T - V_D)$$

Normal resting $V_A = 12 \times (500 - 150) = 4200$ mL/min

In this case of rapid, shallow breathing,

$$V_A = 24 \times (250 - 150) = 2400 \text{ mL/min}$$

Thus, even though the minute ventilation is the same in both cases, the alveolar ventilation rate is less during rapid, shallow breathing because there is a less effective exchange of gases between the atmosphere and the dead space. Because there is less exchange of gases, the partial pressures of alveolar gases become closer to the partial pressure of blood gases. Consequently, the alveolar partial pressure of O_2 decreases and the alveolar partial pressure of CO_2 increases. This decreases the concentration gradients for gases, resulting in less gas exchange between alveolar air and blood.

2. We expect vital capacity to be greatest when standing because the abdominal organs move inferiorly, thereby allowing greater depression of the diaphragm and a greater inspiratory reserve volume.

3. The hose increases dead space and therefore decreases alveolar ventilation. Ima Diver has to compensate by increasing respiratory rate or tidal volume. If the hose is too long, she will not be able to compensate. Furthermore, with a long hose, air is simply moved back and forth in the hose with little exchange of air between the atmosphere and the lungs taking place. Another consideration is the effect of water pressure on the thorax, which decreases compliance and increases the work of ventilation. In fact, a few feet underwater there is enough pressure on the thorax to prevent the intake of air, even through a short hose connected to the atmosphere.

4. The increase in atmospheric pressure increases the partial pressure of oxygen. According to Henry's law, as the partial pressure of oxygen increases, the amount of oxygen dissolved in the body fluids increases. The increase in dissolved oxygen is detrimental to the gangrene bacteria. The HBO treatment does not increase hemoglobin's ability to pick up oxygen in the lungs because hemoglobin is already saturated with oxygen.

5. Compression causes a decrease in thoracic volume and therefore lung volume. Consequently, pressure in the lungs increases over atmospheric pressure and air moves out of the lungs. Raising the arms expands the thorax and lungs. This results in a lower than atmospheric pressure in the lungs, and air moves into the lungs.

6. The victim's lungs expand because of the pressure generated by the rescuer's muscles of expiration. This fills the lungs with air that has a greater pressure than atmospheric pressure. Air flows out of the victim's lungs as a result of this pressure difference and because of the recoil of the thorax and lungs. Although the partial pressure of oxygen of the rescuer's expired air is less than atmospheric, enough oxygen can be provided to sustain the victim. The lower partial pressure of oxygen can also activate the chemoreceptor reflex and stimulate the victim to breathe. In addition, the rescuer's partial pressure of carbon dioxide is higher than atmospheric and this can activate the chemosensitive area in the medulla.

7. The left side of the diaphragm moves superiorly. During inspiration, thoracic volume increases as the right side of the diaphragm moves inferiorly and the intercostal muscles move the ribs outward. Increased thoracic volume causes a decrease in pressure in the thoracic cavity. As a result, the pressure on the superior surface of the diaphragm is less than on the inferior surface. The paralyzed left side of the diaphragm moves superiorly because of this pressure difference.

8. At the end of expiration, the thoracic wall is not moving. Therefore, the forces causing the thoracic wall to move inward and outward must be equal. At the end of expiration, the lungs are adhering to the thoracic wall through the pleurae, and recoil of the lungs is pulling the thoracic wall inward. As a result of the pneumothorax, air enters the pleural cavity and the visceral and parietal pleurae separate from each other. The recoil of the lungs causes them to collapse. Without the inward force produced by lung recoil, the thoracic wall expands outward.

9. All else being equal (i.e., the thickness of the respiratory membrane, the diffusion coefficient of the gas, and the surface area of the respiratory membrane), diffusion is a function of the partial pressure difference of the gas across the respiratory membrane. The greater the difference in partial pressure, the greater the rate of diffusion. The greatest rate of oxygen diffusion should therefore occur at the end of inspiration when the partial pressure of oxygen in the alveoli is at its highest. The greatest rate of carbon dioxide diffusion should occur at the end of inspiration when the partial pressure of carbon dioxide in the alveoli is at its lowest.

10. Cutting the vagus nerves eliminates the Hering-Breuer reflex and results in a greater than normal inspiration. This increases tidal volume. Cutting the phrenic nerves eliminates contraction of the diaphragm. Tidal volume decreases drastically, and death probably results. Cutting the intercostal nerves eliminates raising of the ribs and sternum and decreases tidal volume, unless the diaphragm compensates.

11. While hyperventilating and making ready to leave your instructor behind, you might make the following arguments:
    • Hyperventilation increases the oxygen content of the air in the lungs; therefore, you would have more oxygen to use when holding your breath.
    • It is hemoglobin that is saturated. Hyperventilation increases the amount of oxygen dissolved in the blood plasma.
    • Hyperventilation decreases the amount of carbon dioxide in the blood. This makes it possible to hold one's breath for a longer time because of a decreased urge to take a breath.

Chapter 21

1. With the loss of the swallowing reflex, the vocal folds and vocal cords no longer occlude the glottis. Consequently, vomit can enter the larynx and block the respiratory tract.

2. Without adequate amounts of hydrochloric acid, the pH in the stomach is not low enough for the activation of pepsin. This loss of pepsin function results in inadequate protein digestion. If the food is well chewed, however, proteolytic enzymes in the small intestine (e.g., trypsin, chymotrypsin) can still digest the protein. If the stomach secretion of intrinsic factor decreases, the absorption of vitamin B_{12} is hindered. Inadequate amounts of vitamin B_{12} can result in decreased red blood cell production (pernicious anemia).

3. Even though ulcers are apparently ultimately caused by bacteria, overproduction of hydrochloric acid due to stress is a possible contributing factor. Reducing hydrochloric acid production is recommended. In addition to antibiotic therapy, commonly recommended solutions include relaxation, drugs that reduce stomach acid secretion, and antacids to neutralize the hydrochloric acid. Smaller meals are also advised because distention of the stomach stimulates acid production. Proper diet is also important. The patient is also advised to avoid alcohol, caffeine, and large amounts of protein because they stimulate acid production. Ingestion of fatty acids is recommended because they inhibit acid production by causing the release of cholecystokinin. Stress also stimulates the sympathetic nervous system, which inhibits duodenal gland secretion. As a result, the duodenum has less of a mucous coating and is more susceptible to gastric acid and enzymes. Relaxing after a meal helps decrease sympathetic activities and increase parasympathetic activities.

4. Lack of bile due to blockage of the common bile duct can result in jaundice (due to an accumulation of bile pigments in the blood) and clay-colored stools (due to lack of bile pigments in the feces). Blockage of the bile duct causes abdominal pain, nausea, and vomiting. Fat absorption is impaired because of the absence of bile salts in the duodenum and a loose, bulky stool results. Lack of fat absorption reduces the absorption of fat-soluble vitamins, such as vitamin K, resulting in a lack of normal clotting function.

5. The accumulation of materials above the site of impaction and the action of bacteria on the material would result in an increase in osmotic pressure in the area. Water would move by osmosis into the colon above the site of impaction. Bowel impaction is very dangerous and must be treated quickly. The increased volume and distention of the digestive tract above the site of impaction causes compression of the mucosa. This compression can occlude blood vessels in the mucosa and lead to necrosis. Necrosis of the mucosa results in increased permeability of the mucosa, allowing toxic organisms and substances in the digestive tract to enter the circulation, resulting in septic shock.

6. Cholera toxin irreversibly activates a G protein that causes persistent activation of the chloride channel. The activated channel allows excessive movements of chloride from the cells into the digestive tract. Water follows the osmotic gradient generated by the chloride, which leads to diarrhea. Conversely, in cystic fibrosis, mutations in the channel reduce the movement of chloride into the digestive tract. A thick mucus builds up on the surface of the epithelial cells. Mucus can block the common bile duct and pancreatic duct.

7. Oral rehydration therapy relies on the principle of osmosis. Water follows solutes as they are absorbed across the intestinal epithelium. The combination of sodium and glucose is optimal, since the two molecules are cotransported by a symporter that is driven by a Na^+ gradient established by the Na^+-K^+ pump. Hence, the presence of sodium aids glucose absorption. Fructose is absorbed by a facilitated diffusion and is not coupled to a Na^+ gradient.

Chapter 22

1. In figure 25.2, the Daily Value for saturated fat is listed as less than 20 g for a 2000 kcal/day diet. The % Daily Values appearing on food labels are based on a 2000 kcal/day diet. Therefore, the % Daily Value for saturated fat for one serving of this food is 10% (2/20 = .10, or 10%).

2. According to the Daily Value guidelines, total fats should be no more than 30% of total

kilocaloric intake. For someone consuming 3000 kcal/day, this is 900 kcal (3000 kcal × 0.30). There are 9 kcal in a gram of fat. Therefore, the maximum amount (weight) of fats the active teenage boy should consume is 100 g (900/9).

3. The % Daily Value is the amount of the nutrient in one serving divided by its Daily Value. Therefore, the % Daily Value is 10% (10/100 = .10, or 10%).

4. The % Daily Value for one serving of the food is 10% (see answer to question 3). Since there are four servings in the package, if the teenager eats half of the food in the package, he consumes two servings. Thus, he eats 20% (10% × 2) of the recommended maximum total fat.

5. The protein in meat contains all of the essential amino acids and is a complete protein food. Although plants contain proteins, a variety of plants must be consumed to ensure that all the essential amino acids are included in adequate amounts. Also, plants contain less protein per unit weight than meat, so a larger quantity of plants must be consumed to get the same amount of protein.

6. Copper is necessary for the proper functioning of the electron-transport chain. Inadequate copper in the diet results in reduced ATP production—that is, not enough energy.

7. Fasting can be damaging because proteins are used to produce glucose. The glucose enters the blood and provides an energy source for the brain. This breakdown of proteins can damage tissues, such as muscle, and disrupt chemical reactions regulated by enzyme systems. A single day without food, however, is unlikely to cause permanent harm.

8. Weight is lost when kilocalories used per day exceeds kilocalories ingested per day. About 60% of the kilocalories used per day is due to basal metabolic rate. A person with a high basal metabolic rate loses weight faster than a person with a low basal metabolic rate, all else being equal. Another factor to consider is the amount of physical activity, which accounts for about 30% of kilocalories used per day. An active person loses more weight than a sedentary person does.

9. Amino acids, derived from ingested proteins, are necessary to build muscles. As Lotta and her friend discovered, excess proteins do not accelerate this process. Excess proteins can be used as an energy source in oxidative deamination, for the formation of the intermediate molecules of carbohydrate metabolism, or in gluconeogenesis. Excess proteins are also converted into storage molecules through glycogenesis or lipogenesis. Lotta is in positive nitrogen balance because the amount of nitrogen she gains from her diet is greater than the amount she loses by excretion. Some of the nitrogen in the amino acids she ingests is incorporated into the proteins of her muscles as they enlarge.

10. No, this approach does not work because he is not losing stored energy from adipose tissue. In the sauna, he gains heat, primarily by convection from the hot air and by radiation from the hot walls. The evaporating sweat is removing heat gained from the sauna. The loss of water will make him thirsty, and he will regain the lost weight from fluids he drinks and food he eats.

11. As ATP breakdown increases, more ATP is produced to replace that used. Over an extended time, the ATP must be produced through aerobic respiration. Therefore, oxygen consumption and basal metabolic rate increase. The production of ATP requires the metabolism of carbohydrates, lipids, or proteins. As these molecules are used at a faster than normal rate, body weight decreases. Increased appetite and increased food consumption resist the loss in body weight. As ATP is produced and used, heat is released as a by-product. The heat raises body temperature, which is resisted by the dilation of blood vessels in the skin and by sweating.

12. During fever production, the body produces heat by shivering. The body also conserves heat by the constriction of blood vessels in the skin (producing pale skin) and by a reduction in sweat loss (producing dry skin). When the fever breaks—that is, "the crisis is over"—heat is lost from the body to lower body temperature to normal. This is accomplished by the dilation of blood vessels in the skin (producing flushed skin) and increased sweat loss (producing wet skin).

Chapter 23

1. There are several ways to decrease glomerular filtration rate:
   a. Decrease hydrostatic pressure in the glomerulus.
      1. Decrease systemic arterial blood pressure.
         a. Decrease extracellular fluid volume.
         b. Decrease peripheral resistance.
         c. Decrease cardiac output.
      2. Constrict or occlude the afferent arteriole.
      3. Relax the efferent arteriole.
   b. Increase glomerular capsule pressure.
   c. Increase the colloid osmotic pressure of the plasma.
   d. Decrease the permeability of the filtration barrier.
   e. Decrease the total area of the glomeruli available for filtration.

2. Urea is partially responsible for the high osmolality of the interstitial fluid in the medulla of the kidney. Since a high osmolality of the interstitial fluid must exist for the kidney to produce a concentrated urine, a small amount of urea in the kidney results in the production of dilute urine by the kidney.

3. As the loops of Henle become longer, the mechanisms that increase the concentration of the interstitial fluid of the medulla become more efficient, thus raising the concentration of the interstitial fluid. The maximum concentration for urine is determined by the concentration of the interstitial fluid deep in the medulla of the kidneys. The higher the concentration of interstitial fluid in the medulla of the kidney, the greater the concentration of the urine the kidney is able to produce.

4. Assume that the ascending limb of the loop of Henle and the distal convoluted tubules are impermeable to sodium and other ions but actively pump out water. Other characteristics of the kidney are assumed to be unchanged. As the urine moves up the ascending limb, it becomes hyperosmotic, because sodium remains behind as water is pumped out. Assuming that the collecting ducts are impermeable to sodium, on reaching the collecting ducts the presence or absence of ADH determines the final concentration of the urine. If ADH is absent, there is little or no exchange of water as the urine passes down the collecting ducts and a hyperosmotic urine will be produced. On the other hand, if ADH is present, water moves from the interstitial fluid into the collecting ducts, thus diluting the urine and producing a hyposmotic urine.

5. Two shots of vodka would not significantly change blood volume or osmolality. Inhibition of ADH, however, decreases water reabsorption, which increases urine volume. The urine is less concentrated than normal because it is diluted by the water.

6. The large volume of hyposmotic fluid ingested increases blood volume and causes blood osmolality to decrease. The increased blood volume increases blood pressure, which is detected by baroreceptors, and the decreased blood osmolality is detected by osmoreceptors in the hypothalamus. The response to these stimuli is an inhibition of ADH secretion and decreased water reabsorption. Consequently, a large volume of dilute urine is produced and blood volume decreases. The increased blood volume caused by ingesting the beer increases blood pressure, which inhibits the renin–angiotensin–aldosterone mechanism, which in turn inhibits aldosterone secretion. The increased volume also stimulates ANH secretion. Decreased aldosterone and increased ANH promote decreased Na^+ reabsorption. Thus, water remains in the filtrate and is lost in the urine. The loss of water eventually restores blood volume and osmolality.

7. A low-salt diet tends to reduce the osmolality of the blood. Consequently, ADH secretion is inhibited, producing dilute urine and thus eliminating water. This in turn reduces blood volume and blood pressure.

8. When excess glucose is not reabsorbed, it osmotically obligates water to remain in the nephron. This results in a large production of urine, called polyuria, with a consequent loss of water, salts, and glucose. The loss of water can be compensated for by increasing fluid intake. The intense thirst that stimulates increased fluid intake is called polydipsia. The loss of salts can be compensated for by increasing the salt intake. The high glucose levels in the blood would increase the blood osmolality, thus stimulating the secretion of ADH. This increases the permeability of the distal convoluted tubule and collecting duct to water. Normally, this would allow reabsorption of water from the collecting ducts and thus conserve water. If glucose levels in the urine are high enough, however, water loss increases even with high levels of ADH being present.

9. She was expecting urinary volume to be low because ADH promotes water reabsorption. When ADH levels first increase, the reabsorption of water increases and urinary output is reduced. This also causes an increase in blood volume and therefore an increase in blood pressure. The increased blood pressure increases glomerular filtration rate, which increases urinary output toward normal levels. In addition, the increased blood volume inhibits the renin–angiotensin–aldosterone mechanism, inhibits aldosterone secretion, and stimulates atrial natriuretic hormone secretion. These responses also increase urinary output.

10. Ammonium ion (NH_4^+) levels in the urine increase when pH decreases because decreased pH stimulates increased production of NH_4^+ from glutamine and increased secretion of H^+, which combine with ammonia (NH_3) to form NH_4^+. Thus, increased NH_4^+ could occur in metabolic or respiratory acidosis.

The most logical conclusion is that the condition is metabolic acidosis because an elevated respiratory rate increases blood pH. The observed increase in respiration rate compensates for the metabolic acidosis by lowering H⁺ levels.

11. Diarrhea is one of the most common causes of metabolic acidosis, resulting from the loss of bicarbonate ions. Increasing the respiration rate and producing an acidic urine help increase the blood pH.

12. Breathing through the glass tube increases the dead air space and decreases the efficiency of gas exchange. Consequently, blood carbon dioxide levels increase and produce a decrease in blood pH. Compensatory responses include an increased respiration rate and the production of acidic urine.

13. A major effect of alkalosis is hyperexcitability of the nervous system. If the girl is prone to having convulsions, inducing alkalosis might result in a seizure. This can be accomplished by having the girl hyperventilate. The resulting loss of carbon dioxide from the blood causes an increase in blood pH.

14. At high altitudes, we expect stimulation of the chemoreceptor reflex and an increase in respiration rate. This can result in hyperventilation, a decrease in blood carbon dioxide, and respiratory alkalosis. The increased secretion of hydrochloric acid into the stomach can also increase blood pH and contribute to the problem. The kidney produces a more alkaline urine.

Chapter 24

1. Removing the testes would eliminate the major source of testosterone. Blood levels of testosterone would therefore decrease. Because testosterone has a negative-feedback effect on the hypothalamus and pituitary gland, GnRH, FSH, and LH secretion would increase and the blood levels of these hormones would increase. An adult male's primary and secondary sexual characteristics are already developed. Lack of testosterone causes decreased sperm cell production, decreased sex drive, and decreased muscular strength. The development of baldness, however, would be less.

2. Prior to puberty, the levels of GnRH are very low because the hypothalamus is very sensitive to the inhibitory effects of testosterone. Since GnRH levels are low, so are FSH and LH levels. Loss of the testes and testosterone production would result in an increase in GnRH, FSH, and LH levels. Because little testosterone is produced, the boy would not develop sexually and would have no sex drive. Small amounts of androgens would be produced because the adrenal cortex produces some androgens. He would be taller than normal as an adult, with thin bones and weak musculature. His voice would not deepen and the normal masculine distribution of hair would not develop.

3. Ideally, the pill would inhibit spermatogenesis. Using the same approach as in females, the inhibition of FSH and LH secretion should work. Chronic administration of GnRH suppresses FSH and LH levels enough to cause infertility. Lack of LH can also result in reduced testosterone levels and a loss of sex drive, however. Some evidence indicates that administration of testosterone in the proper amounts reduces FSH and LH secretion, thus

leading to a reduced sperm cell production. The testosterone, however, maintains normal sex drive. The technique appears to work for a large percentage of males, resulting in a sperm concentration in the semen that is too low to result in fertilization. The technique is not sufficiently precise, however, to be used as a standard birth-control technique.

4. The removal of the ovaries from a 20-year-old woman eliminates the major site of estrogen and progesterone production, thereby causing an increase in GnRH, FSH, and LH levels due to lack of negative feedback. One expects to see the symptoms of menopause, such as cessation of menstruation and reduction in the size of the uterus, vagina, and breasts. There may also be a temporary reduction in sex drive.

5. In a postmenopausal woman, the ovaries have stopped producing estrogen and progesterone. Without the negative-feedback effect of these hormones, the levels of GnRH, FSH, and LH increase. Removal of the nonfunctioning ovaries in a postmenopausal woman does not change the level of any of these hormones or produce any symptoms not already occurring due to the lack of ovarian function.

6. It is clear that estrogen and progesterone administration resulted in a large decrease in the amount of LH in the plasma the day of ovulation. The differences in plasma LH levels between the groups at other times are very small. The incidence of pregnancies suggests that the reduced plasma LH levels may result in no ovulation.

7. The progesterone inhibits GnRH in the hypothalamus. Consequently, the anterior pituitary is not stimulated to produce LH and FSH. Lack of LH prevents ovulation and lack of FSH prevents development of the follicles. LH also is required for the maturation of follicles prior to ovulation. Without follicle development, there is inadequate estrogen production, which causes the hot flash symptoms.

8. High progesterone levels after menses inhibit GnRH secretion from the hypothalamus and therefore FSH and LH secretion from the anterior pituitary. Without FSH and LH, follicle development and estrogen production is inhibited. Thickening of the endometrium is not expected because estrogen causes proliferation of the endometrium. Also, estrogen increases the synthesis of uterine progesterone receptors, and without estrogen the secretory response of the endometrium to the elevated progesterone is inhibited.

Chapter 25

1. Polyspermy, the fertilization of one oocyte with two sperm cells, can result in triploidy. Polyspermy is usually prevented by depolarization of the oocyte plasma membrane when a sperm cell penetrates the membrane. If depolarization of the oocyte membrane does not occur, the zona pellucida will not be denatured and expand and other sperm cells can attach to the oocyte membrane, leading to polyspermy.

2. Postovulatory age, the approximate length of time the embryo has been developing, is 14 days less than the time since the last menstrual period (LMP). In this case, the postovulatory age is 30 days (44 − 14). By this time, the neural tube has closed, the somites have formed, the digestive

tract is developing, the limb buds have appeared, a tubular beating heart is present, and the lungs are developing. Based on reproductive structures, which are just forming, male and female embryos are indistinguishable at this age.

3. The fever would have occurred on days 21–31 of development, which is during part of the time of neural tube closure (days 18–25). If the fever prevented neural tube closure, the child might be born with anencephalus or spina bifida.

4. The limb buds develop in a proximal-to-distal sequence. If the limb bud is damaged during embryonic development when it is about one-half grown, the proximal structures, the arm and forearm, develop normally but the distal structures, the wrist and hand, do not form normally. Depending on the degree of damage, the wrist and hand might be completely absent or underdeveloped.

5. The mesonephric duct system develops, because of testosterone, to form portions of the male reproductive duct system. Without the production of müllerian-inhibiting hormone, the paramesonephric duct system also develops to form the uterus and uterine tubes. Although ovaries are present, the clitoris may be enlarged because of testosterone to produce somewhat the appearance of male external genitalia. The amount of masculinization depends on the levels of testosterone and how long it was administered. High levels of testosterone over an extended period completely masculinize the external genitalia.

6. Progesterone inhibits uterine smooth muscle contraction. Progesterone also reduces the release of oxytocin, the hormone that stimulates uterine contractions. Also, progesterone reduces the number of oxytocin receptors in the uterus. All of these effects are expected to reduce labor contractions and can prevent preterm delivery.

7. Suckling the breast stimulates the release of oxytocin from the posterior pituitary. Once the oxytocin is in the blood, it travels to both breasts and causes milk letdown.

8. Yes. Because dimpled cheeks are dominant, a person with dimpled cheeks may have a genotype of Dd ("D" being the dominant gene for dimples and "d" being the normal gene). If two Dd, dimpled people have children, it may be expected that approximately 1/4 of their children would not have dimples (dd).

9. No, not without additional information. To be able to roll their tongue into a tube, both mother and son need to have only one dominant gene. The son could inherit this dominant gene from his mother, so it is possible that his father is a nontongue roller. It is also possible that his father is a tongue roller, but there is no proof for either hypothesis.

10. A person who is AB blood type has the genotype I^A I^B but has no i allele. A person of blood type O has a genotype of ii. As the AB parent has no i allele, an AB individual cannot be the parent of a child with blood type O.

11. Hemophilia is an X-linked recessive gene. A female must have the genotype X^h X^h to have hemophilia, whereas, because males have only one X chromosome, a single hemophilia gene will cause them to have the disorder (X^h Y). For half their children to have hemophilia, the mother's genotype must be XX^h (she does not have hemophilia) and the father's genotype must be X^h Y (he has hemophilia). If the woman did not carry the hemophilia gene, none of her daughters or sons could have the disorder.

Appendix G

Answers to Predict Questions

Chapter 1

1. The chemical level is the level at which correction is currently being accomplished. Insulin can be purchased and injected into the circulation to replace the insulin normally produced by the pancreas. Therapy at the cellular level includes drugs that stimulate pancreatic cells to increase insulin production or make cells more responsive to insulin. Transplantation of pancreatic cells, pancreatic tissue, and even the entire pancreas are also possibilities.

2. Donating a pint of blood results in a decrease in blood pressure. Negative-feedback mechanisms, such as an increase in heart rate, return blood pressure toward a normal value. When a negative-feedback mechanism fails to return a value to its normal level, the value can continue to deviate from its normal range. Homeostasis is not maintained in this situation, and the person's health or life can be threatened.

3. a. The normal response to a decrease in blood pressure is for receptors to detect the decrease and for the brain to stimulate the heart to contract faster, which increases blood pressure and maintains homeostasis.
   b. Just before Molly fainted, her heart rate increased. However, it did not increase enough to accommodate her change in position, and her blood pressure dropped below normal. The decreased blood pressure dropped below normal. The decreased blood pressure resulted in the delivery of less blood to the brain, homeostasis of brain tissue was disrupted, and Molly fainted. Note that the regulation of blood pressure involves more than changes in heart rate and is discussed more thoroughly in chapters 20 and 21.
   c. When Molly fell to the floor, she returned to a lying down position. This eliminated the pooling effect of the blood in the veins below the heart because of gravity. Therefore, blood return to the heart increased, blood pressure increased, blood flow to the brain increased, homeostasis of brain tissue was restored, and she regained consciousness.

4. Negative-feedback mechanisms work to control respiratory rates so that body cells have adequate oxygen and are able to eliminate carbon dioxide. The greater the respiratory rate, the greater the exchange of gases between the body and the air. When a person is at rest, there is less of a demand for oxygen, and less carbon dioxide is produced than during exercise. At rest, homeostasis can be maintained with a low respiration rate. During exercise, there is a greater demand for oxygen, and more carbon dioxide must be eliminated. Consequently, to maintain homeostasis during exercise, the respiratory rate increases.

5. The sensation of thirst is involved in a negative-feedback mechanism that maintains body fluids. The sensation of thirst increases with a decrease in body fluids. The thirst mechanism causes a person to drink fluids, which returns body fluid levels to normal, thereby maintaining homeostasis.

6. Your kneecap is both proximal and superior to the heel. It is also anterior to the heel because it is on the anterior side of the lower limb, whereas the heel is on the posterior side.

7. After passing through the left thoracic wall, the first membrane to be encountered is the parietal pleura. Continuing through the pleural cavity, the visceral peritoneum and the left lung are pierced. Leaving the lung, the bullet penetrates the visceral pleura, the pleural cavity, and the parietal pleura (remember that the lung is surrounded by a double membrane sac). Next the parietal pericardium, the pericardial cavity, the visceral pericardium, and the heart are encountered.

Chapter 2

1. The mass (amount of matter) of the astronaut on the surface of the earth and in outer space does not change. In outer space, where the force of gravity from the earth is very small, the astronaut is "weightless," compared with his or her weight on the earth's surface.

2. Fluorine has 9 protons (the atomic number), 10 neutrons (the mass number minus the atomic number), and 9 electrons (equal to the number of protons).

3. Because atoms are electrically neutral, the iron (Fe) atom has the same number of protons and electrons. The loss of three electrons results in an iron ion that has three more protons than electrons and therefore a charge of $+3$. The correct symbol is Fe^{3+}.

4. A decrease in blood CO_2 decreases the amount of H_2CO_3 and therefore the blood H^+ level. Because CO_2 and H_2O are in equilibrium with H^+ and HCO_3^-, with H_2CO_3 as an intermediate, a decrease in CO_2 causes some H^+ and HCO_3^- to join together to form H_2CO_3, which then forms CO_2 and H_2O. Consequently, the H^+ concentration decreases.

5. During exercise, muscle contractions increase, which requires energy. This energy is obtained from the energy in the chemical bonds of ATP. As ATP is broken down, energy is released. Some of the energy is used to drive muscle contractions, and some becomes heat. Because the rate of these reactions increases during exercise, more heat is produced than when at rest, and body temperature increases.

6. A base added to a solution combines with H^+, which decreases the H^+ concentration. By definition, a decrease in H^+ concentration is an increase in pH. Note that buffers take up H^+ when an acid is added to a buffered solution. It is reasonable to expect a buffer to release H^+ when a base is added to a buffered solution. The released H^+ can combine with the base to prevent an increase in pH. For example, H^+ can combine with OH^- to form H_2O.

Chapter 3

1. Urea is continually produced by cells and diffuses from the cells into the blood. If the kidneys stop eliminating urea, it begins to accumulate in the blood. As urea concentration increases in the blood, the concentration gradient for urea from the cells into the blood decreases. Therefore, less urea diffuses from the cells and the concentration of urea in them increases. The urea finally reaches concentrations high enough to be toxic to cells, thereby causing cell damage followed by cell death.

2. Water moves by osmosis from solution B into solution A. Because solution A is hyperosmotic to solution B, solution A has more solutes and less water than does solution B. Water therefore moves from solution B (with more water) to solution A (with less water).

3. Glucose transported by facilitated diffusion across the plasma membrane moves from a higher to a lower concentration. If glucose molecules are quickly converted to some other molecule as they enter the cell, a steep concentration gradient is maintained. The rate of glucose transport into the cell is directly proportional to the magnitude of the concentration gradient.

4. Digitalis should increase the force of heart contractions. By interfering with Na^+ transport, digitalis decreases the concentration gradient for Na^+ because fewer Na^+ are pumped out of cells by active transport. Consequently, fewer Na^+ diffuse into cells, and fewer Ca^{2+} move out of the cells by antiport. The higher intracellular levels of Ca^{2+} promote more forceful contractions.

5. a. Cells highly specialized to synthesize and secrete proteins have large amounts of rough endoplasmic reticulum (ribosomes attached to endoplasmic reticulum) because these organelles are important for protein synthesis. Golgi apparatuses are well developed because they package materials for release in secretory vesicles. Also, numerous secretory vesicles exist in the cytoplasm.
   b. Cells highly specialized to actively transport substances into the cell have a large surface area exposed to the fluid from which substances are actively transported, and numerous mitochondria are present near the membrane across which active transport occurs.
   c. Cells highly specialized to phagocytize foreign substances have numerous lysosomes in their cytoplasm and evidence of phagocytic vesicles.

6. The mRNA sequence is GCAUGC. The complementary strand sequence is GCATGC. The difference between the complementary strand and the mRNA sequence is the presence of thymine (T) in the complementary strand and the presence of uracil (U) in the mRNA sequence.

7. Because adenine pairs with thymine (no uracil exists in DNA) and cytosine pairs with guanine, the sequence of DNA replicated from strand 1 is TACGAT. This sequence is also the sequence of DNA in the original strand 1. A replicate of strand 2 is therefore ATGCTA, which is the same as the original strand 1.

Chapter 4

1. a. If nonkeratinized stratified squamous epithelium lined the digestive tract, it would provide protection against abrasion but would hinder the secretion of digestive enzymes and the absorption of digested food. Nonkeratinized stratified squamous epithelium is not specialized to absorb or secrete because the many layers of cells hinder the movement of materials across the epithelium. Squamous cells, as opposed to cuboidal or columnar cells, do not contain the number of organelles necessary to support the production and transport of large numbers of complex molecules.
   b. Keratinized stratified epithelium forms a tough layer that is a barrier to the movement of water. Replacing the epithelium of skin with moist stratified squamous epithelium increases the loss of water across the skin because water can diffuse through nonkeratinized stratified squamous epithelium, and it is more delicate and provides less protection than keratinized stratified squamous epithelium.
2. The tight junctions prevent substances from passing between the epithelial cells. NaCl must pass across the epithelial layer. NaCl could pass by facilitated diffusion across one membrane of the epithelial cells and by active transport across the other epithelial membrane. If the membranes have a high permeability to water, the water will move in the same direction as NaCl because of osmosis.
3. When a muscle contracts, the pull it exerts is transmitted along the length of its tendons. The tendons need to be very strong in that direction but not as strong in others. The collagen fibers, which are like microscopic ropes, are therefore all arranged in the same direction to maximize their strength. In the skin, collagen fibers are oriented in many directions because the skin can be pulled in many directions. The collagen fibers can be somewhat randomly oriented, or they can be organized into alternating layers. The fibers within a layer run in the same direction, but the fibers of different layers run in different directions.
4. Hyaline cartilage provides a smooth surface so that bones in joints can move easily. When the smooth surface provided by hyaline cartilage is replaced by dense fibrous connective tissue, the smooth surface is replaced by a less smooth surface, and the movement of bones in joints is much more difficult. The increased friction helps increase the inflammation and pain that occur in the joints of people who have rheumatoid arthritis.
5. In severely damaged tissue in which cells are killed and blood vessels are destroyed, the usual symptoms of inflammation cannot occur. Surrounding these areas of severe tissue damage, however, where blood vessels are still intact and cells are still living, the classic signs of inflammation do develop. The signs of inflammation therefore appear around the periphery of severely injured tissues.

6. Suturing large wounds brings the edges of the wounds close together. Healing is therefore more rapid, there is less danger of infections, less scar tissue is formed, and wound contracture is greatly reduced.

Chapter 5

1. Because the permeability barrier is composed mainly of lipids surrounding the epidermal cells, substances that are lipid-soluble easily pass through, whereas water-soluble substances have difficulty.
2. a. The lips are pinker or redder than the palms of the hand. Several explanations for this are possible: There might be more blood vessels in the lips, increased blood flow might occur in the lips, or the blood vessels might be easier to see through the epidermis of the lips. The last possibility explains most of the difference in color between the lips and the palms. The epidermis of the lips is thinner and not as heavily keratinized as that of the palms. In addition, the papillae containing the blood vessels in the lips are "high" and closer to the surface.
   b. A person who does manual labor has a thicker stratum corneum on the palms (and possibly calluses) than a person who does not perform manual labor. The thicker epidermis masks the underlying blood vessels, and the palms do not appear as pink. In addition, carotene accumulating in the lipids of the stratum corneum might impart a yellowish cast to the palms.
   c. The posterior surface of the forearm appears darker because of the tanning effect of ultraviolet light from the sun.
   d. The genitals normally have more melanin and appear darker than the soles of the feet.
3. The story is not true. Hair color results from the transfer of melanin from melanocytes to keratinocytes in the hair matrix as the hair grows. The hair itself is dead. To turn white, the hair must grow out without the addition of melanin, a process that takes weeks.
4. a. Billy's ears and nose turned pale because of decreased blood flow. As ambient temperature decreases, body temperature decreases. Constriction of blood vessels in the skin reduces heat loss and helps maintain body temperature.
   b. Billy's ears and nose turned red because of increased blood flow. As skin temperature decreases and blood vessels constrict, tissues can be damaged by the lack of blood flow and ice crystal formation. Cold-induced vasodilation periodically increases blood flow and prevents or slows the rate of ice crystal formation. This strategy for maintaining tissue homeostasis is beneficial as long as body temperature can be maintained.
   c. At colder temperatures, cold-induced vasodilation stops and tissues are no longer protected by an infusion of warm blood. Damaged tissues are white because blood no longer flows through them.
   d. One function of the skin is to prevent the entry of microorganisms. If frostbite results in the destruction of skin cells, this function can be compromised, resulting in infections.

Chapter 6

1. In the absence of a good blood supply, nutrients, chemicals, and cells involved in

tissue repair enter cartilage tissue very slowly. As a result, the ability of cartilage to undergo repair is poor. Within a joint, the articular cartilage of one bone presses against and moves against the articular cartilage of another bone. If the articular cartilages were covered by perichondrium, or contained blood vessels and nerves, the resulting pressure and friction could damage these structures.
2. In the elderly, the bone matrix contains proportionately less collagen than hydroxyapatite, compared with the bones of younger people. Collagen provides bone with flexible strength, and a reduction in collagen results in brittle bones. In addition, the elderly have less dense bones with less matrix. The combination of reduced matrix that is more brittle results in a greater likelihood of bones breaking.
3. Cancellous bone consists of trabeculae with spaces between them. Blood vessels can pass through these spaces. In compact bone, the blood vessels pass through the perforating and central canals. The trabeculae in cancellous bone are thin enough that nutrients and gases can diffuse from blood vessels around the trabeculae to the osteocytes through the canaliculi.
4. Chondroblasts are surrounded by cartilage matrix and receive oxygen and nutrients by diffusion through the matrix. When the matrix becomes calcified, diffusion is reduced to the point that the cells die. When osteoblasts form bone matrix, they connect to one another by their cell processes. Thus, when the matrix is laid down, canaliculi are formed. Even though the ossified bone matrix is dense and prevents significant diffusion, the osteocytes can receive gases and nutrients through the canaliculi or by movement from one osteocyte to another.
5. Interstitial growth of cartilage results from the division of chondrocytes within the cartilage, followed by the addition of new cartilage matrix between the chondrocytes. The resulting expansion of the cartilage matrix is possible because cartilage matrix is not too rigid. Bones cannot undergo interstitial growth because bone matrix is rigid and cannot expand from within. New bone must therefore be added to the surface by apposition.
6. Growth of articular cartilage results in an increase in the size of epiphyses. This is only one of the functions of articular cartilage; it also forms a smooth, resilient covering over the ends of the epiphyses within joints. Ossified articular cartilage could not perform that function.
7. Her growth for the next few months increases, and she may be taller than a typical 12-year-old female. Because the epiphyseal plates ossify earlier than normal, however, her height at age 18 will be less than otherwise expected.
8. a. Henry's bone density is less than normal for a man his age. Less dense bone is more likely to break.
   b. Henry's eating habits have resulted in inadequate dietary intake of Ca^{2+} and vitamin D. Therefore, the absorption of Ca^{2+} from his intestine into his blood has been inadequate.
   c. One might expect that Henry's blood Ca^{2+} would be low because of his diet. Low blood Ca^{2+} levels, however, stimulate increased PTH secretion. An increase in PTH maintains normal blood Ca^{2+} levels by increasing the number of osteoclasts,

which break down bone and release Ca^{2+} into the blood. Thus, Henry's blood Ca^{2+} levels are maintained at the expense of his bones, which become less dense as more matrix than usual is broken down. Increased PTH levels also promote increased Ca^{2+} reabsorption from the urine.

d. Normally, exposure to ultraviolet light, especially in sunlight, activates a precursor molecule in the skin that eventually becomes activated vitamin D in the kidneys (see chapter 5). Henry produces few, if any, precursor molecules because of his nocturnal lifestyle. Even though PTH normally increases the formation of active vitamin D in the kidneys, Henry's elevated PTH levels have not resulted in adequate vitamin D levels because of his lack of precursor molecules. Low levels of vitamin D result in less absorption of Ca^{2+} from the small intestine, contributing to Henry's low bone density.

e. Exercise increases the mechanical stress on bones, which increases osteoblast activity. Lack of exercise has resulted in decreased osteoblast activity. Bone density has decreased because less bone matrix has been produced by osteoblasts at the same time that more bone matrix has been broken down by osteoclasts under the influence of PTH.

Chapter 7

1. The sagittal suture is so named because it is in line with the midsagittal plane of the head. The coronal suture is so named because it is in line with the coronal plane (see chapter 1).
2. The bones most often broken in a "broken nose" are the nasals, ethmoid, vomer, and maxillae.
3. The lumbar vertebrae support a greater weight than the other vertebrae. The vertebrae are more massive because of the greater weight they support.
4. The atlas. It has no body.
5. The anterior support of the scapula is lost with a broken clavicle, and the shoulder is located more inferiorly and anteriorly than normal. In addition, since the clavicle normally holds the upper limb away from the body, the upper limb moves medially and rests against the side of the body.
6. The olecranon moves into the olecranon fossa as the elbow is straightened. The coronoid process moves into the coronoid fossa and the head of radius into the radial fossa as the elbow is bent.
7. The dried skeleton seems to have longer "fingers" than the hand with soft tissue intact because the soft tissue fills in the space between the metacarpals. With the soft tissue gone, the metacarpals seem to be an extension of the fingers, which appear to extend from the most distal phalanx to the carpals.
8. The femoral neck is commonly injured in elderly people because it is the smallest portion of the femur, which supports the weight of the body. It also forms an angle between the pelvis and the shaft of the femur, so the downward force of gravity on the body places an enormous amount of pressure on this part of the femur. That pressure is usually resisted in younger people with strong bones. The bone matrix begins to deteriorate in the elderly (osteoporosis), however. Because of hormonal differences between men and women, osteoporosis is much more common among elderly women.

9. The joint between the metacarpals and the phalanges is the metacarpophalangeal joint.
10. Premature sutural synostosis can result in abnormal skull shape, interfere with normal brain growth, and result in brain damage if not corrected. Such an abnormality is usually corrected surgically by removing some of the bone around the suture and creating an artificial fontanel, which then undergoes normal synostosis.
11. The synovial membrane is very thin and delicate. A considerable amount of pressure is exerted on the articular cartilages within a joint, and the articular cartilage is very tough, yet flexible, to withstand the pressure. If the synovial membrane covered the articular cartilage, it would be easily damaged during movement.
12. The movements required are abduction of the arm and flexion of the elbow. Flexion of the shoulder and elbow followed by medial rotation of the arm also works.

Chapter 8

1. When a muscle changes length, the I bands and the H zones change in width, but the A band does not. When a muscle is stretched, the I bands and the H zones increase in width as the length of the sarcomere increases. When a muscle contracts, cross-bridges form and cause the actin myofilaments to slide over the myosin myofilaments. The result is that the I bands and H zones decrease in width as the sarcomeres shorten. When a muscle relaxes, cross-bridges release, and actin myofilaments slide past myosin myofilaments as the sarcomeres lengthen. The I bands and H zones increase in width.
2. a. If Na^+ cannot enter the muscle fiber, no action potentials are produced in the muscle fiber because the influx of Na^+ causes the depolarization phase of the action potential. Without action potentials, the muscle fiber cannot contract at all. The result is flaccid paralysis.
   b. If ATP levels are low in a muscle fiber before stimulation, the following events occur. Energy from the breakdown of ATP already is stored in the heads of the myosin molecules. After stimulation, cross-bridges form. If not enough additional ATP molecules are in the muscle cells to bind to the myosin molecules to allow for cross-bridge release, however, the muscle becomes stiff without contracting or relaxing.
   c. If ATP levels in the muscle fiber are adequate but the action potential frequency is so high that Ca^{2+} accumulate around the myofilaments, the muscle contracts continuously without relaxing. As long as the ions are numerous within the sarcoplasm in the area of the myofilaments, cross-bridge formation is possible. If ATP levels are adequate, cross-bridge formation, release, and formation can proceed again, resulting in a continuously contracting muscle.
3. A decrease occurs in muscle control when reinnervation of muscle fibers occurs after poliomyelitis because the number of motor units in the muscle is decreased. Reinnervation results in a greater number of muscle fibers per motor unit. Control is reduced because the number of motor units that can be recruited is decreased. The greater the number of motor

units in a muscle, the greater the ability to have fine gradations of muscle contraction as motor units are recruited. A smaller number of motor units means that gradations of muscle contraction are not as fine.

4. a. Organophosphate poisons inhibit the activity of acetylcholinesterase, which breaks down acetylcholine at the neuromuscular junction. Consequently, acetylcholine accumulates in the synaptic cleft and continuously stimulates the muscle fiber, producing spastic contractions. Death can occur if multiple wave summation occurs to the point of complete tetanus of the respiratory muscles. Or the muscles can be so continuously stimulated that they fatigue and cannot contract at all.
   b. The problem with organophosphate poisoning is one of prolonged communication between nerves and target organs that utilize ACh as a neurotransmitter. Possible solutions involve administration of drugs that (1) block the receptors for ACh so that the neurotransmitter can no longer bind to the receptors, (2) break down ACh, or (3) bind to ACh so that it is unable to bind to its receptors.
5. As a weight is lifted, the muscle contractions are concentric contractions. When a weight lifter lifts a heavy weight above the head, most of the muscle groups contract with a force while the muscle is shortening. Concentric contractions are a category of isotonic contractions in which tension in the muscle increases or remains about the same while the muscle shortens. While the weight is held above the head, the contractions are isometric contractions, because the length of the muscles does not change. While the weight is lowered, unless the weight lifter simply drops the weight, the length of the muscles increases as the weight is lowered for most of the muscle groups. Eccentric contractions are contractions in which tension is maintained in a muscle while the muscle increases in length. The major muscle groups are therefore contracting eccentrically while the weight is lowered.
6. John has a greater recovery oxygen consumption than Eric. During the race up the stairs, John relied more on anaerobic respiration for the ATP required for his muscle contractions than did Eric, even though they did approximately the same amount of exercise. Eric's aerobic (cross-country) training has improved his ability to generate ATP.
7. Relaxation is initiated when Ca^{2+} are actively transported back into the sarcoplasmic reticulum by Ca^{2+} pumps. Slow-twitch fibers have slower Ca^{2+} pumps than fast-twitch fibers, so they relax more slowly.
8. Chicken breasts, which are responsible for moving the wings during flying, are white meat. Chicken thighs, which are responsible for standing and walking, are dark meat. The color of the meat depends on the number of capillaries (blood is red) within the muscle and is based on its myoglobin content (myoglobin is also red). After cooking, the tissues look darker, not red, because the blood and myoglobin have been broken down by the heat. Thus, the dark meat is darker because it contains more capillaries and more myoglobin. This is consistent with SO fibers, which are used for the maintenance of posture and slow movements, such as walking. The white muscle, with fewer capillaries and

lower myoglobin content, is consistent with FG fibers, which are fatigable fibers. Thus, chickens cannot fly very long. Other birds, such as ducks, that fly long distances have dark breast meat.

9. Susan was unable to keep pace during the finishing sprint because her muscles were not trained for quick movements, such as sprinting. She had trained them only for aerobic, steady-state activity, such as the activity occurring during most of the race. As her coach, you would advise Susan to incorporate high-intensity sprinting activities into her training. Such activities will cause her fast-twitch muscle fibers to increase their performance capacity, which will increase her ability to sprint at the end of the race.

Chapter 9

1. The biceps brachii is the antagonist for the triceps brachii.
2. Raising eyebrows—occipitofrontalis; winking—orbicularis oculi and then levator palpebrae superioris; whistling—orbicularis oris and buccinator; smiling—levator anguli oris, risorius, zygomaticus major, and zygomaticus minor; frowning—corrugator supercilii and procerus; flaring nostrils—levator labii superioris alaeque nasi and nasalis.
3. Shortening the right sternocleidomastoid muscle rotates the head to the left. It also slightly elevates the chin.
4. The trapezius can elevate and depress the scapula and the pectoralis major and the deltoid can flex and extend the shoulder.
5. Two arm muscles are involved in flexion of the elbow: the brachialis and the biceps brachii. The brachialis only flexes, whereas the biceps brachii both flexes the elbow and supinates the forearm. With the forearm supinated, both muscles can flex the elbow optimally; when pronated, the biceps brachii does less to flex the elbow. Chin-ups with the elbow supinated are therefore easier because both muscles flex the forearm optimally in this position. Bodybuilders who wish to build up the brachialis muscle perform chin-ups with the forearms pronated.

Chapter 10

1. When the axon of a neuron is severed, the proximal portion of the axon remains attached to the neuron cell body. The distal portion is detached, however, and has no way to replenish the enzymes and other proteins essential to its survival. Because the DNA in the nucleus provides the information that determines the structure of proteins by directing mRNA synthesis, the distal portion of the axon has no source of new proteins. Consequently, it degenerates and dies. On the other hand, the proximal portion of the axon is still attached to the nucleus and therefore has a source of new proteins. It remains alive and, in many cases, grows to replace the severed distal axon.
2. Tissue A has the larger resting membrane potential. There is a greater tendency for K^+ to diffuse out of the cell because it has significantly more K^+ leak channels. As a result, a greater negative charge develops on the inside of the plasma membrane, resulting in a larger resting membrane potential.
3. If the Na^+–K^+ pump stopped, the normal concentration gradients for Na^+ and K^+ would no longer be maintained. Without a concentration gradient for K^+, there would be no resting membrane potential.

4. If the intracellular concentration of K^+ is increased, the concentration gradient from the inside to the outside of the plasma membrane increases. This situation is similar to decreasing the extracellular concentration of K^+. The greater concentration gradient for K^+ increases their tendency to diffuse out of the cell across the plasma membrane. A greater negative charge then develops inside the cell (hyperpolarization).
5. Calcium ions bind to gating proteins that regulate the voltage-gated Na^+ channels. Low concentrations of Ca^{2+} cause the voltage-gated Na^+ channels to open, and high concentrations of Ca^{2+} cause the voltage-gated Na^+ channels to close. If the extracellular concentration of Ca^{2+} decreases, the resting membrane potential depolarizes because voltage-gated Na^+ channels open and Na^+ diffuse into the cell.
6. When the cells are stimulated, there is an increase in the permeability of their plasma membranes to Na^+. These ions diffuse into the cells down their concentration gradients and cause depolarization of the plasma membranes. If the concentration gradient for Na^+ is reduced, the tendency for Na^+ to diffuse into the cell decreases. In cell A, with the reduced Na^+ concentration gradient, the graded potential (depolarization) is of a smaller magnitude than in cell B because fewer ions are able to diffuse into the cell in response to the stimulus.
7. If the extracellular concentration of Na^+ decreases, the magnitude of the action potential is reduced. The smaller extracellular concentration of Na^+ reduces the tendency for Na^+ to diffuse into the cell when the Na^+ channels are open during an action potential. Consequently, the inside of the plasma membrane does not become as positive as it does in cells with a high extracellular concentration of Na^+. Even though the magnitude of action potentials is reduced when the extracellular Na^+ concentration is reduced, all of the action potentials are the same magnitude (all-or-none principle).
8. A prolonged stronger-than-threshold stimulus produces more action potentials than a prolonged threshold stimulus of the same duration. A prolonged stronger-than-threshold stimulus can stimulate more action potentials because the permeability of the membrane to Na^+ is increased. A very strong stimulus can even stimulate action potentials during the relative refractory period, whereas a prolonged threshold stimulus stimulates a low frequency of action potentials. Thus, when a prolonged stronger-than-threshold stimulus is applied, less time elapses between the production of one action potential and the next, resulting in the production of a greater number of action potentials.
9. If the duration of the absolute refractory period is 1 ms, that means action potentials can be generated no faster than every millisecond. The maximal frequency of action potentials is 1000 per second because there are 1000 milliseconds in a second.
10. Action potentials are transmitted fastest by electrical synapses because the ionic currents can quickly flow through the connexons. In contrast, chemical synapses are slower because the synaptic vesicles must be stimulated to release neurotransmitter, which diffuses across the synaptic cleft. The neurotransmitter must

stimulate ligand-gated Na^+ channels to open. The resulting movement of Na^+ into the cell can produce an action potential. All of these events take time.

Chapter 11

1. Dorsal root ganglia contain neuron cell bodies, which are larger in diameter than the axons of the dorsal roots. Action potentials are propagated in both directions in spinal nerves, toward the spiral cord in the dorsal root and away from the spinal cord in the ventral root.
2. Reflexive removal of the hand better protects the hand from heat damage because it is much faster than conscious removal. For conscious removal to occur, action potentials resulting from the pain stimulus have to be conducted up an ascending tract to reach the brain, where awareness of the stimulus occurs. Then the brain decides to remove the hand and through descending tracts stimulates the motor neurons to skeletal muscles. All of these activities take time.
3. Nerves C5–T1, which innervate the left arm, forearm, and hand, are damaged.
4. Pain, tingling, and numbness radiating from the elbow, down the posterolateral portion of the forearm, and to the lateral side of the hand and fingers (4 and 5) are consistent with damage to the C8–T1 dermatome (see figure 11.12). Ulnar nerve damage results in the same symptoms in the lateral hand and fingers, but the symptoms do not radiate into the forearm and elbow (see figure 11.16). Careful examination of Sarah's right-upper limb to map the extent of the pain and numbness allows for a tentative differential diagnosis between ulnar nerve damage and damage to the C8–T1 brachial plexus roots. A radiograph of the neck can indicate the presence of an extra rib, which most commonly affects roots C8–T1. As the ulnar nerve arises from the C8–T1 roots of the brachial plexus (see figure 11.14), all muscles innervated by that nerve can be affected. Because the radial and median nerves also obtain some axons from the C8–T1 roots, some of their more distal-lateral muscles may also be affected.
5. Damage to the right phrenic nerve results in the absence of muscular contraction in the right half of the diaphragm. Because the phrenic nerves originate from C3–C5, damage to the upper cervical region of the spinal cord eliminates their functions; damage in the lower cord below the point where the spinal nerves originate does not affect the nerves to the diaphragm. Breathing is affected, however, because the intercostal nerves to the intercostal muscles, which move the ribs, are paralyzed.
6. a. Often when one part of the head suffers a heavy blow, the brain moves within the cranial cavity and hits the opposite side of the cranial cavity. In this case, the blow to the back of the head forced the brain anteriorly and the frontal lobes struck the frontal bones with enough force to tear blood vessels between the brain and the dura. Subsequent bleeding from these vessels into the subdural space created the subdural hematoma. This type of injury is a contrecoup brain injury because it occurred on the opposite side of the brain from the point of traumatic impact.
   b. The subdural hematoma forcing the medulla oblongata to herniate into the vertebral canal caused the tissues of the medulla to become compressed and fail to function. The medulla

oblongata contains the centers for the control of respiration and heart rate. Thus, damage to these areas of the brain interrupted the woman's ability to regulate her breathing and heart rate, which was most likely a major contributing factor in her death.

7. The tongue is protruded by contraction of the geniohyoid muscle, which pulls the back of the tongue forward, thereby pushing the muscle mass of the tongue forward. With one side pushed forward and unopposed by muscles of the opposite side, the tongue deviates toward the nonfunctional side. In the example, therefore, the right hypoglossal nerve is damaged.

Chapter 12

1. Lesions on one side of the spinal cord that interrupt the spinothalamic tract (anterolateral system) eliminate pain and temperature sensation below that level on the opposite side of the body. This occurs because the pathway is contralateral (ascending on the opposite side of the spinal cord) to the area of skin innervated by the receptors.

2. The damage to Bill's spinal cord would be on the right side. The dorsal column conveys sensations of proprioception, fine touch, and vibration through the spinal cord on the same side of the body as the sensory receptors. The damage to Mary's brainstem would be on the left side of the pons. The damage occurred above the medulla oblongata where the secondary neurons of the medial lemniscus decussate.

3. Action potentials reaching the auditory cortex is where the sound is "perceived." In the auditory association area the auditory information is compared with past sounds. On the basis of that comparison, the auditory association categorizes the sound as new or as a recognized sound and passes judgement on the importance of the sound.

4. a. Damage to the dorsal column on the left side results in the loss of two-point discrimination, proprioception, vibration, and pressure sensations from the left side of the body because the dorsal column/medial lemniscal system ascends ipsilaterally in the spinal cord. It crosses over in the medulla oblongata. Damage to the anterolateral system (spinothalamic tract) on the left side results in the loss of pain and temperature sensations from the right (contralateral) side of the body because fibers entering the anterolateral system cross over where they enter the spinal cord. Damage to the spinocerebellar tracts on the left side results in loss of unconscious proprioception to the left cerebellum because the spinocerebellar tracts ascend ipsilaterally.

   b. Spinal cord level L2 and below supplies skin of the lower limb.

   c. Damage to the lateral corticospinal tract and to the indirect pathways on the left side of the spinal cord results in loss of voluntary control on the left side because these tracts descend ipsilaterally in the spinal cord. The right cerebral cortex controls the left (contralateral) side of the body because motor tracts cross from one side to the other in the brainstem. The anterior spinothalamic tracts extend to the midthorax and are unaffected by an injury at level L2.

   d. Spinal cord level L2 and below supplies the muscles of the lower limb. All of the nerves arising from the lumbosacral plexus are affected: obturator, femoral, tibial, and common fibular.

5. The right half of the cerebellum receives input from, and projects to, the left cerebral motor cortex. If the right half of the cerebellum is injured, the left cerebral cortex does not have the normal feedback and input from the cerebellum necessary for normal movements. The right side of the body is affected because the left cerebral cortex controls the muscles on the right side of the body.

6. If a person holds an object in her right hand, tactile sensations of various types travel up the spinal cord to the brain, where they reach the somatic sensory cortex of the left hemisphere and the object is recognized. Action potentials then travel to Wernicke's area, where the object is given a name. From there, action potentials travel to Broca's area, where the spoken word is formulated. Action potentials from Broca's area travel to the premotor area and primary motor cortex, where action potentials are initiated that stimulate the muscles necessary to form the word.

Chapter 13

1. Sniffing is the rapid, repeated intake of air. Sniffing brings in more air containing odorants, which increases the likelihood of detecting an odor. Inhaling slowly and deeply is effective for the same reason.

2. Eye drops placed into the eye tend to drain through the nasolacrimal duct into the nasal cavity. Recall that much of what is considered taste is actually smell. The medication is detected by the olfactory neurons and is interpreted by the brain as taste sensation. Crying produces extra tears, which are conducted to the nasal cavity, causing a "runny" nose.

3. Inflammation of the cornea involves edema, the accumulation of fluid. Fluid accumulation in the cornea increases its water content and, because water causes the proteoglycans to expand, the transparency of the lens decreases, interfering with normal vision.

4. Eyestrain, or eye fatigue, occurs primarily in the ciliary muscles. It occurs because close vision requires accommodation. Accommodation occurs as the ciliary muscles contract, releasing the tension of the suspensory ligaments and allowing the lens to become more rounded. Continued close vision requires the maintenance of accommodation, which requires that the ciliary muscles remain contracted for a long time, resulting in their fatigue.

5. Rhodopsin breakdown is associated with adaptation to bright light and occurs rapidly, whereas rhodopsin production is associated with adaptation to conditions of little light and occurs slowly. Eyes adapt rather quickly to bright light but quite slowly to very dim light.

6. Rod cells distributed over most of the retina are involved in both peripheral vision (out of the corner of eye) and vision under conditions of very dim light. When attempting to focus directly on an object, however, a person relies on the cones within the macula and fovea centralis; although the cones are involved in visual acuity, they do not function well in dim light; thus, the object may not be seen at all.

7. A light flashing off to the right side stimulates the left visual cortex.

8. A lesion in the right optic tract at C results in loss of vision in the left temporal and right nasal visual fields.

9. When a tree falls in the forest it, of course, makes a sound. That sound is "Timbre!"

10. It is much easier to perceive subtle musical tones when music is played somewhat softly, as opposed to very loudly, because loud sounds have sound waves with a greater amplitude, which causes the basilar membrane to vibrate more violently over a wider range. The spreading of the wave in the basilar membrane makes it more difficult to hear subtle tone differences.

Chapter 14

1. Terminal ganglia are found near or embedded within the wall of organs supplied by the parasympathetic division and contribute to the enteric nervous system. Postganglionic parasympathetic axons from the terminal ganglia also contribute to the enteric nervous system. Chain ganglia and collateral ganglia contain the cell bodies of sympathetic neurons. They are not embedded within the walls of organs supplied by the sympathetic division. Instead, postganglionic neurons extend from them to organs. Thus, postganglionic sympathetic axons are found in the enteric nervous system.

2. For a sensory axon running alongside sympathetic axons, the sensory axon leaves the wall of the small intestine in a splanchnic nerve. The sensory axon passes through the collateral ganglion of the splanchnic nerve and continues on to a sympathetic chain ganglion. From the sympathetic chain ganglion, the sensory axon passes through a white ramus communicans, the ventral rami of a spinal nerve, a spinal nerve, and the dorsal root of a spinal nerve to a dorsal root ganglion. For a sensory axon running alongside parasympathetic axons, the sensory axon leaves the wall of the small intestine and runs parallel to preganglionic parasympathetic fibers of the vagus nerve to its sensory ganglion.

3. Nicotinic receptors are located within the autonomic ganglia as components of the membranes of the postganglionic neurons of the sympathetic and parasympathetic divisions. Nicotine binds to the nicotinic receptors of the postganglionic neurons, resulting in action potentials. Consequently, the postganglionic neurons stimulate their effector organs. After the consumption of nicotine, structures innervated by both the sympathetic and the parasympathetic divisions are affected. After the consumption of muscarine, only the effector organs that respond to acetylcholine are affected. This includes all the effector organs innervated by the parasympathetic division and the sweat glands, which are innervated by the sympathetic division.

4. a. The radial muscles of the iris cause the pupil to dilate, and the circular muscles cause the pupil to constrict.

   b. The radial muscles are controlled by the sympathetic division, and the circular muscles are controlled by the parasympathetic division.

   c. An adrenergic drug could activate α_1 receptors on the radial muscles of the iris, causing dilation. A muscarinic blocking agent could block muscarinic receptors on the circular muscles, causing less constriction—that is, causing dilation.

d. An inability to see close-up objects indicates that the ciliary muscles of the eye are affected by the drug. A muscarinic blocking agent could produce this effect by blocking the contraction of the ciliary muscles. Usually, pupil-dilating eye drops contain a muscarinic blocking agent and an adrenergic agent.

e. Contraction of the radial muscles of the iris causes dilation of the pupil. An adrenergic blocking agent could block α_1 receptors on the radial muscles, causing less dilation. A muscarinic agent could stimulate muscarinic receptors on the circular muscles of the iris, causing pupil constriction.

f. Sympathetic stimulation of blood vessels normally keeps them in a state of partial contraction. Decreased sympathetic stimulation of blood vessels in the conjunctiva of the eye can result in dilation of the blood vessels and bloodshot eyes. The pupil-dilating eye drops contain an adrenergic blocking agent, which dilates the pupils and, as a side effect, produces bloodshot eyes.

5. The frequency of action potentials in sympathetic neurons to the sweat glands increases as the body temperature increases. The increasing body temperature is detected by the hypothalamus, which activates the sympathetic neurons. Sweating cools the body by evaporation. As the body temperature declines, the frequency of action potentials in sympathetic neurons to the sweat glands decreases. A lack of sweating helps prevent heat loss from the body.

6. In response to an increase in blood pressure, information is transmitted in the form of action potentials along sensory neurons to the medulla oblongata. From the medulla oblongata, the frequency of action potentials delivered along sympathetic nerve fibers to blood vessels decreases. As a result, blood vessels dilate, causing the blood pressure to decrease. In response to a decrease in blood pressure, fewer action potentials are transmitted along sensory neurons to the medulla oblongata, which responds by increasing the frequency of action potentials delivered along sympathetic nerves to blood vessels. As a result, blood vessels constrict, causing blood pressure to increase.

7. The parasympathetic division releases acetylcholine, which binds to muscarinic receptors on organs. Bethanechol chloride produces effects similar to the stimulation of organs by the parasympathetic division. Thus, this drug should stimulate the urinary bladder to contract. Side effects can be produced by the stimulation of muscarinic receptors elsewhere in the body. The stimulation of smooth muscle in the digestive tract can produce abdominal cramps. The stimulation of air passageways can cause an asthmatic attack. Decreased tear production, salivation, and dilation of the pupils are not expected side effects because parasympathetic stimulation causes increased tear production, salivation, and constriction of the pupils. Sweat glands are innervated by the sympathetic division but have muscarinic receptors. Bethanechol chloride can increase sweating.

Chapter 15

1. The tumor causes overproduction of TSH, which stimulates the thyroid gland. Thyroid hormones levels would be above normal. Through negative feedback, the elevated thyroid hormone levels inhibit TRH secretion, so TRH levels would be below normal.

2. A major function of plasma proteins, to which hormones bind, is to increase the length of time that hormones persist in the blood. If the concentration of a plasma protein decreases, the hormone that binds to it is removed from the blood more rapidly than normal and the concentration of the hormone in the blood decreases. It is possible, however, that the concentration of the hormone does not change. This can occur if the secretion rate of the hormone increases, compensating for the increased rate of removal.

3. A drug can increase the cAMP concentration in a cell by stimulating its synthesis or by inhibiting its breakdown. Drugs that bind to a receptor that increases adenylate cyclase activity will increase cAMP synthesis. Because phosphodiesterase normally causes the breakdown of cAMP, an inhibitor of phosphodiesterase decreases the rate of cAMP breakdown and causes cAMP to increase in the smooth muscle cells of the airway and produces relaxation.

4. Membrane-bound receptors that increase the synthesis of intracellular mediators, such as cAMP, normally activate enzymes already existing in the cytoplasm of the cell. The synthesis of cAMP occurs quickly, but the duration is short because cAMP is broken down quickly, and the activated enzymes are then deactivated. Membrane-bound receptor mechanisms are therefore better adapted to short-term and rapid responses. Intracellular receptor mechanisms result in the synthesis of new proteins, which can take time. Intracellular receptors are therefore better adapted for mediating responses that last a relatively long time (i.e., for many minutes, hours, or longer).

5. The cell bodies of the neurons cells that produce ADH are in the hypothalamus, and their axons extend into the posterior pituitary, where ADH is stored and secreted. Removing the posterior pituitary severs the axons, resulting in a temporary reduction in secretion. The cell bodies still produce ADH, however; as the ADH accumulates at the ends of severed axons, ADH secretion resumes.

6. If GH is administered to young people before the growth of their long bones is complete, it causes their long bones to grow and they will grow taller. To accomplish this, however, GH has to be administered over a considerable length of time. It is likely that some symptoms of acromegaly will develop. In addition to undesirable changes in the skeleton, nerves frequently are compressed as a result of the proliferation of connective tissue. Because GH spares glucose usage, chronic hyperglycemia results, frequently leading to diabetes mellitus and the development of severe atherosclerosis. Mr. Hoops' doctor would therefore not prescribe GH.

7. a. TSI is structurally similar to TSH. TSI can bind to TSH receptors in the thyroid gland and stimulate thyroid hormone secretion. High levels of TSI results in hypersecretion of thyroid hormones.

b. TRH and TSH levels are lower than normal because the elevated levels of thyroid hormones inhibit their secretion.

c. The 31I treatment destroys thyroid gland tissue. So much tissue was destroyed that the thyroid gland was no longer able to produce normal amounts of thyroid hormones.

d. TRH and TSH levels would be elevated because the low levels of thyroid hormones do not inhibit the hypothalamus and anterior pituitary.

e. Even though Josie's production of TRH and TSH is unaffected by the 31I treatment, they can no longer regulate thyroid hormone secretion within a normal range because there is not enough viable thyroid tissue left. Josie will have to maintain normal thyroid hormone levels by taking thyroid hormone pills. If Josie maintains normal thyroid hormone levels by taking thyroid hormones, TRH and TSH levels will be normal.

8. In response to a reduced dietary intake of Ca^{2+}, the blood Ca^{2+} levels begin to decline. In response to the decline in blood Ca^{2+} levels, an increase of PTH secretion from the parathyroid glands occurs. The PTH increases the breakdown of bone and the release of Ca^{2+} from bone into blood, reduces Ca^{2+} loss in urine, and increases Ca^{2+} absorption from the small intestine. However, if dietary intake of Ca^{2+} is too low, normal blood Ca^{2+} levels are maintained by removing Ca^{2+} from bone. Severe dietary calcium deficiency results in bones that become soft and eaten away because of the decrease in Ca^{2+} content.

9. High aldosterone levels result in increased K^+ secretion and low blood K^+ levels. The effect of low blood K^+ levels is hyperpolarization of muscle and neurons. The hyperpolarization results from the lower levels of K^+ in the extracellular fluid and a greater tendency for K^+ to diffuse from the cell. As a result, a greater than normal stimulus is required to cause the cells to depolarize to threshold and generate an action potential. Symptoms of low serum K^+ levels therefore include lethargy and muscle weakness. The major effect of a low rate of aldosterone secretion is elevated blood K^+ levels. As a result, nerve and muscle cells partially depolarize. They produce action potentials spontaneously or in response to very small stimuli because their resting membrane potential is closer to threshold. The result is muscle spasms, or tetanus.

10. Large doses of cortisone can damage the adrenal cortex because cortisone inhibits ACTH secretion from the anterior pituitary. ACTH is required to keep the adrenal cortex from undergoing atrophy. Prolonged use of large doses of cortisone can cause the adrenal gland to atrophy to the point at which it cannot recover if ACTH secretion does increase again.

11. An increase in insulin secretion in response to parasympathetic stimulation and gastrointestinal hormones is consistent with the maintenance of homeostasis because parasympathetic stimulation and increased gastrointestinal hormones result from conditions such as eating a meal. Insulin levels therefore increase just before large amounts of glucose and amino acids enter the circulatory system. The elevated insulin levels prevent a large increase in blood glucose and the loss of glucose in the urine.

12. Sympathetic stimulation during exercise inhibits insulin secretion. Blood glucose levels are not high because skeletal muscle tissue continues to take up some glucose and metabolizes it. Muscle contraction depends on glucose stored in the form of glycogen in muscles and fatty acid metabolism. During a long run, glycogen levels are depleted. The "kick" at the end of the race results from increased energy production through anaerobic respiration, which uses glucose or glycogen as an energy source. Because blood glucose levels and glycogen levels are low, the source of energy is insufficient for greatly increased muscle activity.

Chapter 16

1. An elevated reticulocyte count indicates that erythropoiesis and the demand for red blood cells are increased and that immature red blood cells (reticulocytes) are entering the circulation in large numbers. An elevated reticulocyte count can occur for a number of reasons, including loss of blood; therefore, after a person donates a unit of blood, the reticulocyte count increases.

2. Carbon monoxide binds to the iron of hemoglobin and prevents the transport of oxygen. The decreased oxygen stimulates the release of erythropoietin from the kidneys. Erythropoietin increases red blood cell production in red bone marrow, thereby causing the number of red blood cells in the blood to increase.

3. The white blood cells shown in figure 16.8 are (a) lymphocyte, (b) basophil, (c) monocyte, (d) neutrophil, and (e) eosinophil.

4. People with type AB blood were called universal recipients because they could receive type A, B, AB, or O blood with little likelihood of a transfusion reaction. Type AB blood does not have antibodies against type A or B antigens; therefore, the transfusion of these antigens in type A, B, or AB blood does not cause a transfusion reaction in a person with type AB blood. The term is misleading, however, for two reasons. First, other blood groups can cause a transfusion reaction. Second, antibodies in the donor's blood can cause a transfusion reaction. For example, type O blood contains anti-A and anti-B antibodies that can react against the A and B antigens in type AB blood.

5. a. HDN causes a decrease in the number of red blood cells. A transfusion treats this anemia by increasing the number of red blood cells, which promotes oxygen and carbon dioxide transport.
b. An exchange transfusion not only increases the number of red blood cells but also decreases bilirubin and anti-Rh antibody levels by removing them in the withdrawn blood. Fewer anti-Rh antibodies decrease the agglutination and lysis of red blood cells.
c. Erythropoietin stimulates red blood cell production. Thus, Billy has two sources of red blood cells: the donor blood and himself.
d. The destruction of red blood cells in a fetus with HDN results in lower than normal numbers of red blood cells. The resulting decrease in oxygen transport stimulates increased erythropoietin production.
e. After birth, Billy breathes on his own. The ability to oxygenate the blood using the lungs is greater than the ability to oxygenate the blood across the placenta. Billy's blood oxygen levels increase and erythropoietin levels decrease. Thus, his production of red blood cells decreases and his anemia gets worse.
f. The donor blood used in exchange transfusions for the treatment of HDN should be Rh-negative, even though the newborn is Rh-positive. Rh-negative red blood cells do not have Rh antigens. Therefore, any anti-Rh antibodies in the newborn's blood do not react with the transfused Rh-negative red blood cells. Note, however, that Rh-positive blood is often used because supplies of Rh-negative blood are limited. A judgment is made on the severity of the HDN. In severe cases, Rh-negative blood is used.
g. Giving Rh-negative blood to an Rh-positive newborn does not change the blood type of the newborn because blood type is genetically determined. Eventually, all of the Rh-negative red blood cells die and the newborn produces only Rh-positive red blood cells.

6. A white blood count (WBC) should be done. An elevated WBC, leukocytosis, can be an indication of bacterial infections. A differential WBC should also be done. An increase in the number of neutrophils supports the diagnosis of a bacterial infection. Coupled with other symptoms, this can mean appendicitis. If these tests are normal, appendicitis is still a possibility and the physician must rely on other clinical signs. Diagnostic accuracy for appendicitis is approximately 75%–85% for experienced physicians.

Chapter 17

1. The anterior interventricular artery supplies blood to the anterior wall, most of the left ventricle, and the anterior part of the right ventricle. A blocked anterior interventricular artery reduces the oxygen supply to the portion of the heart that is supplied by that artery, and the cardiac muscle in that area dies. If the clot forms in the anterior interventricular artery just after it branches from the coronary artery, a large part of the heart dies and the heart stops pumping blood. If the clot forms at the apex of the heart, a much smaller part of the heart is affected. The pumping capability of the heart may be reduced, but the heart can still pump blood. Thus, the amount of cardiac muscle tissue affected by a blocked artery is critical.

2. The heart must continue to function under all conditions and requires energy in the form of ATP. During heavy exercise, lactic acid is produced in skeletal muscle as a by-product of anaerobic respiration. The ability to use lactic acid provides the heart with an additional energy source.

3. It is important to prevent tetanic contractions in cardiac muscle because the cycle of contraction and relaxation stops during tetanic contractions. This would cause the pumping action of the heart to stop. In skeletal muscle, the cycle of contraction and relaxation is not important as a pump, but it is important to maintain a static contracted state. This is essential to maintaining posture or to holding a limb in a specific position.

4. If the normal blood supply is reduced in a small area of the heart through which the left bundle branch passes, conduction of action potentials through that side of the heart is slowed or blocked. As a consequence, the left side of the heart contracts more slowly. The right side of the heart contracts more normally. The reduced rate of contraction of the left ventricle reduces the pumping effectiveness of the left ventricle.

5. When the AV and semilunar valves are closed, ventricular volume does not change, even though the ventricles are contracting. If the volume does not change, then cardiac muscle fibers are not changing length and the contraction is isometric (see chapter 8).

6. The left ventricle has the thickest wall. The pressure produced by the left ventricle is much higher than the pressure produced by the right ventricle, when the ventricles contract. It is important for each ventricle to pump the same amount of blood because, with two connected circulation loops, the blood flowing into one must equal the blood flowing into the other so that one does not become overfilled with blood at the expense of the other. For example, if the right ventricle pumps less blood than the left ventricle, blood must accumulate in the systemic blood vessels. If the left ventricle pumps less blood than the right ventricle, blood accumulates in the pulmonary blood vessels.

7. Fibrillation makes cardiac muscle an ineffective pump. The ventricles are the primary pumping chambers of the heart. Ventricular fibrillation results in death because of the heart's inability to pump blood to the body and lungs. The atria function primarily as reservoirs. Their pumping action is most important during exercise. Therefore, atrial fibrillation does not destroy the ventricles' ability to pump blood.

8. A lubb–duppshhh is an abnormal heart sound immediately following the second heart sound. An abnormal heart sound immediately following a normal heart sound indicates an incompetent valve. A semilunar valve is incompetent because the second heart sound is caused by closing of the semilunar valves. A schhlubb-dubb is an abnormal sound immediately before the first heart sound. An abnormal heart sound immediately before a valve closes indicates a stenosed valve. An AV valve is stenosed because the second heart sound is caused by closing of the AV valves.

9. Sympathetic stimulation increases heart rate. If venous return remains constant, stroke volume decreases as the number of beats per minute increases because the ventricles have less time to fill.

10. In response to severe hemorrhage, blood pressure decreases, which is detected by baroreceptors. A reduced frequency of action potentials is sent from the baroreceptors to the medulla oblongata. This causes the cardioregulatory center to increase sympathetic stimulation of the heart and increase the heart rate. Sympathetic stimulation of the heart also increases stroke volume, as long as the volume of blood returned to the heart is adequate. Following hemorrhage, however, the blood volume in the body is reduced, which reduces venous return. The decreased venous return results in decreased preload and decreased force of contraction of the ventricles (Starling's law). Thus, stroke volume is reduced.

11. a. The stenosed valve makes it more difficult for contractions of the left ventricle to force blood to flow into the aorta because of the narrowed opening. When the resistance is sufficiently high, the stroke volume decreases because the left ventricle is unable to overcome the increased resistance (afterload) to blood flow through the stenosed valve.

b. The left ventricle must generate a greater force during ventricular systole to cause blood to flow into the aorta because of the increased resistance caused by the stenosed aortic semilunar valve. Consequently, the left ventricular pressure is very high during ventricular systole, even though the systolic pressure is low. Just as skeletal muscle hypertrophies in response to the resistance of moving heavy weights, cardiac muscle hypertrophies in response to the resistance to blood flow through the stenosed valve.

c. The mean arterial pressure is equal to cardiac output (heart rate × stroke volume) × peripheral resistance. The mean arterial pressure decreases when stroke volume decreases, unless there is a compensating increase in heart rate or peripheral resistance.

d. When Norma rises to a standing position, gravity affects blood in arteries and veins, and the blood tends to settle in blood vessels below the heart. Consequently, there is decreased venous return to the heart, which results in decreased cardiac output (Starling's law of the heart). The decreased cardiac output results in decreased blood pressure, decreased blood flow to the brain, and dizziness.

e. When Norma stands, her blood pressure decreases, activating the baroreceptor reflex. Her heart rate increases. Although Norma has a normal baroreceptor reflex, the heart is not able to pump enough blood past the stenosed valve to maintain blood pressure.

f. Angina results from inadequate blood delivery to heart muscle through the coronary circulation. Norma's stenosed valve has reduced blood pressure in the aorta, which reduces blood delivery to the body and to the coronary blood vessels. At rest, there is sufficient blood delivery to cardiac muscle, and there is no pain; however, when Norma starts exercising, the cardiac muscle requires greater delivery of oxygen by the blood, which is not possible with the stenosed valve. This results in an increase in anaerobic respiration and a decrease in pH, which are responsible for the pain of angina pectoris.

Chapter 18

1. Arteriosclerosis slowly reduces blood flow through the carotid arteries and therefore the amount of blood that flows to the brain. As the resistance to flow increases in the carotid arteries during the late stages of arteriosclerosis, the blood flow to the brain is compromised, resulting in reduced oxygen delivery. Confusion, loss of memory, and loss of the ability to perform other normal brain functions occur.

2. a. Vasoconstriction of blood vessels in the skin in response to exposure to cold results in a decreased flow of blood through the skin and in a dramatic increase in resistance (Poiseuille's law). Vasoconstriction makes the skin appear pale.

b. Vasodilation of blood vessels in the skin results in increased blood flow through the skin. Vasodilation makes the skin appear flushed or red in color.

c. In a patient with erythrocytosis, the hematocrit increases dramatically. As a result, the viscosity of the blood increases, which increases resistance to flow. Consequently, flow decreases or a greater pressure is needed to maintain the same flow.

d. Dehydration results in reduced plasma and increased blood viscosity. Resistance to blood flow increases and blood flow decreases.

3. Vasoconstriction causes increased peripheral resistance and increased mean arterial pressure. Vasodilation causes decreased peripheral resistance and decreased mean arterial pressure.

4. Premature beats of the heart result in contraction of the heart muscle before the heart has had time to fill to its normal capacity. Consequently, a reduced stroke volume occurs, which results in a weak pulse in response to that premature contraction. Other contractions and the resulting pulses are normal. Weak pulses occur in response to hemorrhagic shock because of a decreased venous return. The heart does not fill with blood between contractions (decreased preload); the stroke volume is therefore reduced and the pulse is weak. Strong pulses in a person who received too much saline solution in an intravenous transfusion result from an increase in venous return to the heart because of the increased volume of fluid in the circulatory system. The increased venous return (increased preload) causes increased stroke volume, which increases the pulse pressure. Exercise increases stroke volume and pulse pressure.

5. Returning to a horizontal position removes the effect of hydrostatic pressure on the blood vessels of the lower limbs. As the pressure in the lower limbs decreases, the distensible (compliant) veins are less stretched. Venous volume decreases and venous return increases, which increases blood pressure.

6. These conditions decrease plasma protein concentrations. Thus, the movement of water by osmosis from the interstitial fluid into the blood decreases. Less water removal results in edema. Keeping the legs elevated reduces the hydrostatic pressure in the capillaries of the legs. A major effect is that the pressure force that moves fluid out of the capillary is decreased. As a result, the net movement of fluid out of the arterial ends of capillaries decreases and the net movement into the venous ends of the capillaries increases. Therefore, excess interstitial fluid is carried away from the legs.

7. When a blood vessel is blocked (when the legs are crossed, for instance), oxygen and nutrients are depleted, and waste products accumulate in tissue supplied by the blocked blood vessel. The reduced supply of oxygen and nutrients and the accumulated waste products all cause vasodilation of arterioles and relaxation of the precapillary sphincters. When the block to blood flow is removed, a large volume of blood rapidly flows through the arterioles and capillaries, which gives the skin a red appearance.

8. Severe vasoconstriction that results from Raynaud syndrome causes the digits to appear white because of the lack of blood flow through the capillary beds in the digits. The precapillary sphincters, which are not controlled by sympathetic stimulation, would be relaxed because of local control mechanisms, such as reduced oxygen and increased carbon dioxide levels in the tissues.

9. A rapid loss of blood decreases blood volume and venous return. Consequently, stroke volume and blood pressure decrease. Vasoconstriction of veins could oppose this change by increasing venous return.

10. During a headstand, hydrostatic pressure increases blood pressure in the head. As blood pressure in the lower limbs decreases because of decreased hydrostatic pressure, the compliant veins are less stretched and venous return increases. The increased venous return increases stroke volume and blood pressure. Thus, the blood pressure in the area of the aortic arch and carotid sinus baroreceptors increases. The increased pressure activates the baroreceptor reflexes, increasing parasympathetic stimulation of the heart and decreasing sympathetic stimulation. Thus, the heart rate decreases.

11. a. The blocked vein in Harry's right leg caused edema and led to tissue ischemia. Edema developed inferior to the blocked vein. Blockage of the vein increased the capillary hydrostatic pressure in the capillary beds drained by the blocked vein. The increased capillary hydrostatic pressure increased the amount of fluid that flowed from the capillaries into the tissue spaces and reduced the amount of fluid that returned to the capillaries. Consequently, fluid accumulated in the tissue spaces and caused edema (see figure 21.34). The ischemia resulted in pain, much the way ischemia of the heart causes pain during myocardial infarctions (see chapter 20).

b. Emboli that originate in the posterior tibial vein pass through the following parts of the circulatory system before they lodge in the pulmonary arteries of the lungs: posterior tibial vein, popliteal vein, femoral vein, external iliac vein, common iliac vein, inferior vena cava, right atrium, right ventricle, pulmonary trunk, right or left pulmonary artery. Emboli will lodge in branches of the pulmonary arteries. Emboli are most likely to lodge in the lungs because the pulmonary arteries branch many times before they deliver blood to the pulmonary capillaries and, as they branch, their diameters decrease. Even small emboli will eventually lodge in the smaller branches of the pulmonary arteries. The other parts of the circulatory system through which the emboli pass have much larger diameters and emboli can pass readily through them.

c. When emboli are large enough or numerous enough to block blood flow through a significant part of the lungs, resistance to blood flow through the lungs increases. The increased resistance to flow increases the pulmonary venous pressure, which increases the afterload for the right ventricle of the heart. If the right ventricle is unable to overcome the increased afterload, failure of the right side of the heart can occur, and blood flow through the lungs is reduced.

d. Pulmonary emboli large enough to significantly reduce blood flow through the lungs reduces the lungs' ability to carry out gas exchange with blood, and blood oxygen levels decrease. If blood flow through the lungs to the left side of the heart is reduced significantly, hypotension can develop. Reduced blood flow to the left side of the heart will result in a reduced cardiac output because of decreased venous return (Starling's law of the heart). As a result, blood pressure falls and the mechanisms that regulate blood pressure are activated (see "Regulation of Mean Arterial Pressure"). Manifestations of hypotension, such as an increased heart rate, a weak pulse, and pallor, may be present.

  e. Heparin and coumadin are anticoagulants. They are prescribed because they decrease the rate at which coagulation proceeds (see chapter 19). Heparin is administered intravenously, whereas coumadin can be taken orally, which makes home use possible. Prothrombin time is checked periodically because enough anticoagulant must be administered to prevent enlargement of the thrombus in the deep vein of Harry's leg and to prevent additional emboli from forming. Enzymes continually break down coagulated blood. Clots are removed because the slower rate of coagulation allows them to be broken down faster than they can be formed. It is also important to monitor Harry's prothrombin time because excess blockage of coagulation can result in bleeding (see chapter 19). Coumadin is prescribed for a substantial amount of time after a venous thrombosis or a pulmonary embolus has formed to prevent it from reoccurring.

12. The baroreceptor reflex, ADH, and renin–angiotensin–aldosterone mechanisms function similarly in both cases. The fluid shift mechanism, however, is important when the loss of blood occurs over several hours, but it does not operate within a short period. The fluid shift mechanism plays a very important role in the maintenance of blood volume when blood loss or dehydration develops over several hours. When the blood pressure decreases, interstitial fluids pass into the capillaries, which prevents a further decline in blood pressure. The fluid shift mechanism is a powerful method through which blood pressure is maintained because the interstitial fluid acts as a fluid reservoir.

Chapter 19

1. a. The lymphatic vessels remove fluid from tissues. Cancer cells that break free from a tumor can spread, or metastasize, to other parts of the body by entering lymphatic or blood capillaries (see chapter 4). If they enter lymphatic capillaries, they are carried by lymph to lymph nodes, which filter lymph. The first lymph nodes in which cancer cells are likely to become trapped are the sentinel lymph nodes.

  b. All of Cindy's sentinel lymph nodes contained cancer cells, indicating that the cancer cells have spread into her lymphatic system. In order to minimize the risk of further metastasis, her axillary lymph nodes were removed because they drain the superficial thorax and upper limb. If none of the sentinel lymph nodes had contained cancer cells, however, an option would have been to not remove the axillary lymph nodes.

  c. When lymph nodes and their attached lymphatic vessels are removed, it disrupts the normal removal of fluid from tissues. Recall from chapter 18 that approximately one-tenth of the fluid entering tissues is normally removed by the lymphatic system. A reduction in this fluid removal can result in an accumulation of fluid in the tissues.

  d. Exercise increases skeletal muscle contractions and increases breathing, both of which increase lymph flow. Consequently, more fluid can enter lymphatic capillaries, reducing edema.

  e. Compression bandages or garments produce an external pressure on tissues and reduce the movement of fluid into tissues from blood capillaries (see chapter 18). It is recommended that compression bandages or garments be worn during daily activities, especially during exercise.

  f. Lymphatic vessels periodically contract and relax, moving lymph in one direction because of the valves within lymphatic vessels. The compression pump mimics this effect through external pressure changes. Increased pressure compresses the lymphatic vessels, moving lymph. Decreased pressure allows the lymphatic vessels to expand and fill with lymph. The sequential pressure changes help move lymph in a distal-to-proximal direction.

2. The T cells transferred to mouse B do not respond to the antigen. The T cells are MHC-restricted and must have the MHC proteins of mouse A as well as antigen X to respond.

3. When the antigen is eliminated, it is no longer available for processing and combining with MHC class II molecules. Consequently, no signal takes place to cause lymphocytes to proliferate and produce antibodies.

4. The first exposure to the disease-causing agent (antigen) evokes a primary response. Gradually, however, antibodies degrade and memory cells die. If, before all the memory cells are eliminated, a second exposure to the antigen occurs, a secondary response results. The memory cells produced then can provide immunity until the next exposure to the antigen.

5. With the depression of helper T-cell activity, the ability of antigens to activate effector T cells is greatly decreased. The depression of cell-mediated immunity results in an inability to resist intracellular microorganisms and cancer.

6. The booster shot stimulates a secondary (memory) response, resulting in the formation of large amounts of antibodies and memory cells. Consequently, there is better, longer-lasting immunity.

Chapter 20

1. When you sleep with your mouth open, less air passes through the nasal passages. This is especially true when the nasal passages are plugged because you have a cold. As a consequence, inspired air is not humidified and warmed. The dry air dries and irritates the throat. Air breathed through the mouth while running in very cold weather is not warmed and humidified, which can irritate the throat, larynx, and trachea.

2. When food moves down the esophagus, the normally collapsed esophagus expands. If the cartilage rings were solid, expansion of the esophagus, and therefore swallowing, would be more difficult.

3. The conducting zone consists of the nose (i.e., external nose, nostrils, nasal cavity, and choanae), pharynx (i.e., nasopharynx, oropharynx, and laryngopharynx), larynx, and most of the tracheobronchial tree (i.e., trachea, main bronchi, lobar bronchi, segmental bronchi, bronchioles, and terminal bronchioles). The respiratory zone consists of the parts of the tracheobronchial tree distal to the terminal bronchioles—the respiratory bronchioles (with a few alveoli) and alveolar ducts (with many alveoli).

4. During respiratory movements, the parietal and visceral pleurae slide over each other. Normally, the pleural fluid in the pleural cavities lubricates the surfaces of these membranes. When the pleural membranes are inflamed, their surfaces become roughened. The rough surfaces rub against each other and create an intense pain. The pain is increased when a person takes a deep breath because the movement of the membranes is greater than during normal breaths.

5. Relaxation of the abdominal muscles allows the abdominal organs to move inferiorly. Thus, it is easier for the diaphragm to increase the volume of the thoracic cavity.

6. Contraction of the muscles of inspiration causes the volume of the thoracic cavity to increase. According to Boyle's law, as volume increases, pressure decreases.

7. If resistance to airflow increases, the only way to maintain the same airflow is to increase the pressure difference between the alveoli and outside the body. According to Boyle's law, the greater the change in volume, the greater the change in pressure. Increased work by the muscles of respiration can cause greater changes in thoracic volume, resulting in greater pressure differences.

8. The tube should apply suction. In order for the lung to expand, pressure in the alveoli must be greater than the pressure in the pleural cavity. This can be accomplished by lowering the pressure in the pleural cavity through suction. Applying air under pressure would make the pressure in the pleural cavity greater than the pressure in the alveoli, which would keep the alveoli collapsed.

9. The alveolar ventilation is 4200 mL/min ($12 \times [500 - 150]$). During exercise, the alveolar ventilation is 88,800 mL/min ($24 \times [4000 - 300]$), a 21-fold increase. The increased air exchange increases P_{O_2} and decreases P_{CO_2} in the alveoli, thus increasing gas exchange between the alveoli and the blood.

10. The air the diver is breathing has a greater total pressure than atmospheric pressure at sea level. Consequently, the partial pressure of each gas in the air increases. According to Henry's law, as the partial pressure of a gas increases, the amount (concentration) of gas dissolved in the liquid (e.g., body fluids) with which the gas is in contact increases. When the diver suddenly ascends, the partial pressure of gases in the body returns toward sea level atmospheric pressure. As a result, the amount (concentration) of gas that can be dissolved in body fluids suddenly decreases. When the fluids can no longer hold all the gas, the gas bubbles form.

11. During exercise, skeletal muscle cells increase oxygen use in order to produce the ATP molecules required for muscle contraction. The P_{O_2} inside the cells therefore decreases, which increases the diffusion gradient for oxygen, resulting in increased movement of oxygen into the cells. The aerobic production of ATP also produces carbon dioxide. The P_{CO_2} inside the

cell therefore increases, which increases the diffusion gradient for carbon dioxide, resulting in increased movement of carbon dioxide out of cells.

12. Blood pH decreases because of increased CO_2 production by the actively working skeletal muscles. As carbon dioxide levels increase, carbonic anhydrase catalyzes the conversion of CO_2 to carbonic acid, which in turn dissociates into H^+ and HCO_3^-. Therefore, pH declines. In addition, pH declines because some of the skeletal muscle fibers produce lactic acid as a result of anaerobic respiration. Oxygen delivery to the skeletal muscles increases during the race. Under acidic conditions, hemoglobin's affinity for oxygen decreases and more oxygen is released into the tissues (the oxygen–hemoglobin dissociation curve shifts to the right).

13. Remember that the oxygen–hemoglobin dissociation curve normally shifts to the right in tissues. The shift of the curve to the left caused by CO reduces hemoglobin's ability to release oxygen to tissues, which contributes to the detrimental effects of CO poisoning. In the lungs, the shift to the left may slightly increase the hemoglobin's ability to pick up oxygen, but this effect is offset by its decreased ability to release oxygen to tissues.

14. Hyperventilation decreases blood carbon dioxide levels, causing an increase in blood pH. Holding one's breath increases blood carbon dioxide levels and decreases blood pH.

15. a. Normal arterial Po_2 is 95 mm Hg and Pco_2 is 40 mm Hg. A Po_2 of 60 mm Hg and a Pco_2 of 30 mm Hg are below normal.
    b. Asthma causes constriction of the terminal bronchioles, which reduces alveolar ventilation and alveolar Po_2. Consequently, oxygen exchange across the respiratory membrane decreases and blood Po_2 decreases. The carotid and aortic body chemoreceptors detect the decrease in blood Po_2, resulting in increased stimulation of the respiratory centers and an increased respiratory rate.
    c. Constriction of the terminal bronchioles increases resistance to air flow. Will attempted to compensate for the increased resistance by breathing more forcefully.
    d. No, Will's arterial Po_2 and Po_2 are abnormally low. Despite the forceful breathing, Will's tidal volume was so low that his increased respiratory rate did not increase alveolar ventilation enough to restore homeostasis.
    e. Will's arterial Po_2 was lower than normal because of the low alveolar Po_2. The low arterial Po_2 stimulated him to hyperventilate. Despite the hyperventilation, alveolar ventilation was insufficient to restore normal arterial Po_2. The hyperventilation, however, effectively removed carbon dioxide from the blood because the diffusion coefficient for carbon dioxide is 20 times greater than that of oxygen. As a result, arterial Pco_2 is lower than normal.
    f. As Will's arterial Pco_2 decreased, his blood pH would increase (become more alkaline). Normally, if just blood pH increased, that would be detected by the medullary chemoreceptors, leading to a decrease in ventilation (see figure 23.22). However, respiration results from the integration of many stimuli (see figure 20.18). During Will's asthma attack, the effect of the low arterial

Po_2 on the respiratory centers was greater than the effect of the decreased blood pH.
    g. Inhaled β-adrenergic agents cause relaxation of the smooth muscle in the bronchi and bronchioles. Bronchodilation improves airflow and facilitates the return of blood gases to homeostatic levels. Inhaled glucocorticoids help reduce the inflammatory responses and mucus buildup associated with asthma.

16. When a person hyperventilates, Pco_2 in the blood decreases. Consequently, carbon dioxide moves out of cerebrospinal fluid into the blood. As carbon dioxide levels in cerebrospinal fluid decrease, H^+ and HCO_3^- combine to form carbonic acid, which forms carbon dioxide. The result is a decrease in H^+ concentration in cerebrospinal fluid and decreased stimulation of the respiratory center by the chemosensitive area. Until blood Pco_2 levels increase, the chemosensitive area is not stimulated and apnea results.

17. Through touch, thermal, or pain receptors, the respiratory center can be stimulated to cause a sudden inspiration of air.

Chapter 21

1. The moist stratified squamous epithelium of the oropharynx and the laryngopharynx protects these regions from abrasive food when it is first swallowed. The ciliated pseudostratified epithelium of the nasopharynx helps move mucus produced in the nasal cavity and the nasopharynx into the oropharynx and esophagus. It is not as necessary to protect the nasopharynx from abrasion because food does not normally pass through this cavity.

2. Sipping a liquid is the voluntary phase of swallowing. The movement of fluid into the pharynx initiates the involuntary pharyngeal phase of swallowing. As the fluid is moved into the esophagus, the esophageal phase of swallowing is initiated, resulting in peristalsis, which moves the "stuck" food into the stomach.

3. Peptides stimulate chief cells to secrete pepsinogen and parietal cells to secrete hydrochloric acid. The acid converts the inactive pepsinogen into pepsin, which digests proteins (peptides). Thus, peptides stimulate the stomach to digest peptides. The homeostasis "set point" for peptides is no peptides in the stomach. When peptides appear, the negative-feedback mechanism causes them to be digested, which returns the system to its set point.

4. After a heavy meal, blood pH may increase because, as bicarbonate ions pass from the cells of the stomach into the extracellular fluid, the pH of the extracellular fluid increases. As the extracellular fluid exchanges ions with the blood, the blood pH also increases.

5. As long as bile salts are reabsorbed in the ileum, they stimulate additional bile secretion in the liver. This process stops when the duodenum empties, which causes secretin and cholecystokinin levels to decrease. Bile secretion decreases and the sphincters regulating the movement of bile into the duodenum close. Consequently, bile no longer enters the duodenum, which "breaks" the positive-feedback loop of bile salts stimulating additional bile secretion.

6. Secretin production and its stimulation of bicarbonate ion secretion constitute a negative-feedback mechanism because, as the pH of the chyme in the duodenum decreases as a result of

the presence of acid, secretin causes an increase in bicarbonate ion secretion, which increases the pH and restores the proper pH balance in the duodenum.

7. a. The spinal cord injury has disrupted ascending and descending nerve tracts. Dan has no awareness of the need to defecate, and he has lost the ability to relax or contract the external anal sphincter.
    b. The defecation reflex center in the conus medullaris is still functional. Distention caused by an enema can activate the defecation reflex.
    c. After breakfast, the gastrocolic and duodenocolic reflexes can cause mass movement of feces into the rectum, causing distention and activation of the defecation reflex. After breakfast may be better than after other meals because there is more time for feces to have moved into the colon.
    d. Straining increases abdominal pressure, which stimulates reflexive contraction of the external anal sphincter. This reflex is still functional in Dan. During straining in a noninjured person, the brain is able to inhibit this reflex through a descending nerve tract. Because of the disruption of Dan's descending nerve tracts, however, his brain no longer can inhibit the reflex, so, when Dan strains, the external anal sphincter contracts.
    e. Even if the defecation reflex center is damaged, the local reflexes of the enteric nervous system are still functional. Although this local component of the defecation reflex is weak and lacks voluntary control, defecation can still occur with assistance from large enemas and strong cathartics. Cathartics function by increasing fecal volume, softening feces, or stimulating intestinal muscle contractions.

Chapter 22

1. If vitamins are broken down during the process of digestion, their structures are destroyed and, as a result, their ability to function is lost.

2. The Daily Value for carbohydrate is 300 g/day. One serving of food with 30 g of carbohydrate has a % Daily Value of 10% (30/300 = .10, or 10%).

3. On an 1800 kcal/day diet, the total percentage of Daily Values for energy-producing nutrients should add up to no more than 90%, because 1800/2000 = 0.9, or 90%.

4. If the electron of the electron-transport chain cannot be donated to oxygen, the entire electron-transport chain and the citric acid cycle stop, no ATP can be produced aerobically, and death results because too little energy is available for the body to perform vital functions. Anaerobic respiration is not adequate to provide all the energy needed to maintain human life, except for a short time.

5. The kilocalories in a beer or cola is about 145 kcal. It takes about 1.5 h to burn off these kilocalories while watching TV (145/95 = 1.5 h) and about 15 min while jogging (145/580 = 0.25 h). Although it may be difficult to burn off kilocalories through exercise, it is clear that exercise can significantly increase kilocalorie use.

6. When muscles contract, they use ATP. As a result of the chemical reactions necessary to synthesize ATP, heat is also produced. During exercise, the large amounts of heat can raise body temperature, and we feel warm. Shivering consists of small, rapid muscle contractions that produce heat in an effort to prevent a decrease in body temperature in the cold.

7. Vasoconstriction reduces blood flow to the skin, which reduces skin temperature because less warm blood from the deeper parts of the body reaches the skin. As the difference in temperature between the skin and the environment decreases, less loss of heat occurs. If the skin temperature decreases too much, however, dilation of blood vessels to the skin occurs, which prevents the skin from becoming so cold that it is damaged.

Chapter 23

1. Even though hemoglobin is a smaller molecule than albumin, it does not normally enter the filtrate because hemoglobin is contained within red blood cells and these cells cannot pass through the filtration membrane. Conditions that cause red blood cells to rupture (hemolysis) result in large amounts of hemoglobin entering the filtrate.

2. The skin normally prevents water loss because it is relatively impermeable to water (see chapter 5). A large surface area burn results in fluid loss because of the loss of skin. Fluid loss results in decreased blood volume and blood pressure. In response, sympathetic activity increases, causing increased heart rate and pale skin (see chapter 18). Blood pressure is restored to the low normal range. The increased sympathetic activity also causes intense vasoconstriction, including vasoconstriction of arteries and arterioles supplying the kidneys. Constriction of the renal arteries and the renal afferent arterioles decreases the blood pressure in the glomeruli. As a consequence, the glomerular capillary pressure, filtration pressure, glomerular filtration rate, and urine production decrease. After an IV is administered and blood volume increases, the blood pressure returns to normal and renal arteries and arterioles once again dilate; thus, glomerular capillary pressure, filtration pressure, and urine production increase to normal levels.

3. Without the normal active transport of Na$^+$, the concentration of Na$^+$ and ions symported with them remains elevated in the nephron. This increases the osmotic concentration of the filtrate and less water moves out of the filtrate. The result is an increased volume of urine.

4. a. Billy has defective ADH receptors. When ADH binds to the receptor, the G protein mechanism responsible for inserting aquaporin-2 molecules is not activated. Therefore, the distal convoluted tubules and collecting ducts are not permeable to water, resulting in a large volume of dilute urine. Billy's blood Na$^+$ levels are high because he is not reabsorbing normal amounts of water from the filtrate.

   b. The infants' plasma levels of ADH will increase during the water deprivation test. Water deprivation results in increased blood osmolality, which stimulates ADH secretion. The hallmark of NDI is a normal secretion of ADH from the pituitary followed by an abnormal response in the kidney. If Billy had CDI, his ADH plasma levels of ADH would be abnormally low to nondetectable because central diabetes insipidus results from a deficiency in ADH secretion.

   c. The abnormal aquaporin-2 molecules would inhibit water reabsorption. The water would stay in the distal convoluted tubules and collecting ducts to exit the body as part of urine.

   d. Water that moves into the collecting duct cells would diffuse into the interstitial fluid at a reduced rate because of the reduced number of aquaporins. Urine volume would increase because of water retention in the distal convoluted tubule and collecting ducts. Urine concentration would decrease because of dilution by the water. ADH can cause an increase in the number of aquaporin-2 water channel proteins in the apical membrane of the distal convoluted tubules and collecting ducts, but not in aquaporin-3 or aquaporin-4 channels in the basal membranes. These channels determine the permeability of the basal membranes to water. Therefore, ADH would not be an effective treatment. The net effect would be the production of a large volume of dilute urine as in nephrogenic diabetes insipidus caused by an abnormal aquaporin-2 water channel protein.

5. A decrease in blood pressure is most likely caused by a decrease in blood volume. Mechanisms that increase blood volume help restore blood pressure. Increased aldosterone secretion increases Na$^+$ reabsorption in the kidneys. Increased ADH secretion increases water reabsorption from the kidneys. Together, the increased movement of Na$^+$ and water into the blood increases blood volume while maintaining blood osmolality. An increased sensation of thirst results in increased fluid intake, and increased salt appetite increases Na$^+$ intake. This increases blood volume while maintaining blood osmolality.

6. Ingesting the salty potato chips increases blood osmolality, which stimulates the sensation of thirst and increases ADH secretion. The increased ADH promotes water reabsorption by the kidneys in order to maintain normal blood osmolality. Therefore, urine volume decreases. The increased reabsorption of water increases blood volume, which increases blood pressure. The increased blood pressure inhibits renin secretion and, therefore, aldosterone secretion. Decreased aldosterone secretion results in less Na$^+$ reabsorption and more Na$^+$ are lost in the urine. The increased blood pressure also stimulates increased ANH secretion, which decreases the reabsorption of Na$^+$ from the filtrate. The retention of Na$^+$ in the filtrate increases the concentration of Na$^+$ in the urine. The final result is the production of a small volume of concentrated urine.

7. The urethra of females is much shorter than the urethra of males. In addition, the opening of the urethra in females is closer to the anus, which is a potential source of bacteria. The female urinary bladder is therefore more accessible to bacteria from the exterior. This accessibility is a major reason that urinary bladder infections are more common in females than in males.

8. During exercise, the amount of water lost is increased because of increased evaporation from the respiratory system, increased insensible perspiration, and increased sweat. The amount of water lost in the form of sweat can increase substantially. The amount of urine formed decreases during exercise.

9. During hemorrhagic shock, blood pressure decreases and visceral blood vessels constrict, including those supplying the kidneys. As a consequence, blood flow to the kidneys decreases and the glomerular capillary pressure decreases dramatically. Therefore, filtration pressure decreases, glomerular filtration rate

decreases, and urine production decreases, resulting in the production of a small volume of concentrated urine. The decreased blood pressure stimulates the ADH mechanism, resulting in increased ADH secretion; the renin–angiotensin–aldosterone, resulting in increased aldosterone secretion, and the ANH mechanism, resulting in decreased ANH secretion. Increased aldosterone and decreased ANH cause increased Na$^+$ reabsorption, and increased ADH causes increased water reabsorption. The movement of Na$^+$ and water into the blood helps maintain homeostasis by increasing blood volume and blood pressure.

10. Hyperventilation results in a greater than normal rate of carbon dioxide loss from the circulatory system. Because carbon dioxide is lost from the circulatory system, H$^+$ concentration decreases and the pH of body fluids increases. Breathing into a paper bag corrects for the effects of hyperventilation because the person rebreathes air that has a higher concentration of carbon dioxide. The result is an increase in carbon dioxide in the body. Consequently, the H$^+$ concentration increases and pH decreases toward normal levels.

11. Elevated blood carbon dioxide levels cause an increase in H$^+$ and a decrease in blood pH due to the following reaction:

$$CO_2 + H_2O \rightleftharpoons H_2CO_3 \rightleftharpoons H^+ + HCO_3^-$$

However, the kidney plays an important role in the regulation of blood pH. The kidney's rate of H$^+$ secretion into the urine and reabsorption of HCO$_3^-$ increase. This helps prevent high blood H$^+$ levels and low blood pH in Mr. Puffer. The blood levels of HCO$_3^-$ would be greater than normal because of the increased reabsorption of HCO$_3^-$.

Chapter 24

1. The prostate gland is anterior to the wall of the rectum. A finger inserted into the rectum can palpate the prostate gland through the rectal wall.

2. Clotting helps keep the sperm cells within the female reproductive tract. Once the semen is in contact with the uterus, breaking down the clotting proteins facilitates the movement of sperm cells into the uterus.

3. Primary and secondary sexual characteristics develop in response to testosterone. With inadequate testosterone, the male genitalia do not develop and remain juvenile. Normal adult sexual behavior does not develop, nor do the adult pattern of hair growth, skin changes, laryngeal changes, increased metabolism, and muscle and bone development.

4. a. Progesterone inhibits GnRh release from the hypothalamus. Consequently, secretion of LH and FSH from the anterior pituitary decreases. The inhibition of LH lowers LH levels and prevents the LH surge, which is necessary for ovulation. Decreased FSH inhibits follicle development, which decreases estrogen production by the ovaries. Decreased estrogen inhibits the LH surge because normally increased estrogen around the time of ovulation is part of a positive-feedback mechanism that produces the LH surge. Inhibition of follicle development and ovulation prevents pregnancy.

   b. After 21 days, not taking progesterone and estrogen results in a decrease in hormone level, which induces menstruation.

c. Taking the pill every day at the same time reinforces the habit and makes it unnecessary for a woman to remember when to resume taking the pills containing the hormones.

d. The pill should be started at or near the beginning of the cycle because this is the best time to prevent follicle maturation. Later in the cycle, there is a risk that a follicle has already matured and pregnancy could result. It is recommended for the first 2 weeks of taking the pill that other methods of contraception be used when having sex. Starting the pill on Sunday is a way of determining when a woman has her period. As a matter of preference, many women want to have their period during the week rather than the weekend. Therefore, a Sunday-starter would stop taking the pill containing hormones 3 weeks after she starts and would have her period in the following week.

5. After menopause, estrogen and progesterone are produced by the ovary in only small amounts. Consequently, estrogen and progesterone levels in the blood are low. These hormones have a negative-feedback effect on the secretion of GnRH, FSH, and LH. Therefore, in the absence of estrogen and progesterone, GnRH, FSH, and LH are secreted in greater amounts and their blood levels increase. However, there is no LH or FSH surge. The average concentration of LH and FSH is greater than the levels that occur either before or after ovulation.

Chapter 25

1. Two primitive streaks on one embryonic disk result in the development of two embryos. If the two primitive streaks are touching or are very close to each other, the embryos may be joined. This condition is called conjoined (Siamese) twins.

2. Because the early embryonic heart is a simple tube, blood must be forced through the heart in almost a peristaltic fashion, and the contraction begins in the sinus venosus. If the sinus venosus did not contract first, blood could flow in the opposite direction.

3. The lack of müllerian-inhibiting hormone allows the paramesonephric duct system to develop into the internal female reproductive structures, particularly the uterus and the uterine tube. Assuming that testosterone secretion is normal, the internal male reproductive system will also develop. In the presence of testosterone, the external male genitalia also develop.

4.

	Clinical Age	Developmental Age
Fertilization	14 days	0 days
Implantation	21 days	7 days
Fetal period	70 days	56 days
Parturition	280 days	266 days

5. Elevation of Ca^{2+} levels might cause the uterine muscles to contract tetanically. This tetanic contraction might compress blood vessels and cut the blood supply to the fetus. Hypercalcemia can also result in arrhythmias and muscle weakness (see chapter 23). The doctor would therefore not administer Ca^{2+} to the woman in labor but could give oxytocin, which strengthens contractions but is less likely to produce tetany.

6. Nursing stimulates the release of oxytocin from the mother's posterior pituitary gland, which is responsible for milk letdown. Oxytocin can also cause uterine contractions and cramps.

7. Genotype *DD* (homozygous dominant) has the polydactyly phenotype, genotype *Dd* (heterozygous) has the polydactyly phenotype, and genotype *dd* (homozygous recessive) has the normal phenotype.

8. None. One in two will be homozygous normal and one in two will be normal heterozygous carriers.

9. The probability of a girl with Turner syndrome (XO; that is, with one X chromosome and no other sex chromosome) having hemophilia if her mother is a carrier for hemophilia is the same as for a male because she has only one X chromosome. If her mother is a carrier for hemophilia ($X^H X^h$), the daughter will be either $X^H O$ (normal) or $X^h O$ (hemophiliac). The probability is 1/2, or 50%.

Glossary

Many of the words in this glossary and the text are followed by a simplified phonetic spelling showing pronunciation. The pronunciation key reflects standard clinical usage as presented in *Stedman's Medical Dictionary* (27th edition), a leading reference volume in the health sciences.

The phonetic system used is a basic one and has only a few conventions:

- Two diacritical marks are used; the macron (ˉ) for long vowels; and the breve (˘) for short vowels.
- Principal stressed syllables are followed by a prime ('); monosyllables do not have a stress mark.
- Other syllables are separated by hyphens.

The following pronunciation key provides examples and consonant sounds encountered in the phonetic system. No attempt has been made to accommodate the slurred sounds common in speech or regional variations in speech sounds. Note that a vowel with a breve (˘) is used for the indefinite vowel sound of the schwa (ə). Native pronunciation of foreign words is approximated as closely as possible.

Pronunciation Key

Vowels

ā	day, care, gauge	ah	father
a	mat, damage	aw	fall, cause, raw
ă	about, para	ē	be, equal, ear
ĕ	taken, genesis	k	kept
e	term, learn	ks	tax
ī	pie	kw	quit
ĭ	pit, sieve, build	l	law
ō	note, for so,	m	me
o	not, oncology, ought	n	no
oo	food	ng	ring
ow	cow, out	p	pan
oy	troy, void	r	rot
ū	unit, curable	s	so, miss
ŭ	cut	sh	should

Consonants

b	bad	t	ten
ch	child	th	thin, with
d	dog	v	very
dh	this, smooth	w	we
f	fit	y	yes
g	got	z	zero
h	hit	zh	azure,
j	jade		measure

In some words the initial sound is not that of the initial letter(s), or the initial letter(s) is not sounded or has a different sound, as in the following examples:

aerobe (ar'ob)	phthalein (thal'e-in)
eimuria (ime're-a)	pneumonia (nu-mo'ne-a)
gnathic (nath'ik)	psychology (si-kol'o-je)
knuckle (nuk-l)	ptosis (to'sis)
oedipism (ed'i-pizm)	xanthoma (zan-tho'ma)

A

A band Length of the myosin myofilament in a sarcomere.

abdomen (ab-dō'men, ab'dō-men) Belly, between the thorax and the pelvis.

abduction (ab-dŭk'shŭn) [L., *abductio*, take away] Movement away from the midline.

absolute refractory period (ab'sō-loot rē-frak'tōr-ē) Portion of the action potential during which the membrane is insensitive to all stimuli, regardless of their strength.

absorptive cell (ab-sōrp'tiv) Cell on the surface of villi of the small intestines and the luminal surface of the large intestine that is characterized by having microvilli; secretes digestive enzymes and absorbs digested materials on its free surface.

absorptive state Immediately after a meal when nutrients are being absorbed from the intestine into the circulatory system.

accommodation (ă-kom'ŏ-dā'shŭn) [L., *ac* + *commodo*, to adapt] Ability of electrically excitable tissues, such as nerve or muscle cells, to adjust to a constant stimulus so that the magnitude of the local potential decreases through time; also called adaptation.

acetabulum (as-ĕ-tab'ū-lŭm) [L., shallow vinegar vessel or cup] Cup-shaped depression on the external surface of the coxa.

acetylcholine (ACh) (as-e-til-kō'lēn) Neurotransmitter substance released from motor neurons, all preganglionic neurons of the parasympathetic and sympathetic divisions, all postganglionic neurons of the parasympathetic division, some postganglionic neurons of the sympathetic division, and some central nervous system neurons.

acetylcholinesterase (as'ē-til-kō-lin-es'ter-ās) Enzyme found in the synaptic cleft that causes the breakdown of acetylcholine to acetic acid and choline, thus limiting the stimulatory effect of acetylcholine.

Achilles tendon *See* calcaneal tendon.

acid (as'id) Molecule that is a proton donor; any substance that releases hydrogen ions (H⁺).

acidic Solution containing more than 10^{-27} mol of hydrogen ions per liter; has a pH less than 7.

acinus; pl. **acini** (as'i-nŭs, as'ĭ-nī) [L., berry, grape] Grape-shaped secretory portion of a gland. The terms *acinus* and *alveolus* are sometimes used interchangeably. Some authorities differentiate the terms: Acini have a constricted opening into the excretory duct, whereas alveoli have an enlarged opening.

acromion (ă-krō'mē-on) [Gr., *akron*, extremity + *omos*, shoulder] Bone comprising the tip of the shoulder.

acrosome (ak'rō-sōm) [Gr., *akron*, extremity + *soma*, body] Cap on the head of the spermatozoon, with hydrolytic enzymes that help the spermatozoon penetrate the ovum.

actin myofilament (ak'tin) Thin myofilament within the sarcomere; composed of two F actin molecules, tropomyosin, and troponin molecules.

action potential [L., *potentia*, power, potency] Change in membrane potential in an excitable tissue that acts as an electric signal and is propagated in an all-or-none fashion.

activation energy (ak-ti-vā'shŭn) Energy that must be added to molecules to initiate a reaction.

active site Portion of an enzyme in which reactants are brought into close proximity and that plays a role in reducing activation energy of the reaction.

active tension Tension produced by the contraction of a muscle.

active transport Carrier-mediated process that requires ATP and can move substances against a concentration gradient.

adaptive immunity Immune status in which there is an ability to recognize, remember, and destroy a specific antigen.

adenohypophysis (ad'ĕ-nō-hī-pof'ĭ-sis) Portion of the hypophysis derived from the oral ectoderm; also called anterior pituitary.

adenosine diphosphate (ADP) (ă-den'ō-sēn) Adenosine, an organic base, with two phosphate groups attached to it. Adenosine diphosphate combines with a phosphate group to form adenosine triphosphate.

adenosine triphosphate (ATP) Adenosine, an organic base, with three phosphate groups attached to it. Energy stored in ATP is used in nearly all of the endergonic reactions in cells.

adenylate cyclase (ad'e-nil sīklās) An enzyme acting on ATP to form 3',5'-cyclic AMP plus pyrophosphate (two phosphate groups). A crucial step in the regulation and formation of the intracellular chemical signal 3',5'-cyclic AMP.

adipocyte (ad'i-pō-sīt) Fat cell.

adipose (ad'i-pōs) [L., *adeps*, fat] Fat.

adrenal gland (ă-drē'năl) [L., *ad,* to + *ren,* kidney] Located near the superior pole of each kidney, it is composed of a cortex and a medulla. The adrenal medulla is a highly modified sympathetic ganglion that secretes the hormones epinephrine and norepinephrine; the cortex secretes aldosterone and cortisol as its major secretory products. Also called suprarenal gland.

adrenergic neuron (ad-rĕ-ner′jik) Nerve fiber that secretes norepinephrine (or epinephrine) as a neurotransmitter substance.

adrenergic receptor (ad-rĕ-ner′jik) Receptor molecule that binds to adrenergic agents, such as epinephrine and norepinephrine.

adrenocorticotropic hormone (ACTH) (ă-drē′nō-kōr′ti-kō-trō′pik) Hormone of the adenohypophysis that governs the nutrition and growth of the adrenal cortex, stimulates it to functional activity, and causes it to secrete cortisol.

adventitia (ad-ven-tish′ă) [L., *adventicius,* coming from abroad, foreign] Outermost covering of any organ or structure that is properly derived from outside the organ and does not form an integral part of the organ.

aerobic respiration (ār-ō′bik) Breakdown of glucose in the presence of oxygen to produce carbon dioxide, water, and approximately 38 ATPs; includes glycolysis, the citric acid cycle, and the electron-transport chain.

afferent arteriole (af′er-ent) Branch of an interlobular artery of the kidney that conveys blood to the glomerulus.

afferent division Nerve fibers that send impulses from the periphery to the central nervous system.

agglutination (ă-gloo-ti-nā-shŭn) [L., *ad,* to + *gluten,* glue] Process by which blood cells, bacteria, or other particles are caused to adhere to one another and form clumps.

agglutinin (ă-gloo′ti-nin) Antibody that binds to an antigen and causes agglutination.

agglutinogen (ă-gloo-tin′ō-jen) Antigen on surface of red blood cells that can stimulate the production of antibodies (agglutinins) that combine with the antigen and cause agglutination.

agranulocyte (ă-gran′ū-lō-sīt) Nongranular leukocyte (monocyte or lymphocyte).

ala; pl. **alae** (ā′lă, ā′lē) [L., a wing] Wing-shaped structure.

aldosterone (al-dos′ter-ōn) Steroid hormone produced by the zona glomerulosa of the adrenal cortex that facilitates potassium exchange for sodium in the distal renal tubule, causing sodium reabsorption and potassium and hydrogen secretion.

alkaline (al′kă-līn) Solution containing less than 10^{-7} mol of hydrogen ions per liter; has a pH greater than 7.0.

alkalosis (al-kă-lō′sis) Condition characterized by blood pH of 7.45 or above.

allantois (ă-lan′tō-is) Tube extending from the embryonic hindgut into the umbilical cord; forms the urinary bladder.

allele (ă-lēl′) [Gr., *allelon,* reciprocally] Any one of a series of two or more different genes that may occupy the same position or locus on a specific chromosome.

all-or-none principle When a stimulus is applied to a cell, an action potential is either produced or not. In muscle cells, the cell either contracts to the maximum extent possible (for a given condition) or does not contract.

alternative pathway Part of the nonspecific immune system for activation of complement.

alveolar duct (al-vē′ō-lăr) Part of the respiratory passages beyond a respiratory bronchiole; from it arise alveolar sacs and alveoli.

alveolar gland Gland in which the secretory unit has a saclike form and an obvious lumen.

alveolar sac Two or more alveoli that share an opening.

alveolus; pl. **alveoli** (al-vē′ō-lŭs, al-vē′ō-lī) Cavity. Examples include the sockets into which teeth fit, the endings of the respiratory system, and the terminal endings of secretory glands.

amino acid (ă-mē′nō) Class of organic acids that constitute the building blocks for proteins.

amplitude-modulated signal (am′pli-tood) Signal that varies in magnitude or intensity, such as with large versus small concentrations of hormones.

ampulla (am-pul′lă, am-pul′lē) [L., two-handled bottle] Saclike dilatation of a semicircular canal; contains the crista ampullaris. Wide portion of the uterine tube between the infundibulum and the isthmus.

amygdala (a-mig′da-la, -lē) [L. fr. Gr., *amygdale,* almond] Nucleus in the temporal lobe of the brain, amygdaloid nucleus; also called amydaloid nuclear complex.

amylase (am′il-ās) One of a group of starch-splitting enzymes that cleave starch, glycogen, and related polysaccharides.

anabolism (ă-nab′ō-lizm) [Gr., *anabole,* a raising up] All of the synthesis reactions that occur within the body; requires energy.

anaerobic respiration (an-ār-ō′bik) Breakdown of glucose in the absence of oxygen to produce lactic acid and two ATPs; consists of glycolysis and the reduction of pyruvic acid to lactic acid.

anal canal Terminal portion of the digestive tract.

anal triangle Posterior portion of the perineal region through which the anal canal opens.

analgesic (an-al-jē′zik) Compound capable of producing analgesia, without producing anesthesia or loss of consciousness, characterized by reduced response to painful stimuli.

anaphase (an′ă-fāz) Time during cell division when chromatids divide (or, in the case of first meiosis, when the chromosome pairs divide).

anastomoses (ă-nas′tō-mō′sez) Natural communication, direct or indirect, between two blood vessels or other tubular structures. An opening created by surgery, trauma, or disease between two or more normally separate spaces or organs.

anatomical dead air space Volume of the conducting airways from the external environment down to the terminal bronchioles.

androstenedione (an-drō-stēn-dī′ōn) Andro-genic steroid of weaker potency than testosterone; secreted by the testis, ovary, and adrenal cortex.

anencephaly (an′en-sef′ă-lē) [Gr., *an* + *enkephalos,* no brain] Defective development of the brain and absence of the bones of the cranium. Only a rudimentary brainstem and a trace of basal ganglia are present.

aneurysm (an′ū-rizm) [Gr., *eurys,* wide] Dilated portion of an artery.

angiotensin I (an-jē-ō-ten′sin) Peptide derived when renin acts on angiotensinogen.

angiotensin II Peptide derived from angiotensin I; stimulates vasoconstriction and aldosterone secretion; also called active angiotensin.

anion (an′ī-on) Ion carrying a negative charge.

antagonist (an-tag′ŏ-nist) Muscle that works in opposition to another muscle.

anterior chamber Chamber of the eye between the cornea and the iris.

anterior interventricular sulcus Groove on the anterior surface of the heart, marking the location of the septum between the two ventricles.

anterior pituitary *See* adenohypophysis.

antibody (an′tē-bod-ē) Protein found in the plasma that is responsible for humoral immunity; binds specifically to antigen.

antibody-mediated immunity Immunity due to B cells and the production of antibodies.

anticoagulant (an′tē-kō-ag′ū-lant) Agent that prevents coagulation.

antidiuretic hormone (ADH) (an′tē-dī-ū-ret′ik) Hormone secreted from the neurohypophysis that acts on the kidney to reduce the output of urine; also called vasopressin because it causes vasoconstriction.

antigen (an′ti-jen) [anti(body) + Gr., *gen,* producing] Substance that induces a state of sensitivity or resistance to infection or toxic substances after a latent period; substance that stimulates the specific immune system; also called epitope.

antigenic determinant (an-ti-jen′ik) Specific part of an antigen that stimulates an immune system response by binding to receptors on the surface of lymphocytes; also called epitope.

antithrombin (an-tē-throm′bin) Substance that inhibits or prevents the effects of thrombin so that blood does not coagulate.

antrum (an′trŭm) [Gr., *antron,* a cave] Cavity of an ovarian follicle filled with fluid containing estrogen.

anulus fibrosus (an′ū-lŭs fī-brō′sus) [L., fibrous ring] Fibrous material forming the outer portion of an intervertebral disk.

anus (ā′nŭs) Lower opening of the digestive tract through which fecal matter is extruded.

aorta (ā-ōr′tă) [Gr., *aorte* from *aeiro,* to lift up] Large, elastic artery that is the main trunk of the systemic arterial system; carries blood from the left ventricle of the heart and passes through the thorax and abdomen.

aortic arch [L., bow] Curve between the ascending and descending portions of the aorta.

aortic body One of the smallest bilateral structures, similar to the carotid bodies, attached to a small branch of the aorta near its arch; contains chemoreceptors that respond primarily to decreases in blood oxygen; less sensitive to decreases in blood pH or increases in carbon dioxide.

apex (ā′peks) [L., summit or tip] Extremity of a conical or pyramidal structure. The apex of the heart is the rounded tip directed anteriorly and slightly inferiorly.

Apgar score Named for U.S. anesthesiologist Virginia Apgar (1909–1974). Evaluation of a newborn infant's physical status by assigning numerical values to each of five criteria: appearance (skin color), pulse (heart rate), grimace (response to stimulation), activity (muscle tone), and respiratory effort; a score of 10 indicates the best possible condition.

apical ectodermal ridge Layer of surface ectodermal cells at the lateral margin of the embryonic limb bud; stimulates growth of the limb.

apical foramen [L., aperture] Opening at the apex of the root of a tooth, gives passage to the nerve and blood vessels.

apocrine gland (ap′ō-krin) [Gr., *apo,* away from + *krino,* to separate] Gland whose cells contribute cytoplasm to its secretion (e.g., mammary glands). Sweat glands that produce organic secretions traditionally are called apocrine. These sweat glands, however, are actually merocrine glands.

appendicular skeleton (ap′en-dik′ū-lăr) The portion of the skeleton consisting of the upper limbs and the lower limbs and their girdles.

appositional growth (ap-ō-zish′ŭn-al) [L., *ap* + *pono,* to put or place] To place one layer of bone, cartilage, or other connective tissue against an existing layer.

aqueous humor (ak′wē-ŭs, ā′kwē-ŭs) Watery, clear solution that fills the anterior and posterior chambers of the eye.

arachnoid (ă-rak′noyd) [Gr., *arachne*, spider, cobweb] Thin, cobweb-appearing meningeal layer surrounding the brain; the middle of the three layers.

arcuate artery (ar′kū-āt) Artery that originates from the interlobar arteries of the kidney and forms an arch between the cortex and medulla of the kidney.

areola (ă-rē′ō-lă, -lē) [L., area] Circular, pigmented area surrounding the nipple; its surface is dotted with little projections caused by the presence of the areolar glands beneath.

areolar gland (ă-rē′ō-lăr) Gland forming small, rounded projections from the surface of the areola of the mamma.

arrectores pilorum; pl. arrector pili (ă-rek-tō′rez pī-lōr′um, ă-rek′tōr pī′lī) [L., that which raises; hair] Smooth muscle attached to the hair follicle and dermis that raises the hair when it contracts.

arterial capillary (ar-tē′rē-ăl) Capillary opening from an arteriole or a metarteriole.

arteriole (ar-tēr′ē-ōl) Minute artery with all three tunics that transports blood to a capillary.

arteriosclerosis (ar-tēr′ē-ō-skler-ō′sis) [L., *arterio* + Gr., *sklerosis*, hardness] Hardening of the arteries.

arteriovenous anastomosis (ar-tēr′ē-ō-vē′nŭs ă-nas′tō-mō′sis) Vessel through which blood is shunted from an arteriole to a venule without passing through the capillaries.

artery (ar′ter-ē) Blood vessel that carries blood away from the heart.

articular cartilage (ar-tik′ū-lăr kar′ti-lij) Hyaline cartilage covering the ends of bones within a synovial joint.

articulation (ar-tik-ū-lā′shŭn) Place where two bones come together; also called joint.

arytenoid cartilages (ar-i-tē′noyd) Small, pyramidal laryngeal cartilages that articulate with the cricoid cartilage.

ascending aorta Part of the aorta from which the coronary arteries arise.

ascending colon (kō′lon) Portion of the colon between the small intestine and the right colic flexure.

asthma (az′mă) Condition of the lungs in which widespread narrowing of airways occurs, caused by contraction of smooth muscle, edema of the mucosa, and mucus in the lumen of the bronchi and bronchioles.

astrocyte (as′trō-sīt) [Gr., *astron*, star + *kytos*, a hollow, a cell] Star-shaped neuroglia cell involved with forming the blood–brain barrier.

atherosclerosis (ath′er-ō-skler-ō′sis) Arteriosclerosis characterized by irregularly distributed lipid deposits in the intima of large and medium-sized arteries.

atomic number (ă-tom′ik) Number of protons in each type of atom.

atrial diastole (ā′trē-ăl dī-as′tō-lē) Dilation of the heart's atria.

atrial natriuretic hormone (ANH) (ā′trē-ăl nă′trē-ū-ret′ik) Peptide released from the atria when atrial blood pressure is increased; lowers blood pressure by increasing the rate of urinary production, thus reducing blood volume.

atrial systole (ā′trē-ăl sis′tō-lē) Contraction of the atria.

atrioventricular (AV) bundle (ā′trē-ō-ven-trik′ū-lar) Bundle of modified cardiac muscle fibers that projects from the AV node through the interventricular septum.

atrioventricular (AV) node Small node of specialized cardiac muscle fibers that gives rise to the atrioventricular bundle of the conduction system of the heart.

atrioventricular valve One of two valves closing the openings between the atria and ventricles.

atrium; pl. atria (ā′trē-ŭm, ā′trē-ă) [L., entrance hall] One of two chambers of the heart into which veins carry blood.

auditory cortex (aw′di-tōr-ē kōr′teks) Portion of the cerebral cortex that is responsible for the conscious sensation of sound; in the dorsal portion of the temporal lobe within the lateral fissure and on the superolateral surface of the temporal lobe.

auditory ossicle (os′i-kl) Bone of the middle ear; includes the malleus, incus, and stapes.

auricle (aw′ri-kl) [L., *auris*, ear] Part of the external ear that protrudes from the side of the head; also called pinna. Small pouch projecting from the superior, anterior portion of each atrium of the heart.

auscultatory (aws-kŭltă-tō-rē) Relating to auscultation, listening to the sounds made by the various body structures as a diagnostic method.

autoimmune disease (aw-tō-i-mūn′di-zēz′) Disease resulting from a specific immune system reaction against self-antigens.

autonomic ganglia (aw-tō-nom′ik gang′glē-ă) Ganglia containing the nerve cell bodies of the autoimmune division of the nervous system.

autonomic nervous system (ANS) Nervous system composed of nerve fibers that send impulses from the central nervous system to smooth muscle, cardiac muscle, and glands.

autophagia (aw-tō-fā′jē-ă) [Gr., *auto*, self + *phagein*, to eat] Segregation and disposal of organelles within a cell.

autoregulation (aw′tō-reg-ŭ-lā′shŭn) Maintenance of a relatively constant blood flow through a tissue despite relatively large changes in blood pressure; maintenance of a relatively constant glomerular filtration rate despite relatively large changes in blood pressure.

autorhythmic Spontaneous and periodic—for example, in smooth muscle, spontaneous (without nervous or hormonal stimulation) and periodic contractions.

autosome (aw′tō-sōm) [Gr., *auto*, self + *soma*, body] Any chromosome other than a sex chromosome; normally exist in pairs in somatic cells and singly in gametes.

axial skeleton (ak′sē-ăl) Skull, vertebral column, and rib cage.

axillary (ak′sil-ār-ē) Relating to the axilla; the space below the shoulder joint, bounded by the pectoralis major anteriorly, the latissimus dorsi posteriorly, the serratus anterior medially, and the humerus laterally.

axolemma (ak′sō-lem′ă) [Gr., *axo* + *lemma*, husk] Plasma membrane of the axon.

axon (ak′son) [Gr., axis] Main central process of a neuron that normally conducts action potentials away from the neuron cell body.

axon hillock Area of origin of the axon from the nerve cell body.

axoplasm (ak′sō-plazm) Neuroplasm or cytoplasm of the axon.

B

baroreceptor (bar′ō-rē-sep′ter, bar′ō-rē-sep′tōr) Sensory nerve ending in the walls of the atria of the heart, venae cavae, aortic arch, and carotid sinuses; sensitive to stretching of the wall caused by increased blood pressure; also called pressoreceptor.

baroreceptor reflex Detects changes in blood pressure and produces changes in heart rate, heart force of contraction, and blood vessel diameter that return blood pressure to homeostatic levels.

basal ganglia (bā′săl gang′glē-ă) Nuclei at the base of the cerebrum involved in controlling motor functions.

base (bās) Molecule that is a proton acceptor; any substance that binds to hydrogen ions. Lower part or bottom of a structure; the base of the heart is the flat portion directed posteriorly and superiorly; veins and arteries project into and out of the base, respectively.

basement membrane (bās′ment mem′brān) Specialized extracellular material located at the base of epithelial cells and separating them from the underlying connective tissues.

basilar membrane (bas′i-lăr mem′brān) Wall of the membranous labyrinth bordering the scala tympani; supports the organ of Corti.

basophil (bā′sō-fil) [Gr., *basis*, baso + *phileo*, to love] White blood cell with granules that stain specifically with basic dyes; promotes inflammation.

B cell Type of lymphocyte responsible for antibody-mediated immunity.

belly (bel′ē) Largest portion of muscle between the origin and insertion.

beta-oxidation (bā′tă ok-si-dā′shŭn) Metabolism of fatty acids by removing a series of two-carbon units to form acetyl-CoA.

bicarbonate ion (bī-kar′bon-āt) Anion (HCO_3^-) remaining after the dissociation of carbonic acid.

bicuspid valve (bī-kŭs′pid) Valve closing the orifice between left atrium and left ventricle of the heart; also called mitral valve.

bile (bīl) Fluid secreted from the liver into the duodenum; consists of bile salts, bile pigments, bicarbonate ions, cholesterol, fats, fat-soluble hormones, and lecithin.

bile canaliculus (bīl kan′ă-lik′ū-lŭs) One of the intercellular channels approximately 1 μm or less in diameter that occurs between liver cells into which bile is secreted; empties into the hepatic ducts.

bile salt Organic salt secreted by the liver that functions as an emulsifying agent.

bilirubin (bil-i-roo′bin) [L., *bili* + *ruber*, red] Bile pigment derived from hemoglobin during the destruction of red blood cells.

biliverdin (bil-i-ver′din) Green bile pigment formed from the oxidation of bilirubin.

binocular vision (bin-ok′ū-lăr) [L., *bini*, paired + *oculus*, eye] Vision using two eyes at the same time; responsible for depth perception when the visual fields of the eyes overlap.

bipolar neuron (bī-pō′ler) One of the three categories of neurons consisting of a neuron with two processes—one dendrite and one axon—arising from opposite poles of the cell body.

blastocele (blas′tō-sēl) [Gr., *blastos*, germ + *koilos*, hollow] Cavity in the blastocyst.

blastocyst (blas′tō-sist) [Gr., *blastos*, germ + *kystis*, bladder] Stage of mammalian embryos that consists of the inner cell mass and a thin trophoblast layer enclosing the blastocele.

bleaching In response to light, retinal separates from opsin.

blind spot (blīnd) Point in the retina where the optic nerve penetrates the fibrous tunic; contains no rods or cones and therefore does not respond to light.

blood clot Coagulated phase of blood.

blood colloid osmotic pressure Osmotic pressure due to the concentration difference of proteins across a membrane that does not allow passage of the proteins.

blood groups Classification of blood based on the type of antigen found on the surface of red blood cells.

blood island   Aggregation of mesodermal cells in the embryonic yolk sac that forms vascular endothelium and primitive blood cells.

blood pressure (BP) [L., *pressus*, to press]   Tension of the blood within the blood vessels; commonly expressed in units of millimeters of mercury (mm Hg).

blood–brain barrier   Permeability barrier controlling the passage of most large-molecular compounds from the blood to the cerebrospinal fluid and brain tissue; consists of capillary endothelium and may include the astrocytes.

blood–thymic barrier   Layer of reticular cells that separates capillaries from thymic tissue in the cortex of the thymus gland; prevents large molecules from leaving the blood and entering the cortex.

Bohr effect   Named for Danish physiologist Christian Bohr (1855–1911). Shift of the oxygen–hemoglobin dissociation curve to the right or left because of changes in blood pH. The definition sometimes is extended to include shifts caused by changes in blood carbon dioxide levels.

bony labyrinth (lab′i-rinth)   Part of the inner ear; contains the membranous labyrinth that forms the cochlea, vestibule, and semicircular canals.

Boyle's law   The pressure of a gas is equal to the volume of a container times the constant, K, for a given temperature. Assuming a constant temperature, the pressure of a gas is inversely proportional to its volume.

brachial (brā′kē-ăl) [L., *brachium*, arm]   Relating to the arm.

branchial arch   Typically, six arches in vertebrates; in the lower vertebrates, they bear gills, but they appear transiently in the higher vertebrates and give rise to structures in the head and neck.

broad ligament   Peritoneal fold passing from the lateral margin of the uterus to the wall of the pelvis on each side.

bronchiole (brong′kē-ōl)   One of the finer subdivisions of the bronchial tubes, less than 1 mm in diameter; has no cartilage in its wall but does have relatively more smooth muscle and elastic fibers.

brush border   Epithelial surface consisting of microvilli.

buffer (bŭf′er)   Mixture of an acid and a base that reduces any changes in pH that would otherwise occur in a solution when acid or base is added to the solution.

bulb of the penis   Expanded posterior part of the corpus spongiosum of the penis.

bulb of the vestibule   Mass of erectile tissue on each side of the vagina.

bulbar conjunctiva (bŭl′bar kon-jŭnk-tī′vă)   Conjunctiva that covers the surface of the eyeball.

bulbourethral gland (bŭl′bō-ū-rē′thrăl)   One of two small compound glands that produce a mucoid secretion; it discharges through a small duct into the spongy urethra.

bulbus cordis (bŭl′bŭs) [L., plant bulb]   End of the embryonic cardiac tube where blood leaves the heart; becomes part of the ventricle.

bursa; pl. **bursae** (ber′să, ber′sē) [L., purse]   Closed sac or pocket containing synovial fluid, usually found in areas where friction occurs.

bursitis (ber-sī′tis) [L., *purse* + Gr., *ites*, inflammation]   Inflammation of a bursa.

C

calcaneal tendon (kal-kā′nē-al)   Common tendon of the gastrocnemius, soleus, and plantaris muscle that attaches to the calcaneus; also called Achilles tendon.

calcitonin (kal-si-tō′nin)   Hormone released from parafollicular cells that acts on tissues to cause a decrease in blood levels of calcium ions.

calmodulin (kal-mod′ū-lin) [*calcium* + *modulate*]   Protein receptor for Ca²⁺ that plays a role in many Ca^{2+}-regulated processes, such as smooth muscle contraction.

calorie (cal) (kal′ō-rē) [L., *calor*, heat]   Unit of heat content or energy. The quantity of energy required to raise the temperature of 1 g of water 1°C.

calpain (kal′pān)   Enzyme involved in changing the shape of dendrites; involved with long-term memory.

calyx; pl. **calyces** (kā′liks, kal′i-sēz) [Gr., cup of a flower]   Flower-shaped or funnel-shaped structure; specifically, one of the branches or recesses of a renal pelvis into which the tips of the renal pyramids project.

cancellous bone (kan′sĕ-lŭs) [L., grating or lattice]   Bone with a latticelike appearance; also called spongy bone.

cancer (kan′ser)   Any of various types of malignant neoplasms, most of which invade surrounding tissues, may metastasize to several sites, and are likely to recur after attempted removal and to cause death of the patient unless adequately treated.

canine (kā′nīn)   Referring to the cuspid tooth.

cannula (kan′ū-lă) [L., *canna*, reed]   Tube; often inserted into an artery or a vein.

capacitation (kă-pas′i-tā′shŭn) [L., *capax*, capable of]   Process whereby spermatozoa acquire the ability to fertilize ova; occurs in the female genital tract.

capitulum (kă-pit′ū-lŭm) [L., *caput*, head]   Head-shaped structure.

carbaminohemoglobin (kar-bam′i-nō-hē-mō-glō′bin)   Carbon dioxide bound to hemoglobin by means of a reactive amino group on the hemoglobin.

carbohydrate (kar-bō-hī′drāt)   Monosaccharide (simple sugar) or the organic molecules composed of monosaccharides bound together by chemical bonds—for example, glycogen. For each carbon atom in the molecule, there are typically one oxygen molecule and two hydrogen molecules.

carbonic acid/bicarbonate buffer system   One of the major buffer systems in the body; major components are carbonic acid and bicarbonate ions.

carbonic anhydrase   Enzyme that catalyzes the reaction between carbon dioxide and water to form carbonic acid.

carcinoma (kar-si-nō′mă)   Malignant neoplasm derived from epithelial tissue.

cardiac [Gr., *kardia*, heart]   Related to the heart.

cardiac cycle [Gr., *kyklos*, circle]   Complete round of cardiac systole and diastole.

cardiac nerve   Nerve that extends from the sympathetic chain ganglia to the heart.

cardiac output (CO)   Volume of blood pumped by the heart per minute; also called minute volume.

cardiac part   Region of the stomach near the opening of the esophagus.

cardiac reserve [L., *re* + *servo*, to keep back, reserve]   Work that the heart is able to perform beyond that required during ordinary circumstances of daily life.

carotid body (ka-rot′id)   One of the small organs near the carotid sinuses; contains chemoreceptors that respond primarily to decreases in blood oxygen; less sensitive to decreases in blood pH or increases in carbon dioxide.

carotid sinus   Enlargement of the internal carotid artery near the point where the internal carotid artery branches from the common carotid artery; contains baroreceptors.

carpal (kar′păl) [Gr., *karpos*, wrist]   Bone of the wrist.

carrier   Person in apparent health whose chromosomes contain a pathologic mutant gene, which may be transmitted to his or her children.

cartilage (kar′ti-lij) [L., *cartilage*, gristle]   Firm, smooth, resilient, nonvascular connective tissue.

cartilaginous joint (kar-ti-laj′i-nŭs)   Bones connected by cartilage; includes synchondroses and symphyses.

catabolism (kă-tab′ō-lizm) [Gr., *katabole*, a casting down]   All of the decomposition reactions that occur in the body; releases energy.

catalyst (kat′ă-list)   Substance that increases the rate at which a chemical reaction proceeds without being changed permanently.

cataract (kat′a-rakt)   Complete or partial opacity of the lens of the eye.

cations (kat′ī-on) [Gr., *kation*, going down]   Ions carrying a positive charge.

caveola; pl. **caveolae** (kav-ē-ō′lă, kav-ē-ō′lē) [L., small pocket]   Shallow invagination in the membranes of smooth muscle cells that may perform a function similar to that of both the T tubules and sarcoplasmic reticulum of skeletal muscle.

cecum (sē′kŭm, sē′kă) [L., *caecus*, blind]   Cul-de-sac forming the first part of the large intestine.

cell-mediated immunity   Immunity due to the actions of T cells and null cells.

celom (sē′lom, sē-lō′mă) [Gr., *koilo* + *amma*, a hollow]   Principal cavities of the trunk—for example, the pericardial, pleural, and peritoneal cavities. Separate in the adult, they are continuous in the embryo.

cementum (se-men′tŭm) [L., *caementum*, rough quarry stone]   Layer of modified bone covering the dentin of the root and neck of a tooth; blends with the fibers of the periodontal membrane.

central nervous system (CNS)   Major subdivision of the nervous system, consisting of the brain and spinal cord.

central vein   Terminal branches of the hepatic veins that lie centrally in the hepatic lobules and receive blood from the liver sinusoids.

centrosome (sen′trō-sōm)   Specialized zone of cytoplasm close to the nucleus and containing two centrioles.

cerebellum (ser-e-bel′ŭm) [L., little brain]   Separate portion of the brain attached to the brainstem at the pons; important in maintaining muscle tone, balance, and coordination of movement.

cerebrospinal fluid (CSF) (ser′ĕ-brō-spī′năl)   Fluid filling the ventricles and surrounding the brain and spinal cord.

ceruminous glands (sĕ-roo′mi-nŭs)   Modified sebaceous glands in the external acoustic meatus that produce cerumen (earwax).

cervical canal (ser′vĭ-kal)   Canal extending from the isthmus of the uterus to the opening of the uterus into the vagina.

cervix; pl. **cervices** (ser′viks, ser-vī′sēz) [L., neck]   Lower part of the uterus extending from the isthmus of the uterus into the vagina.

chalazion (ka-lā′zē-on)   Chronic inflammation of a meibomian gland; also called meibomian cyst.

cheek (chēk)   Side of the face forming the lateral wall of the mouth.

chemical signal   Molecule that binds to a macromolecule, such as receptors or enzymes, and alters their function; also called ligand.

chemoreceptor (kē′mō-rē-sep′tor)   Sensory cell that is stimulated by a change in the concentration of chemicals to produce action potentials. Examples include taste receptors, olfactory receptors, and carotid bodies.

chemoreceptor reflex Chemoreceptors detect a decrease in blood oxygen, an increase in carbon dioxide, or a decrease in pH and produce an increased rate and depth of respiration and, by means of the vasomotor center, vasoconstriction.

chemosensitive area (kem-ō-sen′si-tiv, kē-mō-sen′si-tiv) Chemosensitive neurons in the medulla oblongata detect changes in blood, carbon dioxide, and pH.

chemotactic factor (kē-mō-tak′tik) Part of a microorganism or chemical released by tissues and cells that act as chemical signals to attract leukocytes.

chemotaxis (kē-mo-tak′sis) [Gr., *chemo* + *taxis*, orderly arrangement] Attraction of living protoplasm (cells) to chemical stimuli.

chief cell Cell of the parathyroid gland that secretes parathyroid hormone. Cell of a gastric gland that secretes pepsinogen.

chloride (klōr′īd) Compound containing chlorine—for example, salts of hydrochloric acid.

chloride shift Diffusion of chloride ions into red blood cells as bicarbonate ions diffuse out; maintains electrical neutrality inside and outside the red blood cells.

choana; pl. choanae (kō′an-ă, kō-ā′nē) *See* internal naris.

cholecystokinin (kō′lē-sis-tō-kī′nin) Hormone liberated by the upper intestinal mucosa on contact with gastric contents; stimulates the contraction of the gallbladder and the secretion of pancreatic juice high in digestive enzymes.

cholinergic neuron (kol-in-er′jik) Nerve fiber that secretes acetylcholine as a neurotransmitter substance.

chondroblast (kon′drō-blast) [Gr., *chondros*, gristle, cartilage + *blastos*, germ] Cartilage-producing cell.

chondrocyte (kon′drō-sīt) [Gr., *chondros*, gristle, cartilage + *kytos*, a cell] Mature cartilage cell.

chorda tympani; pl. chordae (kōr′dă tim′pan-ē, kōr′dē) Branch of the facial nerve that conveys taste sensation from the front two-thirds of the tongue.

chordae tendineae (kōr′dă ten′di-nē-ē) [L., cord] Tendinous strands running from the papillary muscles to the atrioventricular valves.

choroid (ko′royd) Portion of the vascular tunic associated with the sclera of the eye.

choroid plexus [Gr., *chorioeides*, membrane-like] Specialized plexus located within the ventricles of the brain that secretes cerebrospinal fluid.

chromatid (krō′mă-tid) One-half of a chromosome; separates from its partner during cell division.

chromatin (krō′ma-tin) Colored material; the genetic material in the nucleus.

chromosome (krō′mō-sōm) Colored body in the nucleus, composed of DNA and proteins and containing the primary genetic information of the cell; 23 pairs in humans.

chronic pain Prolonged pain.

chylomicron (kī-lō-mī′kron) [Gr., *chylos*, juice + *micros*, small] Microscopic particle of lipid surrounded by protein; in chyle and blood.

chymotrypsin (kī-mō-trip′sin) Proteolytic enzyme formed in the small intestine from the pancreatic precursor chymotrypsinogen.

ciliary body (sil′ē-ar-ē) Structure continuous with the choroid layer at its anterior margin that contains smooth muscle cells; functions in accommodation.

ciliary gland Modified sweat gland that opens into the follicle of an eyelash, keeping it lubricated.

ciliary muscle Smooth muscle in the ciliary body of the eye.

ciliary process Portion of the ciliary body of the eye that attaches by suspensory ligaments to the lens.

ciliary ring Portion of the ciliary body of the eye that contains smooth muscle cells.

circumduction (ser-kŭm-dŭk′shŭn) [L., around + *ductus*, to draw] Movement in a circular motion.

circumferential lamellae (ser-kŭm-fer-en′shē-al lă-mel′ē) Lamellae covering the surface of and extending around compact bone inside the periosteum.

circumvallate papilla (ser-kŭm-val′āt pă-pil′ă) Type of papilla on the surface of the tongue surrounded by a groove.

cisterna; pl. cisternae (sis-ter′nă, sis-ter′nē) Interior space of the endoplasmic reticulum.

cisterna chyli (kīl′ē) [L., tank + Gr., *chylos*, juice] Enlarged inferior end of the thoracic duct that receives chyle from the intestine.

citric acid cycle (sit′rik) Series of chemical reactions in which citric acid is converted into oxaloacetic acid, carbon dioxide is formed, and energy is released. The oxaloacetic acid can combine with acetyl-CoA to form citric acid and restart the cycle. The energy released is used to form NADH, FADH, and ATP.

classical pathway Part of the specific immune system for activation of complement.

clavicle (klav′i-kl) The collarbone, between the sternum and scapula.

cleavage furrow (klēv′ij) Inward pinching of the plasma membrane that divides a cell into two halves, which separate from each other to form two new cells.

cleft palate (kleft) Failure of the embryonic palate to fuse along the midline, resulting in an opening through the roof of the mouth.

clinical age (klin′i-kl) Age of the developing fetus from the time of the mother's last menstrual period before pregnancy.

clinical perineum (klin′i-kl per′i-nē′ŭm) Portion of the perineum between the vaginal and anal openings.

clitoris (klit′ō-ris) Small, cylindrical, erectile body, rarely exceeding 2 cm in length, situated at the most anterior portion of the vulva and projecting beneath the prepuce.

cloaca (klō-ā′kă) [L., sewer] In early embryos, the endodermally lined chamber into which the hindgut and allantois empty.

cloning (klōn′ing) Growing a colony of genetically identical cells or organisms.

clot retraction Condensation of a clot into a denser, compact structure; caused by the elastic nature of fibrin.

coagulation (kō-ag-ū-lā′shŭn) Process of changing from liquid to solid, especially of blood; formation of a blood clot.

cochlear duct (kok′lē-ăr) Interior of the membranous labyrinth of the cochlea; also called cochlear canal or scala media.

cochlear nerve Nerve that carries sensory impulses from the organ of Corti to the vestibulocochlear nerve.

cochlear nucleus Neurons from the cochlear nerve synapse within the dorsal or ventral cochlear nucleus in the superior medulla oblongata.

codon (kō′don) Sequence of three nucleotides in mRNA or DNA that codes for a specific amino acid in a protein.

cofactor (kō′fak′ter, kō-fak′tōr) Nonprotein component of an enzyme, such as coenzymes and inorganic ions essential for enzyme action.

collagen fibers (kol′lă-jen) [Gr., *koila*, glue + *gen*, producing] Ropelike protein of the extracellular matrix.

collateral ganglia (ko-lat′er-ăl gang′glē-ă) Sympathetic ganglia at the origin of large abdominal arteries; include the celiac, superior, and inferior mesenteric arteries; also called prevertebral ganglia.

collecting duct Straight tubule that extends from the cortex of the kidney to the tip of the renal pyramid. Filtrate from the distal convoluted tubes enters the collecting duct and is carried to the calyces.

colloid (kol′oyd) [Gr., *kolla*, glue + *eidos*, appearance] Atoms or molecules dispersed in a gaseous, liquid, or solid medium that resist separation from the liquid, gas, or solid.

colloidal solution (ko-loyd′ăl) Fine particles suspended in a liquid; resistant to sedimentation or filtration.

colon (kō′lon) Division of the large intestine that extends from the cecum to the rectum.

colostrum (kō-los′trŭm) Thin, white fluid; the first milk secreted by the breast at the termination of pregnancy; contains less fat and lactose than the milk secreted later.

columnar Shaped like a column.

commissure (kom′i-shŭr) [L., *commissura*, a joining together] Connection of nerve fibers between the cerebral hemispheres or from one side of the spinal cord to the other.

common bile duct Duct formed by the union of the common hepatic and cystic ducts; it empties into the small intestine.

common hepatic duct Part of the biliary duct system formed by the joining of the right and left hepatic ducts.

compact bone Bone that is denser and has fewer spaces than cancellous bone.

competition Similar molecules binding to the same carrier molecule or receptor site.

complement (kom′plě-ment) Group of serum proteins that stimulates phagocytosis and inflammation.

complement cascade Series of reactions in which each component activates the next component, resulting in activation of complement proteins.

compliance (kom-plī′ans) Change in volume (e.g., in lungs or blood vessels) caused by a given change in pressure.

compound (kom′pownd) Substance composed of two or more different types of atoms that are chemically combined.

concha; pl. conchae (kon′kă, kon′kē) [L., shell] Structure comparable to a shell in shape—for example, the three bony ridges on the lateral wall of the nasal cavity.

conduction (kon-dŭk′shŭn) [L., *con* + *ductus*, to lead, conduct] Transfer of energy, such as heat, from one point to another without evident movement in the conducting body.

cone (kōn) Photoreceptor in the retina of the eye; responsible for color vision.

congenital (kon-jen′i-tăl) [L., *congenitus*, born with] Occurring at birth; may be genetic or due to some influence (e.g., drugs) during development.

conjunctiva (kon-jŭnk-tī′vă) [L., *conjungo*, to bind together] Mucous membrane covering the anterior surface of the eyeball and lining the lids.

conjunctival fornix (kon-jŭnk-tī′văl fōr′niks) Area in which the palpebral and bulbar conjunctiva meet.

constant region Portion of an antibody that does not combine with an antigen and is the same in different antibodies.

continuous capillary [L., *capillaris*, relating to hair] Capillary in which pores are absent; is less permeable to large molecules than are other types of capillaries.

contraction phase (kon-trak′shŭn) One of the three phases of muscle contraction; the time during which tension is produced by the contraction of muscle.

convection (kon-vek′shŭn) [L., *con* + *vectus*, to carry or bring together] Transfer of heat in liquids or gases by movement of the heated particles.

coracoid (kōr′ă-koyd) [Gr., *korakodes*, crow's beak] Resembling a crow's beak—for example, a process on the scapula.

Cori cycle Named for Czech-U.S. biochemist and Nobel laureate Carl F. Cori (1896–1984). Lactic acid, produced by skeletal muscle, is carried in the blood to the liver, where it is aerobically converted into glucose. The glucose may return through the blood to skeletal muscle or may be stored as glycogen in the liver.

cornea (kōr′nē-ă) Transparent portion of the fibrous tunic that makes up the outer wall of the anterior portion of the eye.

corniculate cartilage (kōr-nik′ū-lāt) Conical nodule of elastic cartilage surmounting the apex of each arytenoid cartilage.

corona radiata Single layer of columnar cells derived from the cumulus mass, which anchor on the zona pellucida of the oocyte in a secondary follicle.

coronary (kōr′o-nār-ē) [L., *coronarius*, a crown] Resembling a crown; encircling.

coronary artery One of two arteries that arise from the base of the aorta and carry blood to the muscle of the heart.

coronary ligament Peritoneal reflection from the liver to the diaphragm at the margins of the bare area of the liver.

coronary sinus Short trunk that receives most of the veins of the heart and empties into the right atrium.

coronoid (kōr′ŏ-noyd) [Gr., *korone*, a crow] Shaped like a crow's beak—for example, a process on the mandible.

corpus; pl. **corpora** (kōr′pŭs, -pōr-ă) [L., *body*] Any body or mass; the main part of an organ.

corpus albicans (al′bi-kanz) Atrophied corpus luteum, leaving a connective tissue scar in the ovary.

corpus callosum (kăl-lō′sŭm) [L., *body* + *callous*] Largest commissure of the brain, connecting the cerebral hemispheres.

corpus cavernosum; pl. **corpora cavernosa** One of two parallel columns of erectile tissue forming the dorsal part of the body of the penis or the body of the clitoris.

corpus luteum (lū′tē-ŭm) Yellow endocrine body formed in the ovary in the site of a ruptured vesicular follicle immediately after ovulation; secretes progesterone and estrogen.

corpus luteum of pregnancy Large corpus luteum in the ovary of a pregnant female; secretes large amounts of progesterone and estrogen.

corpus spongiosum (spŭn′jē-ō′sŭm) Median column of erectile tissue located between and ventral to the two corpora cavernosa in the penis; posteriorly it forms the bulb of the penis, and anteriorly it terminates as the glans penis; it is traversed by the urethra. In the female, it forms the bulb of the vestibule.

corpus striatum (strī-ā′tŭm) [L., *corpus*, body + *striatus*, striated or furrowed] Caudate nucleus, putamen, and globus pallidus; so-named because of the striations caused by intermixing of gray and white matter, which result from the number of tracts crossing the anterior portion of the corpus striatum.

cortex; pl. **cortices** (kōr′teks, kōr′ti-sēz) [L., bark] Outer portion of an organ (e.g., adrenal cortex or cortex of the kidney).

corticotropin-releasing hormone (kōr′ti-kō-trō′pin) Hormone from the hypothalamus that stimulates the anterior pituitary gland to release adrenocorticotropic hormone.

cortisol (kōr′ti-sol) Steroid hormone released by the zona fasciculata of the adrenal cortex; increases blood glucose and inhibits inflammation.

covalent bond (kō-vāl′ent) Chemical bond characterized by the sharing of electrons.

coxal bone (kok′să) Hipbone.

cranial nerve (krā′nē-ăl) Nerve that originates from a nucleus within the brain; there are 12 pairs of cranial nerves.

cranial vault Eight skull bones that surround and protect the brain; braincase.

craniosacral division (krā′nē-ō-sā′krăl) Parasympathetic division of the autonomic nervous system.

cranium (krā′nē-ŭm) [Gr., *kranion*, skull] Skull; in a more limited sense, the braincase.

cremaster muscle (krē-mas′ter) Extension of abdominal muscles originating from the internal oblique muscles; in the male, raises the testicles; in the female, envelops the round ligament of the uterus.

crenation (krē-nā′shŭn) [L., *crena*, notched] Denoting the outline of a shrunken cell.

cricoid cartilage (krī′koyd) Most inferior laryngeal cartilage.

cricothyrotomy (krī′kō-thī-rot′ō-mē) Incision through the skin and cricothyroid membrane for relief of respiratory obstruction.

crista, cristae (kris′tă, kris′tē) [L., crest] Shelflike infolding of the inner membrane of a mitochondrion.

crista ampullaris (kris′tă am-pul′ăr-ĭs) [L., crest] Elevation on the inner surface of the ampulla of each semicircular duct for dynamic or kinetic equilibrium.

critical closing pressure Pressure in a blood vessel below which the vessel collapses, occluding the lumen and preventing blood flow.

crown Part of a tooth that is covered with enamel.

cruciate (kroo′shē-āt) [L., *cruciatus*, cross] Resembling or shaped like a cross.

crus of the penis (krŭs) Posterior portion of the corpus cavernosum of the penis attached to the ischiopubic ramus.

crypt (kript) Pitlike depression or tubular recess.

cryptorchidism (krip-tōr′ki-dizm) Failure of the testis to descend.

crystalline (kris′tă-lēn) Protein that fills the epithelial cells of the lens in the eye.

cuboidal Resembling a cube.

cumulus oophorus (ō-of′ōr-ŭs) [L., a heap] Mass of epithelial cells surrounding the oocyte; also called cumulus mass.

cuneiform cartilage (kū′nē-i-fōrm) Small rod of elastic cartilage above each corniculate cartilage in the larynx.

cupula; pl. **cupulae** (koo′poo-lă, kū′pū-lă, koo′poo-lē) [L., *cupa*, tub] Gelatinous mass that overlies the hair cells of the cristae ampullares of the semicircular ducts.

cuticle (kū′ti-kl) [L., *cutis*, skin] Outer, thin layer, usually horny—for example, the outer covering of hair or the growth of the stratum corneum onto the nail.

cystic duct (sis′tik) Duct leading from the gallbladder; joins the common hepatic duct to form the common bile duct.

cytokine (sī′tō-kīn) Protein or peptide secreted by a cell that regulates the activity of neighboring cells.

cytokinesis (sī′tō-ki-nē′sis) [Gr., *cyto*, cell + *kinsis*, movement] Division of the cytoplasm during cell division.

cytology (sī-tol′ō-jē) [Gr., *kytos*, a hollow (cell) + *logos*, study] Study of anatomy, physiology, pathology, and chemistry of the cell.

cytoplasm (sī′tō-plazm) Protoplasm of the cell surrounding the nucleus.

cytoplasmic inclusion (sī-tō-plaz′mik) Any foreign or other substance contained in the cytoplasm of a cell.

cytotoxic reaction (sī′tō-tok′sik) [Gr., *cyto*, cell + L., *toxic*, poison] Antibodies (IgG or IgM) combine with cells and activate complement, and cell lysis occurs.

cytotrophoblast (sī′tō-trof′ō-blast) Inner layer of the trophoblast, composed of individual cells.

D

Daily Values Dietary reference values useful for planing a healthy diet. The Daily Values are taken from the Reference Daily Intakes (RDIs) and the Daily Reference Values.

Daily Reference Values (DRVs) Recommended amounts in the diet for total fat, saturated fat, cholesterol, total carbohydrate, dietary fiber, sodium, potassium, and protein. The values for total fat, saturated fat, cholesterol, and sodium are the uppermost limits considered desirable because of their link to certain diseases.

Dalton's law Named for English chemist John Dalton (1766–1844). In a mixture of gases, the portion of the total pressure resulting from each type of gas is determined by the percentage of the total volume represented by each gas type.

dartos muscle (dar′tōs) Layer of smooth muscle in the skin of the scrotum; contracts in response to lower temperature and relaxes in response to higher temperature; raises and lowers testes in the scrotum.

deciduous tooth (dē-sid′ū-ŭs) Tooth of the first set of teeth; also called primary tooth.

decussate (dē′kŭ-sāt, dē-kŭs′āt) [L., *decusso*, X-shaped, from *decussis*, ten (X)] To cross.

deep inguinal ring (ing′gwi-năl) Opening in the transverse fascia through which the spermatic cord (or round ligament in the female) enters the inguinal canal.

defecation (def-ē-kā′shŭn) [L., *defaeco*, to remove the dregs, purify] Discharge of feces from the rectum.

defecation reflex Combination of local and central nervous system reflexes initiated by distension of the rectum and resulting in the movement of feces out of the lower colon.

deglutition (dē-gloo-tish′ŭn) [L., *de* + *glutio*, to swallow] Act of swallowing.

dendrite (den′drīt) [Gr., *dendrites*, tree] Branching processes of a neuron; receives stimuli and conducts potentials toward the cell body.

dendritic cell (den-drit′ik) Large cells with long, cytoplasmic extensions that are capable of taking up and concentrating antigens, leading to the activation of B or T lymphocytes.

dendritic spine Extension of nerve cell dendrites where axons form synapses with the dendrites; also called gemmule.

dental arch (den′tăl) [L., *arcus*, bow] Curved maxillary or mandibular arch in which the teeth are located.

dentin (den′tin) Bony material forming the mass of the tooth.

deoxyhemoglobin (dē-oks′ē-hē-mō-glō′bin) Hemoglobin without oxygen bound to it.

deoxyribonuclease (de-oks′ē-rī-bō-noo′klē-ās) Enzyme that splits DNA into its component nucleotides.

deoxyribonucleic acid (DNA) (dē-oks′ē-rī′bō-noo-klē′ic) Type of nucleic acid containing deoxyribose as the sugar component, found principally in the nuclei of cells; constitutes the genetic material of cells.

depolarization (dē-pō′lăr-i-zā′shŭn) Change in the electric charge difference across the plasma membrane that causes the difference to be smaller or closer to 0 mV; phase of the action potential in which the membrane potential moves toward zero, or becomes positive.

depression (dē-presh′ŭn) Movement of a structure in an inferior direction.

depth perception (per-sep′shun) Ability to distinguish between near and far objects and to judge their distance.

dermatome (der′mă-tōm) Area of skin supplied by a spinal nerve.

dermis (der′mis) [Gr., *derma*, skin] Dense irregular connective tissue that forms the deep layer of the skin.

descending aorta Part of the aorta, further divided into the thoracic aorta and abdominal aorta.

descending colon Part of the colon extending from the left colonic flexure to the sigmoid colon.

desmosome (dez′mō-sōm) [Gr., *desmos*, a band + *soma*, body] Point of adhesion between cells. Each contains a dense plate at the point of adhesion and a cementing extracellular material between the cells.

desquamate (des′kwă-māt) [L., *desquamo*, to scale off] Peeling or scaling off of the superficial cells of the stratum corneum.

diabetes insipidus (dī-ă-bē′tēz in-sip′ĭ-dŭs) Chronic excretion of large amounts of urine of low specific gravity accompanied by extreme thirst; results from inadequate output of antidiuretic hormone.

diabetes mellitus (me-lī′tŭs) Metabolic disease in which carbohydrate use is reduced and that of lipid and protein enhanced; caused by a deficiency of insulin or an inability to respond to insulin and is characterized, in more severe cases, by hyperglycemia, glycosuria, water and electrolyte loss, ketoacidosis, and coma.

diapedesis (dī′ă-pĕ-dē′sis) [Gr., *dia*, through + *pedesis*, a leaping] Passage of blood or any of its formed elements through the intact walls of blood vessels.

diaphragm (dī′ă-fram) Musculomembranous partition between the abdominal and thoracic cavities.

diaphysis (dī-af′i-sis) [Gr., growing between] Shaft of a long bone.

diastole (dī-as′tō-lē) [Gr., *diastole*, dilation] Relaxation of the heart chambers, during which they fill with blood; usually refers to ventricular relaxation.

diencephalon (dī-en-sef′ă-lon) [Gr., *dia*, through + *enkephalos*, brain] Second portion of the embryonic brain; in the inferior core of the adult cerebrum.

diffuse lymphatic tissue Dispersed lymphocytes and other cells with no clear boundary; found beneath mucous membranes, around lymph nodules, and within lymph nodes and spleen.

diffusion (di-fū′zhŭn) [L., *diffundo*, to pour in different directions] Tendency for solute molecules to move from an area of high concentration to an area of low concentration in solution; the product of the constant random motion of all atoms, molecules, or ions in a solution.

diffusion coefficient Measure of how easily a gas diffuses through a liquid or tissue.

digestive tract (di-jes′tiv, dī-jes′tiv) Mouth, oropharynx, esophagus, stomach, small intestine, and large intestine.

digit (dij′it) Finger, thumb, or toe.

dilator pupillae (dī′lă-tĕr pū-pil′ē) Radial smooth muscle cells of the iris diaphragm that cause the pupil of the eye to dilate.

diploid (2*n*) number (dip′loyd) Normal number of chromosomes (in humans, 46 chromosomes) in somatic cells.

disaccharide (dī-sak′ă-rīd) Condensation product of two monosaccharides by the elimination of water.

dissociate (di-sō′sē-āt) [L., *dis* + *socio*, to disjoin, separate] Ionization in which ions are dissolved in water and the cations and anions are surrounded by water molecules.

distal convoluted tubule Convoluted tubule of the nephron that extends from the ascending limb of the loop of Henle and ends in a collecting duct.

distributing artery Medium-sized artery with a tunica media composed principally of smooth muscle; regulates blood flow to different regions of the body.

dominant (dom′i-nant) [L., *dominus*, a master] Gene that is expressed phenotypically to the exclusion of a contrasting recessive gene.

dorsal root (dōr′săl) Sensory (afferent) root of a spinal nerve.

dorsal root ganglion (gang′glē-on) Collection of sensory neuron cell bodies within the dorsal root of a spinal nerve; also called spinal ganglion.

ductus arteriosus (dŭk′tŭs ar-tēr′ē-ō-sŭs) Fetal vessel connecting the left pulmonary artery with the descending aorta.

ductus deferens (def′er-enz) Duct of the testicle, running from the epididymis to the ejaculatory duct; also called vas deferens.

duodenal gland (doo′ō-dē′năl, doo-od′ĕ-năl) Small gland that opens into the base of intestinal glands; secretes a mucoid alkaline substance.

duodenocolic reflex (doo-ō-dē′nō-kō-lik) Local reflex resulting in a mass movement of the contents of the colon; produced by stimuli in the duodenum.

duodenum (doo-ō-dē′nŭm, doo-od′ĕ-nŭm) [L., *duodeni*] First division of the small intestine; connects to the stomach.

dura mater (doo′ră mā′ter) [L., hard mother] Tough, fibrous membrane forming the outer covering of the brain and spinal cord.

E

eardrum (ēr′drŭm) Cellular membrane that separates the external from the middle ear; vibrates in response to sound waves; also called tympanic membrane.

ectoderm (ek′tō-derm) Outermost of the three germ layers of an embryo.

ectopic focus; pl. foci (ek-top′ik fō′kŭs, fō′sī) Any pacemaker other than the sinus node of the heart; abnormal pacemaker; an ectopic pacemaker.

edema (e-dē′mă) [Gr., *oidema*, a swelling] Excessive accumulation of fluid within or around calls, usually causing swelling.

effector T cell (ē-fek′tŏr, ē-fek′tōr) Subset of T lymphocytes that is responsible for cell-mediated immunity.

efferent arteriole (ef′er-ent ar-tēr′ē-ōl) Vessel that carries blood from the glomerulus to the peritubular capillaries.

efferent division Nerve fibers that send impulses from the central nervous system to the periphery.

efferent ductule (ef′er-ent dŭk′tool) [L., *ductus*, duct] One of a number of small ducts leading from the testis to the head of the epididymis.

ejaculation (ē-jak-ū-lā′shŭn) Reflexive expulsion of semen from the penis.

ejaculatory duct (ē-jak′ū-lă-tōr-ē) Duct formed by the union of the ductus deferens and the excretory duct of the seminal vesicle; opens into the prostatic urethra.

ejection period (ē-jek′shŭn) Time in the cardiac cycle when the semilunar valves are open and blood is being ejected from the ventricles into the arterial system.

elastin (ĕ-las′tin) Yellow, elastic, fibrous mucoprotein that is the major connective tissue protein of elastic structures (e.g., large blood vessels and elastic ligaments).

electrocardiogram (ECG, EKG) (ē-lek-trō-kar′dē-ō-gram) [Gr., *elektron*, amber + *kardia*, heart + *gramma*, a drawing] Graphic record of the heart′s electric currents obtained with an electrocardiograph.

electrolyte (ē-lek′trō-līt) [Gr., *electro* + *lytos*, soluble] Cation or anion in solution that conducts an electric current.

electron (ē-lek′tron) Negatively charged subatomic particle in an atom.

electron-transport chain Series of electron carriers in the inner mitochondrial membrane; they receive electrons from NADH and FADH$_2$, using the electrons in the formation of ATP and water.

element (el′ĕ-ment) [L., *elementum*, rudiment, beginning] Substance composed of atoms of only one kind.

elevation (el-ĕ-vā′shŭn) Movement of a structure in a superior direction.

embolism (em′bō-lizm) [Gr., *embolisma*, a piece of patch, literally something thrust in] Obstruction or occlusion of a vessel by a transported clot, a mass of bacteria, or other foreign material.

embolus; pl. emboli (em′bō-lŭs, em′bō-lī) [Gr., *embolos*, plug, wedge, or stopper] Plug, composed of a detached clot, a mass of bacteria, or another foreign body, occluding a blood vessel.

embryo (em′brē-ō) Developing human from the first to the eighth week of development.

embryonic disk (em-brē-on′ik) Point in the inner cell mass at which the embryo begins to form.

embryonic period From approximately the second to the eighth week of development, during which the major organ systems are organized.

emission (ē-mish′ŭn) [L., *emissio*, to send out] Discharge; accumulation of semen in the urethra prior to ejaculation. A nocturnal emission is a discharge of semen while asleep.

emmetropia (em-ĕ-trō′pē-ă) [Gr., *emmetros*, according to measure + *ops*, eye] In the eye, the state of refraction in which parallel rays are focused exactly on the retina; no accommodation is necessary.

emulsify (ē-mŭl′si-fī) To form an emulsion.

enamel (ē-nam′ĕl) Hard substance covering the exposed portion of the tooth.

endocardium; pl. endocardia (en-dō-kar′dē-ŭm, en-dō-kar′dē-ă) Innermost layer of the heart, including endothelium and connective tissue.

endocrine gland (en′dō-krin) [Gr., *endon*, inside + *krino*, to separate] Ductless gland that secretes a hormone internally, usually into the circulation.

endocytosis (en′dō-sī-tō′sis) Bulk uptake of material through the cell membrane.

endoderm (en′dō-derm) Innermost of the three germ layers of an embryo.

endolymph (en'dō-limf) [Gr., *endo* + L., *lympha*, clear fluid] Fluid found within the membranous labyrinth of the inner ear.

endometrium; pl. **endometria** (en'dō-mē'trē-ŭm, en'dō-mē'trē-ă) Mucous membrane composing the inner layer of the uterine wall; consists of a simple columnar epithelium and a lamina propria that contains simple tubular uterine glands.

endomysium (en'dō-miz'ē-ŭm, en'dō-mis'ē-ŭm) [Gr., *endo*, within + *mys*, muscle] Fine connective tissue sheath surrounding a muscle fiber.

endoneurium (en-dō-noo'rē-ŭm) [Gr., *endo*, within + *neuron*, nerve] Delicate connective tissue surrounding individual nerve fibers within a peripheral nerve.

endoplasmic reticulum; pl. **reticula** (en'dō-plas'mik re-tik'ū-lŭm, re-tik'ū-lă) Double-walled membranous network inside the cytoplasm; rough has ribosomes attached to the surface; smooth does not have ribosomes attached.

endorphin (en-dōr'fin) Opiate-like polypeptide found in the brain and other parts of the body; binds in the brain to the same receptors that bind exogenous opiates.

endosteum (en-dos'tē-ŭm) [Gr., *endo*, within + *osteon*, bone] Membranous lining of the medullary cavity and the cavities of spongy bone.

endothelium; pl. **endothelia** (en-dō-thē'lē-ŭm, en-dō-thē'lē-ă) [Gr., *endo*, within + *thele*, nipple] Layer of flat cells lining blood and lymphatic vessels and the chambers of the heart.

enkephalin (en-kef'ă-lin) Pentapeptide found in the brain; binds to specific receptor sites, some of which may be pain-related opiate receptors.

enteric nervous system Complex network of neuron cell bodies and axons within the wall of the digestive tract; capable of controlling the digestive tract independently of the central nervous system through local reflexes.

enterokinase (en'tēr-ō-kī'nās) Intestinal proteolytic enzyme that converts trypsinogen into trypsin.

enzyme (en'zīm) [Gr., *en*, in + *zyme*, leaven] Protein that acts as a catalyst.

eosinophil (ē-ō-sin'ō-fil) [Gr., *eos*, dawn + *philos*, fond] White blood cell that stains with acidic dyes; inhibits inflammation.

epicardium (ep-i-kar'dē-ŭm) [Gr., *epi*, on + *kardia*, heart] Serous membrane covering the surface of the heart; also called visceral pericardium.

epidermis (ep-i-derm'is) [Gr., *epi*, on + *derma*, skin] Outer portion of the skin formed of epithelial tissue that rests on or covers the dermis.

epididymis; pl. **epididymides** (ep-i-did'i-mis, -di-dim'i-dēz) [Gr., *epi*, on + *didymos*, twin] Elongated structure connected to the posterior surface of the testis, which consists of the head, body, and tail; site of storage and maturation of the spermatozoa.

epiglottis (ep-i-glot'is) [Gr., *epi*, on + *glottis*, mouth of the windpipe] Plate of elastic cartilage covered with mucous membrane; serves as a valve over the glottis of the larynx during swallowing.

epimysium (ep-i-mis'ē-ŭm) [Gr., *epi*, on + *mys*, muscle] Fibrous envelope surrounding a skeletal muscle.

epinephrine (ep'i-nef'rin) Hormone (amino acid derivative) similar in structure to the neurotransmitter norepinephrine; major hormone released from the adrenal medulla; increases cardiac output and blood glucose levels; also called adrenaline.

epineurium (ep-i-noo'rē-ŭm) [Gr., *epi*, on + *neuron*, nerve] Connective tissue sheath surrounding a nerve.

epiphyseal line (ep-i-fiz'ē-ăl) Dense plate of bone in a bone that is no longer growing, indicating the former site of the epiphyseal plate.

epiphyseal plate Site at which bone growth in length occurs; located between the epiphysis and diaphysis of a long bone; area of hyaline cartilage where cartilage growth is followed by endochondral ossification; also called metaphysis or growth plate.

epiphysis; pl. **epiphyses** (e-pif'i-sis, e-pif'i-sēz) [Gr., *epi*, on + *physis*, growth] Portion of a bone developed from a secondary ossification center and separated from the remainder of the bone by the epiphyseal plate.

epiploic appendage (ep'i-plō'ik) One of a number of little processes of peritoneum projecting from the serous coat of the large intestine except the rectum; they are generally distended with fat.

epithelium (ep-i-thē'lē-ŭm) [Gr., *epi*, on + *thele*, nipple] One of the four primary tissue types. *Nipple* refers to the tiny capillary-containing connective tissue in the lips, which is where the term was first used. The use of the term was later expanded to include all covering and lining surfaces of the body.

epitope (ep'i-tōp) [Gr., *epi*, on + *top*, place] *See* antigenic determinant.

eponychium (ep-ō-nik'ē-ŭm) [Gr., *epi*, on + *onyx*, nail] Outgrowth of the skin that covers the proximal and lateral borders of the nail; also called cuticle.

erection (ē-rek'shŭn) [L., *erectio*, to set up] Condition of erectile tissue when filled with blood; tissue becomes hard and unyielding; especially refers to this state of the penis.

erythroblastosis fetalis (ē-rith'rō-blas-tō'sis fē-tă'lis) [*erythroblast* + *osis*, condition] Destruction of erythrocytes in the fetus or newborn caused by antibodies produced in an Rh-negative mother acting on the Rh-positive blood of the fetus or newborn.

erythrocyte (ē-rith'rō-sīt) [Gr., *erythros*, red + *kytos*, cell] Red blood cell; biconcave disk containing hemoglobin.

erythropoiesis (ē-rith'rō-poy-ē'sis) [*erythrocyte* + Gr., *poiesis*, a making] Production of erythrocytes.

erythropoietin (ē-rith'rō-poy'ē-tin) Protein that enhances erythropoiesis by stimulating the formation of proerythroblasts and the release of reticulocytes from bone marrow.

esophagus; pl. **esophagi** (ē-sof'ă-gŭs, ē-sof'ă-gī, ē-sof'ă-jī) [Gr., *oisophagos*, gullet] Portion of the digestive tract between the pharynx and stomach.

essential amino acid Amino acid, required by animals, that must be supplied in the diet.

estrogen (es'trō-jen) Substance that exerts biologic effects characteristic of estrogen hormone, such as stimulating female secondary sexual characteristics, growth, and maturation of long bones, and help controlling the menstrual cycle.

eustachian tube (ū-stā'shŭn, ū-stā'kē-ăn) Named for Italian anatomist Bartolommeo Eustachio (1524–1574). Auditory canal; extends from the middle ear to the nasopharynx.

evagination (ē-vaj-i-nā'shŭn) [L., *e*, out + *vagina*, sheath] Protrusion of some part or organ from its normal position.

evaporation (ē-vap-ŏ-ra'shŭn) [L., *e*, out + *vaporare*, to emit vapor] Change from liquid to vapor form.

eversion (ē-ver'zhŭn) [L., *everto*, to overturn] Turning outward.

excitation–contraction coupling (ek-sī-tā'shŭn kon-trak'shŭn kŭp'ling) Stimulation of a muscle fiber produces an action potential that results in contraction of the muscle fiber.

excitatory postsynaptic potential (EPSP) (ek-sī'tă-tō-rē pōst-si-nap'tik pō-ten'shăl) Depolarization in the postsynaptic membrane that brings the membrane potential close to threshold.

exocrine gland (ek'sō-krin) [Gr., *exo*, outside + *krino*, to separate] Gland that secretes to a surface or outward through a duct.

exocytosis (ek'sō-sī-to'sis) Elimination of material from a cell through the formation of vacuoles.

expiratory reserve volume Maximum volume of air that can be expelled from the lungs after a normal expiration.

extension (eks-ten'shŭn) [L., *extensio*, to stretch out] To stretch out.

external acoustic meatus (mē-ā'tŭs) Short canal that opens to the exterior environment and terminates at the eardrum; part of the external ear.

external anal sphincter Ring of striated muscular fibers surrounding the anus.

external ear Portion of the ear that includes the auricle and external acoustic meatus; terminates at the eardrum.

external nose Nostril; anterior or external opening of the nasal cavity.

external spermatic fascia Outer fascial covering of the spermatic cord.

external urethral orifice Slitlike opening of the urethra in the glans penis.

external urinary sphincter Sphincter skeletal muscle around the base of the urethra external to the internal urinary sphincter.

exteroceptor (eks'ter-ō-sep'ter, eks'ter-ō-sep'tōr) [L., *exterus*, external + *receptor*, receiver] Sensory receptor in the skin or mucous membranes that responds to stimulation by external agents or forces.

extracellular (eks-tră-sel'ū-lăr) Outside the cell.

extracellular matrix; pl. **matrices** (eks-tră-sel'ū-lăr mā'triks, mā'tri-sēz) Nonliving chemical substances located between connective tissue cells.

extrinsic clotting pathway (eks-trin'sik) Series of chemical reactions resulting in clot formation; begins with chemicals (e.g., tissue thromboplastin) found outside the blood.

extrinsic muscle Muscle located outside the structure being moved.

eyebrow Short hairs on the bony ridge above the eyes.

eyelash Hair at the margins of the eyelids.

eyelid Movable fold of skin in front of the eyeball; also called palpebra.

F

F actin (ak'tin) Fibrous actin molecule that is composed of a series of globular actin molecules (G actin).

facilitated diffusion Carrier-mediated process that does not require ATP and moves substances into or out of cells from a high to a low concentration.

falciform ligament (fal'si-fŏrm lig'ă-ment) Fold of peritoneum extending to the surface of the liver from the diaphragm and anterior abdominal wall.

fallopian tube (fa-lō'pē-an) *See* uterine tube.

false pelvis Portion of the pelvis superior to the pelvic brim; composed of the bone on the posterior and lateral sides and by muscle on the anterior side; also called greater pelvis.

falx cerebelli (falks ser-ĕ-bel'ī) Dural fold between the two cerebellar hemispheres.

falx cerebri (falx ser'ĕ-brī) Dural fold between the two cerebral hemispheres.

far point of vision Distance from the eye where accommodation is not needed to have the image focused on the retina.

fascia; pl. **fasciae** (fash′ē-ă, fash′ē-ē) [L., band or fillet]   Loose areolar connective tissue found beneath the skin (hypodermis) or dense connective tissue that encloses and separates muscles.

fasciculus (fă-sik′ū-lŭs) [L., *fascis,* bundle]   Band or bundle of nerve or muscle fibers bound together by connective tissue.

fat [A.S., *faet*]   Greasy, soft-solid material found in animal tissues and many plants; composed of two types of molecules: glycerol and fatty acids.

fatigue (fă-tēg′) [L., *fatigo,* to tire]   Period characterized by a reduced capacity to do work.

fat-soluble vitamin   Vitamin, such as A, D, E, and K, that is soluble in lipids and absorbed from the intestine along with lipids.

fauces (faw′sēz) [L., throat]   Space between the cavity of the mouth and the pharynx.

female climacteric   Period of life occurring in women, encompassing termination of the reproductive period and characterized by endocrine, somatic, and transitory psychologic changes and ultimately menopause; also called perimenopause.

female pronucleus   Nuclear material of the ovum after the ovum has been penetrated by the spermatozoon. Each pronucleus carries the haploid number of chromosomes.

fertilization (fer′til-i-zā′shŭn)   Process that begins with the penetration of the secondary oocyte by the spermatozoon and is completed with the fusion of the male and female pronuclei.

fetus (fē′tus)   Developing human following the embryonic period (after 8 weeks of development).

fibrin (fī′brin)   Elastic filamentous protein derived from fibrinogen by the action of thrombin, which releases peptides from fibrinogen in coagulation of the blood.

fibroblast (fī′brō-blast) [L., *fibra,* fiber + Gr., *blastos,* germ]   Spindle-shaped or stellate cells that form connective tissue.

fibrocyte (fī′brō-sīt)   Mature cell of fibrous connective tissue.

fibrous joint (fī′brŭs)   Bones connected by fibrous tissue with no joint cavity; includes sutures, syndesmoses, and gomphoses.

fibrous layer   Outer layer of the eye; composed of the sclera and the cornea.

filiform (fil′i-fōrm)   Filament-shaped.

filtrate (fil′trāt)   Liquid that has passed through a filter—for example, fluid that enters the nephron through the filtration membrane of the glomerulus.

filtration (fil-trā′shŭn)   Movement, due to a pressure difference, of a liquid through a filter that prevents some or all of the substances in the liquid from passing through.

filtration fraction   Fraction of the plasma entering the kidney that filters into Bowman's capsule. Normally, it is around 19%.

filtration membrane   Membrane formed by the glomerular capillary endothelium, the basement membrane, and the podocytes of Bowman's capsule.

filtration pressure   Pressure gradient that forces fluid from the glomerular capillary through the filtration membrane into Bowman's capsule; glomerular capillary pressure minus glomerular capsule pressure minus colloid osmotic pressure.

fimbria; pl. **fimbriae** (fim′brē-ă, fim′brē-ē) [L., fringe]   Fringelike structure located at the ostium of the uterine tube.

first messenger   *See* intercellular chemical signal.

fixator (fik-sā′ter)   Muscle that stabilizes the origin of a prime mover.

flagellum; pl. **flagella** (flă-jel′ŭm, flă-jel′ă) [L., whip]   Whiplike locomotory organelle of constant structural arrangement consisting of double peripheral microtubules and two single central microtubules.

flatus (flā′tŭs) [L., a blowing]   Gas or air in the gastrointestinal tract that may be expelled through the anus.

flexion (flek′shŭn) [L., *flectus*]   Bending.

focal point   Point at which light rays cross after passing through a concave lens, such as the lens of the eye.

foliate (fō′lē-āt)   Leaf-shaped.

follicle-stimulating hormone (FSH) (fol′i-kl)   Hormone of the adenohypophysis that, in females, stimulates the Graafian follicles of the ovary and assists in follicular maturation and the secretion of estrogen; in males, FSH stimulates the epithelium of the seminiferous tubules and is partially responsible for inducing spermatogenesis.

follicular phase (fō-lik′ū-lăr)   Time between the end of menses and ovulation, characterized by rapid division of endometrial cells and development of follicles in the ovary; also called proliferative phase.

foramen; pl. **foramina** (fō-rā′men, fō-ram′i-nă)   Hole.

foramen ovale (o-val′ē)   In the fetal heart, the oval opening in the septum secundum; the persistent part of septum primum acts as a valve for this interatrial communication during fetal life; postnatally, the septum primum becomes fused to the septum secundum to close the foramen ovale, forming the fossa ovale.

force   That which produces a motion in the body; pull.

foregut   Cephalic portion of the primitive digestive tube in the embryo.

foreskin   *See* prepuce.

formed elements   Cells (i.e., red and white blood cells) and cell fragments (i.e., platelets) of blood.

formula unit   Relative number of cations and ions in an ionic compound.

fornix (fōr′niks) [L., arch, vault]   Recess at the cervical end of the vagina. Recess deep to each eyelid where the palpebral and bulbar conjunctivae meet.

fovea centralis (fō′vĕ-ă)   Depression in the middle of the macula where there are only cones and no blood vessels.

free energy   Total amount of energy that can be liberated by the complete catabolism of food.

frenulum (fren′ū-lŭm) [L., *frenum,* bridle]   Fold extending from the floor of the mouth to the midline of the undersurface of the tongue.

frequency-modulated signals   Signals, all of which are identical in amplitude, that differ in their frequency—for example, strong stimuli may initiate a high frequency of action potentials and weak stimuli may initiate a low frequency of action potentials.

FSH surge   Increase in plasma follicle-stimulating hormone (FSH) levels before ovulation.

fulcrum (ful′krŭm)   Pivot point.

fundus (fŭn′dŭs) [L., bottom]   Bottom, or rounded end, of a hollow organ—for example, the fundus of the stomach or uterus.

fungiform (fŭn′ji-fōrm)   Mushroom-shaped.

G

G actin (jē ak′tin)   Globular protein molecules that, when bound together, form fibrous actin (F actin).

gallbladder (gawl′blad-er)   Pear-shaped receptacle on the inferior surface of the liver; serves as a storage reservoir for bile.

gamete (gam′ēt)   Ovum or spermatozoon.

gamma globulin (gam′ă glob′ū-lin) [L., *globulus,* globule]   Plasma proteins that include the antibodies.

ganglion; pl. **ganglia** (gang′glē-on, gang′glē-ă) [Gr., swelling, or knot]   Any group of nerve cell bodies in the peripheral nervous system.

gap junction   Small channel between cells that allows the passage of ions and small molecules between cells; provides means of intercellular communication.

gastric gland (gas′trik)   Gland located in the mucosa of the fundus and body of the stomach.

gastric inhibitory polypeptide   Hormone secreted by the duodenum that inhibits gastric acid secretion.

gastric pit   Small pit in the mucous membrane of the stomach, at the bottom of which are the mouths of the gastric glands that secrete mucus, hydrochloric acid, intrinsic factor, pepsinogen, and hormones.

gastrin (gas′trin)   Hormone secreted in the mucosa of the stomach and duodenum that stimulates the secretion of hydrochloric acid by the parietal cells of the gastric glands.

gastrocolic reflex (gas′trō-kol′ik)   Local reflex resulting in mass movement of the contents of the colon, which occurs after the entrance of food into the stomach.

gastroesophageal opening (gas′trō-ē-sof′ă-jē′ăl)   Opening of the esophagus into the stomach; also called cardiac opening.

gene (jēn) [Gr., *genos,* birth, descent]   Functional unit of heredity. Each gene occupies a specific place, or locus, on a chromosome; it is capable of reproducing itself exactly at each cell division; it often is capable of directing the formation of an enzyme or another protein.

genetics (jĕ-net′iks) [Gr., *genesis,* origin or production]   Branch of science that deals with heredity.

genital fold (jen′i-tăl)   Paired longitudinal ridges developing in the embryo on each side of the urogenital orifice. In the male, they form part of the penis; in the female, they form the labia minora.

genital tubercle (jen′i-tăl)   Median elevation just cephalic to the urogenital orifice of an embryo; gives rise to the penis of the male or the clitoris of the female.

genotype (jen′ō-tīp) [Gr., *genos,* birth, descent + *typos,* type]   Genetic makeup of an individual.

germ cell (jerm)   Spermatozoon or ovum.

germ layer   One of three layers in the embryo (ectoderm, endoderm, or mesoderm) from which the four primary tissue types arise.

germinal center   Lighter-staining center of a lymphatic nodule; area of rapid lymphocyte division.

gingiva (jin′ji-vă)   Dense fibrous tissue, covered by mucous membrane, that covers the alveolar processes of the upper and lower jaws and surrounds the necks of the teeth.

girdle (ger′dl)   Belt or zone; the bony region where the limbs attach to the body.

gland [L., *glans,* acorn]   Secretory organ from which secretions may be released into the blood, into a cavity, or onto a surface.

glans penis [L., *glans,* acorn]   Conical expansion of the corpus spongiosum that forms the head of the penis.

globin (glō′bin)   Protein portion of hemoglobin.

glomerular capillary pressure (glō-mār′ū-lăr)   Blood pressure within the glomerulus.

glomerular filtration rate (GFR)   Amount of plasma (filtrate) that filters into Bowman's capsules per minute.

glomerulus (glō-mār′ū-lŭs) [L., *glomus,* ball of yarn]   Mass of capillary loops at the beginning of each nephron, nearly surrounded by Bowman's capsule.

glottis (glot′is) [Gr., aperture of the larynx] Vocal apparatus; includes vocal folds and the cleft between them.

glucocorticoid (gloo-kō-kōr′ti-koyd) Steroid hormone (e.g., cortisol) released by zonula fasciculata of the adrenal cortex; increases blood glucose and inhibits inflammation.

gluconeogenesis (gloo′kō-nē-ō-jen′ē-sis) [Gr., glykys, sweet + neos, new + genesis, production] Formation of glucose from noncarbohydrates, such as proteins (amino acids) or lipids (glycerol).

glycogenesis (glī′kō-jĕ-nō′-sis) Formation of glycogen from glucose molecules.

glycolysis (glī-kol′i-sis) [Gr., glykys, sweet + lysis, a loosening] Anaerobic process during which glucose is converted to pyruvic acid; net of two ATP molecules is produced during glycolysis.

goblet cell Mucus-producing epithelial cell that has its apical end distended with mucin.

Golgi apparatus (gol′jē) Named for Camillo Golgi, Italian histologist and Nobel laureate (1843–1926). Specialized endoplasmic reticulum that concentrates and packages materials for secretion from the cell.

Golgi tendon organ Proprioceptive nerve ending in a tendon.

gomphosis (gom-fō′sis) [Gr., gomphos, bolt, nail + osis, condition] Fibrous joint in which a peglike process fits into a hole.

gonad (gō′nad) [Gr., gone, seed] Organ that produces sex cells; testis of a male or ovary of a female.

gonadal ridge (gō-nad′ăl) Elevation on the embryonic mesonephros; primordial germ cells become embedded in it, establishing it as a testis or an ovary.

gonadotropin (gō′nad-ō-trō′pin) Hormone capable of promoting gonadal growth and function. Two major gonadotropins are luteinizing hormone (LH) and follicle stimulating hormone (FSH).

gonadotropin-releasing hormone (GnRH) Hypothalamic-releasing hormone that stimulates the secretion of gonadotropins (LH and FSH) from the adenohypophysis; also called luteinizing hormone-releasing hormone (LHRH).

granulocyte (gran′ū-lō-sīt) Mature granular white blood cell (neutrophil, basophil, or eosinophil).

granulosa cell (gran-ū-lō′sǎ) Cell in the layer surrounding the primary follicle.

gray matter Collection of nerve cell bodies, their dendritic processes, and associated neuroglial cells within the central nervous system.

gray ramus communicans; pl. rami communicantes (rā′mŭs kō-mū′nī-kans, rā′mī kō-mū-nī-kan′tēz) Connection between a spinal nerve and a sympathetic chain ganglion through which unmyelinated postganglionic axons project.

greater omentum Peritoneal fold passing from the greater curvature of the stomach to the transverse colon, hanging like an apron in front of the intestines.

greater vestibular gland One of two mucus-secreting glands on each side of the lower part of the vagina. The equivalent of the bulbourethral glands in the male.

growth hormone Hormone that stimulates general growth of the individual; stimulates cellular amino acid uptake and protein synthesis; also called somatotropin.

gubernaculum (goo′ber-nak′ū-lŭm) [L., helm] Column of tissue that connects the fetal testis to the developing scrotum; involved in testicular descent.

gustatory (gŭs′tă-tōr-ē) Associated with the sense of taste.

gustatory hair Microvillus of gustatory cell in a taste bud.

gynecomastia (gī′nē-kō-mas′tē-ǎ) [Gr., gyne, woman + mastos, breast] Excessive development of the male mammary glands, which sometimes secrete milk.

H

H zone Area in the center of the A band in which there are no actin myofilaments; contains only myosin.

hair [A.S., hear] Columns of dead keratinized epithelial cells.

hair follicle Invagination of the epidermis into the dermis; contains the root of the hair and receives the ducts of sebaceous and apocrine glands.

Haldane effect Named for Scottish physiologist John S. Haldane (1860–1936). Hemoglobin that is not bound to carbon dioxide binds more readily to oxygen than hemoglobin that is bound to carbon dioxide.

half-life Time it takes for one-half of an administered substance to be lost through biologic processes.

haploid (n) number (hap′loyd) One set of chromosomes, in contrast to diploid; characteristic of gametes.

hapten (hap′ten) [Gr., hapto, to fasten] Small molecule that binds to a large molecule; together they stimulate the specific immune system.

hard palate Floor of the nasal cavity that separates the nasal cavity from the oral cavity; composed of the palatine processes of the maxillary bones and the horizontal plates of the palatine bones; also called bony palate.

haustra (haw′strǎ) [L., machine for drawing water] Sacs of the colon, caused by contraction of the taeniae coli, which are slightly shorter than the gut so that the latter is thrown into pouches.

haversian canal (ha-ver′shan) Named for seventeenth-century English anatomist Clopton Havers (1650–1702). Canal containing blood vessels, nerves, and loose connective tissue and running parallel to the long axis of the bone.

heart skeleton Fibrous connective tissue that provides a point of attachment for cardiac muscle cells, electrically insulates the atria from the ventricles, and forms the fibrous rings around the valves.

heat energy Energy that results from the random movement of atoms, ions, or molecules; the greater the amount of heat energy in an object, the higher the object's temperature.

helicotrema (hel′i-kō-trē′mǎ) [Gr., helix, spiral + traema, hole] Opening at the apex of the cochlea through which the scala vestibuli and the scala tympani of the cochlea connect.

helper T cell Subset of T lymphocytes that increases the activity of B cells and T cells.

hematocrit (hē′mǎ-tō-krit) [Gr., hemato, blood + krin, to separate] Percentage of blood volume occupied by erythrocytes.

hematoma (he-ma-tō′ma) Localized mass of blood released from blood vessels but confined within an organ or a space; the blood is usually clotted.

heme (hēm) Oxygen-carrying, color-furnishing part of hemoglobin.

hemidesmosome (hem-ē-des′mō-sōm) Similar to half a desmosome, attaching epithelial cells to the basement membrane.

hemocytoblast (he′mo-si′to-blast) Blood stem cell derived from mesenchyme that can give rise to red and white blood cells and platelets.

hemoglobin (hē-mō-glō′bin) Red, respiratory protein of red blood cells; consists of 6% heme and 94% globin; transports oxygen and carbon dioxide.

hemolysis (hē-mol′i-sis) [Gr., haima + lysis, destruction] Destruction of red blood cells in such a manner that hemoglobin is released.

hemopoiesis (hē′mō-poy-ē′sis) [Gr., haima, blood + poiesis, a making] Formation of the formed elements of blood—that is, red blood cells, white blood cells, and platelets; also called hematopoiesis.

hemopoietic tissue (hē′mō-poy-et′ik) [Gr., haima, blood + poiesis, to make] Blood-forming tissue.

hemostasis (hē′mō-stā-sis) Arrest of bleeding.

Henry's law Named for English chemist William Henry (1775–1837). The concentration of a gas dissolved in a liquid is equal to the partial pressure of the gas over the liquid times the solubility coefficient of the gas.

heparin (hep′ǎ-rin) Anticoagulant that prevents platelet agglutination and thus prevents thrombus formation.

hepatic artery (he-pa′tik) Branch of the aorta that delivers blood to the liver.

hepatic cord Plate of liver cells that radiates away from the central vein of a liver lobule.

hepatic portal system System of portal veins that carries blood from the intestines, stomach, spleen, and pancreas to the liver.

hepatic portal vein Portal vein formed by the superior mesenteric and splenic veins and entering the liver.

hepatic sinusoid (si′nŭ-soyd) Terminal blood vessel having an irregular and larger caliber than an ordinary capillary within the liver lobule.

hepatic vein Vein that drains the liver into the inferior vena cava.

hepatocyte (hep′ǎ-tō-sīt) Liver cell.

hepatopancreatic ampulla Dilation within the major duodenal papilla that normally receives both the common bile duct and the main pancreatic duct.

hepatopancreatic ampullar sphincter Smooth muscle sphincter of the hepatopancreatic ampulla; also called sphincter of Oddi.

Hering-Breuer reflex (her′ing broy′er) Named for German physiologist Heinrich Ewald Hering (1866–1948) and Austrian internist Josef Breuer (1842–1925). Sensory impulses from stretch receptors in the lungs arrest inspiration; expiration then occurs.

heterozygous (het′er-ō-zī′gŭs) [Gr., heteros, other + zygon, yoke] Having different allelic genes at one or more paired loci in homologous chromosomes.

hiatus (hī-ā′tŭs) [L., aperture, to yawn] Opening.

hilum (hī′lŭm) [L., small bit or trifle] Indented surface on many organs, serving as a point where nerves and vessels enter or leave.

hindgut Caudal or terminal part of the embryonic gut.

histamine (his′tă-mēn) Amine released by mast cells and basophils that promotes inflammation.

histology (his-tol′ō-jē) [Gr., histo, web (tissue) + logos, study] Science that deals with the microscopic structure of cells, tissues, and organs in relation to their function.

holocrine gland (hol′ō-krin) [Gr., holos, complete + krino, to separate] Gland whose secretion is formed by the disintegration of entire cells (e.g., sebaceous gland).

homeostasis (hō′mē-ō-stā′sis) [Gr., homoio, like + stasis, a standing] State of equilibrium in the body with respect to functions and composition of fluids and tissues.

homologous (hŏ-mol′ō-gŭs) [Gr., ratio or relation] Alike in structure or origin.

homozygous (hō-mō-zī′gŭs) [Gr., homos, the same + zygon, yoke] State of having identical allelic genes at one or more paired loci in homologous chromosomes.

hormone (hōr′mōn) [Gr., *hormon,* to set into motion] Substance secreted by endocrine tissues into the blood that acts on a target tissue to produce a specific response.

hormone receptor Protein or glycoprotein molecule of cells that specifically binds to hormones and produces a response.

horn Subdivision of gray matter in the spinal cord. The axons of sensory neurons synapse with neurons in the posterior horn, the cell bodies of motor neurons are in the anterior horn, and the cell bodies of autonomic neurons are in the lateral horn.

human chorionic gonadotropin (HCG) Hormone produced by the placenta; stimulates the secretion of testosterone by the fetus; during the first trimester, it stimulates ovarian secretion from the corpus luteum of the estrogen and progesterone required for the maintenance of the placenta. In a male fetus, it stimulates the secretion of testosterone by the fetal testis.

humoral immunity (hū′mōr-ăl) [L., *humor,* a fluid] Immunity due to antibodies in serum.

hyaline cartilage (hī′ă-lin) [Gr., *hyalos,* glass] Gelatinous, glossy cartilage tissue consisting of cartilage cells and their matrix; contains collagen, proteoglycans, and water.

hyaluronic acid (hī′ă-loo-ron′ik) Mucopoly-saccharide made up of alternating β-(1,4)-linked residues of hyalobiuronic acid, forming a gelatinous material in the tissue spaces and acting as a lubricant and shock absorbant generally throughout the body.

hydrochloric acid (HCl) (hī-drō-klōr′ik) Acid of gastric juice.

hydrogen bond (hī′drŏ-jen) Hydrogen atoms bound covalently to either N or O atoms have a small positive charge that is weakly attracted to the small negative charge of other atoms, such as O or N; it can occur within a molecule or between different molecules.

hydrophilic (hi-dro-fil′ik) [Gr., *hydro,* water + *philos,* love] Denoting the property of attracting or associating with water molecules, possessed by polar molecules and ions; the opposite of hydrophobic.

hydrophobic (hi-dro-fob′ik) [Gr., *hydro,* water + *phobos,* fear] Lacking an attraction to water, possessed by nonpolar molecules; the opposite of hydrophilic.

hydroxyapatite (hī-drok′sē-ap′ă-tīt) Mineral with the empiric formula $3 \ Ca_3(PO_4)_2 \cdot Ca(OH)_2$; the main mineral of bone and teeth.

hymen (hī′men) [Gr., membrane] Thin, membranous fold partly occluding the vaginal external orifice; normally disrupted by sexual intercourse or other mechanical phenomena.

hyoid bone (hī′oyd)(Gr., *hyoeides,* shaped like the Greek letter epsilon [ε]) U-shaped bone between the mandible and larynx.

hypercalcemia (hī′per-kal-sē′mē-ă) Abnormally high levels of calcium in the blood.

hypercapnia (hī′per-kap′nē-ă) Higher than normal levels of carbon dioxide in the blood or tissues.

hyperkalemia (hī′per-kă-lē′mē-ă) Greater than normal concentration of potassium ions in the circulating blood.

hypernatremia (hī′per-nă-trē′mē-ă) Abnormally high plasma concentration of sodium ions.

hyperosmotic (hī′per-oz-mot′ik) [Gr., *hyper,* above + *osmos,* an impulsion] Having a greater osmotic concentration or pressure than a reference solution.

hyperpolarization (hī′per-pō′lăr-i-zā′shŭn) Increase in the charge difference across the plasma membrane; causes the charge difference to move away from 0 mV.

hypertonic (hī-per-ton′ik) [Gr., *hyper,* above + *tonos,* tension] Solution that causes cells to shrink.

hypertrophy (hī-per′trŏ-fē) [Gr., *hyper,* above + *trophe,* nourishment] Increase in bulk or size; not due to an increase in number of individual elements.

hypocalcemia (hī-pō-kal-sē′mē-ă) Abnormally low levels of calcium in the blood.

hypocapnia (hī′pō-kap′nē-ă) Lower than normal levels of carbon dioxide in the blood or tissues.

hypodermis (hī′pō-der′mis) [Gr., *hypo,* under + *dermis,* skin] Loose areolar connective tissue found deep to the dermis that connects the skin to muscle or bone.

hypokalemia (hī′pō-ka-lē′mē-ă) Abnormally small concentration of potassium ions in the blood.

hyponatremia (hī′pō-nă-trē′mē-ă) Abnormally low plasma concentration of sodium ions.

hyponychium (hī-pō-nik′ē-ŭm) [Gr., *hypo,* under + *onyx,* nail] Thickened portion of the stratum corneum under the free edge of the nail.

hypopolarization Change in the electric charge difference across the plasma membrane that causes the charge difference to be smaller or move closer to 0 mV.

hyposmotic (hī′pos-mot′ik) [Gr., *hypo,* under + *osmos,* an impulsion] Having a lower osmotic concentration or pressure than a reference solution.

hypospadias (hī′pō-spā′dē-ăs) [Gr., one having the orifice of the penis too low; *hypospao,* to draw away from under] Developmental anomaly in the wall of the urethra so that the canal is open for a greater or lesser distance on the undersurface of the penis; a similar defect in the female in which the urethra opens into the vagina.

hypothalamohypophysial portal system (hī′pō-thal′ă-mō-hī′pō-fiz′ē-ăl) Series of blood vessels that carries blood from the area of the hypothalamus to the anterior pituitary gland; originates from capillary beds in the hypothalamus and terminates as a capillary bed in the anterior pituitary gland.

hypothalamohypophyseal tract Nerve tract consisting of the axons of neurosecretory cells and extending from the hypothalamus into the posterior pituitary gland. Hormones produced in the neurosecretory cell bodies in the hypothalamus are transported through the hypothalamohypophyseal tract to the posterior pituitary gland, where they are stored for later release.

hypothalamus (hī′pō-thal′ă-mŭs) [Gr., *hypo,* under + *thalamus,* bedroom] Important autonomic and neuroendocrine control center beneath the thalamus.

hypothenar (hī-pō-thē′nar) [Gr., *hypo,* under + *thenar,* palm of the hand] Fleshy mass of tissue on the medial side of the palm; contains muscles responsible for moving the little finger.

hypotonic (hī-pō-ton′ik) [Gr., *hypo,* under + *tonos,* tension] Solution that causes cells to swell.

I

I band Area between the ends of two adjacent myosin myofilaments within a myofibril; Z disk divides the I band into two equal parts.

ileocecal sphincter (il′ē-ō-sē′kăl) Thickening of circular smooth muscle between the ileum and the cecum, forming the ileocecal valve.

ileocecal valve Valve formed by the ileocecal sphincter between the ileum and the cecum.

ileum (il′ē-ŭm) [Gr., *eileo,* to roll up; twist] Third portion of the small intestine, extending from the jejunum to the ileocecal opening into the large intestine; the posterior inferior bone of the coxal bone.

immunity (i-mū′ni-tē) [L., *immunis,* free from service] Resistance to infectious disease and harmful substances.

immunization (im-mū′ni-zā′shun) Process by which a subject is rendered immune by deliberately introducing an antigen or antibody into the subject.

immunoglobulin (IG) (im′ū-nō-glob′ū-lin) Antibody found in the gamma globulin portion of plasma.

implantation (im-plan-tā′shŭn) Attachment of the blastocyst to the endometrium of the uterus, occurring 6 or 7 days after fertilization of the ovum.

impotence (im′pō-tens) Inability to accomplish the male sexual act; caused by psychic or physical factors; also called erectile dysfunction (ED).

incisor (in-sī′zŏr) [L., *incido,* to cut into] One of the anterior, cutting teeth.

incisura (in′sī-soo′ră) [L., a cutting into] Notch or indentation at the edge of any structure.

incus (ing′kus) [L., anvil] Middle of the three ossicles in the middle ear.

inferior colliculus (ko-lik′ū-lŭs) [L., *collis,* hill] One of two rounded eminences of the midbrain; involved with hearing.

inferior vena cava Vein that returns blood from the lower limbs and the greater part of the pelvic and abdominal organs to the right atrium.

inflammatory response (in-flam′ă-tōr-ē) Complex sequence of events involving chemicals and immune cells that results in the isolation and destruction of antigens and tissues near the antigens.

infundibulum (in-fŭn-dib′ū-lŭm) [L., funnel] Funnel-shaped structure or passage—for example, the infundibulum that attaches the hypophysis to the hypothalamus or the funnel-like expansion of the uterine tube near the ovary.

Infusion Introduction of a fluid other than blood, such as a saline or glucose solution, into the blood.

inguinal canal (ing′gwi-năl) Passage through the lower abdominal wall that transmits the spermatic cord in the male and the round ligament in the female.

inhibin (in-hib′in) Polypeptide secreted from the testes that inhibits FSH secretion.

inhibitory neuron (in-hib′i-tōr-ē) Neuron that produces IPSPs and has an inhibitory influence.

inhibitory postsynaptic potential (IPSP) Hyperpolarization in the postsynaptic membrane, which causes the membrane potential to move away from threshold.

innate immunity (i′nāt, i-nāt′) Immune system response that is the same with each exposure to an antigen; there is no ability for the system to remember a previous exposure to the antigen.

inner cell mass Group of cells at one end of the blastocyst, part of which forms the body of the embryo.

inner ear Part of the ear that contains the sensory organs for hearing and balance; contains the bony and membranous labyrinth.

insensible perspiration [L., *per,* through + *spiro,* to breathe everywhere] Perspiration that evaporates before it is perceived as moisture on the skin; the term sometimes includes evaporation from the lungs.

insertion (in-ser′shŭn) More movable attachment point of a muscle; usually, the lateral or distal end of a muscle associated with the limbs; also called mobile end.

inspiratory capacity (in-spī′ră-tō-rē) Volume of air that can be inspired after a normal expiration; the sum of the tidal volume and the inspiratory reserve volume.

inspiratory reserve volume   Maximum volume of air that can be inspired after a normal inspiration.

insulin (in′sŭ-lin)   Protein hormone secreted from the pancreas that increases the uptake of glucose and amino acids by most tissues.

interatrial septum (in-ter-ā′trē-ăl) [L., *saeptum*, partition]   Wall between the atria of the heart.

intercalated disk (in-ter′kă-lā-ted)   Cell-to-cell attachment with gap junctions between cardiac muscle cells.

intercalated duct   Minute duct of glands, such as the salivary gland, and the pancreas; leads from the acini to the interlobular ducts.

intercellular (in-ter-sel′ū-lăr)   Between cells.

intercellular chemical signal   Chemical that is released from cells and passes to other cells; acts as a signal that allows cells to communicate with each other; also called first messenger.

interferons (in-ter-fēr′onz)   Proteins that prevent viral replication.

interlobar artery (in-ter-lō′bar)   Branch of the segmental arteries of the kidney; runs between the renal pyramids and gives rise to the arcuate arteries.

interlobular artery (in-ter-lob′ū-lăr)   Artery that passes between lobules of an organ; branches of the interlobar arteries of the kidney pass outward through the cortex from the arcuate arteries and supply the afferent arterioles.

interlobular duct   Any duct leading from a lobule of a gland and formed by the junction of the intercalated ducts draining the acini.

interlobular vein   Vein that parallels the interlobular arteries; in the kidney, it drains the peritubular capillary plexus, emptying into arcuate veins.

intermediate olfactory area   Part of the olfactory cortex responsible for the modulation of olfactory sensations.

internal anal sphincter [Gr., *sphinkter*, band or lace]   Smooth muscle ring at the upper end of the anal canal.

internal naris; pl. nares (nā′ris, nā′rēs)   Opening from the nasal cavity into the nasopharynx.

internal spermatic fascia   Inner connective tissue covering of the spermatic cord.

internal urinary sphincter   Traditionally recognized as a sphincter composed of a thickening of the middle smooth muscle layer of the bladder around the urethral opening.

interphase (in′ter-fāz)   Period between active cell divisions when DNA replication occurs.

interstitial (in-ter-stish′ăl) [L., *inter*, between + *sisto*, to stand]   Space within tissue. Interstitial growth is growth from within.

interstitial cell   Cell between the seminiferous tubules of the testes; secretes testosterone; also called Leydig cell.

interventricular septum (in-ter-ven-trik′ū-lăr)   Wall between the ventricles of the heart.

intestinal gland (in-tes′ti-năl)   Tubular gland in the mucous membrane of the small and large intestines; also called crypt.

intracellular (in-tră-sel′ū-lăr)   Inside a cell.

intracellular mediator   Molecule produced within a cell that binds to a macromolecule, such as receptors or enzymes inside that cell, that regulates their activities; also called second messenger.

intramural plexus (in′tră-mū′răl plek′sus)   Combined submucosal and myenteric plexuses.

intrinsic clotting pathway (in-trin′sik)   Series of chemical reactions resulting in clot formation that begins with chemicals (e.g., plasma factor XII) within the blood.

intrinsic factor   Factor secreted by the parietal cells of gastric glands and required for adequate absorption of vitamin B_{12}.

intrinsic muscles   Muscles located within the structure being moved.

intubation (in-too-ba′shun)   Insertion of a tube into an opening, a canal, or a hollow organ.

inversion (in-ver′zhŭn) [L., *inverto*, to turn about]   Turning inward.

ion (ī′on) [Gr., *ion*, going]   Atom or group of atoms carrying a charge of electricity by virtue of having gained or lost one or more electrons.

ion channel   Pore in the plasma membrane through which ions, such as sodium and potassium, move.

ionic bonding (ī-on′ik)   Chemical bond that is formed when one atom loses an electron and another accepts that electron.

iris (ī′ris)   Specialized portion of the vascular tunic; the "colored" portion of the eye that can be seen through the cornea.

ischemia (is-kē′mē-ă) [Gr., *ischo*, to keep back + *haima*, blood]   Reduced blood supply to an area of the body.

ischium (is′kē-ŭm)   Superior bone of the coxal bone.

isomers (ī′sō-merz) [Gr., *isos*, equal + *meros*, part]   Molecules having the same number and types of atoms but differing in their three-dimensional arrangement.

isometric contraction (ī-sō-met′rik) [Gr., *isos*, equal + *metron*, measure]   Muscle contraction in which the length of the muscle does not change but the tension produced increases.

isosmotic (ī′sō-os-mot′ik) [Gr., *isos*, equal + *osmos*, an impulsion]   Having the same osmotic concentration or pressure as a reference solution.

isotonic (ī′sō-ton′ik) [Gr., *isos*, equal + *tonos*, tension]   Type of solution that causes cells to neither shrink nor swell.

isotope (ī′sō-tōp) [Gr., *isos*, equal + *topos*, part, place]   Either of two or more atoms that have the same atomic number but a different number of neutrons.

isthmus (is′mŭs)   Constriction connecting two larger parts of an organ, such as the constriction between the body and the cervix of the uterus or the portion of the uterine tube between the ampulla and the uterus.

J

jaundice (jawn′dis) [Fr., *jaune*, yellow]   Yellowish staining of the integument, the sclerae, and the other tissues with bile pigments.

jejunum (jě-joo′nŭm) [L., *jejunus*, empty]   Second portion of the small intestine; located between the duodenum and the ileum.

juxtaglomerular apparatus (jŭks′tă-glŏ-mer′ū-lăr)   Complex consisting of juxtaglomerular cells of the afferent arteriole and macular densa cells of the distal convoluted tubule near the renal corpuscle; secretes renin.

juxtaglomerular cell   Modified smooth muscle cell of the afferent arteriole located at the renal corpuscle; a component of the juxtaglomerular apparatus.

juxtamedullary nephron (jŭks′tă-med′ŭ-lăr-ē)   Nephron located near the junction of the renal cortex and medulla.

K

karyotype (kar′ē-ō-tīp)   Display of chromosomes arranged by pairs.

keratinization (ker′ă-tin-i-zā′shŭn)   Production of keratin and changes in the chemical and structural character of epithelial cells as they move to the skin surface.

keratinized (ker′ă-ti-nīzd) [Gr., *keras*, horn]   Having become a structure that contains keratin, a protein found in skin, hair, nails, and horns.

keratinocyte (ke-rat′i-nō-sīt) [Gr., *keras*, horn + *kytos*, cell]   Epidermal cell that produces keratin.

keratohyalin (ker′ă-tō-hī′ă-lin)   Nonmembrane-bound protein granule in the cytoplasm of stratum granulosum cells of the epidermis.

ketogenesis (kē-tō-jen′ě-sis)   Production of ketone bodies, such as from acetyl-CoA.

ketone body (kē′tōn)   One of a group of ketones, including acetoacetic acid, β-hydroxybutyric acid, and acetone.

ketosis (kē-tō′sis) [*ketone* + *osis*, condition]   Condition characterized by the enhanced production of ketone bodies, as in diabetes mellitus or starvation.

kidney (kid′nē) [A.S., *cwith*, womb, belly + *neere*, kidney]   One of the two organs that excrete urine. The kidneys are bean-shaped organs approximately 11 cm long, 5 cm wide, and 3 cm thick lying on each side of the spinal column, posterior to the peritoneum, approximately opposite the twelfth thoracic and first three lumbar vertebrae.

kilocalorie (Kcal) (kil′ō-kal-ō-rē)   Quantity of energy required to raise the temperature of 1 kg of water 1°C; 1000 calories. Equal to one dietary calorie.

kinetic energy (ki-net′ik)   Motion energy or energy that can do work.

kinetic labyrinth (lab′i-rinth)   Part of the membranous labyrinth composed of the semicircular canals; detects dynamic or kinetic equilibrium, such as movement of the head.

kinetochore (ki-nē′tō-kōr, ki-net′o-) [Gr., *kinēto*, moving + Gr., *chōra*, space]   Structural portion of the chromosome to which microtubules attach.

Korotkoff sounds (kō-rot′kof)   Named for Russian physician Nikolai S. Korotkoff (1874–1920). Sounds heard over an artery when blood pressure is determined by the auscultatory method; caused by turbulent flow of blood.

L

labium majus; pl. labia majora (lā′bē-ŭm, lā′bē-ă)   One of two rounded folds of skin surrounding the labia minora and vestibule; homolog of the scrotum in males.

labium minus; pl. labia minora   One of two narrow longitudinal folds of mucous membrane enclosed by the labia majora and bounding the vestibule; anteriorly they unite to form the prepuce.

lacrimal apparatus (lak′ri-măl)   Lacrimal, or tear, gland in the superolateral corner of the orbit of the eye and a duct system that extends from the eye to the nasal cavity.

lacrimal canaliculus   Canal that carries excess tears away from the eye; located in the medial canthus and opening on the lacrimal papilla.

lacrimal gland   Tear gland located in the superolateral corner of the orbit.

lacrimal papilla   Small lump of tissue in the medial canthus or corner of the eye; the lacrimal canal opens within the lacrimal papilla.

lacrimal sac   Enlargement in the lacrimal canal that leads into the nasolacrimal duct.

lactation (lak-tā′shŭn) [L., *lactatio*, suckle]   Period after childbirth during which milk is formed in the breasts.

lacteal (lak′tē-ăl)   Lymphatic vessel in the wall of the small intestine that carries chyle from the intestine and absorbs fat.

lactiferous duct (lak-tif′er-ŭs)   One of 15–20 ducts that drain the lobes of the mammary gland and open onto the surface of the nipple.

lactiferous sinus   Dilation of the lactiferous duct just before it enters the nipple.

lacuna; pl. lacunae (lă-koo′nă, -koo′nē) [L., *lacus,* a hollow, a lake]   Small space or cavity; potential space within the matrix of bone or cartilage normally occupied by a cell that can be visualized only when the cell shrinks away from the matrix during fixation; space containing maternal blood within the placenta.

lag phase   One of the three phases of muscle contraction; time between the application of the stimulus and the beginning of muscular contraction. Also called latent phase.

lamella; pl. lamellae (lă-mel′ă, lă-mel′ē)   Thin sheet or layer of bone.

lamellated corpuscle (lam′ĕ-lāt-ed)   Oval receptor found in the deep dermis or hypodermis (responsible for deep cutaneous pressure and vibration) and in tendons (responsible for proprioception); also called Pacinian corpuscle.

lamina; pl. laminae (lam′i-nă, lam′i-nē) [L., *lamina,* plate, leaf]   Thin plate—for example, the thinner portion of the vertebral arch.

lamina propria (prō′prē-ă)   Layer of connective tissue underlying the epithelium of a mucous membrane.

laminar flow (lam′i-nar)   Relative motion of layers of a fluid along smooth, concentric, parallel paths.

Langerhans cell   Named for German anatomist Paul Langerhans (1847–1888); dendritic cell found in the skin.

lanugo (lă-noo′gō) [L., *lana,* wool]   Fine, soft, unpigmented fetal hair.

Laplace's law   Named for French mathematician Pierre S. de Laplace (1749–1827); the force that stretches the wall of a blood vessel is proportional to the radius of the vessel times the blood pressure.

large intestine   Portion of the digestive tract extending from the small intestine to the anus.

laryngitis (lar-in-jī′tis)   Inflammation of the mucous membrane of the larynx.

laryngopharynx (lă-ring′gō-far-ingks)   Part of the pharynx lying posterior to the larynx.

larynx; pl. larynges (lar′ingks, lă-rin′jēz)   Organ of voice production located between the pharynx and the trachea; it consists of a framework of cartilages and elastic membranes housing the vocal folds and the muscles that control the position and tension of these elements.

last menstrual period (LMP)   Beginning of the last menstruation before pregnancy; used clinically to time events during pregnancy.

lateral geniculate nucleus (je-nik′ū-lāt)   Nucleus of the thalamus where fibers from the optic tract terminate.

lateral olfactory area (ol-fak′tŏ-rē)   Part of the olfactory cortex involved in the conscious perception of olfactory stimuli.

lens   Transparent biconvex structure lying between the iris and the vitreous humor.

lens fiber   Epithelial cell that makes up the lens of the eye.

lesser omentum (ō-men′tŭm) [L., membrane that encloses the bowels]   Peritoneal fold passing from the liver to the lesser curvature of the stomach and to the upper border of the duodenum for a distance of approximately 2 cm beyond the pylorus.

lesser vestibular gland (ves-tib′ū-lăr)   Number of minute mucous glands opening on the surface of the vestibule between the openings of the vagina and urethra; also called paraurethral gland.

leukocyte (loo′kō-sīt)   White blood cell.

leukocytosis (loo′kō-sī-tō′sis)   Abnormally large number of white blood cells in the blood.

leukopenia (loo-kō-pē′nē-ă)   Lower than normal number of white blood cells in the blood.

leukotriene (loo-kō-trī′ēn)   Specific class of physiologically active fatty acid derivatives present in many tissues.

lever   Rigid shaft capable of turning about a fulcrum or pivot point.

LH surge   Increase in plasma luteinizing hormone (LH) levels before ovulation and responsible for initiating it.

ligamentum arteriosum (lig′ă-men′tŭm)   Remains of the ductus arteriosus.

ligamentum venosum   Remnant of the ductus venosus.

ligand   *See* chemical signal.

ligand-gated ion channel   Ion channel in a plasma membrane caused to either open or close by a ligand binding to a receptor.

limbic system (lim′bik) [L., *limbus,* border]   Part of the brain involved with emotions and olfaction; includes the cingulate gyrus, hippocampus, habenular nuclei, parts of the basal ganglia, the hypothalamus (especially the mammillary bodies, the olfactory cortex, and various nerve tracts).

lingual tonsil (ling′gwăl)   Collection of lymphoid tissue on the posterior portion of the dorsum of the tongue.

lipase (lip′ās)   Any fat-splitting enzyme.

lipid (li′pid) [Gr., *lipos,* fat]   Substance composed principally of carbon, oxygen, and hydrogen; contains a lower ratio of oxygen to carbon and is less polar than carbohydrates; generally soluble in nonpolar solvents.

lipid bilayer   Double layer of lipid molecules forming the plasma membrane and other cellular membranes.

lipochrome (lip′ō-krōm)   Lipid-containing pigment that is metabolically inert.

lipotropin (li-pō-trō′pin)   One of the peptide hormones released from the adenohypophysis; increases lipolysis in fat cells.

liver (liv′er)   Largest gland of the body, lying in the upper-right quadrant of the abdomen just inferior to the diaphragm; secretes bile and is of great importance in carbohydrate and protein metabolism and in detoxifying chemicals.

lobar bronchus (brong′kus)   Branch from a primary bronchus that conducts air to each lobe of the lungs. There are two branches in the left lung and three branches from the primary bronchus in the right lung. Also called secondary bronchus.

lobe (lōb)   Rounded, projecting part, such as the lobe of a lung, the liver, or a gland.

lobule (lob′ūl)   Small lobe or subdivision of a lobe, such as a lobule of the lung or a gland.

local inflammation   Inflammation confined to a specific area of the body. Symptoms include redness, heat, swelling, pain, and loss of function.

local potential   Depolarization that is not propagated and that is graded or proportional to the strength of the stimulus.

local reflex   Reflex of the intramural plexus of the digestive tract that does not involve the brain or spinal cord.

locus; pl. loci (lō′kŭs, lō′sī)   Place; usually a specific site.

loop of Henle   Named for German anatomist Friedrich G. J. Henle (1809–1885). U-shaped part of the nephron extending from the proximal to the distal convoluted tubule and consisting of descending and ascending limbs. Some of the loops of Henle extend into the renal pyramids.

lower respiratory tract   Larynx, trachea, and lungs.

lung recoil   Decrease in the size of an expanded lung as a result of a decrease in the size (volume) of its alveoli; due to elastic recoil of elastic fibers surrounding alveoli and water surface tension of a thin film of water within alveoli.

lunula; pl. lunulae (loo′noo-lă, loo′noo-lē) [L., *luna,* moon]   White, crescent-shaped portion of the nail matrix visible through the proximal end of the nail.

luteal phase (loo′tē-ăl)   Portion of the menstrual cycle extending from the time of formation of the corpus luteum after ovulation to the time when menstrual flow begins; usually 14 days in length; also called secretory phase.

luteinizing hormone (LH) (loo′tē-ĭ-nīz-ing)   In females, hormone stimulating the final maturation of the follicles and the secretion of progesterone by them, with their rupture releasing the ovum, and the conversion of the ruptured follicle into the corpus luteum; in males, stimulates the secretion of testosterone in the testes.

lymph (limf) [L., *lympha,* clear spring water]   Clear or yellowish fluid derived from interstitial fluid and found in lymph vessels.

lymph node   Encapsulated mass of lymph tissue found among lymph vessels.

lymphatic capillary   Beginning of the lymphatic system of vessels; lined with flattened endothelium lacking a basement membrane.

lymphatic nodule   Small accumulation of lymph tissue lacking a distinct boundary.

lymphatic sinus   Channels in a lymph node crossed by a reticulum of cells and fibers.

lymphatic vessel   One of the system of vessels carrying lymph from the lymph capillaries to the veins.

lymphedema (limf′e-dē′mă)   Swelling of tissues resulting from the excessive accumulation of fluid caused by the removal, damage, or blockage of lymphatic vessels or lymph nodes; usually results in swelling of the arm or leg.

lymphoblast (lim′fō-blast)   Cell that matures into a lymphocyte.

lymphocyte (lim′fō-sīt)   Nongranulocytic white blood cell formed in lymphoid tissue.

lymphokine (lim′fō-kīn)   Chemical produced by lymphocytes that activates macrophages, attracts neutrophils, and promotes inflammation.

lysis (lī′sis) [Gr., *lysis,* a loosening]   Process by which a cell swells and ruptures.

lysosome (lī′sō-sōm) [Gr., *lysis,* loosening + *soma,* body]   Membrane-bounded vesicle containing hydrolytic enzymes that function as intracellular digestive enzymes.

lysozyme (lī′sō-zīm)   Enzyme that is destructive to the cell walls of certain bacteria; present in tears and some other fluids of the body.

M

M line   Line in the center of the H zone made of delicate filaments that holds the myosin myofilaments in place in the sarcomere of muscle fibers.

macrophage (mak′rō-făj) [Gr., *makros,* large + *phagein,* to eat]   Any large, mononuclear phagocytic cell.

macula; pl. maculae (mak′ū-lă, mak′ū-lē) [L., a spot]   Sensory structure in the utricle and saccule, consisting of hair cells and a gelatinous mass embedded with otoliths.

macula densa   Cells of the distal convoluted tubule located at the renal corpuscle and forming part of the juxtaglomerular apparatus.

main bronchus; pl. **bronchi** (brong′kŭs, brong′kī) One of two tubes arising at the inferior end of the trachea; each primary bronchus extends into one of the lungs; also called primary bronchus.

major duodenal papilla Point of opening of the common bile duct and pancreatic duct into the duodenum.

major histocompatibility complex (MHC) molecules Genes that control the production of major histocompatibility complex proteins, which are glycoproteins found on the surfaces of cells. The major histocompatibility proteins serve as self-markers for the immune system and are used by antigen-presenting cells to present antigens to lymphocytes.

male pronucleus Nuclear material of the sperm cell after the ovum has been penetrated by the sperm cell.

malignant (mă-lig′nănt) Resistant to treatment; occurring in severe form and frequently fatal; having locally invasive and destructive growth and metastasis.

malleus; pl. **mallei** (mal′ē-ŭs, mal′ē-ī) [L., hammer] Largest of the three auditory ossicles; attached to the tympanic membrane.

mamillary bodies (mam′i-lār-ē) [L., breast- or nipple-shaped] Nipple-shaped structures at the base of the hypothalamus.

mamma; pl. **mammae** (mam′ă, mam′ē) Breast; the organ of milk secretion; one of two hemispheric projections of variable size situated in the subcutaneous layer over the pectoralis major muscle on each side of the chest; it is rudimentary in the male.

mammary ligaments (mam′ă-rē) Well-developed ligaments that extend from the overlying skin to the fibrous stroma of mammary gland; also called Cooper's ligaments.

manubrium; pl. **manubria** (mă-noo′brē-ŭm, mă-noo′bre-ă) [L., handle] Part of a bone representing the handle, such as the manubrium of the sternum representing the handle of a sword.

mass movement Forcible peristaltic movement of short duration, occurring only three or four times a day, which moves the contents of the large intestine.

mass number Number of protons plus the number of neutrons in each atom.

mastication (mas-ti-kā′shŭn) [L., mastico, to chew] Process of chewing.

mastication reflex Repetitive cycle of relaxation and contraction of the muscles of mastication.

mastoid (mas′toyd) [Gr., mastos, breast] Resembling a breast.

mastoid air cells Spaces within the mastoid process of the temporal bone connected to the middle ear by ducts.

mature follicle Ovarian follicle in which the oocyte attains its full size. The follicle contains a fluid-filled antrum and is surrounded by the theca interna and externa. Also called Graafian follicle.

maximal stimulus Stimulus resulting in a local potential just large enough to produce the maximum frequency of action potentials.

meatus (mē-ā′tŭs) [L., to go, pass] Passageway or tunnel.

mechanoreceptor (mek′ă-nō-rē-sep′tŏr) Sensory receptor that responds to mechanical pressures—for example, pressure receptors in the carotid sinus or touch receptors in the skin.

meconium (mē-kō′nē-ŭm) [Gr., mekon, poppy] First intestinal discharges of the newborn infant, greenish in color and consisting of epithelial cells, mucus, and bile.

medial olfactory area Part of the olfactory cortex responsible for the visceral and emotional reactions to odors.

medulla oblongata (me-dool′ă ob-long-gah′tă) Inferior portion of the brainstem that connects the spinal cord to the brain and contains autonomic centers controlling functions such as heart rate, respiration, and swallowing.

medullary cavity (med′ul-er-ē, med′oo-lār-ē) Large, marrow-filled cavity in the diaphysis of a long bone.

medullary ray Extension of the kidney medulla into the cortex, consisting of collecting ducts and loops of Henle.

megakaryoblast (meg-ă-kar′ē-ō-blast) [Gr., mega + karyon, nut, nucleus + blastos, germ] Cell that gives rise to platelets or thrombocytes.

meibomian cyst (mī-bō′mē-an) Named for German anatomist Hendrik Meibom (1638–1700). A chronic inflammation of a meibomian gland.

meibomian gland Sebaceous gland near the inner margins of the eyelid; secretes sebum that lubricates the eyelid and retains tears.

meiosis (mī-ō′sis) [Gr., a lessening] Cell division that results in the formation of gametes. Consists of two divisions, which result in one (female) or four (male) gametes, each of which contains one-half the number of chromosomes in the parent cell.

Meissner corpuscle (mīs′nerz kōr′pŭs-l) Named for German histologist Georg Meissner (1829–1905). See tactile corpuscle.

melanin (mel′ă-nin) [Gr., melas, black] Group of related molecules responsible for skin, hair, and eye color. Most melanins are brown to black pigments; some are yellowish or reddish.

melanocyte (mel′ă-nō-sīt) [Gr., melas, black + kytos, cell] Cell found mainly in the stratum basale; produces the brown or black pigment melanin.

melanocyte-stimulating hormone (MSH) Peptide hormone secreted by the anterior pituitary; increases melanin production by melanocytes, making the skin darker in color.

melanosome (mel′ă-nō-sōm) [Gr., melas, black + soma, body] Membranous organelle containing the pigment melanin.

melatonin (mel-ă-tōn′in) Hormone (amino acid derivative) secreted by the pineal body; inhibits the secretion of gonadotropin-releasing hormone from the hypothalamus.

membrane attack complex (MAC) Channel through a plasma membrane produced by activated complement proteins, primarily complement protein C9; in nucleated cells, water enters the channel, causing lysis of cells.

membrane-bound receptor Receptor molecule, such as a hormone receptor, that is bound to the plasma membrane of the target cell.

membranous labyrinth (mem′bră-nŭs lab′i-rinth) Membranous structure within the inner ear consisting of the cochlea, vestibule, and semicircular canals.

membranous urethra (ū-rē′thră) Portion of the male urethra, approximately 1 cm in length, extending from the prostate gland to the beginning of the penile urethra.

memory cell Small lymphocyte that is derived from a B cell or T cell and that rapidly responds to a subsequent exposure to the same antigen.

menarche (me-nar′kē) [Gr., mensis, month + arche, beginning] Establishment of menstrual function; the time of the first menstrual period or flow.

meninx; pl. **meninges** (mē′ninks, mĕ-nin′jes) [Gr., membrane] Connective tissue membrane surrounding the brain.

meniscus; pl. **menisci** (me-nis′kus, me-nis′sī) Crescent-shaped intraarticular fibrocartilage found in certain joints, such as the crescent-shaped fibrocartilaginous structure of the knee.

menopause (men′ō-pawz) [Gr., mensis, month + pausis, cessation] Permanent cessation of the menstrual cycle.

menses (men′sēz) [L., mensis, month] Periodic hemorrhage from the uterine mucous membrane, occurring at approximately 28-day intervals.

menstrual cycle (men′stroo-ăl) Series of changes that occur in sexually mature, nonpregnant women and result in menses. Specifically refers to the uterine cycle but is often used to include both the uterine and ovarian cycles.

Merkel (tactile) disk (mer′kelz) Named for German anatomist Friedrich Merkel (1845–1919). See tactile disk.

merocrine gland (mer′ō-krin) [Gr., meros, part + krino, to separate] Gland that secretes products with no loss of cellular material—for example, water-producing sweat glands.

mesencephalon (mez-en-sef′ă-lon) [Gr., mesos, middle + enkephalos, brain] Midbrain in both the embryo and adult; consists of the cerebral peduncle and the corpora quadrigemini.

mesentery (mes′en-ter-ē) [Gr., mesos, middle + enteron, intestine] Double layer of peritoneum extending from the abdominal wall to the abdominal viscera, conveying to it its vessels and nerves.

mesoderm (mez′ō-derm) Middle of the three germ layers of an embryo.

mesonephros (mez′ō-nef′ros) One of three excretory organs appearing during embryonic development; forms caudal to the pronephros as the pronephros disappears. It is well developed and is functional for a time before the establishment of the metanephros, which gives rise to the kidney. It undergoes regression as an excretory organ, but its duct system is retained in the male as the efferent ductule and epididymis.

mesosalpinx (mez′ō-sal′pinks) [Gr., mesos, middle + salpinx, trumpet] Part of the broad ligament supporting the uterine tube.

mesothelium (mez-ō-thē′lē-ŭm) Single layer of flattened cells forming an epithelium that lines serous cavities, such as peritoneum, pleura, pericardium.

mesovarium (mez′ō-vā′rē-ŭm) Short peritoneal fold connecting the ovary with the broad ligament of the uterus.

messenger ribonucleic acid (mRNA) Type of RNA that moves out of the nucleus and into the cytoplasm, where it is used as a template to determine the structure of proteins.

metabolism (mĕ-tab′ō-lizm) [Gr., metabole, change] Sum of all the chemical reactions that take place in the body, consisting of anabolism and catabolism. Cellular metabolism refers specifically to the chemical reactions within cells.

metacarpal (met′ă-kar′păl) Relating to the fine bones of the hand between the carpus (wrist) and the phalanges.

metanephros (met-ă-nef′ros) Most caudally located of the three excretory organs appearing during embryonic development; becomes the permanent kidney of mammals. In mammalian embryos, it is formed caudal to the mesonephros and develops later as the mesonephros undergoes regression.

metaphase (met′ă-fās) Time during cell division when the chromosomes line up along the equator of the cell.

metarteriole (met′ar-tēr′ē-ōl) One of the small peripheral blood vessels that contain scattered groups of smooth muscle fibers in their walls; located between the arterioles and the true capillaries.

metastasis (mĕ-tas′tă-sis) Shifting of a disease or its local manifestations or the spread of a disease from one part of the body to another, as in a malignant neoplasm.

metatarsal (met′ă-tar′sal) [Gr., *meta*, after + *tarsos*, sole of the foot] Distal bone of the foot.

metencephalon (met′en-sef′ă-lon) [Gr., *meta*, after + *enkephalos*, brain] Second most posterior division of the embryonic brain; becomes the pons and cerebellum in the adult.

micelle (mi-sel′, mī-sel′) [L., *micella*, small morsel] Droplets of lipid surrounded by bile salts in the small intestine.

microfilament (mī-krō-fil′ă-ment) Small fibril forming bundles, sheets, or networks in the cytoplasm of cells; provides structure to the cytoplasm and mechanical support for microvilli and stereocilia.

microglia (mī-krog′lē-ă) [Gr., *micro* + *glia*, glue] Small neuroglial cells that become phagocytic and mobile in response to inflammation; considered to be macrophages within the central nervous system.

microtubule (mī-krō-too′būl) Hollow tube composed of tubulin, measuring approximately 25 nm in diameter and usually several micrometers long. Helps provide support to the cytoplasm of the cell and is a component of certain cell organelles, such as centrioles, spindle fibers, cilia, and flagella.

microvillus; pl. microvilli (mī′krō-vil′ŭs, mī′krō-vil′ī) Minute projection of the cell membrane that greatly increases the surface area.

micturition reflex (mik-choo-rish′ŭn) Contraction of the urinary bladder stimulated by stretching of the bladder wall; results in emptying of the bladder.

middle ear Air-filled space within the temporal bone; contains auditory ossicles; between the external and internal ear.

milk letdown Expulsion of milk from the alveoli of the mammary glands; stimulated by oxytocin.

mineral Inorganic nutrient necessary for normal metabolic functions.

mineralocorticoid (min′er-al-ō-kōr′ti-koyd) Steroid hormone (e.g., aldosterone) produced by the zona glomerulosa of the adrenal cortex; facilitates exchange of potassium for sodium in the distal renal tubule, causing sodium reabsorption and potassium and hydrogen ion secretion.

minor duodenal papilla Site of the opening of the accessory pancreatic duct into the duodenum.

minute ventilation Product of tidal volume times the respiratory rate.

mitochondrion; pl. mitochondria (mī-tō-kon′drē-on, mī-tō-kon′drē-ă) [Gr., *mitos*, thread + *chandros*, granule] Small, spherical, rod-shaped or thin filamentous structure in the cytoplasm of cells that is a site of ATP production.

mitosis (mī-tō′sis) [Gr., thread] Cell division resulting in two daughter cells with exactly the same number and type of chromosomes as the mother cell.

modiolus (mō-dī′ō′lŭs) [L., nave of a wheel] Central core of spongy bone about which turns the spiral canal of the cochlea.

molar (mō′lăr) Tricuspid tooth; the three posterior teeth of each dental arch.

molecule (mol′ĕ-kūl) Substance composed of two or more atoms chemically combined to form a structure that behaves as an independent unit.

monoblast (mon′ō-blast) Cell that matures into a monocyte.

monocyte (mon′o-sīt) Type of white blood cell; large phagocytic white blood cell that moves from the blood into tissues and becomes a macrophage.

mononuclear phagocytic system (mon-ō-noo′klē-ăr fag-ō-sit′ik) Phagocytic cells, each with a single nucleus; derived from monocytes; also called reticuloendothelial system.

monosaccharide (mon-ō-sak′ă-rīd) Simple sugar carbohydrate that cannot form any simpler sugar by hydrolysis.

mons pubis (monz pū′bis) [L., mountain; the genitals] Prominence caused by a pad of fatty tissue over the symphysis pubis in the female.

morula (mōr′oo-lă, mōr′ū-lă) [L., *morus*, mulberry] Mass of 12 or more cells resulting from the early cleavage divisions of the zygote.

motor neuron Neuron that innervates skeletal, smooth, or cardiac muscle fibers.

motor unit Single neuron and the muscle fibers it innervates.

mucosa (mū-kō′să) [L., *mucosus* mucous] Mucous membrane consisting of epithelium and lamina propria. In the digestive tract, there is also a layer of smooth muscle.

mucous membrane (mū′kŭs) Thin sheet consisting of epithelium and connective tissue (lamina propria) that lines cavities that open to the outside of the body; many contain mucous glands that secrete mucus.

mucous neck cell One of the mucous-secreting cells in the neck of a gastric gland.

mucus (mū′kŭs) Viscous secretion produced by and covering mucous membranes; lubricates mucous membranes and traps foreign substances.

multiple motor unit summation Increased force of contraction of a muscle due to recruitment of motor units.

multiple-wave summation Increased force of contraction of a muscle due to increased frequency of stimulation.

multipolar neuron One of three categories of neurons consisting of a neuron cell body, an axon, and two or more dendrites.

muscarinic receptor (mŭs′kă-rin′ik) Class of cholinergic receptor that is specifically activated by muscarine in addition to acetylcholine.

muscle fiber Muscle cell.

muscle spindle Three to 10 specialized muscle fibers supplied by gamma motor neurons and wrapped in sensory nerve endings; detects stretch of the muscle and is involved in maintaining muscle tone.

muscle tone Relatively constant tension produced by a muscle for long periods as a result of asynchronous contraction of motor units.

muscle twitch Contraction of a whole muscle in response to a stimulus that causes an action potential in one or more muscle fibers.

muscular fatigue Fatigue due to a depletion of ATP within the muscle fibers.

muscularis (mŭs-kū-lā′ris) [Modern L., muscular] Muscular coat of a hollow organ or tubular structure.

muscularis mucosa Thin layer of smooth muscle found in most parts of the digestive tube; located outside the lamina propria and adjacent to the submucosa.

musculi pectinati (mŭs-kū-lī pek′tĭ-nă′tē) Prominent ridges of atrial myocardium located on the inner surface of much of the right atrium and both auricles.

mutation (mū-tā′shŭn) Change in the number or kinds of nucleotides in the DNA of a gene.

myelencephalon (mī′el-en-sef′ă-lon) [Gr., *myelos*, medulla, marrow + *enkephalos* brain] Most caudal portion of the embryonic brain; also called medulla oblongata.

myelin sheath (mī′ĕ-lin) Envelope surrounding most axons; formed by Schwann cell membranes being wrapped around the axon.

myelinated axon (mī′ĕ-li-nāt-ed ak′son) Nerve fiber having a myelin sheath.

myeloblast (mī′ĕ-lō-blast) Immature cell from which the different granulocytes develop.

myenteric plexus (mī′en-ter′ik) Plexus of unmelinated fibers and postganglionic autonomic cell bodies lying in the muscular coat of the esophagus, stomach, and intestines; communicates with the submucosal plexuses.

myoblast (mī′ō-blast) [Gr., *mys*, muscle + *blastos*, germ] Primitive, multinucleated cell with the potential to develop into a muscle fiber.

myofilament (mī-ō-fil′ă-ment) Extremely fine molecular thread helping form the myofibrils of muscle; thick myofilaments are formed of myosin, and thin myofilaments are formed of actin.

myometrium (mī′ō-mē′trē-ŭm) Muscular wall of the uterus; composed of smooth muscle; also called muscular layer.

myosin myofilament (mī′ō-sin mī-ō-fil′ă-ment) Thick myofilament of muscle fibrils; composed of myosin molecules.

N

nail (nāl) [A.S., naegel] Several layers of dead epithelial cells containing hard keratin on the ends of the digits.

nail bed Epithelial tissue resting on dermis under the nail between the nail matrix and hyponychium; contributes to the formation of the nail.

nail matrix Epithelial tissue resting on dermis under the proximal end of a nail; produces most of the nail.

nasal cavity (nā′zăl) Cavity between the external nares and the pharynx. It is divided into two chambers by the nasal septum and is bounded inferiorly by the hard and soft palates.

nasal septum Bony partition that separates the nasal cavity into left and right parts; composed of the vomer, the perpendicular plate of the ethmoid, and hyaline cartilage.

nasolacrimal duct (nā-zō-lak′ri-măl) Duct that leads from the lacrimal sac to the nasal cavity.

nasopharynx (nā-zō-far′ingks) Part of the pharynx that lies above the soft palate; anteriorly it opens into the nasal cavity.

near point of vision Closest point from the eye at which an object can be held without appearing blurred.

neck Slightly constricted part of a tooth, between the crown and the root.

neoplasm (nĕ′ō-plazm) Abnormal tissue that grows by cellular proliferation more rapidly than normal and continues to grow after the stimuli that initiated the new growth ceases.

nephron (nef′ron) [Gr., *nephros*, kidney] Functional unit of the kidney, consisting of the renal corpuscle, the proximal convoluted tubule, the loop of Henle, and the distal convoluted tubule.

nerve tract Bundles of parallel axons with their associated sheaths in the central nervous system.

neural crest (noor′ăl) Edge of the neural plate as it rises to meet at the midline to form the neural tube.

neural crest cells Cells derived from the crests of the forming neural tube in the embryo; together with the mesoderm, they form the mesenchyme of the embryo; they give rise to part of the skull, the teeth, melanocytes, sensory neurons, and autonomic neurons.

neural layer   Portion of the retina containing rods and cones.

neural plate   Region of the dorsal surface of the embryo that is transformed into the neural tube and neural crest.

neural tube   Tube formed from the neuroectoderm by the closure of the neural groove; develops into the spinal cord and brain.

neuroectoderm (noor-ō-ek′tō-derm)   Part of the ectoderm of an embryo giving rise to the brain and spinal cord.

neurohormone (noor′ō-hōr′mōn)   Hormone secreted by a neuron.

neuromodulator   Substance that influences the sensitivity of neurons to neurotransmitters but neither strongly stimulates nor strongly inhibits neurons by itself.

neuromuscular junction (noor-ō-mŭs′kū-lăr)   Specialized synapse between a motor neuron and a muscle fiber.

neuron (noor′on) [Gr., nerve]   Morphologic and functional unit of the nervous system, consisting of the nerve cell body, the dendrites, and the axon; also called nerve cell.

neuron cell body   Enlarged portion of the neuron containing the nucleus and other organelles; also called nerve cell body.

neurotransmitter (noor′ō-trans-mit′er) [Gr., neuro, nerve + L., transmitto, to send across]   Any specific chemical agent released by a presynaptic cell on excitation that crosses the synaptic cleft and stimulates or inhibits the postsynaptic cell.

neutral solution (noo′trăl)   Solution, such as pure water, that has 10^{-7} mol of hydrogen ions per liter and an equal concentration of hydroxide ions; has a pH of 7.

neutron (noo′tron) [L., neuter, neither]   Electrically neutral particle in the nuclei of atoms (except hydrogen).

neutrophil (noo′trō-fil) [L., neuter, neither + Gr., philos, fond]   Type of white blood cell; small phagocytic white blood cell with a lobed nucleus and small granules in the cytoplasm.

nicotinic receptor (nik-ō-tin′ik)   Class of cholinergic receptor molecule that is specifically activated by nicotine and by acetylcholine.

nipple (nip′l)   Projection at the apex of the mamma, on the surface of which the lactiferous ducts open; surrounded by a circular pigmented area, the areola.

Nissl substance (nis′l)   Named after German neurologist Franz Nissl (1860–1919). Areas in the neuron cell body containing rough endoplasmic reticulum.

nociceptor (nō-si-sep′ter) [L., noceo, to injure + capio, to take]   Sensory receptor that detects painful or injurious stimuli; also called pain receptor.

nonelectrolyte (non-ē-lek′trō-līt) [Gr., electro + lytos, soluble]   Molecules that do not dissociate and do not conduct electricity.

norepinephrine (nōr′ep-i-nef′rin)   Neurotransmitter substance released from most of the postganglionic neurons of the sympathetic division; hormone released from the adrenal cortex that increases cardiac output and blood glucose levels; also called noradrenaline.

nose, or **nasus** (nōz, nā′sŭs)   Visible structure that forms a prominent feature of the face; nasal cavities.

notochord (nō′tō-kōrd) [Gr., notor, back + chords, cord]   Small rod of tissue lying ventral to the neural tube. A characteristic of all vertebrates, in humans it becomes the nucleus pulposus of the intervertebral disks.

nuchal (noo′kăl)   Back of the neck.

nuclear envelope (noo′klē-er)   Double membrane structure surrounding and enclosing the nucleus.

nuclear pores   Porelike openings in the nuclear envelope where the inner and outer membranes fuse.

nucleic acid (noo-klē′ik, noo-klā′ik)   Polymer of nucleotides, consisting of DNA and RNA, forms a family of substances that comprise the genetic material of cells and control protein synthesis.

nucleolus; pl. nucleoli (noo-klē′ō-lŭs, noo-klē′ō-lī)   Somewhat rounded, dense, well-defined nuclear body with no surrounding membrane; contains ribosomal RNA and protein.

nucleotide (noo′klē-ō-tīd)   Basic building block of nucleic acids consisting of a sugar (either ribose or deoxyribose) and one of several types of organic bases.

nucleus; pl. nuclei (noo′klē-ŭs, noo′klē-ī) [L., inside of a thing]   Cell organelle containing most of the genetic material of the cell; collection of nerve cell bodies within the central nervous system; center of an atom consisting of protons and neutrons.

nucleus pulposus (pŭl-pō′sŭs) [L., central pulp]   Soft central portion of the intervertebral disk.

nutrient (noo′trē-ent) [L., nutriens, to nourish]   Chemicals taken into the body that are used to produce energy, provide building blocks for new molecules, or function in other chemical reactions.

O

olecranon process (ō-lek′ră-non, ō′lē-krā′non)   Process on the distal end of the ulna, forming the point of the elbow.

olfaction (ol-fak′shŭn) [L., olfactus, smell]   Sense of smell.

olfactory area   Extreme superior region of the nasal cavity.

olfactory bulb (ol-fak′tō-rē)   Ganglion-like enlargement at the rostral end of the olfactory tract that lies over the cribriform plate; receives the olfactory nerves from the nasal cavity.

olfactory cortex   Termination of the olfactory tract in the cerebral cortex within the lateral fissure of the cerebrum.

olfactory epithelium   Epithelium of the olfactory recess containing olfactory receptors.

olfactory tract   Nerve tract that projects from the olfactory bulb to the olfactory cortex.

oligodendrocyte (ol′i-gō-den′drō-sīt)   Neuroglial cell that has cytoplasmic extensions that form myelin sheaths around axons in the central nervous system.

oncogene (ong′kō-jēn)   Gene that can change or be activated to cause cancer.

oncology (ong-kol′ō-jē)   Study of neoplasms.

oocyte (ō′ō-sīt) [Gr., oon, egg + kytos, a hollow (cell)]   Immature ovum.

oogenesis (ō-ō-jen′ĕ-sis)   Formation and development of a secondary oocyte or ovum.

oogonium (ō-ō-gō′nē-ŭm) [Gr., oon, egg + gone, generation]   Primitive cell from which oocytes are derived by meiosis.

opposition   Movement of the thumb and little finger toward each other; movement of the thumb toward any of the fingers.

opsin (op′sin)   Protein portion of the rhodopsin molecule; a class of proteins that bind to retinal to form the visual pigments of the rods and cones of the eye.

opsonin (op′sŏ-nin) [Gr., opsonein to prepare food]   Substance, such as antibody or complement, that enhances phagocytosis.

optic chiasma (op′tik kī′az′mă) [Gr., two crossing lines; chi the letter χ]   Point of crossing of the optic tracts.

optic disc   Point at which axons of ganglion cells of the retina converge to form the optic nerve, which then penetrates through the fibrous tunic of the eye.

optic nerve   Nerve carrying visual signals from the eye to the optic chiasm.

optic stalk   Constricted proximal portion of the optic vesicle in the embryo; develops into the optic nerve.

optic tract   Tract that extends from the optic chiasma to the lateral geniculate nucleus of the thalamus.

optic vesicle   One of the paired evaginations from the walls of the embryonic forebrain from which the retina develops.

oral cavity (ōr′ăl)   The mouth; consists of the space surrounded by the lips, cheeks, teeth, and palate; limited posteriorly by the fauces.

orbit (ōr′bit)   Eye socket; formed by seven skull bones that surround and protect the eye.

organ of Corti (ōr′găn)   Named for Italian anatomist Marquis Alfonso Corti (1822–1888). Spiral organ; rests on the basilar membrane and supports the hair cells that detect sounds.

organelle (or′gă-nel) [Gr., organon, tool]   Specialized part of a cell with one or more specific individual functions.

orgasm (ōr′gazm) [Gr., orgao, to swell, be excited]   Climax of the sexual act, associated with a pleasurable sensation.

origin (ōr′i-jin)   Less movable attachment point of a muscle; usually the medial or proximal end of a muscle associated with the limbs; also called fixed end.

oropharynx (ōr′ō-far′ingks)   Portion of the pharynx that lies posterior to the oral cavity; it is continuous above with the nasopharynx and below with the laryngopharynx.

oscillating circuit   Neuronal circuit arranged in a circular fashion that allows action potentials produced in the circuit to keep stimulating the neurons of the circuit.

osmolality (os-mō-lal′i-tē)   Osmotic concentration of a solution; the number of moles of solute in 1 kg of water times the number of particles into which the solute dissociates.

osmoreceptor cell (os-mō-rē-sep′ter, os′mō-rē-sep′tōr) [Gr., osmos, impulsion]   Receptor in the central nervous system that responds to changes in the osmotic pressure of the blood.

osmosis (os-mō′sis) [Gr., osmos, thrusting or an impulsion]   Diffusion of solvent (water) through a membrane from a less concentrated solution to a more concentrated solution.

osmotic pressure (os-mot′ik)   Force required to prevent the movement of water across a selectively permeable membrane.

ossification (os′i-fi-kā′shŭn) [L., os, bone + facio, to make]   Bone formation; also called osteogenesis.

osteoblast (os′tē-ō-blast) [Gr., osteon, bone + blastos, germ]   Bone-forming cell.

osteoclast (os′tē-ō-klast) [Gr., osteon, bone + klastos, broken]   Large, multinucleated cell that absorbs bone.

osteocyte (os′tē-ō-sīt) [Gr., osteon, bone + kytos, cell]   Mature bone cell surrounded by bone matrix.

osteomalacia (os′tē-ō-mă-lā′shē-ă)   Softening of bones due to calcium depletion; adult rickets.

osteon (os′tē-on)   Central canal containing blood capillaries and the concentric lamellae around it; occurs in compact bone; also called haversian system.

osteoporosis (os′tē-ō-pō-rō′sis) [Gr., *osteon*, bone + *poros*, pore + *osis*, condition]   Reduction in quantity of bone, resulting in porous bone.

ostium (os′tē-ŭm) [L., door, entrance, mouth]   Small opening—for example, the opening of the uterine tube near the ovary or the opening of the uterus into the vagina.

otolith (ō′tō-lith)   Crystalline particles of calcium carbonate and protein embedded in the maculae.

oval window (ō′văl)   Membranous structure to which the stapes attaches; transmits vibrations to the inner ear.

ovarian cycle (ō-var′ē-an)   Series of events that occur in a regular fashion in the ovaries of sexually mature, nonpregnant females; results in ovulation and the production of the hormones estrogen and progesterone.

ovarian epithelium   Peritoneal covering of the ovary; also called germinal epithelium.

ovarian ligament   Bundle of fibers passing to the uterus from the ovary.

ovary (ō′vă-rē)   One of two female reproductive glands located in the pelvic cavity; produces the secondary oocyte, estrogen, and progesterone.

oviduct (ō′vi-dŭkt)   *See* uterine tube.

ovulation (ov′ū-lā′shun)   Release of an ovum, or secondary oocyte, from the vesicular follicle.

oxidation (ok-si-dā′shŭn)   Loss of one or more electrons from a molecule.

oxidation–reduction reaction   Reaction in which one molecule is oxidized and another is reduced.

oxidative deamination (ok-si-dā′tiv)   Removal of the amine group of an amino acid to form a keto acid, ammonia, and NADH.

oxygen deficit (ok′sē-jen)   Oxygen necessary for the synthesis of the ATP required to remove lactic acid produced by anaerobic respiration.

oxygen–hemoglobin dissociation curve   Graph describing the relationship between the percentage of hemoglobin saturated with oxygen and a range of oxygen partial pressures.

oxyhemoglobin (ox′sē-hē-mō-glō′bin)   Oxygenated hemoglobin.

P

P wave   First complex of the electrocardiogram representing depolarization of the atria.

Pacinian corpuscle (pa-sin′ē-an)   Named for Italian anatomist Filippo Pacini (1812–1883). *See* lamellated corpuscle.

palate (pal′ăt) [L., *palatum*, palate]   Roof of the mouth.

palatine tonsil (pal′ă-tīn)   One of two large, oval masses of lymphoid tissue embedded in the lateral wall of the oral pharynx.

palpebra; pl. palpebrae (pal-pē′bră, pal-pē′brē) [L., eyelid]   Eyelid.

palpebral conjunctiva (pal-pē′brăl kon-jŭnk-tī′vă)   Conjunctiva that covers the inner surface of the eyelids.

palpebral fissure   Space between the upper and lower eyelids.

pancreas (pan′krē-as) [Gr., *pankreas*, sweetbread]   Abdominal gland that secretes pancreatic juice into the intestine and insulin and glucagon from the pancreatic islets into the bloodstream.

pancreatic duct (pan-krē-at′ik)   Excretory duct of the pancreas that extends through the gland from tail to head, where it empties into the duodenum at the greater duodenal papilla.

pancreatic islet   Cellular mass varying from a few to hundreds of cells lying in the interstitial tissue of the pancreas; composed of different cell types that make up the endocrine portion of the pancreas and are the source of insulin and glucagon; also called islets of Langerhans.

pancreatic juice [L., *jus*, broth]   External secretion of the pancreas; clear, alkaline fluid containing several enzymes.

papilla (pă-pil′ă) [L., nipple]   Small, nipplelike process; projection of the dermis, containing blood vessels and nerves, into the epidermis; projections on the surface of the tongue.

papillary muscle (pap′i-lăr′ē)   Nipplelike, conical projection of myocardium within the ventricle; the chordae tendineae are attached to the apex of the papillary muscle.

parafollicular cell (par-ă-fo-lik′ū-lăr)   Endocrine cell in the thyroid gland; secretes the hormone calcitonin.

paramesonephric duct (par-ă-mes-ō-nef′rik)   One of two embryonic tubes extending along the mesonephros and emptying into the cloaca; in the female, the duct forms the uterine tube, the uterus, and part of the vagina; in the male, it degenerates.

paranasal sinus (par-ă-nā′săl)   Air-filled cavities within certain skull bones that connect to the nasal cavity; located in the frontal, maxillary, sphenoid, and ethmoid bones.

parasympathetic division (par-ă-sim-pa-thet′ik)   Subdivision of the autonomic nervous system; characterized by having the cell bodies of its preganglionic neurons located in the brainstem and the sacral region of the spinal cord (craniosacral division); usually involved in activating vegetative functions, such as digestion, defecation, and urination.

parathyroid gland (par-ă-thī′royd)   One of four glandular masses embedded in the posterior surface of the thyroid gland; secretes parathyroid hormone.

parathyroid hormone (PTH)   Peptide hormone produced by the parathyroid gland; increases bone breakdown and blood calcium levels.

parietal (pă-rī′ĕ-tăl) [L., *paries*, wall]   Relating to the wall of any cavity.

parietal cell   Gastric gland cell that secretes hydrochloric acid.

parietal pericardium   Serous membrane lining the fibrous portion of the pericardial sac.

parietal peritoneum   Layer of peritoneum lining the abdominal walls.

parietal pleura   Serous membrane that lines the different parts of the wall of the pleural cavity.

parotid gland (pă-rot′id)   Largest of the salivary glands; situated anterior to each ear.

partial pressure   Pressure exerted by a single gas in a mixture of gases.

passive tension   Tension applied to a load by a muscle without contracting; produced when an external force stretches the muscle.

patella (pa-tel′ă) [L., *patina*, shallow disk]   Kneecap.

pectoral girdle (pek′tŏ-răl)   Site of attachment of the upper limb to the trunk; consists of the scapula and the clavicle; also called shoulder girdle.

pedicle (ped′i-kl) [L., *pes*, feet]   Stalk or base of a structure, such as the pedicle of the vertebral arch.

pelvic brim (pel′vik)   Imaginary plane passing from the sacral promontory to the pubic crest.

pelvic girdle   Site of attachment of the lower limb to the trunk; ring of bone formed by the sacrum and the coxal bones.

pelvic inlet   Superior opening of the true pelvis.

pelvic outlet   Inferior opening of the true pelvis.

pelvis; pl. pelves (pel′vis, pel′vēz) [L., basin]   Any basin-shaped structure; cup-shaped ring of bone at the lower end of the trunk, formed from the ossa coxal bones, sacrum, and coccyx.

pennate (bipennate) (pen′āt) [L., *penna*, feather]   Muscles with fasciculi arranged like the barbs of a feather along a common tendon.

pepsin (pep′sin) [Gr., *pepsis*, digestion]   Principal digestive enzyme of the gastric juice, formed from pepsinogen; digests proteins into smaller peptide chains.

pepsinogen (pep-sin′ō-jen) [*pepsin* + Gr. *gen*, producing]   Proenzyme formed and secreted by the chief cells of the gastric mucosa; the acidity of the gastric juice and pepsin itself converts pepsinogen into pepsin.

peptidase (pep′ti-dās)   Enzyme capable of hydrolyzing one of the peptide links of a peptide.

peptide bond (pep′tīd)   Chemical bond between amino acids.

Percent Daily Value (% Daily Value)   Percent of the recommended daily value of a nutrient found in one serving of a particular food.

perforating canal   Canal containing blood vessels and nerves and running through bone perpendicular to the haversian canals; also called Volkmann's canal.

periarterial sheath (per′ē-ar-tē′rē-ăl)   Dense accumulation of lymphocytes (white pulp) surrounding arteries within the spleen.

pericapillary cell   One of the slender connective tissue cells in close relationship to the outside of the capillary wall; it is relatively undifferentiated and may become a fibroblast, macrophage, or smooth muscle cell.

pericardial cavity (per-i-kar′dē-ăl)   Space within the mediastinum in which the heart is located.

pericardial fluid   Viscous fluid contained within the pericardial cavity between the visceral and parietal pericardium; functions as a lubricant.

pericardium (per-i-kar′dē-ŭm) [Gr., *pericardion*, membrane around the heart]   Membrane covering the heart; also called pericardial sac.

perichondrium (per-i-kon′drē-ŭm) [Gr., *peri*, around + *chondros*, cartilage]   Double-layered connective tissue sheath surrounding cartilage.

perilymph (per′i-limf) [Gr., *peri*, around + L., *lympha*, clear fluid (lymph)]   Fluid contained within the bony labyrinth of the inner ear.

perimetrium (per-i-mē′trē-ŭm)   Outer serous coat of the uterus; also called serous layer.

perimysium (per-i-mis′ē-ŭm, per-i-miz′ē-ŭm) [Gr., *peri*, around + *mys*, muscle]   Fibrous sheath enveloping a bundle of skeletal muscle fibers (muscle fascicle).

perineum (per′i-nē′ŭm)   Area inferior to the pelvic diaphragm between the thighs; extends from the coccyx to the pubis.

perineurium (per-i-noo′rē-ŭm) [L., *peri*, around + Gr., *neuron*, nerve]   Connective tissue sheath surrounding a nerve fascicle.

periodontal ligament (per′ē-ō-don′tăl)   Connective tissue that surrounds the tooth root and attaches it to its bony socket.

periosteum (per-ē-os′tē-ŭm) [Gr., *peri*, around + *osteon*, bone]   Thick, double-layered connective tissue sheath covering the entire surface of a bone, except the articular surface, which is covered with cartilage.

peripheral nervous system (PNS) (pē-rif′ē-răl)   Major subdivision of the nervous system consisting of nerves and ganglia.

peripheral resistance   Resistance to blood flow in all the blood vessels.

peristaltic wave (per-i-stal′tik)   Contraction in a tube, such as the intestine, characterized by a wave of contraction in smooth muscle preceded by a wave of relaxation that moves along the tube.

peritubular capillary Capillary network located in the cortex of the kidney; associated with the distal and proximal convoluted tubules.

permanent tooth One of the 32 teeth belonging to the permanent dentition; also called secondary tooth.

peroneal (per-ō-nē′ăl) [Gr., *perone*, fibula] Associated with the fibula.

peroxisome (per-ok′si-sōm) Membrane-bounded body similar to a lysosome in appearance but often smaller and irregular in shape; contains enzymes that either decompose or synthesize hydrogen peroxide.

Peyer's patch Named for Swiss anatomist Johann K. Peyer (1653–1712). Lymph nodule found in the lower half of the small intestine and the appendix.

phagocyte (fag′ō-sīt) Cell that ingests bacteria, foreign particles, and other cells.

phagocytosis (fag′ō-sī-tō′sis) [Gr., *phagein*, to eat + *kytos*, cell + *osis*, condition] Cells' ingestion of solid substances, such as other cells, bacteria, bits of necrosed tissue, and foreign particles.

phalange; pl. phalanges (fă-lanj′, fă-lan′jēz) [Gr., *phalanx*, line of soldiers] Bone of a finger or toe.

pharyngeal pouch (fă-rin′jē-ăl) Paired evagination of embryonic pharyngeal endoderm between the brachial arches that gives rise to the thymus, thyroid gland, tonsils, and parathyroid glands.

pharyngeal tonsil (fă-rin′jē-ăl) One of two collections of aggregated lymphoid nodules on the posterior wall of the nasopharynx.

pharynx (far′ingks) [Gr., *pharynx*, throat, the joint opening of the gullet and windpipe] Upper expanded portion of the digestive tube between the esophagus below and the oral and nasal cavities above and in front.

phenotype (fē′nō-tīp) [Gr., *phaino*, to display, show forth + *typos*, model] Characteristic observed in an individual due to the expression of his or her genotype.

phosphodiesterase (fos′fō-dī-es′ter-ās) Enzyme that splits phosphodiester bonds—that is, that breaks down cyclic AMP to AMP.

phospholipid (fos-fō-lip′id) Lipid with phosphorus, resulting in a molecule with a polar end and a nonpolar end; main component of the lipid bilayer.

phosphorylation (fos′fōr-i-lā′shŭn) Addition of phosphate to an organic compound.

photoreceptor (fō′tō-rē-sep′ter, fō′tō-rē-sep′tōr) [L., *photo*, light + *ceptus*, to receive] Sensory receptor that is sensitive to light—for example, rods and cones of the retina.

phrenic nerve (fren′ik) Nerve derived from spinal nerves C3–C5; supplies the diaphragm.

physiologic contracture (fiz-ē-ō-loj′ik kontrak′chūr) Temporary inability of a muscle to either contract or relax because of a depletion of ATP so that active transport of calcium ions into the sarcoplasmic reticulum cannot occur.

physiologic dead space Sum of anatomical dead air space plus the volume of any nonfunctional alveoli.

physiologic shunt Deoxygenated blood from the alveoli plus deoxygenated blood from the bronchi and bronchioles.

pia mater (pī′ă mā′ter, pē′a mah′ter) [L., tender mother] Delicate membrane forming the inner covering of the brain and spinal cord.

pigmented layer Pigmented portion of the retina.

pineal body (pin′ē-ăl) [L., *pineus*, relating to pine trees] A small, pine cone–shaped structure that projects from the epiphysis of the diencephalon; produces melatonin; also called pineal gland.

pinna (pin′ă) [L., *pinna* or *penna*, feather] *See* auricle.

pinocytosis (pin′ō-sī-tō′sis, pī′no-sī-tō′sis) [Gr., *pineo*, to drink + *kytos*, cell + *osis*, condition] Cell drinking; uptake of liquid by a cell.

pituitary gland Endocrine gland attached to the hypothalamus by the infundibulum; also called hypophysis.

plane (plān) [L., *planus*, flat] Flat surface; an imaginary surface formed by extension through any axis or two points—for example, a midsagittal plane, a coronal plane, and a transverse plane.

plasma (plaz′mă) [Gr., something formed] Fluid portion of blood.

plasma cell Cell derived from B cells; produces antibodies.

plasma clearance Volume of plasma per minute from which a substance can be completely removed by the kidneys.

plasmin (plaz′min) Enzyme derived from plasminogen; dissolves clots by converting fibrin into soluble products.

plateau phase of action potential Prolongation of the depolarization phase of a cardiac muscle cell membrane; results in a prolonged refractory period.

platelet (plāt′let) Irregularly shaped disk found in blood; contains granules in the central part and clear protoplasm peripherally but has no definite nucleus; also called thrombocyte.

platelet plug Accumulation of platelets that stick to each other and to connective tissue; prevents blood loss from damaged blood vessels.

pleural cavity (ploor′ăl) Potential space between the parietal and visceral layers of the pleura.

plexus; pl. plexuses (plek′sŭs, plek′sŭs-ez) [L., a braid] Intertwining of nerves or blood vessels.

plicae circulares (plī′kă, plī′sē) Numerous folds of the mucous membrane of the small intestine.

pluripotent (ploo-rip′ō-tent) [L., *pluris*, more + *potentia*, power] In development, a cell or group of cells that have not yet become fixed or determined as to what specific tissues they are going to become.

podocyte cells (pod′ō-sīt) [Gr., *pous, podos*, foot + *kytos*, a hollow (cell)] Epithelial cell of Bowman's capsule attached to the outer surface of the glomerular capillary basement membrane by cytoplasmic foot processes.

Poiseuille's law (pwah-zuh′yez) Named for French physiologist and physicist Jean Léonard Marie Poiseuille (1797–1869). The volume of a fluid passing per unit of time through a tube is directly proportional to the pressure difference between its ends and to the fourth power of the internal radius of the tube and inversely proportional to the tube's length and the viscosity of the fluid.

polar body (pō′lăr) One of the two small cells formed during oogenesis because of unequal division of the cytoplasm.

polar covalent bond Covalent bond in which atoms do not share their electrons equally.

polarization Development of differences in potential between two points in living tissues, as between the inside and outside of the plasma membrane.

polycythemia (pol′ē-sī-thē′mē-ă) Increase in red blood cell number above the normal.

polygenic (pol-ē-jen′ik) Relating to a hereditary disease or normal characteristic controlled by the interaction of genes at more than one locus.

polysaccharide (pol-ē-sak′ă-rīd) Carbohydrate containing a large number of monosaccharide molecules.

polyunsaturated fat Fatty acid that contains two or more double covalent bonds between its carbon atoms.

pons (ponz) [L., bridge] Portion of the brainstem between the medulla and midbrain.

popliteal (pop-lit′ē-ăl, pop-li-tē′ăl) [L., ham] Posterior region of the knee.

porta (pōr′tă) [L., gate] Fissure on the inferior surface of the liver where the portal vein, hepatic artery, hepatic nerve plexus, hepatic ducts, and lymphatic vessels enter or exit the liver.

portal system (pōr′tal) System of vessels in which blood, after passing through one capillary bed, is conveyed through a second capillary network.

portal triad Branches of the portal vein, hepatic artery, and hepatic duct bound together in the connective tissue that divides the liver into lobules.

postabsorptive state State following the absorptive state; blood glucose levels are maintained because of the conversion of other molecules to glucose.

posterior chamber (pos-tēr′ē-ŏr) Chamber of the eye between the iris and the lens.

posterior interventricular sulcus Groove on the diaphragmatic surface of the heart, marking the location of the septum between the two ventricles.

posterior pituitary Portion of the hypophysis derived from the brain. Major secretions include antidiuretic hormone and oxytocin. Also called neurohypophysis.

postganglionic neuron (pōst′gang-glē-on′ik) Autonomic neuron that has its cell body located within an autonomic ganglion and sends its axon to an effector organ.

postovulatory age Age of the developing fetus based on the assumption that fertilization occurs 14 days after the last menstrual period before the pregnancy.

postsynaptic (pōst-si-nap′tik) Relating to the membrane of a nerve, muscle, or gland that is in close association with a presynaptic terminal. The postsynaptic membrane has receptor molecules within it that bind to neurotransmitter molecules.

potential difference (pō-ten′shăl) Difference in electrical potential, measured as the charge difference across the plasma membrane.

potential energy [Gr., *en*, in + *ergon*, work] Energy in a chemical bond that is not being exerted or used to do work.

PQ interval Time elapsing between the beginning of the P wave and the beginning of the QRS complex in the electrocardiogram; also called PR interval.

precapillary sphincter (prē-kap′i-lār-ē sfingk′ter) Smooth muscle sphincter that regulates blood flow through a capillary.

preganglionic neuron (prē′gang-glē-on′ik) Autonomic neuron that has its cell body located within the central nervous system and sends its axon through a nerve to an autonomic ganglion, where it synapses with postganglionic neurons.

premolar (prē-mō′lăr) Bicuspid tooth.

prenatal (prē-nā′tal) [*pre* + L., *natus*, born] Preceding birth.

prepuce (prē′poos) In males, the free fold of skin that more or less completely covers the glans penis; the foreskin. In females, the external fold of the labia minora that covers the clitoris.

pressoreceptor (pres′ō-rē-sep′ter, pres′ō-rē-sep′tōr) *See* baroreceptor.

presynaptic terminal (prē′si-nap′tik) Enlarged axon terminal or terminal bouton.

primary palate In the early embryo, the structure that gives rise to the upper jaw and lips.

primary response Immune response that occurs as a result of the first exposure to an antigen.

primary spermatocyte (sper′mă-tō-sīt) Spermatocyte arising by a growth phase from a spermatogonium; gives rise to secondary spermatocytes after the first meiotic division.

prime mover Muscle that plays a major role in accomplishing a movement.

primitive streak (prim'i-tiv) Ectodermal ridge in the midline of the embryonic disk, from which arises the mesoderm by the inward and then lateral migration of cells.

primordial germ cell (prī-mōr'dē-ăl) Most primitive undifferentiated sex cell, found initially outside the gonad on the surface of the yolk sac.

PR interval *See* PQ interval.

process (pros'es, prō'ses) Projection on a bone.

processus vaginalis (prō-ses'ŭs vaj'i-năl-ŭs) Peritoneal outpocketing in the embryonic lower anterior abdominal wall that traverses the inguinal canal; in the male, it forms the tunica vaginalis testis and normally loses its connection with the peritoneal cavity.

proerythroblast Cell that matures into an erythrocyte.

progeria (prō-jēr'ē-ă) [Gr., *pro,* before + *ge* + *amras,* old age] Severe retardation of growth after the first year accompanied by a senile appearance and death at an early age.

progesterone (prō-jĕs'tē-rōn) Steroid hormone secreted by the corpus luteum and one of the hormones secreted by the placenta.

prolactin (prō-lak'tin) Hormone of the adenohypophysis that stimulates the production of milk.

prolactin-inhibiting hormone (PIH) Neurohormone released from the hypothalamus that inhibits prolactin release from the adenohypophysis.

prolactin-releasing hormone (PRH) Neurohormone released from the hypothalamus that stimulates prolactin release from the adenohypophysis.

proliferative phase (prō-lif'er-ă-tiv) *See* follicular phase.

pronation (prō-nā'shŭn) [L., *pronare,* to bend forward] Rotation of the forearm so that the anterior surface is down (prone).

pronephros (prō-nef'ros) In the embryos of higher vertebrates, a series of tubules emptying into the celomic cavity. It is a temporary structure in the human embryo, followed by the mesonephros and still later by the metanephros, which gives rise to the kidney.

prophase (prō'fāz) First stage in cell division when chromatin strands condense to form chromosomes.

proprioception (prō-prē-ō-sep'shun) [L., *proprius,* one's own + *capio,* to take] Information about the position of the body and its various parts.

proprioceptor (prō'prē-ō-sep'ter) Sensory receptor associated with joints and tendons.

prostaglandin (pros'tă-glan'din) Class of physiologically active substances present in many tissues; among its effects are vasodilation, stimulation and contraction of uterine smooth muscle and the promotion of inflammation and pain.

prostate gland (pros'tāt) [Gr., *prostates,* one standing before] Gland that surrounds the beginning of the urethra in the male. The secretion of the gland is a milky fluid that is discharged by 20–30 excretory ducts into the prostatic urethra as part of the semen.

prostatic urethra (pros-tat'ik) Part of the male urethra, approximately 2.5 cm in length, that passes through the prostate gland.

protease (prō'tē-ās) Enzyme that breaks down proteins.

protein (prō'tēn, prō'tē-ĭn) [Gr., *proteios,* primary] Macromolecule consisting of long sequences of amino acids linked together by peptide bonds.

protein kinase (kin-āz) Class of enzymes that phosphorylates other proteins. Many of these kinases are responsive to other chemical signals (e.g., cAMP, cGMP, insulin, epidermal growth factor, calcium and calmodulin).

proteoglycan (prō'tē-ō-glī'kan) Macromolecule consisting of numerous polysaccharides attached to a common protein core.

prothrombin (prō-throm'bin) Glycoprotein present in blood that, in the presence of prothrombin activator, is converted to thrombin.

proton (prō'ton) [Gr., *protos,* first] Positively charged particle in the nuclei of atoms.

protraction (prō-trak'shŭn) [L., *protractus,* to draw forth] Movement forward or in the anterior direction.

provitamin (prō-vī'tă-min) Substance that may be converted into a vitamin.

proximal convoluted tubule (prok'si-măl) Part of the nephron that extends from the glomerulus to the descending limb of the loop of Henle.

pseudostratified epithelium Epithelium consisting of a single layer of cells but having the appearance of multiple layers.

psychologic fatigue (sī-kō-loj'-ik) Fatigue caused by the central nervous system.

ptosis (tō'sis) [G., *ptosis,* a falling] Falling down of an organ—for example, the drooping of the upper eyelid.

puberty (pū'ber-tē) [L., *pubertas,* grown up] Series of events that transform a child into a sexually mature adult; involves an increase in the secretion of GnRH.

pubis (pū'bis) Anterior inferior bone of the coxal bone.

pudendal cleft (pū-den'dăl) Cleft between the labia majora.

pudendum (pū-den'dŭm) *See* vulva.

pulmonary artery (pŭl'mō-nār-ē) One of the arteries that extend from the pulmonary trunk to the right or left lungs.

pulmonary capacity Sum of two or more pulmonary volumes.

pulmonary trunk Large, elastic artery that carries blood from the right ventricle of the heart to the right and left pulmonary arteries.

pulmonary vein One of the veins that carry blood from the lungs to the left atrium of the heart.

pulp (pŭlp) [L., *pulpa,* flesh] Soft tissue within the pulp cavity of the tooth, consisting of connective tissue containing blood vessels, nerves, and lymphatics.

pulse pressure (pŭls) Difference between systolic and diastolic pressure.

pupil (pū'pĭl) Circular opening in the iris through which light enters the eye.

Purkinje fiber (pŭr-kĭn'jē) Named for Bohemian anatomist Johannes E. von Purkinje (1787–1869). Modified cardiac muscle cells found beneath the endocardium of the ventricles. Specialized to conduct action potentials.

pus (pŭs) Fluid product of inflammation; contains white blood cells, the debris of dead cells, and tissue elements liquefied by enzymes.

pyloric opening (pī-lōr'ik) Opening between the stomach and the superior part of the duodenum.

pyloric sphincter Thickening of the circular layer of the gastric musculature encircling the junction between the stomach and duodenum; also called pylorus.

pyrogen (pī'rō-jen) Chemical released by microorganisms, neutrophils, monocytes, and other cells that stimulates fever production by acting on the hypothalamus.

Q

QRS complex Principal deflection in the electrocardiogram, representing ventricular depolarization.

QT interval Time elapsing from the beginning of the QRS complex to the end of the T wave, representing the total duration of electrical activity of the ventricles.

R

radial pulse (rā'dē-ăl) Pulse detected in the radial artery.

radiation (rā'dē-ā'shŭn) [L., *radius,* ray, beam] Sending forth of light, short radiowaves, ultraviolet or x-rays, or any other rays for treatment or diagnosis or for other reasons; radiant heat.

radioactive isotope (rā'dē-ō-ak'tiv) Isotope with a nuclear composition that is unstable from which subatomic particles and electromagnetic waves are emitted.

ramus; pl. rami (rā'mŭs, rā'mī) [L., branch] One of the primary subdivisions of a nerve or blood vessel; the part of a bone that forms an angle with the main body of the bone.

raphe (rā'fē) [Gr., *rhaphe,* suture, seam] Central line running over the scrotum from the anus to the root of the penis.

receptor Structural protein or glycoprotein molecule on the cell surface or within the cytoplasm that binds to a specific factor (chemical signal).

recessive (rē-ses'iv) Gene that may not be expressed because of suppression by a contrasting dominant gene.

Recommended Dietary Allowances (RDAs) Guide for estimating the nutritional needs of groups of people based on their age, sex, and other factors; first established in 1941.

Recommended Daily Intake (RDI) Generally, the highest RDA value in each of four categories: infants, toddlers, people over 4 years of age, and pregnant or lactating women. The RDIs used to determine the Daily Values are based on the 1968 RDAs.

rectum (rek'tŭm) [L., *rectus,* straight] Portion of the digestive tract that extends from the sigmoid colon to the anal canal.

red marrow (mar'o) Soft, pulpy connective tissue filling the cavities of bones; consists of reticular fibers and the development stages of blood cells and platelets; gradually replaced by yellow marrow in long bones and the skull.

red pulp [L., *pulpa,* flesh] Reddish brown substance of the spleen consisting of venous sinuses and the tissues intervening between them, called pulp cords.

reduction (rē-dŭk'shŭn) Gain of one or more electrons by a molecule.

refraction (rē-frak-shŭn) Bending of a light ray when it passes from one medium into another of different density.

refractory period (rē-frak'tōr-ē) [Gr., *periodos,* a way around, a cycle] Period following effective stimulation during which excitable tissue, such as heart muscle, fails to respond to a stimulus of threshold intensity.

regeneration (rē'jen-er-ā'shŭn) Reproduction or reconstruction of a lost or injured part.

regulatory gene Gene involved with controlling the activity of structural genes.

relative refractory period Portion of the action potential following the absolute refractory period during which another action potential can be produced with a greater-than-threshold stimulus strength.

relaxation phase (rē-lak-sā'shŭn) Phase of muscle contraction following the contraction phase; the time from maximal tension production until tension decreases to its resting level.

relaxin Polypeptide hormone secreted by the corpus luteum and placenta during pregnancy; facilitates the birth process by causing a softening and lengthening of the pubic symphysis and cervix.

renal artery (rē'nŭl) Artery that originates from the aorta and delivers blood to the kidney.

renal blood flow rate Volume at which blood flows through the kidneys per minute; an average of approximately 1200 mL/min.

renal column Cortical substance separating the renal pyramids.

renal corpuscle Glomerulus and Bowman's capsule that encloses it.

renal fascia (făsh'i-ă) Connective tissue surrounding the kidney that forms a sheath or capsule for the organ.

renal fat pad Fat layer that surrounds the kidney and functions as a shock-absorbing material.

renal fraction Portion of the cardiac output that flows through the kidneys; averages 21%.

renal pelvis Funnel-shaped expansion of the upper end of the ureter that receives the calyces.

renal pyramid One of a number of pyramidal masses seen on longitudinal section of the kidney; they contain part of the loops of Henle and the collecting tubules.

renin (rē'nin) Enzyme secreted by the juxtaglomerular apparatus that converts angiotensinogen to angiotensin I.

renin-angiotensin-aldosterone mechanism Renin, released from the kidneys in response to low blood pressure, converts angiotensinogen to angiotensin I. Angiotensin I is converted by angiotensin-converting enzyme to angiotensin II, which causes vasoconstriction, resulting in increased blood pressure. Angiotensin II also increases aldosterone secretion, which increases blood pressure by increasing blood volume.

repolarization (rē'pō-lăr-i-zā'shŭn) Phase of the action potential in which the membrane potential moves from its maximum degree of depolarization toward the value of the resting membrane potential.

reposition Return of a structure to its original position.

residual volume (rē-zid'ū-ăl) Volume of air remaining in the lungs after a maximum expiratory effort.

resolution (rez-ō-loo'shŭn) [L., resolutio, a slackening] Phase of the male sexual act after ejaculation during which the penis becomes flaccid; feeling of satisfaction; inability to achieve erection and second ejaculation. Last phase of the female sexual act, characterized by an overall sense of satisfaction and relaxation.

respiration (res-pi-ră'shŭn) [L., respiratio, to exhale, breathe] Process of life in which oxygen is used to oxidize organic fuel molecules, providing a source of energy, carbon dioxide, and water; movement of air into and out of the lungs, the exchange of gases with blood, the transportation of gases in the blood, and gas exchange between the blood and the tissues.

respiratory bronchiole (res'pi-ră-tōr-ē, rē-spīr'ă-tōr-ē) Smallest bronchiole (0.5 mm in diameter) that connects the terminal bronchiole to the alveolar duct.

respiratory membrane Membrane in the lungs across which gas exchange occurs with blood.

resting membrane potential (RMP) Electric charge difference inside a plasma membrane, measured relative to just outside the plasma membrane.

reticular (re-tik'ū-lăr) [L., rete, net] Relating to a fine network of cells or collagen fibers.

reticular cell Cell with processes making contact with those of other similar cells to form a cellular network; along with the network of reticular fibers, the reticular cells form the framework of bone marrow and lymphatic tissues.

reticulocyte (re-tik'ū-lō-sīt) Young red blood cell with a network of basophilic endoplasmic reticulum occurring in larger numbers during the process of active red blood cell synthesis.

reticuloendothelial system (re-tik'ū-lō-en-dō-thē'lē-ăl) See mononuclear phagocytic system.

retina (ret'i-nă) Nervous tunic of the eyeball.

retinaculum (ret-i-nak'ū-lŭm) [L., band, halter, to hold back] Dense regular connective tissue sheath holding down the tendons at the wrist, ankle, or other sites.

retraction (rē-trak'shŭn) [L., retractio, a drawing back] Movement in the posterior direction.

retroperitoneal (re'trō-per'i-tō-nē'ăl) Behind the peritoneum.

rhodopsin (rō-dop'sin) Light-sensitive substance found in the rods of the retina; composed of opsin loosely bound to retinal.

ribonuclease (rī-bō-nū'klē-ās) Enzyme that splits RNA into its component nucleotides.

ribonucleic acid (RNA) (rī'bō-noo-klē'ik) Nucleic acid containing ribose as the sugar component; found in all cells in both nuclei and cytoplasm; helps direct protein synthesis.

ribosomal RNA (rRNA) (rī'bō-sōm-ăl) RNA that is associated with certain proteins to form ribosomes.

ribosome (rī'bō-sōm) Small, spherical, cytoplasmic organelle where protein synthesis occurs.

right lymphatic duct Lymphatic duct that empties into the right subclavian vein; drains the right side of the head and neck, the right-upper thorax, and the right-upper limb.

rigor mortis (rig'er mōr'tĭs) Increased rigidity of muscle after death due to cross-bridge formation between actin and myosin as calcium ions leak from the sarcoplasmic reticulum.

rod Photoreceptor in the retina of the eye; responsible for noncolor vision in low-intensity light.

root Part below the neck of a tooth covered by cementum rather than enamel and attached by the periodontal ligament to the alveolar bone.

root of the penis Proximal attached part of the penis, including the two crura and the bulb.

rotation (rō-tā'shŭn) Movement of a structure about its axis.

rotator cuff (rō-tā'ter, rō-tā'tor) Four deep muscles that attach the humerus to the scapula.

round ligament Fibromuscular band that is attached to the uterus on each side in front of and below the opening of the uterine tube; it passes through the inguinal canal to the labium majus.

round ligament of the liver Remains of the umbilical vein.

round window Membranous structure separating the scala tympani of the inner ear from the middle ear.

Ruffini end organ (roo-fē'nēz) Named for Italian histologist Angelo Ruffini (1864–1929); receptor located deep in the dermis and responding to continuous touch or pressure.

ruga; pl. **rugae** (roo'gă, roo'gē) [L., a wrinkle] Fold or ridge; fold of the mucous membrane of the stomach when the organ is contracted; transverse ridge in the mucous membrane of the vagina.

S

saccule (sak'yūl) Part of the membranous labyrinth; contains a sensory structure, the macula, that detects static equilibrium.

salivary amylase (sal'i-văr-ē am'il-ās) Enzyme secreted in the saliva that breaks down starch to maltose and isomaltose.

salivary gland Gland that produces and secretes saliva into the oral cavity. The three major pairs of salivary glands are the parotid, submandibular, and sublingual glands.

salt Molecule consisting of a cation other than hydrogen and an anion other than hydroxide.

sarcolemma (sar'kō-lem'ă) [Gr., sarco, muscle + lemma, husk] Plasma membrane of a muscle fiber.

sarcoma (sar-kō'mă) Malignant neoplasm derived from connective tissue.

sarcomere (sar'kō-mēr) [Gr., sarco, muscle + meros, part] Part of a myofibril between adjacent Z disks.

sarcoplasm (sar'kō-plazm) [Gr., sarco, muscle + plasma, a thing formed] Cytoplasm of a muscle fiber, excluding the myofilaments.

sarcoplasmic reticulum (sar'kō-plaz'mik) [Gr., sarco, muscle + plasma, a thing formed + reticulum, net] Endoplasmic reticulum of muscle.

satellite cell (sat'ě-līt) Specialized cell that surrounds the cell bodies of neurons within ganglia.

saturated (satch'ŭ-rāt-ěd) Fatty acid in which the carbon chain contains only single bonds between carbon atoms.

saturation (satch-ŭ-rā'shun) Point when all carrier molecules or enzymes are attached to substrate molecules and no more molecules can be transported or reacted.

scala tympani (skā'lă tim'pă-nī) [L., stairway] Division of the spiral canal of the cochlea lying below the spiral lamina and basilar membrane.

scala vestibuli (skā'lă ves-tib'ū-lī) Division of the cochlea lying above the spiral lamina and vestibular membrane.

scapula (skap'ū-lă) Bone forming the shoulder blade.

scar (skar) [Gr., eschara, scab] Fibrous tissue replacing normal tissue; also called cicatrix.

sciatic nerve (sī-at'ik) Tibial and common peroneal nerves bound together; also called ischiadic nerve.

sclera (sklēr'ă) White of the eye; white, opaque portion of the fibrous tunic of the eye.

scleral venus sinus Series of veins at the base of the cornea that drain excess aqueous humor from the eye.

scrotum; pl. **scrota, scrotums** (skrō'tŭm, skrō'tă, skrō'tŭmz) Musculocutaneous sac containing the testes.

sebaceous gland (sē-bā'shŭs) [L., sebum, tallow] Gland of the skin, usually associated with a hair follicle, that produces sebum.

second messenger See intracellular mediator.

secondary follicle Follicle in which the secondary oocyte is surrounded by granulosa cells at the periphery; contains fluid-filled antral spaces.

secondary oocyte (ō'ō-sīt) Oocyte in which the second meiotic division stops at metaphase II unless fertilization occurs.

secondary palate Roof of the mouth in the early embryo that gives rise to the hard and the soft palates.

secondary response Immune response that occurs when the immune system is exposed to an antigen against which it has already produced a primary response; also called memory response.

secondary spermatocyte (sper′mă-tō-sīt) Spermatocyte derived from a primary spermatocyte by the first meiotic division; each secondary spermatocyte gives rise by the second meiotic division to two spermatids.

secretin (se-krē′tin) Hormone formed by the epithelial cells of the duodenum; stimulates secretion of pancreatic juice high in bicarbonate ions.

secretion (se-krē′shŭn) Substance produced inside a cell and released from the cell.

secretory phase (se-krēt′ĕ-rē, sē′krĕ-tōr-ē) *See* luteal phase.

segmental artery One of five branches of the renal artery, each supplying a segment of the kidney.

segmental bronchus (brong′kŭs) Extends from the secondary bronchus and conducts air to each lobule of the lungs.

self-antigen Antigen produced by the body that is capable of initiating an immune response against the body.

semen (sē′men) [L., seed (of plants, men, animals)] Penile ejaculate; thick, yellowish white, viscous fluid containing spermatozoa and secretions of the testes, seminal vesicles, prostate, and bulbourethral glands.

semicircular canal (sem′ē-sir′kū-lăr) Canal in the petrous portion of the temporal bone that contains sensory organs that detect kinetic or dynamic equilibrium. Three semicircular canals are within each inner ear.

seminal vesicle (sem′i-năl) One of two glandular structures that empty into the ejaculatory ducts; its secretion is one of the components of semen.

seminiferous tubule (sem′i-nif′er-ŭs) Tubule in the testis in which spermatozoa develop.

sensible perspiration (sen′si-bl pers-pi-rā′shŭn) Perspiration excreted by the sweat glands that appears as moisture on the skin; produced in large quantity when there is much humidity in the atmosphere.

septum primum (sep′tŭm prī′mŭm) First septum in the embryonic heart that arises on the wall of the originally single atrium of the heart and separates it into right and left chambers.

septum secundum (sek′ŭn-dŭm) Second of two major septal structures involved in the partitioning of the atrium, arising later than the septum primum and located to the right of it; it remains an incomplete partition until after birth, with its unclosed area constituting the foramen ovale.

serosa (se-rō′să) [L., serosus, serous] Outermost covering of an organ or a structure that lies in a body cavity.

serous fluid (ser′ŭs) Fluid similar to lymph that is produced by and covers serous membrane; it lubricates the serous membrane.

serous membrane Thin sheet composed of epithelial and connective tissues; it lines cavities that do not open to the outside of the body or contain glands but do secrete serous fluid.

serous pericardium Lining of the pericardial sac composed of a serous membrane.

Sertoli cell (ser-tō′lē) Named for Italian histologist Enrico Sertoli (1842–1910). Elongated cell in the wall of the seminiferous tubules to which spermatids are attached during spermatogenesis.

serum (sēr′ŭm) [L., whey] Fluid portion of blood after the removal of fibrin and blood cells.

sesamoid bone (ses′ă-moyd) [Gr., sesamoceies, like a sesame seed] Bone found within a tendon, such as the patella.

sex chromosomes Pair of chromosomes responsible for sex determination, XX in female and XY in male.

sex-linked trait Characteristic resulting from the expression of a gene on a sex chromosome.

sigmoid colon (sig′moyd) Part of the colon between the descending colon and the rectum.

sigmoid mesocolon Fold of peritoneum attaching the sigmoid colon to the posterior abdominal wall.

simple epithelium Epithelium consisting of a single layer of cells.

sinoatrial (SA) node (si′nō-ā′trē-ăl) Mass of specialized cardiac muscle fibers; acts as the "pacemaker" of the cardiac conduction system.

sinus (sī′nŭs) [L., cavity] Hollow in a bone or other tissue; enlarged channel for blood or lymph.

sinus venosus End of the embryonic cardiac tube where blood enters the heart; becomes a portion of the right atrium, including the SA node.

sinusoid (si′nŭ-soyd) [L., sinus + Gr., eidos, resemblance] Terminal blood vessel having a larger diameter than an ordinary capillary.

sinusoidal capillary (si-nŭ-soy′dăl) Capillary with caliber of 10–20 μm or more; lined with a fenestrated type of endothelium.

small intestine [L., intestinus, entrails] Portion of the digestive tube between the stomach and the cecum; consists of the duodenum, jejunum, and ileum.

sodium–potassium (Na⁺–K⁺) pump Biochemical mechanism that uses energy derived from ATP to achieve the active transport of potassium ions opposite to that of sodium ions; also called sodium potassium ATP-ase.

soft palate Posterior muscular portion of the palate, forming an incomplete septum between the mouth and the oropharynx and between the oropharynx and the nasopharynx.

solute (sol′ūt, sō′loot) [L., solutus, dissolved] Dissolved substance in a solution.

solution (sō-loo′shŭn) [L., solutio] Homogenous mixture formed when a solute is dissolved in a solvent.

solvent (sol′vent) [L., solvens, to dissolve] Liquid that holds another substance in solution.

soma (sō′mă) [Gr., body] Neuron cell body or the enlarged portion of the neuron containing the nucleus and other organelles.

somatic (sō-mat′ik) [Gr., somatikos, bodily] Relating to the body; the cells of the body except the reproductive cells.

somatic nervous system Composed of nerve fibers that send impulses from the central nervous system to skeletal muscle.

somatomedin (sō′mă-tō-mē′din) Peptide synthesized in the liver capable of stimulating certain anabolic processes in bone and cartilage, such as synthesis of DNA, RNA, and protein.

somatotropin (sō′mă-tō-trō′pin) Protein hormone of the anterior pituitary gland; it promotes body growth, fat mobilization, and inhibition of glucose utilization.

somite (sō′mīt) [Gr., soma, body + ite] One of the paired segments consisting of cell masses formed in the early embryonic mesoderm on each side of the neural tube.

somitomere (sō′mīt-ō-mēr) Indistinct somite in the head region of the embryo.

spatial summation Summation of the local potentials in which two or more action potentials arrive simultaneously at two or more presynaptic terminals that synapse with a single neuron.

specific heat Heat required to raise the temperature of any substance 1°C compared with the heat required to raise the same volume of water 1°C.

speech Use of the voice in conveying ideas.

spermatic cord (sper-mat′ik) Cord formed by the ductus deferens and its associated structures; extends through the inguinal canal into the scrotum.

spermatid (sper′mă-tid) [Gr., sperma, seed + id] Cell derived from the secondary spermatocyte; gives rise to a spermatozoon.

spermatogenesis (sper′mă-tō-jen′ĕ-sis) Formation and development of the spermatozoon.

spermatogonium (sper′mă-tō-gō′nē-ŭm) [Gr., sperma, seed + gone, generation] Cell that divides by mitosis to form primary spermatocytes.

spermatozoon; pl. spermatozoa (sper′mă-tō-zō′on, sper′mă-to-zō′ă) [Gr., sperma, seed + zoon, animal] Male gamete or sex cell, composed of a head and a tail. The spermatozoon contains the genetic information transmitted by the male. Also called sperm cell.

sphenoid (sfē′noyd) [Gr., shen, wedge] Wedge-shaped.

sphincter pupillae (sfingk′ter pū-pil′ē) Circular smooth muscle fibers of the iris's diaphragm that constrict the pupil of the eye.

sphygmomanometer (sfig′mō-mă-nom′ĕ-ter) [Gr., sphygmos, pulse + manos, thin, scanty + metron, measure] Instrument for measuring blood pressure.

spinal nerve (spī′năl) One of 31 pairs of nerves formed by the joining of the dorsal and ventral roots that arise from the spinal cord.

spindle fiber (spin′dl) Specialized microtubule that develops from each centrosome and extends toward the chromosomes during cell division.

spiral artery (spī′răl) One of the corkscrewlike arteries in premenstrual endometrium; most obvious during the secretory phase of the uterine cycle.

spiral ganglion Cell bodies of sensory neurons that innervate hair cells of the organ of Corti are located in the spiral ganglion.

spiral lamina Attached to the modiolus and supports the basilar and vestibular membranes.

spiral ligament Attachment of the basilar membrane to the lateral wall of the bony labyrinth.

spiral organ Organ of Corti; rests on the basilar membrane and consists of the hair cells that detect sound; also called organ of Corti.

spiral tubular gland Well-developed simple or compound tubular gland; spiral in shape; within the endometrium of the uterus; prevalent in the secretory phase of the uterine cycle.

spirometer (spī-rom′ĕ-ter) [L., spiro, to breathe + Gr., metron, measure] Gasometer used for measuring the volume of respiratory gases; usually understood to consist of a counterbalanced cylindrical bell sealed by dipping into a circular trough of water.

spirometry (spī-rom′ĕ-trē) Making pulmonary measurements with a spirometer.

spleen (splēn) Large lymphatic organ in the upper part of the abdominal cavity on the left side between the stomach and diaphragm, composed of white and red pulp. It responds to foreign substances in the blood, destroys worn-out red blood cells, and is a storage site for blood cells.

spongy urethra Portion of the male urethra, approximately 15 cm in length, that traverses the corpus spongiosum of the penis.

squamous (skwā′mŭs) [L., squama, a scale] Scalelike, flat.

stapedius (stă-pē′dē-ŭs) Small skeletal muscles attached to the stapes.

stapes (stā′pēz) [L., stirrup] Smallest of the three auditory ossicles; attached to the oval window.

Starling's law of the heart Named for English physiologist Ernest H. Starling (1866–1927). Force of contraction of cardiac muscle is a function of the length of its muscle fibers at the end of diastole; the greater the ventricular filling, the greater the stroke volume produced by the heart.

sternum (ster´nŭm) [L., *sternon*, chest] Breastbone.

steroid (stēr´oyd, ster´oyd) Large family of lipids, including some reproductive hormones, vitamins, and cholesterol.

stomach (stŭm´ŭk) Large sac between the esophagus and the small intestine, lying just beneath the diaphragm.

stratified epithelium (strat´i-fīd ep-i-thē´lē-ŭm) Epithelium consisting of more than one layer of cells.

stratum basale (strat-ŭm băh-săl´ē) [L., layer; basal] Basal, or deepest, layer of the epidermis; also called stratum germinativum.

stratum corneum (kōr´nē-ŭm) [L., layer + *corneus*, horny] Most superficial layer of the epidermis consisting of flat, keratinized, dead cells.

stratum germinativum *See* stratum basale.

stratum granulosum (gran´ū-lō´sŭm) [L., layer; granulum, a small grain] Layer of cells in the epidermis filled with granules of keratohyalin.

stratum lucidum (lū´sid-ŭm) [L., layer + *lucidus*, clear] Clear layer of the epidermis found in thick skin between the stratum granulosum and the stratum corneum.

stratum spinosum (spī´nōs-ŭm) [L., layer + *spina*, spine] Layer of many-sided cells in the epidermis with intercellular connections (desmosomes) that give the cells a spiny appearance.

stria; pl. striae (strī´ă, strī´ē) [L., channel] Line or streak in the skin that is a different texture or color from the surrounding skin; also called stretch mark.

striated (strī´āt-ēd) [L., *striatus*, furrowed] Striped; marked by stripes or bands.

stroke volume (SV) [L., *volumen*, something rolled up, scroll, from *volvo*, to roll] Volume of blood pumped out of one ventricle of the heart in a single beat.

structural gene Gene that determines the structure of a specific protein or peptide.

sty (stī) Inflamed ciliary gland of the eye.

subcutaneous (sŭb´-koo-tā´nē-ŭs) [L., *sub*, under + *cutis*, skin] Under the skin; same tissue as the hypodermis.

sublingual gland (sŭb-ling´gwăl) One of two salivary glands in the floor of the mouth beneath the tongue.

submandibular gland (sŭb-man-dib´ū-lăr) One of two salivary glands in the neck, located in the space bounded by the two bellies of the digastric muscle and the angle of the mandible.

submucosa (sŭb-moo-kō´să) Layer of tissue beneath a mucous membrane.

submucosal plexus (sŭb-mū-kō´săl) [L., a braid] Ganglionated plexus of unmyelinated nerve fibers in the intestinal submucosa.

substantia nigra (sŭb-stan´shē-ă nī´gră) [L., substance; black] Black nuclear mass in the midbrain; involved in coordinating movement and maintaining muscle tone.

subthreshold stimulus Stimulus resulting in a local potential so small that it does not reach threshold and produce an action potential.

sucrose (soo´krōs) Disaccharide composed of glucose and fructose; table sugar.

sulcus; pl. sulci (sool´kŭs, sŭl´sī) [L., furrow or ditch] Furrow or groove on the surface of the brain between the gyri; may also refer to a fissure.

superficial inguinal ring (ing´gwi-năl) Slitlike opening in the aponeurosis of the external oblique muscle of the abdominal wall through which the spermatic cord (round ligament in the female) emerges from the inguinal canal.

superior colliculus (ko-lik´ū-lŭs) [L., *collis*, hill] One of two rounded eminences of the midbrain; aids in coordination of eye movements.

superior vena cava (vē´nă cā´vă) Vein that returns blood from the head and neck, upper limbs, and thorax to the right atrium.

supination (soo´pi-nā´shŭn) [L., *supino*, to bend backward, place on back] Rotation of the forearm (when the forearm is parallel to the ground) so that the anterior surface is up (supine).

supramaximal stimulus Stimulus of greater magnitude than a maximal stimulus; however, the frequency of action potentials is not increased above that produced by a maximal stimulus.

suppressor T cell Subset of T lymphocytes that decreases the activity of B cells and T cells.

surfactant (ser-fak´tănt) Lipoproteins forming a monomolecular layer over pulmonary alveolar surfaces; stabilizes alveolar volume by reducing surface tension and the tendency for the alveoli to collapse.

suspension (sŭs-pen´shŭn) Liquid through which a solid is dispersed and from which the solid separates unless the liquid is kept in motion.

suspensory ligament (sŭs-pen´sŏ-rē) Band of peritoneum that extends from the ovary to the body wall; contains the ovarian vessels and nerves. Small ligament attached to the margin of the lens in the eye and the ciliary body to hold the lens in place.

suture (soo´choor) [L., *sutura*, a seam] Junction between flat bones of the skull.

sweat (swet) [A.S., *swat*] Perspiration; secretions produced by the sweat glands of the skin; also called sensible perspiration.

sweat gland Usually means structure that produces a watery secretion called sweat. Some sweat glands, however, produce viscous organic secretions. Also called sudoriferous gland.

sympathetic chain ganglion (sim-pă-thet´ik) Collection of sympathetic postganglionic neurons that are connected to each other to form a chain along both sides of the spinal cord; also called paravertebral ganglion.

sympathetic division Subdivision of the autonomic division of the nervous system characterized by having the cell bodies of its preganglionic neurons located in the thoracic and upper lumbar regions of the spinal cord (thoracolumbar division); usually involved in preparing the body for physical activity; also called thoracolumbar division.

symphysis; pl. symphyses (sim´fi-sis, sim´fă-sēz) [Gr., a growing together] Fibrocartilage joint between two bones.

synapse; pl. synapses (sin´aps, sĭ-naps´, sĭ-nap´sēz) [Gr., *syn*, together + *haptein*, to clasp] Functional membrane-to-membrane contact of a nerve cell with another nerve cell, muscle cell, gland cell, or sensory receptor; functions in the transmission of action potentials from one cell to another; also called neuromuscular junction.

synaptic cleft (si-nap´tik) Space between the presynaptic and the postsynaptic membranes.

synaptic fatigue Fatigue due to depletion of neurotransmitter vesicles in the presynaptic terminals.

synaptic vesicle Secretory vesicle in the presynaptic terminal containing neurotransmitter substances.

synchondrosis; pl. synchondroses (sin´kon-drō´sis, -sēz) [Gr., *syn*, together + *chondros*, cartilage + *osis*, condition] Union between two bones formed by hyaline cartilage.

syncytiotrophoblast (sin-sish´ē-ō-trō´fō-blast) Outer layer of the trophoblast composed of multinucleated cells.

syndesmosis; pl. syndesmoses (sin´dez-mō´sis, sin´dez-mō´sēz) [Gr., *syndeo*, to bind + *osis*, condition] Form of fibrous joint in which opposing surfaces that are some distance apart are united by ligaments.

synergist (sin´er-jist) Muscle that works with other muscles to cause a movement.

synovial (si-nō´vē-ăl) [Gr., *syn*, together + *oon*, egg] Relating to or containing synovia (a substance that serves as a lubricant in a joint, tendon sheath, or bursa).

synovial fluid Slippery fluid found inside synovial joints and bursae; produced by the synovial membranes.

systemic inflammation (sis-tem´ik) Inflammation that occurs in many areas of the body. In addition to symptoms of local inflammation, increased neutrophil numbers in the blood, fever, and shock can occur.

systole (sis´tō-lē) [Gr., *systole*, a contracting] Contraction of the heart chambers during which blood leaves the chambers; usually refers to ventricular contraction.

T

T cell Thymus-derived lymphocyte of immunologic importance; it is of long life and is responsible for cell-mediated immunity.

T tubule (tū´bul) Tubelike invagination of the sarcolemma that conducts action potentials toward the center of the cylindrical muscle fibers.

T wave Deflection in an electrocardiogram following the QRS complex, representing ventricular repolarization.

tactile corpuscle (tak´til) Oval receptor found in the papillae of the dermis; responsible for fine, discriminative touch; also called Meissner corpuscle.

tactile disk Cuplike receptor found in the epidermis; responsible for light touch and superficial pressure; also called Merkel's disk.

talus (tā´lŭs) [L., ankle bone, heel] Tarsal bone contributing to the ankle.

target tissue Tissue on which a hormone acts.

tarsal bone (tar´săl) [Gr., *tarsos*, sole of foot] One of seven ankle bones.

tarsal plate (tar´săl) Crescent-shaped layer of connective tissue that helps maintain the shape of the eyelid.

taste (tāst) Sensations created when a chemical stimulus is applied to the taste receptors in the tongue.

taste bud Sensory structure, mostly on the tongue, that functions as a taste receptor.

tectum (tek´tŭm) Roof of the midbrain.

tegmentum (teg-men´tŭm) Floor of the midbrain.

telencephalon (tel-en-sef´ă-lon) [Gr., *telos*, end + *enkephalos*, brain] Anterior division of the embryonic brain from which the cerebral hemispheres develop.

telophase (tel´ō-fāz) Time during cell division when the chromosomes are pulled by spindle fibers away from the cell equator and into the two halves of the dividing cell.

temporal summation (tem´pŏ-răl) Summation of the local potential that results when two or more action potentials arrive at a single synapse in rapid succession.

tendon (ten´dŏn) Band or cord of dense connective tissue that connects a muscle to a bone or another structure.

tensor tympani (ten′sŏr tim′pa-nī) Small skeletal muscle attached to the malleus.

tentorium cerebelli (ten-tō′rē-ŭm ser′ĕ-bel′ī) Dural folds between the cerebrum and the cerebellum.

terminal bouton (bū-ton′) [Fr., button] Enlarged axon terminal or presynaptic terminal.

terminal cisterna (sis-ter′nă) [L., *terminus*, limit + *cista*, box] Enlarged end of the sarcoplasmic reticulum in the area of the T tubules.

terminal hair [L., *terminus*, a boundary, limit] Long, coarse, usually pigmented hair found in the scalp, eyebrows, and eyelids and replacing vellus hair.

terminal sulcus (sŭl′kŭs) [L., furrow or ditch] V-shaped groove on the surface of the tongue at the posterior margin.

testis; pl. testes (tes′tis, tes′tēz) One of two male reproductive glands located in the scrotum; produces spermatozoa, testosterone, and inhibin.

testosterone (tes-tos′tĕ-rōn) Steroid hormone secreted primarily by the testes; aids in spermatogenesis, maintenance and development of male reproductive organs, secondary sexual characteristics, and sexual behavior.

tetraiodothyronine (T₄) (tet′ră-ī-ō′dō-thī′rō-nēn) One of the iodine-containing thyroid hormones; also called thyroxine.

thalamus (thal′ă-mŭs) [Gr., *thalamos*, bed, bedroom] Large mass of gray matter that forms the larger dorsal subdivision of the diencephalon.

theca (thē′kă) [Gr., *theke*, box] Sheath or capsule.

theca externa External fibrous layer of the theca of a vesicular follicle.

theca interna Inner vascular layer of the theca of the secondary and mature follicle; produces estrogen and contributes to the formation of the corpus luteum after ovulation.

thenar eminence (thē′nar) [Gr., palm of the hand] Fleshy mass of tissue at the base of the thumb; contains muscles responsible for thumb movements.

thick skin Skin in the palms, soles, and tips of the digits; has all five epidermal strata.

thin skin Skin over most of the body, usually without a stratum lucidum; has fewer layers of cells than thick skin.

thoracic cavity (thō-ras′ik) Space within the thoracic walls, bounded below by the diaphragm and above by the neck.

thoracic duct Largest lymph vessel in the body, beginning at the cisterna chyli and emptying into the left subclavian vein; drains the left side of the head and neck, the left-upper thorax, the left-upper limb, and the inferior half of the body.

thoracolumbar division (thōr′ă-kō-lŭm′bar) See sympathetic division.

thoroughfare channel Channel for blood through a capillary bed from an arteriole to a venule.

threshold potential (thresh′ōld) Value of the membrane potential at which an action potential is produced as a result of depolarization in response to a stimulus.

threshold stimulus Stimulus resulting in a local potential just large enough to reach threshold and produce an action potential.

thrombocyte (throm′bō-sīt) Platelet.

thrombocytopenia (throm′bō-sī′tō-pē′nē-ă) [*thrombocyte* + Gr., *penia*, poverty] Condition in which there is an abnormally small number of platelets in the blood.

thromboxane (throm-bok′sān) Specific class of physiologically active fatty acid derivatives present in many tissues.

thrombus; pl. thrombi (throm′bŭs, throm′bī) [Gr., *thrombos*, a clot] Clot in the cardiovascular system formed from constituents of blood; may be occlusive or attached to the vessel or heart wall without obstructing the lumen.

thymus (thī′mŭs) [Gr., *thymos*, sweetbread] Bilobed lymph organ located in the inferior neck and superior mediastinum; secretes the hormone thymosin.

thyroid cartilage (thī′royd) Largest laryngeal cartilage. It forms the laryngeal prominence, or Adam's apple.

thyroid gland [Gr., *thyreoeides*, shield] Endocrine gland located inferior to the larynx and consisting of two lobes connected by the isthmus; secretes the thyroid hormones triiodothyronine (T₃) and tetraiodothyronine (T₄).

thyroid-stimulating hormone (TSH) Glycoprotein hormone released from the hypothalamus; stimulates thyroid hormone secretion from the thyroid gland; also called thyrotropin.

thyrotropin (thī-rot′rō-pin, thī-rō-trō′pin) See thyroid-stimulating hormone (TSH).

thyroxine (thi-rok′sēn, thi-rok′sin) See tetraiodothyronine.

tidal volume (tī′dăl) Volume of air that is inspired or expired in a single breath during regular, quiet breathing.

tissue repair (tish′ū) Substitution of viable cells for damaged or dead cells by regeneration or replacement.

tolerance (tol′er-ăns) Failure of the specific immune system to respond to an antigen.

tongue (tŭng) Muscular organ occupying most of the oral cavity when the mouth is closed; major attachment is through its posterior portion.

tonicity (tō-nis′i-tē) [Gr., *tonos*, tone] Osmotic pressure or tension of a solution, usually relative to that of blood; a state of continuous activity or tension caused by muscle contraction beyond the tension related to physical properties of muscle.

tonsil; pl. tonsils (ton′sil, ton′silz) [L., *tonsilla*, stake] Collection of lymphoid tissue; usually refers to a large collection of lymphatic tissue beneath the mucous membrane of the oral cavity and pharynx; lingual, pharyngeal, and palatine tonsils.

total lung capacity Volume of air contained in the lungs at the end of a maximum inspiration; equals vital capacity plus residual volume.

total tension Sum of active and passive tension.

trabecula; pl. trabeculae (tră-bek′ū-lă, tră-bek′ū-lē) [L., *trabs*, beam] One of the supporting bundles of fibers traversing the substance of a structure, usually derived from the capsule or one of the fibrous septa, such as trabeculae of lymph nodes, testes; a beam or plate of cancellous bone.

trachea (trā′kē-ă) [Gr., *tracheia arteria*, rough artery] Air tube extending from the larynx into the thorax, where it divides to form the bronchi; composed of 16–20 rings of hyaline cartilage.

tracheostomy (trā′ke-os′to-me) [*tracheo, trachea* + Gr., *stoma*, mouth] Operation to make an opening into the trachea; usually the opening is intended to be permanent, and a tube is inserted into the trachea to allow airflow.

tracheotomy (trā′ke-ot′o-me) [*tracheo, trachea* + Gr., *tome*, incision] Act of cutting into the trachea.

transcription (tran-skrip′shŭn) Process of forming RNA from a DNA template.

transfer RNA (tRNA) RNA that attaches to individual amino acids and transports them to the ribosomes, where they are connected to form a protein polypeptide chain.

transfusion (trans-fū′zhŭn) [L., *trans*, across + *fundo*, to pour from one vessel to another] Transfer of blood from one person to another.

transitional epithelium (tran-sish′ŭn-ăl) Stratified epithelium that may be either cuboidal or squamouslike, depending on the presence or absence of fluid in the organ (as in the urinary bladder).

translation (trans-lā′shŭn) Synthesis of polypeptide chains at the ribosome in response to information contained in mRNA molecules.

transverse colon (trans-vers′ kō′lon) Part of the colon between the right and left colic flexures.

transverse mesocolon (mez′ō-kō′lon) Fold of peritoneum attaching the transverse colon to the posterior abdominal wall.

transverse (T) tubule [L., *tubus*, tube] Tubule that extends from the sarcolemma to a myofibril of striated muscles.

treppe (trep′eh) [Ger., staircase] Series of successively stronger contractions that occur when a rested muscle fiber receives closely spaced stimuli of the same strength but with a sufficient stimulus interval to allow complete relaxation of the fiber between stimuli.

triacylglycerol (trī-as′il-glis′er-ol) See triglyceride.

triad (trī′ad) Two terminal cisternae and a T tubule between them.

tricuspid valve (trī-kŭs′pid) Valve closing the orifice between the right atrium and the right ventricle of the heart.

triglyceride (tri-glis′er-id) Three-carbon glycerol molecule with a fatty acid attached to each carbon; constitute approximately 95% of the fats in the human body. Also called triacylglycerol.

trigone (trī′gōn) [Gr., *trigonon*, triangle] Triangular, smooth area at the base of the bladder between the openings of the two ureters and that of the urethra.

triiodothyronine (T₃) (trī-ī′ō-dō-thī′rō-nēn) One of the iodine-containing thyroid hormones.

trochlea (trok′lē-ă) [L., pulley] Structure shaped like or serving as a pulley or spool.

trochlear nerve (trok′lē-ar) [L., *trochlea*, pulley] Cranial nerve IV, to the muscle (superior oblique) turning around a pulley.

trophoblast (trof′ō-blast) [Gr., *trophe*, nourishment + *blastos*, germ] Cell layer forming the outer layer of the blastocyst, which erodes the uterine mucosa during implantation; the trophoblast does not become part of the embryo but contributes to the formation of the placenta.

tropomyosin (trō-pō-mī′ō-sin) Fibrous protein found as a component of the actin myofilament.

troponin (trō′pō-nin) Globular protein component of the actin myofilament.

true pelvis Portion of the pelvis inferior to the pelvic brim.

true rib (ver-tĕ′brō-ster′năl) Rib that attaches by an independent costal cartilage directly to the sternum; also called vertebrosternal rib.

trypsin (trip′sin) Proteolytic enzyme formed in the small intestine from the inactive pancreatic precursor trypsinogen.

tubercle (too′ber-kl) Lump on a bone.

tubular load (too′bū-lăr) Amount of a substance per minute that crosses the filtration membrane into Bowman's capsule.

tubular maximum Maximum rate of secretion or reabsorption of a substance by the renal tubules.

tubular reabsorption Movement of materials, by means of diffusion, active transport, or symport, from the filtrate within a nephron to the blood.

tubular secretion Movement of materials, by means of active transport, from the blood into the filtrate of a nephron.

tumor (too′mŏr) Any swelling or growth; a neoplasm.

tunic (too′nik) [L., coat] One of the enveloping layers of a part; one of the coats of a blood vessel;

one of the coats of the eye; one of the coats of the digestive tract.

tunica adventitia (too′ni-kă ad-ven-tish′ă) Outermost fibrous coat of a vessel or an organ that is derived from the surrounding connective tissue.

tunica albuginea (al-bū-jin′ē-ă) Dense, white, collagenous tunic surrounding a structure, such as the capsule around the testis.

tunica intima (in′ti-mă) Innermost coat of a blood vessel; consists of endothelium, a lamina propria, and an inner elastic membrane.

tunica media Middle, usually muscular, coat of an artery or another tubular structure.

turbulent flow Flow characterized by eddy currents exhibiting nonparallel blood flow.

tympanic membrane (tim-pan′ik) Eardrum; cellular membrane that separates the external from the middle ear; vibrates in response to sound waves.

U

unipolar neuron (oo-ni-pō′lar) One of the three categories of neurons consisting of a nerve cell body with a single axon projecting from it; also called a pseudounipolar neuron.

unmyelinated axon (ŭn-mī′ĕ-li-nā-ted) Nerve fibers lacking a myelin sheath.

unsaturated (ŭn-sach′ŭr-āt-ed) Carbon chain of a fatty acid that possesses one or more double or triple bonds.

upper respiratory tract Nasal cavity, pharynx, and associated structures.

up-regulation Increase in the concentration of receptors in response to a signal.

ureter (ū-rē′ter, ŭ′rē-ter) [Gr., oureter, urinary canal] Tube conducting urine from the kidney to the urinary bladder.

urethral gland (ū-rē′thrăl) One of numerous mucous glands in the wall of the spongy urethra in the male.

urogenital triangle Anterior portion of the perineal region containing the openings of the urethra and vagina in the female and the urethra and root structures of the penis in the male.

uterine cycle (ū′ter-in, ū′ter-īn) Series of events that occur in a regular fashion in the uterus of sexually mature, nonpregnant females; prepares the uterine lining for implantation of the embryo.

uterine part Portion of the uterine tube that passes through the wall of the uterus.

uterine tube One of the tubes leading on each side from the uterus to the ovary; consists of the infundibulum, ampulla, isthmus, and uterine parts; also called fallopian tube or oviduct.

uterus (ū′ter-ŭs) Hollow, muscular organ in which the fertilized ovum develops into a fetus.

utricle (oo′tri-kl) Part of the membranous labyrinth; contains a sensory structure, the macula, that detects static equilibrium.

uvula (ū′vū-lă) [L., uva, grape] Small, grapelike appendage at the posterior margin of the soft palate.

V

vaccination (vak′si-nā′shŭn) Deliberate introduction of an antigen into a subject to stimulate the immune system and produce immunity to the antigen.

vaccine (vak′sēn, vak-sēn′) [L., vaccinus, relating to a cow] Preparation of killed microbes, altered microbes, or derivatives of microbes or microbial products intended to produce immunity. The method of administration is usually inoculation, but ingestion is preferred in some instances, and nasal spray is used occasionally.

vagina (vă-jī′nă) [L., sheath] Genital canal in the female, extending from the uterus to the vulva.

vapor pressure Partial pressure exerted by water vapor.

variable region Part of an antibody that combines with an antigen.

vas deferens (vas def′er-enz) See ductus deferens.

vasa recta (vā′să rek′tă) Specialized capillary that extends from the cortex of the kidney into the medulla and then back to the cortex.

vasa vasorum (vā′sor-ŭm) [L., vessel, dish] Small vessels distributed to the outer and middle coats of larger blood vessels.

vascular layer (vas′kū-lăr) Middle layer of the eye; contains many blood vessels.

vasoconstriction (vā′sō-kon-strik′shŭn, vas′ō-kon-strik′shŭn) Decreased diameter of blood vessels.

vasodilation (vā′sō-dī-lā′shŭn) Increased diameter of blood vessels.

vasomotion (vā-sō-mō′shŭn) Periodic contraction and relaxation of the precapillary sphincter, resulting in cyclic blood flow through capillaries.

vasomotor center (vā-sō-mō′ter, vas-ō-mō′ter) Area within the medulla oblongata that regulates the diameter of blood vessels by way of the sympathetic nervous system.

vasomotor tone Relatively constant frequency of sympathetic impulses that keep blood vessels partially constricted in the periphery.

vasopressin (vā-sō-pres′in, vas-ō-pres′in) See antidiuretic hormone.

vellus hair (vel′ŭs) [L., fleece] Short, fine, usually unpigmented hair that covers the body except for the scalp, eyebrows, and eyelids. Much of the vellus is replaced at puberty by terminal hairs.

venous capillary (vē′nŭs) Capillary opening into a venule.

venous return Volume of blood returning to the heart.

venous sinus Endothelium-lined venous channel in the dura mater that receives cerebrospinal fluid from the arachnoid granulations.

ventilation (ven-ti-lā′shŭn) [L., ventus, the wind] Movement of gases into and out of the lungs.

ventral root (ven′trăl) Motor (efferent) root of a spinal nerve.

ventricle (ven′tri-kl) [L., venter, belly] Chamber of the heart that pumps blood into arteries (i.e., the left and right ventricles); in the brain, a fluid-filled cavity.

ventricular diastole (ven-trik′ū-lăr) Dilation of the heart ventricles.

ventricular systole Contraction of the ventricles.

venule (ven′ool, vē′nool) Minute vein, consisting of endothelium and a few scattered smooth muscles, that carries blood away from capillaries.

vermiform appendix (ver′mi-fōrm) [L., vermis, worm + forma, form; appendage] Wormlike sac extending from the blind end of the cecum.

vesicle (ves′i-kl) [L., vesica, bladder] Small sac containing a liquid or gas, such as a blister in the skin or an intracellular, membrane-bounded sac.

vestibular fold (ves-tib′ū-lăr) One of two folds of mucous membrane stretching across the laryngeal cavity from the angle of the thyroid cartilage to the arytenoid cartilage superior to the vocal cords; helps close the glottis; also called false vocal cord.

vestibular membrane Membrane separating the cochlear duct and the scala vestibuli.

vestibule (ves′ti-bool) [L., antechamber, entrance court] Anterior part of the nasal cavity just inside the external nares that is enclosed by cartilage; space between the lips and the alveolar processes and teeth; middle region of the inner ear containing the utricle and saccule; space behind the labia minora containing the openings of the vagina, urethra, and vestibular glands.

vestibulocochlear nerve (ves-tib′ū-lō-kok′lē-ăr) Nerve formed by the cochlear and vestibular nerves; extends to the brain.

villus; pl. villi (vil′ŭs, vil′ī) [L., shaggy hair (of beasts)] Projection of the mucous membrane of the intestine; leaf-shaped in the duodenum; becomes shorter, more finger-shaped, and sparser in the ileum.

visceral (vis′er-ăl) Relating to the internal organs.

visceral pericardium (per′i-kar′dē-ŭm) Serous membrane covering the surface of the heart; also called epicardium.

visceral peritoneum (per′i-tō-nē′ŭm) [Gr., periteino, to stretch over] Layer of peritoneum covering the abdominal organs.

visceral pleura (vis′er-ăl plūr′ă) Serous membrane investing the lungs and dipping into the fissures between the several lobes.

visceroreceptor (vis′er-ō-rē-sep′tŏr) Sensory receptor associated with the organs.

viscosity (vis-kos′i-tē) [L., viscosus, viscous] Resistance to flow or alteration of shape by any substance as a result of molecular cohesion.

visual cortex (vizh′oo-ăl) Area in the occipital lobe of the cerebral cortex that integrates visual information and produces the sensation of vision.

visual field Area of vision for each eye.

vital capacity (vīt-ăl) Greatest volume of air that can be exhaled from the lungs after a maximum inspiration.

vitamin (vīt′ă-min) [L., vita, life + amine] One of a group of organic substances present in minute amounts in natural foodstuffs that are essential to normal metabolism; insufficient amounts in the diet may cause deficiency diseases.

vitamin D Fat-soluble vitamin produced from precursor molecules in skin exposed to ultraviolet light; increases calcium and phosphate uptake from the intestines.

vitreous humor (vit′rē-ŭs) Transparent, jellylike material that fills the space between the lens and the retina.

Volkmann's canal Named for German surgeon Richard Volkmann (1830–1889). Canal in bone containing blood vessels; not surrounded by lamellae; runs perpendicular to the long axis of the bone and the haversian canals, interconnecting the latter with each other and the exterior circulation.

vulva (vŭl′vă) [L., wrapper or covering, seed covering, womb] External genitalia of the female, composed of the mons pubis, the labia majora and minora, the clitoris, the vestibule of the vagina and its glands, and the opening of the urethra and of the vagina. Also called pudendum.

W

water-soluble vitamin Vitamin, such as B complex and C, that is absorbed with water from the intestinal tract.

white matter Bundles of parallel axons with their associated sheath in the central nervous system.

white pulp Part of the spleen consisting of lymphatic nodules and diffuse lymphatic tissue; associated with arteries.

white ramus communicans; pl. rami communicantes (rā′mŭs kō-mū′nĭ-kans, rā′mī kō-mū-nĭ-kan′tēz) Connection between a spinal nerve and a sympathetic chain ganglion through which myelinated preganglionic axons project.

wisdom tooth Third molar tooth on each side in each jaw.

X

xiphoid (zi′foyd) [Gr., *xiphos,* sword] Sword-shaped, with special reference to the sword tip; the inferior part of the sternum.

X-linked Gene located on an X chromosome.

Y

yellow marrow (mar′o) Connective tissue filling the cavities of bones; consists primarily of reticular fibers and fat cells; replaces red marrow in long bones and the skull.

Y-linked Gene located on a Y chromosome.

yolk sac (yōk, yōlk) Highly vascular layer surrounding the yolk of an embryo.

Z

Z disk Delicate, membranelike structure found at each end of a sarcomere to which actin myofilaments attach.

zona fasciculata (zō′nǎ fa-sik′ū-lǎ′tǎ) [L., *zone,* a girdle, one of the zones of the sphere] Middle layer of the adrenal cortex that secretes cortisol.

zona glomerulosa (glō-mǎr-ū-lōs-ǎ) Outer layer of the adrenal cortex that secretes aldosterone.

zona pellucida (pe-lū′sǐ-dǎ) Layer of viscous fluid surrounding the oocyte.

zona reticularis (rē-tik′ū-lar′is) Inner layer of the adrenal cortex that secretes androgens and estrogens.

zonula adherens (zō′nū-lǎ ad-her′enz) [L., a small zone; adhering] Small zone holding or adhering cells together.

zonula occludens (ō-klūd′enz) [L., occluding] Junction between cells in which the plasma membranes may be fused; occludes or blocks off the space between the cells.

zygomatic (zī-gō-mat′ik) [Gr., *zygon,* yoke] Yoking or joining; bony arch created by the junction of the zygomatic and temporal bones.

zygote (zī′gōt) [Gr., *zygotos,* yoked] Diploid cell resulting from the union of a sperm cell and an oocyte.

Credits

13.29a&b: © Trent Stephens; **13.31:** © Jerry Wachter/ Photo Researchers, Inc.

Chapter 14
Opener: © Innerspace Imaging/ Science Photo Library/Photo Researchers, Inc.

Chapter 15
Opener: © Dr. L. Orci, University of Geneva/Science Photo Library/ Photo Researchers, Inc.; **Figure A:** © Ewing Galloway, Inc./ Index Stock; **15.18d:** © Victor Eroschenko; **Figure Ba&b:** © Ken Greer/Visuals Unlimited; **15.22:** © Victor Eroschenko; **Figure C, 15.25c:** © Bio-Photo Assocs/ Photo Researchers, Inc.; **Figure D (Boy):** © Mark Clark/Science Photo Library/Photo Researchers, Inc.; **Figure D (Kit):** © Visuals Unlimited.

Chapter 16
Opener: © Dennis Kunkel Microscopy, Inc.; **16.1:** © The McGraw-Hill Companies, Inc./Eric Wise, photographer; **16.3a:** © National Cancer Institute/Science Photo Library/ Photo Researchers, Inc.; **16.7:** © Ed Reschke; **16.8:** © Victor Eroschenko.

Chapter 17
Opener: © David M. Phillips/ Photo Researchers, Inc.; **17.2:** © Terry Cockerham/Cynthia Alexander/Synapse Media Productions; **17.5b:** © R. T. Hutchings; **17.7a&b:** © McMinn & Hutchings, Color Atlas of Human Anatomy/Mosby; **17.11b:** © Ed Reschke; **Figure A:** © Hank Morgan/Science Source/Photo Researchers, Inc. **17.17:** © Terry Cockerham/Cynthia Alexander/ Synapse Media Productions.

Chapter 18
Opener: © Prof. P. Motta/Dept. of Anatomy/University "La Sapienza", Rome/Science Photo Library/Photo Researchers, Inc. 18.1: © Visuals Unlimited.

Chapter 19
Opener: © Dennis Kunkel Microscopy, Inc.; **19.3:** © Victor Eroschenko; **19.7b:** © Trent Stephens.

Chapter 20
Opener: © David Phillips/Photo Researchers, Inc.; **20.2b:** © The McGraw-Hill Companies, Inc./ Photo and dissection by Christine Eckel; **20.4b:** © CNRI/Phototake. com; **20.6:** © Branislav Vidic.

Chapter 21
Opener: © Dennis Kunkel Microscopy, Inc.; **21.3b&c:** © The McGraw-Hill Companies, Inc./Rebecca Gray, photographer/ Don Kincaid, dissections; **21.10c:** © Victor Eroschenko; **21.15e:** © David M. Phillips/Visuals Unlimited; **Figure A:** © SIU Biomedical/Photo Researchers, Inc.; **21.24b:** © CNRI/Science Photo Library/Photo Researchers, Inc.

Chapter 22
Opener: © Professors P. Motta & T. Naguro/Science Photo Library/ Photo Researchers, Inc; **22.17:** © Kyle Rothenborg/Pacific Stock.

Chapter 23
Opener: © Dennis Kunkel Microscopy, Inc.; **23.3b:** © The McGraw-Hill Companies, Inc./ Rebecca Gray, photographer/Don Kincaid, dissections.

Chapter 24
Opener: © Prof. P. Motta & S. Makabe/Science Photo Library/ Photo Researchers, Inc.; **Figure Aa-c:** © The McGraw-Hill Companies, Inc./Jill Braaten, photographer; **Figure Ad:** © The McGraw-Hill Companies, Inc.; **Figure Ae:** © The McGraw-Hill Companies, Inc./Bob Coyle, photographer.

Chapter 25
Opener: © Dr. Yorgus Nikas/Jason Burns/Phototake.com; **Figure A:** © NMSB/Custom Medical Stock Photo; **25.9:** © John Giannicchi/Photo Researchers, Inc.; **25.17a&b:** © Petit Format/Nestle/Photo Researchers, Inc.; **25.23:** © CNRI/Science Photo Library/ Photo Researchers, Inc.; **25.25:** © Norman Lightfoot/Photo Researchers, Inc. **25.26:** © Dr. M.A. Ansary/Photo Researchers, Inc.

Index

Selected Abbreviations

α alpha
ACE angiotensin-converting enzyme
acetyl-CoA acetyl coenzyme A
ACh acetylcholine
ADH antidiuretic hormone
ADP adenosine diphosphate
ANH atrial natriuretic hormone
ANS autonomic nervous system
apo E apolipoprotein E
ATP adenosine triphosphate
AV atrioventricular beta
BCOP blood colloid osmotic pressure
BMI body mass index
BMR basal metabolic rate
BP blood pressure
BPG 2,3-bisphosphoglycerate
bpm beats per minute
BUN blood urea nitrogen
$C_6H_{12}O_6$ glucose
$Ca_{10}(PO_4)_6(OH)_2$ hydroxyapatite
Ca^{2+} calcium ion
cal calorie
cAMP cyclic adenosine monophosphate
CBC complete blood count
cGMP cyclic guanosine monophosphate
CH_3COOH acetic acid
Cl^- chloride ion
CNS central nervous system
CO cardiac output
CO carbon monoxide
CO_2 carbon dioxide
—COOH carboxyl group
COX-1 cyclooxygenase-1
CP capsule pressure
CRH corticotrophin-releasing hormone
CSF cerebrospinal fluid
DAG diacylglycerol
DNA deoxyribonucleic acid
ECG or EKG electrocardiogram
EEG electroencephalogram
EGF epidermal growth factor
ENS enteric nervous system
EPSP excitatory postsynaptic potential
FAD flavin adenine dinucleotide
$FADH_2$ reduced flavin adenine diphosphate
Fe^{2+} iron ion
FEV_1 forced expiratory volume in one second
FGF fibroblast growth factor
FSH follicle-stimulating hormone gamma
$\bar{g}$ gram
GABA gamma-aminobutyric acid

GCP glomerular capillary pressure
GDP guanosine diphosphate
GFR glomerular filtration rate
GH growth hormone
GHIH growth hormone-inhibiting hormone
GHRH growth hormone-releasing hormone
GnRH gonadotropin-releasing hormone
GTP guanosine triphosphate
H^+ hydrogen ion
H_2CO_3 carbonic acid
H_2O water
H_2O_2 hydrogen peroxide
$H_2PO_4^-$ dihydrogen phosphate ion
HCG human chorionic gonadotropin
HCl hydrochloric acid
HCO_3^- bicarbonate ion
HDL high-density lipoprotein
Hg mercury
HIV human immunodeficiency virus
HLA human leukocyte antigen
HPO_4^{2-} monohydrogen phosphate ion
HR heart rate
I^- iodide ion
ICSH interstitial cell-stimulating hormone
IFP interstitial fluid pressure
Ig immunoglobulin
IP_3 inositol triphosphate
IPSP inhibitory postsynaptic potential
IU international units
K^+ potassium ion
kcal kilocalorie
kg kilogram
L liter
LDL low-density lipoprotein
LH luteinizing hormone
LHRH luteinizing hormone-releasing hormone
MAC membrane attack complex
MALT mucosa-associated lymphoid tissue
MAO monoamine oxidase
MAP mean arterial pressure
mEq milliequivalent
Mg^{2+} magnesium ion
MHC major histocompatibility complex
mOsm milliosmole
mRNA messenger ribonucleic acid
mV millivolt
Na^+ sodium ion
NaCl sodium chloride

NAD^+ nicotinamide adenine dinucleotide
NADH reduced nicotinamide adenine dinucleotide
$NaHCO_3$ sodium bicarbonate
NaOH sodium hydroxide
NFP net filtration pressure
—NH_2 amine group
NH_3 ammonia
NH_4^+ ammonium ion
NK cells natural killer cells
NO nitric oxide
O_2 oxygen
OH^- hydroxide ion
PAH *para*-aminohippuric acid
P_{ALV} alveolar pressure
P_B barometric air pressure
PGE prostaglandin E
PGF prostaglandin F
P_i inorganic phosphate
PIF prolactin-inhibiting factor
PIH prolactin-inhibiting hormone
PIP_2 phosphoinositol
PMNs polymorphonuclear neutrophils
PNS peripheral nervous system
PO_4^{3-} phosphate ion
P_{PL} pleural pressure
PR peripheral resistance
PRF prolactin-releasing factor
PRH prolactin-releasing hormone
PTH parathyroid hormone
RANKL receptor for activation of nuclear factor kappa B ligand
RAS reticular activating system
RBC red blood count
RDA recommended daily allowances
RDI reference daily intake
RhoGAM Rh_o (D) immune globulin
RMP resting membrane potential
RNA ribonucleic acid
SA sinoatrial
SV stroke volume
T_3 triiodothyronine
T_4 tetraiodothyronine
TF tissue factor
TGF-_ transforming growth factor beta
TMJ temporomandibular joint
TNF tumor necrosis factor
TRH thyroid-releasing hormone
TSH thyroid-stimulating hormone
V_A alveolar ventilation
VLDL very low-density lipoprotein
vWF von Willebrand factor
WBC white blood count

Prefixes, Suffixes, and Combining Forms

The ability to break down medical terms into separate components or to recognize a complete word depends on mastery of the combining forms (roots or stems) and the prefixes and suffixes that alter or modify their meanings. Common prefixes, suffixes, and combining forms are listed below in boldface type, followed by the meaning of each form and an example illustrating its use.

a-, an-   without, lack of: *aphasia* (lack of speech), *anaerobic* (without oxygen)

ab-   away from: *ab*ductor (leading away from)

-able   capable: vi*able* (capable of living)

acou-   hearing: *acou*stics (science of sound)

acr-   extremity: *acr*omegaly (large extremities)

ad-   to, toward, near to: *ad*renal (near the kidney)

adeno-   gland: *adeno*ma (glandular tumor)

-al   expressing relationship: neur*al* (referring to nerves)

-algia   pain: gastr*algia* (stomach pain)

angio-   vessel: *angio*graphy (radiography of blood vessels)

ante-   before, forward: *ante*cubital (before elbow)

anti-   against, reversed: *anti*peristalsis (reversed peristalsis)

arthr-   joint: *arthr*itis (inflammation of a joint)

-ary   associated with: urin*ary* (associated with urine)

-asis   condition, state of: homeost*asis* (state of staying the same)

auto-   self: *auto*lysis (self breakdown)

bi-   twice, double: *bi*cuspid (two cusps)

bio-   live: *bio*logy (study of living)

-blast   bud, germ: fibro*blast* (fiber-producing cell)

brady-   slow: *brady*cardia (slow heart rate)

-c   expressing relationship: cardia*c* (referring to heart)

carcin-   cancer: *carcin*ogenic (causing cancer)

cardio-   heart: *cardio*pathy (heart disease)

cata-   down, according to: *cata*bolism (breaking down)

cephal-   head: *cephal*ic (toward the head)

-cele   hollow: blasto*cele* (hollow cavity inside a blastocyst)

cerebro-   brain: *cerebro*spinal (referring to brain and spinal cord)

chol-   bile: a*chol*ic (without bile)

cholecyst-   gallbladder: *cholecyst*okinin (hormone causing the gallbladder to contract)

chondr-   cartilage: *chondr*ocyte (cartilage cell)

-cide   kill: bacteri*cide* (agent that kills bacteria)

circum-   around, about: *circum*duction (circular movement)

-clast   smash, break: osteo*clast* (cell that breaks down bone)

co-, com-, con-   with, together: *co*enzyme (molecule that functions with an enzyme), *com*misure (coming together), *con*vergence (to incline together)

contra-   against, opposite: *contra*lateral (opposite side)

crypto-   hidden: *crypto*rchidism (undescended or hidden testes)

cysto-   bladder, sac: *cysto*cele (hernia of a bladder)

-cyte-, cyto-   cell: erythro*cyte* (red blood cell), *cyto*skeleton (supportive fibers inside a cell)

de-   away from: *de*hydrate (remove water)

derm-   skin: *derm*atology (study of the skin)

di-   two: *di*ploid (two sets of chromosomes)

dia-   through, apart, across: *dia*pedesis (ooze through)

dis-   reversal, apart from: *dis*sect (cut apart)

-duct   leading, drawing: ab*duct* (lead away from)

-dynia   pain: masto*dynia* (breast pain)

dys-   difficult, bad: *dys*mentia (bad mind)

e-   out, away from: *e*viscerate (take out viscera)

ec-   out from: *ec*topic (out of place)

ecto-   on outer side: *ecto*derm (outer skin)

-ectomy   cut out: append*ectomy* (cut out the appendix)

-edem   swell: myo*edem*a (swelling of a muscle)

em-, en-   in: *em*pyema (pus in), *en*cephalon (in the brain)

-emia   blood: an*emia* (deficiency of blood)

endo-   within: *endo*metrium (within the uterus)

entero-   intestine: *entero*itis (inflammation of the intestine)

epi-   upon, on: *epi*dermis (on the skin)

erythro-   red: *erythro*cyte (red blood cell)

eu-   well, good: *eu*phoria (well-being)

ex-   out, away from: *ex*halation (breathe out)

exo-   outside, on outer side: *exo*genous (originating outside)

extra-   outside: *extra*cellular (outside the cell)

-ferent   carry: af*ferent* (carrying to the central nervous system)

-form   expressing resemblance: fusi*form* (resembling a fusion)

gastro-   stomach: *gastro*dynia (stomach ache)

-genesis   produce, origin: patho*genesis* (origin of disease)

gloss-   tongue: hypo*gloss*al (under the tongue)

glyco-   sugar, sweet: *glyco*lysis (breakdown of sugar)

-gram   a drawing: myo*gram* (drawing of a muscle contraction)

-graph   instrument that records: myo*graph* (instrument for measuring muscle contraction)

hem-   blood: *hem*opoiesis (formation of blood)

hemi-   half: *hemi*plegia (paralysis of half of the body)

hepato-   liver: *hepato*titis (inflammation of the liver)

hetero-   different, other: *hetero*zygous (different genes for a trait)

hist-   tissue: *hist*ology (study of tissues)

homeo-, homo-   same: *homeo*stasis (state of staying the same), *homo*logous (alike in structure or origin)

hydro-   wet, water: *hydro*cephalus (fluid within the head)

hyper-   over, above, excessive: *hyper*trophy (overgrowth)

hypo-   under, below, deficient: *hypo*tension (low blood pressure)

-ia, -id   expressing condition: neuralg*ia* (pain in nerve), flacc*id* (state of being weak)

-iatr   treat, cure: ped*iatr*ics (treatment of children)

-im   not: *im*permeable (not permeable)

in-   in, into: *in*jection (forcing fluid into)

infra-   below, beneath: *infra*orbital (below the eye)

inter-   between: *inter*costal (between the ribs)

intra-   within: *intra*ocular (within the eye)

-ism   condition, state of: dimorph*ism* (condition of two forms)